Master Resource Book for

JEE Main

Physics

Revised Edition

Master Resource Book for

JEE Main

Physics

5000+

Specially Prepared Questions for JEE Main with

Complete Theory

2 Levels Exercises

Exams Questions

DB Singh

arihant

ARIHANT PRAKASHAN (Series), MEERUT

Master Resource Book for

JEE Main

arihant

ARIHANT PRAKASHAN (Series), MEERUT

卐 **Administrative & Production Offices**

Regd. Office

'Ramchhaya' 4577/15, Agarwal Road, Darya Ganj, New Delhi -110002
Tele: 011- 47630600, 43518550

卐 **Head Office**

Kalindi, TP Nagar, Meerut (UP) - 250002
Tel: 0121-7156203, 7156204

卐 **Sales & Support Offices**

Agra, Ahmedabad, Bengaluru, Bareilly, Chennai, Delhi, Guwahati, Hyderabad, Jaipur, Jhansi, Kolkata, Lucknow, Nagpur & Pune.

卐 **ISBN** 978-93-25792-45-6

卐 **PRICE** ₹875.00

PO No : TXT-XX-XXXXXXX-X-XX

Published by Arihant Publications (India) Ltd.

For further information about the books published by Arihant, log on to www.arihantbooks.com or e-mail at info@arihantbooks.com

Follow us on

PREFACE

In sync with the recent changes in the test pattern and format of **JEE Main** (Joint Engineering Entrance), it is my pleasure to introduce **Master Resource Book in Physics for JEE Main,** for the Students aspiring a seat in a reputed Engineering College. JEE Main is a gateway examination for candidates expecting to seek admission in Bachelor in Engineering (BE), Bachelor of Technology (B.Tech) and Bachelor of Architecture (B.Arch) at Indian Institutes of Information Technology (IIITs), National Institutes of Technology (NITs), Delhi Technological University and other Centrally Funded Technical Institutes (CFTIs).

JEE Main is also an examination which is like screening examination for **JEE Advanced** (The gateway examination to India's most reputed Technical Institutes, Indian Institutes of Technology IITs). Only the top 2.2 lacs students passed in JEE Main will be able to attempt JEE Advanced.

Gradually, the number of students aspiring for the seat in the Engineering College has increased rapidly in the last 5 Years or so. This year nearly 10 lacs students appeared for JEE Main and only a few were able to reserve a seat in the college of their choice, so there is a cut throat competition among the aspirants. Thus, it calls for a systematic mastery of all the subjects of the test with paramount importance to problem-solving. Most of the books now in the market have become repetitive with scant respect to the needs of true and effective learning. This book has been designed to fulfill the perceived needs of the students as such.

- This book comprehensively covers all the topics of JEE Main Physics syllabus. The chapters have been sequenced according to the **syllabus of class 11th & 12th**. Each chapter has essential theoretical discussion of the related concepts with sufficient number of solved examples and practice problems. In each chapter previous years' questions of AIEEE and JEE Main have been included to help students know the difficulty levels and nature of questions asked in competitive exams at this level.
- All types of questions have been included in this book: **Single Correct Answer Types and Numerical Value Questions.**
- This is the only book which has its subject matter divided as per class 11th & 12th syllabus.

It is hoped this new effort will immensely benefit the students in their goal to secure a seat in the prestigious engineering college, and would be convenient to teachers in planning their teaching programmes. Suggestions for further improvement are welcomed from the students and teachers.

DB Singh

CONTENTS

PART I

Chapters from Class 11th Syllabus

PART II

Chapters from Class 12th Syllabus

SYLLABUS

NOTE The syllabus contains two Sections - A & B. Section A pertains to the Theory Part, having 80% weightage, while Section B contains Practical Component (Experimental Skills) having 20% weightage.

SECTION- A

UNIT 1 Physics and Measurement

Physics, technology and society, SI units, Fundamental and derived units. Least count, accuracy and precision of measuring instruments, Errors in measurement, Significant figures. Dimensions of Physical quantities, dimensional analysis and its applications.

UNIT 2 Kinematics

Frame of reference. Motion in a straight line: Position-time graph, speed and velocity. Uniform and non-uniform motion, average speed and instantaneous velocity.

Uniformly accelerated motion, velocity-time, position time graphs, relations for uniformly accelerated motion.

Scalars and Vectors, Vector addition and Subtraction, Zero Vector, Scalar and Vector products, Unit Vector, Resolution of a Vector. Relative Velocity, Motion in a plane, Projectile Motion, Uniform Circular Motion.

UNIT 3 Laws of Motion

Force and Inertia, Newton's First Law of motion; Momentum, Newton's Second Law of motion; Impulse; Newton's Third Law of motion. Law of conservation of linear momentum and its applications, Equilibrium of concurrent forces.

Static and Kinetic friction, laws of friction, rolling friction.

Dynamics of uniform circular motion: Centripetal force and its applications.

UNIT 4 Work, Energy and Power

Work done by a constant force and a variable force; kinetic and potential energies, work-energy theorem, power.

Potential energy of a spring, conservation of mechanical energy, conservative and nonconservative forces; Elastic and inelastic collisions in one and two dimensions.

UNIT 5 Rotational Motion

Centre of mass of a two-particle system, Centre of mass of a rigid body; Basic concepts of rotational motion; moment of a force, torque, angular momentum, conservation of angular momentum and its applications; moment of inertia, radius of gyration. Values of moments of inertia for simple geometrical objects, parallel and perpendicular axes theorems and their applications.

Rigid body rotation, equations of rotational motion.

UNIT 6 Gravitation

The universal law of gravitation.

Acceleration due to gravity and its variation with altitude and depth.

Kepler's laws of planetary motion.

Gravitational potential energy; gravitational potential.

Escape velocity. Orbital velocity of a satellite.
Geo-stationary satellites.

UNIT 7 Properties of Solids & Liquids

Elastic behaviour, Stress-strain relationship, Hooke's. Law, Young's modulus, bulk modulus, modulus of rigidity.

Pressure due to a fluid column; Pascal's law and its applications.

Viscosity, Stokes' law, terminal velocity, streamline and turbulent flow, Reynolds number. Bernoulli's principle and its applications.

Surface energy and surface tension, angle of contact, application of surface tension - drops, bubbles and capillary rise.

Heat, temperature, thermal expansion; specific heat capacity, calorimetry; change of state, latent heat.

Heat transfer-conduction, convection and radiation, Newton's law of cooling.

UNIT 8 Thermodynamics

Thermal equilibrium, zeroth law of thermo-dynamics, concept of temperature. Heat, work and internal energy. First law of thermodynamics.

Second law of thermodynamics: reversible and irreversible processes. Camot engine and its efficiency.

UNIT 9 Kinetic Theory of Gases

Equation of state of a perfect gas, work done on compressing a gas.

Kinetic theory of gases - assumptions, concept of pressure. Kinetic energy and temperature: rms speed of gas molecules; Degrees of freedom, Law of equipartition of energy, applications to specific heat capacities of gases; Mean free path, Avogadro's number.

UNIT 10 Oscillations and Waves

Periodic motion - period, frequency, displacement as a function of time. Periodic functions. Simple harmonic motion (S.H.M.) and its equation; phase; oscillations of a spring - restoring force and force constant; energy in S.H.M. - kinetic and potential energies; Simple pendulum - derivation of expression for its time period; Free, forced and damped oscillations, resonance.

Wave motion Longitudinal and transverse waves, speed of a wave. Displacement relation for a progressive wave. Principle of superposition of waves, reflection of waves, Standing waves in strings and organ pipes, fundamental mode and harmonics, Beats, Doppler effect in sound.

UNIT 11 Electrostatics

Electric charges Conservation of charge, Coulomb's law-forces between two point charges, forces between multiple charges; superposition principle and continuous charge distribution.

Electric field Electric field due to a point charge, Electric field lines, Electric dipole, Electric field due to a dipole, Torque on a dipole in a uniform electric field.

Electric flux, Gauss's law and its applications to find field due to infinitely long, uniformly charged straight wire, uniformly charged infinite plane sheet and uniformly charged thin spherical shell.

Electric potential and its calculation for a point charge, electric dipole and system of charges; Equipotential surfaces, Electrical potential energy of a system of two point charges in an electrostatic field.

Conductors and insulators, Dielectrics and electric polarization, capacitor, combination of capacitors in series and in parallel, capacitance of a parallel plate capacitor with and without dielectric medium between the plates, Energy stored in a capacitor.

UNIT 12 Current Electricity

Electric current, Drift velocity, Ohm's law, Electrical resistance, Resistances of different materials, V-I characteristics of Ohmic and nonohmic conductors, Electrical energy and power, Electrical resistivity, Colour code for resistors; Series and parallel combinations of resistors; Temperature dependence of resistance.

Electric Cell and its Internal resistance, potential difference and emf of a cell, combination of cells in series and in parallel.

Kirchhoff's laws and their applications. Wheatstone bridge, Metre bridge.

Potentiometer - principle and its applications.

UNIT 13 Magnetic Effects of Current and Magnetism

Biot-Savart law and its application to current carrying circular loop. Ampere's law and its applications to infinitely long current carrying straight wire and solenoid. Force on a moving charge in uniform magnetic and electric fields Cyclotron.

Force on a current-carrying conductor in a uniform magnetic field. Force between two parallel current-carrying conductors-definition of ampere. Torque experienced by a current loop in uniform magnetic field, Moving coil galvanometer, its current sensitivity and conversion to ammeter and voltmeter.

Current loop as a magnetic dipole and its magnetic dipole moment. Bar magnet as an equivalent solenoid, magnetic field lines; Earth's magnetic field and magnetic elements. Para, dia and ferro-magnetic substances

Magnetic susceptibility and permeability, Hysteresis, Electromagnets and permanent magnets.

UNIT 14 Electromagnetic Induction and Alternating Currents

Electromagnetic induction; Faraday's law, induced emf and current; Lenz's Law, Eddy currents. Self and mutual inductance.

Alternating currents, peak and rms value of alternating current/ voltage; reactance and impedance; LCR series circuit, resonance; Quality factor, power in AC circuits, wattless current. AC generator and transformer.

UNIT 15 Electromagnetic Waves

Electromagnetic waves and their characteristics. Transverse nature of electromagnetic waves.

Electromagnetic spectrum (radio waves, microwaves, infrared, visible, ultraviolet, X-rays, gamma rays). Applications of e.m. waves.

UNIT 16 Optics

Reflection and refraction of light at plane and spherical surfaces, mirror formula, Total internal reflection and its applications, Deviation and Dispersion of light by a prism, Lens Formula, Magnification, Power of a Lens, Combination of thin lenses in contact, Microscope and Astronomical Telescope (reflecting and refracting) and their magnifying powers.

Wave optics wave front and Huygens' principle, Laws of reflection and refraction using Huygen's principle. Interference, Young's double slit experiment and expression for fringe width, coherent sources and sustained interference of light. Diffraction due to a single slit, width of central maximum. Resolving power of microscopes and astronomical telescopes, Polarisation, plane polarized light; Brewster's law, uses of plane polarized light and Polaroids.

UNIT 17 Dual Nature of Matter and Radiation

Dual nature of radiation. Photoelectric effect, Hertz and Lenard's observations; Einstein's photoelectric equation; particle nature of light. Matter waves-wave nature of particle, de Broglie relation. Davisson-Germer experiment.

UNIT 18 Atoms and Nuclei

Alpha-particle scattering experiment; Rutherford's model of atom; Bohr model, energy levels, hydrogen spectrum.

Composition and size of nucleus, atomic masses, isotopes, isobars; isotones. Radioactivity-alpha, beta and gamma particles/rays and their properties; radioactive decay law. Mass-energy relation, mass defect; binding energy per nucleon and its variation with mass number, nuclear fission and fusion.

UNIT 19 Electronic Devices

Semiconductors; semiconductor diode: I-V characteristics in forward and reverse bias; diode as a rectifier; I-V characteristics of LED, photodiode, solar cell, and Zener diode; Zener diode as a voltage regulator. Junction transistor, transistor action, characteristics of a transistor transistor as an amplifier (common emitter configuration) and oscillator. Logic gates (OR, AND, NOT, NAND & NOR). Transistor as a switch.

UNIT 20 Communication Systems

Propagation of electromagnetic waves in the atmosphere; Sky and space wave propagation, Need for modulation, Amplitude and Frequency Modulation, Bandwidth of signals, Bandwidth of Transmission medium, Basic Elements of a Communication System (Block Diagram only)

SECTION- B

UNIT 21 Experimental Skills

Familiarity with the basic approach and observations of the experiments and activities

1. Vernier callipers - its use to measure internal and external diameter and depth of a vessel.
2. Screw gauge - its use to determine thickness/ diameter of thin sheet/wire.
3. Simple Pendulum - dissipation of energy by plotting a graph between square of amplitude and time.
4. Metre Scale - mass of a given object by principle of moments.
5. Young's modulus of elasticity of the material of a metallic wire.
6. Surface tension of water by capillary rise and effect of detergents.
7. Coefficient of Viscosity of a given viscous liquid by measuring terminal velocity of a given spherical body.
8. Plotting a cooling curve for the relationship between the temperature of a hot body and time.
9. Speed of sound in air at room temperature using a resonance tube.
10. Specific heat capacity of a given (i) solid and (ii) liquid by method of mixtures.
11. Resistivity of the material of a given wire using metre bridge.
12. Resistance of a given wire using Ohm's law.
13. Potentiometer
    (i) Comparison of emf of two primary cells.
    (ii) Internal resistance of a cell.
14. Resistance and figure of merit of a galvanometer by half deflection method.
15. Focal length of
    (i) Convex mirror
    (ii) Concave mirror
    (iii) Convex lens
16. Using parallax method. Plot of angle of deviation vs angle of incidence for a triangular prism.
17. Refractive index of a glass slab using a travelling microscope.
18. Characteristic curves of a p-n junction diode in forward and reverse bias.
19. Characteristic curves of a Zener diode and finding reverse break down voltage.
20. Characteristic curves of a transistor and finding current gain and voltage gain.
21. Identification of Diode, LED, Transistor, IC, Resistor, Capacitor from mixed collection of such items.
22. Using multimeter to
    (i) Identify base of a transistor.
    (ii) Distinguish between npn and pnp type transistor.
    (iii) See the unidirectional flow of current in case of a diode and an LED.
    (iv) Check the correctness or otherwise of a given electronic component (diode, transistor or IC).

01

Units and Measurements

Physics is a quantitative science, based on measurement of certain physical quantities. Measurement of any physical quantity involves comparison with some standard unit of that quantity. In this chapter, we shall study about the units and measurement.

IN THIS CHAPTER ....

- Physics, Technology and Society
- Physical Quantities
- Fundamental and Derived Units
- Least Count and Percentage Uncertainty
- Accuracy and Precision of Measuring Instruments
- Errors in Measurement
- Significant Figures
- Dimensions of Physical Quantities
- Dimensional Analysis and Its Applications

Physics, Technology and Society

Physics is the study of matter and its motion, as well as space and time using concepts such as energy, force, mass and charge. It is an experimental science, creating theories that are tested against observation.

The connection between physics, technology and society can be seen in many examples like working of heat engines gave rise to thermodynamics. Wireless communication technology arose from basic laws of electricity and magnetism. Lately discovery of silicon chip triggered the computer revolution.

Physical Quantities

All the quantities which are used to describe the laws of physics are called physical quantities. To measure a physical quantity, some standard unit of that quantity is required, *e.g.* if length of some metal rod is measured to be 20 m, then m is the unit of length and 20 is the numerical value. So,

$$\text{Physical quantity} = \text{Numerical value} \times \text{Unit}$$

Note (i) If the numerical value of any physical quantity in different units u_1 and u_2 are n_1 and n_2 respectively, then $n_1u_1 = n_2u_2$.

(ii) As the unit will change, numerical value will also change, *e.g.* acceleration due to gravity, $g = 32 \text{ fts}^{-2} = 9.8 \text{ ms}^{-2}$.

Fundamental and Derived Units

Those physical quantities which are independent of other physical quantities and not defined in terms of other physical quantities, are called fundamental or *base quantities*. The quantities which can be expressed in terms of the fundamental quantities are called *derived quantities*.

Units of fundamental and derived quantities are respectively, known as the *fundamental units* and *derived units*. A complete set of these units, both fundamental and derived units, is known as the *system of units*.

System of Units

The common system of units are

(i) **FPS system** The units of length, mass and time are respectively foot, pound and second.

(ii) **CGS system** The units of length, mass and time are respectively centimetre, gram and second.

(iii) **MKS system** The units of length, mass and time are respectively metre, kilogram and second.

SI System (International System of Units)

The SI is based on the following seven fundamental units and two supplementary units.

Units and Symbol of Quantities

S. No.	Quantity	Unit	Symbol
Fundamental Units			
1.	Mass	kilogram	kg
2.	Length	metre	m
3.	Time	second	s
4.	Electric current	ampere	A
5.	Temperature	kelvin	K
6.	Amount of substance	mole	mol
7.	Luminous intensity	candela	cd
Supplementary Units			
1.	Plane angle	radian	rad
2.	Solid angle	steradian	Sr

Note (i) Angle and solid angle are considered supplementary base units because although these have units but they are both dimensionless.

(ii) 2π radians = 360°

Practical Units

A large number of units are used in general life for measurement of different quantities in comfortable manner. But they are neither fundamental units nor derived units.

Practical Units of Length, Mass and Time

Practical Units of Length	Practical Units of Mass	Practical Units of Time
1 light year $=9.46\times10^{15}$ m	1 quintal = 10^2 kg	1 year = $365\frac{1}{4}$ solar days
1 astronomical unit or 1 AU $=1.5\times10^{11}$ m	1 metric ton $=10^3$ kg	1 lunar month = 27.3 solar days
1 parsec = 3.26 light year	1 atomic mass unit (amu) $=1.66\times10^{-27}$kg	1 solar day=86400 s
1 seamile = 6020 ft	1 pound = 0.4537 kg	**Tropical year** It is that year in which solar eclipse occurs.
1 micron $=1\ \mu m=10^{-6}$ m	1 chandrasekhar limit = 1.4 times the mass of sun $=2.8\times10^{30}$ kg	**Leap year** It is that year in which the month of February has 29 days.
1 angstrom $=10^{-10}$ m	1 slug = 14.59 kg	1 shake = 10^{-8} s
1 fermi = 10^{-15} m		

Example 1. *What is the SI unit of surface tension?*

(a) Nm^{-1} (b) Nm^{-2} (c) Nm (d) N

Sol. (a) Surface tension $=\dfrac{\text{Force}}{\text{Length}}=\dfrac{N}{m}=Nm^{-1}$

Least Count and Percentage Uncertainty

The smallest value of a physical quantity which can be measured accurately with an instrument is called the Least Count (LC) of the measuring instrument.

$$LC=\frac{\text{Value of 1 main scale division}}{\text{Total number of vernier scale divisions}}$$

The instrument with the least uncertainty is taken to measure objects, as all measurements consider accuracy.

The percentage uncertainty is calculated with the following formula

$$=\frac{\text{Maximum possible error}}{\text{Measurement of object in question}}\times100$$

The smaller the measurement, the larger the percentage uncertainty.

Least Count of Certain Measuring Instruments

- Vernier calliper,

  Least count (LC) $=\dfrac{1\text{ mm}}{10\text{ divisions}}=0.1$ mm
- Screw gauge,

  Least count $=\dfrac{1\text{ mm}}{100\text{ divisions}}=0.01$ mm
- Travelling microscope,

  Least count $=\dfrac{0.5\text{ mm}}{50\text{ divisions}}=0.01$ mm
- Spectrometer,

  Least count $=\dfrac{0.5\text{ degree}}{30\text{ divisions}}=\dfrac{30^\circ}{30\text{ divisions}}=1^\circ$

Accuracy and Precision of Measuring Instruments

In our real world, two terms, *i.e.* accuracy and precision are often used as interchangeably, but they have specific meaning.

Accuracy An instrument is said to be accurate, if the physical quantity measured by it resembles very closely to its true value.

Precision An instrument is said to have high degree of precision, if the measured value remains unchanged, how so ever, large number of times it may have been repeated.

Note (i) Resolution stands for least count or the minimum reading which an instrument can read.

(ii) The least count of an instrument is indirectly proportional to the precision of the instrument.

Example 2. *In an experiment, the angles are required to be measured using an instrument. 29 divisions of the main scale exactly coincide with the 30 divisions of the vernier scale. If the smallest division of the main scale is half a degree (= 0.5°), then the least count of the instrument is* **[AIEEE 2009]**

(a) one minute (b) half minute
(c) one degree (d) half degree

Sol. (a) Least count $= \dfrac{\text{Value of main scale division}}{\text{Number of divisions on vernier scale}}$

$$= \frac{1}{30}\text{ MSD} = \frac{1}{30} \times \frac{1^\circ}{2} = \frac{1^\circ}{60} = 1 \text{ min}$$

Example 3. *A student measured the length of a rod and wrote it as 3.50 cm. Which instrument did he use to measure it?* **[JEE Main 2014]**

(a) A meter scale
(b) A vernier calliper where the 10 divisions in vernier scale matches with 9 divisions in main scale and main scale has 10 divisions in 1 cm
(c) A screw gauge having 100 divisions in the circular scale and pitch as 1 mm
(d) A screw gauge having 50 divisions in the circular scale and pitch as 1 mm

Sol. (b) If student measure 3.50 cm, it means that there is an uncertainty of order 0.01 cm.

For vernier scale with 1 MSD = 1mm and 9 MSD = 10 VSD

∴ LC of VC = 1 MSD − 1 VSD

$$= \frac{1}{10}\left(1 - \frac{9}{10}\right) = \frac{1}{100} \text{ cm}$$

Errors in Measurement

The difference between the measured value and true value (mean value) of a quantity is called error of measurement. Different types of error are given below.

(i) **Absolute error** The difference between the true value and the measured value of a quantity is called an absolute error. Usually the mean value a_m is taken as the true value. So, if

$$a_{\text{mean}} = a_m = \frac{a_1 + a_2 + \ldots + a_n}{n}$$

then absolute errors are

$$\Delta a_1 = a_m - a_1, \Delta a_2 = a_m - a_2,$$
$$\Delta a_3 = a_m - a_3, \ldots, \Delta a_n = a_m - a_n$$

The absolute error may be positive or negative.

(ii) **Mean absolute error** It is the arithmetic mean of the magnitudes of different values of absolute errors.

∴ Mean absolute error,

$$\Delta a_{\text{mean}} = \frac{|\Delta a_1| + |\Delta a_2| + |\Delta a_3| + \ldots + |\Delta a_n|}{n}$$

The final result of measurement can be written as $a = a_m \pm \Delta\bar{a}$. This implies that value of a is likely to lie between $a_m + \Delta\bar{a}$ and $a_m - \Delta\bar{a}$.

(iii) **Relative or fractional error**

$$\therefore \text{Relative error} = \frac{\text{Mean absolute error}}{\text{Mean value of measurement}} = \frac{\Delta a_{\text{mean}}}{a_{\text{mean}}} = \frac{\Delta a_{\text{mean}}}{a_m}$$

(iv) **Percentage error**

$$\therefore \text{Percentage error} = \frac{\Delta a_{\text{mean}}}{a_{\text{mean}}} \times 100\%$$

Example 4. *The average speed of a train is measured by 5 students. The results of measurements are given below*

Number of Students	*Speed (m/s)*
1	10.2
2	10.4
3	9.8
4	10.6
5	10.8

The percentage error in the measurement of average speed is

(a) 2.6% (b) 3.5% (c) 4.5% (d) 5.5%

Sol. (a) Mean value, $v_m = \dfrac{10.2 + 10.4 + 9.8 + 10.6 + 10.8}{5}$

$$= \frac{51.8}{5} = 10.4 \text{ ms}^{-1}$$

$$\Delta v_1 = v_m - v_1 = 10.4 - 10.2 = 0.2$$
$$\Delta v_2 = v_m - v_2 = 10.4 - 10.4 = 0.0$$
$$\Delta v_3 = v_m - v_3 = 10.4 - 9.8 = 0.6$$
$$\Delta v_4 = v_m - v_4 = 10.4 - 10.6 = -0.2$$
$$\Delta v_5 = v_m - v_5 = 10.4 - 10.8 = -0.4$$

Mean absolute error,

$$\overline{\Delta v} = \frac{|\Delta v_1| + |\Delta v_2| + |\Delta v_3| + |\Delta v_4| + |\Delta v_5|}{5}$$

$$= \frac{0.2 + 0.0 + 0.6 + 0.2 + 0.4}{5} = \frac{1.4}{5} = 0.28 \text{ ms}^{-1}$$

$$\text{Relative error} = \pm\frac{\overline{\Delta v}}{v_m} = \pm\frac{0.28}{10.4}$$

$$\text{Percentage error} = \pm\frac{\overline{\Delta v}}{v_m} \times 100 = \pm\frac{0.28}{10.4} \times 100 = \pm 2.6\%$$

Least Count Error

It is an instrumental or random error associated with the resolution of the instrument.

The least count error occurs with both systematic and random error. Instruments of higher precision and improving experimental techniques can reduce the least count error.

Combination of Errors

In Addition If $Z = A + B$, then $\Delta Z = \pm(\Delta A + \Delta B)$, maximum fractional error in this case $\frac{\Delta Z}{Z} = \frac{\Delta A + \Delta B}{A + B}$, *i.e.* when two physical quantities are added, then the maximum absolute error in the result is the sum of the absolute errors of the individual quantities.

In Difference If $Z = A - B$, the maximum absolute error is $\Delta Z = \pm(\Delta A + \Delta B)$ and maximum fractional error in this case

$$\frac{\Delta Z}{Z} = \frac{\Delta A + \Delta B}{A - B}$$

Example 5. *The volumes of two bodies are measured to be $V_1 = (10.2 \pm 0.02)\,cm^3$ and $V_2 = (6.4 \pm 0.01)\,cm^3$. The sum and difference in volumes with error limits is*

(a) $(16.6 \pm 0.03)\ cm^3$ and $(3.8 \pm 0.03)\ cm^3$
(b) $(16.6 \pm 0.01)\ cm^3$ and $(3.8 \pm 0.01)\ cm^3$
(c) $(16.2 \pm 0.03)\ cm^3$ and $(3.6 \pm 0.03)\ cm^3$
(d) $(16.2 \pm 0.01)\ cm^3$ and $(3.6 \pm 0.01)\ cm^3$

Sol. (a) Given, $V_1 = (10.2 \pm 0.02)\ cm^3$

and $V_2 = (6.4 \pm 0.01)\ cm^3$

$$\Delta V = \pm(\Delta V_1 + \Delta V_2)$$
$$= \pm(0.02 + 0.01)\ cm^3 = \pm 0.03\ cm^3$$
$$V_1 + V_2 = (10.2 + 6.4)\ cm^3 = 16.6\ cm^3$$

and $V_1 - V_2 = (10.2 - 6.4)\ cm^3 = 3.8\ cm^3$

Hence, sum of volume $= (16.6 \pm 0.03)\ cm^3$

and difference of volume $= (3.8 \pm 0.03)\ cm^3$

In product If $Z = AB$, then maximum fractional error is

$$\frac{\Delta Z}{Z} = \pm\left(\frac{\Delta A}{A} + \frac{\Delta B}{B}\right)$$

In division If $Z = A/B$, then maximum fractional error is

$$\frac{\Delta Z}{Z} = \pm\left(\frac{\Delta A}{A} + \frac{\Delta B}{B}\right)$$

Note Maximum fractional error in product or division of two (or more) quantities is equal to sum of fractional errors in the individual quantities.

Example 6. *The temperature of two bodies measured by a thermometer are $t_1 = 20°C + 0.5°C$ and $t_2 = 50°C \pm 0.5°C$. The maximum absolute error in temperature difference is*

(a) $\pm 1°C$ (b) $\pm 2°C$ (c) $\pm 3°C$ (d) $\pm 4°C$

Sol. (a) The temperature difference is given by

$$t' = t_2 - t_1 = (50°\ C \pm 0.5°\ C) - (20°\ C \pm 0.5°\ C)$$
$$t' = 30°\ C \pm 1°C$$

Example 7. *Two resistors of resistances $R_1 = 100 \pm 3\ \Omega$ and $R_2 = 200 \pm 4\ \Omega$ are connected in parallel, then the equivalent resistance in parallel is (in ohm)*

$\left(\text{Use } \frac{1}{R'} = \frac{1}{R_1} + \frac{1}{R_2} \text{ and } \frac{\Delta R'}{\Delta R'^2} = \frac{\Delta R_1}{R_1^2} + \frac{\Delta R_2}{R_2^2}\right)$

(a) 66.7 ± 1.8 (b) 300 ± 7
(c) 150.8 ± 2 (d) 92.3 ± 3

Sol. (a) The equivalent resistance of parallel combination is

$$R' = \frac{R_1R_2}{R_1 + R_2} = \frac{200}{3} = 66.7\ \Omega$$

From, $\frac{1}{R'} = \frac{1}{R_1} + \frac{1}{R_2}$, we get

$$\frac{\Delta R'}{R'^2} = \frac{\Delta R_1}{R_1^2} + \frac{\Delta R_2}{R_2^2}$$
$$\Delta R' = (R'^2)\frac{\Delta R_1}{R_1^2} + (R'^2)\frac{\Delta R_2}{R_2^2}$$
$$= \left(\frac{R'}{R_1}\right)^2 \Delta R_1 + \left(\frac{R'}{R_2}\right)^2 \Delta R_2$$
$$= \left(\frac{66.7}{100}\right)^2 3 + \left(\frac{66.7}{200}\right)^2 4 = 1.8\ \Omega$$

Hence, $R' = (66.7 \pm 1.8)\ \Omega$

Example 8. *The following observations were taken for determining surface tension T of water by capillary method. Diameter of capillary, $d = 1.25 \times 10^{-2}$ m, rise of water, $h = 1.45 \times 10^{-2}$m. Using $g = 9.80\ m/s^2$ and the simplified relation $T = \frac{rhg}{2} \times 10^3 N/m$, the possible error in surface tension is closest to* **[JEE Main 2017]**

(a) 1.5% (b) 2.4%
(c) 10% (d) 0.15%

Sol. (a) By ascent formula, we have surface tension,

$$T = \frac{rhg}{2} \times 10^3\ \frac{N}{m}$$
$$= \frac{dhg}{4} \times 10^3\ \frac{N}{m} \qquad \left(\because r = \frac{d}{2}\right)$$
$$\Rightarrow \frac{\Delta T}{T} = \frac{\Delta d}{d} + \frac{\Delta h}{h} \qquad \text{[given, } g \text{ is constant]}$$

So, percentage error $= \frac{\Delta T}{T} \times 100 = \left(\frac{\Delta d}{d} + \frac{\Delta h}{h}\right) \times 100$

$$= \left(\frac{0.01 \times 10^{-2}}{1.25 \times 10^{-2}} + \frac{0.01 \times 10^{-2}}{1.45 \times 10^{-2}}\right) \times 100$$
$$= 1.5\%$$
$$\therefore \frac{\Delta T}{T} \times 100 = 1.5\%$$

Example 9. *In a simple pendulum, experiment for determination of acceleration due to gravity (g), time taken for 20 oscillations is measured by using a watch of 1 second least count. The mean value of time taken comes out to be 30 s. The length of pendulum is measured by using a meter scale of least count 1 mm and the value obtained 55.0 cm. The percentage error in the determination of g is close to*

[JEE Main 2019]

(a) 0.7% (b) 6.8%
(c) 3.5% (d) 0.2%

Sol. (b) Relation used for finding acceleration due to gravity by using a pendulum is

$$g = \frac{4\pi^2 l}{T^2}$$

So, fractional error in value of g is

$$\frac{\Delta g}{g} = \frac{\Delta l}{l} + \frac{2\Delta T}{T} \quad \text{...(i)}$$

Given, $\Delta l = 0.1\,\text{cm}, l = 55\,\text{cm}, \Delta T = 1\,\text{s}$ and T for 20 oscillations $= 30$ s

Substituting above values in Eq. (i), we get

$$\frac{\Delta g}{g} = \frac{0.1}{55} + 2 \times \frac{1}{30}$$

Hence, percentage error in g is

$$= \frac{\Delta g}{g} \times 100$$

$$= \frac{10}{55} + \frac{20}{3} = 6.8\%$$

Significant Figures

"The significant figures are those number of digits in a quantity that are known reliably plus one digit that is uncertain." Larger the number of significant figures obtained in a measurement, greater is the accuracy of the measurement.

Rules for Significant Figures

(i) All non-zero digits are significant figures.

Number	Significant figures
16	2
1683	4
16835	5

(ii) All zeros occurring between non-zero digits are significant figures.

Number	Significant figures
802	3
80004	5

(iii) All zeros to the right of the last non-zero digits are not significant.

Number	Significant number
40	1
410	2
40240	4

(iv) All zeros to the right of a decimal point and to the left of a non-zero digit are not significant.

Number	Significant number
0.08	1
0.008	1
0.0846	3

(v) All zeros to the right of a decimal point and to the right of a non-zero digit are significant.

Number	Significant number
0.40	2
0.430	3

(vi) The powers of ten are not counted as significant digits *e.g.*, 1.4×10^{-7} has only two significant figures 1 and 4.

Rounding off the Digits

Certain rules are applied in order to round off the measurements

(i) If the number lying to the right of digit to be rounded is less than 5, then the rounded digit is retained as such. However, if it is more than 5, then the digit to be rounded is increased by 1.

For example, $x = 6.24$ is rounded off to 6.2 to two significant digits and $x = 8.356$ is rounded off to 8.36 to three significant digits.

(ii) If the digit to be dropped is 5 followed by digits other than zero, then the preceding digit is increased by 1.

For example, $x = 14.252$ is rounded off to $x = 14.3$ to three significant digits.

(iii) If the digit to be dropped is simply 5 or 5 followed by zeros, then the preceding digit is left unchanged if it is even and it is raised by one if it is odd.

For example, $x = 6.250$ or $x = 6.25$ becomes $x = 6.2$ after rounding off to two significant digits and $x = 6.350$ or $x = 6.35$ becomes $x = 6.4$ after rounding off to two significant digits.

Algebraic Operations with Significant Figures

The following rules for algebraic operations with significant figures, that make final result more consistent with the precision of the measured values.

Addition and Subtraction

The number of decimal places in the final result of any of these operations has to be equal to the smallest number of decimal places in any of the terms involved in calculation *e.g.*, sum of terms 2.29 and 62.7 is 64.99. After rounding off to one place of decimal, it will become 65.0. Subtraction of 62.7 from 82.27 gives 19.57. After rounding off to one place of decimal, it will become 19.6.

Multiplication and Division

In these operations, the number of significant figures in the result is same as the smallest number of significant figures in any of the factors.

e.g. $1.2 \times 1.3 = 1.56$. After rounding off to two significant figures, it becomes 1.6.

Similarly, if $\frac{1100}{10.2}$ gives 107.84. Then, the result when rounded off to three significant digits becomes 108.

Example 10. *The respective number of significant figures for the numbers 23.023, 0.0003 and 2.1×10^{-3} are* *[AIEEE 2010]*

(a) 5, 1, 2 (b) 5, 1, 5
(c) 5, 5, 2 (d) 4, 4, 2

Sol. (a) The reliable digit plus the first uncertain digit is known as significant figures.

For the number 23.023, all the non-zero digits are significant, hence 5.

For the number 0.0003, number is less than 1, the zero(s) on the right of decimal point but to the left of the first non-zero digit are not significant, hence 1.

For the number 2.1×10^{-3}, significant figures are 2.

Example 11. *The area enclosed by a circle of diameter 1.06 m to correct number of significant figures is*

(a) $0.88\ m^2$ (b) $0.088\ m^2$
(c) $0.882\ m^2$ (d) $0.530\ m^2$

Sol. (c) Here, $r = \frac{1.06}{2} = 0.530$ m

Area enclosed $= \pi r^2 = 3.14(0.53)^2$

$= 0.882026\ m^2 = 0.882\ m^2$

(rounded to three significant figures)

Example 12. *A body of mass m = 3.513 kg is moving along the X-axis with a speed of $5.00\ ms^{-1}$. The magnitude of its momentum is recorded as* *[AIEEE 2008]*

(a) $17.6\ kg\ ms^{-1}$ (b) $17.565\ kg\ ms^{-1}$
(c) $17.56\ kg\ ms^{-1}$ (d) $17.57\ kg\ ms^{-1}$

Sol. (a) So, momentum, $p = mv = 17.565$ kg ms^{-1}

where $m = 3.513$ kg and $v = 5.00$ ms^{-1}

As the number of significant digits in m is 4 and in v is 3, so, p must have 3 (minimum) significant digits.

Hence, $p = 17.6$ kg ms^{-1}

Example 13. *5.74 g of a substance occupies 1.2 cm^3. Keeping the significant figure in view, its density is given by*

(a) $4.8\ g\ cm^{-3}$ (b) $1.5\ g\ cm^{-3}$
(c) $2.1\ g\ cm^{-3}$ (d) $9.2\ g\ cm^{-3}$

Sol. (a) There are 3 significant figures in the measured mass whereas there are only 2 significant figures in the measured volume. Hence, the density should be expressed to only 2 significant figures.

$\because$ Density $= \frac{\text{Mass}}{\text{Volume}}$

$\therefore$ Density $= \frac{5.74}{1.2}$ g cm^{-3} $= 4.8$ g cm^{-3}

Dimensions of Physical Quantities

Dimensions of a physical quantity are the powers to which the fundamental quantities are raised to represent that quantity. In mechanics, all physical quantities can be expressed in terms of mass [M], length [L] and time [T].

For example, Force = Mass × Acceleration

$$= \text{Mass} \times \frac{\text{Velocity}}{\text{Time}} = \frac{m \times s}{t \times t}$$

$$= \frac{m \times s}{t^2} = [M]\ [L]\ [T^{-2}]$$

So, the dimensions of force are 1 in mass, 1 in length and −2 in time.

Dimensional Formula and Dimensional Equations

The expression which shows how and which of the fundamental quantities represent the dimension of physical quantity is called the *dimensional formula* of the given physical quantity.

For example, as deduced above, $[M^1L^1T^{-2}]$ is the dimensional formula of force. It reveals that unit of force depends on [M], [L] and [T].

Further, if we represent force by [F], then $[F] = [M^1L^1T^{-2}]$ is called the *dimensional equation* of force.

Dimensionless Quantity

In the equation $[M^aL^b\ T^c]$, if $a = b = c = 0$, then the quantity is called dimensionless.

For example, strain, specific gravity and angle are dimensionless quantity because they are ratio of two similar quantities.

Some dimensionless quantities

Angle, solid angle, relative density, specific gravity, Poisson's ratio, Reynold's number, all trigonometric ratios, refractive index, relative permittivity, dielectric constant, magnetic susceptibility. A dimensionless quantity has same numeric value in all system of units.

Principle of Homogeneity

According to this principle, a correct dimensional equation must be homogeneous, *i.e.* dimensions of all the terms in a physical expression must be same.

LHS (dimension) = RHS (dimension)

Dimensions of Important Physical Quantities

Physical Quantity	SI Unit	Dimensional Formula
Power	Watt (W)	$[ML^2T^{-3}]$
Pressure, stress, coefficient of elasticity (ρ, σ, η)	Pascal (Pa) or Nm^{-2}	$[ML^{-1}T^{-2}]$
Frequency, angular frequency	Hz or s^{-1}	$[T^{-1}]$
Angular momentum	$kg\,m^2\,s^{-1}$	$[ML^2T^{-1}]$
Torque	Nm	$[ML^2T^{-2}]$
Gravitational constant (G)	$N\,m^2\,kg^{-2}$	$[M^{-1}L^3T^{-2}]$
Moment of inertia	$kg\,m^2$	$[ML^2]$
Acceleration, acceleration due to gravity	ms^{-2}	$[LT^{-2}]$
Force, thrust, tension, weight	Newton (N)	$[MLT^{-2}]$
Linear momentum, impulse	$kg\,ms^{-1}$ or Ns	$[MLT^{-1}]$
Work, energy, KE, PE, thermal energy, internal energy, etc.	Joule (J)	$[ML^2T^{-2}]$
Surface area, area of cross-section	m^2	$[L^2]$
Electric conductivity	Sm^{-1}	$[M^{-1}L^{-3}T^3A^2]$
Young's modulus, Bulk modulus	Pa	$[ML^{-1}T^{-2}]$
Compressibility	$m^2\,N^{-1}$	$[M^{-1}LT^2]$
Magnetic flux	Wb	$[ML^2T^{-2}A^{-1}]$
Magnetic flux density (σ)	Wb/m^2	$[MT^{-2}A^{-1}]$
Intensity of a wave	Wm^{-2}	$[MT^{-3}]$
Photon flux density	$m^{-2}\,s^{-1}$	$[L^{-2}T^{-1}]$
Luminous energy	Lm s	$[ML^2T^{-2}]$
Luminance	Lux	$[MT^{-3}]$
Specific heat capacity	$Jkg^{-1}K^{-1}$	$[L^2T^{-2}K^{-1}]$
Latent heat of vaporisation	Jkg^{-1}	$[L^2T^{-2}]$
Coefficient of thermal conductivity	$Wm^{-1}K^{-1}$	$[MLT^{-3}K^{-1}]$
Electric voltage	JC^{-1}	$[ML^2T^{-3}A^{-1}]$
Magnetisation	Am^{-1}	$[L^{-1}A]$
Magnetic induction	T	$[MT^{-2}A^{-1}]$
Planck's constant	J-s	$[ML^2T^{-1}]$
Radioactive decay constant	Bq	$[T^{-1}]$
Binding energy	MeV	$[ML^2T^{-2}]$

Dimensional Analysis and Its Applications

There are three applications of dimensional analysis

1. To check the correctness of a given physical equation

As per principle of homogeneity, if the dimensions of each term on both sides of a physical relation are same, then the relation is dimensionally correct otherwise wrong.

Example 14. *Is the given expression of velocity of sound given by $v = \sqrt{\left(\frac{E}{\rho}\right)}$ is dimensionally correct?*

Here, E = coefficient of elasticity,
ρ = density of medium

(a) Yes (b) No
(c) Cannot be predicted (d) The correct expression is $\frac{E}{\rho}$

Sol. (a) $[LHS] = [v] = [LT^{-1}]$

$$[RHS] = \left[\left(\frac{E}{\rho}\right)\right]^{1/2} = \left[\left(\frac{ML^{-1}T^{-2}}{ML^{-3}}\right)\right]^{1/2} = [LT^{-1}]$$

$$[LHS] = [RHS]$$

Hence, equation is dimensionally correct.

Example 15. *Given equation $\frac{1}{2}mv^2 = mgh$, where m is the mass of the body, v is velocity, g is the acceleration due to gravity and h is the height. Then the given equation is*

(a) dimensionally incorrect
(b) dimensionally correct
(c) wrong
(d) None of the above

Sol. (b) Given, $\frac{1}{2}mv^2 = mgh$

The dimensions of LHS are

$$[M][LT^{-1}]^2 = [M][L^2T^{-2}] = [ML^2T^{-2}]$$

The dimensions of RHS are

$$[M][LT^{-2}][L] = [M][L^2T^{-2}] = [ML^2T^{-2}]$$

The dimensions of LHS and RHS are the same and hence the equation is dimensionally correct.

Example 16. *The SI unit of energy is $J = kg\,m^2s^{-2}$, that of speed v is ms^{-1} and of acceleration a is ms^{-2}, which one of the formula for kinetic energy given below is correct on the basis of dimensional arguments?*
(Given, m stands for the mass of body)

(a) $K = m^2v^2$ (b) $K = ma$
(c) $K = \frac{1}{2}mv^2 + ma$ (d) $K = \frac{1}{2}mv^2$

Sol. (a) Every correct formula or equation must have the same dimensions on both sides of the equation. Also, only quantities with the same physical dimensions can be added or subtracted. The dimensions of the quantity on the right side for (a) is

$$K = m^2v^2$$

Putting, $m = [M], v = [LT^{-1}]$

$\therefore$ dimensions are $[M^2L^2T^{-2}]$

for (b), $K = ma$

putting $m = [M], a = [LT^{-2}]$

$\therefore$ dimensions are $[MLT^{-2}]$

option (c) has no proper dimensions,

option (d), $K = \frac{1}{2}mv^2$, putting $m = [M]$

and $V = [LT^{-1}]$, we have

$$K = [M][LT^{-1}]^2 = [ML^2T^{-2}]$$

in units it is written as kg m^2 s^{-2}.

2. Derivation of formula

The method of dimensions is used to deduce the relation among the physical quantities. We should know the dependence of the physical quantity on the other quantities.

We explain the process in following examples.

Example 17. *The time period T of simple pendulum depends upon length l of the pendulum and gravitational acceleration. The formula for time period of simple pendulum is given by*

(a) $T = 2\pi\sqrt{\frac{l}{g}}$ (b) $T = 2\pi\sqrt{\frac{g}{l}}$

(c) $T = \frac{1}{2\pi}\sqrt{lg}$ (d) $T = \frac{2\pi l}{g}$

Sol. (a) Let $T \propto l^a$ and $T \propto g^b$

where a and b are dimensionless constants

$$T = kl^a g^b$$

where, k is dimensionless constant.

$$[LHS] = [T] = [M^0L^0T^1]$$

and $[RHS] = (l^a g^b) = [L]^a [LT^{-2}]^b$

$$= [L^{a+b}T^{-2b}] = [M^0L^{a+b}T^{-2b}]$$

According to homogeneity principle,

or $[M^0L^0T] = [M^0L^{a+b}T^{-2b}]$

For dimensional balance, dimensions on both sides should be same.

$\therefore$ $a + b = 0$ and $-2b = 1$

$\therefore$ $b = -\frac{1}{2}$ and $a = \frac{1}{2}$

$\therefore$ $T = 2\pi\sqrt{\frac{l}{g}}$

[since, numerical value of k in case of simple pendulum is 2π]

Example 18. *Which of the following combinations has the dimension of electrical resistance (ε_0 is the permittivity of vacuum and μ_0 is the permeability of vacuum)?* **[JEE Main 2019]**

(a) $\sqrt{\frac{\mu_0}{\varepsilon_0}}$ (b) $\frac{\mu_0}{\varepsilon_0}$

(c) $\sqrt{\frac{\varepsilon_0}{\mu_0}}$ (d) $\frac{\varepsilon_0}{\mu_0}$

Sol. (a) Let dimensions of resistance R, permittivity ε_0 and permeability μ_0 are $[R]$, $[\varepsilon_0]$ and $[\mu_0]$, respectively.

So, $[R] = [\varepsilon_0]^\alpha [\mu_0]^\beta$ ...(i)

$$[R] = [M^1 L^2 T^{-3} A^{-2}],$$

$$[\varepsilon_0] = [M^{-1} L^{-3} T^4 A^2],$$

$$[\mu_0] = [M^1 L^1 T^{-2} A^{-2}]$$

Now, from Eq. (i), we get

$$[M^1 L^2 T^{-3} A^{-2}] = [M^{-1} L^{-3} T^4 A^2]^\alpha [M^1 L^1 T^{-2} A^{-2}]^\beta$$

$$[M^1 L^2 T^{-3} A^{-2}] = [M^{-\alpha+\beta} L^{-3\alpha+\beta} T^{4\alpha-2\beta} A^{2\alpha-2\beta}]$$

On comparing both sides, we get

$-\alpha + \beta = 1$ ...(ii)

$-3\alpha + \beta = 2$ ...(iii)

$4\alpha - 2\beta = -3$ ...(iv)

$2\alpha - 2\beta = -2$ ...(v)

Value of α and β can be found using any two Eqs. from (ii) to (v),

on subtracting Eq. (iii) from Eq. (ii), we get

$$(-\alpha + \beta) - (-3\alpha + \beta) = 1 - 2$$

$\Rightarrow$ $2\alpha = -1$

or $\alpha = \frac{-1}{2}$

Put the value of α in Eq. (ii), we get

$$\beta = +\frac{1}{2}$$

$\therefore$ $[R] = [\varepsilon_0]^{-1/2} [\mu_0]^{1/2} = \sqrt{\frac{\mu_0}{\varepsilon_0}}$

3. To convert a physical quantity from one system to the other

Let dimensional formula of a given physical quantity be $[M^aL^bT^c]$. If in a system having base units $[M_1L_1T_1]$ the numerical value of given quantity (Q) be n_1 and numerical value n_2 in another unit system having the base units $[M_2, L_2, T_2]$, then

$$Q = n_1u_1 = n_2u_2$$

$$n_1 [M_1^aL_1^bT_1^c] = n_2 [M_2^aL_2^bT_2^c]$$

$\Rightarrow$ $n_2 = n_1\left[\frac{M_1}{M_2}\right]^a\left[\frac{L_1}{L_2}\right]^b\left[\frac{T_1}{T_2}\right]^c$

Example 19. *The density of a material in SI units is 128 kg m^{-3}. In certain units in which the unit of length is 25 cm and the unit of mass is 50 g, the numerical value of density of the material is* **[JEE Main 2019]**

(a) 40 (b) 16
(c) 640 (d) 410

Sol. (a) To convert a measured value from one system to another system, we use

$$n_1 u_1 = n_2 u_2$$

where, n is numeric value and u is unit.

We get $$128 \frac{\text{kg}}{\text{m}^3} = n_2 \frac{50 \text{ g}}{(25 \text{ cm})^3}$$

$$\Rightarrow \frac{128 \times 1000 \text{ g}}{100 \times 100 \times 100 \text{ cm}^3} = \frac{n_2 \times 50 \text{ g}}{25 \times 25 \times 25 \text{ cm}^3}$$

$$\left[\because \text{Density} = \frac{\text{Mass}}{\text{Volume}}\right]$$

$$\Rightarrow n_2 = \frac{128 \times 1000 \times 25 \times 25 \times 25}{50 \times 100 \times 100 \times 100} = 40$$

Limitations of Theory of Dimensions

Although dimensional analysis is very useful but it is not universal, it has some limitations as given below

(i) This method gives no information about dimensional constants. Such as universal constant of gravitation (G) or Planck's constant (h).

(ii) Numerical constant (k), having no dimensions such as 3/4, e, 2π, etc., cannot be deduced by the method of dimensions.

(iii) This technique is useful only for deducing and verifying power relations. Relationship involving exponential, trigonometric functions, etc., cannot be obtained or studied by this method.

(iv) We cannot use this method to obtain the required relation, if the quantity of interest depends upon more parameters than the number of fundamental quantities used.

(v) Even if a physical quantity depends on three physical quantities, out of which two have same dimensions, the formula cannot be derived by theory of dimensions.

Practice Exercise

ROUND I Topically Divided Problems

Physical Quantities and Units

1. The SI unit of electrochemical equivalent is
(a) kg C (b) $C\ kg^{-1}$ (c) $kg\ C^{-1}$ (d) $kg^2\ C^{-1}$

2. SI unit of intensity of wave is
(a) $J\ m^{-2}s^{-1}$ (b) $J\ m^{-1}s^{-2}$ (c) $W\ m^{-2}$ (d) $J\ m^{-2}$

3. Which one of the following pairs of quantities and their unit is properly matched?
(a) Electric field - coulomb/m
(b) Magnetic flux - Weber/m^2
(c) Power - Farad
(d) Capacitance - Henry

4. Farad is not equivalent to
(q = coulomb, V = volt and J = joule)
(a) $\frac{q}{V}$ (b) qV^2 (c) $\frac{q^2}{J}$ (d) $\frac{J}{V^2}$

5. On which of the following factor universal time depends?
(a) Rotation of earth on its axis
(b) Oscillations of quartz crystal
(c) Vibrations of cesium atom
(d) Earth's orbital motion around the sun

6. A sextant is a double reflecting navigation instrument. Which of the following physical quantity, it can measure?
(a) Area of hill
(b) Angular distance between two visible objects
(c) Breadth of a tower
(d) Volume of the building

7. The 'rad' is the correct unit used to report the measurement of [AIEEE 2006]
(a) the ability of a beam of gamma ray photons to produce ions in a target
(b) the energy delivered by radiation to a target
(c) the biological effect of radiation
(d) the rate of decay of a radioactive source

8. A pressure of 10^6 dyne cm^{-2} is equivalent to
(a) $10^5\ Nm^{-2}$ (b) $10^4\ Nm^{-2}$ (c) $10^6\ Nm^{-2}$ (d) $10^7\ Nm^{-2}$

9. How many wavelengths of Kr^{86} are there in one metre?
(a) 1553164.13 (b) 1650763.73
(c) 652189.63 (d) 2348123.73

10. Young modulus of steel is $1.9 \times 10^{11}\ N/m^2$. When expressed in CGS units of dyne/cm^2, it will be equal to ($1\ N = 10^5$ dyne, $1\ m^2 = 10^4 cm^2$) [NCERT]
(a) 1.9×10^{10} (b) 1.9×10^{11}
(c) 1.9×10^{12} (d) 1.9×10^{13}

11. 1 light year is defined as the distance travelled by light in one year. The speed of light $3 \times 10^8\ ms^{-1}$. The same in metre is
(a) 3×10^{12}m (b) 9.461×10^{15} m
(c) 3×10^{15} m (d) None of these

12. 1 slug is equivalent to 14.6 kg. A force of 10 pound is applied on a body of 1 kg. The acceleration of the body is
(a) 43.8 ms^{-2} (b) 4.448 ms^{-2}
(c) 44.4 ms^{-2} (d) None of these

13. The time taken by an electron to go from ground state to excited state is one shake (1 shake = 10^{-8}s). This time in nanosecond will be
(a) 10 (b) 4 (c) 2 (d) 25

14. The concorde is the fastest airlines used for commercial service. It can cruise at 1450 mile per hour (about two times the speed of sound or in other words mach 2). What is it in m/s?
(a) 644.4 m/s (b) 80 m/s
(c) 40 m/s (d) None of these

15. The value of universal gas constant is $R = 8.3$ J/k-mol. The value of R in atmosphere litre per kelvin per mol
(a) 8.12 (b) 0.00812
(c) 81.2 (d) 0.0812

Least Count and Accuracy of Measuring Instruments

16. One main scale division of a vernier callipers is a cm and nth division of the vernier scale coincide with $(n-1)$th division of the main scale. The least count of the callipers (in mm) is [JEE Main 2021]
(a) $\frac{10na}{(n-1)}$ (b) $\frac{10a}{(n-1)}$
(c) $\left(\frac{n-1}{10n}\right)a$ (d) $\frac{10a}{n}$

17. A student measures the thickness of a human hair by looking at it through a microscope of magnification 100. He makes 20 observations and finds that the average width of the hair in the field of view of the microscope is 3.5 mm. The thickness of hair is **[NCERT]**

(a) 0.035 mm (b) 0.04 mm
(c) 0.35 mm (d) 0.40 mm

18. A vernier callipers has 1 mm marks on the main scale. It has 20 equal divisions on the vernier scale which match with 16 main scale divisions. For this vernier callipers, the least count is **[IIT JEE]**

(a) 0.02 mm (b) 0.05 mm
(c) 0.1 mm (d) 0.2 mm

19. A spectrometer gives the following reading when used to measure the angle of a prism.

Main scale reading : 58.5 degree.
Vernier scale reading : 9 division.

Given that, 1 division on main scale corresponds to 0.5 degree. Total division on the vernier scale is 30 and match with 29 divisions of the main scale. The angle of the prism from the above data is **[AIEEE 2012]**

(a) 58.59° (b) 59.77°
(c) 58.65° (d) 59°

20. A screw gauge gives the following reading when used to measure the diameter of a wire.

Main scale reading : 0 mm, Circular scale reading : 52 divisions

Given that, 1 mm on main scale corresponds to 100 divisions of the circular scale.

The diameter of wire from the above data is

(a) 0.052 cm (b) 0.026 cm
(c) 0.005 cm (d) 0.52 cm

21. Two full turns of the circular scale of a screw gauge cover a distance of 1 mm on its main scale. The total number of divisions on the circular scale is 50. Further, it is found that the screw gauge has a zero error of – 0.03 mm. While measuring the diameter of a thin wire, a student notes the main scale reading of 3 mm and the number of circular scale divisions in line with the main scale as 35. The diameter of the wire is

(a) 3.32 mm
(b) 3.73 mm
(c) 3.67 mm
(d) 3.38 mm

22. The least count of the main scale of a screw gauge is 1 mm. The minimum number of divisions on its circular scale required to measure 5 μm diameter of a wire is **[JEE Main 2019]**

(a) 50 (b) 200
(c) 500 (d) 100

23. A student measures the time period of 100 oscillations of a simple pendulum four times. The data set is 90s, 91s, 92s and 95s. If the minimum division in the measuring clock is 1s, then the reported mean time should be **[JEE Main 2016]**

(a) (92 ± 2) s (b) (92 ± 5) s
(c) (92 ± 1.8) s (d) (92 ± 3) s

24. Which of the following is the most precise device for measuring length? **[NCERT]**

(a) A vernier callipers with 20 divisions on the sliding scale.
(b) A screw gauge of pitch 1 mm and 100 divisions on the circular scale.
(c) An optical instrument that can measure length to within a wavelength of light.
(d) All are equally precise device for measuring length.

Errors and Significant Figures

25. The density of a cube is measured by measuring its mass and length of its sides. If the maximum errors in the measurement of mass and length are 3% and 2% respectively, then the maximum error in the measurement of density is

(a) 7% (b) 5%
(c) 1% (d) 9%

26. A physical quantity is represented by $X = M^a L^b T^{-c}$. If percentage errors in the measurements of M, L and T are α%, β% and γ% respectively, then total percentage error is

(a) $(\alpha a + \beta b - \gamma c)$ % (b) $(\alpha a + \beta b + \gamma c)$ %
(c) $(\alpha a - \beta b - \gamma c)$ % (d) 0%

27. If there is a positive error of 50% in the measurement of speed of a body, then the error in the measurement of kinetic energy is

(a) 25% (b) 50%
(c) 100 % (d) 125%

28. The radius of the sphere is (4.3 ± 0.1) cm. The percentage error in its volume is

(a) $\frac{0.1}{4.3} \times 100$ (b) $3 \times \frac{0.1 \times 100}{4.3}$
(c) $\frac{1}{3} \times \frac{0.1 \times 100}{4.3}$ (d) $3 + \frac{0.1 \times 100}{4.3}$

29. The time period of a simple pendulum is given by $T = 2\pi\sqrt{\frac{l}{g}}$. The measured value of the length of pendulum is 10 cm known to a 1 mm accuracy. The time for 200 oscillations of the pendulum is found to be 100 s using a clock of 1 s resolution. The percentage accuracy in the determination of g using this pendulum is x. The value of x to the nearest integer is **[JEE Main 2021]**

(a) 2% (b) 3% (c) 5% (d) 4%

30. The velocity of transverse wave in a string is $v = \sqrt{\frac{T}{M}}$ where T is the tension in the string and M is mass per unit length. If $T = 3.0$ kgf, mass of string is 2.5 g and length of string is 1.000 m, then the percentage error in the measurement of velocity is
(a) 0.5 (b) 0.7 (c) 2.3 (d) 3.6

31. The initial temperature of a liquid is $(80.0 \pm 0.1)°$ C. After it has been cooled, its temperature is $(10.0 \pm 0.1)°$ C. The fall in temperature in degree centigrade is
(a) 70.0 (b) 70.0 ± 0.3
(c) 70.0 ± 0.2 (d) 70.0 ± 0.1

32. A public park, in the form of a square, has an area of $(100 \pm 0.2)\,\text{m}^2$. The side of park is
(a) (10 ± 0.01)m (b) (10 ± 0.1)m
(c) (10.0 ± 0.1)m (d) (10.0 ± 0.2)m

33. Given, potential difference $V = (8 \pm 0.5)$ V and current $I = (2 \pm 0.2)$A. The value of resistance R (in Ω) is
(a) $4 \pm 16.25\%$ (b) $4 \pm 6.25\%$
(c) $4 \pm 10\%$ (d) $4 \pm 8\%$

34. The length, breadth and thickness of a block are measured to be 50 cm, 2.0 cm and 1.00 cm. The percentage error in the measurement of volume is
(a) 0.8% (b) 8%
(c) 10% (d) 12.5 %

35. One side of a cubical block is measured with the help of a vernier callipers of vernier constant 0.01 cm. This side comes out to be 1.23 cm. What is the percentage error in the measurement of area?
(a) $\frac{1.23}{0.01} \times 100$ (b) $\frac{0.01}{1.23} \times 100$
(c) $2 \times \frac{0.01}{1.23} \times 100$ (d) $3 \times \frac{0.01}{1.23} \times 100$

36. A physical quantity z depends on four observables a, b, c and d, as $z = \frac{a^2 b^{2/3}}{\sqrt{c} d^3}$. The percentages of error in the measurement of a, b, c and d are 2%, 1.5%, 4% and 2.5%, respectively. The percentage of error in z is **[JEE Main 2020]**
(a) 13.5% (b) 16.5%
(c) 14.5% (d) 12.25%

37. The relative density of the material of a body is the ratio of its weight in air and the loss of its weight in water. By using a spring balance, the weight of the body in air is measured to be (5.00 ± 0.05) N. The weight of the body in water is measured to be (4.00 ± 0.05) N, then the maximum possible percentage error in relative density is
(a) 11% (b) 10% (c) 9% (d) 7%

38. The diameter and height of a cylinder are measured by a meter scale to be 12.6 ± 0.1 cm and 34.2 ± 0.1 cm, respectively. What will be the value of its volume in appropriate significant figures? **[JEE Main 2019]**
(a) $4300 \pm 80\ \text{cm}^3$
(b) $4260 \pm 80\ \text{cm}^3$
(c) $4264.4 \pm 81.0\ \text{cm}^3$
(d) $4264 \pm 81\ \text{cm}^3$

39. The current voltage relation of diode is given by $I = (e^{1000V/T} - 1)$ mA, where the applied voltage V is in volt and the temperature T is in kelvin. If a student makes an error measuring ± 0.01V while measuring the current of 5 mA at 300K, what will be the error in the value of current (in mA)? **[JEE Main 2013]**
(a) 0.2 mA (b) 0.02 mA
(c) 0.5 mA (d) 0.05 mA

40. The result after adding 3.8×10^{-6} to 4.2×10^{-5} with due regard to significant figures is
(a) 4.58×10^{-5} (b) 0.458×10^{-4}
(c) 4.6×10^{-5} (d) 45.8×10^{-6}

41. The value of π^2 with due regard for significant figures is (Given $\pi = 3.14$)
(a) 9.86 (b) 9.859
(c) 9.8596 (d) 9.85960

42. You measure two quantities as $A = 1.0\ \text{m} \pm 0.2\ \text{m}, B = 2.0\ \text{m} \pm 0.2\ \text{m}$. We should report correct value for $\sqrt{AB}$ as **[NCERT Exemplar]**
(a) $1.4\ \text{m} \pm 0.4\ \text{m}$ (b) $1.41\ \text{m} \pm 0.15\ \text{m}$
(c) $1.4\ \text{m} \pm 0.3\ \text{m}$ (d) $1.4\ \text{m} \pm 0.2\ \text{m}$

43. The area of a square is $5.29\ \text{cm}^2$. The area of 7 such squares taking into account the significant figures is **[JEE Main 2019]**
(a) $37.030\ \text{cm}^2$ (b) $37.0\ \text{cm}^2$
(c) $37.03\ \text{cm}^2$ (d) $37\ \text{cm}^2$

44. For the four sets of three measured physical quantities as given below. Which of the following options is correct? **[JEE Main 2020]**
(i) $A_1 = 24.36, B_1 = 0.0724, C_1 = 256.2$
(ii) $A_2 = 24.44, B_2 = 16.082, C_2 = 240.2$
(iii) $A_3 = 25.2, B_3 = 19.2812, C_3 = 236.183$
(iv) $A_4 = 25, B_4 = 236.191, C_4 = 19.5$
(a) $A_1 + B_1 + C_1 < A_3 + B_3 + C_3 < A_2 + B_2 + C_2 < A_4 + B_4 + C_4$
(b) $A_4 + B_4 + C_4 < A_1 + B_1 + C_1 = A_2 + B_2 + C_2 = A_3 + B_3 + C_3$
(c) $A_4 + B_4 + C_4 < A_1 + B_1 + C_1 = A_3 + B_3 + C_3 < A_2 + B_2 + C_2$
(d) $A_1 + B_1 + C_1 = A_2 + B_2 + C_2 = A_3 + B_3 + C_3 = A_4 + B_4 + C_4$

Dimensions

45. The damping force of an oscillating particle is observed to be proportional to velocity. The constant of proportionality can be measured in

(a) $kg\ s^{-1}$ (b) kg s
(c) $kg\ ms^{-1}$ (d) $kg\ m^{-1}\ s^{-1}$

46. The fundamental unit, which has the same power in the dimensional formulae of surface tension and viscosity is

(a) mass (b) length
(c) time (d) None of these

47. If the units of M and L are increased three times, then the unit of energy will be increased by

(a) 3 times (b) 6 times
(c) 27 times (d) 81 times

48. If L denotes the inductance of an inductor through which a current I is flowing, then the dimensional formula of LI^2 is

(a) $[MLT^{-2}]$
(b) $[ML^2T^{-2}]$
(c) $[M^2L^2T^{-2}]$
(d) not expressible in terms of M, L, T

49. The equation of alternating current is $I = I_0e^{-t/CR}$ where t is time, C is capacitance and R is resistance of coil, then the dimensions of CR is

(a) $[MLT^{-1}]$ (b) $[M^0LT]$
(c) $[M^0L^0T]$ (d) None of these

50. Which of the following pairs has same dimensions?

(a) Current density and charge density
(b) Angular momentum and momentum
(c) Spring constant and surface energy
(d) Force and torque

51. Dimensions of $1/\mu_0\varepsilon_0$, where symbols have their usual meaning, are **[AIEEE 2003]**

(a) $[L^{-1}T]$ (b) $[L^2T^2]$ (c) $[L^2T^{-2}]$ (d) $[LT^{-1}]$

52. The physical quantities not having same dimensions are **[AIEEE 2003]**

(a) torque and work
(b) momentum and Planck's constant
(c) stress and Young's modulus
(d) speed and $(\mu_0\varepsilon_0)^{-1/2}$

53. The dimensions of magnetic field in M, L, T and C is given as **[AIEEE 2008]**

(a) $[MLT^{-1}C^{-1}]$
(b) $[MT^2C^{-2}]$
(c) $[MT^{-1}C^{-1}]$
(d) $[MT^{-2}C^{-1}]$

54. The time dependence of a physical quantity P is given by $P = P_0e^{-\alpha t^2}$, where α is a constant and t is time. Then constant α is

(a) dimensionless (b) dimension of t^{-2}
(c) dimensions of P (d) dimension of t^2

55. The dimensional formula of magnetic permeability is

(a) $[M^0L^{-1}T]$ (b) $[M^0L^2T^{-1}]$
(c) $[M^0L^2T^{-1}A^2]$ (d) $[MLT^{-2}A^{-2}]$

56. $[ML^{-2}T^{-2}]$ represents dimensional formula of which of the following physical quantities?

(a) Energy (b) Pressure
(c) Torque (d) Pressure gradient

57. Which of the following units denotes the dimensions $[ML^2/Q^2]$, where Q denotes the electric charge? **[AIEEE 2006]**

(a) Wb/m^2 (b) henry (H)
(c) H/m^2 (d) weber (Wb)

58. Amount of solar energy received on the earth's surface per unit area per unit time is defined as solar constant. Dimensional formula of solar constant is **[JEE Main 2020]**

(a) $[MLT^{-2}]$ (b) $[ML^0T^{-3}]$
(c) $[M^2L^0T^{-1}]$ (d) $[ML^2T^{-2}]$

59. The dimension of $\frac{B^2}{2\mu_0}$, where B is magnetic field and μ_0 is the magnetic permeability of vacuum, is **[JEE Main 2020]**

(a) $[ML^{-1}T^{-2}]$ (b) $[MLT^{-2}]$
(c) $[ML^2T^{-1}]$ (d) $[ML^2T^{-2}]$

60. In SI units, the dimensions of $\sqrt{\frac{\varepsilon_0}{\mu_0}}$ is **[JEE Main 2019]**

(a) $[A^{-1}TML^3]$ (b) $[AT^2M^{-1}L^{-1}]$
(c) $[AT^{-3}ML^{3/2}]$ (d) $[A^2T^3M^{-1}L^{-2}]$

61. In the relation $y = r\sin(\omega t - kx)$, the dimensional formula of ω/k are

(a) $[M^0L^0T^0]$ (b) $[M^0L^1T^{-1}]$
(c) $[M^0L^0T^1]$ (d) $[M^0L^1T^0]$

62. Let $[\varepsilon_0]$ denotes the dimensional formula of the permittivity of vacuum. If M = mass, L = length, T = time and A = electric current, then **[JEE Main 2013]**

(a) $[\varepsilon_0] = [M^{-1}L^{-3}T^2A]$ (b) $[\varepsilon_0] = [M^{-1}L^{-3}T^4A^2]$
(c) $[\varepsilon_0] = [M^{-2}L^2T^{-1}A^{-2}]$ (d) $[\varepsilon_0] = [M^{-1}L^2T^{-1}A^2]$

63. Let l, r, c and v represent inductance, resistance, capacitance and voltage, respectively. The dimension of $\frac{l}{rcv}$ in SI units will be **[JEE Main 2019]**

(a) $[LT^2]$ (b) $[LTA]$ (c) $[A^{-1}]$ (d) $[LA^{-2}]$

64. $\int \frac{dx}{\sqrt{2ax - x^2}} = a^n \sin^{-1}\left(\frac{x}{a} - 1\right)$. The value of n is

(a) 0 (b) -1
(c) 1 (d) None of these

(You may use dimensional analysis to solve the problem.)

65. The period of a body under SHM is represented by $T = p^a D^b S^c$, where p is pressure, D is density and S is surface tension. The value of a, b and c are

(a) $-\frac{3}{2}, \frac{1}{2}, 1$ (b) $-1, -2, 3$
(c) $\frac{1}{2}, \frac{-3}{2}, \frac{-1}{2}$ (d) $1, 2, \frac{1}{3}$

66. Taking frequency f, velocity v and density ρ to be the fundamental quantities, then the dimensional formula for momentum will be

(a) $[\rho v^4 f^{-3}]$ (b) $[\rho v^3 f^{-1}]$
(c) $[\rho v f^2]$ (d) $[\rho^2 v^2 f^2]$

67. If p represents radiation pressure, v represents speed of light and q represents radiation energy striking a unit area per second, then non-zero integers a, b and c are such that $p^a q^b v^c$ is dimensionless, then

(a) $a = 1$, $b = 1$, $c = -1$
(b) $a = 1$, $b = -1$, $c = 1$
(c) $a = -1$, $b = 1$, $c = 1$
(d) $a = 1$, $b = 1$, $c = 1$

68. The wavelength associated with a moving particle depends upon power p of its mass m, qth power of its velocity v and rth power of Planck's constant h. Then the correct set of values of p, q and r is

(a) $p = 1$, $q = -1$, $r = 1$
(b) $p = 1$, $q = 1$, $r = 1$
(c) $p = -1$, $q = -1$, $r = -1$
(d) $p = -1$, $q = -1$, $r = 1$

69. If speed V, area A and force F are chosen as fundamental units, then the dimensional formula of Young's modulus will be **[JEE Main 2020]**

(a) $[FA^2V^{-3}]$ (b) $[FA^{-1}V^0]$
(c) $[FA^2 V^{-2}]$ (d) $[FA^2V^{-1}]$

70. If momentum p, area A and time T are taken to be the fundamental quantities, then the dimensional formula for energy is **[JEE Main 2020]**

(a) $[p^2AT^{-2}]$ (b) $[pA^{-1}T^{-2}]$
(c) $[pA^{1/2}T^{-1}]$ (d) $[p^{1/2}AT^{-1}]$

71. If speed (v), acceleration (A) and force (F) are considered as fundamental units, the dimension of Young's modulus will be **[JEE Main 2019]**

(a) $[v^{-4}A^{-2}F]$ (b) $[v^{-2}A^2 F^2]$
(c) $[v^{-2}A^2 F^{-2}]$ (d) $[v^{-4}A^2 F]$

72. The dimension of stopping potential V_0 in photoelectric effect in units of Planck's constant h, speed of light c, gravitational constant G and ampere A is **[JEE Main 2020]**

(a) $h^{-2/3}c^{-1/3}G^{4/3} A^{-1}$ (b) $h^{1/3}G^{2/3}c^{1/3} A^{-1}$
(c) $h^0c^5G^{-1} A^{-1}$ (d) $h^{1/3}G^{2/3}c^{1/3} A^{-1}$

73. If surface tension (S), moment of inertia (I) and Planck's constant (h), were to be taken as the fundamental units, the dimensional formula for linear momentum would be **[JEE Main 2019]**

(a) $S^{1/2}I^{1/2}h^{-1}$ (b) $S^{3/2}I^{1/2}h^0$
(c) $S^{1/2}I^{1/2}h^0$ (d) $S^{1/2}I^{3/2}h^{-1}$

ROUND II Mixed Bag

Only One Correct Option

1. The vernier scale used for measurement has a positive zero error of 0.2 mm. If while taking a measurement it was noted that 0 on the vernier scale lies between 8.5 cm and 8.6 cm, vernier coincidence is 6, then the correct value of measurement is .... cm. (Given, least count = 0.01 cm) **[JEE Main 2021]**

(a) 8.36 cm
(b) 8.54 cm
(c) 8.58 cm
(d) 8.56 cm

2. If 1 g cm s^{-1} = x N-s, then the number x is equal to

(a) 1×10^{-3} (b) 3.6×10^{-3}
(c) 1×10^{-5} (d) 6×10^{-4}

3. If muscle times speed equals power, what is the ratio of the SI units and the CGS unit of muscle?

(a) 10^5 (b) 10^3
(c) 10^7 (d) 10^{-5}

4. The pitch of a screw gauge 15 mm and there are 100 divisions on the circular scale. While measuring diameter of a thick wire. The pitch scale reads 1 mm and 63 rad division on the circular scale coincides with the reference. The length of the wire is 5.6 cm, then which one of the following option is correct?

(a) The least count of screw gauge is 0.002 cm
(b) The volume of the wire is 0.117 cm^3
(c) The diameter of the wire is 1.33 m
(d) The cross-section area of the wire is 0.0209 cm^3

5. Two full turns of the circular scale of a screw gauge cover a distance of 1 mm on its main scale. The total number of divisions on the circular scale is 50. Further, it is found that the screw gauge has a zero error of – 0.03 mm. While measuring the diameter of a thin wire, a student notes the main scale reading of 3 mm and the number of circular scale divisions in line with the main scale as 35. The diameter of the wire is **[AIEEE 2008]**

(a) 3.32 mm (b) 3.37 mm
(c) 3.67 mm (d) 3.38 mm

6. The dimensions of a rectangular block measured with callipers having least count of 0.01 cm are 5 mm × 10 mm × 5 mm. The maximum percentage error in the measurement of the volume of the block is
(a) 5% (b) 10 %
(c) 15 % (d) 20%

7. A resistor of 10 kΩ having tolerance 10% is connected in series with another resistor of 20 kΩ having tolerance 20%. The tolerance of the combination will be approximately
(a) 10% (b) 13%
(c) 17% (d) 20%

8. The following observations were taken for determining surface tension of water by capillary tube method. Diameter of capillary, $D = 1.25\times10^{-2}$ m and rise of water in capillary, $h = 1.46 \times 10^{-2}$ m. Taking $g = 9.80\ \text{ms}^{-2}$ and using the relation $T = (rgh/2)\times10^3\ \text{Nm}^{-1}$, what is the possible error in surface tension T?
(a) 2.4% (b) 15 %
(c) 1.6% (d) 0.15%

9. Resistance of a given wire is obtained by measuring the current flowing in it and the voltage difference applied across it. If the percentage errors in the measurement of the current and the voltage difference are 3% each, then error in the value of resistance of the wire is **[AIEEE 2012]**
(a) 6% (b) zero (c) 1% (d) 3%

10. A wire has a mass (0.3 ± 0.003) g, radius (0.5 + 0.005) mm and length (6 ± 0.06) cm. The maximum percentage error in the measurement of its density is **[IIT JEE 2004]**
(a) 1 (b) 2 (c) 3 (d) 4

11. In an experiment to measure the height of a bridge by dropping stone into water underneath, if the error in measurement of time is 0.1s at the end of 2 s, then the error in estimation of height of bridge will be **[Kerala CEE 2004]**
(a) 0.49 m (b) 0.98 m (c) 1.96 m (d) 2.12 m

12. A cube has a side of length 1.2×10^{-2} m. Calculate its volume. **[IIT JEE 2003]**
(a) $1.7 \times 10^{-6}\ \text{m}^3$ (b) $1.73 \times 10^{-6}\ \text{m}^3$
(c) $1.70 \times 10^{-6}\ \text{m}^3$ (d) $1.732 \times 10^{-6}\ \text{m}^3$

13. In the experiment of Ohm's law, a potential difference of 5.0 V is applied across the end of a conductor of length 10.0 cm and diameter of 5.00 mm. The measured current in the conductor is 2.00 A. The maximum permissible percentage error in the resistivity of the conductor is **[JEE Main 2021]**
(a) 3.9 (b) 8.4 (c) 7.5 (d) 3.0

14. A resistor of 4 kΩ with tolerance 10% is connected in parallel with a resistor of 6 kΩ with tolerance 100%. The tolerance of the parallel combination is nearly
(a) 10 % (b) 20 %
(c) 30 % (d) 40 %

15. What is the unit of k in the relation where, $U = \dfrac{ky}{y^2 + a^2}$ where U represents the potential energy, y represents the displacement and a represents amplitude?
(a) m s^{-1} (b) m s
(c) J m (d) J s^{-1}

16. In the equation $X = 3\,YZ^2$, X and Z have dimensions of capacitance and magnetic induction respectively. In MKSQ system, the dimensional formula of Y is
(a) $[M^{-3}L^{-2}T^{-2}Q^{-4}]$ (b) $[ML^{-2}]$
(c) $[M^{-3}L^{-2}Q^4T^8]$ (d) $[M^{-3}L^{-2}Q^4T^4]$

17. The dimensional formula of $\dfrac{1}{\varepsilon_0}\dfrac{e^2}{hc}$ is
(a) $[M^0L^0T^0A^0]$ (b) $[M^{-1}L^3T^2A]$
(c) $[ML^3T^{-4}A^{-2}]$ (d) $[M^{-1}L^{-3}T^4]$

18. Given $X = (Gh/c^3)^{1/2}$, where G, h and c are gravitational constant, Planck's constant and the velocity of light respectively. Dimensions of X are the same as those of
(a) mass (b) time
(c) length (d) acceleration

19. The thrust developed by a rocket-motor is given by $F = mv + A(p_1 - p_2)$, where m is the mass of the gas ejected per unit time, v is velocity of the gas, A is area of cross-section of the nozzle, p_1, p_2 are the pressures of the exhaust gas and surrounding atmosphere. The formula is dimensionally
(a) correct
(b) wrong
(c) sometimes wrong, sometimes correct
(d) data is not adequate

20. When a wave traverses a medium the displacement of a particle located at x at a time t is given by $y = a\sin(bt - cx)$, where a, b and c are constants of the wave. Which of the following is not dimensionless?
(a) $\dfrac{y}{a}$ (b) bt
(c) cx (d) $\dfrac{b}{c}$

21. In a new system of units, unit of mass is 10 kg, unit of length is 1 km and unit of time is 1 min. The value of 1 J in this new hypothetical system is
(a) 3.6×10^{-4} new units (b) 6×10^7 new units
(c) 10^{11} new units (d) 1.67×10^4 new units

22. The number of particles given by $n = D\dfrac{n_2 - n_1}{x_2 - x_1}$ are crossing a unit area perpendicular to X-axis in unit time, where n_1 and n_2 are the number of particles per unit volume for the values x_1 and x_2 of x respectively. Then the dimensional formula of diffusion constant D is

(a) $[M^0LT^0]$ (b) $[M^0L^2T^{-4}]$
(c) $[M^0LT^{-3}]$ (d) $[M^0L^2T^{-1}]$

23. A calorie is a unit of heat and equals 4.2 J. Suppose we employ a system of units in which the unit of mass is α kg, the unit of length is β metre and the unit of time is γ s. In this new system, 1 calorie =

(a) $\alpha^{-1}\beta^{-2}\gamma^2$ (b) $4.2\,\alpha\beta^2\gamma^{-2}$
(c) $\alpha\beta^2\gamma^2$ (d) $4.2\,\alpha^{-1}\beta^{-2}\gamma^2$

24. An important milestone in the evolution of the universe just after the Big Bang is the Planck time t_p, the value of which depends on three fundamental constants speed c of light in vacuum, gravitational constant G and Planck's constant h. Then, $t_p \propto$

(a) Ghc^5 (b) $\dfrac{c^5}{Gh}$
(c) $\dfrac{Gh}{c^5}$ (d) $\left(\dfrac{Gh}{c^5}\right)^{1/2}$

25. Which of the following combinations have the dimensions of time? L-C-R represent inductance, capacitance and resistance, respectively.

(a) RC (b) LC (c) R/C (d) C/L

26. Photon is quantum of radiation with energy $E = h\nu$ where ν is frequency and h is Planck's constant. The dimensions of h are the same as that of [NCERT Exemplar]

(a) linear impulse (b) angular impulse
(c) linear momentum (d) None of these

27. In order to determine the Young's modulus of a wire of radius 0.2 cm (measured using a scale of least count = 0.001 cm) and length 1 m (measured using a scale of least count = 1 mm), a weight of mass 1 kg (measured using a scale of least count = 1 g) was hanged to get the elongation of 0.5 cm (measured using a scale of least count = 0.001 cm). What will be the fractional error in the value of Young's modulus determined by this experiment? [JEE Main 2021]

(a) 0.14% (b) 0.9%
(c) 9% (d) 1.4%

28. What is dimensional formula of thermal conductivity?

(a) $[MLT^{-1}\theta^{-1}]$ (b) $[MLT^{-3}\theta^{-1}]$
(c) $[M^2LT^{-3}\theta^{-2}]$ (d) $[ML^2T^{-2}\theta]$

29. A quantity x is given by $x = \dfrac{IFv^2}{WL^4}$ where, I is moment of inertia, F is force, v is work and L is length. The dimensional formula for x is same as that of [JEE Main 2020]

(a) Planck's constant
(b) force constant
(c) coefficient of viscosity
(d) energy density

30. The quantities $x = \dfrac{1}{\sqrt{\mu_0\varepsilon_0}}$, $y = \dfrac{E}{B}$ and $z = \dfrac{l}{CR}$ are defined, where C is capacitance, R is resistance, l is length, E is electric field, B is magnetic field, ε_0 is free space permittivity and μ_0 is permeability, respectively. Then, [JEE Main 2020]

(a) x, y and z have the same dimension
(b) Only x and z have the same dimension
(c) Only x and y have the same dimension
(d) Only y and z have the same dimension

31. Dimensional formula for thermal conductivity is (Here, K denotes the temperature) [JEE Main 2020]

(a) $[MLT^{-2}K]$ (b) $[MLT^{-2}K^{-2}]$
(c) $[MLT^{-3}K^{-1}]$ (d) $[MLT^{-3}K]$

32. A quantity f is given by $f = \sqrt{\dfrac{hc^5}{G}}$, where c is speed of light, G universal gravitational constant and h is the Planck's constant. Dimension of f is that of [JEE Main 2020]

(a) area (b) volume
(c) momentum (d) energy

33. The force of interaction between two atoms is given by $F = \alpha\beta\exp\left(-\dfrac{x^2}{\alpha kT}\right)$; where x is the distance, k is the Boltzmann constant and T is temperature and α and β are two constants. The dimension of β is [JEE Main 2019]

(a) $[MLT^{-2}]$ (b) $[M^0L^2T^{-4}]$
(c) $[M^2LT^{-4}]$ (d) $[M^2L^2T^{-2}]$

34. In the formula $X = 5YZ^2$, X and Z have dimensions of capacitance and magnetic field, respectively. What are the dimensions of Y in SI units? [JEE Main 2019]

(a) $[M^{-1}L^{-2}T^4A^{-2}]$ (b) $[M^{-2}L^0T^{-4}A^{-2}]$
(c) $[M^{-3}L^{-2}T^8A^4]$ (d) $[M^{-2}L^{-2}T^6A^3]$

Numerical Value Questions

35. If the unit of velocity is run, the unit of time is second and unit of force is strength in a hypothetical system of unit. In this system of unit, the unit of mass is $(\text{strength})^x\,(\text{second})^y\,(\text{run})^z$. The value of $\dfrac{y}{x}$ is ....

36. The edge of a cube is measured using a vernier calliper (9 divisions of the main scale is equal to 10 divisions of vernier scale and 1 main scale division is 1 mm). The main scale division reading is 10 and 1 division of vernier scale was found to be coinciding with the main scale. The mass of the cube is 2.736 g. The density (in gcm^{-3}) upto correct significant figures is .... **[IIT JEE 2005]**

37. The density of a solid metal sphere is determined by measuring its mass and its diameter. The maximum error in the density of the sphere is $\left(\frac{x}{100}\right)$%. If the relative errors in measuring the mass and the diameter are 6.0% and 1.5% respectively, the value of x is .... **[JEE Main 2020]**

38. The focal length of a spherical mirror is given by $\frac{1}{f}=\frac{1}{v}+\frac{1}{u}$, where u is the position of object from pole of the mirror and v is position of image from pole of the mirror.

If $u = (30 \pm 0.3)$ cm and $v = (60 \pm 0.6)$ cm. The maximum percentage error in the measurement of focal length of the mirror is n%, then the value of n is .... **[JEE Main 2021]**

39. If the speed of light (c), acceleration due to gravity (g) and pressure (p) are taken as base units. The dimension of g in the dimensional formula of universal gravitational constant is ....

Answers

Round I

1. (c)	2. (a)	3. (b)	4. (b)	5. (a)	6. (b)	7. (c)	8. (a)	9. (b)	10. (c)
11. (b)	12. (a)	13. (a)	14. (a)	15. (d)	16. (d)	17. (a)	18. (d)	19. (c)	20. (a)
21. (d)	22. (b)	23. (a)	24. (c)	25. (d)	26. (b)	27. (d)	28. (b)	29. (b)	30. (d)
31. (c)	32. (a)	33. (a)	34. (b)	35. (c)	36. (c)	37. (a)	38. (b)	39. (a)	40. (c)
41. (a)	42. (d)	43. (c)	44. (*)	45. (a)	46. (a)	47. (c)	48. (b)	49. (c)	50. (c)
51. (c)	52. (b)	53. (c)	54. (b)	55. (d)	56. (d)	57. (b)	58. (b)	59. (a)	60. (d)
61. (b)	62. (b)	63. (c)	64. (a)	65. (a)	66. (a)	67. (b)	68. (d)	69. (b)	70. (c)
71. (d)	72. (c)	73. (c)							

Round II

1. (b)	2. (c)	3. (a)	4. (b)	5. (d)	6. (a)	7. (c)	8. (c)	9. (a)	10. (d)
11. (c)	12. (a)	13. (a)	14. (c)	15. (c)	16. (d)	17. (a)	18. (c)	19. (a)	20. (d)
21. (a)	22. (d)	23. (d)	24. (d)	25. (a)	26. (b)	27. (d)	28. (b)	29. (d)	30. (a)
31. (c)	32. (d)	33. (c)	34. (c)	35. 1	36. 2.66	37. 1050	38. 1	39. 2	

Solutions

Round I

1. According to Faraday's first law of electrolysis, $m = Zq$ or $Z = \frac{m}{q}$. So, SI unit of Z is kg C^{-1}.

2. Intensity $(I) = \frac{\text{Energy}}{\text{Area} \times \text{Time}} = \frac{\text{J}}{\text{m}^2\text{ s}}$ or $\text{Jm}^{-2}\text{s}^{-1}$

3. Magnetic flux has the unit as weber/ m^2.

4. Capacitance, $C = \frac{\text{Charge}}{\text{Potential}} = \frac{q}{V}$

Also, potential $= \frac{\text{work}}{\text{charge}}$ $\qquad \left(\because V = \frac{W}{q}\right)$

$\therefore \qquad C = \frac{q^2}{J}$ as well as $C = \frac{J}{V^2}$

Thus, (a), (c) and (d) are equivalent to farad but (b) is not equivalent to Farad.

5. Time defined in terms of rotation of the earth is called Universal Time (UT).

6. The primary use of a sextant is to measure the angle between an astronomical object and the horizon.

7. 'rad' is used to measure biological effect of radiation.

8. 1 Newton $= 10^5$ dyne and 1 m = 100 cm

$10^6 \text{ dyne cm}^{-2} = 10^6 \times 10^{-5} \text{ N} \times (10^{-2}\text{ m})^{-2} = 10^5 \text{ Nm}^{-2}$

9. According to definition, metre is the distance containing 1650763.73 wavelength in vacuum of radiation corresponding to orange red light emitted by an atom of Kr-86.

10. Young modulus, $Y = 1.9 \times 10^{11} \text{ N/m}^2$

$\because \qquad 1 \text{ N} = 10^5 \text{ dyne}, 1 \text{ m}^2 = 10^4 \text{ cm}^2$

So, $\quad Y \text{ (in CGS)} = \frac{1.9 \times 10^{11} \times 10^5}{10^4} \text{ dyne/cm}^2$

$= 1.9 \times 10^{12} \text{ dyne/cm}^2$

11. One light year $= 3 \times 10^8 \text{m/s} \times 1 \text{ yr}$

$= \frac{3 \times 10^8 \text{ m}}{\text{s}} \times 365 \times 24 \times 60 \times 60 \text{s}$

$= 3 \times 10^8 \times 365 \times 24 \times 60 \times 60 \text{ m}$

$= 9.461 \times 10^{15} \text{ m}$

12. Force, $F = ma$

$\therefore \qquad a = \frac{F}{m} = \frac{10 \text{ pound}}{1 \text{ kg}}$

$= 10 \frac{\text{pound}}{\text{kg}} = \frac{10 \text{ slug ft}}{\text{kg s}^2}$

$= 10 \times 14.6 \text{ kg} \frac{\text{ft}}{\text{kg s}^2} = 146 \frac{\text{ft}}{\text{s}^2}$

$= 146 \times 0.30 \text{ ms}^{-2}$

$= 43.8 \text{ ms}^{-2}$

13. As, $\qquad n_1 u_1 = n_2 u_2$

$\Rightarrow \qquad n_2 = \frac{1 \text{ shake}}{1 \text{ ns}} = \frac{10^{-8} \text{ s}}{10^{-9} \text{ s}}$

$\therefore \qquad n_2 = 10$

14. As, $\qquad n_1 u_1 = n_2 u_2$

(n_1 and n_2 are numerical values and u_1 and u_2 are the unit in proper system)

$\therefore \qquad n_2 = \frac{n_1 u_1}{u_2}$

$= \frac{1450 \text{ mile/h}}{\text{m/s}} = \frac{1450 \text{ s/mile}}{\text{mh}}$

$= \frac{1450 \text{ s} \times 1.6 \text{ km}}{10^{-3} \text{ km} \times 60 \times 60 \text{ s}} = 644.4$

1450 mile/h = 644.4 m/s

15. As, $R = 8.3$ J/K-mol

Now, $\qquad n_1 u_1 = n_2 u_2$

(u_1 and u_2 are units while n_1 and n_2 are the numerical values)

$\therefore \qquad n_2 = \frac{n_1 u_1}{u_2} = \frac{8.3 \text{ J/K-mol}}{\text{atm L/K-mol}}$

$= \frac{8.3 \text{ J/K-mol}}{(1.013 \times 10^5 \text{ N/m}^2)(10^{-3}\text{m}^3)/\text{K-mol}}$

$= \frac{8.12}{10^2} = 0.0812$

$\therefore \quad$ 8.3 J/K mol = 0.0812 atm L/K mol

16. $na' = (n-1)a$

$\Rightarrow a' = \left(\frac{n-1}{n}\right)a$

$\therefore$ Least Count (LC) = 1MSD − 1VSD $= a - a'$

$= a - \left(\frac{n-1}{n}\right)a = \frac{a}{n} \text{ cm} = \frac{10a}{n} \text{ mm}$

17. Magnification of microscope = 100

Observed width of the hair = 3.5 mm

Magnification $= \frac{\text{Observed width}}{\text{Real width}}$

Real width $= \frac{\text{Observed width}}{\text{Magnification width}} = \frac{3.5}{100} = 0.035 \text{ mm}$

18. Least count of vernier callipers

LC = 1 MSD − 1 VSD

$= \frac{\text{Smallest division on main scale}}{\text{Number of divisions on vernier scale}}$

As, 20 divisions of vernier scale = 16 divisions of main scale

$\therefore \quad 1 \text{ VSD} = \frac{16}{20} \text{ mm} = 0.8 \text{ mm}$

$\text{LC} = 1 \text{ MSD} - 1 \text{ VSD}$

$= 1 \text{ mm} - 0.8 \text{ mm} = 0.2 \text{ mm}$

19. Here, least count $= \frac{0.5}{30}$ degree

Now, total reading = main scale reading + (vernier scale reading) × (least count)

$= 58.5 \text{ degrees} + (9)\left(\frac{0.5}{30}\right) \text{ degrees}$

$= \left(58.5 + \frac{1.5}{10}\right)$

$= 58.5 + 0.15$

$= 58.65$ degree

$= 58.65°$

20. Diameter of wire, $d = \text{MSR} + \text{CSR} \times \text{LC}$

$= 0 + 52 \times \frac{1}{100}$

$= 0.52 \text{ mm} = 0.052 \text{ cm}$

21. The diameter = main scale reading + circular scale reading × LC − zero error

$= 3 + 35 \times \frac{1}{2 \times 50} - (-0.03)$

$= 3.38$ mm

22. In a screw gauge,

Least count

$= \frac{\text{Measure of 1 main scale division (MSD)}}{\text{Number of division on circular scale}}$

Here, minimum value to be measured/least count is

$5 \,\mu\text{m} = 5 \times 10^{-6}$ m

∴ According to the given values,

$5 \times 10^{-6} = \frac{1 \times 10^{-3}}{N}$

or $\quad N = \frac{10^{-3}}{5 \times 10^{-6}} = \frac{1000}{5} = 200$ divisions

23. Arithmetic mean time of a oscillating simple pendulum

$= \frac{\Sigma x_i}{N} = \frac{90 + 91 + 92 + 95}{4} = 92 \text{ s}$

Mean deviation of a simple pendulum

$= \frac{\Sigma |\bar{x} - x_i|}{N}$

$= \frac{2 + 1 + 3 + 0}{4} = 1.5$

Given, minimum division in the measuring clock, i.e. simple pendulum = 1 s. Thus, the reported mean time of a oscillating simple pendulum = (92 ± 2) s.

24. The instrument whose least count is minimum, is called the most precise device.

(a) Number of division (MSD) = 20

Main Scale Division (MSD) = 1 mm

As 20 divisions on vernier scale will be equal to the 19 divisions on main scale.

∴ Vernier Scale Division (VSD) $= \frac{19}{20}$ MSD

Least count of vernier callipers = 1 MSD − 1 VSD

$= 1 \text{ MSD} - 1 \text{ VSD}$

$= 1 \text{ MSD} - \frac{19}{20} \text{ MSD}$

$= \frac{1}{20} \text{ MSD}$

$= \frac{1}{20} \text{ mm} = \frac{1}{200} \text{ cm}$

$= 0.005$ cm

(b) Pitch of screw guage = 1 mm

Number of divisions on circular scale = 100

Least count of screw gauge

$= \frac{\text{Pitch}}{\text{Number of divisions on circular scale}}$

$\frac{1}{100} \text{ mm} = \frac{1}{1000} \text{ cm} = 0.001 \text{ cm}$

(c) Wavelength of light

$(\lambda) \approx 10^{-7} \text{m} = 10^{-5} \text{ cm} = 0.00001 \text{ cm}$

∴ As the given optical instrument can measure length to within a wavelength of light, therefore least count of the given optical instrument

= Wavelength of light

= 0.00001 cm

The least count is minimum for the given optical instrument, therefore the given optical instrument is the most precise.

25. We know that, density, $\rho = \frac{M}{V} = \frac{[\text{M}]}{[\text{L}^3]}$

$\therefore \quad \frac{\Delta \rho}{\rho} \times 100 = \frac{\Delta M}{M} \times 100 + 3 \frac{\Delta L}{L} \times 100$

$= 3\% + 3\,(2\%) = 9\%$

26. Given, $\quad X = [\text{M}^a \text{L}^b \text{T}^{-c}]$

$\therefore \quad \frac{\Delta X}{X} = \pm \left[a \frac{\Delta M}{M} + b \frac{\Delta L}{L} + c \frac{\Delta T}{T} \right]$

$= \pm [a\alpha + \beta b + \gamma c]\%$

27. Kinetic energy, $E = \frac{1}{2} m v^2$

$\therefore \quad \frac{\Delta E}{E} \times 100 = \frac{\Delta v^2 - v^2}{v^2} \times 100$

$= [(1.5)^2 - 1] \times 100 = 125\%$

28. Percentage error in radius is $\frac{0.1}{4.3} \times 100$

Again, $\quad V \propto R^3$

$\therefore$ Error in volume $= \frac{3 \times 0.1}{4.3} \times 100$

29. Given, $T = 2\pi\sqrt{\dfrac{l}{g}} \Rightarrow g = \dfrac{4\pi^2 l}{T^2}$

$$\therefore \frac{\Delta g}{g} = \frac{\Delta l}{l} + \frac{2\Delta T}{T} = \frac{0.1}{10} + 2\left(\frac{1}{0.5 \times 200}\right)$$

$$\frac{\Delta g}{g} \times 100 = \frac{1}{100} \times 100 + \frac{1}{50} \times 100 = 3\%$$

30. $v = \sqrt{\dfrac{T}{m}} = \left[\dfrac{m'g}{M/l}\right]^{1/2} = \left[\dfrac{m'lg}{M}\right]^{1/2}$

It follows from here, $\dfrac{\Delta v}{v} = \dfrac{1}{2}\left[\dfrac{\Delta m'}{m} + \dfrac{\Delta l}{l} + \dfrac{\Delta M}{M}\right]$

$$= \frac{1}{2}\left[\frac{0.1}{3.0} + \frac{0.01}{1.000} + \frac{0.1}{2.5}\right]$$

$$= \frac{1}{2}[0.03 + 0.001 + 0.04]$$

$$= 0.036$$

Percentage error in the measurement = 3.6

31. When quantities are subtracted, their maximum absolute errors are added up.

$\therefore$ Result $= (80 - 10) \pm (0.1 + 0.1)$

$= 70 \pm 0.2$

32. Percentage error in side of public park

$$= \frac{1}{2}\left[\frac{0.2}{100} \times 100\right] = 0.1 \text{ m}$$

Absolute error in side of public park

$$= \frac{0.1}{100} \times 10 = 0.01 \text{ m}$$

Side of public park $= \sqrt{100} = 10$ m

$\therefore$ Side $= (10 \pm 0.01)$ m

33. As, $V = (8 + 0.5)$ V

and $I = (2 + 0.2)$ A

$\therefore$ $R = \dfrac{8}{2} = 4\,\Omega$ (R = resistance)

$\Rightarrow$ $\dfrac{\Delta R}{R}\% = \left(\dfrac{\Delta V}{V} + \dfrac{\Delta I}{I}\right)$

$$= \left(\frac{0.5}{8} + \frac{0.2}{2}\right) \times 100 = 16.25\%$$

$\therefore$ $R = (4 \pm 16.25\%)\,\Omega$

34. Percentage error in length $= \dfrac{1}{50} \times 100 = 2$

Percentage error in breadth $= \dfrac{0.1}{2.0} \times 100 = 5$

Percentage error in thickness $= \dfrac{0.1}{1.00} \times 100 = 1$

Percentage error in volume $= 2 + 5 + 1 = 8$

35. Percentage error in measurement of a side $= \dfrac{0.01}{1.23} \times 100$

$\therefore$ Percentage error in measurement of area

$$= 2 \times \frac{0.01}{1.23} \times 100$$

36. Given, $z = \dfrac{a^2 b^{2/3}}{\sqrt{c}\, d^3}$

According to question,

% error in z = (2)% error in $a + \left(\dfrac{2}{3}\right)$% error in $b + \left(\dfrac{1}{2}\right)$% error in c + (3)% error in d

$$\frac{\Delta z}{z} = 2\frac{\Delta a}{a} + \frac{2}{3}\frac{\Delta b}{b} + \frac{1}{2}\frac{\Delta c}{c} + 3\frac{\Delta d}{d}$$

$$= 2 \times 2\% + \frac{2}{3} \times 1.5\% + \frac{1}{2} \times 4\% + 3 \times 2.5\%$$

$$= 14.5\%$$

37. Weight in air = (5.00 ± 0.05) N

Weight in water = (4.00 ± 0.05) N

Loss of weight in water

= [(5.00 − 4.00) ± (0.05 + 0.05)] N

= (1.00 ± 0.1) N

Now, relative density $= \dfrac{\text{Weight in air}}{\text{Weight loss in water}}$

i.e. $\text{RD} = \dfrac{5.00 \pm 0.05}{1.00 \pm 0.1}$

Now, relative density with maximum permissible error

$$= \frac{5.00}{1.00} \pm \left(\frac{0.05}{5.00} + \frac{0.1}{1.00}\right) \times 100$$

$$= 5.0 \pm (1 + 10)\% = 5 \pm 11\%$$

So, maximum possible percentage error in relative density is ± 11%.

38. Volume of a cylinder of radius r and height h is given by

$$V = \pi r^2 h$$

or $V = \dfrac{1}{4}\pi D^2 h$, where D is the diameter of circular surface. Here, $D = 12.6$ cm and $h = 34.2$ cm

$\Rightarrow$ $V = \dfrac{\pi}{4} \times (12.6)^2 \times (34.2)$

$V = 4262.22 \text{ cm}^3$

$V = 4260$ (in three significant numbers)

Now, error calculation can be done as

$$\frac{\Delta V}{V} = 2\left(\frac{\Delta D}{D}\right) + \frac{\Delta h}{h} = \frac{2 \times 0.1}{12.6} + \frac{0.1}{34.2}$$

$\Rightarrow$ $\dfrac{\Delta V}{V} = 0.0158 + 0.0029$

$\Rightarrow$ $\Delta V = (0.01879) \times (4262.22)$

$\Rightarrow$ $\Delta V = 79.7 \simeq 80 \text{ cm}^3$

$\therefore$ For proper significant numbers, volume reading will be $V = 4260 \pm 80 \text{ cm}^3$.

39. Given, $I = (e^{1000V/T} \quad 1)$ mA

$dV = \pm 0.01$ V, $T = 300$ K

$I = 5$ mA

$$\Rightarrow \quad I = e^{1000V/T} - 1$$

$$I + 1 = e^{1000V/T}$$

Taking log on both sides, we get

$$\log (I+1) = \frac{1000V}{T}$$

$$\Rightarrow \quad \frac{d(I+1)}{I+1} = \frac{1000}{T} dV$$

$$\frac{dI}{I+1} = \frac{1000}{T} dV$$

$$\Rightarrow \quad dI = \frac{1000}{T} \times (I+1)\, dV$$

$$dI = \frac{1000}{300} \times (5+1) \times 0.01 = 0.2 \text{ mA}$$

So, error in the value of current is 0.2 mA.

40. $3.8 \times 10^{-6} + 4.2 \times 10^{-5}$

$$= (3.8 \times 10^{-1} + 4.2) \times 10^{-5}$$
$$= (0.38 + 4.2) \times 10^{-5}$$
$$= (4.58) \times 10^{-5}$$

Rounding off to one place of decimal.

The sum $= 4.6 \times 10^{-5}$

41. As $\pi = 3.14$

$$\therefore \quad \pi^2 = (3.14)^2 = 9.8596$$

On rounding off $\pi^2 = 9.86$

42. Given $A = 1.0 \text{ m} \pm 0.2 \text{ m}$

$$B = 2.0 \text{ m} \pm 0.2 \text{ m}$$

$$x = \sqrt{AB} = \sqrt{1.0 \times 2.0} = 1.414 \text{ m}$$

Rounding-off to two significant digits,

$$x = \sqrt{AB} = 1.4 \text{ m}$$

Now, $$\frac{\Delta x}{x} = \frac{1}{2}\left[\frac{\Delta A}{A} + \frac{\Delta B}{B}\right]$$

$$= \frac{1}{2}\left[\frac{0.2}{1.0} + \frac{0.2}{0.2}\right]$$

$$= \frac{0.6}{2 \times 2.0}$$

$$\Delta x = \frac{0.6 \times x}{2 \times 2.0} = 0.15 \times 1.414 = 0.2121$$

Rounding-off to one significant digits, $\Delta x = 0.2$ m

Thus, $\sqrt{AB} = 1.4\text{m} \pm 0.2\text{m}$

43. Area of 1 square $= 5.29 \text{ cm}^2$

Area of seven such squares

$= 7$ times addition of area of 1 square

$= 5.29 + 5.29 + 5.29 + \ldots 7 \text{ times} = 37.03 \text{ cm}^2$

As we know that, if in the measured values to be added/subtracted the least number of significant digits after the decimal is n.

Then, in the sum or difference also, the number of significant digits after the decimal should be n.

Here, number of digits after decimal in 5.29 is 2, so our answer also contains only two digits after decimal point.

$\therefore$ Area required $= 37.03 \text{ cm}^2$

44. Given, $A_1 = 24.36$, $B_1 = 0.0724$, $C_1 = 256.2$

$$\therefore \quad A_1 + B_1 + C_1 = 280.6324$$

As, sum contains same number of digits after decimal as present in the number having the least number of decimal places.

So, $A_1 + B_1 + C_1 = 280.6$

Similarly, $A_2 + B_2 + C_2 = 280.7$

$A_3 + B_3 + C_3 = 280.7$ (rounded off)

$A_4 + B_4 + C_4 = 280.7$ (rounded off)

So, $A_1 + B_1 + C_1 < A_2 + B_2 + C_2$

$= A_3 + B_3 + C_3 = A_4 + B_4 + C_4$

None of the options is matching with result.

45. Force, $F = kv$, $[k] = \frac{[F]}{[v]} = \frac{[\text{MLT}^{-2}]}{[\text{LT}^{-1}]} = [\text{MT}^{-1}]$.

So, unit is kg s^{-1}.

46. [Surface tension] $= [\text{ML}^0\text{T}^{-2}]$, [viscosity] $= [\text{ML}^{-1}\text{T}^{-1}]$.

Clearly, mass has the same exponent in these physical quantities.

47. [Energy] $= [\text{ML}^2\text{T}^{-2}]$. Increasing M and L by a factor of 3, energy is increased 27 times.

48. LI^2 represents energy, *i.e.* $[\text{ML}^2\text{T}^{-2}]$.

49. CR is known as time constant

$$CR = [\text{T}]$$

50. Spring constant $= \frac{F}{L} = [\text{ML}^0\text{T}^{-2}]$

Surface energy $= \frac{\text{Energy}}{\text{Area}} = [\text{ML}^0\text{T}^{-2}]$

Spring constant and surface energy has same dimensions.

51. As we know that, formula of velocity is

$$v = \frac{1}{\sqrt{\mu_0 \varepsilon_0}}$$

$$\Rightarrow \quad v^2 = \frac{1}{\mu_0 \varepsilon_0} = [\text{LT}^{-1}]^2$$

$$\therefore \quad \frac{1}{\mu_0 \varepsilon_0} = [\text{L}^2\text{T}^{-2}]$$

52. Planck's constant (in terms of unit)

$$h = \text{J-s} = [\text{ML}^2\text{T}^{-2}][\text{T}] = [\text{ML}^2\text{T}^{-1}]$$

Momentum (p) = kg - ms^{-1}

$$= [\text{M}][\text{L}][\text{T}^{-1}] = [\text{MLT}^{-1}]$$

53. From the relation $F = qvB$

$$\Rightarrow \quad [\text{MLT}^{-2}] = [\text{C}][\text{LT}^{-1}][B]$$

$$\Rightarrow \quad [B] = [\text{MC}^{-1}\text{T}^{-1}]$$

54. Here αt^2 is a dimensionless, therefore $\alpha = \frac{1}{t^2}$ and has the dimension of $[t^{-2}]$.

55. As, $\frac{F}{l} = \frac{\mu_0 I_1 I_2}{2\pi r}$

or $[\mu_0] = \frac{[F]}{[I_1 I_2]} = \frac{[MLT^{-2}]}{[A^2]} = [MLT^{-2}A^{-2}]$

56. As, $[ML^{-2}T^{-2}] = \frac{[MLT^{-2}]}{[L][L^2]} = \frac{\text{Force}}{\text{Distance} \times \text{Area}}$

$= \frac{\text{Pressure}}{\text{Distance}} = \text{Pressure gradient}$

57. Magnetic energy $= \frac{1}{2} LI^2 = \frac{LQ^2}{2t^2}$ $\left[\text{as, } I = \frac{Q}{t}\right]$

where, L = inductance and I = current.

Energy has the dimensions = $[ML^2T^{-2}]$

Equate the dimensions, we have

$$[ML^2T^{-2}] = [\text{henry}] \times \frac{[Q^2]}{[T^2]}$$

$\Rightarrow$ $[\text{henry}] = \frac{[ML^2]}{[Q^2]}$

58. Solar constant $= \frac{\text{Solar energy}}{\text{Area} \times \text{Time}} = \frac{[ML^2T^{-2}]}{[L^2][T]} = [ML^0T^{-3}]$

59. As, $\frac{B^2}{2\mu_0}$ = energy density of magnetic field

$= \frac{\text{Energy}}{\text{Volume}}$

So, $\left[\frac{B^2}{2\mu_0}\right] = [\text{Energy /Volume}] = \frac{[ML^2T^{-2}]}{[L^3]}$

$= [ML^{-1}T^{-2}]$

60. Dimensions of ε_0 (permittivity of free space) are

$$[\varepsilon_0] = [M^{-1}L^{-3}T^4A^2]$$

As, c = speed of light.

$\therefore$ Dimension of $[c] = [LT^{-1}]$

So, dimensions of $\sqrt{\frac{\varepsilon_0}{\mu_0}}$ are

$$\left[\sqrt{\frac{\varepsilon_0}{\mu_0}}\right] = \left[\sqrt{\frac{\varepsilon_0^2}{\varepsilon_0\mu_0}}\right] = [\varepsilon_0 c] \qquad \left[\because c^2 = \frac{1}{\mu_0\varepsilon_0}\right]$$

$= [M^{-1}L^{-3}T^4A^2][LT^{-1}] = [M^{-1}L^{-2}T^3A^2]$

61. Given, $y = r\sin(\omega t - kx)$

where, ωt = angle

$\therefore$ $\omega = \frac{1}{T} = [T^{-1}]$ ($\because$ angle is dimensionless)

Similarly, kx = angle

$\therefore$ $k = \frac{1}{x} = [L^{-1}]$

$\therefore$ $\frac{\omega}{k} = \frac{[T^{-1}]}{[L^{-1}]} = [LT^{-1}]$

62. Electrostatic force between two charges,

$$F = \frac{1}{4\pi\varepsilon_0}\frac{q_1q_2}{R^2} \Rightarrow \varepsilon_0 = \frac{q_1q_2}{4\pi FR^2}$$

Substituting the units.

Hence, $\varepsilon_0 = \frac{C^2}{N\text{-}m^2} = \frac{[AT]^2}{[MLT^{-2}][L^2]} = [M^{-1}L^{-3}T^4A^2]$

63. Dimensions of given quantities are

l = inductance = $[M^1L^2T^{-2}A^{-2}]$

r = resistance = $[M^1L^2T^{-3}A^{-2}]$

c = capacitance = $[M^{-1}L^{-2}T^4A^2]$

v = voltage = $[M^1L^2T^{-3}A^{-1}]$

So, dimensions of $\frac{l}{rcv}$ are

$$\left[\frac{l}{rcv}\right] = \frac{[M^1L^2T^{-2}A^{-2}]}{[M^1L^2T^{-2}A^{-1}]} = [A^{-1}]$$

64. Trigonometric function has no dimension, so

$\frac{x}{a}$ = dimensionless

Thus, a has the dimensions as x in equation.

LHS of given equation is dimensionless and hence, a^n is dimensionless

$\Rightarrow$ $n = 0$

65. By substituting the dimensions of each quantity, we get

$$T = [ML^{-1}T^{-2}]^a [L^3M]^b [MT^{-2}]^c$$

By solving, we get $a = -\frac{3}{2}$, $b = \frac{1}{2}$ and $c = 1$

66. Momentum, $p \propto f^a v^b \rho^c$

$[MLT^{-1}] = [T^{-1}]^a[LT^{-1}]^b[ML^{-3}]^c$

$[MLT^{-1}] = [M^cL^{b-3c}T^{-a-b}]$

$\Rightarrow$ $c = 1$

$b - 3c = 1$

$\Rightarrow$ $b = 4$

$-a - b = -1$

$a + b = 1, a = -3$

$[p] = [f^{-3}v^4\rho]$

67. Here, $[M^0L^0T^0] = [ML^{-1}T^{-2}]^a[ML^{-3}]^b[LT^{-1}]^c$

or $[M^0L^0T^0] = [M^{a+b}L^{-a+c}T^{-2a-3b-c}]$

Comparing powers of M, L and T, we get

$a + b = 0, -a + c = 0, -2a - 3b = 0$

Solving, $a = 1, b = -1, c = 1$

68. $\lambda = m^p v^q h^r$

$[M^0LT^0] = [M^p][LT^{-1}]^q[ML^2T^{-2}]^r$

$[M^0LT^0] = [M^{p+r}L^{q+2r}T^{-q-r}]$

$\therefore$ $p + r = 0, q + 2r = 1, -q - r = 0$

After solving, we get

$p = -1, q = -1, r = 1$

69. Let Young's modulus is related to speed, area and force, as $Y = F^x A^y V^z$

Substituting dimensions, we have

$$[ML^{-1}T^{-2}] = [MLT^{-2}]^x [L^2]^y [LT^{-1}]^z$$

Comparing power of similar quantities, we have

$$x = 1,\ x + 2y + z = -1 \text{ and } -2x - z = -2$$

Solving these, we get

$$x = 1,\ y = -1,\ z = 0$$

So, $[Y] = [FA^{-1}V^0]$

70. Let dimensions of energy E in terms of momentum p, area A and time T are

$$[E] = [p]^x [A]^y [T]^z$$

Substituting dimensions of fundamental quantities for E, p, A and T, we have

$$[ML^2T^{-2}] = [MLT^{-1}]^x [L^2]^y [T]^z$$

$$[ML^2T^{-2}] = [M^x L^{x+2y} T^{-x+z}]$$

Equating powers of same physical quantities on both sides, we have

$$x = 1,\ x + 2y = 2 \text{ and } -x + z = -2$$

So, $x = 1,\ y = \frac{1}{2},\ z = -1$

$\therefore$ Dimensional formula of $[E] = [pA^{1/2}\,T^{-1}]$

71. Dimensions of speed are $[v] = [LT^{-1}]$

Dimensions of acceleration are $[A] = [LT^{-2}]$

Dimensions of force are $[F] = [MLT^{-2}]$

Dimension of Young modulus is, $[Y] = [ML^{-1}T^{-2}]$

Let dimensions of Young's modulus is expressed in terms of speed, acceleration and force as

$$[Y] = [v]^\alpha [A]^\beta [F]^\gamma \quad \text{...(i)}$$

Then substituting dimensions in terms of M, L and T, we get

$$[ML^{-1}T^{-2}] = [LT^{-1}]^\alpha [LT^{-2}]^\beta [MLT^{-2}]^\gamma$$

$$= [M^\gamma L^{\alpha+\beta+\gamma}\ T^{-\alpha-2\beta-2\gamma}]$$

Now comparing powers of basic quantities on both sides, we get $\gamma = 1$

$$\alpha + \beta + \gamma = -1$$

and $-\alpha - 2\beta - 2\gamma = -2$

Solving these, we get

$$\alpha = -4,\ \beta = 2,\ \gamma = 1$$

Substituting α, β, and γ in Eq. (i), we get

$$[Y] = [v^{-4}A^2F^1]$$

72. Let $V_0 = (h)^a \cdot (c)^b \cdot (G)^c \cdot (A)^d \quad \text{...(i)}$

Then, $[V_0] = [\text{potential}] = \left[\dfrac{\text{potential energy}}{\text{charge}}\right]$

$$= \frac{[ML^2T^{-2}]}{[AT]} = [ML^2T^{-3}A^{-1}]$$

$$[h] = \left[\frac{\text{Energy}}{\text{Frequency}}\right] = \frac{[ML^2T^{-2}]}{[T^{-1}]} = [ML^2T^{-1}]$$

$$[c] = [\text{Speed}] = [LT^{-1}]$$

$$[G] = \left[\frac{\text{Force} \times (\text{Distance})^2}{(\text{Mass})^2}\right]$$

$$= \frac{[MLT^{-2}][L^2]}{[M^2]}$$

$$= [M^{-1}L^3T^{-2}]$$

Substituting the dimensions of V_0, h, C, G and A in Eq. (i) and equating dimension on both sides, we get

$$[ML^2T^{-3}A^{-1}] = [ML^2T^{-1}]^a \times [LT^{-1}]^b \times [M^{-1}L^3T^{-2}]^c \times [A]^d$$

$$\Rightarrow \quad a - c = 1 \quad \text{...(ii)}$$

$$2a + b + 3c = 2 \quad \text{...(iii)}$$

$$-a - b - 2c = -3 \quad \text{...(iv)}$$

$$d = -1 \quad \text{...(v)}$$

On solving above equations, we get

$$a = 0,\ b = 5,\ c = -1,\ d = -1$$

Substituting these values in Eq. (i), we get

$$V_0 = h^0 \cdot c^5 \cdot G^{-1} \cdot A^{-1}$$

73. Suppose, linear momentum (p) depends upon the Planck's constant (h) raised to the power (a), surface tension(S) raised to the power (b) and moment of inertia (I) raised to the power (c).

Then, $p \propto (h)^a (S)^b (I)^c$ or $p = kh^aS^bI^c$

where, k is a dimensionless proportionality constant.

Thus, $[p] = [h]^a [S]^b [I]^c \quad \text{...(i)}$

Then, the respective dimensions of the given physical quantities, *i.e.*

$$[p] = [\text{mass} \times \text{velocity}] = [MLT^{-1}]$$

$$[I] = [\text{mass} \times (\text{distance})^2] = [ML^2T^0]$$

$$[S] = [\text{force} \times \text{length}] = [ML^0T^{-2}]$$

$$[h] = [ML^2T^{-1}]$$

Then, substituting these dimensions in Eq. (i), we get

$$[MLT^{-1}] = [ML^2T^{-1}]^a [MT^{-2}]^b [ML^2]^c$$

For dimensional balance, the dimensions on both sides should be same.

Thus, equating dimensions, we have

$$a + b + c = 1$$

$$2(a + c) = 1 \quad \text{or} \quad a + c = \frac{1}{2}$$

$$-a - 2b = -1 \quad \text{or} \quad a + 2b = 1$$

Solving these three equations, we get

$$a = 0,\ b = \frac{1}{2},\ c = \frac{1}{2}$$

$$\therefore \quad p = h^0 S^{\frac{1}{2}} I^{\frac{1}{2}}$$

or $$p = S^{\frac{1}{2}} I^{\frac{1}{2}} h^0$$

Round II

1. Vernier scale reading (VSR) $= 6 \times 0.01 = 0.06$ cm

Correct value $= 8.5 + 0.06 - 0.02 = 8.54$ cm

2. As, $x = \frac{1 \text{ g cm s}^{-1}}{\text{N-s}} = \frac{1 \text{ g cm s}^{-1}}{1 \text{ kg} \times 1 \text{ ms}^{-1} \times 1 \text{ s}}$

$$= \frac{1 \text{ g cm s}^{-1}}{10^3 \text{ g} \times 10^2 \text{ cm s}^2 \times 1 \text{ s}} = 10^{-5}$$

3. Muscle × Speed = Power

$$\therefore \quad \text{Muscle} = \frac{\text{Power}}{\text{Speed}} = \frac{\text{Work}}{\text{Time} \times \text{Speed}}$$

$$= \frac{[\text{ML}^2\text{T}^{-2}]}{[\text{T}]\,[\text{LT}^{-1}]} = [\text{MLT}^{-2}]$$

= Mass × acceleration = Force

Hence, $\frac{\text{SI unit of force}}{\text{CGS unit of force}} = \frac{\text{kg} \times \text{m} \times \text{s}^{-2}}{\text{g} \times \text{cm} \times \text{s}^{-2}}$

$$= 10^3 \times 10^2 = 10^5$$

4. Least count $1 = \frac{1}{100}$ mm $= 0.01 = 0.001$ cm

Diameter of wire $D = 1$ mm $+ 63 \times 0.01$ mm

$= 1.63$ mm or 0.163 cm

Volume of wire $= \frac{\pi D^2 l}{4} = \frac{3.14 \times (0.163)^2 \times 5.6}{4}$

$= 0.117 \text{ cm}^3$

5. As, pitch $= \frac{1}{2}$ mm $= 0.5$ mm

Least count $= \frac{0.5}{50}$ mm $= 0.01$ mm

Zero error $= -0.03$ mm

Zero correction $= +\,0.03$ mm

Observed diameter of wire $= 3 + 35 \times 0.01$

$= 3.35$ mm

Corrected diameter of wire $= 3.35$ mm $+ 0.03$ mm

$= 3.38$ mm

6. Given, $l = 10$ mm $= 1$ cm

$b = 5$ mm $= 0.5$ cm and

$h = 5$ mm $= 0.5$ cm

Error in measurements,

$\Delta l = \Delta b = \Delta h = 0.01$ cm

∴ Maximum percentage error in volume,

$$\frac{\Delta V}{V} \times 100 = \frac{\Delta l}{l} \times 100 + \frac{\Delta b}{b} \times 100 + \frac{\Delta h}{h} \times 100$$

$$= \frac{0.01}{1} \times 100 + \frac{0.01}{0.5} \times 100 + \frac{0.01}{0.5} \times 100$$

$$= 1\% + 2\% + 2\% = 5\%$$

7. As, $\Delta R_s = \Delta R_1 + \Delta R_2 = \left[\frac{10}{100} \times 10 + \frac{20}{100} \times 20\right] \text{k}\Omega = 5 \text{ k}\Omega$

$$\frac{\Delta R_s}{R_s} \times 100 = \frac{5}{30} \times 100 = \frac{50}{3} = 17\%$$

8. $T = \frac{rgh}{2} = \frac{dgh}{4}$ $\quad \left[\because r = \frac{d}{2}\right]$

$$\therefore \quad \frac{\Delta T}{T} \times 100 = \frac{\Delta d}{d} \times 100 + \frac{\Delta g}{g} \times 100 + \frac{\Delta h}{h} \times 100$$

$$= \frac{0.01}{1.26} \times 100 + \frac{0.01}{9.80} \times 100 + \frac{0.01}{1.45} \times 100$$

$$= 0.8 + 0.1 + 0.7 = 1.6\%$$

9. $R = \frac{V}{i}$

$$\therefore \quad \log R = \log V - \log i$$

$$\frac{\Delta R}{R} = \frac{\Delta V}{V} + \frac{\Delta i}{i} = 3\% + 3\% = 6\%$$

10. Here, $\frac{\Delta m}{m} = \frac{0.003}{0.3}$,

$$\frac{\Delta r}{r} = \frac{0.005}{0.5} \cdot \frac{\Delta L}{L} = \frac{0.06}{6}$$

As $\rho = \frac{m}{(\pi r^2) L}$

$$\therefore \left(\frac{\Delta \rho}{\rho}\right) \times 100 = \left(\frac{\Delta m}{m} + \frac{2\,\Delta r}{r} + \frac{\Delta L}{L}\right) \times 100$$

$$= \left(\frac{0.003}{0.3} + \frac{2 \times 0.005}{0.5} + \frac{0.06}{6}\right) \times 100$$

$$= 1 + 2 + 1 = 4\%$$

11. From, $s = ut + \frac{1}{2}at^2$

$$h = 0 + \frac{1}{2} \times 9.8\,(2)^2 = 19.6 \text{ m}$$

$$\frac{\Delta h}{h} = \pm \left(\frac{\Delta t}{t}\right) \quad (\because a = g = \text{constant})$$

$$= \pm 2\left(\frac{0.1}{2}\right) = \pm \frac{1}{10}$$

$$\Delta h = \frac{h}{10} = \frac{19.6}{10} = 1.96 \text{ m}$$

12. Here, $L = 1.2 \times 10^{-2}$ m, $V = ?$

$$V = L^3 = (1.2 \times 10^{-2})^3 = 1.728 \times 10^{-6} \text{ m}^3$$

As the result can have only two significant digits, therefore rounding off, we get

$$V = 1.7 \times 10^{-6} \text{ m}^3$$

13. Resistance, $R = \frac{\rho l}{A}$

Also, $R = \frac{V}{I}$

$$\Rightarrow \quad \rho = \frac{AV}{Il} = \frac{\pi D^2 V}{4Il} \quad \left(\because A = \frac{\pi D^2}{4}\right)$$

$$\therefore \quad \frac{\Delta \rho}{\rho} = 2\frac{\Delta D}{D} + \frac{\Delta V}{V} + \frac{\Delta I}{I} + \frac{\Delta l}{l}$$

$$= 2\left(\frac{0.01}{5.00}\right) + \left(\frac{0.1}{5.0}\right) + \left(\frac{0.01}{2.00}\right) + \left(\frac{0.1}{10.0}\right)$$

$$\frac{\Delta \rho}{\rho} \times 100 = 0.4 + 2 + 0.5 + 1 = 3.9\%$$

14. Equivalent, $R_s = \dfrac{R_1 R_2}{R_1 + R_2}$

$$\frac{\Delta R_s}{R_s}\times 100 = \frac{\Delta R_1}{R_1}\times 100 + \frac{\Delta R_2}{R_2}\times 100 + \frac{\Delta (R_1+R_2)}{R_1+R_2}\times 100$$

Now, $\Delta R_1 = \dfrac{10}{100}\times 4\text{ k}\Omega = 0.4\text{ k}\Omega$

$\Delta R_2 = \dfrac{10}{100}\times 6\text{ k}\Omega = 0.6\text{ k}\Omega$

Again, $\dfrac{\Delta R_s}{R_s}\times 100 = \dfrac{0.4}{4}\times 100 + \dfrac{0.6}{6}\times 100 + \dfrac{0.4+0.6}{10}\times 100$

$= 10 + 10 + 10 = 30\%$

15. The right hand side of the given relation is basically $\dfrac{k}{\text{metre}}$. But, since the left hand side is joule, therefore k should be Jm.

16. Capacitance $X = [\text{M}^{-1}\text{L}^{-2}\text{T}^2\text{Q}^2]$

Magnetic induction $Z = [\text{MT}^{-1}\text{Q}^{-1}]$

$[Z^2] = [\text{M}^2\text{T}^{-2}\text{Q}^{-2}]$

Given, $X = 3YZ^2$

or $Y = \dfrac{X}{3Z^2}$ or $[Y] = \dfrac{[X]}{[Z]^2}$

$\therefore$ $[Y] = \dfrac{[\text{M}^{-1}\text{L}^{-2}\text{T}^2\text{Q}^2]}{[\text{M}^2\text{T}^{-2}\text{Q}^{-2}]} = [\text{M}^{-3}\text{L}^{-2}\text{T}^4\text{Q}^4]$

17. The formula for fine structure constant $= \dfrac{e^2}{4\pi\varepsilon_0\left(\dfrac{h}{2\pi}\right)c}$

It is dimensionless.

18. $[X] = \left[\dfrac{\text{M}^{-1}\text{L}^3\text{T}^{-2}\times \text{ML}^2\text{T}^{-1}}{\text{L}^3\text{T}^{-3}}\right]^{1/2} = [\text{L}]$

19. Each of the three terms in the given equation has the dimensional formula of force.

20. Given equation, $y = a\sin(bt - cx)$

Comparing the given equation with general wave equation

$$y = a\sin\left(\frac{2\pi t}{T} - \frac{2\pi x}{\lambda}\right)$$

we get, $b = \dfrac{2\pi}{T}, c = \dfrac{2\pi}{\lambda}$

Dimension of $\dfrac{b}{c} = \dfrac{2\pi/T}{2\pi/\lambda} = [\text{LT}^{-1}]$ and other three quantities are dimensionless.

21. We know that, the dimensional formula of energy is $[\text{ML}^2\text{T}^{-2}]$.

$$n_2 = 1\left[\frac{1\text{ kg}}{10\text{ kg}}\right]^1\left[\frac{1\text{ m}}{1\text{ km}}\right]^2\left[\frac{1\text{ s}}{1\text{ min}}\right]^2$$

$$= \frac{1}{10}\times\frac{1}{10^6}\times\frac{1}{(60)^{-2}}$$

$$= \frac{3600}{10^7} = 3.6\times 10^{-4}$$

22. From the given relation, $D = \dfrac{n(x_2 - x_1)}{n_2 - n_1}$

Here, $[n] = \left[\dfrac{1}{\text{area}\times\text{time}}\right] = \dfrac{1}{[\text{L}^2\text{T}]} = [\text{L}^{-2}\text{T}^{-1}]$

$x_2 - x_1 = [\text{L}]$

and $n_2 - n_1 = \left[\dfrac{1}{\text{volume}}\right] = \left[\dfrac{1}{\text{L}^3}\right] = [\text{L}^{-3}]$

So, $[D] = \dfrac{[\text{L}^{-2}\text{T}^{-1}\text{L}]}{[\text{L}^{-3}]} = [\text{L}^2\text{T}^{-1}]$

23. $[\text{Calorie}] = [\text{ML}^2\text{T}^{-2}]$

Comparing with general dimensional formula $[\text{M}^a\text{L}^b\text{T}^c]$, we get

$a = 1, b = 2, c = 2$

$$n_2 = 4.2\left[\frac{1\text{ kg}}{\alpha\text{ kg}}\right]^1\left[\frac{1\text{ m}}{\beta\text{m}}\right]^2\left[\frac{1s}{\gamma\text{s}}\right]^{-2} = 4.2\,\alpha^{-1}\beta^{-2}\gamma^2$$

24. Note carefully that every alternative has Gh and c^5.

$[Gh] = [\text{M}^{-1}\text{L}^3\text{T}^{-2}][\text{ML}^2\text{T}^{-1}] = [\text{M}^0\text{L}^5\text{T}^{-3}]$

$[c] = [\text{LT}^{-1}]$

$\therefore$ $\left(\dfrac{Gh}{c^5}\right)^{1/2} = [\text{T}]$

25. We know that,

$R = [\text{M}^1\text{L}^2\text{T}^{-3}\text{A}^{-2}]$

$C = [\text{M}^{-1}\text{L}^{-2}\text{T}^4\text{A}^2]$

$L = [\text{ML}^2\text{T}^{-2}\text{A}^{-2}]$

$\therefore$ $RC = [\text{T}]$ and $\sqrt{LC} = [\text{T}]$

26. From $E = h\nu$

$h = \dfrac{E}{\nu} = \dfrac{[\text{ML}^2\text{T}^{-2}]}{[\text{T}^{-1}]} = [\text{ML}^2\text{T}^{-1}]$

Angular impulse $= \tau\times t = [\text{ML}^2\text{T}^{-2}\times\text{T}] = [\text{ML}^2\text{T}^{-1}]$

27. $Y = \dfrac{\text{Stress}}{\text{Strain}} = \dfrac{mgL}{\pi R^2 l}$

$\therefore$ $\dfrac{\Delta Y}{Y} = \dfrac{\Delta m}{m} + \dfrac{\Delta L}{L} + 2\dfrac{\Delta R}{R} + \dfrac{\Delta l}{l}$

$$\frac{\Delta Y}{Y}\times 100 = \left[\frac{1}{1000} + \frac{1}{1000} + 2\left(\frac{0.001}{0.2}\right) + \frac{0.001}{0.5}\right]\times 100$$

$= 0.1 + 0.1 + 1 + 0.2 = 1.4\%$

28. Heat ΔQ transferred through a rod of length L and area A in time Δt is $\Delta Q = KA\left(\dfrac{T_1 - T_2}{L}\right)\Delta t$

where, K = coefficient of thermal conductivity,

$T_1 - T_2$ = temperature difference.

$\therefore$ $K = \dfrac{\Delta Q\times L}{A(T_1 - T_2)\Delta t}$ ...(i)

Substituting dimensions for corresponding quantities in Eq. (i), we have

$$[K] = \frac{[\text{ML}^2\text{T}^{-2}][\text{L}]}{[\text{L}^2][\theta][\text{T}]} = [\text{MLT}^{-3}\theta^{-1}]$$

29. Given that, $x = \frac{IF v^2}{WL^4}$

Dimensionally,

$$[x] = \frac{[I]\,[F]\,[v]^2}{[W]\,[L]^4} = \frac{[M^1L^2]\,[M^1L^1T^{-2}]\,[L^1T^{-1}]^2}{[M^1L^2T^{-2}]\,[L^1]^4}$$

$$= [M^1L^2]\frac{[M^1L^1T^{-2}][L^2T^{-2}]}{[M^1L^2T^{-2}][L^4]}$$

$$= [M^1L^{-1}T^{-2}] \quad \text{...(i)}$$

On checking the alternatives

(a) Planck's constant $\Rightarrow [h] = [M^1L^2T^{-1}]$ doesn't match with dimensional formula of x.

(b) Force constant $\Rightarrow [K] = [M^1T^{-2}]$ doesn't match with dimensional formula of x.

(c) Coefficient of viscosity $\Rightarrow [\eta] = [M^1L^{-1}T^{-1}]$ doesn't match with dimensional formula of x.

(d) Energy density $\Rightarrow [E_d] = [M^1L^{-1}T^{-2}]$ matches with dimensional formula of x.

30. $x = \frac{1}{\sqrt{\mu_0\varepsilon_0}}$ = speed of light in vacuum

$\therefore$ Dimension of x, $[x] = [M^0L^1T^{-1}]$

$$y = \frac{E}{B} = \text{speed of EM wave}$$

$\therefore$ Dimension of y, $[y] = [M^0L^1T^{-1}]$

$$z = \frac{l}{RC} = \frac{l}{\tau} = \frac{\text{length}}{\text{time}}$$

$\therefore$ Dimension of z, $[z] = [M^0L^1T^{-1}]$

Thus, all quantities have same dimensions, *i.e.* of velocity.

31. For conduction of heat,

$$\frac{dQ}{dt} = KA\frac{dT}{dx}$$

$$K = \frac{\left(\frac{dQ}{dt}\right)}{A\left(\frac{dT}{dx}\right)} = \frac{dQ \times dx}{A \times dt \times dT}$$

$$= \frac{\text{joule} \times \text{metre}}{(\text{metre})^2 \times \text{second} \times \text{kelvin}}$$

$$= \frac{\frac{\text{kilogram} \times (\text{metre})^2}{(\text{second})^2} \times \text{metre}}{(\text{metre})^2 \times \text{second} \times \text{kelvin}}$$

$$= \frac{\text{kg}\cdot\text{m}^2\cdot\text{m}}{\text{s}^2\cdot\text{m}^2\cdot\text{s}\cdot\text{K}} = \frac{\text{kg}\cdot\text{m}}{\text{s}^3\cdot\text{K}}$$

$$[K] = \frac{[M^1][L^1]}{[T^3][K^1]}$$

$$\Rightarrow \quad [K] = [M^1L^1T^{-3}K^{-1}]$$

32. Dimensions of quantity f are

$$[f] = \frac{[h]^{\frac{1}{2}}\,[c]^{\frac{5}{2}}}{[G]^{\frac{1}{2}}} \quad \text{...(i)}$$

As, $h = \frac{E}{\nu}$; $[h] = [ML^2T^{-2}]\,[T] = [ML^2T^{-1}]$

$$c = [LT^{-1}] \text{ and } G = \frac{F \cdot r^2}{m^2}$$

$$\Rightarrow \quad [G] = \frac{[MLT^{-2}]\,[L^2]}{[M^2]} = [M^{-1}L^3T^{-2}]$$

So, dimensions of f using Eq. (i),

$$[f] = \frac{[ML^2T^{-1}]^{\frac{1}{2}}\,[LT^{-1}]^{\frac{5}{2}}}{[M^{-1}L^3T^{-2}]^{\frac{1}{2}}}$$

$$= \left[M^{\frac{1}{2}+\frac{1}{2}}, L^{\frac{5}{2}-\frac{3}{2}+1}, T^{-\frac{1}{2}-\frac{5}{2}+\frac{2}{2}}\right] = [ML^2T^{-2}]$$

Thus, it is the dimensions of energy.

33. Force of interaction between two atoms is given as

$$F = \alpha\beta \exp(-x^2/\alpha kT)$$

As we know, exponential terms are always dimensionless, so

$$\text{dimensions of}\left(\frac{-x^2}{\alpha kT}\right) = [M^0L^0T^0]$$

$\Rightarrow$ Dimensions of α = Dimension of (x^2/kT)

Now, substituting the dimensions of individual term in the given equation, we get

$$= \frac{[M^0L^2T^0]}{[M^1L^2T^{-2}]}$$

{$\because$ Dimensions of kT equivalent to the dimensions of energy $= [M^1L^2T^{-2}]$}

$$= [M^{-1}L^0T^2] \quad \text{...(i)}$$

Now from given equation, we have dimensions of

F = dimensions of α × dimensions of β

$\Rightarrow$ Dimensions of β = Dimensions of $\left(\frac{F}{\alpha}\right)$

$$= \frac{[M^1L^1T^{-2}]}{[M^{-1}L^0T^2]} \quad [\because \text{using Eq. (i)}]$$

$$= [M^2L^1T^{-4}]$$

34. To find dimensions of capacitance in the given relation, we can use formula for energy.

Capacitor energy, $U = \frac{1}{2}CV^2$

So, dimensionally,

$$\Rightarrow \quad [C] = \left[\frac{U}{V^2}\right]$$

As, $\quad V = \text{potential} = \frac{\text{potential energy}}{\text{charge}}$

We have, $\quad [C] = \frac{[U]}{\left[\frac{U^2}{q^2}\right]} = \frac{[q^2]}{[U]} = \left[\frac{A^2T^2}{ML^2T^{-2}}\right]$

$$\therefore \quad X = [M^{-1}L^{-2}A^2T^4]$$

To get dimensions of magnetic field, we use force on a current carrying conductor in magnetic field,

$$F = BIl \Rightarrow [B] = \frac{[F]}{[I][l]} = \left[\frac{MLT^{-2}}{AL}\right]$$

$$\therefore \quad Z = [M^1L^0T^{-2}A^{-1}]$$

Now, using given relation,

$$X = 5YZ^2$$

$$[Y] = \frac{[X]}{[Z^2]} = \left[\frac{M^{-1}L^{-2}A^2T^4}{(M^1L^0T^{-2}A^{-1})^2}\right] = \frac{M^{-1}L^{-2}A^2T^4}{M^2T^{-4}A^{-2}}$$

$$\therefore \quad [Y] = [M^{-3}L^{-2}T^8A^4]$$

35. $\because F = ma$

$$\Rightarrow \quad m = \frac{F}{a} = (\text{Force}) \div \frac{\text{Change in velocity}}{\text{Time}}$$

$$= \frac{\text{Force} \times \text{Time}}{\text{Change in velocity}}$$

$$= (\text{strength})\ (\text{second})\ (\text{run})^{-1}$$

Thus, $x = 1$, $y = 1$ and $z = -1$

$$\therefore \quad \frac{y}{x} = 1$$

36.

$$1 \text{ MSD} = 1 \text{ mm}$$

$$9 \text{ MSD} = 10 \text{ VSD}$$

$$\therefore \quad \text{Least count, LC} = 1 \text{ MSD} - \text{VSD}$$

$$= 1 \text{ mm} - \frac{9}{10} \text{ mm}$$

$$= \frac{1}{10} \text{ mm}$$

Measure reading of edge = MSR + VSR (LC)

$$= 10 + 1 \times \frac{1}{10}$$

$$= 10.1 \text{ mm}$$

Volume of cube, $V = (1.01)^3 \text{ cm}^3 = 1.03 \text{ cm}^3$

[after rounding off upto 3 significant digits, as edge length is measured upto 3 significant digits]

$$\therefore \quad \text{Density of cube} = \frac{2.736}{1.03} = 2.6563 \text{ gcm}^{-3}$$

$$= 2.66 \text{ gcm}^{-3}$$

(after rounding off to 3 significant digits)

37. Given, relative error in mass, $\frac{\Delta m}{m} \times 100 = 6\%$

Relative error in diameter, $\frac{\Delta d}{d} \times 100 = 1.5\%$

Density of sphere, $\rho = \frac{\text{Mass}}{\text{Volume}} = \frac{M}{\frac{4}{3}\pi r^3} = \frac{M}{\frac{4}{3}\pi\left(\frac{d}{2}\right)^3}$

$$\rho = \frac{6}{\pi} Md^{-3} \text{ or } \rho \propto Md^{-3}$$

For maximum error in density,

$$\frac{\Delta\rho}{\rho} \times 100 = \frac{\Delta M}{M} \times 100 + 3 \times \frac{\Delta d}{d} = 6\% + 3 \times 1.5\%$$

$$\frac{\Delta\rho}{\rho} \times 100 = 10.5\%$$

$$\Rightarrow \quad \frac{1050}{100}\% = \frac{x}{100}\% \qquad \text{(given)}$$

$$\therefore \quad x = 1050$$

38. $\because \frac{1}{f} = \frac{1}{v} + \frac{1}{u}$

$$\text{or} \quad \frac{1}{f} = \frac{1}{30} + \frac{1}{60}$$

$$\text{or} \quad \frac{1}{f} = \frac{2+1}{60} = \frac{3}{60}$$

$$\therefore \quad f = 20 \text{ cm}$$

$$\because \quad \frac{1}{f} = \frac{1}{v} + \frac{1}{u}$$

$$\text{or} \quad \frac{-1}{f^2}\frac{df}{dx} = -\frac{1}{v^2}\frac{dv}{dx} - \frac{1}{u^2}\frac{du}{dx}$$

$$\Rightarrow \quad \frac{\Delta f}{f^2} = \frac{\Delta v}{v^2} + \frac{\Delta u}{u^2}$$

$$\Rightarrow \quad \frac{\Delta f}{400} = \frac{0.6}{3600} + \frac{0.3}{900}$$

$$\Rightarrow \quad \frac{\Delta f}{20} = \frac{0.6}{180} + \frac{0.3}{45} = \frac{1}{300} + \frac{1}{150}$$

$$\Rightarrow \quad \Delta f\left(\frac{2+1}{300}\right)20 = \frac{60}{300} = 0.2$$

Percentage error in focal length

$$= \frac{\Delta f}{f} \times 100 = \frac{0.2}{20} \times 100 = 1\%$$

$$\therefore \quad n = 1$$

39. Dimensions of $[c] = [LT^{-1}]$

$$\Rightarrow \quad [L] = [cT]$$

The dimensions of $[g] = [LT^{-2}] = [cT\, T^{-2}]$

$$\therefore \quad [T] = [cg^{-1}]$$

The dimensions of $[p] = [ML^{-1}T^{-2}]$

$$= [M\,(cT)^{-1}\,(cg^{-1})^{-2}]$$

$$= [Mc^{-3}\,g^2\,T^{-1}]$$

$$= [M\,c^{-3}\,g^2(cg^{-1})^{-1}]$$

$$= [M\,c^{-4}\,g^3]$$

$$\therefore \quad [M] = [p\,c^4\,g^{-3}]$$

The dimensions of universal gravitational constant,

$$[G] = [M^{-1}L^3T^{-2}]$$

$$= [(p\,c^4g^{-3}0]^{-1}\,(cT)^3\,T^{-2}]$$

$$= [p^{-1}\,c^{-4}g^3\,c^3T]$$

$$= [p^{-1}\,c^{-1}\,g^3\,(c^{-1}g^{-1})]$$

$$= [p^{-1}\,c^0\,g^2]$$

Thus, the dimensions of g in the dimensional formula of universal gravitational constant is 2.

02

Vector Analysis

IN THIS CHAPTER ....

Scalars and Vectors

On the basis of magnitude, direction and rules of addition, all physical quantities are classified into two groups as scalars and vectors.

A *scalar quantity* is one whose specification is completed with its magnitude only. For example, mass, distance, speed, energy, electric flux, current electricity, etc.

A *vector quantity* is one whose specification is completed with its magnitude and direction both. For example, displacement, velocity, acceleration, force, electric field intensity, current density, etc.

Types of Vectors

Some important types of vectors are given in the table below

S.No.	Types of Vectors	Description
(i)	**Polar vectors**	These are the vectors which have a starting point or a point of application. *e.g.* Displacement, force, etc.
(ii)	**Axial vectors**	These are the vectors which represent rotational effect and act along the axis of rotation in accordance with right hand screw rule. *e.g.* Angular velocity, angular acceleration, etc.
(iii)	**Equal vectors**	These are the vectors which have equal magnitude and same direction.
(iv)	**Like parallel vectors**	These are the vectors which have same direction and magnitude may equal or different.
(v)	**Anti-parallel vectors**	These are the vectors which have opposite direction and magnitude may equal or different.
(vi)	**Collinear vectors**	These are the vectors which act along same line, *i.e.* vectors lying in the same line. Angle between them can be zero or 180°.
(vii)	**Zero vector**	A vector having zero magnitude is known as zero vector. Its direction is not specified and hence it is arbitrary. It is represented by 0.

S.No.	Types of Vectors	Description
(viii)	Unit vector	A vector whose magnitude is unit and points in a particular direction is called unit vector. It is represented by $\hat{\mathbf{A}}$ (*A* cap or *A* hat). The unit vector along the direction of **A** is $\hat{\mathbf{A}} = \frac{\mathbf{A}}{\lvert\mathbf{A}\rvert}$, $\therefore$ $\mathbf{A} = \hat{\mathbf{A}}\lvert\mathbf{A}\rvert$, where $\mathbf{A} = A_x\hat{\mathbf{i}} + A_y\hat{\mathbf{j}} + A_z\hat{\mathbf{k}}$ and $\lvert\mathbf{A}\rvert = \sqrt{A_x^2 + A_y^2 + A_z^2}$.
(ix)	Coplanar vectors	These are the vectors which always lie in the same plane.
(x)	Negative vector	If the direction of a vector is reversed, the sign of the vector is reversed. It is called negative vector of the original vector.

Example 1. *A vector may change, if*

(a) *frame of reference is translated*
(b) *vector is rotated*
(c) *frame of reference is rotated*
(d) *vector is translated parallel to itself*

Sol. (b) Vector will change, if it is rotated because its direction changes.

Example 2. *The expression* $\frac{1}{\sqrt{2}}(\hat{\mathbf{i}} + \hat{\mathbf{j}})$ *is a*

(a) *unit vector*
(b) *null vector*
(c) *vector of magnitude* $\sqrt{2}$
(d) *scalar*

Sol. (a) We have $|\mathbf{R}| = \left[\left(\frac{1}{\sqrt{2}}\right)^2 + \left(\frac{1}{\sqrt{2}}\right)^2\right]^{1/2} = 1$

Multiplication of a Vector by a Real Number

The multiplication of a vector by a scalar quantity n gives a new vector whose magnitude is n times the magnitude of the given vector. Its direction is same as that of the given vector, if n is a positive real number and reverses, if n is a negative real number.

Example 3. *If* $\mathbf{A} = 2\hat{\mathbf{i}} - 3\hat{\mathbf{j}} + 4\hat{\mathbf{k}}$, *is multiplied by a number 5, then the vector along y-direction is*

(a) $-15\hat{\mathbf{j}}$
(b) $5\hat{\mathbf{j}}$
(c) $-5\hat{\mathbf{j}}$
(d) $15\hat{\mathbf{j}}$

Sol. (a) As, $n \times \mathbf{A} = n\mathbf{A}$,

So, $5 \times (2\hat{\mathbf{i}} - 3\hat{\mathbf{j}} + 4\hat{\mathbf{k}}) = 10\hat{\mathbf{i}} - 15\hat{\mathbf{j}} + 20\hat{\mathbf{k}}$

$\therefore$ Vector along y-direction $= -15\,\hat{\mathbf{j}}$

Addition of Vectors

Vectors can be added by following laws

1. Triangle Law of Vector Addition

If two vectors are represented both in magnitude and direction by the two sides of a triangle taken in the same order, then the resultant of these vectors is represented both in magnitude and direction by the third side of the triangle taken in reverse order as shown below

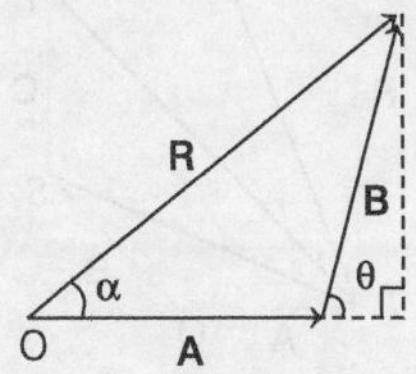

or $\mathbf{R} = \mathbf{A} + \mathbf{B}$

If θ is the angle between **A** and **B**, then the magnitude of resultant, **R** can be given by

$$R = |\mathbf{A} + \mathbf{B}| = \sqrt{A^2 + B^2 + 2AB\cos\theta}$$

If **R** makes an angle α with **A**, then $\tan\alpha = \frac{B\sin\theta}{A + B\cos\theta}$.

2. The Parallelogram Law

If two non-zero vectors **A** and **B** are represented by the two adjacent sides of a parallelogram then, the resultant is given by the diagonal of the parallelogram in magnitude and direction both passing through the point of intersection of the two vectors.

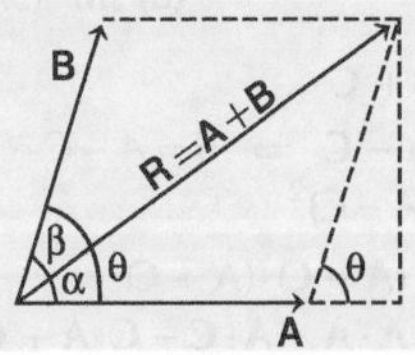

The magnitude of **R** is

$$R = |\mathbf{R}| = \sqrt{A^2 + B^2 + 2AB\cos\theta} \quad \text{...(i)}$$

where, θ is the angle between **A** and **B**.

Here, $\tan\alpha = \frac{B\sin\theta}{A + B\cos\theta}$

and $\tan\beta = \frac{A\sin\theta}{B + A\cos\theta}$ ...(ii)

where, $\alpha + \beta = \theta$.

Special Cases

If $\theta = 0°$, $R_{max} = A + B$

If $\theta = 180°$, $R_{min} = A - B$ and if $\theta = 90°$, $R = \sqrt{A^2 + B^2}$

In all other cases, Eqs. (i) and (ii) can be used to calculate magnitude and direction of **R**.

3. Polygon Law of Vector Addition

If a number of vectors are represented both in magnitude and direction by the sides of a polygon taken in the same order, then the resultant vector is represented both in magnitude and direction by the closing side of the polygon taken in the opposite order.

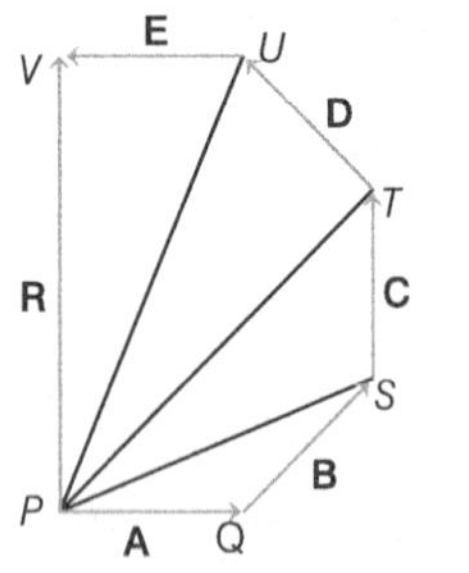

$$\mathbf{R} = \mathbf{A} + \mathbf{B} + \mathbf{C} + \mathbf{D} + \mathbf{E}$$

Properties of Vector Addition

(i) Vector addition is commutative,
i.e. $\mathbf{A} + \mathbf{B} = \mathbf{B} + \mathbf{A}$

(ii) Vector addition is associative,
i.e. $(\mathbf{A} + \mathbf{B}) + \mathbf{C} = \mathbf{A} + (\mathbf{B} + \mathbf{C})$

(iii) Vector addition is distributive,
i.e. $\lambda(\mathbf{A} + \mathbf{B}) = \lambda\mathbf{A} + \lambda\mathbf{B}$

Example 4. *If $\mathbf{A} = \mathbf{B} + \mathbf{C}$ and the magnitude of $\mathbf{A}$, $\mathbf{B}$ and $\mathbf{C}$ are 5, 4 and 3 minutes respectively, then angle between $\mathbf{A}$ and $\mathbf{C}$ is*

(a) $\cos^{-1}(4/5)$ (b) $\cos^{-1}(3/5)$
(c) $\tan^{-1}(3/4)$ (d) $\sin^{-1}(3/5)$

Sol. (b) Given, $\mathbf{A} = \mathbf{B} + \mathbf{C}$

$\Rightarrow \mathbf{B} = \mathbf{A} - \mathbf{C} \Rightarrow B = A - C$

$\Rightarrow B^2 = (A - C)^2$

$\Rightarrow \mathbf{B}\cdot\mathbf{B} = (\mathbf{A} - \mathbf{C})\cdot(\mathbf{A} - \mathbf{C})$

$\Rightarrow \mathbf{B}\cdot\mathbf{B} = \mathbf{A}\cdot\mathbf{A} - \mathbf{A}\cdot\mathbf{C} - \mathbf{C}\cdot\mathbf{A} + \mathbf{C}\cdot\mathbf{C}$

$\Rightarrow B^2 = A^2 - AC\cos\theta - AC\cos\theta + C^2$

$\Rightarrow 2AC\cos\theta = A^2 + C^2 - B^2$

$\Rightarrow 2\times 5\times 3\times\cos\theta = 5^2 + 3^2 - 4^2$

$\Rightarrow \cos\theta = \dfrac{18}{30} = \dfrac{3}{5}$

$\Rightarrow \theta = \cos^{-1}\left(\dfrac{3}{5}\right)$

Example 5. *The resultant of two forces acting at an angle of 150° is 10 N and is perpendicular to one of the forces. One of the two other forces is*

(a) $20/\sqrt{3}$ N (b) $10/\sqrt{3}$ N
(c) 20 N (d) $20/\sqrt{3}$ N

Sol. (c) We have, $R^2 = A^2 + B^2 + 2AB\cos\theta$

$$(10)^2 = A^2 + B^2 + 2AB\cos 150^\circ$$
$$= A^2 + B^2 + 2AB(-\sqrt{3}/2)$$

or $\quad 100 = A^2 + B^2 - \sqrt{3}AB \quad$ ...(i)

Again $\quad \tan 90^\circ = \dfrac{B\sin 150^\circ}{A + B\cos 150^\circ}$

$$= \frac{B\times 1/2}{A + B(-\sqrt{3}/2)}$$

$$= \frac{B}{2A - \sqrt{3}B}$$

or $\quad \infty = \dfrac{B}{2A - \sqrt{3}B}$

or $\quad 2A - \sqrt{3}B = 0$

or $\quad A = \dfrac{\sqrt{3}}{2}B$

Putting the value of A in Eq. (i), we get

$$100 = \frac{3}{4}B^2 + B^2 - \sqrt{3}\times\frac{\sqrt{3}}{2}B\times B = \frac{1}{4}B^2$$

or $\quad B^2 = 4\times 100$

or $\quad B = 20$ N

and $\quad A = \dfrac{\sqrt{3}}{2}20 = 10\sqrt{3}$ N

Example 6. *Three vectors each of magnitude A are acting at a point such that angle between any two vectors is 60°. The magnitude of their resultant is*

(a) zero (b) $\sqrt{2}A$
(c) $\sqrt{3}A$ (d) $\sqrt{6}A$

Sol. (d) We have, $R = |\mathbf{A} + \mathbf{B} + \mathbf{C}|$

$$= [A^2 + B^2 + C^2 + 2\mathbf{A}\cdot\mathbf{B} + 2\mathbf{B}\cdot\mathbf{C} + 2\mathbf{C}\cdot\mathbf{A}]^{1/2}$$

Given, $\quad A = B = C$

and $\quad \theta = 60^\circ$

$$R = [3A^2 + 2\mathbf{A}\cdot\mathbf{B} + 2\mathbf{B}\cdot\mathbf{C} + 2\mathbf{C}\cdot\mathbf{A}]^{1/2}$$
$$= [3A^2 + (2A\cdot A\cos 60^\circ)\times 3]^{1/2}$$
$$= \sqrt{6}A$$

Example 7. *A particle has two equal accelerations in two given directions. If one of the accelerations is halved, then the angle which the resultant makes with the other is also halved. The angle between the accelerations is*

(a) 120° (b) 90°
(c) 60° (d) 45°

Sol. (a) $\quad \tan\beta = \dfrac{B\sin\theta}{A + B\cos\theta}$

$= \dfrac{A\sin\theta}{A + A\cos\theta} \quad (\because A = B)$

$= \dfrac{\sin\theta}{(1 + \cos\theta)} \quad$ ...(i)

$\tan\dfrac{\beta}{2} = \dfrac{(A/2)\sin\theta}{A + (A/2)\cos\theta}$

$= \dfrac{\sin\theta}{2 + \cos\theta} \quad$ ...(ii)

We know that, $\tan\beta = \dfrac{2\tan\frac{\beta}{2}}{1-\tan^2\frac{\beta}{2}}$

$$\Rightarrow \quad \frac{\sin\theta}{1+\cos\theta} = \frac{2\sin\theta}{(2+\cos\theta)\left[1-\left(\dfrac{\sin\theta}{2+\cos\theta}\right)^2\right]}$$

$$\Rightarrow \quad \frac{1}{1+\cos\theta} = \frac{2(2+\cos\theta)}{(2+\cos\theta)^2-\sin^2\theta}$$

$$= \frac{2(2+\cos\theta)}{4+\cos^2\theta+4\cos\theta-(1-\cos^2\theta)}$$

$$\Rightarrow \quad 2\cos^2\theta+4\cos\theta+3 = 2(2+\cos\theta)(1+\cos\theta)$$

$$\Rightarrow \quad 2\cos^2\theta+4\cos\theta+3 = 2\cos^2\theta+6\cos\theta+4$$

$$\Rightarrow \quad \cos\theta = -\frac{1}{2} \Rightarrow \cos\theta = \cos 120° \Rightarrow \theta = 120°$$

Example 8. *The sum of magnitudes of two forces acting at a point is 16 and magnitude of their resultant is $8\sqrt{3}$. If the resultant is at 90° with the force of smaller magnitude, their magnitudes are*

(a) 3, 13 (b) 2, 14
(c) 5, 11 (d) 4, 12

Sol. (b) Given $A + B = 16$

$$R = (A^2+B^2+2AB\cos\theta)^{1/2}$$

$$8\sqrt{3} = (A^2+B^2+2AB\cos\theta)^{1/2}$$

and $\tan 90° = \dfrac{B\sin\theta}{A+B\cos\theta}$ or $\infty = \dfrac{B\sin\theta}{A+B\cos\theta}$

or $A + B\cos\theta = 0$

or $B\cos\theta = -A$

$\therefore$ $8\sqrt{3} = [A^2+B^2+2A(-A)]^{1/2}$

or $192 = B^2 - A^2 = (B-A)(B+A) = (B-A)\times 16$

or $B - A = 192/16 = 12$

On solving, $A = 2$ and $B = 14$

Subtraction of Vectors

Negative of a vector (−A) is a vector of the same magnitude as vector **A** but pointing in a direction opposite to that of **A**.

Therefore, **A** − **B** = **A** + (−**B**).

Let the angle between vectors **A** and **B** be θ, then the angle between **A** and − **B** will be 180° − θ.

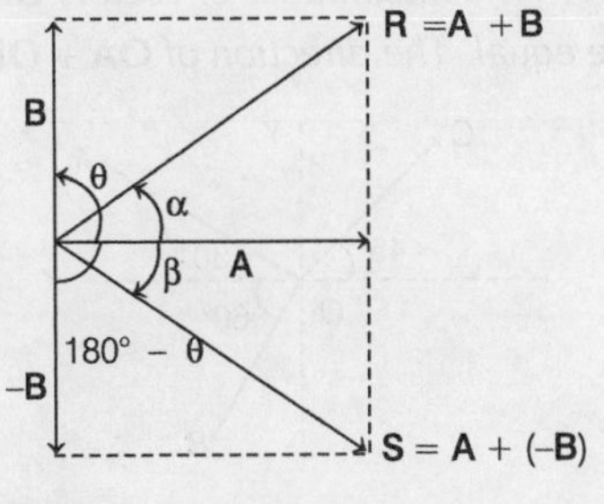

Magnitude of **S** = **A** − **B** will be given by

$$|\mathbf{S}| = |\mathbf{A}-\mathbf{B}|$$

$$= \sqrt{A^2+B^2+2AB\cos(180°-\theta)}$$

or $$S = \sqrt{A^2+B^2-2AB\cos\theta}$$

For direction of **S**, we will either calculate angle α or β,

$$\tan\alpha = \frac{B\sin(180°-\theta)}{A+B\cos(180°-\theta)} = \frac{B\sin\theta}{A-B\cos\theta}$$

$$\tan\beta = \frac{A\sin(180°-\theta)}{B+A\cos(180°-\theta)} = \frac{A\sin\theta}{B-A\cos\theta}$$

Note (i) **A** − **B** or **B** − **A** can also be found by making triangles as shown in figure.

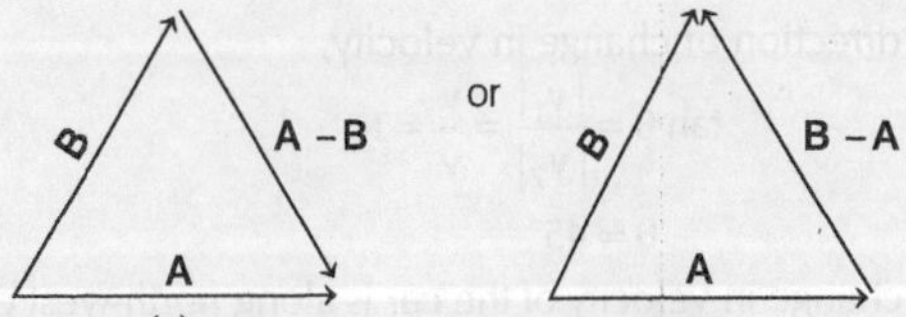

(a) (b)

(ii) Change in velocity of a particle moving along circular path with a constant speed

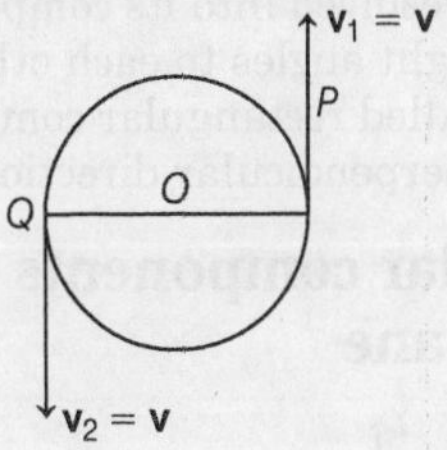

When a particle moves along a circular path with a constant speed, then its velocity changes due to change in direction.

$\therefore$ Change in velocity, $\Delta\mathbf{v} = \mathbf{v}_1 - \mathbf{v}_2 = \mathbf{v} - (-\mathbf{v}) = 2\mathbf{v}$

Example 9. *Find **A** − **B** from the diagram shown in figure. Given A = 4 units and B = 3 units*

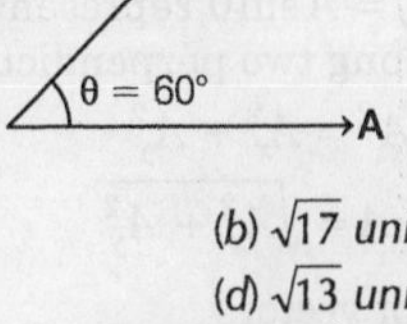

(a) $\sqrt{18}$ *units* (b) $\sqrt{17}$ *units*
(c) $\sqrt{14}$ *units* (d) $\sqrt{13}$ *units*

Sol. (d) Subtraction, $S = \sqrt{A^2+B^2-2AB\cos\theta}$

$$= \sqrt{16+9-2\times4\times3\cos 60°}$$

$$= \sqrt{13} \text{ units}$$

Example 10. *A car moving towards south changes its direction towards west moving with the same speed. Find the change in the direction of velocity of the car.*

(a) *North-West* (b) *North-East*
(c) *South-East* (d) *South-West*

Sol. (a)

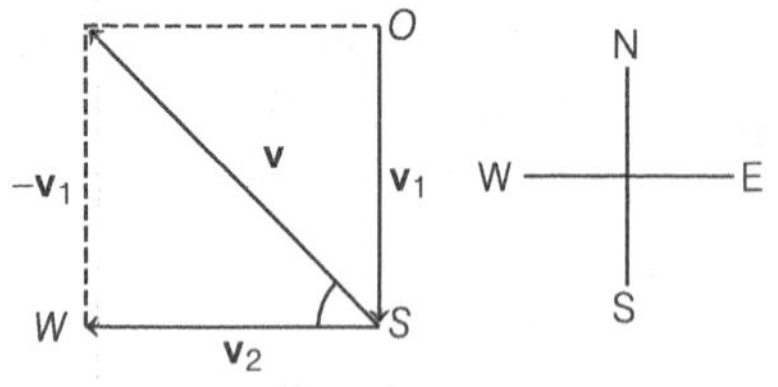

Here, $|\mathbf{v}_1| = |\mathbf{v}_2| = v$ (say)

$\therefore$ Change in velocity of car, $\Delta\mathbf{v} = \mathbf{v}_2 - \mathbf{v}_1$

Magnitude of the change in velocity,

$$|\Delta\mathbf{v}| = \sqrt{v_1^2 + v_2^2 - 2v_1v_2\cos 90^\circ}$$

$$= \sqrt{v^2 + v^2 - 0} = \sqrt{2v^2} = \sqrt{2}\,v$$

The direction of change in velocity,

$$\tan\theta = \frac{|\mathbf{v}_1|}{|\mathbf{v}_2|} = \frac{v}{v} = 1$$

or $\theta = 45^\circ$

The change in velocity of the car is along north-west direction.

Resolution of Vectors

When a vector is resolved into its components and the components are right angles to each other, then such components are called rectangular components of given vector along two perpendicular directions.

The rectangular components of a vector lying in the plane

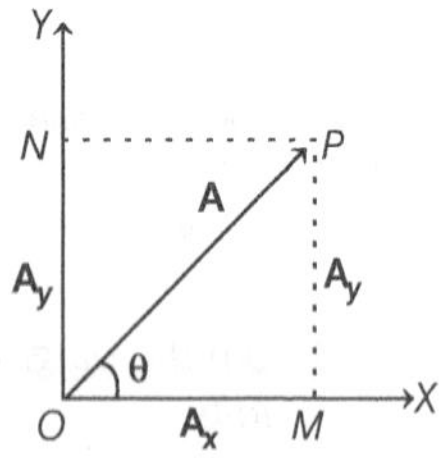

$$\mathbf{A} = A_x\hat{\mathbf{i}} + A_y\hat{\mathbf{j}}$$

If θ is the angle subtended by vector $\mathbf{A}$ with X-axis, then $A_x = A\cos\theta$ and $A_y = A\sin\theta$ represented the rectangular components of A along two perpendicular directions.

$\therefore$ $$A^2 = A_x^2 + A_y^2$$

or $$A = \sqrt{A_x^2 + A_y^2}$$

For the directions of vectors

$$\tan\theta = \frac{A_y}{A_x}$$

or $$\theta = \tan^{-1}\left(\frac{A_y}{A_x}\right)$$

The rectangular components of a vector lying in the space

Suppose there is a vector $\mathbf{A}$ in space as shown in the figure. Let the rectangular components of $\mathbf{A}$ along X-axis, Y-axis and Z-axis be $\mathbf{A}_x$, $\mathbf{A}_y$ and $\mathbf{A}_z$ respectively.

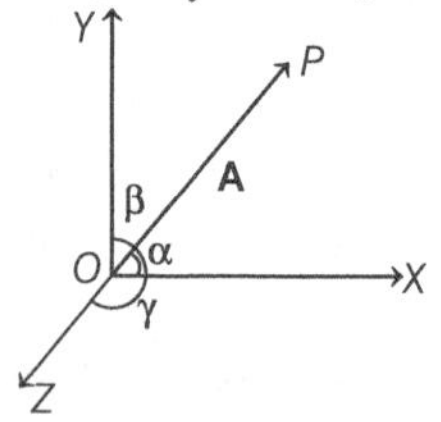

$\therefore$ $$\mathbf{A} = A_x\hat{\mathbf{i}} + A_y\hat{\mathbf{j}} + A_z\hat{\mathbf{k}}$$

$\therefore$ $$|A| = \sqrt{A_x^2 + A_y^2 + A_z^2}$$

This vector makes angle

$$\alpha = \cos^{-1}\left(\frac{A_x}{A}\right) \text{ with } X\text{-axis}$$

$$\beta = \cos^{-1}\left(\frac{A_y}{A}\right) \text{ with } Y\text{-axis}$$

$$\gamma = \cos^{-1}\left(\frac{A_z}{A}\right) \text{ with } Z\text{-axis}$$

Example 11. *A force of 8N makes an angle 30° with X-axis. Find the x and y components of the force.*

(a) $F_x = 4\sqrt{3}\,N, F_y = 4\,N$ (b) $F_x = 4\,N, F_y = 4\sqrt{3}\,N$
(c) $F_x = 2\,N, F_y = 2\sqrt{3}\,N$ (d) $F_x = 2\sqrt{3}\,N, F_y = 2\,N$

Sol. (a) Here, $F = 8$ N, $\theta = 30^\circ$

x-component of force, $F_x = F\cos 30^\circ = 8 \times \frac{\sqrt{3}}{2} = 4\sqrt{3}$ N

y-component of force, $F_y = F\sin 30^\circ = 8 \times \frac{1}{2} = 4$ N

Example 12. *A force of 10.5 N acts on a particle along a direction making an angle of 37° with the vertical. The component of force in the vertical direction is*

(a) 4.5 *N* (b) 8.4 *N*
(c) 6.9 *N* (d) 3.2 *N*

Sol. (*b*) The component of the force in the vertical direction will be

$$F_v = F\cos\theta = 10.5\cos 37^\circ = 10.5 \times \frac{4}{5} = 8.4\text{ N}$$

Example 13. *The magnitudes of vectors* **OA**, **OB** *and* **OC** *as in figure are equal. The direction of* **OA** + **OB** − **OC** *is*

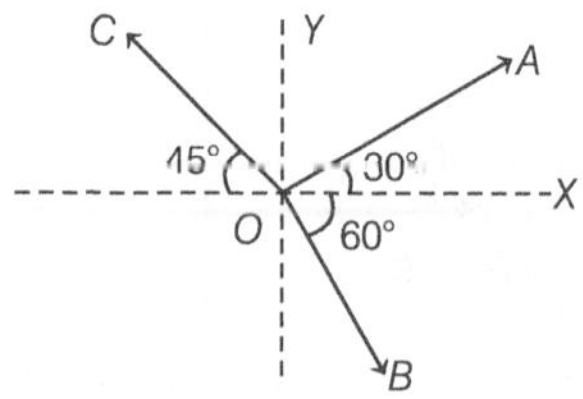

(a) $\tan^{-1}\left[\frac{(1-\sqrt{3}-\sqrt{2})}{(1+\sqrt{3}+\sqrt{2})}\right]$ (b) $\tan^{-1}\left[\frac{(1+\sqrt{3}+\sqrt{2})}{(1-\sqrt{3}-\sqrt{2})}\right]$

(c) $\cot^{-1}\left[\frac{(1-\sqrt{3}-\sqrt{2})}{(1+\sqrt{3}+\sqrt{2})}\right]$ (d) $\cot^{-1}\left[\frac{(1+\sqrt{3}+\sqrt{2})}{(1-\sqrt{3}-\sqrt{2})}\right]$

Sol. (a) Let, $OA = OB = OC = F$

x-component of $\mathbf{OA} = F\cos 30° = F\frac{\sqrt{3}}{2}$

x-component of $\mathbf{OB} = F\cos 60° = \frac{F}{2}$

x-component of $\mathbf{OC} = F\cos 135° = -\frac{F}{\sqrt{2}}$

$\therefore$ x-component of $\mathbf{OA}+\mathbf{OB}-\mathbf{OC}$

$$= \left(\frac{F\sqrt{3}}{3}\right)+\left(\frac{F}{2}\right)-\left(-\frac{F}{\sqrt{2}}\right)$$

$$= \frac{F}{2}(\sqrt{3}+1+\sqrt{2})$$

y-component of $\mathbf{OA} = F\sin 30° = \frac{F}{2}$

y-component of $\mathbf{OB} = -F\sin 60° = -\frac{F\sqrt{3}}{2}$

y-component of $\mathbf{OC} = F\sin 45° = \frac{F}{\sqrt{2}}$

$\therefore$ y-component of $\mathbf{OA}+\mathbf{OB}-\mathbf{OC}$

$$= \left(\frac{F}{2}\right)+\left(-\frac{F\sqrt{3}}{2}\right)-\left(\frac{F}{\sqrt{2}}\right)=\frac{F}{2}(1-\sqrt{3}-\sqrt{2})$$

Angle of $\mathbf{OA}+\mathbf{OB}-\mathbf{OC}$ with x-axis

$$= \tan^{-1}\left[\frac{\frac{F}{2}(1-\sqrt{3}-\sqrt{2})}{\frac{F}{2}(1+\sqrt{3}+\sqrt{2})}\right] = \tan^{-1}\left[\frac{(1-\sqrt{3}-\sqrt{2})}{(1+\sqrt{3}+\sqrt{2})}\right]$$

Scalar Product or Dot Product

The scalar product of two vectors **A** and **B** is defined as the product of magnitude of **A** and **B** multiplied by the cosine of the smaller angle between them.

i.e. $\mathbf{A}\cdot\mathbf{B} = AB\cos\theta$

or $\mathbf{A}\cdot\mathbf{B} = A(B\cos\theta)$ = (magnitude of **A**) (component of **B** in the direction of **A**).

Dot product or scalar product of two vectors gives the scalar quantity.

Some Important Properties of Dot Product

- Dot product is commutative $\mathbf{A}\cdot\mathbf{B} = AB\cos\theta$ and $\mathbf{B}\cdot\mathbf{A} = BA\cos\theta = AB\cos\theta$

  $\therefore \mathbf{A}\cdot\mathbf{B} = \mathbf{B}\cdot\mathbf{A}$, which is commutative law.
- Dot product is distributive over the addition of vectors

  i.e., $\mathbf{A}\cdot(\mathbf{B}+\mathbf{C}) = \mathbf{A}\cdot\mathbf{B}+\mathbf{A}\cdot\mathbf{C}$
- Dot product of two like parallel vectors

  Here, $\theta = 0°$

  $\therefore A\cdot B = AB\cos 0° = AB$ $(\because \cos 0° = 1)$
- Dot product of two antiparallel vectors

  Here, $\theta = 180°$

  $\therefore \mathbf{A}\cdot\mathbf{B} = AB\cos 180° = -AB$ $(\because \cos 180° = -1)$
- Dot product of orthogonal unit vectors

  $$\hat{\mathbf{i}}\cdot\hat{\mathbf{i}} = \hat{\mathbf{j}}\cdot\hat{\mathbf{j}} = \hat{\mathbf{k}}\cdot\hat{\mathbf{k}} = 1$$

  $$\hat{\mathbf{i}}\cdot\hat{\mathbf{j}} = \hat{\mathbf{j}}\cdot\hat{\mathbf{k}} = \hat{\mathbf{k}}\cdot\hat{\mathbf{i}} = 0$$
- Dot product of two vectors in terms of their components

  $$\mathbf{A}\cdot\mathbf{B} = (x_1\hat{\mathbf{i}}+y_1\hat{\mathbf{j}}+z_1\hat{\mathbf{k}})\cdot(x_2\hat{\mathbf{i}}+y_2\hat{\mathbf{j}}+z_2\hat{\mathbf{k}})$$

  or $\mathbf{A}\cdot\mathbf{B} = x_1x_2+y_1y_2+z_1z_2$

Example 14. *If* **A** *and* **B** *are perpendicular vectors, where* $\mathbf{A} = 5\hat{\mathbf{i}}+7\hat{\mathbf{j}}-3\hat{\mathbf{k}}$ *and* $\mathbf{B} = 2\hat{\mathbf{i}}+2\hat{\mathbf{j}}-a\hat{\mathbf{k}}$, *then the value of a is*

(a) -2 (b) 8 (c) -7 (d) -8

Sol. (d) For perpendicular vectors, $\mathbf{A}\cdot\mathbf{B} = 0$

So, $(5\hat{\mathbf{i}}+7\hat{\mathbf{j}}-3\hat{\mathbf{k}})\cdot(2\hat{\mathbf{i}}+2\hat{\mathbf{j}}-a\hat{\mathbf{k}}) = 0$

or $10+14+3a = 0$ or $3a = -24$

or $a = -8$

Example 15. *The angle between the two vectors* $\mathbf{A} = 3\hat{\mathbf{i}}+4\hat{\mathbf{j}}+5\hat{\mathbf{k}}$ *and* $\mathbf{B} = 3\hat{\mathbf{i}}+4\hat{\mathbf{j}}-5\hat{\mathbf{k}}$ *will be*

(a) zero (b) 45° (c) 90° (d) 180°

Sol. (c) $\mathbf{A}\cdot\mathbf{B} = (3\hat{\mathbf{i}}+4\hat{\mathbf{j}}+5\hat{\mathbf{k}})\cdot(3\hat{\mathbf{i}}+4\hat{\mathbf{j}}-5\hat{\mathbf{k}}) = 9+16-25 = 0$

or $AB\cos\theta = 0$

or $\cos\theta = 0$

or $\theta = 90°$

Example 16. *The vectors* $\mathbf{P} = a\hat{\mathbf{i}}+a\hat{\mathbf{j}}+3\hat{\mathbf{k}}$ *and* $\mathbf{Q} = a\hat{\mathbf{i}}-2\hat{\mathbf{j}}-\hat{\mathbf{k}}$ *are perpendicular to each other. The positive value of a is*

(a) 3 (b) 2 (c) 1 (d) 0

Sol. (a) For perpendicular vectos $\mathbf{P}\cdot\mathbf{Q} = 0$

So, $(a\hat{\mathbf{i}}+a\hat{\mathbf{j}}+3\hat{\mathbf{k}})\cdot(a\hat{\mathbf{i}}-2\hat{\mathbf{j}}-\hat{\mathbf{k}}) = 0$

or $a^2-2a-3 = 0$

On solving $a = 3$ or -1

Example 17. *What is the angle* ϕ *between* $\mathbf{a} = 3.0\,\hat{\mathbf{i}}-4.0\,\hat{\mathbf{j}}$ *and* $\mathbf{b} = 2.0\,\hat{\mathbf{i}}-3.0\,\hat{\mathbf{k}}$?

(a) 60° (b) 53°
(c) 71° (d) 75°

Sol. (c) Angle between **a** and **b** is given as

$$\cos\phi = \frac{\mathbf{a}\cdot\mathbf{b}}{ab} = \frac{3\times2+(-4\times0)+0\times(-3)}{\sqrt{3^2+(-4)^2+0^2}\cdot\sqrt{2^2+0^2+(-3)^2}}$$

$$\Rightarrow \cos\phi = \frac{6}{5\sqrt{13}} \Rightarrow \cos\phi = 0.333$$

$$\Rightarrow \phi = \cos^{-1}(0.333) = 70.55° \simeq 71°$$

Vector Product or Cross Product

The vector product or cross product of two vectors is a single vector whose magnitude is equal to the product of the magnitudes of two given vectors multiplied by the sine of the smaller angle between the two given vectors.

i.e. $\mathbf{A} \times \mathbf{B} = (AB \sin\theta)\,\hat{\mathbf{n}}$

The unit vector normal to the plane containing vectors A and **B**, is given by $\hat{\mathbf{n}} = \dfrac{\mathbf{A}\times\mathbf{B}}{AB\sin\theta} = \dfrac{\mathbf{A}\times\mathbf{B}}{|\mathbf{A}\times\mathbf{B}|}$

Some Important Properties of Cross Product

- Cross product of two vectors does not obey the commutative law.

  i.e. $\mathbf{A}\times\mathbf{B} \neq \mathbf{B}\times\mathbf{A}$

  Here, $\mathbf{A}\times\mathbf{B} = -\mathbf{B}\times\mathbf{A}$

- Cross product of two vectors is distributive over vector addition

  i.e. $\mathbf{A}\times(\mathbf{B}+\mathbf{C}) = \mathbf{A}\times\mathbf{B} + \mathbf{A}\times\mathbf{C}$

- Cross product of two like parallel vectors is zero. In this case, the angle between vectors will be zero degree.

  $\therefore\quad \mathbf{A}\times\mathbf{B} = (AB\sin 0^\circ)\,\hat{\mathbf{n}} = 0 \qquad [\because \sin 0^\circ = 0]$

- Cross product of two perpendicular vectors : In this case, $\theta = 90^\circ$

  $$\mathbf{A}\times\mathbf{B} = (AB\sin 90^\circ)\,\hat{\mathbf{n}} = (AB)\,\hat{\mathbf{n}}$$

- Cross product of orthogonal unit vectors

  $$\hat{\mathbf{i}}\times\hat{\mathbf{i}} = (1\times 1\times \sin 0^\circ)\,\hat{\mathbf{n}} = 0,$$

  Similarly, $\hat{\mathbf{j}}\times\hat{\mathbf{j}} = 0$ and $\hat{\mathbf{k}}\times\hat{\mathbf{k}} = 0$

  Also, $\hat{\mathbf{i}}\times\hat{\mathbf{j}} = 1\times 1\times\sin 90^\circ\,\hat{\mathbf{k}} = \hat{\mathbf{k}}$

  Similarly, $\hat{\mathbf{j}}\times\hat{\mathbf{k}} = \hat{\mathbf{i}}, \hat{\mathbf{k}}\times\hat{\mathbf{i}} = \hat{\mathbf{j}}$

  Now, $\hat{\mathbf{j}}\times\hat{\mathbf{i}} = -\hat{\mathbf{k}}, \hat{\mathbf{i}}\times\hat{\mathbf{k}} = -\hat{\mathbf{j}}$ and $\hat{\mathbf{k}}\times\hat{\mathbf{j}} = -\hat{\mathbf{i}}$

- Cross product of two vectors in terms of their rectangular components

  $$\mathbf{A}\times\mathbf{B} = (x_1\hat{\mathbf{i}} + y_1\hat{\mathbf{j}} + z_1\hat{\mathbf{k}})\times(x_2\hat{\mathbf{i}} + y_2\hat{\mathbf{j}} + z_2\hat{\mathbf{k}})$$

  $$= \begin{vmatrix} \hat{\mathbf{i}} & \hat{\mathbf{j}} & \hat{\mathbf{k}} \\ x_1 & y_1 & z_1 \\ x_2 & y_2 & z_2 \end{vmatrix}$$

- Magnitude of cross product of two vectors **A** and **B** represents the area of the parallelogram.

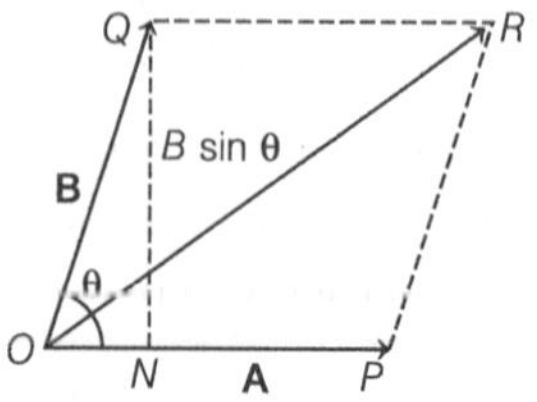

$\therefore |\mathbf{A}\times\mathbf{B}| = AB\sin\theta = A(B\sin\theta) = OP\times QN =$ Area of the parallelogram.

Triple Product

Scalar triple product of three vectors is given by

$$\mathbf{A}\cdot(\mathbf{B}\times\mathbf{C}) = \begin{vmatrix} a_1 & a_2 & a_3 \\ b_1 & b_2 & b_3 \\ c_1 & c_2 & c_3 \end{vmatrix}$$

where $\mathbf{A} = a_1\hat{\mathbf{i}} + a_2\hat{\mathbf{j}} + a_3\hat{\mathbf{k}}$,

$$\mathbf{B} = b_1\hat{\mathbf{i}} + b_2\hat{\mathbf{j}} + b_3\hat{\mathbf{k}} \text{ and } \mathbf{C} = c_1\hat{\mathbf{i}} + c_2\hat{\mathbf{j}} + c_3\hat{\mathbf{k}}$$

It gives volume of parallelopiped formed with **A, B** and **C** as adjacent sides.

Vector triple product is given by

$$\mathbf{A}\times(\mathbf{B}\times\mathbf{C}) = \mathbf{B}(\mathbf{A}\cdot\mathbf{C}) - \mathbf{C}(\mathbf{A}\cdot\mathbf{B})$$

It is worth noting that

(i) $\mathbf{A}\cdot(\mathbf{B}\times\mathbf{C}) = 0$ implies that vectors are coplanar.

(ii) In scalar triple product, dot and cross can be interchanged provided that their cyclic order is maintained.

(iii) Four points *A, B, C* and *D* are coplanar, if $\mathbf{AB}\cdot(\mathbf{BC}\times\mathbf{CD}) = 0$.

Example 18. *The vector* **A** *has a magnitude of 5 unit,* **B** *has a magnitude of 6 unit and the cross product* **A** *and* **B** *has the magnitude of 15 unit. The angle between* **A** *and* **B** *is*

(a) 90° (b) 60° (c) 30° (d) 120°

Sol. (c) If the angle between **A** and **B** is θ, the cross product will have a magnitude,

$$|\mathbf{A}\times\mathbf{B}| = AB\sin\theta \text{ or } 15 = 5\times 6\sin\theta \text{ or } \sin\theta = \frac{1}{2}$$

$$\therefore\quad \theta = 30^\circ$$

Example 19. *If* $\mathbf{A} = \hat{\mathbf{i}} + 2\hat{\mathbf{k}}$ *and* $\mathbf{B} = \hat{\mathbf{j}} - \hat{\mathbf{k}}$*, then,* $\mathbf{A}\times\mathbf{B}$ *is equal to*

(a) $2\hat{\mathbf{i}} - \hat{\mathbf{j}} - \hat{\mathbf{k}}$ (b) $-2\hat{\mathbf{i}} + \hat{\mathbf{j}} + \hat{\mathbf{k}}$ (c) $\hat{\mathbf{i}} + \hat{\mathbf{j}} - \hat{\mathbf{k}}$ (d) $2\hat{\mathbf{i}} + \hat{\mathbf{j}} - \hat{\mathbf{k}}$

Sol. (b) We have, $\mathbf{A}\times\mathbf{B} = (\hat{\mathbf{i}} + 2\hat{\mathbf{k}})\times(\hat{\mathbf{j}} - \hat{\mathbf{k}})$

$$= \begin{vmatrix} \hat{\mathbf{i}} & \hat{\mathbf{j}} & \hat{\mathbf{k}} \\ 1 & 0 & 2 \\ 0 & 1 & -1 \end{vmatrix}$$

$$= \hat{\mathbf{i}}(0-2) + \hat{\mathbf{j}}(0+1) + \hat{\mathbf{k}}(1-0) = -2\hat{\mathbf{i}} + \hat{\mathbf{j}} + \hat{\mathbf{k}}$$

Example 20. *The area of a parallelogram whose adjacent sides are* $\mathbf{P} = 2\hat{\mathbf{i}} + 3\hat{\mathbf{j}}$ *and* $\mathbf{Q} = \hat{\mathbf{i}} + 4\hat{\mathbf{j}}$ *is*

(a) 5 square units (b) 15 square units
(c) 20 square units (d) 25 square units

Sol. (a) $\mathbf{P}\times\mathbf{Q} = (2\hat{\mathbf{i}} + 3\hat{\mathbf{j}})\times(\hat{\mathbf{i}} + 4\hat{\mathbf{j}})$

$$= \begin{vmatrix} \hat{\mathbf{i}} & \hat{\mathbf{j}} & \hat{\mathbf{k}} \\ 2 & 3 & 0 \\ 1 & 4 & 0 \end{vmatrix} = \hat{\mathbf{i}}(0-0) - \hat{\mathbf{j}}(0-0) + \hat{\mathbf{k}}(8-3) = 5\hat{\mathbf{k}}$$

Area of parallelogram $= |\mathbf{P}\times\mathbf{Q}| = 5$ square units

Practice Exercise

ROUND I Topically Divided Problems

Addition and Subtraction of Vectors

1. Consider the quantities, pressure, power, energy, impulse, gravitational potential, electrical charge, temperature, area. Out of these, the only vector quantities are **[NCERT Exemplar]**

(a) Impulse, pressure and area
(b) Impulse and area
(c) Area and gravitational potential
(d) Impulse and pressure

2. What is the numerical value of the vector $3\hat{\mathbf{i}}+4\hat{\mathbf{j}}+5\hat{\mathbf{k}}$?

(a) $3\sqrt{2}$ (b) $5\sqrt{2}$
(c) $7\sqrt{2}$ (d) $9\sqrt{2}$

3. If, $0.5\,\hat{\mathbf{i}}+0.8\,\hat{\mathbf{j}}+c\,\hat{\mathbf{k}}$ is a unit vector, then the value of c is

(a) $\sqrt{0.11}$ (b) $\sqrt{0.22}$
(c) $\sqrt{0.33}$ (d) $\sqrt{0.89}$

4. Which one of the following statements is true? **[NCERT Exemplar]**

(a) A scalar quantity is the one that is conserved in a process.
(b) A scalar quantity is the one that can never take negative values.
(c) A scalar quantity is the one that does not vary from one point to another in space.
(d) A scalar quantity has the same value for observers with different orientations of the axes.

5. Given, vector, $\mathbf{A}=\hat{\mathbf{i}}-\hat{\mathbf{j}}+2\,\hat{\mathbf{k}}$ and vector $\mathbf{B}=3\,\hat{\mathbf{i}}-3\,\hat{\mathbf{j}}+6\,\hat{\mathbf{k}}$, then which one of the following statements is true?

(a) A is perpendicular to B.
(b) A is parallel to B.
(c) Magnitude of A is half of that of B.
(d) Magnitude of B is equal to that of A.

6. What is the unit vector along $\hat{\mathbf{i}}+\hat{\mathbf{j}}$?

(a) $\dfrac{\hat{\mathbf{i}}+\hat{\mathbf{j}}}{\sqrt{2}}$ (b) $\sqrt{2}\,(\hat{\mathbf{i}}+\hat{\mathbf{j}})$
(c) $\hat{\mathbf{i}}+\hat{\mathbf{j}}$ (d) $\hat{\mathbf{k}}$

7. Figure shows the orientation of two vectors **u** and **v** in the XY-plane. **[NCERT Exemplar]**

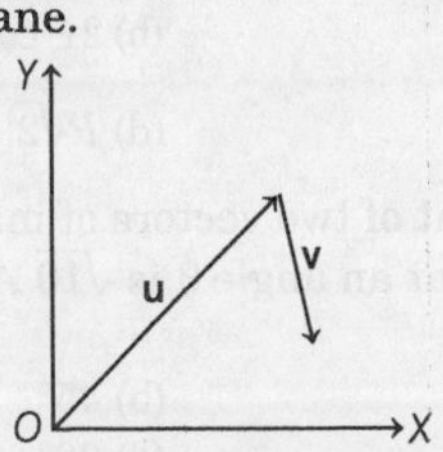

If $\mathbf{u}=a\hat{\mathbf{i}}+b\hat{\mathbf{j}}$ and $\mathbf{v}=p\hat{\mathbf{i}}+q\hat{\mathbf{j}}$, which of the following is correct?

(a) a and p are positive while b and q are negative
(b) a, p and b are positive while q is negative
(c) a, q and b are positive while p is negative
(d) a, b, p and q are all positive

8. In the cube of side 'a' shown in the figure, the vector from the central point of the face $ABOD$ to the central point of the face $BEFO$ will be **[JEE Main 2019]**

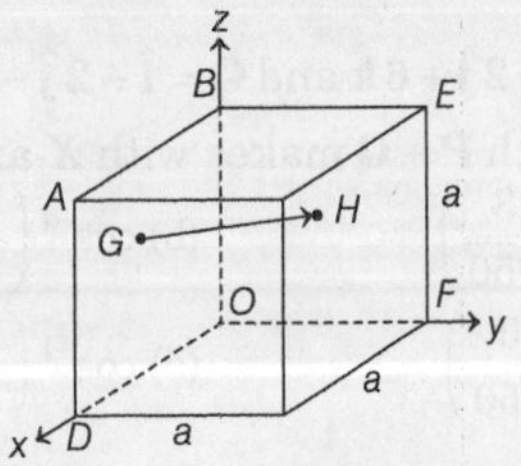

(a) $\dfrac{1}{2}a\,(\hat{\mathbf{i}}-\hat{\mathbf{k}})$ (b) $\dfrac{1}{2}a\,(\hat{\mathbf{j}}-\hat{\mathbf{i}})$
(c) $\dfrac{1}{2}a\,(\hat{\mathbf{j}}-\hat{\mathbf{k}})$ (d) $\dfrac{1}{2}a\,(\hat{\mathbf{k}}-\hat{\mathbf{i}})$

9. Two forces, each equal to $\dfrac{p}{2}$ act at right angles. Their effect may be neutralised by a third force acting along their bisector in the opposite direction with a magnitude of

(a) p (b) $\dfrac{p}{2}$ (c) $\dfrac{p}{\sqrt{2}}$ (d) $\sqrt{2}\,p$

10. Given $\mathbf{A}=\hat{\mathbf{i}}+2\hat{\mathbf{j}}-3\,\hat{\mathbf{k}}$. When a vector **B** is added to **A**, we get a unit vector along X-axis, then **B** is

(a) $-2\hat{\mathbf{j}}+3\,\hat{\mathbf{k}}$ (b) $-\hat{\mathbf{i}}-2\hat{\mathbf{j}}$
(c) $-\hat{\mathbf{i}}+3\,\hat{\mathbf{k}}$ (d) $2\hat{\mathbf{j}}-3\,\hat{\mathbf{k}}$

11. Two forces $\mathbf{F}_1$ and $\mathbf{F}_2$ are acting at right angles to each other, then their resultant is

(a) $F_1 + F_2$ (b) $\sqrt{F_1^{\ 2} + F_2^{\ 2}}$
(c) $\sqrt{F_1^{\ 2} - F_2^{\ 2}}$ (d) $\dfrac{F_1 + F_2}{2}$

12. Given, $\mathbf{R} = \mathbf{A} + \mathbf{B}$ and $R = A = B$. The angle between $\mathbf{A}$ and $\mathbf{B}$ is

(a) 60° (b) 90°
(c) 120° (d) 180°

13. The resultant of two forces, each $\mathbf{P}$, acting at an angle θ is

(a) $2P\sin\dfrac{\theta}{2}$ (b) $2P\cos\dfrac{\theta}{2}$
(c) $2P\cos\theta$ (d) $P\sqrt{2}$

14. The resultant of two vectors of magnitudes $2A$ and $\sqrt{2}\,A$ acting at an angle θ is $\sqrt{10}\,A$. The correct value of θ is

(a) 30° (b) 45°
(c) 60° (d) 90°

15. Two forces, each equal to F, act as shown in figure. Their resultant is

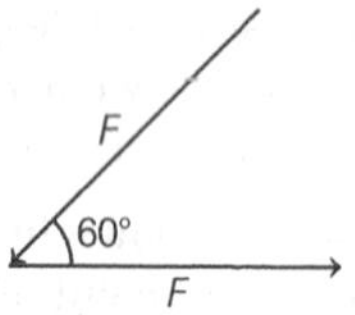

(a) $\dfrac{F}{2}$ (b) F (c) $\sqrt{3}\,F$ (d) $\sqrt{5}\,F$

16. If $\mathbf{P} = 4\hat{\mathbf{i}} - 2\hat{\mathbf{j}} + 6\hat{\mathbf{k}}$ and $\mathbf{Q} = \hat{\mathbf{i}} - 2\hat{\mathbf{j}} - 3\hat{\mathbf{k}}$, then the angle which $\mathbf{P} + \mathbf{Q}$ makes with X-axis is

(a) $\cos^{-1}\left(\dfrac{3}{\sqrt{50}}\right)$ (b) $\cos^{-1}\left(\dfrac{4}{\sqrt{50}}\right)$
(c) $\cos^{-1}\left(\dfrac{5}{\sqrt{50}}\right)$ (d) $\cos^{-1}\left(\dfrac{12}{\sqrt{50}}\right)$

17. If $\mathbf{A} + \mathbf{B} = \mathbf{C}$ and $A = \sqrt{3}, B = \sqrt{3}$ and $C = 3$, then the angle between $\mathbf{A}$ and $\mathbf{B}$ is

(a) 0° (b) 30° (c) 60° (d) 90°

18. If the magnitude of the sum of the two vectors is equal to the difference of their magnitudes, then the angle between vectors is

(a) 0° (b) 45° (c) 90° (d) 180°

19. If, the resultant of two forces $(A + B)$ and $(A - B)$ is $\sqrt{A^2 + B^2}$, then the angle between these forces is

(a) $\cos^{-1}\left[-\dfrac{(A^2 - B^2)}{A^2 + B^2}\right]$ (b) $\cos^{-1}\left[-\dfrac{(A^2 + B^2)}{(A^2 - B^2)}\right]$
(c) $\cos^{-1}\left[-\dfrac{A^2 + B^2}{2(A^2 - B^2)}\right]$ (d) $\cos^{-1}\left[-\dfrac{2(A^2 + B^2)}{A^2 - B^2}\right]$

20. If the resultant of $\mathbf{A}$ and $\mathbf{B}$ makes angle α with $\mathbf{A}$ and β with $\mathbf{B}$, then

(a) $\alpha < \beta$, always (b) $\alpha < \beta$, if $A < B$
(c) $\alpha < \beta$, if $A > B$ (d) $\alpha < \beta$, if $A = B$

21. If the resultant of the vectors $(\hat{\mathbf{i}} + 2\hat{\mathbf{j}} - \hat{\mathbf{k}}), (\hat{\mathbf{i}} - \hat{\mathbf{j}} + 2\hat{\mathbf{k}})$ and $\mathbf{C}$ is a unit vector along the y-direction, then $\mathbf{C}$ is

(a) $-2\hat{\mathbf{i}} - \hat{\mathbf{k}}$ (b) $-2\hat{\mathbf{i}} + \hat{\mathbf{k}}$ (c) $2\hat{\mathbf{i}} - \hat{\mathbf{k}}$ (d) $-2\hat{\mathbf{i}} + \hat{\mathbf{k}}$

22. $(\mathbf{P} + \mathbf{Q})$ is a unit vector along X-axis. If $\mathbf{P} = \hat{\mathbf{i}} - \hat{\mathbf{j}} + \hat{\mathbf{k}}$, then what is the value of vector $\mathbf{Q}$?

(a) $\hat{\mathbf{i}} + \hat{\mathbf{j}} - \hat{\mathbf{k}}$ (b) $\hat{\mathbf{j}} - \hat{\mathbf{k}}$ (c) $\hat{\mathbf{i}} + \hat{\mathbf{j}} + \hat{\mathbf{k}}$ (d) $\hat{\mathbf{j}} + \hat{\mathbf{k}}$

23. What vector must be added to the sum of two vectors $2\hat{\mathbf{i}} - \hat{\mathbf{j}} + 3\hat{\mathbf{k}}$ and $3\hat{\mathbf{i}} - 2\hat{\mathbf{j}} - 2\hat{\mathbf{k}}$ so that the resultant is a unit vector along Z-axis?

(a) $5\hat{\mathbf{i}} + \hat{\mathbf{k}}$ (b) $-5\hat{\mathbf{i}} + 3\hat{\mathbf{j}}$ (c) $3\hat{\mathbf{j}} + 5\hat{\mathbf{k}}$ (d) $-3\hat{\mathbf{j}} + 2\hat{\mathbf{k}}$

24. Two vectors $\mathbf{a}$ and $\mathbf{b}$ are at an angle of 60° with each other. Their resultant makes an angle of 45° with $\mathbf{a}$. If $|\mathbf{b}| = 2$ units, then $|\mathbf{a}|$ is

(a) $\sqrt{3}$ (b) $\sqrt{3} - 1$ (c) $\sqrt{3} + 1$ (d) $\sqrt{3}/2$

25. A vector $\mathbf{A}$ when added to the vector $\mathbf{B} = 3\hat{\mathbf{i}} + 4\hat{\mathbf{j}}$ yields a resultant vector that is in the positive y-direction and has a magnitude equal to that of $\mathbf{B}$. Find the magnitude of $\mathbf{A}$.

(a) $\sqrt{10}$ (b) 10 (c) 5 (d) $\sqrt{15}$

26. It is found that $|\mathbf{A} + \mathbf{B}| = |\mathbf{A}|$. This necessarily implies, **[NCERT Exemplar]**

(a) $\mathbf{B} = 0$ (b) $\mathbf{A}, \mathbf{B}$ are parallel
(c) $\mathbf{A}, \mathbf{B}$ are perpendicular (d) $\mathbf{A} \cdot \mathbf{B} \le 0$

27. The simple sum of two co-initial vectors is 16 units. Their vector sum is 8 units. The resultant of the vectors is perpendicular to the smaller vector. The magnitudes of the two vectors are

(a) 2 units and 14 units (b) 4 units and 12 units
(c) 6 units and 10 units (d) 8 units and 8 units

28. The resultant of two forces at right angle is 5N. When the angle between them is 120°, the resultant is $\sqrt{13}$. Then, the forces are

(a) $\sqrt{12}$ N, $\sqrt{13}$ N (b) $\sqrt{20}$ N, $\sqrt{5}$ N
(c) 3 N, 4 N (d) $\sqrt{40}$ N, $\sqrt{15}$ N

29. Given, $\mathbf{P} = \mathbf{A} + \mathbf{B}$ and $P = A + B$. The angle between $\mathbf{A}$ and $\mathbf{B}$ is

(a) 0° (b) $\dfrac{\pi}{4}$ (c) $\dfrac{\pi}{2}$ (d) π

30. What is the angle between $\mathbf{P}$ and the resultant of $(\mathbf{P} + \mathbf{Q})$ and $(\mathbf{P} - \mathbf{Q})$?

(a) zero (b) $\tan^{-1}(P/Q)$
(c) $\tan^{-1}(Q/P)$ (d) $\tan^{-1}(P - Q)/(P + Q)$

31. Two vectors **a** and **b** are such that |**a** + **b**| = |**a** − **b**|. What is the angle between **a** and **b**?
(a) 0° (b) 90° (c) 60° (d) 180°

32. The vectors $\mathbf{A}_1$ and $\mathbf{A}_2$ each of magnitude A are inclined to each other such that their resultant is equal to $\sqrt{3}A$. Then the resultant of $\mathbf{A}_1$ and $-\mathbf{A}_2$ is
(a) $2A$ (b) $\sqrt{3}A$ (c) $\sqrt{2}A$ (d) A

33. If $\mathbf{A} = \hat{\mathbf{i}} + \hat{\mathbf{j}} - 2\hat{\mathbf{k}}$ and $\mathbf{B} = 2\hat{\mathbf{i}} - \hat{\mathbf{j}} + \hat{\mathbf{k}}$, then the magnitude of $2\mathbf{A} - 3\mathbf{B}$ is
(a) $\sqrt{90}$ (b) $\sqrt{50}$
(c) $\sqrt{190}$ (d) $\sqrt{30}$

34. The vector which can give unit vector along X-axis with $\mathbf{A} = 2\hat{\mathbf{i}} - 4\hat{\mathbf{j}} + 7\hat{\mathbf{k}}$, $\mathbf{B} = 7\hat{\mathbf{i}} + 2\hat{\mathbf{j}} - 5\hat{\mathbf{k}}$ and $\mathbf{C} = -4\hat{\mathbf{i}} + 7\hat{\mathbf{j}} + 3\hat{\mathbf{k}}$ is
(a) $4\hat{\mathbf{i}} + 5\hat{\mathbf{j}} + 5\hat{\mathbf{k}}$ (b) $-5\hat{\mathbf{i}} - 5\hat{\mathbf{j}} + 5\hat{\mathbf{k}}$
(c) $-5\hat{\mathbf{i}} - 5\hat{\mathbf{j}} - 5\hat{\mathbf{k}}$ (d) $4\hat{\mathbf{i}} - 5\hat{\mathbf{j}} - 5\hat{\mathbf{k}}$

Resolution of Vectors

35. The x and y components of a force are 2 N and − 3N. The force is
(a) $2\hat{\mathbf{i}} - 3\hat{\mathbf{j}}$ (b) $2\hat{\mathbf{i}} + 3\hat{\mathbf{j}}$
(c) $-2\hat{\mathbf{i}} - 3\hat{\mathbf{j}}$ (d) $3\hat{\mathbf{i}} + 2\hat{\mathbf{j}}$

36. The magnitude of the X and Y components of **A** are 7 and 6. Also the magnitudes of X and Y components of **A** + **B** are 11 and 9, respectively. What is the magnitude of **B**?
(a) 5 (b) 6
(c) 8 (d) 9

37. One of the rectangular components of a velocity of 60 kmh^{-1} is 30 kmh^{-1}. The other rectangular component is
(a) 30 km h^{-1} (b) $30\sqrt{3}$ km h^{-1}
(c) $30\sqrt{2}$ km h^{-1} (d) zero

38. A force is inclined at 60° to the horizontal. If its rectangular component in the horizontal direction is 50 N, then magnitude of the force in the vertical direction is
(a) 25 N (b) 75 N
(c) 87 N (d) 100 N

39. The component of a vector **r** along X-axis will have maximum value, if **[NCERT Exemplar]**
(a) **r** is along positive Y-axis
(b) **r** is along positive X-axis
(c) **r** makes an angle of 45° with the X-axis
(d) **r** is along negative Y-axis

40. Consider a vector $\mathbf{F} = 4\hat{\mathbf{i}} - 3\hat{\mathbf{j}}$. Another vector that is perpendicular to **F** is
(a) $4\hat{\mathbf{i}} + 3\hat{\mathbf{j}}$ (b) $6\hat{\mathbf{j}}$ (c) $7\hat{\mathbf{k}}$ (d) $3\hat{\mathbf{i}} - 4\hat{\mathbf{j}}$

41. The angle subtended by the vector, $\mathbf{A} = 4\hat{\mathbf{i}} + 3\hat{\mathbf{j}} + 12\hat{\mathbf{k}}$ with the X-axis is
(a) $\sin^{-1}\left(\frac{3}{13}\right)$ (b) $\sin^{-1}\left(\frac{4}{13}\right)$
(c) $\cos^{-1}\left(\frac{4}{13}\right)$ (d) $\cos^{-1}\left(\frac{3}{13}\right)$

Scalar and Vector Product of Vectors

42. The angle between the Z-axis and the vector $\hat{\mathbf{i}} + \hat{\mathbf{j}} + \sqrt{2}\hat{\mathbf{k}}$ is
(a) 30° (b) 45°
(c) 60° (d) 90°

43. The angle between $\mathbf{A} = \hat{\mathbf{i}} + \hat{\mathbf{j}}$ and $\mathbf{B} = \hat{\mathbf{i}} - \hat{\mathbf{j}}$ is **[NCERT Exemplar]**
(a) 45° (b) 90° (c) −45° (d) 180°

44. Given, $\mathbf{P} = 3\hat{\mathbf{j}} + 4\hat{\mathbf{k}}$ and $\mathbf{Q} = 2\hat{\mathbf{i}} + 5\hat{\mathbf{k}}$. The magnitude of the scalar product of these vectors, is
(a) 20 (b) 23
(c) 26 (d) $5\sqrt{33}$

45. If $\mathbf{P} = 2\hat{\mathbf{i}} - 3\hat{\mathbf{j}} + \hat{\mathbf{k}}$ and $\mathbf{Q} = 3\hat{\mathbf{i}} + 2\hat{\mathbf{j}}$, then $\mathbf{P} \cdot \mathbf{Q}$ is
(a) zero (b) 6
(c) 12 (d) 15

46. If **AB** = AB, then the angle between **A** and **B** is
(a) 0° (b) 45°
(c) 90° (d) 180°

47. Projection of **P** on **Q** is
(a) $\mathbf{P} \cdot \hat{\mathbf{Q}}$ (b) $\hat{\mathbf{P}} \cdot \mathbf{Q}$
(c) $\mathbf{P} \times \hat{\mathbf{Q}}$ (d) $\mathbf{P} \times \mathbf{Q}$

48. **A** and **B** are two vectors given by $\mathbf{A} = 2\hat{\mathbf{i}} + 3\hat{\mathbf{j}}$ and $\mathbf{B} = \hat{\mathbf{i}} + \hat{\mathbf{j}}$. The magnitude of the components of **A** along **B** is
(a) $\frac{5}{\sqrt{2}}$ (b) $\frac{3}{\sqrt{2}}$
(c) $\frac{7}{\sqrt{2}}$ (d) $\frac{1}{\sqrt{2}}$

49. If, $\mathbf{A} = 2\hat{\mathbf{i}} + 3\hat{\mathbf{j}} + 4\hat{\mathbf{k}}$ and $\mathbf{B} = 4\hat{\mathbf{i}} + 3\hat{\mathbf{j}} + 2\hat{\mathbf{k}}$, then angle between **A** and **B** is
(a) $\sin^{-1}\left(\frac{25}{29}\right)$ (b) $\sin^{-1}\left(\frac{29}{25}\right)$
(c) $\cos^{-1}\left(\frac{25}{29}\right)$ (d) $\cos^{-1}\left(\frac{29}{25}\right)$

50. What is the angle between $(\hat{\mathbf{i}} + 2\hat{\mathbf{j}} + 2\hat{\mathbf{k}})$ and $\hat{\mathbf{i}}$?
(a) 0° (b) $\pi/6$
(c) $\pi/3$ (d) None of these

51. If **A** = **B**, then which of the following is not correct?
(a) $\hat{\mathbf{A}} = \hat{\mathbf{B}}$ (b) $\hat{\mathbf{A}} \cdot \hat{\mathbf{B}} = AB$
(c) $|\mathbf{A}| = |\mathbf{B}|$ (d) $A\hat{\mathbf{B}} \,||\, B\hat{\mathbf{A}}$

52. For what value of a, $\mathbf{A} = 2\hat{\mathbf{i}} + a\hat{\mathbf{j}} + \hat{\mathbf{k}}$ will be perpendicular to $\mathbf{B} = 4\hat{\mathbf{i}} - 2\hat{\mathbf{j}} - \hat{\mathbf{k}}$?
(a) 4 (b) zero
(c) 3 (d) 1

53. The sum of two vectors **A** and **B** is at right angles to their difference, then
(a) $A = B$
(b) $A = 2B$
(c) $B = 2A$
(d) **A** and **B** have the same direction

54. Given, θ is the angle between $\hat{\mathbf{A}}$ and $\hat{\mathbf{B}}$. Then, $|\hat{\mathbf{A}} \times \hat{\mathbf{B}}|$ is equal to
(a) $\sin\theta$ (b) $\cos\theta$
(c) $\tan\theta$ (d) $\cot\theta$

55. If $\mathbf{P} \cdot \mathbf{Q} = 0$, then $|\mathbf{P} \times \mathbf{Q}|$ is
(a) $|\mathbf{P}||\mathbf{Q}|$ (b) zero
(c) 1 (d) $\sqrt{PQ}$

56. Given, $\mathbf{C} = \mathbf{A} \times \mathbf{B}$. The angle which **A** makes with **C** is
(a) 0° (b) 45°
(c) 90° (d) 180°

57. The adjacent sides of a parallelogram are represented by co-initial vectors $2\hat{\mathbf{i}} + 3\hat{\mathbf{j}}$ and $\hat{\mathbf{i}} + 4\hat{\mathbf{j}}$. The area of the parallelogram is
(a) 5 units along Z-axis (b) 5 units in XY-plane
(c) 3 units in XZ-plane (d) 3 units in YZ-plane

58. The magnitudes of the two vectors **a** and **b** are a and b, respectively. The vector product of **a** and **b** cannot be
(a) equal to zero (b) less than ab
(c) equal to ab (d) greater than ab

59. Given, $\mathbf{A} = 4\hat{\mathbf{i}} + 6\hat{\mathbf{j}}$ and $\mathbf{B} = 2\hat{\mathbf{i}} + 3\hat{\mathbf{j}}$. Which of the following is correct?
(a) $\mathbf{A} \times \mathbf{B} = 0$
(b) $\mathbf{A} \cdot \mathbf{B} = 24$
(c) $\frac{|\mathbf{A}|}{|\mathbf{B}|} = \frac{1}{2}$
(d) **A** and **B** are anti-parallel

60. If $\mathbf{A} \cdot \mathbf{B} = 0$ and $\mathbf{A} \times \mathbf{B} = 1$, then **A** and **B** are
(a) perpendicular unit vectors
(b) parallel unit vectors
(c) parallel
(d) perpendicular

61. The area of a parallelogram formed by the vectors $\mathbf{A} = \hat{\mathbf{i}} - 2\hat{\mathbf{j}} + 3\hat{\mathbf{k}}$ and $\mathbf{B} = 3\hat{\mathbf{i}} - 2\hat{\mathbf{j}} + \hat{\mathbf{k}}$ as its adjacent sides, is
(a) $8\sqrt{3}$ units (b) 64 units
(c) 32 units (d) $4\sqrt{6}$ units

62. The magnitude of the vectors product of two vectors is $\sqrt{3}$ times their scalar product. The angle between the two vectors is
(a) 90° (b) 60°
(c) 45° (d) 30°

63. Three vectors **A**, **B** and **C** satisfy the relation $\mathbf{AB} = 0$ and $\mathbf{AC} = 0$. If **B** and **C** are not lying in the same plane, then **A** is parallel to
(a) **B** (b) **C** (c) $\mathbf{B} \times \mathbf{C}$ (d) **BC**

64. A vector $\mathbf{F}_1$ is along the positive Y-axis. If its vector product with another vector $\mathbf{F}_2$ is zero, then $\mathbf{F}_2$ could be
(a) $4\hat{\mathbf{j}}$ (b) $\hat{\mathbf{j}} + \hat{\mathbf{k}}$
(c) $\hat{\mathbf{j}} - \hat{\mathbf{k}}$ (d) $-4\hat{\mathbf{i}}$

65. The area of the parallelogram represented by the vectors, $\mathbf{A} = 4\hat{\mathbf{i}} + 3\hat{\mathbf{j}}$ and $\mathbf{B} = 2\hat{\mathbf{i}} + 4\hat{\mathbf{j}}$ is
(a) 14 units (b) 7.5 units
(c) 10 units (d) 5 units

66. If **A** and **B** denote the sides of a parallelogram and its area is $\frac{1}{2}AB$ (A and B are the magnitude of **A** and **B** respectively), the angle between **A** and **B** is
(a) 30° (b) 60° (c) 45° (d) 120°

67. Two vectors **A** and **B** are inclined to each other at an angle θ. Which of the following is the unit vector perpendicular to both **A** and **B**?
(a) $\frac{\mathbf{A} \times \mathbf{B}}{\mathbf{AB}}$ (b) $\frac{\hat{\mathbf{A}} \times \hat{\mathbf{B}}}{\sin\theta}$
(c) $\frac{\mathbf{A} \times \mathbf{B}}{AB\sin\theta}$ (d) $\frac{\mathbf{A} \times \mathbf{B}}{AB\cos\theta}$

68. Angle between **A** and **B** is θ. What is the value of $\mathbf{A} \cdot (\mathbf{B} \times \mathbf{A})$?
(a) $A^2B\cos\theta$ (b) $A^2B\sin\theta\cos\theta$
(c) $A^2B\sin\theta$ (d) zero

69. Given, $\mathbf{C} = \mathbf{A} \times \mathbf{B}$ and $\mathbf{D} = \mathbf{B} \times \mathbf{A}$. What is the angle between **C** and **D**?
(a) 30° (b) 60°
(c) 90° (d) 180°

ROUND II Mixed Bag

Only One Correct Option

1. Three forces of magnitudes 6 N, 6 N and $\sqrt{72}$ N act at a corner of a cube along three sides as shown in figure. Resultant of these forces is

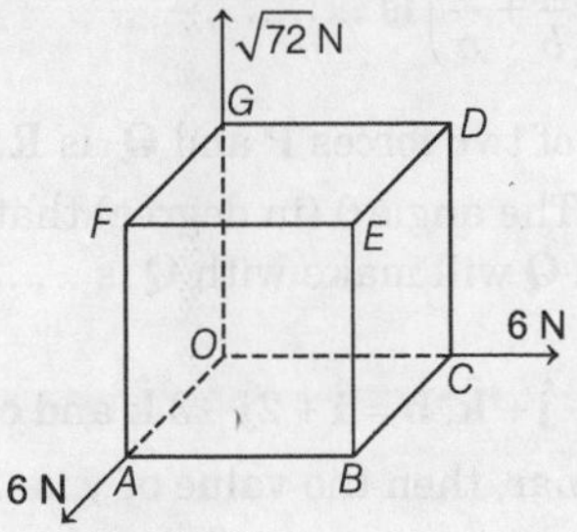

(a) 12 N along OB (b) 18 N along OA
(c) 18 N along OC (d) 12 N along OE

2. The sum of the magnitudes of two forces acting at a point is 16 N. The resultant of these forces is perpendicular to the smaller force has a magnitude of 8 N. If the smaller force has magnitude x, then the value of x is

(a) 2 N (b) 4 N
(c) 6 N (d) 7 N

3. The magnitude of resultant of three vectors of magnitude 1, 2 and 3 whose directions are those of the sides of an equilateral triangle taken in order is

(a) zero (b) $2\sqrt{2}$ units
(c) $4\sqrt{3}$ units (d) $\sqrt{3}$ units

4. Two vectors **A** and **B** have equal magnitudes. The magnitude of $(\mathbf{A}+\mathbf{B})$ is n times the magnitude of $(\mathbf{A}-\mathbf{B})$. The angle between **A** and **B** is **[JEE Main 2019]**

(a) $\sin^{-1}\left(\frac{n^2-1}{n^2+1}\right)$ (b) $\sin^{-1}\left(\frac{n-1}{n+1}\right)$
(c) $\cos^{-1}\left(\frac{n^2-1}{n^2+1}\right)$ (d) $\cos^{-1}\left(\frac{n-1}{n+1}\right)$

5. There are two forces each of magnitude 10 units. One inclined at an angle of 30° and the other at an angle of 135° to the positive direction of X-axis. The x and y components of the resultant are respectively,

(a) $1.59\,\hat{\mathbf{i}}$ and $12.07\,\hat{\mathbf{j}}$ (b) $10\,\hat{\mathbf{i}}$ and $10\,\hat{\mathbf{j}}$
(c) $1.59\,\hat{\mathbf{i}}$ (d) $15.9\,\hat{\mathbf{i}}$ and $12.07\,\hat{\mathbf{j}}$

6. What is the angle between $\hat{\mathbf{i}}+\hat{\mathbf{j}}+\hat{\mathbf{k}}$ and $\hat{\mathbf{i}}$?

(a) 0° (b) $\pi/6$
(c) $\pi/3$ (d) None of these

7. If $\mathbf{A}=2\hat{\mathbf{i}}+3\hat{\mathbf{j}}+6\hat{\mathbf{k}}$ and $\mathbf{B}=3\hat{\mathbf{i}}-6\hat{\mathbf{j}}+2\hat{\mathbf{k}}$, then vector perpendicular to both **A** and **B** has magnitude k times that of $(6\hat{\mathbf{i}}+2\hat{\mathbf{j}}-3\hat{\mathbf{k}})$. The value of k is equal to

(a) 1 (b) 4
(c) 7 (d) 9

8. If $\mathbf{A}_1$ and $\mathbf{A}_2$ are two non-collinear unit vectors and if $|\mathbf{A}_1+\mathbf{A}_2|=\sqrt{3}$, then the value of $(\mathbf{A}_1-\mathbf{A}_2)\cdot(2\mathbf{A}_1+\mathbf{A}_2)$ is

(a) 1 (b) 1/2
(c) 3/2 (d) 2

9. If vectors **A** and **B** are given by $\mathbf{A}=5\hat{\mathbf{i}}+6\hat{\mathbf{j}}+3\hat{\mathbf{k}}$ and $\mathbf{B}=6\hat{\mathbf{i}}-2\hat{\mathbf{j}}-6\hat{\mathbf{k}}$. Which is/are of the following correct?

(a) **A** and **B** are mutually perpendicular
(b) Product of $\mathbf{A}\times\mathbf{B}$ is the same $\mathbf{B}\times\mathbf{A}$
(c) The magnitude of **A** and **B** are equal
(d) The magnitude of $\mathbf{A}\cdot\mathbf{B}$ is non-zero

10. Which of the following statement(s) is/are correct?

(a) The magnitude of the vector $3\hat{\mathbf{i}}+4\hat{\mathbf{j}}$ is 10.
(b) A force $(3\hat{\mathbf{i}}+4\hat{\mathbf{j}})$ N acting on a particle causes a displacement $6\hat{\mathbf{j}}$. The work done by the force is 30 N.
(c) If **A** and **B** represent two adjacent sides of a parallelogram, then $|\mathbf{A}\times\mathbf{B}|$ give the area of that parallelogram.
(d) A force has magnitude 20 N. Its component in a direction making an angle 60° with the force is $10\sqrt{3}$ N.

11. The component of vector $\mathbf{A}=a_x\hat{\mathbf{i}}+a_y\hat{\mathbf{j}}+a_z\hat{\mathbf{k}}$ along the direction of $(\hat{\mathbf{i}}-\hat{\mathbf{j}})$ is

(a) $(a_x-a_y+a_z)$ (b) (a_x+a_y)
(c) $(a_x-a_y)/\sqrt{2}$ (d) $(a_x-a_y+a_z)$

12. If $\mathbf{A}\times\mathbf{B}=\mathbf{B}\times\mathbf{A}$, then the angle between A and B is

(a) π (b) $\pi/3$
(c) $\pi/2$ (d) $\pi/4$

13. Given that, $\mathbf{A}+\mathbf{B}+\mathbf{C}=0$. Out of three vectors, two are equal in magnitude and the magnitude of third vector is $\sqrt{2}$ times that of either of the two having equal magnitude. Then, the angles between vectors are given by

(a) 45°, 45°, 90° (b) 90°, 135°, 135°
(c) 30°, 60°, 90° (d) 45°, 60°, 90°

14. If the vectors $\mathbf{A}=2\hat{\mathbf{i}}+4\hat{\mathbf{j}}$ and $\mathbf{B}=5\hat{\mathbf{i}}-p\hat{\mathbf{j}}$ are parallel to each other, the magnitude of **B** is

(a) $5\sqrt{5}$ (b) 10
(c) 15 (d) $2\sqrt{5}$

15. Two vectors **A** and **B** are inclined at an angle θ. Now if the vectors are interchanged, then the resultant turns through an angle α. Which of the following relation is true?

(a) $\tan\frac{\alpha}{2} = \left(\frac{A-B}{A+B}\right)^2 \tan\frac{\theta}{2}$ (b) $\tan\frac{\alpha}{2} = \left(\frac{A-B}{A+B}\right)\tan\frac{\theta}{2}$

(c) $\tan\frac{\alpha}{2} = \left(\frac{A-B}{A+B}\right)\cot\frac{\theta}{2}$ (d) $\tan\frac{\alpha}{2} = \left(\frac{A-B}{A+B}\right)\cot\frac{\theta}{2}$

16. Let $|\mathbf{A}_1| = 3$, $|\mathbf{A}_2| = 5$ and $|\mathbf{A}_1 + \mathbf{A}_2| = 5$. The value of $(2\mathbf{A}_1 + 3\mathbf{A}_2)\cdot(3\mathbf{A}_1 - 2\mathbf{A}_2)$ is **[JEE Main 2019]**

(a) −106.5 (b) −112.5
(c) −99.5 (d) −118.5

17. The vectors $2\hat{\mathbf{i}} + 3\hat{\mathbf{j}} - 2\hat{\mathbf{k}}$, $5\hat{\mathbf{i}} + a\hat{\mathbf{j}} + \hat{\mathbf{k}}$ and $-\hat{\mathbf{i}} + 2\hat{\mathbf{j}} + 3\hat{\mathbf{k}}$ are coplanar when a is

(a) −9 (b) 9
(c) −18 (d) 18

18. Three vectors **A**, **B** and **C** add up to zero. Find which is false? **[NCERT Exemplar]**

(a) $(\mathbf{A}\times\mathbf{B})\times\mathbf{C}$ is not zero unless **B**, **C** are parallel
(b) $(\mathbf{A}\times\mathbf{B})\cdot\mathbf{C}$ is zero unless **B**, **C** are parallel
(c) If **A**, **B**, **C** defined a plane, $(\mathbf{A}\times\mathbf{B})\times\mathbf{C}$ is in that plane
(d) $(\mathbf{A}\times\mathbf{B})\cdot\mathbf{C} = |\mathbf{A}||\mathbf{B}||\mathbf{C}| \to C^2 = A^2 + B^2$

Numerical Value Questions

19. When two non-zero vectors **a** and **b** are perpendicular to each other, the magnitude of their resultant is R. When they are opposite to each other, the magnitude of their resultant is $\frac{R}{\sqrt{2}}$. The value of $\left(\frac{a}{b} + \frac{b}{a}\right)$ is ……… .

20. The sum of two forces **P** and **Q** is **R** such that $|\mathbf{R}| = |\mathbf{P}|$. The angle θ (in degree) that the resultant of 2**P** and **Q** will make with **Q** is ……… . **[JEE Main 2020]**

21. If $\mathbf{a} = 2\hat{\mathbf{i}} - \hat{\mathbf{j}} + \hat{\mathbf{k}}$, $\mathbf{b} = \hat{\mathbf{i}} + 2\hat{\mathbf{j}} - 3\hat{\mathbf{k}}$ and $\mathbf{c} = 3\hat{\mathbf{i}} - y\hat{\mathbf{j}} + 5\hat{\mathbf{k}}$ are coplanar, then the value of y is ……… .

22. A triangle ABC has its vertices at $A(2, 1, 1)$, $B(1, -3, -5)$ and $C(4, -4, -4)$. If $\angle BAC = \theta$ and $\cos\theta = \frac{8n}{\sqrt{1537}}$. The value of n is ……… .

23. The volume of a parallelopiped, whose edges are represented by $\mathbf{a} = (2\hat{\mathbf{i}} - 3\hat{\mathbf{j}} + 4\hat{\mathbf{k}})$ m, $\mathbf{b} = (\hat{\mathbf{i}} + 2\hat{\mathbf{j}} - \hat{\mathbf{k}})$ m and $\mathbf{c} = (3\hat{\mathbf{i}} - \hat{\mathbf{j}} + 2\hat{\mathbf{k}})$ m is ……… m^3.

Answers

Round I

1. (b)	2. (b)	3. (a)	4. (d)	5. (b)	6. (a)	7. (b)	8. (b)	9. (c)	10. (a)
11. (b)	12. (c)	13. (b)	14. (b)	15. (b)	16. (c)	17. (c)	18. (d)	19. (c)	20. (c)
21. (a)	22. (b)	23. (b)	24. (b)	25. (a)	26. (a)	27. (c)	28. (c)	29. (a)	30. (a)
31. (b)	32. (d)	33. (a)	34. (c)	35. (a)	36. (a)	37. (b)	38. (c)	39. (b)	40. (c)
41. (c)	42. (b)	43. (b)	44. (c)	45. (a)	46. (a)	47. (a)	48. (a)	49. (c)	50. (d)
51. (b)	52. (c)	53. (a)	54. (a)	55. (a)	56. (c)	57. (a)	58. (d)	59. (a)	60. (a)
61. (d)	62. (b)	63. (c)	64. (a)	65. (c)	66. (a)	67. (b)	68. (d)	69. (d)	

Round II

1. (d)	2. (c)	3. (d)	4. (c)	5. (a)	6. (d)	7. (c)	8. (b)	9. (a)	10. (c)
11. (c)	12. (a)	13. (b)	14. (a)	15. (b)	16. (d)	17. (d)	18. (d)	19. 4	20. 90
21. 4	22. 6	23. 7							

Solutions

Round I

1. Out of the given quantities, impulse and area are vector quantities and other are scalar quantities.

2. Required numerical value is $\sqrt{3^2+4^2+5^2}$, *i.e.* $\sqrt{50}$ or $5\sqrt{2}$.

3. Clearly, $(0.5)^2+(0.8)^2+c^2=1$

$$0.25+0.64+c^2=1$$

or $\quad c^2=1-0.25-0.64=0.11$

or $\quad c=\sqrt{0.11}$

4. A scalar quantity has the same value for observers with different orientations of the axes.

5. A vector **A** is parallel to that of vector **B**, if it can be written as

$$A=mB$$

Here, $\quad A=(\hat{\mathbf{i}}-\hat{\mathbf{j}}+2\hat{\mathbf{k}})=\frac{1}{3}(3\hat{\mathbf{i}}-3\hat{\mathbf{j}}+6\hat{\mathbf{k}})$

$$A=\frac{1}{3}B$$

This implies A is parallel to B and magnitude of A is $\frac{1}{3}$ times the magnitude of B.

6. We have $\mathbf{A}=A\hat{\mathbf{A}}$ or $\hat{\mathbf{A}}=\frac{\mathbf{A}}{A}$

$\therefore$ Required unit vector is $\frac{\hat{\mathbf{i}}+\hat{\mathbf{j}}}{|\hat{\mathbf{i}}+\hat{\mathbf{j}}|}=\frac{\hat{\mathbf{i}}+\hat{\mathbf{j}}}{\sqrt{2}}$

7. As per figure, in $\mathbf{u}=a\hat{\mathbf{i}}+b\hat{\mathbf{j}}$, both a and b are positive. In $\mathbf{v}=p\hat{\mathbf{i}}+q\hat{\mathbf{j}}$, p is positive and q is negative.

Thus, a, p and b are positive and q is negative.

8. In the given cube, coordinates of point G(centre of face $ABOD$) are $x_1=\frac{a}{2}, y_1=0, z_1=\frac{a}{2}$

where, a = side of cube

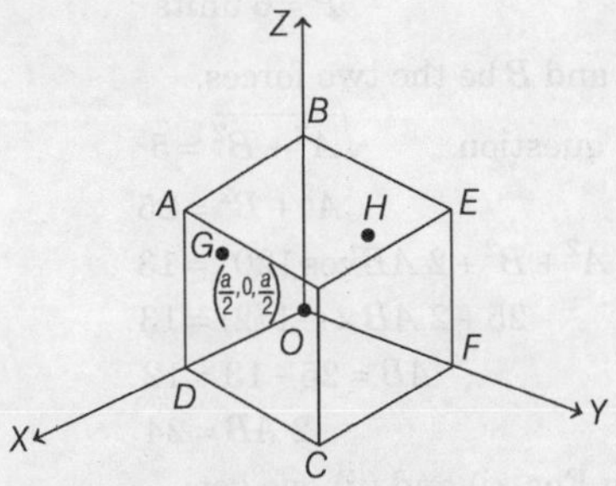

and coordinates of point H are

$$x_2=0,\ y_2=\frac{a}{2},\ z_2=\frac{a}{2}$$

So, vector GH is

$$GH=(x_2-x_1)\hat{\mathbf{i}}+(y_2-y_1)\hat{\mathbf{j}}+(z_2-z_1)\hat{\mathbf{k}}$$

$$=-\frac{a}{2}\hat{\mathbf{i}}+\frac{a}{2}\hat{\mathbf{j}}=\frac{a}{2}(\hat{\mathbf{j}}-\hat{\mathbf{i}})$$

9. As, $\quad R=\sqrt{a^2+b^2+2ab\cos\theta}$

$\Rightarrow \quad R=\sqrt{\left(\frac{p}{2}\right)^2+\left(\frac{p}{2}\right)^2+2\left(\frac{p}{2}\right)\left(\frac{p}{2}\right)\cos 90^\circ}$

$\Rightarrow \quad R=\sqrt{2}\cdot\frac{p}{2}=\frac{p}{\sqrt{2}}$

10. As, $\mathbf{B}+(\hat{\mathbf{i}}+2\hat{\mathbf{j}}-3\hat{\mathbf{k}})=\hat{\mathbf{i}}$ or $\mathbf{B}=-2\hat{\mathbf{j}}+3\hat{\mathbf{k}}$

11. As, resultant of two vectors is given by,

$$R=\sqrt{A^2+B^2+2AB\cos\theta}$$

$\therefore \quad F^2=F_1^2+F_2^2+2F_1F_2\cos 90^\circ$

Given, $\theta=90^\circ$

or $\quad F^2=F_1^2+F_2^2 \Rightarrow F=\sqrt{F_1^2+F_2^2}$

12. For the resultant,

$$R^2=R^2+R^2+2R^2\cos\theta$$

or $\quad R^2=2R^2+2R^2\cos\theta$

$$\frac{1}{2}=1+\cos\theta$$

or $\quad \cos\theta=-\frac{1}{2}$ or $\theta=120^\circ$

13. As, $R^2=P^2+P^2+2P^2\cos\theta$

or $\quad R^2=2P^2+2P^2\cos\theta$

or $\quad R^2=2P^2(1+\cos\theta)$

or $\quad R^2=2P^2\left(2\cos^2\frac{\theta}{2}\right)$

or $\quad R^2=4P^2\cos^2\frac{\theta}{2}$

or $\quad R=2P\cos\frac{\theta}{2}$

14. Resultant, $R=\sqrt{A^2+B^2+2AB\cos\theta}$

$\Rightarrow \quad 10A^2=4A^2+2A^2+2\times 2A\times\sqrt{2}A\times\cos\theta$

or $\quad 4A^2=4\sqrt{2}A^2\cos\theta$

or $\quad \cos\theta=\frac{1}{\sqrt{2}}$

$\Rightarrow \quad \theta=45^\circ$

15. Note that the angle between two forces is 120° and not 60°.

$$R^2=F^2+F^2+2\cdot F\cdot F\cos 120^\circ$$

$\Rightarrow \quad R^2=F^2+F^2-F^2$

$\Rightarrow \quad R^2=F^2$

$\Rightarrow \quad R=F$

16. As, $\mathbf{P}+\mathbf{Q}=5\hat{\mathbf{i}}-4\hat{\mathbf{j}}+3\hat{\mathbf{k}}$

$\Rightarrow \quad \cos\alpha=\frac{5}{\sqrt{5^2+4^2+3^2}}=\frac{5}{\sqrt{50}}$

or $\quad \alpha=\cos^{-1}\left(\frac{5}{\sqrt{50}}\right)$

17. As, $\mathbf{A}+\mathbf{B}=\mathbf{C}$ (Given)

So, it is given that **C** is the resultant of **A** and **B**

$\therefore \quad C^2 = A^2 + B^2 + 2AB\cos\theta$

or $\quad 3^2 = 3+3+2\times3\times\cos\theta$

or $\quad 3 = 6\cos\theta$

or $\quad \cos\theta = \frac{1}{2} \Rightarrow \theta = 60°$

18. According to question, $\sqrt{P^2+Q^2+2PQ\cos\theta} = (P-Q)$

$\Rightarrow \quad P^2+Q^2+2PQ\cos\theta = P^2+Q^2-2PQ$

$\Rightarrow \quad 2PQ(1+\cos\theta)=0$

But, $\quad 2PQ \neq 0$

$\therefore \quad 1+\cos\theta = 0$

or $\quad \cos\theta = -1$

or $\quad \theta = 180°$

19. Here, $\quad P=(A+B),\ Q=(A-B)$

and $\quad R=\sqrt{A^2+B^2}$

$$\Rightarrow \quad \cos\theta = \frac{R^2-P^2-Q^2}{2PQ}$$

$$= \frac{(A^2+B^2)-(A+B)^2-(A-B)^2}{2(A+B)(A-B)}$$

$$= -\left[\frac{A^2+B^2}{2(A^2-B^2)}\right]$$

$$\therefore \quad \theta = \cos^{-1}\left[-\frac{A^2+B^2}{2(A^2-B^2)}\right]$$

20. We can make the diagram as below

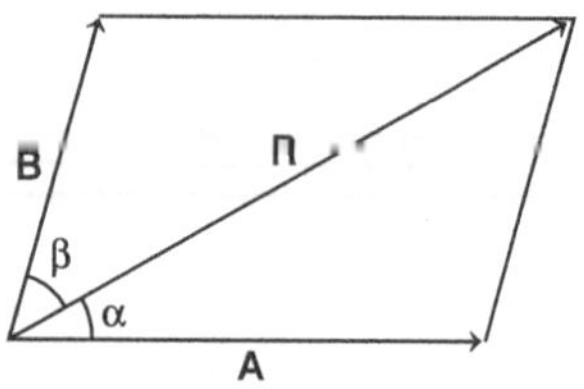

Clearly, $\tan\alpha = \frac{B\sin\theta}{A+B\cos\theta}$ and $\tan\beta = \frac{A\sin\theta}{B+A\cos\theta}$

From the above equation, it is clear that when $\alpha<\beta$, then $B<A$.

21. Given, $(\hat{\mathbf{i}}+2\hat{\mathbf{j}}-\hat{\mathbf{k}})+(\hat{\mathbf{i}}-\hat{\mathbf{j}}+2\hat{\mathbf{k}})+\mathbf{C}=\hat{\mathbf{j}}$

$\therefore \quad \mathbf{C}=\hat{\mathbf{j}}-(\hat{\mathbf{i}}-2\hat{\mathbf{j}}-\hat{\mathbf{k}})-(\hat{\mathbf{i}}-\hat{\mathbf{j}}+2\hat{\mathbf{k}})$

$\quad = -2\hat{\mathbf{i}}-\hat{\mathbf{k}}$

22. Given, $\mathbf{P}=\hat{\mathbf{i}}-\hat{\mathbf{j}}+\hat{\mathbf{k}}$, then we have

$\mathbf{P}+\mathbf{Q}=\hat{\mathbf{i}}$

$\mathbf{Q}=\hat{\mathbf{i}}-\hat{\mathbf{i}}+\hat{\mathbf{j}}-\hat{\mathbf{k}}=\hat{\mathbf{j}}-\hat{\mathbf{k}}$

23. As, $\mathbf{A}=2\hat{\mathbf{i}}-\hat{\mathbf{j}}+3\hat{\mathbf{k}}$, $\mathbf{B}=3\hat{\mathbf{i}}-2\hat{\mathbf{j}}-2\hat{\mathbf{k}}$ and $\mathbf{C}=?$

$\mathbf{R}=\hat{\mathbf{k}}=\mathbf{A}+\mathbf{B}+\mathbf{C}$

$\hat{\mathbf{k}}=(2\hat{\mathbf{i}}-\hat{\mathbf{j}}+3\hat{\mathbf{k}})+(3\hat{\mathbf{i}}-2\hat{\mathbf{j}}-2\hat{\mathbf{k}})+\mathbf{C}$

$\quad =5\hat{\mathbf{i}}-3\hat{\mathbf{j}}+\hat{\mathbf{k}}+\mathbf{C}$

$\therefore \quad \mathbf{C}=-5\hat{\mathbf{i}}+3\hat{\mathbf{j}}$

24. Here, $\tan 45° = \frac{2\sin 60°}{a+2\cos 60°} = \frac{\sqrt{3}}{a+1}$

$1 = \frac{\sqrt{3}}{a+1}$

or $\quad a+1=\sqrt{3}$

$\quad a=\sqrt{3}-1$

25. Given, $\mathbf{C}=|\mathbf{B}|\hat{\mathbf{j}} \Rightarrow \mathbf{C}=5\hat{\mathbf{j}}$

Let, $\quad \mathbf{C}=\mathbf{A}+\mathbf{B}=A+3\hat{\mathbf{i}}+4\hat{\mathbf{j}}$

$\quad 5\hat{\mathbf{j}}=A+3\hat{\mathbf{i}}+4\hat{\mathbf{j}}$

$\Rightarrow \quad \mathbf{A}=-3\hat{\mathbf{i}}+\hat{\mathbf{j}}$

$\quad |\mathbf{A}|=\sqrt{(3)^2+(1)^2}=\sqrt{10}$

26. If $|\mathbf{A}+\mathbf{B}|=|\mathbf{A}|$, then either $|\mathbf{B}|=0$ or **A** and **B** will be antiparallel, where $|\mathbf{B}|=2|\mathbf{A}|$.

27. According to question, $P+Q=16$ ...(i)

$\Rightarrow \quad P^2+Q^2+2PQ\cos\theta=64$ ...(ii)

$\therefore \quad \tan 90° = \frac{Q\sin\theta}{P+Q\cos\theta}$

or $\quad \infty = \frac{Q\sin\theta}{P+Q\cos\theta}$

$\Rightarrow \quad P+Q\cos\theta=0$

or $\quad Q\cos\theta=-P$ ...(iii)

From Eqs. (ii) and (iii), we get

$P^2+Q^2+2P(-P)=64$

or $\quad Q^2-P^2=64$

or $\quad (Q-P)(Q+P)=64$ ...(iv)

Now from Eqs. (i) and (iv), we get

or $\quad Q-P=\frac{64}{16}=4$ ...(v)

Adding, Eqs. (i) and (v), we get

$2Q=20$

or $\quad Q=10$ units

From Eq. (i), $\quad P+10=16$

or $\quad P=6$ units

28. Let, A and B be the two forces.

As per question $\quad \sqrt{A^2+B^2}=5$

or $\quad A^2+B^2=25$ ...(i)

and $\quad A^2+B^2+2AB\cos 120°=13$

or $\quad 25+2AB\times(-1/2)=13$

or $\quad AB=25-13=12$

or $\quad 2AB=24$ ...(ii)

Solving Eqs. (i) and (ii), we get

$A=3\text{ N}$ and $B=4\text{ N}$

29. Given, $|\mathbf{P}|=A+B$

$\Rightarrow \quad |\mathbf{P}|^2=(A+B)^2$

or $\quad |\mathbf{A}+\mathbf{B}|^2=(A+B)^2$

or $\quad A^2+B^2+2AB\cos\theta=A^2+B^2+2AB$

or $\quad \cos\theta=1\Rightarrow\theta=0°$

30. Resultant, $\mathbf{R} = (\mathbf{P}+\mathbf{Q}) + (\mathbf{P}-\mathbf{Q}) = 2\,\mathbf{P}$. Thus, angle between **R** and **P** is 0°.

31. From the condition given in question,

$$a^2 + b^2 + 2\,ab\cos\theta = a^2 + b^2 - 2\,ab\cos\theta$$

or $\quad 4\,ab\cos\theta = 0$

But $\quad 4\,ab \neq 0$

$\therefore \quad \cos\theta = 0$

or $\quad \theta = 90°$

32. Let θ be the angle between $\mathbf{A}_1$ and $\mathbf{A}_2$, then

$$A_1^2 + A_2^2 + 2A_1A_2\cos\theta = R^2$$

or $\quad A^2 + A^2 + 2AA\cos\theta = 3\,A^2$

or $\quad \cos\theta = \dfrac{1}{2} = \cos 60°$

or $\quad \theta = 60°$

The angle between $\mathbf{A}_1$ and $-\mathbf{A}_2$ is $(180° - 60°) = 120°$

$\therefore$ Resultant of $\mathbf{A}_1$ and $-\mathbf{A}_2$ is

$$R' = [A_1^2 + A_2^2 + 2A_1A_2\cos(180° - 60°)]^{1/2}$$
$$= [A^2 + A^2 + 2AA\cos 120°]^{1/2} = A$$

33. Clearly, $2\,\mathbf{A} - 3\,\mathbf{B} = 2\,(\hat{\mathbf{i}} + \hat{\mathbf{j}} - 2\,\hat{\mathbf{k}}) - 3\,(2\,\hat{\mathbf{i}} - \hat{\mathbf{j}} + \hat{\mathbf{k}})$

$$= -4\,\hat{\mathbf{i}} + 5\,\hat{\mathbf{j}} - 7\,\hat{\mathbf{k}}$$

$\therefore$ Magnitude of $\;2\mathbf{A} - 3\mathbf{B} = \sqrt{(-4)^2 + (5)^2 + (-7)^2}$

$$= \sqrt{16 + 25 + 49} = \sqrt{90}$$

34. The vector is $\hat{\mathbf{i}} - (\mathbf{A} + \mathbf{B} + \mathbf{C})$

$$= \hat{\mathbf{i}} - [(2\hat{\mathbf{i}} - 4\hat{\mathbf{j}} + 7\,\hat{\mathbf{k}}) + (7\,\hat{\mathbf{i}} + 2\,\hat{\mathbf{j}} - 5\,\hat{\mathbf{k}}) + (-4\,\hat{\mathbf{i}} + 7\hat{\mathbf{j}} + 3\,\hat{\mathbf{k}})]$$
$$= -5\,\hat{\mathbf{i}} - 5\hat{\mathbf{j}} - 5\,\hat{\mathbf{k}}$$

35. Here, $\mathbf{F} = F_x\hat{\mathbf{i}} + F_y\hat{\mathbf{j}}$ or $\mathbf{F} = 2\,\hat{\mathbf{i}} - 3\,\hat{\mathbf{j}}$

36. Let, $\mathbf{A} + \mathbf{B} = \mathbf{R}$. Given, $A_x = 7$ and $A_y = 6$

Also, $\quad R_x = 11$ and $R_y = 9$

Therefore, $\quad B_x = R_x - A_x = 11 - 7 = 4$

and $\quad B_y = R_y - A_y = 9 - 6 = 3$

Hence, magnitude of $\;\mathbf{B} = \sqrt{B_x^2 + B_y^2} = \sqrt{4^2 + 3^2} = 5$

37.

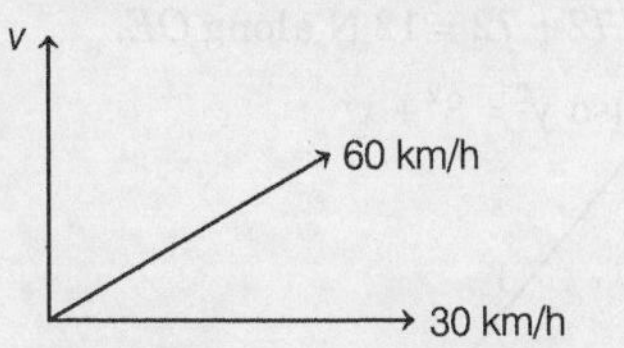

$$60^2 = 30^2 + v^2$$

or $\quad v^2 = 90 \times 30$

or $\quad v = 30\sqrt{3}\ \text{kmh}^{-1}$

38. Given, $A_x = 50$ and $\theta = 60°$

Then $\quad \tan\theta = A_y / A_x$

or $\quad A_y = A_x\tan\theta$

or $\quad A_y = 50\tan 60° = 50 \times \sqrt{3} = 87\ \text{N}$

39. If **r** makes an angle θ with X-axis, then component of **r** along X-axis $= r\cos\theta$.
It will be maximum, if $\cos\theta = \max = 1$ or $\theta = 0°$. *i.e.* **r** is along positive X-axis.

40. Since $\mathbf{F} = 4\,\hat{\mathbf{i}} - 3\,\hat{\mathbf{j}}$ is lying in XY-plane, hence the vector perpendicular to **F** must be lying perpendicular to XY-plane or along Z-axis, *i.e.* $7\hat{\mathbf{k}}$.

41. Let θ be the angle which **A** make with X-axis, then

$$\cos\theta = \frac{A_x}{A} = \frac{4}{\sqrt{4^2 + 3^2 + 12^2}} = \frac{4}{13}$$

$\therefore \quad \theta = \cos^{-1}\left(\dfrac{4}{13}\right)$

42. From, $\mathbf{A}\cdot\mathbf{B} = AB\cos\theta$

$\Rightarrow \quad \cos\theta = \dfrac{\mathbf{A}\cdot\mathbf{B}}{AB}$

or $\quad \cos\theta = \dfrac{(\hat{\mathbf{i}} + \hat{\mathbf{j}} + \sqrt{2}\,\hat{\mathbf{k}})\cdot\hat{\mathbf{k}}}{1\sqrt{1^2 + 1^2 + (\sqrt{2})^2}}$

or $\quad \cos\theta = \dfrac{\sqrt{2}}{2} = \dfrac{1}{\sqrt{2}} \Rightarrow \theta = 45°$

43. $\cos\theta = \dfrac{\mathbf{A}\cdot\mathbf{B}}{AB} = \dfrac{(\hat{\mathbf{i}} + \hat{\mathbf{j}})\cdot(\hat{\mathbf{i}} - \hat{\mathbf{j}})}{\sqrt{1^2 + 1^2} \times \sqrt{1^2 + (-1)^2}} = \dfrac{1 - 1}{2} = 0 = \cos 90°$

$\therefore \quad \theta = 90°$

44. $|\mathbf{P}\cdot\mathbf{Q}| = (3\hat{\mathbf{j}} + 4\,\hat{\mathbf{k}})\cdot(2\hat{\mathbf{i}} + 5\,\hat{\mathbf{k}}) = 6 + 20 = 26$

45. Here, $\mathbf{P}\cdot\mathbf{Q} = (2\hat{\mathbf{i}} - 3\hat{\mathbf{j}} + \hat{\mathbf{k}})\cdot(3\hat{\mathbf{i}} + 2\hat{\mathbf{j}}) = 6\,(\hat{\mathbf{i}}\cdot\hat{\mathbf{i}}) - 6\,(\hat{\mathbf{j}}\cdot\hat{\mathbf{j}}) = 0$

46. We know that, $[\hat{\mathbf{i}}\cdot\hat{\mathbf{j}} = \hat{\mathbf{j}}\cdot\hat{\mathbf{k}} = \hat{\mathbf{k}}\cdot\hat{\mathbf{i}} = 0,\ \hat{\mathbf{i}}\cdot\hat{\mathbf{i}} = \hat{\mathbf{j}}\cdot\hat{\mathbf{j}} = \hat{\mathbf{k}}\cdot\hat{\mathbf{k}} = 1]$

As, $\quad AB\cos\theta = AB$

or $\quad \cos\theta = 1$

or $\quad \theta = 0°$

47. Projection of **P** on **Q** is $P\cos\theta$

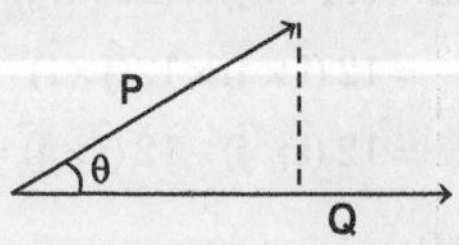

Here, $\quad P\cos\theta = \dfrac{PQ\cos\theta}{Q} = \dfrac{\mathbf{P}\cdot\mathbf{Q}}{Q} = \mathbf{P}\cdot\hat{\mathbf{Q}}$

48. Magnitude of component of **A** along **B**

$$= \frac{\mathbf{A}\cdot\mathbf{B}}{|\mathbf{B}|} = \frac{(2\hat{\mathbf{i}} + 3\hat{\mathbf{j}})\cdot(\hat{\mathbf{i}} + \hat{\mathbf{j}})}{\sqrt{1^2 + 1^2}} = \frac{5}{\sqrt{2}}$$

49. As, $\mathbf{A}\cdot\mathbf{B} = AB\cos\theta$

$\Rightarrow \quad \cos\theta = \dfrac{\mathbf{A}\cdot\mathbf{B}}{AB} = \dfrac{(2\hat{\mathbf{i}} + 3\hat{\mathbf{j}} + 4\,\hat{\mathbf{k}})\cdot(4\,\hat{\mathbf{i}} + 3\hat{\mathbf{j}} + 2\,\hat{\mathbf{k}})}{\sqrt{4 + 9 + 16}\cdot\sqrt{16 + 9 + 4}}$

$$= \frac{8 + 9 + 8}{29} = \frac{25}{29}$$

$\Rightarrow \quad \theta = \cos^{-1}\left(\dfrac{25}{29}\right)$

50. Using, $\mathbf{A}\cdot\mathbf{B} = AB\cos\theta$

$$\Rightarrow \quad \cos\theta = \frac{\mathbf{A}\cdot\mathbf{B}}{AB}$$

or $$\cos\theta = \frac{(\hat{\mathbf{i}}+2\hat{\mathbf{j}}+2\hat{\mathbf{k}})\cdot\hat{\mathbf{i}}}{(1^2+2^2+2^2)^{1/2}} = \frac{1}{\sqrt{9}} = \frac{1}{3} = 0.33$$

$$\Rightarrow \quad \theta = 70°30'$$

51. Here $\hat{\mathbf{A}}\cdot\hat{\mathbf{B}} = (1)\,(1)\cos 0° = 1 \neq AB.$

52. $\mathbf{A}\perp\mathbf{B}$, if $\mathbf{A}\cdot\mathbf{B} = AB\cos 90° = 0$

$$(2\hat{\mathbf{i}} + a\hat{\mathbf{j}} + \hat{\mathbf{k}})\cdot(4\hat{\mathbf{i}} - 2\hat{\mathbf{j}} - 2\hat{\mathbf{k}}) = 0$$

or $$8 - 2a - 2 = 0 \text{ or } a = 3$$

53. Using, $\mathbf{A}\cdot\mathbf{B} = AB\cos\theta$, given $\theta = 90° \Rightarrow \cos 90° = 0$

Then, $$(\mathbf{A}+\mathbf{B})\cdot(\mathbf{A}-\mathbf{B}) = 0$$
$$A^2 - B^2 = 0 \text{ or } A = B$$

54. The vector product,

$$|\mathbf{A}\times\mathbf{B}| = (1)\,(1)\sin\theta = \sin\theta$$

55. Since, $\mathbf{P}\cdot\mathbf{Q} = PQ\cos\theta$, then

$$\mathbf{P}\cdot\mathbf{Q} = 0$$

$$\Rightarrow \quad \mathbf{P}\perp\mathbf{Q} \text{ or } \theta = 90°$$

$$|\mathbf{P}\times\mathbf{Q}| = PQ\sin 90° = PQ \text{ or } |\mathbf{P}||\mathbf{Q}|$$

56. The direction of the vector given by the cross product of the two vectors is perpendicular to the plane containing the two vectors, *i.e.* $\mathbf{A}\times\mathbf{B} = (AB\sin\theta) = \mathbf{C}$. Therefore, the angle which $\mathbf{A}$ makes with $\mathbf{C}$ is 90°.

57. The required area, $\mathbf{A}\times\mathbf{B} = (2\hat{\mathbf{i}} + 3\hat{\mathbf{j}})\times(\hat{\mathbf{i}} + 4\hat{\mathbf{j}})$

$$= 8\,(\hat{\mathbf{i}}\times\hat{\mathbf{j}}) + 3(\hat{\mathbf{j}}\times\hat{\mathbf{i}}) = 8\,\hat{\mathbf{k}} - 3\hat{\mathbf{k}} = 5\,\hat{\mathbf{k}}$$

58. As, $|\mathbf{a}\times\mathbf{b}| = ab\sin\theta$

Since, $\sin\theta$ cannot be greater than 1.

$\therefore |\mathbf{a}\times\mathbf{b}|$ cannot be greater than ab.

59. We have $\mathbf{A}\times\mathbf{B} = (4\,\hat{\mathbf{i}} + 6\hat{\mathbf{j}})\times(2\,\hat{\mathbf{i}} + 3\,\hat{\mathbf{j}})$

$$= 12\,(\hat{\mathbf{i}}\times\hat{\mathbf{j}})\times 12(\hat{\mathbf{j}}\times\hat{\mathbf{i}})$$
$$= 12\,(\hat{\mathbf{i}}\times\hat{\mathbf{j}}) - 12\,(\hat{\mathbf{i}}\times\hat{\mathbf{j}}) = 0$$

60. Given, $\mathbf{A}\cdot\mathbf{B} = 0$

$\Rightarrow \quad \mathbf{A}\perp\mathbf{B} \quad$ (Since, $\mathbf{A}\cdot\mathbf{B} = AB\cos\theta$)

Now, $$\mathbf{A}\times\mathbf{B} = 1$$

or $$AB\sin\theta = 1$$
$$AB\sin 90° = 1 \text{ or } AB = 1$$

$$\Rightarrow \quad A = 1 \text{ and } B = 1$$

So, $\mathbf{A}$ and $\mathbf{B}$ are perpendicular unit vectors.

61. Required area of parallelogram,

$$\mathbf{A}\times\mathbf{B} = (\hat{\mathbf{i}} - 2\hat{\mathbf{j}} + 3\,\hat{\mathbf{k}})\times(3\hat{\mathbf{i}} - 2\hat{\mathbf{j}} + \hat{\mathbf{k}})$$
$$= -2\hat{\mathbf{k}} - \hat{\mathbf{j}} - 6(-\hat{\mathbf{k}}) - 2\,\hat{\mathbf{i}} + 9\,\hat{\mathbf{j}} - 6\,(-\hat{\mathbf{i}})$$
$$= 4\,\hat{\mathbf{i}} + 8\hat{\mathbf{j}} + 4\,\hat{\mathbf{k}}$$
$$|\mathbf{A}\times\mathbf{B}| = \sqrt{4^2 + 8^2 + 4^2} = \sqrt{32 + 64}$$
$$= \sqrt{96} = 4\sqrt{6} \text{ units}$$

62. Given, $$|\mathbf{A}\times\mathbf{B}| = \sqrt{3}\mathbf{A}\cdot\mathbf{B}$$

$$\Rightarrow \quad AB\sin\theta = \sqrt{3}AB\cos\theta$$

or $$\tan\theta = \sqrt{3}$$

$$\Rightarrow \quad \theta = 60°$$

63. As, $\mathbf{A}\cdot\mathbf{B} = 0$ so, $\mathbf{A}$ is perpendicular to $\mathbf{B}$. Also $\mathbf{A}\cdot\mathbf{C} = 0$ means $\mathbf{A}$ is perpendicular to $\mathbf{C}$. Since $\mathbf{B}\times\mathbf{C}$ is perpendicular to $\mathbf{B}$ and $\mathbf{C}$, then clearly $\mathbf{A}$ is parallel to $\mathbf{B}\times\mathbf{C}$.

64. As, $\mathbf{F}_1 = F_1\,\hat{\mathbf{j}}$; $\mathbf{F}_1\times\mathbf{F}_2$ is equal to zero only, if angle between $\mathbf{F}_1$ and $\mathbf{F}_2$ is either 0° or 180°. So, $\mathbf{F}_2$ will be $4\,\hat{\mathbf{j}}$. (In direction of Y-axis) $\quad (\because \hat{\mathbf{i}}\times\hat{\mathbf{i}} = \hat{\mathbf{j}}\times\hat{\mathbf{j}} = 0)$

65. Area $= |\mathbf{A}\times\mathbf{B}| = |(4\hat{\mathbf{i}} + 3\hat{\mathbf{j}})\times(2\hat{\mathbf{i}} + 4\hat{\mathbf{j}}) = |10\,\hat{\mathbf{k}}| = 10$ units

66. Area of parallelogram $= |\mathbf{A}\times\mathbf{B}|$

$$AB\sin\theta = \frac{1}{2}AB$$

$$\therefore \quad \sin\theta = \frac{1}{2}$$

$$\Rightarrow \quad \theta = 30°$$

67. The required unit vector should be

$$\hat{\mathbf{n}} = \frac{\mathbf{A}\times\mathbf{B}}{AB\sin\theta}$$
$$= \frac{A\hat{\mathbf{A}}\times B\hat{\mathbf{B}}}{AB\sin\theta} = \frac{\hat{\mathbf{A}}\times\hat{\mathbf{B}}}{\sin\theta}$$

68. $\mathbf{A}\cdot(\mathbf{B}\times\mathbf{A}) = 0$ (According to rules for scalar triple product)

69. Since, $(\mathbf{A}\times\mathbf{B}) = -(\mathbf{B}\times\mathbf{A})$, so $\mathbf{C} = \mathbf{D}$ *i.e.*, $\mathbf{C}$ and $\mathbf{D}$ are antiparallel to each other, *i.e.* $\theta = 180°$.

Round II

1. The resultant of 6 N along OC and 6 N along OA is

$$R = \sqrt{6^2 + 6^2} = \sqrt{72} \text{ N along } OB$$

The resultant of $\sqrt{72}$ N along OB and $\sqrt{72}$ N along OG is

$$R' = \sqrt{72 + 72} = 12 \text{ N along } OE.$$

2. Given, $x + y = 16$, Also $y^2 = 8^2 + x^2$

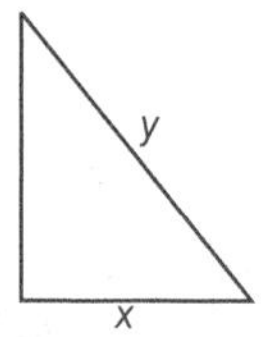

or $$y^2 = 64 + (16 - y)^2 \quad (\because x = 16 - y)$$

or $$y^2 = 64 + 256 + y^2 - 32\,y$$

or $$32\,y = 320$$

or $$y = 10 \text{ N}$$

$$\therefore \quad x + 10 = 16$$

or $$x = 6 \text{ N}$$

3. The three vectors **A**, **B** and **C** are represented as shown in figure (a) where $A = 1, B = 2$ and $C = 3$. Here the sides of the equilateral triangle represent only the directions and not the magnitudes of the vectors.

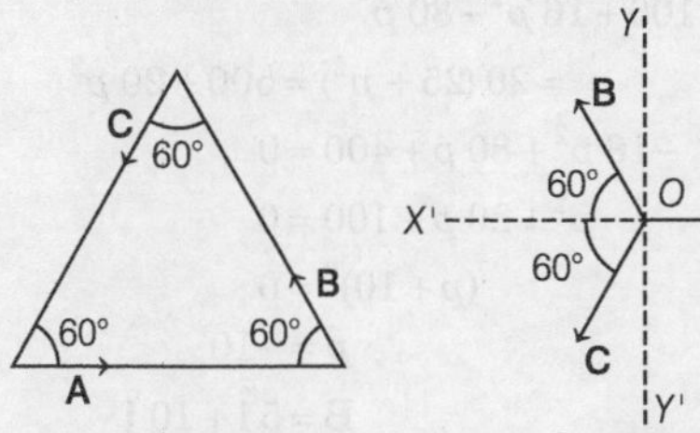

(a) (b)

In figure (b), these vector are drawn from a common point, O and they are lying in XY-plane. Resolving these vectors into two rectangular components along XY-axis and Y-axis, we have, the X-component of resultant vector as

$$R_X = |\mathbf{A}| + |\mathbf{B}| \cos(180° - 60°) + |\mathbf{C}| \cos(180° + 60°)$$
$$= 1 - 2\cos 60° - 3\cos 60°$$
$$= 1 - 2 \times \frac{1}{2} - 3 \times \frac{1}{2} = -\frac{3}{2}$$

Y-component of resultant vector is

$$R_Y = 0 + |\mathbf{B}| \sin(180° - 60°) + |\mathbf{C}| \sin(180° + 60°)$$
$$= 0 + 2\sin 60° - 3\sin 60° = -\sin 60° = -\sqrt{3}/2$$

Magnitude of resultant vector,

$$R = \sqrt{R_X^2 + R_Y^2} = \sqrt{\left(-\frac{3}{2}\right)^2 + \left(-\frac{\sqrt{3}}{2}\right)^2} = \sqrt{3} \text{ units}$$

4. Given, $|\mathbf{A}| = |\mathbf{B}|$ or $A = B$ ...(i)

Let magnitude of $(\mathbf{A} + \mathbf{B})$ is R and for $(\mathbf{A} - \mathbf{B})$ is R'.

Now, $\mathbf{R} = \mathbf{A} + \mathbf{B}$

and $R^2 = A^2 + B^2 + 2AB\cos\theta$

$$R^2 = 2A^2 + 2A^2\cos\theta \quad \text{...(ii)}$$

[∵ using Eq. (i)]

Again, $\mathbf{R'} = \mathbf{A} - \mathbf{B}$

$\Rightarrow$ $R'^2 = A^2 + B^2 - 2AB\cos\theta$

$$R'^2 = 2A^2 - 2A^2\cos\theta \quad \text{...(iii)}$$

[∵ using Eq. (i)]

Given, $R = nR'$ or $\left(\frac{R}{R'}\right)^2 = n^2$

Dividing Eq. (ii) by Eq. (iii), we get

$$\frac{n^2}{1} = \frac{1 + \cos\theta}{1 - \cos\theta}$$

or $$\frac{n^2 - 1}{n^2 + 1} = \frac{(1 + \cos\theta) - (1 - \cos\theta)}{(1 + \cos\theta) + (1 - \cos\theta)}$$

$\Rightarrow$ $$\frac{n^2 - 1}{n^2 + 1} = \frac{2\cos\theta}{2} = \cos\theta$$

or $$\theta = \cos^{-1}\left(\frac{n^2 - 1}{n^2 + 1}\right)$$

5. Here, $\mathbf{A} = \mathbf{OP} = 10$ units along OP

$\mathbf{B} = \mathbf{OQ} = 10$ units along OQ

$\angle XOP = 30°$ and $\angle XOQ = 135°$

$\therefore$ $\angle QOX' = 180° - 135° = 45°$

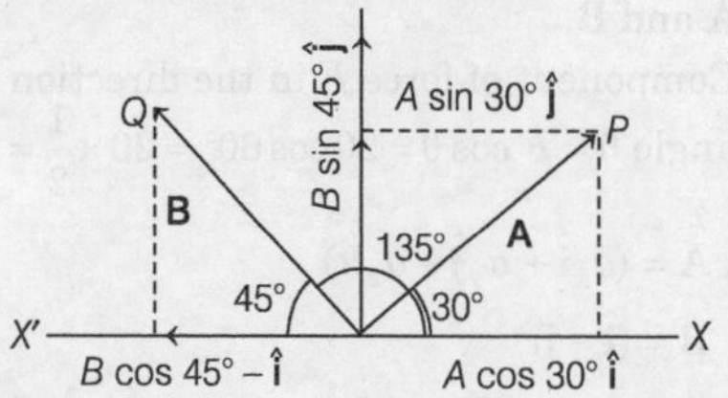

Resolving **A** and **B** into two rectangular components we have $A\cos 30°$ along OX and $A\sin 30°$ along OY. While $B\cos 45°$ along OX' and $B\sin 45°$ along OY'.

Resultant of components of forces along X-axis.

$$= (A\cos 30° - B\cos 45°)\,\hat{\mathbf{i}}$$
$$= (10 \times \sqrt{3}/2 - 10 \times 1/\sqrt{2})\,\hat{\mathbf{i}} = 1.59\,\hat{\mathbf{i}}$$

Resultant of components forces along Y-axis

$$= (A\sin 30° + B\sin 45°)\,\hat{\mathbf{j}}$$
$$= \left(10 \times \frac{1}{2} + 10\frac{1}{\sqrt{2}}\right)\hat{\mathbf{j}} = 12.07\,\hat{\mathbf{j}}$$

6. Using $\mathbf{A} \cdot \mathbf{B} = AB\cos\theta$

$\because$ $\mathbf{A} = \hat{\mathbf{i}} + \hat{\mathbf{j}} + \hat{\mathbf{k}}$,

$\therefore$ $A = \sqrt{1^2 + 1^2 + 1^2} = \sqrt{3}$

$\Rightarrow$ $$\cos\theta = \frac{\mathbf{A} \cdot \hat{\mathbf{i}}}{\sqrt{3}} = \frac{1}{\sqrt{3}} = \frac{\sqrt{3}}{3}$$

$$= \frac{1.732}{3} = 0.5773 = \cos 54°44'$$

or $\theta = 54°44' \simeq 55°$

7. Let, **C** be a vector perpendicular to **A** and **B**, then as per question, $k\mathbf{C} = \mathbf{A} \times \mathbf{B}$

or $$k = \frac{(\mathbf{A} \times \mathbf{B})}{\mathbf{C}} = \frac{(2\hat{\mathbf{i}} + 3\hat{\mathbf{j}} + 6\hat{\mathbf{k}}) \times (3\hat{\mathbf{i}} - 6\hat{\mathbf{j}} + 2\hat{\mathbf{k}})}{(6\hat{\mathbf{i}} + 2\hat{\mathbf{j}} - 3\hat{\mathbf{k}})}$$

$$= \frac{(42\hat{\mathbf{i}} + 14\hat{\mathbf{j}} - 21\hat{\mathbf{k}})}{(6\hat{\mathbf{i}} + 2\hat{\mathbf{i}} - 3\hat{\mathbf{k}})} = 7$$

8. Here, $A_1 = A_2 = 1$

and $A_1^2 + A_2^2 + 2A_1A_2\cos\theta = (\sqrt{3})^2 = 3$

or $1 + 1 + 2 \times 1 \times 1 \times \cos\theta = 3$

or $\cos\theta = \frac{1}{2}$

Now, $(\mathbf{A}_1 - \mathbf{A}_2) \cdot (2\mathbf{A}_1 + \mathbf{A}_2) = 2A_1^2 - A_2^2 - A_1A_2\cos\theta$

$$= 2 \times 1^2 - 1^2 - 1 \times 1 \times \frac{1}{2} = \frac{1}{2}$$

9. Here, $\mathbf{A} \cdot \mathbf{B} = (5\hat{\mathbf{i}} + 6\hat{\mathbf{j}} + 3\hat{\mathbf{k}}) \cdot (6\hat{\mathbf{i}} - 2\hat{\mathbf{j}} - 6\hat{\mathbf{k}}) = 0$

Hence, **A** and **B** are mutually perpendicular to each other. Vector product of two vectors is not commutative, hence $\mathbf{A} \times \mathbf{B} \neq \mathbf{B} \times \mathbf{A}$.

10. (a) If $\mathbf{A} = 3\hat{\mathbf{i}} + 4\hat{\mathbf{j}}$, then $|\mathbf{A}| = \sqrt{3^2+4^2} = 5$

(b) $W = (3\hat{\mathbf{i}} + 4\hat{\mathbf{j}}) \cdot 6\hat{\mathbf{j}} = 24\ \mathbf{J}$

(c) $|\mathbf{A} \times \mathbf{B}|$ = Area of parallelogram whose two adjacent sides are represented by two vectors **A** and **B**.

(d) Component of force F in the direction making an angle $\theta = F\cos\theta = 20\cos 60° = 20 \times \frac{1}{2} = 10$ N

11. Here, $\mathbf{A} = (a_x\hat{\mathbf{i}} + a_y\hat{\mathbf{j}} + a_z\hat{\mathbf{k}})$

Let, $\mathbf{B} = (\hat{\mathbf{i}} - \hat{\mathbf{j}})$

Then, $\hat{\mathbf{B}} = \frac{\mathbf{B}}{B} = \frac{\hat{\mathbf{i}} - \hat{\mathbf{j}}}{\sqrt{(1)^2 + (-1)^2}} = \frac{\hat{\mathbf{i}} - \hat{\mathbf{j}}}{\sqrt{2}}$

Component of **A** along the direction of **B** is

$$\mathbf{A} \cdot \hat{\mathbf{B}} = (a_x\hat{\mathbf{i}} + a_y\hat{\mathbf{j}} + a_z\hat{\mathbf{k}}) \cdot \frac{(\hat{\mathbf{i}} - \hat{\mathbf{j}})}{\sqrt{2}} = \frac{(a_x - a_y)}{\sqrt{2}}$$

12. As, $\mathbf{A} \times \mathbf{B} = \mathbf{B} \times \mathbf{A}$ or $(\mathbf{A} \times \mathbf{B}) - (\mathbf{B} \times \mathbf{A}) = 0$

or $(\mathbf{A} \times \mathbf{B}) + (\mathbf{A} \times \mathbf{B}) = 0$

or $2(\mathbf{A} \times \mathbf{B}) = 0$

or $2AB\sin\theta\hat{\mathbf{n}} = 0$

As, $A \neq 0$

nor $B \neq 0$

So, $\sin\theta = 0$

Hence, $\theta = 0°$ or π

13. If $|\mathbf{A}| = |\mathbf{B}| = x$, then $|\mathbf{C}| = \sqrt{2}\,x$.

Now, $\mathbf{A} + \mathbf{B} = -\mathbf{C}$

or $(\mathbf{A} + \mathbf{B}) \cdot (\mathbf{A} + \mathbf{B}) = (-\mathbf{C}) \cdot (-\mathbf{C})$

$\Rightarrow A^2 + B^2 + 2AB\cos\theta = C^2$

$\Rightarrow x^2 + x^2 + 2x^2\cos\theta = 2x^2$

or $\cos\theta = 0$ or $\theta = 90°$

Again, $\mathbf{A} + \mathbf{C} = -\mathbf{B}$

$\Rightarrow (\mathbf{A} + \mathbf{C}) \cdot (\mathbf{A} + \mathbf{C}) = -\mathbf{B} \cdot \mathbf{B}$

or $\mathbf{A} \cdot \mathbf{A} + \mathbf{C} \cdot \mathbf{C} + 2\mathbf{A} \cdot \mathbf{C} = B^2$

or $x^2 + 2x^2 + 2x^2\sqrt{2}\cos\theta = x^2$

or $\cos\theta = -\frac{1}{\sqrt{2}}$

$\Rightarrow \theta = 135°$

Again, $\mathbf{B} + \mathbf{C} = -\mathbf{A}$

or $(\mathbf{B} + \mathbf{C}) \cdot (\mathbf{B} + \mathbf{C}) = (-\mathbf{A}) \cdot (-\mathbf{A})$

or $x^2 + 2x^2 + 2x^2\sqrt{2}\cos\theta = x^2$

or $\cos\theta = \frac{-2x^2}{2x^2\sqrt{2}\cos\theta} = -\frac{1}{\sqrt{2}}$

$\Rightarrow \theta = 135°$

14. As, $\mathbf{A} = 2\hat{\mathbf{i}} + 4\hat{\mathbf{j}}$ and $\mathbf{B} = 5\hat{\mathbf{i}} - p\hat{\mathbf{j}}$

$\therefore A = \sqrt{2^2 + 4^2} = \sqrt{20}$

and $B = \sqrt{5^2 + p^2}$

Now, $\mathbf{A} \cdot \mathbf{B} = 10 - 4p$

If $\mathbf{A} \parallel \mathbf{B}$, then

$$\mathbf{A} \cdot \mathbf{B} = AB\cos 0° = AB$$

$$10 - 4p = \sqrt{20}\sqrt{25 + p^2}$$

Squaring, $100 + 16p^2 - 80p$

$= 20(25 + p^2) = 500 + 20p^2$

or $20p^2 - 16p^2 + 80p + 400 = 0$

or $p^2 + 20p + 100 = 0$

or $(p + 10)^2 = 0$

$\therefore p = -10$

$\therefore \mathbf{B} = 5\hat{\mathbf{i}} + 10\hat{\mathbf{j}}$

$B = \sqrt{5^2 + (10)^2} = \sqrt{125} = 5\sqrt{5}$

15. As, $\mathbf{A} = A\hat{\mathbf{A}}$ and $\mathbf{B} = B\hat{\mathbf{B}}$. Let θ be the angle between **A** and **B** as per question,

$$\cos\alpha = \frac{(A\hat{\mathbf{A}} + B\hat{\mathbf{B}}) \cdot (A\hat{\mathbf{B}} + B\hat{\mathbf{A}})}{|A\hat{\mathbf{A}} + B\hat{\mathbf{B}}||A\hat{\mathbf{B}} + B\hat{\mathbf{A}}|}$$

or $\cos\alpha = \frac{2AB + (A^2 + B^2)\cos\theta}{\sqrt{(A^2 + B^2 + 2AB\cos\theta)^2}}$

or $2AB + (A^2 + B^2)\cos\theta = (A^2 + B^2)\cos\alpha + 2AB\cos\theta\cos\alpha$

or $2AB(1 - \cos\alpha\cos\theta) = (A^2 + B^2)(\cos\alpha - \cos\theta)$

or $\frac{2AB}{A^2 + B^2} = \frac{\cos\alpha - \cos\theta}{1 - \cos\alpha\cos\theta}$

or $\frac{2AB + (A^2 + B^2)}{(A^2 + B^2) - 2AB} = \frac{(\cos\alpha - \cos\theta) + (1 - \cos\alpha\cos\theta)}{(1 - \cos\alpha\cos\theta) - (\cos\alpha - \cos\theta)}$

or $\frac{(A + B)^2}{(A - B)^2} = \frac{(1 + \cos\alpha)(1 - \cos\theta)}{(1 + \cos\theta)(1 - \cos\alpha)} = \frac{\tan^2\theta/2}{\tan^2\alpha/2}$

or $\tan\frac{\alpha}{2} = \left(\frac{A - B}{A + B}\right)\tan\frac{\theta}{2}$

16. For vector $\mathbf{A}_1 + \mathbf{A}_2$, we have

$|\mathbf{A}_1 + \mathbf{A}_2|^2 = (\mathbf{A}_1 + \mathbf{A}_2) \cdot (\mathbf{A}_1 + \mathbf{A}_2)$ $[\because \mathbf{x} \cdot \mathbf{x} = |\mathbf{x}|^2]$

$\Rightarrow |\mathbf{A}_1 + \mathbf{A}_2|^2 = |\mathbf{A}_1|^2 + |\mathbf{A}_2|^2 + 2\mathbf{A}_1 \cdot \mathbf{A}_2$

Given, $|\mathbf{A}_1| = 3$, $|\mathbf{A}_2| = 5$ and $|\mathbf{A}_1 + \mathbf{A}_2| = 5$

So, we have, $(5)^2 = 9 + 25 + 2\mathbf{A}_1 \cdot \mathbf{A}_2$

$\Rightarrow \mathbf{A}_1 \cdot \mathbf{A}_2 = -\frac{9}{2}$

Now, $(2\mathbf{A}_1 + 3\mathbf{A}_2) \cdot (3\mathbf{A}_1 - 2\mathbf{A}_2)$

$= 6|\mathbf{A}_1|^2 - 4\mathbf{A}_1 \cdot \mathbf{A}_2 + 9\mathbf{A}_1 \cdot \mathbf{A}_2 - 6|\mathbf{A}_2|^2$

$= 6|\mathbf{A}_1|^2 - 6|\mathbf{A}_2|^2 + 5\mathbf{A}_1 \cdot \mathbf{A}_2$

Substituting values, we have

$(2\mathbf{A}_1 + 3\mathbf{A}_2) \cdot (3\mathbf{A}_1 - 2\mathbf{A}_2)$

$$= 6(9) - 6(25) + 5\left(-\frac{9}{2}\right) = -118.5$$

17. If the three vectors are coplanar, then their scalar triple product is zero. So, $(\mathbf{A} \times \mathbf{C}) \cdot \mathbf{B} = 0$

or $[(2\hat{\mathbf{i}} + 3\hat{\mathbf{j}} - 2\hat{\mathbf{k}}) \times (-\hat{\mathbf{i}} + 2\hat{\mathbf{j}} + 3\hat{\mathbf{k}})] \cdot [5\hat{\mathbf{i}} + a\hat{\mathbf{j}} + \hat{\mathbf{k}}] = 0$

or $[(13\hat{\mathbf{i}} - 4\hat{\mathbf{j}} + 7\hat{\mathbf{k}}] \cdot [5\hat{\mathbf{i}} + a\hat{\mathbf{j}} + \hat{\mathbf{k}}] = 0$

or $65 - 4a + 7 = 0$ or $a = 18$

18. Given, $\mathbf{A} + \mathbf{B} + \mathbf{C} = 0$, then $\mathbf{A}$, $\mathbf{B}$ and $\mathbf{C}$ are in one plane and are represented by the three sides of a triangle taken in one order.

(a) $\therefore \quad \mathbf{B} \times (\mathbf{A} + \mathbf{B} + \mathbf{C}) = \mathbf{B} \times 0 = 0$

or $\quad \mathbf{B} \times \mathbf{A} + \mathbf{B} \times \mathbf{B} + \mathbf{B} \times \mathbf{C} = 0$

or $\quad \mathbf{B} \times \mathbf{A} + 0 + \mathbf{B} \times \mathbf{C} = 0$

or $\quad \mathbf{A} \times \mathbf{B} = \mathbf{B} \times \mathbf{C} \qquad \ldots \text{(i)}$

$\therefore \quad (\mathbf{A} \times \mathbf{B}) \times \mathbf{C} = (\mathbf{B} \times \mathbf{C}) \times \mathbf{C};$

It cannot be zero.

If $\mathbf{B} \| \mathbf{C}$, then $\mathbf{B} \times \mathbf{C} = 0$, then $(\mathbf{B} \times \mathbf{C}) \times \mathbf{C} = 0$

Thus, option (a) is correct.

(b) $(\mathbf{A} \times \mathbf{B}) \cdot \mathbf{C} = (\mathbf{B} \times \mathbf{C}) \cdot \mathbf{C} = 0$

If $\mathbf{B} \| \mathbf{C}$, then $\mathbf{B} \times \mathbf{C} = 0$, then $(\mathbf{B} \times \mathbf{C}) \times \mathbf{C} = 0$

Thus, option (b) is correct.

(c) $(\mathbf{A} \times \mathbf{B}) = \mathbf{D} = AB \sin \theta\, \mathbf{D}$. The direction of $\mathbf{D}$ is perpendicular to the plane containing $\mathbf{A}$ and $\mathbf{B}$.

$(\mathbf{A} \times \mathbf{B}) \times \mathbf{C} = \mathbf{D} \times \mathbf{C}$. Its direction is in the plane of $\mathbf{A}$, $\mathbf{B}$ and $\mathbf{C}$. Thus, option (c) is correct.

(d) If $C^2 = A^2 + B^2$, then the angle between $\mathbf{A}$ and $\mathbf{B}$ is $90°$.

$$\therefore \quad (\mathbf{A} \times \mathbf{B}) \cdot \mathbf{C} = (AB \sin 90°\ \mathbf{D}) \cdot \mathbf{C} = AB\,(\mathbf{D} \cdot \mathbf{C})$$
$$= ABC \cos 90° = 0.$$

Thus, option (d) is false.

19. If $\mathbf{a}$ and $\mathbf{b}$ are perpendicular to each other.

$$R = \sqrt{a^2 + b^2}$$

If a and b are opposite to each other,

$$\frac{R}{\sqrt{2}} = a - b$$

$$\Rightarrow \quad \frac{R^2}{2} = a^2 + b^2 - 2ab$$

$$\Rightarrow \quad a^2 + b^2 = 2a^2 + 2b^2 - 4ab$$

$$\Rightarrow \quad a^2 + b^2 - 4ab = 0$$

$$\Rightarrow \quad \frac{a^2 + b^2}{ab} = 4$$

$$\text{or} \quad \frac{a}{b} + \frac{b}{a} = 4$$

20. Given, sum of $\mathbf{P}$ and $\mathbf{Q}$ is $\mathbf{R}$. Let angle between $\mathbf{P}$ and $\mathbf{Q}$ is β, then resultant of $\mathbf{P}$ and $\mathbf{Q}$,

$$|\mathbf{R}| = \sqrt{|\mathbf{P}|^2 + |\mathbf{Q}|^2 + 2|\mathbf{P}||\mathbf{Q}| \cos \beta}$$

As, $\quad |\mathbf{R}| = |\mathbf{P}| \qquad$ (given)

So, $\quad |\mathbf{P}|^2 = |\mathbf{P}|^2 + |\mathbf{Q}|^2 + 2|\mathbf{P}||\mathbf{Q}| \cos \beta$

or $\quad |\mathbf{P}| \cos \beta = -\dfrac{Q}{2} \qquad \ldots \text{(i)}$

If resultant of $2\mathbf{P}$ and $\mathbf{Q}$ makes angle θ with $\mathbf{Q}$, then angle θ is given by

$$\tan \theta = \frac{|2\mathbf{P}| \sin \beta}{|\mathbf{Q}| + |2\mathbf{P}| \cos \beta}$$

Substituting the value of $|\mathbf{P}| \cos \beta$ from Eq. (i) in above equation, we get

$$\tan \theta = \infty \Rightarrow \theta = \frac{\pi}{2} = 90°$$

21. If three vectors are coplanar, then $\mathbf{a} \cdot (\mathbf{b} \times \mathbf{c}) = 0$.

$$\Rightarrow \quad (2\hat{\mathbf{i}} - \hat{\mathbf{j}} + \hat{\mathbf{k}}) \cdot \begin{vmatrix} \hat{\mathbf{i}} & \hat{\mathbf{j}} & \hat{\mathbf{k}} \\ 1 & 2 & -3 \\ 3 & -y & 5 \end{vmatrix} = 0$$

$$\Rightarrow \quad (2\hat{\mathbf{i}} - \hat{\mathbf{j}} + \hat{\mathbf{k}}) \cdot \{\hat{\mathbf{i}}\,(10 - 3y) - \hat{\mathbf{j}}(5 + 9) + \hat{\mathbf{k}}\,(-y - 6)\} = 0$$

$$\Rightarrow \quad (2\hat{\mathbf{i}} - \hat{\mathbf{j}} + \hat{\mathbf{k}}) \cdot \{(10 - 3y)\hat{\mathbf{i}} - 14\hat{\mathbf{j}} - (6 + y)\,\hat{\mathbf{k}}\} = 0$$

$$\Rightarrow \quad 2\,(10 - 3y) + 14 - (6 + y) = 0$$

$$\Rightarrow \quad 20 - 6y + 14 - 6 - y = 0$$

$$\Rightarrow \quad 28 - 7y = 0$$

$$\therefore \quad y = 4$$

22. Here, $\mathbf{AB}$ = position vector of B – position vector of A

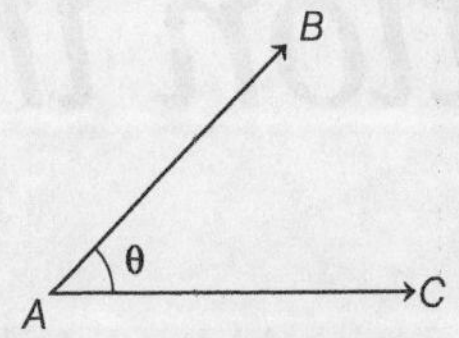

$$= (\hat{\mathbf{i}} - 3\hat{\mathbf{j}} - 5\,\hat{\mathbf{k}}) - (2\hat{\mathbf{i}} + \hat{\mathbf{j}} + \hat{\mathbf{k}})$$
$$= -\hat{\mathbf{i}} - 4\hat{\mathbf{j}} - 6\,\hat{\mathbf{k}}$$

and $\mathbf{AC}$ = position vector of C – position vector of A

$$= (4\hat{\mathbf{i}} - 4\hat{\mathbf{j}} - 4\,\hat{\mathbf{k}}) - (2\hat{\mathbf{i}} + \hat{\mathbf{j}} + \hat{\mathbf{k}}) = 2\hat{\mathbf{i}} - 5\hat{\mathbf{j}} - 5\,\hat{\mathbf{k}}$$

$$\because \quad \mathbf{AC} \cdot \mathbf{AB} = |\mathbf{AC}|\,|\mathbf{AB}| \cos \theta$$

$$\therefore \quad \cos \theta = \frac{\mathbf{AC} \cdot \mathbf{AB}}{|\mathbf{AC}||\mathbf{AB}|}$$

$$= \frac{(-2 + 20 + 30)}{\sqrt{4 + 25 + 25}\,\sqrt{1 + 16 + 36}}$$

$$= \frac{48}{\sqrt{29}\,\sqrt{53}}$$

$$= \frac{48}{\sqrt{1537}} = \frac{8n}{\sqrt{1537}}$$

$$\therefore \quad n = 6$$

23. The volume of parallelopiped is $|\mathbf{a} \cdot (\mathbf{b} \times \mathbf{c})|$

$$= |(2\hat{\mathbf{i}} - 3\hat{\mathbf{j}} + 4\,\hat{\mathbf{k}}) \cdot \{(\hat{\mathbf{i}} + 2\hat{\mathbf{j}} - \hat{\mathbf{k}}) \times (3\hat{\mathbf{i}} - \hat{\mathbf{j}} + 2\hat{\mathbf{k}})\}|$$

$$= \left|(2\hat{\mathbf{i}} - 3\hat{\mathbf{j}} + 4\,\hat{\mathbf{k}}) \cdot \left\{\begin{vmatrix} \hat{\mathbf{i}} & \hat{\mathbf{j}} & \hat{\mathbf{k}} \\ 1 & 2 & -1 \\ 3 & -1 & 2 \end{vmatrix}\right\}\right|$$

$$= |(2\hat{\mathbf{i}} - 3\hat{\mathbf{j}} + 4\,\hat{\mathbf{k}}) \cdot \{\hat{\mathbf{i}}\,(4 - 1) - \hat{\mathbf{j}}\,(2 + 3) + \hat{\mathbf{k}}\,(-1 - 6)\}|$$

$$= |(2\hat{\mathbf{i}} - 3\hat{\mathbf{j}} + 4\,\hat{\mathbf{k}}) \cdot (3\hat{\mathbf{i}} - 5\hat{\mathbf{j}} - 7\,\hat{\mathbf{k}})|$$

$$= |6 + 15 - 28|$$

$$= 7\ \text{m}^3$$

03

Kinematics I (Motion in 1-D)

IN THIS CHAPTER ....

The branch of Physics in which we study the motion of objects or particles and their equilibrium under the action of external forces is known as *mechanics*.

Kinematics is one of the branch of Mechanics. It describes the motion of objects without looking at the cause of the motion. Here, time factor plays an important role.

Frame of Reference

The frame of reference is a suitable coordinate system involving space and time used as a reference to study the motion of different bodies. The most common reference frame is the cartesian frame of reference involving (x, y, z and t).

(i) **Inertial Frame of Reference** A frame of reference which is either at rest or moving with constant velocity is known as inertial frame of reference. In this frame of reference, Newton's first law of motion is applicable.

(ii) **Non-Inertial Frame of Reference** A frame of reference moving with some acceleration is known as non-inertial frame of reference. In this frame of reference, Newton's law of motion is not applicable.

Rest and Motion

An object is said to be at rest, if it does not change its position with time and in a state of motion, if it continuously changes its position with time.

On the basis of number of coordinates required to specify the object, motion of object can be classified as

(i) **One-dimensional motion** The motion of an object is considered as 1-D (one-dimensional), if only one coordinate is needed to specify the position of the object at any time.

(ii) **Two-dimensional motion** The motion of an object is considered as 2-D (two-dimensional), if two coordinates are needed to specify the position of the object at any time.

In 2-D motion, the object moves in a plane.

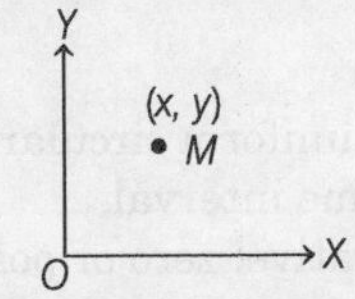

(iii) **Three-dimensional motion** The motion of an object is considered as 3-D (three-dimensional), if all the three coordinates are needed to specify the position of the object.

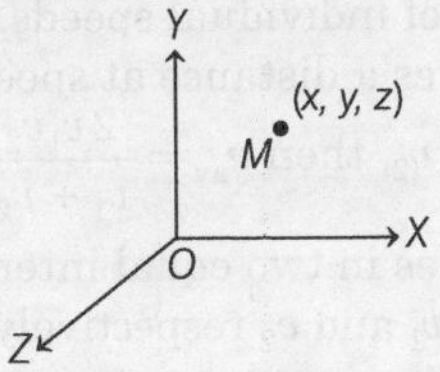

Basic Terms Related to Motion

Here, we will consider the motion of a point object in a straight line in one dimension, therefore, the motion of the point object can be described by specifying the distance x of the point object and the corresponding instant of time t. Mathematically, the position of the object in one dimensional motion can be expressed as follows

$$x = x(t)$$

or $$x = f(t)$$

Here, the distance x is the function of the time t.

Position Vector

It describes the instantaneous position of a particle with respect to the chosen frame of reference. It is a vector joining the origin to the particle. If at any time (x, y, z) be coordinates of the particle, then its position vector is given by $\mathbf{r} = x\hat{\mathbf{i}} + y\hat{\mathbf{j}} + z\hat{\mathbf{k}}$.

In one dimensional motion, position vector may be given by $\mathbf{r} = x\hat{\mathbf{i}}$, $y = z = 0$ (along X-axis). In two dimensional motion, $\mathbf{r} = x\hat{\mathbf{i}} + y\hat{\mathbf{j}}$ (in xy-plane $z = 0$).

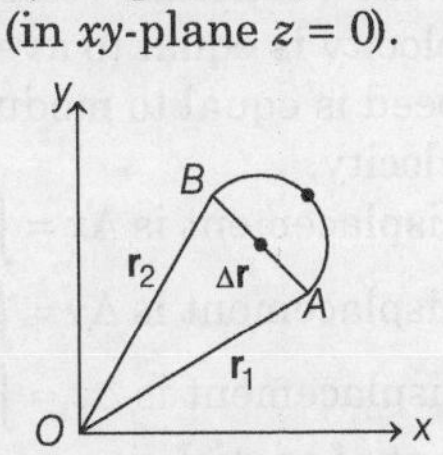

Distance and Displacement

Distance is the total length of the path travelled by the particle in a given interval of time.

Displacement is a vector joining the initial position of the particle to its final position in a given interval of time. Mathematically, it is equal to the change in position vectors, *i.e.* $\Delta\mathbf{r} = \mathbf{r}_2 - \mathbf{r}_1$

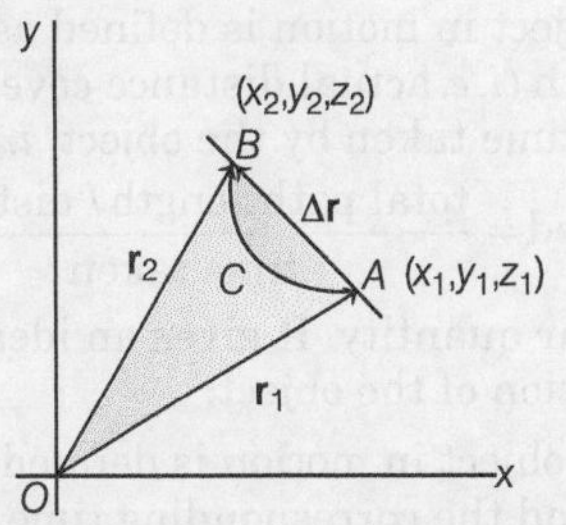

The direction of displacement is directed from A to B through the straight line AB and the magnitude of displacement is

$$|\mathbf{AB}| = \sqrt{(x_2 - x_1)^2 + (y_2 - y_1)^2 + (z_2 - z_1)^2}$$

Distance *vs* Displacement

- Distance is a scalar quantity and displacement is a vector quantity.
- For motion between two points, displacement is single valued while distance depends on actual path and so can have many values.
- Path length or distance is a positive scalar quantity which does not decrease with time and can never be zero for a moving body. Displacement of a body can be zero or negative.
- Magnitude of displacement can never be greater than distance travelled by the body.
- When a body returns to its initial position, its displacement is zero but distance or path length is non-zero.

Example 1. *A particle moves along a circle of radius R. It starts from A and moves in an anti-clockwise direction. What is the distance and displacement of the particle from A to D?*

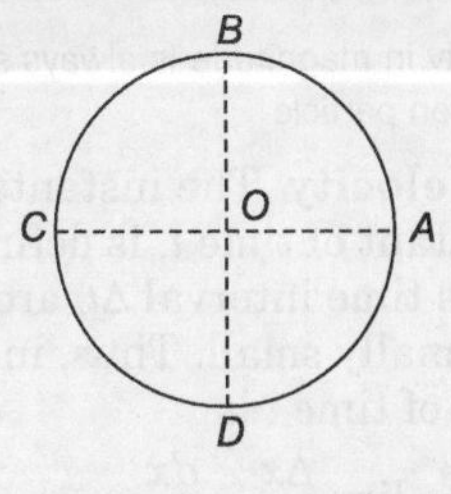

(a) $\dfrac{\pi R}{2}, \sqrt{2R}$ (b) $\pi R, 2R$

(c) $\dfrac{3}{2}\pi R, \sqrt{2}R$ (d) $2\pi R$, zero

Sol. (c) For the motion from A to D,

$$\text{Distance travelled} = \frac{2\pi R \times 3}{4} = \frac{3}{2}\pi R$$

$$\therefore \quad \text{Displacement} = |\mathbf{AD}| = \sqrt{(OA)^2 + (OD)^2}$$

$$= \sqrt{R^2 + R^2} = \sqrt{2}R$$

Speed and Velocity

Speed of an object in motion is defined as the ratio of total path length (*i.e.* actual distance covered) and the corresponding time taken by the object, *i.e.*

$$\text{speed} = \frac{\text{total path length / distance}}{\text{time taken}}$$

Speed is a scalar quantity. It gives an idea about the direction of motion of the object.

Velocity of an object in motion is defined as the ratio of displacement and the corresponding time interval taken by the object, *i.e.*

$$\text{velocity} = \frac{\text{displacement}}{\text{time interval}}$$

Velocity is a vector quantity as it has both the magnitude (speed) and direction.

Average Velocity & Speed and Instantaneous Velocity & Speed

Average Velocity Average velocity $\bar{v}$ of an object moving through a displacement (Δx) during a time interval (Δt) is given by

$$\bar{v} = \frac{\Delta x}{\Delta t} = \frac{x_2 - x_1}{t_2 - t_1}$$

If the beginning and ending velocities for the given motion are known and the acceleration is constant, then the average velocity can also be expressed as

$$v_{\text{average}} = \bar{v} = \frac{v_1 + v_2}{2}$$

Average Speed Average speed is a measurement of the total distance travelled in a given period of time. It is sometimes referred as the distance per time ratio.

$$\text{Average Speed, } v_{\text{av}} = \frac{\text{Total distance travelled}}{\text{Total time taken}}$$

Note Average velocity in magnitude is always smaller than or equal to average speed of a given particle.

Instantaneous Velocity The instantaneous velocity of an object at an instant of time t, is defined as the limit of average velocity as time interval Δt, around time t becomes infinitesimally small. Thus, instantaneous velocity at instant of time t is

$$v_i = \lim_{\Delta t \to 0} \frac{\Delta x}{\Delta t} = \frac{dx}{dt}$$

where, dx/dt = the differential coefficient of x w.r.t. time.

Instantaneous Speed The instantaneous speed is defined as the limit of the average speed as the time interval (Δt) becomes infinitesimally small or approaches to zero.

Thus, instantaneous speed at instant of time t is

$$s_i = \lim_{\Delta t \to 0} \frac{\Delta s}{\Delta t} = \frac{ds}{dt}$$

where, ds is distance for time dt.

Important Points about Speed and Velocity

- The velocity in the uniform circular motion does not depend upon the time interval.
- Velocity can be negative, zero or positive, but speed is never negative.
- If motion takes place in the same direction, then the average speed and average velocity are the same.
- If a particle travels equal distances at speeds $v_1, v_2, v_3, \ldots$ etc respectively, then the average speed is harmonic mean of individual speeds.
- If a particle moves a distance at speed v_1 and comes back with speed v_2, then $v_{\text{av}} = \frac{2v_1v_2}{v_1 + v_2}$ but $\mathbf{v}_{\text{av}} = 0$.
- If a particle moves in two equal intervals of time at different speeds v_1 and v_2 respectively, then $v_{\text{av}} = \frac{v_1 + v_2}{2}$.
- The average velocity between two points in a time interval can be obtained from a position *versus* time graph by calculating the slope of the straight line joining the coordinates of the two points.

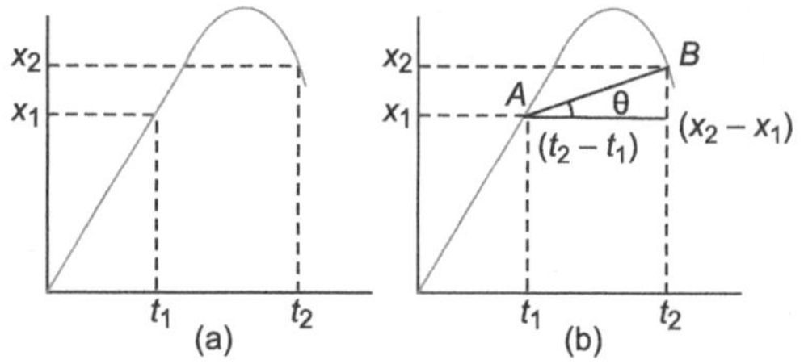

The graph, describes the motion of a particle moving along X-axis (along a straight line).

Suppose, we wish to calculate the average velocity between $t = t_1$ and $t = t_2$. The slope of chord AB [shown in Fig. (b)] gives the average velocity.

Mathematically, $\quad v_{\text{av}} = \tan\theta = \frac{x_2 - x_1}{t_2 - t_1}$

- If a body moves with a constant velocity, the instantaneous velocity is equal to average velocity. The instantaneous speed is equal to modulus of instantaneous velocity.
- x-component of displacement is $\Delta x = \int v_x \, dt$

  y-component of displacement is $\Delta y = \int v_y \, dt$

  z-component of displacement is $\Delta z = \int v_z \, dt$

  Thus, displacement of particle is

  $$\Delta \mathbf{r} = \Delta x \hat{\mathbf{i}} + \Delta y \hat{\mathbf{j}} + \Delta z \hat{\mathbf{k}}$$
- The magnitude of instantaneous velocity is equal to the instantaneous speed at a given instant.
- If during motion, velocity remains constant throughout a given interval of time, then the motion is said to be uniform.

For uniform motion, instantaneous velocity = average velocity = uniform velocity.
However, converse may or may not be true, *i.e.* if average velocity = instantaneous velocity, the motion may or may not be uniform.

Example 2. *A car travels a distance A to B at a speed of 40 kmh^{-1} and returns to A at a speed of 30 kmh^{-1}. What is the average speed for the whole journey?*

(a) Zero (b) 34.3 kmh^{-1} (c) 68.6 kmh^{-1} (d) 120 kmh^{-1}

Sol. (b) Let $AB = s$, time taken to go from A to B, $t_1 = \frac{s}{40}$ h

and time taken to go from B to A, $t_2 = \frac{s}{30}$ h

$$\therefore \quad \text{Total time taken} = t_1 + t_2 = \frac{s}{40} + \frac{s}{30} = \frac{(3+4)s}{120} = \frac{7s}{120} \text{ h}$$

Total distance travelled $= s + s = 2s$

$$\therefore \quad \text{Average speed} = \frac{\text{Total distance travelled}}{\text{Total time taken}} = \frac{2s}{7s/120} = \frac{120 \times 2}{7} = 34.3 \text{ kmh}^{-1}$$

Example 3. *A man walks on a straight road from his home to a market 3 km away with a speed of 6 kmh^{-1} finding the market closed, he instantly turns and walks back with a speed of 9 kmh^{-1}. What is the magnitude of average velocity and average speed of the man, over the interval of time 0 to 40 min ?*

(a) 2.25 kmh^{-1}, 6.75 kmh^{-1} (b) 6.75 kmh^{-1}, 2.25 kmh^{-1}
(c) zero, 2.25 kmh^{-1} (d) zero, 6.75 kmh^{-1}

Sol. (a) Time taken by man to go from his home to market,

$$t_1 = \frac{\text{distance}}{\text{speed}} = \frac{3 \text{ km}}{6 \text{ kmh}^{-1}} = \frac{1}{2} \text{ h} = 30 \text{ min}$$

Time taken by man to go from market to home,

$$t_2 = \frac{3 \text{ km}}{9 \text{ kmh}^{-1}} = \frac{1}{3} \text{ h} = 20 \text{ min}$$

Distance moved in 30 min (from home to market) = 3.0 km
Distance moved in 10 min (from market to home) with speed

$$9 \text{ kmh}^{-1} = 9 \times \frac{1}{6} = 1.5 \text{ km}$$

So, displacement = 3.0 − 1.5 = 1.5 km
Total path length travelled = 3.0 + 1.5 = 4.5 km

$$\text{Average velocity} = \frac{1.5 \text{ km}}{(40/60) \text{ h}} = 2.25 \text{ kmh}^{-1}$$

$$\text{Average speed} = \frac{4.5 \text{ km}}{(40/60) \text{ h}} = 6.75 \text{ kmh}^{-1}$$

Example 4. *A particle travels half the distance with velocity u. The remaining part of the distance is covered with velocity v_1 for the first half time and v_2 for the remaining half time.*

The average velocity of the particle during the complete motion is

(a) $\frac{2u + v_1 + v_2}{2u(v_1 + v_2)}$ (b) $\frac{2u(v_1 + v_2)}{2u + v_1 + v_2}$

(c) $\frac{v_1 + v_2}{v_1 - v_2}$ (d) $\frac{v_1 - v_2}{v_1 + v_2}$

Sol. (b) When time intervals are equal, then distances are equal

$$v_{av} = \frac{2uv_m}{u + v_m} = \frac{2u\left(\frac{v_1 + v_2}{2}\right)}{u + \left(\frac{v_1 + v_2}{2}\right)} = \frac{2u(v_1 + v_2)}{2u + v_1 + v_2}$$

Example 5. *The coordinates of a moving particle at any time t are given by $x = \alpha t^3$ and $y = \beta t^3$. The speed of the particle at time t is given by*

(a) $3t\sqrt{\alpha^2 + \beta^2}$ (b) $3t^2\sqrt{\alpha^2 + \beta^2}$

(c) $t^2\sqrt{\alpha^2 + \beta^2}$ (d) $\sqrt{\alpha^2 + \beta^2}$

Sol. (b) The given co-ordinate

$$x = \alpha t^3, \quad y = \beta t^3$$

Then, $v_x = \frac{dx}{dt} = 3\alpha t^2$ and $v_y = \frac{dy}{dt} = 3\beta t^2$

$$\therefore \text{ Resultant velocity, } v = \sqrt{v_x^2 + v_y^2} = \sqrt{9\alpha^2 t^4 + 9\beta^2 t^4} = 3t^2\sqrt{\alpha^2 + \beta^2}$$

Example 6. *A particle is moving according to graph is shown in figure. What is the average velocity in the interval of 3 s to 8 s?*

(a) 10 ms^{-1} (b) 10 ms^{-2} (c) 8 ms^{-1} (d) 3 ms^{-1}

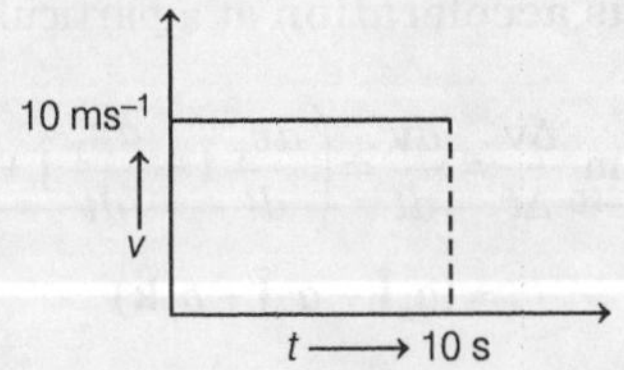

Sol. (a) In uniform motion, average velocity equals to instantaneous velocity. Hence, average velocity will be 10 ms^{-1}.

Example 7. *A particle is moving according to the equation $x = 5t^2 - 20t + 4$. What is the average velocity between time $t_1 = 0$ s to $t_2 = 4$ s? (where, x = displacement, t = time)*

(a) 20t ms^{-1} (b) 20 ms^{-2}
(c) 10 ms^{-1} (d) None of these

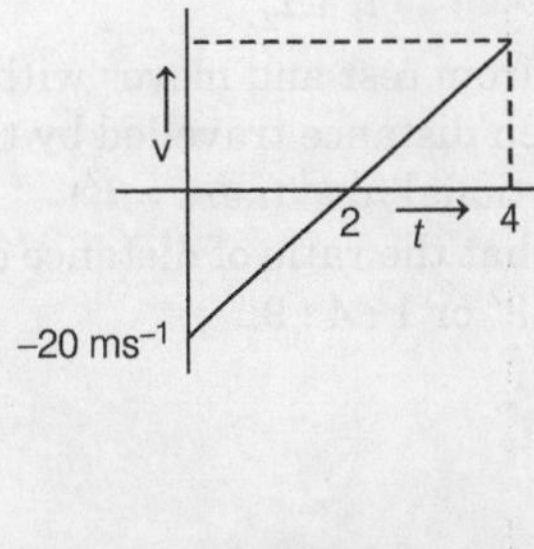

Sol. (c) Given that, $x = 5t^2 - 20t + 4$

$$v = \frac{dx}{dt} = 10t - 20$$

At $t = 0$, $v_i = -20 \text{ ms}^{-1}$

$t = 2$ s, $v_f = 0 \text{ ms}^{-1}$

$t = 4$ s, $v = 20 \text{ ms}^{-1}$

Area of *v-t* graph gives displacement and distance.

Average velocity = 0

Average speed $= \frac{40}{4} = 10 \text{ ms}^{-1}$

Uniform and Non-uniform Motion

Uniform Motion An object is said to be in uniform motion if its velocity is uniform, *i.e.* it undergoes equal displacement in equal intervals of time, howsoever small these intervals may be. For a uniform motion along a straight line in a given direction, the magnitude of displacement is equal to the actual distance covered by the object.

Non-uniform Motion An object is said to be in non-uniform motion if it undergoes equal displacement in unequal intervals of time, howsoever small these intervals may be. Clearly, in non-uniform motion, the velocity of an object is different at different instants.

Acceleration

Acceleration of an object is defined as rate of change of velocity. It is a vector quantity and its SI unit is m/s².

Average acceleration for a given time

$$\mathbf{a}_{av} = \frac{\mathbf{v}_2 - \mathbf{v}_1}{t_2 - t_1} = \frac{\Delta \mathbf{v}}{\Delta t}$$

Instantaneous acceleration at a particular instant is defined as

$$\mathbf{a}_{ins} = \lim_{\Delta t \to 0} \frac{\Delta \mathbf{v}}{\Delta t} = \frac{d\mathbf{v}}{dt} = \left(\frac{dv_x}{dt} \hat{\mathbf{i}} + \frac{dv_y}{dt} \hat{\mathbf{j}} + \frac{dv_z}{dt} \hat{\mathbf{k}} \right)$$

$$= (a_x \hat{\mathbf{i}} + a_y \hat{\mathbf{j}} + a_z \hat{\mathbf{k}})$$

Uniformly Accelerated Motion

In an accelerated motion, if the change in velocity of an object in each unit of time is constant, the object is said to be moving with **constant acceleration** and such a motion is called uniformly accelerated motion.

- If a particle is accelerated for a time t_1 with acceleration a_1 and for time t_2 with acceleration a_2, then average acceleration is

$$a_{av} = \frac{a_1 t_1 + a_2 t_2}{t_1 + t_2}$$

- If a body starts from rest and moves with uniform acceleration, then distance travelled by the body in t second is proportional to t^2 (*i.e.* $s \propto t^2$)

  So, we can say that the ratio of distance covered in 1*s*, 2*s*, 3*s*, is $1^2 : 2^2 : 3^2$ or 1 : 4 : 9.

- A particle moving with a uniform acceleration from A to B along a straight line has velocities v_1 and v_2 at A and B respectively. If C is the mid-point between A and B, then velocity of the particle at C is equal to $v = \sqrt{\frac{v_1^2 + v_2^2}{2}}$.

Example 8. *A particle travels according to the equation $a = A - Bv$, where a is acceleration, A and B are constants and v is velocity of the particle. It's velocity as a function of time is*

(a) $\frac{B}{A}(1 - e^{-At})$ (b) $\frac{B}{A}(1 - e^{-Bt})$ (c) $\frac{A}{B}(1 - e^{-Bt})$ (d) $\frac{A}{B}(1 - e^{-At})$

Sol. (c) Given acceleration, $a = \frac{dv}{dt}$

$$\therefore \quad \frac{dv}{dt} = A - Bv$$

$$\Rightarrow \quad \frac{dv}{A - Bv} = dt$$

$$\int_0^V \frac{dv}{A - Bv} = \int_0^t dt$$

$$-\frac{1}{B} \log_e (A - Bv) \Big|_0^V = t$$

$$\Rightarrow \quad \log_e \frac{A - Bv}{A} = -Bt$$

$$\Rightarrow \quad A - Bv = Ae^{-Bt}$$

$$\therefore \quad v = \frac{A}{B}(1 - e^{-Bt})$$

Example 9. *A body moves along a straight line with an acceleration 3 ms^{-2} for 2 s and then with an acceleration 4 ms^{-2} for 3 s. What is his average acceleration?*

(a) 3.4 ms^{-2} (b) 3.5 ms^{-2}

(c) 3.6 ms^{-2} (d) 3.7 ms^{-2}

Sol. (c) Average acceleration,

$$a_{av} = \frac{a_1 t_1 + a_2 t_2}{t_1 + t_2} = \frac{3 \times 2 + 4 \times 3}{2 + 3} = 3.6 \text{ ms}^{-2}$$

Kinematics Equations for Uniformly Accelerated Motion

We can establish the relation between velocity, acceleration and the distance travelled by the body in a particular time interval by a set of equations. These equations are known as *kinematics equations* or *equations of motion*.

If u be the initial velocity, v the final velocity and distance covered by the body in time t is s, then the equations of motion are as under

(i) $v = u + at$

(ii) $s = ut + \frac{1}{2}at^2$

(iii) $v^2 - u^2 = 2as$ and

(iv) $s_{nth} = u + \frac{a}{2}(2n - 1)$

For motion in a plane, we may consider motion of an object along X-axis and Y-axis independently and then combine the two motions so as to get the net motion of the particle. Thus, we have

(i) $\mathbf{v} = \mathbf{u} + \mathbf{a}\,t$,

$$v_x = u_x + a_x t,$$
$$v_y = u_y + a_y t$$

and $|\mathbf{v}| = \sqrt{v_x^2 + v_y^2}$

$$= (v_x\hat{\mathbf{i}} + v_y\hat{\mathbf{j}}) = (u_x\hat{\mathbf{i}} + u_y\hat{\mathbf{j}}) + (a_x\hat{\mathbf{i}} + a_y\hat{\mathbf{j}})\,t$$

(ii) $\mathbf{s} = \mathbf{u}\,t + \frac{1}{2}\mathbf{a}t^2$,

$$s_x = u_x t + \frac{1}{2}a_x t^2,\; s_y = u_y t + \frac{1}{2}a_y t^2$$

and $|\mathbf{s}| = \sqrt{s_x^2 + s_y^2}$

(iii) $v_x^2 - u_x^2 = 2\,a_x \cdot s_x,\; v_y^2 - u_y^2 = 2\,a_y \cdot s_y$

Equations of motion in free space are similar to those for motion in a plane.

Example 10. *Starting from the origin at time $t = 0$, with initial velocity $5\hat{\mathbf{j}}\ ms^{-1}$, a particle moves in the xy-plane with a constant acceleration of $(10\hat{\mathbf{i}} + 4\hat{\mathbf{j}})\ ms^{-2}$. At time t, its co-ordinates are (20 m, y_0 m). The values of t and y_0 respectively, are* **[JEE Main 2020]**

(a) 2 s and 18 m (b) 5 s and 25 m
(c) 2 s and 24 m (d) 4 s and 52 m

Sol. (a) $\mathbf{u} = (5\hat{\mathbf{j}})$ m/s $= (0\hat{\mathbf{i}} + 5\hat{\mathbf{j}})$ m/s

$$\mathbf{a} = (10\hat{\mathbf{i}} + 4\hat{\mathbf{j}})\ \text{m/s}^2$$
$$\mathbf{s} = [(20 - 0)\hat{\mathbf{i}} + (y_0 - 0)\hat{\mathbf{j}}]\ \text{m} = (20\hat{\mathbf{i}} + y_0\hat{\mathbf{j}})\ \text{m}$$

Using second equation of motion in x-direction,

$$s_x = u_x t + \frac{1}{2}a_x t^2$$
$$20 = (0)t + \frac{1}{2}(10)\,(t)^2$$
$$20 = 0 + 5t^2$$
$$t^2 = 4$$
$$t = 2\ \text{s}$$

Using second equation of motion in y-direction,

$$s_y = u_y t + \frac{1}{2}a_y t^2 = 5\times2 + \frac{1}{2}\times4\times(2)^2 = 10 + 8 = 18\ \text{m}$$

Hence, correct option is (a).

Example 11. *A particle starts from the origin at $t = 0$ with an initial velocity of $3.0\hat{\mathbf{i}}$ m/s and moves in the xy-plane with a constant acceleration $(6.0\hat{\mathbf{i}} + 4.0\hat{\mathbf{j}})\ m/s^2$. The x-coordinate of the particle at the instant when its y-coordinate is 32 m is D metres. The value of D is* **[JEE Main 2020]**

(a) 50 (b) 60 (c) 40 (d) 32

Sol. (b) Given,

At $t = 0$, initial velocity of particle is $\mathbf{x} = 3\hat{\mathbf{i}}\,\text{ms}^{-1}$

Acceleration of particle, $\mathbf{a} = (6\hat{\mathbf{i}} + 4\hat{\mathbf{j}})\text{ms}^{-2}$

By $s = ut + \frac{1}{2}at^2$, we have

$$\mathbf{r} = (3\hat{\mathbf{i}})t + \frac{1}{2}(6\hat{\mathbf{i}} + 4\hat{\mathbf{j}})t^2$$
$$\Rightarrow \mathbf{r} = (3t + 3t^2)\hat{\mathbf{i}} + 2t^2\hat{\mathbf{j}}$$

So, at time t, x-coordinate of particle, $x = 3t + 3t^2$

and y-coordinate of particle, $y = 2t^2$

When $y = 32$ m

$\Rightarrow$ $2t^2 = 32$

$\Rightarrow$ $t^2 = 16$ or $t = 4$s

Value of x-coordinate at $t = 4$s,

$$x = (3t + 3t^2)_{t=4} = 12 + 48 = 60\ \text{m}$$

Hence, $D = 60$ m

Motion Under Gravity

When an object falls towards the earth, an acceleration is involved, this acceleration is due to earth's gravitational pull and is called gravitation acceleration/acceleration due to gravity (g).

The magnitude of gravitational acceleration near the surface of earth is $g = 9.8\ \text{ms}^{-2} = 32\ \text{fts}^{-2}$.

Cases of Motion Under Gravity

***Case* I If particle is moving upwards**

In this case, applicable kinematics relations are

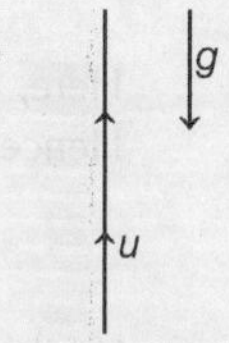

$$v = u - gt \quad \text{... (i)}$$
$$h = ut - \frac{1}{2}gt^2 \quad \text{... (ii)}$$
$$v^2 = u^2 - 2gh \quad \text{... (iii)}$$

Here, h is the vertical height of the particle in upward direction.

Note For maximum height attained by a particle

$$h = h_{max}, v = 0$$

i.e. $(0)^2 = u^2 - 2gh_{max}$

$\therefore$ $h_{max} = \dfrac{u^2}{2g}$

***Case* II If particle is moving vertically downwards.**

In this is case,

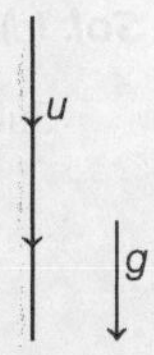

$$v = u + gt \quad \text{... (i)}$$
$$h = ut + \frac{1}{2}gt^2 \quad \text{... (ii)}$$
$$v^2 = u^2 + 2gh \quad \text{... (iii)}$$

Here, h is the vertical height of particle in downward direction.

Note Sign Conventions Normally, vertically upward motion is taken as negative and vertically downward motion is taken as positive. Similarly, for horizontally rightward motion is taken positive and leftward motion is taken negative.

Example 12. *A ball is dropped from a high tower such that distance covered by it in last second of its motion is same as the distance covered by it during first three seconds. Find the height of tower. (Take, g = 10 ms^{-2})*

(a) 125 m (b) 125 cm (c) 120 m (d) 120 cm

Sol. (a) Let ball takes t seconds to reach the ground and h be the height of tower. Then,

$$h = 0 + \frac{1}{2}gt^2 \quad \text{...(i)}$$

and $$h_{nth} = 0 + \frac{g}{2}(2t-1) = 0 + \frac{1}{2}g(3)^2$$

or $$2t - 1 = 9 \quad \text{...(ii)}$$

$\Rightarrow$ $$t = 5\text{ s}$$

and $$h = \frac{1}{2} \times 10 \times (5)^2 = 125\text{m}$$

Example 13. *A particle is projected vertically upwards with velocity 40 ms^{-1}. Find the displacement and distance travelled by the particle in 6 s. (Take g =10 ms^{-2})*

(a) 60 m, 100 m (b) 60 m, 120 m
(c) 40 m, 100 m (d) 40 m, 80 m

Sol. (a) Here, u is positive (upwards) and a is negative). So, first we will find t_0, the time when velocity becomes zero, *i.e.* when the particle is the highest point.

$$t_0 = \left|\frac{u}{a}\right| = \frac{40}{10} = 4\text{ s}$$

Here, $t > t_0$
Hence, distance > displacement

$$s = 40 \times 6 - \frac{1}{2} \times 10 \times 36 = 60\text{ m}$$

While, $$d = \left|\frac{u^2}{2a}\right| + \frac{1}{2}|a(t-t_0)^2|$$

$$= \frac{(40)^2}{2 \times 10} + \frac{1}{2} \times 10 \times (6-4)^2 = 100\text{ m}$$

Example 14. *A ball is thrown upwards from the ground with an initial speed u. The ball is at a height of 80 m at two times, the time interval being 6s. Then, the value of u is (Take, g = 10 ms^{-2})*

(a) 20 ms^{-1} (b) 30 ms^{-1} (c) 40 ms^{-1} (d) 50 ms^{-1}

Sol. (d) Here, $a = g = -10$ ms^{-2} and $s = 80$ m

Substituting the values in $s = ut + \frac{1}{2}at^2$, we have

$$80 = ut - 5t^2$$

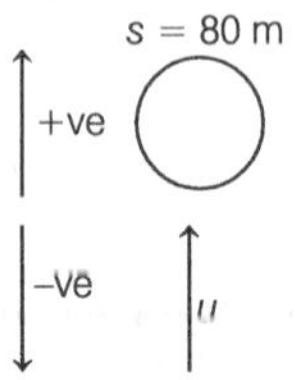

or $$5t^2 - ut + 80 = 0$$

or $$t = \frac{u + \sqrt{u^2 - 1600}}{10}$$

and $$t = \frac{u - \sqrt{u^2 - 1600}}{10}$$

Now, it is given that

$$\frac{u + \sqrt{u^2 - 1600}}{10} - \frac{u - \sqrt{u^2 - 1600}}{10} = 6$$

or $$\frac{\sqrt{u^2 - 1600}}{5} = 6$$

or $$\sqrt{u^2 - 1600} = 30$$

or $$u^2 - 1600 = 900$$

$\therefore$ $$u^2 = 2500$$

or $$u = \pm 50\text{ ms}^{-1}$$

Ignoring the negative sign, we have

$$u = 50\text{ ms}^{-1}$$

Example 15. *A ball is thrown vertically upwards with a velocity of 20 ms^{-1} from the top of a multistorey building. The height of the point from where the ball is thrown is 25 m from the ground. How long will it be before the ball hits the ground? (Take, g = 10 ms^{-2})*

(a) 1 s (b) 3 s
(c) 5 s (d) 7 s

Sol. (c) We split the motion of ball in two parts, the upward motion (*A* to *B*) and the downward motion (*B* to *C*) and calculate the corresponding time taken t_1 and t_2. Since, the velocity at *B* is zero, we have

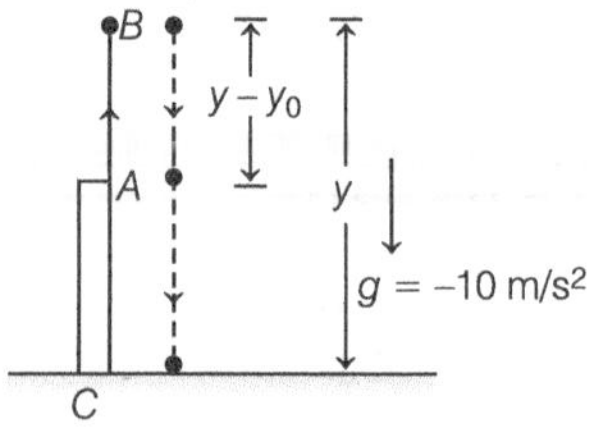

$$v = v_0 + gt$$
$$0 = 20 - 10t_1$$
$$t_1 = 2\text{ s}$$

This is the time in going from *A* to *B*. From *B* or the point of the maximum height the ball falls freely under the acceleration due to gravity. The ball is moving in negative *y*-direction. We use equation

$$y = y_0 + v_0 t + \frac{1}{2}at^2$$

We have, $y_0 = 45$ m, $y = 0, v_0 = 0$, $a = -g = -10\text{ms}^{-2}$

$$0 = 45 + \frac{1}{2}(-10)t_2^2$$

$\Rightarrow$ $$t_2 = 3\text{ s}$$

Total time taken by the ball before it hits the ground

$$= t_1 + t_2 = 2\text{ s} + 3\text{ s} = 5\text{ s}$$

Non-Uniformly Accelerated Motion

When motion of a particle is not uniform, *i.e.* acceleration of particle is not constant or acceleration is a function of time, then following relations hold for one dimensional motion.

(i) $v = \frac{ds}{dt}$

(ii) $a = \frac{dv}{dt} = v\frac{dv}{ds}$

(iii) $ds = v\,dt$ and

(iv) $dv = a\,dt$ or $v\,dv = a\,ds$

Example 16. *A particle is moving with a velocity of* $v = (3 + 6t + 9t^2)\ cms^{-1}$ *The displacement of the particle in time interval* $t = 5$ *s to* $t = 8$ *s is*

(a) 1352 *cm* (b) 1287 *cm*
(c) 1182 *cm* (b) 11000 *cm*

Sol. (*b*) Given, $v = (3 + 6t + 9t^2)$ cms^{-1}

or $\frac{ds}{dt} = (3 + 6t + 9t^2)$

or $ds = (3 + 6t + 9t^2)\,dt$

$\therefore \int_0^s ds = \int_0^t (3 + 6t + 9t^2)\,dt$

$\therefore s = [3t + 3t^2 + 3t^3]_5^8$

or $s = 1287$ cm

Example 17. *The motion of a particle along a straight line is described by the function* $x = (2t - 3)^2$, *where x is in metres and t is in seconds. The acceleration at t = 2 s is*

(a) 5 *ms*$^{-2}$ (b) 6 *ms*$^{-2}$ (c) 7 *ms*$^{-2}$ (d) 8 *ms*$^{-2}$

Sol. (*d*) Position, $x = (2t - 3)^2$

Velocity, $v = \frac{dx}{dt} = 4(2t - 3)\ \text{ms}^{-1}$

and acceleration, $a = \frac{dv}{dt} = 8\ \text{ms}^{-2}$

At $t = 2$ s, $a = 8\ \text{ms}^{-2}$

Graphs in One Dimensional Motion

The tabular forms of *s-t* and *v-t* graphs are given for one dimensional motion with uniform velocity or with constant acceleration.

Position-Time Graph

(i) Position-time graph gives instantaneous value of displacement at any instant.

(ii) The slope of tangent drawn to the graph at any instant of time gives the instantaneous velocity at that instant.

(iii) The *s-t* graph cannot make sharp turns.

Different Cases of Position-Time Graph

Different Cases	*s-t* Graph	Features of Graph
At rest	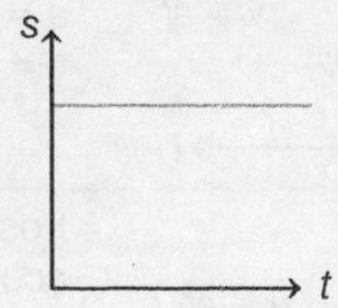	Slope = v = 0
Uniform motion	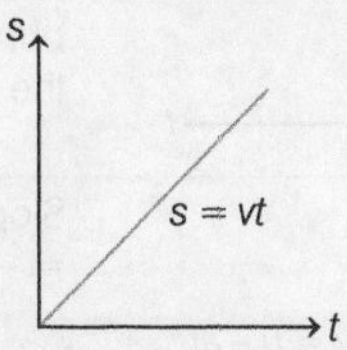	Slope = constant, v = constant, a = 0
Uniformly accelerated motion with $u = 0$, $s = 0$ at $t = 0$	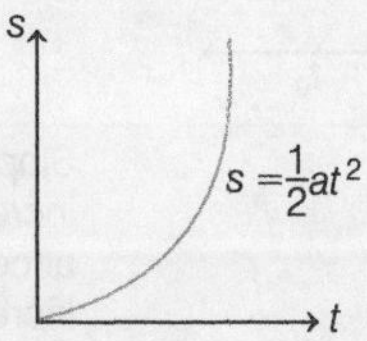	u=0, *i.e.* Slope of *s-t* graph at t=0, should be zero.
Uniformly accelerated motion with $u \neq 0$ but $s = 0$ at $t = 0$	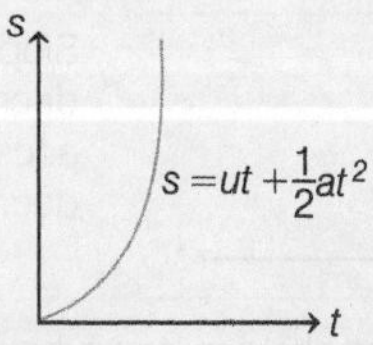	Slope of *s-t* graph gradually goes on increasing
Uniformly retarded motion	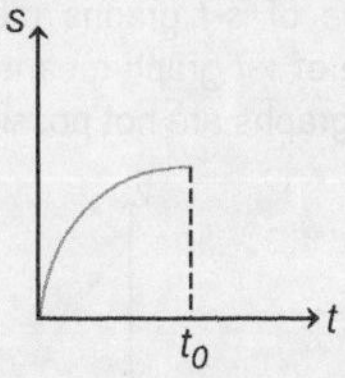	Slope of graph is decreasing with time so, v is decreasing, a is negative

Velocity-Time Graph

(i) Velocity-time graph gives the instantaneous value of velocity at any instant.

(ii) The slope of tangent drawn on graph gives instantaneous acceleration.

(iii) Area under *v-t* graph with time axis gives the value of displacement covered in given time.

(iv) The *v-t* curve cannot take sharp turns.

Different Cases of Velocity-Time Graph

Different Cases	*v-t* Graph	Features of Graph
Uniform motion	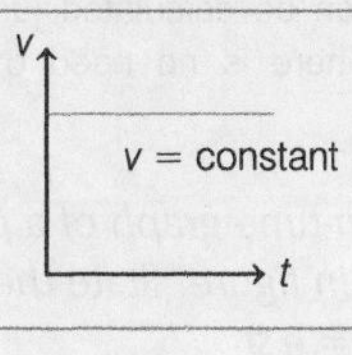	(i) $\theta = 0^\circ$ (ii) v = constant (iii) Slope of *v-t* graph = a = 0

Different Cases	v-t Graph	Features of Graph
Uniformly accelerated motion with $u = 0$ and $s = 0$ at $t = 0$	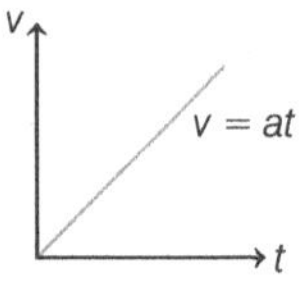	So slope of v-t graph is constant. *i.e.* $u = 0$, so a = constant. Also $v = 0$ at $t = 0$.
Uniformly accelerated motion with $u \neq 0$ but $s = 0$ at $t = 0$	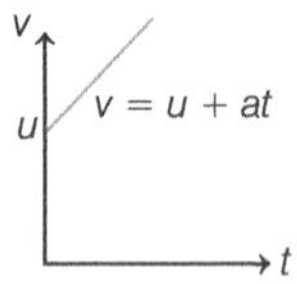	Positive constant acceleration because θ is constant and $<90^\circ$ but the initial velocity of the particle is positive.
Uniformly decelerated motion	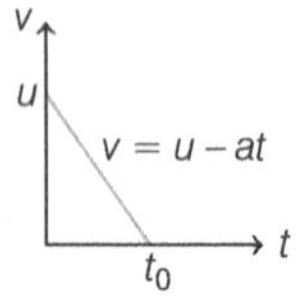	Slope of v-t graphs $= -a$ (retardation)
Non-uniformly accelerated motion	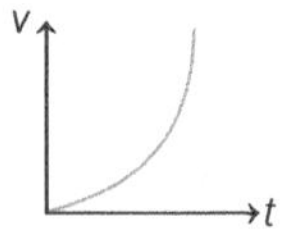	Slope of v-t graph increases with time, so acceleration is increasing.
Non-uniformly decelerating motion	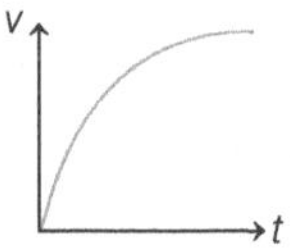	Slope of v-t graph is decreasing, so acceleration is decreasing.

Note (i) Slope of s-t or v-t graphs can never be infinite at any point, because infinite slope of s-t graphs means infinite velocity. Similarly infinity slope of v-t graph means infinite acceleration. Hence, the following graphs are not possible.

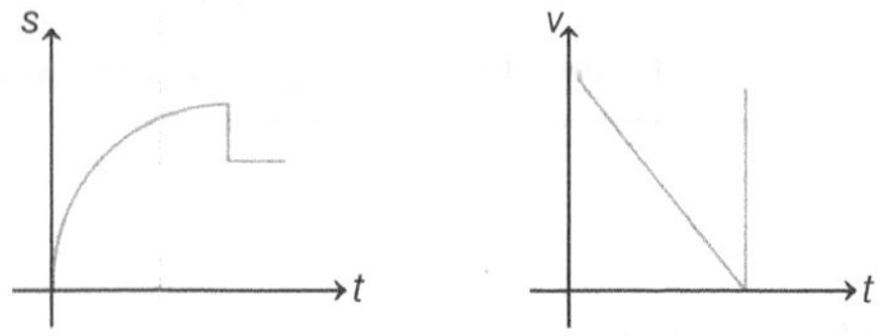

(ii) At one time, two values of velocity or displacement are not possible. Hence, the following graphs are not acceptable

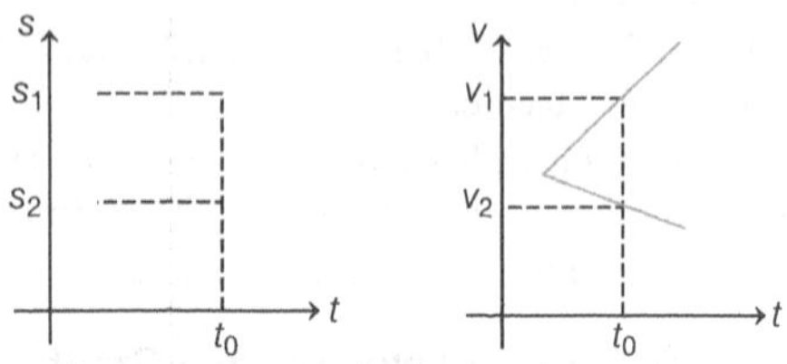

(iii) Different values of displacements in s-t graph corresponding to given v-t graphs can be calculated just by calculating areas under v-t graph. There is no need of using equations like $v = u + at$, etc.

Example 18. *Position-time graph of a particle moving in a straight line is as shown in figure. State the type of motion. (Given, $s_0 = 20$ m and $t_0 = 4$ s)*

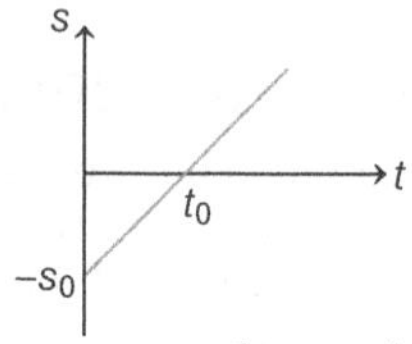

(a) Accelerated (b) Decelerated
(c) Non-accelerated (d) None of these

Sol. (c) Slope of s-t graph is constant. Hence, velocity of particle is constant. Further at time $t = 0$, displacement of the particle from the mean position is $-s_0$ or -20 m. Velocity of particle,

$$v = \text{slope} = \frac{s_0}{t_0} = \frac{20}{4} = 5\ \text{ms}^{-1}$$

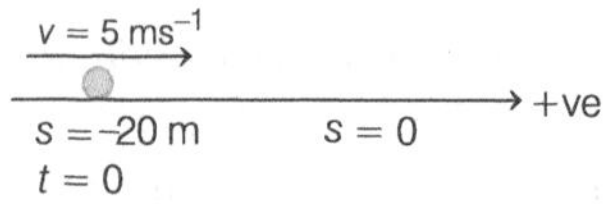

Motion of the particle is as shown in figure. At $t = 0$, particle is at -20 m and has a constant velocity of 5 ms^{-1}. At $t_0 = 4$ s, particle will pass through its mean position. Hence, motion is non-accelerated or uniform.

Relative Velocity

The time rate of change of relative position of one object with respect to another is called relative velocity.

Let two objects A and B are moving along the +ve direction of X-axis. At time t, their displacement from the origin be x_A and x_B.

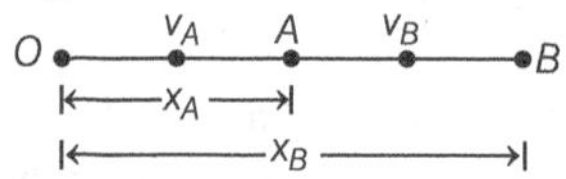

$\therefore$ Their velocities are $v_A = \dfrac{dx_A}{dt}$ and $v_B = \dfrac{dx_B}{dt}$

The displacement of B relative to A,

$$x_{BA} = x_B - x_A$$

Rate of change of relative displacement w.r.t. time is

$$\frac{d(\mathbf{x}_{BA})}{dt} = \frac{d}{dt}(\mathbf{x}_B - \mathbf{x}_A)$$

$$\Rightarrow \quad \frac{d\mathbf{x}_{BA}}{dt} = \frac{d\mathbf{x}_B}{dt} - \frac{d\mathbf{x}_A}{dt}$$

$$\therefore \quad \mathbf{v}_{BA} = \mathbf{v}_B - \mathbf{v}_A$$

Different Cases

***Case* I** When the two objects move with equal velocities, *i.e.* $v_A = v_B$ or $v_B - v_A = 0$. It means, the two objects stay at constant distance apart during the whole journey.

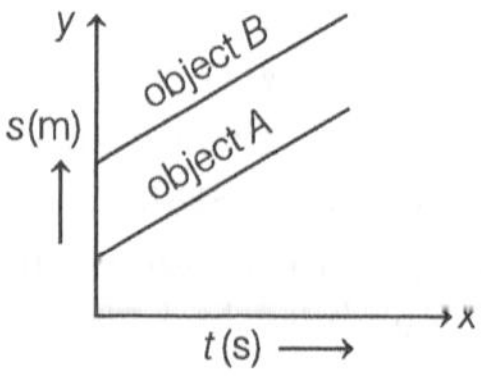

In this case, the position-time graph of two objects are parallel straight lines.

***Case* II** If both objects A and B move along parallel straight lines in the opposite direction, then relative velocity of B w.r.t. A is given as

$$\mathbf{v}_{BA} = \mathbf{v}_B - (-\mathbf{v}_A) = \mathbf{v}_B + \mathbf{v}_A$$

and the relative velocity of A w.r.t. B is given by

$$\mathbf{v}_{AB} = \mathbf{v}_A - \mathbf{v}_B$$

Examples of Relative Motion

1. Relative Velocity of Rain

Consider a man walking East with velocity $\mathbf{v}_m$, represented by **OA**. Let the rain be falling vertically downwards with velocity $\mathbf{v}_r$, represented by **OB**.

To find the relative velocity of rain w.r.t. man (*i.e.*, $\mathbf{v}_{rm}$) bring the man at rest by imposing a velocity $-\mathbf{v}_m$ on man and apply this velocity on rain also.

Now the relative velocity of rain w.r.t. man will be the resultant velocity of $\mathbf{v}_r (= OB)$ and $-\mathbf{v}_m (= OA)$, which will be represented by diagonal **OC** of rectangle $OACB$

$$\therefore \quad v_{rm} = \sqrt{v_r^2 + v_m^2 + 2v_r v_m \cos 90^\circ} = \sqrt{v_r^2 + v_m^2}$$

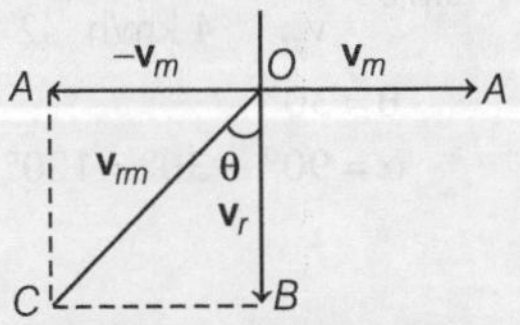

If θ is the angle which $\mathbf{v}_{rm}$ makes with the vertical direction, then

$$\tan\theta = \frac{BC}{OB} = \frac{\mathbf{v}_m}{\mathbf{v}_r} \quad \text{or} \quad \theta = \tan^{-1}\left(\frac{\mathbf{v}_m}{\mathbf{v}_r}\right)$$

Here, angle θ is from vertical towards west and is written as θ, west of vertical.

Note In the above problem if the man wants to protect himself from the rain, he should hold his umbrella in the direction of relative velocity of rain w.r.t. man *i.e.*, the umbrella should be held making an angle $\theta (= \tan^{-1} v_m / v_r)$ west of vertical.

2. Crossing the River

To cross the river over shortest distance, *i.e.* to cross the river straight, the man should swim upstream making an angle θ with **OB** such that, **OB** gives the direction of resultant velocity ($\mathbf{v}_{mR}$) of velocity of swimmer and velocity of river water as shown in figure. Let us consider

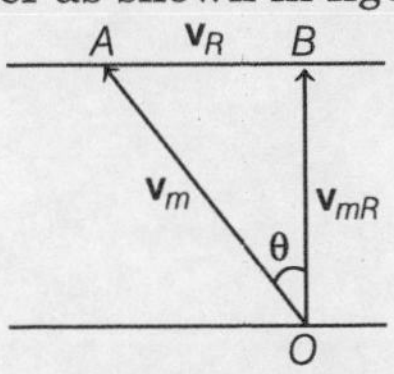

$\mathbf{AB} = \mathbf{v}_R$ (velocity of river water)

$\mathbf{OA} = \mathbf{v}_m$ (velocity of man in still river water)

$\mathbf{OB} = \mathbf{v}_{mR}$ (relative velocity of man w.r.t. river)

$$\mathbf{v}_{mR} = \sqrt{\mathbf{v}_m^2 - \mathbf{v}_R^2}$$

In ΔOAB, $\sin\theta = \dfrac{\mathbf{v}_R}{\mathbf{v}_m}$

where θ is the angle made by man with shortest distance OB,

$$\tan\theta = \frac{\mathbf{v}_R}{\mathbf{v}_m} = \frac{\mathbf{v}_R}{\sqrt{\mathbf{v}_m^2 - \mathbf{v}_R^2}}$$

(a) **Time taken to cross the river** If d be the width of the river, then time taken cross to the river given by

$$t_1 = \frac{d}{\mathbf{v}_{mR}} = \frac{d}{\sqrt{\mathbf{v}_m^2 - \mathbf{v}_R^2}}$$

(b) **To cross the river in possible shortest time** The man should go along OA. Now the swimmer will be going along OB, which is the direction of resultant velocity of $\mathbf{v}_m$ and $\mathbf{v}_R$

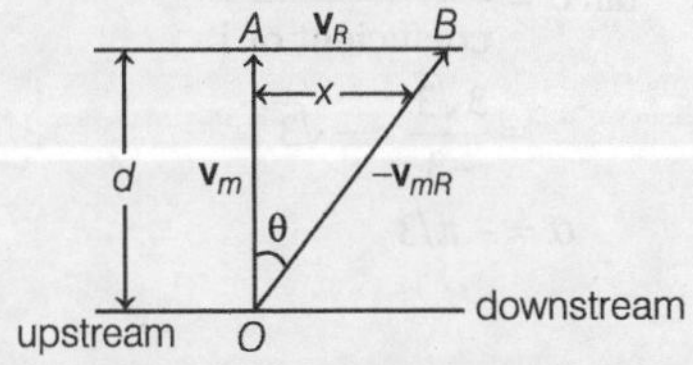

In ΔOAB, $\quad \tan\theta = \dfrac{AB}{OA} = \dfrac{\mathbf{v}_R}{\mathbf{v}_m}$

and $\quad \mathbf{v}_{mR} = \sqrt{\mathbf{v}_m^2 + \mathbf{v}_R^2}$

Time of crossing the river $t = \dfrac{d}{\mathbf{v}_m}$

$$= \frac{OB}{\mathbf{v}_{mR}} = \frac{\sqrt{x^2 + d^2}}{\sqrt{\mathbf{v}_m^2 + \mathbf{v}_R^2}}$$

The boat will be reading the point B instead of point A. If $BD = x$,

if $AB = x$, then

$$\tan\theta = \frac{\mathbf{v}_R}{\mathbf{v}_m} = \frac{x}{d}$$

$$\Rightarrow \quad x = \frac{d\mathbf{v}_R}{\mathbf{v}_m}$$

In this case, the man will reach the opposite bank at a distance AB downstream.

Example 19. *A man A moves due to east with velocity 6 ms^{-1} and another man B moves in N-30°E with 6 ms^{-1}. Find the velocity of B w.r.t. A.*

(a) $3\ ms^{-1}$ (b) $\sqrt{3}\ ms^{-1}$

(c) $\sqrt{6}\ ms^{-1}$ (d) $6\ ms^{-1}$

Sol. (d) Given, $\mathbf{v}_A = 6\,\hat{\mathbf{i}}$

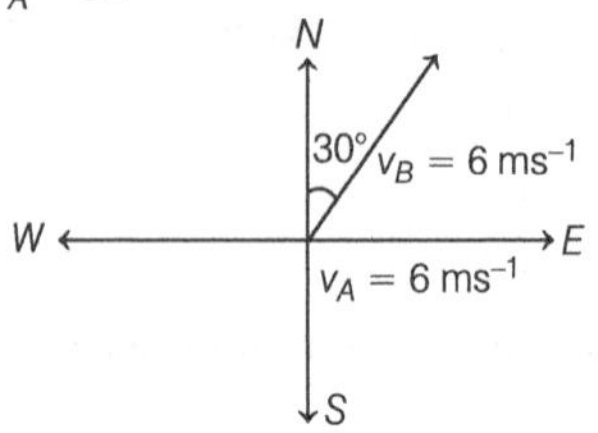

$$\mathbf{v}_B = v_B \cos 60°\,\hat{\mathbf{i}} + v_B \sin 60°\,\hat{\mathbf{j}}$$

$$= 6\left(\frac{1}{2}\right)\hat{\mathbf{i}} + 6\left(\frac{\sqrt{3}}{2}\right)\hat{\mathbf{j}} = 3\,\hat{\mathbf{i}} + 3\sqrt{3}\,\hat{\mathbf{j}}$$

To find the velocity,

$$\mathbf{v}_{BA} = \mathbf{v}_B - \mathbf{v}_A = (3\hat{\mathbf{i}} + 3\sqrt{3}\,\hat{\mathbf{j}}) - 6\,\hat{\mathbf{i}}$$

$$= -3\,\hat{\mathbf{i}} + 3\sqrt{3}\,\hat{\mathbf{j}}$$

$$|\mathbf{v}_{BA}| = \sqrt{(-3)^2 + (3\sqrt{3})^2} = \sqrt{9 + 27}$$

$$= \sqrt{36} = 6\text{ ms}^{-1}$$

Here, $\hat{\mathbf{i}}$ is –ve and $\hat{\mathbf{j}}$ is +ve. So, second quadrant is possible.

Direction, $\tan\alpha = \dfrac{\text{coefficient of }\hat{\mathbf{j}}}{\text{coefficient of }\hat{\mathbf{i}}}$

$$= \frac{3\sqrt{3}}{-3} = -\sqrt{3}$$

$\Rightarrow$ $\alpha = -\pi/3$

Example 20. *The stream of a river is flowing with a speed of 2 km/h. A swimmer can swim at a speed of 4 km/h. What should be the direction of the swimmer with respect to the flow of the river to cross the river straight?*

(a) 60°
(b) 120°
(c) 90°
(d) 150°

Sol. (b) Let the velocity of the swimmer is

$$v_s = 4\text{ km/h}$$

and velocity of river is $v_r = 2$ km/h

Also, angle of swimmer with the flow of the river (down stream) is α as shown in the figure below.

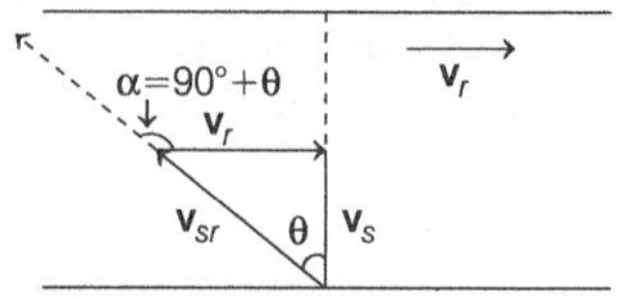

From diagram, angle θ is

$$\sin\theta = \frac{v_r}{v_{sr}} = \frac{2\text{ km/h}}{4\text{ km/h}} = \frac{1}{2}$$

$\Rightarrow$ $\theta = 30°$

Clearly, $\alpha = 90° + 30° = 120°$

Practice Exercise

ROUND I Topically Divided Problems

Distance and Displacement

1 An aeroplane flies 400 m from North and then flies 300 m South and then flies 1200 m upwards, then net displacement is

(a) 1200 m (b) 1300 m
(c) 1400 m (d) 1500 m

2. A clerk starts from his house with a speed of 2 kmh^{-1} and reaches the office 3 min late. Next day he increases speed 1 kmh^{-1} and reaches the office 3 min earlier. Find the distance between his house and office.

(a) 1 km (b) 0.6 km
(c) 2 km (d) 3 km

3. A particle starts from rest at the point $A(1, 2, -3)$ m and reaches at the point $B(3, 4, 5)$ m. The magnitude of displacement of man is

(a) 7.1 m (b) 3.7 m
(c) 3.4 m (d) 8.5 m

4. A body starts from rest and moves with a constant acceleration. The ratio of distance covered in the nth second to the distance covered in n second is

(a) $\frac{2}{n} - \frac{1}{n^2}$
(b) $\frac{1}{n^2} - \frac{1}{n}$
(c) $\frac{2}{n^2} - \frac{1}{n}$
(d) $\frac{2}{n} + \frac{1}{n^2}$

5. A wheel of radius 1 m rolls forward half a revolution on a horizontal ground. The magnitude of the displacement of the point on the wheel initially in contact with the ground is

(a) 2π (b) $\sqrt{2}\,\pi$
(c) $\sqrt{\pi^2 + 4}$ (d) π

6. A body is moving along a straight line path with constant velocity. At an instant of time the distance of time the distance travelled by it is s and its displacement is D, then

(a) $D < s$ (b) $D > s$
(c) $D = s$ (d) $D \leq s$

Speed, Velocity and Acceleration

7. A particle moves with constant acceleration and v_1, v_2 and v_3 denote the average velocities in the three successive intervals t_1, t_2 and t_3 of time. Which of the following relations is correct?

(a) $\frac{v_1 - v_2}{v_2 - v_3} = \frac{t_1 - t_2}{t_2 + t_3}$ (b) $\frac{v_1 - v_2}{v_2 - v_3} = \frac{t_1 - t_2}{t_1 - t_2}$
(c) $\frac{v_1 - v_2}{v_2 - v_3} = \frac{t_1 - t_2}{t_2 - t_3}$ (d) $\frac{v_1 - v_2}{v_2 - v_3} = \frac{t_1 + t_2}{t_2 + t_3}$

8. A bee flies in a line from a point A to another point B in 4 s with a velocity of $|t - 2|$ ms^{-1}. The distance between A and B in metre is

(a) 2 (b) 4
(c) 6 (d) 8

9. A 2 m wide truck is moving with a uniform speed $v_0 = 8$ ms^{-1} along a straight horizontal road. A pedestrian starts to cross the road with a uniform speed v when the truck is 4 m away from him. The minimum value of v, so that he can cross the road safely is

(a) 2.62 ms^{-1} (b) 4.6 ms^{-1}
(c) 3.57 ms^{-1} (d) 1.414 ms^{-1}

10. A particle starts from rest and travels a distance s with uniform acceleration, then it travels a distance $2s$ with uniform speed, finally it travels a distance $3s$ with uniform retardation and comes to rest. If the complete motion of the particle in a straight line then the ratio of its average velocity to maximum velocity is

(a) 6/7 (b) 4/5
(c) 3/5 (d) 2/5

11. Three particles start from the origin at the same time, one with a velocity v_1 along x-axis, the second along the y-axis with a velocity v_2 and the third along $x = y$ line. The velocity of the third so that the three may always lie on the same line is

(a) $\frac{v_1 v_2}{v_1 + v_2}$ (b) $\frac{\sqrt{2}\, v_1 v_2}{v_1 + v_2}$
(c) $\frac{\sqrt{3}\, v_1 v_2}{v_1 + v_2}$ (d) zero

12. A body moves in a plane, so that the displacement along X and Y-axes is $x = 3t^2$ and $y = 4t^2$, then velocity is
(a) $15\,t^2$
(b) $20\,t^2$
(c) $5\,t^2$
(d) None of the above

13. The displacement of a body along X-axis depends on time as $\sqrt{x} = t + 1$. Then, the velocity of body
(a) increase with time (b) decrease with time
(c) independent of time (d) None of these

14. The velocity of a particle is $v = v_0 + gt + Ft^2$. Its position is $x = 0$ at $t = 0$, then its displacement after time $(t = 1)$ is [JEE Main 2021]
(a) $v_0 + g + F$ (b) $v_0 + \frac{g}{2} + \frac{F}{3}$
(c) $v_0 + \frac{g}{2} + F$ (d) $v_0 + 2g + 3F$

15. In one dimensional motion, instantaneous speed v satisfies $0 \le v < v_0$. [NCERT Exemplar]
(a) The displacement in time T must always take non-negative values
(b) The displacement x in time T satisfies $-v_0 T < x < v_0 T$
(c) The acceleration is always a non-negative number
(d) The motion has no turning points

16. A particle located at $x = 0$ at time $t = 0$, starts moving along the positive x-direction with a velocity v that varies as $v = \alpha\sqrt{x}$. The displacement of the particle varies with time as
(a) t^2 (b) t
(c) $t^{1/2}$ (d) t^3

17. A particle moving with a uniform acceleration along a straight line covers distance a and b in successive intervals of p and q second. The acceleration of the particle is
(a) $\frac{pq(p+q)}{2(bp-aq)}$ (b) $\frac{2(aq-bp)}{pq(p-q)}$
(c) $\frac{bp-aq}{pq(p-q)}$ (d) $\frac{2(bp-aq)}{pq(p-q)}$

18. The position vector of particle changes with time according to the relation $\mathbf{r}(t) = 15t^2\hat{\mathbf{i}} + (4 - 20t^2)\hat{\mathbf{j}}$. What is the magnitude of the acceleration (in ms^{-2}) at $t = 1$? [JEE Main 2019]
(a) 50 (b) 100
(c) 25 (d) 40

19. The relation between time t and distance x is $t = ax^2 + bx$, where a and b are constants. The acceleration is
(a) $-2abv^2$ (b) $2bv^3$
(c) $-2av^3$ (d) $2av^2$

20. A particle starts from the origin and moves along the X-axis such that the velocity at any instant is given by $4t^3 - 2t$, where t is in second and velocity is in ms^{-1}. What is the acceleration of the particle when it is 2 m from the origin?
(a) $10\ ms^{-2}$ (b) $12\ ms^{-2}$
(c) $22\ ms^{-2}$ (d) $28\ ms^{-2}$

21. A lift is coming from 8th floor and is just about to reach 4th floor. Taking ground floor as origin and positive direction upwards for all quantities, which one of the following is correct? [NCERT Exemplar]
(a) $x < 0, v < 0, a > 0$
(b) $x > 0, v < 0, a < 0$
(c) $x > 0, v < 0, a > 0$
(d) $x > 0, v > 0, a < 0$

22. The position of a particle as a function of time t, is given by $x(t) = at + bt^2 - ct^3$ where a, b and c are constants. When the particle attains zero acceleration, then its velocity will be [JEE Main 2019]
(a) $a + \frac{b^2}{2c}$ (b) $a + \frac{b^2}{4c}$
(c) $a + \frac{b^2}{3c}$ (d) $a + \frac{b^2}{c}$

23. The driver of a car moving with a speed of $10\ ms^{-1}$ sees a red light ahead, applies brakes and stops after covering 10 m distance. If the same car were moving with a speed of $20\ ms^{-1}$, the same driver would have stopped the car after covering 30 m distance. Within what distance the car can be stopped if travelling with a velocity of $15\ ms^{-1}$? Assume the same reaction time and the same deceleration in each case.
(a) 18.75 m (b) 20.75 m
(c) 22.75 m (d) 25 m

24. A particle moving in a straight line with uniform acceleration is observed to be at a distance a from a fixed point initially. It is at distances b, c, d from the same point after n, $2n$, $3n$ second. The acceleration of the particle is
(a) $\frac{c - 2b + a}{n^2}$ (b) $\frac{c + b + a}{9n^2}$
(c) $\frac{c + 2b + a}{4n^2}$ (d) $\frac{c - b + a}{n^2}$

25. The retardation experienced by a moving motor boat, after its engine is cut-off, is given by $\frac{dv}{dt} = -kv^3$, where k is a constant. If v_0 is the magnitude of the velocity at cut-off, the magnitude of the velocity at time t after the cut off is
(a) v_0 (b) $\frac{v_0}{2}$
(c) $v_0 e^{-kt}$ (d) $\frac{v_0}{\sqrt{2v_0^2 kt + 1}}$

26. The engine of a train can impart a maximum acceleration of 1 ms^{-2} and the brakes can give a maximum retardation of 3 ms^{-2}. The least time during which a train can go from one place to the other place at a distance of 1.2 km is nearly

(a) 108 s (b) 191 s
(c) 56.6 s (d) time is fixed

27. The acceleration of a particle increasing linearly with time t is bt. The particle starts from the origin with an initial velocity v_0. The distance travelled by the particle in time t will be

(a) $v_0 t + \frac{1}{6} bt^3$ (b) $v_0 t + \frac{1}{6} bt^2$
(c) $v_0 t + \frac{1}{3} bt^3$ (d) $v_0 t + \frac{1}{3} bt^2$

Uniform and Non-uniform Motions

28. A car is moving along a straight road with uniform acceleration. It passes through two points P and Q separated by a distance with velocities 30 kmh^{-1} and 40 kmh^{-1} respectively. The velocity of car midway between P and Q is

(a) 33.3 kmh^{-1} (b) 1 kmh^{-1}
(c) $25\sqrt{2}$ kmh^{-1} (d) 35.35 kmh^{-1}

29. A particle moves from the point $(2.0\hat{i} + 4.0\hat{j})$ m at $t = 0$ with an initial velocity $(5.0\hat{i} + 4.0\hat{j})\ ms^{-1}$. It is acted upon by a constant force which produces a constant acceleration $(4.0\hat{i} + 4.0\hat{j})\ ms^{-2}$. What is the distance of the particle from the origin at time 2 s? **[JEE Main 2019]**

(a) 5 m (b) $20\sqrt{2}$ m (c) $10\sqrt{2}$ m (d) 15 m

30. A particle covers 4 m, 5 m, 6 m and 7 m in 3rd, 4th, 5th and 6th second respectively. The particle starts

(a) with an initial non-zero velocity and moves with uniform acceleration
(b) from rest and moves with uniform velocity
(c) with an initial velocity and moves with uniform velocity
(d) from rest and moves with uniform acceleration

31. A mosquito is moving with a velocity $\mathbf{v} = 0.5t^2\hat{\mathbf{i}} + 3t\hat{\mathbf{j}} + 9\mathbf{k}$ m/s and accelerating in uniform conditions. What will be the direction of mosquito after 2s? **[JEE Main 2021]**

(a) $\tan^{-1}\left(\frac{2}{3}\right)$, from X-axis
(b) $\tan^{-1}\left(\frac{2}{3}\right)$, from Y-axis
(c) $\tan^{-1}\left(\frac{5}{2}\right)$, from Y-axis
(d) $\tan^{-1}\left(\frac{5}{2}\right)$, from X-axis

32. A particle is moving with speed $v = b\sqrt{x}$ along positive X-axis. Calculate the speed of the particle at time $t = \tau$ (assume that the particle is at origin at $t = 0$). **[JEE Main 2019]**

(a) $\frac{b^2\tau}{4}$ (b) $\frac{b^2\tau}{2}$
(c) $b^2\tau$ (d) $\frac{b^2\tau}{\sqrt{2}}$

33. Two balls A and B are thrown simultaneously from the top of a tower. A is thrown vertically up with a speed of 4 ms^{-1}. B is thrown vertically down with a speed of 4 ms^{-1}. The ball A and B hit the ground with speed v_A and v_B respectively, then

(a) $v_A < v_B$ (b) $v_A > v_B$
(c) $v_A \geq v_B$ (d) $v_A = v_B$

34. A stone is allowed to fall from the top of a tower 100 m high and at the same time another stone is projected vertically upwards from the ground with a velocity of 25 ms^{-1}. The two stones will meet after

(a) 4 s (b) 0.4 s
(c) 0.04 s (d) 40 s

35. A boy released a ball from the top of a building. It will clear a window 2 m high at a distance 10 m below the top in nearly

(a) 1 s (b) 1.3 s
(c) 0.6 s (d) 0.13 s

36. A ball P is dropped vertically and another ball Q is thrown horizontally with the same velocity from the same height and at the same time. If air resistance is neglected, then

(a) ball P reaches the ground first
(b) ball Q reaches the ground first
(c) both reach the ground at the same time
(d) the respective masses of the two balls will decide the time

37. A ball A is thrown up vertically with a speed u and at the same instant another ball B is released from a height h. At time t, the speed of A relative to B is

(a) u (b) $2u$
(c) $u - gt$ (d) $\sqrt{(u^2 - gt)}$

38. From a balloon rising vertically upwards at 5 m/s a stone is thrown up at 10 m/s relative to the balloon. Its velocity with respect to ground after 2 s is (assume $g = 10\ m/s^2$)

(a) 0 (b) 20 m/s
(c) 10 m/s (d) 5 m/s

39. A body thrown vertically upward with an initial velocity u reaches maximum height in 6 second. The ratio of the distances travelled by the body in the first second and seventh second is

(a) 1 : 1 (b) 11 : 1 (c) 1 : 2 (d) 1 : 11

40. A particle of mass m is initially situated at the point P inside a hemispherical surface of radius r as shown in figure. A horizontal acceleration of magnitude a_0 is suddenly produced on the particle in the horizontal direction. If gravitational acceleration is neglected, the time taken by particle to touch the sphere again is

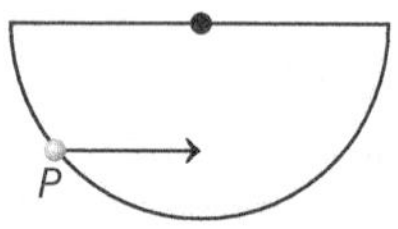

(a) $\sqrt{\frac{4r \sin \alpha}{a_0}}$ (b) $\sqrt{\frac{4r \tan \alpha}{a_0}}$

(c) $\sqrt{\frac{4r \cos \alpha}{a_0}}$ (d) None of these

41. A ball is thrown vertically upwards. It is observed that it remains at a height h twice with a time interval Δt, then the initial velocity of the ball is

(a) $\sqrt{8gh + g^2 (\Delta t)^2}$ (b) $\sqrt{8gh + \left(\frac{g\Delta t}{2}\right)^2}$

(c) $1/2\sqrt{8gh + g^2 (\Delta t)^2}$ (d) $\sqrt{8gh + 4g^2 (\Delta t)^2}$

42. A frictionless wire AB is fixed on a sphere of radius R. A very small spherical ball slips on this wire. The time taken by this ball to slip from A to B is

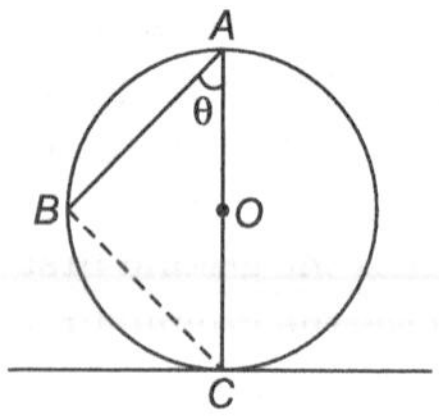

(a) $\frac{2\sqrt{gR}}{g \cos \theta}$ (b) $2\sqrt{gR}\,\frac{\cos \theta}{g}$

(c) $2\sqrt{\frac{R}{g}}$ (d) $\frac{gR}{\sqrt{g \cos \theta}}$

43. A body is thrown vertically up with a velocity u. It passes three points A, B and C in its upward journey with velocities $\frac{u}{2}, \frac{u}{3}$ and $\frac{u}{4}$ respectively. The ratio of the separations between points A and B and between B and C, *i.e.* $\frac{AB}{BC}$ is

(a) 1 (b) 2 (c) $\frac{10}{7}$ (d) $\frac{20}{7}$

44. A juggler keeps on moving four balls in the air throws the balls in regular interval of time. When one ball leaves his hand (speed = 20 ms^{-1}), the position of other ball will be (Take, $g = 10\ ms^{-2}$)

(a) 10 m, 20 m, 10 m (b) 15 m, 20 m, 15 m
(c) 5 m, 15 m, 20 m (d) 5 m, 10 m, 20 m

45. A body freely falling from rest has a velocity v after it falls through distance h. The distance it has to fall down further for its velocity to become double is

(a) h (b) $2h$
(c) $3h$ (d) $4h$

46. A ball is thrown vertically upwards from the top of a tower of height h with velocity v. The ball strikes the ground after

(a) $\frac{v}{g}\left[1+\sqrt{1+\frac{2gh}{v^2}}\right]$ (b) $\frac{v}{g}\left[1+\sqrt{1-\frac{2gh}{v^2}}\right]$

(c) $\frac{v}{g}\left(1+\frac{2gh}{v^2}\right)^{1/2}$ (d) $\frac{v}{g}\left(1-\frac{2gh}{v^2}\right)^{1/2}$

47. From an elevated point A, a stone is projected vertically upwards. When the stone reaches a distance h below A, its velocity is double of what was at a height above A? The greatest height attained by the stone is

(a) $\frac{h}{3}$ (b) $\frac{2h}{2}$ (c) $\frac{h}{2}$ (d) $\frac{5h}{3}$

48. A particle starting from rest falls from a certain height. Assuming that the value of acceleration due to gravity remains the same throughout motion, its displacements in three successive half second intervals are s_1, s_2, s_3. Then,

(a) $s_1 : s_2 : s_3 = 1:5:9$ (b) $s_1 : s_2 : s_3 = 1:2:3$
(c) $s_1 : s_2 : s_3 = 1:1:1$ (d) $s_1 : s_2 : s_3 = 1:3:5$

49. A body released from a great height falls freely towards the earth. Another body is released from the same height exactly one second later. The separation between the two bodies two second after the release of the second body is

(a) 9.8 m (b) 4.9 m
(c) 24.5 m (d) 19.6 m

50. A ball released from the top of a tower travels $\frac{11}{36}$ of the height of the tower in the last second of its journey. The height of the tower is (Take, $g = 10\ ms^{-2}$)

(a) 11 m (b) 36 m (c) 47 m (d) 180 m

51. A stone thrown vertically upwards attains a maximum height of 45 m. In what time, the velocity of stone become equal to one-half the velocity of throw? (Take, $g = 10\ ms^{-2}$)

(a) 2 s (b) 1.5 s (c) 1 s (d) 0.5 s

52. From a tower of height H, a particle is thrown vertically upwards with a speed u. The time taken by the particle to hit the ground, is n times that taken by it to reach the highest point of its path. The relation between H, u and n is **[JEE Main 2014]**

(a) $2gH = n^2 u^2$ (b) $gH = (n-2)^2 u^2$

(c) $2gH = nu^2 (n-2)$ (d) $gH = (n-2)^2 u$

53. Water drops fall from a tap on the floor 5 m below at regular intervals of time, the first drop striking the floor when the fifth drop begins to fall. The height at which the third drop will be, from ground, at the instant when first drop strikes the ground, will be (Take, $g = 10\ ms^{-2}$)

(a) 1.25 m (b) 2.15 m
(c) 2.73 m (d) 3.75 m

54. A ball thrown upward from the top of a tower with speed v reaches the ground in t_1 second. If this ball is thrown downward from the top of the same tower with speed v it reaches the ground in t_2 second. In what time the ball shall reach the ground if it is allowed to falls freely under gravity from the top of the tower?

(a) $\frac{t_1 + t_2}{2}$ (b) $\frac{t_1 - t_2}{2}$
(c) $\sqrt{t_1 t_2}$ (d) $t_1 + t_2$

55. A ball is dropped on the floor from a height of 10 m. It rebounds to a height of 2.5 m. If the ball is in contact with the floor for 0.01 s, the average acceleration during contact is nearly (Take, $g = 10\,ms^{-2}$)

(a) $500\sqrt{2}\ ms^{-2}$ upwards
(b) $1800\ ms^{-2}$ downwards
(c) $1500\sqrt{5}\ ms^{-2}$ upwards
(d) $1500\sqrt{2}\ ms^{-2}$ upwards

56. A helicopter rises from rest on the ground vertically upwards with a constant acceleration g. A food packet is dropped from the helicopter when it is at a height h. The time taken by the packet to reach the ground is close to (Here, g is the acceleration due to gravity.) **[JEE Main 2020]**

(a) $t = \frac{2}{3}\sqrt{\frac{h}{g}}$ (b) $t = 1.8\sqrt{\frac{h}{g}}$
(c) $t = \sqrt{\frac{2h}{3g}}$ (d) $t = 3.4\sqrt{\frac{h}{g}}$

Problem Related to Graph

57. A body is thrown vertically upwards. Which one of the following graph correctly represent the velocity *versus* time? **[JEE Main 2017]**

(a) 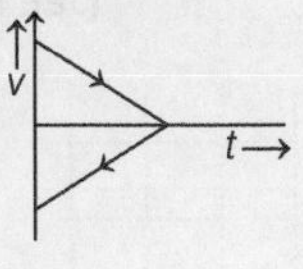
(b) 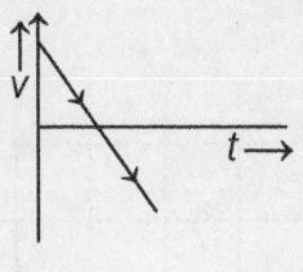
(c) 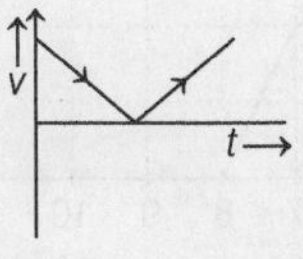
(d) 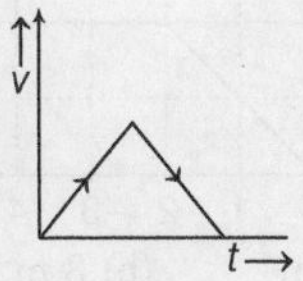

58. Among the four graphs, there is only one graph for which average velocity over the time interval $(0, T)$ can vanish for a suitably chosen T. Which one is it? **[NCERT Exemplar]**

(a) 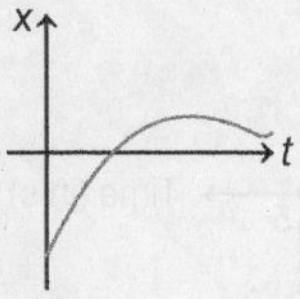
(b) 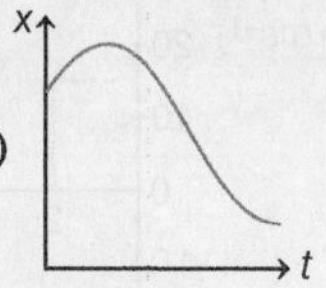
(c) 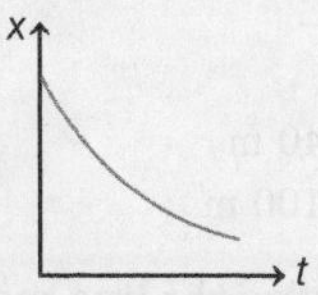
(d) 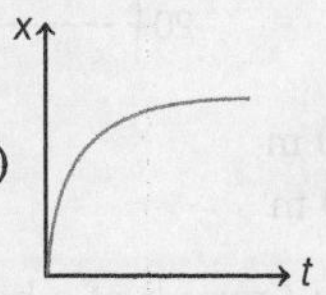

59. A rocket is fired upwards. Its engine explodes fully is 12 s. The height reached by the rocket as calculated from its velocity-time graph is

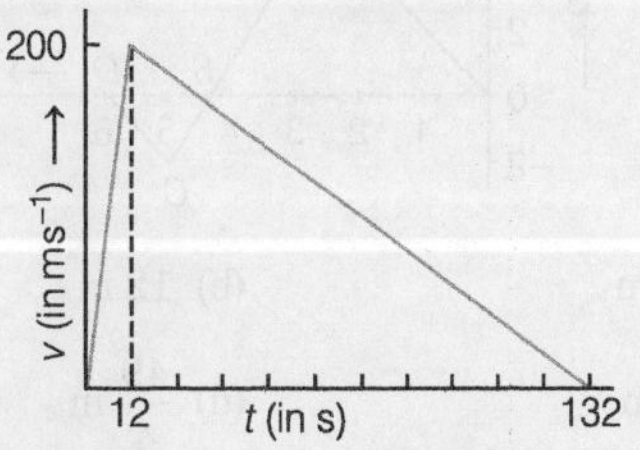

(a) 1200×66 m
(b) 1200×132 m
(c) $\frac{1200}{12}$ m
(d) 1200×12^2 m

60. Figure shows the acceleration-time graphs of a particle. Which of the following represents the corresponding velocity-time graphs?

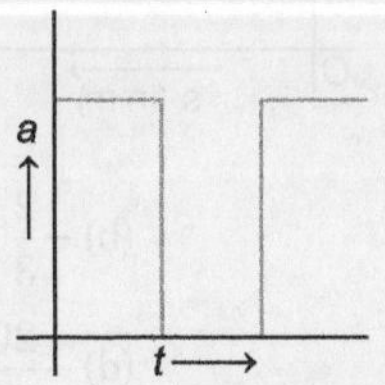

(a) 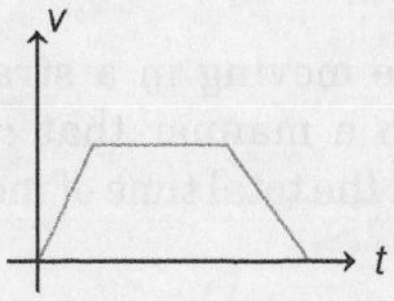
(b) 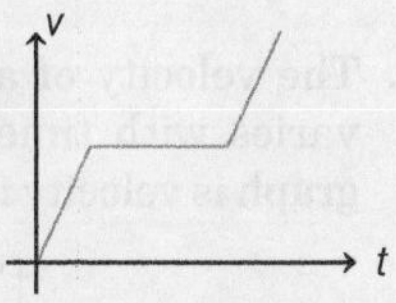
(c) 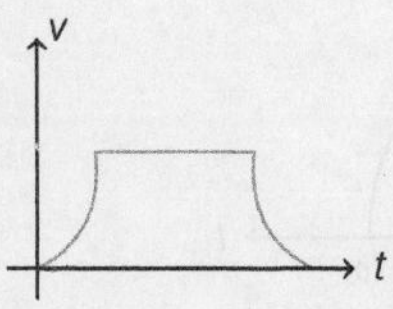
(d) 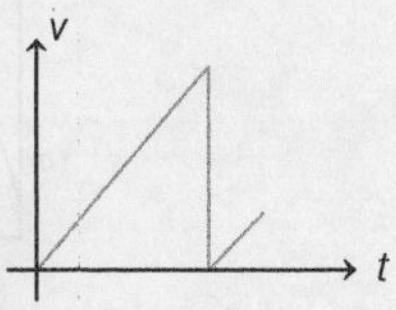

61. In the given v-t graph, the distance travelled by the body in 5 s will be

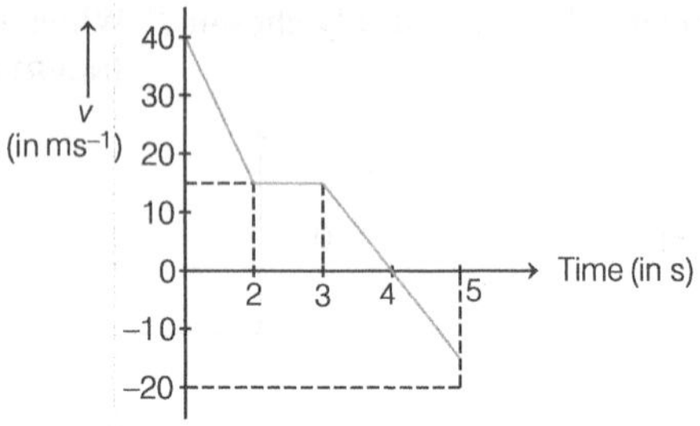

(a) 20 m (b) 40 m
(c) 80 m (d) 100 m

62. The v-t graph of a body in a straight line motion is shown in the figure. The point S is at 4.333 s. The total distance covered by the body in 6 s is **[JEE Main 2020]**

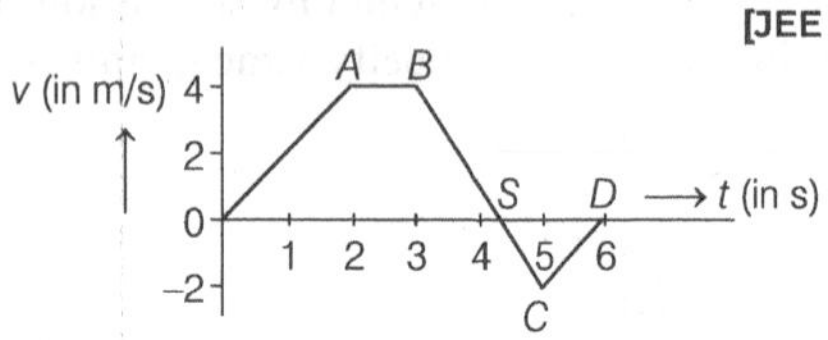

(a) $\frac{37}{3}$ m (b) 12 m
(c) 11 m (d) $\frac{49}{4}$ m

63. If the velocity v of a particle moving along a straight line decreases linearly with its displacement s from 20 ms^{-1} to a value approaching zero at $s = 30$ m, then acceleration of the particle at $s = 15$ m is

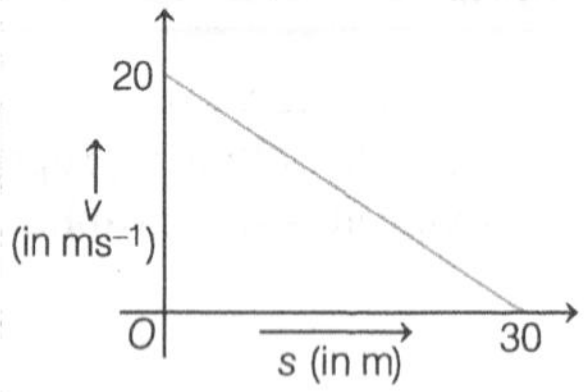

(a) $\frac{2}{3}$ ms^{-2} (b) $-\frac{2}{3}$ ms^{-2}
(c) $\frac{20}{3}$ ms^{-2} (d) $-\frac{20}{3}$ ms^{-2}

64. The velocity of a particle moving in a straight line varies with time in such a manner that v *versus* t graph is velocity is v_m and the total time of motion is t_0

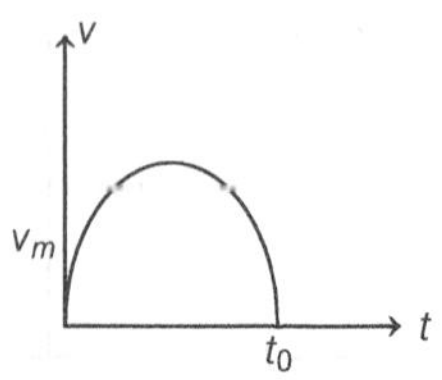

(i) Average velocity of the particle is $\frac{\pi}{4} v_m$

(ii) Such motion cannot be realised in practical terms

(a) Only (i) is correct
(b) Only (ii) is correct
(c) Both (i) and (ii) are correct
(d) Both (i) and (ii) are wrong

65. A body is at rest at $x = 0$. At $t = 0$, it starts moving in the positive x-direction with a constant acceleration. At the same instant, another body passes through $x = 0$ moving in the positive x-direction with a constant speed. The position of the first body is given by $x_1(t)$ after time t and that of the second body by $x_2(t)$ after the same time interval. Which of the following graphs correctly describes $(x_1 - x_2)$ as a function of time?

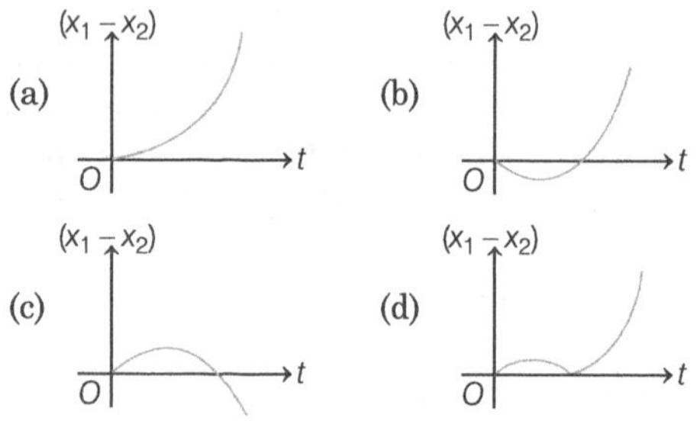

66. All the graphs below are intended to represent the same motion. One of them does it incorrectly. Pick it up. **[JEE Main 2018]**

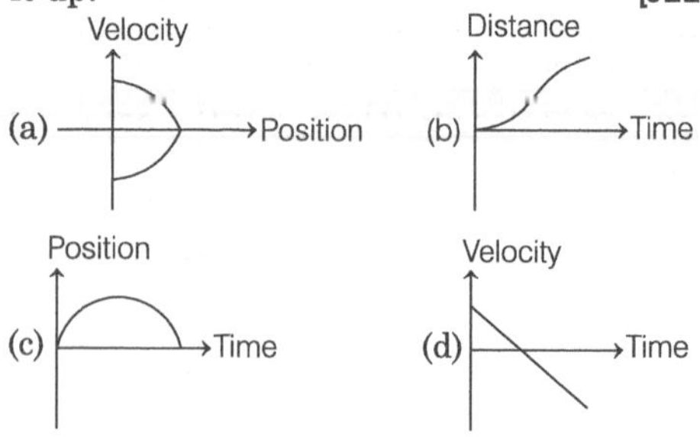

67. A particle starts from the origin at time $t = 0$ and moves along the positive X-axis. The graph of velocity with respect to time is shown in figure. What is the position of the particle at time $t = 5$ s? **[JEE Main 2019]**

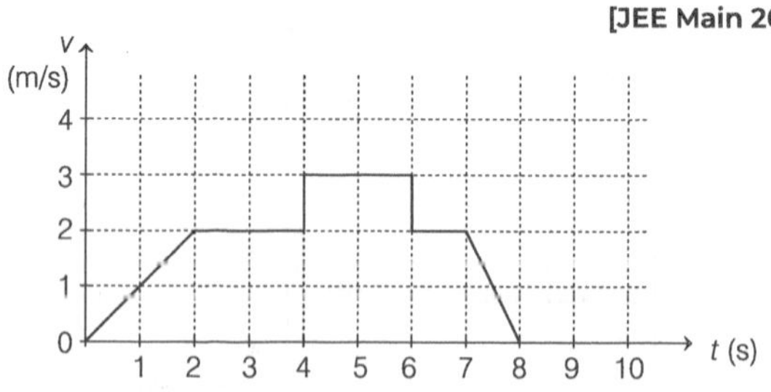

(a) 6 m (b) 3 m (c) 10 m (d) 9 m

68. A particle starts from origin O from rest and moves with a uniform acceleration along the positive X-axis. Identify all figures that correctly represent the motion qualitatively.
(where, a = acceleration, v = velocity, x = displacement, t = time.) [JEE Main 2019]

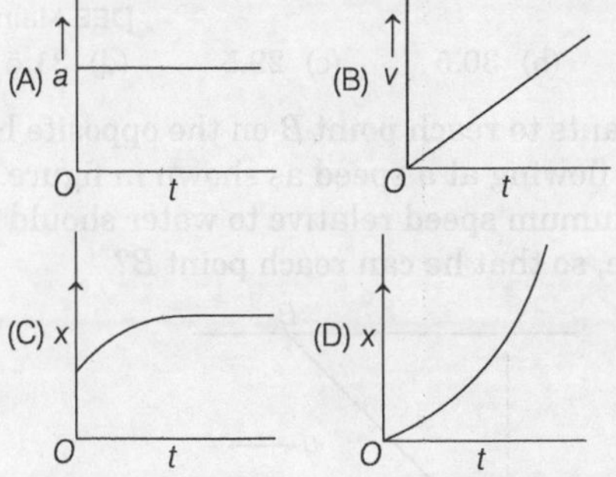

(a) (A) (b) (A), (B), (C)
(c) (B), (C) (d) (A), (B), (D)

69. The velocity-displacement graph of a particle is shown in the figure. The acceleration-displacement graph of the same particle is represented by [JEE Main 2021]

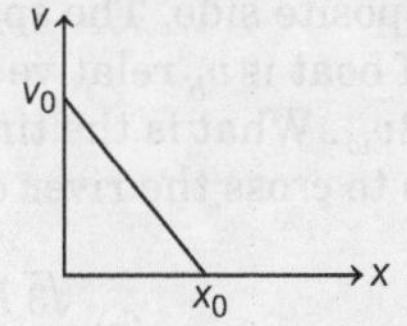

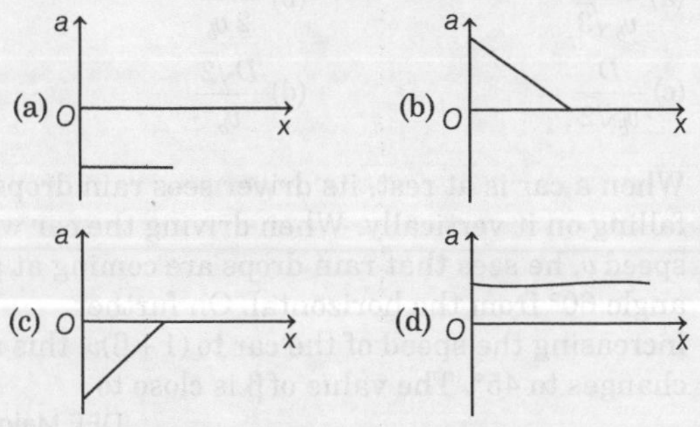

70. The velocity-time graph of a particle in one-dimensional motion is shown in figure. Which of the following formulae are correct for describing the motion of the particle over the time interval t_1 to t_2?

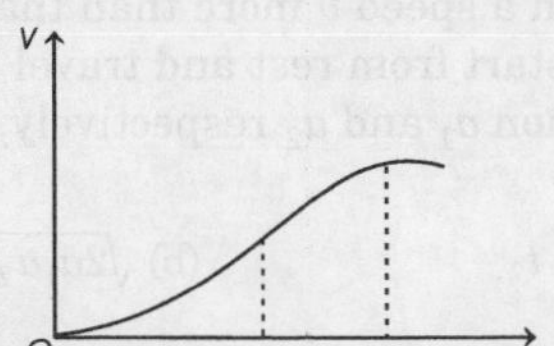

(i) $x(t_2) = x(t_1) + v(t_1)(t_2 - t_1) + \frac{1}{2}a(t_2 - t_1)^2$

(ii) $v(t_2) = v(t_1) + a(t_2 - t_1)$

(iii) $v_{av} = \left[\frac{x(t_2) - x(t_1)}{(t_2 - t_1)}\right]$

(iv) $a_{av} = \frac{[v(t_2) - v(t_1)]}{(t_2 - t_1)}$

(v) $x(t_2) = x(t_1) + v_{av}(t_2 - t_1) + \frac{1}{2}a_{av}(t_2 - t_1)^2$

(vi) $x(t_2) - x(t_1)$ = Area under v-t curve bounded by the t-axis and the dotted line shown

(a) (iii) and (vi)
(b) (iii), (iv) and (vi)
(c) (ii), (iii) and (iv)
(d) (iv) and (vi)

Relative Motion

71. An express train is moving with a velocity v_1 and its driver finds another train is moving on the same track in the same direction with velocity v_2. To avoid collision, driver applies a retardation a on the train. The minimum time of avoiding collision will be

(a) $t = \frac{v_1 - v_2}{a}$ (b) $t = \frac{v_1^2 - v_2^2}{2}$
(c) Neither (a) nor (b) (d) Both (a) and (b)

72. At a metro station, a girl walks up a stationary escalator in time t_1. If she remains stationary on the escalator, then the escalator take her up in time t_2. The time taken by her to walk up on the moving escalator will be [NCERT Exemplar]

(a) $(t_1 + t_2)/2$ (b) $t_1 t_2/(t_2 - t_1)$
(c) $t_1 t_2/(t_2 + t_1)$ (d) $t_1 - t_2$

73. Two cars move in the same direction along parallel roads. One of them is a 100 m long travelling with a velocity of 7.5 ms^{-1}. How long will it take for the first car to overtake the second car?

(a) 24 s (b) 40 s
(c) 60 s (d) 80 s

74. A 210 m long train is moving due North at a speed of 25 m/s. A small bird is flying due South, a little above the train with speed 5 m/s. The time taken by the bird to cross the train is

(a) 6 s (b) 7 s
(c) 9 s (d) 10 s

75. Rain drops fall vertically at a speed of 20 ms^{-1}. At what angle do they fall on the wind screen of a car moving with a velocity of 15 ms^{-1}, if the wind screen velocity inclined at an angle of 23° to the vertical?

$$\left[\cot^{-1}\left(\frac{4}{3}\right) \approx 36^\circ\right]$$

(a) 60° (b) 30° (c) 45° (d) 90°

76. Two trains travelling on the same track are approaching each other with equal speeds of $40\ ms^{-1}$. The drivers of the trains begin to decelerate simultaneously when they are just 2 km apart. If the decelerations are both uniform and equal, then the value of deceleration to barely avoid collision should be

(a) $0.8\ ms^{-2}$ (b) $2.1\ ms^{-2}$
(c) $11.0\ ms^{-2}$ (d) $13.2\ ms^{-2}$

77. A passenger train of length 60 m travels at a speed of 80 km/h. Another freight train of length 120 m travels at a speed of 30 km/h. The ratio of times taken by the passenger train to completely cross the freight train when : (i) they are moving in the same direction and (ii) in the opposite direction is **[JEE Main 2019]**

(a) $\frac{3}{2}$ (b) $\frac{25}{11}$
(c) $\frac{11}{5}$ (d) $\frac{5}{2}$

78. The distance between two particles moving towards each other is decreasing at the rate of 6 m/s. If these particles travel with same speed and in the same direction, then the separation increase at the rate of 4 m/s. The particles have speed as

(a) 5 m/s; 1 m/s (b) 4 m/s; 1 m/s
(c) 4 m/s; 2 m/s (d) 5 m/s; 2 m/s

79. A police jeep is chasing with velocity of 45 km/h a thief in another jeep moving with velocity of 153 km/h. Police fires a bullet with muzzle velocity of 180 m/s towards thief's car. The velocity with which the bullet will strike the car of the thief is

(a) 150 m/s (b) 27 m/s
(c) 450 m/s (d) 250 m/s

80. A train is moving towards east and a car is along north, both with same speed. The observed direction of a car to the passenger in the train is

(a) east-north direction
(b) west-north direction
(c) south-east direction
(d) None of the above

81. A steam boat goes across a lake and comes back (i) on a quiet day when the water is still and (ii) on a rough day when there is a uniform current so as to help the journey onwards and to impede the journey back. If the speed of the launch on both days was same, the time required for complete journey on the rough day, as compared to the quiet day will be

(a) more (b) less
(c) same (d) None of these

82. Trains A and B are running on parallel tracks in the opposite directions with speeds of 36 km/h and 72 km/h, respectively. A person is walking in train A in the opposite direction to its motion with a speed of 1.8 km/h. Speed (in ms^{-1}) of this person as observed from train B will be close to (Take, the distance between the tracks as negligible) **[JEE Main 2020]**

(a) 28.5 (b) 30.5 (c) 29.5 (d) 31.5

83. A man wants to reach point B on the opposite bank of a river flowing at a speed as shown in figure. What minimum speed relative to water should the man have, so that he can reach point B?

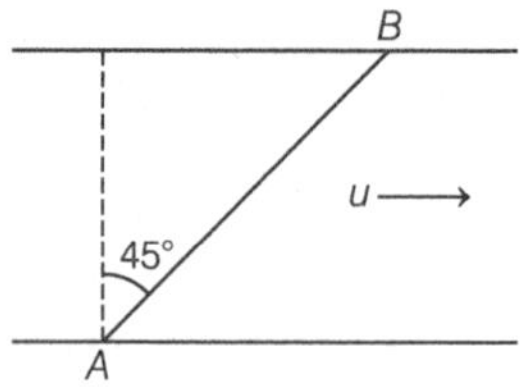

(a) $u\sqrt{2}$ (b) $u/\sqrt{2}$
(c) $2u$ (d) $u/2$

84. A boat crosses a river from part A to part B which are just on opposite side. The speed of the water is v_w and that of boat is v_b relative to still water. Assume $v_b = 2v_w$. What is the time taken by the boat? If it has to cross the river directly on the AB line.

(a) $\frac{2D}{v_b\sqrt{3}}$ (b) $\frac{\sqrt{3}D}{2v_b}$
(c) $\frac{D}{v_b\sqrt{2}}$ (d) $\frac{D\sqrt{2}}{v_b}$

85. When a car is at rest, its driver sees rain drops falling on it vertically. When driving the car with speed v, he sees that rain drops are coming at an angle 60° from the horizontal. On further increasing the speed of the car to $(1+\beta)v$, this angle changes to 45°. The value of β is close to **[JEE Main 2020]**

(a) 0.50 (b) 0.41
(c) 0.37 (d) 0.73

86. In a car race on a straight path, car A takes a time t less than car B at the finish and passes finishing point with a speed v more than that of car B. Both the cars start from rest and travel with constant acceleration a_1 and a_2 respectively, then v is equal to **[JEE Main 2019]**

(a) $\frac{2a_1a_2}{a_1+a_2}t$ (b) $\sqrt{2a_1a_2}\,t$
(c) $\sqrt{a_1a_2}\,t$ (d) $\frac{a_1+a_2}{2}t$

ROUND II Mixed Bag

Only One Correct Option

1. A particle moving in a straight line covers half the distance with speed of 3 m/s. The other half of the distance is covered in two equal time intervals and with speeds of 4.5 m/s and 7.5 m/s, respectively. The average speed of the particle during this motion is

(a) 4.0 m/s (b) 5.0 m/s
(c) 5.5 m/s (d) 4.8 m/s

2. In a race for 100 m distance, the first and the second runners have a gap of one metre at the mid way stage. Assuming the first runner goes steady, by what percentage should the second runner increases his speed just to win the race.

(a) 2% (b) 4%
(c) more than 4% (d) less than 4%

3. A car accelerates from rest at a constant rate α for some time after which it decelerates at a constant rate β to come to rest. If the total time elapsed is t seconds, the total distance travelled is **[JEE Main 2021]**

(a) $\frac{4\alpha\beta}{(\alpha+\beta)}t^2$ (b) $\frac{2\alpha\beta}{(\alpha+\beta)}t^2$
(c) $\frac{\alpha\beta}{2(\alpha+\beta)}t^2$ (d) $\frac{\alpha\beta}{4(\alpha+\beta)}t^2$

4. A particle is moving with a velocity $\mathbf{v} = k(y\hat{\mathbf{i}} + x\hat{\mathbf{j}})$, where k is a constant. The general equation for its path is

(a) $y = x^2$ + constant
(b) $y^2 = x$ + constant
(c) xy = constant
(d) $y^2 = x^2$ + constant

5. An automobile travelling with a speed of 60 kmh^{-1} can brake to stop with a distance of 20 m. If the car is going twice as fast, *i.e.* 120 kms^{-1}, the stopping distance will be

(a) 20 m (b) 40 m
(c) 60 m (d) 80 m

6. A point initially at rest moves along x-axis. Its acceleration varies with time as $a = (6t + 5)\ m/s^2$. If it starts from origin, the distance covered in 2 s is

(a) 20 m (b) 18 m
(c) 16 m (d) 25 m

7. The motion of a body is given by the equation $\frac{dv(t)}{dt} = 6.0 - 3v(t)$, where $v(t)$ is speed in ms^{-1} and t in second. If body was at rest at $t = 0$ and the acceleration is half the initial value, then find its speed.

(a) 2.0 ms^{-1} (b) 3.0 ms^{-1}
(c) 1.0 ms^{-1} (d) 6.0 ms^{-1}

8. An elevator ascends with an upward acceleration of 2.0 ms^{-2}. At the instant its upward speed is 2.5 ms^{-1}, a loose bolt is dropped from the ceiling of the elevator 3.0 m from the floor. If $g = 10\ ms^{-2}$, then find the distance covered by the bolt during the free fall relative to the elevator shaft.

(a) 0.11 m (b) 0.75 m (c) 1.38 m (d) 2.52 m

9. A particle is moving with a uniform acceleration along a straight line AB. Its speed at A and B are 2 ms^{-1} and 14 ms^{-1} respectively. Then

(a) its speed at mid-point of AB is 20 ms^{-1}
(b) its speed at a point P such that $AP : PB = 1:5$ is 4 ms^{-1}
(c) the time to go from A to mid-point of AB is double of that to go from mid-point of B
(d) None of the above

10. A bullet emerges from a barrel of length 1.2 m with a speed of 640 ms^{-1}. Assuming constant acceleration, the approximate time that it spends in the barrel after the gun is fired is

(a) 4 ms (b) 40 ms (c) 400 ms (d) 1 s

11. A train accelerated uniformly from rest attains a maximum speed of 40 ms^{-1} in 20 s. It travels at this speed for 20 s and is brought to rest with uniform retardation in 40 s. The average velocity during this period is

(a) (80/3) ms^{-1} (b) 30 ms^{-1}
(c) 25 ms^{-1} (d) 40 ms^{-1}

12. A car starting from rest, accelerates at the rate f through a distance S, then continues at constant speed for time t and then decelerates at the rate $\frac{f}{2}$ to come to rest. If the total distance traversed in 15 s, then

(a) $S = \frac{1}{4}ft^2$ (b) $S = \frac{1}{72}ft^2$
(c) $S = \frac{1}{6}ft^2$ (d) $S = ft$

13. A particle is dropped vertically from rest from a height. The time taken by it to fall through successive distances of 1 m each, will then be

(a) all being equal to $\sqrt{2/9}$ s
(b) in the ratio of the square roots of the integers 1, 2, 3, ...
(c) in the ratio of the difference in the square roots of the integers is $(\sqrt{2}-\sqrt{1}), (\sqrt{3}-\sqrt{2}), (\sqrt{4}-\sqrt{3})$
(d) in the ratio of the reciprocal of the square roots of the integers is $\left(\frac{1}{\sqrt{1}}, \frac{1}{\sqrt{2}}, \frac{1}{\sqrt{3}}, \frac{1}{\sqrt{4}}\right)$

14. A man throws balls with the same speed vertically upwards one after the other at an interval of 2 s. What should be the speed of the throw so that more than two balls are in the sky at any time? (Take, $g = 9.8\ \text{m/s}^2$)

(a) At least 0.8 m/s
(b) Any speed less than 19.6 m/s
(c) Only with speed 19.6 m/s
(d) More than 19.6 m/s

15. From the top of a tower of height 50 m, a ball is thrown vertically upwards with a certain velocity. It hits the ground 10 s after it is thrown up. How much time does it take to cover a distance AB where A and B are two points 20 m and 40 m below the edge of the tower? (Take, $g = 10\ \text{ms}^{-2}$)

(a) 2.0 s (b) 1.0 s
(c) 0.5 s (d) 0.4 s

16. From the top of a tower, a stone is thrown up and reaches the ground in time $t_1 = 9$ s. A second stone is thrown down with the same speed and reaches the ground in time $t_2 = 4$ s. A third stone is released from rest and reaches the ground in time t_3, which is equal to

(a) 6.5 s (b) 6.0 s
(c) $\frac{5}{36}$ s (d) 65 s

17. A ball which is at rest, is dropped from a height h metre. As it bounces off the floor, its speed is 80% of what it was just before touching the ground. The ball will then rise nearly to a height

(a) 0.94 h (b) 0.80 h
(c) 0.75 h (d) 0.64 h

18. A parachutist after alling out falls 50 m without friction. When parachute opens, it decelerates at $2\ \text{ms}^{-2}$. He reaches the ground with a speed of $3\ \text{ms}^{-1}$. At what height, did he fallen out?

(a) 111 m (b) 293 m
(c) 182 m (d) 91 m

19. A tennis ball is released from a height h and after freely falling on a wooden floor, it rebounds and reaches height $\frac{h}{2}$. The velocity *versus* height of the ball during its motion may be represented graphically by

(Graphs are drawn schematically and on not to scale) **[JEE Main 2020]**

(a)
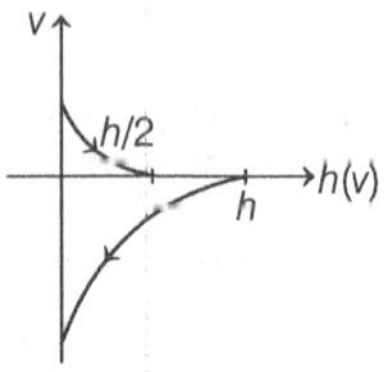

(b)
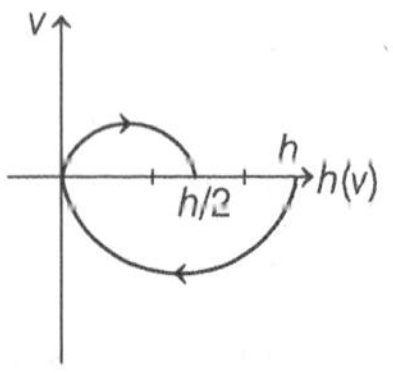

(c)
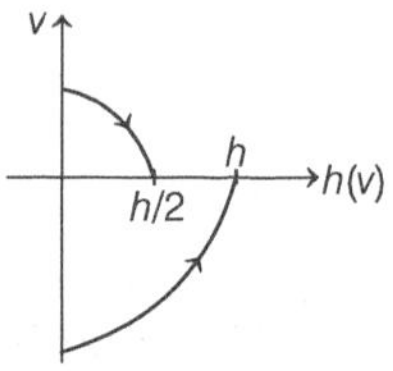

(d)
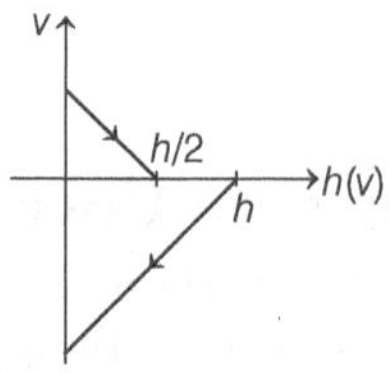

20. A motorboat covers a given distance in 6 h moving downstream on a river. It covers the same distance in 10 h moving upstream. The time it takes to cover the same distance in still water is

(a) 9 h (b) 7.5 h
(c) 6.5 h (d) 8 h

21. Two trains are moving with equal speed in opposite directions along two parallel railway tracks. If the wind is blowing with speed u along the track so that the relative velocities of the trains w.r.t. the wind are in the ratio 1 : 2, then the speed of each train must be

(a) $3u$ (b) $2u$
(c) $5u$ (d) $4u$

22. A boat crosses a river of width 1 km by shortest path in 15 min. If the speed of boat in still water is $5\ \text{kmh}^{-1}$, then what is the speed of the river?

(a) $5\ \text{kmh}^{-1}$
(b) $12\ \text{kmh}^{-1}$
(c) $3\ \text{kmh}^{-1}$
(d) $4\ \text{kmh}^{-1}$

23. Two buses of equal 5 m lengths are moving with the same velocity in the same direction on a highway. The first bus is 40 m ahead of the second bus. The driver of the second bus thinks to overtake the first bus and gives an acceleration of $1\ \text{ms}^{-2}$ to the bus. After what time the second bus just passes the first bus?

(a) 5 s (b) 10 s
(c) 15 s (d) 20 s

24. Two cars A and B are travelling in the same direction with velocities v_A and v_B $(v_A > v_B)$. When the car A is at a distance s behind car B, the driver of the car A applies the brakes producing a uniform retardation a, there will be no collision when

(a) $s < \frac{(v_A - v_B)^2}{2a}$ (b) $s = \frac{(v_A - v_B)^2}{2a}$
(c) $s \geq \frac{(v_A - v_B)^2}{2a}$ (d) $s \leq \frac{(v_A - v_B)^2}{2a}$

25. Two stones are thrown up simultaneously from the edge of a cliff 240 m high with initial speed of 10 m/s and 40 m/s respectively. Which of the following graph best represents the time variation of relative position of the second stone with respect to the first? (Assume stones do not rebound after hitting the ground and neglect air resistance, take $g = 10 \text{m/s}^2$) [JEE Main 2015]

(a) 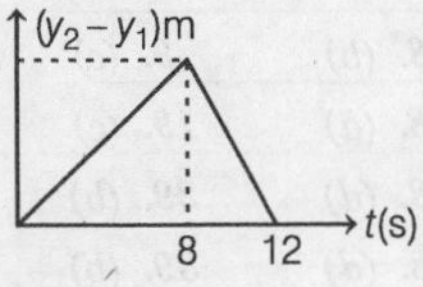

(b) 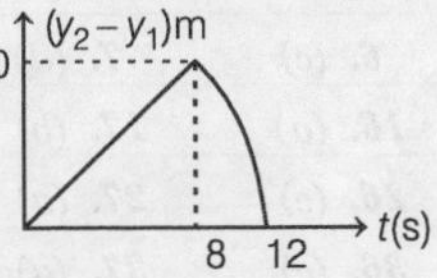

(c) 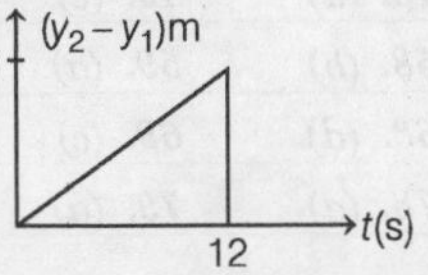

(d) 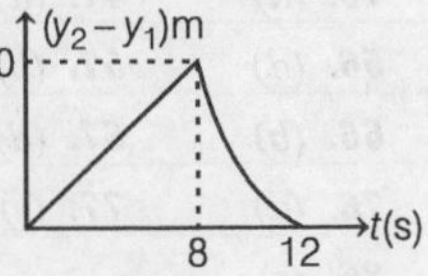

26. Ship A is sailing towards north-east with velocity $\mathbf{v} = 30\hat{\mathbf{i}} + 50\hat{\mathbf{j}}$ km/h, where $\hat{\mathbf{i}}$ points east and $\hat{\mathbf{j}}$ north. Ship B is at a distance of 80 km east and 150 km north of Ship A and is sailing towards west at 10 km/h. A will be at minimum distance from B in [JEE Main 2019]

(a) 4.2 h
(b) 2.6 h
(c) 3.2 h
(d) 2.2 h

Numerical Value Questions

27. The speed *versus* time graph for a particle is shown in the figure. The distance travelled (in m) by the particle during the time interval $t = 0$ s to $t = 5$ s will be ....

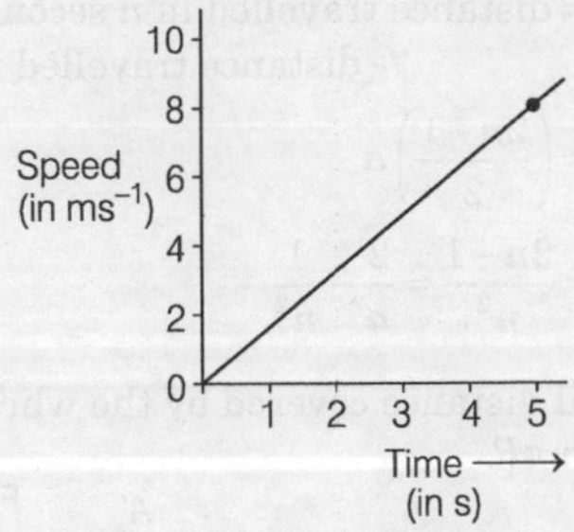

28. The distance x covered by a particle in one dimensional motion varies with time t as $x^2 = at^2 + 2bt + c$. If the acceleration of the particle depends on x as x^{-n}, where n is an integer, the value of n is ....

29. A particle moving on straight line whose velocity-time graph is shown in the figure. The average speed from $t = 0$ to $t = 6$ s is $v = \frac{15}{n}$ ms^{-1}, the value of n is ....

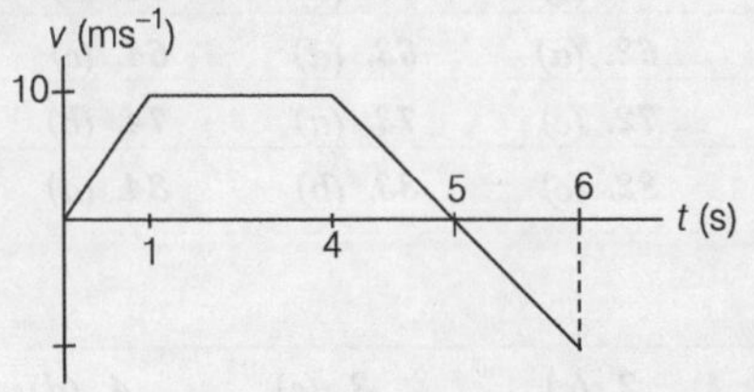

30. A particle is moving on a straight line whose velocity as function of time is $(t - 2)$ ms^{-1}. The distance travelled by particle (in m) in 4 s is ......

31. A particle starts from rest to move along X-axis. The acceleration of the particle is $a = (t - x)$ ms^{-2}. During motion, maximum acceleration of the particle is $a_0 = 2$ ms^{-2}. The velocity (in ms^{-1}) of the particle at $t = \frac{\pi}{3}$ s is ....

32. A person is swimming with a speed of 10 m/s at an angle of 120° with the flow and reaches to a point directly opposite on the other side of the river. The speed of the flow is x m/s. The value of x to the nearest integer is .... [JEE Main 2021]

33. A particle is moving along the X-axis with its coordinate with the t given by $x(t) = 10 + 8t - 3t^2$. Another particle is moving along the Y-axis with its coordinate as a function of time given by $y(t) = 5 - 8t^3$. At $t = 1$ s, the speed of the second particle as measured in the frame of the first particle is given as $\sqrt{v}$. Then v (in m/s) is ....

Answers

Round I

1. (a)	*2. (b)*	*3. (d)*	*4. (a)*	*5. (c)*	*6. (c)*	*7. (d)*	*8. (b)*	*9. (c)*	*10. (c)*
11. (b)	*12. (d)*	*13. (a)*	*14. (b)*	*15. (b)*	*16. (a)*	*17. (b)*	*18. (a)*	*19. (c)*	*20. (c)*
21. (a)	*22. (c)*	*23. (a)*	*24. (a)*	*25. (d)*	*26. (c)*	*27. (a)*	*28. (d)*	*29. (b)*	*30. (a)*
31. (b)	*32. (b)*	*33. (d)*	*34. (a)*	*35. (d)*	*36. (c)*	*37. (a)*	*38. (d)*	*39. (b)*	*40. (c)*
41. (c)	*42. (c)*	*43. (d)*	*44. (c)*	*45. (c)*	*46. (a)*	*47. (d)*	*48. (d)*	*49. (c)*	*50. (d)*
51. (b)	*52. (c)*	*53. (d)*	*54. (c)*	*55. (d)*	*56. (d)*	*57. (b)*	*58. (b)*	*59. (a)*	*60. (b)*
61. (c)	*62. (a)*	*63. (d)*	*64. (c)*	*65. (b)*	*66. (b)*	*67. (d)*	*68. (d)*	*69. (c)*	*70. (b)*
71. (a)	*72. (c)*	*73. (a)*	*74. (b)*	*75. (a)*	*76. (a)*	*77. (c)*	*78. (a)*	*79. (a)*	*80. (b)*
81. (a)	*82. (c)*	*83. (b)*	*84. (a)*	*85. (d)*	*86. (c)*				

Round II

1. (a)	*2. (c)*	*3. (c)*	*4. (d)*	*5. (d)*	*6. (b)*	*7. (c)*	*8. (c)*	*9. (c)*	*10. (a)*
11. (c)	*12. (b)*	*13. (c)*	*14. (d)*	*15. (d)*	*16. (b)*	*17. (d)*	*18. (b)*	*19. (c)*	*20. (b)*
21. (a)	*22. (c)*	*23. (b)*	*24. (c)*	*25. (b)*	*26. (b)*	*27. 20*	*28. 3*	*29. 2*	*30. 4*
31. 1	*32. 5*	*33. 580*							

Solutions

Round I

1. Displacement along North $= 400 - 300 = 100$ m

Upward displacement $= 1200$ m

$\therefore$ Net displacement $= \sqrt{(100)^2 + (1200)^2}$

$= 1204.15 \text{ m} \approx 1200 \text{ m}$

2. If distance between his house and office is s.

$$\therefore \quad 2 = \frac{s}{t + \frac{3}{60}} \text{ (for 1st event)} \quad \ldots\text{(i)}$$

$$\text{and} \quad 2 + 1 = \frac{s}{\left(t - \frac{3}{60}\right)} \text{ (for 2nd event)} \quad \ldots\text{(ii)}$$

On solving Eqs. (i) and (ii), we get

$s = 0.6$ km and $t = 0.25$ h

3. Here, initial position vector of particle is

$$\mathbf{r}_i = \hat{\mathbf{i}} + 2\hat{\mathbf{j}} - 3\hat{\mathbf{k}}$$

and final position vector of particle is

$$\mathbf{r}_f = 3\hat{\mathbf{i}} + 4\hat{\mathbf{j}} + 5\hat{\mathbf{k}}$$

$\therefore$ The displacement of particle is

$$\mathbf{s} = \mathbf{r}_f - \mathbf{r}_i$$
$$= (3\hat{\mathbf{i}} + 4\hat{\mathbf{j}} + 5\hat{\mathbf{k}}) - (\hat{\mathbf{i}} + 2\hat{\mathbf{j}} - 3\hat{\mathbf{k}})$$
$$= 2\hat{\mathbf{i}} + 2\hat{\mathbf{j}} + 8\hat{\mathbf{k}}$$

$\therefore$ Magnitude of displacement, $|\mathbf{s}| = \sqrt{2^2 + 2^2 + 8^2}$

$= \sqrt{72} \text{ m} = 8.5 \text{ m}$

4. Here, $S_n = \frac{1}{2} a n^2$

$S_{n\text{th}}$ = distance travelled in n second − distance travelled in $(n-1)$ second

$$= \left(\frac{2n-1}{2}\right) a$$

$$\therefore \quad \frac{S_{n\text{th}}}{S_n} = \frac{2n-1}{n^2} = \frac{2}{n} - \frac{1}{n^2}$$

5. Horizontal distance covered by the wheel in half revolution πR.

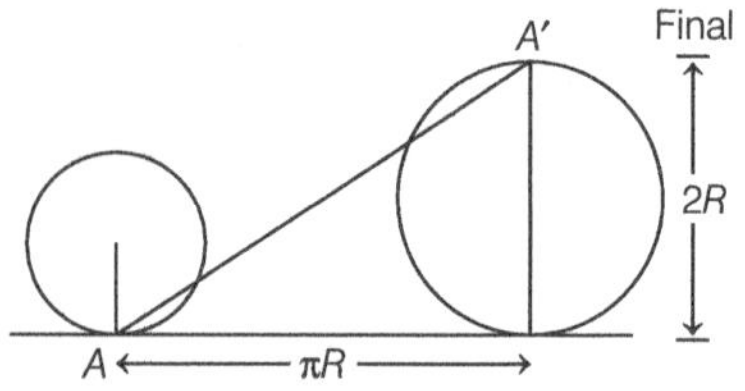

So, the displacement of the point which was initially in contact with ground

$$= AA' = \sqrt{(\pi R)^2 + (2R)^2}$$
$$= R\sqrt{\pi^2 + 4} = \sqrt{\pi^2 + 4} \quad (\because R = 1 \text{ m})$$

6. A body is moving on straight line with constant velocity. Between A and B, the straight path length is the shortest distance. This is the distance travelled. The particle starts at A and reaches B along the straight line. Therefore displacement is also AB, *i.e.* $D = s$.

7. As, $v_1 = \dfrac{u + v_1'}{2} = \dfrac{u + u + at_1}{2} = u + \dfrac{1}{2}at_1$

$$v_2 = \frac{v_1' + v_2'}{2} = \frac{(u + at_1) + u + a(t_1 + t_2)}{2}$$

$$= u + at_1 + \frac{1}{2}at_2$$

$$v_3 = \frac{v_2' + v_3'}{2} = \frac{(u + at_1 + at_2) + u + a(t_1 + t_2 + t_3)}{2}$$

$$= u + at_1 + at_2 + \frac{1}{2}at_3$$

Then, $v_1 - v_2 = -\dfrac{1}{2}a(t_1 + t_2)$

$$v_2 - v_3 = -\frac{1}{2}a(t_2 + t_3)$$

$\therefore$ $\dfrac{v_1 - v_2}{v_2 - v_3} = \dfrac{t_1 + t_2}{t_2 + t_3}$

8. Here, $v = |t - 2|\ \text{ms}^{-1}$

$v = t - 2$, when $t > 2\text{ s}$

$v = 2 - t$, when $t < 2\text{ s}$

$\therefore$ $a = \dfrac{dv}{dt} = 1\ \text{ms}^{-2}$ when $t > 2\text{ s}$

$a = -1\ \text{ms}^{-2}$ when $t < 2\text{ s}$

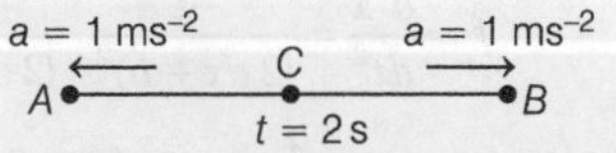

In the direction of motion from A to C, bee decelerates but for C to B, bee accelerates.

Let $AC = s_1, BC = s_2$

$u_A = 2\ \text{ms}^{-1}, t = 0$

$u_C = 0$ at $t = 4\text{ s}$

$\therefore$ $s_1 = \left(\dfrac{u_A + u_C}{2}\right)t_1$

$$s_2 = \left(\frac{u_C + u_B}{2}\right)t_2$$

$\therefore$ $s = s_1 + s_2 = \left(\dfrac{2+0}{2}\right)2 + \left(\dfrac{0+2}{2}\right)2 = 4\text{ m}$

9. Let the man starts crossing the road at an angle θ with the roadside. For safe crossing, the condition is that the man must cross the road by the time truck describes the distance $(4 + 2\cot\theta)$,

So, $\dfrac{4 + 2\cot\theta}{8} = \dfrac{2\sin\theta}{v}$

or $v = \dfrac{8}{2\sin\theta + \cos\theta}$

For minimum v, $\dfrac{dv}{d\theta} = 0$

or $\dfrac{-8(2\cos\theta - \sin\theta)}{(2\sin\theta + \cos\theta)^2} = 0$

or $2\cos\theta - \sin\theta = 0$

or $\tan\theta = 2$

So, $\sin\theta = \dfrac{2}{\sqrt{5}}$

$$\cos\theta = \frac{1}{\sqrt{5}}$$

$\therefore$ $v_{\min} = \dfrac{8}{2\left(\dfrac{2}{\sqrt{5}}\right) + \dfrac{1}{\sqrt{5}}}$

$$= \frac{8}{\sqrt{5}} = 3.57\ \text{ms}^{-1}$$

10. When a particle is moving with uniform acceleration, let v be the velocity of particle at a distance s,

then average velocity $= \dfrac{0 + v}{2} = \dfrac{v}{2}$

Time taken, $t_1 = \dfrac{s}{(v/2)} = \dfrac{2s}{v}$

When particle moves with uniform velocity, time taken,

$$t_2 = \frac{2s}{v}$$

When particle moves with uniform acceleration, time taken,

$$t_3 = \frac{3s}{(0 + v)/2} = \frac{6s}{v}$$

$$\text{Total time} = t_1 + t_2 + t_3$$

$$= \frac{2s}{v} + \frac{2s}{v} + \frac{6s}{v} = \frac{10s}{v}$$

$\therefore$ $v_{\text{av}} = \dfrac{s + 2s + 3s}{10s/v} = \dfrac{6v}{10}$

or $\dfrac{v_{\text{av}}}{v} = \dfrac{6}{10} = \dfrac{3}{5}$

11. Let time interval be chosen as 1 s

$$\frac{PA}{PB} = \frac{OA}{OB} = \frac{v_x}{v_y}$$

So, $P(x, y)$ divides AB in the ratio of $v_x : v_y$.

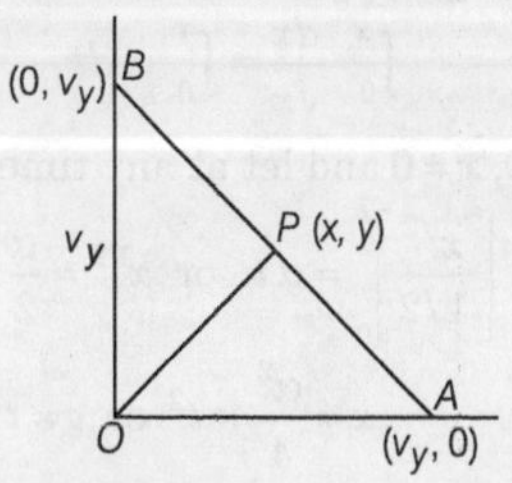

Using section formula,

$$x = \frac{v_x \times 0 + v_y \times v_x}{v_x + v_y} = \frac{v_x v_y}{v_x + v_y}$$

$$y = \frac{v_x v_y + v_y \times 0}{v_x + v_y} = \frac{v_x v_y}{v_x + v_y}$$

$$v = \sqrt{x^2 + y^2} = \sqrt{2}\frac{v_x v_y}{v_x + v_y}$$

Now, replace v_x by v_1 and v_y by v_2.

$$v = \frac{\sqrt{2}\,v_1 v_2}{v_1 + v_2}$$

12. Given, $x = 3t^2 \Rightarrow v_x = \frac{dx}{dt} = 3(2t) = 6t$

and $y = 4t^2 \Rightarrow v_y = \frac{dy}{dt} = 4(2t) = 8t$

$\therefore$ Velocity, $\mathbf{v} = v_x\hat{\mathbf{i}} + v_y\hat{\mathbf{j}} = 6t\,\hat{\mathbf{i}} + 8t\,\hat{\mathbf{j}}$

$$|\mathbf{v}| = \sqrt{(6t)^2 + (8t)^2} = 10\,t$$

13. $\sqrt{x} = t + 1$

Squaring both sides, we get

$$x = (t+1)^2 = t^2 + 1 + 2t$$

Differentiating it w.r.t. t, we get

$$\frac{dx}{dt} = 2t + 2$$

$$\text{Velocity, } v = \frac{dx}{dt} = 2t + 2$$

So, velocity increases with time.

14. Given, $v = v_0 + gt + Ft^2$

$$\Rightarrow \frac{dx}{dt} = v_0 + gt + Ft^2 \qquad \left[\because v = \frac{dx}{dt}\right]$$

$$\Rightarrow dx = (v_0 + gt + Ft^2)\,dt$$

So, $\int_0^x dx = \int_0^1 (v_0 + gt + Ft^2)\,dt$

$$\Rightarrow x = v_0 + \frac{g}{2} + \frac{F}{3}$$

Since, vertical component of velocity is zero.

15. The maximum distance covered in time $T = v_0 T$.
Therefore, for the object having one dimensional motion, the displacement x in time T satisfies $-v_0 T < x < v_0 T$.

16. Given, $v = \alpha\sqrt{x}$ or $\frac{dx}{dt} = \alpha\sqrt{x}$ $\left[\because v = \frac{dx}{dt}\right]$

or $$\frac{dx}{\sqrt{x}} = \alpha\,dt$$

On integrating, we get

$$\int_0^x \frac{dx}{\sqrt{x}} = \int_0^t \alpha\,dt$$

[$\because$ at $t = 0, x = 0$ and let at any time t, particle be at x]

$$\Rightarrow \left[\frac{x^{1/2}}{1/2}\right]_0^x = \alpha t \text{ or } x^{1/2} = \frac{\alpha}{2}t$$

or $$x = \frac{\alpha^2}{4} \times t^2 \text{ or } x \propto t^2$$

17. According to problem, when $s = a, t = p$

$$\because \quad s = ut + \frac{1}{2}ft^2 \text{ (here, } f = \text{acceleration)}$$

$$\therefore \quad a = up + \frac{fp^2}{2} \qquad \ldots\text{(i)}$$

For $s = b, t = q$

$$b = uq + \frac{fq^2}{2} \qquad \ldots\text{(ii)}$$

After solving Eqs. (i) and (ii), we get

$$f = \frac{2(aq - bp)}{pq(p - q)}$$

18. Position vector of particle is given as

$$\mathbf{r} = 15t^2\hat{\mathbf{i}} + (4 - 20t^2)\hat{\mathbf{j}}$$

Velocity of particle is

$$\mathbf{v} = \frac{d\mathbf{r}}{dt} = \frac{d}{dt}[15t^2\hat{\mathbf{i}} + (4 - 20t^2)\hat{\mathbf{j}}] = 30t\,\hat{\mathbf{i}} - 40t\,\hat{\mathbf{j}}$$

Acceleration of particle is

$$\mathbf{a} = \frac{d}{dt}(\mathbf{v}) = \frac{d}{dt}(30t\hat{\mathbf{i}} - 40t\hat{\mathbf{j}}) = 30\hat{\mathbf{i}} - 40\hat{\mathbf{j}}$$

So, magnitude of acceleration at $t = 1$ s is

$$|\mathbf{a}|_{t=1\text{s}} = \sqrt{a_x^2 + a_y^2} = \sqrt{30^2 + 40^2} = 50\text{ ms}^{-2}$$

19. Given, $t = ax^2 + bx$

Differentiating it w.r.t. t, we get

$$\frac{dt}{dt} = 2ax\frac{dx}{dt} + b\frac{dx}{dt}$$

$$v = \frac{dx}{dt} = \frac{1}{(2ax + b)}$$

Again, differentiating w.r.t. t, we get

$$\frac{d^2x}{dt^2} = \frac{-2a}{(2ax + b)^2} \cdot \frac{dx}{dt}$$

$$\therefore \quad f = \frac{d^2x}{dt^2} = \frac{-1}{(2ax + b)^2} \cdot \frac{2a}{(2ax + b)}$$

or $$f = \frac{-2a}{(2ax + b)^3} \Rightarrow f = -2av^3$$

20. $$\frac{dx}{dt} = 4t^3 - 2t$$

or $$dx = 4t^3\,dt - 2t\,dt$$

Integrating, $$x = \frac{4t^4}{4} - \frac{2t^2}{2} = t^4 - t^2$$

When $x = 2$,

$$\therefore \quad t^4 - t^2 - 2 = 0$$

$$t^2 = \frac{-(-1) \pm \sqrt{1+8}}{2}$$

or $$t^2 = \frac{1 \pm 3}{2} = 2 \qquad \text{(Ignoring –ve sign)}$$

Again, $$\frac{d^2x}{dt^2} = 12t^2 - 2$$

When $t^2 = 2$, acceleration $= 12 \times 2 - 2 = 22\text{ ms}^{-2}$

21. As lift is coming from 8th to 4th floor, the value of x becomes negative, *i.e.* $x < 0$. Velocity is downwards (*i.e.* negative). So, $v < 0$. Before reaching 4th floor lift is retarded, *i.e.* acceleration is upwards. Hence, $a > 0$.

22. Position of particle is, $x(t) = at + bt^2 - ct^3$

So, its velocity is, $v = \frac{dx}{dt} = a + 2bt - 3ct^2$

and acceleration is, $a = \frac{dv}{dt} = 2b - 6ct$

Acceleration is zero, then $2b - 6ct = 0$

$$\Rightarrow \quad t = \frac{2b}{6c} = \frac{b}{3c}$$

Substituting this t in expression of velocity, we get

$$v = a + 2b\left(\frac{b}{3c}\right) - 3c\left(\frac{b}{3c}\right)^2$$

$$= a + \frac{2b^2}{3c} - \frac{b^2}{3c} = a + \frac{b^2}{3c}$$

23. If t_0 is the reaction time, then the distance covered during decelerated motion is $10 - 10\,t_0$.

Now, in the first case,

$$10^2 = 2a\,(10 - 10\,t_0) \quad \text{...(i)}$$

Similarly, in the second case,

$$20^2 = 2a\,(30 - 20\,t_0) \quad \text{...(ii)}$$

Again, in the third case,

$$15^2 = 2a\,(x - 5\,t_0) \quad \text{...(iii)}$$

Dividing Eq. (ii) by Eq. (i), we get

$$\frac{20^2}{10^2} = \frac{30 - 20\,t_0}{10 - 10\,t_0}$$

or $\quad 40 - 40\,t_0 = 30 - 20\,t_0$

or $\quad 20\,t_0 = 10$

or $\quad t_0 = \frac{1}{2}\,\text{s}$

Dividing Eq. (iii) by Eq. (i), we get

$$\frac{225}{100} = \frac{x - 15\,t_0}{10 - 10\,t_0}$$

or $\quad \frac{9}{4} = \frac{x - 15 \times \frac{1}{2}}{10 - 10 \times \frac{1}{2}}$

$$45 = 4x - 30$$

or $\quad 4x = 75$

or $\quad x = \frac{75}{4}\,\text{m} = 18.75\,\text{m}$

24. As, $b - a = un + \frac{1}{2}An^2$

$\therefore \quad 2b - 2a = 2\,un + An^2 \quad \text{...(i)}$

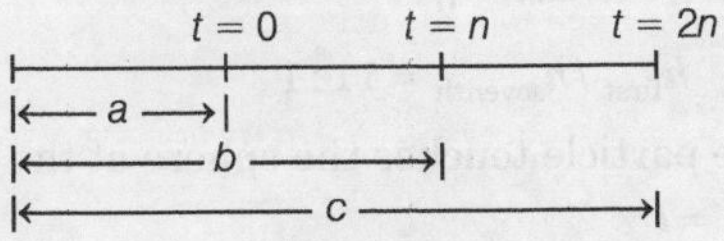

Again, $\quad c - a = u\,(2\,n) + \frac{1}{2}A\,(2\,n)^2 \quad \text{...(ii)}$

Subtracting Eq. (i) from Eq. (ii), we get

$$c - a - 2b + 2a = An^2$$

$$A = \frac{c - 2\,b + a}{n^2}$$

25. As, $\frac{dv}{dt} = -kv^3$

$\Rightarrow \quad \int_{v_0}^{v} \frac{dv}{v^3} = -k\int_0^t dt$

$\Rightarrow \quad -\frac{1}{k}\int_{v_0}^{v} v^{-3}\,dv = t$

or $\quad -\frac{1}{k}\left|\frac{v^{-3+1}}{-3+1}\right|_{v_0}^{v} = t$

or $\quad \frac{1}{2k}\left[\frac{1}{v^2} - \frac{1}{v_0^2}\right] = t$

or $\quad \frac{1}{v^2} - \frac{1}{v_0^2} = 2\,kt$

or $\quad \frac{1}{v^2} = \frac{1}{v_0^2} + 2\,kt$

or $\quad \frac{1}{v^2} = \frac{1 + 2v_0^2 kt}{v_0^2}$

or $\quad v = \frac{v_0}{\sqrt{2v_0^2 kt + 1}}$

26. As, $1 = \frac{v}{t_1}$ and $3 = \frac{v}{t_2}$

$\therefore \quad 1200 = \frac{1}{2}(t_1 + t_2)\,v, \quad \text{...(i)}$

$$1200 = \frac{1}{2}\left(v + \frac{v}{3}\right)v = \frac{1}{2}\,\frac{4\,v^2}{3} = \frac{2\,v^2}{3}$$

or $\quad v^2 = 1800$

From Eq. (i), we get

$\therefore \quad 1200 = \frac{1}{2}(t_1 + t_2) \times \sqrt{1800}$

$$(t_1 + t_2) = \frac{2400}{\sqrt{1800}}\,\text{s} = \frac{2400}{42.43}\,\text{s} = 56.6\,\text{s}$$

27. Given, $\frac{dv}{dt} = bt$ or $dv = bt\,dt$

$\Rightarrow \quad \int_{v_0}^{v} dv = \int_0^t bt\,dt$

or $\quad v - v_0 = \frac{bt^2}{2}$

or $\quad v = v_0 + \frac{bt^2}{2}$

or $\quad dx = v_0 dt + \frac{bt^2}{2}\,dt$

$$\int_0^x dx = \int_0^t v_0 dt + \frac{b}{2}\int_0^t t^2\,dt$$

or $\quad x = v_0 t + \frac{1}{2}\,\frac{bt^3}{3}$

$$= v_0 t + \frac{bt^3}{6}$$

28. $40^2 - 30^2 = 2\,as$, and $v^2 - 30^2 = 2a\,\frac{s}{2}$

or $\quad 2\,(v^2 - 30^2) = 2\,as$

Comparing, we get

$$2\,(v^2 - 900) = 1600 - 900 = 700$$

or $\quad v^2 = 900 + 350 = 1250$

or $\quad v = 35.35\,\text{kmh}^{-1}$

29. Given, initial position of particle, $\mathbf{r}_0 = (2\,\hat{\mathbf{i}} + 4\,\hat{\mathbf{j}})\,\text{m}$,

Initial velocity of particle, $\mathbf{u} = (5\,\hat{\mathbf{i}} + 4\,\hat{\mathbf{j}})\,\text{m/s}$

Acceleration of particle, $\mathbf{a} = (4\,\hat{\mathbf{i}} + 4\,\hat{\mathbf{j}})\,\text{m/s}^2$

According to second equation of motion,
position of particle at time t is, $\mathbf{r} = \mathbf{r}_0 + \mathbf{u}t + \frac{1}{2}\mathbf{a}t^2$
At $t = 2$s, position of particle,

$$\mathbf{r} = (2\hat{\mathbf{i}} + 4\hat{\mathbf{j}}) + (5\hat{\mathbf{i}} + 4\hat{\mathbf{j}}) \times 2 + \frac{1}{2}(4\hat{\mathbf{i}} + 4\hat{\mathbf{j}}) \times 4$$

or $\mathbf{r} = (2 + 10 + 8)\hat{\mathbf{i}} + (4 + 8 + 8)\hat{\mathbf{j}}$

$\Rightarrow \mathbf{r} = 20\hat{\mathbf{i}} + 20\hat{\mathbf{j}}$

$\therefore$ Distance of particle from origin, $|\mathbf{r}| = 20\sqrt{2}$ m

30. $4 = u + \frac{a}{2}(2 \times 3 - 1)$

or $4 = u + \frac{5a}{2}$

$5 = u + \frac{a}{2}(2 \times 4 - 1)$

or $5 = u + \frac{7a}{2}$

Subtracting, $1 = \frac{7a}{2} - \frac{5a}{2} = \frac{2a}{2} = a$

Again, $4 = u + \frac{5}{2}$

or $u = 4 - \frac{5}{2} = 1.5 \text{ ms}^{-1}$

So, the initial velocity is non-zero and acceleration is uniform.

31. $\mathbf{v} = 0.5t^2\hat{\mathbf{i}} + 3t\hat{\mathbf{j}} + 9\hat{\mathbf{k}}$

$\therefore \mathbf{a} = \frac{d\mathbf{v}}{dt} = t\hat{\mathbf{i}} + 3\hat{\mathbf{j}}$

At $t = 2$ s, $\mathbf{a} = 2\hat{\mathbf{i}} + 3\hat{\mathbf{j}}$

$\therefore \tan\theta = \frac{2}{3}$

or $\theta = \tan^{-1}\left(\frac{2}{3}\right)$, from Y-axis

32. Given, speed, $v = b\sqrt{x}$
Now, differentiating it with respect to time, we get

$$\frac{dv}{dt} = \frac{d}{dt}b\sqrt{x}$$

Now, acceleration,

$\Rightarrow a = \frac{b}{2\sqrt{x}} \cdot \frac{dx}{dt}$ $\left[\because \frac{dv}{dt} = a\right]$

$\Rightarrow a = \frac{b}{2\sqrt{x}} \cdot v = \frac{b}{2\sqrt{x}} \cdot b\sqrt{x} = \frac{b^2}{2}$

As acceleration is constant, we use

$v = u + at$ ...(i)

Now, it is given that $x = 0$ at $t = 0$.
So, initial speed of particle is

$$u = b\sqrt{x}\Big|_{x=0} = b \times 0 = 0$$

Hence, when time $t = \tau$, speed of the particle using Eq. (i) is

$$v = u + at = 0 + \frac{b^2}{2} \cdot \tau = \frac{b^2}{2} \cdot \tau$$

33. When A returns to the level of top of tower, its downward velocity is 4 ms^{-1}. This velocity is the same as that of B. So, both A and B hit the ground with the same velocity.

34. As, $x = \frac{1}{2}gt^2$ and $100 - x = 25x - \frac{1}{2}gt^2$,

Adding $25t = 100$

or $t = 4$ s

35. $\Delta t = \sqrt{\frac{2 \times 12}{10}} - \sqrt{\frac{2 \times 10}{10}}$ $\left(\because t = \sqrt{\frac{2H}{g}}\right)$

$= 1.549 \text{ s} - 1.414 \text{ s} = 0.135 \text{ s} \simeq 0.13 \text{ s}$

36. Vertical component of velocities of both the balls are same and equal to zero, so $t = \sqrt{\frac{2h}{g}}$ which is same for both the balls.

37. At time t

Velocity of A, $v_A = u - gt$ (upward)
Velocity of B, $v_B = gt$ (downward)

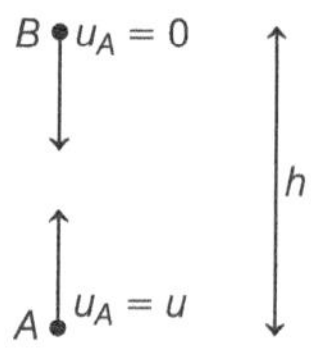

If we assume that height h is smaller than or equal to maximum height reached by A, then at every instant v_A and v_B are in opposite direction.

$\therefore v_{AB} = v_A + v_B$

$= u - gt + gt$

(Speeds in opposite directions get added)

$= u$

38. Initial velocity of balloon with respect to ground

$v = 10 + 5 = 15$ m/s (upward)

After 2 s its velocity $v = u - gt$

$= 15 - 10 \times 2 = -5$ m/s

$= 5$ m/s (downward)

39. Time of ascent $= \frac{u}{g} = 6 \text{ s} \Rightarrow u = 60$ m/s

Distance in first second,

$$h_{\text{first}} = 60 - \frac{g}{2}(2 \times 1 - 1) = 55 \text{ m}$$

Distance in seventh second will be equal to the distance in first second of vertical downward motion.

$$h_{\text{seventh}} = \frac{g}{2}(2 \times 1 - 1) = 5 \text{ m}$$

$\Rightarrow h_{\text{first}} / h_{\text{seventh}} = 11 : 1$

40. Let the particle touches the sphere at the point A.
Let $PA = l$

$\therefore PB = \frac{l}{2}$

In ΔOPB, $\cos\alpha = \frac{PB}{r}$

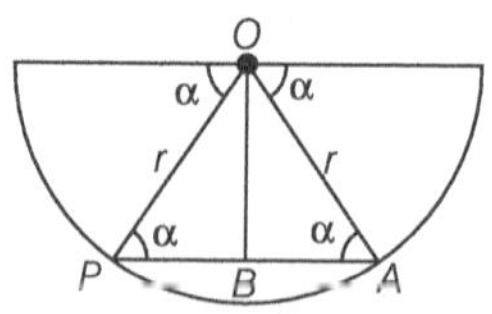

$\therefore PB = r\cos\alpha$

or $\frac{l}{2} = r\cos\alpha$

$l = 2r\cos\alpha$

But $$l = \frac{1}{2} a_0 t^2$$

$\therefore$ $$t = \sqrt{\left(\frac{2l}{a_0}\right)} = \sqrt{\left(\frac{2 \times 2r \cos \alpha}{a_0}\right)} = \sqrt{\left(\frac{4r \cos \alpha}{a_0}\right)}$$

41. Let the ball be at height h at time t and $(t + \Delta t)$, then

$$h = ut - \frac{1}{2} gt^2 \quad \text{...(i)}$$

and $$h = u(t + \Delta t) - \frac{1}{2} g (t + \Delta t)^2 \quad \text{...(ii)}$$

Equating Eqs. (i) and (ii), we get

$$t = \frac{2u - g\Delta t}{2g}$$

Substituting Eq. (ii) in Eq. (i), we get,

$$h = \frac{4u^2 - g^2 (\Delta t)^2}{8g}$$

$\Rightarrow$ $$u = \frac{1}{2} \sqrt{8gh + g^2 (\Delta t)^2}$$

42. Acceleration of body along AB is $g \cos \theta$

Distance travelled in time t sec $= AB = \frac{1}{2} (g \cos \theta) t^2$

From ΔABC, $AB = 2R \cos \theta$

$$2R \cos \theta = \frac{1}{2} g \cos \theta \, t^2$$

$$t^2 = \frac{4R}{g}$$

$$t = 2\sqrt{\frac{R}{g}}$$

43. Here, $$OA = \frac{u^2 - \frac{u^2}{4}}{2g} = \frac{u^2}{2g} \cdot \frac{3}{4}$$

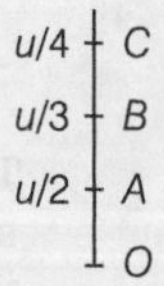

$$OB = \frac{u^2 - \frac{u^2}{9}}{2g} = \frac{u^2}{2g} \cdot \frac{8}{9}$$

and $$OC = \frac{u^2 - \frac{u^2}{16}}{2g} = \frac{u^2}{2g} \cdot \frac{15}{16}$$

$$\therefore \ AB = OB - OA = \frac{u^2}{2g}\left\{\frac{8}{9} - \frac{3}{4}\right\} = \frac{4^2}{2g} \cdot \frac{5}{36}$$

$$BC = OC - OB = \frac{u^2}{2g}\left\{\frac{15}{16} - \frac{8}{9}\right\} = \frac{u^2}{2g} \cdot \frac{7}{144}$$

$$\therefore \ \frac{AB}{BC} = \frac{5}{36} \times \frac{144}{7} = \frac{20}{7}$$

44.

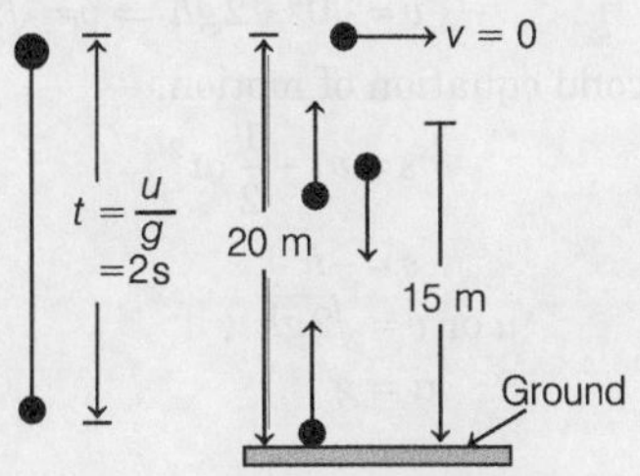

Position of balls

$$h_1 = \frac{1}{2} gt^2 = \frac{1}{2} \times 10 \times 1^2 = 5 \text{ m}$$

$$h_2 = \frac{1}{2} gt^2 = \frac{1}{2} \times 10 \times 2^2 = 20 \text{ m}$$

From ground, 5 m, 20 m, 15 m (shown in figure)

45. Now, retain only the positive sign.

$$(2v)^2 - v^2 = 2gh'$$

or $$4v^2 - v^2 = 2gh' \text{ or } 3v^2 = 2gh'$$

or $$3 \times 2gh = 2gh' \text{ or } h' = 3h$$

46. As, $$h = -vt + \frac{1}{2} gt^2$$

or $gt^2 - 2vt - 2h = 0$

$\Rightarrow$ $$t = \frac{-(-2v) \pm \sqrt{4v^2 + 8gh}}{2g} = \frac{2v \pm \sqrt{v^2 + 2gh}}{2g}$$

$$= \frac{v}{g} \pm \frac{(v^2 + 2gh)^{1/2}}{g} = \frac{v}{g}\left[1 \pm \sqrt{1 + \frac{2gh}{v^2}}\right]$$

$$= \frac{v}{g}\left[1 + \sqrt{1 + \frac{2gh}{v^2}}\right]$$ [taking +ve sign]

47. Let u be the velocity with which the stone is projected vertically upwards.

Given that, $$v_{-h} = 2v_h$$

$$(v_{-h})^2 = 4v_h^2$$

$\therefore$ $$u^2 - 2g(-h) = 4(u^2 - 2gh)$$

$\therefore$ $$u^2 = \frac{10gh}{3}$$

Now, $$h_{max} = \frac{u^2}{2g} = \frac{5h}{3}$$

48. As, $s' \propto t^2$

Now, $$s_1' : s_2' : s_3' = \frac{1}{4} : 1 : \frac{9}{4} \text{ or } 1:4:9$$

For successive intervals,

$$s_1 : s_2 : s_3 = 1 : (4-1) : (9-4)$$

or $$s_1 : s_2 : s_3 = 1 : 3 : 5$$

49. As, $$\Delta x = \frac{1}{2} gt^2 - \frac{1}{2} g(t-1)^2$$

$$= \frac{1}{2} g[t^2 - (t-1)^2] = \frac{1}{2} g(2t - 1)$$

$$= \frac{1}{2} \times 9.8 \times 5 \text{ m} = 24.5 \text{ m}$$

50. Clearly, $$\frac{11h}{36} = \frac{9.8}{2}(2n-1)$$

or $$\frac{11}{36} \times \frac{1}{2} \times 9.8 n^2 = \frac{9.8}{2}(2n-1)$$

or $$2n - 1 = \frac{11}{36} n^2$$

or $$11n^2 = 72n - 36$$

or $$11n^2 - 72n + 36 = 0$$

or $$11n^2 - 66n - 6n + 36 = 0$$

or $$11n(n-6) - 6(n-6) = 0 \Rightarrow n = 6$$

(Rejecting fractional value)

$$h = \frac{1}{2} \times 10 \times 6 \times 6 = 180 \text{ m}$$

51. Let us solve the problem in terms of relative initial velocity, relative acceleration and relative displacement of the coin with respect to floor of the lift.

Given, $u = 0 \text{ ms}^{-1}, a = 9.8 \text{ ms}^{-2}, s = 4.9 \text{ m}, t = ?$

As, $$4.9 = 0 \times t + \frac{1}{2} \times 9.8 \times t^2$$

or $$4.9\,t^2 = 4.9$$

or $$t = 1 \text{ s}$$

$$15 = 30 - 10\,t$$

or $$10\,t = 15$$

or $$t = 1.5 \text{ s}$$

52. Time taken to reach the maximum height

$$t_1 = \frac{u}{g}$$

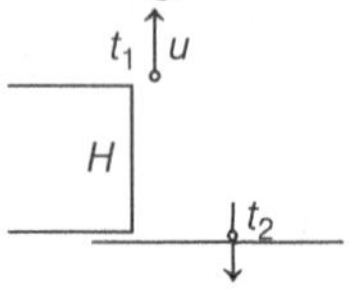

If t_2 is the time taken to hit the ground

i.e. $$-H = ut_2 - \frac{1}{2} gt_2^2$$

But $$t_2 = nt_1 \quad \text{[given]}$$

So, $$-H = u\frac{nu}{g} - \frac{1}{2} g \frac{n^2u^2}{g^2}$$

$$-H = \frac{nu^2}{g} - \frac{1}{2}\frac{n^2u^2}{g}$$

$$H = \frac{1}{2}\frac{n^2u^2}{g} - \frac{nu^2}{g} = \frac{n^2u^2 - 2nu^2}{2g}$$

$$2gH = n^2u^2 - 2nu^2$$

$$2gH = nu^2\,(n-2)$$

53. By the time 5th water drop starts falling, the first water drop reaches the ground.

As $u = 0$, $$h = \frac{1}{2} gt^2$$

$$= \frac{1}{2} \times 10 \times t^2$$

or $$5 = \frac{1}{2} \times 10 \times t^2$$

or $$t = 1 \text{ s}$$

Hence, the interval of each water drop $= \dfrac{1 \text{ s}}{4} = 0.25$ s

When the 5th drop starts its journey towards ground, the third drop travels in air for

$$t_1 = 0.25 + 0.25 = 0.5 \text{ s}$$

∴ Height (distance) covered by 3rd drop in air is

$$h_1 = \frac{1}{2} gt_1^2 = \frac{1}{2} \times 10 \times (0.5)^2$$

$$= 5 \times 0.25 = 1.25 \text{ m}$$

So, third water drop will be at a height of

$$= 5 - 1.25 = 3.75 \text{ m}$$

54. As, $h = -vt_1 + \frac{1}{2} gt_1^2$

or $$\frac{h}{t_1} = -v + \frac{1}{2} gt_1 \quad \text{...(i)}$$

and $$h = vt_2 + \frac{1}{2} gt_2^2$$

or $$\frac{h}{t_2} = v + \frac{1}{2} gt_2 \quad \text{...(ii)}$$

∴ $$\frac{h}{t_1} + \frac{h}{t_2} = \frac{1}{2} g\,(t_1 + t_2)$$

or $$h = \frac{1}{2} gt_1t_2$$

For fall under gravity from the top of the tower,

$$h = \frac{1}{2} gt^2$$

∴ $$\frac{1}{2} gt_1t_2 = \frac{1}{2} gt^2$$

⇒ $$t = \sqrt{t_1\,t_2}$$

55. Average acceleration $= \dfrac{\Delta v}{\Delta t}$

$$= \frac{\sqrt{2\,gh'} - (-\sqrt{2\,gh})}{\Delta t}$$

$$= \frac{\sqrt{2\,gh'} + \sqrt{2\,gh}}{\Delta t}$$

$$= \frac{\sqrt{2 \times 10 \times 2.5} + \sqrt{2 \times 10 \times 10}}{0.01} \text{ ms}^{-2}$$

$$= \frac{\sqrt{50} + \sqrt{200}}{0.01} \text{ ms}^{-2}$$

$$= \frac{5\sqrt{2} + 10\sqrt{2}}{0.01} \text{ ms}^{-2}$$

$$= \frac{15\sqrt{2}}{0.01} = 1500\sqrt{2} \text{ ms}^{-2}$$

The upward velocity has been taken as positive. Since, average acceleration is positive, therefore its direction is vertically upward.

56. Let t be the time taken by the packet to reach the ground. As, the helicopter rises from rest in upward direction, its final velocity is

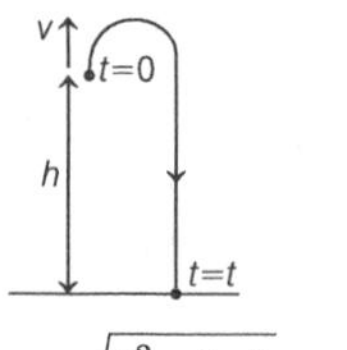

$$v = \sqrt{0^2 + 2gh} \Rightarrow v = \sqrt{2gh}$$

From second equation of motion,

$$s = ut + \frac{1}{2} at^2$$

Here, $$s = -h$$

$$u \text{ or } v = \sqrt{2gh}$$

⇒ $$a = g$$

Substituting all these values in above equation, we get

$$-h = \sqrt{2gh}\, t + \frac{1}{2}(-g)t^2$$

$$\Rightarrow \quad \frac{1}{2}gt^2 \sqrt{2gh}\, t - h = 0$$

This is a quadratic equation in t.

$$\therefore \quad t = \frac{\sqrt{2gh} \pm \sqrt{(\sqrt{2gh})^2 - 4 \times \frac{g}{2}(-h)}}{2 \times \frac{g}{2}}$$

$$= \frac{\sqrt{2gh} \pm \sqrt{2gh + 2gh}}{g}$$

$$= \frac{\sqrt{2gh}}{g}(1 + \sqrt{2})$$

$$= \sqrt{\frac{2h}{g}}(1 + \sqrt{2})$$

$$= (2 + \sqrt{2})\sqrt{\frac{h}{g}} = 3.4\sqrt{\frac{h}{g}}$$

57. Initially velocity keeps on decreasing at a constant rate, then it increases in negative direction with same rate.

58. In graph (b), for one value of displacement, there are two timings. As a result of it, for one time, the average velocity is positive and for other time is equivalent negative. Due to it, the average velocity for the two timings (equal to time period) can vanish.

59. Height reached $= \frac{1}{2} \times 132 \times 1200 \text{ m} = 66 \times 1200 \text{ m}$

60. Since acceleration is constant, therefore there is a uniform increase in velocity. So, the v-t graph is a straight line slopping upward to the right. When acceleration becomes zero, velocity is constant. So, v-t graph is a straight line parallel to the time-axis.

61. Area between v-t graph and time axis gives the distance.

$$\therefore \quad \Delta = \frac{1}{2} \times 2 \times 20 + 15 \times 3 + 2 \times \frac{1}{2} \times 15 \times 1$$

$$= 80 \text{ m}$$

62. The given v-t graph is shown below

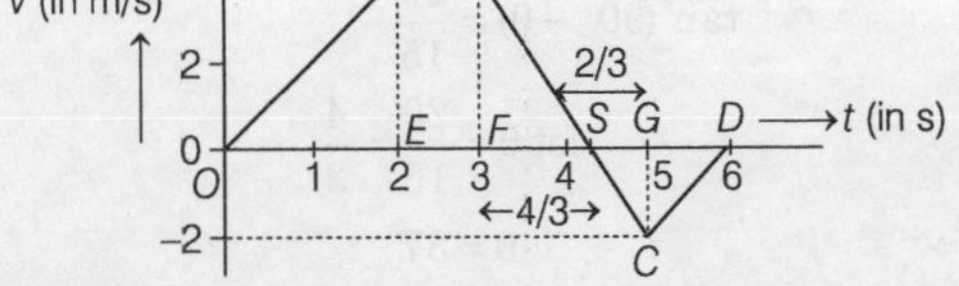

Distance covered by the body in 6s = Sum of magnitudes of area of each part in v-t curve

s = Area of ΔAOE + Area of rectangle $ABFE$ + Area of ΔBSF + Area of ΔSCG + Area of ΔGCD

$$= \frac{1}{2}(OE \times AE) + (EF \times AE) + \frac{1}{2}(SF \times BF) + \frac{1}{2}(SG \times GC) + \frac{1}{2}(GD \times GC)$$

$$= \frac{1}{2}(2 \times 4) + 1 \times 4 + \frac{1}{2} \times 4 \times \left(\frac{4}{3}\right) + \frac{1}{2} \times 2 \times \left(\frac{2}{3}\right) + \frac{1}{2} \times 2 \times 1$$

$$= 4 + 4 + \frac{8}{3} + \frac{2}{3} + 1 = \frac{37}{3} \text{ m}$$

63. Slope of line $= -\frac{2}{3}$

Equation of line is $(v - 20) = -\frac{2}{3}(s - 0)$

$$\Rightarrow \quad v = 20 - \frac{2}{3}s \qquad \text{...(i)}$$

Velocity at $s = 15$ m, *i.e.*

$$v = \left.\frac{ds}{dt}\right|_{s=15\text{ m}} = 20 - \frac{2}{3}(15) = 10 \text{ ms}^{-1}$$

Differentiate Eq. (i) with respect to time,

$$\text{acceleration} = \frac{dv}{dt} = -\frac{2}{3}\frac{ds}{dt}$$

$$\therefore \quad a = \left.\frac{dv}{dt}\right|_{s=15\text{ m}} = -\frac{2}{3}\left.\frac{ds}{dt}\right|_{s=15\text{ m}} = -\frac{20}{3}\text{ms}^{-2}$$

64. The displacement of the particle is determined by the area bounded by the curve, this area is $s = \frac{\pi}{4}v_m t_0$.

The average velocity is $\langle v \rangle = \frac{s}{t_0} = \frac{\pi}{4}v_m$

Such motion cannot be realised in practical terms since at the initial and final moments, the acceleration (which is slope of v-t graph) is infinitely large. Hence, both (i) and (ii) are correct.

65. Here, $\quad x_2 = vt \quad$ and $\quad x_1 = \frac{at^2}{2}$

$$\therefore \quad x_1 - x_2 = -\left(vt - \frac{at^2}{2}\right)$$

From the above expression, it is clear that at $t = 0$, $x_1 - x_2 = 0$, further for increasing values, the graph is as follows

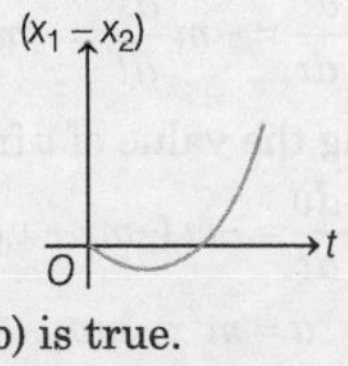

Hence, option (b) is true.

66. In option (a), (c) and (d), velocity of the body becomes zero after sometime or travelling some distance at least once. Hence, these graphs represent the same motion. In option (b), distance increases with time and becomes constant, hence velocity is not zero after starting motion, hence it represents different motion.

67. To get exact position at $t = 5$ s, we need to calculate area of the shaded part in the curve as shown below

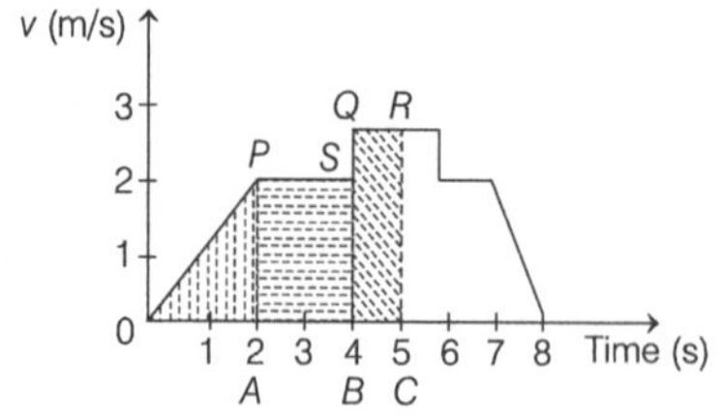

$\therefore$ Displacement of particle = Area of OPA + Area of $PABSP$ + Area of $QBCRQ$

$$= \left(\frac{1}{2} \times 2 \times 2\right) + (2 \times 2) + (3 \times 1)$$

$$= 2 + 4 + 3 = 9 \text{ m}$$

68. Since, the particle starts from rest, this means, initial velocity, $u = 0$.

Also, it moves with uniform acceleration along positive X-axis. This means, its acceleration (a) is constant.

$\therefore$ Given, a-t graph in (A) is correct.

As we know, for velocity-time graph, slope = acceleration.

Since, the given v-t graph in (B) represents that its slope is constant and non-zero.

$\therefore$ Graph in (B) is also correct.

Also, the displacement of such a particle w.r.t. time is given by

$$x = ut + \frac{1}{2}at^2 = 0 + \frac{1}{2}at^2 \Rightarrow x \propto t^2$$

So, x *versus* t graph would be a parabola with starting from origin.

This is correctly represented in displacement time graph given in (D).

69. Given line have positive intercept but negative slope so its equation can be written as

$$v = -mx + v_0 \quad \ldots\text{(i)}$$

$$\left[\text{where, } m = \tan\theta = \frac{v_0}{x_0}\right]$$

By differentiating w.r.t. time, we get

$$\frac{dv}{dt} = -m\frac{dx}{dt} = -mv$$

Now substituting the value of v from Eq. (i), we get

$$\frac{dv}{dt} = -m(-mx + v_0) = m^2x - mv_0$$

$$\therefore \quad a = m^2x - mv_0$$

The graph between a and x should have positive slope but negative intercept on a-axis. So, graph (c) is correct.

70. The slope of the given graph over the time interval t_1 to t_2 is not constant and is not uniform. It means acceleration is not constant and is not uniform, therefore relations (i), (ii) and (v) are not correct which is for uniform accelerated motion, but relations (iii), (iv) and (vi) are correct, because these relations are true for both uniform or non-uniform accelerated motion.

71. As the trains are moving in the same direction, so the initial relative speed $(v_1 - v_2)$ and by applying retardation, final relative speed becomes zero.

From $v = u - at$, we have

$$\Rightarrow \quad 0 = (v_1 - v_2) - at$$

$$\Rightarrow \quad t = \left(\frac{v_1 - v_2}{a}\right)$$

72. Velocity of girl, $v_g = \frac{L}{t_1}$

Velocity of escalator, $v_e = \frac{L}{t_2}$

Effective velocity of girl on escalator $= v_g + v_e$

$$= \frac{L}{t_1} + \frac{L}{t_2}$$

If t is the time taken, then $\frac{L}{t} = \frac{L}{t_1} + \frac{L}{t_2}$

or $\quad t = \frac{t_1 t_2}{t_1 + t_2}$

73. From the figure, the relative displacement is

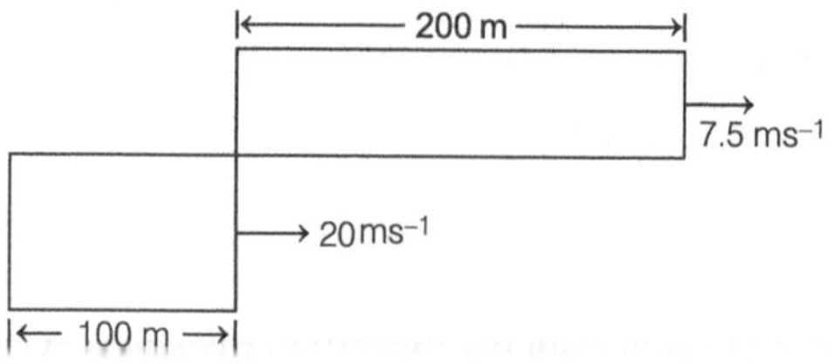

$$s_{rel} = (200 + 100) \text{ m} = 300 \text{ m}$$

$$v_{rel} = v_1 - v_2 = (20 - 7.5) \text{ ms}^{-1} = 12.5 \text{ ms}^{-1}$$

$$\therefore \quad t = \frac{s_{rel}}{v_{rel}} = \frac{300}{12.5} = 24 \text{ s}$$

74. Relative velocity of bird w.r.t. train = 25 + 5 = 30 m/s

Time taken by the bird to cross the train

$$t = \frac{210}{30} = 7 \text{ s}$$

75. Let the required angle is θ.

$$\tan(90° - \theta) = \frac{20}{15}$$

$$\therefore \quad \cot\theta = \frac{20}{15} = \frac{4}{3}$$

$$\Rightarrow \quad \theta = 37°$$

$\therefore$ Required angle $= \theta + 23° = 37° + 23° = 60°$

76. Let us calculate relative deceleration by considering relative velocity.

Using, $\quad v^2 - u^2 = 2as$

$$0^2 - 80^2 = 2 \times a \times 2000$$

or $\quad a = -\dfrac{80 \times 80}{4000} = -\dfrac{64}{40}\,\text{ms}^{-2} = -1.6\ \text{ms}^{-2}$

Deceleration of each train is $\dfrac{1.6}{2}\,\text{ms}^{-2}$, *i.e.* $0.8\ \text{ms}^{-2}$.

77. When trains are moving in same direction, then relative speed $= |v_1 - v_2|$ and in opposite direction, relative speed $= |v_1 + v_2|$.

Hence, ratio of time when trains move in same direction with time when trains move in opposite direction is

$$\frac{t_1}{t_2} = \frac{\left(\dfrac{l_1 + l_2}{|v_1 - v_2|}\right)}{\left(\dfrac{l_1 + l_2}{|v_1 + v_2|}\right)} = \frac{|v_1 + v_2|}{|v_1 - v_2|}$$

where, $l_1 + l_2 =$ sum of lengths of trains which is same as distance covered by trains to cross each other

So, $\quad \dfrac{t_1}{t_2} = \dfrac{80 + 30}{80 - 30} = \dfrac{110}{50} = \dfrac{11}{5}$

78. When two particles moves towards each other, then

$$v_1 + v_2 = 6 \quad \text{...(i)}$$

When these particles moves in the same direction, then

$$v_1 - v_2 = 4 \quad \text{...(ii)}$$

By solving Eqs. (i) and (ii), we get $v_1 = 5$ and $v_2 = 1$ m/s.

79. Effective speed of the bullet

= Speed of bullet + Speed of police jeep

= 180 m/s + 45 km/h

$= (180 + 12.5)$ m/s $= 192.5$ m/s

Speed of thief's jeep = 153 km/h = 42.5 m/s

Velocity of bullet w.r.t. thief's car

$= 192.5 - 42.5 = 150$ m/s

80. $\mathbf{v}_{ct} = \mathbf{v}_c - \mathbf{v}_t$

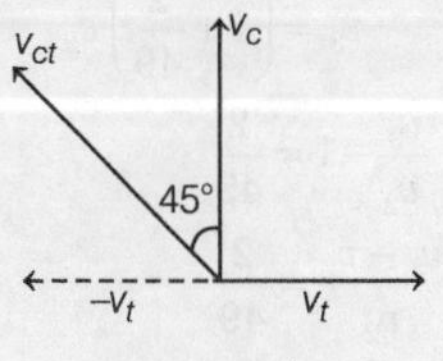

$$\mathbf{v}_{ct} = \mathbf{v}_c + (-\mathbf{v}_t)$$

Velocity of car w.r.t. train (v_{ct}) is towards west-north.

81. Here, $t_1 = \dfrac{x}{v} + \dfrac{x}{v} = \dfrac{2x}{v}$

$$t_2 = \frac{x}{v + \omega} + \frac{x}{v - \omega} = \frac{2xv}{v^2 - \omega^2}$$

or $\quad t_2 = \dfrac{2xv}{v^2\left(1 - \dfrac{\omega^2}{v^2}\right)} = \dfrac{2x}{v\left(1 - \dfrac{\omega^2}{v^2}\right)}$

or $\quad t_2 = \dfrac{t_1}{1 - \dfrac{\omega^2}{v^2}}$ or $t_2 > t_1$

82. Condition given in question is as shown,

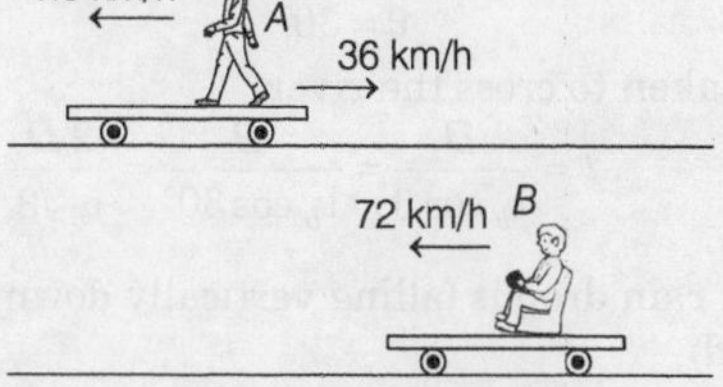

Speed of person on train A with respect to ground $= 36 - 1.8 = 34.2$ km/h in direction of A

So, the situation becomes as shown

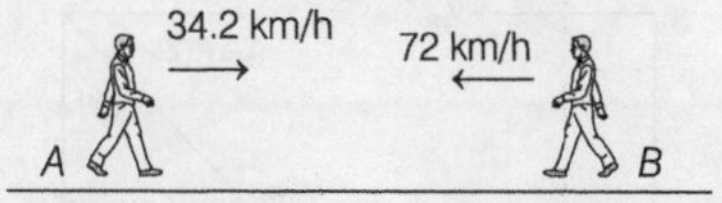

Hence, speed of person on train A as observed by an observer on train B is

$$v_{AB} = 34.2 + 72 = 106.2\ \text{km/h}$$

$$= 106.2 \times \frac{5}{18}\frac{\text{m}}{\text{s}} = 29.5\ \text{ms}^{-1}$$

83. Let v be the speed of boatman in still water.

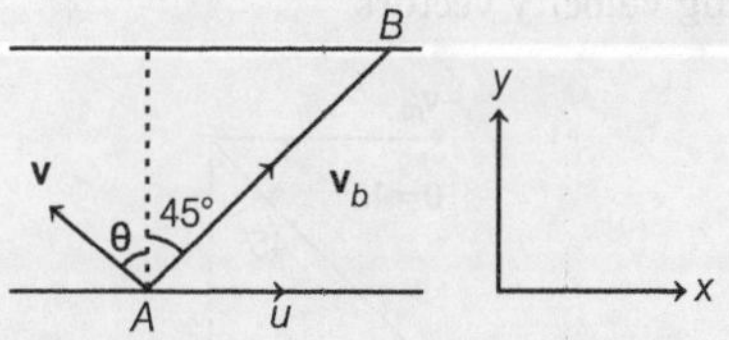

Resultant of v and u should be along AB. Components of $\mathbf{v}_b$ (absolute velocity of boatman) along x and y directions are

$$v_x = u - v\sin\theta$$

and $\quad v_y = v\cos\theta$

Further, $\tan 45° = \dfrac{v_y}{v_x}$

or $\quad 1 = \dfrac{v\cos\theta}{u - v\sin\theta}$

$$v = \frac{u}{\sin\theta + \cos\theta} = \frac{u}{\sqrt{2}\sin(\theta + 45°)}$$

v is minimum, at

$$\theta + 45° = 90° \quad \text{or} \quad \theta = 45°$$

and $\quad v_{\min} = \dfrac{u}{\sqrt{2}}$

84. From figure,

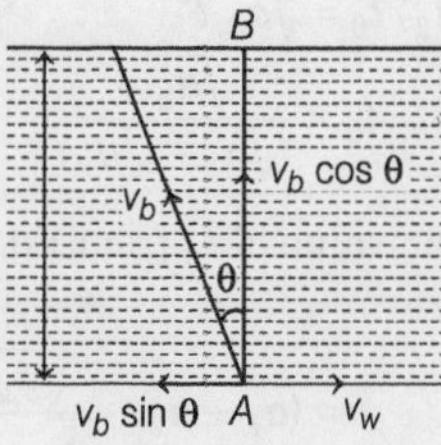

$$v_b \sin\theta = v_\omega$$

$$\sin\theta = \frac{v_\omega}{v_b} = \frac{1}{2}$$

$\Rightarrow \qquad \theta = 30^\circ$

Time taken to cross the river,

$$t = \frac{D}{v_b \cos\theta} = \frac{D}{v_b \cos 30^\circ} = \frac{2D}{v_b\sqrt{3}}$$

85. Given, rain drop is falling vertically downwards (w.r.t. ground).

So, velocity of rain $\mathbf{v}_r = v_r(-\hat{\mathbf{j}})$, velocity of man $\mathbf{v}_m = v\hat{\mathbf{i}}$.

Velocity of rain w.r.t to man $\mathbf{v}_{r/m} = \mathbf{v}_r - \mathbf{v}_m$

Drawing velocity vectors

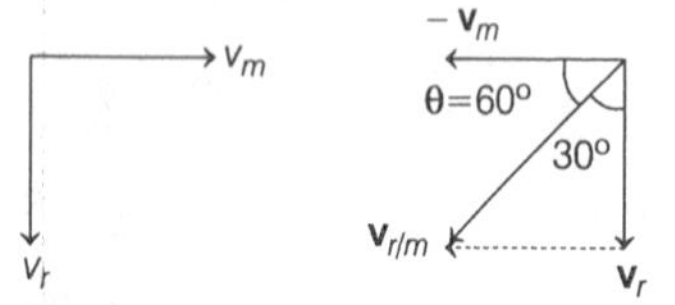

From diagram,

$$\tan 30^\circ = \frac{v_m}{v_r} \Rightarrow \frac{1}{\sqrt{3}} = \frac{v_m}{v_r} \text{ or } v_r = \sqrt{3}v$$

Now, man increases velocity of car to $(1+\beta)v$ in same direction, then $\mathbf{v}'_m = (1+\beta)\,v\hat{\mathbf{i}}$

Drawing velocity vectors

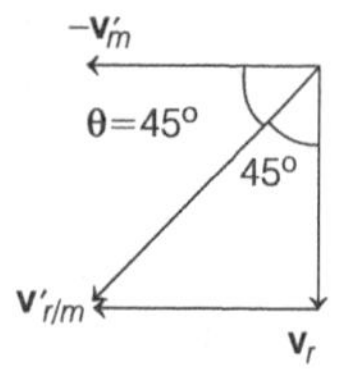

$$\tan 45^\circ = \frac{v'_m}{v_r} \Rightarrow 1 = \frac{(1+\beta)v}{\sqrt{3}v}$$

$$\Rightarrow \qquad 1+\beta = \sqrt{3} \text{ or } \beta = \sqrt{3}-1 = 0.73$$

86. Let car B takes time $(t_0 + t)$ and car A takes time t_0 to finish the race.

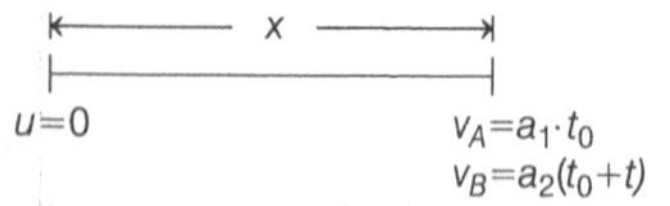

Given, $\qquad v_A - v_B = v = (a_1 - a_2)t_0 - a_2 t \qquad \ldots$(i)

$$s_B = s_A = \frac{1}{2}a_1 t_0^2 = \frac{1}{2}a_2(t_0+t)^2$$

or $\qquad \sqrt{a_1}\, t_0 = \sqrt{a_2}(t_0 + t)$

or $\qquad \sqrt{a_1} t_0 = \sqrt{a_2} t_0 + \sqrt{a_2} t$

or $\qquad (\sqrt{a_1} - \sqrt{a_2})\, t_0 = \sqrt{a_2}\, t$

or $\qquad t_0 = \dfrac{\sqrt{a_2}.\, t}{(\sqrt{a_1} - \sqrt{a_2})} \qquad \ldots$(ii)

Substituting the value of t_0 from Eq. (ii) into Eq. (i), we get

$$v = (a_1 - a_2)\frac{\sqrt{a_2}\, t}{\sqrt{a_1} - \sqrt{a_2}} - a_2 t$$

$$= (\sqrt{a_1} - \sqrt{a_2})(\sqrt{a_1} + \sqrt{a_2}) \cdot \frac{\sqrt{a_2} t}{(\sqrt{a_1} - \sqrt{a_2})} - a_2 t$$

or $\qquad v = (\sqrt{a_1} + \sqrt{a_2}) \cdot \sqrt{a_2}\, t - a_2 t$

$\qquad = \sqrt{a_1 a_2} \cdot t + a_2 t - a_2 t$

or $\qquad v = \sqrt{a_1 \cdot a_2}\; t$

Round II

1. If t_1 and $2\,t_2$ are the time taken by particle to cover first and second half distance, respectively

$$t_1 = \frac{x/2}{3} = \frac{x}{6}$$

Clearly, $\qquad x_1 = 4.5\, t_2$

and $\qquad x_2 = 7.5\, t_2$

So, $\qquad x_1 + x_2 = \dfrac{x}{2} \Rightarrow 4.5\, t_2 + 7.5\, t_2 = \dfrac{x}{2}$

$$t_2 = \frac{x}{24}$$

Total time $t = t_1 + 2\,t_2 = \dfrac{x}{6} + \dfrac{x}{12} = \dfrac{x}{4}$

So, $\qquad$ average speed = 4 m/s

2. Let v_1 and v_2 be the initial speeds of first and second runners respectively. Let t be time taken by them when the first runner has completed 50 m. During this time, the second runner has covered a distance $= 50 - 1 = 49$ m.

So, $\qquad t = \dfrac{50}{v_1} = \dfrac{49}{v_2} \qquad \ldots$(i)

Suppose, the second runner increases his speed to v_3 so that he covers the remaining distance (= 51 m) in time t.

So $\qquad t = \dfrac{51}{v_3} = \dfrac{49}{v_2}$

or $\qquad v_3 = \dfrac{51}{49} v_2$

or $\qquad v_3 = \left(1 + \dfrac{2}{49}\right) v_2$

or $\qquad \dfrac{v_3}{v_2} - 1 = \dfrac{2}{49}$

or $\qquad \dfrac{v_3 - v_2}{v_2} = \dfrac{2}{49}$

or $\qquad$ % increase $= \dfrac{2}{49} \times 100 = 4.1\%$

3. $v_1 = \alpha t_1$, $v_2 = \beta t_2$ and $t_1 + t_2 = t$

$\therefore \qquad v = \dfrac{\alpha\beta t}{(\alpha+\beta)}$

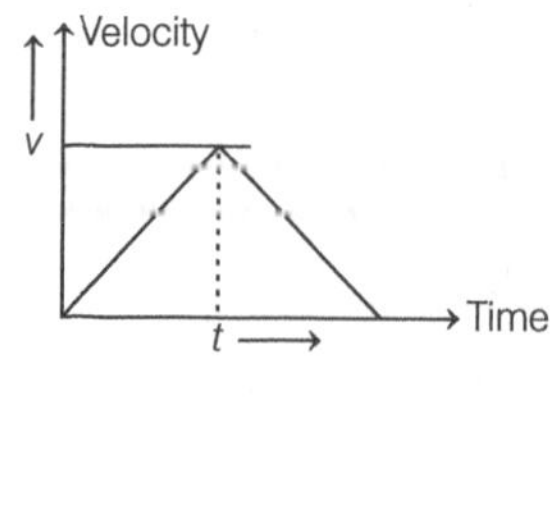

Distance = Area of v-t graph

$$= \frac{1}{2} \times v \times t = \frac{1}{2} \frac{\alpha\beta t^2}{\alpha+\beta}$$

4. Given, velocity of a particle is $\mathbf{v} = k(y\,\hat{\mathbf{i}} + x\,\hat{\mathbf{j}})$ ...(i)

Suppose, it's position is given as $\mathbf{r} = x\,\hat{\mathbf{i}} + y\,\hat{\mathbf{j}}$

$$\therefore \quad \mathbf{v} = \frac{d\mathbf{r}}{dt} = \frac{d}{dt}(x\,\hat{\mathbf{i}} + y\,\hat{\mathbf{j}}) = \frac{dx}{dt}\hat{\mathbf{i}} + \frac{dy}{dt}\hat{\mathbf{j}} \quad \text{...(ii)}$$

Comparing Eqs. (i) and (ii), we get

$$\frac{dx}{dt} = ky \quad \text{...(iii)}$$

and $$\frac{dy}{dt} = kx \quad \text{...(iv)}$$

Dividing Eq. (iii) and Eq. (iv), we get

$$\frac{\frac{dx}{dt}}{\frac{dy}{dt}} = \frac{y}{x}$$

$$\Rightarrow \quad x\frac{dx}{dt} = y\frac{dy}{dt}$$

or $$xdx = ydy$$

Integrating both sides, we get

$$\int xdx = \int ydy$$

or $$\frac{x^2}{2} + \frac{c_1}{2} = \frac{y^2}{2} + \frac{c_2}{2}$$

where, c_1 and c_2 are the constants of integration.

$$\Rightarrow \quad x^2 + c = y^2 \quad [\text{here, } c \text{ (constant)} = c_1 - c_2]$$

or $$y^2 = x^2 + \text{constant}$$

5. Let a be the retardation in both the cases. Using the relation, $v^2 = u^2 + 2\,as$, when automobile is stopped, $v = 0$.

So, $$0 = u^2 - 2\,as$$

or $$s \propto u^2$$

$$\therefore \quad s_2 = 4\,s_1 = 4 \times 20 = 80 \text{ m}$$

6. Given, acceleration $a = (6\,t + 5)\text{ m/s}^2$

$$\Rightarrow \quad a = \frac{dv}{dt} = (6\,t + 5),$$

$$dv = (6\,t + 5)\,dt$$

Integrating it, we have $\int dv = \int (6t + 5)\,dt$

$$v = 3\,t^2 + 5\,t + c$$

where c is constant of integration

When $t = 0$, $v = 0$ so $c = 0$

$$\therefore \quad v = 3\,t^2 + 5\,t$$

$$\Rightarrow \quad ds = (3\,t^2 + 5\,t)\,dt \quad \left(\text{as } v = \frac{ds}{dt}\right)$$

Integrating it within the condition of motion is as t changes from 0 to 2 s, S changes from 0 to S, we have

$$\int_0^S ds = \int_0^2 (3\,t^2 + 5\,t)\,dt$$

$$\therefore \quad S = \left[t^3 + \frac{5}{2}t^2\right]_0^2 = 8 + 10 = 18 \text{ m}$$

7. Given, $$\frac{dv(t)}{dt} = 6 - 3\,v(t) \quad \text{...(i)}$$

or $$\frac{dv(t)}{6 - 3\,v(t)} = dt$$

Integrating it, we have

$$\left[-\frac{1}{3}\log(6 - 3\,v(t))\right] = t + K \quad \text{...(ii)}$$

At $$t = 0, v(t) = 0$$

$$\therefore \quad K = -\frac{1}{3}\log 6$$

Putting this value in Eq. (ii), we have

$$-\frac{1}{3}\log(6 - 3\,v(t)) = t - \frac{1}{3}\log 6$$

or $$\log\left(\frac{6 - 3\,v(t)}{6}\right) = -3\,t$$

or $$\frac{6 - 3\,v(t)}{6} = e^{-3t}$$

or $$1 - \frac{v(t)}{2} = e^{-3t}$$

or $$v(t) = 2\,(1 - e^{-3t})$$

when $$t = 0, v(t) = 2\,(1 - e^{-3t})$$

Initially, $v = 0$, From Eq. (i) acceleration,

$$a_0 = \frac{dv}{dt} = 6 - 3 \times 0 = 6 \text{ ms}^{-2}$$

When $a = \frac{a_0}{2} = \frac{6}{2} = 3 \text{ ms}^{-2}$ then from Eq. (i);

$$3 = 6 - 3\,v$$

or $$3\,v = 6 - 3 = 3$$

or $$v = 1 \text{ ms}^{-1}$$

8. Velocity of bolt relative to elevator $= 2.5 - 2.5 = 0$

Acceleration of bolt relative to elevator,

$$a = 10 - (-2) = 12 \text{ ms}^{-2} \quad (\because g = 10 \text{ m/s}^2)$$

Using the relation, $s = ut + \frac{1}{2}at^2$

we have, $$3.0 = 0 \times t + \frac{1}{2} \times 12 \times t^2$$

or $$t = \frac{1}{\sqrt{2}}\text{ s} = 0.707 \text{ s} = 0.7 \text{ s}$$

$$\text{Displacement} = -2.5 \times 0.71 + \frac{1}{2} \times 10 \times (0.71)^2$$

$$= -1.775 + 2.521 = 0.746 = 0.75 \text{ m}$$

$$\text{Distance covered} = 2 \times \frac{u^2}{2\,g} + \text{displacement}$$

$$= 2 \times \frac{(2.5)^2}{2 \times 10} + 0.75$$

$$= 0.63 + 0.75 = 1.38 \text{ m}$$

9. Here, $u = 2 \text{ ms}^{-1}$, $v = 14 \text{ ms}^{-1}$

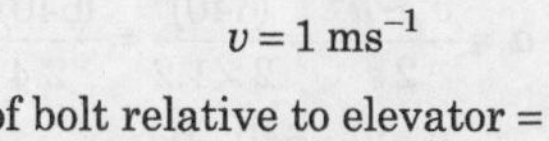

Distance between A and $B = s$

Then acceleration, $$a = \frac{v^2 - u^2}{2\,s} = \frac{14^2 - 2^2}{2\,s} = \frac{194}{2\,s} = \frac{97}{s}$$

The speed at mid-point C,

$$v^2 = u^2 + 2a\frac{s}{2}$$

$$= 2^2 + 2\times\frac{97}{s}\times\frac{s}{2} = 101$$

$$v = \sqrt{101} \simeq 10 \text{ ms}^{-1}$$

As per question, $AP = \frac{1}{6}[AB] = \frac{1}{6}s$

When, $s = \frac{s}{6}$,

then, $v^2 = 2^2 + 2\times\frac{97}{s}\times\frac{s}{6}$

or $v = 3 + 2\times\frac{97}{3} = 36.3 \text{ ms}^{-1}$

$\therefore$ $v = \sqrt{36.3} \approx 6 \text{ ms}^{-1}$

Since velocity at mid-point C is 10 ms^{-1}.

$\therefore$ Taking motion from A to C, we have

$$10 = 2 + a \times t_1$$

or $t_1 = \frac{10-2}{a} = \frac{8}{a}$

Taking motion from C to B, we have

$$14 = 10 + 1 \times t_2$$

or $t_2 = \frac{14-10}{a} = \frac{4}{a}$

$\therefore$ $\frac{t_1}{t_2} = 2$ or $t_2 = \frac{t_1}{2}$

10. Here, $u = 0, v = 640 \text{ ms}^{-1}, s = 1.2 \text{ m}, a = ?$ and $t = ?$

As, $v^2 = u^2 + 2as$

or $a = \frac{v^2 - u^2}{2s} = \frac{(640)^2}{2\times1.2} = \frac{(640)^2}{2.4} \text{ms}^{-1}$

Also, $t = \frac{v-u}{a} = \frac{640-0}{\frac{(640)^2}{2.4}}$

$$= \frac{2.4}{640} = 3.75\times10^{-3} \text{ s} \simeq 4 \text{ ms}$$

11. The acceleration of train in 20 s is given by

$$a = \frac{40-0}{20} = 2 \text{ ms}^{-2}$$

[from the formula $v = u + at$ (here, $u = 0$)].

Now the distance travelled is given by equation of motion,

$$v^2 = u^2 + 2as$$

So, $s_1 = \left(\frac{v^2 - u^2}{2}\right)\Big/a = \frac{40\times40-0}{2\times2} = 400 \text{ m}$

Now distance travelled with constant speed of 40 ms^{-1} in $t = 20\text{s}$ is

$$s_2 = 40\times20 = 800 \text{ m}$$

Again the distance covered in 3rd case is given by

$$s_3 = \frac{40\times40-0}{2\times1} = 800 \text{ m}$$

Therefore, average speed of the train is given by

$$v_{av} = \frac{400+800+800}{20+20+40} = \frac{2000}{80} = 25 \text{ ms}^{-1}$$

12.

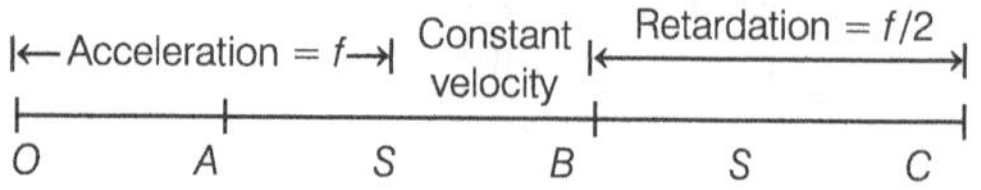

Taking motion of car from 0 to A,

Here, $u = 0, a = f, s = S, v = ?$

As $v^2 = u^2 + 2as$

So $v^2 = 0 + 2\times f\times S$

or $v = \sqrt{2fS}$

The velocity of car at A = velocity of car at $B = (2fS)^{1/2}$

As magnitude of retardation of the car from B to C is half of that of acceleration from O to A when velocity changes by v, so distance $BC = 2S$

Distance, $AB = 15S - (S + 2S) = 12S$

As distance AB is covered with constant velocity in time t

So, $12S = vt = (2fS)^{1/2}\times t$

or $144S^2 = 2fSt^2$

or $S = \frac{1}{72}ft^2$

13. As, $h = ut + \frac{1}{2}gt^2$

$\Rightarrow$ $1 = 0\times t_1 + \frac{1}{2}gt_1^2$

$\Rightarrow$ $t_1 = \sqrt{2/g}$

Velocity after travelling 1m distance

$$v^2 = u^2 + 2gh$$

$\Rightarrow$ $v^2 = (0)^2 + 2g\times1$

$\Rightarrow$ $v = \sqrt{2g}$

For second 1 m distance

$$1 = \sqrt{2g}\times t_2 + \frac{1}{2}gt_2^2$$

$\rightarrow$ $gt_2^2 + 2\sqrt{2}gt_2 - 2 = 0$

$$t_2 = \frac{-2\sqrt{2g}\pm\sqrt{8g+8g}}{2g}$$

$$= \frac{-\sqrt{2}\pm2}{\sqrt{g}}$$

Taking +ve sign, $t_2 = (2-\sqrt{2}t)/\sqrt{g}$

$\therefore \frac{t_1}{t_2} = \frac{\sqrt{2/g}}{(2-\sqrt{2})/\sqrt{g}} = \frac{1}{\sqrt{2}-1}$ or $\frac{t_2}{t_1} = (\sqrt{2}-\sqrt{1})$ and so on.

14. Interval of all ball throw = 2 s

If we want that minimum three (more than two) ball remain in air then, time of flight of first ball must be greater than 4 s.

$$t > 4 \text{ s}$$

$$\frac{2u}{g} > 4 \text{ s}$$

$\Rightarrow$ $u > 19.6$ m/s

For $u = 19.6$, then first ball will just about to strike the ground (in air).

Second ball will be highest point (in air).

Third ball will beat point of projection or at ground (not in air).

15. Let the body be projected upwards with velocity u from top of tower. Taking vertical downward motion of boy from top of tower to ground, we have

$$u = -u, a = g = 10 \text{ ms}^{-2}, s = 50 \text{ m}, t = 10 \text{ s}$$

As $$s = ut + \frac{1}{2}at^2$$

So, $$50 = -u \times 10 + \frac{1}{2} \times 10 \times 10^2$$

On solving, $u = 45 \text{ ms}^{-1}$

If t_1 and t_2 are the time taken by the ball to reach points A and B respectively, then

$$20 = 45\, t_1 + \frac{1}{2} \times 10 \times t_1^2$$

and $$40 = -45\, t_2 + \frac{1}{2} \times 10 \times t_2^2$$

On solving, we get, $t_1 = 9.4 \text{s}$ and $t_2 = 9.8 \text{s}$

Time taken to cover the distance AB

$$= (t_2 - t_1) = 9.8 - 9.4 = 0.4 \text{ s}$$

16. Let u be the initial upward velocity of the ball from top of the tower and h be the height of the tower.

Taking the downward motion of the first stone from A to the ground, we have

$$h = -ut_1 + \frac{1}{2}gt_1^2 \quad \text{...(i)}$$

Taking the downward motion of the second stone from top of the tower to the ground, we have

$$h = ut_2 + \frac{1}{2}gt_1^2 \quad \text{...(ii)}$$

Multiplying Eq. (i) t_2 and Eq. (ii) by t_1 and adding, we get

$$h\,(t_1 + t_2) = \frac{1}{2}gt_1 t_2 (t_1 + t_2)$$

So, $$h = \frac{1}{2}gt_1 t_2 \quad \text{...(iii)}$$

For falls under gravity from the top of the tower

$$h = \frac{1}{2}gt_3^2 \quad \text{...(iv)}$$

From Eqs. (iii) and (iv),

$$t_3^2 = t_1 t_2$$

or $$t_3 = \sqrt{t_1\, t_2} = \sqrt{9 \times 4} = 6 \text{s}$$

17. Clearly, $$v = \sqrt{2\,gh} \quad \text{...(i)}$$

After rebounce, $v^2 = u^2 - 2gh$

$$\Rightarrow \quad u^2 = v^2 + 2gh$$

and $$u^2 = 2gh' \quad \text{...(ii)}$$

$$\therefore \quad \frac{v^2}{u^2} = \frac{2gh}{2gh'}$$

$$\Rightarrow \quad h' = h \times \frac{u^2}{v^2} = h \times \left(\frac{80}{100}\right)^2$$

$$= 0.64\text{ h}$$

18. First 50 m fall is under the effect of gravity only. The velocity acquired, $u = \sqrt{2\,gh} = \sqrt{2 \times 9.8 \times 50} \text{ ms}^{-1}$.

Taking onward motion of parachutist with retardation 2 ms^{-2}, we have, $u = 10\sqrt{9.8} \text{ ms}^{-1}$.

$$d = -2 \text{ ms}^{-2}, v = 3 \text{ ms}^{-1}$$

$$s = \frac{v^2 - u^2}{2\,a} = \frac{(3)^2 - (2 \times 9.8 \times 50)}{2 \times (-2)} = 243 \text{ m}$$

$\therefore$ Total height $= 50 + 243 = 293$ m

19.

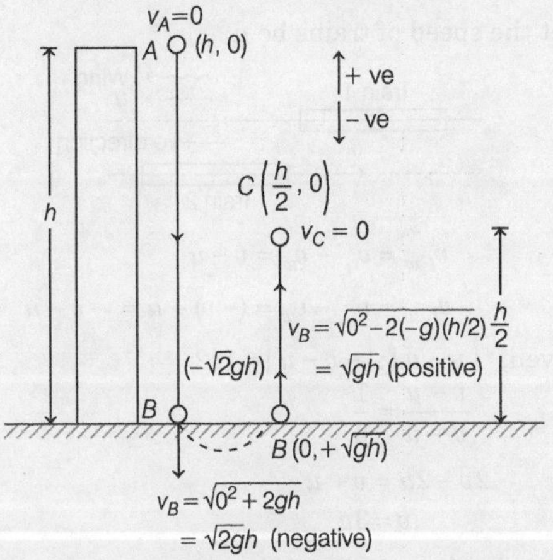

Using third equation of motion, the relation between v and h is given by,

$$v^2 = u^2 + 2gh = 0^2 + 2gh = 2gh$$

$$\Rightarrow \quad v^2 \propto h$$

It is the equation of a parabola. So, the shape of v *versus* h graph will be parabolic.

Now, as we have some co-ordinate points from diagram.

On plotting them and tracing with a parabolic locus, we will get the following graph

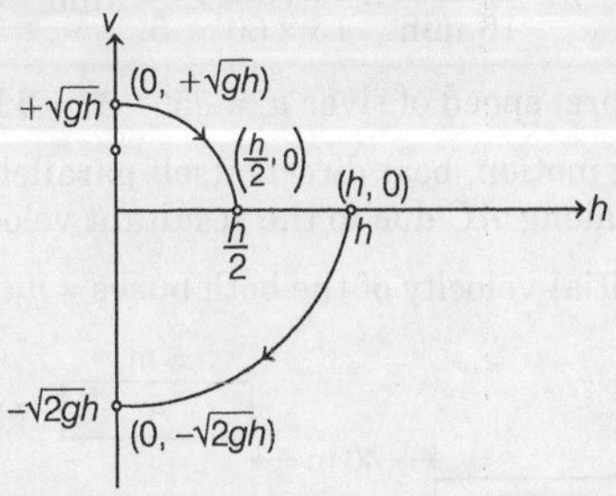

Hence, correct option is (c).

20. If v_w be the velocity of water and v_b be the velocity of motorboat in still water.

The distance covered by motorboat in moving downstream in 6 h is given by

$$x = (v_b + v_w) \times 6 \quad \text{...(i)}$$

Same distance covered by motorboat in moving upstream in 10 h is

$$x = (v_b - v_w) \times 10 \quad \text{...(ii)}$$

From Eqs. (i) and (ii), we have

$$(v_b + v_w) \times 6 = (v_b - v_w) \times 10$$

$\Rightarrow$ $v_w = \dfrac{v_b}{4}$

$\therefore$ $x = (v_b + v_w) \times 6 = 7.5\, v_b$

Time taken by the motorboat to cover the same distance in still water is

$$t = \frac{x}{v_b} = \frac{7.5\, v_b}{v_b} = 7.5 \text{ h}$$

21. Let the speed of trains be v.

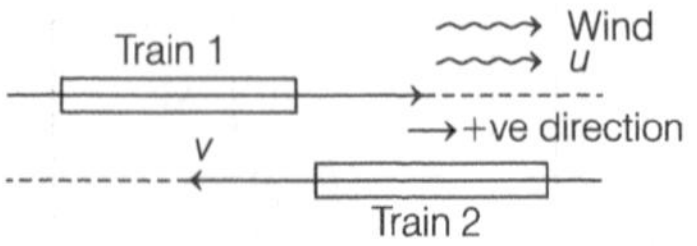

$$v_{t_1 w} = v_{t_1} - v_w = v - u$$

$$v_{t_2 w} = v_{t_2} - v_w = (-v) - u = -v - u$$

Given, $|v - u| : |-v - u| = 1:2$

$\therefore$ $\dfrac{v-u}{v+u} = \dfrac{1}{2}$

or $2v - 2u = v + u$

or $v = 3u$

22. Let v_r be the velocity of river.

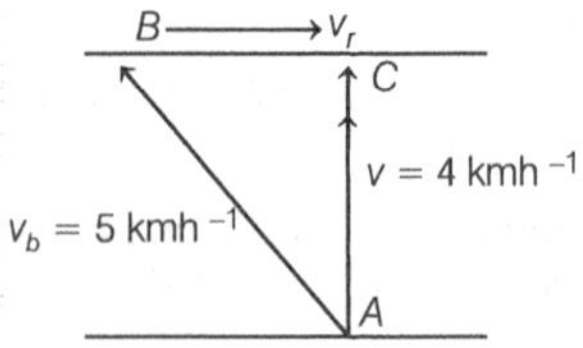

Velocity of boat when crosses river along the shortest path

$$v = \frac{1 \text{ km}}{15 \text{ min}} = \frac{1000}{15 \times 60} \times \frac{18}{5} \text{ kmh}^{-1} = 4 \text{ kmh}^{-1}$$

Therefore, speed of river $v_r = \sqrt{5^2 - 4^2} = 3 \text{ kmh}^{-1}$

During motion, boat directs itself parallel to AB, moves along AC due to the resultant velocity.

23. The initial velocity of the both buses = v_0

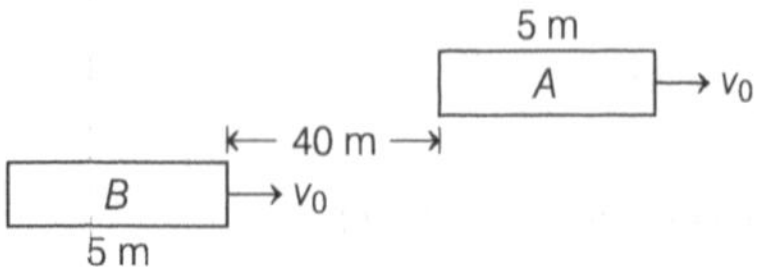

$\therefore$ $u_{rel} = \mathbf{u}_{BA} = v_0 - v_0 = 0$

$\because$ $s_{rel} = u_{rel} t + \dfrac{1}{2} a_{rel} t^2$

$$5 + 40 + 5 = 0 + \frac{1}{2}(1 - 0)\, t^2$$

$\therefore$ $t^2 = 100$

$\Rightarrow$ $t = 10$ s

24. For no collision, the speed of car A should be reduced to v_B before the cars meet, *i.e.*, final relative velocity of car A with respect to car B is zero, *i.e.* $v_r = 0$

Here, initial relative velocity, $u_r = v_A - v_B$

Relative acceleration, $a_r = -a - 0 = -a$

Let relative displacement = s_r

The equation,

$$v_r^2 = u_r^2 + 2\, a_r s_r$$

$$(0)^2 = (v_A - v_B)^2 - 2\, a s_r$$

$$s_r = \frac{(v_A - v_B)^2}{2a}$$

For no collision, $s_r \le s$

i.e. $\dfrac{(v_A - v_B)^2}{2a} \le s$

25. Consider the stones thrown up simultaneously as shown in the diagram below.

Considering motion of the second particle with respect to the first we have relative acceleration

$|\mathbf{a}_{21}| = |\mathbf{a}_2 - \mathbf{a}_1| = g - g = 0$

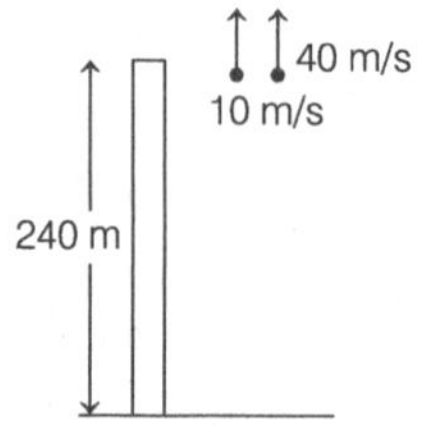

Thus, motion of first particle is straight line with respect to second particle till the first particle strikes ground at a time given by

$$-240 = 10\,t - \frac{1}{2} \times 10 \times t^2$$

or $t^2 - 2t - 48 = 0$

or $t^2 - 8t + 6t - 48 = 0$

or $t = 8, -6$ (not possible)

Thus, distance covered by second particle with respect to first particle in 8 s is

$$s_{12} = (v_{21})\, t$$

$$= (40 - 10)\,(8\text{ s})$$

$$= 30 \times 8 = 240 \text{ m}$$

Similarly, time taken by second particle to strike the ground is given by

$$-240 = 40t - \frac{1}{2} \times 10 \times t^2$$

or $-240 = 40t - 5t^2$

or $5t^2 - 40t - 240 = 0$

or $t^2 - 8t - 48 = 0$

$t^2 - 12t + 4t - 48 = 0$

or $t(t - 12) + 4(t - 12) = 0$

or $t = 12, -4$ (not possible)

Thus, after 8 s, magnitude of relative velocity will increase upto 12 s when second particle strikes the ground.

So, option (b) is correct.

26. Considering the initial position of ship A as origin, so the velocity and position of ship will be

$$\mathbf{v}_A = (30\,\hat{\mathbf{i}} + 50\,\hat{\mathbf{j}})$$

and $$\mathbf{r}_A = (0\,\hat{\mathbf{i}} + 0\,\hat{\mathbf{j}})$$

Now, as given in the question, velocity and position of ship B will be,

$$\mathbf{v}_B = -10\,\hat{\mathbf{i}}$$

and $$\mathbf{r}_B = (80\,\hat{\mathbf{i}} + 150\,\hat{\mathbf{j}})$$

Hence, the given situation can be represented graphically as

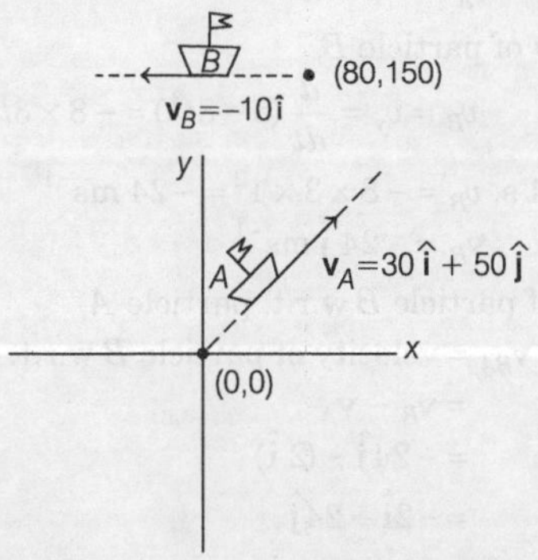

After time t, coordinates of ships A and B are $(80-10t, 150)$ and $(30t, 50t)$.

So, distance between A and B after time t is

$$d = \sqrt{(x_2 - x_1)^2 + (y_2 - y_1)^2}$$

$$d = \sqrt{(80 - 10t - 30t)^2 + (150 - 50t)^2}$$

$$\Rightarrow \quad d^2 = (80 - 40t)^2 + (150 - 50t)^2$$

Distance is minimum when $\frac{d}{dt}(d^2) = 0$

After differentiating, we get

$$\Rightarrow \quad \frac{d}{dt}[(80 - 40t)^2 + (150 - 50t)^2] = 0$$

$$\Rightarrow \quad 2(80 - 40t)(-40) + 2(150 - 50t)(-50) = 0$$

$$\Rightarrow \quad -3200 + 1600t - 7500 + 2500t = 0$$

$$\Rightarrow \quad 4100t = 10700$$

$$\Rightarrow \quad t = \frac{10700}{4100} = 2.6 \text{ h}$$

27. Distance travelled = Area under speed-time graph

Distance travelled (from $t = 0$ s to $t = 5$ s)

$$= \text{Area of } \Delta\, OAB$$

$$= \frac{1}{2} \times \text{Base} \times \text{Height}$$

$$= \frac{1}{2} \times 5 \times 8 = 20 \text{ m}$$

28. Given, displacement (x) and time (t) relation,

$$x^2 = at^2 + 2bt + c \quad \text{...(i)}$$

On differentiating Eq. (i) w.r.t. t, we get

$$\Rightarrow \quad 2x\frac{dx}{dt} = 2at + 2b$$

or $$xv = at + b \quad \text{...(ii)}$$

Differentiating again w.r.t. t, we have

$$x \cdot \frac{dv}{dt} + v \cdot \frac{dx}{dt} = a \quad \left(\because \frac{dx}{dt} = v \text{ and } \frac{dv}{dt} = A\right)$$

$$\Rightarrow \quad xA + v^2 = a$$

$$\Rightarrow \quad xA = a - v^2$$

$$\Rightarrow \quad xA = a - \left(\frac{at + b}{x}\right)^2 \quad \left[\text{from Eq. (ii), } v = \frac{at + b}{x}\right]$$

$$\Rightarrow \quad A = \frac{ac - b^2}{x^3}$$

$$\Rightarrow \quad A \propto x^{-3}$$

Thus, $$n = 3$$

29. The corresponding velocity time graph is

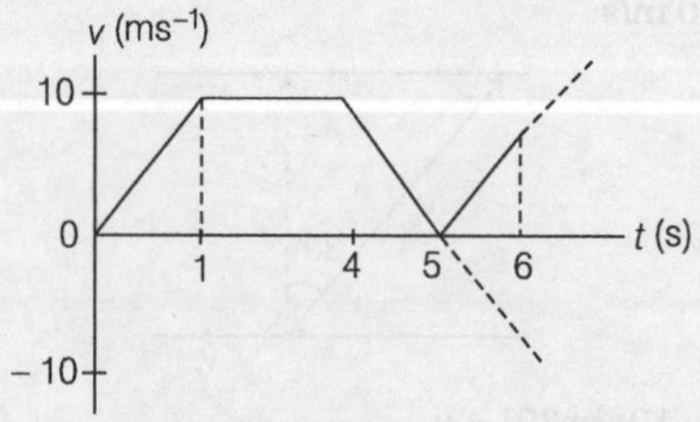

Distance travelled = area of velocity time graph

$$= \frac{1}{2} \times 10\,(3 + 5) + \frac{1}{2} \times 10 \times 1$$

$$= 40 + 5 = 45 \text{ m}$$

$$\therefore \quad \langle v \rangle = \frac{\text{Total distance}}{\text{Total time}}$$

$$= \frac{45}{6} = \frac{15}{2} \text{ ms}^{-1}$$

$$\therefore \quad n = 2$$

30. $\because \; v = t - 2$

or $$0 = t - 2$$

$$\therefore \quad t = 2 \text{ s}$$

$$\therefore \quad |v| = t - 2 \text{ for } t > 2 \text{ s}$$

and $$|v| = 2 - t \text{ for } 0 < t \le 2 \text{ s}$$

$$\therefore \quad s = \int_0^4 |v|\, dt = \int_0^2 (2 - t)\, dt + \int_2^4 (t - 2)\, dt$$

$$= \left[2t - \frac{t^2}{2}\right]_0^2 + \left[\frac{t^2}{2} - 2t\right]_2^4$$

$$= (4 - 2) + 2 = 4 \text{ m}$$

31. $\because \; a = t - x$

or $$\frac{da}{dt} = 1 - \frac{dx}{dt}$$

or $$\frac{da}{dt} = 1 - v$$

or $$\frac{d^2a}{dt^2} = -\frac{dv}{dt} = -a$$

The solution of this differential equation is

$$a = a_0 \sin(\omega t + \phi)$$

where, $\omega = 1$ unit

At $$t = 0, n = 0 \Rightarrow a = 0$$

$\therefore$ $$0 = a_0 \sin(\omega \times 0 + \phi)$$

$\therefore$ $$\phi = 0$$

$\therefore$ $$a = a_0 \sin t$$

or $$\frac{dv}{dt} = a_0 \sin t \quad \text{or} \quad \int_0^v dv = a_0 \int_0^t \sin t \, dt$$

or $$v = a_0[-\cos t]_0^t$$
$$= -a_0 [\cos t - 1]$$
$$v = a_0(1 - \cos t)$$
$$= 2\left(1 - \cos\frac{\pi}{3}\right)$$
$$= \left(2 \times \frac{1}{2}\right) = 1 \text{ ms}^{-1}$$

32. $v_p = 10$ m/s

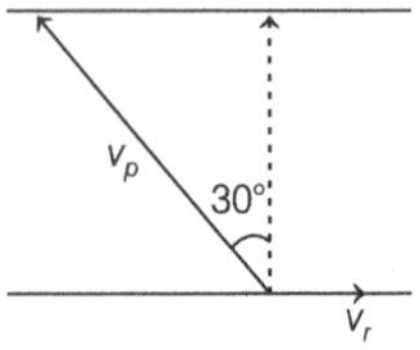

$\Rightarrow$ $$10 \sin 30° = v_r$$

$\Rightarrow$ $$v_r = 5 \text{ m/s}$$

$\therefore$ $$x = 5$$

33. Let particle A moves along X-axis such that

$$x = 10 + 8t - 3t^2$$

and let particle B moves over Y-axis such that

$$y = 5 - 8t^3$$

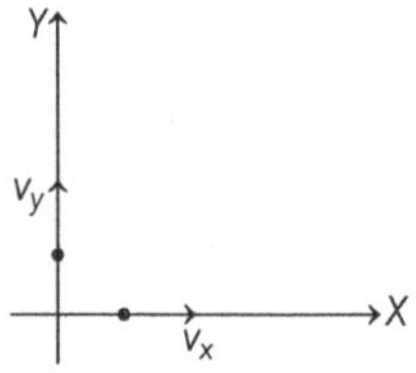

Velocities of particles A and B are

$$v_A = v_x = \frac{dx}{dt} = \frac{d}{dt}(10 + 8t - 3t^2)$$
$$= 8 - 6t$$

or v_A at $t = 1$ s,

$$v_A = 8 - 6 \times 1 = 2 \text{ ms}^{-1}$$

$\Rightarrow$ $$\mathbf{v}_A = +2\hat{\mathbf{i}} \text{ ms}^{-1}$$

Velocity of particle B,

$$v_B = v_y = \frac{d}{dt}(5 - 8t^3) = -8 \times 3t^2$$

At $t = 1$ s, $v_B = -8 \times 3 \times 1^2 = -24 \text{ ms}^{-1}$

$\Rightarrow$ $$\mathbf{v}_B = -24\hat{\mathbf{j}} \text{ ms}^{-1}$$

Speed of particle B w.r.t. particle A,

$$\mathbf{v}_{BA} = \text{velocity of particle } B \text{ w.r.t. } A$$
$$= \mathbf{v}_B - \mathbf{v}_A$$
$$= -24\hat{\mathbf{j}} - (2\hat{\mathbf{i}})$$
$$= -2\hat{\mathbf{i}} - 24\hat{\mathbf{j}}$$

$\Rightarrow$ Magnitude of velocity,

$$|\mathbf{v}_{BA}| = \sqrt{(2)^2 + (24)^2}$$

But given $|\mathbf{v}_{BA}| = \sqrt{v}$

So, $$\sqrt{v} = \sqrt{24^2 + 2^2}$$

or $$v = 580 \text{ ms}^{-1}$$

04

Kinematics II (Projectile Motion)

An object which is projected in the air with some velocity and makes a random angle with the horizontal surface is known *projectile*. The path followed by the projectile (called *trajectory*) is a parabola and the motion exhibited by it is called *projectile motion*. It is a two-dimensional motion.

IN THIS CHAPTER ....

- Equation of Trajectory
- Important Terms in Projectile Motion
- Special Cases of Projectile Motion
- Projectile Motion on an Inclined Plane

Equation of Trajectory

Let us consider a projectile launched, so that its initial velocity u makes an angle θ with the horizontal. In the following diagram, horizontal direction is taken as X-axis and vertical direction is taken as the Y-axis.

$$\mathbf{u} = u_x \hat{\mathbf{i}} + u_y \hat{\mathbf{j}}$$

$$\Rightarrow \quad \mathbf{u} = u \cos\theta \,\hat{\mathbf{i}} + u \sin\theta \,\hat{\mathbf{j}}$$

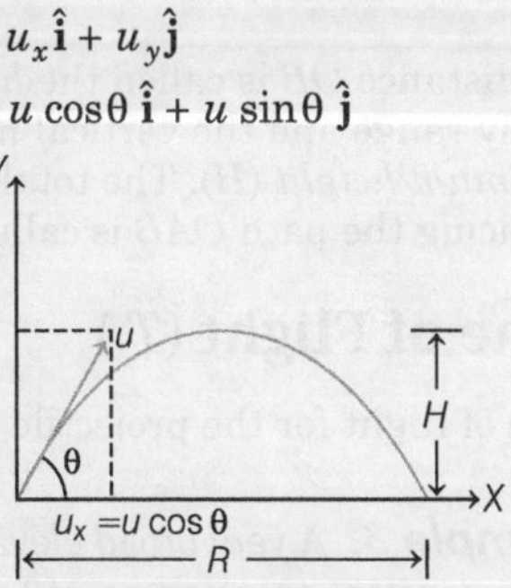

It can be seen that the X-axis is parallel to the horizontal, Y-axis is parallel to the vertical and u lies in the XY-plane. The constant acceleration a is given as

$$a = a_x \hat{\mathbf{i}} + a_y \hat{\mathbf{j}}$$

where, $a_x = 0$ (Since, there is no acceleration along X-axis)

$a_y = -g$ (Since, the acceleration is downwards and equal to g)

Velocity after time t can be given as

$$\mathbf{v}_t = v \cos\theta \hat{\mathbf{i}} + (u \sin\theta - gt) \hat{\mathbf{j}}$$

Speed of the projectile at any time t is $v = \sqrt{v_x^2 + v_y^2}$ and displacement at time t will be

$$= ut\cos\theta\hat{\mathbf{i}} + \left(ut\sin\theta - \frac{1}{2}gt^2\right)$$

The direction of **v** with the X-axis is given by $\theta = \tan^{-1}\left(\frac{v_y}{v_x}\right)$.

Coordinates of the projectile after time t is given by

$$\Rightarrow \quad x = u\cos\theta\, t \quad \text{...(i)}$$

$$\Rightarrow \quad y = u\sin\theta\, t - \frac{1}{2}gt^2 \quad \text{...(ii)}$$

From Eqs. (i) and (ii), eliminating t, we get

$$y = u\sin\theta\frac{x}{u\cos\theta} - \frac{1}{2}g\frac{x^2}{u^2\cos^2\theta}$$

$$\Rightarrow \quad y = x\tan\theta - \frac{gx^2}{2u^2\cos^2\theta} \quad \text{...(iii)}$$

The equation between x and y represents the path of the projectile known as **trajectory**. The Eq. (iii) shows that it is the equation of parabola of the form

$$y = bx + cx^2$$

where, $b = \tan\theta$ = constant and

$$c = \frac{g}{2u^2\cos^2\theta} = \text{constant.}$$

Example 1. *A particle moves according to the equation* $v = a\hat{\mathbf{i}} + b\hat{\mathbf{j}}$

At the initial point, $x = 0 = y$, *the radius of the trajectory is*

(a) $\frac{bx^2}{2a}$ (b) $\frac{2a}{bx^2}$ (c) $\frac{x^2}{a}$ (d) $\frac{a}{x^2}$

Sol. (a) Distance = speed × time

Given, $v = a\hat{\mathbf{i}} + b\hat{\mathbf{j}}$

At the initial point, $x = at, \frac{dy}{dt} = bx$

$$\Rightarrow \quad t = \frac{x}{a} \text{ or } dt = \frac{dx}{a}$$

$$\Rightarrow \quad dy = bx\,dt = bx\frac{dx}{a}$$

On integrating the above expression, using $\int x^n dx = \frac{x^{n+1}}{n+1}$

$$dy = \frac{b}{a}\int x\,dx$$

Thus, $$y = \frac{bx^2}{2a}$$

Example 2. *A biker stands on the edge of a cliff 490 m above the ground and throws a stone horizontally with an initial speed of 15 ms^{-1}. Neglecting air resistance, the speed with which the stone hits the ground is*

(a) 10 ms^{-1} (b) 99 ms^{-1} (c) 15 ms^{-1} (d) 30 ms^{-1}

Sol. (b) We choose the origin of the X-axis and Y-axis at the edge of the cliff. At $t = 0$ s, the stone is thrown. Consider the positive direction of X-axis to be along the initial velocity and the positive direction of Y-axis to be the vertically upward direction. The x and y- components of the motion can be treated independently. The equations of motion are

$$x = x_0 + u_x t$$

$$y = y_0 + u_y t + \frac{1}{2}a_y t^2$$

Here, $x_0 = y_0 = 0, u_y = 0, a_y = -g = -9.8 \text{ ms}^{-2}, u_x = 15 \text{ ms}^{-1}$

The stone hits the ground when, $y = -490$ m

$$-490 = -\frac{1}{2}(9.8)t^2$$

$$\Rightarrow \quad t = 10 \text{ s}$$

The velocity components are, $u_x = u_{0x}$ and $u_y = u_{0y} - gt$, so that when the stone hits the ground

$$u_x = 15 \text{ ms}^{-1}$$

$$u_y = 0 - 9.8 \times 10 = -98 \text{ ms}^{-1}$$

∴ The speed of the stone is

$$\sqrt{u_x^2 + u_y^2} = \sqrt{15^2 + 98^2} = 99 \text{ ms}^{-1}$$

Important Terms in Projectile Motion

Consider a body is projected from the point O with an initial velocity u at an angle θ with the horizontal and it hits the ground at same level at point B.

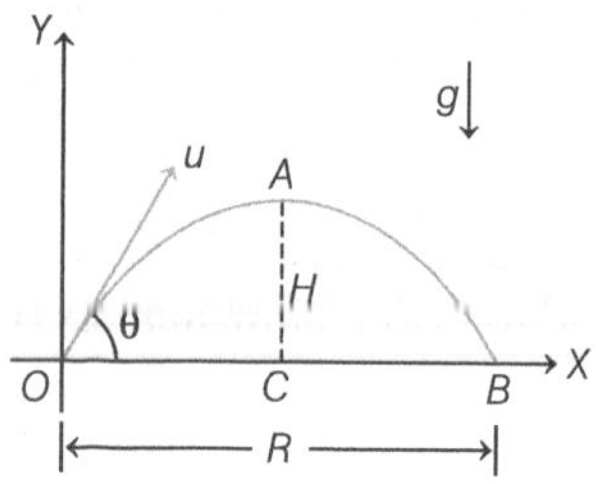

The distance OB is called the *horizontal range* (R) or simply range and the vertical height AC is known as the *maximum height* (H). The total time taken by the particle in tracing the path OAB is called the *time of flight* (T).

Time of Flight (T)

Time of flight for the projectile is given as $T = \frac{2u\sin\theta}{g}$

Example 3. *A very broad elevator is going up vertically with a constant acceleration of 2 ms^{-2}. At the instant, when its velocity is 4 ms^{-1}, a ball is projected from the floor of the lift with a speed of 4 ms^{-1} relative to the floor at an elevation of 30°. If g = 10 ms^{-2}, then what is the time taken by the ball to return to the floor?*

(a) $\frac{1}{2}$ s (b) $\frac{1}{3}$ s

(c) $\frac{1}{4}$ s (d) 1 s

Sol. (b) Here, $u = 4\ ms^{-1}, \theta = 30°$

Acceleration of the ball relative to the lift $= 10 + 2 = 12\ ms^{-2}$ acting in the negative *y*-direction or vertically downwards. It means, here $g' = 12\ ms^{-2}$.

Time of flight, $T = \dfrac{2u\sin\theta}{g'} = \dfrac{2\times 4\times\sin 30°}{12} = \dfrac{1}{3}$ s

Horizontal Range (*R*)

The horizontal range *R* is the horizontal distance covered by the projectile is given as,

$$\Rightarrow \quad R = \frac{u^2\sin 2\theta}{g}$$

Range will be maximum, if

$$\sin 2\theta = 1$$
$$\sin 2\theta = \sin 90°$$
$$2\theta = 90°$$

or $\theta = 45°$

$$R_{max} = \frac{u^2}{g} \quad (\text{at } \theta = 45°)$$

Example 4. *A shell is fired from a fixed artillery gun with an initial speed u such that it hits the target on the ground at a distance R from it. If t_1 and t_2 are the values of the time taken by it to hit the target in two possible ways, the product t_1t_2 is* **[JEE Main 2019]**

(a) $\dfrac{R}{4g}$ (b) $\dfrac{R}{g}$

(c) $\dfrac{R}{2g}$ (d) $\dfrac{2R}{g}$

Sol. (d)

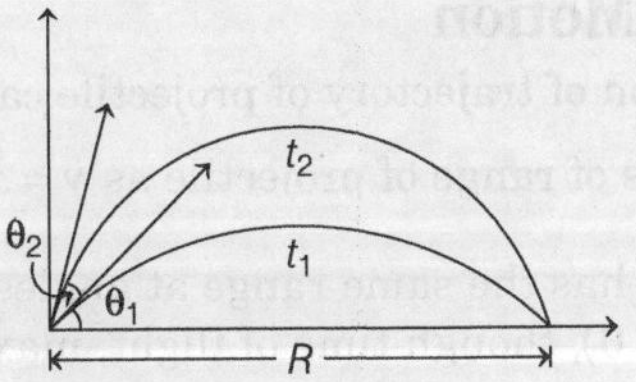

Given, range of the fired shell, $R = R$ and time of flights are t_1 and t_2.

Let θ_1 and θ_2 be the two angles at which shell is fired. As, range in both cases is same, *i.e.*

$$R_1 = R_2 = R$$

Here, $R_1 = \dfrac{u^2\sin 2\theta_1}{g}$

and $R_2 = \dfrac{u^2\sin 2\theta_2}{g}$

$$\Rightarrow \quad R = \frac{u^2\sin 2\theta_1}{g} = \frac{u^2\sin 2\theta_2}{g} \quad \text{...(i)}$$

$$\Rightarrow \quad \sin 2\theta_1 = \sin 2\theta_2$$

$$\Rightarrow \quad \sin 2\theta_1 = \sin(180° - 2\theta_2) \quad [\because \sin(180° - \theta) = \sin\theta]$$

$$\Rightarrow \quad 2(\theta_1 + \theta_2) = 180° \text{ or } \theta_1 + \theta_2 = 90°$$

$$\Rightarrow \quad \theta_2 = 90° - \theta_1 \quad \text{...(ii)}$$

So, time of flight in first case,

$$t_1 = \frac{2u\sin\theta_1}{g} \quad \text{...(iii)}$$

and time of flight in second case,

$$t_2 = \frac{2u\sin\theta_2}{g} = \frac{2u\sin(90° - \theta_1)}{g} = \frac{2u\cos\theta_1}{g} \quad \text{...(iv)}$$

From Eqs. (iii) and (iv), we get

$$t_1t_2 = \frac{2u\sin\theta_1}{g}\times\frac{2u\cos\theta_1}{g}$$

$$\Rightarrow \quad t_1t_2 = \frac{4u^2\sin\theta_1\cos\theta_1}{g^2}$$

$$\Rightarrow \quad t_1t_2 = \frac{2u^2\sin 2\theta_1}{g^2} \quad (\because \sin 2\theta = 2\sin\theta\cos\theta) \text{ ...(v)}$$

From Eq. (i), we get

$$\therefore \quad t_1t_2 = \frac{2R}{g} \quad \left(\because R = \frac{u^2\sin 2\theta_1}{g}\right)$$

Example 5. *A body is projected at $t = 0$ with a velocity $10\ ms^{-1}$ at an angle of $60°$ with the horizontal. The radius of curvature of its trajectory at $t = 1$ s is R. Neglecting air resistance and taking acceleration due to gravity $g = 10\ ms^{-2}$, the value of R is* **[JEE Main 2019]**

(a) 10.3 m (b) 2.8 m

(c) 5.1 m (d) 2.5 m

Sol. (b) Components of velocity at an instant of time *t* of a body projected at an angle θ is

$$v_x = u\cos\theta + g_xt \quad \text{and} \quad v_y = u\sin\theta + g_yt$$

Here, components of velocity at $t = 1$ s, is

$$v_x = u\cos 60° + 0 \quad [\text{as } g_x = 0]$$
$$= 10\times\frac{1}{2} = 5 \text{ m/s}$$

and $v_y = u\sin 60° + (-10)\times(1)$

$$= 10\times\frac{\sqrt{3}}{2} + (-10)\times(1)$$
$$= 5\sqrt{3} - 10$$

$$\Rightarrow \quad |v_y| = |10 - 5\sqrt{3}| \text{ m/s}$$

Now, angle made by the velocity vector at time of $t = 1$ s

$$|\tan\alpha| = \left|\frac{v_y}{v_x}\right| = \frac{|10 - 5\sqrt{3}|}{5}$$

$$\Rightarrow \quad \tan\alpha = |2 - \sqrt{3}|$$

or $\alpha = 15°$

$\therefore$ Radius of curvature of the trajectory of the projected body,

$$R = v^2 / g\cos\alpha$$
$$= \frac{(5)^2 + (10 - 5\sqrt{3})^2}{10\times 0.97}$$
$$[\because v^2 = v_x^2 + v_y^2 \text{ and } \cos 15° = 0.97]$$

$$\Rightarrow \quad R = 2.77 \text{ m} \approx 2.8 \text{ m}$$

Maximum Height (H)

Maximum height attained by the projectile is given as,

$$H = \frac{u^2 \sin^2 \theta}{2g}$$

Example 6. *Two stones having different masses m_1 and m_2 are projected at angles θ and $(90° - \theta)$ with same velocity from the same point. The ratio of their maximum heights is*

(a) 1 : 1 (b) 1 : $\tan\theta$ (c) $\tan\theta$: 1 (d) $\tan^2\theta$: 1

Sol. (d) Maximum height, $H_1 = \frac{u^2 \sin^2 \theta}{2g}$ ...(i)

and $H_2 = \frac{u^2 \sin^2(90° - \theta)}{2g}$ ...(ii)

From Eqs. (i) and (ii), we get

$$\frac{H_1}{H_2} = \frac{\sin^2 \theta}{\sin^2(90° - \theta)}$$

$$\frac{H_1}{H_2} = \frac{\tan^2 \theta}{1}$$

$\therefore$ $H_1 : H_2 = \tan^2\theta : 1$

Example 7. *Two particles are projected from the same point with the same speed u such that they have the same range R, but different maximum heights h_1 and h_2. Which of the following is correct?* **[JEE Main 2019]**

(a) $R^2 = 4h_1h_2$ (b) $R^2 = 16h_1h_2$ (c) $R^2 = 2h_1h_2$ (d) $R^2 = h_1h_2$

Sol. (b)

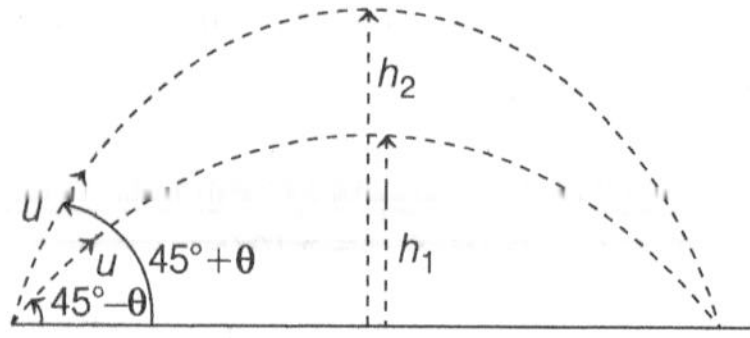

As maximum range occurs at $\theta = 45°$ for a given initial projection speed, we take angles of projection of two particles as

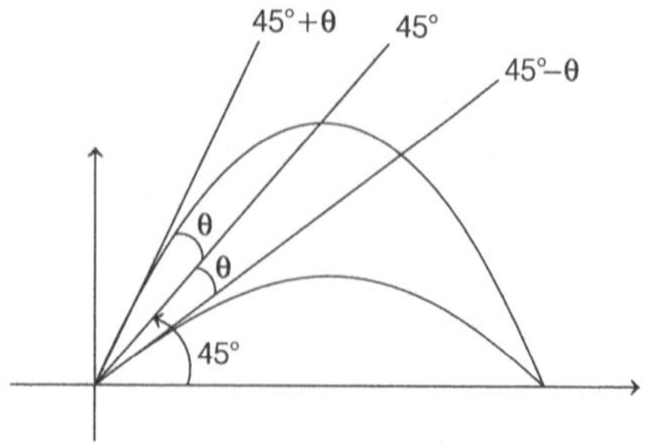

$$\theta_1 = 45° + \theta, \theta_2 = 45° - \theta$$

where, θ is angle of projectiles with 45° line. So, range of projectiles will be

$$R = R_1 = R_2 = \frac{u^2 \sin 2(\theta_1)}{g}$$

$\Rightarrow$ $R = \frac{u^2 \sin 2(45° + \theta)}{g}$

$\Rightarrow$ $R = \frac{u^2 \sin(90° + 2\theta)}{g}$

$\Rightarrow$ $R = \frac{u^2 \cos 2\theta}{g}$

$\Rightarrow$ $R^2 = \frac{u^4 \cos^2 2\theta}{g^2}$ ...(i)

Maximum heights achieved in two cases are

$$h_1 = \frac{u^2 \sin^2(45° + \theta)}{2g}$$

and $h_2 = \frac{u^2 \sin^2(45° - \theta)}{2g}$

So, $h_1h_2 = \frac{u^4 \sin^2(45° + \theta) \sin^2(45° - \theta)}{4g^2}$

Using $2 \sin A \cdot \sin B = \cos(A - B) - \cos(A + B)$, we have

$$\sin(45° + \theta) \sin(45° - \theta) = \frac{1}{2}(\cos 2\theta - \cos 90°)$$

$\Rightarrow \sin(45° + \theta) \sin(45° - \theta) = \frac{\cos 2\theta}{2}$ [$\because \cos 90° = 0$]

So, we have

$$h_1h_2 = \frac{u^4\left(\frac{\cos 2\theta}{2}\right)^2}{4g^2} \Rightarrow h_1h_2 = \frac{u^4 \cos^2 2\theta}{16g^2}$$ ...(ii)

From Eqs. (i) and (ii), we get

$\Rightarrow$ $h_1h_2 = \frac{R^2}{16} \Rightarrow R^2 = 16\, h_1h_2$

Important Points Regarding Projectile Motion

- The equation of trajectory of projectile can be written in the terms of range of projectile as $y = x\left(1 - \frac{x}{R}\right)\tan\theta$.
- A projectile has the same range at angles of projection θ and $(90° - \theta)$, though time of flight, maximum height and trajectories are different. This is also true for a range of projectile for $\theta_1 = (45° - \alpha)$ and $\theta_2 = (45° + \alpha)$ and is equal to $\frac{u^2 \cos 2\alpha}{g}$.
- There are two unique times at which the projectile is at the same height $h\,(< H)$ and the sum of these two times equals the time of flight T. Since, $h = (u \sin\theta)t - \frac{1}{2}gt^2$ is quadratic in time, so it has two unique roots t_1 and t_2 (say) such that sum of roots $(t_1 + t_2)$ is $\frac{2u \sin\theta}{g}$ and product (t_1t_2) is $\frac{2h}{g}$. The time lapse $(t_1 - t_2)$ between these two events is $(t_1 - t_2)^2 = (t_1 + t_2)^2 - 4t_1t_2$

$$t_1 - t_2 = \sqrt{\frac{4u^2 \sin^2\theta}{g^2} - \frac{8h}{g}}$$

- In case of projectile motion, range R is n times the maximum height H.

i.e. $R = nH$

Then, $\dfrac{u^2 \sin 2\theta}{g} = n \dfrac{u^2 \sin^2 \theta}{2g}$

i.e. $\tan\theta = \dfrac{4}{n}$ or $\theta = \tan^{-1}\left(\dfrac{4}{n}\right)$

- If air resists the projectile motion, then
  - (i) Time taken by projectile during upward motion < Time taken during downward motion.
  - (ii) The values of height attained and range of a projectile decrease both.
  - (iii) The projectile returns to the ground with less speed. At its trajectory, its horizontal velocity also decreases.
  - (iv) Time of flight also decreases.
  - (v) The angle which the projectile makes with the ground, increases.
- If K' is the kinetic energy at the point of launch, then kinetic energy at the highest point is

$$K' = \frac{1}{2} mv_x^2 = \frac{1}{2} mu^2 \cos^2\theta \Rightarrow K' = K \cos^2\theta$$

- For complementary angles ϕ and $90° - \phi$, if T_ϕ and $T_{90°-\phi}$ are the times of flight and R is the range, then

$$T_\phi T_{90°-\phi} = \frac{2R_\phi}{g} = \frac{2R_{90°-\phi}}{g} = \frac{2R}{g}$$

e.g. $T_{1°} T_{89°} = \dfrac{2R_{1°}}{g} = \dfrac{2R_{89°}}{g}$

- The velocity of the projectile is minimum at the highest point ($= u\cos\theta$) and is maximum at the point of projection or at the point of striking the ground.
- At the maximum point of projectile motion, the velocity is not zero, but is $u\cos\theta$ which acts in the horizontal direction. The angle between velocity and acceleration varies from $0° < \theta < 180°$.
- Path of a projectile w.r.t. other projectile is a straight line.
- In oblique projection of a projectile, the following physical quantities remains constant during motion.
  - (i) horizontal component of velocity ($u\cos\theta$)
  - (ii) acceleration due to gravity (g)
  - (iii) total energy of the projectile.

  The following physical quantities which change during the motion are
  - (i) speed and velocity
  - (ii) direction of motion
  - (iii) linear momentum
  - (iv) kinetic energy and potential energy
  - (v) angle between direction of motion and acceleration due to gravity. (It changes from $90° + \theta$ to $90° - \theta$).

Special Cases of Projectile Motion

Projection from a Height

Case I Projectile Projected in Horizontal Direction

Let us consider that a projectile is projected with a velocity **u**. The following observations are taken from point O at a height h from the ground.

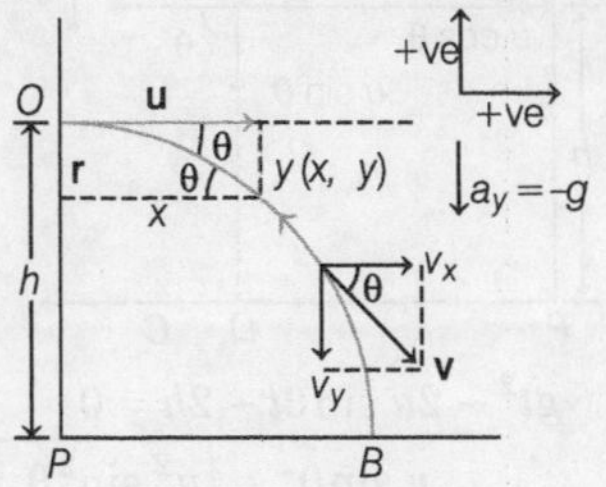

Here, $u_x = u$,
$u_y = 0,\ a_x = 0$
$a_y = -g$

(a) Horizontal motion, $x = ut$ ...(i)

Vertical motion, $-h = 0(t) - \dfrac{1}{2}gt^2$ ...(ii)

or $t = \sqrt{\dfrac{2h}{g}}$

(b) Horizontal range, $(R) = u \times t = u\sqrt{2h/g}$

(c) Let at time t, the coordinates of the position of projectile be (x, y), then

$$x = ut \text{ and } y = 0 - \frac{1}{2}gt^2$$

Therefore, at time t, position vector

$$\mathbf{r} = x\hat{\mathbf{i}} + y\hat{\mathbf{j}} = ut\hat{\mathbf{i}} - \frac{1}{2}gt^2\hat{\mathbf{j}}$$

$$|\mathbf{r}| = \sqrt{x^2 + y^2} = \sqrt{(ut)^2 + \left(-\frac{1}{2}gt^2\right)^2}$$

and $\tan\theta = y/x$

(d) Let at time t, the horizontal and vertical velocities of projectile be v_x and v_y, respectively.

Hence, $v_x = u$

and $v_y = 0 + (-gt) = -gt$

$$\mathbf{r} = v_x\hat{\mathbf{i}} + v_y\hat{\mathbf{j}} = u\hat{\mathbf{i}} + (-gt)\hat{\mathbf{j}}$$

and $v = \sqrt{v_x^2 + v_y^2} = \sqrt{u^2 + (-gt)^2}$

and $\tan\theta = \dfrac{v_y}{v_x}$

Case II Projectile Projected at an Angle θ above Horizontal

$$u_x = u\cos\theta,\ a_x = 0$$
$$u_y = u\sin\theta,\ a_y = -g$$

From equation of horizontal motion,

$$x = u\cos\theta t \quad \text{...(i)}$$

Equation of vertical motion will be

$$-h = u\sin\theta t - \frac{1}{2}gt^2 \quad \text{...(ii)}$$

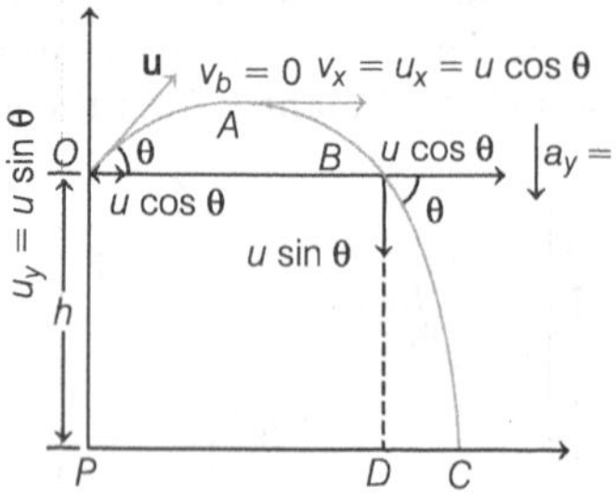

$$gt^2 - 2u\sin\theta t - 2h = 0 \quad \text{...(iii)}$$

or $$t = \frac{u\sin\theta}{g} \pm \sqrt{\frac{u^2\sin^2\theta}{g^2} \times \frac{2h}{g}}$$

Horizontal distance covered in time T,

$$PC = (u\cos\theta)T$$

and horizontal distance covered during this time,

$$OB = \frac{u^2\sin 2\theta}{g}$$

In such case for range PC to become maximum, θ should be 45°.

***Case* III Projection at an Angle θ below Horizontal**

$$u_x = u\cos\theta, a_x = 0$$

$$u_y = -u\sin\theta, a_y = -g$$

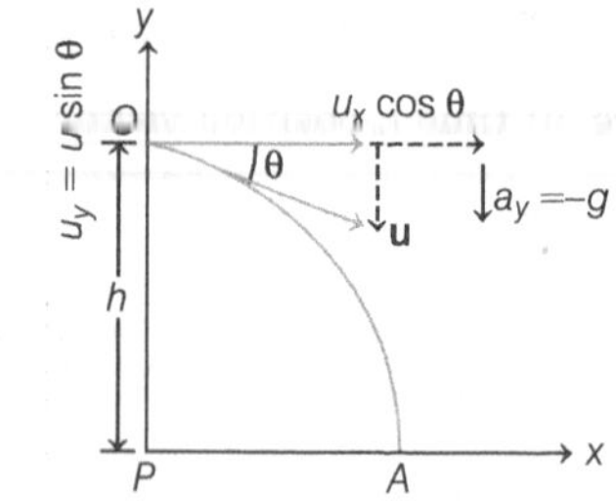

From equation of motion,

$$s = ut + \frac{1}{2}at^2$$

$$-h = (-u\sin\theta)t + \frac{1}{2}(-g)t^2$$

or $$gt^2 + (2u\sin\theta)t - 2h = 0$$

On solving this equation, value of t can be obtained

$$t = \frac{-2u\sin\theta}{2g} \pm \frac{\sqrt{4u^2\sin^2\theta + 8gh}}{2g}$$

Neglect –ve root of t.

In this time, the horizontal distance covered on the Earth

$$PA = (u\cos\theta)t$$

Example 8. *A boy playing on the roof of a 10 m high building throws a ball with a speed of 10 m/s at an angle of 30° with the horizontal. How far from the throwing point will the ball be at the height of 10 m from the ground?*

[$g = 10$ m/s^2, $\sin 30° = \frac{1}{2}$, $\cos 30° = \sqrt{3}/2$]

(a) 5.20 m (b) 4.33 m (c) 2.60 m (d) 8.66 m

Sol. (d) The ball will be at point P when it is at a height of 10 m from the ground. So, we have to find the distance OP, which can be calculated directly by considering it as a projectile on a levelled plane (OX).

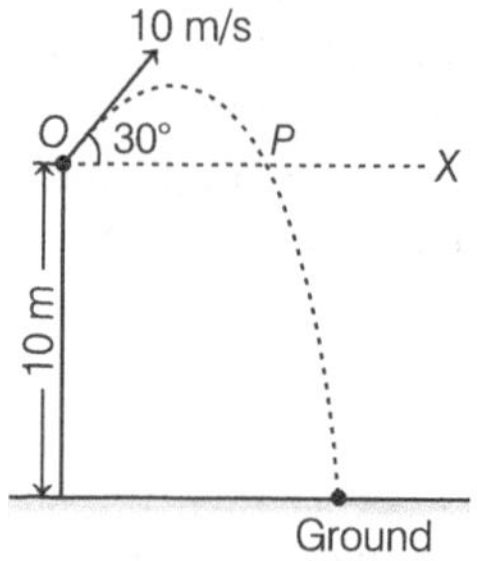

$$OP = R = \frac{u^2\sin 2\theta}{g} = \frac{10^2 \times \sin(2 \times 30°)}{10}$$

$$= \frac{10\sqrt{3}}{2} = 5\sqrt{3} = 8.66 \text{ m}$$

Projectile Motion on an Inclined Plane

When a projectile is projected from an inclined plane, we consider two axes x and y, *i.e.* along and perpendicular to the inclined plane. Different cases of projectile motion on an inclined plane are shown below

***Case* I Motion up the Plane** In xy-plane,

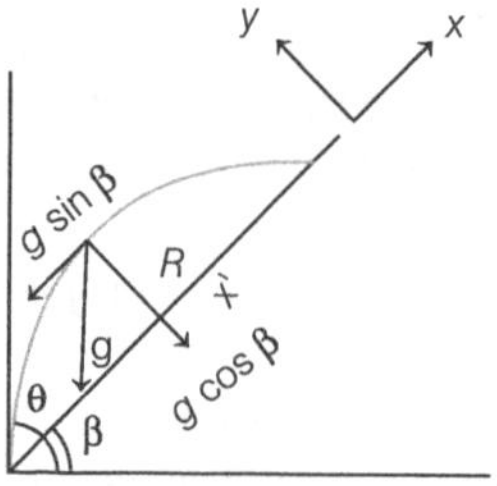

$$u_x = u\cos(\theta - \beta)$$

$$u_y = u\sin(\theta - \beta)$$

$$a_x = -g\sin\beta,$$

$$a_y = -g\cos\beta$$

Time of flight,

$\Rightarrow$ $$T = \frac{2u\sin(\theta - \beta)}{g\cos\beta}$$

Range of inclined plane,

$$R = \frac{2u^2\cos\theta\sin(\theta - \beta)}{g\cos^2\beta}$$

Maximum range on inclined plane,

$$R_{\max} = \frac{u^2}{g(1+\sin\beta)}$$

***Case* II Motion down the Plane** Let the particle be thrown with a velocity u at an angle θ with the horizontal as shown in the figure.

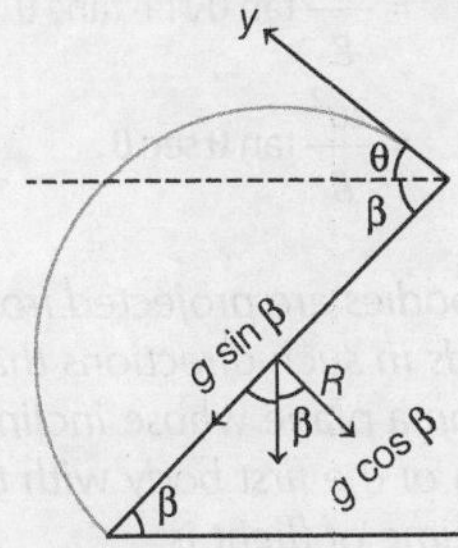

Time of flight (T) and Range (R) on inclined plane is given as,

$$\Rightarrow \quad T = \frac{2\,u\sin(\theta+\beta)}{g\cos\beta}$$

$$R = \frac{u^2}{g}\left[\frac{\sin(2\theta+\beta)+\sin\beta}{1-\sin^2\beta}\right]$$

Since, θ is variable and the maximum value of sine function is 1.

For R to be maximum, $\sin(2\theta+\beta) = 1$

and $\quad R_{\max} = \frac{u^2}{g}\left(\frac{1+\sin\beta}{1-\sin^2\beta}\right)$

$$= \frac{u^2}{g(1-\sin\beta)} \text{ down the plane}$$

Example 9. *A plane is inclined at an angle $\alpha = 30°$ with respect to the horizontal. A particle is projected with a speed $u = 2\ ms^{-1}$, from the base of the plane, making an angle $\theta = 15°$ with respect to the plane as shown in the following figure. The distance from the base, at which the particle hits the plane is close to [Take, $g = 10\ ms^{-2}$]*

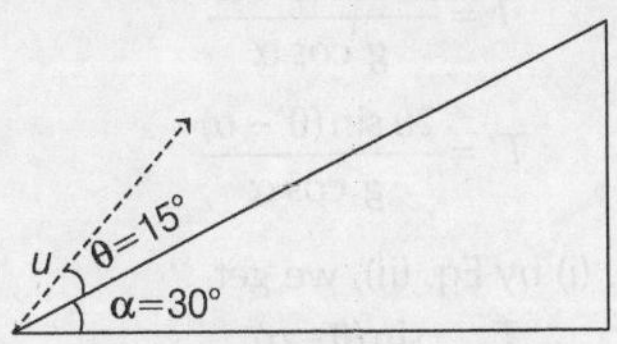

(a) 26 cm (b) 20 cm
(c) 18 cm (d) 14 cm

Sol. (b) When a projectile is projected at an angle θ with an inclined plane making angle α with the horizontal, then

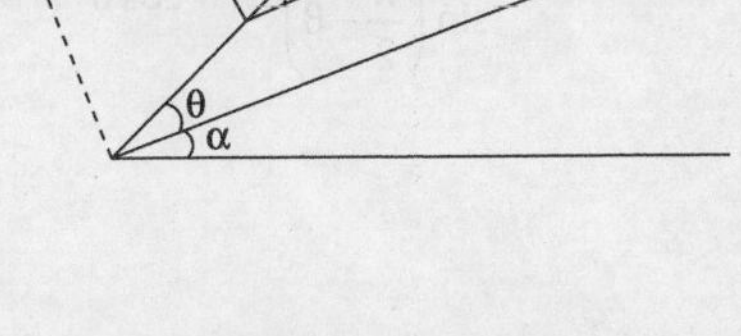

Components of u along and perpendicular to plane are

$$u_x = u\cos\theta \text{ and } u_y = u\sin\theta$$

We can also resolve acceleration due to gravity into its components along and perpendicular to plane as shown below.

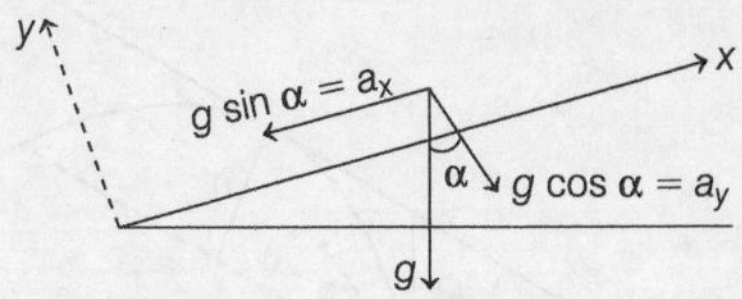

So, we can now apply formula for range, *i.e.* net horizontal displacement of the particle as

$$R = u_xT + \frac{1}{2}a_xT^2 \quad \text{...(i)}$$

where, T = time of flight.

Using formula for time of flight, we have

$$T = \frac{2u_y}{a_y} = \frac{2u\sin\theta}{g\cos\alpha} \quad \text{...(ii)}$$

From Eqs. (i) and (ii), we have

Range up the inclined plane is

$$R = u_xT + \frac{1}{2}a_xT^2 = u\cos\theta\left(\frac{2u\sin\theta}{g\cos\alpha}\right) - \frac{1}{2}g\sin\alpha\left(\frac{2u\sin\theta}{g\cos\alpha}\right)^2$$

Here, $u = 2\ ms^{-1}$, $g = 10\ ms^{-2}$, $\theta = 15°$, $\alpha = 30°$

So, $\quad T = \frac{2u\sin\theta}{g\cos\alpha} = \frac{2\times2\sin15°}{10\times\cos30°}$

$$= \frac{2\times2\times0.258\times2}{10\times1.732} = 0.1191$$

Now, $R = 2\times\cos15°\times0.1191 - \frac{1}{2}\times10\sin30°\,(0.1191)^2$

$$= 2\times0.965\times0.1191 - \frac{5}{2}(0.1191)^2$$

$$= 0.229 - 0.0354 = 0.1936 \text{ m}$$

$$\approx 0.20 \text{ m} = 20 \text{ cm}$$

Example 10. *A cannon fired from under a shelter inclined at an angle α to the horizontal. The cannon is at point A at distant R from the base (B) of the shelter. The initial velocity of the cannon is v_0 and its trajectory lies in the plane. The maximum range R_{max} of the shell is*

(a) $\frac{v_0^2}{g}\sin2\alpha$

(b) $\frac{g}{v_0^2}\sin2(\phi-\alpha)$

(c) $\frac{v_0^2}{g}\sin2\left(\alpha+\sin^{-1}\frac{\sqrt{gR\sin2\alpha}}{v_0}\right)$

(d) $\frac{v_0^2}{2g}\sin2\left(\alpha+\sin^{-1}\frac{\sqrt{R\sin2\alpha}}{g}\right)$

Sol. (c) For h' to be maximum,

$$h' = R\sin\alpha = \frac{v_0^2\sin^2(\phi-\alpha)}{2g\cos\alpha}$$

$$gR \sin 2\alpha = v_0^2 \sin^2(\phi - \alpha)$$

$$\Rightarrow \qquad \phi = \alpha + \sin^{-1} \frac{\sqrt{gR \sin 2\alpha}}{g}$$

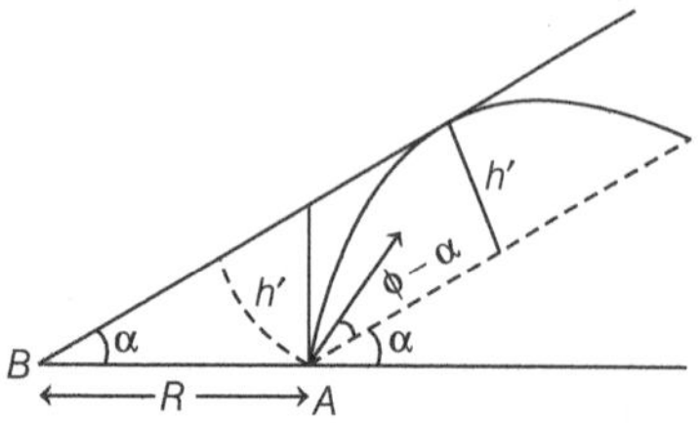

Range, $R_{max} = \dfrac{v_0^2 \sin 2\phi}{g}$

$$= \frac{v_0^2}{g} \sin 2\left(\alpha + \sin^{-1} \frac{\sqrt{gR \sin 2\alpha}}{v_0}\right)$$

Example 11. *A particle is projected horizontally with a speed u from the top of a plane inclined at an angle θ with the horizontal. How far from the point of projection will the particle strike the plane?*

(a) $\dfrac{2u^2}{g} \tan\theta \sec\theta$ (b) $\dfrac{2u}{g} \tan^2\theta \sec\theta$

(c) $\dfrac{2u^2}{g} \tan\theta \cos\theta$ (d) $\dfrac{2u}{g} \tan\theta \cos^2\theta$

Sol. (a) Take X and Y-axes as shown in figure below. Suppose that the particle strikes the plane at a point P with coordinates (x, y). Consider the motion between A and P.

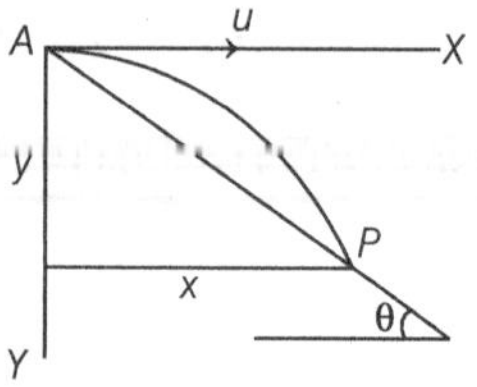

Motion in x-direction,

Initial velocity $= u$

Acceleration $= 0$

$$x = ut \qquad \text{...(i)}$$

Motion in y- direction,

Initial velocity $= 0$

Acceleration $= g$

$$y = \frac{1}{2} gt^2 \qquad \text{...(ii)}$$

Eliminating t from Eqs. (i) and (ii), we get

$$y = \frac{1}{2} g \frac{x^2}{u^2}$$

Also, $\qquad y = x \tan\theta$

Thus, $\dfrac{g x^2}{2u^2} = x \tan\theta$ giving $x = 0$ or $\dfrac{2u^2 \tan\theta}{g}$

Clearly, the point P corresponds to $x = \dfrac{2u^2 \tan\theta}{g}$, then

$$y = x \tan\theta - \frac{2u^2 \tan^2\theta}{g}$$

The distance, $AP = l = \sqrt{x^2 + y^2}$

$$= \frac{2u^2}{g} \tan\theta \sqrt{1 + \tan^2\theta}$$

$$= \frac{2u^2}{g} \tan\theta \sec\theta$$

Example 12. *Two bodies are projected from the same point with equal speeds in such directions that they both strike the same point on a plane whose inclination is α. If θ be the angle of projection of the first body with the horizontal, then the ratio of their time of flight is*

(a) $\dfrac{\cos(\theta - \alpha)}{\cos\beta}$ (b) $\dfrac{\sin(\theta - \alpha)}{\cos\theta}$

(c) $\dfrac{\sin(\theta - \alpha)}{\sin\beta}$ (d) $\dfrac{\cos(\theta - \alpha)}{\sin\theta}$

Sol. (b) Range, $R = \dfrac{u^2}{g \cos^2\theta}[\sin(2\theta - \alpha) - \sin\alpha]$

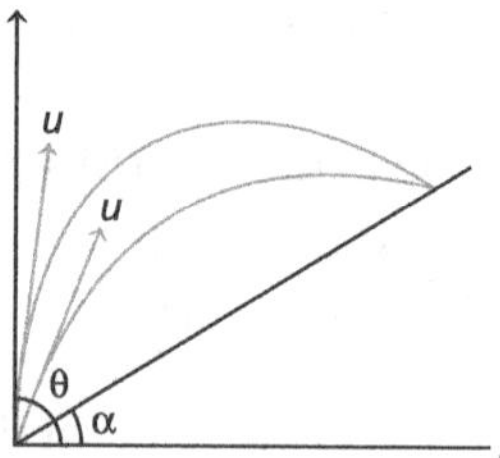

Range of both the bodies is same

Therefore,

$$\sin(2\theta - \alpha) = \sin(2\theta' - \alpha)$$

$$2\theta' - \alpha = \pi - (2\theta - \alpha)$$

$$\theta' = \frac{\pi}{2} - (\theta - \alpha)$$

Now, $\qquad T = \dfrac{2u \sin(\theta - \alpha)}{g \cos\alpha} \qquad \text{...(i)}$

and $\qquad T' = \dfrac{2u \sin(\theta' - \alpha)}{g \cos\alpha} \qquad \text{...(ii)}$

Dividing Eq. (i) by Eq. (ii), we get

$$\frac{T}{T'} = \frac{\sin(\theta - \alpha)}{\sin(\theta' - \alpha)}$$

$$= \frac{\sin(\theta - \alpha)}{\sin\left\{\frac{\pi}{2} - (\theta - \alpha) - \alpha\right\}}$$

$$= \frac{\sin(\theta - \alpha)}{\sin\left(\frac{\pi}{2} - \theta\right)} = \frac{\sin(\theta - \alpha)}{\cos\theta}$$

Practice Exercise

ROUND I Topically Divided Problems

Equation of Trajectory

1. The height y and distance x along the horizontal for a body projected in the xy-plane are given by $y = 8t - 5t^2$ and $x = 6t$. The initial speed of projection is

(a) 8 m/s (b) 9 m/s
(c) 10 m/s (d) (10/3) m/s

2. A particle moves in the xy-plane with velocity $v_x = 8t - 2$ and $v_y = 2$. If it passes through the point $x = 14$ and $y = 4$ at $t = 2$ s, find the equation (x-y relation) of the path.

(a) $x = y^2 - y + 2$
(b) $x = 2y^2 + 2y - 3$
(c) $x = 3y^2 + 5$
(d) Cannot be found from above data

3. A particle slides down a frictionless parabolic ($y = x^2$) track ($A - B - C$) starting from rest at point A. Point B is at the vertex of parabola and point C is at a height less than that of point A. After C, the particle moves freely in air as a projectile. If the particle reaches highest point at P, then [NCERT Exemplar]

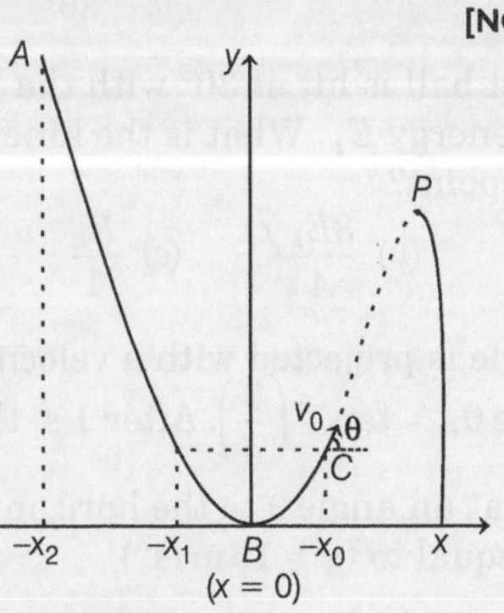

(a) KE at P = KE at B
(b) height at P = height at A
(c) total energy at P = total energy at A
(d) time of travel from A to B = time of travel from B to P.

4. A projectile A is thrown at an angle of 30° to the horizontal from point P. At the same time, another projectile B is thrown with velocity v^2 upwards from the point Q vertically below the highest point. For B to collide with A, $\frac{v_2}{v_1}$ should be

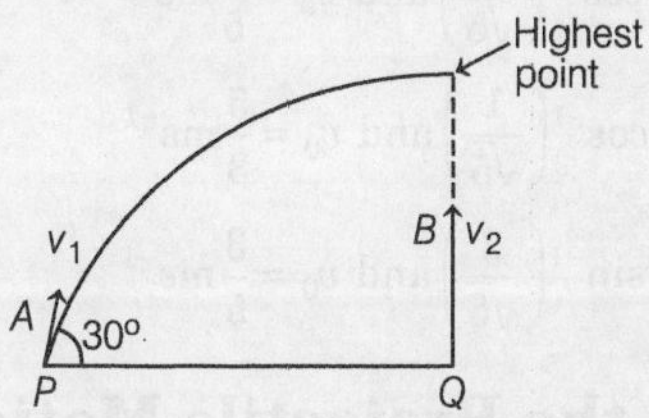

(a) 1 (b) 2
(c) $\frac{1}{2}$ (d) 4

5. A projectile is fired with a velocity v at an angle θ with the horizontal. The speed of the projectile when its direction of motion makes an angle β with the horizontal is

(a) $v \cos\theta$ (b) $v \cos\theta \cos\beta$
(c) $v \cos\theta \sec\beta$ (d) $v \cos\theta \tan\beta$

6. A ball is projected with velocity u at an angle α with horizontal plane. Its speed when it makes an angle β with the horizontal is

(a) $u \cos\alpha$ (b) $\frac{u}{\cos\beta}$
(c) $u \cos\alpha \cos\beta$ (d) $\frac{u \cos\alpha}{\cos\beta}$

7. A projectile is thrown with a velocity of 10 m/s at an angle 60° with horizontal. The interval between the moment when speed is $\sqrt{5g}$ m/s, is ($g = 10\,\text{m/s}^2$)

(a) 1 s (b) 3 s
(c) 2 s (d) 4 s

8. A body of mass m is thrown upward at an angle θ with the horizontal with velocity v. While rising up, the velocity of the mass after t second will be

(a) $\sqrt{(v\cos\theta)^2 + (v\sin\theta)^2}$
(b) $\sqrt{(v\cos\theta - v\sin\theta)^2 - gt}$
(c) $\sqrt{v^2 + g^2t^2 - (2v\sin\theta)\,gt}$
(d) $\sqrt{v^2 + g^2t^2 - (2v\cos\theta)\,gt}$

9. The equation of motion of a projectile are given by $x = 36t$ and $2y = 96t - 9.8t^2$ m. The angle of projectile will be

(a) $\sin^{-1}\left(\frac{4}{5}\right)$ (b) $\sin^{-1}\left(\frac{3}{5}\right)$

(c) $\sin^{-1}\left(\frac{4}{3}\right)$ (d) $\sin^{-1}\left(\frac{3}{4}\right)$

10. The trajectory of a projectile near the surface of the earth is given as $y = 2x - 9x^2$. If it were launched at an angle θ_0 with speed v_0, then (Take, $g = 10\text{ ms}^{-2}$) **[JEE Main 2019]**

(a) $\theta_0 = \sin^{-1}\left(\frac{1}{\sqrt{5}}\right)$ and $v_0 = \frac{5}{3}\text{ms}^{-1}$

(b) $\theta_0 = \cos^{-1}\left(\frac{2}{\sqrt{5}}\right)$ and $v_0 = \frac{3}{5}\text{ms}^{-1}$

(c) $\theta_0 = \cos^{-1}\left(\frac{1}{\sqrt{5}}\right)$ and $v_0 = \frac{5}{3}\text{ms}^{-1}$

(d) $\theta_0 = \sin^{-1}\left(\frac{2}{\sqrt{5}}\right)$ and $v_0 = \frac{3}{5}\text{ms}^{-1}$

Terms in the Projectile Motion

11. The horizontal range of a projectile fired at an angle of 15° is 50 m. If it is fired with the same speed at an angle of 45°, its range will be **[NCERT Exemplar]**

(a) 60 m (b) 71 m (c) 100 m (d) 141 m

12. A particle is projected from horizontal making an angle 60° with initial velocity 40 ms^{-1}. The time taken by the particle to make angle 45° from horizontal, is

(a) 15 s (b) 2.0 s (c) 20 s (d) 1.5 s

13. A particle is projected from the ground at an angle of 60° with horizontal with speed $u = 20\text{ ms}^{-1}$. The radius of curvature of the path of the particle, when its velocity makes an angle of 30° with horizontal is ($g = 10\text{ ms}^{-2}$)

(a) 10.6 m (b) 12.8 m (c) 15.4 m (d) 24.2 m

14. A bomb is dropped on an enemy post by an aeroplane flying horizontally with a velocity of 60 kmh^{-1} and at a height of 490 m. At the time of dropping the bomb, how far the aeroplane should be from the enemy post so that the bomb may directly hit the target?

(a) $\frac{400}{3}$ m (b) $\frac{500}{3}$ m (c) $\frac{1700}{3}$ m (d) 498 m

15. A stone is thrown at an angle θ to be the horizontal reaches a maximum height H, then the time of flight of stone will be

(a) $\sqrt{\frac{2H}{g}}$ (b) $2\sqrt{\frac{2H}{g}}$

(c) $\frac{2\sqrt{2H}\sin\theta}{g}$ (d) $\frac{\sqrt{2H}\sin\theta}{g}$

16. An arrow is shot into air. Its range is 200 m and its time of flight is 5 s. If $g = 10\text{ m/s}^2$, then horizontal component of velocity and the maximum height will be respectively

(a) 20 m/s, 62.50 m (b) 40 m/s, 31.25 m

(c) 80 m/s, 62.5 m (d) None of these

17. A ball is thrown up with a certain velocity at an angle θ to the horizontal. The kinetic energy (KE) of the ball varies in the horizontal displacement x as

(a)
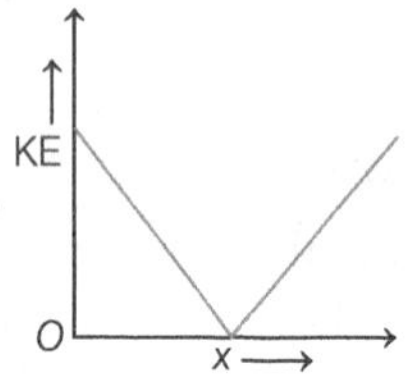

(b)
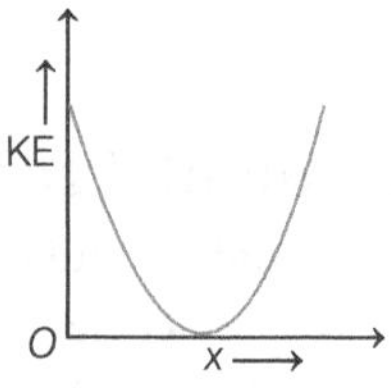

(c)
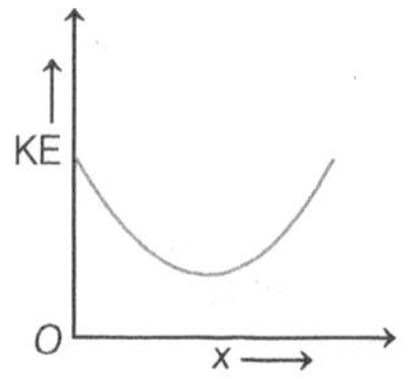

(d)
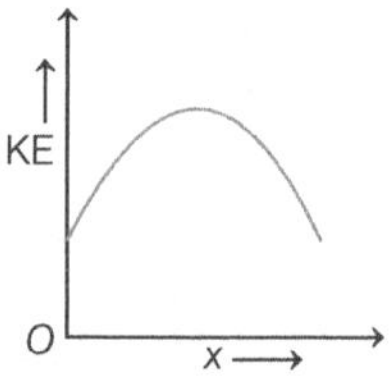

18. It was calculated that a shell when fired from a gun with a certain velocity and at an angle of elevation $\frac{5\pi}{36}$ rad should strike a given target. In actual practice, it was found that a hill just prevented the trajectory. At what angle of elevation, should the gun be to hit the target?

(a) $\frac{5\pi}{36}$ rad (b) $\frac{11\pi}{36}$ rad

(c) $\frac{7\pi}{36}$ rad (d) $\frac{13\pi}{36}$ rad

19. A cricket ball is hit at 30° with the horizontal with kinetic energy E_k. What is the kinetic energy at the highest point?

(a) $\frac{E_k}{2}$ (b) $\frac{3E_k}{4}$ (c) $\frac{E_k}{4}$ (d) Zero

20. A particle is projected with a velocity of 30 m/s, at an angle $\theta_0 = \tan^{-1}\left(\frac{3}{4}\right)$. After 1 s, the particle is moving at an angle θ to the horizontal, where $\tan\theta$ will be equal to ($g = 10\text{ m/s}^2$)

(a) 1 (b) 2 (c) $\frac{1}{2}$ (d) $\frac{1}{3}$

21. Two stones thrown at different angles have same initial velocity and same range. If H is the maximum height attained by one stone thrown at an angle of 30°, then the maximum height attained by the other stone is

(a) $\frac{H}{2}$ (b) H (c) $2H$ (d) $3H$

22. A projectile shot into air at some angle with the horizontal has a range of 200 m. If the time of flight is 5 s, then the horizontal component of the velocity of the projectile at the highest point of trajectory is
(a) 40 ms^{-1}
(b) 0 ms^{-1}
(c) 9.8 ms^{-1}
(d) equal to the velocity of projection of the projectile

23. The kinetic energy of a projectile at the highest, point is half of the initial kinetic energy. What is the angle of projection with the horizontal?
(a) 30° (b) 45° (c) 60° (d) 90°

24. A ball is projected from a certain point on the surface of a planet at a certain angle with the horizontal surface. The horizontal and vertical displacement x and y vary with time t in second as $x = 10\sqrt{3}\, t$ and $y = 10\, t - t^2$. The maximum height attained by the ball is
(a) 100 m (b) 75 m (c) 50 m (d) 25 m

25. For a projectile thrown into space with a speed v, the horizontal range is $\frac{\sqrt{3}v^2}{2g}$. The vertical range is $\frac{v^2}{8g}$. The angle which the projectile makes with the horizontal initially is
(a) 15° (b) 30°
(c) 45° (d) 60°

26. A projectile of mass m is thrown with a velocity v making an angle of 45° with the horizontal. The change in momentum from departure to arrival along vertical direction, is
(a) $2mv$ (b) $\sqrt{2}\, mv$
(c) mv (d) $\frac{mv}{2}$

27. A body is projected at an angle θ to the horizontal with kinetic energy E_k. The potential energy at the highest point of the trajectory is
(a) E_k (b) $E_k \cos^2 \theta$
(c) $E_k \sin^2 \theta$ (d) $E_k \tan^2 \theta$

28. The horizontal range of an oblique projectile is equal to the distance through which a projectile has to fall freely from rest to acquire a velocity equal to the velocity of projection in magnitude. The angle of projection is
(a) 15° (b) 60° (c) 45° (d) 30°

29. A projectile is fired at an angle of 30° to the horizontal such that the vertical component of its initial velocity is 80 ms^{-1}. Its time of flight is T. Its velocity at $t = \frac{T}{4}$ has a magnitude of nearly
(a) 200 ms^{-1} (b) 300 ms^{-1}
(c) 140 ms^{-1} (d) 100 ms^{-1}

30. Two projectiles A and B are thrown with velocities v and $\frac{v}{2}$, respectively. They have the same range. If projectile B is thrown at an angle of 15° to the horizontal, then projectile A must have been thrown at an angle
(a) $\sin^{-1}\left(\frac{1}{16}\right)$ (b) $\sin^{-1}\left(\frac{1}{4}\right)$
(c) $2\sin^{-1}\left(\frac{1}{4}\right)$ (d) $\frac{1}{2}\sin^{-1}\left(\frac{1}{8}\right)$

31. The velocity of projection of an oblique projectile is $(6\hat{i} + 8\hat{j})\, ms^{-1}$. The horizontal range of the projectile is
(a) 4.9 m (b) 9.6 m (c) 19.6 m (d) 14 m

32. A projectile is thrown with velocity v making an angle θ with the horizontal. It just crosses the tops of two poles, each of height h, after 1 s and 3 s respectively. The time of flight of the projectile is
(a) 1 s (b) 3 s (c) 4 s (d) 7.8 s

33. Two particles are simultaneously projected in opposite directions horizontally from a given point in space whose gravity g is uniform. If u_1 and u_2 be their initial speeds, then the time t after which their velocites are mutually perpendicular is given by
(a) $\frac{\sqrt{u_1 u_2}}{g}$ (b) $\frac{\sqrt{u_1^2 + u_2^2}}{g}$
(c) $\frac{\sqrt{u_1(u_1 + u_2)}}{g}$ (d) $\frac{\sqrt{u_2(u_1 + u_2)}}{g}$

34. A cricketer can throw a ball to a maximum horizontal distance of 100 m. How much high above the ground can the cricketer throw the same ball? **[NCERT Exemplar]**
(a) 40 m (b) 45 m (c) 500 m (d) 50 m

35. A piece of marble is projected from Earth's surface with velocity of 50 ms^{-1}. 2 s later, it just clears a wall 5 m high. What is the angle of projection?
(a) 45° (b) 30°
(c) 60° (d) None of these

36. The ceiling of a long hall is 25 m high. Then, the maximum horizontal distance that a ball thrown with a speed of 40 m/s can go without hitting the ceiling of the hall, is **[NCERT Exemplar]**
(a) 95.5 m (b) 105.5 m
(c) 100 m (d) 150.5 m

37. A particle leaves the origin with an initial velocity $\mathbf{v} = (3.00\, \hat{i})\, ms^{-1}$ and a constant acceleration $\mathbf{a} = (-1.00\, \hat{i} - 0.50\, \hat{j})\, ms^{-2}$. When the particle reaches its maximum x-coordinate, what is its y-component of velocity?
(a) −2.0 ms^{-1} (b) −1.0 ms^{-1}
(c) −1.5 ms^{-1} (d) 1.0 ms^{-1}

38. Two stones are projected so as to reach the same distance from the point of projection on a horizontal surface. The maximum height reached by one exceeds the other by an amount equal to half the sum of the height attained by them, then angle of projection of the stone which attains smaller height is

(a) 45° (b) 60°
(c) 30° (d) $\tan^{-1}(3/4)$

39. The angle of projection of a projectile for which the horizontal range and maximum height are equal to

(a) $\tan^{-1}(2)$ (b) $\tan^{-1}(4)$
(c) $\cot^{-1}(2)$ (d) 60°

40. A particle is projected with velocity $2\sqrt{gh}$, so that it just clears two walls of equal height h, which are at a distance of $2h$ from each other. What is the time interval of passing between the two walls?

(a) $\frac{2h}{g}$ (b) $\sqrt{\frac{gh}{g}}$ (c) $\sqrt{\frac{h}{g}}$ (d) $2\sqrt{\frac{h}{g}}$

41. A particle is projected from the ground with an initial speed of v at an angle θ with horizontal. The average velocity of the particle between its point of projection and highest point of trajectory is

(a) $\frac{v}{2}\sqrt{1+2\cos^2\theta}$ (b) $\frac{v}{2}\sqrt{1+\cos^2\theta}$
(c) $\frac{v}{2}\sqrt{1+3\cos^2\theta}$ (d) $v\cos\theta$

42. The maximum range of a bullet fired from a toy pistol mounted on a car at rest is $R_0 = 40$ m. What will be the acute angle of inclination of the pistol for maximum range when the car is moving in the direction of firing with uniform velocity $v = 20$ m/s, on a horizontal surface? $(g = 10\text{ m/s}^2)$ [JEE Main 2013]

(a) 30° (b) 60°
(c) 75° (d) 45°

43. A projectile is thrown at an angle θ such that it is just able to cross a vertical wall as its highest point as shown in the figure. The angle θ at which the projectile is thrown is given by

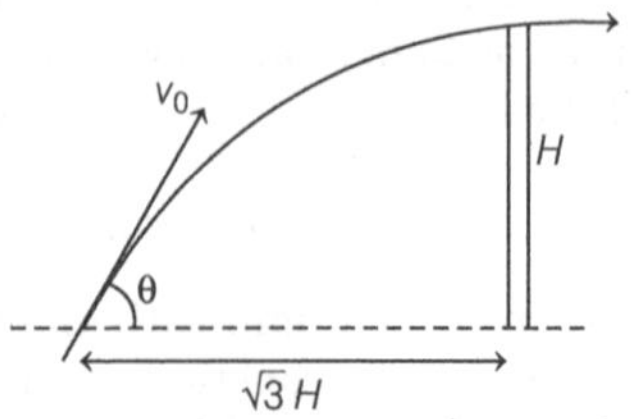

(a) $\tan^{-1}\left(\frac{1}{\sqrt{3}}\right)$ (b) $\tan^{-1}\sqrt{3}$
(c) $\tan^{-1}\left(\frac{2}{\sqrt{3}}\right)$ (d) $\tan^{-1}\left(\frac{\sqrt{3}}{2}\right)$

44. Trajectories of two projectiles are shown in figure. Let T_1 and T_2 be the time periods and u_1 and u_2 their speeds of projection, then

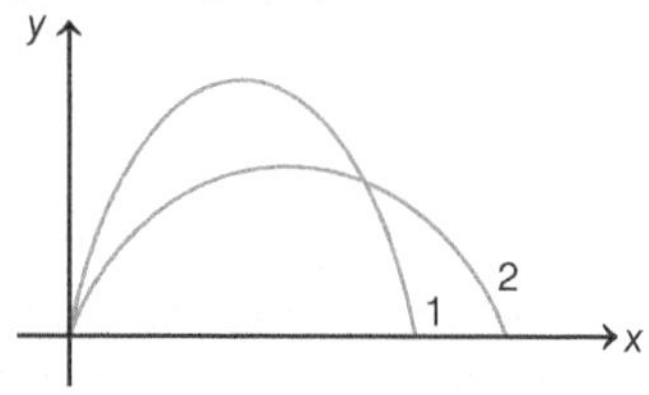

(a) $T_2 > T_1$ (b) $T_1 = T_2$
(c) $u_1 > u_2$ (d) $u_1 < u_2$

Special Cases of Projectile Motion

45. A body of mass m thrown horizontally with velocity v, from the top of tower of height h touches the level ground at distance of 250 m from the foot of the tower. A body of mass 2 m thrown horizontally with velocity $\frac{v}{2}$, from the top of tower of height $4h$ will touch the level ground at a distance x from the foot of the tower. The value of x is

(a) 250 m (b) 500 m
(c) 125 m (d) $250\sqrt{2}$ m

46. A man standing on a hill top projects a stone horizontally with speed v_0 as shown in figure. Taking the coordinate system as given in the figure. The coordinates of the point where the stone will hit the hill surface

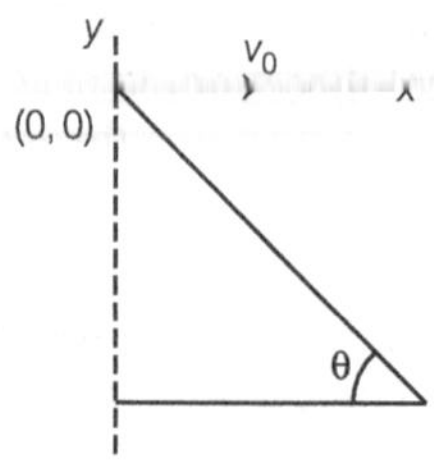

(a) $\left(\frac{2v_0^2\tan\theta}{g}, \frac{-2v_0^2\tan^2\theta}{g}\right)$

(b) $\left(\frac{2v_0^2}{g}, \frac{2v_0^2\tan^2\theta}{g}\right)$

(c) $\left(\frac{2v_0^2\tan\theta}{g}, \frac{2v_0^2}{g}\right)$

(d) $\left(\frac{2v_0^2\tan^2\theta}{g}, \frac{2v_0^2\tan\theta}{g}\right)$

47. A fighter plane enters inside the enemy territory, at time $t = 0$ with velocity $v_0 = 250\text{ ms}^{-1}$ and moves horizontally with constant acceleration $a = 20\text{ ms}^{-2}$ (see figure). An enemy tank at the border, spot the plane and fire shots at an angle $\theta = 60°$ with the horizontal and with velocity $u = 600\ \text{ms}^{-1}$. At what altitude H of the plane, it can be hit by the shot?

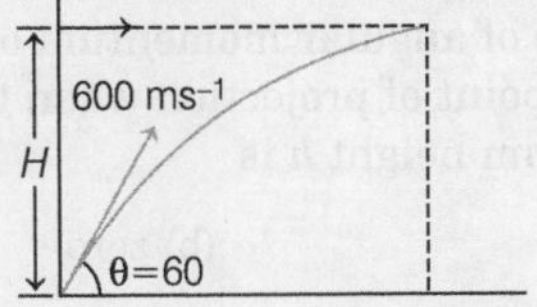

(a) $1500\sqrt{3}$ m (b) 125 m
(c) 1400 m (d) 2473 m

48. A particle A is projected from the ground with an initial velocity of $10\ \text{ms}^{-1}$ at an angle of 60° with horizontal. From what height h should an another particle B be projected horizontally with velocity $5\ \text{ms}^{-1}$, so that both the particles collide with velocity $5\ \text{ms}^{-1}$ on the ground at point C, if both are projected simultaneously? ($g = 10\ \text{ms}^{-2}$)

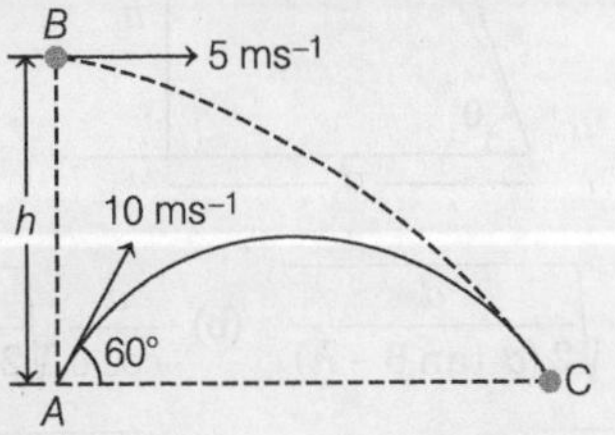

(a) 10 m (b) 30 m
(c) 15 m (d) 25 m

49. A particle is projected with speed v at an angle $\theta \left(0 < \theta < \frac{\pi}{2}\right)$ above the horizontal from a height H above the ground. If v = speed with which particle hits the ground and t = time taken by particle to reach ground, then

(a) as θ increases, v decreases and t increases
(b) as θ increases, v increases and t increases
(c) as θ increases, v remains same and t increases
(d) as θ increases, v remains same and t decreases

50. A projectile A is thrown at an angle 30° to the horizontal from point P. At the same time, another projectile B is thrown with velocity v_2 upwards from the point Q vertically below the highest point A. If projectile B collides with projectile A, then the ratio $\frac{v_2}{v_1}$ should be

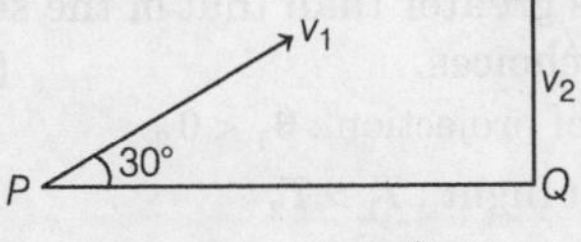

(a) $\frac{\sqrt{3}}{2}$ (b) 2 (c) $\frac{1}{2}$ (d) $\frac{2}{\sqrt{3}}$

51. An aircraft, diving at an angle of 53.0° with the vertical releases a projectile at an altitude of 730 m. The projectile hits the ground 5.00 s after being released. What is the speed of the aircraft?

(a) $282\ \text{ms}^{-1}$
(b) $202\ \text{ms}^{-1}$
(c) $182\ \text{ms}^{-1}$
(d) $102\ \text{ms}^{-1}$

52. A bomber plane moves horizontally with a speed of $500\ \text{ms}^{-1}$ and a bomb releases from it, strikes the ground in 10 s. Angle at which it strikes the ground will be ($g = 10\ \text{ms}^{-2}$)

(a) $\tan^{-1}\left(\frac{1}{5}\right)$
(b) $\tan\left(\frac{1}{5}\right)$
(c) $\tan^{-1}(1)$
(d) $\tan^{-1}(5)$

53. A plane surface is inclined making an angle θ with the horizontal. From the bottom of this inclined plane, a bullet is fired with velocity v. The maximum possible range of the bullet on the inclined plane is

(a) $\frac{v^2}{g}$ (b) $\frac{v^2}{g(1+\sin\theta)}$
(c) $\frac{v^2}{g(1-\sin\theta)}$ (d) $\frac{v^2}{g(1+\sin\theta)^2}$

54. An aeroplane is flying in a horizontal direction with a velocity $600\ \text{kmh}^{-1}$ at a height of 1960 m. When it is vertically above the point A on the ground, a body is dropped from it. The body strikes the ground at point B. Calculate the distance AB.

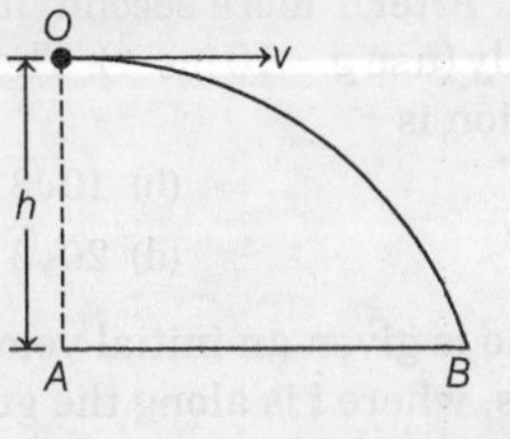

(a) 3.33 km (b) 333 km
(c) 33.3 km (d) 3330 km

55. A ball is projected up an incline of 30° with a velocity of $30\ \text{ms}^{-1}$ at an angle of 30° with reference to the inclined plane from the bottom of the inclined plane. If $g = 10\ \text{ms}^{-2}$, then the range on the inclined plane is

(a) 12 m (b) 60 m
(c) 120 m (d) 600 m

ROUND II Mixed Bag

Only One Correct Option

1. A ball projected from ground at an angle of 45° just clears a wall in front. If point of projection is 4 m from the foot of wall and ball strikes the ground at a distance of 6 m on the other side of the wall, the height of the wall is **[JEE Main 2013]**

(a) 4.4 m (b) 2.4 m
(c) 3.6 m (d) 1.6 m

2. A shell is fired from a cannon with a velocity v at angle θ with horizontal. At the highest point, it explodes into two pieces of equal mass. One of the pieces retraces its path to the cannon. The speed of the other piece just after explosion is

(a) $3v\cos\theta$ (b) $2v\cos\theta$
(c) $\frac{3}{2}v\cos\theta$ (d) $\frac{\sqrt{3}}{2}v\cos\theta$

3. Figure shows four paths for a kicked football ignoring the effects of air on the flight. Rank the paths according to the initial horizontal velocity component highest first.

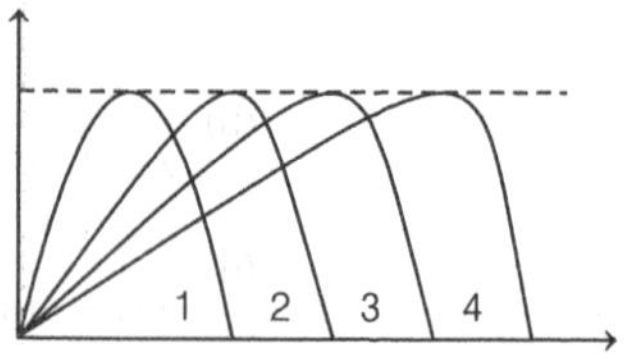

(a) 1, 2, 3, 4 (b) 2, 3, 4, 1
(c) 3, 4, 1, 2 (d) 4, 3, 2, 1

4. Two second after projection, a projectile is travelling in a direction inclined at 30° to the horizontal. After 1 more second, it is travelling horizontally (use $g = 10\,\text{ms}^{-2}$). The initial velocity of its projection is

(a) $10\,\text{ms}^{-1}$ (b) $10\sqrt{3}\,\text{ms}^{-1}$
(c) $20\,\text{ms}^{-1}$ (d) $20\sqrt{3}\,\text{ms}^{-1}$

5. A projectile is given an initial velocity of $(\hat{\mathbf{i}} + 2\hat{\mathbf{j}})$ m/s, where $\hat{\mathbf{i}}$ is along the ground and $\hat{\mathbf{j}}$ is along the vertical. If $g = 10\,\text{m/s}^2$, the equation of its trajectory is **[JEE Main 2013]**

(a) $y = x - 5x^2$ (b) $y = 2x - 5x^2$
(c) $4y = 2x - 5x^2$ (d) $4y = 2x - 25x^2$

6. A boy can throw a stone upto a maximum height of 10 m. The maximum horizontal distance that the boy can throw the same stone upto will be **[AIEEE 2012]**

(a) $20\sqrt{2}$ m (b) 10 m
(c) $10\sqrt{2}$ (d) 20 m

7. A particle of mass m is projected with a velocity v making an angle of 30° with the horizontal. The magnitude of angular momentum of the projectile about the point of projection when the particle is at its maximum height h is **[AIEEE 2011]**

(a) $\frac{\sqrt{3}}{2}\frac{mv^2}{g}$ (b) zero
(c) $\frac{mv^3}{\sqrt{2}g}$ (d) $\frac{\sqrt{3}}{16}\frac{mv^3}{g}$

8. If a stone is to hit at a point which is at a distance d away and at a height h above the point from where the stone starts, then what is the value of initial speed u, if the stone is launched at an angle θ?

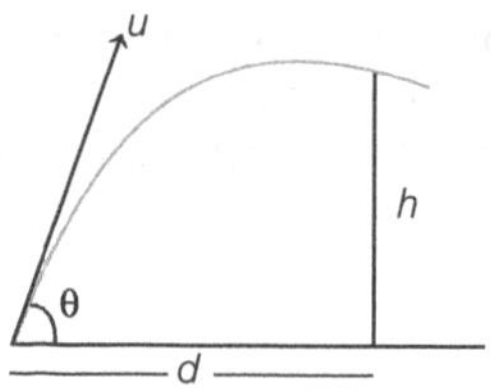

(a) $\frac{g}{\cos\theta}\sqrt{\frac{d}{2(d\tan\theta - h)}}$ (b) $\frac{d}{\cos\theta}\sqrt{\frac{g}{2(d\tan\theta - h)}}$
(c) $\sqrt{\frac{gd^2}{h\cos^2\theta}}$ (d) $\sqrt{\frac{gd^2}{(d-h)}}$

9. Two projectiles A and B are projected with same speed at angles 15° and 75° respectively to the maximum height and have same horizontal range. If h be the maximum height and T total time of flight of a projectile, then

(a) $h_A > h_B$ (b) $h_A = h_B$
(c) $T_A < T_B$ (d) $T_A > T_B$

10. A projectile has the same range R for two angles of projections. If T_1 and T_2 be the times of flight in the two cases, then (using θ as the angle of projection corresponding to T_1)

(a) $T_1T_2 \propto R$ (b) $T_1T_2 \propto R^2$
(c) $T_1/T_2 = \cot\theta$ (d) $T_1/T_2 = 1$

11. Two particles are projected in air with speed v_0 at angles θ_1 and θ_2 (both acute) to the horizontal, respectively. If the height reached by the first particle is greater than that of the second, then tick the right choices. **[NCERT Exemplar]**

(a) Angle of projection : $\theta_1 < \theta_2$
(b) Time of flight : $T_1 > T_2$
(c) Horizontal range : $R_1 > R_2$
(d) Total energy : $U_1 > U_2$

12. The trajectory of a projectile in vertical plane is $y = ax - bx^2$, where a and b are constants and x and y are respectively horizontal and vertical distances of the projectile from the point of projection. The maximum height attained by the particle and the angle of projection from the horizontal are

(a) $\frac{b^2}{4b}$, $\tan^{-1}(b)$ (b) $\frac{a^2}{b}$, $\tan^{-1}(2b)$

(c) $\frac{a^2}{4b}$, $\tan^{-1}(a)$ (d) $\frac{2a^2}{b}$, $\tan^{-1}(a)$

13. The speed of projection of a projectile is increased by 10%, without changing the angle of projection. The percentage increase in the range will be

(a) 10% (b) 20%

(c) 15% (d) 5%

14. A body of mass 1 kg is projected with velocity 50 m/s at an angle of 30° with the horizontal. At the highest point of its path, a force 10 N starts acting on body for 5 s vertically upward besides gravitational force, what is horizontal range of the body? (Take, $g = 10 \, m/s^2$)

(a) $125\sqrt{3}$ m (b) $200\sqrt{3}$ m

(c) 500 m (d) $250\sqrt{3}$ m

15. A particle is projected with a velocity 200 ms^{-1} at an angle of 60°. At the highest point, it explodes into three particles of equal masses. One goes vertically upwards with a velocity 100 ms^{-1}, the second particle goes vertically downwards. What is the velocity of third particle?

(a) 120 ms^{-1} making 60° angle with horizontal

(b) 200 ms^{-1} making 30° angle with horizontal

(c) 300 ms^{-1}

(d) 200 ms^{-1}

16. Two stones are projected with the same velocity in magnitude but making different angles with the horizontal. Their ranges are equal. If the angle of projection of one is $\frac{\pi}{3}$ and its maximum height is y_1, the maximum height of the other will be

(a) $3y_1$ (b) $2y_1$

(c) $\frac{y_1}{2}$ (d) $\frac{y_1}{3}$

17. A car is travelling at a velocity of 10 kmh^{-1} on a straight road. The driver of the car throws a parcel with a velocity of $10\sqrt{2}$ kmh^{-1} when the car is passing by a man standing on the side of the road. If the parcel is to reach the man, the direction of throw makes the following angle with direction of the car

(a) 135° (b) 45°

(c) $\tan^{-1}(\sqrt{2})$ 60° (d) $\tan\left(\frac{1}{\sqrt{2}}\right)$

18. A particle of mass m is projected with a velocity v at an angle of 60° with horizontal. When the particle is at its maximum height, the magnitude of its angular momentum about the point of projection is

(a) zero (b) $\frac{3mv^3}{16g}$

(c) $\frac{\sqrt{3}mv^2}{16g}$ (d) $\frac{3mv^2}{3g}$

19. A stone is projected with a velocity $20\sqrt{2}$ ms^{-1} at an angle of 45° to the horizontal. The average velocity of stone during its motion from starting point to its maximum height is (Take, $g = 10 \, ms^{-2}$)

(a) $5\sqrt{5}$ ms^{-1} (b) $10\sqrt{5}$ ms^{-1}

(c) 20 ms^{-1} (d) $20\sqrt{5}$ ms^{-1}

20. Average torque on a projectile of mass m, initial speed u and angle of projection θ, between initial and final position P and Q as shown in figure about the point of projection is

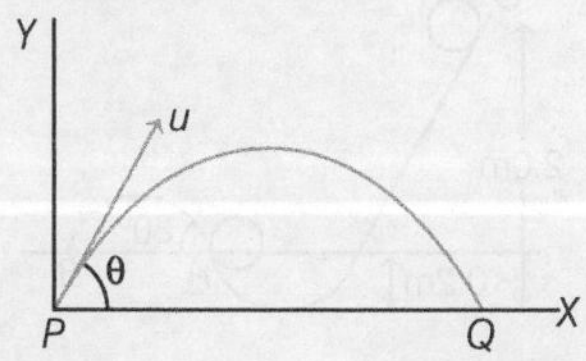

(a) $mu^2 \sin\theta$ (b) $mu^2 \cos\theta$

(c) $\frac{1}{2}mu^2 \sin 2\theta$ (d) $\frac{1}{2}mu^2 \cos 2\theta$

21. After one second, the velocity of a projectile makes an angle of 45° with the horizontal. After another one second, it is travelling horizontally. The magnitude of its initial velocity and angle of projection are ($g = 10 \, ms^{-2}$)

(a) 14.62 ms^{-1}, $\tan^{-1}(2)$ (b) 22.36 ms^{-1}, $\tan^{-1}(2)$

(c) 14.62 ms^{-1}, 60° (d) 22.36 ms^{-1}, 60°

22. A body is projected up over a smooth inclined plane with a velocity v_0 from the point A as shown in the figure. The angle of inclination is 45° and top B of the plane is connected to a well of diameter 40 m. If the body just manages to cross the well, what is the value of v_0? Length of the inclined plane is $20\sqrt{2}$ m, and $g = 10 \, ms^{-2}$.

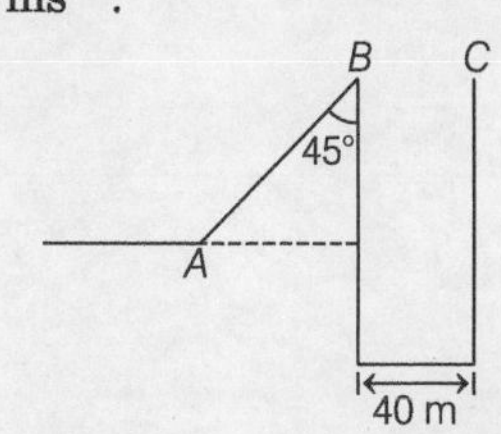

(a) 20 ms^{-1} (b) $20\sqrt{2}$ ms^{-1}

(c) 40 ms^{-1} (d) $40\sqrt{2}$ ms^{-1}

23. A projectile is launched with a speed of 10 m/s at an angle 60° with the horizontal from a sloping surface of inclination 30°. The range R is. (Take, $g = 10\,\text{m/s}^2$)

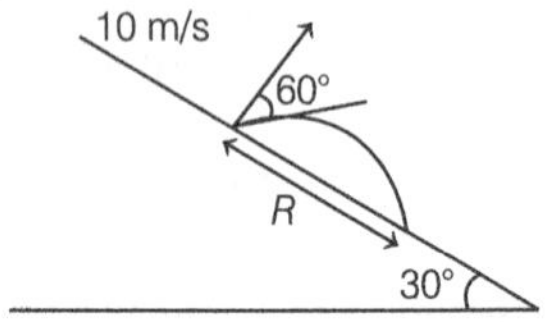

(a) 4.9 m (b) 13.3 m (c) 9.1 m (d) 12.6 m

24. A tennis ball (treated as hollow spherical shell) starting from O rolls down a hill. At point A, the ball becomes air borne leaving at an angle of 30° with the horizontal. The ball strikes the ground at B. What is the value of the distance AB?

(Moment of inertia of a spherical shell of mass m and radius R about its diameter $= \frac{2}{3} mR^2$) [JEE Main 2013]

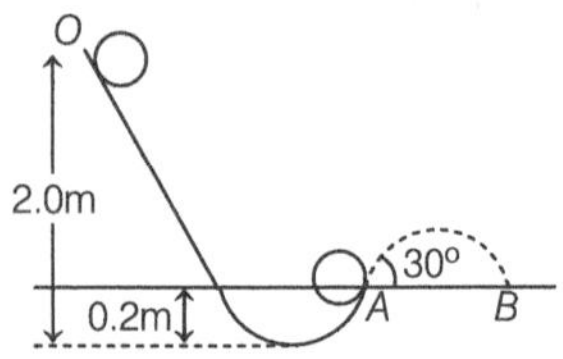

(a) 1.87 m (b) 2.08 m (c) 1.57 m (d) 1.77 m

25. A projectile is fired with a velocity v at right angle to the slope which is inclined at an angle θ with the horizontal. What is the range covered by the projectile?

(a) $\frac{2v^2}{g}\tan\theta$ (b) $\frac{v^2}{g}\tan\theta$

(c) $\frac{2v^2}{g}\sec\theta$ (d) $\frac{2v^2}{g}\tan\theta\sec\theta$

Numerical Value Questions

26. An aeroplane in a level flight at 144 kmh^{-1} is at an altitude of 1000 m. The value of distance (in m) from which the body should be released to hit the given target will be ....

27. A ball thrown by one player reaches the other in 2s. The maximum value of height (in m) attained by the ball above the point of projection will be (Take, $g = 10\,\text{ms}^{-2}$) ....

28. Two stones are projected with the same velocity but making different angles with the horizontal. Their ranges are equal. If angle of projection of one is 30° and its maximum height is y, then the maximum height of other will be ny, where value of n will be ....

29. When the angle of projection is 75°, a ball falls 10 m shorter of the target. When the angle of projection is 45°, it falls 10 m ahead of the target. Both are projected from the same point with the same speed in the same direction, the distance of the target (in m) from the point of projection is ....

30. A marble starts falling from rest on a smooth inclined plane of inclination α. After covering a distance h, the ball rebounds off the plane. The distance from the impact point where the ball rebounds for the second time $nh\sin\alpha$. Here, the value of n is ....

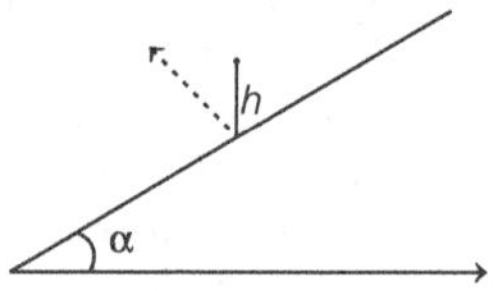

31. The projectile motion of a particle of mass 5g is shown in the figure.

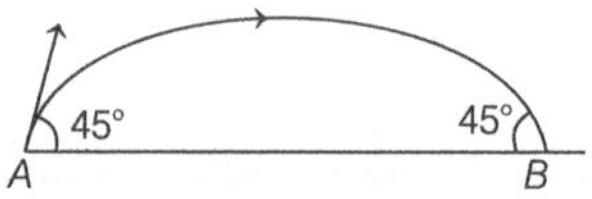

The initial velocity of the particle is $5\sqrt{2}\,\text{ms}^{-1}$ and the air resistance is assumed to be negligible. The magnitude of the change in momentum between the points A and B is $x \times 10^{-2}$ kg-ms^{-1}. The value of x, to the nearest integer,is .... [JEE Main 2021]

Answers

Round I

1. (c)	2. (a)	3. (c)	4. (c)	5. (c)	6. (d)	7. (a)	8. (c)	9. (a)	10. (c)
11. (c)	12. (d)	13. (c)	14. (b)	15. (b)	16. (b)	17. (c)	18. (d)	19. (b)	20. (d)
21. (d)	22. (a)	23. (b)	24. (d)	25. (b)	26. (b)	27. (c)	28. (a)	29. (c)	30. (d)
31. (b)	32. (c)	33. (a)	34. (d)	35. (b)	36. (d)	37. (c)	38. (c)	39. (b)	40. (d)
41. (c)	42. (b)	43. (c)	44. (d)	45. (a)	46. (a)	47. (d)	48. (c)	49. (c)	50. (c)
51. (b)	52. (a)	53. (b)	54. (a)	55. (b)					

Round II

1. (c)	2. (a)	3. (d)	4. (d)	5. (b)	6. (d)	7. (d)	8. (b)	9. (c)	10. (a)
11. (b)	12. (c)	13. (b)	14. (d)	15. (c)	16. (d)	17. (b)	18. (b)	19. (b)	20. (c)
21. (b)	22. (b)	23. (b)	24. (b)	25. (a)	26. 571.43	27. 5	28. 3	29. 30	30. 8
31. 5									

Solutions

Round I

1. $v_y = \frac{dy}{dt} = 8 - 10t,\ v_x = \frac{dx}{dt} = 6$

At $t = 0$, $v_y = 8$ m/s and $v_x = 6$ m/s

$$\therefore \quad v = \sqrt{v_x^2 + v_y^2} = 10 \text{ m/s}$$

2. $v_x = 8t - 2$

or $$\frac{dx}{dt} = 8t - 2$$

or $$\int_{14}^{x} dx = \int_{2}^{t} (8t - 2)dt$$

or $$x - 14 = [4t^2 - 2t]_2^t = 4t^2 - 2t - 12$$

or $$x = 4t^2 - 2t + 2 \quad \text{...(i)}$$

Further, $v_y = 2$

or $$\frac{dy}{dt} = 2$$

$$\therefore \quad \int_4^y dy = \int_2^t 2\,dt$$

or $$y - 4 = [2t]_2^t = 2t - 4$$

or $$y = 2t$$

or $$t = \frac{y}{2} \quad \text{...(ii)}$$

Substituting the value of t from Eq. (ii) in Eq. (i), we have

$$x = y^2 - y + 2$$

3. Since $y = x^2$, the motion is in two dimensions. Velocity at B is greater than at P. In the given motion of a particle, the law of conservation of energy is obeyed. Therefore, total energy at P = total energy at A. As vertical distance $AB > BP$, time of travel from A to B is greater than that from B to P.

4. Equating velocities along the vertical,

$$v_2 = v_1 \sin 30^\circ$$

or $$\frac{v_2}{v_1} = \frac{1}{2}$$

5. As, $v' \cos\beta = v\cos\theta$

($\because$ horizontal component of velocities are always equal.)

or $$v' = v\cos\theta\sec\beta$$

6. As, $v\cos\beta = u\cos\alpha$

($\because$ horizontal component of velocities are always equal.)

$$\therefore \quad v = \frac{u\cos\alpha}{\cos\beta}$$

7. $v^2 = v_y^2 + v_x^2$

or $$5g = (u_y - gt)^2 + u_x^2$$

or $$50 = (5\sqrt{3} - 10t)^2 + (5)^2$$

$$\therefore \quad (5\sqrt{3} - 10t) = \pm 5$$

$$t_1 = \frac{5\sqrt{3} - 5}{10}$$

and $$t_2 = \frac{5\sqrt{3} + 5}{10}$$

$$\therefore \quad t_2 - t_1 = 1 \text{ s}$$

8. Instantaneous velocity of rising mass after t s will be

$$v_t = \sqrt{v_x^2 + v_y^2}$$

where, $v_x = v\cos\theta$ = Horizontal component of velocity

$v_y = v\sin\theta - gt$ = Vertical component of velocity

$$v_t = \sqrt{(v\cos\theta)^2 + (v\sin\theta - gt)^2}$$

$$v = \sqrt{v^2 + g^2t^2 - (2v\sin\theta)\,gt}$$

9. $x = 36t$

$\therefore \quad v_x = \frac{dx}{dt} = 36$

$y = 48t - 4.9t^2$

$\therefore \quad v_y = \frac{dy}{dt} = 48 - 9.8t$

At $t = 0$, $\quad v_x = 36$ m/s

and $\quad v_y = 48$ m/s

So, angle of projection, $\theta = \tan^{-1}\left(\frac{v_y}{v_x}\right)$

$= \tan^{-1}\left(\frac{48}{36}\right)$

$= \tan^{-1}\left(\frac{4}{3}\right)$

or $\quad \theta = \sin^{-1}\left(\frac{4}{5}\right)$

10. Given, $g = 10$ m/s^2

Equation of trajectory of the projectile,

$$y = 2x - 9x^2 \quad \text{...(i)}$$

In projectile motion, equation of trajectory is given by

$$y = x\tan\theta_0 - \frac{gx^2}{2v_0^2\cos^2\theta_0} \quad \text{...(ii)}$$

By comparison of Eqs. (i) and (ii), we get

$$\tan\theta_0 = 2 \quad \text{...(iii)}$$

and $\frac{g}{2v_0^2\cos^2\theta_0} = 9$ or $v_0^2 = \frac{g}{9\times 2\cos^2\theta_0}$ ...(iv)

From Eq. (iii), we can get value of $\cos\theta$ and $\sin\theta$

$$\cos\theta_0 = \frac{1}{\sqrt{5}} \text{ and } \sin\theta_0 = \frac{2}{\sqrt{5}} \quad \text{...(v)}$$

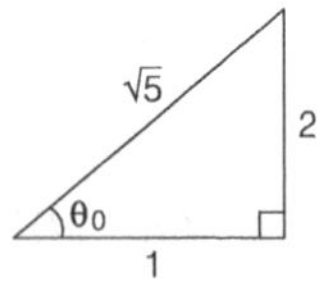

Using value of $\cos\theta_0$ from Eq. (v) to Eq. (iv), we get

$$v_0^2 = \frac{10\times(\sqrt{5})^2}{2\times(1)^2\times 9} = \frac{10\times 5}{2\times 9}$$

$\Rightarrow \quad v_0^2 = \frac{25}{9}$ or $v_0 = \frac{5}{3}$ m/s ...(vi)

From Eq. (v), we get

$$\theta_0 = \cos^{-1}\left(\frac{1}{\sqrt{5}}\right)$$

11. Here, $\quad 50 = \frac{u^2\sin 2\times 15^\circ}{g}$

or $\quad \frac{u^2}{g} = \frac{50}{\sin 30^\circ} = \frac{50}{1/2} = 100$

$$R = \frac{u^2\sin 2\times 45^\circ}{g} = \frac{u^2}{g} = 100\,\text{m}$$

12. At 45°, $v_x = v_y$

or $\quad u_x = u_y - gt$

$\therefore \quad t = \frac{u_y - u_x}{g} = \frac{40(\sin 60^\circ - \sin 30^\circ)}{9.8} = 1.5$ s

13. Let v be the velocity of particle when it makes 30° with horizontal, then

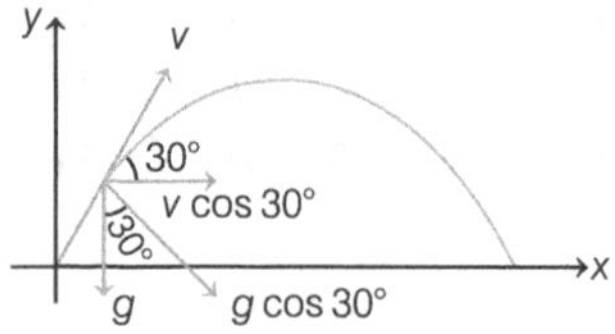

$$v\cos 30^\circ = u\cos 60^\circ$$

or $\quad v = \frac{u\cos 60^\circ}{\cos 30^\circ} = \frac{(20)\left(\frac{1}{2}\right)}{\left(\frac{\sqrt{3}}{2}\right)} = \frac{20}{\sqrt{3}}\,\text{ms}^{-1}$

Now, $\quad g\cos 30^\circ = \frac{v^2}{R}$

or $\quad R = \frac{v^2}{g\cos 30^\circ} = \frac{\left(\frac{20}{\sqrt{3}}\right)^2}{(10)\frac{\sqrt{3}}{2}} = 15.4$ m

14. $t = \sqrt{\frac{2h}{g}} = \sqrt{\frac{2\times 490}{9.8}} = \sqrt{100} = 10$ s

$x = vt = \left(60\times\frac{5}{18}\right)\text{ms}^{-1}\times 10\,\text{s} = \frac{500}{3}$ m

15. $H = \frac{u^2\sin^2\theta}{2g}$ and $T = \frac{2u\sin\theta}{g}$

$\Rightarrow \quad T^2 = \frac{4u^2\sin^2\theta}{g^2}$

$\therefore \quad \frac{T^2}{H} = \frac{8}{g}$

$\Rightarrow \quad T = \sqrt{\frac{8H}{g}} = 2\sqrt{\frac{2H}{g}}$

16. $T = \frac{2u_y}{g}$

$\therefore \quad u_y = \frac{gT}{2} = 25$ m/s

Now, $\quad H = \frac{u_y^2}{2g} = \frac{(25)^2}{20} = 31.25$ m

Further, $\quad R = u_xT$

$\therefore \quad u_x = \frac{R}{T} = 40$ m/s

17. As x increases, kinetic energy initially decreases till the highest point of projectile, because vertical component of velocity is decreasing (horizontal velocity remains same).

At maximum height, kinetic energy becomes minimum but not zero, because of some horizontal velocity. Thus after that, kinetic energy (KE) again increases with horizontal displacement (x) as magnitude of y-component of velocity increases on falling down. This is depicted in option (c).

18. For same target, there are two angle of projection, if one is θ, then other is $\left(\frac{\pi}{2}-\theta\right)$, hence required angle

$$=\frac{\pi}{2}-\frac{5\pi}{36}=\frac{18\pi-5\pi}{36}$$

$$=\frac{13\pi}{36}\text{ rad}$$

19. As, $E_k' = E_k\cos^2 30^\circ = \frac{3E_k}{4}$

20. Given, $u_x = u\cos\theta_0 = 30\times\frac{4}{5} = 24\text{ m/s}$

and $u_y\sin\theta_0 = 30\times\frac{3}{5} = 18\text{ m/s}$

After 1 s, u_x will remain as it is and u_y will decrease by 10 m/s or it will remain 8 m/s.

$$\therefore \quad \tan\theta = \frac{v_y}{u_x} = \frac{8}{24} = \frac{1}{3}$$

21. Since, range is given to be same, therefore the other angle is $(90^\circ - 30^\circ)$, *i.e.* 60°.

$$H = \frac{v^2\sin^2 30^\circ}{2g} = \frac{1}{4}\left[\frac{v^2}{2g}\right]$$

$$H' = \frac{v^2\sin^2 60^\circ}{2g} = \frac{3}{4}\left[\frac{v^2}{2g}\right]$$

$$\frac{H'}{H} = \frac{3}{4}\times\frac{4}{1} = 3 \quad\text{or}\quad H' = 3H$$

22. $R = \frac{v^2\sin 2\theta}{g} = 200,\ T = \frac{2v\sin\theta}{g} = 5$

Dividing, we get

$$\frac{v^2\times 2\sin\theta\cos\theta}{g}\times\frac{g}{2v\sin\theta} = \frac{200}{5} = 40$$

or $$v\cos\theta = 40\text{ ms}^{-1}$$

It may be noted here that the horizontal component of the velocity of projection remains the same during the flight of the projectile.

23. As the kinetic energy of a projectile at the highest, point is equal to half of the initial kinetic energy, so

$$(\text{KE})_H = \frac{1}{2}(\text{KE})_i$$

$$\frac{1}{2}mv^2\cos^2\theta = \frac{1}{2}\left(\frac{1}{2}mv^2\right) = \frac{1}{4}mv^2$$

or $$\cos^2\theta = \frac{1}{2} \quad\text{or}\quad \cos\theta = \frac{1}{\sqrt{2}}$$

or $$\theta = 45^\circ$$

24. $\because y = 10t - t^2$ (Given)

$$v_y\frac{d}{dt}(y) = \frac{d}{dt}(10t) - \frac{d}{dt}(t^2) = 10 - 2t$$

At maximum height, $v_y = 0$

$\therefore$ $$10 - 2t = 0$$

or $$2t = 10 \text{ or } t = 5\text{ s}$$

$\therefore$ $$y = (10\times 5 - 5\times 5) = 25\text{ m}$$

25. As, $\frac{v^2\sin 2\theta}{g} = \frac{\sqrt{3}v^2}{2g}$

or $$\sin 2\theta = \frac{\sqrt{3}}{2}$$

or $$2\theta = 60^\circ$$

or $$\theta = 30^\circ$$

Let us cross check with the help of data for vertical range.

$$\frac{v^2\sin^2\theta}{2g} = \frac{v^2}{8g} \quad\text{or}\quad \sin^2\theta = \frac{1}{4}$$

or $$\sin\theta = \frac{1}{2}$$

or $$\theta = 30^\circ$$

26. Change in momentum is the product of force and time.

$$\Delta p = mg\times\frac{2u\sin\theta}{g} \qquad \left(\because F = \frac{\Delta p}{\Delta t}\right)$$

$$= 2mv\sin\theta = 2mv\sin 45^\circ \qquad [\because \text{Here } u = v]$$

$$= \frac{2mv}{\sqrt{2}} = \sqrt{2}\,mv$$

27. Let v be the velocity of projection and θ the angle of projection.

Kinetic energy at highest point

$$= \frac{1}{2}mv^2\cos^2\theta \quad\text{or}\quad E_k\cos^2\theta$$

Potential energy at highest point

$$= E_k - E_k\cos^2\theta$$

$$= E_k(1-\cos^2\theta)$$

$$= E_k\sin^2\theta$$

28. Using, $v^2 - u^2 = 2as$, we get

$$s = \frac{v^2}{2g}$$

Now, $$\frac{v^2\sin 2\theta}{g} = \frac{v^2}{2g}$$

or $$\sin 2\theta = \frac{1}{2}$$

or $$\sin 2\theta = \sin 30^\circ \qquad \left[\because \sin 30^\circ = \frac{1}{2}\right]$$

or $$\theta = 15^\circ$$

29. $u_y = u\sin 30^\circ$

$$\Rightarrow u = \frac{u_y}{\sin 30^\circ} = \frac{80^\circ}{1/2} = 160\text{ ms}^{-1}$$

$\therefore u_x = u\cos 30° = 160 \times \frac{\sqrt{3}}{2} = 80\sqrt{3}\ \text{ms}^{-1}$

$$T = \frac{2u_y}{g} = \frac{2\times 80}{10} = 16\ \text{s}$$

At $t = \frac{T}{4} = 4\ \text{s},\ v_x = 80\sqrt{3}\ \text{ms}^{-1}$

$$v_y = 80 - 10\times 4 = 40\ \text{ms}^{-1}$$

$\therefore$ $v = \sqrt{(80\sqrt{3})^2 + (40)^2} = 144.2\ \text{ms}^{-1}$

$$\approx 140\ \text{ms}^{-1}$$

30. As, $R = \frac{v^2\sin 2\theta}{g}$

In the given problem, $v^2\sin 2\theta = \text{constant}$

$$v^2\sin 2\theta = \left(\frac{v}{2}\right)^2 \sin 30° = \frac{v^2}{8} \quad \left[\because v = \frac{v}{2} \text{ and } \theta = 15°\right]$$

or $\sin 2\theta = \frac{1}{8}$ or $2\theta = \sin^{-1}\left(\frac{1}{8}\right)$

or $\theta = \frac{1}{2}\sin^{-1}\left(\frac{1}{8}\right)$

31. Here, $\mathbf{v} = 6\hat{\mathbf{i}} + 8\hat{\mathbf{j}}\ \text{ms}^{-1}$

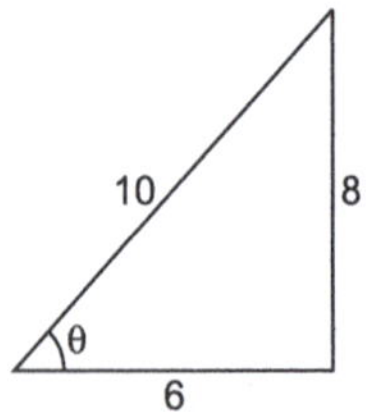

Comparing with $\mathbf{v} = v_x\hat{\mathbf{i}} + v_y\hat{\mathbf{j}}$, we get

$v_x = 6\ \text{ms}^{-1}$

and $v_y = 8\ \text{ms}^{-1}$

Also, $v^2 = v_x^2 + v_y^2$

$$= 36 + 64 = 100$$

or $v = 10\ \text{ms}^{-1}$

$$\sin\theta = \frac{8}{10} \text{ and } \cos\theta = \frac{6}{10}$$

$$R = \frac{v^2\sin 2\theta}{g} = \frac{2v^2\sin\theta\cos\theta}{g}$$

$$R = 2\times 10\times 10\times \frac{8}{10}\times\frac{6}{10}\times\frac{1}{10}\ \text{m} = 9.6\ \text{m}$$

32. $h = v\sin\theta t - \frac{1}{2}gt^2$

or $\frac{1}{2}gt^2 - v\sin\theta t + h = 0$

$$t_1 + t_2 = -\frac{-v\sin\theta}{\frac{1}{2}g}$$

or $t_1 + t_2 = \frac{2v\sin\theta}{g} = T$

$$T = (1+3)\ \text{s} = 4\ \text{s}$$

33. Since, $v_1 \perp v_2$

$\therefore$ $v_1 \cdot v_2 = 0$

or $(u_1\hat{\mathbf{i}} - gt\hat{\mathbf{j}})\cdot(-u_2\hat{\mathbf{i}} - gt\hat{\mathbf{j}}) = 0$

$\therefore$ $g^2t^2 = u_1u_2$

or $t = \frac{\sqrt{u_1u_2}}{g}$

34. Horizontal range of a projectile is given by

$$R = \frac{u^2\sin 2\theta}{g}$$

If $\theta = 45°$, then R is maximum and is equal to

$$R_{\text{max}} = \frac{u^2}{g}$$

Given, $R_{\text{max}} = 100\ \text{m}$

$\therefore$ $100 = \frac{u^2}{g}$ ...(i)

When cricketer throws the ball vertically upward, then ball goes upto height H.

Using equation of motion,

$$v^2 = u^2 + 2as$$

$$(0)^2 = u^2 + 2(-g)H$$

or $H = \frac{u^2}{2g} = \frac{1}{2}\left(\frac{u^2}{g}\right)$

$= \frac{1}{2}\times 100$ [using Eq. (i)]

$$= 50\ \text{m}$$

35. Horizontal component $= u\cos\theta$

Vertical component $= u\sin\theta$

$$g = -10\ \text{ms}^{-2},\ u = 50\ \text{ms}^{-1},\ h = 5\ \text{m},\ t = 2\ \text{s}$$

$$h = u_y t + \frac{1}{2}gt^2$$

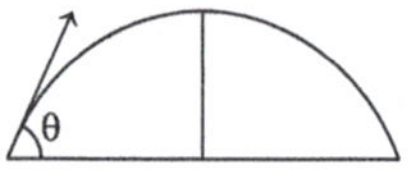

$\therefore$ $5 = 50\sin\theta - \frac{1}{2}\times 10\times 4$

or $5 = 50\sin\theta - 20$

or $\sin\theta = \frac{25}{50} = \frac{1}{2}$

$\therefore$ $\theta = 30°$

36. Given, initial velocity $(u) = 40$ m/s

Height of the hall $(H) = 25$ m

Let the angle of projection of the ball be θ, when maximum height attained by it is 25 m.

Maximum height attained by the ball,

$$H = \frac{u^2\sin^2\theta}{2g}$$

$$25 = \frac{(40)^2\sin^2\theta}{2\times 9.8}$$

or $\quad \sin^2\theta = \dfrac{25\times 2\times 9.8}{1600} = 0.3063$

or $\quad \sin\theta = 0.5534 = \sin 33.6^\circ$

or $\quad \theta = 33.6^\circ$

$\therefore$ Horizontal range $(R) = \dfrac{u^2\sin 2\theta}{g}$

$$= \frac{(40)^2\sin 2\times 33.6^\circ}{9.8} = \frac{1600\times\sin 67.2^\circ}{9.8} = \frac{1600\times 0.9219}{9.8} = 150.5\text{ m}$$

37. The velocity of the particle at any time t

$$\mathbf{v} = \mathbf{v}_0 + \mathbf{a}\,t$$

The x-component is $v_x = v_{ox} + a_x t$

The y-component is $v_y = v_{oy} + a_x = (-0.5\,t)\text{ ms}^{-1}$

When the particle reaches its maximum x-coordinate, $v_x = 0$.

i.e. $\quad 3 - t = 0 \Rightarrow t = 3\text{ s}$

The y-component of the velocity of this time is

$$v_y = -0.5\times 3 = -1.5\text{ ms}^{-1}$$

38. As, $H_1 - H_2 = \dfrac{H_1 + H_2}{2}$

or $\quad H_1 = 3H_2$

$\therefore \quad \dfrac{u^2\sin^2\theta}{2g} = 3\left\{\dfrac{u^2\sin^2(90^\circ-\theta)}{2g}\right\}$

$\tan^2\theta = 3$

$\therefore \quad \tan\theta = \sqrt{3}$

or $\quad \theta = 60^\circ$

Therefore, the other angle is $(90^\circ - \theta)$ or 30°.

Hence, when angle of projection is 30°, then stone attains smaller height.

39. Given, $R = H$

$$\frac{u^2\sin 2\alpha}{g} = \frac{u^2\sin^2\alpha}{2g}$$

or $\quad 2\sin\alpha\cos\alpha = \dfrac{\sin^2\alpha}{2}$

or $\quad \dfrac{\sin\alpha}{\cos\alpha} = 4$

or $\quad \tan\alpha = 4$

$\therefore \quad \alpha = \tan^{-1}(4)$

40. Let Δt be the time interval, then

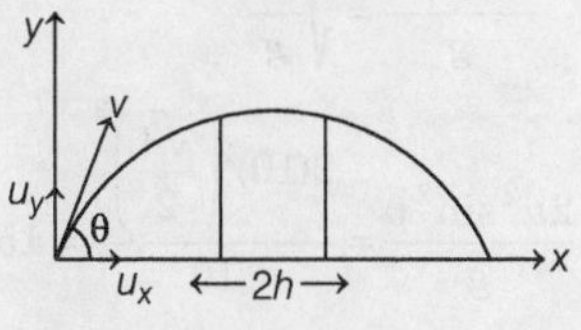

$$2h = (u_x)(\Delta t)$$

or $\quad u_x = \dfrac{2h}{\Delta t} \quad$ ...(i)

Further, $\quad h = u_y t - \dfrac{1}{2}gt^2$

or $\quad gt^2 - 2u_y t + 2h = 0$

$\therefore \quad t_1 = \dfrac{2u_y + \sqrt{4u_y^2 - 8gh}}{2g}$

and $\quad t_2 = \dfrac{2u_y - \sqrt{4u_y^2 - 8gh}}{2g}$

$$\Delta t = t_1 - t_2 = \frac{\sqrt{4u_y^2 - 8gh}}{g}$$

or $\quad u_y^2 = \dfrac{g^2(\Delta t)^2}{4} + 2gh$

Given, $\quad u_x^2 + u_y^2 = (2\sqrt{gh})^2$

$\therefore \quad \dfrac{4h^2}{(\Delta t)^2} + \dfrac{g^2(\Delta t)^2}{4} + 2gh = 4gh$

$$\frac{g^2}{4}(\Delta t)^4 - 2gh(\Delta t)^2 + 4h^2 = 0$$

$$(\Delta t)^2 = \frac{2gh \pm \sqrt{4g^2h^2 - 4g^2h^2}}{g^2/2} = \frac{4h}{g}$$

or $\quad \Delta t = 2\sqrt{\dfrac{h}{g}}$

41. Average velocity $= \dfrac{\text{Displacement}}{\text{Time}}$

$$v_{av} = \frac{\sqrt{H^2 + \dfrac{R^2}{4}}}{T/2} \quad \text{...(i)}$$

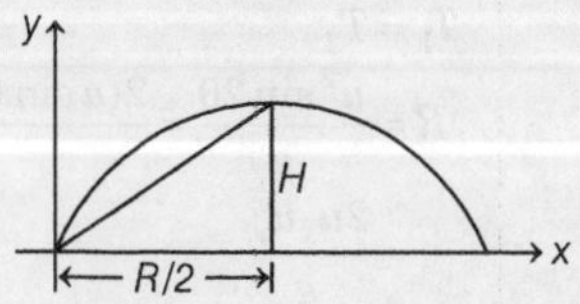

Here, $\quad H = \text{maximum height} = \dfrac{v^2\sin^2\theta}{2g}$,

$R = \text{range} = \dfrac{v^2\sin 2\theta}{g}$,

and $\quad T = \text{time of light} = \dfrac{2v\sin\theta}{g}$

Substituting in Eq. (i), we get

$$v_{av} = \frac{v}{2}\sqrt{1 + 3\cos^2\theta}$$

42. According to question,

$$R_{max} = R_0 = \frac{u^2}{g} = 40$$

$\Rightarrow \quad u^2 = 40\times g = 40\times 10 = 400$

$\Rightarrow \quad u = 20\text{ m/s}$

When the car is moving : We will take ground as a frome of reference,

In ground frame; Range

$$R = \frac{2u_x u_y}{g}$$

$$\Rightarrow \quad R = \frac{2\,(20 + u\cos\theta)\,u\sin\theta}{g}$$

Range R will be maximum, if $\frac{dR}{d\theta} = 0$

$$\Rightarrow \quad \frac{2[20u\cos\theta + u^2(\cos^2\theta - \sin^2\theta)]}{g} = 0$$

$$\Rightarrow \quad 20\cos\theta + u(\cos^2\theta - \sin^2\theta) = 0$$

$$\Rightarrow \quad 20\cos\theta + (2\cos^2\theta - 1)u = 0$$

$$\Rightarrow \quad 20\cos\theta + 40\cos^2\theta - 20 = 0 \qquad [\because u = 20 \text{ m/s}]$$

$$\Rightarrow \quad 2\cos^2\theta + \cos\theta - 1 = 0$$

$$\Rightarrow \quad \cos\theta = -1, \frac{1}{2}$$

$$\Rightarrow \quad \cos\theta = \frac{1}{2}$$

$$\Rightarrow \quad \theta = 60^\circ$$

43. $\frac{R/2}{H} = \frac{\sqrt{3}\,H}{H} = \sqrt{3}$

or $$\frac{(v_0^2 \sin\theta\cos\theta)/g}{(v_0^2\sin^2\theta)/2g} = \sqrt{3}$$

$$2\cot\theta = \sqrt{3}$$

$$\tan\theta = \frac{2}{\sqrt{3}}$$

or $$\theta = \tan^{-1}\left(\frac{2}{\sqrt{3}}\right)$$

44. Maximum height and time of flight depend on the vertical component of initial velocity.

$$H_1 = H_2 \Rightarrow u_{y_1} = u_{y_2}$$

Here $$T_1 = T_2$$

Range $$R = \frac{u^2\sin 2\theta}{g} = \frac{2(u\sin\theta)\,(u\cos\theta)}{g} = \frac{2u_x u_y}{g}$$

$$R_2 > R_1$$

$\therefore$ $$u_{x_2} > u_{x_1} \text{ or } u_2 > u_1$$

45. $t = \sqrt{\frac{2h}{g}}$

Distance from the foot of the tower

$$d = vt = v\sqrt{\frac{2h}{g}} = 250 \text{ m}$$

When velocity $= \frac{v}{2}$ and height of tower $= 4h$, then

distance $x - \frac{v}{2}\sqrt{\frac{2(4h)}{g}}$

$$x = v\sqrt{\frac{2h}{g}} = 250 \text{ m}$$

46. Range of the projectile on an inclined plane (down the plane) is,

$$R = \frac{u^2}{g\cos^2\beta}[\sin(2\alpha + \beta) + \sin\beta]$$

Here, $$u = v_0, \alpha = 0 \text{ and } \beta = \theta$$

$\therefore$ $$R = \frac{2v_0^2\sin\theta}{g\cos^2\theta}$$

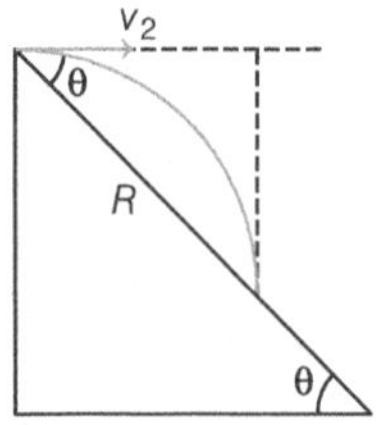

Now, $$x = R\cos\theta = \frac{2v_0^2\tan\theta}{g}$$

and $$y = -R\sin\theta = -\frac{2v_0^2\tan^2\theta}{g}$$

47. If it is being hit, then

$$d = v_0 t + \frac{1}{2}at^2 = (u\cos\theta)t$$

($\because$ acceleration in horizontal direction is zero)

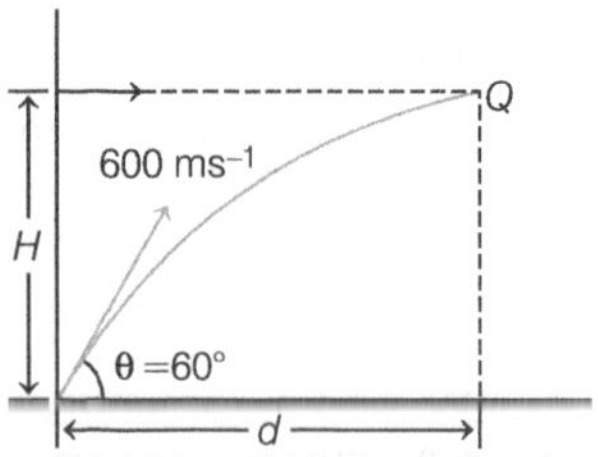

or $$t = \frac{u\cos\theta - v_0}{a/2}$$

$\therefore$ $$t = \frac{600 \times \frac{1}{2} - 250}{10} = 5 \text{ s}$$

$$H = (u\sin\theta)t - \frac{1}{2} \times gt^2$$

$$= 600 \times \frac{\sqrt{3}}{2} \times 5 - \frac{1}{2} \times 10 \times 25$$

$$H = 2473 \text{ m}$$

48. Horizontal component of velocity of A is $10\cos 60^\circ$ or 5 ms^{-1} which is equal to the velocity of B in horizontal direction. They will collide at C, if time of flight of the particles are equal or $t_A = t_B$

$$\frac{2u\sin\theta}{g} = \sqrt{\frac{2h}{g}} \qquad \left(\because h = \frac{1}{2}gt_B^2\right)$$

or $$h = \frac{2u^2\sin^2\theta}{g} = \frac{2(10)^2\left(\frac{\sqrt{3}}{2}\right)^2}{10} = 15 \text{ m}$$

49. From figure,

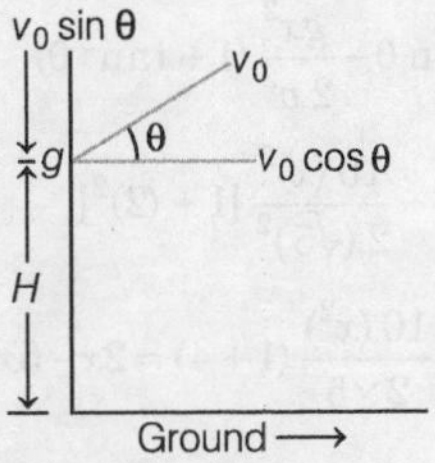

$$H = (-v_0 \sin\theta)t + \frac{1}{2}gt^2$$

$$v_x = v_0 \cos\theta$$

$$v_y^2 = (v_0 \sin\theta)^2 + 2gH$$

$$v = \sqrt{v_x^2 + v_y^2} \text{ at ground}$$

$$v = \sqrt{v_0^2 + 2gH}$$

It means speed is independent of angle of projection.

Also, $\frac{1}{2}gt^2 = H + t\, v_0 \sin\theta$

From this we can say that as θ increases, t increases.

50. Vertical component of velocity of A should be equal to vertical velocity of B.

or $v_1 \sin 30^\circ = v_2$

or $\frac{v_1}{2} = v_2$

$\therefore$ $\frac{v_2}{v_1} = \frac{1}{2}$

51. Since, the projectile is released, therefore, its initial velocity is same as the velocity of the plane at the time of release.

Take the origin at the point of release.

Let x and $y(= -730 \text{ m})$ be the coordinates of the point on the ground where the projectile hits and let t be the time when it hits. Then,

$$y = -v_0\, t \cos\theta - \frac{1}{2}gt^2$$

where, $\theta = 53.0^\circ$

This equation gives $v_0 = -\dfrac{y + \frac{1}{2}gt^2}{t\cos\theta}$

$$= \frac{-730 + \frac{1}{2}(9.8)(5)^2}{5\cos 53^\circ} = 202 \text{ ms}^{-1}$$

52. Horizontal component of velocity, $v_x = 500 \text{ ms}^{-1}$ and vertical component of velocity while striking the ground

$$v_y = 0 + 10 \times 10 = 100 \text{ ms}^{-1}$$

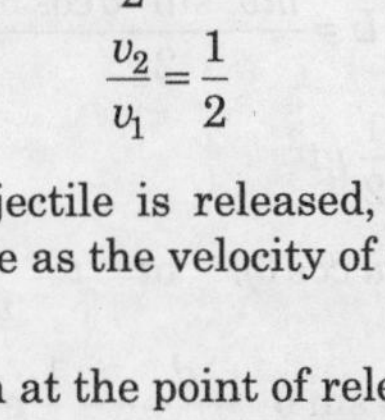

$\therefore$ Angle with which it strikes the ground,

$$\theta = \tan^{-1}\left(\frac{v_y}{v_x}\right) = \tan^{-1}\left(\frac{100}{500}\right)$$

$$\theta = \tan^{-1}\left(\frac{1}{5}\right)$$

53. As, $v_x = v\cos(\alpha - \theta);\ v_y = v\sin(\alpha - \theta)$

$$a_x = -g\sin\theta;\ a_y = -g\cos\theta$$

If T is the time of flight, then

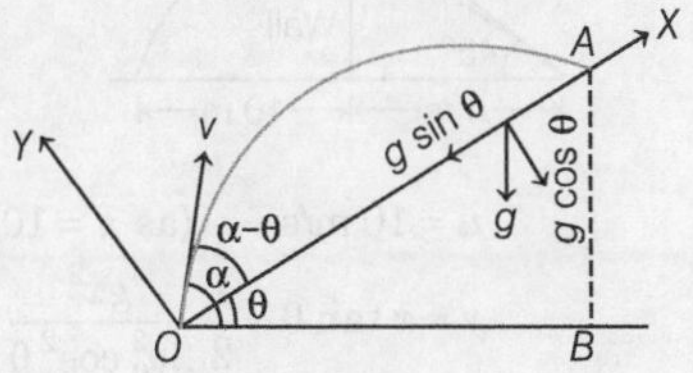

$$T = \frac{2v\sin(\alpha - \theta)}{g\cos\theta}$$

Again, $OB = v\cos\alpha \times T$

Now, $\cos\theta = \frac{OB}{OA}$

or $OA = \frac{OB}{\cos\theta}$

or $OA = \frac{v\cos\alpha \cdot T}{\cos\theta}$

or $OA = v\cos\alpha \times \frac{2v\sin(\alpha-\theta)}{g\cos\theta} \times \frac{1}{\cos\theta}$

or $OA = \frac{v^2}{g\cos^2\theta}[2\sin(\alpha-\theta)\cos\alpha]$

or $OA = \frac{v^2}{g\cos^2\theta}[\sin(2\alpha-\theta) + \sin(-\theta)]$

or $OA = \frac{v^2}{g\cos^2\theta}[\sin(2\alpha-\theta) - \sin\theta]$

Clearly, the range $R (= OA)$ will be maximum when $\sin(2\alpha - \theta)$ is maximum, *i.e.* 1. This would mean

$$2\alpha - \theta = \frac{\pi}{2} \quad \text{or} \quad \alpha = \frac{\theta}{2} + \frac{\pi}{4}$$

Maximum range up the inclined plane,

$$R_{max} = \frac{v^2}{g\cos^2\theta}(1 - \sin\theta) = \frac{v^2(1-\sin\theta)}{g(1-\sin^2\theta)}$$

$$= \frac{v^2(1-\sin\theta)}{g(1-\sin\theta)(1+\sin\theta)} = \frac{v^2}{g(1+\sin\theta)}$$

54. From $h = \frac{1}{2}gt^2$,

We have $t_{OB} = \sqrt{\frac{2h_{OA}}{g}} = \sqrt{\frac{2 \times 1960}{9.8}} = 20 \text{ s}$

Horizontal distance, $AB = vt_{OB}$

$$= \left(600 \times \frac{5}{18}\right)(20) = 3333.33 \text{ m} = 3.33 \text{ km}$$

55. $R = \dfrac{2 \times 30 \times 30 \sin 30^\circ \cos 60^\circ}{10 \cos^2 30^\circ}$

$= 180 \times \dfrac{1}{2} \times \dfrac{1}{2} \times \dfrac{2 \times 2}{3} \text{ m} = 60 \text{ m}$

Round II

1. As range $= 10 = \dfrac{u^2 \sin 2\theta}{g} \Rightarrow u^2 = 10g$

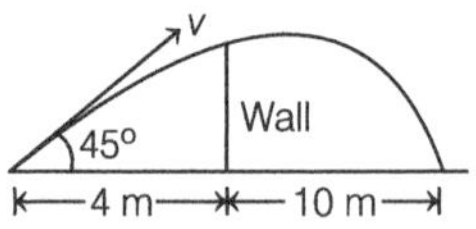

$\therefore \quad u = 10 \text{ m/s} \quad (\text{as } g = 10 \text{ m/s}^2)$

$$y = x \tan \theta - \frac{1}{2} \frac{gx^2}{2v_0^2 \cos^2 \theta}$$

$$= 4 \tan 45^\circ - \frac{1}{2} \frac{g \times 16}{2 v_0^2 \cos^2 45^\circ}$$

$$= 4 \times 1 - \frac{1}{2} \frac{10 \times 16}{2 \times 10 \times 10 \times \frac{1}{2}}$$

$$= 4 - 0.8$$

$$= 3.2 \approx 3.6 \text{ m}$$

2. According to law of conservation of linear momentum at the highest point,

$$mv \cos \theta = \frac{m}{2}(-v \cos \theta) + \frac{m}{2} v_1$$

or $\quad v_1 = 3v \cos \theta$

3. $R = \dfrac{u^2 \sin 2\theta}{g} = \dfrac{2u_x u_y}{g}$

$\therefore$ Range $\propto$ horizontal initial velocity component (u_x)

In path 4, range is maximum of football and has maximum horizontal velocity component in this path.

4. Time of flight, $T = \dfrac{2u_y}{g}$

$$u_y = \frac{gT}{2} = \frac{10 \times 6}{2} = 30 \text{ ms}^{-1}$$

Vertical velocity after 2s, $v_y = u_y - gT$

$$= 30 - 10 \times 2$$

$$= 30 - 20 = 10 \text{ ms}^{-1}$$

According to question,

$$\tan 30^\circ = \frac{v_y}{u_x} = \frac{10}{u_x}$$

or $\quad u_x = \dfrac{10}{\tan 30^\circ} = 10\sqrt{3} \text{ ms}^{-1}$

$\therefore \quad u = \sqrt{u_x^2 + u_y^2} = \sqrt{(10\sqrt{3})^2 + (30)^2}$

$$= 20\sqrt{3} \text{ ms}^{-1}$$

5. Initial velocity, $\mathbf{v} = (\hat{\mathbf{i}} + 2\hat{\mathbf{j}})$ m/s

Magnitude of velocity, $\mathbf{v} = \sqrt{(1)^2 + (2^2)} = \sqrt{5}$ m/s

Equation of trajectory of projectile,

$$y = x \tan \theta - \frac{gx^2}{2v^2}(1 + \tan^2 \theta) \quad \left(\tan \theta = \frac{y}{x} = \frac{2}{1} = 2\right)$$

$\therefore \quad y = x \times 2 - \dfrac{10(x)^2}{2(\sqrt{5})^2}[1 + (2)^2]$

$$= 2x - \frac{10(x^2)}{2 \times 5}(1 + 4) = 2x - 5x^2$$

6. Maximum speed with which the boy can throw stone is

$$u = \sqrt{2gh} = \sqrt{2 \times 10 \times 10} = 10\sqrt{2} \text{ m/s}$$

Range is maximum when projectile is thrown at an angle of 45°, thus

$$R_{\text{max}} = \frac{u^2}{g} = \frac{(10\sqrt{2})^2}{10} = 20 \text{ m}$$

7. Angular momentum of the projectile as given by

$$L = mv_h r_\perp = m(v \cos \theta) h$$

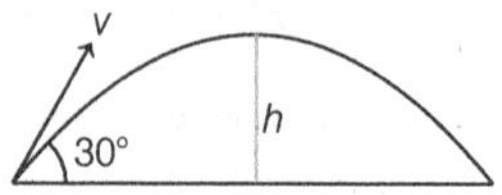

where, h is the maximum height.

$$= m(v \cos \theta)\left(\frac{v^2 \sin^2 \theta}{2g}\right)$$

$$L = \frac{mv^3 \sin^2 \theta \cos \theta}{2g} = \frac{\sqrt{3}}{16} \frac{mv^3}{g}$$

8. $h = (u \sin \theta)t - \dfrac{1}{2}gt^2$

$$d = (u \cos \theta)t \quad \text{or} \quad t = \frac{d}{u \cos \theta}$$

$$h = u \sin \theta \cdot \frac{d}{u \cos \theta} - \frac{1}{2} g \cdot \frac{d^2}{u^2 \cos^2 \theta}$$

$$u = \frac{d}{\cos \theta} \sqrt{\frac{g}{2(d \tan \theta - h)}}$$

9. For $\theta_A = 15^\circ$ and $\theta_B = 75^\circ$, $R_A = R_B$

$$u_A = u_B,$$

But $\quad \dfrac{h_A}{h_B} = \dfrac{u_A^2 \sin^2 \theta_A}{u_B^2 \sin^2 \theta_B} = \left(\dfrac{\sin 15^\circ}{\sin 75^\circ}\right)^2 < 1$

or $\quad h_A < h_B$

Again, $\quad \dfrac{T_A}{T_B} = \dfrac{u_A \sin \theta_A}{u_B \sin \theta_B} = \dfrac{\sin 15^\circ}{\sin 75^\circ} < 1$

or $\quad T_A < T_B$

10. Horizontal range is same when angle of projection is θ and $(90^\circ - \theta)$.

$\therefore \quad R = \dfrac{u^2 \sin 2\theta}{g} = \dfrac{2u^2 \sin \theta \cos \theta}{g}$

When angle of projection is θ,

$$T_1 = \frac{2u \sin \theta}{g}$$

When angle of projection is $(90° - \theta)$,

$$T_2 = \frac{2u\sin(90° - \theta)}{g} = \frac{2u\cos\theta}{g}$$

$$\therefore \quad T_1T_2 = \frac{4u^2\sin\theta\cos\theta}{g^2} = \left(\frac{2u^2\sin\theta\cos\theta}{g}\right)\left(\frac{2}{g}\right) = \frac{2R}{g}$$

$$\therefore \quad T_1T_2 \propto R$$

and $$\frac{T_1}{T_2} = \frac{2u\sin\theta/g}{2u\cos\theta/g} = \tan\theta$$

11. Height, $h = \frac{v_0^2\sin^2\theta}{2g}$, i.e. $h \propto \sin^2\theta$

$$\therefore \quad \frac{h_1}{h_2} = \frac{\sin^2\theta_1}{\sin^2\theta_2} > 1$$

So, $$\sin^2\theta_1 > \sin^2\theta_2$$

or $$\theta_1 > \theta_2$$

Time of flight, $T = \frac{2v_0\sin\theta}{g}$

or $$T \propto \sin\theta$$

$$\therefore \quad \frac{T_1}{T_2} = \frac{\sin\theta_1}{\sin\theta_2} > 1$$

or $$T_1 > T_2$$

Horizontal range, $R = \frac{u^2\sin 2\theta}{g}$

or $$R \propto \sin 2\theta$$

$$\therefore \quad \frac{R_1}{R_2} = \frac{\sin 2\theta_1}{\sin 2\theta_2} \le 1$$

or $$R_1 \le R_2$$

Total energy of each particle will be equal to KE of each particle at the time of its projection.

12. $y = ax - bx^2$

For height of y to be maximum,

$$\frac{dy}{dx} = 0$$

or $$a - 2bx = 0$$

or $$x = \frac{a}{2b}$$

$$\therefore \quad y_{max} = a\left(\frac{a}{2b}\right) - b\left(\frac{a}{2b}\right)^2 = \frac{a^2}{4b}$$

and $$\left(\frac{dy}{dx}\right)_{x=0} = a = \tan\theta$$

where, θ = angle of projection

$$\therefore \quad \theta = \tan^{-1}(a).$$

13. $H = \frac{u^2\sin^2\theta}{2g}$

$$dH = \frac{2u\sin^2\theta}{2g}du$$

$$\therefore \quad \frac{dH}{H} = \frac{2du}{u} = 2 \times \frac{1}{10}$$

$\therefore$ % increase in $H = \frac{dH}{H} \times 100 = \frac{2}{10} \times 100 = 20\%$

14. For 5 s, weight of the body is balanced by the given force. Hence, it will move in a straight line as shown below.

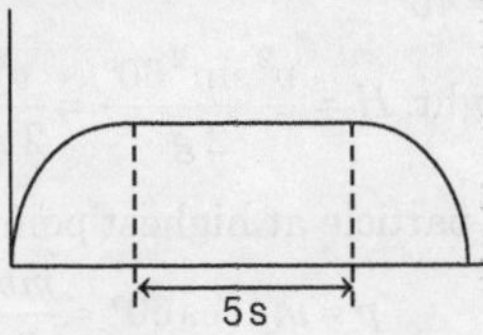

$$R = \frac{u^2\sin 2\theta}{g} + (u\cos\theta)(5)$$

$$= \frac{(50)^2 \cdot \sin 60°}{10} + (50 \times \cos 30°)(5) = 250\sqrt{3}\text{ m}$$

15. If a particle is projected with velocity u at an angle θ with the horizontal, the velocity of the particle at the highest point is

$$v = u\cos\theta = 200\cos 60° = 100\text{ ms}^{-1}$$

If m is the mass of the particle, then its initial momentum at highest point in the horizontal direction $= mv = m \times 100$. It means at the highest point, initially the particle has no momentum in vertically upward or downward direction. Therefore, after explosion, the final momentum of the particles going upward and downward must be zero.

Hence, the final momentum after explosion is the momentum of the third particle, in the horizontal direction. If the third particle moves with velocity v', then its momentum becomes $\frac{mv'}{3}$. According to law of conservation of linear momentum, we have

$$\frac{mv'}{3} = m \times 100 \text{ or } v' = 300\text{ ms}^{-1}.$$

16. Given, $\theta_1 = \pi/3 = 30°$

Horizontal range is same if $\theta_1 + \theta_2 = 90°$

$$\therefore \quad \theta_2 = 90° - 30° = 60°$$

$$y_1 = \frac{u^2\sin^2 30°}{2g}$$

and $$y_2 = \frac{u^2\sin^2 60°}{2g}$$

$$\therefore \quad \frac{y_2}{y_1} = \frac{\sin^2 30°}{\sin^2 60°} = \left(\frac{\frac{1}{2}}{\frac{\sqrt{3}}{2}}\right)^2 = \frac{1}{3}$$

or $$y_2 = \frac{y_1}{3}$$

17. Let v_1 be the velocity of the car and v_2 be the velocity of the parcel. The parcel is thrown at an angle θ from O, it reaches the man at M.

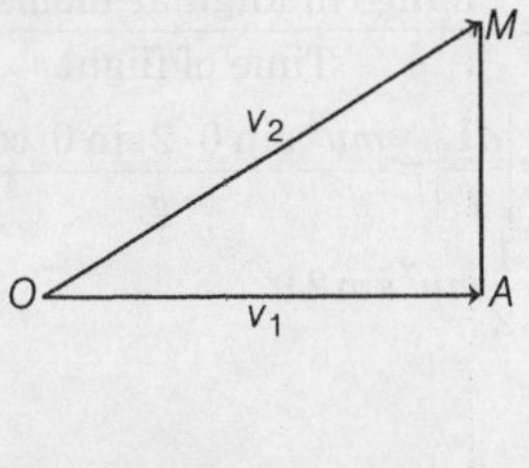

$\therefore \quad \cos\theta = \frac{v_1}{v_2} = \frac{10}{10\sqrt{2}} = \frac{1}{\sqrt{2}} = \cos 45°$

So, $\theta = 45°$

18. Maximum height, $H = \frac{v^2 \sin^2 60°}{2g} = \frac{v^2}{2g} \times \frac{3}{4} = \frac{3v^2}{8g}$

Momentum of particle at highest point,

$$p = mv\cos 60° = \frac{mv}{2}$$

Angular momentum $= pH = \frac{mv}{2} \times \frac{3v^2}{8g} = \frac{3mv^3}{16g}$

19. When projectile is at A, the given situation can be shown as

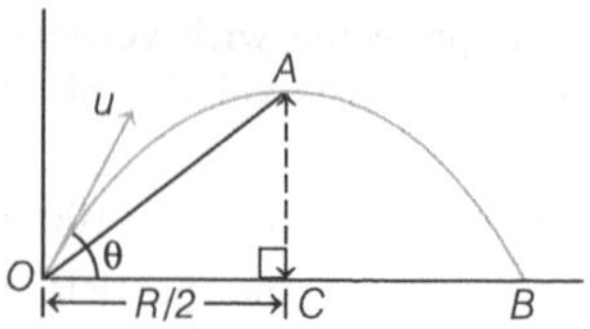

$$OC = \frac{R}{2} = \frac{1}{2}\frac{u^2}{g}\sin 2\theta$$

$$= \frac{1}{2} \times \frac{(20\sqrt{2})^2}{10}\sin 2 \times 45° = 40 \text{ m}$$

$$AC = H = \frac{u^2 \sin^2\theta}{2g} = \frac{(20\sqrt{2})^2}{2 \times 10}\sin^2 45° = 20 \text{ m}$$

$\therefore$ Displacement, $OA = \sqrt{OC^2 + CA^2} = \sqrt{40^2 + 20^2}$

Time of projectile from O to A

$$= \frac{1}{2}\left(\frac{2u\sin\theta}{g}\right) = \frac{u\sin\theta}{2g}$$

$$= \frac{(20\sqrt{2})\sin 45°}{10} = 2 \text{ s}$$

$\therefore$ Average velocity $= \frac{\text{Displacement}}{\text{Time}} = \frac{\sqrt{40^2 + 20^2}}{2}$

$$= 10\sqrt{5} \text{ ms}^{-1}$$

20. Time of flight, $t = \frac{2u\sin\theta}{g}$

Horizontal range, $R = \frac{u^2 \sin 2\theta}{g}$

Change in angular momentum,

$|d\mathbf{L}| = (\mathbf{L}_f - \mathbf{L}_i)$ about point of projection

$$= (mu\sin\theta) \times \frac{u^2 \sin 2\theta}{g}$$

$$= \frac{mu^3 \sin\theta \sin 2\theta}{g}$$

Torque $|\tau| = \frac{\text{Change in angular momentum}}{\text{Time of flight}}$

$$= \left|\frac{d\mathbf{L}}{T}\right| = \frac{mu^3 \sin\theta \; 2\sin\theta \; \cos\theta}{g} \times \frac{g}{2u\sin\theta}$$

or $\quad \tau = \frac{1}{2}mu^2 \sin 2\theta$

21. Time of flight of this particle, $T = 4$ s. If u is its initial speed and θ is the angle of projection, then

$$T = 4 = \frac{2u\sin\theta}{g}$$

or $\quad u\sin\theta = 2g \quad$ ...(i)

After 1 s, the velocity vector particle makes an angle of 45° with horizontal, so $v_x = v_y$

i.e., $\quad u\cos\theta = (u\sin\theta) - gt$

or $\quad u\cos\theta = 2g - g \quad (\because t = 1 \text{ s})$

or $\quad u\cos\theta = g \quad$ ...(ii)

Squaring and adding Eqs. (i) and (ii), we have

$$u^2 = 5g^2 = 5(10)^2 = 500$$

or $\quad u = \sqrt{500} = 22.36 \text{ ms}^{-1}$

Dividing Eq. (i) by Eq. (ii), we have

$$\tan\theta = 2 \text{ or } \theta = \tan^{-1}(2)$$

22. Let v be the velocity acquired by the body at B which will be moving and making an angle 45° with the horizontal direction. As the body just crosses the well, so $\frac{v^2}{g} = 40$

or $\quad v^2 = 40g = 40 \times 10 = 400$

or $\quad v = 20 \text{ ms}^{-1}$

Taking motion of the body from A to B along the inclined plane, we have

$$u = v_0,\ a = -g\sin 45° = -\frac{10}{\sqrt{2}} \text{ ms}^{-2},$$

$$s = 20\sqrt{2} \text{ m},\ v = 20 \text{ ms}^{-1}$$

As $\quad v^2 = u^2 + 2as$

$\therefore \quad 400 = v_0^2 + 2\left(-\frac{10}{\sqrt{2}}\right) \times 20\sqrt{2}$

or $\quad v_0^2 = 400 + 400 = 800$

or $\quad v = 20\sqrt{2} \text{ ms}^{-1}$

23. At $B, S_y = 0$

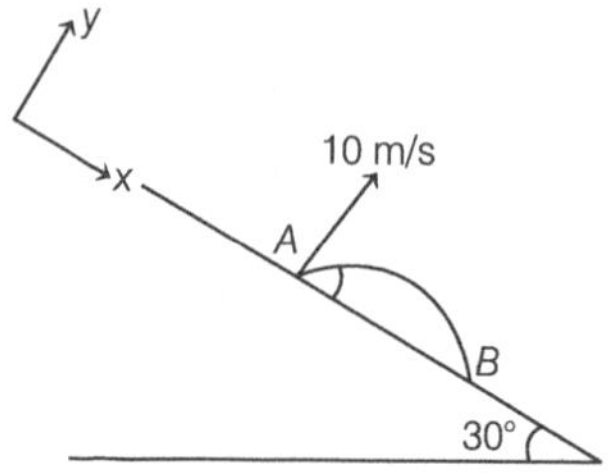

$\therefore \quad u_y t + \frac{1}{2}a_y t^2 = 0$

or $\quad t = -\frac{2u_y}{a_y} = \frac{-2(10)}{-10 \times \sqrt{3}/2} = \frac{4}{\sqrt{3}} \text{ s}$

Now, $\quad AB = R - \frac{1}{2}a_x t^2$

$$= \frac{1}{2}\left(10 \times \frac{1}{2}\right)\left(\frac{16}{3}\right) = 13.33 \text{ m}$$

24. Total mechanical energy at $O = mgh = 2\,mg$

Now, total mechanical energy at A

$$= \frac{1}{2}mv^2\left(1 + \frac{k^2}{R^2}\right) + 0.2\,mgh$$

Since, total mechanical energy is conserved

$$\Rightarrow \quad \frac{1}{2}mv^2\left(1 + \frac{k^2}{R^2}\right) + 0.2\,mgh = 2\,mg$$

$$\Rightarrow \quad \frac{1}{2}v^2\left(1 + \frac{\left(R\sqrt{\frac{2}{5}}\right)^2}{R^2}\right) + 0.2\,g = 2\,g$$

$$\left[\because k = \sqrt{\frac{2}{5}}\,R \text{ for spherical sphere}\right]$$

$$\Rightarrow \quad \frac{1}{2}v^2\left[1 + \frac{2}{5}\right] + 2 = 20 \Rightarrow v^2\left[\frac{7}{5}\right] = 36$$

$$\Rightarrow \quad v = \sqrt{\frac{36 \times 5}{7}} = \sqrt{\frac{180}{7}} = 5.1 \text{ m/s}$$

Now, range of projectile

$$\text{Range} = \frac{u^2 \sin 2\theta}{g} = \frac{(5.1)^2 \sin 60^\circ}{10} \approx 2.08 \text{ m}$$

25. We know that the range of projectile projected with velocity u, making an angle θ with the horizontal direction up the inclined plane, whose inclination with the horizontal direction is θ_0, is

$$R = \frac{u^2}{g\cos^2\theta_0}[\sin(2\theta - \theta_0) - \sin\theta_0]$$

Here, $\quad u = v,\ \theta = (90^\circ + \theta),\ \theta_0 = \theta$

$$\therefore \quad R = \frac{v^2}{g\cos^2\theta_0}\{\sin[2(90^\circ + \theta)] - \theta\} - \sin\theta\}$$

$$= \frac{v^2}{g\cos^2\theta_0}[\sin(180^\circ + \theta) - \sin\theta]$$

$$= -\frac{v^2}{g\cos^2\theta_0}2\sin\theta = -\frac{2v^2}{g}\tan\theta\sec\theta$$

$$\Rightarrow \quad R = \frac{2v^2}{g}\tan\theta \text{ (in magnitude)}$$

26. Horizontal velocity, $u = \dfrac{144 \times 1000}{60 \times 60} = 40 \text{ ms}^{-1}$

Time of flight, $T = \sqrt{2h/g} = \sqrt{2 \times 1000/9.8}$

Horizontal range $= 40\sqrt{2 \times 1000/9.8} = 571.43 \text{ m}$

27. Given, $T = \dfrac{2u\sin\theta}{g} = 2$ or $\dfrac{u\sin\theta}{g} = 1$

Maximum height,

$$H = \frac{u^2\sin^2\theta}{2g} = \frac{g}{2}\left(\frac{u^2\sin^2\theta}{g^2}\right) = \frac{10}{2} \times 1^2 = 5 \text{ m}$$

28. As horizontal range of the two stones is same. So the sum of angles of projection of two stones must be 90°, $30^\circ + \theta = 90^\circ$ or $\theta = 60^\circ$.

According to question, $y = \dfrac{u^2\sin^2 30^\circ}{2g} = \dfrac{u^2(1/2)^2}{2g}$

And $\quad y' = \dfrac{u^2\sin^2 60^\circ}{2g} = \dfrac{u^2(3\sqrt{3}/2)^2}{2g}$

$$\therefore \quad \frac{y'}{y} = 3 \text{ or } y' = 3y$$

29. If range is R then, $\dfrac{u^2 \sin 2 \times 75^\circ}{g} = R - 10$ ...(i)

and $\quad \dfrac{u^2\sin 2 \times 45^\circ}{g} = R + 10$

or $\quad \dfrac{u^2}{g} = R + 10$

From Eq. (i), $(R + 10)\sin 150^\circ = R - 10$

or $\quad (R + 10)\dfrac{1}{2} = R - 10$

or $\quad R = 30 \text{ m}$

30. Velocity before strike, $u = \sqrt{2gh}$

Component of acceleration along the inclined plane $= g\sin\alpha$ and the perpendicular component $= g\cos\alpha$

Using $\quad s = ut + \dfrac{1}{2}at^2$,

For vertical direction, we get

$$0 = v\cos\alpha t - \frac{1}{2}g\cos\alpha t^2$$

and for horizontal direction,

$$x = u\sin\alpha t + \frac{1}{2}g\sin\alpha t^2$$

$$= u\sin\alpha\frac{2u}{g} + \frac{1}{2}g\sin\alpha\left(\frac{2u}{g}\right)^2 \quad \left(\because t = \frac{2u}{g}\right)$$

$$= \frac{2u^2\sin\alpha}{g} + \frac{2u^2\sin\alpha}{g}$$

$$= \frac{4u^2\sin\alpha}{g} = 4 \times \frac{2gh \times \sin\alpha}{g} = 8h\sin\alpha$$

31.

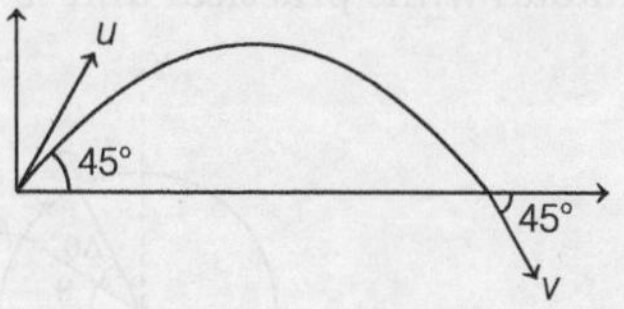

$$|\mathbf{u}| = |\mathbf{v}|$$

$$\mathbf{u} = u\cos 45\hat{\mathbf{i}} + u\sin 45\hat{\mathbf{j}} \quad \text{...(i)}$$

$$\mathbf{v} = v\cos 45\hat{\mathbf{i}} - v\sin 45\hat{\mathbf{j}} \quad \text{...(ii)}$$

$$|\Delta\mathbf{p}| = |m(\mathbf{v} - \mathbf{u})|$$

$$\Delta\mathbf{p} = 2mu\sin 45^\circ \quad \text{[from Eqs. (i) and (ii)]}$$

$$= 2 \times 5 \times 10^{-3} \times 5\sqrt{2} \times \frac{1}{\sqrt{2}}$$

$$= 50 \times 10^{-3} = 5 \times 10^{-2}$$

$$\therefore \quad x = 5$$

05

Circular Motion

When a particle moves in two dimensions or in a plane such that its distance from a fixed (or moving) point remains constant, then its motion is called *circular motion*. This fixed point is called the *centre* and the distance is called *radius of the circular path*. When the speed of the particle performing circular motion is constant, its motion is said to be *uniform circular motion*, if the speed of the particle performing circular motion changes with respect to time, its motion is said to be *non-uniform circular motion*.

IN THIS CHAPTER ....

- Kinematics of Circular Motion
- Dynamics of Circular Motion
- Circular Turning of Roads
- Conical Pendulum
- Motion in a Vertical Circle

Kinematics of Circular Motion

For a particle in circular motion, following variables are required to describe its motion.

Angular Displacement

The angle between initial and final positions of particle in a given interval of time, is called angular displacement. In the given figure **OP** is the initial position and **OP′** is the final position of the particle. Then, angular displacement

$$\angle P'OP = \Delta\theta$$

It is a dimensionless quantity, as $\theta = \Delta s / r$.

Its SI unit is radian while practical unit is degree.

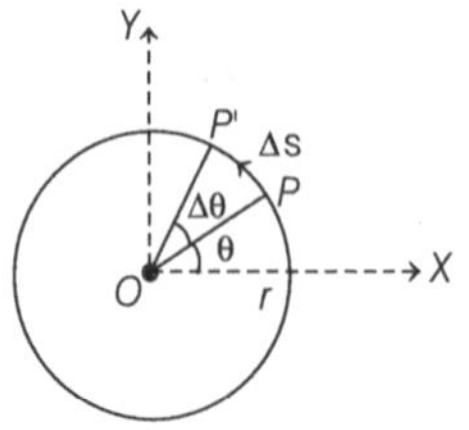

If a body makes n revolutions, its angular displacement $\theta = 2\pi n$ radian.

Note Angular displacement is a vector quantity, provided $\Delta\theta$ is small. Commutative law of vector addition is not valid for large $\Delta\theta$.

Angular Velocity

If the angular position of a particle changes with time, it is said to have angular velocity. If θ_1 and θ_2 are the angular positions of a particle at time t_1 and t_2 respectively, the average angular velocity ω_{av} for this time interval is defined as

$$\omega_{av} = \frac{\theta_2 - \theta_1}{t_2 - t_1} = \frac{\Delta\theta}{\Delta t}$$

and instantaneous angular velocity is defined as the limiting value of this ratio

i.e. $$\omega = \lim_{\Delta t \to 0} \frac{\Delta\theta}{\Delta t} = \frac{d\theta}{dt}$$

In case of uniform circular motion, $\omega = \omega_{av}$

Its unit is rads^{-1} and its dimensional formula is $[M^0L^0T^{-1}]$.

Important points regarding the angular velocity are given below.

(i) Average angular velocity is a scalar quantity but, instantaneous angular velocity is a vector quantity.

(ii) If a particle makes n rotations in t second.

$$\omega_{av} = \left[\frac{2\pi n}{t}\right] \text{rad s}^{-1}$$

Hence, if T is the time period and f is the frequency of uniform circular motion.

$$\omega = \frac{2\pi \times 1}{T} = 2\pi f \qquad \left[\because f = \frac{1}{T}\right]$$

or $\qquad T = 2\pi/\omega$

(iii) Angular velocity depends on the point about which rotation is considered. For the figure given below,

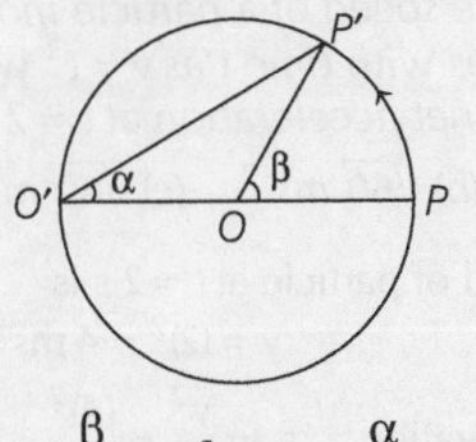

$$\omega_o = \frac{\beta}{t} \text{ and } \omega_{o'} = \frac{\alpha}{t}$$

Relation between angular velocity and linear velocity

If linear velocity of particle performing circular motion is v and angular velocity is ω, then

$$v = r\omega$$

where, r is the radius of the circular path.

In vector form, $\mathbf{v} = \boldsymbol{\omega} \times \mathbf{r}$

Note Linear velocity is always along the tangent to the circular path.

Relative angular velocity

Relative angular velocity for two important cases are as follows

(i) When two particles are moving on same circle or different coplanar concentric circles in same direction with different angular velocities ω_1 and ω_2, respectively, the angular velocity of particle 2 relative to particle 1 for an observer at the centre will be

$$\omega_{rel} = \omega_2 - \omega_1$$

(ii) When two particles are moving on two different concentric circles with different angular velocities, the angular velocity of particle 2 relative to particle 1 as observed by 1 will depend on their positions and velocities.

If 1 and 2 are closest to each other and moving in the same direction, then

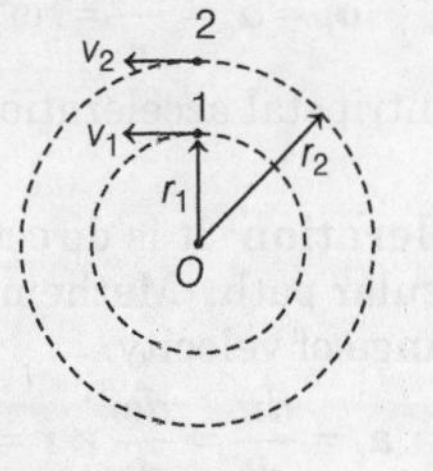

$$v_{rel} = v_2 - v_1$$

$$r_{rel} = r_2 - r_1$$

So, $$\omega_{rel} = \frac{v_{rel}}{r_{rel}} = \frac{v_2 - v_1}{r_2 - r_1}$$

Angular Acceleration (α)

If the angular speed of a particle is variable, the body is said to have an angular acceleration. Let ω_1 and ω_2 be the instantaneous angular speeds at times t_1 and t_2 respectively, then the *average angular acceleration* α_{av} is defined as

$$\alpha_{av} = \frac{\omega_2 - \omega_1}{t_2 - t_1} = \frac{\Delta\omega}{\Delta t}$$

The instantaneous angular acceleration is the limit of this ratio as Δt approaches to zero, *i.e.*

$$\alpha_{inst} = \lim_{\Delta t \to 0} \frac{\Delta\omega}{\Delta t} = \frac{d\omega}{dt} = \frac{d^2\theta}{dt^2}$$

The SI unit of angular acceleration is rad s^{-2} and its dimensional formula is $[M^0 L^0 T^{-2}]$.

If $\alpha = 0$, circular motion is said to be uniform.

It has same characteristics as that of angular velocity.

Kinematics Equations for Circular Motion

If the angular acceleration (α) is constant, then the kinematic equations of motion are as follows:

(i) $\omega = \omega_0 + \alpha t$

(ii) $\theta = \omega_0 t + \frac{1}{2}\alpha t^2$

(iii) $\omega^2 = \omega_0^2 + 2\alpha\theta$

(iv) $\theta = \left(\frac{\omega + \omega_0}{2}\right) t$

(v) $\theta_t = \omega_0 + \frac{1}{2}\alpha(2t - 1)$

where, ω_0 and ω are the initial and final angular velocities, respectively.

Radial and Tangential acceleration

In circular motion, acceleration of the particle can be resolved into two components

Radial acceleration It is directed along the radial direction towards the centre of the circular path. It is expressed as,

$$a_r = a_c = \frac{v^2}{r} = r\omega^2$$

It is also called centripetal acceleration or normal acceleration.

Tangential acceleration It is directed along the tangent to the circular path. Mathematically, it is equal to time rate of change of velocity.

It is expressed as, $\mathbf{a}_t = \frac{d\mathbf{v}}{dt} = \frac{d\omega}{dt} \times \mathbf{r} = \alpha \times \mathbf{r} \Rightarrow a_t = \alpha r$

Net acceleration The two components of acceleration are mutually perpendicular. So, the net acceleration is given as,

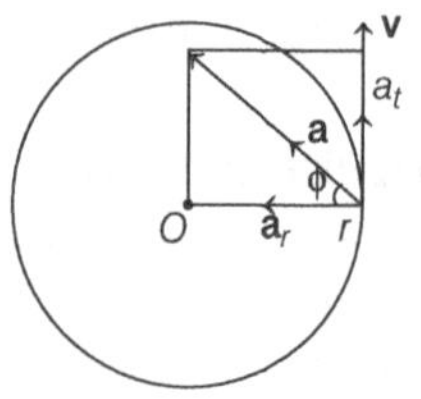

$$\mathbf{a} = \frac{d\mathbf{v}}{dt} = \mathbf{a}_t + \mathbf{a}_r, \text{ where } \mathbf{a}_t = \alpha \times \mathbf{r} \text{ and } \mathbf{a}_r = \omega \times \mathbf{v}$$

$$\Rightarrow \quad a = \sqrt{a_t^2 + a_r^2}$$

If a (acceleration) makes an angle ϕ with the radius, then $\tan\phi = \frac{a_t}{a_r}$.

Regarding circular motion, following possibilities exist

(a) If $a_r = 0$ and $a_t = 0$, then $a = 0$ and motion is uniform translatory.

(b) If $a_r = 0$ and $a_t \neq 0$, then $a = a_t$ and motion is accelerated translatory.

(c) If $a_r \neq 0$ but $a_t = 0$, then $a = a_r$ and motion is uniform circular.

(d) If $a_r \neq 0$ and $a_t \neq 0$, then $a = \sqrt{a_t^2 + a_r^2}$ and motion is non-uniform circular.

Example 1. *A clock has a continuously moving second's hand of 0.1 m length. The average acceleration of the tip of the hand (in ms^{-2}) is of the order of* **[JEE Main 2020]**

(a) 10^{-3}
(b) 10^{-4}
(c) 10^{-2}
(d) 10^{-1}

Sol. (a) Given that, radius, $R = 0.1$ m

Angular frequency, $\omega = \frac{2\pi}{T} = \frac{2\pi}{60} = 0.105$ rad/s

$\therefore$ Average acceleration, $a = \omega^2 R = (0.105)^2(0.1)$

$$= 1.102 \times 10^{-3} \text{ m/s}^2$$

and it is of the order of 10^{-3}.

Example 2. *A particle moves in a circle of radius $\left(\frac{20}{\pi}\right)$ m with constant tangential acceleration. If the speed of the particle is 80 m/s at the end of the second revolution after motion has begun, then the tangential acceleration is*

(a) $160\,\pi$ m/s^2 (b) $40\,\pi$ m/s^2
(c) 40 m/s^2 (d) $640\,\pi$ m/s^2

Sol. (c) Given, radius $r = \frac{20}{\pi}$ m, tangential acceleration $a_t =$ constant, velocity $v = 80$ m/s and initial angular velocity $\omega_0 = 0$,

Final angular velocity, $\omega_f = \frac{v}{r} = \frac{80}{20/\pi} = 4\pi$ rad/s

Angular displacement, $\theta = 2\pi \times 2 = 4\pi$

From 3rd equation of motion,

$$\omega^2 = \omega_0^2 + 2\alpha\theta$$

$$\Rightarrow \quad (4\pi)^2 = 0^2 + 2 \times \alpha \times (4\pi)$$

$$\alpha = 2\pi \text{ rad/s}^2$$

Also, $a = \alpha r = 2\pi \times \frac{20}{\pi} = 40$ m/s^2

Example 3. *The speed of a particle moving in a circle of radius $r = 2$ m varies with time t as $v = t^2$ where, t is in second and v in ms^{-1}. The net acceleration at $t = 2$ s is*

(a) $\sqrt{40}$ ms^{-2} (b) $\sqrt{60}$ ms^{-2} (c) $\sqrt{80}$ ms^{-2} (d) 10 ms^{-2}

Sol. (c) Linear speed of particle at $t = 2$ s is

$$v = (2)^2 = 4 \text{ ms}^{-1}$$

$\therefore$ Radial acceleration, $a_r = \frac{v^2}{r} = \frac{(4)^2}{2} = 8 \text{ ms}^{-2}$

The tangential acceleration is

$$a_t = \frac{dv}{dt} = 2t$$

At $t = 2$ s $\quad a_t = (2)(2) = 4 \text{ ms}^{-2}$

Net acceleration of the particle at $t = 2$ s is

$$a = \sqrt{(a_r)^2 + (a_t)^2} = \sqrt{(8)^2 + (4)^2}$$

or $\quad a = \sqrt{80}$ ms^{-2}

Example 4. *In a circular motion of a body, which amongst the following relation between tangential acceleration and linear velocity is correct?*

(a) $a_t = \mathbf{a} \cdot \mathbf{v}$ (b) $a_t = \mathbf{v}/\mathbf{a}$ (b) $a_t = \mathbf{a}/\mathbf{v}$ (d) $a_t = \frac{\mathbf{a} \cdot \mathbf{v}}{|\mathbf{v}|}$

Sol. (d) Let velocity of the particle be, $\mathbf{v} = v_x\hat{\mathbf{i}} + v_y\hat{\mathbf{j}}$

Acceleration, $\mathbf{a} = \frac{dv_x}{dt}\hat{\mathbf{i}} + \frac{dv_y}{dt}\hat{\mathbf{j}}$

Component of $\mathbf{a}$ along $\mathbf{v}$ will be, $\frac{\mathbf{a} \cdot \mathbf{v}}{|\mathbf{v}|} = \frac{v_x \frac{dv_x}{dt} + v_y \cdot \frac{dv_y}{dt}}{\sqrt{v_x^2 + v_y^2}}$ ...(i)

Further, tangential acceleration of particle is rate of change of speed.

or $$a_t = \frac{dv}{dt} = \frac{d}{dt}\left(\sqrt{v_x^2 + v_y^2}\right)$$

or $$a_t = \frac{1}{2\sqrt{v_x^2 + v_y^2}}\left[2v_x \cdot \frac{dv_x}{dt} + 2v_y \frac{dv_y}{dt}\right]$$

or $$a_t = \frac{v_x \cdot \frac{dv_x}{dt} + v_y \cdot \frac{dv_y}{dt}}{\sqrt{v_x^2 + v_y^2}} \quad \text{...(ii)}$$

From Eqs. (i) and (ii), we can see that

$$a_t = \frac{\mathbf{a} \cdot \mathbf{v}}{|\mathbf{v}|}$$

or Tangential acceleration = component of acceleration along velocity.

Dynamics of Circular Motion

In circular motion of a particle, there are two kinds of forces that occur. They are described below

Centripetal Force

When a particle perform circular motion, it is acted upon by a force directed along the radius towards the centre of the circle. This force is called the centripetal force.

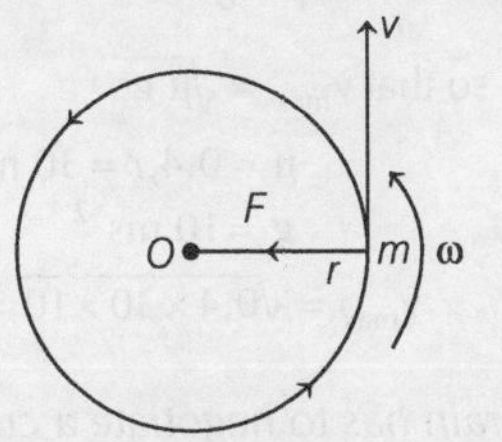

If m is the mass of the particle, then the centripetal force is given by

$$\mathbf{F} = \frac{mv^2}{r}\,\hat{\mathbf{r}} = -mr\omega^2\hat{\mathbf{r}}$$

where $\hat{\mathbf{r}}$ is the unit vector acting along $\mathbf{r}$.

Important Points

- In non-uniform circular motion, the particle simultaneously possesses two forces

Centripetal force, $F_c = ma_c = \frac{mv^2}{r} = mr\omega^2$

Tangential force, $F_t = ma_t$

$\therefore$ Net force, $F_{\text{net}} = ma = m\sqrt{a_c^2 + a_t^2}$

- If a moving particle comes to stand still, *i.e.* the particle will move along the radius towards the centre and if radial acceleration a_r is zero, the body will fly off along the tangent. So, a tangential velocity and a radial acceleration (hence force) is always present in uniform circular motion.

Centrifugal Force

It can be defined as the radially directed outward force acting on a body in circular motion, as observed by a person moving with the body.

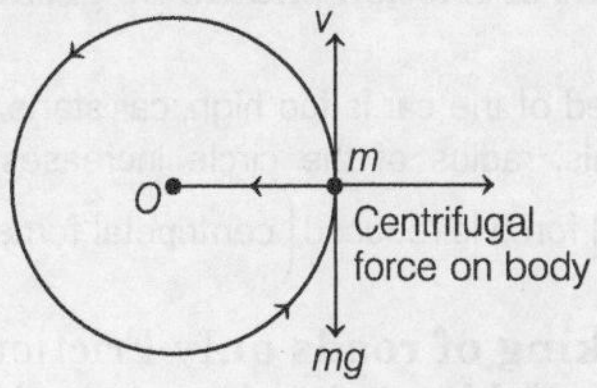

Centrifugal force is given as, $\mathbf{F} = \frac{mv^2}{r}\,\hat{\mathbf{r}} = mr\omega^2\hat{\mathbf{r}}$

Also,

(i) In an inertial frame, the centrifugal force does not act on the object.

(ii) In non-inertial rotating frames, pseudo force arises as centrifugal force and need to be considered.

Circular Turning of Roads

When vehicles go through turns, they travel along a nearly circular arc. There must be some force which will produce the required centripetal acceleration. If the vehicles travel on a horizontal circular path, this resultant force is also horizontal. The necessary centripetal force is being provided to the vehicles by following three ways:

(i) **By friction only** For a car of mass m moving with a speed v in a horizontal circular arc of radius r, the necessary centripetal force to the car will be provided by force of friction f acting towards centre.

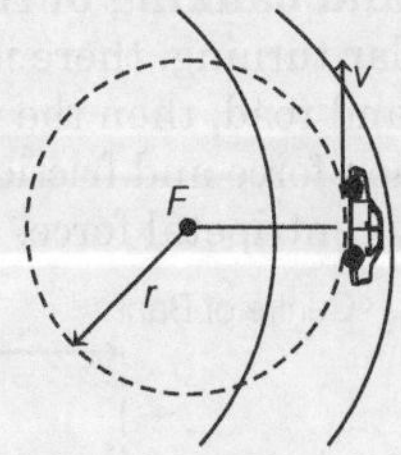

Thus, $$f = \frac{mv^2}{r}$$

Further, limiting value of f,

$$f_L = \mu N = \mu\, mg$$

Therefore, for a safe turn without sliding

$$\frac{mv^2}{r} \le f_L$$

or $$\frac{mv^2}{r} \le \mu\, mg$$

$$\mu \ge \frac{v^2}{rg} \text{ or } v \le \sqrt{\mu rg}$$

Here, two situations may arise. If μ and r are known to us, the speed of the vehicle should not exceed $\sqrt{\mu rg}$ and if v and r are known to us, the coefficient of friction should be greater than $\frac{v^2}{rg}$.

Note If the speed of the car is too high, car starts skidding outwards. Due to this, radius of the circle increases or the necessary centripetal force is reduced $\left(\text{centripetal force} \propto \frac{1}{r}\right)$.

(ii) **By banking of roads only** Friction is not always reliable at circular turns, if high speed and sharp turns are involved. To avoid dependence on friction, the roads are banked at the turn so that the outer part of the road is somewhat raised compared to the inner part.

∴ For safe turn without sliding,

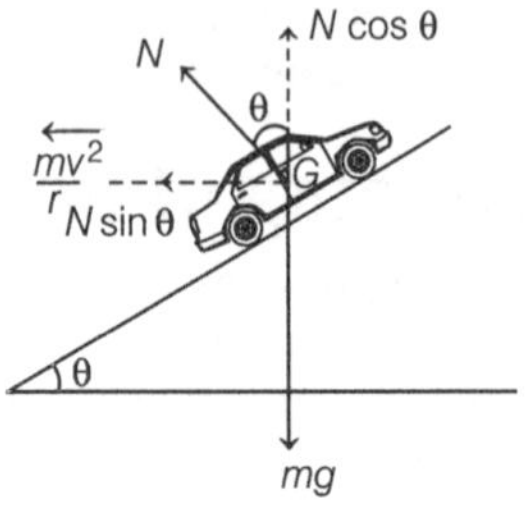

$$N \sin\theta = \frac{mv^2}{r} \text{ and } N\cos\theta = mg$$

From these two equations, we get

$$\tan\theta = \frac{v^2}{rg} \text{ or } v = \sqrt{rg\tan\theta}$$

(iii) **By friction and banking of roads both** If on a banked circular turning, there is a frictional force between car and road, then the vector sum of normal reaction force and frictional force provides the necessary centripetal force.

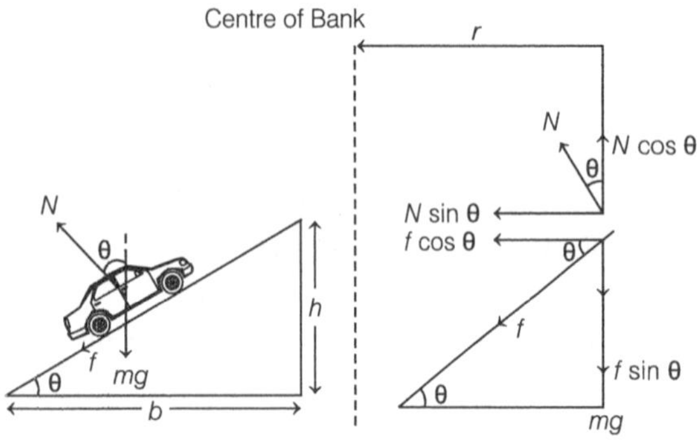

$$N \sin\theta + f\cos\theta = \frac{mv^2}{r} \quad \text{...(i)}$$

$$N\cos\theta = mg + f\sin\theta \quad \text{...(ii)}$$

(∵ vertical force is balanced)

Taking limiting condition, we can write

$$f = \mu_s N \quad \text{...(iii)}$$

To obtain the value of N, Solve above three equations properly.

$$N = \frac{mg}{\cos\theta - \mu_s \sin\theta}$$

After putting the value of N in Eq. (i), we get

∴ For a car moving upward on inclined road, the maximum speed for no skidding is given as,

$$v_{\text{max}} = \left[\frac{rg(\sin\theta + \mu_s\cos\theta)}{\cos\theta - \mu_s\sin\theta}\right]^{1/2} = \left[\frac{rg(\mu_s + \tan\theta)}{1 - \mu_s\tan\theta}\right]^{1/2}$$

Example 5. *Find the maximum speed at which a car can turn round a curve of 30 m radius on a level road, if the coefficient of friction between the tyres and the road is 0.4. (acceleration due to gravity = 10 ms^{-2})*

(a) 12 ms^{-1} (b) 10 ms^{-1} (c) 11 ms^{-1} (d) 15 ms^{-1}

Sol. (c) We know that for turning, centripetal force is provided by friction, so

$$\frac{mv^2}{r} \le f_L$$

As, $f_L = \mu N = \mu mg$ (where, $N = mg$)

Thus, $\frac{mv^2}{r} \le \mu mg$

i.e., $v \le \sqrt{\mu gr}$, so that $v_{\text{max}} = \sqrt{\mu gr}$

Here, $\mu = 0.4, r = 30$ m

and $g = 10$ ms^{-2}

So, $v_{\text{max}} = \sqrt{0.4 \times 30 \times 10} = 10.95 \simeq 11 \text{ms}^{-1}$

Example 6. *A train has to negotiate a curve of radius 2000 m. By how much should the outer rail be raised with respect to inner rail for a speed of 72 km h^{-1}. The distance between the rails is 1 m.*

(a) 4 cm (b) 5 cm
(c) 2 cm (d) 1 cm

Sol. (c) Given, $v = 72 \text{ km h}^{-1} = 72 \times \frac{5}{18} = 20 \text{ ms}^{-1}$

$l = 1$ m, $r = 2000$ m, $g = 10$ ms^{-2}

We have, $\tan\theta = \frac{v^2}{rg}$

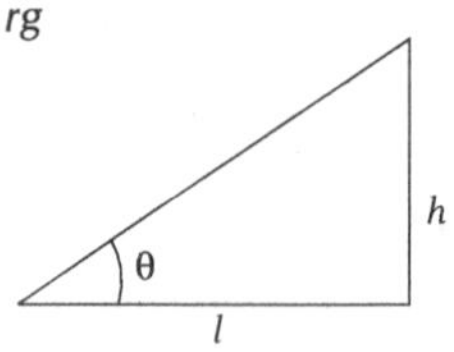

Also, $\tan\theta = \frac{h}{l}$

$\Rightarrow \quad \frac{v^2}{rg} = \frac{h}{l} \Rightarrow h = \frac{v^2 l}{rg}$

$\therefore \quad h = \frac{(20)^2 \times 1}{2000 \times 10} = \frac{1}{50} \text{ m} = \frac{100}{50} = 2 \text{ cm}$

Example 7. *A circular race track of radius 300 m is banked at angle of 15°. If the coefficient of friction between the wheels of a race-car and the road is 0.2, then the maximum permissible speed to avoid slipping is*

(a) 28.1 ms^{-1} (b) 50 ms^{-1} (c) 38.1 ms^{-1} (d) 42 ms^{-1}

Sol. (c) Here, $r = 300$ m, $\theta = 15°$, $\mu_s = 0.2$, $g = 10\ ms^{-2}$,

hence, maximum permissible speed to avoid slipping,

$$V_{max} = \left[\frac{rg(\mu_s + \tan\theta)}{1 - \mu_s \tan\theta}\right]^{1/2} = \left[\frac{300 \times 10[0.2 + \tan 15°]}{1 - 0.2 \tan 15°}\right]$$

We obtain, $v_{max} = 38.1\ ms^{-1}$

Conical Pendulum

It consists of a string OA, whose upper end O is fixed and bob of mass m is tied at the other free end which is whirled in a horizontal circle. As the string traces the surface of the cone, and such arrangement is called a conical pendulum.

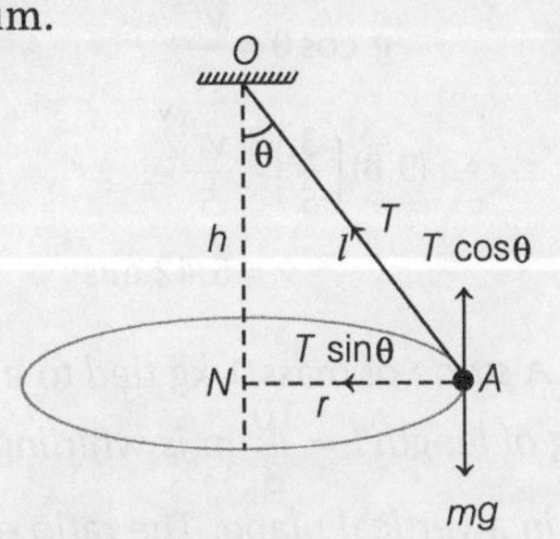

In this case, vertical component of tension T balances the weight of the bob and horizontal component provides the necessary centripetal force.

∴ Angular speed, $\omega = \sqrt{\frac{g \tan\theta}{r}}$

and time period of pendulum,

$$\tau = 2\pi\sqrt{\frac{l\cos\theta}{g}} = 2\pi\sqrt{\left(\frac{h}{g}\right)}$$

where, $h = ON = l\cos\theta$

Example 8. *A hemispherical bowl of radius R is rotating about its axis of symmetry which is kept vertical. A small ball kept in the bowl rotates with the bowl without slipping on its surface. If the surface of the bowl is smooth and the angle made by the radius through the ball with the vertical is α, find the angular speed at which bowl is rotating.*

(a) $\omega = \sqrt{\frac{R}{g\cos\alpha}}$ (b) $\omega = \sqrt{\frac{g}{\cos\alpha}}$

(c) $\sqrt{\frac{g}{R\cos\alpha}}$ (d) *None of these*

Sol. (c) Let ω be the angular speed of rotation of the bowl. Two forces are acting on the ball.

(i) Normal reaction, N (ii) Weight, mg

The ball is rotating in a circle of radius, $r = (R \sin\alpha)$ with centre at A with an angular speed, ω

Hence, $N \sin\alpha = mr\omega^2$

⇒ $N \sin\alpha = m(R\sin\alpha)\omega^2$ [∵ $r = R\sin\alpha$]

⇒ $N = mR\omega^2$ ...(i)

and $N\cos\alpha = mg$ ...(ii)

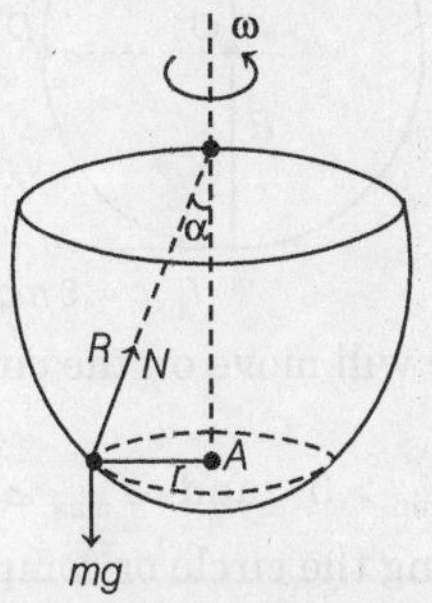

Dividing Eq. (i) by Eq. (ii), we get

$$\frac{1}{\cos\alpha} = \frac{\omega^2 R}{g}$$

∴ $\omega = \sqrt{\frac{g}{R\cos\alpha}}$

Motion in a Vertical Circle

This is an example of non-uniform circular motion. Consider a particle of mass m attached to a string of length R to be whirled in a vertical circle about a fixed point O.

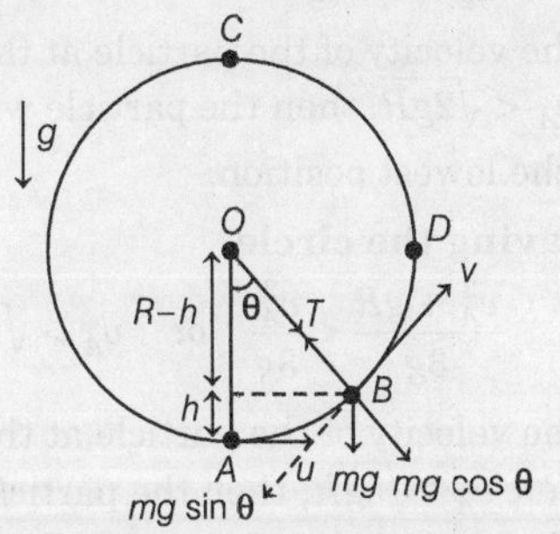

(i) **Velocity at any point on vertical loop**

At the lowest point A, it is imparted a velocity u in the horizontal direction.

Let v be its velocity at any point B at height h, is given by

$$v = \sqrt{u^2 - 2gh} = \sqrt{u^2 - 2gR(1 - \cos\theta)}$$

(ii) **Tension at any point on the vertical loop,**

At any point B, the necessary centripetal force is provided by the resultant of tension T and $mg\cos\theta$, so

$$T_B = mg\cos\theta + \frac{mv^2}{R} = \frac{m}{R}(u^2 + gR - 3gh)$$

(iii) **Tension at the lowest point A and highest point C,**

$$T_A = T_{max}$$
$$T_C = T_{min} = 0$$

Velocity at the highest point C

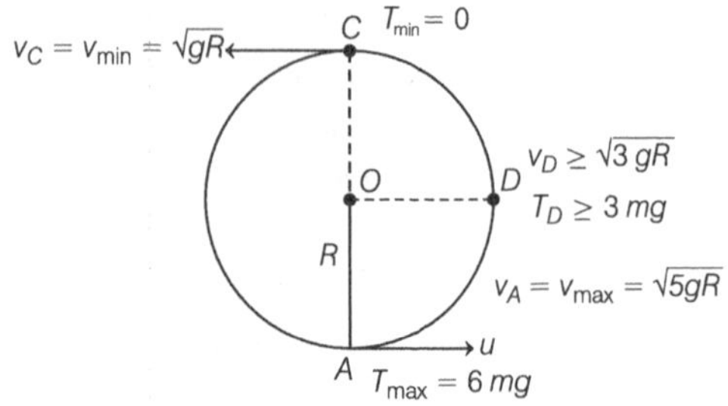

(iv) The particle will move on the circular path only and only if,

$$T_{min} > 0 \quad \text{and} \quad T_{max} \geq 6\,mg$$

So, for looping the circle or completing the loop, velocity at the lowest point should be $v_A \geq \sqrt{5gR}$ and at the highest point $v_C \geq \sqrt{gR}$

Note If $T_{min} \leq 0$, the string will slack and the particle will fall down instead of moving on the circle.

(v) In case of looping the circle for horizontal position of string,

$$v_D \geq \sqrt{3gR} \quad \text{and} \quad T_D \geq 3\,mg$$

(vi) **For oscillations,**

$$\frac{v_A^2}{2g} < \frac{v_A^2 + gR}{3g} \quad \text{or} \quad v_A < \sqrt{2gR}$$

So, if the velocity of the particle at the lowest point is $0 > v_A < \sqrt{2gR}$, then the particle will oscillate about the lowest position.

(vii) **For leaving the circle,**

$$\frac{v_A^2 + gR}{3g} < \frac{v_A^2}{3g} \quad \text{or} \quad v_A > \sqrt{2gR}$$

So, if the velocity of the particle at the lowest point is $\sqrt{2gR} < v_A > \sqrt{5gR}$, then the particle will move along the circle for $\theta > 90°$ and will not reach the point C but will leave the circle somewhere between $90° < \theta < 180°$.

Note (i) Instead of string, if a light rod is used, then the particle will complete the circular path if the velocity given at the lowest position is $\sqrt{4gR}$.

(ii) Oscillation of a pendulum is a part of circular motion. At points A and C, since velocity is zero, net centripetal force will be zero. Only tangential force is present. From A to B or C to B, speed of the bob increases. Therefore, tangential force is parallel to velocity. From B to A or B to C, speed of the bob decreases. Hence, tangential force is antiparallel to velocity.

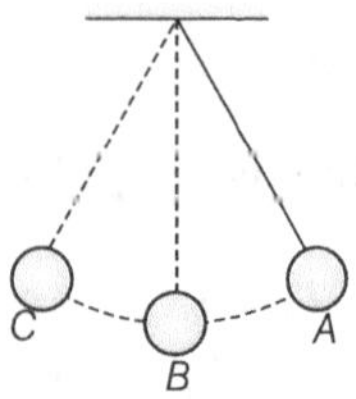

Example 9. *A particle is suspended from a fixed point by a string of length 5 m. It is projected from the equilibrium position with such a velocity that the string slackens after the particle has reached a height 8 m above the lowest point. Find the velocity of the particle, just before the string slackens.*

(a) 5.42 m/s (b) 6.24 m/s (c) 10.26 m/s (d) 9.28 m/s

Sol. (a) At P,

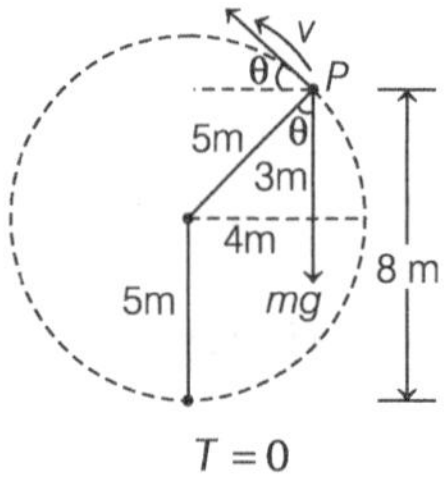

$$T = 0$$

$\therefore$ $$mg\cos\theta = \frac{mv^2}{R}$$

or $$g\cos\theta = \frac{v^2}{R}$$

or $$(9.8)\left(\frac{3}{5}\right) = \frac{v^2}{5}$$

$\therefore$ $$v = 5.42 \text{ m/s}$$

Example 10. *A stone of mass 1 kg tied to a light in extensible string of length $L = \frac{10}{3}$ m is whirling in a circular path of radius L in a vertical plane. The ratio of the maximum tension in the string to the minimum tension in the string is 4 and if g is taken to be 10 m/s^2. The speed of stone at the highest point of the circular is* **[JEE Main 2019]**

(a) 20 m/s (b) $10\sqrt{3}$ m/s (c) $5\sqrt{2}$ m/s (d) 10 m/s

Sol. (d) Since the maximum tension T_B in the string moving in the vertical circle is at the bottom and minimum tension T_T is at the top.

$\therefore$ $$T_B = \frac{mv_B^2}{L} + mg$$

and $$T_T = \frac{mv_T^2}{L} - mg$$

$\therefore$ $$\frac{T_B}{T_T} = \frac{\frac{mv_B^2}{L} + mg}{\frac{mv_T^2}{L} - mg} = \frac{4}{1}$$

or $$\frac{v_B^2 + gL}{v_T^2 - gL} = \frac{4}{1}$$

or $$v_B^2 + gL = 4v_T^2 - 4gL$$

but $$v_B^2 = v_T^2 + 4gL$$

$\therefore$ $$v_T^2 + 4gL + gL = 4v_T^2 - 4gL$$

$\Rightarrow$ $$3v_T^2 = 9gl$$

$\therefore$ $$v_T^2 = 3 \times g \times L = 3 \times 10 \times \frac{10}{3}$$

or $$v_T = 10 \text{ m/s}$$

Practice Exercise

ROUND I Topically Divided Problems

Kinematics of Circular Motion

1. A wheel rotates with a constant angular velocity of 300 rpm. The angle with which the wheel rotates in one second is

(a) π rad (b) 5π rad

(c) 10π rad (d) 20π rad

2. A fan is making 600 revolutions per minute. If after some time, it makes 1200 revolution per minute, then increase in its angular velocity is

(a) 10π rad/s (b) 20π rad/s

(c) 40π rad/s (d) 60π rad/s

3. The radius of the Earth's orbit around the Sun is 1.5×10^{11} m. The linear velocity of the Earth is

(a) 1.99×10^{-7} ms^{-1} (b) 2.99×10^{-7} ms^{-1}

(c) 1.99×10^{4} ms^{-1} (d) 2.99×10^{4} ms^{-1}

4. The length of second's hand in a watch is 1 cm. The change in velocity of its tip in 15 s is

(a) zero (b) $\frac{\pi}{30\sqrt{2}}$ cm/s

(c) $\frac{\pi}{30}$ cm/s (d) $\frac{\pi\sqrt{2}}{30}$ cm/s

5. A particle is moving along a circular path with a constant speed of 10 ms^{-1}. What is the magnitude of the change in velocity of the particle, when it moves through an angle of 60° around the centre of the circle? [JEE Main 2019]

(a) $10\sqrt{2}$ m/s (b) 10 m/s

(c) $10\sqrt{3}$ m/s (d) Zero

6. Two particles A and B are moving on two concentric circles of radii R_1 and R_2 with equal angular speed ω. At $t = 0$, their positions and direction of motion are shown in the figure.

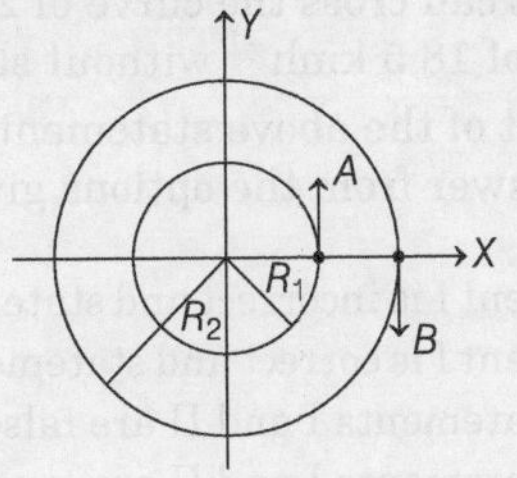

The relative velocity $\mathbf{v}_A - \mathbf{v}_B$ at $t = \frac{\pi}{2\omega}$ is given by [JEE Main 2019]

(a) $\omega(R_1 + R_2)\hat{\mathbf{i}}$ (b) $-\omega(R_1 + R_2)\hat{\mathbf{i}}$

(c) $\omega(R_1 - R_2)\hat{\mathbf{i}}$ (d) $\omega(R_2 - R_1)\hat{\mathbf{i}}$

7. A car-wheel is rotated to uniform angular acceleration about its axis, initially its angular velocity is zero. It rotates through an angle θ_1 in the first 2 s, in the next 2 s, it rotates through an additional angle θ_2, the ratio of $\frac{\theta_2}{\theta_1}$ is

(a) 1 (b) 2 (c) 3 (d) 5

8. The speed of revolution of a particle moving round a circle is doubled and its angular speed is halved. Then, centripetal acceleration will be

(a) unchanged

(b) halved

(c) doubled

(d) 4 times

9. A particle is undergoing uniformly accelerated circular motion with angular retardation π rad/s^2. If the angular velocity of the particle at $t = 0$ is 2π rad/s, the velocity and acceleration vectors of the body at $t = 0$ s are best represented by

(a) 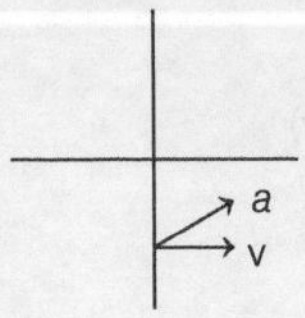

(b) 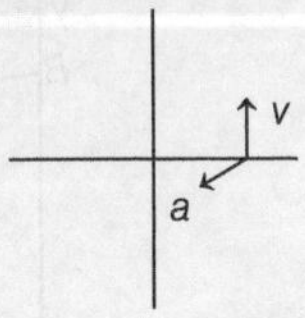

(c) 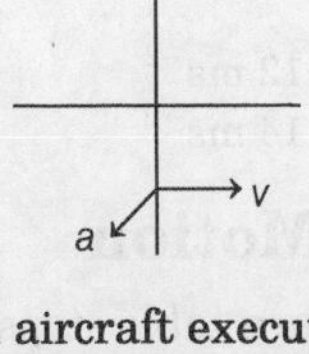

(d) 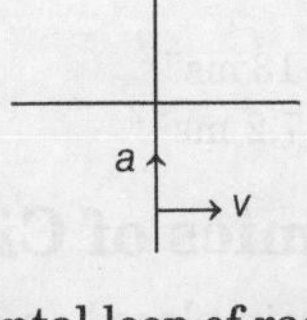

10. An aircraft executes a horizontal loop of radius 1 km with a speed of 900 km/h, then the ratio of its centripetal acceleration and the acceleration due to gravity will be

(a) 6 (b) 7

(c) 8 (d) 5

11. A stone tied to the end of a string 80 cm long is whirled in a horizontal circle with a constant speed. If the stone makes 14 revolutions in 25 s, what is the magnitude and direction of acceleration of the stone?
(a) 9.9 m/s^2 along the tangent
(b) 7.9 m/s^2 along the radius
(c) 9.9 m/s^2 along the radius
(d) None of the above

12. Read the following statements and choose the correct option given below.
(i) The net acceleration of a particle in the circular motion is always along the radius of the circle towards the centre.
(ii) The velocity vector of a particle at a point is always along the tangent to the path of the particle at that point.
(iii) The acceleration vector of a particle in uniform circular motion averaged over one cycle is a null vector.
(a) (i) and (iii) (b) (ii) and (iii)
(c) (iii) Only (d) All the three

13. A car is moving along a circular path of radius 500 m with a speed of 30 ms^{-1}. If at some instant, its speed increases at the rate of 2 ms^{-1}, then at that instant, the magnitude of resultant acceleration will be
(a) 4.7 ms^{-2} (b) 3.8 ms^{-2}
(c) 3 ms^{-2} (d) 2.7 ms^{-2}

14. A point P moves in counter-clockwise direction on a circular path as shown in the figure. The movement of P is such that it sweeps out a length $s = t^3 + 5$, where s is in metre and t is in second. The radius of the path is 20 m. The acceleration of P when $t = 2$ s is nearly

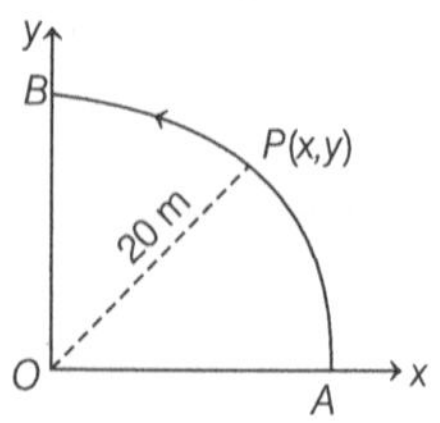

(a) 13 ms^{-2} (b) 12 ms^{-2}
(c) 7.2 ms^{-2} (d) 14 ms^{-2}

Dynamics of Circular Motion

15. A coin placed on a rotation turn table stops when it is placed at a distance of 9 cm from the centre. If the angular velocity of the turn table is trippled. It will just slip, if its distance from the centre is
(a) 27 cm (b) 9 cm
(c) 3 cm (d) 1 cm

16. Three identical particles are joined together by a thread as shown in figure. All the three particles are moving in a horizontal plane. If the velocity of the outermost particle is v_0, then the ratio of tensions in the three sections of the string is

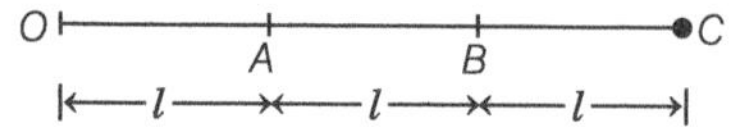

(a) 3 : 5 : 7 (b) 3 : 4 : 5 (c) 7 : 11 : 6 (d) 3 : 5 : 6

17. Two particles of equal mass are connected to a rope AB of negligible mass such that one is at end A and other dividing the length of rope in the ratio 1 : 2 from B. The rope is rotated about end B in a horizontal plane. Ratio of tensions in the smaller part to the other is (ignore the effect of gravity)
(a) 4 : 3 (b) 1 : 4 (c) 1 : 2 (d) 1 : 3

18. A coin is placed on a gramophone record rotating at a speed of 45 rpm. It flies away when the rotational speed is 50 rpm. If two such coins are placed over the other on the same record, both of them will fly away when rotational speed is
(a) 100 rpm (b) 25 rpm (c) 12.5 rpm (d) 50 rpm

19. A body moves along a circular path of radius 5 m. The coefficient of friction between the surface of path and the body is 0.5. The angular velocity, in rad/s, with which the body should move so that it does not leave the path is (Take, $g = 10$ ms^{-2})
(a) 4 (b) 3 (c) 2 (d) 1

20. A motorcycle moving with a velocity of 72 kmh^{-1} on a flat road takes a turn on the road at a point where the radius of curvature of the road is 20 m. The acceleration due to gravity is 10 ms^{-2}. In order to avoid skidding, he must not bent with respect to the vertical plane by an angle greater than
(a) $\theta = \tan^{-1}(2)$ (b) $\theta = \tan^{-1}(6)$
(c) $\theta = \tan^{-1}(4)$ (d) $\theta = \tan^{-1}(25.92)$

21. **Statement I** A cyclist is moving on an unbanked road with a speed of 7 kmh^{-1} and takes a sharp circular turn along a path of radius of 2m without reducing the speed. The static friction coefficient is 0.2. The cyclist will not slip and pass the curve (Take, $g = 9.8\ m/s^2$)
Statement II If the road is banked at an angle of 45°, cyclist can cross the curve of 2m radius with the speed of 18.5 kmh^{-1} without slipping.
In the light of the above statements, choose the correct answer from the options given below.
[JEE Main 2021]
(a) Statement I is incorrect and statement II is correct.
(b) Statement I is correct and statement II is incorrect.
(c) Both statements I and II are false.
(d) Both statements I and II are true.

22. A curved road of 50 m radius is banked at correct angle for a given speed. If the speed is to be doubled keeping the same banking angle, the radius of curvature of the road should be changed to

(a) 25 m (b) 100 m
(c) 150 m (d) 200 m

23. A vehicle is moving with a velocity v on a curved road of width b and radius of curvature R. For counteracting the centrifugal force on the vehicle, the difference in elevation required in between the outer and inner edges of the rod is

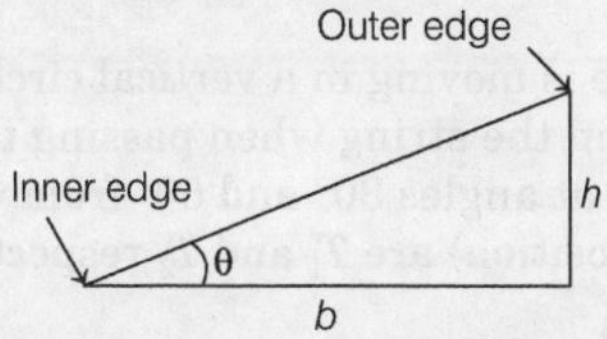

(a) v^2b/Rg (b) vb/Rg
(c) vb^2/Rg (d) vb/R^2g

24. An object is being weighed on a spring balance moving around a curve of radius 100 m at a speed 7 ms^{-1}. The object has a weight of 60 kg-wt. The reading registered on the spring balance would be

(a) 60.075 kg-wt (b) 60.125 kg-wt
(c) 60.175 kg-wt (d) 60.225 kg-wt

25. A small block is shot into each of the four tracks shown below. Each of the tracks rises to the same height. The speed with which the block enters the track is the same in all cases. At the highest point of the track, the normal reaction is the maximum in the case

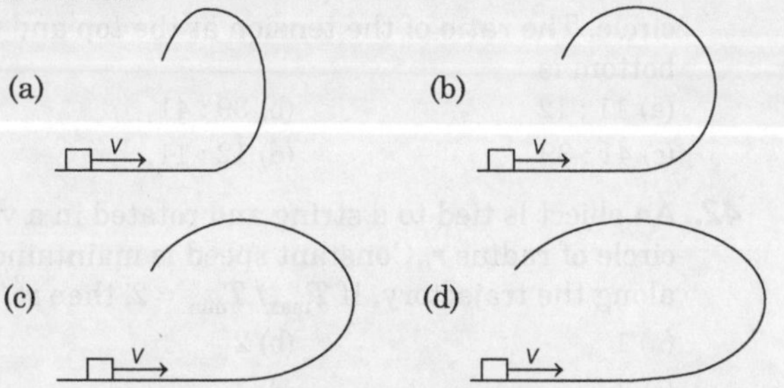

26. A long horizontal rod has a bead, which can slide along its length and initially placed at a distance L from one end A of the rod. The rod is set in angular acceleration α. If the coefficient of friction, between the rod and the bead is μ and gravity is neglected, then the time after which the bead starts slipping is

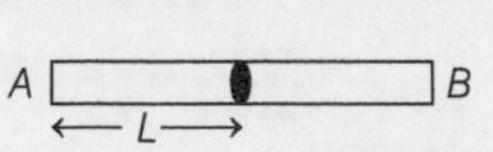

(a) $\sqrt{\mu/\alpha}$ (b) $\mu/\sqrt{\alpha}$
(c) $1/\sqrt{\mu\alpha}$ (d) $\mu\alpha$

27. A self-propelled vehicle (assume it as a point mass) runs on a track with constant speed v. It passes through three positions A, B and C on the circular part of the track. Suppose N_A, N_B and N_C are the normal forces exerted by the track on the vehicle when it is passing through points A, B and C respectively, then

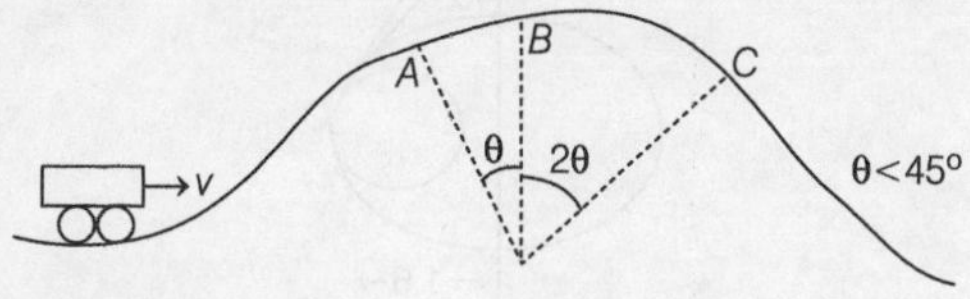

(a) $N_A = N_B = N_C$
(b) $N_B > N_A > N_C$
(c) $N_C > N_A > N_B$
(d) $N_B > N_C > N_A$

28. A car moving on a circular path and takes a turn. If R_1 and R_2 be the reactions on the inner and outer wheels respectively, then

(a) $R_1 = R_2$ (b) $R_1 < R_2$
(c) $R_1 > R_2$ (d) $R_1 \geq R_2$

Conical Pendulum and Motion in Vertical Circle

29. A sphere of mass 0.2 kg is attached to an inextensible string of length 0.5 m whose upper end is fixed to the ceilling. The sphere is made to describe a horizontal circle of radius 0.3 m. The speed of the sphere will be

(a) 1.5 ms^{-1} (b) 2.5 ms^{-1}
(c) 3.2 ms^{-1} (d) 4.7 ms^{-1}

30. A particle describes a horizontal circle in a conical funnel whose inner surface is smooth with speed of 0.5 m/s. What is height of the plane of circle from vertex of the funnel?

(a) 0.25 m (b) 2 cm
(c) 4 cm (d) 2.5 cm

31. How many revolutions per minute must the apparatus shown in figure make about a vertical axis, so that the cord makes an angle of 45° with the vertical? (Given, $l = \sqrt{2}$ m, $a = 0.2$ m, $m = 5$ kg)

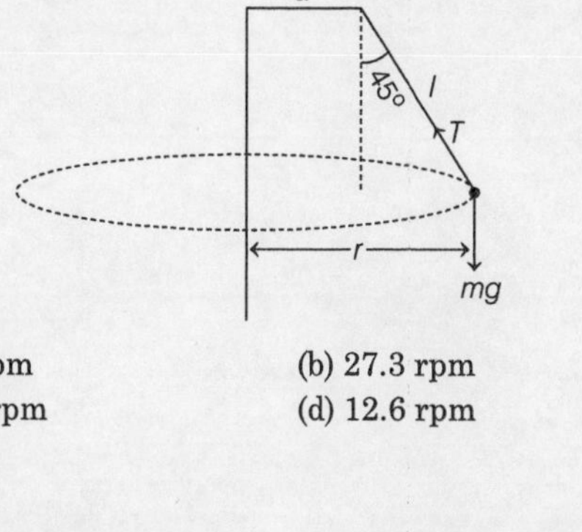

(a) 6 rpm (b) 27.3 rpm
(c) 29 rpm (d) 12.6 rpm

32. Two wires AC and BC are tied at C of small sphere of mass 5 kg, which revolves at a constant speed v in the horizontal circle of radius 1.6 m. The minimum value of v is

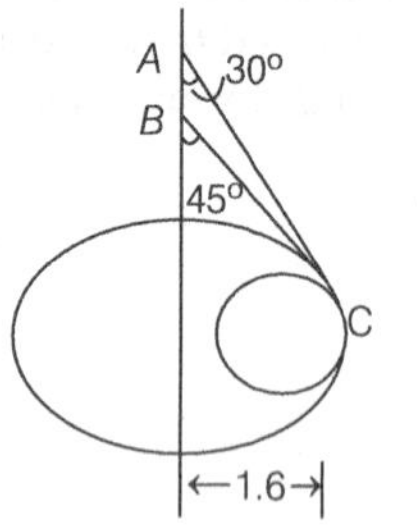

(a) 3.01 ms^{-1} (b) 4.01 ms^{-1}
(c) 8.2 ms^{-1} (d) 3.96 ms^{-1}

33. The string of a pendulum of length l is displaced through 90° from the vertical and released. Then, the minimum strength of the string in order to withstand the tension as the pendulum passes through the mean position is
(a) mg (b) $6mg$ (c) $3mg$ (d) $5mg$

34. An insect is at the bottom of a hemispherical ditch of radius 1 m. It crawls up the ditch but starts slipping after it is at height h from the bottom. If the coefficient of friction between the ground and the insect is 0.75, then h is (Take, $g = 10 \text{ ms}^{-2}$) [JEE Main 2020]
(a) 0.20 m (b) 0.45 m (c) 0.60 m (d) 0.80 m

35. A fighter plane is pulling out for a dive at 900 kmh^{-1} in a vertical circle of radius 2 km. Its mass is 5000 kg. Find the force exerted by the air on it at the lowest point.
(a) 2.0625×10^4 N upward
(b) 2.0625×10^5 N downward
(c) 2.0625×10^5 N upward
(d) 2.0625×10^4 N downward

36. A stone of mass m is tied to a string and is moved in a vertical circle of radius r making n revolution per minute. The total tension in the string when the stone is at its lowest point is
(a) mg (b) $m(g + \pi n r^2)$
(c) $m(g + \pi n r)$ (d) $m\{g + (\pi^2 n^2 r)/900\}$

37. A 2 kg stone at the end of a string 1 m long is whirled in a vertical circle at a constant speed. The speed of the stone is 4 m/s. The tension in the string will be 52 N, when the stone is
(a) at the top of the circle
(b) at the bottom of the circle
(c) halfway down
(d) None of the above

38. A body crosses the topmost point of a vertical circle with critical speed. What will be its acceleration when the string is horizontal?
(a) g (b) $2g$
(c) $3g$ (d) $6g$

39. A particle is moving in a vertical circle. The tensions in the string when passing through two positions at angles 30° and 60° from vertical (lowest position) are T_1 and T_2 respectively.
(a) $T_1 = T_2$
(b) $T_2 > T_1$
(c) $T_1 > T_2$
(d) Tension in the string always remains the same

40. A stone tied to a string of length L is whirled in a vertical circle, with the other end of the string at the centre. At a certain instant of time, the stone is at its lowest position, and has a speed u. The magnitude of change in its velocity as it reaches a position, where the string is horizontal is
(a) $\sqrt{u^2 - 2gL}$ (b) $\sqrt{2gL}$
(c) $\sqrt{u^2 - gL}$ (d) $\sqrt{2(u^2 - gL)}$

41. A stone of mass 1 kg is tied to a string 4 m long and is rotated at constant speed of 40 ms^{-1} in a vertical circle. The ratio of the tension at the top and the bottom is
(a) 11 : 12 (b) 39 : 41
(c) 41 : 39 (d) 12 : 11

42. An object is tied to a string and rotated in a vertical circle of radius r. Constant speed is maintained along the trajectory. If $T_{max}/T_{min} = 2$, then v^2/rg is
(a) 1 (b) 2
(c) 3 (d) 4

ROUND II Mixed Bag

Only One Correct Option

1. A block of 200 g mass moves with a uniform speed in a horizontal circular groove, with vertical side walls of radius 20 cm. If the block takes 40 s to complete one round, the normal force by the side walls of the groove is **[JEE Main 2021]**

(a) 0.0314 N (b) 9.859×10^{-2} N
(c) 6.28×10^{-3} N (d) 9.859×10^{-4} N

2. The distance r from the origin of a particle moving in xy-plane varies with time as $r = 2t$ and the angle made by the radius vector with positive X-axis is $\theta = 4t$. Here, t is in second, r in metre and θ in radian. The speed of the particle at $t = 1$ s is

(a) 10 ms^{-1} (b) 16 ms^{-1}
(c) 20 ms^{-1} (d) 12 ms^{-1}

3. A particle moves along a circle of radius $\left(\frac{20}{\pi}\right)$ m with constant tangential acceleration. If the velocity of the particle is 80 ms^{-1}, at the end of seconds revolution after motion has begun, the tangential acceleration is

(a) 40 ms^{-2} (b) 640 π ms^{-2}
(c) 1609 π ms^{-2} (d) 40 π ms^{-2}

4. A particle moves such that its position vector $\mathbf{r}(t) = \cos \omega t \hat{\mathbf{i}} + \sin \omega t \hat{\mathbf{j}}$, where ω is a constant and t is time. Then, which of the following statements is true for the velocity $\mathbf{v}(t)$ and acceleration $\mathbf{a}(t)$ of the particle? **[JEE Main 2020]**

(a) $\mathbf{v}$ and $\mathbf{a}$ both are parallel to $\mathbf{r}$.
(b) $\mathbf{v}$ is perpendicular to $\mathbf{r}$ and $\mathbf{a}$ is directed away from the origin.
(c) $\mathbf{v}$ and $\mathbf{a}$ both are perpendicular to $\mathbf{r}$.
(d) $\mathbf{v}$ is perpendicular to $\mathbf{r}$ and $\mathbf{a}$ is directed towards the origin.

5. For a particle in uniform circular motion, the acceleration $\mathbf{a}$ at a point $P(R, \theta)$ on the circle of radius R is (here θ is measured from the X-axis)

(a) $-\frac{v^2}{R}\cos\theta\,\hat{\mathbf{i}} + \frac{v^2}{R}\sin\theta\,\hat{\mathbf{j}}$ (b) $-\frac{v^2}{R}\sin\theta\,\hat{\mathbf{i}} + \frac{v^2}{R}\sin\theta\,\hat{\mathbf{j}}$
(c) $-\frac{v^2}{R}\cos\theta\,\hat{\mathbf{i}} - \frac{v^2}{R}\sin\theta\,\hat{\mathbf{j}}$ (d) $\frac{v^2}{R}\hat{\mathbf{i}} + \frac{v^2}{R}\hat{\mathbf{j}}$

6. When a ceiling fan is switched on, it makes 10 rotations in the first 4 s. How many rotations will it make in the next 4 s? (Assuming uniform angular acceleration).

(a) 10 (b) 20
(c) 40 (d) 30

7. A particle is moving in a circle of radius R in such a way that at any instant the normal and tangential components of its acceleration are equal. If its speed at $t = 0$ is v_0, the time taken to complete the first revolution is

(a) $\frac{R}{v_0}$ (b) $\frac{R}{v_0}(1 - e^{-2\pi})$ (c) $\frac{R}{v_0}e^{-2\pi}$ (d) $\frac{2\pi R}{v_0}$

8. An aeroplane flying at a velocity of 900 kmh^{-1} loops the loop. If the maximum force pressing the pilot against the seat is five times its weight, the loop radius should be

(a) 1594 m (b) 1402 m
(c) 1315 m (d) 1167 m

9. A string of length L is fixed at one end and the string makes $\frac{2}{\pi}$ rev/s around the vertical axis through the fixed and as shown in the figure, then tension in the string is

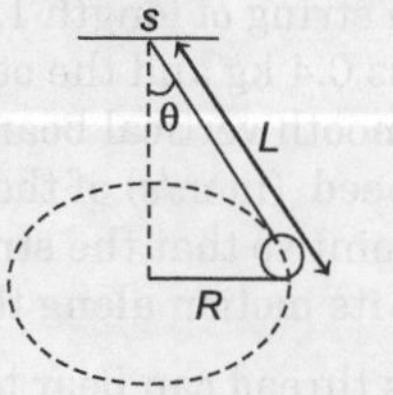

(a) ML (b) $2\,ML$
(c) $4\,ML$ (d) $16\,ML$

10. An object of mass 10 kg is whirled round a horizontal circle of radius 4 m by a revolving string inclined 30° to the vertical. If the uniform speed of the object is 5 ms^{-1}, the tension in the string (approximately) is

(a) 720 N (b) 960 N
(c) 114 N (d) 125 N

11. When the road is dry and coefficient of friction is μ, the maximum speed of a car in a circular path is 10 ms^{-1}. If the road becomes wet and $\mu' = \mu/2$, what is the maximum speed permitted?

(a) 5 ms^{-1} (b) 10 ms^{-1}
(c) $10\sqrt{2}$ ms^{-1} (d) $5\sqrt{2}$ ms^{-1}

12. A heavy small-sized sphere is suspended by a string of length l. The sphere rotates uniformly in a horizontal circle with the string making an angle θ with the vertical, then the time period of this conical pendulum is

(a) $t = 2\pi\sqrt{\frac{l}{g}}$ (b) $t = 2\pi\sqrt{\frac{l\sin\theta}{g}}$
(c) $t = 2\pi\sqrt{\frac{l\cos\theta}{g}}$ (d) $t = 2\pi\sqrt{\frac{l}{g\cos\theta}}$

13. A smooth wire of length $2\pi r$ is bent into a circle and kept in a vertical plane. A bead can slide smoothly on the wire. When the circle is rotating with angular speed ω about the vertical diameter AB, as shown in figure, the bead is at rest with respect to the circular ring at position P as shown. Then, the value of ω^2 is equal to **[JEE Main 2019]**

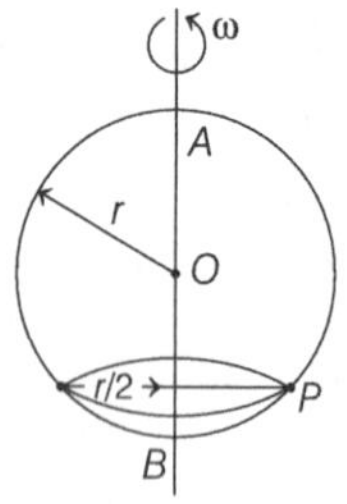

(a) $\frac{\sqrt{3}g}{2r}$ (b) $2g/(r\sqrt{3})$
(c) $(g\sqrt{3})/r$ (d) $2g/r$

Numerical Value Questions

14. One end of a string of length 1.5 m is tied to the stone of mass 0.4 kg and the other end to a small pivot on a smooth vertical board. What is the minimum speed (in m/s) of the stone required at its lowermost point so that the string does not slack at any point in its motion along the vertical circle?

15. A weightless thread can bear tension upto 3.7 kg-wt. A stone of mass 500 g is tied to it and revolved in a circular path of radius 4 m in a vertical plane. If $g = 10\,\text{ms}^{-2}$, then the maximum angular velocity (in rad/s) of the stone will be …… .

16. A circular disc rotates at 60 rpm. A coin of 18 g is placed at a distance of 8 cm, from the centre. The centrifugal force on the coin becomes $q \times 10^{-2}$ N. The value of q is ……… .

17. A stone of mass 2 kg is tied to a string of length 0.5 m. If the breaking tension of the string is 900 N, then the maximum angular velocity (in rad/s), that stone can have in uniform circular motion is ……… .

18. In figure, a particle is placed at the highest point A of a smooth sphere of radius r. It is given slight push, and it leaves the sphere at B, at a depth h vertically below A such that h is equal to r / n, where the value of n is ……… .

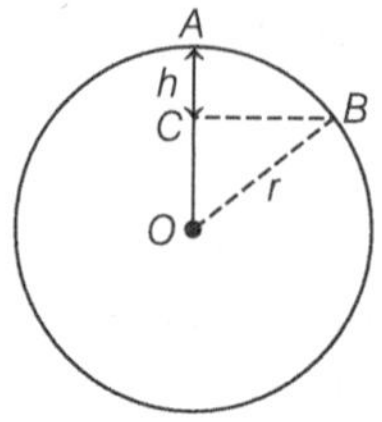

19. Consider a 20 kg uniform circular disc of radius 0.2 m. It is pin supported at its centre and is at rest initially. The disc is acted upon by a constant force $F = 20$ N through a massless string wrapped around its periphery as shown in the figure.

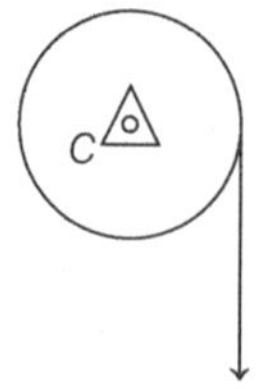

Suppose the disc makes n number of revolutions to attain an angular speed of 50 rad s^{-1}. The value of n, to the nearest integer, is ……… . (Given, in one complete revolution, the disc rotates by 6.28 rad) **[JEE Main 2021]**

20. The angular speed of truck wheel is increased from 900 rpm to 2460 rpm in 26 s. The number of revolutions by the truck engine during this time is ……… . (Assuming the acceleration to be uniform) **[JEE Main 2021]**

Answers

Round I

1. (c)	2. (b)	3. (d)	4. (d)	5. (b)	6. (d)	7. (c)	8. (a)	9. (b)	10. (a)
11. (c)	12. (b)	13. (d)	14. (d)	15. (d)	16. (d)	17. (a)	18. (d)	19. (d)	20. (a)
21. (d)	22. (d)	23. (a)	24. (a)	25. (a)	26. (a)	27. (b)	28. (b)	29. (a)	30. (d)
31. (b)	32. (d)	33. (c)	34. (a)	35. (d)	36. (d)	37. (b)	38. (c)	39. (c)	40. (d)
41. (b)	42. (c)								

Round II

1. (d)	2. (b)	3. (a)	4. (d)	5. (c)	6. (d)	7. (b)	8. (a)	9. (d)	10. (d)
11. (d)	12. (c)	13. (b)	14. 8.6	15. 4	16. 5.689	17. 30	18. 3	19 20	20. 728

Solutions

Round I

1. Frequency of wheel, $\nu = \frac{300}{60} = 5$ rps. Angle rotated by wheel in one rotation $= 2\pi$ rad. Therefore, angle rotated by wheel in 1 s $= 2\pi \times 5$ rad $= 10\pi$ rad.

2. Increase in angular velocity, $\omega = 2\pi\,(n_2 - n_1)$

$$\omega = 2\pi\,(1200 - 600)\,\frac{\text{rad}}{\text{min}}$$

$$= \frac{2\pi \times 600}{60}\,\frac{\text{rad}}{\text{s}} = 20\pi\,\frac{\text{rad}}{\text{s}}$$

3. Here, $r = 1.5 \times 10^{11}$ m; time period of revolution of Earth around the Sun is 1 yr, *i.e.*

$$T = 1\text{ yr} = 365 \times 24 \times 60 \times 60 \text{ s}$$

$\therefore$ Angular velocity, $\omega = \frac{2\pi}{T}$

$$= \frac{2 \times (22/7)}{365 \times 24 \times 60 \times 60}$$

$$= 1.99 \times 10^{-7} \text{ rad s}^{-1}$$

Linear velocity, $v = \omega r$

$$= 1.99 \times 10^{-7} \times 1.5 \times 10^{11}$$

$$= 2.99 \times 10^{4} \text{ ms}^{-1}$$

4. In 15 s, second's hand rotate through 90°

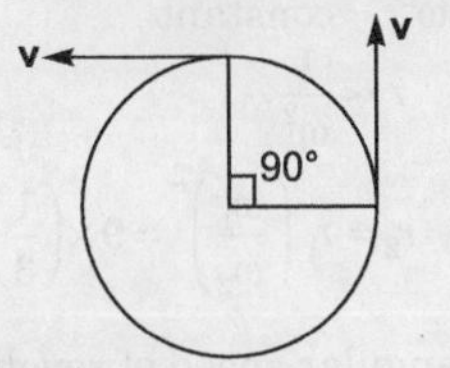

$\therefore$ Change in velocity,

$$|\Delta \mathbf{v}| = 2\,v \sin\left(\frac{\theta}{2}\right)$$

$$= 2\,(r\omega)\sin\left(\frac{90^\circ}{2}\right)$$

$$= 2 \times 1 \times \frac{2\pi}{T} \times \frac{1}{\sqrt{2}}$$

$$= \frac{4\pi}{60\sqrt{2}} = \frac{\pi\sqrt{2}}{30}\,\frac{\text{cm}}{\text{s}} \qquad [\text{As, } T = 60 \text{ s}]$$

5. Let v_1 be the velocity of the particle moving along the circular path initially, v_1 and v_2 be the velocity when it moves through an angle of 60° as shown below.

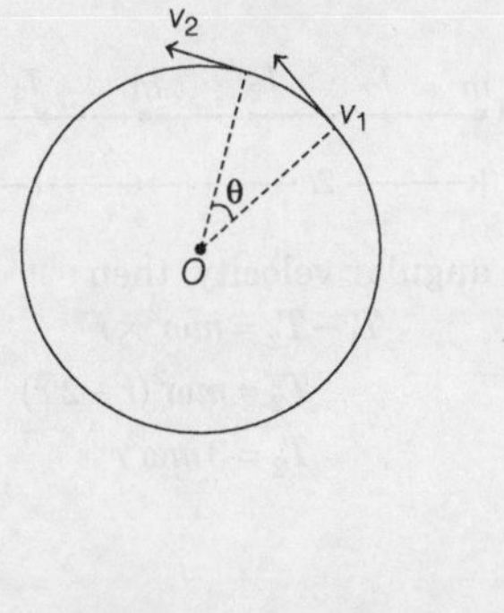

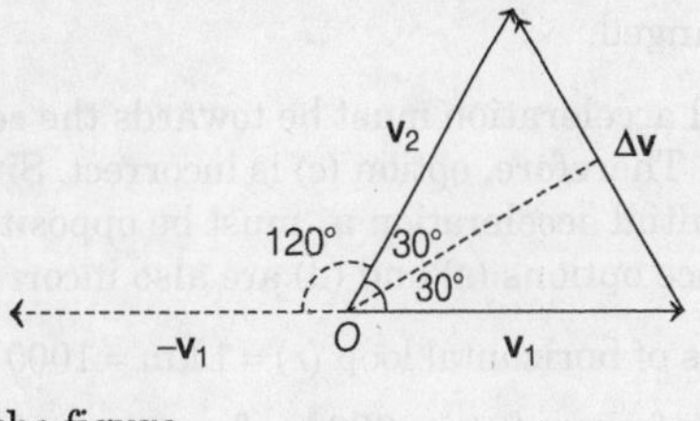

From the figure,

$$\because \quad \Delta\mathbf{v} = \mathbf{v}_2 - \mathbf{v}_1 = \mathbf{v}_2 + (-\mathbf{v}_1)$$

$$\therefore \quad |\Delta\mathbf{v}|^2 = v_1^2 + v_2^2 + 2v_1v_2\cos 120^\circ$$

$$= v^2 + v^2 + 2v \times v \times \left(-\frac{1}{2}\right)$$

$$\Rightarrow \quad |\Delta\mathbf{v}| = v = 10 \text{ m/s}$$

6. Angle covered by each particle in time duration 0 to $\frac{\pi}{2\omega}$ is

$$\theta = \omega \times t = \omega \times \frac{\pi}{2\omega} = \frac{\pi}{2} \text{ rad}$$

So, positions of particles at $t = \frac{\pi}{2\omega}$ is as shown below;

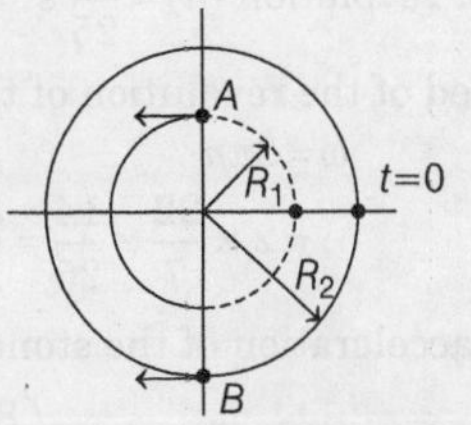

Velocities of particles at $t = \frac{\pi}{2\omega}$ are

$$\mathbf{v}_A = -\omega R_1\hat{\mathbf{i}} \quad \text{and} \quad \mathbf{v}_B = -\omega R_2\hat{\mathbf{i}}$$

The relative velocity of particles is

$$\mathbf{v}_A - \mathbf{v}_B = -\omega R_1\hat{\mathbf{i}} - (-\omega R_2\hat{\mathbf{i}})$$

$$= -\omega(R_1 - R_2)\hat{\mathbf{i}} = \omega(R_2 - R_1)\hat{\mathbf{i}}$$

7. $\alpha = \frac{\omega}{t}$ and $\omega = \frac{\theta}{t}$

$$\therefore \quad \alpha = \frac{\theta}{t^2}$$

but $\quad \alpha = \text{constant}$

So, $\quad \frac{\theta_1}{\theta_1 + \theta_2} = \frac{(2)^2}{(2+2)^2}$

or $\quad \frac{\theta_1}{\theta_1 + \theta_2} = \frac{1}{4}$

or $\quad \frac{\theta_1 + \theta_2}{\theta_1} = \frac{4}{1}$

or $\quad 1 + \frac{\theta_2}{\theta_1} = \frac{4}{1}$

$$\therefore \quad \frac{\theta_2}{\theta_1} = 3$$

8. Centripetal acceleration $a_c = \frac{v^2}{r} = v\omega$

When v is doubled, ω is halved, $v\omega$ shall remain unchanged.

9. Radial acceleration must be towards the centre of the circle. Therefore, option (c) is incorrect. Since, tangential acceleration $\mathbf{a}_t$ must be opposite to velocity **v**, hence options (a) and (d) are also incorrect.

10. Radius of horizontal loop $(r) = 1\text{ km} = 1000\text{ m}$

Speed of aircraft $(v) = 900\text{ km/h}$

$$= 900 \times \frac{5}{18}\text{ m/s} \left(\because 1\text{ km/h} = \frac{5}{18}\text{ m/s}\right)$$

$$= 250\text{ m/s}$$

Centripetal acceleration of the aircraft

$$a = \frac{v^2}{r} = \frac{(250)^2}{1000} = \frac{62500}{1000} = 62.5\text{ m/s}^2$$

Acceleration due to gravity $(g) = 9.8\text{ m/s}^2$

$$\therefore \frac{\text{Centripetal acceleration }(a)}{\text{Acceleration due to gravity }(g)} = \frac{62.5}{9.8} = 6.38 \simeq 6$$

11. Radius of the horizontal circle = Length of the string

$$= 80\text{ cm} = 0.80\text{ m}$$

Frequency of revolution $(n) = \frac{14}{25}\text{ s}^{-1}$

Angular speed of the revolution of the stone

$$\omega = 2\pi n$$

$$= 2 \times \frac{22}{7} \times \frac{14}{25} = \frac{88}{25}\text{ rad/s}$$

Centripetal acceleration of the stone,

$$a = r\omega^2 = 0.80 \times \left(\frac{88}{25}\right)^2$$

$$= 0.80 \times \frac{88}{25} \times \frac{88}{25} = 9.91\text{ m/s}^2$$

The direction of the acceleration is towards the centre of the horizontal circle, *i.e.* along its radius.

12. (i) False, because in uniform circular motion, the centripetal acceleration is along the radius of the circle towards the centre, but in non-uniform circular motion, the direction of the resultant acceleration is not along the radius of the circle.

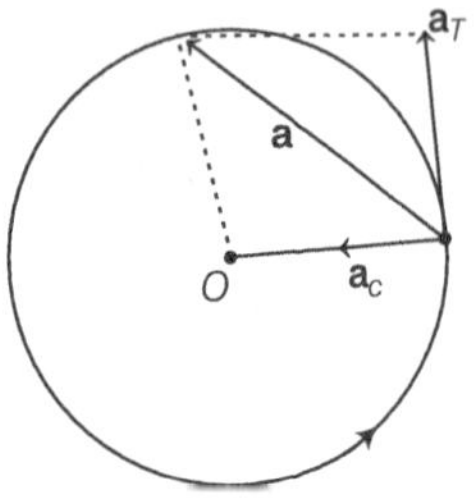

(ii) True, the velocity vector of a particle is always along the tangent to the path of the particle either it is in rectilinear, circular or curvilinear motion.

(iii) True, the direction of acceleration vector in uniform circular motion is directed towards the centre of the circular path, which is continuously changing with time. Therefore, the resultant of all these vectors over one cycle will be a null vector.

13. Centripetal acceleration, $a_c = \frac{v^2}{r} = \frac{(30)^2}{500} = 1.8\text{ ms}^{-2}$

Tangential acceleration, $a_t = 2\text{ ms}^{-1}$

$\therefore$ Resultant acceleration, $a = \sqrt{a_t^2 + a_c^2}$

$$= \sqrt{(2)^2 + (1.8)^2} = 2.7\text{ ms}^{-2}$$

14. Given, $s = t^3 + 5$

$\therefore$ Speed, $v = \frac{ds}{dt} = 3t^2$

and rate of change of speed, $a_t = \frac{dv}{dt} = 6t$

$\therefore$ Tangential acceleration at $t = 2$ s,

$$a_t = 6 \times 2 = 12\text{ ms}^{-2}$$

and at $t = 2$ s, $v = 3(2)^2 = 12\text{ ms}^{-1}$

$\therefore$ Centripetal acceleration, $a_c = \frac{v^2}{R} = \frac{144}{20}\text{ ms}^{-2}$

$\therefore$ Net acceleration $= \sqrt{a_t^2 + a_c^2} \approx 14\text{ ms}^{-2}$

15. In the given condition friction provides the required centripetal force and that is constant. *i.e.*

$$m\omega^2 r = \text{constant}$$

$$\Rightarrow \quad r \propto \frac{1}{\omega^2}$$

$$\therefore \quad r_2 = r_1\left(\frac{\omega_1}{\omega_2}\right)^2 = 9 \times \left(\frac{1}{3}\right)^2 = 1\text{ cm}$$

16. Let ω be the angular speed of revolution

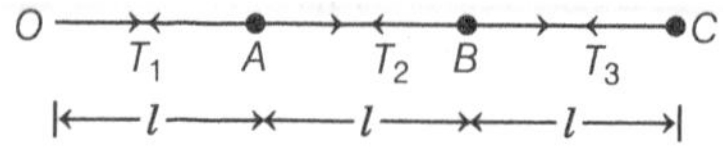

$$T_3 = m\omega^3 3l$$

$$T_2 - T_3 = m\omega^2 2l$$

$$\Rightarrow \quad T_2 = m\omega^2 5l$$

$$T_1 - T_2 = m\omega^2 l$$

$$\Rightarrow \quad T_1 = m\omega^2 6l$$

$$\Rightarrow \quad T_3 : T_2 : T_1 = 3 : 5 : 6$$

17. Tension in the respective parts are shown in figure.

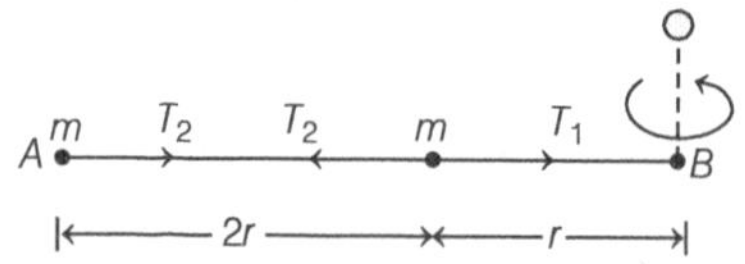

Let ω be angular velocity, then

$$T_1 - T_2 = m\omega^2 \times r \quad \text{...(i)}$$

and $$T_2 = m\omega^2(r + 2r)$$

$$T_2 = 3m\omega^2 r \quad \text{...(ii)}$$

From Eqs. (i) and (ii), we get

$$T_1 = 4m\omega^2 r$$

$$\Rightarrow \quad \frac{T_1}{T_2} = \frac{4}{3}$$

18. A coin flies off when centrifugal force just exceeds the force of friction, *i.e.* $mr\omega^2 \geq \mu mg$

or $$\omega \geq \sqrt{\frac{\mu g}{r}}$$

Thus, ω does not depend upon mass and will remain the same (*i.e.* 50 rpm).

19. Here, $r = 5\text{ m}, \mu = 0.5, \omega = ?, g = 10\text{ ms}^{-2}$

As, $mr\omega^2 = F = \mu R = \mu\, mg$

$$\omega = \sqrt{\frac{\mu g}{r}} = \sqrt{\frac{0.5 \times 10}{5}} = 1\text{ rad s}^{-1}$$

20. Using the formula for motor cycle not to skid,

$$\theta = \tan^{-1}\left(\frac{v^2}{rg}\right)$$

where, $r = 20\text{ m}$

$$v = 72\text{ km h}^{-1} = 72 \times \frac{5}{18} = 20\text{ ms}^{-1}$$

$$\therefore \quad \theta = \tan^{-1}\left(\frac{20 \times 20}{20 \times 10}\right)$$

or $$\theta = \tan^{-1}(2)$$

21. **Statement I**

$$v_{max} = \sqrt{\mu Rg} = \sqrt{(0.2) \times 2 \times 9.8}$$

$$v_{max} = 1.97\text{ m/s}$$

$$7\text{ km/h} = 1.944\text{ m/s}$$

Speed is lower than v_{max}, hence it can take safe turn.

Statement II

$$V_{max} = \sqrt{Rg\left(\frac{\tan\theta + \mu}{1 - \mu\tan\theta}\right)}$$

$$= \sqrt{2 \times 9.8\left(\frac{1 + 0.2}{1 - 0.2}\right)} = 5.42\text{ m/s}$$

$$18.5\text{ km/h} = 5.14\text{ m/s}$$

Speed is lower than v_{max}, hence it can take safe turn.

22. As, $\tan\theta = \frac{v^2}{rg}$, therefore, when speed v is doubled; r must be made 4 times, if θ remains the same.

∴ New radius of curvature,

$$r' = 4r = 4 \times 50\text{ m} = 200\text{ m}$$

23. As, $$\tan\theta = \frac{v^2}{Rg}$$

From the given figure, we get

$$\frac{h}{b} = \frac{v^2}{Rg} \Rightarrow h = \frac{v^2 b}{Rg}$$

24. Here, $r = 100\text{ m}, v = 7\text{ ms}^{-1}, m = 60\text{ kg}$

Reading registered = resultant force = ?

Two forces are acting as weight (mg) and centripetal force, $\left(\frac{mv^2}{r}\right)$ are at 90° to each other.

$$\therefore \text{ Resultant force} = \sqrt{(mg)^2 + \left(\frac{mv^2}{r}\right)^2}$$

$$= mg\left[1 + \left(\frac{v^2}{rg}\right)^2\right]^{1/2}$$

$$= 60 \times 9.8\left[1 + \left(\frac{7 \times 7}{100 \times 9.8}\right)^2\right]^{1/2}$$

$$= 60.075 \times 9.8\text{ N}$$

$$= 60.075\text{ kg-wt}$$

25. The blocks will have the same speed, say, equal to v, at the highest point of each track, as they all rise to the same height. If R be the radius of curvature of a track and N be the normal reaction of the track at the highest point of the track, then

$$\text{centripetal force} = N + mg = \frac{mv^2}{R}$$

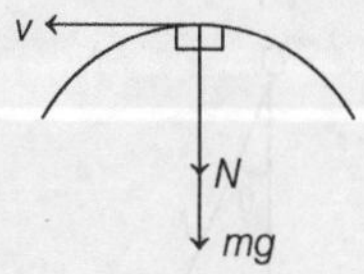

$\Rightarrow N$ will be the maximum when R is the minimum. This occurs when the track is most sharply curved, *i.e.* in option (a).

26. Tangential acceleration, $a = L\alpha$

∴ Normal reaction, $N = Ma = ML\alpha$

∴ Frictional force, $F = mN = \mu\, ML\alpha$

For no sliding along the length, frictional force ≥ centripetal force.

i.e. $\mu\, ML\alpha \geq ML\omega^2$

As $\omega = \omega_0 + \alpha t = \alpha t$

∴ $\mu\, ML\alpha \geq ML(\alpha t)^2$

$$\Rightarrow \quad t = \sqrt{\frac{\mu}{\alpha}}$$

27. As we know, $mg\cos\theta - N = \frac{mv^2}{R}$

$$\Rightarrow \quad N = mg\cos\theta - \frac{mv^2}{R}$$

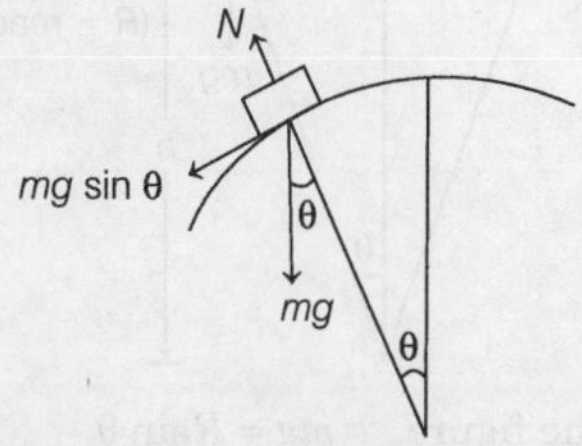

Hence, N decrease as θ increases.

So $N_B > N_A > N_C$

28. The given situation is shown below.

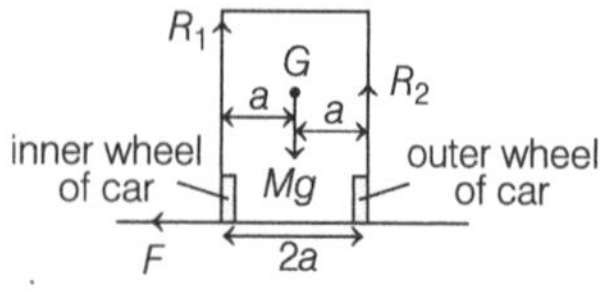

Since, $R_1 + R_2 = Mg$

If F be the horizontal force which provides the necessary centripetal force for motion in a circle, then

$$F = \frac{mv^2}{r}$$

Taking moment about point G, there should be no resultant turning effect about the centre of gravity.

$\therefore \quad Fh + R_1 a = R_2 a$

$\therefore \quad R_2 = \frac{Fh + R_1 a}{a}$

$$R_2 = R_1 + \frac{Fh}{a}$$

$\therefore$ Clearly, $R_2 > R_1$

29. In figure, $T \sin\theta = \frac{mv^2}{r}$; $T\cos\theta = mg$;

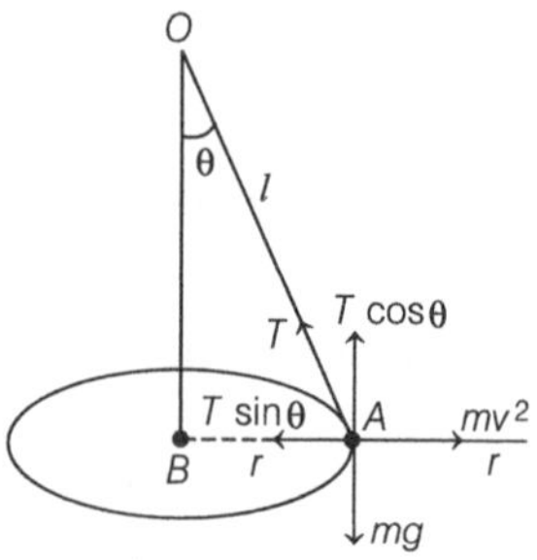

So, $\tan\theta = \frac{v^2}{rg} = \frac{r}{\sqrt{l^2 - r^2}}$

$$v = \left[\frac{r^2 g}{(l^2 - r^2)^{1/2}}\right]^{1/2} = \left[\frac{0.09 \times 10}{(0.25 - 0.09)^{1/2}}\right]^{1/2}$$

$= 1.5 \text{ ms}^{-1}$

30. The particle is moving in circular path as shown below

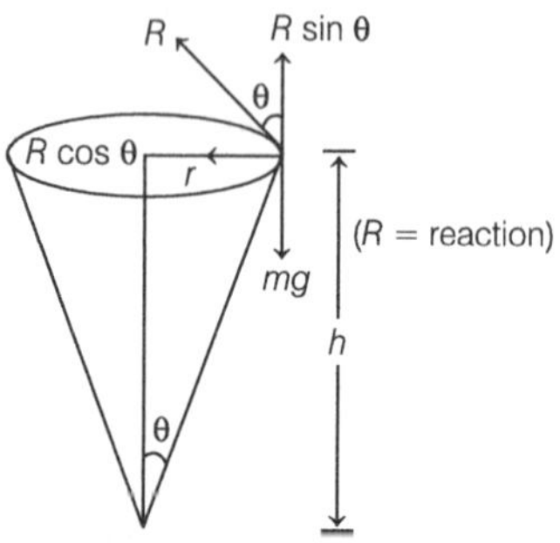

From the figure, $mg = R\sin\theta$ ...(i)

$\frac{mv^2}{r} = R\cos\theta$ ...(ii)

From Eqs. (i) and (ii), we get

$$\tan\theta = \frac{rg}{v^2}$$

and $\tan\theta = \frac{r}{h}$

$$h = \frac{v^2}{g} = \frac{(0.5)^2}{10} = 0.25 \text{ m} = 2.5 \text{ cm}$$

31. $r = a + l\sin 45° = (0.2) + (\sqrt{2})\left(\frac{1}{\sqrt{2}}\right) = 1.2 \text{ m}$

Now, $T\cos 45° = mg$ ...(i)

and $T\sin 45° = mr\omega^2$ ...(ii)

From Eqs. (i) and (ii), we have $\omega = 2n\pi = \sqrt{\frac{g}{r}}$

$\therefore \quad n = \frac{1}{2\pi}\sqrt{\frac{g}{r}} = \frac{60}{2\pi}\sqrt{\frac{9.8}{1.2}}$ rpm $= 27.3$ rpm

32. From free body diagram shown in figure,

$T_1 \cos 30° + T_2 \cos 45° = mg$ ...(i)

$T_1 \sin 30° + T_2 \sin 45° = \frac{mv^2}{r}$ ...(ii)

After solving Eqs. (i) and (ii), we get

$$T_1 = \frac{mg - \frac{mv^2}{r}}{\left(\frac{\sqrt{3}-1}{2}\right)}$$

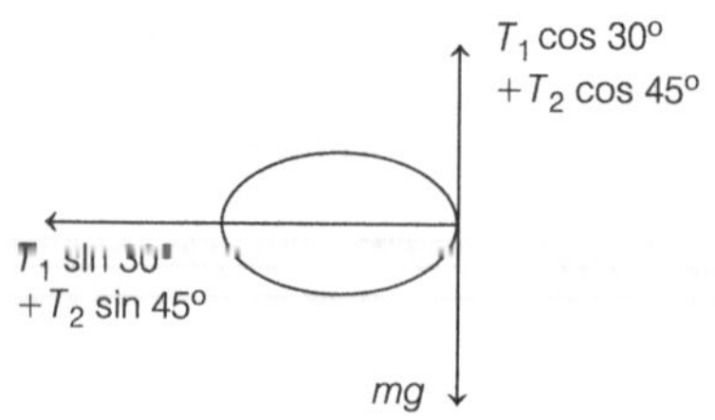

But $T_1 \geq 0$

$\therefore \quad \frac{mg - \frac{mv^2}{r}}{\frac{\sqrt{3}-1}{2}} \geq 0$

or $mg \geq \frac{mv^2}{r}$

or $v \leq \sqrt{rg}$

$\therefore \quad v_{max} = \sqrt{rg} = \sqrt{1.6 \times 9.8} = 3.96 \text{ ms}^{-1}$

33. Velocity at the lowest point,

$$v = \sqrt{2gl}$$

At the lowest point, the tension in the string,

$$T = mg + \frac{mv^2}{l}$$

$$= mg + \frac{m}{l}(2gl) = 3mg$$

34. Let h be maximum height up to which insect crawls up the ditch. The free body diagram is as shown

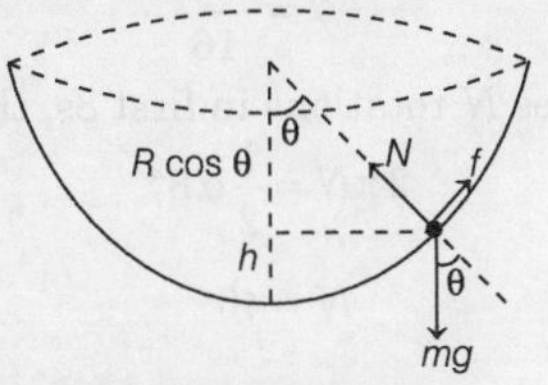

Resolving the components of force along tangential and radial direction

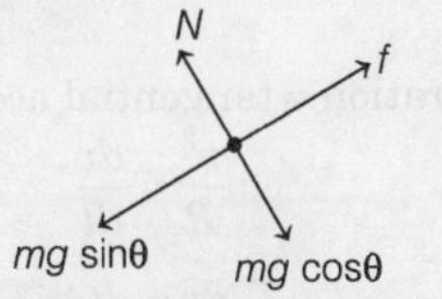

For balancing, $mg\cos\theta = N$

$$mg\sin\theta = f_{\max} = \mu N$$

$$\Rightarrow \quad mg\sin\theta = \mu mg\cos\theta$$

$$\frac{\sin\theta}{\cos\theta} = \frac{\mu mg}{mg}$$

$$\tan\theta = \mu = 0.75 = \frac{3}{4}$$

$$\Rightarrow \quad \cos\theta = \frac{4}{5}$$

From diagram, $h = R - R\cos\theta = R - R\left(\frac{4}{5}\right) = \frac{R}{5}$

or $\quad h = \frac{1}{5} = 0.20$ m

35. Force, $F = \frac{mv^2}{r} + mg = \frac{5\times10^3\times(250)^2}{2\times10^3} + 5\times10^4$

$= 2.0625\times10^4$ N downward

36. Here, $T = mg + m\omega^2 r = m(g + 4\pi^2 n^2 r)$

$$= m\left\{g + \left(4\pi^2\left(\frac{n}{60}\right)^2 r\right)\right\} = m\left\{g + \left(\frac{\pi^2 n^2 r}{900}\right)\right\}$$

37. Here, $mg = 2\times10 = 20$ N and $\frac{mv^2}{r} = \frac{2\times(4)^2}{1} = 32$ N

It is clear that, 52 N tension will be at the bottom of the circle because we know that

$$T_{\text{bottom}} = mg + \frac{mv^2}{r}$$

38. The body crosses the topmost position of a vertical circle with critical velocity, so the velocity at the lowest point of vertical circle, $u = \sqrt{5\,gr}$.

Velocity of the body when string is horizontal, is

$$v^2 = u^2 - 2\,gr = 5\,gr - 2\,gr = 3\,gr$$

$\therefore$ Centripetal acceleration, $a_c = \frac{v^2}{r} = \frac{3\,gr}{r} = 3\,g$

39. Tension, $T = \frac{mv^2}{r} + mg\cos\theta$

For $\theta = 30°$, $\quad T_1 = \frac{mv^2}{r} + mg\cos30°$

For $\theta = 60°$, $\quad T_2 = \frac{mv^2}{r} + mg\cos60°$

Since, $\cos30° > \cos60°$

$\therefore \quad T_1 > T_2$

40. The velocity at B is v, where $v^2 = u^2 - 2\,gL$, in vertically upward direction.

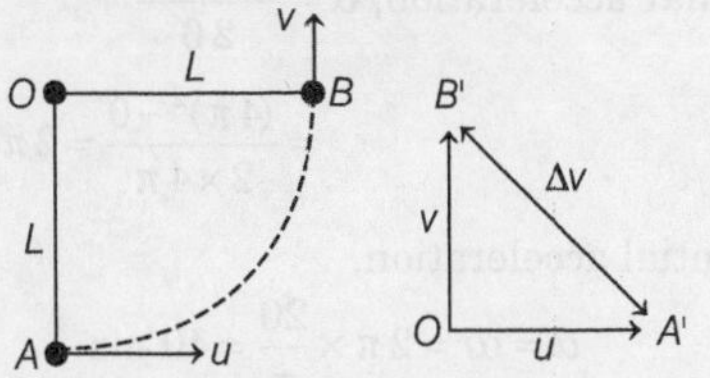

$$\therefore \quad \Delta v = \sqrt{u^2 + v^2} = \sqrt{u^2 + (u^2 - 2\,gL)} = \sqrt{2\,(u^2 - gL)}$$

41. As, $\quad T_{\text{top}} = \frac{mv^2}{r} - mg \quad$ ...(i)

and $\quad T_{\text{bottom}} = \frac{mv^2}{r} + mg \quad$ ...(ii)

From Eqs. (i) and (ii), we get

$$\Rightarrow \quad \frac{T_{\text{top}}}{T_{\text{bottom}}} = \frac{\frac{v^2}{r} - g}{\frac{v^2}{r} + g} = \frac{\frac{40\times40}{4} - 10}{\frac{40\times40}{4} + 10}$$

$$= \frac{400-10}{400+10} = \frac{390}{410} = \frac{39}{41}$$

42. At the lowest point, $\quad \frac{mv^2}{r} = T_L - mg \quad$ ...(i)

At the highest point, $\quad \frac{mv^2}{r} = T_H + mg \quad$ ...(ii)

As $\quad \frac{T_{\max}}{T_{\min}} = \frac{T_L}{T_H} = 2$

$\therefore \quad T_L = 2\,T_H$

From Eqs. (i) and (ii), we have

$$2\,T_H - mg = T_H + mg$$

$$T_H = 2\,mg$$

From Eq. (ii), $\quad \frac{mv^2}{r} = 3\,mg \quad$ or $\quad \frac{v^2}{rg} = 3$

Round II

1. $N = m\omega^2 R$

$$N = m\left[\frac{4\pi^2}{T^2}\right]R \quad \text{...(i)}$$

Given, $m = 0.2$ kg, $T = 40$ s, $R = 0.2$ m

Putting the values in Eq. (i), we get

$$N = 9.859\times10^{-4}\ \text{N}$$

2. Here, $r = 2\,t, \theta = 4\,t$

As, $\quad l = r\theta = (2\,t)(4\,t) = 8\,t^2$

$$\therefore \quad v = \frac{dl}{dt} = \frac{d}{dt}(8\,t^2) = 16\,t$$

$\Rightarrow \quad v_t = 16\ \text{ms}^{-1} \quad$ (at $t = 1$ s)

3. Initial angular velocity, $\omega_0 = 0$.

Final angular velocity,

$$\omega = \frac{v}{r} = \frac{80}{(20/\pi)} = 4\pi \text{ rads}^{-1}$$

Angle described, $\theta = 4\pi$ rad

$\therefore$ Angular acceleration, $\alpha = \dfrac{\omega^2 - \omega_0^2}{2\theta}$

$$= \frac{(4\pi)^2 - 0}{2 \times 4\pi} = 2\pi \text{ rads}^{-2}$$

Tangential acceleration,

$$a = \alpha r = 2\pi \times \frac{20}{\pi} = 40 \text{ ms}^{-2}$$

4. Position vector of a particle moving around a unit circle ($r = 1$) in xy-plane is given by

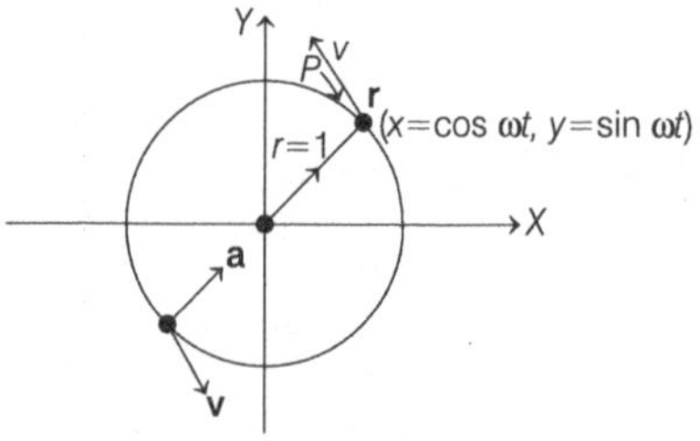

$$\mathbf{r}(t) = \cos \omega t\, \hat{\mathbf{i}} + \sin \omega t\, \hat{\mathbf{j}} \quad \text{...(i)}$$

Velocity, $\mathbf{v}(t) = \dfrac{d\mathbf{r}(t)}{dt} = \omega(-\sin \omega t\, \hat{\mathbf{i}} + \cos \omega t\, \hat{\mathbf{j}})$

$$\Rightarrow \quad \mathbf{v}(t) = \omega[\cos(\omega t + \pi/2)\, \hat{\mathbf{i}} + \sin(\omega t + \pi/2)\, \hat{\mathbf{j}}] \quad \text{...(ii)}$$

$$\mathbf{a}(t) = \frac{d\mathbf{v}(t)}{dt} = -\omega^2 \cos \omega t\, \hat{\mathbf{i}} - \omega^2 \sin \omega t\, \hat{\mathbf{j}}$$

$$= -\omega^2(\cos \omega t\, \hat{\mathbf{i}} + \sin \omega t\, \hat{\mathbf{j}})$$

$$\mathbf{a}(t) = -\omega^2\, \mathbf{r}(t) \quad \text{...(iii)}$$

From Eqs. (i), (ii) and (iii), it is clear that **v** is perpendicular to **r** and **a** is directed towards the origin.

5. For a particle in uniform circular motion,

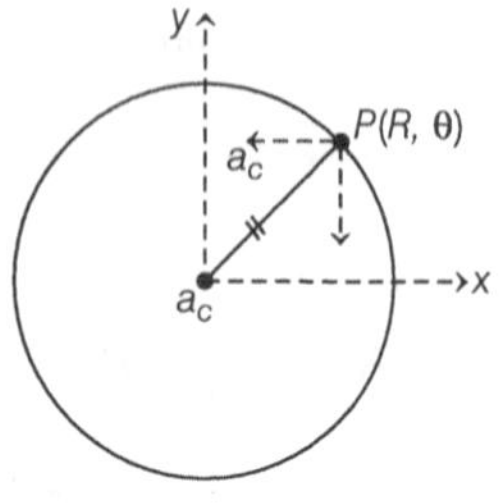

$$\mathbf{a} = \frac{v^2}{R} \text{ towards centre of circle}$$

$$\mathbf{a} = \frac{v^2}{R}(-\cos\theta\, \hat{\mathbf{i}} - \sin\theta\, \hat{\mathbf{j}})$$

or $$\mathbf{a} = -\frac{v^2}{R}\cos\theta\, \hat{\mathbf{i}} - \frac{v^2}{R}\sin\theta\, \hat{\mathbf{j}}$$

6. As, $\theta = 2\pi n = \omega_0 t + \dfrac{1}{2}\alpha t^2$

$$\Rightarrow \quad 2\pi \times 10 = \frac{1}{2}\alpha (4)^2$$

or $$\alpha = \frac{40\pi}{16}$$

Let it makes N rotations in first 8s, then

$$2\pi N = \frac{1}{2}\alpha 8^2$$

$$\Rightarrow \quad N = 40 \quad \left(\text{as } \alpha = \frac{40\pi}{16}\right)$$

$\therefore$ Required number of rotations $= 40 - 10 = 30$

7. According to question, in circular motion of particle, at any instant,

radial acceleration = tangential acceleration

$$\Rightarrow \quad \frac{v^2}{R} = \frac{dv}{dt}$$

Hence, $$\frac{R}{v^2}dv = dt$$

$$\Rightarrow \quad \int \frac{R}{v^2}dv = dt$$

$$\Rightarrow \quad \frac{-R}{v} = t + c \quad \text{...(i)}$$

When $t = 0$, $v = v_0$

$$\therefore \quad \frac{-R}{v_0} = c \quad \text{...(ii)}$$

Hence, from Eqs. (i) and (ii), we get

$$\frac{-R}{v} = t - \frac{R}{v_0}$$

$$\Rightarrow \quad t = \frac{R}{v_0} - \frac{R}{v}$$

$$\Rightarrow \quad t = \frac{R}{v_0} - \frac{R}{(ds/dt)} \quad \left[\because v = \frac{ds}{dt}\right]$$

$$\Rightarrow \quad ds\left(\frac{R}{v_0} - t\right) = R\,dt$$

$$\Rightarrow \quad \int \frac{ds}{R} = \int \frac{dt}{\frac{R}{v_0} - t}$$

$$\Rightarrow \quad \frac{s}{R} = -\ln\left(\frac{R}{v_0} - t\right) + c$$

When $t = 0$, $s = 0$,

$$\therefore \quad c = \frac{\ln R}{v_0}$$

$$\therefore \quad \frac{s}{R} = \frac{\ln R}{v_0} - \ln\left(\frac{R}{v_0 - t}\right) = \ln\left[\frac{R}{R - v_0 t}\right]$$

For complete revolution, $t = T$ and $s = 2\pi R$

$$\therefore \quad \frac{2\pi R}{R} = \ln\left(\frac{R}{R - v_0 T}\right)$$

$$\Rightarrow \quad T = \frac{R}{v_0}(1 - e^{-2\pi})$$

8. Here, $v = 900 \text{ km h}^{-1} = \dfrac{900 \times 1000}{60 \times 60} \text{ms}^{-1} = 250 \text{ ms}^{-1}$

Maximum force is at bottom of the vertical circle

$$F_{max} = \frac{mv^2}{r} + mg = 5\,mg$$

$\therefore$ $$v^2 = 4\,gr$$

or $$r = \frac{v^2}{4\,g} = \frac{250 \times 250}{4 \times 9.8} = 1594.4 \simeq 1594 \text{ m}$$

9. From figure in question,

$$T \sin\theta = M\omega^2 R \qquad [\because R = L\sin\theta]$$

$$\Rightarrow \quad T\sin\theta = M\omega^2 L \sin\theta$$

$$\Rightarrow \quad T = M\omega^2 L = M \cdot 4\pi^2 n^2 L$$

$$= M \cdot 4\pi^2 \left(\frac{2}{\pi}\right)^2 L$$

$$= 16\,ML$$

10. In figure, $\sin 30° = \dfrac{AB}{OA}$

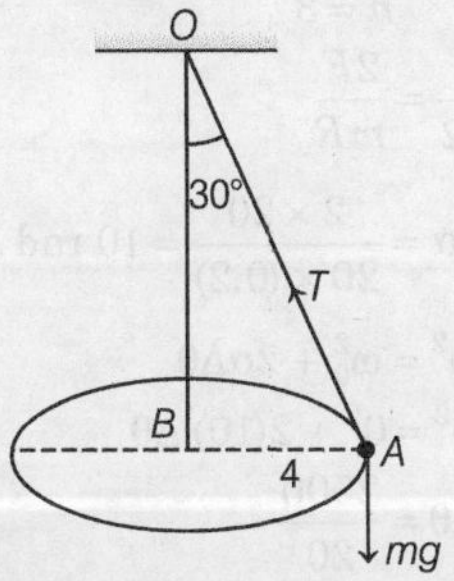

or $$OA = \frac{AB}{\sin 30°} = \frac{4}{1/2} = 8 \text{ m}$$

$$\frac{T}{AO} = \frac{F}{AB} = \frac{mg}{OB}$$

$$F = \frac{AO}{AB} \times F$$

$$= \frac{AO}{AB}\frac{mv^2}{r}$$

$$= \frac{8}{4} \times 10 \times \frac{5^2}{4} = 125 \text{ N}$$

11. The maximum speed without skidding is

$$v = \sqrt{\mu\, rg}$$

$\therefore$ $$\frac{v_2}{v_1} = \sqrt{\frac{\mu_1}{\mu_1}} = \sqrt{\frac{\mu/2}{\mu}} = \frac{1}{\sqrt{2}} \quad (\text{for } rg = \text{constant})$$

$$v_2 = \frac{v_1}{\sqrt{2}} = 5\sqrt{2} \text{ ms}^{-1} \qquad (\because v_1 = 10 \text{ ms}^{-1})$$

12. Radius of circular path in the horizontal plane

$$r = l\sin\theta$$

Resolving T along the vertical and horizontal directions, we get,

$$T\cos\theta = Mg \qquad \text{...(i)}$$

$$T\sin\theta = Mr\omega^2 = M\,(l\sin\theta)\,\omega^2 \qquad [\because r = l\sin\theta]$$

or $$T = Ml\omega^2 \qquad \text{...(ii)}$$

Dividing Eq. (ii) by Eq. (i), we get

$$\frac{1}{\cos\theta} = \frac{l\omega^2}{g}$$

or $$\omega^2 = \frac{g}{l\cos\theta}$$

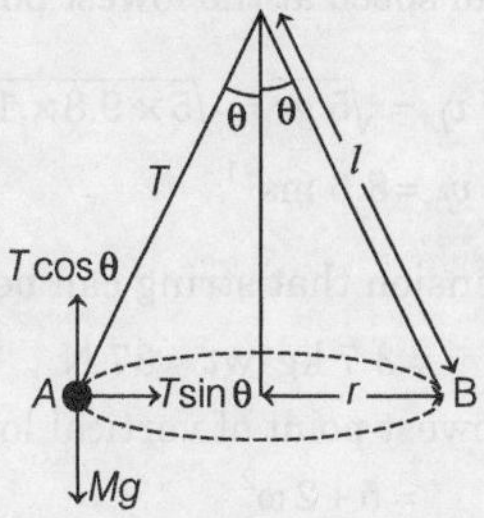

$\therefore$ Time period, $$t = \frac{2\pi}{\omega} = 2\pi\sqrt{\frac{l\cos\theta}{g}}$$

13. Let N = normal reaction of wire loop acting towards centre.

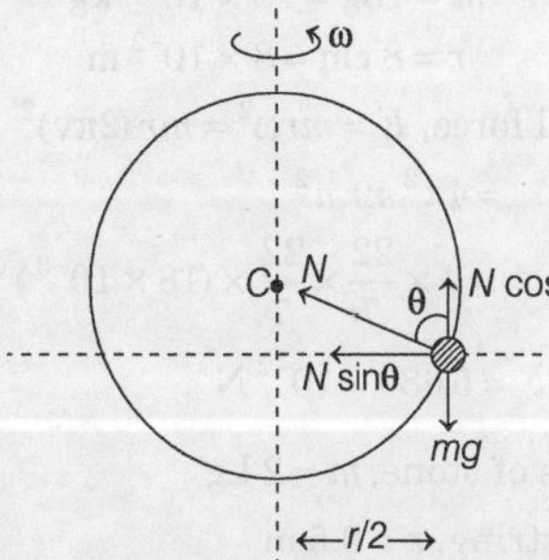

Then, component $N\cos\theta$ balances weight of bead,

$$\Rightarrow \quad N\cos\theta = mg \qquad \text{...(i)}$$

and component $N\sin\theta$ provides necessary centripetal pull on the bead,

$$\Rightarrow \quad N\sin\theta = m\left(\frac{r}{2}\right)\omega^2 \qquad \text{...(ii)}$$

From Eqs. (i) and (ii), we have

$$\tan\theta = \frac{r\omega^2}{2g} \qquad \text{...(iii)}$$

Now, from geometry of figure,

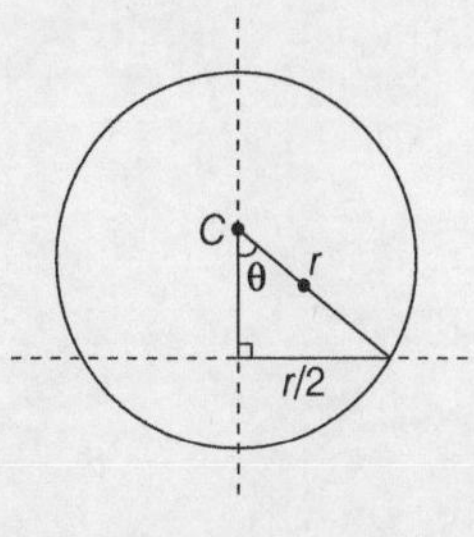

$$\tan\theta = \frac{\frac{r}{2}}{\sqrt{r^2 - \left(\frac{r}{2}\right)^2}} = \frac{r}{2\left(\frac{\sqrt{3}}{2}\right)r} = \frac{1}{\sqrt{3}} \qquad \text{...(iv)}$$

Put this value in Eq. (iii), we get

$$\omega^2 = \frac{2g}{\sqrt{3}\,r}$$

14. Here, $r = 1.5$ m, $m = 0.4$ kg; $g = 9.8\ \text{ms}^{-2}$

The minimum speed at the lowest point of the vertical circle is

$$v_L = \sqrt{5\,gr} = \sqrt{5 \times 9.8 \times 1.5}$$

$$v_L = 8.6\ \text{ms}^{-1}$$

15. Maximum tension that string can bear

$$= 3.7\ \text{kg-wt} = 37\ \text{N}$$

Tension at lowest point of vertical loop $= mg + m\omega^2 r$

$$= 5 + 2\,\omega^2$$

$\therefore$ $$37 = 5 + 2\,\omega^2$$

or $$\omega = 4\ \text{rad/s}$$

16. Here, $n = 60\ \text{rpm} = \dfrac{60}{60} = 1$ rps

$$m = 18\text{g} = 18 \times 10^{-3}\ \text{kg}$$

$$r = 8\ \text{cm} = 8 \times 10^{-2}\ \text{m}$$

Centrifugal force, $F = mr\omega^2 = mr\,(2\pi\nu)^2$

$$= 4\pi^2 mr\,n^2$$

$$= 4 \times \frac{22}{7} \times \frac{22}{7} \times (18 \times 10^{-3}) \times (8 \times 10^{-2}) \times 1^2$$

$$F = 5.689 \times 10^{-2}\ \text{N}$$

17. Here, mass of stone, $m = 2$ kg

Length of string, $r = 0.5$ m

Breaking tension, $T = 900$ N

At $$T = mr\omega^2$$

or $$\omega^2 = \frac{T}{mr} = \frac{900}{2 \times 0.5} = 900$$

$\Rightarrow$ $$\omega = \sqrt{900} = 30\ \text{rad/s}$$

18. If v is velocity acquired at B, then

$$v^2 = 2\,gh$$

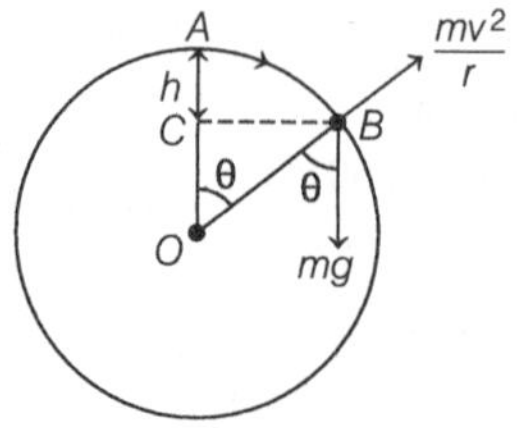

The particle will leave the sphere at B, when

$$\frac{mv^2}{r} \geq mg\cos\theta$$

$$\frac{2\,gh}{r} = \frac{g\,(r-h)}{r}$$

which gives, $$h = \frac{r}{3}$$

$\therefore$ $$n = 3$$

19. $$\alpha = \frac{\tau}{I} = \frac{FR}{mR^2/2} = \frac{2F}{mR}$$

$$\alpha = \frac{2 \times 20}{20 \times (0.2)} = 10\ \text{rad/s}^2$$

$$\omega^2 = \omega_0^2 + 2\alpha\Delta\theta$$

$$(50)^2 = 0^2 + 2(10)\,\Delta\theta$$

$\Rightarrow$ $$\Delta\theta = \frac{2500}{20}$$

$$= 125\ \text{rad}$$

Number of revolutions $n = \dfrac{125}{2\pi} \approx 20$

20. We know, $\theta = \left(\dfrac{\omega_1 + \omega_2}{2}\right)t$

Let number of revolutions be N.

$\therefore$ $$2\pi N = 2\pi\left(\frac{900 + 2460}{60 \times 2}\right) \times 26$$

$$N = 728$$

06

Laws of Motion and Friction

IN THIS CHAPTER ....

- Force
- Inertia
- Newton's Laws of Motion
- Principle of Conservation of Linear Momentum
- Equilibrium of Concurrent Forces
- Common Forces in Mechanics
- Connected Motion
- Friction

Force

Force is a push or pull which

(i) generates or tends to generate motion in a body at rest.

(ii) stops or tends to stop a body in motion.

(iii) increases or decreases the magnitude of velocity of the moving body.

(iv) changes or tries to change the direction of a moving body.

(v) tends to change the shape of the body.

Based on the nature of interaction between two bodies, forces may be broadly classified into two types

(a) **Non-contact Forces** These are the forces that act between two bodies separated by a distance without any actual contact, e.g. gravitational force, electrostatic force.

(b) **Contact Forces** These are the forces that act between two bodies in contact, e.g. tension, normal reaction, friction, etc.

Inertia

The inability of a body to change by itself its state of rest or state of uniform motion along a straight line is called *inertia* of the body.

As inertia of a body is measured by the mass of the body. Heavier the body, greater the force required to change its state and hence, greater is its inertia. It is of three types (i) inertia of rest (ii) inertia of motion (iii) inertia of direction.

Newton's Laws of Motion

According to Newton, there are three laws' of motion

First Law of Motion (Law of Inertia)

It states that a body continues to be in a state of rest or of uniform motion along a straight line, unless it is acted upon by some external force to change of state. This is also called law of inertia.

If $F = 0 \Rightarrow v = \text{constant} \Rightarrow a = 0$

- This law defines force.
- The body opposes any external change in its state of rest or of uniform motion.

e.g. When a stationary vehicle suddenly moves, then the passengers inside the vehicle fall backward.

Linear Momentum

It is measured as the product of the mass of the body and its velocity.

The momentum of a body of mass m moving with a velocity $\mathbf{v}$ is given by $\mathbf{p} = m\mathbf{v}$.

Its unit is kg-ms^{-1} and dimensional formula is $[\text{ML T}^{-1}]$

Second Law of Motion

The rate of change of momentum of a body is directly proportional to the applied force and takes place in the direction in which the force acts. According to second law,

$$F \propto \frac{dp}{dt} \quad \text{or} \quad F = k\frac{dp}{dt}$$

where, k is constant.

Second law can be written as $F = \dfrac{dp}{dt} = ma$

The SI unit of force is newton (N) and in CGS system is dyne.

$$1\,\text{N} = 10^5 \text{ dyne}$$

Impulse

It is defined as the product of the average force and the time for which the force acts.

Impulse, $\mathbf{I} = \mathbf{F}_{\text{av}}\, t$

Impulse is also equal to the total change in momentum of the body during the impact.

Impulse, $\mathbf{I} = \mathbf{p}_2 - \mathbf{p}_1$

Impulse = Change in momentum

Third Law of Motion

To every action, there is an equal and opposite reaction.

i.e. $\mathbf{F}_{12} = -\mathbf{F}_{21}$

Action and reaction are mutually opposite and act on two different bodies.

e.g. Jumping of a man from a boat onto the bank of a river, Jerk is produced in a gun when bullet is fired from it.

Example 1. *A block of metal weighing 2 kg is resting on a frictionless plane. It is struck by a jet releasing water at the rate of 1 kgs^{-1} and with a speed of 5 ms^{-1}. The initial acceleration of the block is*

(a) 2 ms^{-2} (b) 2.5 ms^{-2} (c) 3 ms^{-2} (d) 3.5 ms^{-2}

Sol. (b) The water releasing from jet striking the block at the rate of 1 kgs^{-1} with a speed of 5 ms^{-1} will exert a force on the block,

$$F = v\frac{dm}{dt} = 5 \times 1 = 5\text{ N}$$

Under the action of this force of 5 N, the block of mass 2 kg will move with an acceleration given by

$$a = \frac{F}{m} = \frac{5}{2} = 2.5\text{ ms}^{-2}$$

Example 2. *A bullet of mass 0.04 kg moving with a speed of 90 ms^{-1} enters a having wooden block and is stopped after a distance of 60 cm. The average resistive force exerted by the block on the bullet is*

(a) 100 N (b) 270 N (c) 50 N (d) 298 N

Sol. (b) The retardation a of the bullet is given by

$$a = -\frac{u^2}{2s} = \frac{-90 \times 90}{2 \times 0.6}\text{ms}^{-2} = -6750\text{ ms}^{-2}$$

The retarding force by the second law of motion is

$$F = ma = 0.04 \times 6750 = 270\text{ N}$$

Note The actual resistive force and therefore, retardation of the bullet may not be uniform. The answer therefore only indicates the average resistive force.

Example 3. *A batsman hits back a ball straight in the direction of the bowler without changing its initial speed of 12 ms^{-1}. If the mass of the ball is 0.15 kg, the impulse imparted to the ball is* **[NCERT Exemplar]**

(a) 1.5 Ns (b) 3.6 Ns (c) 7.2 Ns (d) 10.2 Ns

Sol. (b) Change in momentum $= m\Delta v$

Given, $m = 0.15$ kg

$\therefore \quad \Delta p = 0.15 \times 12 - (-0.15 \times 12) = 3.6$ N s

In the direction from the batsman to the bowler.

Example 4. *Three forces start acting simultaneously on a particle moving with velocity v. These forces are represented in magnitude and direction by the three sides of a ΔABC (as shown). The particle will now move with velocity* **[AIEEE 2003]**

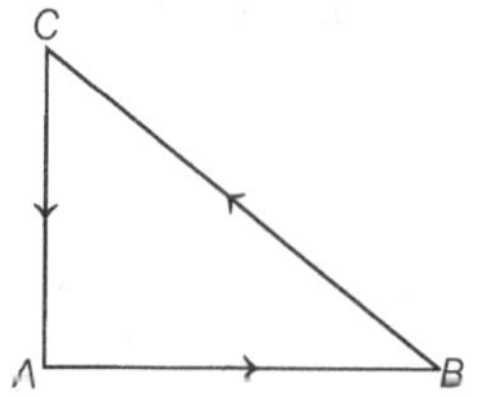

(a) less than v

(b) greater than v

(c) $|v|$ in the direction of largest force BC

(d) v, remaining unchanged

Sol. (*d*) Resultant force is zero, as three forces acting on the particle can be represented in magnitude and direction by three sides of a triangle in same order. Hence, by Newton's second law,

$$\mathbf{F} = m\frac{d\mathbf{v}}{dt} \Rightarrow m\frac{dv}{dt} = 0 \Rightarrow \frac{dv}{dt} = 0$$

$\Rightarrow$ v = constant, hence v remains unchanged.

Principle of Conservation of Linear Momentum

It states that if no external force is acting on a system, the momentum of the system remains constant.

If no force is acting, then $\mathbf{F} = 0$

$$\therefore \quad \frac{d\mathbf{p}}{dt} = 0 \Rightarrow \mathbf{p} = \text{constant}$$

or $$m_1 v_1 = m_2 v_2 = \text{constant}$$

Equilibrium of Concurrent Forces

- If all the forces working on a body are acting on the same point, then they are said to be concurrent.
- A body under the action of concurrent forces, is said to be in equilibrium, when there is no change in the state of rest or of uniform motion along a straight line.
- When a particle is at rest or moving with constant velocity in an inertial frame of reference, the net force on it, i.e. the vector sum of all the forces acting on it must be zero.

  i.e. $\mathbf{F}_1 + \mathbf{F}_2 + \mathbf{F}_3 = 0$ or $\Sigma \mathbf{F} = 0$

  In component form, $\Sigma F_x = 0$, $\Sigma F_y = 0$, $\Sigma F_z = 0$

  Lami's Theorem Three forces P, Q and R are acting on a body are in equilibrium, if

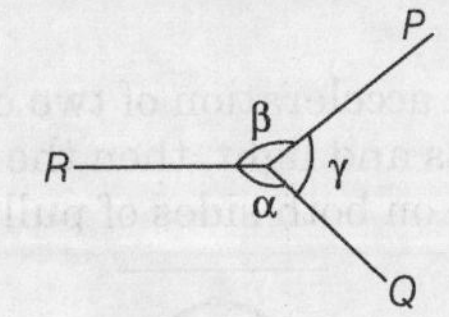

$$\frac{P}{\sin\alpha} = \frac{Q}{\sin\beta} = \frac{R}{\sin\gamma}$$

Common Forces in Mechanics

(i) Tension (T)

When a body is connected through a string or rope, a force may act on the body by the string or rope due to the tendency of extension. This force is called tension.

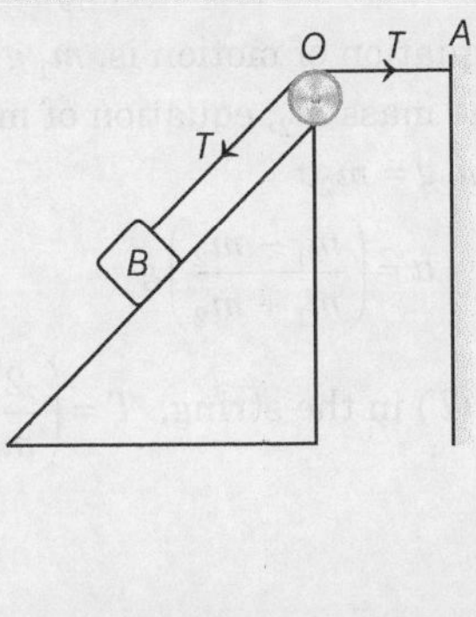

If the string is massless, then

(i) the tension T has the same magnitude at all points throughout the string.

(ii) the magnitude of acceleration of any number of masses connected through string is always same.

(iii) if there is friction between string and pulley, tension is different on two sides of the pulley but if there is no friction between pulley and string, tension will be same on both sides of the pulley.

Note (i) If string slacks, tension in string becomes zero.

(ii) The direction of tension on a body or pulley is always away from the point of contact.

The direction of tension (T) in some cases are shown below

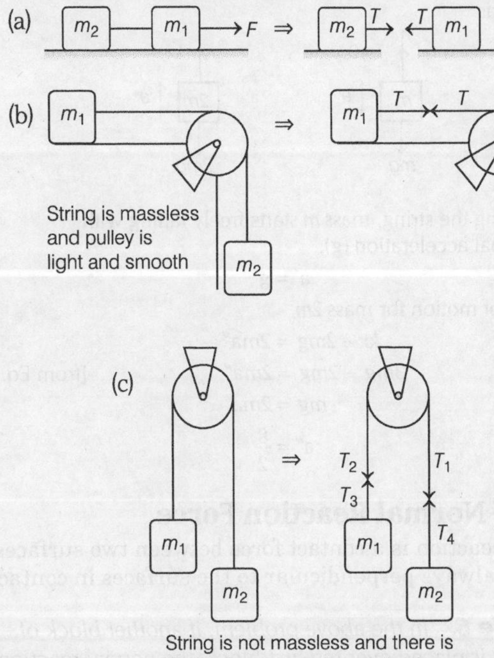

Example 5. *System shown in figure is in equilibrium and at rest. The spring and string are massless, now the string is cut. The acceleration of masses 2m and m, just after the string is cut will be*

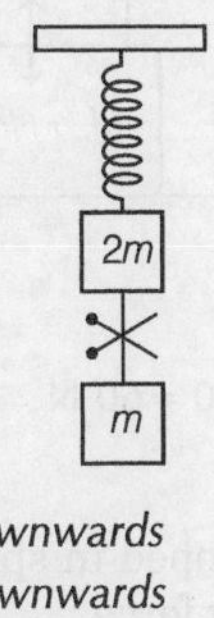

(a) g/2 upwards, g downwards

(b) g upwards, g/2 downwards

(c) g upwards, 2g downwards

(d) 2g upwards, g downwards

Sol. (b) Before cutting the string, free body diagrams of masses m and $2m$ are given as

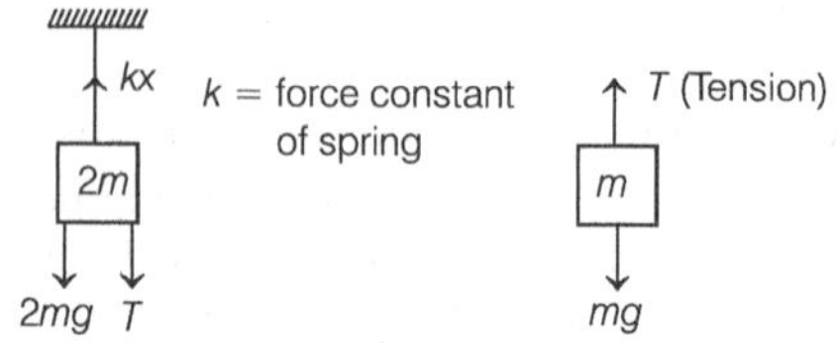

$$\Rightarrow \quad T + 2mg = kx \quad \text{...(i)}$$
$$T = mg \quad \text{...(ii)}$$

From Eqs. (i) and (ii), we get

$$mg + 2mg = kx$$
$$\Rightarrow \quad kx = 3mg \quad \text{...(iii)}$$

After cutting the string, free body diagrams of masses m and $2m$ are given as

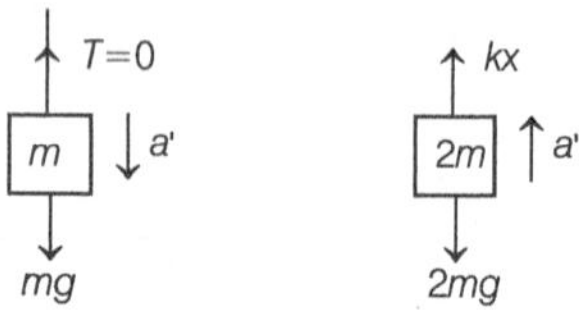

After cutting the string, mass m starts freely falling with gravitational acceleration (g).

i.e. $\quad a' = g$

Equation of motion for mass $2m$,

$$kx - 2mg = 2ma''$$
$$\Rightarrow \quad 3mg - 2mg = 2ma'' \quad \text{[from Eq. (i)]}$$
$$\Rightarrow \quad mg = 2ma''$$
$$\Rightarrow \quad a'' = \frac{g}{2}$$

(ii) The Normal Reaction Force

Normal reaction is a contact force between two surfaces, which is always perpendicular to the surfaces in contact.

Example 6. *In the above problem, if another block of mass 2 kg is placed over the first block, the normal reaction between the first block and the ground surface in new case is*

(a) 40 N (b) 60 N (c) 20 N (d) 10 N

Sol. (b)

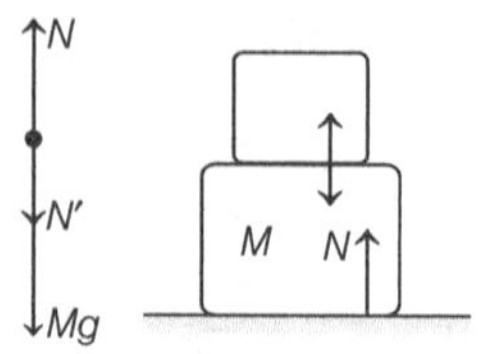

$$N - N' = Mg$$
$$\Rightarrow \quad N = (M + m)\, g$$
$$\Rightarrow \quad N = (4 + 2) \times 10 = 60 \text{ N}$$

(iii) Spring Force

The resistive force developed in spring, when its length is changed, is called spring force.

$$F \propto -x$$

Spring force, $\quad F = -kx$

where, k = spring constant and x = change in length of the spring.

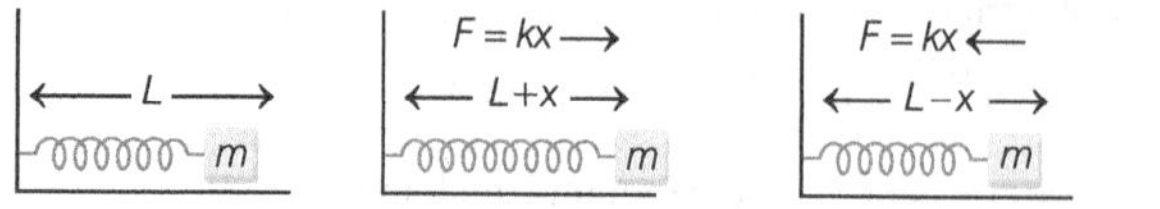

Spring in natural length (a) | Spring in the condition of elongation (b) | Spring in the condition of compression (c)

(iv) Weight of a Body in a Lift

Earth attracts every body towards its centre. The force of attraction exerted by the earth on the body is called *gravity force*. If m be the mass of the body, then the gravity force on it is mg. Normally, the weight of a body is equal to the gravity force, $w = mg$.

- If lift is accelerating upward with acceleration a, then apparent weight of the body is $R = m(g + a)$.
- If lift is accelerating downward at the rate of acceleration a, then apparent weight of the body is $R = m(g - a)$.
- If lift is moving upward or downward with constant velocity, then apparent weight of the body is equal to actual weight, *i.e.* $R = mg$.
- If the lift is falling freely under the effect of gravity ($g = a$), then it is called weightlessness condition ($R = 0$).

Connected Motion

When two objects of masses m_1 and m_2 tied at the ends of an inextensible string which passes over the light and frictionless pulley. Suppose $m_1 > m_2$, the heavier object m_1 moves downward and lighter object m_2 moves upward.

If a be the common acceleration of two objects, since pulley is frictionless and light, then the tension in the string will be same on both sides of pulley.

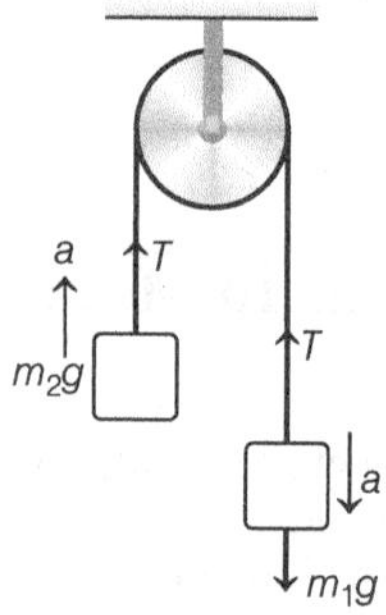

For mass m_1, equation of motion is, $m_1 g - T = m_1 a$

Similarly for the mass m_2, equation of motion is

$$T - m_2 g = m_2 a$$

$$\therefore \quad a = \left(\frac{m_1 - m_2}{m_1 + m_2}\right) g$$

Hence, tension (T) in the string, $T = \left(\dfrac{2\, m_1 m_2}{m_1 + m_2}\right) g$

Example 7. *Two masses $m_1 = 5$ kg and $m_2 = 4.8$ kg tied to a string are hanging over a light frictionless pulley. What is the acceleration of the masses when lift is free to move ? ($g = 9.8\ ms^{-2}$)* **[AIEEE 2004]**

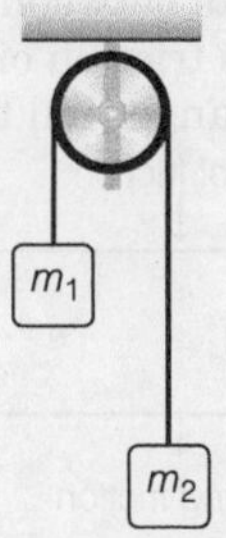

(a) $0.2\ ms^{-2}$ (b) $9.8\ ms^{-2}$ (c) $5\ ms^{-2}$ (d) $4.8\ ms^{-2}$

Sol. (a) On releasing, the motion of the system will be according to figure,

$$m_1 g - T = m_1 a \quad \text{...(i)}$$

and $$T - m_2 g = m_2 a \quad \text{...(ii)}$$

On solving Eqs. (i) and (ii), we get

$$a = \left(\frac{m_1 - m_2}{m_1 + m_2}\right) g \quad \text{...(iii)}$$

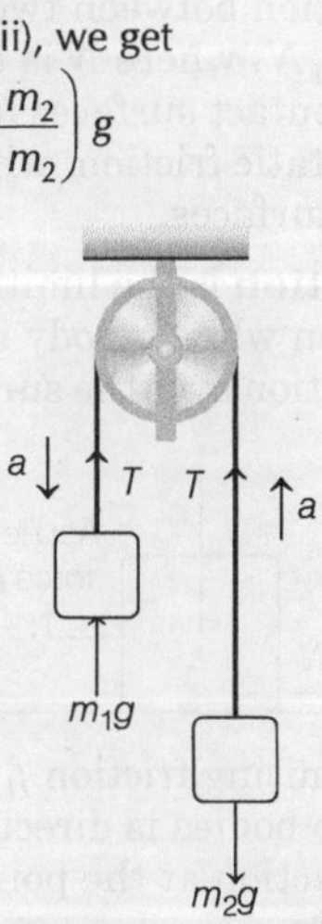

Here, $m_1 = 5$ kg, $m_2 = 4.8$ kg, $g = 9.8\ \text{ms}^{-2}$

$$\therefore \quad a = \left(\frac{5 - 4.8}{5 + 4.8}\right) \times 9.8 = \frac{0.2}{9.8} \times 9.8 = 0.2\ \text{ms}^{-2}$$

Some Important Cases of Connected Motion

(i) Motion of blocks in contact

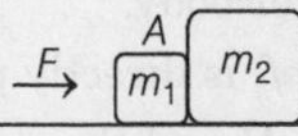

Acceleration, $$a = \frac{F}{m_1 + m_2}$$

(ii) Motion of blocks connected by massless string

Acceleration, $$a = \frac{F}{m_1 + m_2}$$

and tension, $$T = \frac{m_1 F}{m_1 + m_2}$$

(iii) Pulley-mass system

(a) Body accelerated on a horizontal surface by a falling body,

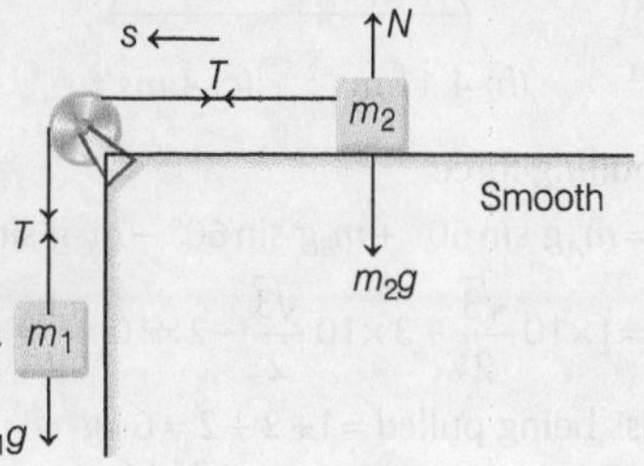

Acceleration, $$a = \left(\frac{m_1}{m_1 + m_2}\right) g$$

Tension, $$T = \left(\frac{m_1 m_2}{m_1 + m_2}\right) g$$

(b) Motion on a smooth inclined plane

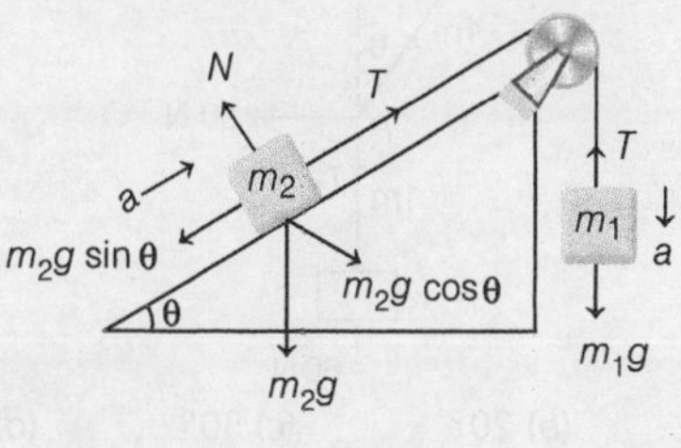

$$\therefore \quad a = \left(\frac{m_1 - m_2 \sin\theta}{m_1 + m_2}\right) g$$

and $$T = \frac{m_1 m_2 (1 + \sin\theta) g}{(m_1 + m_2)}$$

(c) Body accelerated on a wedge due to a another falling body ($m_2 > m_1$)

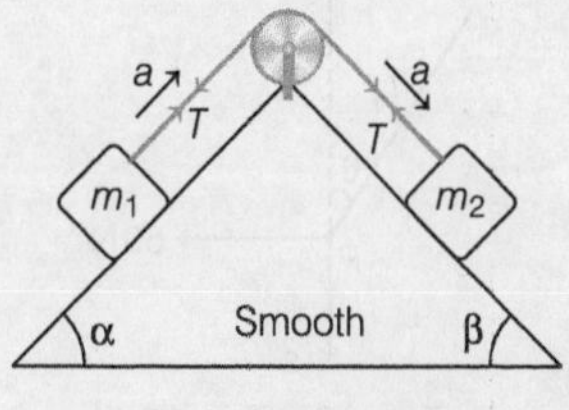

$$T = \frac{m_1 m_2}{m_1 + m_2} (\sin\alpha + \sin\beta) g$$

and $$a = \left(\frac{m_2 \sin\beta - m_1 \sin\alpha}{m_2 + m_1}\right) g$$

Example 8. *As shown in figure A, B and C are 1 kg, 3 kg and 2 kg, respectively. The acceleration of the system is*

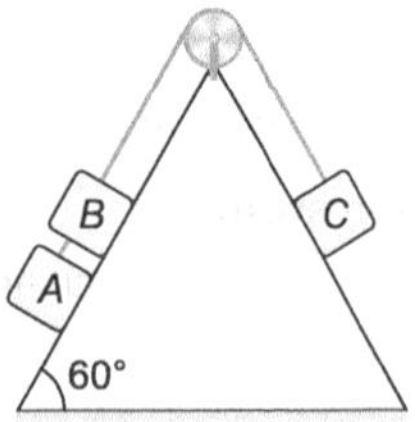

(a) 5 ms^{-2} (b) 4.11 ms^{-2} (c) 4 ms^{-2} (d) 5.11 ms^{-2}

Sol. (b) Net pulling force

$$= m_A g \sin 60^\circ + m_B g \sin 60^\circ - m_C g \sin 30^\circ$$

$$= 1\times 10 \frac{\sqrt{3}}{2} + 3\times 10 \frac{\sqrt{3}}{2} - 2\times 10\times \frac{1}{2} = 24.66\,\text{N}$$

Total mass being pulled $= 1 + 3 + 2 = 6$ kg

$\therefore$ Acceleration of the system, $a = \frac{24.66}{6} = 4.11\,\text{ms}^{-2}$

Example 9. *A mass of 6 kg is suspended by a rope of length 2 m from the ceiling. A force of 50 N in the horizontal direction is applied at the mid-point P of the rope as shown. The angle that rope makes with the vertical in equilibrium is*

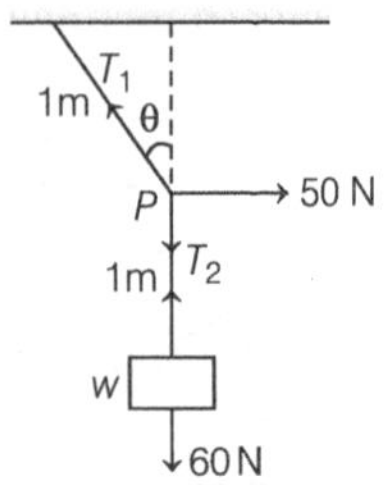

(a) 10° (b) 20° (c) 30° (d) 40°

Sol. (d) Making the free body diagram of *P* and *w*, we consider the point of equilibrium of the weight *w*.

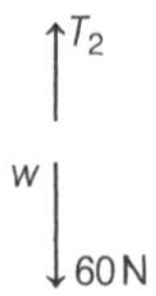

$\therefore$ $$T_2 = 6\times 10 = 60\,\text{N}$$

Consider the equilibrium of the point *P* under the action of three forces, the tensions T_1 and T_2 and the horizontal force 50 N. The horizontal and vertical components of the resultant force must vanish separately.

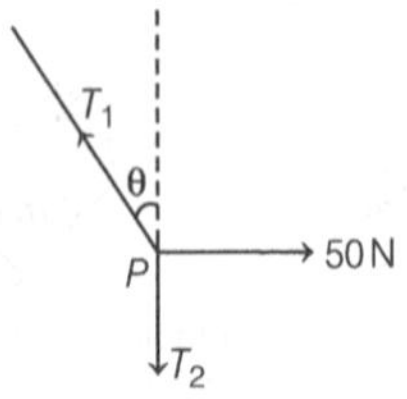

$$T_1 \cos\theta = T_2 = 60\,\text{N}$$
$$T_1 \sin\theta = 50\,\text{N}$$

which gives that

$$\tan\theta = \frac{5}{6}$$

or $$\theta = \tan^{-1}\left(\frac{5}{6}\right) = 40^\circ$$

Friction

Whenever an object actually slides or rolls over the surface of another body or tends to do, a force opposing the relative motion acts between these two surfaces in contact. It is known as friction or force due to friction. Force of friction acts tangential to the surfaces in contact.

Types of friction
- External friction
  - Static friction
  - Limiting friction
  - Kinetic friction
    - Sliding friction
    - Rolling friction
- Internal friction

(i) **Static friction** is a self-adjusting force and is always equal and opposite to the applied force.

The static friction between two contact surfaces is given by $f_s \le \mu_s N$, where N is the normal force between the contact surfaces and μ_s is the coefficient of static friction, which depends on the nature of the surfaces.

(ii) **Limiting friction** is the limiting (maximum) value of static friction when a body is just on the verge of starting its motion over the surface of another body.

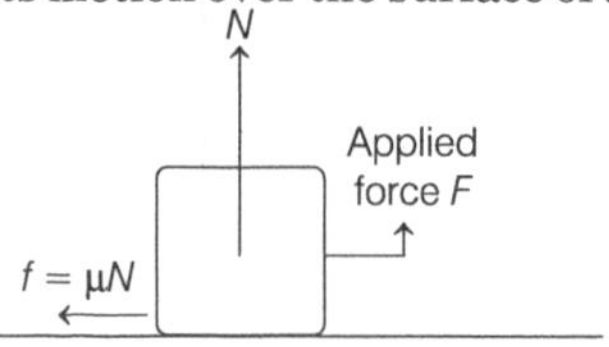

The force of limiting friction f_l between the surfaces of two bodies is directly proportional to the normal reaction at the point of contact. Mathematically,

$$f_l \propto N \text{ or } f_l = \mu_l N \Rightarrow \mu_l = \frac{f_l}{N}$$

where, μ_l is the *coefficient of limiting friction* for the given surfaces in contact.

(iii) **Kinetic friction** is the opposing force that comes into play, when one body actually slides over the surface of another body.

Force of friction f_k is directly proportional to the normal reaction N and the ratio $\frac{f_k}{N}$ is called *coefficient of kinetic* μ_k.

Whenever limiting friction is converted into kinetic friction, body starts its motion with an abrupt uncontrolled movement.

- **Sliding friction** is opposing force that comes into play when one body actually slides over the surface of the other body is called *sliding friction*.

  e.g. A book is moving over a horizontal table.

- **Rolling friction** is the opposing force that comes into play, when one body of symmetric shape (wheel or cylinder or disc, etc.) rolls over the surface of another body.

  Force of rolling friction f_r is directly proportional to the normal reaction N and inversely proportional to the radius (r) of wheel. Thus,

$$f_r \propto \frac{N}{r} \text{ or } f_r = \mu_r \frac{N}{r}$$

The constant μ_r is known as the *coefficient of rolling friction*, μ_r has the unit and dimensions of length.

Magnitude wise $\mu_r << \mu_k$ or μ_l

Angle of Friction

It is defined as the angle θ at which the resultant R of the force of limiting friction f_l and normal reaction N, subtends with the normal reaction.

The tangent of the angle of friction is equal to the coefficient of friction, i.e. $\mu = \tan\theta$.

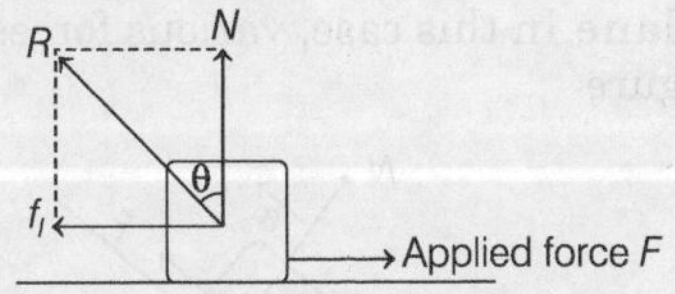

Angle of Repose (α)

If a body is placed on an inclined plane and it is just on the point of sliding down, then the angle of inclination of the plane with the horizontal is called the *angle of repose* (α).

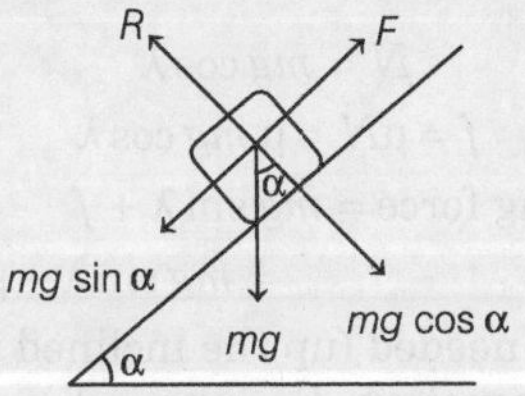

In limiting condition, $F = mg\sin\alpha$ and $R = mg\cos\alpha$

So, $\frac{F}{R} = \tan\alpha$

$\therefore \frac{F}{R} = \mu_s = \tan\alpha$

Thus, the coefficient of limiting friction is equal to tangent of the angle of repose.

As well as, $\alpha = \theta$

i.e. Angle of repose = Angle of friction

Example 10. *A block rests on a rough inclined plane making an angle of 30° with the horizontal. The coefficient of static friction between the block and the plane is 0.8. If the frictional force on the block is 10 N, the mass of the block (in kg) is ($g = 10$ m/s^2)* **[AIEEE 2004]**

(a) 2.0 (b) 4.0 (c) 1.6 (d) 2.5

Sol. (a) Let mass of the block be m.

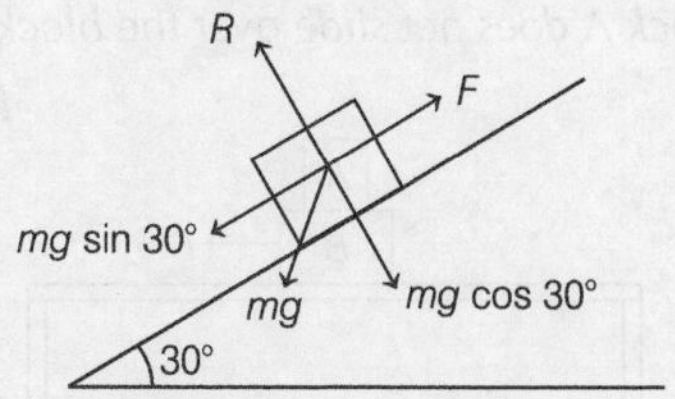

Frictional force in rest position,

$$F = mg \sin 30°$$

(this is static frictional force and may be less than the limiting frictional force)

$$\therefore \quad 10 = m \times 10 \times \frac{1}{2}$$

$$\text{or} \quad m = \frac{2 \times 10}{10} = 2 \text{ kg}$$

Example 11. *A block of mass m is placed on a surface with a vertical cross-section given by $y = x^3/6$. If the coefficient of friction is 0.5, the maximum height above the ground at which the block can be placed without slipping is* **[JEE Main 2014]**

(a) $\frac{1}{6}$ m (b) $\frac{2}{3}$ m (c) $\frac{1}{3}$ m (d) $\frac{1}{2}$ m

Sol. (a) A block of mass m is placed on a surface with a vertical cross-section, then

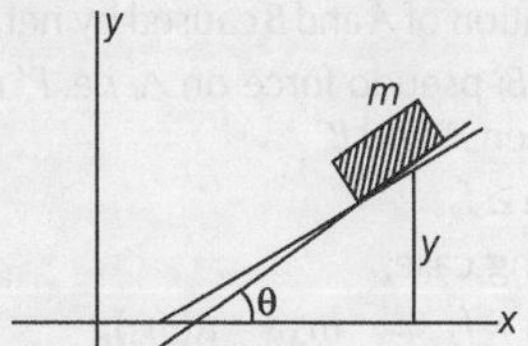

$$\tan\theta = \frac{dy}{dx} = \frac{d\left(\frac{x^3}{6}\right)}{dx} = \frac{x^2}{2}$$

At limiting equilibrium, we get

$$\mu = \tan\theta, \; 0.5 = x^2/2$$

$$\Rightarrow \quad x^2 = 1 \Rightarrow x = \pm 1$$

Now, putting the value of x in $y = x^3/6$, we get

When $x = 1$	When $x = -1$
$\therefore \; y = \frac{(1)^3}{6} = \frac{1}{6}$	$y = \frac{(-1)^3}{6} = \frac{-1}{6}$

So, the maximum height above the ground at which the block can be placed without slipping is $\frac{1}{6}$ m.

Example 12. *Two blocks A and B of masses $m_A = 1$ kg and $m_B = 3$ kg are kept on the table as shown in figure. The coefficient of friction between A and B is 0.2 and between B and the surface of the table is also 0.2.*

The maximum force F that can be applied on B horizontally, so that the block A does not slide over the block B is (Take, $g = 10\ m/s^2$) *[JEE Main 2019]*

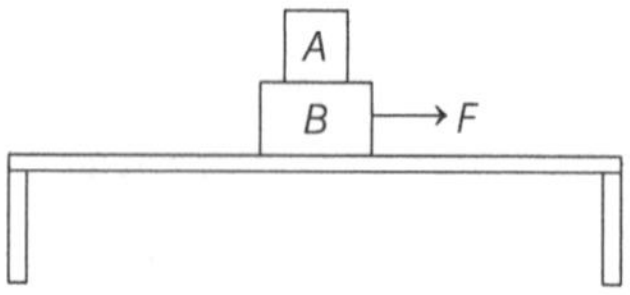

(a) 12 N (b) 16 N (c) 8 N (d) 40 N

Sol. (b) Acceleration a of system of blocks A and B is

$$a = \frac{\text{Net force}}{\text{Total mass}} = \frac{F - f_1}{m_A + m_B}$$

where, f_1 = friction between B and the surface

$$= \mu(m_A + m_B)g$$

So, $$a = \frac{F - \mu(m_A + m_B)g}{(m_A + m_B)} \quad \text{...(i)}$$

Here, $\mu = 0.2$, $m_A = 1\text{ kg}$, $m_B = 3\text{ kg}$, $g = 10\text{ ms}^{-2}$

Substituting the above values in Eq. (i), we have

$$a = \frac{F - 0.2(1+3)\times 10}{1+3}$$

$$a = \frac{F - 8}{4} \quad \text{...(ii)}$$

Due to acceleration of block B, a pseudo force F' acts on A. This force F' is given by $F' = m_A a$.

where, a is acceleration of A and B caused by net force acting on B.

For A to slide over B; pseudo force on A, i.e. F' must be greater than friction between A and B.

$$\Rightarrow \quad m_A a \geq f_2$$

We consider limiting case,

$$m_A a = f_2 \Rightarrow m_A a = \mu(m_A)g$$

$$\Rightarrow \quad a = \mu g = 0.2 \times 10 = 2\text{ ms}^{-2} \quad \text{...(iii)}$$

Putting the value of a from Eq. (iii) into Eq. (ii), we get

$$\frac{F-8}{4} = 2$$

$$\therefore \quad F = 16\text{ N}$$

Some Special Cases of Friction

***Case* I Acceleration of a block on a horizontal surface** As shown in figure, we have

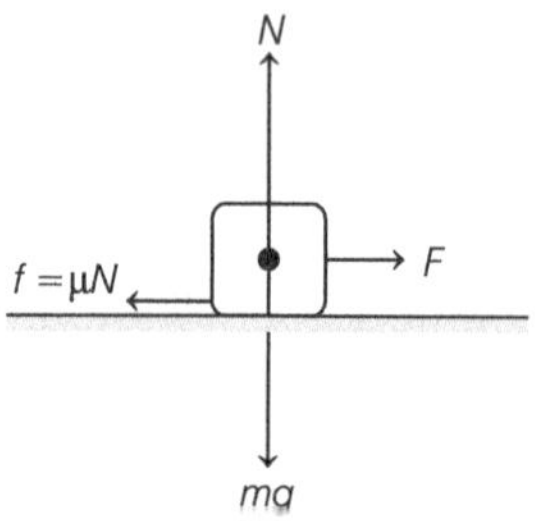

$$N = mg \text{ and } f = \mu N = \mu\, mg$$

$\therefore$ Net acceleration produced,

$$a = \frac{F - f}{m} = \frac{F - \mu\, mg}{m} = \frac{F}{m} - \mu g$$

where, μ = coefficient of (kinetic) friction between the two surfaces in contact.

***Case* II Acceleration of a block sliding down a rough inclined plane** Let there be an inclined plane having angle of inclination λ, which is more than the angle of repose α. Then, as shown in figure

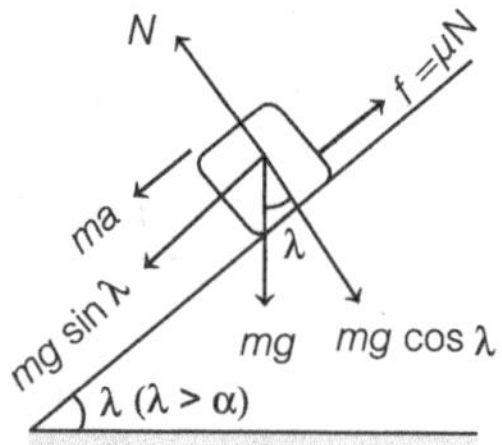

$$N = mg\cos\lambda \text{ and } f = \mu N = \mu\, mg\cos\lambda$$

$\therefore$ Net accelerating force down the inclined plane,

$$mg\sin\lambda - f = mg\sin\lambda - \mu\, mg\cos\lambda = ma$$

Acceleration, $a = g(\sin\lambda - \mu\cos\lambda)$

***Case* III Retardation of a block sliding up a rough inclined plane** In this case, various forces have been shown in figure

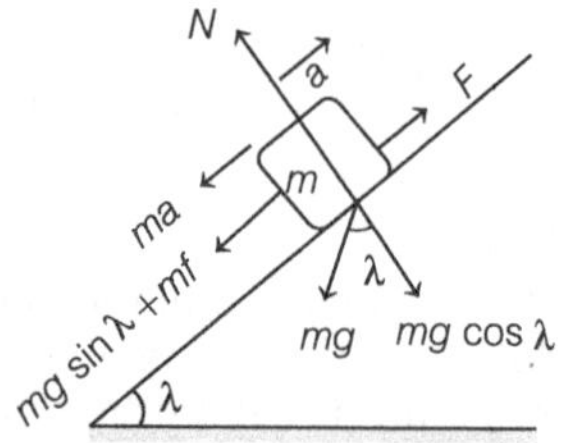

$$N = mg\cos\lambda$$

Force of friction, $f = \mu N = \mu\, mg\cos\lambda$

and net retarding force $= mg\sin\lambda + f$

$$= mg\sin\lambda + \mu\, mg\cos\lambda$$

$\therefore$ External force needed (up the inclined plane) to maintain sliding motion $F = mg(\sin\lambda + \mu\cos\lambda)$

In the absence of external force, the motion of given block will be retarded and the value of retardation will be

$$a = g(\sin\lambda + \mu\cos\lambda)$$

***Case* IV Maximum length of hung chain** A uniform chain of length l is placed on the table in such a manner that its l' part is hanging over the edge of table without sliding. Since, the chain has uniform linear density, therefore coefficient of friction,

$$\mu = \frac{m_2}{m_1} = \frac{\text{mass hanging from the table}}{\text{mass lying on the table}}$$

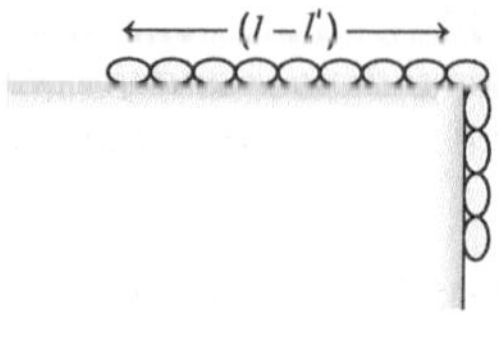

∴ For this case, we can rewrite the above expression in the following manner,

$$\mu = \frac{\text{length hanging from the table}}{\text{length lying on the table}}$$

(as chain has uniform linear density)

$$\therefore \quad \mu = \frac{l'}{l - l'}$$

By solving, $l' = \dfrac{\mu l}{(\mu + l)}$

Case V Motion of an insect in the rough bowl The insect crawls up the bowl upto a certain height h only till the component of its weight along the bowl is balanced by limiting frictional force.

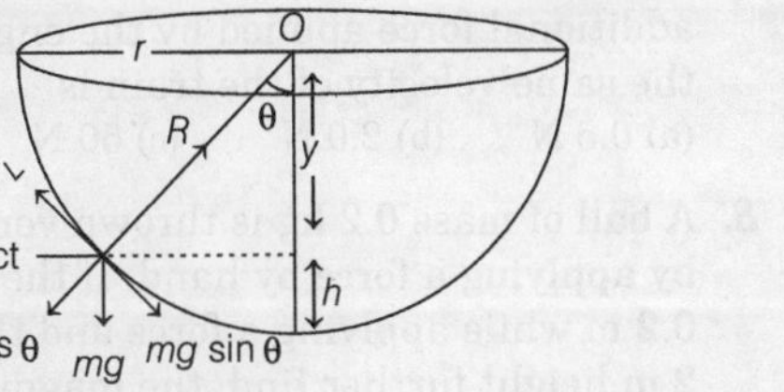

Let m = mass of the insect,
r = radius of the bowl
and μ = coefficient of friction.
For limiting condition at point A,

$$R = mg\cos\theta \qquad \text{...(i)}$$

and

$$F_L = mg\sin\theta \qquad \text{...(ii)}$$

Dividing Eq. (ii) by Eq. (i), we get

$$\tan\theta = \frac{F_L}{R} = \mu$$

$$\therefore \quad \frac{\sqrt{r^2 - y^2}}{y} = \mu$$

or

$$y = \frac{r}{\sqrt{1 + \mu^2}}$$

Example 13. *A man standing stationary with respect to a horizontal conveyor belt that is accelerating with 1 ms^{-2}. If the coefficient of static friction between the man's shoes and the belt is 0.2, upto what acceleration of the belt can the man continue to be stationary relative to the belt? (Mass of the man = 65 kg)*

(a) 1.25 ms^{-2} (b) 1.96 ms^{-2} (c) 2.5 ms^{-2} (d) 3.6 ms^{-2}

Sol. (b) As the man is standing stationary w.r.t. the horizontal conveyor belt, he is also accelerating at 1 ms^{-2}, the acceleration of the belt. Thus,

Acceleration of the man, $a = 1\text{ms}^{-2}$

Mass of the man, $M = 65$ kg

Therefore, net force on the man, $ma = 65 \times 1 = 65$ N

The limiting friction between the shoes of the man and the belt is given by $F = \mu Mg = 0.2 \times 65 \times 9.8$ N

If the man can remain stationary upto an acceleration say a', then

$$Ma' = F$$

or

$$a' = \frac{F}{M} = \frac{0.2 \times 65 \times 9.8}{65} = 1.96 \text{ ms}^{-2}$$

Example 14. *A block of mass 10 kg is kept on a rough inclined plane as shown in the figure. A force of 3 N is applied on the block. The coefficient of static friction between the plane and the block is 0.6. What should be the minimum value of force F, such that the block does not move downward)? (Take, g = 10ms^{-2})* ***[JEE Main 2019]***

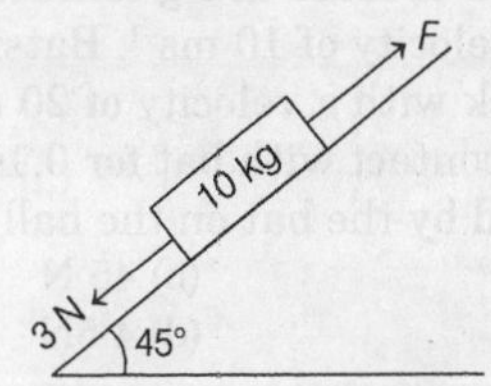

(a) 32 N (b) 25 N (c) 23 N (d) 18 N

Sol. (a) Free body diagram, for the given figure is as follows:

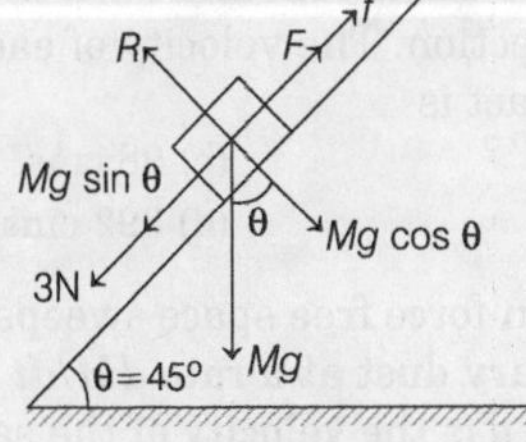

For the block to be in equilibrium, i.e. so that it does not move downward, then $\Sigma f_x = 0$

$$\therefore \quad 3 + Mg\sin\theta - F - f = 0 \text{ or } 3 + Mg\sin\theta = F + f$$

As, frictional force, $f = \mu R$

$$\therefore \quad 3 + Mg\sin\theta = F + \mu R \qquad \text{...(i)}$$

Similarly, $\Sigma f_y = 0$

$$-Mg\cos\theta + R = 0 \text{ or } Mg\cos\theta = R \qquad \text{...(ii)}$$

Substituting the value of R from Eq. (ii) to Eq. (i), we get

$$3 + Mg\sin\theta = F + \mu(Mg\cos\theta) \qquad \text{...(iii)}$$

Here, $M = 10$ kg, $\theta = 45°$, $g = 10$ m/s^2

and $\mu = 0.6$

Substituting these values in Eq. (iii), we get

$$3 + (10 \times 10\sin 45°) - (0.6 \times 10 \times 10\cos 45°) = F$$

$$\Rightarrow \quad F = 3 + \frac{100}{\sqrt{2}} - \frac{60}{\sqrt{2}} = 3 + \frac{40}{\sqrt{2}}$$

$$= 3 + 20\sqrt{2} = 31.8 \text{ N or } F \simeq 32 \text{ N}$$

Practice Exercise

ROUND I Topically Divided Problems

Newton's Laws of Motion

1. A cricket ball of mass 150 g collides straight with a bat with a velocity of 10 ms^{-1}. Batsman hits it straight back with a velocity of 20 ms^{-1}. If ball remains in contact with bat for 0.1s, then average force exerted by the bat on the ball is
(a) 15 N (b) 45 N
(c) 150 N (d) 4.5N

2. A disc of mass 10 g is kept floating horizontally in air by firing bullets, each of mass 5 g, with the same velocity at the same rate of 10 bullets per second. The bullets rebound with the same speed in positive direction. The velocity of each bullet at the time of impact is
(a) 196 cms^{-1} (b) 98 cms^{-1}
(c) 49 cms^{-1} (d) 392 cms^{-1}

3. A satellite in force free space sweeps stationary interplanetary dust at a rate $dM/dt = \alpha v$, where M is the mass, v is the velocity of the satellite and α is a constant. What is the deacceleration of the satellite?
(a) $-2\,\alpha v^2/M$ (b) $-\alpha v^2/M$
(c) $+\,\alpha v^2/M$ (d) $-\,\alpha v^2$

4. The engine of a car produces an acceleration of 6 ms^{-2} in the car. If this car pulls another car of the same mass, then the acceleration would be
(a) 6 ms^{-2} (b) 12 ms^{-2}
(c) 3 ms^{-2} (d) 1.5 ms^{-2}

5. A body of mass 2 kg travels according to law $x(t) = pt + qt^2 + rt^3$, where $p = 3\,ms^{-1}$, $q = 4\,ms^{-2}$ and $r = 5\,ms^{-3}$. The force acting on the body at $t = 2$ s is **[NCERT Exemplar]**
(a) 136 N (b) 134 N
(c) 158 N (d) 68 N

6. In a rocket of mass 1000 kg fuel is consumed at a rate of 40 kg/s. The velocity of the gases ejected from the rocket is 5×10^4 m/s. The thrust on the rocket is
(a) 2×10^3 N (b) 5×10^4 N
(c) 2×10^6 N (d) 2×10^9 N

7. An open carriage in a goods train is moving with a uniform velocity of 10 ms^{-1}. If the rain adds water with zero velocity at the rate of 5 kgs^{-1}, then the additional force applied by the engine to maintain the same velocity of the train is
(a) 0.5 N (b) 2.0 N (c) 50 N (d) 25 N

8. A ball of mass 0.2 kg is thrown vertically upwards by applying a force by hand. If the hand moves 0.2 m while applying a force and the ball goes upto 2 m height further find, the magnitude of the force. (Take, $g = 10$ m/s^2)
(a) 16 N (b) 20 N
(c) 22 N (d) 44 N

9. A constant force acting on a body of mass 3.0 kg changes its speed from 2.0 m/s to 3.5 m/s. The direction of motion of the body remains unchanged. What is the magnitude and direction of the force?
(a) 0.18 N, along the direction of motion
(b) 0.18 N, opposite to the direction of motion
(c) 0.28 N, along the direction of motion
(d) 0.28 N, opposite to the direction of motion

10. A body of mass 5 kg is acted upon by two perpendicular forces 8 N and 6 N. Give the magnitude and direction of the acceleration of the body
(a) 2 m/s^2 at an angle 37° to force
(b) 2 m/s^2 at an angle 57° to force
(c) 4 m/s^2 at angle 37° to force
(d) 4 m/s^2 at an angle 57° to force

11. A particle of mass m is moving in a straight line with momentum p. Starting at time $t = 0$, a force $F = kt$ acts in the same direction on the moving particle during time interval T, so that its momentum changes from p to $3p$. Here, k is a constant.
The value of T is **[JEE Main 2019]**
(a) $\sqrt{\frac{2p}{k}}$ (b) $2\sqrt{\frac{p}{k}}$
(c) $\sqrt{\frac{2k}{p}}$ (d) $2\sqrt{\frac{k}{p}}$

Conservation of Linear Momentum and Impulse

12. If a force of 250 N act on body the momentum acquired is 125 kg-m/s. What is the period for which force acts on the body?
(a) 0.5 s (b) 0.2 s (c) 0.4 s (d) 0.25 s

13. 100 g of an iron ball having velocity 10 ms^{-1} collides with wall at an angle 30° and rebounds with the same angle. If the period of contact between the ball and wall is 0.1s, then the average force experienced by the wall is
(a) 10 N (b) 100 N
(c) 1.0 N (d) 0.1 N

14. A player caught a cricket ball of mass 150 g moving at a rate of 20 m/s. If the catching process is completed in 0.1 s, the force of the blow exerted by the ball on the hand of the player is equal to **[AIEEE 2006]**
(a) 150 N (b) 3 N
(c) 30 N (d) 300 N

15. Conservation of momentum in a collision between particles can be understood from **[NCERT Exemplar]**
(a) conservation of energy
(b) Newton's first law only
(c) Newton's second law only
(d) Both Newton's second and third laws

16. A bag of sand of mass m is suspended by a rope. A bullet of mass $\frac{m}{20}$ is fired at it with a velocity v and gets embedded into it. The velocity of the bag finally is
(a) $\frac{v}{20}\times 21$ (b) $\frac{20\,v}{21}$
(c) $\frac{v}{20}$ (d) $\frac{v}{21}$

17. The figure shows the position-time (x-t) graph of one-dimensional motion of a body of mass 0.4 kg. The magnitude of each impulse is **[AIEEE 2010]**

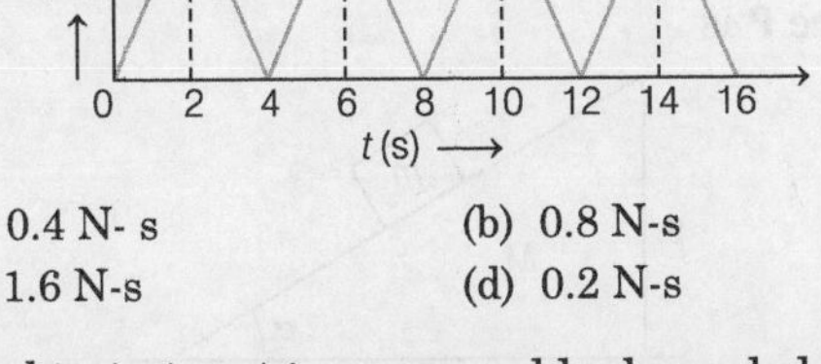

(a) 0.4 N- s (b) 0.8 N-s
(c) 1.6 N-s (d) 0.2 N-s

18. An object at rest in space suddenly explodes into three parts of same mass. The momentum of the two parts are $2p\hat{\mathbf{i}}$ and $p\hat{\mathbf{j}}$. The momentum of the third part
(a) will have a magnitude $p\sqrt{3}$
(b) will have a magnitude $p\sqrt{5}$
(c) will have a magnitude p
(d) will have a magnitude $2p$

19. A machine gun fires a bullet of mass 40 g with a velocity 1200 ms^{-1}. The man holding it, can exert a maximum force of 144 N on the gun. How many bullets can he fire per second at the most? **[AIEEE 2004]**
(a) 1 (b) 4
(c) 2 (d) 3

20. A man weighing 60 kg is standing on a trolley weighing 240 kg. The trolley is resting on frictionless horizontal rails. If the man starts walking on the trolley with a velocity of 1 ms^{-1}, then after 4 s, his displacement relative to the ground is
(a) 6 m (b) 4.8 m
(c) 3.2 m (d) 2.4 m

21. A ball is thrown vertically up (taken as + Z-axis) from the ground. The correct momentum-height (p-h) diagram is **[JEE Main 2019]**

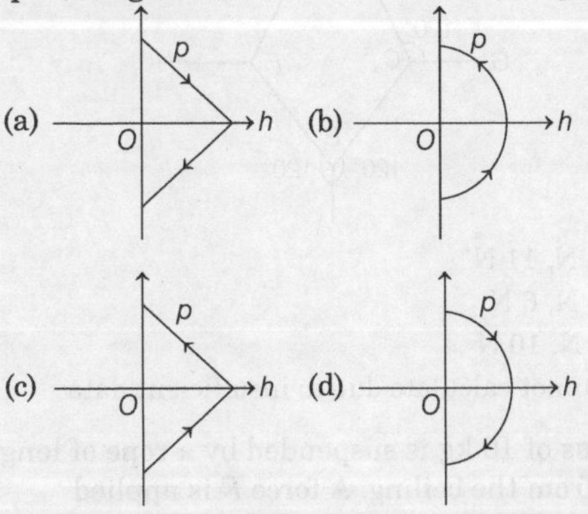

22. A batsman deflects a ball by an angle of 45° without changing its initial speed which is equal to 54 km/h. What is the impulse imparted to the ball? (Mass of the ball is 0.15 kg)
(a) 4 kg-m/s
(b) 6 kg-m/s
(c) 2 kg-m/s
(d) 5 kg-m/s

23. A shell is fired from a cannon with velocity v ms^{-1} at an angle θ with the horizontal direction. At the highest point in its path it explodes into two pieces of equal mass. One of the pieces retraces its path to the cannon and the speed in m/s of the piece immediately after the explosion is
(a) $3\,v\cos\theta$ (b) $2\,v\cos\theta$
(c) $\frac{3\,v}{2}\cos\theta$ (d) $\frac{\sqrt{3}\,v\cos\theta}{2}$

Equilibrium of Forces

24. When a force F acts on a body of mass m, the acceleration produced in the body is a. If three equal forces $F_1 = F_2 = F_3 = F$ act on the same body as shown in figure, the acceleration produced is

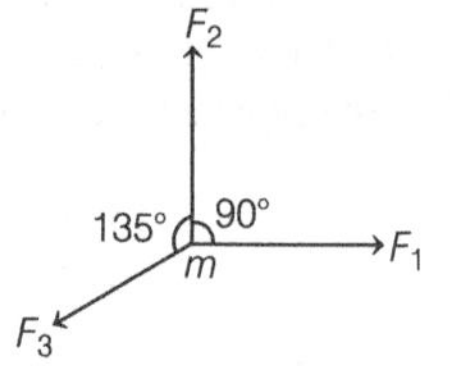

(a) $(\sqrt{2}-1)\,a$ (b) $(\sqrt{2}+1)\,a$
(c) $\sqrt{2}\,a$ (d) a

25. The following figure is the part of a horizontally stretched net section AB which is stretched with a force of 10 N. The tension in the section BC and BF are

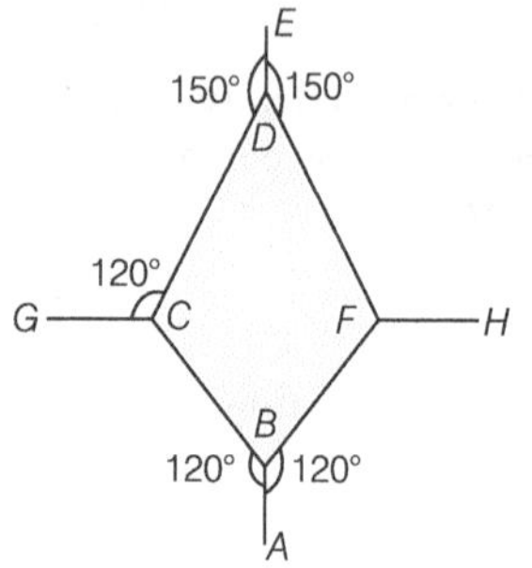

(a) 10 N, 11 N
(b) 10 N, 6 N
(c) 10 N, 10 N
(d) Cannot calculate due to insufficient data

26. A mass of 10 kg is suspended by a rope of length 4 m, from the ceiling. A force F is applied horizontally at the mid-point of the rope such that the top half of the rope makes an angle of 45° with the vertical. Then, F equals (Take, $g = 10\ \text{ms}^{-2}$ and the rope to be massless) **[JEE Main 2020,19]**

(a) 75 N (b) 70 N (c) 100 N (d) 90 N

27. A body of weight 2 kg is suspended as shown in figure. The tension T_1 in the horizontal string (in kg-wt) is

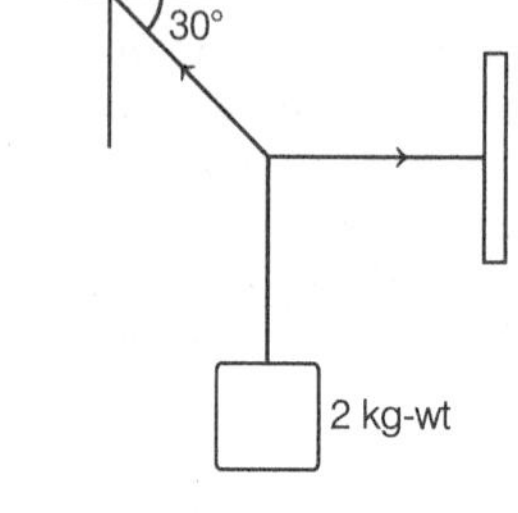

(a) $2\sqrt{3}$ (b) $\sqrt{3}/2$ (c) $\sqrt{3}$ (d) 2

28. A ball of mass 1 kg hangs in equilibrium from two strings OA and OB as shown in figure. What are the tensions in strings OA and OB? (Take, $g = 10\ \text{ms}^{-2}$)

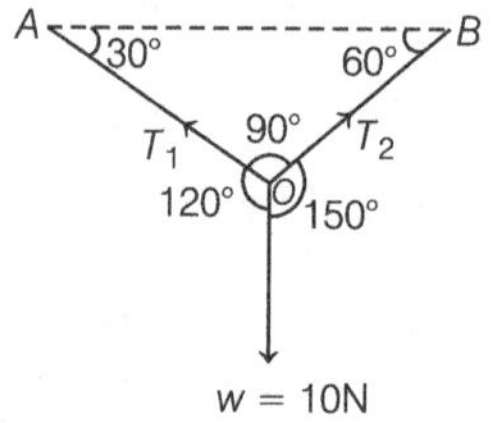

(a) 5 N, zero (b) Zero, N
(c) 5 N, $5\sqrt{3}$ N (d) $5\sqrt{3}$ N, 5 N

29. A piece of wire is bent in the shape of a parabola $y = kx^2$ (Y-axis vertical) with a bead of mass m on it. The bead can slide on the wire without friction. It stays at the lowest point of the parabola when the wire is at rest. The wire is now accelerated parallel to the X-axis with a constant acceleration a. The distance of the new equilibrium position of the bead, where the bead can stay at rest with respect to the wire, from the Y-axis is

(a) $\frac{a}{gk}$ (b) $\frac{a}{2gk}$ (c) $\frac{2a}{gk}$ (d) $\frac{a}{4gk}$

Motion of Connected Bodies

30. Two blocks are in contact on a frictionless table. One has mass m and other $2m$. A force f is applied on $2m$ as shown in figure. Next the same force F is applied from the right on m. In the two cases respectively, the force of contact between the two blocks will be

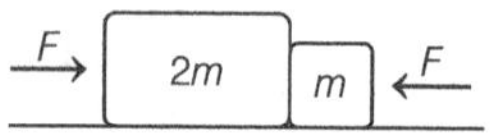

(a) 2 : 1 (b) 1 : 3
(c) 1 : 2 (d) 3 : 1

31. A wooden wedge of mass M and inclination angle α rests on a smooth floor. A block of mass m is kept on wedge. A force **P** is applied on the wedge as shown in figure, such that a block remains stationary with respect to wedge. The magnitude of force **P** is

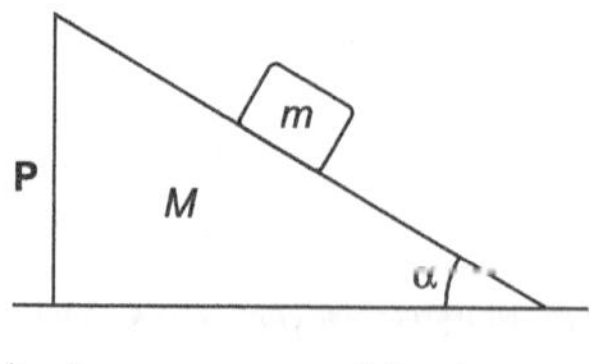

(a) $(M+m)\,g\tan\alpha$ (b) $g\tan\alpha$
(c) $mg\cos\alpha$ (d) $(M+m)\,g\,\text{cosec}\,\alpha$

32. Two blocks are connected by a string as shown in the diagram. The upper block is hung by another string. A force applied on the upper string produces an acceleration of 2 m/s^2 in the upward direction in both the blocks. If T and T' be the tensions in the two parts of the string, then (Take, $g = 9.8$ m/s^2)

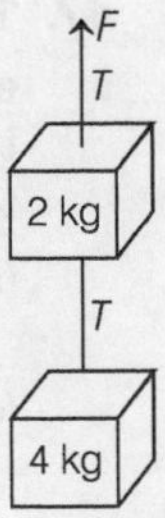

(a) $T = 70.8$ N and $T' = 47.2$ N
(b) $T = 58.8$ N and $T' = 47.2$ N
(c) $T = 70.8$ N and $T' = 58.8$ N
(d) $T = 70.8$ N and $T' = 0$

33. A light spring balance hangs from the hook of the other light spring balance and a block of mass M kg hangs from the former one. Then, the true statement about the scale reading is **[AIEEE 2003]**
(a) both the scales read M kg each
(b) the scale of the lower one reads M kg and of the upper one zero
(c) the reading of the two scales can be anything but the sum of the readings will be M kg
(d) both the scales read $M/2$ kg

34. Three blocks are placed at rest on a smooth inclined plane with force acting on m_1 parallel to the inclined plane. Find the contact force between m_2 and m_3.

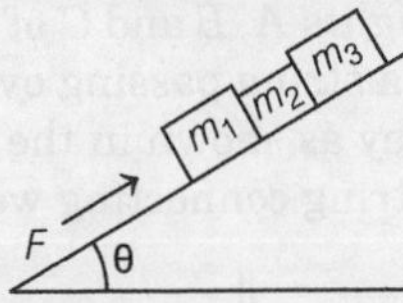

(a) $\frac{(m_1 + m_2 + m_3) F}{m_3}$ (b) $\frac{m_3 F}{m_1 + m_2 + m_3}$
(c) $F - (m_1 + m_2) g$ (d) None of these

35. Given in the figure are two blocks A and B of weight 20 N and 100 N, respectively. These are being pressed against a wall by a force F as shown in figure. If the coefficient of friction between the blocks is 0.1 and between block B and the wall is 0.15, the frictional force applied by the wall in block B is **[JEE Main 2015]**

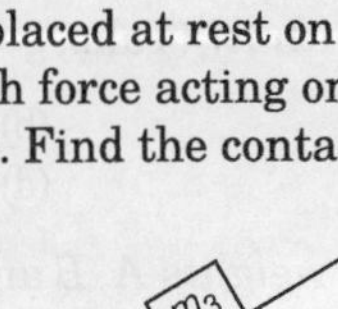

(a) 100 N (b) 80 N
(c) 120 N (d) 150 N

36. A block is kept on a frictionless inclined surface with angle of inclination α. The incline is given an acceleration a to keep the block stationary. Then, a is equal to **[AIEEE 2005]**

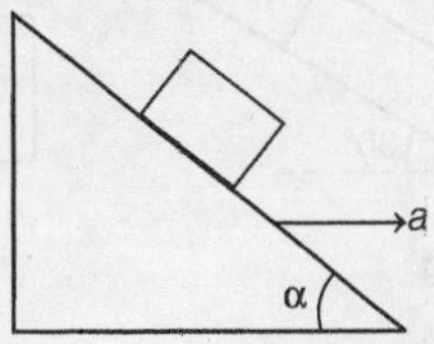

(a) $\frac{g}{\tan \alpha}$ (b) g cosec
(c) g (d) $g \tan \alpha$

37. An elevator and its load have a total mass of 800 kg. The elevator is originally moving downwards at 10 ms^{-1}, it slows down to stop with constant acceleration in a distance of 25 m. Find the tension T in the supporting cable while the elevator is being brought to rest. (Take, $g = 10$ ms^{-2})
(a) 8000 N (b) 1600 N
(c) 9600 N (d) 6400 N

38. Two blocks of masses $m_1 = 4$ kg and $m_2 = 2$ kg are connected to the ends of a string which passes over a massless, frictionless pulley. The total downward thrust on the pulley is nearly
(a) 27 N (b) 54 N
(c) 0.8 N (d) zero

39. A man wants to slide down a rope. The breaking load for the rope $\frac{2}{3}$rd of the weight of the man. With what minimum acceleration should fireman slide down?
(a) $\frac{g}{4}$ (b) $\frac{g}{3}$
(c) $\frac{2g}{3}$ (d) $\frac{g}{6}$

40. A sphere is accelerated upward by a cord whose breaking strength is four times its weight. The maximum acceleration with which the sphere can move up without breaking the cord is
(a) g (b) $3g$
(c) $2g$ (d) $4g$

41. A body of mass 0.05 kg is observed to fall with an acceleration of 9.5 ms^{-2}. The opposite force of air on the body is (Take, $g = 9.8$ ms^{-2})
(a) 0.015 N
(b) 0.15 N
(c) 0.030 N
(d) zero

42. The acceleration of the 500 g block in figure is

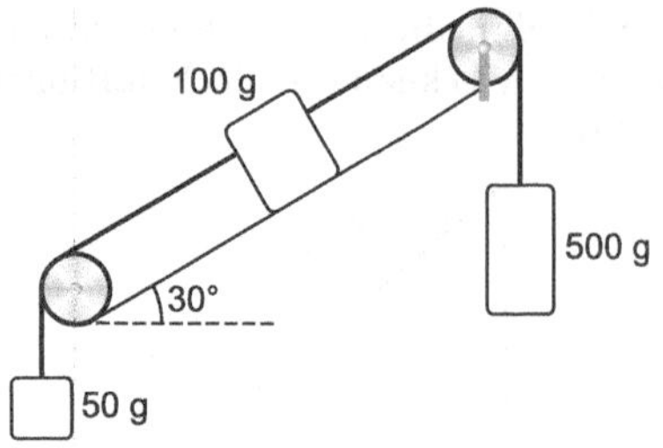

(a) $\frac{6g}{13}$ downwards (b) $\frac{7g}{13}$ downwards
(c) $\frac{8g}{13}$ downwards (d) $\frac{9g}{13}$ upwards

43. A frictionless inclined plane of length l having inclination θ is placed inside a lift which is accelerating downward with an acceleration $a (< g)$. If a block is allowed to move down the inclined plane from rest, then the time taken by the block to slide from top of the inclined plane to the bottom of the inclined plane is

(a) $\sqrt{\frac{2l}{g}}$ (b) $\sqrt{\frac{2l}{g-a}}$
(c) $\sqrt{\frac{2l}{g+a}}$ (d) $\sqrt{\frac{2l}{(g-a)\sin\theta}}$

44. For the system shown in figure, the pulleys are light and frictionless. The tension in the string will be

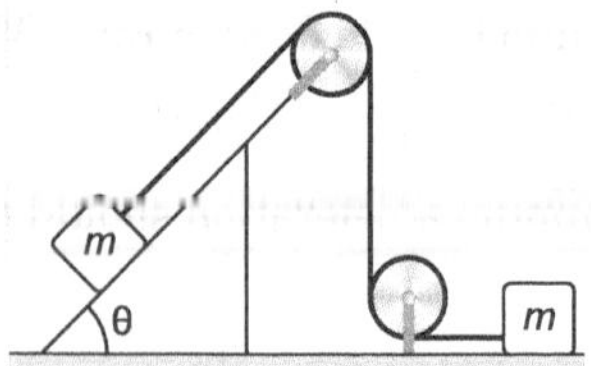

(a) $\frac{2}{3}mg\sin\theta$ (b) $\frac{3}{2}mg\sin\theta$
(c) $\frac{1}{2}mg\sin\theta$ (d) $2mg\sin\theta$

45. A spring balance is attached to the ceiling of a lift. A man hangs his bag on the spring and the spring reads 49 N, when the lift is stationary. If the lift moves downward with an acceleration of 5 ms^{-2}, the reading of the spring balance will be **[AIEEE 2003]**

(a) 24 N (b) 74 N
(c) 15 N (d) 49 N

46. A block of mass M is pulled along a horizontal frictionless surface by a rope of mass m. If a force P is applied at the free end of the rope, the force exerted by the rope on the block is **[AIEEE 2003]**

(a) $\frac{Pm}{M+m}$ (b) $\frac{Pm}{M-m}$ (c) P (d) $\frac{PM}{M+m}$

47. Two weights w_1 and w_2 are suspended from the ends of a light string over a smooth fixed pulley. If the pulley is pulled up with acceleration g, the tension in the string will be

(a) $\frac{4w_1w_2}{w_1+w_2}$ (b) $\frac{2w_1w_2}{w_1+w_2}$
(c) $\frac{w_1-w_2}{w_1+w_2}$ (d) $\frac{w_1w_2}{2(w_1+w_2)}$

48. A lift is moving down with acceleration a. A man in the lift drops a ball inside the lift. The acceleration of the ball as observed by the man in the lift and a man standing stationary on the ground are respectively **[AIEEE 2002]**

(a) g, g (b) $g-a, g-a$
(c) $g-a, g$ (d) a, g

49. The monkey B shown in figure is holding on to the tail of the monkey A which is climbing up a rope. The masses of the monkeys A and B are 5 kg and 2 kg respectively. If A can tolerate a tension of 30 N in its tail, what force should it apply on the rope in order to carry the monkey B with it? (Take, $g = 10\ ms^{-2}$)

(a) 105 N (b) 108 N
(c) 10.5 N (d) 100 N

50. Three equal weights A, B and C of mass 2 kg each are hanging on a string passing over a fixed frictionless pulley as shown in the figure. The tension in the string connecting weight B and C is

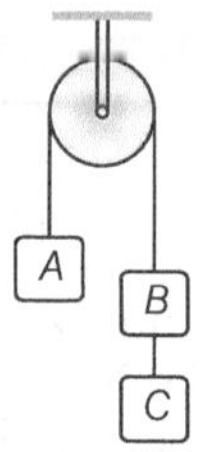

(a) zero (b) 13 N
(c) 3.3 N (d) 19.6 N

51. Two block of masses 7 kg and 5 kg are placed in contact with each other on a smooth surface. If a force of 6 N is applied on a heavier mass the force on the lighter mass is

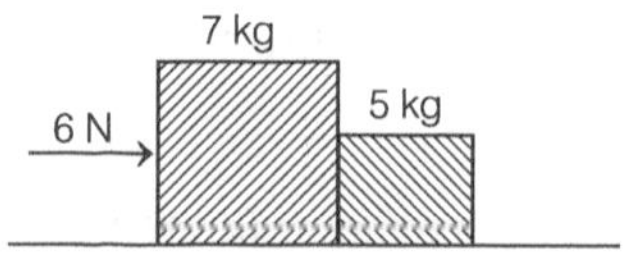

(a) 3.5 N (b) 2.5 N
(c) 7 N (d) 5 N

52. Refer to the system shown in figure. The acceleration of the masses is

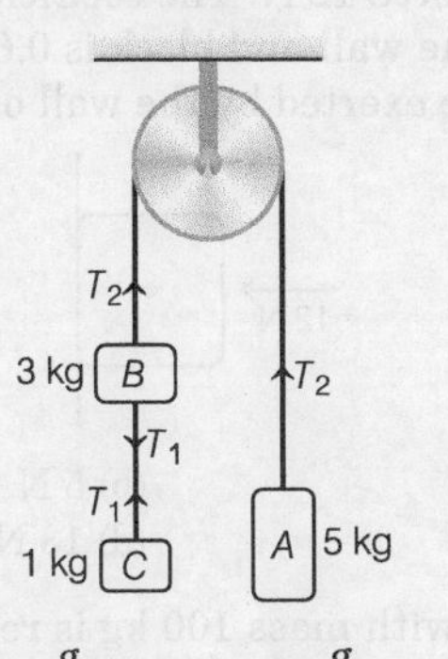

(a) $\frac{g}{3}$ (b) $\frac{g}{6}$ (c) $\frac{g}{9}$ (d) $\frac{g}{12}$

53. A block is dragged on a smooth horizontal plane with the help of a light rope which moves with a velocity v as shown in figure. The horizontal velocity of the block is

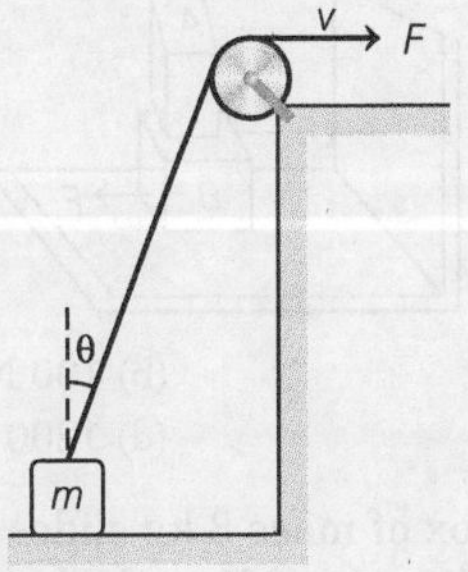

(a) v (b) $v \sin \theta$
(c) $\frac{v}{\sin \theta}$ (d) $\frac{v}{\cos \theta}$

54. Three identical blocks of masses $m = 2$ kg are drawn by a force $F = 10.2$ N with an acceleration of 0.6 ms^{-2} on a frictionless surface, then what is the tension (in newton) in the string between the blocks B and C? **[AIEEE 2002]**

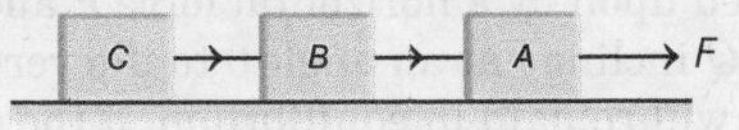

(a) 9.2 (b) 7.8
(c) 4 (d) 9.8

55. A blumb bob is hung from the ceiling of a train compartment. The train moves on an inclined track of inclination 30° with horizontal. Acceleration of train up the plane is $a = 9/2$. The angle which the string supporting the bob makes with normal to the ceiling in equilibrium is

(a) 30° (b) $\tan^{-1} \frac{2}{\sqrt{3}}$
(c) $\tan^{-1} \frac{\sqrt{3}}{2}$ (d) $\tan^{-1} 2$

56. Two fixed frictionless inclined plane making the angles 30° and 60° with the vertical are shown in the figure. Two blocks A and B are placed on the two planes. What is the relative vertical acceleration of A with respect to B? **[AIEEE 2010]**

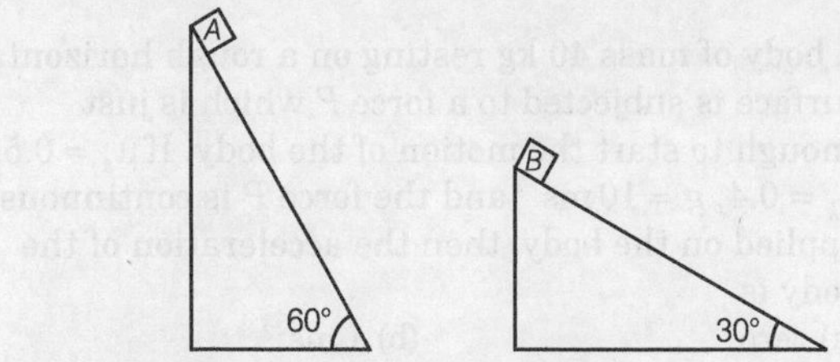

(a) 4.9 ms^{-2} in horizontal direction
(b) 9.8 ms^{-2} in vertical direction
(c) zero
(d) 4.9 ms^{-2} in vertical direction

Friction and Motion on Inclined Surface

57. A chain lies on a rough horizontal table. It starts sliding when one-fourth of its length hangs over the edge of the table. The coefficient of static friction between the chain and the surface of the table is

(a) $\frac{1}{2}$ (b) $\frac{1}{3}$
(c) $\frac{1}{4}$ (d) $\frac{1}{5}$

58. The coefficient of friction between a body and the surface of an inclined plane at 45° is 0.5 if $g = 9.8 \text{ m/s}^2$. The acceleration of the body downwards (in m/s^2) is

(a) $\frac{4.9}{\sqrt{2}}$ (b) $4.9\sqrt{2}$
(c) $19.2\sqrt{2}$ (d) 4.9

59. A fireman of mass 60 kg slides down a pole. He is pressing the pole with a force of 600 N. The coefficient of friction between the hands and the pole is 0.5 with what acceleration with the fireman slide down? (Take, $g = 10 \text{ m/s}^2$)

(a) 1 m/s^2 (b) 2.5 m/s^2
(c) 10 m/s^2 (d) 5 m/s^2

60. A marble block of mass 2 kg lying on ice when given a velocity of 6 ms^{-1} is stopped by friction in 10 s. Then, the coefficient of friction is **[AIEEE 2003]**

(a) 0.02 (b) 0.03
(c) 0.06 (d) 0.01

61. A box of mass m kg is placed on the rear side of an open truck accelerating at 4 ms^{-2}. The coefficient of friction between the box and the surface below it is 0.4.

The net acceleration of the box with respect to the truck is zero. The value of m is
(Take, $g = 10 \text{ ms}^{-2}$)
(a) 4 kg (b) 8 kg
(c) 9.78 kg (d) It could be any value

62. A body of mass 40 kg resting on a rough horizontal surface is subjected to a force P which is just enough to start the motion of the body. If $\mu_s = 0.5$, $\mu_k = 0.4$, $g = 10 \text{ ms}^{-2}$ and the force P is continuously applied on the body, then the acceleration of the body is
(a) zero (b) 1 ms^{-2}
(c) 2 ms^{-2} (d) 2.4 ms^{-2}

63. A 40 kg slab rests on a frictionless floor. A 10 kg block rests on top of the slab. The static coefficient of friction between the block and the slab is 0.60 while the kinetic coefficient of friction is 0.40. The 10 kg block is acted upon by a horizontal force of 100 N. If $g = 9.8 \text{ ms}^{-2}$, the resulting acceleration of the slab will be

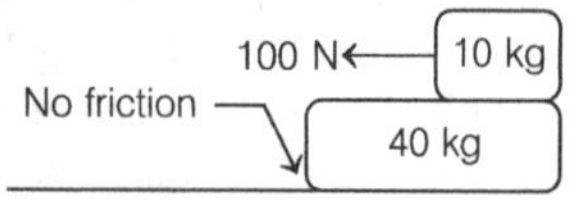

(a) 1.47 ms^{-2} (b) 1.69 ms^{-2}
(c) 9.8 ms^{-2} (d) 0.98 ms^{-2}

64. A horizontal force of 10 N is necessary to just hold a block stationary against a wall. The coefficient of friction between the block and the wall is 0.2. The weight of the block is

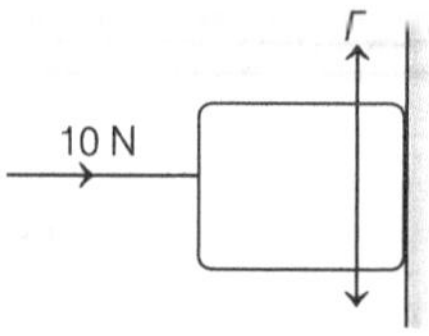

(a) 20 N (b) 50 N
(c) 100 N (d) 2 N

65. Consider a car moving on a straight road with a speed of 100 ms^{-1}. The distance at which car can be stopped, is (Take, $\mu_k = 0.5$)
(a) 800 m (b) 1000 m
(c) 100 m (d) 400 m

66. A block of mass 1 kg is at rest on a horizontal table. The coefficient of static friction between the block and the table is 0.5. If $g = 10 \text{ ms}^{-2}$, then the magnitude of the force acting upwards at an angle of 60° from the horizontal that will just start the block moving is
(a) 5 N (b) 5.36 N
(c) 74.6 N (d) 10 N

67. A block of weight 5 N is pushed against a vertical wall by a force 12 N. The coefficient of friction between the wall and block is 0.6. The magnitude of the force exerted by the wall on the block is

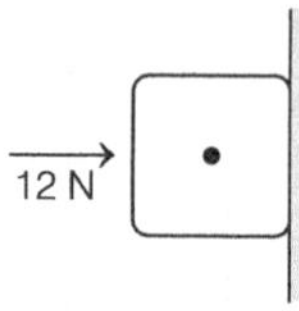

(a) 12 N (b) 5 N
(c) 7.2 N (d) 13 N

68. A block A with mass 100 kg is resting on another block B of mass 200 kg. As shown in figure, a horizontal rope tied to a wall holds it. The coefficient of friction between A and B is 0.2 while coefficient of friction between B and the ground is 0.3. The minimum required force F to start moving B will be

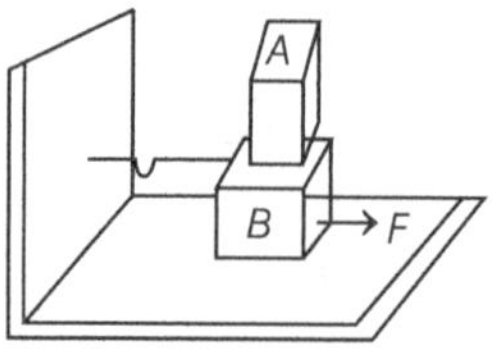

(a) 900 N (b) 100 N
(c) 1100 N (d) 1200 N

69. A wooden box of mass 8 kg slides down an inclined plane of inclination 30° to the horizontal with a constant acceleration of 0.4 ms^{-2}. What is the force of friction between the box and inclined plane? (Take, $g = 10 \text{ ms}^{-2}$)
(a) 36.8 N
(b) 76.8 N
(c) 65.6 N
(d) 97.8 N

70. A block of mass m lying on a rough horizontal plane is acted upon by a horizontal force P and another force Q inclined at an angle θ to the vertical. The block will remain in equilibrium, if the coefficient of friction between it and the surface is

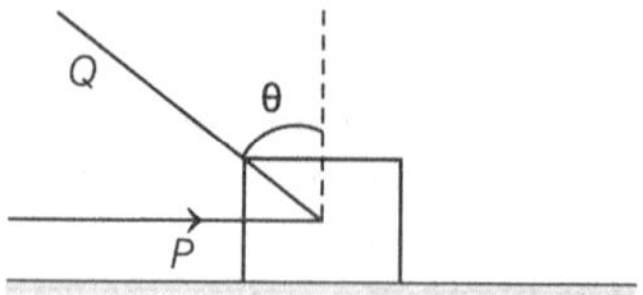

(a) $\dfrac{P+Q\sin\theta}{mg+Q\cos\theta}$ (b) $\dfrac{P\cos\theta+Q}{mg-Q\sin\theta}$

(c) $\dfrac{P+Q\cos\theta}{mg+Q\sin\theta}$ (d) $\dfrac{P\sin\theta-Q}{mg-Q\cos\theta}$

71. A partly hanging uniform chain of length L is resting on a rough horizontal table. l is the maximum possible length that can hang in equilibrium. The coefficient of friction between the chain and table is

(a) $\frac{l}{L-l}$ (b) $\frac{L}{l}$ (c) $\frac{l}{L}$ (d) $\frac{lL}{L+l}$

72. Two masses $m_1 = 5$ kg and $m_2 = 10$ kg connected by an inextensible string over a frictionless pulley, are moving as shown in the figure. The coefficient of friction of horizontal surface is 0.15. The minimum weight m that should be put on top of m_2 to stop the motion is [JEE Main 2018]

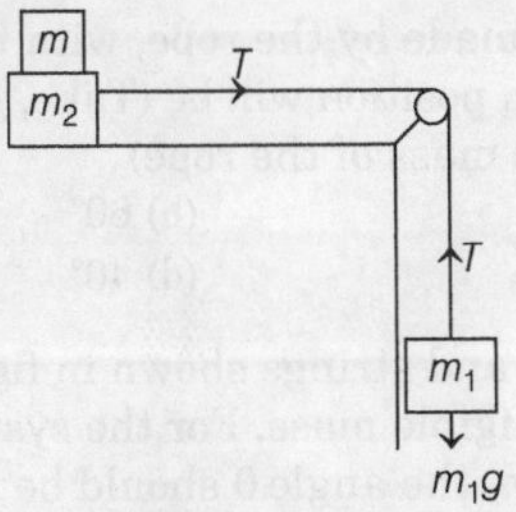

(a) 18.3 kg (b) 27.3 kg (c) 43.3 kg (d) 10.3 kg

73. The minimum force required to start pushing a body up a rough (frictional coefficient μ) inclined plane is F_1 while the minimum force needed to prevent it from sliding down is F_2. If the inclined plane makes an angle θ from the horizontal such that $\tan\theta = 2\mu$, then the ratio $\frac{F_1}{F_2}$ is

(a) 4 (b) 1 (c) 2 (d) 3

74. A point particle of mass m, moves along the uniformly rough track PQR as shown in the figure. The coefficient of friction between the particle and the rough track equals μ. The particle is released, from rest , from the point P and it comes to rest at a point R. The energies, lost by the ball, over the parts, PQ and QR, of the track, are equal to each other, and no energy is lost when particle changes direction from PQ to QR. The values of the coefficient of friction μ and the distance $x(= QR)$, are respectively close to [JEE Main 2016]

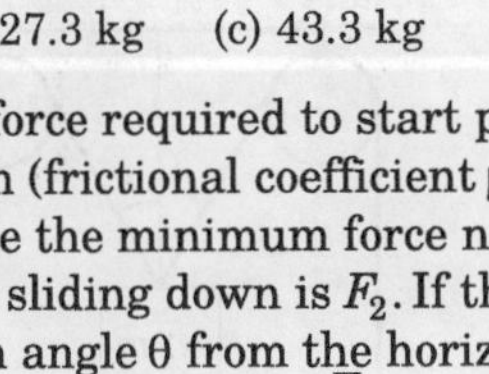

(a) 0.2 and 6.5 m (b) 0.2 and 3.5 m
(c) 0.29 and 3.5 m (d) 0.29 and 6.5 m

75. A block of mass $\sqrt{3}$ kg resting on a horizontal surface. A force F is applied on the block as shown in figure. If coefficient of friction between the block be $\frac{1}{2\sqrt{3}}$ what can be the maximum value of force F, so that block does not start moving? (Take, $g = 10\ \text{ms}^{-2}$)

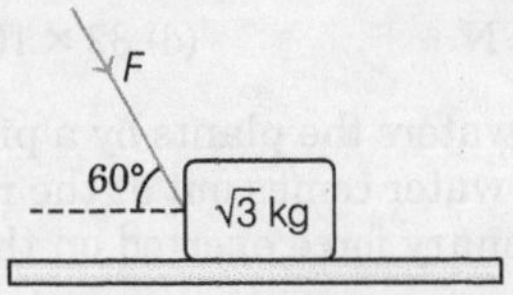

(a) 20 N (b) 10 N
(c) 12 N (d) 15 N

76. A heavy uniform chain lies on horizontal table top. If the coefficient of friction between the chain and the table surface is 0.25, then the maximum fraction of the length of the chain that can hang over one edge of the table is

(a) 20% (b) 25%
(c) 35% (d) 15%

77. A car starts from rest to cover a distance s. The coefficient of friction between the road and the tyres is m. The minimum time in which the car can cover the distance is proportional to

(a) μ (b) $\sqrt{\mu}$
(c) $1/\mu$ (d) $1/\sqrt{\mu}$

78. The coefficient of kinetic friction between a 20 kg box and the floor is 0.40. How much work does a pulling force do on the box in pulling it 8.0 m across the floor at constant speed? The pulling force is directed 37° above the horizontal

(a) 343 J (b) 482 J
(c) 14.4 J (d) None of these

79. A block moves down a smooth inclined plane of inclination θ. Its velocity on reaching the bottom is v. If it slides down a rough inclined plane of same inclination, its velocity on reaching the bottom is v/n, where n is a number greater than 1. The coefficient of friction is given by

(a) $\mu = \tan\theta\left(1-\frac{1}{n^2}\right)$

(b) $\mu = \cot\theta\left(1-\frac{1}{n^2}\right)$

(c) $\mu = \tan\theta\left(1-\frac{1}{n^2}\right)^{1/2}$

(d) $\mu = \cot\theta\left(1-\frac{1}{n^2}\right)^{1/2}$

ROUND II Mixed Bag

Only One Correct Option

1. 80 railway wagons all of same mass 5×10^3 kg are pulled by an engine with a force of 4×10^5 N. The tension in the coupling between 30th and 31st wagon from the engine is

(a) 25×10^4 N
(b) 40×10^4 N
(c) 20×10^4 N
(d) 32×10^4 N

2. A gardner waters the plants by a pipe of diameter 1 mm. The water comes out at the rate of 10 cm^3 s^{-1}. The reactionary force exerted on the hand of the gardner is

(a) zero
(b) 1.27×10^{-2} N
(c) 1.27×10^{-4} N
(d) 0.127 N

3. When forces F_1, F_2, F_3 are acting on a particle of mass m such that F_2 and F_3 are mutually perpendicular, then the particle remains stationary. If the force F_1 is now removed, then the acceleration of the particle is

(a) F_1/m
(b) F_2F_3/mF_1
(c) $(F_2 - F_3)/m$
(d) F_2/m

4. A block of mass m slides along a floor while a force of magnitude F is applied to it at an angle θ as shown in figure. The coefficient of kinetic friction is μ_k, then the block's acceleration a is given by (g is acceleration due to gravity) [JEE Main 2021]

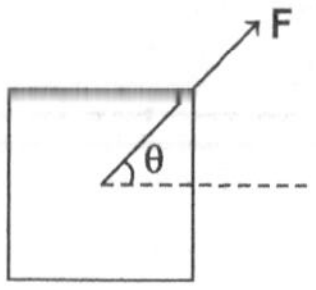

(a) $-\frac{F}{m}\cos\theta - \mu_K\left(g - \frac{F}{m}\sin\theta\right)$

(b) $\frac{F}{m}\cos\theta - \mu_K\left(g - \frac{F}{m}\sin\theta\right)$

(c) $\frac{F}{m}\cos\theta - \mu_K\left(g + \frac{F}{m}\sin\theta\right)$

(d) $\frac{F}{m}\cos\theta + \mu_K\left(g - \frac{F}{m}\sin\theta\right)$

5. A 5 kg stationary bomb is exploded in three parts having mass 1 : 1 : 3 respectively. Parts having same mass move in perpendicular directions with velocity 39 ms^{-1}, then the velocity of bigger part will be

(a) $10\sqrt{2}$ ms^{-1}
(b) $\frac{10}{\sqrt{2}}$ ms^{-1}
(c) $13\sqrt{2}$ ms^{-1}
(d) $\frac{15}{\sqrt{2}}$ ms^{-1}

6. Two persons are holding a rope of negligible weight tightly at its ends so that it is horizontal. A 15 kg weight is attached to rope at the mid-point which now no more remains horizontal. The minimum tension required to completely straighten the rope is

(a) 15 kg
(b) (15/2) kg
(c) 5 kg
(d) infinitely large

7. A mass of 6 kg is suspended by a rope of length 2 m from a ceiling. A force of 50 N is applied in the horizontal direction at the mid-point of the rope. The angle made by the rope, with the vertical, in equilibrium position will be (Take, $g = 10\,\text{ms}^{-2}$, neglect the mass of the rope)

(a) 90°
(b) 60°
(c) 50°
(d) 40°

8. The pulley and strings shown in figure are smooth and of negligible mass. For the system to remain in equilibrium, the angle θ should be

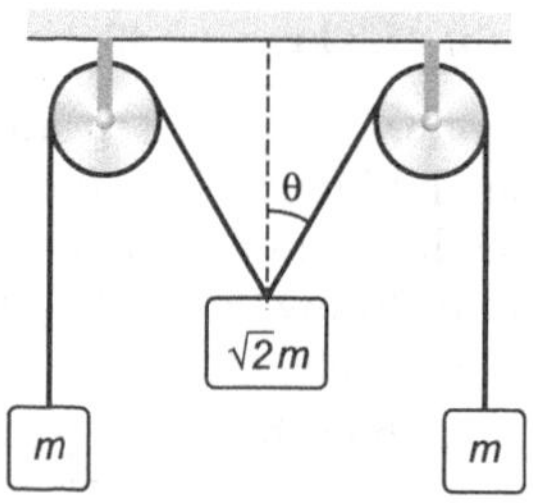

(a) 0°
(b) 30°
(c) 45°
(d) 60°

9. A 24 kg block resting on a floor has a rope tied to its top. The maximum tension, the rope can withstand without breaking is 310 N. The minimum time in which the block can be lifted a vertical distance of 4.6 m by pulling on the rope is

(a) 1.2 s
(b) 1.3 s
(c) 1.7 s
(d) 2.3 s

10. A rope of mass 0.1 kg is connected at the same height of two opposite walls. It is allowed to hang under its own weight. At the constant point between the rope and the wall, the rope makes an angle $\theta = 10°$ with respect to horizontal. The tension in the rope at its mid-point between the wall is

(a) 2.78 N
(b) 2.56 N
(c) 2.82 N
(d) 2.71 N

11. Block A of mass m and block B of mass $2m$ are placed on a fixed triangular wedge by means of a massless, inextensible string and a frictionless at 45° to the horizontal on both the side. If the coefficient of friction between the block A and the wedge is 2/3 and that between the block B and the wedge is 1/3 and both blocks A and B are released from rest, the acceleration of A will be **[UP SEE 2008]**

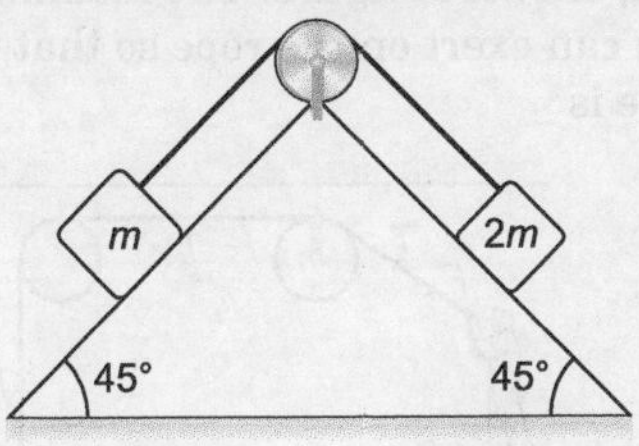

(a) $-1\ m/s^2$ (b) $1.2\ m/s^2$
(c) $0.2\ m/s^2$ (d) $0\ m/s^2$

12. A block of mass 5 kg is moving horizontally at a speed of $1.5\ ms^{-1}$. A vertically upward force 5 N acts on it for 4 s. What will be the distance of the block from the point where the force starts acting?
(a) 2 m (b) 6 m
(c) 8 m (d) 10 m

13. A particle of mass m is at rest at the origin at time $g = 0$. It is subjected to a force $F(t) = f_0 e^{-bt}$ in the x-direction. Its speed $v(t)$ is depicted by which of the following curves?

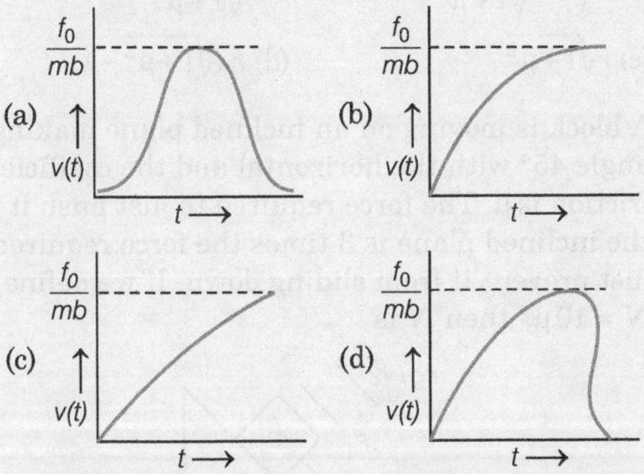

14. A mass m hangs with the help of a string wrapped around a pulley on a frictionless bearing. The pulley has mass m and radius R. Assuming pulley to be the perfect uniform circular disc, the acceleration of the mass m. If the string does not slip on the pulley, is
(a) g (b) $\frac{2}{3}g$
(c) $\frac{g}{3}$ (d) $\frac{3}{2}g$

15. A ball of mass 0.2 kg is thrown vertically upwards by applying a force by hand. If the hand moves 0.2 m while applying the force and the ball goes upto 2 m height further, find the magnitude of the force. (Take, $g = 10\ m/s^2$)
(a) 4 N (b) 16 N
(c) 20 N (d) 22 N

16. A block at rest slides down a smooth inclined plane which makes an angle 60° with the vertical and it reaches the ground in t_1 seconds. Another block is dropped vertically from some point and reaches the ground in t_2 seconds, then the ratio of $t_1 : t_2$ is
(a) 1 : 2 (b) 2 : 1 (c) 1 : 3 (d) $1 : \sqrt{2}$

17. The acceleration of system of two bodies over the wedge as shown in figure.

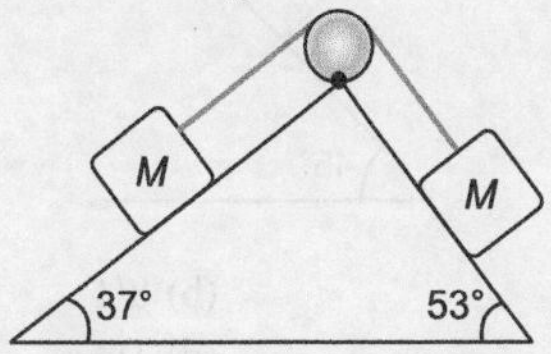

(a) $1\ ms^{-2}$ (b) $2\ ms^{-2}$ (c) $0.5\ ms^{-2}$ (d) $10\ ms^{-2}$

18. One end of massless rope, which passes over a massless and frictionless pulley P is tied to a hook C while the other end is free. Maximum tension that the rope can bear is 360 N. With what value of maximum safe acceleration (in ms^{-2}) can a man of 60 kg climb on the rope?

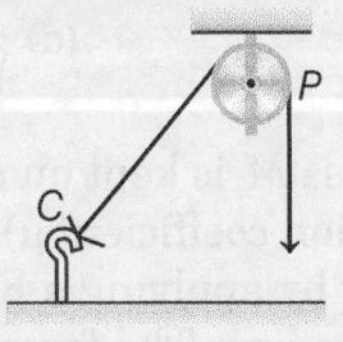

(a) 16 (b) 6 (c) 4 (d) 80

19. A small ball of mass m is thrown upward with velocity u from the ground. The ball experiences a resistive force mkv^2, where v is its speed. The maximum height attained by the ball is **[JEE Main 2020]**

(a) $\frac{1}{k}\tan^{-1}\left(\frac{ku^2}{2g}\right)$ (b) $\frac{1}{2k}\ln\left(1+\frac{ku^2}{g}\right)$

(c) $\frac{1}{k}\ln\left(1+\frac{ku^2}{2g}\right)$ (d) $\frac{1}{2k}\tan^{-1}\left(\frac{ku^2}{g}\right)$

20. A circular disc with a groove along its diameter is placed horizontally. A block of mass 1 kg is placed as shown. The coefficient of friction between the block and all surfaces of groove in contact is $\mu = \frac{2}{5}$, the disc has an acceleration of $25\ m/s^2$. Find the acceleration of block with respect to disc. **[JEE Main 2021]**

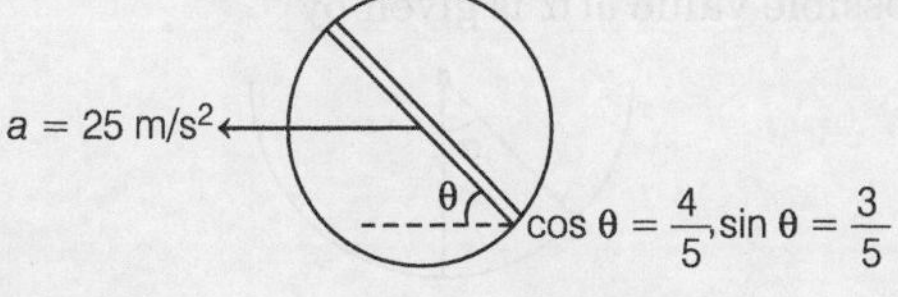

(a) $10\ m/s^2$ (b) $5\ m/s^2$
(c) $20\ m/s^2$ (d) $1\ m/s^2$

21. A block of mass 15 kg is resting on a rough inclined plane as shown in figure. The block is tied by a horizontal string which has a tension of 50 N. The coefficient of friction between the surfaces of contact is (Take, $g = 10\ \text{ms}^{-2}$)

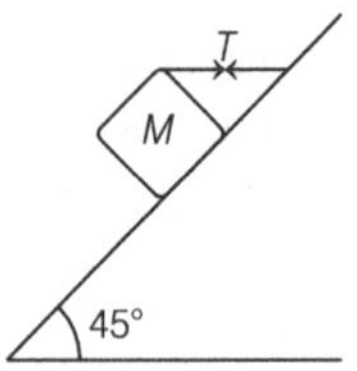

(a) 1/2 (b) 3/4
(c) 2/3 (d) 1/4

22. A given object takes n times more time to slide down a 45° rough inclined plane as it takes to slide down a perfectly smooth 45° incline. The coefficient of kinetic friction between the object and the incline is

(a) $\dfrac{1}{1-n^2}$ (b) $1-\dfrac{1}{n^2}$
(c) $\sqrt{1-\dfrac{1}{n^2}}$ (d) $\sqrt{\dfrac{1}{1-n^2}}$

23. A body of mass M is kept on a rough horizontal surface (friction coefficient μ). A person is trying to pull the body by applying a horizontal force but the body is not moving. The force by the surface on the body is F, where
(a) $F = Mg$
(b) $F = \mu MgF$
(c) $Mg \le f \le Mg\sqrt{1+\mu^2}$
(d) $Mg \ge f \ge Mg\sqrt{1+\mu^2}$

24. The upper half of an inclined plane with inclination ϕ is perfectly smooth, while the lower half is rough. A body starting from rest at the top will again come to rest at the bottom if coefficient of friction for the lower half is given by
(a) $2\sin\phi$ (b) $2\cos\phi$
(c) $2\tan\phi$ (d) $\tan\phi$

25. An insect crawls up a hemispherical surface very slowly, figure. The coefficient of friction between the insect and the surface is 1/3. If the line joining the centre of the hemispherical surface to the insect makes an angle α with the vertical, the maximum possible value of α is given by

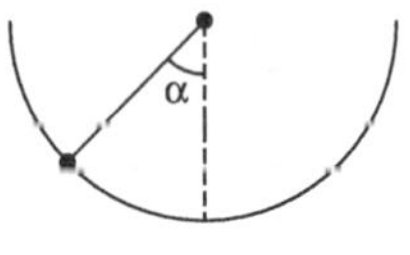

(a) $\cot\alpha = 3$ (b) $\sec\alpha = 3$
(c) $\operatorname{cosec}\alpha = 3$ (d) None of these

26. A man of mass M is standing on a board of mass m. The friction coefficient between the board and the floor is μ, shown in figure. The maximum force that the man can exert on the rope so that the board does not move is

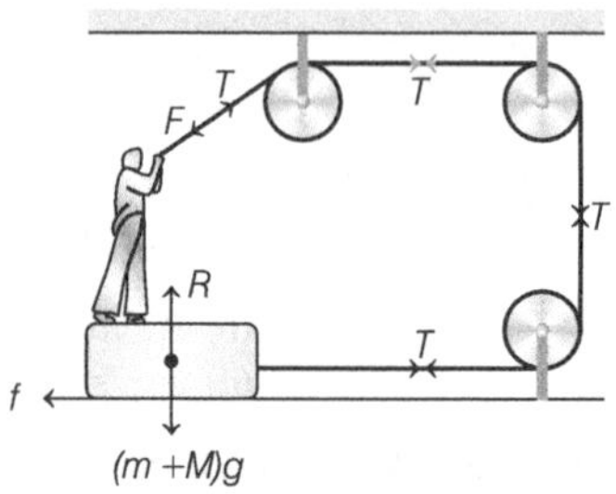

(a) $\mu(m+M)g$ (b) $\dfrac{\mu(m+M)g}{\mu+1}$
(c) $\dfrac{\mu(m+M)g}{\mu-1}$ (d) None of these

27. If coefficient of friction between an insect and bowl is μ and radius of the bowl is r, the maximum height to which the insect can crawl in the bowl is

(a) $r\left(1-\dfrac{1}{\sqrt{1+\mu^2}}\right)$ (b) $\dfrac{r}{\sqrt{1+\mu^2}}$
(c) $r\sqrt{1+\mu^2}$ (d) $r(\sqrt{1+\mu^2}-1)$

28. A block is moving on an inclined plane making an angle 45° with the horizontal and the coefficient of friction is μ. The force required to just push it up the inclined plane is 3 times the force required to just prevent it from sliding down. If we define $N = 10\mu$, then N is

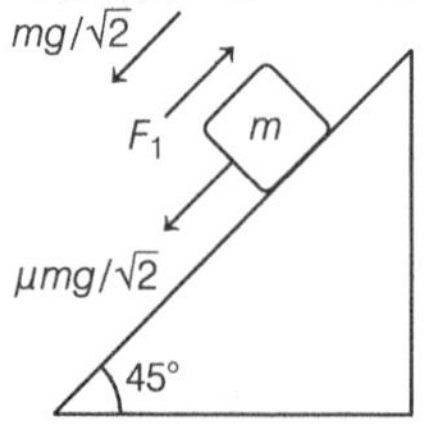

(a) 2 (b) 4 (c) 5 (d) 6

29. A block of mass 200 kg is being pulled up by men on an inclined plane at angle of 45° as shown in figure. The coefficient of static friction is 0.5. Each man can only apply a maximum force of 500 N. Calculate the number of men required for the block to just start moving up the plane.

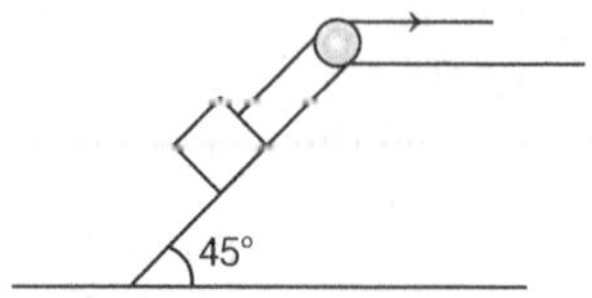

(a) 10 (b) 15 (c) 5 (d) 3

30. A block kept on a rough inclined plane, as shown in the figure, remains at rest upto a maximum force 2 N down the inclined plane. The maximum external force up the inclined plane that does not move the block is 10 N. The coefficient of static friction between the block and the plane is (Take, $g = 10\text{ m/s}^2$) **[JEE Main 2019]**

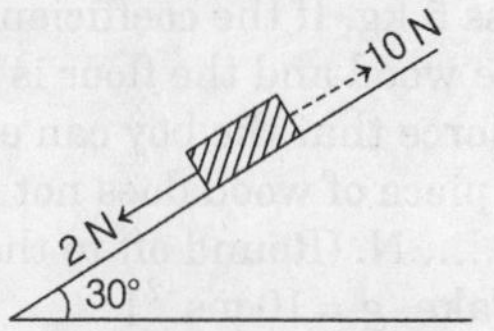

(a) $\frac{2}{3}$ (b) $\frac{\sqrt{3}}{2}$

(c) $\frac{\sqrt{3}}{4}$ (d) $\frac{1}{2}$

31. A smooth block is released at rest on a 45° incline and then slides a distance d. The time taken to slide is n times as much to slide on rough incline than on a smooth incline. The coefficient of friction is

(a) $\mu_k = 1 - \frac{1}{n^2}$

(b) $\mu_k = \sqrt{1 - \frac{1}{n^2}}$

(c) $\mu_s = 1 - \frac{1}{n^2}$

(d) $\mu_s = \sqrt{1 - \frac{1}{n^2}}$

32. A block of mass 5 kg is (i) pushed in case (A) and (ii) pulled in case (B), by a force $F = 20$ N, making an angle of 30° with the horizontal, as shown in the figures. The coefficient of friction between the block, the floor is $\mu = 0.2$. The difference between the accelerations of the block, in case (B) and case (A) will be (Take, $g = 10\text{ ms}^{-2}$) **[JEE Main 2019]**

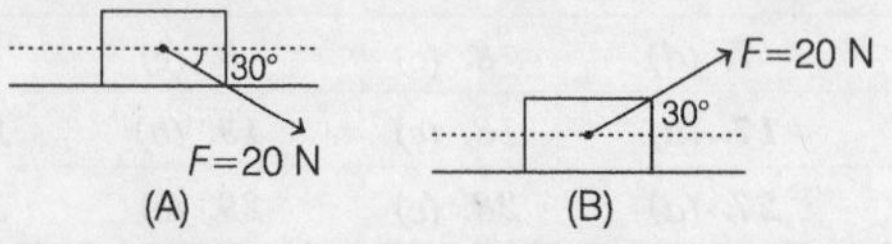

(a) 0.4 ms^{-2} (b) 3.2 ms^{-2}

(c) 0.8 ms^{-2} (d) 0 ms^{-2}

Numerical Value Questions

33. It is found that when 0.5 m was cut-off from the muzzle of a gun firing block of 50 kg, the velocity of the block was changed from 700 m/s to 600 m/s. The force exerted on the block by the powder of gas at the muzzle when expanded in the bore was 6.5×10^n N. The value of n is ....

34. A block B of mass 1 kg is placed on a light plank shown in figure. A force F of 10 N is applied on the plank horizontally. The acceleration (in m/s^2) of block B is ....

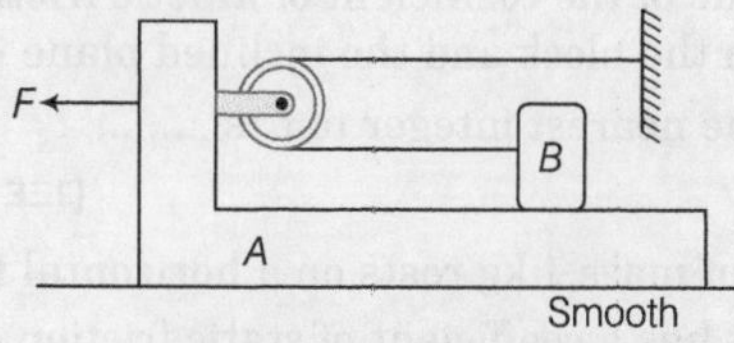

35. For the situation shown in the figure, pulleys are light and smooth. Pulleys C and D are fixed, but A and B are movable. The value of m (in kg), if m is in equilibrium is .... (Take, $m_1 = 2$ kg and $m_2 = 1$ kg)

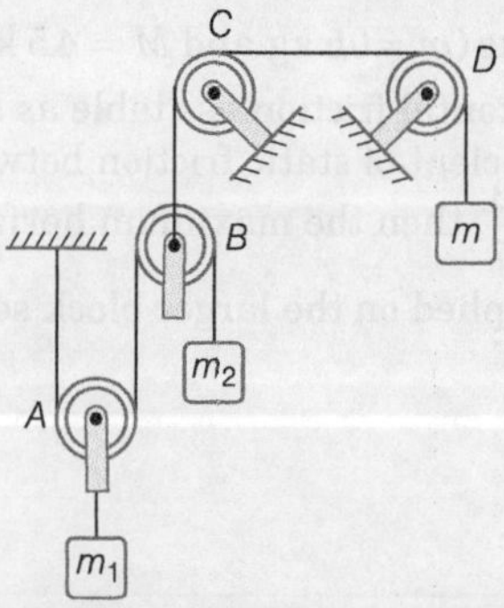

36. A small block starts slipping down from a point B on an inclined plane AB, which is making an angle θ with the horizontal section BC which is smooth and the remaining section CA is rough with a coefficient of friction μ. It is found that the block comes to rest as it reaches the bottom (point A) of the inclined plane. If $BC = 2AC$, the coefficient of friction is given by $\mu = k \tan\theta$. Then, the value of k is......... .

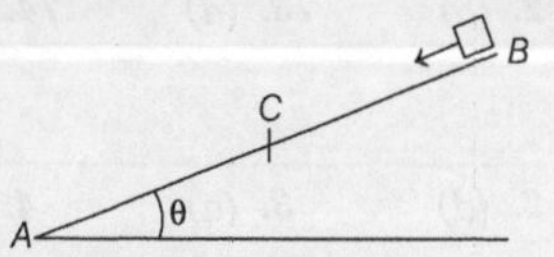

[JEE Main 2020]

37. Consider a uniform cubical box of side a on a rough floor that is to be moved by applying minimum possible force F at a point b above its centre of mass (see figure). If the coefficient of friction is $\mu = 0.4$, the maximum possible value of $100 \times \frac{b}{a}$ for box not to topple before moving is ....

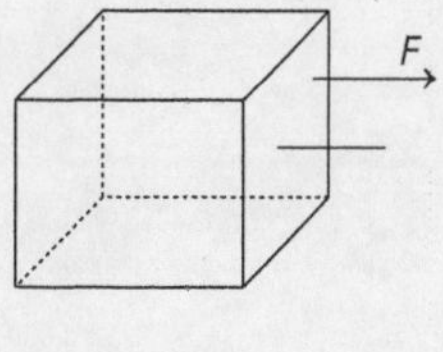

[JEE Main 2020]

38. A block starts moving up an inclined plane of inclination 30° with an initial velocity of v_0. It comes back to its initial position with velocity $\frac{v_0}{2}$. The value of the coefficient of kinetic friction between the block and the inclined plane is close to $\frac{I}{1000}$, the nearest integer to I is .... **[JEE Main 2020]**

39. A body of mass 1 kg rests on a horizontal floor with which it has a coefficient of static friction $\frac{1}{\sqrt{3}}$. It is desired to make the body move by applying the minimum possible force F N. The value of F will be .... (Round off to the nearest integer)

(Take, $g = 10\ \text{ms}^{-2}$) **[JEE Main 2021]**

40. Two blocks ($m = 0.5$ kg and $M = 4.5$ kg) are arranged on a horizontal frictionless table as shown in figure. The coefficient of static friction between the two blocks is $\frac{3}{7}$, then the maximum horizontal force that can be applied on the larger block so that the blocks move together is .... N. (Round off to the nearest integer) (Take, $g = 9.8\ \text{ms}^{-2}$)

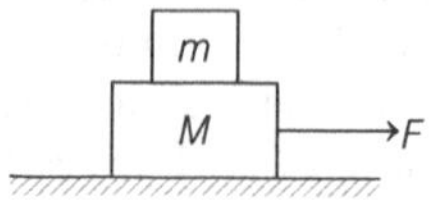

[JEE Main 2021]

41. A boy of mass 4 kg is standing on a piece of wood having mass 5 kg. If the coefficient of friction between the wood and the floor is 0.5, the maximum force that the boy can exert on the rope so that the piece of wood does not move from its place is .... N. (Round off to the nearest integer). [Take, $g = 10\ \text{ms}^{-2}$] **[JEE Main 2021]**

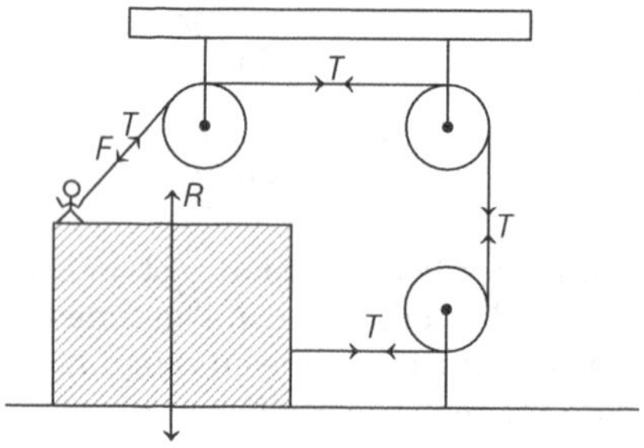

Answers

Round I

1. (b)	2. (b)	3. (b)	4. (c)	5. (a)	6. (c)	7. (c)	8. (c)	9. (a)	10. (a)
11. (b)	12. (a)	13. (a)	14. (c)	15. (d)	16. (d)	17. (b)	18. (b)	19. (d)	20. (c)
21. (d)	22. (a)	23. (a)	24. (a)	25. (c)	26. (c)	27. (a)	28. (c)	29. (b)	30. (c)
31. (a)	32. (a)	33. (a)	34. (b)	35. (c)	36. (d)	37. (c)	38. (b)	39. (b)	40. (b)
41. (a)	42. (c)	43. (d)	44. (c)	45. (a)	46. (d)	47. (a)	48. (c)	49. (a)	50. (b)
51. (b)	52. (c)	53. (c)	54. (b)	55. (b)	56. (d)	57. (b)	58. (a)	59. (d)	60. (c)
61. (d)	62. (b)	63. (d)	64. (d)	65. (b)	66. (b)	67. (d)	68. (c)	69. (a)	70. (a)
71. (a)	72. (b)	73. (d)	74. (c)	75. (a)	76. (a)	77. (d)	78. (b)	79. (a)	

Round II

1. (a)	2. (d)	3. (a)	4. (b)	5. (c)	6. (d)	7. (d)	8. (c)	9. (c)	10. (c)
11. (d)	12. (d)	13. (a)	14. (b)	15. (d)	16. (b)	17. (a)	18. (c)	19. (b)	20. (a)
21. (a)	22. (b)	23. (c)	24. (c)	25. (a)	26. (b)	27. (a)	28. (c)	29. (c)	30. (b)
31. (a)	32. (c)	33. 6	34. 5	35. 2	36. 3	37. 75	38. 346	39. 5	40. 21
41. 30									

Solutions

Round I

1. $F = \dfrac{m(v-u)}{t} = \dfrac{0.15\,[20-(-10)]}{0.1} = \dfrac{0.15 \times 30}{0.1} = 45\text{ N}$

2. $2mnv = Mg$

$\therefore \quad v = \dfrac{Mg}{2mn}$

or $\quad u = \dfrac{10 \times 980}{2 \times 5 \times 10}$

$= \dfrac{9800}{100}\text{ cms}^{-1} = 98\text{ cms}^{-1}$

3. The force acting on the satellite is given by

$$F = \frac{d}{dt}(Mv) = M\frac{dv}{dt} + v\frac{dM}{dt}$$

$$= M\frac{dv}{dt} + v(\alpha v) \qquad \left(\because \frac{dM}{dt} = \alpha v\right)$$

We know that, net force on satellite is zero,

i.e. $M\dfrac{dv}{dt} + v(\alpha v) = 0$

$\Rightarrow$ Acceleration, $a = \dfrac{dv}{dt} = \dfrac{-v^2\alpha}{M}$

4. Force applied by engine $= 6\,m$

When two cars are pulled,

$(m+m)\,a = 6m$

or $\quad 2ma = 6m$

or $\quad a = 3\text{ ms}^{-2}$

5. Given, $x(t) = pt + qt^2 + rt^3$ and $p = 3\text{ ms}^{-1}$, $q = 4\text{ ms}^{-2}$, $r = 5\text{ ms}^{-3}$, then $x(t) = 3t + 4t^2 + 5t^3$

$$a = \frac{d^2x(t)}{dt^2} = 8 + 30t$$

$\therefore \quad t = 2\text{ s}$

$\therefore \quad a = 8 + 30 \times 2 = 68$

Now, $\quad F = m \times a = 2 \times 68 = 136\text{ N}$

6. Thrust, $F = u\left(\dfrac{dm}{dt}\right) = 5 \times 10^4 \times 40 = 2 \times 10^6\text{ N}$

7. $F = v\dfrac{dm}{dt} = 10 \times 5\text{ N} = 50\text{ N}$

8. Let the ball starts moving with velocity u and it reaches upto maximum height $H_{\max}$, then from

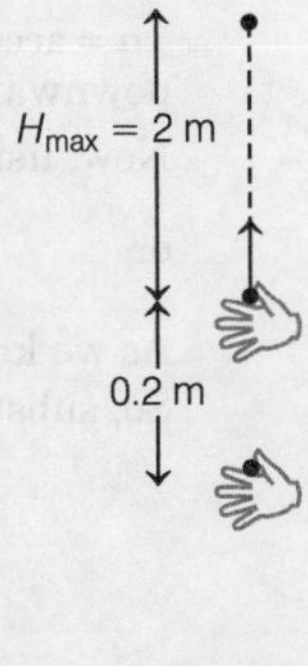

$$H_{\max} = \frac{u^2}{2g}$$

$\Rightarrow \quad u = \sqrt{2g(H_{\max})}$

$= \sqrt{2 \times 10 \times 2}$

$= 2\sqrt{10}\text{ m/s}$

This velocity is supplied to the ball by the hand and initially the hand was at rest, so it requires this velocity during travelled distance of 0.2 m,

$\therefore \quad a = \dfrac{u^2}{2s} = \dfrac{40}{2 \times 0.2} = 100\text{ m/s}^2$

So, upward force on the ball,

$F = m(g+a)$

$= 0.2\,(10 + 100)$

$= 0.2 \times 110 = 22\text{ N}$

9. Mass of the body, $m = 3.0$ kg

Initial speed, $\quad u = 2.0$ m/s

Final speed, $\quad v = 3.5$ m/s

Time, $\quad t = 25$ s

Force, $\quad F = ?$

Using the first equation of motion, $v = u + at$

$\therefore \quad 3.5 = 2.0 + a \times 25$

or $\quad a = \dfrac{3.5 - 2.0}{25}\text{ m/s}^2$

Acceleration, $a = \dfrac{1.5}{25}\text{ m/s}^2$

$\therefore$ Force acting on the body,

$$F = ma = 3.0 \times \frac{1.5}{25} = \frac{4.5}{25}\text{ N} = 0.18\text{ N}$$

As direction of motion of the body remains unchanged, therefore the direction of force acting on the body is along the direction of motion.

10. Mass of the body, $m = 5$ kg

Force acting on body, $F_1 = 8$ N

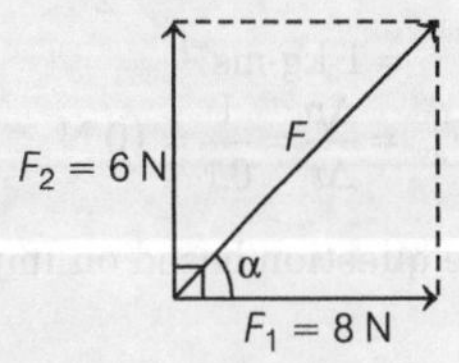

Force perpendicular to force F_1 on the body,

$F_2 = 6$ N

Angle between two forces, $\theta = 90°$

Resultant force acting on the body,

$$F = \sqrt{F_1^2 + F_2^2 + 2F_1F_2\cos\theta}$$

$$= \sqrt{(8)^2 + (6)2 + 2 \times 8 \times 6 \times \cos 90°}$$

$$= \sqrt{64 + 36} \qquad (\because \cos 90° = 0)$$

$= 10$ N

If resultant force F makes an angle α with force F_1, then

$$\tan\alpha = \frac{F_2}{F_1} = \frac{6}{8} = 0.75 = \tan 36°\,33'$$

$\alpha = 36°53'$

Using relation, $F = ma$

$\Rightarrow$ Acceleration, $a = \frac{F}{m} = \frac{10}{5} = 2 \text{ m/s}^2$

$\therefore$ An acceleration of 2 m/s^2 is acting on body at an angle of $36°33'$ from the direction of force $F_1 = 8$ N.

11. Here, $F = kt$

When $t = 0$, linear momentum $= p$

When $t = T$, linear momentum $= 3p$

According to Newton's second law of motion,

applied force, $F = \frac{dp}{dt}$

or $dp = F \cdot dt$

or $dp = kt \cdot dt$

Now, integrate both side with proper limit,

$$\int_p^{3p} dp = k\int_0^T t\,dt \text{ or } [p]_p^{3p} = k\left[\frac{t^2}{2}\right]_0^T$$

or $$(3p - p) = \frac{1}{2} k [T^2 - 0]$$

or $$T^2 = \frac{4p}{k}$$

or $$T = 2\sqrt{\frac{p}{k}}$$

12. Change in momentum = Impulse

$\Rightarrow$ $\Delta p = F \times \Delta t$

$\Rightarrow$ $\Delta t = \frac{\Delta p}{F} = \frac{125}{250} = 0.5$ s

13. Change in momentum,

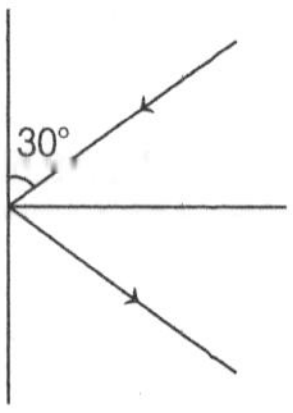

$$\Delta p = 2\,mu \sin 30°$$
$$= 2 \times 0.1 \times 10 \times \frac{1}{2}$$
$$= 1 \text{ kg-ms}^{-1}$$

$\therefore$ $F_{av} = \frac{\Delta p}{\Delta t} = \frac{1}{0.1} = 10$ N

14. This is the question based on impulse-momentum theorem,

$$|F \cdot \Delta t| = |\text{Change in momentum}|$$

$\Rightarrow$ $F \times 0.1 = |p_f - p_i|$

As the ball will stop after catching

$$p_i = mv_i = 0.15 \times 20 = 3, p_f = 0$$

$\Rightarrow$ $F \times 0.1 = 3$

$\Rightarrow$ $F = 30$ N

15. Conservation of momentum in a collision between particles can be understood from both, Newton's second and third laws.

16. Applying law of conservation of momentum,

$$\frac{m}{20} v = \left(m + \frac{m}{20}\right) V$$

or $$V = \frac{v}{20} \times \frac{20}{21} = \frac{v}{21}$$

17. From the graph, it is a straight line, so it is showing uniform motion. Because of impulse direction of velocity changes as can be seen from the slope of the graph.

Initial velocity, $v_1 = \frac{2}{2} = 1 \text{ ms}^{-1}$

Final velocity, $v_2 = -\frac{2}{2} = -1 \text{ ms}^{-1}$

$$\mathbf{p}_i = mv_1 = 0.4 \text{ N-s}$$
$$\mathbf{p}_f = mv_2 = -0.4 \text{ N-s}$$
$$\mathbf{J} = \mathbf{p}_f - \mathbf{p}_i = -0.4 - 0.4 = -0.8 \text{ N-s}$$

(where, $\mathbf{J}$ = impulse)

$\therefore$ $|\mathbf{J}| = 0.8$ N-s

18. Combined momentum $= 2p\hat{\mathbf{i}} + p\hat{\mathbf{j}}$

Magnitude of combined momentum

$$= \sqrt{(2p)^2 + p^2} = \sqrt{5p^2} = \sqrt{5}p$$

This must be equal to the momentum of the third part.

19. Force exerted by machine gun on man's hand in firing a bullet = Change in momentum per second on a bullet or rate of change of momentum

$$= \left(\frac{40}{1000}\right) \times 1200 = 48 \text{ N}$$

Force exerted by man on machine gun

= Force exerted on man by machine gun

= 144 N

Hence, number of bullets fired $= \frac{144}{48} = 3.$

20. The trolley shall move backwards to conserve momentum. The backward momentum would be shared by both the trolley and man.

Applying conservation of momentum,

$$60 \times 1 = (240 + 60)\, v$$

or $$60 = 300\, v$$

or $$v = \frac{60}{300}$$
$$= \frac{1}{5} \text{ ms}^{-1} = 0.2 \text{ ms}^{-1}$$

Speed of man w.r.t. ground $= (1 - 0.2) \text{ ms}^{-1} = 0.8 \text{ ms}^{-1}$

Displacement of man $= 0.8 \times 4$ m $= 3.2$ m

21. When a ball is thrown vertically upward, then the acceleration of the ball,

a = acceleration due to gravity (g) (acting in the downward direction).

Now, using the equation of motion, $v^2 = u^2 - 2gh$

or $$h = \frac{-v^2 + u^2}{2g} \quad \ldots(i)$$

As we know, momentum, $p = mv$ or $v = p/m$

So, substituting the value of v in Eq. (i), we get

$$h = \frac{u^2 - (p/m)^2}{2g}$$

As we know that, at the maximum height, velocity of the ball thrown would be zero.

So, for the flight when the ball is thrown till it reaches the maximum height (h).

$v \rightarrow$ changes from u to 0

$\Rightarrow \quad p \rightarrow$ changes from mu to 0

Similarly, when it reacher it's initial point, then

$h \rightarrow$ changes from h_{max} to 0

Also, $p \rightarrow$ changes from 0 to some values.

Thus, these conditions are only satisfied in the plot given in option (d).

22. Mass of the ball, $m = 0.15$ kg

Velocity of the ball, $v = 54 \text{ km/h} = 54 \times \frac{5}{18} \text{ m/s} = 15 \text{ m/s}$

Let the ball be incident along path PO. Batsman deflects the ball by an angle of 45° along both OQ.

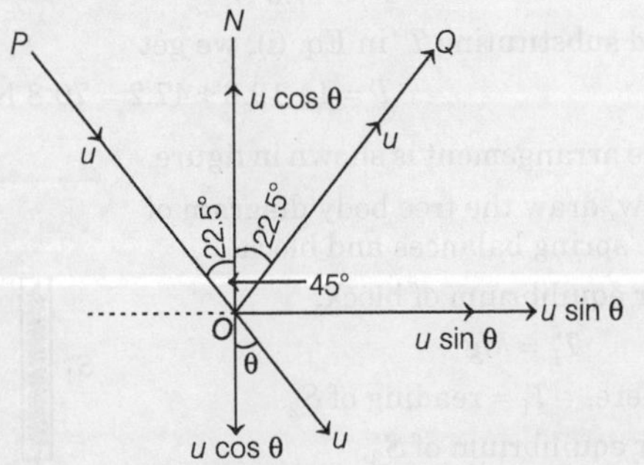

$$\angle PON = \angle NOQ = \frac{45°}{2} = 22.5°$$

The horizontal component of velocity $u \sin \theta$ remains unchanged while vertical component of velocity is just reversed.

$\therefore$ Impulse imparted to the ball

$= $ Change in linear momentum of the ball

$= mu \cos \theta - (-mu \cos \theta)$

$= 2\, mu \cos \theta = 2 \times 0.15 \times 15 \times \cos 22.5°$

$= 4.5 \times 0.9239 \text{ kg-m/s} = 4.16 \text{ kg-m/s}$

$\simeq 4$ kg-m/s

23. In case of projectile motion at the highest point,

$$(v)_{\text{vertical}} = 0$$

and $\quad (v)_{\text{horizontal}} = v \cos \theta$

The initial linear momentum of the system will be $mv \cos \theta$. Now, as force of blasting is internal and force of gravity is vertical.

So, linear momentum of the system along horizontal is conserved.

$$p_1 + p_2 = mv \cos \theta$$

$$m_1 v_1 + m_2 v_2 = mv \cos \theta$$

But it is given that $m_1 = m_2 = \frac{m}{2}$ and as one part retraces its path, $v_1 = -v \cos \theta$

$\therefore \quad \frac{1}{2} m(-v \cos \theta) + \frac{1}{2} mv_2 = mv \cos \theta$

or $\quad v_2 = 3\, v \cos \theta$

24. Since, acceleration acts in the direction of force, hence according to diagram,

resultant acceleration $(a)_R = \sqrt{2}a - a$

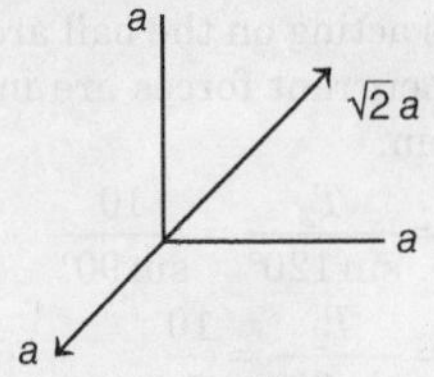

$$(a)_R = (\sqrt{2} - 1)\, a$$

25. By drawing the free body diagram of point B. Let the tension in the section BC and BF be T_1 and T_2, respectively.

From Lami's theorem,

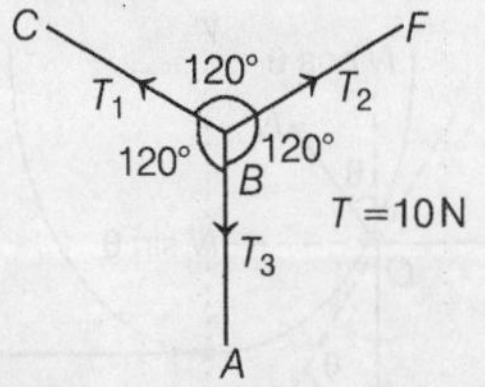

$$\frac{T_1}{\sin 120°} = \frac{T_2}{\sin 120°} = \frac{T}{\sin 120°}$$

$$\Rightarrow \quad T = T_1 = T_2 = 10 \text{ N}$$

26. Given situation is as shown below.

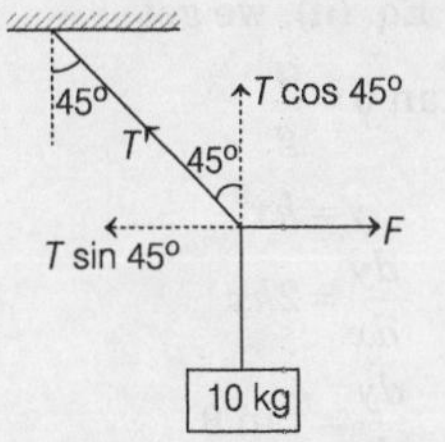

We resolve tension T in string into vertical and horizontal components.

For equilibrium, $\quad F = T \sin 45° \quad$ ...... (i)

and $\quad Mg = T \cos 45° \quad$ .... (ii)

On dividing Eq. (i) by Eq. (ii), we get

$$\frac{F}{Mg} = \tan 45° \text{ or } F = Mg = 10 \times 10 = 100 \text{ N}$$

27. According to equilibrium of forces at point O,

$T \sin 30° = 2$ kg-wt

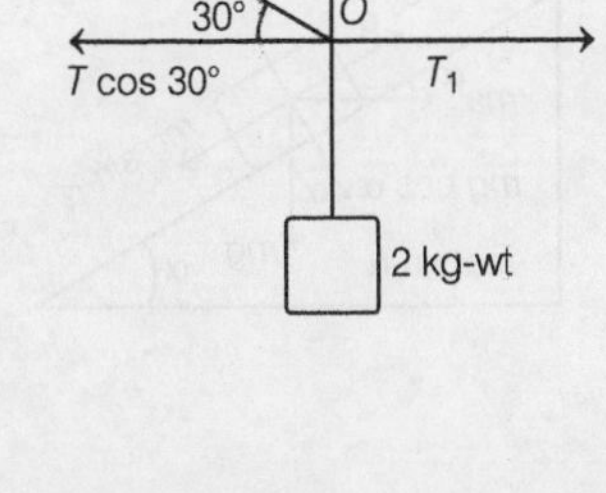

$$\Rightarrow \qquad T = 4 \text{ kg-wt}$$
$$T_1 = T \cos 30°$$
$$= 4 \cos 30° = 2\sqrt{3}$$

28. Various forces acting on the ball are as shown in figure. The three concurrent forces are in equilibrium. Using Lami's theorem.

$$\frac{T_1}{\sin 150°} = \frac{T_2}{\sin 120°} = \frac{10}{\sin 90°}$$
$$\frac{T_1}{\sin 30°} = \frac{T_2}{\sin 60°} = \frac{10}{1}$$
$$\therefore \qquad T_1 = 10 \sin 30° = 10 \times 0.5 = 5 \text{ N}$$
$$T_2 = 10 \sin 60° = 10 \times \frac{\sqrt{3}}{2} = 5\sqrt{3} \text{ N}$$

29. The given situation is shown in the figure

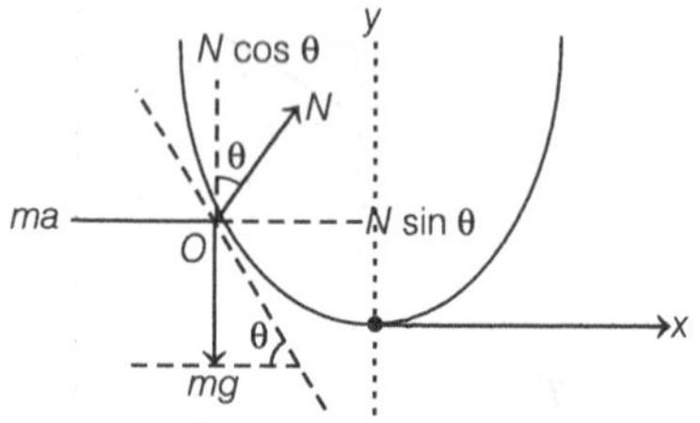

At equilibrium condition at point O.

$$N \sin \theta = ma \qquad \ldots\text{(i)}$$
$$N \cos \theta = mg \qquad \ldots\text{(ii)}$$

Dividing Eq. (i) by Eq. (ii), we get

$$\Rightarrow \qquad \tan \theta = \frac{a}{g} \qquad \ldots\text{(iii)}$$

Given, $\quad y = kx^2$

$$\frac{dy}{dx} = 2kx$$

But $\quad \dfrac{dy}{dx} = \tan \theta$

Hence, $\tan \theta = 2kx \Rightarrow \dfrac{a}{g} = 2kx \Rightarrow x = \dfrac{a}{2gk}$

30. When force F is applied on $2m$ from left, contact force,

$$F_1 = \frac{m}{m + 2m} F = \frac{F}{3}$$

When force F is applied on m from right, contact force,

$$F_2 = \frac{2m}{m + 2m} F = \frac{2F}{3}$$
$$\therefore \qquad F_1 : F_2 = 1 : 2$$

31. Since, $P = (M + m) a$

Now, as in free body diagram of block,

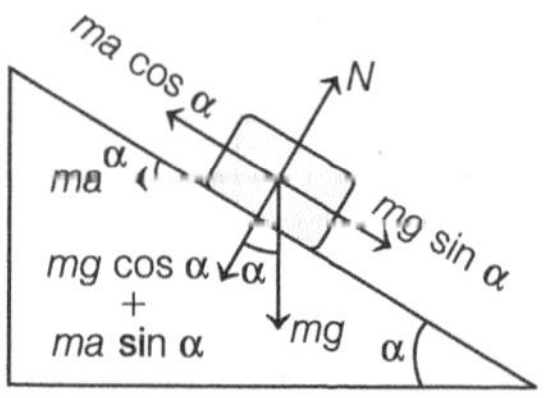

$$ma \cos \alpha = mg \sin \alpha$$
$$\therefore \qquad a = g \frac{\sin \alpha}{\cos \alpha} = g \tan \alpha$$
$$\text{or} \qquad P = (M + m) g \tan \alpha$$

32. FBD of mass 2 kg FBD of mass 4 kg

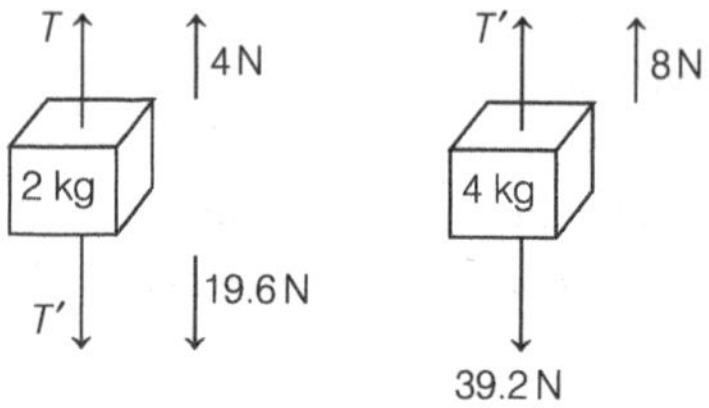

$$T - T' = 19.6 = 4 \qquad \ldots\text{(i)}$$
$$T' - 39.2 = 8 \qquad \ldots\text{(ii)}$$

From Eq. (ii), we get

$$T' = 47.2 \text{ N}$$

and substituting T' in Eq. (i), we get

$$T = 4 + 19.6 + 47.2 = 70.8 \text{ N}$$

33. The arrangement is shown in figure.

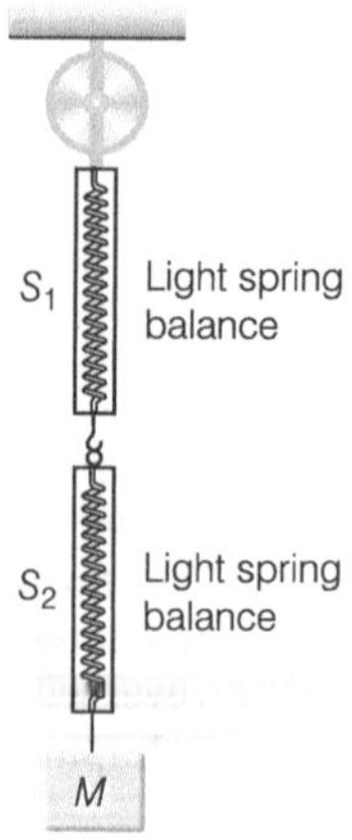

Now, draw the free body diagram of the spring balances and block.

For equilibrium of block,

$$T_1 = Mg$$

where, T_1 = reading of S_2.

For equilibrium of S_2,

$$T_2 = T_1$$

where, T_2 = reading of S_1.

For equilibrium of S_1,

$$T_3 = T_2$$

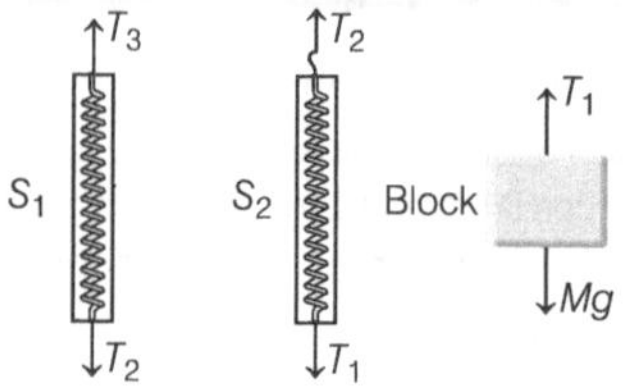

Hence, $\quad T_1 = T_2 = Mg$

So, both scales read M kg.

34. Acceleration of system, $a = \dfrac{\text{Net pushing force}}{\text{Total mass}}$

$$\text{or} \qquad a = \frac{F - (m_1 + m_2 + m_3) g \sin \theta}{(m_1 + m_2 + m_3)}$$

Equation of motion for m_3,

$$N - m_3 g \sin \theta = m_3 a$$

$$\text{or } N = m_3 g \sin \theta + m_3 \left[\frac{F - (m_1 + m_2 + m_3) g \sin \theta}{(m_1 + m_2 + m_3)}\right]$$
$$= \frac{m_3 F}{m_1 + m_2 + m_3}$$

35. In vertical direction, weights are balanced by frictional forces.

Consider FBD of block A and B as shown in figure.

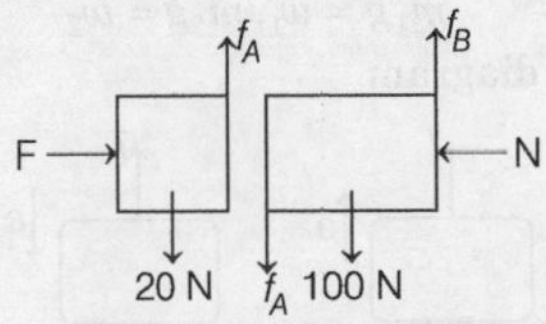

As the blocks are in equilibrium, balance forces are in horizontal and vertical direction.

For the system of blocks $(A + B)$,

$$F = N$$

For block A, $f_A = 20$ N and for block B,

$$f_B = f_A + 100 = 120 \text{ N}$$

36. In the frame of wedge, the force diagram of block is shown in figure. From free body diagram of wedge,

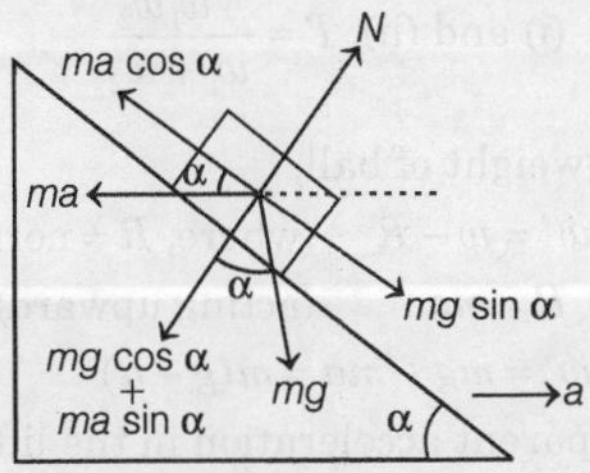

For block to remain stationary,

$$ma \cos \alpha = mg \sin \alpha \text{ or } a = g \tan \alpha$$

37. As the elevator is going down with decreasing speed, so acceleration is in upward direction. Let it be a,

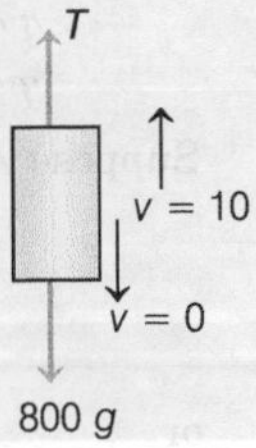

$$T - 800\, g = 800\, a$$
$$T = 800\, (g + a)$$

From $\quad v^2 = u^2 - 2as$

$\Rightarrow \quad 10^2 = 0 - 2 \times a \times 25$

$\therefore \quad a = 2 \text{ ms}^{-2}$

$\therefore \quad T = 800\, (10 + 2),$

$\therefore \quad T = 9600$ N

38. $T = \dfrac{2 m_1 m_2}{m_1 + m_2} g = \dfrac{2 \times 4 \times 2 \times 10}{4 + 2}$

$$= \frac{160}{6} = 26.6 \approx 27 \text{ N}$$

Total downward thrust on the pulley

$$= 2T = 2 \times 27 = 54 \text{ N}$$

39. Tension in rope, $T <$ Breaking load, $\dfrac{2}{3} mg$

$\therefore \quad m\,(g - a) < \dfrac{2}{3} mg$

or $\quad a > \dfrac{g}{3}$

40. Here, the tension in the cord is given by

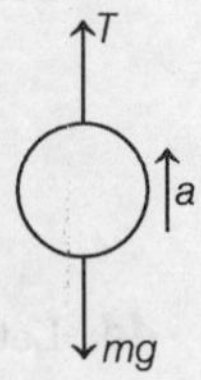

$$T = mg + ma$$

Here, upward acceleration $= a$

Mass of sphere $= M$

$$T = 4\,mg$$

$\Rightarrow \quad 4\,mg = mg + ma$

$$3\,mg = ma$$

$\Rightarrow \quad a = 3\,g$

41. Here, mass of the body,

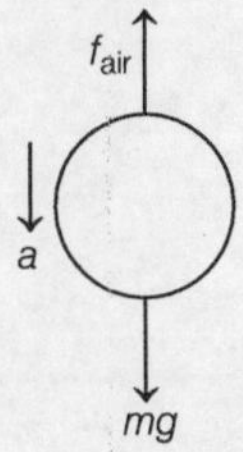

$$M = 0.05 \text{ kg}$$

Acceleration, $g = 9.8 \text{ ms}^{-2}$, $a = 9.5 \text{ ms}^{-2}$

$\therefore \quad mg - f_{air} = ma$

$\Rightarrow \quad f_{air} = m\,(g - a)$

$$= 0.05\,(9.8 - 9.5)$$
$$= 0.015 \text{ N}$$

42. The given situation is shown below.

Let a be the common acceleration.

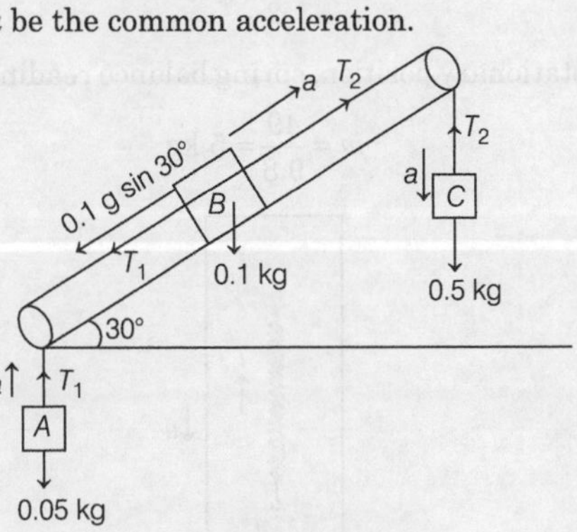

$$m_B = 100 \text{ g} = 0.1 \text{ kg}$$
$$m_C = 500 \text{ g} = 0.5 \text{ kg}$$
$$m_A = 50 \text{ g} = 0.05 \text{ kg}$$

Equation of motion for the mass A,

$$T_1 - 0.05\, g = 0.05\, a \quad \text{...(i)}$$

Equation of motion for the mass B,

$$T_2 = 0.1\, g \sin 30° - T_1 = 0.1\, a$$
$$T_2 - T_1 - 0.05\, g = 0.1\, a \quad \text{...(ii)}$$

Equation of motion for the mass C,

$$0.5\, g - T_2 = 0.5\, a \quad \text{...(iii)}$$

Adding Eqs. (i) and (ii), we have

$$T_2 - 0.1\, g = 0.15\, a$$

$\Rightarrow \quad T_2 = 0.1\, g + 0.15\, a \quad$ ...(iv)

Putting the value of T_2 from Eq. (iv) in Eq. (iii), we get

$$0.5\, g - 0.1\, g - 0.15\, a = 0.5\, a$$

$\Rightarrow \quad 0.4\, g = 0.65\, a$

$\Rightarrow \quad a = \dfrac{0.4g}{0.65} = \dfrac{8g}{13}$ (downwards)

43. Effective value of acceleration due to gravity in the lift $= g - a$

Using $s = ut + \frac{1}{2}at^2$, we get

$$l = \frac{1}{2}(g-a)\sin\theta t^2, \text{ we get}$$

$$t = \sqrt{\frac{2l}{(g-a)\sin\theta}}$$

44. Let force in downward to the incline $mg\sin\theta - T = ma$

$$T = ma$$

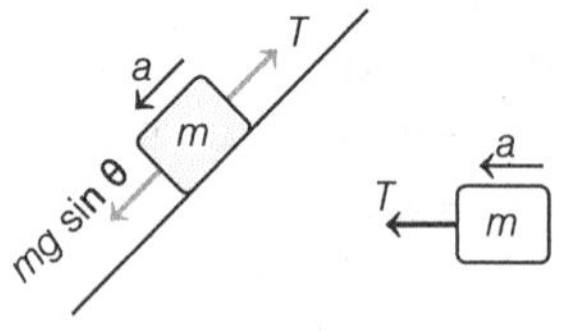

$\therefore \quad mg\sin\theta - T = T$

or $\quad 2T = mg\sin\theta$

or $\quad T = \frac{1}{2}mg\sin\theta$

45. In stationary position, spring balance reading $= mg = 49$

or $\quad m = \frac{49}{9.8} = 5$ kg

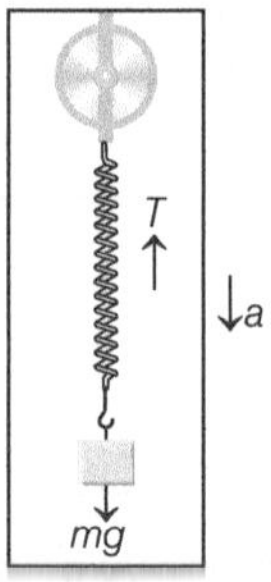

When lift moves downward, $mg - T = ma$

Reading of balance,

$$T = mg - ma = 5\,(9.8 - 5)$$
$$= 5 \times 4.8 = 24.0 \text{ N}$$

46. Let acceleration of system (rope + block) be a along the direction of applied force, then $a = \frac{P}{M+m}$.

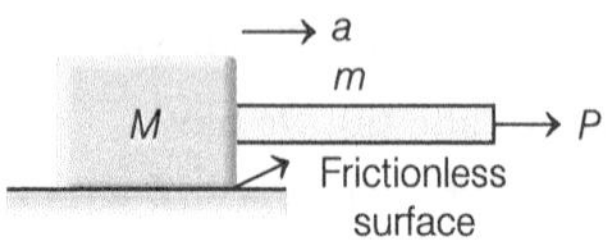

Draw the FBD of block and rope as shown in the following figure

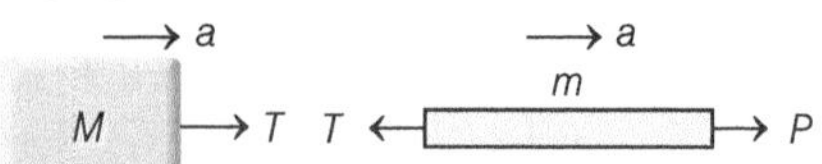

where, T is the required parameter.

For block, $\quad T = Ma$

$\Rightarrow \quad T = \frac{MP}{M+m}$

47. For solving the problem, we assume that observer is situated in the frame of pulley (non-inertial reference frame),

$$m_1 g = w_1,\ m_2 g = w_2$$

From force diagram,

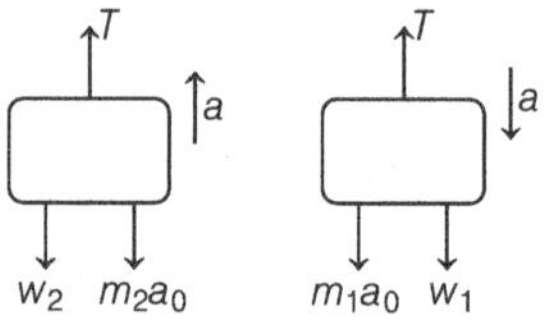

$$T - m_2 a_0 - w_2 = m_2 a$$

or $\quad T - m_2 g - w_2 = m_2 a \quad (\because a_0 = g)$

or $\quad T - 2w_2 = m_2 a \quad$ ...(i)

From force diagram, $m_1 a_0 + w_1 - T = m_1 a$

or $\quad m_1 a_0 + w_1 - T = m_1 a$

or $\quad 2w_1 - T = m_1 a \quad$ ...(ii) $(\because a_0 = g)$

From Eqs. (i) and (ii), $T = \frac{4w_1 w_2}{w_1 + w_2}$

48. Apparent weight of ball,

$$w' = w - R \quad \text{(where, } R = \text{normal reaction)}$$
$$R = ma \quad \text{(acting upward)}$$
$$w' = mg - ma = m(g-a)$$

Hence, apparent acceleration in the lift is $g - a$. Now, if the man is standing stationary on the ground, then the apparent acceleration of the falling ball is g.

49. If A is climbing with constant velocity, then

$$T' = 5g + T \text{ and } T = 2g$$
$$T' = 5g + 2g = 7g$$

Suppose A is climbing with acceleration a such that

$$T = 30 \text{ N}$$
$$T - 2a = 2a$$
$$30 - 2 \times 10 = 2a$$

or $\quad a = 5 \text{ ms}^{-2}$

Again, $\quad T' - T - 5g = 5a$

or $\quad T' = T + 5g + 5a$

or $\quad T' = (30 + 50 + 25)$ N

$\quad = 105$ N

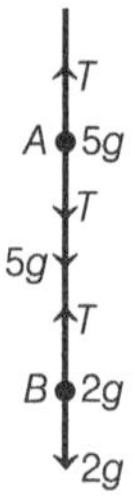

50. Tension between m_2 and m_3 is given by

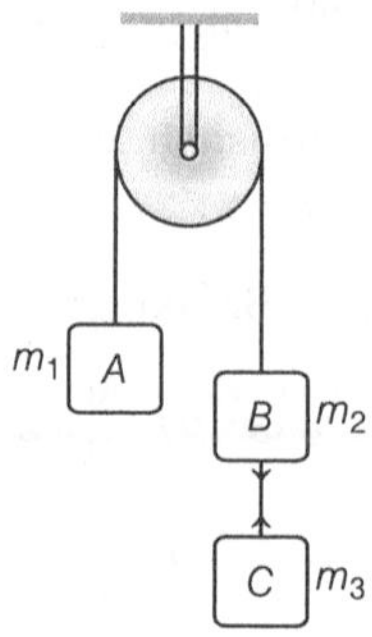

$$T = \frac{2m_1m_2}{m_1+m_2+m_3} \times g$$

$$= \frac{2\times2\times2}{2+2+2} \times 9.8 = 13 \text{ N}$$

51. From Newton's second law,

$$F = ma$$

$\Rightarrow$ $6 = (7+5)\,a$

$$a = \frac{1}{2}\,\text{m/s}^2$$

$$f' = 5 \text{ kg}$$

Now $f' = 5 \times \frac{1}{2} = 2.5 \text{ N}$

52. From diagram, $5g - T_2 = 5a$ ...(i)

$T_2 - T_1 = 3a$ ...(ii)

$T_1 - g = a$ ...(iii)

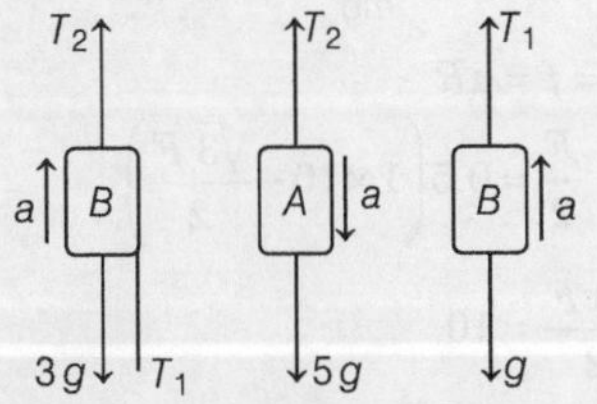

Solving Eqs. (i), (ii) and (iii), we get

$$g = 9a$$

or $a = \frac{g}{9}$

53. From geometry, $l^2 = x^2 + y^2$ but y is constant, hence on differentiating, we have $2l\frac{dl}{dt} = 2x\frac{dx}{dt}$

But $\frac{dl}{dt} = v$. Hence, horizontal velocity of block,

$$v_x = \frac{dx}{dt}$$

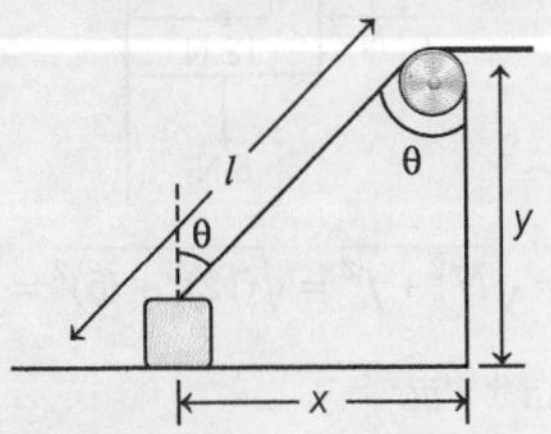

$\Rightarrow$ $lv = x \cdot v_x$

or $v_x = \frac{l \cdot v}{x} = \frac{v}{\sin\theta}$

54. The system of masses is shown below

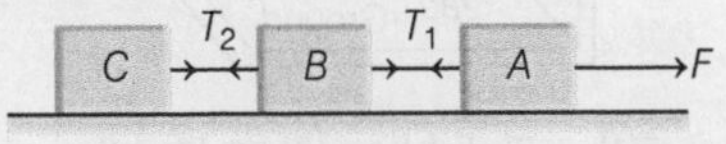

From the figure,

$F - T_1 = ma$ ...(i)

and $T_1 - T_2 = ma$ ...(ii)

Eq. (i) gives,

$$10.2 - T_1 = 2 \times 0.6$$

$\Rightarrow$ $T_1 = 10.2 - 1.2 = 9\text{N}$

Again, from Eq. (ii), we get

$$9 - T_2 = 2 \times 0.6$$

$\Rightarrow$ $T_2 = 9 - 1.2 = 7.8 \text{ N}$

55.

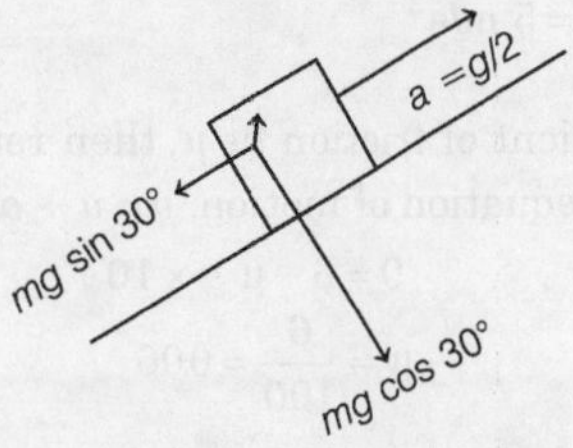

From diagram, $T\sin\theta - mg\sin\theta = ma$

$T\sin\theta = mg\sin\theta + \frac{mg}{2}$ ...(i)

$T\cos\theta = mg\cos\theta$ ...(ii)

Dividing Eq. (i) by Eq. (ii), we get

$$\tan\theta = \frac{2}{\sqrt{3}} \Rightarrow \theta = \tan^{-1}\frac{2}{\sqrt{3}}$$

56. Force applying on the block,

$$F = mg\sin\theta$$

or $mg\sin\theta = ma$

$\therefore$ $a = g\sin\theta$

where, a is acceleration along the inclined plane.

$\therefore$ Vertical component of acceleration

$$= a\sin\theta = g\sin\theta\cdot\sin\theta = g\sin^2\theta$$

$\therefore$ Relative vertical acceleration of A with respect to B is

$$g\,(\sin^2 60° - \sin^2 30°\,] = \frac{g}{2} = 4.9\,\text{ms}^{-2}$$

(in vertical direction)

57. Weight of chain on table $= mg - \frac{1}{4}mg = \frac{3}{4}mg$

For maximum possible friction, $\frac{3}{4}\mu mg = \frac{1}{4}mg$

$\therefore$ $\mu = \frac{1}{3}$

58. Net acceleration,

$$a = g\,(\sin\theta - \mu\cos\theta) = 9.8\,(\sin 45° - 0.5\cos 45°\,)$$

$$= \frac{4.9}{\sqrt{2}}\,\text{m/s}^2$$

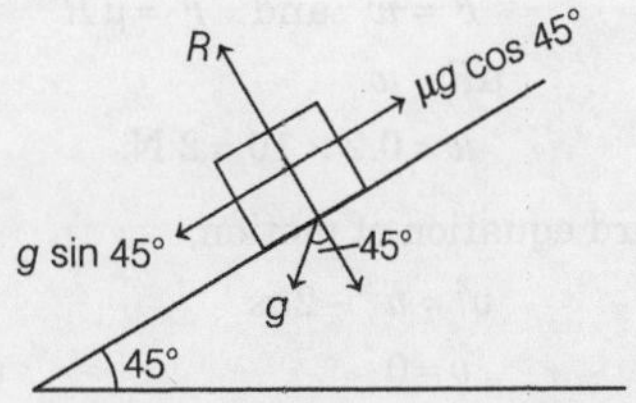

59. Net downward acceleration

$$= \frac{\text{Weight} - \text{Frictional force}}{\text{Mass}} = \frac{mg - \mu R}{m} = \frac{60 \times 10 - 0.5 \times 600}{60} = \frac{300}{60} = 5 \text{ m/s}^2$$

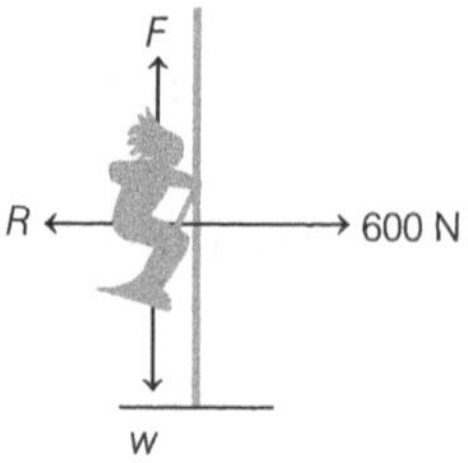

60. Let coefficient of friction be μ, then retardation will be μg. From equation of motion, $v = u + at$

$\Rightarrow \quad 0 = 6 - \mu g \times 10$

or $\quad \mu = \frac{6}{100} = 0.06$

61. Pseudo force on the block $= m \times 4$ N (backward)

Force of friction $= 0.4 \times m \times 10$ N (forward)

Equating, $m \times 4 = 0.4 \times m \times 10 = 4m$

Clearly, the equation holds good for all values of m.

62. Force, $P = f_{ms} = \mu_s mg$ (when body is at rest)

When the body starts moving with acceleration a, then

$$P - f_k = ma$$

$$\mu_s mg - \mu_k mg = ma$$

or $\quad a = (\mu_0 - \mu_k) g$

or $\quad a = (0.5 - 0.4)\ 10$

$\quad = 0.1 \times 10 \text{ ms}^{-2} = 1 \text{ ms}^{-2}$

63. $f_{ms} = 0.6 \times 10 \times 9.8 \text{ N} = 58.8 \text{ N}$

Since, the applied force is greater than f_{ms}, therefore the block will be in motion. So, we should consider f_k.

$$f_k = 0.4 \times 10 \times 9.8 \text{ N}$$

or $\quad f_k = 4 \times 9.8$ N

This would cause acceleration of 40 kg block.

$$\text{Acceleration} = \frac{4 \times 9.8 \text{ N}}{40 \text{ kg}} = 0.98 \text{ ms}^{-2}$$

64. Let R be the normal contact force by wall on the block.

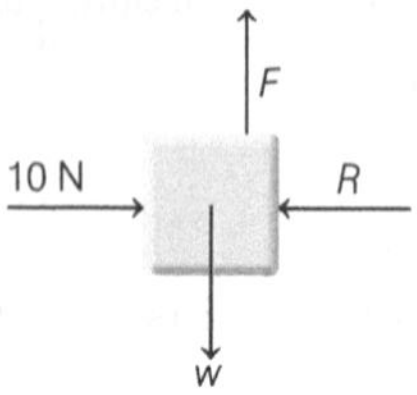

$$R = 10 \text{ N}$$

$$F = w \quad \text{and} \quad F = \mu R$$

$\therefore \quad \mu R = w$

or $\quad w = 0.2 \times 10 = 2$ N

65. From third equation of motion,

$$v^2 = u^2 - 2as$$

Given, $\quad v = 0 \quad$ (car is stopped)

As friction provide the retardation,

$$a = \mu g, \; v = 100 \text{ ms}^{-1}$$

$\therefore \quad (100)^2 = 2\mu gs$

$\Rightarrow \quad s = \frac{100 \times 100}{2 \times 0.5 \times 10} = \frac{100 \times 100}{5 \times 2} = 1000 \text{ m}$

66. $R + F\sin 60° = mg$

or $\quad R = mg - \frac{\sqrt{3}F}{2}$

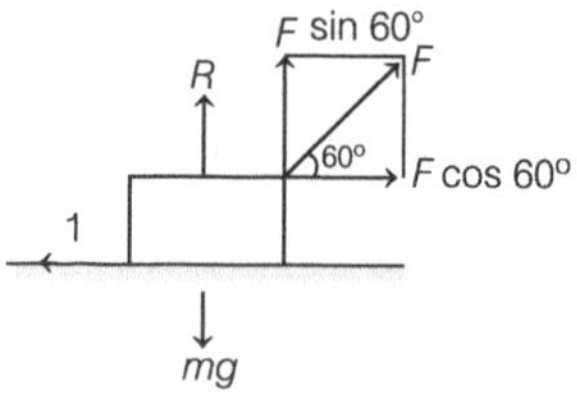

$F\cos 60° = f = \mu R$

or $\quad \frac{F}{2} = 0.5\left(1 \times 10 - \frac{\sqrt{3}F}{2}\right)$

or $\quad F + \frac{\sqrt{3}F}{2} = 10$

or $\quad F = \frac{20}{2 + \sqrt{3}}$

or $\quad F = \frac{20}{3.732} = 5.36 \text{ N}$

67. Wall applies 2 forces of the block (i) normal reaction, $R = 12$ N, and (ii) frictional force, $f_2 = mg = 5$ N tangentially upward.

$\therefore$ Total force exerted by wall on block

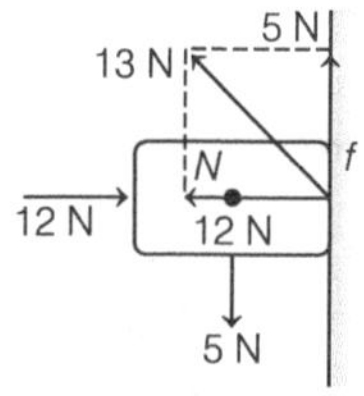

$$F = \sqrt{N^2 + f_s^2} = \sqrt{(12)^2 + (5)^2} = 13 \text{ N}$$

68. As, $F = F_{AB} + F_{BG}$

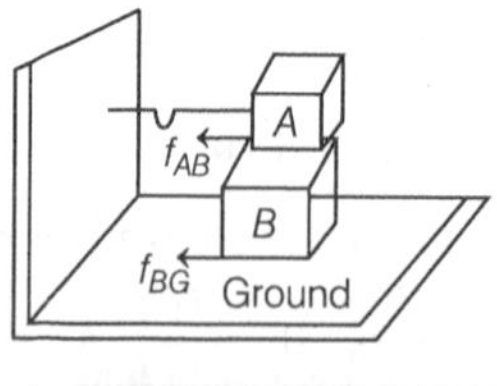

$$= \mu_{AB} m_a g + \mu_{BG}(m_A + m_B) g$$

$$= 0.2 \times 100 \times 10 + 0.3 \times (300) \times 10$$

$$= 200 + 900 = 1100 \text{ N}$$

69. $ma = mg\sin\theta - f$

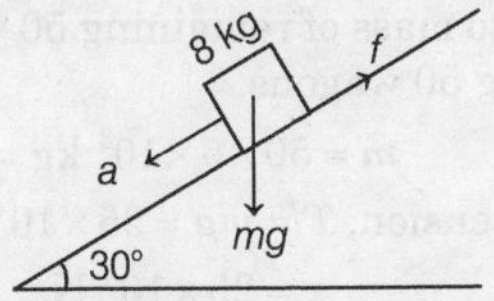

or $\quad f = mg\sin\theta - ma = m(g\sin\theta - a)$

$$= 8\left(10\times\frac{1}{2} - 0.4\right)\text{N} = 8\times 4.6\text{ N} = 36.8\text{ N}$$

70. For equilibrium of the block,

$$N = mg + Q\cos\theta$$

$$Q\sin\theta + P = \mu N$$

$$Q\sin\theta + P = \mu\,(mg + Q\cos\theta)$$

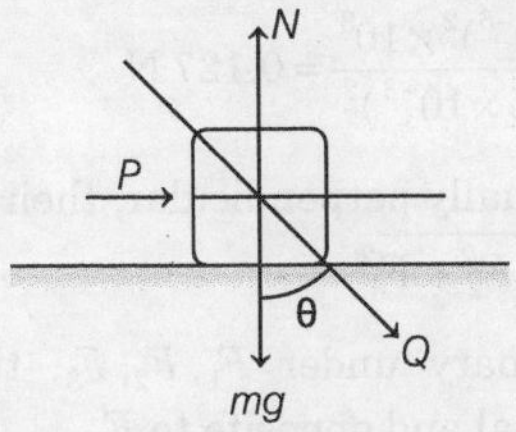

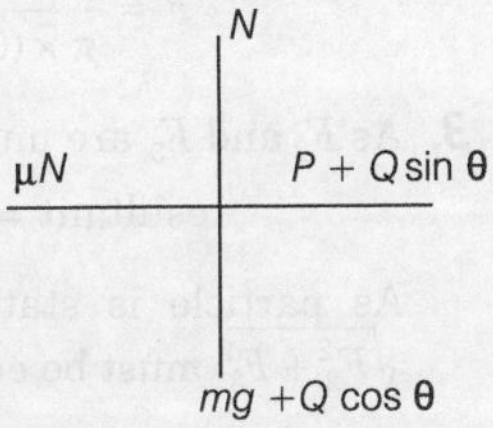

$$\therefore \quad \mu = \left(\frac{Q\sin\theta + P}{mg + Q\cos\theta}\right)$$

71. If μ is the mass/length, then

weight of hanging length $= \mu\, lg$

Weight of chain on table $= \mu\,(L - l)\,g$

$\because \quad R = \mu\,(L - l)\,g$

$\therefore \quad f = \mu_s R = \mu_s \mu\,(L - l)\,g$

Equating $\quad \mu_s \mu\,(L - l)\,g = \mu l g$

or $\quad \mu_s = \dfrac{l}{L - l}$

72. Motion stops when pull due to $m_1 \le$ force of friction between m and m_2 and surface.

$\Rightarrow \quad m_1 g \le \mu(m_2 + m)g$

$\Rightarrow \quad 5\times 10 \le 0.15(10 + m)\times 10$

$\Rightarrow \quad m \ge 23.33$ kg

Here, nearest value is 27.3 kg.

So, $\quad m_{\min} = 27.3$ kg

73. $F_1 = mg\,(\sin\theta + \mu\cos\theta)$ (as body just in position to move up, friction force downward)

$F_2 = mg(\sin\theta - \mu\cos\theta)$ (as body just in position to slide down, friction upward)

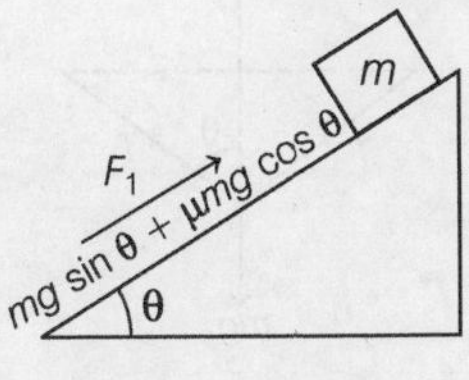

$$\frac{F_1}{F_2} = \frac{\sin\theta + \mu\cos\theta}{\sin\theta - \mu\cos\theta}$$

$$= \frac{\tan\theta + \mu}{\tan\theta - \mu} = \frac{2\mu + \mu}{2\mu - \mu} = 3$$

74. Energy lost over path $PQ = \mu\, mg\cos\theta\times 4$

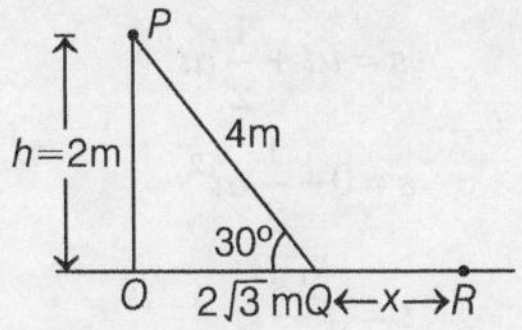

Energy lost over path $QR = \mu mgx$

i.e. $\mu\, mg\cos 30°\times 4 = \mu mgx \qquad (\because \theta = 30°)$

$$x = 2\sqrt{3} = 3.45\text{ m}$$

From Q to R energy loss is half of the total energy loss.

i.e. $\quad \mu mgx = \dfrac{1}{2}\times mgh$

$\Rightarrow \quad \mu = 0.29$

The values of the coefficient of friction μ and the distance $x(= QR)$ are 0.29 and 3.5.

75. From acting on block are shown in adjoining figure

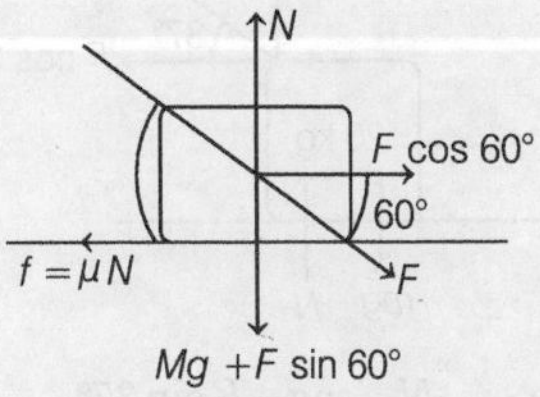

As the block does not move, hence

$$F\cos 60° = f = \mu N = \mu\,(Mg + F\sin 60°)$$

$$\therefore \quad F\frac{1}{2} = \frac{1}{2\sqrt{3}}\left(\sqrt{3}0\times 10 + F\frac{\sqrt{3}}{2}\right)$$

On simplification, we get $F = 20$ N

76. Let the length of chain be l and mass m. Let a part x of chain can hang over one edge of table having coefficient of friction.

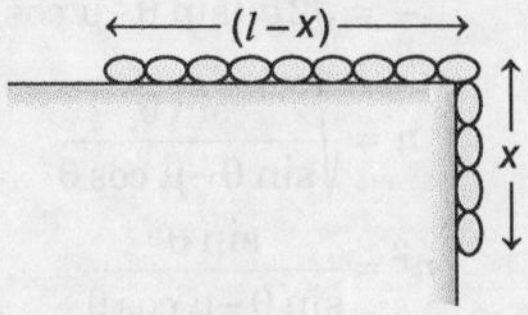

$\therefore$ Pulling force, $F = \dfrac{mx}{l}\,g$

and friction force, $f = \mu N = \mu\dfrac{m}{l}\,(l - x)\,g$

For equilibrium, $\quad F = f$, hence

$$\frac{mx}{l}\cdot g = \mu\frac{m}{l}\,(l - x)\,g = 0.25\frac{m}{l}\,(l - x)\,g$$

$\Rightarrow \quad x = \dfrac{l}{5}$

$\therefore$ % change in $\dfrac{x}{l} = \dfrac{1}{5}\times 100 = 20\%$

77. Force on the car,

$$F = \mu R$$

or $\quad ma = \mu mg \qquad (\because R = mg)$

or $\quad a = \mu g$

Now, from second equation of motion,

$$s = ut + \frac{1}{2}at^2$$

or $\quad s = 0 + \frac{1}{2}at^2 \qquad (\because u = 0)$

or $\quad t = \sqrt{\frac{2s}{\mu g}}$

$\therefore \quad t = \sqrt{\frac{2s}{\mu g}}$

or $\quad t \propto \frac{1}{\sqrt{\mu}}$

78. The work done by the force is $F \cos 37°$,

where $\quad F \cos 37° = f = \mu N$

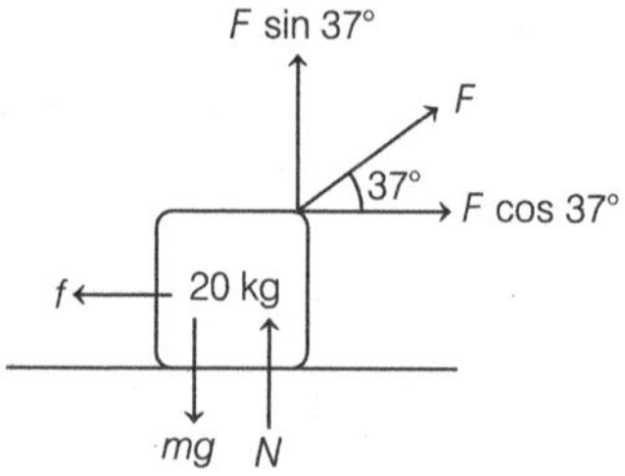

In this case, $\quad N = mg - F \sin 37°$,

So that, $\quad F = \frac{\mu mg}{(\cos 37° + \mu \sin 37°)}$

Here, $\mu = 0.40$ and $m = 20$ kg

$\therefore \quad F = 75.4$ N

Hence, $\quad W = (75.4 \cos 37°)(8.0) = 482$ J

79. For a smooth plane, $v = \sqrt{2g \sin \theta \cdot s}$ and for a rough plane,

$$\frac{v}{n} = \sqrt{2g(\sin \theta - \mu \cos \theta) s}$$

$\therefore \quad n = \sqrt{\frac{\sin \theta}{\sin \theta - \mu \cos \theta}}$

or $\quad n^2 = \frac{\sin \theta}{\sin \theta - \mu \cos \theta}$

$\Rightarrow \quad (n^2 - 1) \sin \theta = n^2 \mu \cos \theta$

or $\quad \mu = \left(\frac{n^2 - 1}{n^2}\right) \tan \theta = \tan \theta \left(1 - \frac{1}{n^2}\right)$

Round II

1. Total mass of 80 wagons = $80 \times 5 \times 10^3 = 4 \times 10^5$ kg

Acceleration, $\quad a = \frac{F}{M} = \frac{4 \times 10^5}{4 \times 10^5} = 1 \text{ ms}^{-2}$

Tension in the coupling between 30th and 31st wagon will be due to mass of remaining 50 wagons. Now, mass of remaining 50 wagons.

$$m = 50 \times 5 \times 10^3 \text{ kg} = 25 \times 10^4 \text{ kg}$$

$\therefore$ Required tension, $T = mg = 25 \times 10^4 \times 1$

$$= 25 \times 10^4 \text{ N}$$

2. Rate of flow water, $\frac{v}{t} = 10 \text{ cm}^3\text{s}^{-1}$

Density of water, $\rho = 10^3$ kg/m^3

Cross-sectional area of pipe $OA = \pi (0.5 \times 10^{-3})^2$

$$\text{Force} = m\frac{dv}{dt} = \frac{mv}{t} = \frac{V\rho v}{t} = \frac{\rho v}{t} \times \frac{v}{At}$$

$$= \left(\frac{v}{t}\right)^2 \frac{l}{A} \qquad \left(\because V = \frac{v}{At}\right)$$

$$F = \frac{(10 \times 10^{-6})^2 \times 10^3}{\pi \times (0.5 \times 10^{-3})^2} = 0.127 \text{ N}$$

3. As F_2 and F_3 are mutually perpendicular, their resultant $= \sqrt{F_2^2 + F_3^2}$

As particle is stationary under F_1, F_2, F_3, therefore $\sqrt{F_2^2 + F_3^2}$ must be equal and opposite to F_1.

The force F_1 is now removed. So, the resultant of F_2 and F_3 will now make the particle move with force equal to F_1. Hence, the acceleration is F_1/m.

4.

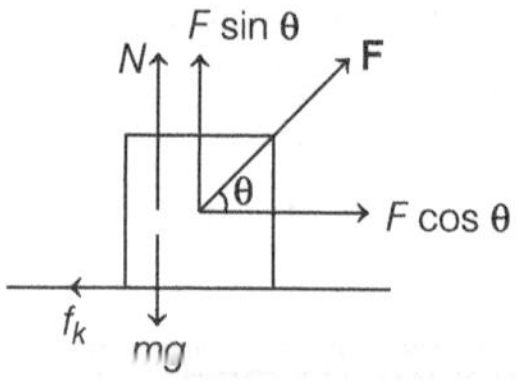

$$N = mg - F \sin \theta$$

$$F \cos \theta - \mu_K N = ma$$

$$F \cos \theta - \mu_K (mg - F \sin \theta) = ma$$

$$a = \frac{F}{m} \cos \theta - \mu_K \left(g - \frac{F}{m} \sin \theta\right)$$

5. As $m_1 : m_2 : m_3 = 1 : 1 : 3$

and momentum is conserved, $\sqrt{p_1^2 + p_2^2 + p_3^2} = 3v_3$

$$\sqrt{1 \times 39^2 + 1 \times 39^2} = 3v_3$$

$\Rightarrow \quad 39\sqrt{2} = 3v_3$

$\Rightarrow \quad v_3 = \frac{39\sqrt{2}}{3} = 13\sqrt{2} \text{ ms}^{-1}$

6. Let T be the tension in the string. Since, the system is in equilibrium, therefore from figure,

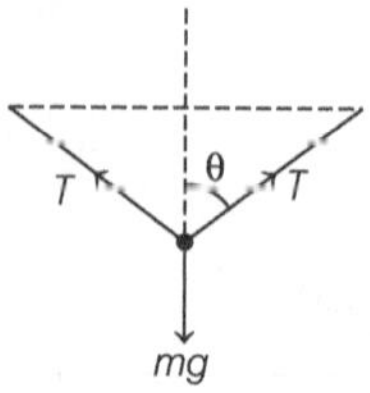

$$2T\cos\theta = mg$$

or $\quad T = mg/2\cos\theta$

The string will be straight, if $\theta = 90°$

$\therefore \quad T = mg/2\cos 90° = mg/2(0) = \infty$

7. The three forces acting on the mass at location A have been shown in figure. Since, the mass is in equilibrium, therefore the three forces acting on the mass must be represented by the three sides of a triangle in one order. Hence,

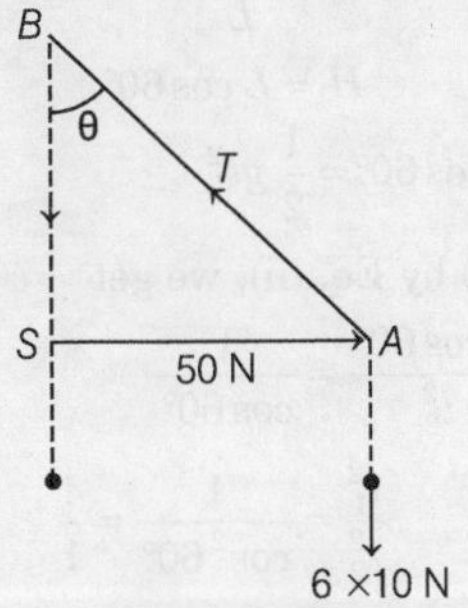

In ΔSBA, $\quad \dfrac{50}{SA} = \dfrac{6\times 10}{SB}$

or $\quad \dfrac{SA}{SB} = \dfrac{50}{60} = \dfrac{5}{6}$

or $\quad \tan\theta = \dfrac{SA}{SB} = \dfrac{5}{6} = 0.8333 = \tan 40°$

$\therefore \quad \theta = 40°$

8. If T is tension in each part of the string holding mass $\sqrt{2}\,m$, then in equilibrium,

$$T\cos\theta + T\cos\theta = \sqrt{2}\,mg$$

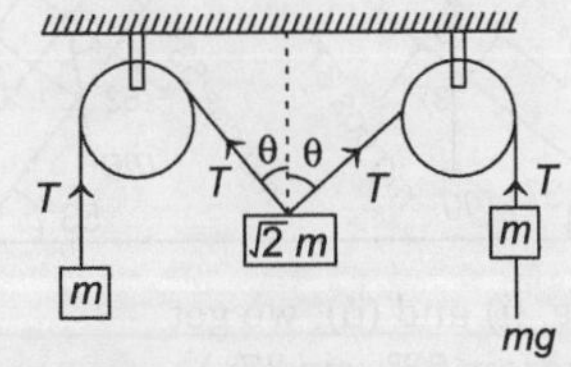

$$2T\cos\theta = \sqrt{2}\,mg$$

But $\quad T = mg$

$\therefore \quad 2mg\cos\theta = \sqrt{2}\,mg$

$$\cos\theta = \frac{1}{\sqrt{2}}$$

$$\theta = 45°$$

9. Effective upward force $= 310 - mg$

$$= 310 - 24\times 9.8 = 74.8\text{ N}$$

Upward acceleration,

$$a = 74.8\ /\ 24 = 3.12\text{ ms}^{-2}$$

As, $\quad s = ut + \dfrac{1}{2}at^2$

$\Rightarrow \quad 4.6 = 0 + \dfrac{1}{2}\times 3.12\times t^2$

$\Rightarrow \quad t^2 = \dfrac{4.6}{1.56} = 2.95$

$\Rightarrow \quad t = \sqrt{2.95} \approx 1.7\text{s}$

10. Mass of rope, $M = 0.1$ kg, $\theta = 10°$

From figure,

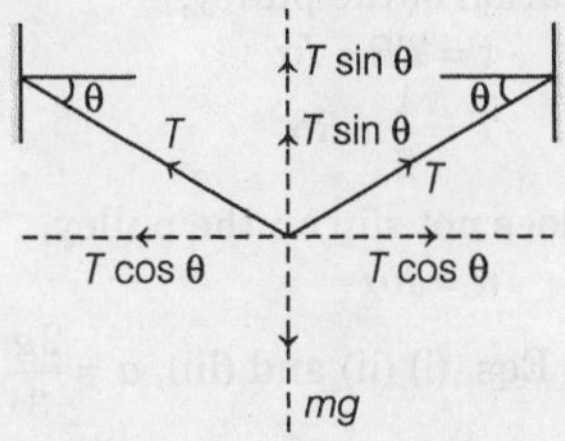

$$2T\sin\theta = mg$$

$$T = \frac{mg}{2\sin\theta} = \frac{0.1\times 9.8}{2\sin 10°} = 2.82\text{ N}$$

11. The situation is as shown in the figure

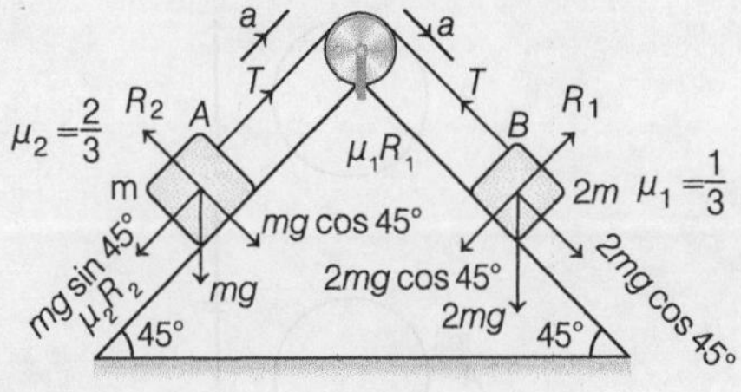

The equation of motion for body B,

$$2mg\sin 45° - \mu_1 R_1 - T_2 = 2ma$$

$$2mg\sin 45° - \frac{1}{3}2mg\cos 45° - T = 2ma$$

$$\Rightarrow 2mg\times\frac{1}{\sqrt{2}} - \frac{1}{3}2mg\times\frac{1}{\sqrt{2}} - T = 2ma$$

In this problem as $(m_B = m_A)g\sin\theta = (mg/\sqrt{2})$ is lesser than $(\mu_B m_B + \mu_A m_A)g\cos\theta = (4mg/3\sqrt{2})$, the masses will not move and hence,

acceleration of B = acceleration of $A = 0$

12. $\quad a = \dfrac{5}{5} = 1\text{ ms}^{-2}$

Upward distance covered in 4 s,

$$y = \frac{1}{2}at^2 = \frac{1}{2}\times 1\times (4)^2 = 8\text{ m}$$

Horizontal distance covered in 4s,

$$x = vt = 1.5\times 4 = 6\text{ m}$$

$$s = \sqrt{x^2 + y^2} = \sqrt{6^2 + 8^2}$$

$$= \sqrt{36 + 64} = 10\text{ m}$$

13. $\quad m\dfrac{dv}{dt} = f_0 e^{-bt}$

$\Rightarrow \quad \dfrac{dv}{dt} = \dfrac{f_0}{m}e^{-bt}$

$\Rightarrow \quad \displaystyle\int_0^v dv = \frac{f_0}{m}\int_0^t e^{-bt}\,dt$

$\Rightarrow \quad v = \dfrac{f_0}{m}\left(\dfrac{e^{-bt}}{-b}\right)_0^t$

$\Rightarrow \quad v = \dfrac{f_0}{mb}(1 - e^{bt})$

Hence, option (a) is correct.

14. For the motion of the block,

$$mg - T = ma$$

For the rotation of the pulley,

$$\tau = TR = I\alpha$$

$$T = \frac{1}{2}mR\alpha$$

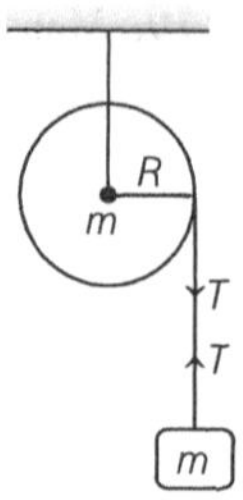

As string does not slip on the pulley,

$$a = R\alpha$$

On solving Eqs. (i),(ii) and (iii), $a = \frac{2g}{3}$

15. The situation is shown in figure. At initial time, the ball is at P, then under the action of a force (exerted by hand) from P to A and then from A to B. Let acceleration of ball during PA be a ms^{-2} (assumed to be constant) in upward direction and velocity of ball be v m/s.

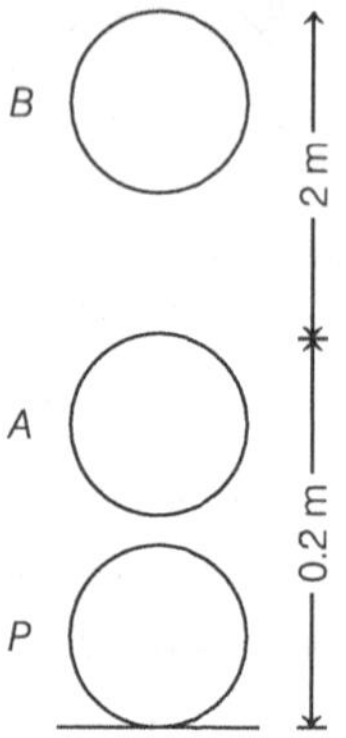

Then, for PA, $v^2 = 0^2 + 2a \times 0.2$

For AB, $0 = v^2 - 2 \times g \times 2$

$\Rightarrow$ $v^2 = 2g \times 2$

From above equations,

$$a = 10g = 100 \text{ ms}^{-2}$$

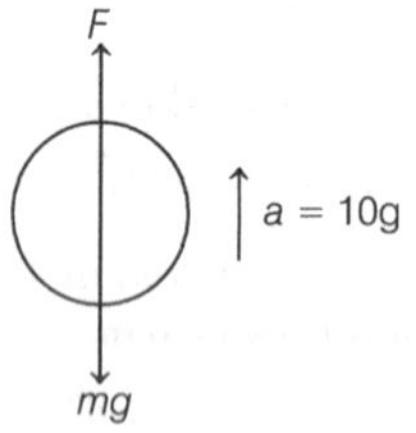

Then, for PA, FBD of ball is

$F - mg = ma$ (F is the force exerted by hand on ball)

$\Rightarrow$ $F = m(g + a) = 0.2\,(10 + 100) = 22$ N

16. Let L be the length and H be the height of the inclined plane, respectively. Acceleration of the block slide down the smooth inclined plane is

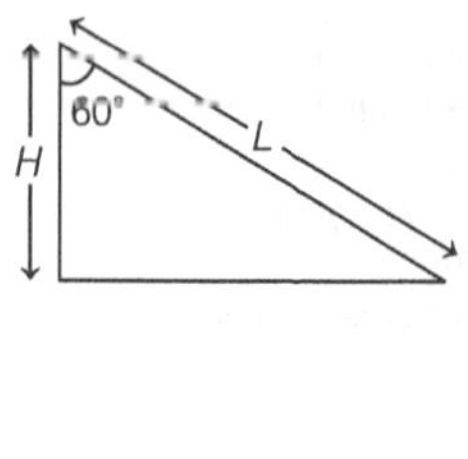

$$a = g\cos 60^\circ$$

$\therefore$ $L = \frac{1}{2}g\cos 60^\circ\, t_1^2$ ...(i) ($\because u = 0$)

Acceleration to another block dropped vertically down from the same inclined plane is

$$a = g$$

$\therefore$ $H = \frac{1}{2}at_2^2 = \frac{1}{2}gt_2^2$ ($\because u = 0$)

From figure, $\cos 60^\circ = \frac{H}{L}$

$\Rightarrow$ $H = L\cos 60^\circ$

$\therefore$ $L\cos 60^\circ = \frac{1}{2}gt_2^2$ ...(ii)

Divide Eq. (i) by Eq. (ii), we get

$$\frac{t_1^2 \cos 60^\circ}{t_2^2} = \frac{1}{\cos 60^\circ}$$

$\Rightarrow$ $\frac{t_1^2}{t_2^2} = \frac{1}{\cos^2 60^\circ} = \frac{4}{1}$

$\Rightarrow$ $\frac{t_1}{t_2} = \frac{2}{1}$

17. Let T be the tension in the string. Let a be the acceleration of the system. The equation of motion are

$Ma = Mg\sin 53^\circ - T$ ...(i)

and $Ma = T - mg\sin 37^\circ$ ...(ii)

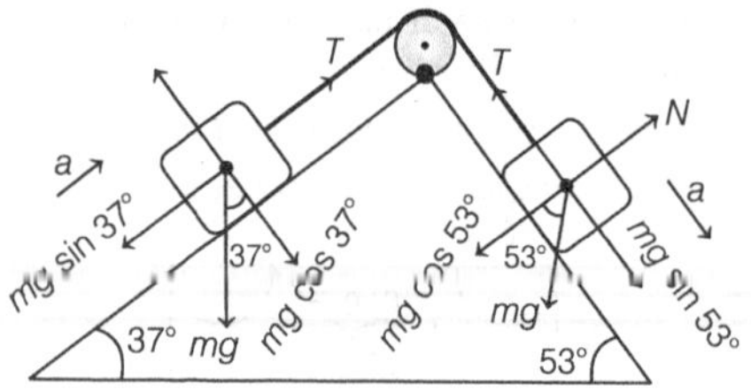

Adding Eqs. (i) and (ii), we get

$$a = \frac{mg\,(\sin 53^\circ - \sin 37^\circ)}{2m}$$

$$= g\cos 45^\circ \sin 8^\circ$$

$$\left[\because \sin A - \sin B = 2\cos\left(\frac{A+B}{2}\right)\sin\left(\frac{A-B}{2}\right)\right]$$

$$= 10 \times \frac{1}{\sqrt{2}} \times 0.139$$

$$= 0.98 \text{ ms}^{-1} \cong 1 \text{ ms}^{-2}$$

18. The free body diagram of the person can be drawn as

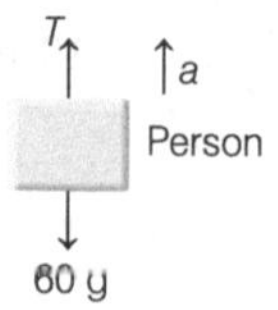

Let the person move up with an acceleration a, then

$$T - 60g = 60a$$

$$\Rightarrow \quad a_{max} = \frac{T_{max} - 60g}{60}$$

or $\quad a_{max} = \dfrac{360 - 60g}{60} \rightarrow -\text{ve value}$

That means, it is not possible to climb up on the rope.

Even in this problem, it is not possible to remain at rest on rope.

Hence, no option is correct.

But, if they will ask for the acceleration of climbing down, then

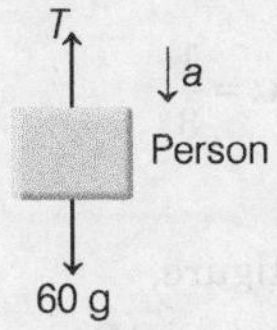

$$60g - T = 60a$$

$$\Rightarrow \quad 60g - T_{max} = 60a_{min}$$

or $\quad a_{min} = \dfrac{60g - 360}{60} = 4\ \text{ms}^{-2}$

19.

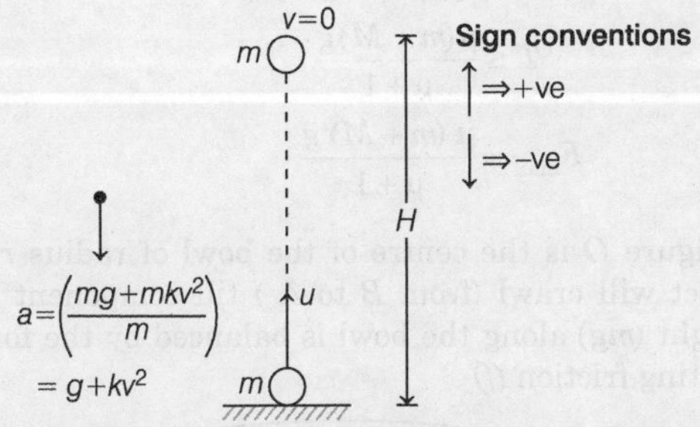

Net force on ball = Weight of ball + Resistive force

i.e. $\quad F_{net} = w + F_{resistive} = (-mg) + (-mkv^2)$

$$= -m(g + kv^2)$$

So, net acceleration of ball,

$$a = \frac{F_{net}}{m} = \frac{-m(g+kv^2)}{m} = -(g + kv^2)$$

$$\Rightarrow \quad v\frac{dv}{dy} = -(g+kv^2) \Rightarrow \frac{vdv}{(g+kv^2)} = -dy$$

Integrating both sides, we get

$$\int_u^0 \frac{v}{(g+kv^2)}dv = -\int_0^H dy \quad \text{...(i)}$$

Let $\quad g + kv^2 = t \Rightarrow 0 + 2kvdv = dt$

$$\Rightarrow \quad vdv = \frac{1}{2k}dt$$

Lower limit

If $v = u$, then $g + ku^2 = t \Rightarrow t = g + ku^2$

Upper limit

If $v = 0$, then $g + k(0)^2 = t \Rightarrow t = g$

Putting these values in Eq. (i), we get

$$\int_{g+ku^2}^{g} \frac{\frac{1}{2k}dt}{t} = -\int_0^H dy$$

$$\Rightarrow \quad \frac{1}{2k}\int_{g+ku^2}^{g} \frac{1}{t}dt = -[y]_0^H$$

$$\Rightarrow \quad \frac{1}{2k}[\ln(t)]_{g+ku^2}^{g} = -[H - 0]$$

$$\Rightarrow \quad \frac{1}{2k}[\ln(g) - \ln(g+ku^2)] = -[H]$$

$$\Rightarrow \quad -\frac{1}{2k}[\ln(g) - \ln(g+ku^2)] = H$$

$$\Rightarrow \quad H = \frac{1}{2k}[\ln(g+ku^2) - \ln(g)]$$

$$= \frac{1}{2k}\left[\ln\left(\frac{g+ku^2}{g}\right)\right] = \frac{1}{2k}\ln\left(1 + \frac{ku^2}{g}\right)$$

20. The free body diagram of the given system is given as

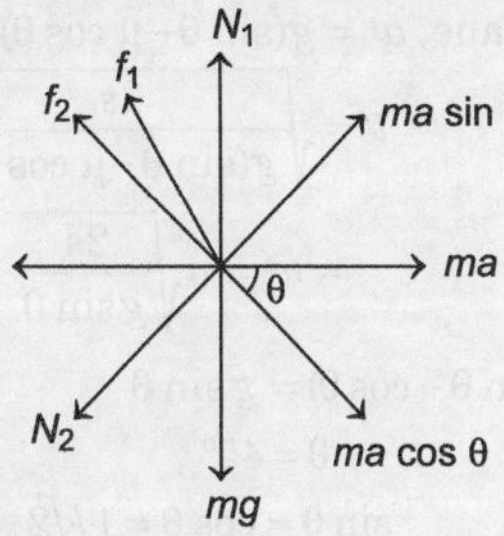

Given, $m = 1$ kg,

$$\mu = \frac{2}{5},\ a = 25\ \text{ms}^{-2},\ \cos\theta = \frac{4}{5},\ \sin\theta = \frac{3}{5}$$

Here, f_1 and f_2 are the two frictional forces corresponding to the two points of contact.

If a' be the acceleration of the block with respect to disc, then from above figure,

$$ma\cos\theta - f_1 - f_2 = ma'$$

$$\Rightarrow \quad ma\cos\theta - \mu N_1 - \mu N_2 = ma'$$

$$\Rightarrow \quad ma\cos\theta - \mu mg - \mu ma\sin\theta = ma'$$

$$\Rightarrow \quad a' = a\cos\theta - \mu g - \mu a\sin\theta$$

$$= 25 \times \frac{4}{5} - \frac{2}{5} \times 10 - \frac{2}{5} \times 25 \times \frac{3}{5}$$

$$= 20 - 4 - 6 = 10\ \text{ms}^{-2}$$

21. Figure shows free body diagram of the block,

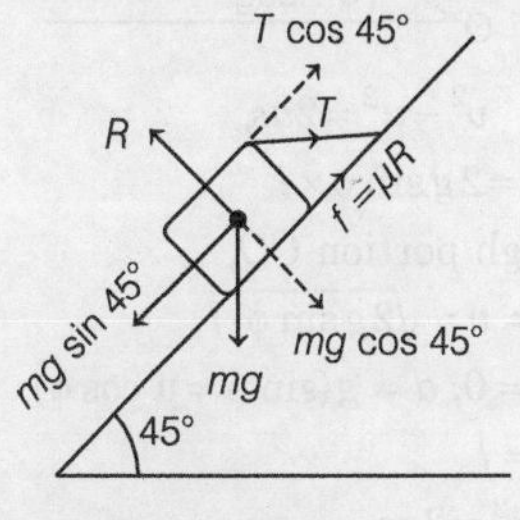

For equilibrium, along the plane,

$$\mu R + T\cos 45° = mg\sin 45°$$

$$\mu R + \frac{T}{\sqrt{2}} = \frac{mg}{\sqrt{2}} \quad \text{...(i)}$$

For equilibrium, in direction perpendicular to inclined plane,

$$R = T \sin 45° = mg \cos 45°$$

$$= \frac{T}{\sqrt{2}} = \frac{mg}{\sqrt{2}}$$

Put in Eq. (i), $\frac{\mu}{\sqrt{2}}(T + mg) = -\frac{1}{\sqrt{2}}(mg - T)$

$$\mu(50 + 15 \times 10) = (15 \times 10 - 50)$$

$$\mu = \frac{100}{200} = \frac{1}{2}$$

22. From $s = ut + \frac{1}{2}at^2 = 0 + \frac{1}{2}at^2 \Rightarrow t = \sqrt{\frac{2s}{a}}$

For smooth plane, $a = g \sin \theta$

For rough plane, $a' = g(\sin \theta - \mu \cos \theta)$

$$\therefore \quad t' = \sqrt{\frac{2s}{g(\sin \theta - \mu \cos \theta)}}$$

$$= nt = n\sqrt{\frac{2s}{g \sin \theta}}$$

$$\therefore \quad n^2 g(\sin \theta - \cos \theta) = g \sin \theta$$

When $\theta = 45°$,

$$\sin \theta = \cos \theta = 1/\sqrt{2}$$

Solving, we get $\mu = 1 - \frac{1}{n^2}$

23. Maximum force by surface when friction works,

$$F = \sqrt{f^2 + R^2} = \sqrt{\mu R^2 + R^2} = R\sqrt{\mu^2 + 1}$$

Maximum force $= R$ when there is no friction

Hence, ranging from R to $R\sqrt{\mu^2 + 1}$, we get

$$Mg \le f \le Mg\sqrt{\mu^2 + 1}$$

24. For the smooth portion BC,

$u = 0,\ s = l,\ g \sin \phi,\ u = ?$

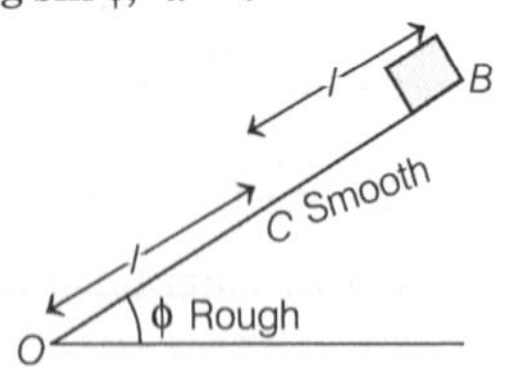

From $\quad v^2 - u^2 = 2as,$

$$v^2 - 0 = 2g \sin \phi \times l$$

For the rough portion CO,

$$u = v = \sqrt{2g \sin \phi \cdot l}$$

$$v = 0,\ a = g(\sin \phi = \mu \cos \phi)$$

$$s = l$$

From $v^2 - u^2 = 2as,$

$$0 - 2gl \sin \phi = 2g(\sin \phi - \mu \cos \phi)l$$

$$-\sin \phi = \sin \phi - \mu \cos \phi$$

$$\mu \cos \phi = 2 \sin \phi$$

$$\mu = 2 \tan \phi$$

25. As is from figure.

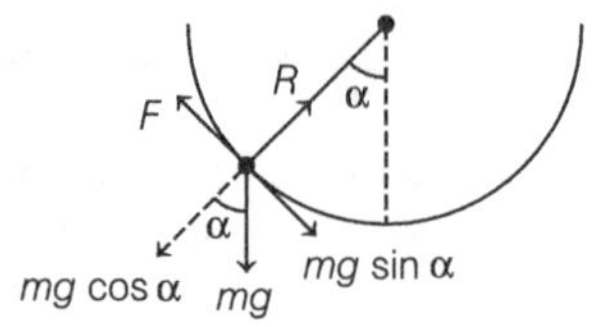

$$F = mg \sin \alpha$$

$$R = mg \cos \alpha$$

$$\frac{F}{R} = \tan \alpha$$

i.e. $\quad \mu = \tan \alpha = \frac{1}{3}$

$\therefore \quad \cot \alpha = 3$

26. As is clear from figure,

$$R + T = (m + M)g$$

$$R = (m + M)g - T$$

The system will not move till,

$$T \le F \text{ or } (T \le \mu R)$$

$$T \le \mu[(m + M)g = T]$$

$$T \le \frac{\mu(m + M)g}{\mu + 1}$$

$$\therefore \quad F_{max} = \frac{\mu(m + M)g}{\mu + 1}$$

27. In figure O is the centre of the bowl of radius r. The insect will crawl (from B to A) till component of its weight (mg) along the bowl is balanced by the force of limiting friction (f)

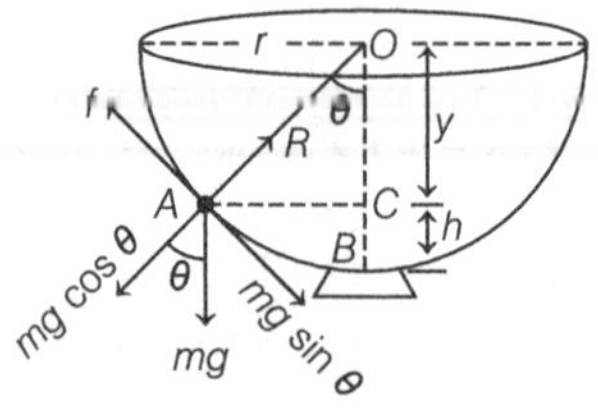

i.e $\quad mg \sin \theta = f = \mu R = \mu mg \cos \theta$

or $\quad \mu = \tan \theta = \frac{AC}{OC}$

$$= \sqrt{\frac{OA^2 - OC^2}{OC}} = \frac{\sqrt{r^2 - y^2}}{y}$$

or $\quad \mu^2 = \frac{r^2 - y^2}{y^2}$

$$\mu^2 y^2 + y^2 = r^2$$

$$y = \frac{r}{\sqrt{\mu^2 + 1}}$$

$$h = BC = OB - OC = r - y$$

$$= r - \frac{r}{\sqrt{\mu^2 + 1}} = r\left(1 - \frac{1}{\sqrt{\mu^2 + 1}}\right)$$

28. $$f_1 = \frac{mg}{\sqrt{2}} + \frac{\mu mg}{\sqrt{2}}$$

$$f_2 = \frac{mg}{\sqrt{2}} - \frac{\mu mg}{\sqrt{2}}$$

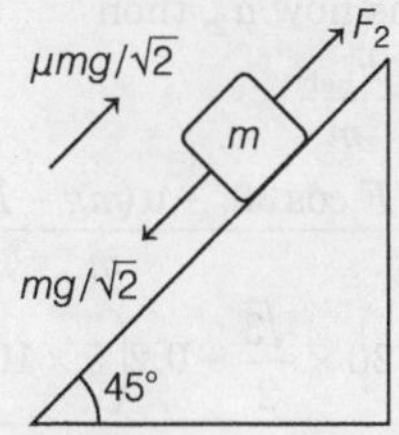

$$F_1 = 3\sqrt{2}$$

$$1 + \mu = 3 - 3\mu$$

$$\Rightarrow \quad 4\mu = 2$$

$$\Rightarrow \quad \mu = \frac{1}{2}$$

$$N = 10\mu$$

$$\Rightarrow \quad N = 5$$

29. Here, mass of the block, $m = 200$ kg, coefficient of static friction, $\mu_s = 0.5 = \frac{1}{2}$

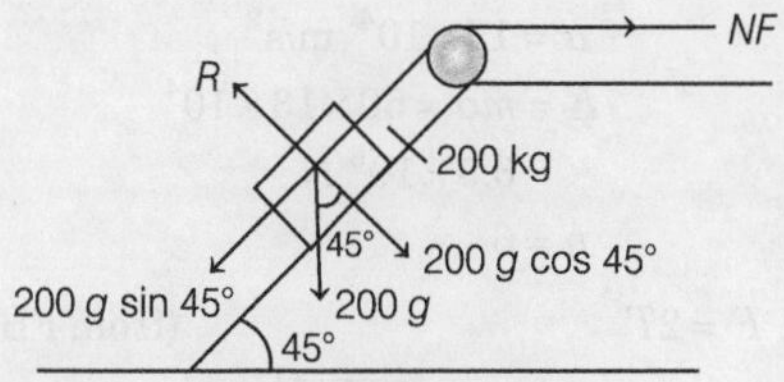

Angle to inclined plane, $\theta = 45°$

Maximum force that each man can apply, $F = 500$ N

Let N number of men are required for the block to just start moving up the plane,

$$NF = mg \sin\theta + f = mg \sin\theta + \mu_s R$$

$$= mg \sin\theta + \mu_s mg \cos\theta$$

$$= mg[\sin\theta + \mu_s \cos\theta]$$

$$NF = 200 \times 10 \left(\sin 45° + \frac{1}{2} \cos 45° \right)$$

$$= \frac{200 \times 10 \times 3}{2\sqrt{2}}$$

$$N = \frac{200 \times 10 \times 3}{2\sqrt{2} \times 500} = 4.2 \simeq 5$$

30. Block does not move upto a maximum applied force of 2 N down the inclined plane.

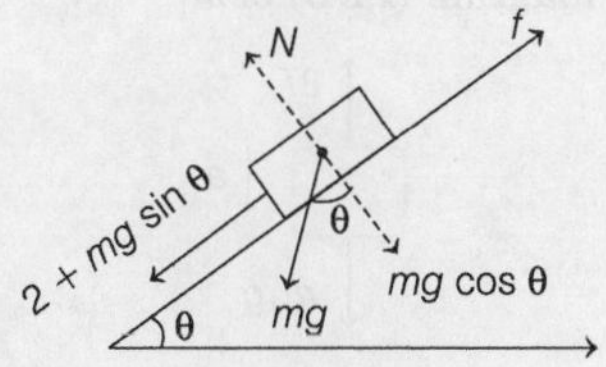

So, equating forces, we have

$$2 + mg \sin\theta = f$$

or $$2 + mg \sin\theta = \mu\, mg \cos\theta \quad \text{...(i)}$$

Similarly, block also does not move upto a maximum applied force of 10 N up the plane.

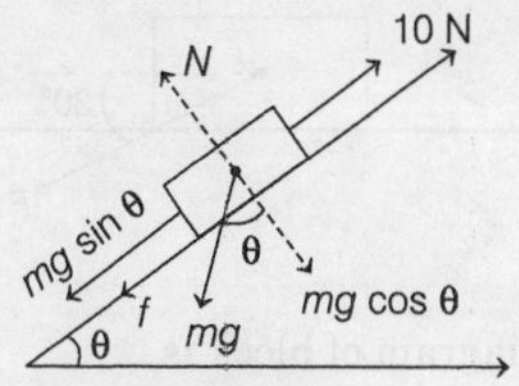

Now, equating forces, we have

$$mg \sin\theta + f = 10 \text{ N}$$

or $$mg \sin\theta + \mu mg \cos\theta = 10 \quad \text{...(ii)}$$

Now, solving Eqs. (i) and (ii), we get

$$mg \sin\theta = 4 \quad \text{...(iii)}$$

and $$\mu\, mg \cos\theta = 6 \quad \text{...(iv)}$$

Dividing Eq. (iii) by Eq. (iv), we get

$$\mu \cot\theta = \frac{3}{2}$$

$$\Rightarrow \quad \mu = \frac{3 \tan\theta}{2} = \frac{3 \tan 30°}{2}$$

$$\Rightarrow \quad \mu = \frac{\sqrt{3}}{2}$$

31. When friction is absent, $ma_1 = mg \sin\theta$

$$a = g \sin\theta$$

$$\therefore \quad s_1 = \frac{1}{2} a_1 t_1^2 \quad \text{...(i)}$$

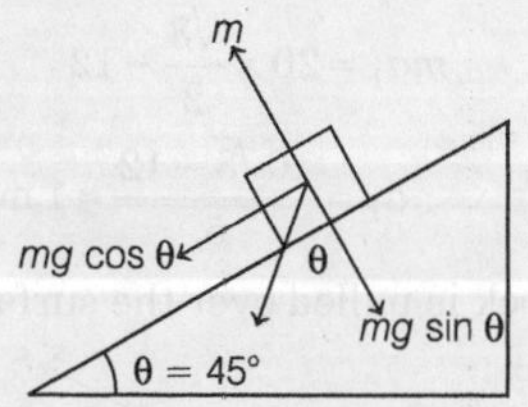

When friction is present, friction is in opposite to the direction of motion,

$$a_2 = g \sin\theta - \mu_k\, g \cos\theta$$

$$\therefore \quad s_2 = \frac{1}{2} a_2 t_2^2 \quad \text{...(ii)}$$

From Eqs. (i) and (ii), we get

$$\frac{1}{2} a_1 t_1^2 = \frac{1}{2} a_2 t_2^2$$

$$\Rightarrow \quad a_1 t_1^2 = a_2 (n\, t_1)^2 \qquad (\because\ t_2 = n\, t_1)$$

or $$a_1 = n^2 a_2$$

$$\Rightarrow \quad \frac{a_2}{a_1} = \frac{g \sin\theta - \mu_k g \cos\theta}{g \sin\theta} = \frac{1}{n^2}$$

or $$\frac{g \sin 45° - \mu_k g \cos 45°}{g \sin 45°} = \frac{1}{n^2}$$

or $$1 - \mu_k = \frac{1}{n^2}$$

or $$\mu_k = 1 - \frac{1}{n^2}$$

32. ***Case*** **I** Block is pushed over surface

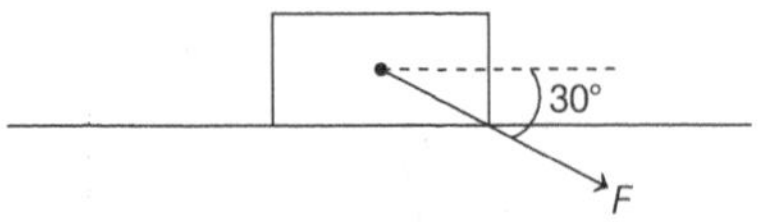

Free body diagram of block is

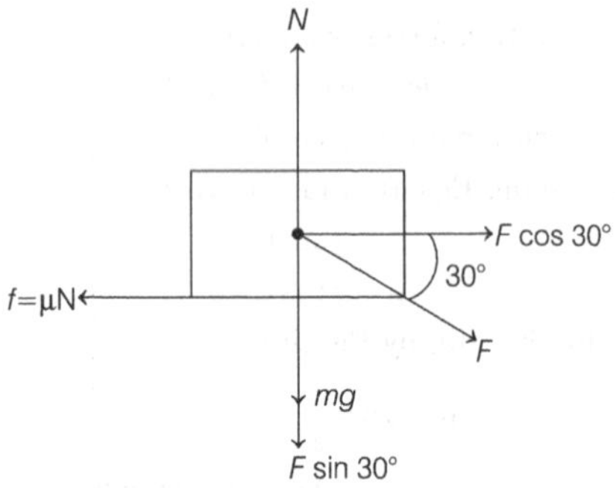

In this case, normal reaction,

$N = mg + F\sin 30° = 5 \times 10 + 20 \times \frac{1}{2} = 60$ N

(Given, $m = 5$ kg, $F = 20$ N)

Force of friction, $f = \mu N = 0.2 \times 60$ $(\because \mu = 0.2)$

$= 12$ N

So, net force causing acceleration (a_1) is

$$F_{net} = ma_1 = F\cos 30° - f$$

$$\Rightarrow \quad ma_1 = 20 \times \frac{\sqrt{3}}{2} - 12$$

$$\therefore \quad a_1 = \frac{10\sqrt{3} - 12}{5} \approx 1 \text{ ms}^{-2}$$

Case **II** Block is pulled over the surface

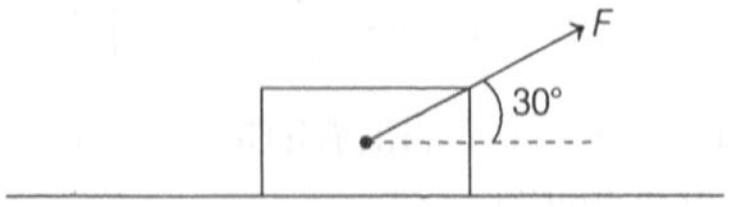

Free body diagram of block is,

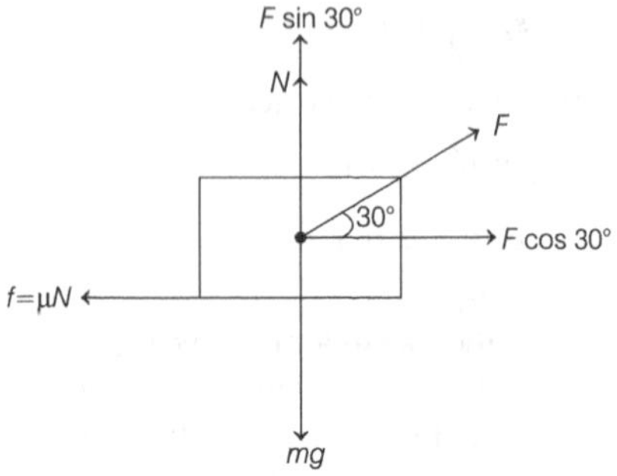

Net force causing acceleration is

$$F_{net} = F\cos 30° - f$$

$$= F\cos 30° - \mu N$$

$$\Rightarrow \quad F_{net} = F\cos 30° - \mu(mg - F\sin 30°)$$

If acceleration is now a_2, then

$$a_2 = \frac{F_{net}}{m}$$

$$= \frac{F\cos 30° - \mu(mg - F\sin 30°)}{m}$$

$$= \frac{20 \times \frac{\sqrt{3}}{2} - 0.2\left(5 \times 10 - 20 \times \frac{1}{2}\right)}{5}$$

$$= \frac{10\sqrt{3} - 8}{5}$$

$$\Rightarrow \quad a_2 \approx 1.8 \text{ ms}^{-2}$$

So, difference $= a_2 - a_1 = 1.8 - 1 = 0.8 \text{ ms}^{-2}$

33. With full barrel length, the muzzle velocity of block is 700 m/s and when 0.5 m is cut-off from the barrel , the muzzle velocity falls to 600 m/s. Hence, due to pressure of powder, the velocity increases from 600 m/s to 700 m/s over a distance of 0.5 m.

$$\therefore \quad 700^2 = 600^2 + 2 \times a \times 0.5$$

$$\therefore \quad a = 13 \times 10^4 \text{ m/s}^2$$

$$\therefore \quad F = ma = 50 \times 13 \times 10^4$$

$$= 6.5 \times 10^6 \text{ N}$$

$$\therefore \quad n = 6$$

34. Here, $F = 2T$ (from FBD of pulley)

$$\therefore \quad T = 5 \text{ N}$$

But, $$T = ma$$

$$\therefore \quad a = \frac{T}{m} = \frac{5}{1} = 5 \text{ m/s}^2$$

35. For m_2,

$$T - m_2 g = m_2(2a) \quad ...(i)$$

Free body diagram (FBD) of m_2,

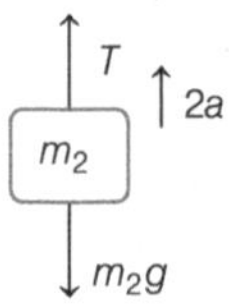

For m_1, $$m_1 g - 2m_2 = m_1 (a) \quad ...(ii)$$

Free body diagram (FBD) of m_1,

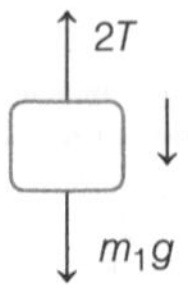

From Eqs. (i) and (ii), we get

$$m_1 g - 2m_2 g = 4m_2 a + m_1 a$$

$\Rightarrow \quad a = 0 \text{ m/s}^2$

or $\quad T = m_2 g = 10 \text{ N}$

For m, $2T = mg$

$\Rightarrow \quad m = \frac{20}{10} = 2 \text{ kg}$

Free body diagram of m,

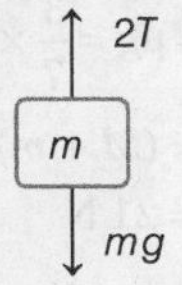

$$mg = 2T$$

$\Rightarrow \quad m = \frac{2 \times 10}{10} = 2 \text{ kg}$

36. Different forces acting on the inclined plane are shown below.

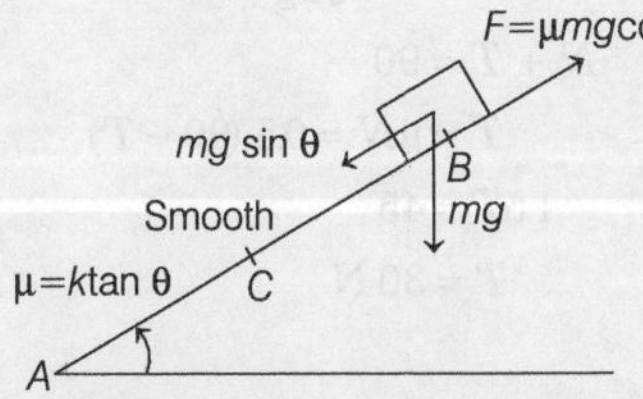

As block stops at point A, this means work done by component of weight down the plane is dissipated in doing work against friction.

$\Rightarrow \quad mg \sin\theta (AB) = \mu mg \cos\theta (AC)$

$\Rightarrow \quad mg \sin\theta (3AC) = \mu mg \cos\theta (AC)$

$\Rightarrow \quad 3 \tan\theta = \mu$

Given, $\mu = k \tan\theta$

Comparing both, we get

$\Rightarrow \quad k = 3$

37. According to the given situation,

When the minimum force F capable to topple the block is applied, the block will be on the verge of toppling.

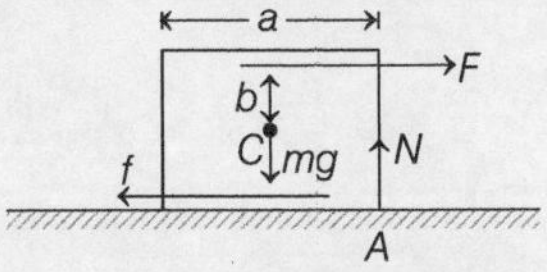

As the block is not moving, we have friction f such that

$$f = F \quad \text{... (i)}$$

Also, note that reaction N acts from point A as block is at the verge of toppling.

To maintain the equilibrium, net torque about centre of mass C is zero.

$\Rightarrow \quad F \cdot b + f\left(\frac{a}{2}\right) = N\left(\frac{a}{2}\right)$

Using result of Eq. (i), we get

$$f\left(\frac{a}{2} + b\right) = N\left(\frac{a}{2}\right)$$

Now, $f = \mu mg$ and $N = mg$

$\therefore \quad \mu mg\left(\frac{a}{2} + b\right) = mg\left(\frac{a}{2}\right)$

$\Rightarrow \quad \mu\left(\frac{a}{2} + b\right) = \frac{a}{2}$

As, $\quad \mu = 0.4$

$\therefore \quad \frac{b}{a} = \frac{3}{4}$

Hence, $\quad 100 \times \frac{b}{a} = 100 \times \frac{3}{4} = 75$

38. Let μ be the coefficient of kinetic friction between the block and the inclined plane.

The free body diagram of the given situation is shown below.

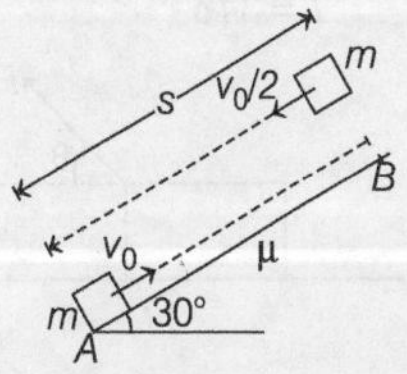

While going from A to B, the acceleration of the block is

$$a_1 = g \sin 30° + \mu g \cos 30°$$

$\Rightarrow \quad a_1 = \frac{g}{2} + \frac{\mu g\sqrt{3}}{2}$

$\Rightarrow \quad a_1 = 5 + 5\sqrt{3}\mu \quad \text{...(i)} \; (g = 10 \text{ ms}^{-2})$

Let s be the distance between A and B.

From third equation of motion,

$$v_0^2 - 0 = 2a_1 s$$

$\Rightarrow \quad s = \frac{v_0^2}{2a_1} = \frac{v_0^2}{2(5 + 5\sqrt{3}\mu)} \quad \text{...(ii)}$

[using Eq. (i)]

When the block comes back to its initial position (i.e. from B to A), its velocity is $\frac{v_0}{2}$.

So, the acceleration of the block while coming from B to A is

$$a_2 = g \sin 30° - \mu g \cos 30°$$

$\Rightarrow \quad a_2 = \frac{g}{2} - \frac{\mu g\sqrt{3}}{2}$

$\Rightarrow \quad a_2 = 5 - 5\sqrt{3}\mu \quad \text{...(iii)}$

Again, using third equation of motion, we get

$$\left(\frac{v_0}{2}\right)^2 - 0 = 2a_2 s$$

$\Rightarrow \quad \frac{v_0^2}{4} = 2a_2 s$

$$\Rightarrow \quad s = \frac{v_0^2}{8a_2}$$

$$\Rightarrow \quad s = \frac{v_0^2}{8(5 - 5\sqrt{3}\mu)} \quad \ldots\text{(iv)}$$

[using Eq. (iii)]

Equating Eqs. (ii) and (iv), we get

$$\frac{v_0^2}{2(5 + 5\sqrt{3}\mu)} = \frac{v_0^2}{8(5 - 5\sqrt{3}\mu)}$$

$$\Rightarrow \quad 4(5 - 5\sqrt{3}\mu) = 5 + 5\sqrt{3}\mu$$

$$\Rightarrow \quad 20 - 20\sqrt{3}\mu = 5 + 5\sqrt{3}\mu$$

$$\Rightarrow \quad 25\sqrt{3}\mu = 15$$

$$\Rightarrow \quad \mu = \frac{15}{25\sqrt{3}} = \frac{3}{5\sqrt{3}}$$

$$\Rightarrow \quad \mu = \frac{\sqrt{3}}{5}$$

$$\Rightarrow \quad \mu = 0.346 = \frac{346}{1000}$$

$$\therefore \quad I = 346$$

39.

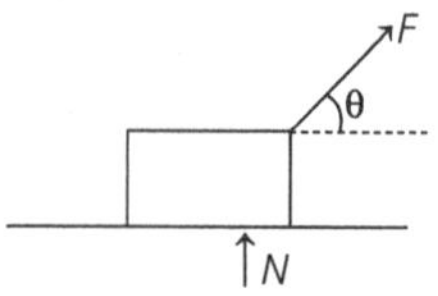

$$F\cos\theta = \mu N$$

$$F\sin\theta + N = mg$$

$$\Rightarrow \quad F = \frac{\mu mg}{\cos\theta + \mu\sin\theta}$$

$$F_{\text{min}} = \frac{\mu mg}{\sqrt{1 + \mu^2}} = \frac{\frac{1}{\sqrt{3}} \times 10}{\frac{2}{\sqrt{3}}} = 5\text{ N}$$

40.

$$a_{\text{max}} = \mu g = \frac{3}{7} \times 9.8$$

$$F = (M + m)\, a_{\text{max}} = 5\, a_{\text{max}}$$

$$= 21\text{ N}$$

41.

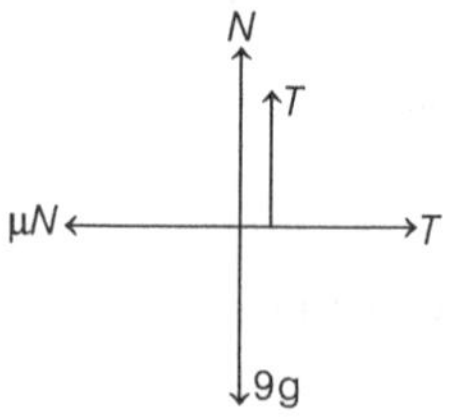

$$N + T = 90$$

$$T = \mu N = 0.5\,(90 - T)$$

$$1.5T = 45$$

$$T = 30\text{ N}$$

07

Work, Energy and Power

The terms work, energy and power are frequently used in everyday language. In physics, however the word *work* converse a definite and precise meaning. *Energy* is our capacity to do work and it is related to work. In other words, an object with lots of energy can do lots of work. The term *power* is usually associated with the time in which the work is done, *i.e.* the rate at which work is done.

IN THIS CHAPTER ....

- Work
- Conservative and Non-Conservative Forces
- Energy
- Work-Energy Theorem
- Conservation of Mechanical Energy
- Power

Work

The scientific meaning of work is transfer of energy by mechanical means. Work is said to be done by a force, when a body a displaced actually through some distance in the direction of applied force. The SI unit of work is joule (J) and in CGS, it is erg.

$$1 \text{ joule (J)} = 10^7 \text{ erg}$$

Work Done by a Constant Force

The work done by a constant force F on a particle which undergoes displacement s is given by

$W = \mathbf{F} \cdot \mathbf{s} = F\, s \cos\theta$, where θ is the angle between **F** and **s**.

Work done is a scalar or dot product of **F** and **s**.

Work can be *positive*, *negative* or even *zero*, also depending on the angle (θ) between the force vector **F** and displacement vector **s**. Work done by a force is zero when $\theta = 90°$, it is positive when $\theta < 90°$ and negative when $\theta > 90°$.

For example, when a person lifts a body, the work done by the lifting force is positive (as $\theta = 0°$) but work done by the force of gravity is negative (as $\theta = 180°$). Similarly, work done by centripetal force is always zero (as $\theta = 90°$).

Work Done by a Variable Force

The force is said to be variable force, if it changes its direction or magnitude or both. The work done by a variable force can be calculated as

$$W = \int_{r_1}^{r_2} \mathbf{F} \cdot d\mathbf{r}$$

where integration is performed along the path of particle and $d\mathbf{r}$ is the position vector of the particle.

If the particle moves from $\mathbf{r}_1(x_1, y_1, z_1)$ to $\mathbf{r}_2(x_2, y_2, z_2)$, then the work done by force $\mathbf{F}$ is given by

$$W = \int dW = \int_{x_1}^{x_2} F_x dx + \int_{y_1}^{y_2} F_y dy + \int_{z_1}^{z_2} F_z dz$$

where, F_x, F_y and F_z are the rectangular components of force in x, y and z-directions, respectively.

If the motion is one dimensional, $W = \int_{x_i}^{x_f} F_x dx$ in which F_x is the component of force along motion.

Example 1. *When a force is applied on a moving body, its motion is retarded, then the work done is*

(a) positive (b) negative
(c) zero (d) positive and negative

Sol. (b) The angle between the displacement and the applied retarded force is 180°. $\{\because \cos 180° = -1\}$

$\therefore$ Work done $= Fs \cos 180° = -Fs$
= negative

Example 2. *A force $\mathbf{F} = -(y\hat{\mathbf{i}} + x\hat{\mathbf{j}})$ acts on a particle moving in the XY-plane starting from the origin, the particle is taken along the positive X-axis to the point $(2a, 0)$ and then parallel to the Y-axis to the point $(2a, 2a)$. The total work done on the particle is*

(a) $-4a^2$ (b) $-2a^2$ (c) $4a^2$ (d) $2a^2$

Sol. (a) Here, $\mathbf{F} = -(y\hat{\mathbf{i}} + x\hat{\mathbf{j}})$

The given situation is shown in the figure. According to question, particle is taken from origin point O to A and then point A to B.

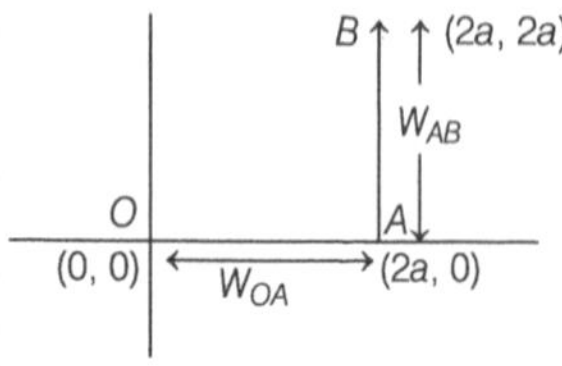

Hence, total work done,

$$W = W_{OA} + W_{AB} \quad \text{...(i)}$$

$\therefore$ $W_{OA} = \int_0^{2a} \mathbf{F} \cdot \mathbf{dx}$ [only x-coordinate is varying]

$$= \int_0^{2a} -(0 \cdot \hat{\mathbf{i}} + x\hat{\mathbf{j}})\, dx\, \hat{\mathbf{i}}$$

$$= \int_0^{2a} x dx (\hat{\mathbf{j}} \cdot \hat{\mathbf{i}})$$

$W_{OA} = 0$ $[\because \hat{\mathbf{j}} \cdot \hat{\mathbf{i}}] = 0$

$\because$ $W_{AB} = \int_0^{2a} -(y\hat{\mathbf{i}} + x\hat{\mathbf{j}})\,(dy\hat{\mathbf{j}})$

[x-coordinate is constant $\therefore x = 2a$]

$$= \int_0^{2a} -(y\hat{\mathbf{i}} + 2a\hat{\mathbf{j}})\, dy\, \hat{\mathbf{j}}$$

$$= \int_0^{2a} (-y\, \hat{\mathbf{i}} \cdot \hat{\mathbf{j}})\, dy - \int_0^{2a} 2a\, dy\, (\hat{\mathbf{j}} \cdot \hat{\mathbf{j}})$$

$\Rightarrow$ $W_{AB} = 0 - 2a\,[y]_0^{2a} = -2a \cdot 2a - 4a^2$

$\therefore$ $W = W_{OA} + W_{AB} = 0 - 4a^2 = -4a^2$

Example 3. *A force $F = (2 + x)$ acts on a particle in x-direction, where F is in newton and x in metre. The work done by this force during a displacement from $x = 1.0$ m to $x = 2.0$ m is*

(a) 2 J (b) 3.5 J
(c) 4.5 J (d) None of these

Sol. (b) The work done in small displacement from x to $x + dx$ is

$$dW = Fdx = (2 + x)\, dx$$

Hence, $W = \int_1^2 dW = \int_1^2 (2 + x)\, dx = \int_1^2 2\, dx + \int_1^2 x dx$

$$= \left[2x + \frac{x^2}{2}\right]_1^2 = 3.5 \text{ J}$$

Example 4. *A body of mass 2kg initially at rest moves under the action of an applied horizontal force of 7 N on a table with coefficient of kinetic friction = 0.1. The work done by applied force in 10 s will be equal to*

(a) 800 J (b) 825 J (c) 882 J (d) 856 J

Sol. (c) Here, $m = 2$ kg, $u = 0$, $F = 7$ N, $\mu = 0.1$, $t = 10$ s

Acceleration produced by applied force,

$$a_1 = \frac{F}{m} = \frac{7}{2} = 3.5 \text{ ms}^{-2}$$

Force of friction, $f = \mu R = \mu mg = 0.1 \times 2 \times 9.8 = 1.96$ N

Retardation produced by friction,

$$a_2 = \frac{-f}{m} = -\frac{1.96}{2} = -0.98 \text{ ms}^{-2}$$

Net acceleration with which body moves,

$$a = a_1 + a_2 = 3.5 - 0.98 = 2.52 \text{ ms}^{-2}$$

Distance moved by the body in 10 s,

$$s = ut + \frac{1}{2}at^2 = 0 + \frac{1}{2} \times 2.52 \times 10^2 = 126 \text{ m}$$

$\therefore$ Work done by the applied force, $W = F \times s = 7 \times 126 = 882$ J

Conservative and Non-Conservative Forces

Conservative Forces

A force is said to be conservative, if work done by or against the force in moving a body depends only on the initial and final positions of the body and not the nature of path followed between the final and initial positions.

For example, gravitational force, electrostatic force, etc., are conservative forces.

Non-Conservative Forces

A force is said to be non-conservative, if work done by or against the force in moving a body from one position to another, depends on the path followed between these two positions.

For example, force of friction and viscous force are non-conservative forces.

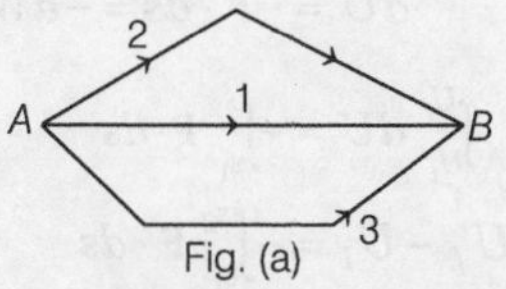

Fig. (a)

Fig. (a) Let W_1, W_2, W_3 denote the amounts of work done moving a body from A to B along three different paths 1, 2, 3, respectively. If the force is non-conservative, then $W_1 \neq W_2 \neq W_3$.

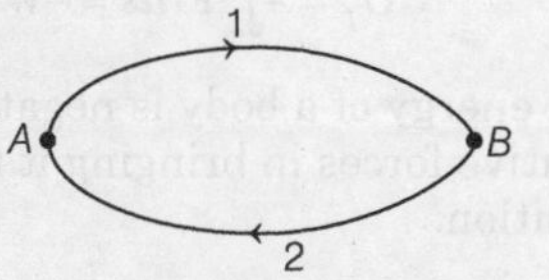

Fig. (b)

Fig. (b) Shows that a particle moving a closed path. $A \to 1 \to B \to 2 \to A$. If W_1 is work done in moving the particle from $A \to 1 \to B$ and W_2 is work done in moving the particle from $B \to 2 \to A$, then for a non-conservative force $|W_1| \neq |W_2|$.

$\therefore$ Net work done along the closed path, $A \to B \to A$ is not zero.

i.e. $$\oint \mathbf{F} \cdot d\mathbf{s} \neq 0$$

Example 5. *A particle is taken from point A to point B via the path ACB and then come back to point A via the path BDA. What is the work done by gravity on the body over this closed path, if the motion of the particle is in the vertical plane?*

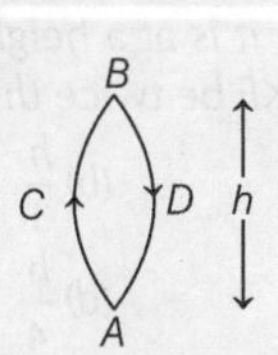

(a) mgh (b) $-mgh$

(c) $\frac{1}{2}mgh$ (d) zero

Sol. (d) Here, displacement of the particle is **AB**, gravity is acting vertically downwards. The vertical component of **AB** is h upwards, hence

$$W_{(ACB)} = -mgh$$

For the path BDA, component of the displacement acting along vertical direction is h (downward).

In this case, $W_{(BDA)} = mgh$

Total work done, $W_{ACB} + W_{BDA} = 0$

Energy

Capacity or ability of a body to do work is called its energy. Like work, energy is a scalar quantity. The units of measurement of energy are same as the units of work. In SI, the unit of energy of joule (J) and in CGS, the unit of energy is erg.

There are so many types of energy. *e.g.* Kinetic, potential, electrostatic, magnetic, geothermal, elastic, solar, etc. In this chapter, we will discuss only mechanical energy. Mechanical energy consists of kinetic energy and potential energy.

$$\text{ME} = \text{KE} + \text{PE}$$

Note KE is always positive but PE may be positive, negative or zero. Infact, when forces involved are repulsive, PE is positive and when forces involved are attractive, PE is negative.

Kinetic Energy

Kinetic energy (KE) is the capacity of a body to do work by virtue of its motion. The faster the object moves, the greater is the kinetic energy. When the object is stationary, its kinetic energy is zero.

An object of mass m moving with velocity $\mathbf{v}$, has a kinetic energy,

$$\text{KE} = \frac{1}{2}mv^2 \qquad \left[\frac{1}{2}mv^2 = \frac{1}{2}m(|\mathbf{v}|)^2\right]$$

- Kinetic energy is corelated with momentum as

$$K = \frac{p^2}{2m}$$

or $$p = \sqrt{2mK}$$

- Kinetic energy for a system of particle will be

$$K = \frac{1}{2}\sum_i m_i v_i^2$$

Example 6. *A 120 g mass has a velocity* $\mathbf{v} = (2\,\hat{\mathbf{i}} + 5\,\hat{\mathbf{j}})\,ms^{-1}$ *at a certain instant, KE of the body at that instant is*

(a) 3.0 J (b) 1.74 J

(c) 4.48 J (d) 5.84 J

Sol. (b) Here, $m = 120\text{ g} = 0.12\text{ kg}$,

$$\mathbf{v} = (2\hat{\mathbf{i}} + 5\hat{\mathbf{j}})\text{ ms}^{-1}$$

$$v = |\mathbf{v}| = \sqrt{2^2 + 5^2} = \sqrt{29}\text{ ms}^{-1}$$

$$\therefore \quad \text{KE} = \frac{1}{2}mv^2 = \frac{1}{2} \times 0.12 \times 29 = 1.74\text{ J}$$

Example 7. *If the linear momentum of a body is increased by 50%, then the kinetic energy of that body increases by*

(a) 100% (b) 125%

(c) 225% (d) 25%

Sol. (b) Kinetic energy of the body,

$$K = \frac{p^2}{2m}$$

Since the mass remains constant, so $K \propto p^2$.

So, $$\frac{K_2}{K_1} = \frac{p_2^2}{p_1^2} = \left[\frac{150}{100}\right]^2 = \frac{9}{4}$$

Thus, $$\left(\frac{K_2}{K_1} - 1\right) \times 100 = \left(\frac{9}{4} - 1\right) \times 100 = 125\%$$

Example 8. *A body is projected horizontally with a velocity of u ms^{-1} at an angle β with the horizontal. The kinetic energy at the highest point is (3/4)th of the initial kinetic energy. The value of β is*

(a) 30° (b) 45°
(c) 60° (d) 120°

Sol. (a) The kinetic energy at the highest point would be equal to $\frac{1}{2}m(u\cos\beta)^2$ as the vertical component of the velocity is zero.

$$\Rightarrow \quad \text{KE} = \frac{1}{2}mu^2\cos^2\beta$$

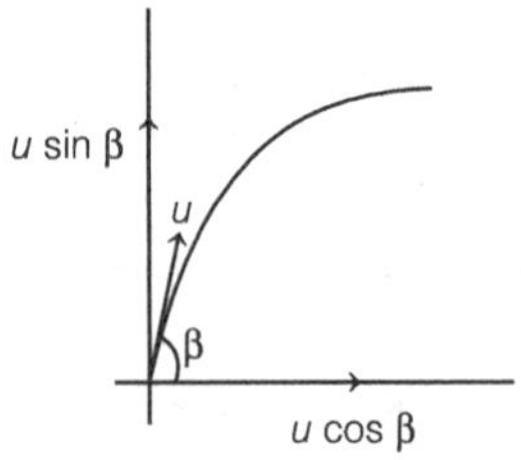

The initial kinetic energy (K) is the maximum kinetic energy.

So, $$\text{KE} = K\cos^2\beta$$

Given, $$K\cos^2\beta = \frac{3}{4}K$$

$$\Rightarrow \quad \cos\beta = \frac{\sqrt{3}}{2}$$

So, $$\beta = 30°$$

Example 9. *At time $t = 0$, particle starts moving along the X-axis. If its kinetic energy increases uniformly with time t, the net force acting on it must be proportional to*

(a) $\sqrt{t}$ (b) constant
(c) t (d) $1/\sqrt{t}$

Sol. (d) Given, $K \propto t$

$$\Rightarrow \quad \frac{dK}{dt} = \text{constant}$$

$$\Rightarrow \quad K \propto t$$

$$\frac{1}{2}mv^2 \propto t$$

$$\Rightarrow \quad v \propto \sqrt{t}$$

Also, $P = Fv = \frac{dK}{dt} = \text{constant}$

$$\Rightarrow \quad F \propto \frac{1}{v}$$

$$\Rightarrow \quad F \propto \frac{1}{\sqrt{t}}$$

Potential Energy

Potential energy (PE) of a body is the energy stored in a body or a system by virtue of its position of configuration in a field.

The change in potential energy (dU) of a system corresponding to a conservative force is

$$dU = -\mathbf{F}\cdot d\mathbf{s} = -dW \qquad \left(\because F = -\frac{dU}{ds}\right)$$

$$\int_{U_i}^{U_f} dU = -\int_{\mathbf{s}_1}^{\mathbf{s}_2} \mathbf{F}\cdot d\mathbf{s}$$

or $$U_f - U_i = -\int_{\mathbf{s}_1}^{\mathbf{s}_2} \mathbf{F}\cdot d\mathbf{s}$$

We generally choose the reference point at infinity and assume potential energy to be zero at that point.

i.e. $\mathbf{s}_1 = \infty$ and $U_i = 0$

then $$U_f = -\int_{\infty}^{\mathbf{s}} \mathbf{F}\cdot d\mathbf{s} = -W$$

Thus, potential energy of a body is negative of work done by the conservative forces in bringing it from infinity to the present position.

Potential Energy of a Spring

Whenever an elastic body (say a spring) is either stretched or compressed, work is being done against the elastic spring force. The work done is $W = \frac{1}{2}kx^2$, where k is spring constant and x is the displacement.

Elastic potential energy, $U = \frac{1}{2}kx^2$

If spring is stretched from initial position x_1 to final position x_2, then

Work done = Increment in elastic potential energy

$$= \frac{1}{2}k(x_2^2 - x_1^2)$$

Example 10. *When a body is projected vertically up, its PE is twice its KE, when it is at a height h above the ground. At what height, will its KE be twice the KE?*

(a) $2h$ (b) $\frac{h}{3}$
(c) $\frac{h}{2}$ (d) $\frac{h}{4}$

Sol. (c) Total energy,

$$E_1 = \text{PE} + \text{KE}$$

$$= \text{PE} + \frac{1}{2}\text{PE} = \frac{3}{2}\text{PE} = \frac{3}{2}mgh$$

$$E_2 = \text{PE} + \text{KE} = \text{PE} + 2\text{PE} + 3\text{PE} = 3mgh'$$

As, $$E_2 = E_1$$

$$3mgh' = \frac{3}{2}mgh$$

$$\Rightarrow \quad h' = \frac{h}{2}$$

Example 11. *A person trying to lose weight by burning fat lifts a mass of 10 kg upto a height of 1 m, 1000 times. Assume that, the potential energy lost each time he lowers the mass is dissipated. How much fat will he use up considering the work done only when the weight is lifted up? Fat supplies 3.8×10^7 J of energy per kg which is converted into mechanical energy with a 20% of efficiency rate. (Take, $g = 9.8\ ms^{-2}$)* **[JEE Main 2016]**

(a) 2.45×10^{-3} kg (b) 6.45×10^{-3} kg
(c) 9.89×10^{-3} kg (d) 12.89×10^{-3} kg

Sol. (d) Given, potential energy burnt by lifting weight

$$= mgh = 10 \times 9.8 \times 1 \times 1000 = 9.8 \times 10^4$$

If mass lost by a person be m, then energy dissipated

$$= m \times \frac{2}{10} \times 3.8 \times 10^7\ \text{J}$$

$$\Rightarrow \quad 9.8 \times 10^4 = m \times \frac{1}{5} \times 3.8 \times 10^7$$

$$\Rightarrow \quad m = \frac{5}{3.8} \times 10^{-3} \times 9.8 = 12.89 \times 10^{-3}\ \text{kg}$$

Example 12. *A spring of spring constant 5×10^3 N/m is stretched initially by 5 cm from the unstretched position. Then, the work required to stretch it further by another 5 cm is*

(a) 12.50 N-m (b) 18.75 N-m
(c) 25.00 N-m (d) 6.25 N-m

Sol. (b) Work done to stretch the spring by 5 cm from mean position,

$$W_1 = \frac{1}{2} k x_1^2$$

$$= \frac{1}{2} \times 5 \times 10^3 \times (5 \times 10^{-2})^2 = 6.25\ \text{J}$$

Work done to stretch the spring by 10 cm from mean position,

$$W_2 = \frac{1}{2} k(x_1 + x_2)^2$$

$$= \frac{1}{2} \times 5 \times 10^3\ (5 \times 10^{-2} + 5 \times 10^{-2})^2 = 25\ \text{J}$$

Net work done to stretch the spring from 5 cm to 10 cm

$$= W_2 - W_1 = 25 - 6 \cdot 25$$
$$= 18.75\ \text{J}$$
$$= 18.75\ \text{N-m}$$

Work-Energy Theorem

Work done by all the forces (conservative or non-conservative, external or internal) acting on a particle or an object is equal to the change in its kinetic energy of the particle. Thus, we can write,

$$W = \Delta K = K_f - K_i = \frac{1}{2} m(v_f^2 - v_i^2)$$

We can also write, $K_f = K_i + W$

Note In a non-inertial frame, work done by all the forces (including the pseudo forces) = change in kinetic energy in non-inertial frame.

Example 13. *A particle of mass 0.5 kg travels in a straight line with velocity $v = ax^{3/2}$, where $a = 5 m^{-1/2} s^{-1}$. The work done by the net force during its displacement from $x = 0$ to $x = 2$ m is*

(a) 50 J (b) 45 J
(c) 25 J (c) None of these

Sol. (a) Here, $m = 0.5$ kg,

$$v = ax^{3/2},\ a = 5\ m^{-1/2}\ s^{-1}, W = ?$$

Initial velocity (at $x = 0$), $v_1 = a \times 0 = 0$

Final velocity (at $x = 2$), $v_2 = 5 \times 2^{3/2}$

$$\text{Work done} = \text{Increase in KE}$$

$$= \frac{1}{2} m(v_2^2 - v_1^2)$$

$$W = \frac{1}{2} \times 0.5\ [(5 \times 2^{3/2})^2 - 0] = 50\text{J}$$

Example 14. *A particle which is experiencing a force, is given by $\mathbf{F} = 3\hat{\mathbf{i}} - 12\hat{\mathbf{j}}$, undergoes a displacement of $\mathbf{d} = 4\hat{\mathbf{i}}$. If the particle had a kinetic energy of 3 J at the beginning of the displacement, what is its kinetic energy at the end of the displacement?* **[JEE Main 2019]**

(a) 9 J (b) 15 J (c) 12 J (d) 10 J

Sol. (b) We know that, work done in displacing a particle at displacement **d** under force **F** is given by

$$\Delta W = \mathbf{F} \cdot \mathbf{d}$$

By substituting given values, we get

$$\Rightarrow \quad \Delta W = (3\hat{\mathbf{i}} - 12\hat{\mathbf{j}}) \cdot (4\hat{\mathbf{i}})$$

$$\Rightarrow \quad \Delta W = 12\ \text{J} \quad \text{... (i)}$$

Now, using work-energy theorem, we get

work done (ΔW) = change in kinetic energy (ΔK)

$$\text{or} \quad \Delta W = K_2 - K_1 \quad \text{... (ii)}$$

Comparing Eqs. (i) and (ii), we get

$$K_2 - K_1 = 12\ \text{J}$$

$$\text{or} \quad K_2 = K_1 + 12\ \text{J}$$

Given, initial kinetic energy, $K_1 = 3$ J

$\therefore$ Final kinetic energy, $K_2 = 3\ \text{J} + 12\ \text{J} = 15\ \text{J}$

Example 15. *A block of mass $m = 1$ kg moving on a horizontal surface with speed $v = 2\ ms^{-1}$ enters a rough patch ranging from $x = 0.10$ m to $x = 2.01$ m. The retarding force F_r on the block in this range inversely proportional to x over this range*

$$F_r = -\frac{k}{x} \quad \text{for } 0.1 < x < 2.01\ m$$

$$= 0 \quad \text{for } x < 0.1\ m \text{ and } x > 2.01\ m$$

where, $k = 0.5$ J. The final kinetic energy of the block as it crosses this patch is

(a) 5 J (b) 50 J
(c) 0.5 J (d) 500 J

Sol. (c) If K_i and K_f are initial and final kinetic energies corresponding to x_i and x_f, then

$$K_f - K_i = \int_{x_i}^{x_f} F\,dx$$

$$K_f = K_i + \int_{0.1}^{2.01} \frac{(-k)}{x}\,dx$$

$$K_f = \frac{1}{2}mv_i^2 - k\ln(x)\Big|_{0.1}^{2.01}$$

$$= 2 - 0.5\ln(20.1)$$

$$= 2 - 1.5$$

$$= 0.5\text{ J}$$

Conservation of Mechanical Energy

The mechanical energy E of a system is the sum of its kinetic energy K and its potential energy U.

$$E = K + U$$

When the forces acting on the system are conservative in nature, the mechanical energy of the system remains constant.

$$K + U = \text{constant}$$

$$\Rightarrow \quad \Delta K + \Delta U = 0$$

There are physical situations, where one or more non-conservative force act on the system but net work done by them is zero, then the mechanical energy of the system remains constant.

If $\Sigma W_{net} = 0$, then mechanical energy, E = constant.

Example 16. *In the arrangement shown in figure, string is light and inextensible and friction is absent everywhere.*

The speed of both the blocks after the block A has ascend a height of 1 m will be

(Given that, $m_A = 1$ kg and 2 kg)

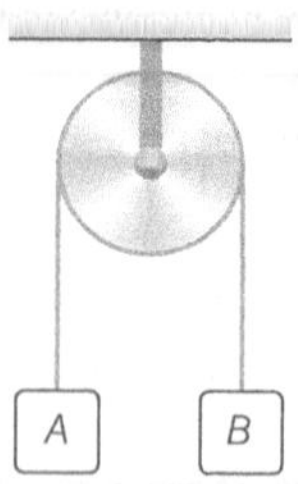

(a) 2 ms^{-1} (b) 2.58 ms^{-1}
(c) 3 ms^{-1} (c) 3.58 ms^{-1}

Sol. (b) Since, there is no friction anywhere, so mechanical energy will be conserved.

Here, the speed of both the blocks are same. Let the speed be $v\text{ ms}^{-1}$. Since block of 2 kg is coming down, hence the gravitational potential energy is decreasing while the gravitational potential energy of 1 kg block is increasing.

So, kinetic energy of both the blocks will increase.

Hence, $\quad m_B gh = m_A gh + \frac{1}{2}m_A v^2 + \frac{1}{2}m_B v^2$

or $\quad 2\times10\times1 = 1\times10\times1 + \frac{1}{2}\times1\times v^2 + \frac{1}{2}\times2v^2$

or $\quad 20 = 10 + 0.5v^2 + v^2$

or $\quad 1.5v^2 = 10$

$\therefore \quad v^2 = \frac{10}{1.5} = 6.67$

or $\quad v = 2.58\text{ ms}^{-1}$

Example 17. *The potential energy of a 1 kg particle free to move along the X-axis is given by*

$$V(x) = \left(\frac{x^4}{4} - \frac{x^2}{2}\right)\text{ J}$$

The total mechanical energy of the particle is 2 J, then the maximum speed (in ms^{-1}) is [AIEEE 2006]

(a) $\frac{3}{\sqrt{2}}$ (b) $\sqrt{2}$ (c) $\frac{1}{\sqrt{2}}$ (d) 2

Sol. (a) Given, $V(x) = \left(\frac{x^4}{4} - \frac{x^2}{2}\right)$

For minimum value of V, $\frac{dV}{dx} = 0$

$$\Rightarrow \quad \frac{4x^3}{4} - \frac{2x}{2} = 0$$

$$\Rightarrow \quad x = 0, \quad x = \pm 1$$

So, $\quad V_{min}(x = \pm 1) = \frac{1}{4} - \frac{1}{2} = \frac{-1}{4}\text{ J}$

Now, $K_{max} + V_{min}$ = Total mechanical energy

$$\Rightarrow \quad K_{max} = \left(\frac{1}{4}\right) + 2 \quad \text{or} \quad K_{max} = \frac{9}{4}$$

or $\quad \frac{mv^2}{2} = \frac{9}{4} \quad \text{or} \quad v = \frac{3}{\sqrt{2}}\text{ ms}^{-1}$

Example 18. *A spherical ball of mass 20 kg is stationary at the top of a hill of height 100 m. It rolls down a smooth surface to the ground, then climbs up another hill of height 30 m and finally rolls down to a horizontal base at a height of 20 m above the ground. The velocity attained by the ball is* [AIEEE 2005]

(a) 40 m/s (b) 20 m/s (c) 10 m/s (d) $10\sqrt{30}$ m/s

Sol. (a) According to conservation of energy, potential energy at height H is sum of kinetic energy and potential energy at h_2.

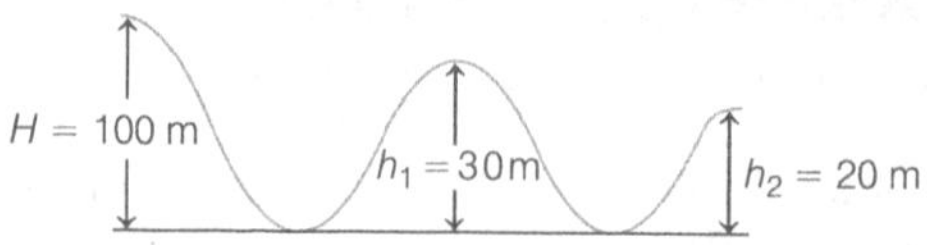

$$\therefore \quad mgH = \frac{1}{2}mv^2 + mgh_2$$

$$\Rightarrow \quad mg(H - h_2) = \frac{1}{2}mv^2$$

or $$v = \sqrt{2g(100 - 20)}$$

or $$v = \sqrt{2 \times 10 \times 80}$$

$$= 40 \text{ m/s}$$

Power

The time rate of doing work is called power. If an external force is applied to an object and if the work done by this force is ΔW in the time interval Δt, then the average power during this interval is defined as

$$P = \frac{\Delta W}{\Delta t}$$

The work done on the object contributes to increasing the energy of the object.

The instantaneous power is the limiting value of the **average power** as Δt approaches zero.

i.e. $$P = \lim_{\Delta t \to 0} \frac{\Delta W}{\Delta t} = \frac{dW}{dt}$$

where we have represented the infinitesimal value of the work done by dW.

$$P = \frac{dW}{dt} = \mathbf{F}\frac{ds}{dt} = \mathbf{F} \cdot \mathbf{v} \quad \left[\text{as, } \frac{ds}{dt} = \mathbf{v}\right]$$

- Power is equal to the scalar product of force and velocity.
- Power is a scalar with dimensions $[ML^2T^{-3}]$. The SI unit of power is Js^{-1} and is called **watt (W)**. Practical unit of power is **horse power (HP)**.

$$1 \text{ HP} = 746 \text{ W}$$

- Kilowatt-hour or watt-day are units of work or energy but not of power.

$$1 \text{ kWh} = 10^3 \times Js^{-1} \times (60 \times 60 \text{ s})$$
$$= 3.6 \times 10^6 \text{ J}$$

- The slope of work-time curve gives the instantaneous power as $P = \frac{dW}{dt} = \tan\theta$ [from Fig. (a)] while the area under P-t curve gives the work done.

Since, $P = \frac{dW}{dt}$, which means $W = \int P dt$ = area under P-t curve [as shown in Fig. (b)].

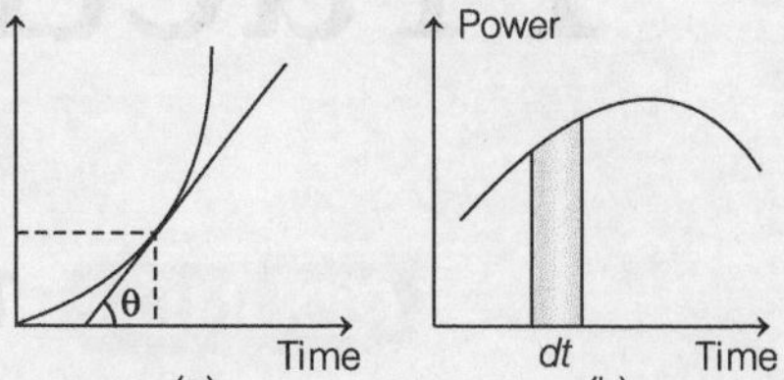

(a) (b)

Example 19. *An advertisement claims that a certain 1200 kg car can accelerate from rest to a speed of 25 ms^{-1} in a time of 8 s. What average power must the motor produce to cause this acceleration? (Ignore friction)*

(a) 45 kW (b) 45.9 kW
(c) 46.9 kW (d) None of these

Sol. (c) The work done in accelerating the car is given by

$$W = \Delta K = \frac{1}{2}m(v_f^2 - v_i^2) = \frac{1}{2}(1200)[(25)^2 - 0^2]$$

or $$W = 375 \text{ kJ}$$

$$\text{Power} = \frac{W}{t} = \frac{375}{8} = 46.9 \text{ kW}$$

Example 20. *A body of mass m is accelerated uniformly from rest to a speed v in a time T. The instantaneous power delivered to the body as a function of time, is given by* **[AIEEE 2005, 04]**

(a) $\frac{mv^2}{T^2}t$ (b) $\frac{mv^2}{T^2}t^2$
(c) $\frac{1}{2}\frac{mv^2}{T^2}t$ (d) $\frac{1}{2}\frac{mv^2}{T^2}t^2$

Sol. (a) A body of mass m with uniform acceleration, then force,

$$F = ma = \frac{mv}{T} \quad \left[\because a = \frac{v - 0}{T}\right]$$

$$\text{Instantaneous power} = Fv = mav = \frac{mv}{T} \cdot at$$

$$= \frac{mv}{T} \cdot \frac{v}{T} \cdot t = \frac{mv^2}{T^2}t$$

Practice Exercise

ROUND I Topically Divided Problems

Work Done by Variable and Constant Forces

1. The work done in pulling up a block of wood weighing 2 kN for a length of 10 m on a smooth plane inclined at an angle of 15° with the horizontal is [$\sin 15° = 0.2588$]

(a) 4.36 kJ (b) 5.13 kJ
(c) 8.91 kJ (d) 9.82 kJ

2. A block of mass 10 kg slides down a rough slope which is inclined at an angle of 45° to the horizontal. The coefficient of sliding friction is 0.30. When the block has slide 5 m, the work done on the block by the force of friction is nearly

(a) 115 J (b) $-75\sqrt{2}$ J
(c) 321.4 J (d) – 321.4 J

3. A position-dependent force $F = 3x^2 - 2x + 7$ acts on a body of mass 7 kg and displaces it from $x = 0$ m to $x =$ 5m. The work done on the body is x' joule. If both F and x' are measured in SI units, the value of x' is

(a) 135 (b) 235
(c) 335 (d) 935

4. Water is drawn from a well in a 5 kg drum of capacity 55 L by two ropes connected to the top of the drum. The linear mass density of each rope is 0.5 kgm^{-1}. The work done in lifting water to the ground from the surface of water in the well 20 m below is ($g = 10$ ms^{-2})

(a) 1.4×10^4 J (b) 1.5×10^4 J
(c) $9.8 \times 10 \times 6$ J (d) 18 J

5. Under the action of a force, a 2 kg body moves such that its position x as a function of time t is given by $x = t^3 / 3$, where x is in metre and t in second. The work done by the force in the first two seconds is

(a) 1.6 J (b) 16 J
(c) 160 J (d) 1600 J

6. A mass M is lowered with the help of a string by a distance h at a constant acceleration $g/2$. The work done by the string will be

(a) $\frac{Mgh}{2}$ (b) $\frac{-Mgh}{2}$
(c) $\frac{3Mgh}{2}$ (d) $\frac{-3Mgh}{2}$

7. A 5 kg brick of 20 cm × 10 cm × 8 cm dimension lying on the largest base. It is now made to stand with length vertical. If $g = 10$ ms^{-2}, then the amount of work done is

(a) 3 J (b) 5 J (c) 7 J (d) 9 J

8. In a children's park, there is a slide which has a total length of 10 m and a height of 8.0 m. A vertical ladder is provided to reach the top. A boy weighing 200 N climbs up the ladder to the top of the slide and slides down to the ground. The average friction offered by the slide is three-tenth of his weight. The work done by the slide on the boy as he comes down is

(a) zero (b) + 600 J (c) –600 J (d) +1600 J

9. A ball is released from the top of a tower. The ratio of work done by force of gravity in Ist second, 2nd second and 3rd second of the motion of ball is

(a) 1 : 2 : 3 (b) 1 : 4 : 16 (c) 1 : 3 : 5 (d) 1 : 9 : 25

10. A plate of mass m, length b and breadth a is initially lying on a horizontal floor with length parallel to the floor and breadth perpendicular to the floor. The work done to erect it on its breadth is

(a) $mg\left[\frac{b}{2}\right]$ (b) $mg\left[a + \frac{b}{2}\right]$
(c) $mg\left[\frac{b-a}{2}\right]$ (d) $mg\left[\frac{b+a}{2}\right]$

11. A uniform chain of length L and mass M overhangs a horizontal table with its two-third part on the table. The friction coefficient between the table and the chain is μ. The work done by the friction during the period the chain slips off the table is

(a) $-\frac{1}{4}\mu MgL$ (b) $-\frac{2}{9}\mu MgL$
(c) $-\frac{4}{9}\mu MgL$ (d) $-\frac{6}{7}\mu MgL$

12. A uniform chain of length L and mass M is lying on a smooth table and one-third of its length is hanging vertically down over the edge of the table. If g is acceleration due to gravity, the work required to pull the hanging part on the table is

(a) MgL (b) $MgL/3$
(c) $MgL/9$ (d) $MgL/18$

13. A rod AB of mass 10 kg and length 4 m rests on a horizontal floor with end A fixed, so as to rotate it in vertical. Work done on the rod is 100 J. The height to which the end B be raised vertically above the floor is

(a) 1.5 m (b) 2.0 m
(c) 1.0 m (d) 2.5 m

14. Consider a force $\mathbf{F} = -x\hat{\mathbf{i}} + y\hat{\mathbf{j}}$. The work done by this force in moving a particle from point $A(1, 0)$ to $B(0, 1)$ along the line segment is (all quantities are in SI units) **[JEE Main 2020]**

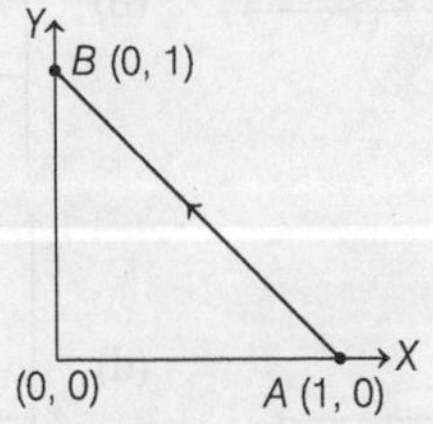

(a) $\frac{3}{2}$ (b) 2 (c) 1 (d) $\frac{1}{2}$

15. A cord is used to lower vertically a block of mass M by a distance d with constant downward acceleration $g/4$. Work done by the cord on the block is

(a) $Mg\frac{d}{4}$ (b) $3\,Mg\frac{d}{4}$
(c) $-3\,Mg\frac{d}{4}$ (d) Mgd

16. Two inclined frictionless tracks, one gradual and the other steep meet at A from where two stones are allowed to slide down from rest, one on each track as shown in figure.

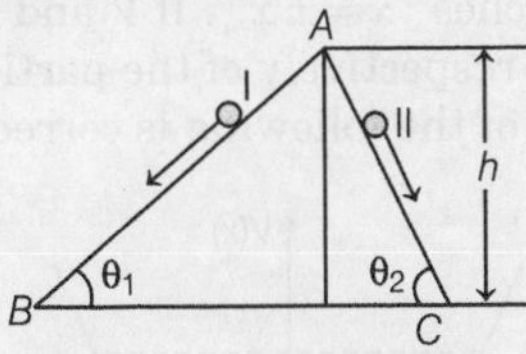

Which of the following statement is correct? **[NCERT Exemplar]**

(a) Both the stones reach the bottom at the same time but not with the same speed.
(b) Both the stones reach the bottom with the same speed and stone I reaches the bottom earlier than stone II.
(c) Both the stones reach the bottom with the same speed and stone II reaches the bottom earlier than stone I.
(d) Both the stones reach the bottom at different times and with different speeds.

17. A 10 kg brick moves along an X-axis. Its acceleration as a function of its position is shown in figure. What is the net work performed on the brick by the force causing the acceleration as the brick moves from $x = 0$ to $x = 8.0$ m ?

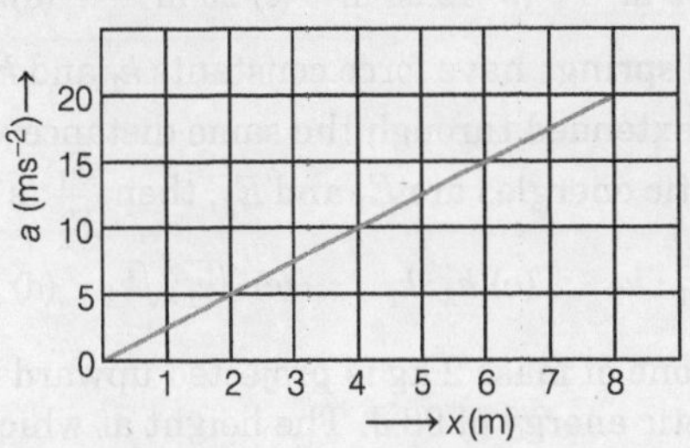

(a) 4 J (b) 8 J (c) 2 J (d) 1 J

18. A block of mass m is kept on a platform which starts from rest with constant acceleration $\frac{g}{2}$ upwards as shown in figure. Work done by normal reaction on block in time t is **[JEE Main 2019]**

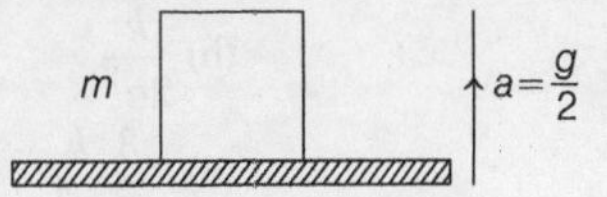

(a) $\frac{mg^2t^2}{8}$ (b) $\frac{3mg^2t^2}{8}$ (c) 0 (d) $-\frac{mg^2t^2}{8}$

Energy

19. A stone is dropped from the top of a tall tower. The ratio of the kinetic energy of the stone at the end of three seconds to the increase in the kinetic energy of the stone during the next three seconds is

(a) 1 : 1 (b) 1 : 2 (c) 1 : 3 (d) 1 : 9

20. An engine pumps water continuously through a hole. Speed with which water passes through the hole nozzle is v and k is the mass per unit length of the water jet as it leaves the nozzle. Find the rate at which kinetic energy is being imparted to the water.

(a) $\frac{1}{2}kv^2$ (b) $\frac{1}{2}kv^3$
(c) $\frac{v^2}{2k}$ (d) $\frac{v^3}{2k}$

21. Two masses of 1 g and 4 g are moving with equal kinetic energies. The ratio of the magnitudes of their linear momenta is

(a) 4 : 1 (b) $\sqrt{2}$: 1
(c) 1 : 2 (d) 1 : 16

22. A mass of 5 kg is moving along a circular path of radius 1m. If the mass moves with 300 revolutions per minute, its kinetic energy (in J) would be [NCERT Exemplar]

(a) $250\pi^2$ (b) $100\pi^2$ (c) $5\pi^2$ (d) 0

23. A body of mass 2 kg is thrown up vertically with kinetic energy of 490 J. The height at which the kinetic energy of the body becomes half of its original value is

(a) 50 m (b) 12.25 m (c) 25 m (d) 10 m

24. Two springs have force constants k_1 and k_2. These are extended through the same distance x. If their elastic energies are E_1 and E_2, then $\frac{E_1}{E_2}$ is equal to

(a) $k_1 : k_2$ (b) $k_2 : k_1$ (c) $\sqrt{k_1} : \sqrt{k_2}$ (d) $k_1^2 : k_2^2$

25. A stone of mass 2 kg is projected upward with kinetic energy of 98 J. The height at which the kinetic energy of the body becomes half of its original value, is given by (Take, $g = 10\ ms^{-2}$)

(a) 5 m (b) 2.5 m (c) 1.5 m (d) 0.5 m

26. A particle is moving in a circular path of radius a under the action of an attractive potential $U = -\frac{k}{2r^2}$. Its total energy is [JEE Main 2018]

(a) $-\frac{k}{4a^2}$ (b) $\frac{k}{2a^2}$

(c) zero (d) $-\frac{3}{2}\frac{k}{a^2}$

27. The potential energy of a particle of mass 5 kg moving in the XY-plane is given by $U = (-7x + 24y)$ J, x and y being in metre. If the particle starts from rest from origin, then speed of particle at $t = 2$ s is

(a) $5\ ms^{-1}$ (b) $0.1\ ms^{-1}$

(c) $17.5\ ms^{-1}$ (d) $10\ ms^{-1}$

28. An elastic string of unstretched length L and force constant k is stretched by a small length x. It is further stretched by another small length y. The work done in the second stretching is

(a) $\frac{1}{2}ky^2$ (b) $\frac{1}{2}k(x^2 + y^2)$

(c) $\frac{1}{2}k(x + y)^2$ (d) $\frac{1}{2}ky(2x + y)$

29. If the potential energy between two molecules is given by $U = -\frac{A}{r^6} + \frac{B}{r^{12}}$, then at equilibrium, separation between molecules and the potential energy are [JEE Main 2020]

(a) $\left(\frac{B}{2A}\right)^{1/6}, -\frac{A^2}{2B}$ (b) $\left(\frac{B}{A}\right)^{1/6}, 0$

(c) $\left(\frac{2B}{A}\right)^{1/6}, -\frac{A^2}{4B}$ (d) $\left(\frac{2B}{A}\right)^{1/6}, -\frac{A^2}{2B}$

30. A bomb of mass 3.0 kg explodes in air into two pieces of masses 2.0 kg and 1.0 kg. The smaller mass goes at a speed of 80 m/s. The total energy imparted to the two fragments is

(a) 1.07 kJ (b) 2.14 kJ

(c) 2.4 kJ (d) 4.8 kJ

31. A uniform chain of length 2 m is kept on a table such that a length of 60 cm hangs freely from the edge of the table. The total mass of the chain is 4 kg. What is the work done in pulling the entire chain on the table? [AIEEE 2004]

(a) 7.2 J (b) 3.6 J

(c) 120 J (d) 1200 J

32. Which of the diagrams shown in figure represents variation of total mechanical energy of a pendulum oscillating in air as function of time? [NCERT Exemplar]

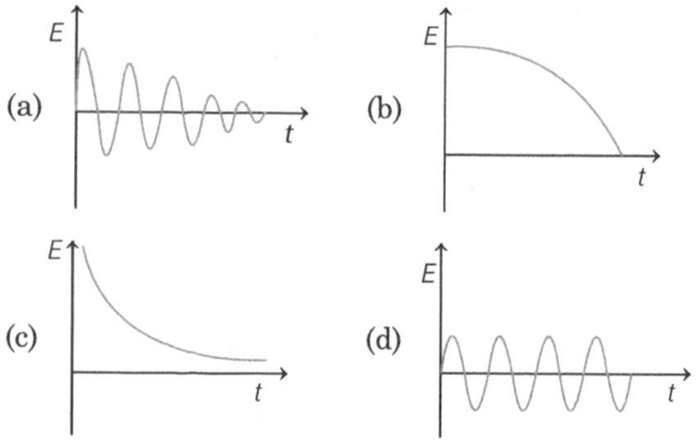

33. A ball is projected vertically upwards with a certain initial speed. Another ball of the same mass is projected at an angle of 60° with the vertical with the same initial speed. At highest point of their journey, the ratio of their potential energies will be

(a) 1 : 1 (b) 2 : 1 (c) 3 : 2 (d) 4 : 1

34. The potential energy function for a particle executing linear SHM is given by $V(x) = \frac{1}{2}kx^2$, where k is the force constant of the oscillator. For $k = 0.5$ N/m, the graph of $V(x)$ *versus* x is shown in the figure. A particle of total energy E turns back when it reaches $x = \pm x_m$. If V and K indicate the PE and KE respectively of the particle at $x = \pm x_m$, then which of the following is correct? [NCERT Exemplar]

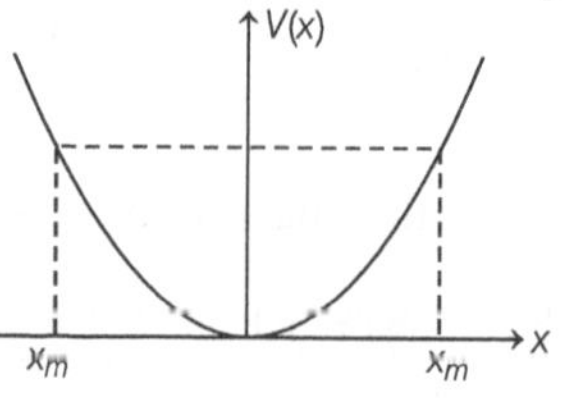

(a) $V = 0, K = E$ (b) $V = E\ K = 0$

(c) $V < E, K = 0$ (d) $V = 0, K < 0$

35. The block of mass M moving on the frictionless horizontal surface collides with the spring of spring constant k and compresses it by length L. The maximum momentum of the block after collision is **[AIEEE 2005]**

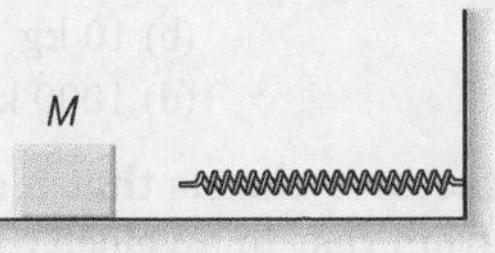

(a) $\sqrt{MkL}$ (b) $\frac{kL^2}{2M}$ (c) zero (d) $\frac{ML^2}{k}$

Work-Energy Theorem

36. A spring of spring constant 5×10^3 Nm^{-1} is stretched initially by 5 cm from the unstretched position, then the work required to stretch it further by another 5 cm is

(a) 12.50 N-m (b) 18.75 N-m
(c) 25.00 N-m (d) 6.25 N-m

37. A body of mass 4 kg is moving with momentum of 8 kg-ms^{-1}. A force of 0.2 N acts on it in the direction of motion of the body for 10 s. The increase in kinetic energy (in J) is

(a) 10 (b) 8.5
(c) 4.5 (d) 4

38. A force acts on a 30 g particle in such a way that the position of the particle as function of time is given by $x = 3t - 4t^2 + t^3$, where x is in metre and t is in second. The work done during the first 4 s is

(a) 5.28 J (b) 450 mJ
(c) 490 mJ (d) 530 mJ

39. The upper half of an inclined plane with inclination ϕ is perfectly smooth, while the lower half is rough. A body starting from rest at the top will again come to rest at the bottom, if the coefficient of friction for the lower half is given by **[AIEEE 2005]**

(a) $2 \sin \phi$ (b) $2 \cos \phi$
(c) $2 \tan \phi$ (d) $\tan \phi$

40. A time dependent force $F = 6t$ acts on a particle of mass 1 kg. If the particle starts from rest, the work done by the force during the first 1 s will be **[JEE Main 2017]**

(a) 22 J (b) 9 J (c) 18 J (d) 4.5 J

41. When a rubber band is stretched by a distance x, it exerts a restoring force of magnitude $F = ax + bx^2$, where a and b are constants. The work done in stretching the unstretched rubber band by L is **[JEE Main 2014]**

(a) $aL^2 + bL^3$ (b) $\frac{1}{2}(aL^2 + bL^3)$
(c) $\frac{aL^2}{2} + \frac{bL^3}{3}$ (d) $\frac{1}{2}\left(\frac{aL^2}{2} + \frac{bL^3}{3}\right)$

42. A bullet fired into a fixed target loses half of its velocity after penetrating 3 cm. How much further it will penetrate before coming to rest, assuming that it faces constant resistance to motion? **[AIEEE 2005]**

(a) 3.0 cm (b) 2.0 cm
(c) 1.5 cm (d) 1.0 cm

43. A 2 kg block slides on a horizontal floor with a speed of 4 m/s. It strikes a uncompressed spring and compresses it till the block is motionless. The kinetic friction force is 15 N and spring constant is 10000 N/m. The spring compresses by **[AIEEE 2007]**

(a) 5.5 cm (b) 2.5 cm
(c) 11.0 cm (d) 8.5 cm

44. A body of mass M is dropped from a height h on a sand floor. If the body penetrates x cm into the sand, the average resistance offered by the sand to the body is

(a) $Mg\left(\frac{h}{x}\right)$ (b) $Mg\left(1+\frac{h}{x}\right)$
(c) $Mgh + Mgx$ (d) $Mg\left(1-\frac{h}{x}\right)$

45. A stone tied to a string of length L is whirled in a vertical circle with the other end of the string at the centre. At a certain instant of time, the stone is at its lowest position and has a speed u. The magnitude of the change in its velocity as it reaches a position where the string is horizontal is

(a) $\sqrt{u^2 - 2gl}$
(b) $\sqrt{2gl}$
(c) $\sqrt{u^2 - gl}$
(d) $\sqrt{2(u^2 - gL)}$

46. A uniform cable of mass M and length L is placed on a horizontal surface such that its $\left(\frac{1}{n}\right)$th part is hanging below the edge of the surface. To lift the hanging part of the cable upto the surface, the work done should be **[JEE Main 2019]**

(a) $\frac{2MgL}{n^2}$ (b) $nMgL$
(c) $\frac{MgL}{n^2}$ (d) $\frac{MgL}{2n^2}$

47. In a shotput event, an athlete throws the shotput of mass 10 kg with an inital speed of 1 m s^{-1} at 45° from a height 1.5 m above ground. Assuming, air resistance to be negligible and acceleration due to gravity to be 10 ms^{-2}, the kinetic energy of the shotput when it just reaches the ground will be **[NCERT Exemplar]**

(a) 2.5 J (b) 5.0 J
(c) 52.5 J (d) 155.0 J

48. The graph between the resistive force F acting on a body and the distance covered by the body is shown in the figure. The mass of the body is 25 kg and initial velocity is 2 m/s. When the distance covered by the body is 4 m, its kinetic energy would be

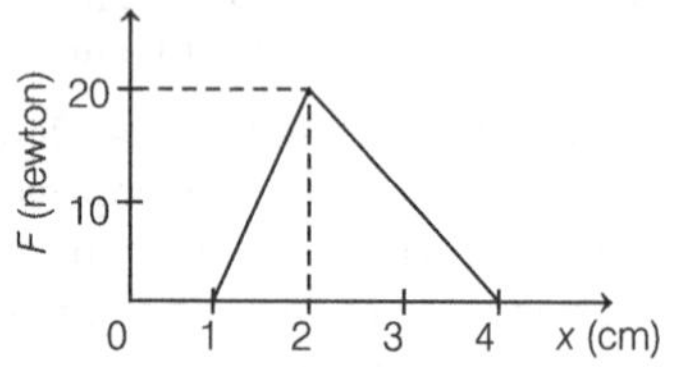

(a) 50 J (b) 40 J (c) 20 J (d) 10 J

49. A 0.5 kg ball is thrown up with an initial speed 14 m/s and reaches a maximum height of 8.0 cm. How much energy is dissipated by air drag acting on the ball during the time of ascent?
(a) 19.6 J (b) 4.9 J (c) 10 J (d) 9.8 J

50. A simple pendulum is released from A as shown in figure. If M and l represent the mass of the bob and length of the pendulum respectively, the gain in kinetic energy at B is

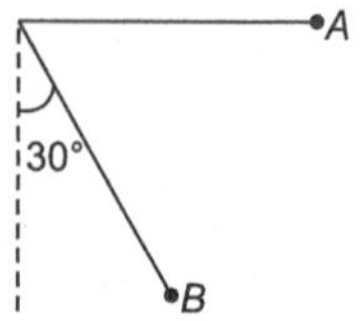

(a) $\frac{Mgl}{2}$ (b) $\frac{\sqrt{3}}{2}Mgl$ (c) $\frac{Mgl}{\sqrt{2}}$ (d) $\frac{2}{\sqrt{3}}mgl$

51. In the given curved road, if particle is released from A, then

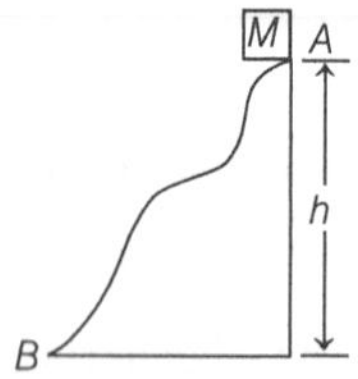

(a) kinetic energy at B must be mgh
(b) kinetic energy at B may be zero
(c) kinetic energy at B must be less than mgh
(d) kinetic energy at B must not be equal to zero

52. A 50 g bullet moving with a velocity of 10 ms^{-1} gets embedded into a 950 g stationary body. The loss in kinetic energy of the system will be
(a) 95% (b) 100% (c) 5% (d) 50%

53. Given that, the position of the body in metre is a function of time as follows

$$x = 2t^4 + 5t + 4$$

The mass of the body is 2 kg. What is the increase in its kinetic energy, one second after the start of motion?
(a) 168 J (b) 169 J (c) 32 J (d) 144 J

Power

54. A one kilowatt motor is used to pump water from a well 10 m deep. The quantity of water pumped out per second is nearly
(a) 1 kg (b) 10 kg
(c) 100 kg (d) 1000 kg

55. A car manufacturer claims that his car can be accelerated from rest to a velocity of 10 ms^{-1} in 5 s. If the total mass of the car and its occupants is 1000 kg, then the average horse power developed by the engine is
(a) $\frac{10^3}{746}$ (b) $\frac{10^4}{746}$
(c) $\frac{10^5}{746}$ (d) 8

56. A 10 HP motor pump take out water from a well of depth 20 m and falls a water tank of volume 22380 L at a height of 10 m from the ground. The running time of the motor to fill the empty water tank is (Take, $g = 10\,ms^{-2}$)
(a) 5 min (b) 10 min
(c) 15 min (d) 20 min

57. Ten litre of water per second is lifted from well through 20 m and delivered with a velocity of 10 ms^{-1}, then the power of the motor is
(a) 1.5 kW (b) 2.5 kW
(c) 3.5 kW (d) 4.5 kW

58. A dam is situated at a height of 550 m above sea level and supplies water to a power house which is at a height of 50 m above sea level. 2000 kg of water passes through the turbines per second. What would be the maximum electrical power output of the power house, if the whole system were 80% efficient?
(a) 8 MW (b) 10 MW
(c) 12.5 MW (d) 16 MW

59. An automobile weighing 120 kg climbs up a hill that rises 1 m in 20 s. Neglecting frictional effects, the minimum power developed by the engine is 9000 W. If $g = 10\,ms^{-2}$, then the velocity of the automobile is
(a) 36 kmh^{-1}
(b) 54 kmh^{-1}
(c) 72 kmh^{-1}
(d) 90 kmh^{-1}

60. A 60 HP electric motor lifts an elevator having a maximum total load capacity of 2000 kg. If the frictional force on the elevator is 4000 N, the speed of the elevator at full load is close to (Take, 1 HP = 746 W and $g = 10\,ms^{-2}$) **[JEE Main 2020]**
(a) 2.0 ms^{-1} (b) 1.5 ms^{-1}
(c) 1.9 ms^{-1} (d) 1.7 ms^{-1}

61. A 500 kg car, moving with a velocity of 36 kmh^{-1} on a straight road unidirectionally, doubles its velocity in one minute. The power delivered by the engine for doubling the velocity is

(a) 750 W (b) 1050 W
(c) 1150 W (d) 1250 W

62. A quarter horse power motor runs at a speed of 600 rpm. Assuming 40% efficiency, the work done by the motor in one rotation will be

(a) 7.46 J (b) 7400 J
(c) 7.46 erg (d) 74.6 J

63. A particle of mass m is moving in a circular path of constant radius r such that its centripetal acceleration a_c is varying with time t as $a_c = k^2rt^2$. The power is

(a) $2\pi mk^2r^2t$ (b) mk^2r^2t
(c) $\frac{mk^4r^2t^5}{3}$ (d) zero

64. The power supplied by a force acting on a particle moving in a straight line is constant. The velocity of the particle varies with the displacement x as

(a) $x^{1/2}$ (b) x (c) x^2 (d) $x^{1/3}$

65. A body is moved along straight line by a machine delivering a constant power. The distance moved by the body in time t is proportional to

(a) $t^{3/4}$ (b) $t^{3/2}$ (c) $t^{1/4}$ (d) $t^{1/2}$

66. An elevator in a building can carry a maximum of 10 persons with the average mass of each person being 68 kg. The mass of the elevator itself is 920 kg and it moves with a constant speed of 3 m/s. The frictional force opposing the motion is 6000 N.

If the elevator is moving up with its full capacity, the power delivered by the motor to the elevator must be at least (Take, $g = 10\ m/s^2$) **[JEE Main 2020]**

(a) 62360 W (b) 48000 W
(c) 56300 W (d) 66000 W

67. A particle is moving unidirectionally on a horizontal plane under the action of a constant power supplying energy source. The displacement (s)-time (t) graph that describes the motion of the particle is (graphs are drawn schematically and are not to scale) **[JEE Main 2020]**

(a)
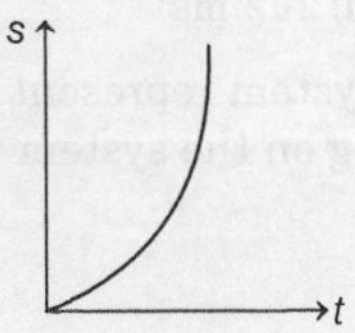

(b)
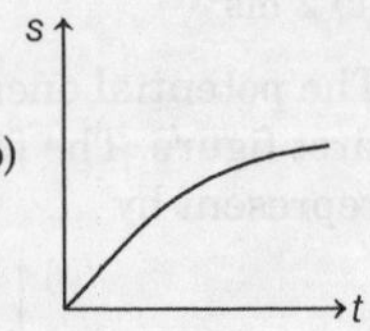

(c)
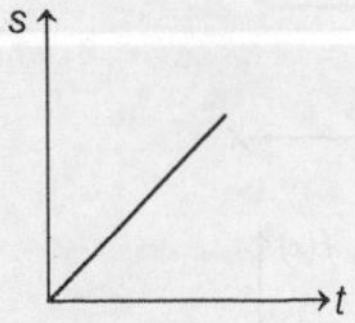

(d)
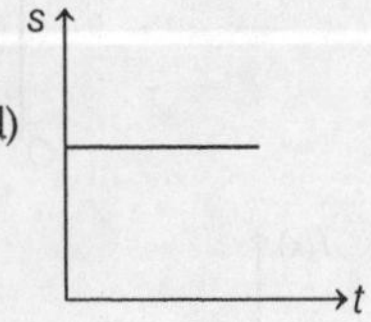

ROUND II Mixed Bag

Only One Correct Option

1. A body of mass 3 kg is under a force which causes a displacement is given by $s = \frac{t^3}{3}$ (in m). Find the work done by the force in first 2 s.

(a) 2 J (b) 3.8 J (c) 5.2 J (d) 24 J

2. A man of mass m, standing at the bottom of the staircase of height L climbs it and stands at its top. **[NCERT Exemplar]**

(a) Work done by all forces on man is equal to the rise in potential energy mgL
(b) Work done by all forces on man is zero
(c) Work done by the gravitational force on man is mgL
(d) The reaction force from a step works as the point of application of the force does not move while the force exists

3. A force acts on a 2 kg object, so that its position is given as a function of time as $x = 3t^2 + 5$. What is the work done by this force in first 5 s? **[JEE Main 2019]**

(a) 850 J (b) 900 J (c) 950 J (d) 875 J

4. A person pushes a box on a rough horizontal platform surface. He applies a force of 200 N over a distance of 15 m. Thereafter, he gets progressively tired and his applied force reduces linearly with distance to 100 N. The total distance through which the box has been moved is 30 m. What is the work done by the person during the total movement of the box? **[JEE Main 2020]**

(a) 5250 J (b) 2780 J
(c) 3280 J (d) 5690 J

5. If a man speeds up by 1 ms^{-1}, his kinetic energy increases by 44%. His original speed in ms^{-1} is

(a) 1 (b) 2
(c) 5 (d) 4

6. A boy is rolling a 0.5 kg ball on the frictionless floor with the speed of 20 ms^{-1}. The ball gets deflected by an obstacle on the way. After deflection, it moves with 5% of its initial kinetic energy. What is the speed of the ball now? **[JEE Main 2021]**

(a) 19.0 ms^{-1} (b) 4.47 ms^{-1}
(c) 14.41 ms^{-1} (d) 1.00 ms^{-1}

7. Two blocks of mass m each are connected to a spring of spring constant k as shown in figure. The maximum displacement in the block is

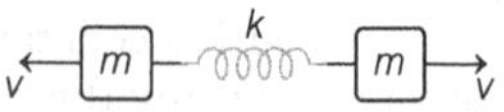

(a) $\sqrt{\dfrac{2mv^2}{k}}$ (b) $\sqrt{\dfrac{mv^2}{k}}$

(c) $2\sqrt{\dfrac{mv^2}{k}}$ (d) $2\sqrt{\dfrac{k}{mv^2}}$

8. Power supplied to a particle of mass 2 kg varies with time as $P = \dfrac{3t^2}{2}$ watt, here t is in second. If the velocity of particle at $t = 0$ is $v = 0$, the velocity of particle at time $t = 2$ s will be

(a) 1 ms^{-1} (b) 4 ms^{-1}

(c) 2 ms^{-1} (d) $2\sqrt{2}$ ms^{-1}

9. The potential energy of a system represent in the first figure. The force acting on the system will be represent by

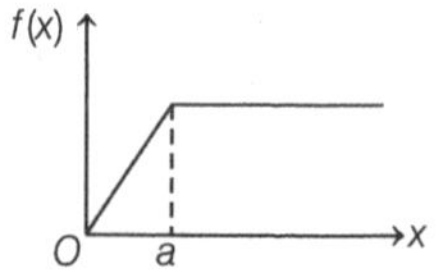

(a)

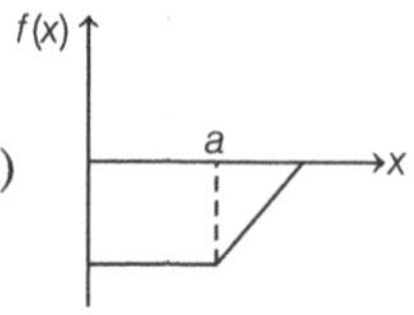

(b)

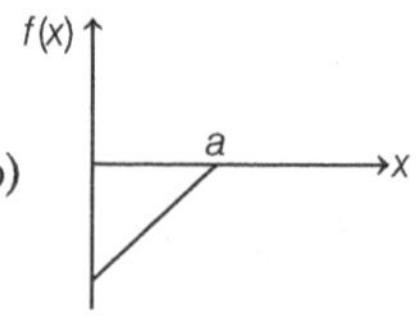

(c)

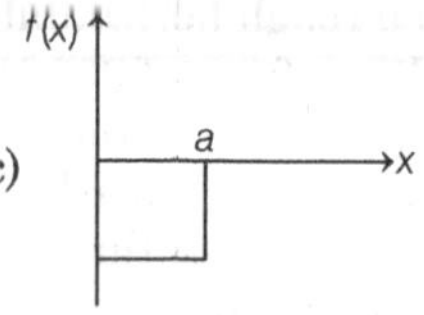

(d)

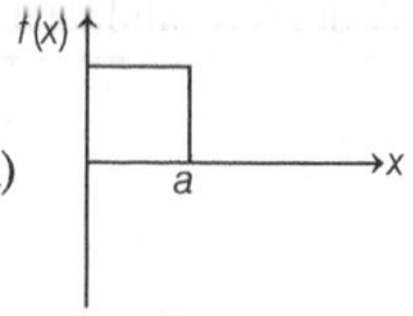

10. A gun of mass 20 kg has bullet of mass 0.1 kg in it. The gun is free to recoil 804 J of recoil energy are released on firing the gun. The speed of bullet (ms^{-1}) is

(a) $\sqrt{804 \times 2010}$ (b) $\sqrt{\dfrac{2010}{804}}$

(c) $\sqrt{\dfrac{804}{2010}}$ (d) $\sqrt{804 \times 4 \times 10^3}$

11. A particle moves on a rough horizontal ground with some initial velocity v_0. If (3/4)th of its kinetic energy is lost due to friction in time t_0, the coefficient of friction between the particle and the ground is

(a) $\dfrac{v_0}{2gt_0}$ (b) $\dfrac{v_0}{4gt_0}$ (c) $\dfrac{3v_0}{4gt_0}$ (d) $\dfrac{v_0}{gt_0}$

12. An ideal spring with spring constant k is hung from the ceiling and a block of mass M is attached to its lower end. The mass is released with the spring initially unstretched, then the maximum extension in the spring is

(a) $\dfrac{4Mg}{k}$ (b) $\dfrac{2Mg}{k}$

(c) $\dfrac{Mg}{k}$ (d) $\dfrac{Mg}{2k}$

13. A bullet when fired at a target with velocity of 100 ms^{-1} penetrates 1 m into it. If the bullet is fired at a similar target with a thickness 0.5 m, then it will emerge from it with a velocity of

(a) $50\sqrt{2}$ m/s (b) $\dfrac{50}{\sqrt{2}}$ m/s

(c) 50 m/s (d) 10 m/s

14. Velocity-time graph of a particle of mass 2 kg moving in a straight line is as shown in figure. Work done by all forces on the particle is

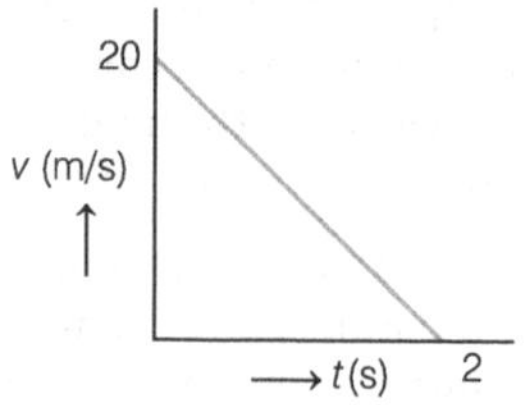

(a) 400 J (b) – 400 J

(c) – 200 J (d) 200 J

15. A box of mass 50 kg is pulled up on an incline 12 m long and 2 m high by a constant force of 100 N from rest. It acquires a velocity of 2 ms^{-1} on reaching the top. Work done against friction ($g = 10$ ms^{-2}) is

(a) 50 J (b) 100 J

(c) 150 J (d) 200 J

16. A particle is released from a height s. At certain height, its kinetic energy is three times its potential energy. The height and speed of the particle at that instant are respectively

(a) $\dfrac{s}{4}, \dfrac{3gs}{2}$ (b) $\dfrac{s}{4}, \sqrt{\dfrac{3gs}{2}}$

(c) $\dfrac{s}{2}, \sqrt{\dfrac{3gs}{2}}$ (d) $\dfrac{s}{8}, \sqrt{\dfrac{3gs}{2}}$

17. A man running has half the kinetic energy of a boy of half his mass. The man speeds up by 1 ms^{-1} and then has kinetic energy as that of the boy. What were the original speeds of man and the boy?

(a) $\sqrt{2}$ ms^{-1}; $2\sqrt{2} - 1$ ms^{-1}

(b) $(\sqrt{2} - 1)$ ms^{-1}, $2(\sqrt{2} - 1)$ ms^{-1}

(c) $(\sqrt{2} + 1)$ ms^{-1}; $2(\sqrt{2} + 1)$ ms^{-1}

(d) None of the above

18. If the kinetic energy of a body is directly proportional to time t, the magnitude of the force acting on the body is
(a) directly proportional to $\sqrt{t}$
(b) inversely proportional to $\sqrt{t}$
(c) directly proportional to the speed of the body
(d) inversely proportional to the root of speed of the body

19. The kinetic energy k of a particle moving along a circle of radius R depends upon the distance s as $k = as^2$. The force acting on the particle is
(a) $2a\frac{s^2}{R}$ (b) $2as\left[1+\frac{s^2}{R^2}\right]^{1/2}$
(c) $2as$ (d) $2a$

20. A particle is placed at the origin and a force $F = kx$ is acting on it (where k is a positive constant). If $U(0) = 0$, the graph of $U(x)$ *versus* x will be (where U is the potential energy function) **[UP SEE 2004]**

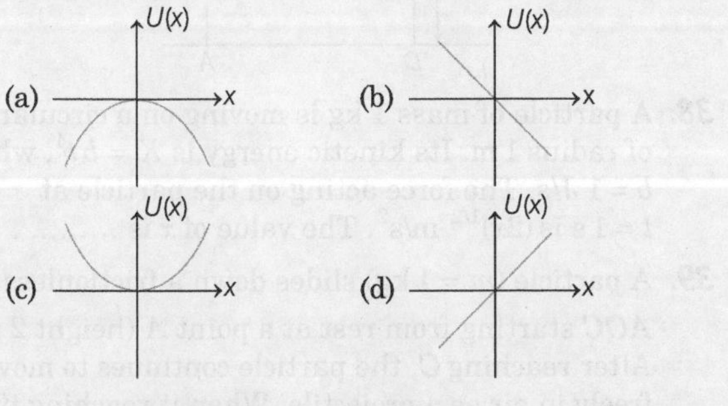

21. A block of mass 5 kg is resting on a smooth surface. At what angle can a force of 20 N be acted on the body so that it will acquired a kinetic energy of 40 J after moving 4 m
(a) 30° (b) 45° (c) 60° (d) 120°

22. A variable force given by the two-dimensional vector $\mathbf{F} = (3x^2\,\hat{\mathbf{i}} + 4\hat{\mathbf{j}})$ acts on a particle. The force is in newton and X is in metre. What is the change in the kinetic energy of the particle as it moves from the point with coordinates (2, 3) to (3, 0)? (The coordinates are in metres)
(a) –7 J (b) zero (c) +7 J (d) +19 J

23. A block of mass 2 kg is free to move along the X-axis its is at rest and form $t = 0$ onwards it is subjected to a time dependent force $F(t)$ in the x-direction. The force $F(t)$ varies with t as shown in the figure. The kinetic energy of the block after 4.5 s is

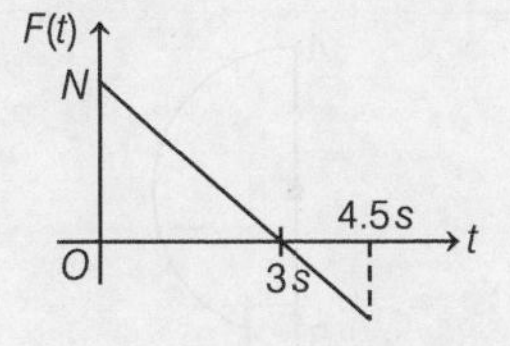

(a) 4.50 J (b) 7.50 J (c) 5.06 J (d) 14.06 J

24. A particle moves in a straight line with retardation proportional to its displacement. Its loss of kinetic energy for any displacement x is proportional to
(a) x^2 (b) e^x
(c) x (d) $\log_e x$

25. A particle moves in one dimension from rest under the influence of a force that varies with the distance travelled by the particle as shown in the figure. The kinetic energy of the particle after it has travelled 3 m is **[JEE Main 2019]**

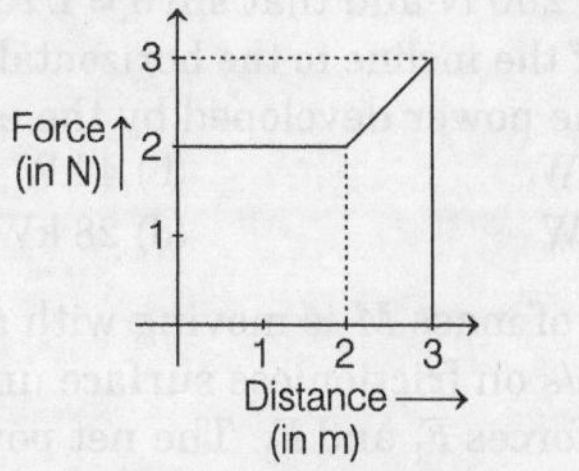

(a) 4 J (b) 2.5 J (c) 6.5 J (d) 5 J

26. A block of mass m lying on a smooth horizontal surface is attached to a spring (of negligible mass) of spring constant k. The other end of the spring is fixed as shown in the figure. The block is initially at rest in its equilibrium position. If now the block is pulled with a constant force F, the maximum speed of the block is **[JEE Main 2019]**

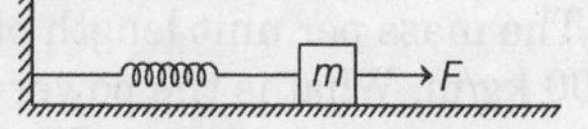

(a) $\frac{\pi F}{\sqrt{mk}}$ (b) $\frac{F}{\sqrt{mk}}$ (c) $\frac{2F}{\sqrt{mk}}$ (d) $\frac{F}{\pi\sqrt{mk}}$

27. A car of mass m is driven with an acceleration a along a straight level road against a constant external resistive force R. When the velocity of the car is v, the rate at which engine of the car is doing work, will be
(a) $R \cdot v$ (b) $ma \cdot v$
(c) $(R + ma) \cdot v$ (d) $(ma - R) \cdot v$

28. A motor drives a body along a straight line with a constant force. The power P developed by the motor must vary with time t as shown in figure

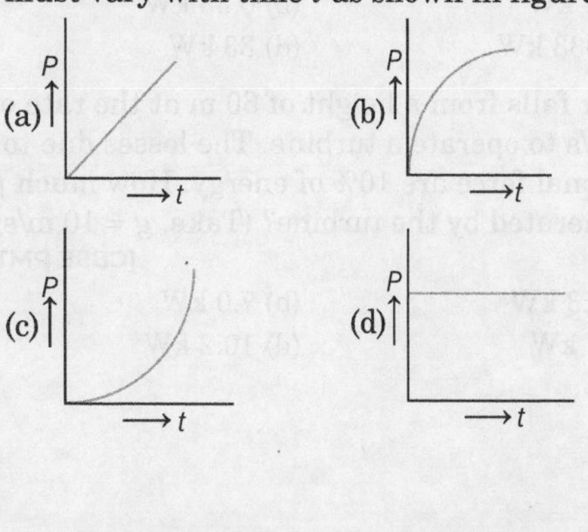

29. A bob of mass m accelerates uniformly from rest to v_1 in time t_1. As a function of t, the instantaneous power delivered to the body is

(a) $\frac{mv_1 t}{t_2}$ (b) $\frac{mv_1 t}{t_1}$

(c) $\frac{mv_1 t^2}{t_1}$ (d) $\frac{mv_1^2 t}{t_1^2}$

30. A car of mass 1000 kg moves at a constant speed of 20 ms^{-1} up an incline. Assume that the frictional force is 200 N and that $\sin\theta = 1/20$, where, θ is the angle of the incline to the horizontal. The $g = 10\ ms^{-2}$. Find the power developed by the engine.

(a) 14 kW (b) 4 kW

(c) 10 kW (d) 28 kW

31. A body of mass M is moving with a uniform speed of 10 m/s on frictionless surface under the influence of two forces F_1 and F_2. The net power of the system is

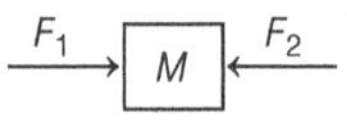

(a) $10\,F_1F_2M$

(b) $10\,(F_1 + F_2)\,M$

(c) $(F_1 + F_2)\,M$

(d) zero

32. An engine pumps water through a hose pipe. Water passes through the pipe and leaves with a velocity of 2 m/s. The mass per unit length of water in the pipe is 100 kg/m. What is the power of the engine?

(a) 800 W (b) 400 W

(c) 200 W (d) 100 W

33. Power applied to a particle varies with time as $P = (3t^2 - 2t + 1)$ watt, where t is in second. Find the change in its kinetic energy between $t = 2$ s and $t = 4$ s.

(a) 32 J (b) 46 J

(c) 61 J (d) 100 J

34. An engine pumps up 100 kg of water through a height of 10 m in 5 s. Given that, the efficiency of the engine is 60%. If $g = 10\,m/s^2$, the power of the engine is

(a) 3.3 kW (b) 0.33 kW

(c) 0.033 kW (d) 33 kW

35. Water falls from a height of 60 m at the rate of 15 kg/s to operate a turbine. The losses due to frictional force are 10% of energy. How much power is generated by the turbine? (Take, $g = 10\,m/s^2$) **[CBSE PMT 2008]**

(a) 12.3 kW (b) 7.0 kW

(c) 8.1 kW (d) 10.2 kW

Numerical Value Questions

36. A cricket ball of mass 0.15 kg is thrown vertically up by a bowling machine, so that it rises to a maximum height of 20 m after leaving the machine. If the part pushing the ball applies a constant force F on the ball and moves horizontally a distance of 0.2 m, while launching the ball, the value of F (in N) is .... (Take, $g = 10\,ms^{-2}$) **[JEE Main 2020]**

37. A boy wishes to move the block slowly shown in the figure, a distance of 2 m to the right by either sliding or by tiping over one corner with the least amount of work. The mass of block is 100 kg and coefficient of friction is 0.3. If the least amount of work is 207 nJ. The value of n is ....

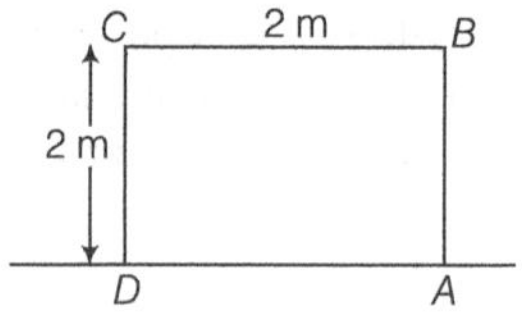

38. A particle of mass 1 kg is moving on a circular path of radius 1 m. Its kinetic energy is $K = bt^4$, where $b = 1$ J/s. The force acting on the particle at $t = 1$ s is $(2x)^{1/2}$ m/s^2. The value of x is ....

39. A particle ($m = 1$ kg) slides down a frictionless track AOC starting from rest at a point A (height 2 m). After reaching C, the particle continues to move freely in air as a projectile. When it reaching its highest point P (height 1 m), the kinetic energy of the particle (in J) is .... (Figure drawn is schematic and not to scale) (Take, $g = 10\,ms^{-2}$) **[JEE Main 2020]**

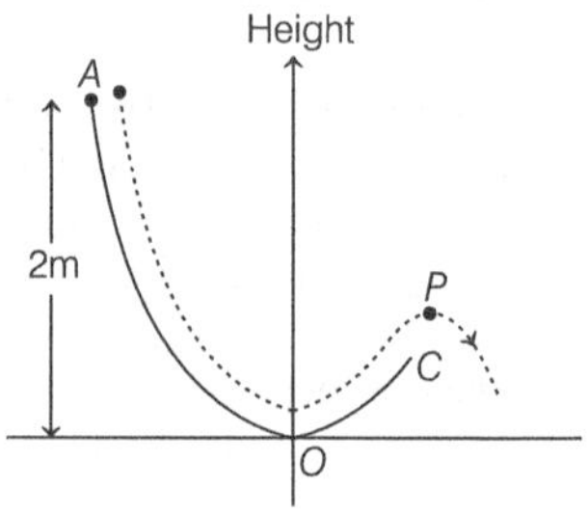

40. A sleeve of mass m is moving along perimeter by applying a constant force of 10N always directed toward point A. The radius of semi-circle is 10 cm. The work done (in J) by applied force F during motion of sleeve from B to A is ....

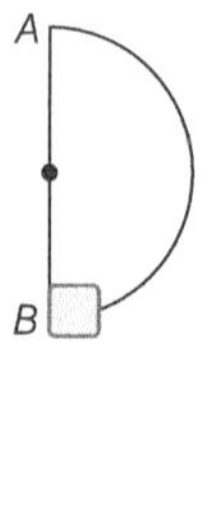

41. A body of mass 2 kg is driven by an engine delivering a constant power of 1 J/s. The body starts from rest and moves in a straight line. After 9 s, the body has moved a distance (in m) .... **[JEE Main 2020]**

42. A ball of mass 4 kg is moving with a velocity of $10\,ms^{-1}$, collides with a spring of length 8 m and force constant $100\ Nm^{-1}$. The length of the compressed spring is x m. The value of x, to the nearest integer, is .... **[JEE Main 2021]**

43. Two blocks A and B of equal masses ($m = 10$ kg) are connected by a light spring of spring constant $k = 150$ N/m. The system is in equilibrium. The minimum value of initial downward velocity v_0 of the block B for which the block A bounce up is $\frac{20}{\sqrt{3\,n}}$ m/s. The value of n is ....

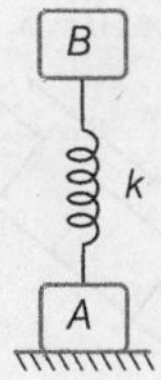

Answers

Round I

1. (b)	2. (b)	3. (a)	4. (a)	5. (b)	6. (b)	7. (a)	8. (c)	9. (c)	10. (c)
11. (b)	12. (d)	13. (b)	14. (c)	15. (c)	16. (c)	17. (b)	18. (b)	19. (c)	20. (b)
21. (c)	22. (a)	23. (b)	24. (a)	25. (b)	26. (c)	27. (d)	28. (d)	29. (c)	30. (d)
31. (b)	32. (c)	33. (d)	34. (b)	35. (a)	36. (b)	37. (c)	38. (a)	39. (c)	40. (d)
41. (c)	42. (d)	43. (a)	44. (b)	45. (d)	46. (d)	47. (d)	48. (d)	49. (d)	50. (b)
51. (b)	52. (a)	53. (d)	54. (b)	55. (b)	56. (a)	57. (b)	58. (a)	59. (b)	60. (c)
61. (d)	62. (a)	63. (b)	64. (d)	65. (b)	66. (d)	67. (a)			

Round II

1. (d)	2. (d)	3. (b)	4. (a)	5. (c)	6. (b)	7. (a)	8. (c)	9. (c)	10. (d)
11. (a)	12. (b)	13. (a)	14. (b)	15. (b)	16. (b)	17. (c)	18. (b)	19. (b)	20. (a)
21. (c)	22. (c)	23. (c)	24. (a)	25. (c)	26. (b)	27. (c)	28. (a)	29. (d)	30. (a)
31. (d)	32. (a)	33. (b)	34. (a)	35. (c)	36. 150	37. 2	38. 6	39. 10	40. 2
41. 18	42. 6	43. 5							

Solutions

Round I

1. $W = mg\sin\theta \times s$

$= 2\times10^3 \times \sin 15^\circ \times 10 = 5.13$ kJ

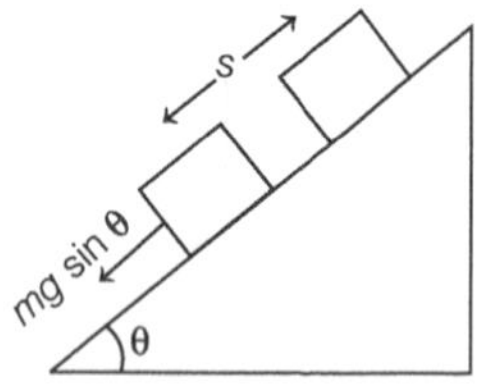

2. As, $F = \mu mg\cos\theta$

or $F = 0.30\times10\times10\cos45^\circ$

or $F = \frac{30}{\sqrt{2}}$ N

$W = F\times s\times\cos180^\circ = -\frac{30}{\sqrt{2}}\times5$

$= -\frac{150}{\sqrt{2}}\times\frac{\sqrt{2}}{\sqrt{2}} = -75\sqrt{2}$ J

3. This is the case of work done by a variable force

$$W = \int_0^5 (3x^2 - 2x + 7)\,dx$$

$$W = |x^3 - x^2 + 7x|_0^5$$

or $W = (5\times5\times5 - 5\times5 + 7\times5)$

or $W = (125 - 25 + 35) = 135$ J

$\therefore$ $x' = 135$

4. Work done in lifting water and drum $= (60\times10\times20)$ J

$= 12000$ J

Total mass of ropes $= (4\times0.5)$ kg $= 20$ kg

Work done in the case of ropes $= 20\times10\times10 = 2000$ J

Total work done $= 12000 + 2000 = 14000$ J $\simeq 1.4\times10^4$ J

5. $v = \frac{dx}{dt} = \frac{d}{dt}\left(\frac{t^3}{3}\right) = t^2$

When $t = 0$, then $v = 0$ and when $t = 2$, then

$v = 4$ m/s

According to work-energy theorem,

$$W = \frac{1}{2}m[(4)^2 - (0)^2] = \frac{1}{2}\times2\times16 = 16\text{J}$$

6. Tension in the string,

$$T = M(g-a) = M\left(g - \frac{g}{2}\right) = \frac{Mg}{2}$$

$$W = T\times h\times\cos180^\circ$$

$$= -\frac{Mgh}{2}$$

7. Initial height of CG = 4 cm

Final height of CG = 10 cm

Increase in height = (10 − 4) cm = 6 cm = 0.06 m

Work done $= 5\times10\times0.06 = 3$ J

8. Average friction, $F = \frac{3}{10}mg$

As, $W = -Fs$ or $W = -\frac{3}{10}mgs$

or $W = \left(-\frac{3}{10}\times200\times10\right)$ J $= -600$ J

9. Initial velocity of ball is zero, *i.e.* $u = 0$.

$\therefore$ Displacement of ball in t^{th} second,

$$s = gt - \frac{1}{2}g = g\left(t - \frac{1}{2}\right)$$

$$s \propto \left(t - \frac{1}{2}\right)$$

or $s_1 : s_2 : s_3 = \left(1 - \frac{1}{2}\right) : \left(2 - \frac{1}{2}\right) : \left(3 - \frac{1}{2}\right) = 1:3:5$

Now, $W = mgs,\ W \propto s$

$\therefore$ $W_1 : W_2 : W_3 = 1:3:5$

10. Initial height of CG $= \frac{a}{2}$

Final height of CG $= \frac{b}{2}$

Work done $= mg\left[\frac{b}{2} - \frac{a}{2}\right] = mg\left(\frac{b-a}{2}\right)$

11. $dW = -\mu\left[\frac{M}{L}\right]gl\,dl$

$$W = \int_0^{\frac{2L}{3}} -\frac{\mu Mg}{L}\,l\,dl$$

or $W = -\frac{\mu Mg}{L}\left[\frac{l^2}{2}\right]_0^{\frac{2L}{3}}$

or $W = -\frac{\mu Mg}{2L}\frac{4L^2}{9} - 0$ or $W = -\frac{2}{9}\mu MgL$

12. The weight of hanging part $\left(\frac{L}{3}\right)$ of chain is $\left(\frac{1}{3}Mg\right)$.

This weight acts at centre of gravity of the hanging part which is at a depth of $L/6$ from the table.

As, work done = force × distance

$\therefore$ $W = \frac{Mg}{3}\times\frac{L}{6} = \frac{MgL}{18}$

13. Work done $= mg\left(\frac{h}{2}\right)$

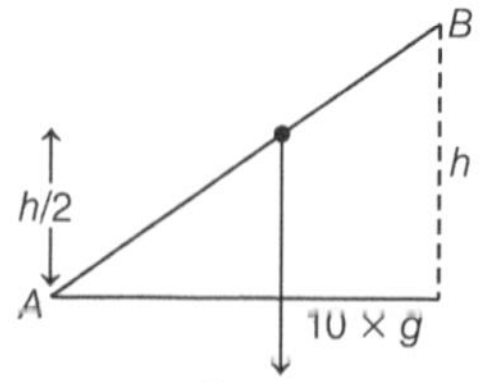

$$100 = \frac{10\times10\times h}{2}$$

$h = 2.0$ m

14. Work done by a variable force on the particle,

$$W = \int \mathbf{F} \cdot d\mathbf{r}$$
$$= \int \mathbf{F} \cdot (dx\hat{\mathbf{i}} + dy\hat{\mathbf{j}})$$

$\therefore$ In two dimension, $d\mathbf{r} = dx\hat{\mathbf{i}} + dy\hat{\mathbf{j}}$

and it is given $\mathbf{F} = -x\hat{\mathbf{i}} + y\hat{\mathbf{j}}$

$$\therefore \quad W = \int (-x\hat{\mathbf{i}} + y\hat{\mathbf{j}}) \cdot (dx\hat{\mathbf{i}} + dy\hat{\mathbf{j}})$$
$$= \int -x\,dx + y\,dy$$
$$= \int -x\,dx + \int y\,dy$$

As particle is displaced from $A(1, 0)$ to $B(0, 1)$, so x varies from 1 to 0 and y varies from 0 to 1.

So, with limits, work will be

$$W = \int_1^0 -x\,dx + \int_0^1 y\,dy$$
$$= \left[\frac{-x^2}{2}\right]_1^0 + \left[\frac{y^2}{2}\right]_0^1$$
$$= \frac{1}{2}[(0-(-1)^2] + [(1)^2 - 0)]$$
$$= 1\text{ J}$$

15. When the block moves vertically downward with acceleration $\frac{g}{4}$, then tension in the cord,

$$T = M\left(g - \frac{g}{4}\right) = \frac{3}{4}Mg$$

Work done by the cord $\mathbf{F} \cdot \mathbf{s} = Fs\cos\theta = Td\cos 180°$

$$= \left(-\frac{3}{4}Mg\right) \times d \qquad [\because \cos 180° = -1]$$
$$= -3\,Mg\frac{d}{4}$$

16. As both surfaces I and II are frictionless and two stones slide from the same height, therefore both the stones reach the bottom with same speed $\left(\frac{1}{2}mv^2 = mgh\right)$. As acceleration down the plane II is larger ($a_2 = g\sin\theta_2$ greater than $a_1 = g\sin\theta_1$), therefore stone II reaches the bottom earlier than stone I.

17. According to the graph, the acceleration a varies linearly with the coordinate x. We may write $a = \alpha x$, where α is the slope of the graph.

$$\alpha = \frac{20}{8} = 2.5\text{ s}^{-2}$$

The force on the brick is in the positive x- direction and according to Newton's second law, its magnitude is given by

$$F = \frac{a}{m} = \frac{\alpha}{m}x$$

If x_f is the final coordinate, the work done by the force is

$$W = \int_0^{x_f} F\,dx = \frac{\alpha}{m}\int_0^{x_f} x\,dx$$
$$= \frac{\alpha}{2m}x_f^2 = \frac{2.5}{2 \times 10} \times (8)^2 = 8\text{ J}$$

18. Normal reaction force on the block,

$$N = ma_{\text{net}}$$

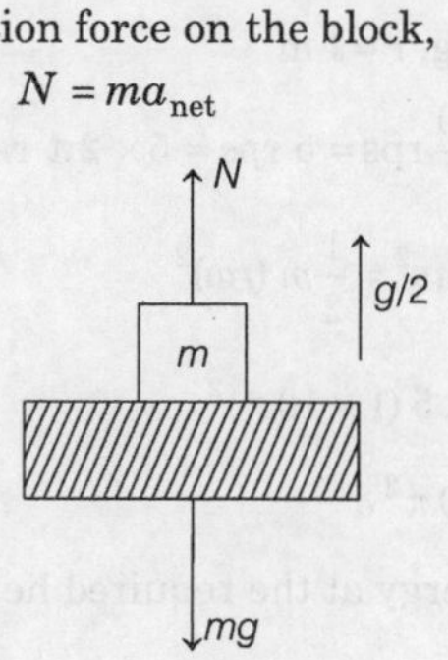

where, a_{net} = net acceleration of block.

$$= g + a$$
$$= g + \frac{g}{2} = \frac{3g}{2}$$
$$\Rightarrow \quad N = m\left(g + \frac{g}{2}\right)$$
$$= \frac{3mg}{2}$$

Now, in time t, block moves by a displacement s given by

$$s = 0 + \frac{1}{2}at^2 = \frac{1}{2}\left(\frac{g}{2}\right)t^2 \qquad (\because u = 0)$$

Here, $\quad a = \frac{g}{2}$ (given)

$\therefore$ Work done = Force $\times$ Displacement

$$\Rightarrow \quad W = \frac{3mg}{2} \times \frac{gt^2}{4}$$
$$= \frac{3mg^2t^2}{8}$$

19. As, $\quad E = \frac{1}{2}mv^2$

$$= \frac{1}{2}m(gt)^2 \qquad [\because v = 0 + gt = gt]$$
$$= \frac{1}{2}mg^2t^2$$
$$\frac{E_1}{E_2} = \frac{\frac{1}{2}mg^2 \times 3^2}{\frac{1}{2}mg^2(6^2 - 3^2)} = \frac{9}{9 \times 3} = \frac{1}{3}$$

20. Velocity of water = v

Mass flowing per unit length = k

Mass flowing per second = kv

$\therefore$ Rate of kinetic energy or KE per second

$$= \frac{1}{2} \times \text{mass flowing per second} \times v^2$$
$$= \frac{1}{2} \times kv \times v^2 = \frac{1}{2}kv^3$$

21. $p = \sqrt{2mE_K}$

or $\quad p \propto \sqrt{m} \qquad [\because E_k$ is given to be constant$]$

$$\therefore \quad \frac{p_1}{p_2} = \sqrt{\frac{m_1}{m_2}} = \sqrt{\frac{1}{4}} = \frac{1}{2}$$

22. Here, $m = 5$ kg, $r = 1$ m

$$\omega = \frac{300}{60} \text{ rps} = 5 \text{ rps} = 5 \times 2\pi \text{ rad s}^{-1}$$

$$\text{KE} = \frac{1}{2}mv^2 = \frac{1}{2}m(r\omega)^2$$

$$= \frac{1}{2} \times 5(1 \times 10\pi)^2$$

$$= 250\pi^2 \text{ J}$$

23. Potential energy at the required height $= \frac{490}{2} = 245$J

Again, $245 = 2 \times 10 \times h$

or $h = \frac{245}{20} \text{ m} = 12.25 \text{ m}$

24. $E = \frac{1}{2}kx^2$

$$F \propto k$$

$$\frac{E_1}{E_2} = \frac{k_1}{k_2}$$

25. As, $K = \frac{1}{2}mv^2$

$$v^2 = \frac{98 \times 2}{2} = 98$$

$$h = \frac{v^2}{2g} = \frac{98}{2 \times 9.8} = 5$$

$$K_1 = \frac{1}{2}mv^2 = \frac{1}{2}m \times 2gh$$

$\therefore$ $\frac{K_2}{K_1} = \frac{h_2}{h_1}$

Given $K_2 = \frac{K_1}{2} = \frac{K_1}{2K_1} = \frac{h_2}{5}$

$\therefore$ $h_2 = 2.5$ m

26. $\therefore$ Force $= -\frac{dU}{dr}$

$\Rightarrow$ $F = -\frac{d}{dr}\left(\frac{-k}{2r^2}\right) = -\frac{k}{r^3}$

As particle is on circular path, this force must be centripetal force.

$\Rightarrow$ $|F| = \frac{mv^2}{r}$

So, $\frac{k}{r^3} = \frac{mv^2}{r}$

$\Rightarrow$ $\frac{1}{2}mv^2 = \frac{k}{2r^2}$

$\therefore$ Total energy of particle = KE + PE

$$= \frac{k}{2r^2} - \frac{k}{2r^2} = 0$$

Total energy = 0

27. $F = \frac{\partial U}{\partial x}\hat{\mathbf{i}} - \frac{\partial U}{\partial y}\hat{\mathbf{j}} = -7\hat{\mathbf{i}} + 24\hat{\mathbf{j}}$

$\therefore$ $a_x = \frac{F_X}{m} = -\frac{7}{5} = -1.4 \text{ ms}^{-2}$ along negative X- axis

$a_y = \frac{F_Y}{m} = \frac{24}{5} = 4.8 \text{ ms}^{-2}$ along positive Y-axis

$\therefore$ $v_x = a_x t = 1.4 \times 2 = 2.8 \text{ ms}^{-2}$

and $v_y = 4.8 \times 2 = 9.6 \text{ ms}^{-1}$

$\therefore$ $v = \sqrt{v_x^2 + v_y^2} = 10 \text{ ms}^{-1}$

28. Elastic force in string is conservative in nature.

$$W = -\Delta V_1$$

where, W = work done by elastic force of string

$$W = -(V_f - V_i) = V_i - V_f$$

or $W = \frac{1}{2}kx^2 - \frac{1}{2}k(x+y)^2$

or $W = \frac{1}{2}kx^2 - \frac{1}{2}k(x^2 + y^2 + 2xy)$

$$= \frac{1}{2}kx^2 - \frac{1}{2}kx^2 - \frac{1}{2}ky^2 - \frac{1}{2}k(2xy)$$

$$= -kxy - \frac{1}{2}ky^2 = \frac{1}{2}ky(-2x - y)$$

The work done against elastic force,

$$W_{\text{ext}} = -W = \frac{ky}{2}(2x + y)$$

29. Potential energy between two molecules is given by

$$U = -\frac{A}{r^6} + \frac{B}{r^{12}}$$

From the relation between force and potential energy,
Force acting between them, $F = -\frac{dU}{dr}$

$\therefore$ $F = -\frac{d}{dr}\left(-\frac{A}{r^6} + \frac{B}{r^{12}}\right)$

$$= -\left(\frac{6A}{r^7} - \frac{12B}{r^{13}}\right)$$

$$= \frac{6}{r^7}\left(-A + \frac{2B}{r^6}\right)$$

At equilibrium, $F = 0$

$\Rightarrow$ $\frac{6}{r^7}\left(-A + \frac{2B}{r^6}\right) = 0$

$\Rightarrow$ $-A + \frac{2B}{r^6} = 0$ $(\because r \neq 0)$

$\Rightarrow$ $r^6 = \frac{2B}{A}$

or $r = \left(\frac{2B}{A}\right)^{\frac{1}{6}}$

The above calculated value of r is the separation between molecules at equilibrium.
Now, putting this value in the expression of potential energy, we get

$$U = -\frac{A}{\left\{\left(\frac{2B}{A}\right)^{\frac{1}{6}}\right\}^6} + \frac{B}{\left\{\left(\frac{2B}{A}\right)^{\frac{1}{6}}\right\}^{12}} = -\frac{A^2}{2B} + \frac{A^2}{4B}$$

$\therefore$ $U = -\frac{A^2}{4B}$

30. Both fragments will possess the equal linear momentum.

i.e. $m_1 v_1 = m_2 v_2$

$\Rightarrow \quad 1 \times 80 = 2 \times v_2$

$\Rightarrow \quad v_2 = 40$ m/s

$\therefore$ Total energy of system $= \frac{1}{2} m_1 v_1^2 + \frac{1}{2} m_2 v_2^2$

$$= \frac{1}{2} \times 1 \times (80)^2 + \frac{1}{2} \times 2 \times (40)^2$$

$= 4800$ J

$= 4.8$ kJ

31. Mass per unit length $= \frac{M}{L} = \frac{4}{2} = 2$ kg/m

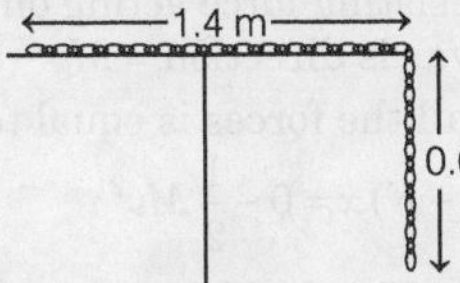

The mass of 0.6 m of chain $= 0.6 \times 2 = 1.2$ kg

The height of centre of mass of hanging part,

$$h = \frac{0.6 + 0}{2} = 0.3 \text{ m}$$

Hence, work done in pulling the chain on the table = Work done against gravity force

i.e. $W = mgh = 1.2 \times 10 \times 0.3 = 3.6$ J

32. When a pendulum oscillates in air, it loses energy continuously in overcoming resistance due to air. Therefore, total mechanical energy of the pendulum decreases continuously with time. The variation of total mechanical energy E with time t is shown correctly by curve (c).

33. For first ball, $mgh_1 = \frac{1}{2} mu^2$

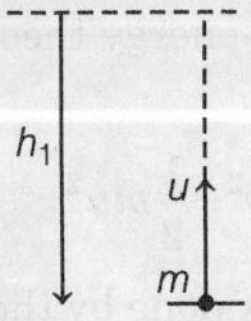

i.e. $h_1 = \frac{u^2}{2g}$

For second ball,

$$mgh_2 = mg \frac{u^2 \cos^2 \theta}{2g}$$

$$= \frac{1}{2} mu^2 \cos^2 \theta$$

$$= \frac{1}{2} mu^2 \cos^2 60°$$

$$= \frac{1}{2} mu^2 \left(\frac{1}{2}\right)^2$$

$$= \frac{1}{2} mu^2 \left(\frac{1}{4}\right)$$

$\Rightarrow \quad h_2 = \frac{u^2}{8g}$

$$\frac{h_1}{h_2} = \frac{u^2}{2g} \times \frac{8g}{u^2}$$

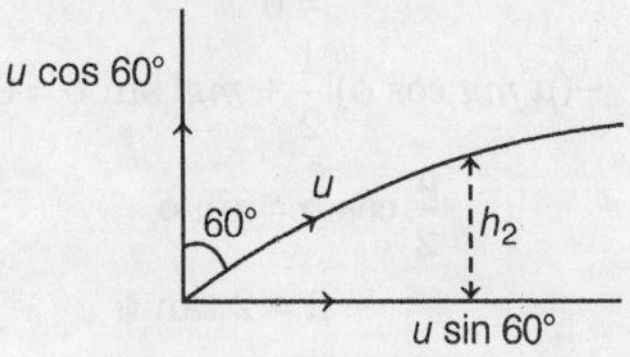

$\Rightarrow \quad \frac{h_1}{h_2} = \frac{4}{1}$

$\therefore$ At highest point, $U_1 : U_2 = h_1 : h_2 = 4 : 1$

34. At $x = +x_m$, the particle turns back. Therefore, its velocity at this point is zero.Therefore, kinetic energy $K = 0$. The total energy E is in the form of potential energy, *i.e.* $V = E$.

35. Momentum would be maximum when KE would be maximum and this is the case when total elastic PE is converted into KE.

According to conservation of energy,

$$\frac{1}{2} kL^2 = \frac{1}{2} Mv^2$$

$\Rightarrow \quad kL^2 = \frac{(Mv)^2}{M}$

or $\quad MkL^2 = p^2 \quad [\because p = Mv]$

$\Rightarrow \quad p = L\sqrt{Mk}$

36. As, $W_1 = \frac{1}{2} kx_1^2 = \frac{1}{2} \times 5 \times 10^3 \times (5 \times 10^{-2})^2 = 6.25$ J

$$W_2 = \frac{1}{2} k (x_1 + x_2)^2$$

$$\frac{1}{2} \times 5 \times 10^3 (5 \times 10^{-2} + 5 \times 10^{-2})^2 = 25 \text{ J}$$

Net work done $= W_2 - W_1 = 25 - 6.25$

$= 18.75$ J $= 18.75$ N-m

37. Initially, $4u = 8 \Rightarrow u = 2$ m/s

Now, $\quad mv - mu = Ft$

$mv - 8 = 0.2 \times 10$

or $\quad v = (5/2)\ \text{ms}^{-1}$

Increase in KE $= \frac{1}{2} m(v^2 - u^2)$

$$= \frac{1}{2} \times 4 \left[\left(\frac{5}{2}\right)^2 - (2)^2\right] = 4.5 \text{ J}$$

38. As, $v = \frac{dx}{dt} = 3 - 8t + 3t^2$

At $t = 0$, $v_0 = 3 - 8 \times 0 + 3 \times 0^2 = 3$ m/s

and at $t = 4$ s, $v_4 = 3 - 8 \times 4 + 3 \times 4^2 = 19$ m/s

$$W = \frac{1}{2} m (v_4^2 - v_0^2)$$

(According to work-energy theorem)

$$= \frac{1}{2} \times 0.03 \times (19^2 - 3^3) = 5.28 \text{ J}$$

39. According to work-energy theorem,

Work done = Change in kinetic energy

$$W = \Delta K = 0$$

$\Rightarrow$ Work done by friction + Work done by gravity $= 0$

$$\Rightarrow \quad -(\mu\, mg \cos\phi)\frac{1}{2} + mgl \sin\phi = 0$$

or $$\frac{\mu}{2}\cos\phi = \sin\phi$$

or $$\mu = 2\tan\phi$$

40. From Newton's second law,

$$\frac{\Delta p}{\Delta t} = F$$

$$\Rightarrow \quad \Delta p = F\Delta t$$

$$\therefore \quad p = \int dp = \int_0^1 F\,dt$$

$$\Rightarrow \quad p = \int_0^1 6t\,dt = 3\,\text{kg}\left(\frac{\text{m}}{\text{s}}\right)$$

Also, change in kinetic energy,

$$\Delta K = \frac{\Delta p^2}{2m} = \frac{3^2}{2\times 1} = 4.5\text{ J}$$

From work-energy theorem,

Work done = Change in kinetic energy

So, work done $= \Delta K = 4.5$ J

41. Given, $F = ax + bx^2$

According to work-energy theorem, work done in stretching the rubber band by t is

$$|dW| = Fdx$$

$$|W| = \int_0^L (ax + bx^2)\,dx = \left[\frac{ax^2}{2}\right]_0^L + \left[\frac{bx^3}{3}\right]_0^L$$

$$= \left[\frac{aL^2}{2} - \frac{a\times(0)^2}{2}\right] + \left[\frac{b\times L^3}{3} - \frac{b\times(0)^3}{3}\right]$$

$$\therefore \quad |W| = \frac{aL^2}{2} + \frac{bL^3}{3}$$

42. According to work-energy theorem,

Total work done = Change in kinetic energy

$$W = \Delta K$$

***Case* I** $$F\times 3 = \frac{1}{2}m\left(\frac{v_0}{2}\right)^2 - \frac{1}{2}mv_0^2$$

where, F is resistive force and v_0 is initial speed.

***Case* II** Let the further distance travelled by the bullet before coming to rest be s.

$$\therefore \quad -F(3+s) = K_f - K_i = -\frac{1}{2}mv_0^2$$

$$\Rightarrow \quad -\frac{1}{8}mv_0^2(3+s) = -\frac{1}{2}mv_0^2$$

or $$\frac{1}{4}(3+s) - 1$$

or $$\frac{3}{4} + \frac{s}{4} = 1$$

or $$s = 1\text{ cm}$$

43. According to work-energy theorem,

Loss in kinetic energy = Work done against friction + Potential energy of spring

$$\frac{1}{2}mv^2 = f\,x + \frac{1}{2}kx^2$$

$$\Rightarrow \quad \frac{1}{2}\times 2\,(4)^2 = 15\,x + \frac{1}{2}\times 10000\,x^2$$

$$\Rightarrow \quad 5000\,x^2 + 15x - 16 = 0$$

$$\therefore \quad x = 0.055\text{ m} = 5.5\text{ cm}$$

44. If the body strikes the sand floor with a velocity v, then

$$Mgh = \frac{1}{2}mv^2$$

With this velocity v, when body passes through the sand floor it comes to rest after travelling a distance x. Let F be the resisting force acting on the body. Net force in downwards direction $= Mg - F$

Work done by all the forces is equal to change in KE

$$(Mg - F)\,x = 0 - \frac{1}{2}Mv^2$$

$$(Mg - F)\,x = -Mgh$$

or $$Fx = Mgh + Mgx$$

or $$F = Mg\left(1 + \frac{h}{x}\right)$$

45. In this case, motion of stone is in vertical circle of radius L and centre at O.

The change in velocity is

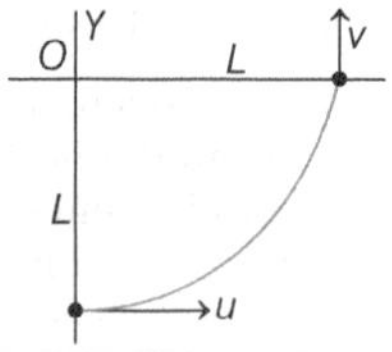

$$\Delta\mathbf{v} = \mathbf{v} - \mathbf{u} = v\mathbf{j} - u\mathbf{i}$$

$$|\Delta\mathbf{v}| = \sqrt{(v)^2 + (-u)^2} = \sqrt{v^2 + u^2}$$

According to work-energy theorem,

$$W = \Delta K$$

or $$W_T + W_g = \frac{1}{2}mv^2 - \frac{1}{2}mu^2 \quad \ldots\text{(i)}$$

W_T = work done by the force of tension = 0

W_g = work done by the force of gravity = mgL (path independent)

From Eq. (i), $0 - mgL = \frac{1}{2}mv^2 - \frac{1}{2}mu^2$

$$\because \quad v^2 = u^2 - 2gL$$

$$\therefore \quad |\Delta\mathbf{v}| = \sqrt{v^2 + u^2} = \sqrt{2\,(u^2 - gl)}$$

46.

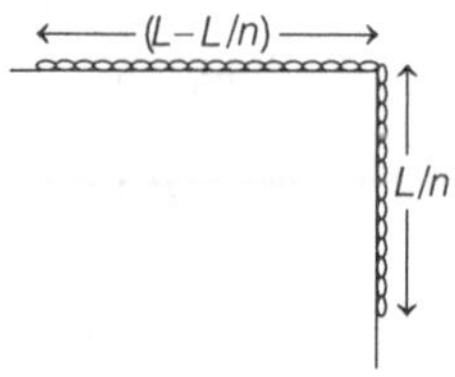

Given, mass of the cable is M.

So, mass of $\frac{1}{n}$th part of the cable, *i.e.* hanged part of the cable is M/n. ...(i)

Now, centre of mass of the hanged part will be its middle point.

So, its distance from the top of the table will be $L/2n$.

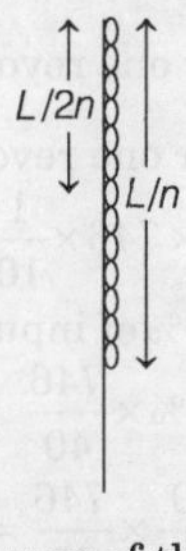

$\therefore$ Initial potential energy of the hanged part of cable,

$$U_i = \left(\frac{M}{n}\right)(-g)\left(\frac{L}{2n}\right)$$

$$\Rightarrow \quad U_i = -\frac{MgL}{2n^2} \quad ...(ii)$$

When the whole cable is on the table, its potential energy will be zero.

$$\therefore \quad U_f = 0 \quad ...(iii)$$

Now, using work-energy theorem,

$$W_{net} = \Delta U = U_f - U_i$$

$$\Rightarrow \quad W_{net} = 0 - \left(-\frac{MgL}{2n^2}\right)$$

[using Eqs. (ii) and (iii)]

$$\Rightarrow \quad W_{net} = \frac{MgL}{2n^2}$$

47. As the shotput reaches the ground, its KE

= PE of shotput when it is thrown + KE given

$$= mgh + \frac{1}{2}mv^2$$

$$= 10 \times 10 \times 1.5 + \frac{1}{2} \times 10\,(1)^2$$

$$= 150 + 5 = 155\text{ J}$$

48 Initial KE of the body $= \frac{1}{2}mv^2 = \frac{1}{2} \times 25 \times 4 = 50\text{ J}$

Work done against resistive force

= Area between F-x graph

$$= \frac{1}{2} \times 4 \times 20 = 40\text{ J}$$

Final KE = Initial KE − Work done against resistive force

$$= 50 - 40 = 10\text{ J}$$

49. If there is no air drag, then maximum height,

$$H = \frac{u^2}{2g} = \frac{14 \times 14}{2 \times 9.8} = 10\text{ m}$$

But due to air drag ball reaches upto height 8 m only, so loss of energy

$$= Mg\,(10 - 8) = 0.5 \times 9.8 \times 2 = 9.8\text{ J}$$

50. Vertical height, $h = l\cos\theta = l\cos 30^\circ$

Loss of potential energy = Mgh

$$= Mgl\cos 30^\circ = \frac{\sqrt{3}}{2}Mgl$$

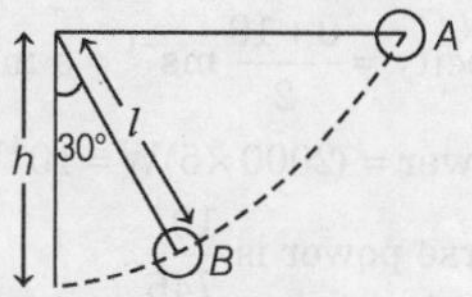

$\therefore$ Kinetic energy gained $= \frac{\sqrt{3}}{2}Mgl$

51. (a) If the surface is smooth, then the kinetic energy at B never be zero.

(b) If the surface is rough, the kinetic energy at B be zero. Because, work done by force of friction is negative. If work done by friction is equal to mgh, then net work done on body will be zero. Hence, net change in kinetic energy is zero. Hence, (b) is correct.

(c) If the surface is rough, the kinetic energy at B must be lesser than mgh. If surface is smooth, the kinetic energy at B is equal to mgh.

(d) The reason is same as in (a) and (b)

52. Applying principle of conservation of linear momentum, velocity of the system (v) is

$$m_1 v_1 = (m_1 + m_2)v$$

$$\Rightarrow \quad v = \frac{m_1 v_1}{m_1 + m_2} = \frac{50 \times 10 \times 100}{(50 + 950) \times 100} = \frac{1}{2}\text{ ms}^{-1}$$

Initial KE, $\quad E_1 = \frac{1}{2}m_1 v_1^2$

$$= \frac{1}{2} \times \left(\frac{50}{1000}\right) \times 10^2 = 2.5\text{ J}$$

Final KE, $\quad E_2 = \frac{1}{2}(m_1 + m_2)v^2$

$$= \frac{1}{2}\frac{(50 + 950)}{1000} \times \left(\frac{1}{2}\right)^2 = 0.125\text{J}$$

Percentage loss in KE,

$$\frac{E_1 - E_2}{E_1} \times 100 = \frac{2.5 - 0.125}{2.5} \times 100 = 95\%$$

53. $x = 2t^4 + 5t + 4 \Rightarrow v = \frac{dx}{dt} = 8t^3 + 5$

At $t = 0$, $v = 5$ m/s

At $t = 1$ s, $v = 8 \times 1 + 5 = 13$ m/s

Increase in KE $= \frac{1}{2}m\,[(13)^2 - (5)^2]$

$$= \frac{1}{2} \times 2 \times 144 = 144\text{ J}$$

54. As, $\quad P = \frac{mgh}{t}$

or $\quad \frac{m}{t} = \frac{P}{gh}$

or $\quad \frac{m}{t} = \frac{1000}{10 \times 10}\text{ kg} = 10\text{ kg}$

55. As, $a = \frac{10-0}{5}\ ms^{-2} = 2\ ms^{-2}$

$\because \quad F = ma$

$\therefore \quad F = (1000 \times 2)\ N = 2000\ N$

Average velocity $= \frac{0+10}{2} ms^{-1} = 5\ ms^{-1}$

Average power $= (2000 \times 5)W = 10^4 W$

Required horse power is $\frac{10^4}{746}$.

56. Volume of water of raise $= 22380\ L = 22380 \times 10^{-3}\ m^3$

$$P = \frac{mgh}{t} = \frac{V\rho gh}{t}$$

$$\Rightarrow \quad t = \frac{V\rho gh}{p}$$

$$t = \frac{22380 \times 10^{-3} \times 10^3 \times 10 \times 10}{10 \times 746} = 5\ min$$

57. As power, $P = \frac{\text{total energy}}{t} = \frac{mgh + \frac{1}{2}mv^2}{t}$

$$= \frac{10 \times 10 \times 20 + \frac{1}{2} \times 10 \times 10 \times 10}{1}$$

$= 2000 + 500 = 2500\ W$

$= 2.5\ kW$

58. Given, $h = 500\, m$, $\frac{dm}{dt} = 2000\ kgs^{-1}$

$\therefore$ Power output $= \frac{80}{100} \times \frac{dm}{dt} gh$

$= \frac{4}{5} \times 2000 \times 10 \times 500\ W$

$= 8 \times 10^6 W = 8\ MW$

59. Minimum force $mg\sin\theta$, so minimum power is given by

$$P = mg\sin\theta\, v \qquad (\because F = mg\sin\theta)$$

or $\quad v = \frac{P}{mg\sin\theta}$

or $\quad v = \frac{9000 \times 2}{1200 \times 10 \times 1} = 15\ ms^{-1}$

$= 15 \times \frac{18}{5} = 54\ kmh^{-1}$

60. At maximum load, force provided by motor to pull the lift,

F = weight carried + friction $= mg + f$

$= (2000 \times 10) + 4000 = 24000\ N$

Power delivered by motor at speed v of load, $P = F \times v$

$$\Rightarrow \quad v = \frac{P}{F} = \frac{60 \times 746}{24000} = 1.865 = 1.9\ ms^{-1}$$

61. Given, $u = 36\ km/h = 36 \times \frac{5}{18}\ ms^{-1}$

$= 10\, ms^{-1}$ and $v = 20\ ms^{-1}$

$\therefore$ Work done = Increase in kinetic energy

$= \frac{1}{2} \times 500\ [20^2 - 10^2]$

$= \frac{500 \times 30 \times 10}{2}$

Power $= \frac{500 \times 30 \times 10}{2 \times 60} = 1250\ W$

62. Motor makes 600 revolution per minute,

$\therefore \quad n = 600 \frac{\text{revolution}}{\text{min}} = 10 \frac{\text{rev}}{\text{s}}$

$\therefore$ Time required for one revolution $= \frac{1}{10}$ s

Energy required for one revolution = Power $\times$ Time

$= \frac{1}{4} \times 746 \times \frac{1}{10} = \frac{746}{40}$ J

But work done = 40% of input

$= 40\% \times \frac{746}{40}$

$= \frac{40}{100} \times \frac{746}{40} = 7.46$ J

63. Here $a_c = \frac{v^2}{r} = k^2rt^2$

$\therefore \quad v = krt$

The tangential acceleration, $a_t = \frac{dv}{dt}$

$= \frac{d(krt)}{dt} = kr$

The work done by centripetal force will be zero.

So, power is delivered to the particle by only tangential force which acts in the same direction of instantaneous velocity.

$\therefore \quad$ Power $= F_t v = ma_t krt$

$= m\,(kr)(krt)$

$= mk^2r^2t$

64. Since, power, $P = \frac{W}{t} = F \times v$

$\Rightarrow \quad P = F \times v = ma \times v \qquad [\because F = ma]$

$\Rightarrow \quad P = mav \qquad \ldots(i)$

Again, by equation of motion,

$v^2 - u^2 = 2ax$

$\Rightarrow \quad v^2 - 0 = 2ax$

$\Rightarrow \quad a = \frac{v^2}{2x} \qquad \ldots(ii)$

From Eqs. (i) and (ii), we have

$$P = m \cdot \frac{v^2}{2x} \cdot v$$

$$P = \frac{mv^3}{2x}$$

$$\Rightarrow \quad v^3 = \left(\frac{2P}{m}\right)x$$

Since, P is constant, hence

$v^3 \propto x$

$\Rightarrow \quad v \propto x^{1/3}$

65. Delivering power of a machine,

P = constant

$$P = F \cdot v \qquad \left[\because \text{Power} = \frac{W}{t} = F \cdot \frac{s}{t} = F \cdot v\right]$$

$$= mav = m\frac{dv}{dt}v$$

$$P = mv\frac{dv}{dt}$$

$$vdv = \frac{P}{m}\,dt$$

Integrating on both sides, we get

$$\int_0^v v\,dv = \int_0^t \frac{P}{m}\,dt$$

$$\frac{v^2}{2} = \frac{Pt}{m} \Rightarrow v = \left(\frac{2Pt}{m}\right)^{1/2}$$

$$v = \frac{ds}{dt} = \left(\frac{2Pt}{m}\right)^{1/2}$$

$$ds = \left(\frac{2Pt}{m}\right)^{1/2} dt$$

$$\int_0^s ds = \sqrt{\frac{2P}{m}} \int_0^t t^{1/2}\,dt$$

$$s = \sqrt{\frac{2P}{m}}\,\frac{t^{3/2}}{3/2}$$

$$\therefore \qquad s \propto t^{3/2}$$

66. Mass of elevator, $M = 920$ kg

Mass of all 10 passengers carried by elevator $= 10 \times m$
$= 10 \times 68 = 680$ kg

Total weight of elevator and passengers
$= (M + 10m)g$
$= (920 + 680) \times 10$
$= 16000$ N

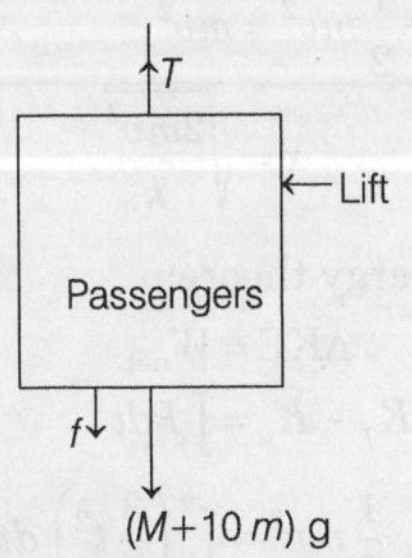

Force of friction $= 6000$ N

Total force (T) applied by the motor of elevator
$= 16000 + 6000 = 22000$ N

Power delivered by elevator's motor,

$$P = F \cdot v = 22000 \times 3 \qquad [\because v = 3\text{ms}^{-1}]$$
$$= 66000 \text{ W}$$

67. As, power $= \dfrac{d}{dt}$ (KE) $=$ constant

$$\Rightarrow \qquad \frac{d}{dt}\left(\frac{1}{2}mv^2\right) = P$$

$$\Rightarrow \qquad \frac{1}{2}m \cdot 2v\frac{dv}{dt} = P$$

$$\Rightarrow \qquad vdv = \frac{P}{m} \cdot dt$$

Integrating both sides, we get

$$\int v\,dv = \frac{P}{m}\int dt$$

$$\Rightarrow \qquad \frac{v^2}{2} = \frac{P}{m} \cdot t$$

$$\Rightarrow \qquad v = \sqrt{\frac{2P}{m}} \cdot t^{1/2}$$

$$\Rightarrow \qquad \frac{ds}{dt} = \sqrt{\frac{2P}{m}} \cdot t^{1/2}$$

$$\Rightarrow \qquad ds = \sqrt{\frac{2P}{m}} \cdot t^{1/2} dt$$

Again integrating both sides, we get

$$s = \int ds = \int \sqrt{\frac{2P}{m}} \cdot t^{1/2} dt$$

$$\Rightarrow \qquad s = C \cdot t^{3/2}$$

where, C is a constant $= \dfrac{2}{3}\sqrt{\dfrac{2P}{m}}$.

Hence, displacement (s)-time (t) graph is correctly represented in option (a).

Round II

1. Given, $\qquad s = \dfrac{t^3}{3}$

$$\therefore \qquad ds = t^2\,dt$$

$$\Rightarrow \qquad a = \frac{d^2 s}{dt^2} = \frac{d^2}{dt^2}\left[\frac{t^3}{3}\right] = 2\,t \text{ m/s}^2$$

Now work done by the force,

$$W = \int_0^2 F \cdot ds = \int_0^2 m \cdot a\,ds$$

$$\int_0^2 3 \times 2\,t \times t^2\,dt = \int_0^2 6\,t^3\,dt = \frac{3}{2}[t^4]_0^2 = 24 \text{ J}$$

2. When a man of mass m climbs up the staircase of height L, work done by the gravitational force on the man is $-mgL$ and work done by muscular force is mgL. If we ignore air resistance and friction, then the work done by all forces on man is equal to $-mgL + mgL =$ zero. Further, force from a step does not do work because the point of application of force does not move while the force exists.

3. Here, the displacement of an object is given by

$$x = (3\,t^2 + 5) \text{ m}$$

Therefore, velocity $(v) = \dfrac{dx}{dt} = \dfrac{d(3\,t^2 + 5)}{dt}$

or $\qquad v = 6\,t$ m/s $\qquad$ ...(i)

The work done in moving the object from $t = 0$ to $t = 5$ s,

$$W = \int_{x_0}^{x_5} F \cdot dx \qquad \text{...(ii)}$$

The force acting on this object is given by

$$F = ma = m \times \frac{dv}{dt}$$

$$= m \times \frac{d(6t)}{dt} \qquad \text{[using Eq. (i)]}$$

$$F = m \times 6 = 6\,m = 12\text{ N}$$

Also, $x_0 = 3t^2 + 5 = 3 \times (0)^2 + 5 = 5\text{ m}$

and at $t = 5$ s,

$$x_5 = 3 \times (5)^2 + 5 = 80\text{ m}$$

Put the values in Eq. (ii), we get

$$W = 12 \times \int_{x_0}^{x_5} dx = 12\,[80 - 5]$$

$$W = 12 \times 75 = 900\text{ J}$$

Alternate Solution

To using work-kinetic energy theorem,

$$W = \Delta\text{KE} = \frac{1}{2}m(v_f^2 - v_i^2)$$

$$= \frac{1}{2}m \times (30^2 - 0^2)$$

$$= \frac{1}{2} \times 2 \times 900 = 900\text{ J}$$

4. For $0 \le x \le 15$ m, $F = 200$ N ...(i)

For 15 m $< x \le 30$ m, force F is linearly decreasing from 200 N to 100 N.

So, using two-point form of straight line, we have

$$(y - y_1) = \left(\frac{y_2 - y_1}{x_2 - x_1}\right)(x - x_1)$$

$$(F - F_1) = \left(\frac{F_2 - F_1}{x_2 - x_1}\right)(x - x_1)$$

Here, $x_1 = 15$ m, $F_1 = 200$ N

$x_2 = 30$ m, $F_2 = 100$ N

So, $$(F - 200) = \left(\frac{100 - 200}{30 - 15}\right)(x - 15)$$

$$F - 200 = \frac{-100}{15}(x - 15)$$

$$F = 200 - \frac{20}{3}(x - 15)$$

$$F = 200 - \frac{20}{3}x + 100$$

$$F = \left(300 - \frac{20}{3}x\right)\text{ N} \qquad \text{...(ii)}$$

Therefore, $$F = \begin{cases} 200\text{ N}; & 0 \le x \le 15\text{ m} \\ \left(300 - \frac{20}{3}x\right)\text{ N}; & 15\text{ m} < x \le 30\text{ m} \end{cases}$$

Now, work done during the complete movement of the box,

$$W = \int_0^{30} F\,dx = \int_0^{15} 200\,dx + \int_{15}^{30}\left(300 - \frac{20x}{3}\right)dx$$

$$= 200\,[x]_0^{15} + \left[300x - \frac{20}{3}\frac{x^2}{2}\right]_{15}^{30}$$

$$= 200\,[15 - 0] + \left[300x - \frac{10}{3}x^2\right]_{15}^{30}$$

$$= 200 \times 15 + \left[\left\{300(30) - \frac{10}{3}(30)^2\right\} - \left\{300\,(15) - \frac{10}{3}(15)^2\right\}\right]$$

$$= 3000 + [\{9000 - 3000\} - \{4500 - 750\}]$$

$$= 3000 + [6000 - 3750]$$

$$= 3000 + 2250 = 5250\text{ J}$$

5. As, $E_1 = \frac{1}{2}mv^2$

$$E_2 = \frac{1}{2}m\,(v + 1)^2$$

$$\frac{E_2 - E_1}{E_1} = \frac{\frac{1}{2}m\,[(v+1)^2 - v^2]}{\frac{1}{2}mv^2} = \frac{44}{100}$$

On solving, we get $v = 5\text{ ms}^{-1}$.

6. Given, $m = 0.5$ kg and $u = 20$ m/s

Initial kinetic energy, $(K_i) = \frac{1}{2}mu^2$

$$= \frac{1}{2} \times 0.5 \times 20 \times 20 = 100\text{ J}$$

After deflection it moves with 5% of K_i

$\therefore$ $$K_f = \frac{5}{100} \times K_i \Rightarrow \frac{5}{100} \times 100$$

$\Rightarrow$ $$K_f = 5\text{ J}$$

Now, let the final speed be v m/s, then

$$K_f = 5 = \frac{1}{2}mv^2$$

$\Rightarrow$ $$v^2 = 20$$

$\Rightarrow$ $$v = \sqrt{20} = 4.47\text{ m/s}$$

7. $$\frac{1}{2}kx^2 = \frac{1}{2}mv^2 + \frac{1}{2}mv^2 = mv^2$$

$$x = \sqrt{\frac{2mv^2}{k}}$$

8. From work-energy theorem,

$$\Delta\text{KE} = W_{\text{net}}$$

or $$K_f - K_i = \int P\,dt$$

or $$\frac{1}{2}mv^2 = \int_0^2 \left(\frac{3}{2}t^2\right)dt$$

$$v^2 = \left[\frac{t^3}{2}\right]_0^2$$

$$v = 2\text{ ms}^{-1}$$

9. As slope of problem graph is positive and constant upto certain distance and then it becomes zero.

So, from $F = \frac{-dU}{dx}$, upto distance F = constant (negative) and become zero suddenly.

10. Here, $m_1 = 20$ kg

$m_2 = 0.1$ kg

where, v_1 = velocity of recoil of gun

and v_2 = velocity of bullet.

As, $m_1 v_1 = m_2 v_2$ ($\because$ momentum is conserved)

$$v_1 = \frac{m_2}{m_1} v_2 = \frac{0.1}{20} v_2 = \frac{v_2}{200}$$

$$\text{Recoil energy of gun} = \frac{1}{2} m_1 v_1^2 = \frac{1}{2} \times 20 \left(\frac{v_2}{200}\right)^2$$

$$804 = \frac{10\, v_2^2}{4 \times 10^4} = \frac{v_2^2}{4 \times 10^3}$$

$$v_2 = \sqrt{804 \times 4 \times 10^3} \text{ ms}^{-1}$$

11. KE lost is $\frac{3}{4}$ th, therefore KE left is $\frac{1}{4}$ th. Hence, velocity of particle reduces from v_0 to $\frac{v_0}{2} = v_0 - \mu g\, t_0$

or $$\mu = \frac{v_0}{2 g t_0}$$

12. Let x be the maximum extension of the spring as shown in figure. From conservation of mechanical energy, decreases in gravitational potential energy = increase in elastic potential energy

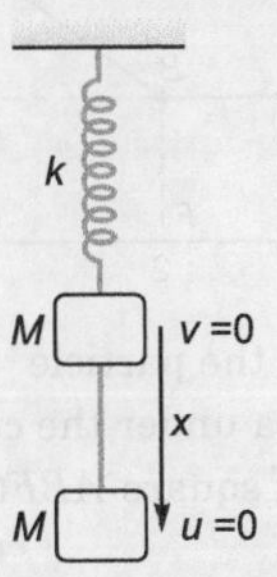

$$\therefore \quad Mgx = \frac{1}{2} kx^2$$

$$\therefore \quad x = \frac{2\,Mg}{k}$$

13. Let v be the velocity with which the bullet will emerge. Now, change in kinetic energy = work done

For first case, $\frac{1}{2} m\,(100)^2 - \frac{1}{2} m \times 0 = F$ ...(i)

For second case, $\frac{1}{2} m (100)^2 - \frac{1}{2} m v^2 = F \times 0.5$ ...(ii)

Dividing Eq. (ii) by Eq.(i), we get

$$\frac{(100)^2 - (v)^2}{(100)^2} = \frac{0.5}{1} = \frac{1}{2}$$

or $$v = \frac{100}{\sqrt{2}} = 50\sqrt{2} \text{ ms}^{-1}$$

14. Initial velocity of particle, $v_i = 20 \text{ ms}^{-1}$

Final velocity of the particle, $v_f = 0$

According to work-energy theorem,

$$W_{\text{net}} = \Delta \text{KE} = K_f - K_i$$

$$= \frac{1}{2} m (v_f^2 - v_i^2) = \frac{1}{2} \times 2 (0^2 - 20^2)$$

$$= -400 \text{ J}$$

15. If W_1 = work done by applied force

W_2 = work done against friction then applying work-energy theorem

$$W_1 - W_2 = \text{PE} + \text{KE (at the top)}$$

$$F \times s - W_2 = mgh + \frac{1}{2} \times mv^2$$

$$100 \times 12 - W_2 = 50 \times 10 \times 2 + \frac{1}{2} \times 50 \times 2^2$$

$$1200 - W_2 = 1100$$

$$W_2 = 100 \text{ J}$$

16. Velocity at B when dropped from A

where, $AC = s$

$$v^2 = u^2 + 2g\,(s - x) \quad \text{...(i)}$$

$$v^2 = 2g(s - x) \quad \text{...(ii)}$$

Potential energy at $B = mgx$

$\therefore$ kinetic energy = 3 × potential energy

$$\frac{1}{2} m \times 2g(s - x) = 3 \times mgx$$

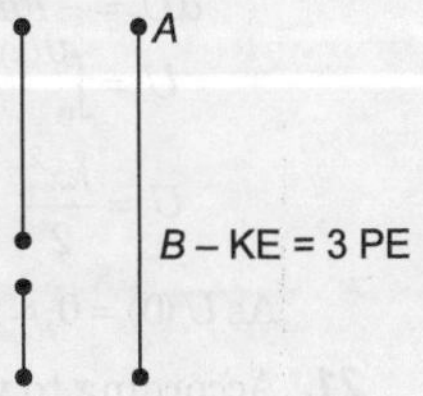

or $(s - x) = 3x$

or $s = 4x$

or $x = \frac{s}{4}$

From Eq. (i),

$$v^2 = 2g\,(s - x)$$

$$= 2g\left(s - \frac{s}{4}\right)$$

$$= \frac{2g \times 3s}{4} = \frac{3gs}{2}$$

$$\therefore \quad x = \frac{s}{4} \text{ and } v = \sqrt{\frac{3gs}{2}}$$

17. Let mass of boy be m, therefore mass of man = $2m$.

As, KE of man = $\frac{1}{2}$ KE of boy

$$\therefore \quad \frac{1}{2}(2m)\,u^2 = \frac{1}{2} \times \frac{1}{2} m\,u'^2$$

$$u^2 = \frac{u'^2}{4}, \; u = \frac{u'}{2}$$

When man speeds up to 1 m/s

KE of man = KE of boy

$$\frac{1}{2}(2m)\,(u+1)^2 = \frac{1}{2} m u'^2 = \frac{1}{2} m\,(2\,u)^2$$

$$(u+1)^2 = 2\,u^2$$

$$u + 1 = \sqrt{2}\,u$$

$$u = \frac{1}{\sqrt{2} - 1} = \frac{\sqrt{2} + 1}{(\sqrt{2} - 1)(\sqrt{2} + 1)}$$

$$u = (\sqrt{2} + 1) \text{ ms}^{-1}$$

$$u' = 2\,u = 2\,(\sqrt{2} + 1) \text{ ms}^{-1}$$

18. $\frac{1}{2}mv^2 = a(t)$

or $\quad v \propto t^{1/2}$

$$a = \frac{dv}{dt} \propto t^{-1/2}$$

or $\quad F \propto \frac{1}{\sqrt{t}}$ and $F \propto \frac{1}{v}$

19. Here $K = \frac{1}{2}mv^2 = as^2$

$\therefore \quad mv^2 = 2as^2$

Differentiating w.r.t. time t, we get

$$2mv\frac{dv}{dt} = 4as\frac{ds}{dt} = 4\,asv \Rightarrow m\frac{dv}{dt} = 2as$$

This is tangential force, $F_t = 2as$

Centripetal force, $F_c = \frac{mv^2}{R} = \frac{2as^2}{R}$

$\therefore$ Force acting on the particle,

$$F = \sqrt{F_t^2 + F_c^2} = \sqrt{(2as)^2 + \left(\frac{2as^2}{R}\right)^2} = 2as\sqrt{1 + s^2/R^2}$$

20. From $F = -\frac{dU}{dx}$

$$dU = -Fdx$$

$$U = \int_0^{U(x)} dU = -\int_0^x Fdx = -\int_0^x kx\,dx$$

$$U = \frac{kx^2}{2}$$

As $U(0) = 0, \propto x^2$ and U is negative.

21. According to work-energy theorem,

W = change in kinetic energy

$$Fs\cos\theta = \frac{1}{2}mv^2 - \frac{1}{2}mv^2$$

Substituting the given values, we get

$$20 \times 4 \times \cos\theta = 40 - 0$$

$$\cos\theta = \frac{40}{80} = \frac{1}{2}$$

$$\theta = \cos^{-1}\left(\frac{1}{2}\right) = 60^\circ$$

22. Given, $\quad \mathbf{F} = 3x^2\hat{\mathbf{i}} + 4\hat{\mathbf{j}}, \mathbf{r} = x\hat{\mathbf{i}} + y\hat{\mathbf{k}}$

Let, $\quad d\mathbf{r} = dx\hat{\mathbf{i}} + dy\hat{\mathbf{j}}$

Work done, $W = \int \mathbf{F}\,d\mathbf{r} = \int_{(23)}^{(3,0)} (3x^2\hat{\mathbf{i}} + 4\hat{\mathbf{j}})\,(dx\hat{\mathbf{i}} + dy\hat{\mathbf{j}})$

$$= \int_{(2,3)}^{(3,0)} (3x^2\,dx + 4\,dy) = (x^3 + 4\,y)_{(2,3)}^{(3,0)}$$

$$= 3^3 + 4 \times 0 - (2^3 + 4 \times 3)$$

$$= 27 + 0 - 20 = +7\text{ J}$$

23. $\int F\,dt = \Delta p$

$$\Rightarrow \quad \frac{1}{2} \times 4 \times 3 - \frac{1}{2} \times 1.5 \times 2 = p_f - 0$$

$$\Rightarrow \quad p_f = 6 - 1.5 = \frac{9}{2}$$

$$\text{KE} = \frac{p^2}{2m} = \frac{81}{4 \times 2 \times 2}$$

$$\text{KE} = 5.06\text{ J}$$

24. From given information $a = -kx$, where a is acceleration, x is displacement and k is a proportionality constant.

$$\frac{vdv}{dx} = -kx \Rightarrow v\,dv = -kx\,dx$$

Let for any displacement from 0 to x, the velocity changes from v_0 to v.

$$\Rightarrow \quad \int_{v_0}^{v} v\,dv = -\int_0^x kx\,dx$$

$$\Rightarrow \quad \left[\frac{v^2}{2}\right]_{v_0}^{v} = -k\left[\frac{x^2}{2}\right]_0^x$$

$$\Rightarrow \quad \frac{v^2 - v_0^2}{2} = -\frac{kx^2}{2}$$

$$\Rightarrow \quad m\left(\frac{v^2 - v_0^2}{2}\right) = -\frac{mkx^2}{2}$$

$$\frac{1}{2}mv^2 - \frac{1}{2}mv_0^2 = -\frac{mkx^2}{2}$$

$\Rightarrow \quad \Delta K \propto x^2 \quad$ [ΔK is loss in KE]

25. Area under force-displacement graph gives the value of work done.

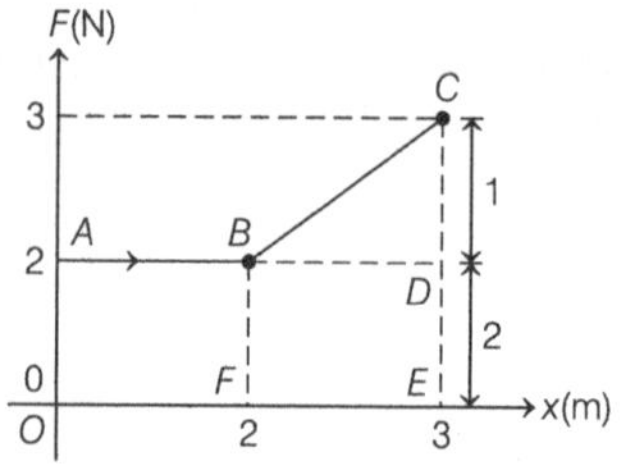

$\therefore$ Work done on the particle

= Area under the curve ABC

W = Area of square $ABFO$ + Area of ΔBCD + Area of rectangle $BDEF$

$$= 2 \times 2 + \frac{1}{2} \times 1 \times 1 + 2 \times 1 = 6.5\text{ J}$$

Now, from work-energy theorem,

$$\Delta W = K_f - K_i$$

$\Rightarrow \quad K_f = \Delta W = 6.5\text{ J} \quad [\because K_i = 0]$

26. According to the work-energy theorem,

Net work done = Change in the kinetic energy

Here, net work done = work done due to external force (W_{ext}) + work done due to the spring (W_{spr}).

As, $\quad W_{\text{ext}} = F \cdot x$

and $\quad W_{\text{spr}} = \frac{-1}{2}kx^2$

$$\Rightarrow \quad \Delta\text{KE} = F \cdot x + \left(-\frac{1}{2}kx^2\right)$$

$$(\Delta\text{KE})_f - (\Delta\text{KE})_i = F \cdot x - \frac{1}{2}kx^2$$

$$\Rightarrow \quad \frac{1}{2}mv_{\max}^2 - \frac{1}{2}m(0)^2 = F \cdot \left(\frac{F}{k}\right) - \frac{1}{2}k\left(\frac{F}{k}\right)^2$$

[using Eq. (i)]

$$\Rightarrow \quad \frac{1}{2}mv_{max}^2 = \frac{F^2}{k} - \frac{F^2}{2k} = \frac{F^2}{2k}$$

or $$v_{max}^2 = \frac{F^2}{km}$$

$$\Rightarrow \quad v_{max} = F/\sqrt{km}$$

27. From the diagram,

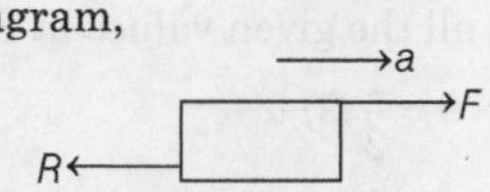

$$F - R = ma \quad \text{or} \quad F = R + ma$$

Rate of doing work = Power = $F \cdot v = (R + ma) \cdot v$

28. We know that, $P = F \cdot v = F \cdot \frac{l}{t}$

As F = constant

$\therefore \quad l \propto t^2$

$$\therefore \quad P = F \cdot \frac{l}{t} = F \cdot \frac{t^2}{t} = F \cdot T$$

or $P \propto t$

29. From $v = u + at$, $v_1 = 0 + at_1$ $\quad \left(\because a = \frac{v_1}{t_1}\right)$

$$F = ma = \frac{mv_1}{t_1}$$

Velocity acquired in t seconds $= at = \frac{v_1}{t_1} t$

$$\text{Power} = F \cdot v = \frac{mv_1}{t_1} \times \frac{v_1 t}{t_1} = \frac{mv_1^2 t}{t_1^2}$$

30. As, $P = (mg \sin\theta + F)v$

$$= \left(1000 \times 10 \times \frac{1}{20} + 200\right) \times 20$$

$$= 14000 \text{ W} = 14 \text{kW}$$

31. $\because$ Speed is constant.

$\therefore$ Work done by force = 0

$$\therefore \quad \text{Power} = \frac{\text{Work}}{\text{Time}} = 0$$

32. We know that,

$$P = F \cdot v \quad \text{...(i)}$$

But $$F = \frac{\Delta p}{\Delta t} = \frac{mv}{t} \quad [\because \Delta p = mv]$$

$$= \frac{mv}{(l/v)} \quad \left[\because l = v \cdot t \Rightarrow t = \frac{l}{v}\right]$$

$$= \frac{mv^2}{l}$$

$$\Rightarrow \quad F = \left(\frac{m}{l}\right) \cdot v^2 \quad \text{...(ii)}$$

From Eqs. (i) and (ii), we have

$$P = \left(\frac{m}{l}\right) \cdot v^2 \cdot v$$

$$= \left(\frac{m}{l}\right) \cdot v^3 = 100 \times 2^3 = 800 \text{ W}$$

33. Given, $P = 3t^2 - 2t + 1 = \frac{dE}{dt}$

$$\therefore \quad dE = (3t^2 - 2t + 1)dt$$

$$\Rightarrow \quad E = \int_{t=2s}^{t=4s} (3t^2 - 2t + 1)\, dt$$

$$= \left[\frac{3t^3}{3} - \frac{2t^2}{2} + t\right]_{t=2s}^{t=4s}$$

$$= [(4^3 - 2^3) - (4^2 - 2^2) + (4 - 2)]$$

or $$E = 56 - 12 + 2 = 46 \text{ J}$$

34. Work output of engine $= mgh = 100 \times 10 \times 10 = 10^4$ J

$$\text{Efficiency } (\eta) = \frac{\text{output}}{\text{input}} = \frac{10^4}{60} \times 100 = \frac{10^5}{6} \text{ J}$$

$$\therefore \quad \text{Power} = \frac{\text{input energy}}{\text{time}} = \frac{10^5/6}{5}$$

$$= \frac{10^5}{30} = 3.3 \text{ kW}$$

35. Power given to turbine $= \frac{mgh}{t}$

$$P_{in} = \left(\frac{m}{t}\right) gh = P_{in} = 15 \times 10 \times 60$$

$$P_{in} = 9000 \text{ W}$$

$$\Rightarrow \quad P_{in} = 9 \text{ kW}$$

As efficiency of turbine is 90%, therefore power generate = 90% of 9 kW

$$\Rightarrow \quad P_{out} = 9 \times \frac{90}{100}$$

$$\Rightarrow \quad P_{out} = 8.1 \text{ kW}$$

36. Work done by bowling machine = Initial kinetic energy of ball = Final potential energy of ball.

$\Rightarrow$ Force $\times$ displacement $= mgh$

$$\Rightarrow \quad F(0.2) = (0.15)(10)(20)$$

$$F = 150 \text{ N}$$

37. To slide the block, the boy has to overcome the force of friction $\mu mg = 300$ N

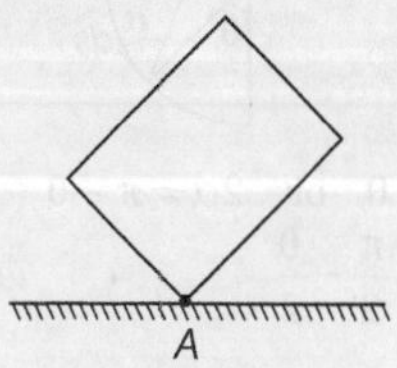

$$\therefore \quad W_1 = 300 \times 2 = 600 \text{ J}$$

To displace the block by tipping about corner A, the change in the height of centre of gravity is

$$h = \sqrt{2} - 1 = 1.414 - 1 = 0.414$$

$$\therefore \quad W_2 = mgh = 414 = 207 \times 2 \quad \text{(least)}$$

$$\therefore \quad n = 2$$

38. Here, $\frac{1}{2}mv^2 = bt^4$ or $mv^2 = 2bt^4$

The centripetal force on the body,

$$F_c = \frac{mv^2}{R} = \frac{2bt^4}{R}$$

and $$a_t = \frac{dv}{dt} = \sqrt{\frac{2b}{m}} \frac{d(t)^2}{dt} = \sqrt{\frac{8b}{m}}\, t$$

$\therefore$ Tangential force,

$$F_t = ma_t = m\sqrt{\frac{8b}{m}}\,t$$

The force on the particle,

$$F = \sqrt{\left(\frac{2bt^4}{R}\right)^2 + 8mbt^2}$$

$$= \sqrt{4+8} = \sqrt{12} = 2\sqrt{3}\text{ N}$$

$\therefore \quad 2\sqrt{3} = (2x)^{1/2}$

$\Rightarrow \quad 4\times 3 = 2x \Rightarrow x = 6$

39. Following is the situation given

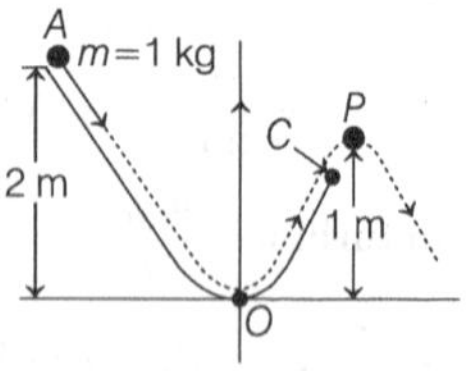

Since, the given path AOC is frictionless.

So, it starting from point A, the particle during the path AOC will attain maximum height at P.

Energy conservation at A and P gives

PE at A = (PE + KE) at P

$$\Rightarrow \quad U_A = U_P + K_P$$

$$mgh_A = mgh_P + K_P$$

$$1\times 10\times 2 = 1\times 10\times 1 + K_P$$

$$\therefore \quad K_P = 10\text{ J}$$

40. Here, $\frac{\pi}{2} - \alpha + \frac{\pi}{2} - \alpha = \theta$

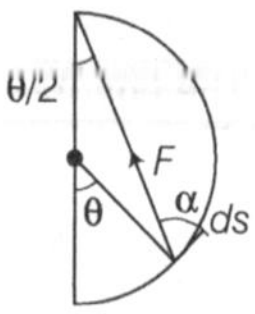

or $\quad \pi - 2\alpha = \theta$ or $2\alpha = \pi - \theta$

or $\quad \alpha = \frac{\pi}{2} - \frac{\theta}{2}$

$$dW = Fds\cos\alpha = F(r\,d\theta)\cos\left(\frac{\pi}{2} - \frac{\theta}{2}\right)$$

$$= Fr\sin\frac{\theta}{2}\,d\theta$$

Total work done, $W = \int dW = Fr\int_0^{\pi}\sin\frac{\theta}{2}\,d\theta$

$$= 2Fr\left[-\cos\frac{\theta}{2}\right]_0^{\pi} = 2Fr\,[0+1]$$

$$= 2Fr = 2\times 10\times 0.1 = 2\text{ J}$$

41. Let s be the required distance.

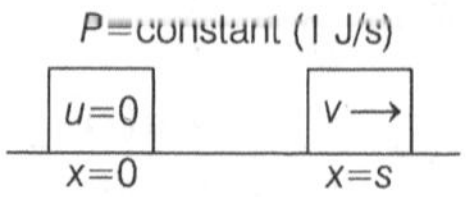

From work-energy theorem,

Work = Change in kinetic energy

$\Rightarrow \quad$ Power × Time = ΔK

i.e. $\quad Pt = \Delta K \Rightarrow Pt = \frac{1}{2}mv^2 \quad \ldots$(i)

Given, $\quad P = 1\text{ Js}^{-1}, t = 9\text{ s}, m = 2\text{ kg}$

Substituting all the given values in Eq. (i), we get

$$1\times 9 = \frac{1}{2}(2)\,v^2$$

$$v^2 = 9 \Rightarrow v = 3\text{ m/s} \quad (\text{at } t = 9\text{ s})$$

As, $\quad Fv = P \Rightarrow (ma)v = P \quad [\because F = ma]$

$$\Rightarrow \quad m\left[\frac{dv}{dt}\right]v = P$$

$$\Rightarrow \quad m\left[\frac{ds}{dt}\frac{dv}{ds}\right]v = P$$

$$\Rightarrow \quad m\left[v\frac{dv}{ds}\right]v = P$$

$$\Rightarrow \quad 2v^2dv = ds \quad \{\because P = 1\text{ J/s and } m = 2\text{ kg}\}$$

Integrating both sides, we get

$$\int_0^3 2v^2dv = \int_0^s ds \Rightarrow \frac{2}{3}[v^3]_0^3 = s$$

$$\frac{2}{3}[27-0] = s \Rightarrow s = 18\text{ m}$$

Hence, after 9 s, the body has moved a distance of 18 m.

42. Let's say the compression in the spring by y, so, by work-energy theorem, we have

$$\Rightarrow \quad \frac{1}{2}mv^2 = \frac{1}{2}ky^2$$

$$\Rightarrow \quad y = \sqrt{\frac{m}{k}}\cdot v \Rightarrow y = \sqrt{\frac{4}{100}}\times 10$$

$$\Rightarrow \quad y = 2\text{ m}$$

Final length of spring = 8 − 2 = 6 m

43. Let for bouncing of the block A, the elongation in the spring is x_1.

$$kx_1 > m_A g \quad \text{or} \quad kx_1 > 100$$

$$\therefore \quad x_1 > \frac{100}{150} \quad \text{or} \quad x_1 > \frac{2}{3}$$

$$\therefore \quad x_{1\,\min} = \frac{2}{3}$$

For $x_{1\,\min}$, the velocity of block B will be zero.

According to conservation principle of mechanical energy,

$$U_i + T_i = U_f + T_f$$

$$\frac{1}{2}kx_0^2 + \frac{1}{2}mv_{0\,\min}^2 = mg(x_0 + x_{1\,\min}) + \frac{1}{2}kx_{1\,\min}^2$$

Here, $\quad x_0 = \frac{mg}{k}$

By putting the value, we get

$$v_{0\,\min} = \frac{20}{\sqrt{15}}\text{ m/s}$$

$$\therefore \quad n = 5$$

08

Centre of Mass

Centre of mass of a system of particles is the point that behaves as whole mass of the system is concentrated on it and all the external forces are acting on it.

For rigid bodies, centre of mass is independent of the state of the body, *i.e.* whether it is in rest or in accelerated motion, centre of mass will remain same.

IN THIS CHAPTER ....

- Position of Centre of Mass
- Motion of Centre of Mass
- Collision

Position of Centre of Mass

For different particles system, the position of centre of mass is as given below,

(i) **Two Particles System** If a system consists two particles of masses m_1, m_2 and respective position vectors $\mathbf{r}_1, \mathbf{r}_2$, then the position of centre of mass is given by

$$\mathbf{r}_{CM} = \frac{m_1\mathbf{r}_1 + m_2\mathbf{r}_2}{m_1 + m_2}$$

If $m_1 = m_2 = m$(say), then $\mathbf{r}_{CM} = \frac{\mathbf{r}_1 + \mathbf{r}_2}{2}$.

(ii) **System of n Particles** If a system consists of n particles, of masses $m_1, m_2, \dots m_n$ with $\mathbf{r}_1, \mathbf{r}_2, \dots, \mathbf{r}_n$ as their respective position vectors, at a given instant of time, then the position vector of CM, *i.e.* $\mathbf{r}_{CM}$ of the system at that instant is given by

$$\mathbf{r}_{CM} = \frac{m_1\mathbf{r}_1 + m_2\mathbf{r}_2 + \dots m_n\mathbf{r}_n}{m_1 + m_2 + \dots + m_n} = \frac{\sum_{i=1}^{n} m_i\mathbf{r}_i}{\sum_{i=1}^{n} m_i} = \frac{\sum_{i=1}^{n} m_i\mathbf{r}_i}{M}$$

In terms of coordinates, $\mathbf{x}_{CM} = \frac{\sum_{i=1}^{n} m_i\mathbf{x}_i}{M}$

Further, $\mathbf{y}_{CM} = \frac{\sum_{i=1}^{n} m_i\mathbf{y}_i}{M}$

$$\mathbf{z}_{CM} = \frac{\sum_{i=1}^{n} m_i\mathbf{z}_i}{M}$$

Note The centre of mass of a body may lie within or outside the body. It is not at all necessary that some mass has to be present at the centre of mass.

Example 1. *The position vectors of three particles of masses $m_1 = 1$ kg, $m_2 = 2$ kg and $m_3 = 3$ kg are $\mathbf{r}_1 = (\hat{\mathbf{i}} + 4\hat{\mathbf{j}} + \hat{\mathbf{k}})$ m, $\mathbf{r}_2 = (\hat{\mathbf{i}} + \hat{\mathbf{j}} + \hat{\mathbf{k}})$ m and $\mathbf{r}_3 = (2\hat{\mathbf{i}} - \hat{\mathbf{j}} - 2\hat{\mathbf{k}})$ m respectively. Find position vector of their centre of mass.*

(a) $\frac{1}{3}(3\hat{\mathbf{i}}+2\hat{\mathbf{j}}-4\hat{\mathbf{k}})$ (b) $\frac{1}{2}(3\hat{\mathbf{i}}+\hat{\mathbf{j}}-\hat{\mathbf{k}})$

(c) $\frac{1}{3}(3\hat{\mathbf{i}}+\hat{\mathbf{j}}-\hat{\mathbf{k}})$ (d) $\frac{1}{2}(3\hat{\mathbf{i}}+2\hat{\mathbf{j}}-4\hat{\mathbf{k}})$

Sol. (b) The position vector of CM of the three parts will be given by

$$\mathbf{r}_{CM} = \frac{m_1\mathbf{r}_1 + m_2\mathbf{r}_2 + m_3\mathbf{r}_3}{m_1 + m_2 + m_3}$$

Substituting the values, we get

$$\mathbf{r}_{CM} = \frac{1(\hat{\mathbf{i}}+4\hat{\mathbf{j}}+\hat{\mathbf{k}}) + 2(\hat{\mathbf{i}}+\hat{\mathbf{j}}+\hat{\mathbf{k}}) + 3(2\hat{\mathbf{i}}-\hat{\mathbf{j}}-2\hat{\mathbf{k}})}{1+2+3}$$

$$= \frac{9\hat{\mathbf{i}}+3\hat{\mathbf{j}}-3\hat{\mathbf{k}}}{6}$$

$$\therefore \quad \mathbf{r}_{CM} = \frac{1}{2}(3\hat{\mathbf{i}}+\hat{\mathbf{j}}-\hat{\mathbf{k}}) \text{ m}$$

Example 2. *Three point particles of masses 1.0 kg, 1.5 kg and 2.5 kg are placed at three corners of a right angle triangle of sides 4.0 cm, 3.0 cm and 5.0 cm as shown in the figure. The centre of mass of the system is at a point*

[JEE Main 2020]

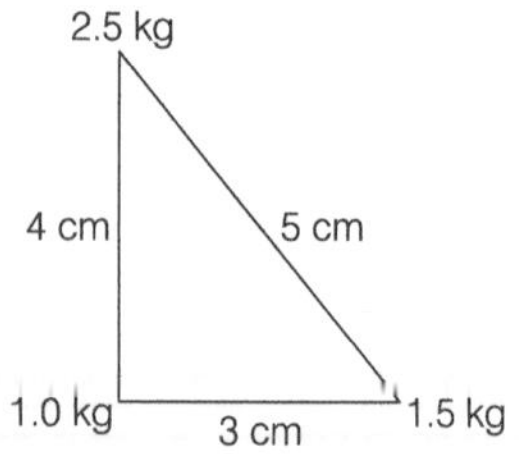

(a) 2.0 cm right and 0.9 cm above 1 kg mass
(b) 0.6 cm right and 2.0 cm above 1 kg mass
(c) 1.5 cm right and 1.2 cm above 1 kg mass
(d) 0.9 cm right and 2.0 cm above 1 kg mass

Sol. (d) We choose origin as shown in the figure.

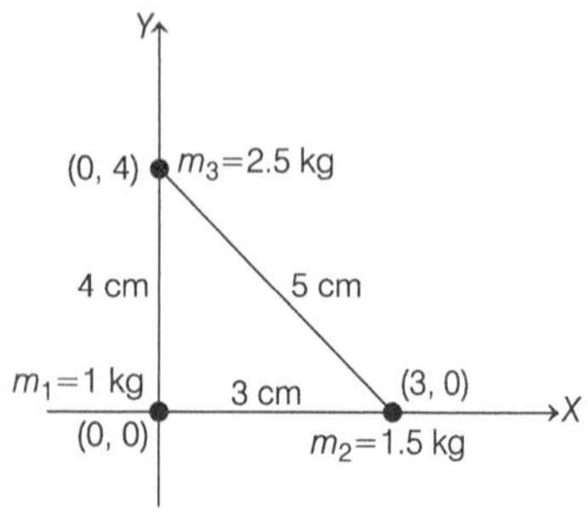

Using $x_{CM} = \frac{m_1x_1 + m_2x_2 + m_3x_3}{m_1 + m_2 + m_3}$, we have

$$x_{CM} = \frac{1\times 0 + 1.5\times 3 + 2.5\times 0}{1 + 1.5 + 2.5} = 0.9 \text{ cm}$$

Similarly,

$$y_{CM} = \frac{1\times 0 + 1.5\times 0 + 2.5\times 4}{1 + 1.5 + 2.5} = 2.0 \text{ cm}$$

So, centre of mass (CM) is 0.9 cm right and 2.0 cm above 1 kg mass.

Example 3. *The centre of mass of a uniform L-shaped lamina (a thin flat plate) with dimensions as shown is (g mass of lamina is 3 kg)*

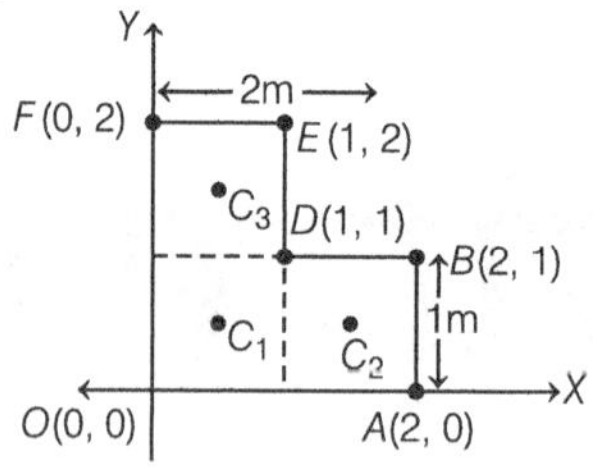

(a) $\left(\frac{6}{5}, \frac{5}{6}\right)$ (b) $\left(\frac{5}{6}, \frac{5}{6}\right)$ (c) $\left(\frac{5}{6}, \frac{6}{5}\right)$ (d) $\left(\frac{7}{5}, \frac{6}{5}\right)$

Sol. (b) Taking the L-shape to consist of 3 squares each of length 1m. The mass of each square is 1kg, since the lamina is uniform. The centre of mass C_1, C_2 and C_3 of the squares are, by symmetry their geometric centres and have coordinates $\left(\frac{1}{2}, \frac{1}{2}\right), \left(\frac{3}{2}, \frac{1}{2}\right), \left(\frac{1}{2}, \frac{3}{2}\right)$ respectively. We take the masses of the squares to be concentrated at these points. The centre of mass of the whole L shape (x, y) is the centre of mass of these mass points.

Hence, $$X = \frac{\left[1\left(\frac{1}{2}\right)+1\left(\frac{3}{2}\right)+1\left(\frac{1}{2}\right)\right] \text{kg-m}}{(1+1+1)\text{ kg}} = \frac{5}{6}\text{m}$$

and $$Y = \frac{\left[1\left(\frac{1}{2}\right)+1\left(\frac{1}{2}\right)+1\left(\frac{3}{2}\right)\right] \text{kg-m}}{(1+1+1)\text{ kg}} = \frac{5}{6}\text{m}$$

Centre of Mass of Rigid Continuous Bodies

A body is said to be a rigid body when it has perfectly definite shape and volume.

The distance between all points of particles of such a body do not change, while applying any force on it.

Since, a real rigid body contains so many particles (atoms) that we can treat it as a continuous distribution of matter. The particles then become differential mass elements dm, and the coordinates of the centre of mass are defined as

$$x_{CM} = \frac{1}{M}\int x\, dm$$

$$y_{CM} = \frac{1}{M}\int y\, dm$$

$$z_{CM} = \frac{1}{M}\int z\, dm$$

For uniform objects which have uniform density or mass per unit volume, we have

$$\rho = \frac{dm}{dV} = \frac{M}{V}$$

where dV is the volume occupied by a mass element dm and V is the total volume of the object. Thus, we find that

$$x_{CM} = \frac{1}{V}\int x\,dV$$
$$y_{CM} = \frac{1}{V}\int y\,dV$$
$$z_{CM} = \frac{1}{V}\int z\,dV$$

Centre of Mass of Different Rigid Bodies

- For a *uniform rectangular, square* or *circular plate*, CM lies at its centre.

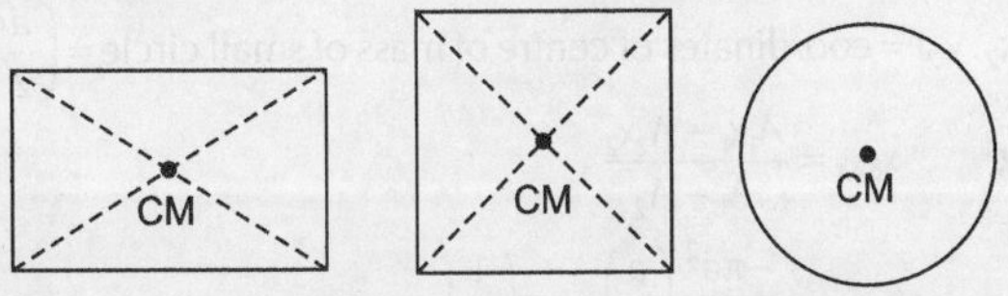

- Centre of mass of a **uniform semicircular** ring lies at a distance of $h = \frac{2R}{\pi}$ from its centre, on the axis of symmetry, where R is the radius of the ring.

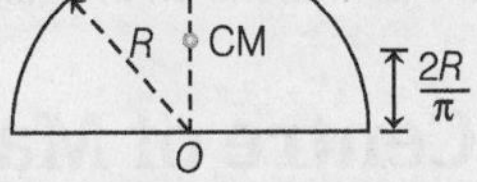

- For a **uniform semicircular disc** of radius R, CM lies at a distance of $h = \frac{4R}{3\pi}$ from the centre on the axis of symmetry as shown in the following figure.

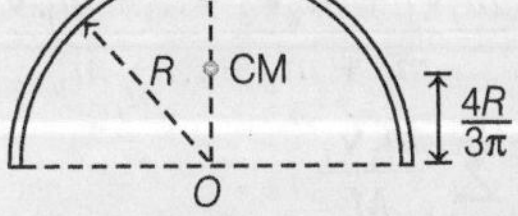

- Centre of mass of a **hemispherical shell** of radius R lies at a distance of $h = R/2$ from its centre on the axis of symmetry as shown in the following figure.

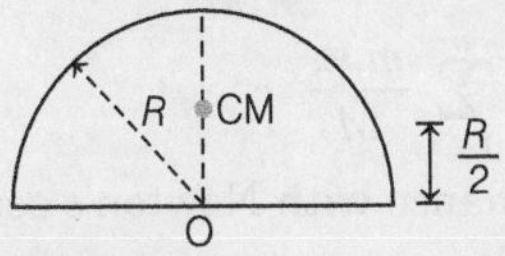

- For a **solid hemisphere** of radius R, CM lies at a distance of $h = \frac{3R}{8}$ from its centre on the axis of symmetry.

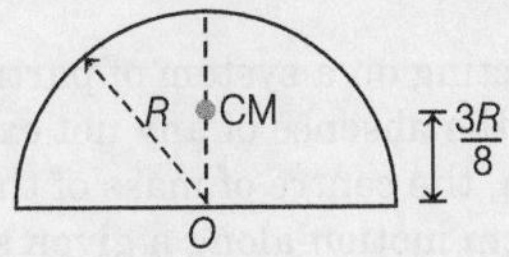

- Centre of mass of a uniform rod is located at its mid-point.

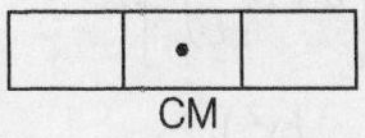

Example 4. *Find the centre of mass of a uniform semicircular ring of radius R and mass M.*

(a) $0, 2\pi R$ (b) $0, \frac{2R}{\pi}$ (c) $0, \frac{3R}{\pi}$ (d) $0, \pi R$

Sol. (b) Consider the centre of the ring as origin. Let a differential element of length dl of the ring whose radius vector makes an angle θ with the X-axis. If the angle subtended by the length dl is $d\theta$ at the centre, then $dl = R\,d\theta$.

Let λ be the mass per unit length.

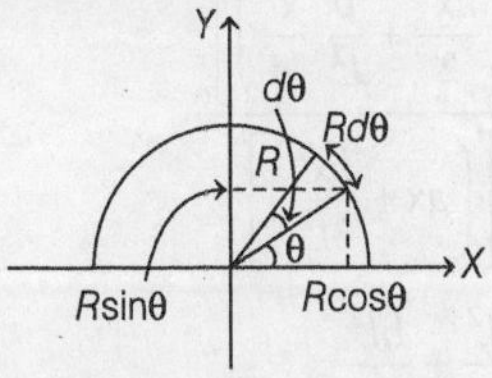

Then, mass of this element is $dm = \lambda R\,d\theta$

$$X_{CM} = \frac{1}{m}\int_0^{\pi} (R\cos\theta)\lambda R\,d\theta = 0$$

$\Rightarrow$ $$Y_{CM} = \frac{1}{m}\int_0^{\pi} (R\sin\theta)\cdot\lambda R\,d\theta$$

or $$= \frac{\lambda R^2}{m}\int_0^{\pi}\sin\theta\,d\theta = \frac{\lambda R^2}{\lambda\pi R}[-\cos\theta]_0^{\pi}$$

$\Rightarrow$ $$Y_{CM} = \frac{2R}{\pi}$$

As, $$m = \int_0^{\pi}\lambda R\,d\theta = 2\pi R$$

Example 5. *A rod of length L has non-uniform linear mass density given by* $\rho(x) = a + b\left(\frac{x}{L}\right)^2$, *where a and b are constants and* $0 \le x \le L$. *The value of x for the centre of mass of the rod is at* **[JEE Main 2020]**

(a) $\frac{3}{4}\left(\frac{2a+b}{3a+b}\right)L$ (b) $\frac{4}{3}\left(\frac{a+b}{2a+3b}\right)L$

(c) $\frac{3}{2}\left(\frac{a+b}{2a+b}\right)L$ (d) $\frac{3}{2}\left(\frac{2a+b}{3a+b}\right)L$

Sol. (a) For a continuous mass distribution,

$$x_{CM} = \frac{\int x\,dm}{\int dm}$$

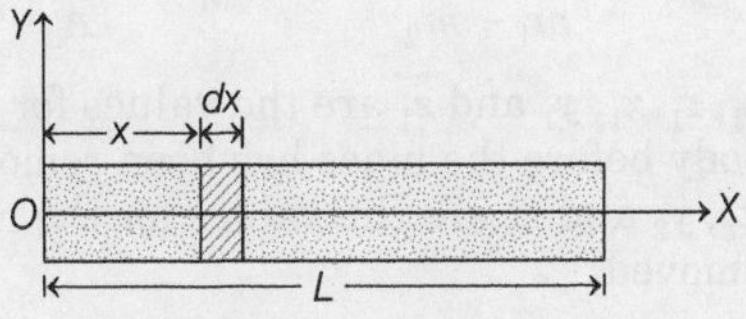

Here, mass of an elemental length dx of rod,

$$dm = \rho dx = \left[a + b\left(\frac{x}{L}\right)^2\right]dx$$

So, $$x_{CM} = \frac{\int_0^L x\left(a + \frac{bx^2}{L^2}\right)dx}{\int_0^L \left(a + \frac{bx^2}{L^2}\right)dx}$$

$$= \frac{\int_0^L \left(ax + \frac{bx^3}{L^2}\right)dx}{\int_0^L \left(a + \frac{bx^2}{L^2}\right)dx}$$

$$= \frac{\left[\left(\frac{ax^2}{2} + \frac{b}{L^2}\cdot\frac{x^4}{4}\right)\right]_0^L}{\left[\left(ax + \frac{bx^3}{3L^2}\right)\right]_0^L}$$

$$= \frac{\frac{aL^2}{2} + \frac{bL^2}{4}}{aL + \frac{bL}{3}}$$

$$= \left(\frac{\frac{a}{2} + \frac{b}{4}}{a + \frac{b}{3}}\right)L$$

So, $$x_{CM} = \frac{3}{4}\left(\frac{2a + b}{3a + b}\right)L$$

Position of Centre of Mass after Removal of a Part from a Rigid Body

If some mass or area is removed from a rigid body, then the position of centre of mass of the remaining portion is obtained from the following formulae.

$$\mathbf{r}_{CM} = \frac{m_1\mathbf{r}_1 - m_2\mathbf{r}_2}{m_1 - m_2}$$

or $$\mathbf{r}_{CM} = \frac{A_1\mathbf{r}_1 - A_2\mathbf{r}_2}{A_1 - A_2}$$

In terms of coordinates,

$$x_{CM} = \frac{m_1x_1 - m_2x_2}{m_1 - m_2} \text{ or } x_{CM} = \frac{A_1x_1 - A_2x_2}{A_1 - A_2}$$

$\therefore$ $$y_{CM} = \frac{m_1y_1 - m_2y_2}{m_1 - m_2} \text{ or } y_{CM} = \frac{A_1y_1 - A_2y_2}{A_1 - A_2}$$

and $$z_{CM} = \frac{m_1z_1 - m_2z_2}{m_1 - m_2} \text{ or } z_{CM} = \frac{A_1z_1 - A_2z_2}{A_1 - A_2}$$

Here, $m_1, A_1, \mathbf{r}_1, x_1, y_1$ and z_1 are the values for the mass of the whole body before the mass has been removed while $m_2, A_2, \mathbf{r}_2, x_2, y_2$ and z_2 are the values for the mass which has been removed.

Example 6. *Find the position of centre of mass of the uniform lamina shown in figure is*

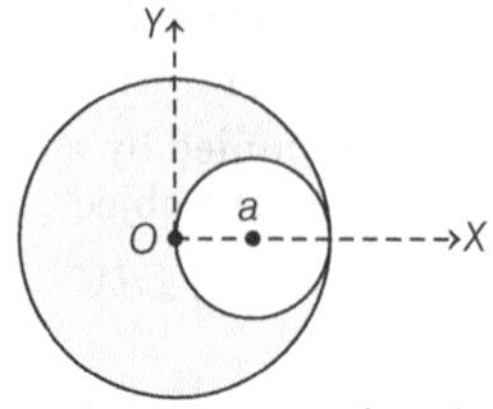

(a) $\left(-\frac{a}{6}, 0\right)$ (b) (0, 0) (c) $\left(\frac{a}{6}, 0\right)$ (d) $\left(\frac{a}{6}, -\frac{a}{6}\right)$

Sol. (a) Here, A_1 = area of complete circle = πa^2

and A_2 = area of small circle = $\pi\left(\frac{a}{2}\right)^2 = \frac{\pi a^2}{4}$

(x_1, y_1) = coordinates of centre of mass of large circle = (0, 0)

and (x_2, y_2) = coordinates of centre of mass of small circle = $\left(\frac{a}{2}, 0\right)$.

Using $$x_{CM} = \frac{A_1x_1 - A_2x_2}{A_1 - A_2}$$

We get, $$x_{CM} = \frac{\frac{-\pi a^2}{4}\left(\frac{a}{2}\right)}{\pi a^2 - \frac{\pi a^2}{4}} = \frac{-\left(\frac{1}{8}\right)}{\left(\frac{3}{4}\right)}a = -\frac{a}{6}$$

and $y_{CM} = 0$ as y_1 and y_2 both are zero. Therefore, the coordinates of centre of mass of the lamina shown in figure are $\left(-\frac{a}{6}, 0\right)$.

Motion of Centre of Mass

Consider the motion of a system of n particles of individual masses $m_1, m_2, \ldots, m_n$ and total mass M. It is assumed that no mass enters or leaves the system during its motion, so that M remains constant. Then, velocity of centre of mass is

$$\mathbf{v}_{CM} = \frac{m_1\mathbf{v}_1 + m_2\mathbf{v}_2 + \ldots + m_n\mathbf{v}_n}{m_1 + m_2 + \ldots + m_n}$$

or $$\mathbf{v}_{CM} = \sum_{i=1}^{n} \frac{m_i\mathbf{v}_i}{M}$$

Acceleration of the centre of mass is

$$\mathbf{a}_{CM} = \frac{m_1\mathbf{a}_1 + m_2\mathbf{a}_2 + \ldots + m_n\mathbf{a}_n}{M}$$

or $$\mathbf{a}_{CM} = \sum_{i=1}^{n} \frac{m_i\mathbf{a}_i}{M}$$

Further, in accordance with Newton's second law of motion, $\mathbf{F} = m\mathbf{a}$.

We can write, force on centre of mass is

$$\mathbf{F}_{CM} = \sum_{i=1}^{n} F_i$$

If total net force acting on a system of particles is zero, then $\mathbf{a}_{CM} = 0$. Hence, in the absence of any net external force acting on a system, the centre of mass of the system is either at rest or in uniform motion along a given straight line.

Example 7. *Four particles A, B, C and D with masses $m_A = m$, $m_B = 2m$, $m_C = 3m$ and $m_D = 4m$ are at the corners of a square. They have accelerations of equal magnitude with directions as shown in the figure. The acceleration of the centre of mass of the particles (in ms^{-2}) is* **[JEE Main 2019]**

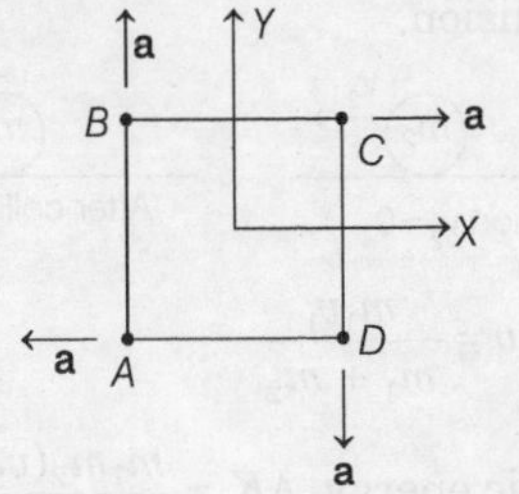

(a) $\frac{a}{5}(\hat{\mathbf{i}} - \hat{\mathbf{j}})$ (b) $a(\hat{\mathbf{i}} + \hat{\mathbf{j}})$

(c) zero (d) $\frac{a}{5}(\hat{\mathbf{i}} + \hat{\mathbf{j}})$

Sol. (a) For a system of discrete masses, acceleration of centre of mass (CM) is given by

$$\mathbf{a}_{CM} = \frac{m_A\mathbf{a}_A + m_B\mathbf{a}_B + m_C\mathbf{a}_C + m_D\mathbf{a}_D}{m_A + m_B + m_C + m_D}$$

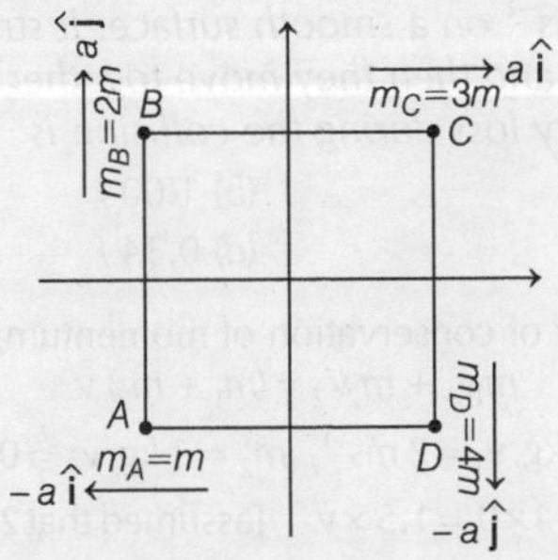

where, $m_A = m, m_B = 2m, m_C = 3m$ and $m_D = 4m$,

$|\mathbf{a}_A| = |\mathbf{a}_B| = |\mathbf{a}_C| = |\mathbf{a}_D| = a$ (according to the question)

$$a_{CM} = \frac{-ma\hat{\mathbf{i}} + 2ma\hat{\mathbf{j}} + 3ma\hat{\mathbf{i}} - 4ma\hat{\mathbf{j}}}{m + 2m + 3m + 4m}$$

$$= \frac{2a\hat{\mathbf{i}} - 2a\hat{\mathbf{j}}}{10}$$

$$= \frac{a}{5}\cdot(\hat{\mathbf{i}} - \hat{\mathbf{j}})\ ms^{-2}$$

Note Centre of mass is not the geometric centre.

Collision

The physical interaction of two or more bodies in which equal and opposite forces act upon each other causing the exchange of energy and momentum is called collision.

Types of Collision

Collision between two bodies may be classified in two ways

(i) **Elastic Collision** A collision is said to be elastic, if total linear momentum and kinetic energy remains conserved before and after collision.

(ii) **Inelastic Collision** A collision is said to be inelastic, if only linear momentum remains conserved and not the kinetic energy, the collision is said to be perfectly inelastic, if approaching particles permanently stick to each other and move with a common velocity.

Elastic Collision in One Dimension

Let the two balls of mass m_1 and m_2, collide each other elastically with velocities u_1 and u_2 in the directions shown as in Fig. (a). Their velocities become v_1 and v_2 after the collision along the same line.

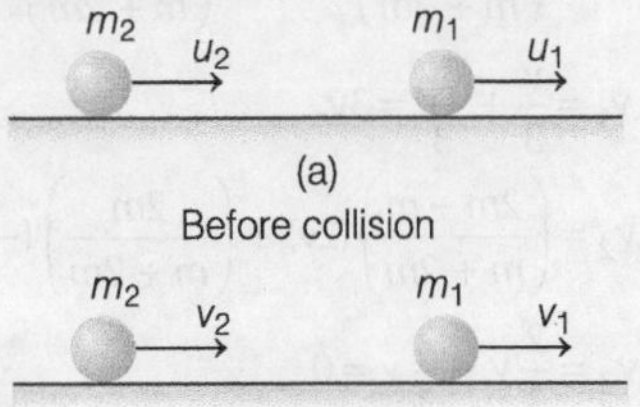

(a) Before collision

(b) After collision

Since, in a perfectly elastic collision, total energy and total linear momentum of colliding particles remains conserved, then

Relative velocity of approach = Relative velocity of separation, *i.e.* $u_1 - u_2 = v_2 - v_1$

$$\Rightarrow \quad v_1 = \left(\frac{m_1 - m_2}{m_1 + m_2}\right)u_1 + \left(\frac{2m_2}{m_1 + m_2}\right)u_2 \qquad \text{...(i)}$$

$$\text{and} \quad v_2 = \left(\frac{2m_1}{m_1 + m_2}\right)u_1 + \left(\frac{m_2 - m_1}{m_1 + m_2}\right)u_2 \qquad \text{...(ii)}$$

Special Cases of Elastic Collision in One Dimension

- If $m_1 = m_2$, then from Eqs. (i) and (ii), we get

$$v_1 = u_2 \text{ and } v_2 = u_1$$

i.e. when two particles of equal mass collide elastically and the collision is head on, they exchange their velocities.

- If $m_1 >> m_2$ and $u_1 = 0$, then

$$\frac{m_2}{m_1} \approx 0$$

Now, we get

$$v_1 \approx 0 \text{ and } v_2 \approx -u_2$$

i.e. The particle of mass m_1 remains at rest while the particle of mass m_2 bounces back with same velocity.

- If $m_2 >> m_1$ and $u_1 = 0$

Now, we get, $v_1 \approx 2u_2$ and $v_2 \approx u_2$

i.e. The mass m_1 (lighter particle) moves with velocity $2u_2$, while the velocity of mass m_2 (heavier particle) remains same.

Example 8. *Two particles of mass m and 2m moving in opposite directions collide elastically with velocities v and 2v. Find their velocities after collision.*

(a) 0, 3v (b) 3v, 0 (c) 3v, 3v (d) 0, 0

Sol. (b) Here, $u_1 = -v, u_2 = 2v, m_1 = m$ and $m_2 = 2m$

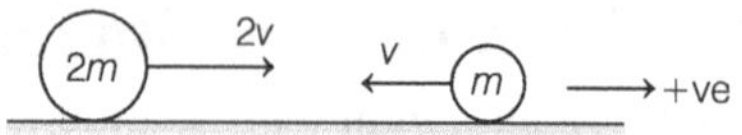

Substituting these values in Eqs. (i) and (ii), we get

$$v_1 = \left(\frac{m-2m}{m+2m}\right)(-v) + \left(\frac{4m}{m+2m}\right)(2v)$$

or $$v_1 = \frac{v}{3} + \frac{8v}{3} = 3v$$

and $$v_2 = \left(\frac{2m-m}{m+2m}\right)(2v) + \left(\frac{2m}{m+2m}\right)(-v)$$

or $$v_2 = \frac{2}{3}v - \frac{2}{3}v = 0$$

i.e. The second particle (of mass 2m) comes to a rest while the first (of mass m) moves with velocity 3v in the direction shown in figure.

Example 9. *In a collinear collision, a particle with an initial speed v_0 strikes a stationary particle of the same mass. If the final total kinetic energy is 50% greater than the original kinetic energy, the magnitude of the relative velocity between the two particles after collision, is* **[JEE Main 2018]**

(a) $\frac{v_0}{4}$ (b) $\sqrt{2}\,v_0$ (c) $\frac{v_0}{2}$ (d) $\frac{v_0}{\sqrt{2}}$

Sol. (b) Final kinetic energy is 50% more than initial kinetic energy

$$\Rightarrow \quad \frac{1}{2}mv_2^2 + \frac{1}{2}mv_1^2 = \frac{150}{100} \times \frac{1}{2}mv_0^2 \quad \text{...(i)}$$

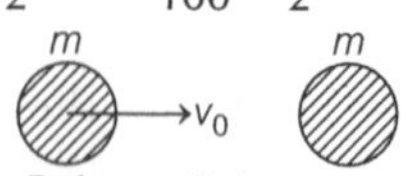

Before collision

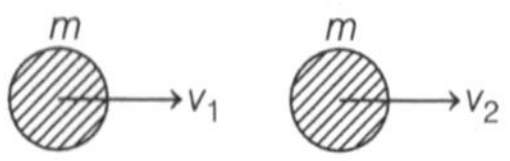

After collision

Conservation of momentum gives

$$mv_0 = mv_1 + mv_2$$

$$v_0 = v_2 + v_1 \quad \text{...(ii)}$$

From Eqs. (i) and (ii), we have

$$v_1^2 + v_2^2 + 2v_1v_2 = v_0^2$$

$$\Rightarrow \quad 2v_1v_2 = \frac{-v_0^2}{2}$$

$$\therefore \quad (v_1 - v_2)^2 = (v_1 + v_2)^2 - 4v_1v_2 = 2v_0^2$$

or $$v_{rel} = \sqrt{2}v_0$$

Inelastic Collision in One Dimension

In an inelastic collision, the total linear momentum as well as total energy remain conserved but total kinetic energy after collision is not equal to kinetic energy before collision.

For inelastic collision,

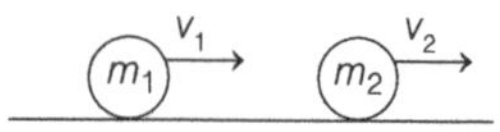

Before collision $v_2 = 0$

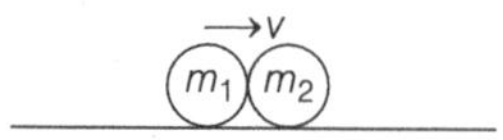

After collision

Common speed, $v = \dfrac{m_1 v_1}{m_1 + m_2}$

and loss of kinetic energy, $\Delta K = \dfrac{m_1 m_2 (v_1 - v_2)^2}{2(m_1 + m_2)}$

Here, $v_2 = 0$

$$\therefore \quad \Delta K = \frac{m_1 m_2 v_1^2}{2(m_1 + m_2)}$$

which is a positive quantity. Therefore, kinetic energy is lost mainly in the form of light, sound and heat.

Example 10. *A block of mass 0.50 kg is moving with a speed of 2.00 ms^{-1} on a smooth surface. It strikes another mass of 1.00 kg and then they move together as a single body. The energy loss during the collision is*

(a) 0.16 J (b) 1.00 J
(c) 0.67 J (d) 0.34 J

Sol. (c) From law of conservation of momentum, we have

$$m_1v_1 + m_2v_2 = (m_1 + m_2)\,v$$

Given, $m_1 = 0.50$ kg, $v_1 = 2\ ms^{-1}$, $m_2 = 1$ kg, $v_2 = 0$ [at rest]

$0.5 \times 2 + 1 \times 0 = 1.5 \times v$ [assumed that 2nd body is at rest]

$$\Rightarrow \quad v = \frac{2}{3}$$

$$\therefore \quad \Delta K = K_f - K_i$$

$$= \frac{1.5 \times \left(\frac{2}{3}\right)^2}{2} - (0.5) \times \frac{2^2}{2} = -\frac{2}{3}\,\text{J} = -0.67\,\text{J}$$

So, energy lost is 0.67 J.

Newton's Law of Restitution

According to this law, the ratio of relative velocity of separation after collision to the relative velocity of approach before collision remains constant,

$$e = \frac{v_2 - v_1}{u_2 - u_1}$$

where, e is called the **coefficient of restitution** and it is constant for two particular objects.

Following are few important points related to coefficient of restitution.

- In general, $0 \le e \le 1$
- $e = 0$, for completely inelastic collision, as both the objects stick together and $e = 1$, for an elastic collision.
- If $0 < e < 1$, the collision is said to be partially elastic.

- For head-on collision, the final velocities of the colliding bodies can be written as

$$v_1 = \left(\frac{m_1 - em_2}{m_1 + m_2}\right)u_1 + \frac{(1+e)m_2}{(m_1+m_2)}u_2$$

and $$v_2 = \frac{(1+e)m_1}{(m_1+m_2)}u_1 + \left(\frac{m_2 - em_1}{m_1 + m_2}\right)u_2$$

Putting, $e = 1$, we will get formulae of v_1 and v_2 for an elastic collision.

Similarly, putting $e = 0$, we get formulae for inelastic collision.

- The loss in KE during an inelastic collision can be given as,

$$\Delta K = \frac{1}{2}\frac{m_1 m_2}{m_1 + m_2}(1 - e^2)(u_1 - u_2)^2$$

However, if the target is massive (*i.e.* $m_2 >> m_1$) and $u_2 = 0$, then the lighter body loses all its kinetic energy.

Example 11. *A ball of mass m moving with a speed v makes a head-on collision with an identical ball at rest. The kinetic energy after collision of the balls is three-fourth the original kinetic energy. The coefficient of restitution (e) is*

(a) $\frac{1}{2}$ (b) $\frac{1}{3}$

(c) $\frac{1}{\sqrt{2}}$ (d) $\frac{1}{\sqrt{3}}$

Sol. (c) From the law of conservation of momentum,

$$mv = mv_1 + mv_2$$

or $$v = v_1 + v_2 \quad \text{...(i)}$$

which gives, $$e = \frac{v_2 - v_1}{v - 0}$$

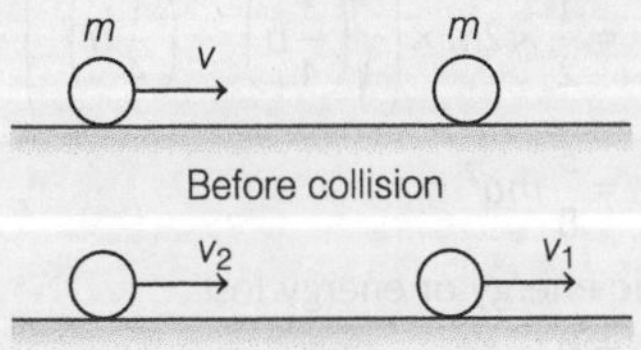

or $$v_2 - v_1 = ev \quad \text{...(ii)}$$

Adding Eqs. (i) and (ii), we get

$$v_2 = \frac{v + ev}{2}$$

and $$v_1 = \frac{(1-e)v}{2}$$

$$\frac{3}{4}\frac{mv^2}{2} = \frac{m}{2}\left[\frac{v^2(1+e)^2}{4} + \frac{(1-e)^2 v^2}{4}\right]$$

$\Rightarrow$ $$3 = (1+e)^2 + (1-e)^2 = 2(1+e^2)$$

$\Rightarrow$ $$e^2 = \frac{1}{2}$$

or $$e = \frac{1}{\sqrt{2}}$$

Rebounding of a Ball on Collision with the Floor

- Speed of the ball after the nth rebound

$$v_n = e^n v_0 = e^n\sqrt{2gh_0}$$

- Height covered by the ball after the nth rebound

$$h_n = e^{2n} h_0$$

- Total distance (vertical) covered by the ball before it stops bouncing

$$H = h_0 + 2h_1 + 2h_2 + 2h_3 + \ldots = h_0\left(\frac{1+e^2}{1-e^2}\right)$$

- Total time taken by the ball before it stops bouncing

$$T = t_0 + t_1 + t_2 + t_3 + \ldots$$

$$= \sqrt{\frac{2h_0}{g}} + 2\sqrt{\frac{2h_1}{g}} + 2\sqrt{\frac{2h_2}{g}} + \ldots = \sqrt{\frac{2h_0}{g}}\left(\frac{1+e}{1-e}\right)$$

Two Dimensional or Oblique Collision

If the initial and final velocities of colliding bodies do not lie along the same line, then the collision is called two dimensional or oblique collision.

In horizontal direction,

$$m_1u_1\cos\alpha_1 + m_2u_2\cos\alpha_2 = m_1v_1\cos\beta_1 + m_2v_2\cos\beta_2$$

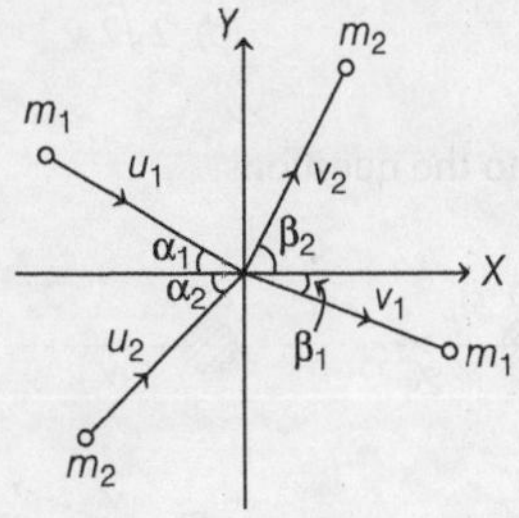

In vertical direction,

$$m_1u_1\sin\alpha_1 - m_2u_2\sin\alpha_2 = m_1v_1\sin\beta_1 - m_2v_2\sin\beta_2$$

If $m_1 = m_2$ and $\alpha_1 + \alpha_2 = 90°$, then $\beta_1 + \beta_2 = 90°$.

If a particle A of mass m_1 moving along X-axis with a speed u and makes an elastic collision with another stationary body B of mass m_2. Then,

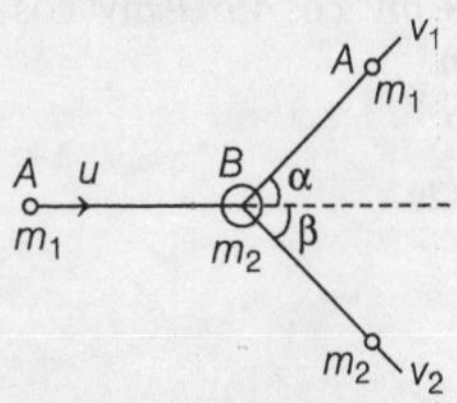

$$m_1v_1\sin\alpha = m_2v_2\sin\beta$$

Example 12. *After perfectly inelastic collision between two identical particles moving with same speed in different directions, the speed of the particles becomes half the initial speed. Find the angle between the two before collision.*

(a) 120° (b) 140° (c) 90° (d) 100°

Sol. (a) Let θ be the desired angle. Linear momentum of the system will remain conserved. Hence,

$$p^2 = p_1^2 + p_2^2 + 2p_1p_2 \cos\theta$$

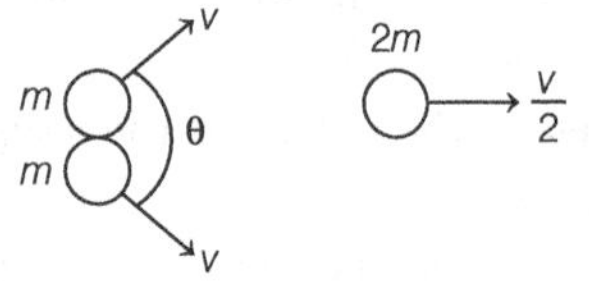

or $$\left\{2m\left(\frac{v}{2}\right)\right\}^2 = (mv)^2 + (mv)^2 + 2(mv)(mv)\cos\theta$$

or $$1 = 1 + 1 + 2\cos\theta$$

or $$\cos\theta = -\frac{1}{2}$$

$\therefore$ $$\theta = 120^\circ$$

Example 13. *A particle of mass m is moving with speed 2v and collides with a mass 2m moving with speed v in the same direction. After collision, the first mass is stopped completely while the second one splits into two particles each of mass m, which move at angle 45° with respect to the original direction. The speed of each of the moving particle will be* [JEE Main 2019]

(a) $\sqrt{2}\,v$ (b) $\dfrac{v}{\sqrt{2}}$

(c) $\dfrac{v}{(2\sqrt{2})}$ (d) $2\sqrt{2}\,v$

Sol. (d) According to the questions,

Initial condition,

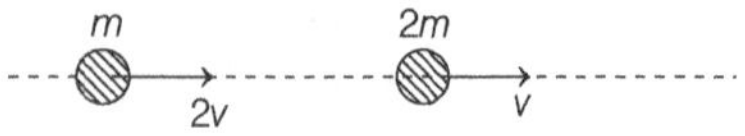

Final condition,

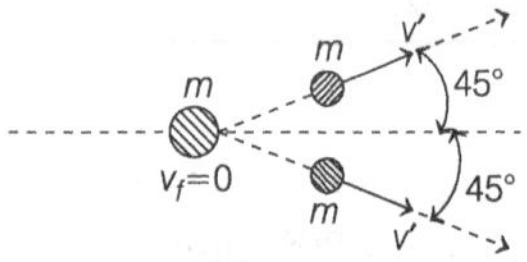

As we know that, in collision, linear momentum is conserved in both x and y directions separately.

So, $$(p_x)_{\text{initial}} = (p_x)_{\text{final}}$$

$$m(2v) + 2m(v) = 0 + mv'\cos 45^\circ + mv'\cos 45^\circ$$

$\Rightarrow$ $$4mv = \frac{2m}{\sqrt{2}}v'$$

$\Rightarrow$ $$v' = 2\sqrt{2}v$$

Example 14. *Two particles of equal mass m have respective initial velocities $u\hat{\mathbf{i}}$ and $u\left(\frac{\hat{\mathbf{i}}+\hat{\mathbf{j}}}{2}\right)$. They collide completely inelastically. The energy lost in the process is* [JEE Main 2020]

(a) $\frac{3}{4}mu^2$ (b) $\sqrt{\frac{2}{3}}mu^2$ (c) $\frac{1}{3}mu^2$ (d) $\frac{1}{8}mu^2$

Sol. (d) Collision between particles are as shown in the figure.

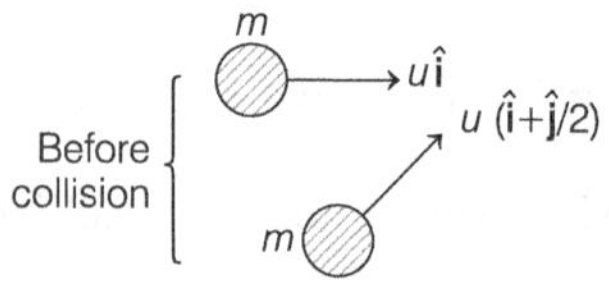

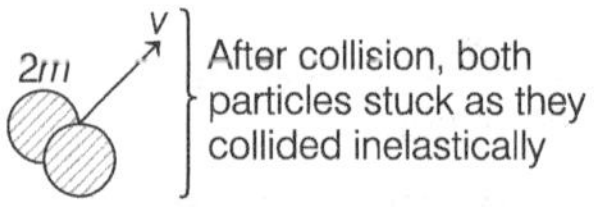

From momentum conservation, we have

$$m(u\hat{\mathbf{i}}) + mu\left(\frac{\hat{\mathbf{i}}+\hat{\mathbf{j}}}{2}\right) = 2m\mathbf{v}$$

$\Rightarrow$ $$\mathbf{v} = \frac{u}{2}\hat{\mathbf{i}} + u\left(\frac{\hat{\mathbf{i}}+\hat{\mathbf{j}}}{4}\right) = \frac{3}{4}u\hat{\mathbf{i}} + \frac{1}{4}u\hat{\mathbf{j}}$$

Initial kinetic energy of particles,

$$K_1 = \frac{1}{2}mu^2 + \frac{1}{2}m\left(\frac{u}{\sqrt{2}}\right)^2$$

$$= \frac{1}{2}mu^2 + \frac{1}{4}mu^2 = \frac{3}{4}mu^2$$

Final kinetic energy of combined particles,

$$K_2 = \frac{1}{2}(2m)v^2$$

$$= \frac{1}{2} \times 2m \times \left(\sqrt{\left(\frac{3}{4}u\right)^2 + \left(\frac{1}{4}u\right)^2}\right)^2$$

$$= \frac{5}{8}mu^2$$

Change in kinetic energy or energy lost

$$= K_1 - K_2$$

$$= \frac{3}{4}mu^2 - \frac{5}{8}mu^2$$

$$= \frac{1}{8}mu^2$$

Practice Exercise

ROUND I Topically Divided Problems

Position of Centre of Mass

1. A cricket bat is cut at the location of its centre of mass as shown. Then,

(a) the two pieces will have the same mass
(b) the bottom piece will have larger mass
(c) the handle piece will have larger mass
(d) mass of handle piece is double the mass of bottom piece

2. Which of the following does the centre of mass lie outside the body? **[NCERT Exemplar]**

(a) A pencil (b) A shotput
(c) A dice (d) A bangle

3. The centre of mass of two particles with masses 4 kg and 2 kg located at (1,0,1) and (2,2,0) respectively has coordinates.

(a) (1/3, 2/3, 2/3) (b) (4/3, 1/3, 1/3)
(c) (2/3, 1/3, 1/3) (d) (4/3, 2/3, 2/3)

4. Three rods of the same mass are placed as shown in figure. What will be the coordinates of centre of mass of the system?

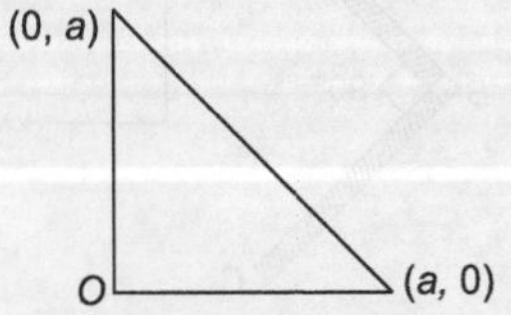

(a) $\left(\frac{a}{2}, \frac{a}{2}\right)$ (b) $\left(\frac{a}{\sqrt{2}}, \frac{a}{\sqrt{2}}\right)$ (c) $(\sqrt{2}\,a, \sqrt{2}\,a)$ (d) $\left(\frac{a}{3}, \frac{a}{3}\right)$

5. Which of the following points is the likely position of the centre of mass of the system shown in figure? **[NCERT Exemplar]**

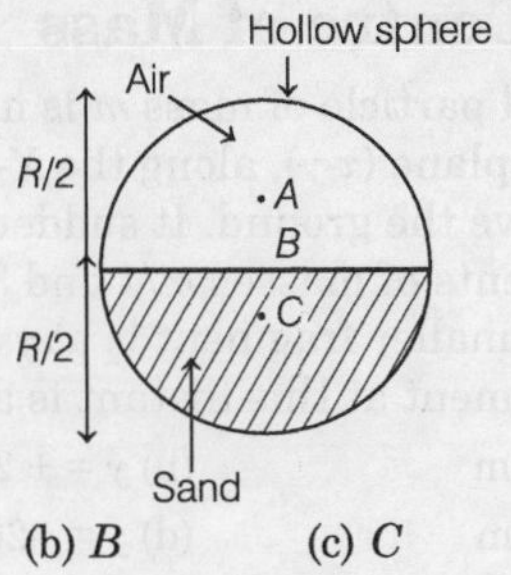

(a) A (b) B (c) C (d) D

6. Four particles of mass 1 kg, 2 kg, 3 kg and 4 kg are placed at the corners A, B, C and D respectively of a square ABCD of edge X-axis and edge AD is taken along Y-axis, the coordinates of centre of mass in SI unit is

(a) (1, 1) m
(b) (5, 7) m
(c) (0.5, 0.7) m
(d) None of the above

7. In the HCl molecule, the separation between the nuclei of the two atoms is about 1.27 Å (1 Å = 10^{-10} m). Find the approximate location of the centre of mass of the molecule, given that a chlorine atom is about 35.5 times as massive as a hydrogen atom and nearly all the mass of an atom is concentrated in its nucleus. **[NCERT]**

(a) r_{CM} = 1.24 Å
(b) r_{CM} = 2.24 Å
(c) r_{CM} = 0.24 Å
(d) r_{CM} = 3.24 Å

8. The centre of mass of three particles of masses 1 kg, 2 kg and 3 kg is at (2, 2, 2). The position of the fourth mass of 4 kg to be placed in the system as that the new centre of mass is at (0, 0, 0) is

(a) (−3, −3, −3) (b) (−3, 3, −3) **[EAMCET]**
(c) (2, 3, −3) (d) (2, −2, 3)

9. Three particles of masses 50 g, 100 g and 150 g are placed at the vertices of an equilateral triangle of side 1 m (as shown in the figure). The (x, y) coordinates of the centre of mass will be **[JEE Main 2019]**

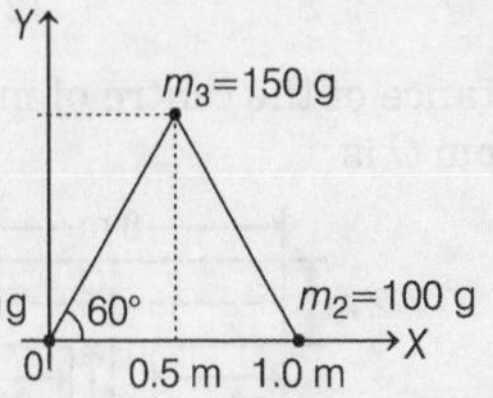

(a) $\left(\frac{\sqrt{3}}{4}\text{ m}, \frac{5}{12}\text{ m}\right)$ (b) $\left(\frac{7}{12}\text{ m}, \frac{\sqrt{3}}{8}\text{ m}\right)$
(c) $\left(\frac{7}{12}\text{ m}, \frac{\sqrt{3}}{4}\text{ m}\right)$ (d) $\left(\frac{\sqrt{3}}{8}\text{ m}, \frac{7}{12}\text{ m}\right)$

10. Three bricks each of length L and mass M are arranged from the wall as shown. The distance of the centre of mass of the system from the wall is

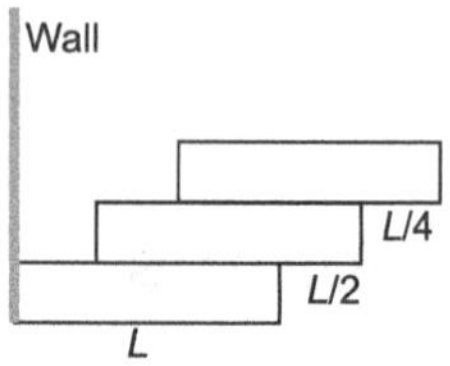

(a) $\frac{L}{4}$ (b) $\frac{L}{2}$ (c) $\frac{3}{2}L$ (d) $\frac{11}{12}L$

11. The coordinates of centre of mass of a uniform flag shaped lamina (thin flat plate) of mass 4 kg. (The coordinates of the same are shown in figure) are **[JEE Main 2020]**

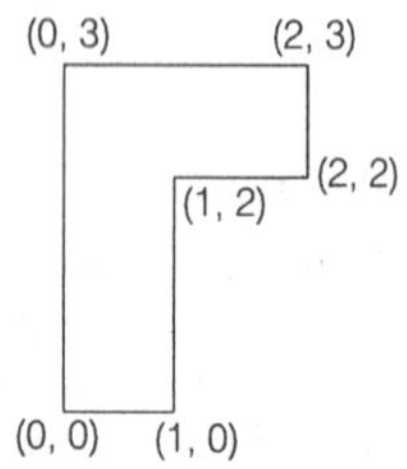

(a) (1.25 m, 1.50 m) (b) (1 m, 1.75 m)
(c) (0.75 m, 0.75 m) (d) (0.75 m, 1.75 m)

12. Look at the drawing given in the figure, which has been drawn with ink of uniform line thickness. The mass of ink used to draw each of the two inner circles and each of the two line segments is m. The mass of the ink used to draw the outer circle is $6m$. The coordinates of the centres of the different parts are outer circle (0, 0) left inner circle $(-a, a)$, right inner circle (a, a), vertical line (0, 0) and horizontal line $(0, -a)$. The y-coordinate of the centre of mass of the ink in this drawing is

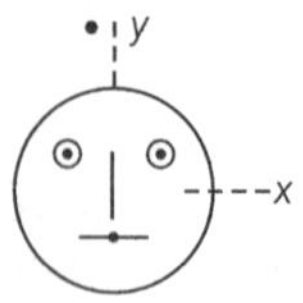

(a) $\frac{a}{10}$ (b) $\frac{a}{8}$ (c) $\frac{a}{12}$ (d) $\frac{a}{3}$

13. The distance of the centre of mass of the T-shaped plate from O is

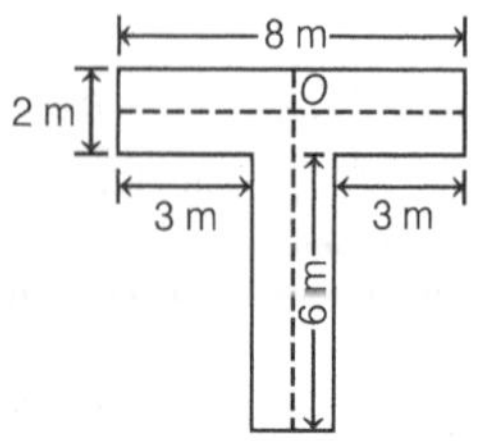

(a) 7 m (b) 2.7 m (c) 4 m (d) 1 m

14. A circular hole of radius 1 cm is cut-off from a disc of radius 6 cm. The centre of hole is 3 cm from the centre of the disc. The position of centre of mass of the remaining disc from the centre of disc is

(a) $-\frac{3}{35}$ cm (b) $\frac{1}{35}$ cm
(c) $\frac{3}{10}$ cm (d) None of these

15. The density of a non-uniform rod of length 1m is given by $\rho(x) = a(1+bx^2)$ where a and b are constants and $0 \le x \le 1$. The centre of mass of the rod will be at **[NCERT Exemplar]**

(a) $\frac{3(2+b)}{4(3+b)}$ (b) $\frac{4(2+b)}{3(3+b)}$
(c) $\frac{3(2+b)}{4(2+b)}$ (d) $\frac{4(3+b)}{3(2+b)}$

16. A non-uniform thin rod of length L is placed along X-axis as such its one of ends is at the origin. The linear mass density of rod is $\lambda = \lambda_0 x$. The distance of centre of mass of rod from the origin is

(a) $\frac{L}{2}$ (b) $\frac{2L}{3}$
(c) $\frac{L}{4}$ (d) $\frac{L}{5}$

17. An L-shaped object made of thin rods of uniform mass density is suspended with a string as shown in figure. If $AB = BC$ and the angle is made by AB with downward vertical is θ, then **[JEE Main 2019]**

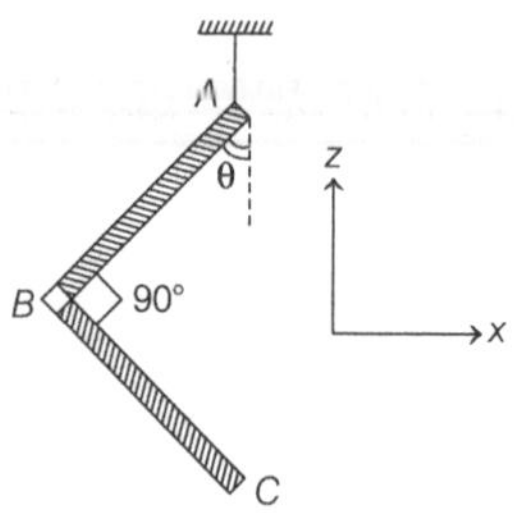

(a) $\tan\theta = \frac{2}{\sqrt{3}}$ (b) $\tan\theta = \frac{1}{2\sqrt{3}}$
(c) $\tan\theta = \frac{1}{2}$ (d) $\tan\theta = \frac{1}{3}$

Motion of Centre of Mass

18. An isolated particle of mass m is moving in a horizontal plane $(x\text{-}y)$, along the X-axis, at a certain height above the ground. It suddenly explodes into two fragments of masses $m/4$ and $3m/4$. An instant later, the smaller fragment is at $y = +15$ cm. The larger fragment at this instant is at

(a) $y = -5$ cm (b) $y = +20$ m
(c) $y = +5$ cm (d) $y = -20$ cm

19. Two identical particles move towards each other with velocity $2v$ and v, respectively. The velocity of centre of mass is

(a) v (b) $\frac{v}{3}$

(c) $\frac{v}{2}$ (d) zero

20. Find the velocity of centre of the system shown in the figure.

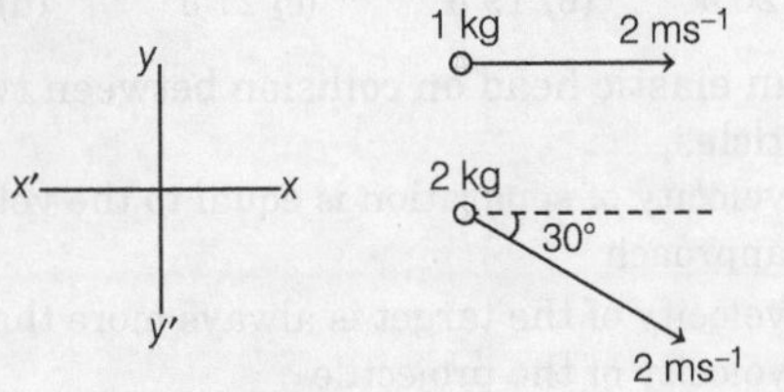

(a) $\left(\frac{2+2\sqrt{3}}{3}\right)\hat{i}-\frac{2}{3}\hat{j}$ (b) $4\hat{i}$

(c) $\left(\frac{2-2\sqrt{3}}{3}\right)\hat{i}-\frac{2}{3}\hat{j}$ (d) None of these

21. Consider a two particle system with particles having masses m_1 and m_2. If the first particles is pushed towards the centre of mass through a distance d, by what distance should the second particle be moved, so as to keep the centre of mass at the same position?

(a) $\frac{m_2}{m_1}d$ (b) $\frac{m_1}{m_1+m_2}d$

(c) $\frac{m_1}{m_2}d$ (d) d

22. Two blocks A and B are connected by a massless string (shown in figure). A force of 30 N is applied on block B. The distance travelled by centre of mass in 2 s starting from rest is

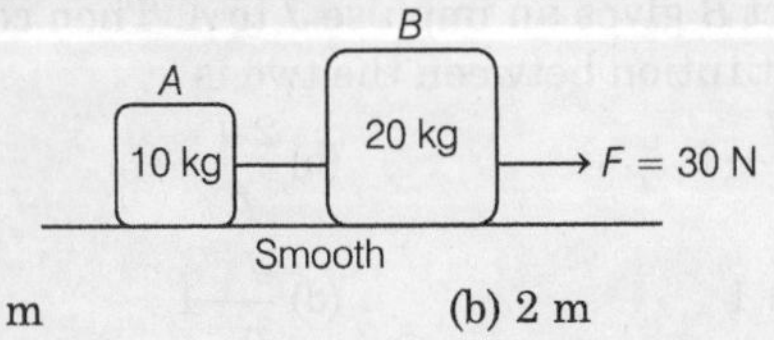

(a) 1 m (b) 2 m

(c) 3 m (d) None of these

23. A particle of mass m moving with a velocity $(3\hat{i}+2\hat{j})\,\text{ms}^{-1}$ collides with a stationary body of mass M and finally moves with a velocity $(-2\hat{i}+\hat{j})\,\text{ms}^{-1}$. If $\frac{m}{M}=\frac{1}{13}$, then

(a) the impulse received by M is $m(5\hat{i}+\hat{j})$

(b) the velocity of the M is $\frac{1}{13}(5\hat{i}+\hat{j})$

(c) the coefficient of restitution $\frac{11}{17}$

(d) All of the above are correct

Collision in One Dimension

24. Two equal masses m_1 and m_2 moving along the same straight line with velocities +3 m/s and −5 m/s respectively collide elastically. Their velocities after the collision will be respectively

(a) + 4 m/s for both

(b) − 3 m/s and + 5 m/s

(c) − 4 m/s and + 4 m/s

(d) − 5 m/s and + 3 m/s

25. A large block of wood of mass $M = 5.99$ kg is hanging from two long massless cords. A bullet of mass $m = 10$ g is fired into the block and gets embedded in it. The (block + bullet) then swing upwards, their centre of mass rising a vertical distance $h = 9.8$ cm before the (block + bullet) pendulum comes momentarily rest at the end of its arc. The speed of the bullet just before collision is (Take, $g = 9.8\ \text{ms}^{-2}$) **[JEE Main 2021]**

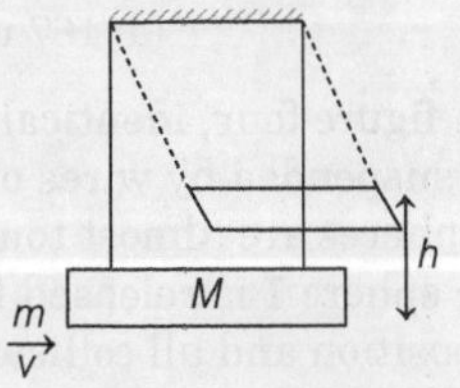

(a) 841.4 m/s (b) 811.4 m/s

(c) 831.4 m/s (d) 821.4 m/s

26. A body of mass 2 kg makes an elastic collision with a second body at rest and continues to move in the original direction but with one-fourth of its original speed. What is the mass of the second body? **[JEE Main 2019]**

(a) 1.5 kg (b) 1.2 kg

(c) 1.8 kg (d) 1.0 kg

27. A small block of mass M moves with velocity 5 ms^{-1} towards an another block of same mass M placed at a distance of 2 m on a rough horizontal surface. Coefficient of friction between the blocks and ground is 0.25. Collision between the two blocks is elastic, the separation between the blocks, when both of them come to rest, is (g = 10 ms^{-2})

(a) 3 m (b) 4 m

(c) 2 m (d) 1.5 m

28. A body of mass m_1 moving with an unknown velocity of $v_1\hat{i}$, undergoes a collinear collision with a body of mass m_2 moving with a velocity $v_2\hat{i}$. After collision, m_1 and m_2 move with velocities of $v_3\hat{i}$ and $v_4\hat{i}$, respectively. If $m_2 = 0.5m_1$ and $v_3 = 0.5v_1$, then v_1 is **[JEE Main 2019]**

(a) $v_4 + v_2$ (b) $v_4 - \frac{v_2}{4}$

(c) $v_4 - \frac{v_2}{2}$ (d) $v_4 - v_2$

29. An α-particle of mass m suffers one-dimensional elastic collision with a nucleus at rest of unknown mass. It is scattered directly backwards losing 64% of its initial kinetic energy. The mass of the nucleus is **[JEE Main 2019]**

(a) $1.5\,m$ (b) $4\,m$
(c) $3.5\,m$ (d) $2\,m$

30. A thick uniform bar lies on a frictionless horizontal surface and is free to move in any way on the surface. Its mass is 0.16 kg and length is 1.7 m. Two particles each of mass 0.08 kg are moving on the same surface and towards the bar in the direction perpendicular to the bar, one with a velocity of 10 ms^{-1} and other with velocity 6 ms^{-1}. If collision between particles and bar is completely inelastic, both particles strike with the bar simultaneously. The velocity of centre of mass after collision is

(a) 2 ms^{-1} (b) 4 ms^{-1}
(c) 10 ms^{-1} (d) 167 ms^{-1}

31. In the given figure four, identical spheres of equal mass m are suspended by wires of equal length l_0, so that all spheres are almost touching to one other. If the sphere 1 is released from the horizontal position and all collisions are elastic, the velocity of sphere 4 just after collision is

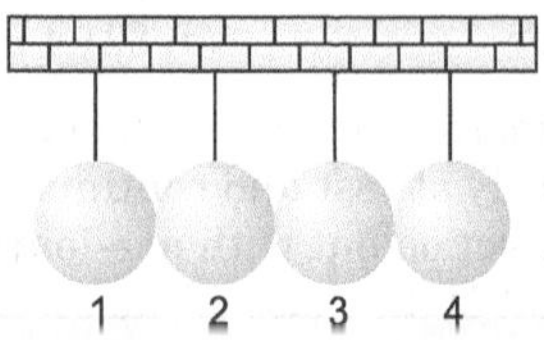

(a) $\sqrt{2\,gl_0}$ (b) $\sqrt{3\,gl_0}$
(c) $\sqrt{gl_0}$ (d) $\frac{\sqrt{g\,l_0}}{2}$

32. A body of mass 1 kg falls freely from a height of 100 m on a platform of mass 3 kg which is mounted on a spring having spring constant $k = 1.25 \times 10^6$ N/m. The body sticks to the platform and the spring's maximum compression is found to be x. Given that $g = 10\,ms^{-2}$, the value of x will be close to **[JEE Main 2019]**

(a) 8 cm (b) 4 cm
(c) 40 cm (d) 80 cm

33. A man (mass = 50 kg) and his son (mass = 20 kg) are standing on a frictionless surface facing each other. The man pushes his son, so that he starts moving at a speed of 0.70 ms^{-1} with respect to the man. The speed of the man with respect to the surface is **[JEE Main 2019]**

(a) 0.28 ms^{-1} (b) 0.20 ms^{-1}
(c) 0.47 ms^{-1} (d) 0.14 ms^{-1}

34. A block of mass 1.9 kg is at rest at the edge of a table of height 1 m. A bullet of mass 0.1 kg collides with the block and sticks to it. If the velocity of the bullet is 20 m/s in the horizontal direction just before the collision, then the kinetic energy just before the combined system strikes the floor, is (Take, $g = 10\,m/s^2$ and assume there is no rotational motion and loss of energy after the collision is negligible) **[JEE Main 2020]**

(a) 20 J (b) 19 J (c) 21 J (d) 23 J

35. In an elastic head on collision between two particles,

(a) velocity of separation is equal to the velocity of approach
(b) velocity of the target is always more than the velocity of the projectile
(c) the maximum velocity of the target is double to that of the projectile
(d) maximum transfer of kinetic energy occurs when masses of both projectile and target are equal

36. A sphere of mass m moving with a constant velocity v hits another stationary sphere of same mass. If e is the coefficient of restitution, then the ratio of velocity of two spheres after collision will be

(a) $\frac{1-e}{1+e}$ (b) $\frac{1+e}{1-e}$ (c) $\frac{e+1}{e-1}$ (d) $\frac{e-1}{e+1}t^2$

37. A particle of mass m collides with another stationary particle of mass M. If the particle m stops just after collision, the coefficient of restitution for collision is equal to

(a) 1 (b) $\frac{m}{M}$ (c) $\frac{M-m}{M+m}$ (d) $\frac{m}{M+m}$

38. In a one dimensional collision between two identical particles A and B, where B is stationary and A has momentum p before impact. During impact B gives an impulse J to A. Then coefficient of restitution between the two is

(a) $\frac{2J}{p}-1$ (b) $\frac{2J}{p}+1$
(c) $\frac{J}{p}+1$ (d) $\frac{J}{p}-1$

39. Three identical blocks A, B and C are placed on horizontal frictionless surface. The blocks A and C are at rest. But A is approaching towards B with a speed 10 ms^{-1}. The coefficient of restitution for all collisions is 0.5. The speed of the block C just after collision is

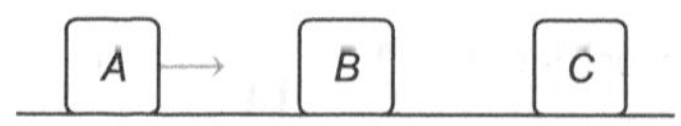

(a) 11.25 ms^{-1} (b) 6 ms^{-1}
(c) 8 ms^{-1} (d) 10 ms^{-1}

40. A bullet of mass m hits a target of mass M hanging by a string and gets embedded in it. If the block rises to a height h as a result of this collision, the velocity of the bullet before collision is
(a) $v=\sqrt{2gh}$
(b) $v=\sqrt{2gh}\left(1+\frac{m}{M}\right)$
(c) $v=\left(1+\frac{M}{m}\right)\sqrt{2gh}$
(d) $v=\sqrt{2gh}\left(1-\frac{m}{M}\right)$

41. If a ball is dropped from rest, its bounces from the floor. The coefficient of restitution is 0.5 and the speed just before the first bounce is 5 ms^{-1}. The total time taken by the ball to come to rest is
(a) 2 s
(b) 1 s
(c) 0.5 s
(d) 0.25 s

42. A tennis ball bounces down flight of stairs striking each step in turn and rebounding to the height of the step above. The coefficient of restitution has a value
(a) 1/2 (b) 1
(c) $1/\sqrt{2}$ (d) $1/2\sqrt{2}$

43. Two bodies A and B of definite shape (dimensions of bodies are not ignored). A is moving with speed of 10 ms^{-1} and B is in rest, collides elastically. The
(a) body A comes to rest and B moves with speed of 10 ms^{-1}
(b) they may move perpendicular to each other
(c) A and B may come to rest
(d) they must move perpendicular to each other

Collision in Two Dimensions

44. A ball moving with a certain velocity hits another identical ball at rest. If the plane is frictionless and collision is elastic, the angle between the directions in which the balls move after collision, will be
(a) 30° (b) 60°
(c) 90° (d) 120°

45. A body at rest breaks up into 3 parts. If 2 parts having equal masses fly off perpendicularly each after with a velocity of 12 m/s, then the velocity of the third part which has 3 times mass of each part is
(a) $4\sqrt{2}$ m/s at an angle of 45° from each body
(b) $24\sqrt{2}$ m/s at an angle of 135° from each body
(c) $6\sqrt{2}$ m/s at 135° from each body
(d) $4\sqrt{2}$ m/s at 135° from each body

46. A particle of mass m with an initial velocity $u\hat{i}$ collides perfectly elastically with a mass $3m$ at rest. It moves with a velocity $v\,\hat{j}$ after collision, then v is given by **[JEE Main 2020]**
(a) $v=\sqrt{\frac{2}{3}}u$ (b) $v=\frac{1}{\sqrt{6}}u$ (c) $v=\frac{u}{\sqrt{2}}$ (d) $v=\frac{u}{\sqrt{3}}$

47. A smooth steel ball strikes a fixed smooth steel plate at an angle θ with the vertical. If the coefficient of restitution is e, the angle at which the rebounce will take place is
(a) θ (b) $\tan^{-1}\left(\frac{\tan\theta}{e}\right)$
(c) $e\tan\theta$ (d) $\tan^{-1}\left(\frac{e}{\tan\theta}\right)$

48. Particle A of mass m_1 moving with velocity $(\sqrt{3}\,\hat{i}+\hat{j})\,ms^{-1}$ collides with another particle B of mass m_2 which is at rest initially. Let $\mathbf{v}_1$ and $\mathbf{v}_2$ be the velocities of particles A and B after collision, respectively. If $m_1=2m_2$ and after collision $\mathbf{v}_1=(\hat{i}+\sqrt{3}\,\hat{j})\,ms^{-1}$, then the angle between $\mathbf{v}_1$ and $\mathbf{v}_2$ is **[JEE Main 2020]**
(a) 15° (b) 60° (c) −45° (d) 105°

49. A mass m moves with a velocity v and collides inelastically with another identical mass. After collision, the 1st mass moves with velocity $\frac{v}{\sqrt{3}}$ in a direction perpendicular to the initial direction of motion. Find the speed of the second mass after collision.
(a) v (b) $\sqrt{3}\,v$ (c) $\frac{2}{\sqrt{3}}v$ (d) $\frac{v}{\sqrt{3}}$

50. A particle of mass m moving in the x-direction with speed $2v$ is hit by another particle of mass $2m$ moving in the y-direction with speed v. If the collision is perfectly inelastic, the percentage loss in the energy during the collision is close to **[JEE Main 2015]**
(a) 44% (b) 50% (c) 56% (d) 62%

51. A particle of mass m is projected with a speed u from the ground at an angle $\theta=\frac{\pi}{3}$ w.r.t. horizontal (X-axis). When it has reached its maximum height, it collides completely inelastically with another particle of the same mass and velocity $u\hat{i}$. The horizontal distance covered by the combined mass before reaching the ground is **[JEE Main 2020]**
(a) $\frac{3\sqrt{3}}{8}\frac{u^2}{g}$ (b) $\frac{3\sqrt{2}}{4}\frac{u^2}{g}$
(c) $\frac{5}{8}\frac{u^2}{g}$ (d) $2\sqrt{2}\frac{u^2}{g}$

ROUND II Mixed Bag

Only One Option is Correct

1. A 10 kg object collides with stationary 5 kg object and after collision they stick together and move forward with velocity 4 ms^{-1}. What is the velocity with which the 10 kg object hit the second one?

(a) 4 ms^{-1} (b) 6 ms^{-1}
(c) 10 ms^{-1} (d) 12 ms^{-1}

2. A bullet of mass M hits a block of mass M'. The energy transfer is maximum, when

(a) $M' = M$ (b) $M' = 2M$
(c) $M' << M$ (d) $M' >> M$

3. Two bodies having masses m_1 and m_2 and velocities $\mathbf{u}_1$ and $\mathbf{u}_2$ collide and form a composite system of $m_1\mathbf{v}_1 + m_2\mathbf{v}_2 = 0 (m_1 \neq m_2)$. The velocity of the composite system is

(a) zero (b) $\mathbf{u}_1 + \mathbf{u}_2$
(c) $\mathbf{u}_1 - \mathbf{u}_2$ (d) $\frac{\mathbf{u}_1 + \mathbf{u}_2}{2}$

4. A gas molecule of mass m strikes the wall of the container with a speed v at an angle θ with the normal to the wall at the point of collision. The impulse of the gas molecule has a magnitude

(a) $3mv$ (b) $2mv\cos\theta$
(c) mv (d) zero

5. A system of three particles having masses $m_1 = 1$ kg, $m_2 = 2$ kg and $m_3 = 4$ kg respectively is connected by two light springs. The acceleration of the three particles at any instant are 1 ms^{-2}, 2 ms^{-2} and 0.5 ms^{-2} respectively directed as shown in the figure. The net external force acting on the system is

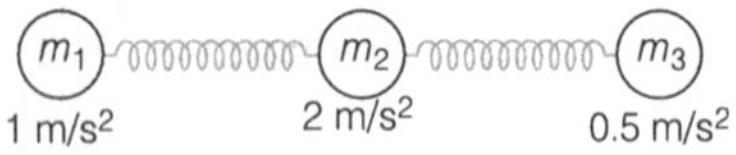

(a) 1 N (b) 7 N (c) 3 N (d) 6 N

6. A loaded spring gun of mass M fires a shot of mass m with a velocity v at an angle of elevation θ. The gun was initially at rest on a horizontal frictionless surface. After firing, the centre of mass of gun-shot system

(a) moves with a velocity $\frac{mv}{M}$
(b) moves with a velocity $\frac{mv}{M\cos\theta}$ in the horizontal direction
(c) remains at rest
(d) moves with velocity $\frac{(M-m)v}{(M+m)}$ in the horizontal direction

7. Three identical balls A, B and C are lying on a horizontal frictionless table as shown in figure. If ball A is imparted a velocity v towards B and C and the collisions are perfectly elastic, then finally

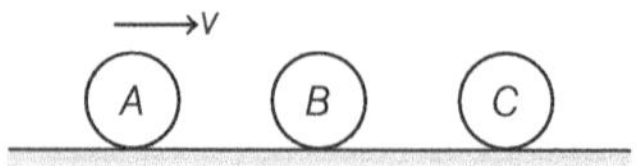

(a) ball A comes to rest and balls B and C roll out with speed $v/2$ each
(b) balls A and B are at rest and ball C rolls out with speed v
(c) all the three balls roll out with speed $v/3$ each
(d) all the three balls come to rest

8. In the figure shown below, the magnitude of acceleration of centre of mass of the system is (Take, $g = 10 ms^{-2}$)

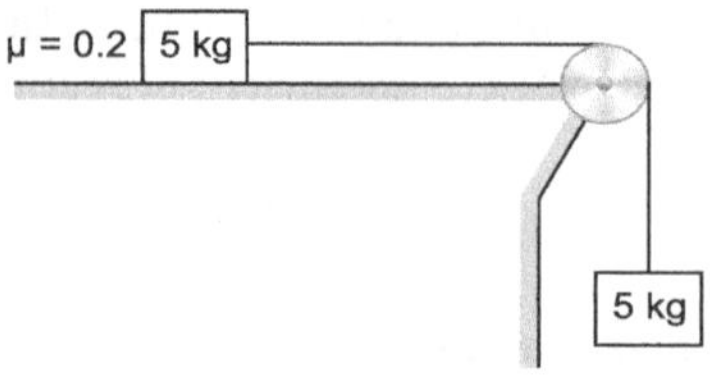

(a) 4 ms^{-2} (b) 10 ms^{-2}
(c) $2\sqrt{2}$ ms^{-2} (d) 5 ms^{-2}

9. A piece of wood of mass 0.03 kg is dropped from the top of a 100 m height building. At the same time, a bullet of mass 0.02 kg is fired vertically upward with a velocity 100 ms^{-1} from the ground. The bullet gets embedded in the wood. Then, the maximum height to which the combined system reaches above the top of the building before falling below is (Take, $g = 10 ms^{-2}$) **[JEE Main 2019]**

(a) 20 m (b) 30 m
(c) 10 m (d) 40 m

10. Two blocks of masses 10 kg and 4 kg are connected by a spring of negligible mass and placed on a frictionless horizontal surface. An impulsive force gives a velocity of 14 ms^{-1} to the heavier block in the direction of the lighter block. The velocity of centre of mass of the system at that moment is

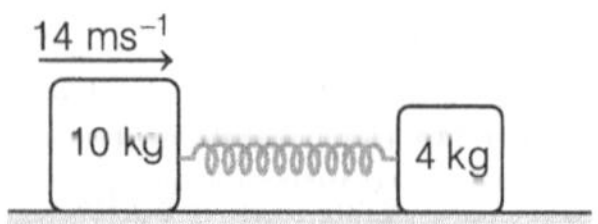

(a) 30 ms^{-1} (b) 20 ms^{-1}
(c) 10 ms^{-1} (d) 5 ms^{-1}

11. A particle is projected with 200 ms^{-1}, at an angle of 60º. At the highest point, it explodes into three particles of equal masses. One goes vertically upward with velocity 100 ms^{-1}, the second particle goes vertically downward with the same velocity as the first. Then, what is the velocity of the third particle?

(a) 120 ms^{-1} with 60º angle
(b) 200 ms^{-1} with 30º angle
(c) 50 ms^{-1} vertically upwards
(d) 300 ms^{-1} horizontally

12. An object of mass m_1 collides with another object of mass m_2, which is at rest. After the collision the objects move with equal speeds in opposite direction. The ratio of the masses $m_2 : m_1$ is **[JEE Main 2021]**

(a) 3 : 1 (b) 2 : 1 (c) 1 : 2 (d) 1 : 1

13. Two bodies of masses m_1 and m_2 $(m_1 > m_2)$ respectively are tied to the ends of a massless string, which passes over a light and frictionless pulley. The masses are initially at rest and then released. Then acceleration of the centre of mass of the system is

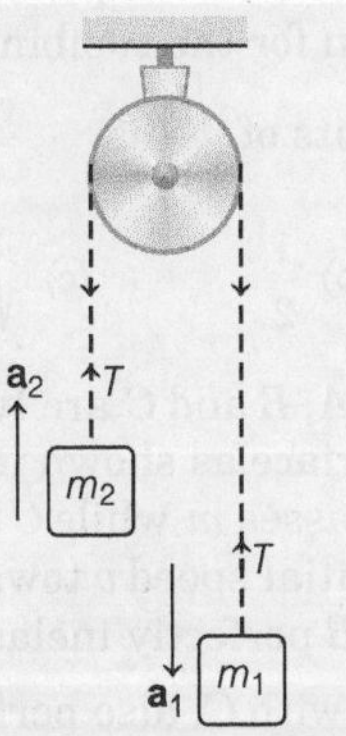

(a) $\left(\frac{m_1 - m_2}{m_1 + m_2}\right)^2 g$ (b) $\left(\frac{m_1 - m_2}{m_1 + m_2}\right)^2$
(c) g (d) zero

14. Three identical sphere lie at rest along a line on a smooth horizontal surface. The separation between any two adjacent spheres is L. The first sphere is moved with a velocity u towards the second sphere at time $t = 0$. The coefficient of restitution for collision between any two blocks is 1/3. Then choose the correct statement.

(a) The third sphere will start moving at $t = \frac{5L}{2u}$
(b) The third sphere will start moving at $t = \frac{4L}{u}$
(c) The centre of mass of the system will have a final speed $u/3$.
(d) The centre of mass of the system will have a final speed u

15. Two carts on horizontal straight rails are pushed apart by an explosion of a powder charge Q placed between the carts. Suppose the coefficient of friction between carts and rails are identical. If the 200 kg cart travels a distance of 36 m and stops, the distance covered by the cart weighing 300 kg is

(a) 32 m (b) 24 m (c) 16 m (d) 12 m

16. The masses of five balls at rest and lying at equal distances in a straight line are in geometrical progression with ratio 2 and their coefficients of restitution are each 2/3. If the first ball be started towards the second with velocity u, then the velocity communicated to 5th ball is

(a) $\frac{5}{9}u$ (b) $\left(\frac{5}{9}\right)^2 u$ (c) $\left(\frac{5}{9}\right)^3 u$ (d) $\left(\frac{5}{9}\right)^4 u$

17. Consider a rubber ball freely falling from a height $h = 4.9$ m onto a horizontal elastic plate. Assume that the duration of collision is negligible and the collision with the plate is totally elastic. Then the velocity as a function of time and the height as function of time will be

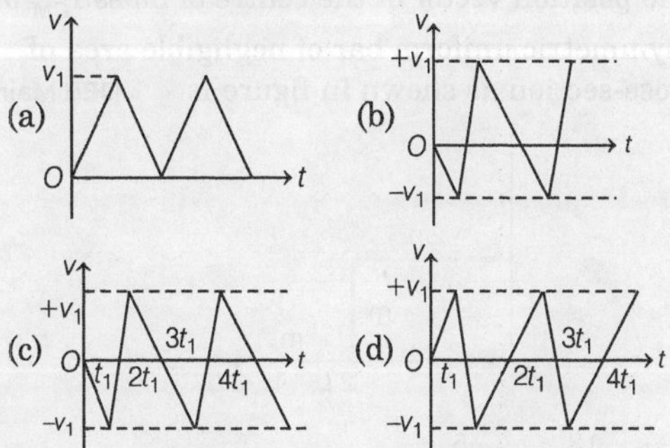

18. A ball strikes a horizontal floor at an angle $\theta = 45°$. The coefficient of restitution between the ball and the floor is $e = 1/2$. The fraction of its kinetic energy lost in collision is

(a) 5/8 (b) 3/8
(c) 3/4 (d) 1/4

19. A ball falls freely from a height of 45 m. When the ball is at a height of 25m, it explodes into two equal pieces. One of them moves horizontally with a speed of 10 ms^{-1}. The distance between the two pieces when both strike the ground is

(a) 10 m (b) 20 m
(c) 15 m (d) 30 m

20. A set on n identical cubical blocks lies at rest parallel to each other along a line on a smooth horizontal surface. The separation between the near surface of any two adjacent block is L. The block at one end is given a speed v towards the next one at time $t = 0$. All collisions are completely inelastic, then

(a) the last block starts moving at $t = \frac{(n-1)}{v} L$

(b) the last block starts moving at $t = \frac{n(n-1)L}{2v}$

(c) the centre of mass of the system will have a final speed v

(d) the centre of mass of the system will have a final speed zero

21. A block C of mass m is moving with velocity v_0 and collides elastically with block A of mass m and connected to another block B of mass $2m$ through spring constant k. What is k? If x_0 is compression of spring when velocity of A and B is same.

(a) $\frac{mv_0^2}{x_0^2}$ (b) $\frac{mv_0^2}{2x_0^2}$

(c) $\frac{3}{2}\frac{mv_0^2}{x_0^2}$ (d) $\frac{2}{3}\frac{mv_0^2}{x_0^2}$

22. The position vector of the centre of mass $\mathbf{r}_{CM}$ of an asymmetric uniform bar of negligible area of cross-section as shown in figure is **[JEE Main 2019]**

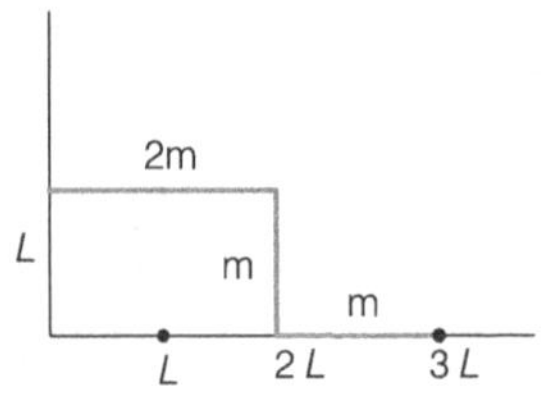

(a) $\mathbf{r} = \frac{13}{8} L\,\hat{\mathbf{x}} + \frac{5}{8} L\,\hat{\mathbf{y}}$

(b) $\mathbf{r} = \frac{11}{8} L\,\hat{\mathbf{x}} + \frac{3}{8} L\,\hat{\mathbf{y}}$

(c) $\mathbf{r} = \frac{3}{8} L\,\hat{\mathbf{x}} + \frac{11}{8} L\,\hat{\mathbf{y}}$

(d) $\mathbf{r} = \frac{5}{8} L\,\hat{\mathbf{x}} + \frac{13}{8} L\,\hat{\mathbf{y}}$

23. Blocks of masses m, $2m$, $4m$ and $8m$ are arranged in a line on a frictionless floor. Another block of mass m, moving with speed v along the same line (see figure) collides with mass m in perfectly inelastic manner. All the subsequent collisions are also perfectly inelastic. By the time, the last block of mass $8m$ starts moving, the total energy loss is p% of the original energy. Value of p is close to **[JEE Main 2020]**

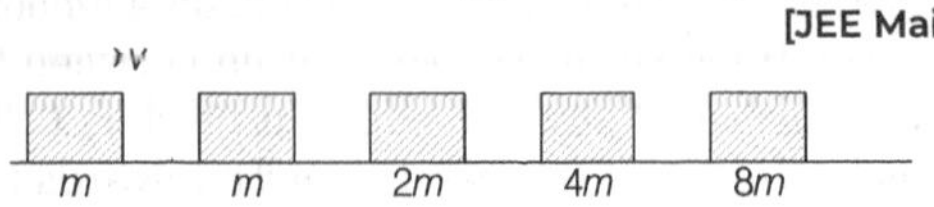

(a) 77 (b) 87 (c) 94 (d) 37

24. A uniform rectangular thin sheet $ABCD$ of mass M has length a and breadth b, as shown in the figure. If the shaded portion $HBGO$ is cut-off, the coordinates of the centre of mass of the remaining portion will be **[JEE Main 2019]**

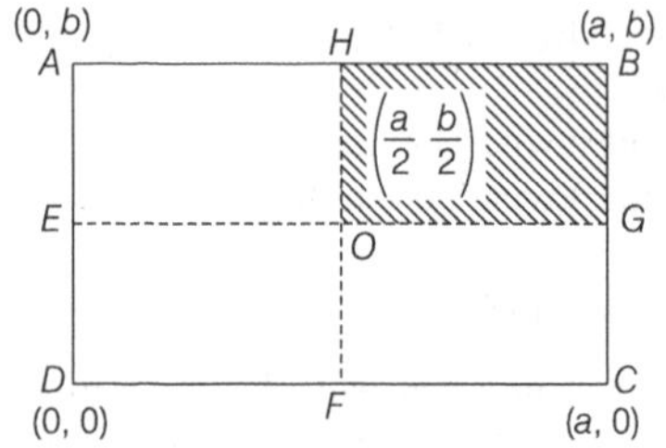

(a) $\left(\frac{2a}{3}, \frac{2b}{3}\right)$ (b) $\left(\frac{5a}{12}, \frac{5b}{12}\right)$

(c) $\left(\frac{3a}{4}, \frac{3b}{4}\right)$ (d) $\left(\frac{5a}{3}, \frac{5b}{3}\right)$

25. A particle of mass m is dropped from a height h above the ground. At the same time another particle of the same mass is thrown vertically upwards from the ground with a speed of $\sqrt{2gh}$. If they collide head-on completely inelastically, then the time taken for the combined mass to reach the ground, in units of $\sqrt{\frac{h}{g}}$ is **[JEE Main 2020]**

(a) $\sqrt{\frac{1}{2}}$ (b) $\frac{1}{2}$ (c) $\sqrt{\frac{3}{2}}$ (d) $\sqrt{\frac{3}{4}}$

26. Three blocks A, B and C are lying on a smooth horizontal surface as shown in the figure. A and B have equal masses m while C has mass M. Block A is given an initial speed v towards B due to which it collides with B perfectly inelastically. The combined mass collides with C, also perfectly inelastically $\frac{5}{6}$th of the initial kinetic energy is lost in whole process. What is value of $\frac{M}{m}$? **[JEE Main 2019]**

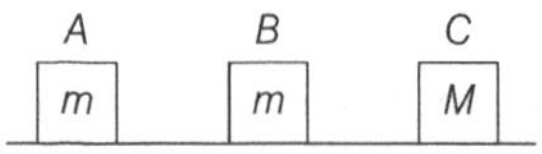

(a) 4 (b) 2 (c) 3 (d) 5

27. A simple pendulum is made of a string of length l and a bob of mass m, is released from a small angle θ_0. It strikes a block of mass M, kept on a horizontal surface at its lowest point of oscillations, elastically. It bounces back and goes up to an angle θ_1. Then, M is given by **[JEE Main 2019]**

(a) $m\left(\frac{\theta_0 + \theta_1}{\theta_0 - \theta_1}\right)$ (b) $\frac{m}{2}\left(\frac{\theta_0 - \theta_1}{\theta_0 + \theta_1}\right)$

(c) $m\left(\frac{\theta_0 - \theta_1}{\theta_0 + \theta_1}\right)$ (d) $\frac{m}{2}\left(\frac{\theta_0 + \theta_1}{\theta_0 - \theta_1}\right)$

28. Two particles of masses M and $2M$, moving as shown, with speeds of 10 m/s and 5 m/s, collide elastically at the origin. After the collision, they move along the indicated directions with speed v_1 and v_2 are nearly **[JEE Main 2019]**

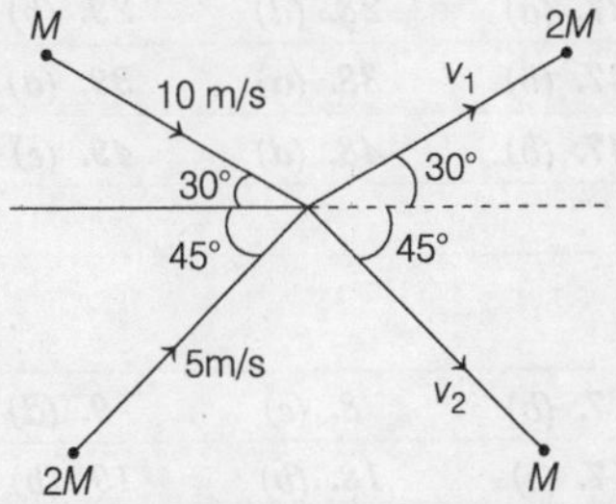

(a) 6.5 m/s and 3.2 m/s (b) 3.2 m/s and 6.3 m/s
(c) 3.2 m/s and 12.6 m/s (d) 6.5 m/s and 6.3 m/s

29. As shown in figure, when a spherical cavity (centred at O) of radius 1 is cut out of a uniform sphere of radius R (centred at C), the centre of mass of remaining (shaded) part of sphere is at G, *i.e.* on the surface of the cavity. R can be determined by the equation

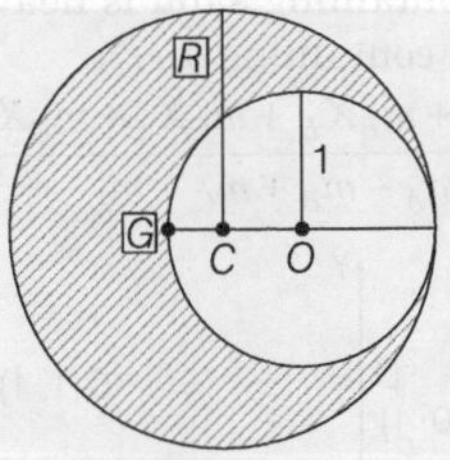

[JEE Main 2020]

(a) $(R^2 + R + 1)(2 - R) = 1$
(b) $(R^2 + R - 1)(2 - R) = 1$
(c) $(R^2 - R - 1)(2 - R) = 1$
(d) $(R^2 - R + 1)(2 - R) = 1$

Numerical Value Questions

30. The centre of mass of a solid hemisphere of radius 8 cm is x cm from the centre of the flat surface. Then, value of x is .... **[JEE Main 2020]**

31. A body A, of mass $m = 0.1$ kg has an initial velocity of $3\hat{\mathbf{i}}\ \text{ms}^{-1}$. It collides elastically with another body, B of the same mass which has an initial velocity of $5\hat{\mathbf{j}}\ \text{ms}^{-1}$. After collision, A moves with a velocity $\mathbf{v} = 4(\hat{\mathbf{i}} + \hat{\mathbf{j}})$. The energy of B after collision is written as $\frac{x}{10}$ J. The value of x is .... **[JEE Main 2020]**

32. A square shaped hole of side $l = \frac{a}{2}$ is curved out at a distance $d = \frac{a}{2}$ from the centre O of a uniform circular disc of radius a. If the distance of the centre of mass of the remaining portion from O is $-\frac{a}{X}$, value of X (to the nearest integer) is .... **[JEE Main 2020]**

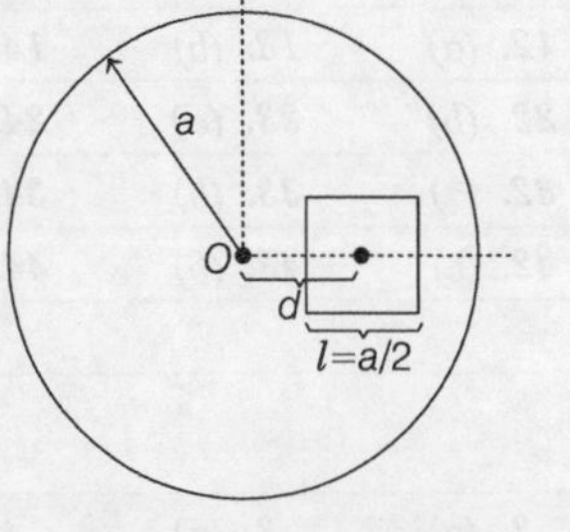

33. Two bodies of the same mass are moving with the same speed, but in different directions in a plane. They have a completely inelastic collision and move together thereafter with a final speed which is half of their initial speed. The angle between the initial velocities of the two bodies (in degree) is .... **[JEE Main 2020]**

34. A particle of mass m is moving along the X-axis with initial velocity $u\hat{\mathbf{i}}$. It collides elastically with a particle of mass 10 m at rest and then moves with half its initial kinetic energy (see figure). If $\sin\theta_1 = \sqrt{n}\sin\theta_2$, then value of n is .... **[JEE Main 2020]**

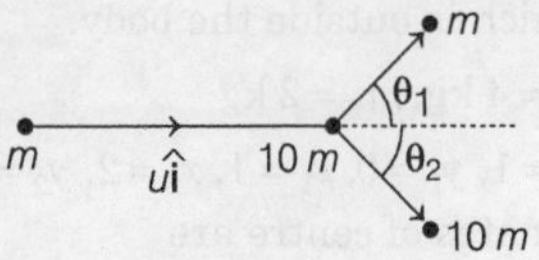

35. A ball of mass 10 kg moving with a velocity $10\sqrt{3}\ \text{ms}^{-1}$ along X-axis, hits another ball of mass 20 kg which is at rest. After collision, the first ball comes to rest and the second one disintegrates into two equal pieces. One of the pieces starts moving along Y-axis at a speed of 10 m/s. The second piece starts moving at a speed of 20 m/s at an angle θ (degree) with respect to the X-axis. The configuration of pieces after collision is shown in the figure. The value of θ to the nearest integer is ..... **[JEE Main 2021]**

After collision

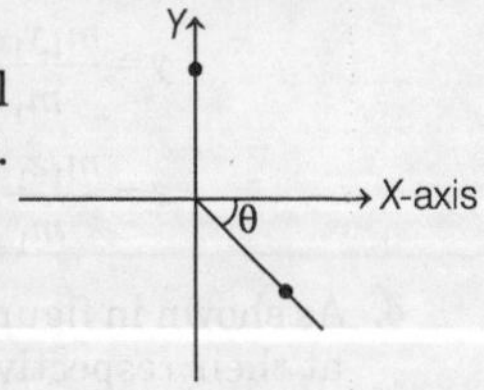

36. The disc of mass M with uniform surface mass density σ is shown in the figure. The centre of mass of the quarter disc (the shaded area) is at the position $\frac{x}{3}\frac{a}{\pi}, \frac{x}{3}\frac{a}{\pi}$ where x is ....
(Round off to the nearest integer)
(Here, a is an area as shown in the figure.) **[JEE Main 2021]**

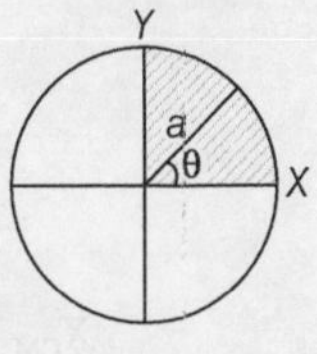

Answers

Round I

1. (b)	2. (d)	3. (d)	4. (d)	5. (c)	6. (c)	7. (a)	8. (a)	9. (c)	10. (d)
11. (d)	12. (a)	13. (b)	14. (a)	15. (a)	16. (b)	17. (d)	18. (a)	19. (c)	20. (a)
21. (c)	22. (b)	23. (d)	24. (d)	25. (c)	26. (b)	27. (a)	28. (d)	29. (b)	30. (b)
31. (a)	32. (*)	33. (b)	34. (c)	35. (d)	36. (a)	37. (b)	38. (a)	39. (a)	40. (c)
41. (c)	42. (c)	43. (b)	44. (c)	45. (d)	46. (c)	47. (b)	48. (d)	49. (c)	50. (c)
51. (a)									

Round II

1. (b)	2. (a)	3. (a)	4. (b)	5. (c)	6. (c)	7. (b)	8. (c)	9. (d)	10. (c)
11. (d)	12. (a)	13. (a)	14. (a)	15. (c)	16. (d)	17. (c)	18. (b)	19. (b)	20. (b)
21. (d)	22. (a)	23. (c)	24. (b)	25. (c)	26. (a)	27. (a)	28. (d)	29. (a)	30. 3
31. 1	32. 23	33. 120	34. 10	35. 30	36. 4				

Solutions

Round I

1. Centre of mass is closer to massive part of the body, therefore the bottom piece of bat has larger mass.

2. Centre of mass of a bangle lies at the centre of the bangle, which is outside the body.

3. Here, $m_1 = 4$ kg, $m_2 = 2$ kg

$$x_1 = 1, y_1 = 0, z_1 = 1, x_2 = 2, y_2 = 2, z_2 = 0$$

The coordinates of centre are

$$x = \frac{m_1x_1 + m_2x_2}{m_1 + m_2} = \frac{4\times1+2\times2}{4+2} = \frac{4}{3}$$

$$y = \frac{m_1y_1 + m_2y_2}{m_1 + m_2} = \frac{4\times0+2\times2}{4+2} = \frac{2}{3}$$

$$z = \frac{m_1z_1 + m_2z_2}{m_1 + m_2} = \frac{4\times1+2\times0}{4+2} = \frac{2}{3}$$

4. As shown in figure, centre of mass of respective rods are at their respective mid-points.

Hence centre of mass of the system has coordinates (X_{CM}, Y_{CM}), then

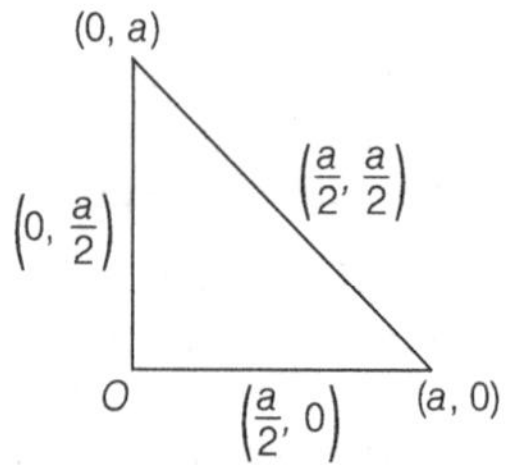

$$X_{CM} = \frac{m\times\frac{a}{2} + m\times\frac{a}{2} + m\times0}{3m} = \frac{a}{3}$$

$$Y_{CM} = \frac{m\times0 + m\times\frac{a}{2} + m\times\frac{a}{2}}{3m} = \frac{a}{3}$$

5. The position of centre of mass of the system shown in figure is likely to be at C. This is because lower part of the sphere containing sand is heavier than upper part of the sphere containing air.

6. $X_{CM} = \dfrac{m_AX_A + m_BX_B + m_CX_C + m_DX_D}{m_A + m_B + m_C + m_D}$

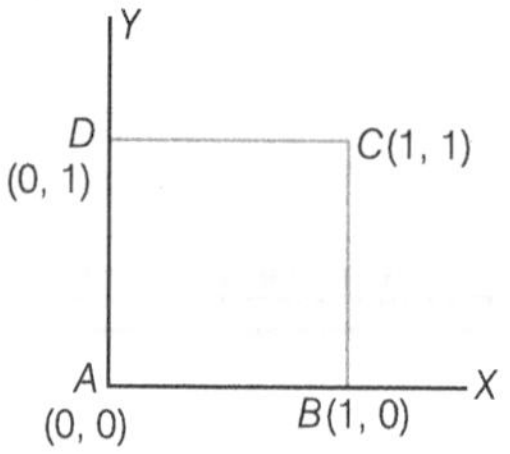

$$= \frac{1\times0+2\times1+3\times1+4\times0}{1+2+3+4}$$

$$= \frac{2+3}{10} = \frac{1}{2} = 0.5 \text{ m}$$

Similarly, $Y_{CM} = \dfrac{m_AY_A + m_BY_B + m_CY_C + m_DY_D}{m_A + m_B + m_C + m_D}$

$$= \frac{1\times0+2\times0+3\times1+4\times1}{1+2+3+4}$$

$$= \frac{7}{10} = 0.7 \text{ m}$$

7. Let mass of hydrogen atom $= m$

$\therefore$ Mass of chlorine atom $= 35.5m$

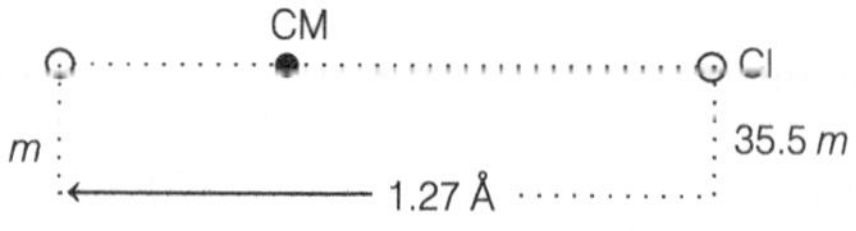

Let hydrogen atom be at origin, *i.e.* position vector of it, $r_1 = 0$.

$\therefore$ Position vector of chlorine atom, $r_2 = 1.27\times10^{-10}$ m

Position vector to centre of mass is given by

$$r_{CM} = \frac{m_1 r_1 + m_2 r_2}{m_1 + m_2}$$

$$= \frac{m \times 0 + 35.5 m \times 1.27\times 10^{-10}}{m + 35.5\, m}$$

$$= \frac{35.5\times 1.27\times 10^{-10}}{36.5}$$

$$= 1.235\times 10^{-10}\text{ m} = 1.24\text{ Å}$$

8. $m_1 = 1$ kg, $m_2 = 2$ kg, $m_3 = 3$ kg

Position of centre of mass (2, 2, 2).

$$m_4 = 4\text{ kg}$$

New position of centre of mass (0, 0 ,0).

For initial position,

$$x_{CM} = \frac{m_1 x_1 + m_2 x_2 + m_3 x_3}{m_1 + m_2 + m_3}$$

$$2 = \frac{m_1 x_1 + m_2 x_2 + m_3 x_3}{1+2+3}$$

$$m_1 x_1 + m_2 x_2 + m_3 x_3 = 12$$

Similarly, $m_1 y_1 + m_2 y_2 + m_3 y_3 = 12$

and $m_1 z_1 + m_2 z_2 + m_3 z_3 = 12$

For new position,

$$x'_{CM} = \frac{m_1 x_1 + m_2 x_2 + m_3 x_3 + m_4 x_4}{m_1 + m_2 + m_3 + m_4}$$

$$0 = \frac{12 + 4\times x_4}{1+2+3+4}$$

$\Rightarrow \quad 4x_4 = -12 \Rightarrow x_4 = -3$

Similarly, $y_4 = -3 \Rightarrow z_4 = -3$

$\therefore$ Position of fourth mass $(-3, -3, -3)$.

9. The height of equilateral Δ is

$$h = y_3 = \sqrt{(1)^2 - (0.5)^2} = \sqrt{3}/2\text{ m}$$

Thus, coordinates of three masses are (0, 0), (1, 0) and $\left(0.5, \frac{\sqrt{3}}{2}\right)$.

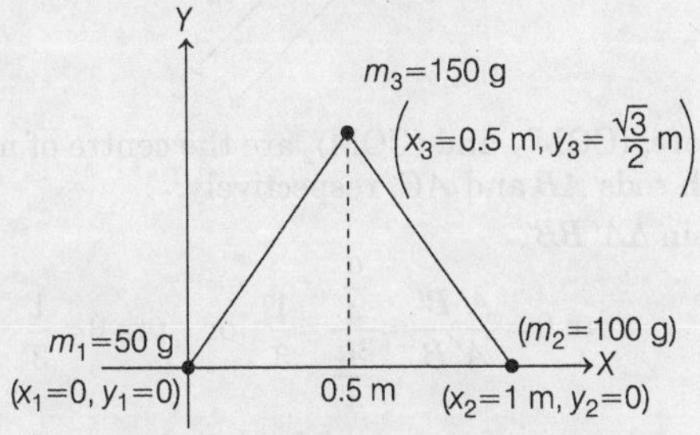

Using, $X_{CM} = \frac{m_1 x_1 + m_2 x_2 + m_3 x_3}{m_1 + m_2 + m_3}$,

$$= \frac{50\times 0 + 100\times 1 + 150\times 0.5}{50 + 100 + 150}$$

$$= \frac{175}{300} = \frac{7}{12}\text{ m}$$

Similarly, $Y_{CM} = \frac{m_1 y_1 + m_2 y_2 + m_3 y_3}{m_1 + m_2 + m_3}$

$$= \frac{50\times 0 + 100\times 0 + 150\times \frac{\sqrt{3}}{2}}{50 + 100 + 150} = \frac{\sqrt{3}}{4}\text{ m}$$

10. From figure,

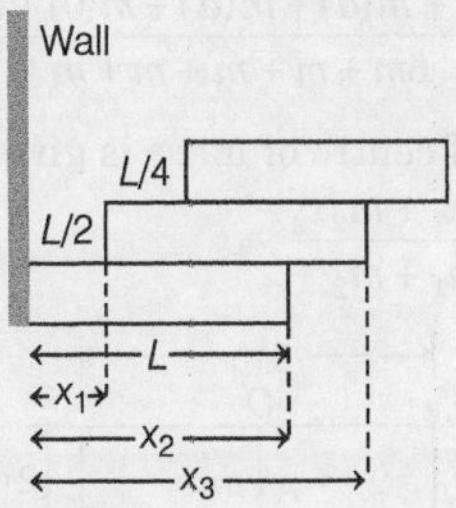

$$x_1 = \frac{L}{2},\ x_2 = \frac{L}{2} + \frac{L}{2} = L$$

$$x_3 = \frac{L}{2} + \frac{L}{4} + \frac{L}{2} = \frac{5L}{4}$$

$$\therefore \quad x_{CM} = \frac{M_1 x_1 + M_2 x_2 + M_3 x_3}{M_1 + M_2 + M_3}$$

$$= \frac{M\times\frac{L}{2} + M\times L + M\times\frac{5L}{4}}{M + M + M}$$

$$= \frac{\frac{11}{4} ML}{3M} = \frac{11}{12} L$$

11. Given lamina consists of 2 parts I and II as shown in the figure.

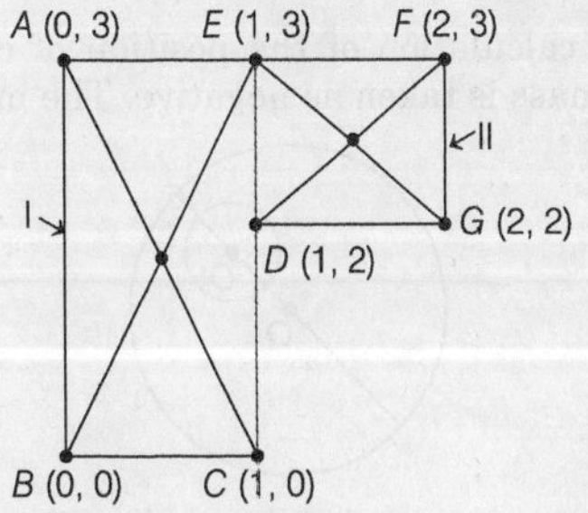

As mass of uniform lamina is 4 kg, mass of part I is $m_1 = 3$ kg and mass of part II is $m_2 = 1$ kg.

These masses can be assumed to be concentrated at geometrical centres of sections I and II.

So, $m_1 = 3$ kg has coordinates $x_1 = 0.5$ m, $y_1 = 1.5$ m

and $m_2 = 1$ kg has coordinates $x_2 = 1.5$ m, $y_2 = 2.5$ m

Now, we use formula of centre of mass (CM) to find X_{CM} and Y_{CM}.

So, $X_{CM} = \frac{m_1 x_1 + m_2 x_2}{m_1 + m_2}$

$$= \frac{(3\times 0.5) + (1\times 1.5)}{4} = 0.75\text{ m}$$

and $Y_{CM} = \dfrac{m_1 y_1 + m_2 y_2}{m_1 + m_2}$

$= \dfrac{(3\times 1.5) + (1\times 2.5)}{4} = 1.75\text{ m}$

12. $y_{CM} = \dfrac{m_1 y_1 + m_2 y_2 + m_3 y_3 + m_4 y_4 + m_5 y_5}{m_1 + m_2 + m_3 + m_4 + m_5}$

$= \dfrac{(6m)(0) + m(a) + m(a) + m(0) + m(-a)}{6m + m + m + m + m} = \dfrac{a}{10}$

13. Coordinate of centre of mass is given by

$$X_{CM} = \frac{m_1 x_1 + m_2 x_2}{m_1 + m_2}$$

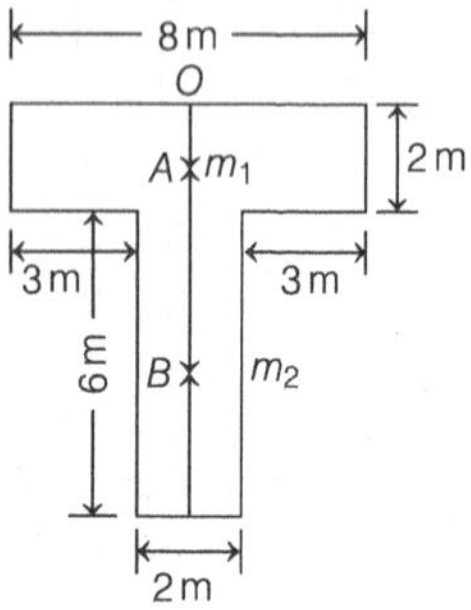

Taking parts A and B as two bodies of same system.

$m_1 = l \times b \times \sigma = 8 \times 2 \times \sigma = 16\sigma$

$m_2 = l \times b \times \sigma = 6 \times 2 \times \sigma = 12\sigma$

Choosing O as origin,

$x_1 = 1$ m, $x_2 = 2 + 3 = 5$ m

$\therefore \quad X_{CM} = \dfrac{16\sigma \times 1 + 12\sigma \times 5}{16\sigma + 12\sigma} = \dfrac{19}{7}$

$= 2.7$ m from O

14. For the calculation of the position of centre of mass, cut-off mass is taken as negative. The mass of disc is

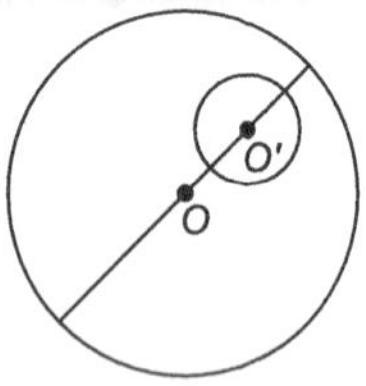

$m_1 = \pi r_1^2 \sigma$

$= \pi (6)^2 \sigma = 36\pi\sigma$

where σ is surface mass density.

The mass of cutting portion is

$m_2 = \pi (1)^2 \sigma = \pi\sigma$

$x_{CM} = \dfrac{m_1 x_1 - m_2 x_2}{m_1 - m_2}$

Taking origin at the centre of disc,

$x_1 = 0, x_2 = 3$ cm

$x_{CM} = \dfrac{36\pi\sigma \times 0 - \pi\sigma \times 3}{36\pi\sigma - \pi\sigma}$

$= \dfrac{-3\pi\sigma}{35\pi\sigma} = -\dfrac{3}{35}$ cm

15. Here, $\rho(x) = a(1 + bx^2)$

When $b \to 0$, $(x) = a =$ constant, *i.e.* density of rod of length 1 m is constant. In that event, centre of mass of rod would lie at 0.5 m, (*i.e.* at the centre of rod.)

When we try $b \to 0$ in all the four given options, we find choice (a) alone given $x = \dfrac{3(2+b)}{4(3+b)} = \dfrac{6}{12} = 0.5$

16. The mass of considered element is

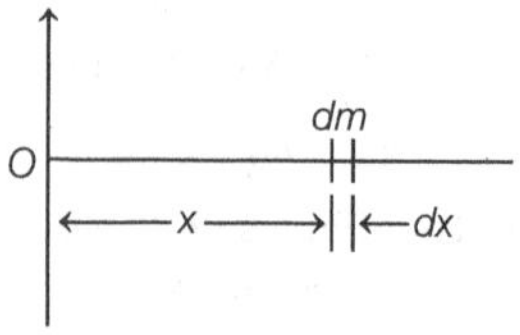

$dm = \lambda\, dx = \lambda_0 x\, dx$

$\therefore \quad x_{CM} = \dfrac{\int_0^1 x\, dm}{\int dm} = \dfrac{\int_0^1 x(\lambda_0 x\, dx)}{\int_0^1 \lambda_0 x\, dx}$

$= \dfrac{\lambda_0 \left[\dfrac{x^3}{3}\right]_0^L}{\lambda_0 \left[\dfrac{x^2}{2}\right]_0^L} = \dfrac{\lambda_0 \dfrac{L^3}{3}}{\lambda_0 \dfrac{L^2}{2}} = \dfrac{2}{3}L$

17. The given system of rods can be drawn using geometry as,

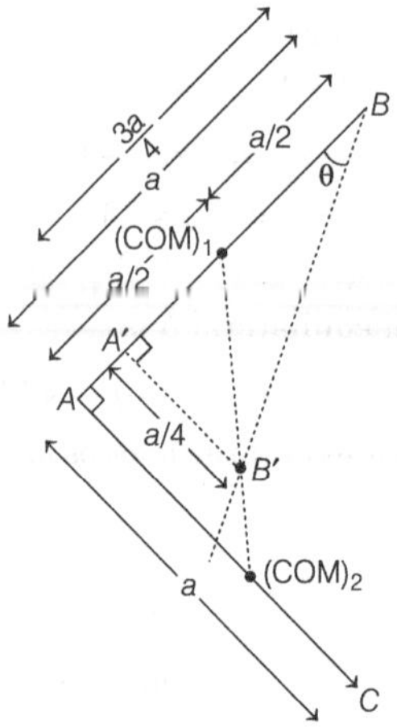

where, $(COM)_1$ and $(COM)_2$ are the centre of mass of both rods AB and AC, respectively.

So, in $\Delta A'BB'$,

$\tan\theta = \dfrac{A'B'}{A'B} = \dfrac{\frac{a}{4}}{\frac{3a}{4}} = \dfrac{1}{3}$ or $\tan\theta = \dfrac{1}{3}$

18. Since there is no external force acting on the particle, hence

$y_{CM} = \dfrac{m_1 y_1 + m_2 y_2}{m_1 + m_2} = 0,$

Hence, $\left(\dfrac{m}{4}\right) \times (15) + \left(\dfrac{3m}{4}\right)(y_2) = 0$

$\Rightarrow \quad y_2 = -5$ cm

19. Let mass of each body be m.

$$\Rightarrow \quad v_{CM} = \frac{m \times 2v - mv}{m + m} = \frac{v}{2}$$

20. Here, $m_1 = 1$ kg, $\mathbf{v}_1 = 2\,\hat{\mathbf{i}}$

$$m_2 = 2 \text{ kg}, \mathbf{v}_2 = 2\cos 30^\circ\,\hat{\mathbf{i}} - 2\sin 30^\circ\,\hat{\mathbf{j}}$$

$$\mathbf{v}_{CM} = \frac{m_1\mathbf{v}_1 + m_2\mathbf{v}_2}{m_1 + m_2}$$

$$= \frac{1 \times 2\,\hat{\mathbf{i}} + 2\,(2\cos 30^\circ\,\hat{\mathbf{i}} - 2\sin 30^\circ\,\hat{\mathbf{j}})}{1+2}$$

$$= \frac{2\,\hat{\mathbf{i}} + 2\sqrt{3}\,\hat{\mathbf{i}} - 2\,\hat{\mathbf{j}}}{3} = \left(\frac{2+2\sqrt{3}}{3}\right)\hat{\mathbf{i}} - \frac{2}{3}\,\hat{\mathbf{j}}$$

21. To keep applied centre of mass at the same position, velocity of centre of mass is zero, so

$$\frac{m_1\mathbf{v}_1 + m_2\,\mathbf{v}_2}{m_1 + m_2} = 0$$

where, $\mathbf{v}_1$ and $\mathbf{v}_2$ are velocities of particles 1 and 2 respectively.

$$\Rightarrow \quad m_1\,\frac{d\mathbf{r}_1}{dt} + m_2\,\frac{d\mathbf{r}_2}{dt} = 0 \quad \left[\because v_1 = \frac{d\mathbf{r}_1}{dt} \text{ and } v_2 = \frac{d\mathbf{r}_2}{dt}\right]$$

$$\Rightarrow \quad m\,d\mathbf{r}_1 + m_2\,d\mathbf{r}_2 = 0$$

Let 2nd particle has been displaced by distance x.

$$\Rightarrow \quad m_1\,(d) + m_2\,(x) = 0 \quad \Rightarrow \quad x = -\frac{m_1 d}{m_2}$$

(negative sign shows that both the particles have to move in opposite directions.)

22. The acceleration of centre of mass is

$$a_{CM} = \frac{F}{m_A + m_B} = \frac{30}{10+20} = 1 \text{ ms}^{-2}$$

$$\therefore \quad s = \frac{1}{2}\,a_{CM}\,t^2 = \frac{1}{2} \times 1 \times 2^2 = 2 \text{ m}$$

23. (a) Impulse received by m

$$\mathbf{J} = m\,(\mathbf{v}_f - \mathbf{v}_i)$$
$$= m\,(-2\,\hat{\mathbf{i}} + \hat{\mathbf{j}} - 3\,\hat{\mathbf{i}} - 2\,\hat{\mathbf{j}})$$
$$= m\,(-5\,\hat{\mathbf{i}} - \hat{\mathbf{j}})$$

and impulse received by M

$$= -\mathbf{J} = m\,(5\hat{\mathbf{i}} + \hat{\mathbf{j}})$$

(b) $mv = m\,(5\,\hat{\mathbf{i}} + \hat{\mathbf{j}})$

or $\quad v = \frac{m}{M}\,(5\,\hat{\mathbf{i}} + \hat{\mathbf{j}}) = \frac{1}{13}\,(5\,\hat{\mathbf{i}} + \hat{\mathbf{j}})$

(c) e = (relative velocity of separation/relative velocity of approach) in the direction of $-\hat{\mathbf{j}} = 11/17$

24.

As $m_1 = m_2$, therefore after elastic collision, velocities of masses get interchanged.

i.e. velocity of mass, $m_1 = -5$ m/s

and velocity of mass, $m_2 = +3$ m/s

25. From energy conservation

$\left[\text{after bullet gets embedded till the system comes momentarily at rest}\right]$

$$(M + m)\,g\,h = \frac{1}{2}\,(M + m)v_1^2$$

(where, v_1 is velocity after collision)

$$\therefore \quad v_1 = \sqrt{2gh}$$

Applying momentum conservation, (just before and just after collision)

$$mv = (M + m)v_1$$

$$v = \left(\frac{M + m}{m}\right)v_1$$

$$= \frac{6}{10 \times 10^{-3}} \times \sqrt{2 \times 9.8 \times 9.8 \times 10^{-2}}$$

$$\approx 831.55 \text{ m/s}$$

26. Given situation is as shown

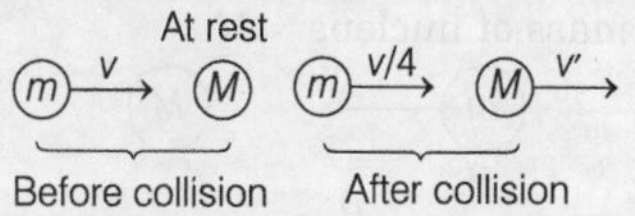

Using momentum conservation law for the given system,

$$(\text{Total momentum})_{\text{before collision}} = (\text{Total momentum})_{\text{after collision}}$$

$$\Rightarrow \quad m(v) + M\,(0) = m\left(\frac{v}{4}\right) + M(v') \qquad \ldots\text{(i)}$$

$\because \quad e = 1$ and

we know that, $\quad e = -\frac{v_2 - v_1}{u_2 - u_1}$

$$\Rightarrow \quad 1 = -\frac{v' - v/4}{0 - v}$$

$$\Rightarrow \quad v = v' - v/4$$

or $\quad v' = 5v/4 \qquad \ldots\text{(ii)}$

Using value from Eq. (ii) into Eq. (i), we get

$$mv = \frac{mv}{4} + M\left(\frac{5v}{4}\right)$$

$$M = \frac{3}{5}\,m = \frac{3}{5} \times 2 = 1.2 \text{ kg}$$

27. Retardation due to friction

$$a = \mu g = (0.25)\,(10) = 2.5 \text{ ms}^{-2}$$

Collision is elastic, *i.e.* after collision, first block comes to rest and the second block acquires the velocity of first block or we can understand it is this manner that second block is permanently at rest while only the first block moves. Distance travelled by it will be,

$$s = \frac{v^2}{2\,a} = \frac{(5)^2}{(2)\,(2.5)} = 5 \text{ m}$$

$\therefore$ Final separation will be $(s - 2) = 3$ m

28. Given, $m_2 = 0.5m_1$

$\Rightarrow \quad m_1 = 2m_2$

Let $m_2 = m$, then, $m_1 = 2m$

Also, $v_3 = 0.5v_1$

Given situation of collinear collision is as shown below.

Before collision,

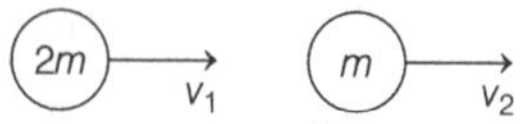

After collision,

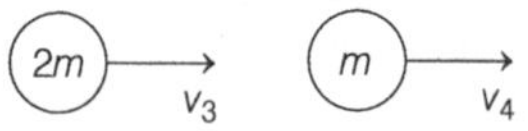

$\therefore$ According to the conservation of linear momentum,
Initial momentum = Final momentum

$$m_1 v_1 \hat{\mathbf{i}} + m_2 v_2 \hat{\mathbf{i}} = m_1 v_3 \hat{\mathbf{i}} + m_2 v_4 \hat{\mathbf{i}}$$

$$\Rightarrow \quad 2mv_1 \hat{\mathbf{i}} + mv_2 \hat{\mathbf{i}} = 2m(0.5v_1)\hat{\mathbf{i}} + mv_4 \hat{\mathbf{i}}$$

$$\Rightarrow \quad v_4 = v_1 + v_2$$

$$\Rightarrow \quad v_1 = v_4 - v_2$$

29. We have following collision, where mass of α particle $= m$ and mass of nucleus $= M$

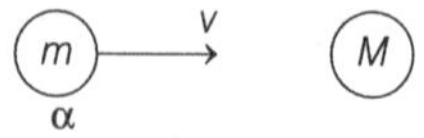

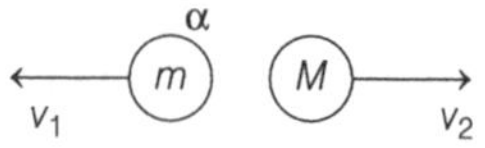

Let α-particle rebounds with velocity v_1, then
Given; final energy of α = 36% of initial energy

$$\Rightarrow \quad \frac{1}{2}mv_1^2 = 0.36 \times \frac{1}{2}mv^2$$

$$\Rightarrow \quad v_1 = 0.6\, v \qquad \ldots\text{(i)}$$

As unknown nucleus gained 64% of energy of α, we have

$$\frac{1}{2}Mv_2^2 = 0.64 \times \frac{1}{2}mv^2$$

$$\Rightarrow \quad v_2 = \sqrt{\frac{m}{M}} \times 0.8\, v \qquad \ldots\text{(ii)}$$

From momentum conservation, we have

$$mv = Mv_2 - mv_1$$

Substituting values of v_1 and v_2 from Eqs. (i) and (ii), we have

$$mv = M\sqrt{\frac{m}{M}} \times 0.8\, v - m \times 0.6\, v$$

$$\Rightarrow \quad M = 4m$$

30. Here, $m = 0.08$ kg, $m_0 = 0.16$ kg

According to conservation principle of momentum,

$$mv_1 + mv_2 = (2m + m_0)\, v_{CM}$$

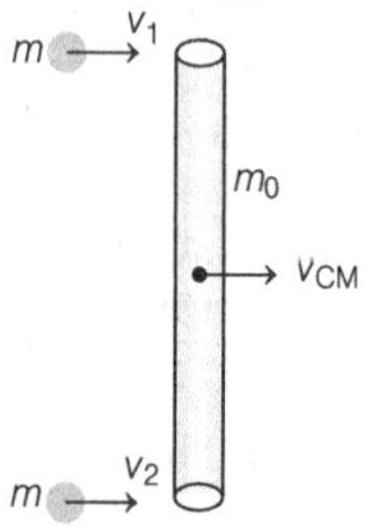

$$\therefore \quad v_{CM} = \frac{mv_1 + mv_2}{2m + m_0}$$

$$= \frac{0.08 \times 16}{0.16 + 0.16} = \frac{1.28}{1.32} = \frac{128}{32} = 4 \text{ ms}^{-1}$$

31. When the sphere 1 is released from horizontal position, from energy conservation, potential energy at height l_0 = kinetic energy at bottom

or $\quad mgl_0 = \frac{1}{2}mv^2$ or $v = \sqrt{2\,gl_0}$

Since, all collisions are elastic, so velocity of sphere 1 is transferred to sphere 2, then from 2 to 3 and finally from 3 to 4. Hence, just after collision, the sphere 4 attains a velocity to $\sqrt{2\,gl_0}$.

32. Initial compression of the spring,

$$mg = k\left(\frac{x_0}{100}\right) \qquad (x_0 \text{ in cm})$$

$$\Rightarrow \quad x_0 = \frac{3 \times 10 \times 100}{1.25 \times 10^6} = \frac{3}{1250}$$

which is very small and can be neglected.

Applying conservation of momentum before and after the collision, *i.e.* momentum before collision = momentum after collision.

$$m \times \sqrt{2gh} = (m + M)\, v$$

($\because$ Velocity of the block just before the collision is $v^2 - 0^2 = 2gh$ or $v = \sqrt{2gh}$)

After substituting the given values, we get

$$1 \times \sqrt{2 \times 10 \times 100} = 4v$$

or $\quad 4v = 20\sqrt{5}$

So, $\quad v = 5\sqrt{5}$ m/s

Let this be the maximum velocity, then for the given system, using

$$\frac{1}{2}mv^2 = \frac{1}{2}kx^2$$

$$\therefore \quad \frac{1}{2} \times 4 \times 125 = \frac{1}{2} \times 1.25 \times 10^6 \times \left(\frac{x}{100}\right)^2$$

$$\Rightarrow \quad 4 = 10^4 \times \frac{x^2}{10^4}$$

or $\quad x = 2$ cm

$\therefore$ No option given is correct.

33. The given situation can be shown as below

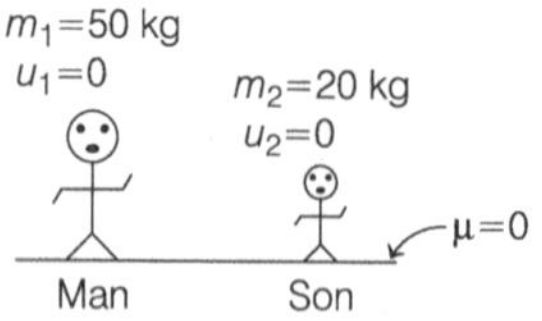

Before collision

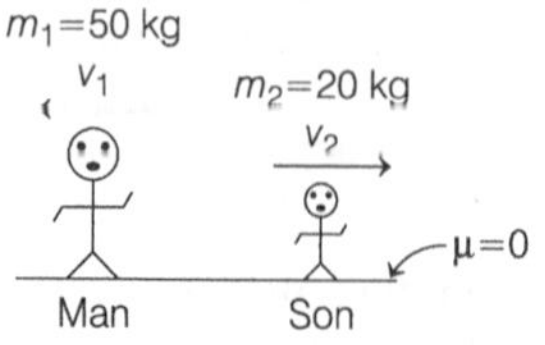

After collision

Using momentum conservation law,

(Total momentum)$_{\text{before collision}}$

$$= (\text{Total momentum})_{\text{after collision}}$$

$$(m_1 \times 0) + (m_2 \times 0) = m_1 \mathbf{v}_1 + m_2 \mathbf{v}_2$$

$$0 = m_1(-v_1)\hat{\mathbf{i}} + m_2 v_2 \hat{\mathbf{i}}$$

$$\Rightarrow \quad m_1 v_1 = m_2 v_2$$

$$\Rightarrow \quad 50 v_1 = 20 v_2$$

$$\Rightarrow \quad v_2 = 2.5 v_1 \quad \text{...(i)}$$

Again, relative velocity = 0.70 m/s

But from figure, relative velocity = $v_1 + v_2$

$$\therefore \quad v_1 + v_2 = 0.7 \quad \text{...(ii)}$$

From Eqs. (i) and (ii), we get

$$v_1 + 2.5 v_1 = 0.7$$

$$\Rightarrow \quad v_1 (3.5) = 0.7$$

$$v_1 = \frac{0.7}{3.5} = 0.20 \text{ m/s}$$

34. When the bullet undergoes an inelastic collision with block, a part of KE of bullet is lost.

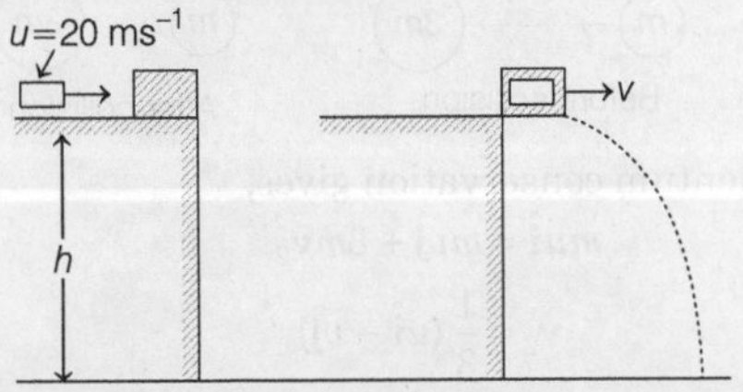

When bullet + block system falls from height h, its total energy (kinetic + potential) becomes kinetic energy, so kinetic energy of bullet + block system at bottom just before collision is equal to total energy just after collision.

Now, by law of conservation of momentum, we have

$$mu = (m + M)v$$

$$\Rightarrow \quad v = \frac{mu}{m+M} = \frac{0.1 \times 20}{(0.1 + 1.9)} = 1 \text{ ms}^{-1}$$

Total energy of bullet and block just after collision

$$= \text{KE} + \text{PE} = \frac{1}{2}(m+M)v^2 + (m+M)gh$$

$$= \frac{1}{2} \times 2 \times 1^2 + 2 \times 10 \times 1$$

$$= 1 + 20 = 21 \text{ J}$$

35. For elastic collision $e = 1$ and velocity of separation is equal to velocity of approach.

The velocity of the target may be more, equal or less than that of projectile depending on their masses.

The maximum velocity of target is double to that of projectile, when projectile is extremely massive as compared to the target.

Maximum kinetic energy is transferred from projectile to target when their masses are exactly equal.

36. Given, $m_1 = m_2 = m$, $u_1 = v$ and $u_2 = 0$

$$v_1 = \frac{v}{2}(1 - e)$$

$$v_2 = \frac{v}{2}(1 + e)$$

$$\therefore \quad \frac{v_1}{v_2} = \left(\frac{1-e}{1+e}\right)$$

37. As net horizontal force acting on the system is zero, hence momentum must remain conserved.

Hence, $\quad mu + 0 = 0 + Mv_2$

$$\Rightarrow \quad v_2 = \frac{mu}{M}$$

As per definition,

$$e = \frac{(v_2 - v_1)}{(u_2 - u_1)} = \frac{v_2 - 0}{0 - u}$$

$$= \frac{v_2}{u} = \frac{\frac{mu}{M}}{u} = \frac{m}{M}$$

38. Let p_1 and p_2 be the momenta of A and B after collision.

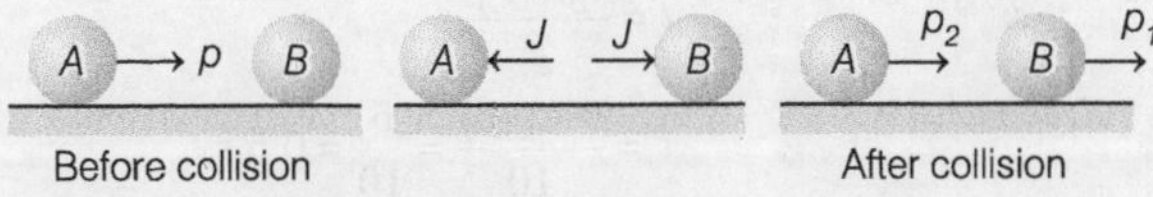

Then applying impulse = change in linear momentum for the two particles

For B $\quad J = p_1$

For A $\quad J = p - p_2$

or $\quad p_2 = p - J$

Coefficient of restitution, $e = \dfrac{p_1 - p_2}{p}$

$$= \frac{p_1 - p + J}{p}$$

$$= \frac{J - p + J}{p}$$

$$= \frac{2J}{p} - 1$$

39. For collision between block A and B,

$$e = \frac{v_B - v_A}{u_A - u_B}$$

$$= \frac{v_B - v_A}{10 - 0} = \frac{v_B - v_A}{10}$$

$$\therefore \quad v_B - v_A = 10e = 10 \times 0.5 = 5 \quad \text{...(i)}$$

From principle of momentum conservation,

$$m_A u_A + m_B u_B = m_A v_A + m_B v_B$$

or $\quad m \times 10 + 0 = mv_A + mv_B$

$$\therefore \quad v_A + v_B = 10 \quad \text{...(ii)}$$

Adding Eqs. (i) and (ii), we get

$$v_B = 7.5 \text{ ms}^{-1} \quad \text{...(iii)}$$

Similarly, for collision between B and C,

$$v_C - v_B = 7.5e = 7.5 \times 0.5 = 3.75$$

$$\therefore \quad v_C - v_B = 3.75 \text{ ms}^{-1} \quad \text{...(iv)}$$

Adding Eqs. (iii) and (iv), we get

$$v_C = 11.25 \text{ ms}^{-1}$$

40. If initial velocity of bullet be v, then after collision combined velocity of bullet and target is

$$v' = \frac{mv}{(M+m)}$$

and $$h = \frac{v'^2}{2g} \text{ or } v' = \sqrt{2gh}$$

$$\therefore \quad \frac{mv}{(M+m)} = \sqrt{2gh}$$

$$\Rightarrow \quad v = \left(\frac{M+m}{m}\right) \cdot \sqrt{2gh} = \left(1 + \frac{M}{m}\right)\sqrt{2gh}$$

41. Acceleration, $g = \dfrac{v - v_0}{t}$

$$\therefore \quad v = 0$$

Speed before first bounce

$$v_0 = -5 \text{ ms}^{-1}$$

$$\therefore \quad t = \frac{v_B - v_A}{g}$$

$$= \frac{0-(-5)}{10} = \frac{5}{10} = 0.5 \text{ s}$$

42. As shown in the following figure, ball is falling from height $2h$ and rebounding to a height h only. It means that velocity of ball just before collision.

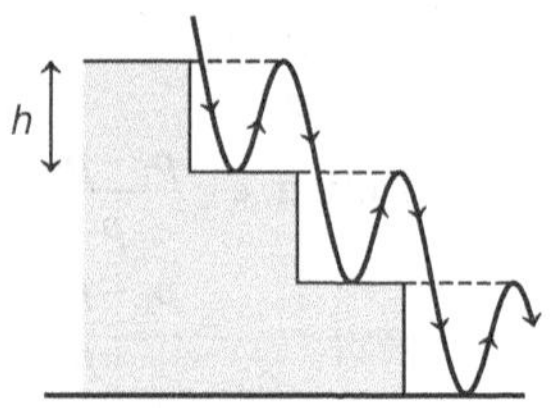

$$u = \sqrt{\frac{2(2h)}{g}} = \sqrt{\frac{4h}{g}}$$

and velocity just after collision.

$$v = -\sqrt{\frac{2h}{g}}$$

$$\therefore \quad e = \frac{-v}{u} = \frac{\sqrt{\frac{2h}{g}}}{\sqrt{\frac{4h}{g}}} = \frac{1}{\sqrt{2}}$$

43. (a) This is only possible when collision is head-on elastic.

(b) When collision is oblique elastic, then in this case, both bodies move perpendicular to each other after collision.

(c) Since, in elastic collision, kinetic energy of system remains constant so, this is not possible.

(d) The same reason as (b)

44. This is an example of elastic oblique collision. When a moving body collides obliquely with another identical body at rest, then during elastic collision, the angle of divergence will be 90°.

45. The momentum of third part will be equal and opposite of the resultant of momentum of rest two equal parts.

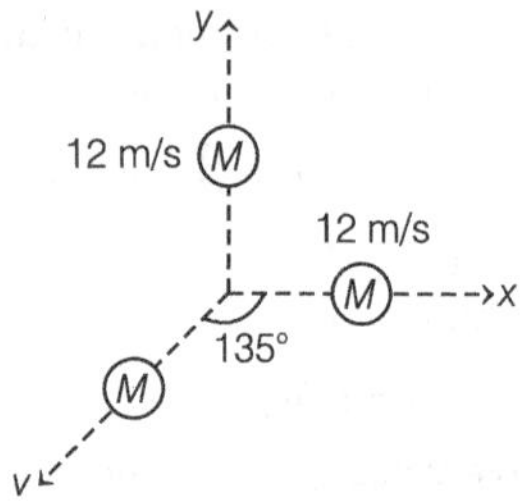

Let v be the velocity of third part.

By the conservation of linear momentum,

$$3m \times v = m \times 12\sqrt{2}$$

$$\Rightarrow \quad v = 4\sqrt{2} \text{ m/s}$$

46. As collision is elastic as shown below, both momentum and KE are conserved.

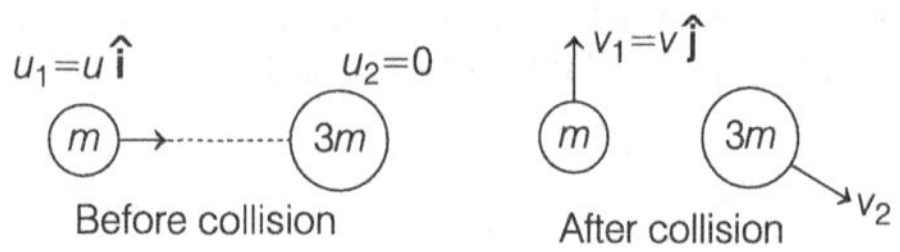

Momentum conservation gives,

$$mu\hat{\mathbf{i}} = mv\hat{\mathbf{j}} + 3m\mathbf{v}_2$$

$$\Rightarrow \quad \mathbf{v}_2 = \frac{1}{3}(u\hat{\mathbf{i}} - v\hat{\mathbf{j}})$$

$$\Rightarrow \quad |\mathbf{v}_2| = \sqrt{\frac{u^2 + v^2}{9}}$$

or $$v_2^2 = (u^2 + v^2)/9 \qquad \text{...(i)}$$

Kinetic energy conservation gives,

$$\Rightarrow \quad \frac{1}{2}mu^2 = \frac{1}{2}mv^2 + \frac{1}{2}3mv_2^2$$

$$\Rightarrow \quad u^2 = v^2 + 3v_2^2 \qquad \text{...(ii)}$$

Substituting value of v_2 from Eq (i) into Eq (ii), we get

$$v = \frac{u}{\sqrt{2}}$$

47. Since, no force is present along the surface, so momentum conservation principle for ball is applicable along the surface of plate.

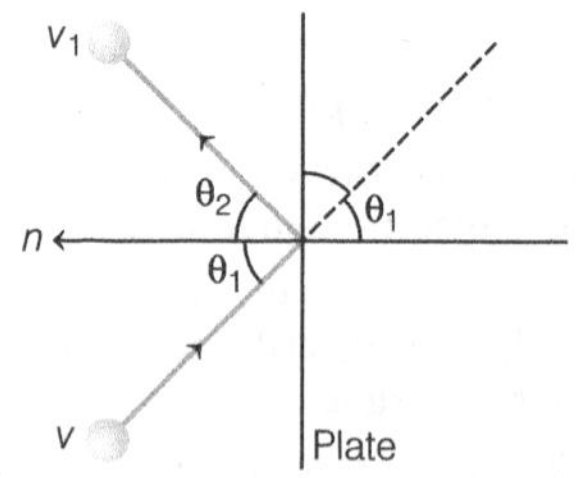

$$mv\sin\theta_1 = mv_1\sin\theta_2$$

or $$v\sin\theta_1 = v_1\sin\theta_2$$

$$e = \frac{v_1\cos\theta_2}{v\cos\theta_1} = \frac{v_1\cos\theta_2}{v\cos\theta}$$

$\therefore \quad v_1 = \cos\theta_2 = ev\cos\theta$

$\therefore \quad \dfrac{v_1 \sin\theta_2}{v_1 \cos\theta_2} = \dfrac{v\sin\theta}{ev\cos\theta} = \dfrac{\tan\theta}{e}$

$\therefore \quad \tan\theta = \dfrac{\tan\theta}{e}$

$\therefore \quad \theta_2 = \tan^{-1}\left(\dfrac{\tan\theta}{e}\right)$

48. Given that, $\mathbf{u}_1 = (\sqrt{3}\,\hat{\mathbf{i}} + \hat{\mathbf{j}})$ m/s, $\mathbf{u}_2 = 0$

$\mathbf{v}_1 = (\hat{\mathbf{i}} + \sqrt{3}\,\hat{\mathbf{j}})$ m/s and $m_1 = 2m_2$

Using conservation of linear momentum,

$$\mathbf{p}_i = \mathbf{p}_f$$

$$m_1\mathbf{u}_1 + m_2\mathbf{u}_2 = m_1\mathbf{v}_1 + m_2\mathbf{v}_2$$

$$2m_2(\sqrt{3}\,\hat{\mathbf{i}} + \hat{\mathbf{j}}) + m_2(0) = 2m_2(\hat{\mathbf{i}} + \sqrt{3}\,\hat{\mathbf{j}}) + m_2\mathbf{v}_2$$

$$\mathbf{v}_2 = 2(\sqrt{3}\,\hat{\mathbf{i}} + \hat{\mathbf{j}}) - 2(\hat{\mathbf{i}} + \sqrt{3}\,\hat{\mathbf{j}})$$

$$= 2(\sqrt{3} - 1)(\hat{\mathbf{i}} - \hat{\mathbf{j}}) \text{ m/s}$$

Let the angle between $\mathbf{v}_1$ and $\mathbf{v}_2$ be θ, then

$$\cos\theta = \frac{\mathbf{v}_1 \cdot \mathbf{v}_2}{v_1 v_2}$$

$$= \frac{(\hat{\mathbf{i}} + \sqrt{3}\,\hat{\mathbf{j}})\cdot 2(\sqrt{3} - 1)(\hat{\mathbf{i}} - \hat{\mathbf{j}})}{2\cdot 2\sqrt{2}(\sqrt{3} - 1)}$$

$$(\because v_1 = 2 \text{ m/s}, \ v_2 = 2\sqrt{2}(\sqrt{3} - 1) \text{ m/s})$$

$$= \frac{2(\sqrt{3} - 1) - 2\sqrt{3}(\sqrt{3} - 1)}{4\sqrt{2}(\sqrt{3} - 1)}$$

$$= \frac{2(\sqrt{3} - 1)(1 - \sqrt{3})}{4\sqrt{2}(\sqrt{3} - 1)}$$

$$\cos\theta = \frac{1 - \sqrt{3}}{2\sqrt{2}} = -0.259$$

$\Rightarrow \quad \theta = 105°$

Alternate solution

Directly observing the direction of $\mathbf{v}_1$ and $\mathbf{v}_2$.

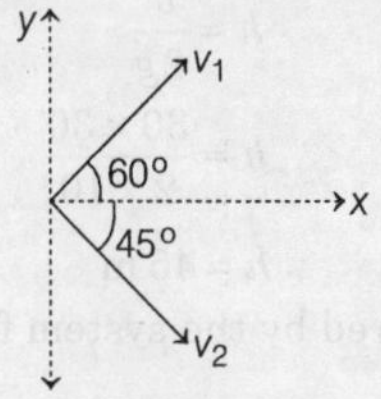

$$\theta_1 = \tan^{-1}\left(\frac{1}{\sqrt{3}}\right) = 60°$$

$\Rightarrow \quad \theta_2 = \tan^{-1}(-1) = 45°$

$\therefore \quad \theta = \theta_1 + \theta_2 = 60° + 45° = 105°$

49. In x-direction, apply conservation of momentum, we get

$$mu_1 + 0 = mv_x$$

$\Rightarrow \quad mv = mv_x$

$\Rightarrow \quad v_x = v$

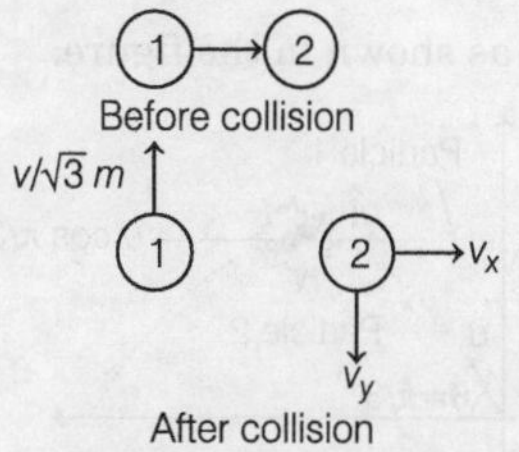

In y-direction, apply conservation of momentum, we get

$$0 + 0 = m\left(\frac{v}{\sqrt{3}}\right) - mv_y$$

$\Rightarrow \quad v_y = \dfrac{v}{\sqrt{3}}$

Velocity of second mass after collision,

$$v' = \sqrt{\left(\frac{v}{\sqrt{3}}\right)^2 + v^2} = \sqrt{\frac{4}{3}v^2} \quad \text{or} \quad v' = \frac{2}{\sqrt{3}}v$$

50. Consider the movement of two particles as shown below.

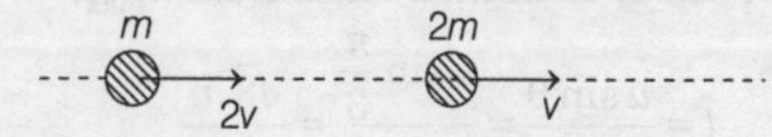

Conserving linear momentum in x-direction

$$(p_i)x = (p_f)x$$

or $\quad 2mv = (2m + m)\,v_x$

or $\quad v_x = \dfrac{2}{3}v$

Conserving linear momentum in y-direction

$$(p_i)y = (p_f)y$$

or $\quad 2mv = (2m + m)\,v_y$

or $\quad v_y = \dfrac{2}{3}v$

Initial kinetic energy of the two particles system is

$$E_i = \frac{1}{2}m(2v)^2 + \frac{1}{2}(2m)(v)^2$$

$$= \frac{1}{2}\times 4mv^2 + \frac{1}{2}\times 2mv^2$$

$$= 2mv^2 + mv^2 = 3mv^2$$

Final energy of the combined two particles system is

$$E_f = \frac{1}{2}(3m)(v_x^2 + v_y^2)$$

$$= \frac{1}{2}(3m)\left[\frac{4v^2}{9} + \frac{4v^2}{9}\right]$$

$$= \frac{3m}{2}\left[\frac{8v^2}{9}\right] = \frac{4mv^2}{3}$$

Loss in the energy $\Delta E = E_i - E_f$

$$= mv^2\left[3 - \frac{4}{3}\right] = \frac{5}{3}mv^2$$

Percentage loss in the energy during the collision

$$\frac{\Delta E}{E_i}\times 100 = \frac{(5/3)\,mv^2}{3mv^2}\times 100 = \frac{5}{9}\times 100 \simeq 56\%$$

51. Collision is as shown in the figure.

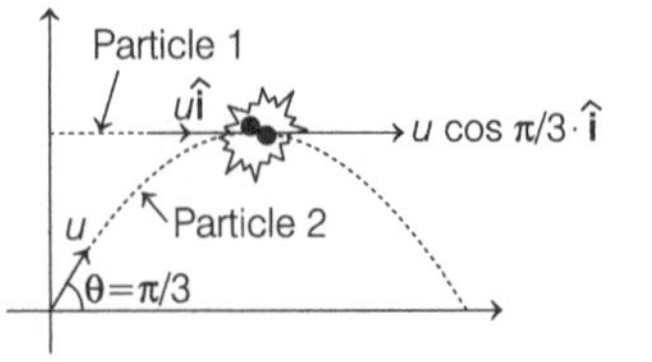

Velocity of the particle projected from origin at its topmost point,

$$\mathbf{u}_2 = u\cos\frac{\pi}{3}\cdot\hat{\mathbf{i}} = \frac{u}{2}\hat{\mathbf{i}}$$

By conservation of momentum (velocity of combined mass after collision (v)), we have

$$mu\hat{\mathbf{i}} + m\frac{u}{2}\hat{\mathbf{i}} = 2m\mathbf{v}$$

$$\Rightarrow \quad \mathbf{v} = \frac{3}{4}u\hat{\mathbf{i}}$$

Time of fall of combined mass from h_{max},

$$t = \frac{u\sin\theta}{g} = \frac{u\sin\frac{\pi}{3}}{g} = \frac{\sqrt{3}}{2}\frac{u}{g}$$

During this time, combined particle keeps on moving with a horizontal speed of $|\mathbf{v}| = \frac{3}{4}u$.

So, horizontal distance covered by combined mass before reaching the ground,

$$R = \text{speed} \times \text{time} = \frac{3}{4}u \times \frac{\sqrt{3}}{2}\frac{u}{g} = \frac{3\sqrt{3}}{8}\cdot\frac{u^2}{g}$$

Round II

1. As, $m_1u_1 + m_2u_2 = (m_1 + m_2)v$

$$\Rightarrow \quad 10\times u_1 + 5\times 0 = (10+5)\times 4$$

$$\Rightarrow \quad u_1 = \frac{15\times 4}{10} = 6\ \text{ms}^{-1}$$

2. If $M = M'$, then bullet will transfer whole of its velocity (and consequently 100% of its KE) to block and will itself come to rest as per theory of collision.

3. Since net momentum of the composite system is zero, hence resultant velocity of the composite system should also be zero.

4. In the following figure, it can be seen that the component of momentum along X-axis (parallel to the wall of container) remains unchanged even after the collision.

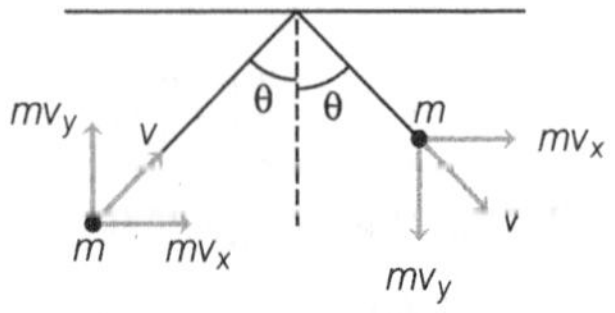

$\therefore$ Impulse = change in momentum of gas molecule along Y-axis, *i.e.* in a direction normal to the wall $= 2mv\cos\theta$

5. $\because \quad \mathbf{a}_{CM} = \dfrac{\mathbf{F}_{eq}}{(m_1+m_2+m_3)} = \dfrac{m_1\mathbf{a}_1 + m_2\mathbf{a}_2 + m_3\mathbf{a}_3}{(m_1+m_2+m_3)}$

$$\therefore \quad \mathbf{F}_{eq} = m_1\mathbf{a}_1 + m_2\mathbf{a}_2 + m_3\mathbf{a}_3$$

$$= 1\times 1 + 2\times 2 + 4\times(-0.5) = 1 + 4 - 2 = 3\ \text{N}$$

6. Since gun-shot system is an isolated closed system, its centre of mass must remain at rest.

7. When two identical balls collide head-on elastically, they exchange their velocities. Hence when A collides with B, A transfers its whole velocity to B. When B collides with C, B transfers its whole velocity to C. Hence, finally A and B will be at rest and only C will be moving forward with a speed v.

8. $a_{system} = \dfrac{5g - 5\mu g}{5+5} = 4\ \text{m/s}^2$

$$a_{CM} = \frac{m_1\mathbf{a}_1 + m_2\mathbf{a}_2}{m_1+m_2} = \frac{5(4\hat{\mathbf{i}}) + 5(4\hat{\mathbf{j}})}{10}$$

$$= \frac{5\sqrt{4^2+4^2}}{10} = 2\sqrt{2}\ \text{m/s}^2$$

9. Velocity of bullet is very high compared to velocity of wooden block so, in order to calculate time for collision, we take relative velocity nearly equal to velocity of bullet.

So, time taken for particles to collide is

$$t = \frac{d}{v_{rel}} = \frac{100}{100} = 1\ \text{s}$$

Speed of block just before collision is,

$$v_1 = gt = 10\times 1 = 10\ \text{ms}^{-1}$$

Speed of bullet just before collision is

$$v_2 = u - gt = 100 - 10\times 1 = 90\ \text{ms}^{-1}$$

Let v = velocity of bullet + block system, then by conservation of linear momentum, we get

$$-(0.03\times 10) + (0.02\times 90) = (0.05)\,v$$

$$\Rightarrow \quad v = 30\ \text{ms}^{-1}$$

Now, maximum height reached by bullet and block is

$$h = \frac{v^2}{2g}$$

$$\Rightarrow \quad h = \frac{30\times 30}{2\times 10}$$

$$\Rightarrow \quad h = 45\ \text{m}$$

$\therefore$ Height covered by the system from point of collision = 45 m

Now, distance covered by bullet before collision in 1 s.

$$= 100\times 1 - \frac{1}{2}\times 10\times 1^2 = 95\ \text{m}$$

Distance of point of collision from the top of the building

$$= 100 - 95 = 5\ \text{m}$$

$\therefore$ Maximum height to which the combined system reaches above the top of the building before falling below $= 45 - 5 = 40$ m.

10. At the time of applying the impulsive force on block of 10 kg pushes the spring forward but 4 kg mass is at rest. Hence,

$$v_{CM} = \frac{m_1 v_1 + m_2 v_2}{m_1 + m_2} = \frac{10 \times 14 + 4 \times 0}{10 + 4}$$

$$= \frac{140}{14} = 10 \text{ ms}^{-1}$$

11. At the highest point momentum of particle before explosion

$$\mathbf{p} = mv \cos 60^\circ$$

$$= m \times 200 \times \frac{1}{2} = 100\, m \text{ horizontally.}$$

Now, as there is no external force during explosion, hence

$$\mathbf{p} = \mathbf{p}_1 + \mathbf{p}_2 + \mathbf{p}_3 = \text{constant}$$

However, since velocities of two fragments, of masses $m/3$ each, are 100 ms^{-1} downward and 100 ms^{-1} upward.

Hence, $\quad \mathbf{p}_1 = -\mathbf{p}_2$

or $\quad \mathbf{p}_1 + \mathbf{p}_2 = 0$

$$\mathbf{p}_3 = \frac{m}{3} \cdot \mathbf{v}_3 = \mathbf{p} = 100\, m \text{ horizontally}$$

$$v_3 = 300 \text{ ms}^{-1} \text{ horizontally}$$

12.

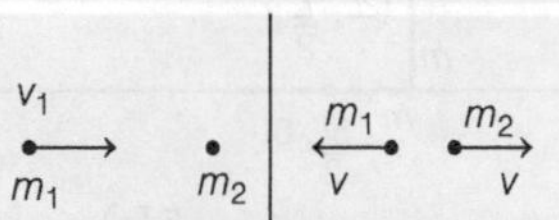

$$m_1 v_1 = -m_1 v + m_2 v$$

$$v_1 = -v + \frac{m_2}{m_1} v$$

$$\frac{(v_1 + v)}{v} = \frac{m_2}{m_1}$$

$$e = \frac{2v}{v_1} = 1$$

$$v = \frac{v_1}{2}$$

$$\frac{v_1 + v_1/2}{v_1/2} = \frac{m_2}{m_1} \Rightarrow 3 = \frac{m_2}{m_1}$$

Hence, the ratio of masses $m_2 : m_1$ is 3 : 1.

13. In the pulley arrangement, $|\mathbf{a}_1| = |\mathbf{a}_2| = a = \left(\frac{m_1 - m_2}{m_1 + m_2}\right) g$

but $\mathbf{a}_1$ is in downward direction and $\mathbf{a}_2$ in the upward direction, *i.e.* $\mathbf{a}_2 = -\mathbf{a}_1$.

$\therefore$ Acceleration of centre of mass

$$\mathbf{a}_{CM} = \frac{m_1 \mathbf{a}_1 + m_2 \mathbf{a}_2}{m_1 + m_2}$$

$$= \frac{m_1 \left(\frac{m_1 - m_2}{m_1 + m_2}\right) g - m_2 \left(\frac{m_1 - m_2}{m_1 + m_2}\right) g}{(m_1 + m_2)}$$

$$= \left(\frac{m_1 - m_2}{m_1 + m_2}\right)^2 g$$

14. First sphere will take a time t_1 to start motion in second sphere on colliding with it, where $t_1 = \frac{L}{u}$.

Now speed of second sphere will be

$$v_2 = \frac{u}{2}(1 + e) = \frac{2}{3} u \qquad \left(\because e = \frac{1}{3}\right)$$

Hence, time taken by second sphere to start motion in third sphere $t_2 = \frac{L}{2/3\, u} = \frac{3\,L}{2\,u}$.

$$\therefore \text{Total time} \quad t = t_1 + t_2 = \frac{L}{u} + \frac{3\,L}{2\,u} = \frac{5\,L}{2\,u}$$

15. Consider the two cart system as a single system. Due to explosion of power, total momentum of system remains unchanged, *i.e.* $\mathbf{p}_1 + \mathbf{p}_2 = 0$ or $m_1 v_1 = m_2 v_2$, hence

$$\frac{v_1}{v_2} = \frac{m_2}{m_1}$$

As coefficient of friction between carts and rails are identical, hence $a_1 = a_2$ and at the time of stopping, final velocity of cart is zero. Using equation $v^2 - u^2 = 2\,as$, we have

$$\frac{s_1}{s_2} = \frac{v_1^2}{v_2^2} = \frac{m_2^2}{m_1^2}$$

$$\Rightarrow \qquad s_2 = \frac{s_1 m_1^2}{m_2^2} = \frac{36 \times (200)^2}{(300)^2} = 16 \text{ m}$$

16. We know that velocity of 2nd ball after collision is given by

$$v_2 = \frac{u_1 (1 + e) m_1}{(m_1 + m_2)} + u_2 \frac{(m_2 - m_1 e)}{(m_1 + m_2)}$$

In present problem $u_2 = 0, m_2 = 2 m_1$ and $e = 2/3$, hence

$$v_2 = \frac{u\left(1 + \frac{2}{3}\right) m_1}{(m_1 + 2 m_1)} = \frac{5}{9} u$$

As four exactly similar type of collisions are taking place successively, hence velocity communicated to fifth ball

$$v_5 = \left(\frac{5}{9}\right)^4 u$$

17. $h = \frac{1}{2} g t^2$ (parabolic)

$v = -gt$ and after the collision $v = gt$ (straight line). Collision is perfectly elastic, then ball reaches to same height again and again with same velocity.

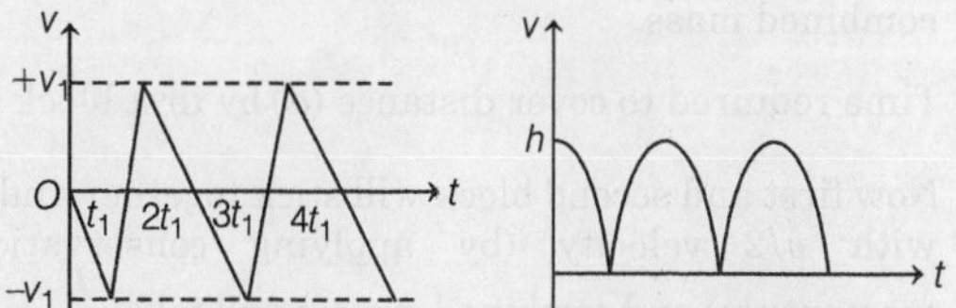

18. Let ball strikes at a speed u, the $K_1 = \frac{1}{2} m u^2$.

Due to collision, tangential component of velocity remains unchanged at $u \sin 45^\circ$, but the normal component of velocity change to $u \sin 45^\circ = \frac{1}{2} u \cos 45^\circ$

$\therefore$ Final velocity of ball after collision

$$v = \sqrt{(u \sin 45^\circ) + \left(\frac{1}{2} u \cos 45^\circ\right)^2}$$

$$= \sqrt{\left(\frac{u}{\sqrt{2}}\right)^2 + \left(\frac{u}{2\sqrt{2}}\right)^2} = \sqrt{\frac{5}{8}}\, u$$

Hence, final kinetic energy

$$K_2 = \frac{1}{2} mv^2 = \frac{5}{16} mu^2$$

$\therefore$ Fractional loss in KE

$$= \frac{K_1 - K_2}{K_1}$$

$$= \frac{\frac{1}{2} mu^2 - \frac{5}{16} mu^2}{\frac{1}{2} mu^2} = \frac{3}{8}$$

19. Let at the time of explosion velocity of one piece of mass $m/2$ is $(10\,\hat{\mathbf{i}})$. If velocity of other be $\mathbf{v}_2$, then from conservation law of momentum (since there is no force in horizontal direction), horizontal component of $\mathbf{v}_2$, must be $-10\,\hat{\mathbf{i}}$.

$\therefore$ Relative velocity of two parts in horizontal direction $= 20\ \text{ms}^{-1}$.

Time taken by ball to fall through 45 m,

$$= 20 = \sqrt{\frac{2h}{g}} = \sqrt{\frac{2 \times 45}{10}} = 3\,\text{s}$$

and time taken by ball to fall through first 20 m,

$$t' = \sqrt{\frac{2h'}{g}} = \sqrt{\frac{2 \times 20}{10}} = 2\,\text{s}$$

Hence time taken by ball pieces to fall from 25 m height to ground $= t - t' = 3 - 2 = 1\,\text{s}$.

$\therefore$ Horizontal distance between the two pieces at the time of striking on ground $= 20 \times 1 = 20$ m.

20.

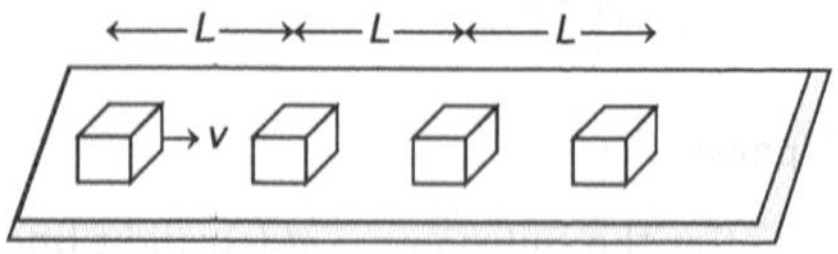

Since, collision is perfectly inelastic, so all the block will stick together one by one and move in a form of combined mass.

Time required to cover distance (d) by first block $= \dfrac{L}{v}$.

Now first and second block will stick together and move with $v/2$ velocity (by applying conservation of momentum) and combined system will take $\dfrac{L}{v/2} = \dfrac{2L}{v}$ to reach upto block third.

Now, these three blocks will move with velocity $v/3$ and combined system will take time $\dfrac{L}{v/3} = \dfrac{3L}{v}$ to reach upto the fourth block.

So, total time $\dfrac{L}{v} + \dfrac{2L}{v} + \dfrac{3L}{v} + \ldots \dfrac{(n-1)L}{v} = \dfrac{n(n-1)L}{2v}$

Final velocity of the centre of mass of the system will be v/n.

21. Using law of conservation of linear momentum, we have

$$mv_0 = mv + 2mv$$

$$\Rightarrow \quad v = \frac{v_0}{3}$$

Using conservation of energy, we have

$$\frac{1}{2} mv_0^2 = \frac{1}{2} x_0^2 + \frac{1}{2} (3m) v^2$$

where, x_0 = compression in the spring.

$$\therefore \quad mv_0^2 = kx_0^2 + (3m) \frac{v_0^2}{9}$$

$$\Rightarrow \quad k = \frac{2mv_0^2}{3x_0^2}$$

22. For given system of rods, masses and coordinates of centre of rods are as shown in the following figure.

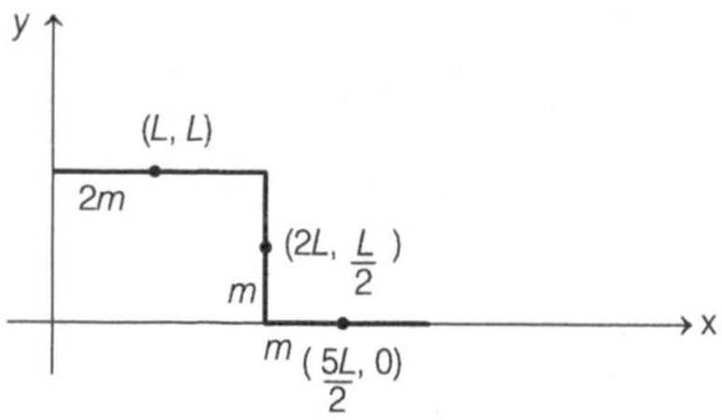

$$\text{So, } X_{\text{COM}} = \left(\frac{2mL + m2L + m\frac{5L}{2}}{4m}\right) = \frac{13}{8} L$$

$$\text{and } Y_{\text{COM}} = \frac{2mL + m \times \frac{L}{2} + m \times 0}{4m} = \frac{5L}{8}$$

So, position vector of COM is

$$\mathbf{r}_{\text{COM}} = X_{\text{COM}} \hat{\mathbf{x}} + Y_{\text{COM}} \hat{\mathbf{y}}$$

$$= \frac{13}{8} L\hat{\mathbf{x}} + \frac{5}{8} L\hat{\mathbf{y}}$$

23.

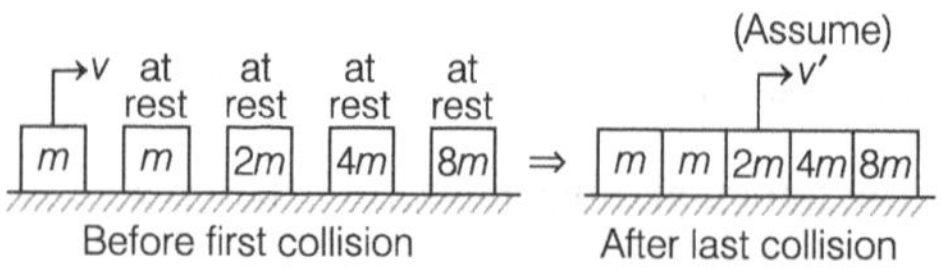

Since, all the collisions are perfectly inelastic, so after the final collision, all blocks will be moving together. Let their final velocity be v'.

By law of conservation of linear momentum,

$$(\mathbf{p}_{\text{sys}})_i = (\mathbf{p}_{\text{sys}})_f$$

$$\Rightarrow mv + m(0) + 2m(0) + 4m(0) + 8m(0) = (m + m + 2m + 4m + 8m)v'$$

$$\Rightarrow \quad mv = 16mv'$$

$$\Rightarrow \quad v' = \frac{v}{16} \qquad \ldots\text{(i)}$$

Now, initial kinetic energy of system,

$$(K_{sys})_i = \frac{1}{2}mv^2 + 0 + 0 + 0 + 0 = \frac{1}{2}mv^2$$

And final kinetic energy of system,

$$(K_{sys})_f = \frac{1}{2}(m + m + 2m + 4m + 8m)(v')^2$$

$$= \frac{1}{2} \times 16m \times \left(\frac{v}{16}\right)^2$$

$$= \frac{1}{2} \times 16m \times \frac{v^2}{256}$$

$$= \frac{1}{32}mv^2$$

Loss in kinetic energy,

$$(\Delta K_{sys})_{loss} = (K_{sys})_i - (K_{sys})_f$$

$$= \frac{1}{2}mv^2 - \frac{1}{32}mv^2$$

$$= \frac{15}{32}mv^2$$

% loss in kinetic energy,

$$\%(\Delta K_{sys})_{loss} = \frac{(\Delta K_{sys})_{loss}}{(K_{sys})_i} \times 100\%$$

$$= \frac{\frac{15}{32}mv^2}{\frac{1}{2}mv^2} \times 100\%$$

$$= \frac{15}{16} \times 100\%$$

$$= 93.75\%$$

Given that, $\% (\Delta K_{sys})_{loss} = p\%$

so, $$p = 93.75 \simeq 94$$

24. The given rectangular thin sheet $ABCD$ can be drawn as shown in the figure below,

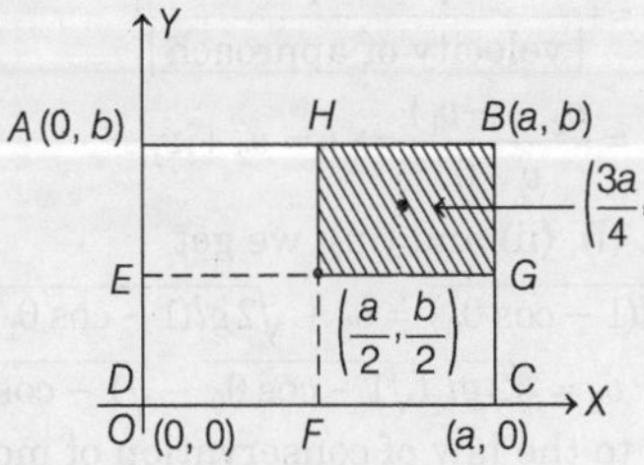

Here,

Area of complete lamina, $A_1 = ab$

Area of shaded part of lamina $= \frac{a}{2} \times \frac{b}{2} = \frac{ab}{4}$

(x_1, y_1) = coordinates of centre of mass of complete lamina $= \left(\frac{a}{2}, \frac{b}{2}\right)$

(x_2, y_2) = coordinates of centre of mass of shaded part of lamina $= \left(\frac{3a}{4}, \frac{3b}{4}\right)$

$\therefore$ Using formula for centre of mass, we have

$$X_{CM} = \frac{A_1x_1 - A_2x_2}{A_1 - A_2}$$

$$= \frac{ab\left(\frac{a}{2}\right) - \frac{ab}{4}\left(\frac{3a}{4}\right)}{ab - \frac{ab}{4}}$$

$$= \frac{\frac{8a^2b - 3a^2b}{16}}{\frac{3ab}{4}} = \frac{5a}{12}$$

Similarly, $$Y_{CM} = \frac{A_1y_1 - A_2y_2}{A_1 - A_2}$$

$$= \frac{ab\left(\frac{b}{2}\right) - \frac{ab}{4}\left(\frac{3b}{4}\right)}{ab - \frac{ab}{4}} = \frac{5b}{12}$$

25. Let particles collide at some distance h' from top at time t_0. Then,

$$h' = \frac{1}{2}gt_0^2 \quad \text{... (i)}$$

(for particle A)

and $$h - h' = \sqrt{2gh} \cdot t_0 - \frac{1}{2}gt_0^2 \quad \text{.... (ii)}$$

(for particle B)

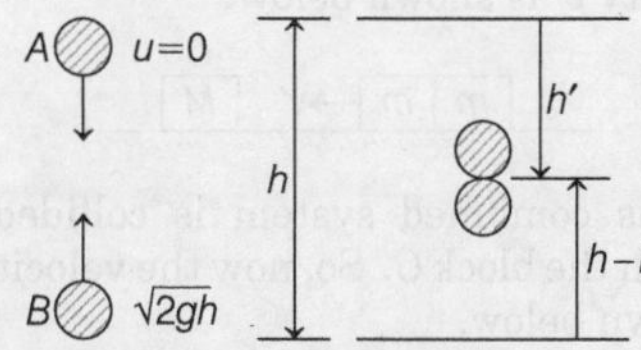

From these equations, particles meet after time t_0 given by

$$t_0 = \frac{h}{\sqrt{2gh}} = \sqrt{\frac{h}{2g}}$$

Velocities of particles A and B at instant of collision are $v_A = gt_0$ and $v_B = \sqrt{2gh} - gt_0$.

Hence,

$$v_A = g \times \sqrt{\frac{h}{2g}} = \sqrt{\frac{1}{2}gh} = \frac{1}{\sqrt{2}}\sqrt{gh}$$

and $$v_B = \sqrt{2gh} - g\sqrt{\frac{h}{2g}}$$

$$= \left(\sqrt{2} - \frac{1}{\sqrt{2}}\right)\sqrt{gh} = \frac{1}{\sqrt{2}} \times \sqrt{gh}$$

So, particles collide as shown in the figure.

From momentum conservation, we can see that particles stuck, $p_{initial} = p_{final}$.

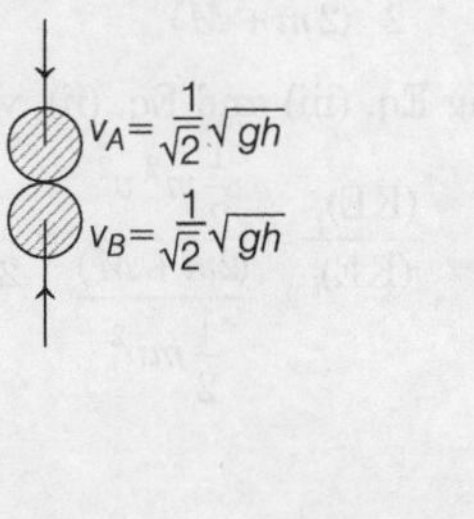

This means the combined system of particles comes to rest ($v_{\text{combined mass}} = 0$) instantaneously.

Now, we have to calculate time of fall of combined mass.

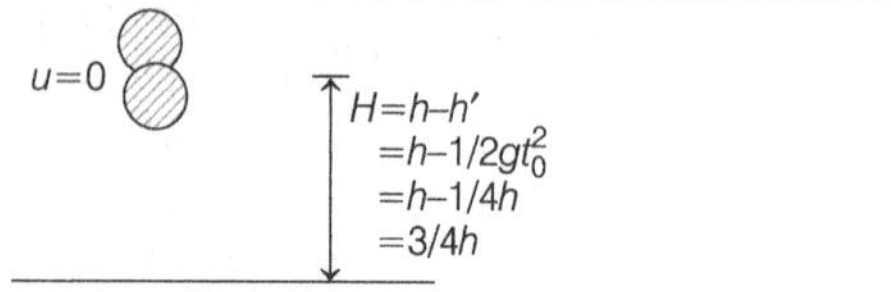

Combined mass starts with $u = 0$

and its height above earth's surface is $H = \frac{3}{4}h$.

So, time taken by combined mass to reach ground is given by

$$H = ut + \frac{1}{2}at^2$$

$$\Rightarrow \quad \frac{3}{4}h = \frac{1}{2}g \times t^2 \Rightarrow \sqrt{\frac{3}{2}}\sqrt{\frac{h}{g}} = t$$

26. Initially, block A is moving with velocity v as shown in the figure below,

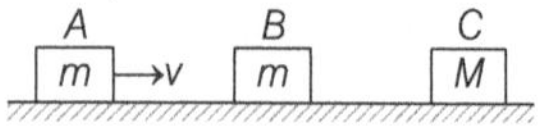

Now, A collides with B such that they collide inelastically. Thus, the combined mass (say) move with the velocity v as shown below.

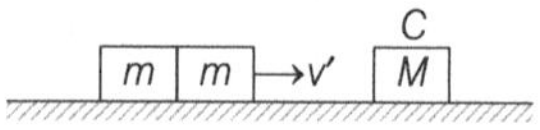

Then, this combined system is collided inelastically again with the block C. So, now the velocity of system be v'' as shown below.

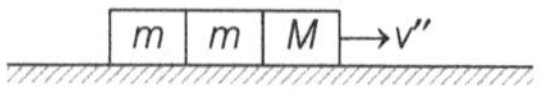

Thus, according to the principle of conservation of momentum,

initial momentum of the system = final momentum of the system

$$\Rightarrow \quad m\,v = (2m + M)\,v''$$

$$\text{or} \quad v'' = \left(\frac{mv}{2m + M}\right) \quad \ldots \text{(i)}$$

Initial kinetic energy of the system,

$$(\text{KE})_i = \frac{1}{2}mv^2 \quad \ldots \text{(ii)}$$

Final kinetic energy of the system, $(\text{KE})_f$

$$= \frac{1}{2}(2m + M)(v'')^2 = \frac{1}{2}(2m + M)\left(\frac{mv}{2m + M}\right)^2$$

[$\because$ using Eq. (i)]

$$= \frac{1}{2} \cdot \frac{v^2 m^2}{(2m + M)} \quad \ldots \text{(iii)}$$

Dividing Eq. (iii) and Eq. (ii), we get

$$\frac{(\text{KE})_f}{(\text{KE})_i} = \frac{\frac{\frac{1}{2}m^2v^2}{(2m + M)}}{\frac{1}{2}mv^2} = \frac{m}{2m + M} \quad \ldots \text{(iv)}$$

It is given that $\frac{5}{6}$th of $(\text{KE})_i$ is lost in this process.

$$\Rightarrow \quad (\text{KE})_f = \frac{1}{6}(\text{KE})_i$$

$$\Rightarrow \quad \frac{(\text{KE})_f}{(\text{KE})_i} = \frac{1}{6} \quad \ldots \text{(v)}$$

Comparing Eq. (iv) and Eq. (v), we get

$$\frac{m}{2m + M} = \frac{1}{6} \Rightarrow 6m = 2m + M$$

$$4m = M \Rightarrow \frac{M}{m} = 4$$

27. Initially, when pendulum is released from angle θ_0 as shown in the figure below,

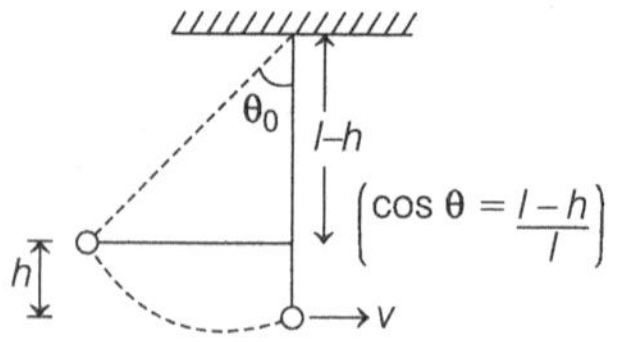

We have, $mgh = \frac{1}{2}mv^2$

Here, $h = l - l\cos\theta_0$

So, $v = \sqrt{2gl\,(1 - \cos\theta_0)}$ ...(i)

With velocity v, bob of pendulum collides with block. After collision, let v_1 and v_2 are final velocities of masses m and M respectively as shown.

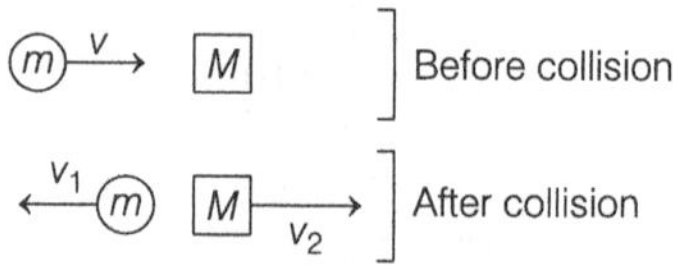

Then, if pendulum is deflected back upto angle θ_1, then

$$v_1 = \sqrt{2gl\,(1 - \cos\theta_1)} \quad \ldots \text{(ii)}$$

Using definition of coefficient of restitution to get

$$e = \frac{|\text{velocity of separation}|}{|\text{velocity of approach}|}$$

$$1 = \frac{v_2 - (-v_1)}{v - 0} \Rightarrow v = v_2 + v_1 \quad \ldots \text{(iii)}$$

From Eqs. (i), (ii) and (iii), we get

$$\Rightarrow \quad \sqrt{2gl(1 - \cos\theta_0)} = v_2 + \sqrt{2gl(1 - \cos\theta_1)}$$

$$\Rightarrow \quad v_2 = \sqrt{2gl}\,(\sqrt{1 - \cos\theta_0} - \sqrt{1 - \cos\theta_1}) \quad \ldots \text{(iv)}$$

According to the law of conservation of momentum,

initial momentum of the system = final momentum of the system

$$\Rightarrow \quad mv = Mv_2 - mv_1$$

$$Mv_2 = m\sqrt{2gl}\,(\sqrt{1 - \cos\theta_0} + \sqrt{1 - \cos\theta_1}) \quad \ldots \text{(v)}$$

Dividing Eq. (v) by Eq. (iv), we get

$$\Rightarrow \quad \frac{M}{m} = \frac{\sqrt{1 - \cos\theta_0} + \sqrt{1 - \cos\theta_1}}{\sqrt{1 - \cos\theta_0} - \sqrt{1 - \cos\theta_1}}$$

$$\frac{M}{m} = \frac{\sin\left(\frac{\theta_0}{2}\right) + \sin\left(\frac{\theta_1}{2}\right)}{\sin\left(\frac{\theta_0}{2}\right) - \sin\left(\frac{\theta_1}{2}\right)}$$

For small θ_0, we have

$$\frac{M}{m} = \frac{\frac{\theta_0}{2} + \frac{\theta_1}{2}}{\frac{\theta_0}{2} - \frac{\theta_1}{2}} \quad \text{or} \quad M = m\left(\frac{\theta_0 + \theta_1}{\theta_0 - \theta_1}\right)$$

28. The given condition can be drawn as shown below.

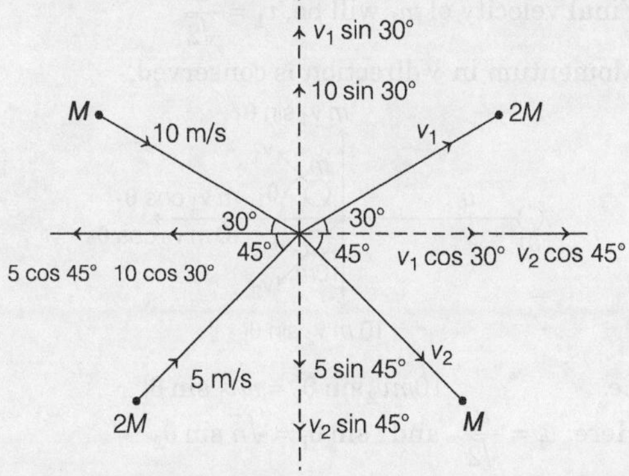

Applying linear momentum conservation law in x-direction, we get

Initial momentum = Final momentum

$$(M \times 10 \cos 30°) + (2M \times 5 \cos 45°)$$
$$= (M \times v_2 \cos 45°) + (2M \times v_1 \cos 30°)$$

$$\Rightarrow \quad 5\sqrt{3} + 5\sqrt{2} = \frac{v_2}{\sqrt{2}} + v_1\sqrt{3} \quad \ldots \text{(i)}$$

Similarly, applying linear momentum conservation law in y-direction, we get

$$(M \times 10 \sin 30°) - (2M \times 5 \sin 45°)$$
$$= (M \times v_2 \sin 45°) - (2M \times v_1 \sin 30°)$$

$$\Rightarrow \quad 5 - 5\sqrt{2} = \frac{v_2}{\sqrt{2}} - v_1 \quad \ldots \text{(ii)}$$

Subtracting Eq. (ii) from Eq. (i), we get

$$(5\sqrt{3} + 5\sqrt{2}) - (5 - 5\sqrt{2}) = \left(\frac{v_2}{\sqrt{2}} + v_1\sqrt{3}\right) - \left(\frac{v_2}{\sqrt{2}} - v_1\right)$$

$$\Rightarrow \quad v_1 = 6.516 \text{ m/s} \approx 6.5 \text{ m/s} \quad \ldots \text{(iii)}$$

Substituting the value from Eq. (iii) in Eq. (i), we get

$$\Rightarrow \quad v_2 \approx 6.3 \text{ m/s}$$

29. If we combine the remaining and removed spherical parts at their initial relative positions, we get complete sphere as shown in the figure.

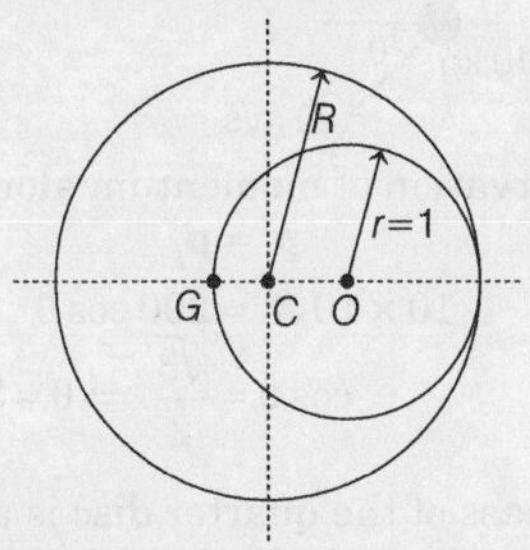

So, for complete sphere,

$$X_{CM} = \frac{m_{\text{remaining}} \times x_{\text{remaining}} + m_{\text{removed}} \times x_{\text{removed}}}{m_{\text{remaining}} + m_{\text{removed}}}$$

As, X_{CM} is at C which is taken at origin of our chosen reference axis, so $X_{CM} = 0$.

$$\Rightarrow \quad m_{\text{remaining}} \times x_{\text{remaining}} + m_{\text{removed}} \times x_{\text{removed}} = 0 \quad \ldots \text{(i)}$$

Here, $m_{\text{remaining}} = \left(\frac{4}{3}\pi R^3 - \frac{4}{3}\pi 1^3\right)\rho$

where, ρ = density of sphere,

$$x_{\text{remaining}} = 2 - R,$$

$$M_{\text{removed}} = -\left(\frac{4}{3}\pi 1^3\right)\rho.$$

Here, mass removed is negative

and $\quad x_{\text{removed}} = R - 1$.

So, from Eq. (i), we get

$$\left(\frac{4}{3}\pi R^3 - \frac{4}{3}\pi\right)\rho \times (2 - R) = \left(\frac{4}{3}\pi\right)\rho \times (R - 1)$$

$$\Rightarrow \quad (R^3 - 1)(2 - R) = R - 1$$

$$\Rightarrow \quad (R^2 + R + 1)(2 - R) = 1$$

30. The distance of centre of mass of a solid hemisphere from the centre of the flat surface is given by

$$d = \frac{3R}{8}$$

Here, $\quad R = 8$ cm

$\therefore \quad d = \frac{3 \times 8}{8} = 3$ cm

Given, $\quad d = x$ cm

$\therefore \quad x = 3$

31. Given situation is shown in the figure.

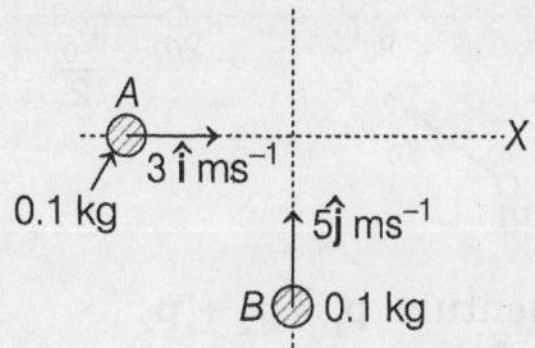

Total initial momentum, $\mathbf{p}_i = m_A v_A + m_B v_B$

$$= 0.1 \times 3\hat{i} + 0.1 \times 5\hat{j}$$
$$= 0.3\,\hat{i} + 0.5\,\hat{j} \text{ (kg-ms}^{-1})$$

Final velocity of A, $\quad v_A = 4\,(\hat{i} + \hat{j})$

Let final velocity of B be v_B.

Then, final momentum after collision,

$$\mathbf{p}_f = m_A v_A + m_B v_B$$
$$= 0.1 \times 4(\hat{i} + \hat{j}) + 0.1 \times v_B$$

Now, by conservation of momentum, we have

$$\mathbf{p}_i = \mathbf{p}_f$$

$$\Rightarrow \quad 0.3\hat{i} + 0.5\hat{j} = 0.4\hat{i} + 0.4\hat{j} + 0.1 v_B$$

$$\Rightarrow \quad v_B = -\hat{i} + \hat{j}$$

Kinetic energy of B after collision will be

$$K = \frac{1}{2}mv^2 = \frac{1}{2} \times 0.1 \times |(-\hat{i} + \hat{j})|^2$$

$$= \frac{1}{2} \times 0.1 \times 2 = 0.1 \text{ J}$$

It is given, energy of B after collision is $\frac{x}{10}$.

So, $\frac{x}{10} = 0.1$ or $x = 1$.

32. Centre of mass of remaining portion is given by

$$X_{CM} = \frac{m_1x_1 - m_2x_2}{m_1 - m_2}$$

Here note that, removed mass (m_2) is treated as a negative mass.

Let x = mass per unit area of uniform disc.

Then, m_1 = mass of complete disc $= x \cdot \pi a^2$

and m_2 = mass of square portion $= x \cdot \frac{a^2}{4}$.

Also, x_1 = x-coordinate of centre of mass of $m_1 = 0$

and x_2 = x- coordinate of centre of mass of $m_2 = d$.

$$\Rightarrow \quad X_{CM} = \frac{x\pi a^2 \times 0 - x\frac{a^2}{4} \times d}{x\pi a^2 - x\frac{a^2}{4}} = \frac{-a}{2(4\pi - 1)} \quad \ldots\text{(i)}$$

$$\left(\text{Since, } d = \frac{a}{2}\right)$$

Given, $X_{CM} = -\frac{a}{X}$ ...(ii)

Comparing the Eqs. (i) and (ii), we get

$\therefore \quad X = 2(4\pi - 1) = 23.12$ or $X \approx 23$.

33. Let initial velocities of two bodies are making angle θ_1 and θ_2 with horizontal direction as shown in figure.

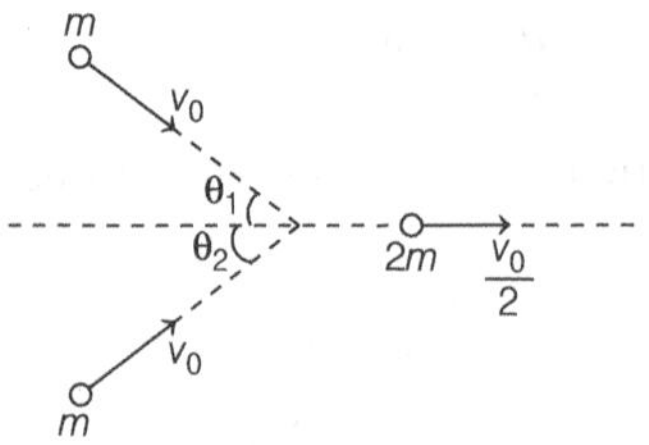

Initial momentum, $\mathbf{p}_i = \mathbf{p}_1 + \mathbf{p}_2$

$= \{mv_0 \cos\theta_1 \hat{\mathbf{i}} + mv_0 \sin\theta_1 \hat{\mathbf{j}}\} + \{mv_0 \cos\theta_2 \hat{\mathbf{i}} + mv_0 \sin\theta_2 - \hat{\mathbf{j}}\}$

$= mv_0(\cos\theta_1 + \cos\theta_2)\hat{\mathbf{i}} + mv_0(\sin\theta_1 - \sin\theta_2)\hat{\mathbf{j}}$

Final momentum, $\mathbf{p}_f = (2m)\left(\frac{v_0}{2}\right)\hat{\mathbf{i}}$

$\Rightarrow \quad \mathbf{p}_f = mv_0\hat{\mathbf{i}}$

In collision momentum remains conserved, so, applying momentum conservation,

$$\mathbf{p}_f = \mathbf{p}_i$$

or $\quad mv_0\hat{\mathbf{i}} = mv_0(\cos\theta_1 + \cos\theta_2)\hat{\mathbf{i}} + mv_0(\sin\theta_1 - \sin\theta_2)\hat{\mathbf{j}}$

$\Rightarrow \quad \sin\theta_1 - \sin\theta_2 = 0 \Rightarrow \theta_1 = \theta_2$

and $\quad mv_0 = 2mv_0\cos\theta$ or $\cos\theta = \frac{1}{2}$ or $\theta = 60°$

But angle between initial velocities is $\theta_1 + \theta_2$ which is equal to $60° + 60° = 120°$.

34. Given, impact is shown below,

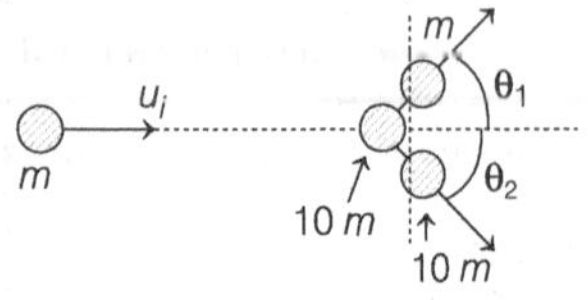

For particle 1, final KE is equals to half of its initial value,

$$K_f = \frac{1}{2}K_i$$

$$\Rightarrow \quad \frac{1}{2}m_1v_1^2 = \left(\frac{1}{2}m_1u^2\right) \times \frac{1}{2}$$

Final velocity of m_1 will be, $v_1 = \frac{u}{\sqrt{2}}$

Momentum in y-direction is conserved,

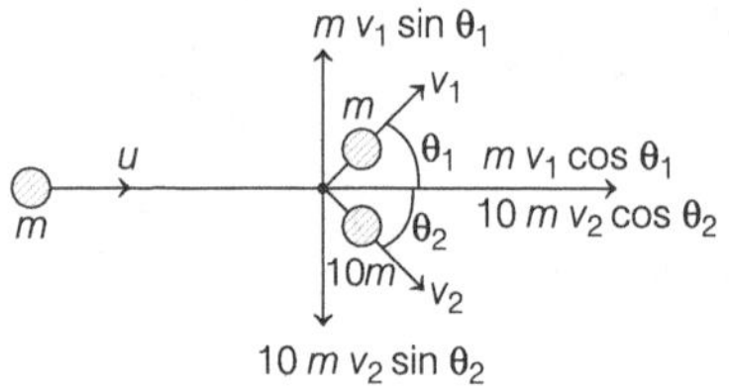

i.e. $\quad 10mv_2\sin\theta_2 = mv_1\sin\theta_1$

Here, $v_1 = \frac{u}{\sqrt{2}}$ and $\sin\theta_1 = \sqrt{n}\sin\theta_2$

So, we have

$$10mv_2 \cdot \sin\theta_2 = \frac{mu}{\sqrt{2}} \cdot \sqrt{n}\sin\theta_2$$

$$\Rightarrow \quad v_2 = \frac{u\sqrt{n}}{10\sqrt{2}} \quad \ldots\text{(i)}$$

Also collision is elastic, so KE is conserved.

$$\Rightarrow \quad \frac{1}{2}mu^2 = \frac{1}{2}mv_1^2 + \frac{1}{2}(10m)v_2^2$$

Substituting values of v_1 and v_2, we have

$$u^2 = \left(\frac{u}{\sqrt{2}}\right)^2 + 10\left(\frac{u^2n}{100 \times 2}\right)$$

$$\Rightarrow \quad \frac{u^2}{2} = \frac{u^2n}{10 \times 2} \Rightarrow n = 10$$

35. Before collision

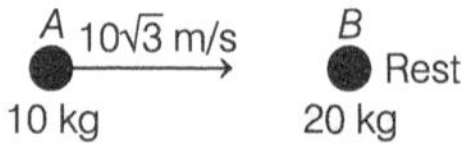

After collision

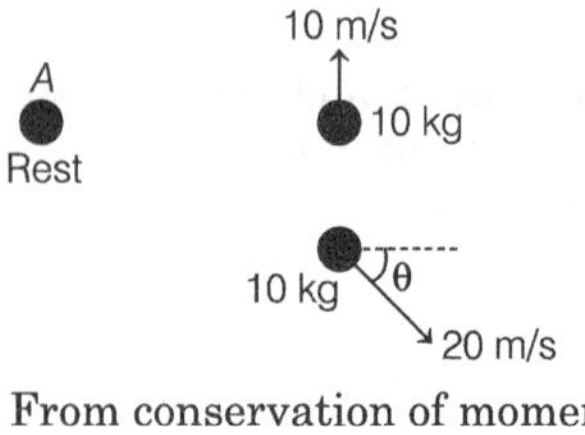

From conservation of momentum along X-axis,

$$\mathbf{p}_i = \mathbf{p}_f$$

$$10 \times 10\sqrt{3} = 200\cos\theta$$

$$\cos\theta = \frac{\sqrt{3}}{2} \Rightarrow \theta = 30°$$

36. Centre of mass of the quarter disc is at $\frac{4a}{3\pi}, \frac{4a}{3\pi}$

According to the centre of mass of the quarter disc (the shaded area) is at $\frac{x}{3}\frac{a}{\pi} \cdot \frac{x}{3}\frac{a}{\pi}$, So, $x = 4$

09

Rotational Motion

If all particles of a rigid body perform circular motion and the centres of these circles are steady on a definite straight line, then the motion of the rigid body is called *rotational motion*.

IN THIS CHAPTER ....

Basic Concepts of Rotational Motion

In rotation of a rigid body about a fixed axis, every particle of the body moves in a circle, which lies in a plane perpendicular to the axis and has its centre on the axis. For *e.g.* A disc (rigid body) of radius R and mass M rotating about a fixed axis passing through its centre as shown in figure.

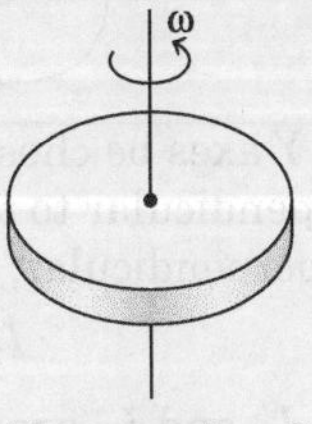

Rotational motion is characterised by angular displacement $d\theta$ and angular velocity $\omega = \dfrac{d\theta}{dt}$.

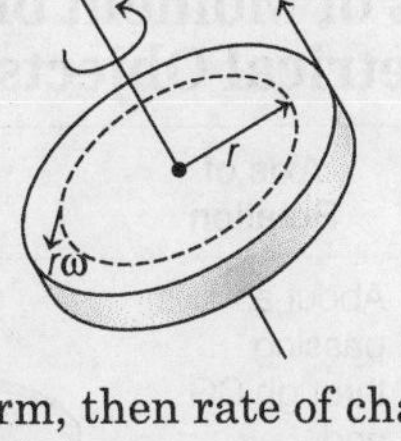

If angular velocity is not uniform, then rate of change of angular velocity is known as *angular acceleration* (α).

Angular acceleration, $\alpha = \dfrac{\Delta\omega}{\Delta t}$ (unit of α = rad s^{-2})

If the angular velocity of rotation varies with the magnitude of tangential velocity as $v = r\omega$. The rate of change of tangential velocity is known as *tangential acceleration* (a_t) given by

$$a_t = \frac{\Delta v}{\Delta t} = r \cdot \frac{\Delta \omega}{\Delta t}$$

$$\Rightarrow \quad a_t = r\alpha$$

Moment of Inertia

Moment of inertia of a rotating body is its property to oppose any change in its state of uniform rotation.

The moment of inertia of a rigid body about any axis of rotation is the sum of the product of masses of the particles and the square of their respective distances from axis of rotation.

$$I = MR^2$$

$$I = m_1 r_1^2 + m_2 r_2^2 + \ldots + m_n r_n^2$$

or

$$I = \sum_{i=1}^{n} m_i r_i^2$$

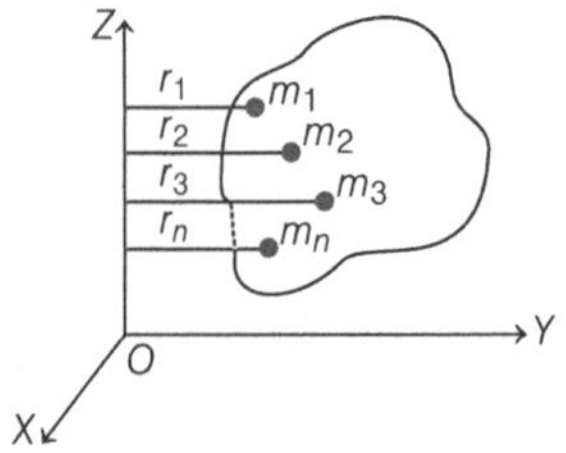

where, $m_1, m_2, \ldots, m_n$ are the masses of n particles and $r_1, r_2, \ldots, r_n$ be their distances from axis of rotation.

The unit of moment of inertia in SI system is kg-m². It is neither a scalar nor a vector, *i.e.* it is a tensor.

Radius of Gyration

The radius of gyration of a body about a given axis is the perpendicular distance of a point from the axis at which the whole mass of the body could be concentrated without any change in the moment of inertia of the body about that axis.

If a body has mass M and radius of gyration is K, then moment of inertia,

$$I = MK^2$$

$$K = \sqrt{\frac{I}{M}}$$

Radius of gyration is also defined as the root mean square distance of all the particles about the axis of rotation.

i.e. $$K = \sqrt{\frac{r_1^2 + r_2^2 + r_3^2 + \ldots + r_n^2}{n}}$$

Note Radius of gyration depends upon shape and size of the body, position and configuration of the axis of rotation.

Theorems on Moment of Inertia

There are two theorems based on moment of inertia are given below

Theorem of Parallel Axes

It states that the moment of inertia of a rigid body about any axis is equal to the sum of the moment of inertia about a parallel axis through its centre of mass (I_{CM}), the product of the mass of the body (M) and the square of the perpendicular distance between the two axes.

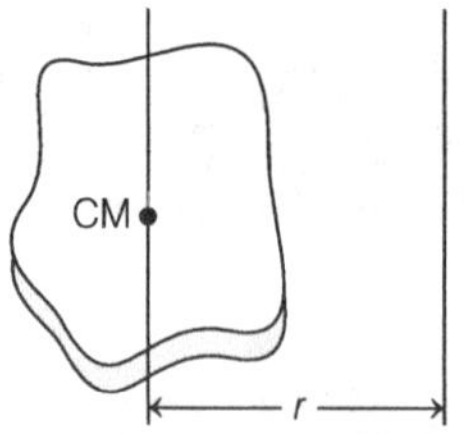

$$I = I_{CM} + Mr^2$$

Theorem of Perpendicular Axes

It states that the moment of inertia of a plane lamina about an axis perpendicular to its plane is equal to the sum of the moments of inertia of the lamina about any two mutually perpendicular axes in its plane and intersecting each other at the point, where the perpendicular axis passes through it.

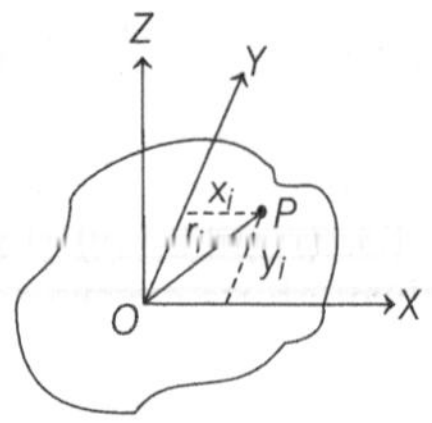

Let X and Y axes be chosen in the plane of the body and Z-axis perpendicular to this plane, three axes being mutually perpendicular, then according to the theorem

$$I_Z = I_X + I_Y$$

where, I_X, I_Y and I_Z are the moments of inertia about the X, Y, Z axes respectively.

Values of Moment of Inertia for Simple Geometrical Objects

Body	Axis of Rotation	Figure	Moment of Inertia	K	K^2/R^2
Thin circular ring, radius R	About an axis passing through CG and perpendicular to its plane	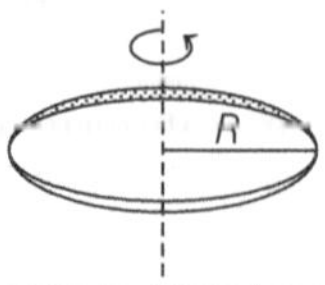	MR^2	R	1

Body	Axis of Rotation	Figure	Moment of Inertia	K	K^2/R^2
Thin circular ring, radius R	About its diameter	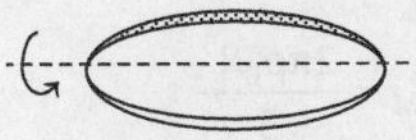	$\frac{1}{2}MR^2$	$\frac{R}{\sqrt{2}}$	$\frac{1}{2}$
Thin rod length, L	Perpendicular to rod at mid-point	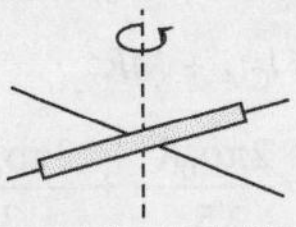	$\frac{1}{12}ML^2$	$\frac{L}{\sqrt{12}}$	
Thin rod length, L	About an axis passing through its edge and perpendicular to the rod	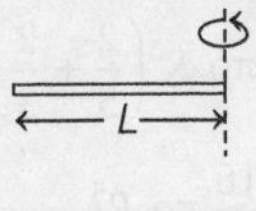	$\frac{ML^2}{3}$	$\frac{L}{\sqrt{3}}$	
Circular disc, radius R	Perpendicular to disc at centre	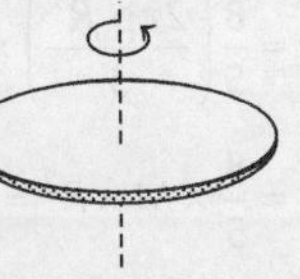	$\frac{1}{2}MR^2$	$\frac{R}{2}$	$\frac{1}{2}$
Circular disc, radius R	Diameter	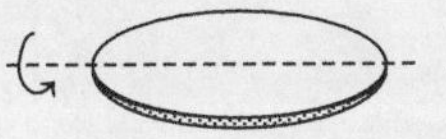	$\frac{1}{4}MR^2$	$\frac{R}{4}$	$\frac{1}{4}$
Hollow cylinder, radius R	About its own axis	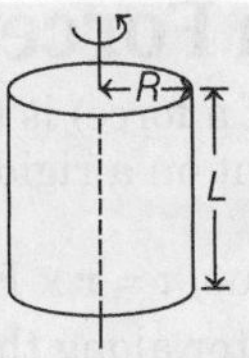	MR^2	R	1
Solid cylinder, radius R	About its own axis		$\frac{MR^2}{2}$	$\frac{R}{2}$	$\frac{1}{2}$
Solid sphere, radius R	About its diametric axis	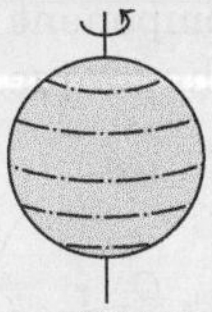	$\frac{2}{5}MR^2$	$\sqrt{\frac{2}{5}}R$	$\frac{2}{5}$

Example 1. *Consider a uniform square plate of side a and mass m. The moment of inertia of this plate about an axis perpendicular to its plane and passing through one of its corners is* **[AIEEE 2008]**

(a) $\frac{5}{6}ma^2$ (b) $\frac{1}{12}ma^2$ (c) $\frac{7}{12}ma^2$ (d) $\frac{2}{3}ma^2$

Sol. (d)

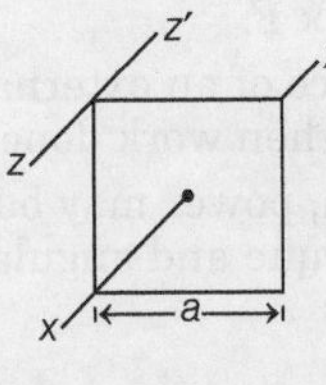

Moment of inertia of square plate about xx' is $\frac{ma^2}{6}$, therefore moment of inertia about $z z'$ can be computed using parallel axes theorem,

$$I_{zz'} = I_{xx'} + m\left(\frac{a}{\sqrt{2}}\right)^2 = \frac{ma^2}{6} + \frac{ma^2}{2} = \frac{2ma^2}{3}$$

Example 2. *From a uniform circular disc of radius R and mass 9 M, a small disc of radius $\frac{R}{3}$ is removed as shown in the figure. The moment of inertia of the remaining disc about an axis perpendicular to the plane of the disc and passing through centre of disc is* **[JEE Main 2018]**

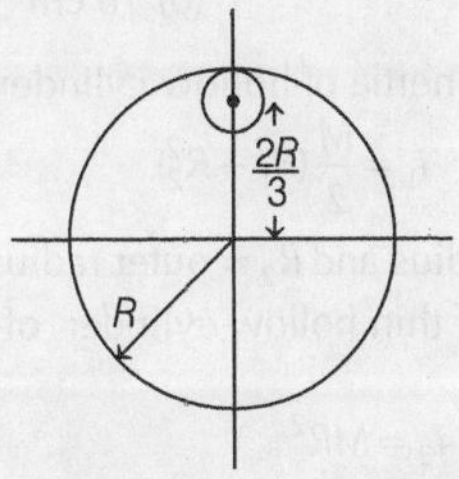

(a) $4MR^2$ (b) $\frac{40}{9}MR^2$ (c) $10MR^2$ (d) $\frac{37}{9}MR^2$

Sol. (a) Moment of inertia of remaining solid
= Moment of inertia of complete solid
− Moment of inertia of removed portion

$$\therefore \quad I = \frac{9MR^2}{2} - \left[\frac{M(R/3)^2}{2} + M\left(\frac{2R}{3}\right)^2\right] \Rightarrow I = 4MR^2$$

Example 3. *Four spheres, each of diameter 2a and mass M are placed with their centres on the four corners of a square of side b. Then, moment of inertia of the system about one side of the square taken as the axis is*

(a) $\frac{2}{5}M(4a^2+5b^2)$ (b) $\frac{2}{3}M(4a^2+5b^2)$

(c) $\frac{1}{3}M(2a^2+5b^2)$ (d) $\frac{1}{4}M(2a^2+5b^2)$

Sol. (a) *ABCD* is a square of side *b*. Four spheres, each of mass *M* and radius a are placed at the four corners of the square.

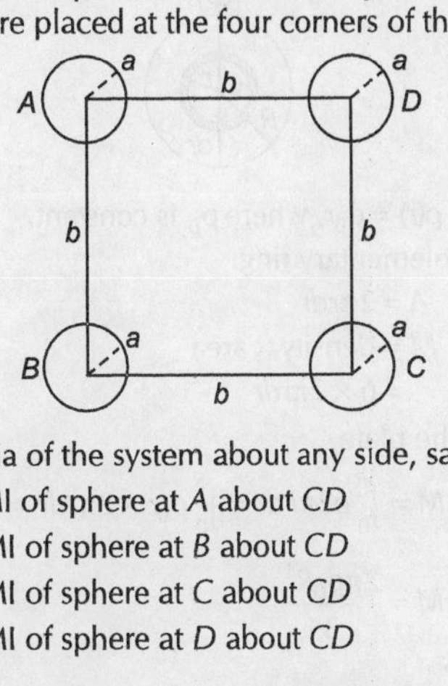

Moment of inertia of the system about any side, say *CD*
= MI of sphere at *A* about *CD*
+ MI of sphere at *B* about *CD*
+ MI of sphere at *C* about *CD*
+ MI of sphere at *D* about *CD*

$$= \left(\frac{2}{5}Ma^2 + Mb^2\right) + \left(\frac{2}{5}Ma^2 + Mb^2\right) + \frac{2}{5}Ma^2 + \frac{2}{5}Ma^2$$
$$= \frac{8}{5}Ma^2 + 2Mb^2$$
$$= \frac{2}{5}M(4a^2 + 5b^2)$$

Example 4. *Let the moment of inertia of a hollow cylinder of length 30 cm (inner radius 10 cm and outer radius 20 cm) about its axis be I. The radius of a thin cylinder of the same mass such that its moment of inertia about its axis is also I, is* **[JEE Main 2019]**

(a) 16 cm (b) 14 cm
(c) 12 cm (d) 18 cm

Sol. (a) Moment of inertia of hollow cylinder about its axis is

$$I_1 = \frac{M}{2}(R_1^2 + R_2^2)$$

where, R_1 = inner radius and R_2 = outer radius.
Moment of inertia of thin hollow cylinder of radius R about its axis is

$$I_2 = MR^2$$

Given, $I_1 = I_2$ and both cylinders have same mass (M). So, we have

$$\frac{M}{2}(R_1^2 + R_2^2) = MR^2$$
$$(10^2 + 20^2)/2 = R^2$$
$$\Rightarrow \quad R^2 = 250$$
$$R \approx 16 \text{ cm}$$

Example 5. *A thin circular plate of mass M and radius R has its density varying as $\rho(r) = \rho_0 r$ with ρ_0 as constant and r is the distance from its centre. The moment of inertia of the circular plate about an axis perpendicular to the plate and passing through its edge is $I = aMR^2$. The value of the coefficient a is* **[JEE Main 2019]**

(a) $\frac{1}{2}$ (b) $\frac{3}{5}$ (c) $\frac{8}{5}$ (d) $\frac{3}{2}$

Sol. (c) Consider an elementary ring of thickness dx and radius r as shown below

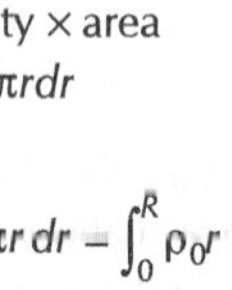

Varying density, $\rho(r) = \rho_0 r$, where ρ_0 is constant.
Area of circular elementary ring.

$$A = 2\pi r dr$$

$\therefore$ Mass of ring, M = density × area

$$= \rho \times 2\pi r dr$$

$\therefore$ Total mass of the plate,

$$M = \int_0^R \rho 2\pi r\, dr = \int_0^R \rho_0 r \times 2\pi r dr = \int_0^R \rho_0 \times 2\pi \times r^2 dr$$
$$\Rightarrow \quad M = \frac{2\pi\rho_0 R^3}{3}$$

$\therefore$ Moment of inertia about centre of mass,

$$I_{CM} = \int_0^R Mr^2 = \int_0^R \rho_0 r \times 2\pi r\, dr \times r^2$$
$$\Rightarrow \quad I_{CM} = \frac{2\pi\rho_0 R^5}{5}$$

Now, using parallel axes theorem, we know that

$$\Rightarrow \quad I = I_{CM} + MR^2$$
$$= \frac{2\pi\rho_0 R^5}{5} + \frac{2\pi\rho_0 R^3}{3} \times R^2$$
$$= \pi\rho_0 R^5\left(\frac{2}{5} + \frac{2}{3}\right)$$
$$\Rightarrow \quad I = \frac{16}{15}\pi\rho_0 R^5$$
$$= \frac{8}{5}\left(\frac{2\pi\rho_0 R^3}{3}\right) \times R^2$$
$$= \frac{8}{5} \times M \times R^2 = a(MR)^2 \quad \text{(given)}$$

Hence, $a = \frac{8}{5}$

Moment of a Force or Torque

Torque (or moment of a force) is the turning effect of a force applied at a point on a rigid body about the axis of rotation.
Mathematically, torque, $\tau = \mathbf{r} \times \mathbf{F} = |\mathbf{r} \times \mathbf{F}|\, \hat{\mathbf{n}} = rF\sin\theta\hat{\mathbf{n}}$
where, $\hat{\mathbf{n}}$ is a unit vector along the axis of rotation.
Torque is an axial vector and its SI unit is newton-metre (N-m).

- The torque about axis of rotation is independent of choice of origin O, so long as it is chosen on the axis of rotation AB.
- Only normal component of force contributes towards the torque. Radial component of force does not contribute towards the torque.

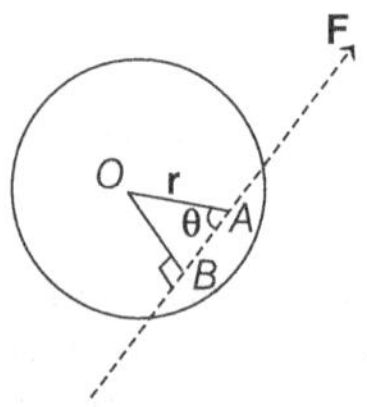

- A torque produces angular acceleration in a rotating body. Thus, torque, $\tau = I\alpha$.
- Moment of a couple (or torque) is given by product of position vector $\mathbf{r}$ between the two forces and either force $\mathbf{F}$. Thus, $\tau = \mathbf{r} \times \mathbf{F}$.
- If under the influence of an external torque, τ the given body rotates by $d\theta$, then work done, $dW = \tau \cdot d\theta$.
- In rotational motion, power may be defined as the scalar product of torque and angular velocity, *i.e.* Power $P = \tau \cdot \omega$.

Example 6. *Let **F** be the force acting on a particle having position vector **r** and τ be the torque of this force about the origin. Then,* **[AIEEE 2003]**

(a) $\mathbf{r}\cdot\tau=0$ and $\mathbf{F}\cdot\tau\neq0$ (b) $\mathbf{r}\cdot\tau\neq0$ and $\mathbf{F}\cdot\tau=0$
(c) $\mathbf{r}\cdot\tau\neq0$ and $\mathbf{F}\cdot\tau\neq0$ (d) $\mathbf{r}\cdot\tau=0$ and $\mathbf{F}\cdot\tau=0$

Sol. (d) As $\tau=\mathbf{r}\times\mathbf{F}$

τ is perpendicular to both **r** and **F**, so $\mathbf{r}\cdot\tau$ as well as $\mathbf{F}\cdot\tau$ has to be zero.

Example 7. *A force of $-F\hat{\mathbf{k}}$ acts on O, the origin of the coordinate system. The torque about the point (1, – 1) is* **[AIEEE 2006]**

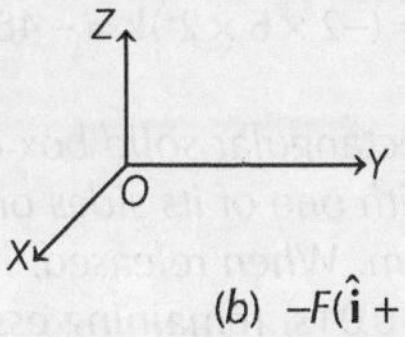

(a) $F(\hat{\mathbf{i}}-\hat{\mathbf{j}})$ (b) $-F(\hat{\mathbf{i}}+\hat{\mathbf{j}})$
(c) $F(\hat{\mathbf{i}}+\hat{\mathbf{j}})$ (d) $-F(\hat{\mathbf{i}}-\hat{\mathbf{j}})$

Sol. (c) The torque about the given position, $\tau=\mathbf{r}\times\mathbf{F}$

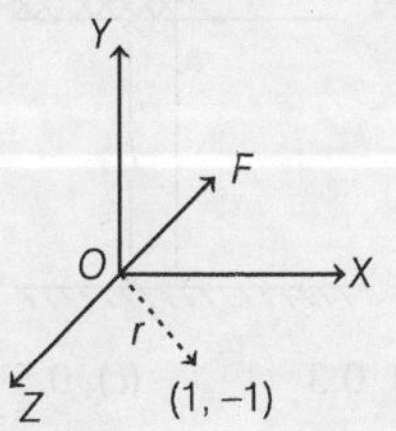

Here, $\mathbf{r}=\hat{\mathbf{i}}-\hat{\mathbf{j}}$ and $\mathbf{F}=-F\hat{\mathbf{k}}$

$$\therefore\quad \tau=(\hat{\mathbf{i}}-\hat{\mathbf{j}})\times(-F\hat{\mathbf{k}})$$
$$=F[(-\hat{\mathbf{i}}\times\hat{\mathbf{k}})+(\hat{\mathbf{j}}\times\hat{\mathbf{k}})]$$
$$=F(\hat{\mathbf{j}}+\hat{\mathbf{i}})=F(\hat{\mathbf{i}}+\hat{\mathbf{j}})$$

Example 8. *The magnitude of torque on a particle of mass 1 kg is 2.5 N-m about the origin. If the force acting on it is 1 N and the distance of the particle from the origin is 5 m, then the angle between the force and the position vector is (in rad)* **[JEE Main 2019]**

(a) $\frac{\pi}{8}$ (b) $\frac{\pi}{4}$
(c) $\frac{\pi}{3}$ (d) $\frac{\pi}{6}$

Sol. (d) Given, $m=1\,\text{kg}$

$|\tau|=2.5$ N-m, $F=1$ N and $r=5$ m

We know that,

torque, $|\tau|=rF\sin\theta$

$$\Rightarrow\quad 2.5=5\times1\times\sin\theta$$
$$\Rightarrow\quad \sin\theta=\frac{1}{2}$$
$$\text{or}\quad \theta=\frac{\pi}{6}\text{ rad}$$

Example 9. *A slab is subjected to two forces $\mathbf{F}_1$ and $\mathbf{F}_2$ of same magnitude F as shown in the figure. Force $\mathbf{F}_2$ is in XY-plane while force $\mathbf{F}_1$ acts along Z-axis at the point $(2\hat{\mathbf{i}}+3\hat{\mathbf{j}})$. The moment of these forces about point O will be* **[JEE Main 2019]**

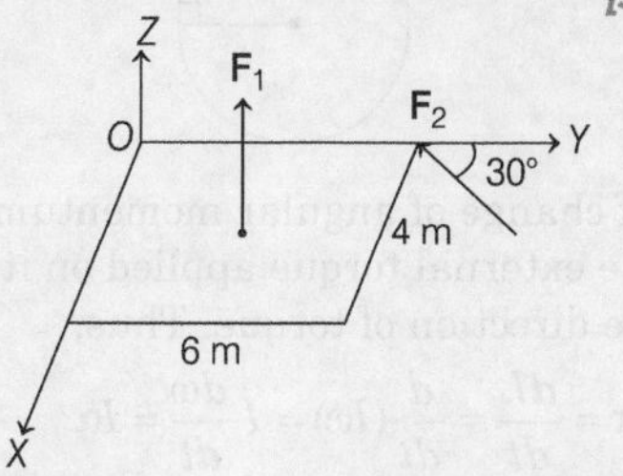

(a) $(3\hat{\mathbf{i}}+2\hat{\mathbf{j}}-3\hat{\mathbf{k}})F$ (b) $(3\hat{\mathbf{i}}-2\hat{\mathbf{j}}+3\hat{\mathbf{k}})F$
(c) $(3\hat{\mathbf{i}}-2\hat{\mathbf{j}}-3\hat{\mathbf{k}})F$ (d) $(3\hat{\mathbf{i}}+2\hat{\mathbf{j}}+3\hat{\mathbf{k}})F$

Sol. (b) According to the question, as shown in the figure below

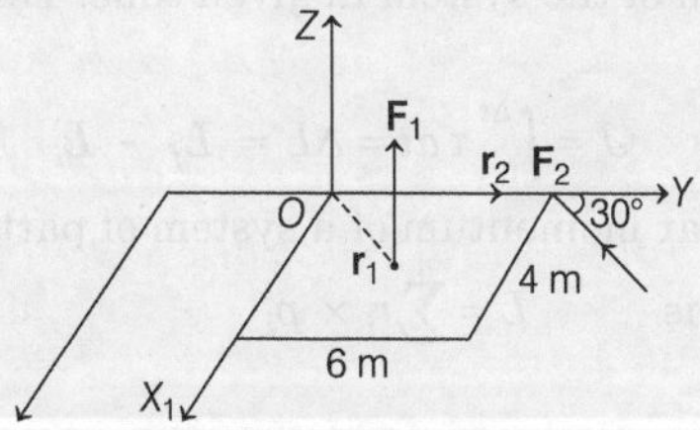

$$\mathbf{r}_1=2\hat{\mathbf{i}}+3\hat{\mathbf{j}}\text{ and }\mathbf{r}_2=6\hat{\mathbf{j}}$$
$$\mathbf{F}_1=F\hat{\mathbf{k}}$$
and
$$\mathbf{F}_2=(-\sin30^\circ\,\hat{\mathbf{i}}-\cos30^\circ\,\hat{\mathbf{j}})F$$

Moment of force is given as $\tau=\mathbf{r}\times\mathbf{F}$

where, **r** is the perpendicular distance and **F** is the force.

∴ Moment due to $\mathbf{F}_1$,

$$\tau_1=(2\hat{\mathbf{i}}+3\hat{\mathbf{j}})\times(F\hat{\mathbf{k}})$$
$$=-2F\hat{\mathbf{j}}+3F\hat{\mathbf{i}}\qquad\text{...(i)}$$

Moment due to $\mathbf{F}_2$,

$$\tau_2=(6\hat{\mathbf{j}})\times(-\sin30^\circ\,\hat{\mathbf{i}}-\cos30\,\hat{\mathbf{j}})F$$
$$=6\sin30^\circ F\,\hat{\mathbf{k}}=3F\,\hat{\mathbf{k}}\qquad\text{...(ii)}$$

∴ Resultant torque,

$$\tau=\tau_1+\tau_2=3F\,\hat{\mathbf{i}}-2F\,\hat{\mathbf{j}}+3F\,\hat{\mathbf{k}}$$
$$=(3\hat{\mathbf{i}}-2\hat{\mathbf{j}}+3\hat{\mathbf{k}})F$$

Angular Momentum

The moment of linear momentum of a given body about an axis of rotation is called its angular momentum.

If $\mathbf{p}=m\mathbf{v}$ be the linear momentum of a particle and **r** is its position vector from the point of rotation, then

Angular momentum, $\mathbf{L}=\mathbf{r}\times\mathbf{p}=r\,p\sin\theta\,\hat{\mathbf{n}}=mvr\sin\theta\,\hat{\mathbf{n}}$,

where $\hat{\mathbf{n}}$ is a unit vector in the direction of rotation. Angular momentum is an axial vector and its SI unit is kg-m^2s^{-1} or J-s.

- For rotational motion of a rigid body, angular momentum is equal to the product of angular velocity and moment of inertia of the body about the axis of rotation.

Mathematically, $L = I\omega$

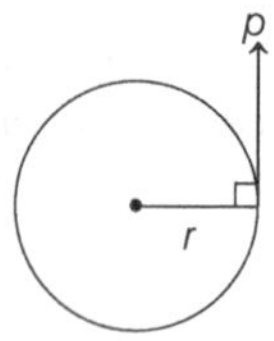

- The rate of change of angular momentum of a body is equal to the external torque applied on it and takes place in the direction of torque. Thus,

$$\tau = \frac{dL}{dt} = \frac{d}{dt}(I\omega) = I\frac{d\omega}{dt} = I\alpha \qquad \left[\because \alpha = \frac{d\omega}{dt}\right]$$

- Total effect of a torque applied on a rotating body in a given time is called angular impulse. **Angular impulse** is equal to total change in angular momentum of the system in given time. Thus, angular impulse,

$$J = \int_0^{\Delta t} \tau\, dt = \Delta L = L_f - L_i$$

- The angular momentum of a system of particles about the origin is $\quad L = \sum^{n} r_i \times p_i$

Law of Conservation of Angular Momentum

According to the law of conservation of angular momentum, if no external torque is acting on a system, then total vector sum of angular momentum of different particles of the system remains constant.

We know that, $\tau_{ext} = \frac{d\mathbf{L}}{dt}$

If $\tau_{ext} = 0$, then $\frac{d\mathbf{L}}{dt} = 0 \Rightarrow L = \text{constant}$.

Therefore, in the absence of any external torque, the total angular momentum of a system must remain conserved.

As $\mathbf{L} = I\omega$, the law of conservation of momentum leads us to the conclusion.

For an isolated system, $I\omega = \text{constant}$ or $I_1\omega_1 = I_2\omega_2$

This principle is often made use by gymnast, swimmers, circus acrobats and ballet dancers etc.

Example 10. *The time dependence of the position of a particle of mass $m = 2$ kg is given by $\mathbf{r}(t) = 2t\hat{\mathbf{i}} - 3t^2\hat{\mathbf{j}}$. Its angular momentum, with respect to the origin, at time $t = 2$ s is* [JEE Main 2019]

(a) $36\,\hat{\mathbf{k}}$ (b) $-34\,(\hat{\mathbf{k}} - \hat{\mathbf{i}})$

(c) $-48\,\hat{\mathbf{k}}$ (d) $48\,(\hat{\mathbf{i}} + \hat{\mathbf{j}})$

Sol. (c) Position of particle, $\mathbf{r} = 2t\hat{\mathbf{i}} - 3t^2\hat{\mathbf{j}}$

where, t is instantaneous time.

Velocity of particle, $\mathbf{v} = \frac{d\mathbf{r}}{dt} = 2\hat{\mathbf{i}} - 6t\hat{\mathbf{j}}$

Now, angular momentum of particle with respect to origin is given by

$$\mathbf{L} = m(\mathbf{r} \times \mathbf{v})$$
$$= m\{(2t\hat{\mathbf{i}} - 3t^2\hat{\mathbf{j}}) \times (2\hat{\mathbf{i}} - 6t\hat{\mathbf{j}})\}$$
$$= m(-12t^2(\hat{\mathbf{i}} \times \hat{\mathbf{j}}) - 6t^2(\hat{\mathbf{j}} \times \hat{\mathbf{i}}))$$

As, $\hat{\mathbf{i}} \times \hat{\mathbf{i}} = \hat{\mathbf{j}} \times \hat{\mathbf{j}} = 0$

$\Rightarrow \mathbf{L} = m(-12t^2\hat{\mathbf{k}} + 6t^2\hat{\mathbf{k}})$

As, $\hat{\mathbf{i}} \times \hat{\mathbf{j}} = \hat{\mathbf{k}}$ and $\hat{\mathbf{j}} \times \hat{\mathbf{i}} = -\hat{\mathbf{k}}$

$\Rightarrow \mathbf{L} = -m(6t^2)\hat{\mathbf{k}}$

So, angular momentum of particle of mass 2 kg at time $t = 2$ s is

$$\mathbf{L} = (-2 \times 6 \times 2^2)\hat{\mathbf{k}} = -48\,\hat{\mathbf{k}}$$

Example 11. *A rectangular solid box of length 0.3 m is held horizontally, with one of its sides on the edge of a platform of height 5 m. When released, it slips off the table in a very short time $\tau = 0.01$ s, remaining essentially horizontal. The angle by which it would rotate when it hits the ground will be (in rad) close to* [JEE Main 2019]

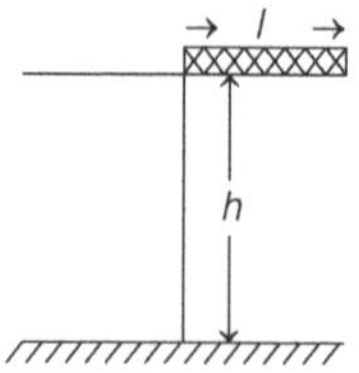

(a) 0.02 (b) 0.3 (c) 0.5 (d) 0.28

Sol. (c) Now, angular impulse of weight = change in angular momentum

$$\therefore \quad mg\frac{l}{2} \times \tau = \frac{ml^2}{3}\omega \Rightarrow \omega = \frac{3g \times \tau}{2 \times l}$$

Substituting the given values, we get

$$\omega = \frac{3 \times 10 \times 0.01}{2 \times 0.3} = 0.5 \text{ rad s}^{-1}$$

Time of fall of box, $t = \sqrt{\frac{2h}{g}} = \sqrt{\frac{2 \times 5}{10}} \approx 1$ s

So, angle turned by box in reaching ground,

$$\theta = \omega t = 0.5 \times 1 = 0.5 \text{ rad}$$

Example 12. *Four point masses, each of mass m are fixed at the corners of a square of side l. The square is rotating with angular frequency ω, about an axis passing through one of the corners of the square and parallel to its diagonal, as shown in the figure. The angular momentum of the square about this axis is* [JEE Main 2020]

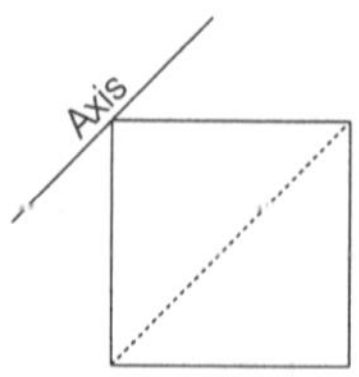

(a) $ml^2\omega$ (b) $4\,ml^2\omega$ (c) $3\,ml^2\omega$ (d) $2\,ml^2\omega$

Sol. (c) The angular momentum of the square about the given axis will be the sum of angular momentum due to each point masses.

i.e. $$L = L_1 + L_2 + L_3 + L_4$$
$$= I_1\omega_1 + I_2\omega_2 + I_3\omega_3 + I_4\omega_4$$

Here, angular frequency will be same.

i.e., $$\omega_1 = \omega_2 = \omega_3 = \omega_4 = \omega \text{ (say)}$$

$\therefore$ $$L = (I_1 + I_2 + I_3 + I_4)\omega$$

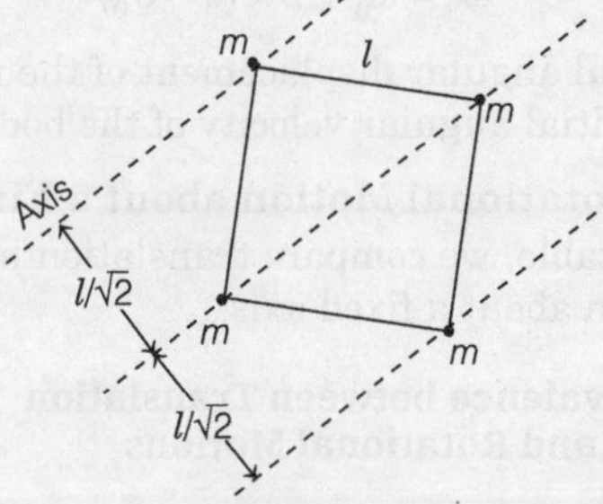

$$L = \left\{ m(0)^2 + 2 \times m\left(\frac{l}{\sqrt{2}}\right)^2 + m(\sqrt{2}l)^2 \right\}\omega = 3ml^2\omega$$

Example 13. *A particle of mass m is moving along the side of a square of side a, with a uniform speed v in the XY-plane as shown in the figure.*

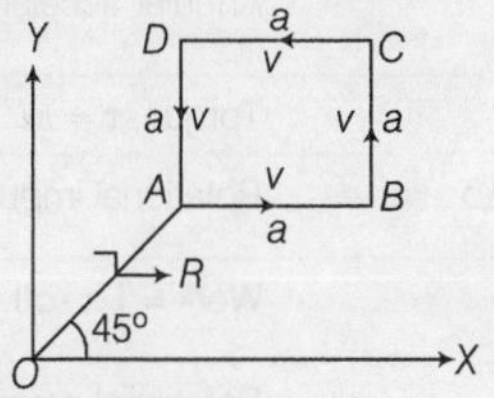

Which of the following statements is false, for the angular momentum **L** *about the origin?* **[JEE Main 2016]**

(a) $\mathbf{L} = -\frac{mv}{\sqrt{2}}R\,\hat{\mathbf{k}}$, when the particle is moving from A to B.

(b) $\mathbf{L} = mv\left(\frac{R}{\sqrt{2}} + a\right)\hat{\mathbf{k}}$, when the particle is moving from B to C.

(c) $\mathbf{L} = mv\left(\frac{R}{\sqrt{2}} - a\right)\hat{\mathbf{k}}$, when the particle is moving from C to D.

(d) $\mathbf{L} = \frac{mv}{\sqrt{2}}R\,\hat{\mathbf{k}}$, when the particle is moving from D to A.

Sol. (c, d) For a particle of mass m is moving along the side of a square of side a such that

Angular momentum **L** about the origin, $\mathbf{L} = \mathbf{r} \times \mathbf{p} = rp \sin\theta\,\hat{\mathbf{n}}$

or $$\mathbf{L} = r(p)\,\hat{\mathbf{n}}$$

When a particle is moving from D to A, $\mathbf{L} = \frac{R}{\sqrt{2}}mv(-\hat{\mathbf{k}})$

A particle is moving from A to B, $\mathbf{L} = \frac{R}{\sqrt{2}}mv(-\hat{\mathbf{k}})$

and it moves from C to D, $\mathbf{L} = \left(\frac{R}{\sqrt{2}} + a\right)mv\,(\hat{\mathbf{k}})$

For B to C, we have

$$\mathbf{L} = \left(\frac{R}{\sqrt{2}} + a\right)mv\,(\hat{\mathbf{k}})$$

Example 14. *A thin smooth rod of length L and mass M is rotating freely with angular speed ω_0 about an axis perpendicular to the rod and passing through its centre. Two beads of mass m and negligible size are at the centre of the rod initially. The beads are free to slide along the rod. The angular speed of the system, when the beads reach the opposite ends of the rod, will be* **[JEE Main 2019]**

(a) $\frac{M\omega_0}{M+3m}$ (b) $\frac{M\omega_0}{M+m}$

(c) $\frac{M\omega_0}{M+2m}$ (d) $\frac{M\omega_0}{M+6m}$

Sol. (d) As there is no external torque on system.

$\therefore$ Angular momentum of system is conserved.

$\Rightarrow$ $$I_i\omega_i = I_f\omega_f$$

Initially,

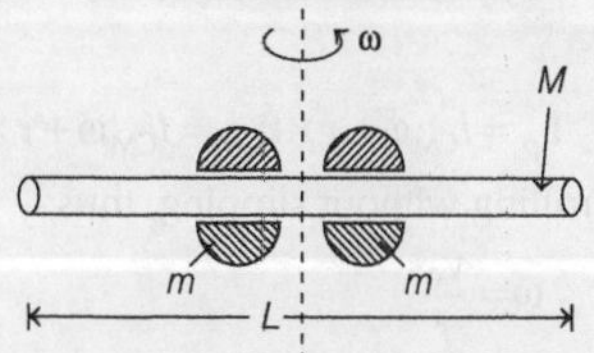

Finally,

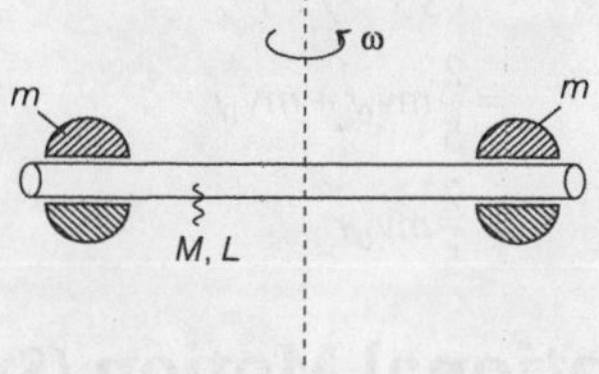

$\Rightarrow$ $$\frac{ML^2}{12}\cdot\omega_0 + 0 = \left(\frac{ML^2}{12} + 2(m)\left(\frac{L}{2}\right)^2\right)\omega$$

So, final angular speed of system is

$\Rightarrow$ $$\omega = \frac{\frac{ML^2}{12}\cdot\omega_0}{\left(\frac{ML^2 + 6mL^2}{12}\right)} = \frac{M\omega_0}{M+6m}$$

Angular Momentum of a Rigid Body

When a rigid body describes pure rotational motion, all its constituent particles describe circular motion about the axis of rotation. In such case, angular momentum of the rigid body,

$$L = \Sigma mvr = \Sigma mr^2\omega = (\Sigma mr^2)\cdot\omega = I\omega$$

Direction of **L** is same as ω. Hence, ω angular momentum is an axial vector.

When a rigid body is rotating about its centre of mass axis with angular velocity ω_{CM}, simultaneously moving translationally with a linear velocity v, the angular momentum of that body about a point P in the laboratory frame is given by

$$\mathbf{L}_p = I_{CM}\,\omega + r_X m\,\mathbf{v}_{CM}$$

Example 15. *A solid sphere rolls without slipping on a rough surface with centre of mass having a constant speed v_0. If mass of the sphere be m and r be its radius, then what is the value of angular momentum of the sphere about the point of contact with rough surface?*

(a) $\frac{7}{5}mv_0 r$ (b) $\frac{5}{7}mv_0 r$

(c) $\frac{3}{5}mv_0 r$ (d) $\frac{5}{3}mv_0 r$

Sol. (a) The angular momentum of the sphere about the point of contact P will be

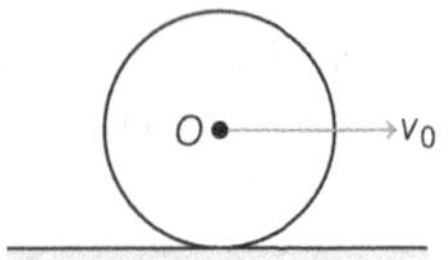

P

$$\mathbf{L}_p = I_{CM}\omega + \mathbf{r}\times\mathbf{P}_{CM} = I_{CM}\omega + \mathbf{r}\times m\mathbf{v}_{CM}$$

As sphere is rolling without slipping, thus

$$\omega = \frac{v_0}{r}$$

$$\mathbf{L}_p = \left(\frac{2}{3}mr^2\right)\left(\frac{v_0}{r}\right) + r\times m\times v_0$$

$$= \frac{2}{5}mv_0 r + mv_0 r$$

$$= \frac{7}{5}mv_0 r$$

Pure Rotational Motion (Spinning)

When the body rotates in such a manner that its axis of rotation does not move, then its motion is called spinning motion.

In spinning, rotational kinetic energy is given by

$$K_R = \frac{1}{2}I\omega^2$$

$\because$ $$I = mK^2$$

and $$v = \omega R$$

$\therefore$ $$K_R = \frac{1}{2}mK^2\frac{v^2}{R^2}$$

$$= \frac{1}{2}mv^2\left(\frac{K^2}{R^2}\right)$$

Here, $\frac{K^2}{R^2}$ is a constant for different bodies. Value of $\frac{K^2}{R^2} = 1$ for ring and cylindrical shell and $\frac{K^2}{R^2} = \frac{1}{2}$ for disc and solid cylinder and $\frac{K^2}{R^2} = \frac{2}{5}$ for a solid sphere.

Equations of Rotational Motion

The kinematical quantities in rotational motion, angular displacement (θ), angular velocity (ω) and angular acceleration (α) respectively, then the kinematic equations for rotational motion with uniform angular acceleration are

$$\omega = \omega_0 t + \alpha t,$$

$$\theta = \theta_0 + \omega_0 t + \frac{1}{2}\alpha t^2$$

and $$\omega^2 = \omega_0^2 + 2\times(\theta - \theta_0)$$

where, θ_0 = initial angular displacement of the rotating body and ω_0 = initial angular velocity of the body.

Dynamics of Rotational Motion about a Fixed Axis

From the given table, we compare translation motion and rotational motion about a fixed axis.

Equivalence between Translation and Rotational Motions

Translation Motion	Rotational Motion
Displacement, x	Angular displacement, θ
Linear velocity, $v = \frac{dx}{dt}$	Angular velocity, $\omega = \frac{d\theta}{dt}$
Linear momentum, $p = mv$	Angular momentum, $L = I\omega$
Acceleration, $a = \frac{dv}{dt}$	Angular acceleration, $\alpha = \frac{d\omega}{dt}$
Force, $F = ma$	Torque, $\tau = I\alpha$
Impulse, $I = F\Delta t = \Delta p$	Rotational impulse, $J = \int \tau dt = \Delta L$
Work, $W = \int F\cdot dS$	Work $= \int \tau\cdot d\theta$
Power, $P = F\cdot v$	Rotational power, $P = \tau\cdot\omega$

Example 16. *A disc of mass 5 kg and radius 50 cm rolls on the ground at the rate of 10 ms^{-1}. The kinetic energy of the disc is (Given, $I = \frac{1}{2}MR^2$)*

(a) 300 J (b) 325 J (c) 350 J (d) 375 J

Sol. (d) Here, mass of the disc, $M = 5$ kg

Radius of the disc, $R = 50\text{ cm} = \frac{1}{2}\text{ m}$

Linear velocity of the disc, $v = 10\text{ ms}^{-1}$

As, $$v = R\omega$$

$\therefore$ $$10 = \frac{1}{2}\omega$$

or $$\omega = 10\times 2 = 20\text{ rad s}^{-1}$$

Also, moment of inertia of disc about an axis through its centre,

$$I = \frac{1}{2}MR^2 = \frac{1}{2}\times 5\times\left(\frac{1}{2}\right)^2 = \frac{5}{8}\text{kgm}^2$$

$\therefore$ KE of translation $= \frac{1}{2}Mv^2 = \frac{1}{2}\times 5\times(10)^2 = 250$ J and KE of rotation $= \frac{1}{2} I\omega^2 = \frac{1}{2} \times \frac{5}{8} \times (20)^2 = 125$ J

Total KE $= 250 + 125 = 375$ J

Example 17. *A rod of length 50 cm is pivoted at one end. It is raised such that, if makes an angle of 30° from the horizontal as shown and released from rest. Its angular speed when it passes through the horizontal (in rad s^{-1}) will be (Take, $g = 10\ ms^{-2}$)* ***[JEE Main 2019]***

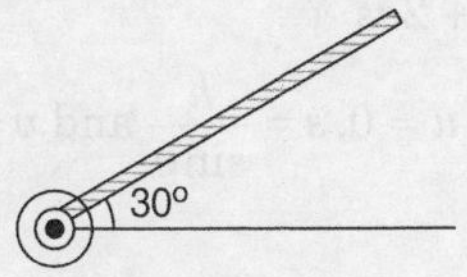

(a) $\frac{\sqrt{30}}{2}$ (b) $\sqrt{30}$

(c) $\frac{\sqrt{20}}{3}$ (d) $\sqrt{\frac{30}{2}}$

Sol. (b) Loss of potential energy of rod = Gain of kinetic energy

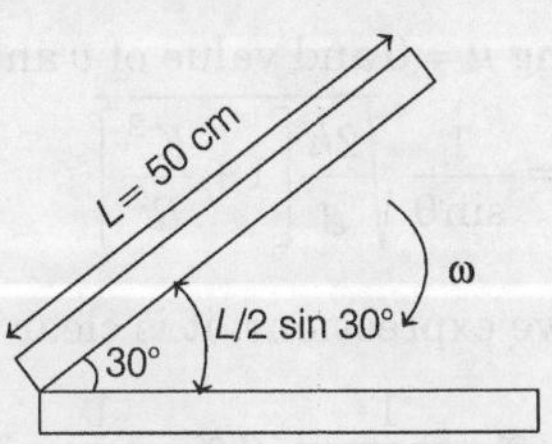

$\therefore \quad \Delta PE = \frac{1}{2} I\omega^2$

(where, I = moment of inertia of rod and ω = angular frequency of rod)

$\Rightarrow \quad Mg \times \frac{L}{2} \sin 30° = \frac{1}{2} \times I \times \omega^2$

$\Rightarrow \quad Mg \frac{L}{2} \times \frac{1}{2} = \frac{1}{2} \times I \times \omega^2$

$\Rightarrow \quad \frac{Mg\ L \times 2}{4 \times I} = \omega^2$

$\Rightarrow \quad \sqrt{\frac{MgL}{4 \times \frac{ML^2}{3}} \times \frac{2}{1}} = \omega \quad \left[\because I = \frac{ML^2}{3} \text{ for rod}\right]$

$\omega = \sqrt{30}$ rad/s

Example 18. *A stationary horizontal disc is free to rotate about its axis. When a torque is applied on it, its kinetic energy as a function of θ, where θ is the angle by which it has rotated, is given as $k\theta^2$. If its moment of inertia is I, then the angular acceleration of the disc is* ***[JEE Main 2019]***

(a) $\frac{k}{2I}\theta$ (b) $\frac{k}{I}\theta$ (c) $\frac{k}{4I}\theta$ (d) $\frac{2k}{I}\theta$

Sol. (d) Given, kinetic energy $= k\theta^2$

We know that, kinetic energy of a rotating body about its axis $= \frac{1}{2} I\omega^2$

where, I is moment of inertia and ω is angular velocity.

So, $\frac{1}{2} I\omega^2 = k\theta^2$

or $\omega^2 = \frac{2k\theta^2}{I}$

$\Rightarrow \quad \omega = \sqrt{\frac{2k}{I}}\,\theta \quad \ldots \text{(i)}$

Differentiating the above equation w.r.t. time on both sides, we get

$\frac{d\omega}{dt} = \sqrt{\frac{2k}{I}} \cdot \frac{d\theta}{dt} = \sqrt{\frac{2k}{I}} \cdot \omega \quad \left[\because \omega = \frac{d\theta}{dt}\right]$

$\therefore$ Angular acceleration,

$\alpha = \frac{d\omega}{dt} = \sqrt{\frac{2k}{I}} \cdot \omega = \sqrt{\frac{2k}{I}} \cdot \sqrt{\frac{2k}{I}}\,\theta \quad$ [using Eq. (i)]

or $\alpha = \frac{2k}{I}\theta$

Rolling Motion

When a body performs translatory motion as well as rotational motion, then this type of motion is known as rolling motion or combined rotational and translational motion. *e.g.* Motion of football rolling on a surface.

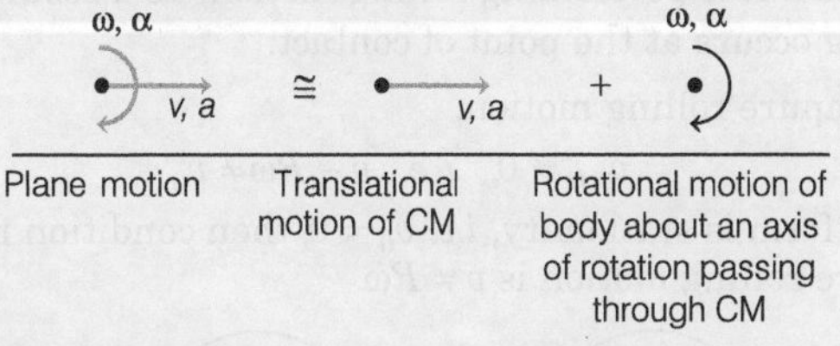

The kinetic energy of rolling body is the sum of kinetic energy of a translation motion and kinetic energy of rotation motion.

Net kinetic energy,

$$K_N = K_T + K_R = \frac{1}{2} mv^2 + \frac{1}{2} I\omega^2$$

$$= \frac{1}{2} mv^2 + \frac{1}{2} mv^2 \frac{K^2}{R^2}$$

$$\therefore \quad K_N = \frac{1}{2} mv^2 \left(1 + \frac{K^2}{R^2}\right)$$

Classification of Rolling Motion

Depending on the fact that relative velocity of point of contact of the body undergoing rolling motion, with the platform (on which the body is performing plane motion) is zero or non-zero, rolling motion is classified into two categories, which are as follows

(i) Pure rolling or rolling without slipping/sliding or perfect rolling motion.

(ii) Impure rolling or rolling with slipping/sliding or imperfect rolling motion.

Pure Rolling Motion

If the relative velocity of the point of contact (between body and platform) is zero, then the rolling motion is said to be pure rolling motion.

For pure rolling motion, $v_{AB} = 0$

i.e. $(v - R\omega) - v_0 = 0 \Rightarrow v - R\omega = v_0$

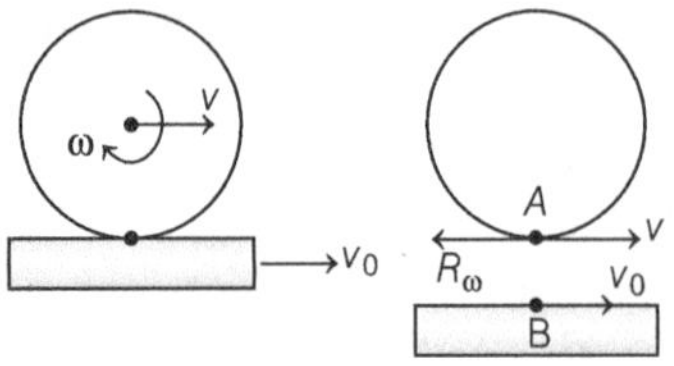

If the platform is stationary, *i.e.* $v_0 = 0$, then for pure rolling motion $v = R\omega$.

Note Friction is responsible for pure rolling motion. In rolling motion, friction is non-dissipative in nature, *i.e.* work done by friction force is zero because point of contact is relatively at rest.

Impure Rolling Motion

In impure rolling motion, the point of contact of the body with the platform is not relatively at rest w.r.t. platform on which it is performing rolling motion, as a result sliding occurs at the point of contact.

For impure rolling motion,

$$v_{AB} \neq 0, \quad i.e. \quad v - R\omega \neq v_0$$

If platform is stationary, *i.e.* $v_0 = 0$, then condition for impure rolling motion is $v \neq R\omega$.

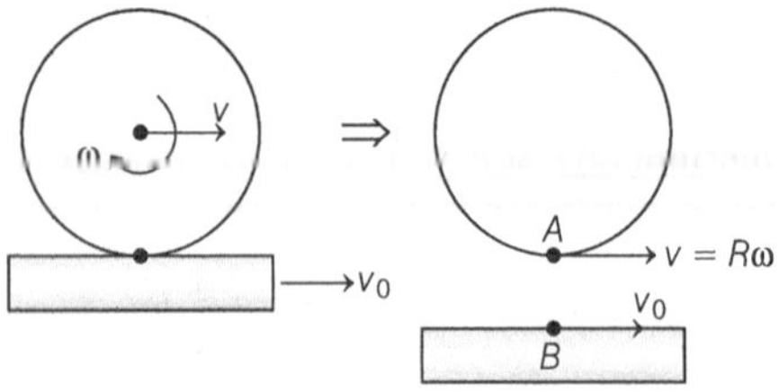

Here as, $v \neq R\omega$,

So, $a \neq R\alpha$

Note In this case, friction is opposing the relative motion, *i.e.* rolling motion and is dissipative in nature, *i.e.* work done by friction force is non-zero.

Rolling Motion on an Inclined Plane

When a body of mass m and radius R rolls down on inclined plane of height h and angle of inclination θ, it loses potential energy. However, it acquires both linear and angular speeds and hence gain kinetic energy of translation and that of rotation.

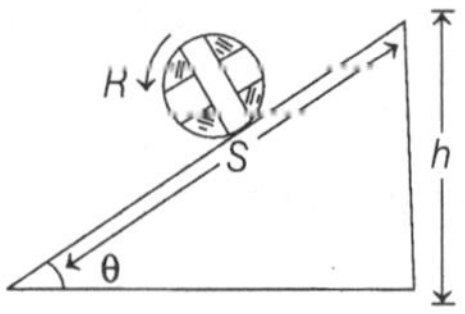

By conservation of mechanical energy,

$$mgh = \frac{1}{2}mv^2\left(1 + \frac{K^2}{R^2}\right)$$

- **Velocity at the lowest point,** $v = \sqrt{\dfrac{2gh}{1 + \dfrac{K^2}{R^2}}}$
- **Acceleration in motion** From second equation of motion, $v^2 = u^2 + 2as$

  By substituting $u = 0, s = \dfrac{h}{\sin\theta}$ and $v = \sqrt{\dfrac{2gh}{1 + \dfrac{K^2}{R^2}}}$, we get

$$a = \frac{g\sin\theta}{1 + \dfrac{K^2}{R^2}}$$

- **Time of descent** From first equation of motion,

$$v = u + at$$

  By substituting $u = 0$ and value of v and a from above expressions $t = \dfrac{1}{\sin\theta}\sqrt{\dfrac{2h}{g}\left[1 + \dfrac{K^2}{R^2}\right]}$

  From the above expressions, it is clear that,

$$v \propto \frac{1}{\sqrt{1 + \dfrac{K^2}{R^2}}}; a \propto \frac{1}{1 + \dfrac{K^2}{R^2}}; t \propto \sqrt{1 + \frac{K^2}{R^2}}$$

Note (i) Factor $\left(\dfrac{K^2}{R^2}\right)$ is a measure of moment of inertia of a body and its value is constant for given shape of the body and it does not depend on the mass and radius of a body.

(ii) Velocity, acceleration and time of descent (for a given inclined plane) all depends on $\dfrac{K^2}{R^2}$. Lesser the moment of inertia of the rolling body, lesser will be the value of $\dfrac{K^2}{R^2}$. So, greater will be its velocity and acceleration and lesser will be the time of descent.

Example 19. *A uniform sphere of mass 500 g rolls without slipping on a plane horizontal surface with its centre moving at a speed of 5.00 cm/s. Its kinetic energy is* **[JEE Main 2020]**

(a) 6.25×10^{-4} J (b) 1.13×10^{-3} J

(c) 8.75×10^{-4} J (d) 8.75×10^{-3} J

Sol. (c) The given situation is as shown in the figure below.

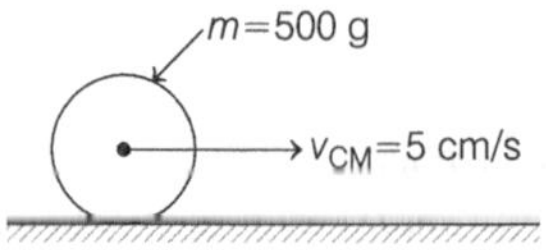

In pure rolling, object possesses both translational kinetic energy and rotational kinetic energy.

So, $\quad KE_{sphere} = \frac{1}{2}mv_{CM}^2 + \frac{1}{2}I\omega^2$

Here, $\quad I = \frac{2}{5}mr^2$ and $\omega = \frac{v}{r} \qquad [\because v_{CM} = v]$

So, $\quad KE_{sphere} = \frac{1}{2}mv^2 + \frac{1}{2}\times\frac{2}{5}mr^2\times\frac{v^2}{r^2}$

$$= \left(\frac{1}{2}+\frac{1}{5}\right)mv^2$$

$$= \frac{7}{10}\times 500\times10^{-3}\times25\times10^{-4}$$

$$= 8.75\times10^{-4}\text{ J}$$

Example 20. *A homogeneous solid cylindrical roller of radius R and mass m is pulled on a cricket pitch by a horizontal force. Assuming rolling without slipping, angular acceleration of the cylinder is* **[JEE Main 2019]**

(a) $\frac{F}{2\,mR}$ (b) $\frac{2F}{3\,mR}$ (c) $\frac{3F}{2\,mR}$ (d) $\frac{F}{3\,mR}$

Sol. (b) When force F is applied at the centre of roller of mass m as shown in the figure below

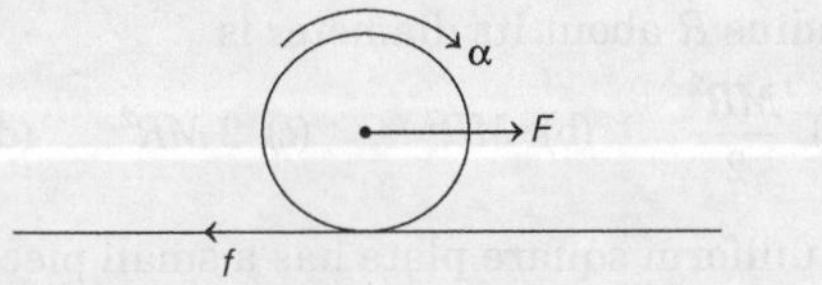

Its acceleration is given by

$$\frac{(F-f)}{m} = \alpha \qquad \text{... (i)}$$

where, f = force of friction and m = mass of roller.

Torque on roller is provided by friction f and it is

$$\tau = fR = I\alpha \qquad \text{...(ii)}$$

where, $\quad I$ = moment of inertia of solid cylindrical roller $= mR^2/2$

and $\quad \alpha$ = angular acceleration of cylinder $= a/R$.

Hence, $\tau = \frac{mR^2}{2}\cdot\frac{a}{R} = \frac{maR}{2}$

From Eq. (ii), $(\tau = fR)$

$$f = \frac{ma}{2} \qquad \text{...(iii)}$$

From Eqs. (i) and (iii), we get

$$F = \frac{3}{2}ma$$

$\Rightarrow \quad a = \frac{2F}{3m}$

So, $\quad \alpha = \frac{2F}{3mR} \qquad \left[\because \alpha = \frac{a}{R}\right]$

Example 21. *Three bodies, a ring, a solid cylinder and a solid sphere roll down the same inclined plane without slipping. They start from rest. The radii of the bodies are identical. Which of the body reaches the ground with maximum velocity?*

(a) *Ring*

(b) *Cylinder*

(c) *Sphere*

(d) *All have the same velocity*

Sol. (c) We assume conservation of energy of the rolling body, *i.e.* there is no loss of energy due to friction etc. The potential energy lost by the body in rolling down the inclined plane (= mgh) must, therefore be equal to kinetic energy gained.

Since, the bodies start from rest. Therefore, the kinetic energy gained is equal to the final kinetic energy of the bodies.

$$K = \frac{1}{2}mv_{CM}^2\left(1+\frac{k^2}{R^2}\right)$$

where, v_{CM} is the final velocity of (the centre of mass) of the body. Equating K and mgh, we have

$$mgh = \frac{1}{2}mv^2\left(1+\frac{k^2}{R^2}\right)$$

or $\quad v^2 = \left(\frac{2\,gh}{1+k^2/R^2}\right)$

Note It is independent of the mass of the rolling body.

For a ring, $\quad k^2 = R^2$

$$v_{ring} = \sqrt{\frac{2\,gh}{1+1}} = \sqrt{gh}$$

For a solid cylinder, $\quad k^2 = \frac{R^2}{2}$

$$v_{disc} = \sqrt{\frac{2\,gh}{1+1/2}} = \sqrt{\frac{4}{3}gh}$$

For a solid sphere, $\quad k^2 = \frac{2R^2}{5}$

$$v_{sphere} = \sqrt{\frac{2\,gh}{1+\frac{2}{5}}} = \sqrt{\frac{10}{7}gh}$$

From the results obtained it is clear that among the three bodies, the sphere has the greatest and the ring has the least velocity of the centre of mass at the bottom of the inclined plane.

Practice Exercise

ROUND I Topically Divided Problems

Moment of Inertia

1. Moment of inertia of a thin rod of mass M and length L about an axis passing through its centre is $\frac{ML^2}{12}$. Its moment of inertia about a parallel axis at a distance of $\frac{L}{4}$ from this axis is given by

(a) $\frac{ML^2}{48}$ (b) $\frac{ML^3}{48}$ (c) $\frac{ML^2}{12}$ (d) $\frac{7\,ML^2}{48}$

2. Moment of inertia of a solid cylinder of length L and diameter D about an axis passing through its centre of gravity and perpendicular to its geometric axis is

(a) $M\left(\frac{D^2}{4}+\frac{L^2}{12}\right)$ (b) $M\left(\frac{L^2}{16}+\frac{D^2}{8}\right)$ (c) $M\left(\frac{D^2}{4}+\frac{L^2}{6}\right)$ (d) $M\left(\frac{L^2}{12}+\frac{D^2}{16}\right)$

3. Three identical thin rods each of length l and mass M are joined together to form a ladder H. What is the moment of inertia of the system about one of the sides of H?

(a) $\frac{Ml^2}{4}$ (b) $\frac{Ml^2}{3}$ (c) $\frac{2Ml^2}{3}$ (d) $\frac{4Ml^2}{3}$

4. Three particles each of mass m gram, are situated at the vertices of an equilateral triangle ABC of side l cm (as shown in figure). The moment of inertia of the system about a line AX perpendicular to AB and in the plane of ABC in g-cm^2 unit will be

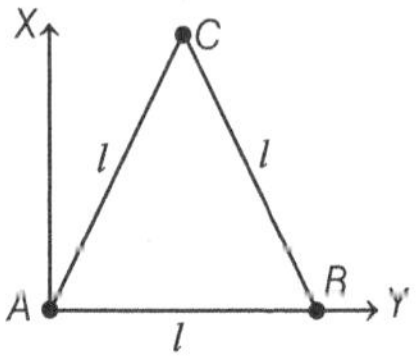

(a) $\frac{3}{4}ml^2$ (b) $2\,ml^2$ (c) $\frac{5}{4}ml^2$ (d) $\frac{3}{2}ml^2$

5. Two thin uniform circular rings each of radius 10 cm and mass 0.1 kg are arranged such that they have a common centre and their planes are perpendicular to each other. The moment of inertia of this system about an axis passing through their common centre and perpendicular to the plane of one of the rings in kg-m^2 is

(a) 15×10^{-3} (b) 5×10^{-3} (c) 1.5×10^{-3} (d) 18×10^{-4}

6. Moment of inertia of a circular wire of mass M and radius R about its diameter is **[AIEEE 2002]**

(a) $\frac{MR^2}{2}$ (b) MR^2 (c) $2\,MR^2$ (d) $\frac{MR^2}{4}$

7. A uniform square plate has a small piece Q of an irregular shape removed and glued to the centre of the plate leaving a hole behind. The moment of inertia about the Z-axis is **[NCERT Exemplar]**

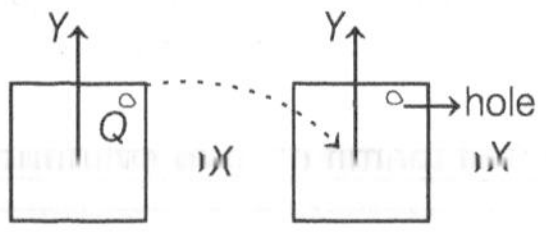

(a) increased
(b) decreased
(c) the same
(d) changed in unpredicted manner

8. Two discs have same mass and thickness and their materials have densities d_1 and d_2. The ratio of their moments of inertia about central axis will be

(a) $d_1 : d_2$ (b) $d_1 d_2 : 1$ (c) $1 : d_1 d_2$ (d) $d_2 : d_1$

9. A thin rod of length L and mass M is bent at the middle point O at an angle of 60°. The moment of inertia of the rod about an axis passing through O and perpendicular to the plane of the rod will be

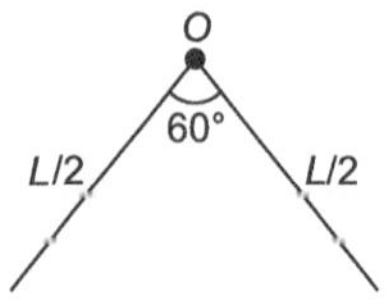

(a) $\frac{ML^2}{6}$ (b) $\frac{ML^2}{12}$ (c) $\frac{ML^2}{24}$ (d) $\frac{ML^2}{3}$

10. Four point masses, each of value m, are placed at the corners of a square $ABCD$ of side l. The moment of inertia of this system about an axis passing through A and parallel to BD is **[AIEEE 2006]**

(a) $2ml^2$ (b) $\sqrt{3}\,ml^2$
(c) $3ml^2$ (d) ml^2

11. The moment of inertia of uniform semi-circular disc of mass M and radius r about a line perpendicular to the plane of the disc through the centre is **[AIEEE 2005]**

(a) $\frac{1}{4}Mr^2$ (b) $\frac{2}{5}Mr^2$
(c) Mr^2 (d) $\frac{1}{2}Mr^2$

12. If I_1 is the moment of inertia of a thin rod about an axis perpendicular to its length and passing through its centre of mass and I_2 is the moment of inertia of ring about an axis perpendicular to plane of ring and passing through its centre formed by bending the rod, then

(a) $\frac{I_1}{I_2}=\frac{3}{\pi^2}$ (b) $\frac{I_1}{I_2}=\frac{2}{\pi^2}$
(c) $\frac{I_1}{I_2}=\frac{\pi^2}{2}$ (d) $\frac{I_1}{I_2}=\frac{\pi^2}{3}$

13. One solid sphere A and another hollow sphere B are of same mass and same outer radii. Their moment of inertia about their diameters are respectively I_A and I_B such that **[AIEEE 2004]**

(a) $I_A = I_B$ (b) $I_A > I_B$
(c) $I_A < I_B$ (d) $\frac{I_A}{I_B}=\frac{d_A}{d_B}$

14. Seven identical circular planar discs, each of mass M and radius R are welded symmetrically as shown in the figure. The moment of inertia of the arrangement about the axis normal to the plane and passing through the point P is **[JEE Main 2018]**

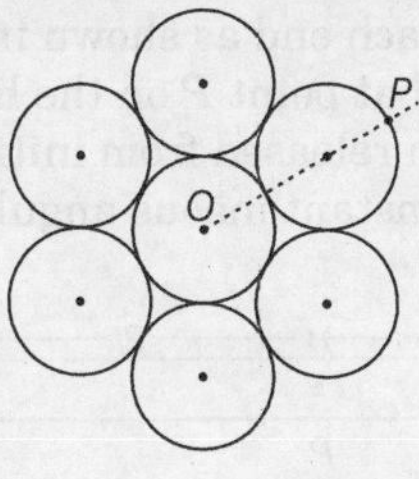

(a) $\frac{19}{2}MR^2$
(b) $\frac{55}{2}MR^2$
(c) $\frac{73}{2}MR^2$
(d) $\frac{181}{2}MR^2$

15. The moment of inertia of a solid sphere, about an axis parallel to its diameter and at a distance of x from it, is $I(x)$. Which one of the graphs represents the variation of $I(x)$ with x correctly? **[JEE Main 2019]**

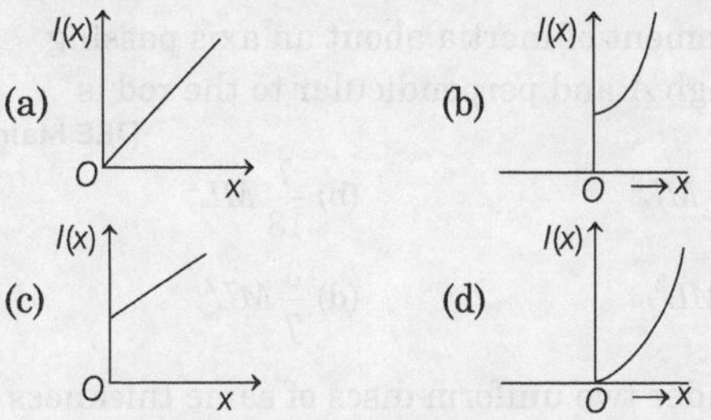

16. A circular disc D_1 of mass M and radius R has two identical discs D_2 and D_3 of the same mass M and radius R attached rigidly at its opposite ends (see figure). The moment of inertia of the system about the axis OO' passing through the centre of D_1, as shown in the figure will be **[JEE Main 2019]**

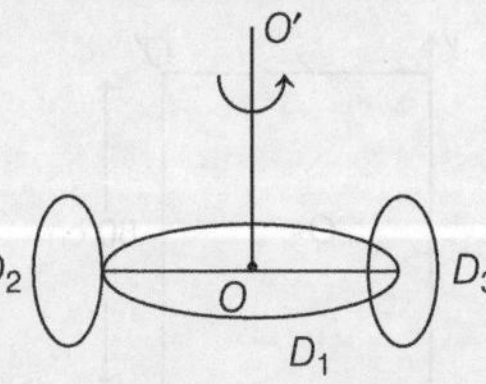

(a) $\frac{2}{3}MR^2$ (b) $\frac{4}{5}MR^2$ (c) $3MR^2$ (d) MR^2

17. A thin disc of mass M and radius R has mass per unit area $\sigma(r) = kr^2$, where r is the distance from its centre. Its moment of inertia about an axis going through its centre of mass and perpendicular to its plane is **[JEE Main 2019]**

(a) $\frac{MR^2}{2}$ (b) $\frac{MR^2}{6}$ (c) $\frac{MR^2}{3}$ (d) $\frac{2MR^2}{3}$

18. The radius of gyration of a uniform rod of length l, about an axis passing through a point $\frac{l}{4}$ away from the centre of the rod and perpendicular to it, is **[JEE Main 2020]**

(a) $\frac{1}{8}l$ (b) $\sqrt{\frac{3}{8}}\,l$ (c) $\sqrt{\frac{7}{48}}\,l$ (d) $\frac{1}{4}l$

19. Mass per unit area of a circular disc of radius a depends on the distance r from its centre as $\sigma(r) = A + Br$. The moment of inertia of the disc about the axis, perpendicular to the plane and passing through its centre is **[JEE Main 2020]**

(a) $2\pi a^4\left(\frac{A}{4}+\frac{B}{5}\right)$ (b) $2\pi a^4\left(\frac{aA}{4}+\frac{B}{5}\right)$
(c) $\pi a^4\left(\frac{A}{4}+\frac{aB}{5}\right)$ (d) $2\pi a^4\left(\frac{A}{4}+\frac{aB}{5}\right)$

20. The linear mass density of a thin rod AB of length L varies from A to B as $\lambda(x) = \lambda_0\left(1+\frac{x}{L}\right)$, where x is the distance from A. If M is mass of the rod, then its moment of inertia about an axis passing through A and perpendicular to the rod is [JEE Main 2020]

(a) $\frac{5}{12}ML^2$ (b) $\frac{7}{18}ML^2$
(c) $\frac{2}{5}ML^2$ (d) $\frac{3}{7}ML^2$

21. Consider two uniform discs of same thickness and different radii $R_1 = R$ and $R_2 = \alpha R$ made of the same material. If the ratio of their moments of inertia I_1 and I_2 respectively, about their axes is $I_1 : I_2 = 1:16$, then the value of α is [JEE Main 2020]

(a) 2 (b) $2\sqrt{2}$
(c) 4 (d) $\sqrt{2}$

22.

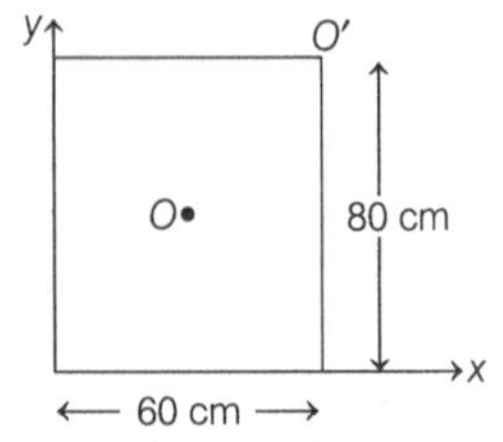

For a uniform rectangular sheet, shown in the above figure, the ratio of moments of inertia about the axes perpendicular to the sheet and passing through O (the centre of mass) and O' (corner point) is [JEE Main 2020]

(a) 1/8 (b) 2/3
(c) 1/4 (d) 1/2

23. Moment of inertia of a cylinder of mass M, length L and radius R about an axis passing through its centre and perpendicular to the axis of the cylinder is $I = M\left(\frac{R^2}{4}+\frac{L^2}{12}\right)$. If such a cylinder is to be made for a given mass of a material. To have minimum possible moment of inertia, the ratio L/R for cylinder is [JEE Main 2020]

(a) $\frac{2}{3}$ (b) $\frac{3}{2}$
(c) $\sqrt{\frac{3}{2}}$ (d) $\sqrt{\frac{2}{3}}$

24. An equilateral triangle ABC is cut from a thin solid sheet of wood (see figure). D, E and F are the mid points of its sides as shown and G is the centre of the triangle. The moment of inertia of the triangle about an axis passing through G and perpendicular to the plane of the triangle is I_0. If the smaller triangle DEF is removed from ABC, the moment of inertia of the remaining figure about the same axis is I, then [JEE Main 2019]

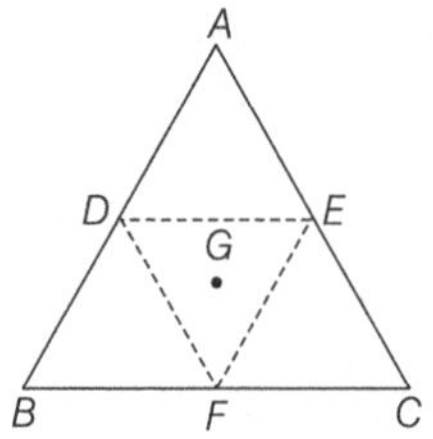

(a) $I' = \frac{3}{4}I_0$ (b) $I' = \frac{15}{16}I_0$
(c) $I' = \frac{I_0}{4}$ (d) $I' = \frac{9}{16}I_0$

Torque, Angular Momentum and Conservation of Angular Momentum

25. Four 2 kg masses are connected by $\frac{1}{4}$ m spokes to an axle. A force of 24 N acts on a lever (1/2) m long to produce angular acceleration α. The magnitude of α (in rad s^{-2}) is

(a) 24 (b) 12 (c) 6 (d) 3

26. A horizontal force F is applied such that the block remains stationary, then which of the following statement is false?

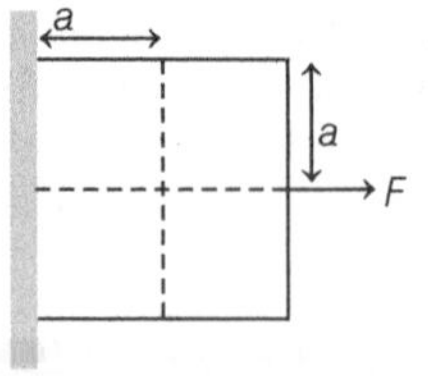

(a) $F = mg$ (where F is the frictional force).
(b) $F = N$ (where N is the normal force).
(c) F will not produce torque.
(d) N will not produce torque.

27. A rigid massless rod of length $3l$ has two masses attached at each end as shown in the figure. The rod is pivoted at point P on the horizontal axis (see figure). When released from initial horizontal position, its instantaneous angular acceleration will be [JEE Main 2019]

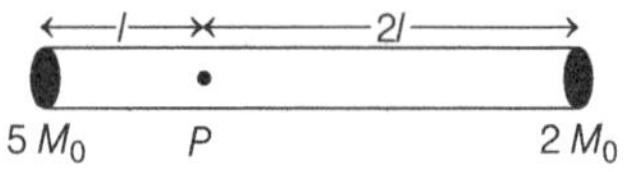

(a) $\frac{g}{13l}$ (b) $\frac{g}{2l}$ (c) $\frac{7g}{3l}$ (d) $\frac{g}{3l}$

28. A slender uniform rod of mass M and length l is pivoted at one end, so that it can rotate in a vertical plane (see the figure). There is negligible friction at the pivot. The free end is held vertically above the pivot and then released. The angular acceleration of the rod when it makes an angle θ with the vertical, is **[JEE Main 2017]**

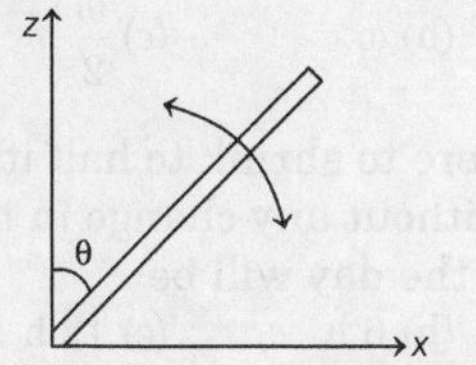

(a) $\frac{2g}{3l}\sin\theta$ (b) $\frac{3g}{2l}\cos\theta$

(c) $\frac{2g}{3l}\cos\theta$ (d) $\frac{3g}{2l}\sin\theta$

29. A particle of mass m is moving along a trajectory given by $x = x_0 + a\cos\omega_1 t$ and $y = y_0 + b\sin\omega_2 t$. The torque acting on the particle about the origin at $t = 0$ is **[JEE Main 2019]**

(a) zero (b) $m(-x_0 b + y_0 a)\,\omega_1^2\,\hat{\mathbf{k}}$

(c) $-m(x_0 b\omega_2^2 - y_0 a\omega_1^2)\,\hat{\mathbf{k}}$ (d) $+my_0 a\,\omega_1^2\,\hat{\mathbf{k}}$

30. A uniform cylinder of mass M and radius R is to be pulled over a step of height a $(a < R)$ by applying a force F at its centre O perpendicular to the plane through the axes of the cylinder on the edge of the step (see figure). The minimum value of F required is **[JEE Main 2020]**

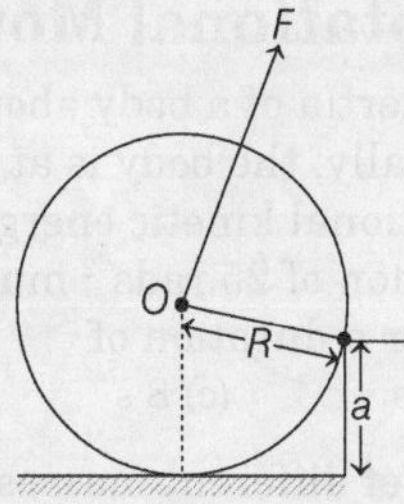

(a) $Mg\frac{a}{R}$ (b) $Mg\sqrt{1-\frac{a^2}{R^2}}$

(c) $Mg\sqrt{\left(\frac{R}{R-a}\right)^2 - 1}$ (d) $Mg\sqrt{1-\left(\frac{R-a}{R}\right)^2}$

31. A ring of diameter 0.4 m and of mass 10 kg is rotating about its axis at the rate of 1200 rpm. The angular momentum of the ring is

(a) 60.28 kg-m^{2}s^{-1} (b) 55.26 kg-m^{2}s^{-1}

(c) 40.28 kg-m^{2}s^{-1} (d) 50.28 kg-m^{2}s^{-1}

32. A mass M hangs on a massless rod of length l which rotates at a constant angular frequency. The mass M moves with steady speed in a circular path of constant radius. Assume that, the system is in steady circular motion with constant angular velocity ω. The angular momentum of M about point A in L_A which lies in the position z-direction and the angular momentum of M about B is L_B.

The correct statement for this system is

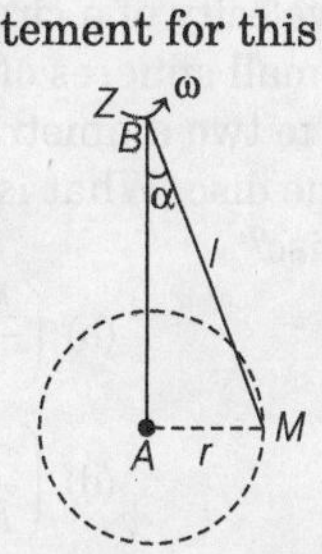

[JEE Main 2021]

(a) L_A and L_B both are constant in magnitude and direction

(b) L_B is constant in direction with varying magnitude

(c) L_B is constant, both in magnitude and direction

(d) L_A is constant, both in magnitude and direction

33. A particle of mass M moves along the line PC with velocity v as shown. What is the angular momentum of the particle about O?

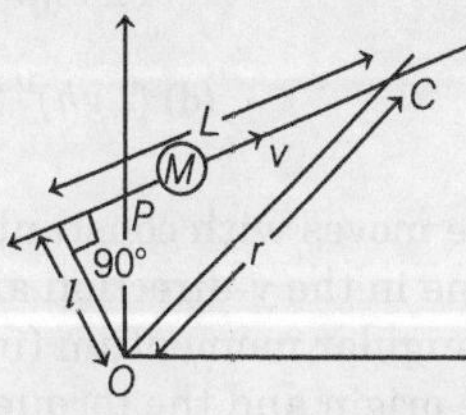

(a) mvL (b) mvl (c) mvr (d) Zero

34. A particle of mass 20 g is released with an initial velocity 5 m/s along the curve from the point A, as shown in the figure. The point A is at height h from point B. The particle slides along the frictionless surface. When the particle reaches point B, its angular momentum about O will be (Take, $g = 10$ m/s^2) **[JEE Main 2019]**

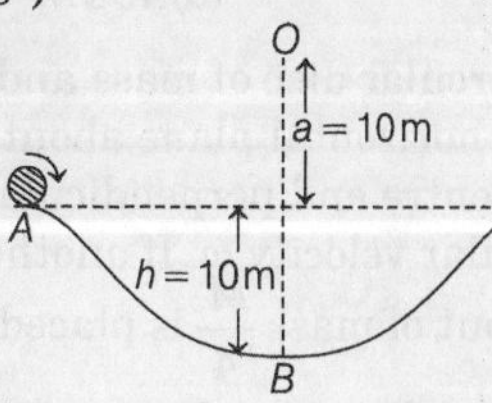

(a) 8 kg-m^2/s (b) 3 kg-m^2/s

(c) 2 kg-m^2/s (d) 6 kg-m^2/s

35. A triangular plate is shown. A force $\mathbf{F} = 4\hat{\mathbf{i}} - 3\hat{\mathbf{j}}$ is applied at point P. The torque at point P with respect to points O and Q are

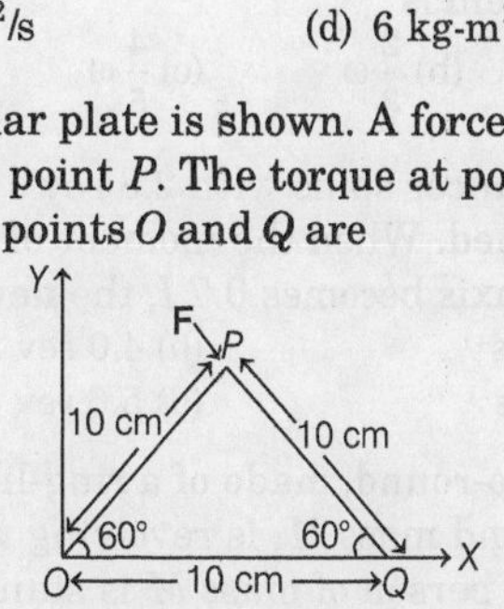

[JEE Main 2021]

(a) $-15-20\sqrt{3}, 15-20\sqrt{3}$ (b) $15+20\sqrt{3}, 15-20\sqrt{3}$

(c) $15-20\sqrt{3}, 15+20\sqrt{3}$ (d) $-15+20\sqrt{3}, 15+20\sqrt{3}$

36. Initial angular velocity of a circular disc of mass M is ω_1, then two small spheres of mass m are attached gently to two diametrically opposite points on the edge of the disc. What is the final angular velocity of the disc? [AIEEE 2002]

(a) $\left(\frac{M+m}{M}\right)\omega_1$ (b) $\left(\frac{M+m}{m}\right)\omega_1$

(c) $\left(\frac{M}{M+4m}\right)\omega_1$ (d) $\left(\frac{M}{M+2m}\right)\omega_1$

37. A cord is wound round the circumference of a wheel of radius r. The axis of the wheel of horizontal and moment of inertia about it is I. A weight mg is attached to the end of the cord and falls from rest. After falling through a distance h, the angular velocity of the wheel will be

(a) $\left(\frac{2gh}{1+mr}\right)^{1/2}$ (b) $\left(\frac{2mgh}{I+mr^2}\right)^{1/2}$

(c) $\left(\frac{2mgh}{1+2m}\right)^{1/2}$ (d) $(2gh)^{1/2}$

38. A 3 kg particle moves with constant speed of 2 ms^{-1} in the XY-plane in the y-direction along the line $x = 4$ m. The angular momentum (in kg-m^2s^{-1}) relative to the origin and the torque about the origin needed to maintain this motion are respectively

(a) 12, 0 (b) 24, 0
(c) 0, 24 (d) 0, 12

39. If the earth suddenly changes its radius x times the present value, the new period of rotation would be

(a) $6x^2h$ (b) $12x^2h$
(c) $24x^2h$ (d) $48x^2h$

40. A thin and circular disc of mass and radius R is rotating in a horizontal plane about axis passing through its centre and perpendicular of its plane with an angular velocity ω. If another disc of same dimensions but of mass $\frac{M}{4}$ is placed gently on the first disc co-axially, then the new angular velocity of the system is

(a) $\frac{5}{4}\omega$ (b) $\frac{2}{3}\omega$ (c) $\frac{4}{5}\omega$ (d) $\frac{3}{2}\omega$

41. A ballet dancer spins with 2.8 rev s^{-1} with her arms out stretched. When the moment of inertia about the same axis becomes $0.7\,I$, the new rate of spin is

(a) 3.2 rev s^{-1} (b) 4.0 rev s^{-1}
(c) 4.8 rev s^{-1} (d) 5.6 rev s^{-1}

42. A merry-go-round, made of a ring-like platform of radius R and mass M, is revolving with angular speed ω. A person of mass M is standing on it. At one instant, the person jumps off the round, radially away from the centre of the round (as seen from the round). The speed of the round afterward is [NCERT Exemplar]

(a) 2ω (b) ω (c) $\frac{\omega}{2}$ (d) 0

43. If earth where to shrink to half its present diameter without any change in its mass, the duration of the day will be

(a) 48 h (b) 6 h (c) 12 h (d) 24 h

44. A man of 80 kg mass is standing on the rim of a circular platform of mass 200 kg rotating about its axis. The mass of the platform with the man on it rotates at 12.0 rpm. If the man now moves to centre of the platform, the rotational speed would become

(a) 16.5 rpm (b) 25.7 rpm
(c) 32.3 rpm (d) 31.2 rpm

45. Consider a uniform rod of mass $M = 4m$ and length l pivoted about its centre. A mass m moving with velocity v making angle $\theta = \frac{\pi}{4}$ to the rod's long axis collides with one end of the rod and sticks to it. The angular speed of the rod-mass system just after the collision is [JEE Main 2020]

(a) $\frac{3\sqrt{2}}{7}\frac{v}{l}$ (b) $\frac{4}{7}\frac{v}{l}$ (c) $\frac{3}{7}\frac{v}{l}$ (d) $\frac{3}{7\sqrt{2}}\frac{v}{l}$

Dynamics of Rotational Motion

46. The moment of inertia of a body about a given axis is 1.2 kg-m^2. Initially, the body is at rest. In order to produce a rotational kinetic energy of 1500 J and angular acceleration of 25 $rads^{-2}$ must be applied about that axis for a duration of

(a) 4 s (b) 2 s (c) 8 s (d) 10 s

47. A ring and a disc of different masses are rotating with the same kinetic energy. If we apply a retarding torque τ on the ring stops after making n revolutions, then in how many revolutions will the disc stop under the same retarding torque?

(a) n (b) $2n$ (c) $4n$ (d) $n/2$

48. A thin metal disc of radius 0.25 m and mass 2 kg starts from rest and rolls down an inclined plane. If its rotational kinetic energy is 4 J at the foot of the inclined plane, then its linear velocity at the same point is

(a) 1.2 ms^{-1} (b) $2\sqrt{2}$ ms^{-1}
(c) 20 ms^{-1} (d) 2 ms^{-1}

49. The motor of an engine is rotating about its axis with an angular velocity of 100 rev min^{-1}. It come to rest in 15s, after being switched off. Assuming constant angular deceleration. What are the number of revolutions made by it before coming to rest?

(a) 15.6 (b) 32.6 (c) 12.5 (d) 40

50. A thin uniform rod of length l and mass m is swinging freely about a horizontal axis passing through its end. Its maximum angular speed is ω. Its centre of mass rises to a maximum height of **[AIEEE 2009]**

(a) $\frac{1}{3}\frac{l^2\omega^2}{g}$ (b) $\frac{1}{6}\frac{l\omega}{g}$ (c) $\frac{1}{2}\frac{l^2\omega^2}{g}$ (d) $\frac{1}{6}\frac{l^2\omega^2}{g}$

51. The oxygen molecule has a mass of 5.30×10^{-26} kg and a moment of inertia of 1.94×10^{-46} kg-m^{-2} about an axis through its centre perpendicular to the lines joining the two atoms. Suppose the mean speed of such a molecule in a gas is 500 m/s and that is KE of rotation is $\frac{2}{3}$ of its KE translation.

Find the average angular velocity of the molecule.

(a) 3.75×10^{12} rad/s (b) 5.75×10^{12} rad/s
(c) 9.75×10^{12} rad/s (d) 6.75×10^{12} rad/s

52. A circular disc rolls down an inclined plane. The ratio of the rotational kinetic energy to total kinetic energy is

(a) $\frac{1}{2}$ (b) $\frac{1}{3}$ (c) $\frac{2}{3}$ (d) $\frac{3}{4}$

53. Moment of inertia of a body about a given axis is 1.5 kg-m^2. Initially, the body is at rest. In order to produce a rotational kinetic energy of 1200 J, the angular acceleration of 20 rad/s^2 must be applied about the axis for a duration of **[JEE Main 2019]**

(a) 5 s (b) 2 s (c) 3 s (d) 2.5 s

54. Two discs of moment of inertia I_1 and I_2 about their respective axes and rotating with angular speed ω_1 and ω_2 are brought into contact face-to-face with their axes of rotation coincident. Then the loss of kinetic energy of the system in the process is

(a) $\frac{I_1 I_2}{2(I_1 + I_2)}(\omega_1 - \omega_2)^2$ (b) $-\frac{I_1 I_2}{2(I_1 + I_2)}(\omega_1 - \omega_2)^2$

(c) $\frac{I_1 I_2}{(I_1 + I_2)}(\omega_1 - \omega_2)^2$ (d) zero

55. A thin uniform rod AB of mass m and length L is hinged at one end A to the level floor. Initially, it stands vertically and is allowed to fall freely to the floor in the vertical plane. The angular velocity of the rod, when its end B strikes the floor is (g is acceleration due to gravity)

(a) $\left(\frac{mg}{L}\right)^{1/2}$ (b) $\left(\frac{mg}{3L}\right)^{1/2}$ (c) $\left(\frac{g}{L}\right)$ (d) $\left(\frac{3g}{L}\right)^{1/2}$

56. A particle performs uniform circular motion with an angular momentum L. If the frequency of a particle's motion is doubled and its kinetic energy is halved, the angular momentum becomes

(a) $2L$ (b) $4L$ (c) $L/2$ (d) $L/4$

57. A wheel is rotating freely with an angular speed ω on a shaft. The moment of inertia of the wheel is I and the moment of inertia of the shaft is negligible. Another wheel of moment of inertia $3I$ initially at rest is suddenly coupled to the same shaft. The resultant fractional loss in the kinetic energy of the system is **[JEE Main 2020]**

(a) $\frac{3}{4}$ (b) $\frac{5}{6}$ (c) 0 (d) $\frac{1}{4}$

58. Two uniform circular discs are rotating independently in the same direction around their common axis passing through their centres. The moment of inertia and angular velocity of the first disc are 0.1 kg-m^2 and 10 rad s^{-1} respectively, while those for the second one are 0.2 kg-m^2 and 5 rad s^{-1}, respectively. At some instant, they get stuck together and start rotating as a single system about their common axis with some angular speed. The kinetic energy of the combined system is **[JEE Main 2020]**

(a) $\frac{10}{3}$ J (b) $\frac{20}{3}$ J (c) $\frac{5}{3}$ J (d) $\frac{2}{3}$ J

59. Two coaxial discs, having moments of inertia I_1 and $\frac{I_1}{2}$ are rotating with respective angular velocities ω_1 and $\frac{\omega_1}{2}$, about their common axis. They are brought in contact with each other and thereafter they rotate with a common angular velocity. If E_f and E_i are the final and initial total energies, then $(E_f - E_i)$ is **[JEE Main 2019]**

(a) $-\frac{I_1\omega_1^2}{24}$ (b) $\frac{I_1\omega_1^2}{12}$ (c) $\frac{3}{8}I_1\omega_1^2$ (d) $\frac{I_1\omega_1^2}{6}$

60. A uniformly thick wheel with moment of inertia I and radius R is free to rotate about its centre of mass (see figure). A massless string is wrapped over its rim and two blocks of masses m_1 and $m_2 (m_1 > m_2)$ are attached to the ends of the string.

The system is released from rest. The angular speed of the wheel when m_1 descents by a distance h is **[JEE Main 2020]**

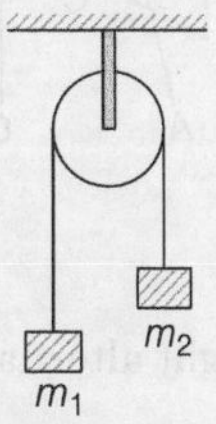

(a) $\left[\frac{2(m_1 + m_2)gh}{(m_1 + m_2)R^2 + I}\right]^{\frac{1}{2}}$ (b) $\left[\frac{m_1 + m_2}{(m_1 + m_2)R^2 + I}\right]^{\frac{1}{2}} gh$

(c) $\left[\frac{m_1 - m_2}{(m_1 + m_2)R^2 + I}\right]^{\frac{1}{2}} gh$ (d) $\left[\frac{2(m_1 - m_2)gh}{(m_1 + m_2)R^2 + I}\right]^{\frac{1}{2}}$

Rolling Motion

61. A solid sphere, a hollow sphere and a ring are released from top of an inclined plane (frictionless), so that they slide down the plane. Then, maximum acceleration down the plane is for (no rolling) [AIEEE 2002]

(a) solid sphere (b) hollow sphere
(c) ring (d) All same

62. A sphere and a hollow cylinder roll without slipping down two separate inclined planes and travel the same distance in the same time. If the angle of the plane down which the sphere rolls is 30°, the angle of the other plane is

(a) 60° (b) 53°
(c) 37° (d) 45°

63. A sphere of mass 2 kg and radius 0.5 m is rolling with an initial speed of 1ms^{-1} goes up an inclined plane which makes an angle of 30° with the horizontal plane, without slipping. How long will the sphere take to return to the starting point A? [JEE Main 2021]

(a) 0.60 s (b) 0.52 s
(c) 0.57 s (d) 0.80 s

64. A solid sphere rolls down without slipping on an inclined plane at angle 60° over a distance of 10 m. The acceleration (in ms^{-2}) is

(a) 4 (b) 5
(c) 6.06 (d) 7

65. A roller is made by joining together two corners at their vertices O. It is kept on two rails AB and CD which are placed a symmetrically (see the figure), with its axis perpendicular to CD and its centre O at the centre of line joining AB and CD (see the figure). It is given a light path, so that it starts rolling with its centre O moving parallel to CD in the direction shown. As it moves, the roller will tend to [JEE Main 2016]

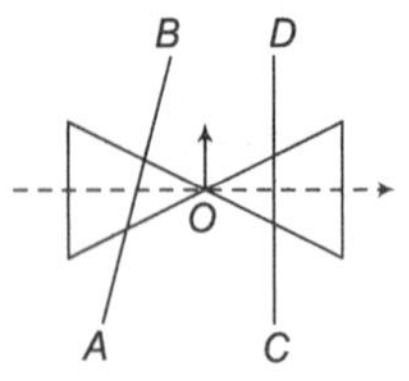

(a) turn left
(b) turn right
(c) go straight
(d) turn left and right alternately

66. A solid cylinder on moving with constant speed v_0 reaches the bottom of an incline of 30°. A hollow cylinder of same mass and radius moving with the same constant speed v_0 reaches the bottom of a different incline of inclination θ. There is no slipping and both of them go through the same distance in the same time; θ is then equal to

(a) 37° (b) 30°
(c) 42° (d) 45°

67. A hemispherical bowl or radius R is kept on a horizontal table. A small sphere of radius r ($r << R$) is placed at the highest point at the inside of the bowl and let go. The sphere rolls without slipping. Its velocity at the lowest point is

(a) $\sqrt{5\,gR/7}$ (b) $\sqrt{3\,gR/2}$
(c) $\sqrt{4\,gR/3}$ (d) $\sqrt{10\,gR/7}$

68. A marble and a cube have the same mass starting from rest, the marble rolls and the cube slides down a frictionless ramp. When they arrive at the bottom, the ratio of speed of the cube to the centre of mass and speed of the marble is

(a) 7 : 5 (b) $\sqrt{7}:\sqrt{5}$
(c) $\sqrt{2}:1$ (d) 5 : 2

69. A string is wound around a hollow cylinder of mass 5 kg and radius 0.5 m. If the string is now pulled with a horizontal force of 40 N and the cylinder is rolling without slipping on a horizontal surface (see figure), then the angular acceleration of the cylinder will be (Neglect the mass and thickness of the string) [JEE Main 2019]

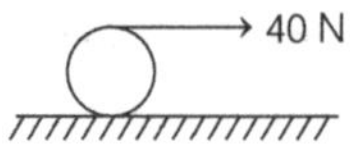

(a) 10 rad/s^2 (b) 16 rad/s^2
(c) 20 rad/s^2 (d) 12 rad/s^2

70. A tennis ball (treated as hollow spherical shell) starting from O rolls down a hill. At point A, the ball becomes air borne leaving at an angle of 30° with the horizontal. The ball strikes the ground at B, then what is the value of the distance AB?

(Moment of inertia of a spherical shell of mass m and radius R about its diameter $= \frac{2}{3}mR^2$) [JEE Main 2013]

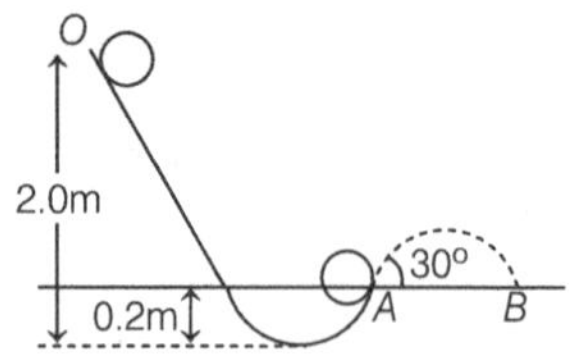

(a) 1.87 m (b) 2.08 m
(c) 1.57 m (d) 1.77 m

ROUND II **Mixed Bag**

Only One Correct Option

1. If the moment of inertia of a disc about an axis tangential and parallel to its surface be I, then what will be the moment of inertia about the axis tangential but perpendicular to the surface?

(a) $\frac{6}{5}I$ (b) $\frac{3}{4}I$ (c) $\frac{3}{2}I$ (d) $\frac{5}{4}I$

2. A ring of radius R is first rotated with an angular velocity ω_0 and then carefully placed on a rough horizontal surface. The coefficient of friction between the surface and the ring is μ. Time after which its angular speed is reduced to half is

(a) $\frac{\omega_0 \mu R}{2g}$ (b) $\frac{2\omega_0 R}{\mu g}$

(c) $\frac{\omega_0 R}{2\mu g}$ (d) $\frac{\omega_0 g}{2\mu R}$

3. The curve between $\log_e L$ and $\log_e p$ is (L is angular momentum and p is linear momentum)

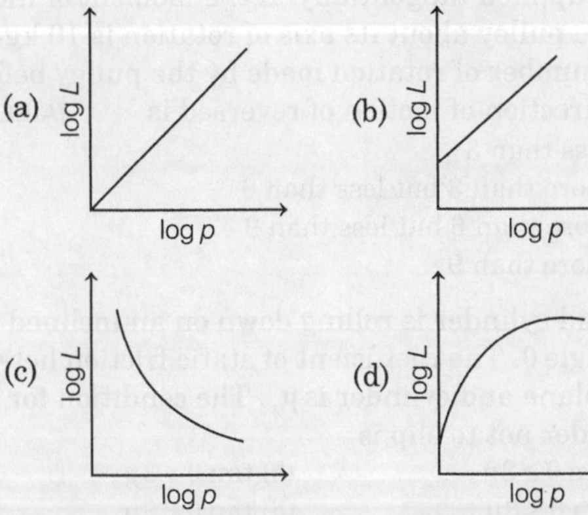

4. Two thin discs each of mass M and radius R are placed at either end of a rod of mass m, length l and radius r. Moment of inertia of the system about an axis passing through the centre of rod and perpendicular to its length is

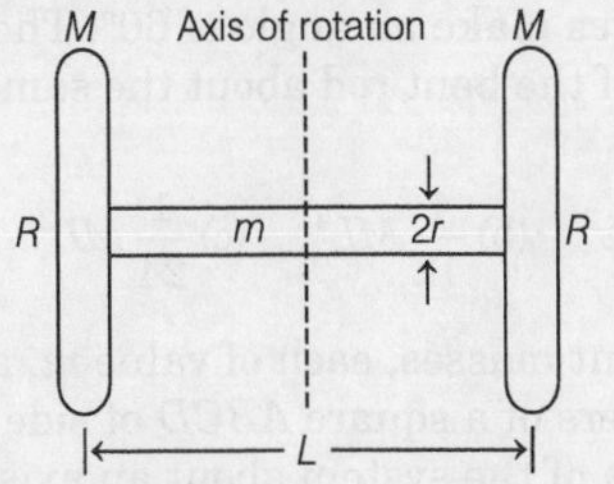

(a) $\frac{mL^2}{12}+\frac{1}{4}MR^2+\frac{1}{4}ML^2$ (b) $\frac{ML^2}{12}+\frac{1}{2}mR^2+\frac{1}{2}mL^2$

(c) $\frac{1}{2}mL^2+\frac{mR^2}{2}+\frac{ML^2}{12}$ (d) $\frac{mL^2}{12}+MR^2+\frac{1}{2}ML^2$

5. A cubical block of side a is moving with velocity v on a horizontal smooth plane as shown. It hits a ridge at point O. The angular speed of the block after is hits O is

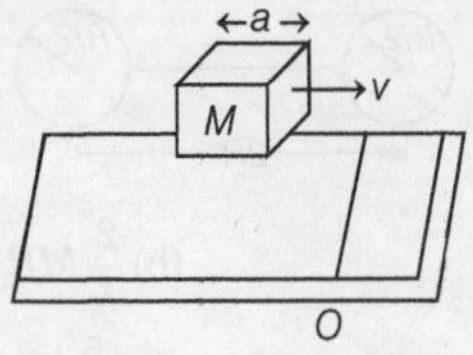

(a) $\frac{3v}{4a}$ (b) $\frac{3v}{2a}$

(c) $\frac{\sqrt{3}v}{\sqrt{2}a}$ (d) zero

6. Four particles each of mass m are lying symmetrically on the rim of a disc of mass M and radius R. Moment of inertia of this system about an axis passing through one of the particles and perpendicular to plane of disc is

(a) $16mR^2$ (b) $3(M+16m)\frac{R^2}{2}$

(c) $(3M+12m)\frac{R^2}{2}$ (d) zero

7. If the radius r of earth suddenly changes to x times the present values, the new period of rotation would be

(a) $dT/dt = (T/r)\,(dr/dt)$ (b) $dT/dt = (2T/r)\,(dr/dt)$

(c) $dT/dt = (r/T)\,(dr/dt)$ (d) $dT/dt = \left(\frac{1}{2}T/r\right)\left(\frac{dr}{dt}\right)$

8. The mass of the earth is increasing at the rate of 1 part in 5×10^{19} per day by the attraction of meteors falling normally on the earth's surface. Assuming that the density of earth is uniform, the rate of change of the period of rotation of the earth is

(a) 2.0×10^{-20} (b) 2.66×10^{-19}

(c) 4.33×10^{-18} (d) 5.66×10^{-17}

9. For the given uniform square lamina $ABCD$, whose centre is O as shown in figure. **[AIEEE 2007]**

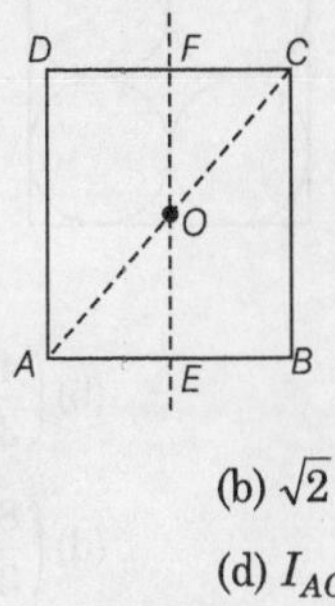

(a) $I_{AC}=\sqrt{2}\,I_{EF}$ (b) $\sqrt{2}\,I_{AC}=I_{EF}$

(c) $I_{AD}=3\,I_{EF}$ (d) $I_{AC}=I_{EF}$

10. Two spheres each of mass M and radius $R/2$ are connected with a massless rod of length $2R$ as shown in the figure. What will be the moment of inertia of the system about an axis passing through the centre of one of the sphere and perpendicular to the rod?

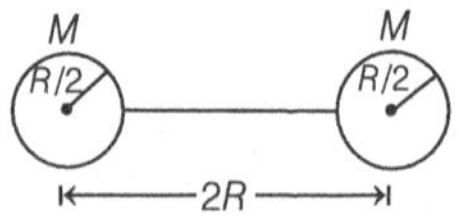

(a) $\frac{21}{5}MR^2$ (b) $\frac{2}{5}MR^2$

(c) $\frac{5}{2}MR^2$ (d) $\frac{5}{21}MR^2$

11. With referene to figure of a cube of edge a and mass m, state whether the following is true.
(O is the centre of the cube).

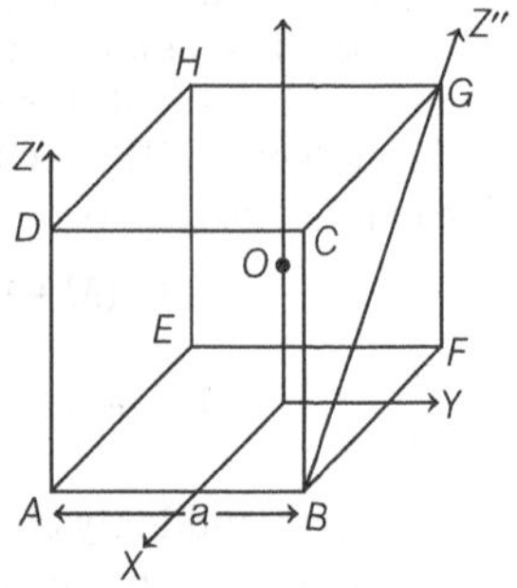

(a) The moment of inertia of cube about Z-axis is $I_Z = I_X + I_Y$

(b) The moment of inertia of cube about Z' is $I'_Z = I_Z + \frac{ma^2}{2}$

(c) The moment of inertia of cube about Z'' is $I''_Z = I_Z + \frac{mz^2}{2}$

(d) $I_X \neq I_Y$

12. Four holes of radius R are cut from a thin square plate of side $4R$ and mass M. The moment of inertia of the remaining portion about z-axis is

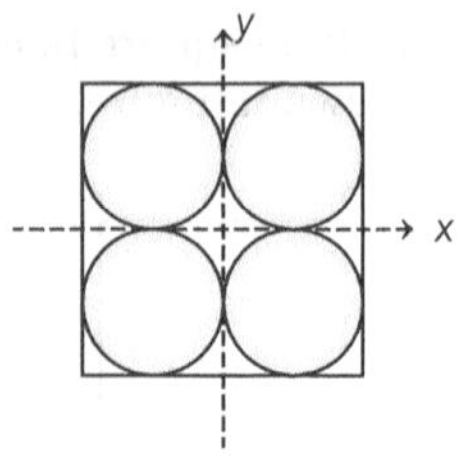

(a) $\frac{\pi}{12}MR^2$ (b) $\left(\frac{4}{3}-\frac{\pi}{4}\right)MR^2$

(c) $\left(\frac{4}{3}-\frac{\pi}{6}\right)MR^2$ (d) $\left(\frac{8}{3}-\frac{10\pi}{16}\right)MR^2$

13. A circular platform is free to rotate in a horizontal plane about a vertical axis passing through its centre. A tortoise is sitting at the edge of the platform. Now, the platform is given an angular velocity ω_0. When the tortoise moves along a chord of the platform with a constant velocity (w.r.t. the platform), the angular velocity of the platform will vary with the time t as

(a) 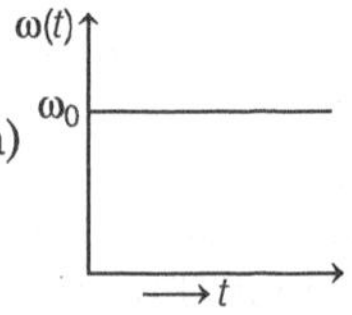

(b) 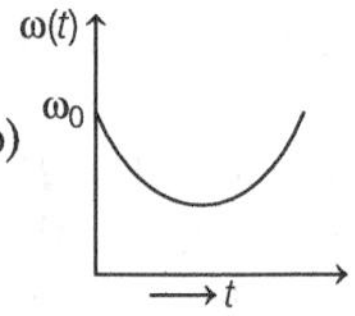

(c) 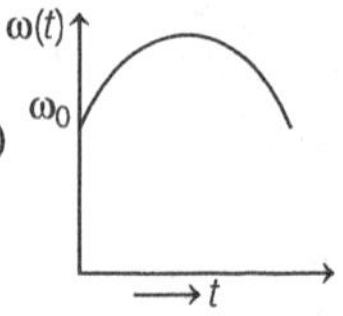

(d) 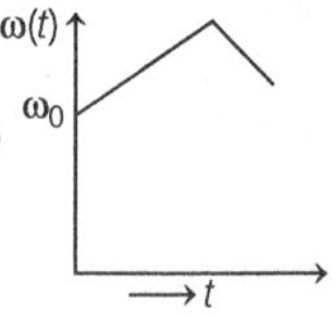

14. A pulley of radius 2m is rotated about its axis by a force $F = (20t - 5t^2)$ newton (where t is measured in s) applied tangentially. If the moment of inertia of the pulley about its axis of rotation is 10 kg-m^2, the number of rotation made by the pulley before its direction of motion of reversed is **[AIEEE 2011]**

(a) less than 3
(b) more than 3 but less than 6
(c) more than 6 but less than 9
(d) more than 9

15. A solid cylinder is rolling down on an inclined plane of angle θ. The coefficient of static friction between the plane and cylinder is μ_s. The condition for the cylinder not to slip is

(a) $\tan\theta \geq 3\mu_s$ (b) $\tan\theta > 3\mu_s$

(c) $\tan\theta \leq 3\mu_s$ (d) $\tan\theta < 3\mu_s$

16. The moment of inertia of a rod about an axis through its centre and perpendicular to it is $\frac{1}{12}ML^2$ (where, M is the mass and L the length of the rod). The rod is bent in the middle to that the two halves make an angle of 60°. The moment of inertia of the bent rod about the same axis would be

(a) $\frac{1}{48}ML^2$ (b) $\frac{1}{12}ML^2$ (c) $\frac{1}{24}ML^2$ (d) $\frac{ML^2}{8\sqrt{3}}$

17. Four point masses, each of value m, are placed at the corners of a square $ABCD$ of side l. The moment of inertia of the system about an axis passing through A and parallel to BD is **[AIEEE 2006]**

(a) $\sqrt{3}\,ml^2$ (b) $3\,ml^2$

(c) ml^2 (d) $2\,ml^2$

18. From a circular disc of radius R and mass $9M$, a small disc or radius $R/3$ is removed. The moment of inertia of the remaining disc about an axis perpendicular to the plane of the disc and passing through O is **[IIT JEE 2005]**

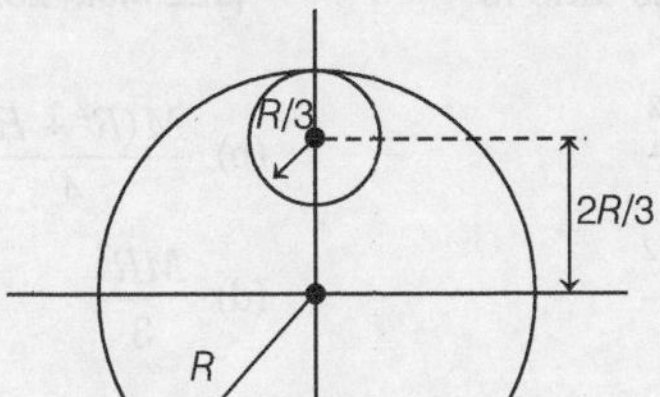

(a) $4MR^2$ (b) $\frac{40}{9}MR^2$ (c) $10MR^2$ (d) $\frac{37}{9}MR^2$

19. As shown in the figure, a bob of mass m is tied by a massless string whose other end portion is wound on a flywheel (disc) of radius r and mass m. When released from rest the bob starts falling vertically. When it has covered a distance of h, the angular speed of the wheel will be **[JEE Main 2020]**

(a) $\frac{1}{r}\sqrt{\frac{2gh}{3}}$ (b) $r\sqrt{\frac{3}{4gh}}$ (c) $\frac{1}{r}\sqrt{\frac{4gh}{3}}$ (d) $r\sqrt{\frac{3}{2gh}}$

20. A bead of mass m stays at point $P(a, b)$ on a wire bent in the shape of a parabola $y = 4Cx^2$ and rotating with angular speed ω (see figure). The value of ω is (neglect friction) **[JEE Main 2020]**

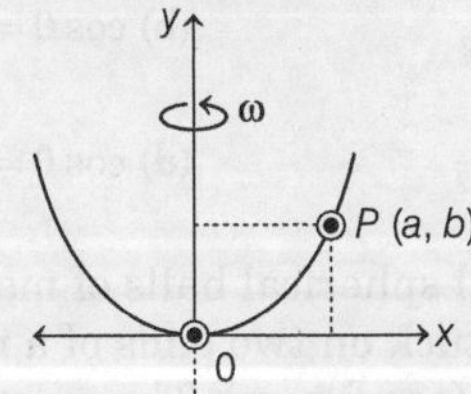

(a) $2\sqrt{2gC}$ (b) $\sqrt{\frac{2g}{C}}$

(c) $2\sqrt{gC}$ (d) $\sqrt{\frac{2gC}{ab}}$

21. From a solid sphere of mass M and radius R, a cube of maximum possible volume is cut. Moment of inertia of cube about an axis passing through its centre and perpendicular to one of its faces is **[JEE Main 2015]**

(a) $\frac{MR^2}{32\sqrt{2}\pi}$ (b) $\frac{MR^2}{16\sqrt{2}\pi}$ (c) $\frac{4MR^2}{9\sqrt{3}\pi}$ (d) $\frac{4MR^2}{3\sqrt{3}\pi}$

22. A uniform rod of length l is being rotated in a horizontal plane with a constant angular speed about an axis passing through one of its ends. If the tension generated in the rod due to rotation is $T(x)$ at a distance x from the axis, then which of the following graphs depicts it most closely? **[JEE Main 2019]**

(a)

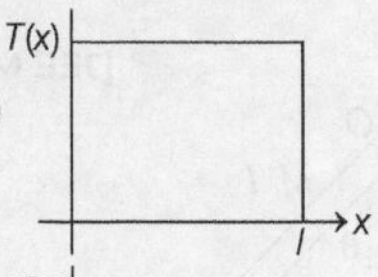

(b)

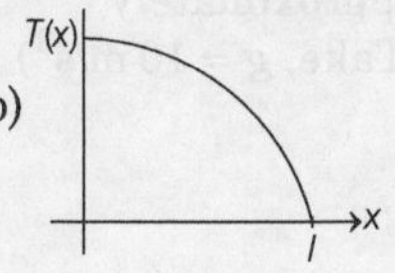

(c)

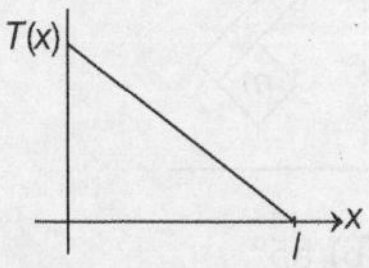

(d)

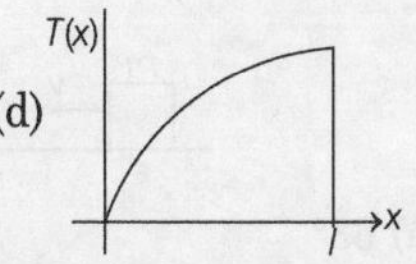

23. A ball rolls without slipping. The radius of gyration of the ball about an axis passing through its centre of mass is k. If radius of the ball be R, then the fraction of total energy associated with its rotational energy will be

(a) $\frac{k^2}{k^2 + R^2}$ (b) $\frac{R^2}{k^2 + R^2}$ (c) $\frac{k^2 + R^2}{R^2}$ (d) $\frac{k^2}{R^2}$

24.

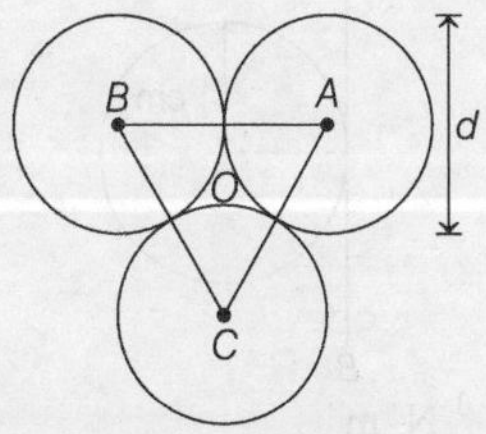

Three solid sphere each of mass m and diameter d are stuck together such that the lines connecting the centres form an equilateral triangle of side of length d. The ratio I_0 / I_A of moment of inertia I_0 of the system about an axis passing the centroid and about centre of any of the spheres I_A and perpendicular to the plane of the triangle is **[JEE Main 2020]**

(a) $\frac{15}{13}$ (b) $\frac{13}{15}$ (c) $\frac{13}{23}$ (d) $\frac{23}{13}$

25. A solid sphere of mass M and radius R is divided into two unequal parts. The first part has a mass of $\frac{7M}{8}$ and is converted into a uniform disc of radius $2R$. The second part is converted into a uniform solid sphere. Let I_1 be the moment of inertia of the disc about its axis and I_2 be the moment of inertia of the new sphere about its axis. The ratio I_1 / I_2 is given by **[JEE Main 2019]**

(a) 285 (b) 185

(c) 65 (d) 140

26. A block of mass $m = 1$ kg slides with velocity $v = 6$ m/s on a frictionless horizontal surface and collides with a uniform vertical rod and sticks to it as shown. The rod is pivoted about O and swings as a result of the collision, making angle θ before

momentarily coming to rest. If the rod has mass $M = 2$ kg and length $l = 1$ m, then the value of θ is approximately
(Take, $g = 10\ m/s^2$) [JEE Main 2020]

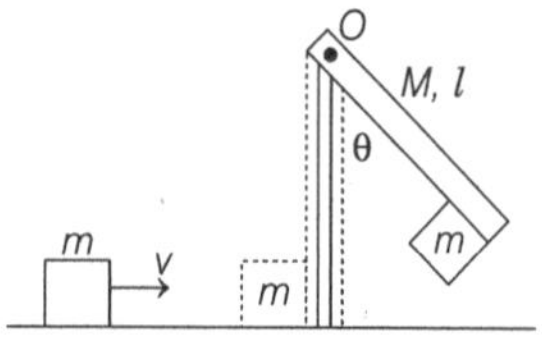

(a) 63° (b) 55°
(c) 69° (d) 49°

27. A metal coin of mass 5g and radius 1 cm is fixed to a thin stick AB of negligible mass as shown in the figure. The system is initially at rest. The constant torque, that will make the system rotate about AB at 25 rotations per second in 5s, is close to [JEE Main 2019]

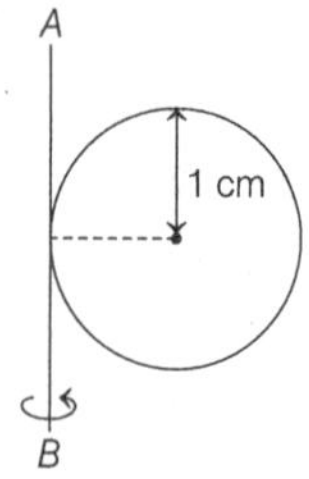

(a) 4.0×10^{-6} N-m
(b) 2.0×10^{-5} N-m
(c) 1.6×10^{-5} N-m
(d) 7.9×10^{-6} N-m

28. A person of mass M is sitting on a swing to length L and swinging with an angular amplitude θ_0. If the person stands up when the swing passes through its lowest point, the work done by him, assuming that his centre of mass moves by a distance $l (l << L)$, is close to [JEE Main 2019]

(a) $Mgl\,(1 - \theta_0^2)$
(b) $Mgl\,(1 + \theta_0^2)$
(c) Mgl
(d) $Mgl\left(1 + \frac{\theta_0^2}{2}\right)$

29. A round uniform body of radius R, mass M and moment of inertia I, rolls down (without slipping) an inclined plane making an angle θ with the horizontal. Then, its acceleration is [AIEEE 2007]

(a) $\frac{g \sin \theta}{1 + I/MR^2}$ (b) $\frac{g \sin \theta}{1 + MR^2/I}$
(c) $\frac{g \sin \theta}{1 - I/MR^2}$ (d) $\frac{g \sin \theta}{1 - MR^2/I}$

30. Shown in the figure is a hollow icecream cone (it is open at the top). If its mass is M, radius of its top R and height H, then its moment of inertia about its axis is [JEE Main 2020]

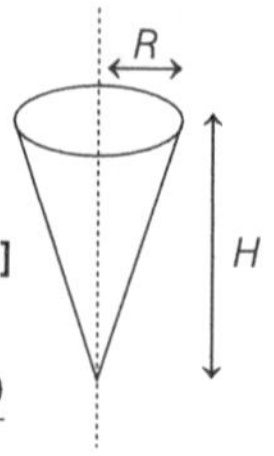

(a) $\frac{MR^2}{2}$ (b) $\frac{M(R^2 + H^2)}{4}$
(c) $\frac{MH^2}{3}$ (d) $\frac{MR^2}{3}$

31. A uniform rod of length l is pivoted at one of its ends on a vertical shaft of negligible radius. When the shaft rotates at angular speed ω, the rod makes an angle θ with it (see figure). To find θ, equate the rate of change of angular momentum (direction going into the paper) $\frac{ml^2}{12}\omega^2 \sin\theta\cos\theta$ about the centre of mass to the torque provided by the horizontal and vertical forces F_H and F_V about the centre of mass. The value of θ is then such that [JEE Main 2020]

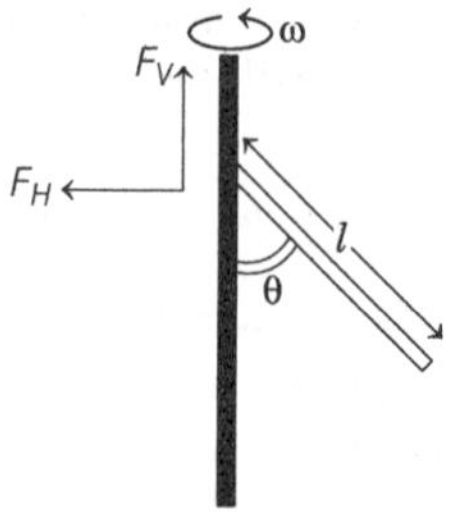

(a) $\cos\theta = \frac{g}{l\omega^2}$ (b) $\cos\theta = \frac{g}{2l\omega^2}$
(c) $\cos\theta = \frac{2g}{3l\omega^2}$ (d) $\cos\theta = \frac{3g}{2l\omega^2}$

32. Two identical spherical balls of mass M and radius R each are stuck on two ends of a rod of length $2R$ and mass M (see figure). The moment of inertia of the system about the axis passing perpendicularly through the centre of the rod is [JEE Main 2019]

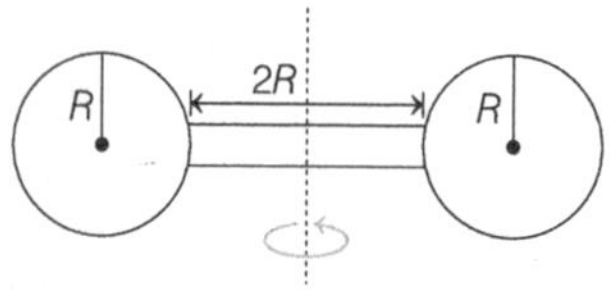

(a) $\frac{137}{15}MR^2$ (b) $\frac{209}{15}MR^2$ (c) $\frac{17}{15}MR^2$ (d) $\frac{152}{15}MR^2$

33. A circular disc of radius b has a hole of radius a at its centre (see figure). If the mass per unit area of the disc varies as $\left(\frac{\sigma_0}{r}\right)$, then the radius of gyration of the disc about its axis passing through the centre is [JEE Main 2019]

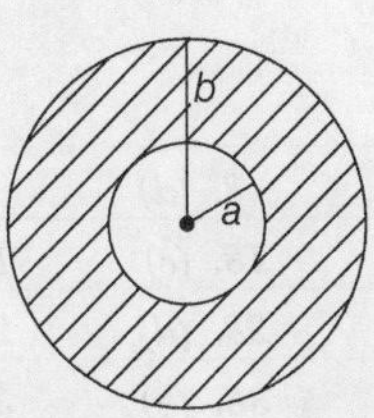

(a) $\sqrt{\dfrac{a^2+b^2+ab}{2}}$ (b) $\dfrac{a+b}{2}$

(c) $\sqrt{\dfrac{a^2+b^2+ab}{3}}$ (d) $\dfrac{a+b}{3}$

34. A solid sphere and solid cylinder of identical radii approach an incline with the same linear velocity (see figure). Both roll without slipping all throughout. The two climb maximum heights h_{sph} and h_{cyl} on the incline. The ratio $\dfrac{h_{sph}}{h_{cyl}}$ is given by [JEE Main 2019]

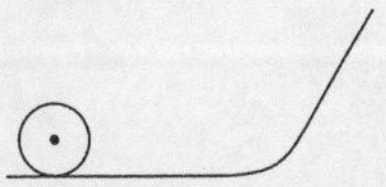

(a) $\dfrac{2}{\sqrt{5}}$ (b) $\dfrac{14}{15}$ (c) 1 (d) $\dfrac{4}{5}$

35. The following bodies are made to roll up (without slipping) the same inclined plane from a horizontal plane : (i) a ring of radius R, (ii) a solid cylinder of radius $R/2$ and (iii) a solid sphere of radius $R/4$. If in each case, the speed of the centre of mass at the bottom of the incline is same, the ratio of the maximum height they climb is

(a) 10 : 15 : 7 (b) 4 : 3 : 2

(c) 20 : 15 : 14 (d) 2 : 3 : 4

Numerical Value Questions

36. A massless equilateral triangle EFG of side a (as shown in figure) has three particles of mass m situated at its vertices. The moment of inertia of the system about the line EX perpendicular to EG in the plane of EFG is $\dfrac{N}{20}ma^2$, where N is an integer. The value of N is .... [JEE Main 2020]

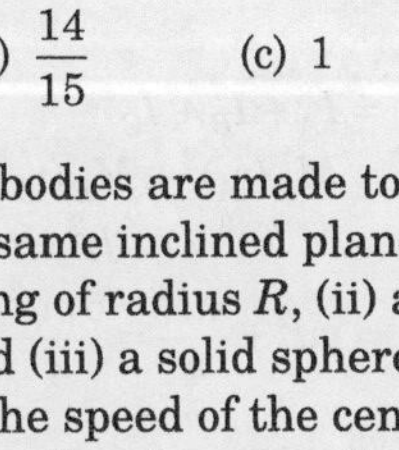

37. A force $\mathbf{F} = (\hat{\mathbf{i}} + 2\hat{\mathbf{j}} + 3\hat{\mathbf{k}})$ N acts at a point $(4\hat{\mathbf{i}} + 3\hat{\mathbf{j}} - \hat{\mathbf{k}})$ m. Then, the magnitude of torque about the point $(\hat{\mathbf{i}} + 2\hat{\mathbf{j}} + \hat{\mathbf{k}})$m will be $\sqrt{x}$ N-m. The value of x is .... [JEE Main 2020]

38. A person of 80 kg mass is standing on the rim of a circular platform of mass 200 kg rotating about its axis at 5 rpm. The person now starts moving towards the centre of the platform. What will be the rotational speed (in rpm) of the platform, when the person reaches its centre? [JEE Main 2020]

39. One end of a straight uniform 1 m long bar is pivoted on horizontal table. It is released from rest when it makes an angle 30° from the horizontal (see figure). Its angular speed when its hits the table is given as $\sqrt{n}\ s^{-1}$, where n is an integer. The value of n is .... [JEE Main 2020]

40. A thin rod of mass 0.9 kg and length 1 m is suspended at rest from one end, so that it can freely oscillate in the vertical plane.
A particle of mass 0.1 kg moving in a straight line with velocity 80 m/s hits the rod at its bottom-most point and sticks to it (see figure). The angular speed (in rad/s) of the rod immediately after the collision will be .... [JEE Main 2020]

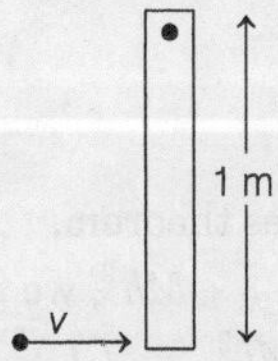

41. A circular disc of mass M and radius R is rotating about its axis with angular speed ω_1. If another stationary disc having radius $\dfrac{R}{2}$ and same mass M is dropped co-axially on to the rotating disc. Gradually, both discs attain constant angular speed ω_2. The energy lost in the process is $p\%$ of the initial energy. Value of p is .... [JEE Main 2020]

42. ABC is a plane lamina of the shape of an equilateral triangle. D, E are mid-points of AB, AC and G is the centroid of the lamina. Moment of inertia of the lamina about an axis passing through G and perpendicular to the plane ABC is I_0. If part ADE is removed, the moment of inertia of the remaining part about the same axis is $\dfrac{NI_0}{16}$, where N is an integer. Value of N is .... [JEE Main 2020]

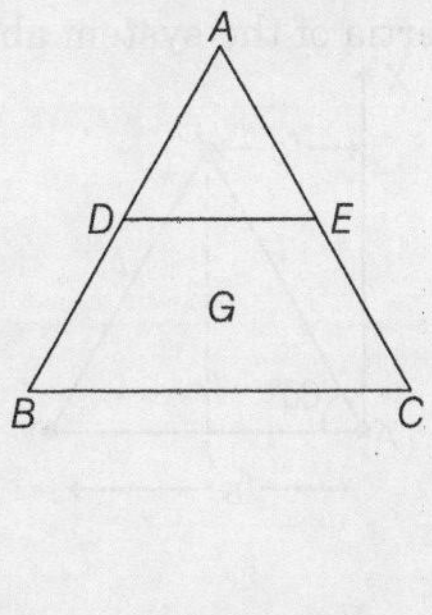

Answers

Round I

1. (d)	2. (d)	3. (d)	4. (c)	5. (c)	6. (a)	7. (b)	8. (d)	9. (b)	10. (c)
11. (d)	12. (d)	13. (c)	14. (d)	15. (b)	16. (c)	17. (d)	18. (c)	19. (d)	20. (b)
21. (a)	22. (c)	23. (c)	24. (b)	25. (b)	26. (d)	27. (a)	28. (d)	29. (d)	30. (d)
31. (d)	32. (d)	33. (b)	34. (d)	35. (a)	36. (c)	37. (b)	38. (b)	39. (c)	40. (c)
41. (b)	42. (a)	43. (b)	44. (d)	45. (a)	46. (b)	47. (b)	48. (b)	49. (c)	50. (d)
51. (d)	52. (b)	53. (b)	54. (b)	55. (d)	56. (d)	57. (a)	58. (b)	59. (a)	60. (d)
61. (d)	62. (d)	63. (c)	64. (c)	65. (a)	66. (c)	67. (d)	68. (b)	69. (b)	70. (b)

Round II

1. (a)	2. (c)	3. (b)	4. (a)	5. (a)	6. (b)	7. (b)	8. (a)	9. (d)	10. (a)
11. (b)	12. (d)	13. (c)	14. (b)	15. (c)	16. (b)	17. (b)	18. (a)	19. (c)	20. (a)
21. (c)	22. (b)	23. (a)	24. (c)	25. (d)	26. (a)	27. (b)	28. (b)	29. (a)	30. (a)
31. (d)	32. (a)	33. (c)	34. (b)	35. (c)	36. 25	37. 195	38. 9	39. 15	40. 20
41. 20	42. 11								

Solutions

Round I

1. Apply parallel axes theorem,

$$I = I_{CM} + Mh^2, \text{ we get}$$

$$\Rightarrow \quad = \frac{ML^2}{12} + M\left(\frac{L}{4}\right)^2 = \frac{7\,ML^2}{48}$$

2. Moment of inertia,

$$I = M\left(\frac{L^2}{12} + \frac{r^2}{4}\right) = M\left(\frac{L^2}{12} + \frac{D^2}{16}\right)$$

3. Moment of inertia of the system about rod x shown in the figure

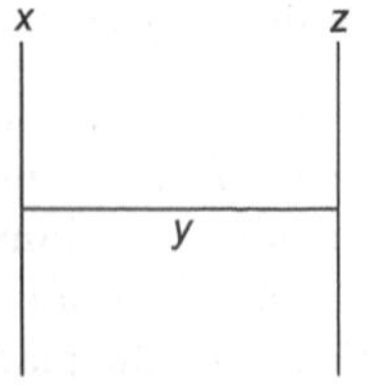

$$I = I_x + I_y + I_z$$

$$= 0 + \left(\frac{Ml^2}{12} + \frac{Ml^2}{4}\right) + Ml^2 = \frac{4}{3}Ml^2$$

4. Moment of inertia of the system about axis AX

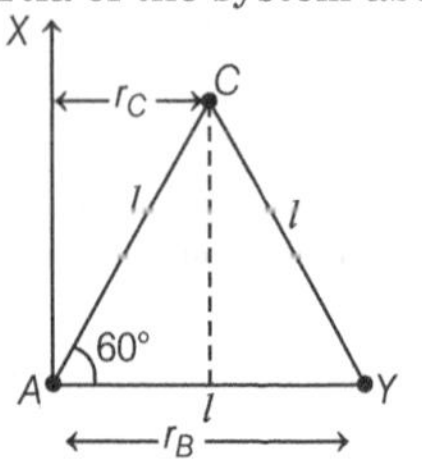

$$= I_A + I_B + I_C$$

$$= M_A(r_A)^2 + M_B(r_B)^2 + M_C(r_C)^2$$

$$= M(0)^2 + m(l)^2 + m(l\cos 60^\circ)^2$$

$$= ml^2 + \frac{ml^2}{4} = \frac{5\,ml^2}{4} \qquad \left(\because \cos 60^\circ = \frac{1}{2}\right)$$

5. Here, $m_1 = m_2 = 0.1$ kg

$$r_1 = r_2 = 10 \text{ cm} = 0.1 \text{ m}$$

and $$I = I_1 + I_2 = m_1 r_1^2 + \frac{1}{2} m_2 r_2^2 = \frac{3}{2} m_1 r_1^2 \qquad (\because m_1 = m_2)$$

$$= \frac{3}{2} \times 0.1\,(0.1)^2 = 1.5 \times 10^{-3} \text{ kg-m}^2$$

6. Moment of inertia of circular wire about its axis is MR^2. Consider two diameters XX' and YY'. Moment of inertia about any of these diameters is same, let us say I.

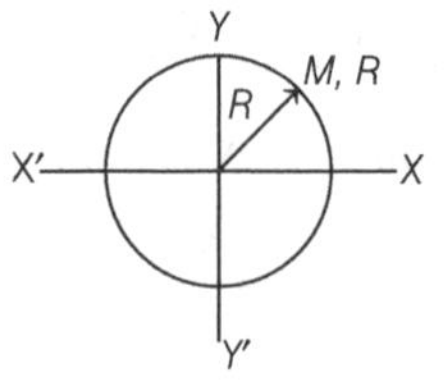

From perpendicular axes theorem,

$$I + I = MR^2 \text{ or } I = \frac{MR^2}{2}$$

7. According to the theorem of perpendicular axes, $I_Z = I_X + I_Y$.

With the hole, I_X and I_Y both decrease gluing the removed piece at the centre of square plate does not affect I_Z. Hence, I_Z decreases, overall.

8. As, $m_1 = m_2$

$$\Rightarrow \quad \pi R_1^2 x d_1 = \pi R_2^2 x d_2$$

$$\frac{R_1^2}{R_2^2} = \frac{d_2}{d_1}$$

Now, $$\frac{I_1}{I_2} = \frac{\frac{1}{2} m R_1^2}{\frac{1}{2} m R_2^2} = \frac{R_1^2}{R_2^2} = \frac{d_2}{d_1}$$

9. Moment of inertia of a uniform rod about one end

$$= \frac{mL^2}{3}$$

$\therefore$ Moment of inertia of the system, which rod is bent

$$= 2 \times \left(\frac{M}{2}\right) \frac{(L/2)^2}{3} = \frac{ML^2}{12}$$

10. The situation is shown in figure

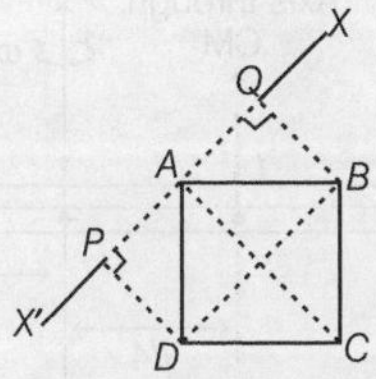

$$I_{XX'} = m \times DP^2 + m \times BQ^2 + m \times CA^2$$

$$= m \times 2 \times \left(\frac{\sqrt{2}\, l}{2}\right)^2 + m \times (\sqrt{2}\, l)^2 = 3ml^2$$

11. The mass of complete (circular) disc is

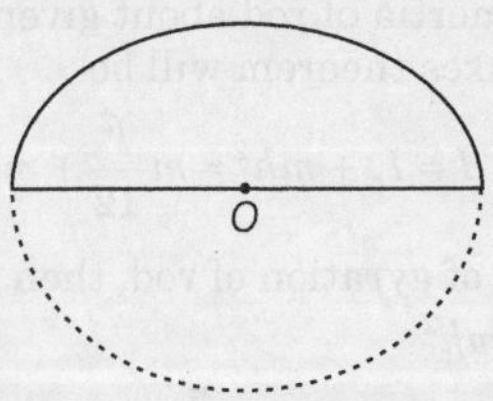

The moment of inertia of disc about the given axis is

$$I = \frac{2\,Mr^2}{2} = Mr^2$$

Let the moment of inertia of semi-circular disc be I_1. The disc may be assumed as combination of two semi-circular parts.

Thus, $\quad I_1 = I - I_1 \quad$ or $\quad I_1 = \frac{I}{2} = \frac{Mr^2}{2}$

12. Here, as ring is made by bending the rod $l = 2\pi R$.

$$R = \frac{l}{2\pi}$$

$$I_1 = \frac{ml^2}{12},$$

$$I_2 = mR^2 = \frac{ml^2}{4\pi^2}$$

$$\frac{I_1}{I_2} = \frac{ml^2}{12} / \frac{ml^2}{4\pi^2} = \frac{\pi^2}{3}$$

13. Let same mass and same outer radii of solid sphere and hollow sphere be M and R, respectively.

The moment of inertia of solid sphere A about its diameter,

$$I_A = \frac{2}{5} MR^2 \quad \ldots\text{(i)}$$

Similarly, the moment of inertia of hollow sphere (spherical shell) B about its diameter,

$$I_B = \frac{2}{3} MR^2 \quad \ldots\text{(ii)}$$

It is clear from Eqs. (i) and (ii), $I_A < I_B$.

14. First we found moment of inertia (MI) of system using parallel axes theorem about centre of mass, then we use it to find moment of inertia about given axis.

Moment of inertia of an outer disc about the axis through centre

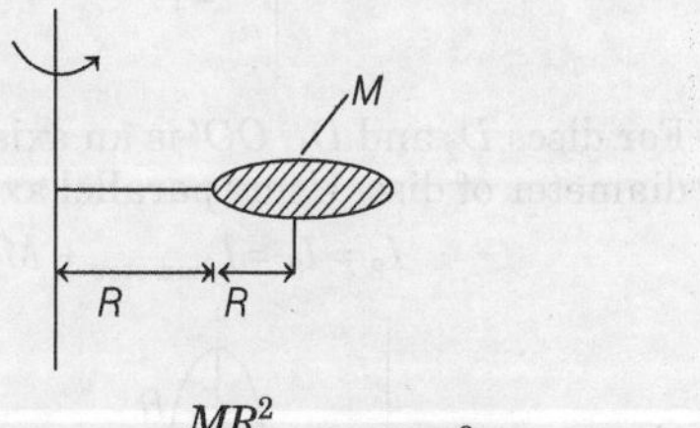

$$= \frac{MR^2}{2} + M(2R)^2$$

$$= MR^2\left(\frac{1}{2} + 4\right) = \frac{9}{2} MR^2$$

For 6 such discs,

$$\text{Moment of inertia} = 6 \times \frac{9}{2} MR^2 = 27MR^2$$

So, moment of inertia of system

$$= \frac{MR^2}{2} + 27MR^2 = \frac{55}{2} MR^2$$

Hence, $$I_P = \frac{55}{2} MR^2 + (7M \times 9R^2)$$

$$\Rightarrow \quad I_P = \frac{181}{2} MR^2$$

$$I_{\text{system}} = \frac{181}{2} MR^2$$

15. Moment of inertia of a solid sphere about an axis through its centre of mass is $I_C = \frac{2}{5} MR^2$

Moment of inertia about a parallel axis at a distance x from axis through its COM is

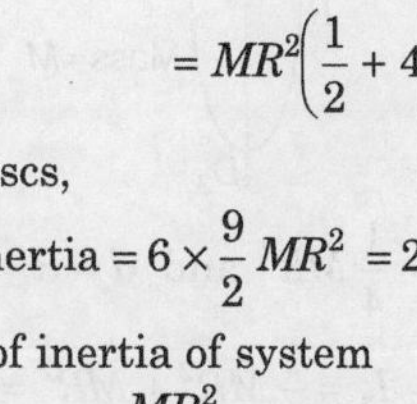

$$I = I_C + Mx^2 \quad \text{(by parallel axes theorem)}$$

So graph of I *versus* x is parabolic as shown

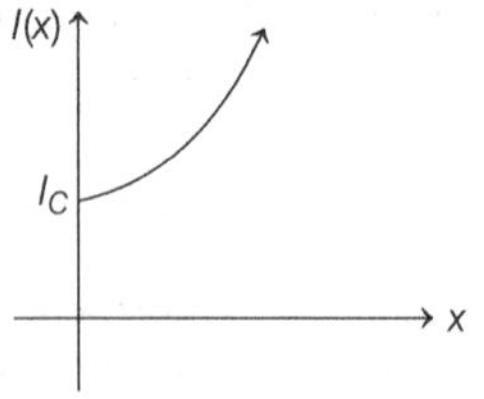

16. For disc D_1, moment of inertia across axis OO' will be

$$I_1 = \frac{1}{2}MR^2 \quad \ldots(i)$$

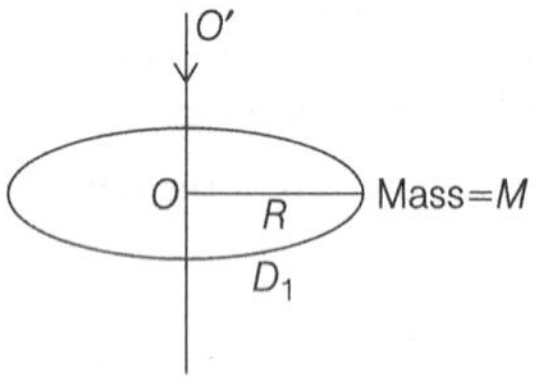

For discs D_2 and D_3, OO' is an axis parallel to the diameter of disc. Using parallel axes theorem,

$$I_2 = I_3 = I_{\text{diameter}} + Md^2 \quad \ldots(ii)$$

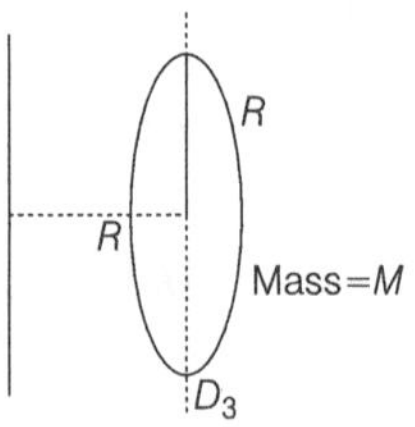

Here, $I_{\text{diameter}} = \frac{1}{4}MR^2$ and $d = R$

$$\therefore \quad I_2 = I_3 = \frac{1}{4}MR^2 + MR^2 = \frac{5}{4}MR^2$$

Now, total moment of inertia of the system,

$$I = I_1 + I_2 + I_3$$
$$= \frac{1}{2}MR^2 + 2 \times \frac{5}{4}MR^2$$
$$= 3MR^2$$

17. Given, surface mass density,

$$\sigma = kr^2$$

So, mass of the disc can be calculated by considering small element of area $2\pi rdr$ on it and then integrating it for complete disc, i.e.

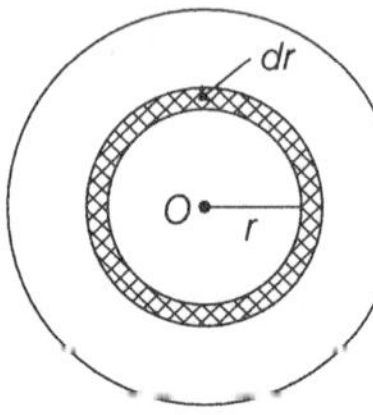

$$dm = \sigma\ dA = \sigma \times 2\pi rdr$$
$$\int dm = M = \int_0^R (kr^2)\, 2\pi rdr$$

$$\Rightarrow \quad M = 2\pi k\frac{R^4}{4} = \frac{1}{2}\pi kR^4$$

$$\Rightarrow \quad k = \frac{2M}{\pi R^4} \quad \ldots(i)$$

Moment of inertia about the axis of the disc,

$$I = \int dI = \int dmr^2 = \int \sigma dAr^2$$
$$= \int_0^R kr^2\,(2\pi rdr)\,r^2$$

$$\Rightarrow \quad I = 2\pi k\int_0^R r^5\ dr = \frac{2\pi kR^6}{6} = \frac{\pi kR^6}{3} \quad \ldots(ii)$$

From Eqs. (i) and (ii), we get

$$I = \frac{2}{3}MR^2$$

18. To calculate radius of gyration, first we calculate moment of inertia of rod about given axis.

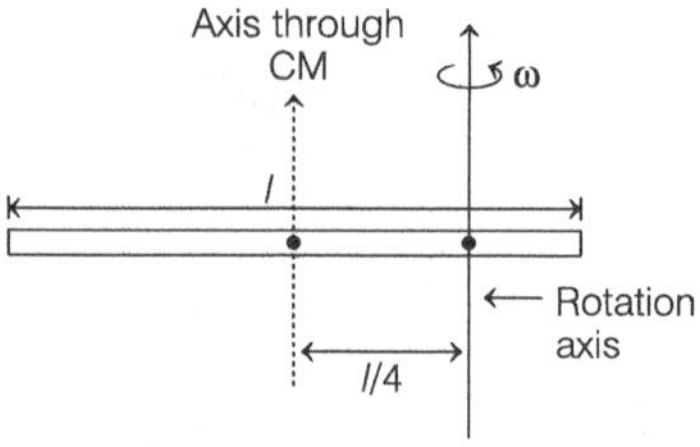

Moment of inertia of rod about an axis through centre of mass, $I_c = \frac{ml^2}{12}$

where, $m =$ mass and $l =$ length of rod.

Moment of inertia of rod about given axis of rotation by parallel axes theorem will be

$$I = I_c + mh^2 = m\frac{l^2}{12} + m\left(\frac{l}{4}\right)^2 = \frac{7}{48}ml^2$$

If $k =$ radius of gyration of rod, then moment of inertia of rod, $I = mk^2$

$$\Rightarrow \quad mk^2 = \frac{7}{48}ml^2$$

$\therefore$ Radius of gyration of rod, $k = \sqrt{\frac{7}{48}}l$

19. Consider an elemental ring of thickness d at a distance from the centre.

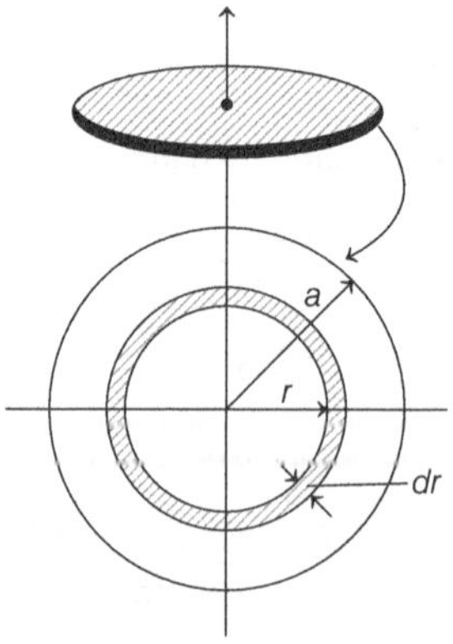

Area of this elemental ring, $dA = 2\pi r dr$

Given, mass per unit area, $\sigma(r) = A + Br$

Mass of elemental ring,

$$dm = \sigma(r) \cdot dA = (A + Br) 2\pi r\, dr$$

Moment of inertia of this elemental ring about axis,

$$dI = dm \cdot r^2$$

$$= 2\pi r^3 \ (A + Br) \cdot dr$$

Moment of inertia of complete disc of radius a,

$$I = \int_0^a dI = \int_0^a 2\pi r^3 (A + Br)\, dr = \left[2\pi A \frac{r^4}{4} + 2\pi B \frac{r^5}{5} \right]_0^a$$

$$= \frac{2\pi A a^4}{4} + \frac{2\pi B a^5}{5} = 2\pi a^4 \left(\frac{A}{4} + \frac{aB}{5} \right)$$

20. According to question,

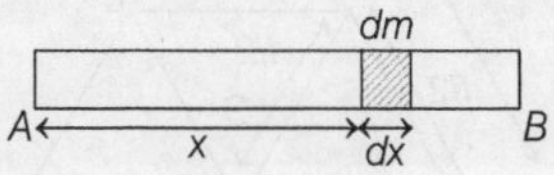

Linear mass density, $\lambda(x) = \lambda_0 \left(1 + \frac{x}{L} \right)$

Moment of inertia of small element about given axis,

$$dI = (dm) x^2 = (\lambda dx) x^2$$

$$\Rightarrow \quad dI = \lambda_0 \left(1 + \frac{x}{L} \right) x^2 dx$$

Total moment of inertia about same axis due to the complete rod,

$$I = \int dI = \int_0^L \lambda_0 \left(1 + \frac{x}{L} \right) x^2 dx = \lambda_0 \left[\frac{x^3}{3} + \frac{x^4}{4L} \right]_0^L$$

$$= \frac{7L^3 \lambda_0}{12} \quad \text{...(i)}$$

Now, mass of complete rod,

$$M = \int dm = \int \lambda dx$$

$$= \int_0^L \lambda_0 \left(1 + \frac{x}{L} \right) dx$$

$$= \lambda_0 \left[x + \frac{x^2}{2L} \right]_0^L$$

$$= \lambda_0 \left[L + \frac{L^2}{2L} - 0 \right] = \frac{3\lambda_0 L}{2}$$

$$\Rightarrow \quad \lambda_0 = \frac{2M}{3L}$$

Putting the value of λ_0 in Eq. (i), we get

$$I = \frac{7}{18} ML^2$$

21. Both the discs are made up of same material.

Let the density of material used for both discs be ρ and their thickness be t.

Now, moment of inertia of first disc,

$$I_1 = \frac{M_1 R_1^2}{2} = \frac{\rho A_1 t R_1^2}{2}$$

$$= \frac{\rho (\pi R_1^2) t R_1^2}{2} = \frac{\pi \rho R_1^4 t}{2}$$

$$\because \quad R_1 = R \quad \text{(given)}$$

$$\therefore \quad I_1 = \frac{\pi \rho R^4 t}{2} \quad \text{...(i)}$$

and moment of inertia of second disc,

$$I_2 = \frac{M_2 R_2^2}{2} = \frac{\rho A_2 t R_2^2}{2}$$

$$= \frac{\rho (\pi R_2^2) t R_2^2}{2} = \frac{\pi \rho R_2^4 t}{2}$$

$$\because \quad R_2 = \alpha R \quad \text{(given)}$$

$$\therefore \quad I_2 = \frac{\pi \rho (\alpha R)^4 t}{2} = \frac{\alpha^4 \pi \rho R^4 t}{2} \quad \text{...(ii)}$$

Dividing Eq. (i) by Eq. (ii), we get

$$\frac{I_1}{I_2} = \frac{\left(\frac{\pi \rho R^4 t}{2} \right)}{\left(\frac{\alpha^4 \pi \rho R^4 t}{2} \right)}$$

$$\Rightarrow \quad \frac{I_1}{I_2} = \frac{1}{\alpha^4} \quad \text{...(iii)}$$

It is given that,

$$\frac{I_1}{I_2} = \frac{1}{16} \quad \text{...(iv)}$$

On comparing the Eqs. (iii) and (iv), we get

$$\frac{1}{\alpha^4} = \frac{1}{16}$$

$$\Rightarrow \quad \alpha = (16)^{1/4}$$

$$= (2^4)^{1/4} = 2$$

22.

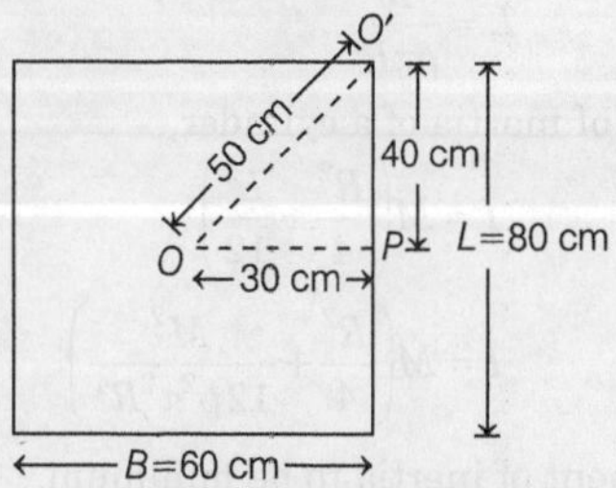

Let the mass of this rectangular sheet be M.

From figure,

$$O'P = \frac{L}{2} = \frac{80}{2} \text{ cm} = 40 \text{ cm}$$

$$OP = \frac{B}{2} = \frac{60}{2} \text{ cm} = 30 \text{ cm}$$

$$OO' = \sqrt{(OP)^2 + (O'P)^2}$$

(by Pythagoras' theorem)

$$= \sqrt{(30)^2 + (40)^2}$$

$$= \sqrt{900 + 1600}$$

$$= \sqrt{2500} = 50 \text{ cm}$$

Now, moment of inertia about the axis perpendicular to the plane of sheet and passing through point O is

$$I_O = \frac{M}{12}[L^2 + B^2] = \frac{M}{12}[(80)^2 + (60)^2]$$
$$= \frac{M}{12}[6400 + 3600]$$
$$= \frac{M}{12}[10000] \quad \text{...(i)}$$

and moment of inertia about the axis perpendicular to the plane of sheet and passing through point O' is

$I_{O'} = I_O + Md^2$ (using theorem of parallel axes)

Here, $d = OO' = 50$ cm

So, $$I_{O'} = \frac{M}{12}[10000] + M[50]^2$$
$$= \frac{M}{12}[10000] + M[2500]$$
$$= \frac{M[10000] + M[30000]}{12}$$
$$= \frac{M}{12}[40000] \quad \text{...(ii)}$$

Dividing Eq. (i) by Eq. (ii), we get

$$\frac{I_O}{I_{O'}} = \frac{\frac{M}{12}[10000]}{\frac{M}{12}[40000]} = \frac{1}{4}$$

23. 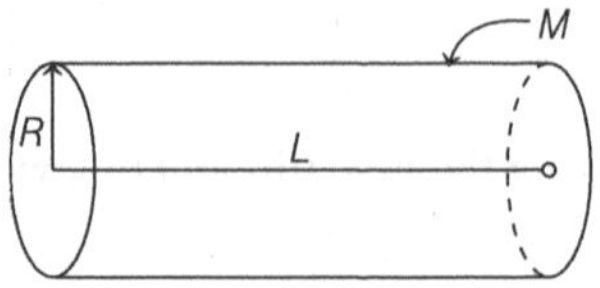

Mass of cylinder, $M = \rho\pi R^2 L$ ...(i)

(where, ρ = density of material)

$$\Rightarrow \quad L = \frac{M}{\rho\pi R^2} \quad \text{...(ii)}$$

Moment of inertia of a cylinder,

$$I = M\left(\frac{R^2}{4} + \frac{L^2}{12}\right)$$

$$\Rightarrow \quad I = M\left(\frac{R^2}{4} + \frac{M^2}{12\rho^2\pi^2R^4}\right) \quad \text{[from Eq. (ii)]}$$

For moment of inertia to be minimum,

$$\frac{dI}{dR} = 0$$

$$\Rightarrow \quad \frac{d}{dR}\left[M\left(\frac{R^2}{4} + \frac{M^2}{12\rho^2\pi^2R^4}\right)\right] = 0$$

$$\Rightarrow \quad \frac{2R}{4} + \frac{M^2}{12\rho^2\pi^2}\cdot(-4R^{-5}) = 0$$

[$\because M$ = constant]

$$\Rightarrow \quad \frac{2R}{4} = \frac{4M^2}{12\rho^2\pi^2R^5}$$

$$\Rightarrow \quad R^6 = \frac{2}{3}\frac{M^2}{\rho^2\pi^2}$$

$$\Rightarrow \quad R^6 = \frac{2}{3}\cdot\frac{\rho^2\pi^2R^4L^2}{\rho^2\pi^2} \quad \text{[from Eq. (i)]}$$

$$\Rightarrow \quad \left(\frac{R}{L}\right)^2 = \frac{2}{3}$$

$$\Rightarrow \quad \frac{L}{R} = \sqrt{\frac{3}{2}}$$

24. Suppose the mass of the ΔABC be M and length of the side be l.

When the ΔDEF is being removed from it, then the mass of the removed Δ will be $M/4$ and length of its side will be $l/2$ as shown below

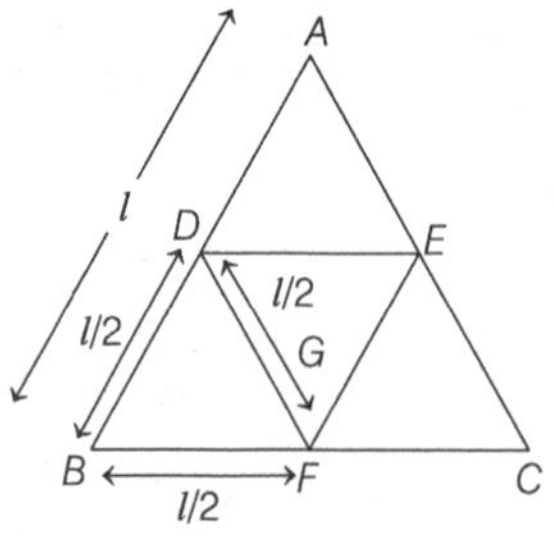

Since we know, moment of inertia of the triangle about the axis, passing through its centre of gravity is $I = kml^2$, where k is a constant.

Then for ΔDEF, moment of inertia of the triangle about the axis,

$$I = k\left(\frac{M}{4}\right)\left(\frac{l}{2}\right)^2 = \frac{kMl^2}{16} \quad \text{...(i)}$$

Moment of inertia of ΔABC is

$$I_0 = kMl^2 \quad \text{...(ii)}$$

The moment of inertia of the remaining part will be

$$I' = I_0 - I = kMl^2 - \frac{kMl^2}{16}$$

[$\because$ using Eqs. (i) and (ii)]

$$= \frac{15kMl^2}{16}$$

or $$I' = \frac{15}{16}I_0$$

25. As, force $\times$ distance $= \tau = I\alpha$

$$\Rightarrow \quad F\sin 30° \times \frac{1}{2} = 4\left[2\times\left(\frac{1}{4}\right)^2\right]\alpha$$

or $$24\times\frac{1}{2}\times\frac{1}{2} = \frac{\alpha}{2}$$

$\therefore$ $$\alpha = 12 \text{ rad s}^{-2}$$

26. As the block remains stationary, therefore for translatory equilibrium,

$$\Sigma f_x = 0 \quad \therefore \quad F = N$$

and $$\Sigma f_y = 0 \quad \therefore \quad F = mg$$

For rotational equilibrium, $\Sigma\tau = 0$.

By taking the torque of different force about point O,

$$\tau_f + \tau_f + \tau_N + \tau_{mg} = 0$$

As f and mg passing through point O

$$\therefore \quad \tau_f + \tau_N = 0$$

As $\tau_f \neq 0$

$\therefore \tau_N \neq 0$ and torque by friction and normal reaction will be in opposite direction or N will not produce torque.

27. The given condition can be drawn in the figure below

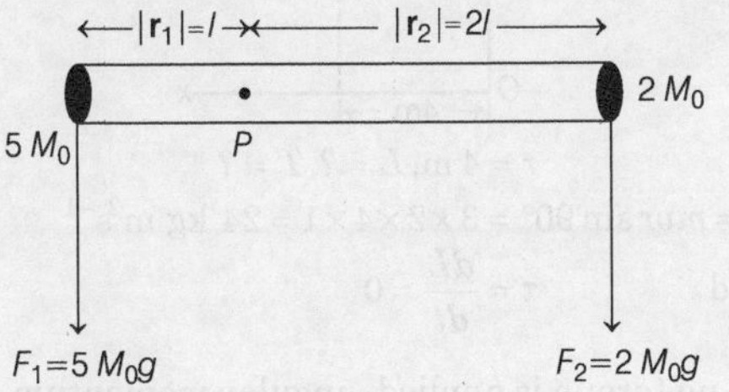

Torque (τ) about $P = \mathbf{r}_1 \times \mathbf{F}_1 + \mathbf{r}_2 \times \mathbf{F}_2$ ... (i)

$\Rightarrow \quad \tau = l \times 5M_0 g$ (outwards) $- 2l \times 2M_0 g$ (inwards)

$\Rightarrow \quad \tau = 5M_0 gl - 4M_0 gl$ (outwards)

$\Rightarrow \quad \tau = M_0 gl$ (outwards) ...(ii)

Now we know that, torque is also given by

$$\tau = I\alpha \quad \text{...(iii)}$$

Here, I = moment of inertia (w.r.t. point P) of rod and α = angular acceleration.

For point P, $\quad I = (5M_0) \times l^2 + (2M_0)(2l)^2 \quad [\because I = MR^2]$

$\Rightarrow \quad I = 13M_0 l^2$ ...(iv)

Putting value of I from Eq. (iv) in Eq. (iii), we get

$$\tau = (13M_0 l^2)\,\alpha \quad \text{...(v)}$$

From Eqs. (ii) and (v), we get

$$M_0 gl = 13M_0 l^2 \alpha$$

$$\Rightarrow \quad \alpha = \frac{g}{13l}$$

28. As the rod rotates in vertical plane, so a torque is acting on it, which is due to the vertical component of weight of rod.

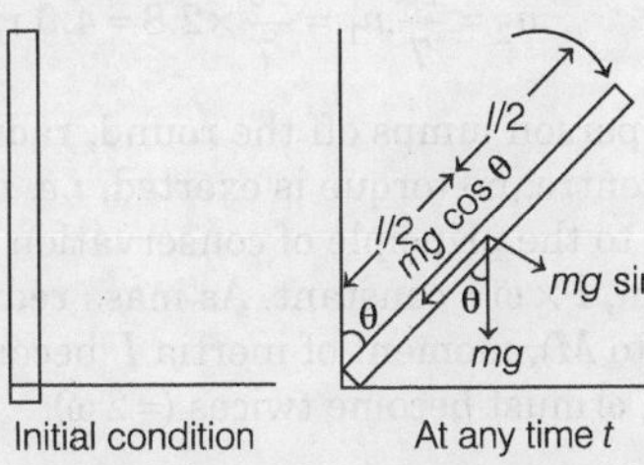

Now, torque τ = force × perpendicular distance of line of action of force from axis of rotation

$$= mg \sin\theta \times \frac{l}{2}$$

Again, torque, $\tau = I\alpha$

where, I = moment of inertia $= \dfrac{ml^2}{3}$

[force and torque frequency along axis of rotation passing through in end]

α = angular acceleration

$$\therefore \quad mg \sin\theta \times \frac{l}{2} = \frac{ml^2}{3}\alpha$$

$$\therefore \quad \alpha = \frac{3g \sin\theta}{2l}$$

29. Given equations of trajectory are

$$x = x_0 + a \cos\omega_1 t$$
$$y = y_0 + b \sin\omega_2 t$$

Now, $v_x = dx/dt = -a\omega_1 \sin\omega_1 t$

and $\quad v_y = \dfrac{dy}{dt} = b\omega_2 \cos\omega_2 t$

Similarly, $a_x = \dfrac{dv_x}{dt} = -a\omega_1^2 \cos(\omega_1 t)$

Similarly, $a_y = \dfrac{dv_y}{dt} = -b\omega_2^2 \sin(\omega_2 t)$

Now, at $t = 0$; $x = x_0 + a$; $y = y_0$ and

$$a_x = -a\omega_1^2;\ a_y = 0 \quad \text{... (i)}$$

Torque acting on the particle is given as

$$\tau = \mathbf{r} \times \mathbf{F} = m(\mathbf{r} \times \mathbf{a})$$

Here, $\mathbf{r} = (x_0 + a)\hat{\mathbf{i}} + y_0\hat{\mathbf{j}} + 0\hat{\mathbf{k}}$

and $\quad \mathbf{a} = -a\omega_1^2\hat{\mathbf{i}} + 0\hat{\mathbf{j}} + 0\hat{\mathbf{k}}$

So, torque at $t = 0$,

$$\tau = m(-a\omega_1^2) \times y_0(-\hat{\mathbf{k}}) \quad \text{[using Eq. (i)]}$$

$$\Rightarrow \quad \tau = +my_0 a\omega_1^2\hat{\mathbf{k}}$$

30. Force F is applied to cylinder as shown

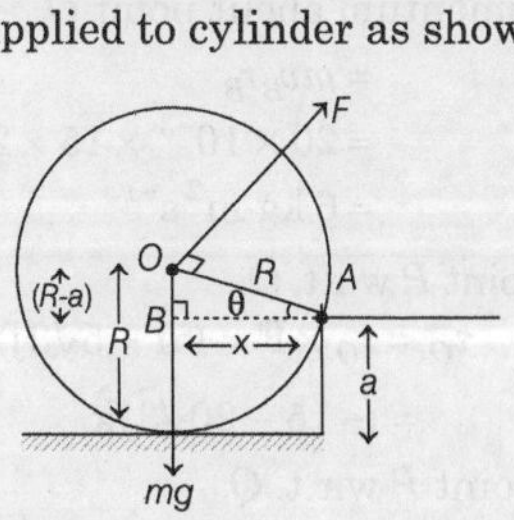

Force F must be large enough to create a torque about point A, so that it is able to counteract the torque due to weight about point A.

$\therefore$ Torque due to force about point A = Torque due to weight about point A

$$\Rightarrow \quad F \cdot R = Mgx$$

From ΔOAB, $x = \sqrt{R^2 - (R-a)^2}$

$$\Rightarrow \quad F \cdot R = Mg\sqrt{R^2 - (R-a)^2}$$

$$\Rightarrow \quad F = \frac{Mg}{R}\sqrt{R^2 - (R-a)^2}$$

$$\Rightarrow \quad F = Mg\sqrt{1 - \left(\frac{R-a}{R}\right)^2}$$

31. Here, $r = 0.2$ m, $M = 10$ kg,

$$n = 1200 \text{ rpm} = 20 \text{ rps}$$
$$L = I\omega = (Mr^2)(2\pi n)$$
$$= 10 \times (0.2)^2 \times 2 \times \frac{22}{7} \times 20$$
$$= 50.28 \text{ kg-m}^2\text{s}^{-1}$$

32. Angular momentum, $\mathbf{L} = m(\mathbf{r} \times \mathbf{v})$

The direction of $\mathbf{L}$ with respect to A is along the Z-axis and magnitude is mvr. So, L_A is constant, both in magnitude and direction.

But the direction of $\mathbf{L}$ with respect to B is continuously changing with time with constant magnitude.

33. Angular momentum

= linear momentum × perpendicular distance of line of action of linear momentum from the axis of rotation

$= mv \times l$

34. The given figure is shown below as

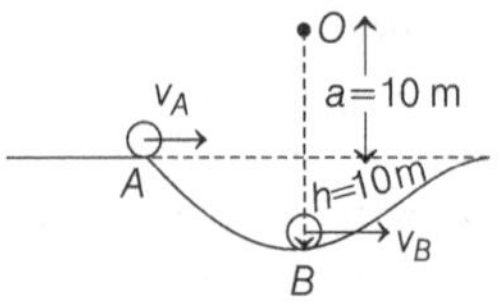

As friction is absent, energy at A = energy at B

$$\Rightarrow \quad \frac{1}{2}mv_A^2 + mgh = \frac{1}{2}mv_B^2$$
$$\Rightarrow \quad v_A^2 + 2gh = v_B^2$$
or $$v_B^2 = (5)^2 + 2 \times 10 \times 10 = 225$$
$$\Rightarrow \quad v_B = 15 \text{ ms}^{-1}$$

Angular momentum about point O,

$$= mv_B r_B$$
$$= 20 \times 10^{-3} \times 15 \times 20$$
$$= 6 \text{ kg-m}^2\text{s}^{-1}$$

35. Torque at point P w.r.t. O,

$$\tau_O = \mathbf{r}_O \times \mathbf{F} = (5\hat{\mathbf{i}} + 5\sqrt{3}\hat{\mathbf{j}}) \times (4\hat{\mathbf{i}} - 3\hat{\mathbf{j}})$$
$$= (-15 - 20\sqrt{3})\hat{\mathbf{k}}$$

Torque at point P w.r.t. Q,

$$\tau_Q = \mathbf{r}_Q \times \mathbf{F} = (-5\hat{\mathbf{i}} + 5\sqrt{3}\hat{\mathbf{j}}) \times (4\hat{\mathbf{i}} - 3\hat{\mathbf{j}})$$
$$= (15 - 20\sqrt{3})\hat{\mathbf{k}}$$

36. Conservation of angular momentum gives

$$\frac{1}{2}MR^2\omega_1 = \left(\frac{1}{2}MR^2 + 2mR^2\right)\omega_2$$
$$\Rightarrow \quad \frac{1}{2}MR^2\omega_1 = \frac{1}{2}R^2(M + 4m)\omega_2$$
or $$\omega_2 = \left(\frac{M}{M + 4m}\right)\omega_1$$

37. By conservation of energy,

$$mgh = \frac{1}{2}I\omega^2 + \frac{1}{2}mv^2$$
$$= \frac{1}{2}I\omega^2 + \frac{1}{2}mr^2\omega^2 = \frac{\omega^2}{2}[I + mr^2]$$
$$\omega = \left(\frac{2mgh}{I + mr^2}\right)^{1/2}$$

38. Here, $m = 3$ kg, $v = 2$ ms^{-1}

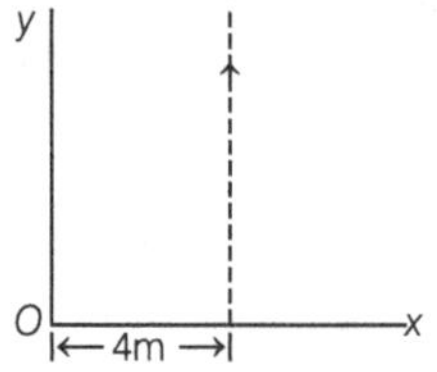

$$r = 4 \text{ m}, L = ?, T = ?$$
$$L = mvr\sin 90° = 3 \times 2 \times 4 \times 1 = 24 \text{ kg m}^2\text{s}^{-1}$$
and $$\tau = \frac{dL}{dt} = 0$$

39. As no torque is applied, angular momentum

$$L = I\omega = \text{constant} = \left(\frac{2}{5}MR^2\right)\left(\frac{2\pi}{T}\right) = \text{constant. } i.e.$$
$$\Rightarrow \quad \frac{R^2}{T} = \text{constant}$$
$$\Rightarrow \quad \frac{R_1^2}{T_1} = \frac{R_2^2}{T_2}$$
$$\therefore \quad T_2 = \left(\frac{R_2}{R_1}\right)^2 T_1 = \left(\frac{xR_1}{R_1}\right)^2 \times 24\,h = 24x^2h$$

40. According to conservation of angular momentum,

$$I_1\omega_1 = I_2\omega_2$$
$$\Rightarrow \quad \frac{1}{2}MR^2\omega = \left(\frac{1}{2}MR^2 + \frac{1}{2}\left\{\frac{M}{4}\right\}R^2\right)\omega_2$$
$$\therefore \quad \omega_2 = \frac{4}{5}\omega$$

41. Here, $n_1 = 2.8$ rps, $n_2 = ?$

$$I_2 = 0.7\,I_1$$
As, $$\frac{\omega_2}{\omega_1} = \frac{I_1}{I_2} = \frac{1}{0.7}$$
$$\therefore \quad \frac{n_2}{n_1} = \frac{10}{7} \qquad (\because \omega = 2\pi n)$$
$$n_2 = \frac{10}{7}n_1 = \frac{10}{7} \times 2.8 = 4.0 \text{ rps}$$

42. When the person jumps off the round, radially away from the centre, no torque is exerted, *i.e.* $\tau = 0$. According to the principle of conservation of angular momentum, $I \times \omega$ = constant. As mass reduces to half (from $2M$ to M), moment of inertia I becomes half. Therefore, ω must become twices (= 2ω).

43. As $L - I\omega -$ constant

$$\therefore \quad \frac{2}{5}MR^2 \times \left(\frac{2\pi}{7}\right) = \text{constant}$$
i.e. $$\frac{R^2}{T} = \text{constant}$$

where R is halved, R^2 becomes (1/4)th. Therefore, T becomes 1/4th, *i.e.* 6 h.

44. Here, mass of man, $m = 80$ kg
Mass of platform, $M = 200$ kg
Let R be the radius of platform.
When man is standing on the rim,

$$I_1 = M(R/2)^2 + mR^2 = \left(\frac{R}{2}\right)^2 (M + 4m)$$

When man reaches the centre of platform,

$$I_2 = M(R/2)^2 + m \times 0 = M(R/2)^2$$

As angular momentum is conserved,

$$\frac{I_1}{I_2} = \frac{\omega_2}{\omega_1} = \frac{2\pi n_2}{2\pi n_1} = \frac{n_2}{n_1}$$

$$n_2 = \frac{I_1}{I_2} \times n_2 = \frac{(M+4m)(R/2)^2}{M(R/2)^2} \times 12 = \frac{(200 + 4 \times 80)}{200} \times 12 = \frac{520 \times 12}{200}$$

$$n_1 = 31.2 \text{ rpm}$$

45. Collision of particle and rod occurs as shown in the figure

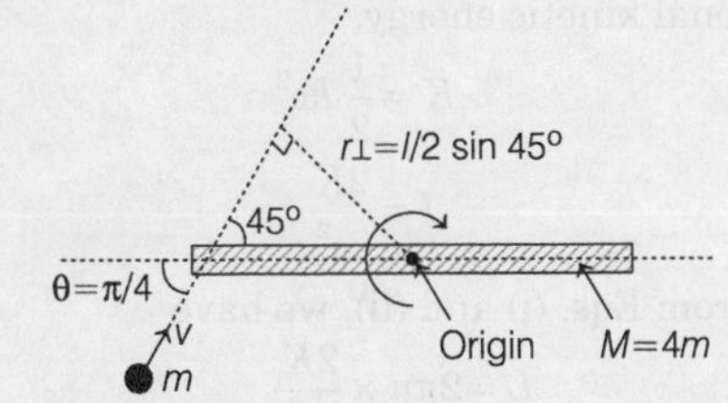

Angular momentum of particle about origin (centre of rod) before collision,

$$L_i = m \cdot v \cdot r_\perp$$

where, $r_\perp$ = perpendicular distance of direction of velocity from centre of rod.
From geometry of figure,

$$r_\perp = \frac{l}{2} \sin 45° = \frac{l}{2\sqrt{2}}$$

So, initial angular momentum of system,

$$L_i = m \cdot v \cdot \frac{l}{2\sqrt{2}}$$

Now, when mass m stuck to one of the ends of rod, let system rotates about origin with initial angular speed ω, then final angular momentum of system just after collision,

$$L_f = I_{\text{system}} \cdot \omega = (I_{\text{rod}} + I_{\text{particle}})\,\omega = \left(\frac{4m \cdot l^2}{12} + m\left(\frac{l}{2}\right)^2\right)\omega = \frac{7}{12} m l^2 \omega$$

By angular momentum conservation,

$$L_i = L_f$$

$$\Rightarrow \quad \frac{mvl}{2\sqrt{2}} = \frac{7}{12} m l^2 \omega$$

$$\Rightarrow \quad \omega = \frac{3\sqrt{2}}{7}\left(\frac{v}{l}\right)$$

46. As, $I = 1.2$ kg-m², $E_r = 1500$ J, $\alpha = 25$ rad s^{-2} $\omega_1 = 0$, $t = ?$

Now, $E_r = \frac{1}{2} I\omega^2$

$$\Rightarrow \quad \omega = \sqrt{\frac{2E_r}{I}} = \sqrt{\frac{2 \times 1500}{1.2}} = 50 \text{ rad s}^{-1}$$

From, $\omega_2 = \omega_1 + \alpha t$

$$50 = 0 + 25t$$

$$\Rightarrow \quad t = 2 \text{ s}$$

47. Work done in stopping = Change in KE = Final KE – Initial KE, *i.e.* $\tau\theta = K$ = constant. As τ is same in the two cases, θ must be same, *i.e.* number of revolution must be same.
Then, total number of revolutions made by disc when it stops = $n + n = 2n$.

48. Here, $r = 0.25$ m and $m = 2$ kg

$$\text{Rotational KE} = \frac{1}{2} I\omega^2 = \frac{1}{2} \times \left(\frac{1}{2} mr^2\right)\omega^2$$

$$\Rightarrow \quad 4 = \frac{1}{4} mv^2 = \frac{1}{4} \times 2v^2$$

$$\therefore \quad v = \sqrt{8} = 2\sqrt{2} \text{ ms}^{-1}$$

49. As $\omega_f = \omega_i + \alpha t$
Here, $0 = \omega_0 - \alpha t$

$$\therefore \quad \alpha = \frac{\omega_0}{t} = \frac{(100 \times 2\pi)/60}{15} = 0.7 \text{ rad s}^{-2}$$

Angle rotated before coming to rest,

$$\theta = \frac{\omega_0^2}{2\alpha}$$

or

$$\theta = \frac{\left(\frac{100 \times 2\pi}{60}\right)^2}{2 \times 0.7} = 78.33 \text{ rad}$$

Hence, number of rotations, $n = \frac{\theta}{2\pi} = \frac{78.33}{2\pi} = 12.5$

50. By conservation of energy,
Initial total energy = Final total energy

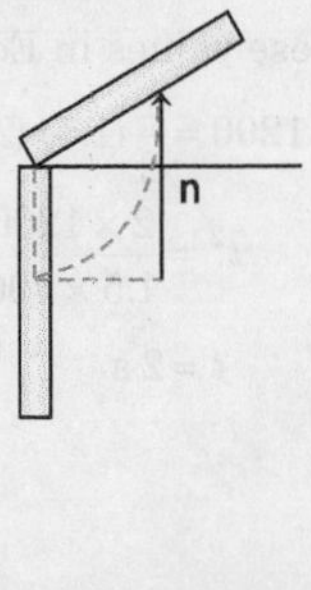

$$(TE)_i = (TE)_f$$
$$\frac{1}{2}I\omega^2 = mgh$$
$$\frac{1}{2}\times\frac{1}{3}ml^2\omega^2 = mgh$$
$$\Rightarrow \quad h = \frac{1}{6}\frac{l^2\omega^2}{g}$$

51. Mass of oxygen molecule, $M = 5.30\times 10^{-26}$ kg

Moment of inertia, $I = 1.94\times 10^{-46}$ kg-m^2

Mean speed of the molecule, $v = 500$ m/s

Given, KE of rotation $= \frac{2}{3}\times$ KE of translation

$$\frac{1}{2}I\omega^2 = \frac{2}{3}\times\frac{1}{2}Mv^2$$

or
$$\omega = \sqrt{\frac{2Mv^2}{I}}$$
$$= \sqrt{\frac{2\times 5.30\times 10^{-26}\times(500)^2}{1.94\times 10^{-46}}}$$
$$= 1.35\times 10^{10}\times 500$$
$$= 6.75\times 10^{12}\text{ rad/s}$$

52. Rotational kinetic energy, $K_R = \frac{1}{2}I\omega^2$

$$K_R = \frac{1}{2}\times\frac{MR^2}{2}\times\omega^2 = \frac{1}{4}Mv^2 \qquad (\because v = R\omega)$$

Translational kinetic energy, $K_T = \frac{1}{2}Mv^2$

Total kinetic energy $= K_T + K_R = \frac{1}{2}Mv^2 + \frac{1}{4}Mv^2$

$$= \frac{3}{4}Mv^2$$

$$\therefore \quad \frac{\text{Rotational kinetic energy}}{\text{Total kinetic energy}} = \frac{\frac{1}{4}Mv^2}{\frac{3}{4}Mv^2} = \frac{1}{3}$$

53. Rotational kinetic energy of a body is given by

$$KE_{\text{rotational}} = \frac{1}{2}I\omega^2$$

where, $\omega = \omega_0 + \alpha t$

So, $KE_{\text{rotational}} = \frac{1}{2}I(\omega_0 + \alpha t)^2$ ... (i)

Here, $I = 1.5$ kgm^2,

KE $= 1200$ J

and $\alpha = 20$ rad/s^2 and $\omega_0 = 0$

Substituting these values in Eq.(i), we get

$$1200 = \frac{1}{2}(1.5)(20\times t)^2$$
$$\Rightarrow \quad t^2 = \frac{2\times 1200}{1.5\times 400} = 4$$
$$\therefore \quad t = 2\text{ s}$$

54. Here, $K_f = \frac{1}{2}(I_1 + I_2)\left[\frac{(I_1\omega_1 + I_2\omega_2)^2}{(I_1 + I_2)^2}\right]$

$$= \frac{1}{2}\frac{(I_1\omega_1 + I_2\omega_2)^2}{I_1 + I_2}$$

and $K_i = \frac{1}{2}(I_1\omega_1^2 + I_2\omega_2^2)$

$$\Rightarrow \quad \Delta K = K_f - K_i$$
$$= -\frac{I_1 I_2}{2(I_1 + I_2)}(\omega_1 - \omega_2)^2$$

55. As, the rod is hinged at one end, its moment of inertia about this end is $I = \frac{ML^2}{3}$.

Total energy in upright position

= Total energy on striking the floor

$$0 + \frac{MgL}{2} = \frac{1}{2}I\omega^2 + 0 = \frac{1}{2}\frac{ML^2}{3}\omega^2$$
$$\Rightarrow \quad g = \frac{L\omega^2}{3}$$
$$\text{or} \quad \omega = \sqrt{\frac{3g}{L}}$$

56. Given, angular momentum,

$$L = I\omega$$
$$\Rightarrow \quad L = I(2\pi n) \qquad [\because \omega = 2\pi n]$$
$$\Rightarrow \quad L = 2\pi n I \qquad \text{...(i)}$$

Rotational kinetic energy,

$$K = \frac{1}{2}I\omega^2$$
$$\Rightarrow \quad I = \frac{2K}{\omega^2} \qquad \text{...(ii)}$$

Now, from Eqs. (i) and (ii), we have

$$L = 2\pi n\times\frac{2K}{\omega^2}$$
$$= 2\pi n\times\frac{2K}{(2\pi n)^2}$$
$$L = \frac{K}{\pi n}$$
$$\Rightarrow \quad \frac{L_2}{L_1} = \frac{K_2}{K_1}\times\frac{n_1}{n_2}$$
$$= \frac{K_1/2}{K}\times\frac{n_1}{2n_1} \qquad \left[\because K_2 = \frac{K_1}{2} \text{ and } n_2 = 2n_1\right]$$
$$\Rightarrow \quad \frac{L_2}{L_1} = \frac{1}{4}$$
$$\Rightarrow \quad L_2 = \frac{L_1}{4} = \frac{L}{4} \qquad [\because L_1 = L]$$

57. Let ω_c be the final angular speed.

According to law of conservation of angular momentum,

Initial angular momentum = Final angular momentum

$$I\omega + 0 = I\omega_c + 3I\omega_c$$

$$I\omega = 4I\omega_c$$

$$\Rightarrow \quad \omega_c = \frac{\omega}{4}$$

Loss of rotational kinetic energy,

$$(\Delta KE)_{loss} = \text{Initial KE} - \text{Final KE}$$

$$\Rightarrow \quad (\Delta KE)_{loss} = \frac{1}{2}I\omega^2 - \frac{1}{2}(I + 3I)\omega_c^2$$

$$= \frac{1}{2}I\omega^2 - \frac{1}{2}(I + 3I)\left(\frac{\omega}{4}\right)^2$$

$$= \frac{1}{2}I\omega^2 - \frac{1}{2}(4I)\left(\frac{\omega}{4}\right)^2$$

$$= \frac{1}{2}I\omega^2\left(1 - \frac{1}{4}\right)$$

$$= \frac{3}{8}I\omega^2$$

The resultant fractional loss in KE of the system

$$= \frac{(\Delta KE)_{loss}}{(KE)_{initial}}$$

$$= \frac{(3/8)I\omega^2}{(1/2)I\omega^2} \quad \left[\because (KE)_{initial} = \frac{1}{2}I\omega^2\right]$$

$$= \frac{3}{4}$$

58. The given situation is shown in figure

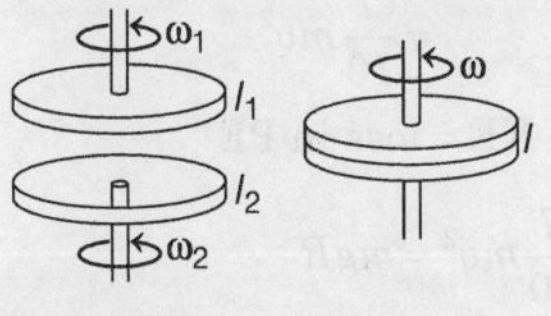

As there is no external torque, angular momentum of system is conserved.

i.e. $L_1 + L_2 = L_{system}$

$$\Rightarrow \quad I_1\omega_1 + I_2\omega_2 = (I_1 + I_2)\omega$$

$$\Rightarrow \quad \omega = \frac{I_1\omega_1 + I_2\omega_2}{I_1 + I_2}$$

$$\omega = \frac{0.1 \times 10 + 0.2 \times 5}{(0.1 + 0.2)} \text{ or } \omega = \frac{20}{3} \text{ rad/s}$$

Kinetic energy of combined system is

$$K = \frac{1}{2}I_{system}\omega^2$$

$$= \frac{1}{2}(0.1 + 0.2) \times \left(\frac{20}{3}\right)^2 = \frac{20}{3} \text{ J}$$

59. Initial kinetic energy of the given system,

$$(KE)_i = \frac{1}{2}I_1\omega_1^2 + \frac{1}{2}\left(\frac{I_1}{2}\right)\left(\frac{\omega_1}{2}\right)^2$$

$$= \left(\frac{1}{2} + \frac{1}{16}\right)I_1\omega_1^2 = \frac{9}{16}I_1\omega_1^2 \quad \text{...(i)}$$

Now, using angular momentum conservation law (assuming angular velocity after contact is ω)

Initial angular momentum = Final angular momentum

$$I_1\omega_1 + \left(\frac{I_1}{2}\right)\left(\frac{\omega_1}{2}\right) = I_1\omega_2' + \frac{I_1}{2}\omega_2'$$

$$\Rightarrow \quad \frac{5}{4}\omega_1 = \frac{3}{2}\omega_2'$$

$$\text{or} \quad \omega_2' = \frac{5}{6}\omega_1 \quad \text{... (ii)}$$

Now, final kinetic energy (after contact) is

$$(KE)_f = \frac{1}{2}I_1\omega_2'^2 + \frac{1}{2}\left(\frac{I_1}{2}\right)\omega_2'^2$$

$$= \frac{1}{2}I_1\left(\frac{5}{6}\omega_1\right)^2 + \frac{1}{4}I_1\left(\frac{5}{6}\omega_1\right)^2 \quad \text{[using Eq. (ii)]}$$

$$= \left(\frac{25}{72} + \frac{25}{144}\right)I_1\omega_1^2$$

$$= \frac{25}{48}I_1\omega_1^2 \quad \text{...(iii)}$$

Hence, change in KE,

$$\Delta KE = (KE)_f - (KE)_i$$

$$= \frac{25}{48}I_1\omega_1^2 - \frac{9}{16}I_1\omega_1^2 \quad \text{[using Eq. (i)]}$$

$$\Delta KE = -\frac{1}{24}I_1\omega_1^2$$

60. Loss of gravitational potential energy of system appears in the form of kinetic energy of system.

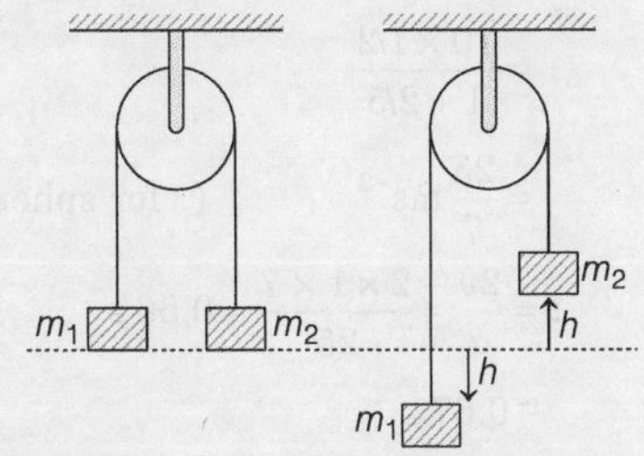

Loss of potential energy of masses
= Kinetic energy of masses + Kinetic energy of rotation of wheel

$$\Rightarrow \quad m_1gh - m_2gh = \frac{1}{2}m_1v^2 + \frac{1}{2}m_2v^2 + \frac{1}{2}I\omega^2$$

Assuming no slip, we have

$$v = \omega R$$

$$\therefore \quad (m_1 - m_2)gh = \frac{1}{2}(m_1 + m_2)(\omega^2R^2) + \frac{1}{2}I\omega^2$$

$$\Rightarrow \quad \omega = \sqrt{\frac{2(m_1 - m_2)gh}{\left(m_1 + m_2 + \frac{I}{R^2}\right)R^2}}$$

$$\text{or} \quad \omega = \sqrt{\frac{2(m_1 - m_2)gh}{(m_1 + m_2)R^2 + I}}$$

$$\text{or} \quad \omega = \left[\frac{2(m_1 - m_2)gh}{(m_1 + m_2)R^2 + I}\right]^{\frac{1}{2}}$$

61. Since, the inclined plane is frictionless, then there will be no rolling and the mass will only slide down. Hence, acceleration is same for all the given bodies.

62. For rolling, $t = \sqrt{\dfrac{2l(1 + K^2/R^2)}{g\sin\theta}}$ = same (given in question)

$$\therefore \quad \frac{2l(1 + K_1^2/R^2)}{g\sin\theta_1} = \frac{2l(1 + K_2^2/R^2)}{g\sin\theta_2}$$

$$\Rightarrow \quad \frac{1 + K_1^2/R^2}{\sin\theta_1} = \frac{1 + K_2^2/R^2}{\sin\theta_2} \quad \text{...(i)}$$

For sphere, $K_1^2 = \dfrac{2}{5}R^2, \ \theta_1 = 30°$

For hollow cylinder, $\dfrac{K_2^2}{R^2} = 1, \theta_2 = ?$

$\therefore$ From Eq. (i), we get

$$\Rightarrow \quad \frac{1+\frac{2}{5}}{\sin 30°} = \frac{1+1}{\sin\theta_2}$$

or $$\sin\theta_2 = \frac{5}{7} = 0.7143$$

or $$\theta = 45.58° \simeq 45°$$

63. For rolling, $a = \dfrac{g\sin\theta}{\left(1 + \dfrac{I}{MR^2}\right)}$

$$= \frac{10 \times 1/2}{1 + 2/5}$$

$$= \frac{25}{7}\,\text{ms}^{-2} \quad \left(\because \text{for sphere, } I = \frac{2}{5}MR^2\right)$$

$$\Rightarrow \quad t = \frac{2v}{a} = \frac{2 \times 1 \times 7}{25} = 0.56\text{ s}$$

$$\simeq 0.57\text{ s}$$

64. Here, $\theta = 60°, l = 10\text{ m}, a = ?$

For solid sphere, $K^2 = \dfrac{2}{5}R^2$

$$\therefore \quad a = \frac{9.8\sin 60°}{1 + \frac{2}{5}} \quad \left(\because a = \frac{g\sin\theta}{1 + \frac{K^2}{R^2}}\right)$$

or $$a = \frac{5}{7} \times 9.8 \times \frac{\sqrt{3}}{2} = 6.06\text{ ms}^{-2}$$

65. As, the wheel rolls forward the radius of the wheel decreases along *AB*, hence for the same number of rotations, it moves less distance along *AB*. So, it turns left.

66. For solid cylinder, $\theta = 30°, \dfrac{K_1^2}{R_1^2} = \dfrac{1}{2}$

For hollow cylinder, $\theta = ?, \dfrac{K_2^2}{R_2^2} = 1$

For rolling on inclined plane,

$$\frac{1 + \frac{K_1^2}{R_1^2}}{\sin\theta_1} = \frac{1 + \frac{K_2^2}{R_2^2}}{\sin\theta_2}$$

Hence, $$\frac{\left(1 + \frac{1}{2}\right)}{\sin 30°} = \frac{1+1}{\sin\theta}$$

$$\therefore \quad \sin\theta = \frac{2}{3} = 0.6667$$

$$\theta = 41.8° \simeq 42°$$

67. As, it is clear from figure,

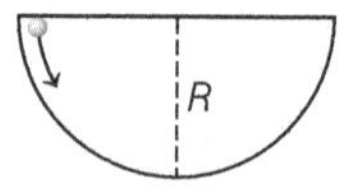

On reaching the bottom of the bowl, loss in PE = mgR, and gain in

$$\text{KE} = \frac{1}{2}mv^2 + \frac{1}{2}I\omega^2$$

$$\Rightarrow \quad |\Delta K| = \frac{1}{2}mv^2 + \frac{1}{2} \times \left(\frac{2}{5}mr^2\right)\omega^2$$

$$= \frac{1}{2}mv^2 + \frac{1}{5}mv^2$$

$$= \frac{7}{10}mv^2$$

As, gain in KE = loss in PE

$$\therefore \quad \frac{7}{10}mv^2 = mgR$$

$$v = \sqrt{\frac{10\,gR}{7}}$$

68. If h is height of the ramp, then in rolling of marble, speed

$$v = \sqrt{\frac{2\,gh}{1 + K^2/R^2}}$$

The speed of the cube to the centre of mass,

$$v' = \sqrt{2\,gh}$$

$$\therefore \quad \frac{v'}{v} = \sqrt{1 + \frac{K^2}{R^2}}$$

For marble sphere, $K^2 = \dfrac{2}{5}R^2$

$$\therefore \quad \frac{v'}{v} = \sqrt{1 + \frac{2}{5}}$$

$$= \sqrt{\frac{7}{5}} = \sqrt{7} : \sqrt{5}$$

69. Given, $m = 5$ kg, $R = 0.5$ m.

Horizontal force, $F = 40$ N

As, cylinder is rolling without slipping.

Hence, torque is producing rotation about centre O.

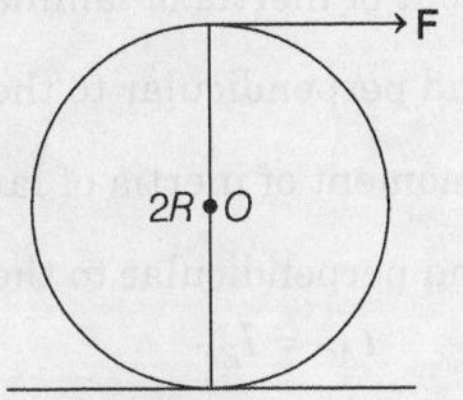

So, $\tau = \mathbf{r} \times \mathbf{F}$ (Here, $r = R$)

$\theta = 90°$

So, $\tau = \mathbf{r} \times \mathbf{F} = rF$

or $\tau = 0.5 \times 40 = 20$ N-m ...(i)

If α is acceleration of centre of mass 'O' then torque is,

$$\tau = I\alpha$$

where, $I = MR^2$

$\therefore$ $\tau = MR^2\alpha$ ...(ii)

Comparing Eq. (i) with Eq. (ii),

$$MR^2\alpha = 20$$

$$\Rightarrow \quad \alpha = \frac{20}{5 \times (0.5)^2}$$

or $\alpha = 16 \text{ rad/s}^2$

70. Total mechanical energy at O

$$= mgh = 2\,mg$$

Now, total mechanical energy at A

$$= \frac{1}{2} mv^2 \left(1 + \frac{K^2}{R^2}\right) + 0.2\,mgh$$

Since, total mechanical energy is conserved

$$\Rightarrow \quad \frac{1}{2} mv^2 \left(1 + \frac{K^2}{R^2}\right) + 0.2\,mgh = 2\,mg$$

$$\Rightarrow \quad \frac{1}{2} v^2 \left(1 + \frac{\left(R\sqrt{\frac{2}{5}}\right)^2}{R^2}\right) + 0.2\,g = 2\,g$$

$$\left[\because K = \sqrt{\frac{2}{5}}\,R \text{ for spherical sphere}\right]$$

$$\Rightarrow \quad \frac{1}{2} v^2 \left[1 + \frac{2}{5}\right] + 2 = 20$$

$$\Rightarrow \quad v^2 \left[\frac{7}{5}\right] = 36$$

$$\Rightarrow \quad v = \sqrt{\frac{36 \times 5}{7}} = \sqrt{\frac{180}{7}} = 5.1 \text{ m/s}$$

Now, range of projectile

$$= \frac{u^2 \sin 2\theta}{g}$$

$$= \frac{(5.1)^2 \sin 60°}{10} \approx 2.08 \text{ m}$$

Round II

1. MI of disc about tangent in a plane $= \frac{5}{4} MR^2 = I$

$$\therefore \quad MR^2 = \frac{4}{5} I$$

MI of disc about tangent I to plane, $I' = \frac{3}{2} MR^2$

$$\therefore \quad I' = \frac{3}{2}\left(\frac{4}{5} I\right) = \frac{6}{5} I$$

2. Angular retardation, $\alpha = \frac{\tau}{I} = \frac{\mu\, mgR}{mR^2} = \frac{\mu g}{R}$

As, $\omega = \omega_0 - \alpha t$

$$\therefore \quad t = \frac{\omega_0 - \omega}{\alpha}$$

$$= \frac{\omega_0 - \omega_0/2}{\mu g/R}$$

$$= \frac{\omega_0 R}{2\mu g}$$

3. As, $L = rP$

$$\Rightarrow \quad \log_e L = \log_e P + \log_e r$$

If graph is drawn between $\log_e L$ and $\log_e P$, then it will be straight line which will not pass through the origin because of presence of constant term ($\log_e r$) in the equation.

4. Moment of inertia of rod about the given axis $= \frac{mL^2}{12}$

Moment of inertia of each disc about disc about its diameter

$$= \frac{MR^2}{4}$$

Using theorem of parallel axes, moment of inertia of each disc about the given axis $= \frac{MR^2}{4} + M\left(\frac{L}{2}\right)^2$

$$= \frac{MR^2}{4} + \frac{ML^2}{4}$$

$\therefore$ For theorem of parallel axes, moment of inertia about the given axis is

$$I = \frac{mL^2}{12} + \left(\frac{MR^2}{4} + \frac{ML^2}{4}\right)$$

$$I = \frac{mL^2}{12} + \frac{MR^2}{4} + \frac{ML^2}{4}$$

5. Angular momentum of block w.r.t. O before collision at $O = Mv\frac{a}{2}$. On collision, the block will rotate about the side passing through O. Now its angular momentum $= I\omega$

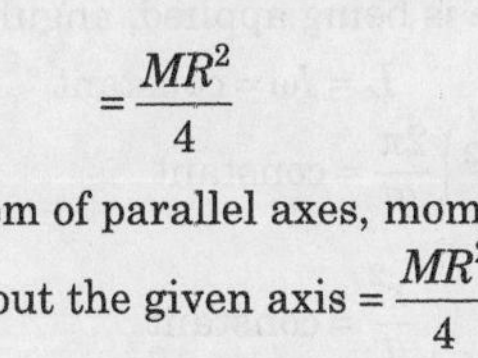

By law of conservation of angular momentum,

$$Mv\frac{a}{2} = I\omega$$

$$\Rightarrow \quad Mv\frac{a}{2} = \left(\frac{Ma^2}{6} + \frac{Ma^2}{2}\right)\omega$$

$$\Rightarrow \quad \omega = \frac{3v}{4a}$$

where I is moment of inertia of the block about the axis perpendicular to the plane passing through O.

6. According to the theorem of parallel axes, moment of inertia of disc about an axis passing through K and perpendicular to plane of disc,

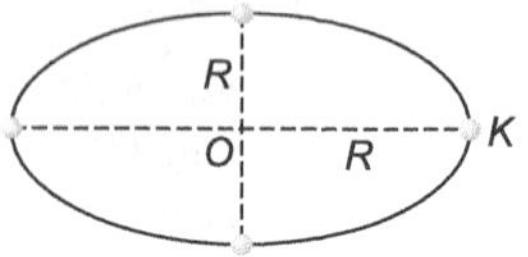

$$= \frac{1}{2}MR^2 + MR^2$$

$$= \frac{3}{2}MR^2$$

Total moment of inertia of the system

$$= \frac{3}{2}MR^2 + m(2R)^2 + m(\sqrt{2}R)^2 + m(\sqrt{2}R)^2$$

$$= 3(M + 16m)\frac{R^2}{2}$$

7. As no torque is being applied, angular momentum,

$$L = I\omega = \text{constant}$$

$$\Rightarrow \quad \left(\frac{2}{5}Mr^2\right)\frac{2\pi}{T} = \text{constant}$$

or $$\frac{r^2}{T} = \text{constant}$$

Differentiating w.r.t. time t, we get

$$\frac{T\cdot 2r\frac{dr}{dt} - r^2\frac{dT}{dt}}{T^2} = 0$$

or $$2Tr\frac{dr}{dt} = r^2\frac{dT}{dt}$$

or $$\frac{dT}{dt} = \frac{2T}{r}\frac{dr}{dt}$$

8. As angular momentum is conserved in the absence of a torque, therefore

$$I_0\omega_0 = I\omega$$

$$\left(\frac{2}{3}MR^2\right)\left(\frac{2\pi}{T_0}\right) = \left[\frac{2}{5}MR^2 + \frac{2}{5}\frac{MR^2}{5\times10^{19}}\right]\frac{2\pi}{T}$$

$$\frac{T}{T_0} = 1 + \frac{1}{5\times10^{19}}$$

$$\frac{T}{T_0} - 1 = \frac{1}{5\times10^{19}} = 2\times10^{-20}$$

9. From symmetry considerations,

$I_{AC} = \frac{1}{2}$ moment of inertia of lamina about an axis through O and perpendicular to the plane $ABCD$.

and $I_{EF} = \frac{1}{2}$ moment of inertia of lamina about an axis through O and perpendicular to the plane $ABCD$

$$\therefore \quad I_{AC} = I_{EF}$$

10.

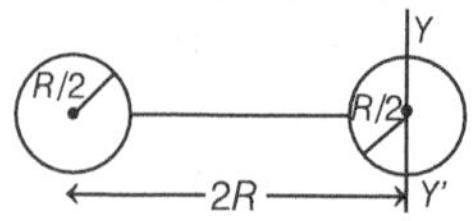

Moment of inertia of the system about YY'

$I_{yy'}$ = Moment of inertia of sphere P about YY' + Moment of inertia of sphere Q about YY'

Moment of inertia of sphere P about YY'

$$= \frac{2}{5}M\left(\frac{R}{2}\right)^2 + M(x)^2$$

$$= \frac{2}{5}M\left(\frac{R}{2}\right)^2 + M(2R)^2$$

$$= \frac{MR^2}{10} + 4MR^2$$

Moment of inertia of sphere Q about YY' is $\frac{2}{5}M\left(\frac{R}{2}\right)^2$

$$\text{Now, } I_{YY'} = \frac{MR^2}{10} + 4MR^2 + \frac{2}{5}M\left(\frac{R}{2}\right)^2 = \frac{21}{5}MR^2$$

11. Choice (a) is false, as theorem of perpendicular axes applies only to a plane lamina.

Now, Z axis parallel to Z' axis and distance between them $= \frac{a\sqrt{2}}{2} = \frac{a}{\sqrt{2}}$. Therefore, according to the theorem of parallel axes,

$$I'_Z = I_Z + m\left(\frac{a}{\sqrt{2}}\right)^2 = I_Z + \frac{ma^2}{2}$$

Choice (b) is true.

Again, choice (c) is false as Z'' axis is not parallel to Z-axis.

Choice (d) is also false as from symmetry, we find that $I_X = I_Y$.

12. If M is mass of the square plate before cutting the holes, then mass of portion of each hole,

$$m = \frac{M}{16R^2}\times\pi R^2 = \frac{\pi}{16}M$$

$\therefore$ Moment of inertia of remaining portion,

$$I = I_{\text{square}} - 4I_{\text{hole}}$$

$$= \frac{M}{12}(16R^2 + 16R^2) - 4\left[\frac{mR^2}{2} + m(\sqrt{2}R)^2\right]$$

$$= \frac{M}{12} \times 32R^2 - 10mR^2$$

$$= \frac{8}{3}MR^2 - \frac{10\pi}{16}MR^2$$

$$= \left(\frac{8}{3} - \frac{10\pi}{16}\right)MR^2$$

13. As there is no external torque, angular momentum will remain constant. When the tortoise moves from A to C, moment of inertia of the platform and tortoise decreases. Therefore, angular velocity of the system increases. When the tortoise moves from C to B, moment of inertia increases. Therefore, angular velocity decreases.

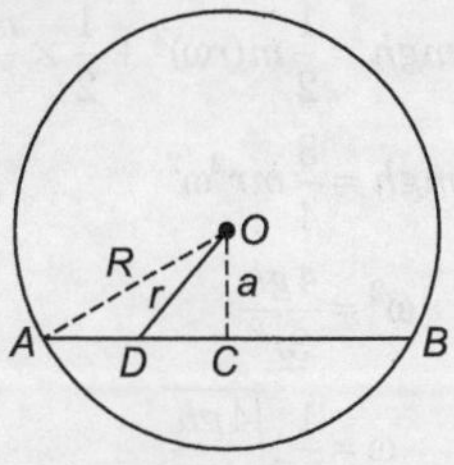

If M = mass of platform,

R = radius of platform,

m = mass of tortoise moving along the chord AB

and a = perpendicular distance of O from AB.

Initial angular momentum, $I_1 = mR^2 + \frac{MR^2}{2}$

At any time t, let the tortoise reach D moving with velocity v.

$\therefore \quad AD = vt$

$AC = \sqrt{R^2 - a^2}$

As $\quad DC = AC - AD = (\sqrt{R^2 - a^2} - vt)$

$\therefore \quad OD = r = a^2 + [\sqrt{R^2 - a^2} - vt]^2$

Angular momentum at time t

$$I_2 = mr^2 + \frac{MR^2}{2}$$

As angular momentum is conserved

$\therefore \quad I_1\omega_0 = I_2\omega(t)$

This shows that variation of $\omega(t)$ with time is non-linear.

14. To reverse the direction,

$$\int \tau \, d\theta = 0 \quad \text{(work done is zero)}$$

As, $\quad \tau = F \times r = (20t - 5t^2) \times 2 = 40t - 10t^2$

$$\alpha = \frac{\tau}{I} = \frac{40t - 10t^2}{10} = 4t - t^2 = \frac{d\omega}{dt}$$

$$\omega = \int_0^t \alpha \, dt = 2t^2 - \frac{t^3}{3}$$

ω is zero at

$$2t^2 - \frac{t^3}{3} = 0$$

$\Rightarrow \quad t^3 = 6t^2$

$\Rightarrow \quad t = 6 \text{ s}$

As $\quad \frac{d\theta}{dt} = \omega$

$\Rightarrow \quad \theta = \int_0^6 \omega \, dt = \int_0^6 \left(2t^2 - \frac{t^3}{3}\right) dt$

$$= \left[\frac{2t^3}{3} - \frac{t^4}{12}\right]_0^6$$

$$= 216\left[\frac{2}{3} - \frac{1}{2}\right] = 36 \text{ rad}$$

Number of rotation $= \frac{\theta}{2\pi} = \frac{36}{2\pi} = 5.73 < 6$

15. Linear acceleration for rolling, $a = \frac{g\sin\theta}{\sqrt{1 + \frac{K^2}{R^2}}}$

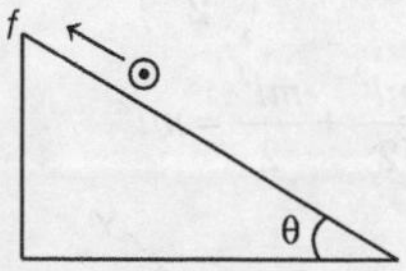

For cylinder, $\frac{K^2}{R^2} = \frac{1}{2}$

$\therefore \quad a_{\text{cylinder}} = \frac{2}{3} g\sin\theta$

For rotation, the torque

$fR = I\alpha = (MR^2\alpha)/2$

(where, f = force of friction)

But $R\alpha = a$

$\therefore \quad f = \frac{M}{2}a$

$\therefore \quad f = \frac{M}{2}\frac{2}{3}g\sin\theta$

$$= \frac{M}{3}g\sin\theta$$

$\mu_s = f/N$, where N is normal reaction,

$\therefore \quad \mu_s = \frac{\frac{M}{3}g\sin\theta}{Mg\cos\theta} = \frac{\tan\theta}{3}$

$\therefore$ For rolling without slipping of a roller down the inclined plane, $\tan\theta \le 3\mu_s$.

16. Since, rod is bent at the middle, so each part of it will have same length $\left(\frac{L}{2}\right)$ and mass $\left(\frac{M}{2}\right)$ as shown

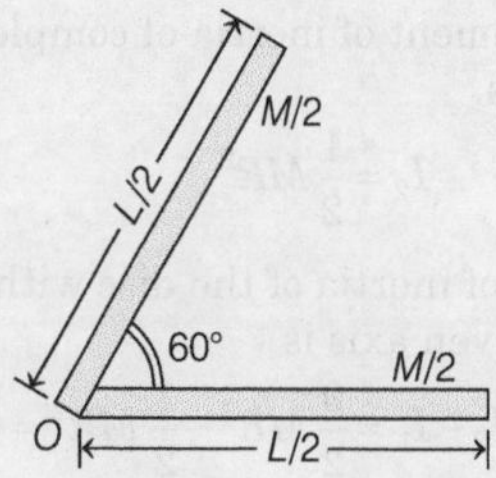

Moment of inertia of each part about an axis passing through its one end

$$= \frac{1}{3}\left(\frac{M}{2}\right)\left(\frac{L}{2}\right)^2$$

Hence, net moment of inertia about an axis passing through its middle point O is

$$I = \frac{1}{3}\left(\frac{M}{2}\right)\left(\frac{L}{2}\right)^2 + \frac{1}{3}\left(\frac{M}{2}\right)\left(\frac{L}{2}\right)^2$$

$$= \frac{1}{3}\left[\frac{ML^2}{8} + \frac{ML^2}{8}\right] = \frac{ML^2}{12}$$

17. As, it is clear from figure,

$$AC = BD = \sqrt{l^2 + l^2} = l\sqrt{2}$$

Moment of inertia of four point masses about BD,

$$I_{BD} = m\left(\frac{l\sqrt{2}}{2}\right)^2 + m \times 0 + m\left(\frac{l\sqrt{2}}{2}\right)^2 + m \times 0$$

$$= \frac{ml^2}{2} + \frac{ml^2}{2} = ml^2$$

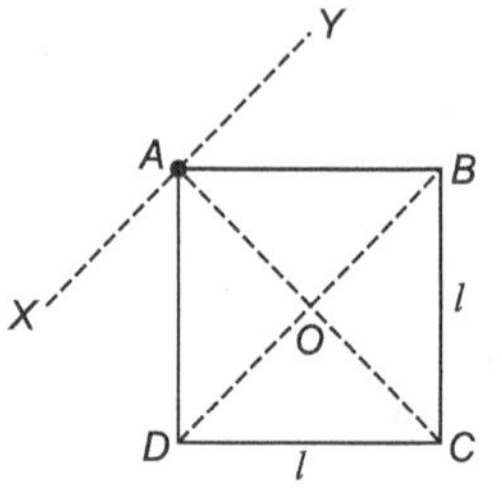

Applying the theorem of parallel axes,

$$I_{XY} = I_{BD} + M\,(AO)^2$$

$$= ml^2 + 4m\left(\frac{l}{\sqrt{2}}\right)^2$$

$$= 3\,ml^2$$

18. Mass per unit area of disc $= \dfrac{9\,M}{\pi R^2}$

Mass of removed portion of disc $= \dfrac{9\,M}{\pi R^2} \times \pi\left(\dfrac{R}{3}\right)^2 = M$

Moment of inertia of removed portion about an axis passing through centre of disc and perpendicular to the plane of disc, using theorem of parallel axes is

$$I_1 = \frac{M}{2}\left(\frac{R}{3}\right)^2 + M\left(\frac{2\,R}{3}\right)^2 = \frac{1}{2}MR^2$$

When portion of disc would not have been removed, then the moment of inertia of complete disc about the given axis is

$$I_2 = \frac{1}{2}MR^2$$

So moment of inertia of the disc with removed portion, about the given axis is

$$I = I_2 - I_1 = \frac{9}{2}MR^2 - \frac{1}{2}MR^2 = 4\,MR^2$$

19. When bob falls, its potential energy is converted into kinetic energy of bob and rotational kinetic energy of disc.

So, decrease in potential energy of bob = kinetic energy of bob + kinetic energy of disc

$$\Rightarrow \quad mgh = \frac{1}{2}mv^2 + \frac{1}{2}I\omega^2 \quad \text{...(i)}$$

If there is no slip, then $v = r\omega$

Also, moment of inertia of disc,

$$I = \frac{mr^2}{2}$$

Substituting this value in Eq. (i), we get

$$mgh = \frac{1}{2}m(r\omega)^2 + \frac{1}{2} \times \frac{mr^2}{2} \times \omega^2$$

$$\Rightarrow \quad mgh = \frac{3}{4}mr^2\omega^2$$

$$\Rightarrow \quad \omega^2 = \frac{4gh}{3r^2}$$

$$\text{or} \quad \omega = \frac{1}{r}\sqrt{\frac{4gh}{3}}$$

20. Various forces acting on bead at point P can be resolved as shown in figure

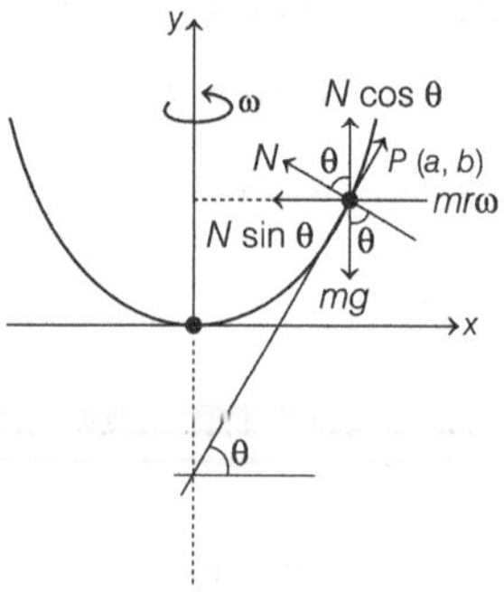

Normal force of rotating wire are

$$N\cos\theta = mg \text{ and } N\sin\theta = mx\omega^2$$

$$\Rightarrow \quad \tan\theta = \frac{dy}{dx} = \frac{x\omega^2}{g} \quad \text{...(i)}$$

Given, $\quad y = 4\,Cx^2$

$$\Rightarrow \quad \frac{dy}{dx} = 8\,Cx$$

$$\left.\frac{dy}{dx}\right|_{\text{at } x = a,\, y = b} = 8\,Ca \quad \text{...(ii)}$$

From Eqs. (i) and (ii), we get

$$\frac{x\omega^2}{g} = 8\,Ca$$

At $x = a$, we have

$$\frac{a\omega^2}{g} = 8\,Ca$$

$$\Rightarrow \quad \omega = \sqrt{8\,Cg}$$

$$= 2\sqrt{2g\,C}$$

21. Use geometry of the figure to calculate mass and side length of the cube interms of M and R, respectively.

Consider the cross-sectional view of a diametric plane as shown in the following diagram.

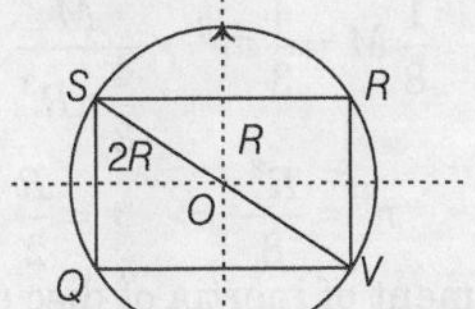

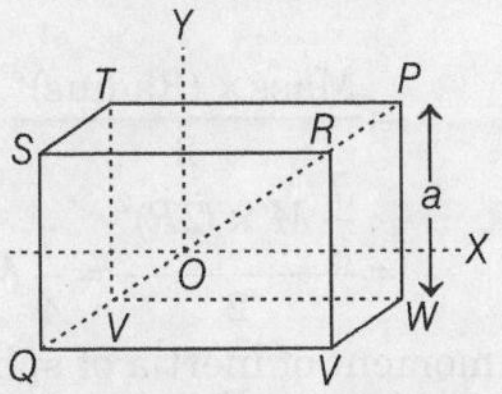

Using geometry of the cube,

$$PQ = 2R = (\sqrt{3})\,a$$

or
$$a = \frac{2R}{\sqrt{3}}$$

Volume density of the solid sphere

$$\rho = \frac{M}{(4/3)\,\pi R^3} = \frac{3}{4\pi}\left(\frac{M}{R^3}\right)$$

Mass of cube $(m) = (\rho)\,(a)^3$

$$= \left(\frac{3}{4\pi} \times \frac{M}{R^3}\right)\left[\frac{2R}{\sqrt{3}}\right]^3$$

$$= \frac{3M}{4\pi R^3} \times \frac{8R^3}{3\sqrt{3}} = \frac{2M}{\sqrt{3}\pi}$$

Moment of inertia of the cube about given axis is

$$I_Y = \frac{ma^2}{12}(a^2 + a^2) = \frac{ma^2}{6}$$

$$\Rightarrow \quad I_Y = \frac{ma^2}{6}$$

$$= \frac{2M}{\sqrt{3}\pi} \times \frac{1}{6} \times \frac{4R^2}{3} = \frac{4MR^2}{9\sqrt{3}\pi}$$

22. In a uniform rod, mass per unit length remains constant. If it is denoted by λ, then

$\lambda = \dfrac{m}{l}$ = constant for all segments of rod.

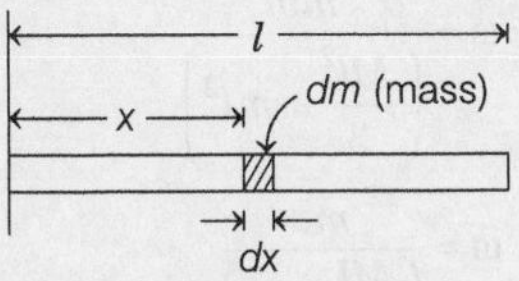

To find tension at x distance from fixed end, let us assume an element of dx length and dm mass. Tension on this part due to rotation is

$$dT = Kx \qquad \text{...(i)}$$

As,
$$K = m\omega^2$$

For this element,
$$K = (dm)\,\omega^2 \qquad \text{...(ii)}$$

$$\therefore \quad dT = (dm)\,\omega^2 x \qquad \text{...(iii)}$$

To find complete tension in the rod, we need to integrate Eq. (iii),

$$\int_0^T dT = \int_0^m (dm)\,\omega^2 x \qquad \text{...(iv)}$$

Using linear mass density,

$$\lambda = \frac{m}{l} = \frac{dm}{dx}$$

$$\Rightarrow \quad dm = \frac{m}{l} \cdot dx \qquad \text{...(v)}$$

Putting the value of Eq. (v) in Eq. (iv), we get

$$T = \int_x^l \frac{m}{l} \cdot \omega^2 x \cdot dx = \frac{m}{l} \cdot \omega^2 \left[\frac{x^2}{2}\right]_x^l$$

$$\Rightarrow \quad T = \frac{m\omega^2}{2l}[l^2 - x^2]$$

Now, if we observe the equation, it will look like a parabolic equation and when we see from given options, the correct option is (b).

23. Kinetic energy of rotation,

$$K_{\text{rot}} = \frac{1}{2} I\omega^2 = \frac{1}{2} Mk^2 \frac{v^2}{R^2} \qquad \left(\because \omega = \frac{v}{R}\right)$$

where, k is radius of gyration.

Kinetic energy of translation,

$$K_{\text{trans}} = (1/2)Mv^2$$

Thus, total energy,

$$E = K_{\text{rot}} + K_{\text{trans}}$$

$$= \frac{1}{2} Mk^2 \frac{v^2}{R^2} + \frac{1}{2} Mv^2$$

$$= \frac{1}{2} Mv^2\left(\frac{k^2}{R^2} + 1\right)$$

$$= \frac{1}{2} \frac{Mv^2}{R^2}(k^2 + R^2)$$

Hence,
$$\frac{K_{\text{rot}}}{K_{\text{total}}} = \frac{\frac{1}{2} Mk^2 \frac{v^2}{R^2}}{\frac{1}{2} \frac{Mv^2}{R^2}(k^2 + R^2)}$$

$$= \frac{k^2}{k^2 + R^2}$$

24. In given arrangement, mass of each sphere $= m$

Radius of each sphere $= \dfrac{d}{2}$

Distance of centre of each sphere from axis of rotation or centroidal axis $= \dfrac{2}{3} \times$ Altitude of equilateral triangle of side $d = \dfrac{2}{3} \times \dfrac{\sqrt{3}}{2} \times d = \dfrac{d}{\sqrt{3}}$

Now, moment of inertia of each sphere along an axis through its centre,

$$I_1 = \frac{2}{5}mr^2 = \frac{2}{5}m\left(\frac{d}{2}\right)^2 = \frac{1}{10}md^2$$

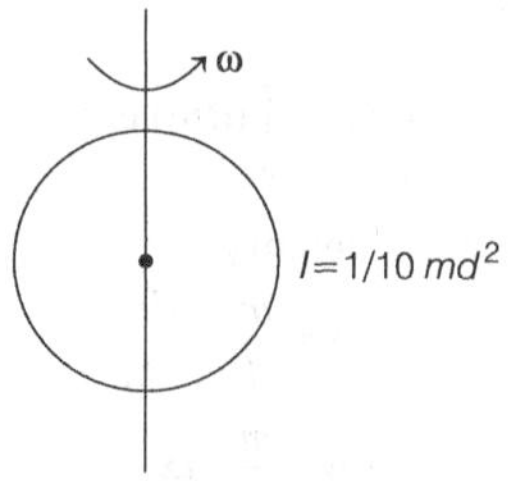

Now, by using parallel axes theorem, moment of inertia of each sphere around axis through O,

$$I' = I_1 + m(h^2)$$

$$= I_1 + m\left(\frac{d}{\sqrt{3}}\right)^2 = \frac{1}{10}md^2 + \frac{md^2}{3} = \frac{13}{30}md^2$$

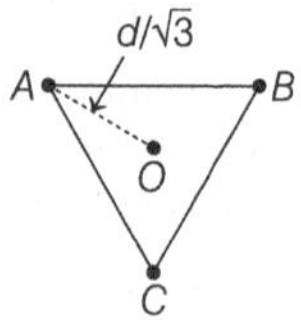

So, moment of inertia of system of spheres about O,

$$I_0 = 3I' = 3 \times \frac{13}{30}md^2 = \frac{13}{10}md^2 \quad \text{...(i)}$$

Again by using parallel axes theorem, moment of inertia of system about point A,

$$I_A = I_0 + m_{\text{system}} \times h^2$$

$$= \frac{13}{10}md^2 + 3m \times \left(\frac{d}{\sqrt{3}}\right)^2$$

$$= \frac{23}{10}md^2 \quad \text{...(ii)}$$

Hence, required ratio from Eqs. (i) and (ii), we get

$$\frac{I_0}{I_A} = \frac{13}{10}md^2 \times \frac{10}{23\,md^2}$$

$$\Rightarrow \quad \frac{I_0}{I_A} = \frac{13}{23}$$

25. The given situation is shown in the figure given below

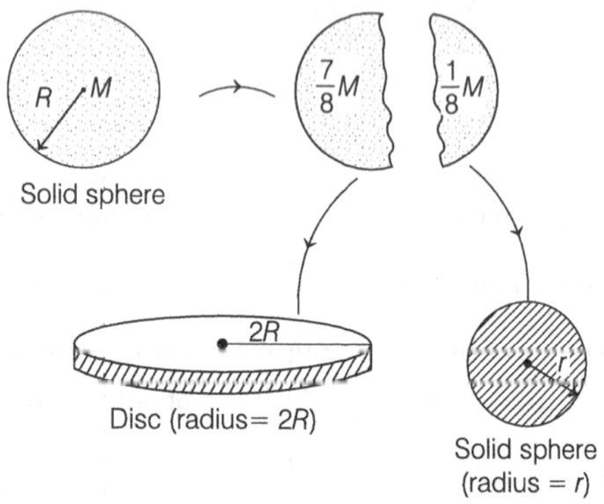

Density of given sphere of radius R,

$$\rho = \frac{\text{Mass}}{\text{Volume}} = \frac{M}{\frac{4}{3}\pi R^3}$$

Let radius of sphere formed from second part is r, then mass of second part = volume × density

$$\frac{1}{8}M = \frac{4}{3}\pi r^3 \times \frac{M}{\frac{4}{3}\pi R^3}$$

$$\therefore \quad r^3 = \frac{R^3}{8} \Rightarrow r = \frac{R}{2}$$

Now, I_1 = moment of inertia of disc (radius $2R$ and mass $\frac{7}{8}M$) about its axis

$$= \frac{\text{Mass} \times (\text{Radius})^2}{2}$$

$$= \frac{\frac{7}{8}M \times (2R)^2}{2} = \frac{7}{4}MR^2$$

and $\quad I_2$ = moment of inertia of sphere (radius $\frac{R}{2}$ and mass $\frac{1}{8}M$) about its axis

$$= \frac{2}{5} \times \text{Mass} \times (\text{Radius})^2$$

$$= \frac{2}{5} \times \frac{1}{8}M \times \left(\frac{R}{2}\right)^2 = \frac{MR^2}{80}$$

$$\therefore \text{ Ratio } \frac{I_1}{I_2} = \frac{\frac{7}{4}MR^2}{\frac{1}{80}MR^2} = 140$$

26.

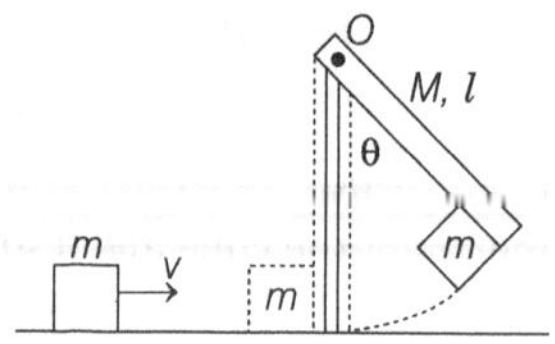

As there is no external torque, so angular momentum about pivot O is conserved.

$$\Rightarrow \quad L_i = L_f$$

$$\Rightarrow \quad mvl = (I_{\text{system}})\,\omega$$

$$\Rightarrow \quad mvl = (I_{\text{rod}} + I_{\text{block}})\,\omega$$

$$\Rightarrow \quad mvl = \left(\frac{Ml^2}{3} + ml^2\right)\omega$$

(where, ω = initial angular speed of rod and block)

$$\omega = \frac{mvl}{\left(\frac{Ml^2}{3} + ml^2\right)}$$

$$\Rightarrow \quad \omega = \frac{mv}{\left(\frac{Ml}{3} + ml\right)}$$

$$\Rightarrow \quad \omega = \frac{1 \times 6}{\left(\frac{2 \times 1}{3} + 1 \times 1\right)} = \frac{6 \times 3}{5} = \frac{18}{5} \text{ rad/s}$$

Now, by law of conservation of energy,

Initial rotational kinetic energy of rod and block = Final potential energy at angle θ of rod and block

$$\Rightarrow \quad \frac{1}{2}(I_{\text{system}}) \times \omega^2 = Mgh_1 + mgh_2$$

where, h_1 = height raised by centre of mass of rod

$$= \frac{l}{2}(1 - \cos\theta)$$

and h_2 = height raised by block = $l(1 - \cos\theta)$.

Hence,

$$\frac{1}{2}(I_{\text{rod}} + I_{\text{block}})\omega^2 = Mg\frac{l}{2}(1 - \cos\theta) + mgl(1 - \cos\theta)$$

$$\Rightarrow \quad \frac{1}{2}\left(\frac{Ml^2}{3} + ml^2\right)\omega^2 = gl\left(\frac{M}{2} + m\right) \times (1 - \cos\theta)$$

$$\Rightarrow \quad \frac{\frac{1}{2}\left(\frac{M}{3} + m\right) l\omega^2}{\left(\frac{M}{2} + m\right) g} = 1 - \cos\theta$$

$$\Rightarrow \quad \frac{\frac{1}{2}\left(\frac{2}{3} + 1\right) \times 1 \times \left(\frac{18}{5}\right)^2}{\left(\frac{2}{2} + 1\right) \times 10} = 1 - \cos\theta$$

$$\Rightarrow \quad \frac{27}{50} = 1 - \cos\theta$$

$$\Rightarrow \quad \cos\theta = \frac{23}{50} = 0.46$$

$$\Rightarrow \quad \theta = \cos^{-1}(0.46) = 62.61° \simeq 63°$$

The nearest answer is 63°.

27. Moment of inertia (MI) of a disc about a tangential axis in the plane of disc can be obtained as below.

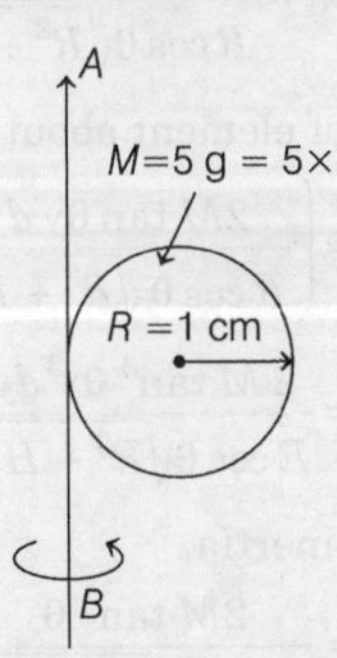

Moment of inertia of disc about its axis,

$$I_1 = \frac{MR^2}{2}$$

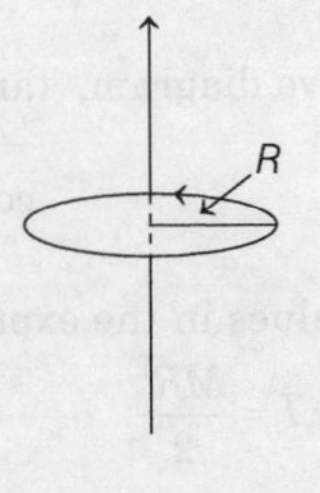

From perpendicular axes theorem, moment of inertia of disc about an axis along its diameter is

$$I_X + I_Y = I_Z \Rightarrow 2I_2 = I_1$$

$$\Rightarrow \quad I_2 = \frac{I_1}{2} = \frac{MR^2}{4}$$

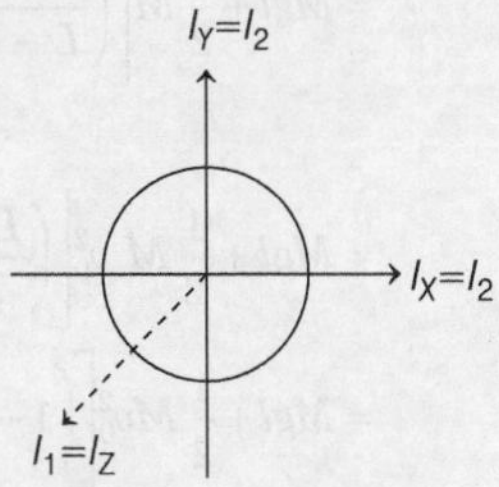

So, moment of inertia about a tangential axis from parallel axes theorem is

$$I = I_2 + MR^2 = \frac{MR^2}{4} + MR^2 = \frac{5}{4}MR^2$$

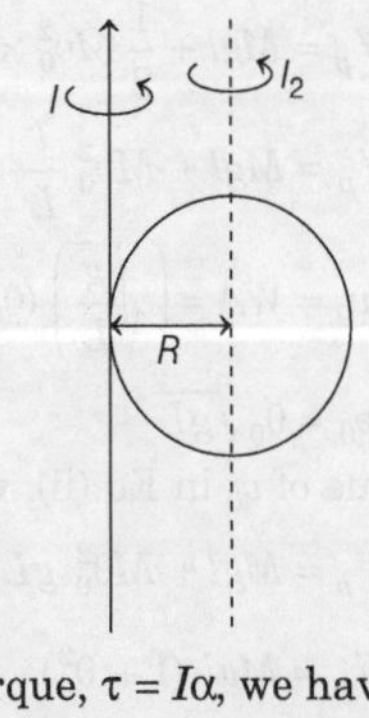

Now, using torque, $\tau = I\alpha$, we have

$$\tau = I\alpha = \frac{5}{4}MR^2\left(\frac{\omega_f - \omega_i}{\Delta t}\right)$$

Here, $M = 5 \times 10^{-3}$ kg, $R = 1 \times 10^{-2}$ m

$$\omega_f = 25 \text{ rps} = 25 \times 2\pi \frac{\text{rad}}{\text{s}} = 50\pi \frac{\text{rad}}{\text{s}},$$

$$\omega_i = 0, \ \Delta t = 5 \text{ s}$$

So,
$$\tau = \frac{\frac{5}{4} \times 5 \times 10^{-3} \times (10^{-2})^2 \times 50\pi}{5}$$

$$\approx 2 \times 10^{-5} \text{ N-m}$$

28. Initially, centre of mass is at distance L from the top end of the swing. It shifts to $(L - l)$ distance when the person stands up on the swing.

∴ Using angular momentum conservation law, if v_0 and v_1 are the velocities before standing and after standing of the person, then

$$Mv_0L = Mv_1(L - l)$$

$$\Rightarrow \quad v_1 = \left(\frac{L}{L - l}\right)v_0 \quad \text{...(i)}$$

Now, total work done by (person + gravitation) system will be equal to the change in kinetic energy of the person, *i.e.*

$$W_g + W_p = \text{KE}_1 - \text{KE}_0$$

$$\Rightarrow \quad -Mgl + W_p = \frac{1}{2}Mv_1^2 - \frac{1}{2}Mv_0^2$$

$$\Rightarrow \quad W_p = Mgl + \frac{1}{2}M(v_1^2 - v_0^2)$$

$$= Mgl + \frac{1}{2}M\left[\left(\frac{L}{L-l}\right)^2 v_0^2 - v_0^2\right]$$

[from Eq. (i)]

$$= Mgl + \frac{1}{2}M\,v_0^2\left[\left(\frac{L-l}{L}\right)^{-2} - 1\right]$$

$$= Mgl + \frac{1}{2}Mv_0^2\left[\left(1 - \frac{l}{L}\right)^{-2} - 1\right]$$

$$= Mgl + \frac{1}{2}Mv_0^2\left[\left(1 + \frac{2l}{L}\right) - 1\right]$$

[using $(1 + x)^n = 1 + nx$ as higher terms can be neglected, if $n << 1$]

$$\Rightarrow \quad W_p = Mgl + \frac{1}{2}Mv_0^2 \times \frac{2l}{L}$$

or $$W_p = Mgl + Mv_0^2\,\frac{l}{L} \quad \ldots\text{(ii)}$$

Here, $$v_0 = WA = \left(\sqrt{\frac{g}{L}}\right)(\theta_0 L)$$

$$\Rightarrow \quad v_0 = \theta_0\sqrt{gL}$$

$\therefore$ Using this value of v_0 in Eq.(ii), we get

$$W_p = Mgl + M\theta_0^2\, gL \cdot \frac{l}{L}$$

$$\Rightarrow \quad W_p = Mgl\,(1 + \theta_0^2)$$

29. Assuming that no energy is used up against friction, the total loss in potential energy is equal to the total gain in the kinetic energy.

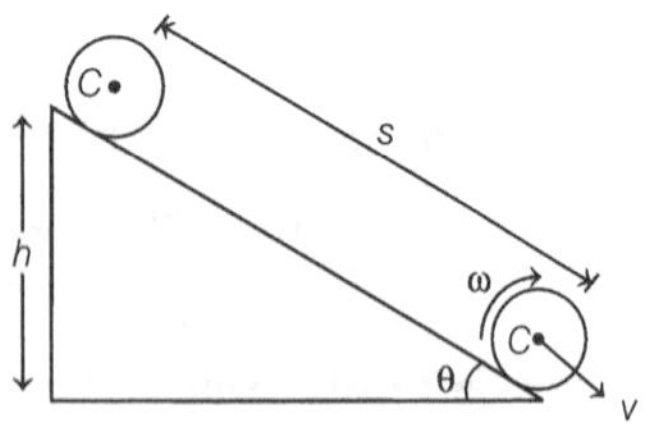

$$Mgh = \frac{1}{2}I\omega^2 + \frac{1}{2}Mv^2 \quad [\text{as, } v = r\omega]$$

i.e. $$Mgh = \frac{1}{2}I\left(\frac{v^2}{R^2}\right) + \frac{1}{2}Mv^2$$

or $$\frac{1}{2}v^2\left(M + \frac{I}{R^2}\right) = Mgh$$

or $$v^2 = \frac{2\,Mgh}{M + I/R^2} = \frac{2\,gh}{1 + I/MR^2}$$

If s is the distance covered along the plane, then

$$h = s\sin\theta$$

$$\therefore \quad v^2 = \frac{2gs\sin\theta}{1 + I/MR^2}$$

Now, $$v^2 = 2\,as$$

$$\therefore \quad 2\,as = \frac{2\,g\,s\sin\theta}{1 + I/MR^2}$$

or $$a = \frac{g\sin\theta}{1 + I/MR^2}$$

30. The hollow icecream cone can be assumed to be formed of several parts of rings having different radii. In general, we shall consider a ring (element) which is at distance y away from vertex O, of thickness dy, having radius r as shown in figure.

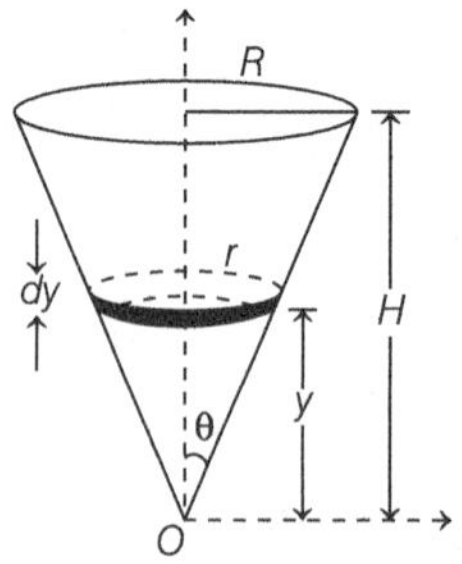

Area of cone, $A = \pi Rl = \pi R\sqrt{R^2 + H^2}$

Area of element, $dA = 2\pi r dl = 2\pi r\,\dfrac{dy}{\cos\theta}$

Mass of element, dm = Surface density × Area

$$= \frac{M}{\pi R\sqrt{R^2 + H^2}} \times \frac{2\pi r\,dy}{\cos\theta}$$

From above diagram, $\tan\theta = \dfrac{r}{y}$ or $r = y\tan\theta$

$$\therefore \quad dm = \frac{2M\tan\theta\, y\,dy}{R\cos\theta\sqrt{R^2 + H^2}}$$

Moment of inertia of element about its axis,

$$dI = (dm)r^2 = \left(\frac{2M\tan\theta y\,dy}{R\cos\theta\sqrt{R^2 + H^2}}\right)(y\tan\theta)^2$$

$$= \frac{2\,M\tan^3\theta y^3\,dy}{R\cos\theta\sqrt{R^2 + H^2}}$$

$\therefore$ Total moment of inertia,

$$I = \int dI = \frac{2M\tan^3\theta}{R\cos\theta\sqrt{R^2 + H^2}}\int_0^H y^3\,dy$$

$$= \frac{2M\tan^3\theta}{R\cos\theta\sqrt{R^2 + H^2}} \cdot \frac{H^4}{4}$$

Again, from above diagram, $\tan\theta = \dfrac{R}{H}$

$$\cos\theta = \frac{H}{\sqrt{R^2 + H^2}}$$

Putting these values in the expression of I, we get

$$I = \frac{MR^2}{2}$$

31. Rod rotates around vertical axis as

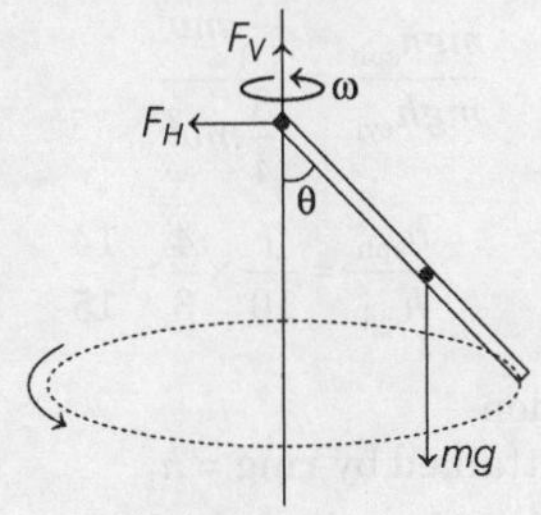

Given that,

Rate of change of angular momentum of rod = Torque provided by horizontal and vertical forces about centre of mass

$$\Rightarrow \quad \frac{ml^2}{12}\omega^2 \sin\theta\cos\theta = \text{Net torque of } F_H \text{ and } F_V \quad \text{...(i)}$$

Here, $F_V = mg$ and its perpendicular distance from axis is $\frac{l}{2}\sin\theta$.

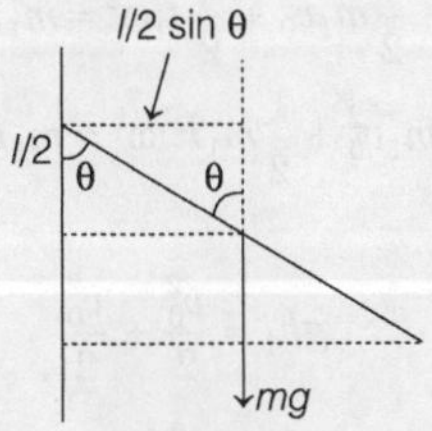

$$\therefore \text{ Torque of } F_V = mg\cdot\frac{l}{2}\sin\theta$$

and $\quad F_H = \text{centripetal force} = m\omega^2\frac{l}{2}\sin\theta$

Also, distance of F_H from axis is also $\frac{l}{2}\cos\theta$.

$$\therefore \text{ Torque of } F_H = m\omega^2\frac{l}{2}\sin\theta.\frac{l}{2}\cos\theta$$

So, from Eq. (i), we have

$$mg\frac{l}{2}\sin\theta - m\omega^2\frac{l}{2}\sin\theta.\frac{l}{2}\cos\theta = \frac{ml^2}{12}\omega^2\sin\theta\cos\theta$$

$$\Rightarrow \quad mg\frac{l}{2}\sin\theta = ml^2\omega^2\left(\frac{1}{12}+\frac{1}{4}\right)\sin\theta\cos\theta$$

$$\Rightarrow \quad \cos\theta = \frac{3}{2}\left(\frac{g}{\omega^2 l}\right)$$

32. We know that, moment of inertia (MI) about the principal axis of the sphere is given by

$$I_{\text{sphere}} = \frac{2}{5}MR^2 \quad \text{... (i)}$$

Using parallel axes theorem, moment of inertia about the given axis in the figure below will be

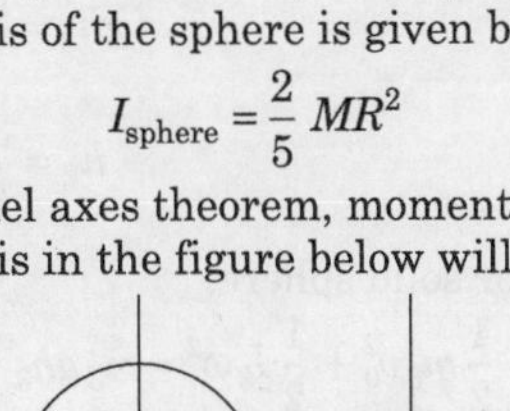

$$I_1 = \frac{2}{5}MR^2 + M(2R)^2$$

$$I_1 = \frac{22}{5}MR^2 \quad \text{... (ii)}$$

Considering both spheres at equal distance from the axis, moment of inertia due to both spheres about this axis will be

$$2I_1 = 2\times(22/5)\,MR^2$$

Now, moment of inertia of rod about its perpendicular bisector axis is given by

$$I_2 = \frac{1}{12}ML^2$$

Here, given that $\quad L = 2R$

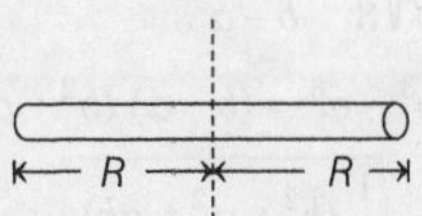

$$\therefore \quad I_2 = \frac{1}{12}M(2R)^2 = \frac{1}{3}MR^2 \quad \text{... (iii)}$$

So, total moment of inertia of the system is

$$I = 2I_1 + I_2 = 2\times\frac{22}{5}MR^2 + \frac{1}{3}MR^2$$

$$\Rightarrow \quad I = \left(\frac{44}{5}+\frac{1}{3}\right)MR^2 = \frac{137}{15}MR^2$$

33. Given, variation in mass per unit area (surface mass density),

$$\sigma = \frac{\sigma_0}{r} \quad \text{...(i)}$$

Calculation of Mass of Disc

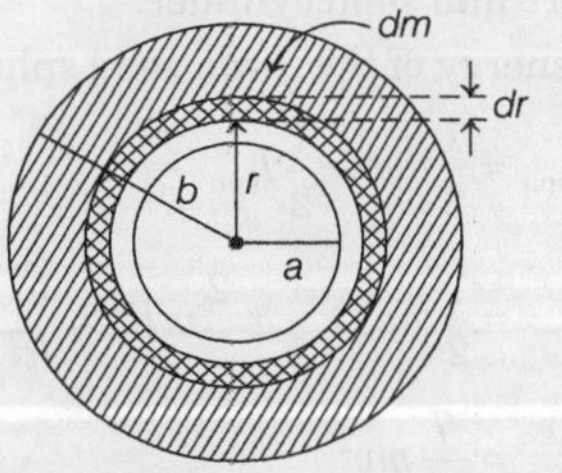

Let us divide whole disc in small area elements, one of them shown at r distance from the centre of the disc with its width as dr.

Mass of this element is $dm = \sigma\cdot dA$

$$\Rightarrow \quad dm = \frac{\sigma_0}{r}\times 2\pi r dr \quad \text{[from Eq. (i)]} \quad \text{...(ii)}$$

Mass of the disc can be calculated by integrating it over the given limits of r,

$$\int_0^M dm = \int_a^b \sigma_0\times 2\pi\times dr$$

$$\Rightarrow \quad M = \sigma_0 2\pi(b-a) \quad \text{...(iii)}$$

Calculation of Moment of Inertia

$$I = \int_0^M r^2 dm = \int_a^b r^2.\frac{\sigma_0}{r}\times 2\pi r dr \quad \text{[from Eq. (ii)]}$$

$$= \sigma_0 2\pi \int_a^b r^2 dr$$

$$= \sigma_0 2\pi \left[\frac{r^3}{3}\right]_a^b$$

$$\Rightarrow \quad I = \frac{1}{3}\sigma_0 2\pi\,[b^3 - a^3] \quad \text{...(iv)}$$

Now, radius of gyration,

$$K = \sqrt{\frac{I}{M}} = \sqrt{\frac{\frac{2\pi\sigma_0}{3}(b^3 - a^3)}{2\pi\sigma_0 (b-a)}}$$

$$\Rightarrow \quad K = \sqrt{\frac{1}{3}\frac{(b^3 - a^3)}{b-a}}$$

As we know, $b^3 - a^3 = (b-a)(b^2 + a^2 + ab)$

$$\therefore \quad K = \sqrt{\frac{1}{3}(b^2 + a^2 + ab)}$$

$$\text{or} \quad K = \sqrt{\frac{a^2 + b^2 + ab}{3}}$$

34. When a spherical/circular body of radius r rolls without slipping, its total kinetic energy is

$$K_{\text{total}} = K_{\text{translation}} + K_{\text{rotation}}$$

$$= \frac{1}{2}mv^2 + \frac{1}{2}I\omega^2$$

$$= \frac{1}{2}mv^2 + \frac{1}{2}I \cdot \frac{v^2}{r^2} \quad \left[\because \omega = \frac{v}{r}\right]$$

Let v be the linear velocity and R be the radius for both solid sphere and solid cylinder.

$\therefore$ Kinetic energy of the given solid sphere will be

$$K_{\text{sph}} = \frac{1}{2}mv^2 + \frac{1}{2}I_{\text{sph}}\frac{v^2}{R^2}$$

$$= \frac{1}{2}mv^2 + \frac{1}{2} \times \frac{2}{5}mR^2 \times \frac{v^2}{R^2}$$

$$= \frac{7}{10}mv^2 \quad \text{...(i)}$$

Similarly, kinetic energy of the given solid cylinder will be $K_{\text{cyl}} = \frac{1}{2}mv^2 + \frac{1}{2}I_{\text{cyl}}\frac{v^2}{R}$

$$= \frac{1}{2}mv^2 + \frac{1}{2} \times \frac{mR^2}{2} \times \frac{v^2}{R^2} = \frac{3}{4}mv^2 \quad \text{...(ii)}$$

Now, from the conservation of mechanical energy,

$$mgh = K_{\text{total}}$$

$\therefore$ For solid sphere,

$$mgh_{\text{sph}} = \frac{7}{10}mv^2 \quad \text{...(iii) [using Eq. (i)]}$$

Similarly, for solid cylinder,

$$mgh_{\text{cyl}} = \frac{3}{4}mv^2 \quad \text{...(iv) [using Eq. (ii)]}$$

Taking the ratio of Eqs. (iii) and (iv), we get

$$\frac{mgh_{\text{sph}}}{mgh_{\text{cyl}}} = \frac{\frac{7}{10}mv^2}{\frac{3}{4}mv^2}$$

$$\Rightarrow \quad \frac{h_{\text{sph}}}{h_{\text{cyl}}} = \frac{7}{10} \times \frac{4}{3} = \frac{14}{15}$$

35. From question,
let height attained by ring $= h_1$
Height attained by cylinder $= h_2$
Height attained by sphere $= h_3$
As we know that for a body which is rolling up an inclined plane (without slipping), follows the law of conservation of energy.

$\therefore$ For ring, using energy conservation law at its height h_1,

$$(\text{KE})_{\text{linear}} + (\text{KE})_{\text{rotational}} = (\text{PE})$$

$$\Rightarrow \quad \frac{1}{2}m_1 v_0^2 + \frac{1}{2}I_1\omega^2 = m_1 g h_1$$

$$\Rightarrow \quad \frac{1}{2}m_1 v_0^2 + \frac{1}{2}m_1 R^2\omega^2 = m_1 g h_1$$

$$(\because I = mR^2 \text{ for ring})$$

$$\Rightarrow \quad gh_1 = \frac{v_0^2}{2} + \frac{v_0^2}{2} \quad (\because v_0 = \omega R)$$

$$\Rightarrow \quad h_1 = v_0^2 / g \quad \text{...(i)}$$

Similarly, for solid cylinder, applying the law of conservation of energy,

$$\frac{1}{2}m_2 v_0^2 + \frac{1}{2}I_2\omega^2 = m_2 g h_2$$

$$\Rightarrow \quad \frac{1}{2}m_2 v_0^2 + \frac{1}{2}\left(\frac{1}{2} \times m_2 \times \left(\frac{R}{2}\right)^2\right)\omega^2 = m_2 g h_2$$

$$\left[\because I = \frac{1}{2}mR^2 \text{ for cylinder and } R = \frac{R}{2}\right]$$

$$\Rightarrow \quad \frac{1}{2}m_2 v_0^2 + \frac{1}{2} \times \frac{1}{8}m_2 R^2 \times \frac{v_0^2}{(R/2)^2} = m_2 g h_2$$

$$\Rightarrow \quad \frac{1}{2}v_0^2 + \frac{1}{4}v_0^2 = gh_2$$

$$\Rightarrow \quad gh_2 = \frac{3}{4}v_0^2$$

$$\Rightarrow \quad h_2 = \frac{3}{4}\left(\frac{v_0^2}{g}\right) \quad \text{...(ii)}$$

Similarly, for solid sphere,

$$\frac{1}{2}m_3 v_0^2 + \frac{1}{2}I_3\omega^2 = m_3 g h_3$$

$$\Rightarrow \quad \frac{1}{2}m_3 v_0^2 + \frac{1}{2}\left[\frac{2}{5}m_3\left(\frac{R}{4}\right)^2\right]\omega^2 = m_3 g h_3$$

$$\left[\because I = \frac{2}{5}mR^2 \text{ for solid sphere and } R = \frac{R}{4}\right]$$

$$\Rightarrow \quad \frac{1}{2}m_3 v_0^2 + \frac{1}{2}\times\frac{2}{5}\times m_3 \frac{R^2}{16}\times\frac{v_0^2}{(R/4)^2} = m_3 g h_3$$

$$\Rightarrow \quad \frac{1}{2}v_0^2 + \frac{1}{5}v_0^2 = gh_3$$

$$\Rightarrow \quad gh_3 = \frac{7}{10}v_0^2$$

or $$h_3 = \frac{7}{10}\left(\frac{v_0^2}{g}\right) \quad \text{...(iii)}$$

∴ Taking the ratio of h_1, h_2 and h_3 by using Eqs. (i), (ii) and (iii), we get

$$h_1 : h_2 : h_3 = \frac{v_0^2}{g} : \frac{3}{4}\frac{v_0^2}{g} : \frac{7}{10}\frac{v_0^2}{g} = 1 : \frac{3}{4} : \frac{7}{10}$$

$$\Rightarrow \quad h_1 : h_2 : h_3 = 40:30:28 = 20:15:14$$

∴ The most appropriate option is (c).

36.

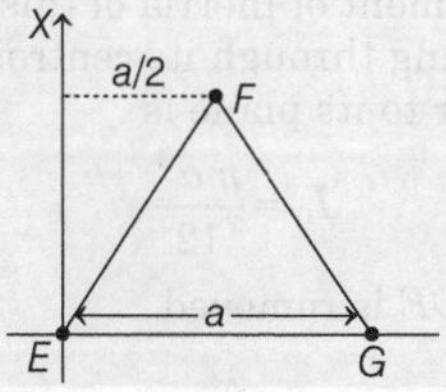

Moment of inertia of system

$$= \sum_{i=1}^{3} m_i r_i^2$$

$$= m_1 r_1^2 + m_2 r_2^2 + m_3 r_3^2$$

$$= m\times 0 + m\times a^2 + m\times\left(\frac{a}{2}\right)^2$$

$$= \frac{5}{4}ma^2 = \frac{25}{20}ma^2$$

$$= \frac{N}{20}ma^2$$

$$\therefore \quad N = 25$$

37. Given, $\mathbf{F} = (\hat{\mathbf{i}} + 2\hat{\mathbf{j}} + 3\hat{\mathbf{k}})$ N

$$\mathbf{r} = (4\hat{\mathbf{i}} + 3\hat{\mathbf{j}} - \hat{\mathbf{k}}) - (\hat{\mathbf{i}} + 2\hat{\mathbf{j}} + \hat{\mathbf{k}})$$

$$= (4-1)\hat{\mathbf{i}} + (3-2)\hat{\mathbf{j}} + (-1-1)\hat{\mathbf{k}}$$

$$= (3\hat{\mathbf{i}} + \hat{\mathbf{j}} - 2\hat{\mathbf{k}}) \text{ m}$$

Torque, $\tau = \mathbf{r}\times\mathbf{F}$

$$= \begin{vmatrix} \hat{\mathbf{i}} & \hat{\mathbf{j}} & \hat{\mathbf{k}} \\ 3 & 1 & -2 \\ 1 & 2 & 3 \end{vmatrix} = (3+4)\hat{\mathbf{i}} - (9+2)\hat{\mathbf{j}} + (6-1)\hat{\mathbf{k}}$$

$$= (7\hat{\mathbf{i}} - 11\hat{\mathbf{j}} + 5\hat{\mathbf{k}}) \text{ N-m}$$

Magnitude of torque, $|\tau| = \sqrt{(7)^2 + (-11)^2 + (5)^2}$

$$= \sqrt{49 + 121 + 25}$$

$$= \sqrt{195} \text{ N-m}$$

So, the value of $x = 195$.

38.

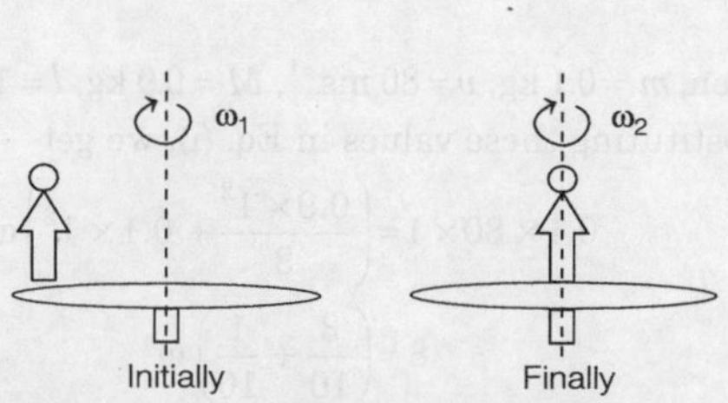

According to conservation of angular momentum,

$$I_1\omega_1 = I_2\omega_2$$

$$\Rightarrow \quad \left(80R^2 + \frac{200R^2}{2}\right)\omega_1 = \left(\frac{200R^2}{2}\right)\omega_2$$

$$\Rightarrow \quad \frac{360R^2}{2}\times 5 = \frac{200R^2}{2}\times\omega_2$$

$$\omega_2 = 9 \text{ rpm}$$

39. When bar is released, its potential energy appears in form of rotational kinetic energy.

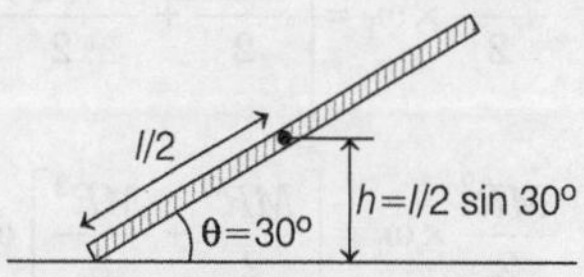

Now, from geometry of figure, height of centre of mass above surface,

$$h = \frac{l}{2}\cdot\sin 30^\circ = \frac{l}{4}$$

If angular speed of bar as it hits the table is ω, then Gravitational potential energy = Rotational kinetic energy

$$\Rightarrow \quad mgh = \frac{1}{2}I\omega^2$$

$$\Rightarrow \quad mg\frac{l}{4} = \frac{1}{2}\left(\frac{ml^2}{3}\right)\cdot\omega^2$$

$$\Rightarrow \quad \omega^2 = \frac{3}{2}\frac{g}{l} = \frac{3}{2}\times\frac{10}{1} = 15$$

$$\Rightarrow \quad \omega = \sqrt{15} \text{ s}^{-1}$$

Thus, $\quad n = 15$

40. The given situation is shown in the following figure

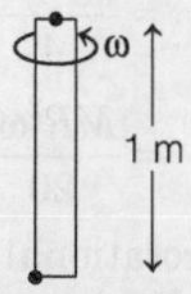

Applying law of conservation of angular momentum about pivotal point,

$$L_i = L_f$$

$$\Rightarrow \quad mvl = I\omega$$

$$\Rightarrow \quad mvl = \left(\frac{1}{3}ML^2 + mL^2\right)\omega \quad \text{...(i)}$$

Given, $m = 0.1$ kg, $v = 80\ \text{ms}^{-1}$, $M = 0.9$ kg, $l = 1$ m

Substituting these values in Eq. (i), we get

$$0.1 \times 80 \times 1 = \left(\frac{0.9 \times 1^2}{3} + 0.1 \times 1^2\right)\omega$$

$$8 = \left(\frac{3}{10} + \frac{1}{10}\right)\omega$$

$$\Rightarrow \quad 8 = \frac{4}{10}\omega$$

$$\Rightarrow \quad \omega = 20 \text{ rad/s}$$

41. In this process, no external torque is applied, so the angular momentum will remain conserved.

From law of conservation of angular momentum,

$$(L_{\text{sys}})_i = (L_{\text{sys}})_f$$

$$\Rightarrow \quad I_1\omega_1 + 0 = (I_1 + I_2)\,\omega_2$$

$$\Rightarrow \quad \frac{MR^2}{2} \times \omega_1 = \left[\frac{MR^2}{2} + \frac{M\left(\frac{R}{2}\right)^2}{2}\right]\omega_2$$

$$\Rightarrow \quad \frac{MR^2}{2} \times \omega_1 = \left[\frac{MR^2}{2} + \frac{MR^2}{8}\right]\omega_2$$

$$\Rightarrow \quad \frac{MR^2}{2} \times \omega_1 = \frac{5\,MR^2}{8} \times \omega_2$$

$$\Rightarrow \quad \omega_1 = \frac{5}{4}\omega_2$$

$$\Rightarrow \quad \omega_2 = \frac{4}{5}\omega_1 \qquad \ldots\text{(i)}$$

Now, the loss in rotational kinetic energy,

$$(\Delta K_{\text{sys}})_{\text{loss}} = (K_{\text{sys}})_i - (K_{\text{sys}})_f$$

$$= \left[\frac{1}{2} I_1\omega_1^2 + 0\right] - \left[\frac{1}{2}(I_1 + I_2)\omega_2^2\right]$$

$$= \left[\frac{1}{2} \times \frac{MR^2}{2} \times \omega_1^2 + 0\right] - \left[\frac{1}{2} \times \left(\frac{MR^2}{2} + \frac{MR^2}{8}\right) \times \left(\frac{4}{5}\omega_1\right)^2\right]$$

$$= \frac{MR^2\omega_1^2}{4} - \frac{1}{2} \times \frac{5MR^2}{8} \times \frac{16\,\omega_1^2}{25}$$

$$= \frac{MR^2\omega_1^2}{4} - \frac{MR^2\omega_1^2}{5}$$

$$= \frac{MR^2\omega_1^2}{20}$$

Now, the % loss in rotational kinetic energy,

$$\%(\Delta K_{\text{sys}})_{\text{loss}} = \frac{(\Delta K_{\text{sys}})_{\text{loss}}}{(K_{\text{sys}})_i} \times 100\%$$

$$= \frac{\left(\frac{MR^2\omega_1^2}{20}\right)}{\left(\frac{MR^2\omega_1^2}{4}\right)} \times 100\%$$

$$= \frac{4}{20} \times 100\%$$

$$= 20\%$$

and given that $\%\,(\Delta K_{\text{sys}})_{\text{loss}} = p\%$

On comparing, we get $\quad p = 20$

42. Let the mass of plane lamina having a shape of equilateral triangle be m and its side be a.

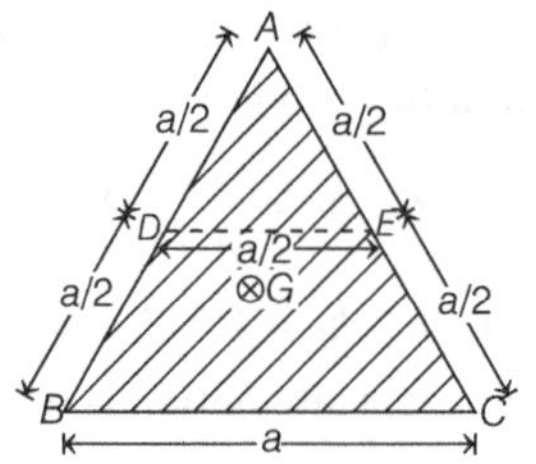

Then, the moment of inertia of this lamina ABC about the axis passing through its centroid G and perpendicular to its plane is

$$I_0 = \frac{ma^2}{12} \qquad \ldots\text{(i)}$$

Now, part ADE is removed.

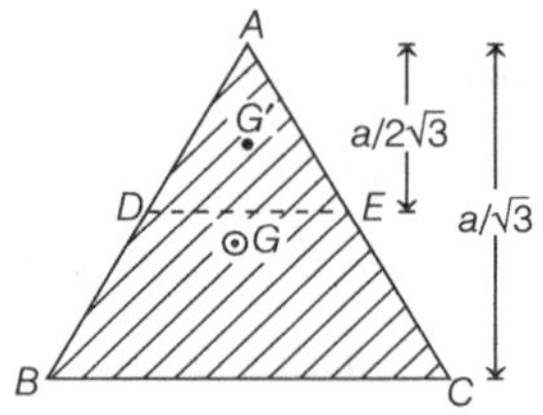

The side of the new triangular lamina ADE is $\frac{a}{2}$.

It is also equilateral in shape.

If its centroid is G', so $AG' = \frac{a/2}{\sqrt{3}} = \frac{a}{2\sqrt{3}}$

We know that for triangular lamina ABC, the centroid is G.

So, $\quad AG = \frac{a}{\sqrt{3}}$

Now, side $G'G = AG - AG'$

$$= \frac{a}{\sqrt{3}} - \frac{a}{2\sqrt{3}}$$

$$= \frac{a}{2\sqrt{3}}$$

Now, the moment of inertia of lamina ADE about the axis passing through point G and perpendicular to its plane (using parallel axes theorem),

$$I_1 = \frac{m_1\left(\frac{a}{2}\right)^2}{12} + m_1(G'G)^2 \qquad \ldots\text{(ii)}$$

Let m_1 be the mass of lamina ADE.

$$\therefore \quad m_1 = \frac{\text{mass of lamina } ABC}{\text{area of lamina } ABC} \times \text{area of lamina}$$

$$ADE = \frac{m}{\frac{\sqrt{3}}{4}(a)^2} \times \frac{\sqrt{3}}{4}\left(\frac{a}{2}\right)^2 = \frac{m}{4}$$

So, putting all the values in Eq. (ii), we get

$$I_1 = \frac{\left(\frac{m}{4}\right)\left(\frac{a}{2}\right)^2}{12} + \left(\frac{m}{4}\right)\left(\frac{a}{2\sqrt{3}}\right)^2$$

$$= \frac{ma^2}{192} + \frac{ma^2}{48}$$

$$= \frac{5ma^2}{192} \qquad \ldots\text{(iii)}$$

So, the moment of inertia of remaining part about the axis passing through point G and perpendicular to its plane is given by

$$I_{\text{net}} = I_0 - I_1$$

$$= \frac{ma^2}{12} - \frac{5ma^2}{192}$$

$$= \frac{11\,ma^2}{192}$$

$$= \frac{11}{16} \times \frac{ma^2}{12} = \frac{11}{16} I_0$$

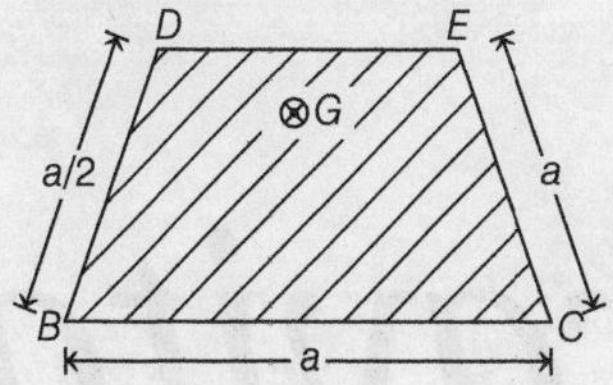

It is given that, $I_{\text{net}} = \frac{N}{16} I_0$

On comparing, we get

$$N = 11$$

10

Gravitation

IN THIS CHAPTER ....

- The Universal Law of Gravitation
- Acceleration due to Gravity
- Gravitational Field
- Gravitational Potential (V)
- Gravitational Potential Energy
- Escape Velocity
- Motion of a Satellite
- Geostationary Satellite
- Kepler's Laws of Planetary Motion

The Universal Law of Gravitation

In this universe, each body attracts other body with a force that is directly proportional to the product of their masses and inversely proportional to the square of the distance between them. Let m_1 and m_2 be the masses of two bodies and r be the separation between them.

$$F \propto \frac{m_1 m_2}{r^2} \quad \Rightarrow \quad F = \frac{G m_1 m_2}{r^2}$$

Here, G is the constant of proportionality which is called *universal gravitational constant*. The value of G is 6.67×10^{-11} N-m^{2}kg^{-2}.

Example 1. *Two particles of equal mass m go around a circle of radius R under the action of their mutual gravitational attraction. The speed of each particle with respect to their centre of mass is* ***[AIEEE 2011]***

(a) $\sqrt{\frac{Gm}{R}}$ (b) $\sqrt{\frac{Gm}{4R}}$ (c) $\sqrt{\frac{Gm}{3R}}$ (d) $\sqrt{\frac{Gm}{2R}}$

Sol. (b) Given, masses of the two particles are $m_1 = m_2 = m$ and radius of the two particles are $r_1 = r_2 = R$. So, the gravitational force of attraction between the particles,

$$F = \frac{G \times m \times m}{R^2} \quad \text{...(i)}$$

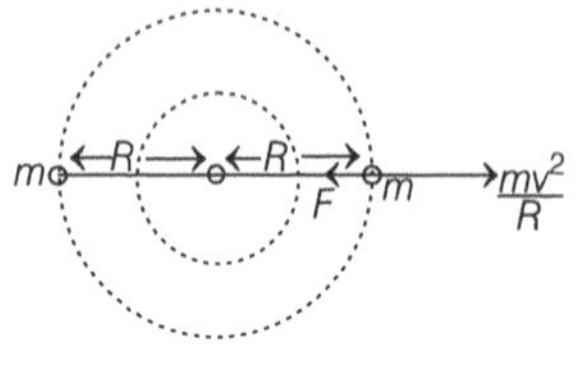

Centripetal force, $F = \frac{mv^2}{R}$ ...(ii)

From Eqs. (i) and (ii), we get

$$\frac{Gm^2}{(2R)^2} = \frac{mv^2}{R}$$

So, the speed of the each particle is $v = \sqrt{\frac{Gm}{4R}}$.

Acceleration due to Gravity

The acceleration of an object during its free fall towards the earth is called acceleration due to gravity.

If M is the mass of earth and R is the radius, the earth attracts a mass m on its surface with a force F given by

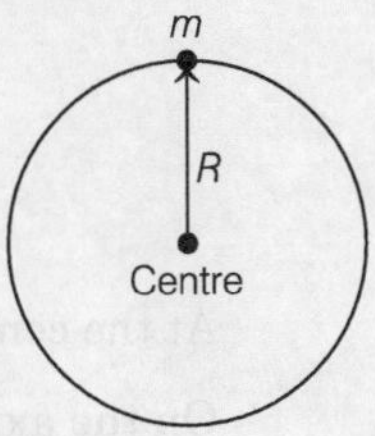

$$F = \frac{GMm}{R^2}$$

This force imparts an acceleration to the mass m which is known as acceleration due to gravity (g).

By Newton's law, we have

$$\text{Acceleration, } g = \frac{F}{m} = \frac{\frac{GMm}{R^2}}{m} = \frac{GM}{R^2}$$

On the surface of earth, $g = \frac{GM}{R^2}$

Substituting the values of G, M and R, we get

$$g = 9.81 \text{ ms}^{-2}$$

Mass of the earth, $M = 6 \times 10^{24}$ kg and radius of the earth, $R = 6.4 \times 10^6$ m.

Variation in g with Altitude and Depth

The value of g is variable and can vary in some cases as mentioned below

(i) **Value of acceleration due to gravity (g) at a height (h) from the surface of the earth is given** by $g' = \frac{gR^2}{(R+h)^2}$

If $h << R$, then $g' = g\left[1 - \frac{2h}{R}\right]$.

(ii) **Value of acceleration due to gravity (g) at a depth (d) from the surface of the earth is given by**

$$g' = g\left(1 - \frac{d}{R}\right)$$

At the centre of the earth, $d = R$ and hence $g' = 0$.

where, g = acceleration due to gravity on the surface.

Note It should be noted that the value of g decreases, if we move above the surface or below the surface of the earth.

Variation in the Value of g due to Rotation of the Earth

Due to rotation of the earth, the value of g decreases as the speed of rotation of the earth increases. The value of acceleration due to gravity at a latitude is

$$g'_\lambda = g - R\omega^2 \cos^2 \lambda$$

Following conclusions can be drawn from the above discussion

(i) The effect of centrifugal force due to rotation of the earth is to reduce the effective value of g.

(ii) The effective value of g is not truely in vertical direction.

(iii) At the equators, $\lambda = 0°$

Therefore, $g' = g - R\omega^2$ (minimum value)

(iv) At the poles, $\lambda = 90°$

Therefore, $g' = g$ (maximum value)

Example 2. *The acceleration due to gravity becomes $g/2$ (g = acceleration due to gravity on the surface of the earth) at a height equal to*

(a) $\frac{R}{4}$ (b) $\frac{R}{2}$ (c) $\frac{R}{3}$ (d) $\frac{R}{5}$

Sol. (a) The acceleration due to gravity,

$$g = \frac{GM}{R^2} \quad \text{...(i)}$$

At a height h above the surface of the earth, the acceleration due to gravity is

$$g' = \frac{GM}{(R+h)^2} \quad \text{...(ii)}$$

From Eqs. (i) and (ii), we get

$$\frac{g}{g'} = \left(\frac{R+h}{R}\right)^2 = \left(1 + \frac{h}{R}\right)^2$$

$$\Rightarrow \quad \frac{g'}{g} = \left(1 + \frac{h}{R}\right)^{-2} = \left(1 - \frac{2h}{R}\right)$$

Here, $g' = \frac{g}{2}$

$$\therefore \quad \frac{g/2}{g} = \left(1 - \frac{2h}{R}\right)$$

$$\Rightarrow \quad \frac{2h}{R} = \frac{1}{2} \text{ or } h = \frac{R}{4}$$

Example 3. *The value of the acceleration due to gravity is g_1 at a height $h = \frac{R}{2}$ (where, R = radius of the earth) from the surface of the earth. It is again equal to g_1 at a depth d below the surface of the earth. The ratio $\left(\frac{d}{R}\right)$ equals* **[JEE Main 2020]**

(a) $\frac{7}{9}$ (b) $\frac{1}{3}$ (c) $\frac{4}{9}$ (d) $\frac{5}{9}$

Sol. (d) Given that, acceleration due to gravity at height h from the surface of earth = acceleration due to gravity at depth d below the surface of earth = g_1

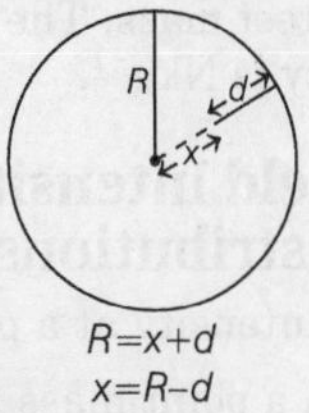

$R = x + d$

$x = R - d$

$$\Rightarrow \quad g_1 = \frac{GM}{(R+h)^2} = \frac{GM}{R^3}(R-d)$$

$$\Rightarrow \quad \frac{GM}{\left(R+\frac{R}{2}\right)^2} = \frac{GM}{R^3}(R-d) \qquad \left[\because h = \frac{R}{2}\right]$$

$$\Rightarrow \quad \frac{4GM}{9R^2} = \frac{GM}{R^2}\left(1-\frac{d}{R}\right)$$

$$\Rightarrow \quad 1-\frac{d}{R} = \frac{4}{9}$$

$$\Rightarrow \quad \frac{d}{R} = 1-\frac{4}{9} = \frac{5}{9}$$

Example 4. *A box weighs 196 N on a spring balance at the north pole. Its weight recorded on the same balance, if it is shifted to the equator is close to (Take, $g = 10\ ms^{-2}$ at the north pole and the radius of the earth = 6400 km)* **[JEE Main 2020]**

(a) 195.66 N (b) 195.32 N (c) 194.66 N (d) 194.32 N

Sol. (b) Acceleration due to gravity at poles, $g_p = 10\ \text{ms}^{-2}$

Weight of box at poles, $w_p = 196$ N

So, mass of box, $m = \frac{w_p}{g_p} = 19.6$ kg

Now, due to rotation of earth acceleration due to gravity at equator,

$$g_e = g_p - R\omega^2 \qquad \text{...(i)}$$

Here, $g_p = 10\ \text{ms}^{-2}$, $R = 6400 \times 10^3$ m,

$$\omega = \frac{2\pi}{T} \text{ and } T = 24 \times 3600\ \text{s}$$

Now, substituting the given values in Eq. (i), we get

$$g_e = 10 - 0.034 = 9.966\ \text{ms}^{-2}$$

So, weight of box at equator,

$$w_e = g_e \times m = 9.966 \times 19.6 = 195.32\ \text{N}$$

Gravitational Field

The space surrounding a material body in which its gravitational force of attraction can be experienced is called its gravitational field.

Gravitational Field Intensity (E)

Gravitational field intensity at any point is defined as the gravitational force experienced by any test mass divided by the magnitude of test mass when placed at the desired point.

$$\mathbf{E} = \frac{\mathbf{F}_r}{m_0}$$

where, m_0 is a small test mass. The SI unit of gravitational intensity is Nkg^{-1}.

Gravitational Field Intensity due to Various Mass Distributions

(i) Gravitational intensity at a point P situated at a distance r from a point mass M is given by $\mathbf{E} = \frac{GM}{r^2}$

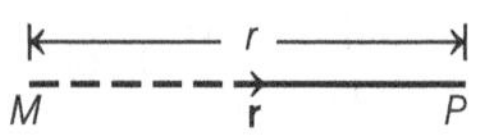

(ii) Due to a ring having a uniform mass distribution.

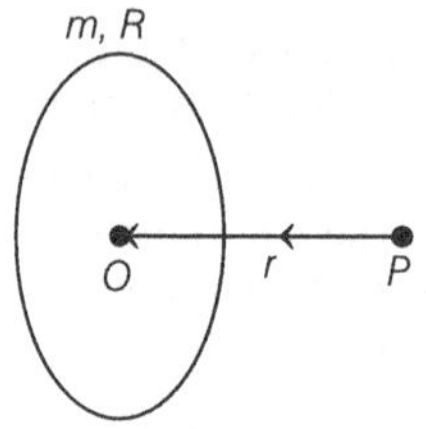

At the centre, $\mathbf{E} = 0$

On the axis, $\mathbf{E} = \frac{Gmr}{(R^2 + r^2)^{3/2}}$ [along P to O]

(iii) Due to hollow sphere having a uniform mass distribution. For inside point ($r < R$), $E = 0$. For outside point ($r \geq R$),

$$E = -\frac{GM}{r^2}$$

where, r is the distance of point from the centre.

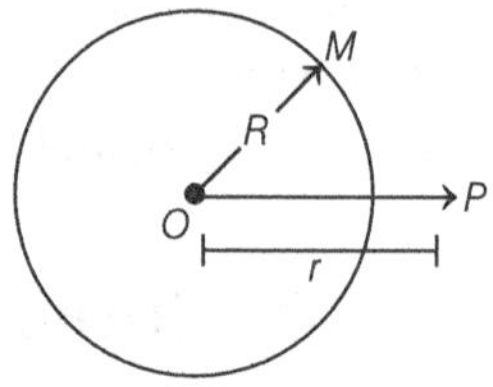

(iv) Due to a solid sphere having a uniform mass distribution.

For inside point ($r < R$),

$$E = -\frac{GMr}{R^3}$$

For outside point ($r \geq R$),

$$E = -\frac{GM}{r^2}$$

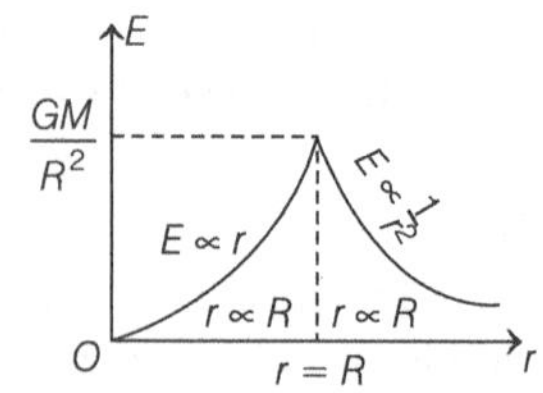

Gravitational Potential (V)

Gravitational potential at any point in a gravitational field is defined as the work done in bringing a unit mass from infinity to that point.

Gravitational potential, $V = \lim_{m_0 \to 0} \frac{W}{m_0}$

Gravitational potential due to a point mass is

$$V = -\frac{GM}{r}$$

Gravitational potential is always negative and it is maximum at infinity. It is a scalar term and its SI unit is J kg^{-1}.

Gravitational Potential due to Various Mass Distributions

(i) Due to ring having a uniform mass distribution.

At the centre, $V = -\frac{GM}{R}$

On the axis, $V = -\frac{GM}{\sqrt{R^2 + r^2}}$

(ii) Due to a hollow sphere of radius R having a uniform mass distribution. For inside points ($r \le R$),

$$V = -\frac{GM}{R}$$

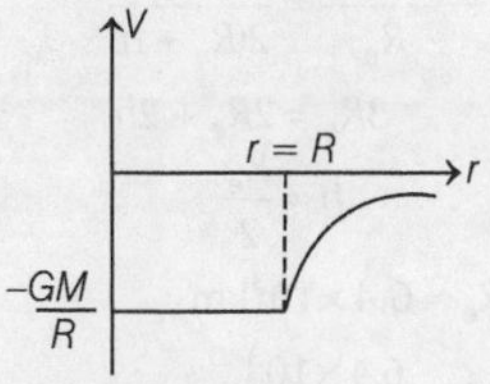

For outside points ($r > R$),

$$V = -\frac{GM}{r}$$

(iii) Due to a solid sphere of radius R having a uniform mass distribution.

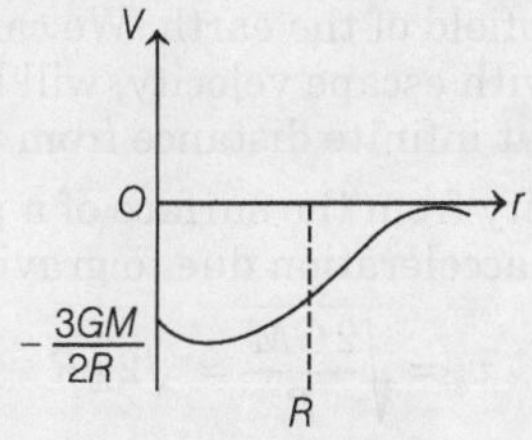

For inside points ($r < R$),

$$V = \frac{-GM(3R^2 - r^2)}{2R^3}$$

For outside points ($r \ge R$),

$$V = -\frac{GM}{r}$$

Example 5. *Three particles, each of mass m, are placed at the vertices of an equilateral triangle of side a. What is the gravitational field at the centroid of the triangle?*

(a) Zero (b) $-3\sqrt{3}\frac{Gm}{a^2}$ (c) $\frac{3Gm}{a^2}$ (d) $\frac{2Gm}{a^2}$

Sol. (a) In the figure, O is the centroid of triangle ABC,

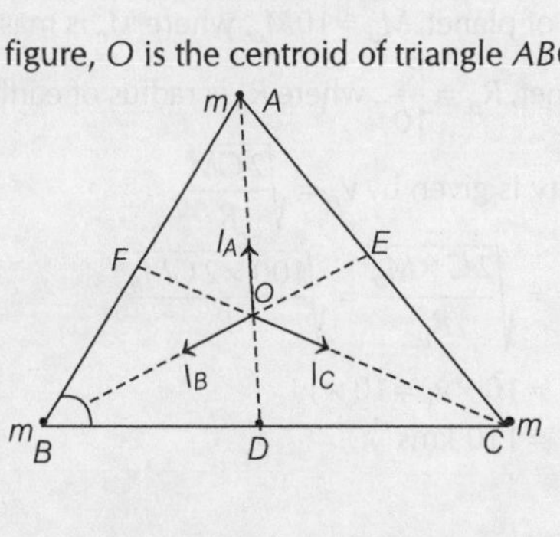

where, $OA = \frac{2}{3}AD = \frac{2}{3}(AB\sin 60°)$

$$= \frac{2}{3} \times a \times \frac{\sqrt{3}}{2} = \frac{a}{\sqrt{3}}$$

Thus, $OA = OB = OC = \frac{a}{\sqrt{3}}$

The gravitational intensity at O due to mass m at A is

$$E_A = \frac{Gm}{(OA)^2} = \frac{Gm}{(a/\sqrt{3})^2} \text{ along } OA$$

Similarly, the gravitational intensity at O due to mass m at B is

$$E_B = \frac{Gm}{(OB)^2} = \frac{Gm}{(a/\sqrt{3})^2} \text{ along } OB$$

and gravitational intensity at O due to mass m at C is

$$E_C = \frac{Gm}{(OC)^2} = \frac{Gm}{(a/\sqrt{3})^2} \text{ along } OC$$

As E_A, E_B and E_C are equal in magnitude and equally inclined to each other, the resultant gravitational intensity at O is zero.

Example 6. *Two bodies of masses m and 4 m are placed at a distance r. The gravitational potential at a point on the line joining them, where the gravitational field is zero, is* **[AIEEE 2011]**

(a) $-\frac{4Gm}{r}$ (b) $-\frac{6Gm}{r}$ (c) $-\frac{9Gm}{r}$ (d) zero

Sol. (c) Let gravitational field be zero at P as shown in figure.

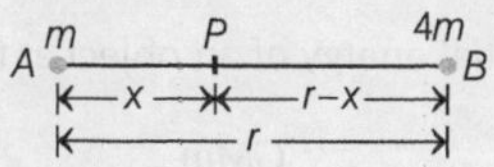

$$\therefore \quad \frac{Gm}{x^2} = \frac{G(4m)}{(r-x)^2}$$

$$\Rightarrow \quad 4x^2 = (r-x)^2$$

$$\Rightarrow \quad 2x = r - x$$

$$\Rightarrow \quad x = \frac{r}{3}$$

$$\therefore \quad V_P = -\frac{Gm}{x} - \frac{G(4m)}{r-x}$$

$$= -\frac{Gm}{r/3} - \frac{4Gm}{r - r/3} \qquad \left[\because x = \frac{r}{3}\right]$$

$$= -\frac{3Gm}{r} - \frac{6Gm}{r} = -\frac{9Gm}{r}$$

Gravitational Potential Energy

Gravitational potential energy of a body or system is negative of work done by the conservative gravitational force F in bringing it from infinity to the present position. Mathematically, gravitational potential energy,

$$U = -W = -\int_{\infty}^{\mathbf{r}} \mathbf{F} \cdot d\mathbf{r}$$

- The gravitational potential energy of two particles of masses m_1 and m_2 separated by a distance r is given by

$$U = -\frac{Gm_1 m_2}{r}$$

- The gravitational potential energy of mass m at the surface of the earth is

$$U = -\frac{GMm}{R}$$

- Difference in potential energy of mass m at a height h from the earth's surface and at the earth's surface is

$$U_{(R+h)} - U_R = \frac{mgh}{1+\frac{h}{R}} \simeq mgh \qquad \text{(if } h << R\text{)}$$

- For three particles system,

$$U = -\left[\frac{Gm_1m_2}{r_{12}} + \frac{Gm_1m_3}{r_{13}} + \frac{Gm_2m_3}{r_{23}}\right]$$

- For n-particles system, $\frac{n(n-1)}{2}$ pairs form and total potential energy of the system is sum of potential energies of all such pairs.

Example 7. *If g is acceleration due to gravity on earth's surface, the gain of the potential energy of an object of mass m raised from the surface of the earth to a height equal to the radius R of the earth, is*

(a) $2mgR$ (b) mgR
(c) $\frac{1}{2}mgR$ (d) $\frac{1}{4}mgR$

Sol. (c) The potential energy of an object at the surface of the earth is

$$U_1 = -\frac{GMm}{R}$$

The potential energy of the object at a height ($h = R$) from the surface of the earth is $U_2 = -\frac{GMm}{R+h} = -\frac{GMm}{R+R}$

Gain in potential energy of the object is $\Delta U = U_2 - U_1$

$$\Delta U = -\frac{GMm}{R+R} + \frac{GMm}{R}$$

$$\Delta U = \frac{1}{2}\frac{GMm}{R}$$

As, $gR^2 = GM$

Hence, $\Delta U = \frac{1}{2}g\frac{R^2m}{R}$

$\Rightarrow$ $\Delta U = \frac{1}{2}mgR$

Example 8. *The energy required to take a satellite to a height h above earth surface (where, radius of earth $= 6.4 \times 10^3$ km) is E_1 and kinetic energy required for the satellite to be in a circular orbit at this height is E_2. The value of h for which E_1 and E_2 are equal, is* **[JEE Main 2019]**

(a) 3.2×10^3 km (b) 1.28×10^4 km
(c) 6.4×10^3 km (d) 1.6×10^3 km

Sol. (a) The energy required for taking a satellite upto a height h from earth's surface is the difference between the energy at h height and energy at surface, then

$\Rightarrow$ $E_1 = U_f - U_i$

$$E_1 = -\frac{GM_em}{R_e+h} + \frac{GM_em}{R_e} \qquad \text{...(i)}$$

(where, U = potential energy)

$\therefore$ Orbital velocity of satellite,

$$v_o = \sqrt{\frac{GM_e}{(R_e+h)}} \qquad \text{(where, } M_e = \text{ mass of earth)}$$

So, energy required to perform circular motion,

$$\Rightarrow \quad E_2 = \frac{1}{2}mv_o^2 = \frac{GM_em}{2(R_e+h)} \qquad \text{...(ii)}$$

According to the question, $E_1 = E_2$

$$\therefore \quad \frac{-GM_em}{R_e+h} + \frac{GM_em}{R_e} = \frac{GM_em}{2(R_e+h)}$$

$$\Rightarrow \quad 3R_e = 2R_e + 2h$$

$$h = \frac{R_e}{2}$$

As, radius of earth, $R_e \approx 6.4\times 10^3$ km

Hence, $h = \frac{6.4\times 10^3}{2}$ km or 3.2×10^3 km

Escape Velocity

It is the minimum velocity with which a body must be projected from the surface of the earth so that it escapes the gravitational field of the earth. We can also say that a body, projected with escape velocity, will be able to go to a point which is at infinite distance from the earth.

The escape velocity from the surface of a planet of mass M, radius R and acceleration due to gravity g is

$$\Rightarrow \quad v_e = \sqrt{\frac{2GM}{R}} = \sqrt{2gR}$$

Substituting the values of $g = 9.81 \text{ ms}^{-2}$ and $R = 6400$ km, we get

$$v_e = 11.2 \text{ kms}^{-1}$$

Example 9. *A planet in a distant solar system is 10 times more massive than the earth and its radius is 10 times smaller. Given that the escape velocity from the earth is 11 kms^{-1}, the escape velocity from the surface of the planet would be* **[AIEEE 2008]**

(a) 1.1 kms^{-1} (b) 11 kms^{-1}
(c) 110 kms^{-1} (d) 0.11 kms^{-1}

Sol. (c) Mass of planet, $M_p = 10M_e$, where M_e is mass of earth.

Radius of planet, $R_p = \frac{R_e}{10}$, where R_e is radius of earth.

Escape velocity is given by $v_e = \sqrt{\frac{2GM}{R}}$

For planet, $v_p = \sqrt{\frac{2G\times M_p}{R_p}} = \sqrt{\frac{100\times 2GM_e}{R_e}}$

$= 10\times v_e = 10\times 11$

$= 110 \text{ kms}^{-1}$

Motion of a Satellite

Any body that revolves around earth or any planet is called *satellite*. These can be natural (*e.g.* moon) or artifical. The artifical satellites are man made satellites launched from the earth. The path of these satellites are elliptical with the centre of earth at a focus.

Orbital Velocity

Orbital velocity is the velocity required to put the satellite into its orbit around the earth and given as

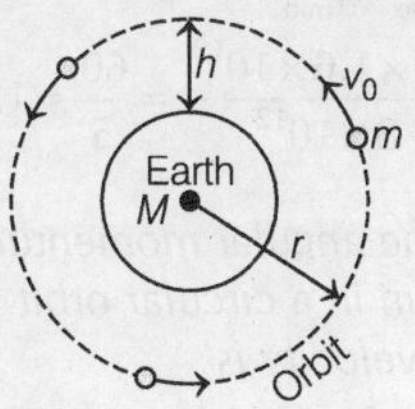

$$\Rightarrow \quad v_o = \sqrt{\frac{GM}{r}}$$

$$\text{or} \quad v_o = \sqrt{\frac{GM}{R+h}}$$

where, h = height of satellite above the earth's surface.

For a satellite very close to the earth's surface, *i.e.* $h << R$ or $r \approx R$

$$v_o = \sqrt{\frac{GM}{r}} = \frac{GM}{R} = \sqrt{gR}$$

$$\therefore \quad v_e = \sqrt{2}v_o$$

Time Period

The time taken by a satellite to complete one revolution around the earth is called the *time period.* It is given by

$$T = \frac{2\pi r}{v_o} = 2\pi r\sqrt{\frac{r}{GM}}$$

$$\Rightarrow \quad T = \frac{2\pi r^{3/2}}{\sqrt{GM}}$$

If $r \approx R$, then $T = 2\pi\sqrt{\frac{R}{g}}$.

Total Energy of the Satellite

Kinetic energy of satellite, $K = \frac{1}{2}mv_0^2 = \frac{GMm}{2r}$

Potential energy of satellite, $U = -\frac{GMm}{r}$

and total energy of satellite,

$$E = K + U = -\frac{GMm}{2r} = -K$$

Binding Energy of Satellite

It is the energy required to remove the satellite from its orbit and take it to infinity.

Binding energy $= -E = +\frac{GMm}{2r}$

Geostationary Satellite

If an artificial satellite revolves around the earth in an equatorial plane with a time period of 24 h in the same sense as that of the earth, then it will appear stationary to the observer on the earth. Such a satellite is known as a *geostationary satellite* or *parking satellite.*

Example 10. *A satellite is revolving in a circular orbit at a height h from the earth surface such that h << R, where R is the radius of the earth. Assuming that, the effect of earth's atmosphere can be neglected the minimum increase in the speed required, so that the satellite could escape from the gravitational field of earth is* **[JEE Main 2019]**

(a) $\sqrt{\frac{gR}{2}}$ (b) $\sqrt{gR}$

(c) $\sqrt{2gR}$ (d) $\sqrt{gR}(\sqrt{2}-1)$

Sol. (*d*) Orbital velocity of the satellite is given as $v_o = \sqrt{\frac{GM}{R+h}}$

Since, $R >> h$

$$\therefore \quad v_o = \sqrt{\frac{GM}{R}} = \sqrt{gR} \quad \left[\because g = \frac{GM}{R^2}\right]$$

Escape velocity of the satellite,

$$v_e = \sqrt{\frac{2GM}{R+h}} = \sqrt{\frac{2GM}{R}} = \sqrt{2gR}$$

Since, we know that in order to escape the earth's gravitational field a satellite must get escape velocity.

$\therefore$ Change in velocity,

$$\Delta v = v_e - v_o = \sqrt{gR}(\sqrt{2}-1)$$

Example 11. *A 400 kg satellite is in a circular orbit of radius 2 R_E about the earth. Amount of energy required to transfer it to a circular orbit of radius 4R_E is*

(a) 3.13×10^9 J

(b) 5.29×10^2 J

(c) 3.13×10^3 J

(d) 5.29×10^9 J

Sol. (a) Initially, $E_i = -\frac{GM_Em}{4R_E}$

While finally, $E_f = -\frac{GM_Em}{8R_E}$

The change in the total energy is

$$\Delta E = E_f - E_i = \frac{GM_Em}{8R_E} = \left(\frac{GM_E}{R_E^2}\right)\frac{mR_E}{8}$$

$$\Delta E = \frac{gmR_E}{8} = \frac{9.81\times400\times6.37\times10^6}{8}$$

$$\Delta E = 3.13\times10^9 \text{ J}$$

Kepler's Laws of Planetary Motion

Kepler discovered three empirical laws which accurately describe the motion of the planets. These laws are

Kepler's First Law or Law of Orbits

All the planets move around the sun in an elliptical orbit with sun at one of the focus of ellipse.

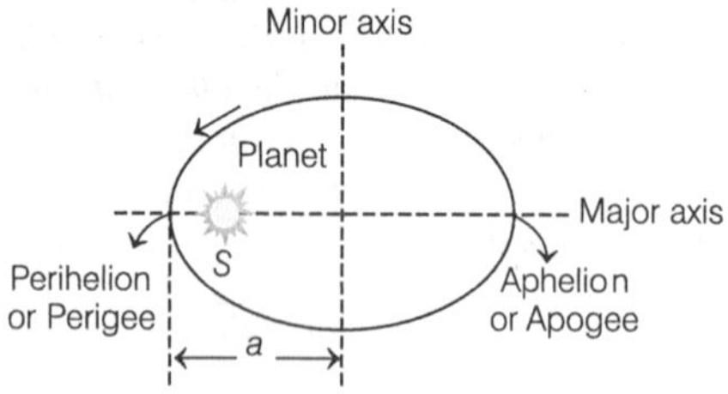

The point when the planet is nearest to the sun is termed as **perihelion** and the farthest one is known as **aphelion**.

Kepler's Second Law or Law of Areas

The line joining the sun to the planet sweeps out equal areas in equal intervals of time, *i.e.* areal velocity of the planet w.r.t. sun is constant.

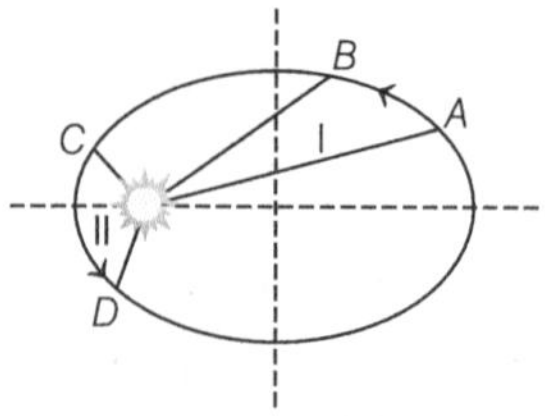

This law indicates that a planet moves faster near the sun and slowly when away from the sun.

According to second law, area of region I = area of region II, where the planet takes same time to move from A to B and from C to D.

Kepler's Third Law or Law of Periods

The square of the planet's time period of revolution is directly proportional to the cube of semi-major axis of its orbit.

$$T^2 \propto a^3$$

where, a is the semi-major axis.

Example 12. *The maximum and minimum distances of a comet from the sun are 8×10^{12} m and 1.6×10^{12} m. If its velocity when nearest to the sun is 60 ms^{-1}. What will be its velocity (in ms^{-1}) when it is farthest?*

(a) 12 (b) 60
(c) 112 (d) 6

Sol. (a) By conservation of angular momentum,

$$mvr = \text{constant}$$

i.e. $$v_{min} \times r_{max} = v_{max} \times r_{min}$$

$\therefore$ $$v_{min} = \frac{60 \times 1.6 \times 10^{12}}{8 \times 10^{12}} = \frac{60}{5} = 12 \text{ ms}^{-1}$$

Example 13. *If the angular momentum of a planet of mass m, moving around sun in a circular orbit is L about the centre of the sun , its areal velocity is* **[JEE Main 2019]**

(a) $\frac{4L}{m}$ (b) $\frac{2L}{m}$ (c) $\frac{L}{2m}$ (d) $\frac{L}{m}$

Sol. (c) According to Kepler's second law, the line joining the planet to sun sweeps out equal areas in equal interval of time. This means the rate of change of area with time is constant.

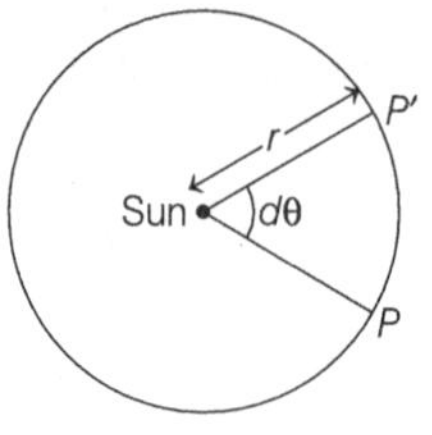

The area covered from P to P' is dA, which is given by

$$dA = \frac{d\theta}{2\pi} \times \pi r^2$$

$\Rightarrow$ $$dA = \frac{1}{2} r^2 d\theta \text{ or } \frac{dA}{dt} = \frac{1}{2} r^2 \frac{d\theta}{dt}$$

where, $\frac{dA}{dt}$ = areal velocity.

$$\frac{dA}{dt} = \frac{1}{2} r^2 \omega \quad \text{or} \quad \frac{dA}{dt} = \frac{1}{2} r^2 \cdot \frac{L}{mr^2}$$

(because angular momentum, $L = mr^2\omega$)

$$\frac{dA}{dt} = \frac{L}{2m}$$

Practice Exercise

ROUND I Topically Divided Problems

Universal Law of Gravitation

1. Imagine a light planet revolving around a very massive star in a circular orbit of radius r with a period of revolution T. If the gravitational force of attraction between the planet and the star is proportional to $R^{-3/2}$, then T^2 is proportional to

(a) R^3 (b) $R^{5/2}$
(c) $R^{3/2}$ (d) $R^{7/2}$

2. If a planet of given density were made larger its force of attraction for an object on its surface would increase because of planet's greater mass but would decrease because of the greater distance from the object to the centre of the planet. Which effect predominate?

(a) Increase in mass
(b) Increase in radius
(c) Both affect attraction equally
(d) None of the above

3. A uniform ring of mass M and radius r is placed directly above a uniform sphere of mass 8 M and of same radius R. The centre of the ring is at a distance of $d = \sqrt{3}\,R$ from the centre of the sphere. The gravitational attraction between the sphere and the ring is

(a) $\dfrac{GM^2}{R^2}$ (b) $\dfrac{3\,GM^2}{2\,R^2}$
(c) $\dfrac{2\,GM^2}{\sqrt{2}\,R^2}$ (d) $\dfrac{\sqrt{3}\,GM^2}{R^2}$

4. A solid sphere of uniform density and radius R applies a gravitational force of attraction equal to F_1 on a particle placed at P, distance $2R$ from the centre O of the sphere. A spherical cavity of radius $R/2$ is now made in the sphere as shown in figure. The sphere with cavity now applies an gravitational force F_2 on same particle placed at P. The ratio F_2/F_1 will be

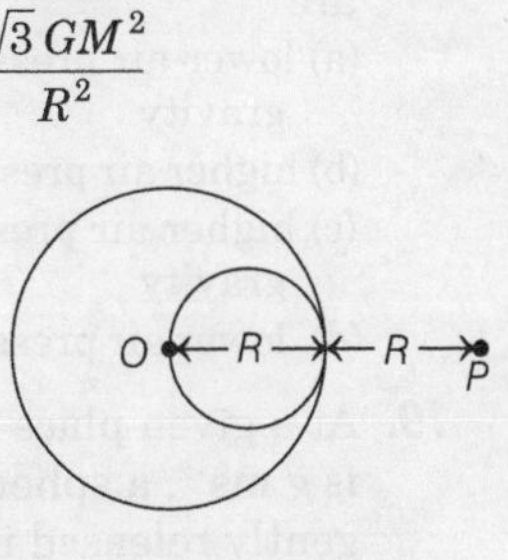

(a) 1/2 (b) 7/9
(c) 3 (d) 7

5. Both earth and moon are subject to the gravitational force of the sun. As observed from the sun, the orbit of the moon **[NCERT Exemplar]**

(a) will be elliptical
(b) will not be strictly elliptical because the total gravitational force on it is not central
(c) is not elliptical but will necessarily be a closed curve
(d) deviates considerably from being elliptical due to influence of planets other than earth

6. Different points in earth are at slightly different distances from the sun and hence experience different forces due to gravitation. For a rigid body, we know that if various forces act at various points in it, the resultant motion is as if a net force acts on the CM (centre of mass) causing translation and a net torque at the CM causing rotation around an axis through the CM for the earth-sun system (approximating the earth as a uniform density sphere) **[NCERT Exemplar]**

(a) the torque is zero
(b) the torque causes the earth to spin
(c) the rigid body result is not applicable since the earth is not even approximately a rigid body
(d) the torque causes the earth to move around the sun

7. Two astronauts have deserted their spaceship in a region of space far from the gravitational attraction of any other body. Each has a mass of 100 kg and they are 100 m apart. They are initially at rest relative to one another. How long will it be before the gravitational attraction brings them 1 cm closer together?

(a) 2.52 days (b) 1.41 days (c) 0.70 days (d) 0.41 days

8. Particles of masses $2M$, m and M are respectively at points A, B and C with $AB = \dfrac{1}{2}\,(BC)$. m is much-much smaller than M and at time $t = 0$, they are all at rest

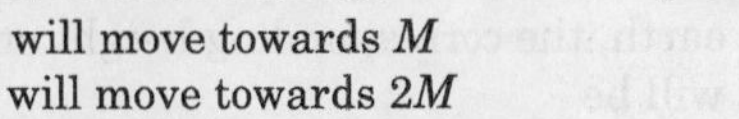

[NCERT Exemplar]

(a) m will remain at rest
(b) m will move towards M
(c) m will move towards $2M$
(d) m will have oscillatory motion

9. A spherical hollow is made in a lead sphere of radius R such that its surface touches the outside surface of the lead sphere and passes through the centre. The mass of the lead sphere before hollowing was M. The force of attraction that this sphere would exert on a particle of mass m which lies at a distance d (> R) from the centre of the lead sphere on the straight line joining the centres of the sphere and the hollow is

(a) $\frac{GMm}{d^2}$ (b) $\frac{GMm}{8d^2}$

(c) $\frac{GMm}{d^2}\left[1+\frac{1}{8\left(1+\frac{R}{2d}\right)}\right]$ (d) $\frac{GMm}{d^2}\left[1-\frac{1}{8\left(1-\frac{R}{2d}\right)^2}\right]$

10. A straight rod of length L extends from $x = a$ to $x = L + a$. The gravitational force it exerts on a point mass m at $x = 0$, if the mass per unit length of the rod is $A + Bx^2$, is given by **[JEE Main 2019]**

(a) $Gm\left[A\left(\frac{1}{a+L}-\frac{1}{a}\right)-BL\right]$

(b) $Gm\left[A\left(\frac{1}{a+L}-\frac{1}{a}\right)+BL\right]$

(c) $Gm\left[A\left(\frac{1}{a}-\frac{1}{a+L}\right)+BL\right]$

(d) $Gm\left[A\left(\frac{1}{a}-\frac{1}{a+L}\right)-BL\right]$

11. Four identical particles of mass M are located at the corners of a square of side a. What should be their speed, if each of them revolves under the influence of other's gravitational field in a circular orbit circumscribing the square? **[JEE Main 2019]**

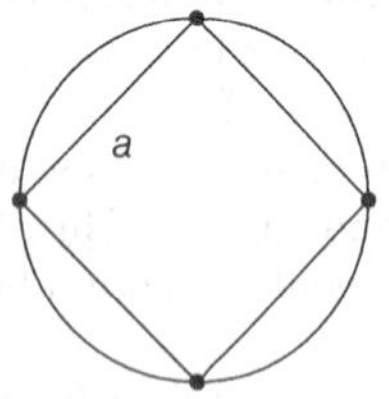

(a) $1.35\sqrt{\frac{GM}{a}}$ (b) $1.16\sqrt{\frac{GM}{a}}$

(c) $1.21\sqrt{\frac{GM}{a}}$ (d) $1.41\sqrt{\frac{GM}{a}}$

Acceleration due to Gravity

12. The acceleration due to gravity on a planet is 1.96 ms^{-2}. If it is safe to jump from a height of 3 m on the earth, the corresponding height on the planet will be

(a) 3 m (b) 6 m (c) 9 m (d) 15 m

13. A thief stole a box full of valuable articles of weight w and while carrying it on his head jumped down from a wall of height h from the ground. Before he reaches the ground, he experienced a load

(a) zero (b) $w/2$ (c) w (d) $2w$

14. The mass of the moon is 1/8 of the earth but the gravitational pull is 1/6 of the earth. It is due to the fact that

(a) moon is the satellite of the earth
(b) the radius of the earth is 8.6 the moon
(c) the radius of the earth is $\sqrt{8/6}$ of the moon
(d) the radius of the moon is 6/8 of the earth

15. Average density of the earth **[AIEEE 2005]**

(a) does not depend on g
(b) is a complex function of g
(c) is directly proportional to g
(d) is inversely proportional to g

16. The change in the value of g at a height h above the surface of the earth is the same as at a depth d below the surface of earth. When both d and h are much smaller than the radius of earth, then which one of the following is correct? **[AIEEE 2005]**

(a) $d = \frac{h}{2}$ (b) $d = \frac{3h}{2}$ (c) $d = 2h$ (d) $d = h$

17. The earth is an approximate sphere. If the interior contained matter which is not of the same density everywhere, then on the surface of the earth, the acceleration due to gravity **[NCERT Exemplar]**

(a) will be directed towards the centre but not the same everywhere
(b) will have the same value everywhere but not directed towards the centre
(c) will be same everywhere in magnitude directed towards the centre
(d) cannot be zero at any point

18. There is a mine of depth about 2.0 km. In this mine the conditions as compared to those at the surface are

(a) lower air pressure, higher acceleration due to gravity
(b) higher air pressure, lower acceleration due to gravity
(c) higher air pressure, higher acceleration due to gravity
(d) lower air pressure, lower acceleration due to gravity

19. At a given place where, acceleration due to gravity is $g \text{ ms}^{-2}$, a sphere of lead of density $d \text{ kgm}^{-3}$ is gently released in a column of liquid of density $\rho \text{ kgm}^{-3}$. If $d > \rho$, the sphere will

(a) fall vertically with an acceleration of $g \text{ ms}^{-2}$
(b) fall vertically with no acceleration
(c) fall vertically with an acceleration $g\left(\frac{d-\rho}{d}\right)$
(d) fall vertically with an acceleration ρ/d

20. At a distance 320 km above the surface of earth, the value of acceleration due to gravity will be lower than its value on the surface of the earth by nearly (radius of earth = 6400 km)
(a) 2% (b) 6%
(c) 10% (d) 14%

21. The acceleration due to gravity at a height (1/20)th of the radius of the earth above the earth surface is 9 ms^{-2}. Its value at a point at an equal distance below the surface of the earth (in ms^{-2}) is about
(a) 8.5 (b) 9.5
(c) 9.8 (d) 11.5

22. If the value of g acceleration due to gravity at earth surface is 10 ms^{-2}, its value in ms^{-2} at the centre of the earth, which is assumed to be a sphere of radius R metre and uniform mass density is
(a) 5 (b) $10/R$
(c) $10/2R$ (d) zero

23. The height at which the acceleration due to gravity decreases by 36% of its value on the surface of the earth. (The radius of the earth is R)
(a) $\frac{R}{6}$ (b) $\frac{R}{4}$
(c) $\frac{R}{2}$ (d) $\frac{2}{3}R$

24. Two spherical planets A and B have same mass but densities in the ratio 8 : 1. For these planets, the ratio of acceleration due to gravity at the surface of A to its value at the surface of B is
(a) 1 : 4 (b) 1 : 2
(c) 4 : 1 (d) 8 : 1

25. If the radius of the earth were to shrink by 1% and its mass remaining same, the acceleration due to gravity on the earth's surface would
(a) decrease by 2%
(b) remain unchanged
(c) increase by 2%
(d) become zero

26. If the force inside the earth surface varies as r^n, where r is the distance of body from the centre of earth, then the value of n will be
(a) –1 (b) –2
(c) 1 (d) 2

27. The height at which the acceleration due to gravity becomes $\frac{g}{9}$ (where, g is the acceleration due to gravity on the surface of the earth) in terms of R (the radius of the earth) is **[AIEEE 2009]**
(a) $2R$ (b) $\frac{R}{\sqrt{3}}$
(c) $\frac{R}{2}$ (d) $\sqrt{2}R$

28. The variation of acceleration due to gravity g with distance d from centre of the Earth is best represented by (R = Earth's radius) **[JEE Main 2017]**

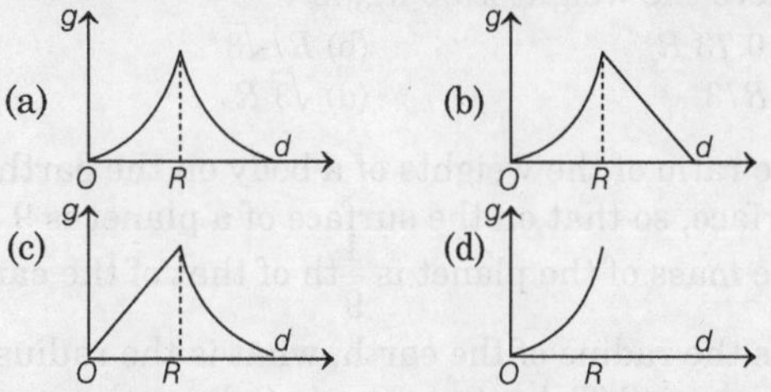

29. The depth from the surface of the earth of radius R at which the acceleration due to gravity will be 75% of the value on the surface of the earth is
(a) $R/4$ (b) $R/2$ (c) $3R/4$ (d) $R/8$

30. Two equal masses m and m are hung from a balance whose scale pans differ in height by h. If ρ is the mean density of earth, then the error in weighing machine is
(a) zero (b) $4\pi G\rho mh/3$
(c) $8\pi G\rho mh/3$ (d) $2\pi G\rho mh/3$

31. The speed of earth's rotation about its axis is ω. Its speed is increased to x times to make the effective acceleration due to gravity equal to zero at the equator, then value of x is around
(g = 10 ms^{-2}, R = 6400 km)
(a) 1 (b) 8.5 (c) 17 (d) 34

32. The bodies situated on the surface of earth at its equator, becomes weightless, when the earth has KE about it axis
(a) mgR (b) $2\,mgR/5$ (c) $MgR/5$ (d) $5\,MgR/2$

33. The height h at which the weight of a body will be the same as that at the same depth h from the surface of the earth is
(Radius of the earth is R and effect of the rotation of the earth is neglected) **[JEE Main 2020]**
(a) $\frac{\sqrt{5}}{2}(R-R)$ (b) $\frac{R}{2}$
(c) $\frac{\sqrt{5}R-R}{2}$ (d) $\frac{\sqrt{3}R-R}{2}$

34. At what height above the earth's surface, does the force of gravity decrease by 10%? The radius of the earth is 6400 km?
(a) 345.60 km (b) 687.20 km
(c) 1031.8 km (d) 12836.80 km

35. The ratio of acceleration due to gravity at a height h above the surface of the earth and at a depth h below the surface of the earth for h < radius of earth
(a) is constant
(b) increases linearly with h
(c) decreases linearly with h
(d) decreases parabolically with h

36. If a man weights 90 kg on the surface of earth, the height above the surface of the earth of radius R, where the weight is 30 kg, is

(a) $0.73\ R$ (b) $R/\sqrt{3}$
(c) $R/3$ (d) $\sqrt{3}\ R$

37. The ratio of the weights of a body on the earth's surface, so that on the surface of a planet is 9 : 4. The mass of the planet is $\frac{1}{9}$th of that of the earth. If R is the radius of the earth, what is the radius of the planet? (Take, the planets to have the same mass density) **[JEE Main 2019]**

(a) $\frac{R}{3}$ (b) $\frac{R}{4}$ (c) $\frac{R}{9}$ (d) $\frac{R}{2}$

38. The value of acceleration due to gravity at earth's surface is $9.8\ \text{ms}^{-2}$. The altitude above its surface at which the acceleration due to gravity decreases to $4.9\ \text{ms}^{-2}$, is close to (Take, radius of earth $= 6.4 \times 10^6$ m) **[JEE Main 2019]**

(a) 9.0×10^6 m (b) 2.6×10^6 m
(c) 6.4×10^6 m (d) 1.6×10^6 m

Gravitational Field, Potential and Potential Energy

39. The gravitational field due to a mass distribution is $E = k/x^3$ in the x-direction (k is a constant). Taking the gravitational potential to be zero at infinity, its value at a distance $x/\sqrt{2}$ is

(a) k/x (b) $k/2x$
(c) k/x^2 (d) $k/2x^2$

40. Which of the following graphs represents correctly the variation of the intensity of gravitational field (I) with the distance (r) from the centre of a spherical shell of mass M and radius a?

(a)
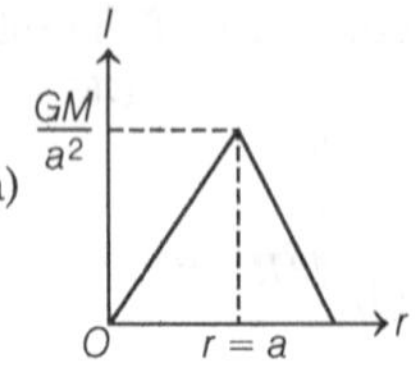

(b)
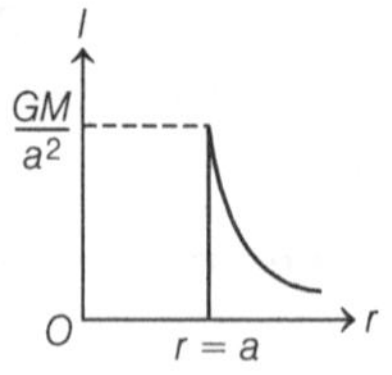

(c)
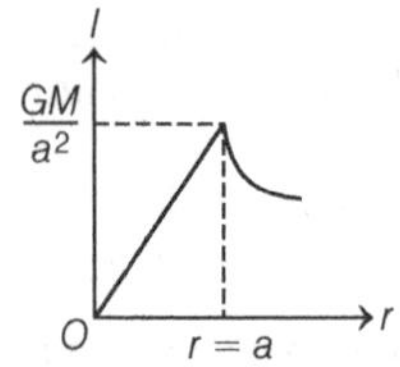

(d)
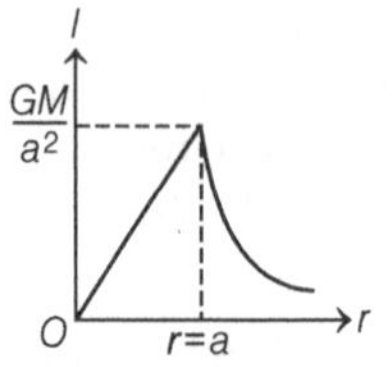

41. Two bodies of masses 2 kg and 8 kg are separated by a distance of 9 m. The point where the resultant gravitational field intensity is zero is at a distance of

(a) 4.5 m from each mass (b) 6 m from 2 kg
(c) 6 m from 8 kg (d) 2.5 m from 2 kg

42. There are two bodies of masses 100000 kg and 1000 kg separated by a distance of 1 m. At what distance (in metre) from the smaller body, the intensity of gravitational field will be zero?

(a) 1/9 (b) 1/10
(c) 1/11 (d) 10/11

43. Consider two solid spheres of radii $R_1 = 1$ m, $R_2 = 2$ m and masses M_1 and M_2, respectively. The gravitational field due to spheres 1 and 2 are shown. The value of $\frac{M_1}{M_2}$ is **[JEE Main 2020]**

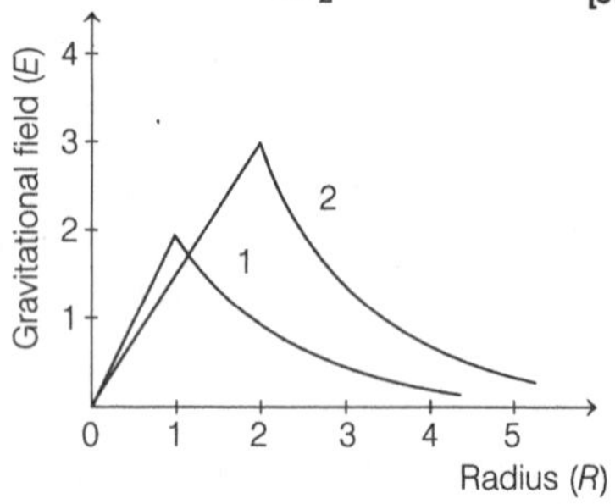

(a) $\frac{1}{3}$ (b) $\frac{1}{2}$ (c) $\frac{1}{6}$ (d) $\frac{2}{3}$

44. In a certain region of space, the gravitational field is given by $-k/r$, where r is the distance and k is a constant. If the gravitational potential at $r = r_0$ be V_0, then what is the expression for the gravitational potential V?

(a) $k \log\left(\frac{r}{r_0}\right)$ (b) $k \log\left(\frac{r_0}{r}\right)$
(c) $V_0 + k \log\left(\frac{r}{r_0}\right)$ (d) $V_0 + k \log\left(\frac{r_0}{r}\right)$

45. A particle of mass m is placed at the centre of a uniform spherical shell of mass 3 m and radius R. The gravitational potential on the surface of the shell is

(a) $-\frac{Gm}{R}$ (b) $-\frac{3Gm}{R}$
(c) $-\frac{4Gm}{R}$ (d) $-\frac{2Gm}{R}$

46. The kinetic energy needed to project a body of mass m from the earth's surface (radius R) to infinity is **[AIEEE 2002]**

(a) $\frac{mgR}{2}$ (b) $2\,mgR$ (c) mgR (d) $\frac{mgR}{4}$

47. Energy required to move a body of mass m from an orbit of radius $2R$ to $3R$ is **[AIEEE 2002]**

(a) $\frac{GMm}{12R^2}$ (b) $\frac{GMm}{3R^2}$
(c) $\frac{GMm}{8R}$ (d) $\frac{GMm}{6R}$

48. A mass m is placed at a point B in the gravitational field of mass M. When the mass m is brought from B to near point A, its gravitational potential energy will
(a) remain unchanged (b) increase
(c) decrease (d) become zero

49. The gravitational potential difference between the surface of a planet and a point 20 m above it is 14 J kg^{-1}. The work done in moving a 2.0 kg mass by 8.0 m on a slope of 60° from the horizontal, is equal to
(a) 7 J (b) 9.6 J
(c) 16 J (d) 32 J

50. If W_1, W_2 and W_3 represent the work done in moving a particle from A to B along three different paths 1, 2 and 3 respectively (as shown) in a gravitional field of point mass m, then

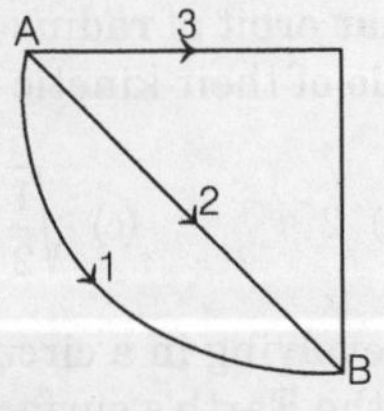

(a) $W_1 = W_2 = W_3$
(b) $W_1 > W_2 > W_3$
(c) $W_1 > W_2 < W_3$
(d) $W_1 < W_2 < W_3$

51. The mass of the earth is 6.00×10^{22} kg. The constant of gravitation $G = 6.67 \times 10^{-11} \text{Nm}^2\text{kg}^{-2}$. The potential energy of the system is -7.79×10^{28} J. The mean distance between earth and moon is
(a) 3.80×10^8 m
(b) 3.37×10^6 m
(c) 7.60×10^4 m
(d) 1.90×10^2 m

52. If g is the acceleration due to gravity on the earth's surface, the gain in the potential energy of an object of mass m raised from the surface of the earth to a height equal to the radius R of the earth, is **[AIEEE 2004]**
(a) $2\,mgR$ (b) $\frac{1}{2}mgR$
(c) $\frac{1}{4}mgR$ (d) mgR

53. The change in potential energy when a body of mass m is raised to a height nR from the centre of earth (R = radius of earth)
(a) $mgR\frac{(n-1)}{n}$ (b) $nmgR$
(c) $mgR\frac{n^2}{n^2+1}$ (d) $mgR\frac{n}{n+1}$

54. A particle of mass 10 g is kept on the surface of a uniform sphere of mass 100 kg and radius 10 cm. Find the work to be done against the gravitational force between them, to take the particle far away from the sphere. (Take, $G = 6.67 \times 10^{-11} \text{ Nm}^2/\text{kg}^2$)
(a) 13.34×10^{-10} J **[AIEEE 2005]**
(b) 3.33×10^{-10} J
(c) 6.67×10^{-9} J
(d) 6.67×10^{-10} J

55. A body of mass m rises to a height $h = R/5$ from the surface of earth, where R is the radius of earth. If g is the acceleration due to gravity at the surface of earth, the increase in potential energy is
(a) $(4/5)\,mgh$ (b) $(5/6)\,mgh$
(c) $(6/7)\,mgh$ (d) mgh

Motion of Satellite and Escape Velocity

56. What is the minimum energy required to launch a satellite of mass m from the surface of a planet of mass M and radius R in a circular orbit at an altitude of $2R$? **[JEE Main 2013]**
(a) $\frac{5GmM}{6R}$ (b) $\frac{2GmM}{3R}$
(c) $\frac{GmM}{2R}$ (d) $\frac{GmM}{3R}$

57. A solid sphere of mass M and radius a is surrounded by a uniform concentric spherical shell of thickness $2a$ and $2M$. The gravitational field at distance $3a$ from the centre will be **[JEE Main 2019]**
(a) $\frac{GM}{9a^2}$ (b) $\frac{2GM}{9a^2}$
(c) $\frac{GM}{3a^2}$ (d) $\frac{2GM}{3a^2}$

58. Two planets have masses M and $16\,M$ and their radii are a and $2a$, respectively. The separation between the centres of the planets is $10a$. A body of mass m is fired from the surface of the larger planet towards the smaller planet along the line joining their centres. For the body to be able to reach at the surface of smaller planet, the minimum firing speed needed is **[JEE Main 2020]**
(a) $2\sqrt{\frac{GM}{a}}$ (b) $4\sqrt{\frac{GM}{a}}$
(c) $\sqrt{\frac{GM^2}{ma}}$ (d) $\frac{3}{2}\sqrt{\frac{5GM}{a}}$

59. The ratio of the radii of the planets P_1 and P_2 is a. The ratio of their acceleration due to gravity is b. The ratio of the escape velocities from them will be
(a) ab (b) $\sqrt{ab}$
(c) $\sqrt{a/b}$ (d) $\sqrt{b/a}$

60. The escape velocity for a body projected vertically upwards from the surface of the earth is 11.2 kms^{-1}. If the body is projected in a direction making an angle of 45° with the vertical, the escape velocity will be
(a) 11.2 kms^{-1}
(b) $11.2 \times \sqrt{2}$ kms^{-1}
(c) 11.2×2 kms^{-1}
(d) $11.2/\sqrt{2}$ kms^{-1}

61. The escape velocity of a body depends upon mass as [AIEEE 2002]
(a) m^0
(b) m^1
(c) m^2
(d) m^3

62. The time period of an earth satellite in circular orbit is independent of [AIEEE 2004]
(a) the mass of the satellite
(b) radius of its orbit
(c) both the mass and radius of the orbit
(d) neither the mass of the satellite nor the radius of its orbit

63. If suddenly the gravitational force of attraction between earth and a satellite revolving around it becomes zero, then the satellite will [AIEEE 2002]
(a) continue to move in its orbit with same velocity
(b) move tangentially to the original orbit with the same velocity
(c) become stationary in its orbit
(d) move towards the earth

64. A satellite S is moving in an elliptical orbit around earth. The mass of the satellite is very small compared to the mass of the earth?
(a) The acceleration of S is always directed towards the centre of the earth
(b) The angular momentum of S about the centre of the earth changes in direction but its magnitude remains constant
(c) The total mechanical energy of S varies periodically with time
(d) The linear momentum of S remains constant in magnitude

65. A satellite of mass M is in a circular orbit of radius R about the centre of the earth. A meteorite of the same mass falling towards the earth collides with the satellite completely inelastically. The speeds of the satellite and the meteorite are the same just before the collision. The subsequent motion of the combined body will be [JEE Main 2019]
(a) in the same circular orbit of radius R
(b) in an elliptical orbit
(c) such that it escapes to infinity
(d) in a circular orbit of a different radius

66. Satellites orbiting the earth have finite life and sometimes debris of satellites fall to the earth. This is because, [NCERT Exemplar]
(a) the solar cells and batteries in satellites run out
(b) the laws of gravitation predict a trajectory spiralling inwards
(c) of viscous forces causing the speed of satellite and hence height to gradually decrease
(d) of collisions with other satellites

67. The escape velocity from the earth is 11 kms^{-1}. The escape velocity from a planet having twice the radius and the same mean density as the earth would be
(a) 5.5 kms^{-1}
(b) 11 kms^{-1}
(c) 15.5 kms^{-1}
(d) 22 kms^{-1}

68. The mass of the moon is 1/81 of earth's mass and its radius 1/4th that of the earth. If the escape velocity from the earth's surface is 11.2 kms^{-1}, its value for the moon will be
(a) 0.15 kms^{-1}
(b) 5 kms^{-1}
(c) 2.5 kms^{-1}
(d) 0.5 kms^{-1}

69. Two satellites A and B have masses m and $2m$, respectively. A is in a circular orbit of radius R and B is in a circular orbit of radius $2R$ around the earth. The ratio of their kinetic energies, T_A / T_B is [JEE Main 2019]
(a) $\frac{1}{2}$
(b) 2
(c) $\sqrt{\frac{1}{2}}$
(d) 1

70. A satellite is revolving in a circular orbit at a height h from the Earth's surface (radius of Earth $R, h << R$). The minimum increase in its orbital velocity required, so that the satellite could escape from the Earth's gravitational field, is close to (Neglect the effect of atmosphere) [JEE Main 2017]
(a) $\sqrt{2gR}$
(b) $\sqrt{gR}$
(c) $\sqrt{gR/2}$
(d) $\sqrt{gR}\,(\sqrt{2}-1)$

71. Planet A has mass M and radius R and planet B has half the mass and half the radius of planet A. If the escape velocities from the planets A and B are v_A and v_B respectively, then $\frac{v_A}{v_B} = \frac{n}{4}$. The value of n is [JEE Main 2020]
(a) 1
(b) 2
(c) 3
(d) 4

72. A satellite is in an elliptical orbit around a planet P. It is observed that the velocity of the satellite when it is farthest from the planet is 6 times less than that when it is closest to the planet. The ratio of distances between the satellite and the planet at closest and farthest points is [JEE Main 2020]
(a) 1 : 6
(b) 1 : 3
(c) 1 : 2
(d) 3 : 4

73. A body is moving in a low circular orbit about a planet of mass M and radius R. The radius of the orbit can be taken to be R itself. Then, the ratio of the speed of this body in the orbit to the escape velocity from the planet is [JEE Main 2020]
(a) 1
(b) $\frac{1}{\sqrt{2}}$
(c) $\sqrt{2}$
(d) 2

74. A satellite of mass m revolves around the earth of radius R at a height x from its surface. If g is the acceleration due to gravity on the surface of the earth, the orbital speed of the satellite is **[AIEEE 2004]**

(a) gx (b) $\dfrac{gR}{R-x}$

(c) $\dfrac{gR^2}{R+x}$ (d) $\left(\dfrac{gR^2}{R+x}\right)^{1/2}$

75. A satellite is moving with a constant speed v in circular orbit around the earth. An object of mass m is ejected from the satellite such that it just escapes from the gravitational pull of the earth. At the time of ejection, the kinetic energy of the object is **[JEE Main 2019]**

(a) $\dfrac{3}{2}mv^2$ (b) $2mv^2$

(c) mv^2 (d) $\dfrac{1}{2}mv^2$

Kepler's Laws

76. If the radius of earth's orbit is made (1/4)th, then duration of an year will become

(a) 8 times (b) 4 times

(c) (1/8) times (d) (1/4) times

77. In our solar system, the inter-planetary region has chunks of matter (much smaller in size compared to planets) called asteroids. They **[NCERT Exemplar]**

(a) will not move around the sun since they have very small masses compared to sun

(b) will move in an irregular way because of their small masses and will drift away into outer space

(c) will move around the sun in closed orbits but not obey Kepler's laws

(d) will move in orbits like planets and obey Kepler's laws

78. The time period of a satellite of earth is 5 h. If the separation between the earth and the satellite is increased to 4 times the previous value, the new time period will become **[AIEEE 2003]**

(a) 10 h (b) 80 h (c) 40 h (d) 20 h

79. A geostationary satellite is orbiting around an arbitrary planet P at a height of $11R$ above the surface of P, R being the radius of P. The time period of another satellite (in h) at a height of $2R$ from the surface of P is .... P has the time period of 24 h. **[JEE Main 2021]**

(a) $6\sqrt{2}$ (b) $\dfrac{6}{\sqrt{2}}$ (b) 3 (d) 5

80. The period of revolution of planet A around the sun is 8 times that B. The distance of A from the sun is how many times greater than that of B from the sun?

(a) 2 (b) 3 (c) 4 (d) 5

81. The largest and the shortest distance of the earth from the sun are r_1 and r_2, its distance from the sun when it is perpendicular to the major axis of the orbit drawn from the sun, is

(a) $\dfrac{r_1+r_2}{4}$ (b) $\dfrac{r_1r_2}{r_1+r_2}$ (c) $\dfrac{2r_1r_2}{r_1+r_2}$ (d) $\dfrac{r_1+r_2}{3}$

82. A comet of mass m moves in a highly elliptical orbit around the sun of mass M. The maximum and minimum distances of the comet from the centre of the sun are r_1 and r_2 respectively. The magnitude of angular momentum of the comet with respect to the centre of sun is

(a) $\left[\dfrac{GMr_1}{(r_1+r_2)}\right]^{1/2}$ (b) $\left[\dfrac{GMmr_1}{(r_1+r_2)}\right]^{1/2}$

(c) $\left[\dfrac{2Gm^2r_1r_2}{r_1+r_2}\right]^{1/2}$ (d) $\left[\dfrac{2GMm^2r_1r_2}{(r_1+r_2)}\right]^{1/2}$

ROUND II Mixed Bag

Only One Correct Option

1. A rocket of mass M is launched vertically from the surface of the earth with an initial speed v. Assuming the radius of the earth to be R and negligible air resistance. The maximum height attained by the rocket above the surface of the earth,

(a) $R\Big/\left(\dfrac{gR}{2v^2}-1\right)$ (b) $R\left(\dfrac{gR}{2v^2}-1\right)$

(c) $R\Big/\left(\dfrac{2gR}{v^2}-1\right)$ (d) $R\left(\dfrac{2gR}{v^2}-1\right)$

2. A spherical symmetric gravitational system of particles has a mass density $\rho = \begin{cases}\rho_0 \text{ for } r \le R\\ 0 \text{ for } r > R\end{cases}$ where ρ_0 is a constant. A test mass can undergo circular motion under the influence of the gravitational field of particles. Its speed v as a function of distance $r (0 < r < \infty)$ from the centre of the system is represent by

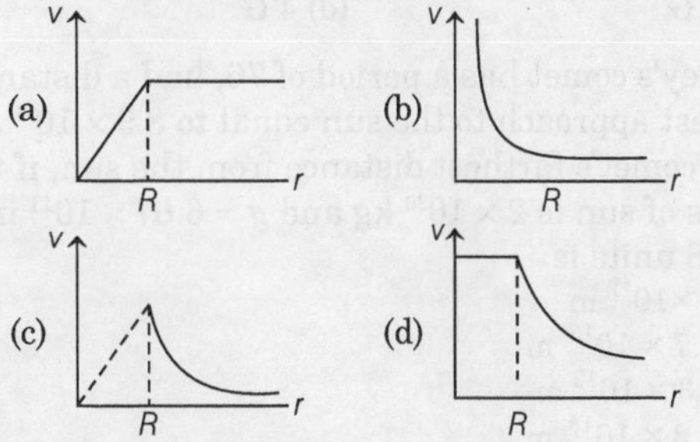

3. The correct graph representing the variation of total energy (E), kinetic energy (K) and potential energy (U) of a satellite with its distance from the centre of earth is

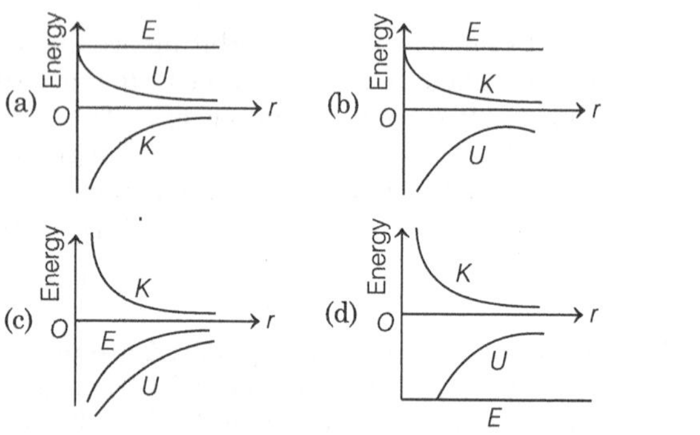

4. A simple pendulum has a time period T_1 when on the earth's surface and T_2 when taken to a height $2R$ above the earth's surface when R is $2R$ above earth's surface where R is the radius of the earth. The value of (T_1/T_2) is **[Kerala CET 2006]**

(a) 1/9 (b) 1/3
(c) $\sqrt{3}$ (d) 9

5. The escape velocity on the surface of earth is 11.2 kms^{-1}. If mass and radius of a planet is 4 and 2 times respectively, than that of earth, the escape velocity on the planet **[BVP Engg. 2006]**

(a) 11.2 kms^{-1}
(b) 1.12 kms^{-1}
(c) 22.4 kms^{-1}
(d) 15.8 kms^{-1}

6. The time period of a geostationary satellite at a height 36000 km is 24 h. A spy satellite orbits very close to earth surface (R = 6400 km). What will be its time period? **[Orissa JEE 2008]**

(a) 4 h (b) 1 h
(c) 2 h (d) 1.5 h

7. Infinite number of masses, each of 1 kg are placed along the x-axis at $x = +1\,m, \pm 2\,m_1, \pm 4\,m, \pm 8\,m, \pm 16\,m\ldots$ The magnitude of the resultant gravitational potential in terms of gravitational constant G at the origin ($x = 0$) is **[Kerala CET 2008]**

(a) $G/2$ (b) G
(c) $2\,G$ (d) $4\,G$

8. Halley's comet has a period of 76, had a distance of closest approach to the sun equal to 8.9×10^{10} m. The comet's farthest distance from the sun, if the mass of sun is 2×10^{30} kg and $g = 6.67 \times 10^{11}$ in MKS units is

(a) 2×10^{12} m
(b) 2.7×10^{13} m
(c) 5.3×10^{12} m
(d) 5.3×10^{13} m

9. The potential energy of gravitational interaction of a point mass m and a thin uniform rod of mass M and length l, if they are located along a straight line at distance a from each other is

(a) $U = \dfrac{GMm}{a}\log_e\left(\dfrac{a+l}{a}\right)$

(b) $U = GMm\left(\dfrac{1}{a} - \dfrac{1}{a+l}\right)$

(c) $U = \dfrac{GMm}{l}\log_e\left(\dfrac{a+l}{a}\right)$

(d) $U = -\dfrac{GMm}{a}$

10. The work that must be done in lifting a body of weight P from the surface of the earth to a height h is

(a) $\dfrac{PRh}{R-h}$ (b) $\dfrac{R+h}{PRh}$
(c) $\dfrac{PRh}{R+h}$ (d) $\dfrac{R-h}{PRh}$

11. A spaceship is launched into a circular orbit close to earth's surface. The additional velocity that should be imparted to the spaceship in the orbit to overcome the gravitational pull is (Radius of earth = 6400 km and g = 9.8 ms^{-2})

(a) 11.2 kms^{-1} (b) 8 kms^{-1}
(c) 3.2 kms^{-1} (d) 1.5 kms^{-1}

12. Two identical thin rings each of radius R are coaxially placed at a distance R. If the rings have a uniform mass distribution and each has mass m_1 and m_2 respectively, then the work done in moving a mass m from centre of one ring to that of the other is

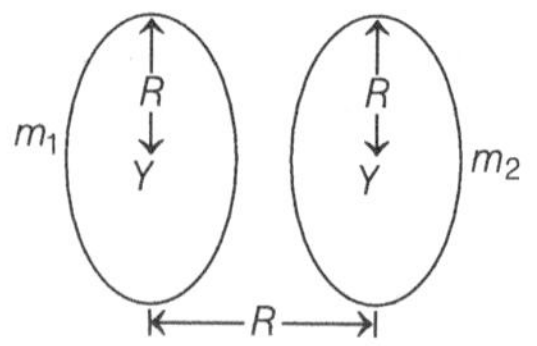

(a) $\dfrac{Gmm_1(\sqrt{2}+1)}{m_2R}$ (b) $\dfrac{Gm(m_1-m_2)(\sqrt{2}-1)}{\sqrt{2}R}$
(c) $\dfrac{Gm\sqrt{2}(m_1+m_2)}{R}$ (d) zero

13. Four particles, each of mass M and equidistant from each other, move along a circle of radius R under the action of their mutual gravitational attraction, the speed of each particle is **[JEE Main 2014]**

(a) $\sqrt{\dfrac{GM}{R}}$ (b) $\sqrt{2\sqrt{2}\dfrac{GM}{R}}$
(c) $\sqrt{\dfrac{GM}{R}(1+2\sqrt{2})}$ (d) $\dfrac{1}{2}\sqrt{\dfrac{GM}{R}(1+2\sqrt{2})}$

14. A thin uniform annular disc (see figure) of mass M has outer radius $4R$ and inner radius $3R$. The work required to take a unit mass from point P on its axis to infinity is

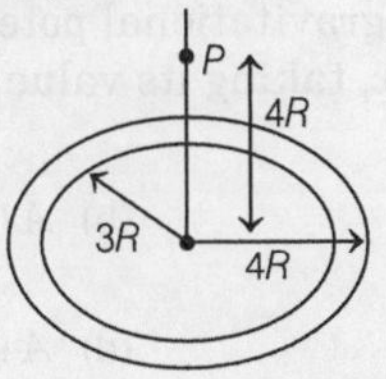

(a) $\frac{2GM}{7R}(4\sqrt{2}-5)$ (b) $-\frac{2GM}{7R}(4\sqrt{2}-5)$

(c) $\frac{GM}{4R}$ (d) $\frac{2GM}{5R}(\sqrt{2}-1)$

15. If satellite is revolving around a planet of mass M in an elliptical orbit of semi-major axis a, find the orbital speed of the satellite when it is at a distance r from the focus.

(a) $v^2 = GM\left[\frac{2}{r}-\frac{1}{a}\right]$ (b) $v^2 = GM\left[\frac{2}{r^2}-\frac{1}{a}\right]$

(c) $v^2 = GM\left[\frac{2}{r^2}-\frac{1}{a^2}\right]$ (d) $v^2 = GM\left[\frac{2}{r}-\frac{1}{a^2}\right]$

16. The gravitational force between a point like mass M and an infinitely long, thin rod of linear mass density perpendicular to distance L from M is

(a) $\frac{MG\lambda}{L}$ (b) $\frac{1}{2}\frac{MG\lambda}{L}$

(c) $\frac{2MG\lambda}{L^2}$ (d) infinite

17. The acceleration due to gravity on the earth's surface at the poles is g and angular velocity of the earth about the axis passing through the pole is ω. An object is weighed at the equator and at a height h above the poles by using a spring balance. If the weights are found to be same, then h is ($h << R$, where R is the radius of the earth) **[JEE Main 2020]**

(a) $\frac{R^2\omega^2}{2g}$ (b) $\frac{R^2\omega^2}{g}$ (c) $\frac{R^2\omega^2}{4g}$ (d) $\frac{R^2\omega^2}{8g}$

18. Two equal mases m and m are hung from balance whose scale pans differ in vertical height by h. Calculate the error in weighing, if any, in terms of density of earth ρ.

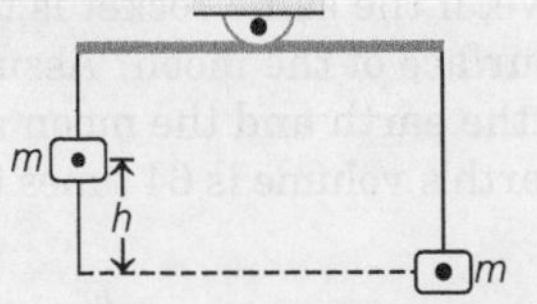

(a) $\frac{2}{3}\pi\rho R^3 Gm$ (b) $\frac{8}{3}\pi\rho Gmh$

(c) $\frac{8}{3}\pi\rho R^3 Gm$ (d) $\frac{4}{3}\pi\rho Gm^2 h$

19. How will you weight the sun, *i.e.* estimate its mass? You will need to know the period of one of its planets and the radius of the planetary orbit. The mean orbital radius of the earth around the sun is 1.5×10^8 km, then the mass of the sun is **[NCERT Exemplar]**

(a) 4×10^{30} kg (b) 5×10^{30} kg

(c) 2×10^{30} kg (d) 3×10^{30} kg

20. Two satellites S_1 and S_2 revolve around a planet in coplanar circular orbits in the same sense. Their periods of revolution are 1 h and 8 h respectively. The radius of orbit of S_1 is 10^4 km. When S_2 is closest to S_1, the speed of S_2 relative to S_1 is

(a) $\pi\times10^4$ km h^{-1}

(b) $2\pi\times10^4$ km h^{-1}

(c) $3\pi\times10^4$ km h^{-1}

(d) $4\pi\times10^4$ km h^{-1}

21. If a planet was suddenly stopped in its orbit, k suppose to be circular, find how much time will it take in falling onto the sun.

(a) $\sqrt{2}/8$ times the period of the planet's revolution

(b) $4\sqrt{2}$ times the period of the planet's revolution

(c) $3\sqrt{2}$ times the period of the planet's revolution

(d) 9 times the period of the planet's revolution

22. Distance between the centres of two stars is $10a$. The masses of these stars are M and $16M$ and their radii a and $2a$ respectively. A body of mass m is fired straight from the surface of the larger star towards the smaller star. The minimum initial speed for the body to reach the surface of smaller star is

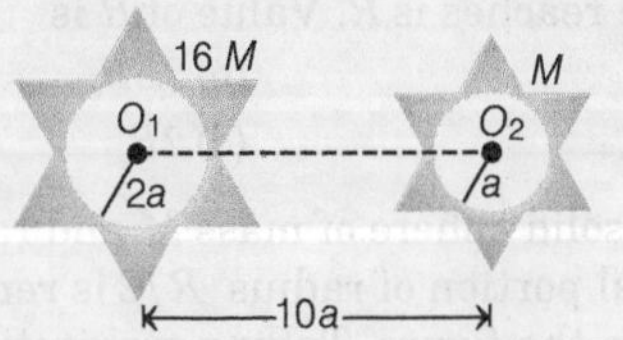

(a) $\frac{2}{3}\sqrt{\frac{GM}{a}}$ (b) $\frac{3}{2}\sqrt{\frac{5GM}{a}}$

(c) $\frac{2}{3}\sqrt{\frac{5GM}{a}}$ (d) $\frac{3}{2}\sqrt{\frac{GM}{a}}$

23. The mass density of a planet of radius R varies with the distance r from its centre as $\rho(r) = \rho_0\left(1-\frac{r^2}{R^2}\right)$, then the gravitational field is maximum at **[JEE Main 2020]**

(a) $r = \sqrt{\frac{3}{4}}R$ (b) $r = R$

(c) $r = \sqrt{\frac{5}{9}}R$ (d) $r = \frac{1}{\sqrt{3}}R$

24. A body A of mass m is moving in a circular orbit of radius R about a planet. Another body B of mass $\frac{m}{2}$ collides with A with a velocity which is half $\left(\frac{v}{2}\right)$, the instantaneous velocity v of A. The collision is completely inelastic, then the combined body [JEE Main 2020]

(a) escapes from the planet's gravitational field
(b) starts moving in an elliptical orbit around the planet
(c) falls vertically downward towards the planet
(d) continues to move in a circular orbit

25. Two stars of masses 3×10^{31} kg each and at distance 2×10^{11} m rotate in a plane about their common centre of mass O. A meteorite passes through O moving perpendicular to the star's rotation plane. In order to escape from the gravitational field of this double star, the minimum speed that meteorite should have at O is (Take, gravitational constant, ($G = 6.67 \times 10^{-11}$ N-m^{2}kg^{-2}) [JEE Main 2019]

(a) 2.8×10^5 m/s (b) 3.8×10^4 m/s
(c) 2.4×10^4 m/s (d) 1.4×10^5 m/s

26. A satellite is moving in a low nearly circular orbit around the earth. Its radius is roughly equal to that of the earth's radius R_e. By firing rockets attached to it, its speed is instantaneously increased in the direction of its motion, so that it becomes $\sqrt{\frac{3}{2}}$ times larger. Due to this, the farthest distance from the centre of the earth that the satellite reaches is R. Value of R is [JEE Main 2020]

(a) $4R_e$ (b) $2.5R_e$
(c) $3R_e$ (d) $2R_e$

27. From a solid sphere of mass M and radius R, a spherical portion of radius $R/2$ is removed as shown in the figure. Taking gravitational potential $V = 0$ at $r = \infty$, the potential at the centre of the cavity thus formed is (G = gravitational constant) [JEE Main 2015]

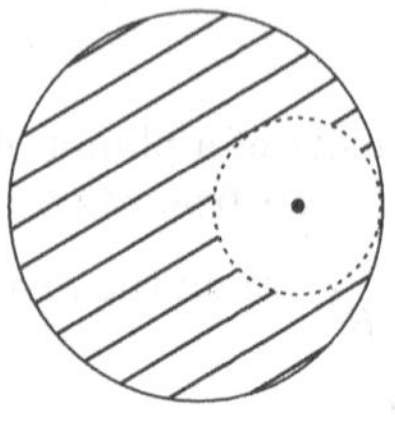

(a) $\frac{-GM}{2R}$ (b) $\frac{-GM}{R}$
(c) $\frac{-2GM}{3R}$ (d) $\frac{-2GM}{R}$

28. On the X-axis and at a distance x from the origin, the gravitational field due to a mass distribution is given by $\frac{Ax}{(x^2 + a^2)^{3/2}}$ in the x-direction. The magnitude of gravitational potential on the X-axis at a distance x, taking its value to be zero at infinity, is [JEE Main 2020]

(a) $\frac{A}{(x^2 + a^2)^{1/2}}$ (b) $A(x^2 + a^2)^{3/2}$
(c) $\frac{A}{(x^2 + a^2)^{3/2}}$ (d) $A(x^2 + a^2)^{1/2}$

29. The mass density of a spherical galaxy varies as $\frac{K}{r}$ over a large distance r from its centre. In that region, a small star is in a circular orbit of radius R. Then, the period of revolution T depends on R as [JEE Main 2020]

(a) $T \propto R$ (b) $T^2 \propto R^3$
(c) $T^2 \propto R$ (d) $T^2 \propto \frac{1}{R^3}$

30. A test particle is moving in a circular orbit in the gravitational field produced by mass density $\rho(r) = \frac{K}{r^2}$. Identify the correct relation between the radius R of the particle's orbit and its period T [JEE Main 2019]

(a) $\frac{T^2}{R^3}$ is a constant (b) $\frac{T}{R^2}$ is a constant
(c) TR is a constant (d) $\frac{T}{R}$ is a constant

31. A spaceship orbits around a planet at a height of 20 km from its surface. Assuming that only gravitational field of the planet acts on the spaceship, what will be the number of complete revolutions made by the spaceship in 24 h around the planet?
(Take, mass of planet = 8×10^{22} kg, radius of planet = 2×10^6 m and gravitational constant $G = 6.67 \times 10^{-11}$ N-m^2/kg^2) [JEE Main 2019]

(a) 11 (b) 17 (c) 13 (d) 9

32. A rocket has to be launched from earth in such a way that it never returns. If E is the minimum energy delivered by the rocket launcher, what should be the minimum energy that the launcher should have, if the same rocket is to be launched from the surface of the moon? Assume that, the density of the earth and the moon are equal and that the earth's volume is 64 times the volume of the moon. [JEE Main 2019]

(a) $\frac{E}{64}$ (b) $\frac{E}{16}$
(c) $\frac{E}{32}$ (d) $\frac{E}{4}$

33. A satellite of mass m is launched vertically upwards with an initial speed u from the surface of the earth. After it reaches height R (R = radius of the earth), it ejects a rocket of mass $\frac{m}{10}$, so that subsequently the satellite moves in a circular orbit. The kinetic energy of the rocket is (G is the gravitational constant, M is the mass of the earth) **[JEE Main 2020]**

(a) $\frac{3m}{8}\left(u+\sqrt{\frac{5\,GM}{6R}}\right)^2$ (b) $\frac{m}{20}\left(u-\sqrt{\frac{2GM}{3R}}\right)^2$

(c) $\frac{m}{20}\left(u^2+\frac{113}{200}\frac{GM}{R}\right)$ (d) $5m\left(u^2-\frac{119}{200}\frac{GM}{R}\right)$

Numerical Value Questions

34. The radius in kilometre to which the present radius of earth (R = 6400 km) to be compressed, so that the escape velocity is increased to ten time is ...... km. **[JEE Main 2021]**

35. If one wants to remove all the mass of the earth to infinity in order to break it up completely. The amount of energy that needs to be supplied will be $\frac{x}{5}\frac{GM^2}{R}$, where x is .... (Round off to the nearest integer)
(Here, M is the mass of earth, R is the radius of earth and G is the gravitational constant). **[JEE Main 2021]**

36. An artificial satellite is moving in a circular orbit around the earth with a speed equal to half the magnitude of escape velocity from the earth. If the satellite is stopped suddenly in its orbit and allowed to fall freely onto the earth, then speed (in ms^{-1}) with which it hits the surface of the earth is $\sqrt{kgR}$. The value of k is .... (Take, $g = 9.8\ ms^{-2}$ and $R = 6.4\times 10^6$ m)

37. An asteroid is moving directly towards the centre of the earth. When at a distance of 10 R (R is the radius of the earth) from the earth's centre, it has a speed of 12 km/s. Neglecting the effect of earth's atmosphere, what will be the speed of the asteroid when it hits the surface of the earth (escape velocity from the earth is 11.2 km/s)? Give your answer to the nearest integer (in km/s) .... **[JEE Main 2020]**

38. Inside a fixed sphere of radius R and uniform density ρ, there is a spherical cavity of radius $\frac{R}{2}$ such that, the surface of cavity passes through the centre of the sphere as shown in the figure. A particle of mass m_0 is released from rest at centre B of the cavity. The velocity with which particle strikes the centre A of sphere is $\frac{\sqrt{2\pi G\rho R^2}}{\sqrt{\beta}}$. The value of β is ....

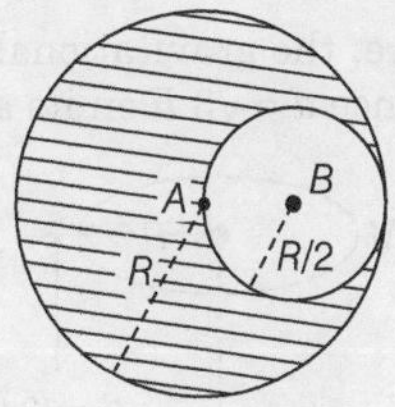

Answers

Round I

1. (b)	2. (a)	3. (d)	4. (b)	5. (b)	6. (a)	7. (b)	8. (c)	9. (d)	10. (c)
11. (b)	12. (d)	13. (a)	14. (c)	15. (c)	16. (c)	17. (d)	18. (b)	19. (c)	20. (c)
21. (b)	22. (d)	23. (b)	24. (c)	25. (c)	26. (c)	27. (a)	28. (c)	29. (a)	30. (c)
31. (c)	32. (c)	33. (c)	34. (a)	35. (c)	36. (a)	37. (d)	38. (b)	39. (c)	40. (b)
41. (c)	42. (c)	43. (c)	44. (c)	45. (c)	46. (c)	47. (d)	48. (c)	49. (b)	50. (a)
51. (a)	52. (b)	53. (a)	54. (d)	55. (b)	56. (a)	57. (c)	58. (d)	59. (b)	60. (a)
61. (a)	62. (a)	63. (b)	64. (a)	65. (b)	66. (c)	67. (d)	68. (c)	69. (d)	70. (d)
71. (d)	72. (a)	73. (b)	74. (d)	75. (c)	76. (c)	77. (d)	78. (c)	79. (c)	80. (b)
81. (c)	82. (d)								

Round II

1. (c)	2. (c)	3. (c)	4. (b)	5. (d)	6. (d)	7. (d)	8. (c)	9. (c)	10. (c)
11. (c)	12. (b)	13. (d)	14. (a)	15. (a)	16. (c)	17. (a)	18. (b)	19. (c)	20. (a)
21. (a)	22. (b)	23. (c)	24. (b)	25. (a)	26. (d)	27. (b)	28. (a)	29. (c)	30. (d)
31. (a)	32. (b)	33. (d)	34. 64	35. 3	36. 1	37. 16	38. 3		

Solutions

Round I

1. Gravitational force $= \left(\dfrac{GMm}{R^{3/2}}\right)$ provides the necessary centripetal force (*i.e.* $mR\omega^2$)

So, $$\frac{GMm}{R^{3/2}} = mR\omega^2 = mR\left(\frac{2\pi}{T}\right)^2 = \frac{4\pi^2 mR}{T^2}$$

or $$T^2 = \frac{4\pi^2 R^{5/2}}{GM}$$

i.e. $$T^2 \propto R^{5/2}$$

2. Let R be the original radius of a planet. Then attraction on a body of mass m placed on its surface will be

$$F = \frac{GMm}{R^2}$$

If radius of the planet is made double, *i.e.* $R' = 2R$, then mass of the planet becomes

$$M' = \frac{4}{3}\pi(2R)^3\rho = 8\times\frac{4}{3}\pi R^3\rho = 8M$$

New force, $$F' = \frac{GM'm}{R'^2} = \frac{G\times 8M\times m}{(2R)^2} = 2F$$

i.e. Force of attraction increases due to the increase in mass of the planet.

3. From the figure, the gravitational intensity due to the ring at a distance $d = \sqrt{3}R$ on its axis is

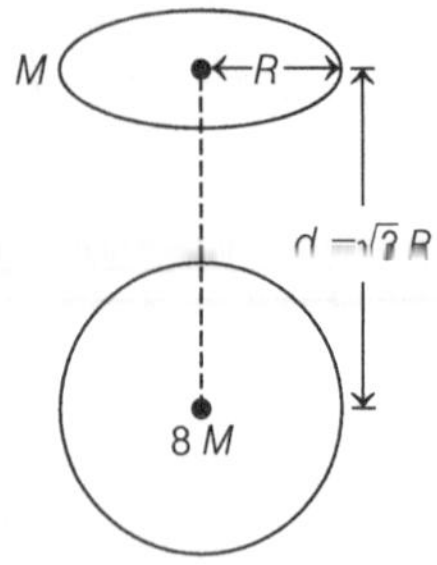

$$E = \frac{GMd}{(d^2+R^2)^{3/2}} = \frac{GM\times\sqrt{3}R}{(3R^2+R^2)^{3/2}} = \frac{\sqrt{3}GM}{8R^2}$$

Force on sphere $= (8M)\times E = (8M)\times\dfrac{\sqrt{3}GM}{8R^2} = \dfrac{\sqrt{3}GM^2}{R^2}$

4. Gravitational force due to solid sphere, $F_1 = \dfrac{GMm}{(2R)^2}$, where M and m are masses of the solid sphere and particle respectively and R is the radius of the sphere. The gravitational force on particle due to sphere with cavity = force due to solid sphere creating cavity, assumed to be present above at that position.

i.e. $$F_2 = \frac{GMm}{4R^2} - \frac{G(M/8)m}{(3R/2)^2} = \frac{7}{36}\frac{GMm}{R^2}$$

So, $$\frac{F_2}{F_1} = \frac{7GMm}{36R^2}\Big/\left(\frac{GMm}{4R^2}\right) = \frac{7}{9}$$

5. Moon is revolving around earth in almost circular orbit. Sun exerts gravitational pull on both, earth and moon. When observed from sun, the orbit of the moon will not be strictly elliptical because the total gravitational force (*i.e.* force due to earth on moon and force due to sun on moon) is not central.

6. The earth is revolving on circular orbit around sun due to gravitational force (F) which acts along the radius of circular path, towards the sun, *i.e.* angle between **r** and **F** is zero. As,

Torque, $|\tau| = |\mathbf{r}\times\mathbf{F}| = rF\sin 0° = 0$

Therefore, torque is zero.

7. Here, $m_1 = m_2 = 100$ kg, $r = 100$ m

Acceleration of first astronaut,

$$a_1 = \frac{Gm_1m_2}{r^2}\times\frac{1}{m_1} = \frac{Gm_2}{r^2}$$

Acceleration of second astronaut,

$$a_2 = \frac{Gm_1m_2}{r^2}\times\frac{1}{m_2} = \frac{Gm_1}{r^2}$$

Net acceleration of approach,

$$a = a_1 + a_2 = \frac{Gm_2}{r^2} + \frac{Gm_1}{r^2} = \frac{2Gm_1}{r^2} \quad [\because m_1 = m_2]$$

$$= \frac{2\times(6.67\times10^{-11})\times100}{(100)^2}$$

$$= 2\times6.67\times10^{-13}\ \text{ms}^{-2}$$

As, $$s = \frac{1}{2}at^2$$

$\therefore$ $$t = \left(\frac{2s}{a}\right)^{1/2} = \left[\frac{2\times(1/100)}{2\times6.67\times10^{-13}}\right]^{1/2}\ \text{second}$$

On solving, we get $t = 1.41$ days

8. Resultant force on mass m due to masses at A and B is $F = \dfrac{G2M\times m}{(AB)^2} - \dfrac{GMm}{(BC)^2}$ towards BA. Therefore, m will move towards $2M$.

9. Gravitational force between sphere of mass M and the particle of mass m at B is

$$F_1 = \frac{GMm}{d^2}$$

If M_1 is the mass of the removed part of sphere, then

$$M_1 = \frac{4}{3}\pi\left(\frac{R}{2}\right)^3\rho = \frac{1}{8}\left(\frac{4}{3}\pi R^3\rho\right) = \frac{M}{8} \qquad \left(\because \rho = \frac{M}{\frac{4}{3}\pi R^3}\right)$$

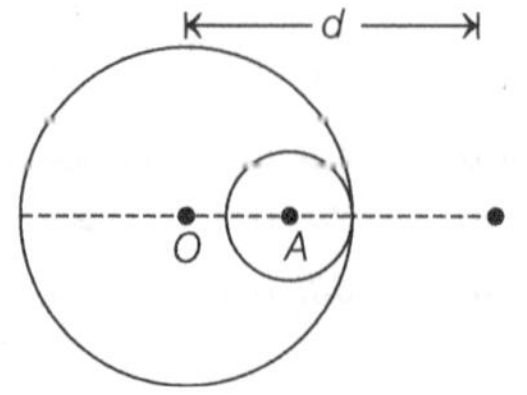

Gravitational force between the removed part and the particle of mass m at B is

$$F_2 = \frac{GM_1 m}{(d-R/2)^2} = \frac{G(M/8)m}{(d-R/2)^2} = \frac{GMm}{8(d-R/2)^2}$$

$\therefore$ Required force,

$$F = F_1 - F_2 = \frac{GMm}{d^2} - \frac{GMm}{8[d-(R/2)]^2}$$

$$= \frac{GMm}{d^2}\left[1 - \frac{1}{8\left(1-\frac{R}{2d}\right)^2}\right]$$

10. Given, situation is

Force of attraction between mass m and an elemental mass dm of rod,

$$dF = \frac{Gmdm}{x^2} = \frac{Gm(A+Bx^2)\,dx}{x^2}$$

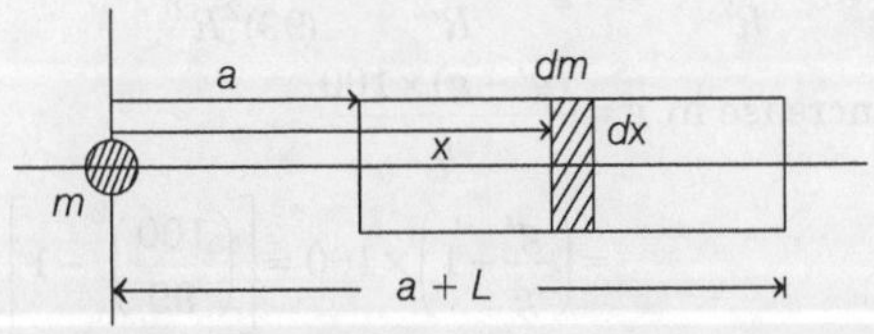

Total attraction force is sum of all such differential forces produced by elemental parts of rod from $x = a$ to $x = a + L$.

$$\therefore\ F = \int dF = \int_{x=a}^{x=a+L} \frac{Gm(A+Bx^2)}{x^2}\,dx$$

$$= Gm\int_{x=a}^{x=a+L}\left(\frac{A}{x^2}+B\right)dx$$

$$= Gm\left[-\frac{A}{x}+Bx\right]_{x=a}^{x=a+L}$$

$$= Gm\left(\frac{-A}{a+L}+B(a+L)+\frac{A}{a}-Ba\right)$$

$$= Gm\left(\frac{A}{a}-\frac{A}{a+L}+BL\right)$$

$$= Gm\left[A\left(\frac{1}{a}-\frac{1}{a+L}\right)+BL\right]$$

11. In given configuration of masses, net gravitational force provides the necessary centripetal force for rotation.

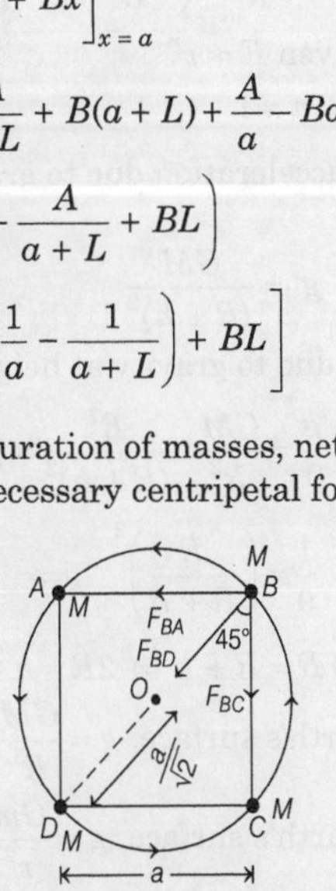

Net force on mass M at position B towards centre of circle is

$$F_{BO(\text{net})} = F_{BD} + F_{BA}\sin 45° + F_{BC}\cos 45°$$

$$= \frac{GM^2}{(\sqrt{2}a)^2} + \frac{GM^2}{a^2}\left(\frac{1}{\sqrt{2}}\right) + \frac{GM^2}{a^2}\left(\frac{1}{\sqrt{2}}\right)$$

[where, diagonal length BD is $\sqrt{2}a$]

$$= \frac{GM^2}{2a^2} + \frac{GM^2}{a^2}\left(\frac{2}{\sqrt{2}}\right) = \frac{GM^2}{a^2}\left(\frac{1}{2}+\sqrt{2}\right)$$

This force will act as centripetal force.

Distance of particle from centre of circle is $\frac{a}{\sqrt{2}}$.

Here, $F_{\text{centripetal}} = \frac{Mv^2}{r} = \frac{Mv^2}{\frac{a}{\sqrt{2}}} = \frac{\sqrt{2}Mv^2}{a}$ $\quad\left(\because r = \frac{a}{\sqrt{2}}\right)$

So, for rotation about the centre, $F_{\text{centripetal}} = F_{BO(\text{net})}$

$$\Rightarrow \quad \sqrt{2}\frac{Mv^2}{a} = \frac{GM^2}{a^2}\left(\frac{1}{2}+\sqrt{2}\right)$$

$$\Rightarrow \quad v^2 = \frac{GM}{a}\left(1+\frac{1}{2\sqrt{2}}\right) = \frac{GM}{a}(1.35)$$

$$\Rightarrow \quad v = 1.16\sqrt{\frac{GM}{a}}$$

12. A person is safe, if his velocity while reaching the surface of moon from a height h' is equal to its velocity while falling from height h on earth. So,

$$\sqrt{2g'h'} = \sqrt{2gh}$$

or $\quad h' = \frac{gh}{g'} = 9.8 \times \frac{3}{1.96} = 15\text{ m}$

13. When the thief with box on his head jumped down from a wall, he along with box is falling down with acceleration due to gravity, so the apparent weight of box becomes zero, (because, $R = mg - mg = 0$), so he experiences no load till he reaches the ground.

14. As, $\quad \frac{g_m}{g_e} = \frac{\frac{GM}{8R_m^2}}{GM/R_e^2} = \frac{R_e^2}{8R_m^2}$ ...(i)

Given, $\quad \frac{mg_m}{mg_e} = \frac{1}{6}$

$\therefore \quad \frac{g_m}{g_e} = \frac{1}{6}$ ...(ii)

From Eqs. (i) and (ii), we get

$$\frac{R_e^2}{8R_m^2} = \frac{1}{6}$$

or $\quad R_e = R_m\sqrt{\frac{8}{6}}$

15. Mass of earth = Volume × Density

As, $\quad g = \frac{GM}{R^2}$; $M = \left(\frac{4}{3}\pi R^3\right)\rho$

$\therefore \quad g = \frac{4G}{3}\frac{\pi R^3}{R^2}\rho$ or $g = \left(\frac{4G\pi R}{3}\right)\rho$

[where, ρ = average density]

$\Rightarrow \quad g \propto \rho$ or $\rho \propto g$

16. Acceleration due to gravity at height h,

$$g_h = g\left(1 - \frac{2h}{R}\right) \quad [\text{ if } h << R] \ldots\text{(i)}$$

Acceleration due to gravity at depth d,

$$g_d = g\left(1 - \frac{d}{R}\right) \quad \ldots\text{(ii)}$$

As per statement of the problem,

$$g_h = g_d$$

i.e. $g\left(1 - \frac{2h}{R}\right) = g\left(1 - \frac{d}{R}\right)$

$\Rightarrow \quad 2h = d$

17. If the earth is an approximate sphere of non-uniform density, then the centre of gravity of earth will not be situated at the centre of earth. The distance of different points on earth will be at different distances from the centre of gravity of earth. As, $g = \frac{GM}{r^2}$ or $g \propto \frac{1}{r^2}$, so g is different on different points on the surface of earth but never zero.

18. Below the sea level the pressure is increasing with depth in mine due to presence of atmospheric air there. The acceleration due to gravity below the surface of the earth decreases with the distance from the surface of the earth, as $g' = g\left(1 - \frac{d}{R}\right)$.

19. When a sphere of mass m is released in a liquid, it falls vertically down with acceleration $= \frac{mg - F_B}{m}$

$$= \frac{\frac{4}{3}\pi r^3 dg - \frac{4}{3}\pi r^3 \rho g}{\frac{4}{3}\pi r^3 d} = \frac{(d - \rho)g}{d}$$

20. As, $\frac{g'}{g} = 1 - \frac{2h}{R} = 1 - \frac{2 \times 320}{6400} = 1 - \frac{1}{10} = \frac{9}{10}$

$\therefore$ % decrease in $g = \left(\frac{g - g'}{g}\right) \times 100 = \frac{1}{10} \times 100 = 10\%$

21. Given, $g_h = 9 = \frac{gR^2}{(R + R/20)^2} = \frac{20 \times 20}{21 \times 21} g$

or $g = \frac{9 \times 21 \times 21}{20 \times 20}$

Now, $g_d = g\left(1 - \frac{d}{R}\right)$

$$= \frac{9 \times 21 \times 21}{20 \times 20}\left(1 - \frac{R/20}{R}\right)$$

$$= 9.5 \text{ ms}^{-2}$$

22. As, value of g at internal point which is at a distance x from the centre is given by

$$g' = \frac{GM}{R^3} x$$

When $x = 0$, then $g' = 0$.

23. The value of acceleration due to gravity at a height h reduces to $= 100 - 36 = 64\% = \frac{64}{100} g$

$\therefore \quad \frac{64}{100} g = \frac{gR^2}{(R + h)^2}$

or $\quad \frac{8}{10} = \frac{R}{R + h}$ or $h = \frac{R}{4}$

24. Mass of two planets is same, so

$$\frac{4}{3}\pi R_1^3 \rho_1 = \frac{4}{3}\pi R_2^3 \rho_2$$

or $\quad \frac{R_1}{R_2} = \left(\frac{\rho_2}{\rho_1}\right)^{1/3} = \left(\frac{1}{8}\right)^{1/3} = \frac{1}{2}$

$$\frac{g_1}{g_2} = \frac{GM/R_1^2}{GM/R_2^2} = \left(\frac{R_2}{R_1}\right)^2 = (2)^2 = 4$$

$\Rightarrow \quad g_1 : g_2 = 4 : 1$

25. As, $g = \frac{GM}{R^2} \Rightarrow g' = \frac{GM}{R'^2} = \frac{GM\,(100)^2}{(99)^2 R^2}$

% Increase in $g = \frac{(g' - g) \times 100}{g}$

$$= \left(\frac{g'}{g} - 1\right) \times 100 = \left[\left(\frac{100}{99}\right)^2 - 1\right] \times 100$$

$$= \left[\left(1 + \frac{1}{99}\right)^2 - 1\right] \times 100 \approx 2\%$$

26. Force acting on a body of mass M at a point at depth d inside the earth is

$$F = mg' = mg\left(1 - \frac{d}{R}\right)$$

$$= \frac{mGM}{R^2}\left(\frac{R - d}{R}\right) = \frac{GMm}{R^3} r, \quad (\because R - d = r)$$

So, $F \propto r$; Given $F \propto r^n$

Clearly, $\quad n = 1$

27. As, value of acceleration due to gravity above the earth surface is g.

$$g' = \frac{GM}{(R + h)^2}$$

Acceleration due to gravity at height h

$\Rightarrow \quad \frac{g}{9} = \frac{GM}{R^2} \cdot \frac{R^2}{(R + h)^2} = g\left(\frac{R}{R + h}\right)^2$

$\Rightarrow \quad \frac{1}{9} = \left(\frac{R}{R + h}\right)^2 \Rightarrow \frac{R}{R + h} = \frac{1}{3}$

$\Rightarrow \quad 3R = R + h \Rightarrow 2R = h$

28. Inside the earth's surface, $g = \frac{GM}{R^3} r$, i.e. $g \propto r$

Outside the earth's surface, $g = \frac{Gm}{r^2}$, i.e. $g \propto \frac{1}{r^2}$

So, till earth's surface g increases linearly with distance r, shown only in graph (c).

29. At depth d from the surface of the earth,

$$g' = g\left(1 - \frac{d}{R}\right)$$

Given, $g' = \frac{75}{100}g = \frac{3}{4}g$

Then, $\frac{3g}{4} = g\left(1 - \frac{d}{R}\right)$

On solving, we get

$$d = \frac{R}{4}$$

30. Error in weighing $= mg - mg' = mg - mg\left(1 - \frac{2h}{R}\right)$

$$= \frac{mg\,2h}{R} = \frac{m2\,hg}{R}$$

$$= \frac{m\,2h}{R} \times \frac{G\frac{4}{3}\pi R^3\rho}{R^2} = \frac{8\pi G\rho mh}{3}$$

31. At equator, $g' = g - R\omega^2$

When angular velocity be $\omega' (= x\omega)$, then $g' = 0$

Since, at equator,

$$g' = g - \omega'^2 R$$

or $\omega' = \sqrt{g/R} = x\omega$

or $x = (\sqrt{g/R})/\omega$

or $x = \frac{\sqrt{10/(6.4 \times 10^6)}}{2\pi} \times 24 \times 60 \times 60 = 17$

32. When there is a weightlessness in the body at the equator, then $g' = r - R\omega^2 = 0$

or $\omega = \sqrt{g/R}$ and

linear velocity

$$= \omega R = (\sqrt{g/R})\,R = \sqrt{gR} \quad \text{...(i)}$$

$\therefore$ KE of rotation of earth $= \frac{1}{2}I\omega^2 = \frac{1}{2} \times \frac{2}{5}MR^2 \times \omega^2$

$= \frac{1}{5}MgR$ [from Eq. (i)]

33. Acceleration due to gravity at height h above earth's surface is

$$g_h = g\left(\frac{R^2}{(R+h)^2}\right)$$

Also, acceleration at depth $d = h$ below surface is

$$g_d = g\left(1 - \frac{d}{R}\right) = g\left(1 - \frac{h}{R}\right)$$

Given, weight of body at height h = weight of body at depth h

So, $mg_h = mg_d \Rightarrow g_h = g_d$

$$\Rightarrow g\left(\frac{R^2}{(R+h)^2}\right) = g\left(1 - \frac{h}{R}\right) \Rightarrow \frac{R^2}{(R+h)^2} = 1 - \frac{h}{R}$$

$$\Rightarrow R^3 = (R-h)(R+h)^2$$

$$\Rightarrow R^3 = (R^3 - hR^2 + h^2R - h^3 + 2R^2h - 2Rh^2)$$

$$\Rightarrow h^2 + Rh - R^2 = 0$$

$$\Rightarrow h = \frac{-R \pm \sqrt{R^2 - 4(1)(-R^2)}}{2(1)}$$

$$\Rightarrow h = \frac{-R \pm \sqrt{5}R}{2}$$

As h is a positive number, $h = \frac{-R + \sqrt{5}R}{2}$ or $\frac{\sqrt{5}R - R}{2}$

34. As, $g' = g - \frac{10\,g}{100} = \frac{90}{100}g$

$\therefore$ $g' = g\frac{R^2}{(R+h)^2}$

or $\frac{9}{10} = \frac{R^2}{(R+h)^2}$

or $\frac{3}{\sqrt{10}} = \frac{R}{R+h}$

or $h = (\sqrt{10} - 3)R/3$

$$= \frac{(\sqrt{10} - 3) \times 6400}{3} = 345.60 \text{ km}$$

35. Acceleration due to gravity at height h,

$$g_1 = g\left(1 - \frac{2h}{R}\right)$$

Acceleration due to gravity at depth h,

$$g_2 = g\left(1 - \frac{h}{R}\right)$$

$$\therefore \frac{g_1}{g_2} = \frac{1 - 2h/R}{1 - h/R} = \left(1 - \frac{2h}{R}\right)\left(1 - \frac{h}{R}\right)^{-1} = \left(1 - \frac{h}{R}\right)$$

$\therefore \frac{g_1}{g_2}$ decreases linearly with h.

36. Given, $\frac{mg'}{mg} = \frac{30}{90}$ or $\frac{g'}{g} = \frac{1}{3}$

Now, $g' = g\frac{R^2}{(R+h)^2}$

or $\frac{g'}{g} = \frac{R^2}{(R+h)^2} = \frac{1}{3}$ or $\frac{R}{R+h} = \frac{1}{\sqrt{3}}$

or $(R+h) = \sqrt{3}R$

or $h = (\sqrt{3} - 1)R = 0.73R$

37. Let mass of given body is m, then it's weight on earth's surface $= mg_e$

where, g_e = acceleration due to gravity on earth's surface and weight on the surface of planet $= mg_p$, where g_p = acceleration due to gravity on planet's surface.

Given, $\frac{mg_e}{mg_p} = \frac{9}{4} \Rightarrow \frac{g_e}{g_p} = \frac{9}{4}$

But $g = \frac{GM}{R^2}$, so we have $\frac{\left(\frac{GM}{R^2}\right)}{\left(\frac{GM_p}{R_p^2}\right)} = \frac{9}{4}$

where, M = mass of earth, R = radius of earth,

M_p = mass of planet $= \frac{M}{9}$ (given)

and R_p = radius of planet.

$\Rightarrow \quad \frac{M}{M_p} \cdot \frac{R_p^2}{R^2} = \frac{9}{4} \Rightarrow 9 \cdot \left(\frac{R_p}{R}\right)^2 = \frac{9}{4}$

$\Rightarrow \quad \frac{R_p}{R} = \frac{1}{2} \Rightarrow R_p = \frac{R}{2}$

38. The relation of gravitational acceleration at an altitude h above the earth's surface is given as

$$g_h = g\left(1 + \frac{h}{R_e}\right)^{-2}$$

where, g is the acceleration due to gravity at earth's surface and R_e is the radius of the earth.

Given that at some height h, acceleration due to gravity,

$$g_h = 4.9 \text{ m/s}^2 \approx \frac{g}{2} \quad \text{... (i)}$$

$\therefore$ The ratio of acceleration due to gravity at earth's surface and at some altitude h is

$$\left(1 + \frac{h}{R_e}\right) = \sqrt{\frac{g}{g_h}} = \sqrt{2} \quad \text{[from Eq. (i)]}$$

$\therefore \quad \frac{h}{R_e} = \sqrt{2} - 1$ or $h = 0.414 \times R_e$

$h = 0.414 \times 6400$ km

($\because$ Given, radius of earth, $R_e = 6400 \text{ km} = 6.4 \times 10^6$ m)

or $\quad h = 2649.6 \text{ km} = 2.6 \times 10^6$ m

Thus, at 2.6×10^6 m above the earth's surface, acceleration due to gravity decreases to 4.9 m/s^2.

39. As, $\quad dV = -E\,dx$

or $\quad V = -\int_{\infty}^{x/\sqrt{2}} E\,dx = -\int_{\infty}^{x/\sqrt{2}} kx^{-3}\,dx = k/x^2$

40. Intensity of gravitational field at a point inside the spherical shell is zero

$$I = 0 \text{ upto } r = a$$

outside the shell, $I \propto \frac{1}{r^2}$ where $r > a$

41. If x is the distance of point on the line joining the two masses from mass m_2 where gravitational field intensity is zero, then

$$\frac{Gm_1}{(r-x)^2} = \frac{Gm_2}{x^2}$$

or $\quad \frac{2}{(9-x)^2} = \frac{8}{x^2}$ or $\frac{1}{9-x} = \frac{2}{x}$

On solving, $x = 6$ m

42. Let x be the distance of point from the smaller body where gravitational intensity is zero.

$\therefore \quad \frac{Gm_1}{(1-x)^2} = \frac{Gm_2}{x^2}$

or $\quad \frac{x}{1-x} = \sqrt{\frac{m_2}{m_1}} = \sqrt{\frac{1000}{100000}} = \frac{1}{10}$

or $\quad 10x = 1 - x$ or $x = (1/11)$ m

43. Gravitational field of a solid sphere is maximum at its surface ($r = R$) and its value at surface,

$$E = \frac{GM}{R^2}$$

From graph given in the question,

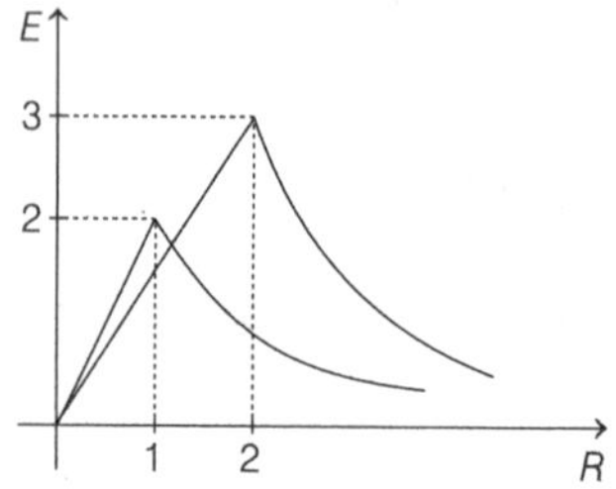

We can observe that,

$E_1 = \frac{GM_1}{(1)^2} = 2$ and $E_2 = \frac{GM_2}{(2)^2} = 3$

or $\quad GM_1 = 2 \quad \text{...(i)}$

and $\quad GM_2 = 12 \quad \text{...(ii)}$

On dividing Eq. (i) by Eq. (ii), we get

$$\frac{M_1}{M_2} = \frac{2}{12} = \frac{1}{6}$$

44. Here, $I = -\frac{dV}{dr} = -k/r$ or $dV = k\frac{dr}{r}$

Integrating it, we get

$$\int_{V_0}^{V} dV = \int_{r_0}^{r} k\frac{dr}{r}$$

or $\quad V = V_0 + k\log(r/r_0)$

45. Gravitational potential on the surface of the shell is

V = Gravitational potential due to particle (V_1)

+ Gravitational potential due to shell particle (V_2)

$$= -\frac{Gm}{R} + \left(-\frac{G3m}{R}\right) = -\frac{4Gm}{R}$$

46. The minimum kinetic energy required to project a body of mass m from earth's surface to infinity is known as escape energy. Therefore,

$$\text{KE} = \frac{GM_e m}{R} = mgR \quad \left[\because gR = \frac{GM_e}{R}\right]$$

47. Gravitational potential energy of body will be

$$E = -\frac{GMm}{r}$$

where, $\quad M$ = mass of earth,

m = mass of the body

and $\quad r$ = radius of earth.

At $\quad r = 2R$,

$$E_1 = -\frac{GMm}{(2R)}$$

At $\quad r = 3R, E_2 = -\frac{GMm}{(3R)}$

Energy required to move a body of mass m from an orbit of radius $2R$ to $3R$ is

$$\Delta E = \frac{GMm}{R}\left[\frac{1}{2}-\frac{1}{3}\right] = \frac{GMm}{6R}$$

48. Gravitational potential energy of a body in the gravitational field, $E = \frac{-GMm}{r}$. When r decreases negative value of E increases, *i.e.* E decreases.

49. Gravitational intensity, $I = \frac{dV}{dx} = \frac{14}{20} = 0.7 \text{ Nkg}^{-1}$

Acceleration due to gravity, $g = I = 0.7 \text{ N kg}^{-1}$

Work done under this field in displacing a body on a slope of 60° through a distance s

$$= m\,(g \sin 60°)\,s$$
$$= 2 \times (0.7 \times \sqrt{3}/2) \times 8$$
$$= 9.6 \text{ J}$$

50. Since the gravitational field is conservative field, hence the work done in taking a particle from one point to another in a gravitational field is path independent.

Hence, $W_1 = W_2 = W_3$.

51. As, $U = \frac{-GMm}{r}$

or $r = \frac{-GMm}{U}$

$$r = \frac{-6.67 \times 10^{-11} \times 6 \times 10^{24} \times 7.4 \times 10^{22}}{-7.79 \times 10^{28}}$$
$$= 3.8 \times 10^{8} \text{ m}$$

52. Gravitational potential energy of body on the earth's surface is

$$U = -\frac{GM_e m}{R}$$

At a height h from earth's surface, its value is

$$U_h = -\frac{GM_e m}{(R+h)} = -\frac{GM_e m}{2R} \quad [\text{as } h = R]$$

where, M_e = mass of earth, m = mass of body

and R = radius of earth.

$\therefore$ Gain in potential energy

$$= U_h - U = -\frac{GM_e m}{2R} - \left(-\frac{GM_e m}{R}\right)$$
$$= -\frac{GM_e m}{2R} + \frac{GM_e m}{R}$$
$$= \frac{GM_e m}{2R} = \frac{gR^2 m}{2R} \quad \left[\text{as, } g = \frac{GM_e}{R^2}\right]$$
$$= \frac{1}{2} mgR$$

53. The change in potential energy in gravitational field is given by

$$\Delta E = GMm\left(\frac{1}{r_1} - \frac{1}{r_2}\right)$$

In this problem; $r_1 = R$ and $r_2 = nR$

$$\Delta E = GMm\left(\frac{1}{R} - \frac{1}{nR}\right)$$
$$= \frac{GMm}{R}\left(\frac{n-1}{n}\right)$$
$$= mgR\left(\frac{n-1}{n}\right) \quad \left(\because g = \frac{Gm}{R^2}\right)$$

54. The initial potential energy of the particle,

$$U_i = -\frac{GMm}{r}$$
$$U_i = -\frac{6.67 \times 10^{-11} \times 100 \times 10^{-2}}{0.1}$$
$$U_i = -\frac{6.67 \times 10^{-11}}{0.1} = -6.67 \times 10^{-10} \text{ J}$$

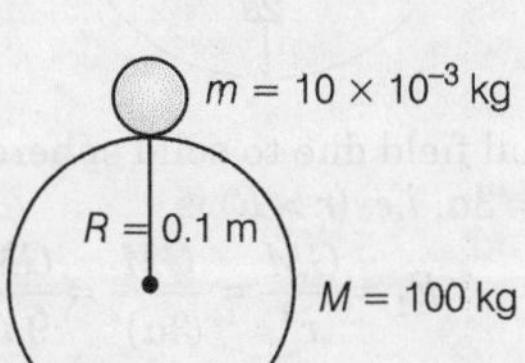

We know that, work done = difference in potential energy

$$\therefore \quad W = \Delta U = U_f - U_i$$
$$\Rightarrow \quad W = -U_i \quad [\because U_f = 0]$$
$$= 6.67 \times 10^{-10} \text{ J}$$

55. Gravitational force on a body at a distance x from the centre of earth, $F = \frac{GMm}{x^2}$

$\therefore$ Work done,

$$W = \int_R^{R+h} F\,dx = \int_R^{R+h} \frac{GMm}{x^2}\,dx$$
$$= GMm\left[-\frac{1}{x}\right]_R^{R+h}$$
$$= mgR^2\left(\frac{1}{R} - \frac{1}{R+h}\right)$$

This work done appears as increase in potential energy

$$\therefore \quad \Delta U = mgR^2\left[\frac{1}{R} - \frac{1}{R+h}\right]$$
$$= mg\,(5h)^2\left[\frac{1}{5h} - \frac{1}{6h}\right] = \frac{5}{6} mgh$$

56. From conservation of energy,

Total energy at the planet = Total energy at altitude

$$-\frac{GMm}{R} + (\text{KE})_{\text{surface}} = -\frac{GMm}{3R} + \frac{1}{2} mv_A^2 \quad \text{...(i)}$$

In the orbit of planet, the necessary centripetal force is obtained by gravitational force.

$\therefore \quad \dfrac{m v_A^2}{R+2R} = \dfrac{GMm}{(R+2R)^2}$

$\Rightarrow \quad v_A = \dfrac{GM}{3R}$ ...(ii)

From Eqs. (i) and (ii), we get

$$(\text{KE})_{\text{surface}} = \frac{5}{6}\frac{GMm}{R}$$

57. According to question, diagram will be as follows

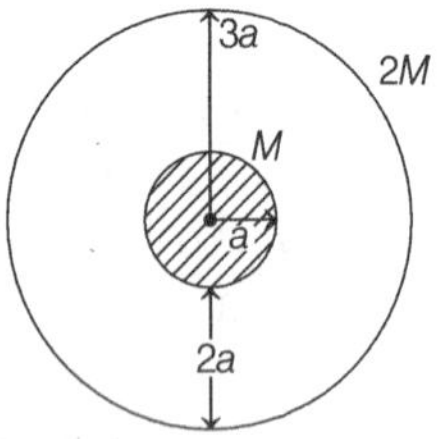

Gravitational field due to solid sphere of radius a at a distance, $r = 3a$, *i.e.* $(r > R)$ is

$$E_1 = \frac{GM}{r^2} = \frac{GM}{(3a)^2} = \frac{GM}{9a^2}$$

Similarly, gravitational field due to spherical shell at a distance, $r = 3a$,

i.e. $\quad E_2 = \dfrac{GM}{R^2} = \dfrac{G(2M)}{(3a)^2} = \dfrac{2GM}{9a^2}$

Both fields are attractive in nature, so direction will be same.

Net electric field, $E_{\text{net}} = \dfrac{GM}{9a^2} + \dfrac{2GM}{9a^2}$

$$E_{\text{net}} = \frac{GM}{3a^2}$$

58. For the body to be able to reach at the surface of smaller planet, it just has to overcome the gravitational field of bigger planet and at that point (let say P in given figure) net force will be zero. After crossing this P point, body will move towards smaller planet itself and will reach with minimum speed.

Let the distance of point P from bigger planet be x.

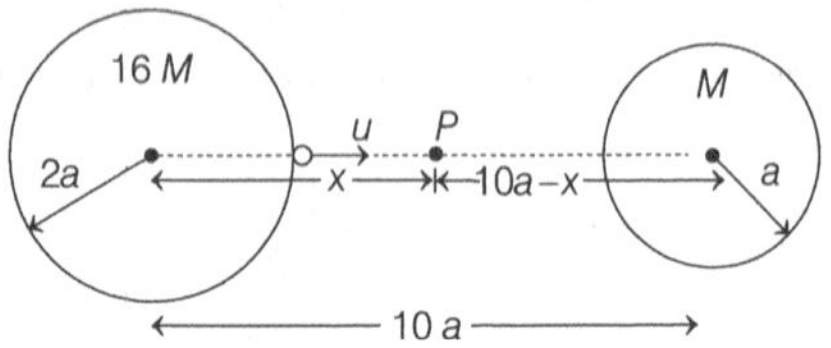

At point P, $\mathbf{F}_{\text{net}} = 0 \Rightarrow F_{16M} = F_M$

$\Rightarrow \quad \dfrac{G(16M)m}{x^2} = \dfrac{G(M)(m)}{(10a - x)^2} \Rightarrow x = 8a$ ...(i)

Since, there is no net external force on system, so we can apply conservation of mechanical energy from the surface of bigger planet to point P.

i.e. $\quad K_i + U_i = K_f + U_f$

$$\frac{1}{2}mu^2 + \left(-\frac{G(16M)m}{2a} - \frac{GMm}{8a}\right) = 0 + \left(\frac{-G(16M)m}{8a} - \frac{GMm}{2a}\right)$$

$$\frac{1}{2}mu^2 - \frac{65}{8}\frac{GMm}{a} = -\frac{5}{2}\frac{GMm}{a}$$

$$\frac{1}{2}mu^2 = \frac{45}{8}\frac{GMm}{a}$$

$\Rightarrow \quad u = \dfrac{3}{2}\sqrt{\dfrac{5GM}{a}}$

59. As, $\dfrac{(v_e)_{P_1}}{(v_e)_{P_2}} = \dfrac{\sqrt{2g_1 R_1}}{\sqrt{2g_2 R_2}} = \sqrt{\dfrac{g_1}{g_2} \times \dfrac{R_1}{R_2}} = \sqrt{ab}$

60. Escape velocity of a body from the surface of earth is 11.2 kms^{-1}, which is independent of the angle of projection.

61. Escape velocity $= \sqrt{2gR_e}$

So, escape velocity is independent of m.

62. Time period of satellite, $T = 2\pi\sqrt{\dfrac{(R+h)^3}{GM_e}}$

where, $R + h$ = orbital radius of satellite

and $\quad M_e$ = mass of earth.

Thus, time period does not depend on mass of the satellite.

63. When gravitational force becomes zero, the centripetal force required cannot be provided.

So, the satellite will move with the velocity as it has at the instant when gravitational force becomes zero, *i.e.* it moves tangentially to the original orbit.

64. When a satellite is moving in an elliptical orbit, it's angular momentum $(= \mathbf{r} \times \mathbf{p})$ about the centre of earth does not change its direction. The linear momentum $(= m\mathbf{v})$ does not remain constant as velocity of satellite is not constant. The total mechanical energy of S is constant at all locations.

The acceleration of S (= centripetal acceleration) is always directed towards the centre of earth.

65. According to the given condition in the question, after collision the mass of combined system is doubled. Also, this system would be displaced from its circular orbit.

So, the combined system revolves around centre of mass of 'system + earth' under action of a central force. Hence, orbit must be elliptical.

66. Orbital velocity of satellite at distance r from the centre of earth, $v = \sqrt{\dfrac{GM}{r}}$

Total energy of satellite = PE + KE $= -\dfrac{GMm}{r} + \dfrac{1}{2}mv^2$

$$= -\frac{GMm}{r} + \frac{1}{2}m\frac{GM}{r}$$

$$= -\frac{GMm}{2r}$$

The viscous force acting on satellite decreases the energy of satellite. As a result of it, the value of r gradually decreases, consequently the height of satellite gradually decreases.

67. Escape velocity, $v_e = \sqrt{\dfrac{2GM}{R}}$

$$\Rightarrow \quad v_e = \sqrt{\frac{2G\frac{4}{3}\pi R^3 \times d}{R}}$$

$$= \sqrt{2G\frac{4}{3}\pi R^2 \times d} = R\sqrt{\frac{8}{3}\pi Gd}$$

(Here, d = mean density of earth)

Now, $\quad v_e \propto R\sqrt{d}$

$$\therefore \quad \frac{v_e}{v_p} = \frac{R_e}{R_p}\sqrt{\frac{d_e}{d_p}} = \frac{R_e}{2R_e}\sqrt{\frac{d_e}{d_e}}$$

$$v_p = 2v_e = 2 \times 11 = 22 \text{ kms}^{-1}$$

68. Here, $\dfrac{g_m}{g_e} = \dfrac{M_m}{M_e} \times \left(\dfrac{R_e}{R_m}\right)^2 = \dfrac{1}{81} \times (4)^2 = \dfrac{16}{81}$

$$\Rightarrow \quad g_m = \frac{16}{81} g_e$$

Also, $\quad v_e = \sqrt{2g_e R_e} = 11.2 \text{ kms}^{-1}$

$$v_m = \sqrt{2g_m R_m} = \sqrt{2 \times \frac{16}{81} g_e \times \frac{1}{4} R_e}$$

$$= \frac{2}{9}\sqrt{2g_e R_e} = \frac{2}{9} \times 11.2$$

$$= 2.48 \approx 2.5 \text{ kms}^{-1}$$

69. Orbital speed of a satellite in a circular orbit is

$$v_o = \sqrt{\left(\frac{GM}{r_o}\right)}$$

where, r_o is the radius of the circular orbit.

So, kinetic energies of satellites A and B are

$$T_A = \frac{1}{2} m_A v_{OA}^2 = \frac{GMm}{2R}$$

$$T_B = \frac{1}{2} m_B v_{OB}^2 = \frac{GM(2m)}{2(2R)} = \frac{GMm}{2R}$$

So, ratio of their kinetic energies is $\dfrac{T_A}{T_B} = 1$.

70. Given, a satellite is revolving in a circular orbit at a height h from the Earth's surface having radius of earth R, i.e. $h << R$.

Orbital velocity of a satellite, $v = \sqrt{\dfrac{GM}{R+h}} = \sqrt{\dfrac{GM}{R}}$

(as, $h << R$)

Velocity required to escape,

$$\frac{1}{2} mv'^2 = \frac{GMm}{R+h}$$

$$v' = \sqrt{\frac{2GM}{R+h}} = \sqrt{\frac{2GM}{R}} \quad (h << R)$$

$\therefore$ Minimum increase in its orbital velocity required to escape from the Earth's gravitational field,

$$v' - v = \sqrt{\frac{2GM}{R}} - \sqrt{\frac{GM}{R}}$$

$$= \sqrt{2gR} - \sqrt{gR} = \sqrt{gR}(\sqrt{2} - 1) \quad \left(\because g = \frac{GM}{R^2}\right)$$

71. Escape velocity is given by

$$v_e = \sqrt{\frac{2GM}{R}}$$

where, m = mass and R = radius of planet.

So, $\quad v_A = \sqrt{\dfrac{2GM}{R}}$

and $\quad v_B = \sqrt{\dfrac{2G\left(\frac{M}{2}\right)}{\left(\frac{R}{2}\right)}} = \sqrt{\dfrac{2GM}{R}}$

$$\Rightarrow \quad \frac{v_A}{v_B} = 1 \text{ or } \frac{n}{4} = 1 \Rightarrow n = 4$$

72. The motion of a satellite in an elliptical orbit is shown in figure

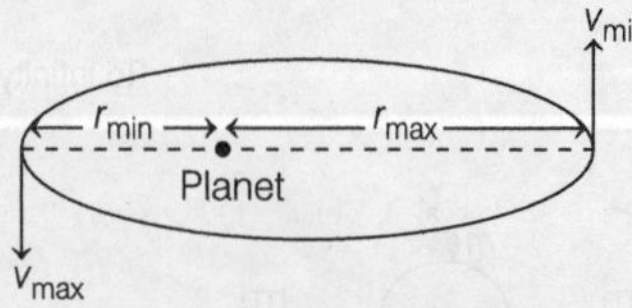

Applying, angular momentum conservation about planet at closest and farthest points,

$$L_{closest} = L_{farthest}$$

$$mv_{max}r_{min} = mv_{min}r_{max} \quad \text{or} \quad \frac{r_{min}}{r_{max}} = \frac{v_{min}}{v_{max}}$$

Given that, $v_{max} = 6v_{min}$ or $\dfrac{v_{min}}{v_{max}} = \dfrac{1}{6}$

$$\therefore \quad \frac{r_{min}}{r_{max}} = \frac{1}{6}$$

73. Orbital velocity of a body revolving around a planet,

$$v_o = \sqrt{\frac{GM}{r}} \quad \text{...(i)}$$

where, M = mass of the planet,

r = orbital radius of revolving body = R (given)

and G = universal gravitational constant.

Escape velocity of any body from the surface of a planet,

$$v_e = \sqrt{\frac{2GM}{R}} \quad \text{...(ii)}$$

where, M = mass of the planet,

R = radius of the planet and G = universal gravitational constant. Ratio of body's orbital velocity to the escape velocity from the planet,

$$\frac{v_o}{v_e} = \frac{\sqrt{\frac{GM}{R}}}{\sqrt{\frac{2GM}{R}}} = \frac{1}{\sqrt{2}}$$

74. The gravitational force exerted on satellite at a height x is

$$F_G = \frac{GM_e m}{(R+x)^2}$$

where, M_e = mass of earth.

Since, gravitational force provides the necessary centripetal force, so

$$\frac{GM_e m}{(R+x)^2} = \frac{mv_o^2}{(R+x)}$$

where, v_o is orbital speed of satellite.

$$\Rightarrow \quad \frac{GM_e m}{(R+x)} = mv_o^2$$

$$\text{or} \quad \frac{gR^2 m}{(R+x)} = mv_o^2 \qquad \left[\because g = \frac{GM_e}{R^2}\right]$$

$$\text{or} \quad v_o = \sqrt{\left[\frac{gR^2}{(R+x)}\right]} = \left[\frac{gR^2}{(R+x)}\right]^{1/2}$$

75. Kinetic energy must be given to an object, so that it just escapes to infinity is as shown in the figure below

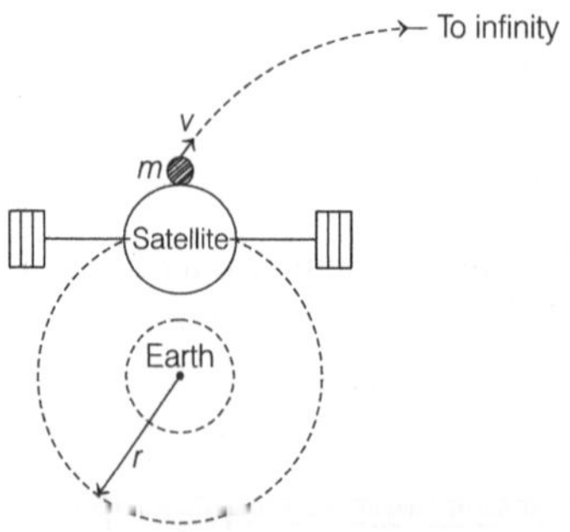

Applying energy-conservation for mass m, we get

Initial energy of m at satellite = Final energy of m at infinity

$$\Rightarrow \quad U_i + K_i = U_f + K_f \qquad \ldots\text{(i)}$$

Here, U_i = initial PE of

$$m = \frac{-G \cdot Mm}{r} = -(v^2)m$$

As, $v = \sqrt{\frac{GM}{r}}$ = orbital speed of the satellite,

K_i = initial kinetic energy of m,

U_f = final potential energy of m which is zero

and K_f = final kinetic energy of m which is also zero.

So, from Eq. (i), we have

$-v^2 m + K_i = 0 + 0$ or $K_i = mv^2$

76. As, $T_2 = T_1\left(\frac{r_2}{r_1}\right)^{3/2} = T_1\left(\frac{1}{4}\right)^{3/2} = \frac{1}{8}T_1 = \frac{1}{8}$ times

77. Asteroids move in circular orbits like planets under the action of central forces and obey Kepler's laws.

78. According to Kepler's law,

$$T^2 \propto r^3 \quad \Rightarrow \quad 5^2 \propto r^3 \qquad \ldots\text{(i)}$$

$$\text{and} \quad (T')^2 \propto (4r)^3 \qquad \ldots\text{(ii)}$$

From Eqs. (i) and (ii), we have

$$\frac{25}{(T')^2} = \frac{r^3}{64r^3}$$

$$\text{or} \quad T' = \sqrt{1600} \text{ or } T' = 40 \text{ h}$$

79. From Kepler's law,

$$T^2 \propto R^3 \Rightarrow T \propto R^{3/2}$$

$$\therefore \quad \frac{T_1}{T_2} = \left(\frac{12R}{3R}\right)^{3/2} = 8$$

$$\Rightarrow \quad T_2 = \frac{T_1}{8} = \frac{24}{8} = 3 \text{ h}$$

80. As, $T^2 \propto r^3$

So, $$\frac{T_A^2}{T_B^2} = \frac{r_A^3}{r_B^3}$$

or $$\frac{r_A}{r_B} = \left(\frac{T_A}{T_B}\right)^{2/3} = (8)^{2/3} = 4 \text{ or } r_A = 4r_B$$

So, $$r_A - r_B = 4r_B - r_B = 3r_B$$

81. The earth moves around the sun in elliptical path. So, by using the properties of ellipse

$r_1 = (1+e)a$ and $r_2 = (1-e)a$

$\Rightarrow \quad a = \frac{r_1 + r_2}{2}$ and $r_1 r_2 = (1-e^2)a^2$

where, a = semi-major axis,

b = semi-minor axis

and e = eccentricity.

Now, required distance = semi-latus rectum = $\frac{b^2}{a}$

$$= \frac{a^2(1-e^2)}{a} = \frac{r_1 r_2}{(r_1+r_2)/2} = \frac{2r_1 r_2}{r_1 + r_2}$$

82. The gravitational force of sun on comet is radial, hence angular momentum is constant over the entire orbit. Using law of conservation of angular momentum, at locations A and B, we get

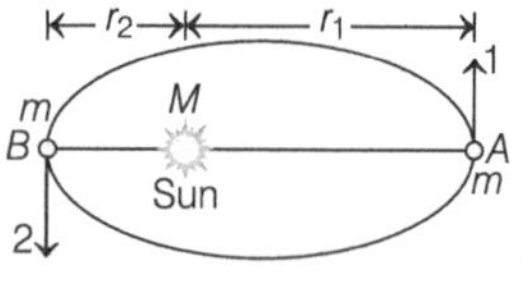

$$L = mv_1 r_1 = mv_2 r_2$$

$$\text{or} \quad v_2 = \frac{v_1 r_1}{r_2} \qquad \ldots\text{(i)}$$

Using the principle of conservation of total energy at A and B,

$$\frac{1}{2}mv_1^2 - \frac{GMm}{r_1} = \frac{1}{2}mv_2^2 - \frac{GMm}{r_2}$$

$$\text{or} \quad v_2^2 - v_1^2 = 2GM\left(\frac{1}{r_2} - \frac{1}{r_1}\right) \qquad \ldots\text{(ii)}$$

Putting the values from Eq. (i) in Eq. (ii) and solving, we get

$$v_1 = \left[\frac{2GMr_2}{r_1(r_1+r_2)}\right]^{1/2}$$

$$\therefore \quad L = mv_1r_1 = \left[\frac{2GMm^2r_1r_2}{(r_1+r_2)}\right]^{1/2}$$

Round II

1. As, $\Delta \text{KE} = \Delta U$

$$\Rightarrow \quad \frac{1}{2}mv^2 = GM_em\left(\frac{1}{R} - \frac{1}{R+h}\right) \quad \text{...(i)}$$

Also, $$g = \frac{GM_e}{R^2} \quad \text{...(ii)}$$

On solving Eqs. (i) and (ii), we have

$$h = \frac{R}{\left(\frac{2gR}{v^2} - 1\right)}$$

2. For $r \le R$,

$$\frac{mv^2}{r} = \frac{GmM}{r^2} \quad \left(\because M = \frac{4}{3}\pi r^3\rho_0\right)$$

$\therefore \quad v \propto r$

i.e. v-r graph is a straight line passing through the origin.

For $r \ge R$,

$$\frac{mv^2}{r} = \frac{GM\left(\frac{4}{3}\pi R^3\right)\rho_0}{r^2} \Rightarrow v \propto \frac{1}{\sqrt{r}}$$

The corresponding v-r graph will be as shown in option (c).

3. We know that, total energy (E), kinetic energy (K) and potential energy (U) of moving satellite is given as

$$E = -\frac{GMm}{2r}, K = \frac{GMm}{2r} \text{ and } U = -\frac{GMm}{r}$$

For satellite U, K and E vary with r and also U and E remain negative whereas K remains always positive.

Hence, from above expressions, graph represented in option (c) is correct.

4. As, $T = 2\pi\sqrt{\frac{l}{g}}$

i.e. $T \propto 1/\sqrt{g}$ (for fixed value of l)

$$\therefore \quad \frac{T_1}{T_2} = \sqrt{\frac{g'}{g}} = \left[\frac{gR^2/(R+2R)^2}{g}\right]^{1/2} = \frac{1}{3}$$

5. Escape velocity from the surface of earth is given by

$$v_e = \sqrt{\frac{2GM_e}{R_e}} \quad \text{...(i)}$$

Here, $v_e = 11.2 \text{ kms}^{-1}, M_p = 4M_e$

$$v_p = \sqrt{\frac{2GM_p}{R_p}} = \sqrt{\frac{2G \times 4M_e}{2R_e}} \quad \text{...(ii)}$$

Dividing Eq. (ii) by Eq. (i), we get

$$v_e' = \sqrt{2} \times v_e = 1.414 \times 11.2 = 15.8 \text{ kms}^{-1}$$

6. As, $T = \frac{2\pi r}{v} = \frac{2\pi r}{[GM/r]^{1/2}} = 2\pi\left[\frac{r^3}{GM}\right]^{1/2}$

As per question, $24 = 2\pi\left[\frac{(6400+36000)^3}{GM}\right]^{1/2}$

and $T' = 2\pi\left[\frac{(6400)^3}{GM}\right]^{1/2}$

$$\therefore \quad \frac{T'}{24} = \left[\frac{(6400)^3}{(6400+36000)^3}\right]^{1/2} = (0.4)^3$$

or $T' = (0.4)^3 \times 24 = 1.53 \text{ h}$

7. $V = \left[-\frac{Gm}{1} + \left(-\frac{Gm}{2}\right) + \left(\frac{Gm}{4}\right) + \ldots\right] \times 2$

$$= -2Gm\left[1 + \frac{1}{2} + \frac{1}{4} + \ldots\right] = -2Gm\left[\frac{1}{1-\frac{1}{2}}\right]$$

$$= -4Gm = -4G \times 1 = -4G$$

In magnitude, $V = 4G$

8. It is self-evident that, the orbit of the comet is elliptic with sun begin at one of the focus. Now, as for elliptic orbits, according to Kepler's third law,

$$T^2 = \frac{4\pi^2a^3}{GM} \Rightarrow a = \left(\frac{T^2GM}{4\pi^2}\right)^{1/3}$$

or $$a = \left[\frac{(76 \times 3.14 \times 10^7) \times 6.67 \times 10^{-11} \times 2 \times 10^{30}}{4\pi^2}\right]^{1/3}$$

$$a = 2.7 \times 10^{12}$$

But in case of ellipse,

$$2a = r_{\min} + r_{\max}$$

$$\therefore \quad r_{\max} = 2a - r_{\min} = 2 \times 2.7 \times 10^{12} - 8.9 \times 10^{10}$$
$$\simeq 5.3 \times 10^{12} \text{ m}$$

9.

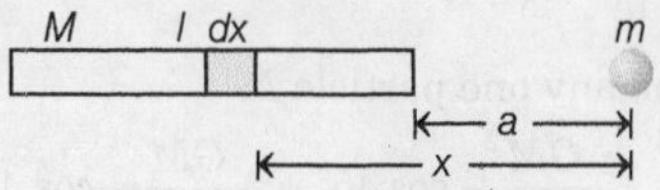

$$\Rightarrow \quad dU = \frac{Gm\left(\frac{M}{l}dx\right)}{x}$$

Integrating, we get

$$\Rightarrow \quad \int dU = \frac{GmM}{l}\int_a^{a+l}\frac{dx}{x}$$

$$\Rightarrow \quad U = \frac{GmM}{l}\log_e\left(\frac{a+l}{a}\right)$$

10. Force on the body $= \frac{GMm}{x^2}$

To move it by a small distance dx,

Work done $= F\,dx = \frac{GMm}{x^2}dx$

Total work done

$$= \int F\,dx = GMm\int_R^{R+h}\frac{dx}{x^2} = \left[\frac{-GMm}{x}\right]_R^{R+h}$$

$$= GMm\left[\frac{1}{R}-\frac{1}{R+h}\right]$$

$$= GMm\left[\frac{(R+h)-R}{R(R+h)}\right] = \frac{GMmh}{R(R+h)}$$

$$= \frac{GM}{R^2}\times\frac{mhR}{R+h} = \frac{gmhR}{R+h} = \frac{PRh}{R+h}$$

11. For orbiting spaceship close to earth's surface,

$$\frac{mv_o^2}{R} = \frac{GMm}{R^2}$$

i.e. $\quad v_o = \sqrt{\frac{GM}{R}} = \sqrt{gR}$

$\therefore \quad v_o = \sqrt{(9.8\times 6.4\times 10^6)} \approx 8\ \text{kms}^{-1}$

For escaping from closed to the surface of earth,

$$\frac{GMm}{R} = \frac{1}{2}mv_e^2$$

$$v_e = \sqrt{\frac{2GM}{R}} = \sqrt{2\,gR}$$

$\Rightarrow \quad v_e = \sqrt{2}\times v_o = 1.41\times 8 = 11.2\ \text{km/s}$

$\therefore$ The additional velocity to be imparted to the orbiting satellite for escaping is $11.2-8 = 3.2\ \text{kms}^{-1}$.

12. V_A = (Potential at A due to m_1) + (Potential at A due to m_2)

$\Rightarrow \quad V_A = -\frac{Gm_1}{R} - \frac{Gm_2}{\sqrt{2}\,R}$

Similarly, V_B = (Potential at B due to m_1) + (Potential at B due to m_2)

$\Rightarrow \quad V_B = -\frac{Gm_2}{R} - \frac{Gm_1}{\sqrt{2}\,R}$

Since, $W_{A\to B} = m(V_B - V_A) \Rightarrow W_{A\to B}$

$$= \frac{Gm\,(m_1-m_2)(\sqrt{2}-1)}{\sqrt{2}\,R}$$

13. Net force acting on any one particle M

$$= \frac{GM^2}{(2R)} + \frac{GM^2}{(R\sqrt{2})^2}\cos 45° + \frac{GM}{(R\sqrt{2})^2}\cos 45°$$

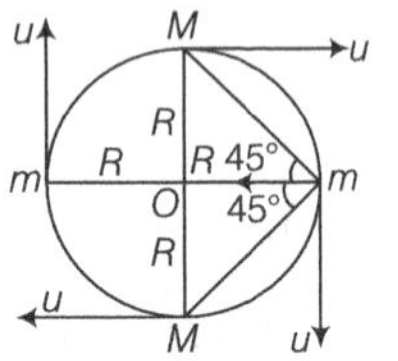

$$= \frac{GM^2}{R^2}\left(\frac{1}{4}+\frac{1}{\sqrt{2}}\right)$$

This force will be equal to centripetal force.

So, $\quad \frac{Mu^2}{R} = \frac{GM^2}{R^2}\left(\frac{1}{4}+\frac{1}{\sqrt{2}}\right)$

$\Rightarrow \quad u = \sqrt{\frac{GM}{4R}(1+2\sqrt{2})}$

$$= \frac{1}{2}\sqrt{\frac{GM}{R}(2\sqrt{2}+1)}$$

Hence, speed of each particle in a circular motion is $\frac{1}{2}\sqrt{\frac{GM}{R}(2\sqrt{2}+1)}$.

14. As, $W_{\text{ext}} = U_\infty - U_\rho$

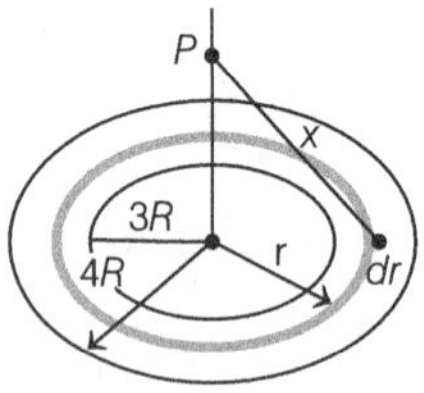

$$W_{\text{ext}} = 0 - \int -\left(-\frac{G\rho M}{x}\cdot dx\right)$$

$$W_{\text{ext}} = G\int \frac{M}{\pi\times 7R^2}\cdot\frac{2\pi r dx}{\sqrt{16R^2+r^2}}$$

$$= \frac{2GM}{7R^2}\int\frac{r\,dr}{\sqrt{16R^2+r^2}}$$

$$W_{\text{ext}} = \frac{2GM}{7R^2}\int\frac{2\,dz}{2} = \frac{2GM}{7R^2}[z]$$

$$= \frac{2GM}{7R^2}\left[\sqrt{16R^2+r^2}\right]_{3R}^{4R}$$

$$= \frac{2GM}{7R^2}[4\sqrt{2}R - 5R]$$

$$= \frac{2GM}{7R}(4\sqrt{2}-5)$$

15. As in case of elliptic orbit of a satellite, mechanical energy $E = -\left(\frac{GMm}{2a}\right)$ remains constant, at any position of satellite in the orbit.

i.e. $\quad \text{KE} + \text{PE} = -\frac{GMm}{2a} \quad$ ...(i)

Now, if at position r, v is the orbital speed of satellite

$$\text{KE} = \frac{1}{2}mv^2$$

and $\quad \text{PE} = -\frac{GMm}{r} \quad$ ...(ii)

So, from Eqs. (i) and (ii), we have

$$\frac{1}{2}mv^2 - \frac{GMm}{r} = -\frac{GMm}{2a},$$

i.e. $\quad v^2 = GM\left[\frac{2}{r}-\frac{1}{a}\right]$

16.

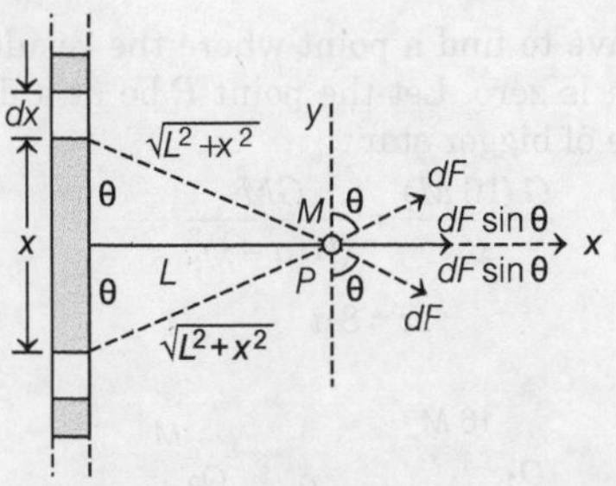

Let the mass M be placed symmetrically.

$$\Rightarrow \quad F_{\text{net}} = \int_{-\infty}^{\infty} dF \sin\theta \quad (dF\cos\theta \text{ are cancelled out})$$

$$= \int_{-\infty}^{\infty} \frac{GM(\lambda\, dx)}{X^2+L^2} \frac{L}{\sqrt{X^2+L^2}}$$

$$\Rightarrow \quad F_{\text{net}} = GM\lambda L \int_{-\infty}^{\infty} \frac{dx}{(X^2+L^2)^{3/2}}$$

$$\Rightarrow \quad F_{\text{net}} = \frac{GM\lambda L}{L^2}(2) \Rightarrow F_{\text{net}} = \frac{2GM\lambda}{L^2}$$

17. Weight of the object near the equator $= m(g - R\omega^2)$ and at height h near the poles

$$= mg\frac{R^2}{(R+h)^2}$$

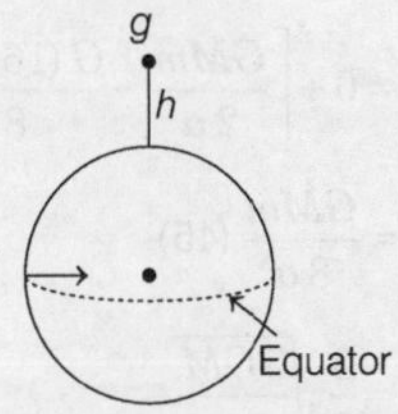

According to question, both the weights are same.

$$\Rightarrow \quad m(g - R\omega^2) = \frac{mgR^2}{(R+h)^2}$$

$$(g - R\omega^2) = gR^2\left[\frac{1}{R^2+h^2+2Rh}\right]$$

$$= \frac{gR^2}{R^2}\left[\frac{1}{1+\frac{h^2}{R^2}+\frac{2Rh}{R^2}}\right]$$

$$= g\left[\frac{1}{1+\frac{2h}{R}}\right] \qquad (\because h << R)$$

$$= g\left[1-\frac{2h}{R}\right]$$

$$\Rightarrow \quad R\omega^2 = \frac{2gh}{R}$$

$$\Rightarrow \quad h = \frac{R^2\omega^2}{2g}$$

18. As with height g varies as

$$g' = \frac{g}{[1+h/R]^2} = g\left[1-\frac{2h}{R}\right]$$

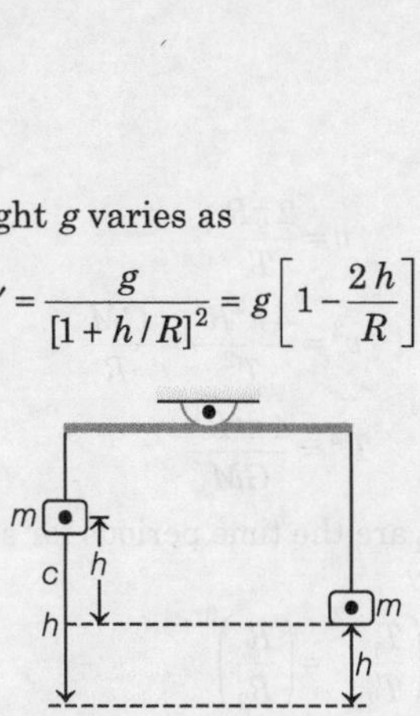

and in according with figure $h_1 > h_2$, W_1 will be lesser than W_2 and

$$i.e., \quad W_2 - W_1 = mg_2 - mg_1 = 2mg\left[\frac{h_1}{R} - \frac{h_2}{R}\right]$$

$$\text{or} \quad W_2 - W_1 = 2m\frac{GM}{R^2}\frac{h}{R}$$

$$\left[\text{As, } g = \frac{GM}{R^2} \text{ and } (h_1 - h_2) = h\right]$$

$$\text{or} \quad W_2 - W_1 = \frac{2mhG}{R^3}\left(\frac{4}{3}\pi R^3\rho\right)$$

$$= \frac{8}{3}\pi\rho\, Gmh \qquad \left[\text{as, } M = \frac{4}{3}\pi R^3\rho\right]$$

19. Mean orbital radius of the earth around the sun

$r = 1.5 \times 10^8$ km $= 1.5 \times 10^{11}$ m

Time period of earth around the sun = 1 yr

$= 365$ days $= 365 \times 24 \times 60 \times 60$ s

As required centripetal force is obtained from the gravitational force, therefore

Centripetal force = Gravitational force

$$\frac{mv^2}{r} = \frac{GM_s m}{r^2}$$

$$v^2 = \frac{GM_s}{r}$$

$$(r\omega)^2 = \frac{GM_s}{r} \qquad (\because v = r\omega)$$

$$\text{or} \quad \omega^2 = \frac{GM_s}{r^3}$$

$$\text{But} \quad \omega = \frac{2\pi}{T}$$

$$\therefore \quad \left(\frac{2\pi}{T}\right)^2 = \frac{GM_s}{r^3}$$

$$\text{or} \quad M_s = \frac{4\pi^2 r^3}{GT^2}$$

$$= \frac{4 \times (3.14)^2 \times (1.5 \times 10^{11})^3}{6.67 \times 10^{-11}(365 \times 24 \times 60 \times 60)^2}$$

$$\approx 2 \times 10^{30} \text{ kg}$$

20. For satellite, $\frac{mv^2}{R} = \frac{GMm}{R^2}$

$$\Rightarrow \quad v^2 = \frac{GM}{R}$$

$$v = \frac{2\pi R}{T}$$

$$\Rightarrow \quad v^2 = \frac{4\pi^2 R^2}{T^2} = \frac{GM}{R}$$

$$\therefore \quad T^2 = \frac{4\pi^2 R^3}{GM}$$

If T_1 and T_2 are the time periods for satellites S_1 and S_2 respectively.

$$\left(\frac{T_1}{T_2}\right)^2 = \left(\frac{R_1}{R_2}\right)^3$$

$$\Rightarrow \quad R_2 = \left(\frac{T_2}{T_1}\right)^{2/3} x R_1$$

$$T_1 = 1\text{ h}, T_2 = 8\text{ h}, R_1 = 10^4\text{ km}$$

$$R_2 = \left(\frac{8}{1}\right)^{2/3} \times 10^4\text{ km} = 4 \times 10^4\text{ km}$$

$$v_1 = \frac{2\pi R_1}{T_1} = \frac{2\pi \times 10^4}{1} = 2\pi \times 10^4\text{ kmh}^{-1}$$

$$v_2 = \frac{2\pi R_2}{T_2} = \frac{2\pi \times 4 \times 10^4}{8} = \pi \times 10^4\text{ kmh}^{-1}$$

Relative velocity of S_2 with respect to S_1 is

$$v = v_2 - v_1 = (\pi \times 10^4 - 2\pi \times 10^4)\text{ kmh}^{-1}$$

$$|v| = \pi \times 10^4\text{ kmh}^{-1}$$

21. If the mass of sun is M and radius of the planet's orbit is r, then as $v_o = \sqrt{GM/r}$

$$T = \frac{2\pi r}{v_0} = 2\pi r \times \sqrt{\frac{r}{GM}}$$

i.e. $$T^2 = \frac{4\pi^2 r^3}{GM} \quad \text{...(i)}$$

Now, if the planet (When stopped in the orbit) has velocity v when it is at a distance x from the sun, by conservation of mechanical energy, we get

$$\frac{1}{2}mv^2 + \left(-\frac{GMm}{x}\right) = 0 - \frac{GMm}{r}$$

or $$\left(-\frac{dx}{dt}\right) = \frac{2GM}{r}\left[\frac{r-x}{x}\right]$$

i.e. $$-\frac{dx}{dt} = \sqrt{\frac{2GM}{r}}\sqrt{\frac{(r-x)}{x}}$$

or $$\int_0^t dt = -\sqrt{\frac{r}{2GM}}\int_r^0 \left[\frac{x}{(r-x)}\right] dx$$

Substituting $x = r\sin^2\theta$ and solving the RHS, we get

$$t = \sqrt{\frac{r}{2GM}} \times \left(\frac{\pi r}{2}\right) \quad \text{(After integrating)}$$

Now, Eq. (i) reduces to

$$t = \frac{1}{\sqrt{4\sqrt{2}}} T, \text{ i.e. } t = \left(\frac{\sqrt{2}}{8}\right) T$$

22. First we have to find a point where the resultant field due to both is zero. Let the point P be at a distance x from centre of bigger star.

$$\Rightarrow \quad \frac{G(16M)}{x^2} = \frac{GM}{(10a - x)^2}$$

$$\Rightarrow \quad x = 8a \quad \text{(from } O_1\text{)}$$

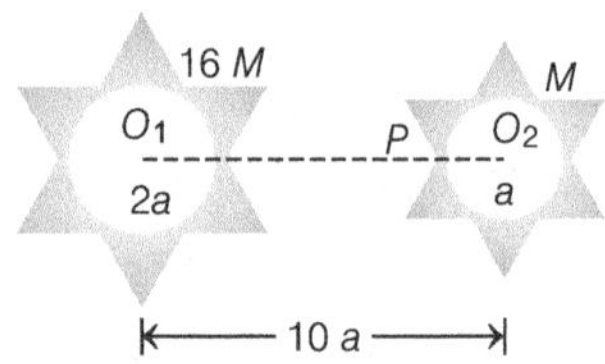

i.e. Once the body reaches P, the gravitational pull of attraction due to M takes the lead to make m move towards it automatically as the gravitational pull of attraction due to 16 M vanishes i.e., a minimum KE or velocity has to be imparted to m from surface of 16 M such that it is just able to overcome the gravitational pull of 16 M. By law of conservation of energy,

(Total mechanical energy at A)

= (Total mechanical energy at P)

$$\Rightarrow \quad \frac{1}{2}mv_{\min}^2 + \left[\frac{G(16M)m}{2a} - \frac{G(16M)m}{8a}\right] = 0 + \left[\frac{GMm}{2a} - \frac{G(16M)m}{8a}\right]$$

$$\Rightarrow \quad \frac{1}{2}mv_{\min}^2 = \frac{GMm}{8a}(45)$$

$$\Rightarrow \quad v_{\min} = \frac{3}{2}\sqrt{\frac{5GM}{a}}$$

23. Gravitational field intensity at a distance x from centre of planet is $E = \frac{GM}{x^2}$, where M = mass enclosed within a spherical volume of radius x.

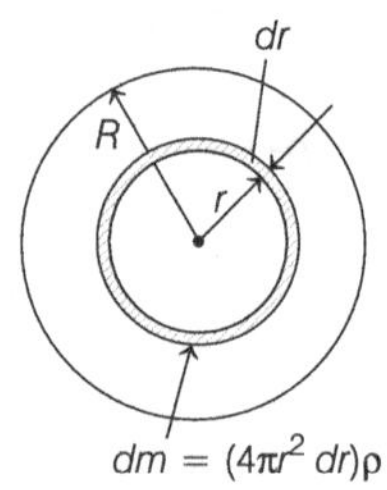

So, $$E = \frac{G}{x^2}\int_0^x (4\pi r^2 dr)[\rho(r)] \quad \text{...(i)}$$

Here, $M = \int_0^x dm = \int_0^x 4\pi r^2 dr \cdot \rho(r)$

So, from Eq. (i), we get

$$E = \frac{G}{x^2}\int_0^x 4\pi r^2 \cdot \rho_0 \left(1 - \frac{r^2}{R^2}\right) dr$$

$$= \frac{G}{x^2}\int_0^x 4\pi\rho_0\left(r^2 - \frac{r^4}{R^2}\right)dr$$

$$= \frac{4\pi G\rho_0}{x^2}\left[\frac{r^3}{3} - \frac{r^5}{5R^2}\right]_0^x$$

$$= \frac{4\pi G\rho_0}{x^2}\left(\frac{x^3}{3} - \frac{x^5}{5R^2}\right)$$

$$= 4\pi G\rho_0\left(\frac{x}{3} - \frac{x^3}{5R^2}\right)$$

For gravitational field to be maximum, $\frac{dE}{dx} = 0$

$$\Rightarrow \quad \frac{d}{dx}\left[4\pi G\rho_0\left(\frac{x}{3} - \frac{x^3}{5R^2}\right)\right] = 0$$

$$\Rightarrow \quad 4\pi G\rho_0\left(\frac{1}{3} - \frac{3x^2}{5R^2}\right) = 0$$

$$\Rightarrow \quad x = \sqrt{\frac{5}{9}}\,R$$

So, E is maximum at a distance of $\sqrt{\frac{5}{9}}\,R$ from centre of planet.

24. For a particle or satellite with speed v_p around a planet whose escape speed is v_e and orbital speed of particle v_o, we have following possibilities

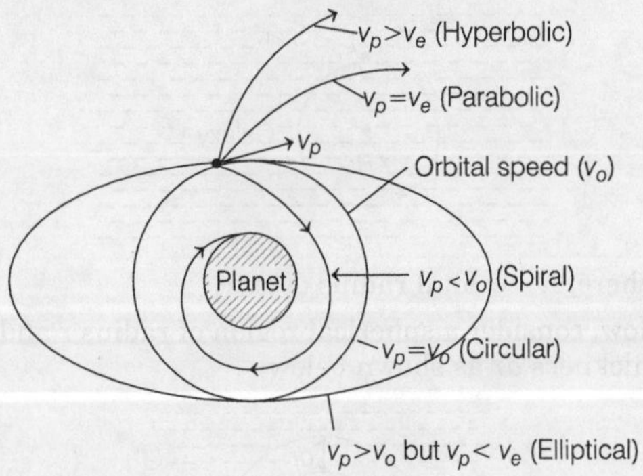

Now, by momentum conservation in given collision, we have, $p_i = p_f$

$$\Rightarrow \quad mv_A + \frac{m}{2}\left(\frac{v_A}{2}\right) = \left(m + \frac{m}{2}\right)v_p \qquad \text{...(i)}$$

Body A is orbiting in orbit of radius $R + h$, hence for a small value of h,

$$v_A = \sqrt{\frac{GM}{r}} = \sqrt{\frac{GM}{R}} = \frac{v_e}{\sqrt{2}}$$

So, final speed of combined mass system,

$$v_p = \frac{5}{6}v_o = \frac{5}{6} \times \frac{1}{\sqrt{2}}v_e$$

As v_p is less than v_o, so it must spiral towards planet. Most nearest option is (b), if we take spiral close to elliptical.

25. Let us assume that, stars are moving in XY-plane with origin as their centre of mass as shown in the figure below

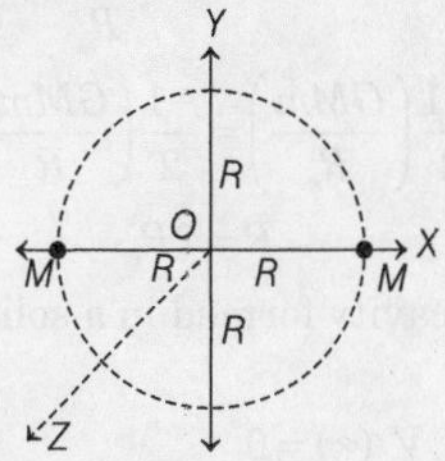

According to question, mass of each star,

$$M = 3 \times 10^{31}\ \text{kg}$$

and diameter of circle, $2R = 2 \times 10^{11}$ m

$$\Rightarrow \quad R = 10^{11}\ \text{m}$$

Potential energy of meteorite at O,

$$U_{\text{total}} = -\frac{2\,GMm}{r}$$

If v is the velocity of meteorite at O, then kinetic energy K of the meteorite is $K = \frac{1}{2}mv^2$.

To escape from this dual star system, total mechanical energy of the meteorite at infinite distance from stars must be at least zero.

By conservation of energy, we have

$$\frac{1}{2}mv^2 - \frac{2\,GMm}{R} = 0$$

$$\Rightarrow \quad v^2 = \frac{4\,GM}{R} = \frac{4 \times 6.67 \times 10^{-11} \times 3 \times 10^{31}}{10^{11}}$$

$$\Rightarrow \quad v = 2.83 \times 10^5 \simeq 2.8 \times 10^5\ \text{m/s}$$

26.

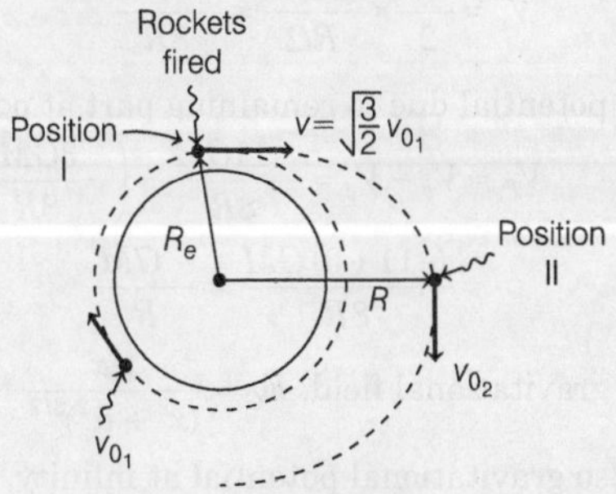

Using law of conservation of energy for position I and position II of satellite, we have

$$(\text{PE} + \text{KE})_{\text{Position I}} = (\text{PE} + \text{KE})_{\text{Position II}}$$

$$\Rightarrow \quad \frac{-GMm}{R_e} + \frac{1}{2}mv_1^2 = \frac{-GMm}{R} + \frac{1}{2}mv_2^2$$

$$\Rightarrow \quad \frac{-GMm}{R_e} + \frac{1}{2}m\left(\frac{3}{2}v_{0_1}^{\,2}\right) = \frac{-GMm}{R} + \frac{1}{2}mv_{0_2}^{\,2}$$

As, orbital speed, $v_0 = \sqrt{\frac{GM}{R}}$

(where, R = orbital radius)

Therefore, $\frac{-GMm}{R_e}+\frac{1}{2}m\left(\frac{3}{2}\frac{GM}{R_e}\right)$

$$= -\frac{GMm}{R}+\frac{1}{2}m\left(\frac{GM}{R}\right)$$

$$\Rightarrow \quad \frac{-1}{4}\left(\frac{GMm}{R_e}\right)=\frac{-1}{2}\left(\frac{GMm}{R}\right)$$

$$\Rightarrow \quad R = 2R_e$$

27. Consider the cavity formed in a solid sphere as shown in figure.

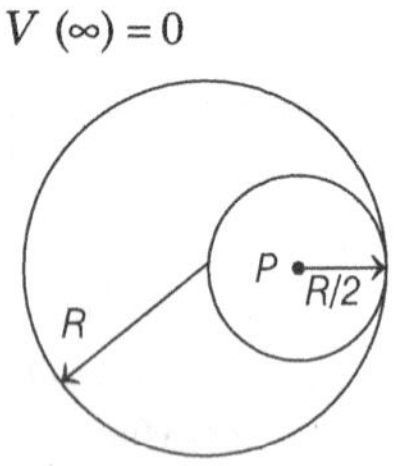

According to the question, we can write potential at an internal point P due to complete solid sphere,

$$V_s = -\frac{GM}{2R^3}\left[3R^2-\left(\frac{R}{2}\right)^2\right]$$

$$= \frac{-GM}{2R^3}\left[3R^2-\frac{R^2}{4}\right]$$

$$= \frac{-GM}{2R^3}\left[\frac{11R^2}{4}\right] = \frac{-11GM}{8R}$$

Mass of removed part $= \frac{M}{(4/3)\times\pi R^3}\times\frac{4}{3}\pi\left(\frac{R}{2}\right)^3=\frac{M}{8}$

Potential at point P due to removed part,

$$V_c = \frac{-3}{2}\times\frac{GM/8}{R/2}=\frac{-3GM}{8R}$$

Thus, potential due to remaining part at point P,

$$V_P = V_s - V_c = \frac{-11GM}{8R}-\left(-\frac{3GM}{8R}\right)$$

$$= \frac{(-11+3)GM}{8R} = \frac{-GM}{R}$$

28. Here, gravitational field, $E_G = \frac{Ax}{(x^2+a^2)^{3/2}}$

and also gravitational potential at infinity, $V_\infty = 0$

Now, $\quad dV = -\mathbf{E}_G\cdot d\mathbf{x}$

$$\int_{V_\infty}^{V_x} dV = -\int_\infty^x \mathbf{E}_G\cdot d\mathbf{x}$$

$$[V]_{V_\infty}^{V_x} = -\int_\infty^x \frac{Ax}{(x^2+a^2)^{3/2}}\,dx\cos 0^\circ$$

$$V_x - V_\infty = -\int_\infty^x \frac{Ax}{(x^2+a^2)^{3/2}}\,dx\times 1$$

$$V_x - 0 = -\int_\infty^x \frac{Ax}{(x^2+a^2)^{3/2}}\,dx$$

$$V_x = -A\int_\infty^x \frac{x}{(x^2+a^2)^{3/2}}\,dx \qquad \text{...(i)}$$

Let $x^2+a^2=t$, then $2xdx+0=dt$

$$x\,dx = \frac{dt}{2}$$

At $x=\infty$, $(\infty)^2+a^2=t$

$t=\infty$ (lower limit)

Again, at $x=x$

$x^2+a^2=t \Rightarrow t=x^2+a^2$ (upper limit)

From Eq. (i), we get

$$V_x = -A\int_\infty^{x^2+a^2}\frac{dt/2}{(t)^{3/2}}$$

$$= -\frac{A}{2}\int_\infty^{x^2+a^2}(t)^{-3/2}dt$$

$$= -\frac{A}{2}\left[\frac{(t)^{-\frac{3}{2}+1}}{-\frac{3}{2}+1}\right]_\infty^{x^2+a^2} = -\frac{A}{2}\left[\frac{(t)^{-\frac{1}{2}}}{-\frac{1}{2}}\right]_\infty^{x^2+a^2}$$

$$= A\left[\frac{1}{(t)^{\frac{1}{2}}}\right]_\infty^{x^2+a^2} = A\left[\frac{1}{(x^2+a^2)^{\frac{1}{2}}}-\frac{1}{(\infty)^{\frac{1}{2}}}\right]$$

$$= A\left[\frac{1}{(x^2+a^2)^{\frac{1}{2}}}-0\right] = \frac{A}{(x^2+a^2)^{\frac{1}{2}}}$$

29. Force on star is provided by the mass enclosed inside region of radius R as shown below

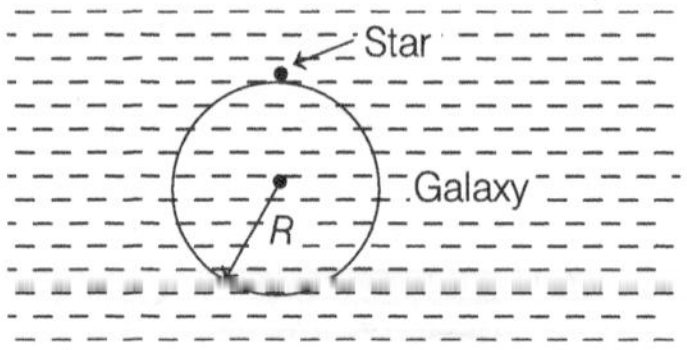

where, R = orbital radius of star.

Now, consider a spherical region of radius r and thickness dr as shown below

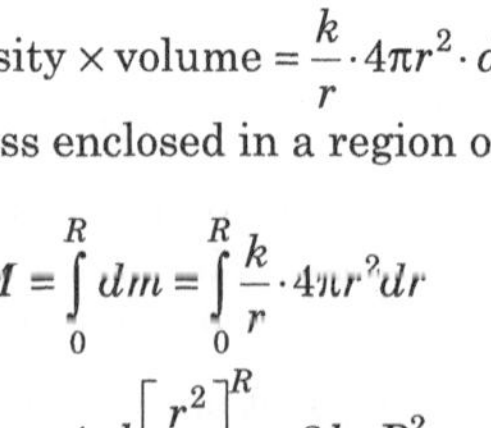

Mass enclosed in this region is given by

$$dm = \text{density}\times\text{volume} = \frac{k}{r}\cdot 4\pi r^2\cdot dr$$

So, total mass enclosed in a region of radius R is given by

$$M = \int_0^R dm = \int_0^R \frac{k}{r}\cdot 4\pi r^2 dr$$

$$= 4\pi k\left[\frac{r^2}{2}\right]_0^R = 2k\pi R^2$$

Gravitational pull of this mass provides necessary centripetal force for rotation of star.

So, $\dfrac{GMm}{R^2} = \dfrac{mv^2}{R} \Rightarrow v^2 = \dfrac{GM}{R} = \dfrac{G2k\pi R^2}{R}$

$\Rightarrow \quad v^2 = 2Gk\pi R \Rightarrow v = \sqrt{2Gk\pi R}$

Time period of rotation,

$$T = \frac{2\pi R}{v} = \frac{2\pi R}{\sqrt{2Gk\pi R}} \Rightarrow T \propto \sqrt{R} \text{ or } T^2 \propto R$$

30. Let the mass of the test particle be m and its orbital linear speed be v.

Force of gravity of the mass-density would provide the necessary centripetal pull on test particle.

$\Rightarrow$ Gravitational force (F_g) = Centripetal force (F_c)

Let us now assume an elementary ring at a distance r from the centre of the mass density, such that mass of this elementary ring be M.

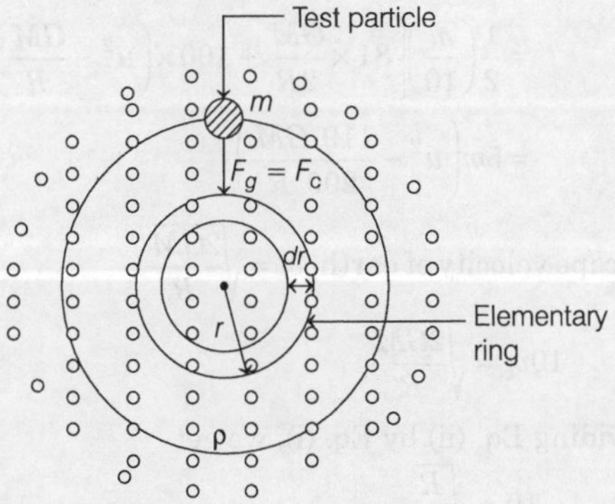

$\Rightarrow \quad \dfrac{GMm}{R^2} = \dfrac{mv^2}{R}$

where, $M = \int_0^R \rho \cdot 4\pi r^2 \cdot dr$

$\Rightarrow \quad \dfrac{G\left(\int_0^R \frac{K}{r^2} \cdot 4\pi r^2 \cdot dr\right)}{R^2} = \dfrac{v^2}{R} \quad \left[\because \text{ given, } \rho(r) = \dfrac{K}{r^2}\right]$

$\Rightarrow \quad G \cdot 4\pi K = v^2$

$\therefore$ Orbital speed of mass, $m = v = \sqrt{G4\pi K}$ ... (i)

Time period of rotation of the test particle,

$$T = \frac{2\pi R}{v} = \frac{2\pi R}{\sqrt{G \cdot 4\pi K}} \quad [\because \text{ using Eq. (i)}]$$

Hence, $\dfrac{T}{R} = \sqrt{\dfrac{\pi}{GK}}$ = a constant

31. A satellite or spaceship in a circular orbit at a distance $(R + h)$ from centre of a planet experiences a gravitational force given by

$$F_g = \frac{GmM}{(R+h)^2}$$

where, M = mass of planet,

m = mass of spaceship,

R = radius of planet

and h = height of spaceship above surface.

This gravitational pull provides necessary centripetal pull for orbital motion of spaceship.

So, $F_g = F_{\text{centripetal}} \Rightarrow \dfrac{GmM}{(R+h)^2} = \dfrac{mv^2}{(R+h)}$

where, v = orbital speed of spaceship.

Orbital speed of spaceship,

$$v = \sqrt{\frac{GM}{(R+h)}} \quad \text{...(i)}$$

Here, $G = 6.67 \times 10^{-11}$ N-m^2/kg^2,

$M = 8 \times 10^{22}$ kg

$R + h = (2 \times 10^6 + 20 \times 10^3)$ m

$= 2.02 \times 10^6$ m

So, substituting these values in Eq. (i), we get

$$v = \sqrt{\frac{6.67 \times 10^{-11} \times 8 \times 10^{22}}{2.02 \times 10^6}} = 1.6 \times 10^3 \text{ ms}^{-1}$$

Time period of rotation of spaceship will be

$$T = \frac{2\pi(R+h)}{v} \Rightarrow T = \frac{2\pi \times 2.02 \times 10^6}{1.6 \times 10^3}$$

$$\approx 8 \times 10^3 \text{ s} = \frac{8 \times 10^3}{60 \times 60} \text{ (h)} = 2.2 \text{ h}$$

So, number of revolutions made by spaceship in 24 h,

$$n = \frac{24}{T} = \frac{24}{2.2} \approx 11 \text{ rev}$$

32. Given, volume of earth (V_e) is 64 times of volume of moon (V_m), *i.e.*

$$\frac{V_e}{V_m} = 64 = \frac{\frac{4}{3}\pi R_e^3}{\frac{4}{3}\pi R_m^3}$$

where, R_e and R_m are the radii of earth and moon, respectively.

Then, $\dfrac{R_e}{R_m} = 4$ ...(i)

Also, since the density of moon and earth are equal, *i.e.* $\rho_m = \rho_e$

$\Rightarrow \dfrac{M_e}{V_e} = \dfrac{M_m}{V_m}$, where M_e and M_m are the mass of the earth and moon, respectively.

$\Rightarrow \quad \dfrac{M_e}{M_m} = \dfrac{V_e}{V_m} = 64$ ...(ii)

The minimum energy or escape energy delivered by the rocket launcher, so that the rocket never returns to earth is $E_e = \dfrac{GM_e m}{R_e} = E$

where, m is the mass of the rocket.

Similarly, minimum energy that a launcher should have to escape or to never return, if rocket is launched from surface of the moon is

$$E_m = \frac{GM_m m}{R_m}$$

$\therefore$ Ratio of escape energies E_e and E_m is

$$\frac{E_e}{E_m} = \frac{\left(\frac{GM_e m}{R_e}\right)}{\left(\frac{GM_m m}{R_m}\right)} = \frac{M_e}{M_m} \cdot \frac{R_m}{R_e}$$

$$= 64 \times \frac{1}{4} = 16$$

[using Eqs. (i) and (ii)]

$$\therefore \quad E_m = \frac{E_e}{16} = \frac{E}{16}$$

33. Let us assume that, when satellite reaches in orbit at height R, it has a velocity v_r radially away from earth.

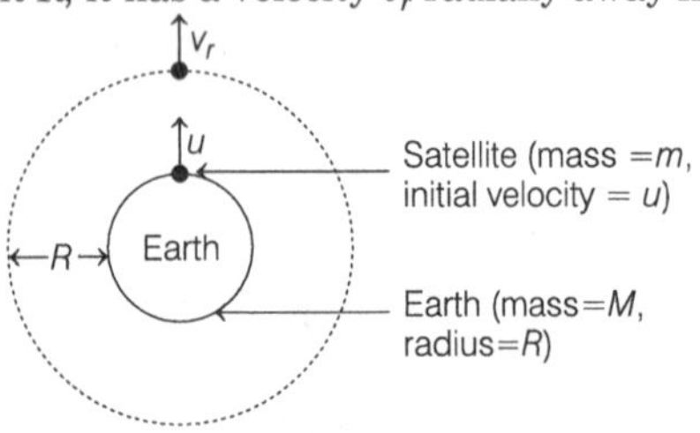

According to conservation of energy,
Total energy of satellite at surface of earth = Total energy at height R

$\Rightarrow (\text{KE} + \text{PE})_{\text{initial}} = (\text{KE} + \text{PE})_{\text{final}}$

$$\Rightarrow \frac{1}{2}mu^2 + \left(\frac{-GMm}{R}\right) = \frac{1}{2}mv_r^2 + \left(\frac{-GMm}{2R}\right)$$

So, radial velocity of satellite at height R,

$$v_r = \left(u^2 - \frac{GM}{R}\right)^{1/2} \quad \ldots\text{(i)}$$

To provide satellite an orbital velocity $\left(v_o = \sqrt{\frac{GM}{2R}}\right)$, a rocket $\left(\text{mass} = \frac{m}{10}\right)$ is fired.

Rocket must be given a velocity v such that it destroyed or opposes radial momentum of satellite and also provides orbital velocity to the satellite.

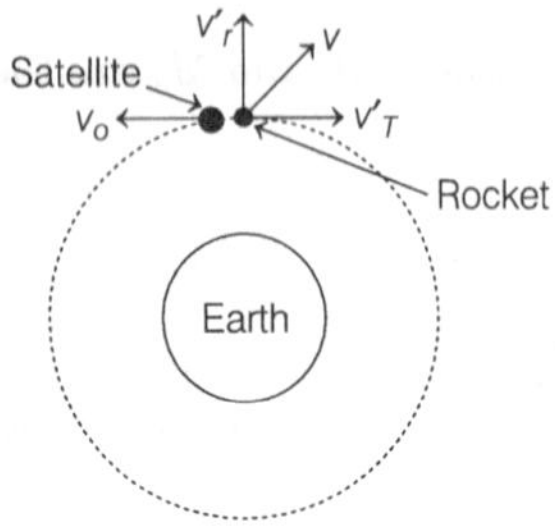

Now, we apply conservation of momentum to calculate radial (v_r') and tangential (v_T') components of rocket's velocity.

When rocket is ejected from satellite, mass of satellite will be

$$m - \frac{m}{10} = \frac{9m}{10}$$

By applying conservation of tangential momentum of rocket and satellite system, we have

$$\frac{m}{10} \times v_T' + \frac{9m}{10}(-v_o) = 0 \quad \ldots\text{(ii)}$$

Similarly, by applying conservation of radial momentum of rocket and satellite system, we have

$$\frac{m}{10} \times v_r' + \frac{9m}{10}(v_r) = 0 \quad \ldots\text{(iii)}$$

From Eqs. (i), (ii) and (iii), we have

$$v_T' = 9\sqrt{\frac{GM}{2R}} \text{ and } v_r' = 10\sqrt{u^2 - \frac{GM}{R}}$$

So, kinetic energy of the rocket,

$$K = \frac{1}{2}\left(\frac{m}{10}\right) \cdot v^2 = \frac{1}{2}\left(\frac{m}{10}\right)\left(\sqrt{v_T'^2 + v_r'^2}\right)^2$$

$$= \frac{1}{2}\left(\frac{m}{10}\right)(v_T'^2 + v_r'^2)$$

$$= \frac{1}{2}\left(\frac{m}{10}\right)\left[81 \times \frac{GM}{2R} + 100 \times \left(u^2 - \frac{GM}{R}\right)\right]$$

$$= 5m\left(u^2 - \frac{119}{200}\frac{GM}{R}\right)$$

34. Escape velocity of earth, $v_e = \sqrt{\frac{2GM}{R}}$ ...(i)

$$\Rightarrow \quad 10v_e = \sqrt{\frac{2GM}{R'}} \quad \ldots\text{(ii)}$$

Dividing Eq. (ii) by Eq. (i), we get

$$10 = \sqrt{\frac{R}{R'}}$$

$$\Rightarrow \quad R' = \frac{R}{100} = \frac{6400}{100} = 64 \text{ km}$$

35. Initial potential energy, $U_i = -\frac{3}{5}\frac{GM^2}{R}$

Final potential energy, $U_f = 0$

$\therefore$ Energy supplied $= U_f - U_i$

$$= 0 - \left(-\frac{3}{5}\frac{GM^2}{R}\right)$$

$$= \frac{3}{5}\frac{GM^2}{R}$$

$\therefore \quad x = 3$

36. For a satellite in circular orbit,

$$\frac{mv_0^2}{r} = \frac{GMm}{r^2}$$

or $$v_0^2 = \frac{GM}{R+h}$$

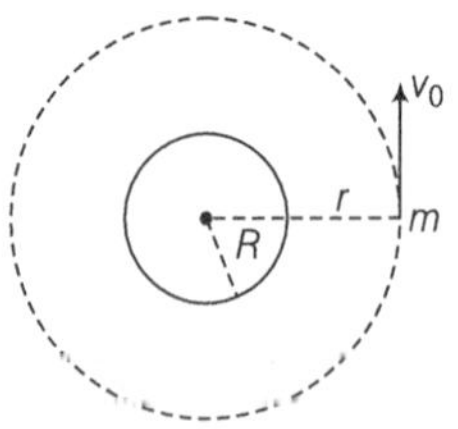

Given, $$v_0 = \frac{v_e}{2}$$

$$\Rightarrow \sqrt{\frac{GM}{R+h}} = \frac{1}{2}\sqrt{\frac{2GM}{R}}$$

$$\frac{GM}{R+h} = \frac{GM}{2R}$$

$$\Rightarrow \quad h = R$$

When the satellite stops, $K = 0$

$$\therefore \quad U_i - U_f = K_f$$

$$\Rightarrow \quad -\frac{GMm}{2R} - \left(-\frac{GMm}{R}\right) = \frac{1}{2}mv^2$$

$$v = \sqrt{\frac{GM}{R}} = \sqrt{gR}$$

$$\therefore \quad k = 1$$

37.

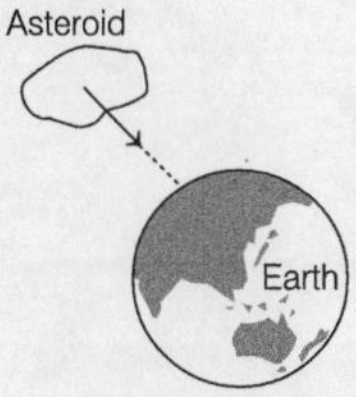

Applying energy conservation, we have

Total energy of asteroid at 12 km = Total energy of asteroid at surface of earth

$$\Rightarrow \quad U_1 + K_1 = U_2 + K_2$$

$$\Rightarrow \quad \frac{-GM_e m}{10R_e} + \frac{1}{2}mv_0^2 = \frac{-GM_e m}{R_e} + \frac{1}{2}mv^2$$

$$\Rightarrow \quad \frac{9}{10}\frac{GM_e m}{R_e} + \frac{1}{2}mv_0^2 = \frac{1}{2}mv^2$$

$$\Rightarrow \quad \frac{9}{10} \times \frac{2GM_e}{R_e} + v_0^2 = v^2$$

$$\Rightarrow \quad \frac{9}{10} \times (\sqrt{v_e})^2 + v_0^2 = v^2$$

$$\Rightarrow \quad \frac{9}{10} \times (11.2)^2 + (12)^2 = v^2$$

$$\Rightarrow \quad v^2 = \sqrt{\frac{9}{10} \times (11.2)^2 + (12)^2} \Rightarrow v \approx 16 \text{ kms}^{-1}$$

38. $V_A = \begin{bmatrix}\text{Potential due to the}\\ \text{complete sphere at } A\end{bmatrix} - \begin{bmatrix}\text{Potential due to the}\\ \text{spherical cavity at } A\end{bmatrix}$

$$= -\frac{3}{2}\frac{GM}{R} - \left(-\frac{GM'}{r}\right)$$

$$M = \frac{4}{3}\pi R^3 \rho$$

$$M' = \frac{4}{3}\pi\left(\frac{R}{2}\right)^3 \rho = \frac{1}{8}\left(\frac{4}{3}\pi R^3 \rho\right), r = \frac{R}{2}$$

$$\therefore \quad V_A = \frac{G}{R}\left[\frac{\pi \rho R^3}{3} - 2\pi \rho R^3\right] = -\frac{5}{3}\pi G \rho R^2$$

Now, $V_B = \begin{bmatrix}\text{Potential due to the}\\ \text{complete sphere at } B\end{bmatrix} - \begin{bmatrix}\text{Potential due to the}\\ \text{spherical cavity at } B\end{bmatrix}$

$$= -\frac{GM}{2R^3}(3R^2 - r^2) - \left[-\frac{3}{2}\frac{GM'}{r}\right]$$

or
$$V_B = -\frac{11}{8}\frac{GM}{R} + \frac{3GM'}{R}$$

$$= \frac{G}{R}\left[\pi \rho R^3 - \frac{11\pi \rho R^3}{6}\right]$$

$$= -\frac{4}{3}\pi G \rho R^2$$

Now, $V_B - V_A = \frac{1}{3}\pi G \rho R^2$

$$\frac{1}{2}m_0 v^2 = U_B - U_A = m_0 (V_B - V_A)$$

$$\Rightarrow \quad v = \sqrt{2(V_B - V_A)}$$

or
$$v = \frac{\sqrt{2\pi G \rho R^2}}{\sqrt{3}}$$

$$\Rightarrow \quad \beta = 3$$

11

Properties of Solids

IN THIS CHAPTER ....

- Elastic Behaviour
- Stress
- Strain
- Hooke's Law
- Poisson's Ratio (σ)
- Stress-Strain Relationship
- Work Done or Potential Energy Stored in a Stretched Wire
- Thermal Stresses and Strains

A body is said to be *elastic*, if it regains its original shape and size on the removal of deforming force. The property of matter by virtue of which a body tends to regain its original shape and size after the removal of deforming forces is called *elasticity*.

If a body completely gains its natural shape after the removal of the deforming forces it is called *perfectly elastic body*. If a body remains in the deformed state or does not even partially regain its original shape after removal of deforming forces it is called *perfectly inelastic* or *plastic body*.

Elastic Behaviour

The elastic behaviour can be understood by taking the microscopic nature of solids. It is shown in the following figure, in which we can see that the balls represent the atoms or molecules and the springs represent the interatomic or intermolecular forces.

In this system originally, the ball is in the position of its stable equilibrium, if any ball is displaced a little from its equilibrium position, the springs attached to that ball will either be stretched or compressed.

Therefore, the restoring forces are developed in the springs and they will bring the ball back to its natural position. This is known as the elastic behaviour of the solid body.

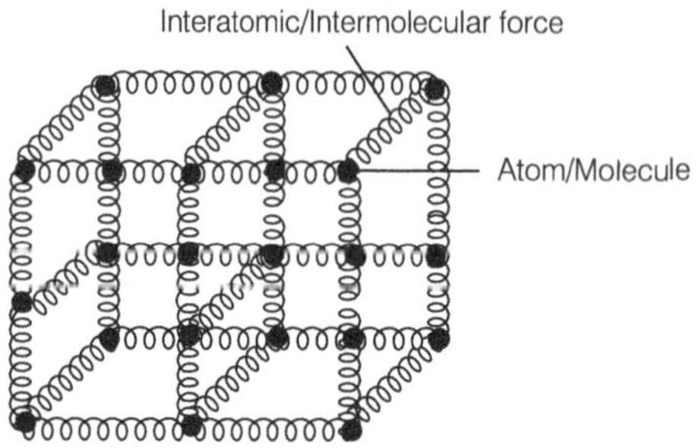

Stress

When an external force is applied to a body, an internal restoring force is set up at each cross-section of the body which tends to restore the body back to its original state. The restoring force set up inside the body per unit area is known as stress.

$$\text{Stress} = \frac{\text{Restoring force}}{\text{Area}}$$

In SI system, unit of stress is Nm^{-2} or pascal (denoted by Pa) and in CGS system, it is dyne/cm^2. Mainly, there are three types of stress. They are listed as follows.

(i) Normal or Longitudinal Stress

If area of cross-section of a rod is A and a deforming force F is applied along the length of the rod and perpendicular to its cross-section, then in this case, stress produced in the rod is known as normal or longitudinal stress. It is also known as axial stress.

$$\text{Longitudinal stress} = \frac{F_n}{A}$$

Longitudinal stress is of two types

(a) **Tensile stress** A type of longitudinal stress in which length of the rod is increased on application of deforming force over it, the stress produced in rod is called tensile stress.

(b) **Compressive stress** A type of longitudinal stress in which length of the rod is decreased on application of deforming force, the stress produced is called compressive stress.

(ii) Volumetric Stress

In this type of stress, when a force is applied on a body such that it produces a change in volume and density without change in its shape and

(a) at any point, the force is perpendicular to its surface.

(b) at any small area, the magnitude of force is directly proportional to its area.

Then, force per unit area is called volumetric stress.

$$\therefore \quad \text{Volumetric stress} = \frac{F_V}{A}$$

This is the case when small solid body is immersed in fluid.

(iii) Shearing or Tangential Stress

In shearing or tangential stress, the force is applied tangentially or parallel to a surface of a body, which produces a change in shape without change in its volume.

$$\text{Tangential stress} = \frac{F_T}{A}$$

Example 1. *The ratio of radii of two wires of same material is 2 : 1. If these wires are stretched by equal force, what is the ratio of stress produced in them?*

(a) 1 : 2 (b) 1 : 3 (c) 1 : 4 (d) 1 : 5

Sol. (c) Here, $r_1 : r_2 = 2:1, F_1 = F_2 = F$

$$\text{Stress } (S) = \frac{\text{Force}}{\text{Area}} = \frac{F}{\pi r^2}$$

$$\therefore \quad S \propto \frac{1}{r^2}$$

$$\therefore \quad \frac{S_1}{S_2} = \frac{r_2^2}{r_1^2} = \left(\frac{1}{2}\right)^2 = \frac{1}{4}$$

Example 2. *The elastic limit of brass is 379 MPa. What should be the minimum diameter of a brass rod, if it is to support a 400 N load without exceeding its elastic limit?*

(a) 0.90 mm (b) 1.00 mm
(c) 1.16 mm (d) 1.36 mm **[JEE Main 2019]**

Sol. (c) Let d_{min} be the minimum diameter of brass.

Then, stress in brass rod is given by

$$\sigma = \frac{F}{A} = \frac{4F}{\pi d_{min}^2} \qquad \left[\because A = \frac{\pi d^2}{4}\right]$$

For stress not to exceed elastic limit, we have $\sigma \le 379$ MPa

$$\Rightarrow \quad \frac{4F}{\pi d_{min}^2} \le 379 \times 10^6$$

Here, $F = 400$ N

$$\therefore \quad d_{min}^2 = \frac{1600}{\pi \times 379 \times 10^6}$$

$$\Rightarrow \quad d_{min} = 1.16 \times 10^{-3} \text{ m} = 1.16 \text{ mm}$$

Example 3. *A steel wire having a radius of 2.0 mm, carrying a load of 4 kg, is hanging from a ceiling. Given that $g = 3.1\pi \, ms^{-2}$, what will be the tensile stress that would be developed in the wire?* **[JEE Main 2019]**

(a) $6.2 \times 10^6 Nm^{-2}$ (b) $5.2 \times 10^6 Nm^{-2}$
(c) $3.1 \times 10^6 Nm^{-2}$ (d) $4.8 \times 10^6 Nm^{-2}$

Sol. (c) Given, radius of wire, $r = 2$ mm $= 2 \times 10^{-3}$ m

Weight of load, $m = 4$ kg, $g = 3.1\pi$ ms^{-2}

$$\therefore \quad \text{Tensile stress} = \frac{\text{Force } (F)}{\text{Area } (A)} = \frac{mg}{\pi r^2}$$

$$= \frac{4 \times 3.1 \times \pi}{\pi \times (2 \times 10^{-3})^2}$$

$$= 3.1 \times 10^6 Nm^{-2}$$

Strain

When the size or shape of a body is changed under an external force, the body is said to be strained. The change occurred in the unit size of the body is called strain. Usually, it is denoted by ε.

$$\text{Strain} = \frac{\text{Change in dimension}}{\text{Original dimension}} = \frac{\Delta x}{x}$$

Here, Δx is the change (may be in length, in volume etc.) and x is the original value of quantity in which change has occurred. It has no dimension as it is a pure number.

Since, a body may have three types of deformation, *i.e.* in length, in volume or in shape, likewise there are following three types of strains

(i) Longitudinal Strain

It is associated with longitudinal stress. The change in length per unit original length of the body under deformation produced by the external force is known as longitudinal strain.

$$\text{Longitudinal strain} = \frac{\text{Change in length}}{\text{Original length}} \quad \text{or} \quad \varepsilon_l = \frac{\Delta l}{l}$$

It is of two types

(a) **Tensile strain** If on applying a deforming force, there is an increase of Δl in length of a rod, then strain produced in the rod is called tensile strain.

(b) **Compressive strain** If on applying a deforming force, there is decrease of Δl in length of a rod, then strain produced in the rod is called compressive strain.

(ii) Volumetric Strain

It is produced when there's volumetric stress in the body. The change in volume per unit original volume of the body under deformation produced by the external force is known as volumetric strain.

$$\text{Volumetric strain} = \frac{\text{Change in volume}}{\text{Original volume}}$$

or $$\varepsilon_V = \frac{\Delta V}{V}$$

(iii) Shearing Strain

This type of strain is produced when a shearing stress is present.

It is defined as the angle in radians through which a plane perpendicular to the fixed surface of the cubical body is turned under the effect of tangential force.

$$\text{Shearing strain, } \tan\phi = \frac{x}{L}$$

or $$\phi = \frac{x}{L} \quad \text{(for very small deformation)}$$

Hooke's Law

It states that for a body within elastic limit, stress applied to a body is proportional to the resulting strain, *i.e.*

$$\text{Stress} \propto \text{Strain} \quad \text{or} \quad \frac{\text{Stress}}{\text{Strain}} = E = \text{constant}$$

where, E is constant and is known as *modulus of elasticity* or *coefficient of elasticity* of the material of the body.

Types of Modulus of Elasticity

Depending on the type of stress applied on a body and resulting strain produced, we have following three modulii of elasticity and they are as follows

(i) Young's modulus of elasticity
(ii) Bulk modulus of elasticity
(iii) Modulus of rigidity

Young's Modulus of Elasticity (*Y*)

It is defined as the ratio of normal stress (either tensile or compressive stress) to the longitudinal strain within an elastic limit.

Thus, $$Y = \frac{\text{Normal stress}}{\text{Longitudinal strain}}$$

Consider a metal wire, PQ of length l, radius r and the uniform area of cross-section A. Let it be suspended from a rigid support at P, a stretching force F be applied normally at the free end Q and let its length increase by $\Delta l (= QQ')$.

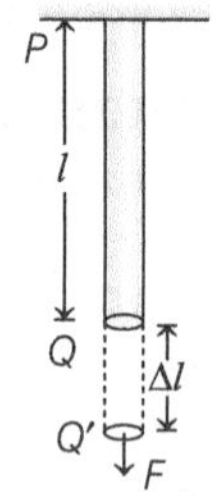

Then, longitudinal strain $= \dfrac{\Delta l}{l}$

$$\text{Normal stress} = \frac{F}{A} = \frac{F}{\pi r^2} \qquad [\because A = \pi r^2]$$

$$\text{Young's modulus } (Y) = \frac{\text{Normal stress}}{\text{Longitudinal strain}} = \frac{F/\pi r^2}{\Delta l / l} = \frac{Fl}{\pi r^2 \Delta l}$$

Since strain is a dimensionless quantity, therefore the unit of Young's modulus is same as that of stress. Young's modulus has unit of pressure.

Example 4. *A structural steel rod has a radius of 10 mm and a length of 1m. A 100 kN force stretches it along its length and Young's modulus of structural steel is 2×10^{11} Nm^{-2}. The strain on the rod is*

(a) 1% (b) 0.04% (c) 0.08% (d) 0.16%

Sol. (d) We assume that the rod is held by a clamp at one end and the force F is applied at the other end, parallel to the length of the rod. Then,

$$\text{Stress} = \frac{F}{A} = \frac{F}{\pi r^2}$$

Given, $F = 100 \text{ kN} = 100 \times 10^3 \text{ N}, r = 10^{-2} \text{m}$

$$= \frac{F}{A} = \frac{100 \times 10^3 \text{ N}}{3.14 \times (10^{-2})^2} = 3.18 \times 10^8 \text{ Nm}^{-2}$$

Elongation, $$\Delta L = \frac{(F/A)\, L}{Y}$$

$$\Delta L = \frac{(3.18 \times 10^8)(1)}{2 \times 10^{11}} = 1.59 \times 10^{-3} \text{ m}$$

$$\Delta L = 1.59 \text{ mm}$$

$$\text{Strain} = \frac{\Delta L}{L} = \frac{1.59 \times 10^{-3} \text{ m}}{1 \text{m}}$$

$$= 1.59 \times 10^{-3} = 0.16\%$$

Example 5. *Young's modulii is of two wires A and B are in the ratio 7 : 4. Wire A is 2 m long and has radius R. Wire B is 1.5 m long and has radius 2 mm. If the two wires stretch by the same length for a given load, then the value of R is close to* **[JEE Main 2019]**

(a) 1.3 mm (b) 1.5 mm
(c) 1.9 mm (d) 1.7 mm

Sol. (d) When a wire is stretched, the change in length of wire is $\Delta l = \frac{Fl}{\pi r^2 Y}$, where Y is its Young's modulus.

Here, for wires A and B,

$$l_A = 2\text{ m}, l_B = 1.5\text{ m},$$
$$\frac{Y_A}{Y_B} = \frac{7}{4}, r_B = 2\text{ mm} = 2\times10^{-3}\text{ m and } \frac{F_A}{F_B} = 1$$

As, it is given that $\Delta l_A = \Delta l_B$

$$\Rightarrow \frac{F_A l_A}{\pi r_A^2 Y_A} = \frac{F_B l_B}{\pi r_B^2 Y_B}$$
$$\Rightarrow r_A^2 = \frac{F_A}{F_B}\cdot\frac{l_A}{l_B}\cdot\frac{Y_B}{Y_A}\cdot r_B^2 = 1\times\frac{2}{1.5}\times\frac{4}{7}\times4\times10^{-6}\text{ m}$$
$$= 3.04\times10^{-6}\text{ m}$$
$$\Rightarrow r_A = 1.7\times10^{-3}\text{ m}$$
$$\text{or } r_A = 1.7\text{ mm}$$

Example 6. *A copper wire of length 2.2 m and a steel wire of length 1.6 m, both of diameter 3.0 mm are connected end-to-end. When stretched by a load, the net elongation is found to be 0.70 mm. The load applied (in N) is (Take, $Y_C = 1.1\times10^{11}\text{ Nm}^{-2}, Y_S = 2\times10^{11}\text{ Nm}^{-2}$)* **[NCERT]**

(a) 1.8×10^2 N (b) 2.5×10^6 N
(c) 3.8×10^8 N (d) 6.1×10^{-2} N

Sol. (a) The copper and steel wires are under a tensile stress because they have the same tension (equal to the load W) and the same area of cross-section A.

Stress = Strain × Young's modulus

$$\therefore \frac{W}{A} = \left(\frac{\Delta L_C}{L_C}\right)\times Y_C = \left(\frac{\Delta L_S}{L_S}\right)\times Y_S$$

where the subscripts C and S refer to copper and stainless steel, respectively.

$$\frac{\Delta L_C}{\Delta L_S} = \left(\frac{Y_S}{Y_C}\right)\times\left(\frac{L_C}{L_S}\right) \quad \text{...(i)}$$

Given, $L_C = 2.2\text{ m}, L_S = 1.6\text{ m}$

$$\frac{\Delta L_C}{\Delta L_S} = \frac{2\times10^{11}}{1.1\times10^{11}}\times\frac{2.2}{1.6} = 2.5$$

The total elongation is given to be

$$\Delta L_C + \Delta L_S = 7\times10^{-4}\text{ m} \quad \text{...(ii)}$$

Solving Eqs. (i) and (ii), we get

$$\Delta L_C = 5\times10^{-4}\text{ m},$$
$$\Delta L_S = 2\times10^{-4}\text{ m}$$

$$W = \frac{(A\times Y_C\times\Delta L_C)}{L_C}$$
$$= \frac{\pi(1.5\times10^{-3})^2\times(1.1\times10^{11}\times5\times10^{-4})}{2.2}$$
$$= 1.8\times10^2\text{ N}$$

Example 7. *A thin uniform metallic rod of length 0.5 m and radius 0.1 m with an angular velocity 400 rad s^{-1} in a horizontal plane about a vertical axis passing through one of its ends. Elongation in the rod (in m) is (Take, density of material of the rod is 10^4 kgm^{-3} and $Y = 2\times10^{11}$ Nm^{-2})* **[NCERT]**

(a) $\frac{10^{-3}}{3}$ (b) $\frac{3}{10^{-3}}$ (c) $\frac{2}{10^2}$ (d) $\frac{10^2}{2}$

Sol. (a) Consider an element of length dx at a distance x from the axis of rotation,

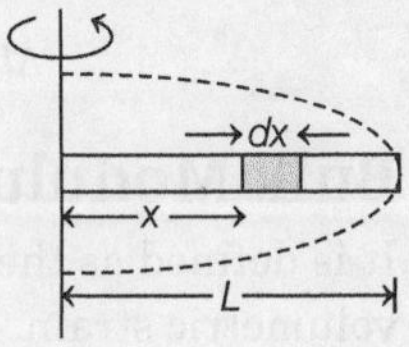

$$dF = dm\, x\omega^2$$
$$= \rho A dx\cdot x\cdot\omega^2$$
$$F = \rho A\omega^2\int_x^L x\,dx$$
$$= \frac{1}{2}\rho A\omega^2(L^2 - x^2)$$

If dy is the elongation in the element of length dx, then

$$\frac{dy}{dx} = \frac{F}{AY}$$
$$dy = \left.\frac{F\,dx}{A\cdot Y}\right|_{\Delta l} = \frac{\rho\omega^2}{2Y}\int_0^L (L^2 - x^2)\,dx$$
$$\Delta l = \frac{\rho\omega^2\cdot L^2}{3Y}$$
$$\Delta l = \frac{10^4\times(400)^2\times\left(\frac{1}{2}\right)^3}{3\times2\times10^{11}} = \frac{10^{-3}}{3}\text{ m}$$

Example 8. *A body of mass 3.14 kg is suspended from one end of a wire of length 10 m. The radius of the wire is changed uniformly from 9.8×10^{-4} m at one end to 5×10^{-4} m at the other end. The change in length of the wire is (Take, $Y = 2\times10^{11}$ Nm^{-2})*

(a) 10^{-2} m (b) 10^{-1} m (c) 10^{-3} m (d) 10^{-5} m

Sol. (c) Total change in length of the wire is

$$\tan\theta = \frac{b-a}{L} = \frac{r}{x}$$
$$\Rightarrow r = x\tan\theta$$

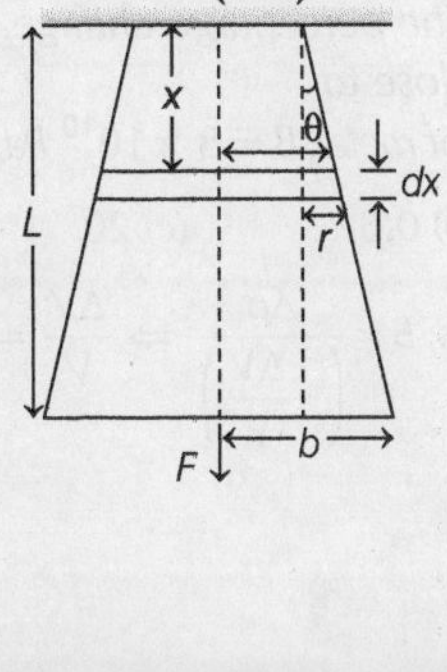

Change in dx length of the wire is $dy = \frac{F}{Ry} \cdot \frac{dx}{(a+r)^2}$

or $$dy = \frac{Fdx}{Ry(a + x\tan\theta)^2}$$

So, change in overall length of the wire,

$$\int_0^{\Delta L} dy = \frac{F}{\pi Y}\int_0^L \frac{dx}{(a + x\tan\theta)^2}$$

$$\Delta l = \frac{F}{\pi Y \tan\theta}\left(\frac{1}{a + x\tan\theta}\right)_0^L$$

$$= \frac{FL}{\pi a(a + L\tan\theta)Y}$$

$$= \frac{FL}{\pi abY} \qquad (\because a + L\tan\theta = b)$$

$\therefore$ $$\Delta l = \frac{3.14 \times 9.8 \times 10}{3.14 \times (9.8 \times 10^{-4}) \times 5 \times 10^{-4} \times 2 \times 10^{11}}$$

$$\Delta l = 10^{-3}\text{ m}$$

Bulk Modulus of Elasticity (K)

It is defined as the ratio of the volumetric stress to the volumetric strain. It is denoted by K. Thus,

$$K = \frac{\text{Volumetric stress}}{\text{Volumetric strain}}$$

Suppose a force F acts uniformly over the whole surface of the sphere (shown), decreasing its volume by ΔV. Then,

$$K = \frac{F/A}{-\Delta V/V}$$

or $$K = \frac{-FV}{A\Delta V} = -\frac{pV}{\Delta V}$$

The negative sign indicates that on increasing stress, the volume of the sphere decreases. The units of bulk modulus are Pa or Nm^{-2} in SI system.

Compressibility (C) The reciprocal of the bulk modulus of the material of the body is called the compressibility of the material. Thus,

$$\text{Compressibility } (C) = \frac{1}{K} = \frac{-1}{V}\left(\frac{\Delta V}{\Delta p}\right)$$

Its unit is N^{-1}m^2 or Pa^{-1} in SI system.

Note Young's modulus and Bulk modulus for a perfectly rigid body is infinity.

Example 9. *A cube of metal is subjected to a hydrostatic pressure of 4 GPa. The percentage change in the length of the side of the cube is close to*
(Take, Bulk modulus of metal, $B = 8 \times 10^{10}$ Pa) **[JEE Main 2020]**

(a) 1.67 (b) 0.6 (c) 20 (d) 5

Sol. (a) Bulk modulus, $B = \frac{\Delta p}{\left(-\frac{\Delta V}{V}\right)} \Rightarrow \frac{\Delta V}{V} = -\frac{\Delta p}{B}$

$$\frac{\Delta V}{V} \times 100\% = -\frac{\Delta p}{B} \times 100\%$$

$$= \frac{-4 \times 10^9}{8 \times 10^{10}} \times 100\%$$

$$= -\frac{1}{2} \times 10\%$$

$$= -0.5 \times 10\% = -5\%$$

Now, $V = l^3$ (for cube)

$\Rightarrow$ $l = (V)^{1/3}$

$\Rightarrow$ $$\frac{\Delta l}{l} \times 100\% = \frac{1}{3}\left(\frac{\Delta V}{V} \times 100\%\right)$$

$$= \frac{1}{3}(-5\%) = -1.67\ \%$$

So, length of cube will be decreased by 1.67%.

Modulus of Rigidity

The ratio of tangential stress to shearing strain is known as modulus of rigidity. It is also called *shear modulus*. It is denoted by Greek letter η (eta). Thus,

$$\eta = \frac{\text{Tangential stress}}{\text{Shearing strain}}$$

Here, a body (shown) is acted upon by an external force tangential to the surface of the body, the opposite face being kept fixed, its volume remains unchanged. Then,

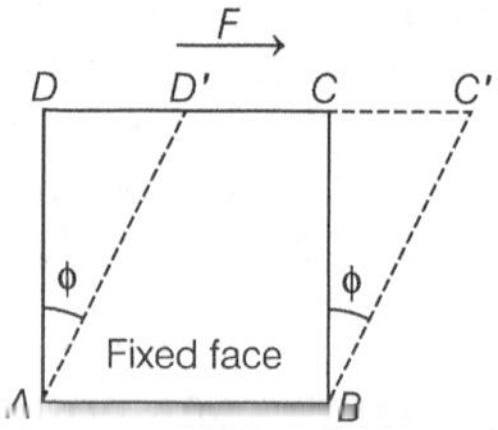

$$\eta = \frac{F/A}{DD'/AD}$$

Here, $$\tan\phi \approx \phi = \frac{DD'}{AD} \qquad \text{[for small angle } \phi]$$

$$\eta = \frac{F/A}{\phi}$$

or $$\eta = \frac{F}{A\phi}$$

The units of modulus of rigidity are Pa or Nm^{-2} in SI system.

Note Modulus of rigidity (or shear modulus) is involved with solids only. Modulus of rigidity for a solid is generally less than its Young's modulus.

Example 10. *A square lead slab of side 50 cm and thickness 10 cm is subject to a shaping force (on its narrow face) of 9×10^4 N. The lower edge is riveted to the floor. The upper edge is displaced by (Take, G = 5.6 GPa)* **[NCERT]**

(a) 0.30 mm (b) 0.16 mm
(c) 0.28 mm (d) 0.92 mm

Sol. (b) The lead slab is fixed and the force is applied parallel to the narrow face. The area of the face parallel to which this force is applied is

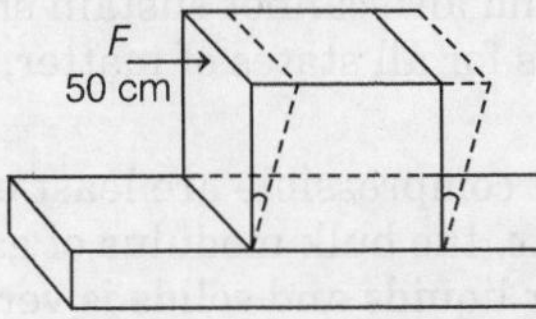

$$A = 50\text{ cm} \times 10\text{ cm} = 0.5\text{ m} \times 0.1\text{m} = 0.05\text{ m}^2$$

$$\therefore \text{ Stress applied} = (9.4 \times 10^4 \text{ N}/0.05\text{ m}^2)$$
$$= 1.80 \times 10^6 \text{ Nm}^{-2}$$

We know that, shearing strain $= \dfrac{\Delta x}{L} = \dfrac{\text{stress}}{G}$

$$\therefore \text{ Displacement, } \Delta x = \frac{\text{Stress} \times L}{G}$$
$$= \frac{(1.8 \times 10^6 \times 0.5)}{5.6 \times 10^9}$$
$$= 1.6 \times 10^{-4}\text{ m} = 0.16\text{ mm}$$

Example 11. *A 5 cm cube has its upper face displaced by 0.2 cm by a tangential force of 8 N. The modulus of rigidity of the material of cube is*

(a) 5×10^4 Nm^{-2}

(b) 6×10^4 Nm^{-2}

(c) 7×10^4 Nm^{-2}

(d) 8×10^4 Nm^{-2}

Sol. (d) Here, $l = 5\text{ cm} = 5 \times 10^{-2}\text{ m}$,
$\Delta l = 0.2\text{ cm} = 0.2 \times 10^{-2}\text{ m}, F = 8\text{ N}$

Modulus of rigidity, $\eta = \dfrac{\text{Shearing stress}}{\text{Shearing strain}}$

Hence, shearing stress $= \dfrac{F}{A} = \dfrac{F}{l^2} = \dfrac{8}{(5 \times 10^{-2})^2} = 3200\text{ Nm}^{-2}$

Shearing strain $= \dfrac{\Delta l}{l} = \dfrac{0.2}{5} = 0.04$

$$\therefore \quad \eta = \frac{3200}{0.04} = 80000\text{ Nm}^{-2}$$
$$= 8 \times 10^4\text{ Nm}^{-2}$$

Poisson's Ratio (σ)

The ratio of lateral strain to the longitudinal strain is constant for a given material. This constant is called as Poisson's ratio.

$$\sigma = \frac{\text{Lateral strain}}{\text{Longitudinal strain}} = \left(\frac{-\Delta R}{R}\right) \bigg/ \frac{\Delta l}{l}$$

Here, negative sign shows that if the length increases, then the radius of wire decreases. Poisson's ratio (σ) has no units and dimensions.

Theoretically, $\quad -1 < \sigma = \dfrac{1}{2}$

Practically, $0 < \sigma < \dfrac{1}{2}$ while practically no substance has been found for which σ is negative.

Example 12. *A tension of 20 N is applied to a wire of cross-sectional area 0.01 cm^2. The decrease in cross-sectional area is (Take, Young's modulus of Cu $= 1.1 \times 10^{11}$ Nm^{-2} and Poisson's ratio = 0.32)*

(a) 1.81×10^{-4} cm^2 (b) 1.16×10^{-6} cm^2

(c) 2.81×10^{-8} cm^2 (d) 5.23×10^{-3} cm^2

Sol. (b) As, $\dfrac{\Delta l}{l} = \dfrac{F}{AY} = \dfrac{20}{10^{-6} \times 1.1 \times 10^{11}} = 1.81 \times 10^{-4}$

$$\frac{\Delta r}{r} = \sigma \times \frac{\Delta l}{l} = 0.32 \times 1.81 \times 10^{-4}$$

$$\frac{\Delta A}{A} = \frac{2\,\Delta r}{r} = 2 \times 0.32 \times 1.81 \times 10^{-4}$$
$$= 1.16 \times 10^{-4}$$

and $\quad \Delta A = A(1.16 \times 10^{-4})$
$$= 1.16 \times 10^{-6}\text{ cm}^2 \qquad [\because A = 0.01\text{ cm}^2]$$

Stress-Strain Relationship

For a small deformation (say < 0.01) the longitudinal stress is proportional to the longitudinal strain. But when the deformation is not small, the relation of stress and strain is complicated.

Figure below shows qualitatively relation between the stress and strain of a stretched metal wire when the load is gradually increased in it.

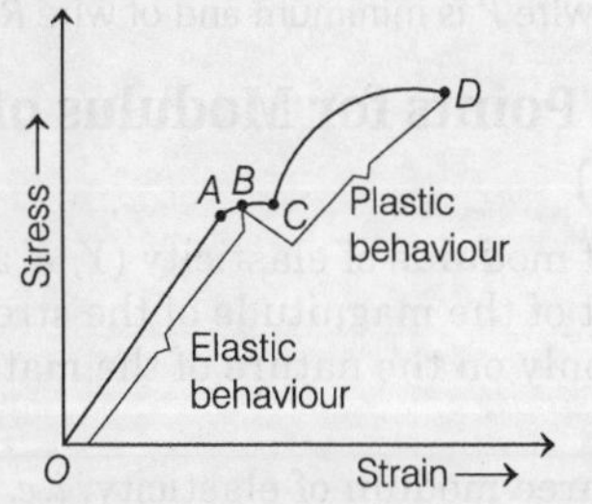

In figure, A = proportional limit,
B = elastic limit,
C = yield point (lower)
and D = fracture point.

- Point A is *limit of proportionality* and beyond which linear variation of stress and strain ceases. Hooke's law is valid in region OA and it is also called *linear elastic region.*
- Point B is *elastic limit, i.e.* the maximum stress upto which a metal wire regains its original shape (length) after removal of applied load. Region AB is called *non-linear elastic region.*
- Point C is called *yield point,* at this point yielding of the wires begins, *i.e.* even if the stretching force is removed the wire does not come back to its original length, some permanent increase in length takes place. The behaviour of wire is now plastic.

- Point D is known as a *fracture point*. The stress corresponding to this point is called breaking stress, region BD represents *plastic region*.

If large deformation takes place between elastic limit and the fracture point, the material is called *ductile*. If it breaks soon after the elastic limit is crossed, it is called *brittle*.

Note Elastic limit is the property of a body whereas elasticity is the property of material of a body.

Example 13. *The strain-stress curves of three wires of different materials are shown in the figure. P, Q and R are the elastic limits of the wires, the figure shows that*

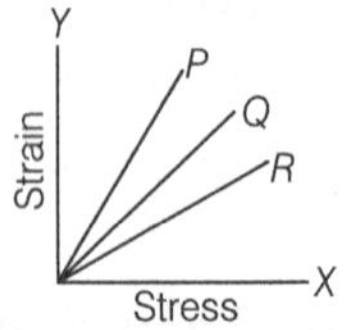

(a) *elasticity of wire P is maximum*
(b) *elasticity of wire Q is maximum*
(c) *elasticity of wire R is maximum*
(d) *None of the above*

Sol. (c) As stress is shown on X-axis and strain on Y-axis.

So, we can say that $y = \cot\theta = \dfrac{1}{\tan\theta} = \dfrac{1}{\text{slope}}$

So, elasticity of wire P is minimum and of wire R is maximum.

Important Points for Modulus of Elasticity (Y, K and η)

- The value of modulus of elasticity (Y, K and η) is independent of the magnitude of the stress and strain. It depends only on the nature of the material of the body.
- There are three modulii of elasticity, *i.e.* Y, K and η while elastic constants are four, *i.e.* Y, K, η and σ. Poisson's ratio σ is not modulus of elasticity as it is the ratio of two strains and not of stress to strain.

  Elastic constants are found to depend on each other through the relations

  $$Y = 3K(1 - 2\sigma) \quad \text{and } Y = 2\eta(1 + \sigma)$$

  Eliminating σ or Y between these, we get

  $$Y = \frac{9K\eta}{3K + \eta} \quad \text{and} \quad \sigma = \frac{3K - 2\eta}{6K + 2\eta}$$

- The modulii of elasticity has same dimensional formula and units as that of stress, since strain is dimensionless, *i.e.* the dimensional formula for Y, K or η is $[ML^{-1}T^{-2}]$, while unit is dyne cm^{-2} or Nm^{-2}.
- Greater the value of modulii of elasticity, more elastic is the material. As $Y \propto \left(\dfrac{1}{\Delta l}\right)$, $K \propto \dfrac{1}{\Delta V}$ and $\eta \propto \left(\dfrac{1}{\phi}\right)$.
- The modulii of elasticity Y and η exist only for solids as liquids and gases cannot be deformed along one dimension only and also cannot sustain shear strain. However, K exists for all states of matter, *i.e.* solid, liquid and gas.
- Gases being most compressible are least elastic while solids are most, *i.e.* the bulk modulus of gases is very low while that for liquids and solids is very high. *i.e.*

  $$E_{\text{solid}} > E_{\text{liquid}} > E_{\text{gas}}$$

- If a liquid of density ρ, volume V and Bulk modulus K is compressed, then its density increases.

  As density, $\rho = \dfrac{M}{V}$

  So, $$\frac{\Delta\rho}{\rho} = -\frac{\Delta V}{V} \quad \text{...(i)}$$

  But by definition of Bulk modulus,

  $$K = -\frac{-V\,\Delta p}{\Delta V}$$

  $$\Rightarrow \quad \frac{\Delta V}{V} = \frac{\Delta p}{K} \quad \text{...(ii)}$$

  From Eqs. (i) and (ii), we get

  $$\frac{\Delta\rho}{\rho} = \frac{\rho' - \rho}{\rho} = \frac{\Delta p}{K}$$

  or $$\rho' = \rho\left(1 + \frac{\Delta p}{K}\right) = \rho\,(1 + C\,\Delta p)$$

Work Done or Potential Energy Stored in a Stretched Wire

When a wire is stretched, work is done against the interatomic forces. This work is stored in the wire in the form of elastic potential energy.

Let us consider a wire of length l and the cross-sectional area A. If a force F acts along the length of the wire and stretches it by x, then

$$Y = \frac{\text{Stress}}{\text{Strain}} = \frac{F/A}{x/l} = \frac{Fl}{Ax}$$

$$F = \frac{YA}{l}x$$

So, work done for an additional small increase dx in length,

$$dW = F\,dx = (YA/l)\,x\,dx$$

So, total work done in increasing the length by Δl,

$$W = \int_0^{\Delta l} \frac{YA}{l} x\,dx$$

$$W = \frac{1}{2}\frac{YA}{l}(\Delta l)^2$$

$\therefore$ Work done per unit volume,

$$\frac{W}{V} = \frac{1}{2}Y\left(\frac{\Delta l}{l}\right)^2 = \frac{1}{2}Y\,(\text{strain})^2$$

$$\left(\because V = Al \text{ and strain} = \frac{\Delta l}{l}\right)$$

or $$\frac{W}{V} = \frac{1}{2} Y \times \text{Strain} \times \text{Strain}$$

$$\left(\because Y = \frac{\text{Stress}}{\text{Strain}} \text{ or } Y \times \text{Strain} = \text{Stress} \right)$$

or $$\frac{W}{V} = \frac{1}{2} \times \text{Stress} \times \text{Strain}$$

or $$\frac{W}{Al} = \frac{1}{2} \frac{F}{A} \times \frac{\Delta l}{l}$$

or $$W = \frac{1}{2} F \Delta l$$

$$W = \frac{1}{2} \text{Load} \times \text{Elongation}$$

Thus, stored elastic energy is

$$U = \frac{1}{2} \text{Stress} \times \text{Strain} \times \text{Volume}$$

$$U = \frac{1}{2} Y (\text{Strain})^2 \times \text{Volume}$$

$$U = \frac{1}{2} \text{Load} \times \text{Elongation}$$

Example 14. *Two steel wires having same length are suspended from a ceiling under the same load. If the ratio of their energy stored per unit volume is 1 : 4, the ratio of their diameters is* **[JEE Main 2020]**

(a) $\sqrt{2} : 1$ (b) $1 : \sqrt{2}$

(c) $2 : 1$ (d) $1 : 2$

Sol. (a) Elastic potential energy stored in a loaded wire,

$$U = \frac{1}{2} (\text{Stress} \times \text{Strain} \times \text{Volume})$$

$\therefore$ Energy stored per unit volume,

$$u = \frac{U}{\text{Volume}} = \frac{1}{2} \times \text{Stress} \times \text{Strain}$$

$$= \frac{1}{2} \left(\frac{F}{A} \right)^2 \times \frac{1}{Y}$$

Here, both wires are of same material and under same load, so the ratio of stored energies per unit volume, for both the wires will be

$$\frac{u_A}{u_B} = \frac{\frac{1}{2Y} \cdot \frac{F^2}{A_A^2}}{\frac{1}{2Y} \cdot \frac{F^2}{A_B^2}} = \frac{A_B^2}{A_A^2}$$

$\Rightarrow$ $$\frac{u_A}{u_B} = \frac{d_B^4}{d_A^4} \qquad \left(\because A = \pi \frac{d^2}{4} \right)$$

Here, $$\frac{u_A}{u_B} = \frac{1}{4}$$

So, $$\frac{d_B^4}{d_A^4} = \frac{1}{4} \text{ or } \frac{d_B}{d_A} = \frac{1}{\sqrt{2}}$$

$\Rightarrow$ $$\frac{d_A}{d_B} = \sqrt{2} : 1$$

Example 15. *A boy's catapult is made of rubber cord which is 42 cm long, with 6 mm diameter of cross-section and of negligible mass. The boy keeps a stone weighing 0.02 kg on it and stretches the cord by 20 cm by applying a constant force. When released the stone flies off with a velocity of 20 ms^{-1}. Neglect the change in the area of cross-section of the cord while stretched. The Young's modulus of rubber is closest to* **[JEE Main 2019]**

(a) 10^6Nm^{-2} (b) 10^4Nm^{-2} (c) 10^8Nm^{-2} (d) 10^3Nm^{-2}

Sol. (a) When rubber cord is stretched, it stores potential energy and when released, this potential energy is given to the stone as kinetic energy.

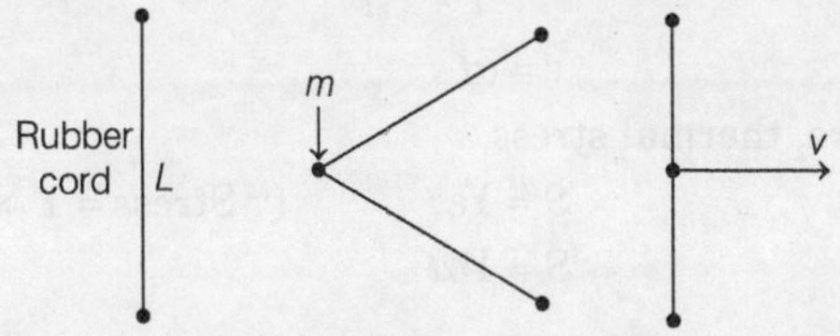

So, potential energy of stretched cord
= kinetic energy of stone

$\Rightarrow$ $$\frac{1}{2} Y \left(\frac{\Delta L}{L} \right)^2 A \cdot L = \frac{1}{2} m v^2$$

Here, $\Delta L = 20 \text{ cm} = 0.2 \text{ m}$, $L = 42 \text{ cm} = 0.42 \text{ m}$,
$v = 20 \text{ ms}^{-1}$, $m = 0.02 \text{ kg}$
$d = 6 \text{ mm} = 6 \times 10^{-3} \text{m}$

$\therefore$ $$A = \pi r^2 = \pi \left(\frac{d}{2} \right)^2 = \pi \left(\frac{6 \times 10^{-3}}{2} \right)^2$$

$$= \pi (3 \times 10^{-3})^2$$

$$= 9\pi \times 10^{-6} \text{ m}^2$$

On substituting values, we get

$$Y = \frac{mv^2 L}{A (\Delta L)^2} = \frac{0.02 \times (20)^2 \times 0.42}{9\pi \times 10^{-6} \times (0.2)^2}$$

$$\approx 3.0 \times 10^6 \text{Nm}^{-2}$$

So, the closest value of Young's modulus is 10^6 Nm^{-2}.

Thermal Stresses and Strains

When a body is allowed to expand or contract with increasing temperature or decreasing temperature, no stresses are induced in the body.

But if the deformation of the body is prevented, some stresses are induced in the body. Such stresses are called *thermal stresses* or *temperature stresses*. The corresponding strains are called *thermal strains* or *temperature strains.*

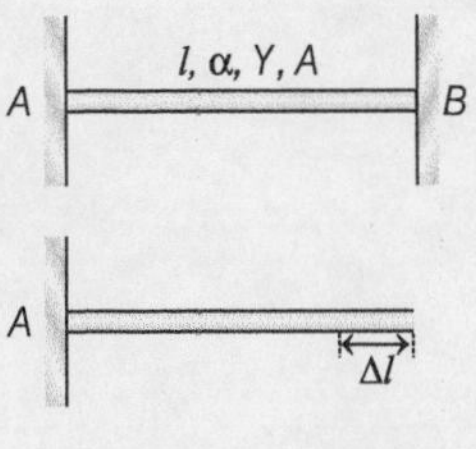

A body having linear dimensions is shown in above figure.

Let the temperature of the rod be increased by an amount t. The length of the rod would increase by an amount Δl, if it is not fixed at two supports. Here,

$$\Delta l = l\alpha t$$

But since the rod is fixed at the supports, a compressive strain will be produced in the rod.

Because at the increased temperature, the natural length of the rod is $l + \Delta l$, while being fixed at two supports, its actual length is l. Hence, thermal strain,

$$\varepsilon = \frac{\Delta l}{l} = \frac{l\alpha t}{l} = \alpha t$$

or $\varepsilon = \alpha t$

Therefore, thermal stress,

$$S = Y\varepsilon \qquad (\because \text{Stress} = Y \times \text{Strain})$$

or $S = Y\alpha t$

Example 16. *A uniform cylindrical rod of length L and radius r, is made from a material whose Young's modulus of elasticity equals Y. When this rod is heated by temperature T and simultaneously subjected to a net longitudinal compressional force F, its length remains unchanged. The coefficient of volume temperature expansion of the material of the rod, is (nearly) equal to* **[JEE Main 2019]**

(a) $9F/(\pi r^2 YT)$

(b) $6F/(\pi r^2 YT)$

(c) $3F/(\pi r^2 YT)$

(d) $F/(3\pi r^2 YT)$

Sol. (c) As length of rod remains unchanged,

Strain caused by compressive forces is equal and opposite to the thermal strain.

Now, compressive strain is obtained by using formula for Young's modulus,

$$Y = \frac{F/A}{\Delta l / l}$$

Compressive strain, $\frac{\Delta l}{l} = \frac{F}{AY} = \frac{F}{\pi Y r^2}$ ...(i)

Also, thermal strain in rod is obtained by using formula for expansion in rod,

$$\Delta l = l\,\alpha\,\Delta T$$

Thermal strain, $\frac{\Delta l}{l} = \alpha\,\Delta T$ ...(ii)

From Eqs. (i) and (ii), we get

$$\frac{F}{\pi r^2 Y} = \alpha T \qquad [\because \Delta T = T]$$

$$\Rightarrow \quad \alpha = \frac{F}{\pi r^2 YT}$$

Hence, coefficient of volume temperature expansion of rod is

$$\gamma = 3\alpha = \frac{3F}{\pi r^2 YT}$$

Example 17. *A rod of length L at room temperature and uniform area of cross-section A, is made of a metal having coefficient of linear expansion α /°C. It is observed that an external compressive force F, is applied on each of its ends, prevents any change in the length of the rod, when its temperature rises by ΔT K. Young's modulus Y for this metal is* **[JEE Main 2019]**

(a) $\frac{F}{2A\alpha\,\Delta T}$

(b) $\frac{F}{A\alpha(\Delta T - 273)}$

(c) $\frac{2F}{A\alpha\Delta T}$

(d) $\frac{F}{A\alpha\Delta T}$

Sol. (d) If a rod of length L and coefficient of linear expansion α/° C, then with the rise in temperature by ΔT K, its change in length is given as

$$\Delta L = L\,\alpha\,\Delta T$$

$$\Rightarrow \quad \frac{\Delta L}{L} = \alpha\,\Delta T \qquad \text{...(i)}$$

Also, when a rod is subjected to some compressive force (F), then its' Young's modulus is given as

$$Y = \frac{\text{Stress}}{\text{Strain}} = \frac{F/A}{\Delta L / L}$$

$$\frac{\Delta L}{L} = \frac{F}{YA} \qquad \text{...(ii)}$$

Since, it is given that the length of the rod does not change. So, from Eqs. (i) and (ii), we get

$$\alpha\,\Delta T = \frac{F}{YA}$$

$$\Rightarrow \quad Y = \frac{F}{A\alpha\,\Delta T}$$

Practice Exercise

ROUND I Topically Divided Problems

Stress, Strain and Hooke's Law

1. A steel cable with a radius of 1.5 cm supports a chair lift at a ski area. If the maximum stress is not to exceed 10^8 N/m^2, what is the maximum load the cable can support? **[NCERT]**
(a) 7×10^5 N (b) 7×10^6 N
(c) 7×10^4 N (d) 9×10^5 N

2. A rectangular frame is to be suspended symmetrically by two strings of equal length on two supports (Fig.) It can be done in one of the following three ways **[NCERT Exemplar]**

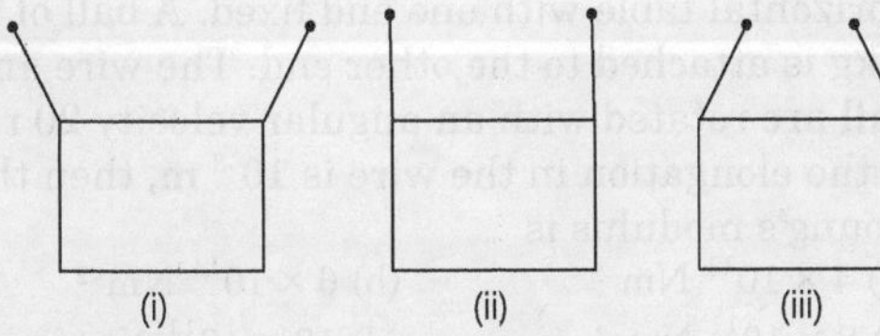

The tension in the strings will be
(a) the same in all cases
(b) least in (i)
(c) least in (ii)
(d) least in (iii)

3. Two identical wires of rubber and iron are stretched by the same weight, then the number of atoms in unit volume of iron wire will be
(a) equal to that of rubber
(b) less than that of the rubber
(c) more than that of the rubber
(d) None of the above

4. A force F is required to break a wire of length l and radius r. What force is required to break a wire, of the same material, having twice the length and six times the radius?
(a) F (b) $3F$
(c) $9F$ (d) $36F$

5. A substance breaks down by a stress of 10^6 Nm^{-2}. If the density of the material of the wire is 3×10^3 kg m^{-3}, then the length of the wire of the substance which will break under its own weight when suspended vertically is
(a) 66.6 m (b) 60.0 m
(c) 33.3 m (d) 30.0 m

6. A body of mass $m = 10$ kg is attached to a wire of length 0.3 m. The maximum angular velocity with which it can be rotated in a horizontal circle is (Breaking stress of wire = 4.8×10^7 Nm^{-2} and area of cross-section of a wire = 10^{-6} m^2)
(a) 4 rads^{-1} (b) 8 rads^{-1}
(c) 1 rads^{-1} (d) 2 rads^{-1}

7. A mild steel wire of length $2L$ and cross-sectional area A is stretched, well within elastic limit, horizontally between two pillars as shown in figure. A mass m is suspended from the mid-point of the wire. Strain in the wire is **[NCERT Exemplar]**

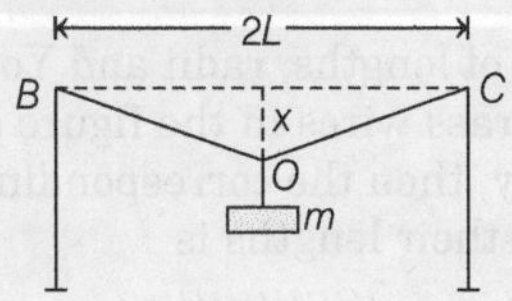

(a) $\dfrac{x^2}{2L^2}$ (b) $\dfrac{x}{L}$ (c) $\dfrac{x^2}{L}$ (d) $\dfrac{x^2}{2L}$

8. In the figure, three identical springs are shown. From spring A, a mass of 4 kg is hung and spring shows elongation of 1 cm. But when a weight of 6 kg is hung on B & C as shown in figure, the hook descends through

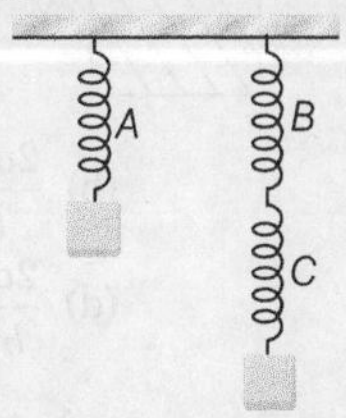

(a) 1 cm (b) 2 cm (c) 3 cm (d) 4 cm

9. A load suspended by a massless spring produces an extension of x cm, in equilibrium. When it is cut into two unequal parts, the same load produces an extension of 7.5 cm when suspended by the larger part of length 60 cm. When it is suspended by the smaller part, the extension is 5.0 cm. Then
(a) $x = 12.5$
(b) $x = 3.0$
(c) the length of the original spring is 90 cm
(d) the length of the original spring is 80 cm

10. The length of an elastic string is a metre when the tension is 4 N and b metre when the tension is 5 N. The length in metre when the tension is 9 N, is

(a) $4a-5b$ (b) $5b-4a$
(c) $9b-9a$ (d) $a+b$

Types of Modulus of Elasticity

11. The ratio of diameters of two wires of same materials is $n:1$. The length of each wire is 4 m. On applying the same load, the increase in length of thin wire will be $(n>1)$

(a) n^2 times (b) n times
(c) $2n$ times (d) $(2n+1)$ times

12. A rigid bar of mass M is supported symmetrically by three wires each of length l. Those at each end are of copper and the middle one is of iron. The ratio of their diameter, if each is to have the same tension, is equal to [NCERT Exemplar]

(a) Y_{copper}/Y_{iron} (b) $\sqrt{\dfrac{Y_{iron}}{Y_{copper}}}$

(c) $\dfrac{Y_{iron}^2}{Y_{copper}^2}$ (d) $\dfrac{Y_{iron}}{Y_{copper}}$

13. If the ratio of lengths, radii and Young's moduli of steel and brass wires in the figure are a, b and c respectively, then the corresponding ratio of increase in their lengths is [JEE Main 2013]

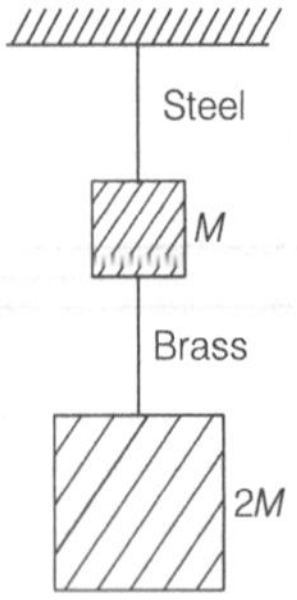

(a) $\dfrac{3c}{2ab^2}$ (b) $\dfrac{2a^2c}{b}$

(c) $\dfrac{3a}{2b^2c}$ (d) $\dfrac{2ac}{b^2}$

14. In an experiment, brass and steel wires of length 1 m each with areas of cross-section 1 mm^2 are used. The wires are connected in series and one end of the combined wire is connected to a rigid support and other end is subjected to elongation. The stress required to produce a net elongation of 0.2 mm is

[Take, the Young's modulus for steel and brass as $120\times10^9\ Nm^{-2}$ and $60\times10^9\ Nm^{-2}$, respectively] [JEE Main 2019]

(a) $1.2\times10^6\ Nm^{-2}$ (b) $0.2\times10^6\ Nm^{-2}$
(c) $8\times10^6\ Nm^{-2}$ (d) $4.0\times10^6\ Nm^{-2}$

15. Speed of a transverse wave on a straight wire (mass 6.0 g, length 60 cm and area of cross-section 1.0 mm^2) is 90 ms^{-1}. If the Young's modulus of wire is $16\times10^{11}\ Nm^{-2}$, the extension of wire over its natural length is [JEE Main 2020]

(a) 0.01 mm (b) 0.04 mm
(c) 0.03 mm (d) 0.02 mm

16. A thick rope of rubber of density $1.5\times10^3\ kgm^{-3}$ and Young's modulus $5\times10^6\ Nm^{-2}$, 8 m in length is hung from the ceiling of a room, the increase in its length due to its own weight is

(a) 9.6×10^{-2} m (b) 19.2×10^{-2} m
(c) 9.6×10^{-3} m (d) 9.6 m

17. A substance breaks down by a stress of $10^6\ Nm^{-2}$. If the density of the material of the wire is $3\times10^3\ kgm^{-3}$, then the length of the wire of that substance which will break under its own weight when suspended vertically is nearly

(a) 3.4 m (b) 34 m (c) 340 m (d) 3400 m

18. A copper wire of negligible mass, 1 m length and cross-sectional area 10^{-6} is kept on a smooth horizontal table with one end fixed. A ball of mass 1 kg is attached to the other end. The wire and the ball are rotated with an angular velocity 20 rad s^{-1}. If the elongation in the wire is 10^{-3} m, then the Young's modulus is

(a) $4\times10^{11}\ Nm^{-2}$ (b) $6\times10^{11}\ Nm^{-2}$
(c) $8\times10^{11}\ Nm^{-2}$ (d) $10\times10^{11}\ Nm^{-2}$

19. Two wires of the same material and length are stretched by the same force. Their masses are in the ratio 3:2. Their elongations are in the ratio

(a) 3 : 2 (b) 9 : 4
(c) 2 : 3 (d) 4 : 9

20. The dimensions of four wires of the same material are given below. In which wire, the increase in length will be maximum?

(a) Length 100 cm, Diameter 1 mm
(b) Length 200 cm, Diameter 2 mm
(c) Length 300 cm, Diameter 3 mm
(d) Length 50 cm, Diameter 0.5 mm

21. Two wires of the same length and same material but radii in the ratio of 1 : 2 are stretched by unequal forces to produce equal elongation. The ratio of the two forces is

(a) 1 : 1 (b) 1 : 2
(c) 2 : 3 (d) 1 : 4

22. Two wires of the same material have lengths in the ratio 1 : 2 and their radii are in the ratio $1:\sqrt{2}$. If they are stretched by applying equal forces, the increase in their lengths will be in the ratio of

(a) $\sqrt{2}:2$ (b) $2:\sqrt{2}$ (c) 1:1 (d) 1:2

23. When a weight of 5 kg is suspended from a copper wire of length 30 m and diameter 0.5 mm, the length of the wire increases by 2.4 cm. If the diameter is doubled, the extension produced is
(a) 1.2 cm (b) 0.6 cm (c) 0.3 cm (d) 0.15 cm

24. A wire of length L and radius r is clamped rigidly at one end. When the other end of the wire is pulled by a force F, its length increases by l. Another wire of the same material of length $4L$, radius $4r$ is pulled by a force $4F$. The increase in length will be
(a) $\frac{l}{2}$ (b) l
(c) $2l$ (c) $4l$

25. A wire extends by 1 mm when a force is applied. Double the force is applied to another wire of same material and length but half the radius of cross-section. The elongation of the wire (in mm) will be
(a) 8 (b) 4
(c) 2 (d) 1

26. A steel ring of radius r and cross-sectional area A is fitted on a wooden disc of radius $R (R > r)$. If Young's modulus be E, then the force with which the steel ring is expanded is
(a) $AE\frac{R}{r}$ (b) $AE\frac{(R-r)}{r}$
(c) $\frac{E}{A}\left(\frac{R-r}{A}\right)$ (d) $\frac{Er}{AR}$

27. A wire is stretched 1 mm by a force of 1 kN. How far would a wire of the same material and length but of four times that diameter be stretched by the same force ?
(a) $\frac{1}{2}$ mm (b) $\frac{1}{4}$ mm
(c) $\frac{1}{8}$ mm (d) $\frac{1}{16}$ mm

28. Two bars A and B of circular cross-section and of same volume and made of the same material are subjected to tension. If the diameter of A is half that of B and if the force applied to both the rods is the same and it is in the elastic limit, the ratio of extension of A to that of B will be
(a) 16 : 1 (b) 8 : 1
(c) 4 : 1 (d) 2 : 1

29. A steel wire has length 2 m, radius 1 mm and $Y = 2 \times 10^{11}\,\text{Nm}^{-2}$. A 1 kg sphere is attached to one end of the wire and whirled in a vertical circle with an angular velocity of 2 revolutions per second. When the sphere is at the lowest point of the vertical circle, the elongation of the wire is nearly (Take, $g = 10\,\text{ms}^{-2}$)
(a) 1 mm (b) 2 mm
(c) 0.1 mm (d) 0.01 mm

30. Two wires of equal cross-section but one made of steel and the other of copper are joined end-to-end. When the combination is kept under tension, the elongations in the two wires are found to be equal. What is the ratio of the lengths of the two wires? (Given, Y for steel $= 2 \times 10^{11}\text{Nm}^{-2}$ and Y for copper $= 1.1 \times 10^{11}\,\text{Nm}^{-2}$)
(a) 2 : 11 (b) 11 : 2
(c) 20 : 11 (d) 11 : 20

31. The Young's modulus of brass and steel are $10 \times 10^{10}\,\text{Nm}^{-2}$ and $2 \times 10^{11}\,\text{Nm}^{-2}$, respectively. A brass wire and a steel wire of the same length are extended by 1 mm under the same force. The radii of the brass and steel wires are R_B and R_S respectively, then
(a) $R_S = \sqrt{2}R_B$ (b) $R_S = \frac{R_B}{\sqrt{2}}$
(c) $R_S = 4R_B$ (d) $R_S = \frac{R_B}{4}$

32. When the tension in a metal wire is T_1, its length is l_1. When the tension is T_2, its length is l_2. The natural length of wire is
(a) $\frac{T_2}{T_1}(l_1 + l_2)$ (b) $T_1l_1 + T_2l_2$
(c) $\frac{l_1T_2 - l_2T_1}{T_2 - T_1}$ (d) $\frac{l_1T_2 + l_2T_1}{T_2 + T_1}$

33. Two wires, one made of copper and other of steel are joined end-to-end (as shown in figure). The area of cross-section of copper wire is twice that of steel wire.

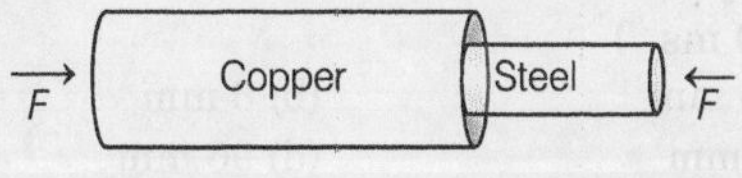

They are placed under compressive force of magnitudes F. The ratio for their lengths such that change in lengths of both wires are same is ($Y_S = 2 \times 10^{11}\,\text{Nm}^{-2}$ and $Y_C = 1.1 \times 10^{11}\,\text{Nm}^{-2}$)
(a) 2 : 1 (b) 1.1 : 1
(c) 1 : 2 (d) 2 : 3

34. A uniform slender rod of length L, cross-sectional area A and Young's modulus Y is acted upon by the forces shown in the figure. The elongation of the rod is

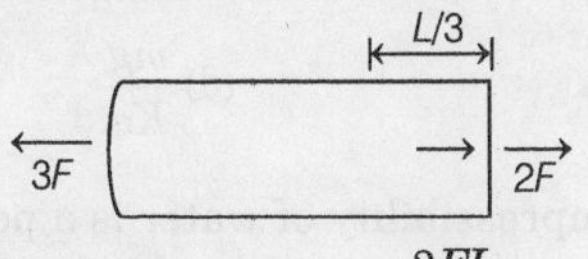

(a) $\frac{3FL}{5AY}$ (b) $\frac{2FL}{5FY}$
(c) $\frac{3FL}{8AY}$ (d) $\frac{8FL}{3AY}$

35. The stress *versus* strain graphs for wires of two materials A and B are as shown in the figure. If Y_A and Y_B are the Young's modulus of the materials, then

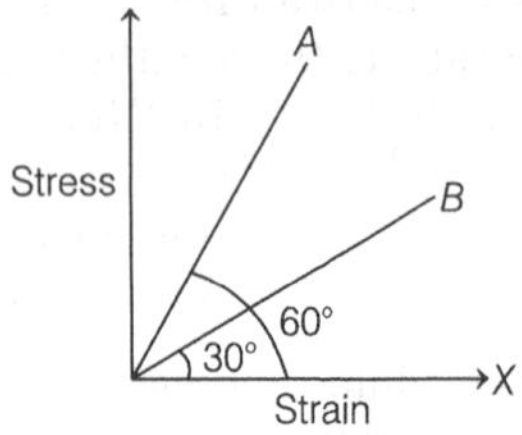

(a) $Y_B = 2Y_A$ (b) $Y_A = Y_B$
(c) $Y_B = 3Y_A$ (d) $Y_A = 3Y_B$

36. One end of steel wire is fixed to ceiling of an elevator moving up with an acceleration 2 ms^{-2} and a load of 10 kg hangs from other end. Area of cross-section of the wire is 2 cm^2. The longitudinal strain in the wire is (Take, $g = 10\ ms^{-2}$ and $Y = 2 \times 10^{11}\ Nm^{-2}$)
(a) 4×10^{11} (b) 3×10^{-6}
(c) 8×10^{-6} (d) 2×10^{-6}

37. The upper end of a wire of radius 4 mm and length 10 cm is clamped and its other end is twisted through an angle of 30°. Then angle of shear is
(a) 12° (b) 0.12°
(c) 1.2° (d) 0.012°

38. A rubber rope of length 8 m is hung from the ceiling of a room. What is the increase in length of the rope due to its own weight? (Given, Young's modulus of elasticity of rubber = $5 \times 10^6\ Nm^{-2}$, density of rubber = $1.5 \times 10^3\ kgm^{-3}$ and $g = 10\ ms^{-2}$)
(a) 1.5 mm (b) 6 mm
(c) 24 mm (d) 96 mm

39. A solid sphere of radius r made of a soft material of bulk modulus K is surrounded by a liquid in a cylindrical container. A massless piston of area a floats on the surface of the liquid, covering entire cross-section of cylindrical container. When a mass m is placed on the surface of the piston to compress the liquid, the fractional decrement in the radius of the sphere $\left(\frac{dr}{r}\right)$ is **[JEE Main 2018]**
(a) $\frac{Ka}{mg}$ (b) $\frac{Ka}{3mg}$
(c) $\frac{mg}{3Ka}$ (d) $\frac{mg}{Ka}$

40. If the compressibility of water is σ per unit atmospheric pressure, then the decrease in volume V due to atmospheric pressure p will be
(a) $\sigma\ p/V$ (b) σpV
(c) σ/pV (d) $\sigma V/p$

41. A cube is compressed at 0°C equally from all sides by an external pressure p. By what amount should the temperature be raised to bring it back to the size it had before the external pressure was applied? (Take, K as bulk modulus of elasticity of the material of the cube and α as the coefficient of linear expansion)
(a) $\frac{p}{K\alpha}$ (b) $\frac{p}{3K\alpha}$
(c) $\frac{3\pi\alpha}{p}$ (d) $\frac{K}{3p}$

42. A cube is shifted to a depth of 100 m in a lake. The change in volume is 0.1%. The bulk modulus of the material is nearly
(a) 10 Pa (b) 10^4 Pa
(c) 10^7 Pa (d) 10^9 Pa

43. A cube is subjected to a uniform volume compression. If the side of the cube decreases by 1% the bulk strain is
(a) 0.01 (b) 0.02
(c) 0.03 (d) 0.06

44. A copper bar of length L and area of cross-section A is placed in a chamber at atmospheric pressure. If the chamber is evacuated, the percentage change in its volume will be (Take, compressibility of copper is $8 \times 10^{-12}\ m^2\ N^{-1}$ and 1 atm = $10^5\ Nm^{-2}$)
(a) 8×10^{-7} (b) 8×10^{-5}
(c) 1.25×10^{-4} (d) 1.25×10^{-5}

45. The compressibility of water is $6 \times 10^{-10}\ N^{-1}m^2$. If one litre is subjected to a pressure of $4 \times 10^7\ Nm^{-2}$. The decrease in its volume is
(a) 2.4 cc (b) 10 cc
(c) 24 cc (d) 15 cc

46. Forces of 100 N each are applied in opposite directions on the upper and lower faces of a cube of side 20 cm. The upper face is shifted parallel to itself by 0.25 cm. If the side of the cube were 10 cm, then the displacement would be
(a) 0.25 cm (b) 0.5 cm
(c) 0.75 cm (d) 1 cm

47. The Young's modulus of the material of a wire is $6 \times 10^{12}\ Nm^{-2}$ and there is no transverse strain in it, then its modulus of rigidity will be
(a) $3 \times 10^{12}\ Nm^{-2}$ (b) $2 \times 10^{12}\ Nm^{-2}$
(c) $10^{12}\ Nm^{-2}$ (d) None of these

48. Equal torsional torques act on two rods x and y having equal length. The diameter of rod y is twice the diameter of rod x. If θ_x and θ_y are the angles of twist, then $\frac{\theta_x}{\theta_y}$ is equal to
(a) 1 (b) 2 (c) 4 (d) 16

49. Two rods A and B of the same material and length have their radii r_1 and r_2, respectively. When they are rigidly fixed at one end and twisted by the same couple applied at the other end, the ratio of the angle of twist at the end of A and the angle of twist at the end of B is

(a) $\frac{r_2^4}{r_1^4}$ (b) $\frac{r_1^4}{r_2^4}$

(c) $\frac{r_2^2}{r_1^2}$ (d) $\frac{r_1^2}{r_2^2}$

Thermal Stress, Strain and Poisson's Ratio

50. A 1 m long wire is stretched without tension at 30°C between two rigid supports. What strain will be produced in the wire if the temperature falls to 0°C ? (Given, $\alpha = 12 \times 10^{-6}\ K^{-1}$)

(a) 36×10^{-5} (b) 64×10^{-5}

(c) 0.78 (d) 0.32

51. Two different wires having lengths L_1 and L_2 and respective temperature coefficients of linear expansion α_1 and α_2, are joined end-to-end. Then the effective temperature coefficient of linear expansion is **[JEE Main 2020]**

(a) $\frac{\alpha_1 L_1 + \alpha_2 L_2}{L_1 + L_2}$ (b) $2\sqrt{\alpha_1 \alpha_2}$

(c) $\frac{\alpha_1 + \alpha_2}{2}$ (d) $4\frac{\alpha_1 \alpha_2}{\alpha_1 + \alpha_2}\frac{L_2 L_1}{(L_2 + L_1)^2}$

52. At 40°C, a brass wire of 1 mm radius is hung from the ceiling. A small mass M is hung from the free end of the wire. When the wire is cooled down from 40°C to 20°C, it regains its original length of 0.2 m. The value of M is close to (Coefficient of linear expansion and Young's modulus of brass are 10^{-5}/°C and 10^{11} N/m^2 respectively, $g = 10\ ms^{-2}$) **[JEE Main 2019]**

(a) 9 kg (b) 0.5 kg (c) 1.5 kg (d) 0.9 kg

53. A uniform wire (Young's modulus $2 \times 10^{11}\ Nm^{-2}$) is subjected to longitudinal tensile stress of $5 \times 10^7\ Nm^{-2}$. If the overall volume change in the wire is 0.02%, the fractional decrease in the radius of the wire is close to **[JEE Main 2013]**

(a) 1.0×10^{-4} (b) 1.5×10^{-4}

(c) 0.25×10^{-4} (d) 5×10^{-4}

54. The temperature of a rod of length 1 m and area of cross-section 1 cm² is increased from 0°C to 100°C. If the rod is not allowed to increase in length, the force required will be (Take, $\alpha = 10^{-5}$/°C and $Y = 10^{11}$ N/m²)

(a) 10^3 N (b) 10^4 N

(c) 10^5 N (d) 10^9 N

55. For a given material, the Young's modulus is 2.4 times that of modulus of rigidity. Its Poisson's ratio is

(a) 0.1 (b) 0.2 (c) 0.3 (d) 0.4

56. When a rubber cord is stretched, the change in volume with respect to change in its linear dimensions is negligible. The Poisson's ratio for rubber is

(a) 1 (b) 0.25 (c) 0.5 (d) 0.75

Potential Energy Stored in Stretched Wire

57. Modulus of rigidity of ideal liquids is **[NCERT Exemplar]**

(a) infinity

(b) zero

(c) unity

(d) some finite small non-zero constant value

58. A spring is extended by 30 mm when a force of 1.5 N is applied to it. Calculate the energy stored in the spring when hanging vertically supporting a mass of 0.20 kg, if the spring was relaxed before applying the mass.

(a) 0.01 J (b) 0.02 J (c) 0.04 J (d) 0.08 J

59. Two wires of the same material (Young's modulus Y) and same length L but radii R and $2R$ respectively are joined end-to-end and a weight w is suspended from the combination as shown in the figure. The elastic potential energy in the system is

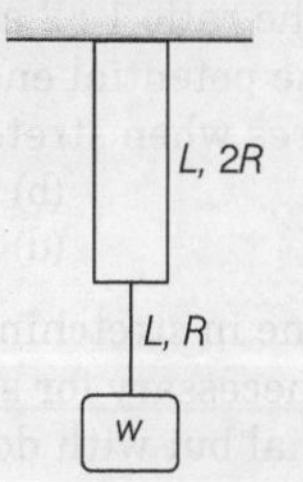

(a) $\frac{3w^2L}{4\pi R^2 Y}$ (b) $\frac{3w^2L}{8\pi R^2 Y}$ (c) $\frac{5w^2L}{8\pi R^2 Y}$ (d) $\frac{w^2L}{\pi R^2 Y}$

60. A wire suspended vertically from one of its ends is stretched by attaching a weight of 200 N to the lower end. The weight stretches the wire by 1 mm. Then, the elastic energy stored in the wire is

(a) 0.2 J (b) 10 J

(c) 20 J (d) 0.1 J

61. A height spring extends 40 mm when stretched by a force of 10 N and for tensions upto this value, the extension is proportional to the stretching force. Two such springs are joined end-to-end and the double length spring is stretched 40 mm beyond its natural length. The total strain energy (in joule) stored in the double spring is

(a) 0.05 (b) 0.10 (c) 0.80 (d) 0.40

62. The force constant of a wire is k and that of another wire of the same material is $2k$. When both the wires are stretched, then work done is
(a) $W_2 = 1.5 W_1$ (b) $W_2 = 2 W_1$
(c) $W_2 = W_1$ (d) $W_2 = 0.5 W_1$

63. If the shear modulus of a wire material is 5.9×10^{11} dyne cm^{-2}, then the potential energy of a wire of 4×10^3 cm in diameter and 5 cm long twisted through an angle of 10°, is
(a) 1.253×10^{-12} J (b) 2.00×10^{-12} J
(c) 1.00×10^{-12} J (d) 0.8×10^{-12} J

64. Two cylinders of same material and of same length are joined end-to-end as shown in figure. The upper end of A is rigidly fixed and their radii are in ratio of 1 : 2. If the lower end of B is twisted by an angle θ, the angle of twist of cylinder A is

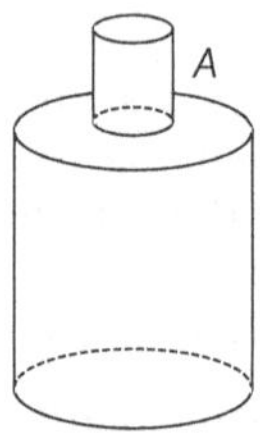

(a) $\frac{15}{16}\theta$ (b) $\frac{16}{15}\theta$ (c) $\frac{16}{17}\theta$ (d) $\frac{17}{16}\theta$

65. Two wires of the same material and length but diameters in the ratio 1 : 2 are stretched by the same force. The potential energy per unit volume for the two wires when stretched will be in the ratio
(a) 16 : 1 (b) 4 : 1
(c) 2 : 1 (d) 1 : 1

66. If the work done in stretching a wire by 1 mm is 2 J, the work necessary for stretching another wire of same material but with double radius of cross-section and half the length by 1 mm is
(a) $\frac{1}{4}$ J (b) 4 J
(c) 8 J (d) 16 J

67. A copper wire 2 m long is stretched by 1 mm. If the energy stored in the stretched wire is converted to heat, calculate the rise in temperature of the wire. (Take, $Y = 12 \times 10^{11}$ dyne cm^{-2}, density of copper = 9 gcm^{-3} and specific heat of copper = 0.1 cal g^{-1}°C^{-1})
(a) 252°C (b) (1/252)°C
(c) 1000°C (d) 2000°C

68. A wire ($Y = 2 \times 10^{11}$ Nm^{-2}) has length 1 m and cross-sectional area 1 mm^2. The work required to increase the length by 2 mm is
(a) 0.4 J (b) 4 J
(c) 40 J (d) 400 J

69. The increase in length on stretching a wire is 0.05%. If its Poisson's ratio is 0.4, the diameter is reduced by
(a) 0.01% (b) 0.02%
(c) 0.03% (d) 0.04%

70. The following data were obtained when a wire was stretched within the elastic region
Force applied to wire = 100 N
Area of cross-section of wire = 10^{-6} m^2
Extension of wire = 2×10^{-3}m
Which of the following deductions can be correctly made from this data?
I. The value of Young's modulus is 10^{11} Nm^{-2}.
II. The strain is 10^{-3}.
III. The energy stored in the wire when the load is applied is 10 J.
(a) I, II and III (b) I and II
(c) Only I (d) Only III

71. When a 4 kg mass is hung vertically on a light spring that obeys Hooke's law, the spring stretches by 2 cm. The work required to be done by an external agent in stretching this spring by 5 cm will be
(a) 4.9 J (b) 2.45 J
(c) 0.495 J (d) 0.245 J

72. Consider two cylindrical rods of identical dimensions, one of rubber and the other of steel. Both the rods are fixed rigidly at one end to the roof. A mass M is attached to each of the free ends at the centre of the rods, then **[NCERT Exemplar]**
(a) Both the rods will elongate but there shall be no perceptible change in shape
(b) the steel rod will elongate and change shape but the rubber rod will only elongate
(c) the steel rod will elongate without any perceptible change in shape, but the rubber rod will elongate and the shape of the bottom edge will change to an ellipse.
(d) the steel rod will elongate, without any perceptible change in shape, but the rubber rod will elongate with the shape of the bottom edge tapered to a tip at the centre

ROUND II Mixed Bag

Only One Correct Option

1. A load of 4.0 kg is suspended from a ceiling through a steel wire of length 2.0 m and radius 2.0 mm. It is found that, the length of the wire increases by 0.031 mm as equilibrium is achieved. Taking, $g = 3.1\,\pi$ ms^{-2}, the Young's modulus of steel is

(a) 2.0×10^8 Nm^{-2} (b) 2.0×10^9 Nm^{-2}
(c) 2.0×10^{11} Nm^{-2} (d) 2.0×10^{13} Nm^{-2}

2. A wire of length $2L$ and radius r is stretched between A and B without the application of any tension. If Y is the Young modulus of the wire and it is stretched like ACB, then the tension in the wire in this position will be

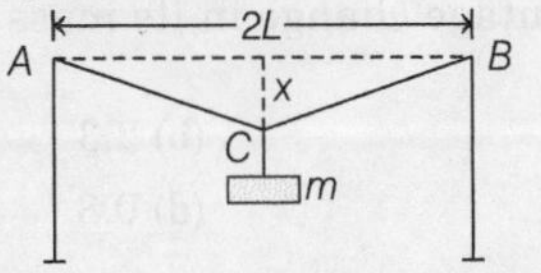

(a) $\dfrac{\pi r^2 Y d^3}{2L^2}$ (b) $\dfrac{\pi r^2 Y d^2}{2L^2}$

(c) $\dfrac{\pi^2 r^2 Y \cdot 2L^2}{d^2}$ (d) $\dfrac{\pi r^2 Y \cdot 2L}{d}$

3. One end of a uniform rod of mass m_1, uniform area of cross-section A is suspended from the roof and mass m_2 is suspended from the other end. What is the stress at the mid-point of the rod?

(a) $(m_1 + m_2)g/A$ (b) $(m_1 - m_2)g/A$

(c) $\left[\dfrac{(m_1/2) + m_2}{A}\right]g$ (d) $\left[\dfrac{m_1 + (m_2/2)}{A}\right]g$

4. A man grows into a giant such that his linear dimensions increase by a factor of 9. Assuming that his density remains the same, the stress in the leg will change by a factor of **[JEE Main 2017]**

(a) $\dfrac{1}{9}$ (b) 81 (c) $\dfrac{1}{81}$ (d) 9

5. The graph shown was obtained from the experimental measurements of the period of oscillation T for different masses M placed in the scale pan on the lower end of the spring balance. The most likely reason for the line not passing through the origin is that

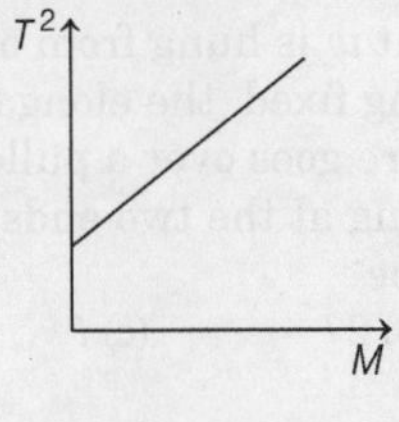

(a) spring did not obey Hooke's law
(b) amplitude of oscillation was too large
(c) clock used needed regulation
(d) mass of the pan was not neglected

6. A rectangular bar 2 cm in breadth and 1 cm in depth and 100 cm in length is supported at its ends and a load of 2 kg is applied at its middle. If Young's modulus of the material of the bar is 20×10^{11} dyne cm^{-2}, the depression in the bar is

(a) 0.2450 cm (b) 0.3675 cm
(c) 0.1225 cm (d) 0.9800 cm

7. Find the ratio of Young's modulus of wire A to wire B.

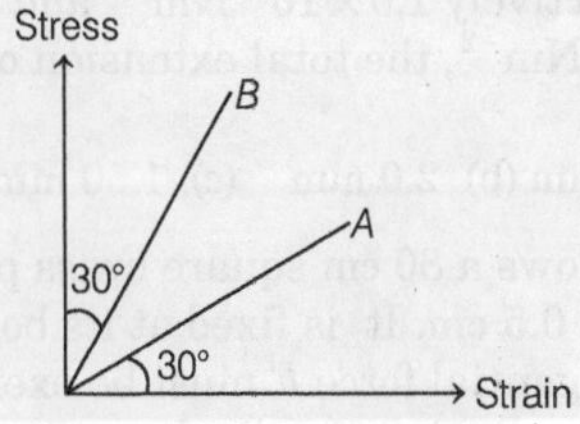

(a) 1 : 1 (b) 1 : 1
(c) 1 : 3 (d) 1 : 4

8. If work done in stretching a wire by 1 mm is 2 J, the work necessary for stretching another wire of same material, but double the radius and half length by 1 mm is

(a) (1/4) J (b) 4 J
(c) 8 J (d) 16 J

9. A steel wire of length 20 cm and uniform cross-section 1 mm² is tied rigidly at both the ends. The temperature of the wire is altered from 40°C to 20°C. Coefficient of linear expansion of steel is $\alpha = 1.1 \times 10^{-5}$ °C^{-1} and Y for steel is 2.0×10^{11} Nm^2, the tension in the wire is

(a) 2.2×10^6 N (b) 16 N
(c) 8 N (d) 44 N

10. The stress-strain graph for a metallic wire is shown at two different temperatures T_1 and T_2. Which temperature is high T_1 or T_2?

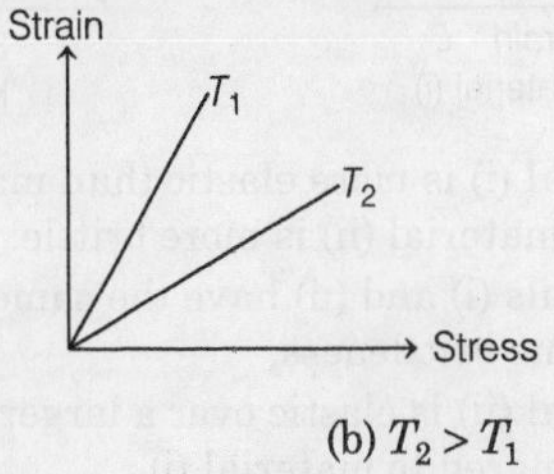

(a) $T_1 > T_2$ (b) $T_2 > T_1$
(c) $T_1 = T_2$ (d) None of these

11. A wire is suspended from the ceiling and stretched under the action of a weight F suspended from its other end. The force exerted by the ceiling on it is equal and opposite to the weight, then **[NCERT Exemplar]**

(a) tensile stress at any cross section A of the wire is F/A
(b) tensile stress at any cross section is zero
(c) tensile stress at any cross section A of the wire is $2F/A$
(d) tension at any cross section A of the wire is O

12. A copper wire of length 1.0 m and a steel wire of length 0.5 m having equal cross-sectional areas are joined end-to-end. The composite wire is stretched by a certain load which stretches the copper wire by 1 mm. If the Young's modulii of copper and steel are respectively $1.0 \times 10^{11} \text{Nm}^{-2}$ and $2.0 \times 10^{11} \text{Nm}^{-2}$, the total extension of the composite wire is **[JEE Main 2013]**

(a) 1.75 mm (b) 2.0 mm (c) 1.50 mm (d) 1.25 mm

13. Figure shows a 80 cm square brass plate of thickness 0.5 cm. It is fixed at its bottom edge. What tangential force F must be exerted on the upper edge, so that the displacement x of this edge in the direction of force is 0.16 mm?

(The shear modulus of brass is 3.5×10^{10} Pa.)

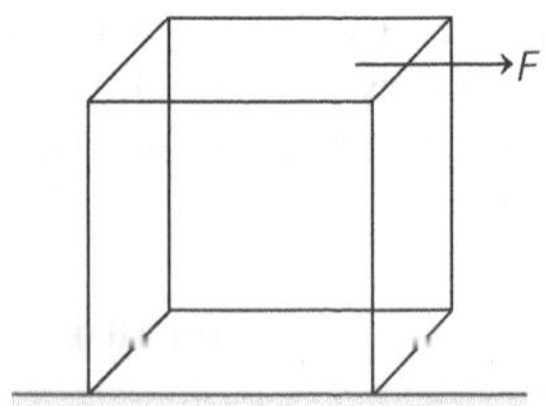

(a) 2.8×10^4 N (b) 3.8×10^{-4} N
(c) 5×10^5 N (d) 4×10^{-5} N

14. The stress-strain graphs for two materials are shown in figure. (assume same scale). **[NCERT Exemplar]**

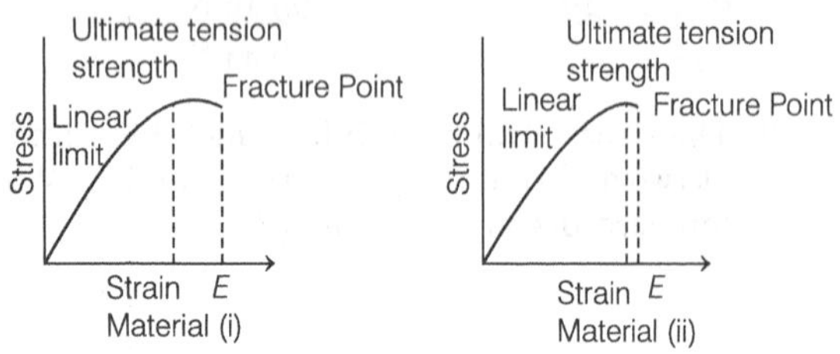

(a) Material (i) is more elastic than material (ii) and hence material (ii) is more brittle.
(b) Materials (i) and (ii) have the same elasticity and the same brittleness.
(c) Material (ii) is elastic over a larger region of strain as compared to material (i).
(d) Material (ii) is less brittle than material (i).

15. Given, initial stretching force on rod = 100 kN and initial elongation = 1.59×10^{-3} m. What is the increase in elastic potential energy when the stretching force is increased by 200 kN?

(a) 238.5 J (b) 636.0 J
(c) 115.5 J (d) 79.5 J

16. The pressure that has to be applied to the ends of a steel rod of length 10 cm to keep its length constant when its temperature is raised by 100°C is (For steel Young's modulus is $2 \times 10^{11} \text{Nm}^{-2}$ and coefficient of thermal expansion is $1.1 \times 10^{-5} \text{K}^{-1}$) **[JEE Main 2014]**

(a) 2.2×10^8 Pa (b) 2.2×10^9 Pa
(c) 2.2×10^7 Pa (d) 2.2×10^6 Pa

17. When the temperature of a metal wire is increased from 0°C to 10°C, its length increases by 0.02%. The percentage change in its mass density will be closest to **[JEE Main 2020]**

(a) 0.06 (b) 2.3
(c) 0.008 (d) 0.8

18. Determine the volume contraction of a solid copper cube, 10 cm on an edge, when subjected to a hydraulic pressure of 7×10^6 Pa. (Take, K for copper = 140×10^9 Pa) **[NCERT]**

(a) 5×10^{-7} m^3 (b) 4×10^{-8} m^3
(c) 5×10^{-8} m^3 (d) 6×10^{-8} m^3

19. The bulk modulus of a metal is 8×10^9 Nm^{-2} and its density is 11 gcm^{-2}. The density of this metal under a pressure of 20000 N cm^{-2} will be (in gcm^{-3})

(a) $\frac{440}{39}$ (b) $\frac{431}{39}$
(c) $\frac{451}{39}$ (d) $\frac{40}{39}$

20. The Poisson's ratio of a material is 0.1. If the longitudinal strain of a rod of this material is 10^{-3}, then the percentage change in the volume of the rod will be

(a) 0.008% (b) 0.08%
(c) 0.8% (d) 8%

21. A solid block of silver with density 10.5×10^3 kg m^{-3} is subjected to an external pressure of 10^7 Nm^{-2}. If the bulk modulus of silver is 17×10^{10} Nm^{-2}, the change in density of silver (in kg m^{-3}) is

(a) 0.61 (b) 1.7
(c) 6.1 (d) 17×10^3

22. When a weight w is hung from one end of the wire other end being fixed, the elongation produced in it be l. If this wire goes over a pulley and two weights w each are hung at the two ends, the elongation of the wire will be

(a) $4l$ (b) $2l$ (c) l (d) $l/2$

23. A sonometer wire of length 1.5 m is made of steel. The tension in it produces an elastic strain of 1%. What is the fundamental frequency of steel, if density and elasticity of steel are 7.7×10^3 kg/m^3 and 2.2×10^{11} N/m^2, respectively? [JEE Main 2013]

(a) 188.5 Hz (b) 178.2 Hz
(c) 200.5 Hz (d) 770 Hz

24. A pendulum made of a uniform wire of cross-sectional area A has time period T. When an additional mass M is added to its bob, the time period changes T_M. If the Young's modulus of the material of the wire is Y, then $\frac{1}{Y}$ is equal to
(g = gravitational acceleration) [JEE Main 2015]

(a) $\left[\left(\frac{T_M}{T}\right)^2 - 1\right]\frac{A}{Mg}$ (b) $\left[\left(\frac{T_M}{T}\right)^2 - 1\right]\frac{Mg}{A}$
(c) $\left[1 - \left(\frac{T_M}{T}\right)^2\right]\frac{A}{Mg}$ (d) $\left[1 - \left(\frac{T}{T_M}\right)^2\right]\frac{A}{Mg}$

25. The twisting couple per unit twist for a solid cylinder of radius 3 cm is 0.1 N-m. The twisting couple per unit twist, for a hollow cylinder of same material with outer and inner radius 5 cm and 4 cm respectively, will be

(a) 0.1 N-m (b) 0.455 N-m
(c) 0.91 N-m (d) 1.82 N-m

26. A rigid bar of mass 15 kg is supported symmetrically by three wires each 2 m long. These at each end are of copper and middle one is of iron. Determine the ratio of their diameters, if each wire is to have the same tension. (Take, Young's modulus of elasticity for copper and steel are 110×10^9 N/m^2 and 190×10^9 N/m^2, respectively.) [NCERT]

(a) 1 : 1.3 (b) 1.3 : 1 (c) 2.3 : 1.3 (d) 2.3 : 1

27. The wires A and B shown in figure, are made of the same material and have radii r_A and r_B, respectively. A block of mass m is connected between them. When a force F is $mg/3$, one of the wires breaks, then which of the following option is correct?

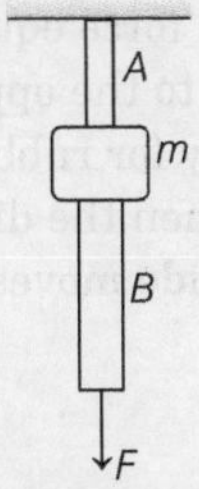

(a) A will break before B if $r_A > r_B$
(b) A will break before B if $r_A = r_B$
(c) Either A or B will break if $r_A = 2r_B$
(d) The length of A and B must be known to decide which wire will break

28. A uniform rectangular bar of area of cross-section A is fixed at one end and on other end force F is applied as shown in figure. Find the shear stress at a plane through the bar making an angle θ with the vertical as shown in figure.

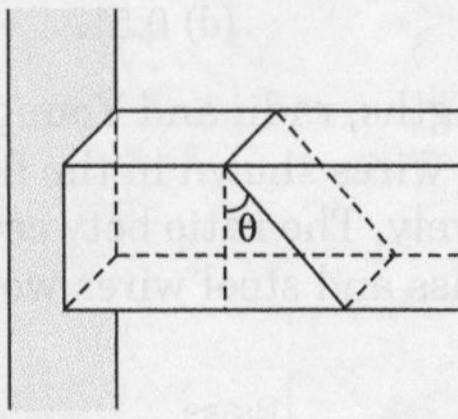

(a) $\frac{F}{2A}(\cos 2\theta)$ (b) $\frac{F}{2A}$
(c) $\frac{F}{2A}(\sin 2\theta)$ (d) $\frac{F}{2A}\cos\theta$

29. A uniform rod of length L and area of cross-section A is subjected to tensile load F. If σ be Poisson's ratio and Y be the Young's modulus of the material of the rod, then find the volumetric strain produced in rod.

(a) $\frac{F}{AY}(1 + 2\sigma)$ (b) $\frac{F}{AY}(1 - 2\sigma)$
(c) Zero (d) None of these

30. A solid sphere of radius r made of a material of Bulk modulus K is surrounded by a liquid in a cylindrical container. A massless piston of area a floats on the surface of the liquid. When a mass m is placed on the piston to compress the liquid, the fractional change in the radius of the sphere $(\Delta r/r)$ is

(a) Ka/mg (b) $Ka/3mg$
(c) $mg/3Ka$ (d) mg/Ka

31. A pendulum clock loses 12 s a day, if the temperature is 40°C and gains 4 s in a day, if the temperature is 20°C. The temperature at which the clock will show correct time and the coefficient of linear expansion (α) of the metal of the pendulum shaft are, respectively [JEE Main 2016]

(a) 25°C, $\alpha = 1.85 \times 10^{-5}$ /°C
(b) 60°C, $\alpha = 1.85 \times 10^{-4}$ /°C
(c) 30°C, $\alpha = 1.85 \times 10^{-3}$ /°C
(d) 55°C, $\alpha = 1.85 \times 10^{-2}$ /°C

32. A steel wire of length 4.7 m and cross-sectional area 3.0×10^{-5} m^2 stretches by the same amount as a copper wire of length 3.5 m and cross-sectional area 4.0×10^{-5} m^2 under a given load. The ratio of the Young's modulus of steel to that of copper is [NCERT]

(a) 1.2 :1 (b) 1.8 : 1
(c) 1.5 : 1 (d) 1.19 : 1

33. The Poisson's ratio of a material is 0.4. If a force is applied to a wire of this material, there is a decrease of cross-sectional area by 2%. The percentage increase in its length is

(a) 3% (b) 2.5%
(c) 1% (d) 0.5%

34. The ratio of lengths, radii and Young's modulus of steel and brass wires shown in the figure are a, b and c, respectively. The ratio between the increase in length of brass and steel wires would be

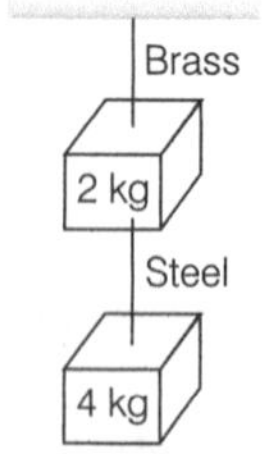

(a) $\frac{b^2a}{2c}$ (b) $\frac{bc}{2a^2}$
(c) $\frac{ba^2}{2c}$ (d) $\frac{c}{2b^2a}$

35. A wire of cross-sectional area A is stretched horizontally between two clamps loaded at a distance $2l$ metres from each other. A weight w kg is suspended from the mid-point of the wire. The strain produced in the wire, (if the vertical distance through which the mid-point of the wire moves down $x < l$) will be

(a) x^2/l^2
(b) $2x^2/l^2$
(c) $x^2/2l^2$
(d) $x/2l$

36. A boy's catapult is made of rubber cord 42 cm long and 6 mm in diameter. The boy keeps a stone weighing 0.02 kg on it and stretches the cord by 20 cm. When released, the stone flies off with a velocity of 20 ms^{-1}. Neglect the change in the cross-section of the cord in stretched position. The stress in the rubber cord is

(a) 1.8×10^6 Nm^{-2}
(b) 1.4×10^6 Nm^{-2}
(c) 2.4×10^5 Nm^{-2}
(d) 1.8×10^5 Nm^{-2}

37. One litre of a gas is maintained at pressure 72 cm of mercury. It is compressed isothermally, so that its volume becomes 900 cm^3. The value of stress and strain will be respectively

(a) 0.106 Nm^{-2} and 0.1 (b) 1.106 Nm^{-2} and 0.1
(c) 106.62 Nm^{-2} and 0.1 (d) 10662.4 Nm^{-2} and 0.1

Numerical Value Questions

38. Wires A and B are made from the same material. A has twice the diameter and three times the length of B. If the elastic limits are not reached, when each is stretched by the same tension, the ratio of energy stored in A to that of B will be ....

39. Two separate wires A and B are stretched by 2 mm and 4 mm respectively, when they are subjected to a force of 2 N. Assume that, both the wires are made up of same material and the radius of wire B is 4 times that of the radius of wire A. The length of the wires A and B are in the ratio of $a : b$, then a/b can be expressed as $1/x$, where x is ....

[JEE Main 2021]

40. A wire of density 9×10^{-3} kg cm^{-3} is stretched between two clamps 1 m apart. The resulting strain in the wire is 4.9×10^{-4}. The lowest frequency (in Hz) of the transverse vibrations in the wire is (Young's modulus of wire, $Y = 9 \times 10^{10}$ Nm^{-2}), (to the nearest integer) ....

[JEE Main 2020]

41. A body of mass $m = 10$ kg is attached to one end of a wire of length 0.3 m. The maximum angular speed (in rad s^{-1}) with which it can be rotated about its other end in space station is (breaking stress of wire $= 4.8 \times 10^7$ Nm^{-2} and area of cross-section of the wire $= 10^{-2}$ cm^2) ....

[JEE Main 2020]

42. A solid ball 3 cm in diameter, is submerged in a lake to a depth where the pressure is 10^3 $kgfm^{-2}$. If Bulk modulus of the material of the ball is 10^7 dyne cm^{-2}, then the change in the volume of ball (in cm^3) will be ....

43. A rubber cube of each side 7 cm has one side fixed, while a tangential force equal to the weight of 300 kgf is applied to the opposite face. If the modulus of rigidity for rubber is 2×10^7 dyne cm^{-2} and $g = 10 ms^{-2}$, then the distance (in cm) through which the strain side moves will be ....

Answers

Round I

1. (c)	2. (c)	3. (c)	4. (d)	5. (c)	6. (a)	7. (a)	8. (c)	9. (a)	10. (b)
11. (a)	12. (b)	13. (c)	14. (c)	15. (c)	16. (a)	17. (b)	18. (a)	19. (c)	20. (d)
21. (d)	22. (c)	23. (b)	24. (b)	25. (a)	26. (b)	27. (d)	28. (a)	29. (a)	30. (c)
31. (b)	32. (c)	33. (b)	34. (d)	35. (d)	36. (b)	37. (c)	38. (d)	39. (c)	40. (b)
41. (b)	42. (d)	43. (c)	44. (b)	45. (c)	46. (b)	47. (a)	48. (d)	49. (a)	50. (a)
51. (a)	52. (a)	53. (c)	54. (b)	55. (b)	56. (c)	57. (b)	58. (c)	59. (c)	60. (d)
61. (b)	62. (b)	63. (a)	64. (c)	65. (a)	66. (d)	67. (b)	68. (a)	69. (b)	70. (b)
71. (b)	72. (d)								

Round II

1. (c)	2. (b)	3. (c)	4. (d)	5. (d)	6. (c)	7. (c)	8. (d)	9. (d)	10. (a)
11. (a)	12. (d)	13. (a)	14. (c)	15. (b)	16. (a)	17. (a)	18. (c)	19. (a)	20. (b)
21. (a)	22. (c)	23. (b)	24. (a)	25. (b)	26. (b)	27. (a)	28. (c)	29. (b)	30. (c)
31. (a)	32. (b)	33. (b)	34. (d)	35. (c)	36. (b)	37. (d)	38. 3 : 4	39. 32	40. 35
41. 4	42. 0.1386	43. 2.1							

Solutions

Round I

1. Given, radius of steel cable, $r = 1.5\text{ cm} = 1.5\times 10^{-2}\text{ m}$

$$\text{Maximum stress} = 10^8\ \text{N/m}^2$$

Area of cross-section of steel cable, $A = \pi r^2$

$$= 3.14\times(1.5\times 10^{-2})^2\ \text{m}^2$$
$$= 3.14\times 2.25\times 10^{-4}\ \text{m}^2$$

$$\text{Maximum stress} = \frac{\text{Maximum force}}{\text{Area of cross-section}}$$

Maximum force = Maximum stress × Area of cross-section

$$= 10^8\times(3.14\times 2.25\times 10^{-4})\ \text{N}$$
$$= 7.065\times 10^4\ \text{N}$$
$$= 7.1\times 10^4\ \text{N} \simeq 7\times 10^4\ \text{N}$$

2. Let m be the mass of rectangular frame and θ be the angle which the tension T in the string make with the horizontal.

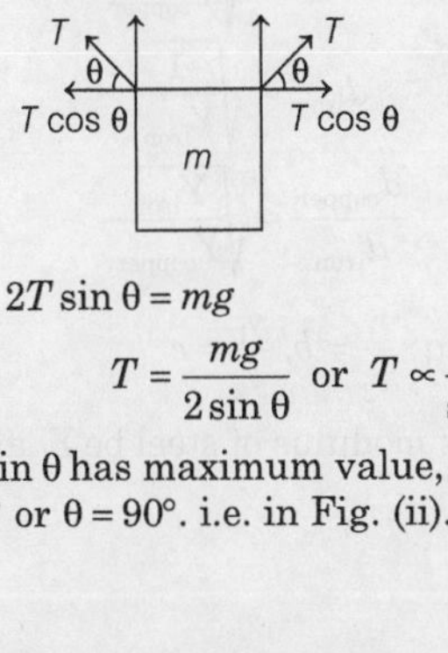

Therefore, $2T\sin\theta = mg$

or $T = \dfrac{mg}{2\sin\theta}$ or $T \propto \dfrac{1}{\sin\theta}$

T is least if $\sin\theta$ has maximum value, *i.e.* $\sin\theta = 1 = 90°$ or $\theta = 90°$. i.e. in Fig. (ii).

3. As, rubber is being more stretched as compare to the iron under the action of same weight. Therefore, the number of atoms in unit volume of iron will be more than that of rubber.

4. Breaking force does not depend upon length. Breaking force = Breaking stress × Area of cross-section. For a given material, breaking stress is constant.

∵ Breaking force ∝ Area of cross-section

$$\Rightarrow \quad \frac{F_2}{F_1} = \frac{A_2}{A_1} = \frac{\pi(6r)^2}{\pi r^2} = 36$$

or $\quad F_2 = 36F_1 = 36F$

5. Length, $L = \dfrac{\text{Breaking stress}}{\rho g}$

$$L = \frac{10^6}{3\times 10^3\times 10} = 33.3\ \text{m}$$

6. Breaking strength = Tension in the wire = $mr\omega^2$ (centrifugal force)

$$\Rightarrow \quad 4.8\times 10^7\times 10^{-6} = 10\times 0.3\times\omega^2$$
$$\omega^2 = \frac{48}{0.3\times 10} = 16$$
$$\omega = 4\ \text{rads}^{-1}$$

7. According to the figure,

Increase in length $= BO + OC - BC$

$$\Delta L = 2\,BO - 2L$$
$$= 2\,(L^2 + x^2)^{1/2} - 2\,L$$

or $$\Delta L = 2L\left[1+\frac{x^2}{L^2}\right]^{1/2} - 2L$$

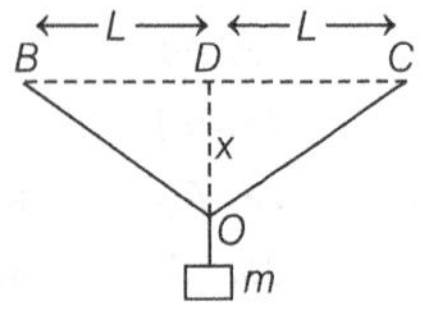

Using binomial theorem, $(1+x)^n = 1 + nx + \ldots +$ higher terms.

$$\Rightarrow \quad \Delta L = 2L\left[1+\frac{x^2}{2L^2}\right] - 2L$$

$$= 2L + 2L \times \frac{x^2}{2L^2} - 2L = \frac{x^2}{L}$$

$$\therefore \quad \text{Strain} = \frac{\Delta L}{2L} = \frac{x^2/L}{2L} = \frac{x^2}{2L^2}$$

8. As, $x = \frac{F}{k}$

If spring constant is k for the first case, it is $\frac{k}{2}$ for second case.

For first case, $$1 = \frac{4}{k} \quad \ldots(i)$$

For second case, $$x' = \frac{6}{k/2} = \frac{12}{k} \quad \ldots(ii)$$

Dividing Eq. (ii) by Eq. (i), we get

$$x' = \frac{12/k}{4/k} = 3 \text{ cm}$$

9. Assume original length of spring = l

As, $$mg = kx$$

$$\Rightarrow \quad k_1(60) = k_2(l-60) = kl$$

According to question,

$$\therefore \quad \frac{mg}{k_1} = 7.5$$

and $$\frac{mg}{k_2} = 5$$

$$\therefore \quad k_1 = \frac{kl}{60}, k_2 = \frac{kl}{(l-60)}$$

$$\frac{k_1}{k_2} = \frac{5.0}{7.5} = \frac{(l-60)}{60}$$

$$\Rightarrow \quad \frac{2}{3} = \frac{(l-60)}{60}$$

$$\therefore \quad l = 100 \text{ cm}$$

and $$kx = k_1 \times 7.5,$$

$$kx = \left(\frac{5k}{3}\right) \times 7.5$$

$$\therefore \quad x = 12.5 \text{ cm}$$

10. As, $$T_1 = K(l - l_1) \quad \ldots(i)$$

and $$T_2 = K(l - l_2) \quad \ldots(ii)$$

So, $$\frac{T_1}{T_2} = \frac{l - l_1}{(l - l_2)}$$

$$\therefore \quad T_1 l - T_1 l_2 = T_2 l - T_2 l_1$$

$$(T_1 - T_2)\, l = T_1 l_2 - T_2 l_1$$

$$l = \frac{T_1 l_2 - T_2 l_1}{(T_1 - T_2)} = \frac{(4b - 5a)}{(4-5)} \quad \ldots(i)$$

$$l = (5a - 4b) \quad \ldots(ii)$$

So from Eqs. (i) and (ii), we get

$$k = \frac{1}{b-a}$$

So, length of wire when tension is 9 N, is given by

$$9 = kl' \quad (l' = \text{change in length})$$

$$9 = \frac{1}{(b-a)} \times l'$$

$$l' = 9b - 9a$$

Hence, final length = $l + l'$

$$= 5a - 4b + 9b - 9a$$

$$l_0 = 5b - 4a$$

11. As, $$Y = \frac{\frac{F}{A}}{\frac{\Delta l}{l}} = \frac{Fl}{A\,\Delta l}$$

or $$Y = \frac{F \times 4}{\pi D^2 \times \Delta l}$$

or $$\Delta l \propto \frac{1}{D^2}$$

or $$\frac{\Delta L_2}{\Delta L_1} = \frac{D_1^2}{D_2^2} = \frac{n^2}{1} \quad \left\{\because \frac{D_2}{D_1} = \frac{n}{1}\right\}$$

On applying the same load, the increase in length of thin wire will be n^2 times.

12. As, $$Y = \frac{mgL}{\pi r^2 l} \quad (Y = \text{Young's modulus})$$

$$\Rightarrow \quad Y = \frac{4mgL}{\pi(2r)^2 l} = \frac{4mgL}{\pi(d)^2 l_1}$$

So under same tension,

$$\Rightarrow \quad Y \propto \frac{1}{d^2}$$

$$\Rightarrow \quad d \propto \sqrt{\frac{1}{Y}}$$

Then, $$d_{\text{copper}} \propto \sqrt{\frac{1}{Y_{\text{copper}}}}$$

and $$d_{\text{iron}} \propto \sqrt{\frac{1}{Y_{\text{iron}}}}$$

So, $$\frac{d_{\text{copper}}}{d_{\text{iron}}} = \sqrt{\frac{Y_{\text{iron}}}{Y_{\text{copper}}}}$$

13. Given $\frac{l_1}{l_2} = a, \frac{r_1}{r_2} = b, \frac{y_1}{y_2} = c$

Let Young's modulus of steel be Y_1 and that of brass be Y_2

$$Y_1 = \frac{F_1 l_1}{A_1 \Delta l_1} \quad \text{...(i)}$$

and $$Y_2 = \frac{F_2 l_2}{A_2 \Delta l_2} \quad \text{...(ii)}$$

Dividing Eq. (i) by Eq. (ii) we get

$$\frac{Y_1}{Y_2} = \frac{F_1 A_2 l_1 \Delta l_2}{F_2 A_1 l_2 \Delta l_1} = \frac{F_1}{F_2} \cdot \frac{\pi r_2^2}{\pi r_1^2} \frac{l_1}{l_2} \cdot \frac{\Delta l_2}{\Delta l_1}$$

From the figure,

Force on steel wire, $F_1 = 2M + M = 3M$

Force on brass wire, $F_2 = 2M$

$$\therefore \quad \frac{\Delta l_1}{\Delta l_2} = \frac{3a}{2b^2c}$$

14. In given experiment, a composite wire is stretched by a force F.

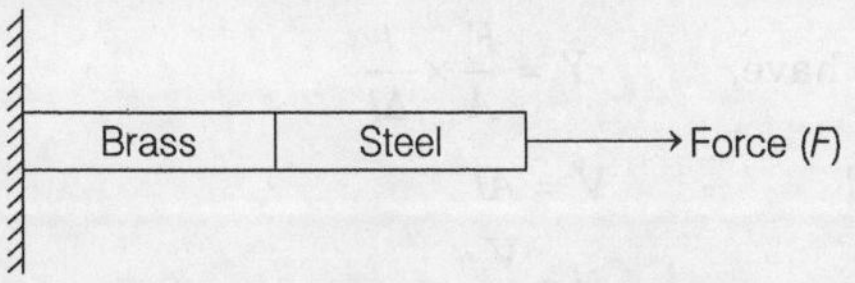

Net elongation in the wire = Elongation in brass wire + Elongation in steel wire ...(i)

Now, Young's modulus of a wire of cross-section (A) when some force (F) is applied, $Y = \frac{Fl}{A\Delta l}$

We have,

$$\Delta l = \text{elongation} = \frac{Fl}{AY}$$

So, from Eq. (i), we have

$$\Delta l_{net} = \Delta l_{brass} + \Delta l_{steel}$$

$$\Rightarrow \quad \Delta l_{net} = \left(\frac{Fl}{AY}\right)_{brass} + \left(\frac{Fl}{AY}\right)_{steel}$$

As wires are connected in series and they are of same area of cross-section, length and subjected to same force, so

$$\Delta l_{net} = \frac{F}{A}\left(\frac{l}{Y_{brass}} + \frac{l}{Y_{steel}}\right) \quad \text{...(ii)}$$

Here,

$$\Delta l_{net} = 0.2 \text{ mm} = 0.2 \times 10^{-3} \text{ m}$$

and $l = 1$ m

$Y_{brass} = 60 \times 10^9 \text{Nm}^{-2}$, $Y_{steel} = 120 \times 10^9 \text{Nm}^{-2}$

On putting the values in Eq (ii), we have

$$0.2 \times 10^{-3} = \frac{F}{A}\left(\frac{1}{60 \times 10^9} + \frac{1}{120 \times 10^9}\right)$$

$$\Rightarrow \quad \text{Stress} = \frac{F}{A} = 8 \times 10^6 \text{Nm}^{-2}$$

15. Speed of transverse wave over a string,

$$v = \sqrt{\frac{T}{\mu}} \quad \text{...(i)}$$

where, T = tension or force on string

and $\mu = \frac{m}{l}$ = mass per unit length.

Also, Young's modulus of string,

$$Y = \frac{Tl}{A\Delta l} \Rightarrow T = \frac{YA\Delta l}{l} \quad \text{...(ii)}$$

From Eqs. (i) and (ii), we have

$$v^2 = \frac{YA\Delta l}{\mu l} \text{ or } \Delta l = \frac{mv^2}{YA} \quad \text{...(iii)}$$

Here, $m = 6 \text{ g} = 6 \times 10^{-3}$ kg, $l = 60$ cm $= 60 \times 10^{-2}$ m, $A = 1 \text{ mm}^2 = 1 \times 10^{-6} \text{ m}^2$, $Y = 16 \times 10^{11} \text{ Nm}^{-2}$ and $v = 90 \text{ ms}^{-1}$

Substituting these given values in Eq. (iii), we get

$$\Delta l = \frac{6 \times 10^{-3} \times (90)^2}{16 \times 10^{11} \times 1 \times 10^{-6}} = 3.03 \times 10^{-5} \text{ m}$$

$$\approx 30 \times 10^{-6} \text{ m} = 0.03 \text{ mm}$$

16. If A is the area of cross-section and l is the length of rope, then the mass of rope, $m = Al\rho$. As, the weight of the rope acts at the mid-point of the rope.

So, $$Y = \frac{mg}{A} \times \frac{(l/2)}{\Delta l}$$

$$\Delta l = \frac{mgl}{2AY} = \frac{Al\rho gl}{2AY} = \frac{g\rho l^2}{2Y}$$

or $$\Delta l = \frac{9.8 \times 1.5 \times 10^3 \times 8^2}{2 \times 5 \times 10^6}$$

$$= 9.6 \times 10^{-2} \text{ m}$$

17. From the question, $10^6 = \frac{LAdg}{A}$

$$\therefore \quad L = \frac{10^6}{3 \times 10^3 \times 9.8} \text{ m}$$

$$= \frac{1000}{3 \times 9.8} = 34.01 \text{ m} \simeq 34 \text{ m}$$

18. As, $$Y = \frac{Fl}{A\Delta l} = \frac{(ml\,\omega^2)l}{A\,\Delta l}$$

or $$Y = \frac{ml^2\omega^2}{A\,\Delta l}$$

or $$Y = \frac{1 \times 1 \times 1 \times 20 \times 20}{10^{-6} \times 10^{-3}}$$

$$= 4 \times 10^{11} \text{ Nm}^{-2}$$

19. As, $Y = \frac{Fl}{A\,\Delta l} \Rightarrow \Delta l \propto \frac{1}{A}$

Again, $m = Al\rho, m \propto A$

$$\therefore \quad \Delta l \propto \frac{1}{m}$$

$$\therefore \quad \frac{\Delta l_1}{\Delta l_2} = \frac{m_2}{m_1} = \frac{2}{3}$$

20. As, $$Y = \frac{F}{A}\frac{L}{\Delta l}$$

$$\Rightarrow \quad \Delta l \propto \frac{L}{A} \propto \frac{L}{\pi d^2}$$

$\therefore \quad \Delta l \propto \dfrac{L}{d^2}$

The ratio of $\dfrac{L}{d^2}$ is maximum for case (d).

21. As, $Y = \dfrac{Fl}{A\Delta l}$

In the given problem, Y, l and Δl are constants.

$\therefore \quad F \propto A$

or $\quad F \propto \pi r^2 \qquad (\because \text{area} = \pi r^2)$

or $\quad F \propto r^2$

or $\quad \dfrac{F_1}{F_2} = \dfrac{r_1^2}{r_2^2} = \dfrac{1}{4}$

22. As, $Y = \dfrac{Fl}{\pi r^2 \Delta l}$ or $\Delta l = \dfrac{F}{\pi r^2 Y}$

$\Rightarrow \quad \Delta l \propto \dfrac{1}{r^2}$ and $\Delta l' \propto \dfrac{2l}{(\sqrt{2}r)^2}$

$\therefore \quad \Delta l' \propto \dfrac{1}{r^2}$

Again, $\quad \dfrac{\Delta l}{\Delta l'} = 1$

23. As, $\quad Y = \dfrac{Mg \times 4 \times l}{\pi D^2 \times \Delta l} \Rightarrow \Delta l \propto \dfrac{1}{D^2}$

When D is doubled, Δl becomes one-fourth, *i.e.*

$$\frac{1}{4} \times 2.4 \text{ cm}, \; \textit{i.e.} \; 0.6 \text{ cm}$$

24. As, $Y = \dfrac{FL}{\pi r^2 l}$ (Here, l = change in length)

or $\quad l = \dfrac{FL}{\pi r^2 Y}$ or $l \propto \dfrac{FL}{r^2}$

$\Rightarrow \quad \dfrac{l_1}{l_2} = \dfrac{\dfrac{FL}{r^2}}{\dfrac{(4F) \times (4L)}{(4r)^2}}$

or $\quad \dfrac{l_1}{l_2} = \dfrac{F \times L}{r^2} \times \dfrac{(4r)^2}{4F \times 4L}$

or $\quad l_1 = l_2 = l$

So, l remains unchanged.

25. As, $Y = \dfrac{Fl}{A\Delta l}$

$\Rightarrow \quad \Delta l \propto \dfrac{F}{r^2}$

$\Rightarrow \quad \dfrac{\Delta l_2}{\Delta l_1} = \dfrac{F_2}{F_1} \times \dfrac{r_1^2}{r_2^2}$

or $\quad \dfrac{\Delta l_2}{\Delta l_1} = 2 \times 2 \times 2 = 8$

or $\quad \Delta l_2 = 8\Delta l_1 = 8 \times 1 \text{ mm} = 8 \text{ mm}$

26. Initial length (circumference) of the ring $= 2\pi r$

Final length (circumference) of the ring $= 2\pi R$

Change in length $= 2\pi R - 2\pi r$

$$\text{Strain} = \frac{\text{Change in length}}{\text{Original length}} = \frac{2\pi(R - r)}{2\pi r} = \frac{R - r}{r}$$

Now Young's modulus,

$$E = \frac{F/A}{l/L} = \frac{F/A}{(R - r)/r}$$

$$\therefore \quad F = AE\left(\frac{R - r}{r}\right)$$

27. As, $Y = \dfrac{Fl}{A\Delta l}$

where, Y, l and F are constants.

$\Rightarrow \quad \Delta l \propto \dfrac{1}{D^2}$

$\Rightarrow \quad \dfrac{\Delta l_2}{\Delta l_1} = \dfrac{D_1^2}{D_2^2} = \dfrac{1}{16}$

$\therefore \quad \Delta l_2 = \dfrac{1}{16}$ mm $\qquad (\because \Delta l_1 = 1 \text{ mm})$

28. We have, $\quad Y = \dfrac{F}{A} \times \dfrac{l}{\Delta l} \qquad$ ...(i)

and $\quad V = Al$

or $\quad l = \dfrac{V}{A} \qquad$ ...(ii)

$\therefore \quad Y = \dfrac{FV}{A^2 \Delta l} \qquad$ [from Eqs. (i) and (ii)]

$\Rightarrow \quad \Delta l \propto \dfrac{1}{A^2}$

or $\quad \Delta l \propto \dfrac{1}{d^4}$

$\Rightarrow \quad \dfrac{\Delta l_A}{\Delta l_B} = \dfrac{d_B^4}{d_A^4} = \dfrac{1^4}{\left(\dfrac{1}{2}\right)^4} = 16$

29. As, $Y = \dfrac{(mg + ml\omega^2)\, l}{\pi r^2 \Delta l}$

(because tension at lowest point $T = (mg + m\omega^2 l)$)

or $\quad \Delta l = \dfrac{m\,(g + l\omega^2)\, l}{\pi r^2 Y}$

or $\quad \Delta l = \dfrac{1\,(10 + 2 \times 4\pi^2 \times 4)^2}{\pi\,(1 \times 10^{-3})^2 \times 2 \times 10^{11}}$

or $\quad \Delta l = \dfrac{(20 + 64 \times 9.88)7}{2 \times 22 \times 10^5}$

$\quad = \dfrac{4566.24}{44 \times 10^5} \times 10^3 \text{ mm} = 1 \text{ mm}$

30. As, $\quad Y = \dfrac{\text{stress}}{\text{strain}}$

or $\quad \text{strain} = \dfrac{\text{stress}}{Y}$

or $\quad \dfrac{\Delta L}{L} = \dfrac{\text{stress}}{Y}$

Since, cross-section are equal and same tension exists in both wires, therefore the stresses developed are equal. Also, ΔL is given to be the same for both the wires.

$$\therefore \quad L \propto Y$$

$$\Rightarrow \quad \frac{L_s}{L_{Cu}} = \frac{Y_s}{Y_{Cu}} = \frac{2 \times 10^{11}}{1.1 \times 10^{11}} = \frac{20}{11}$$

31. As, $Y = \dfrac{F}{\pi R^2} \times \dfrac{l}{\Delta l}$

F, l and Δl are constants.

$$\Rightarrow \quad R^2 \propto \frac{1}{Y}$$

$$\Rightarrow \quad \frac{R_S^2}{R_B^2} = \frac{Y_B}{Y_S} = \frac{10^{11}}{2 \times 10^{11}} = \frac{1}{2}$$

$$\therefore \quad \frac{R_S}{R_B} = \frac{1}{\sqrt{2}}$$

$$\text{or} \quad R_S = \frac{R_B}{\sqrt{2}}$$

32. We have, $Y = \dfrac{Fl}{A\Delta l}$

where, Y, l and A are constants.

$$\therefore \quad \frac{F}{\Delta l} = \text{constant} \Rightarrow \Delta l \propto F$$

Now, $l_1 - l \propto T_1$ and $l_2 - l \propto T_2$

Dividing, we get $\dfrac{l_1 - l}{l_2 - l} = \dfrac{T_1}{T_2}$

$$\text{or} \quad l_1 T_2 - l T_2 = l_2 T_1 - l T_1$$

$$\text{or} \quad l(T_1 - T_2) = l_2 T_1 - l_1 T_2$$

$$\text{or} \quad l = \frac{l_2 T_1 - l_1 T_2}{T_1 - T_2}$$

$$\text{or} \quad l = \frac{l_1 T_2 - l_2 T_1}{T_2 - T_1}$$

33. As, $Y_S = \dfrac{FL_S}{A_S \Delta L_S}$

and $\quad Y_C = \dfrac{FL_C}{A_C \Delta L_C}$

$$\therefore \quad \frac{L_C}{L_S} = \frac{\dfrac{Y_C A_C \Delta L_C}{F}}{\dfrac{Y_S A_S \Delta L_S}{F}} = \left(\frac{Y_C}{Y_S}\right)\left(\frac{A_C}{A_S}\right)\left(\frac{\Delta L_C}{\Delta L_S}\right) \quad \ldots \text{(i)}$$

Here, $\quad \dfrac{A_C}{A_S} = 2, \dfrac{\Delta L_C}{\Delta L_S} = 1, \dfrac{Y_C}{Y_S} = \dfrac{1.1}{2}$

Putting the value of ratios in Eq. (i), we get

$$\therefore \quad \frac{L_C}{L_S} = \frac{1.1}{2} \times 2 \times 1 = 1.1$$

Hence, $L_C : L_S = 1.1 : 1$

34. Net elongation of the rod is

$$l = \frac{3F\left(\dfrac{2L}{3}\right)}{AY} + \frac{2F\left(\dfrac{L}{3}\right)}{AY}$$

$$l = \frac{8FL}{3AY}$$

35. As, $\dfrac{Y_A}{Y_B} = \dfrac{\tan \theta_A}{\tan \theta_B} = \dfrac{\tan 60^\circ}{\tan 30^\circ} = \dfrac{\sqrt{3}}{1/\sqrt{3}} = 3 \Rightarrow Y_A = 3Y_B$

36. Here, $\quad T = m(g + a_0) = 10(10 + 2) = 120 \text{ N}$

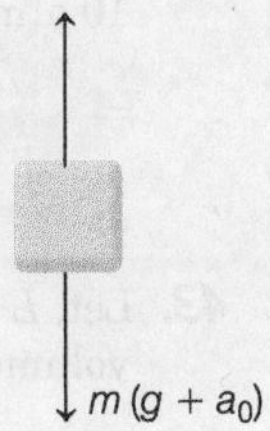

$$\therefore \quad \text{Stress} = \frac{T}{A}$$

$$= \frac{120}{2 \times 10^{-4}} = 60 \times 10^4 \text{ Nm}^{-2}$$

and $\quad Y = \dfrac{\text{stress}}{\text{strain}}$

$$\therefore \quad \text{Strain} = \frac{\text{Stress}}{Y}$$

$$= \frac{60 \times 10^4}{2 \times 10^{11}} = 30 \times 10^{-7} = 3 \times 10^{-6}$$

37. As, $\phi = \dfrac{r\theta}{l}$; so, $\phi = \dfrac{0.4 \times 30^\circ}{10} = 1.2^\circ$

38. As, $Y = \dfrac{Mg}{A} \times \dfrac{L/2}{\Delta L}$

(Length is taken as $\dfrac{L}{2}$ because weight acts at centre of gravity (CG))

Now, $\quad M = AL\rho$

(For the purpose of calculation of mass, the whole of geometrical length L is to be considered.)

$$\therefore \quad Y = \frac{AL\rho gL}{2A\,\Delta L}$$

$$\text{or} \quad \Delta L = \frac{\rho g L^2}{2Y}$$

$$= \frac{1.5 \times 10^3 \times 10 \times 8 \times 8}{2 \times 5 \times 10^6}$$

$$= 9.6 \times 10^{-2} \text{ m}$$

$$= 9.6 \times 10^{-2} \times 10^3 \text{ mm}$$

$$= 96 \text{ mm}$$

39. $\therefore$ Bulk modulus, $K = \dfrac{\text{Volumetric stress}}{\text{Volumetric strain}} = \dfrac{\Delta p}{\dfrac{\Delta V}{V}}$

$$\Rightarrow \quad K = \frac{mg}{a\left(\dfrac{3\Delta r}{r}\right)} \quad \left[\because V = \frac{4}{3}\pi r^3, \text{ so } \frac{\Delta V}{V} = \frac{3\Delta r}{r}\right]$$

$$\Rightarrow \quad \frac{\Delta r}{r} = \frac{mg}{3aK} = \frac{dr}{r}$$

40. As, $K = \dfrac{p}{\dfrac{\Delta V}{V}}$ (Here, K = bulk modulus of elasticity)

$$\text{or} \quad \frac{1}{K} = \frac{\Delta V / V}{p}$$

or $\sigma = \dfrac{\Delta V}{pV}$

or $\Delta V = \sigma p V$

41. As, $K = \dfrac{pV}{\Delta V} = \dfrac{pV}{3V\alpha\Delta T} = \dfrac{p}{3\alpha\Delta T}$

$\Rightarrow \quad \Delta T = \dfrac{p}{3K\alpha}$

42. 10 m column of water exerts nearly 1 atmospheric pressure. So, 100 m column of water exerts nearly 10 atmospheric pressure, *i.e.* 10×10^5 Pa or 10^6 Pa.

$\Rightarrow \quad 10^6 = K\dfrac{\Delta V}{V}$ or $10^6 = K\dfrac{0.1}{100}$

$\Rightarrow \quad K = 10^9$ Pa

43. Let, L be the length of each side of cube, then initial volume $= L^3$. When each side decrease by 1%, then

New length, $L' = L - \dfrac{1}{100} = \dfrac{99L}{100}$

New volume $= L'^3 = \left(\dfrac{99L}{100}\right)^3$,

$\therefore$ Change in volume,

$$\Delta V = L^3 - \left(\frac{99L}{100}\right)^3$$

$$= L^3\left[1 - \left(1 - \frac{3}{100} + \dots\right)\right] = L^3\left[\frac{3}{100}\right] = \frac{3L^3}{100}$$

$\therefore$ Bulk strain $= \dfrac{\Delta V}{V} = \dfrac{3L^3/100}{L^3} = 0.03$

44. As, $\dfrac{1}{K} = \dfrac{\Delta V/V}{\Delta p}$ or $\dfrac{\Delta V}{V} = \Delta p\left[\dfrac{1}{K}\right]$

or $\dfrac{\Delta V}{V} \times 100 = 10^5 \times 8 \times 10^{-10} \times 100 = 8 \times 10^{-3}$

45. Bulk modulus, $K = -\dfrac{p}{\dfrac{\Delta V}{V}}$

Negative sign shows that an increase in pressure and a decrease in volume occur.

Compressibility, $C = \dfrac{1}{K} = \dfrac{-dV}{pV}$

Decrease in volume, $\Delta V = pVC$

$$= 4 \times 10^7 \times 1 \times 6 \times 10^{-10}$$
$$= 24 \times 10^{-3} \text{ L}$$
$$= 24 \times 10^{-3} \times 10^3 \text{ cm}^3 = 24 \text{ cc}$$

46. As, $\eta = \dfrac{Fl}{A\Delta l} = \dfrac{Fl}{l^2\Delta l} = \dfrac{F}{l\,\Delta l}$ or $\Delta l \propto \dfrac{1}{l}$ (for, $F =$ constant)

If l is halved, then Δl is doubled, *i.e.* $0.25 \times 2 = 0.5$ cm.

47. As, $\eta = \dfrac{Y}{2(1+\sigma)}$

Now, $\sigma = 0$

then, $\eta = \dfrac{Y}{2} = \dfrac{6 \times 10^{12}}{2} = 3 \times 10^{12}$ Nm^{-2}

48. As, $\tau_x = \dfrac{\pi\eta r^4}{2l}\theta_x$

and $\tau_y = \dfrac{\pi\eta(2r)^4}{2l}\theta_y$

Since, $\tau_x = \tau_y$

$\therefore \quad \theta_x = 16\theta_y$

or $\dfrac{\theta_x}{\theta_y} = 16$

49. As, torque, $\tau = \dfrac{\pi\eta r^4}{2l}\theta$

In the given problem, $r^4\theta =$ constant

$\Rightarrow \quad \theta \propto \dfrac{1}{r^4}$

$\Rightarrow \quad \dfrac{\theta_A}{\theta_B} = \dfrac{r_2^4}{r_1^4}$

50. Strain = Fractional change in length

$$= \frac{\Delta l}{l} = \frac{l\,\alpha t}{l} = \alpha t = 12 \times 10^{-6} \times 30 = 36 \times 10^{-5}$$

51. Let α be the effective temperature coefficient of linear expansion.

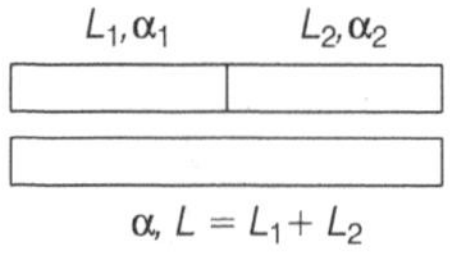

Change in length of equivalent wire = Sum of change in length of each wire

$$\Delta L = \Delta L_1 + \Delta L_2$$

$\Rightarrow \quad L\alpha\Delta T = L_1\alpha_1\Delta T + L_2\alpha_2\Delta T$

$\because \quad L = L_1 + L_2$

$\therefore \quad (L_1 + L_2)\,\alpha\,\Delta T = \Delta T(L_1\alpha_1 + L_2\alpha_2)$

$\Rightarrow \quad \alpha = \dfrac{L_1\alpha_1 + L_2\alpha_2}{L_1 + L_2}$

52. Given, $T_1 = 40°$ C and $T_2 = 20°$ C

$\Rightarrow \quad \Delta T = T_1 - T_2 = (40 - 20)°\text{ C} = 20°$ C

Also, Young's modulus,

$$Y = 10^{11} \text{ N/m}^2$$

Coefficient of linear expansion,

$$\alpha = 10^{-5}/°\text{C}$$

Area of the brass wire, $A = \pi \times (10^{-3})^2$ m^2

Now, contraction in the wire due to fall in temperature is

$$\Delta l = l\,\alpha\Delta T$$

$\Rightarrow \quad \dfrac{\Delta l}{l} = \alpha\Delta T \quad \dots\text{(i)}$

We know that, Young's modulus is defined as

$$Y = \frac{Mgl}{A\Delta l}$$

$\Rightarrow \quad M = \dfrac{YA\Delta l}{gl} \quad \dots\text{(ii)}$

Using Eq. (i), we get

$$M = \frac{YA}{g} \times \alpha \Delta T$$

$$= \frac{10^{11} \times 22 \times 10^{-6} \times 10^{-5} \times 20}{7 \times 10}$$

$$\Rightarrow \quad M = \frac{22 \times 20}{7 \times 10} = \frac{44}{7} = 6.28 \text{ kg}$$

which is closest to 9, so option (a) is nearly correct.

53. Given, $Y = 2 \times 10^{11}$ N/m^2

Stress $= 5 \times 10^7$ N/m^2

As, $\frac{\text{stress}}{\text{strain}} = Y$

$$\Rightarrow \quad \text{Strain} = \frac{5 \times 10^7}{2 \times 10^{11}} = 2.5 \times 10^{-4}$$

It is symmetric strain.

Now, strain of 2.5×10^{-4} is equivalent.

As, $\frac{\Delta V}{V} = 3\left(\frac{\Delta r}{r}\right)$

$$\therefore \quad \frac{2.5 \times 10^{-4}}{3} = \frac{\Delta r}{r} = 0.75 \times 10^{-4}$$

$\therefore$ Fraction decrease in radius

$$= (1.00 - 0.75) 10^{-4} = 0.25 \times 10^{-4}$$

54. F = force developed

$$= YA\, \alpha\, (\Delta\theta)$$

$$= 10^{11} \times 10^{-4} \times 10^{-5} \times 100 = 10^4 \text{ N}$$

55. As, $\eta = \frac{Y}{2(1+\sigma)}$

or $\eta = \frac{2.4\eta}{2(1+\sigma)}$

or $1 + \sigma = 1.2$

or $\sigma = 0.2$

56. As volume, $V = \pi r^2 l$

$$\Rightarrow \quad \frac{\Delta V}{V} = \frac{\Delta(\pi r^2 l)}{\pi r^2 l}$$

or $\frac{\Delta V}{V} = \frac{r^2 \Delta l + 2rl\Delta r}{r^2 l}$

or $\frac{\Delta V}{V} = \frac{\Delta l}{l} + \frac{2\Delta r}{r}$

$$\Rightarrow \quad \frac{\Delta l}{l} = \frac{-2\Delta r}{r} \qquad \left(\because \frac{\Delta V}{V} \approx 0\right)$$

Now, Poisson's ratio,

$$\sigma = -\frac{\Delta r/r}{\frac{\Delta l}{l}} = -\frac{\Delta r/r}{-2\frac{\Delta r}{r}} = 0.5$$

57. Modulus of rigidity of ideal liquids is zero, because a liquid at rest begins to move under the effect to tangential force.

58. $K = \frac{1.5 \text{ N}}{30 \times 10^{-3} \text{ m}} = 50 \text{ Nm}^{-1}$ (as, $mg = kx$)

$$\therefore \quad l = \frac{0.2 \times 10}{50} \text{ m} = 0.04 \text{ m}$$

Now, energy stored $= \frac{1}{2} \times 0.20 \times 10 \times 0.04$ J $= 0.04$ J

59. As, $k_1 = \frac{Y\pi(2R)^2}{L}$ and $k_2 = \frac{Y\pi(R)^2}{L}$

Since, $k_1 x_1 = k_2 x_2 = w$

Elastic potential energy of the system,

$$U = \frac{1}{2}k_1 x_1^2 + \frac{1}{2}k_2 x_2^2$$

$$= \frac{1}{2}k_1\left(\frac{w}{k_1}\right)^2 + \frac{1}{2}k_2\left(\frac{w}{k_2}\right)^2$$

$$= \frac{1}{2}w^2\left\{\frac{1}{k_1} + \frac{1}{k_2}\right\} \quad \text{...(i)}$$

Now, $\frac{1}{k_1} + \frac{1}{k_2} = \frac{L}{4Y\pi R^2} + \frac{L}{Y\pi R^2}$ ...(ii)

$$\therefore \quad U = \frac{1}{2}w^2\left(\frac{5L}{4Y\pi R^2}\right)$$

$$= \frac{5w^2 L}{8\pi Y R^2} \quad \text{[from Eqs. (i) and (ii)]}$$

60. Elastic energy stored in the wire,

$$U = \frac{1}{2} \times \text{stress} \times \text{strain} \times \text{volume}$$

$$= \frac{1}{2} \times \frac{F}{A} \times \frac{\Delta l}{l} \times Al = \frac{1}{2}F\Delta l$$

or $U = \frac{1}{2} \times 200 \times 1 \times 10^{-3} = 0.1$ J

61. As, $mg = kx$

$$k = \frac{10 \text{ N}}{40 \times 10^{-3} \text{ m}} = \frac{1000}{4} \text{ Nm}^{-1} = 250 \text{ Nm}^{-1}$$

Spring constant of combination of two such springs

$$= \frac{250}{2} \text{ Nm}^{-1} = 125 \text{ Nm}^{-1}$$

$$\text{Energy} = \frac{1}{2} \times 125 \times (40 \times 10^{-3})^2 \text{ J} = 0.1 \text{ J}$$

62. Work done in stretching the wire,

$$W = \frac{1}{2} \times \text{force constant} \times x^2$$

For first wire, $W_1 = \frac{1}{2} \times kx^2 = \frac{1}{2}kx^2$

For second wire, $W_2 = \frac{1}{2} \times 2k \times x^2 = kx^2$

Hence, $W_2 = 2W_1$

63. To twist the wire through the angle $d\theta$, it is necessary to do the work

$$dW = \tau d\theta$$

and $\theta = 10' = \frac{10}{60} \times \frac{\pi}{180} = \frac{\pi}{1080}$ rad

$\therefore \quad W = \int dW = \int_0^{\theta} \tau \, d\theta = \int_0^{\theta} \frac{\eta \pi r^4 \theta \, d\theta}{2l} = \frac{\eta \pi r^4 \theta}{4l}$

or $\quad W = \frac{5.9 \times 10^{11} \times 10^{-5} \times \pi (2 \times 10^{-5})^4 \pi^2}{10^{-4} \times 4 \times 5 \times 10^{-2} \times (1080)^2}$

or $\quad W = 1.253 \times 10^{-12}$ J

64. For cylinder A, $\quad \tau = \frac{\pi \eta r^4}{2l} \theta'$

For cylinder B, $\quad \tau = \frac{\pi \eta (2r)^4 (\theta - \theta')}{2l}$

Since, torque acting both of the cylinders will be equal, so

$$\frac{\pi \eta r^4 \theta'}{2l} = \frac{n\eta (2r)^4 (\theta - \theta')}{2l} \text{ (for equal torque)}$$

$\Rightarrow \quad \theta' = \frac{16}{17} \theta$

65. As, energy density $= \frac{1}{2} \times \text{stress} \times \text{strain}$

$$= \frac{1}{2} \times \text{stress} \times \frac{\text{stress}}{Y} = \frac{(\text{stress})^2}{2Y} \propto \frac{1}{D^4}$$

$$\left(\because \text{stress} = \frac{\text{force}}{\text{area}} = \frac{4F}{\pi D^2} \right)$$

Now, $\quad \frac{u_A}{u_B} = \frac{D_B^4}{D_A^4} = (2)^4 = 16$

(where, u_A and u_B are energy densities)

Hence, $u_A : u_B = 16 : 1$

66. As, $W - \frac{1}{2} F \Delta l$ (where, Δl – extension)

$\Rightarrow \quad W = \frac{1}{2} \times \frac{Y \pi r^2 \Delta l}{l} \Delta l \quad \left(\because F = \frac{Y \pi r^2 \Delta l}{l} \right)$

or $\quad W = \frac{Y \pi r^2 \Delta l^2}{2l}$

$\Rightarrow \quad W \propto \frac{r^2}{l}$ and $W' \propto \frac{(2r')^2 2}{l}$

$\Rightarrow \quad \frac{W'}{W} = 8$

or $\quad W' = 8 \times 2 \text{ J} = 16$ J

67. As, $E = \frac{1}{2} \frac{YA \Delta l^2}{l}$

But $\quad m = Ald$

or $\quad A = \frac{m}{ld}$

$\therefore \quad E = \frac{Y m \Delta l^2}{2l^2 d}$

E in calorie $= \frac{Y m \Delta l^?}{2l^2 dJ}$

Now, $\quad mS\theta = \frac{Y m \Delta l^2}{2l^2 dJ}$

or $\quad \theta = \frac{Y \Delta l^2}{2l^2 d \, JS}$

or $\quad \theta = \frac{12 \times 10^{11} \times 10^{-1} \times 10^{-3} \times 10^{-3}}{2 \times 2 \times 2 \times 9 \times 10^3 \times 4.2 \times 0.1 \times 10^3}$

$= \frac{12 \times 10^5}{72 \times 42 \times 10^5}$

$= \frac{1}{252}$ °C

68. As, work done $= \frac{1}{2} F \Delta l = \frac{1}{2} \frac{YA \Delta l^2}{l} \quad \left(\because F = \frac{YA \Delta l}{l} \right)$

$= \frac{2 \times 10^{11} \times 10^{-6} (2 \times 10^{-3})^2}{2 \times 1} = 0.4$ J

69. As, $\sigma = \frac{\text{Lateral strain}}{\text{Longitudinal strain}}$

or $\quad$ Lateral strain $= \sigma \times$ Longitudinal strain

$= 0.4 \times \frac{0.05}{100} = \frac{0.02}{100}$

So, percentage reduction in diameter is 0.02.

70. As, stress $= \frac{100 \text{ N}}{10^{-6} \text{ m}^2} = 10^8 \text{ Nm}^{-2}$ and

Strain $\quad = \frac{2 \times 10^{-3}}{2} = 10^{-3}$

$\therefore$ Young modulus, $Y = \frac{10^8}{10^{-3}} \text{ Nm}^{-2} = 10^{11} \text{ Nm}^{-2}$

Thus, energy stored $= \frac{1}{2} \times 100 \times 2 \times 10^{-3}$ J

$= 10^{-1}$ J $= 0.1$ J

71. As, $mg = kx$

$k = \frac{4 \times 9.8}{2 \times 10^{-2}}$ or $k = 19.6 \times 10^2 \text{ Nm}^{-1}$

$\therefore$ Work done $= \frac{1}{2} \times 19.6 \times 10^2 \times (5 \times 10^{-2})^2$ J $= 2.45$ J

72. When a mass M is attached to each of the free ends at the centre of rods, the steel rod will elongate without any perceptible change in shape, but the rubber rod will elongate with the shape of the bottom edge tapered to a tip at the centre.

Round II

1. As, $Y = \frac{MgL}{\pi r^2 \times l} = \frac{4 \times (3.1 \, \pi) \times (2.0)}{\pi \times (2 \times 10^{-3})^2 \times (0.031 \times 10^{-3})}$

$= 2 \times 10^{11} \text{ Nm}^{-2}$

2. As, $T = \frac{YAl}{L}$ (where, T = force)

Increase in length of one segment of wire,

$$l - \left(L + \frac{1}{2} \frac{d^2}{L} \right) - L - \frac{1}{2} \frac{d^2}{L}$$

So, $\quad T = \frac{Y \pi r^2 \cdot d^2}{2L^2}$

3. As, stress = (weight due to mass m_2 + half of the weight of rod)/area

$$= (m_2 g + m_1 g/2)/A = [(m_1/2) + m_2]g/A$$

4. Stress $= \dfrac{\text{Weight}}{\text{Area}}$

Volume will become (9^3) times.

So weight = volume × density × g will also become $(9)^3$ times.

Area of cross-section will become $(9)^2$ times.

$$= \frac{9^3 \times W_0}{9^2 \times A_0} = 9\left(\frac{W_0}{A_0}\right)$$

Hence, the stress increases by a factor of 9.

5. When no weight is placed in pan and T^2 shows some value, it means the pan is not weightless and hence, the mass of the pan cannot be neglected.

6. Here, $W = 2 \times 1000 \times 980$ dyne; $l = 100$ cm, $b = 2$ cm, $d = 1$ cm, $Y = 20 \times 10^{11}$ dyne cm^{-2},

Now, $\delta = \dfrac{Wl^3}{4Ybd^3}$

$$= \frac{(2 \times 1000 \times 980) \times (100)^3}{4 \times (20 \times 10^{11}) \times 2 \times (1)^3}$$

$$= 0.1225 \text{ cm}$$

7. The graph is given as follows

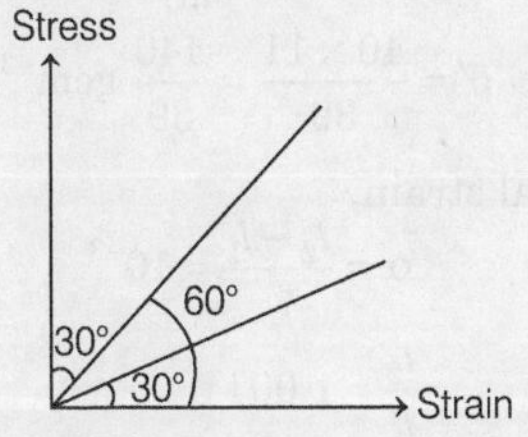

As, $Y = \tan\theta$

$$\therefore \quad \frac{Y_A}{Y_B} = \frac{\tan 30^\circ}{\tan 60^\circ} = \frac{1/\sqrt{3}}{\sqrt{3}} = \frac{1}{3} \quad \left\{\because \tan 30^\circ = \frac{1}{\sqrt{3}} \text{ and } \tan 60^\circ = \sqrt{3}\right\}$$

8. As, work done $= \dfrac{1}{2} \times Y \times (\text{strain})^2 \times \text{volume}$

$$2 = \frac{1}{2} \times Y \times \left(\frac{\Delta L}{L}\right)^2 AL = \frac{YA\Delta L^2}{2L}$$

New work done, $W' = \dfrac{Y(4A)\Delta L^2}{2(L/2)}$

$$= 8\left[\frac{YA\,\Delta L^2}{2L}\right] = 8 \times 2 = 16 \text{ J}$$

9. Increase in length due to rise in temperature,

$$\Delta L = \alpha L \,\Delta T$$

As, $Y = \dfrac{FL}{A\Delta L}$

So, $F = \dfrac{YA\Delta L}{L} = \dfrac{YA \times \alpha L \Delta T}{L} = YA\alpha\Delta T$

$$\therefore \quad F = 2 \times 10^{11} \times 10^{-6} \times 1.1 \times 10^{-5} \times 20 = 44 \text{ N}$$

10. The slope of stress-strain curve with strain axis gives the value of Young's modulus.

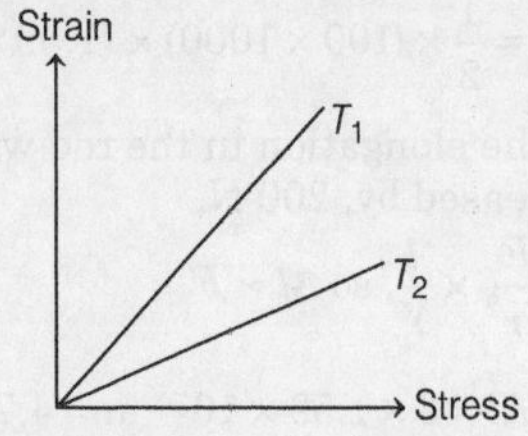

In the above graph, strain is taken along Y-axis. Therefore, the slope of graph at temperature T_1 is less than the slope of graph at temperature T_2.

Now as we know with increase in temperature, the value of modulus of elasticity decreases. Therefore, temperature T_1 is greater than temperature T_2.

11. Tensile stress $= \dfrac{\text{Force}}{\text{Area}} = \dfrac{F}{A}$

Tension = Applied force = F

12. Here, $Y_c = 1 \times 10^{11}$ N/m^2

$Y_s = 2 \times 10^{11}$ N/m^2

$l_c = 1.0$ m, $l_2 = 0.5$ m and $\Delta l_c = 1 \times 10^{-3}$ m

As, $(\text{strain})_c = \dfrac{\text{stress}}{Y_c}$

$$\Rightarrow \quad 1 \times 10^{-3} = \frac{\text{stress}}{1 \times 10^{11}} \Rightarrow \text{stress} = 10^8 \text{ N/m}^2$$

Now, $Y_s = \dfrac{\text{stress}}{\text{strain}}$

$$\Rightarrow \quad \text{strain} = \frac{10^8}{2 \times 10^{11}} = 0.5 \times 10^{-3}$$

or $\dfrac{\Delta l_s}{1/2} = 0.5 \times 10^{-3} \Rightarrow \Delta l_s = 0.25 \times 10^{-3}$

$$\therefore \quad \Delta l = \Delta l_c + \Delta l_s$$

$$= 1 + 0.25 = 1.25 \text{ mm}$$

13. Shearing strain $= \dfrac{x}{h} = \dfrac{1.6 \times 10^{-4} \text{ m}}{0.8 \text{ m}} = 2.0 \times 10^{-4}$

Stress = (Shearing strain) (Shear modulus)

$= (2.0 \times 10^{-4}) \times (3.5 \times 10^{10} \text{ Pa})$

$= 7.0 \times 10^6$ Pa

$\therefore \quad F = \text{Stress} \times \text{Area}$

$= 7.0 \times 10^6 \text{ Pa} \times 0.8 \text{ m} \times 0.005 \text{ m}$

$= 2.8 \times 10^4$ N

14. From graphs, it is clear that ultimate strength of material (ii) is greater than that of material (i). Therefore, the elastic behaviour of material (ii) is elastic over a larger region of strain as compared to material (i).

If the fracture point of a material is closer to ultimate strength point, then the material is a brittle material.

Therefore, the material (ii) is more brittle than material (i).

15. Initial elastic potential energy,

$$U_1 = \frac{1}{2} F\Delta l = \frac{1}{2} \times (100 \times 1000) \times (1.59 \times 10^{-3}) = 79.5 \text{ J}$$

Let, Δl_1 be the elongation in the rod when stretching force is increased by, 200 N.

Since $\Delta l = \frac{F}{\pi r^2} \times \frac{l}{Y}$, so $\Delta l \propto F$

$$\therefore \quad \frac{\Delta l_1}{\Delta l} = 3\Delta l = 3 \times 1.59 \times 10^{-3} \text{ m} = 4.77 \times 10^{-3} \text{ m}$$

Final elastic potential energy,

$$U_1 = \frac{1}{2} F_1 \Delta l_1 = \frac{1}{2} \times (300 \times 10^3) \times (4.77 \times 10^{-3}) = 715.5 \text{ J}$$

Increase in elastic potential energy

$$= 715.5 - 79.5 = 636.0 \text{ J}$$

16. If the deformation is small, then the stress in a body is directly proportional to the corresponding strain.

According to Hooke's law,

$$\text{Young's modulus } (Y) = \frac{\text{Tensile stress}}{\text{Tensile strain}}$$

So, $$Y = \frac{F/A}{\Delta L/L} = \frac{FL}{A\Delta L}$$

If the rod is compressed, then compressive stress and strain appear. Their ratio Y is same as that for tensile case.

Given, length of a steel rod (L) = 10cm,

Temperature (θ) = $100°$ C

As length is constant,

$$\therefore \quad \text{strain} = \frac{\Delta L}{L} = \alpha \Delta \theta$$

Now, pressure = stress = $Y \times$ strain

[Given, $Y = 2 \times 10^{11} \text{N/m}^2$ and $\alpha = 1.1 \times 10^{-5} \text{K}^{-1}$]

$$= 2 \times 10^{11} \times 1.1 \times 10^{-5} \times 100 = 2.2 \times 10^8 \text{ Pa}$$

17. $$\text{Density} = \frac{\text{Mass}}{\text{Volume}} = \frac{\text{Mass}}{\text{Area} \times \text{Length}}$$

$$\Rightarrow \quad \rho = \frac{M}{AL}$$

$$\Rightarrow \quad \frac{\Delta\rho}{\rho} = \frac{\Delta M}{M} + \frac{\Delta A}{A} + \frac{\Delta L}{L}$$

But $\Delta M = 0$, as mass remains constant.

$$\therefore \quad \frac{\Delta\rho}{\rho} = \frac{\Delta A}{A} + \frac{\Delta L}{L} = \beta\Delta T + \alpha\Delta T$$

$$= 2\alpha\Delta T + \alpha\Delta T$$

[$\because$ coefficient of expansion of area, $\beta = 2\alpha$]

$$= 3\alpha\Delta T \quad \ldots\text{(i)}$$

Now, given that,

$$\frac{\Delta L}{L} \times 100 = \alpha\Delta T = 0.02\%$$

So, from Eq (i), we get

$$\frac{\Delta\rho}{\rho} \times 100 = 3 \times 0.02\% = 0.06\%$$

18. Given, each side of cube (l) = 10 cm = 0.1 m

Hydraulic pressure (p) = 7×10^6 Pa

Bulk modulus for copper (K) = 140×10^9 Pa

Volume contraction (ΔV) = ?

Volume of the cube (V) = $l^3 = (0.1)^3 = 1 \times 10^{-3} \text{ m}^3$

$$\text{Bulk modulus for copper } (K) = \frac{p}{\Delta V/V} = \frac{pV}{\Delta V}$$

or $$\Delta V = \frac{pV}{K}$$

$$\Delta V = \frac{7 \times 10^6 \times 1 \times 10^{-3}}{140 \times 10^9} = \frac{1}{20} \times 10^{-6} \text{ m}^3$$

$$= 0.05 \times 10^{-6} \text{ m}^3 = 5 \times 10^{-8} \text{ m}^3$$

19. Here, $p = 20000 \text{ Ncm}^{-2} = 2 \times 10^8 \text{ Nm}^{-2}$

As, $$k = \frac{pV}{\Delta V}$$

or $$\Delta V = \frac{pV}{k} = \frac{2 \times 10^8 \times V}{8 \times 10^9} = \frac{V}{40}$$

New volume of the metal,

$$V' = V - \Delta V = V - \frac{V}{40} = \frac{39V}{40}$$

New mass of the metal

$$= V' \times \rho' = \frac{39V}{40}\rho' = V \times 11$$

or $$\rho' = \frac{40 \times 11}{39} = \frac{440}{39} \text{ gcm}^{-3}$$

20. Longitudinal strain,

$$\Rightarrow \quad \alpha = \frac{l_2 - l_1}{l_1} = 10^{-3}$$

$$\frac{l_2}{l_1} = 1.001$$

$$\text{Poisson's ratio, } \sigma = \frac{\text{lateral strain}}{\text{longitudinal strain}} = \frac{\beta}{\alpha}$$

or $$\beta = \sigma\alpha = 0.1 \times 10^{-3} = 10^{-4} = \frac{r_1 - r_2}{r_1}$$

or $$\frac{r_2}{r_1} = 1 - 10^{-4} = 0.9999$$

$$\% \text{ increase in volume} = \left(\frac{V_2 - V_1}{V_1}\right) \times 100$$

$$= \left(\frac{\pi r_2^2 l_2 - \pi r_1^2 l_1}{\pi r_1^2 l_1}\right) \times 100$$

$$= \left(\frac{r_2^2 l_2}{r_1^2 l_1} - 1\right) \times 100$$

$$= [(0.9999)^2 \times 1.001 - 1] \times 100$$

$$= 0.08\%$$

21. Decrease in volume, $\Delta V = \frac{\Delta p \times V}{K}$

Final volume, $V' = V - \Delta V$

$$= V - \frac{V\Delta p}{K} = V(1 - \Delta p / K)$$

or $$\frac{m}{\rho'} = \frac{m}{\rho}\left(1 - \frac{\Delta p}{K}\right)$$

or $$\rho' = \frac{\rho}{\left(1 - \frac{\Delta p}{K}\right)}$$

or $$\rho = \frac{10.5 \times 10^3}{(1 - 10^7 / 17 \times 10^{10})} = 0.61 \text{ kg m}^{-3}$$

22. As, $Y = \frac{w}{A} \times \frac{L}{l}$ or $l = \frac{wL}{YA}$

When wire goes over a pulley and weight w is attached each free end of wire, then the tension in the wire is doubled, but the original length of wire is reduced to half, so extension in the wire is

$$l' = \frac{2w \times (L/2)}{YA} = \frac{wL}{YA} = l$$

23. Frequency, $f = \frac{v}{2l} = \frac{1}{2l}\sqrt{\frac{T}{\mu}} = \frac{1}{2l}\sqrt{\frac{T}{Ad}}$

where, T is tension in the wire and μ is the mass per unit length of wire.

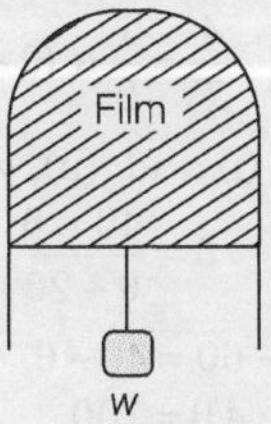

Also, Young's modulus, $Y = \frac{Tl}{A\Delta l}$

$$\Rightarrow \quad \frac{T}{A} = \frac{Y\Delta l}{l}$$

Putting this value in expression of frequency, we have

$$f = \frac{1}{2l}\sqrt{\frac{Y\Delta l}{ld}}$$

Given, $l = 1.5$ m, $\frac{\Delta l}{l} = 0.01$

$d = 7.7 \times 10^3 \text{ kg / m}^3, Y = 2.2 \times 10^{11} \text{ N / m}^2$

Putting these values, we have

$$f = \frac{1}{2l}\sqrt{\frac{2.2 \times 10^{11} \times 0.01}{7.7 \times 10^3}}$$

$$f = \sqrt{\frac{2}{7} \times \frac{10^3}{3}}$$

$$\Rightarrow \quad f \approx 178.2 \text{ Hz}$$

24. We know that, time period,

$$T = 2\pi\sqrt{\frac{L}{g}}$$

When additional mass M is added to its bob

$$T_M = 2\pi\sqrt{\frac{L + \Delta L}{g}}$$

where, ΔL is increase in length.
We know that,

$$Y = \frac{Mg / A}{\Delta L / L} = \frac{MgL}{A\Delta L} \Rightarrow \Delta L = \frac{MgL}{AY}$$

$$\Rightarrow \therefore \quad T_M = 2\pi\sqrt{\frac{L + \frac{MgL}{AY}}{g}}$$

$$\Rightarrow \quad \left(\frac{T_M}{T}\right)^2 = 1 + \frac{Mg}{AY} \text{ or } \frac{Mg}{AY} = \left(\frac{T_M}{T}\right)^2 - 1$$

or $$\frac{1}{Y} = \frac{A}{Mg}\left[\left(\frac{T_M}{T}\right)^2 - 1\right]$$

25. Twisting couple per unit twist for solid cylinder,

$$C_1 = \frac{\pi\eta r^4}{2l}$$

$\therefore$ Hollow cylinder,

$$C_2 = C_1\left(\frac{r_2^4 - r_1^4}{r^4}\right) = \frac{0.1 \times (5^4 - 4^4)}{3^4} = \frac{36.9}{81}$$

$$= 0.455 \text{ N-m}$$

26. Young's modulus of copper $(Y_1) = 110 \times 10^9 \text{ N/m}^2$

Young's modulus of steel $(Y_2) = 190 \times 10^9 \text{ N/m}^2$

Let d_1 and d_2 be the diameters of steel and copper wires, respectively.

Since tension in each wire is same, therefore each wire has same extension. As each wire is of same length, hence each wire has same strain.

Young's modulus $(Y) = \frac{\text{Stress}}{\text{Strain}} = \frac{F / A}{\text{Strain}}$

or $$Y = \frac{F}{\left(\frac{\pi d^2}{4}\right)\text{Strain}} = \frac{4F}{\pi d^2 \times \text{Strain}}$$

$$\therefore \quad Y \propto \frac{1}{d^2} \text{ or } d^2 \propto \frac{1}{Y}$$

$$\therefore \quad \frac{d_1^2}{d_2^2} = \frac{Y_2}{Y_1}$$

or $$\frac{d_1}{d_2} = \sqrt{\frac{Y_2}{Y_1}} = \sqrt{\frac{190 \times 10^9}{110 \times 10^9}} = \sqrt{\frac{19}{11}} = \sqrt{1.73}$$

$$= 1.31$$

$$\therefore \quad d_1 : d_2 = 1.31 : 1$$

27. Here, tension in B, $T_B = F = mg / 3$

Tension in A, $T_A = T_B + mg = \frac{mg}{3} + mg = \frac{4mg}{3}$

$$\therefore \quad T_A = 4T_B$$

A wire will break when the stress is breaking stress (s).

$$S = \frac{\text{Tension}}{\text{Area of cross-section}} = \frac{T}{\pi r^2}$$

For $r_A = 2r_B, S_A = 4S_B$

So, A will break before B.

28. Consider A' be the area of cross-section of plane inclined at an angle θ shown in figure

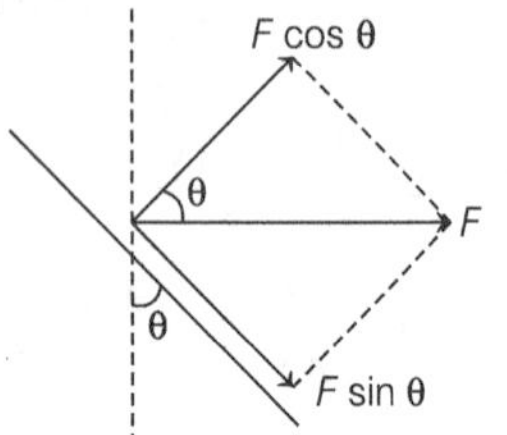

$\therefore \quad A' = \dfrac{A}{\cos\theta}$

Now, the tangential force on plane is $F\sin\theta$.

Shear stress $= \dfrac{F\sin\theta}{\dfrac{A}{\cos\theta}} = \dfrac{F}{A}\sin\theta\cos\theta$

$= \dfrac{F}{2A} 2\sin\theta\cos\theta = \dfrac{F}{2A}(\sin 2\theta)$

29. $Y = \dfrac{F/A}{\Delta L/L}$

$\Rightarrow \quad Y = \dfrac{\Delta L}{L} = \dfrac{F}{AY}$ ...(i)

Now, by definition of Poisson's ratio,

$\sigma = -\dfrac{\Delta r/r}{\Delta L/L}$

$\Rightarrow \quad \dfrac{\Delta r}{r} = -\dfrac{\sigma\Delta L}{L} = -\dfrac{\sigma F}{AY}$ [by using Eq. (i)]

$\Rightarrow \quad \dfrac{\Delta r}{r} = -\dfrac{\sigma F}{AY}$ ...(ii)

Now, $\quad V = \pi r^2 L$

$\Rightarrow \quad \dfrac{\Delta V}{V} = \dfrac{2\Delta r}{r} + \dfrac{\Delta L}{L}$ ...(iii)

By using Eqs. (i), (ii) and (iii), we get

$\dfrac{\Delta V}{V} = 2\left(\dfrac{-\sigma F}{AY}\right) + \dfrac{F}{AY}$

$\Rightarrow \quad \dfrac{\Delta V}{V} = \dfrac{F}{AY}[1 - 2\sigma]$

30. The volume of sphere in liquid,

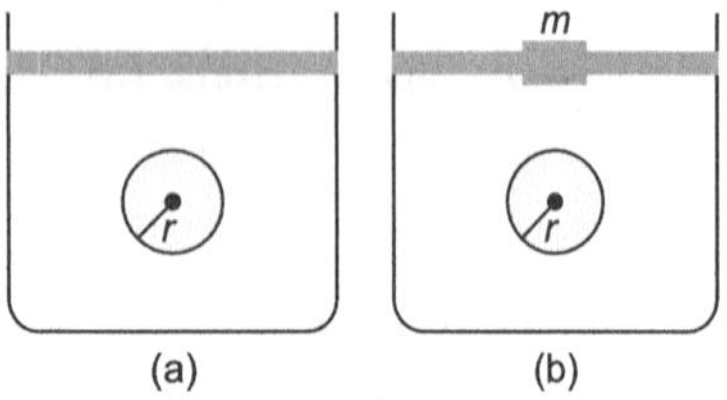

$V = \dfrac{4}{3}\pi r^3$ ...(i)

When mass m is placed on the piston, the increased pressure $p = \dfrac{mg}{a}$.

Since, this increased pressure is equally applicable to all directions on the sphere, so there will be decrease in volume of sphere, due to decrease in its radius.

From Eq. (i), change in volume is

$\Delta V = \dfrac{4}{3}\pi \times 3r^2\Delta r = 4\pi r^2 \Delta r$

$\dfrac{\Delta V}{V} = \dfrac{4\pi r^2\Delta r}{(4/3)\pi r^3} = \dfrac{3\Delta r}{r}$

$\therefore$ Bulk modulus ,

$K = \dfrac{p}{dV/V} = \dfrac{mg}{a} \times \dfrac{r}{3\Delta r}$

$\therefore \quad \dfrac{\Delta r}{r} = \dfrac{mg}{3Ka}$

31. Time period of a pendulum,

$T = 2\pi\sqrt{\dfrac{l}{g}}$

where, l is length of pendulum and g is acceleration due to gravity.

Change in time period of a pendulum is

$\dfrac{\Delta T}{T} = \dfrac{1}{2}\dfrac{\Delta l}{l}$

When clock gains 12 s, we get

$\dfrac{12}{T} = \dfrac{1}{2}\alpha(40 - \theta)$ ...(i)

When clock loses 4 s, we get

$\dfrac{4}{T} = \dfrac{1}{2}\alpha(\theta - 20)$ ...(ii)

Comparing Eqs. (i) and (ii), we get

$3 = \dfrac{40 - \theta}{\theta - 20}$

$\Rightarrow \quad 3\theta - 60 = 40 - \theta$

$\Rightarrow \quad 4\theta = 100$

$\Rightarrow \quad \theta = 25°\text{C}$

Substituting the value of θ in Eq. (i), we have

$\dfrac{12}{T} = \dfrac{1}{2}\alpha(40 - 25)$

$\Rightarrow \quad \dfrac{12}{24 \times 3600} = \dfrac{1}{2}\alpha(15)$

$\Rightarrow \quad \alpha = \dfrac{24}{24 \times 3600 \times 15}$

$\Rightarrow \quad \alpha = 1.85 \times 10^{-5}/°\text{C}$

Thus, the coefficient of linear expansion in a pendulum clock $= 1.85 \times 10^{-5}/°\text{C}$

32. Given, **for steel wire**

Length $(l_1) = 4.7$ m

Area of cross-section $(A_1) = 3.0 \times 10^{-5}$ m^2

For copper wire

Length $(l_2) = 3.5$ m

Area of cross-section $(A_2) = 4.0 \times 10^{-5}$ m^2

Let F be the given load under which steel and copper wires be stretched by the same amount Δl.

Young's modulus $(Y) = \dfrac{F/A}{\Delta l/l} = \dfrac{F \times l}{A \times \Delta l}$

For steel, $\quad Y_s = \dfrac{F \times l_1}{A_1 \times \Delta l}$ ...(i)

For copper, $Y_c = \dfrac{F \times l_2}{A_2 \times \Delta l}$ ...(ii)

Dividing Eq. (i) by Eq. (ii), we get

$$\frac{Y_s}{Y_c} = \frac{F \times l_1}{A_1 \times \Delta l} \times \frac{A_2 \times \Delta l}{F \times l_2}$$

$$= \frac{l_1}{l_2} \times \frac{A_2}{A_1} = \frac{4.7}{3.5} \times \frac{4.0 \times 10^{-5}}{3.0 \times 10^{-5}}$$

$$= \frac{18.8}{10.5} = 1.79 = 1.8$$

Hence, $Y_s : Y_c = 1.8 : 1$

33. Poisson's ratio, $\sigma = 0.4 = \dfrac{\Delta d/d}{\Delta l/l}$

Area, $A = \pi r^2 = \dfrac{\pi d^2}{4}$ or $d^2 = \dfrac{4A}{\pi}$

Differentiating, we get

$$2d\,\Delta d = \frac{4}{\pi} \cdot \Delta A$$

As, $A = \dfrac{\pi d^2}{4}$

So, $\Delta A = \dfrac{2\pi d \Delta d}{4}$

$$\frac{\Delta A}{A} = \frac{\pi \dfrac{d}{2} \Delta d}{\pi d^2/4} = 2\frac{\Delta d}{d}$$

Given $\dfrac{\Delta A}{A} \times 100 = 2\%$

$\Rightarrow$ $\dfrac{2\Delta d}{d} \times 100 = 2\%$

or $\dfrac{\Delta d}{d} \times 100 = 1\%$

Given $\sigma = \dfrac{\Delta d/d}{\Delta l/l} = 0.4$

or $\dfrac{\Delta d}{d} = 0.4\dfrac{\Delta l}{l} = \dfrac{\Delta l}{l} = \dfrac{1}{0.4}\left(\dfrac{\Delta d}{d}\right)$

$\Rightarrow$ $\dfrac{\Delta l}{l} \times 100 = 2.5 \times 1\% = 2.5\%$

34. Given, $\dfrac{l_1}{l_2} = a, \dfrac{r_1}{r_2} = b, \dfrac{y_1}{y_2} = c$

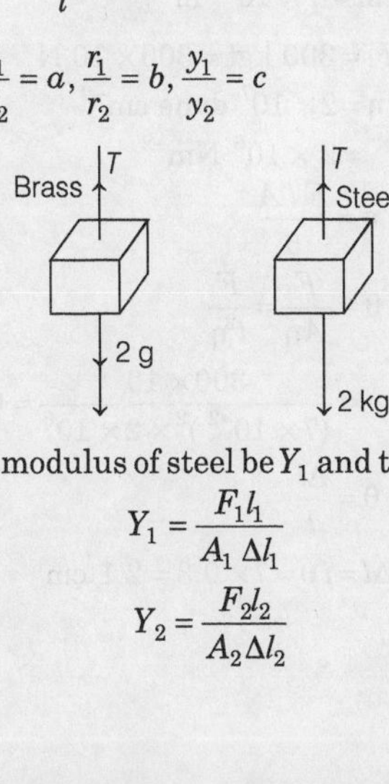

Let Young modulus of steel be Y_1 and that of brass be Y_2

$$Y_1 = \frac{F_1 l_1}{A_1\,\Delta l_1} \quad ...(i)$$

and $$Y_2 = \frac{F_2 l_2}{A_2 \Delta l_2} \quad ...(ii)$$

Dividing Eq. (i) by Eq. (ii), we get

$$\frac{Y_1}{Y_2} = \frac{F_1 A_2 \cdot l_1 \Delta l_2}{F_2 A_1\, l_2 \Delta l_1} \quad ...(iii)$$

Force on steel wire,

$$F_1 = T = 2g$$

Force on brass wire,

$$F_2 = T_1' = T + 2g = 4g$$

Now putting the value of $F_1 \cdot F_2$ in Eq. (iii), we get

$$\frac{Y_2}{Y_1} = \left(\frac{2g}{4g}\right)\left(\frac{\pi r_2^2}{\pi r_1^2}\right)\left(\frac{l_1}{l_2}\right)\left(\frac{\Delta l_2}{\Delta l_1}\right) \quad ...(iv)$$

Now given that, $\dfrac{l_1}{l_2} = a, \dfrac{r_1}{r_2} = b$

and $\dfrac{Y_1}{Y_2} = c$

From Eq. (iv), we get

$$\frac{\Delta l_1}{\Delta l_2} = \frac{c}{2b^2 a}$$

35. From figure, the increase in length,

$$\Delta l = (PR + RQ) - PQ = 2PR - PQ$$

$$= 2(l^2 + x^2)^{1/2} - 2l$$

$$= 2l\left(1 + \frac{x^2}{l^2}\right)^{1/2} - 2l$$

$$= 2l\left[1 + \frac{1}{2}\frac{x^2}{l^2}\right] - 2l$$

$$= x^2/l \quad \text{(By Binomial theorem)}$$

$\therefore$ Strain $= \Delta l/2l = x^2/2l^2$

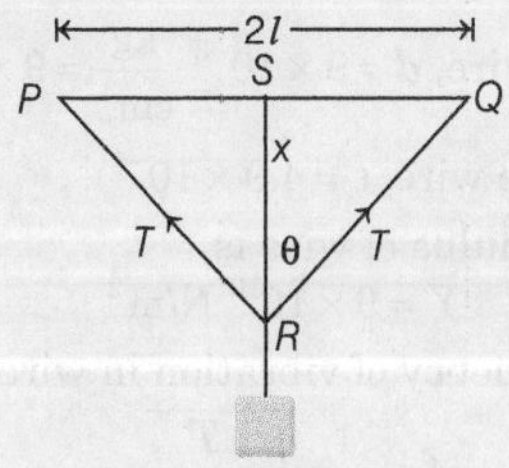

36. Here $m = 0.02$ kg; $v = 20\ \text{ms}^{-1}$;

$$l = 42\ \text{cm} = 0.42\ \text{m}$$

$$\Delta l = 20\ \text{cm} = 0.20\ \text{cm}$$

$$r = 3\ \text{mm} = 3 \times 10^{-3}\ \text{m}$$

Due to extension, energy is stored in the cord. This is converted into kinetic energy when the stone flies off.

$\therefore$ Work done $= \dfrac{1}{2}mv^2 = \dfrac{1}{2}F\Delta l$

or $F = \dfrac{mv^2}{\Delta l} = \dfrac{0.02 \times (20)^2}{0.20} = 40$ N

Stress $= \dfrac{F}{\pi r^2} = \dfrac{40}{(22/7)\,(3 \times 10^{-3})^2}$

$$= 1.4 \times 10^6\ \text{Nm}^{-2}$$

37. According to Boyle's law, $p_2V_2 = p_1V_1$

or $$p_2 = p_1\left(\frac{V_1}{V_2}\right)$$

or $$p_2 = 72 \times 1000/900 = 80 \text{ cm of Hg}$$

Stress = Increase in pressure

$$= p_2 - p_1 = 80 - 72 = 8$$
$$= (8 \times 10^{-2}) \times 13.6 \times 10^3 \times 9.8$$
$$= 10662.4 \text{ Nm}^{-2}$$

$$\text{Volumetric strain} = \frac{V_1 - V_2}{V_1} = \frac{1000 - 900}{1000} = 0.1$$

38. Elastic energy per unit volume,

$$U = \frac{1}{2} F\Delta l = \frac{F^2 l}{2\,AY}$$

$$U \propto \frac{l}{r^2} \text{ (} F \text{ and } Y \text{ are constants)}$$

$$\therefore \quad \frac{U_A}{U_B} = \left(\frac{l_A}{l_B}\right)\left(\frac{r_B}{r_A}\right)^2 = (3) \times \left(\frac{1}{2}\right)^2 = \frac{3}{4}$$

39. For A $$\frac{F}{\pi r^2} = Y\frac{2\text{ mm}}{a} \quad \text{...(i)}$$

For B $$\frac{F}{\pi \cdot 16r^2} = Y\frac{4\text{ mm}}{b} \quad \text{...(ii)}$$

Dividing Eq. (i) by Eq. (ii), we get

$$\therefore \quad 16 = \frac{2b}{4a}$$

$$\frac{a}{b} = \frac{1}{32}$$

$$\therefore \quad x = 32$$

40. Density of wire, $d = 9 \times 10^{-3} \dfrac{\text{kg}}{\text{cm}^3} = 9 \times 10^3 \text{ kg/m}^3$

Strain in the wire, $\varepsilon = 4.9 \times 10^{-4}$

Young's modulus of wire is

$$Y = 9 \times 10^{10} \text{ N/m}^2$$

Lowest frequency of vibration in wire will be

$$f = \frac{1}{2L}\sqrt{\frac{T}{(M/L)}} \quad \text{...(i)}$$

Now, $$\frac{T}{M/L} = \frac{T}{V\rho/L} = \frac{T}{\left(\frac{LA\cdot\rho}{L}\right)} = \frac{T}{A\rho}$$

But $\dfrac{T}{A}$ = stress = Y × strain = $Y \times \varepsilon$

$$\Rightarrow \quad \frac{T}{M/L} = \frac{Y \times \varepsilon}{\rho} \quad \left(\because A = \frac{M}{\rho L}\right)$$

So, from Eq (i), frequency will be

$$f = \frac{1}{2L}\sqrt{\frac{T}{(M/L)}} = \frac{1}{2L}\sqrt{\frac{Y \times \varepsilon}{\rho}}$$

$$= \frac{1}{2 \times 1} \times \sqrt{\frac{9 \times 10^{10} \times 4.9 \times 10^{-4}}{9 \times 10^3}}$$

$$= 35 \text{ Hz}$$

41. Centripetal force is provided by the tension in wire.

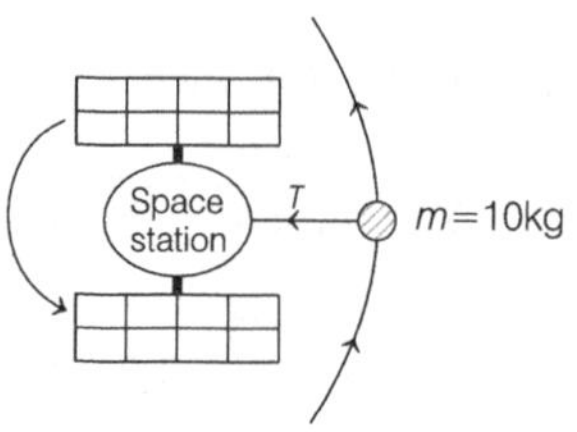

So, $$T = m\omega^2 l$$

Stress in wire, $$\sigma = \frac{T}{A} = \frac{m\omega^2 l}{A} \quad \text{...(i)}$$

Here, $\sigma_{max} = 4.8 \times 10^7 \text{ Nm}^{-2}$,

$A = 10^{-2} \text{ cm}^2 = 10^{-2} \times 10^{-4} \text{ m}^2 = 10^{-6} \text{ m}^2$,

$m = 10$ kg and $l = 0.3$ m

If maximum angular speed of rotation is ω_{max}, then from Eq. (i), we have

$$\omega^2_{max} = \frac{\sigma_{max} A}{ml} = \frac{4.8 \times 10^7 \times 10^{-6}}{10 \times 0.3} = 16$$

or $$\omega_{max} = 4 \text{ rad s}^{-1}$$

42. Here, $2r = 3 \times 10^{-2}$ m

or $r = (3/2) \times 10^{-2}$ m

$\Delta p = 10 \text{ kgfm}^{-2} = 10^3 \times 9.8 \text{ Nm}^{-2}$

$K = 10^7 \text{ dyne cm}^2 = 10^6 \text{ Nm}^{-2}$

Volume of the ball,

$$V = \frac{4}{3}\pi r^3 = \frac{4}{3} \times \frac{22}{7} \times \left(\frac{3}{2} \times 10^{-2}\right)^3 \text{ m}^3$$

Now, $$K = \frac{\Delta p \times V}{\Delta V} \text{ or } \Delta V = \frac{V\Delta p}{K}$$

$$= \frac{\frac{4}{3} \times \frac{22}{7} \times \left(\frac{3}{2} \times 10^{-2}\right)^3 \times 10^3 \times 9.8}{10^6}$$

$$= 0.1386 \times 10^{-6} \text{ m}^3 = 0.1386 \text{ cm}^3$$

43. Given $l = 7 \text{ cm} = 7 \times 10^{-2}$ m

$$F = 300 \text{ kgf} = 300 \times 10 \text{ N}$$
$$\eta = 2 \times 10^7 \text{ dyne cm}^{-2}$$
$$= 2 \times 10^6 \text{ Nm}^{-2}$$

As, $$\eta = \frac{F/A}{\theta}$$

or $$\theta = \frac{F}{A\eta} = \frac{F}{l^2\eta}$$

$$= \frac{300 \times 10}{(7 \times 10^{-2})^2 \times 2 \times 10^6} = 0.3 \text{ rad}$$

$$\theta = \frac{\Delta l}{l}$$

or $$\Delta l = l\theta = 7 \times 0.3 = 2.1 \text{ cm}$$

12

Properties of Fluids

The substances which flow are called *fluids*. Fluids include both liquids and gases. The branch of Science which deals with fluids at rest is called fluid statics or hydro statics and the branch of Science which deals with fluids in motion is called *hydrodynamics*. Fluid statics includes hydrostatic pressure, laws of floatation, Pascal's law and Archimedes' principle, while hydrodynamics includes continuity equation and Bernoulli's principle and Torricelli's theorem.

IN THIS CHAPTER ....

- Density and Relative Density
- Pressure due to a Fluid Column
- Pascal's Law and its Applications
- Archimedes' Principle and Buoyancy
- Flow of Liquids
- Critical Velocity and Reynold's Number
- Energy of a Flowing Liquid
- Bernoulli's Principle
- Torricelli's Theorem
- Venturimeter
- Surface Tension
- Surface Energy
- Angle of Contact
- Capillarity
- Viscosity
- Poiseuille's Formula
- Stoke's Law and Terminal Velocity

Density and Relative Density

Density and relative density are given below

Density

Fluid is characterised by density ρ at every point which is defined as the ratio of the mass of the fluid contained in an infinitesimal volume element around the point to the volume of element.

$$\rho = \lim_{\Delta V \to 0} \frac{\Delta m}{\Delta V} = \frac{dm}{dV}$$

Density is a positive scalar quantity. In case of a homogeneous isotropic substance, density has no directional properties, so it is a scalar. It has dimensions $[ML^{-3}\ T^0]$. Its SI unit in MKS is kgm^{-3}, while in CGS it is gcc^{-1} with $1\ gcc^{-1} = 10^3\ kgm^{-3}$.

Relative Density

It is defined as the ratio of density of substance to the density of water at 4°C.

$$i.e.\ \text{Relative density} = \frac{\text{Density of substance}}{\text{Density of water at } 4^\circ \text{C}}$$

Relative density has no units and no dimensions. It is also known as *specific gravity*.

Important Points Related to Density

- When immiscible liquids of different densities are poured in a container, the liquid of highest density will be at the bottom, while that of lowest density at the top and interfaces will be plane.
- If a liquid of mass m_1 and density ρ_1 and other liquid of mass m_2 and density ρ_2 are mixed.

$$m_1 + m_2 = M$$

and $$V = \frac{m_1}{\rho_1} + \frac{m_2}{\rho_2} \qquad \left[\because V = \frac{m}{\rho}\right]$$

Density of mixed liquids,

$$\rho = \frac{M}{V} = \frac{m_1 + m_2}{\frac{m_1}{\rho_1} + \frac{m_2}{\rho_2}} = \frac{\Sigma m_i}{\Sigma\left(\frac{m_i}{\rho_i}\right)}$$

If $m_1 = m_2$, then $\rho = \frac{2\rho_1 \rho_2}{\rho_1 + \rho_2}$.

- If a liquid of volume V_1 and density ρ_1 and other liquid of volume V_2 and density ρ_2 are mixed.

Then, we have

$$m = \rho_1 V_1 + \rho_2 V_2 \text{ and } V = V_1 + V_2 \qquad \left[\because \rho = \frac{m}{V}\right]$$

$$\rho = \frac{m}{V} = \frac{\rho_1 V_1 + \rho_2 V_2}{V_1 + V_2} = \frac{\Sigma \rho_i V_i}{\Sigma V_i}$$

- With increase in pressure due to a decrease in volume, density will increase.

$$\therefore \quad \frac{\rho}{\rho_0} = \frac{m/V}{m/V_0} = \frac{V_0}{V}$$

Example 1. *A rock of volume 10 cm^3 has mass of 100g, the specific gravity of rock is*

(a) 10 (b) 100
(c) 1000 (d) 10000

Sol. (a) Given, $V = 10 \text{ cm}^3 = 10 \times 10^{-6} = 10^{-5} \text{m}^3$

$$m = 100 \text{ g} = 0.1 \text{kg}$$

$$\therefore \text{ Density of rock, } d_r = \frac{\text{Mass}}{\text{Volume}} = \frac{0.1}{10^{-5}} = 10^4 \text{ kg/m}^3$$

$$\therefore \text{ Specific gravity of rock} = \frac{\text{Density of rock}}{\text{Density of water}} = \frac{10^4}{10^3} = 10$$

Example 2. *When equal volumes of two metals are mixed together, the specific gravity of alloys is 4. When equal masses of the same two metals are mixed together, the specific gravity of alloy is 3. What is the specific gravity of each metal?*

(a) 3 and 6 (b) 5 and 4
(c) 2 and 6 (d) 4 and 4

Sol. (c) In case of two metals,

$$\rho = \frac{(m_1 + m_2)}{V_1 + V_2}$$

If equal volumes are mixed,

$$V_1 = V_2 = V \text{ and}$$
$$m_1 = V\rho_1 \text{ and } m_2 = V\rho_2$$
$$\rho = \frac{V\rho_1 + V\rho_2}{V + V}$$

i.e. $$\rho = \frac{\rho_1 + \rho_2}{2} = 4$$

$$\Rightarrow \quad \rho_1 + \rho_2 = 8 \qquad \text{...(i)}$$

and when equal masses are mixed, $m_1 = m_2 = m$ and

$$V_1 = \frac{m}{\rho_1}$$

and $$V_2 = \frac{m}{\rho_2}$$

$$\therefore \quad \rho = \frac{m + m}{\frac{m}{\rho_1} + \frac{m}{\rho_2}}$$

$$\therefore \quad \frac{2\rho_1\rho_2}{\rho_1 + \rho_2} = 3 \qquad \text{...(ii)}$$

From Eqs. (i) and (ii), we get

$$\frac{2\rho_1\rho_2}{8} = 3$$

$$\Rightarrow \quad \rho_1\rho_2 = 12$$

$$\Rightarrow \quad \rho_1 = \frac{12}{\rho_2} \qquad \text{...(iii)}$$

From Eqs. (i) and (iii), we get

$$\frac{12}{\rho_2} + \rho_2 = 8$$

$$\Rightarrow \quad 12 + \rho_2^2 - 8\rho_2 = 0$$

$$\Rightarrow \quad \rho_2^2 - 8\rho_2 + 12 = 0$$

$$\Rightarrow \quad \rho_2^2 - 6\rho_2 - 2\rho_2 + 12 = 0$$

$$\Rightarrow \rho_2(\rho_2 - 6) - 2(\rho_2 - 6) = 0$$

$$\Rightarrow \quad (\rho_2 - 2)(\rho_2 - 6) = 0$$

$$\Rightarrow \quad \rho_2 = 2 \text{ or } 6 \qquad \text{...(iv)}$$

Solving Eqs. (iii) and (iv) for ρ_1 and ρ_2, we find specific gravities of metals are 2 and 6.

Pressure due to a Fluid Column

The pressure p is defined as the ratio of the magnitude of the force to the area of the contact, *i.e.*

$$p = \frac{\text{magnitude of force}}{\text{area of the contact}} = \frac{F}{A} \qquad \text{...(i)}$$

For a point at a depth h below the surface of a liquid of density ρ, pressure p is given by

$$p = p_0 + h\rho g \qquad \text{...(ii)}$$

where, p_0 is the *atmospheric pressure*.

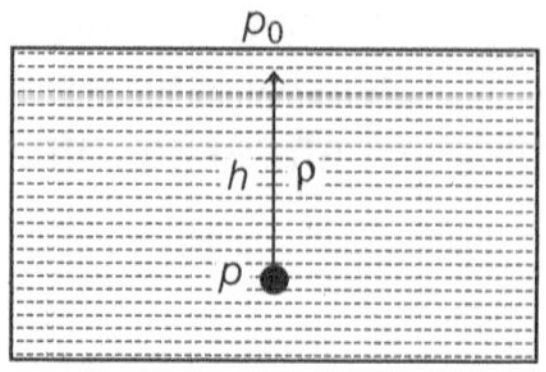

The excess pressure above atmospheric pressure is called *gauge pressure* and total pressure is called *absolute pressure.* Thus,

gauge pressure = absolute pressure – atmospheric pressure

i.e. $h\rho g = p - p_0$ ...(iii)

Some important points related to fluid column pressure are given below

- Forces acting on a fluid in equilibrium have to be perpendicular to its surface, because it cannot sustain the shear stress. *i.e.* fluid at rest has no tangential force acting on its surface.
- In the same liquid, pressure will be same at all points at the same level.

  For example, in the given figure,

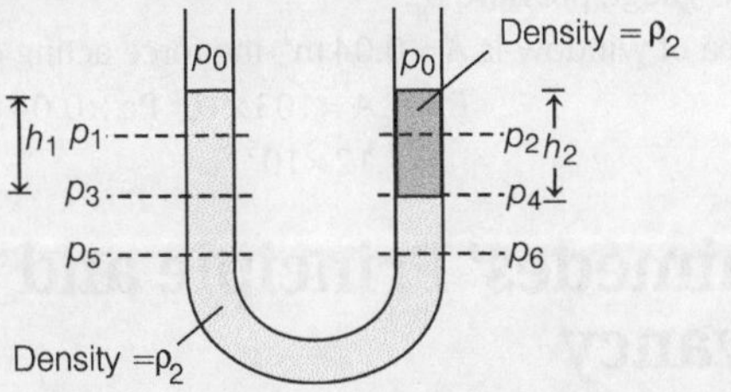

$$p_1 \neq p_2,\ p_3 = p_4 \text{ and } p_5 = p_6$$

Further, $p_3 = p_4$

$$p_0 + \rho_1 g h_1 = p_0 + \rho_2 g h_2$$

or $\rho_1 h_1 = \rho_2 h_2$ or $h \propto \frac{1}{\rho}$

i.e. for a given pressure at a point, the height of liquid column above it is inversly proportional to its density.

- *Barometer* is used to measure atmospheric pressure, while *manometer* measures pressure difference, *i.e.* gauge pressure.

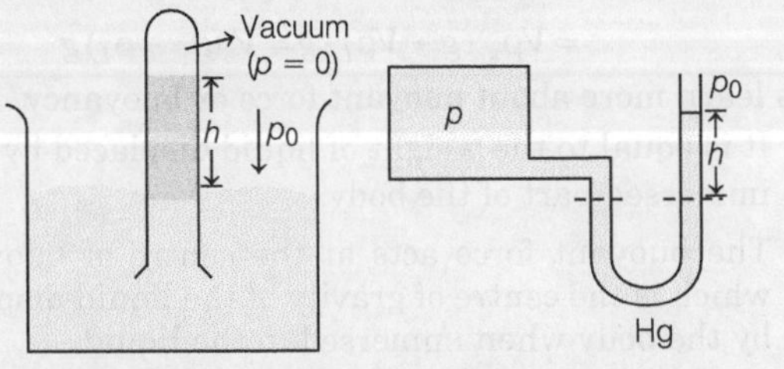

(a) Barometer $p_0 = h\rho g$ (b) Manometer $p - p_0 = h\rho g$

- Pressure at two points within a liquid at vertical separation of h, when the liquid container is accelerating up are related by expression

$$p_2 - p_1 = \rho(g + a)h$$

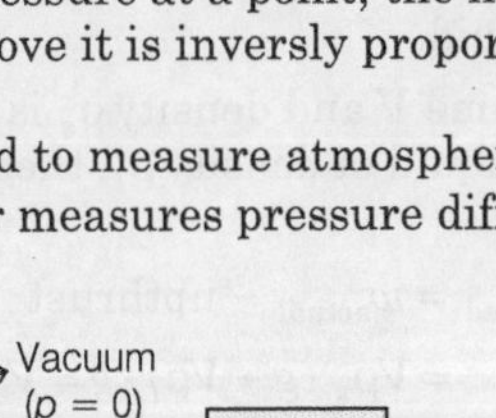

If container is accelerating down, then $p_2 - p_1 = \rho(g - a)h$ and if container is accelerating down with acceleration > g, then the liquid occupies upper part of the vessel.

- Variation of pressure within an accelerating closed container.

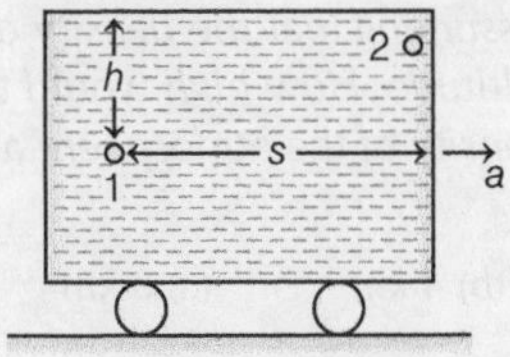

Here, all the points lying on a particular line making an angle of $\tan^{-1}(a/g)$ with the horizontal have the same pressure. In the given scenario, point 2 is the least pressure point, if the vessel is completely closed, we can take its pressure to be zero.

$$p_1 = p_2 + \rho g h + \rho s a$$

as $p_2 = 0$

So, $p_1 = \rho g h + \rho s a$

Pascal's Law and its Applications

Pascal's law states that the increase in pressure at one point of the enclosed liquid in equilibrium at rest is transmitted equally to all other points of liquid provided the gravity effect is neglected.

Some applications of Pascal's law are given as follows

(i) Pascal's law is used in the working of the hydraulic lift which is used to support or lift heavy objects. In hydraulic lift,

$$F_2 = \frac{A_2}{A_1} F_1$$

where, A_1, A_2 = area of cross-section of smaller and larger piston of hydraulic lift, F_1 = force applied on smaller piston.

(ii) Hydraulic lift is a force multiplying device, which is used in dentist's chair, car lifts and jacks, many elevators and hydraulic brakes.

Example 3. *The average depth of Indian Ocean is about 3000 m. Bulk modulus of water is 2.2×10^9 Nm^{-2} and $g = 10$ ms^{-2}, then fractional compression $\left(\frac{\Delta V}{V}\right)$ of water at the bottom of the Indian Ocean is*

(a) 1.36% (b) 20.6%
(c) 13.9% (d) 0.52%

Sol. (a) The pressure exerted by a 3000 m column of water on the bottom layer

$$p = h\rho g = 3000 \times 1000 \times 10$$

($\because$ Density of water, $\rho = 1000$ kg/m^3)

$$= 3 \times 10^7 \text{ kg m}^{-1}\text{s}^{-2}$$
$$= 3 \times 10^7 \text{ Nm}^{-2}$$

Fractional compression, $\left(\frac{\Delta V}{V}\right)=\frac{\text{Stress}}{B}=\frac{3\times10^7}{2.2\times10^9}$

$$=1.36\times10^{-2}$$

$$\frac{\Delta V}{V}=1.36\%$$

Example 4. *Assuming that the density of atmosphere does not change with altitude. How high would the atmosphere extend? (Given, density of the atmosphere at sea level is 1.29 kg/m³).*

(a) 2 km (b) 4 km (c) 8 km (d) 16 km

Sol. (c) Given, $\rho=1.29$ kg/m³, $g=9.8$ ms^{-2}, $p=1.01\times10^5$ Nm^{-2}

$\because$ Pressure $=\rho gh=$ density $\times$ gravity $\times$ height

$\therefore$ $1.01\times10^5=1.29\times9.8\times h$

$\Rightarrow$ $h=\frac{1.01\times10^5}{1.29\times9.8}\approx7989$ m

$\Rightarrow$ $h\approx8$ km

Note The acceleration due to gravity decreases with height and hence the density of air also decreases. The atmospheric cover extends with decreasing pressure over 100 km. We should also note that the sea level atmospheric pressure is not always 760 mm of Hg. A drop in the Hg level by 10 mm or more is a sign of an approaching storm.

Example 5. *A submarine experience a pressure of 5.05×10^6 Pa at a depth of d_1 in a sea. When it goes further to a depth of d_2, it experiences a pressure of 8.08×10^6 Pa, then d_2-d_1 is approximately (density of water $=10^3$ kg/m³ and acceleration due to gravity $=10$ ms^{-2})* **[JEE Main 2019]**

(a) 500 m (b) 400 m (c) 600 m (d) 300 m

Sol. (d) Pressure inside a fluid volume open to atmosphere is

$$p=p_0+h\rho g$$

where, $p=$ pressure at depth h, $h=$ depth,

$\rho=$ density of fluid

and $g=$ acceleration due to gravity.

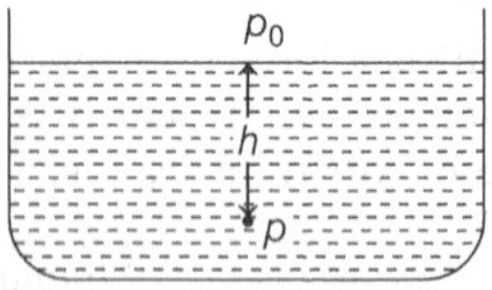

In problem given,
when $h=d_1$, pressure $p_1=5.05\times10^6$ Pa and when $h=d_2$, pressure $p_2=8.08\times10^6$ Pa

So, we have

$$p_1=p_0+d_1\rho g=5.05\times10^6$$

and $$p_2=p_0+d_2\rho g=8.08\times10^6$$

$\Rightarrow$ $$p_2-p_1=(d_2-d_1)\rho g=3.03\times10^6$$

$\Rightarrow$ $$d_2-d_1=\frac{3.03\times10^6}{\rho g}$$

Given, $\rho=10^3$ kgm^{-3}

and $g=10$ ms^{-2}

$\therefore$ $$d_2-d_1=\frac{3.03\times10^6}{10^3\times10}=303\text{ m}\approx300\text{ m}$$

Example 6. *At a depth of 1000 m in an ocean, the force acting on the window of area 20 cm × 20 cm of a submarine at this depth, the interior of which is maintained at sea-level atmospheric pressure is [Given, density of sea water $=1.03\times10^3$ kgm^{-3}, $g=10$ ms^{-2}]*

(a) 3.2×10^8 N (b) 4.12×10^5 N
(c) 8.3×10^2 N (d) 3.1×10^{-5} N

Sol. (b) Given, $h=1000$ m, $\rho=1.03\times10^3$ kgm^{-3}, gauge pressure, $p_g=\rho gh$

$$p_g=1.03\times10^3\times10\times1000$$

$$p_g=103\times10^5\text{ Pa}$$

The pressure outside the submarine is $p=p_a+\rho gh$ and pressure inside the submarine is p_a. Hence, the net pressure acting on the window is gauge pressure p_g.

Since, area of window is $A=0.04$ m², the force acting on it is

$$F=p_gA=103\times10^5\text{ Pa}\times0.04\text{ m}^2$$
$$=4.12\times10^5\text{ N}$$

Archimedes' Principle and Buoyancy

Whenever a body is immersed in a fluid, the fluid exerts an upward force on the body and this upward force is called *buoyant force*. In fact, any body wholly or partially immersed in a fluid is brought up by a force equal to the weight of the displaced fluid. This result is known as *Archimedes' principle*. Thus, buoyant force $=V_i\rho_l g$, where V_i is the volume of immersed part of body and ρ_l is the density of fluid.

If a body of volume V and density ρ_s is completely immersed in a liquid of density ρ_l, then its observed weight,

$$w_{\text{observed}}=w_{\text{actual}}-\text{upthrust} \quad \text{(buoyant force)}$$
$$=V\rho_s\cdot g-V\rho_l\cdot g=V(\rho_s-\rho_l)g$$

Let us learn more about buoyant force or buoyancy

(i) It is equal to the weight of liquid displaced by the immersed part of the body.

(ii) The buoyant force acts at the centre of buoyancy which is the centre of gravity of the liquid displaced by the body when immersed in the liquid.

(iii) The line joining the centre of gravity and centre of buoyancy is called *central line*.

Laws of Floatation

When a body of density ρ_B and volume V is immersed in a liquid of density σ, the forces acting on the body are

(i) The *weight of body* $w=mg=V\rho_B g$ acting vertically downwards through the centre of gravity of the body.

(ii) The *upthrust* $F=V\sigma g$ acting vertically upwards through the centre of gravity of the displaced liquid, *i.e.* centre of buoyancy.

So, the following three cases are possible

***Case* I** The density of body is greater than that of liquid, (*i.e.* $\rho_B > \sigma$). In this case, as weight will be more than upthrust, the body will sink as shown in Fig. (a).

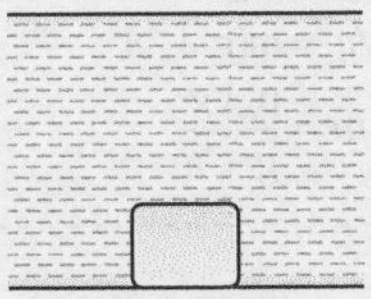

(a) $\rho_B > \sigma$

***Case* II** The density of body is equal to the density of liquid, (*i.e.* $\rho_B = \sigma$). In this case, $w = F$. So, the body will float fully submerged in neutral equilibrium anywhere in the liquid as shown in Fig. (b).

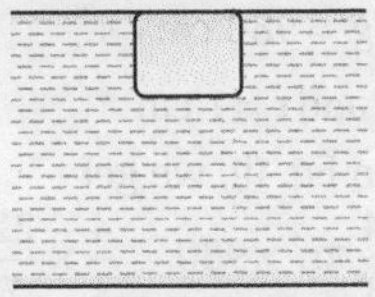

(b) $\rho_B = \sigma$

***Case* III** The density of body is lesser than that of liquid, (*i.e.* $\rho_B < \sigma$). In this case, $w < F$. So, the body will move upwards and in equilibrium will float partially immersed in the liquid such that

$$w = V_{in}\sigma g$$

[V_{in} is the volume of body in the liquid]

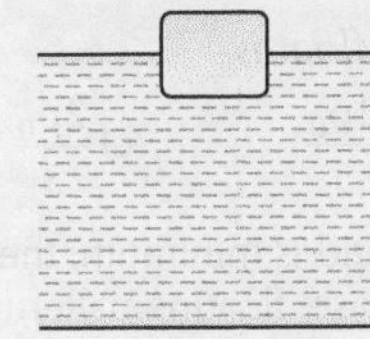

(c) $\rho_B < \sigma$

or $\quad V\rho_B g = V_{in}\sigma g \quad$ [as, $w = mg = V\rho_B g$]

or $\quad V\rho_B = V_{in}\sigma \quad$ ...(i)

Some Particular Cases

- If object is immersed in water, then

$$\text{Relative density, RD} = \frac{\text{Weight of body in air}}{\text{Loss in weight in water}}$$

$$\text{RD} = \frac{\text{Weight of body in air}}{\text{Weight in air} - \text{Weight in water}}$$

So, by weighing a body in air and in water, we can determine the relative density of the body.

- The upthrust on a body immersed in a liquid of density ρ_l in a lift moving downwards with acceleration **a** is

$$F = V\rho_l|\mathbf{g} - \mathbf{a}|$$

- The upthrust on a body immersed in a liquid of density ρ_L in a lift moving upwards with acceleration **a** is

$$F = V\rho_l|\mathbf{g} + \mathbf{a}|$$

- If lift is falling freely, then

$$F = 0$$

Example 7. *A boat having a length of 3 m and breadth 2 m is floating on a lake. The boat sinks by 1 cm, when a man gets on it. The mass of the man is*

(a) 60 kg (b) 62 kg
(c) 72 kg (d) 128 kg

Sol. (a) Mass of the man = Mass of water displaced

$$= \text{Volume} \times \text{Density}$$

$$= 3 \times 2 \times \frac{1}{100} \times 10^3 \text{ kg}$$

$$= 60 \text{ kg}$$

Example 8. *An ice-berg is floating partially immersed in sea water of density 1.03 gcm^{-3}. The density of ice is 0.92 gcm^{-3}. The fraction of the total volume of the ice-berg above the level of sea water is*

(a) 8.1 % (b) 11 %
(c) 34 % (d) 0.8 %

Sol. (b) Let V be the volume of the ice-berg outside the sea water and V be the total volume of ice-berg. According to question,

Weight of iceberg = Weight of liquid displaced

$$0.92 \times V \times g = 1.03\,(V - V')\,g$$

$$0.92\,V = 1.03\,(V - V')$$

or $$\frac{V'}{V} = \frac{1.03 - 0.92}{1.03} = \frac{11}{103}$$

$$\frac{V'}{V} \times 100 = \frac{11}{103} \times 100$$

$$= 10.67\% \approx 11\%$$

Example 9. *A cubical block of side 0.5 m floats on water with 30% of its volume under water. What is the maximum weight that can be put on the block without fully submerging it under water? [Take, density of water = 10^3 kg / m^3]*

[JEE Main 2019]

(a) 30.1 kg (b) 46.3 kg
(c) 87.5 kg (d) 65.4 kg

Sol. (c) When only block is floating on water, 30% of its volume is in water as shown below

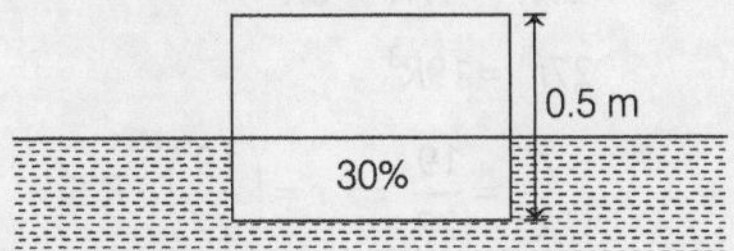

By Archimedes' principle,

weight of block = weight of displaced water

$\Rightarrow \quad V\rho_b g = (30\%V)\,\rho_w g$

where, $\quad \rho_b$ = density of block,

ρ_w = density of water

and $\quad V$ = volume of block.

So, $\quad \rho_b = 0.3\,\rho_w$

Now, let a mass m is placed over block to just keep the cube fully immersed in water.

Now, by Archimedes' principle,

weight of water displaced = weight of block + weight of mass m

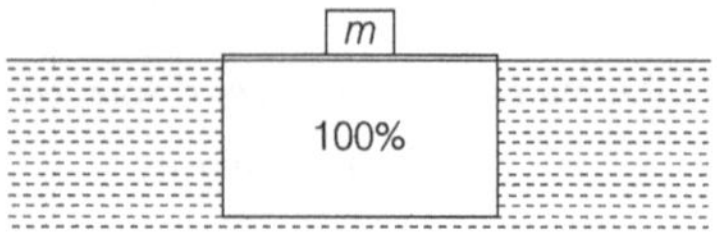

$$\Rightarrow \quad V\rho_w g = V\rho_b g + mg$$

$$\Rightarrow \quad V(\rho_w - \rho_b) = m \Rightarrow V \times 0.7\rho_w = m \quad [\because \rho_b = 0.3\rho_w]$$

Here, $\quad V = (\text{side})^3 = (0.5)^3 \text{m}^3$

$$\rho_w = 10^3 \text{ kg m}^{-3}$$

Substituting these values in the above relation, we get

$$m = (0.5)^3 \times 0.7 \times 10^3 = 87.5 \text{ kg}$$

Example 10. *A hollow spherical shell at outer radius R floats just submerged under the water surface. The inner radius of the shell is r. If the specific gravity of the shell material is $\frac{27}{8}$ with respect to water, then the value of r is* **[JEE Main 2020]**

(a) $\frac{8}{9}R$ (b) $\frac{1}{3}R$ (c) $\frac{2}{3}R$ (d) $\frac{4}{9}R$

Sol. (a) Relative density or specific gravity,

$$\rho_r = \frac{\text{Density of object}}{\text{Density of water}} = \frac{\rho_o}{\rho_w} = \frac{27}{8}$$

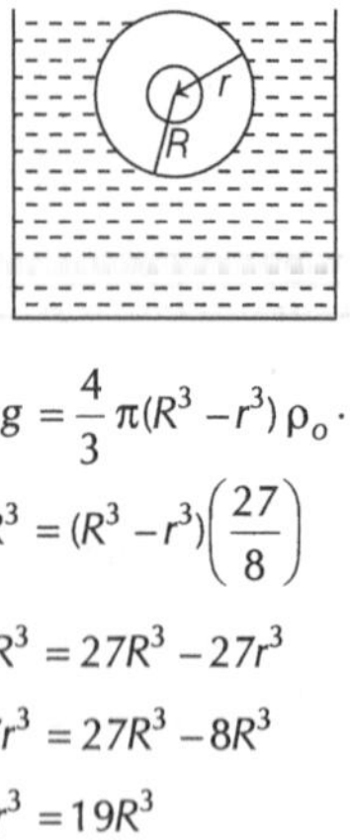

In equilibrium, $\frac{4}{3}\pi R^3 \rho_w g = \frac{4}{3}\pi(R^3 - r^3)\rho_o \cdot g$

$$\Rightarrow \quad R^3 = (R^3 - r^3)\left(\frac{27}{8}\right)$$

$$\Rightarrow \quad 8R^3 = 27R^3 - 27r^3$$

$$\Rightarrow \quad 27r^3 = 27R^3 - 8R^3$$

$$\Rightarrow \quad 27r^3 = 19R^3$$

$$\Rightarrow \quad \frac{r^3}{R^3} = \frac{19}{27} \Rightarrow r = \left(\frac{19}{27}\right)^{1/3} R$$

$$= \frac{2.66}{3}R = 0.88R \Rightarrow r \approx \frac{8}{9}R$$

Example 11. *An air bubble of radius 1 cm in water has an upward acceleration 9.8 cm s^{-2}. The density of water is 1 g/cm^3 and water offers negligible drag force on the bubble. The mass of the bubble is (Take, g = 980 cm/s^2)* **[JEE Main 2020]**

(a) 1.52 g (b) 4.51 g
(c) 4.15 g (d) 3.15 g

Sol. (c) From free body diagram,

$$B - mg = ma \quad \text{(here, } B = \text{buoyant force)}$$

$$\rho_w Vg - mg = ma \quad \text{(here, } V = \text{volume)}$$

$$\rho_w Vg = mg + ma$$

$$\rho_w Vg = m(g + a)$$

$$\Rightarrow \quad m = \frac{\rho_w Vg}{g + a} \quad \ldots\text{(i)}$$

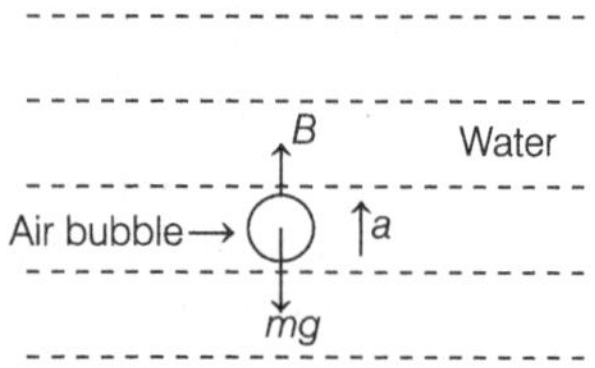

Given, $\rho_w = 1 \text{ g/cm}^3$, $g = 980 \text{ cm/s}^2$,

$a = 9.8 \text{ cm/s}^2$ and $r = 1$ cm

Substituting all values in Eq. (i), we get

$$m = \frac{(1) \times \frac{4}{3}\pi r^3 \times 980}{980 + 9.8} = \frac{\frac{4}{3} \times \frac{22}{7} \times (1)^3 \times 980}{989.8}$$

$$= \frac{4 \times 22 \times 980}{3 \times 7 \times 989.8} = 4.148 \text{ g} \simeq 4.15 \text{ g}$$

Example 12. *Consider a solid sphere of radius R and mass density $\rho(r) = \rho_0\left(1 - \frac{r^2}{R^2}\right)$, $0 < r \le R$. The minimum density of a liquid in which it will float is* **[JEE Main 2020]**

(a) $\frac{\rho_0}{3}$ (b) $\frac{2\rho_0}{5}$ (c) $\frac{2\rho_0}{3}$ (d) $\frac{\rho_0}{5}$

Sol. (b) If density of liquid is minimum, then its buoyant force is also minimum. In this case, floating pattern of sphere is as shown below

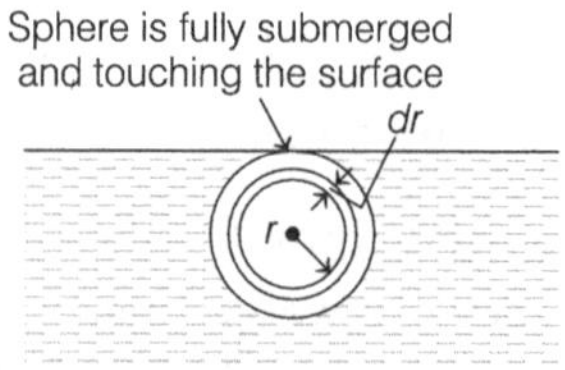

So, weight of sphere = upthrust or weight of fluid displaced

$$\Rightarrow \quad M_{\text{sphere}} \times g = V_{\text{sphere}} \times \rho_l \times g$$

$$\Rightarrow \quad \int_0^R \rho(r) \cdot 4\pi r^2 \cdot dr = \frac{4}{3}\pi R^3 \cdot \rho_l$$

$$\Rightarrow \quad \int_0^R \rho_0\left(1 - \frac{r^2}{R^2}\right) \cdot 4\pi r^2 \cdot dr = \frac{4}{3}\pi R^3 \rho_l$$

$$\Rightarrow \quad \rho_0 \times 4\pi \int_0^R \left(r^2 - \frac{r^4}{R^2}\right) dr = \frac{4}{3}\pi R^3 \cdot \rho_l$$

$$\Rightarrow \quad \rho_0 \times 4\pi \left(\frac{R^3}{3} - \frac{R^3}{5}\right) = \frac{4}{3}\pi R^3 \cdot \rho_l$$

$$\Rightarrow \quad \rho_l = \frac{2}{5}\rho_0$$

Example 13. *A leak proof cylinder of length 1 m, made of a metal which has very low coefficient of expansion is floating vertically in water at 0°C such that its height above the water surface is 20 cm. When the temperature of water is increased to 4°C, the height of the cylinder above the water surface becomes 21 cm. The density of water at T = 4 °C, relative to the density at T = 0 °C is close to* **[JEE Main 2020]**

(a) 1.26 (b) 1.03 (c) 1.01 (d) 1.04

Sol. (c) For a floating body, weight of body

= upthrust or buoyant force = weight of water displaced

Initially at 0°C, cylinder floats as shown in the figure

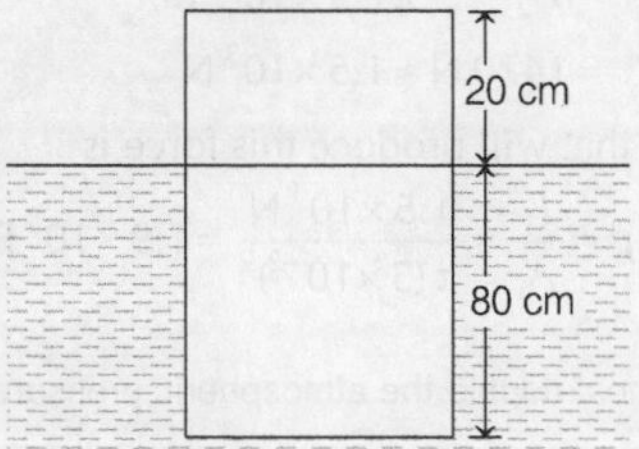

Here, weight of cylinder = weight of water displaced by cylinder of 80 cm length

$\Rightarrow \quad mg = \rho_0 \cdot g \cdot 80\, A$ ...(i)

where, ρ_0 = density of water at 0°C

and A = area of cross-section of cylinder.

When temperature is raised to 4°C, cylinder floats as shown in the figure.

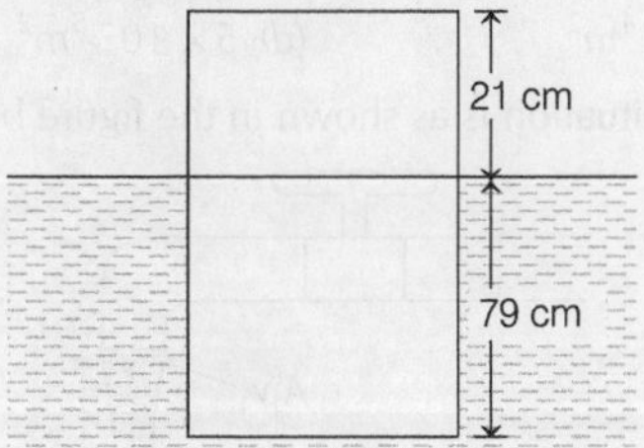

In above case, weight of cylinder = weight of water displaced by cylinder of 79 cm length

$\Rightarrow \quad mg = \rho_4 \cdot g \cdot 79\, A$ ...(ii)

From Eqs. (i) and (ii), we have

$$80\, \rho_0 = \rho_4 \cdot 79$$

$$\Rightarrow \quad \frac{\rho_4}{\rho_0} = \frac{80}{79} = 1.01$$

Flow of Liquids

The different types of fluid flow are

Streamline Flow The flow in which each particles of the liquid passing through a point travels along the same path and with the same velocity as the preceeding particles passing through the same point is known as *streamline flow* of a liquid.

Hence, it is a regular flow. The path followed by each element is called *streamline.*

- The tangent drawn at any point of streamline gives the direction of the flow of liquid at that point.
- The streamlines cannot intersect each other.

Turbulent Flow The flow of liquid with a velocity greater than its critical velocity is disordered and called *turbulent flow.* In case of turbulent flow, maximum part of external energy is spent for producing eddies in the liquid and small part of external energy is available for forward flow.

Laminar Flow If a liquid is flowing over a horizontal surface with a steady flow and moves in the form of layers of different velocities which do not mix with each other, then the flow of liquid is called *laminar flow.* Thus a flow, in which the liquid moves in layers or lamina is called a laminar flow.

Critical Velocity and Reynold's Number

The largest velocity which allows a steady flow is called **critical velocity.**

Value of critical velocity for flow of liquid of density ρ and coefficient of viscosity η flowing through a horizontal tube of diameter D is given by

$$v_c \propto \frac{\eta}{\rho D}$$

Reynold's number is a pure number which determines the nature of flow of liquid through a horizontal tube.

Reynold's number (R_e) is a unitless and dimensionless number given by

$$R_e = \frac{\rho v D}{\eta}$$

If the value of Reynold's number

(i) lies between 0 to 2000, the flow of liquid is streamline or laminar.

(ii) lies between 2000 to 3000, the flow of liquid is unstable and changing from streamline to turbulent flow.

(iii) lies above 3000, the flow of liquid is definitely turbulent.

Example 14. *Water from a pipe is coming at a rate of 100 L/min. If the radius of the pipe is 5 cm, the Reynold's number for the flow is of the order of (density of water = 1000 kg /m^3, coefficient of viscosity of water = 1 mPa-s)* **[JEE Main 2019]**

(a) 10^3 (b) 10^4

(c) 10^2 (d) 10^6

Sol. (b) Reynolds' number for flow of a liquid is given by

$$R_e = \frac{\rho v D}{\eta}$$

where, velocity of flow, $v = \frac{\text{volume flow rate}}{\text{area of flow}} = \frac{V/t}{A}$

So, $R_e = \frac{\rho V D}{\eta A t} = \frac{\rho V 2r}{\eta \times \pi r^2 \times t} = \frac{2\rho V}{\eta \pi r t}$

Here, ρ = density of water $= 1000 \text{ kgm}^{-3}$

$$\frac{V}{t} = \frac{100 \times 10^{-3}}{60} \text{m}^3\text{s}^{-1}$$

where, η = viscosity of water $= 1 \times 10^{-3}$ Pa-s

and r = radius of pipe $= 5 \times 10^{-2}$ m

$$R_e = \frac{2 \times 1000 \times 100 \times 10^{-3}}{1 \times 10^{-3} \times 60 \times 3.14 \times 5 \times 10^{-2}}$$

$$= 212.3 \times 10^2 \approx 2.0 \times 10^4$$

So, order of Reynolds' number is of 10^4.

Equation of Continuity

Let us consider streamline flow of an ideal, non-viscous fluid through a tube of varying cross-section. Let at two sections, the

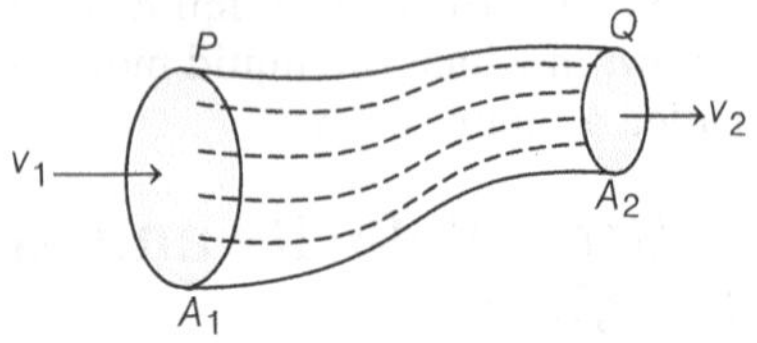

cross-section areas be A_1 and A_2 respectively and fluid flow velocities be v_1 and v_2, then according to equation of continuity

$$A_1 v_1 \rho_1 = A_2 v_2 \rho_2$$

where, ρ_1 and ρ_2 are the respective densities of fluid. Equation of continuity is based on the conservation of mass.

If fluid flowing is incompressible, then

$$\rho_1 = \rho_2$$

and equation of continuity is simplified as

$$A_1 v_1 = A_2 v_2$$

Example 15. *Two syringes of different cross-sections (without needles) filled with water are connected with a tightly fitted rubber tube filled with water. Diameters of the smaller piston and larger piston are 1 cm and 3 cm respectively. If the smaller piston is pushed in through 6 cm, how much does the larger piston move out?*

(a) 0.37 cm (b) 0.67 cm

(c) 37 cm (d) 67 cm

Sol. (b) Water is considered to be perfectly incompressible. Volume covered by the movement of smaller piston inwards is equal to the volume moved outwards due to longer piston.

$$L_1 A_1 = L_2 A_2$$

$$\Rightarrow \quad L_2 = \frac{A_1}{A_2} L_1$$

$$= \frac{\pi \left(\frac{1}{2} \times 10^{-2}\right)^2}{\pi \left(\frac{3}{2} \times 10^{-2}\right)^2} \times 6 \times 10^{-2}$$

$$= 0.67 \times 10^{-2} \text{ m} \approx 0.67 \text{ cm}$$

Note Atmospheric pressure is common to both pistons and has been ignored.

Example 16. *In a car lift compressed air exerts a force F_1 on a small piston having a radius of 5 cm. This pressure is transmitted to the second piston of a radius 15 cm. If the mass of the car to be lifted is 1350 kg, then the pressure necessary to accomplish this task is*

(a) 1.9×10^5 Pa (b) 3×10^6 Pa

(c) 6.5×10^3 Pa (d) 0.23×10^3 Pa

Sol. (a) Since, pressure is transmitted undiminished throughout the fluid

$$F_1 = \frac{A_1}{A_2} F_2 = \frac{\pi (5 \times 10^{-2} \text{m})^2}{\pi (15 \times 10^{-2} \text{m})^2} (1350 \text{ N} \times 9.8 \text{ ms}^{-2})$$

$$= 1470 \text{ N} \approx 1.5 \times 10^3 \text{ N}$$

The air pressure that will produce this force is

$$p = \frac{F_1}{A_1} = \frac{1.5 \times 10^3 \text{ N}}{\pi (5 \times 10^{-2})^2} = 1.9 \times 10^5 \text{ Pa}$$

Note This is almost double the atmospheric pressure.

Example 17. *Water from a tap emerges vertically downwards with an initial speed of 1.0 ms^{-1}. The cross-sectional area of the tap is 10^{-4}m^2. Assume that the pressure is constant throughout the stream of water and that the flow is streamlined. The cross-sectional area of the stream, 0.15 m below the tap would be (Take, $g = 10 \text{ ms}^{-2}$)*

[JEE Main 2019]

(a) $2 \times 10^{-5} \text{m}^2$ (b) $1 \times 10^{-5} \text{m}^2$

(c) $5 \times 10^{-4} \text{m}^2$ (d) $5 \times 10^{-5} \text{m}^2$

Sol. (d) Given situation is as shown in the figure below

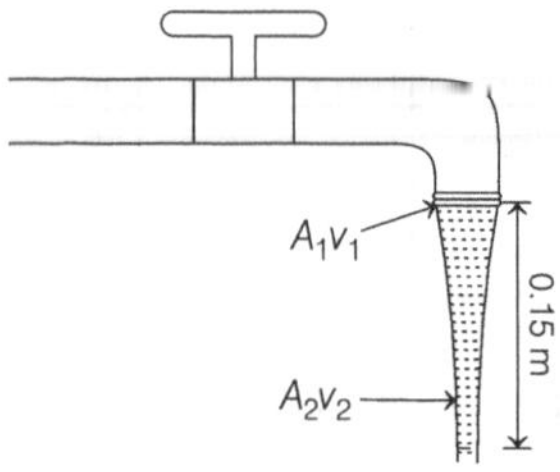

From equation of continuity, $A \propto \frac{1}{v}$

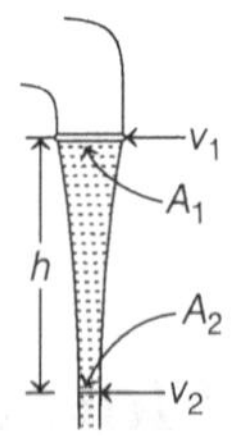

where, A = area of flow

and v = velocity of flow.

$\therefore$ Increase in speed of flow causes a decrease in area of flow.

Here given that height of fall, $h = 0.15$ m

Area, $A = 10^{-4} \text{m}^2$

Initial speed, $v = 1 \text{ ms}^{-1}$

Velocity of water stream below h height is

$$v_2 = \sqrt{v_1^2 + 2gh} \quad [\because v^2 - u^2 = 2gh]$$

Substituting the given values, we get

$$= \sqrt{1^2 + 2 \times 10 \times 0.15}$$

$$= \sqrt{4} = 2 \text{ ms}^{-1}$$

Now, from the equation of continuity, we have

$$A_1v_1 = A_2v_2 \text{ or } A_2 = \frac{A_1v_1}{v_2}$$

$$\therefore \quad A_2 = \frac{10^{-4}\times 1}{2} = 0.5\times 10^{-4} = 5\times 10^{-5}\ m^2$$

Energy of a Flowing Liquid

There are three types of energies in a flowing liquid

Pressure Energy If p is the pressure on the area A of a fluid and the liquid moves through a distance l due to this pressure, then

pressure energy of liquid = work done

$$= \text{force} \times \text{displacement} = pAl$$

The volume of the liquid is Al.

Hence, pressure energy per unit volume of liquid

$$= \frac{pAl}{Al} = p$$

Kinetic Energy If a liquid of mass m and volume V is flowing with velocity v, then the kinetic energy is $\frac{1}{2}mv^2$.

$\therefore$ Kinetic energy per unit volume of liquid

$$= \frac{1}{2}\left(\frac{m}{V}\right)v^2 = \frac{1}{2}\rho v^2$$

Here, ρ is the density of liquid.

Potential Energy If a liquid of mass m is at a height h from the reference line ($h = 0$), then its potential energy is mgh.

$\therefore$ Potential energy per unit volume of the liquid

$$= \left(\frac{m}{V}\right)gh = \rho gh$$

Bernoulli's Principle

If an ideal liquid is flowing in streamlined flow, then total energy, *i.e.* sum of pressure energy, kinetic energy and potential energy per unit volume of the liquid remains constant at every cross-section of the tube.

Mathematically, $p + \frac{1}{2}\rho v^2 + \rho gh = \text{constant}$

It can be expressed as, $\frac{p}{\rho g} + \frac{v^2}{2g} + h = \text{constant}$

where, $\frac{p}{\rho g}$ = pressure head, $\frac{v^2}{2g}$ = velocity head

and h = gravitational head or potential head.

For horizontal flow of liquid, $p + \frac{1}{2}\rho v^2 = \text{constant}$

where, p is called *static pressure* and $\frac{1}{2}\rho v^2$ is called *dynamic pressure*.

Therefore in horizontal flow of liquid, if p increases, v decreases and *vice-versa*.

This theorem is applicable to ideal liquid, *i.e.* a liquid which is non-viscous, incompressible and irrotational.

Applications of Bernoulli's Theorem

- The action of carburetor, paintgun, scent sprayer, atomiser and insect sprayer is based on Bernoulli's theorem.
- The action of Bunsen's burner, gas burner, oil stove and exhaust pump is also based on Bernoulli's theorem.
- Motion of a spinning ball (Magnus effect) is based on Bernoulli's theorem.
- Blowing of roofs by wind storms, attraction between two closely parallel moving boats, fluttering of a flag etc are also based on Bernoulli's theorem.
- Bernoulli's theorem helps in explaining blood flow in artery.
- Working of an aeroplane is based on Bernoulli's theorem.

Example 18. *A fluid is flowing through a horizontal pipe of varying cross-section with speed v ms^{-1} at a point where the pressure is p Pa. At another point, where pressure is $\frac{p}{2}$ Pa its speed is V ms^{-1}. If the density of the fluid is ρ kg m^{-3} and the flow is streamline, then V is equal to* **[JEE Main 2020]**

(a) $\sqrt{\frac{p}{\rho} + v}$ (b) $\sqrt{\frac{2p}{\rho} + v^2}$

(c) $\sqrt{\frac{p}{2\rho} + v^2}$ (d) $\sqrt{\frac{p}{\rho} + v^2}$

Sol. (*d*) Since, pipe is horizontal. Applying Bernoulli's equation at two different cross-section,

$$p_1 + \rho gh_1 + \frac{1}{2}\rho v_1^2 = p_2 + \rho gh_2 + \frac{1}{2}\rho v_2^2$$

For same level,

$$p + \frac{1}{2}\rho v_1^2 = \frac{p}{2} + \frac{1}{2}\rho v_2^2$$

Given, $v_1 = v$ m/s and $v_2 = V$ m/s

$$\frac{p}{2} = \frac{1}{2}\rho(V^2 - v^2) \Rightarrow V^2 - v^2 = \frac{p}{\rho}$$

$$\Rightarrow \quad V = \sqrt{\frac{p}{\rho} + v^2}$$

Example 19. *At what speed, the velocity head of water is equal to pressure head of 40 cm of Hg?*

(a) 10.32 ms^{-1} (b) 2.8 ms^{-1}

(c) 5.5 ms^{-1} (d) 8.4 ms^{-1}

Sol. (*a*) From Bernoulli's equation,

$$\frac{p}{\rho g} + \frac{v^2}{2g} + h = \text{constant}$$

Given that, velocity head = pressure head

$$\frac{v^2}{2g} = \frac{p}{\rho g}$$

$\Rightarrow$ $$v^2 = \frac{2p}{\rho}$$

Given, $p = 40$ cm of Hg $= 40 \times 10^{-2} \times 9.8 \times 13.6 \times 10^3$

$$v^2 = \frac{2 \times 13.6 \times 10^3 \times 40 \times 10^{-2} \times 9.8}{10^3}$$

$\Rightarrow$ $$v = 10.32 \text{ ms}^{-1}$$

Example 20. *Water is flowing with a speed of 2 ms^{-1} in a horizontal pipe with cross-sectional area decreasing from $2 \times 10^{-2} m^2$ to 0.01 m^2 at pressure 4×10^4 Pa. What will be the pressure at small cross-section?*

(a) 2×10^4 Pa (b) 3.4×10^4 Pa

(c) 2.4×10^4 Pa (d) 4×10^4 Pa

Sol. (b) Here, $v_1 = 2 \text{ ms}^{-1}$, $A_1 = 2 \times 10^{-2} \text{m}^2$,

$$p_1 = 4 \times 10^4 \text{ Pa}, A_2 = 0.01 \text{m}^2, p_2 = ?$$

As, $$A_1 v_1 = A_2 v_2$$

or $$v_2 = \frac{A_1 v_1}{A_2}$$

$$= \frac{2 \times 10^{-2} \times 2}{0.01} = 4 \text{ ms}^{-1}$$

Now, $$p_1 + \frac{1}{2}\rho v_1^2 = p_2 + \frac{1}{2}\rho v_2^2$$

or $$p_2 = p_1 + \frac{1}{2}\rho (v_1^2 - v_2^2)$$

or $$p_2 = 4 \times 10^4 + \frac{1}{2} \times 10^3 (2^2 - 4^2)$$

$$= 4 \times 10^4 - 6 \times 10^3$$

$$= 3.4 \times 10^4 \text{ Pa}$$

Example 21. *A long cylindrical vessel is half-filled with a liquid. When the vessel is rotated about its own vertical axis, the liquid rises up near the wall. If the radius of vessel is 5 cm and its rotational speed is 2 rps, then the difference in the heights between the centre and the sides (in cm) will be* **[JEE Main 2019]**

(a) 0.1 (b) 1.2 (c) 0.4 (d) 2.0

Sol. (d) When liquid filled vessel is rotated the liquid profile becomes a paraboloid due to centripetal force, as shown in the figure below

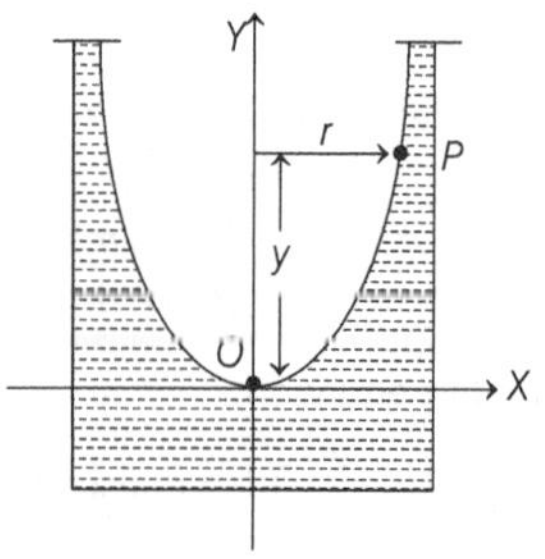

Pressure at any point P due to rotation is given as,

$$P_r = \frac{1}{2}\rho v^2 = \frac{1}{2}\rho \cdot (\omega t)^2 = \frac{1}{2}\rho \omega^2 r^2$$

Gauge pressure at depth y is $p_G = -\rho g y$

If p_0 is atmospheric pressure, then total pressure at point P is

$$p = p_0 + \frac{1}{2}\rho r^2 \omega^2 - \rho g y$$

For any point on surface of rotating fluid, $p = p_0$

Hence, for any surface point.

$$p_0 = p_0 + \frac{1}{2}\rho r^2 \omega^2 - \rho g y$$

or $$\frac{1}{2}\rho r^2 \omega^2 = \rho g y$$

or $$y = \frac{r^2 \omega^2}{2g} \quad \text{... (i)}$$

In the given case,

Angular speed, $\omega = 2$ rps $= 2 \times 2\pi = 4\pi$ rad s^{-1}

Radius of vessel, $r = 5$ cm $= 0.05$ m and $g = 10 \text{ ms}^{-2}$

Hence, substituting these values in Eq. (i), we get

$$y = \frac{\omega^2 r^2}{2g} = \frac{(4\pi)^2 (0.05)^2}{2 \times 10}$$

$$= 0.02 \text{ m} = 2 \text{ cm}$$

Torricelli's Theorem

Velocity of efflux (the velocity with which the liquid flows out of a orifice or narrow hole) is equal to the velocity acquired by a freely falling body through the same vertical distance equal to the depth of orifice below the free surface of liquid.

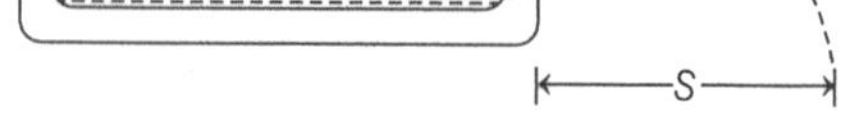

Velocity of efflux, $v = \sqrt{2gh}$

where, h = depth of orifice below the free surface of liquid.

Time taken by the liquid to reach the base-level

$$t = \sqrt{\frac{2(H-h)}{g}}$$

Horizontal range, $S = \sqrt{4h(H-h)}$

where, H – height of liquid column.

Horizontal range is maximum, equal to height of the liquid column H, when orifice is at half of the height of liquid column.

If the hole is at the bottom of the tank, then time required to make the tank empty is $t = \frac{A}{A_0}\sqrt{\frac{2H}{g}}$

where, A is area of the container and A_0 is area of orifice.

Volume of liquid coming out from the orifice per second

$$= VA_0 = A_0\sqrt{2gh}$$

Venturimeter

It is a device which is used to measure the rate of flow of fluids through pipes. Its working is based on Bernoulli's theorem. In the arrangement shown, the rate of flow of fluid V is given by

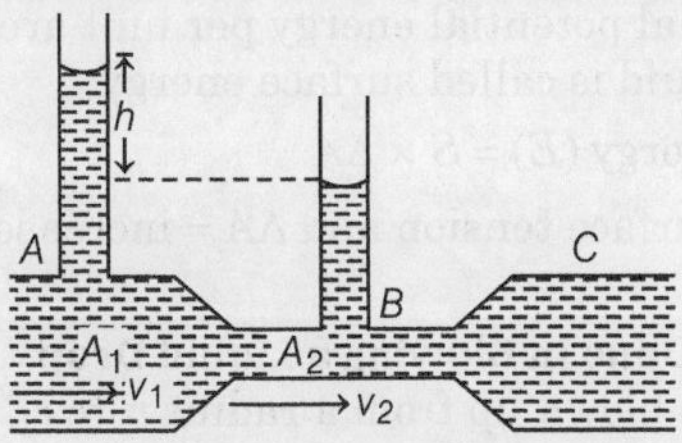

$$V = A_1A_2\sqrt{\frac{2gh}{(A_1^2 - A_2^2)}}$$

$$V = A_1A_2\sqrt{\frac{2(p_1 - p_2)}{\rho(A_1^2 - A_2^2)}}$$

Example 22. *The diameter of a pipe at two points, where a venturimeter is connected is 8 cm and 5 cm and the difference of levels in it is 4 cm. The volume of water flowing through the pipe per second is*

(a) 1889 ccs^{-1} (b) 1520 ccs^{-1}
(b) 1321 ccs^{-1} (d) 1125 ccs^{-1}

Sol. (a) Here, $r_1 = 8/2 = 4.0$ cm,

$r_2 = 5/2 = 2.5$ cm, $h = 4$ cm

Now, $A_1 = \pi r_1^2 = \pi(4)^2 = 16\,\pi\text{cm}^2$

and $A_2 = \pi r_2^2 = \pi(2.5)^2 = 6.25\,\pi\text{ cm}^2$

Here, $\rho = \rho_m$

So, the rate of flow of water in venturimeter is given by

$$V = A_1A_2\sqrt{\frac{2\,gh}{(A_1^2 - A_2^2)}}$$

$$= 6.25\,\pi \times 16\,\pi\sqrt{\frac{2\times980\times4}{(16\,\pi)^2-(6.25\,\pi)^2}}$$

$$= \frac{100\,\pi^2\times28\sqrt{10}}{\sqrt{(16\,\pi-6.25\,\pi)\,(16\,\pi+6.25\,\pi)}}$$

$$= 1889\text{ ccs}^{-1}$$

Example 23. *The top of a water tank is open to air and its water level is maintained. It is giving out 0.74 m^3 water per minute through a circular opening of 2 cm radius in its wall. The depth of the centre of the opening from the level of water in the tank is close to* **[JEE Main 2019]**

(a) 4.8 m (b) 6.0 m (c) 2.9 m (d) 9.6 m

Sol. (a) For the given condition, a water tank is open to air and its water level maintained.

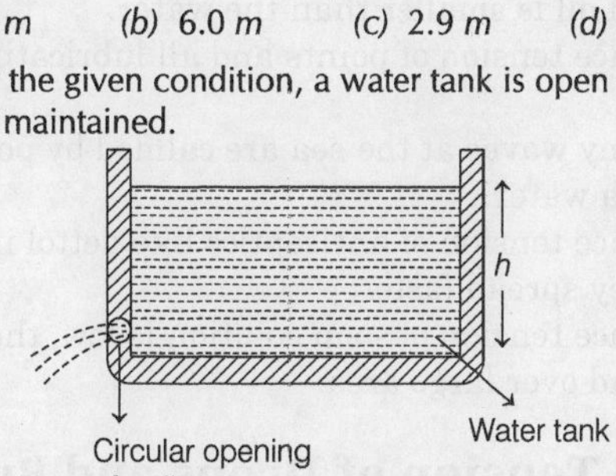

Suppose the depth of the centre of the opening from level of water in tank is h and the radius of opening is r.

According to question, the water per minute through a circular opening

Flow rate $(Q) = 0.74\text{ m}^3/\text{min} = \frac{0.74}{60}\text{ m}^3/\text{s}$

r = radius of circular opening = 2 cm

Here, the area of circular opening $= \pi(r^2)$

$$a = \pi\times(2\times10^{-2})^2 = 4\pi\times10^{-4}\text{ m}^2$$

Now, flow rate through an area is given by

$$Q = \text{Area of circular opening} \times \text{Velocity of water}$$

$$Q = a\times v = \pi(r^2)\times v$$

$$\Rightarrow \quad \frac{0.74}{60} = (4\pi\times10^{-4})\times v \quad \ldots\text{(i)}$$

According to Torricelli's law (velocity of efflux)

$$v = \sqrt{2gh}$$

Putting the value of v in Eq. (i), we get

$$\sqrt{2gh} = \frac{0.74\times10^4}{60\times4\pi}$$

$$\Rightarrow \quad h = \left(\frac{0.74\times10^4}{60\times4\pi}\right)^2\times\frac{1}{2g}$$

$$h \approx 4.8\text{ m}$$

Surface Tension

The property of a liquid at rest by virtue of which its free surface behaves like a stretched membrane under tension and tries to occupy as small area as possible is called **surface tension.**

$$\text{Surface tension, } S = \frac{\text{Force}}{\text{Length}} = \frac{F}{L}$$

The SI unit of surface tension is N/m or J/m². It is a scalar quantity and its dimensional formula is $[\text{MT}^{-2}]$.

Surface tension of a liquid depends only on the nature of liquid and is independent of the surface area of film or length of the line considered. Small liquid drops are spherical due to the property of surface tension.

Surface tension of a liquid decreases with an increase in temperature.

Applications of Surface Tension

- Oil spreads over the water surface because the surface tension of oil is smaller than the water.
- The surface tension of points and all lubricating oils is low.
- The stromy waves at the sea are calmed by pouring oil on the sea water.
- The surface tension of antiseptics like dettol is low and hence they spread faster.
- The surface tension of soap solution is low, therefore it can spread over large area.

Surface Tension of Drops and Bubbles

Due to the property of surface tension, a drop or bubble tends to contract and so compresses the matter enclosed. This in turn increases the internal pressure which prevents further contraction and equilibrium is achieved. So, in equilibrium, the pressure inside a bubble or drop is greater than outside and difference of pressure between two sides of the liquid surface is called **excess pressure**.

Excess pressure in different cases is given below

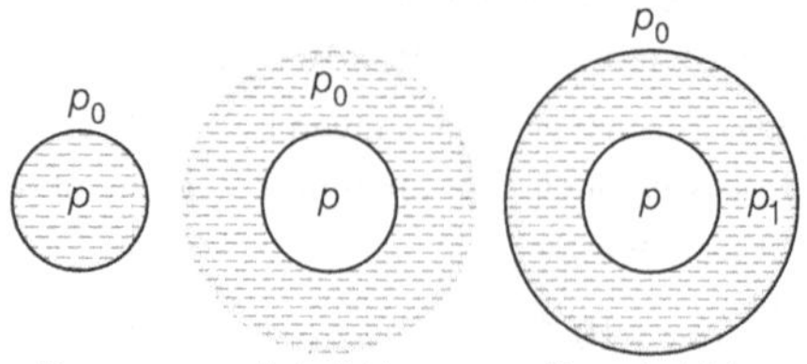

Drop Air bubble Soap bubble

Excess pressure inside a drop, $\Delta p = \dfrac{2S}{r}$

Excess pressure inside an air bubble, $\Delta p = \dfrac{2S}{r}$

Excess pressure inside a soap bubble, $\Delta p = \dfrac{4S}{r}$

- When two soap bubbles of same material having different radii r and $R(> r)$ are combined to form a double bubble, then

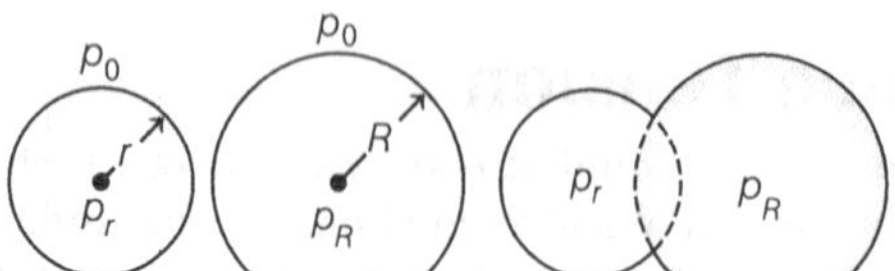

$$p_r - p_0 = \frac{4S}{r}$$

and
$$p_R - p_0 = \frac{4S}{R}$$

$\therefore$
$$p_r - p_R = 4S\left(\frac{1}{r} - \frac{1}{R}\right)$$

So, the radius of the common surface is given as

$$R_0 = \frac{rR}{R-r}.$$

- If two plates are placed in contact with a thin film of liquid in between them to pull them apart, a large force is needed.

Excess pressure in this case is $\dfrac{2S}{d}$, where d is the separation between the plates. Force required to separate two plates, each of area A is given by

$$F = \frac{2\,A \times S}{d}$$

Surface Energy

If we increase the free surface area of a liquid, then work has to be done against the force of surface tension. This work done is stored in liquid surface as potential energy.

This additional potential energy per unit area of free surface of liquid is called surface energy.

Surface energy $(E) = S \times \Delta A$

where, S = surface tension and ΔA = increase in surface area.

(i) **Work Done in Blowing a Liquid Drop** If a liquid drop is blown up from a radius r_1 to r_2, then work done for that is

$$W = S \cdot 4\pi (r_2^2 - r_1^2)$$

(ii) **Work Done in Blowing a Soap Bubble** As a soap bubble has two free surfaces, hence work done in blowing a soap bubble so as to increase its radius from r_1 to r_2 is given by

$$W = S \cdot 8\pi (r_2^2 - r_1^2)$$

(iii) **Work Done in Splitting a Bigger Drop into *n* Smaller Droplets**

If a liquid drop of radius R is split up into n smaller droplets, all of same size, then radius of each droplet

$$r = R \cdot (n)^{-1/3}$$

Work done, $W = 4\pi S(nr^2 - R^2) = 4\pi SR^2 (n^{1/3} - 1)$

(iv) **Coalescance of Drops** If n small liquid drops of radius r each combine together so as to form a single bigger drop of radius $R = n^{1/3} \cdot r$, then in the process energy is released. Release of energy is given by $\Delta U = S \cdot 4\pi (nr^2 - R^2) = 4\pi Sr^2 n(1 - n^{-1/3})$

Example 24. *A rectangular film of liquid is extended from 5 cm × 3 cm to 6 cm × 5 cm. If the work done is 3.0×10^{-4} J. The surface tension of liquid is*

(a) 0.05 Nm^{-1} (b) 0.1 Nm^{-1}
(c) 0.2 Nm^{-1} (d) 2 Nm^{-1}

Sol. (a) Increase in area,

$$\Delta A = 2(6 \times 5 - 5 \times 3) = 2 \times (30 - 15)\ \text{cm}^2$$
$$= 30 \times 10^{-4}\ m^2$$

As, work done, W = surface tension × increase in area

$$3.0 \times 10^{-4} = S \times 2 \times 30 \times 10^{-4} \text{ or } S = 0.05\ \text{Nm}^{-1}$$

Example 25. *Surface tension of a detergent solution is $2.8 \times 10^{-2}\ Nm^{-1}$. What is the work done in blowing a bubble of 2 cm diameter?*

(a) $4 \times 10^{-6}\ J$ (b) $70.3 \times 10^{-6}\ J$
(c) $50.8 \times 10^{-6}\ J$ (d) $60.8 \times 10^{-6}\ J$

Sol. (b) Given that, $S = 2.8 \times 10^{-2}\ Nm^{-1}$

and $R = \frac{2}{2} = 1\ cm = 0.01\ m$

As soap bubble has two free surfaces,

$\therefore$ Work done, $W = S \cdot 8\pi R^2 = 2.8 \times 10^{-2} \times 8 \times 3.14 \times (0.01)^2$

$= 70.3 \times 10^{-6}\ J$

Angle of Contact

The angle subtended between the tangents drawn at liquid surface and at solid surface inside the liquid at the point of contact is called angle of contact (θ).

If liquid molecules is in contact with solid (*i.e.* wall of capillary tube), then forces acting on liquid molecules are

(i) Force of cohesion F_c (acts at an angle 45° to the vertical)

(ii) Force of adhesion F_a (acts outwards at right angle to the wall of the tube)

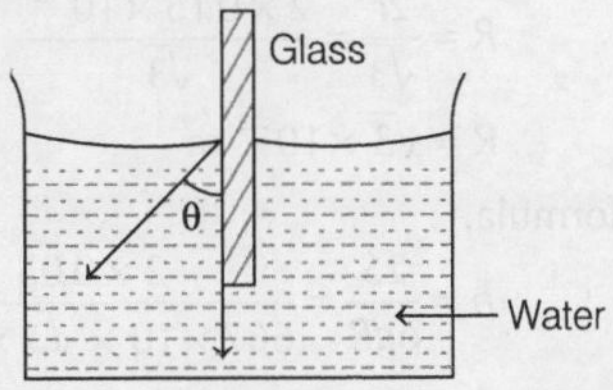

Angle of contact depends upon the nature of the liquid and solid in contact and the medium which exists above the free surface of the liquid.

When wax is coated on a glass capillary tube, it becomes water-proof. The angle of contact increases and becomes obtuse. Water does not rise in it. Rather it falls in the tube by virtue of obtuse angle of contact.

- If θ is acute angle, *i.e.* $\theta < 90°$, then liquid meniscus will be concave upwards.
- If θ is 90°, then liquid meniscus will be plane.
- If θ is obtuse, *i.e.* $\theta > 90°$, then liquid meniscus will be convex upwards.
- If angle of contact is acute angle, *i.e.* $\theta < 90°$, then liquid will wet the solid surface.
- If angle of contact is obtuse angle, *i.e.* $\theta > 90°$, then liquid will not wet the solid surface.

Angle of contact increases with increase in temperature of liquid. Angle of contact decreases on adding soluble impurity to a liquid.

Angle of contact for pure water and glass is zero.

For ordinary water and glass, it is 8°.

For mercury and glass, it is 138°.

For pure water and silver, it is 90°.

For alcohol and clean glass $\theta = 0°$.

Angle of Contact, Meniscus and Shape of liquid surface

Property	Angle of Contact < 90°	Angle of Contact = 90°	Angle of Contact > 90°
Substances	Water and glass	Water and silver	Mercury and glass
Angle of contact	Almost zero, acute angle	Right angle = 90°	Obtuse angle = 138°
Meniscus shape	Concave	Plane	Convex
Capillary action	Liquid rises	No effect	Liquid falls
Sticking to solid	Stick/wets	Does not wet	Does not wet
Relation between cohesive force (F_c) and adhesive force (F_a)	$F_a > \frac{F_c}{\sqrt{2}}$ $F_a > F_c$	$F_a = \frac{F_c}{\sqrt{2}}$	$F_a < \frac{F_c}{\sqrt{2}}$ $F_c > F_a$
Shape of liquid surface	Almost round	Spreads on surface	Flat on interface

Capillarity

The phenomenon of rise or fall of liquid column in a capillary tube is called capillarity.

Ascent of a liquid column in a capillary tube is given by

$$h = \frac{2S\cos\theta}{r\rho g} - \frac{r}{3}$$

If capillary is very narrow, then $h = \frac{2S\cos\theta}{r\rho g}$

where, r = radius of capillary tube, ρ = density of the liquid,

θ = angle of contact and S = surface tension of liquid.

- If $\theta < 90°$, $\cos\theta$ is positive, so h is positive, *i.e.* liquid rises in a capillary tube.
- If $\theta > 90°$, $\cos\theta$ is negative, so h is negative, *i.e.* liquid falls in a capillary tube.
- Rise of liquid in a capillary tube does not violate law of conservation of energy.

Some Practical Examples of Capillarity

(i) The kerosene oil in a lantern and the melted wax in a candle, rise in the capillaries formed in the cotton wick and burns.

(ii) Coffee powder is easily soluble in water because water immediately wets the fine granules of coffee by the action of capillarity.

(iii) The water given to the fields rises in the innumerable capillaries formed in the stems of plants and trees and reaches the leaves.

Example 26. *The lower end of a capillary tube of diameter 2 mm is dipped 8 cm below the surface of water in a beaker. The surface tension of water at temperature of the experiment is $7.3 \times 10^{-2}\ Nm^{-1}$, 1 atmospheric pressure*

$= 1.01 \times 10^5$ *Pa, density of water* $= 1000$ *kg/m*^{3}, $g = 9.8$ ms^{-2}, *then the pressure inside the bubble is*

(a) 2.13×10^3 Pa (b) 1.02×10^5 Pa
(c) 5×10^{-5} Pa (d) 7.3×10^{-3} Pa

Sol. (b) The excess pressure in a bubble of gas in a liquid is given by $\frac{2S}{r}$, where S is the surface tension of the liquid gas interface. The radius of the bubble is r.

The pressure outside the bubble p_0 equals atmospheric pressure plus the pressure due to 8 cm of water column. That is

$$p_0 = (1.01 \times 10^5 \text{ Pa} + 0.08 \text{ m} \times 1000 \text{ kg m}^{-3} \times 9.80 \text{ ms}^{-2})$$
$$p_0 = 1.01784 \times 10^5 \text{ Pa}$$

The pressure inside the bubble is

$$p_i = p_0 + \frac{2S}{r} = 1.01784 \times 10^5 + \frac{(2 \times 7.3 \times 10^{-2})}{10^{-3}}$$
$$= (1.01784 + 0.00146) \times 10^5$$
$$= 1.02 \times 10^5 \text{ Pa}$$

Note This is a 100% increase in pressure from surface level. At a depth of 1 km, the increase in pressure is 100 atm. Submarines are designed to withstand such enormous pressures.

Example 27. *The lower end of a capillary tube is dipped into water and it is seen that water rises through 7.5 cm in the capillary. Given surface tension of water is* 7.5×10^{-2} Nm^{-1} *and angle of contact between water and glass capillary tube is zero. What will be the diameter of the capillary tube? (Given,* $g = 10$ ms^{-2}*)*

(a) 0.2 mm (b) 0.3 mm (c) 0.4 mm (d) 0.5 mm

Sol. (c) Given, $h = 7.5$ cm $= 7.5 \times 10^{-2}$ m, $S = 7.5 \times 10^{-2}$ Nm^{-1}, $\theta = 0°$, $2r = ?$

As, $$h = \frac{2S\cos\theta}{r\rho g}$$

$$2r = \frac{4S\cos\theta}{h\rho g}$$
$$= \frac{4 \times 7.5 \times 10^{-2} \times \cos 0°}{7.5 \times 10^{-2} \times 10^3 \times 10}$$
$$\Rightarrow \quad = 4 \times 10^{-4}$$
$$= 0.4 \text{ mm}$$

Example 28. *A capillary tube made of glass of radius 0.15 mm is dipped vertically in a beaker filled with methylene iodide, which rises to height h in the tube. It is observed that the two tangents drawn from liquid- glass interface (from opposite sides of the capillary) make an angle of* $60°$ *with one another. Then, h is close to (Given, surface tension* $= 0.05$ Nm^{-1}*, density* $= 667$ kg m^{-3} *and* $g = 10$ ms^{-2}*)* **[JEE Main 2020]**

(a) 0.049 m (b) 0.087 m (c) 0.137 m (d) 0.172 m

Sol. (b) It is given that tangents drawn from point of contact of meniscus makes $60°$ angle with each other.

We have following situation, as shown in figure

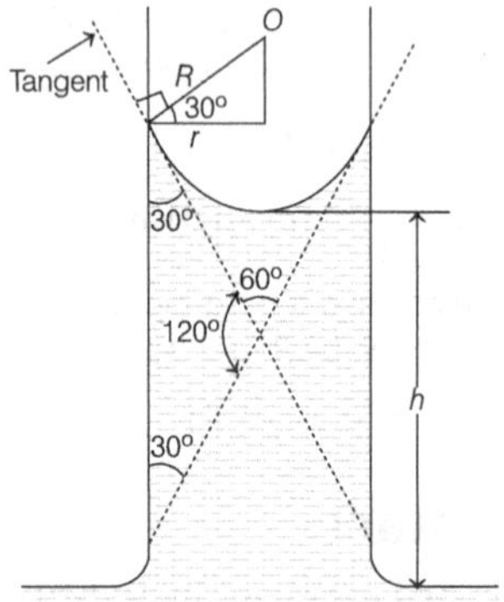

A normal drawn on the tangent, will pass through the centre O of meniscus. Let r be the radius of capillary tube and R be the radius of meniscus.

From geometry of figure, $\frac{\text{radius of capillary}}{\text{radius of meniscus}} = \frac{r}{R} = \cos 30°$

$$\Rightarrow \quad \frac{r}{R} = \frac{\sqrt{3}}{2}$$

or $$R = \frac{2r}{\sqrt{3}} = \frac{2 \times 0.15 \times 10^{-3}}{\sqrt{3}}$$

$$\Rightarrow \quad R = \sqrt{3} \times 10^{-4} \text{ m}$$

Now, by ascent formula,

$$h = \frac{2S}{\rho g R} = \frac{2 \times 0.05}{667 \times 10 \times \sqrt{3} \times 10^{-4}} = 0.087 \text{ m}$$

Viscosity

The property of a fluid due to which it opposes the relative motion between its different layers is called *viscosity* (or fluid friction or internal friction) and the force between the layers opposing the relative motion is called *viscous force.*

According to Newton, the frictional force F (or viscous force) between two layers depends upon the following factors

(i) Force F is directly proportional to the area (A) of the layers in contact, *i.e.* $F \propto A$

(ii) Force F is directly proportional to the velocity gradient $\left(\frac{dv}{dy}\right)$ between the layers, $\frac{dv}{dy}$ is also called *strain rate.*

Combining these two, we have

$$F \propto A\frac{dv}{dy} \text{ or } F = -\eta A\frac{dv}{dy}$$

where, η is a constant called ***coefficient of viscosity.***

The negative sign shows that viscous force on a liquid layer acts in a direction opposite to the relative velocity of flow of fluid.

Its unit is poise or dyne cm^{-2} s in CGS system and poiseuille or deca poiseuille or Newton-s-m^{-2} in SI system. It is a scalar quantity.

1 poiseuille = 1 deca poiseuille = 10 poise

Note In case of a steady flow of a liquid of viscosity η in a capillary tube of length L and radius *i* under a pressure difference p across it, the velocity of flow at a distance x from the axis is given by $v = \frac{p}{4\eta L}(r^2 - x^2)$

Example 29. *A metal block of area 0.10 m^2 is connected to a 0.010 kg mass via a string that passes over an ideal pully (considered massless and frictionless). A liquid with a film thickness of 0.30 mm is placed between the block and the table. When released the block moves to the right with a constant speed of 0.085 ms^{-1}. The coefficient of viscosity of the liquid is*

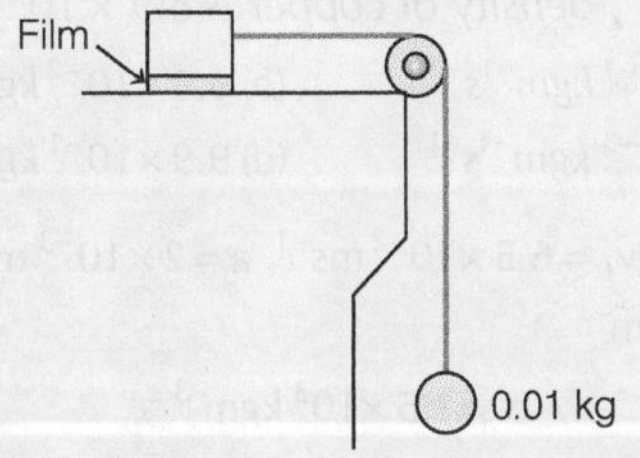

(a) 45.2×10^{2} *Pa-s* (b) 13.4×10^{-4} *Pa-s*
(c) 3.45×10^{-3} *Pa-s* (d) 1.42×10^{-2} *Pa-s*

Sol. (c) The metal block moves to the right because of the tension in the string. The tension T is equal in magnitude to the weight of the suspended mass m. Thus, the shear force

$$F = T = mg = 0.01 \text{ kg} \times 9.8 \text{ ms}^{-2} = 9.8\times10^{-2}\text{N}$$

$$\text{Shear stress on the fluid} = \frac{F}{A} = \frac{9.8\times10^{-2}}{0.10}$$

$$\text{Strain rate} = \frac{v}{t} = \frac{0.085}{0.030}$$

$$\eta = \frac{\text{Stress}}{\text{Strain rate}}$$

$$\eta = \frac{(9.8\times10^{-2}\text{ N})(0.30\times10^{-3}\text{ m})}{(0.085\text{ ms}^{-1})(0.10\text{ m}^2)}$$

$$\eta = 3.45\times10^{-3}\text{ Pa-s}$$

Poiseuille's Formula

In case of steady flow of a liquid of viscosity η in a capillary tube of length L and radius R under a pressure difference p across it, the volume of liquid flowing per second is given by

$$\frac{dQ}{dt} = \frac{\pi p R^4}{8\eta L}$$

This is called *Poiseuille's formula.*

Poiseuille's equation can also be written as

$$Q = \frac{P_1 - P_2}{\left(\frac{8\eta L}{\pi R^4}\right)} = \frac{\Delta p}{X}$$

Here, $$X = \frac{8\eta L}{\pi R^4}$$

Example 30. *Water is flowing through a horizontal tube 8 cm in diameter and 4 km in length at the rate of 20 L/s. Assuming only viscous resistance. The pressure required to maintain the flow in terms of mercury column. (Coefficient of viscosity of water is 0.001 Pa-s) is*

(a) *69.68 cm* (b) *59.68 cm* (c) *49.68 cm* (d) *39.68 cm*

Sol. (b) Here, $2r = 8\text{ cm} = 0.08\text{ m}$

or $r = 0.04\text{ m};\ l = 4\text{ km} = 4000\text{ m};$

$V = 20\text{ L/s} = 20\times10^{-3}\text{ m}^3\text{s}^{-1},$

$\eta = 0.001\text{ Pa-s},\ p = ?$

As, $$V = \frac{\pi p r^4}{8\eta l}$$

or $$p = \frac{8V\eta l}{\pi r^4} = \frac{8\times(20\times10^{-3})\times0.001\times4000}{\left(\frac{22}{7}\right)\times(0.04)^4}$$

$$= 7.954\times10^4\text{ Pa}$$

$\therefore$ Height of mercury column for pressure difference p will be

$$h = \frac{p}{\rho g} = \frac{7.954\times10^4}{(13.6\times10^3)\times9.8}$$

$$= 0.5968\text{ m} = 59.68\text{ cm}$$

Stoke's Law and Terminal Velocity

When a small spherical body falls in a long liquid column, then after sometime it falls with a constant velocity, called *terminal velocity.*

When a small spherical body falls in a liquid column with terminal velocity, then viscous force acting on it is

$$F = 6\pi\eta rv$$

where, r = radius of the body, v = terminal velocity and η = coefficient of viscosity.

This is called *Stoke's law.*

$$\text{Terminal velocity, } v = \frac{2}{9}\frac{r^2(\rho - \sigma)g}{\eta}$$

where, ρ = density of body,

σ = density of liquid,

η = coefficient of viscosity of liquid

and g = acceleration due to gravity.

(i) If $\rho > \sigma$, the body falls downwards.

(ii) If $\rho < \sigma$, the body moves upwards with the constant velocity.

(iii) If $\sigma << \rho$, $v = \frac{2r^2\rho g}{9\eta}$.

- Terminal velocity depends on the radius of the sphere in such a way that, if radius becomes n times, then terminal velocity will become n^2 times.
- Terminal velocity-Time/distance graph

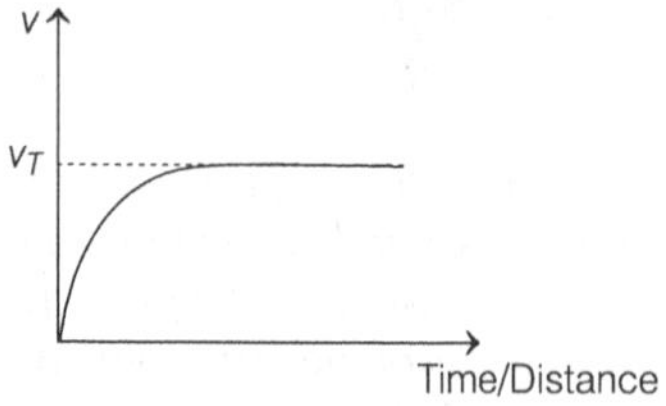

Importance of Stoke's Law

- This law is used in the determination of electronic charge by Millikan in his oil drop experiment.
- This law helps a man coming down with the help of a parachute.
- This law accounts for the formation of clouds.

Example 31. *A rain drop of radius 0.3 mm has a terminal velocity in air 1 ms^{-1}. The viscosity of air is 18×10^{-5} poise. Find the viscous force on the rain drops.*

(a) 2.05×10^{-7} N (b) 1.018×10^{-7} N
(c) 1.05×10^{-7} N (d) 2.058×10^{-7} N

Sol. (b) Here, $r = 0.3\,\text{mm} = 0.3 \times 10^{-3}$ m; $v = 1\,\text{ms}^{-1}$

$$\eta = 18 \times 10^{-5} \text{ poise} = 18 \times 10^{-6} \text{ decapoise}$$

Viscous force, $F = 6\pi\eta rv$

$$= 6 \times \frac{22}{7} \times (18 \times 10^{-6}) \times (0.3 \times 10^{-3}) \times 1$$

$$= 1.018 \times 10^{-7}\,\text{N}$$

Example 32. *In an experiment to verify Stoke's law, a small spherical ball of radius r and density ρ falls under gravity through a distance h in air before entering a tank of water. If the terminal velocity of the ball inside water is same as its velocity just before entering the water surface, then the value of h is proportional to (Ignore viscosity of air)* **[JEE Main 2020]**

(a) r^4 (b) r (c) r^3 (d) r^2

Sol. (a) Let ρ_l be the density of water.

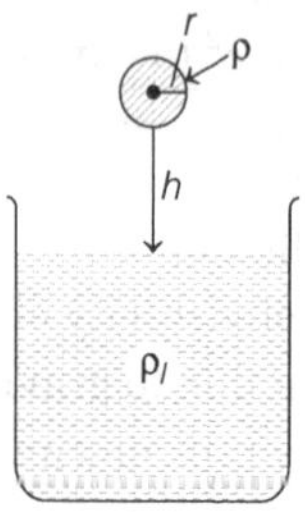

Velocity of the ball after free fall, $v = \sqrt{2gh}$ ...(i)

After falling through distance h, the velocity of the ball will be equal to terminal velocity.

Terminal velocity, $v = \dfrac{2r^2g(\rho_l - \rho)}{9\eta}$ ...(ii)

From Eqs. (i) and (ii), we get

$$\sqrt{2gh} = \frac{2}{9}\frac{r^2g}{\eta}(\rho_l - \rho)$$

$$\Rightarrow \quad 2gh = \frac{4}{81}\frac{r^4g^2}{\eta^2}(\rho_l - \rho)^2$$

$$\Rightarrow \quad h = \frac{2}{81}\frac{r^4g^2}{\eta^2}(\rho_l - \rho)^2 \Rightarrow h \propto r^4$$

Hence, correct option is (a).

Example 33. *The terminal velocity of a copper ball of radius 2 mm falling through a tank of oil at 20°C is 6.5 cms^{-1}. The viscosity of the oil at 20°C is [Given, density of oil is 1.5×10^3 kgm^{-3}, density of copper is 8.9×10^3 kgm^{-3}]*

(a) $3.3 \times 10^{-1}\,kgm^{-1}s^{-1}$ (b) $6.3 \times 10^{-2}\,kgm^{-1}s^{-1}$
(c) $9.2 \times 10^{-3}\,kgm^{-1}s^{-1}$ (d) $9.9 \times 10^{-1}\,kgm^{-1}s^{-1}$

Sol. (d) Given, $v_t = 6.5 \times 10^{-2}\,\text{ms}^{-1}$, $a = 2 \times 10^{-3}$ m, $g = 9.8\,\text{ms}^{-2}$, $\rho = 8.9 \times 10^3\,\text{kgm}^{-3}$

$$\sigma = 1.5 \times 10^3\,\text{kgm}^{-3}$$

$$\therefore \quad \eta = \frac{2a^2(\rho - \sigma)g}{9v_t}$$

$$\eta = \frac{2}{9} \times \frac{(2 \times 10^{-3})^2 \times 9.8}{6.5 \times 10^{-2}} \times 7.4 \times 10^3\,\text{kgm}^{-3}$$

$$\eta = 9.9 \times 10^{-1}\,\text{kgm}^{-1}\text{s}^{-1}$$

Example 34. *Eight spherical rain drops of equal size are falling vertically through air with a terminal velocity of 0.10 ms^{-1}. What should be the velocity, if these drops were to combine to form one large spherical drop?*

(a) 0.1 ms^{-1} (b) 0.2 ms^{-1}
(c) 0.3 ms^{-1} (d) 0.4 ms^{-1}

Sol. (d) Let r be the radius of each of the small rain drop and R be the radius of big rain drop formed.

As, volume of big drop = 8 × volume of each small drop

$$\therefore \quad \frac{4}{3}\pi R^3 = 8 \times \frac{4}{3}\pi r^3$$

$$R = 2r$$

Let terminal velocity of small drop be v_1 and of big drop be v_2.

As, terminal velocity,

$$v = \frac{2r^2(\rho - \sigma)g}{9\eta} \text{ or } v \propto r^2$$

$$\therefore \quad \frac{v_2}{v_1} = \frac{R^2}{r^2}$$

or $$v_2 = v_1\frac{R^2}{r^2} = 0.1\left(\frac{2r}{r}\right)^2 = 0.1 \times 4 = 0.4\,\text{ms}^{-1}$$

Practice Exercise

ROUND I Topically Divided Problems

Density and Relative Density of Substance

1. A beaker containing water is balanced on the pan of a common balance. A solid of specific gravity 1 and mass 5 g is tied to the arm of the balance and immersed in water contained in the beaker. The scale pan with the beaker

(a) goes down
(b) goes up
(c) remains unchanged
(d) None of the above

2. A U-tube contains 10 cm of water and 12.5 cm of methylated spirit separated by mercury. If the 15.0 cm of water and spirit each are further poured into the respective arms of the tube, what is the difference in the levels of mercury in the two arms? (Specific gravity of mercury = 13.6)

(a) 0.221 cm (b) 2.22 cm
(c) 0.02 cm (d) None of these

3. A U-tube contains water and methylated spirit separated by mercury. The mercury columns in the two arms are in level with 10.0 cm of water in one arm and 12.5 cm of spirit in the other. The specific gravity of spirit would be

(a) 0.70 (b) 0.80
(c) 0.90 (d) 0.60

4. The density ρ of water of bulk modulus B at a depth y in the ocean is related to the density at surface ρ_0 by the relation

(a) $\rho = \rho_0\left(1 - \frac{\rho_0 gy}{B}\right)$ (b) $\rho = \rho_0\left(1 + \frac{\rho_0 gy}{B}\right)$

(c) $\rho = \rho_0\left(1 + \frac{B}{\rho_0 hgy}\right)$ (d) $\rho = \rho_0\left(1 - \frac{B}{\rho_0 gy}\right)$

5. An ice block floats in a liquid whose density is less than water. A part of block is outside the liquid. When whole of ice has melted, the liquid level will

(a) rise
(b) go down
(c) remain same
(d) first rise then go down

Pressure due to Fluid Column and Pascal's Law

6. Density of ice is ρ and that of water is σ. What will be the decrease in volume when a mass M of ice melts?

(a) $\frac{M}{\sigma - \rho}$ (b) $\frac{\sigma - \rho}{M}$ (c) $M\left(\frac{1}{\rho} - \frac{1}{\sigma}\right)$ (d) $\frac{1}{M}\left(\frac{1}{\rho} - \frac{1}{\sigma}\right)$

7. A 50 kg girl wearing high heel shoes balances on a single heel. If the heel is circular with a diameter 1.0 cm. What is the pressure exerted on the horizontal floor?

(a) 6.9×10^6 Pa (b) 6.2×10^6 Pa
(c) 9.6×10^6 Pa (d) 9.0×10^6 Pa

8. A uniform tapering vessel shown in figure is filled with liquid of density 900 kgm^{-3}. The force that acts on the base of the vessel due to liquid is (Take, $g = 10\ ms^{-2}$)

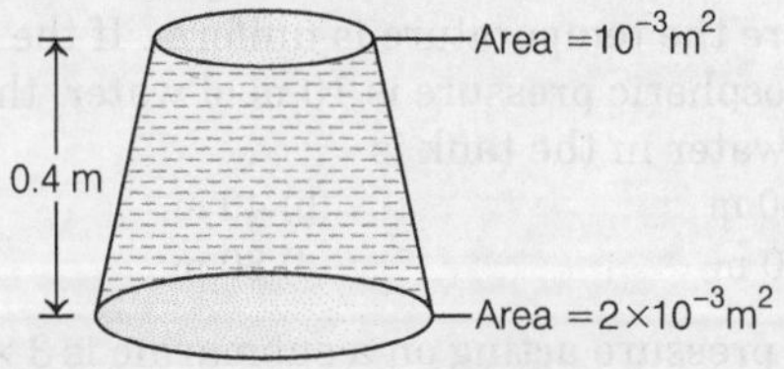

(a) 3.6 N (b) 7.2 N
(c) 9.0 N (d) 12.0 N

9. Figure shows the vertical cross-section of a vessel filled with a liquid of density ρ. The normal thrust per unit area on the walls of the vessel at point P, as shown will be

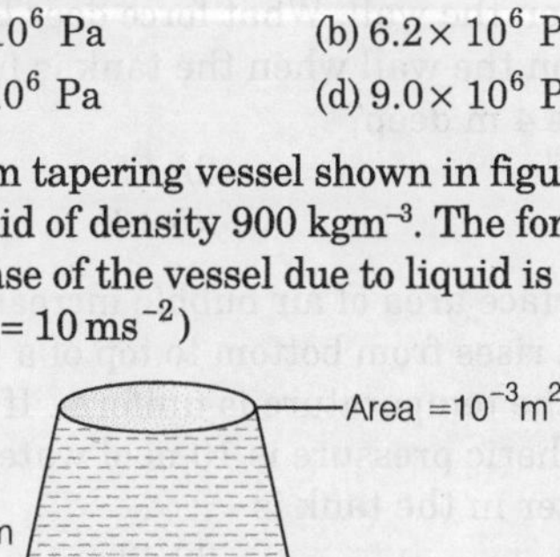

(a) $h\rho g$ (b) $H\rho g$
(c) $(H - h)\rho g$ (d) $(H - h)\rho g \cos\theta$

10. Torricelli's barometer used mercury. Pascal duplicated it using French wine of density 984 kg/m^3. Determine the height of the wine column for normal atmospheric pressure.
(a) 9.5 m (b) 5.5 m
(c) 10.5 m (d) 11.5 m

11. A tank 5m high is half filled with water and then is filled to the top with oil of density 0.85 gcm^{-3}. The pressure at the bottom of the tank, due to these liquids is
(a) 1.85 g dyne cm^{-2} (b) 89.25 g dyne cm^{-2}
(c) 462.5 g dyne cm^{-2} (d) 500 g dyne cm^{-2}

12. A cylindrical vessel is filled with equal amounts of weight of mercury and water. The overall height of the two layers is 29.2 cm, specific gravity of mercury is 13.6. Then the pressure of the liquid at the bottom of the vessel is
(a) 29.2 cm of water (b) 29.2/13.6 cm of mercury
(c) 4 cm of mercury (d) 15.6 cm of mercury

13. An aquarium tank is in the shape of a cube with one side a 4m tall glass wall. When the tank is half filled and the water is 2 m deep, the water exerts a force F on the wall. What force does the water exerts on the wall when the tank is full and the water is 4 m deep?
(a) 1/2 F (b) F
(c) 2 F (d) 4 F

14. The surface area of air bubble increases four times when it rises from bottom to top of a water tank where the temperature is uniform. If the atmospheric pressure is 10 m of water, the depth of the water in the tank is
(a) 30 m (b) 40 m
(c) 70 m (d) 80 m

15. The pressure acting on a submarine is 3×10^5 Pa at a certain depth. If the depth is doubled, the percentage increase in the pressure acting on the submarine would be (Assume that, atmospheric pressure $= 1 \times 10^5$ Pa, density of water $= 10^3$ kg m^{-3}, $g = 10$ ms^{-2}) [JEE Main 2021]
(a) $\frac{200}{3}\%$ (b) $\frac{200}{5}\%$ (c) $\frac{5}{200}\%$ (d) $\frac{3}{200}\%$

16. Two cubes each weighting 22 g exactly are taken. One is of iron ($d = 8 \times 10^3$ kgm^{-3}) and the other is of marble ($D = 3 \times 10^3$ kgm^{-3}). They are immersed in alcohol and then weighted again
(a) iron cube weights less (b) iron cube weights more
(c) both have equal weight (d) nothing can be said

17. Two liquids of densities ρ_1 and $\rho_2 (\rho_2 = 2\rho_1)$ are filled up behind a square wall of side 10 m as shown in figure. Each liquid has a height of 5 m. The ratio of the forces due to these liquids exerted on upper part MN to that at the lower part NO is (assume that the liquids are not mixing)

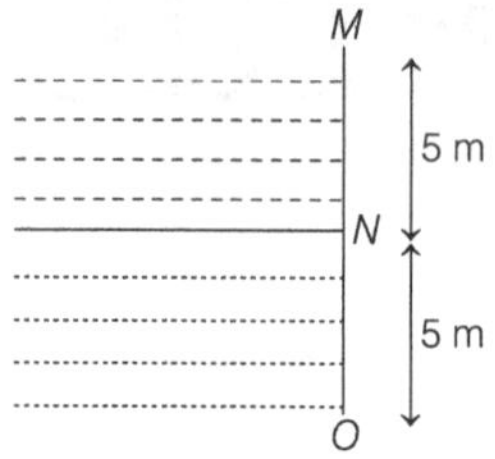

[JEE Main 2020]
(a) 1/2 (b) 2/3 (c) 1/4 (d) 1/3

Archimedes' Principle and Laws of Floatation

18. A cubic block is floating in a liquid with half of its volume immersed in the liquid. When the whole system accelerates upward with acceleration of $g/3$, then the fraction of volume immersed in the liquid will be

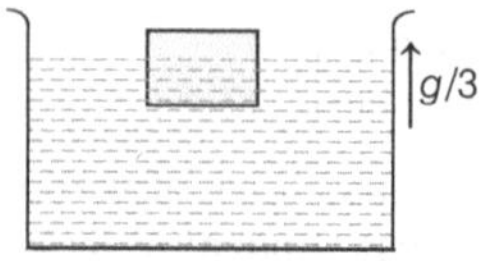

(a) $\frac{1}{2}$ (b) $\frac{3}{8}$ (c) $\frac{2}{3}$ (d) $\frac{3}{4}$

19. The total weight of a piece of wood is 6 kg. In the floating state in water, its $\frac{1}{3}$ part remains inside the water. On this floating solid, what maximum weight is to be put such that the whole of the piece of wood is to be drowned in the water?
(a) 12 kg (b) 10 kg (c) 14 kg (d) 15 kg

20. The spring balance A reads 2 kg with a block of mass m suspended from it. A balance B reads 5 kg when a beaker with liquid is put on the pan of the balance. The two balances are now so arranged that the hanging mass is inside the liquid in a beaker as shown in figure.

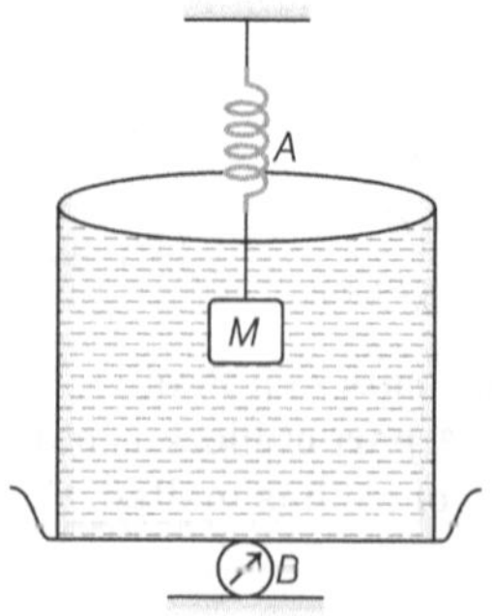

(a) The balance A will read more than 2 kg
(b) The balance B will read less than 5 kg

(c) The balance A will read less than 2 kg and B will read more than 5 kg

(d) The balance A will read more than 2 kg and B will read less than 5 kg

21. A cylinder of mass m and density ρ hanging from a string is lowered into a vessel of cross-sectional area A containing a liquid of density $\sigma\,(<\rho)$ until it is fully immersed. The increase in pressure at the bottom of the vessel is

(a) Zero (b) $\frac{mg}{A}$

(c) $\frac{mg\rho}{\sigma A}$ (d) $\frac{m\sigma g}{\rho A}$

22. A rectangular plate 2m × 3m is immersed in water in such a way that its greatest and least depth are 6 m and 4 m respectively, from the water surface. The total thrust on the plate is

(a) 294×10^3 N (b) 294 N

(c) 100×10^3 N (d) 400×10^3 N

23. A body of density ρ is dropped from rest at a height h into a lake of density σ, where $\sigma > \rho$. Neglecting all dissipative forces, calculate the maximum depth to which the body sinks before returning to float on the surface.

(a) $\frac{h}{\sigma-\rho}$ (b) $\frac{h\rho}{\sigma}$

(c) $\frac{h\rho}{\sigma-\rho}$ (d) $\frac{h\sigma}{\sigma-\rho}$

24. A hemispherical bowl just floats without sinking in a liquid of density 1.2×10^3 kgm^{-3}. If outer diameter and the density of the bowl are 1 m and 2×10^4 kgm^{-3} respectively, then the inner diameter of the bowl will be

(a) 0.94 m (b) 0.96 m

(c) 0.98 m (d) 0.99 m

25. A load of mass M kg is suspended from a steel wire of length 2 m and radius 1.0 mm in Searle's apparatus experiment. The increase in length produced in the wire is 4.0 mm. Now, the load is fully immersed in a liquid of relative density 2. The relative density of the material of load is 8. The new value of increase in length of the steel wire is **[JEE Main 2019]**

(a) zero (b) 5.0 mm

(c) 4.0 mm (d) 3.0 mm

26. A wooden block floating in a bucket of water has $\frac{4}{5}$ of its volume submerged. When certain amount of an oil is poured into the bucket, it is found that the block is just under the oil surface with half of its volume under water and half in oil. The density of oil relative to that of water is **[JEE Main 2019]**

(a) 0.6 (b) 0.8 (c) 0.7 (d) 0.5

27. A cubical block of wooden edge l and a density ρ floats in water of density 2ρ. The lower surface of cube just touches the free end of a massless spring of force constant k fixed at the bottom of the vessel. The weight w put over the block so that it is completely immersed in water without wetting the weight is

(a) $a\,(l\rho g + k)$ (b) $a\,(l^2\rho g + k)$

(c) $a\left(\frac{l\rho g}{2} + 2k\right)$ (d) $l\left(l^2\rho g + \frac{k}{2}\right)$

28. Two cylinders of same cross-section and length L but made of two material of densities ρ_1 and ρ_2 (in CGS units) are cemented together to form a cylinder of length 2 L. If the combination floats in water with a length $L/2$ above the surface of water and $\rho_1 < \rho_2$, then

(a) $\rho_1 > 1$ (b) $\rho_1 < 3/4$

(c) $\rho_1 > 1/2$ (d) $\rho_1 > 3/4$

29. The density of ice is 0.9 gcc^{-1} and that of sea water is 1.1 gcc^{-1}. An ice berg of volume V is floating in sea water. The fraction of ice berg above water level is

(a) 1/11 (b) 2/11 (c) 3/11 (d) 4/11

30. A hollow cylinder of mass m made heavy at its bottom is floating vertically in water. It is tilted from its vertical position through an angle θ and then released. The restoring force acting on it is

(a) $mg\cos\theta$ (b) $mg\sin\theta$

(c) $mg\left[\frac{1}{\cos\theta} - 1\right]$ (d) $mg\left[\frac{1}{\cos\theta} + 1\right]$

Flow of Liquids

31. Fig. (i) and Fig. (ii) refer to the steady flow of a (non-viscous) liquid. Which of the figures is/are incorrect?

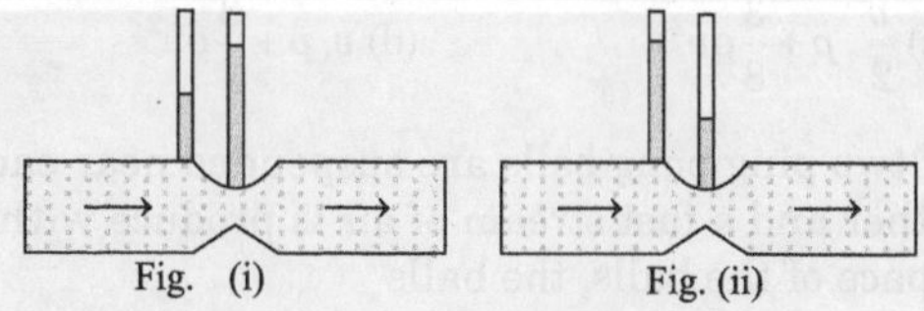

Fig. (i) Fig. (ii)

(a) Fig. (i) (b) Fig. (ii)

(c) both (i) and (ii) (d) None of these

32. An incompressible liquid flows through a horizontal tube as shown in the figure. Then, the velocity v of the fluid is

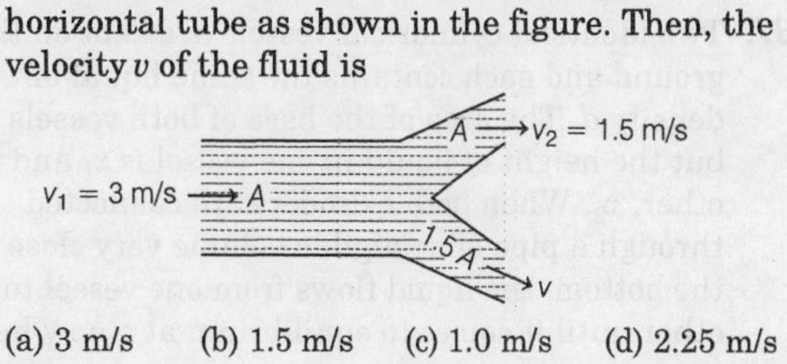

(a) 3 m/s (b) 1.5 m/s (c) 1.0 m/s (d) 2.25 m/s

33. Water flowing out of the mouth of a tap and falling vertically in streamline flow forms a tapering column, *i.e.* the area of cross-section of the liquid column decreases as it moves down. Which of the following is the most accurate explanation for this?

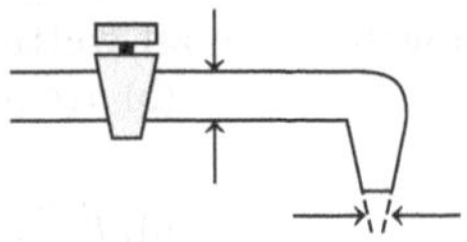

(a) Falling water tries to reach a terminal velocity and hence, reduces the area of cross-section to balance upward and downward forces
(b) As the water moves down, its speed increases and hence, its pressure decreases. It is then compressed by atmosphere
(c) The surface tension causes the exposed surface area of the liquid to decrease continuously
(d) The mass of water flowing out per second through any cross-section must remain constant. As the water is almost incompressible, so the volume of water flowing out per second must remain constant. As this is equal to velocity × area, the area decreases as velocity increases.

34. Two water pipes P and Q having diameter 2×10^{-2} m and 4×10^{-2} m respectively are joined in series with the main supply line of water. The velocity of water flowing in pipe P is
(a) 4 times that of Q (b) 2 times that of Q
(c) (1/2) times that of Q (d) (1/4) times that of Q

35. A fluid flows through a horizontal pipe having two different cross-sections of area A and $2A$. If the pressure at the thin cross-section is p and fluid velocity is v, the velocity and pressure at the thicker cross-section is (take the density of fluid as ρ)
(a) $\frac{v}{2}, p+\frac{1}{2}\rho v^2$ (b) $\frac{v}{4}, p+\frac{3}{8}\rho v^2$
(c) $\frac{v}{2}, p+\frac{3}{8}\rho v^2$ (d) $v, p+\frac{3}{4}\rho v^2$

36. If two ping pong balls are suspended near each other and a fast stream of air is produce within the space of the balls, the balls
(a) come nearer to each other
(b) move away from each other
(c) remain in their original positions
(d) move far away

37. Two identical cylindrical vessels are kept on the ground and each contains the same liquid of density d. The area of the base of both vessels is S but the height of liquid in one vessel is x_1 and in the other, x_2. When both cylinders are connected through a pipe of negligible volume very close to the bottom, the liquid flows from one vessel to the other until it comes to equilibrium at a new height. The change in energy of the system in the process is **[JEE Main 2020]**
(a) $gdS(x_2^2 + x_1^2)$ (b) $\frac{3}{4} gdS(x_2 - x_1)^2$
(c) $gdS(x_2 + x_1)^2$ (d) $\frac{1}{4} gdS(x_2 - x_1)^2$

38. Air is streaming past a horizontal air plane wing such that its speed is 120 ms^{-1} over the upper surface and 90 ms^{-1} at the lower surface. If the density of air is 1.3 kgm^{-3}, what will be the gross lift on the wing? If the wing is 10 m long and has an average width of 2 m,
(a) 81.9 N (b) 8.19 kN (c) 81.9 kN (d) 819 kN

39. A tank is filled with water upto a height H. Water is allowed to come out of a hole P in one of the walls at a depth h below the surface of water (see figure). Express the horizontal distance X in terms of H and h.

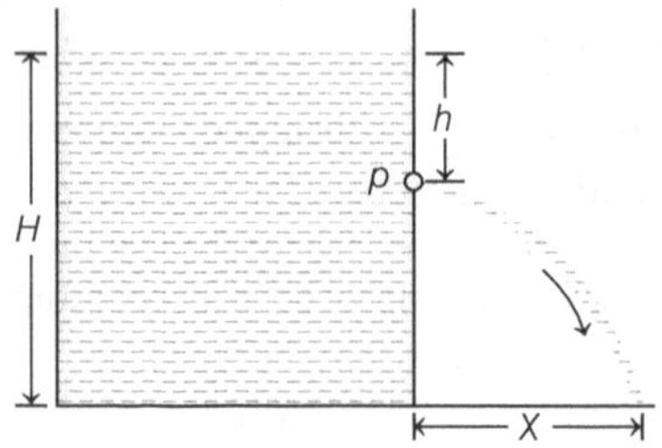

(a) $X = \sqrt{h(H-h)}$ (b) $X = \sqrt{\frac{h}{2}(H-h)}$
(c) $X = 2\sqrt{h(H-h)}$ (d) $X = 4\sqrt{(H-h)}$

40. A cylindrical drum, open at the top, contains 15 L of water. It drains out through a small opening at the bottom. 5 L of water comes out in time t_1, the next 5 L in further time t_2 and the last 5 L in further time t_3. Then
(a) $t_1 < t_2 < t_3$ (b) $t_1 > t_2 > t_3$
(c) $t_1 = t_2 = t_3$ (d) $t_2 > t_1 = t_3$

41. In a test experiment on a model aeroplane in a wind tunnel, the flow speeds on the upper and lower surfaces of the wing are 70 m/s and 63 m/s respectively. What is the lift on the wing, if its area is 2.5 m^2? Take the density of air to be 1.3 kg/m^3.
(a) 5.1×10^2 N (b) 6.1×10^2 N
(c) 1.6×10^3 N (d) 1.5×10^3 N

Surface Tension and Surface Energy

42. A thin metal disc of radius r float on water surface and bends the surface downwards along the perimeter making an angle θ with vertical edge of the disc. If the disc displaces a weight of water w and surface tension of water is T, then the weight of metal disc is
(a) $2\pi rT + w$ (b) $2\pi rT\cos\theta - w$
(c) $2\pi rT\cos\theta + w$ (d) $w - 2\pi rT\cos\theta$

43. A ring is cut from a platinum tube 8.5 cm internal diameter and 8.7 cm external diameter. It is supported horizontally from a pan of a balance so, that it comes in contact with the water in glass vessel. If an extra 3.47 g-wt is required to pull it away from water, surface tension of water is
(a) 62.96 dyne cm^{-1} (b) 70.80 dyne cm^{-1}
(c) 65.35 dyne cm^{-1} (d) 60.00 dyne cm^{-1}

44. What is the radius of the biggest aluminium coin of thickness, t and density ρ, which will still be able to float on the water surface of surface tension S?
(a) $\frac{4S}{3\rho gt}$ (b) $\frac{3S}{4\rho gt}$
(c) $\frac{2S}{\rho gt}$ (d) $\frac{S}{\rho gt}$

45. 8000 identical water drops are combined to form a big drop then the ratio to the final surface energy to the initial surface energy, if all the drops together is
(a) 1 : 10 (b) 1 : 15
(c) 1 : 20 (d) 1 : 25

46. A mercury drop of radius 1 cm is broken into 10^6 droplets of equal size. The work done is (Take, $S = 35 \times 10^{-2}$ Nm^{-1})
(a) 4.35×10^{-2} J (b) 4.35×10^{-3} J
(c) 4.35×10^{-6} J (d) 4.35×10^{-8} J

47. What change in surface energy will be noticed when a drop of radius R splits up into 1000 droplets of radius r, surface tension S ?
(a) $4\pi R^2 S$ (b) $7\pi R^2 S$
(c) $16\pi R^2 S$ (d) $36\pi R^2 S$

48. Let W be the work done, when a bubble of volume V is formed from a given solution. How much work is required to be done to form a bubble of volume 2 V?
(a) W (b) $2W$
(c) $2^{1/3}$ W (d) $4^{1/3}$ W

49. What is the ratio of surface energy of 1 small drop and 1 large drop if 1000 drops combine to form 1 large drop?
(a) 100 : 1 (b) 1000 : 1
(c) 10 : 1 (d) 1 : 100

50. Two spherical soap bubbles of radii a and b in vacuum coalesce under isothermal conditions. The resulting bubble has a radius given by
(a) $\frac{(a+b)}{2}$
(b) $\frac{ab}{a+b}$
(c) $\sqrt{a^2+b^2}$
(d) $a+b$

51. Which graph represent the variation of surface tension with temperature over small temperature ranges for water?

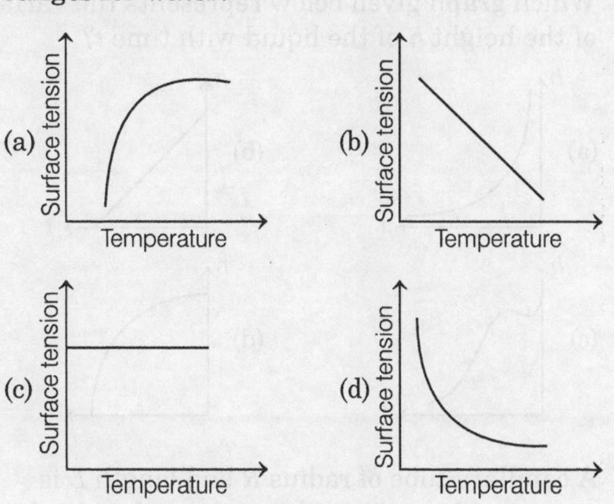

52. The amount of work done in blowing a soap bubble such that its diameter increases from d to D is
(Here, S = surface tension of solution)
(a) $\pi(D^2 - d^2)S$ (b) $2\pi(D^2 - d^2)S$
(c) $4\pi(D^2 - d^2)S$ (d) $8\pi(D^2 - d^2)S$

Excess of Pressure, Shape of Meniscus and Capillarity

53. Pressure inside two soap bubbles are 1.01 atm and 1.02 atm, respectively. The ratio of their volume is **[JEE Main 2020]**
(a) 4 : 1 (b) 0.8 : 1 (c) 8 : 1 (d) 2 : 1

54. The ratio of surface tensions of mercury and water is given to be 7.5 while the ratio of their densities is 13.6. Their contact angles with glass are close to 135° and 0°, respectively. It is observed that mercury gets depressed by an amount h in a capillary tube of radius r_1, while water rises by the same amount h in a capillary tube of radius r_2. The ratio (r_1 / r_2), is then close to **[JEE Main 2019]**
(a) 3/5 (b) 2/3 (c) 2/5 (d) 4/5

55. The following observations were taken for determining surface tension T of water by capillary method. Diameter of capillary, $d = 1.25 \times 10^{-2}$ m rise of water, $h = 1.45 \times 10^{-2}$m. Using $g = 9.80$ m/s^2 and the simplified relation $T = \frac{rhg}{2} \times 10^3$N/m, the possible error in surface tension is closest to **[JEE Main 2017]**
(a) 1.5% (b) 2.4% (c) 10% (d) 0.15%

56. When two soap bubbles of radius r_1 and r_2 ($r_2 > r_1$) coalesce, the radius of curvature of common surface is
(a) $(r_2 - r_1)$ (b) $(r_2 + r_1)$ (c) $\frac{r_2 - r_1}{r_1 r_2}$ (d) $\frac{r_2 r_1}{r_2 - r_1}$

57. Water in a vessel of uniform cross-section escapes through a narrow tube at the base of the vessel. Which graph given below represents the variation of the height h of the liquid with time t?

(a) 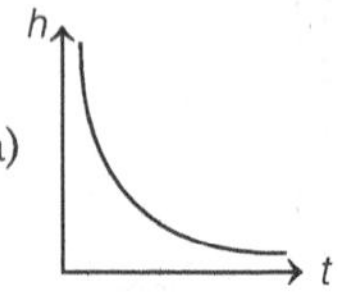

(b) 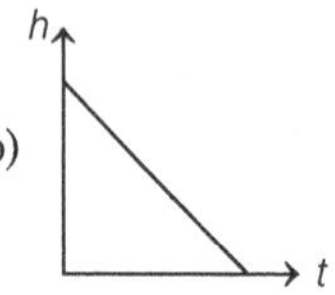

(c) 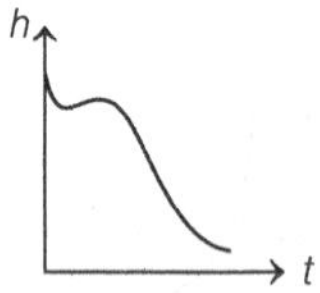

(d) 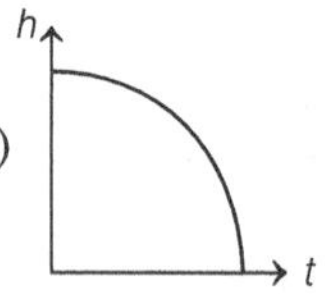

58. A capillary tube of radius R and length L is connected in series with another tube of radius $R/2$ and length $L/4$. If the pressure difference across the two tubes taken together is p, then the ratio of pressure difference across the first tube to that across the second tube is

(a) 1 : 4 (b) 1 : 1
(c) 4 : 1 (d) 2 : 1

59. The rate of steady volume flow of water through a capillary tube of length l and radius r under a pressure difference of p, is V. This tube is connected with another tube of the same length but half the radius in series. Then the rate of steady volume flow through them is (The pressure difference across the combination is p)

(a) $\frac{V}{16}$ (b) $\frac{V}{17}$
(c) $\frac{16V}{17}$ (d) $\frac{17V}{16}$

60. The rate of flow of liquid through a capillary tube of radius r is V, when the pressure difference across the two ends of the capillary is p. If pressure is increased by $3p$ and radius is reduced to $r/2$, then the rate of flow becomes

(a) $V/9$ (b) $3V/8$
(c) $V/4$ (d) $V/3$

61. When two soap bubbles of radii a and b $(b > a)$ coalesce, the radius of curvature of common surface is **[JEE Main 2021]**

(a) $\frac{ab}{b-a}$ (b) $\frac{a+b}{ab}$
(c) $\frac{b-a}{ab}$ (d) $\frac{ab}{a+b}$

62. The angle of contact at the interface of water-glass is 0° Ethylalcohol-glass is 0°, Mercury-glass is 140° and Methyliodide-glass is 30°. A glass capillary is put in a trough containing one of these four liquids. It is observed that the meniscus is convex. The liquid in the trough is **[NCERT Exemplar]**

(a) water (b) ethylalcohol
(c) mercury (d) methyliodide

63. The diagram shows three soap bubbles A, B and C prepared by blowing the capillary tube fitted with stop cocks S, S_1, S_2 and S_3. With stop cock S closed and stop cocks S_1, S_2 and S_3 opened

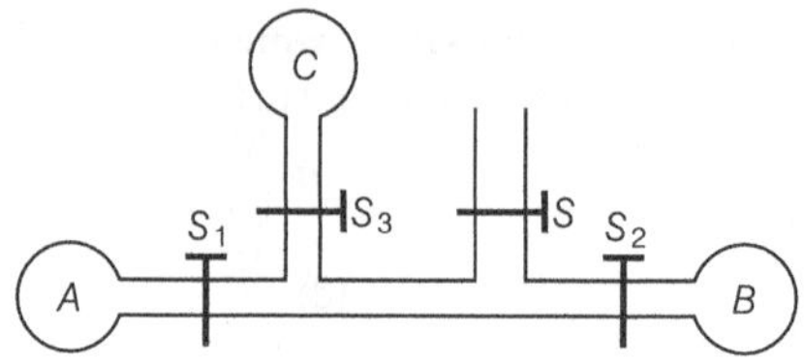

(a) B will start collapsing with volumes of A and C increasing
(b) C will start collapsing with volume of A and B increasing
(c) volume of A, B and C will become equal in equilibrium
(d) C and A will both start collapsing with volume of B increasing

64. Water rises in a capillary tube to a height h. It will rise to a height more than h

(a) on the surface of sun
(b) in a lift moving down with an acceleration
(c) at the poles
(d) in a lift moving up with an acceleration

65. Water rises to a height of 16.3 cm in a capillary of height 18 cm above the water level. If the tube is cut at a height of 12 cm in the capillary tube, then

(a) water will come as a fountain from the capillary tube
(b) water will stay at a height of 12 cm in the capillary tube
(c) the height of water in the capillary tube will be 10.3 cm
(d) water height flow down the sides of the capillary tube

66. Two capillary tubes of radii 0.2 cm and 0.4 cm are dipped in the same liquid. The ratio of heights through which liquid will rise in the tubes is

(a) 1 : 2 (b) 2 : 1
(c) 1 : 4 (d) 4 : 1

67. By inserting a capillary tube upto a depth l in water, the water rises to a height h. If the lower end of the capillary tube is closed inside water and the capillary is taken out and closed end opened, to what height the water will remain in the tube, when $l > h$?

(a) zero (b) $l + h$ (c) $2h$ (d) h

68. Water rises to a height of 10 cm in a capillary tube and mercury falls to a depth of 3.42 cm in the same capillary tube. If the ratio of density of mercury and water is 13.5, then the ratio of surface tension of water and mercury is

(a) 1 : 0.15
(b) 1 : 3
(c) 1 : 6.5
(d) 1.5 : 1

Stoke's Law, Terminal Velocity and Variation of Viscosity

69. The relative velocity of two parallel layers of water is 8 cms^{-1}. If the perpendicular distance between the layers is 0.1 cm, then velocity gradient will be

(a) 40 s^{-1}
(b) 50 s^{-1}
(c) 60 s^{-1}
(d) 80 s^{-1}

70. A small spherical ball of steel falls through a viscous medium with terminal velocity v. If a ball of twice the radius of the first one but of the same mass is dropped through the same method, it will fall with a terminal velocity (neglect buoyancy)

(a) $\frac{v}{2}$ (b) $\frac{v}{\sqrt{2}}$
(c) v (d) $2\,v$

71. The terminal velocity v of a spherical ball of lead of radius R falling through a viscous liquid varies with R such that

(a) $\frac{v}{R}$ = constant (b) vR = constant
(c) v = constant (d) $\frac{v}{R^2}$ = constant

72. A solid sphere of radius R acquires a terminal velocity v_1 when falling (due to gravity) through a viscous fluid having a coefficient of viscosity η. The sphere is broken into 27 identical solid spheres. If each of these spheres acquires a terminal velocity, v_2 when falling through the same fluid, the ratio (v_1 / v_2) equals **[JEE Main 2019]**

(a) 9 (b) 1/27
(c) 1/9 (d) 27

73. A tall cylinder is filled with viscous oil. A round pebble is dropped from the top with zero initial velocity. From the plot shown in figure, indicate the one that represents the velocity (v) of the pebble as a function of time (t).

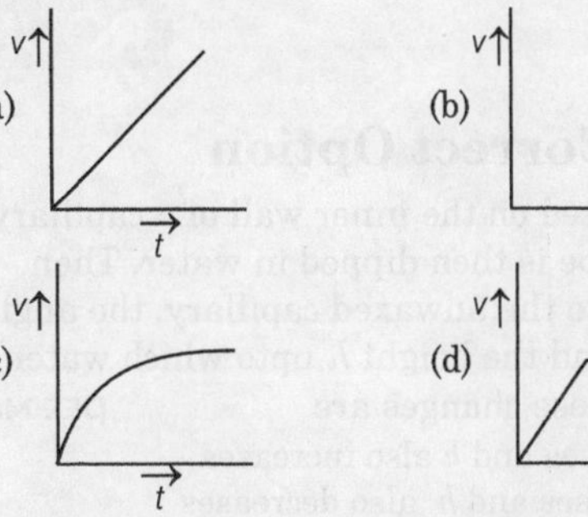

74. A rain drop of radius 1.5 mm, experiences a drag force $F = (2 \times 10^{-5}\, v)$ N, while falling through air from a height 2 km, with a velocity v. The terminal velocity of the rain drop will be nearly (use $g = 10$ ms^{-2})

(a) 200 ms^{-1} (b) 80 ms^{-1}
(c) 7 ms^{-1} (d) 3 ms^{-1}

75. A spherical ball is dropped in a long column of viscous liquid. Which of the following graphs represent the variation of

(i) gravitational force with time
(ii) viscous force with time
(iii) net force acting on the ball with time?

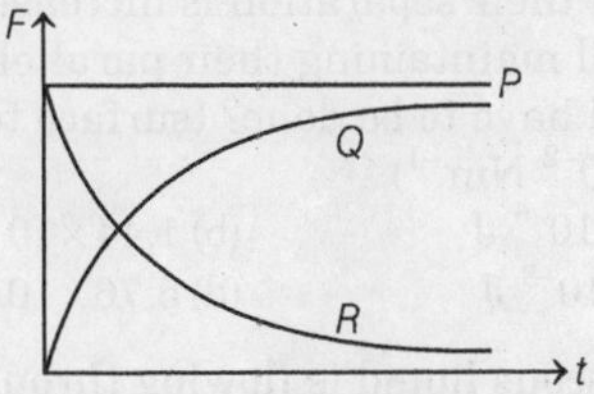

(a) Q, R, P (b) R, Q, P
(c) P, Q, R (d) R, P, Q

76. A small iron sphere is dropped from a great height. It attains its terminal velocity after fallen 32 m. Then, it covers the rest of the path with terminal velocity only. The work done by air friction during the first 32 m of fall is W_1. The work done by air friction during the subsequent 32 m fall is W_2, then

(a) $W_1 > W_2$ (b) $W_1 < W_2$
(c) $W_1 = W_2$ (d) $W_2 = 32\,W_1$

77. A marble of mass x and diameter $2\,r$ is gently released a tall cylinder containing honey. If the marble displaces mass y ($< x$) of the liquid, then the terminal velocity is proportional to

(a) $(x + y)$ (b) $(x - y)$
(c) $\frac{x+y}{r}$ (d) $\frac{(x-y)}{r}$

ROUND II Mixed Bag

Only One Correct Option

1. Wax is coated on the inner wall of a capillary tube and the tube is then dipped in water. Then, compared to the unwaxed capillary, the angle of contact θ and the height h upto which water rises change. These changes are **[JEE Main 2013]**

(a) θ increases and h also increases
(b) θ decreases and h also decreases
(c) θ increases and h decreases
(d) θ decreases and h increases

2. What will be the nature of flow of water from a circular tap, when its flow rate increased from 0.18 L/min to 0.48 L/min? The radius of the tap and viscosity of water are 0.5 cm and 10^{-3} Pa-s, respectively. (Take, density of water = 10^3 kg/m^3)

(a) Unsteady flow to steady flow **[JEE Main 2021]**
(b) Remains steady flow
(c) Remains turbulent flow
(d) Steady flow to unsteady flow

3. A film of water is found between two straight parallel wires of length 10 cm each separated by 0.2 cm. If their separation is increased by 1 mm, while still maintaining their parallelism, how much work will have to be done? (surface tension of water is 7.2×10^{-2} Nm^{-1})

(a) 7.22×10^{-6} J
(b) 1.44×10^{-5} J
(c) 2.88×10^{-8} J
(d) 5.76×10^{-5} J

4. A non-viscous liquid is flowing through a frictionless duct, with cross section varying as shown in figure.

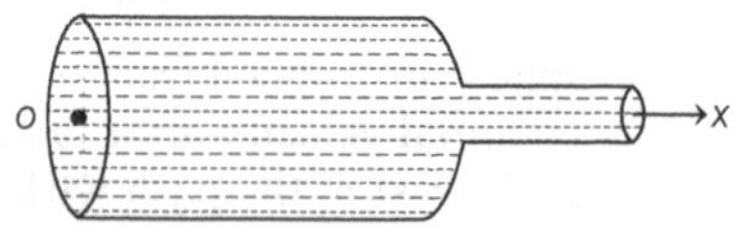

Which of the following graph represents the variation of pressure p along the axis of tube?

(a)
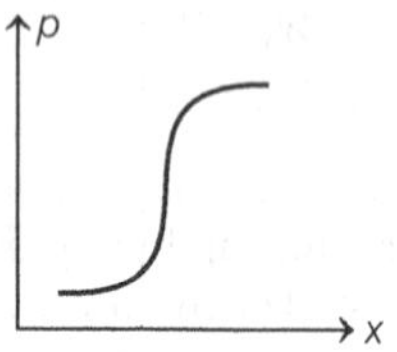

(b)
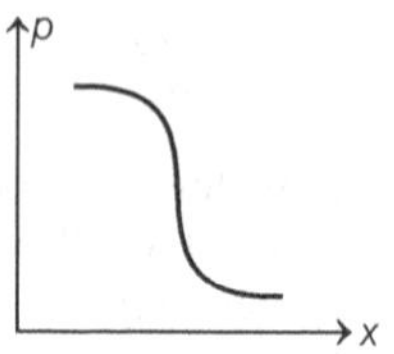

(c)
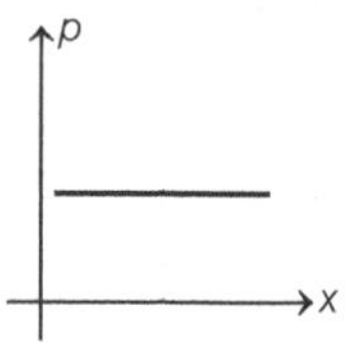

(d)
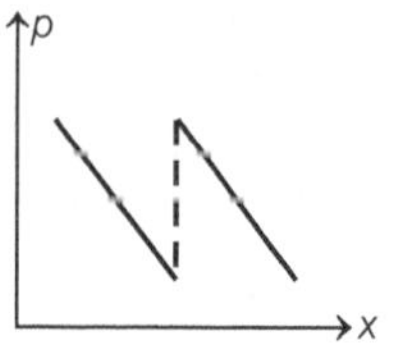

5. A wooden block with a coin placed on its top.floats in water as shown in figure

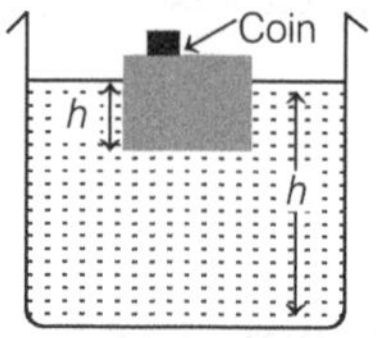

The distance l and h are shown in the figure. After some time the coin falls into the water. Then, **[NCERT Exemplar]**

(a) l decreases
(b) h increases
(c) l increases
(d) None of these

6. Two soap bubbles of radii r_1 and r_2 equal to 4 cm and 5 cm respectively are touching each other over a common surface AB (shown in figure). Its radius will be

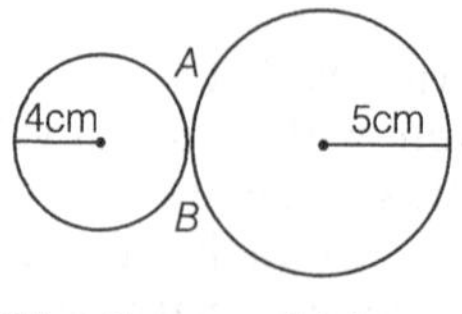

(a) 4 cm (b) 4.5 cm (c) 5 cm (d) 20 cm

7. A spring balance reads w_1 when a ball of mass m is suspended from it. A weighing machine reads w_2 when a beaker of liquid is kept on the pan of balance. When the ball is immersed in liquid, the spring balance reads w_3 and the weighing machine reads w_4. The two balances are now so arranged that the suspended mass is inside the liquid in a beaker. Then,

(a) $w_3 > w_1$
(b) $w_4 > w_2$
(c) $w_3 < w_1$ and $w_4 > w_2$
(d) $w_3 > w_1$ and $w_4 < w_2$

8. Water is filled up to a height h in beaker of radius R as shown in the figure. The density of water ρ the surface tension of water is T and the atmosphere pressure is p_0. Consider a vertical section $ABCD$ of the water column through n diameter of the beaker. The force on water on one side of this section by water on the other side of this section has magnitude.

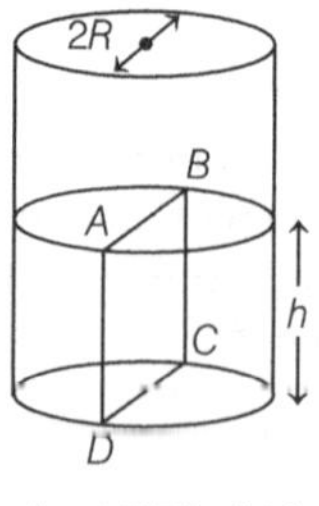

(a) $|2p_0Rh + \pi r^2\rho gh - 2RT|$
(b) $|2p_0Rh + R\rho gh^2 - 2RT|$
(c) $|p_0\pi R^2 + R\rho gh^2 - 2RT|$
(d) $|p_0\pi R^2 + R\rho gh^2 + 2RT|$

9. Water of density ρ at a depth h behind the vertical face of dam whose cross-sectional length is λ and cross-sectional area A. It exerts a horizontal resultant force on the dam tending to slide it along its foundation and a torque tending to overturn the dam about the point O.

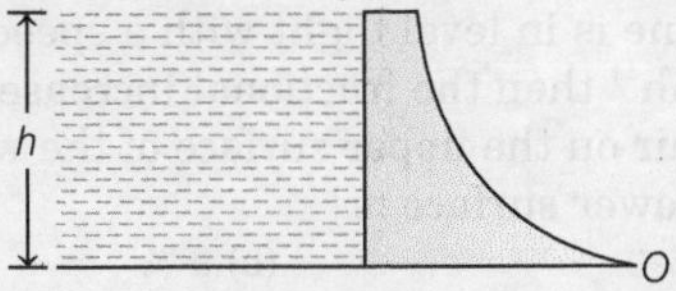

The height at which the resultant force would have to act to the same torque is

(a) $\frac{h}{6}$ (b) $\frac{h}{3}$

(c) $\frac{h}{2}$ (d) $\frac{2h}{3}$

10. Water flows through a vertical tube of variable cross-section. The area of cross-section at A and B are 6 mm^2 and 3 mm^2, respectively. If 12 cc of water enters per second through A, find the pressure difference $p_A - p_B$ ($g = 10\,\text{ms}^{-2}$) The separation between cross-section at A and B is 100 cm.

(a) 1.6×10^5 dyne cm^{-2}

(b) 2.29×10^5 dyne cm^{-2}

(c) 5.9×10^4 dyne cm^{-2}

(d) 3.9×10^5 dyne cm^{-2}

11. A canister has a small hole at its bottom. Water penetrates into the canister when its base is at a depth of 40 cm from the surface of water. If surface tension of water is 73.5 dyne/cm, find the radius of the hole.

(a) 375 mm

(b) 3.75 mm

(c) 0.0375 mm

(d) zero

12. A piece of gold weights 50 g in air and 45 g in water. If there is a cavity inside the piece of gold, then find its volume [Density of gold = 19.3 g/cc].

(a) 2.4 cm^3 (b) 2.4 m^3

(c) 4 .2 m^3 (d) 4.2 mm^3

13. The bottom of a cylindrical vessel has a circular hole of radius r and at depth h below the water level. If the diameter of the vessel is D, then find the speed with which the water level in the vessel drops.

(a) $\frac{4r^2}{D^2}\sqrt{2gh}$ (b) $\frac{4D^2}{r^2}$

(c) $\frac{4D^2}{r^2}\sqrt{2gh}$ (d) None of these

14. A liquid of density ρ is filled in a U-tube is accelerated with an acceleration a so that the height of liquid in its two vertical arms are h_1 and h_2 as shown in the figure. If l is the length of horizontal arm of the tube, the acceleration a is

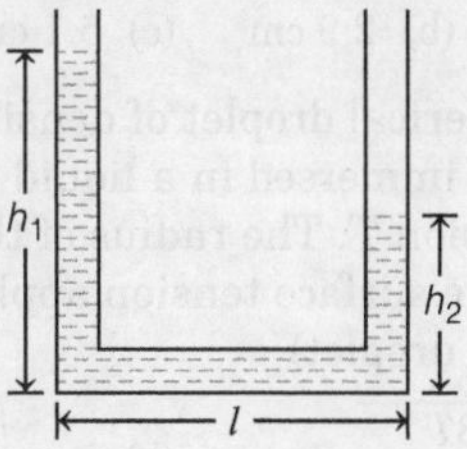

(a) $\frac{g(h_1 - h_2)}{2l}$ towards right

(b) $\frac{g(h_1 - h_2)}{2l}$ towards left

(c) $\frac{g(h_1 - h_2)}{l}$ towards right

(d) $\frac{g(h_1 - h_2)}{l}$ towards left

15. A trough contains mercury to a depth of 3.6 cm. If same amount of mercury is poured in it, then height of mercury in the trough will be

(a) 3.6 cm (b) 7.2 cm

(c) 6 cm (d) None of these

16. A vessel whose bottom has round holes with diameter of 1 mm is filled with water. Assuming that surface tension acts only at holes, then the maximum height to which the water can be filled in vessel without leakage is (Surface tension of water is $75\times10^{-3}\,\text{Nm}^{-1}$ and $g = 10\,\text{ms}^{-2}$)

(a) 3 cm (b) 0.3 cm

(c) 3 mm (d) 3 m

17. A jar shown in figure is filled with a liquid of density ρ. The jar is placed in vacuum. Cross-section of the jar is circular and base is having a radius R. The force exerted by the liquid column on the base of the jar is

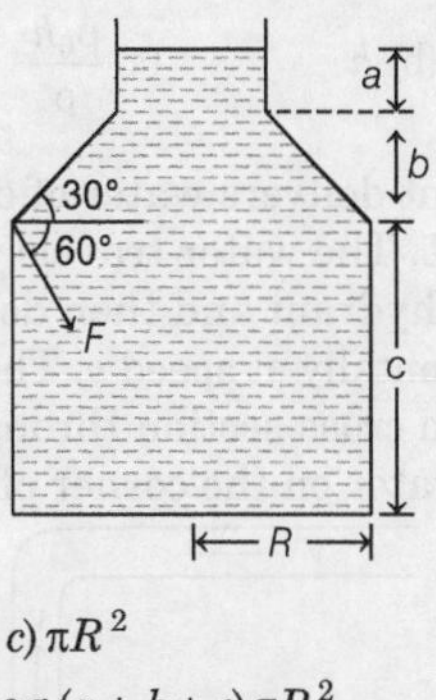

(a) $\rho g(a + b + c)\pi R^2$

(b) less than $\rho g(a + b + c)\pi R^2$

(c) greater than $\rho g(a + b + c)\pi R^2$

(d) $2\rho g(a + b + c)\pi R^2$

18. Water flows into a large tank with flat bottom at the rate of $10^{-4}\ m^3s^{-1}$. Water is also leaking out of a hole of area 1 cm^2 at its bottom. If the height of the water in the tank remains steady, then this height is [JEE Main 2019]

(a) 4 cm (b) 2.9 cm (c) 5.1 cm (d) 1.7 cm

19. A small spherical droplet of density d is floating exactly half immersed in a liquid of density ρ and surface tension T. The radius of the droplet is (take note that the surface tension applies an upward force on the droplet) [JEE Main 2020]

(a) $r = \sqrt{\frac{3T}{(2d-\rho)g}}$ (b) $r = \sqrt{\frac{T}{(d-\rho)g}}$

(c) $r = \sqrt{\frac{T}{(d+\rho)g}}$ (d) $r = \sqrt{\frac{2T}{3(d+\rho)g}}$

20. A uniform cylinder of length L and mass M having cross-sectional area A is suspended, with its length vertical from a fixed point by a massless spring such that it is half submerged in a liquid of density σ at equilibrium position. The extension x_0 of the spring when it is in equilibrium is [JEE Main 2013]

(a) $\frac{Mg}{k}$ (b) $\frac{Mg}{k}\left(1-\frac{LA\sigma}{M}\right)$

(c) $\frac{Mg}{k}\left(1-\frac{LA\sigma}{2M}\right)$ (d) $\frac{Mg}{k}\left(1+\frac{LA\sigma}{M}\right)$

21. The glycerine of density $1.25 \times 10^3\ kgm^{-3}$ is flowing through a conical tube with end radii 0.1 m and 0.04 m respectively. The pressure difference across the ends is 10 Nm^{-2}. The rate of flow of glycerine through the tube is

(a) $6.4\times10^{-2}\ m^3s^{-1}$ (b) $6.4\times10^{-4}\ m^3s^{-1}$

(c) $12.8\times10^{-2}\ m^3s^{-1}$ (d) $12.8\times10^{3}\ m^3s^{-1}$

22. A wooden ball of density ρ is immersed in water of density ρ_0 to depth h and then released. The height H above the surface of water upto which the ball jump out of water is

(a) zero (b) h (c) $\frac{\rho_0 h}{\rho}$ (d) $\left(\frac{\rho_0}{\rho}-1\right)h$

23. A fire hydrant delivers water of density ρ at a volume rate L. The water travels vertically upwards through the hydrant and then does 90° turn to emerge horizontally at speed v. The pipe and nozzle have uniform cross-section throughout. The force exerted by water on the corner of the hydrant is

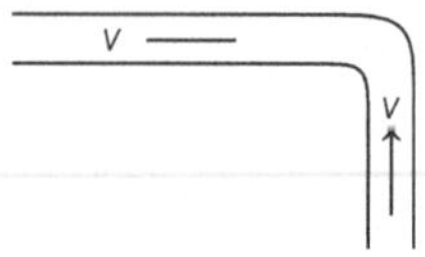

(a) zero (b) ρvL

(c) $2\rho vL$ (d) $\sqrt{2}\rho vL$

24. A plane of mass 3250 g is in level flight at a constant speed and each wing has an area of 25 m^2. During flight the speed of the air is 216 kmh^{-1} over the lower wing surface and 252 kmh^{-1} over the upper wing surface of each wing of aeroplane. Take density of air = 1 kgm^{-3} and $g = 10\ ms^{-2}$.

If a plane is in level flight with a speed of 360 kmh^{-1} then the fractional increase in the speed of the air on the upper surface of the wing relative to the lower surface is

(a) 13 % (b) 9 %

(c) 6.5 % (d) 4.5 %

25. Two capillaries of radii r_1 and r_2, lengths l_1 and l_2 respectively are in series. A liquid of viscosity η is flowing through the combination under a pressure difference p. What is the rate of volume flow of liquid?

(a) $\frac{\pi p}{8\eta}\left(\frac{l_4}{r_1^4}+\frac{l_4}{r_2^4}\right)^{-1}$ (b) $\frac{8\pi p}{\eta}\left(\frac{l_1}{r_1^4}+\frac{l_2}{r_2^4}\right)$

(c) $\frac{\pi p}{8\eta}\left(\frac{r_1^4}{l_1}+\frac{r_2^4}{l_2}\right)^{-1}$ (d) $\frac{\pi p}{8\eta}\left(\frac{l_1}{r_1^4}+\frac{l_2}{r_2^4}\right)^{-1}$

26. A metal ball immersed in alcohol weighs W_1 at 0°C and W_2 at 59°C. The coefficient of cubical expansion of the metal is less than that of alcohol. Assuming that the density of the metal is large compared to that of alcohol, it can be shown that

(a) $W_1 > W_2$

(b) $W_1 < W_2$

(c) $W_1 = W_2$

(d) $W_1 = 2W_2$

27. A soap film is made by dipping a circular frame of radius b in soap solution. A bubble is formed by blowing air with speed v in the form of cylinder. The radius of the bubble formed $R >> b$ so that the air is incident normally on the surface of bubble. Air stops after striking surface of soap bubble. Density of air is ρ. The radius R of the bubble when the soap bubble separates from the ring is (surface tension of liquid is S).

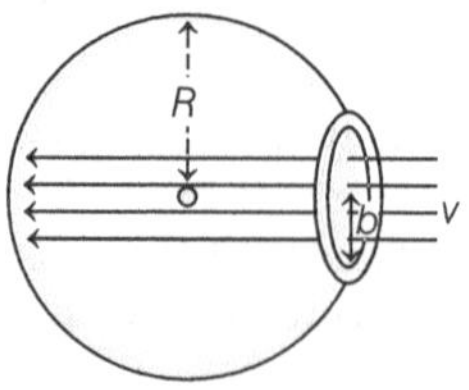

(a) $\frac{S}{\rho v^2}$ (b) $\frac{4S}{\rho v^2}$

(c) $\frac{Sb}{\rho v}$ (d) $\frac{4Sb}{\rho v^2}$

28. A block is submerged in vessel filled with water by a spring attached to the bottom of the vessel. In equilibrium, the spring is compressed. The vessel now moves downward with an acceleration a ($< g$). The spring length

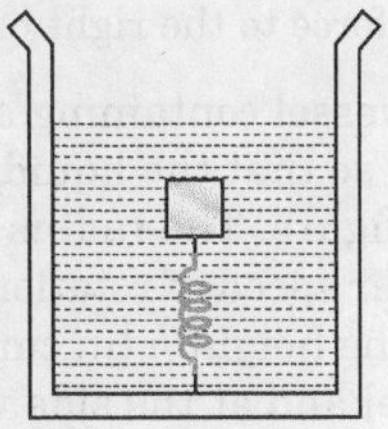

(a) will become zero
(b) will decrease but not zero
(c) will increase
(d) may increase or decrease or remain constant

29. A glass tube 80 cm long and open at both ends is half immersed in mercury. Then the top of the tube is closed and it is taken out of the mercury. A column of mercury 20 cm long then remains in the tube. The atmospheric pressure (in cm of Hg) is

(a) 90 (b) 75
(c) 60 (d) 45

30. There is a hole of area A at the bottom of a cylindrical vessel. Water is filled upto a height h and water flows out in t sec. If water is filled to a height $4h$, then it will flow out in time

(a) $2\ t$ (b) $4\ t$
(c) $16\ t$ (d) $7/4\ t$

31. Two soap bubbles A and B are kept in closed chamber where the air is maintained at pressure $8\ \text{N/m}^2$. The radius of bubbles A and B are 2 cm and 4 cm respectively surface tension of the soap water used to make bubbles is 0.04 N/m. Find the ratio n_B/n_A, where n_A and n_B are the number of moles of air in bubbles A and B respectively
[Neglect the effect of gravity]

(a) 2 (b) 9
(c) 8 (d) 6

32. What is the excess pressure inside a bubble of soap solution of radius 5.00 mm, given that the surface tension of soap solution at the temperature (20° C) is 2.50×10^{-2} N/m? If an air bubble of the same dimension were formed at a depth of 40.0 cm inside a container containing the soap solution (of relative density 1.20), what would be the pressure inside the bubble? (1 atmospheric pressure is 1.01×10^5 Pa.)

(a) 7.06×10^5 Pa
(b) 2.06×10^5 Pa
(c) 1.06×10^5 Pa
(d) 1.86×10^5 Pa

33. A uniform rod of density ρ is placed in a wide tank containing a liquid σ ($\sigma > \rho$). The depth of liquid in the tank is half the length of the rod. The rod is in equilibrium, with its lower end resting on the bottom of the tank. In this position, the rod makes an angle θ with the horizontal. Then, $\sin\theta$ is equal to

(a) $\frac{1}{2}\sqrt{\frac{\sigma}{\rho}}$ (b) $\frac{1}{2}\frac{\sigma}{\rho}$
(c) $\sqrt{\frac{\rho}{\sigma}}$ (d) $\sqrt{\frac{\rho}{\sigma}}$

34. If M is the mass of water that rises in a capillary tube of radius r, then mass of water which will rise in a capillary tube of radius $2r$ is **[JEE Main 2019]**

(a) $2M$ (b) $4M$
(c) $\frac{M}{2}$ (d) M

35. A liquid of density ρ is coming out of a hose pipe of radius a with horizontal speed v and hits a mesh. 50% of the liquid passes through the mesh unaffected 25% losses all of its momentum and, 25% comes back with the same speed. The resultant pressure on the mesh will be **[JEE Main 2019]**

(a) ρv^2 (b) $\frac{1}{2}\rho v^2$
(c) $\frac{1}{4}\rho v^2$ (d) $\frac{3}{4}\rho v^2$

36. On heating water, bubbles beings formed at the bottom of the vessel detach and rise. Take the bubbles to be spheres of radius R and making a circular contact of radius r with the bottom of the vessel. If $r << R$ and the surface tension of water is T, value of r just before bubbles detach is (density of water is ρ) **[JEE Main 2014]**

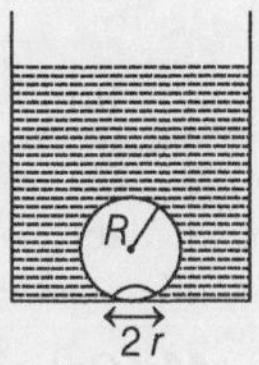

(a) $R^2\sqrt{\frac{2\rho_w g}{3T}}$ (b) $R^2\sqrt{\frac{\rho_w g}{6T}}$
(c) $R^2\sqrt{\frac{\rho_w g}{T}}$ (d) $R^2\sqrt{\frac{3\rho_w g}{T}}$

37. Water flows in a horizontal tube (see figure). The pressure of water changes by $700\ \text{Nm}^{-2}$ between A and B, where the area of cross- section are $40\ \text{cm}^2$ and $20\ \text{cm}^2$, respectively. Find the rate of flow of water through the tube.
(Take, density of water = $1000\ \text{kgm}^{-3}$) **[JEE Main 2020]**

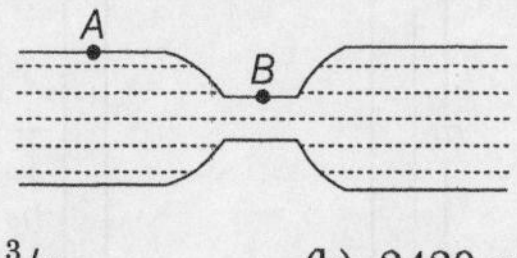

(a) $3020\ \text{cm}^3/\text{s}$ (b) $2420\ \text{cm}^3/\text{s}$
(c) $2720\ \text{cm}^3/\text{s}$ (d) $1810\ \text{cm}^3/\text{s}$

38. An alloy of Zn and Cu (*i.e.* brass) weighs 16.8 g in air and 14.7 g in water. If relative density of Cu and Zn are 8.9 and 7.1 respectively, then the amount of Cu and Zn in the alloy, respectively are
(a) 2g, 4g
(b) 4g, 2g
(c) 9.345g, 7.455 g
(d) 0, 3g

39. There are two identical small holes on the opposite sides of a tank containing a liquid. The tank is open at the top. The difference in height between the two holes is h. As the liquid comes out of the two holes, the tank will experience a net horizontal force proportional to

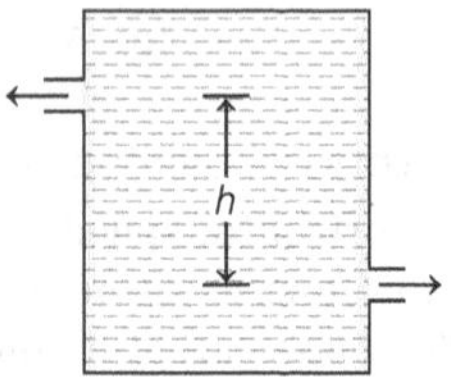

(a) $h^{1/2}$ (b) $h^{3/2}$
(c) h (d) h^2

40. A streamline body with relative density ρ_1 falls into air from a height h_1 on the surface of a liquid of relative density ρ_2, where $\rho_2 > \rho_1$. The time of immersion of the body into the liquid will be

(a) $\sqrt{\frac{2h_1}{g}}$ (b) $\sqrt{\frac{2h}{g}\times\frac{\rho_1}{\rho_2}}$
(c) $\sqrt{\frac{2h_1}{g}\times\frac{\rho_1}{\rho_2}}$ (d) $\sqrt{\frac{2h_1}{g}\times\frac{\rho_1}{(\rho_2-\rho_1)}}$

41. Calculate the force of attraction between two parallel plates separated by a distance 0.2 mm after a water drop of mass 80 mg is introduced between them. The wetting is assumed to be complete. (surface tension of water is 0.07 Nm^{-1})
(a) 0.14 N (b) 0.28 N
(c) 0.42 N (d) 0.56 N

42. The U-tube has a uniform cross-section as shown in figure. A liquid is filled in the two arms upto heights h_1 and h_2 and then the liquid is allowed to move. Neglect viscosity and surface tension. When the level equalize in the two arms, the liquid will

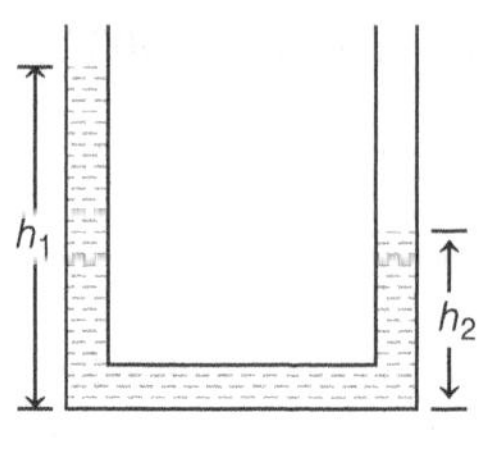

(a) be at rest
(b) be moving with an acceleration of $g\left(\frac{h_1-h_2}{h_1+h_2+2}\right)$
(c) be moving with a velocity of $(h_1-h_2)\sqrt{\frac{g}{2(h_1+h_2+h)}}$
(d) exert a net force to the right on the cube

43. A cylindrical vessel containing a liquid is rotated about its axis, so that the liquid rises at its sides as shown in the figure. The radius of vessel is 5 cm and the angular speed of rotation is ω rad s^{-1}. The difference in the height h (in cm) of liquid at the centre of vessel and at the side will be **[JEE Main 2020I]**

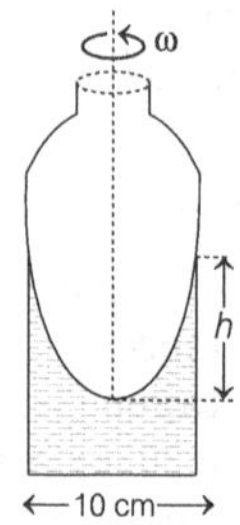

(a) $\frac{25\,\omega^2}{2g}$ (b) $\frac{5\,\omega^2}{2g}$ (c) $\frac{2\,\omega^2}{25g}$ (d) $\frac{2\,\omega^2}{5g}$

Numerical Value Questions

44. Consider a water tank as shown in the figure. Its cross-sectional area is 0.4 m^2. The tank has an opening B near the bottom whose cross-sectional area is 1 cm^2. A load of 24 kg is applied on the water at the top when the height of the water level is 40 cm above the bottom, the velocity of water coming out the opening B is v ms^{-1}. The value of v, to the nearest integer, is .... (Take, $g = 10\ ms^{-2}$) **[JEE Main 2021]**

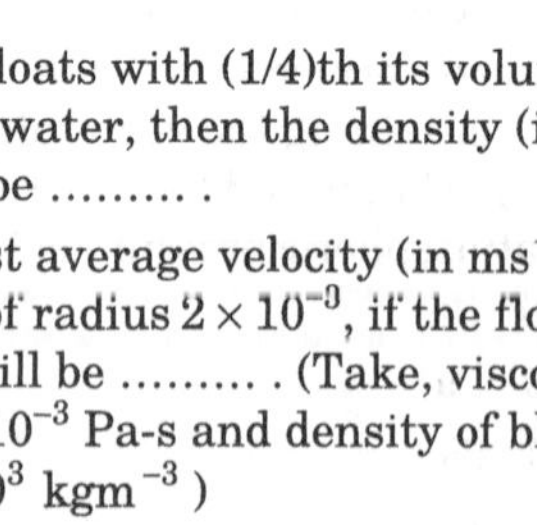

45. If a solid floats with (1/4)th its volume above the surface of water, then the density (in kgm^{-3}) of the solid will be ....

46. The largest average velocity (in ms^{-1}) of blood flow in artery of radius 2×10^{-3}, if the flow must remain laminar will be .... (Take, viscosity of blood $= 2.084\times10^{-3}$ Pa-s and density of blood $= 1.06\times10^{3}\ kgm^{-3}$)

47. A wire ring of 30.0 mm radius resting flat on the surface of the liquid is raised. If the pull required is 3.03 gf force before the film breaks, so the surface tension (in dyne cm^{-1}) of the liquid will be ....

48. A square plate of side 10 cm moves parallel to another plate with a velocity of 10 cm s^{-1}; both plates immersed in water. If the viscous force is 200 dyne and viscosity of water is 0.01 poise, then their separation distance is found to be $x \times 10^{-2}$ m, then the value of x is ....

49. The density of ice is 917 kg m^{-3}. Therefore, the fraction of the volume of a piece of ice while floating above the fresh water will be ....

50. When a long glass capillary tube of radius 0.015 cm is dipped in a liquid, the liquid rises to a height of 15 cm within it. If the contact angle between the liquid and glass is close to 0°, the surface tension of the liquid (in mN m^{-1}) to the nearest integer is (Take, $\rho_{(liquid)} = 900$ kgm^{-3}, $g = 10$ ms^{-2}) .... **[JEE Main 2020]**

51. A bakelite beaker has volume capacity of 500 cc at 30°C. When it is partially filled with V_m volume (at 30°C) of mercury, it is found that the unfilled volume of the beaker remains constant as temperature is varied. If $\gamma_{(beaker)} = 6 \times 10^{-6}$ °C^{-1} and $\gamma_{(mercury)} = 1.5 \times 10^{-4}$ °C^{-1}, where γ is the coefficient of volume expansion, then V_m (in cc) is close to .... **[JEE Main 2020]**

Answers

Round I

1. (c)	2. (a)	3. (b)	4. (b)	5. (b)	6. (c)	7. (b)	8. (b)	9. (c)	10. (c)
11. (c)	12. (c)	13. (d)	14. (c)	15. (a)	16. (b)	17. (c)	18. (a)	19. (a)	20. (c)
21. (d)	22. (a)	23. (c)	24. (c)	25. (d)	26. (a)	27. (d)	28. (b)	29. (b)	30. (c)
31. (a)	32. (c)	33. (d)	34. (a)	35. (c)	36. (a)	37. (d)	38. (c)	39. (c)	40. (a)
41. (d)	42. (c)	43. (a)	44. (c)	45. (c)	46. (a)	47. (d)	48. (d)	49. (d)	50. (c)
51. (b)	52. (b)	53. (c)	54. (c)	55. (a)	56. (d)	57. (a)	58. (a)	59. (b)	60. (c)
61. (a)	62. (c)	63. (b)	64. (b)	65. (b)	66. (b)	67. (c)	68. (c)	69. (d)	70. (a)
71. (d)	72. (a)	73. (c)	74. (c)	75. (c)	76. (b)	77. (c)			

Round II

1. (c)	2. (d)	3. (b)	4. (b)	5. (a)	6. (d)	7. (b)	8. (b)	9. (b)	10. (a)
11. (c)	12. (a)	13. (a)	14. (c)	15. (b)	16. (a)	17. (c)	18. (c)	19. (a)	20. (c)
21. (b)	22. (d)	23. (d)	24. (c)	25. (d)	26. (b)	27. (b)	28. (c)	29. (c)	30. (a)
31. (d)	32. (c)	33. (a)	34. (a)	35. (d)	36. (a)	37. (c)	38. (c)	39. (c)	40. (d)
41. (b)	42. (c)	43. (a)	44. 3	45. 750	46. 0.98	47. 78.76	48. 5	49. 0.083	50. 101
51. 20									

Solutions

Round I

1. Effective weight of solid of specific gravity 1 when immersed in water, will be zero. Hence, the scale pan with the beaker remains unchanged.

2. When 15.0 cm of water is poured in each arm, then

height of water column $(h_1) = 10 + 15 = 25$ cm

Height of spirit column $(h_2) = 12.5 + 15 = 27.5$ cm

Density of water $(\rho_w) = 1 \text{ g/cm}^3$

Density of spirit $(\rho_s) = 0.80 \text{ g/cm}^3$

Density of mercury $(\rho_m) = 13.6 \text{ g/cm}^3$

Let in equilibrium, the difference in the level of mercury in both arms be h cm.

$\therefore \quad h\rho_m g = h_1\rho_w g - h_2\rho_s g$

or $\quad h = \dfrac{h_1\rho_w - h_2\rho_s}{\rho_m} = \dfrac{25\times 1 - 27.5\times 0.80}{13.6}$

$= 0.221$ cm

Therefore, mercury will rise in the arm containing spirit by 0.221 cm.

3.

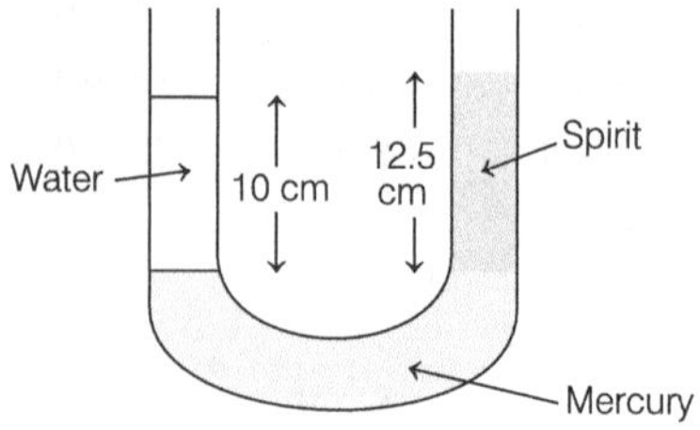

Height of water column $(h_1) = 10.0$ cm

Density of water $(\rho_1) = 1 \text{ g/cm}^3$

Height of spirit column $(h_2) = 12.5$ cm

Density of spirit $(\rho_2) = ?$

The mercury column in both arms of the U-tube are at same level, therefore pressure in both arms will be same.

$\therefore$ Pressure exerted by water column = Pressure exerted by sprit column

$\therefore \quad p_1 = p_2$

$h_1\rho_1 g = h_2\rho_2 g$

or $\quad \rho_2 = \dfrac{h_1\rho_1}{h_2} = \dfrac{10\times 1}{12.5} = 0.80 \text{ g/cm}^3$

Specific gravity of spirit $= \dfrac{\text{Density of spirit}}{\text{Density of water}}$

$= \dfrac{0.80}{1} = 0.80$

4. As, Bulk modulus, $B = -V_0\dfrac{\Delta p}{\Delta V}$

$\Rightarrow \quad \Delta V = -V_0\dfrac{\Delta p}{B}$

$\Rightarrow \quad V = V_0\left(1 - \dfrac{\Delta p}{B}\right)$

$\therefore$ Density, $\quad \rho = \rho_0\left(1 - \dfrac{\Delta p}{B}\right)^{-1} = \rho_0\left(1 + \dfrac{\Delta p}{B}\right)$

where, $\quad \Delta p = p - p_0 = h\rho_0 g$

Pressure difference between depth and surface of ocean.

$\therefore \quad \rho = \rho_0\left(1 + \dfrac{\rho_0 g y}{B}\right) \quad$ (As, $h = y$)

5. Ice is lighter than water. When ice melts, the volume occupied by water is less than that of ice. Due to which the level of water goes down.

6. Volume of ice $= \dfrac{M}{\rho}$, volume of water $= \dfrac{M}{\sigma}$

Change in volume $= \dfrac{M}{\rho} - \dfrac{M}{\sigma} = M\left(\dfrac{1}{\rho} - \dfrac{1}{\sigma}\right)$

7. Given, mass of girl $(m) = 50$ kg

Diameter of circular heel $(2r) = 1.0$ cm

$\therefore \quad$ Radius $(r) = 0.5 \text{ cm} = 5\times 10^{-3}$ m

Area of circular heel $(A) = \pi r^2$

$= 3.14 \times (5 \times 10^{-3})^2 \text{ m}^2$

$= 78.50 \times 10^{-6} \text{m}^2$

$\therefore$ Pressure exerted on the horizontal floor,

$p = \dfrac{F}{A} = \dfrac{mg}{A}$

$= \dfrac{50\times 9.8}{78.50\times 10^{-6}}$

$= 6.24\times 10^6$ Pa

$\simeq 6.2\times 10^6$ Pa

8. Force on the base of the vessel

= Pressure × Area of the base

$= h\rho g \times A$

$= 0.4\times 900\times 10\times 2\times 10^{-3}$

$= 7.2$ N

9. Depth of point P below the free surface of water in the vessel $= (H - h)$. Since, the liquid exerts equal pressure in all direction at one level, hence the pressure at $p = (H - h)\rho g$.

10. Atmospheric pressure $(p) = 1.013\times 10^5$ Pa

Density of French wine $(\rho) = 984 \text{ kg/m}^3$

Let h be the height of the wine column for normal atmospheric pressure.

For normal atmospheric pressure $(p) = h\rho g$

$\therefore \quad h = \dfrac{p}{\rho g} = \dfrac{1.013\times 10^5}{984\times 9.8} = 10.5$ m

11. Pressure at the bottom $p = (h_1 d_1 + h_2 d_2)g$

$$= [250\times 1 + 250\times 0.85]\, g$$
$$= 250\, [1.85]\, g$$
$$= 462.5\, g \text{ dyne/cm}^2$$

12. Let A be the area of cross-section of the cylindrical vessel and x cm be the height of mercury in vessel. The height of water in the vessel = $(29.2 - x)$ cm.

As per question,

$$Ax \times 13.6 = (29.2 - x)\times 1$$

or $\quad x = 2$ cm

$\therefore$ Height of water column = $(29.2 - 2) = 27.2$ cm

$\therefore$ Pressure of the liquids at the bottom

= 27.2 cm of water column + 2 cm of Hg column

$= \frac{27.2}{13.6}$ of Hg column + 2 cm of Hg column

= 4 cm of Hg column

13. Let b be width of the glass wall. When the tank is half filled, then the average force on the glass wall is

$$F = \text{average pressure} \times \text{area}$$
$$= \left[\left(\frac{4}{2}\right)\rho_w g\right]\times\left(\frac{4}{2}\times b\right)$$

When tank is filled up to height 4 m, then

$$F' = (4\rho_w g)\,(4\times b)$$
$$\frac{F'}{F} = \frac{4\times 4}{2\times 2} = 4$$

or $\quad F' = 4F$

14. Surface area, $A = 4\pi r^2$

or $\quad r = (A/4\pi)^{1/2}$

Volume, $V = \frac{4}{3}\pi r^3 = \frac{4}{3}\pi\,(A/4\pi)^{3/2} = kA^{3/2}$

where, $\frac{4\pi}{3}\times\frac{1}{(4\pi)^{3/2}} = k$ = constant.

Using Boyle's law, we have

$$p_1 V_1 = p_2 V_2$$

or $\quad p_2 = \frac{p_1 V_1}{V_2} = \frac{(10+h)\, kA_1^{3/2}}{kA_2^{3/2}}$

or $\quad p_2 = (10+h)\left(\frac{A_1}{A_2}\right)^{3/2}$

As, $\quad p_2 = 10$ m of water, so

$$10 = \frac{10+h}{8}$$

or $\quad 80 = 10 + h$

or $\quad h = 70$ m

15. $p_1 = \rho g d + p_0 = 3\times 10^5$ Pa

$\therefore \quad \rho g d = 2\times 10^5$ Pa

$$p_2 = 2\rho g d + p_0$$
$$= 4\times 10^5 + 10^5 = 5\times 10^5 \text{ Pa}$$

% increase $= \frac{p_2 - p_1}{p_1}\times 100$

$$= \frac{5\times 10^5 - 3\times 10^5}{3\times 10^5}\times 100 = \frac{200}{3}\%$$

16. Since, density of iron is more than that of marble, the volume of iron is less than that of marble for the given mass. The upthrust of water on iron will be less than that on marble. Due to which, iron cube will weight more.

17. Force on a vertical surface of area A and dipped upto height h in a fluid of density ρ is

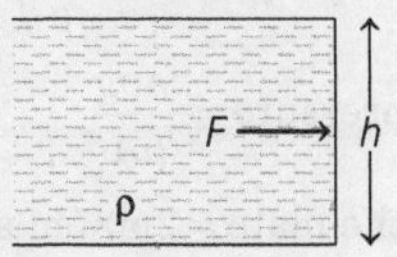

Force, $F = \frac{1}{2}(\rho g h \times A)$

So, in given case,

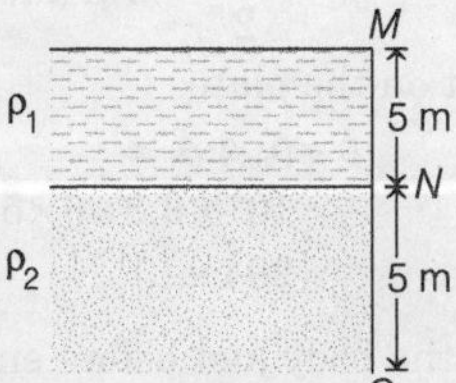

Force on portion $MN = \frac{\rho_1 g h}{2}\times A$

Force on portion NO

= Force due to pressure of liquid in MN portion + Force due to pressure of liquid in NO portion

$$= \rho_1 g h A + \frac{1}{2}(\rho_2 g h A)$$
$$= \rho_1 g h A + \frac{1}{2}\times 2\rho_1 \times ghA \qquad (\because \rho_2 = 2\rho_1)$$
$$= 2\rho_1 g h A$$

Required ratio, $\frac{F_{MN}}{F_{NO}} = \frac{\frac{1}{2}(\rho_1 g h A)}{2\rho_1 g h A} = \frac{1}{4}$

18. Fraction of volume immersed in the liquid, $V_{in} = \left(\frac{\rho}{\sigma}\right)V$

i.e., It depends upon the densities of the block and liquid, so there will be no change in it if system moves upward or downward with constant acceleration, *i.e.* uniform acceleration.

19. Given, $\quad 6\,g = \frac{V}{3}\times 10^3 \times g \qquad$ ...(i)

and $\quad (6+m)\,g = V\times 10^3 \times g \qquad$ ...(ii)

Dividing Eq. (ii) by Eq. (i), we get

or $\quad m = 18 - 6 = 12$ kg

20. The effective weight of the block in liquid will become less than 2 kg due to buoyancy of liquid. As a result of which A will read less than 2 kg.

As, the body immersed in liquid has some effective weight acting downwards so the reading of B will be more than 5 kg.

21. Volume of cylinder $= \dfrac{m}{\rho}$

Upthrust on cylinder $= \left(\dfrac{m}{\rho}\right)\sigma g$

From Newton's third law, the downward force exerted by cylinder on the liquid is $= \left(\dfrac{m}{\rho}\right)\sigma g$

$\therefore$ Increase in pressure $= \dfrac{m\sigma g}{\rho A}$

22. Given, size of the plate = 2m×5m and

Greatest and least depths of the plate are 6m and 4m.

We know that area of the plate $A = 2 \times 3 = 6\ \text{m}^2$

and depth of centre of the plate,

$$x = \frac{6+4}{2} = 5\,\text{m (mean depth)}$$

$\therefore$ Total thrust on the plate,

$$\rho = \rho_w g\, A \cdot x$$
$$= 10^3 \times 9.8 \times 6 \times 5$$
$$= 294 \times 10^3\ \text{N}$$

23. The speed of the body just before entering the liquid is $v = \sqrt{2gh}$. The buoyant force F_B of the lake (*i.e.*, upward thrust of liquid on the body) is greater than the weight of the body w, since $\sigma > \rho$. If V is the volume of the body and a is the acceleration of the body inside the liquid, then $F_B - w = ma$

or $\sigma Vg - \rho Vg = \rho Va$

or $(\sigma - \rho)\, g = \rho a$

or $a = \dfrac{(\sigma - \rho)\, g}{\rho}$

Using the relation, $v^2 = u^2 + 2as$, we have

$$0 = (\sqrt{2gh})^2 - 2g\frac{(\sigma - \rho)}{\rho}s$$

or $s = \dfrac{h\rho}{\sigma - \rho}$

24. Let D_1 be the inner diameter of the hemispherical bowl and D_2 be the outer diameter of the bowl. As, bowl is just floating so

$$\frac{4}{3}\pi\left(\frac{1}{2}\right)^3 \times 1.2 \times 10^3 = \frac{4}{3}\pi\left[\left(\frac{1}{2}\right)^3 - \left(\frac{D_1}{2}\right)^3\right] \times (2 \times 10^4)$$

or $\dfrac{1.2 \times 10^3}{2 \times 10^4} = 1 - D_1^3$

$\Rightarrow$ $D_1 = \left(1 - \dfrac{1.2}{20}\right)^{1/3} = \left(\dfrac{18.8}{20}\right)^{1/3}$

On solving, we get

$$D_1 = 0.98\ \text{m}$$

25. When load M is attached to wire, extension in length of wire is

$$\Delta l_1 = \frac{Mgl}{AY} \qquad \text{...(i)}$$

where, Y is the Young's modulus of the wire.

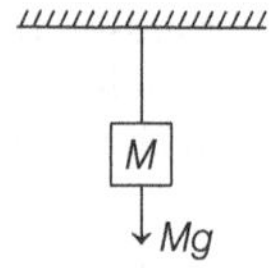

When load is immersed in liquid of relative density 2, increase in length of wire as shown in the figure is

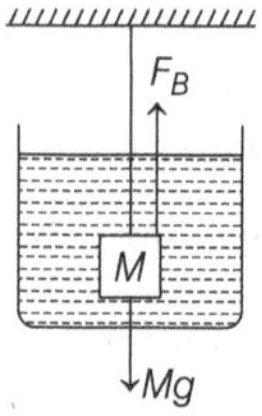

$$\Delta l_2 = \frac{(Mg - F_B)l}{AY}$$

where, F_B = buoyant force.

$$\therefore \quad \Delta l_2 = \frac{\left(Mg - Mg \cdot \dfrac{\rho_l}{\rho_b}\right)l}{AY} \qquad \left[\because F_B = V\rho_l g = \frac{Mg}{\rho_b}\rho_l g\right]$$

Here given that, $\dfrac{\rho_l}{\rho_b} = \dfrac{2}{8} = \dfrac{1}{4}$

So, $$\Delta l_2 = \frac{\left(\dfrac{3}{4}Mg\right)l}{AY} \qquad \text{...(ii)}$$

Dividing Eqs. (ii) by (i), we get

$$\frac{\Delta l_2}{\Delta l_1} = \frac{3}{4}$$

$\Rightarrow$ $\Delta l_2 = \dfrac{3}{4} \times \Delta l_1 = \dfrac{3}{4} \times 4\ \text{mm} = 3\ \text{mm}$

26. For a floating body, upthrust = weight of the part of object, i.e. submerged in the fluid.

In first situation,

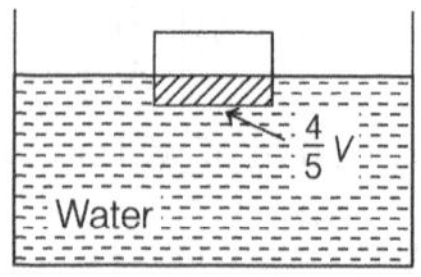

So, weight of block of volume V = weight of water of volume $\dfrac{4}{5}V \Rightarrow V\rho_b g = \dfrac{4}{5}V\rho_w g$

where, ρ_b = density of block

and ρ_w = density of water.

$\Rightarrow$ $\dfrac{\rho_b}{\rho_w} = \dfrac{4}{5}$ ... (i)

In second situation,

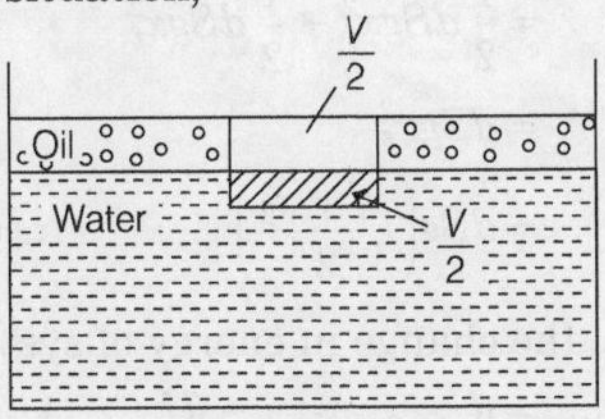

So, weight of block of volume V = weight of oil of volume $\frac{V}{2}$ + weight of water of volume $\frac{V}{2}$.

$$\Rightarrow \quad V\rho_b g = \frac{V}{2}\rho_o g + \frac{V}{2}\rho_w g$$

where, $\quad \rho_o$ = density of oil.

$$\Rightarrow \quad 2\rho_b = \rho_o + \rho_w$$

$$\Rightarrow \quad \frac{2\rho_b}{\rho_w} = \frac{\rho_o}{\rho_w} + 1$$

$$\Rightarrow \quad 2 \times \frac{4}{5} = \frac{\rho_o}{\rho_w} + 1 \quad \text{[using Eq. (i)]}$$

$$\Rightarrow \quad \rho_0 / \rho_w = 8/5 - 1 = 3/5 = 0.6$$

27. Initially the position of wooden block is as shown in Fig. (a). Since, the density of block is half than that of water, hence half of its volume is immersed in water.

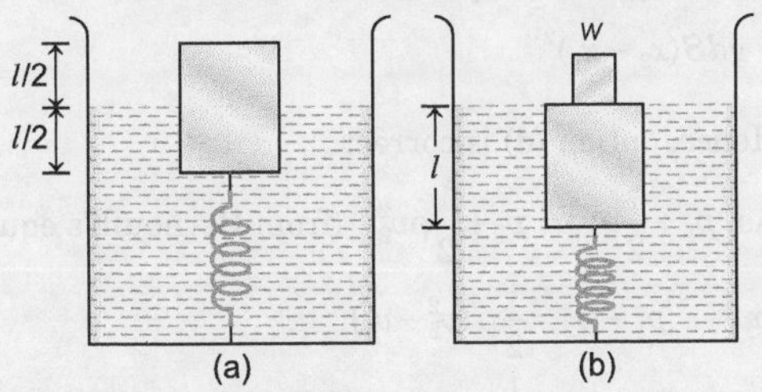

(a) (b)

When weight w is put on the block, the remaining half of the volume of block is immersed in water, figure (b). Therefore,

$$w = \text{additional upthrust} + \text{spring force}$$

$$= l \times l \times \frac{l}{2} \times 2\rho \times g + k\left(\frac{l}{2}\right)$$

$$= l\left(l^2\rho g + \frac{k}{2}\right)$$

28. Mass of the cylinders = $AL(\rho_1 + \rho_2)$. As cylinders float with length $L/2$ outside the water, therefore length of cylinder inside the water = $3L/2$. When cylinders are floating, then, weight of cylinder = weight of water displaced by cylinder.

So, $\quad AL(\rho_1 + \rho_2)\, g = A(3L/2) \times 1 \times g$

or $\quad \rho_1 + \rho_2 = 3/2$

As, $\quad \rho_1 < \rho_2$

so, $\quad \rho_1 < 3/4$

29. Let v be the volume of ice-berg outside the sea water while floating. Therefore, volume of ice-berg inside the sea water = $(V - v)$. As ice-berg is floating, so weight of ice-berg = weight of sea water displaced by ice-berg.

i.e. $\quad V \times 0.9 \times g = (V - v) \times 1.1 \times g$

or $\quad 1.1\, v = 1.1\, V - 0/9\, V$

or $\quad v/V = 0.2/1.1 = 2/11$

30. Let l be the length of the cylinder in water when it is in the vertical position and A be the cross-sectional area of the cylinder. As cylinder is floating, so

weight of cylinder = upward thrust

or $\quad mg = Al\rho g$

or $\quad m = Al\rho$

When the cylinder is tilted through an angle θ, then

length of cylinder in water = $\frac{l}{\cos\theta}$

Weight of water displaced = $\frac{l}{\cos\theta} A\rho g$

$$\therefore \quad \text{Restoring force} = \frac{lA\rho g}{\cos\theta} - lA\rho g$$

$$= lA\rho g\left[\frac{1}{\cos\theta} - 1\right]$$

$$= mg\left[\frac{1}{\cos\theta} - 1\right]$$

31. Fig. (i) is incorrect. From equation of continuity, the speed of liquid is larger at smaller area. According to Bernoulli's theorem, due to larger speed, the pressure will be lower at smaller area (large velocity) and therefore height of liquid column will also be at lesser height, while in Fig. (i) height of liquid column at narrow area in higher.

32. If the liquid is incompressible, then mass of liquid per second entering through left end should be equal to mass of liquid coming out from the right end.

$$M = M_1 + M_2$$

$$\Rightarrow \quad \rho A v_1 = \rho A v_2 + \rho(1.5A)v$$

$$\Rightarrow \quad A v_1 = A v_2 + 1.5\, A v$$

$$A \times 3 = A \times 1.5 + 1.5 \times v \times A$$

$$\Rightarrow \quad 1.5 = 1.5 v$$

$$v = 1 \text{ m/s}$$

33. According to equation of continuity, av = constant. As v increases, a decreases.

34. Using theorem of continuity, we have

$$\pi D_p^2 v_p = \pi D_Q^2 v_Q$$

$$v_p = \left(\frac{D_Q}{D_P}\right)^2 v_Q$$

$$= \left(\frac{4 \times 10^{-2}}{2 \times 10^{-2}}\right) \times v_Q = 4\, v_Q$$

35. As, $Av = 2\,Av'$ or $v' = v/2$

For a horizontal pipe, according to Bernoulli's theorem,

$$p + \frac{1}{2}\rho v^2 = p' + \frac{1}{2}\rho\left(\frac{v}{2}\right)^2$$

or $$p' = p + \frac{1}{2}\rho v^2\left(1 - \frac{1}{4}\right)$$

$\Rightarrow$ $$p' = p + \frac{3}{8}\rho v^2$$

36. When air stream is produced in between two suspended balls, the pressure there becomes less than the pressure on the opposite faces of the balls. Due to which the balls are pushed towards each other.

37.

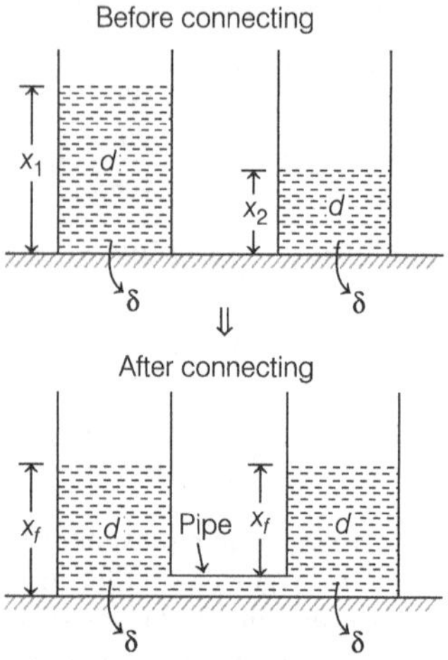

By conservation of volume,

$$(V_{\text{system}})_{\text{initial}} = (V_{\text{system}})_{\text{final}}$$

$$Sx_1 + Sx_2 = Sx_f + Sx_f$$

$$x_1 + x_2 = x_f + x_f$$

$$x_1 + x_2 = 2x_f$$

$$x_f = \frac{x_1 + x_2}{2} \quad \text{...(i)}$$

Now, initial energy of system,

$$(U_{\text{system}})_{\text{initial}} = M_1gh_1 + M_2gh_2$$

$$= dV_1gh_1 + dV_2gh_2$$

$$= dSx_1g\left(\frac{x_1}{2}\right) + dSx_2g\left(\frac{x_2}{2}\right)$$

$$= \frac{1}{2}dSgx_1^2 + \frac{1}{2}dSgx_2^2$$

$$= \frac{1}{2}dSg(x_1^2 + x_2^2) \quad \text{...(ii)}$$

and final energy of system,

$$(U_{\text{system}})_{\text{final}} = M'gh' + M'gh'$$

$$= dV'gh' + dV'gh'$$

$$= dSx_fg\left(\frac{x_f}{2}\right) + dSx_fg\left(\frac{x_f}{2}\right)$$

$$= \frac{1}{2}dSgx_f^2 + \frac{1}{2}dSgx_f^2$$

$$= dSgx_f^2$$

$$= dSg\left(\frac{x_1 + x_2}{2}\right)^2 \quad \text{...(iii) [Using eq. (i)]}$$

Therefore, the change in energy of system,

$$\Delta U_{\text{system}} = (U_{\text{system}})_{\text{final}} - (U_{\text{system}})_{\text{initial}}$$

$$= dSg\left(\frac{x_1 + x_2}{2}\right)^2 - \frac{1}{2}dSg(x_1^2 + x_2^2)$$

$$= dSg\left(\frac{x_1^2 + x_2^2 + 2x_1x_2}{4}\right) - dSg\left(\frac{x_1^2 + x_2^2}{2}\right)$$

$$= dSg\left[\left(\frac{x_1^2 + x_2^2 + 2x_1x_2}{4}\right) - \left(\frac{x_1^2 + x_2^2}{2}\right)\right]$$

$$= dSg\left[\frac{x_1^2 + x_2^2 + 2x_1x_2 - 2(x_1^2 + x_2^2)}{4}\right]$$

$$= \frac{dSg}{4}(x_1^2 + x_2^2 + 2x_1x_2 - 2x_1^2 - 2x_2^2)$$

$$= \frac{dSg}{4}(-x_1^2 - x_2^2 + 2x_1x_2)$$

$$= -\frac{dSg}{4}(x_1^2 + x_2^2 - 2x_1x_2) = -\frac{1}{4}gdS(x_2 - x_1)^2$$

So, energy of the system will be decreased by $\frac{1}{4}gdS(x_2 - x_1)^2$.

Hence, option (d) is correct.

38. As, $p_1 + \frac{1}{2}\rho v_1^2 = p_2 + \frac{1}{2}\rho v_2^2$ (from Bernoulli's equation)

or $$p_1 - p_2 = \frac{1}{2}\rho(v_2^2 - v_1^2)$$

$$= \frac{1}{2} \times 1.3 \times (120^2 - 90^2)$$

$$= 4.095 \times 10^3 \text{ Nm}^{-2}$$

Gross lift on the wing $= (p_1 - p_2) \times \text{Area}$

$$= 4.095 \times 10^3 \times 10 \times 2$$

$$= 81.9 \times 10^3 \text{ N}$$

39. Vertical distance covered by water before striking ground $= (H - h)$. Time taken is, $t = \sqrt{2(H - g)\cdot g}$; Horizontal velocity of water coming out of hole at P, $u = \sqrt{2gh}$

$\therefore$ Horizontal range $= ut$

$$= \sqrt{2gh} \times \sqrt{2(H - g)/g}$$

$$= 2\sqrt{h(H - h)}$$

40. If h is the initial height of liquid in drum above the small opening, then velocity of efflux, $v = \sqrt{2gh}$. As the water drains out, h decreases, hence v decreases. This reduces the rate of drainage of water. Due to which, as the drainage continues, a longer time is required to drain out the same volume of water. So, clearly $t_1 < t_2 < t_3$.

41. Let the lower and upper surface of the wings of the aeroplane be at the same height h and speeds of air on the upper and lower surfaces of the wings be v_1 and v_2.

Speed of air on the upper surface of the wing,
$$v_1 = 70 \text{ m/s}$$
Speed of air on the lower surface of the wing,
$$v_2 = 63 \text{ m/s}$$
Density of the air, $\rho = 1.3 \text{ kg/m}^3$

Area, $A = 2.5 \text{ m}^2$

According to Bernoulli's theorem,
$$p_1 + \frac{1}{2}\rho v_1^2 + \rho gh = p_2 + \frac{1}{2}\rho v^2 + \rho gh$$
or
$$p_2 - p_1 = \frac{1}{2}\rho(v_1^2 - v_2^2)$$
$\therefore$ Lifting force acting on the wings,
$$F = (p_2 - p_1) \times A = \frac{1}{2}\rho(v_1^2 - v_2^2) \times A \quad \left[\because \text{ pressure} = \frac{\text{force}}{\text{area}}\right]$$
$$= \frac{1}{2} \times 1.3 \times [(70)^2 - (63)^2] \times 2.5$$
$$= \frac{1}{2} \times 1.3\ [4900 - 3969] \times 2.5$$
$$= \frac{1}{2} \times 1.3 \times 931 \times 2.5 = 1.51 \times 10^3 \text{ N}$$

42. As, weight of metal disc = total upward force

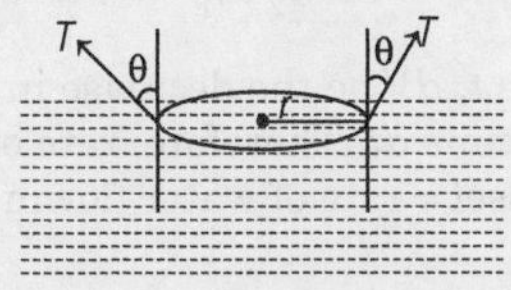

$= $ upthrust force + force due of surface tension

$= $ weight of displaced water $+ T\cos\theta\ (2\pi r)$

$= w + 2\pi rT\cos\theta$

43. Force on the ring due to surface tension of water
$$= (\pi D_1 + \pi D_2) S = mg$$
So,
$$S = \frac{mg}{\pi(D_1 + D_2)} = \frac{3.47 \times 980}{(22/7) \times (8.5 + 8.7)}$$
$$= 62.96 \text{ dyne cm}^{-1}$$

44. Let R be the radius of the biggest aluminium coin which will be supported on the surface of water due to surface tension.

Then,
$$mg = S \times 2\pi R$$
or
$$\pi R^2 t\rho g = S \times 2\pi R$$
or
$$R = 2S/\rho gt$$

45. As volume remains constant, *i.e.* $R^3 = 8000\ r^3$ or $R = 20\ r$

Now,
$$\frac{\text{Surface energy of one big drop}}{\text{Surface energy of 8000 small drops}} = \frac{4\pi R^2 T}{8000 \times 4\pi r^2 T} = \frac{R^2}{8000\ r^2}$$
$$= \frac{(20\ r)^2}{8000\ r^2} = \frac{1}{20}$$

46. If r is the radius of smaller droplet and R is the radius of bigger drop, then according to question,
$$\frac{4}{3}\pi R^3 = 10^6 \times \frac{4}{3}\pi r^3$$
or
$$r = \frac{R}{100} = 0.01\ R$$
$$= 0.01 \times 10^{-2} \text{ m} = 10^{-4} \text{ m}$$
$\therefore$ Work done = Surface tension $\times$ Increase in area
$$= 35 \times 10^{-2} \times [(10^6 \times 4\pi \times (10^{-4})^2 - 4\pi \times (10^{-3})^2]$$
$$= 4.35 \times 10^{-2} \text{ J}$$

47. Increase in surface energy = surface tension $\times$ increase in surface area
$$= S\ (1000 \times 4\pi r^2 - 4\pi R^2)$$
$$\left(100 \times \frac{4}{3}\pi r^3 = \frac{4}{3}R^2 \text{ or } r = \frac{R}{10}\right)$$
$$= S \times 4\pi\left(1000 \times \frac{R^2}{100} - R^2\right) = 36\ \pi R^2 S$$

48. Let R and R' be the radius of bubble of volume V and $2V$ respectively, then
$$\frac{4}{3}\pi R^3 = V \text{ and } \frac{4}{3}\pi R'^3 = 2V$$
So,
$$\frac{R'^3}{R^3} = 2 \quad \text{or} \quad R' = (2)^{1/3} R$$
As
$$W = S \times (4\pi R^2)^2$$
and
$$W' = S \times (4\pi R'^2)^2$$
$$\frac{W'}{W} = \frac{R'^2}{R^2} = 2^{2/3} = (4)^{1/3}$$
or
$$W' = (4)^{1/3}\ W$$

49. As,
$$\frac{4}{3}\pi R^3 = 1000 \times \frac{4}{3}\pi r^3$$
$\Rightarrow$
$$R = 10\ r$$
Surface energy of small drop, $E_1 = S \times 4\pi r^2$

Surface energy of large drop, $E_2 = S \times 4\pi\ (10\ r)^2$

$\therefore$
$$E_1 / E_2 = 1/100$$

50. Since, the bubbles coalesce in vacuum and there is no change in temperature, hence its surface energy does not change. This means that the surface area remains unchanged. Hence,
$$4\pi a^2 + 4\pi b^2 = 4\pi R^2$$
or
$$R = \sqrt{a^2 + b^2}$$

51. As, $T_c = T_0(1 - \alpha t)$, *i.e.* surface tension decreases with increase in temperature. Therefore, graph shown in option (b) is correct.

52. Change in surface area $= 2 \times 4\pi\ [(D/2)^2 - (d/2)^2]$
$$= 2\pi\ (D^2 - d^2)$$
$\therefore$ Work done = surface tension $\times$ change in area
$$= 2\pi S\ (D^2 - d^2)$$

53. Excess of pressure inside a soap bubble is given by
$$\Delta p = p_i - p_o = \frac{4T}{R}$$

Given that, $(p_i)_{\text{I}} = 1.01$ atm
and $(p_i)_{\text{II}} = 1.02$ atm
As, $(p_o)_{\text{I}} = (p_o)_{\text{II}} = 1$ atm
We have, $(\Delta p)_{\text{I}} = 0.01$ atm and
$(\Delta p)_{\text{II}} = 0.02$ atm

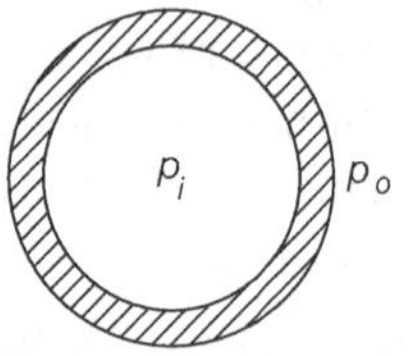

Hence, $\dfrac{(\Delta p)_{\text{I}}}{(\Delta p)_{\text{II}}} = \dfrac{4T/R_1}{4T/R_2} = \dfrac{R_2}{R_1}$

$\Rightarrow \quad \dfrac{0.01}{0.02} = \dfrac{R_2}{R_1}$

$\Rightarrow \quad \dfrac{R_1}{R_2} = \dfrac{2}{1}$

Now, ratio of volumes of soap bubbles is

$$\frac{V_1}{V_2} = \frac{\frac{4}{3}\pi R_1^3}{\frac{4}{3}\pi R_2^3} = \left(\frac{R_1}{R_2}\right)^3 = \left(\frac{2}{1}\right)^3 = 8:1$$

Hence, option (c) is correct.

54. Given, $\dfrac{T_{\text{Hg}}}{T_w} = 7.5, \ \dfrac{\rho_{\text{Hg}}}{\rho_w} = 13.6$

and $\dfrac{\cos\theta_{\text{Hg}}}{\cos\theta_w} = \dfrac{\cos 135°}{\cos 0°} = \dfrac{1}{\sqrt{2}}$

Height of the fluid inside capillary tube is given by

$$h = \frac{2T\cos\theta}{\rho g r}$$

According to given situation, $h_w = h_{\text{Hg}}$

$\therefore \quad \dfrac{2T_w\cos\theta_w}{\rho_w g r_w} = \dfrac{2T_{\text{Hg}}\cos\theta_{\text{Hg}}}{\rho_{\text{Hg}} g r_{\text{Hg}}}$

$\therefore \quad \dfrac{r_{\text{Hg}}}{r_w} = \left(\dfrac{T_{\text{Hg}}}{T_w}\right)\left(\dfrac{\cos\theta_{\text{Hg}}}{\cos\theta_w}\right)\left(\dfrac{\rho_W}{\rho_{\text{Hg}}}\right)$

Given, $r_{\text{Hg}} = r_1$ and $r_w = r_2$, then
Substituting the given values, we get

$$\frac{r_{\text{Hg}}}{r_w} = \frac{r_1}{r_2} = 7.5 \times \frac{1}{\sqrt{2}} \times \frac{1}{13.6}$$
$$= 0.4 = 2/5$$

55. By ascent formula, we have surface tension,

$$T = \frac{rhg}{2} \times 10^3 \ \frac{\text{N}}{\text{m}}$$
$$= \frac{dhg}{4} \times 10^3 \ \frac{\text{N}}{\text{m}} \qquad \left(\because r = \frac{d}{2}\right)$$

$\Rightarrow \quad \dfrac{\Delta T}{T} = \dfrac{\Delta d}{d} + \dfrac{\Delta h}{h}$ [given, g is constant]

So, percentage $= \dfrac{\Delta T}{T} \times 100 = \left(\dfrac{\Delta d}{d} + \dfrac{\Delta h}{h}\right) \times 100$

$$= \left(\frac{0.01 \times 10^{-2}}{1.25 \times 10^{-2}} + \frac{0.01 \times 10^{-2}}{1.45 \times 10^{-2}}\right) \times 100$$
$$= 1.5\%$$

$\therefore \quad \dfrac{\Delta T}{T} \times 100 = 1.5\%$

56. The excess of pressure inside the first bubble of radius r_1 is $p_1 = 4S/r_1$ and in the second bubble of radius r_2 is $p_2 = 4S/r_2$.

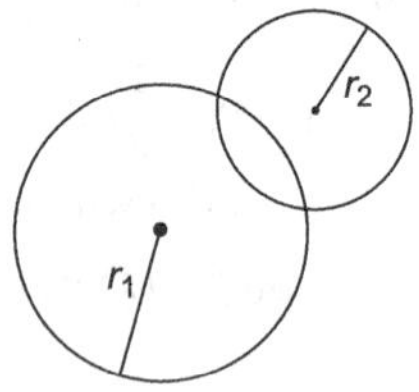

$\therefore$ Excess pressure, $p = \dfrac{4S}{r} = \dfrac{4S}{r_1} - \dfrac{4S}{r_2}$

$\Rightarrow \quad \dfrac{1}{r} = \dfrac{r_2 - r_1}{r_1 r_2}$

$\Rightarrow \quad r = \dfrac{r_1 r_2}{r_2 - r_1}$

57. Let at a time t, dV be the decrease in volume of water in vessel in time dt. Therefore, rate of decrease of water in vessel = rate of water flowing out of narrow tube

So, $\dfrac{dV}{dt} = \dfrac{\pi(p_1 - p_2)r^4}{8\eta l}$

But, $p_1 = p_2 = h\rho g$

$\therefore \quad -\dfrac{dV}{dt} = \dfrac{\pi(h\rho g)r^4}{8\eta l} = \dfrac{(\pi\rho g r^4)}{8\eta l \times A} \times (h \times A)$

where $h \times A$ = volume of water in vessel at a time $t = V$

$\therefore \quad dV = -\left(\dfrac{\pi\rho g r^4}{8\eta l A}\right) \times V\,dt = -\lambda V\,dt$

or $\quad \dfrac{dV}{V} = -\lambda\,dt$

where, $\dfrac{\pi\rho g r^4}{8\eta l A} = \lambda$ = constant

Integrating it within the limits as time changes from 0 to t, volume changes from V_0 to V.

or $\quad \log_e \dfrac{V}{V_0} = -\lambda t$ or $V = V_0 e^{-\lambda t}$

where, V_0 = initial volume of water in vessel $= Ah_0$
Therefore, $h \times A = h_0 A e^{-\lambda t}$ or $h = h_0 e^{-\lambda t}$

Thus, the variation of h and t will be represented by exponential curve as given by (a).

58. Volume of liquid flowing per second through each of the two tubes in series will be the same. So,

$$V = \frac{\pi r_1 R^4}{8\eta L} = \frac{\pi p_2 (R/2)^4}{8\eta\ (L/2)} \text{ or } \frac{p_1}{p_2} = \frac{1}{4}$$

59. Rate of flow of liquid, $V = \frac{p}{R}$

where liquid resistance, $R = \frac{8\eta l}{\pi r^2}$

For another tube, liquid resistance,

$$R' = \frac{8\eta l}{\pi\left(\frac{r}{2}\right)^4} = \frac{8\eta l}{\pi r^4}.16 = 16R$$

For series combination,

$$V_{\text{new}} = \frac{p}{R+R'} = \frac{p}{R+16R} = \frac{p}{17R} = \frac{V}{17}$$

60. As, $V = \frac{\pi p r^4}{8\eta l}$ and $V' = \frac{\pi\,(3p+p)\,(r/2)^4}{8\,\eta l}$

$$\therefore \quad \frac{V'}{V} = 4\times(1/2)^4 = \frac{1}{4}$$

or $$V' = \frac{V}{4}$$

61. Excess pressure at common surface is given by

$$p_{\text{ex}} = 4T\left(\frac{1}{a} - \frac{1}{b}\right) = \frac{4T}{r}$$

$$\therefore \quad \frac{1}{r} = \frac{1}{a} - \frac{1}{b}$$

$$r = \frac{ab}{b-a}$$

62. The meniscus of liquid in a capillary tube will be convex upwards if angle of contact is obtuse. It is so when one end of glass capillary tube is immersed in a trough of mercury.

63. As excess pressure $p \propto 1/r$, therefore pressure inside C is highest and pressure inside B is lowest. The pressure inside A is in between of B and C. Therefore, C starts collapsing with volume of A and B increasing.

64. When lift is accelerated downwards, the observed weight of body in a lift decreases. Hence, to counter balance the upward pull due to surface tension on the liquid meniscus, the height through which the liquid rises must increase.

65. There will be no over flowing of liquid in a tube of insufficient height but there will be adjustment of the radius of curvature of meniscus so that $hR = a$ finite constant.

Here, the maximum height upto which the fluid can rise is 16.3 cm. If the height of the tube is only 12 cm, Water will be able to rise only upto 12 cm because no tube means no adhesion pulling further upwards.

66. Height, $h \propto 1/R$

So, $h_1/h_2 = R_2/R_1 = 0.4/0.2 = 2$

67. Due to surface tension, water rises in the capillary tube upto a height, h with concave meniscus on both the sides. Therefore, the total height of water column in the capillary tube $= h + h = 2h$.

68. As, $h = \frac{2S\cos\theta}{r\rho g}$ (height raised $= h$)

or $$S = \frac{hr\rho g}{2\cos\theta} \text{ or } S \propto \frac{h\rho}{\cos\theta}$$

$$\therefore \quad \frac{S_w}{S_{\text{Hg}}} = \frac{h_1}{h_2}\times\frac{\cos\theta_2}{\cos\theta_1}\times\frac{\rho_1}{\rho_2}$$

$$= \frac{10}{(-3.42)}\times\frac{\cos 135^\circ}{\cos 0^\circ}\times\frac{1}{13.6}$$

$$= \frac{10}{3.42}\times\frac{0.707}{13.6} = \frac{1}{6.5}$$

69. The velocity gradient, $\frac{\Delta V}{\Delta r} = \frac{8}{0.1} = 80\text{ s}^{-1}$

70. Given, $v = \frac{2\,r^2\rho g}{9\eta}$ ...(i)

$$\text{Mass} = \frac{4}{3}\pi r^3\rho = \frac{4}{3}\pi\,(2\,r)^3\rho_1 \text{ or } \rho_1 = \rho/8$$

Terminal velocity of second ball is

$$v_1 = \frac{2\,(2\,r)^2\,(\rho/8)\,g}{8\eta} = \frac{v}{2}$$

71. Terminal velocity, $v = \frac{2R^2\,(\rho - \rho_0)\,g}{9\eta}$

or $$\frac{v}{R^2} = \frac{2\,(\rho-\rho_0)\,g}{9\eta} = \text{constant}$$

72. Terminal speed of a sphere falling in a viscous fluid is

$$v_T = \frac{2}{9}\frac{r^2}{\eta}(\rho_0 - \rho_f)g$$

where, η = coefficient of viscosity of fluid,

ρ_0 = density of falling sphere

and ρ_f = density of fluid.

As we know, if other parameters remains constant, terminal velocity is proportional to square of radius of falling sphere.

i.e. $v_T \propto r^2$ ...(i)

Now, when sphere of radius R is broken into 27 identical solid sphere of radius r, then

Volume of sphere of radius $R = 27 \times$ Volume of sphere of radius r

$$\Rightarrow \quad \frac{4}{3}\pi R^3 = 27\times\frac{4}{3}\pi r^3$$

$$\Rightarrow \quad R = 3r$$

$$\Rightarrow \quad r = \frac{R}{3}$$

So, from Eq. (i), we have

$$\frac{v_1}{v_2} = \frac{R^2}{\left(\frac{R}{3}\right)^2} = 9$$

73. When a round pebble is dropped from the top of a tall cylinder, filled with viscous oil the pebble acquires terminal velocity (*i.e.*, constant velocity) after some time.

Hence, graph in option (c) represents velocity (v) of the pebble as a function of time (t).

74. When terminal velocity v is reaching, then

$$F = 2\times10^{-5}\ v = \frac{4}{3}\pi r^3 \rho g$$

$$\Rightarrow \quad 2\times10^{-5} v = \frac{4}{3}\times\frac{22}{7}\times(1.5\times10^{-3})^3\times10^3\times10$$

On solving, $v = 7.07\ \text{ms}^{-1} \approx 7\ \text{ms}^{-1}$

75. Gravitational force remains constant on the falling spherical ball. It is represented by straight line P. The viscous force ($F = 6\pi\eta rv$) increases as the velocity increases with time. Hence, it is represented by curve Q. Net force = gravitational force – viscous force. As viscous force increases, net force decreases and finally becomes zero. Then the body falls with a constant terminal velocity. It is thus represented by curve R.

76. Work done against air friction is the average gain in kinetic energy before attaining the terminal velocity

$$W_1 = \frac{0+\frac{1}{2}mv_{\text{ter}}^2}{2} = \frac{1}{4}mv_{\text{ter}}^2$$

Work done against air friction after attaining terminal velocity is

$$W_2 = \frac{1}{2}mv_{\max}^2$$

$$\therefore \quad W_2 > W_1$$

77. If v is the terminal velocity, then equation of force,

$$xg - yg = 6\pi\eta rv$$

or $\quad v = \frac{(x-y)}{r}\frac{g}{6\pi\eta}$ or $v \propto \frac{(x-y)}{r}$

Round II

1. Angle of contact increases due to less adhesive force between wax molecule and liquid molecule, inside the capillary. Hence, height is reduced.

2. The nature of flow is determined by Reynolds number.

$$R_e = \frac{\rho v D}{\eta}$$

where, ρ = density of fluid,
η = coefficient of viscosity,
v = velocity of flow
and D = diameter of pipe.

If $R_e < 1000$ = flow is steady

$1000 < R_e < 2000$ = flow becomes unsteady

$R_e > 2000$ = flow is turbulent

$$(R_e)_{\text{initial}} = 10^3\times\frac{0.18\times10^{-3}}{\pi\times(0.5\times10^{-2})^2\times60}\times\frac{1\times10^{-2}}{10^{-3}}$$

$$= 382.16$$

$$(R_e)_{\text{final}} = 10^3\times\frac{0.48\times10^{-3}}{\pi\times(0.5\times10^{-2})^2\times60}\times\frac{1\times10^{-2}}{10^{-3}}$$

$$= 1019.09$$

So, option (d) is correct.

3. As, work done = surface tension × increase in area

$\Rightarrow \quad W$ = surface tension $\times[0.10\times0.006-0.10\times0.005]\times2$

$$= 7.2\times10^{-2}\times0.10\times0.001\times2$$

$$= 1.44\times10^{-5}\ \text{J}$$

4. As we know according to equation of continuity, when cross-section of duct decreases, the velocity of flow of liquid increases and in accordance with Bernoulli's theorem, in a horizontal pipe, the place where speed of liquid is maximum, the value of pressure is minimum. Hence, the 2nd graph correctly represents the variation of pressure.

5. When a coin placed on the top of a wooden box floating in water falls in water, upthrust on the block decreases. Due to it, l decreases as well as h decreases.

6. Excess pressure $\frac{4S}{r_1} - \frac{4S}{r_2} = \frac{4S}{r}$

or $\quad \frac{1}{r} = \frac{1}{r_1} - \frac{1}{r_2} = \frac{1}{4} - \frac{1}{5} = \frac{1}{20}$

or $\quad r = 20$ cm

7. The effective weight of ball in liquid w_3 becomes less then w_1 due to buoyancy of liquid. As, the ball immersed in liquid has some effective weight acting vertically downwards, so, $w_4 > w_2$.

8. As, net force = Averge pressure × Area – $T\times2R$

$$= \left(p_0 + \rho g\frac{h}{2}\right)(2Rh) - T2R$$

$$= |2p_0Rh + R\rho gh^2 - 2RT|$$

9. Let H be the height above O at which the total force F would have to act to produce the given torque. Then

$$H\times F = \tau$$

or $\quad H = \frac{\tau}{F}$

$$H = \frac{\rho glh^3/6}{(\rho glh^2/2)} = \frac{h}{3}$$

10. As, $v_1 = \frac{V}{A_1} = \frac{12\times10^{-6}}{6\times10^{-6}} = 2\ \text{ms}^{-1} = 200\ \text{cms}^{-1}$

and $\quad v_2 = \frac{V}{A_2} = \frac{12\times10^{-6}}{3\times10^{-6}} = 4\ \text{ms}^{-1} = 400\ \text{cms}^{-1}$

Now, $p_A - p_B = \rho g(h_2-h_1) + \frac{\rho}{2}(v_2^2 - v_1^2)$

$$= 1\times1000(100) + \frac{1}{2}(16\times10^4 - 4\times10^4)$$

$$= 10^5 + 6\times10^4$$

$$= 1.6\times10^5\ \text{dyne cm}^{-2}$$

11. As the water tries to enter the hole, it forms a liquid surface through the hole with its concave surface downward. Due to which it can withstand the pressure of the liquid upto which the canister is lowered.

$\therefore$ In equilibrium, $\frac{2T}{r} = h\rho g$

$$\Rightarrow \quad h = \frac{2\sigma}{r\rho g}$$

Putting the given values, we get

$$r = \frac{2 \times 73.5}{40 \times 1 \times 980}$$

$$= 0.00375 \text{ cm}$$

$$= 0.0375 \text{ mm}$$

12. Let V_c be the volume of cavity and V is the actual volume of gold piece [excluding volume of cavity]

$$\therefore \quad V = \frac{50}{19.3} = 2.6 \text{ cm}^3$$

Now, loss in weight of gold in water = Thrust due to water

$$\Rightarrow \quad 50\ g - 45\ g = [V + V_c]\rho_w g$$

$$\Rightarrow \quad 5 = (2.6 + V_c) \times 1$$

$$\Rightarrow \quad V_c = 2.4 \text{ cm}^3$$

13. The velocity of efflux = $\sqrt{2gh}$

The rate of flow of liquid out of hole = $Av = \pi r^2 \sqrt{2gh}$

By using equation of continuity

$$(Av)_{\text{container}} = (Av)_{\text{hole}}$$

$$\pi \frac{D^2}{4} v = \pi r^2 \sqrt{2gh}$$

$$v = \frac{4r^2}{D^2}\sqrt{2gh}$$

$\therefore$ Speed with which water level falls = $\frac{4r^2}{D^2}\sqrt{2gh}$

14. Pressure on left end of horizontal tube, $p_1 = p_0 + h_1\rho g$

Pressure on right end of horizontal tube, $p_2 = p_0 + h_2\rho g$

As $p_1 > p_2$, so acceleration should be towards right hand side. If A is the area of cross-section of the tube in the horizontal portion of U-tube, then

$$p_1 A - p_2 A = (lA\rho)a$$

or $$(h_1 - h_2)\rho g A = lA\rho a$$

or $$a = \frac{g(h_1 - h_2)}{l} \text{ towards right}$$

15. Let A be the area of cross-section of through and ρ be the density of mercury.

Initial mass of mercury in trough = $A \times 3.6 \times \rho$

Final mass of mercury in trough = $Ah'\rho$

$$= (A \times 3.6 \times \rho) \times 2$$

or $$h' = 7.2 \text{ cm}$$

16. As, $h\rho g = \frac{2S}{r}$

or $$h = \frac{2S}{r\rho g}$$

$$= \frac{2 \times 75 \times 10^{-3}}{\left(\frac{1}{2} \times 10^{-3}\right) \times 10^3 \times 10}$$

$$= 0.03 \text{ m} = 3 \text{ cm}$$

17. When jar is placed in vacuum, the liquid level rises up to the top of jar. The force exerted by liquid on the base of jar = force due to vertical column of liquid of height $(a + b + c)$ + vertical downward.

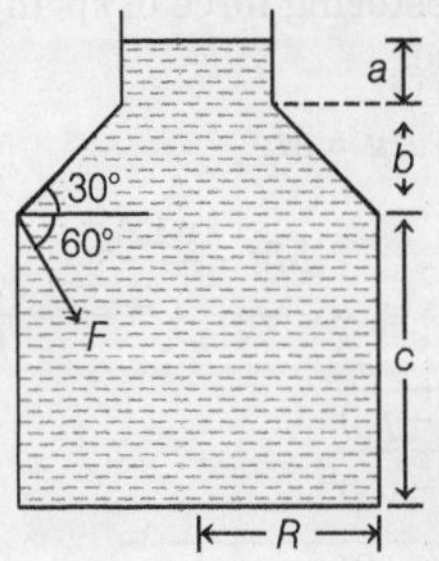

Component of thrust F acting on the portion BC of jar

$$= (a + b + c)\rho g \times \pi R^2 + F \sin 60°$$

$$= \text{greater than } (a + b + c)\rho\, g \times \pi R^2$$

18. As, level of water in tank remains constant with time, so (water inflow rate) = (outflow rate of water)

$\Rightarrow 10^{-4} \text{ m}^3\text{s}^{-1}$ = Area of orifice $\times$ Velocity of outflow

$\Rightarrow 10^{-4} \text{ m}^3\text{s}^{-1} = 10^{-4} \times \sqrt{2gh}$

where, h = Height of water above the orifice or hole.

$$\Rightarrow \quad \sqrt{2gh} = 1 \text{ or } 2 \times 9.8 \times h = 1$$

$$\Rightarrow \quad h = \frac{1}{19.6} \text{ m} = \frac{100}{19.6} \text{ cm or } h = 5.1 \text{ cm}$$

19. Weight of the drop is balanced by upthrust or buoyant force (F_B) and surface tension (T).

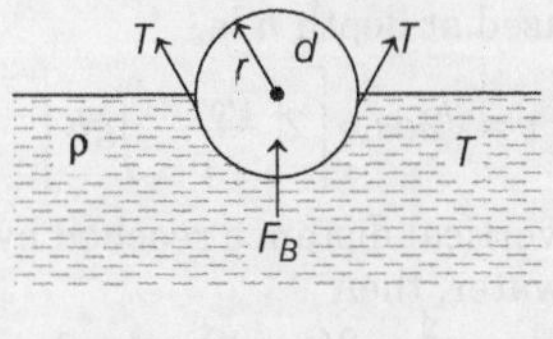

So, weight, $w = F_B + T \times 2\pi r$

$$\Rightarrow \quad d \cdot V \cdot g = \rho \cdot \frac{V}{2} g + T \times 2\pi r$$

$$\left[\because w = dVg \text{ and } F_B = \rho \frac{V}{2} g\right]$$

$$\Rightarrow d \cdot \frac{4}{3}\pi r^3 \cdot g = \rho \cdot \frac{2}{3}\pi r^3 \cdot g + T \times 2\pi r \quad \left[\because V = \frac{4}{3}\pi r^3\right]$$

$$\Rightarrow \quad r = \sqrt{\frac{3T}{(2d - \rho)g}}$$

20. In equilibrium, upward force = downward force

$$kx_0 + F_B = mg$$

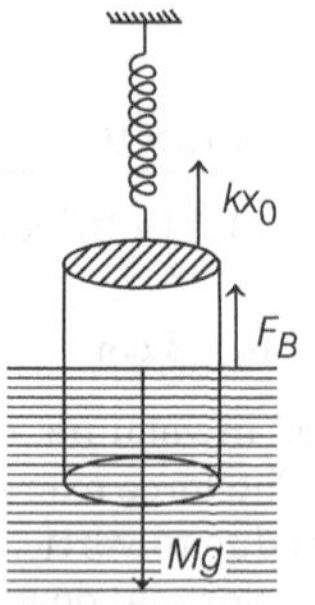

Here, kx_0 is restoring force of spring and F_B is buoyancy force.

$$kx_0 + \sigma \frac{L}{2} Ag = Mg$$

$$x_0 = \frac{Mg - \frac{\sigma LAg}{2}}{k} = \frac{Mg}{k}\left(1 - \frac{\sigma LA}{2M}\right)$$

21. $$v = a_1 a_2 \sqrt{\frac{2(p_1 - p_2)}{\rho(a_1^2 - a_2^2)}}$$

$$= \pi r_1^2 \times \pi r_2^2 \sqrt{\frac{2(p_1 - p_2)}{\rho[(\pi r_1^2)^2 - (\pi r_2^2)^2]}}$$

$$= \pi r_1^2 r_2^2 \sqrt{\frac{2(p_1 - p_2)}{\rho(r_1^4 - r_2^4)}}$$

$$= \frac{22}{7} \times (0.1)^2 \times (0.04)^2 \sqrt{\frac{2 \times 10}{(1.25 \times 10^3)[(0.1)^4 - (0.04)^4]}}$$

$$= 6.4 \times 10^{-4} \text{m}^3\text{s}^{-1}$$

22. Let V be the volume of wooden ball. The mass of ball is $m = V\rho$.

Upward acceleration,

$$a = \frac{\text{upward thrust} - \text{weight of ball}}{\text{mass of ball}}$$

$$\Rightarrow \quad a = \frac{V\rho_0 g - V\rho g}{V\rho} = \frac{(\rho_0 - \rho)g}{\rho}$$

If v' is the velocity of ball on reaching the surface after being released at depth h is

$$v = \sqrt{2as} = \left[2\left(\frac{\rho_0 - \rho}{\rho}\right)gh\right]^{1/2}$$

If h' is the vertical distance reached by ball above the surface of water, then

$$h' = \frac{v^2}{2g} = \frac{2(\rho_0 - \rho)}{\rho} gh \times \frac{1}{2g}$$

$$= \left(\frac{\rho_0 - \rho}{\rho}\right)h = \left(\frac{\rho_0}{\rho} - 1\right)h$$

23. In time Δt, momentum of water entering the hydrant

$$\mathbf{p}_1 = (\rho L \Delta t) v \hat{\mathbf{j}}$$

Momentum of water while leaving the hydrant in time Δt is

$$\mathbf{p}_2 = (\rho L \Delta t) v(-\hat{\mathbf{i}})$$

Change in momentum in time Δt is

$$\Delta \mathbf{p} = \mathbf{p}_2 - \mathbf{p}_1 = \rho L t v(-\hat{\mathbf{i}} - \hat{\mathbf{j}})$$

$$|\Delta \mathbf{p}| = \rho L \Delta t v\sqrt{(-1)^2 + (-1)^2} = \sqrt{2} \rho L \Delta t v$$

Force exerted by water, $F = \frac{|\Delta \mathbf{p}|}{\Delta t} = \sqrt{2}\rho L v$

24. If v_1 and v_2 are the speeds of air on the lower and upper surface of the wings of aeroplane and p_1, p_2 are the pressures there, then

Assume difference,

$$p_1 - p_2 = \frac{1}{2}\rho(v_2^2 - v_1^2)$$

or $$\Delta p = \rho\left(\frac{v_2 + v_1}{2}\right)(v_2 - v_1) = \rho v_{av}(v_2 - v_1)$$

Here, $v_{av} = 360 \text{ kmh}^{-1} = 100 \text{ ms}^{-1}$

$$\Rightarrow \quad \frac{v_2 - v_1}{v_{av}} = \frac{\Delta p}{\rho v_{av}^2} = \frac{mg/A}{\rho v_{av}^2} = \frac{3250 \times 10/50}{1 \times (100)^2}$$

$$= 0.065 = 6.5\%$$

25. The rate of flow of liquid (V) through capillary tube is

$$V = \frac{\pi p r^2}{8\eta h l} = p\left(\frac{\pi r^4}{8\eta l}\right) = \frac{p}{R} = \frac{\text{pressure difference}}{\text{resistance}}$$

where, $$T = \frac{8\eta h l}{\pi r^4}$$

When two tubes are in series,

total resistance, $R = R_1 + R_2$

Rate of flow of liquid, $$V' = \frac{p}{R_1 + R_2}$$

$$= \frac{p}{\frac{8\eta}{\pi}\left[\frac{l_1}{r^4} + \frac{l_2}{r_2^4}\right]}$$

$$= \frac{\pi p}{8\eta}\left[\frac{l_1}{r_1^4} + \frac{l_2}{r_2^4}\right]^{-1}$$

26. Let V_0, V_t = volume of the metal ball at 0° C and t° C respectively, $\rho_0 \rho_2$ = density of alcohol at 0° C and t° C respectively. Then

$$W_1 = W_0 - V_0 \rho_0 g$$
$$W_2 = W_t - V_t \rho_t g$$

where, $V_t = V_0(1 + \gamma_m t)$ and $\rho_t = \frac{\rho_0}{(1 + \gamma_a l)}$

Upthrust at t° C $= V_t \rho_t g$

$$= V_0(1 + \gamma_m t) \times \frac{\rho_0}{(1 + \gamma_a t)} g$$

$$= V_0 \rho_0 \frac{(1 + \gamma_m l)}{(1 + \gamma_a l)} g$$

As $\gamma_m < \gamma_a$, hence upthrust at t° C is less than at 0° C. It means upthrust has been decreased with increase in temperature. Due to which $W_2 > W_1$.

27. In the given figure, total force on the ring due to surface tension of soap film = $(2\pi b) \times 2S \sin\theta$

Mass of air entering per second the bubble

$$= \text{volume} \times \text{density}$$
$$= (Av)\rho = \pi b^2 \times v\rho$$

Momentum of air entering per sec,

$$\pi b^2 v\rho \times v = \pi^2 b^2 v^2 \rho$$

The soap bubble will separate from the ring, when force of surface tension of ring equal to the force

or $$2\pi b \times 2S \times \frac{b}{R} = \pi b^2 v^2 \rho$$

or $$R = \frac{4S}{\rho v^2}$$

28. Let k be the spring constant of spring and it gets compressed by length x in equilibrium position. Let m be the mass of the block and F be the upward thrust of water on block. When the block is at rest,

$$w = kx + F$$

or $$w - F = kx \quad \ldots(i)$$

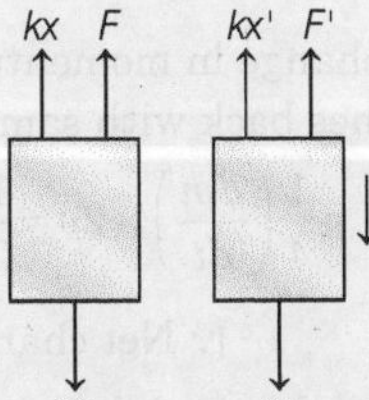

When the vessel moves downward with accleration a ($< g$), the effective downward acceleration $= g - a$. Now, upthrust is reduced and becomes F'

where $$F' = \frac{F}{g}(g-a)$$

In figure, then

$$w - kx' - F' = ma$$

or $$w - kx' - \left(\frac{g-a}{g}\right)F = \frac{wa}{g}$$

or $$(w - F) - kx' + \frac{a}{g}F = \frac{wa}{g}$$

or $$kx - kx' + \frac{a}{g}F = \frac{wa}{g}$$

or $$x' = x + (F - w)\frac{a}{gk}$$

Hence, the spring length will increase.

29. Let p_0 = atmospheric pressure, then

$$p_1 V_1 = p_2 V_2$$

or $$p_2 = p_1 \frac{V_1}{V_2}$$

or $$p_2 = p_0\left(\frac{40}{60}\right) = \frac{2}{3}p_0$$

Now, p_2 + (20cm of Hg) = p_0

or $$\frac{2}{3}p_0 + (20\text{cm of Hg}) = p_0$$

or $$\frac{p_0}{3} = 20 \text{ cm of Hg}$$

$$p_0 = 60 \text{ cm of Hg}$$

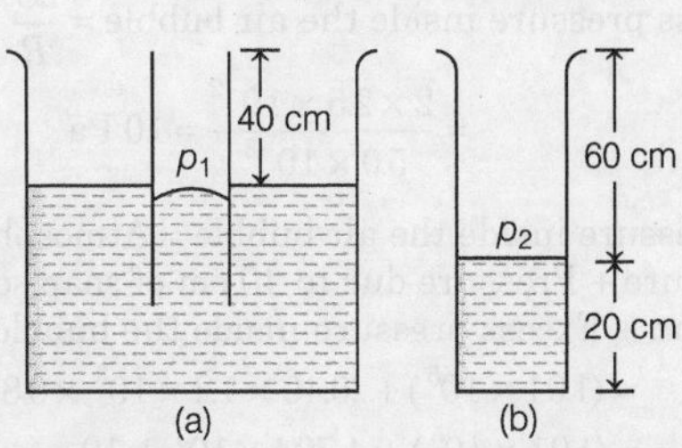

30. Volume of water in the vessel of base area A' and height h is $V = A'h$. Average velocity of out flowing water when height of water changes from h to 0 is

$$v = \frac{\sqrt{2gh} + 0}{2} = \frac{\sqrt{2gh}}{2}$$

$\therefore$ $$V\ A\,v\,t$$

When vessel is filled to height $4\,h$, the volume in vessel

$$= 4V = 4Avt = 4A\frac{\sqrt{2gh}}{2} \times t$$

If t is the time taken for the out flowing liquid and v_1 is the averege velocity of out flowing liquid, then

$$4V = A\,v_1\,t_1$$

or $$t_1 = \frac{4V}{Av_1} = \frac{4A\sqrt{2gh} \times t \times 2}{2 \times A \times \sqrt{2g \times 4h}} = 2t$$

31. Excess pressure inside the soap bubble $= \frac{4S}{r}$. So, the pressure inside the soap bubble $= p_{\text{atm}} + \frac{4S}{r}$

From ideal gas equation, $pV = nRT$

$$\frac{p_A V_A}{p_B V_B} = \frac{n_A}{n_B}$$

$$\Rightarrow \quad \frac{\left(8 + \frac{4S}{r_A}\right)\frac{4}{3}\pi r_A^3}{\left(8 + \frac{4S}{r_B}\right)\frac{4}{3}\pi r_B^2} = \frac{n_A}{n_B} \quad \ldots(i)$$

Substituting, $S = 0.04$ N/m, $r_A = 2$ cm, $r_B = 4$ cm in Eq. (i), we get

$$\frac{n_A}{n_B} = \frac{1}{6}$$

$\therefore$ $$\frac{n_B}{n_A} = 6$$

32. Given, surface tension of soap solution $(S) = 2.5 \times 10^{-2}$ N/m

Density of soap solution $(\rho) = 1.2 \times 10^3$ kg/m^3

Radius of soap bubble $(r) = 5.00$ mm

$$= 5.0 \times 10^{-3}\,\text{m}$$

Atmospheric pressure $(p_0) = 1.01 \times 10^5$ Pa

Excess pressure inside the soap bubble $= \frac{4S}{r}$

$$= \frac{4 \times 2.5 \times 10^{-2}}{5.0 \times 10^{-3}} = 20 \text{ Pa}$$

Excess pressure inside the air bubble $= \frac{2S}{R}$

$$= \frac{2 \times 2.5 \times 10^{-2}}{5.0 \times 10^{-3}} = 10 \text{ Pa}$$

$\therefore$ Pressure inside the air bubble = Atmospheric pressure + Pressure due to 40 cm of soap solution column + Excess pressure inside the bubble

$$= (1.01 \times 10^5) + (0.40 \times 1.2 \times 10^3 \times 9.8) + 10$$
$$= (1.01 \times 10^5) + 4.704 \times 10^3 + 10$$
$$= 1.01 \times 10^5 + 0.04704 \times 10^5 + 0.00010 \times 10^5$$
$$= 1.05714 \times 10^5$$
$$= 1.06 \times 10^5 \text{ Pa}$$

33. As, $AB = L$, $AC = \frac{L}{2}$; $AD = l$ (say)

Let A = area of cross-section of the rod.

Weight of the rod = $AL\rho\, g$ acting vertically downwards at C.

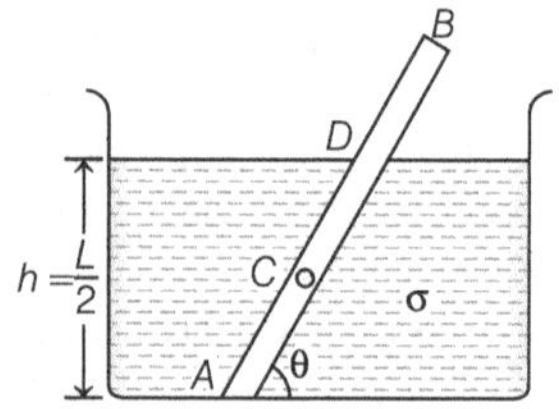

Upthrust of liquid on rod = $A\,l\,\sigma\, g$ acting upwards through the mid-point of AD.

For rotational equilibrium of rod, net torque about point A should be zero. So,

$$(LA\rho\, g)\frac{L}{2}\cos\theta = (lA\sigma g)\frac{l}{2}\cos\theta$$

or $$\frac{l^2}{L^2} = \frac{\rho}{\sigma} \text{ or } \sin\theta = \frac{1}{2}\sqrt{\frac{\sigma}{\rho}}$$

34. Height of liquid rise in capillary tube,

$$h = \frac{2T\cos\theta_c}{\rho rg}$$

$\Rightarrow$ $$h \propto \frac{1}{r}$$

So, when radius is doubled, height becomes half.

$\therefore$ $$h' = h/2$$

Now, density $(\rho) = \frac{\text{mass}(M)}{\text{volume}(V)}$

$\Rightarrow$ $$M = \rho \times \pi r^2 h$$

$\therefore$ $$M' = \rho\pi r'^2 h'$$

$\therefore$ $$\frac{M'}{M} = \frac{r'^2 h'}{r^2 h} = \frac{(2r)^2 (h/2)}{r^2 h} = 2$$

$\Rightarrow$ $$M' = 2M$$

Alternate Solution

According to the given figure, force inside the capillary tube is

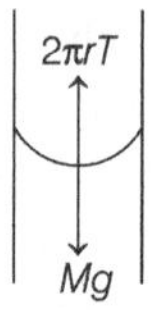

$$2\pi rT = Mg \Rightarrow M \propto r$$

When $r' = 2r$, then

$$M' = 2M$$

35. Mass per unit time of a liquid flow is given by

$$\frac{dm}{dt} = \rho Av$$

where, ρ is density of liquid, A is area through which it is flowing and v is velocity.

$\therefore$ Rate of change in momentum of the 25% of liquid which loses all momentum is

$$\frac{dp_1}{dt} = \frac{1}{4}\left(\frac{dm}{dt}\right)v = \frac{1}{4}\rho Av^2 \quad \text{...(i)}$$

and the rate of change in momentum of the 25% of the liquid which comes back with same speed.

$$\frac{dp_2}{dt} = \frac{1}{4}\left(\frac{dm}{dt}\right) \times 2v = \frac{1}{2}\rho Av^2 \quad \text{...(ii)}$$

[$\because$ Net change in velocity is $= 2v$]

$\therefore$ Net pressure on the mesh is

$$p = \frac{F_{\text{net}}}{A} = \frac{(dp_1/dt + dp_2/dt)}{A} \quad \left[\because F = \frac{dp}{dt}\right]$$

$\therefore$ From Eqs. (i) and (ii), we get

$$p = \frac{3}{4}\rho v^2 A/A = \frac{3}{4}\rho v^2$$

36. The bubble will detach if,

Buoyant force $\geq$ Surface tension force

$$\frac{4}{3}\pi R^3 \rho_w g \geq \int T \times dl \sin\theta$$

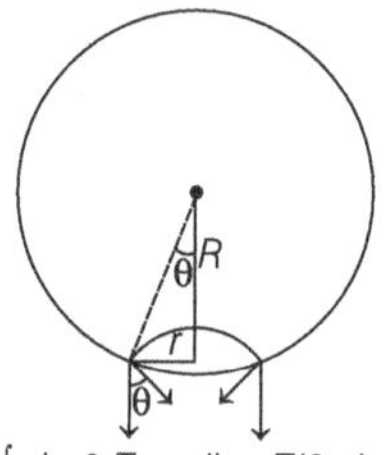

$\int \sin\theta\, T \times dl = T(2\pi r)\sin\theta$

$$(\rho_w)\left(\frac{4}{3}\pi R^3\right) g \geq (T)(2\pi r)\sin\theta$$

$$\sin\theta = \frac{r}{R}$$

Solving, we get $r = \sqrt{\frac{2\rho_w R^4 g}{3T}} = R^2\sqrt{\frac{2\rho_w g}{3T}}$

37. By equation of continuity for sections A and B, we have

$$A_A v_A = A_B v_B$$

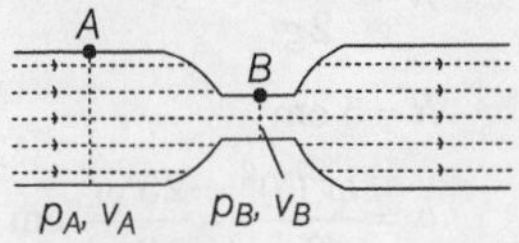

$$\Rightarrow \quad 40\, v_A = 20\, v_B$$

$$\Rightarrow \quad 2\, v_A = v_B \quad \text{...(i)}$$

Now, using Bernoulli's equation (for horizontal tube), we have

$$p_A + \frac{1}{2}\rho v_A^2 = p_B + \frac{1}{2}\rho v_B^2$$

$$\Rightarrow \quad p_A - p_B = \frac{1}{2}\rho(v_B^2 - v_A^2)$$

Here, $p_A - p_B = 700 \text{ Nm}^{-2}$ and $\rho = 1000 \text{ kg m}^{-3}$

$$\Rightarrow \quad 700 = \frac{1}{2} \times 1000\, (v_B^2 - v_A^2)$$

$$\Rightarrow \quad v_B^2 - v_A^2 = 1.4 \quad \text{...(ii)}$$

From Eqs. (i) and (ii), we have

$$3\, v_A^2 = 1.4$$

$$\Rightarrow \quad v_A = \sqrt{0.467} = 0.68 \text{ ms}^{-1} = 68 \text{ cm s}^{-1}$$

So, volume flow rate of water $= A_A \cdot v_A$

$$= 40 \times 68 = 2720 \text{ cm}^3/\text{s}$$

38. Let $m_1 g$ and $m_2 g$ be the mass of Cu and Zn respectively in alloy,

$$\therefore \quad \text{Volume of Cu} = \frac{m_1}{8.9} \text{ cc}$$

$$\text{and} \quad \text{Volume of Zn} = \frac{m_2}{7.1} \text{ cc}$$

$$\therefore \text{Total volume of alloy} = \left[\frac{m_1}{8.9} + \frac{m_2}{7.1}\right] \text{cc}$$

Now, loss of wt. in water = thrust due to water

$$\Rightarrow \quad (16.8 - 14.7)g = \left[\frac{m_1}{8.9} + \frac{m_2}{7.1}\right] \times 1g$$

$$\Rightarrow \quad 2.1 = \frac{m_1}{8.9} + \frac{m_2}{7.1} \quad \text{...(i)}$$

Also, the total mass of alloy

$$m_1 + m_2 = 16.8 \text{ g} \quad \text{...(ii)}$$

Solving Eqs. (i) and (ii), we get

$$m_1 = 9.345 \text{ g}$$

$$\text{and} \quad m_2 = 7.455 \text{ g}$$

39. Here, $v_1 = \sqrt{2g(h+x)}$; $v_2 = \sqrt{2gx}$

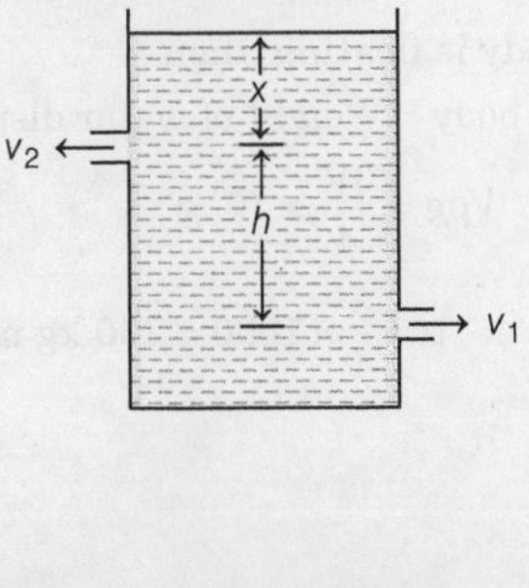

Let, a = area of cross-section of each hole

ρ = density of the liquid

The momentum of the liquid flowing out per second through lower hole = mass × velocity

$$= a v_1 \rho \times v_1 = a\rho v_1^2$$

The force exerted on the lower hole towards left $= a\rho v_1^2$

Similarly, the force exerted on the upper hole towards right

$$= a\rho v_2^2$$

Net force on the tank, $F = a\rho(v_1^2 - v_2^2)$

$$= a\rho[2g(h+x) - 2gx]$$

$$= 2a\rho gh$$

$$\Rightarrow \quad F \propto h$$

40. If V is the volume of the body, its weight $= V\rho_1 g$. Velocity gained by body when it falls from a height $h_1 = \sqrt{2gh_1}$. The weight of liquid displaced by the body as body starts immersing into the liquid $= V\rho_2 g$. The net retarding force on the body when it starts going in the liquid, $F = V(\rho_2 - \rho_1)g$

$$\therefore \text{Retardation,} \quad a = \frac{F}{V\rho_1} = \left[\frac{V(\rho_2 - \rho_1)g}{V\rho_1}\right]$$

The time of immersion of the body is that time in which the velocity of the body becomes zero. Using the relation $v = u + at$, we have $v = 0$, $u = \sqrt{2gh_1}$,

$$a = \frac{v(\rho_2 - \rho_1)g}{V\rho_1} = -\left(\frac{\rho_2 - \rho_1}{\rho_1}\right)g$$

$$\text{we have} \quad 0 = \sqrt{2gh_1} = \left(\frac{\rho_2 - \rho_1}{\rho_1}\right)g$$

$$\text{or} \quad t = \sqrt{\frac{2h_1}{g}} \times \left(\frac{\rho_1}{\rho_2 - \rho_1}\right)$$

41. Let A be the circular area over which the liquid wets the plate and d be the distance between two plates. Mass of liquid drop, $m = Ad\rho$. If S is the force of surface tension of water, then excess of pressure inside the liquid film in excess of atmospheric pressure is given by

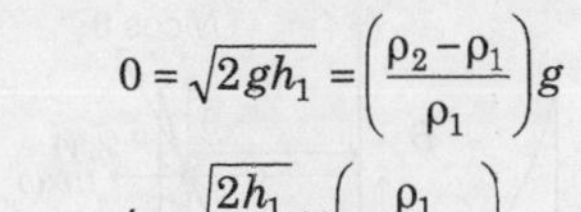

$$p = \frac{S}{r} = \frac{S}{d/2} = \frac{2S}{d}$$

Force of attraction between the plates,

$$F = \frac{2S}{d} A \qquad \left[\because p = \frac{F}{A}\right]$$

$$F = \frac{2S}{\rho d^2} \times A\rho d = \frac{2Sm}{\rho d^2}$$

$$\Rightarrow \quad \frac{2 \times 0.07 \times (80 \times 10^{-6})}{10^3 \times (4 \times 10^{-8})} = 0.28 \text{ N}$$

42. When, there is an equal level of liquid in two arms of U-tube, the height of liquid in each arm of U-tube $= \frac{h_1 + h_2}{2}$. We may consider that a height,

$h_1 - \frac{(h_1 + h_2)}{2} = \frac{h_1 - h_2}{2}$ of the liquid has been transferred from left arm to right arm of U-tube $= \left(\frac{h_1 - h_2}{2}\right) A\rho$

where, A = area of cross-section of tube and ρ = density of liquid.

The decrease in height of this liquid $= \left(\frac{n_1 - n_2}{2}\right)$

Loss in potential energy of this liquid $= \left(\frac{h_1 - h_2}{2}\right)^2 A\rho g$

The mass of the entire liquid in U-tube

$$= (h_1 + h_2 + h)\rho A$$

If this liquid moves with velocity v, then its

$$\text{KE} = \frac{1}{2}(h_1 + h_2 + h)\rho A v^2$$

Using law of conservation of energy, we have

$$\frac{1}{2}(h_1 + h_2 + h)\rho A v^2 = \left(\frac{h_1 - h_2}{2}\right)^2 A\rho g$$

or $$v = (h_1 - h_2)\sqrt{\frac{g}{2(h_1 + h_2 + h)}}$$

43. Let tangent at any point $P(x, y)$ makes angle θ with X-axis as shown below

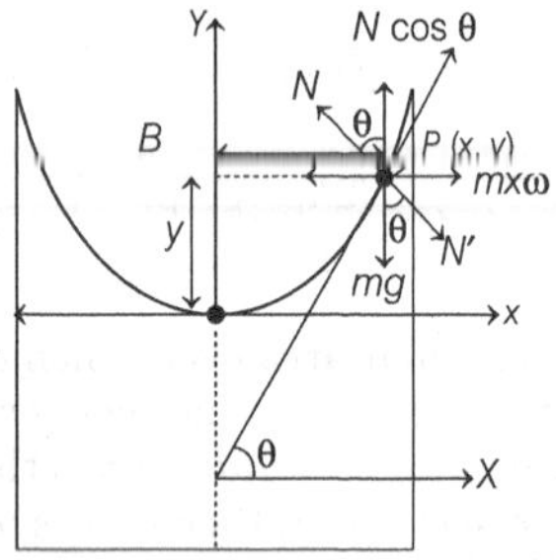

In rotating frame of vessel, let N be normal reaction. Then,

$$N\cos\theta = mg \quad \text{and} \quad N\sin\theta = mx\omega^2$$

$$\Rightarrow \quad \tan\theta = \frac{\omega^2 x}{g}$$

But $\tan\theta = \frac{dy}{dx}$ = slope of tangent

$$\therefore \quad \frac{dy}{dx} = \frac{\omega^2 x}{g}$$

$$\Rightarrow \quad y = \int dy = \int \frac{\omega^2 x}{g} \cdot dx$$

$$\Rightarrow \quad y = \frac{\omega^2}{g} \cdot \frac{x^2}{2}$$

At topmost point of profile of rotating fluid body, $y = h$ and $x = R$

$$\Rightarrow \quad h = \frac{\omega^2 R^2}{2g}$$

Given, $R = 5$ cm

$$\Rightarrow \quad h = \frac{\omega^2 (5)^2}{2g} = \frac{25\,\omega^2}{2g}\text{ cm}$$

Hence, correct option is (a).

44.

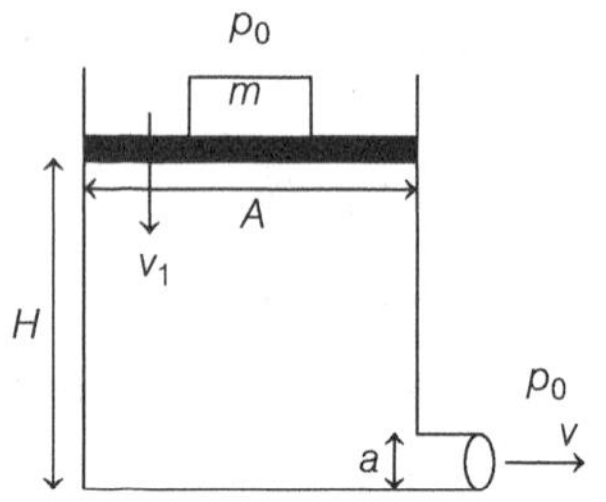

Given, $m = 24$ kg

$A = 0.4\text{ m}^2$

$a = 1\text{ cm}^2$

$H = 40$ cm

Using Bernoulli's equation,

$$\Rightarrow \quad \left(p_0 + \frac{mg}{A}\right) + \rho g H + \frac{1}{2}\rho v_1^2$$

$$\Rightarrow \quad p_0 + 0 + \frac{1}{2}\rho v^2 \qquad \text{...(i)}$$

Neglecting v_1, we get

$$\Rightarrow \quad v = \sqrt{2gH + \frac{2mg}{A\rho}}$$

$$\Rightarrow \quad v = \sqrt{8 + 1.2}$$

$$\Rightarrow \quad v = 3.033\text{ m/s}$$

$$\Rightarrow \quad v = 3\text{ m/s}$$

45. Let V and ρ be volume and density of solid respectively and ρ' be the density of water, *i.e.* $p' = 10^3\text{ kg m}^{-3}$

Weight of body $= V\rho g$

Volume of solid body outside water $= V/4$

$\therefore$ Volume of solid body inside water

$$= V - V/4 = 3V/4$$

Weight of water displaced by solid

$$= \frac{3V}{4} \times 10^3 \times g$$

As solid body is floating, then

Weight of body = Weight of water displaced by it

$$V\rho g = \frac{3V}{4} \times 10^3 g$$

$$\rho = \frac{3}{4} \times 1000 = 750\text{ kg m}^{-3}$$

46. Here, $r = 2 \times 10^{-3}$m, $D = 2r = 2 \times 2 \times 10^{-3} = 4 \times 10^{-3}$ m;

$$\eta = 2.084 \times 10^{-3} \text{ Pa-s};$$

$$\rho = 1.06 \times 10^{3} \text{ kg m}^{-3}$$

For flow to be laminar, $N_R = 2000$

Now, $$v = \frac{N_R \eta}{\rho D} = \frac{2000 \times (2.084 \times 10^{-3})}{(1.06 \times 10^{3}) \times (4 \times 10^{-3})}$$

$$= 0.98 \text{ ms}^{-1}$$

47. Here, r = 30.0 mm = 3 cm, F = 3.03 gf = 3.03 × 980 dyne. Since, the liquid is touching the ring, both inside as well as outside therefore, force acting on the ring due to surface tension is given by

$$F'' = 2\,(S \times \text{circumference of ring})$$

$$= 2(S \times 2\pi r) = 4S\pi r$$

$$= 4 \times S \times \frac{22}{7} \times 3 \text{ dyne}$$

As, $$F' = F = 4 \times S \times \frac{22}{7} \times 3 = 3.03 \times 980$$

$$S = \frac{3.03 \times 980 \times 7}{4 \times 22 \times 3} = 78.76 \text{ dyne cm}^{-1}$$

48. Here, side of the square plate, $l = 10$ cm

Area of the plate = 100 cm^2, dv = 10 cms^{-1},

$$F = 200 \text{ dyne},$$

$$\eta = 0.01 \text{ poise}, dx = ?$$

As, $F = \eta A \dfrac{dv}{dx}$

$$dx = \frac{\eta A dv}{F} = \frac{0.01 \times 100 \times 10}{200} = 0.05 \text{ cm}$$

$$= 5 \times 10^{-2} \text{m}$$

$$= x \times 10^{-2} \text{m (given)}$$

$\therefore \quad x = 5$

49. Here, density of ice, $\rho = 917$ kgm^{-3}

Density of fresh water, $\rho = 1000$ kgm^{-3}

Let V be the total volume of the ice and v be the volume of ice above the water. Then volume of the water displaced by the immersed part of ice = $(V - v)$

According to law of floatation,

weight of ice = weight of the water displaced

$$V \times 917 \times g = (V - v) \times 1000 \times g$$

or $$1000\,v = 1000\,V - 917\,V = 83\,V$$

or $$\frac{v}{V} = \frac{83}{1000} = 0.083$$

50. Height of liquid in capillary tube,

$$h = \frac{2S \cos\theta}{\rho g r}$$

$$\Rightarrow \quad S = \frac{\rho g r h}{2 \cos\theta}$$

Here, $\rho = 900$ kgm^{-3},

$g = 10$ ms^{-2},

$r = 0.015$ cm $= 15 \times 10^{-5}$ m,

$h = 15$ cm $= 15 \times 10^{-2}$ m and $\theta \approx 0°$

$\Rightarrow \cos\theta = 1$

Hence, $$S = \frac{900 \times 10 \times 15 \times 10^{-5} \times 15 \times 10^{-2}}{2 \times 1}$$

$$= 101.25 \times 10^{-3} \text{ N/m} \approx 101 \text{ mN/m}$$

51.

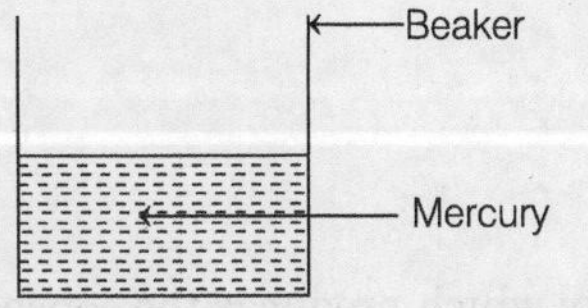

Volume of empty part = Volume of beaker − Volume of mercury

Also it is given that, there is no change in unfilled volume of the beaker with the varying temperature.

So, change in volume of beaker = change in volume of mercury

$$\Rightarrow \quad V_b \gamma_b \Delta T = V_m \gamma_m \Delta T$$

$$\Rightarrow \quad V_b \gamma_b = V_m \gamma_m$$

$$\Rightarrow \quad V_m = \frac{V_b \gamma_b}{\gamma_m} = \frac{500 \times 6 \times 10^{-6}}{1.5 \times 10^{-4}}$$

$$= 20 \text{ cm}^3 = 20 \text{ cc}$$

13

Thermometry, Calorimetry and Heat Transfer

IN THIS CHAPTER ....

Heat

A form of energy which produces the sensation of warmth is known as *heat*. The SI unit of heat is joule (J). It is mainly measured in calorie.

Thermometry

The branch that deals with measurement of temperature is called thermometry and the devices used to measure temperature are called *thermometers*.

To establish the measurement of temperature, a property of a substance is used which changes linearly with temperature. Such property of a substance is called *thermometric property*.

For example, at changing temperature, change in pressure of gas at constant volume takes place. Also, there is change in electric resistance of a metallic wire, etc. Let thermometric properties at temperatures 0°C (ice point), 100°C (steam point) and t°C (unknown temperature) are X_0, X_{100} and X_t, respectively, then

$$\frac{X_t - X_0}{t} = \frac{X_{100} - X_0}{100} \quad \text{or} \quad \frac{X_t - X_0}{X_{100} - X_0} = \frac{t}{100}$$

Thus, $$t = \left(\frac{X_t - X_0}{X_{100} - X_0}\right) \times 100^\circ \text{C}$$

Triple Point

Triple point is a state in which ice, water and water vapour can stay together in equilibrium. It refers to temperature at the equilibrium.

$\therefore$ The temperature scale is given by the equation

$$T = \lim_{p_{tr} \to 0} \frac{p}{p_{tr}} \times 273.16 \text{ K}$$

where, p = pressure and p_{tr} = pressure at equilibrium and K stands for kelvin scale of temperature.

If values of thermometric properties at 0 K, 273.16 K and T K are 0, X_{tr} and X, respectively. Then,

$$\frac{T}{T_{tr}} = \frac{X}{X_{tr}}$$

or $$T = \frac{X}{X_{tr}} T_{tr} = \left[\frac{X}{X_{tr}} \times 273.16\right] \text{K}$$

Types of Thermometers

Some different types of thermometer are discussed below

Constant-Volume Gas Thermometer If p_0 and p_{100} are the pressures of gas at temperatures 0°C and 100°C, then triple point of water and unknown temperature (t°C) respectively, will be

$$t = \left(\frac{p - p_0}{p_{100} - p_0} \times 100\right)° \text{C}$$

or $$T = \left(273.16 \frac{p}{p_{tr}}\right) \text{K}$$

Platinum Resistance Thermometer If R_0 and R_{100} are the resistances of a platinum wire at temperatures 0°C and 100°C, then triple point of water and unknown temperature (t°C) respectively, will be

$$t = \left(\frac{R_t - R_0}{R_{100} - R_0} \times 100\right)° \text{C}$$

or $$T = \left(\frac{R_t}{R_{tr}} \times T_{tr}\right) = \left(\frac{R_t}{R_{tr}} \times 273.16\right) \text{K}$$

Mercury Thermometer In this thermometer, the length of a mercury column from some fixed point is taken as thermometric property. Thus,

$$t = \left(\frac{l_t - l_0}{l_{100} - l_0} \times 100\right)° \text{C}$$

or $$T = \left(\frac{l_t}{l_{tr}} \times 273.16\right) \text{K}$$

Ranges of Different Thermometers

Thermometers	Lower Limit	Upper Limit
Mercury thermometer	– 30°C	300°C
Gas thermometer	– 268°C	1500°C
Platinum resistance thermometer	– 200°C	1200°C
Thermo-couple thermometer	– 200°C	1600°C
Radiation thermometer	– 50°C	6000°C

Saturated and Unsaturated Vapour Pressure

When a space actually contains the maximum possible amount of vapour, the vapour is called *saturated*. If the amount is less than the maximum possible amount of vapour, the vapour is called *unsaturated*.

Dew Point

The temperature at which the saturation vapour pressure is equal to the present vapour pressure is called *dew point*.

If the temperature is decreased below the dew point, source of the vapour condenses.

Humidity and Relative Humidity

The amount of water vapour present in a unit volume of air is called the *absolute humidity* of air. It is always measured in gm^{-3}. The ratio of the amount of water vapour required to saturate the volume at same temperature is called *relative humidity*. Relative humidity is generally expressed as a percentage.

Temperature Scales

The Centigrade (°C), Fahrenheit (°F), Kelvin (K), Reaumer (R) and Rankine (Ra) are commonly used temperature scales.

Various Temperature Scales with LFP and UFP

Temperature Scale	°C	°F	K	°R	°Ra	Another Scale (T)
LFP	0	32	273	0	492	T_L
UFP	100	212	373	80	672	T_U
Number of divisions (N)	100	180	100	80	180	$T_U - T_L$

All these temperatures are related to each other by the following relationship

$$\frac{C-0}{100-0} = \frac{F-32}{212-32} = \frac{K-273}{373-273} = \frac{R-0}{80-0} = \frac{Ra-492}{672-492} = \frac{T-T_L}{T_U-T_L}$$

or $$\frac{C-0}{100} = \frac{F-32}{180} = \frac{K-273}{100} = \frac{R-0}{80} = \frac{Ra-492}{180} = \frac{T-T_L}{T_U-T_L}$$

In general, whenever we have to go from any known scale to any unknown scale, then we follow the equation

$$\frac{\begin{pmatrix}\text{Temperature on}\\ \text{known scale}\end{pmatrix} - \begin{pmatrix}\text{LFP for}\\ \text{known scale}\end{pmatrix}}{(\text{UFP} - \text{LFP})_{\text{known scale}}} = \frac{\begin{pmatrix}\text{Temperature on}\\ \text{unknown scale}\end{pmatrix} - \begin{pmatrix}\text{LFP for}\\ \text{unknown scale}\end{pmatrix}}{(\text{UFP} - \text{LFP})_{\text{unknown scale}}}$$

Example 1. *A thermometer graduated according to a linear scale reads a value x_0, when in contact with boiling water and $x_0/3$, when in contact with ice. What is the temperature of an object (in °C), if this thermometer in the contact with the object reads $x_0/2$?* **[JEE Main 2019]**

(a) 35 (b) 60 (c) 40 (d) 25

Sol. (d) By principle of thermometry, for any linear temperature scale,

$$\frac{T - T_{LFP}}{T_{UFP} - T_{LFP}} = \text{a (constant)}$$

where, T = temperature measured,

T_{LFP} = temperature of melting ice or lower fixed point

and T_{UFP} = temperature of boiling water or upper fixed point.

If T = temperature of given object, then

we have, $$\frac{T - 0° C}{100° C - 0° C} = \frac{\frac{x_0}{2} - \frac{x_0}{3}}{x_0 - \frac{x_0}{3}} \Rightarrow \frac{T}{100} = \frac{1}{4}$$

$$\Rightarrow \quad T = 25° C$$

Example 2. *The resistance of a platinum-resistance thermometer is found to be 11.00 Ω when dipped in a triple point cell. When it is dipped in a bath, resistance is found to be 28.887 Ω. Find the temperature of bath in °C on platinum scale.*

(a) 444.17 °C (b) 333.17 °C (c) 450.17 °C (d) 350.17 °C

Sol. (a) In terms of triple point of water, temperature of bath,

$$T_K = \left[273.16 \frac{R}{R_{tr}}\right] K$$

So, $$T_K = 273.16 \times \frac{28.887}{11.00} = 717.32 \text{ K}$$

Now, as $$T_C = T_K - 273.15$$

$$T_C = 717.32 - 273.15 = 444.17° C$$

Example 3. *A faulty thermometer has its fixed points marked as 5° and 95°. The temperature of a body as measured by the faulty thermometer is 59°. Find the correct temperature of the body on Celsius scale.*

(a) 30°C (b) 45°C (c) 60°C (d) 75°C

Sol. (c) Let θ_0 be the lower fixed point of the faulty thermometer and N be the number of divisions between its lower and the upper fixed points. If a temperature C on Celsius scale corresponds to temperature θ on the scale of faulty thermometer, then

$$\frac{C - 0}{100} = \frac{\theta - \theta_0}{N} \quad \text{...(i)}$$

Here, $\theta_0 = 5°$; $N = 95 - 5 = 90$ and $\theta = 59°$

Therefore, Eq. (i) becomes,

$$\frac{C - 0}{100} = \frac{59 - 5}{95 - 5}$$

or $$C = \frac{54}{90} \times 100 \text{ or } C = 60° C$$

Thermal Expansion

It has been a general observation that heating a body is accompanied by an increase in size of the body. This phenomenon is called *thermal expansion*.

There are three types of thermal expansion

Expansion of Solids

Three types of expansion takes place in solid.

Linear expansion (expansion in length of a solid)

Consider a rod of length l_1 at a temperature θ_1. Let it be heated to a temperature θ_2 and the increased length of the rod be l_2, then

$$l_2 = l_1 (1 + \alpha\theta)$$

where, α = coefficient of linear expansion and $\theta = \theta_2 - \theta_1$.

Superficial expansion (expansion in surface area)

If A_1 is the area of solid at θ_1°C and A_2 is the area at θ_2°C, then

$$A_2 = A_1 (1 + \beta\theta)$$

where, β = coefficient of superficial (areal) expansion and $\theta = \theta_2 - \theta_1$.

Volume expansion (expansion in volume)

If V_1 is the volume of solid at θ_1°C and V_2 is the volume at θ_2°C, then

$$V_2 = V_1 (1 + \gamma\theta)$$

where, γ = coefficient of cubical (volume) expansion and $\theta = \theta_2 - \theta_1$.

Note For isotropic solids, $\beta = 2\alpha, \gamma = 3\alpha$, *i.e.* $\alpha : \beta : \gamma = 1 : 2 : 3$

Expansion of Liquids

Thermal expansion in liquids is identical to that of volume expansion in solids and is governed by the relationship which is $V = V_0(1 + \gamma\,\Delta\theta)$, γ for liquids is generally higher than that of solids.

Liquids have two coefficients of volume expansion

(i) **Coefficient of real expansion** (γ_r), which is due to the actual increase in volume of liquid due to heating.

(ii) **Coefficient of apparent expansion** (γ_a), which is apparent increase in the volume of liquid if expansion of vessel containing the liquid is not taken into account.

$$\therefore \quad \gamma_r = \gamma_a + \gamma_V$$

Here, γ_V = expansion of vessel.

Generally, with increasing temperature, the coefficient of volume expansion of liquids is about ten times greater than that of solids. Water is an exception to this rule. From 0°C to 4°C, water contracts and beyond 4°C, it expands.

Thus, density of water reaches a maximum value of 1000 kgm^{-3} at 4°C.

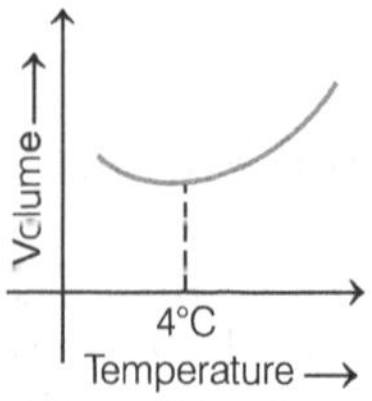

(a)

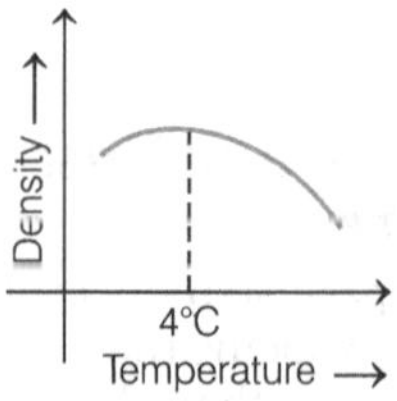

(b)

Expansion of Gases

On heating, gases expand more than solids or liquids and equal volumes of different gases expand equally, when heated by the same amount.

All gases have coefficient of volume expansion γ_V with volume variation given by

$$V = V_0(1 + \gamma_V \Delta\theta)$$

and pressure variation given by

$$p = p_0(1 + \gamma_p \Delta\theta)$$

Applications of Thermal Expansion

- There is a common misconception that, if the object having hole inside it expands, the hole will shrink because material expands into the hole.

  But in actual, if the object expands, the hole will expand too, because every linear dimension of an object changes in the same way when the temperature changes.

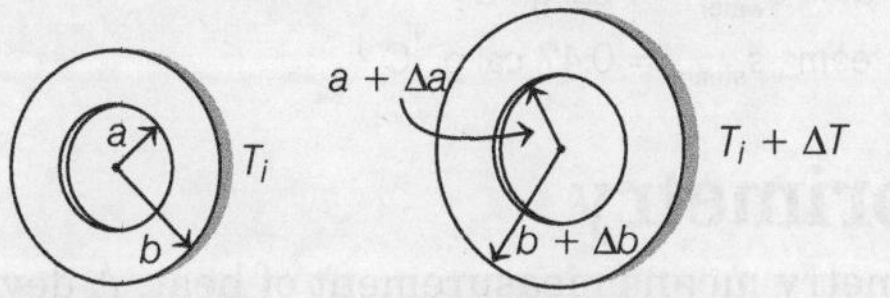

(a) Before heating (b) After heating

- **Expansion of a bimetallic strip** A bimetallic strip based on thermal expansion, is used to convert a temperature change into mechanical displacement. The strip consists of two strips of different metals which expand at different rates as they are heated. The different expansions forces on the flat strip bend in one way, if they heated and in the opposite direction, if cooled below its initial temperature.

  Each strip (made by different substance) has its own average value of coefficient of thermal expansion, therefore length of their bimetallic strip also differs. For example, the brass rod expands more than the steel rod as shown in figure below because brass has a greater average coefficient of expansion than steel.

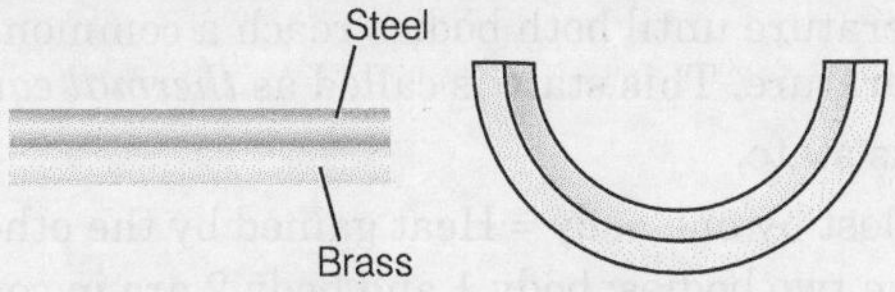

Room temperature Higher temperature

Some Important Points Related to Thermal Expansion

- **Variation of density with temperature** Most substances expand when they are heated, *i.e.*, volume of a given mass of a substance increases on heating, so the density should decrease $\left(\text{as } \rho \propto \frac{1}{V}\right)$. The density ($\rho$) varies with increase in temperature as

$$\therefore \quad \rho' = \frac{\rho}{1 + \gamma\Delta T}$$

This expression can also be written as

$$\rho' = \rho\,(1 + \gamma\Delta T)^{-1}$$

As γ is small, $(1 + \gamma\Delta T)^{-1} \approx 1 - \gamma\Delta T$

$$\therefore \quad \rho' \approx \rho\,(1 - \gamma\Delta T)$$

- **Effect of temperature on upthrust** When a solid body is completely immersed in a liquid, its apparent weight gets decreased due to an upthrust acting on it by the liquid. The apparent weight is given by,

$$w_{\text{app}} = w - F$$

Here, $F = \text{upthrust} = V_S \rho_L g$

where, V_S = volume of solid

and ρ_L = density of liquid.

Now, as the temperature is increased, V_S increases, while ρ_L decreases. So, F may increase or decrease (or may remain constant also) depending upon the condition that which factor dominates on the other.

We can write

$$F \propto V_S \rho_L \text{ or } \frac{F'}{F} = \frac{V_S'}{V_S} \cdot \frac{\rho_L'}{\rho_L}$$

After solving, we get

$$F' = F\left(\frac{1 + \gamma_S \Delta T}{1 + \gamma_L \Delta T}\right)$$

Now, if $\gamma_S > \gamma_L, F' > F$

or $w'_{\text{app}} < w_{\text{app}}$ and *vice-versa.*

and if $\gamma_S = \gamma_L, F' = F$ or $w'_{\text{app}} = w_{\text{app}}$

- **Effect of temperature on the time period of a pendulum** The time period of a simple pendulum is given by

$$T = 2\pi\sqrt{\frac{l}{g}} \text{ or } T \propto \sqrt{l}$$

As the temperature is increased, length of the pendulum is also increased and hence, time period gets increased or a pendulum clock becomes slow and it loses the time.

Time lost in time t (by a pendulum clock whose actual time period is T and the changed time period at some higher temperature is T') is $\Delta t = \left(\frac{\Delta T}{T'}\right) t$

Similarly, if the temperature is decreased, then due to decrease in length, the time period gets decreased. A pendulum clock in this case runs fast and it gains the time and time gained in time t is the same.

i.e. $$\Delta t = \left(\frac{\Delta T}{T'}\right) t$$

- **Induced stress due to change in temperature** When a rod whose ends are rigidly fixed to prevent from expansion or contraction undergoes a change in temperature, thermal stresses are developed in the rod.

This is because, if the temperature is increased, the rod has a tendency to expand but since, it is fixed at two ends, the rod exerts a force on supports.

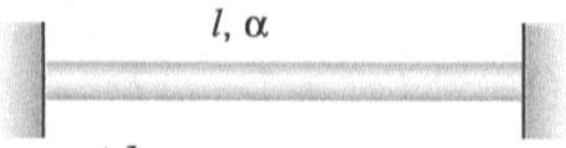

Thermal strain $= \dfrac{\Delta l}{l} = \alpha \cdot \Delta T$

So, thermal stress = (Y) (thermal strain)

$= Y\alpha\Delta T$

or force on supports $F = A$ (stress)

$= YA\alpha\Delta T$

Here, Y = Young's modulus of elasticity of the rod.

Example 4. *A steel ring of 3.000 inch inside diameter at 20°C is to be heated and slipped over a brass shaft measuring 3.002 inch in diameter at 20°C. To what temperature should the ring be heated?* (Take, $\alpha_{steel} = 1.1 \times 10^{-5}\,°C^{-1}$)

(a) 70.6°C (b) 75.6°C
(c) 80.6°C (d) 78.6°C

Sol. (b) Let θ be the temperature to which the ring must be heated.

Final diameter of ring should be 3.002 inch.

$$d_2 = d_1[1 + \alpha\Delta T]$$

$\Rightarrow$ $3.002 = 3[1 + \alpha(\theta - 20)]$ [d_2 = final diameter and d_1 = initial diameter]

$$\Rightarrow \quad \theta = \frac{3.002 - 3}{3\alpha} + 20$$

$$\Rightarrow \quad \theta = 75.6°\,C$$

Example 5. *A pendulum clock loses 12 s a day, if the temperature is 40°C and goes fast by 4 s a day if the temperature is 20°C. Find the temperature at which the clock will show correct time and the coefficient of linear expansion of the metal of the pendulum clock.*

(a) 120°C, $1.85 \times 10^{-5}\,°C^{-1}$ (b) 28°C, $1.85 \times 10^{-6}\,°C^{-1}$
(c) 25 °C, $1.85 \times 10^{-5}\,°C^{-1}$ (d) 27°C, $1.85 \times 10^{-6}\,°C^{-1}$

Sol. (c) A pendulum clock keeps proper time at temperature θ_1 and if temperature is increased to $\theta_2 > \theta_1$, then due to linear expansion, length of pendulum is increased and hence, its time period will increase. Fractional change in time period will be

$$\frac{\Delta T}{T} = \frac{1}{2}\alpha\Delta\theta$$

Let θ be the temperature at which the clock is correct.

Time lost per day = 1/2 α(rise in temperature) × 86400

$$\Rightarrow \quad 12 = 1/2\,\alpha\,(40 - \theta) \times 86400 \quad \text{...(i)}$$

Time gained per day

= 1/2 α (drop in temperature) × 86400

$$4 = \frac{1}{2}\alpha(\theta - 20) \times 86400 \quad \text{...(ii)}$$

On adding Eqs. (i) and (ii), we get

$$32 = 86400\,\alpha\,(40 - 20)$$

$$\Rightarrow \quad \alpha = 1.85 \times 10^{-5}\,°C^{-1}$$

On dividing Eqs. (i) and (ii), we get

$$12(\theta - 20) = 4(40 - \theta)$$

$$\Rightarrow \quad \theta = 25°\,C$$

Clock shows correct time at 25°C.

Specific Heat

The quantity of heat in calorie required to raise the temperature of 1 g of that substance by 1°C is known as specific heat (s) of a substance. Its unit is cal $g^{-1}C^{-1}$.

The heat lost by a body or gained from a body depends upon the difference in the temperature.

The heat lost or gained by a body, $Q = ms\Delta\theta$

where, m = mass of the body, s = specific heat

and $\Delta\theta$ = rise or fall in the temperature of body.

Note Specific heat for ice, $s_{ice} = 0.5$ cal $g^{-1}C^{-1}$

for water, $s_{water} = 1$ cal $g^{-1}C^{-1}$

for steam , $s_{steam} = 0.47$ cal $g^{-1}C^{-1}$

Calorimetry

Calorimetry means measurement of heat. A device in which heat measurement can be made is called a calorimeter.

1 calorie is the quantity of heat required to raise the temperature of 1 g of water by 1°C.

Water Equivalent

It is the quantity of water whose thermal capacity is same as the heat capacity of the body. It is denoted by W.

$W = ms$ = heat capacity of the body

Principle of Calorimetry

When two bodies at different temperatures are placed in contact with each other or mixed with each other (liquid-in-liquid, solid-in-liquid), the heat will pass from the body at higher temperature to the body at lower temperature until both bodies reach a common temperature. This state is called as *thermal equilibrium*.

At this state,

Heat lost by one body = Heat gained by the other body

Let the two bodies; body 1 and body 2 are in contact with each other.

Also, consider that the temperature of body 1 is greater than that of body 2. Therefore, heat will flow from body 1 to body 2. So,

Heat lost by body 1 = Heat gained by body 2

$$m_1 s_1(\theta_1 - \theta) = m_2 s_2(\theta - \theta_2) \qquad (\theta_2 < \theta < \theta_1)$$

where m_1 = mass of body 1,

m_2 = mass of body 2,

s_1 = specific heat of body 1 at temperature θ_1,
s_2 = specific heat of body 2 at temperature θ_2
and θ = common temperature of bodies 1 and 2 at thermal equilibrium state.

Heat Capacity

The quantity of heat required by the body to raise its temperature by 1°C is called heat capacity of a body. It is also known as *thermal capacity*.

Heat capacity = (mass × specific heat)

Change of State

Normally, matter exists in three states : solid, liquid and gas. The conversion of one of these states of matter to another is called the *change of state*.

There are two common changes of states

(i) The change of state from solid to liquid is called *melting* and from liquid to solid is called *fusion*. Both the solid and liquid states of the substance coexist in thermal equilibrium during the change of states from solid to liquid.

(ii) The change of state from liquid to gas (or vapour) is called *vaporisation*. The temperature at which the liquid and the vapour states of the substance coexist is called its *boiling point*. The change of state from solid state to vapour state without passing through the liquid state is called *sublimation*.

Latent Heat

The amount of heat that has to be supplied to (or taken from) the body during the change of state while temperature remaining constant is called *latent heat*. It is measured in terms of calg^{-1} or kcalkg^{-1}. It is given by $Q = mL$, where L is the latent heat.

Latent Heat of Fusion The quantity of heat required to change the unit mass of a solid substance to the liquid state at its melting point is called *latent heat of fusion*. For ice, latent heat of fusion is 80 calg^{-1}.

Latent Heat of Vaporisation The quantity of heat required to convert unit mass of a liquid to gaseous state at the boiling point of the liquid is called *latent heat of vaporisation*. For water, latent heat of vaporisation is 540 calg^{-1}.

Example 6. *A copper ball of mass 100 g is at a temperature T. It is dropped in a copper calorimeter of mass 100 g, filled with 170 g of water at room temperature. Subsequently, the temperature of the system is found to be 75°C. T is (Given, room temperature = 30°C, specific heat of copper = 0.1 cal/g°C)* **[JEE Main 2017]**

(a) 885°C (b) 1250°C
(c) 825°C (d) 800°C

Sol. (a) Heat gained (water + calorimeter) = Heat lost by copper ball

$\Rightarrow \; m_w s_w \Delta T + m_c s_c \Delta T = m_B s_B \Delta T$

$\Rightarrow \; 170 \times 1 \times (75-30) + 100 \times 0.1 \times (75-30) = 100 \times 0.1 \times (T-75)$

$\therefore \; T = 885°\text{C}$

Example 7. *Calculate the heat of fusion of ice from the following data for ice at 0°C added to water. Mass of calorimeter = 60 g, mass of calorimeter + water = 460 g, mass of calorimeter + water + ice = 618 g, initial temperature of water = 38°C, temperature of the mixture = 5°C. The specific heat of calorimeter = 0.10 cal $g^{-1}°C^{-1}$.*

(a) 73.85 calg^{-1} (b) 78.35 calg^{-1}
(c) 88.7 calg^{-1} (d) 84.3 calg^{-1}

Sol. (b) Mass of water = 460 − 60 = 400 g

Mass of ice = 618 − 460 = 158 g

Heat lost by water = Heat gained by ice to melt + Heat gained by (water + calorimeter) to reach 5°C

$\Rightarrow 400 \times 1 \times (38-5) = 158 \times L + 158 \times 1 \times 5 + 60 \times 0.1 \times 5$

where, L is the latent heat of fusion of ice.

$\Rightarrow \; L = 78.35 \text{ calg}^{-1}$

Example 8. *What will be the temperature, when 150 g of ice at 0°C is mixed with 300 g of water at 50°C? Specific heat of water = 1 $\text{calg}^{-1}{}^\circ\text{C}^{-1}$. Latent heat of fusion of ice = 80 calg^{-1}.*

(a) 6.0°C (b) 5.6°C
(c) 6.7°C (d) 17.6°C

Sol. (c) Let us assume that $T > 0°$ C.

Heat lost by water = Heat gained by ice to melt + heat gained by water formed from ice

$300 \times 1 \times (50 - T) = 150 \times 80 + 150 \times 1 \times (T - 0)$

$\Rightarrow \; T = 6.7°\text{ C}$

Hence, our assumption that $T > 0°$C is correct.

Example 9. *A sphere of aluminium of 0.047 kg placed for sufficient time in a vessel containing boiling water, so that the sphere is at 100°C. It is then immediately transferred to 0.14 kg copper calorimeter containing 0.25 kg of water at 20°C. The temperature of water rises and attains a steady state at 23°C. The specific heat capacity of aluminium is* **[NCERT]**

(a) 0.911 $\text{kJ kg}^{-1}\text{K}^{-1}$ (b) 211 $\text{kJ kg}^{-1}\text{K}^{-1}$
(c) 423 $\text{kJ kg}^{-1}\text{K}^{-1}$ (d) 143 $\text{kJ kg}^{-1}\text{K}^{-1}$

Sol. (a) At a steady state, heat given by an aluminium sphere will be equal to the heat absorbed by the water and calorimeter.

Mass of aluminium sphere (m_1) = 0.047 kg

Initial temperatue of aluminium sphere = 100° C

Final temperature = 23° C

Change in temperature (ΔT_1) = 100° C − 23° C = 77° C

Let the specific heat capacity of aluminium be s_{Al}, then the amount of heat gained by the aluminium sphere

$= m_1 s_{Al} \Delta T_1 = 0.047 \times s_{Al} \times 77$

Mass of water (m_2) = 0.25 kg

Mass of calorimeter $(m_3) = 0.14$ kg

Initial temperature of water and calorimeter = 20° C

Final temperature of the mixture = 23° C

Change in temperature $(\Delta T_2) = 23° \text{ C} - 20° \text{ C} = 3° \text{ C}$

Specific heat capacity of water $(s_w) = 4.18 \times 10^3 \text{ Jkg}^{-1} \text{K}^{-1}$

Specific heat capacity of copper calorimeter (s_{Cu})

$= 0.386 \times 10^3 \text{ J kg}^{-1} \text{ K}^{-1}$

The amount of heat gained by water + calorimeter

$= m_2 s_w \Delta T_2 + m_3 s_{Cu} \Delta T_2$

$= \Delta T_2 (m_2 s_w + m_3 s_{Cu})$

$= 3(0.25 \times 4.16 \times 10^3 + 0.14 \times 0.386 \times 10^3)$

In the steady state, heat lost by the aluminium sphere

= heat gained by water + calorimeter

So, $0.047 \times s_{Al} \times 77$

$= 3(0.25 \times 4.18 \times 10^3 + 0.14 \times 0.386 \times 10^3)$

$s_{Al} = 0.911 \text{ kJ kg}^{-1} \text{K}^{-1}$

Example 10. *When 0.15 kg of ice at 0° C is mixed with 0.30 kg of water at 50° C in a container, the resulting temperature is 6.7°C. The heat of fusion of ice is (Take, specific heat capacity of water, $s_w = 4186 \ J \ kg^{-1} K^{-1}$)* ***[NCERT]***

(a) $1.45 \times 10^3 \ J kg^{-1}$ (b) $3.34 \times 10^5 \ J kg^{-1}$

(c) $5.23 \times 10^6 \ J kg^{-1}$ (d) $6.23 \times 10^7 \ J kg^{-1}$

Sol. (b) Heat lost by water $= m_w s_w (T_1 - T_2)$

$= (0.30)(4186)(50 - 6.7)$

$= 54376.14$ J

Heat taken by ice $= m_i L + m_i s_w (T_2 - T_0)$

$= 0.15 \times L + 0.15 \times 4186 \times (6.7 - 0)$

$= 0.15 \, L + 4206.93$ J

Heat lost = Heat gained

$54376.14 = 0.15 \, L + 4206.93$ J

or $L = 3.34 \times 10^5 \text{ Jkg}^{-1}$

Example 11. *Heat required to convert 3 kg of ice at −12° C kept in a calorimeter to steam at 100° C at atmospheric pressure.*

[Given, specific heat capacity of ice $= 2100 \ J kg^{-1} K^{-1}$, specific heat capacity of water $= 4186 \ J kg^{-1} K^{-1}$, latent heat of fusion of ice $= 3.35 \times 10^5 \ J kg^{-1}$ and latent heat of steam $= 2.256 \times 10^6 \ J kg^{-1}$] ***[NCERT]***

(a) 1250 J (b) 9.1×10^6 J (c) 9.1×10^{-6} J (d) 1.250×10^3 J

Sol. (b) We have mass of ice $m = 3$ kg

Specific heat capacity of ice,

$s_{ice} = 2100 \text{ Jkg}^{-1} \text{K}^{-1}$

Specific heat capacity of water,

$s_{water} = 4186 \text{ Jkg}^{-1} \text{K}^{-1}$

Latent heat of fusion of ice,

$L_{ice} = 3.35 \times 10^5 \text{ Jkg}^{-1}$

Latent heat of steam,

$L_{steam} = 2.256 \times 10^6 \text{ Jkg}^{-1}$

Q = heat required to convert 3 kg of ice at −12°C to steam at 100°C

Q_1 = heat required to convert ice at −12°C to ice at 0°C

$= m \, s_{ice} \Delta T_1 = (3)(2100)[0 - (-12)]$

$= 75600$ J

Q_2 = heat required to melt ice at 0°C to water at 0°C

$= m L_{ice} = (3)(3.35 \times 10^5)$

$= 1005000$ J

Q_3 = heat required to convert water at 0°C to water at 100°C

$= m s_w \Delta T_2 = (3)(4186)(100)$

$= 1255800$ J

Q_4 = heat required to convert water at 100°C to steam at 100°C

$= m L_{steam} = (3)(2.256 \times 10^6)$

$= 6768000$ J

So, $Q = Q_1 + Q_2 + Q_3 + Q_4$

$= 75600 + 1005000 + 1255800 + 6768000$

$= 9.1 \times 10^6$ J

Phase Change of Water

For water kept at 1 atm latent heat of fusion is 80.0 cal/g. This simply means 80.0 cal of heat are required to melt 1.0 g of water or 80.0 cal heat is liberated when 1.0 g of water freezes at 0°C. Similarly, latent heat of vaporisation for water at 1 atm is 539 cal/g.

Figure shows how the temperature varies when we add heat continuously to a specimen of ice with an initial temperature below 0°C. Suppose we have taken 1 g of ice at −20° C, therefore the specific heat of ice is 0.53 cal/g-°C.

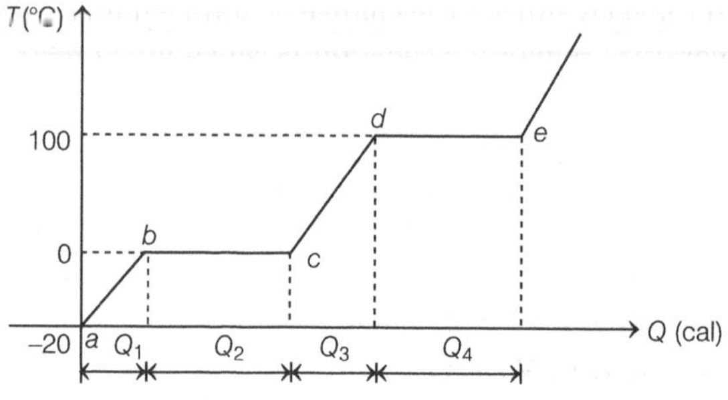

In the above figure,

a* to *b temperature of ice increases until it reaches its melting point 0°C.

$Q_1 = m s_{ice} [0 - (-20)]$

$w = (1)(0.53)(20) = 10.6$ cal

b* to *c temperature remains constant until all the ice has melted.

$Q_2 = m L_f = (1)(80) = 80$ cal

c* to *d temperature of water again rises until it reaches its boiling point 100°C.

$Q_3 = m s_{water} [100 - 0] = (1)(1.0)(100) = 100$ cal

d* to *e temperature is again constant until all the water is transformed into the vapour phase.

$$Q_4 = mL_v = (1)\,(539) = 539 \text{ cal}$$

Thus, the net heat required to convert 1 g of ice at – 20° C into steam at 100°C is

$$Q = Q_1 + Q_2 + Q_3 + Q_4 = 729.6 \text{ cal}$$

Example 12. *How much heat is required to convert 8.0 g of ice at –15° C to steam at 100°C?*
(Given, s_{ice} = 0.53 cal/g-°C, L_f = 80 cal/g and L_v = 539 cal/g and s_{water} = 1 cal/g-°C)

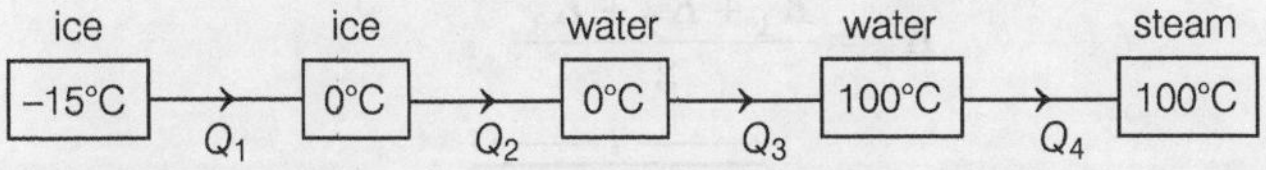

(a) 7895.1 cal (b) 5815.6 cal
(c) 9891.5 cal (d) 9210.5 cal

Sol. (b) $Q_1 = ms_{ice}\,(T_f - T_i) = (8.0)\,(0.53)\,[0 - (-15)] = 63.6$ cal

$Q_2 = mL_f = (8)\,(80) = 640$ cal

$Q_3 = ms_{water}\,(T_f - T_i) = (8.0)\,(1.0)\,[100 - 0] = 800$ cal

$Q_4 = mL_v = (8.0)\,(539) = 4312$ cal

∴ Net heat required

$$Q = Q_1 + Q_2 + Q_3 + Q_4 = 5815.6 \text{ cal}$$

Example 13. *10 g of water at 70°C is mixed with 5 g of water at 30°C. Find the temperature of the mixture in equilibrium.*

(a) 36.67°C (b) 25.98°C
(c) 42.96°C (d) 15.63°C

Sol. (a) Let t° C be the temperature of the mixture.
From energy conservation,
heat given by 10 g of water
= heat taken by 5 g of water

or $m_1 s_{water}\,|\Delta T_1| = m_2 s_{water}\,|\Delta T_2|$

∴ $(10)\,(70 - T) = 5\,(T - 30)$

∴ $T = 36.67°$ C

Example 14. *In a container of negligible mass, 30 g of steam at 100°C is added to 200 g of water that has a temperature of 40°C. The amount of heat required to attain final temperature of the system, masses of water and steam in equilibrium, respectively are (Take, L_v = 539 cal/g and s_{water} = 1 cal/g-° C)*

(a) 12000 cal, 22.6g, 222.6g
(b) 14000 cal, 15.32g, 20.96g
(c) 19000 cal, 20.36g, 15.97g
(d) 23000 cal, 18.52g, 16.56g

Sol. (a) Let Q be the heat required to convert 200 g of water at 40°C into 100°C, then

$$Q = ms\Delta T$$
$$= (200)\,(1.0)\,(100 - 40)$$
$$= 12000 \text{ cal}$$

Now, suppose m_0 mass of steam converts into water to liberate this much amount of heat, then

$$m_0 = \frac{Q}{L} = \frac{12000}{539} = 22.26 \text{ g}$$

Since, it is less than 30 g, the temperature of the mixture = 100°C
mass of steam in the mixture = 30 – 22.26 = 7.74 g
and mass of water in the mixture = 200 + 22.26 = 222.26 g.

Heat Transfer

There are three different ways in which heat can be transferred, *i.e.* conduction, convection and radiation.

Conduction

The process by which the heat is transferred in solid is known as conduction. In conduction, molecules vibrate about a fixed location and transfer the heat by collision.

Thermal Conductivity

In solids, heat is transferred through conduction.
Regarding conduction, following points are worth noting

- The **amount of heat flowing** in a rod of surface area A in time interval Δt is

$$\frac{\Delta Q}{\Delta t} = -KA\frac{\Delta\theta}{\Delta x}$$

  where, K = coefficient of heat conduction,
  $\frac{\Delta\theta}{\Delta x}$ = temperature gradient between faces of a rod

  In this expression, negative sign is used to make $\frac{\Delta Q}{\Delta t}$ a positive quantity, since it is negative.

- The **ratio of thermal and electrical conductivities** is the same for the metals at a particular temperature and is proportional to the absolute temperature of the metal. If T is the absolute temperature, then $\frac{K}{\sigma} \propto T$

  or $\frac{K}{\sigma T}$ = constant.

- Let two rods of thermal conductivities K_1, K_2 lengths l_1, l_2 and cross-sectional area A are connected in series. In steady state, the temperatures of ends of rod are T_1 and T_2 and the temperature of junction is T. Then,

$$T = \frac{K_1T_1l_2 + K_2T_2l_1}{K_1l_2 + K_2l_1}$$

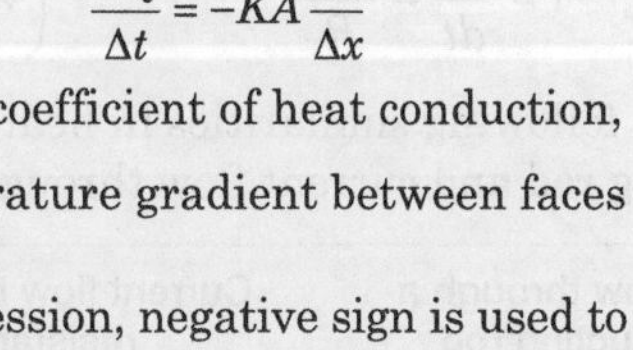

- The **thermoelectric conductivity or diffusivity** is defined as the ratio of the coefficients of thermal conductivity to the product of density and heat capacity at constant pressure. So, diffusivity, $D = \frac{K}{\rho c}$

- The hindrance offered by a body to the flow of heat is called its **thermal resistance**.

$$R = \frac{\text{Temperature difference } (\Delta T)}{\text{Heat current } (H)} = \frac{l}{KA}$$

where, l is length of rod, A is the area and ΔT is the temperature difference across its ends.

If different rods are connected in series, then heat flowing per second is same.

i.e. $H_1 = H_2 = H_3 = \ldots$

$\therefore$ $R_s = R_1 + R_2 + R_3 + \ldots$

If different rods are connected in parallel, then temperature difference is same.

i.e. $\Delta T_1 = \Delta T_2 = \Delta T_3 = \ldots$

$$\therefore \quad \frac{1}{R_p} = \frac{1}{R_1} = \frac{1}{R_2} + \frac{1}{R_3} + \ldots$$

- **Heat current** $H = \dfrac{dQ}{dt} = \dfrac{\Delta T}{R}$ $\left(\text{where, } R = \dfrac{l}{KA}\right)$

Current flow through a resistance,

$$i = \frac{dq}{dt} = \frac{\Delta V}{R} \quad \left(\text{where, } R = \frac{l}{\sigma A}\right)$$

We find the following similarities in heat flow through a conducting rod and current flow through a resistance

Heat flow through a conducting rod	Current flow through a resistance
Heat current, $H = \dfrac{dQ}{dt}$ = rate of heat flow	Electric current, $i = \dfrac{dq}{dt}$ = rate of charge flow
$H = \dfrac{\Delta T}{R} = \dfrac{TD}{R}$	$i = \dfrac{\Delta V}{R} = \dfrac{PD}{R}$
$R = \dfrac{l}{KA}$ Here, K = thermal conductivity.	$R = \dfrac{l}{\sigma A}$ Here, σ = electrical conductivity.

From the above table, it is evident that flow of heat through rods in series and parallel is analogous to the flow of current through resistances in series and parallel. This analogy is of great importance in solving complicated problems of heat conduction.

- In **series combination of rods of different materials**, equivalent conductivity will be given as follows

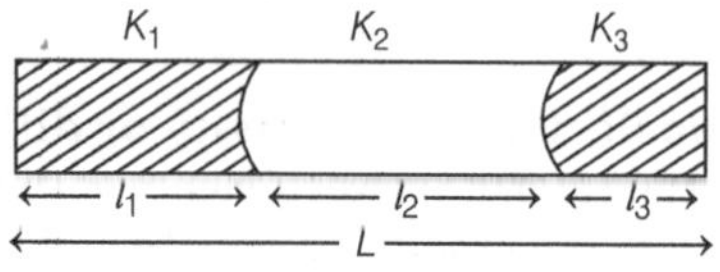

$$\frac{l_1 + l_2 + l_3}{K_s} = \frac{l_1}{K_1} + \frac{l_2}{K_2} + \frac{l_3}{K_3}$$

If lengths of rods are equal, then

$$\frac{1}{K_s} = \frac{\frac{1}{K_1} + \frac{1}{K_2} + \frac{1}{K_3}}{3}$$

- In **parallel combination of slabs of different materials**, equivalent conductivity will be given as follows

$$K_p = \frac{K_1 A_1 + K_2 A_2 + K_3 A_3}{A_1 + A_2 + A_3}$$

If areas of slabs are equal, then

$$K_p = \frac{K_1 + K_2 + K_3}{3}$$

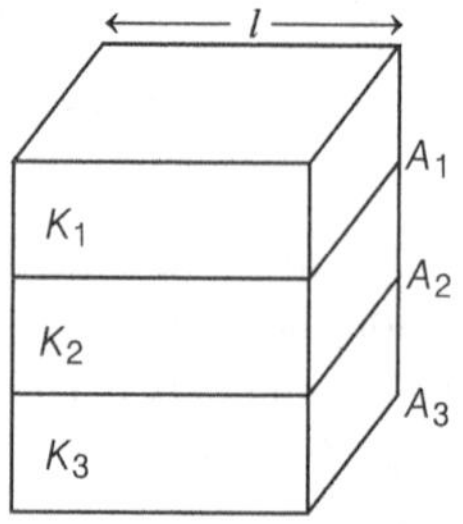

- When the atmospheric pressure falls below 0°C (say $-T$°C), the cold air above water extracts heat from the water. As a result, the water begins to freeze into the ice layers. Let at any time the thickness of ice is x and further, the layer of ice of thickness dx is formed in time dt. If ρ is density of ice and L be the latent heat of fusion, then

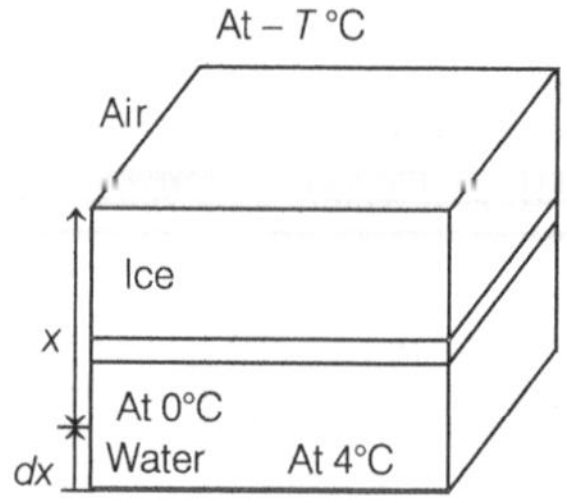

Time taken by ice to grow to a thickness x,

$$t = \frac{\rho L}{K\theta} \int_0^x x\,dx = \frac{\rho L}{2K\theta} x^2$$

If the thickness is from x_1 to x_2, then time taken

$$t = \frac{\rho L}{2KT}(x_2^2 - x_1^2)$$

where, K = coefficient of thermal conductivity of ice.

Example 15. *A cylinder of radius R is surrounded by a cylindrical shell of inner radius R and outer radius 2R. The thermal conductivity of the material of the inner cylinder is K_1 and that of the outer cylinder is K_2. Assuming no loss of heat, the effective thermal conductivity of the system for heat flowing along the length of the cylinder is* **[JEE Main 2019]**

(a) $\dfrac{K_1 + K_2}{2}$ (b) $\dfrac{K_1 + 3K_2}{4}$ (c) $\dfrac{2K_1 + 3K_2}{5}$ (d) $K_1 + K_2$

Sol. (b) Both the given cylinders are in parallel, as heat flow is given along length. In parallel, equivalent thermal conductivity of system is

$$K_{eq} = \frac{K_1A_1 + K_2A_2}{A_1 + A_2}$$

So, in given system

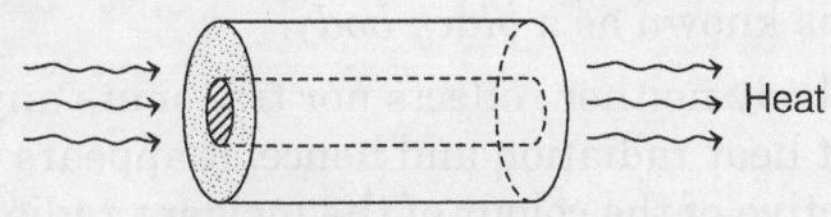

$$K_{eq} = \frac{K_1(\pi R^2) + K_2[\pi(2R)^2 - \pi R^2]}{(\pi R^2) + (4\pi R^2 - \pi R^2)}$$

or $$K_{eq} = \frac{K_1 + 3K_2}{4}$$

Example 16. *A copper rod 2 m long has a circular cross-section of radius 1 cm. One end is kept at 100°C and the other at 0°C and the surface is insulated so that negligible heat is lost through the surface. Find*

(i) the thermal resistance of the bar,
(ii) the thermal current H,
(iii) the temperature gradient $\frac{dT}{dx}$ and
(iv) the temperature 25 cm from the hot end.
Which collection of the answer is correct ?
(Given, thermal conductivity of copper is 401 $Wm^{-1}K^{-1}$)

	(i)	(ii)	(iii)	(iv)
(a)	3 kW^{-1}	6 W	5 Cm^{-1}	8°C
(b)	5.9 kW^{-1}	60 W	−50 Cm^{-1}	87.5°C
(c)	15.9 kW^{-1}	6.3 W	−50 Cm^{-1}	67.5°C
(d)	15.9 kW^{-1}	6.3 W	−50 Cm^{-1}	87.5°C

Sol. (d)

(i) Thermal resistance, $R = \frac{l}{KA} = \frac{l}{K(\pi r^2)}$

or $$R = \frac{(2)}{(401)(\pi)(10^{-2})^2} = 15.9\,kW^{-1}$$

(ii) Thermal current, $H = \frac{\Delta T}{R} = \frac{\Delta\theta}{R} = \frac{100}{15.9}$

or $H = 6.3$ W

(iii) Temperature gradient $\frac{dT}{dx} = \frac{0-100}{2} = -50$

$= -50°\ Cm^{-1}$

(iv) Let θ°C be the temperature at 25 cm from the hot end, then

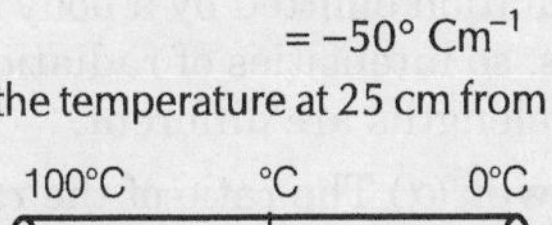

$(\theta - 100) =$ (temperature gradient) × (distance)

or $\theta - 100 = (-50)(0.25)$

$\theta = 87.5°$ C

Example 17. *Two metal cubes with 3 cm edges of copper and aluminium are arranged as shown in figure. Find*

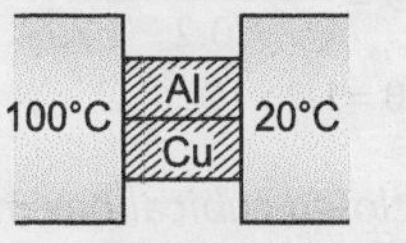

(i) the total thermal current from one reservoir to the other.
(ii) the ratio of the thermal currents carried by the copper cube to that carried by the aluminium cube.
(Take, thermal conductivity of copper = 401 $Wm^{-1}K^{-1}$ and thermal conductivity of aluminium = 237 $Wm^{-1}K^{-1}$).

(a) 0.08 kW^{-1}, 1.75
(b) 0.01 kW^{-1}, 1.05
(c) 0.25 kW^{-1}, 1.32
(d) 0.02 kW^{-1}, 1.02

Sol. (a) Thermal resistance of aluminium cube,

$$R_1 = \frac{l}{k_{Al}A}$$

or $$R_1 = \frac{(3.0\times10^{-2})}{(237)(3.0\times10^{-2})^2} = 0.14\,kW^{-1}$$

and thermal resistance of copper cube, $R_2 = \frac{l}{k_{Cu}A}$

or $$R_2 = \frac{(3.0\times10^{-2})}{(401)(3.0\times10^{-2})^2} = 0.08\,kW^{-1}$$

As, these two resistances are in parallel, their equivalent resistance will be

$$R = \frac{R_1R_2}{R_1 + R_2} = \frac{(0.14)(0.08)}{(0.14) + (0.08)} = 0.05\,kW^{-1}$$

∴ Thermal current, $H = \frac{\text{Temperature difference}}{\text{Thermal resistance}}$

$$= \frac{(100-20)}{0.05} = 1.6\times10^3\ W$$

In parallel, thermal current distributes in inverse ratio of resistances.

Hence, $$\frac{H_{Cu}}{H_{Al}} = \frac{R_{Al}}{R_{Cu}} = \frac{R_1}{R_2} = \frac{0.14}{0.08} = 1.75$$

Example 18. *Water is boiled in flat bottom kettle placed on a stove. The area of the bottom is 3000 cm^2 and the thickness is 2 mm. If the amount of steam produced is1 $gmin^{-1}$, calculate the difference of temperature between the inner and outer surfaces of the bottom. (Given, K for the material of kettle = 0.5 $cal°C^{-1}s^{-1}cm^{-1}$)*

(a) 2.1×10^{-3} °C
(b) 3.1×10^{-3} °C
(c) 1.2×10^{-3} °C
(d) 2.5×10^{-3} °C

Sol. (c) Mass of steam produced = $\frac{dm}{dt} = \frac{1}{60}\,gs^{-1}$

Heat transferred per second = $\frac{dQ}{dt} = L\frac{dm}{dt}$

$$\Rightarrow \frac{dQ}{dt} = 540 \times \frac{1}{60}\ cal°\ C^{-1}s^{-1}cm^{-1}$$

d = thickness = 2 m = 0.2 cm

$$\Rightarrow \quad \frac{dQ}{dt} = \frac{KA\theta}{d} \Rightarrow L\frac{dm}{dt} = \frac{KA\theta}{d}$$

$$\Rightarrow \quad 9 = \frac{0.5 \times 3000 \times \theta}{0.2}$$

$$\Rightarrow \quad \theta = 1.2 \times 10^{-3\circ} C$$

Example 19. *A closed cubical box made of perfectly insulating material has walls of thickness 8 cm and the only way for the heat to enter or leave the box is through the solid cylindrical metallic plugs each of cross-sectional area 12 cm^2 and length 8 cm fixed in the opposite walls of the box as shown in figure. The outer surface A is kept at 100°C while the outer surface B of other plug is kept at 4°C. K of the material of the plugs is 0.5 $cals^{-1\circ}C^{-1}cm^{-1}$. A source of energy generating 36 $cals^{-1}$ is enclosed inside the box. Find the equilibrium of the inner surface of the box, assuming that it is same at all points on the inner surface.*

(a) 76°C (b) 86°C (c) 66°C (d) 56°C

Sol. (a) Let θ be the temperature of inner surface of box. Heat transfer per second through *A* + Heat produced by source per second = Heat transfer per second through *B*

$$\Rightarrow \quad \left(\frac{dQ}{dt}\right)_A + 36 = \left(\frac{dQ}{dt}\right)_B$$

$$\Rightarrow \quad \frac{KA(100-\theta)}{d} + 36 = \frac{KA(\theta-4)}{d}$$

$$\Rightarrow \quad KA(\theta - 4 - 100 + \theta) = 36 \times d$$

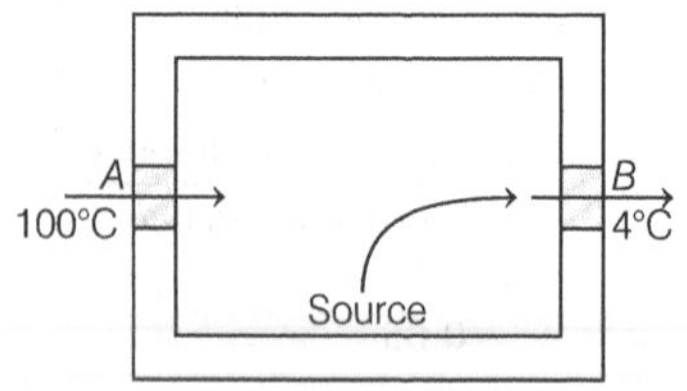

Now, $d = 8$ cm, $A = 12$ cm^2, $K = 0.5$ $cals^{-1\circ}C^{-1}cm^{-1}$

$$\Rightarrow \quad 2\theta - 104 = \frac{36 \times 8}{12 \times 0.5} \Rightarrow \theta = 76^\circ C$$

Convection

The process by which heat is transferred in fluids (liquids and gases) is known as *convection*. In convection, transfer of heat takes place by transport of matter (in form of motion of particles).

When a liquid in a container is heated, the molecules at the lower layers are heated up and their densities decrease. As a result, the molecules rise up and heavier ones come down and hence a continuous movement of molecules takes place giving rise to convection currents. In this way, the whole of liquid gets heated.

Radiation

In radiation, heat is transferred from one body to other or to the surroundings even in the absence of any medium in the intervening space. Heat energy of the Sun is transmitted to Earth through radiations.

Radiation is only a mode of transfer of energy by transverse electromagnetic waves.

Perfectly Black Body

A body which can absorb all radiations (of each wavelength at any temperature), which is incident on its surface and emits the full radiation spectrum on being heated is known as a *black body*.

Such a body neither reflects nor transmits any part of the incident heat radiation and hence, it appears black irrespective of the colour of the incident radiation. This means that a perfectly black body has unit absorptance. Also, a perfectly black body when heated, emits radiation of all possible wavelengths at that temperature.

Absorption, Reflection and Transmission

When radiations are incident on a surface, then three things happen—a part of the *radiation is absorbed*, some is *reflected back* and remaining is *transmitted.*

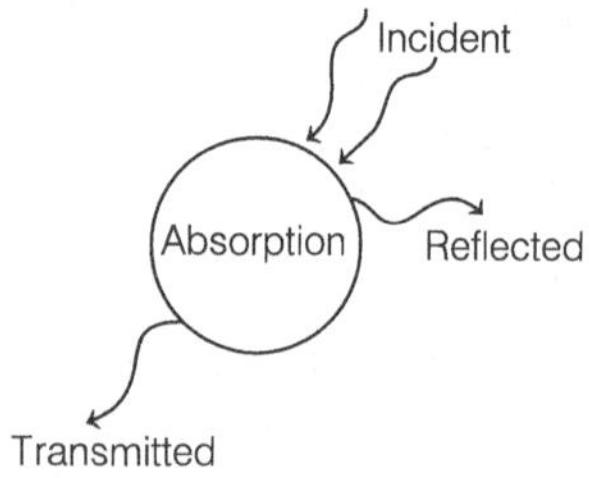

$$Q_{\text{incident}} = Q_{\text{absorbed}} + Q_{\text{reflected}} + Q_{\text{transmitted}}$$

where, Q represents the energy of thermal radiation.

Absorptivity or absorptive power, $a = \frac{Q_{\text{absorbed}}}{Q_{\text{incident}}}$

Reflectivity, $r = \frac{Q_{\text{reflected}}}{Q_{\text{incident}}}$

Transmissivity, $t = \frac{Q_{\text{transmitted}}}{Q_{\text{incident}}}$

$$\frac{Q_s}{Q} + \frac{Q_v}{Q} + \frac{Q_t}{Q} = a + r + t = 1$$

For a perfect black body, $a = 1, r = t = 0$

For a perfect reflector, $a = t = 0, r = 1$

For a perfect transmitter, $a = r = 0, t = 1$

Some Common Terms

The thermal radiation emitted by a body comprises of all the wavelengths, so intensities of radiation corresponding to different wavelengths are different.

Absorptive power (α) The ratio of the radiant energy absorbed by the body in a given time to the total radiant energy incident on it in the same interval of time is known as *absorptive power*.

$$\alpha = \frac{\text{Energy absorbed}}{\text{Energy incident}}$$

As a perfectly black body absorbs all radiations incident on it, the absorptive power of a perfectly black body is maximum and unity.

Spectral Absorptive Power (a_λ) The ratio of radiant energy absorbed by a surface to the radiant energy incident on it for a particular wavelength λ is known as spectral absorptive power. It may have different values for different wavelengths for a given surface.

The spectral absorptive power a_λ is related to absorptive power a through the relation

$$a = \int_0^\infty a_\lambda d\lambda$$

Emissive power (e) For a given surface, it is defined as the radiant energy emitted per second per unit area of the surface.

It is the total amount of energy radiated by a body per second per unit area of surface.

$$e = \frac{1}{A}\frac{\Delta Q}{\Delta t}$$

Spectral emissive power (e_λ) It is emissive power for a particular wavelength λ.

Thus, $$e = \int_0^\infty e_\lambda d\lambda$$

Emissivity (ε) The ratio of the total emissive power of the body (e) to the total emissive power of a perfect black body (E) at given temperature is known as *emissivity* of a body at a given temperature.

i.e. $$\varepsilon = \frac{e}{E}$$

Kirchhoff's Law

Kirchhoff's law states that, "at any temperature and for particular wavelength, the ratio of the emissive power to the absorptive power of all the bodies is same and is equal to the emissive power of a perfectly black body."

i.e. $$\frac{e}{a} = E \quad \text{(constant)}$$

Now, $E = 1$ (for perfectly black body)

So, $e = a$

Kirchhoff's law signifies that good absorbers are always good emitters.

Stefan's Law

According to this law, "the energy emitted per second per unit area of a black body (emissive power = 1) is proportional to the fourth power of the absolute temperature."

i.e. $$E = \sigma T^4$$

Here, σ = Stefan's constant

$$= 5.67 \times 10^{-8}\ \text{Jm}^{-2}\text{s}^{-1}\text{K}^{-4}$$

For any other body, $e = \varepsilon\sigma T^4$

Here, ε = emissivity of body ($\varepsilon = 1$ for a black body).

If Q is the total energy radiated by the ordinary body, then

$$e = \frac{Q}{A \times t} = \varepsilon\sigma T^4$$

$$\Rightarrow \quad Q = A\varepsilon\sigma T^4 t$$

Example 20. *The rate at which the radiant energy reaches the surface of the Earth from the Sun is about $1.4\ kWm^{-2}$. The distance from the Earth to the Sun is about 1.5×10^{11} m and the radius of the Sun is about 0.7×10^9 m. What is the rate of radiation of energy per unit area from the Sun's surface?*

(a) $6.43 \times 10^7\ Wm^{-2}$

(b) $6.43 \times 10^6\ Wm^{-2}$

(c) $5.43 \times 10^7\ Wm^{-2}$

(d) $6.43 \times 10^{-7}\ Wm^{-2}$

Sol. (a) Let D = distance from the Sun to the Earth $= 1.5 \times 10^{11}$ m,

radius of the Sun, $R = 0.7 \times 10^9$ m

and power of the sun, P = energy radiated from the surface of the Sun per second.

Hence, in every one second, P joule of energy are radiated from the surface of the Sun and this energy passes through a big sphere of radius D centred at the Sun.

Hence, at the circumference of this big sphere (*i.e.* near the surface of the Earth), the energy crossing through a unit area per second

$$= \frac{P}{\text{area of big sphere}} = \frac{P}{4\pi D^2}$$

$$= 1.4 \times 10^3\ \text{Wm}^{-2}$$

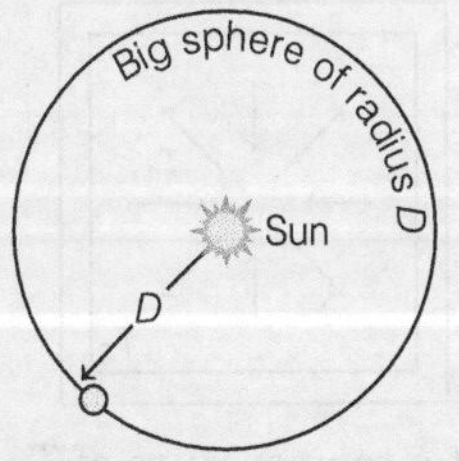

$$\Rightarrow \quad P = 4\pi(1.5 \times 10^{11})^2 \times 1.4 \times 10^3\ \text{W}$$

$$\Rightarrow \quad P = 3.96 \times 10^{26}\ \text{W}$$

Rate of radiation of energy per second per unit area of the Sun's surface is given by

$$\frac{P}{\text{area of big sphere}} = \frac{P}{4\pi R^2}$$

$$= 6.43 \times 10^7\ \text{Wm}^{-2}$$

Example 21. *In the above problem, if the Sun radiates as an ideal black body, what is the temperature of its surface?*

(a) 6803 K (b) 5603 K

(c) 5803 K (d) 5503 K

Sol. (c) If the Sun is an ideal black body, $\varepsilon = 1$

$$\Rightarrow \quad E = \sigma T^4$$

$$\Rightarrow \quad T = \left(\frac{E}{\sigma}\right)^{1/4}$$

$$= \left(\frac{6.43 \times 10^7}{5.67 \times 10^{-8}}\right)^{1/4} = 5803 \text{ K}$$

Example 22. *A black body at 227°C radiates heat at a rate of 7 cal/cm²s. At a temperature of 727°C, the rate of heat radiated in the same units will be*

(a) 112 (b) 105 (c) 101 (d) 89

Sol. (a) According to Stefan's law, $E = \sigma T^4$

$$\therefore \quad 7 = \sigma (227 + 273)^4 = \sigma \times (500)^4$$

and $$x = \sigma (727 + 273)^4 = \sigma \times (1000)^4$$

Hence, $$\frac{x}{7} = \frac{(1000)^4}{(500)^4} = 16$$

$$\Rightarrow \quad x = 16 \times 7 = 112 \text{ cal/cm}^2\text{s}$$

Net Heat Loss from the Surface of a Body

The rate at which a body radiates energy is determined by the temperature of the body and its surroundings. When a body is hotter than its surroundings, the rate of emission exceeds the rate of absorption; there is net loss of energy and the body cools down, unless it is heated by some other means.

When a body is cooler than its surroundings, the rate of absorption is greater than the rate of emission and its temperature rises. At thermal equilibrium, the two rates are equal.

Walls

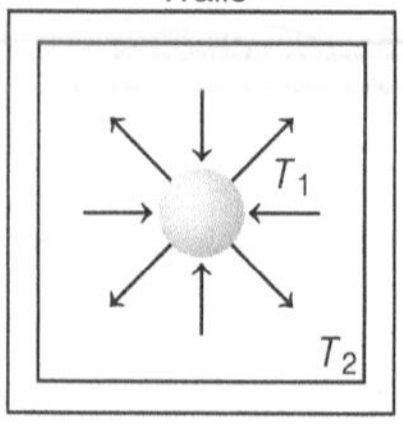

Hence, for a body at a temperature of T_1, surrounded by walls at a temperature T_2 (as shown in figure), the net rate of loss (or gain) of energy per second by radiation is

$$H_{\text{net}} = A\varepsilon\sigma T_1^4 - A\varepsilon\sigma T_2^4$$

$$H_{\text{net}} = A\varepsilon\sigma (T_1^4 - T_2^4)$$

Example 23. *The emissivity of tungsten is approximately 0.35. A tungsten sphere 1 cm in radius is suspended within a large evacuated enclosure whose walls are at 300 K. What power input is required to maintain the sphere at a temperature of 3000 K, if heat conduction along the supports is neglected? (Take, $\sigma = 5.67 \times 10^{-8}$ SI units).*

(a) 2119.8 W (b) 2019.8 W

(c) 2219.8 W (d) 1919.8 W

Sol. (b) Net heat lost by sphere per second

$$H_{\text{net}} = \varepsilon\sigma A(T^4 - T_0^4)$$

where, T = temperature of sphere = 3000 K

and T_0 = temperature of surrounding = 300 K.

Area, $A = 4\pi r^2 = 4\pi (0.01)^2$

To maintain constant temperature,

power input required = net heat loss from the surface

$$P_{\text{input}} = \varepsilon\sigma A(T^4 - T_0^4)$$

$$= 0.35 \times 5.67 \times 10^{-8} \times 4\pi (0.01)^2 \times (3000^4 - 300^4)$$

$$= 2019.8 \text{ W}$$

Newton's Law of Cooling

According to this law, "if the temperature T of the body is not very different from that of the surroundings T_0, then rate of cooling $-\frac{dT}{dt}$ is proportional to the temperature difference between them."

Newton's law of cooling formula can be expressed by

$$T(t) = T_0 + (T_1 - T_0)e^{-kt}$$

where, t = time,

$T(t)$ = temperature of the given body at time t,

T_0 = surrounding temperature,

T_1 = initial temperature of the body and K = constant.

Note Approximate solution for Newton's law of cooling is $\frac{T_1 - T_2}{t} = K\left[\frac{T_1 + T_2}{2} - T_0\right]$, where t is the time in which temperature of body changes from T_1 to T_2.

Example 24. *A body cools down from 60°C to 55°C in 30 s. Using Newton's law of cooling, calculate the approximate time taken by same body to cool down from 55°C to 50°C. Assume that the temperature of surroundings is 45°C.*

(a) 41.28 s

(b) 55.28 s

(c) 51.28 s

(d) 60.28 s

Sol. (c) According to Newton's law of cooling,

$$\frac{\theta_1 - \theta_2}{t} = K\left[\frac{\theta_1 + \theta_2}{2} - \theta_0\right]$$

$$\frac{60 - 55}{30} = K\left[\frac{60 + 55}{2} - 45\right] \quad \text{...(i)}$$

Similarly, for 2nd case,

$$\frac{55 - 50}{t} = K\left[\frac{55 + 50}{2} - 45\right] \quad \text{...(ii)}$$

Dividing Eq. (i) by Eq. (ii), we get

$$t = 51.28 \text{ s}$$

Wien's Displacement Law

According to this law, "the wavelength (λ_m) of maximum intensity of emission of black body radiation is inversely proportional to absolute temperature of the black body."

i.e. $$\lambda_m \propto \frac{1}{T}$$

or $$\lambda_m T = b \text{ (constant)}$$

where, b is Wien's constant and its value 2.89×10^{-3} m-K and λ_m is the wavelength corresponding to maximum intensity (energy constant) of radiation emitted by body at temperature T.

- The thermal radiation emitted by a body at any temperature consists of all wavelengths from small to large values.
  The intensities of all wavelengths are different.

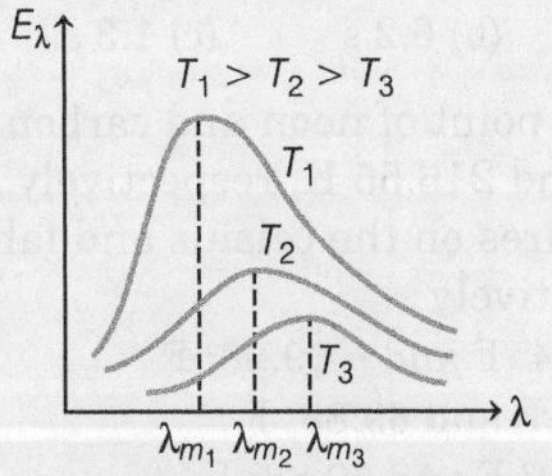

- The given figure shows the experimental curves for radiation emitted by a black body *versus* wavelength for different temperatures.
- The most significant feature of the curves obtained is that they are universal, *i.e.* black-body radiation curves obtained depend only on the temperature and not on the shape, size or material of the black body.
- As the temperature of the body increases, the wavelength at which the spectral intensity (E_λ) is maximum shifts towards left.

Example 25. *Which of the following is ν_m-T graph for perfectly black body? Here, ν_m is the frequency of radiations with maximum intensity and T is the absolute temperature.*

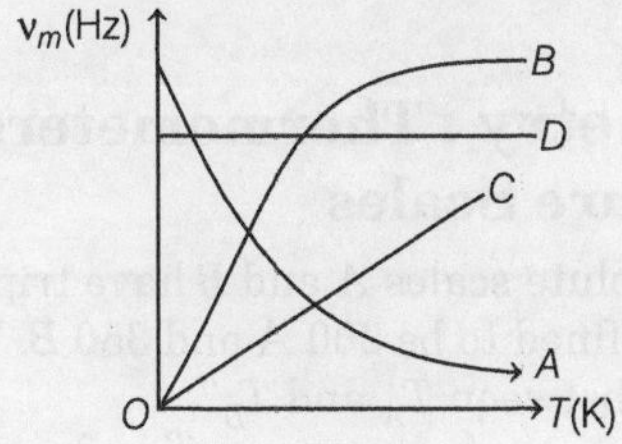

(a) A (b) B
(c) C (d) D

Sol. (c) According to Wien's displacement law,

$$\lambda_m T = b \text{ (Wein's constant)}$$

If ν_m is the frequency corresponding to wavelength λ_m, then

$$\left(\frac{C}{\nu_m}\right) T = b$$

or $$\nu_m = \frac{C}{b} T$$

i.e. $$\nu_m \propto T$$

$\therefore$ $\nu_m - T$ graph is straight line shown by the curve C.

Practice Exercise

ROUND I Topically Divided Problems

Thermometry : Thermometers and Temperature Scales

1. Two absolute scales A and B have triple points of water defined to be 200 A and 350 B. What is the relation between T_A and T_B? [NCERT]

(a) $\frac{T_A}{T_B}=\frac{4}{7}$ (b) $\frac{T_A}{T_B}=\frac{3}{7}$

(c) $\frac{T_A}{T_B}=\frac{7}{3}$ (d) $\frac{T_A}{T_B}=\frac{7}{4}$

2. The Fahrenheit and Kelvin scales of temperature will give the same reading at

(a) – 40 (b) 313
(c) 574.25 (d) 732.75

3. A faulty thermometer has its fixed points marked as 5 and 95. When this thermometer reads 68, the correct temperature (in degree celsius) is

(a) 68 (b) 70 (c) 66 (d) 72

4. The graph between two temperature scales A and B is shown in figure. Between upper fixed point and lower point, there are 150 equal division on scale A and 100 on scale B. The relationship for conversion between the two scales is given by [NCERT Exemplar]

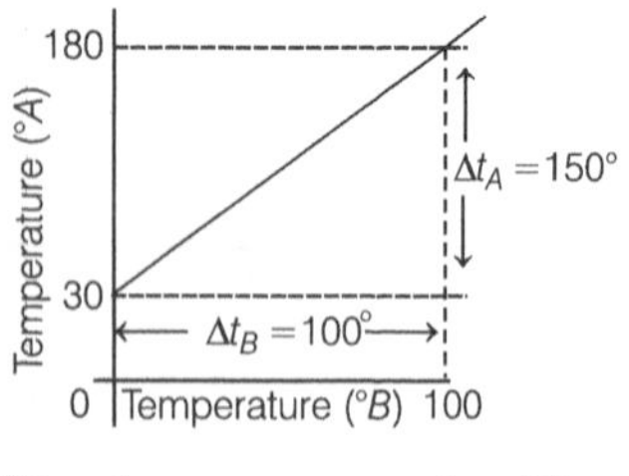

(a) $\frac{t_A-180}{100}=\frac{t_B}{150}$ (b) $\frac{t_A-30}{150}=\frac{t_B}{100}$

(c) $\frac{t_B-180}{150}=\frac{t_A}{100}$ (d) $\frac{t_B-40}{100}=\frac{t_A}{180}$

5. A bimetallic strip is made of aluminium and steel ($\alpha_{Al} > \alpha_{steel}$). On heating, the strip will [NCERT Exemplar]

(a) remain straight
(b) get twisted
(c) will bend with aluminium on concave side
(d) will bend with steel on concave side

6. A clock with an iron pendulum keeps correct time at 15°C. What will be the error in second per day, if the room temperature is 20°C?
(The coefficient of linear expansion of iron is 0.000012°C^{-1})

(a) 2.6 s (b) 6.2 s (c) 1.3 s (d) 3.1 s

7. The triple point of neon and carbon dioxide are 24.57 K and 216.55 K, respectively. These temperatures on the celsius and fahrenheit scales are respectively [NCERT]

(a) – 415.44° F and – 69.88° F
(b) 415.44° F and 69.88° F
(c) – 315.44° F and 69.88° F
(d) – 69.88° F and 415.44° F

Thermal Expansion

8. A metal rod having linear expansion coefficient 2×10^{-5} °C^{-1} has a length of 1 m at 20°C. The temperature at which it is shortened by 1 mm is

(a) –20°C (b) –15°C (c) –30°C (d) –25°C

9. A uniform metallic rod rotates about its perpendicular bisector with constant angular speed. If it is heated uniformly to raise its temperature slightly,

(a) its speed of rotation increases
(b) its speed of rotation decreases
(c) its speed of rotation remains same
(d) its speed increases because its moment of inertia increases

10. A steel wire of uniform cross-sectional area 2 mm^2 is heated upto 50°C and is stretched by clamping its two ends rigidly. The change in tension when the temperature falls from 50°C to 30°C is given by (Take, $Y = 2 \times 10^{11}$ Nm^{-2}, $\alpha = 1.1 \times 10^{-5}$ °C^{-1})

(a) 1.5×10^{10} N (b) 5 N
(c) 88 N (d) 2.5×10^{10} N

11. Density of substance at 0°C is 10 g/cc and at 100°C, its density is 9.7 g/cc. The coefficient of linear expansion of the substance is

(a) 1.03×10^{-4}°C^{-1} (b) 3×10^{-4}°C^{-1}
(c) 19.7×10^{-3}°C^{-1} (d) 10^{-3} °C^{-1}

12. What should be the lengths of a steel and copper rod at 0°C, so that the length of the steel rod is 5 cm longer than the copper rod at any temperature?

[Take, α (steel) = $1.1 \times 10^{-5}\,°C^{-1}$
α (copper) = $1.7 \times 10^{-5}°C^{-1}$]

(a) 14.17 cm; 9.17 cm (b) 9.17 cm, 14.17 cm
(c) 28.34 cm; 18.34 cm (d) 14.17 cm, 18.34 cm

13. When a liquid in a glass vessel is heated, its coefficient of apparent expansion is $10.30 \times 10^{-4}\,°C^{-1}$. When the same liquid is heated in a metal vessel, its coefficient of apparent expansion is $10.06 \times 10^{-4}\,°C^{-1}$. If the coefficient of linear expansion of glass = $9 \times 10^{6}\,°C^{-1}$, what is the coefficient of linear expansion of metal?

(a) $51 \times 10^{-6}\,°C^{-1}$ (b) $17 \times 10^{-6}\,°C^{-1}$
(c) $25 \times 10^{-6}\,°C^{-1}$ (d) $43 \times 10^{-6}\,°C^{-1}$

14. A vertical column 50 cm long at 50°C balances another column of same liquid 60 cm along at 100°C. The coefficient of absolute expansion of the liquid is

(a) 0.005/°C (b) 0.0005/°C
(c) 0.002/°C (d) 0.0002/°C

15. The radius of a metal sphere at room temperature T is R, and the coefficient of linear expansion of the metal is α. The sphere is heated a little by a temperature ΔT, so that its new temperature is $T + \Delta T$. The increase in the volume of the sphere is approximately **[NCERT Exemplar]**

(a) $2\pi R\alpha\,\Delta T$ (b) $\pi R^2 \alpha\,\Delta T$
(c) $4\pi R^3 \alpha\,\Delta T/3$ (d) $4\pi R^3 \alpha \Delta T$

16. Two rods A and B of identical dimensions are at temperature 30°C. If A is heated upto 180°C and B upto T° C, then new lengths are the same. If the ratio of the coefficients of linear expansion of A and B is 4 : 3, then the value of T is **[JEE Main 2019]**

(a) 230°C (b) 270°C
(c) 200°C (d) 250°C

17. The volume of a metal sphere increases by 0.24% when its temperature is raised by 40°C. The coefficient of linear expansion of the metal is

(a) 2×10^{-5} per°C (b) 6×10^{-5} per°C
(c) 2.1×10^{-5} per°C (d) 1.2×10^{-5} per°C

18. The coefficient of apparent expansion of a liquid when determined using two different vessels A and B are γ_1 and γ_2, respectively. If the coefficient of linear expansion of the vessel A is α, the coefficient of linear expansion of vessel B is

(a) $\dfrac{\alpha\gamma_1\gamma_2}{\gamma_1 + \gamma_2}$ (b) $\dfrac{\gamma_1 - \gamma_2}{2\alpha}$
(c) $\dfrac{\gamma_1 - \gamma_2 + \alpha}{3\alpha}$ (d) $\dfrac{\gamma_1 - \gamma_2}{3} + \alpha$

19. A wheel is 80.3 cm in circumference. An iron tyre measures 80.0 cm around its inner face. If the coefficient of linear expansion for iron is $1.2 \times 10^{-5}\,°C^{-1}$, the temperature of the tyre must be raised by

(a) 105°C (b) 417°C (c) 312°C (d) 223°C

20. The coefficient of apparent expansion of mercury in a glass vessel is 153×10^{-6}/°C and in a steel vessel is 144×10^{6}/°C. If α for steel is 12×10^{-6}/°C, then that of glass is

(a) 9×10^{-6}/°C (b) 6×10^{-6} /°C
(c) 36×10^{-6}/°C (d) 27×10^{-6} /°C

21. A bimetallic strip consists of metals A and B. It is mounted rigidly as shown. The metal A has higher coefficient of expansion compared to that of metal B. When the bimetallic strip is placed in a cold both, it will **[JEE Main 2021]**

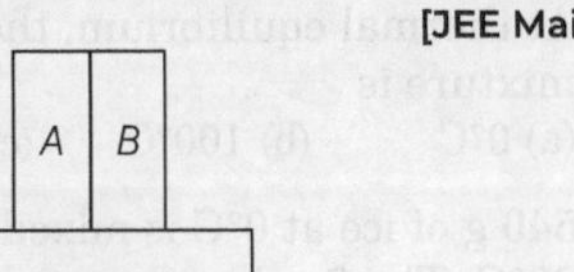

(a) bend towards the right
(b) not bend but shrink
(c) Neither bend nor shrink
(d) bend towards the left

22. The coefficient of linear expansion of crystal in one direction is α_1 and that in every direction perpendicular to it is α_2. The coefficient of cubical expansion is

(a) $\alpha_1 + \alpha_2$ (b) $2\alpha_1 + \alpha_2$
(c) $\alpha_1 + 2\alpha_2$ (d) None of these

23. When the temperature of a rod increases from t to $t + \Delta t$, its moment of inertia increases from I to $I + \Delta I$. If α be the coefficient of linear expansion of the rod, then the value of $\dfrac{\Delta I}{I}$ is

(a) $2\alpha\,\Delta t$ (b) $\alpha\,\Delta t$ (c) $\dfrac{\alpha\,\Delta t}{2}$ (d) $\dfrac{\Delta t}{\alpha}$

24. A piece of metal weighs 46 g in air. When it is immersed in the liquid of specific gravity 1.24 at 27°C, it weighs 30 g. When the temperature of liquid is raised to 42°C, the metal piece weighs 30.5 g, specific gravity of the liquid at 42°C is 1.20, then the linear expansion of the metal will be

(a) 3.316×10^{-5}/°C (b) 2.316×10^{-5}/°C
(c) 4.316×10^{-5}/°C (d) None of these

25. A hole is drilled in a copper sheet. The diameter of the hole is 4.24 cm at 27.0°C. What is the change in the diameter of the hole when the sheet is heated to 227°C? Coefficient of linear expansion of copper is 1.70×10^{-5}/°C. **[NCERT]**

(a) 1.44×10^{-2} cm (b) 2.44×10^{-3} cm
(c) 1.44×10^{-2} mm (d) 2.44×10^{-3} mm

26. The coefficient of volume expansion of glycerine is 49×10^{-5}/K. What is the fractional change in its density for a 30°C rise in temperature? [NCERT]
(a) 1.45×10^{-3} (b) 2.45×10^{-3}
(c) 2.45×10^{-2} (d) 1.45×10^{-2}

Specific Heat and Calorimetry

27. An amount of water of mass 20 g at 0°C is mixed with 40 g of water at 10°C, final temperature of the mixture is
(a) 5°C (b) 0°C (c) 20°C (d) 6.66°C

28. The amount of heat required to convert 10 g of ice at –10°C into steam at 100°C is (in calories)
(a) 6400 (b) 5400 (c) 7200 (d) 7250

29. One gram of ice is mixed with one gram of steam. At thermal equilibrium, the temperature of mixture is
(a) 0°C (b) 100°C (c) 55°C (d) 80°C

30. 540 g of ice at 0°C is mixed with 540 g of water at 80°C. The final temperature of the mixture is
(a) 0°C (b) 53°C
(c) 80°C (d) less than 0°C

31. If the ratio of densities of two substances is 5 : 6 and that of the specific heats is 3 : 5, then the ratio between heat capacities per unit volume is
(a) 1 : 1 (b) 2 : 1
(c) 1 : 2 (d) 1 : 3

32. When the room temperature becomes equal to the dew point, the relative humidity of the room is
(a) 100% (b) 0% (c) 70% (d) 85%

33. Water of volume 2L in a container is heated with a coil of 1 kW at 27°C. The lid of the container is open and energy dissipates at rate of 160 J/s. In how much time temperature will rise from 27°C to 77°C? (Take, specific heat of water is 4.2 kJ/kg)
(a) 8 min 20 s (b) 6 min 2 s
(c) 7 min (d) 14 min

34. Heat is supplied to a certain homogeneous sample of matter, at a uniform rate. Its temperature is plotted against time as shown. Which of the following conclusions can be drawn?

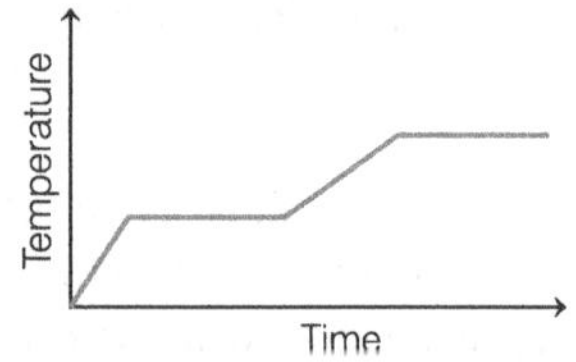

(a) Its specific heat capacity is greater in the solid state than in the liquid state
(b) Its specific heat capacity is greater in the liquid state than in the solid state
(c) Its latent heat of vaporisation is smaller than its latent of fusion
(d) None of the above

35. When two substances at different temperatures are mixed together, exchange of heat occurs between them till they acquire a common temperature. In thermal equilibrium, heat gained by one substance is equal to heat lost by the other substance. This is called principle of calorimetry. 10 g of ice at 0°C and water at 100°C are mixed together. The resultant temperature would be
(a) 10°C (b) 5°C (c) 50°C (d) 40°C

36. A steel ball of mass 0.1 kg falls freely from a height of 10 m and bounces to a height of 5.4 m from the ground. If the dissipated energy in this process is absorbed by the ball, the rise in its temperature is [Take, specific heat of steel = 460 $Jkg^{-1}°C^{-1}$)
(a) 0.01°C (b) 0.1°C
(c) 1.1°C (d) 1°C

37. A lead bullet of 10 g travelling at 300 ms^{-1} strikes against a block of wood comes to rest. Assuming 50% of heat is absorbed by the bullet, the increase in its temperature is
(Take, specific heat of lead = 150 $Jkg^{-1}K^{-1}$)
(a) 100°C (b) 125°C
(c) 150°C (d) 200°C

38. Steam is passed into 22 g of water at 20°C. The mass of water that will be present when the water acquires a temperature of 90°C (Take, latent heat of steam is 540 cal/g) is
(a) 24.8 g (b) 24 g
(c) 36.6 g (d) 30 g

39. Ice at – 20°C is added to 50 g of water at 40°C.
When the temperature of the mixture reaches 0°C, it is found that 20 g of ice is still unmelted. The amount of ice added to the water was close to
(Take, specific heat of water = 4.2 J/g/°C,
specific heat of ice = 2.1 J/g/°C and
heat of fusion of water at 0°C = 334 J/g)
[JEE Main 2019]
(a) 40 g (b) 50 g
(c) 60 g (d) 100 g

40. Water falls from a height of 500 m. What is the rise in temperature of water at the bottom, if whole energy is used up in heating water?
(a) 0.96°C (b) 1.02°C
(c) 1.16°C (d) 0.23°C

41. Which one of the following would raise the temperature of 20 g of water at 30°C most when mixed with it?
(a) 20 g of water at 40°C (b) 40 g of water at 35°C
(c) 10 g of water at 50°C (d) 4 g of water at 80°C

42. A metal ball of mass 0.1 kg is heated upto 500ºC and dropped into a vessel of heat capacity 800 JK^{-1} and containing 0.5 kg water. The initial temperature of water and vessel is 30°C. What is the approximate percentage increment in the temperature of the water?
(Take, specific heat capacities of water and metal are respectively 4200 $Jkg^{-1}K^{-1}$ and 400 $Jkg^{-1}K^{-1}$) **[JEE Main 2019]**

(a) 25% (b) 15%
(c) 30% (d) 20%

43. As the temperature is increased, the time period of a pendulum **[NCERT Exemplar]**

(a) increases as its effective length increases even though its centre of mass still remains at the centre of the bob
(b) decreases as its effective length increases even though its centre of mass still remains at the centre of the bob
(c) increases as its effective length increases due to shifting of centre of mass below the centre of the bob
(d) decreases as its effective length increases remains same but the centre of mass shifts above the centre of the bob

Thermal Conduction and Convection

44. Five rods of same dimensions are arranged as shown in figure. They have thermal conductivities K_1, K_2, K_3, K_4 and K_5. When points A and B are maintained at different temperatures, no heat would flow through central rod, if

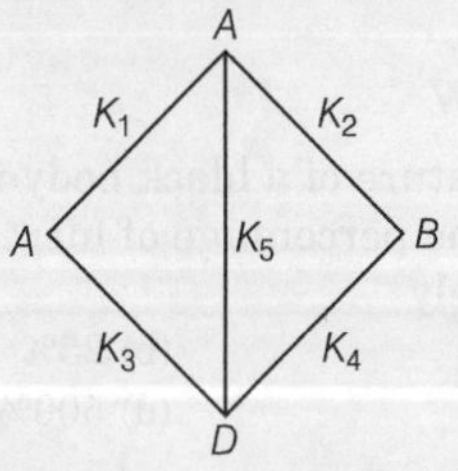

(a) $K_1K_4 = K_2K_3$ (b) $K_1 = K_4$ and $K_2 = K_3$
(c) $\frac{K_1}{K_4} = \frac{K_2}{K_3}$ (d) $K_1 K_2 = K_3 K_4$

45. Two materials having coefficients of thermal conductivity $3K$ and K and thickness d and $3d$ respectively, are joined to form a slab as shown in the figure. The temperatures of the outer surfaces are θ_2 and θ_1 respectively, $(\theta_2 > \theta_1)$. The temperature at the interface is **[JEE Main 2019]**

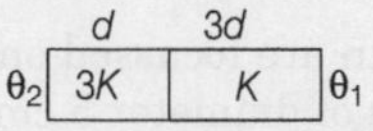

(a) $\frac{\theta_2 + \theta_1}{2}$ (b) $\frac{\theta_1}{3} + \frac{2\theta_2}{3}$
(c) $\frac{\theta_1}{6} + \frac{5\theta_2}{6}$ (d) $\frac{\theta_1}{10} + \frac{9\theta_2}{10}$

46. Two identical square rods of metal are welded end to end as shown in Fig. (i), Q cal of heat flows through this combination in 4 min. If the rods were welded as shown in Fig. (ii), the same amount of heat will flow through the combination in

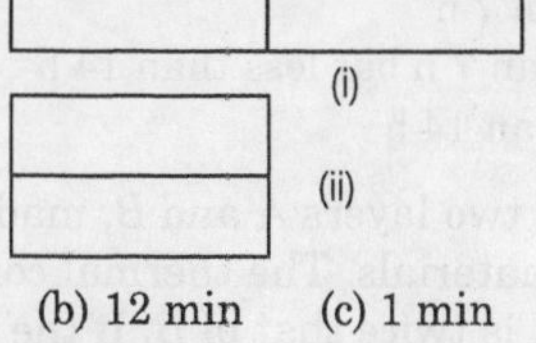

(a) 16 min (b) 12 min (c) 1 min (d) 4 min

47. A cylinder of radius r and thermal conductivity K_1 is surrounded by a cylindrical shell of linear radius r and outer radius $2r$, whose thermal conductivity is K_2. There is no loss of heat across cylindrical surfaces, when the ends of the combined system are maintained at temperatures T_1 and T_2. The effective thermal conductivity of the system, in the steady state is

(a) $\frac{K_1K_2}{K_1 + K_2}$ (b) $K_1 + K_2$ (c) $\frac{K_1 + 3K_2}{4}$ (d) $\frac{3K_1 + K_2}{4}$

48. The coefficient of thermal conductivity of copper is nine times that of steel. In the composite cylindrical bar shown in figure, what will be the temperature at the junction of copper and steel?

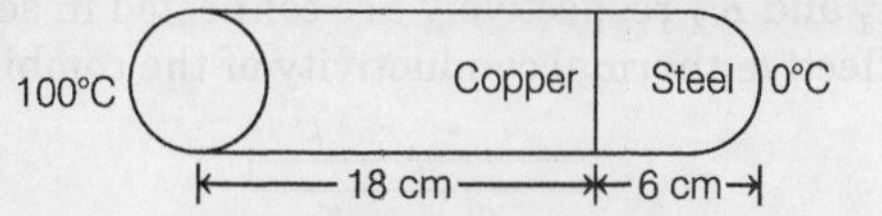

(a) 75°C (b) 67°C (c) 33°C (d) 25°C

49. Two rods of equal length and area of cross-section are kept parallel and lagged between temperatures 20°C and 80°C. The ratio of the effective thermal conductivity to that of the first rod is
[the ratio $\left(\frac{K_1}{K_2}\right) = \frac{3}{4}$]

(a) 7 : 4 (b) 7 : 6 (c) 4 : 7 (d) 7 : 8

50. Two plates of same thickness, of coefficients of thermal conductivity K_1 and K_2 and areas of cross-section A_1 and A_2 are connected as shown in figure. The common coefficient of thermal conductivity K will be

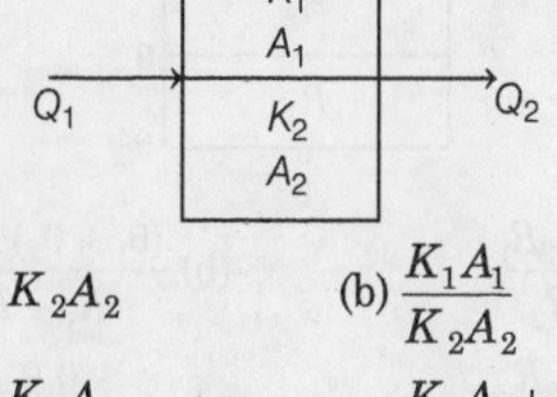

(a) $K_1A_1 + K_2A_2$ (b) $\frac{K_1A_1}{K_2A_2}$
(c) $\frac{K_1A_1 + K_2A_2}{A_1 + A_2}$ (d) $\frac{K_1A_2 + K_2A_1}{K_1 + K_2}$

51. Ice starts forming in a lake with water at 0°C, when the atmospheric temperature is –10°C. If time taken for 1 cm of ice to be formed is 7 h, the time taken for the thickness of ice to increase from 1 cm to 2 cm is
(a) 7 h
(b) less than 7 h
(c) more than 7 h but less than 14 h
(d) more than 14 h

52. A wall has two layers A and B, made of two different materials. The thermal conductivity of material A is twice that of B. If the two layers have same thickness and under thermal equilibrium, the temperature difference across the wall is 48°C, the temperature difference across layer B is
(a) 40°C (b) 32°C
(c) 16°C (d) 24°C

53. A heat source at $T = 10^3$ K is connected to another heat reservoir at $T = 10^2$ K by a copper slab which is 1 m thick. Given that the thermal conductivity of copper is 0.1 $WK^{-1}m^{-1}$, the energy flux through it in the steady state is **[JEE Main 2019]**
(a) 90 Wm^{-2} (b) 65 Wm^{-2}
(c) 120 Wm^{-2} (d) 200 Wm^{-2}

54. Two identical metal wires of thermal conductivities K_1 and K_2 respectively are connected in series. The effective thermal conductivity of the combination is

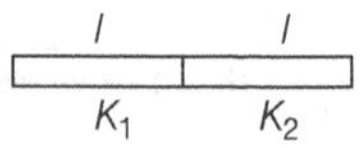

[JEE Main 2021]

(a) $\dfrac{2K_1K_2}{K_1 + K_2}$ (b) $\dfrac{K_1 + K_2}{2K_1K_2}$
(c) $\dfrac{K_1 + K_2}{K_1 K_2}$ (d) $\dfrac{K_1 K_2}{K_1 + K_2}$

55. Four rods of different radii r and length l are used to connect two reservoirs of heat at different temperatures. Which one will conduct heat fastest?
(a) $r = 2$ cm, $l = 0.5$ m (b) $r = 1$ cm, $l = 0.5$ m
(c) $r = 2$ cm, $l = 2$ m (d) $r = 1$ cm, $l = 1$ m

56. Consider two insulating sheets with thermal resistances R_1 and R_2 as shown in figure. The temperature θ is

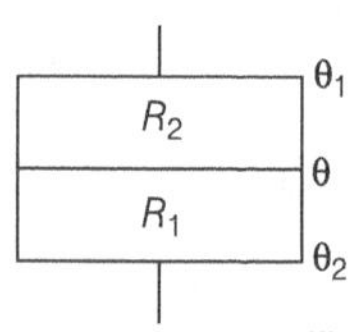

(a) $\dfrac{\theta_1 R_2 + \theta_2 R_1}{R_1 + R_2}$ (b) $\dfrac{(\theta_1 + \theta_2) R_1 R_2}{R_1^2 + R_2^2}$
(c) $\dfrac{\theta_1 R_1 + \theta_2 R_2}{R_1 + R_2}$ (d) $\dfrac{\theta_1 \theta_2 R_1 R_2}{(\theta_1 + \theta_2)(R_1 + R_2)}$

57. Two rods P and Q have equal lengths. Their thermal conductivities are K_1 and K_2 and cross-sectional areas are A_1 and A_2. When the temperature at ends of each rod are T_1 and T_2 respectively, the rate of flow of heat through P and Q will be equal, if
(a) $\dfrac{A_1}{A_2} = \dfrac{K_2}{K_1}$ (b) $\dfrac{A_1}{A_2} = \dfrac{K_2}{K_1} \times \dfrac{T_2}{T_1}$
(c) $\dfrac{A_1}{A_2} = \sqrt{\dfrac{K_1}{K_2}}$ (d) $\dfrac{A_1}{A_2} = \left(\dfrac{K_2}{K_1}\right)^2$

58. The amount of heat conducted out per second through a window, when inside temperature is 10°C and outside temperature is –10°C, is 1000 J. Same heat will be conducted inside through the window, when outside temperature is –23°C and inside temperature is
(a) 23°C (b) 230 K
(c) 270 K (d) 296 K

59. Two identical square rods of metal are welded end to end as shown in Ffig. (i), 20 cal of heat flows through it in 4 min. If the rods are welded as shown in Fig. (ii), the same amount of heat will flow through the rods in **[NCERT Exemplar]**

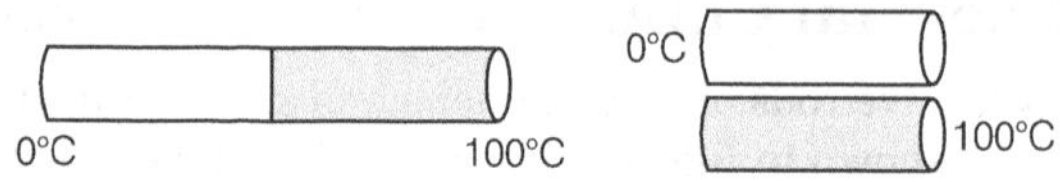

(i) (ii)

(a) 1 min (b) 2 min
(c) 4 min (d) 16 min

Stefan's Law

60. The temperature of a black body is increased by 50%, then the percentage of increase of radiation is approximately
(a) 100% (b) 25%
(c) 400% (d) 500%

61. The rectangular surface of area 8 cm × 4 cm of a black body at a temperature of 127°C emits energy at the rate of E per second. If the length and breadth of the surface are each reduced to half of its initial value, and the temperature is raised to 327°C, the rate of emission of energy will become
(a) $\dfrac{3}{8}E$ (b) $\dfrac{81}{16}E$
(c) $\dfrac{9}{16}E$ (d) $\dfrac{81}{64}E$

62. The rays of Sun are focussed on a piece of ice through a lens of diameter 5 cm, as a result of which 10 g ice melts in 10 min. The amount of heat received from Sun, per unit area per min is
(a) 4 cal cm^{-2} min^{-1} (b) 40 cal cm^{-2} min^{-1}
(c) 4 Jm^{-2} min (d) 400 cal cm^{-2} min^{-1}

63. A black body at a temperature of 327°C radiates 4 cal $cm^{-2}s^{-1}$. At a temperature of 927°C, the rate of heat radiated per unit area (in cal $cm^{-2}s^{-1}$) will be

(a) 16 (b) 32
(c) 64 (d) 128

64. A planet is at an average distance d from the Sun and its average surface temperature is T. Assume that the planet receives energy only from the Sun, and loses energy only through radiation from its surface. Neglect atmospheric effects. If $T \propto d^{-n}$, the value of n is

(a) 2 (b) 1
(c) 1/2 (d) 1/4

65. Two bodies A and B are placed in an evacuated vessel maintained at a temperature of 27°C. The temperature of A is 327°C and that of B is 227°C. The ratio of heat loss from A and B is about

(a) 2 : 1 (b) 4 : 1
(c) 1 : 2 (d) 1 : 4

66. The rates of heat radiation from two patches of skin each of area A, on a patient's chest differ by 2%. If the patch of the lower temperature is at 300 K and emissivity of both the patches is assumed to be unity, the temperature of other patch would be

(a) 306 K (b) 312 K
(c) 308.5 K (d) 301.5 K

67. A surface at temperature T_0 K receives power P by radiation from a small sphere at temperature $T > T_0$ and at a distance d. If both T and d are doubled, the power received by the surface will become

(a) P (b) $2P$
(c) $4P$ (d) $16P$

68. Assuming the Sun to have a spherical outer surface of radius r radiating like a black body at temperature t° C, the power received by a unit surface (normal to the incident rays) at a distance R from the centre of the Sun is (σ is Stefan's constant)

(a) $4\pi r^2 \sigma t^4$ (b) $\frac{r^2\sigma (t+273)^4}{4\pi R^2}$

(c) $\frac{16\pi^2 r^2 \sigma t^4}{R^2}$ (d) $\frac{r^2\sigma (t+273)^4}{R^2}$

Wien's Law

69. The wavelength of maximum intensity of radiation emitted by a star is 289.8 nm and the radiation intensity for the star is
(Stefan's constant $= 5.67 \times 10^{-8}$ $Wm^{-2}K^{-4}$, Wien's constant, $b = 2898$ μmK)

(a) 5.67×10^{8} Wm^{-2} (b) 5.67×10^{-12} Wm^{-2}
(c) 10.67×10^{7} Wm^{-2} (d) 10.67×10^{14} Wm^{-2}

70. A black body radiates at two temperatures T_1 and T_2, such that $T_1 < T_2$. The frequency corresponding to maximum intensity is

(a) less at T_1 (b) more at T_1
(c) equal in the two cases (d) Cannot say

71. A black body is at a temperature of 2880 K. The energy of radiation emitted by this object with wavelength between 499 nm and 500 nm is U_1 and between 999 nm and 1000 nm is U_2. The Wien's constant $= 2.88 \times 10^6$ nm K. Then,

(a) $U_1 = 0$ (b) $U_3 = 0$
(c) $U_1 > U_2$ (d) $U_2 > U_1$

72. Solar radiation emitted by Sun resembles that emitted by a black body at a temperature of 6000 K. Maximum intensity is emitted at a wavelength of about 4800Å. If the Sun were cooled down from 6000 K to 3000 K, then the peak intensity would occur at a wavelength of

(a) 4800 Å (b) 9600 Å
(c) 2400 Å (d) 19200 Å

73. When the temperature of a black body increases, it is observed that the wavelength corresponding of maximum energy changes from 0.26 μm to 0.13 μm. The ratio of the emissive powers of the body at the respective temperatures is

(a) 16/1 (b) 4/1
(c) 1/4 (d) 1/16

74. The temperature of Sun is 5500 K and it emits maximum intensity radiation in the yellow region (5.5×10^{-7} m). The maximum radiation from a furnace occurs at wavelength 11×10^{-7} m. The temperature of furnace is

(a) 2550 K (b) 2750 K
(c) 2650 K (d) 2850 K

Newton's Law of Cooling

75. A body cools from 80°C to 50°C in 5 min. Calculate the time it takes to cool from 60°C to 30°C. The temperature of the surroundings is 20°C. [NCERT]

(a) 9 min (b) 7 min
(c) 8 min (d) 10 min

76. An object is cooled from 75°C to 65°C in 2 min in a room at 30°C. The time taken to cool another identical object from 55°C to 45°C in the same room, in minutes is

(a) 4 (b) 5 (c) 6 (d) 7

77. The total energy radiated from a black body source is collected for one minute and is used to heat a quantity of water. The temperature of water is found to increase from 20°C to 20.5°C. If the absolute temperature of the black body is doubled and the experiment is repeated with the same quantity of water at 20°C, the temperature of water will be

(a) 21°C (b) 22°C
(c) 24°C (d) 28°C

78. A liquid is filled in a container which is kept in a room whose temperature is 20°C. When temperature of liquid is 80°C, it emits heat at the rate of 45 cals^{-1}. When temperature of liquid falls to 40°C, its rate of heat loss will be

(a) 15 cals^{-1} (b) 30 cals^{-1}
(c) 45 cals^{-1} (d) 60 cals^{-1}

79. A metallic sphere cools from 50°C to 40°C in 300 s. If the room temperature is 20°C, then its temperature in next 5 min will be

(a) 38°C (b) 33.3°C (c) 30°C (d) 36°C

80. A solid cube and a solid sphere have equal surface areas. Both are at the same temperature of 120°C. Then,

(a) both of them will cool down at the same rate
(b) the cube will cool down faster than the sphere
(c) the sphere will cool down faster than the cube
(d) whichever of the two is heavier will cool down faster

81. The temperature of a liquid drops from 365 K to 361 K in 2 min. Find the time during which temperature of the liquid drops from 344 K to 342 K. Temperature of room is 292 K.

(a) 84 s
(b) 72 s
(c) 66 s
(d) 60 s

ROUND II Mixed Bag

1. The only possibility of heat flow in a thermos flask is through its cork which is 75 cm^2 in area and 5 cm thick. Its thermal conductivity is 0.0075 cal/cm sec°C. The outside temperature is 40°C and latent heat of ice is 80 cal g^{-1}.

Time taken by 500 g of ice at 0°C in the flask to melt into water at 0°C is

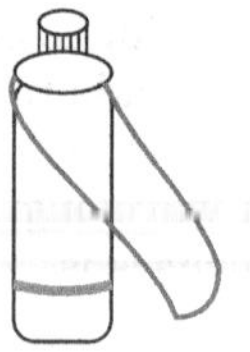

(a) 2.47 h (b) 4.27 h (c) 7.42 h (d) 4.82 h

2. The figure given below shows the cooling curve of pure wax material after heating. It cools from A to B and solidifies along BD. If L and C are respective values of latent heat and the specific heat of the liquid wax, the ratio L/C is

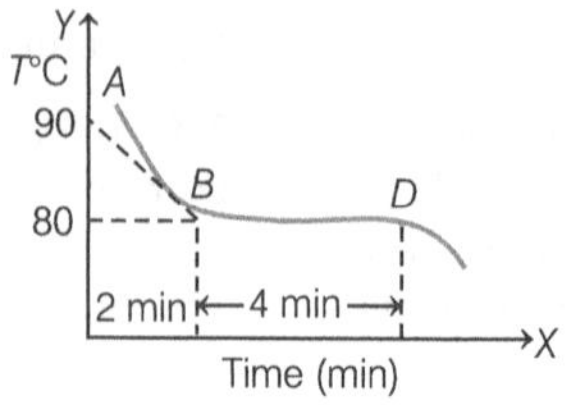

(a) 40 (b) 80 (c) 100 (d) 20

3. The graph shown in the adjacent diagram, represents the variation of temperature (T) of two bodies, x and y having same surface area, with time (t) due to the emission of radiation. Find the correct relation between the emissivity (e) and absorptivity (a) of the two bodies.

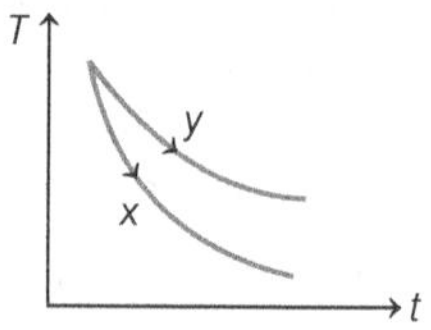

(a) $e_x > e_y$ and $a_x < a_y$ (b) $e_x < e_y$ and $a_x > a_y$
(c) $e_x > e_y$ and $a_x > a_y$ (d) $e_x < e_y$ and $a_x < a_y$

4. If α and λ are coefficients of linear, superficial and volume expansion respectively, then

(a) $\frac{\beta}{\alpha} = \frac{1}{2}$ (b) $\frac{\beta}{\gamma} = \frac{1}{3}$
(c) $\frac{\gamma}{\alpha} = \frac{3}{1}$ (d) $\frac{\beta}{\alpha} = \frac{\gamma}{\beta}$

5. Shown below are the black body radiation curves at temperatures T_1 and T_2 ($T_2 > T_1$). Which of the following plots is correct?

(a)
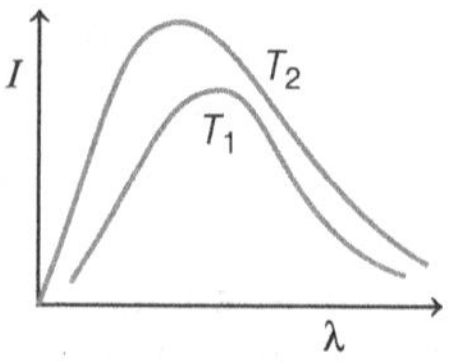

(b)
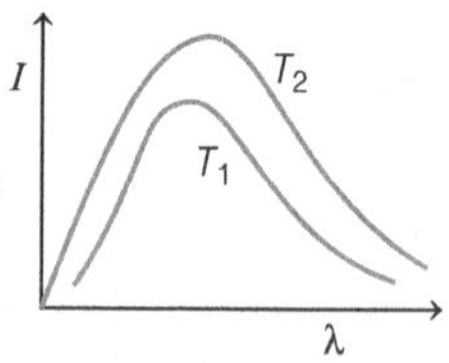

(c)
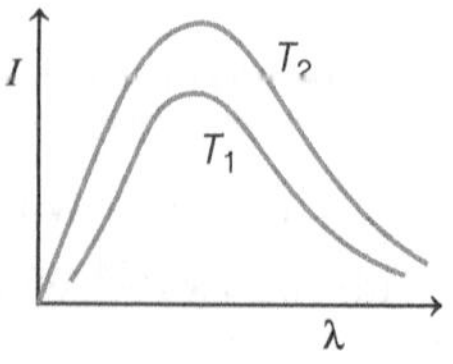

(d)
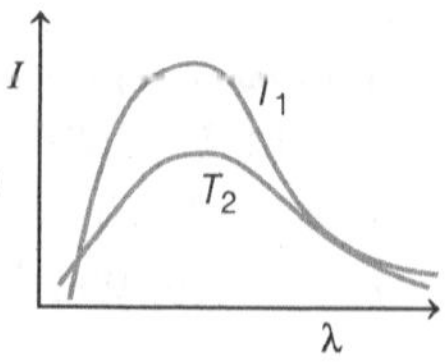

6. Heat is flowing through a conductor of length l from $x = 0$ to $x = l$. If its thermal resistance per unit length is uniform, which of the following graph is correct?

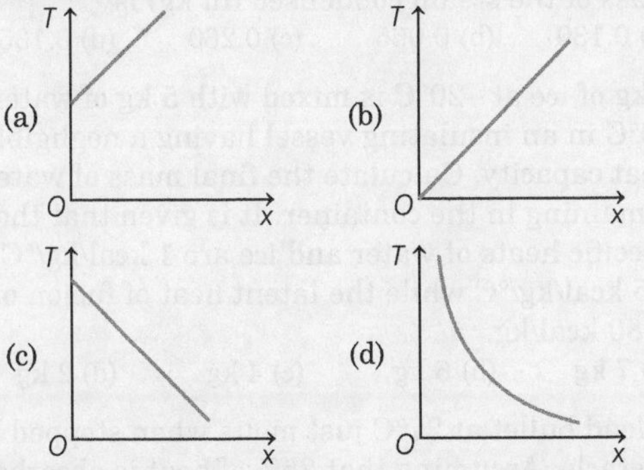

7. Mark the correct options. **[NCERT Exemplar]**

(a) A system X is in thermal equilibrium with Y but not with Z. System Y and Z may be in thermal equilibrium with each other.

(b) A system X is in thermal equilibrium with Y but not with Z. System Y and Z are not in thermal equilibrium with each other.

(c) A system X is neither in thermal equilibrium with Y nor with Z. The systems Y and Z must be in thermal equilibrium with each other.

(d) A system X is neither in thermal equilibrium with Y nor with Z. The system Y and Z may be in thermal equilibrium with each other.

8. A thin copper wire of length l increase its length by 1%, when heated from 0°C to 100°C. If a thin copper plate of area $2l \times l$ is heated from 0°C to 100°C, the percentage increase in its area would be

(a) 1% (b) 4% (c) 3% (d) 2%

9. Which of the following graphs correctly represents the relation between $\ln E$ and $\ln T$, where E is the amount of radiation emitted per unit time from unit area of a body and T is the absolute temperature?

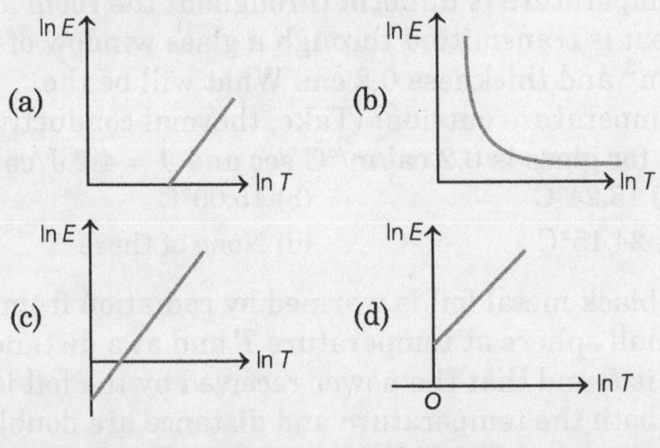

10. 22 g of CO_2 at 27°C is mixed with 16 g of oxygen at 37°C. The temperature of the mixture is

(a) 32°C (b) 27°C (c) 37°C (d) 30°C

11. Two cylindrical conductors A and B of same metallic material have their diameters in the ratio 1 : 2 and lengths in the ratio 2 : 1. If the temperature difference between their ends is same, the ratio of heat conducted respectively by A and B per second is

(a) 1 : 2 (b) 1 : 4
(c) 1 : 16 (d) 1 : 8

12. A hollow copper sphere S and a hollow copper cube C, both of negligible thin walls of same area, are filled with water at 90°C and allowed to cool in the same environment. The graph that correctly represents their coolings is

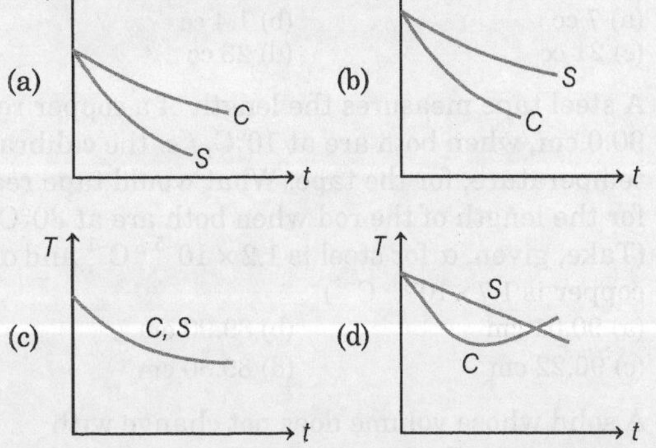

13. The ends of two different materials with their thermal conductivities, radii of cross-section and length all in the ratio of 1:2 maintained at temperature difference. If the rate of the flow of heat in the longer rod is 4 cals^{-1}, that in the shorter rod (in cals^{-1}) will be

(a) 1 (b) 2
(c) 8 (d) 6

14. Two spheres made of same substance have diameters in the ratio 1 : 2. Their thermal capacities are in the ratio of

(a) 1 : 2 (b) 1 : 8
(c) 1 : 4 (d) 2 : 1

15. Three rods made of same material and having same cross-section have been joined as shown in figure. Each rod is of same length. The left and right ends are kept at 0°C and 90°C, respectively. The temperature of the junction of the three rods will be

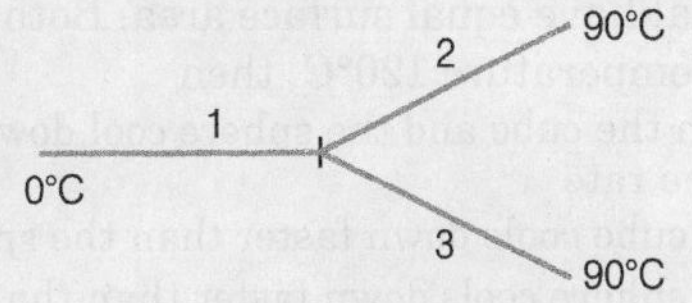

(a) 45°C (b) 60°C
(c) 30°C (d) 20°C

16. 70 cal of heat are required to raise the temperature of 2 mol of an ideal gas at constant pressure from 30°C to 35°C. The amount of heat required to raise the temperature of the same sample of the gas through the same range at constant volume is nearly (Take, gas constant = 1.99 cal K^{-1}-mol^{-1})
(a) 30 cal
(b) 50 cal
(c) 70 cal
(d) 90 cal

17. A flask of volume 10^3 cc is completely filled with mercury at 0°C. The coefficient of cubical expansion of mercury is $1.80 \times 10^{-6} {}^\circ C^{-1}$ and that of glass is $1.4 \times 10^{-6} {}^\circ C^{-1}$. If the flask is now placed in boiling water at 100°C, how much mercury will overflow?
(a) 7 cc (b) 1.4 cc
(c) 21 cc (d) 28 cc

18. A steel tape measures the length of a copper rod as 90.0 cm, when both are at 10°C, *i.e.* the calibration temperature, for the tape. What would tape read for the length of the rod when both are at 30°C? (Take, given, α for steel is $1.2 \times 10^{-5} {}^\circ C^{-1}$ and α for copper is $1.7 \times 10^{-5} {}^\circ C^{-1}$)
(a) 90.01 cm (b) 89.90 cm
(c) 90.22 cm (d) 89.80 cm

19. A solid whose volume does not change with temperature floats in liquid. For two different temperatures t_1 and t_2, the fractions f_1 and f_2 of volume of solid remain submerged. What is the coefficient of volume expansion of liquid?
(a) $\frac{f_1 - f_2}{f_2 t_1 - f_1 t_2}$ (b) $\frac{f_1 - f_2}{f_1 t_1 - f_2 t_2}$
(c) $\frac{f_1 + f_2}{f_2 t_1 - f_1 t_2}$ (d) $\frac{f_1 + f_2}{f_1 t_1 - f_2 t_2}$

20. A steel scale measures the length of a copper wire as 80.0 cm, when both are at 20°C (the calibration temperature for scale). What would be the scale read for the length of the wire when both are at 40°C? (Take, $\alpha_{steel} = 11 \times 10^{-6}$ per°C and $\alpha_{copper} = 17 \times 10^{-6}$ per°C)
(a) 80.0096 cm
(b) 80.0272 cm
(c) 1 cm
(d) 25.2 cm

21. A solid cube and a solid sphere of the same material have equal surface area. Both are at the same temperature 120°C, then
(a) both the cube and the sphere cool down at the same rate
(b) the cube cools down faster than the sphere
(c) the sphere cools down faster than the cube
(d) whichever is having more mass will cool down faster

22. Steam at 100°C is passed into 1.1 kg of water contained in a calorimeter of water equivalent to 0.02 kg at 15°C till the temperature of the calorimeter and its contents rises to 80°C. The mass of the steam condensed (in kg) is
(a) 0.130 (b) 0.065 (c) 0.260 (d) 0.135

23. 2 kg of ice at –20°C is mixed with 5 kg of water at 20°C in an insulating vessel having a negligible heat capacity. Calculate the final mass of water remaining in the container. It is given that the specific heats of water and ice are 1 kcal/kg/°C and 0.5 kcal/kg/°C while the latent heat of fusion of ice is 80 kcal/kg.
(a) 7 kg (b) 6 kg (c) 4 kg (d) 2 kg

24. A lead bullet at 27°C just melts when stopped by an obstacle. Assuming that 25% of heat is absorbed by the obstacle, then the velocity of the bullet at the time of striking (MP of lead = 327°C, specific heat of lead = 0.03 cal/g/°C, latent heat of fusion of lead = 6 cal/g and $J = 4.2$ joule/cal)
(a) 410 m/s (b) 1230 m/s
(c) 307.5 m/s (d) None of these

25. In a vertical U-tube containing a liquid, the two arms are maintained at different temperatures t_1 and t_2. The liquid columns in the two arms have heights l_1 and l_2, respectively. The coefficient of volume expansion of the liquid is equal to

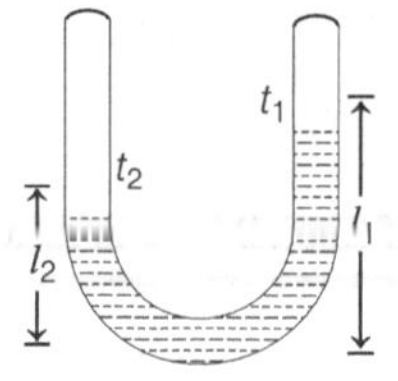

(a) $\frac{l_1 - l_2}{l_2 t_1 - l_1 t_2}$ (b) $\frac{l_1 - l_2}{l_1 t_1 - l_2 t_2}$ (c) $\frac{l_1 + l_2}{l_2 t_1 + l_1 t_2}$ (d) $\frac{l_1 + l_2}{l_1 t_1 + l_2 t_2}$

26. A room is maintained at 20°C by a heater of resistance 20 Ω connected to 200 V mains. The temperature is uniform throughout the room and heat is transmitted through a glass window of area $1 \, m^2$ and thickness 0.2 cm. What will be the temperature outside? (Take, thermal conductivity K for glass is 0.2 cal/m/°C sec and $J = 4.2$ J/cal)
(a) 15.24°C (b) 15.00°C
(c) 24.15°C (d) None of these

27. A black metal foil is warmed by radiation from a small sphere at temperature T and at a distance d. It is found that the power received by the foil is P. If both the temperature and distance are doubled, the power received by the foil will be
(a) $16P$ (b) $4P$
(c) $2P$ (d) P

28. Three discs A, B and C having radii 2 m, 4 m and 6 m respectively are coated with carbon black on their other surfaces. The wavelengths corresponding to maximum intensity are 300 nm, 400 nm and 500 nm, respectively. The power radiated by them are Q_a, Q_b and Q_c respectively, then

(a) Q_a is maximum (b) Q_b is maximum
(c) Q_c is maximum (d) $Q_a = Q_b = Q_c$

29. Four identical rods of same material are joined end to end to form a square. If the temperature difference between the ends of a diagonal is 100°C, then the temperature difference between the ends of other diagonal will be

(a) 0°C
(b) $\frac{100}{l}$°C; where l is the length of each rod
(c) $\frac{100}{2l}$°C
(d) 100°C

30. One end of a copper rod of length 1.0 m and area of cross-section 10^{-3} m^2 is immersed in boiling water and the other end in ice. If the coefficient of thermal conductivity of copper is 92 cal/m-s-°C and the latent heat of ice is 8×10^4 cal/kg, then the amount of ice which will melt in one min is

(a) 9.2×10^{-3} kg (b) 8×10^{-3} kg
(c) 6.9×10^{-3} kg (d) 5.4×10^{-3} kg

31. A solid copper sphere (density ρ and specific heat capacity c) of radius r at an initial temperature 200 K is supended inside a chamber whose walls are at almost 0 K. The time required (in μs) for the temperature of the sphere to drop at 100 K is

(a) $\frac{72}{7}\frac{r\rho c}{\sigma}$ (b) $\frac{7}{72}\frac{r\rho c}{\sigma}$ (c) $\frac{27}{7}\frac{r\rho c}{\sigma}$ (d) $\frac{7}{27}\frac{r\rho c}{\sigma}$

32. An ice box used for keeping eatable cold has a total wall area of 1 m^2 and a wall thickness of 5.0 cm. The thermal conductivity of the ice box is $K = 0.01$ joule/metre-°C. It is filled with ice at 0°C along with eatables on a day when the temperature is 30°C. The latent heat of fusion of ice is 334×10^3 J/kg. The amount of ice melted in one day is (1 day = 86.400 s)

(a) 776 g (b) 7760 g
(c) 11520 g (d) 1552 g

33. A cylindrical rod with one end in a steam chamber and the other end in ice results in melting of 0.1 g of ice per second. If the rod is replaced by another with half the length and double the radius of the first and if the thermal conductivity of material of second rod is 1/4 that of first, the rate at which ice melts (in g/s) will be

(a) 3.2 (b) 1.6 (c) 0.2 (d) 0.1

34. A solid copper cube of edges 1 cm is suspended in an evacuated enclosure. Its temperature is found to fall from 100°C to 99°C in 100 s. Another solid copper cube of edges 2 cm, with similar surface nature, is suspended in a similar manner. The time required for this cube to cool from 100°C to 99°C will be approximately

(a) 25 s (b) 50 s (c) 200 s (d) 400 s

35. Which of the substances A, B or C has the highest specific heat? The temperature *versus* time graph is shown

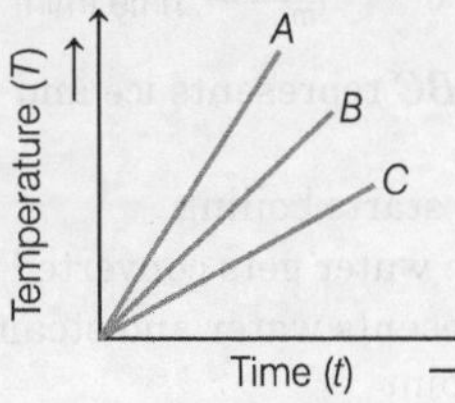

(a) A
(b) B
(c) C
(d) All have equal specific heat

36. The adjoining diagram shows the spectral energy density distribution E_λ of a black body at two different temperatures. If the areas under the curves are in the ratio 16 : 1, the value of temperature T is

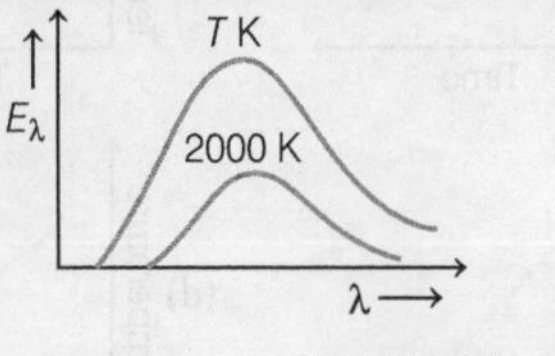

(a) 32000 K (b) 16000 K
(c) 8000 K (d) 4000 K

37. A body cools in a surrounding which is at a constant temperature of θ_0. Assume that it obeys Newton's law of cooling. Its temperature θ is plotted against time t. Tangents are drawn to the curve at the points $P(\theta = \theta_2)$ and $Q(\theta = \theta_1)$. These tangents meet the time axis at angles of ϕ_2 and ϕ_1 as shown, then

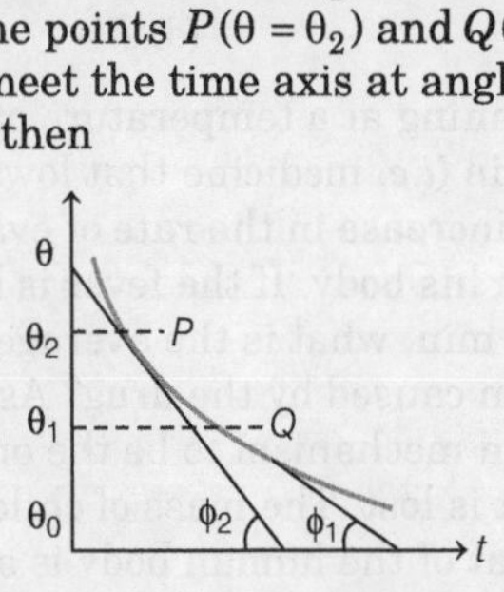

(a) $\frac{\tan\phi_2}{\tan\phi_1} = \frac{\theta_1 - \theta_0}{\theta_2 - \theta_0}$ (b) $\frac{\tan\phi_2}{\tan\phi_1} = \frac{\theta_2 - \theta_0}{\theta_1 - \theta_0}$
(c) $\frac{\tan\phi_1}{\tan\phi_2} = \frac{\theta_1}{\theta_2}$ (d) $\frac{\tan\phi_1}{\tan\phi_2} = \frac{\theta_2}{\theta_1}$

38. Refer to the plot of temperature *versus* time showing the changes in the state of ice on heating (not to scale)
Which of the following is correct? **[NCERT Exemplar]**

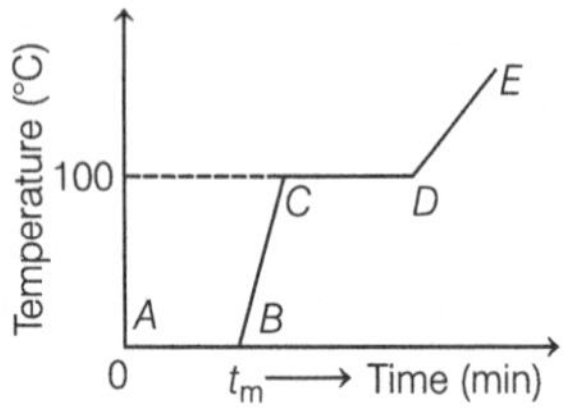

(a) The region BC represents ice and water in thermal equilibrium
(b) At B, water starts boiling
(c) At C, all the water gets converted into steam
(d) C to D represents water and steam in equilibrium at boiling point

39. A block of metal is heated to a temperature much higher than the room temperature and allowed to cool in a room free from air currents. Which of the following curves correctly represents the rate of cooling?

(a) 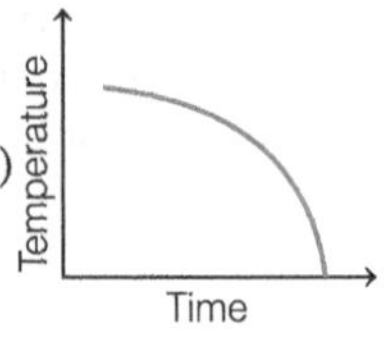

(b) 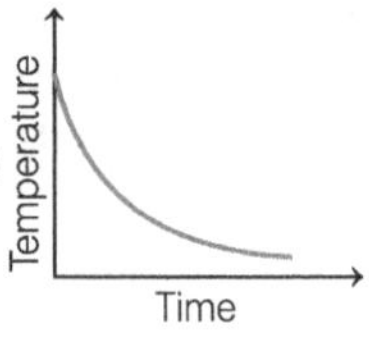

(c) 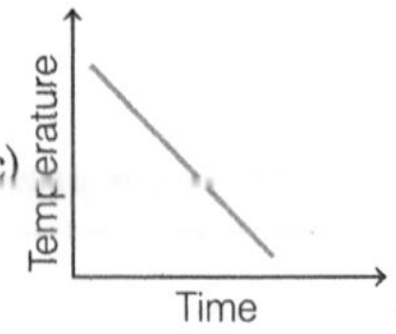

(d) 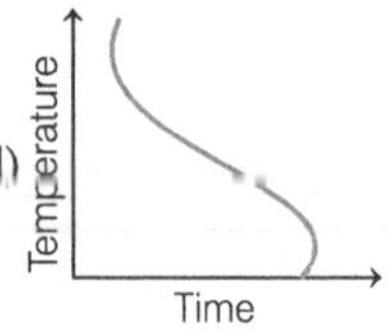

40. A spherical body of radius r radiates power P and its rate of cooling is R.
(a) $P \propto \frac{1}{r}$ (b) $P \propto r^2$
(c) $R \propto r^2$ (d) $R \propto r$

41. A child running at a temperature of 101°F is given an antipyrin (*i.e.* medicine that lowers fever) which causes an increase in the rate of evaporation of sweat from his body. If the fever is brought down to 98°F in 20 min, what is the average rate of extra evaporation caused by the drug? Assume the evaporation mechanism to be the only way by which heat is lost. The mass of child is 30 kg. The specific heat of the human body is approximately the same as that of water and latent heat of evaporation of water at that temperature is about 580 cal/g. **[NCERT]**
(a) 4.31 g/min (b) 4.31 g/s
(c) 2.31 g/min (d) 2.31 g/s

42. Temperature difference of 120°C is maintained between two ends of a uniform rod AB of length $2L$. Another bent rod PQ, of same cross-section as AB and length $\frac{3L}{2}$ is connected across AB (see figure).
In steady state, temperature difference between P and Q will be close to **[JEE Main 2019]**

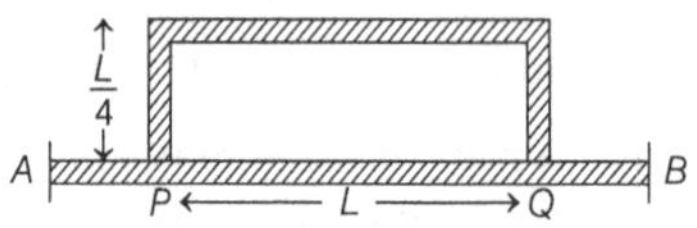

(a) 45°C (b) 35°C
(c) 75°C (d) 60°C

43. Solids expand on heating because
(a) kinetic energy of the atoms increases
(b) potential energy of the atoms increases
(c) total energy of the atoms increases
(d) the potential energy curve is asymmetric about the equilibrium distance between neighbouring atoms

44. An iron tyre is to be fitted on a wooden wheel 1 m in diameter. The diameter of tyre is 6 mm smaller than that of wheel. The tyre should be heated, so that its temperature increases by a minimum of (the coefficient of cubical expansion of iron is $3.6 \times 10^{-5} /°C$)
(a) 167°C (b) 334°C
(c) 500°C (d) 1000°C

45. The power radiated by a black body is P, and it radiates maximum energy around the wavelength λ_0. If the temperature of black body is now changed so that it radiates maximum energy around a wavelength $\lambda_0/4$, the power radiated by it will increase by a factor of
(a) $\frac{4}{3}$ (b) $\frac{16}{9}$
(c) $\frac{64}{27}$ (d) $\frac{256}{81}$

46. A metal ball immersed in alcohol weighs w_1 at 0°C and w_2 at 59°C. The coefficient of cubical expansion of the metal is less than that of alcohol. Assuming that the density of metal is large compared to that of alcohol, it can be shown that
(a) $w_1 > w_2$ (b) $w_1 = w_2$
(c) $w_1 < w_2$ (d) $w_2 = (w_1/2)$

47. The temperature of equal masses of three different liquids A, B and C are 12°C, 19°C and 28°C, respectively. The temperature when A and B are mixed is 16°C and when B and C are mixed is 23°C. The temperature when A and C are mixed, is
(a) 18.2°C (b) 22°C
(c) 20.2°C (d) 25.2°C

48. In an industrial process 10 kg of water per hour is to be heated from 20°C to 80°C. To do this steam at 150°C is passed from a boiler into a copper coil immersed in water. The steam condenses in the coil and is returned to the boiler as water at 90°C. How many kg of steam is required per hour? (Specific heat of steam = 1 calorie per g°C, latent heat of vaporisation = 540 cal/g)

(a) 1 g
(b) 1 kg
(c) 10 g
(d) 10 kg

49. A composite metal bar of uniform section is made up of length 25 cm of copper, 10 cm of nickel and 15 cm of aluminium. Each part being in perfect thermal contact with the adjoining part. The copper end of the composite rod is maintained at 100°C and the aluminium end at 0°C. The whole rod is covered with belt, so that no heat loss occurs at the side. If $K_{Cu} = 2K_{Al}$ and $K_{Al} = 3K_{Ni}$, then what will be the temperatures of Cu-Ni and Ni-Al junctions, repectively?

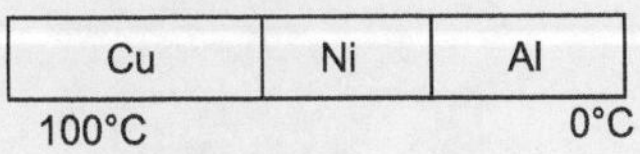

(a) 23.33°C and 78.8°C
(b) 83.33°C and 20°C
(c) 50°C and 30°C
(d) 30°C and 50°C

50. Three rods of identical area of cross-section and made from the same metal form the sides of an isosceles triangle ABC right angled at B. The points A and B are maintained at temperatures T and $\sqrt{2}\,T$, respectively. In the steady state, the temperature of the point C is T_C. Assuming that only heat conduction takes place, $\frac{T_C}{T}$ is equal to

(a) $\frac{1}{(\sqrt{2}+1)}$ (b) $\frac{3}{(\sqrt{2}+1)}$
(c) $\frac{1}{2(\sqrt{2}-1)}$ (d) $\frac{1}{\sqrt{3}(2-1)}$

51. A metal rod AB of length $10x$ has its one end A in ice at 0°C and the other end B in water at 100°C. If a point P on the rod is maintained at 400°C, then it is found that equal amounts of water and ice evaporate and melt per unit time. The latent heat of evaporation of water is 540 cal/g and latent heat of melting of ice is 80 cal/g. If the point P is at a distance of λx from the ice end A, find the value of λ (Neglect any heat loss to the surroundings).

(a) 9 (b) 2
(c) 6 (d) 1

52. A brass boiler has a base area 0.15 m^2 and thickness 1.0 cm. It boils water at the rate of 6.0 kg/min when placed on a gas stove. The temperature of the part of the flame in contact with the boiler will be (Take, thermal conductivity of brass = 109 J/s-m-K, heat of vaporisation of water = 2256×10^3 J/kg) **[NCERT]**

(a) 158°C (b) 208°C
(c) 238°C (d) 264° C

53. A sphere and a cube of same material and same volume are heated upto same temperature and allowed to cool in the same surroundings. The ratio of the amounts of radiation emitted will be

(a) 1 : 1 (b) $\frac{4\pi}{3}$: 1
(c) $\left(\frac{\pi}{6}\right)^{1/3}$: 1 (d) $\frac{1}{2}\left(\frac{4\pi}{3}\right)^{2/3}$: 1

54. A body initially at 80°C cools to 64°C in 5 min and to 52°C in 10 min. The temperature of the body after 15 min will be

(a) 42.7°C (b) 35°C (c) 47°C (d) 40°C

55. Two bodies A and B have thermal emissivities of 0.01 and 0.81, respectively. The outer surface areas of the two bodies are the same. The two bodies emit total radiant power at the same rate. The wavelength λ_B corresponding to maximum spectral radiancy in the radiation from B is shifted from the wavelength corresponding to maximum spectral radiancy in the radiation from A, by 1.00 µm. If the temperature of A is 5802 K, then

(a) the temperature of B is 1836 K
(b) $\lambda_B = 1.5\,\mu m$
(c) the temperature of B is 11604 K
(d) the temperature of B is 2901 K

Numerical Value Questions

56. An unknown metal of mass 192 g heated to a temperature of 100°C was immersed into a brass calorimeter of mass 128 g containing 240 g of water at a temperature of 8.4°C. The specific heat of the unknown metal, if water temperature stabilises at 21.5°C is .... J kg^{-1}K^{-1}.
(Take, specific heat of brass is 394 J kg^{-1}K^{-1}) **[JEE Main 2019]**

57. A cylindrical rod with one end in a steam chamber and the other end in ice, results in melting of 0.1 g of ice per second. If the rod is replaced by another with half the length and double the radius of the first and if the thermal conductivity of the material of the second rod is 1/4 that of the first, the rate at which ice melts in gs^{-1} will be ....

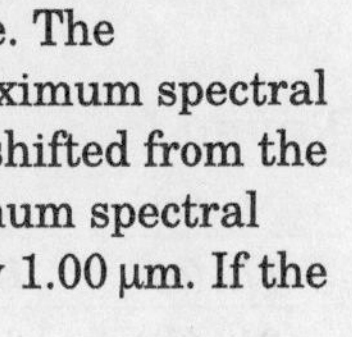

58. The maximum energy in the thermal radiation from a hot source occurs at $\lambda = 11 \times 10^{-5}$ cm. If temperature of another source is n times, for which wavelength of maximum energy is 5.5×10^{-5} cm, then n is ....

59. The latent heat of fusion of ice is 80 calg^{-1} and latent heat of steam is 540 calg^{-1}. Change of state occurs only at melting point or boiling point of the substance. There is no change in temperature during the entire change of state. For rise in temperature (ΔT) heat required $\Delta Q = mc\Delta T$, where c is specific heat of the substance. Heat required (in cal) to convert 1 g of ice at – 5°C to water at 0°C is .... (Take, specific heat of ice = 0.5 $\text{calg}^{-1}\text{°C}^{-1}$)

60. The temperature gradient in the Earth's crust is 32°C km^{-1} and the mean conductivity of earth is 0.008 $\text{cals}^{-1}\text{cm}^{-1}\text{°C}^{-1}$. If considering Earth to be a sphere of radius 6000 km, the loss of heat by Earth everyday is 10^x cal, then find the value of x.

61. 10 g of ice at –20°C is droped into a calorimeter containing 10 g of water at 10°C; the specific heat of water is twice that of ice. When equilibrium is reached, the difference of masses of ice and water in calorimeter is x gram, then the value of x is ....

62. Two metallic spheres S_1 and S_2 are made of the same material and have identical surface finish. The mass of S_1 is three times that of S_2. Both the spheres are heated to the same high temperature and placed in the same room having lower temperature but are thermally insulated from each other. The ratio of the initial rate of cooling of S_1 to that of S_2 is found to be $\left(\frac{1}{3}\right)^{1/m}$, then the value of m is ....

63. Two substances A and B of equal mass m are heated at uniform rate of 6 cal s^{-1} under similar conditions. A graph between temperature and time is shown in figure. Ratio of heat absorbed H_A / H_B by them for complete fusion is ....

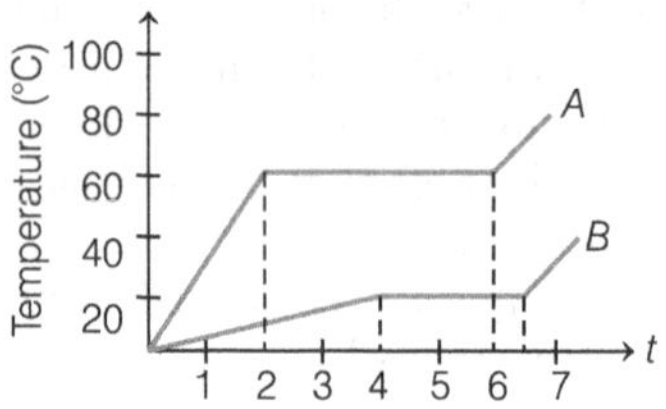

Answers

Round I

1. (a)	2. (c)	3. (b)	4. (b)	5. (d)	6. (a)	7. (a)	8. (c)	9. (b)	10. (c)
11. (a)	12. (a)	13. (b)	14. (a)	15. (d)	16. (a)	17. (a)	18. (d)	19. (c)	20. (a)
21. (d)	22. (c)	23. (a)	24. (b)	25. (a)	26. (d)	27. (d)	28. (d)	29. (b)	30. (b)
31. (c)	32. (a)	33. (a)	34. (a)	35. (a)	36. (b)	37. (c)	38. (a)	39. (a)	40. (c)
41. (d)	42. (d)	43. (a)	44. (a)	45. (d)	46. (c)	47. (c)	48. (a)	49. (b)	50. (c)
51. (d)	52. (b)	53. (a)	54. (a)	55. (a)	56. (a)	57. (a)	58. (c)	59. (a)	60. (c)
61. (d)	62. (a)	63. (c)	64. (c)	65. (a)	66. (d)	67. (c)	68. (d)	69. (a)	70. (a)
71. (d)	72. (b)	73. (d)	74. (b)	75. (a)	76. (a)	77. (d)	78. (a)	79. (b)	80. (a)
81. (a)									

Round II

1. (a)	2. (d)	3. (c)	4. (c)	5. (a)	6. (c)	7. (c)	8. (d)	9. (d)	10. (a)
11. (d)	12. (c)	13. (a)	14. (b)	15. (b)	16. (b)	17. (b)	18. (a)	19. (a)	20. (a)
21. (b)	22. (a)	23. (b)	24. (a)	25. (a)	26. (a)	27. (b)	28. (b)	29. (a)	30. (c)
31. (b)	32. (d)	33. (c)	34. (c)	35. (c)	36. (d)	37. (b)	38. (d)	39. (b)	40. (b)
41. (a)	42. (a)	43. (d)	44. (c)	45. (d)	46. (c)	47. (c)	48. (b)	49. (b)	50. (b)
51. (a)	52. (c)	53. (c)	54. (a)	55. (b)	56. 916	57. 0.2	58. 2	59. 82.5	60. 18
61. 0	62. 3	63. 1.6							

Solutions

Round I

1. Given, triple point of water on scale $A = 200\,A$

Triple point of water on scale $B = 350\,B$

We know that, triple point of water on absolute scale

$$= 273.16\text{ K}$$

$$\therefore \quad 200A = 350B = 273.16\text{ K}$$

$$\therefore \quad 1A = \frac{273.16}{200}\text{ K and } 1B = \frac{273.16}{350}\text{ K}$$

If T_A and T_B are the triple point of water on two scales A and B, then

$$\frac{273.16}{200}T_A = \frac{273.16}{350}T_B$$

$$\therefore \quad \frac{T_A}{T_B} = \frac{200}{350} = \frac{4}{7}$$

2. Let $F = K = x$

As, $$\frac{F-32}{9} = \frac{K-273}{5}$$

$$\therefore \quad \frac{x-32}{9} = \frac{x-273}{5}$$

$$\Rightarrow \quad 9x - 2457 = 5x - 160$$

$$\Rightarrow 4x - 2457 + 160 = 0$$

$$\Rightarrow \quad x = \frac{2297}{4} = 574.25$$

3. $$\frac{C}{100} = \frac{x - \text{lower fixed point}}{\text{upper fixed point} - \text{lower point}} = \frac{68-5}{95-5} = \frac{7}{10}$$

$$\Rightarrow C = \frac{700}{10} = 70°\text{C}$$

4. From graph, we note that for scale A, the lowest fixed point is higher than 0° A and the highest point is 180°A. For scale B, the lowest point is 0°B and the highest point is 100°B. Therefore, the relation $\frac{t_A - 30}{150} = \frac{t_B - 0}{100} = \frac{t_B}{100}$ is correct.

5. Since, $\alpha_{Al} > \alpha_{steel}$, so in bimetallic strip on heating, aluminium strip will expand more than that of steel strip. Due to it, aluminium strip will bend more on convex side and steel strip on concave side.

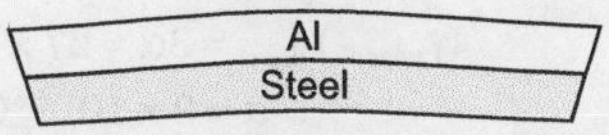

6. Here, $\Delta T = 20°\text{C} - 15°\text{C} = 5°\text{C}$,

$$\alpha = 0.000012°\text{ C}^{-1} = 12 \times 10^{-6}\,°\text{C}^{-1}$$

$$\text{Time lost per day} = \frac{1}{2}\alpha(\Delta T) \times 86400\text{ s}$$

$$= \frac{1}{2} \times 12 \times 10^{-6} \times 5 \times 86400\text{ s}$$

$$= 2.590\text{ s} \simeq 2.6\text{ s}$$

7. Triple point of neon $(T_1) = 24.57$ K

Triple point of CO_2 $(T_2) = 216.55$ K

On celsius scale, °C = $K - 273.15$

Triple point of neon,

$t_1°\text{ C} = 24.57 - 273.15 = -248.58°\text{C}$

Triple point of CO_2, $t_2°\text{ C} = 216.55 - 273.15$

$= -56.60°\text{C}$

On Fahrenheit scale,

$$\frac{K - 273.15}{5} = \frac{F-32}{9}$$

or $$F = (K_1 - 273.15) \times \frac{9}{5} + 32$$

Triple point of neon, $$F_1 = (K_1 - 273.15) \times \frac{9}{5} + 32$$

$$= (24.57 - 273.15) \times \frac{9}{5} + 32$$

$$= -248.58 \times \frac{9}{5} + 32 = -415.44°\text{F}$$

Triple point of CO_2, $$F_2 = (K_2 - 273.15) \times \frac{9}{5} + 32$$

$$= (216.55 - 273.15) \times \frac{9}{5} + 32$$

$$= -56.6 \times \frac{9}{5} + 32 = -69.88°\text{F}$$

8. From $$l_2 = l_1\,[1 + \alpha(t_2 - t_1)]$$

$$\Rightarrow \quad t_2 = t_1 + \frac{l_2 - l_1}{l_1\alpha} = 20 + \frac{-10^3}{1.0 \times 2 \times 10^{-5}}$$

$$= -30°\text{ C}$$

9. When a metallic rod is heated, it expands. Its moment of inertia (I) about a perpendicular bisector increases. According to law of conservation of angular momentum, its angular speed (ω) decreases, since $\omega \propto 1/I$ (according to law of conservation of angular momentum).

10. As, $\Delta L = \alpha L(\Delta T) = \frac{F\,L}{A\,Y}$

$$\therefore \quad F = \alpha(\Delta T)AY$$

$$= 1.1 \times 10^{-5} \times (50 - 30) \times 2 \times 10^{-6} \times 2 \times 10^{11} = 88\text{ N}$$

11. Here, $\rho_0 = 10$ g/cc

and $\rho_{100} = 9.7$ g/cc, $\alpha = ?$

From $\rho_0 = \rho_{100}(1 + \gamma \times 100)$

$$\Rightarrow \quad \gamma = \frac{\rho_0 - \rho_{100}}{\rho_{100} \times 100} = \frac{10 - 9.7}{9.7 \times 100} = 3.09 \times 10^{-4}$$

$$\because \quad \alpha = \frac{\gamma}{3} = \frac{3.09 \times 10^{-4}}{3} = 1.03 \times 10^{-4}\,°\text{ C}^{-1}$$

12. Here, $\alpha\text{ (steel)} = 1.1 \times 10^{-5}\,°\text{ C}^{-1}$

$\alpha\text{ (copper)} = 1.7 \times 10^{-5}\,°\text{ C}^{-1}$

As, $\frac{l_0(S)}{l_0(C)} = \frac{\alpha(C)}{\alpha(S)} = \frac{1.7\times10^{-5}}{1.1\times10^{-5}} = 1.545$

$\therefore \quad l_0(S) = 1.545\, l_0(C)$

Also, $l_0(S) - l_0(C) = 5$

$0.545\, l_0(C) = 5$

$\Rightarrow \quad l_0(C) = \frac{5}{0.545} = 9.17 \text{ cm}$

and $\quad l_0(S) = 1.545 \times 9.17 \text{ cm} = 14.17 \text{ cm}$

13. Here, $\gamma_{ag} = 10.30 \times 10^{-4\circ}\ C^{-1}$,

$\gamma_{am} = 10.06 \times 10^{-4\circ}\ C^{-1}$,

$\alpha_a = 9 \times 10^{-6}\,{}^\circ C^{-1}, \alpha_m = ?$

Now, $\gamma_r = \gamma_{ag} + \gamma_{glass} = \gamma_{am} + \gamma_m$

$\therefore \quad 10.30 \times 10^{-4} + 3 \times 9 \times 10^{-6} = 10.06 \times 10^{-4} + \gamma_m$

$(\because \gamma_{glass} = 3 \times \alpha_a)$

$\therefore \quad \gamma_m = (10.30 + 0.27 - 10.06)\, 10^{-4} = 0.51 \times 10^{-4}$

$\because \quad \alpha_m = \frac{1}{3}\gamma_m = \frac{0.50 \times 10^{-4}}{3}$

$= 0.17 \times 10^{-4} = 17 \times 10^{-6\circ}\ C^{-1}$

14. Given, $h_1 = 50$ cm, $T_1 = 50^\circ$ C,

$h_2 = 60$ cm and $T_2 = 100^\circ$ C.

Let the density of the given liquid at STP is ρ_0.

If both vertical columns balance each other, then their pressure should be equal.

i.e. $\quad p = \rho g h$

$\Rightarrow \quad \rho_1 g h_1 = \rho_2 g h_2$

$\Rightarrow \quad \frac{\rho_1}{\rho_2} = \frac{h_1}{h_2} \quad$ ......(i)

If γ be the coefficient of absolute expansion of liquid, then

$$\rho_1 = \frac{\rho_0}{1+\gamma T_1} \text{ and } \rho_2 = \frac{\rho_0}{1+\gamma T_2}$$

$\therefore$ From Eq. (i), we have

$$\frac{\frac{\rho_0}{1+\gamma T_1}}{\frac{\rho_0}{1+\gamma T_2}} = \frac{h_1}{h_2} = \frac{60}{50}$$

$\Rightarrow \quad \frac{1+\gamma T_1}{1+\gamma T_2} = \frac{6}{5}$

$\Rightarrow \quad 5\gamma T_2 - 6\gamma T_1 = 1$

$\Rightarrow \quad \gamma = \frac{1}{5T_2 - 6T_1} = \frac{1}{5\times100 - 6\times50}$

$\Rightarrow \quad \gamma = \frac{1}{200} = 0.005/^\circ C$

15. Let V be the volume of sphere of radius R at temperature, then $V = \frac{4}{3}\pi R^3$.

Increase in volume of sphere with rise in temperature ΔT,

$$\Delta V = \gamma V \Delta T = 3\alpha \times \frac{4}{3}\pi R^3$$

$$\Delta T = 4\pi R^3 \alpha \Delta T$$

16. Change in length for both rods are same,

$\because \quad \Delta l_1 = \Delta l_2$

$\Delta l = \alpha l \Delta T$

$\Rightarrow \quad l\alpha_1 \Delta T_1 = l\alpha_2 \Delta T_2$

$\Rightarrow \quad \frac{\alpha_1}{\alpha_2} = \frac{\Delta T_2}{\Delta T_1}$

$\therefore \quad \frac{4}{3} = \frac{T - 30}{180 - 30} \quad$ (given, $\alpha_1 : \alpha_2 = 4:3$)

$\frac{4}{3} = \frac{T-30}{150}$

$\Rightarrow \quad T - 30 = 200$

or $\quad T = 200 + 30 = 230^\circ$C

17. Here, $\gamma = \frac{\Delta V}{V \times \Delta T} = \frac{0.24}{100 \times 40} = 6 \times 10^{-5}/^\circ C$

$\therefore$ Coefficient of linear expansion, $\alpha = \frac{\gamma}{3} = 2 \times 10^{-5}/^\circ C$

18. Using $\gamma_{real} = \gamma_{app}$ + vessel

$\therefore \quad \gamma_{vessel} = 3\alpha$

For vessel A, $\gamma_{real} = \gamma_1 + 3\alpha$

For vessel B, $\gamma_{real} = \gamma_2 + 3\alpha_2$

Hence, $\gamma_1 + 3\alpha_1 = \gamma_2 + 3\alpha_2$

$\Rightarrow \quad \gamma_1 - \gamma_2 = 3\alpha_2 - 3\alpha_1$

$\Rightarrow \quad \frac{\gamma_1 - \gamma_2}{3} = \alpha_2 - \alpha_1$

$\Rightarrow \quad \alpha_2 = \frac{\gamma_1 - \gamma_2}{3} + \alpha_1 = \frac{\gamma_1 - \gamma_2}{3} + \alpha$

19. Here, $\Delta l = 80.3 - 80.0 = 0.3$ cm,

$l = 80$ cm, $\alpha = 1.2 \times 10^{-5}\,{}^\circ C^{-1}$

Rise in temperature, $\Delta T = \frac{\Delta l}{l\alpha}$

$\Rightarrow \quad \Delta T = \frac{0.3}{80 \times 1.2 \times 10^{-5}} = 312.5^\circ$ C

20. $\gamma_{real} = \gamma_{app} + \gamma_{vessel}$

So, $\quad (\gamma_{app} + \gamma_{vessel})_{glass} = (\gamma_{app} + \gamma_{vessel})_{steel}$

$\Rightarrow \quad 153 \times 10^{-6} + (\gamma_{vessel})_{glass} = (144 \times 10^{-6} + \gamma_{vessel})_{steel}$

Further, $\quad (\gamma_{vessel})_{steel} = 3\alpha = 3 \times (12 \times 10^{-6})$

$= 36 \times 10^{-6}/^\circ C$

$\Rightarrow \quad 153 \times 10^{-6} + (\gamma_{vessel})_{glass} = 144 \times 10^{-6} + 36 \times 10^{-6}$

$\Rightarrow \quad (\gamma_{vessel})_{glass} = 3\alpha = 27 \times 10^{-6}/^\circ C$

$\Rightarrow \quad \alpha = 9 \times 10^{-6}/^\circ C$

21. Length of both strips will decrease

$\because \quad \alpha_A > \alpha_B$

$\therefore \quad \Delta L_A > \Delta L_B$

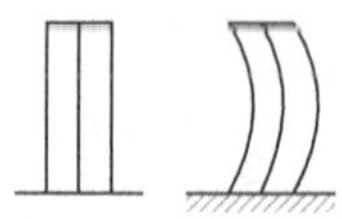

Therefore, it bend towards the left.

22. $V = V_0(1+\gamma\Delta\theta)$

$$L^3 = L_0(1+\alpha_1\Delta\theta)L_0^2(1+\alpha_2\Delta\theta)^2$$
$$= L_0^3(1+\alpha_1\Delta\theta)(1+\alpha_2\Delta\theta)^2$$

Since, $L_0^3 = V_0$ and $L^3 = V$

$\therefore \quad V = V_0(1+\alpha_1\Delta\theta)(1+\alpha_2\Delta\theta)^2$

$\Rightarrow \quad V_0(1+\gamma\Delta\theta) = V_0(1+\alpha_1\Delta\theta)(1+\alpha_2\Delta\theta)^2$

Hence, $1+\gamma\Delta\theta = (1+\alpha_1\Delta\theta)(1+\alpha_2\Delta\theta)^2$

$$\cong (1+\alpha_1\Delta\theta)(1+2\alpha_2\Delta\theta)$$
$$\cong (1+\alpha_1\Delta\theta+2\alpha_2\Delta\theta)$$

$\Rightarrow \quad 1+\gamma\Delta\theta = [1+(\alpha_1+2\alpha_2)\Delta\theta]$

$\Rightarrow \quad \gamma = \alpha_1 + 2\alpha_2$

23. Moment of inertia of a rod,

$$I = \frac{1}{12}ML^2 \quad \text{...(i)}$$

where, M is the mass of the rod and L is the length of the rod.

$\therefore \quad \Delta I = \frac{1}{12}2ML\Delta L \quad (\because M \text{ is a constant}) \text{ ...(ii)}$

Dividing Eq. (ii) by (i), we get

$$\frac{\Delta I}{I} = 2\frac{\Delta L}{L} \quad \text{...(iii)}$$

As, $\Delta L = L\alpha\Delta t$

or $\frac{\Delta L}{L} = \alpha\Delta t$

Substituting the value of $\frac{\Delta L}{L}$ in Eq. (iii), we get

$$\frac{\Delta I}{I} = 2\alpha\Delta t$$

24. Loss of weight at 27°C is

$$= 46 - 30 = 16 = V_1 \times 1.24\,\rho_1 \times g \quad \text{...(i)}$$

Loss of weight at 42°C is

$$= 46 - 30.5 = 15.5 = V_2 \times 1.2\rho_1 \times g \quad \text{...(ii)}$$

Now, dividing Eq. (i) by (ii), we get

$$\frac{16}{15.5} = \frac{V_1}{V_2} \times \frac{1.24}{1.2}$$

$\Rightarrow \quad \frac{V_2}{V_1} = \frac{1.24}{1.2} \times \frac{15.5}{16}$

But $\frac{V_2}{V_1} = 1 + 3\alpha(t_2 - t_1)$

$\Rightarrow \quad \frac{1.24 \times 15.5}{1.2 \times 16} = 1 + 3\alpha(42 - 27)$

$\Rightarrow \quad \alpha = 2.316 \times 10^{-5}/°\text{C}$

25. Given, diameter of the hole $(d_1) = 4.24$ cm

Initial temperature $(T_1) = 27 + 273 = 300$ K

Final temperature $(T_2) = 227 + 273 = 500$ K

Coefficient of linear expansion $(\alpha) = 1.70 \times 10^{-5}/°\text{C}$

Coefficient of superficial expansion $(\beta) = 2\alpha$

$= 3.40 \times 10^{-5}/°\text{C}$

Initial area of hole at 27°C $(A_1) = \pi r^2 = \frac{\pi d_1^2}{4}$

$$= \frac{\pi}{4}(4.24)^2 = 4.494\pi \text{ cm}^2$$

Area of hole at 227°C $(A_2) = A_1(1+\beta\cdot\Delta t)$

$$= 4.494\,\pi[1 + 3.40 \times 10^{-5} \times (227 - 27)]$$
$$= 4.494\pi\,[1 + 3.40 \times 10^{-5} \times 200]$$
$$= 4.495\pi \times 1.0068$$
$$= 4.525\pi \text{ cm}^2$$

If diameter of hole becomes d_2 at 227°C, then $A_2 = \frac{\pi d_2^2}{4}$

$$4.525\pi = \frac{\pi d_2^2}{4}$$

or $d_2^2 = 4.525 \times 4$

or $d_2 = 4.2544$ cm

$\therefore$ Change in diameter $(\Delta d) = d_2 - d_1 = 4.2544 - 4.24$

$$= 0.0144 \text{ cm}$$
$$= 1.44 \times 10^{-2} \text{ cm}$$

26. Given, coefficient of volume expansion $(\gamma) = 49 \times 10^{-5}/\text{K}$

Rise in temperature $(\Delta t) = 30°\text{C}$

Let initial volume of glycerine be V_0.

$\therefore$ Volume of glycerine when temperature is increased by 30°C,

$$V = V_0(1+\gamma\Delta t)$$
$$= V_0(1 + 49 \times 10^{-5} \times 30)$$
$$= V_0(1 + 0.01470) = 1.0147V_0$$

$\therefore \quad \frac{V_0}{V} = \frac{1}{1.0147} \quad \text{...(i)}$

If mass of glycerine is m, then

initial density of glycerine $(\rho_0) = \frac{m}{V_0}$

and final density of glycerine $(\rho) = \frac{m}{V}$

$\therefore \quad \frac{\rho}{\rho_0} = \frac{m/V}{m/V_0} = \frac{V_0}{V} = \frac{1}{1.0147} \quad \text{...(ii)}$

Fractional change in density $= \frac{\Delta\rho}{\rho_0} = \frac{\rho - \rho_0}{\rho_0} = \frac{\rho}{\rho_0} - 1$

Substituting value from Eq. (ii), we get

Fractional change in density $= \frac{1}{1.0147} - 1$

$$= -0.0145 = -1.45 \times 10^{-2}$$

Negative sign shows that the density of glycerine decreases with rise in temperature.

27. Let θ be the temperature of the mixture.

Heat gained by water at 0°C = Heat lost by water at 10°C

$$cm_1(\theta - 0) = cm_2(10 - \theta)$$

$\Rightarrow \quad 20 \times \theta = 40 \times 10$

$$\theta = \frac{400}{60} = 6.66° \text{ C}$$

28. 10 g of ice at –10°C to ice at 0°C,

$$Q_1 = cm\,\Delta\theta = 0.5 \times 10 \times 10 = 50 \text{ cal}$$

10 g of ice at 0°C to water at 0°C,

$$Q_2 = mL = 10 \times 80 = 800 \text{ cal}$$

10 g of water at 0°C to water at 100°C,

$$Q_3 = cm\,\Delta\theta = 1 \times 10 \times 100 = 1000 \text{ cal}$$

10 g of water at 100°C to steam at 100°C,

$$Q_4 = mL = 10 \times 540 = 5400 \text{ cal}$$

Total heat required, $Q = Q_1 + Q_2 + Q_3 + Q_4$

$$= 50 + 800 + 1000 + 5400$$
$$= 7250 \text{ cal}$$

29. Heat required to melt 1 g of ice at 0°C to water at 0°C $= 1 \times 80$ cal

Heat required to raise temperature of 1 g of water from 0°C to 100°C $= 1 \times 1 \times 100 = 100$ cal

Total heat required for maximum temperature of 100° C $= 80 + 100 = 180$ cal

As, one gram of steam gives 540 cal of heat when it is converted to water at 100°C, therefore temperature of the mixture would be 100°C.

30. From the principle of calorimetry,

$$m_1 s_1 \Delta\theta_1 = m_2 s_2 \Delta\theta_2$$

$$\Rightarrow \quad 540 \times s_w \times (80 - \theta) = 540 \times \frac{s_w}{2} \times (\theta - 0)$$

$$\Rightarrow \quad \theta = \frac{160^\circ}{3} \text{C}$$

$$= 53.3 \approx 53^\circ \text{ C}$$

31. Heat capacity per volume $= c \times \frac{m}{V} = c \times \rho$

Desired ratio $= \frac{c_1\rho_1}{c_2\rho_2} = \frac{3}{5} \times \frac{5}{6} = 1\ .2$

32. Relative humidity is the ratio of water vapour pressure and saturated vapour pressure at the present temperature. When the room temperature becomes equal to the dew point, then the air is saturated. Hence, the relative humidity of room is 100%.

33. Heat gained by the water = (Heat supplied by the coil) – (Heat dissipated to environment)

$$\Rightarrow \quad mc\Delta\theta = P_{\text{Coil}}\, t - P_{\text{Loss}}\, t$$

$$\Rightarrow \quad 2 \times 4.2 \times 10^3 \times (77 - 27) = 1000t - 160t$$

$$\Rightarrow \quad t = \frac{4.2 \times 10^5}{840}$$

$$= 500 \text{ s} = 8 \text{ min } 20 \text{ s}$$

34. Slope of graph is greater in the solid state i.e. temperature is rising faster, hence lower heat capacity. The transition from solid to liquid state takes lesser time, hence latent heat is smaller.

Therefore, specific heat capacity is greater in solid state than in the liquid state.

35. As, $Q_{\text{mix}} = \dfrac{m_w Q_w - \dfrac{m_l L_i}{C_w}}{m_i + m_w} = \dfrac{10 \times 100 - \dfrac{10 \times 80}{1}}{10 + 10}$

$$= \frac{1000 - 800}{20}$$

$$= \frac{200}{20} = 10^\circ\text{C}$$

36. Here, $\quad m = 0.1 \text{ kg}, h_1 = 10 \text{ m}, h_2 = 5.4 \text{ m}$

$c = 460 \text{ J kg}^{-1}{}^\circ\text{C}^{-1}, g = 10 \text{ ms}^{-2}, \theta = ?$

Energy dissipated, $Q = mg\,(h_1 - h_2)$

$$= 0.1 \times 10\ (10 - 5.4) = 4.6 \text{ J}$$

From $\quad Q = cm\theta$

$$\Rightarrow \quad \theta = \frac{Q}{cm} = \frac{4.6}{460 \times 0.1} = 0.1^\circ \text{ C}$$

37. Here, $\quad m = 10 \text{ g} = 10^{-2} \text{ kg},$

$v = 300 \text{ ms}^{-1}, \theta = ?, c = 150 \text{ Jkg}^{-1} \text{ K}^{-1}$

$$Q = \frac{50}{100}\left(\frac{1}{2}mv^2\right) = \frac{1}{4} \times 10^{-2}\,(300)^2 = 225 \text{ J}$$

From $\quad Q = cm\,\theta$

$$\Rightarrow \quad \theta = \frac{Q}{cm} = \frac{225}{150 \times 10^{-2}} = 150^\circ \text{ C}$$

38. Let m gram of steam get condensed into water (by heat loss). This happens in following two steps

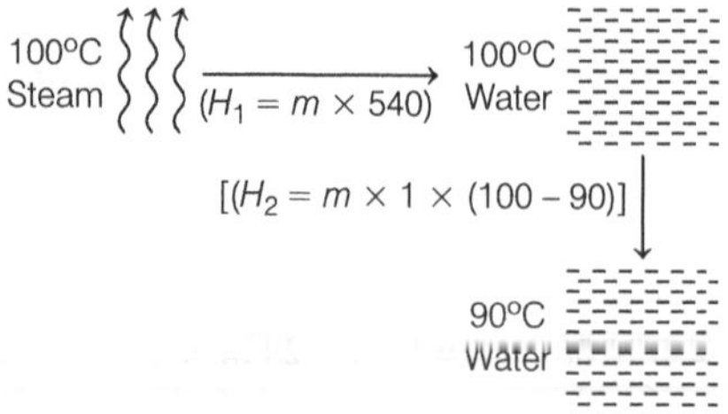

Heat gained by water (20°C) to raise its temperature upto 90° $= 22 \times 1 \times (90 - 20)$

Hence, in equilibrium; heat lost = heat gain

$\Rightarrow m \times 540 + m \times 1 \times (100 - 90) = 22 \times 1 \times (90 - 20)$

$$\Rightarrow \quad m = 2.8 \text{ g}$$

The net mass of the water present in the mixture

$$= 22 + 2.8 = 24.8 \text{ g}$$

39. Let the amount of ice be 'x' gram.

According to the principle of calorimeter,

heat lost by water = heat gained by ice

Here, heat lost by water, $\Delta Q = ms_{\text{water}}\,\Delta T$

$$\Delta Q = 50 \times 4.2 \times 40$$

Heat gained by ice,

$$\Delta Q = x\, s_{\text{ice}}\, \Delta T + (x - 20)\, L$$
$$= x \times 2.1 \times 20 + (x - 20) \times 334$$
$$= 20\, x \times 2.1 + 334x - 6680$$

$\therefore\ 20\, x \times 2.1 + 334x - 6680 = 50 \times 4.2 \times 40$

$$42x + 334x - 6680 = 8400$$

$$\Rightarrow \quad 376\,x = 15080$$

or $\quad x = 40.10 \text{ g}$

$$x \simeq 40 \text{ g}$$

40. Heat absorbed by water = Heat produced

$$mc\Delta T = \frac{mgh}{J}$$

$$\Delta T = \frac{gh}{Jc} = \frac{980 \times 500 \times 100}{4.2 \times 10^7 \times 1}$$

$$= \frac{900}{420} = 1.16° \text{ C}$$

41. Let m gram of water whose temperature is θ ($> 30°$ C) be added to 20 g of water at 30°C. Then,

$$m \times 1\,(\theta - \theta_0) = 20 \times 1\,(\theta_0 - 30)$$

$$\Rightarrow \quad (m + 20)\,\theta_0 = 600 + m\theta$$

$$\Rightarrow \quad \theta_0 = \frac{600 + m\theta}{20 + m}$$

For θ_0 to be maximum, m should be small and θ should be large. So, the correct option is (d).

42. .Using heat lost or gained without change in state is $\Delta Q = ms\Delta T$, where s is specific heat capacity and T = change in temperature.

Let final temperature of ball be T.
Then, heat lost by ball is,

$$\Delta Q = 0.1 \times 400(500 - T) \quad \text{...(i)}$$

This lost heat by ball is gained by water and vessel given as

Heat gained by water,

$$\Delta Q_1 = 0.5 \times 4200(T - 30) \quad \text{...(ii)}$$

and heat gained by vessel,

$$\Delta Q_2 = \text{heat capacity} \times \Delta T$$

$$= 800 \times (T - 30) \quad \text{...(iii)}$$

According to principle of calorimetry,

total heat lost = total heat gained

$$\Rightarrow 0.1 \times 400(500 - T) = 0.5 \times 4200(T - 30) + 800(T - 30)$$

$$\Rightarrow \quad (500 - T) = \frac{(2100 + 800)(T - 30)}{40}$$

$$\Rightarrow \quad 500 - T = 72.5(T - 30)$$

$$\Rightarrow \quad 500 + 217.5 = 72.5\,T \text{ or } T = 36.39 \text{ K}$$

So, percentage increment in temperature of water

$$= \frac{36.39 - 30}{30} \times 100 \approx 20\%$$

43. With increases in temperature, the effective length (l) of simple pendulum increases even though its centre of mass still remains at the centre of the bob. As time period, $T = 2\pi\sqrt{l/g}$ or $T \propto \sqrt{l}$. So, T increases as temperature increases.

44. The equivalent electrical circuit, figure in this case is of Wheatstone bridge. No current would flow through central rod CD when the bridge is balanced. The condition for balanced Wheatstone bridge is $\frac{P}{Q} = \frac{R}{S}$

(in terms of resistances)

$$\frac{1/K_1}{1/K_2} = \frac{1/K_3}{1/K_4} \text{ or } \frac{K_2}{K_1} = \frac{K_4}{K_3}$$

or $\quad K_1K_4 = K_2K_3$

45. Let interface temperature in steady state conduction is θ, then assuming no heat loss through sides;

$$\begin{pmatrix}\text{Rate of heat} \\ \text{flow through} \\ \text{first slab}\end{pmatrix} = \begin{pmatrix}\text{Rate of heat} \\ \text{flow through} \\ \text{second slab}\end{pmatrix}$$

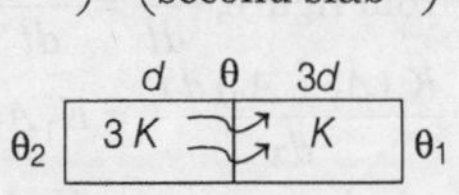

$$\Rightarrow \quad \frac{(3K)\,A(\theta_2 - \theta)}{d} = \frac{KA(\theta - \theta_1)}{3d}$$

$$\Rightarrow \quad 9(\theta_2 - \theta) = \theta - \theta_1$$

$$\Rightarrow \quad 9\,\theta_2 + \theta_1 = 10\,\theta$$

$$\Rightarrow \quad \theta = \frac{9}{10}\theta_2 + \frac{1}{10}\theta_1$$

46. From $\frac{\Delta Q}{\Delta t} = KA\left(\frac{\Delta T}{\Delta x}\right)$

$$\Rightarrow \quad \Delta t = \frac{\Delta Q\,\Delta x}{KA\,(\Delta T)}$$

In arrangement (ii), A is doubled and Δx is halved.

$$\therefore \quad \Delta t \to \frac{1/2}{2} \to \frac{1}{4} \text{ time}$$

i.e. $\quad \frac{1}{4} \times 4 \text{ min} = 1 \text{ min}$

47. As, $\left(\frac{\Delta Q}{\Delta t}\right)_{\text{inner}} + \left(\frac{\Delta Q}{\Delta t}\right)_{\text{outer}} = \left(\frac{\Delta Q}{\Delta t}\right)_{\text{total}}$

$$\frac{K_1\pi r^2(T_2 - T_1)}{l} + \frac{K_2\pi\,[(2\,r)^2 - r^2]\,(T_2 - T_1)}{l}$$

$$= \frac{K\,\pi\,(2\,r)^2\,(T_2 - T_1)}{l}$$

or $\quad (K_1 + 3\,K_2)\frac{\pi r^2\,(T_2 - T_1)}{l} = \frac{K\,\pi 4\,r^2(T_2 - T_1)}{l}$

or $\quad K = \frac{K_1 + 3\,K_2}{4}$

48. Let the temperature of junction be θ.

$$\left(\frac{\Delta Q}{\Delta t}\right)_{\text{copper}} = \left(\frac{\Delta Q}{\Delta t}\right)_{\text{steel}}$$

$$\Rightarrow \quad K_1A\frac{(100 - \theta)}{18} = \frac{K_2\,A\,(\theta - 0)}{6}$$

$$\Rightarrow \quad 9\,K_2\frac{(100 - \theta)}{3} = K_2\theta$$

or $\quad 3\,\theta = 900 - 9\,\theta$

or $\quad 12\,\theta = 900$

or $\quad \theta = 75°\text{C}$

49. For parallel combination of two rods of equal length and equal area of cross-section.

$$K = \frac{K_1 + K_2}{2}$$

$$= \frac{K_1 + \frac{4K_1}{3}}{2} \qquad \left[\therefore K_2 = \frac{4K_1}{3}\right]$$

$$= \frac{7K_1}{6}$$

Hence, $\frac{K}{K_1} = \frac{7}{6}$

50. As it is clear from figure, $\frac{dQ}{dt} = \frac{dQ_1}{dt} + \frac{dQ_2}{dt}$

$$\frac{K(A_1 + A_2)\,dT}{dx} = K_1A_1\frac{dT}{dx} + K_2A_2\frac{dT}{dx}$$

$$\Rightarrow \quad K = \frac{K_1A_1 + K_2A_2}{A_1 + A_2}$$

51. Growth of ice in a pond is conduction process governed by the relation, $t = \frac{\rho L}{K\theta}\frac{y^2}{2}$

The ratio of times for thickness of ice from 0 to y is 1 : 3.

$\therefore$ Time taken to increase the thickness from 1 cm to 2 cm is equal to $3 \times 7 = 21$ h.

52. Here, $K_A = 2K_B\ (dx)_A = d\ (dx)_B$

If θ is temperature of junction, $(dT)_A = \theta_A - \theta$,

$$(dT)_B = (\theta - \theta_B)$$

As, $$\left(\frac{dQ}{dt}\right)_A = \left(\frac{dQ}{dt}\right)_B$$

$$\therefore \quad K_AA\frac{(dT)_A}{(dx)_A} = K_B\frac{A\,(dT)_B}{(dx)_B}$$

$$\Rightarrow \quad 2K_B(\theta_A - \theta) = K_B(\theta - \theta_B)$$

$$\Rightarrow \quad 2\theta_A - 2\theta = \theta - \theta_B$$

$$\Rightarrow \quad 2\theta_A + \theta_B = 3\theta \qquad \ldots(i)$$

As, $$\theta_A - \theta_B = 48°$$

$$\Rightarrow \quad \theta_A = 48 + \theta_B$$

Put in Eq. (i), we get

$$2(48 + \theta_B) + \theta_B = 3\theta$$

$$\Rightarrow \quad 96 + 3\theta_B = 3\theta$$

$$\Rightarrow \quad 96 = 3(\theta - \theta_B)$$

$$\therefore \quad \theta - \theta_B = 96/3 = 32°\text{ C}$$

53. Energy flux is the rate of heat flow per unit area through the rod.

Also, rate of flow of heat per unit time through a material of area of cross-section A and thermal conductivity K between the temperatures T_1 and T_2 $(T_1 > T_2)$ is given as

$$\frac{\Delta Q}{\Delta t} = \frac{KA(T_1 - T_2)}{l} \qquad \ldots(i)$$

Energy flux using Eq. (i), we get

$$= \frac{1}{A}\cdot\frac{\Delta Q}{\Delta t} = \frac{k(T_1 - T_2)}{l}$$

Here, $k = 0.1\text{ W K}^{-1}\text{m}^{-1}$, $l = 1$ m,

$T_1 = 1000$ K and $T_2 = 100$ K

$$\therefore \quad \text{Energy flux} = \frac{0.1(1000 - 100)}{1} = 90\text{ Wm}^{-2}$$

54.

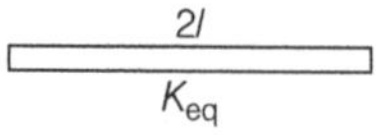

$$R_{eff} = \frac{l}{K_1A} + \frac{l}{K_2A} = \frac{2l}{K_{eq}A}$$

$$K_{eq} = \frac{2K_1K_2}{K_1 + K_2}$$

55. $\frac{\Delta Q}{\Delta t} = KA\left(\frac{\Delta T}{\Delta x}\right) = K(\pi r^2)\frac{\Delta T}{(l)}$

$\therefore \left(\frac{\Delta Q}{\Delta t}\right) \propto \frac{r^2}{l}$, which is maximum in case (a).

56. For the two sheets shown in figure, rate of heat transfer is same, *i.e.*

$$\frac{dQ_1}{dt} = \frac{dQ_2}{dt}$$

$$\therefore \quad \frac{dT_1}{R_1} = \frac{dT_2}{R_2}$$

$$\frac{\theta_1 - \theta}{R_1} = \frac{\theta - \theta_2}{R_2}$$

$$\Rightarrow \quad \theta_1R_2 - \theta R_2 = \theta R_1 - \theta_2R_1$$

$$\text{or} \quad \theta = \frac{\theta_1R_2 + \theta_2R_1}{R_1 + R_2}$$

57. As, $\left(\frac{\Delta Q}{\Delta t}\right)_P = \left(\frac{\Delta Q}{\Delta t}\right)_Q$

$$\Rightarrow \quad K_1A_1\frac{(T_1 - T_2)}{l} = K_2A_2\frac{(T_1 - T_2)}{l}$$

$$\text{or} \quad K_1A_1 = K_2A_2$$

$$\text{or} \quad \frac{A_1}{A_2} = \frac{K_2}{K_1}$$

58. For the same heat to be conducted, temperature difference must be same.

Initial temperature difference

$$= 10 - (-10) = 20°\text{ C} = 20\text{ K}$$

Outside temperature $= -23°\text{ C} = -23 + 273 = 250$ K

Inside temperature $= 250 + 20 = 270$ K

59. As, $\frac{Q}{t} = \frac{KA\Delta\theta}{l} = \frac{\Delta\theta}{(l/KA)} = \frac{\Delta\theta}{R}$ (R = thermal resistance)

$$\Rightarrow \quad t \propto R \Rightarrow \frac{t_p}{t_s} = \frac{R_p}{R_s} = \frac{R/2}{2R} = \frac{1}{4}$$

$$\Rightarrow \quad t_p = \frac{t_s}{4} = \frac{4}{4} = 1\text{ min}$$

60. When temperature of a black body is increased by 50%,

$$T_2 = \frac{150}{100}T_1 = \frac{2}{3}T_1$$

According to Stefan's law,

$$\frac{E_2}{E_1} = \left(\frac{T_2}{T_1}\right)^4 = \left(\frac{3}{2}\right)^4 = \frac{81}{16}$$

Percentage increase in radiation,

$$\frac{(E_2 - E_1)}{E_1} \times 100 = \frac{(81-16)}{16} \times 100 \approx 400\%$$

61. As, energy emitted $\propto AT^4$

$$\therefore \quad \frac{E_2}{E_1} = \frac{E'}{E} = \frac{1}{4}\frac{(327+273)^4}{(127+273)^4} = \frac{1}{4} \times \left(\frac{600}{400}\right)^4 = \frac{81}{64}$$

or $\quad E' = \dfrac{81}{64} E$

62. Mass of ice melted/min $= \dfrac{10}{10} = 1$ g

Quantity of heat used $= 1 \times 80$ cal

Area of lens $= \pi r^2 = 3.14\,(2.5)^2 = 19.625\ \text{cm}^2$

$\therefore$ Amount of heat received/min/cm^2 from Sun

$$= \frac{80}{19.625} \approx 4\ \text{cal cm}^{-2}\text{min}^{-1}$$

63. Here, $T_1 = 327°\text{C} = (327+273)\ \text{K} = 600\ \text{K}$

and $\quad T_2 = 927°\text{C} = (927+273)\ \text{K} = 1200\ \text{K}$

According to Stefan's law, $E \propto T^4$

$$\frac{E_2}{E_1} = \left(\frac{T_2}{T_1}\right)^4 = \left(\frac{1200}{600}\right)^4 = 16$$

$$\Rightarrow \quad E_2 = 16\,E_1 = 16 \times 4 = 64\ \text{cal cm}^{-2}\,\text{s}^{-1}$$

64. Let R = radius of planet

and P = power radiated by the Sun.

Energy received by planet $= \dfrac{P}{4\pi d^2} \times 4\pi R^2$

Energy radiated by planet $= (4\pi R)^2 \sigma T^4$

For thermal equilibrium,

$$4\pi R^2 (\sigma T^4) = \frac{P}{4\pi d^2} \times 4\pi R^2$$

$$T^4 \propto \frac{1}{d^2} \quad \text{or} \quad T \propto d^{-1/2}$$

$$\therefore \quad n = \frac{1}{2}$$

65. As, $\dfrac{E_1}{E_2} = \dfrac{T_1^4 - T_0^4}{T_2^4 - T_0^4}$

$$= \frac{(327+273)^4 - (27+273)^4}{(227+273)^4 - (27+273)^4}$$

$$= \frac{(600)^4 - (300)^4}{(500)^4 - (300)^4}$$

$$= \frac{10^8\,(1296-81)}{10^8\,(625-81)}$$

$$= \frac{1215}{544} = 2.23 \approx 2$$

66. According to Stefan's law,

$$E_1 = \sigma T^4 \text{ and } E_2 = \sigma\,(T+\Delta T)^4$$

$$\therefore \quad \frac{E_2 - E_1}{E_1} = \frac{\sigma\,[(T+\Delta T)^4 - T^4]}{\sigma T^4} = 4\frac{\Delta T}{T}$$

Now, $\quad \dfrac{E_2 - E_1}{E_1} = \dfrac{2}{100} = 4\dfrac{\Delta T}{300}$

$$\therefore \quad \Delta T = \frac{300 \times 2}{4 \times 100} = 1.5\ \text{K}$$

$\therefore$ Temperature of other patch $= T + \Delta T$

$$= 300 + 1.5 = 301.5\ \text{K}$$

67. As, $P \propto T^4$ and $P \propto \dfrac{1}{d^2}$, therefore

$$P \propto \frac{T^4}{d^2} \propto \frac{2^4}{2^2} = 2^2 = 4 \text{ times, } i.e.\ 4P$$

68. Power radiated by Sun at $t°$ C $= \sigma\,(t+273)^4\,4\pi r^2$

Power received by a unit surface $= \dfrac{\sigma\,(t+273)^4\,4\pi r^2}{4\pi R^2}$

$$= \frac{r^2 \sigma\,(t+273)^4}{R^2}$$

69. Here, $\lambda_m = 289.8\ \text{nm} = 289.8 \times 10^{-9}$ m,

$\sigma = 5.67 \times 10^{-8}\ \text{Wm}^{-2}\text{K}^{-4}$,

$b = 2889\ \mu\text{mK} = 2889 \times 10^{-6}$ mK

If T is temperature of star, then according to Wien's law, $\lambda_m T = b$

$$\Rightarrow \quad T = \frac{b}{\lambda_m} = \frac{2898 \times 10^{-6}}{289.8 \times 10^{-9}} = 10^4$$

From Stefan's law, $E = \sigma T^4$

$$= 5.67 \times 10^{-8}\,(10^4)^4 = 5.67 \times 10^8\ \text{Wm}^{-2}$$

70. According to Wien's law, as T increases, λ_m decreases and ν_m increases.

$$\therefore \quad T_1 < T_2$$

Therefore, $\quad \nu_{m_1} < \nu_{m_2}$

71. From Wien's law, $\lambda_m T = b$

$$\Rightarrow \quad \lambda_m = \frac{b}{T} = \frac{2.88 \times 10^6}{2880} = 10^3\ \text{nm}$$

Energy distribution with wavelength will be as follows

From the graph, it is clear that

$U_2 > U_1$ (in fact U_2 is maximum)

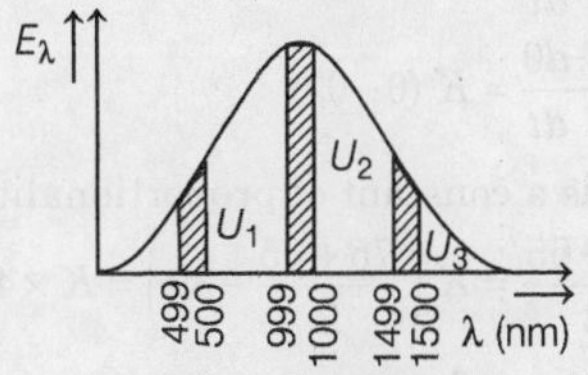

$\lambda_m = 1000$ nm

72. Here, $T_1 = 6000$ K, $\lambda_1 = 4800$ Å,

$T_2 = 3000$ K, $\lambda_2 = ?$

According to Wien's law, $\frac{\lambda_2}{\lambda_1} = \frac{T_1}{T_2}$

$$\Rightarrow \quad \lambda_2 = \frac{T_1}{T_2} \times \lambda_1 = \frac{6000}{3000} \times 4800 = 9600 \text{ Å}$$

73. According to Wien's displacement law,

$$\frac{\lambda_1}{\lambda_2} = \frac{T_2}{T_1} = \frac{0.26}{0.13} = 2$$

$$\therefore \quad \frac{E_1}{E_2} = \left(\frac{T_1}{T_2}\right)^4 = \left(\frac{1}{2}\right)^4 = \frac{1}{16}$$

74. $\lambda_{m_1} T_1 = \lambda_{m_2} T_2$

$$\Rightarrow \quad 5.5 \times 10^{-7} \times 5500 = 11 \times 10^{-7} T$$

$$T = 550 \times 5 \text{ K} = 2750 \text{ K}$$

75. Given, initial temperature, $T_1 = 80°$C

Final temperature, $T_2 = 50°$C

Temperature of the surroundings, $T_0 = 20°$C,

$t_1 = 5$min

According to Newton's law of cooling,

Rate of cooling, $\frac{dT}{dt} = K\left[\frac{T_1 + T_2}{2} - T_0\right]$

$$\Rightarrow \quad \frac{(80-50)}{5} = K\left[\frac{80+50}{2} - 20\right]$$

$$\Rightarrow \quad \frac{30}{5} = K\,(65-20)$$

$$\Rightarrow \quad 6 = K \times 45$$

$$\Rightarrow \quad K = \frac{6}{45} = \frac{2}{15} \quad \text{...(i)}$$

In second condition,

Initial temperature, $T_1' = 60°$C

Final temperature, $T_2' = 30°$C

$t' = ?$

Now, $\frac{(60-30)}{t'} = \frac{2}{15}\left(\frac{60+30}{2} - 20\right)$

$$\Rightarrow \quad \frac{30}{t'} = \frac{2}{15}(45-20)$$

$$\Rightarrow \quad t' = \frac{30 \times 15}{2 \times 25} = 9 \text{ min}$$

76. According to Newton's law of cooling,

$$-\frac{d\theta}{dt} \propto (\theta - \theta_0)$$

$$-\frac{d\theta}{dt} = K\,(\theta - \theta_0)$$

where, K is a constant of proportionality.

Now, $\frac{75-65}{2} = K\left(\frac{75+65}{2} - 30\right) = K \times 40$

$$\Rightarrow \quad K = \frac{1}{8}$$

In case of second identical object,

$$\frac{55-45}{t} = K\left(\frac{55+45}{2} - 30\right)$$

$$\frac{10}{t} = \frac{1}{8} \times 20$$

$$\Rightarrow \quad t = \frac{80}{20} = 4 \text{ min}$$

77. The total energy radiated from a black body per minute,

$$Q \propto T^4$$

$$\Rightarrow \quad \frac{Q_2}{Q_1} = \left(\frac{2T}{T}\right)^4 = 16 \Rightarrow Q_2 = 16 Q_1$$

If m be mass of water taken and s be its specific heat capacity, then $Q_1 = ms(20.5 - 20)$ and $Q_2 = ms(\theta - 20)$

θ°C = Final temperature of water

$$\Rightarrow \quad \frac{Q_2}{Q_1} = \frac{\theta - 20}{0.5}$$

$$\Rightarrow \quad \frac{16}{1} = \frac{\theta - 20}{0.5}$$

$$\Rightarrow \quad \theta = 28°\text{C}$$

78. According to Newton's law of cooling,

Rate of heat loss = Temperature difference

$$\therefore \quad \frac{R_2}{R_1} = \frac{40-20}{80-20} = \frac{1}{3}$$

$$R_2 = \frac{R_1}{3} = \frac{45}{3} = 15 \text{ cals}^{-1}$$

79. According to Newton's law of cooling,

$$\frac{\theta_1 - \theta_2}{t} = K\left(\frac{\theta_1 + \theta_2}{t} - \theta_0\right)$$

$$\Rightarrow \quad \frac{50-40}{300} = K\left(\frac{50+40}{2} - 20\right)$$

or $\quad K = \frac{1}{25 \times 30}$

Now, $\frac{40-\theta}{300} = K\left(\frac{40+\theta}{2} - 20\right) = \frac{K\theta}{2} = \frac{\theta}{1500}$

or $\quad 300\,\theta = 60000 - 1500\,\theta$

$$\theta = \frac{60000}{1800} = 33.3° \text{ C}$$

80. When both have same area and same temperature, they will cool at the same rate.

81. As, $\frac{365-361}{2} = K\left[\frac{365+361}{2} - 293\right]$

(from Newton's law of cooling)

$$\Rightarrow \quad K = \frac{1}{35}$$

Again, $\frac{344-342}{t} = \frac{1}{35}\left[\frac{344+342}{2} - 293\right] = \frac{10}{7}$

$$t = \frac{14}{10} \text{ min} = \frac{14}{10} \times 60 = 84 \text{ s}$$

Round II

1. $mL = \frac{KA\Delta\theta t}{\Delta x}$

$\Rightarrow \quad 500 \times 80 = \frac{0.0075 \times 75 \times (40-0)t}{5}$

$\Rightarrow \quad t = 8.9 \times 10^3 \text{ s} = 2.47 \text{ h}$

2. Let the quantity of heat supplied per minute be Q. Then, quantity of heat supplied in 2 min $= mC(90-80)$ In 4 min, heat supplied $= 2mC(90-80)$

$\therefore \quad 2mC(90-80) = mL \Rightarrow \frac{L}{C} = 20$

3. Rate of cooling $\left(-\frac{dT}{dt}\right) \propto$ Emissivity (e)

From graph, $\left(-\frac{dT}{dt}\right)_x > \left(-\frac{dT}{dt}\right)_y \Rightarrow e_x > e_y$

Further emissivity (e) $\propto$ Absorptive power (a)

$\Rightarrow \quad a_x > a_y$

($\because$ good absorbers are good emitters)

4. As, $\beta = 2\alpha$ and $\gamma = 3\alpha$

$\therefore \quad \frac{\beta}{\gamma} = \frac{2\alpha}{3\alpha} = \frac{2}{3}$

and $\quad \frac{\gamma}{\alpha} = \frac{3\alpha}{\alpha} = \frac{3}{1}$

5. According to Wien's displacement law,

$$\lambda_m \propto \frac{1}{T}$$

$\Rightarrow \quad \lambda_{m_2} < \lambda_{m_1} \qquad (\because T_1 < T_2)$

Therefore, I-λ graph for T_2 has lesser wavelength (λ_m) and so curve for T_2 will shift towards left side.

So, the graph shown in option (a) is correct.

6. $\frac{dQ}{dt} = -KA\frac{d\theta}{dx}$

$\because \frac{dQ}{dt}$, K and A are constants for all points

$\Rightarrow d\theta \propto -dx$, i.e. temperature will decrease linearly with x.

So, the graph shown in option (c) is correct.

7. If $T_x = T_y$ and $T_x \neq T_z$, then $T_y \neq T_z$

If $T_x \neq T_y$ and $T_x \neq T_z$, then $T_y \neq T_z$ but T_y may be equal to T_z.

8. Here, $\Delta T = 100 - 0 = 100°\text{C}$

$$\frac{\Delta l}{l} \times 100 = 1\%$$

$\Rightarrow \quad \frac{\Delta l}{l} = \frac{1}{100} = 0.01$

We know that, $\Delta l = \alpha l \Delta T$

$\Rightarrow \quad \frac{\Delta l}{l} = \alpha \times 100$

$$0.01 = \alpha \times 100$$

$\Rightarrow \quad \alpha = 10^{-4}$

$\because$ Area of copper plate $= 2l \times l = 2l^2$

By area expansion theory,

$\Rightarrow \quad \Delta A = \beta A \Delta T$

$\Rightarrow \quad \frac{\Delta A}{A} = \beta \Delta T$

But $\quad \beta = 2\alpha$

$\therefore \quad \frac{\Delta A}{A} = 2\alpha\Delta T$

$$= 2 \times 10^{-4} \times 100$$

$$= 2 \times 10^{-2}$$

$\therefore \quad \frac{\Delta A}{A} \times 100 = 2 \times 10^{-2} \times 100$

$$= 2\%$$

9. According to Stefan's law, $E = \sigma T^4$

$\Rightarrow \quad \ln E = \ln \sigma + 4 \ln T$

$\Rightarrow \quad \ln E = 4 \ln T + \ln \sigma$

On comparing this equation with $y = mx + c$, we find that graph between $\ln E$ and $\ln T$ will be a straight line, having positive slope ($m = 4$) and intercept on $\ln E$ axis equal to $\ln \sigma$.

Hence, graph shown in option (d) is correct.

10. 22 g of CO_2 is half mole of CO_2, i.e. $n_1 = 0.5$

16 g of O_2 is half mole of O_2, i.e. $n_2 = 0.5$

$\therefore \quad T = \frac{n_1T_1 + n_2T_2}{n_1 + n_2}$

$$= \frac{0.5 \times (27+273) + 0.5\,(37+273)}{0.5+0.5}$$

$$= 305 \text{ K} = 305 - 273 = 32°\text{C}$$

11. Here, $\frac{D_1}{D_2} = \frac{1}{2}$

$\therefore \quad \frac{A_1}{A_2} = \frac{D_1^2}{D_2^2} = \frac{1}{4}$

$\Rightarrow \quad \frac{dx_1}{dx_2} = \frac{2}{1}$

$$\frac{dQ_1}{dt} = \left(KA_1\frac{dT}{dx_1}\right); \frac{dQ_2}{dt}\left(= KA_2\frac{dT}{dx_2}\right)$$

$$\Rightarrow \frac{dQ_1/dt}{dQ_2/dt} = \frac{A_1}{dx_1} \cdot \frac{dx_2}{A_2} = \frac{1}{4} \times \frac{1}{2} = \frac{1}{8}$$

12. $\frac{d\theta}{dt} = \frac{\varepsilon A\sigma}{mc} 4\theta_0^3 \Delta\theta$

For given sphere and cube, $\frac{\varepsilon A\sigma}{mc} 4\theta_0^3\Delta\theta$ is constant, so for both rate of fall temperature $\frac{d\theta}{dt}$ = constant, i.e. same.

13. Here, $\dfrac{K_1}{K_2} = \dfrac{1}{2}, \dfrac{r_1}{r_2} = \dfrac{1}{2}$

$$\therefore \quad \frac{A_1}{A_2} = \frac{1}{4}$$

$$\frac{dx_1}{dx_2} = \frac{1}{2}, \frac{dQ_2}{dt} = 4 \text{ cals}^{-1}, \frac{dQ_1}{dt} = ?$$

$$\because \frac{dQ_2/dt}{dQ_1/dt} = \frac{K_2 A_2 dT/dx_2}{K_1 A_1 dT/dx_1} = \frac{K_2}{K_1}\frac{A_2}{A_1}\frac{dx_1}{dx_2}$$

$$= 2 \times 4 \times \frac{1}{2} = 4$$

$$\Rightarrow \frac{dQ_1}{dt} = \frac{dQ_2/dt}{4} = \frac{4}{4} = 1 \text{ cals}^{-1}$$

14. Thermal capacity = Mass × Specific heat

Due to same material, both spheres will have same specific heat.

Also, mass = volume (V) × density (ρ)

∴ Ratio of thermal capacity

$$= \frac{m_1}{m_2} = \frac{V_1\rho}{V_2\rho} = \frac{\frac{4}{3}\pi r_1^3}{\frac{3}{4}\pi r_2^3} = \left(\frac{r_1}{r_2}\right)^3 = \left(\frac{1}{2}\right)^3 = \frac{1}{8}$$

15. Let the temperature of junction be Q. In equilibrium, rate of flow of heat through rod 1 = sum of rate of flow of heat through rods 2 and 3

$$\left(\frac{dQ}{dt}\right)_1 = \left(\frac{dQ}{dt}\right)_2 + \left(\frac{dQ}{dt}\right)_3$$

$$KA\frac{(\theta - 0)}{l} = \frac{KA(90^\circ - \theta)}{l} + \frac{KA(90^\circ - \theta)}{l}$$

$$\Rightarrow \quad \theta = 2(90^\circ - \theta)$$

or $\quad 3\theta = 180^\circ$

$$\Rightarrow \quad \theta = \frac{180^\circ}{3} = 60°\text{C}$$

16. As, $dQ = C_p m\,\Delta T$

$\therefore \quad 70 = C_p \times 2\,(35 - 30)$

$\Rightarrow \quad C_p = 7 \text{ cal mol}^{-1}{}^\circ\text{C}^{-1}$

As, $\quad C_V = C_p - R$

$\quad = 7 - 1.99 = 5.01 \text{ cal mol}^{-1\circ}\text{ C}^{-1}$

$\therefore \quad dQ' = C_V m\,\Delta T$

$\quad = 5.01 \times 2 \times (35 - 30) = 50.1 \text{ cal} \simeq 50 \text{ cal}$

17. Here, $V_0 = 10^3$ cc,

$\gamma_r = 1.80 \times 10^{-6}{}^\circ\text{C}^{-1}$,

$g = 1.4 \times 10^{-6}{}^\circ\text{C}^{-1}, t = 100°\text{C}$

$\therefore \quad \gamma_a = \gamma_r - g = (1.80 - 4.0)\,10^{-6}$

$V_t = V_0\,(1 + 1.40 \times 10^{-6} \times 10^2)$

$\quad = (10^3 + 1.4)$ cc

∴ Volume of mercury that will overflow

$\quad = V_t - V_0 = 1.4$ cc

18. Initial temperature = 10°C

Final temperature = 30°C

$$\alpha_s = 1.2 \times 10^{-5}{}^\circ\text{C}^{-1}$$

$$\alpha_{Cu} = 1.7 \times 10^{-5}{}^\circ\text{C}^{-1}$$

According to given situation, both the metals will expand, but copper will lengthen more due to greater α_{Cu} than that that of steel. The extra length that copper expands, will be the extra length that the steel scale will measure.

∴ Let the initial length of copper and steel scales be L'_{Cu} and L'_s respectively and their initial length $L_0 = 90$ cm.

∴ Extra expansion of copper = $L'_{Cu} - L'_s$

We know that,

$$L'_{Cu} - L_0(1 + \alpha_{Cu}\Delta T)$$

and $\quad L'_s - L_0(1 + \alpha_s\Delta T)$

$$\therefore \quad L'_{Cu} - L'_s = L_0(\alpha_{Cu} - \alpha_s)\Delta T$$

$$= 90(1.7 - 1.2) \times 10^{-5} \times (30 - 10)$$

$$= 0.009 \text{ cm}$$

∴ Total length measured = 90 + 0.009

= 90.009 cm

≃ 90.01 cm

19. Let V be the volume of solid, d be its density and m be its mass; if γ is coefficient of volume expansion of liquid, then

Density at temperature t_1, $d_1 = \dfrac{d_0}{1 + \gamma t_1}$

Density at temperature t_2, $d_2 = \dfrac{d_0}{1 + \gamma t_2}$

According to Archimedes's principle,

$$f_1 V d_1 g = m = g f_2 V d_2$$

or $\quad \dfrac{d_1}{d_2} = \dfrac{f_2}{f_1} = \dfrac{d_0}{(1 + \gamma t_1)}\dfrac{(1 + \gamma t_2)}{d_0}$

or $\quad f_1 + f_1\gamma t_2 = f_2 + f_2\gamma t_1$

$$f_1 - f_2 = \gamma\,(f_2 t_1 - f_1 t_2)$$

$$\gamma = \frac{(f_1 - f_2)}{f_2 t_1 - f_1 t_2}$$

20. With temperature rise (same 20°C for both), steel scale and copper wire both expand. Hence, length of copper wire w.r.t. steel scale or apparent length of copper after rise in temperature is

$$L_{app} = L'_{Cu} - L'_{steel}$$

$$= [L_0(1 + \alpha_{cu}\Delta\theta) - L_0(1 + \alpha_s\Delta\theta)]$$

$$\Rightarrow \quad L_{app} = L_0(\alpha_{Cu} - \alpha_s)\Delta\theta$$

$$= 80(17 \times 10^{-6} - 11 \times 10^{-6}) \times 20$$

$$= 0.0096 \text{ cm}$$

∴ Length of the wire read = 80.0096 cm

21. Rate of cooling of a body,

$$R = \frac{\Delta\theta}{t} = \frac{A\varepsilon\sigma(T^4 - T_0^4)}{mc}$$

$\Rightarrow \quad R \propto \dfrac{A}{m} \propto \dfrac{\text{Area}}{\text{Volume}}$

For the same surface area, $R \propto \dfrac{1}{\text{Volume}}$

$\because$ Volume of cube < Volume of sphere

$\Rightarrow R_{\text{Cube}} > R_{\text{Sphere}}$, *i.e.* cube cools down with faster rate.

22. Heat is lost by steam in two stages

(i) for change of state from steam at 100°C to water at 100°C is $m \times 540$.

(ii) to change water at 100°C to water at 80°C is $m \times 1 \times (100 - 80)$, where m is the mass of steam condensed.

Total heat lost by steam is $m \times 540 + m \times 20 = 560\, m$(cals.) Heat gained by calorimeter and its contents is

$$= (1.1 + 0.02) \times (80 - 15) = 1.12 \times 65 \text{ cal}$$

Using principle of calorimetry, heat gained = heat lost

$$\therefore \quad 560m = 1.12 \times 65$$
$$m = 0.130 \text{ kg}$$

23. Initially ice will absorb heat to raise its temperature to 0°C, then its melting takes place.

If m_i = initial mass of ice,

m_i' = mass of ice that melts

and m_w = initial mass of water.

By law of mixture,

heat gained by ice = heat lost by water

$$\Rightarrow \quad m_i \times c \times (20) + m_i' \times L = m_w c_w (20)$$
$$\Rightarrow \quad 2 \times 0.5(20) + m_1' \times 80 = 5 \times 1 \times 20$$
$$\Rightarrow \quad m_1' = 1 \text{ kg}$$

So, final mass of water

= initial mass of water + mass of ice that melts

= 5 + 1 = 6 kg

24. If mass of the bullet is m gram, then total heat required for bullet to just melt down

$$Q_1 = mc\Delta\theta + mL = m \times 0.03(327 - 27) + m \times 6$$
$$= 15\, m \text{ cal} = (15m \times 4.2)\text{J}$$

Now, when bullet is stopped by the obstacle, the loss in its mechanical energy $= \dfrac{1}{2}(m \times 10^{-3})v^2\text{J}$

(as $mg = m \times 10^{-3}$kg)

As, 25% of this energy is absorbed by the obstacle.

The energy absorbed by the bullet,

$$Q_2 = \frac{75}{100} \times \frac{1}{2} mv^2 \times 10^{-3} = \frac{3}{8} mv^2 \times 10^{-3} \text{ J}$$

Now, the bullet will melt, if $Q_2 \geq Q_1$

i.e. $\dfrac{3}{8} mv^2 \times 10^{-3} \geq 15\, m \times 4.2$

$$\Rightarrow \quad v_{\min} = 410 \text{ m/s}$$

25. Suppose, height of liquid in each arm before rising the temperature is l.

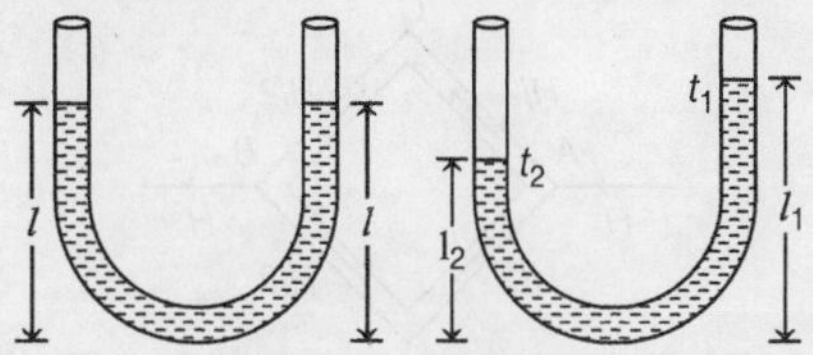

With temperature rise height of liquid in each arm increases, *i.e.* $l_1 > l$ and $l_2 > l$

Also, $$l = \frac{l_1}{1 + \gamma t_1} = \frac{l_2}{1 + \gamma t_2}$$

$$\Rightarrow \quad l_1 + \gamma l_1 t_2 = l_2 + \gamma l_2 t_1$$
$$\Rightarrow \quad \gamma = \frac{l_1 - l_2}{l_2 t_1 - l_1 t_2}$$

26. Heat developed by the heater,

$$H = \frac{V^2}{R} \frac{t}{J} = \frac{(200)^2 \times t}{20 \times 4.2}$$

Heat conducted by the glass

$$H = \frac{0.2 \times 1 \times (20 - \theta)t}{0.002}$$

Hence, $$\frac{(200)^2 \times t}{20 \times 4.2} = \frac{0.2 \times (20 - \theta)\, t}{0.002}$$

$$\Rightarrow \quad \theta = 15.24°\text{C}$$

27. Energy received per second, *i.e.* power, $P \propto (T^4 - T_0^4)$

$$\Rightarrow \quad P \propto T^4 \quad (\because T_0 << T)$$

Also, energy received per sec $(P) \propto \dfrac{1}{d^2}$ (inverse square law)

$$\Rightarrow \quad P \propto \frac{T^4}{d^2}$$
$$\Rightarrow \quad \frac{P_1}{P_2} = \left(\frac{T_1}{T_2}\right)^4 \times \left(\frac{d_2}{d_1}\right)^2$$
$$\Rightarrow \quad \frac{P}{P_2} = \left(\frac{T}{2T}\right)^4 \times \left(\frac{2d}{d}\right)^2 \times \left(\frac{2d}{d}\right)^2 = \frac{1}{4}$$
$$\Rightarrow \quad P_2 = 4P$$

28. Radiated power, $P = A\varepsilon\sigma T^4 \Rightarrow P \propto AT^4$

From Wien's law, $\lambda_m T = \text{constant} \Rightarrow T \propto \dfrac{1}{\lambda_m}$

$$\therefore \quad P \propto \frac{A}{(\lambda_m)^4} \propto \frac{r^2}{(\lambda_m)^4}$$
$$\Rightarrow \quad Q_a : Q_b : Q_c = \frac{2^2}{(300)^4} : \frac{4^2}{(400)^4} : \frac{6^2}{(500)^4}$$

$\therefore$ Q_b will be maximum.

29. Suppose temperature difference between A and B is $100°C$ and $\theta_A > \theta_B$

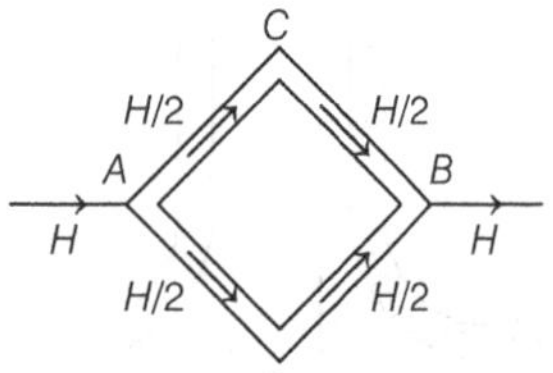

Heat current will flow from A to B *via* path ACB and ADB. Since, all the rods are identical, so

$$(\Delta\theta)_{AC} = (\Delta\theta)_{AD}$$

(because heat current $H = \frac{\Delta\theta}{R}$, here R = same for all)

$$\Rightarrow \quad \theta_A - \theta_C = \theta_A - \theta_D$$

$$\Rightarrow \quad \theta_C = \theta_D$$

i.e. Temperature difference between C and D will be zero.

30. Heat transferred in one minute is utilised in melting the ice, so

$$\frac{KA(\theta_1 - \theta_2)t}{l} = m \times L$$

$$\Rightarrow \quad m = \frac{10^{-3} \times 92 \times (100 - 0) \times 60}{1 \times 8 \times 10^4}$$

$$= 6.9 \times 10^{-3} \text{ kg}$$

31. $\frac{dT}{dt} = \frac{\sigma A}{mcJ}(T^4 - T_0^4)$ [in the given problem, fall in temperature of body, $dT = (200 - 100) = 100\text{K}$, temperature of surrounding, $T_0 = 0\text{K}$, initial temperature of body, $T = 200$ K]

$$\frac{100}{dt} = \frac{\sigma 4\pi r^2}{\frac{4}{3}\pi r^3 \rho c J}(200^4 - 0^4)$$

$$\Rightarrow \quad dt = \frac{r\rho cJ}{48\sigma} \times 10^{-6} \text{ s} = \frac{r\rho c}{\sigma} \times \frac{4.2}{48} \times 10^{-6}$$

$$= \frac{7}{80}\frac{r\rho c}{\sigma}\mu\text{s} \approx \frac{7}{72}\frac{r\rho c}{\sigma}\mu\text{s} \quad (\because J = 4.2)$$

32. $\frac{dQ}{dt} = \frac{KA}{l}d\theta = \frac{0.01 \times 1}{0.05} \times 30 = 6 \text{ J/s}$

Heat transferred in one day (86400 s),

$$Q = 6 \times 86400 = 518400 \text{ J}$$

Now, $Q = mL$

$$\Rightarrow \quad m = \frac{Q}{L} = \frac{518400}{334 \times 10^3}$$

$$= 1.552 \text{ kg} = 1552 \text{ g}$$

33. $\frac{Q}{t} = \frac{KA\Delta\theta}{l} \rightarrow \frac{mL}{t} = \frac{K(\pi r^2)\Delta\theta}{l}$

Rate of melting of ice $\left(\frac{m}{t}\right) \propto \frac{Kr^2}{l}$

Since, for second rod K becomes $\frac{1}{4}$th, radius becomes double and length becomes half, so rate of melting will be twice *i.e.* $\left(\frac{m}{t}\right)_2 = 2\left(\frac{m}{t}\right)_1 = 2 \times 0.1 = 0.2$ g/s

34. Rate of cooling, $\frac{\Delta\theta}{t} = \frac{A\varepsilon\sigma(T^4 - T_0^4)}{mc}$

$$\Rightarrow \quad t \propto \frac{m}{A}$$

($\because \Delta\theta, t, \sigma, (T^4 - T_0^4)$ are constants)

$$\Rightarrow \quad t \propto \frac{m}{A} \propto \frac{\text{Volume}}{\text{Area}} \propto \frac{a^3}{a^2}$$

$$\Rightarrow \quad t \propto a$$

$$\Rightarrow \quad \frac{t_1}{t_2} = \frac{a_1}{a_2}$$

$$\Rightarrow \quad \frac{100}{t_2} = \frac{1}{2}$$

$$\Rightarrow \quad t_2 = 200 \text{ s}$$

35. Substances having more specific heat take longer time to get heated to a higher temperature and longer time to get cooled.

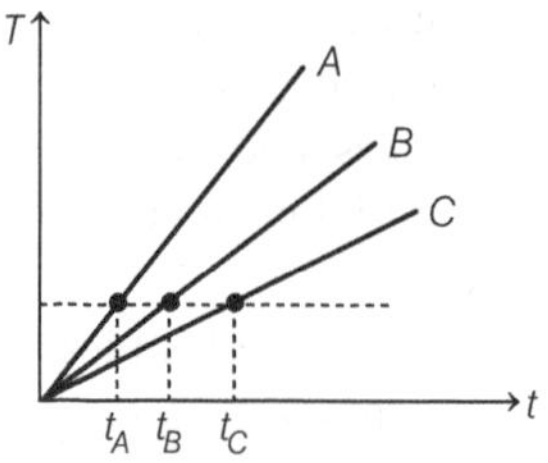

If we draw a line parallel to the time axis, then it cuts the given graphs at three different points. Corresponding points on the times axis shows that

$$t_C > t_B > t_A$$

$$\Rightarrow \quad C_C > C_B > C_A$$

36. $\frac{A_T}{A_{2000}} = \frac{16}{1}$ (given)

Area under E_λ-λ curve represents the emissive power of body and emissive power $\propto T^4$.

(Hence, area under E_λ-λ curve) $\propto T^4$

$$\Rightarrow \quad \frac{A_T}{A_{2000}} = \left(\frac{T}{2000}\right)^4$$

$$\Rightarrow \quad \frac{16}{1} = \left(\frac{T}{2000}\right)^4$$

$$\Rightarrow \quad T = 4000 \text{ K}$$

37. For θ-t plot, rate of cooling $= \frac{d\theta}{dt}$ = slope of the curve.

At P, $\frac{d\theta}{dt} = \tan\phi_2 = k(\theta_2 - \theta_0)$

where, k = constant.

At $Q, \dfrac{d\theta}{dt} = \tan\phi_1 = k(\theta_1 - \theta_0)$

$$\Rightarrow \frac{\tan\phi_2}{\tan\phi_1} = \frac{\theta_2 - \theta_0}{\theta_1 - \theta_0}$$

38. In the given graph, the region AB represents no change in temperature with time. It means ice and water are in thermal equilibrium.
The region BC shows the change in temperature with time. The region CD represents a constant temperature (100°C) with time. It means, water and steam are in thermal equilibrium at boiling point.
Therefore, (d) is the correct option.

39. According to Newton's law of cooling,

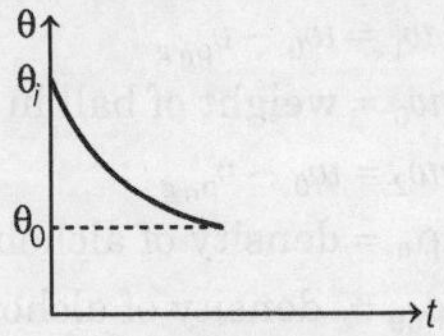

Rate of cooling ∝ Temperature difference

$$\Rightarrow \quad -\frac{d\theta}{dt} \propto (\theta - \theta_0)$$

$$\Rightarrow \quad -\frac{d\theta}{dt} = \alpha(\theta - \theta_0) \qquad (\alpha = \text{constant})$$

$$\Rightarrow \quad \int_{\theta_i}^{\theta} \frac{d\theta}{(\theta - \theta_0)} = -\alpha\int_0^{\tau} dt$$

$$\Rightarrow \quad \theta = \theta_0 + (\theta_i - \theta_0)e^{-\alpha t}$$

This relation tells us that, temperature of the body varies exponentially with time from θ_i to θ_0.

40. The power at which the body radiates is directly proportiional to area, *i.e.* $P \propto A \propto r^2$

Power radiated, $P = (4\pi r^2)\sigma T^4 = ms\left(\dfrac{dT}{dt}\right)$

$$= \frac{3}{4}\pi r^3 \rho s \frac{dT}{dt}$$

where, $\dfrac{dT}{dt} = R =$ rate of cooling

$$\therefore \quad 4\pi r^2 \sigma T^4 = \frac{4}{3}\pi r^3 \rho s (R)$$

$$R = \frac{3\sigma T^4}{r\rho s}$$

$$R \propto \frac{1}{r}$$

41. Given, mass of the child $(m) = 30$ kg

Time taken $(t) = 20$ min

Fall in temperature = (101 − 98)°F

$$\Delta T = 3°\text{F} = 3 \times \frac{5}{9}\,°\text{C} = \frac{5}{3}°\text{C}$$

Specific heat of human body $(s) = 4.2 \times 10^3$ J/kg-°C

Latent heat of evaporation

$$(L) = 580 \text{ cal/g}$$
$$= 580 \times 10^3 \text{ cal/kg}$$
$$= (580 \times 10^3 \times 4.2) \text{ J/kg}$$

Heat given by body during fall in temperature

$$Q_1 = ms\Delta T$$

Let m' be the mass of sweat evaporates from the human body.

Heat taken in evaporation,

$$Q_2 = m'L$$

But $\quad Q_1 = Q_2$

$$\therefore \quad ms\Delta T = m'L$$

or

$$m' = \frac{ms\Delta T}{L} = \frac{30 \times 4.2 \times 10^3 \times 5/3}{580 \times 4.2 \times 10^3}$$

$$= \frac{10}{116} = 0.0862 \text{ kg}$$

$\therefore$ Rate of evaporation of sweat $= \dfrac{0.0862}{20}$

$$= 0.00431 \text{ kg/min}$$
$$= 4.31 \text{ g/min}$$

42. According to the given question, the given figure with its length for each section is given as below

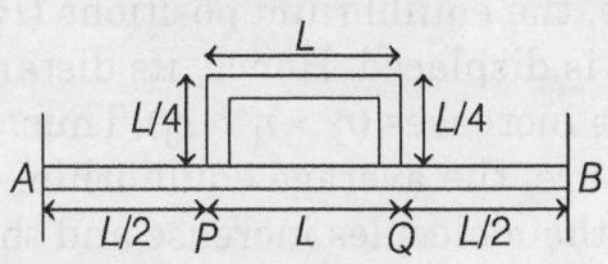

The above figure considering that every section has the same thermal conductivity, then in terms of thermal resistance is shown in the figure below

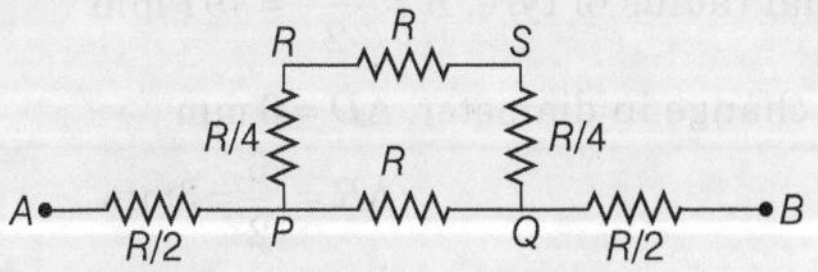

Net resistance of the section $PQSR$ is

$$= \frac{R \times \frac{3R}{2}}{\frac{5R}{2}} = \frac{3R}{5} \qquad \text{...(i)}$$

Total resistance of the net network,

$$R_{\text{net}} = \frac{R}{2} + \frac{R}{2} + \frac{3R}{5} = \frac{8R}{5}$$

$\therefore$ Thermal current, $I = \dfrac{\Delta T_{AB}}{R_{\text{net}}}$

$$\Rightarrow \quad I = \frac{120 - 0}{\left(\frac{8R}{5}\right)} = \frac{120 \times 5}{8R}$$

Thus, the net temperature difference between points P and Q is

$$T_P - T_Q = I \times \frac{3R}{5} \quad \text{[using Eq. (i)]}$$

$$= \frac{120 \times 5}{8R} \times \frac{3R}{5} = 45°C$$

43. The expansion of solids can be well understood by potential energy curve for two adjacent atoms in crystalline solid as a function of their intermolecular separation (r).

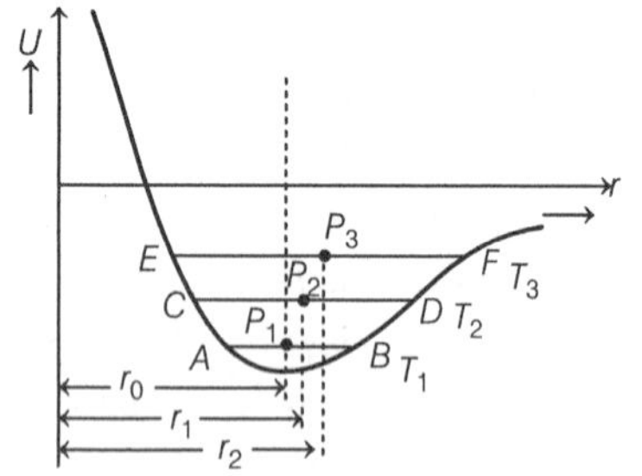

$$T_3 > T_2 > T_1$$

At ordinary temperature Each molecule of the solid vibrates about its equilibrium position P_1 between A and B (r_0 is the equilibrium distance of it from some other molecules)

At high temperature Amplitude of vibration increases ($C \leftrightarrow D$ and $E \leftrightarrow F$). Due to asymmetry in the curve, the equilibrium positions (P_2 and P_3) of the molecule is displaced. Hence, its distance from other molecules increases ($r_2 > r_1 > r_0$). Thus, on raising the temperature, the average equilibrium distance between the molecules increase and the solid as a whole expands.

44. Initial diameter of tyre = (1000 − 6) mm = 994 mm

Initial radius of tyre, $R = \frac{994}{2} = 497$ mm

and change in diameter, $\Delta D = 6$ mm

$$\Delta R = \frac{6}{2} = 3\text{mm}$$

After increasing temperature by $\Delta\theta$, tyre will fit on wheel.

Increment in the length (circumference) of the iron tyre,

$$\Delta L = L \times \alpha \times \Delta\theta = L \times \frac{\gamma}{3} \times \Delta\theta \quad \left(\because \alpha = \frac{\gamma}{3}\right)$$

$$2\pi\Delta R = 2\pi R\left(\frac{\gamma}{3}\right)\Delta\theta$$

$$\Rightarrow \quad \Delta\theta = \frac{3}{\gamma}\frac{\Delta R}{R} = \frac{3 \times 3}{3.6 \times 10^{-5} \times 497}$$

$$\Rightarrow \quad \Delta\theta \cong 500°C$$

45. Let T_0 be the initial temperature of the black body.

$$\therefore \quad \lambda_0 T_0 = b \quad \text{(Wien's law)}$$

Power radiated, $P_0 = CT_0^4$, where C is constant.

If T is new temperature of black body, then

$$\frac{3\lambda_0}{4}T = b = \lambda_0 T_0$$

$$\text{or} \quad T = \frac{4}{3}T_0$$

Power radiated, $P = CT^4 = CT_0^4\left(\frac{4}{3}\right)^4$

$$P = P_0 \times \frac{256}{81} \text{ or } \frac{P}{P_0} = \frac{256}{81}$$

46. As the coefficient of cubical expansion of metal is less as compared to the coefficient of cubical expansion of liquid, we may neglect the expansion of metal ball. So, when the ball is immersed in alcohol at 0°C, it displaces some volume V of alcohol at 0°C and has weight w_1.

$\therefore \quad w_1 = w_0 - v\rho_0 g$

where, $\quad w_0$ = weight of ball in air.

Similarly, $\quad w_2 = w_0 - v\rho_0 g$

where, $\quad \rho_0$ = density of alchohol at 0°C

and $\quad \rho_{59}$ = density of alcohol at 59°C

As $\quad \rho_{59} < \rho_0$,

$\Rightarrow \quad w_2 > w_1$

or $\quad w_1 < w_2$

47. Heat gain = Heat lost

$$C_A(16 - 12) = C_B(19 - 16)$$

$$\Rightarrow \quad \frac{C_A}{C_B} = \frac{3}{4}$$

$$\text{and} \quad C_B(23 - 19) = C_C(28 - 23)$$

$$\Rightarrow \quad \frac{C_B}{C_C} = \frac{5}{4}$$

$$\Rightarrow \quad \frac{C_A}{C_C} = \frac{15}{16} \quad \text{...(i)}$$

If θ is the temperature when A and C are mixed, then

$$C_A(\theta - 12) = C_C(28 - \theta)$$

$$\Rightarrow \quad \frac{C_A}{C_C} = \frac{28 - \theta}{\theta - 12} \quad \text{...(ii)}$$

On solving Eqs. (i) and (ii), we get $\theta = 20.2°C$

48. Suppose m kg steam is required per hour.

Heat is released by steam in following three steps

(i) When 150°C steam $\xrightarrow[Q_1]{}$ 100°C steam

$$Q_1 = mc_{\text{steam}}\,\Delta\theta = m \times 1(150 - 100) = 50m \text{ cal}$$

(ii) When 100°C steam $\xrightarrow[Q_2]{}$ 100°C water

$$Q_2 = mL_v = m \times 540 = 540m \text{ cal}$$

(iii) When 100°C water $\xrightarrow[Q_2]{}$ 90°C water

$$Q_3 = mc_w\Delta\theta = m \times 1 \times (100 - 90) = 10\,m \text{ cal}$$

Hence, total heat given by the steam

$$Q = Q_1 + Q_2 + Q_3 = 600\,m \text{ cal}$$

Heat taken by 10 kg water,

$$Q' = mc_w \, \Delta\theta = 10 \times 10^3 \times 1 \times (80 - 20)$$
$$= 600 \times 10^3 \text{ cal}$$

Hence, $Q = Q'$

$$\Rightarrow \quad 600m = 600 \times 10^3$$
$$\Rightarrow \quad m = 10^3 \text{ g} = 1 \text{ kg}$$

49. If suppose $K_{\text{Ni}} = K$

$$\Rightarrow \quad K_{\text{Al}} = 3K \text{ and } K_{\text{Cu}} = 6K$$

Since, all metal bars are connected in series.

So, $\left(\dfrac{Q}{t}\right)_{\text{Combination}} = \left(\dfrac{Q}{t}\right)_{\text{Cu}} = \left(\dfrac{Q}{t}\right)_{\text{Al}} = \left(\dfrac{Q}{t}\right)_{\text{Ni}}$

and
$$\frac{3}{K_{\text{eq}}} = \frac{1}{K_{\text{Cu}}} + \frac{1}{K_{\text{Al}}} + \frac{1}{K_{\text{Ni}}}$$
$$\frac{3}{K_{\text{eq}}} = \frac{1}{6K} + \frac{1}{3K} + \frac{1}{K} = \frac{9}{6K}$$
$$\Rightarrow \quad K_{\text{eq}} = 2K$$

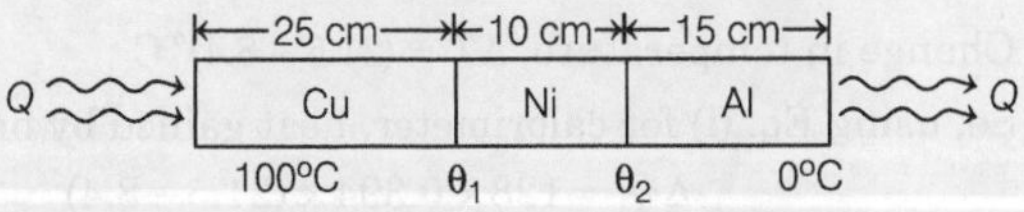

Hence, if $\left(\dfrac{Q}{t}\right)_{\text{Combination}} = \left(\dfrac{Q}{t}\right)_{\text{Cu}}$

$$\Rightarrow \quad \frac{K_{\text{eq}} A (100 - 0)}{l_{\text{Combination}}} = \frac{K_{\text{Cu}} A (100 - \theta_1)}{l_{\text{Cu}}}$$
$$\Rightarrow \quad \frac{2KA(100 - 0)}{(25 + 10 + 15)} = \frac{6KA(100 - \theta_1)}{25}$$
$$\Rightarrow \quad \theta_1 = 83.33°\text{C}$$

Similar, if $\left(\dfrac{Q}{t}\right)_{\text{Combination}} = \left(\dfrac{Q}{t}\right)_{\text{Al}}$

$$\Rightarrow \quad \frac{2KA(100 - 0)}{50} = \frac{3\,KA(\theta_2 - 0)}{15}$$
$$\Rightarrow \quad \theta_2 = 20°\text{C}$$

50. $\because T_B > T_A \Rightarrow$ Heat will flow B to A *via* two paths (i) B to A (ii) and along BCA as shown

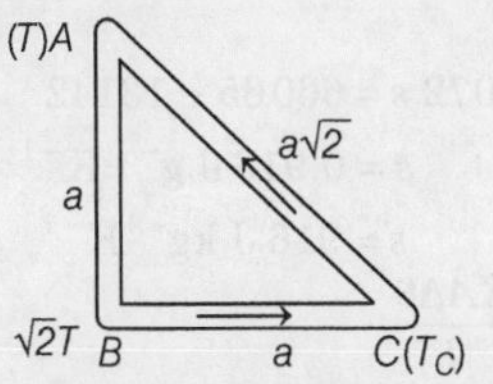

Rate of flow of heat in path BCA will be same,

i.e.
$$\left(\frac{Q}{t}\right)_{BC} = \left(\frac{Q}{t}\right)_{CA}$$
$$\Rightarrow \quad \frac{K(\sqrt{2}T - T_c)A}{a} = \frac{K(T_c - T)A}{\sqrt{2}a}$$
$$\Rightarrow \quad \frac{T_c}{T} = \frac{3}{1 + \sqrt{2}}$$

51.

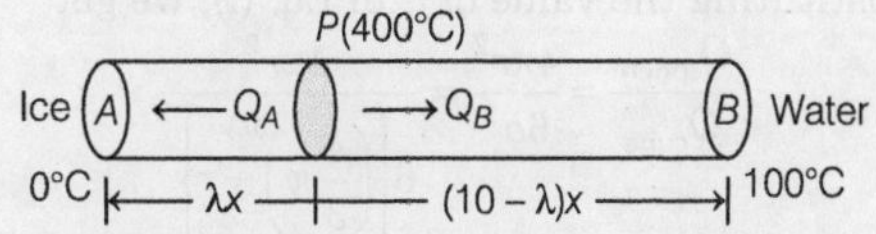

Heat received by end A, for melting of ice,

$$Q_A = \frac{KA(400 - 0)t}{\lambda x} = mL_{\text{ice}} \quad \text{...(i)}$$

Heat received by end B, for vaporisation of water,

$$Q_B = \frac{KA(400 - 100)t}{(10 - \lambda)x} = mL_{\text{vap}} \quad \text{...(ii)}$$

Dividing both equations,
$$\frac{\dfrac{400}{\lambda x}}{\dfrac{300}{(10 - \lambda)x}} = \frac{L_{\text{ice}}}{L_{\text{vap}}}$$
$$\Rightarrow \quad \frac{4}{3}\,\frac{(10 - \lambda)}{\lambda} = \frac{80}{540}$$
$$\Rightarrow \quad \lambda = 9$$

52. Given, base area of boiler $(A) = 0.15 \text{ m}^2$

Thickness $(d) = 1.0 \text{ cm} = 1 \times 10^{-2}$ m

Rate of water boils $= 6.0 \text{ kg/min} = \dfrac{6.0}{60} \text{ kg /s} = 0.1 \text{ kg /s}$

Thermal conductivity of brass $(K) = 109$ J/s-m-K

Latent heat of vaporisation of water (L)
$= 2256 \times 10^3$ J/kg

Let θ_1 be the temperature of the part of the boiler in contact with the stove.

Rate of heat energy supplied

= Rate of heat energy utilised in vaporisation

$$\frac{KA\Delta\theta}{d} = mL$$
$$\frac{KA(\theta_1 - \theta_2)}{d} = mL$$
$$\frac{109 \times 0.15(\theta_1 - 100)}{1 \times 10^{-2}} = 0.1 \times 2256 \times 10^3$$
$$\Rightarrow \quad 1635\,(\theta_1 - 100) = 2256 \times 10^2$$
$$\Rightarrow \quad \theta_1 = \frac{225600}{1635} + 100$$
$$= 137.98 + 100$$
$$= 237.98°\text{C} \approx 238°\text{C}$$

53. $Q = \sigma At(T^4 - T_0^4)$

If T, T_0, σ and t are same for both bodies,

$$\frac{Q_{\text{sphere}}}{Q_{\text{cube}}} = \frac{A_{\text{sphere}}}{A_{\text{cube}}} = \frac{4\pi r^2}{6a^2} \quad \text{...(i)}$$

But according to problem,
volume of sphere = volume of cube

$$\Rightarrow \quad \frac{4}{3}\pi r^3 = a^3$$
$$\Rightarrow \quad a = \left(\frac{4}{3}\pi\right)^{1/3} r$$

Substituting the value of a in Eq. (i), we get

$$\frac{Q_{\text{sphere}}}{Q_{\text{cube}}} = \frac{4\pi r^2}{6a^2} = \frac{4\pi r^2}{6\left\{\left(\frac{4}{3}\pi\right)^{1/3} r\right\}^2}$$

$$= \frac{4\pi r^2}{6\left(\frac{4}{3}\pi\right)^{2/3} r^2} = \left(\frac{\pi}{6}\right)^{1/3} : 1$$

54. According to Newton's law of cooling,

$$\frac{\theta_1 - \theta_2}{t} = K\left[\frac{\theta_1 + \theta_2}{2} - \theta_0\right]$$

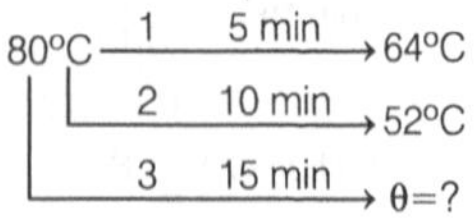

For first process,

$$\frac{(80-64)}{5} = K\left[\frac{80+64}{2} - \theta_0\right] \quad \text{...(i)}$$

For second process,

$$\frac{(80-52)}{10} = K\left[\frac{80+52}{2} - \theta_0\right] \quad \text{...(ii)}$$

For third process,

$$\frac{(80-\theta)}{15} = K\left[\frac{80+\theta}{2} - \theta_0\right] \quad \text{...(iii)}$$

On solving Eqs. (i) and (ii), we get

$$K = \frac{1}{15} \text{ and } \theta_0 = 24° \text{ C}$$

Putting these values in Eq. (iii), we get

$$\theta_0 = 42.7°\text{C}$$

55. According to Stefan's law, $E = eA\sigma T^4$

$\Rightarrow \quad E_A = e_A A\sigma T_A^4$

and $\quad E_B = e_B A\sigma T_B^4$

$\because \quad E_A = E_B$

$\therefore \quad e_A T_A^4 = e_B T_B^4$

$\Rightarrow \quad T_B = \left(\frac{e_A}{e_B} T_A^4\right)^{\frac{1}{4}} = \left(\frac{1}{81} \times (5802)^4\right)^{\frac{1}{4}}$

$\Rightarrow \quad T_B = 1934$

and from Wiens law,

$$\lambda_A \times T_A = \lambda_B \times T_B$$

$\Rightarrow \quad \frac{\lambda_A}{\lambda_B} = \frac{T_B}{T_A}$

$\Rightarrow \quad \frac{\lambda_B - \lambda_A}{\lambda_B} = \frac{T_A - T_B}{T_A}$

$\Rightarrow \quad \frac{1}{\lambda_B} = \frac{5802 - 1934}{5802} = \frac{3868}{5802}$

$\Rightarrow \quad \lambda_B = 1.5\,\mu\text{m}$

56. The principle of calorimetry states that total heat lost by the hotter body equals to the total heat gained by colder body, provided that there is no exchange of heat with the surroundings.

Let specific heat of unknown metal be s and heat lost by this metal be ΔQ.

Heat lost and specific heat of a certain material/substance are related as $\Delta Q = ms\Delta T$ ... (i)

For unknown metal, $m = 192$ g and

$$\Delta T = (100 - 21.5)\ °\text{C}$$

$\therefore \quad \Delta Q' = 192(100 - 21.5) \times s \quad$ ...(ii)

Now, this heat is gained by the calorimeter and water inside it.

As, heat gained by calorimeter can be calculated by Eq. (i).

So, for brass specific heat,

$$s = 394 \text{ J kg}^{-1}\text{ K}^{-1} \quad \text{(given)}$$

$$= 0.394 \text{ J g}^{-1}\text{ K}^{-1}$$

Mass of calorimeter, $m = 128$ g

Change in temperature, $\Delta T = (21.5 - 8.4)°$ C

So, using Eq. (i) for calorimeter, heat gained by brass

$$\Delta Q_1 = 128 \times 0.394 \times (21.5 - 8.4) \quad \text{...(iii)}$$

Heat gained by water can be calculated as follows

mass of water, $m = 240$ g,

specific heat of water, $s = 4.18 \text{ J g}^{-1}\text{K}^{-1}$,

change in temperature, $\Delta T = (21.5 - 8.4)°$ C

Using Eq. (i) for water also, we get

heat gained by water,

$$\Delta Q_2 = 240 \times 4.18 \times (21.5 - 8.4) \quad \text{...(iv)}$$

Now, according to the principle of calorimeter, the total heat gained by the calorimeter and water must be equal to heat lost by unknown metal

$$\Delta Q' = \Delta Q_1 + \Delta Q_2$$

Using Eqs. (ii), (iii) and (iv), we get

$\Rightarrow \quad = 192(100 - 21.5) \times s$

$= 128 \times 0.394 \times (21.5 - 8.4) + 240 \times 4.18 \times (21.5 - 8.4)$

$\Rightarrow \quad 15072\, s = 660.65 + 13142$

$\Rightarrow \quad s = 0.916 \text{ J g}^{-1}\text{ K}^{-1}$

or $\quad s = 916 \text{ J kg}^{-1}\text{K}^{-1}$

57. Since, $\frac{Q}{t} = \frac{KA\Delta\theta}{l}$

$\Rightarrow \quad \frac{mL}{t} = \frac{K(\pi r^2)\Delta\theta}{l}$

$\Rightarrow \quad \frac{m}{t} \propto \frac{Kr^2}{l}$

$\Rightarrow \quad \frac{\left(\frac{m}{t}\right)_2}{\left(\frac{m}{t}\right)_1} = \frac{K_2}{K_1} \times \left(\frac{r_2}{r_1}\right)^2 \times \frac{l_1}{l_2}$

$$= \frac{\frac{1}{4}K_1}{K_1} \times \left(\frac{2r_1}{r_1}\right)^2 \times \frac{l_1}{l_{1/2}} = \frac{1}{4} \times 4 \times 2 = 2$$

$$\Rightarrow \quad \left(\frac{m}{t}\right)_2 = 2\left(\frac{m}{t}\right)_1 = 2 \times 0.1 = 0.2 \text{ gs}^{-1}$$

58. As, $\frac{T_1}{T_2} = n = \frac{\lambda_{m_1}}{\lambda_{m_2}} = \frac{11 \times 10^{-5}}{5.5 \times 10^{-5}} = 2$

59. As, $\Delta Q = cm\,\Delta T + mL$

$= 0.5 \times 1\ (5) + 1 \times 80 = 82.5$ cal

60. As, $\Delta Q = KA\left(\frac{\Delta T}{\Delta x}\right)\Delta t$, where, $A = 4\,\pi r^2$

$$= 0.008 \times 4 \times \frac{22}{7}\ (6 \times 10^8)^2 \times \left(\frac{32}{10^5}\right) \times 86400$$

$= 10^{18}$ cal $= 10^x$ cal (given)

$\therefore \quad x = 18$

61. Heat given by water, $Q_1 = 10 \times 10 = 100$ cal

Heat taken by ice to melt,

$Q_2 = 10 \times 0.5 \times [0 - (-20)] + 10 \times 80 = 900$cal

As $Q_1 < Q_2$, so ice will not completely melt and final temperature = 0°C

As heat given by water in cooling up to 0°C is only just sufficient to increase the temperature of ice from −20° C to 0°C, hence mixture in equilibrium will consist of 10 g ice and 10 g water at 0°C.

62. Rate of cooling $(R) = \frac{\Delta\theta}{t} = \frac{A\varepsilon\sigma(T^4 - T_0^4)}{mc}$

$$\Rightarrow \quad R \propto \frac{A}{m} \propto \frac{\text{area}}{\text{volume}} \propto \frac{r^2}{r^3} \propto \frac{1}{r}$$

$$\Rightarrow \quad \text{Rate } (R) \propto \frac{1}{r} \propto \frac{1}{m^{1/3}} \left[\because m = \rho \times \frac{4}{3}\pi r^3 \Rightarrow r \propto m^{1/3}\right]$$

$$\Rightarrow \quad \frac{R_1}{R_2} = \left(\frac{m_2}{m_1}\right)^{1/3} = \left(\frac{1}{3}\right)^{1/3} = \left(\frac{1}{3}\right)^{1/m} \quad \text{(given)}$$

$\therefore \quad m = 3$

63. From given curve,

Melting point for $A = 60$°C

and melting point for $B = 20$°C

Time taken by A for fusion = (6 − 2) = 4 min

Time taken by B for fusion = (6.5 − 4) = 2.5 min

Then, $\frac{H_A}{H_B} = \frac{6 \times 4 \times 60}{6 \times 2.5 \times 60} = \frac{8}{5} = 1.6$

14

Kinetic Theory of Gases

IN THIS CHAPTER ....

Ideal Gas or Perfect Gas

A gas which strictly obeys the gas laws such as Boyle's law, Charles' law, Gay-Lussac's law, Avogadro's law, etc., is known as an *ideal gas* or *perfect gas*.

These law (gaseous laws) are given below

Boyle's Law According to this law, "for a given mass of an ideal gas at constant temperature (called *isothermal process*), the volume of a gas is inversely proportional to its pressure." *i.e.* $V \propto \frac{1}{p}$ (if m and T = constant)

Graphical forms of such law are shown in following figure

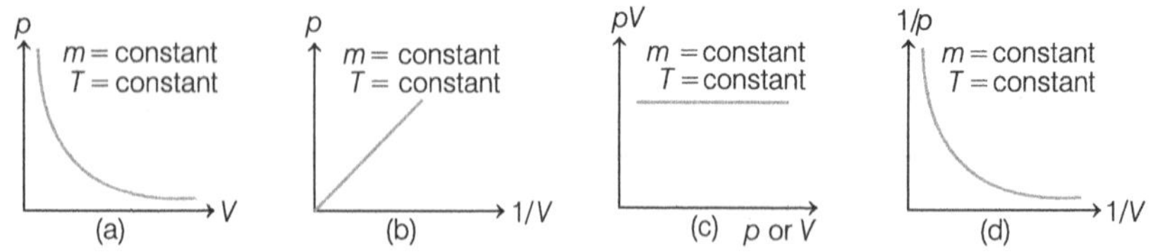

Charles' Law According to this law, "for a given mass of an ideal gas at constant pressure (called **isobaric process**), volume of a gas is directly proportional to its absolute temperature."

i.e. $V \propto T$ (if m and p = constant)

Graphical forms of such law are shown in following figure

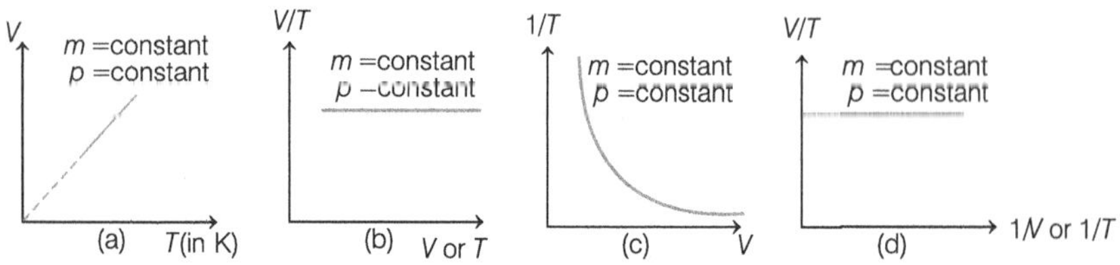

Gay-Lussac's Law or Pressure Law According to this law, "for a given mass of an ideal gas at constant volume (called *isochoric process*), pressure of a gas is directly proportional to its absolute temperature."

i.e. $p \propto T$ (if m and V = constant)

Graphical forms of such law are shown in following figure

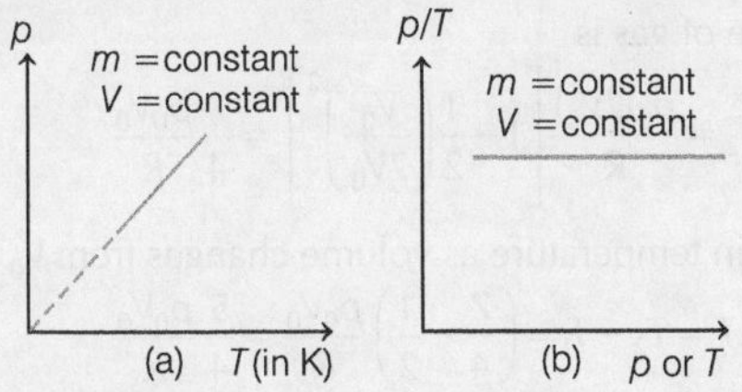

Avogadro's Law According to this law, "at same temperature and pressure, equal volumes of all the gases contain equal number of molecules."

i.e. $N_1 = N_2$

where, N_1 and N_2 are number at molecules at first and second, gas respectively.

Equation of State of a Perfect Gas

In practice, the gases do not obey the gas laws at all values of temperature and pressure. It is because of the intermolecular forces between the gas molecules.

Ideal gas equation is a form of combined effect of above first four laws. Thus, the equation is given by

$$pV = nRT = \frac{m}{M}RT$$

Here, n = number of moles of the gas = $\frac{m}{M}$,

m = total mass of the gas,

M = molecular mass

and R = universal gas constant

$= 8.31 \text{ Jmol}^{-1}\text{ K}^{-1}$

$= 2.0 \text{ cal mol}^{-1}\text{K}^{-1}$.

Above equation is also known as equation of state of an ideal gas.

- In terms of density, the ideal gas equation may be expressed as
$$\frac{p}{\rho T} = \text{constant}$$
- In terms of number of molecules (n) per unit volume of a gas, the ideal gas equation may be expressed as
$$p = nkT$$
- If n_1 mole of a non-reactive gas in thermodynamical state (p_1, V_1, T_1) be mixed with n_2 mole of another non-reactive gas at (p_2, V_2, T_2) and the resultant gas mixture is at a state (p, V, T), then
$$\frac{p_1V_1}{T_1} + \frac{p_2V_2}{T_2} = \frac{pV}{T}$$

Work Done on Compressing a Gas

Work done on compressing a gas can be written as

$$W = \frac{mR(T_2 - T_1)}{1 - n}$$

where, m = total mass of the gas,

R = universal gas constant,

n = number of moles of the gas,

T_1 and T_2 = initial and final temperature of the gas.

Example 1. *During an experiment, an ideal gas is found to obey an additional law Vp^2 = constant. The gas is initially at temperature T and volume V. What will be the temperature of the gas when it expands to a volume 2V?*

(a) $\sqrt{3}\,T$ (b) $\sqrt{1/2\,T}$

(c) $\sqrt{2}\,T$ (d) $\sqrt{3\,T}$

Sol. (c) Here it is given that Vp^2 = constant K (say). Hence, we may write the gas equation as, $pV = nRT$

or $\sqrt{\frac{K}{V}} \cdot V = nRT$ or $\sqrt{V} = \frac{nR}{\sqrt{K}}T$

$\Rightarrow$ $\sqrt{\frac{V_1}{V_2}} = \frac{T_1}{T_2}$

$\therefore$ $T_2 = T_1\sqrt{\frac{V_2}{V_1}} = T\sqrt{\frac{2V}{V}} = \sqrt{2}T$

Example 2. *A closed container of volume 0.02 m^3 contains a mixture of neon and argon gases, at a temperature of 27°C and pressure of 1×10^5 Nm^{-2}. The total mass of the mixture is 28 g. If the gram molecular weights of neon and argon are 20 and 40 respectively, find the masses of the individual gases in the container, assuming them to be ideal.*
(Given, R = 8.31 $Jmol^{-1}K^{-1}$)

(a) 4 g, 24 g (b) 5 g, 25 g

(c) 6 g, 26 g (d) 7 g, 27 g

Sol. (a) Let in the given container mass of neon be m and mass of argon be $(28 - m)$ g, so that

$$n_{Ne} = \frac{m}{20}$$

and $$n_{Ar} = \frac{28 - m}{40}$$

$$n = n_{Ne} + n_{Ar} = \frac{m}{20} + \frac{(28 - m)}{40} = \frac{28 + m}{40} \quad \text{...(i)}$$

and using ideal gas equation for the mixture, we have

$$n = \frac{pV}{RT} = \frac{1 \times 10^5 \times 0.02}{8.314 \times 300} = 0.8 \quad \text{...(ii)}$$

Comparing Eqs.(i) and (ii), we get

$$\frac{28 + m}{40} = 0.8$$

$\Rightarrow$ $m = 4$ g

$\therefore$ $m_{Ne} = 4$ g

and $m_{Ar} = 28 - 4 = 24$ g

Example 3. *Two moles of an ideal gas is contained in a cylinder fitted with a frictionless movable piston, exposed to the atmosphere, at an initial temperature T_0. The gas is slowly heated so that its volume becomes four times the initial value. The work done by the gas is*

(a) zero (b) $2RT_0$ (c) $4RT_0$ (d) $6RT_0$

Sol. (d) Given that gas is slowly heated, which means it remains in equilibrium with the atmosphere, *i.e.* the process takes place at constant pressure.

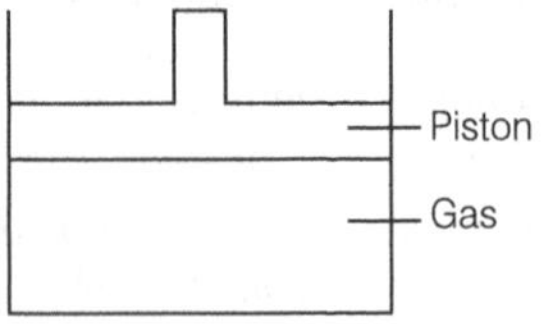

From the equation of ideal gas law,

$$pV = nRT$$

For infinitesimal change,

$$pdV = nR\,dT$$

or $\quad p\Delta V = nR\,\Delta T$

Also, $p\Delta V$ = work done by the gas = ΔW

$\therefore \quad \Delta W = nR\Delta T$

Also, $\quad \Delta V \propto \Delta T$

$\therefore \quad \Delta T \propto \Delta V \propto V_2 - V_1$

Given, $\quad V_2 = 4V_1$

$\therefore \quad \Delta T \propto 4V_1 - V_1 \propto 3V_1 \propto 3T_0$

Also, it is given that, $n = 2$ moles

The expression for work done becomes

$$\Delta W = nR\Delta T$$

$$\Delta W = 2R3T_0 = 6RT_0$$

Example 4. *One mole of an ideal gas passes through a process, where pressure and volume obey the relation $p = p_0\left[1 - \frac{1}{2}\left(\frac{V_0}{V}\right)^2\right]$. Here, p_0 and V_0 are constants. Calculate the change in the temperature of the gas if its volume changes from V_0 to $2V_0$.* **[JEE Main 2019]**

(a) $\frac{1}{2}\frac{p_0V_0}{R}$ (b) $\frac{1}{4}\frac{p_0V_0}{R}$ (c) $\frac{3}{4}\frac{p_0V_0}{R}$ (d) $\frac{5}{4}\frac{p_0V_0}{R}$

Sol. (d) Given process equation for 1 mole of an ideal gas is

$$p = p_0\left[1 - \frac{1}{2}\left(\frac{V_0}{V}\right)^2\right] \qquad \text{...(i)}$$

Also, for 1 mole of ideal gas,

$$pV = RT$$

$$\therefore \quad p = \frac{RT}{V} \qquad \text{...(ii)}$$

So, from Eqs. (i) and (ii), we have

$$\frac{RT}{V} = p_0\left[1 - \frac{1}{2}\left(\frac{V_0}{V}\right)^2\right]$$

$$\therefore \quad T = \frac{p_0V}{R}\left[1 - \frac{1}{2}\left(\frac{V_0}{V}\right)^2\right] \qquad \text{...(iii)}$$

When volume of gas is V_0, then by substituting $V = V_0$ in Eq. (iii), we get

Temperature of gas is

$$T_1 = \frac{p_0V_0}{R}\left[1 - \frac{1}{2}\left(\frac{V_0}{V_0}\right)^2\right] = \frac{p_0V_0}{2R}$$

Similarly, at volume, $V = 2V_0$

Temperature of gas is

$$T_2 = \frac{p_0(2V_0)}{R}\left[1 - \frac{1}{2}\left(\frac{V_0}{2V_0}\right)^2\right] = \frac{7}{4}\frac{p_0V_0}{R}$$

So, change in temperature as volume changes from V_0 to $2V_0$ is

$$\Delta T = T_2 - T_1 = \left(\frac{7}{4} - \frac{1}{2}\right)\frac{p_0V_0}{R} = \frac{5}{4}\frac{p_0V_0}{R}$$

Example 5. *A vessel contains two non-reactive gases neon (monoatomic) and oxygen (diatomic). The ratio of their partial pressure is 3 : 2. The ratio of number of molecules is [Given atomic mass of Ne = 20.2 u, molecular mass of O_2 = 32 u]* **[NCERT]**

(a) $\frac{2}{3}$ (b) $\frac{3}{2}$ (c) $\frac{4}{3}$ (d) $\frac{3}{4}$

Sol. (b) Each gas (assumed ideal) obey the gas laws. Since, V and T are common to the two gases, we have

$$p_1V = \mu_1RT \text{ and } p_2V = \mu_2RT$$

$$\frac{p_1}{p_2} = \frac{\mu_1}{\mu_2}$$

Here, 1 and 2 refer to neon and oxygen gas respectively.

Given, $\quad \frac{p_1}{p_2} = \frac{3}{2}$

$$\frac{\mu_1}{\mu_2} = \frac{3}{2}$$

By definition, $\quad \mu_1 = \frac{N_1}{N_A}$ and $\mu_2 = \frac{N_2}{N_A}$

where, N_1 and N_2 are the number of molecules of two gases and N_A is Avogadro's number.

$$\frac{N_1}{N_2} = \frac{\mu_1}{\mu_2} = \frac{3}{2}$$

Kinetic Theory of Gases

The kinetic theory of gases correlates the macroscopic properties of gases (*e.g.* pressure, temperature, etc) to the microscopic properties of gas molecules (*e.g.* speed, momentum, kinetic energy of molecules, etc). This theory attempts to develop a model of the molecular behaviour of gas, which results in the observed behaviour of an ideal gas. The kinetic theory of gases is based on the following assumptions

(i) A gas consists of a large number of tiny, identical, spherical and electrically neutral, stable elastic particles called *molecules.*

(ii) The space occupied by the molecules of a gas is extremely small as compared to the volume of the gas.

(iii) The molecules of a gas are in a state of continuous, random motion with all possible speeds ranging from zero to infinity in different possible directions. The speed distribution is in accordance with Maxwell's distribution law of molecular speeds and has been shown in figure.

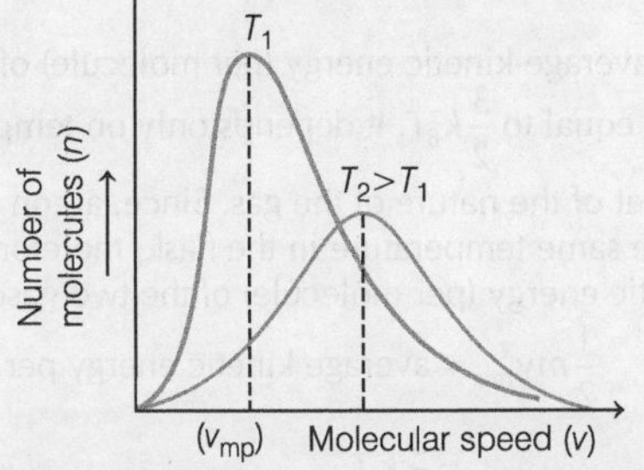

(iv) Each molecule behaves as an independent entity. There is no force of attraction among the molecules. Thus, gas molecules have no potential energy but possess only kinetic energy which is directly proportional to temperature of the gas.

(v) The pressure of a gas is due to elastic collision of gas molecules with the walls of the container.

(vi) The dynamics of the particles is governed by Newton's laws of motion.

(vii) The time of contact of a moving molecule with the container walls at the time of collision is negligible as compared to the time between two successive collisions with the same wall of the container.

Concept of Pressure

On the basis of above assumptions, we can perform mathematical calculations to find expression for pressure exerted by a gas. Accordingly, we find that

$$pV = \frac{1}{3} mN (\bar{v})^2$$

where, m = mass of 1 gas molecule and N = total number of gas molecules and $\bar{v}$ = root mean square velocity.

$$v_{rms} = \bar{v} = \left[\frac{v_1^2 + v_2^2 + \ldots + v_N^2}{N}\right]^{1/2}$$

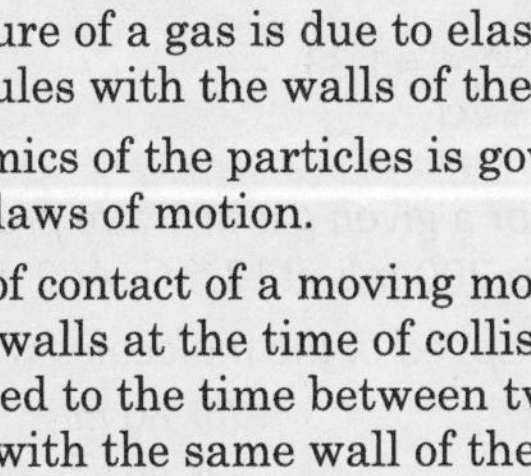

A cubical box with sides of length d containing an ideal gas. The molecule shown moves with velocity v.

Thus, pressure exerted by a gas $p = \frac{1}{3}\frac{mN}{V}\bar{v}^2 = \frac{1}{3}\rho\bar{v}^2$,

where $\rho = \frac{mN}{V}$ = density of given gas.

Now, $$p = \frac{1}{3} m \left(\frac{N}{V}\right) \bar{v}^2$$

$$= \frac{2}{3}\frac{N}{V}\left(\frac{1}{2} m\bar{v}^2\right)$$

or $$pV = \frac{2}{3} N \,\text{KE}$$

Also, $$p = \frac{1}{3}\rho\bar{v}^2 = \frac{2}{3}\left(\frac{1}{2}\rho\bar{v}^2\right)$$

Now, $\frac{1}{2}\rho\bar{v}^2$ = average kinetic energy of the gas per unit volume = E.

$\therefore$ $$p = \frac{2}{3}E$$

Kinetic Energy and Temperature

According to kinetic theory of gases, $pV = \frac{1}{3} mN\bar{v}^2$ ...(i)

but according to equation of state for an ideal gas,

$$pV = nRT \quad \text{...(ii)}$$

From Eqs. (i) and (ii), we get

$$\frac{1}{2} m\bar{v}^2 = \frac{3}{2}\frac{RT}{n} = \frac{3}{2}kT$$

where, k is the Boltzmann's constant. Its value is 1.38×10^{-23} J mol^{-1} K^{-1}.

$\therefore$ **Mean translational kinetic energy** of a gas molecule $= \frac{3}{2}kT$, *i.e.* the mean translational kinetic energy of a gas molecule depends only on its temperature and is independent of its nature or mass, etc.

On this basis, we can define *absolute zero temperature* as the temperature at which translational kinetic energy of a gas molecule becomes zero, *i.e.* at which the molecular motion ceases altogether.

Example 6. *At what temperature does the average translational kinetic energy of a molecule in a gas becomes equal to the kinetic energy of an electron accelerated from rest through 1 V potential difference* ($k = 1.38 \times 10^{-23}$ JK^{-1})

(a) 7330 K (b) 7730 K
(c) 7530 K (d) 7430 K

Sol. (b) Kinetic energy gained by an electron when accelerated by a potential difference of 1 V is 1 eV $= 1.6 \times 10^{-19}$ J.

According to kinetic theory of gases, kinetic energy $= \frac{3}{2}kT$

As, $$\frac{3}{2}kT = 1\text{ eV} = 1.6 \times 10^{-19}\text{ J}$$

$\Rightarrow$ $$T = \frac{2 \times 1.6 \times 10^{-19}}{3k}$$

$$= \frac{2 \times 1.6 \times 10^{-19}}{3 \times 1.38 \times 10^{-23}} = 7730\text{ K}$$

Critical Temperature, Pressure and Volume

Gases cannot be liquified above a temperature called *critical temperature* (T_c) however large the pressure may be. The pressure required to liquify the gas at critical temperature is called *critical pressure* (p_c) and the volume of the gas at critical temperature and pressure is called *critical volume* (V_c).

Value of critical constants in terms of Van der Waals' constants a and b are as under

$$V_c = 3b,$$

$$p_c = \frac{a}{27b^2}$$

and
$$T_c = \frac{8a}{27Rb}$$

Further, $\frac{RT_c}{p_c V_c} = \frac{8}{3}$ is called **critical coefficient** and is same for all gases.

Different Speeds of Gas Molecules

- **rms Speed** The square root of the mean of squares of the speeds of different molecules is known as root mean square (rms) speed of gas molecules.

  i.e.
  $$v_{\text{rms}} = \sqrt{\overline{v}^2}$$
  $$= \sqrt{(v_1^2 + v_2^2 + \ldots + v_N^2)/N}$$

  According to kinetic theory of gases, it is observed that

  $$v_{\text{rms}} = \sqrt{\overline{v}^2} = \sqrt{\frac{3p}{\rho}} = \sqrt{\frac{3pV}{M}}$$
  $$= \sqrt{\frac{3RT}{M}}$$
  $$= \sqrt{\frac{3kT}{m}}$$

  where, M is the molar mass of gas, while m is the mass of a single gas molecule.

- **Average Speed** It is the arithmetic mean of the speeds of molecules in a gas. Thus,

  $$v_{\text{av}} = \overline{v} = \frac{v_1 + v_2 + \ldots + v_N}{N}$$

  On the basis of kinetic theory, it is observed that

  $$v_{\text{av}} = \sqrt{\frac{8RT}{\pi M}} = \sqrt{\frac{8p}{\pi \rho}} = \sqrt{\frac{8kT}{\pi m}}$$

- **Most Probable Speed** It is the speed possessed by maximum number of gas molecules in a given gas. Basis on kinetic theory, it is found that

  $$v_{\text{mp}} = \sqrt{\frac{2RT}{M}} = \sqrt{\frac{2kT}{m}} = \sqrt{\frac{2p}{\rho}}$$

  Thus, we find that for a given gas, $v_{\text{mp}} < v_{\text{av}} < v_{\text{rms}}$ and $v_{\text{mp}} : v_{\text{av}} : v_{\text{rms}} = \sqrt{2} : \sqrt{\frac{8}{\pi}} : \sqrt{3}$.

Example 7. *A flask contains argon and chlorine in the ratio of 2 :1 by mass. The temperature of the mixture is 27°C. Ratio of root mean square speed v_{rms} of the molecules of the two gases is [Given, atomic mass of argon = 39.9 u, molecular mass of chlorine = 70.9 u]* ***[NCERT]***

(a) 0.22 (b) 2.2 (c) 1.33 (d) 3.3

Sol. (c) The average kinetic energy (per molecule) of any (ideal) gas is always equal to $\frac{3}{2}k_BT$. It depends only on temperature and is independent of the nature of the gas. Since, argon and chlorine both have the same temperature in the flask, therefore the ratio of average kinetic energy (per molecule) of the two gases is 1 : 1.

Now, $\frac{1}{2}mv_{\text{rms}}^2$ = average kinetic energy per molecule

$$= \frac{3}{2}k_BT$$

$$\frac{(v_{\text{rms}}^2)_{\text{Ar}}}{(v_{\text{rms}}^2)_{\text{Cl}}} = \frac{(m)_{\text{Cl}}}{(M)_{\text{Ar}}} = \frac{(M)_{\text{Cl}}}{(M)_{\text{Al}}} = \frac{70.9}{39.9} = 1.77$$

where, M denotes the molecular mass of the gas. Taking square root

$$\frac{(v_{\text{rms}})_{\text{Ar}}}{(v_{\text{rms}})_{\text{Cl}}} = 1.33$$

Example 8. *For a given gas at 1 atm pressure, rms speed of the molecules is 200 m/s at 127° C. At 2 atm pressure and at 227° C, the rms speed of the molecules will be*

(a) $100\sqrt{5}$ m/s *(b) 80 m/s* ***[JEE Main 2019]***
(c) 100 m/s (d) $80\sqrt{5}$ m/s

Sol. (a) For a gas molecule, $v_{\text{rms}} = \sqrt{\frac{3RT}{M}}$

$\therefore$ $v_{\text{rms}} \propto \sqrt{T}$

Let unknown rms speed be $v_{\text{rms, 2}}$.

At $T_2 = 227°\text{C}$ (or 500 K)

and at $T_1 = 127°\text{C}$ (or 400 K)

$v_{\text{rms,1}} = 200$ m/s

$\therefore$ Using the relation $v_{\text{rms}} \propto \sqrt{T}$, we can write

$$\frac{v_{\text{rms, 2}}}{v_{\text{rms, 1}}} = \sqrt{\frac{T_2}{T_1}} \quad \ldots \text{(i)}$$

Substituting these given values in Eq. (i), we get

$$\therefore \quad v_{\text{rms, 2}} = \sqrt{\frac{500}{400}} \times 200 \text{ m/s}$$
$$= \frac{1}{2}\sqrt{5} \times 200 \text{ m/s} = 100\sqrt{5} \text{ m/s}$$

Example 9. *The temperature, at which the root mean square velocity of hydrogen molecules equals their escape velocity from the earth, is closest to (Given, Boltzmann constant, $k_B = 1.38 \times 10^{-23}$ J/K, Avogadro number $N_A = 6.02 \times 10^{26}$/kg, radius of earth $= 6.4 \times 10^6$ m and gravitational acceleration on earth $= 10$ ms^{-2}]*

(a) 10^4 K *(b) 650 K* *(c) 3×10^5 K* *(d) 800 K*

Sol. (a) Root mean square velocity of hydrogen molecule is given as

$$v_{rms} = \sqrt{\frac{3k_BT}{m}}$$

Escape velocity of hydrogen molecule from the earth is given as

$$v_e = \sqrt{2gR_e}$$

Given, $\quad v_{rms} = v_e$

or $\quad \sqrt{\frac{3k_BT}{m}} = \sqrt{2gR_e}$

$\Rightarrow \quad T = \frac{2gR_em}{3 \times k_B}$

Substituting the given values, we get

$$T = \frac{2 \times 10 \times 6.4 \times 10^6 \times 2}{3 \times 1.38 \times 10^{-23} \times 6.02 \times 10^{26}} \approx 10^4 \text{ K}$$

Note Question gives Avogadro number, $N_A = 6.02 \times 10^{26}$/kg but we take $N_A = 6.02 \times 10^{26}$/kmol.

Van der Waals' Gas Equation

The gases actually found in nature are called *real gases*. They do not obey the gas laws. A real gas behaves as ideal gas most closely at *low pressure and high temperature.*

Equation of state for real gases is given by Van der Waals' equation with two corrections in ideal gas (i) volume correction (ii) pressure correction.

Van der Waals' gas equation for 1 mole of gas is given by

$$\left(p + \frac{a}{V^2}\right)(V - b) = RT$$

For n moles, $\quad \left(p + \frac{an^2}{V^2}\right)(V - nb) = nRT$

Here, a and b are constants called **Van der Waals' constants.**

Degree of Freedom (f)

The term degree of freedom refers to the number of possible independent ways in which a system can have energy.

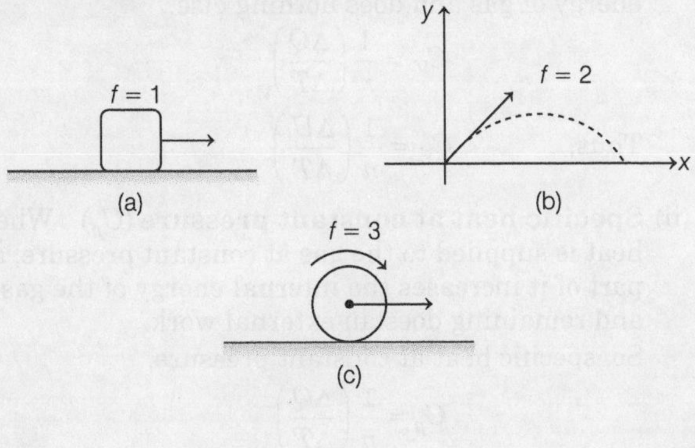

For example In Fig. (a), block has one degree of freedom, because it is confined to move in a straight line and has only one translational degree of freedom.

In Fig. (b), the projectile has two degrees of freedom because it is confined to move in a plane and so it has two translational degrees of freedom.

In Fig. (c), the sphere has two degrees of freedom, *i.e.* one rotational and another translational.

Similarly, a particle free to move in space will have three translational degrees of freedom.

- Degrees of freedom for different gases according to atomicity of gas at low temperature is given below.

Atomicity of Gas	Translational	Rotational	Total	
Monoatomic, *e.g.* Ar, Ne, ideal gas, etc.	3	0	3	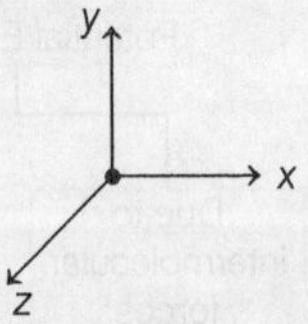
Diatomic, *e.g.* O_2, Cl_2, N_2, etc.	3	2	5	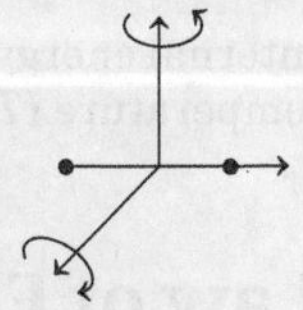
Triatomic (linear), *e.g.* CO_2, C_2H_2, etc.	3	2	5	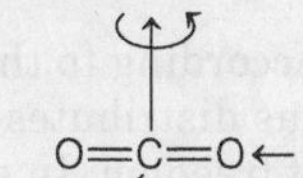
Triatomic (non-linear) or Polyatomic, *e.g.* H_2O, NH_3, CH_4, etc.	3	3	6	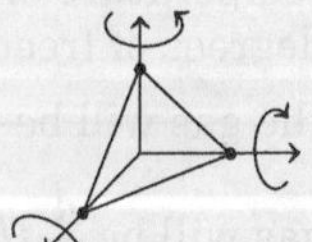

- Degrees of freedom of a diatomic and polyatomic gas depends on temperature and since there is no clear cut demarcation line above which vibrational energy become significant. Moreover, this temperature varies from gas to gas. On the other hand, for a monoatomic gas there is no such confusion.
- Degree of freedom here is 3 at all temperatures. Unless and until stated in the question you can take $f = 3$ for monoatomic gas, $f = 5$ for a diatomic gas and $f = 6$ for a non-linear polyatomic gas.
- When a diatomic or polyatomic gas dissociates into atoms it behaves as a monoatomic gas, whose degrees of freedom are changed accordingly.

Internal Energy of an Ideal Gas

The internal energy (U) of the gas is concerned only with its disordered motion, *i.e.* it is not concerned with its ordered motion. When the temperature of the gas is increased, its disordered motion increases and hence its internal energy is increased.

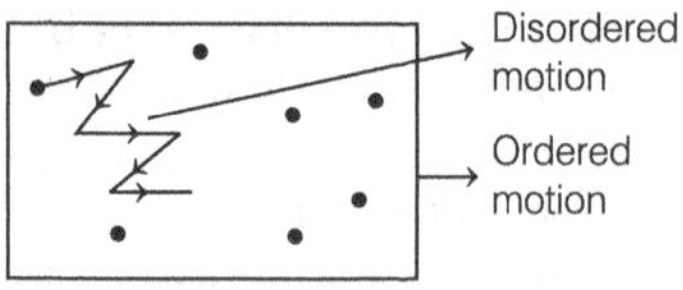

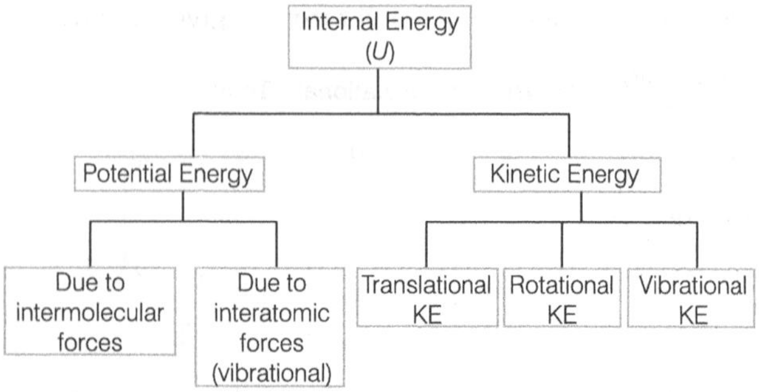

Internal energy of an ideal gas depends only on its absolute temperature (T) and is directly proportional to T.

or $$U \propto T$$

Law of Equipartition of Energy

According to the law of equipartition of energy, an ideal gas distributes its internal energy equally in all degrees of freedom. In each degree of freedom, energy of one mole of an ideal gas is $\frac{1}{2}RT$, where T is the absolute temperature of the gas. Thus, if f be the number of degrees of freedom, then the internal energy of 1 mole of the gas will be $\frac{f}{2}RT$ or internal energy of n moles of the gas will be $\frac{n}{2}fRT$. Thus,

$$U = \frac{n}{2}fRT \quad \text{...(i)}$$

For a monoatomic gas, $f = 3$.

Therefore, $$U = \frac{3}{2}RT$$

(for 1 mole of a monoatomic gas)

For a diatomic and linear polyatomic gas at low temperatures, $f = 5$, so,

$$U = \frac{5}{2}RT \quad \text{(for 1 mole)}$$

and for non-linear polyatomic gas at low temperatures, $f = 6$, so

$$U = \frac{6}{2}RT = 3RT \quad \text{(for 1 mole)}$$

Example 10. *A gas mixture consists of 3 moles of oxygen and 5 moles of argon at temperature T. Assuming the gases to be ideal and the oxygen bond to be rigid, the total internal energy (in units of RT) of the mixture is*

(a) 15 (b) 13
(c) 11 (d) 20

Sol. (a) Internal energy of a gas mixture = Sum of internal energies of its components.

Also, internal energy of a gas is given by

$$U = \frac{f}{2}(nRT)$$

Here, gases are oxygen (diatomic, $f = 5$) and argon (monoatomic, $f = 3$).

So, total internal energy of the mixture is

$$U = \left(\frac{f}{2}nRT\right)_{O_2} + \left(\frac{f}{2}nRT\right)_{Ar}$$

$$= \frac{5}{2} \times 3 \times RT + \frac{3}{2} \times 5 \times RT = 15RT$$

Molar Specific Heat of the Gases

Consider a container containing m gram of gas of molecular mass M. If n is the number of moles of gas in container, ΔQ is the heat supplied and rise in temperature is ΔT, then

$$c = \frac{1}{m}\frac{\Delta Q}{\Delta T}$$

Further, $$n = \frac{m}{M} \quad \text{or} \quad m = nM$$

so $$c = \frac{1}{nM}\frac{\Delta Q}{\Delta T}$$

Thus, molar specific heat,

$$C = Mc = \frac{1}{n}\left(\frac{\Delta Q}{\Delta T}\right)$$

where, C = molar specific heat of the mass
and c = gram specific heat of the gas.

Molar specific heat is of two types

(i) **Specific heat at constant volume (C_V)** When heat is supplied to the gas at constant volume, the entire heat supplied just increases the internal energy of gas and does nothing else.

$$C_V = \frac{1}{n}\left(\frac{\Delta Q}{\Delta T}\right)_V$$

Thus, $$C_V = \frac{1}{n}\left(\frac{\Delta U}{\Delta T}\right)$$

(ii) **Specific heat at constant pressure (C_p)** When heat is supplied to the gas at constant pressure, a part of it increases the internal energy of the gas and remaining does an external work.

So, specific heat at constant pressure,

$$C_p = \frac{1}{n}\left(\frac{\Delta Q}{\Delta T}\right)_p$$

- At constant pressure, to increase the internal energy of the gas by the same amount (as in case of heat supplied at constant volume), more amount of heat is supplied. Hence, we conclude that $C_p > C_V$
- The relation between C_p and C_V is given by Mayer's relation which is $C_p - C_V = R$

  Here, R is gas constant.
- C_p and C_V in terms of degrees of freedom f can be written as $\quad C_V = \frac{f}{2}R$

  and $\quad C_p = C_V + R = \frac{f}{2}R + R = \left(\frac{f}{2}+1\right)R$
- Ratio of specific heats C_P and C_V is $\gamma = \frac{C_p}{C_V} = 1 + \frac{2}{f}$

Specific Heat Capacity of Monoatomic, Diatomic and Polyatomic Gases

Monoatomic Gases The molar specific heat at constant volume C_V is

$$C_V \text{ (monoatomic gas)} = \frac{dU}{dT} = \frac{3}{2}R$$

From Mayer's formula, $C_p - C_V = R$

where, C_p is molar specific heat at constant pressure.

Thus, $\quad C_p = \frac{5}{2}R$

Ratio of specific heat, $\gamma = \frac{C_p}{C_V} = \frac{5}{3}$

Diatomic Gases A diatomic molecule has 5 degrees of freedom, *i.e.* 3 translational and 2 rotational. Using the law of equipartition of energy, the total internal energy of a mole of such a gas is

$$U = \frac{5}{2}k_B T \times N_A = \frac{5}{2}RT$$

The molar specific heats are then given by

$$C_V \text{ (rigid diatomic)} = \frac{5}{2}R, C_p = \frac{7}{2}R$$

$$\gamma \text{ (rigid diatomic)} = \frac{7}{5}$$

If the diatomic molecule is not rigid but has an additional vibrational mode, then

$$U = \left(\frac{5}{2}k_B T + k_B T\right)N_A = \frac{7}{2}RT$$

$$C_V = \frac{7}{2}R, C_p = \frac{9}{2}R, \gamma = \frac{9}{7}R$$

Polyatomic Gases A polyatomic molecule has 3 translational, 3 rotational degrees of freedom and a certain number (f) of vibrational modes.

From law of equipartition of energy, one mole of such a gas has

$$U = \left(\frac{3}{2}\ k_B T + \frac{3}{2}k_B T + f\,k_B T\right)N_A$$

$$C_V = (3+f)R$$

$$C_p = (4+f)R$$

$$\gamma = \frac{(4+f)}{(3+f)}$$

Note Specific heat of lighter elements is higher than that of heavier elements and *vice versa*. Specific heat of the same substance in different states (solid, liquid and vapour) is different. *For example*, specific heat of water is 1 cal $g^{-1o}C^{-1}$ and that of ice is 0.5 cal $g^{-1o}C^{-1}$.

Example 11. *Molecules of an ideal gas are known to have three translational degrees of freedom and two rotational degrees of freedom. The gas is maintained at a temperature of T. The total internal energy U of a mole of this gas and the value of* $\gamma\left(=\frac{C_p}{C_V}\right)$ *are given respectively, by*

[JEE Main 2020]

(a) $U = \frac{5}{2}RT$ and $\gamma = \frac{6}{5}$ (b) $U = 5RT$ and $\gamma = \frac{7}{5}$

(c) $U = \frac{5}{2}RT$ and $\gamma = \frac{7}{5}$ (d) $U = 5RT$ and $\gamma = \frac{6}{5}$

Sol. (c) Given that, translational degree of freedom, $f_{trans} = 3$

Rotational degree of freedom, $f_{rot} = 2$

Total degree of freedom,

$$f = f_{trans} + f_{rot} = 3 + 2 = 5$$

At T temperature, total internal energy of an ideal gas is given by

$U = \frac{f}{2}\mu RT$

$$U = \frac{5}{2}RT \qquad (\because \mu = 1)$$

$$\gamma = \frac{C_p}{C_V} = 1 + \frac{2}{f} = 1 + \frac{2}{5}$$

$$\gamma = \frac{7}{5}$$

Example 12. *A cylinder of fixed capacity 44.8 L contains helium gas at standard temperature and pressure. The amount of heat needed to raise the temperrature of the gas in the cylinder by 15°C is (Given, R = 8.31 J mol*$^{-1}$*K*$^{-1}$*)*

(a) 45 J (b) 374 J

(c) 273 J (d) 432 J

Sol. (b) From ideal gas law, $pV = \mu RT$

1 mole of any (ideal) gas at standard temperature (273 K) and pressure (1 atm $= 1.01\times10^5$ Pa) occupies a volume of 22.4 L. This universal volume is called molar volume. Thus, the cylinder in this example contains 2 mole of helium. Further, since helium is monoatomic, its predicted (and observed) molar specific heat at constant volume $C_V = \frac{3}{2}R$ and molar specific heat at constant pressure.

$$C_p = \frac{3}{2}R + R = \frac{5}{2}R$$

Since, volume of the cylinder is fixed, the heat required is determined by C_V.

$\therefore$ Heat required = Number of moles × Molar specific heat at constant volume × Rise in temperature

$$= 2 \times 1.5R \times 15 = 45R$$

$$= 45 \times 8.31 = 374 \text{ J}$$

Mean Free Path

Every gas consists of a very large number of molecules. These molecules are in a state of continuous rapid and random motion. They undergo perfectly elastic collision against one another. Therefore, path of a single gas molecule consists of a series of short *zig-zag* paths of different lengths. The mean free path of a gas molecule is the average distance between two successive collisions.

Mathematically, it is expressed as $\lambda = \dfrac{1}{\sqrt{2}\,\pi d^2 n_V}$

where, d = diameter of molecules,

n_V = number of molecules per unit volume.

The number of molecules per unit volume can be determined from Avogadro's number and the ideal gas law leading to

$$n_V = \frac{nN_A}{V} = \frac{nN_A}{\dfrac{nRT}{p}} = \frac{N_A p}{RT}$$

$$\therefore \quad \lambda = \frac{RT}{\sqrt{2}\,\pi d^2 N_A p}$$

One mole of an ideal gas will occupy a volume of 22.4L at STP.

Avogadro's number, $N_A = 6.0221367 \times 10^{23}$ mol

Note Standard temperature 0°C = 273.15 K
Standard pressure = 1 atmosphere = 760 mm of Hg = 101.3 kPa
Standard volume of 1 mole of an ideal gas at STP = 22.4 L.

Example 13. *In a dilute gas at pressure p and temperature T, the mean time between successive collisions of a molecule varies with T as*

(a) T (b) $\dfrac{1}{\sqrt{T}}$ (c) $\dfrac{1}{T}$ (d) $\sqrt{T}$

Sol. (b) The expression of mean time (τ) between successive collisions of a molecule at pressure p and temperature T is given by

$$\tau = \frac{\lambda}{v_{avg}}$$

As, mean free path of gas molecule, $\lambda = \dfrac{1}{\sqrt{2}\pi d^2 n_V}$

where, n_V = number of molecules volume.

$\Rightarrow \quad \lambda \propto (T)^0$

Also, average speed, $v_{avg} = \sqrt{\dfrac{8RT}{\pi M}}$

$\Rightarrow \quad v_{avg} \propto \sqrt{T} \therefore \tau \propto \dfrac{1}{\sqrt{T}}$

Example 14. *A vessel contains a mixture of 7 g of nitrogen and 11g of carbon dioxide at temperature T = 300 K. If the pressure of the mixture is 1 atm (1×10^5 N/m²), its density is (Given, gas constant, R = 2513 J/mol K)*

(a) 0.72 kg/m³ (b) 1.44 kg/m³
(c) 2.88 kg/m³ (d) 5.16 kg/m³

Sol. (b) The expression for density of a mixture of gases is given by, $\rho_{mix} = \dfrac{pM_{mix}}{RT}$

where, M_{mix} is the mass of mixture of non-reactive gases, p is pressure, R is gas constant and T is temperature.

$$M_{mix} = \frac{n_1M_1 + n_2M_2}{n_1 + n_2}$$

Given, Mass of nitrogen gas, $m_{N_2} = 7\text{ g} = 7 \times 10^{-3}$ kg

Mass of carbon dioxide, $m_{CO_2} = 11\text{g} = 11 \times 10^{-3}$ kg

Molecular weight of nitrogen gas, $m_{N_2} = 28 \times 10^{-3}$ kg

Molecular weight of carbon dioxide, $M_{CO_2} = 44 \times 10^{-3}$ kg

$$M_{mix} = \frac{n_{N_2}M_{N_2}}{n_{N_2} + n_{CO_2}} = \frac{\dfrac{m_{N_2}}{M_{N_2}}M_{N_2} + \dfrac{m_{CO_2}}{M_{CO_2}}M_{CO_2}}{\dfrac{m_{N_2}}{M_{N_2}} + \dfrac{m_{CO_2}}{M_{CO_2}}}$$

$$M_{mix} = \frac{m_{N_2} + m_{CO_2}}{\dfrac{m_{N_2}}{M_{N_2}} + \dfrac{m_{CO_2}}{M_{CO_2}}} = \frac{(7+11) \times 10^{-3}}{\left(\dfrac{7 \times 10^{-3}}{28 \times 10^{-3}} + \dfrac{11 \times 10^{-3}}{44 \times 10^{-3}}\right)}$$

$$M_{mix} = \frac{18 \times 10^{-3}}{\dfrac{1}{4} + \dfrac{1}{4}} = 36 \times 10^{-3} \text{ kg}$$

$$\therefore \quad \rho = \frac{(1 \times 10^5)\,(36 \times 10^{-3})}{\dfrac{25}{3} \times 300} = 1.44 \text{ kg/m}^3$$

Example 15. *In the arrangement shown in the figure gas is thermally insulated. An ideal gas is filled in the cylinder having pressure p_0 (> atmospheric pressure p_a). The spring of force constant k is initially unstretched. The piston of mass m and area S is frictionless. In equilibrium, the piston rises up a distance x_0, then the decrease in internal energy of the gas is given by*

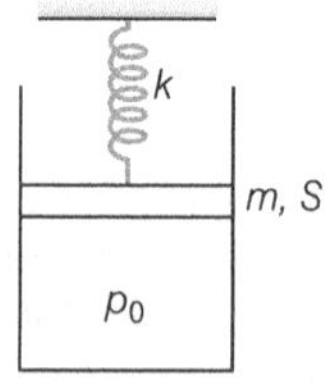

(a) $p_aSx_0 + \dfrac{1}{2}kx_0^2 + mgx_0$ (b) $p_aSx_0 + \dfrac{4}{5}x_0^2 + 2mgx_0$

(c) $2\,p_aSx_0 + \dfrac{4}{5}x_0^2 + \dfrac{2}{3}mgx_0$ (d) $p_aSx_0 + 2kx_0^2 + 2mgx_0$

Sol. (a) Equilibrium of piston gives

$$pS = p_a S + mg + kx_0$$

$$p = p_a + \frac{mg}{S} + \frac{kx_0}{S} \qquad (p = \text{final pressure of gas})$$

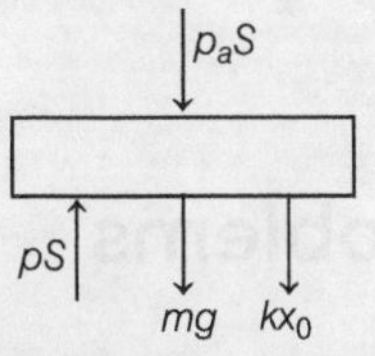

Work done by the gas is equal to work done against atmospheric pressure + elastic potential energy stored in the spring + increase in gravitational potential energy of the piston.

$$= p_a \Delta V + \frac{1}{2}kx_0^2 + mgx_0$$

$$= p_a S x_0 + \frac{1}{2}kx_0^2 + mgx_0$$

This is also the decrease in internal energy of the gas, because the gas is thermally insulated and this work is done at the expense of internal energy of the gas.

Example 16. *Carbon monoxide is carried around a closed cycle abc, in which bc is an isothermal process, as shown.*

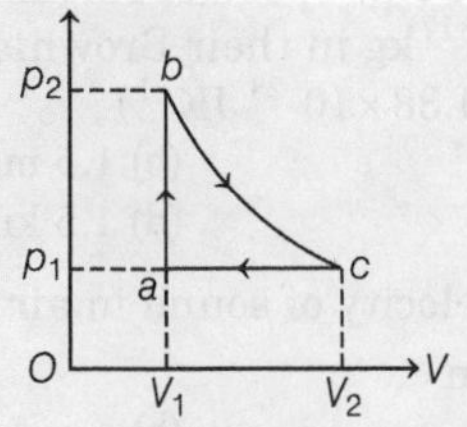

The gas absorbs 7000 J of heat as its temperature is increased from 300 K to 1000 K is going from a to b. The quantity of heat ejected by the gas. process ca is

(a) 4200 J (b) 5000 J

(c) 9800 J (d) 3500 J

Sol. (c) Given that gas absorbs 7000 J of energy, hence

$$(\Delta Q)_{ab} = \mu C_V \Delta T$$

$$\therefore \qquad +7000 = \mu C_V (1000 - 300)$$

For the process ca, $T_a = 300$ K

$$T_c = T_b = 1000 \text{ K}$$

$$(\Delta Q)_{ca} = \mu C_p \Delta T$$

$$= \mu C_p (300 - 1000)$$

$$= -\mu C_p \times 700$$

Also, $\qquad C_p - C_V = R$

$$\therefore \qquad C_p = R + C_V$$

$$(\Delta Q)_{ca} = -\mu (C_V + R) 700$$

For carbon monoxide, $\gamma = 1 + \frac{2}{n} = 1 + \frac{2}{5} = \frac{7}{5}$

$$C_V = \frac{R}{\gamma - 1} = \frac{R}{\frac{7}{5} - 1} = \frac{5R}{2}$$

Hence, we have $(\Delta Q)_{ab} = \mu C_V 700$

$$(\Delta Q)_{ab} = \mu \frac{5R}{2} \times 700 = 7000$$

or $\qquad \mu R = \frac{20}{5} = 4$

$$\therefore \qquad (\Delta Q)_{ca} = -(7000 + 4 \times 700) = -9800 \text{ J}$$

Negative sign shows that heat is ejected.

Practice Exercise

ROUND I Topically Divided Problems

Ideal Gas, Kinetic Theory of Gases and Degree of Freedom

1. An oxygen cylinder of volume 30 L has an initial gauge pressure of 15 atm and a temperature of 27°C. After some oxygen is withdrawn from the cylinder, the gauge pressure drops to 11 atm and its temperature drops to 17°C. Estimate the mass of oxygen taken out of the cylinder.
(Take, $R = 8.31 \text{J mol}^{-1} \text{K}^{-1}$ and molecular mass of $O_2 = 32 \text{ u}$) **[NCERT]**
(a) 0.14 g (b) 0.02 g (c) 0.14 kg (d) 0.014 kg

2. At room temperature (27°C), the rms speed of the molecules of a certain diatomic gas is found to be 1920 ms^{-1}. The gas is
(a) Cl_2 (b) O_2 (c) N_2 (d) H_2

3. The rms velocity of gas molecules is 300 ms^{-1}. The rms velocity of molecules of gas with twice the molecular mass and half the absolute temperature is
(a) 300 ms^{-1} (b) 600 ms^{-1}
(c) 75 ms^{-1} (d) 150 ms^{-1}

4. 8 g of O_2, 14 g of N_2 and 22 g of CO_2 is mixed in a container of 10 L capacity at 27°C. The pressure exerted by the mixture in terms of atmospheric pressure is (R = 0.082 L atm K^{-1} mol^{-1})
(a) 1.4 atm (b) 2.5 atm
(c) 3.7 atm (d) 8.7 atm

5. At a certain temperature, the ratio of the rms velocity of H_2 molecules to O_2 molecule is
(a) 1 : 1 (b) 1 : 4
(c) 4 : 1 (d) 16 : 1

6. Two containers of equal volume contain the same gas at the pressures p_1 and p_2 and absolute temperatures T_1 and T_2 respectively. On joining the vessels, the gas reaches a common pressure p and a common temperature T. The ratio p/T is equal to
(a) $\frac{p_1T_2 + p_2T_2}{T_1 \times T_2}$ (b) $\frac{p_1T_2 + p_2T_1}{T_1 + T_2}$
(c) $\frac{1}{2}\left(\frac{p_1T_2 + p_2T_1}{T_1T_2}\right)$ (d) $\frac{p_1T_2 - p_2T_1}{T_1 \times T_2}$

7. The volume V of an enclosure contains a mixture of three gases : 16g of oxygen, 28 g of nitrogen and 44 g of carbon dioxide at absolute temperature T. Consider R as universal gas constant. The pressure of the mixture of gases is **[JEE Main 2021]**
(a) $\frac{88\,RT}{V}$ (b) $\frac{3\,RT}{V}$ (c) $\frac{5}{2} \cdot \frac{RT}{V}$ (d) $\frac{4\,RT}{V}$

8. rms velocity of a particle is c at pressure p. If pressure is increased two times, then rms velocity becomes
(a) 0.5 c (b) c (c) 2 c (d) 3 c

9. Calculate the rms speed of smoke particles each of mass 5×10^{-17} kg in their Brownian motion in air at NTP ($k = 1.38 \times 10^{-23} \text{ JK}^{-1}$)
(a) 1.5 mm s^{-1} (b) 1.5 ms^{-1}
(c) 1.5 cms^{-1} (d) 1.5 kms^{-1}

10. If c_s is the velocity of sound in air and c is the rms velocity, then
(a) $c_s < c$ (b) $c_s = c$
(c) $c_s = c\left(\frac{\gamma}{3}\right)^{1/2}$ (d) None of these

11. In a diatomic molecule, the rotational energy at a given temperature **[NCERT Exemplar]**
(a) do not obey Maxwell's distribution
(b) have the same value for all molecules
(c) equals the translational kinetic energy for each molecule
(d) is (2/3)rd the translational kinetic energy for each molecule

12. $ABCDEFGH$ is hollow cube made of an insulator. Face $ABCD$ has positive charge on it. Inside the cube, we have ionized hydrogen.

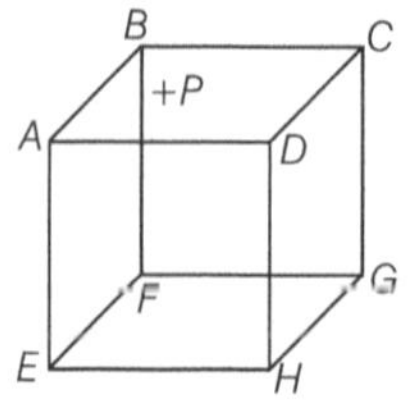

The usual kinetic theory expression for pressure **[NCERT Exemplar]**

(a) will be valid
(b) will be valid since the ions would experience forces larger than due to collisions with the walls
(c) will not be valid since collisions with walls would not be elastic
(d) will not be valid because isotropy is lost

13. Consider a sample of oxygen behaving like an ideal gas. At 300 K, the ratio of root mean square (rms) velocity to the average velocity of gas molecule would be (Take, molecular weight of oxygen = 32 g/mol and $R = 8.3\ JK^{-1}\ mol^{-1}$) **[JEE Main 2021]**

(a) $\sqrt{\frac{3}{8}}$ (b) $\sqrt{\frac{8}{3}}$
(c) $\sqrt{\frac{3\pi}{8}}$ (d) $\sqrt{\frac{8\pi}{3}}$

14. N molecules, each of mass m, of gas A and $2N$ molecules, each of mass $2m$, of gas B are contained in the same vessel which is maintained at a temperature T. The mean square velocity of molecules of B type is denoted by V_2 and the mean square velocity of A type is denoted by V_1, then $\frac{V_1}{V_2}$ is

(a) 2 (b) 1
(c) 1/3 (d) 2/3

15. If 10^{22} gas molecules each of mass 10^{-26} kg collide with a surface (perpendicular to it) elastically per second over an area 1 m^2 with a speed 10^4 m/s, the pressure exerted by the gas molecules will be of the order of **[JEE Main 2019]**

(a) $10^4 N/m^2$ (b) $10^8 N/m^2$
(c) $10^0 N/m^2$ (d) $10^{16} N/m^2$

16. To raise the temperature of a certain mass of gas by 50°C at a constant pressure, 160 cal of heat is required. When the same mass of gas is cooled by 100°C at constant volume, 240 cal of heat is released. How many degrees of freedom does each molecule of this gas have (assume gas to be ideal)? **[JEE Main 2020]**

(a) 5 (b) 7
(c) 6 (d) 3

17. The specific heats, C_p and C_V of a gas of diatomic molecules, A are given (in units of $J\ mol^{-1}\ K^{-1}$) by 29 and 22, respectively. Another gas of diatomic molecules B, has the corresponding values of C_p and C_V given as 30 and 21, respectively. If they are treated as ideal gases, then respectively. **[JEE Main 2019]**

(a) A has a vibrational mode but B has none
(b) both A and B have a vibrational mode each
(c) A has one vibrational mode and B has two
(d) A is rigid but B has a vibrational mode

18. A $25 \times 10^{-3}\ m^3$ volume cylinder is filled with 1 mole of O_2 gas at room temperature (300 K). The molecular diameter of O_2 and its root mean square speed are found to be 0.3 nm and 200 m/s, respectively. What is the average collision rate (per second) for an O_2 molecule? **[JEE Main 2019]**

(a) $\sim 10^{10}$ (b) $\sim 10^{12}$
(c) $\sim 10^{11}0$ (d) $\sim 10^8$

19. Consider two ideal diatomic gases A and B at some temperature T. Molecules of the gas A are rigid and have a mass m. Molecules of the gas B have an additional vibrational mode and have a mass $\frac{m}{4}$.

The ratio of the specific heats (C_V^A and C_V^B) of gas A and B respectively is **[JEE Main 2020]**

(a) 5 : 9 (b) 7 : 9
(c) 3 : 5 (d) 5 : 7

20. The average kinetic energy of a gas molecule at 27°C is 6.21×10^{-21} J. Its average kinetic energy at 127°C will be

(a) 12.2×10^{-21} J
(b) 8.28×10^{-21} J
(c) 10.35×10^{-21} J
(d) 11.35×10^{-21} J

21. The root mean square velocity of the molecules in a sample of helium is 5/7th that of the molecules in a sample of hydrogen. If the temperature of the hydrogen as is 0°C, that of helium sample is about

(a) 0°C (b) 4 K
(c) 273°C (d) 100°C

Internal Energy, Specific Heat Capacity and Mean Free Path

22. The value of molar specific heat at constant volume for 1 mole of polyatomic gas having n number of degrees of freedom at temperature T K is (R = universal gas constant)

(a) $\frac{nR}{2T}$ (b) $\frac{nR}{2}$
(c) $\frac{nRT}{2}$ (d) $2nRT$

23. For a gas, if the ratio of specific heats at constant pressure and constant volume is γ, then the value of degrees of freedom is

(a) $\frac{\gamma+1}{\gamma-1}$ (b) $\frac{\gamma-1}{\gamma+1}$
(c) $\frac{(\gamma-1)}{2}$ (d) $\frac{2}{\gamma-1}$

24. Two moles of monoatomic gas is mixed with three moles of a diatomic gas. The molar specific heat of the mixture at constant volume is

(a) 1.55 R (b) 2.10 R
(c) 1.63 R (d) 2.20 R

25. An ideal gas occupies a volume of $2\ m^3$ at a pressure of 3×10^6 Pa. The energy of the gas is **[JEE Main 2019]**

(a) 6×10^4 J (b) 10^8 J
(c) 9×10^6 J (d) 3×10^2 J

26. Calculate the value of mean free path λ for oxygen molecules at temperature 27°C and pressure 1.01×10^5 Pa. Assume the molecular diameter 0.3 nm and the gas is ideal.
(Take, $k = 1.38 \times 10^{-23}\ JK^{-1}$) **[JEE Main 2021]**

(a) 58 nm (b) 32 nm
(c) 86 nm (d) 102 nm

27. The value of molar specific heat at constant pressure for one mole of triatomic gas (triangular arrangement) at temperature T K is (R = universal gas constant)

(a) $3R$ (b) $\frac{2}{7}R$
(c) $\frac{5}{2}R$ (d) $4R$

28. The diameter of a gas molecule is 2.4×10^{-10} m. The mean free path of gas molecule at NTP is
($k = 1.38 \times 10^{-23}\ JK^{-1}$)

(a) 1.46×10^{-7} m (b) 2.46×10^{-6} m
(c) 1.46×10^{-6} m (d) 2.46×10^{-7} m

29. Match the C_p/C_V ratio for ideal gases with different type of molecules **[JEE Main 2020]**

	Molecule type	C_p/C_V
(A)	Monatomic molecules	I. 7/5
(B)	Diatomic rigid molecules	II. 9/7
(C)	Diatomic non-rigid molecules	III. 4/3
(D)	Triatomic rigid molecules	IV. 5/3

A B C D
(a) IV I II III
(b) III IV II I
(c) II III I IV
(d) IV II I III

30. An ideal gas in a closed container is slowly heated. As its temperature increases, which of the following statements are true? **[JEE Main 2020]**

A. The mean free path of the molecules decreases.
B. The mean collision time between the molecules decreases.
C. The mean free path remains unchanged.
D. The mean collision time remains unchanged.

(a) B and C (b) A and B
(c) C and D (d) A and D

31. The plot that depicts the behaviour of the mean free time τ (time between two successive collisions) for the molecules of an ideal gas, as a function of temperature (T), qualitatively is (graphs are schematic and not drawn to scale)

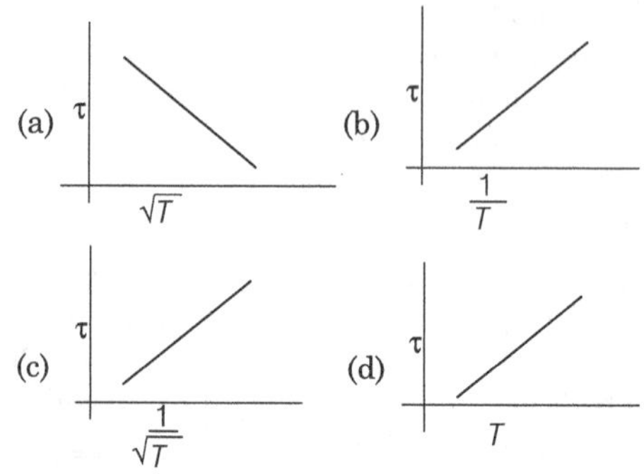

32. Under an adiabatic process, the volume of an ideal gas gets doubled. Consequently, the mean collision time between the gas molecule changes from τ_1 to τ_2. If $\frac{C_p}{C_V} = \gamma$ for this gas, then a good estimate for $\frac{\tau_2}{\tau_1}$ is given by **[JEE Main 2020]**

(a) $\left(\frac{1}{2}\right)^{\frac{\gamma+1}{2}}$ (b) $\frac{1}{2}$
(c) 2 (d) $\left(\frac{1}{2}\right)^{\gamma}$

33. Two gases-argon (atomic radius 0.07 nm, atomic weight 40) and xenon (atomic radius 0.1 nm, atomic weight 140) have the same number density and are at the same temperature. The ratio of their respective mean free times is closest to **[JEE Main 2020]**

(a) 4.67 (b) 2.3
(c) 3.67 (d) 1.09

34. Two moles of an ideal gas with $\frac{C_p}{C_V} = \frac{5}{3}$ are mixed with 3 mol of another ideal gas with $\frac{C_p}{C_V} = \frac{4}{3}$. The value of $\frac{C_p}{C_V}$ for the mixture is **[JEE Main 2020]**

(a) 1.42 (b) 1.47
(c) 1.50 (d) 1.45

35. Number of molecules in a volume of $4\ cm^3$ of a perfect monoatomic gas at some temperature T and at a pressure of 2 cm of mercury is close to
(Given, mean kinetic energy of a molecule at T is 4×10^{-14} erg, $g = 980\ cm/s^2$, density of mercury $= 13.6\ g/cm^3$) **[JEE Main 2020]**

(a) 5.8×10^{16} (b) 4.0×10^{16}
(c) 4.0×10^{18} (d) 5.8×10^{18}

36. Consider a mixture of n moles of helium gas and $2n$ moles of oxygen gas (molecules taken to be rigid) as an ideal gas. Its C_p / C_V value will be **[JEE Main 2020]**

(a) 40/27 (b) 23/15
(c) 19/13 (d) 67/45

37. An HCl molecule has rotational, translational and vibrational motions. If the rms velocity of HCl molecules in its gaseous phase is $\bar{v}$, m is its mass and k_B is Boltzmann constant, then its temperature will be [JEE Main 2019]

(a) $\frac{m\bar{v}^2}{3k_B}$ (b) $\frac{m\bar{v}^2}{7k_B}$ (c) $\frac{m\bar{v}^2}{5k_B}$ (d) $\frac{m\bar{v}^2}{6k_B}$

38. Consider a gas of triatomic molecules. The molecules are assumed to be triangular and made of massless rigid rods whose vertices are occupied by atoms. The internal energy of a mole of the gas at temperature T is [JEE Main 2020]

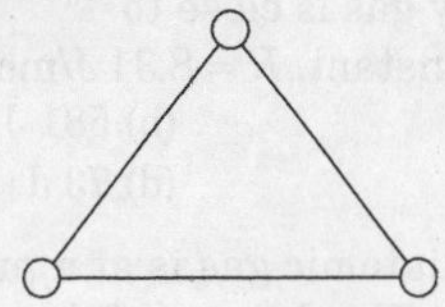

(a) $\frac{5}{2}RT$ (b) $\frac{3}{2}RT$

(c) $\frac{9}{2}RT$ (d) $3RT$

ROUND II Mixed Bag

Only One Correct Option

1. Three perfect gases at absolute temperatures T_1, T_2 and T_3 are mixed. The masses of molecules are m_1, m_2 and m_3 and the number of molecules are n_1, n_2 and n_3 respectively. Assuming no loss of energy, the final temperature of the mixture is [AIEEE 2011]

(a) $\frac{n_1T_1 + n_2T_2 + n_3T_3}{n_1 + n_2 + n_3}$ (b) $\frac{n_1T_1^2 + n_2T_2^2 + n_3T_3^2}{n_1T_1 + n_2T_2 + n_3T_3}$

(c) $\frac{n_1^2T_1^2 + n_2^2T_2^2 + n_3^2T_3^2}{n_1T_1 + n_2T_2 + n_3T_3}$ (d) $\frac{(T_1 + T_2 + T_3)}{3}$

2. The temperature of an open room of volume 30 m^3 increases from 17°C to 27°C due to the sunshine. The atmospheric pressure in the room remains 1×10^5 Pa. If n_i and n_f are the number of molecules in the room before and after heating, then $n_f - n_i$ will be [JEE Main 2017]

(a) 1.38×10^{23} (b) 2.5×10^{25}

(c) -2.5×10^{25} (d) -1.61×10^{23}

3. An insulated container of gas has two chambers separated by an insulating partition. One of the chambers has volume V_1 and contains ideal gas at pressure p_1 and temperature T_1. The other chamber has volume V_2 and contains ideal gas at pressure p_2 and temperature T_2. If the partition is removed without doing any work on the gas, the final equilibrium temperature of the gas in the container will be [AIEEE 2008, 04]

(a) $\frac{T_1T_2(p_1V_1 + p_2V_2)}{p_1V_1T_2 + p_2V_2T_1}$

(b) $\frac{p_1V_1T_1 + p_2V_2T_2}{p_1V_1 + p_2V_2}$

(c) $\frac{p_1V_1T_2 + p_2V_2T_1}{p_1V_1 + p_2V_2}$

(d) $\frac{T_1T_2(p_1V_1 + p_2V_2)}{p_1V_1T_1 + p_2V_2T_2}$

4. 22 g of carbon dioxide at 27°C is mixed in a closed container with 16 g of oxygen at 37°C. If both gases are considered as ideal gases, then the temperature of the mixture is

(a) 24.2°C (b) 28.5°C

(c) 31.5°C (d) 33.5°C

5. A rigid diatomic ideal gas undergoes an adiabatic process at room temperature. The relation between temperature and volume for this process is TV^x = constant, then x is [JEE Main 2019]

(a) $\frac{2}{5}$ (b) $\frac{2}{3}$

(c) $\frac{5}{3}$ (d) $\frac{3}{5}$

6. A cylinder is containing nitrogen at 2 atm and temperature 17°C. The radius of a nitrogen molecule is roughly 1.0A. Molecular mass of nitrogen = 28.0 u, Boltzmann constant, $k = 1.38 \times 10^{-23}$ JK^{-1}. The mean free path of nitrogen molecule is

(a) 1.1×10^{-7} m (b) 2.1×10^{-7} m

(c) 3.1×10^{-7} m (d) 0.8×10^{-7} m

7. A polyatomic ideal gas has 24 vibrational modes. What is the value of γ? [JEE main 2021]

(a) 1.03 (b) 1.30

(c) 1.37 (d) 10.3

8. Inside a cylinder closed at both ends is a movable piston. On one side of the piston is a mass m of a gas, and on the other side a mass $2m$ of the same gas.

What fraction of the volume of the cylinder will be occupied by the larger mass of the gas when the piston is in equilibrium? The temperature is the same throughout.

(a) $\frac{2}{3}$ (b) $\frac{1}{3}$

(c) $\frac{1}{2}$ (d) $\frac{1}{4}$

9. Half-mole of an ideal monoatomic gas is heated at constant pressure of 1 atm from 20° C to 90° C. Work done by gas is close to
(Take, gas constant, $R = 8.31$ J/mol-K) [JEE Main 2019]
(a) 291 J (b) 581 J
(c) 146 J (d) 73 J

10. One kg of a diatomic gas is at a pressure of 8×10^4 Nm^{-2}. The density of the gas is 4 kgm^{-3}. What is the energy of the gas due to its thermal motion? [AIEEE 2009]
(a) 3×10^4 J (b) 5×10^4 J
(c) 6×10^4 J (d) 7×10^4 J

11. The potential energy function for the force between two atoms in a diatomic molecule is approximately given by $U(x) = \frac{a}{x^{12}} - \frac{b}{x^6}$, where a and b are constants and x is the distance between the atoms. If the dissociation energy of the molecule is $D = [U(x = \infty) - U_{\text{at equilibrium}}]$, D is [AIEEE 2010]
(a) $\frac{b^2}{2a}$ (b) $\frac{b^2}{12a}$
(c) $\frac{b^2}{4a}$ (d) $\frac{b^2}{6a}$

12. A 15 g mass of nitrogen gas is enclosed in a vessel at a temperature 27°C. Amount of heat transferred to the gas, so that rms velocity of molecules is doubled is about (Take, $R = 8.3$ J/mol K) [JEE Main 2019]
(a) 10 kJ (b) 0.9 kJ
(c) 14 kJ (d) 6 kJ

13. Two chambers containing m_1 and m_2 gram of a gas at pressures p_1 and p_2 respectively are put in communication with each other, temperature remaining constant. The common pressure reached will be
(a) $\frac{p_1 p_2 (m_1 + m_2)}{p_2 m_1 + p_1 m_2}$ (b) $\frac{p_1 p_2 m_1}{p_2 m_1 + p_1 m_2}$
(c) $\frac{m_1 m_2 (p_1 + p_2)}{p_2 m_1 + p_1 m_2}$ (d) $\frac{m_1 m_2 p_2}{p_2 m_1 + p_1 m_2}$

14. An ideal gas is enclosed in a cylinder at pressure of 2 atm and temperature, 300 K. The mean time between two successive collisions is 6×10^{-8} s. If the pressure is doubled and temperature is increased to 500 K, the mean time between two successive collisions will be close to [JEE Main 2019]
(a) 4×10^{-8} s (b) 3×10^{-6} s
(c) 2×10^{-7} s (d) 0.5×10^{-8} s

15. Two moles of helium gas is mixed with three moles of hydrogen molecules (taken to be rigid). What is the molar specific heat of mixture at constant volume? (Take, $R = 8.3$ J/mol-K) [JEE Main 2019]
(a) 19.7 J/mol-K (b) 15.7 J/mol-K
(c) 17.4 J/mol-K (d) 21.6 J/mol-K

16. Consider an ideal gas confined in an isolated closed chamber. As the gas undergoes an adiabatic expansion, the average time of collision between molecules increases as V^q, where V is the volume of the gas. The value of q is $\left(\gamma = \frac{C_p}{C_V}\right)$ [JEE Main 2015]
(a) $\frac{3\gamma + 5}{6}$ (b) $\frac{3\gamma - 5}{6}$
(c) $\frac{\gamma + 1}{2}$ (d) $\frac{\gamma - 1}{2}$

17. A gaseous mixture consists of 16 g of helium and 16 g of oxygen. The ratio $\frac{C_p}{C_V}$ of the mixture is [AIEEE 2005]
(a) 1.59 (b) 1.62 (c) 1.4 (d) 1.54

18. A vertical closed cylinder is separated into two parts by a frictionless piston of mass m and of negligible thickness. The piston is free to move along the length of the cylinder. The length of the cylinder above the piston is l_1 and that below the piston is l_2, such that $l_1 > l_2$. Each part of the cylinder contains n moles of an ideal gas at equal temperature T. If the piston is stationary, its mass m, will be given by (where, R is universal gas constant and g is the acceleration due to gravity) [JEE Main 2019]
(a) $\frac{nRT}{g}\left[\frac{l_1 - l_2}{l_1 l_2}\right]$ (b) $\frac{nRT}{g}\left[\frac{1}{l_2} + \frac{1}{l_1}\right]$
(c) $\frac{RT}{g}\left[\frac{2l_1 + l_2}{l_1 l_2}\right]$ (d) $\frac{RT}{ng}\left[\frac{l_1 - 3l_2}{l_1 l_2}\right]$

19. One mole of ideal monoatomic gas $\left(\gamma = \frac{5}{3}\right)$ is mixed with one mole of diatomic gas $\left(\gamma = \frac{7}{5}\right)$. What is γ for the mixture ? γ denotes the ratio of specific heat at constant pressure, to that at constant volume. [AIEEE 2004]
(a) $\frac{3}{2}$ (b) $\frac{23}{15}$ (c) $\frac{35}{23}$ (d) $\frac{4}{3}$

20. Two rigid boxes $\frac{1}{2}$ containing different ideal gases are placed on a table. Box A contains one mole of nitrogen at temperature T_0, while box B contains one mole of helium at temperature (7/3) T_0. The boxes are then put into thermal contact with each other and heat flows between them until the gases reach a common final temperature (ignore the heat capacity of boxes). Then, the final temperature of the gases, T_f, in terms of T_0 is [AIEEE 2006]
(a) $T_f = \frac{3}{7} T_0$ (b) $T_f = \frac{7}{3} T_0$
(c) $T_f = \frac{3}{2} T_0$ (d) $T_f = \frac{5}{2} T_0$

21. The number density of molecules of a gas depends on their distance r from the origin as, $n(r) = n_0 e^{-\alpha r^4}$. Then, the total number of molecules is proportional to **[JEE Main 2019]**

(a) $n_0 \alpha^{-3/4}$ (b) $\sqrt{n_0}\ \alpha^{1/2}$
(c) $n_0 \alpha^{1/4}$ (d) $n_0 \alpha^{-3}$

Numerical Value Questions

22. The total number of air molecules (inclusive of oxygen, nitrogen, water vapour and other constituents) in a room of capacity 25.0 m^3 at a temperature of 27°C and 1 atm pressure, is found to be 6.1×10^{13x}, then the value of x is .... (Take, Boltzmann constant $= 1.38 \times 10^{-23}\ JK^{-1}$)

23. A closed vessel contains 0.1 mole of a monatomic ideal gas at 200 K. If 0.05 mole of the same gas at 400 K is added to it, the final equilibrium temperature (in K) of the gas in the vessel will be close to .... **[JEE Main 2020]**

24. If four molecules of a gas have speed 2, 4, 6, 8 kms^{-1} respectively, then their average speed (in kms^{-1}) and root mean square speed (in kms^{-1}) will be ...... and ......, respectively.

25. Initially a gas of diatomic molecules is contained in a cylinder of volume V_1 at a pressure p_1 and temperature 250 K. Assuming that 25% of the molecules get dissociated causing a change in number of moles. The pressure of the resulting gas at temperature 2000 K, when contained in a volume $2V_1$ is given by p_2. The ratio p_2/p_1 is .... **[JEE Main 2020]**

26. From the following data, find the magnitude of Joule's mechanical equivalent of heat. If C_p for hydrogen $= 3.409$ cal $g^{-1}\,°C^{-1}$, C_V for hydrogen $= 2.409$ cal $g^{-1}\,°C^{-1}$ and molecular weight of hydrogen $= 2$, then the magnitude of joule's mechanical equivalent of heat (in $Jcal^{-1}$) will be ....

27. Nitrogen gas is at 300°C temperature. The temperature (in K) at which the rms speed of a H_2 molecule would be equal to the rms speed of a N_2 molecule is ....
(Molar mass of N_2 gas is 28 g.) **[JEE Main 2020]**

28. In an experiment on the specific heat of a metal, a 0.20 kg block of the metal at 150°C is dropped in a copper calorimeter (of water equivalent 0.025 kg) containing 150 cc of water at 27°C. The final temperature is 40ºC. If heat losses to the surroundings is not negligible, then the value of specific heat (in J/g°C) of the metal will be ....

29. A geyser heats water glowing at the rate of 3.0 $Lmin^{-1}$ from 27°C to 77°C. If the geyser operates on a gas burner, then the rate of combustion (in $gmin^{-1}$) of fuel, when its heat of combustion is $4.0 \times 10^4\ Jg^{-1}$, will be ....

30. The molecules of a given mass of a gas have root mean square speeds of 100 ms^{-1} at 27°C and 1.00 atm pressure. If the root mean square speed of the molecules of the gas at 127°C and 2 atm pressure is found to be $\frac{200}{\sqrt{n}}$ m/s, then find the value of n. **[JEE Main 2021]**

31. An air bubble of volume 1.0 cm^3 rises from the bottom of a lake 40 m deep at a temperature of 12°C. The volume of bubble when it reaches the surface, which is at temperature of 3°C, is found to be $5.275 \times 10^{-x}\ m^3$, then the value of x is .... (Given, 1 atm $= 1.01 \times 10^5$ Pa)

Answers

Round I

1. (c)	2. (d)	3. (d)	4. (c)	5. (c)	6. (c)	7. (c)	8. (b)	9. (c)	10. (c)
11. (d)	12. (d)	13. (c)	14. (b)	15. (c)	16. (c)	17. (a)	18. (d)	19. (d)	20. (b)
21. (c)	22. (b)	23. (d)	24. (b)	25. (c)	26. (d)	27. (d)	28. (a)	29. (a)	30. (a)
31. (c)	32. (a)	33. (d)	34. (a)	35. (c)	36. (c)	37. (b)	38. (d)		

Round II

1. (a)	2. (c)	3. (a)	4. (c)	5. (a)	6. (a)	7. (a)	8. (a)	9. (a)	10. (b)
11. (c)	12. (a)	13. (a)	14. (a)	15. (c)	16. (c)	17. (b)	18. (a)	19. (a)	20. (c)
21. (a)	22. 2	23. 266	24. 5, 5.48	25. 5	26. 4.11	27. 41	28. 0.1		
29. 15.75	30. 3	31. 6							

Solutions

Round I

1. Absolute pressure, $p_1 = (15 + 1)$ atm

$$[\because \text{Absolute pressure} = \text{Gauge pressure} + 1 \text{ atm}]$$
$$= 16 \times 1.013 \times 10^5 \text{ Pa}$$
$$V_1 = 30 \text{ L} = 30 \times 10^{-3} \text{ m}^3$$
$$T_1 = 273.15 + 27 = 300.15 \text{ K}$$

Using ideal gas equation, $pV = nRT$

or $$n = \frac{pV}{RT}$$
$$= \frac{p_1V_1}{RT_1} = \frac{16 \times 1.013 \times 10^5 \times 30 \times 10^{-3}}{8.314 \times 300.15}$$
$$= 19.48$$

Final pressure, $p_2 = (11 + 1) = 12$ atm Pa
$$V_2 = 30 \text{ L} = 30 \times 10^{-3} \text{m}^3$$
$$T_2 = 273.15 + 17 = 290.15 \text{ K}$$

Number of moles $= \dfrac{p_2V_2}{RT_2} = \dfrac{12 \times 1.013 \times 10^5 \times 30 \times 10^{-3}}{8.314 \times 290.15}$
$$= 15.12$$

Hence, moles removed $= 19.48 - 15.12 = 4.36$

Mass removed $= 4.36 \times 32 \text{ g} = 0.1396 \text{ kg} \approx 0.14 \text{ kg}$

2. As, $c_{\text{rms}} = \sqrt{\dfrac{3RT}{M}}$

or $$M = \frac{3RT}{c_{\text{rms}}^2}$$
$$= \frac{3 \times 8.31 \times 300}{(1920)^2}$$
$$= 2 \times 10^{-3} \text{ kg} = 2 \text{ g}$$

Since, $M = 2$ for the hydrogen molecule. Hence, the gas is hydrogen.

3. As, $c = \sqrt{\dfrac{3pV}{M}} = \sqrt{\dfrac{3RT}{M}}$

and the new rms speed,
$$c_1 = \sqrt{\frac{3R(T/2)}{(2M)}} = \frac{1}{2}\sqrt{\frac{3RT}{M}}$$
$$= \frac{c}{2} = \frac{300}{2} = 150 \text{ ms}^{-1}$$

4. As, $p = \dfrac{n_1RT + n_2RT + n_3RT}{V} = (n_1 + n_2 + n_3)\dfrac{RT}{V}$
$$= \left(\frac{8}{16} + \frac{14}{28} + \frac{22}{44}\right) \times \frac{0.082 \times 300}{10} = 3.69 \text{ atm} \approx 3.7\text{atm}$$

5. As, $c = \sqrt{\dfrac{3RT}{M}}$

So, $\dfrac{c_H}{c_O} = \sqrt{\dfrac{M_O}{M_H}} = \sqrt{\dfrac{16}{1}} = 4$

6. For a closed system, the total number of moles remains constant.

So, $p_1V = n_1RT_1 \Rightarrow \dfrac{n_1R}{V} = \dfrac{p_1}{T_1}$

and $p_2V = n_2RT_2 \Rightarrow \dfrac{n_2R}{V} = \dfrac{p_2}{T_2}$

$\therefore\ p(2V) = (n_1 + n_2)RT$

where, V is the volume of each vessel. When the vessels are joined.
$$\therefore \quad \frac{p}{T} = \frac{(n_1 + n_2)}{2V}R = \frac{1}{2}\left(\frac{p_1}{T_1} + \frac{p_2}{T_2}\right) = \frac{1}{2}\left(\frac{p_1T_2 + p_2T_1}{T_1T_2}\right)$$

7. $pV = (n_1 + n_2 + n_3)RT$
$$= \left(\frac{16}{32} + \frac{28}{28} + \frac{44}{44}\right)RT$$
$$= \left(\frac{1}{2} + 1 + 1\right)RT$$
$$\Rightarrow \quad pV = \frac{5}{2}RT$$
$$\Rightarrow \quad p = \frac{5}{2} \cdot \frac{RT}{V}$$

8. rms velocity does not change with pressure, till temperature remains constant.

9. As, $c = \sqrt{\dfrac{3kT}{m}} = \sqrt{\dfrac{3 \times 1.38 \times 10^{-23} \times 273}{5 \times 10^{-17}}}$
$$= 15 \times 10^{-3} \text{ ms}^{-1} = 1.5 \text{ cms}^{-1}$$

10. As, $c_s = \sqrt{\dfrac{\gamma p}{\rho}}$

and $c = \sqrt{\dfrac{3pV}{M}} = \sqrt{\dfrac{3p}{\rho}}$
$$\frac{c_s}{c} = \sqrt{\frac{\gamma}{3}} \quad \text{or} \quad c_s = c\sqrt{\frac{\gamma}{3}}$$

11. In a diatomic molecule, the rotational energy at a given temperature obeys Maxwell's energy distribution law. As each such atom has three translational degrees of freedom and two rotational degrees of freedom, therefore, at a given temperature, rotational energy is $\frac{2}{3}$rd the translational KE of each molecule. Energy associated with each molecule per degree of freedom is fixed $\left(= \frac{1}{2}k_B T\right)$.

12. As face $ABCD$ has positive charge on it and the gas consists of ionized hydrogen, therefore, isotropy is lost, The usual expression for pressure on the basis of kinetic theory will not be valid as ions would also experience forces, other than the forces due to collisions with the walls of the container.

13. $$v_{rms} = \sqrt{\frac{3RT}{M}}$$

$$v_{avg} = \sqrt{\frac{8}{\pi} \cdot \frac{RT}{M}}$$

$$\therefore \quad \frac{v_{rms}}{v_{avg}} = \sqrt{\frac{3\pi}{8}}$$

14. As, $p = \frac{1}{3}\frac{M}{V}c^2$

or $$c^2 = \frac{3\,pV}{M} = \frac{3\,RT}{M}$$

For gas A, $$V_1^2 = \frac{3 \times RT}{M} = \frac{3\,RT}{M}$$

For gas B, $$V_2^2 = \frac{3\,RT}{M};$$

So, $$\frac{V_1^2}{V_2^2} = 1 \Rightarrow \frac{V_1}{V_2} = 1$$

15. Momentum imparted to the surface in one collision, $\Delta p = (p_i - p_f) = mv - (-mv) = 2mv$ ...(i)

Force on the surface due to n collision per second,

$$F = \frac{n}{t}(\Delta p) = n\Delta p \qquad (\because t = 1\,\text{s})$$

$$= 2\,mnv \qquad \text{[from Eq. (i)]}$$

So, pressure on the surface, $p = \frac{F}{A} = \frac{2mnv}{A}$

Here, $m = 10^{-26}$ kg, $n = 10^{22}\ \text{s}^{-1}$,

$v = 10^4\ \text{ms}^{-1}$, $A = 1\ \text{m}^2$

$$\therefore \text{Pressure, } p = \frac{2 \times 10^{-26} \times 10^{22} \times 10^4}{1} = 2\ \text{N/m}^2$$

So, pressure exerted is of order of 10^0.

16. Given, in constant pressure process,

$$\Delta Q = 160\ \text{cal}$$
$$\Delta T = 50^\circ\text{C}$$

So, $$\Delta Q = nC_p\Delta T$$

$$\Rightarrow \quad 160 = nC_p(50) \qquad \text{...(i)}$$

Also, in constant volume process,

$$\Delta Q = 240\ \text{cal}$$

$$\Delta T = 100^\circ\text{C} \Rightarrow \Delta Q = nC_V\Delta T$$

$$\Rightarrow \quad 240 = nC_V(100) \qquad \text{...(ii)}$$

Dividing Eq. (i) by Eq. (ii), we have

$$\frac{160}{240} = \frac{nC_p(50)}{nC_V(100)}$$

$$\Rightarrow \quad \frac{C_p}{C_V} = \frac{2 \times 160}{240} = \frac{4}{3}$$

Now, $\frac{C_p}{C_V} = \gamma$ and degrees of freedom of a gas,

$$f = \frac{2}{\gamma - 1}$$

$$\Rightarrow \quad f = \frac{2}{\gamma - 1} = \frac{2}{\frac{4}{3} - 1} \Rightarrow f = 6$$

17. A diatomic gas molecule has 5 degrees of freedom, *i.e.* 3 translational and 2 rotational, at low temperature ranges (~ 250 K to 750 K). At temperatures above 750 K, molecular vibrations occurs and this causes two extra degrees of freedom.

Now, in given case,

For gas A, $C_p = 29$, $C_V = 22$

For gas B, $C_p = 30$, $C_V = 21$

By using $\gamma = \frac{C_p}{C_V} = 1 + \frac{2}{f}$, we have

For gas A,

$$1 + \frac{2}{f} = \frac{29}{22} \approx 1.3 \Rightarrow f = 6.67 \approx 7$$

So, gas A has vibrational mode of degree of freedom.

For gas B,

$$1 + \frac{2}{f} = \frac{30}{21} \approx 1.4 \Rightarrow f = 5$$

Hence, gas B does not have any vibrational mode of degree of freedom.

18. Given, volume, $V = 25 \times 10^{-3}\ \text{m}^3$

$$N = 1 \text{ mole of } O_2 = 6.023 \times 10^{23} \text{ atoms of } O_2,$$

$$T = 300\ \text{K}$$

Root mean square velocity of a gas molecule of O_2, $v_{rms} = 200$ m/s

Radius, $r = \frac{0.3}{2}\ \text{nm} = \frac{0.3}{2} \times 10^{-9}\ \text{m}$

Now, average time, $\frac{1}{\tau} = \frac{v_{av}}{\lambda}$

where, $$\lambda = \frac{RT}{\sqrt{2}\,N\pi r^2 p}$$

As, $$p = \frac{RT}{V}$$

$$\Rightarrow \quad \lambda = \frac{V}{\sqrt{2}\,N\pi r^2}$$

$\therefore$ Average collision per second,

$$\frac{1}{\tau} = \frac{v_{av}}{\lambda} = \frac{\sqrt{\frac{8}{3\pi}} \times v_{rms}}{\lambda}$$

$$= \frac{200 \times \sqrt{2} \times 6.023 \times 10^{23} \times \pi \times \frac{0.09}{4} \times 10^{-18}}{25 \times 10^{-3}}$$

$$\Rightarrow \quad \frac{1}{\tau} = 4.4 \times 10^8 \text{ per second} \approx 10^8$$

19. For a gas, value of specific heat at constant volume,

$$C_V = \frac{1}{2}fR$$

where, f = degrees of freedom

and R = gas constant.

For a diatomic gas molecule, degrees of freedom is in general 5 but when vibration occurs, then degrees of freedom increases to 7.

So, for gas A, specific heat is $C_V^A = \frac{1}{2} 5R$

and for gas B, specific heat is $C_V^B = \frac{1}{2} 7R$

$\therefore$ Ratio of specific heats of gas A and B is

$$\frac{C_V^A}{C_V^B} = \frac{5}{7}$$

20. Given, $T_1 = 27° \text{ C} = 300 \text{ K}$,

$$k_1 = 6.21 \times 10^{-21} \text{ J}$$

and $T_2 = 127° \text{ C} = 400 \text{ K}$.

We know that, $\frac{k_1}{k_2} = \frac{T_1}{T_2}$

$$k_2 = \frac{T_2}{T_1} \times k_1$$

$$= 6.21 \times 10^{-21} \times 400/300$$

$$= 8.28 \times 10^{-21} \text{ J}$$

21. As, $\left(\frac{c_{He}}{c_H}\right) = \sqrt{\frac{\rho_H}{\rho_{He}}} = \sqrt{\frac{1}{4}} = \frac{1}{2}$

$$(c_{He})_t = (c_{He})_0 \sqrt{\frac{T}{T_0}}$$

$$\therefore \quad \frac{(c_{He})_t}{(c_H)_0} = \frac{(c_{He})_0}{(c_H)_0} \sqrt{\frac{T}{T_0}} = \frac{5}{7}$$

or $T \approx 2T_0 = 2 \times 273 = 546° \text{ K}$

$= 273° \text{ C}$

22. According to law of equipartition of energy, average kinetic energy per molecule per degree of freedom at temperature T is $\frac{1}{2} kT$. The average kinetic energy per molecule of polyatomic gas molecule $= \frac{n}{2} kT$

(n = number of mole)

The average kinetic energy per mol of polyatomic gas

$$E = \frac{n}{2} kT \times N = \frac{n}{2} RT$$

$$C_V = \frac{d}{dT}\left(\frac{n}{2} RT\right) = \frac{n}{2} R$$

23. As, $\gamma = 1 + \frac{2}{n}$ or $\frac{2}{n} = \gamma - 1$ or $n = \frac{2}{\gamma - 1}$

24. Molar specific heat of the mixture at constant volume is

$$C_V = \frac{n_1 C_{V_1} + n_2 C_{V_2}}{(n_1 + n_2)}$$

$$= \frac{2\left(\frac{3}{2} R\right) + 3\left(\frac{5}{2} R\right)}{2 + 3} = 2.1 \, R$$

25. Internal energy of 'n' moles of a gas with degree of freedom f ($= 3$ for an ideal gas), at temperature T is

$$E = \frac{f}{2} \cdot n \, RT = \frac{3}{2} nRT$$

For an ideal gas, internal energy, $E = \frac{3}{2} nRT$

$$= \frac{3}{2} p \cdot V \qquad (\because \text{ from } pV = nRT)$$

Substituting the given values, we get

$$= \frac{3}{2} \times 3 \times 10^6 \times 2 = 9 \times 10^6 \text{ J}$$

26. $\lambda = \frac{RT}{\sqrt{2}\pi d^2 N_A P}$

$\lambda = 102 \text{ nm}$

27. Here, $n = 6$,

$$C_p = \left(1 + \frac{n}{2}\right) R = \left(1 + \frac{6}{2}\right) R = 4R$$

28. At NTP, $T = 273 \text{ K}$, $p = 1.01 \times 10^5 \text{ Nm}^{-2}$

Here, $d = 2.4 \times 10^{-10} \text{ m}$

$$\lambda = \frac{kT}{\sqrt{2}\,\pi d^2 p}$$

$$= \frac{(1.38 \times 10^{-23}) \times 273}{1.414 \times 3.14 \times (2.4 \times 10^{-10})^2 \times 1.01 \times 10^5}$$

$$= 1.46 \times 10^{-7} \text{ m}$$

29. $\because \gamma = \frac{C_p}{C_V} = 1 + \frac{2}{f}$

where, f = degree of freedom.

Now,

(A) for monatomic gas, $f = 3$

so, $\gamma = 1 + \frac{2}{3} = \frac{5}{3}$

(B) for diatomic rigid gas, $f = 5$

so, $\gamma = 1 + \frac{2}{5} = \frac{7}{5}$

(C) for diatomic non-rigid gas, $f = 7$

so, $\gamma = 1 + \frac{2}{7} = \frac{9}{7}$

(D) for triatomic rigid gas, $f = 6$

so, $\gamma = 1 + \frac{2}{6} = \frac{8}{6} = \frac{4}{3}$

Therefore,

A-IV, B-I, C-II, D-III.

Hence, correct option is (a).

30. Mean free path, $\lambda = \frac{V}{\sqrt{2}\pi d^2 N} = \frac{kT}{\sqrt{2}\pi d^2 p}$

where, V = volume of conductor,
N = number of molecules,
d = diameter of molecule,
k = Boltzmann's constant,
T = temperature
and p = pressure.

But from gas equation, $\frac{p}{T} = \frac{nR}{V}$

For a constant n and V, $\frac{p}{T}$ = constant.

So, mean free path does not depends on $\frac{p}{T}$ or p and T.

So, in given condition (closed container), mean free path remains constant or unchanged.

$\therefore$ Statement A is incorrect, but C is correct.

Also, mean time between two collisions is

$$\tau = \frac{\text{mean free path}}{v_{\text{avg}}} = \frac{\lambda}{v_{\text{avg}}}$$

As, $\quad v_{\text{avg}} \propto \sqrt{T}$

$\Rightarrow$ Average time, $\tau \propto \frac{1}{\sqrt{T}}$

As, the temperature is increasing, so mean collision time (τ) decreases.

$\therefore$ Statement B is correct.

Hence, option (a) is correct.

31. Mean free time (τ) for a gas molecule is time elapsed between two successive collisions.

$$\therefore \quad \tau = \frac{\text{Mean free path length } (\lambda)}{\text{Average speed } (v_{\text{av}})} = \frac{\left(\frac{1}{\sqrt{2} \cdot \pi d^2 \cdot n}\right)}{\left(\sqrt{\frac{8}{\pi} \frac{R}{M} \cdot T}\right)}$$

So, for an ideal gas, $\tau \propto \frac{1}{\sqrt{T}}$

Thus, graph of τ *versus* $\frac{1}{\sqrt{T}}$ is a straight line as shown below

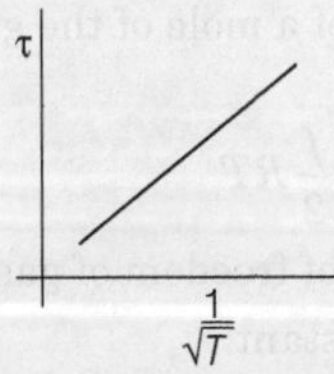

32. Mean free path (λ) of a gas molecule and mean relaxation time (or collision time) τ are related as

$$\tau = \frac{\lambda}{v_{\text{mean}}} = \frac{\frac{1}{\sqrt{2}} \frac{k_B T}{\pi d^2 p}}{\sqrt{\frac{8}{\pi} \cdot \frac{k_B}{m} \cdot T}} = \frac{1}{4d^2}\left(\frac{k_B m}{\pi}\right)^{1/2} \cdot \left(\frac{T^{1/2}}{p}\right)$$

Using $pV = nRT$, we have

$$\tau = \frac{1}{4d^2} \cdot \left(\frac{k_B m}{\pi}\right)^{1/2} \cdot \frac{VT^{1/2}}{nRT}$$

$$= \left\{\frac{1}{4d^2}\left(\frac{k_B m}{\pi}\right)^{1/2} \frac{1}{nR}\right\} \frac{V}{\sqrt{T}}$$

$\Rightarrow \tau \propto \frac{V}{\sqrt{T}}$, as other quantities remains constant during the process.

So, $\quad \frac{\tau_1}{\tau_2} = \frac{V_1}{V_2} \cdot \left(\frac{T_2}{T_1}\right)^{1/2} \quad \ldots \text{(i)}$

As the process is adiabatic,

$$T_1 V_1^{\gamma - 1} = T_2 V_2^{\gamma - 1}$$

$$\Rightarrow \quad \frac{T_2}{T_1} = \left(\frac{V_1}{V_2}\right)^{\gamma - 1}$$

So, Eq. (i) becomes

$$\frac{\tau_1}{\tau_2} = \frac{V_1}{V_2} \cdot \left(\frac{V_1}{V_2}\right)^{\frac{\gamma - 1}{2}}$$

$$\Rightarrow \quad \frac{\tau_1}{\tau_2} = \left(\frac{V_1}{V_2}\right)^{\frac{\gamma + 1}{2}}$$

Now, given $V_2 = 2V_1$, so

$$\frac{\tau_1}{\tau_2} = \left(\frac{1}{2}\right)^{\frac{\gamma + 1}{2}}$$

33. Mean free path for a gas,

$$\lambda = \frac{1}{\sqrt{2} \pi n d^2}$$

where, $\quad n$ = number density

and $\quad d$ = diameter of gas atom.

Mean speed of a gas atom,

$$v = \sqrt{\frac{8}{\pi} \frac{kT}{m}}$$

where, $\quad k$ = Boltzmann's constant,

T = temperature of sample

and $\quad m$ = atomic mass.

Hence, mean relaxation time,

$$T = \frac{\lambda}{v} = \frac{\left(\frac{1}{\sqrt{2} \pi n d^2}\right)}{\sqrt{\frac{8}{\pi} \frac{kT}{m}}}$$

$\tau \propto \frac{\sqrt{m}}{d^2}$, when T and n are same.

Hence, $\frac{\tau_{\text{Ar}}}{\tau_{\text{Xe}}} = \frac{\sqrt{m_{\text{Ar}}}}{\sqrt{m_{\text{Xe}}}} \cdot \frac{d_{\text{Xe}}^2}{d_{\text{Ar}}^2} = \frac{\sqrt{m_{\text{Ar}}}}{\sqrt{m_{\text{Xe}}}} \cdot \frac{r_{\text{Xe}}^2}{r_{\text{Ar}}^2}$

Here, $m_{\text{Ar}} = 40$, $m_{\text{Xe}} = 140$,

$r_{\text{Ar}} = 0.07$ nm, $r_{\text{Xe}} = 0.1$ nm

$$\therefore \quad \frac{\tau_{\text{Ar}}}{\tau_{\text{Xe}}} = \sqrt{\frac{40}{140}} \times \left(\frac{0.1}{0.07}\right)^2 = 1.09$$

34. When n_1 moles of an ideal gas having specific heats C_{p_1} and C_{V_1} is mixed with n_2 moles of another ideal gas with specific heats C_{p_2} and C_{V_2}, specific heats of mixture are given by

$$C_{p(\text{mix})} = \frac{n_1 C_{p_1} + n_2 C_{p_2}}{n_1 + n_2}$$

and $\quad C_{V(\text{mix})} = \frac{n_1 C_{V_1} + n_2 C_{V_2}}{n_1 + n_2}$

So, ratio of specific heats of mixture will be

$$\frac{C_{p(\text{mix})}}{C_{V(\text{mix})}} = \frac{n_1 C_{p_1} + n_2 C_{p_2}}{n_1 C_{V_1} + n_2 C_{V_2}} \quad \ldots(\text{i})$$

Now, using $C_p = \left(\frac{\gamma}{\gamma - 1}\right) R$ and $C_V = \frac{R}{\gamma - 1}$

Note As, $\frac{C_p}{C_V} = \gamma$ and $C_p - C_V = R$

So, $C_p - \frac{C_p}{\gamma} = R$ or $C_p = \frac{\gamma}{\gamma - 1}$ and $C_V = \frac{R}{\gamma - 1}$]

We can write Eq. (i) as

$$\frac{C_{p(\text{mix})}}{C_{V(\text{mix})}} = \frac{n_1\left(\frac{\gamma_1}{\gamma_1 - 1}\right)R + n_2\left(\frac{\gamma_2}{\gamma_2 - 1}\right)R}{n_1\left(\frac{R}{\gamma_1 - 1}\right) + n_2\left(\frac{R}{\gamma_2 - 1}\right)}$$

$$= \frac{n_1\gamma_1(\gamma_2 - 1) + n_2\gamma_2(\gamma_1 - 1)}{n_1(\gamma_2 - 1) + n_2(\gamma_1 - 1)} \quad \ldots(\text{ii})$$

Here, we are given

$$n_1 = 2, \gamma_1 = \frac{5}{3}, n_2 = 3 \text{ and } \gamma_2 = \frac{4}{3}$$

Substituting the above values in Eq.(ii), we have

$$\frac{C_{p(\text{mix})}}{C_{V(\text{mix})}} = \frac{2 \times \frac{5}{3}\left(\frac{4}{3} - 1\right) + 3 \times \frac{4}{3}\left(\frac{5}{3} - 1\right)}{2 \times \left(\frac{4}{3} - 1\right) + 3 \times \left(\frac{5}{3} - 1\right)}$$

$$= \frac{\frac{10}{9} + \frac{24}{9}}{\frac{2}{3} + \frac{6}{3}}$$

$$= \frac{34 \times 3}{9 \times 8} = \frac{17}{12} = 1.42$$

35. Given, height, $h = 2$ cm

Kinetic energy, $U = 4 \times 10^{-14}$ erg

Density of mercury, $\rho = 13.6 \text{ g/cm}^3$

Volume, $V = 4 \text{ cm}^3$ $g = 980 \text{ cm/s}^2$

According to gas equation,

$$pV = NKT \Rightarrow N = \frac{pV}{KT}$$

For monoatomic gas,

Energy, $U = \frac{3}{2}KT \Rightarrow KT = \frac{2}{3}U$

Now, $N = \frac{3pV}{2U}$

So, $N = \frac{3g\rho hV}{2U}$ $\quad (\because \text{ pressure, } p = \rho gh) \ldots(\text{i})$

Substituting the given values in eq. (i), we get

$$N = \frac{3 \times 980 \times 13.6 \times 2 \times 4}{2 \times 4 \times 10^{-14}}$$

$$= 3.99 \times 10^{18}$$

$$\approx 4 \times 10^{18}$$

36. For a mixture of two gases, ratio of specific heats is given by

$$\gamma_{\text{mixture}} = \frac{C_{p_{\text{mixture}}}}{C_{V_{\text{mixture}}}}$$

$$= \frac{n_1 C_{p_1} + n_2 C_{p_2}}{n_1 C_{V_1} + n_2 C_{V_2}} \quad \ldots(\text{i})$$

Here, gas 1 is helium which is monoatomic, so

$$n_1 = n, C_{p_1} = \frac{5}{2}R, C_{V_1} = \frac{3}{2}R$$

and gas 2 is oxygen which is diatomic, so

$$n_2 = 2n, C_{p_2} = \frac{7}{2}R, C_{V_2} = \frac{5}{2}R$$

Hence, from Eq. (i), we get

$$\gamma_{\text{mix}} = \frac{n \times \frac{5R}{2} + 2n \times \frac{7R}{2}}{n \times \frac{3R}{2} + 2n \times \frac{5R}{2}} = \frac{19}{13}$$

37. According to the law of equipartition of energy,

$\frac{1}{2}mv_{\text{rms}}^2 = \frac{n}{2}k_BT$ where, n is the degree of freedom.

Since, HCl is a diatomic molecule that has rotational, translational and vibrational motion.

So, $n = 7$

$$\Rightarrow \frac{1}{2}mv_{\text{rms}}^2 = \frac{7}{2}k_BT$$

Here, $v_{\text{rms}}^2 = \bar{v}$

$$\Rightarrow T = \frac{m\bar{v}^2}{7k_B}$$

38. Internal energy U of a mole of the gas at temperature T is given by

$$U = \frac{f}{2}RT$$

where, f = degree of freedom of gas molecule

and R = gas constant.

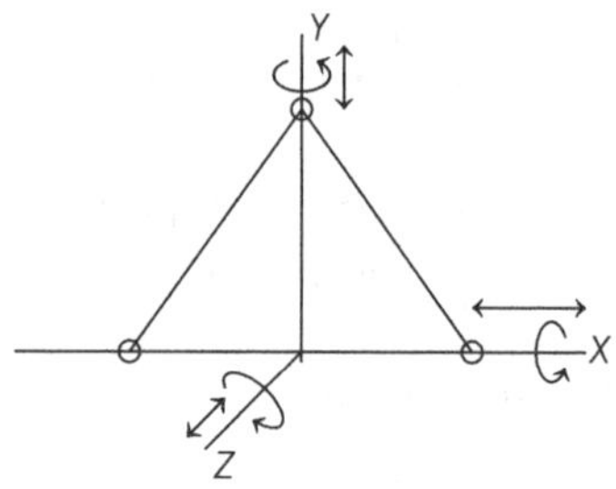

A triatomic triangular gas molecules can have three-translational motions (translation along X, Y or Z-axis) and three-rotational motions (rotation along X, Y or Z-axis). So, it has total 6 degrees of freedom.

So, internal energy, $U = \frac{f}{2}RT = \frac{6}{2}RT = 3RT$

Round II

1. $\frac{F}{2} n_1 k T_1 + \frac{F}{2} n_2 k T_2 + \frac{F}{2} n_3 k T_3 = \frac{F}{2} (n_1 + n_2 + n_3) kT$

[$\because F$ is degree of freedom]

Temperature of mixture, $T = \frac{n_1 T_1 + n_2 T_2 + n_3 T_3}{n_1 + n_2 + n_3}$

2. From $pV = nRT = \frac{N}{N_A} RT$, we have

$$n_f - n_i = \frac{pVN_A}{RT_f} - \frac{pVN_A}{RT_i}$$

$$\Rightarrow \quad n_f - n_i = \frac{10^5 \times 30}{8.3} \times 6.02 \times 10^{23} \cdot \left(\frac{1}{300} - \frac{1}{290}\right)$$

$$= -2.5 \times 10^{25}$$

$\therefore \quad \Delta n = -2.5 \times 10^{25}$

3. As no work is done and system is thermally insulated from surrounding, it means sum of internal energy of gas in two partitions is constant, *i.e.* $U = U_1 + U_2$.

Assuming both gases have same degree of freedom, then

Internal energy, $U = \frac{f(n_1 + n_2) RT}{2}$

and $\quad U_1 = \frac{f n_1 R T_1}{2}, \; U_2 = \frac{f n_2 R T_2}{2}$

Solving, we get

Common temperature, $T = \frac{(p_1 V_1 + p_2 V_2) T_1 T_2}{p_1 V_1 T_2 + p_2 V_2 T_1}$

4. For carbon dioxide, number of mole $(n_1) = \frac{22}{44} = \frac{1}{2}$;

Molar specific heat of CO_2 at constant volume, $C_{V_1} = 3R$

For oxygen, number of moles $(n_2) = \frac{16}{32} = \frac{1}{2}$;

molar specific heat of O_2 at constant volume, $C_{V_2} = \frac{5R}{2}$.

Let T K be the temperature of mixture.

Heat lost by O_2 = Heat gained by CO_2.

$$n_2 C_{V_2} \Delta T_2 = n_1 C_{V_1} \Delta T_1$$

$$\frac{1}{2}\left(\frac{5}{2} R\right)(310 - T) = \frac{1}{2} \times (3R)(T - 300)$$

or $\quad 1550 - 5T = 6T - 1800$ or $T = 304.54 \text{ K} = 31.5° \text{C}$

5. For an ideal gas undergoing an adiabatic process at room temperature,

pV^{γ} = constant or $TV^{\gamma - 1}$ = constant

For a diatomic gas, degree of freedom, $f = 5$

$$\therefore \quad \gamma = 1 + 2/f = 1 + \frac{2}{5} = \frac{7}{5}$$

As for adiabatic process, $TV^{\gamma - 1}$ = constant ...(i)

and it is given that, here TV^{x} = constant ...(ii)

Comparing Eqs. (i) and (ii), we get

$$\gamma - 1 = x \Rightarrow \frac{7}{5} - 1 = x \text{ or } x = 2/5$$

6. Here, $p = 2 \text{ atm} = 2 \times 1.013 \times 105 \text{ Nm}^{-2}$

$$T = 17 + 273 = 290 \text{ K}$$

$$\sigma = 2r = 2 \times 1 \text{ Å} = 2 \times 10^{-10} \text{ m}$$

$$\lambda = \frac{kT}{\sqrt{2}\, m\sigma^2 p}$$

$$= \frac{(1.38 \times 10^{-23}) \times 290}{1.414 \times 3.14 \times (2 \times 10^{-10})^2 \times 2.026 \times 10^5}$$

$$= 1.11 \times 10^{-7} \text{ m}$$

7. Since, each vibrational mode has 2 degrees of freedom, hence total vibrational degrees of freedom $= 24 \times 2 = 48$

$\therefore$ Total number of degree of freedom

$$= 3 + 3 + 48 = 54$$

$$\therefore \quad \gamma = 1 + \frac{2}{f}$$

$$= 1 + \frac{2}{54}$$

$$= \frac{28}{27} = 1.03$$

8. When the piston is in equilibrium, the pressure is same on both the sides of the piston. It is given that temperature and weight of gas on the two sides of piston does not change. From ideal gas equation, $pV = nRT$, we have $V \propto$ mass of the gas.

So, $\quad \frac{V_1}{V_2} = \frac{m_1}{m_2}$

or $\quad \frac{V_1}{V_2} + 1 = \frac{m_1}{m_2} + 1$

or $\quad \frac{V_1 + V_2}{V_2} = \frac{m_1 + m_2}{m_2}$

or $\quad \frac{V_2}{V_1 + V_2} = \frac{m_2}{m_1 + m_2}$

$$= \frac{2m}{m + 2m} = \frac{2}{3}$$

9. Work done by gas during heating process at constant pressure is given by

$$\Delta W = p\Delta V$$

Using ideal gas equation,

$$pV = nRT$$

$\Rightarrow \quad p\Delta V = nR\Delta T$

So, $\quad \Delta W = nR\Delta T$ ... (i)

Now, it is given that, $n = \frac{1}{2}$ and

$$\Delta T = 90° \text{C} - 20° \text{C}$$

$$= 363 \text{ K} - 293 \text{ K} = 70 \text{ K}$$

and R (gas constant) = 8.31 J/mol-K

Substituting these values in Eq. (i), we get

$$\Delta W = \frac{1}{2} \times 8.31 \times 70 = 290.85 \text{ J}$$

$$\Delta W \simeq 291 \text{ J}$$

10. Thermal energy corresponds to internal energy

$$\text{Mass} = 1 \text{ kg}$$

$$\text{Density} = 4 \text{ kgm}^{-3}$$

$\Rightarrow$ $$\text{Volume} = \frac{\text{Mass}}{\text{Density}} = \frac{1}{4} \text{ m}^3$$

$$\text{Pressure} = 8 \times 10^4 \text{ Nm}^{-2}$$

As internal energy $= \frac{f}{2} nRT = \frac{f}{2} pV$

Degree of freedom for diatomic = 5

$\therefore$ Internal energy $= \frac{5}{2} p \times V$

Putting the values of $p = 8 \times 10^4 \text{ Nm}^{-2}$ and $V = \frac{1}{4} \text{m}^3$, we get

Internal energy $= \frac{5}{2} \times 8 \times 10^4 \times \frac{1}{4} = 5 \times 10^4$ J

11. $U(x) = \frac{a}{x^{12}} - \frac{b}{x^6}$

$U(x = \infty) = 0$

As, $F = -\frac{dU}{dx} = -\left[\frac{12a}{x^{13}} + \frac{6\,b}{x^7}\right]$

At equilibrium, $F = 0$

$\therefore$ $$x^6 = \frac{2a}{b}$$

$\therefore$ $$U_{\text{at equilibrium}} = \frac{a}{\left(\frac{2a}{b}\right)^2} - \frac{b}{\left(\frac{2a}{b}\right)} = \frac{-b^2}{4a}$$

$\therefore$ $$D = [U(x = \infty) - U_{\text{at equilibrium}}] = \frac{b^2}{4a}$$

12. We know that, from kinetic theory of gases,

$$v_{\text{rms}} = \sqrt{\frac{3RT}{M}}$$

or $$v_{\text{rms}} \propto \sqrt{T}$$

where, R is gas constant, T is temperature and M is molecular mass of a gas.

Here, to double the v_{rms}, temperature must be 4 times of the initial temperature.

$\therefore$ New temperature, $T_2 = 4 \times T_1$

As, $T_1 = 27° \text{ C} = 300 \text{ K}$

$\Rightarrow$ $T_2 = 4 \times 300 = 1200 \text{ K}$

As, gas is kept inside a closed vessel.

$\therefore$ $$Q = nC_V \cdot dT$$

$$= \frac{15}{28} \times \frac{5R}{2} (1200 - 300)$$

$$\left[\text{As, } n = \frac{m}{M} = \frac{15}{28} \text{ for nitrogen}\right]$$

$$= \frac{15}{28} \times \frac{5}{2} \times 8.3 \times 900$$

[Given, $R = 8.3$ J/K-mol]

or $Q = 10004.4 \text{ J} \simeq 10 \text{ kJ}$

13. According to Boyle's law, $pV = k$ (a constant)

or $$p\frac{m}{\rho} = k \text{ or } \rho = \frac{pm}{k}$$

or $$\rho = \frac{p}{k} \quad \left(\text{where, } \frac{k}{m} = k = \text{constant}\right)$$

So, $$\rho_1 = \frac{p_1}{k}$$

and $$V_1 \frac{p_1}{k} = \frac{m_1}{\rho_1} = \frac{m_1}{p_1/k} = \frac{km_1}{p_1}$$

Similarly, $$V_2 = \frac{km_2}{p_2}$$

Total volume $= V_1 + V_2 = k\left(\frac{m_1}{p_1} + \frac{m_2}{p_1}\right)$

Let p be the common pressure and ρ be the common density of mixture. Then

$$\rho = \frac{m_1 + m_2}{V_1 + V_2} = \frac{m_1 + m_2}{k\left(\frac{m_1}{p_1} + \frac{m_2}{p_2}\right)}$$

$\therefore$ $$p = k\rho = \frac{m_1 + m_2}{\frac{m_1}{p_1} + \frac{m_2}{p_2}} = \frac{p_1 p_2 (m_1 + m_2)}{(m_1 p_2 + m_2 p_1)}$$

14. Mean time elapsed between two successive, collisions is $t = \frac{\lambda}{v}$

where, λ = mean free path length and

v = mean speed of gas molecule

$\therefore$ $$t = \frac{\left(\frac{k_B t}{\sqrt{2}\,\pi\, d^2 p}\right)}{\sqrt{\frac{8}{\pi} \cdot \frac{k_B T}{M}}} \Rightarrow t = \frac{C\sqrt{T}}{p}$$

where, $C = \frac{1}{4d^2}\sqrt{\frac{k_B M}{\pi}}$ = a constant for a gas.

So, $$\frac{t_2}{t_1} = \frac{\sqrt{T_2}}{\sqrt{T_1}} \cdot \left(\frac{p_1}{p_2}\right) \quad \ldots \text{(i)}$$

Here given, $\frac{p_1}{p_2} = \frac{1}{2}, \sqrt{\frac{T_2}{T_1}} = \sqrt{\frac{500}{300}} = \sqrt{\frac{5}{3}}$

and $t_1 = 6 \times 10^{-8}$ s

Substituting there values in Eq. (i), we get

$$t_2 = 6 \times 10^{-8} \times \sqrt{\frac{5}{3}} \times \frac{1}{2}$$

$$= 3.86 \times 10^{-8} \text{ s} \approx 4 \times 10^{-8} \text{ s}$$

15. Let molar specific heat of the mixture is C_V.

Total number of molecules in the mixture = 3 + 2 = 5

$\therefore$ C_V can be determined using

$$n\, C_V\, dT = n_1\, C_{V_1}\, dT + n_2\, C_{V_2} dT$$

or $(n_1 + n_2)\,(C_V)_{\text{mix}} = n_1 C_{V_1} + n_2\, C_{V_2}$

[here, $n = n_1 + n_2$]

Here, $C_{V_1} = \frac{3R}{2}$ (for helium); $n_1 = 2$

$C_{V_2} = \frac{5R}{2}$ (for hydrogen); $n_2 = 3$

[Because for monoatomic gases, $C_V = \frac{3}{2}R$ and for diatomic gases, $C_V = \frac{5}{2}R$]

$$\therefore \quad 5 \times C_V = \left(2 \times \frac{3R}{2}\right) + \left(3 \times \frac{5R}{2}\right) \Rightarrow 5C_V = \frac{21R}{2}$$

or $\quad C_V = \frac{21R}{10} = \frac{21 \times 8.3}{10} = \frac{174.3}{10}$

or $\quad C_V = 17.4$ J/mol-K

16. For an adiabatic process $TV^{\gamma-1}$ = constant.

We know that average time of collision between molecules

$$\tau = \frac{1}{n\pi\sqrt{2}\, v_{\text{rms}}\, d^2}$$

where, n = number of molecules per unit volume

v_{rms} = rms velocity of molecules

As $\quad n \propto \frac{1}{V}$ and $v_{\text{rms}} \propto \sqrt{T}$

$$\tau \propto \frac{V}{\sqrt{T}}$$

Thus, we can write $n = K_1 V^{-1}$ and $v_{\text{rms}} = K_2 T^{1/2}$ where, K_1 and K_2 are constants.

For adiabatic process, $TV^{\gamma-1}$ = constant. Thus, we can write

$$\tau \propto VT^{-1/2} \propto V\,(V^{1-\gamma})^{-1/2}$$

or $\quad \tau \propto V^{\frac{\gamma+1}{2}}$

17. For mixture, $C_V = \frac{n_1 C_{V_1} + n_2 C_{V_2}}{n_1 + n_2}$

For helium,

Number of moles, $n_1 = \frac{16}{4} = 4$ and $\gamma_1 = \frac{5}{3}$

For oxygen,

Number of moles, $n_2 = \frac{16}{32} = \frac{1}{2}$ and $\gamma_2 = \frac{7}{5}$

Then, $C_{V_1} = \frac{R}{\gamma_1 - 1} = \frac{R}{\frac{5}{3} - 1} = \frac{3}{2}R$

and $C_{V_2} = \frac{R}{\gamma_2 - 1} = \frac{R}{\frac{7}{5} - 1} = \frac{5}{2}R$

$\therefore$ For mixture,

$$C_V = \frac{4 \times \frac{3}{2}R + \frac{1}{2} \times \frac{5}{2}R}{4 + \frac{1}{2}} = \frac{6R + \frac{5}{4}R}{\frac{9}{2}}$$

$$= \frac{29R \times 2}{9 \times 4} = \frac{29R}{18}$$

Now, $\quad C_V = \frac{R}{\gamma - 1}$

$\Rightarrow \quad \gamma - 1 = \frac{R}{C_V}$

or $\quad \gamma = \frac{R}{C_V} + 1 = \frac{R}{\frac{29}{18}R} + 1$

$\Rightarrow \quad \frac{C_p}{C_V} = \frac{18}{29} + 1 = \frac{18 + 29}{29} = 1.62$

18. As piston is in equilibrium, so net force on piston is zero.

When the piston is stationary, *i.e.* on equilibrium as shown in the figure below

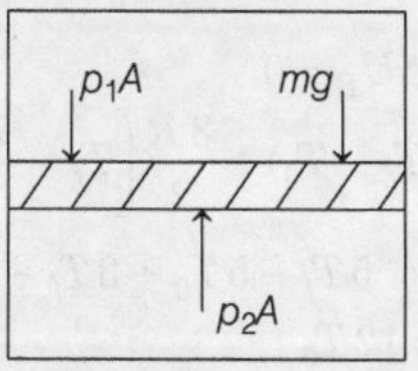

then $\quad p_1A + mg = p_2A$

$\Rightarrow \quad mg = p_2A - p_1A$

or $\quad mg = \left(\frac{nRTA}{V_2} - \frac{nRTA}{V_1}\right)$

[$\because pV = nRT$ (ideal gas equation)]

$$= nRT\left(\frac{A}{Al_2} - \frac{A}{Al_1}\right) = nRT\left(\frac{l_1 - l_2}{l_1 l_2}\right)$$

or $\quad m = \frac{nRT}{g}\left(\frac{l_1 - l_2}{l_1 l_2}\right)$

19. Mayer's formula is given by, $C_p - C_V = R$ and $\gamma = \frac{C_p}{C_V}$

Therefore, using above two relations, we find

$$C_V = \frac{R}{\gamma - 1}$$

For a mole of monatomic gas, $\gamma = \frac{5}{3}$

$$\therefore \quad C_V = \frac{R}{\left(\frac{5}{3}\right) - 1} = \frac{3}{2}R$$

For a mole of diatomic gas, $\gamma = \frac{7}{5}$

$$\therefore \quad C_V = \frac{R}{\left(\frac{7}{5}\right) - 1} = \frac{5}{2}R$$

When these two moles are mixed, then heat required to raise the temperature to 1°C is $C_V = \frac{3}{2}R + \frac{5}{2}R = 4R$

Hence, for one mole, heat required $= \frac{4R}{2} = 2R$

$\therefore \quad C_V = 2R$

$\Rightarrow \quad \frac{R}{\gamma - 1} = 2R$ or $\gamma = \frac{3}{2}$

20. Here, change in internal energy of the system is zero, *i.e.* increase in internal energy of one is equal to decrease in internal energy of other.

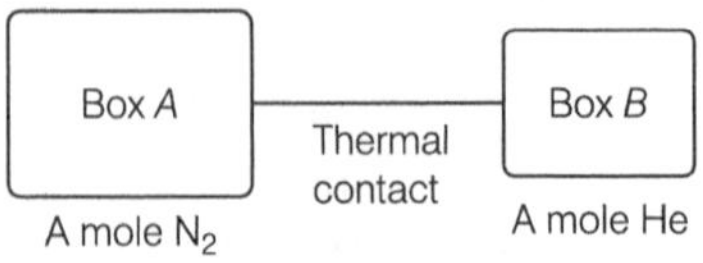

Change in internal energy in box A,

$$\Delta U_A = 1 \times \frac{5}{2} R (T_f - T_0)$$

Change in internal energy in box B

$$\Delta U_B = 1 \times \frac{3R}{2}\left(T_f - \frac{7}{3}T_0\right)$$

Now, $\Delta U_A + \Delta U_B = 0$

$$\Rightarrow \quad \frac{5R}{2}(T_f - T_0) + \frac{3R}{2}\left(T_f - \frac{7T_0}{3}\right) = 0$$

or $\quad 5T_f - 5T_0 + 3T_f - 7T_0 = 0$

or $\quad 8T_f = 12T_0$

or $\quad T_f = \frac{12}{8}T_0 = \frac{3}{2}T_0$

21. Number density of gas molecules,

$$n = \frac{\text{Number of molecules}}{\text{Volume of gas}}$$

$\Rightarrow$ Number of molecules, $N = n \times$ Volume of gas

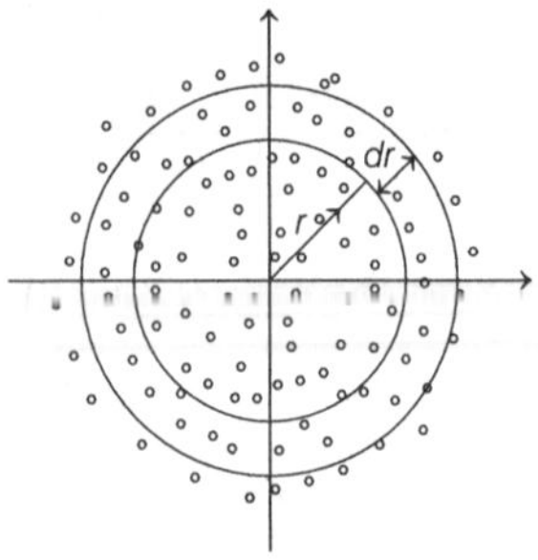

Now, consider a shell of radius r and thickness dr with centre at $r = 0$.

Volume of shell of differentiable thickness (dr), $dV =$ surface area $\times$ thickness $= 4\pi r^2\, dr$

Now, number of molecules in this shell is

$$dN = n(r)\cdot dV = n_0 e^{-\alpha r^4}\cdot 4\pi r^2\, dr$$

So, total number of molecules present in given volume (extending from $r = 0$ to $r = \infty$) is

$$N = \int_0^\infty n(r)\cdot dV = \int_0^\infty n_0\, e^{-\alpha r^4}\cdot 4\pi r^2\, dr$$

$$= \int_0^\infty 4\pi n_0\, e^{-\alpha r^4}\cdot r^2\, dr \qquad \ldots\text{(i)}$$

Here, we take $\alpha r^4 = t$

$$\Rightarrow \quad r = t^{\frac{1}{4}}\cdot \alpha^{\frac{-1}{4}} \Rightarrow 4\alpha r^3 dr = dt$$

$$\Rightarrow \quad r^2\, dr = \frac{dt}{4\alpha r} = \frac{dt}{4\alpha t^{\frac{1}{4}}\cdot \alpha^{-\frac{1}{4}}} = \frac{dt}{4\,\alpha^{3/4}\, t^{\frac{1}{4}}}$$

Also, when $r = 0$, $t = 0$ and when $r = \infty$, $t = \infty$ substituting in Eq. (i), we get

$$N = \int_0^\infty 4\pi\, n_0\, e^{-t}\cdot \frac{dt}{4\,\alpha^{\frac{3}{4}}\, t^{\frac{1}{4}}}$$

$$N = \pi\alpha^{-3/4}\cdot n_0 \cdot \int_0^\infty e^{-t}\cdot t^{-\frac{1}{4}}\, dt$$

As value of definite integral $\int_0^\infty e^{-t}\cdot t^{-\frac{1}{4}}\, dt$ is a constant ($= k$ let), we have

$$N = \pi\, k n_0 \alpha^{-3/4}$$

$$\Rightarrow \quad N \propto n_0 \alpha^{-3/4}$$

22. Here, $V = 25.0\ \text{m}^3$

$$T = 27 + 273 = 300\ \text{K}$$

$$k = 1.38 \times 10^{-23}\ \text{JK}^{-1}$$

Now, $pV = nRT = n(Nk)T$

$= (nN)kT = N'kT$

where, $nN = N' =$ total number of air molcules in the given gas

$$N' = \frac{pV}{kT} = \frac{(1.01 \times 10^5) \times 25}{(1.38 \times 10^{-23}) \times 300}$$

$$= 6.10 \times 10^{26} = 6.10 \times 10^{13 \times 2}$$

$\therefore \quad x = 2$

23. If T be the final equilibrium temperature, then applying conservation of internal energy,

$$U_1 + U_2 = U_1' + U_2'$$

$$n_1 C_V T_1 + n_2 C_V T_2 = (n_1 + n_2) C_V T$$

$$n_1 T_1 + n_2 T_2 = (n_1 + n_2)\, T$$

$$(0.1)(200) + (0.05)(400) = (0.1 + 0.05)\, T$$

$$20 + 20 = (0.15)\, T$$

$$T = \frac{40}{0.15} = 266.66\ \text{K} \simeq 266\ \text{K}$$

24. Here, $c_1 = 2\ \text{kms}^{-1}$; $c_2 = 4\ \text{kms}^{-1}$

$c_3 = 6\ \text{kms}^{-1}$ and $c_4 = 8\ \text{kms}^{-1}$

(i) Average speed,

$$c_{av} = \frac{c_1 + c_2 + c_3 + c_4}{4} = \frac{2 + 4 + 6 + 8}{4} = 5\ \text{kms}^{-1}$$

(ii) Root mean square speed,

$$c = \sqrt{\frac{c_1^2 + c_2^2 + c_3^2 + c_4^2}{4}} = \sqrt{\frac{2^2 + 4^2 + 6^2 + 8^2}{4}}$$

$$= 5.48\ \text{kms}^{-1}$$

25.

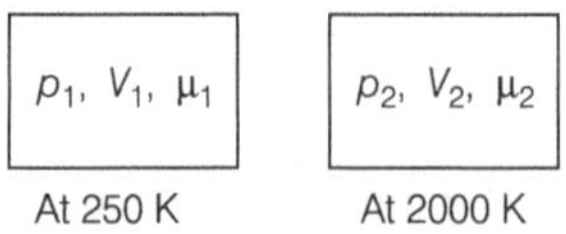

Let $\quad \mu_1 = \mu_0$, $V_2 = 2V_1$ (given)

Total number of moles at 2000 K = Number of moles of diatomic molecules + Number of moles of monoatomic

$$\mu_2 = 0.75\mu_0 + 2 \times 0.25\mu_0 = 1.25\,\mu_0$$

Ideal gas equation for situation 1,

$$p_1V_1 = \mu_0 RT_1$$

$$p_1V_1 = \mu_0 R(250) \quad \text{... (i)}$$

For situation 2,

$$p_2V_2 = (1.25\mu_0) R(2000)$$

$$p_2(2V_1) = (1.25\mu_0) R(2000) \quad \text{... (ii)}$$

Dividing Eq. (i) by Eq. (ii), we get

$$\frac{p_1V_1}{p_2(2V_1)} = \frac{\mu_0 R(250)}{(1.25\mu_0)R(2000)}$$

$$\frac{p_1}{p_2} = \frac{1}{5} \text{ or } \frac{p_2}{p_1} = 5$$

26. Here, $J = ?$

$$C_p = 3.409 \text{ calg}^{-1\circ}\text{C}^{-1}$$

$$C_V = 2.409 \text{ calg}^{-1\circ}\text{C}^{-1}, M = 2$$

$$R = 8.31 \text{ J mol}^{-1}\text{K}^{-1}$$

As, $C_p - C_V = \frac{r}{J} = \frac{R}{MJ}$

$$\therefore \quad 3.409 - 2.409 = \frac{8.31}{2J}$$

$$1 = \frac{4.155}{J}$$

$$J = 4.11 \text{ J cal}^{-1}$$

27. The rms speed is given by $v_{rms} = \sqrt{\frac{3RT}{M}}$

For nitrogen molecule, $(v_{rms})_{N_2} = \sqrt{\frac{3RT_1}{M_1}}$

and for hydrogen molecule, $(v_{rms})_{H_2} = \sqrt{\frac{3RT_2}{M_2}}$

According to question, $(v_{rms})_{N_2} = (v_{rms})_{H_2}$

$$\Rightarrow \sqrt{\frac{3RT_1}{M_1}} = \sqrt{\frac{3RT_2}{M_2}} \Rightarrow \frac{T_1}{M_1} = \frac{T_2}{M_2}$$

$$\Rightarrow \frac{T_1}{T_2} = \frac{M_1}{M_2} \Rightarrow T_2 = \frac{T_1M_2}{M_1} \quad \text{...(i)}$$

Given, $T_1 = 300^\circ \text{C} = 300 + 273 = 573 \text{ K}$

$M_1 = 28$ g and $M_2 = 2$ g

Substituting all these values in Eq. (i), we get

$$T_2 = 573 \times \frac{2}{28} = 40.93 \text{ K} \approx 41 \text{ K}$$

Hence, the required temperature is 41 K.

28. Here, mass of metal, $m = 0.20 \text{ kg} = 200 \text{ g}$

Fall in temperature of metal

$$\Delta T = 150^\circ \text{C} - 40^\circ \text{C} = 110^\circ\text{C}$$

If c is specific heat of the metal, then heat lost by the metal,

$$\Delta Q = mc\Delta T = 200\, s \times 110 \quad \text{...(i)}$$

Volume of water = 150 cc

Mass of water, $m' = 150$ g

Water equivalent of calorimeter

$$w = 0.025 \text{ kg} = 25\text{g}$$

Rise in temperature of water in calorimeter

$$\Delta T' = 40^\circ \text{C} - 27^\circ \text{C} = 13^\circ \text{C}$$

Heat gained by water and calorimeter

$$\Delta Q' = (m' + w)\,\Delta T' = (150 + 25) \times 13$$

$$\Delta Q' = 175 \times 13 \quad \text{...(ii)}$$

As $\Delta Q = \Delta Q'$

$\therefore$ From Eqs. (i) and (ii), we get

$$200 \times s \times 100 = 175 \times 13$$

$$s = \frac{175 \times 13}{200 \times 110} \approx 0.1 \text{ J/g}^\circ\text{C}$$

29. Here, volume of water heated = 3.0 Lmin^{-1}

Mass of water heated, $m = 3000 \text{ gmin}^{-1}$

Rise of temperature, $\Delta T = 77^\circ \text{C} - 27^\circ \text{C} = 50^\circ \text{C}$

Specific heat of water, $s = 4.2 \text{ Jg}^{-1}\,^\circ\text{C}^{-1}$

Amount of heat used, $\Delta Q = ms\Delta T = 3000 \times 4.2 \times 50$

$$= 63 \times 10^4 \text{ Jmin}^{-1}$$

Heat of combustion $= 4 \times 10^4 \text{ Jg}^{-1}$

Rate of combustion of fuel $= \frac{63 \times 10^4}{4 \times 10^4} = 15.75 \text{ gmin}^{-1}$

30. Given, at 27°C, root mean square speed,

$$(v_{rms})_1 = 100 \text{ ms}^{-1}$$

$$(v_{rms})_1 = \sqrt{\frac{3p_1}{d_1}} = \sqrt{\frac{3p_1V_1}{M}} \quad \text{...(i)}$$

According to ideal gas equation,

$$\Rightarrow \frac{V_1}{V_2} = \frac{p_2T_1}{p_1T_2} = \frac{2 \times 300}{400} = \frac{3}{2}$$

$\therefore$ At 127°C, root mean square speed

$$(v_{rms})_2 = \sqrt{\frac{3p_2}{d_2}} = \sqrt{\frac{3p_2V_2}{M}} \quad \text{...(ii)}$$

From Eqs. (i) and (ii), we get

$$\therefore \quad (v_{rms})_2^2 = (v_{rms})_1^2 \times \frac{V_2}{V_1} \times \frac{p_2}{p_1} = (100)^2 \times \frac{2}{3} \times 2$$

or $\quad (v_{rms})_2 = \frac{200}{\sqrt{3}} \text{ ms}^{-1}$

$$\therefore \quad n = 3$$

31. When the air bubble is at 40 m depth, then

$$V_1 = 1 \text{ cm}^3 = 1.0 \times 10^{-6} \text{ m}^3$$

$$T_1 = 12\,^\circ\text{C} = 12 + 273 = 285 \text{ K}$$

$$P_1 = 1\text{atm} + h_1 \rho g = 1.01 \times 10^5 + 40 \times 10^3 \times 9.8$$

$$= 493000 \text{ Pa}$$

When the air bubble reaches at the surface of lake, then, $V_2 = ?, T_2 = 35^\circ \text{C} = 35 + 273 = 308 \text{ K}$

$$P_2 = 1\text{atm} = 1.01 \times 10^5 \text{ Pa}$$

Now, $\frac{p_1V_1}{T_1} = \frac{P_2V_2}{T_2}$ or $V_2 = \frac{p_1V_1T_2}{T_2p_2}$

$$\therefore \quad V_2 = \frac{(493000) \times 1.0 \times 10^{-6} \times 308}{285 \times 1.01 \times 10^5}$$

$$= 5.275 \times 10^{-6} \text{ m}^3$$

15

Thermodynamics

The branch of science which deals with transformation of heat energy into other forms of energy and *vice-versa* is known as *thermodynamics*.

IN THIS CHAPTER ....

Thermodynamic Terms

In order to understand these transformation we need to understand the terms given below.

Thermodynamical System An assembly of an extremely large number of particles which is capable of exchange of energy with its surroundings is called thermodynamic system.

Thermodynamic Parameters The state of thermodynamic system can be described by specifying pressure, volume, temperature, internal energy and number of moles, etc. These are called thermodynamic parameters or coordinates or variables.

The state variables may be extensive (*e.g.* volume, total mass, internal energy) or intensive (*e.g.* pressure, temperature and density.) in nature.

Thermal Equilibrium A thermodynamical system is said to be in thermal equilibrium when macroscopic variables (like pressure, volume, temperature, mass, composition etc) that characterise the system do not change with time

Zeroth Law of Thermodynamics

According to this law, if two systems A and B are each in thermal equilibrium with a third system C, then A and B will also be in thermal equilibrium with each other.

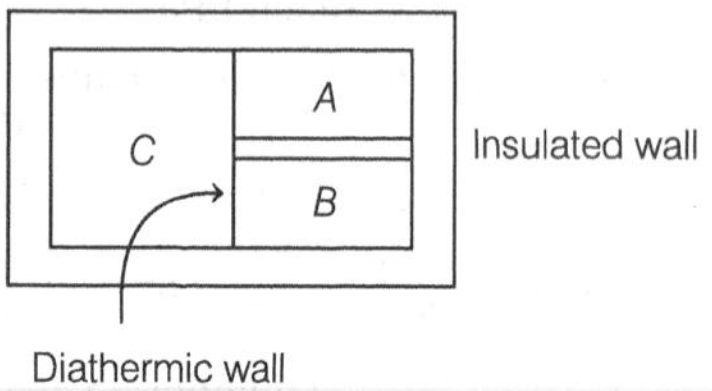

Therefore, there must be a certain scalar physical quantity which is identical for all systems in thermal equilibrium. This quantity (scalar) is the temperature.

Suppose for systems A, B and C are in thermal equilibrium, then $T_A = T_B = T_C$

So for a body, temperature is that physical quantity which decides the degree of hotness or coldness of a body and is responsible for heat flow.

Work

It is defined as the product of force and its displacement in the direction of force. Its unit is joule or N-m, *i.e.*

$$W = \int dW = F\,dx$$

Work Done by a Gas During Expansion

Consider an ideal gas is enclosed in a perfectly insulated cylinder fitted with a non-conducting and frictionless piston. If p is the pressure exerted by the gas, V be the volume of the gas at any particular instant and A be the area of cross-section of the piston, then

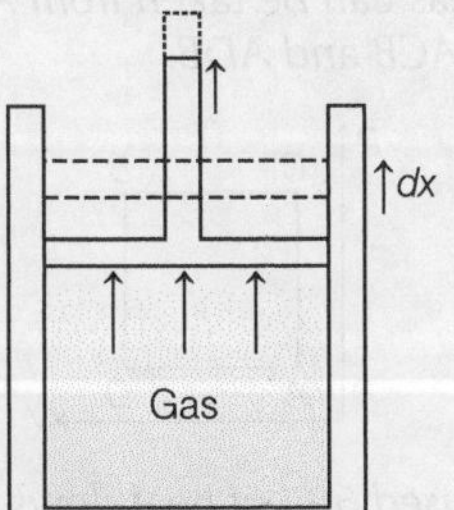

Work done, $dW = (pA)dx = pdV$

$(\because A\,dx = dV = \text{infinitesimal change in volume})$

During expansion of the gas, *work is done by the gas* and is taken as *positive* while during compression *work is done on the gas* and it is taken as *negative*.

Example 1. *When water is boiled under a pressure of 2 atm, the heat of vaporisation is 2.20×10^6 Jkg^{-1} and the boiling point is 120°C. At this pressure, 1 kg of water has a volume of 10^{-3} m^3 and 1 kg of steam has a volume of $0.824\,m^3$. What is the work done when 1 kg of steam is formed at this temperature?*

(a) 166.74 *kJ* (b) 266.74 *kJ* (c) 366.74 *kJ* (d) 466.74 *kJ*

Sol. (a) Work done = $p(\Delta V)$ (at constant pressure)

$\therefore \quad W = 2 \text{ atm} \times (0.824 - 0.001) \text{ m}^3$

$\Rightarrow \quad W = 2 \times 1.013 \times 10^5 \text{ Nm}^{-2} \times 0.823 \text{ m}^3 = 166.74 \text{ kJ}$

Internal Energy

Internal energy of a system is the energy possessed by the system due to molecular motion and molecular configuration. The energy due to molecular motion is called *internal kinetic energy* (U_K) and that due to molecular configuration is called *internal potential energy* (U_P).

i.e. $\quad U = U_K + U_P$

Important points related to internal energy

(i) Change in internal energy is path independent and depends only on the initial and final states of the system.

i.e., $\quad \Delta U = U_f - U_i$

(ii) Change in internal energy in a cyclic process is always zero as for cyclic process, $U_f = U_i$, so that

$$\Delta U = U_f - U_i = 0$$

(iii) In case of ideal gas, as there is no molecular attraction $U_P = 0$, *i.e.* so internal energy of an ideal gas is totally kinetic and is given by

$$U = U_K = \frac{3}{2}nRT$$

with $\quad \Delta U = \frac{3}{2}nR\,\Delta T$

where, n = number of moles

and $\quad R$ = gas constant.

(iv) In case of gases, whatever be the process,

$$\Delta U = nC_V\Delta T = \frac{nR\,\Delta T}{(\gamma - 1)}$$

(v) The change in internal energy (ΔU) of a system in case of gain is taken as positive, while it is taken as negative in case of loss of energy.

First Law of Thermodynamics

When a system changes for a given initial state to a given final state, both the work W and heat Q depend on the nature of the process.

The change in Q and W represent a change in same intrinsic property of the system. This property is the *internal energy* U and we write

$$\Delta U = U_f - U_i = Q - W \quad \text{...(i)}$$

This equation is the *first law of thermodynamics*.

It can be written as $\quad dQ = dU + dW \quad$ ...(ii)

Thus, we can say that, heat supplied to the system is the sum of external work done dW by the system and increase in its internal energy dU.

In Eq. (ii), dQ is change in heat. So, dQ will be positive when heat is given to the system and dQ will be negative when heat is given by the system.

Note First law of thermodynamics is a direct consequence of law of conservation of energy.

Example 2. *When a system goes from state A to state B, it is supplied with 400 J of heat and it does 100 J of work. For this translation, what will be the change in internal energy of the system?*

(a) 250 *J* (b) 300 *J*

(c) 350 *J* (d) 150 *J*

Sol. (b) From the first law of thermodynamics,

$$\Delta U_{AB} = Q_{AB} - W_{AB} = (400 - 100) \text{ J} = 300 \text{ J}$$

Example 3. *In the above example, if the system moves from B to A, what is the change in internal energy?*

(a) 300 J (b) -300 J
(c) 400 J (d) -400 J

Sol. (b) Consider a closed path that passes through the state A and B. Internal energy is a state function, so ΔU is zero for a closed path.

Thus, $\Delta U = \Delta U_{AB} + \Delta U_{BA} = 0$

or $\Delta U_{BA} = -\Delta U_{AB} = -300$ J

Example 4. *A certain amount of an ideal gas passes from state A to B first by means of process 1, then by means of process 2. In which of the process, is the amount of heat absorbed by the gas greater?*

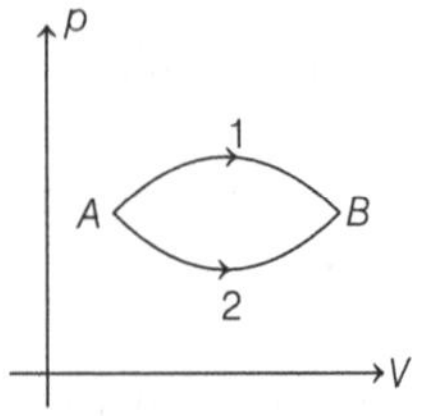

(a) *Process 1* (b) *Process 2*
(c) *Equal in both process* (d) *None of these*

Sol. (a) For process 1, $\Delta Q_1 = \Delta W_1 + \Delta U_1$

and For process 2, $\Delta Q_2 = \Delta W_2 + \Delta U_2$

U is a state function. Hence, ΔU depends only on the initial and final positions. Therefore,

$$\Delta U_1 = \Delta U_2$$

But $W_1 > W_2$

As the area under 1 is greater than area under 2. Hence, $Q_1 > Q_2$

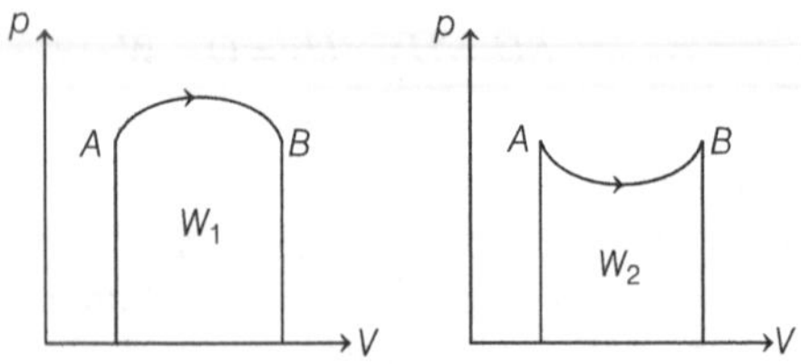

Example 5. *One mole of diatomic ideal gas undergoes a cyclic process ABC as shown in figure. The process BC is adiabatic. The temperatures at A, B and C are 400 K, 800 K and 600 K, respectively.*
Choose the correct statement. **[JEE Main 2014]**

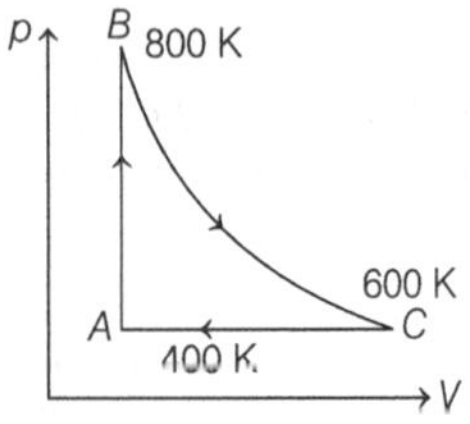

(a) The change in internal energy in whole cyclic process is 250 R.
(b) The change in internal energy in the process CA is 700 R.
(c) The change in internal energy in the process AB is -350 R.
(d) The change in internal energy in the process BC is $-500R$.

Sol. (d) According to first law of thermodynamics, we get

(i) Change in internal energy from A to B, i.e.

$$\Delta U_{AB} = nC_V (T_B - T_A) = 1 \times \frac{5R}{2} (800 - 400) = 1000\ R$$

(ii) Change in internal energy from B to C,

$$\Delta U_{BC} = nC_V (T_C - T_B) = 1 \times \frac{5R}{2} (600 - 800) = -500\ R$$

(iii) $\Delta U_{\text{total}} = 0$

(iv) Change in internal energy from C to A,

$$\Delta U_{CA} = nC_v (T_A - T_C) = 1 \times \frac{5R}{2} (400 - 600) = -500\ R$$

Hence, option (d) is correct.

Example 6. *A gas can be taken from A to B via two different processes ACB and ADB.*

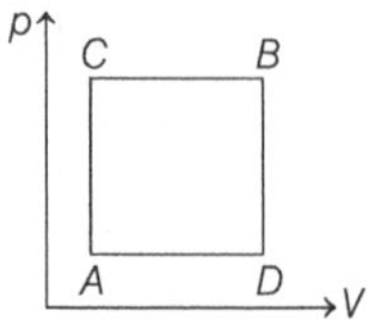

When path ACB is used 60 J of heat flows into the system and 30 J of work is done by the system. If path ADB is used, work done by the system is 10 J, the heat flowing into the system in path ADB is **[JEE Main 2019]**

(a) 80 J (b) 40 J (c) 100 J (d) 20 J

Sol. (b) For the ACB as shown in the figure below

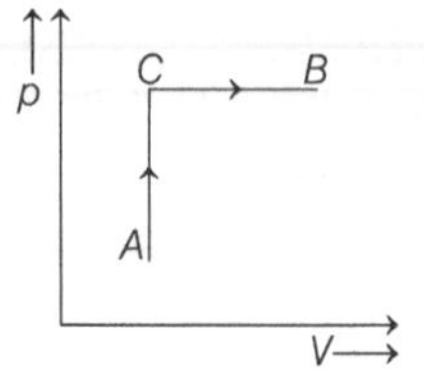

According to the first law of thermodynamics,

Heat supplied, ΔQ = Work done (ΔW) + Internal energy (ΔU)

$\Rightarrow \Delta Q_{CB} = \Delta W_{ACB} + (U_B - U_A)$ [where, $\Delta U = U_B - U_A$]

Substituting the given values, we get

$$U_B - U_A = 60 - 30 = 30\text{ J} \quad \text{...(i)}$$

Similarly for the ADB as shown in the figure below

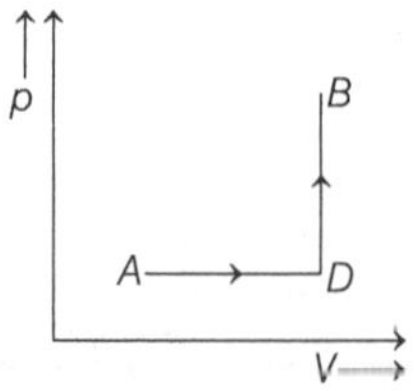

$$\Delta Q_{ADB} = \Delta W_{ADB} + (U_B - U_A)$$

$\Rightarrow \Delta Q_{ADB} = 10 + 30$ [using Eq. (i)]

$= 40$ J

Thermodynamic Process

A thermodynamical process is said to take place when some changes occur in the state of a thermodynamic system *i.e.* the thermodynamic parameters of the system change with time.

Reversible and Irreversible Processes

Reversible Process A process which could be reserved in such a way that the system and its surrounding returns exactly to their initial states with no other changes in the universe is known as reversible process.

Irreversible Process Any process which is not reversible exactly is an irreversible process.

Isothermal Process

In isothermal process, the temperature remains constant. Melting and boiling points are examples of isothermal process. Specific heat in isothermal process is infinity.

Work done in an isothermal process,

$$W = \int p\,dV = nRT\int_{V_1}^{V_2}\frac{dV}{V}$$

$$= nRT \log_e \frac{V_2}{V_1}$$

$$= 2.303\, nRT \log\frac{V_2}{V_1}$$

$$= 2.303\, nRT \log\frac{p_1}{p_2}$$

Isothermal elasticity = p (Bulk modulus)

The pressure *versus* volume curve for isothermal process is as follows

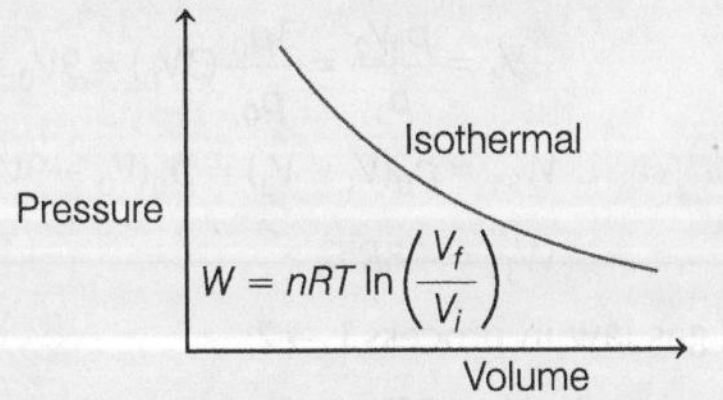

Since for an ideal gas, the internal energy is proportional to temperature, so it follows that there is no change in the internal energy of the gas during an isothermal process. The first law of thermodynamics then becomes

$$\Delta U = 0 = Q - W$$

$$\therefore \quad Q = W$$

All the heat added to the system is used to do work in an isothermal process.

Example 7. *How much work is done by ideal gas in expanding isothermally from an initial volume of 3 L at 20 atm to a final volume of 24 L?*

(a) 1.36×10^5 J (b) 1.26×10^4 J
(c) 1.36×10^4 J (d) 2.36×10^5 J

Sol. (b) In isothermal process at temperature T,

$$W = 2.303\, nRT \log_{10}\frac{V_2}{V_1}$$

$$W = 2.303(p_1V_1)\log_{10}\frac{V_2}{V_1} \quad [\text{using } p_1V_1 = nRT]$$

$$= 2.303(20 \times 3)\log_{10}\frac{24}{3} \text{ L-atm}$$

$$= [2.303 \times 60 \times \log_{10} 8] \times 10^{-3} \times 1.01 \times 10^5$$

$$= 1.26 \times 10^4 \text{ J}$$

Adiabatic Process

In an adiabatic process, heat is neither allowed to enter nor allowed to escape from the system. Specific heat in an adiabatic process is zero.

Since, $dQ = 0$

$\therefore \quad dU = -p\,dV$

So, let us read through the following graphs of p *versus* V that what happens when adiabatic expansion and compression happens.

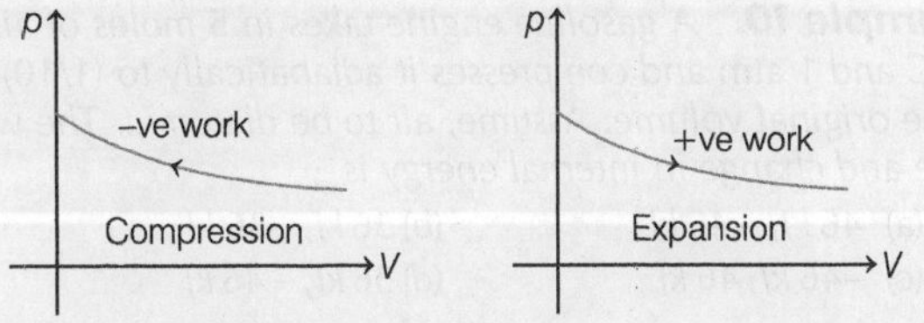

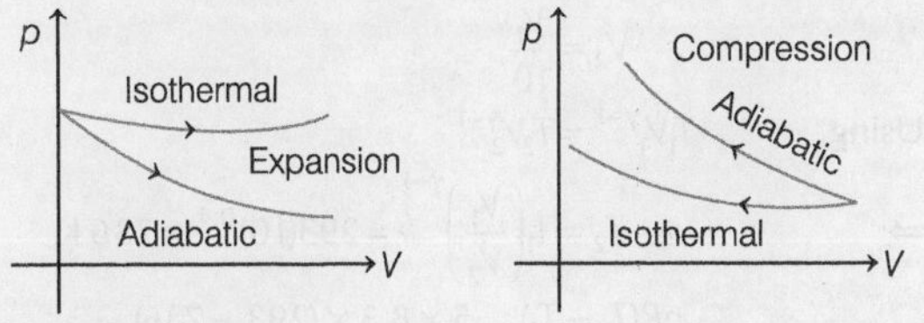

In an adiabatic process,

(a) pV^γ = constant (b) $p^{1-\gamma}T^\gamma$ = constant
(c) $TV^{\gamma-1}$ = constant

Work done in an adiabatic process,

$$W = \frac{p_1V_1 - p_2V_2}{\gamma - 1} = \frac{nR(T_1 - T_2)}{\gamma - 1}$$

where, $\gamma = \frac{C_p}{C_V}$

Adiabatic elasticity (bulk modulus) = γp.

Example 8. *A balloon filled with helium (32°C and 1.7 atm) bursts. Immediately afterwards, the expansion of helium can be considered as* **[JEE Main 2020]**

(a) irreversible, isothermal (b) irreversible, adiabatic
(c) reversible, adiabatic (d) reversible, isothermal

Sol. (b) Expansion after bursting of balloon is a fast process in which helium presses the atmosphere. This expansion occurs at the expense of internal energy of helium molecules.

So, this process is irreversible and adiabatic.

Example 9. *During an adiabatic expansion, the increase in volume is associated with which of the following possibilities w.r.t. pressure and temperature?*

Pressure	*Temperature*
(a) Increase	Increase
(b) Decrease	Decrease
(c) Increase	Decrease
(d) Decrease	Increase

Sol. (b) According to first law of thermodynamics,

$$\Delta Q = \Delta U + \Delta W$$

For an adiabatic process, $\Delta Q = 0$

$\therefore \quad \Delta U = -W$

In adiabatic process, $p \propto \dfrac{1}{V^{\gamma}}$

and $T \propto \dfrac{1}{V^{\gamma-1}}$

$\gamma > 1$, because volume increases.
Then, p and T will decrease.

Example 10. *A gasoline engine takes in 5 moles of air at 20°C and 1 atm and compresses it adiabatically to (1/10) th of the original volume. Assume, air to be diatomic. The work done and change in internal energy is*

(a) 46 kJ, – 46 kJ (b) 36 kJ, –36 kJ
(c) –46 kJ, 46 kJ (d) 36 kJ, – 46 kJ

Sol. (c) Let $p_1 = 1\,\text{atm}$, $n = 5$ mol, $T_1 = 293$ K

$$V_2 = \frac{V_1}{10}$$

Using, $T_1V_1^{\gamma-1} = T_2V_2^{\gamma-1}$

$$\Rightarrow \quad T_2 = T_1\left(\frac{V_1}{V_2}\right)^{\gamma-1} = 293(10)^{0.4} = 736\text{ K}$$

$$\text{Work done} = \frac{nR(T_1 - T_2)}{\gamma - 1} = \frac{5 \times 8.3 \times (293 - 736)}{0.4} = -46\text{ kJ}$$

$$\Delta U = \Delta Q - W = 0 - W = 46\text{ kJ}$$

Cyclic Process

Figure shows a cyclic process $ABCA$, the work done during the cycle can be calculated as

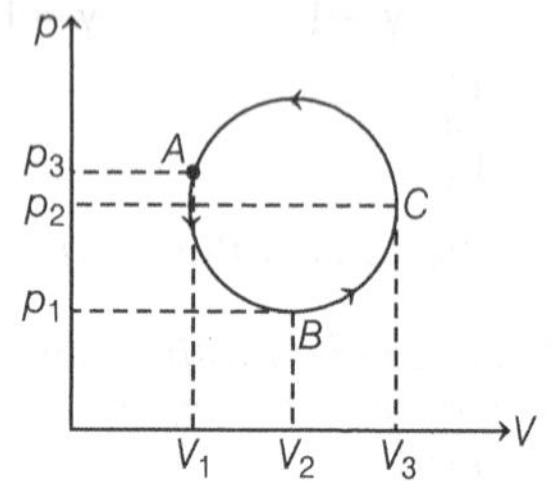

Work done during the process $AB = W_{AB}$.
Work done during the process $BC = W_{BC}$.
Work done during the process $CA = W_{CA}$.
The net work done during the cycle is

$$W = W_{AB} + W_{BC} + W_{CA}$$

Example 11. *One mole of an ideal monoatomic gas undergoes thermodynamic cycle $1 \to 2 \to 3 \to 1$ as shown in the figure. Initial temperature of the gas is $T_0 = 300$ K.*

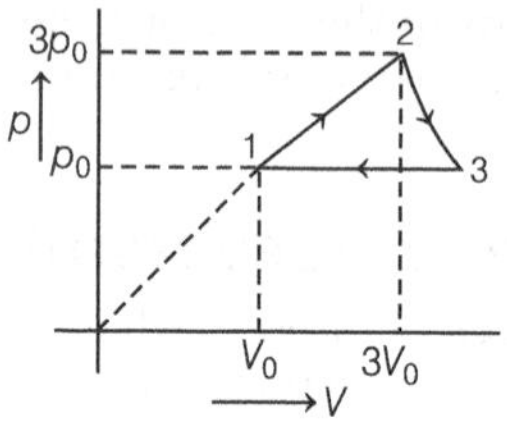

Process $1 \to 2$: $p = \alpha V$
Process $2 \to 3$: pV = constant
Process $3 \to 1$: p = constant (*Take* $\ln|3| = 1.09$)

The net work done by the cycle is

(a) $3.27\,RT_0$ (b) $6.83\,RT_0$
(c) $4.53\,RT_0$ (d) $5.81\,RT_0$

Sol. (d) For process $1 \to 2$,

$$W_{12} = \int_1^2 \alpha V dV = \alpha \int_{V_0}^{3V_0} V dV = \frac{\alpha}{2}(9V_0^2 - V_0^2) = 4\,\alpha V_0^2$$

Using gas law, $\dfrac{p_1V_1}{T_1} = \dfrac{p_2V_2}{T_2}$

$$\Rightarrow \quad T_2 = \frac{p_2V_2}{p_1V_1}T_1 = \frac{V_2^2}{V_1^2}T_1 = \left(\frac{3V_0}{V_0}\right)^2 T_0 = 9T_0$$

For process $2 \to 3$,

$$W_{23} = RT_2 \log\left|\frac{p_2}{p_3}\right| = R(9T_0)\log\left|\frac{3p_0}{p_0}\right|$$

$$= 9RT_0 \log|3| = 9.81RT_0$$

For isothermal process,

$$p_2V_2 = p_3V_3$$

$$\therefore \quad V_3 = \frac{p_2V_2}{p_3} = \frac{3p_0}{p_0}(3V_0) = 9V_0$$

$$W_{31} = p_0(V_1 - V_3) = p_0(V_0 - 9V_0) = -8p_0V_0$$

$$W_{31} = -8RT_0$$

Applying gas law in process $1 \to 2$,

$$p_0V_0 = RT$$

or $\quad \alpha V_0^2 = RT$

The net work,

$$\therefore \quad W_{\text{net}} = W_{12} + W_{23} + W_{31}$$

$$= 4RT_0 + 9.81RT_0 - 8RT_0 = 5.81RT_0$$

Example 12. *A thermodynamic system is taken through the cycle abcda shown in figure. Find the work done by the gas during the parts ab, bc, cd and da.*

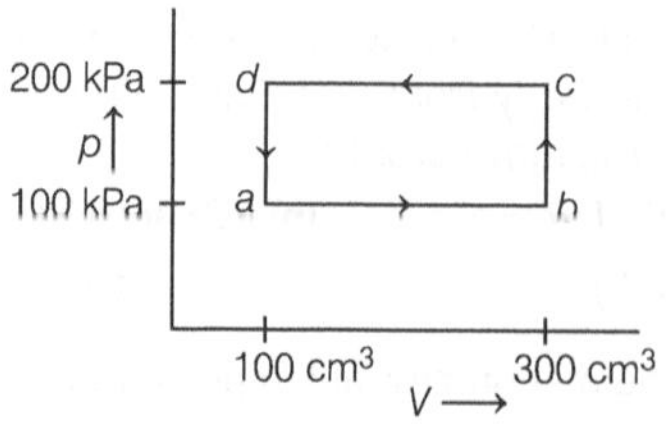

(a) (20 J, 0, – 40 J, 0) (b) (0, 0, – 40 J, 20 J)
(c) (– 40 J, 20 J, 0, 0) (d) None of these

Sol. (a) The work done during the part ab

$$= \int_a^b p\,dV = (100\text{ kPa}) \int_a^b dV$$
$$= (100\text{ kPa})\,(300\text{ cm}^3 - 100\text{ cm}^3)$$
$$= 20\text{ J}$$

The work during bc is zero, as volume does not change.
The work done during cd

$$= \int_c^d p\,dV = (200\text{ kPa}) \int_c^d dV$$
$$= (200\text{ kPa})\,(100\text{ cm}^3 - 300\text{ cm}^3)$$
$$= -40\text{ J}$$

The work done during da is zero, as the volume does not change.

Isochoric Process

A process taking place in a thermodynamic system at constant volume is called an **isochoric process**.

Process equation is $\frac{p}{T}$ = Constant.

$dQ = nC_V\,dT$, molar heat capacity for isochoric process is C_V.

Volume is constant, so $dW = 0$.

From the first law of thermodynamics,

$$dQ = dU = nC_V\,dT$$

p-V curve is a straight line parallel to pressure axis as shown below.

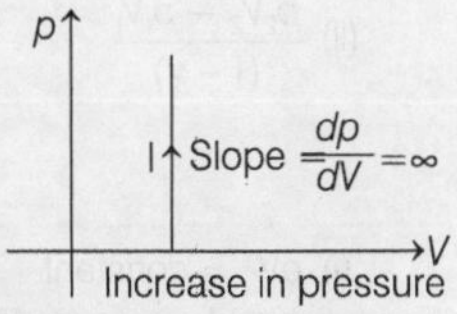

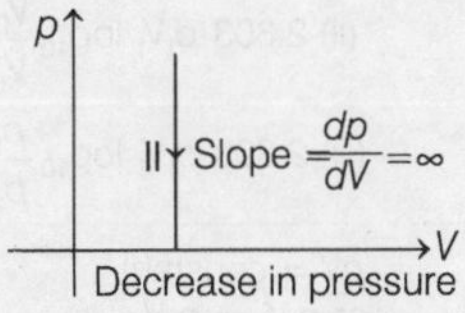

Isobaric Process

An isobaric process is one in which volume and temperature of system may change but pressure remain constant, *i.e.* $\Delta p = 0$.

- For this process, Charles' law is obeyed.

  Hence, $V \propto T \Rightarrow \frac{V_1}{T_1} = \frac{V_2}{T_2}$

- Specific heat of a gas during an isobaric process,

$$C_p = \left(\frac{f}{2} + 1\right) R = \frac{Q}{n\Delta T}$$

- Work done in an isobaric process,

$$W = p(V_f - V_i) = nR(T_f - T_i) = nR\Delta T$$

- From first law of thermodynamics,

$$\Delta Q = \Delta U + \Delta W$$

  Here, $\Delta W = nR\Delta T$

$$\Delta U = nC_V\Delta T$$

$$\therefore \quad \Delta Q = nC_V\Delta T + nR\Delta T = nC_p\Delta T$$

- Bulk modulus of an isobaric process is zero

$$K_{\text{isobaric}} = \frac{\Delta p}{-\frac{\Delta V}{V}} = 0 \qquad (\text{as, } \Delta p = 0)$$

- p-V curve is a straight line parallel to volume axis, slope of p-V curve for an isobaric process, $\frac{dp}{dV} = 0$.

  (a) Graph 1 indicates isobaric expansion. The heat is given to the gas. The volume and temperature of the gas both will rise. The gas expands during positive work.

  (b) Graph 2 indicates isobaric compression. In this process, heat is taken out of the gas. The temperature falls and the gas contracts causing negative work.

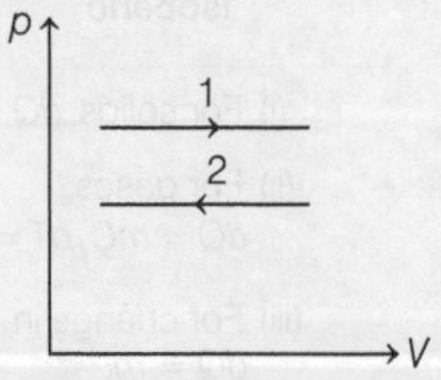

Indication Diagram of p-V Curve

We can determine the work done from area under p-V curve.

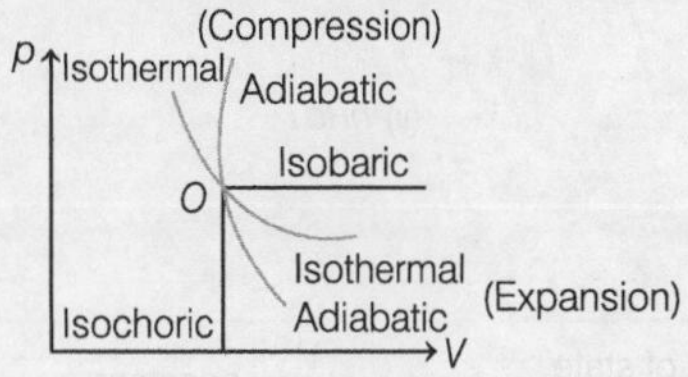

For isothermal process, $\frac{\Delta p}{\Delta V} = -\frac{p}{V}$.

For adiabatic process, $\frac{\Delta p}{\Delta V} = -\frac{\gamma p}{V}$, *i.e.* it means that at a particular point, slope (value) of adiabatic curve is more than that for isotherm or we can say adiabatic curve is more steep than an isotherm for expansion and just reverse for compression.

It is clear from the figure that, for expansion that occurs within same limits.

$$W_{\text{isobaric}} > W_{\text{isothermal}} > W_{\text{adiabatic}} > W_{\text{isochoric}}$$

Important Points Related to Thermodynamic Process

- For a reversible process, the first law of thermodynamics gives the change in the internal energy of the system.

$$dU = dQ - dW$$

Replacing work with a change in volume gives

$$dU = dQ - pdV$$

Since the process is isochoric, $dV = 0$, therefore the above equation becomes

$$dU = dQ$$

- Work done on gas in some process

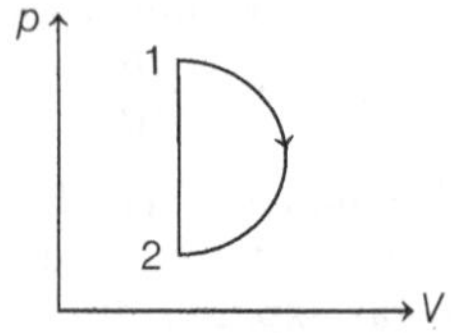

$\Delta V = 0$ but $dV \neq 0$

Work done = + ve

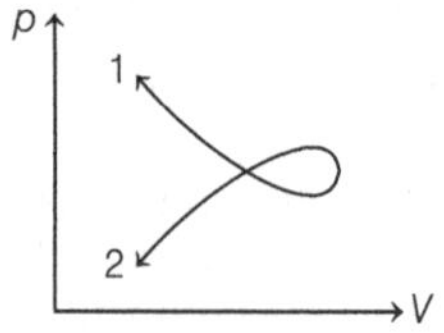

ΔW can be zero

But $dW \neq 0$

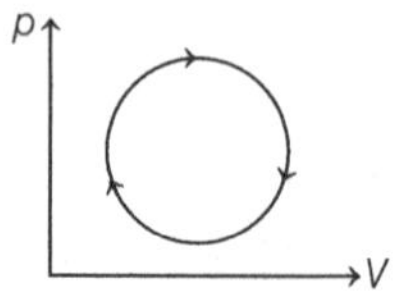

For clockwise, $\Delta W = +$ ve

For anti-clockwise, $\Delta W = -$ve

- Work done is least for monoatomic gas expansion.

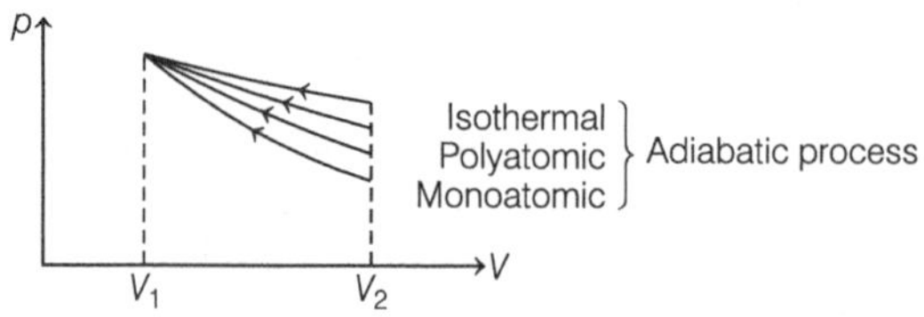

Different Thermodynamic Processes at a Glance

Change or Name of Process	Isobaric	Isochoric	Isothermal	Adiabatic
dQ	(i) For solids, $dQ = mC_p dT$ (ii) For gases, $dQ = mC_p dT = nC_p dT$ (iii) For change in state $dQ = mL$	(i) For solids, $dQ = mC_V dT$ (ii) For gases, $dQ = nC_V dT$	$dQ = dW$	Zero
dU	(i) $dQ - pdV$ (ii) $dQ - nRdT$	dQ	Zero	$- dW$
dW	(i) pdV (ii) $nRdT$	Zero	(i) $2.303\, nRT \log_{10} \frac{V_2}{V_1}$ (ii) $2.303\, p_1V_1 \log_{10} \frac{V_2}{V_1}$ (iii) $2.303\, p_1V_1 \log_{10} \frac{p_1}{p_2}$	(i) $\frac{R(T_2 - T_1)}{(1-\gamma)}$ (ii) $\frac{p_2V_2 - p_1V_1}{(1-\gamma)}$
Equation of state	$\frac{V}{T}$ = constant or $\frac{V_1}{T_1} = \frac{V_2}{T_2}$	$\frac{p}{T}$ = constant or $\frac{p_1}{T_1} = \frac{p_2}{T_2}$	pV = constant or $p_1V_1 = p_2V_2$	(i) pV^{γ} = constant (ii) $TV^{\gamma - 1}$ = constant (iii) $p^{1-\gamma}T^{\gamma}$ = constant
Slope of p-V curve	Zero	∞	$-\frac{p}{V}$	$-\frac{\gamma p}{V}$
Law	Charles' law	Gay-Lussac's law	Boyle's law	Poisson's law
Form of first law	$dQ = dU + dW$ $= nC_p dT + p\,dV$	$dQ = dU = nC_V dT$	$dQ = dW = pdV$	(i) $dW = - dU$ (ii) $dU = - dW$
Bulk modulus	Zero	Infinity	$- p$	$- \gamma p$
Result of maximum work	Maximum	Zero	Less from isobaric process but greater from adiabatic process	Minimum but not zero

Example 13. *A cylinder containing an ideal gas and closed by a movable piston is submerged in an ice-water mixture. The piston is quickly pushed down from position 1 to position 2 (process AB).*

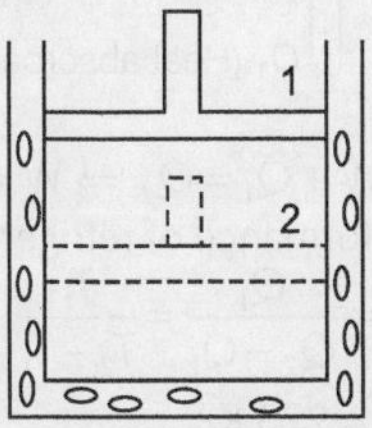

The piston is held at position 2 until the gas is again at 0°C (process BC). Then the piston is slowly raised back to position 1 (process CA). Which one of the following p-V diagrams correctly represent the processes AB, BC and CA and the cycle ABCA?

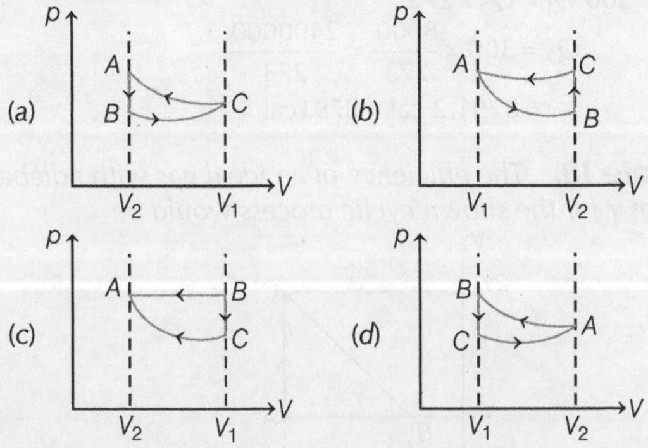

Sol. (d) In an adiabatic process, heat is neither allowed to enter nor allowed to escape the system, the process *AB* is adiabatic compression because piston is pushed very quickly from position 1 to position 2.

The process *BC* is isochoric because in this case volume remains constant, whereas process *CA* is an isothermal expansion because temperature remains constant. These are shown on the *p-V* diagram correctly in option (d).

Second Law of Thermodynamics

The second law of thermodynamics gives a fundamental limitation to the efficiency of a heat engine and the coefficient of performance of a refrigerator. It says that efficiency of a heat engine can never be unity (or 100%). This implies that heat released to the cold reservoir can never be made zero.

Kelvin's Statement

It is impossible for an engine operating in a cyclic process to extract heat from a reservoir and convert it completely into work. In other words, whole of heat can never be converted into work. Heat engine works on this principle.

Clausius Statement

It is impossible for a self-acting machine unaided by any external agency to transfer heat from a colder to a hotter reservoir. In other words, heat by itself can not pass from a colder to a hotter body without an external agency. Refrigerator is based on this statement.

Heat Engine

A heat engine is a device which converts thermal energy into another useful forms of energy and work.

The efficiency of heat engine,

$$\eta = \frac{\text{Work done by working substance}}{\text{Heat given to working substance}}$$

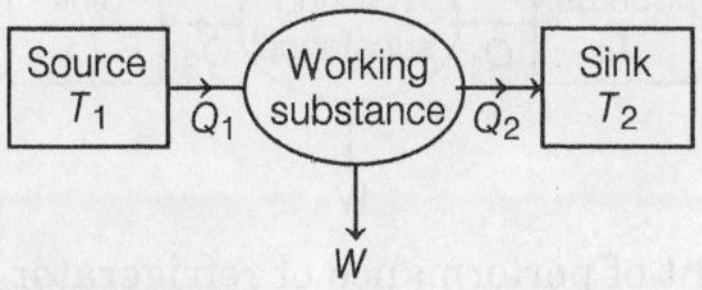

$$\eta = \frac{W}{Q_1} = \frac{Q_1 - Q_2}{Q_1}$$

$$\% \eta = \frac{W}{Q_1} \times 100 = \frac{Q_1 - Q_2}{Q_1} \times 100$$

$$= \left(1 - \frac{Q_2}{Q_1}\right) \times 100 = \left(1 - \frac{T_2}{T_1}\right) \times 100$$

Note If $Q_2 = 0$ or $T_2 = 0$ K, then $\eta = 100\%$, which is impossible.

Example 14. *A heat engine is involved with exchange of heat of 1915 J, – 40 J, +125 J and – QJ, during one cycle achieving an efficiency of 50.0%. The value of Q is* **[JEE Main 2020]**

(a) 640 J (b) 40 J (c) 980 J (d) 400 J

Sol. (c) As processes involved in one cycle of a heat engine must be

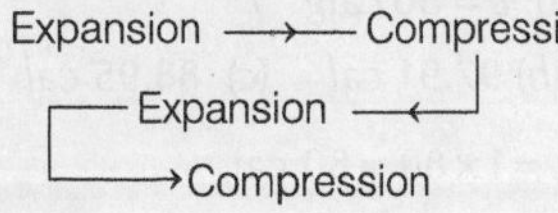

So, Q must be negative.

Now, work done in process = heat supplied – heat rejected
= (1915 + 125) – (40 + Q) J

and heat supplied to cycle = (1915 + 125) J

Given, efficiency = 50% $\Rightarrow \eta = \frac{\text{Work done}}{\text{Heat supplied}}$

$$\Rightarrow \frac{50}{100} = \frac{(1915 + 125) - (40 + Q)}{1915 + 125}$$

$$\Rightarrow Q = 2000 - 1020 = 980 \text{ J}$$

Types of Heat Engines

In practice, heat engine are of two types

(i) **External Combustion Engine** In this engine, heat is produced by burning the fuel in a chamber outside the main body (working substance) of the engine. Steam engine is an external combustion engine. The thermal efficiency of a steam engine varies from 10 to 20%.

(ii) **Internal Combustion Engine** In this engine, heat is produced by burning the fuel inside the main body of the engine. Petrol engine and diesel engines are internal combustion engine.

Refrigerator

A refrigerator or heat pump is basically a heat engine running in reverse direction. It takes heat from the colder body (sink) and after doing some work gives the rest heat to the hotter body (source).

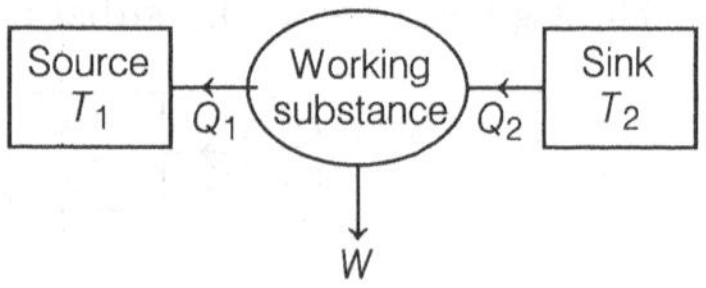

The coefficient of performance of refrigerator,

$$\beta = \frac{\text{Heat extracted}}{\text{Work done}} = \frac{Q_2}{W}$$

$$= \frac{Q_2}{Q_1 - Q_2} = \frac{T_2}{T_1 - T_2} = \frac{1-\eta}{\eta}$$

Example 15. *An automobile engine absorbs 1600 J of heat from a hot reservoir and expels 1000 J to a cold reservoir in each cycle. What maximum work is done in each cycle?*

(a) 400 J (b) 500 J (c) 600 J (d) 450 J

Sol. (c) We know that, $W = Q_H - Q_L = 1600 - 1000 = 600$ J

Example 16. *Calculate the least amount of work that must be done to freeze one gram of water at 0°C by means of a refrigerator. Temperature of surrounding is 27°C. How much heat is passed on the surrounding in this process ? (Take, latent heat of fusion, $l = 80\ \text{cal g}^{-1}$)*

(a) 87.91 cal (b) 97.91 cal (c) 88.95 cal (d) 89.95 cal

Sol. (a) As, $Q_2 = mL = 1 \times 80 = 80$ cal,

$$T_2 = 0°\text{C} = 273 \text{ K}$$

and $$T_1 = 27°\text{C} = 300 \text{ K}$$

$$\frac{Q_2}{W} = \frac{T_2}{T_1 - T_2}$$

$$\Rightarrow \quad W = \frac{Q_2(T_1 - T_2)}{T_2} \text{ cal} = \frac{80(300-273)}{273} = 7.91$$

$$\Rightarrow \quad Q_1 = Q_2 + W = (80 + 7.91) = 87.91 \text{ cal}$$

Example 17. *If minimum possible work is done by a refrigerator in converting 100 g of water at 0°C to ice, how much heat (in cal) is released to the surroundings at temperature 27°C to the nearest integer? (Take, latent heat of ice = 80 cal/g)* **[JEE Main 2020]**

Sol. (8791) To convert 100 g water at 0°C to ice at 0°C, heat absorbed by refrigerator,

$$Q_1 = mL = 100 \times 80 = 8000 \text{ cal}$$

W = work done or energy supplied to refrigerator

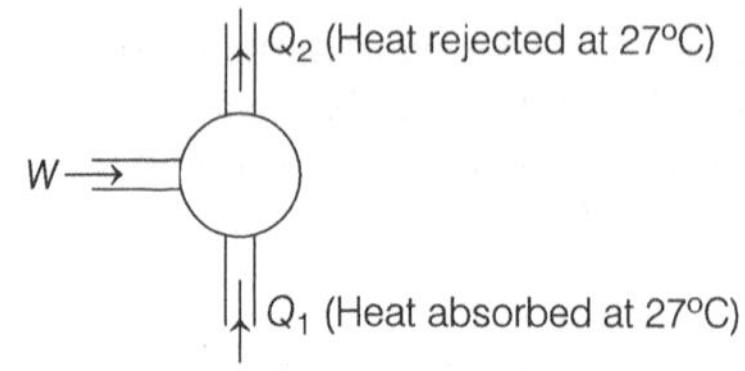

From figure, we have $W + Q_1 = Q_2 \Rightarrow W = Q_2 - Q_1$

Now, coefficient of performance of refrigerator,

$$\beta = \frac{Q_1}{W} = \frac{Q_1}{Q_2 - Q_1} = \frac{T_1}{T_2 - T_1}$$

Here, $T_1 = 0°\text{C} = 273$ K

$T_2 = 27°\text{ C} = 300$ K

So, $$\frac{Q_1}{Q_2 - Q_1} = \frac{273}{300 - 273}$$

$$\Rightarrow \quad 27Q_1 = (Q_2 - Q_1)273$$

$$\Rightarrow \quad 300\ Q_1 = Q_2 \times 273$$

$$\Rightarrow \quad Q_2 = 300 \times \frac{8000}{273} = \frac{2400000}{273}$$

$$= 8791.2 \text{ cal} \approx 8791 \text{ cal}$$

Example 18. *The efficiency of an ideal gas with adiabatic exponent γ for the shown cyclic process would be*

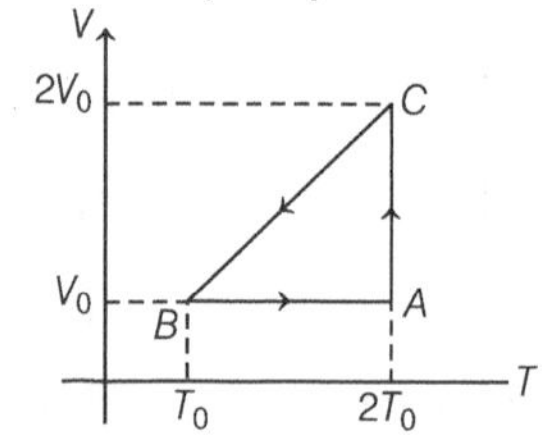

(a) $\frac{(2\ln 2 - 1)}{\gamma/(\gamma-1)}$ (b) $\frac{(1 - 1\ln 2)}{\gamma/(\gamma-1)}$ (c) $\frac{(2\ln 2 + 1)}{\gamma/(\gamma-1)}$ (d) $\frac{(2\ln 2 - 1)}{\gamma/(\gamma+1)}$

Sol. (a) As, $W_{BC} = p\Delta V = nR\Delta T = -nRT_0$

and $$W_{CA} = +\ 2nRT_0 \ln 2$$

Also, $$\Delta Q_{BC} = nC_p\Delta T = \frac{nR_\gamma T_0}{\gamma - 1}$$

$$\therefore \quad \text{Efficiency} = \frac{\text{Work}}{\text{Input heat}} = \frac{(2\ln 2 - 1)}{\gamma/(\gamma-1)}$$

Carnot Cycle

Carnot devised an ideal cycle of operation for a heat engine called Carnot's cycle.

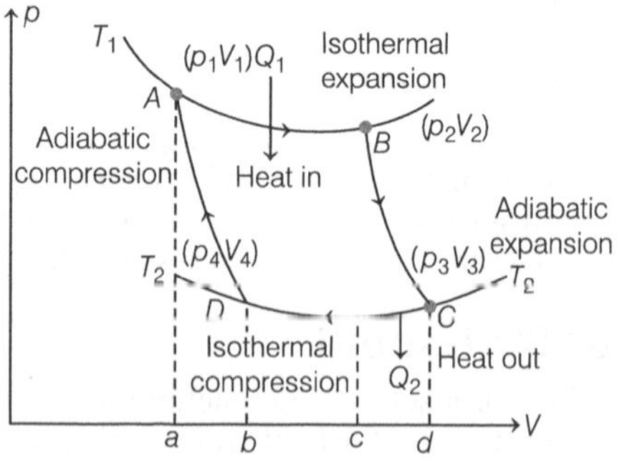

A Carnot's cycle contains the following four processes

(i) Isothermal expansion (AB)

(ii) Adiabatic expansion (BC)

(iii) Isothermal compression (CD)

(iv) Adiabatic compression (DA)

The net work done per cycle by the engine is numerically equal to the area of the loop representing the Carnot's cycle.

After doing the calculations for different processes, we can show that $\frac{Q_2}{Q_1}=\frac{T_2}{T_1}$

Therefore, efficiency of the cycle is $\eta = 1-\frac{T_2}{T_1}$

Efficiency of Carnot engine is maximum (not 100%) for given temperatures T_1 and T_2. But still Carnot engine is not a practical engine because many ideal situations have been assumed while designing this engine which can practically not be obtained.

Carnot Theorem

According to Carnot theorem,

- A heat engine working between the two given temperatures T_1 of hot reservoir *i.e.*, source and T_2 of cold reservoir *i.e.*, sink cannot have efficiency more than that of the Carnot engine.
- The efficiency of the Carnot engine is independent of the nature of working substance.

Carnot's Engine

Carnot engine is an ideal heat engine proposed by *Sadi Carnot* in 1824. The reversible engine which operates between two temperatures of source (T_1) and sink (T_2) is known as *Carnot heat engine.*

The designed engine is a theoretical engine which is free from all the defects of a practical engine. This engine cannot be realised in actual practice, however, this can be taken as a standard against which the performance of an actual engine can be judged.

Example 19. *Carnot's engine takes in a thousand kilo calories of heat from a reservoir at 827°C and exhausts it to a sink at 27°C. How much work does it perform? What is the efficiency of the engine?*

(a) 2.70×10^5 *cal*, 70.70% (b) 2.70×10^5 *cal*, 72.72%

(c) 2.70×10^5 *cal*, 80.70% (d) 3.70×10^5 *cal*, 70.70%

Sol. (b) Given, $Q=10^6$ cal

$$T_1=(827+273)=1100\text{ K}$$

and $$T_2=(27+273)=300\text{ K}$$

As, $$\frac{Q_1}{T_1}=\frac{Q_2}{T_2}$$

$\therefore$ $$Q_2=\frac{T_2}{T_1}Q_1=\left(\frac{300}{1100}\right)(10^6)=2.720\times10^5\text{ cal}$$

Efficiency of the engine,

$$\eta=\left(1-\frac{T_2}{T_1}\right)\times100 \text{ or } \eta=\left(1-\frac{300}{1100}\right)\times100=72.72\%$$

Example 20. *A Carnot engine has an efficiency of 1/6. When the temperature of the sink is reduced by 62°C, its efficiency is doubled. The temperatures of the source and the sink are respectively,* **[JEE Main 2019]**

(a) 62°C, 124°C (b) 99°C, 37°C

(c) 124°C, 62°C (d) 37°C, 99°C

Sol. (b) Efficiency of a Carnot engine working between source of temperature T_1 and sink of temperature T_2 is given by $\eta=1-\frac{T_2}{T_1}$

Here, T_2 and T_1 are absolute temperatures. Initially, $\eta=\frac{1}{6}$

$\therefore$ $$\frac{1}{6}=1-\frac{T_2}{T_1} \Rightarrow \frac{T_2}{T_1}=\frac{5}{6}$$

Finally, efficiency is doubled on reducing sink temperature by 62°C.

$\therefore$ $$\eta=\frac{2}{6},\ T_{\text{sink}}=T_2'=T_2-62$$

So, $$\eta=1-\frac{T'_2}{T'_1}$$

$$\Rightarrow \frac{2}{6}=1-\frac{T_2-62}{T_1} \Rightarrow \frac{T_2-62}{T_1}=\frac{4}{6}$$

$$\Rightarrow \frac{T_2}{T_1}-\frac{62}{T_1}=\frac{4}{6} \Rightarrow \frac{5}{6}-\frac{62}{T_1}=\frac{4}{6} \quad \left(\because \frac{T_2}{T_1}=\frac{5}{6}\right)$$

$$\Rightarrow T_1=6\times62=372\text{ K}=372-273=99^\circ\text{ C}$$

$$\Rightarrow T_2=\frac{5}{6}\times T_1\simeq310\text{ K}=310-273=37^\circ\text{ C}$$

Example 21. *Two ideal Carnot engines operate in cascade (all heat given up by one engine is used by the other engine to produce work) between temperatures T_1 and T_2. The temperature of the hot reservoir of the first engine is T_1 and the temperature of the cold reservoir of the second engine is T_2. T is temperature of the sink of first engine which is also the source for the second engine. How is T related to T_1 and T_2, if both the engines perform equal amount of work?* **[JEE Main 2020]**

(a) $T=0$ (b) $T=\frac{2T_1T_2}{T_1+T_2}$ (c) $T=\sqrt{T_1T_2}$ (d) $T=\frac{T_1+T_2}{2}$

Sol. (d) Carnot engines given are operating in cascade configuration as shown in the figure.

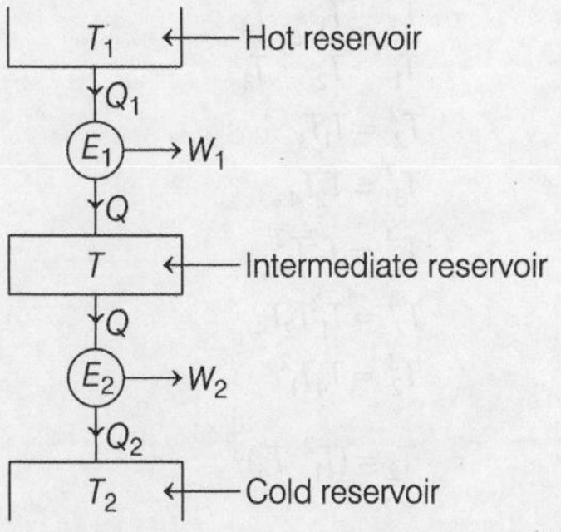

Work done by engine 1, $W_1 = Q_1 - Q = k(T_1 - T)$
Work done by engine 2, $W_2 = Q - Q_2 = k(T - T_2)$

So, $$\frac{W_1}{W_2} = \frac{T_1 - T}{T - T_2}$$

As, $$\frac{W_1}{W_2} = 1 \quad \text{(given)}$$

$$\Rightarrow \quad T - T_2 = T_1 - T$$

$$\Rightarrow \quad T = \frac{T_1 + T_2}{2}$$

Example 22. *Three Carnot engines operate in series between a heat source at a temperature T_1 and a heat sink at temperature T_4 (see figure). There are two other reservoirs at temperatures T_2 and T_3, as shown with $T_1 > T_2 > T_3 > T_4$. The three engines are equally efficient, if* **[JEE Main 2019]**

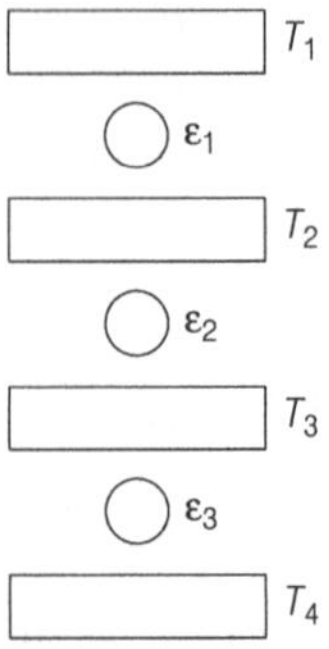

(a) $T_2 = (T_1^3 T_4)^{1/4}$; $T_3 = (T_1 T_4^3)^{1/4}$ (b) $T_2 = (T_1^2 T_4)^{1/3}$; $T_3 = (T_1 T_4^2)^{1/3}$
(c) $T_2 = (T_1 T_4)^{1/2}$; $T_3 = (T_1^2 T_4)^{1/3}$ (d) $T_2 = (T_1 T_4^2)^{1/3}$; $T_3 = (T_1^2 T_4)^{1/3}$

Sol. (b) Given, Carnot engines operates as,

Efficiency of a Carnot's engine is given by $\eta = 1 - \dfrac{T_{sink}}{T_{source}}$

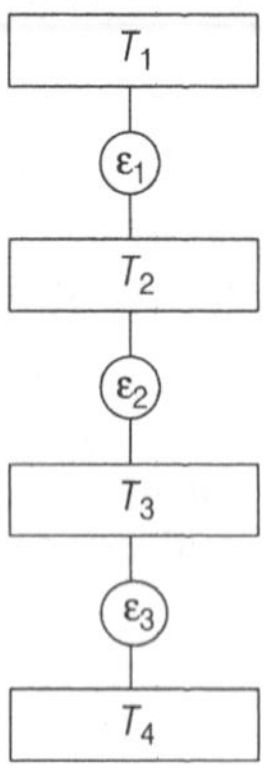

We have, η_1 = efficiency of engine $\varepsilon_1 = 1 - \dfrac{T_2}{T_1}$

η_2 = efficiency of engine $\varepsilon_2 = 1 - \dfrac{T_3}{T_2}$

η_3 = efficiency of engine $\varepsilon_3 = 1 - \dfrac{T_4}{T_3}$

For equal efficiencies, $\eta_1 = \eta_2 = \eta_3$

$$\Rightarrow \quad 1 - \frac{T_2}{T_1} = 1 - \frac{T_3}{T_2} = 1 - \frac{T_4}{T_3}$$

$$\Rightarrow \quad \frac{T_2}{T_1} = \frac{T_3}{T_2} = \frac{T_4}{T_3}$$

$$\Rightarrow \quad T_2^2 = T_1 T_3$$

and $$T_3^2 = T_2 T_4$$

$$\Rightarrow \quad T_2^4 = T_1^2 T_3^2$$

$$T_2^4 = T_1^2 T_2 T_4$$

$$T_2^3 = T_4 T_1^2$$

or $$T_2 = (T_1^2 T_4)^{\frac{1}{3}}$$

Also, $$T_3^4 = T_2^2 T_4^2 = T_1 T_3 T_4^2$$

$$T_3^3 = T_1 T_4^2$$

or $$T_3 = (T_1 T_4^2)^{\frac{1}{3}}$$

Example 23. *Two Carnot engines A and B are operated in series. The first one, A receives heat at T_1(= 600 K) and rejects to a reservoir at temperature T_2. The second engine B receives heat rejected by the first engine and in turn rejects to a heat reservoir at T_3 (= 400 K).*
Calculate the temperature T_2, if the work outputs of the two engines are equal. **[JEE Main 2019]**

(a) 600 K (b) 500 K
(c) 400 K (d) 300 K

Sol. (b) In a Carnot engine, the heat flow from higher temperature source (at T_1) to lower temperature sink (at T_2) and give the work done equal to the $W = Q_1 - Q_2$.

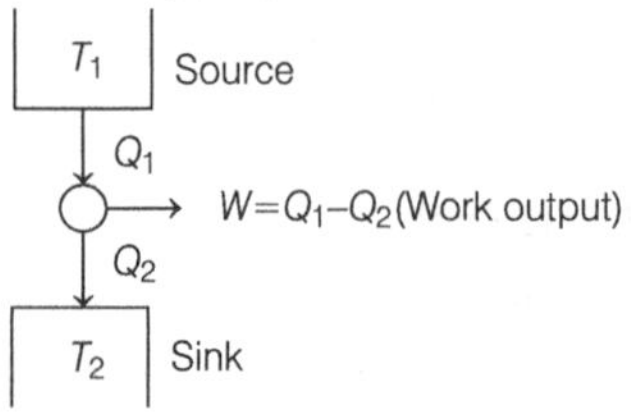

For the given condition, Carnot engines A and B are operated in series as shown below

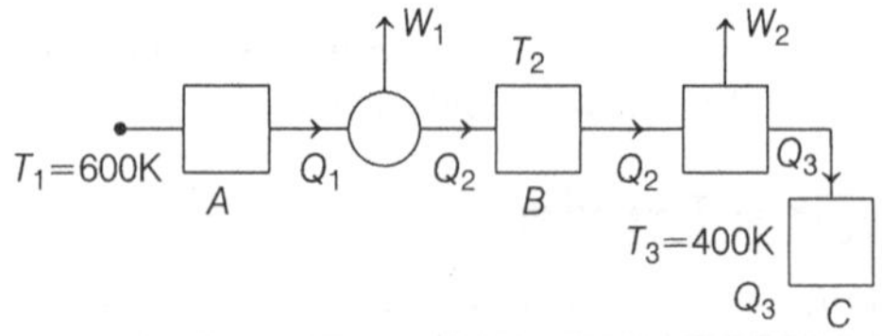

where, Q_1 = heat rejected by engine A at T_1K,
Q_2 = heat received by engine B at T_2K
and Q_3 = heat rejected by engine B to source C at T_3K.

According to Carnot engine principle,
$W_1 = Q_1 - Q_2$ (work output from source A and B)
$W_2 = Q_2 - Q_3$ (work output from source B and C)

As per the given condition, if the work outputs of the two engines are equal, then

$$Q_1 - Q_2 = Q_2 - Q_3$$

$$\Rightarrow \quad Q_1 + Q_3 = 2Q_2$$

$$\frac{Q_1}{Q_2} + \frac{Q_3}{Q_2} = 2$$

Therefore, $$\frac{T_1}{T_2} + \frac{T_3}{T_2} = 2$$

So, $$\frac{T_1 + T_3}{T_2} = 2$$

$$\Rightarrow \quad T_2 = \frac{T_1 + T_3}{2} = \frac{600 + 400}{2}$$

$$T_2 = 500 \text{ K}$$

Practice Exercise

ROUND I Topically Divided Problems

Zeroth Law and First Law of Thermodynamics

1. In a certain process, 400 cal of heat are supplied to a system and at the same time 105 J of mechanical work was done on the system. The increase in its internal energy is

(a) 20 cal (b) 303 cal
(c) 404 cal (d) 425 cal

2. If the heat of 110 J is added to a gaseous system and it acquires internal energy of 40 J, then the amount of internal work done is

(a) 40 J (b) 70 J
(c) 150 J (d) 110 J

3. In a thermodynamic process, pressure of a fixed mass of gas is changed in such a manner that the gas releases 20 J of heat and 8 J of work is done on the gas. If internal energy of the gas was 30 J, then the final internal energy will be

(a) 42 J (b) 18 J
(c) 12 J (d) 60 J

4. What is the nature of change in internal energy in the following three thermodynamic processes shown in figure?

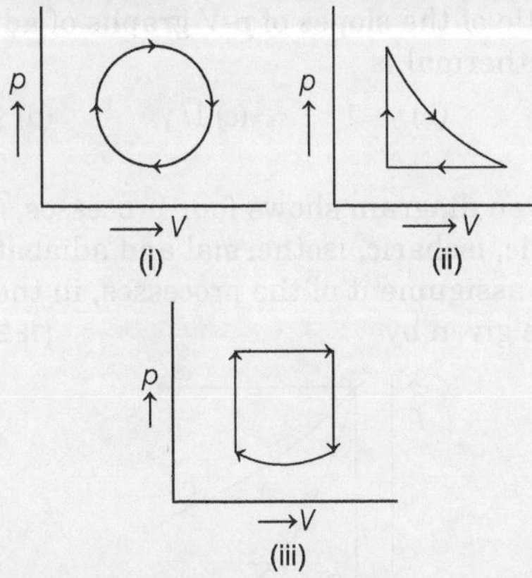

(a) ΔU is positive in all the three cases
(b) ΔU is negative in all the three cases
(c) ΔU is positive for (i), negative for (ii), zero for (iii)
(d) $\Delta U = 0$ in all the cases

5. 1 cm^3 of water at its boiling point absorbs 540 cal of heat to become steam with a volume = 1.013×10^5 Nm^{-2} and the mechanical equivalent of heat = 4.19 $Jcal^{-1}$. The energy spend in this process in overcoming intermolecular forces is

(a) 540 cal (b) 40 cal
(c) 500 cal (d) zero

6. In thermodynamics, heat and work are **[JEE Main 2021]**

(a) path functions
(b) intensive thermodynamic state variables
(c) extensive thermodynamic state variables
(d) point functions

7. Following figure shows two processes A and B for a gas. If ΔQ_A and ΔQ_B are the amount of heat absorbed by the system in two cases and ΔU_A and ΔU_B are changes in internal energies respectively, then **[JEE Main 2019]**

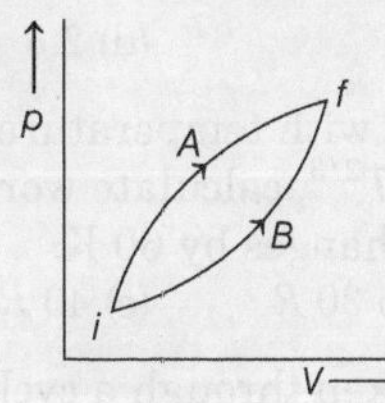

(a) $\Delta Q_A > \Delta Q_B,\ \Delta U_A > \Delta U_B$
(b) $\Delta Q_A < \Delta Q_B,\ \Delta U_A < \Delta U_B$
(c) $\Delta Q_A > \Delta Q_B;\ \Delta U_A = \Delta U_B$
(d) $\Delta Q_A = \Delta Q_B;\ \Delta U_A = \Delta U_B$

8. In a thermodynamic process, pressure of a fixed mass of a gas is changed in such a manner that the gas molecules gives out 20 J of heat and 10 J of work is done on the gas. If the internal energy of gas is 40 J, then the final internal energy will be

(a) 30 J (b) 20 J (c) 60 J (d) 40 J

9. In an isothermal change of an ideal gas, $\Delta U = 0$. The change in heat energy ΔQ is equal to

(a) 0.5 W (b) W (c) 1.5 W (d) 2 W

10. 5 mole of an ideal gas with ($\gamma = 7/5$) initially at STP are compressed adiabatically, so that its temperature becomes 400°C. The increase in the internal energy of gas (in kJ) is

(a) 21.55 (b) 41.55 (c) 65.55 (d) 50.55

11. When an ideal monoatomic gas is heated at constant pressure, fraction of heat energy supplied which increases the internal energy of gas is

(a) 2/5 (b) 3/5
(c) 3/7 (d) 3/4

12. A thermodynamic system is taken through the cycle *PQRSP* process. The net work done by the system is

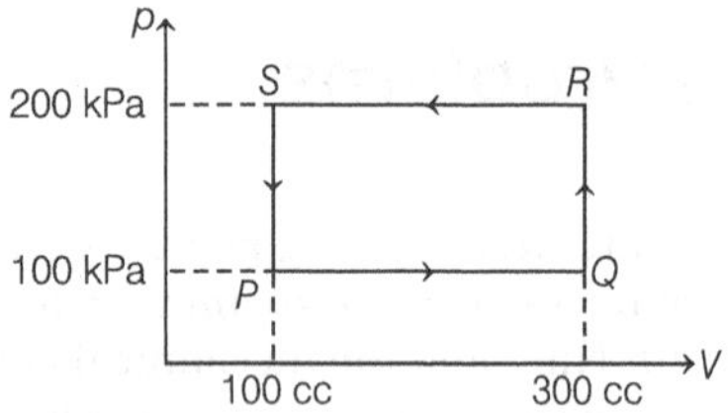

(a) 20 J (b) –20 J
(c) 400 J (d) –374 J

13. One mole of a monoatomic gas is heated at a constant pressure of 1 atm from 0 K to 100 K. If the gas constant $R = 8.32\ \text{Jmol}^{-1}\ \text{K}^{-1}$, the change in internal energy of the gas is approximately

(a) 2.3 J (b) 46 J
(c) 8.67×10^3 J (d) 1.25×10^3 J

14. For the same rise in temperature of one mole of gas at constant volume, heat required for a non-linear triatomic gas is *K* times that required for monoatomic gas. The value of *K* is

(a) 1 (b) 0.5
(c) 2 (d) 2.5

15. A gas expands with temperature according to the relation $V = kT^{2/3}$, calculate work done when the temperature changes by 60 K.

(a) 10 *R* (b) 30 *R* (c) 40 *R* (d) 20 *R*

16. A system is taken through a cyclic process represented by an ellipse as shown. The heat absorbed by the system is

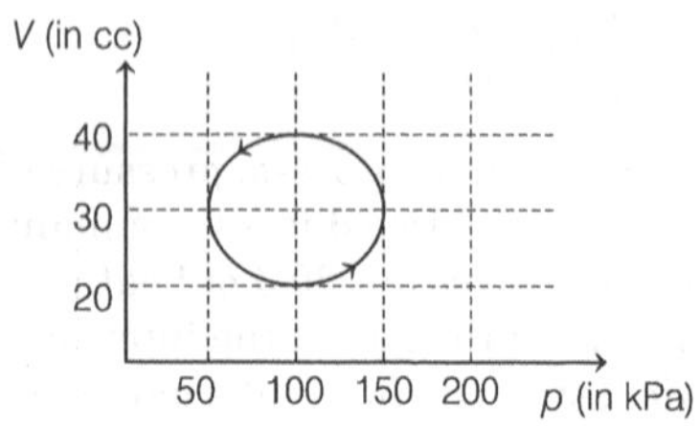

(a) $\pi \times 10^3$ J (b) $\frac{\pi}{2}$ J (c) $4\pi \times 10^2$ J (d) π J

Different Thermodynamic Processes

17. During adiabatic expansion of 10 moles of a gas, the internal energy decreases by 50 J. Work done during the process is

(a) + 50 J (b) – 50 J
(c) zero (d) Cannot say

18. A gas is expanded from volume V_0 to $2V_0$ under three different processes shown in figure.
Process 1 is isobaric process, process 2 is isothermal and process 3 is adiabatic. Let ΔU_1, ΔU_2 and ΔU_3 be the change in internal energy of the gas in these three processes, then

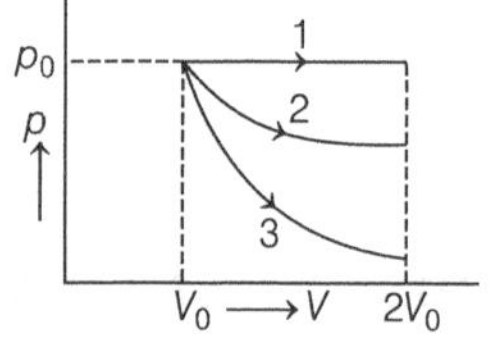

(a) $\Delta U_1 > \Delta U_2 > \Delta U_3$ (b) $\Delta U_1 < \Delta U_2 < \Delta U_3$
(c) $\Delta U_2 < \Delta U_1 > \Delta U_3$ (d) $\Delta U_2 < \Delta U_3 < \Delta U_1$

19. During an adiabatic process, the pressure of a gas is found to be proportional to the cube of its absolute temperature. The ratio C_p/C_V for the gas is

(a) $\frac{3}{2}$ (b) $\frac{4}{3}$ (c) 2 (d) $\frac{5}{3}$

20. The adiabatic elasticity of hydrogen gas ($\gamma = 1.4$) at NTP is

(a) $1 \times 10^5\ \text{N/m}^2$ (b) $1 \times 10^{-8}\ \text{N/m}^2$
(c) $1.4\ \text{N/m}^2$ (d) $1.4 \times 10^5\ \text{N/m}^2$

21. Consider two containers *A* and *B* containing identical gases at the same pressure, volume and temperature. The gas in container *A* is compressed to half of its original volume isothermally while the gas in container *B* is compressed to half of its original value adiabatically. The ratio of final pressure of gas in *B* to that of gas in *A* is **[NCERT Exemplar]**

(a) $2^{\gamma - 1}$ (b) $\left(\frac{1}{2}\right)^{\gamma - 1}$ (c) $\left(\frac{1}{1-\gamma}\right)^2$ (d) $\left(\frac{1}{\gamma - 1}\right)^2$

22. The ratio of the slopes of *p*-*V* graphs of adiabatic and isothermal is

(a) $\frac{\gamma - 1}{\gamma}$ (b) $\gamma - 1$ (c) $1/\gamma$ (d) γ

23. The given diagram shows four processes, *i.e.* isochoric, isobaric, isothermal and adiabatic. The correct assignment of the processes, in the same order is given by **[JEE Main 2019]**

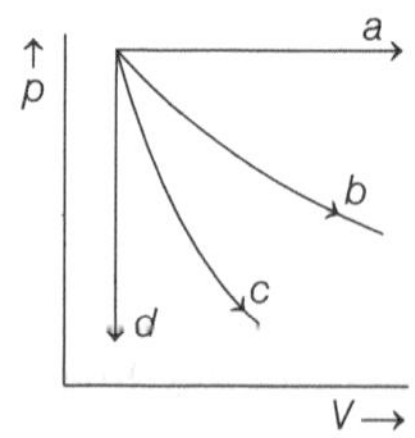

(a) *d a b c* (b) *a d b c*
(c) *d a c b* (d) *a d c b*

24. Two moles of an ideal monoatomic gas occupies a volume V at 27°C. The gas expands adiabatically to a volume $2V$. Calculate (i) the final temperature of the gas and (ii) change in its internal energy. **[JEE Main 2018]**

(a) (i) 189 K (ii) 2.7 kJ
(b) (i) 195 K (ii) −2.7 kJ
(c) (i) 189 K (ii) −2.7 kJ
(d) (i) 195 K (ii) 2.7 kJ

25. An ideal gas at a pressure 1 atm and temperature of 27°C is compressed adiabatically until its pressure becomes 8 times, the initial pressure. Then, the final temperature is $\left(\gamma = \frac{3}{2}\right)$

(a) 627°C (b) 527°C
(c) 427°C (d) 327°C

26. Which one is the correct option for the two different thermodynamic processes?

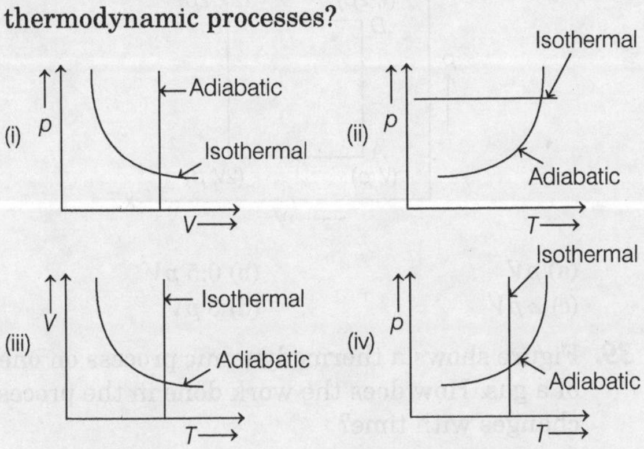

(a) (iii) and (i) (b) (iii) and (iv)
(c) (i) only (d) (ii) and (iii)

27. During an adiabatic process, the pressure p of a fixed mass of an ideal gas changes by Δp and its volume V changes by ΔV. If $\gamma = C_p / C_V$, then $\Delta V/V$ is given by

(a) $-\frac{\Delta p}{p}$ (b) $-\gamma\frac{\Delta p}{p}$
(c) $-\frac{\Delta p}{\gamma p}$ (d) $\frac{\Delta p}{\gamma^2 p}$

28. p-V diagram of an ideal gas is as shown in figure. Work done by the gas in the process $ABCD$ is

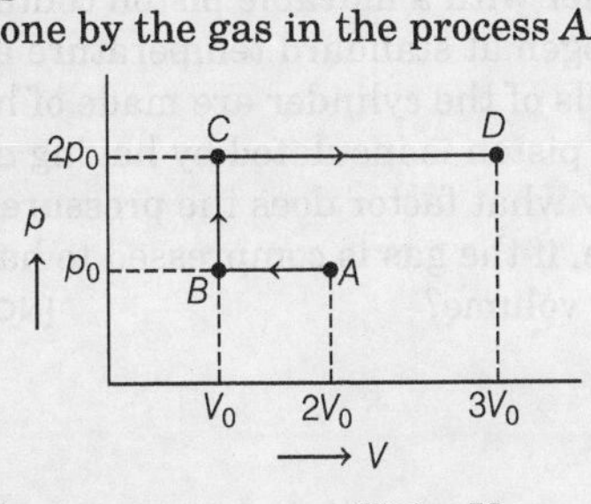

(a) $4pV_0$ (b) $2p_0V_0$
(c) $3p_0V_0$ (d) p_0V_0

29. An ideal gas is taken from state A to state B following three different paths as shown in p-V diagram. Which one of the following is true?

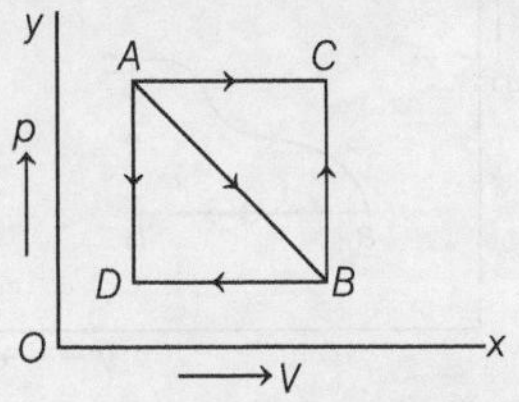

(a) Work done is maximum along AB
(b) Work done is minimum along AB
(c) Work done along ACB = work done along ADB
(d) Work done along ADB is minimum

30. Figure shows four p-V diagrams. Which of these curves represent isothermal and adiabatic process?

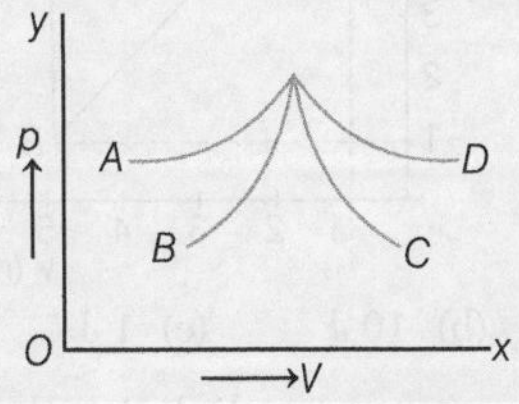

(a) D and C (b) A and C
(c) A and B (d) B and D

31. Which of the following is an equivalent cyclic process corresponding to the thermodynamic cyclic given in the figure? where, $1 \rightarrow 2$ is adiabatic.

(Graphs are schematic and are not to scale) **[JEE Main 2020]**

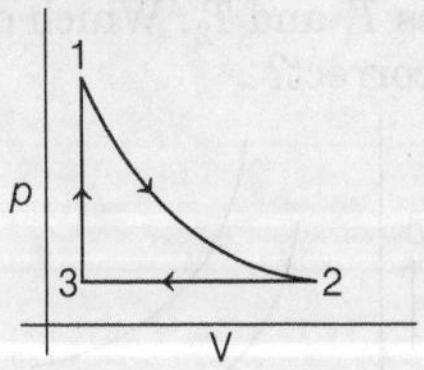

(a)
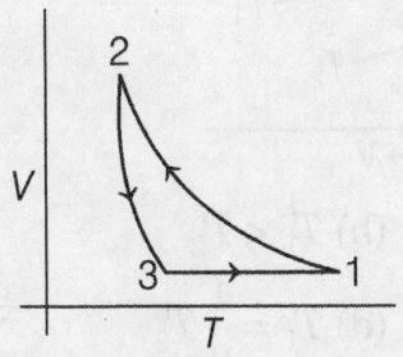

(b)
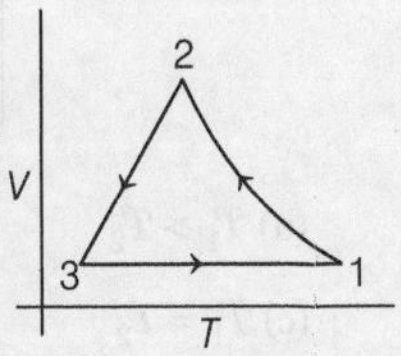

(c)
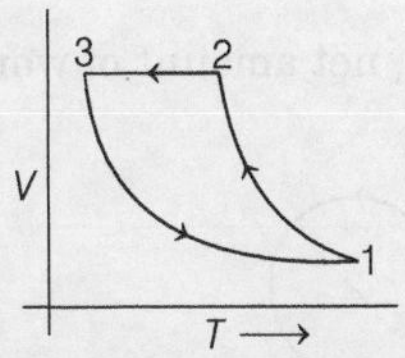

(d)
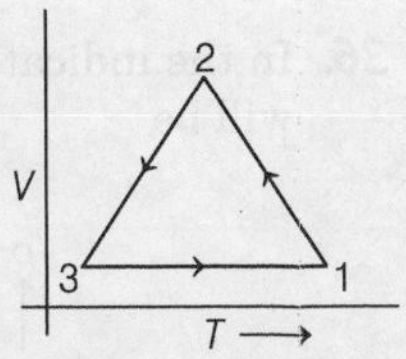

32. A sample of an ideal gas is taken through the cyclic process $abca$ as shown in the figure. The change in the internal energy of the gas along the path ca is −180 J. The gas absorbs 250 J of heat along the path ab and 60 J along the path bc. The work done by the gas along the path abc is **[JEE Main 2019]**

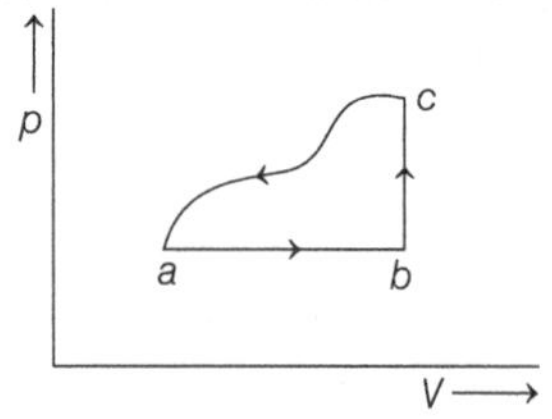

(a) 120 J (b) 130 J (c) 100 J (d) 140 J

33. For the given cyclic process CAB as shown for a gas, the work done is **[JEE Main 2019]**

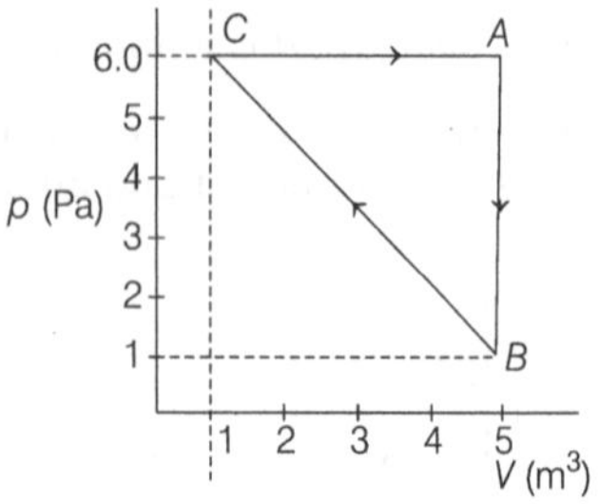

(a) 5 J (b) 10 J (c) 1 J (d) 30 J

34. A gas at pressure p is adiabatically compressed, so that its density becomes twice that of initial value. Given that $\gamma = C_p/C_V = 7/5$, what will be the final pressure of the gas?

(a) $2p$ (b) $\frac{7}{5}p$

(c) $2.63\,p$ (d) p

35. Two isothermal curves are shown in figure at temperatures T_1 and T_2. Which of the following relations is correct?

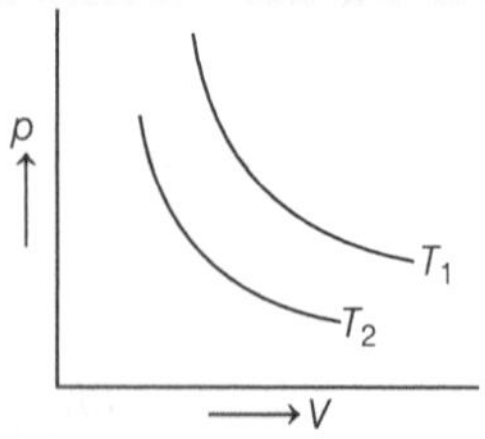

(a) $T_1 > T_2$ (b) $T_1 < T_2$

(c) $T_1 = T_2$ (d) $T_1 = \frac{1}{2}T_2$

36. In the indicator diagram, net amount of work done will be

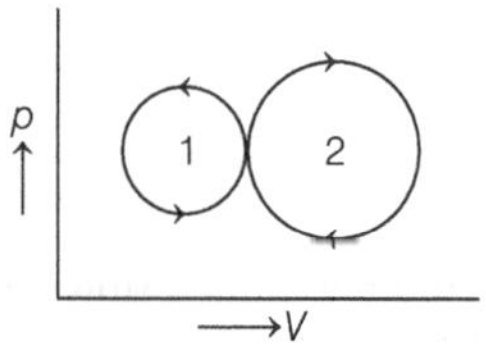

(a) positive (b) zero (c) infinity (d) negative

37. A cyclic process is shown in figure. Work done during isobaric expansion is

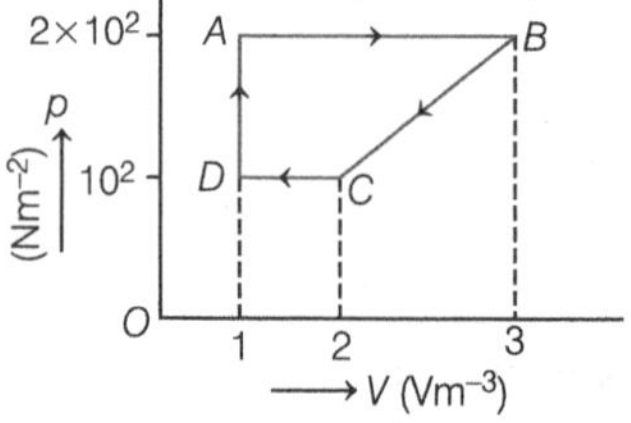

(a) 1600 J (b) 100 J
(c) 400 J (d) 600 J

38. An ideal monoatomic gas is taken around the cycle $ABCD$ as shown in p *versus* V diagram. Work done during the cycle is

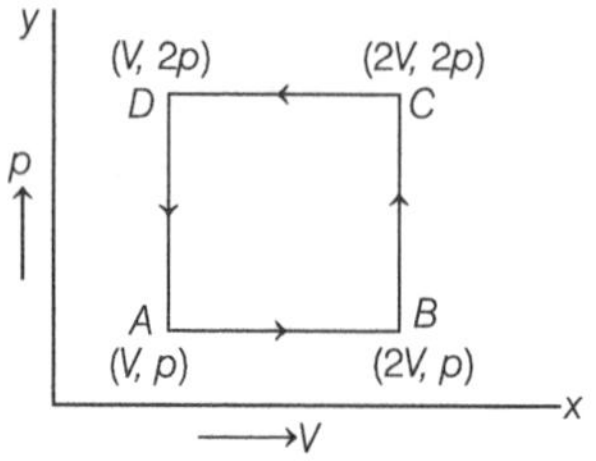

(a) pV (b) $0.5\,pV$
(c) $2\,pV$ (d) $3\,pV$

39. Figure shows a thermodynamic process on one mole of a gas. How does the work done in the process changes with time?

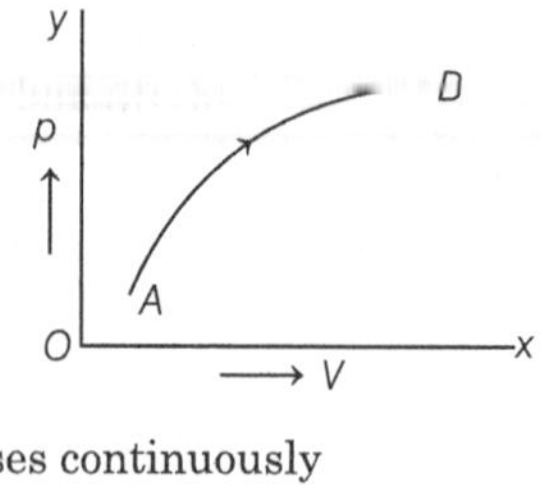

(a) decreases continuously
(b) increases continuously
(c) remains constant
(d) first increases and then decreases

40. A cylinder with a movable piston contains 3 moles of hydrogen at standard temperature and pressure. The walls of the cylinder are made of heat insulator and the piston is insulated by having a pile of sand on it. By what factor does the pressure of the gas increase, if the gas is compressed to half of its original volume? **[NCERT Exemplar]**

(a) 1.40
(b) 1.60
(c) 2.64
(d) 1.94

41. In the following *p-V* diagram, two adiabatic curves and two isothermal curves at T_1 and T_2. The value of V_b / V_c is

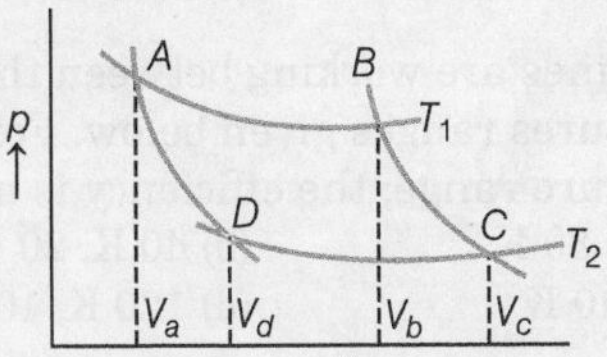

(a) $= V_a / V_d$ (b) $< V_a / V_d$
(c) $> V_a / V_d$ (d) Cannot say

42. For an adiabatic expansion of an ideal gas, the fractional change in its pressure is equal to (where, γ is the ratio of specific heats) **[JEE Main 2021]**

(a) $-\gamma \frac{dV}{V}$ (b) $-\gamma \frac{V}{dV}$
(c) $-\frac{1}{\gamma} \frac{dV}{V}$ (d) $\frac{dV}{V}$

43. One mole of an ideal gas expands adiabatically from an initial temperature T_1 to a final temperature T_2. The work done by the gas would be

(a) $(C_p - C_V)(T_1 - T_2)$
(b) $C_p(T_1 - T_2)$
(c) $C_V(T_1 - T_2)$
(d) $(C_p - C_V)(T_1 + T_2)$

44. A gas at pressure 6×10^5 Nm^{-2} and volume 1 m^3. Its pressure falls to 4×10^5 Nm^{-2} when its volume is 3m^3.Given that the indicator diagram is a straight line, work done by the system is

(a) 6×10^5 J (b) 3×10^5 J
(c) 4×10^5 J (d) 10×10^5 J

45. A thermodynamic system goes from state (i) (*p*, *V*) to (2 *p*, *V*) and (ii) (*p*,*V*) to (*p*, 2*V*). Work done in the two cases is

(a) zero, zero (b) zero, *pV*
(c) *pV*, zero (d) *pV*, *pV*

46. In the indicator diagram, T_a, T_b, T_c, T_d represent temperatures of gas at *A, B, C, D,* respectively. Which of the following is correct relation?

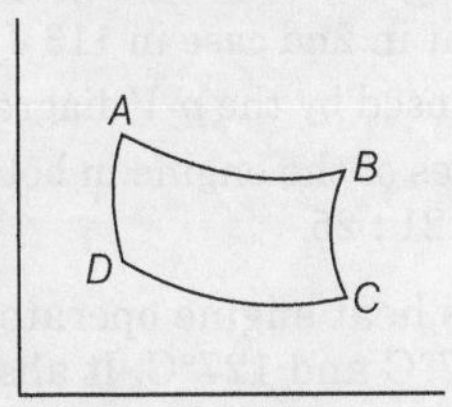

(a) $T_a = T_b = T_c = T_d$ (b) $T_a \neq T_b \neq T_c \neq T_d$
(c) $T_a = T_b$ and $T_c = T_d$ (d) None of these

47. The pressure inside a tyre is 4 atm at 27°C. If the tyre bursts suddenly, new temperature will be (γ = 7/5)

(a) $300\ (4)^{7/2}$ (b) $300\ (4)^{2/7}$
(c) $300\ (2)^{7/2}$ (d) $300\ (4)^{-2/7}$

48. A monoatomic ideal gas, initially at temperature T_1 is enclosed in a cylinder fitted with a frictionless piston. The gas is allowed to expand adiabatically to a temperature T_2 by releasing the piston suddenly. If L_1, L_2 are the lengths of the gas column before and after expansion respectively, then T_1 / T_2 is given by

(a) $(L_1 / L_2)^{2/3}$ (b) (L_1 / L_2)
(c) L_1 / L_2 (d) $(L_2 / L_1)^{2/3}$

49. For adiabatic expansion of a perfect monoatomic gas, when volume increases by 24%. What is the percentage decrease in pressure?

(a) 24% (b) 30%
(c) 48% (d) 71%

50. Starting with the same initial conditions, an ideal gas expands from volume V_1 to V_2 in three different ways. The work done by the gas is W_1 if the process is purely isothermal, W_2 if purely isobaric and W_3 if purely adiabatic. Then,

(a) $W_2 > W_1 > W_3$
(b) $W_2 > W_3 > W_1$
(c) $W_1 > W_2 > W_3$
(d) $W_1 > W_3 > W_2$

51. One litre of dry air at STP is allowed to expand to a volume of 3 L under adiabatic conditions. If γ = 1.40, the work done is ($3^{1.4} = 4.6555$)

(a) 48 J (b) 60.7 J
(c) 90.5 J (d) 100.8 J

52. An ideal gas is heated at constant pressure and absorbs amount of heat *Q*. If the adiabatic exponent is γ, then the fraction of heat absorbed in raising the internal energy and performing the work, is

(a) $1 - \frac{1}{\gamma}$ (b) $1 + \frac{1}{\gamma}$
(c) $1 - \frac{2}{\gamma}$ (d) $1 + \frac{2}{\gamma}$

53. In changing the state of a gas adiabatically from an equilibrium state *A* to another equilibrium state *B*, an amount of work equal to 22.3 J is done on the system. If the gas is taken from state *A* to *B via* a process in which the net heat absorbed by the system is 9.35 cal, how much is the net work done by the system in the latter case?

(a) 15.6 J (b) 11.2 J
(c) 14.9 J (d) 16.9 J

54. During an isothermal expansion, a confined ideal gas does –150 J of work against its surrounding. This implies that
(a) 150 J of heat has been added to the gas
(b) 150 J heat has been removed from the gas
(c) 300 J of heat has been added to the gas
(d) No heat is transferred because the process is isothermal

Second Law of Thermodynamics, Refrigerator and Carnot Engine

55. The efficiency of a Carnot engine working between 800 K and 500 K is
(a) 0.4 (b) 0.625 (c) 0.375 (d) 0.5

56. A Carnot engine whose sink is at 300 K has an efficiency of 40%. By how much should the temperature of source be increased, so as to increase its efficiency by 50% of original efficiency ?
(a) 280 K (b) 275 K (c) 325 K (d) 250 K

57. An engine has an efficiency of 1/3. The amount of work this engine can perform per kilocalorie of heat input is
(a) 1400 cal (b) 700 cal
(c) 700 J (d) 1400 J

58. A Carnot engine whose low temperature reservoir is at 27°C has an efficiency 37.5%. The high temperature reservoir is at
(a) 480°C (b) 327°C
(c) 307°C (d) 207°C

59. The coefficient of performance of a refrigerator working between 10°C and 20°C is
(a) 28.3 (b) 29.3
(c) 2 (d) Cannot be calculated

60. A reversible heat engine converts (1/6)th of heat it absorbs from source into work. When temperature of source is 600 K, temperature at which heat exhausts is
(a) 500 K (b) 100 K
(c) 0 K (d) 600 K

61. A Carnot engine having an efficiency of $\frac{1}{10}$ is being used as a refrigerator. If the work done on the refrigerator is 10 J, then the amount of heat absorbed from the reservoir at lower temperature is: [JEE Main 2020]
(a) 99 J (b) 100 J (c) 90 J (d) 1 J

62. An ideal Carnot engine whose efficiency is 40% receives heat at 500 K. If its efficiency were 50%, then in take temperature for same exhaust temperature would be
(a) 700 K (b) 900 K
(c) 800 K (d) 600 K

63. A Carnot's engine works between a source at a temperature of 27°C and a sink at – 123°C. Its efficiency is
(a) 0.5 (b) 0.25 (c) 0.75 (d) 0.4

64. Four engines are working between the given temperatures ranges given below. For which temperature range, the efficiency is maximum?
(a) 100 K, 80 K (b) 40 K, 20 K
(c) 60 K, 40 K (d) 120 K, 100 K

65. In a refrigerator, the low temperature coil of evaporator is at – 23°C and the compressed gas in the condenser has a temperature of 77°C. How much electrical energy is spent in freezing 1 kg of water already at 0°C ?
(a) 134400 J (b) 1344 J
(c) 80000 J (d) 3200 J

66. A refrigerator absorbs 2000 cal of heat from ice trays. If the coefficient of performance is 4, then work done by the motor is
(a) 2100 J (b) 4200 J
(c) 8400 J (d) 500 J

67. A Carnot engine has same efficiency between (i) 100 K and 500 K, (ii) T K and 900 K. The value of T is
(a) 180 K (b) 90 K (c) 270 K (d) 360 K

68. A refrigerator works between temperature of melting ice and room temperatures (17°C). The amount of energy (in kWh) that must be supplied to freeze 1 kg of water at 0°C is
(a) 1.4 (b) 1.8 (c) 0.058 (d) 2.5

69. A Carnot's engine working between 400 K and 800 K has a work output of 1200 J per cycle. The amount of heat energy supplied to the engine from the source in each cycle is [JEE Main 2021]
(a) 3200 J (b) 1800 J (c) 1600 J (d) 2400 J

70. A Carnot engine used first ideal monoatomic gas and then an ideal diatomic gas, if the source and sink temperatures are 411°C and 69°C, respectively and the engine extracts 1000 J of heat from the source in each cycle, then
(a) area enclosed by the p-V diagram is 10 J
(b) heat energy rejected by engine is 1st case is 600 J while that in 2nd case in 113 J
(c) area enclosed by the p-V diagram is 500 J
(d) efficiencies of the engine in both the cases are in the ratio 21 : 25

71. An ideal gas heat engine operates in Carnot cycle between 227°C and 127°C. It absorbs 6×10^4 cal of heat at higher temperature. Amount of heat converted into work is
(a) 1.2×10^4 cal (b) 2.4×10^4 cal
(c) 6×10^4 cal (d) 4.8×10^4 cal

72. Two heat engines A and B have their sources at 1000 K and 1100 K and their sinks are at 500 K and 400 K, respectively. What is true about their efficiencies?

(a) $\eta_A = \eta_B$ (b) $\eta_A > \eta_B$
(c) $\eta_A < \eta_B$ (d) Cannot say

73. A Carnot engine has the same efficiency between 800 K to 500 K and x K to 600 K. The value of x is

(a) 100 K (b) 960 K (c) 846 K (d) 754 K

74. What is the temperature of source in Carnot cycle of 10% efficiency, when heat exhausts at 270 K?

(a) 400 K (b) 500 K
(c) 300 K (d) 600 K

75. A Carnot engine take 3×10^6 cal of heat from a reservoir at 627°C and gives it to a sink at 27°C. The work done by the engine is

(a) 4.2×10^6 J (b) 8.4×10^6 J
(c) 16.8×10^6 J (d) zero

ROUND II Mixed Bag

Only One Correct Option

1. A given system undergoes a change in which the work done by the system equals the decrease in its internal energy. The system must have undergone an

(a) isothermal change (b) adiabatic change
(c) isobaric change (d) isochoric change

2. One mole of an ideal monoatomic gas is heated at a constant pressure of 1 atm from 0°C to 100°C. Work done by the gas is

(a) 8.31×10^3 J (b) 8.31×10^{-3} J
(c) 8.31×10^{-2} J (d) 8.31×10^2 J

3. The change in internal energy, when a gas is cooled from 927°C to 27°C

(a) 300% (b) 400% (c) 200% (d) 100%

4. A mass of dry air at NTP is compressed to (1/20)th of its original volume suddenly. If $\gamma = 1.4$, the final pressure would be

(a) 20 atm (b) 66.28 atm
(c) 30 atm (d) 150 atm

5. An ideal gas heat engine is operating between 227°C and 127°C. It absorbs 10^4 J of heat at the higher temperature. The amount of heat converted into work is

(a) 2000 J (b) 4000 J (c) 8000 J (d) 5600 J

6. The given p-V diagram represents the thermodynamic cycle of an engine, operating with an ideal monatomic gas. The amount of heat, extracted from the source in a single cycle is **[JEE Main 2013]**

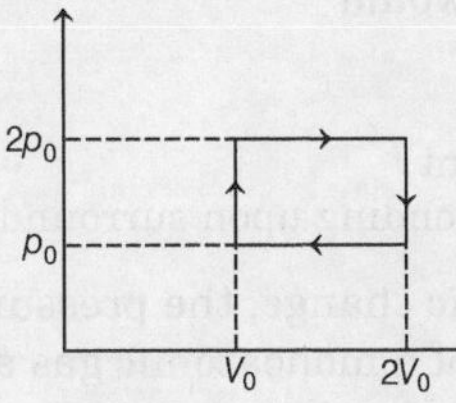

(a) p_0V_0 (b) $\frac{13}{2} p_0V_0$ (c) $\frac{11}{2} p_0V_0$ (d) $4p_0V_0$

7. A solid body of constant heat capacity 1 J/°C is being heated by keeping it in contact with reservoirs in two ways **[JEE Main 2015]**

(i) Sequentially keeping in contact with 2 reservoirs such that each reservoir supplies same amount of heat.

(ii) Sequentially keeping in contact with 8 reservoirs such that each reservoir supplies same amount of heat.

In both the cases, body is brought from initial temperature 100°C to final temperature 200°C. Entropy change of the body in the two cases respectively, is

(a) ln 2, 4 ln 2 (b) ln 2, ln 2
(c) ln 2, 2 ln 2 (d) 2 ln 2, 8 ln 2

8. If AB is an isothermal, BC is an isochoric and AC is an adiabatic curve, which of the graph correctly represents them in figure

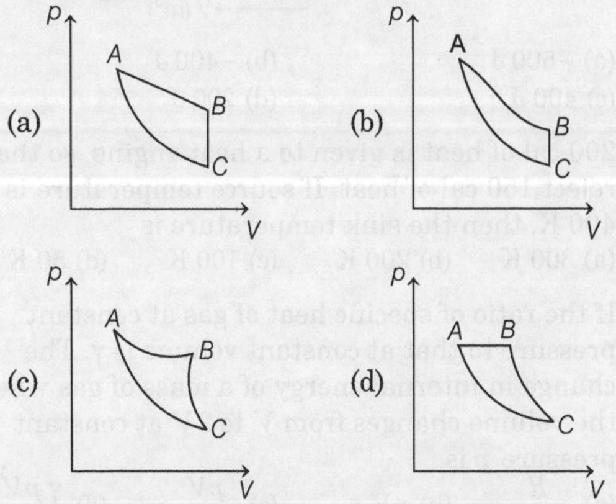

9. In an isothermal reversible expansion, if the volume of 96 g of oxygen at 27°C in increased from 70 L to 140 L, then the work done by the gas will be

(a) $300 R \log_{10} 2$
(b) $81 R \log_e 2$
(c) $900 R \log_{10} 2$
(d) $2.3 \times 900 R \log_{10} 2$

10. An ideal gas at 27°C is compressed adiabatically to $\frac{8}{27}$ of its original volume. If $\gamma = \frac{5}{3}$, then the rise of temperature is
(a) 450 K (b) 375 K
(c) 225 K (d) 405 K

11. A perfect gas goes from state A to state B by absorbing 8×10^5 J of heat and doing 6.5×10^5 J of external work. It is now transferred between the same two states in another process in which it absorbs 10^5 J of heat. In the second process,
(a) work done on gas is 10^5 J
(b) work done on gas is -0.5×10^5 J
(c) work done by gas is 10^5 J
(d) work done by gas is 0.5×10^5 J

12. At NTP, one mole of diatomic gas is compressed adiabatically to half of its volume and $\gamma = 1.41$. The work done on gas will be
(a) 1280 J (b) 1610 J
(c) 1815 J (d) 2025 J

13. The p-V diagram of a diatomic ideal gas system going under cyclic process as shown in figure. The work done during an adiabatic process CD is (Use, $\gamma = 1.4$) [JEE Main 2021]

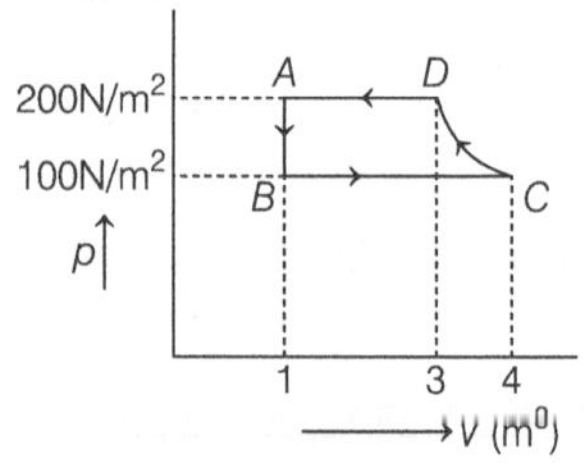

(a) −500 J (b) −400 J
(c) 400 J (d) 200 J

14. 200 cal of heat is given to a heat engine, so that it reject 150 cal of heat. If source temperature is 400 K, then the sink temperature is
(a) 300 K (b) 200 K (c) 100 K (d) 50 K

15. If the ratio of specific heat of gas at constant pressure to that at constant volume is γ. The change in internal energy of a mass of gas when the volume changes from V to $2V$ at constant pressure p is
(a) $\frac{R}{(\gamma-1)}$ (b) pV (c) $\frac{pV}{(\gamma-1)}$ (d) $\frac{\gamma pV}{(\gamma-1)}$

16. Two moles of an ideal monoatomic gas at 27°C occupies a volume of V. If the gas is expanded adiabatically to the volume $2V$, then the work done by the gas will be $\left(\gamma = \frac{5}{3}, R = 8.31 \text{ J/mol K}\right)$
(a) + 2767.23 J (b) 2627.23 J
(c) 2500 J (d) −2500 J

17. One mole of O_2 gas having a volume equal to 22.4 L at 0°C and 1 atmospheric pressure is compressed isothermally, so that its volume reduces to 11.2 L. The work done in this process is
(a) 1672.5 J (b) 1728 J
(c) −1728 J (d) −1572.5 J

18. The volume of an ideal gas is 1 L and its pressure is equal to 72 cm of mercury column. The volume of gas is made 900 cm^3 by compressing it isothermally. The stress of the gas will be
(a) 8 cm (mercury) (b) 7 cm (mercury)
(c) 6 cm (mercury) (d) 4 m (mercury)

19. The pressure and density of a diatomic gas ($\gamma = 7/5$) change adiabatically from (p, d) to (p', d') if $\frac{d'}{d} = 32$, then $\frac{p'}{p}$ should be
(a) 1/128 (b) 32
(c) 128 (d) None of these

20. Ideal gas undergoes an adiabatic change in its state from $(p_1V_1T_1)$ to $(p_2V_2T_2)$. The work done in the process is (μ = number of moles, C_p and C_V are molar specific heats of gas)
(a) $W = \mu(T_1 - T_2)C_p$ (b) $W = \mu(T_1 - T_2)C_V$
(c) $W = \mu(T_1 + T_2)C_p$ (d) $W = \mu(T_1 + T_2)C_V$

21. 2 kg of water is converted into steam by boiling at atmospheric pressure. The volume changes from 2×10^{-3} m^3 to 3.34 m^3. The work done by the system is about
(a) −340 kJ (b) −170 kJ
(c) 170 kJ (d) 340 kJ

22. The changes in pressure and volume of a gas when heat content of the gas remains constant are called adiabatic changes. The equation of such changes is pV^γ = constant. The changes must be sudden and the container must be perfectly insulting to disallow any exchange of heat with the surroundings. In such changes, $dQ = 0$, then as per first law of thermodynamics, $dQ = dU + W = 0$
$\Rightarrow dU = -dW$

A gas in a container is compressed suddenly, its temperature would
(a) increase
(b) decrease
(c) stay constant
(d) change depending upon surrounding temperature.

23. In an adiabatic change, the pressure and temperature of a monoatomic gas are related as $p \propto T^{-c}$, where c equals
(a) $\frac{2}{5}$ (b) $\frac{5}{2}$ (c) $\frac{3}{5}$ (d) $\frac{5}{3}$

24. The efficiency of a Carnot engine working between source temperature T_1 and sink temperature T_2 is $\eta = 1 - \frac{T_2}{T_1}$. The efficiency cannot be 100%, as we cannot maintain $T_2 = 0$. Coefficient of performance of a refrigerator working between the same two temperature is

$$\frac{T_2}{T_1 - T_2} = \frac{1-\eta}{\eta}$$

The efficiency of a Carnot engine is 40%. If temperature of sink is 27°C, what is the source temperature?

(a) 300 K (b) 400 K
(c) 600 K (d) 500 K

25. Figure shows the p-V diagram of an ideal gas undergoing a change of state from A to B. Four different paths I, II, III and IV as shown in the figure may lead to the same changes of state. **[NCERT Exemplar]**

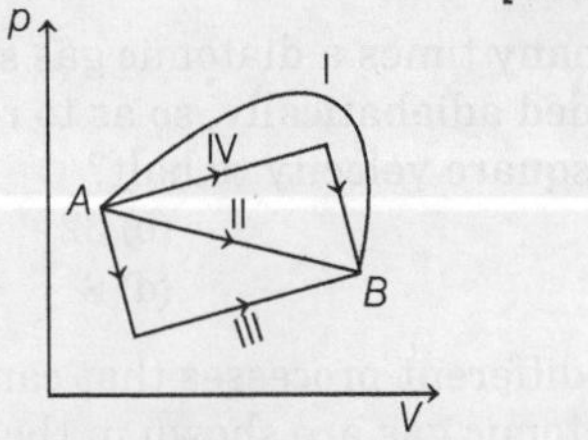

(a) Change in internal energy is same in IV and III cases, but not in I and II
(b) Change in internal energy is different in all the four cases
(c) Work done is maximum in case I
(d) Work done is minimum in case II

26. Consider a heat engine as shown in figure. Q_1 and Q_2 are heat added to heat bath T_1 and heat T_2 is taken from one cycle of engine. W is the mechanical work done on the engine. If $W > 0$, then possibilities are **[NCERT Exemplar]**

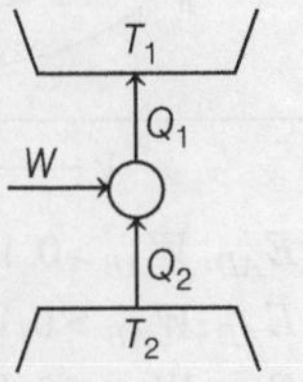

(a) $Q_2 < Q_1 < 0$
(b) $Q_2 > Q_1 > 0$
(c) $Q_1 < Q_2 < 0$
(d) $Q_1 < 0, Q_2 > 0$

27. Consider a cycle followed by an engine (figure)
1 to 2 is isothermal
2 to 3 is adiabatic
3 to 1 is adiabatic

Such a process does not exist because

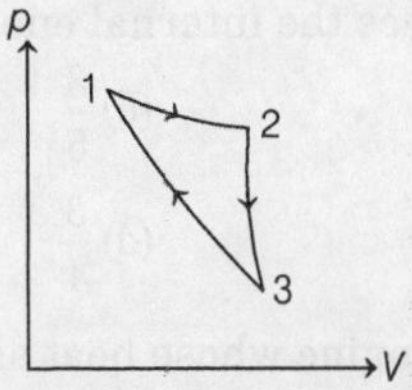

[NCERT Exemplar]

(a) heat is completely converted to mechanical energy in such a process, which is not possible.
(b) mechanical energy is completely converted to heat in this process, which is not possible.
(c) curves representing two adiabatic processes intersect.
(d) curves representing an adiabatic process and an isothermal process don't intersect.

28. p-V plots for two gases during adiabatic processes are shown in figure. Plots 1 and 2 should correspond respectively to

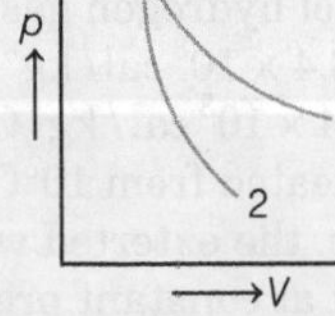

(a) He and O_2 (b) O_2 and He
(c) He and Ar (d) O_2 and N_2

29. Temperature of an ideal gas is 300 K. The change in temperature of the gas when its volume changes from V to $2V$ in the process $p = \alpha V$ (here α is a positive constant) is

(a) 900 K (b) 1200 K
(c) 600 K (d) 300 K

30. A gas under constant pressure of 4.5×10^5 Pa when subjected to 800 kJ of heat changes the volume from 0.5 m³ to 2.0 m³. The change in the internal energy of the gas is

(a) 6.75×10^5 J (b) 5.25×10^5 J
(c) 3.25×10^5 J (d) 1.25×10^5 J

31. A Carnot engine is made to work between 200°C and 0°C first and then between 0°C to −200°C. The ratio of efficiencies of the engine in the two cases is

(a) 1 : 2 (b) 1 : 1
(c) 1.73 : 1 (d) 1 : 1.73

32. Pressure p, volume V and temperature T of a certain material are related by $p = \alpha T^2/V$, where α is constant. Work done by the material when temperature changes from T_0 to $2T_0$ and pressure remains constant is

(a) $3\alpha T_0^2$ (b) $5\alpha T_0^2$ (c) $\frac{3}{2}\alpha T_0^2$ (d) $7\alpha T_0^2$

33. When the ideal monoatomic gas is heated at constant pressure fraction of heat energy supplied which increases the internal energy of gas is

(a) $\frac{2}{5}$ (b) $\frac{3}{5}$
(c) $\frac{3}{7}$ (d) $\frac{3}{4}$

34. If a Carnot engine whose heat sink is at 27°C has an efficiency of 40%. By how many degrees should the temperature of the source be changed to increase the efficiency by 10% of the original efficiency? [JEE Main 2021]

(a) 50.5°C (b) 60.2°C
(c) 35.7°C (d) 72.8°C

35. A refrigerator is to remove heat from the eatable kept inside at 9°C. Calculate the coefficient of performance, if the room temperature is 36°C. [NCERT]

(a) 10.4 (b) 11.5
(c) 9.8 (d) None of these

36. The specific heat of hydrogen gas at constant pressure is $C_p = 3.4 \times 10^3$ cal/ kg° C and at constant volume is $C_V = 2.4 \times 10^3$ cal/ kg° C. If one kilogram hydrogen gas is heated from 10°C to 20°C at constant pressure, the exterted work done on the gas to maintain it at constant pressure is

(a) 10^5 cal (b) 10^4 cal
(c) 10^3 cal (d) 5×10^3 cal

37. In changing the state of a gas adiabatically from an equillibrium state A to another equillibrium state B, an amount of work equal to 22.3 J is done on the system. If the gas is taken from state A to B *via* a process in which the net heat absorbed by the system is 9.35 cal, the net work done by the system in latter case will be

(a) 5.9 J (b) 16.9 J
(c) 9.3 J (d) 4.6 J

38. 540 cal of heat converts 1 cubic centimeter of water at 100°C into 1670 cubic centimeter of steam at 100°C at a pressure of one atmosphere. Then, the work done against the atmospheric pressure is nearly

(a) 540 cal (b) 40 cal
(c) 200 cal (d) 500 cal

39. 5.6 L of helium gas at STP is adiabatically compressed to 0.7 L. Taking the initial temperature to be T_1, the work done in the process is

(a) $\frac{9}{8}RT_1$ (b) $\frac{3}{2}RT_1$
(c) $\frac{15}{8}RT_1$ (d) $\frac{9}{2}RT_1$

40. An ideal gas is taken through the cycle $A \to B \to C \to A$, as shown in figure. If the net heat supplied to the gas in cycle is 5J, work done by the gas in the process $C \to A$

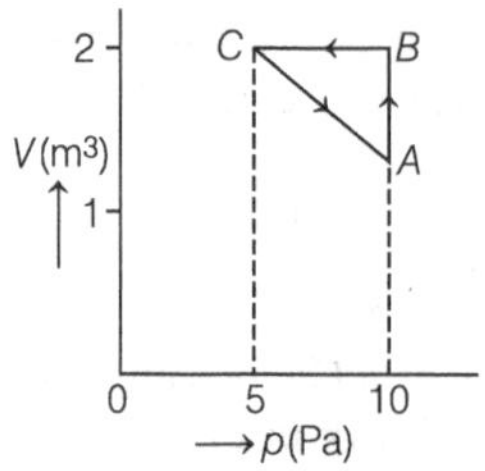

(a) –5 J (b) –10 J
(c) 15 J (d) –20 J

41. For a monoatomic gas, work done at constant pressure is W. The heat supplied at constant volume for the same rise in temperature of the gas is

(a) $W/2$ (b) $3W/2$
(c) $5W/2$ (d) W

42. How many times a diatomic gas should be expanded adiabatically, so as to reduce the root mean square velocity to half?

(a) 64 (b) 32
(c) 16 (d) 8

43. Three different processes that can occur in an ideal monoatomic gas are shown in the p *versus* V diagram. The paths are labelled as $A \to B, A \to C$ and $A \to D$. The change in internal energies during these process are taken as E_{AB}, E_{AC} and E_{AD} and the work done as W_{AB}, W_{AC} and W_{AD}.

The correct relation between these parameters are [JEE Main 2020]

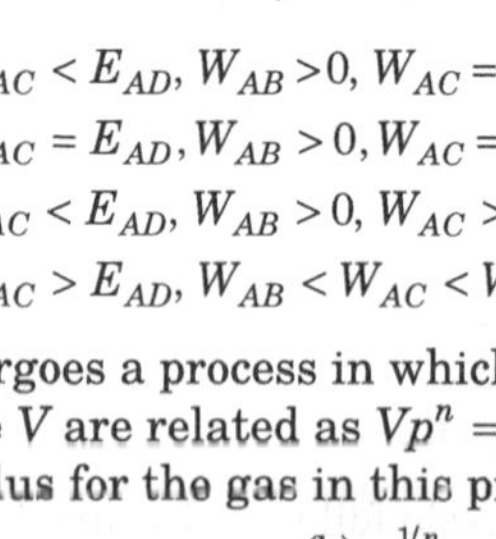

(a) $E_{AB} = E_{AC} < E_{AD}, W_{AB} > 0, W_{AC} = 0, W_{AD} < 0$
(b) $E_{AB} = E_{AC} = E_{AD}, W_{AB} > 0, W_{AC} = 0, W_{AD} < 0$
(c) $E_{AB} < E_{AC} < E_{AD}, W_{AB} > 0, W_{AC} > W_{AD}$
(d) $E_{AB} > E_{AC} > E_{AD}, W_{AB} < W_{AC} < W_{AD}$

44. A gas undergoes a process in which its pressure p and volume V are related as Vp^n = constant. The Bulk modulus for the gas in this process is

(a) np (b) $p^{1/n}$
(c) $\frac{p}{n}$ (d) p^n

45. An ideal gas expands isothermally from a volume V_1 to V_2 and then compressed to original volume V_1 adiabatically. Initial pressure is p_1, final pressure is p_3 and total work done is W. Then,

(a) $p_3 > p_1; W > 0$ (b) $p_3 < p_1; W < 0$
(c) $p_3 > p_1; W < 0$ (d) $p_3 = p_1; W = 0$

46. One litre of dry air at STP expands adiabatically to a volume of 3 L. If $\gamma = 1.40$, the work done by air is ($3^{1.4} = 4.6555$)
(Take, air to be an ideal gas) **[JEE Main 2020]**

(a) 100.8 J (b) 90.5 J
(c) 48 J (d) 60.7 J

47. Certain amount of an ideal gas is contained in a closed vessel. The vessel is moving with a constant velocity v. The rise in temperature of the gas when the vessel is suddenly stopped is (M is molecular mass, $\gamma = C_p/C_V$)

(a) $\frac{Mv^2(\gamma-1)}{2R}$ (b) $\frac{Mv^2(\gamma+1)}{2R}$
(c) $\frac{Mv^2}{2R_\gamma}$ (d) $\frac{Mv^2}{2R(\gamma+1)}$

48. An ideal gas is made to go through a cyclic thermodynamical process in four steps. The amount of heat involved are $Q_1 = 600$ J, $Q_2 = -400$ J, $Q_3 = -300$ J and $Q_4 = 200$ J, respectively. The corresponding work involved are $W_1 = 300$ J, $W_2 = -200$ J, $W_3 = -150$ J and W_4. What is the value of W_4?

(a) –50 J (b) 100 J (c) 150 J (d) 50 J

49. A thermodynamic cycle $xyzx$ is shown on a V-T diagram. The p-V diagram that best describes this cycle is (diagrams are schematic and not to scale) **[JEE Main 2020]**

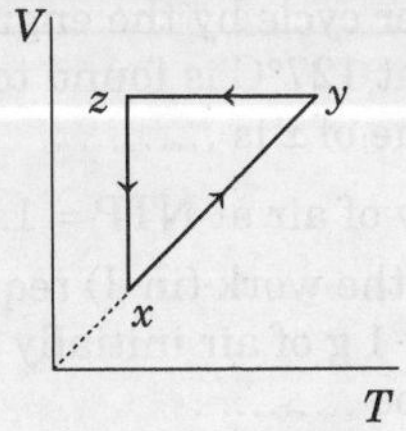

(a) 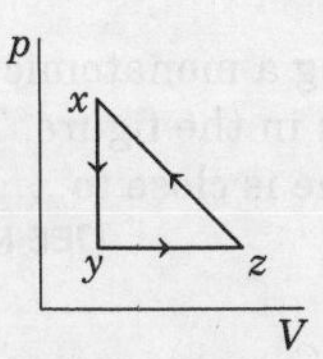 (b) 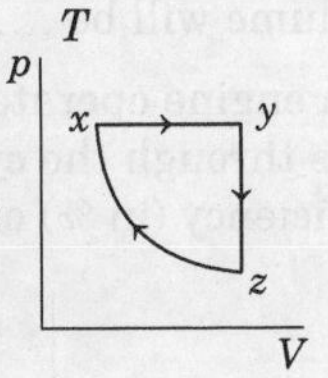

(c) 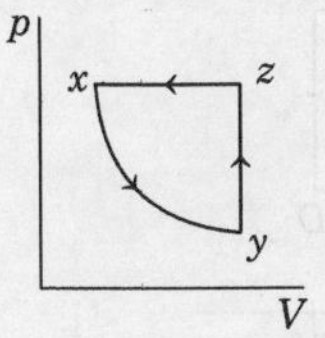 (d) 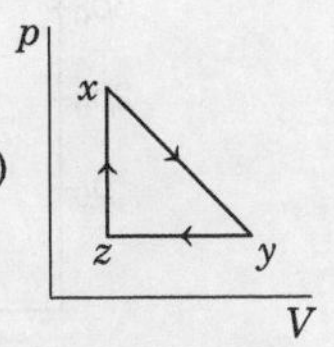

50. An ideal gas is taken from the state A (p, V) to the state B $(p/2, 2V)$ along a straight line path as shown in figure. Select the correct statement from the following.

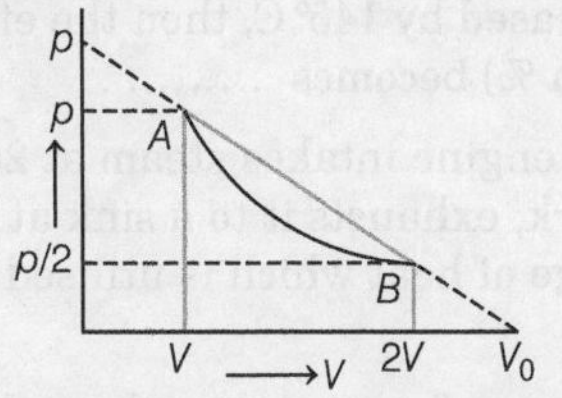

(a) Work done by the gas in going from A to B exceeds the work done in going from A to B under isothermal conditions.

(b) In the T-V diagram, part AB would become a hyperbola.

(c) In the p-T diagram, path AB would be part of hyperbola.

(d) In going from A to B, the temperature T of gas first increases to a maximum value 1 and then decreases.

51. Match the thermodynamic processes taking place in a system with the correct conditions. In the table, ΔQ is heat supplied, ΔW is work done and ΔU is change in internal energy of the system. **[JEE Main 2020]**

Process	Condition
A. Adiabatic	p. $\Delta W = 0$
B. Isothermal	q. $\Delta Q = 0$
C. Isochoric	r. $\Delta U \neq 0, \Delta W \neq 0$ and $\Delta Q \neq 0$
D. Isobaric	s. $\Delta U = 0$

	A	B	C	D
(a)	q	p	s	r
(b)	p	p	q	r
(c)	q	s	p	r
(d)	p	q	s	s

52. n moles of an ideal gas undergoes a process A to B as shown in the figure. The maximum temperature of the gas during the process will be **[JEE Main 2016]**

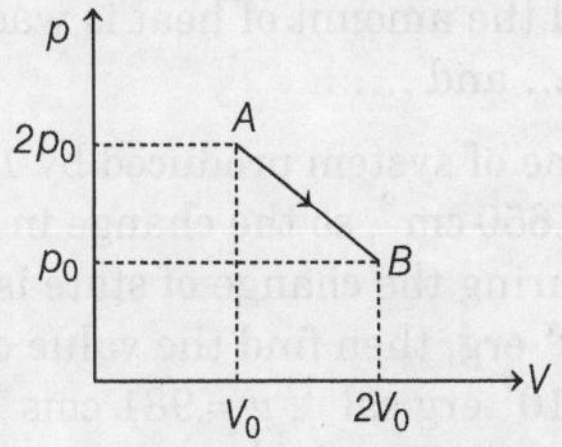

(a) $\frac{9}{4}\frac{p_0V_0}{nR}$ (b) $\frac{3}{2}\frac{p_0V_0}{nR}$
(c) $\frac{9}{2}\frac{p_0V_0}{nR}$ (d) $\frac{9p_0V_0}{nR}$

Numerical Value Questions

53. A Carnot's engine, with its cold body at 17° C has 50% efficiency. If the temperature of its hot body is now increased by 145° C, then the efficiency of engine (in %) becomes ....

54. A Carnot engine intakes steam at 200°C and after doing work, exhausts it to a sink at 100°C, so the percentage of heat which is utilised for doing work is ....

55. A fixed mass of gas is taken through a process $A \to B \to C \to A$, here $A \to B$ is isobaric $B \to C$ is adiabatic and $C \to A$ is isothermal.

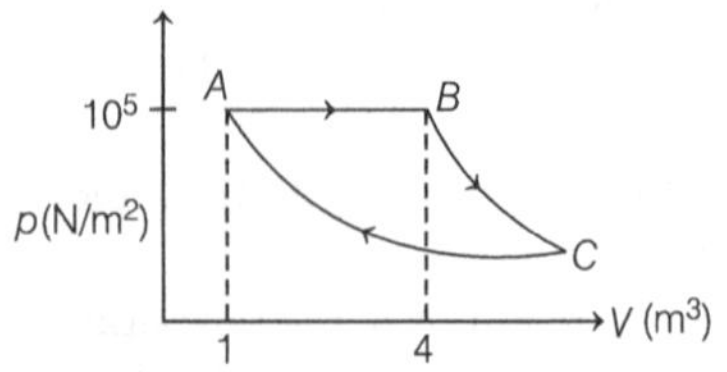

If the pressure at point C is $\frac{10^x}{64}$ N / m², then the value of x is .... (Take, $\gamma = 1.5$)

56. An ideal gas is taken through a cyclic thermodynamic process through four steps. The amount of heat involved in these steps are $Q_1 = 5960$ J, $Q_2 = -5585$ J, $Q_3 = -2980$ J and $Q_4 = 3645$ J, respectively. The efficiency (in %) of the cycle is ....

57. An engine takes in 5 moles of air at 20°C and 1 atm and compresses it adiabatically to (1/10)th of the original volume. Assuming air to be a diatomic ideal gas made up of rigid molecules, the change in its internal energy during this process comes out to be XkJ. The value of X to the nearest integer is .... **[JEE Main 2020]**

58. A steam engine delivers 5.4×10^8 J of work per minute and takes 3.6×10^9 J of heat per minute from the boiler. So, the efficiency of the engine (in %) and the amount of heat is wasted (in J/min) will be ...... and ....

59. The volume of system produced by 1g of water at 100°C is 1650 cm^3, so the change in internal energy (in erg) during the change of state is found to be 21.01×10^x erg, then find the value of x. (Take, $J = 4.2 \times 10^7$ erg cal^{-1}, $g = 981$ cms^{-2} and latent heat of steam = 50 calg^{-1})

60. A Carnot engine operates between two reservoirs of temperatures 900 K and 300 K. The engine performs 1200 J of work per cycle. The heat energy (in J) delivered by the engine to the low temperature reservoir in a cycle, is .... **[JEE Main 2020]**

61. A gram molecule of a gas at 127°C expands isothermally until its volume is doubled, so the amount of heat energy (in cal) absorbed will be ....

62. Starting at temperature 300 K, one mole of an ideal diatomic gas ($\gamma = 1.4$) is first compressed adiabatically from volume V_1 to $V_2 = \frac{V_1}{16}$. It is then allowed to expand isobarically to volume $2V_2$. If all the processes are the quasi-static, then the final temperature of the gas (in K) is (to the nearest integer) .... **[JEE Main 2020]**

63. A cylinder containing one gram molecule of the gas was compressed adiabatically until its temperature raise from 27°C to 97°C. The heat (in cal) produced in the gas is .... (Take, $\gamma = 1.5$)

64. 5 moles of an ideal gas is carried by a quasi-static isothermal process at 500 K to twice its volume as shown in figure.

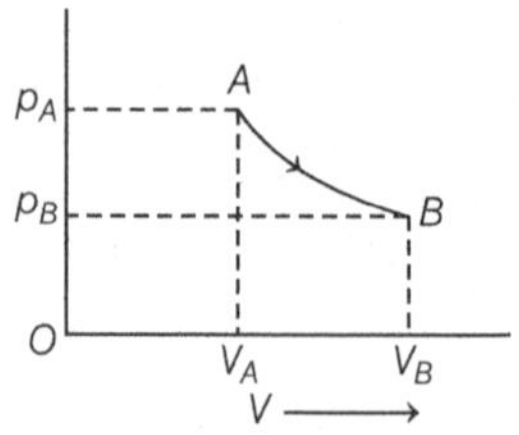

The work done (in J) by the gas along the path AB is .... (Given, $R = 8.31$ J mol^{-1} K^{-1})

65. A Carnot engine absorbs 6×10^5 cal at 227°C. The work done per cycle by the engine, if its sink is maintained at 127°C is found to be 5.04×10^x J, then the value of x is ....

66. If the density of air at NTP = 1.29×10^{-3} gcm^{-3} and $\gamma = 1.4$, then the work (in J) required to compress adiabatically 1 g of air initially at NTP to half of its volume will be ....

67. An engine operates by taking a monatomic ideal gas through the cycle shown in the figure. The efficiency (in %) of the engine is close to .... **[JEE Main 2020]**

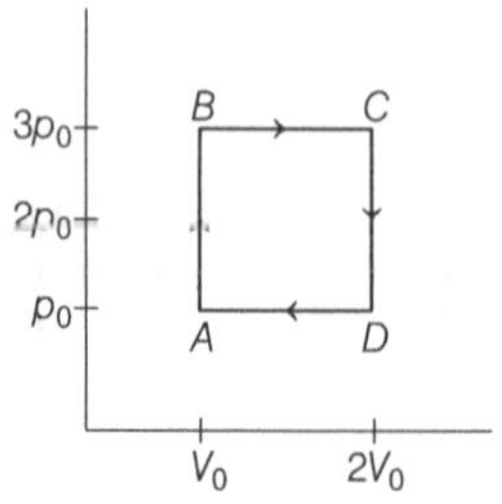

Answers

Round I

1. (d)	**2.** (b)	**3.** (b)	**4.** (d)	**5.** (c)	**6.** (a)	**7.** (c)	**8.** (a)	**9.** (b)	**10.** (b)
11. (b)	**12.** (b)	**13.** (d)	**14.** (c)	**15.** (c)	**16.** (b)	**17.** (a)	**18.** (a)	**19.** (a)	**20.** (d)
21. (a)	**22.** (d)	**23.** (a)	**24.** (c)	**25.** (d)	**26.** (b)	**27.** (c)	**28.** (c)	**29.** (d)	**30.** (a)
31. (b)	**32.** (b)	**33.** (b)	**34.** (c)	**35.** (a)	**36.** (a)	**37.** (c)	**38.** (a)	**39.** (b)	**40.** (c)
41. (a)	**42.** (a)	**43.** (c)	**44.** (d)	**45.** (b)	**46.** (c)	**47.** (d)	**48.** (d)	**49.** (b)	**50.** (a)
51. (c)	**52.** (a)	**53.** (d)	**54.** (b)	**55.** (c)	**56.** (d)	**57.** (d)	**58.** (d)	**59.** (a)	**60.** (a)
61. (c)	**62.** (d)	**63.** (a)	**64.** (b)	**65.** (a)	**66.** (a)	**67.** (a)	**68.** (c)	**69.** (d)	**70.** (c)
71. (a)	**72.** (c)	**73.** (b)	**74.** (c)	**75.** (b)					

Round II

1. (b)	**2.** (d)	**3.** (a)	**4.** (b)	**5.** (a)	**6.** (b)	**7.** (b)	**8.** (b)	**9.** (d)	**10.** (b)
11. (b)	**12.** (c)	**13.** (a)	**14.** (a)	**15.** (c)	**16.** (a)	**17.** (d)	**18.** (a)	**19.** (c)	**20.** (b)
21. (d)	**22.** (a)	**23.** (a)	**24.** (d)	**25.** (c)	**26.** (a)	**27.** (a)	**28.** (a)	**29.** (b)	**30.** (d)
31. (d)	**32.** (a)	**33.** (b)	**34.** (c)	**35.** (a)	**36.** (b)	**37.** (b)	**38.** (b)	**39.** (a)	**40.** (a)
41. (b)	**42.** (b)	**43.** (b)	**44.** (c)	**45.** (c)	**46.** (b)	**47.** (a)	**48.** (c)	**49.** (b)	**50.** (a)
51. (c)	**52.** (a)	**53.** 60	**54.** 21.14	**55.** 5	**56.** 10.82	**57** 4	**58.** 15, 3.1×10^{19}	**59.** 9	
60. 600	**61.** 548	**62.** 1819	**63.** 276.7	**64.** 14401.3	**65.** 5	**66.** – 62.7	**67.** 19		

Solutions

Round I

1. Here, $dQ = 400$ cal, $dW = -105$ J

$$= -105/4.2 \text{ cal}$$
$$= -25 \text{ cal}; \ dU = ?$$

Now, $dU = dQ - dW$

$$dU = 400 - (-25) = 425 \text{ cal}$$

Note dW is negative because work is done on the system.

2. Here, $dQ = 110$ J, $dU = 40$ J, $dW = ?$

From $dQ = dU + dW$

$$dW = dQ - dU = 110 - 40 = 70 \text{ J}$$

3. As, $\Delta U = \Delta Q - \Delta W$

$$\therefore \quad \Delta U = (-20) - (-8) = -12 \text{ J}$$
$$\Delta U = U_f - U_i = -12$$
$$\therefore \quad U_f = -12 + U_i$$
$$= -12 + 30 = 18 \text{ J}$$

4. As indicator diagrams in all the three cases are closed curves, representing cyclic changes, therefore U = constant and $\Delta U = 0$ in all the cases.

5. As, $dU = dQ - dW$

$$= mL - p(dV)$$
$$= 1 \times 540 - \frac{1.013 \times 10^5 \,(1671 - 1)10^{-6}}{4.2}$$
$$= 540 - 40 = 500 \text{ cal}$$

6. Heat and work are treated as path functions in thermodynamics.

$$\Delta Q = \Delta U + \Delta W$$

Since work done by gas depends on type of process, i.e. path and ΔU depends just on initial and final states, so ΔQ i.e. heat, also has to depend on process is path.

7. According to the first law of thermodynamics,
Heat supplied (ΔQ) = Work done (W)
+ Change in internal energy of the system (ΔU)

$$\Delta Q_A = \Delta U_A + W_A$$

Similarly, for process B,

$$\Delta Q_B = \Delta U_B + W_B$$

Now, we know that,
Work done for a process = Area under its p-V curve

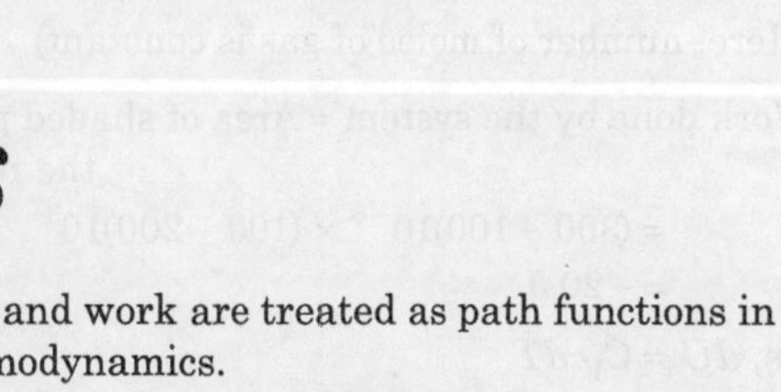

Thus, it is clear from the above graphs,

$$W_A > W_B \qquad \ldots\text{(i)}$$

Also, since the initial and final state are same in both process, so

$$\Delta U_A = \Delta U_B \qquad \ldots\text{(ii)}$$

So, from Eqs. (i) and (ii), we can conclude that

$$\Delta Q_A > \Delta Q_B$$

8. Given, $\Delta Q = -20$ J ; $\Delta W = -10$ J

Now, $\Delta Q = (U_f - U_i) + \Delta W$

$\Rightarrow \quad -20 = (U_f - 40) - 10$

$\Rightarrow \quad U_f = -10 + 40 = 30$ J

9. According to the first law of thermodynamics,

$$\Delta Q = \Delta U + \Delta W$$

$\Rightarrow \quad \Delta Q = 0 + W = W$

10. Here, $n = 5, \gamma = \frac{7}{5}, T_1 = 0° \text{C}, T_2 = 400° \text{C}$

$\therefore \quad dU = \dfrac{nRdT}{\frac{7}{5} - 1} \qquad \left(\because dU = \dfrac{nRdT}{\gamma - 1}\right)$

$$dU = \frac{5 \times 8.31 \times (400 - 0)}{\frac{7}{5} - 1} = 41550 \text{ J}$$

$$dU = 41.55 \text{ kJ}$$

11. As, $\dfrac{dU}{dQ} = \dfrac{C_V dT}{C_p dT} = \dfrac{C_V}{C_p} = \dfrac{(3/2)R}{(5/2)R} = \dfrac{3}{5}$

(Here, number of moles of gas is constant)

12. Work done by the system = Area of shaded portion on the p-V diagram

$$= (300 - 100)10^{-6} \times (100 - 200)10^3$$
$$= -20 \text{ J}$$

13. As, $dU = C_V \, dT$

$$= \left(\frac{3}{2}R\right) dT = \frac{3}{2} \times 8.32 \times 100 = 1.25 \times 10^3 \text{ J}$$

14. For a non linear triatomic gas, $C_V = 3R$

and for monoatomic gas, $C_V' = \frac{3}{2}R$

$\therefore \quad \dfrac{Q}{Q'} = \dfrac{C_V}{C_V'} = K = \dfrac{3R}{\frac{3}{2}R} = 2$

15. As $V = kT^{2/3}$

$\therefore \quad dV = k\frac{2}{3}T^{-1/3}dT \quad$ (after differentiating)

$\therefore \quad \dfrac{dV}{V} = \dfrac{\frac{2}{3}kT^{-1/3}dT}{kT^{2/3}} = \dfrac{2}{3}\dfrac{dT}{T}$

Work done, $W = \int_{T_1}^{T_2} RT \dfrac{dV}{V} = \int_{T_1}^{T_2} RT \dfrac{2}{3}\dfrac{dT}{T}$

$$W = \frac{2}{3}R(T_2 - T_1) = \frac{2}{3}R \times 60 = 40R$$

16. In cyclic process, ΔQ = Work done = Area inside the closed curve treat as an ellipse

$$= \frac{\pi}{4}(p_2 - p_1)(V_2 - V_1)$$

$\Rightarrow \quad \Delta Q = \dfrac{\pi}{4}\{(180 - 50) \times 10^3 \times (40 - 20) \times 10^{-6}\} = \dfrac{\pi}{2}$ J

17. In adiabatic expansion, $dQ = 0$.

$\therefore \quad dW = -dU$
$= -(-50 \text{ J})$
$= 50$ J

18. Process 1 is isobaric (p = constant) expansion. Hence, temperature of gas will increase.

$\therefore \quad \Delta U_1$ = positive

Process 2 is an isothermal process

$\therefore \quad \Delta U_2 = 0$

Process 3 is an adiabatic expansion. Hence, temperature of gas will fall.

$\therefore \quad \Delta U_3$ = negative

$\therefore \quad \Delta U_1 > \Delta U_2 > \Delta U_3$

19. Given, $p \propto T^3$ but we know that, for an adiabatic process the pressure, $p \propto T^{\gamma/\gamma-1}$

So, $\quad \dfrac{\gamma}{\gamma - 1} = 3$

$\Rightarrow \quad \gamma = \dfrac{3}{2} \Rightarrow \dfrac{C_p}{C_V} = \dfrac{3}{2}$

20. We know that, $E_\phi = \gamma p = 1.4 \times (1 \times 10^5) = 1.4 \times 10^5 \text{ N/m}^2$

21. When the compression is isothermal for gas in container A,

$$p_2V_2 = p_1V_1$$
$$p_2 = \frac{p_1V_1}{V_2} = p_1\frac{V_1}{V_1/2} = 2p_1$$

For gas in container B, when compression is adiabatic,

$$p_2'V_2^{\gamma\prime} = p_1V_1^\gamma$$
$$p_2' = p_1\left(\frac{V_1}{V_2'}\right)^\gamma$$
$$= p_1\left(\frac{V_1}{V_1/2}\right)^\gamma = 2^\gamma p_1$$

$\therefore \quad \dfrac{p_2'}{p_2} = \dfrac{2^\gamma p_1}{2p_1} = 2^{\gamma - 1}$

22. Slope of p-V graph of adiabatics = $\gamma \, p/V$

Slope of p-V graph of isothermal = p/V

Required ratio = γ

23. Given processes are

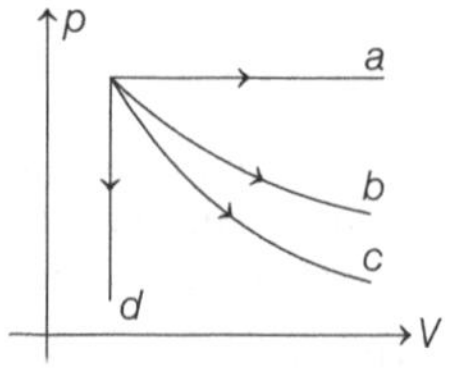

For process a, pressure is constant.

$\therefore a$ is isobaric.

For process d, volume is constant.

$\therefore d$ is isochoric.
Also, as we know that, slope of adiabatic curve in p-V diagram is more than that of isothermal curve.
$\therefore b$ is isothermal and c is adiabatic.

24. For adiabatic process, relation of temperature and volume is

$$T_2V_2^{\gamma-1} = T_1V_1^{\gamma-1}$$

$$\Rightarrow \quad T_2(2V)^{2/3} = 300(V)^{2/3}$$

$$\left[\gamma = \frac{5}{3} \text{ for monoatomic gases}\right]$$

$$\Rightarrow \quad T_2 = \frac{300}{2^{2/3}} \approx 189 \text{ K}$$

Also, in adiabatic process,

$$\Delta Q = 0, \Delta U = -\Delta W$$

or $$\Delta U = \frac{-nR(\Delta T)}{\gamma - 1}$$

$$= -2 \times \frac{3}{2} \times \frac{25}{3}(300 - 189)$$

$$\approx -2.7 \text{ kJ}$$

$$T_2 \approx 189 \text{ K}, \Delta U \approx -2.7 \text{kJ}$$

25. Here, $p_1 = 1$ atm, $T_1 = 27° \text{ C} = 27 + 273 = 300$ K

$p_2 = 8$ atm, $T_2 = ?, \gamma = 3/2$

As change are adiabatic,

$$\therefore \quad p_1^{\gamma-1}T_1^{-\gamma} = p_2^{\gamma-1}T_2^{-\gamma}$$

$$\left(\frac{T_2}{T_1}\right)^{-\gamma} = \left(\frac{p_1}{p_2}\right)^{\gamma-1}$$

$$T_2 = T_1\left(\frac{p_2}{p_1}\right)^{(\gamma-1)/\gamma} = 300 \times (8)^{(1.5-1)/1.5} = 300 \times (8)^{1/3}$$

$$\Rightarrow \quad T_2 = 600 \text{ K} = (600 - 273)° \text{ C} = 327° \text{ C}$$

26. (i) is incorrect ; since in adiabatic process $V \neq$ constant.
(ii) is incorrect, since in isothermal process $T =$ constant.
(iii) and (iv) match isotherms and adiabatic formula

$$TV^{\gamma-1} = \text{constant and } \frac{T^\gamma}{p^{\gamma-1}} = \text{constant}$$

27. In adiabatic process, $pV^\gamma =$ constant

$$\Rightarrow \quad \frac{dp}{dV} \cdot V^\gamma + p \cdot \gamma V^{\gamma-1} = 0$$

$$\Rightarrow \quad \frac{dp}{dV} = -\frac{p\gamma V^{\gamma-1}}{V^\gamma} = \frac{-p\gamma}{V}$$

$$\Rightarrow \quad -\frac{dp}{p\gamma} = \frac{dV}{V}$$

$$\Rightarrow \quad -\frac{\Delta p}{p\gamma} = \frac{\Delta V}{V}$$

28. As, $W_{AB} = -p_0V_0$,

$$W_{BC} = 0 \text{ and } W_{CD} = 4p_0V_0$$

$$\therefore \quad W_{ABCD} = W_{AB} + W_{BC} + W_{CD}$$

$$= -p_0V_0 + 0 + 4p_0V_0 = 3p_0V_0$$

29. As, $W = \int p\,dV =$ area under the p-V curve

$=$ minimum along ADB

30. During an adiabatic process, slope of pV curve is

$$\frac{dp}{dV} = -\gamma\frac{p}{V}$$

During an isothermal process, slope of pV curve is

$$\frac{dp}{dV} = \frac{-p}{V}$$

Both slopes are negative, hence A and B are not possible.
Slope of adiabatic curve is more than the slope of isothermal curve, so D is isothermal and C is adiabatic.

31. In given p-V diagram,

1 to 2 is adiabatic expansion in which temperature reduces.

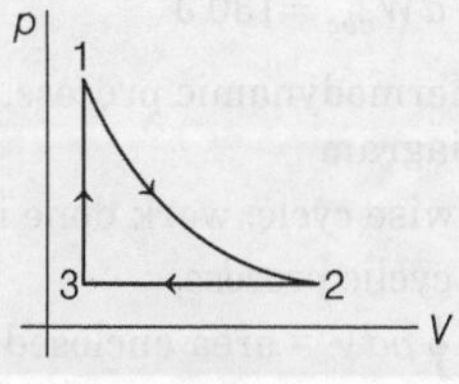

2 to 3 is isobaric compression, so volume decreases and temperature increases in this process.
3 to 1 is isochoric process in which pressure increases. So, it must be isochoric heating in which temperature increases. So, resultant V-T graph will be

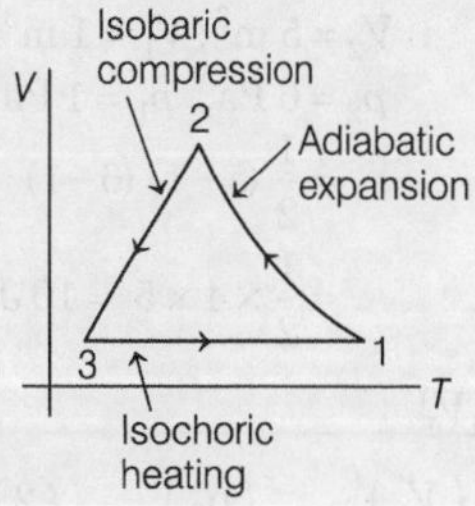

Thus, option (b) is correct.

32. In p-V curve, work done dW, change in internal energy ΔU and heat absorbed ΔQ are connected with first law of thermodynamics, *i.e.*

$$\Delta Q = \Delta U + dW \quad \text{...(i)}$$

and total change in internal energy in complete cycle is always zero. Using this equation in different part of the curve, we can solve the given problem.

In Process $a \to b$

Given, $\Delta Q_{ab} = 250$ J

$$\therefore \quad 250 \text{ J} = \Delta U_{ab} + dW_{ab} \quad \text{...(ii)}$$

In Process $b \to c$

Given, $\Delta Q_{bc} = 60$ J

Also, V is constant, so $dV = 0$.

$$\Rightarrow \quad dW_{bc} = p(dV)_{bc} = 0$$

$\therefore \quad 60\,J = \Delta U_{bc} + 0$

$\Rightarrow \quad \Delta U_{bc} = 60\,J \quad \ldots(iii)$

In Process $c \rightarrow a$

Given, $\Delta U_{ca} = -180\,J \quad \ldots(iv)$

Now, for complete cycle,

$$\Delta U_{abca} = \Delta U_{ab} + \Delta U_{bc} + \Delta U_{ca} = 0 \quad \ldots(v)$$

From Eqs. (iii), (iv) and (v), we get

$$\Delta U_{ab} = -\Delta U_{bc} - \Delta U_{ca}$$

$$\Delta U_{ab} = -60 + 180 = 120\ J \quad \ldots(vi)$$

From Eq. (ii), we get

$$250\,J = 120\,J + dW_{ab}$$

$\Rightarrow \quad dW_{ab} = 130\,J \quad \ldots(vii)$

From Eqs. (i) and (vii), we get

Work done by the gas along the path abc,

$$dW_{abc} = dW_{ab} + dW_{bc}$$

$$= 130\ J + 0\ K$$

$\Rightarrow \quad dW_{abc} = 130\ J$

33. In a cyclic thermodynamic process, work done = area under p-V diagram

Also in clockwise cycle, work done is positive.

In the given cyclic process,

work done $= \oint p\,dV$ = area enclosed by the cycle

$= \frac{1}{2} \times$ base $\times$ height of triangle (CAB) made by cycle

$= \frac{1}{2}(V_2 - V_1)(p_2 - p_1)$

From graph, given

$$V_2 = 5\ m^3,\ V_1 = 1\ m^3,$$

$$p_2 = 6\ Pa,\ p_1 = 1\ Pa$$

$$= \frac{1}{2}(5-1)(6-1)$$

$$= \frac{1}{2} \times 4 \times 5 = 10\,J$$

34. As, $p_2V_2^\gamma = p_1V_1^\gamma$

$$\therefore \quad p_2 = p_1\left(\frac{V_1}{V_2}\right)^\gamma = p_1\left(\frac{\rho_2}{\rho_1}\right)^\gamma = p\left(\frac{2}{1}\right)^{7/5} = 2.63p$$

35. Using, $pV = nRT$, we see the higher value of pV, higher will be T. Since, pV is higher for T_1.

Thus, $T_1 > T_2$.

36. Figure shows that loop 1 is anti-clockwise, therefore W_1 is negative and loop 2 is clockwise, therefore W_2 is positive.

Also, loop 2 is bigger.

$$W_2 > W_1$$

Hence, net work done, $W = -W_1 + W_2 = W_2 - W_1 = (+)$ ve

37. Isobaric expansion is represented by curve AB.

Work done = Area under AB

$$= 2 \times 10^2 \times (3-1) = 4 \times 10^2 = 400\ J$$

38. Work done during the complete cycle

= Area $ABCDA = AD \times AB = p \times V = pV$

39. As work done in a process = area under the curve, which increases continuously.

40. Let initial volume of the gas in the cylinder be V.

$\therefore \quad V_1 = V$

and $\quad V_2 = V/2$

$$\frac{p_2}{p_1} = ?$$

As, hydrogen is a diatomic gas.

$$\therefore \quad \gamma = \frac{7}{5} = 1.4$$

For an adiabatic change, $p_1V_1^\gamma = p_2V_2^\gamma$

$$\therefore \quad \frac{p_2}{p_1} = \left(\frac{V_1}{V_2}\right)^\gamma$$

$$= \left(\frac{V}{V/2}\right)^{1.4}$$

$$= (2)^{1.4} = 2.64$$

41. From symmetry considerations and also from theory,

$$\frac{V_a}{V_d} = \frac{V_b}{V_c}$$

42. pV^γ = constant

Differentiating, we get $\frac{dp}{dV} = \frac{\gamma p}{V}$

$$\frac{dp}{p} = -\frac{\gamma dV}{V}$$

43. As, $dW = dU = \mu C_V \Delta T = -C_V(T_2 - T_1) = C_V(T_1 - T_2)$

$(\because \mu = 1)$

44. According to question, the figure can be drawn as below

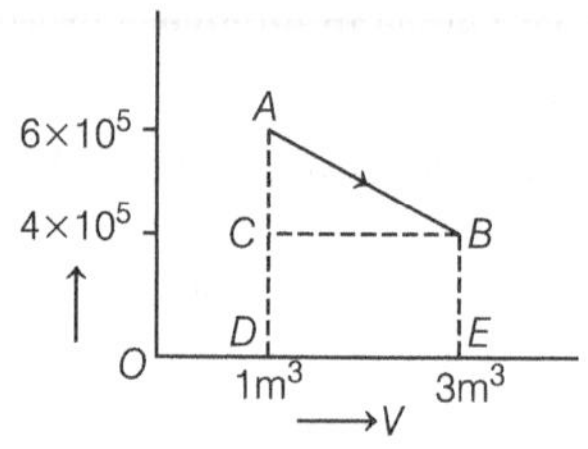

Now, work done by the system

= area under p-V diagram

= area of rectangle $BCDE$ + area of ΔABC

$$= 4 \times 10^5 \times 2 + \frac{2 \times 10^5 \times 2}{2}$$

$\Rightarrow \quad W = 10 \times 10^5\ J$

45. As $dW = p\,dV$

$\therefore$ (i) $dW = p \times 0 = 0 \quad (\because$ change in volume $= 0)$

and (ii) $dW = p(2V - V) = pV$

46. AB and CD are isothermal curves, therefore $T_a = T_b$ and $T_c = T_d$ but all the four temperatures are not equal.

47. In adiabatic operation (*e. g.* bursting of tyre),

$$p_2^{1-\gamma}T_2^{\gamma} = p_1^{1-\gamma}T_1^{\gamma}$$

or
$$T_2 = T_1\left(\frac{p_1}{p_2}\right)^{1-\gamma/\gamma}$$

$$= 300\left(\frac{4}{1}\right)^{\left(\frac{1-7/5}{7/5}\right)} = 300(4)^{-2/7}$$

($\because$ atmospheric pressure = 1 atm)

48. During adiabatic expansion.

$$TV^{\gamma-1} = \text{constant}$$

$$\Rightarrow \quad T_2V_2^{\gamma-1} = T_1V_1^{\gamma-1}$$

or
$$\frac{T_1}{T_2} = \left(\frac{V_2}{V_1}\right)^{\gamma-1}$$

For monoatomic gas, $\gamma = 5/3$

$$\frac{T_1}{T_2} = \left(\frac{AL_2}{AL_1}\right)^{5/3-1} = \left(\frac{L_2}{L_1}\right)^{2/3}$$

49. From, $p_2V_2^{\gamma} = p_1V_1^{\gamma}$

$$\Rightarrow \quad p_2 = p_1\left(\frac{V_1}{V_2}\right)^{\gamma}$$

$$p_2 = \left(\frac{100}{124}\right)^{5/3} p_1$$

$$p_2 = 0.7088 p_1$$

$\therefore$ % decrease in pressure

$$= \frac{p_1 - p_2}{p_1} \times 100\%$$

$$= \frac{p_1 - 0.6985 p_1}{p_1} \times 100$$

$$= \frac{0.2911 p_1}{p_1} \times 100\%$$

$$= 29.11\% \approx 30\%$$

50. The p-V graphs for three given processes are shown in figure.

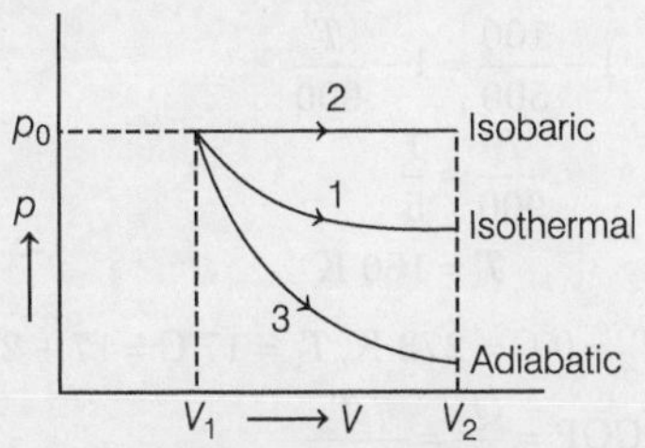

As, work done by the gas = area under the p-V graph (between the curve and V axis)

$$\Rightarrow \quad (\text{Area})_2 > (\text{Area})_1 > (\text{Area})_3$$

$$\therefore \quad W_2 > W_1 > W_3$$

51. Here, $V_1 = 1\text{ L} = 10^{-3}\text{ m}^3$, $V_2 = 3\text{L} = 3\times10^{-3}\text{ m}^3$

$p_1 = 1\text{ atm} = 1.013\times10^5\text{ Nm}^{-2}, \gamma = 1.40, W = ?$

As changes are adiabatic, $p_1V_1^{\gamma} = p_2V_2^{\gamma}$

$$\frac{p_1}{p_2} = \left(\frac{V_2}{V_1}\right)^{\gamma} = (3)^{1.4} = 4.6555$$

$$\therefore \quad p_2 = \frac{p_1}{4.6555} = \frac{1.013\times10^5}{4.6555}$$

$$= 0.217\times10^5\text{ Nm}^{-2}$$

Now, work done $= \dfrac{p_1V_1 - p_2V_2}{\gamma - 1}$

$$= \frac{1.013\times10^5\times10^{-3} - 0.217\times10^5\times3\times10^{-3}}{1.4-1}$$

$$= 90.5\text{ J}$$

52. Heat absorbed by the system at constant pressure,

$Q = nC_p\Delta T$

Change in internal energy, $\Delta U = nC_V\Delta t$

As, $W = Q - \Delta U$

(first law of thermodynamics)

$$\therefore \quad \frac{W}{Q} = \frac{Q-\Delta U}{Q} = 1 - \frac{\Delta U}{Q}$$

$$= 1 - \frac{RC_{\gamma}\Delta T}{RC_{\gamma}\Delta T} = 1 - \frac{C_V}{C_p} = \left(1 - \frac{1}{\gamma}\right)$$

53. Given, work done, $W = -22.3$ J

Work done is taken negative as work is done on the system.

In an adiabatic change, $\Delta Q = 0$

Using first law of thermodynamics,

$$\Delta U = \Delta Q - W$$
$$= 0 - (-22.3)$$
$$= 22.3\text{ J}$$

For another process between state A and state B,

Heat absorbed (ΔQ) = + 9.35 cal

$$= +(9.35\times4.19)\text{ J}$$
$$= +39.18\text{ J}$$

Change in internal energy between two states *via* different paths are equal.

$$\therefore \quad \Delta U = 22.3\text{ J}$$

$\therefore$ From first law of thermodynamics

$$\Delta U = \Delta Q - W$$

or
$$W = \Delta Q - \Delta U$$
$$= 39.18 - 22.3 = 16.88\text{ J}$$
$$\approx 16.9\text{ J}$$

54. $\Delta Q = \Delta U + \Delta W$

$$\Delta Q = 0 - 150\text{ J} = -150\text{ J}$$

So, heat has been given by the system.

i.e. Heat has been removed from gas.

55. As, $\eta = 1 - \dfrac{T_2}{T_1} = 1 - \dfrac{500}{800} = \dfrac{3}{8} = 0.375$

56. As, $\dfrac{T_2}{T_1} = 1 - \eta = 1 - \dfrac{40}{100} = \dfrac{3}{5}$

$$\therefore \quad T_1 = \frac{5}{3}T_2 = \frac{5}{3}\times300 = 500\text{ K}$$

Increase in efficiency = 50% of 40% = 20%

$\therefore$ New efficiency, $\eta' = 40 + 20 = 60\%$

$$\therefore \quad \frac{T_2}{T_1'} = 1 - \eta' = 1 - \frac{60}{100} = \frac{2}{5}$$

$$T_1' = \frac{5}{2} \times 300 = 750 \text{ K}$$

Increase in temperature of source $= T_1' - T_1$
$= 750 - 500 = 250$ K

57. We have, $\eta = \dfrac{W}{Q_1}$

$$\Rightarrow \quad W = \eta Q_1 = \frac{1}{3} \times 1000 \text{ cal}$$
$$= \frac{1000}{3} \times 4.2 = 1400 \text{ J}$$

58. Given, $T_2 = 27 + 273 = 300$ K and $\eta = 37.5\%$

As, $$\eta = 1 - \frac{T_2}{T_1}$$

$$\therefore \quad \frac{37.5}{100} = 1 - \frac{300}{T_1}$$

or $$\frac{300}{T_1} = \frac{62.5}{100} = \frac{5}{8}$$

$$T_1 = \frac{2400}{5} = 480 \text{ K}$$
$$= 480 - 273 = 207°\text{C}$$

59. Given, $T_1 = 273 + 20 = 293$ K, $T_2 = 273 + 10 = 283$ K

$\therefore$ Coefficient of performance

$$= \frac{T_2}{T_1 - T_2} = \frac{283}{293 - 283}$$
$$= \frac{283}{10} = 28.3$$

60. As, $\eta = 1 - \dfrac{T_2}{T_1}$

or $$\frac{T_2}{T_1} = 1 - \eta = 1 - \frac{1}{6} = \frac{5}{6}$$

$$T_2 = \frac{5}{6} T_1 = \frac{5}{6} \times 600 = 500 \text{ K}$$

61. Relation of coefficient of performance β of refrigerator and efficiency η of engine,

$$\beta = \frac{1 - \eta}{\eta}$$

In given case, $\eta = \dfrac{1}{10}$

So, $$\beta = \frac{1 - \frac{1}{10}}{\frac{1}{10}} = \frac{\frac{9}{10}}{\frac{1}{10}} = 9$$

Now, for a Carnot refrigerator, we have

$$\beta = \frac{\text{heat extracted from cold reservoir}}{\text{work done}}$$

$$\Rightarrow \quad 9 = \frac{Q}{10} \Rightarrow Q = 90\text{J}$$

62. From $\eta = 1 - \dfrac{T_2}{T_1}$

$$\Rightarrow \quad \frac{T_2}{T_1} = 1 - \eta = 1 - \frac{40}{100} = \frac{3}{5}$$

$$\therefore \quad T_2 = \frac{3}{5} T_1 = \frac{3}{5} \times 500 = 300 \text{ K}$$

Again $$\frac{T_2}{T_1'} = 1 - \eta$$

or $$\frac{300}{T_1'} = 1 - \frac{50}{100} = \frac{1}{2}$$

or $$T_1' = 600 \text{ K}$$

63. Given, $T_1 = 27°\text{ C} = (27 + 273) \text{ K} = 300$ K,

and $$T_2 = -123 + 273 = 150 \text{ K}$$

$$\therefore \quad \eta = 1 - \frac{T_2}{T_1} = 1 - \frac{150}{300} = 0.5$$

64. As, $\eta = 1 - \dfrac{T_2}{T_1} = \dfrac{T_1 - T_2}{T_1}$

In all the four cases, $T_1 - T_2 = 20$ K. Therefore, η is highest, when T_1 is lowest.
i.e. $T_1 = 40$ K

65. $$\text{COP} = \frac{T_2}{T_1 - T_2} = \frac{273 - 23}{(273 + 77) - (273 - 23)} = \frac{250}{100} = 2.5$$

As, $$\text{COP} = \frac{Q_2}{W}$$

$$\therefore \quad 2.5 = \frac{1000 \times 80 \times 4.2}{W}$$

or $$W = \frac{1000 \times 80 \times 4.2}{2.5} = 134400 \text{ J}$$

66. Here, $Q_2 = 2000$ cal

As, $\text{COP} = \dfrac{Q_2}{W}$

$$\therefore \quad 4 = 2000 / W$$
$$W = 500 \text{ cal} = 500 \times 4.2 = 2100 \text{ J}$$

67. As, $\eta = 1 - \dfrac{T_2}{T_1}$

$$\Rightarrow \quad 1 - \frac{100}{500} = 1 - \frac{T}{900}$$

$$\therefore \quad \frac{T}{900} = \frac{1}{5}$$

or $$T = 180 \text{ K}$$

68. Given, $T_2 = 0°\text{C} = 273$ K, $T_1 = 17°\text{C} = 17 + 273 = 290$ K

$$\text{COP} = \frac{Q_2}{W} = \frac{T_2}{T_1 - T_2}$$

$$\Rightarrow \quad \frac{80 \times 1000 \times 4.2}{W} = \frac{273}{290 - 273} = \frac{273}{17}$$

$$W = \frac{80 \times 1000 \times 4.2 \times 17}{273} \text{ J}$$

$$W = \frac{33.6 \times 17 \times 10^4}{273 \times 3.6 \times 10^5} = 0.058 \text{ kWh}$$

69. $\eta = \frac{T_2}{T_1} = \frac{Q_2}{Q_1} = \frac{Q_1 - W}{Q_1}$ $\quad (\because W = Q_1 - Q_2)$

$$\frac{400}{800} = 1 - \frac{W}{Q_1}$$

$$\frac{W}{Q_1} = 1 - \frac{1}{2} = \frac{1}{2}$$

$$Q_1 = 2W = 2400\ \text{J}$$

70. Here, $T_1 = 411°\text{ C} = (411 + 273)\text{ K} = 684\text{ K}$

$$T_2 = 69°\text{C} = (69 + 273)\text{ K} = 342\text{ K}$$

and $Q_1 = 1000\ \text{J}$

$\because \quad \eta = \frac{W}{Q_1} = 1 - \frac{T_2}{T_1} = 1 - \frac{342}{684} = \frac{1}{2}$

$\Rightarrow \quad W = \eta Q_1 = \frac{1000}{2} = 500\ \text{J}$

71. As, $\frac{Q_2}{Q_1} = \frac{T_2}{T_1}$

$\therefore \quad \frac{Q_2}{6 \times 10^4} = \frac{127 + 273}{227 + 273} = \frac{400}{500}$

$$Q_2 = \frac{4}{5} \times 6 \times 10^4 = 4.8 \times 10^4 \text{cal}$$

$\therefore \quad W = Q_1 - Q_2 = 6 \times 10^4 - 4.8 \times 10^4$

$$= 1.2 \times 10^4\ \text{cal}$$

72. The efficiency of two engines are

$$\eta_A = 1 - \frac{T_2}{T_1} = 1 - \frac{500}{1000} = \frac{1}{2}$$

and $\eta_B = 1 - \frac{T_2}{T_1} = 1 - \frac{400}{1100} = \frac{7}{11}$

Clearly, $\eta_A < \eta_B$

73. As, $\eta = 1 - \frac{T_2}{T_1} = 1 - \frac{500}{800} = 1 - \frac{600}{x}$

$\therefore \quad \frac{3}{8} = 1 - \frac{600}{x}$

$$\frac{600}{x} = 1 - \frac{3}{8} = \frac{5}{8}$$

$$5x = 4800$$

$$x = \frac{4800}{5} = 960\ \text{K}$$

74. As, $\eta = 1 - \frac{T_2}{T_1}$

$\therefore \quad \frac{T_2}{T_1} = 1 - \eta = 1 - \frac{10}{100} = \frac{90}{100}$

or $\quad T_1 = \frac{100 T_2}{90} = \frac{100}{90} \times 270 = 300\ \text{K}$

75. As, $\eta = 1 - \frac{T_2}{T_1} = \frac{W}{Q}$

$\Rightarrow \quad W = \left(1 - \frac{T_2}{T_1}\right) Q = \left\{1 - \frac{(273 + 27)}{(273 + 627)}\right\} \times Q$

$\Rightarrow \quad W = \left(1 - \frac{300}{900}\right) \times 3 \times 10^6$

$$= 2 \times 10^6 \times 4.2\ \text{J} = 8.4 \times 10^6 \text{J}$$

Round II

1. According to first law of thermodynamics,

$$dQ = dU + dW \quad \text{...(i)}$$

As, $\quad dW = -dU \quad \text{...(ii)}$

So from Eqs. (i) and (ii), we get

$\Rightarrow \quad dQ = dU - dU = 0$

The change must be adiabatic.

2. As, $dW = dQ - dU$

$$= C_p (T_2 - T_1) - C_V (T_2 - T_1)$$

$$= R (T_2 - T_1) \quad (\because C_p - C_V = R)$$

$$= 8.31 \times 100$$

$$= 8.31 \times 10^2\ \text{J}$$

3. Here, $T_1 = 927°\text{ C} = (927 + 273)\text{ K} = 1200\text{ K}$

and $\quad T_2 = 27°\text{ C} = (27 + 273)\text{ K} = 300\text{ K}$

As $\quad U \propto T$

$\therefore \quad \frac{\Delta U}{U_2} = \frac{U_1 - U_2}{U_2}$

$$= \frac{1200 - 300}{300} \times 100$$

$$= 300\%$$

4. From $\quad p_2 V_2^\gamma = p_1 V_1^\gamma$

$\Rightarrow \quad p_2 = p_1 \left(\frac{V_1}{V_2}\right)^\gamma = 1 \left(\frac{V_1}{1/20 V_1}\right)^{1.4}$

$$= 66.28\ \text{atm}$$

5. As, $\eta = 1 - \frac{T_2}{T_1} = 1 - \frac{(127 + 273)}{(227 + 273)} = \frac{1}{5}$

$$W = \eta Q_1 = \frac{1}{5} \times 10^4\ \text{J} = 2000\ \text{J}$$

6. Heat supplied $= nC_V\ (2T_0 - T_0) + nC_p\ (2T_0 - T_0)$

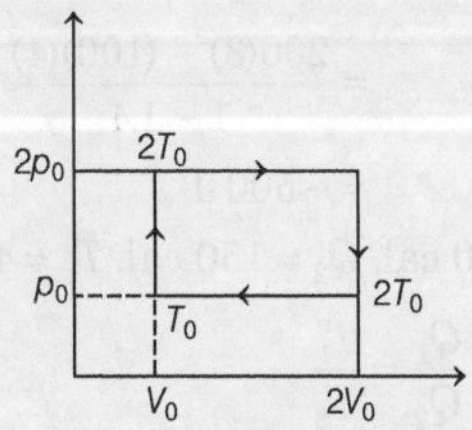

$$= \frac{n 3 R T_0}{2} + \frac{n 5 R}{2} 2T_0$$

$$= \frac{13}{2} nRT_0 = \frac{13}{2} p_0 V_0$$

7. Since, entropy is a state function, therefore change in entropy in both the processes must be same. Therefore, correct option should be (b).

8. As slope of adiabatic AC is more than the slope of isothermal AB and isochoric BC (*i.e.* at constant volume), therefore Fig. (b) represents the curves correctly.

9. As work done in an isothermal process is

$$W = \mu RT \log_e \frac{V_2}{V_1}$$
$$= \left(\frac{m}{M}\right) RT \log_e \frac{V_2}{V_1}$$
$$= 2.3 \times \frac{m}{M} RT \log_{10} \frac{V_2}{V_1}$$
$$= 2.3 \times \frac{96}{32} R(273 + 27) \log_{10} \frac{140}{70}$$
$$= 2.3 \times 900 R \log_{10} 2$$

10. As, $\frac{T_2}{T_1} = \left(\frac{V_1}{V_2}\right)^{\gamma - 1}$

$\Rightarrow \quad T_2 = 300\left(\frac{27}{8}\right)^{\frac{5}{3}-1} = 300\left(\frac{27}{8}\right)^{\frac{2}{3}} = 300\left(\frac{3}{2}\right)^2 = 675 \text{ K}$

$\Rightarrow \quad \Delta T = 576 - 300 = 375 \text{ K}$

11. As, $dU = dQ - dW = 8 \times 10^5 - 6.5 \times 10^5 = 1.5 \times 10^5$ J

In the second process, dU remains the same.

$\therefore \quad dW = dQ - dU = 10^5 - 1.5 \times 10^5$
$= -0.5 \times 10^5$ J

12. As, $T_2 = T_1\left(\frac{V_1}{V_2}\right)^{\gamma-1} = 273(2)^{0.41} = 273 \times 1.328 = 363$ K

Now, for one mole of the gas, the work done,

$$|W| = \frac{R(T_1 - T_2)}{\gamma - 1} = \frac{8.31(273 - 363)}{1.41 - 1} = -1815$$

$\Rightarrow \quad |W| = 1815$ J

13. Adiabatic process is from C to D,

$$\text{Work done} = \frac{p_2V_2 - p_1V_1}{1 - \gamma}$$
$$= \frac{p_DV_D - p_CV_C}{1 - \gamma}$$
$$= \frac{200(3) - (100)(4)}{1 - 1.4}$$
$$= -500 \text{ J}$$

14. Here $Q_1 = 200$ cal, $Q_2 = 150$ cal, $T_1 = 400$ K

As, $\frac{Q_1}{Q_2} = \frac{T_1}{T_2}$

$$T_2 = \frac{Q_2}{Q_1} \times T_1 = \frac{150}{200} \times 400 = 300 \text{ K}$$

15. As, $\Delta U = \mu C_V \Delta T = \mu\left(\frac{R}{\gamma - 1}\right)\Delta T$

$\therefore \quad \Delta U = \frac{p\Delta V}{(\gamma - 1)} = \frac{p(2V - V)}{(\gamma - 1)} = \frac{pV}{(\gamma - 1)}$

16. As, $W = \frac{\mu R(T_1 - T_2)}{(\gamma - 1)} = \frac{\mu RT_1}{(\gamma - 1)}\left[1 - \frac{T_2}{T_1}\right]$

$$= \frac{\mu RT_1}{(\gamma - 1)}\left[1 - \left(\frac{V_1}{V_2}\right)^{\gamma - 1}\right]$$
$$= \frac{2 \times 8.31 \times 300}{\left(\frac{5}{3} - 1\right)}\left[1 - \left(\frac{1}{2}\right)^{\frac{5}{3}-1}\right]$$
$$= + 2767.23 \text{ J}$$

17. As, $W = -\mu RT \log_e \frac{V_2}{V_1} = -1 \times 8.31 \times (273 + 0) \log_e\left(\frac{22.4}{11.2}\right)$

(–ve sign shows compression)

$= -8.31 \times 273 \times \log_e 2$

$= -1572.5$ J $\quad [\because \log_e 2 = 0.693]$

18. For isothermal process, $p_1V_1 = p_2V_2$

$\Rightarrow \quad p_2 = \frac{p_1V_1}{V_2} = \frac{72 \times 1000}{900} = 80$ cm of mercury

$\therefore$ Stress, $\Delta p = p_2 - p_{12} = 80 - 72 = 8$ cm of mercury

19. As, volume of the gas, $V = \frac{M}{d}$

Now, using $pV^\gamma =$ constant, we get

$$\frac{p'}{p} = \left(\frac{V}{V'}\right)^\gamma = (32)^{7/5} = 128$$

20. Work done during an adiabatic change,

$$W = \frac{\mu R(T_1 - T_2)}{(\gamma - 1)}$$

Now, $\gamma = \frac{C_p}{C_V}$

$\therefore \quad \gamma - 1 = \frac{C_p - C_V}{C_V} = \frac{R}{C_V}$

$\therefore \quad W = \frac{\mu R(T_1 - T_2)C_V}{R}$
$= \mu(T_1 - T_2)C_V$

21. As, $W = p\Delta V = 1.01 \times 10^5 (3.34 - 2 \times 10^{-3})$
$= 337 \times 10^3 \text{ J} = 340$ kJ

22. As compression is sudden, changes are adiabatic, $dQ = 0$. Therefore, work done on the gas increases the temperature.

23. In an adiabatic change,

$p^{1-\gamma}T^\gamma =$ constant

or $\quad pT^{\gamma/1-\gamma} =$ constant

or $\quad p \propto T^{-(1-\gamma)/\gamma}$

Thus, $\quad c = -\frac{1 - \gamma}{\gamma}$

For a monoatomic gas, $\gamma = \frac{5}{3}$

$\therefore \quad -c = \frac{1 - 5/3}{5/3} = -\frac{2}{5}$

$\Rightarrow \quad c = \frac{2}{5}$

24. From $\eta = 1 - \frac{T_2}{T_1}$

$$\frac{40}{100} = 1 - \frac{(27 + 273)}{T_1}$$

$$\frac{300}{T_1} = 1 - \frac{40}{100} = \frac{3}{5}$$

$$T_1 = \frac{300 \times 5}{3} = 500 \text{ K}$$

25. From the given initial state A to final state B, change in internal energy is same in all the four cases, as it is independent of the path from A to B.
As work done = area under p-V curve, therefore work done is maximum in case I.

26. Figure represents the working of a refrigerator, where

$$Q_1 = Q_2 + W$$

According to problem, $W > 0$, then

$$W = Q_1 - Q_2 > 0$$

So, there are two possibilities
(i) If both Q_1 and Q_2 are positive,

$$\Rightarrow \quad Q_1 > Q_2 > 0$$

(ii) If both Q_1 and Q_2 are negative,

$$Q_2 < Q_1 < 0$$

27. In the given work, one complete cycle, $1 \to 2 \to 3 \to 1$, the system returns to its initial state.
$\therefore dU = 0$ and $dQ = dW$, *i.e.* heat is completely converted into mechanical energy, which is not possible in such a process. Further, the two adiabatic curves $(2 \to 3)$ and $(3 \to 1)$ cannot intersect each other.

28. As it is clear from figure,
Slope of curve 2 > Slope of curve 1

$$(\gamma p)_2 > (\gamma p)_1$$

$$\gamma_2 > \gamma_1$$

We know that, for monoatomic gas, $\gamma = 1.67$ and for diatomic gas, $\gamma = 1.4$
$\therefore$ Adiabatic curve 2 corresponds to helium (monoatomic) and adiabatic curve 1 corresponds to oxygen (diatomic).

29. The given relation is $p = \alpha V$
Therefore, $\quad p \propto V$
When V changes from V to $2V$, pressure p is also doubled.
For an ideal gas, $\frac{pV}{T}$ = constant

$$\therefore \quad T \propto pV$$

Hence, T becomes $2 \times 2 = 4$ times
i.e. $\quad 4 \times 300 \text{ K} = 1200 \text{ K}$

30. Here, $\quad p = 4.5 \times 10^5$ Pa,

$$dV = (2.0 - 0.5) \text{ m}^3 = 1.5 \text{ m}^3$$

and $\quad dQ = 800 \text{ kJ} = 8 \times 10^5 \text{ J}, dU = ?$

$$dW = pdV = 4.5 \times 10^5 \times 1.5 = 6.75 \times 10^5 \text{ J}$$

$$dU = dQ - dW$$

$$= 8 \times 10^5 - 6.75 \times 10^5$$

$$= 1.25 \times 10^5 \text{ J}$$

31. Given, $\quad T_1 = 200° \text{C} = 200 + 273 = 473 \text{ K}$

$$T_2 = 0° \text{C} = 0 + 273 = 273 \text{ K}$$

$$\therefore \quad \eta_1 = 1 - \frac{T_2}{T_1} = 1 - \frac{273}{473} = \frac{200}{473}$$

Again, $\quad T_1' = 0° \text{C} = (0 + 273) \text{ K} = 273 \text{ K}$
and $\quad T_2' = -200° \text{C} = (-200 + 273) \text{ K} = 73 \text{ K}$

$$\therefore \quad \eta_2 = 1 - \frac{T_2}{T_1'} = 1 - \frac{73}{273} = \frac{200}{273}$$

$$\text{Now,} \quad \frac{\eta_1}{\eta_2} = \frac{200}{473} \times \frac{273}{200} = \frac{273}{473} = \frac{1}{1.732}$$

32. The given relation is $p = \frac{\alpha T^2}{V}$

$$\therefore \quad V = \frac{\alpha T^2}{p}$$

As, pressure is kept constant.

$$\therefore \quad dV = \left(\frac{2\alpha T}{p}\right) dT \quad \text{(after differentiating)}$$

$$\text{Now,} \quad W = \int p\, dV = \int_{T_0}^{2T_0} p \left(\frac{2\alpha T}{p}\right) dT$$

$$= -2\alpha \left[\frac{T^2}{2}\right]_{T_0}^{2T_0} = 3\alpha T_0^2$$

33. For monoatomic gas, $\gamma = \frac{C_p}{C_V} = \frac{5}{3}$

We know that, $\quad \Delta Q = \mu C_p \Delta T$
and $\quad \Delta U = \mu C_V \Delta T$

$$\Rightarrow \quad \frac{\Delta U}{\Delta Q} = \frac{C_V}{C_p} = \frac{3}{5}$$

i.e. Fraction of heat energy required to increases the internal energy will be $\frac{3}{5}$.

34. Given, $T_2 = 27 + 273 = 300 \text{ K}, \eta = 40\%$

$$\text{As,} \quad \eta = 1 - \frac{T_2}{T_1}$$

$$\therefore \quad \frac{T_2}{T_1} = 1 - \eta = 1 - \frac{40}{100} = \frac{60}{100} = \frac{3}{5}$$

$$\Rightarrow \quad T_1 = \frac{5}{3} \times T_2 = \frac{5}{3} \times 300 = 500 \text{ K}$$

$\therefore$ New efficiency, $\eta' = 40 + 4 = 44\%$

($\because$ Increase in efficiency = 10% of 40 = 4%)

Let the new temperature of the source be T_1' K, then

$$\eta' = 1 - \frac{T_2}{T_1'} \Rightarrow \frac{44}{100} = \frac{1 - 300}{T_1'}$$

$$\Rightarrow \quad \frac{300}{T_1'} = 1 - \frac{44}{100} = \frac{56}{100}$$

$$\Rightarrow \quad T_1' = \frac{100 \times 300}{56} = 535.7 \text{ K}$$

$\therefore$ Increase in the temperature of the source,

$$= 535.7 - 500 = 35.7 \text{ K or } 35.7° \text{C}$$

35. Given, temperature of source, $T_1 = (36 + 273)$ K

$$= 309 \text{ K}$$

Temperature of sink, $T_2 = (9 + 273) \text{ K} = 282 \text{ K}$

Coefficient of performance of a refrigerator

$$\beta = \frac{T_2}{T_1 - T_2} = \frac{282}{309 - 282} = \frac{282}{27} = 10.4$$

36. From first law of thermodynamics, $\Delta Q = \Delta U + W$

Work done at constant pressure $(\Delta W)_p = (\Delta Q)_p - \Delta U$

$= (\Delta Q)_p - (\Delta Q)_V$ (as we know, $\Delta Q_V = \Delta U$)

Also $(\Delta Q)_p = MC_p\Delta T$

and $(\Delta Q)_V = MC_V\Delta T$

$\Rightarrow$ $(\Delta W_p) = M(C_p - C_V)\Delta T$

$= 1 \times (3.4 \times 10^3 - 2.4 \times 10^3) \times 10 = 10^4 \text{cal}$

37. According to first law of thermodynamics,

$\Delta Q = \Delta U + \Delta W$, in adiabatic process, $\Delta Q = 0$

$0 = \Delta U - \Delta W$ (work done on the system negative

$\Delta U = + \Delta W = + 22.3 \text{ J}$

In second process, $\Delta Q = \Delta U + \Delta W$

$9.35 \times 4.18 = 22.3 + \Delta W$

Work done by system, $\Delta W = 16.95 \text{ J} \simeq 16.9 \text{ J}$

38. Amount of heat given = 540 cal

Change in volume, $\Delta V = 1670$ cc

and atmospheric pressure, $p = 1.01 \times 10^6 \text{ dyne/cm}^2$

$\therefore$ Work done against atmospheric pressure,

$$W = p\Delta V = \frac{1.01 \times 10^6 \times 1670}{4.2 \times 10^7} = 40 \text{ cal}$$

39. Number of moles of He $= \frac{1}{4}$

Now, $T_1(5.6)^{\gamma-1} = T_2(0.7)^{\gamma-1}$

$$T_1 = T_2\left(\frac{1}{8}\right)^{2/3}$$

$$4T_1 = T_2$$

$\therefore$ Work done, $W = \frac{-nR[T_2 - T_1]}{\gamma - 1}$

$$= \frac{-\frac{1}{4}R[3T_1]}{\frac{2}{3}} = \frac{-9}{8}RT_1$$

$\Rightarrow$ $|W| = \frac{9}{8}RT_1$

40. As, $\Delta W_{AB} = p\Delta V = 10(2 - 1) = 10 \text{ J}$

and $\Delta W_{BC} = 0$, because V is constant

From first law of thermodynamics,

$$\Delta Q = \Delta W + \Delta U$$

As $ABCA$ is a cyclic process, therefore

$$\Delta U = 0$$

$\therefore$ $\Delta Q = \Delta W_{AB} + \Delta W_{BC} + \Delta W_{CA}$

$= \Delta W_{AB} + \Delta W_{CA}$

or $\Delta W_{CA} = \Delta Q - \Delta W_{AB} = 5 - 10 = -5 \text{ J}$

41. For the process at constant pressure,

$$dQ = C_p dT + dW$$

$\therefore$ $dT = \frac{dQ - dW}{C_p}$

For the process at constant volume,

$dQ = C_V dT$ $(\because dW = 0)$

$$= C_V\left(\frac{dQ - dW}{C_p}\right) = \frac{dQ - dW}{C_p / C_V} = \frac{dQ - dW}{\gamma}$$

or $(\gamma - 1)dQ = dW$

$$\left(\frac{5}{3} - 1\right)dQ = W$$

$\therefore$ $dQ = \frac{3}{2}W$

42. As, $v_{\text{rms}} = \sqrt{\frac{3RT}{M}} = \frac{(v_{\text{rms}})_1}{(v_{\text{rms}})_2} = \sqrt{\frac{T_1}{T_2}}$

Now, $T_1V_\gamma^{\gamma-1} = T_2V_2^{\gamma-1}$

$$\frac{T_1}{T_2} = \left(\frac{V_2}{V_1}\right)^{\gamma-1}$$

Thus, $\frac{(V_{\text{rms}})_1}{(V_{\text{rms}})_2} = \left(\frac{V_2}{V_1}\right)^{\frac{\gamma-1}{2}}$

$$= \frac{V}{V/2} = \left(\frac{V_2}{V_1}\right)^{\frac{\frac{7}{5}-1}{2}} = 2$$

$$= \left(\frac{V_2}{V_1}\right)^{\frac{2}{5} \times \frac{1}{2}} = \left(\frac{V_1}{V_1}\right)^{\frac{1}{5}}$$

$$= \left(\frac{V_2}{V_1}\right)^5 = 2^5 = 32$$

43. As, initial and final temperatures for each process are same.

$\therefore$ Change in internal energy for each process will be equal, *i.e.*

$$\Delta U = nC_V\Delta T = \text{same}$$

Thus, $E_{AB} = E_{AC} = E_{AD}$

Now, work done, $W = pdV$

For process AB, volume is increasing

$\Rightarrow$ $W_{AB} > 0$

For process AD, volume is decreasing

$\Rightarrow$ $W_{AD} < 0$

For process AC, volume is constant

$\Rightarrow$ $W_{AC} = 0$

Hence, correct option is (b).

44. Given that, $Vp^n = \text{constant}$

$\therefore$ $Vp^n = (V + \Delta V)(p + \Delta p)^n$

$$= Vp^n\left(1 + \frac{\Delta V}{V}\right)\left(1 + n\frac{\Delta p}{p}\right)$$

$$1 = 1 + \frac{\Delta V}{V} + n\frac{\Delta p}{p} + n\frac{\Delta V}{V}\frac{\Delta p}{p}$$

or $$\frac{\Delta V}{V} = -n\frac{\Delta p}{p} \quad \text{(neglecting the product)}$$

$\Rightarrow$ Bulk modulus, $K = \frac{-\Delta p}{\Delta V / V} = \frac{p}{n}$

45. As, slope of adiabatic process at a given state is more than the slope of isothermal process, therefore in figure, AB is isothermal and BC is an adiabatic.

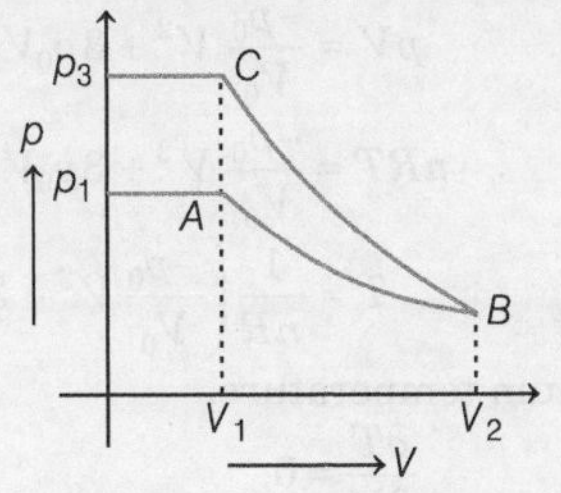

In going from A to B, volume is increasing

$\therefore$ W_{AB} = positive

In going from B to C, volume is decreasing

$\therefore$ W_{BC} = negative

As work done is area under p-V graph, therefore

$$|W_{BC}| > |W_{AB}|$$

$\therefore$ $W = W_{AB} + W_{BC}$ = Negative, *i.e.* $W < 0$.

From the graph, it is clear that $p_3 > p_1$.

46. The given expansion process is adiabatic in nature and we have the following data

Initial volume, $V_1 = 1\text{ L} = 10^{-3}\text{ m}^3$

Final volume, $V_2 = 3\text{ L} = 3 \times 10^{-3}\text{ m}^3$

Initial pressure, $p_1 = 1\text{ atm} = 1.01 \times 10^5$ Pa

Final pressure, p_2 = ?

Using $p_1 V_1^\gamma = p_2 V_2^\gamma$, we have

$$1 \times 1^\gamma = p_2 \times (3)^\gamma$$

Here, $\gamma = 1.40$ and $(3)^{1.4} = 4.6555$

So, $$p_2 = \frac{1}{4.6555} \approx 0.22\text{ atm}$$

$$= 0.22 \times 1.01 \times 10^5\text{ Pa}$$

Work done in an adiabatic expansion is given by

$$W = \frac{(p_1V_1 - p_2V_2)}{\gamma - 1}$$

$$= \frac{(1.01 \times 10^5 \times 1 \times 10^3 - 0.22 \times 1.01 \times 10^5 \times 3 \times 10^{-3})}{(1.40 - 1)}$$

$$= 0.85 \times 1.01 \times 10^5 \times 10^{-3}$$

$$= 85.85\text{ J}$$

Closest value of work done, $W \approx 90.5$ J

47. KE of the vessel $= \frac{1}{2}Mv^2$

When the vessel is suddenly stopped, the ordered motion of the gas molecules is converted into disordered motion of the molecules, thereby increasing the internal energy of the gas. Thus,

$$\Delta U = nC_V\Delta T = \frac{1}{2}mv^2 = \frac{1}{2}(nM)v^2$$

where, n is number of moles of the gas in the vessel and M is molecular weight of the gas.

$\therefore$ $$\Delta T = \frac{Mv^2}{2C_V}$$

As, $$C_V = \frac{R}{\gamma - 1}$$

$\therefore$ $$\Delta T = \frac{Mv^2(\gamma - 1)}{2R}$$

48. From the first law of thermodynamics,

$$\Delta Q = \Delta U + \Delta W \quad \text{..(i)}$$

For a cyclic process, $\Delta U = 0$

$\therefore$ $\Delta Q = \Delta W$

Now, $\Delta Q = Q_1 + Q_2 + Q_3 + Q_4$
$= 600\text{ J} - 400\text{ J} - 300\text{ J} + 200\text{ J} = 100\text{ J}$

and $\Delta W = W_1 + W_2 + W_3 + W_4$

$\Rightarrow$ $\Delta W = 300\text{ J} - 200\text{ J} - 150\text{ J} + W_4$
$= -50\text{ J} + W_4$

Substitute the value of ΔQ and ΔW in Eq. (i), we get

$$100\text{ J} = -50\text{ J} + W_4$$
$$W_4 = 150\text{ J}$$

49. For the given V-T graph of thermodynamic cycle $xyzx$,

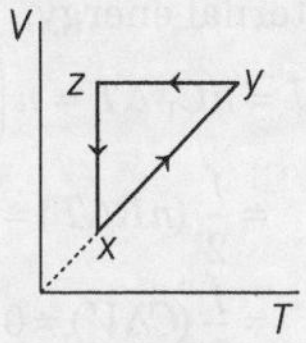

In process xy, $V \propto T \Rightarrow$ pressure is constant.

Process zx is a isothermal compression process, so pressure increases in this process. Process yz is a isochoric process in which temperature decreases, so pressure must decrease in this process. Hence, the p-V graph is as shown in the figure

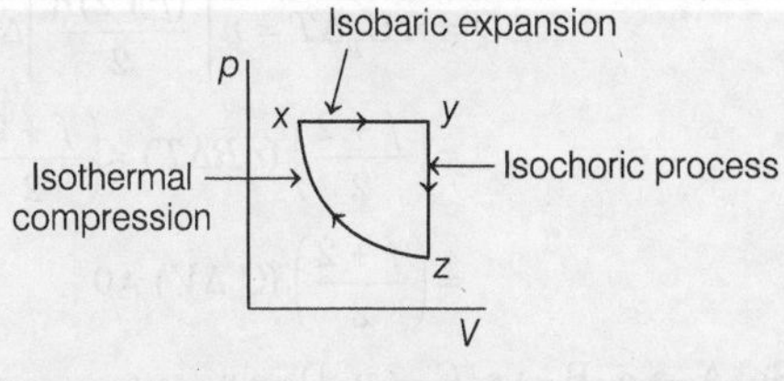

50. Isothermal curve from A to B will be parabolic with lesser area under the curve than the area under straight line AB. Therefore, work done by the gas in going straight from A to B is more.

If p_0, V_0 be the intercepts of curve on p and V axes, then its equation is obtained from $y = mx + c$

i.e. $$p = \frac{p_0}{V_0}V + p_0 \text{ or } \frac{RT}{V} = \frac{p_0 V}{V_0} + p_0$$

or $$T = \frac{p_0}{V_0 R}V^2 + \frac{p_0 V}{R}$$

which is the equation of a parabola. Hence T-V curve is parabolic. Therefore (b) is incorrect.

Also $(p/2) \times (2V) = pV =$ constant, *i.e.* process is isothermal.

Hence, option (a) is correct.

51. **I. Adiabatic process**

No heat is transferred between system and surroundings, therefore

$$\Delta Q = 0$$

II. Isothermal process

Temperature of the system remains constant.

i.e. $T =$ constant $\Rightarrow \Delta T = 0$

So, change in internal energy,

$$\Delta U = nC_V \Delta T = nC_V(0) = 0$$

III. Isochoric process

Volume of system remains constant.

i.e. $V =$ constant $\Rightarrow \Delta V = 0$

So, work done,

$$\Delta W = \int p\, dV = \int p\,(0) = 0$$

IV. Isobaric process

Pressure of system remains constant.

i.e. $p =$ constant (C)

$\Rightarrow \Delta p = 0$

So, change in internal energy,

$$\Delta U = nC_V \Delta T = n\left(\frac{fR}{2}\right)\Delta T$$

$$= \frac{f}{2}(nR\Delta T) = \frac{f}{2}(p\Delta V)$$

$$= \frac{f}{2}(C\Delta V) \neq 0$$

and work done,

$$\Delta W = \int p\, dV = \int C\, dV$$

$$= C\int dV = C\,\Delta V \neq 0$$

and heat transferred,

$$\Delta Q = n\,C_p \Delta T = n\left[\frac{(f+2)R}{2}\right]\Delta T$$

$$= \left(\frac{f+2}{2}\right)(nR\Delta T) = \left(\frac{f+2}{2}\right)(p\Delta V)$$

$$= \left(\frac{f+2}{2}\right)(C\,\Delta V) \neq 0$$

So, A $\rightarrow$ q, B $\rightarrow$ s, C $\rightarrow$ p, D $\rightarrow$ r

52. As, T will be maximum temperature where product of pV is maximum.

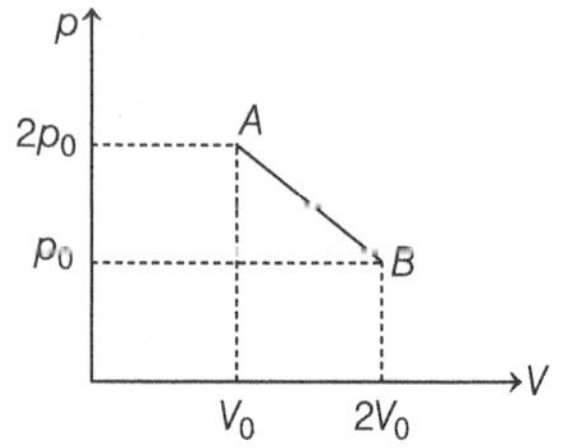

Equation of line AB, we have

$$y - y_1 = \frac{y_2 - y_1}{x_2 - x_1}(x - x_1)$$

$$\Rightarrow \quad p - p_0 = \frac{2p_0 - p_0}{V_0 - 2V_0}(V - 2V_0)$$

$$\Rightarrow \quad p - p_0 = \frac{-p_0}{V_0}(V - 2V_0)$$

$$\Rightarrow \quad p = \frac{-p_0}{V_0}V + 3p_0$$

$$pV = \frac{-p_0}{V_0}V^2 + 3p_0V$$

$$nRT = \frac{-p_0}{V_0}V^2 + 3p_0V$$

$$T = \frac{1}{nR}\left(\frac{-p_0}{V_0}V^2 + 3p_0V\right)$$

For maximum temperature,

$$\frac{\partial T}{\partial V} = 0$$

$$\frac{-p_0}{V_0}(2V) + 3p_0 = 0$$

$$\frac{-p_0}{V_0}(2V) = -3p_0 \Rightarrow V = \frac{3}{2}V_0$$

(condition for maximum temperature)

Thus, the maximum temperature of the gas during the process will be

$$T_{\max} = \frac{1}{nR}\left(\frac{-p_0}{V_0} \times \frac{9}{4}V_0^2 + 3p_0 \times \frac{3}{2}V_0\right)$$

$$= \frac{1}{nR}\left(-\frac{9}{4}p_0V_0 + \frac{9}{2}p_0V_0\right) = \frac{9}{4}\frac{p_0V_0}{nR}$$

Alternate solution

Since, initial and final temperatures are equal, hence maximum temperature is at the middle of line.

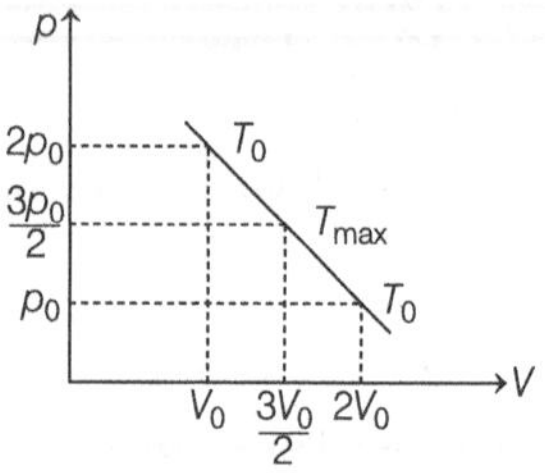

i.e. $pV = nRT$

$$\Rightarrow \quad \frac{\left(\frac{3}{2}p_0\right)\left(\frac{3V_0}{2}\right)}{nR} = T_{\max}$$

$$\frac{9}{4}\frac{p_0V_0}{nR} = T_{\max}$$

53. $1 - \dfrac{T_2}{T_1} = 0.5$ or $T_1 = 2T_2 = 2(17 + 273) = 580\text{ K}$

Temperature of hot body is increased by 145°C or 145 K,

$\therefore \quad T_1' = (580 + 145) = 725\text{ K}$

and $\quad T_2 = (17 + 273) = 290\text{ K}$

$$\therefore \quad \eta = \left(1 - \frac{290}{725}\right) \times 100 = 60\%$$

54. Here, $T_1 = 200°C = 200 + 273 = 473\text{ K}$

$$T_2 = 100°C = 100 + 273 = 373\text{ K}$$

$$\eta = 1 - \frac{T_2}{T_1} = 1 - \frac{373}{473} = \frac{100}{473}$$

$$\eta = \frac{100}{473} \times 100\% = 21.14\%$$

This is the percentage of heat which is utilised for doing work.

55. For adiabatic process BC (pressure constant),

$$p_B = p_C \quad \text{...(i)}$$

For isothermal process CA,

$$p_A V_A = p_C V_C \quad \text{...(ii)}$$

From Eqs. (i) and (ii), we get

$$V_C = \left[\frac{V_B^\gamma}{V_A}\right]^{\frac{1}{\gamma-1}} = \left[\frac{4^\gamma}{1}\right]^{\frac{1}{\gamma-1}}$$

For $\gamma = 1.5, V_C = 4^3 = 64\text{ m}^3$

$\therefore$ $$p_C = \frac{p_A V_A}{V_C} = \frac{10^5}{64}\text{ N/m}^2$$

$$= \frac{10^x}{64}\text{ N/m}^2 \quad \text{(given)}$$

$\therefore$ $x = 5$

56. From the given problem,

$$\Delta Q = Q_1 + Q_2 + Q_3 + Q_4$$

$$= 5960 - 5585 - 2980 + 3645$$

$$\Delta Q = 9605 - 8565 = 1040\text{ J}$$

Efficiency of a cycle is defined as

$$\eta = \frac{\text{Net work}}{\text{Input heat}} = \frac{\Delta W}{Q_1 + Q_4} = \frac{\Delta Q}{Q_1 + Q_4}$$

Putting $\Delta Q = 1040\text{ J}$

and $Q_1 + Q_4 = 5960 + 3645 = 9605\text{ J}$

$\therefore$ $$\eta = \frac{1040}{9605} = 0.1082 = 10.82\%$$

57. Given, air is diatomic gas

$\therefore$ Degrees of freedom, $f = 5$

So, ratio of specific heats of gas, $\gamma = 1 + \frac{2}{f} = \frac{7}{5}$

Also, given that $V_i = V$ and $V_f = \frac{V}{10}$

$$T_i = 20°\text{ C} = 273 + 20 = 293\text{ K}$$

As for adibatic change,

$$TV^{\gamma-1} = \text{constant}$$

$\Rightarrow$ $$T_i V_i^{\gamma-1} = T_f V_f^{\gamma-1}$$

$\Rightarrow$ $$293 \cdot V^{\frac{7}{5}-1} = T_f \cdot \left(\frac{V}{10}\right)^{\frac{7}{5}-1}$$

$\Rightarrow$ $$T_f = 293 \times 10^{2/5}\text{ K}$$

Change in internal energy,

$$\Delta U = \frac{n \cdot f \cdot R(T_f - T_i)}{2}$$

$$= \frac{5 \times 5 \times \frac{25}{3}(293 \times 10^{2/5} - 293)}{2}$$

$$= 4.195 \times 10^3\text{ J} \approx 4\text{kJ}$$

58. Here, Q_1 = heat absorbed per minute

and Q_2 = heat rejected per minute.

We know that, $\eta\% = \frac{W}{Q_1} \times 100$

$\therefore$ $$\eta\% = \frac{5.4 \times 10^8\text{ J}}{3.6 \times 10^9\text{ J}} \times 100$$

$$= \frac{3}{20} \times 100 = 15\%$$

Also using the relation $Q_1 = W + Q_2$, we get

$$Q_2 = Q_1 - W$$

$$= 36 \times 10^8 - 5.4 \times 10^8$$

$$= 30.6 \times 10^8\text{ J/min}$$

$$= 3.06 \times 10^9\text{ J/min}$$

$$= 3.1 \times 10^9\text{ J/min}$$

59. Here, mass of water = 1 g

$\therefore$ Initial volume of water, $V_1 = 1\text{ cm}^3$

Volume of steam, $V_2 = 1650\text{ cm}^3$

Change of internal energy, $dU = ?$

As the state of water is changing,

$\therefore$ $$dQ = mL = 1 \times 540\text{ cal}$$

$$= 540 \times 4.2 \times 10^7\text{ erg}$$

$$= 22.68 \times 10^9\text{ erg}$$

Taking $p = 1$ atm,

$$= 76 \times 13.6 \times 981\text{ dyne cm}^{-2}$$

$$dW = pdV = p(V_2 - V_1)$$

$$= 76 \times 13.6 \times 981\,(1650 - 1)$$

$$= 76 \times 13.6 \times 981 \times 1649\text{ erg}$$

$$= 1.67 \times 10^9\text{ erg}$$

As, $dQ = dU + dW$

$\therefore$ $$dU = dQ - dW = 22.68 \times 10^9 - 1.67 \times 10^9$$

$$dU = 21.01 \times 10^9\text{ erg}$$

$$= 21.01 \times 10^x\text{ erg} \quad \text{(given)}$$

$\therefore$ $x = 9$

60. According to the question,

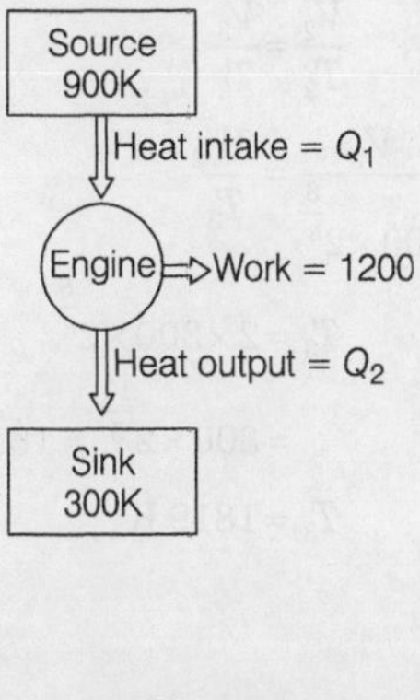

For the given Carnot engine, we have work output, $W = Q_1 - Q_2$ and ratio of heat taken and rejected is

$$\frac{Q_1}{Q_2} = \frac{T_1}{T_2}$$

So, we have

$$1200 = Q_1 - Q_2$$

and $$\frac{Q_1}{Q_2} = \frac{900}{300} = 3$$

On solving these, we get

$$Q_2 = 600 \text{ J}$$

61. Here, temperature of the gas,

$$T = 273 + 127 = 400 \text{ K}$$

Let initial volume of the gas, $V_1 = V$

$\therefore$ Final volume of the gas, $V_2 = 2V$

In an isothermal expansion,

Work done $(W) = 2.3026\, RT \log_{10} \frac{V_2}{V_1}$

$$= 2.3026 \times 8.3 \times 400 \times \log_{10} \frac{2V}{V}$$

$$= 2.3026 \times 8.3 \times 400 \times 0.3010$$

or $$W = 2.30 \times 10^3 \text{ J}$$

The amount of heat absorbed,

$$H = \frac{W}{J} = \frac{2.30 \times 10^3}{4.2} \; 547.6 = 548 \text{ cal}$$

62. Initially gas is compressed adiabatically.

Initial temperature, $T_1 = 300$ K

Number of moles = 1, $\gamma = 1.4$

Initial volume = V_1

Final volume = $\frac{V_1}{16}$

Process equation is

$$T_1V_1^{\gamma-1} = T_2V_2^{\gamma-1}$$

$$\Rightarrow \quad 300V_1^{1.4-1} = T_2\left(\frac{V_1}{16}\right)^{1.4-1}$$

$$\Rightarrow \quad T_2 = 300 \times 2^{\frac{8}{5}} \quad \text{...(i)}$$

In next process, gas is expanded isobarically. Initial volume, $V_2 = \frac{V_1}{16}$

Final volume, $V_3 = 2\left(\frac{V_1}{16}\right) = \frac{V_1}{8}$

Now, for isobaric expansion,

$$\frac{V_2}{T_2} = \frac{V_3}{T_3}$$

$$\Rightarrow \quad \frac{V_2}{300 \times 2^{\frac{8}{5}}} = \frac{2V_2}{T_3}$$

$$\Rightarrow \quad T_3 = 2 \times 300 \times 2^{\frac{8}{5}}$$

$$= 300 \times 2^{\frac{13}{5}} = 1818.85$$

or $$T_3 = 1819 \text{ K}$$

63. Here, initial temperature,

$$T_1 = 27^\circ \text{C} = 273 + 27 = 300 \text{ K}$$

Final temperature, $T_2 = 97^\circ \text{C} = 273 + 97 = 370$ K

When a gas is compressed adiabatically, work done on the gas is given by

$$W = \frac{R}{(1-\gamma)}(T_2 - T_1)$$

$$= \frac{8.3 \times (370 - 300)}{1 - 1.5}$$

or $$W = -11.62 \times 10^2 \text{J}$$

$\therefore$ Heat produced,

$$H = \frac{W}{J} = \frac{11.62 \times 10^2}{4.2} = 276.7 \text{ cal}$$

64. Here, $n = 5$, $T = 500\text{K}$, $V_B = 2\,V_A$,

$$R = 8.31 \text{ mol}^{-1} \text{ K}^{-1}$$

Now, $$W_{\text{iso}} = n \times 2.303 \times RT \log \frac{V_B}{V_A}$$

$$= 5 \times 2.303 \times 8.31 \times 500 \log \frac{2V_A}{V_A}$$

$$= 5 \times 2.303 \times 8.31 \times 500 \log 2$$

$$= 5 \times 2.303 \times 8.31 \times 500 \times 0.3010$$

$$= 14401.3 \text{ J}$$

65. Here, $Q_1 = 6 \times 10^5$ cal

$$T_1 = 227^\circ \text{C} = 227 + 273 = 500 \text{ K}$$

$$T_2 = 127^\circ \text{C} = 127 + 273 = 400 \text{ K}$$

Work done/cycle, $W = ?$

As, $$\frac{Q_2}{Q_1} = \frac{T_2}{T_1}$$

$\therefore$ $$Q_2 = \frac{T_2}{T_1} \times Q_1 = \frac{400}{500} \times 6 \times 10^5$$

$$= 4.8 \times 10^5 \text{ cal}$$

As, $$W = Q_1 - Q_2 = 6 \times 10^5 - 4.8 \times 10^5$$

$$W = 1.2 \times 10^5 \text{ cal}$$

$$= 1.2 \times 10^5 \times 4.2 \text{ J}$$

$$W = 5.04 \times 10^5 \text{ J}$$

$$= 5.04 \times 10^x \text{J} \quad \text{(given)}$$

$\therefore$ $$x = 5$$

66. Here, $T_1 = 0 + 273 = 273$ K;

$$p_1 = 1.013 \times 10^6 \text{ dyne cm}^{-2}; \gamma = 1.4$$

and density of air at NTP, $\rho = 1.29 \times 10^{-3}$ gcm^{-3}

$\therefore$ $$V_1 = \frac{\text{Mass}}{\text{Density}} = \frac{1}{1.29 \times 10^{-3}} = 775.2 \text{ cm}^3$$

and $$V_2 = \frac{V_1}{2} = \frac{775.2}{2} = 387.6 \text{ cm}^3$$

Now, $$p_1V_1^\gamma = p_2V_2^\gamma \text{ or } p_2 = p_1 \times \left(\frac{V_1}{V_2}\right)^\gamma$$

or $$p_2 = 1.013 \times 10^6 \times \left(\frac{V_1}{V_1/2}\right)^{1.4} = 1.013 \times 10^6 \times (2)^{1.4}$$

$$= 1.013 \times 10^6 \times 2.639 = 2.673 \times 10^6 \text{ dyne cm}^{-2}$$

Work done during adiabatic change,

$$W = \frac{p_1V_1 - p_2V_2}{\gamma - 1}$$

$$= \frac{1.013 \times 10^6 \times 775.2 - 2.673 \times 10^6 \times 387.6}{1.4 - 1}$$

$$= \frac{7.853 \times 10^8 - 10.36 \times 10^8}{0.4} = -\ 6.27 \times 10^8 \text{ erg}$$

$$= -\ 62.7 \text{ J}$$

67. From given p-V diagram,

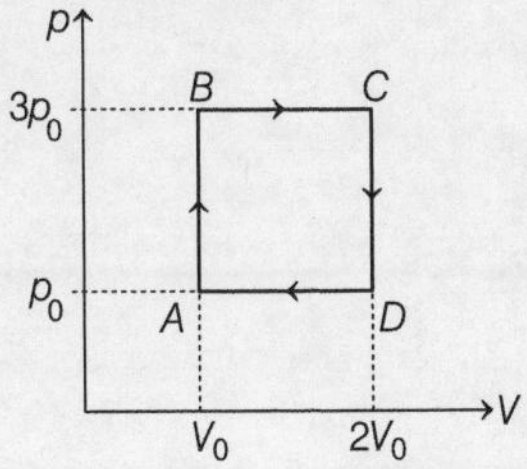

Work done by gas in close loop, W = Area of loop

$$= (3p_0 - p_0) \times (2V_0 - V_0)$$

$$W = 2p_0V_0$$

During the cycle $ABCDA$, processes AB and BC absorb the heat.

Heat absorption during process AB,

$$Q_{AB} = \mu C_V \Delta T \qquad \text{(as } AB = \text{isochoric process)}$$

$$Q_{AB} = \mu\left(\frac{3}{2}R\right)(T_B - T_A)$$

$$\left(\because C_V \text{ for monatomic gas is } \frac{3}{2}R\right)$$

$$= \frac{3}{2}(\mu R T_B - \mu R T_A)$$

$$= \frac{3}{2}(p_BV_B - p_AV_A) \qquad (\because pV = \mu RT)$$

$$= \frac{3}{2}(3p_0V_0 - p_0V_0)$$

$$= 3p_0V_0$$

Heat absorption during process BC,

$$Q_{BC} = \mu C_p \Delta T \qquad \text{(as } BC = \text{isobaric process)}$$

$$= \mu\left(\frac{5}{2}R\right)(T_C - T_B) \qquad (\because C_p = C_V + R)$$

$$= \frac{5}{2}(\mu RT_C - \mu RT_B)$$

$$= \frac{5}{2}(p_CV_C - p_BV_B)$$

$$= \frac{5}{2}(6p_0V_0 - 3p_0V_0)$$

$$= \frac{15}{2}p_0V_0$$

Total heat absorption, $Q = Q_{AB} + Q_{BC}$

$$= 3p_0V_0 + \frac{15}{2}p_0V_0$$

$$= \frac{21}{2}p_0V_0$$

$\therefore$ Efficiency of engine,

$$\eta\% = \frac{W}{Q} \times 100 = \frac{2p_0V_0}{\frac{21}{2}p_0V_0} \times 100 = 19\%$$

16

Oscillations

IN THIS CHAPTER ....

Periodic Motion

A motion that repeats itself over and over again after a regular interval of time is called a periodic motion. The regular interval of time after which the periodic motion is repeated is called its *time period*. Revolution of the Earth around the Sun, rotation of the Earth about its axis are the common examples of periodic motion.

Oscillatory Motion

A special type of periodic motion in which a particle moves to and fro (back and forth or up and down) about a fixed point after regular interval of time is termed as oscillatory or vibratory motion. The fixed point about which the body oscillates is called *mean position* or *equilibrium position*. Thus, a periodic or bounded motion of a body about a fixed point is called an oscillatory or vibratory motion. Examples of oscillatory motion are simple pendulum, spring pendulum, etc.

A body that undergoes oscillatory motion always have a stable equilibrium position (where net force on the body is zero). As the body is displaced from its mean/equilibrium position, a force (torque) comes into existence, which tends to bring the body back to the equilibrium position, this force or torque is termed as the *restoring force* (*torque*).

Note There is no significant difference between oscillations and vibrations. When to and fro motion of the body about a fixed position has small frequency, we call it oscillation such as the oscillation of a simple pendulum.
When to and fro motion of the body about a fixed position has high frequency, we call it as vibrations such as oscillation of a musical instrument.

Periodic Functions

Those functions which are used to represent periodic motion are known as periodic functions.

A function $f(t)$ is said to be periodic, if $f(t) = f(t+T) = f(t+2T)$ ...(i)

$\because$ sine and cosine functions are example of periodic functions.

When T is the period of this periodic motion, then for periodic motion,

$$y = A\sin\omega t = A\sin\omega(t+T) \quad \text{...(ii)}$$

and

$$x = A\cos\omega t = A\cos\omega(t+T) \quad \text{...(iii)}$$

But the value of sine or cosine functions repeat after a period of 2π radian.

$$\therefore \quad \omega(t+T) = \omega t + 2\pi$$

or $$\omega T = 2\pi \quad \text{...(iv)}$$

or $$\omega = 2\pi\nu \quad \text{...(v)}$$

Consider a linear combination of sine and cosine functions are given as

$$x = f(t) = A\sin\omega t + B\cos\omega t$$

Taking, $A = R\cos\phi$ and $B = R\sin\phi$

Then, $x = R\cos\phi\sin\omega t + R\sin\phi\cos\omega t = R\sin(\omega t + \phi)$ ...(vi)

It represents a period function of time period T and amplitude R,

where $R = \sqrt{A^2 + B^2}$ and $\tan\phi = B/A$.

The combination of any number of periodic functions will also be periodic one, whose time period will be minimum of the periodic functions used in the combinations.

Example 1. *A function of time is represented as follows* $\sin\omega t + \cos 2\omega t + \sin 4\omega t$.
The motion represented by it is

(a) *non-periodic*
(b) *periodic*
(c) *both non-periodic and periodic*
(d) *data insufficient*

Sol. (b) This is an example of a periodic motion. It can be noted that each term represents a periodic function with different angular frequency. Since period is the least interval of time after which a function repeats its value, $\sin\omega t$ has a period $T_0 = \frac{2\pi}{\omega}$,

$\cos 2\omega t$ has a period $\frac{\pi}{\omega} = \frac{T_0}{2}$

and $\sin 4\omega t$ has a period $\frac{2\pi}{4\omega} = \frac{T_0}{4}$

The period of the first term is a multiple of the periods of the last two terms. Therefore, the smallest interval of time after which the sum of the three terms repeats is T_0 and thus, the sum is a periodic function with a period $\frac{2\pi}{\omega}$.

Example 2. *Among the following, what is the time period of* $\sin^2\omega t$?

(a) $\frac{\omega}{\pi}$ (b) $\frac{2\omega}{\pi}$ (c) $\frac{\pi}{\omega}$ (d) $\frac{\pi}{2\omega}$

Sol. (c) Given, $\sin^2\omega t$
Using the trigonometric identity,

$$\cos 2\theta = 1 - 2\sin^2\theta$$

We have, $$\sin^2\omega t = \frac{1}{2} - \frac{1}{2}\cos 2\omega t$$

The function is periodic having a period $T = \frac{\pi}{\omega}$. It also represents a harmonic motion with the point of equilibrium occuring at $\frac{1}{2}$ instead of zero.

Simple Harmonic Motion

The type of oscillatory motion in which the particle moves to and fro in a straight line about a fixed point under a restoring force (or torque) whose magnitude is directly proportional to its displacement (or angular displacement), is known as simple harmonic motion.

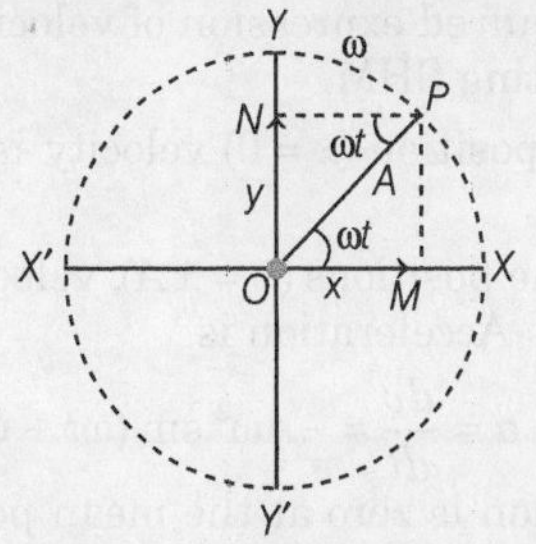

A simple harmonic motion may also be considered as the projection of a uniform circular motion on any diameter of the circle. As shown in figure, let us consider uniform circular motion in a plane with constant angular velocity ω and let at an instant t the particle executing circular motion be at point P making an angle $\theta = \omega t$ from X-axis, then the projection of this circular motion along a diameter YY' is given by,

$$y = A\sin\omega t$$

and projection along diameter XX' is given by

$$x = A\cos\omega t$$

Such a motion is a simple harmonic motion. The radius A of the reference circle is the maximum value of displacement (to and fro motion about the mean position O) and is known as the *amplitude of SHM.*

Terms Related to SHM

- **Displacement** Displacement of a particle in the case of SHM is always measured from the mean position.
  (a) If the particle is at the mean position at $t = 0$, then displacement is given by $x = A\sin\omega t$.
  (b) If the particle is at the extreme position at $t = 0$, then displacement is given by $x = A\cos\omega t$.
  (c) In general displacement is given by $x = A\sin(\omega t + \phi)$, where ϕ is the *initial phase* or *phase constant*.
- **Time period** The time taken by the particle to complete one oscillation is called *time period of oscillation*. It is denoted by T and given by $T = \frac{2\pi}{\omega}$.
- **Frequency** The number of oscillations made by the particle in one second is called *frequency of oscillations.* It is denoted by ν or n. Thus, $\nu = \frac{1}{T} = \frac{\omega}{2\pi}$. ω being the angular frequency of the oscillating particle.
- **Velocity** As, $x = A\sin(\omega t + \phi)$

$$v = \frac{dx}{dt} = A\omega\cos(\omega t + \phi)$$

$$= A\omega\,[1-\sin^2(\omega t+\phi)]^{1/2} \quad \text{(In terms of time)}$$

$$= A\omega\left(1-\frac{x^2}{A^2}\right)^{1/2}$$

So, $$v = \omega\sqrt{A^2-x^2}$$

(In terms of displacement from mean position)

This is the required expression of velocity of the particle executing SHM.

(a) At mean position ($x = 0$) velocity is maximum, *i.e.* $v_{max} = A\omega$.

(b) At extreme positions ($x = \pm A$), velocity is zero.

- **Acceleration** Acceleration is

$$a = \frac{dv}{dt} = -A\omega^2 \sin(\omega t+\phi) = -\omega^2 x$$

(a) Acceleration is zero at the mean position, *i.e.* $a = 0$, at $x = 0$.

(b) Acceleration is maximum at extreme position, *i.e.* $a_{max} = -\omega^2 A$ at $x = A$.

In above expressions negative sign indicate that in SHM, the acceleration is proportional to the displacement but is in opposite.

Also in linear SHM, the force and displacement are related by

$$F = -(\text{ a positive constant })\, x$$

which says that the force is proportional to the displacement but is in opposite direction.

- **Phase relationship between displacement, velocity and acceleration of SHM**

The term $(\omega t+\phi)$ is called phase and ϕ is called phase constant.

As, we have seen that

$$x = A\sin(\omega t+\phi)$$

$$v = A\omega\cos(\omega t+\phi) \quad \text{...(i)}$$

$$= A\omega\sin\left(\omega t+\phi+\frac{\pi}{2}\right)$$

and $$a = -A\omega^2\sin(\omega t+\phi)$$

$$= A\omega^2\sin(\omega t+\phi+\pi) \quad \text{...(ii)}$$

Thus from Eqs. (i) and (ii), we conclude that in SHM, particle velocity is ahead in phase to the displacement by $\pi/2$ and acceleration is further ahead in phase by $\pi/2$.

In figure, x, v and a as functions of time are illustrated.

(i)

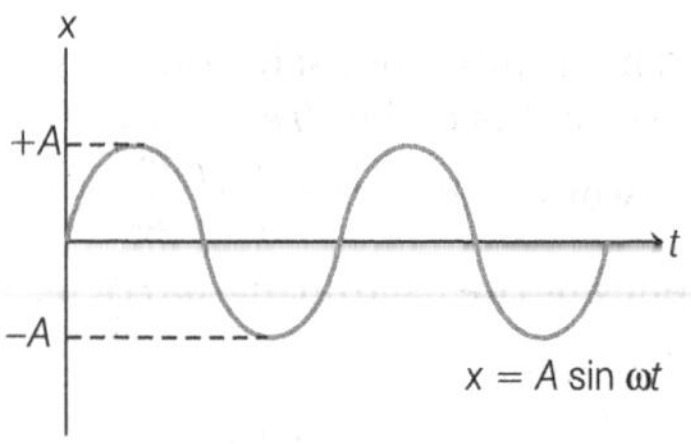

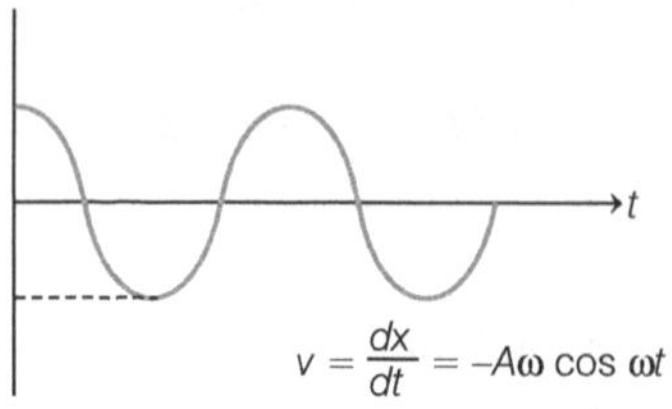

(iii)

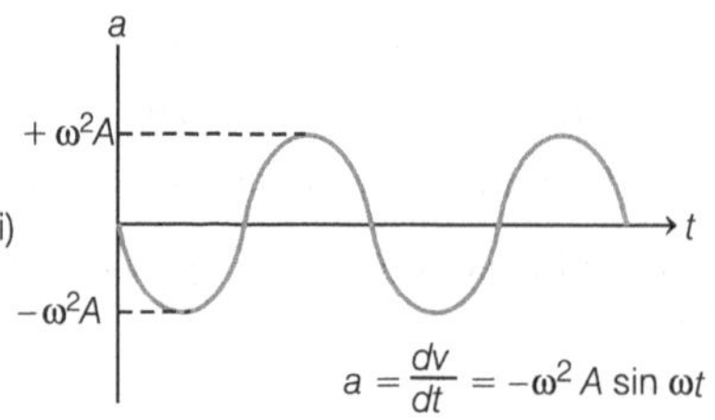

Example 3. *The displacement of a particle executing periodic motion is given by*

$$y = 4\cos^2\left(\frac{t}{2}\right)\sin(1000t)$$

Find independent constituents of SHMs.

(a) $y_1 = 2\sin 1000\,t,\ y_2 = \sin 1001t,\ y_3 = \sin 999t$

(b) $y_1 = 3\sin 1000\,t,\ y_2 = \sin 1000\,t,\ y_3 = \sin 899t$

(c) $y_1 = \sin 1001t,\ y_2 = \sin 999t,\ y_3 = 2\sin 1000\,t$

(d) None of the above

Sol. (a) $y = 4\cos^2\left(\frac{t}{2}\right)\sin(1000t)$

$$= 2(1+\cos t)\sin(1000t)$$

$$= 2\sin(1000t) + 2\cos t\sin(1000t)$$

$$= 2\sin(1000t) + \sin(1001t) + \sin(999t)$$

Thus, the given periodic motion is a combination of three independent SHMs, which are given by

$$y_1 = 2\sin 1000t, y_2 = \sin 10001t$$

and $$y_3 = 999t$$

Example 4. *The following figure depicts circular motion. The radius of the circle, the period of revolution, the initial position and sense of revolution are indicated in the figure.*

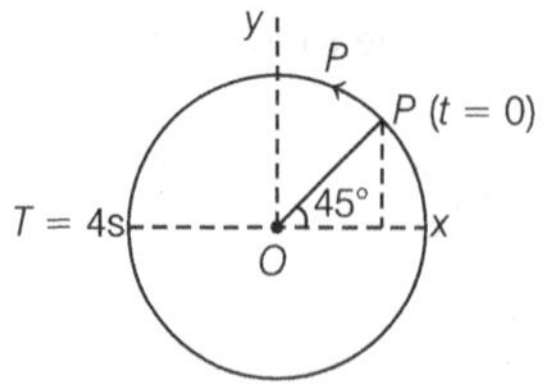

The simple harmonic motion of the x-projection of the radius vector of the rotating particle P is as follows

(a) $x(t) = A\cos\left(\frac{2\pi}{4}t+\frac{\pi}{4}\right)$ (b) $x(t) = A\cos\left(\frac{\pi}{4}t-\frac{2\pi}{4}\right)$

(c) $x(t) = A\sin\left(\frac{2\pi}{4}t+\frac{\pi}{2}\right)$ (d) $x(t) = A\sin\left(\frac{\pi}{4}t-\frac{\pi}{2}\right)$

Sol. (a) At $t = 0$, OP makes an angle of $45° = \frac{\pi}{4}$ rad with the positive direction of X-axis. After time t, it covers an angle $\frac{2\pi}{T}t$ in the anti-clockwise sense, and makes an angle of $\frac{2\pi}{T}t + \frac{\pi}{4}$ with the X-axis.

The projection of OP on the X-axis at time t is given by

$$x(t) = A\cos\left(\frac{2\pi}{T}t + \frac{\pi}{4}\right)$$

For, $T = 4$ s $\quad x(t) = A\cos\left(\frac{2\pi}{4}t + \frac{\pi}{4}\right)$

which is a SHM of amplitude A, period 4 s and an initial phase $= \frac{\pi}{4}$.

Example 5. *The periodic time of a body executing SHM is 2 s. After what interval from $t = 0$, will its displacement be half of its amplitude?*

(a) (1/5) s (b) (1/6) s (c) (1/10) s (d) (1/8) s

Sol. (b) Given, $T = 2$ s, $t = ?$; $x = \frac{A}{2}$

Now, $\quad x = A\sin\omega t = A\sin\frac{2\pi}{T}t$

$\therefore \quad \frac{A}{2} = A\sin\frac{2\pi}{2}t = A\sin\pi t$

$\Rightarrow \quad \sin\pi t = \sin 30° = \sin\frac{\pi}{6}$

or $\quad \pi t = \frac{\pi}{6}$ or $t = \frac{1}{6}$ s

Example 6. *A particle in SHM is described by the displacement function.*

$$x = A\cos(\omega t + \phi), \omega = 2\pi/T$$

If the initial ($t = 0$) position of the particle is 1 cm and its initial velocity is π cm s^{-1}, what is the initial phase angle? (The angular frequency of the particle is π s^{-1}.)

(a) $3\pi/4$ (b) $2\pi/4$ (c) $5\pi/4$ (d) $7\pi/4$

Sol. (a) Here, at $t = 0$, $x = 1$ cm and $v = \pi$ cms^{-1}; $\phi = ?$; $\omega = \pi$ s^{-1}

Given, $\quad x = A\cos(\omega t + \phi)$

$1 = A\cos(\pi \times 0 + \phi)$

$1 = A\cos\phi \quad$ ...(i)

Velocity, $v = \frac{dx}{dt} = -A\omega\sin(\omega t + \phi)$

$\therefore \quad \pi = -A\pi\sin(\pi \times 0 + \phi) = -A\pi\sin\phi$

or $\quad 1 = -A\sin\phi$

or $\quad A\sin\phi = -1 \quad$ ...(ii)

Dividing Eq. (ii) by (i), we get

$$\tan\phi = -1 = -\tan\frac{\pi}{4} = \tan\left(\pi - \frac{\pi}{4}\right) = \tan\frac{3\pi}{4}$$

or $\quad \phi = \frac{3\pi}{4}$

Example 7. *The displacement-time graph of a particle executing SHM is given in figure (sketch is schematic and not to scale).* **[JEE Main 2020]**

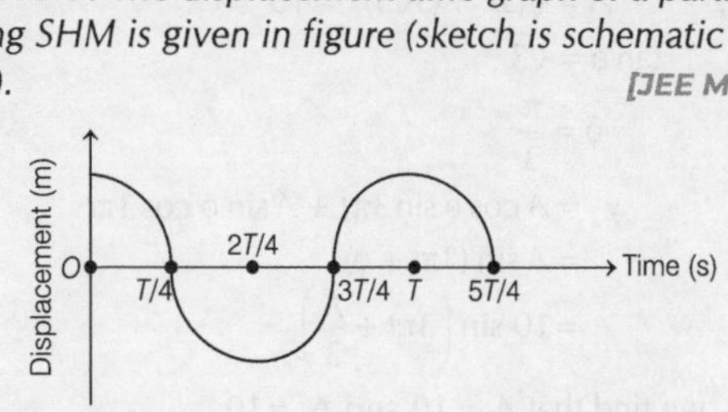

Which of the following statement(s) is/are true for this motion?

A. The force is zero at $t = \frac{3T}{4}$.

B. The acceleration is maximum at $t = T$.

C. The speed is maximum at $t = \frac{T}{4}$.

D. The potential energy is equal to kinetic energy of the oscillation at $t = \frac{T}{2}$.

(a) A, B and D (b) B, C and D
(c) A, B and C (d) A and D

Sol. (c) The displacement-time graph is as shown below,

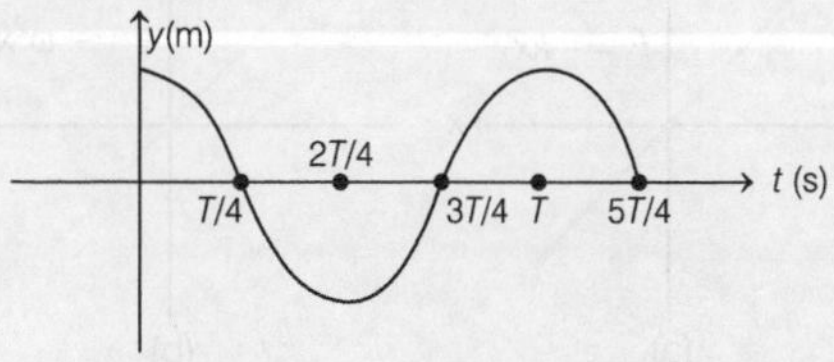

At $t = \frac{3T}{4}$, particle is at mean position. At mean position of SHM, acceleration of particle is zero. *i.e.* force is zero.

Statement (A) is correct.

At $t = T$, particle is at extreme position. As direction is changing, so its acceleration is maximum. Statement (B) is correct.

At $t = \frac{T}{4}$, particle is at mean position, so whole of energy is kinetic. *i.e.* its speed is maximum. Statement (C) is correct.

PE = KE at $t = \frac{T}{8}$, so statement (D) is incorrect.

Hence, option (c) is correct.

Example 8. *If two SHMs are represented by equations*

$$y_1 = 10\sin\left(3\pi t + \frac{\pi}{4}\right) \text{ and } y_2 = 5(\sin 3\pi t + \sqrt{3}\cos 3\pi t),$$

then the ratio of their amplitudes is

(a) 1 : 2 (b) 2 : 1
(c) 1 : 1 (d) 1 : 3

Sol. (c) Here, $y_1 = 10\sin\left(3\pi t + \frac{\pi}{4}\right)$

But $y_2 = 5(\sin 3\pi + \sqrt{3}\cos 3\pi t)$ does not appear to be a single compact function of time t. We therefore, assume that

$$5 = A\cos\phi \quad \text{and} \quad 5\sqrt{3} = A\sin\phi$$

Thus, $A = \sqrt{(5)^2 + (5\sqrt{3})^2} = 10$

and $\tan \phi = \sqrt{3}$

or $\phi = \frac{\pi}{3}$

then, $y_2 = A \cos \phi \sin 3\pi t + A \sin \phi \cos 3\pi t$

$= A \sin (3\pi t + \phi)$

$= 10 \sin \left(3\pi t + \frac{\pi}{3}\right)$

Now, we find that $A_1 = 10$ and $A_2 = 10$

or $A_1 : A_2 = 10 : 10 = 1 : 1$

Differential Equations of SHM

For linear SHM, $\frac{d^2x}{dt^2} + \omega^2 x = 0$

or $a = -\omega^2 x$

In SHM, $F = -kx$ or $a = -\omega^2 x$, *i.e.* F-x graph or a-x graph is a straight line passing through the origin with negative slope. The corresponding graphs are shown below.

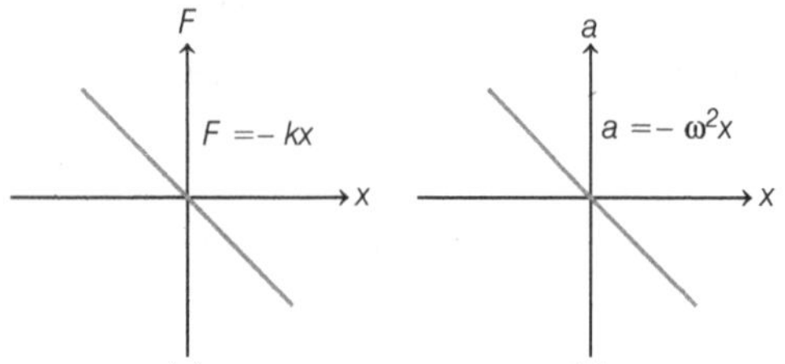

(a) Slope = − k (b) Slope = − ω^2

Energy in SHM

A particle executing SHM possesses two types of energy. If a particle executes SHM, its kinetic energy changes into potential energy and *vice-versa* keeping total energy constant (if friction of air is neglected).

Kinetic energy $K = \frac{1}{2} mv^2$

$$= \frac{1}{2} mA^2\omega^2 \cos^2 (\omega t + \phi) = \frac{1}{2} m\omega^2 (A^2 - x^2)$$

(because $x = A \sin(\omega t + \phi)$)

Potential energy $U = \frac{1}{2} m\omega^2 x^2$

$$= \frac{1}{2} mA^2\omega^2 \sin^2 (\omega t + \phi)$$

Thus, total energy = kinetic energy + potential energy

or $E = \frac{1}{2} m\omega^2 A^2$ (constant)

The total energy is constant and is proportional to the square of amplitude (A) of motion.

Figures show the variations of total energy (E), potential energy (U) and kinetic energy (K) with displacement (x).

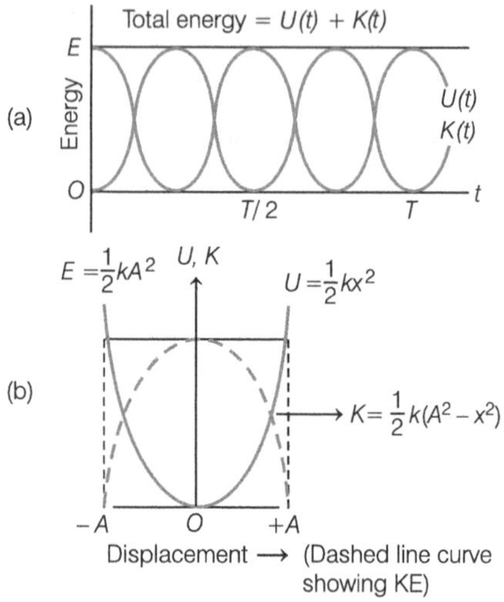

Example 9. *A particle executes SHM of amplitude A. At what distance from the mean position is its KE equal to its PE?*

(a) 0.71 A (b) 0.61 A (c) 0.65 A (d) 0.8 A

Sol. (a) As, $\text{KE} = \frac{1}{2} m\omega^2 (A^2 - x^2)$

and $\text{PE} = \frac{1}{2} m\omega^2 x^2$

As, KE = PE

$\therefore \frac{1}{2} m\omega^2 (A^2 - x^2) = \frac{1}{2} m\omega^2 x^2$

or $A^2 - x^2 = x^2$

or $x^2 = A^2/2$

or $x = A/\sqrt{2} = 0.71A$

Example 10. *A block of mass m attached to a massless spring is performing oscillatory motion of amplitude A on a frictionless horizontal plane. If half of the mass of the block breaks off when it is passing through its equilibrium point, the amplitude of oscillation for the remaining system becomes fA. The value of f is* **[JEE Main 2020]**

(a) $\sqrt{2}$ (b) $\frac{1}{2}$ (c) $\frac{1}{\sqrt{2}}$ (d) 1

Sol. (c) In the equilibrium position, the velocity is maximum and it is equal to $v_{max} = \omega A$

where, $\omega = \sqrt{\frac{k}{m}}$ = angular frequency

and k = spring constant.

Now, kinetic energy in equilibrium position,

$$E = \frac{1}{2} m\omega^2 A^2 = \frac{1}{2} m\left(\frac{k}{m}\right) A^2 \quad \ldots\text{(i)}$$

When half of the mass of the block breaks off, also the kinetic energy of the system will become half, *i.e.* $E/2$.

The angular frequency will become,

$$\omega' = \sqrt{\frac{k}{(m/2)}}$$

Now, the amplitude will become fA.

$\therefore$ The new kinetic energy,

$$E' = \frac{E}{2} = \frac{1}{2}\left(\frac{m}{2}\right)\omega'^2 (fA)^2$$

$$\Rightarrow \quad \frac{E}{2} = \frac{1}{2}\left(\frac{m}{2}\right)\left(\frac{2k}{m}\right)(fA)^2$$

$$\Rightarrow \quad \frac{1}{2}\left[\frac{1}{2}m\left(\frac{k}{m}\right)A^2\right] = \frac{1}{2}\cdot\frac{m}{2}\cdot\frac{2k}{m}\cdot f^2A^2 \quad \text{[Using Eq. (i)]}$$

$$\Rightarrow \quad 1 = 2f^2 \Rightarrow f = \frac{1}{\sqrt{2}}$$

Example 11. *A particle undergoing simple harmonic motion has time dependent displacement given by $x(t) = A \sin\frac{\pi t}{90}$. The ratio of kinetic to potential energy of this particle at $t = 210$ s will be* ***[JEE Main 2019]***

(a) 2 (b) $\frac{1}{3}$ (c) $\frac{1}{9}$ (d) 3

Sol. (b) Here given, displacement, $x(t) = A \sin\frac{\pi t}{90}$

where A is amplitude of SHM, t is time taken by particle to reach a point where its potential energy $U = \frac{1}{2}kx^2$ and kinetic energy $= \frac{1}{2}k(A^2 - x^2)$ here k is force constant and x is position of the particle.

Potential energy (U) at $t = 210$ s is

$$U = \frac{1}{2}kx^2 = \frac{1}{2}kA^2 \sin^2\left(\frac{210}{90}\pi\right)$$

$$= \frac{1}{2}kA^2 \sin^2\left(2\pi + \frac{3}{9}\pi\right) = \frac{1}{2}kA^2 \sin^2\left(\frac{\pi}{3}\right)$$

Kinetic energy at $t = 210$ s, is

$$K = \frac{1}{2}k(A^2 - x^2) = \frac{1}{2}kA^2\left[1 - \sin^2\left(\frac{210\pi}{90}\right)\right]$$

$$= \frac{1}{2}kA^2 \cos^2(210\,\pi/90)$$

$$\Rightarrow \quad K = \frac{1}{2}kA^2 \cos^2(\pi/3)$$

So, ratio of kinetic energy to potential energy is

$$\frac{K}{U} = \frac{\frac{1}{2}kA^2 \cos^2(\pi/3)}{\frac{1}{2}kA^2 \sin^2(\pi/3)} = \cot^2(\pi/3) = \frac{1}{3}$$

Angular Simple Harmonic Motion

If the angular displacement of the body at an instant is θ, the resultant torque acting on the body in angular SHM should be

$$\tau = -k\theta$$

If the moment of inertia is I, the angular acceleration is

$$\alpha = \frac{\tau}{I} = -\frac{k}{I}\theta$$

or

$$\frac{d^2\theta}{dt^2} = -\omega^2\theta \quad \text{...(i)}$$

where

$$\omega = \sqrt{\frac{k}{I}}$$

The Eq. (i) may be integrated in the similar manner and we shall get

$$\theta = \theta_0 \sin(\omega t + \delta) \quad \text{...(ii)}$$

where, θ_0 is the maximum angular displacement on either side.

The angular velocity at time t is

$$\omega = \frac{d\theta}{dt} = \theta_0\omega\cos(\omega t + \delta) \quad \text{...(iii)}$$

The time period of oscillation is

$$T = \frac{2\pi}{\omega} = 2\pi\sqrt{\frac{I}{k}} \quad \text{...(iv)}$$

and the frequency of oscillation is

$$\nu = \frac{1}{T} = \frac{1}{2\pi}\sqrt{\frac{k}{I}} \quad \text{...(v)}$$

The quantity $\omega = \sqrt{\frac{k}{I}}$ is the angular frequency.

Energy in Angular SHM

The potential energy is, $U = \frac{1}{2}k\theta^2 = \frac{1}{2}I\omega^2\theta^2$

The kinetic energy is, $K = \frac{1}{2}I\omega^2$

Total energy, $E = U + K$

$$= \frac{1}{2}I\omega^2\theta^2 + \frac{1}{2}I\omega^2$$

From $\theta = \theta_0 \sin(\omega t + \delta)$ we have,

$$E = \frac{1}{2}I\omega^2\theta_0^2 \sin^2(\omega t + \delta) + \frac{1}{2}I\theta_0^2\omega^2 \cos^2(\omega t + \delta)$$

$$= \frac{1}{2}I\omega^2\theta_0^2 \text{ (Constant)}$$

Identifying Angular SHM

When a system undergoes simple harmonic motion, its angular acceleration α and angular displacement θ are related by

$$\alpha = -\text{(a positive constant)}\,\theta$$

which says that the angular acceleration α is proportional to the angular displacement θ from the equilibrium position but tends to rotate the system in the direction opposite to the displacement.

Also in angular SHM, the torque τ in terms of the angular displacement θ is given by

$$\tau = -\text{(a positive constant)}\,\theta$$

It says that the torque τ is proportional to the angular displacement θ from the equilibrium position but tends to rotate the system in an opposite direction.

Example 12. *Two masses m and $\frac{m}{2}$ are connected at the two ends of a massless rigid rod of length l. The rod is suspended by a thin wire of torsional constant k at the centre of mass of the rod-mass system (see figure). Because of torsional constant k, the restoring torque is $\tau = k\theta$ for angular displacement θ. If the rod is rotated by θ_0 and released, the tension in it when it passes through its mean position will be* **[JEE Main 2019]**

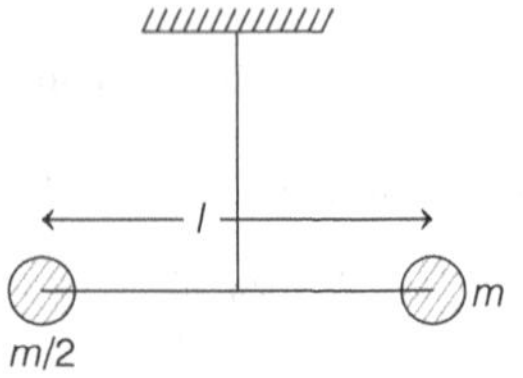

(a) $\frac{2k\theta_0^2}{l}$ (b) $\frac{k\theta_0^2}{l}$ (c) $\frac{3k\theta_0^2}{l}$ (d) $\frac{k\theta_0^2}{2l}$

Sol. (b) Since in the given question, rotational torque, $\tau \propto$ angular displacement.

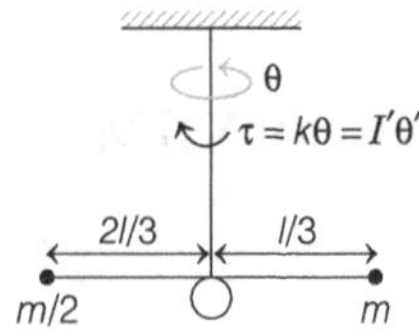

Thus, when it will be released, the system will execute SHM with a time period, $T = 2\pi\sqrt{\frac{I}{k}}$

(Where I is moment of inertia and k is torsional constant)

and the angular frequency is given as, $\omega = \sqrt{\frac{k}{I}}$.

If we now look at the top view of the above figure, we have

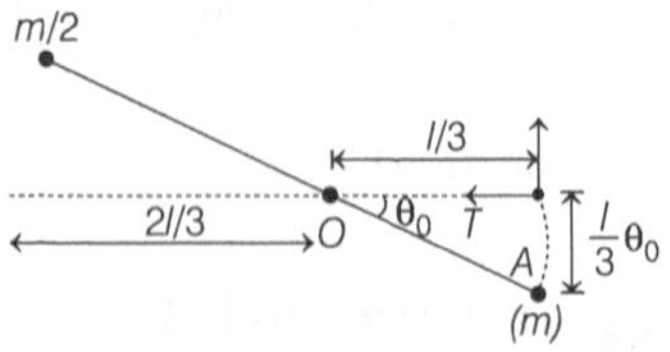

At some angular displacement 'θ_0', at point 'A' the maximum velocity will be

$$v_{max} = \frac{l}{3}\theta_0\omega = \frac{l}{3}\theta_0\sqrt{\frac{k}{I}} \quad \text{...(i)}$$

Then, tension in the rod when it passes through mean position will be

$$T = \frac{m \times v_{max}^2}{\frac{l}{3}} = \frac{ml^2\theta_0^2 k \times 3}{9 \times I \times l} \quad \text{[using Eq. (i)]}$$

$$= \frac{ml\,\theta_0^2 k}{3I}$$

The moment of inertia I at point O,

$$= \frac{m}{2}\left(\frac{2l}{3}\right)^2 + m\left(\frac{l}{3}\right)^2$$

$$= \frac{2l^2m}{9} + \frac{ml^2}{9} = \frac{3ml^2}{9} = \frac{ml^2}{3}$$

$$\Rightarrow \quad T = \frac{ml\,\theta_0^2 k \times 3}{3 \times ml^2} = \frac{\theta_0^2 k}{l} = \frac{k\theta_0^2}{l}$$

Spring Block System

Let a mass m be attached to the free end of a massless spring of spring constant (also known as force constant or spring factor or stiffness) k, with its other end fixed to a rigid support. If the mass be displaced through a distance x and then released, a linear restoring force $F = -kx$ acts on the mass due to elastic nature of the spring.

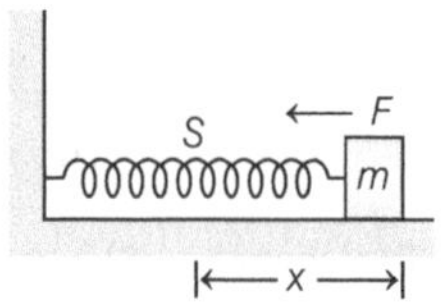

Force Constant for Spring-Block system

From Newton's second law of motion and the expression for acceleration of a particle undergoing SHM, the force acting on a particle of mass m in SHM is

$$F(t) = ma = -m\omega^2 x(t)$$

i.e. $$F(t) = -kx(t)$$

where, $$k = m\omega^2$$

or $$\omega = \sqrt{\frac{k}{m}}$$

Force is always directed towards mean position and is also called *restoring force* and k is called the *spring constant*, its value is governed by the elastic properties of the spring. A stiff spring has large k and a soft spring has small k.

Example 13. *A 5 kg collar is attached to a spring of spring constant 500 Nm^{-1}. It slides without friction over a horizontal rod. The collar is displaced from its equilibrium position by 10 cm and released. The maximum speed is*

(a) 1 ms^{-1} (b) 5 ms^{-1} (c) 10 ms^{-1} (d) 20 ms^{-1}

Sol. (a) The velocity of the collar executing SHM is given by

$$v(t) = -A\omega\sin(\omega t + \phi)$$

The maximum speed is given by

$$v_m = A\omega$$

Given, $$A = 10 \text{ cm} = 0.1 \text{ m},$$

$$\omega = \sqrt{\frac{k}{m}}$$

$$v_m = 0.1 \times \sqrt{\frac{500}{5}}$$

$$v_m = 1 \text{ ms}^{-1}$$

Spring Pendulum

A point mass suspended from a massless (or light) spring constitutes a spring pendulum. If the mass is once pulled downward so as to stretch the spring and then released, the system oscillates up and down about its mean position simple harmonically. Time period and frequency of oscillations are given by

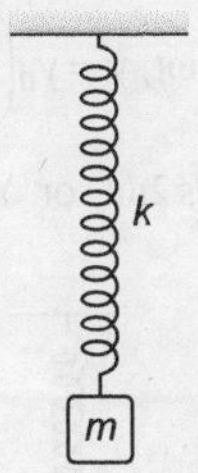

$$T = 2\pi\sqrt{\frac{m}{k}}$$

or $$\nu = \frac{1}{2\pi}\sqrt{\frac{k}{m}}$$

If the spring is not light but has a definite mass m_s, then it can be easily shown that period of oscillation will be

$$T = 2\pi\sqrt{\frac{m + \frac{m_s}{3}}{k}}$$

Oscillations of Spring Combination

For different combinations of spring block system, following cases occurs

- If a spring pendulum is constructed by using two springs in series and a mass m as shown in Fig. (a), the resultant spring constant of the combination is given by

$$\frac{1}{k_s} = \frac{1}{k_1} + \frac{1}{k_2} = \frac{k_1 + k_2}{k_1 k_2}, \text{ and hence}$$

$$T = 2\pi\sqrt{\frac{m}{k_s}} = 2\pi\sqrt{\frac{(k_1 + k_2)}{k_1 k_2}}$$

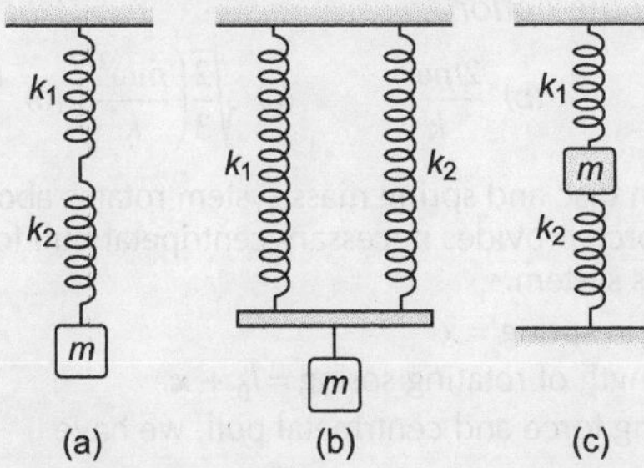

(a) (b) (c)

- If in a spring pendulum, two springs are joined in parallel arrangement as shown in Fig. (b) and (c), then

$$k_p = k_1 + k_2, \text{ and hence, } T = 2\pi\sqrt{\frac{m}{k_1 + k_2}}$$

These rules are applicable for horizontal motion of spring mass systems too.

- The force constant of a spring is inversely proportional to its length, *i.e.* $k \propto \frac{1}{l}$. Thus, if we change the length of a spring, the time of oscillation of spring pendulum will change. As an example, if a spring is cut into two parts of equal length and a pendulum is prepared by using one part of spring and some mass m, then the new time period will be

$$T' = 2\pi\sqrt{\frac{m}{2k}} = \frac{T}{\sqrt{2}}$$

- If two masses m_1 and m_2 are connected by a spring as shown in figure and the arrangement is made to oscillate on a horizontal surface, then time period is given by

$$T' = 2\pi\sqrt{\frac{\mu}{k}}$$

where, $\mu = \frac{m_1 m_2}{m_1 + m_2}$ = reduced mass.

Example 14. *A block with a mass of 2 kg hangs without vibrating at the end of a spring of spring constant 800 Nm^{-1}, which is attached to the ceiling of an elevator. The elevator is going upwards with an acceleration g/3. At a certain instant, the acceleration suddenly ceases and elevator starts moving with constant velocity. What is the angular frequency of oscillation of block when the acceleration ceases?*

(a) 12 *rad* s^{-1} (b) 20 *rad* s^{-1} (c) 21 *rad* s^{-1} (d) 19 *rad* s^{-1}

Sol. (b) Angular frequency, $\omega = \sqrt{\frac{k}{m}} = \sqrt{\frac{800}{2}} = 20 \text{ rads}^{-1}$

Example 15. *In the above example, by what amount is the spring stretched during the time when the elevator is accelerating?*

(a) 2.3 cm (b) 3.3 cm
(c) 5.3 cm (d) 4.4 cm

Sol. (b) When the elevator is accelerating upwards with an acceleration a = g/3, the equation of motion of the block is

$$\left(ky - mg = ma = m\frac{g}{3}\right)$$

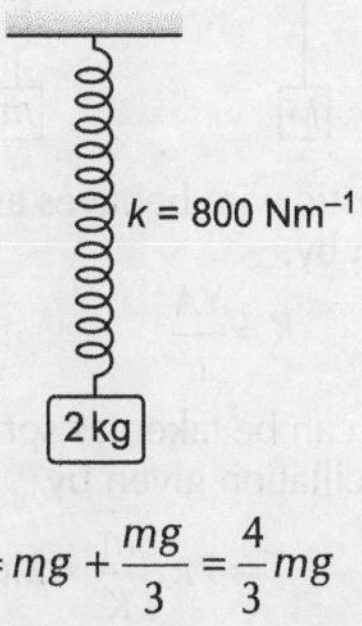

$$\Rightarrow \quad ky = mg + \frac{mg}{3} = \frac{4}{3}mg$$

$$\Rightarrow \quad y = \frac{4mg}{3k} = \frac{4\times2\times10}{3\times800} = 0.033 \text{ m or } 3.3 \text{ cm}$$

Example 16. *In the above example, what is the amplitude of oscillation? (Take, $g = 10\ ms^{-2}$)*

(a) 0.1 cm (b) 0.7 cm (c) 0.8 cm (d) 0.5 cm

Sol. (c) In equilibrium, when the elevator has no acceleration, the equation of motion is

$$ky_0 = mg$$

$$\Rightarrow \quad y_0 = \frac{mg}{k} = \frac{2\times 10}{800} = 0.025\text{ m} \quad \text{or } 2.5\text{ cm}$$

$\therefore$ Amplitude of oscillation,

$$A = y - y_0 = 3.3 - 2.5 = 0.8\text{ cm}$$

Example 17. *A spring whose unstretched length is l has a force constant k. The spring is cut into two pieces of unstretched lengths l_1 and l_2 where, $l_1 = nl_2$ and n is an integer. The ratio k_1/k_2 of the corresponding force constants k_1 and k_2 will be* **[JEE Main 2019]**

(a) n (b) $\frac{1}{n^2}$ (c) $\frac{1}{n}$ (d) n^2

Sol. (c) If parameters like material, number of loops per unit length, area of cross-section, etc., are kept same, then force constant of spring is inversely proportional to its length.

In given case, all other parameters are same for both parts of spring.

So, $$k_1 \propto \frac{1}{l_1} \text{ and } k_2 \propto \frac{1}{l_2}$$

$$\therefore \quad \frac{k_1}{k_2} = \frac{l_2}{l_1} = \frac{l_2}{nl_2} = \frac{1}{n} \qquad [\because l_1 = nl_2]$$

Example 18. *An object of mass m is suspended at the end of a massless wire of length L and area of cross-section A. Young modulus of the material of the wire is Y. If the mass is pulled down slightly its frequency of oscillation along the vertical direction is* **[JEE Main 2020]**

(a) $f = \frac{1}{2\pi}\sqrt{\frac{mL}{YA}}$ (b) $f = \frac{1}{2\pi}\sqrt{\frac{YA}{mL}}$

(c) $f = \frac{1}{2\pi}\sqrt{\frac{mA}{YL}}$ (d) $f = \frac{1}{2\pi}\sqrt{\frac{YL}{mA}}$

Sol. (b)

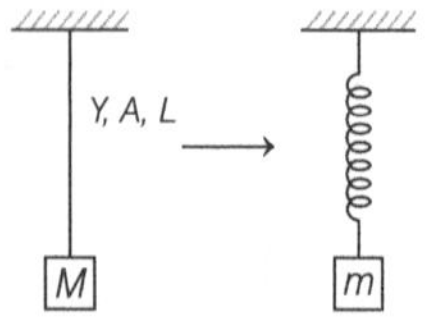

As we know that an elastic wire behaves as an elastic spring with spring constant K given by,

$$K = \frac{YA}{L}$$

This block-wire system can be taken as spring block system with time period of small oscillation given by

$$l = 2\pi\sqrt{\frac{m}{K}} = 2\pi\sqrt{\frac{ml}{YA}}$$

$\therefore$ Frequency of oscillation, $f = \frac{1}{T} = \frac{1}{2\pi}\sqrt{\frac{YA}{mL}}$

Example 19. *When a particle of mass m is attached to a vertical spring of spring constant k and released, its motion is described by $y(t) = y_0 \sin^2 \omega t$, where y is measured from the lower end of unstretched spring. Then ω is* **[JEE Main 2020]**

(a) $\frac{1}{2}\sqrt{\frac{g}{y_0}}$ (b) $\sqrt{\frac{g}{y_0}}$ (c) $\sqrt{\frac{g}{2y_0}}$ (d) $\sqrt{\frac{2g}{y_0}}$

Sol. (c) Given that, displacement of particle at any instant

$$y(t) = y_0 \sin^2 \omega t,\ y = y_0\left(\frac{1-\cos 2\omega t}{2}\right)$$

$$y - \frac{y_0}{2} = -\frac{y_0}{2}\cos 2\omega t \quad \text{or} \quad Y = A\cos 2\omega t$$

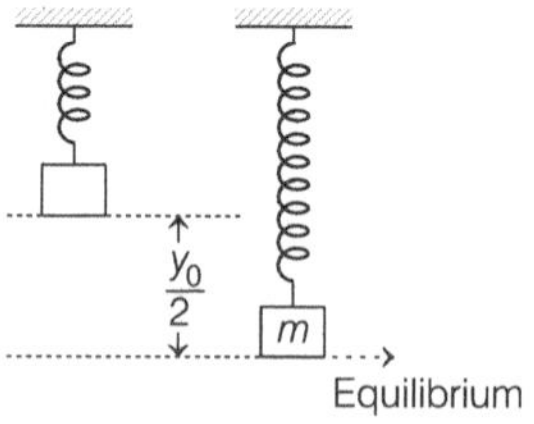

where, amplitude (displacement), $A = \frac{y_0}{2}$

and $$2\omega = \sqrt{\frac{k}{m}} \qquad \ldots(i)$$

Maximum displacement, $\frac{y_0}{2} = \frac{mg}{k}$ (see figure)

or $$k = \frac{2mg}{y_0} \qquad \ldots(ii)$$

Putting the value of k from Eq. (ii) in Eq. (i), we get

$$2\omega = \sqrt{\frac{2mg}{y_0}\times\frac{1}{m}} \quad \text{or} \quad \omega = \sqrt{\frac{g}{2y_0}}$$

Example 20. *A spring mass system (mass m, spring constant k and natural length l) rests in equilibrium on horizontal disc. The free end of the spring is fixed at the centre of the disc. If the disc together with spring mass system, rotates about it's axis with an angular velocity ω, ($k >> m\omega^2$), the relative change in the length of the spring is best given by the option* **[JEE Main 2020]**

(a) $\frac{m\omega^2}{3k}$ (b) $\frac{2m\omega^2}{k}$ (c) $\sqrt{\frac{2}{3}}\left(\frac{m\omega^2}{k}\right)$ (d) $\frac{m\omega^2}{k}$

Sol. (d) When disc and spring mass system rotates about central axis, spring force provides necessary centripetal pull for rotation of spring mass system.

Let extension in spring = x.

Then, total length of rotating spring = $l_0 + x$.

Equating spring force and centripetal pull, we have

$$kx = m\omega^2(l_0 + x)$$

$$\Rightarrow \quad \frac{x}{l_0} = \frac{m\omega^2}{k - m\omega^2}$$

$$\Rightarrow \quad \frac{x}{l_0} = \frac{m\omega^2}{k\left(1 - \frac{m\omega^2}{k}\right)}$$

As $k >> m\omega^2$, so $1 - \frac{m\omega^2}{k} \approx 1$

Hence, $\frac{x}{l_0} = \frac{m\omega^2}{k}$

Example 21. *Two identical springs of spring constant k are attached to a block of mass m and to fixed supports as shown, when the mass is displaced from its equilibrium position on either side, it executes SHM, the period of oscillation is*

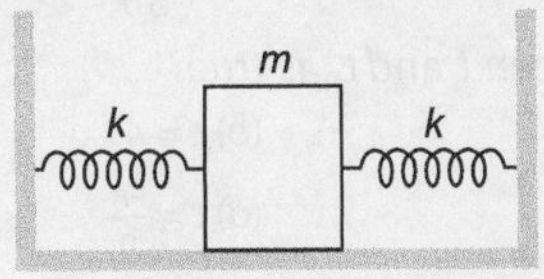

[NCERT]

(a) $2\pi\sqrt{\frac{m}{k}}$ (b) $2\pi\sqrt{\frac{2m}{k}}$ (c) $2\pi\sqrt{\frac{m}{3k}}$ (d) $2\pi\sqrt{\frac{m}{2k}}$

Sol. (*d*) Let the mass be displaced by a small distance *x* to the right side of the equilibrium position. Under this situation, the spring on the left side elongated by a length equal to *x* and that on the right side gets compressed by the same length. The forces acting on the mass are then

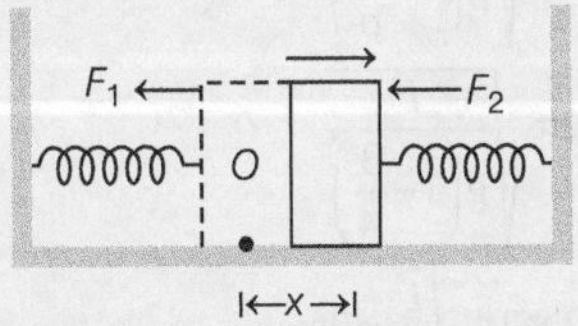

(force exerted by the spring on the left side, trying to pull the mass towards the mean position)

$$F_l = -kx$$

(force exerted by the spring on the right side, trying to push the mass towards the mean position)

$$F_2 = -kx$$

The net force, *F* acting on the mass is then

$$F = -2kx$$

Hence, the force acting on the mass is proportional to the displacement and is directed towards the mean position, therefore the motion executed by the mass is simple harmonics. The time period of oscillation is

$$T = 2\pi\sqrt{\frac{m}{2k}}$$

Example 22. *A block whose mass is 1 kg is fastened to a spring. The spring has a spring constant of 50 Nm^{-1}. The block is pulled to a distance x = 10 cm from its equilibrium position at x = 0 on a frictionless surface from rest at t = 0. The total energy of the block when it is 5 cm away from the mean position is*

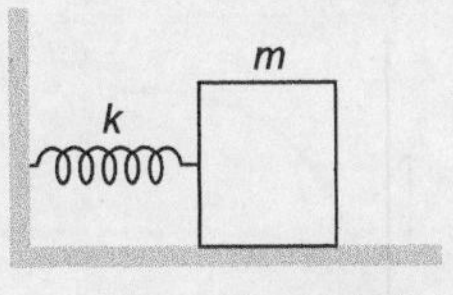

(a) 0.19 *J* (b) 0.0625 *J*
(c) 0.09 *J* (d) 0.25 *J*

Sol. (*d*) The block executes SHM to angular frequency is

$$\omega = \sqrt{\frac{k}{m}} = \sqrt{\frac{50}{1}} = 7.07 \text{ rad s}^{-1}$$

Its displacement at any time *t* is given by

$$x(t) = 0.1\cos(7.07t)$$

Therefore, when the particle is 5 cm away from the mean position, we have

$$0.05 = 0.1\cos(7.07t)$$

or $\cos(7.07t) = 0.5$

$$\sin(7.07t) = \frac{\sqrt{3}}{2} = 0.866$$

Then, the velocity of the block at $x = 5$ cm is

$$= 0.1 \times 7.07 \times 0.866 \text{ ms}^{-1} = 0.61 \text{ms}^{-1}$$

Hence, the kinetic energy of the block

$$\text{KE} = \frac{1}{2}mv^2 = \frac{1}{2}[1 \text{ kg} \times (0.6123)^2] = 0.19 \text{ J}$$

and the potential energy,

$$\text{PE} = \frac{1}{2}kx^2 = \frac{1}{2}(50 \times 0.05 \times 0.05) = 0.0625 \text{ J}$$

∴ Total energy = KE + PE = 0.25 J

Note At maximum displacement kinetic energy is zero and hence the total energy of the system is equal to potential energy. The result is also in conformity with the principle of conservation of energy.

Simple Pendulum

Given figure shows simple pendulum in which a small bob of mass *m* tied to an in extensible massless string of length *L*. The other end of the string is fixed to a support in the ceiling. The bob oscillates in a plane about the vertical line through the support.

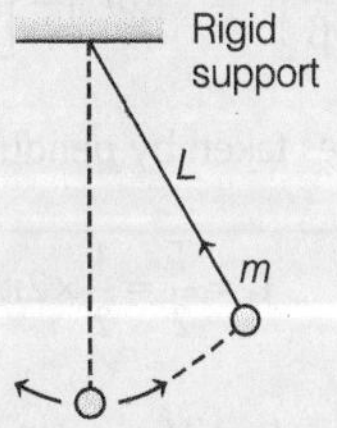

The various force acting on the system are as shown in the diagram. If θ is the angle made by the string with the vertical. When the bob is at the mean position θ = 0, there are only two forces acting on the bob, the tension *T* along the string and the vertical force due to gravity (= *mg*). Resolving force (= *mg*) into the component *mg* cosθ along the string and *mg* sinθ perpendicular to it.

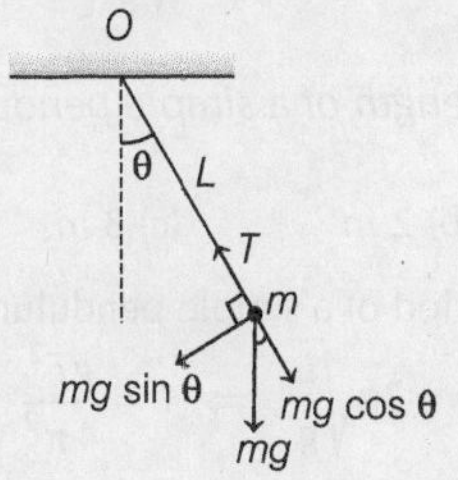

Taking torque about point O.

$$\tau = - mgL \sin\theta$$

$$\tau = - mgL\theta \text{ (for small angular displacement } \sin\theta \simeq \theta)$$

$$I\alpha = - mgL\theta$$

$$\alpha = \frac{- mgL\theta}{I}$$

$$\alpha = \frac{-mgL}{mL^2}\theta = \left(\frac{-g}{L}\right)\theta \Rightarrow \omega = \sqrt{\frac{g}{L}}$$

So, the time period of the simple pendulum can be given as

$$T = 2\pi\sqrt{\frac{L}{g}}$$

Example 23. *A ball is suspended by a thread of length l at the point O on the wall PQ. Another wall OR is inclined to the vertical by α. The thread with the ball is displaced by a small angle β away from the vertical and also away from the wall OR. If the ball is released, assuming the collision to be perfectly elastic the period of oscillation of the pendulum for $\beta > \alpha$ is*

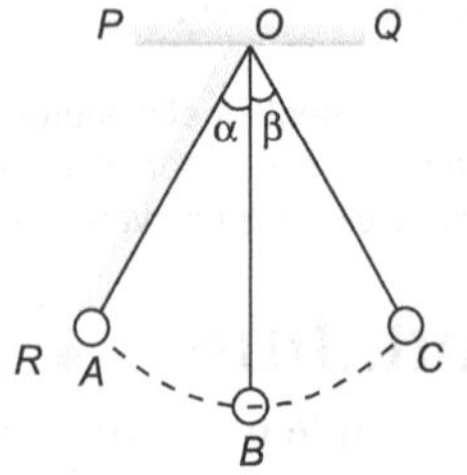

(a) $\sqrt{\frac{L}{g}}\left[\pi + 2\cos^{-1}\frac{\beta}{\alpha}\right]$ (b) $\sqrt{\frac{L}{g}}\left[\pi + 2\sin^{-1}\frac{\alpha}{\beta}\right]$

(c) $\sqrt{\frac{g}{L}}\left[\pi + 2\tan^{-1}\frac{\alpha}{\beta}\right]$ (d) $\sqrt{\frac{g}{L}}\left[2\pi + 2\tan^{-1}\frac{\beta}{\alpha}\right]$

Sol. (b) When $\beta > \alpha$ times taken by pendulum from B to C and C to B is

$$t_1 = \frac{T}{2} = \frac{1}{2}\times 2\pi\sqrt{\frac{L}{g}} = \pi\sqrt{\frac{L}{g}}$$

and $$t_2 = 2t = \frac{2}{\omega}\sin^{-1}\left(\frac{\alpha}{\beta}\right)$$

Using, $$\theta = \theta_0 \sin\omega t$$
$$\alpha = \beta\sin\omega t$$

or $$t = \frac{1}{\omega}\sin^{-1}\left(\frac{\alpha}{\beta}\right)$$

Time period of motion, $T = t_1 + t_2 = \sqrt{\frac{L}{g}}\left[\pi + 2\sin^{-1}\frac{\alpha}{\beta}\right]$

Example 24. *Length of a simple pendulum which ticks seconds is*

(a) 1 m (b) 2 m (c) 3 m (d) 4 m

Sol. (a) The time period of a simple pendulum is given by

$$T = 2\pi\sqrt{\frac{L}{g}} \Rightarrow L = \frac{gT^2}{4\pi^2}$$

The time period of a simple pendulum which ticks seconds is 2 s. Therefore, for $g = 9.8\ \text{ms}^{-2}$ and $T = 2\ \text{s}, L$ is

$$L = \frac{9.8\times 4}{4\pi^2} \approx 1\text{m}$$

Example 25. *The bob of a simple pendulum executes SHM in water with a period t, while the period of oscillation of the bob is t_0 in air. Neglecting frictional force of water and given that the density of the bob is $\left(\frac{4}{3}\right)\times 1000\ kgm^{-3}$. What relationship between t and t_0 is true?*

(a) $t = t_0$ (b) $t = 4t_0$

(c) $t = 2t_0$ (d) $t = \frac{t_0}{2}$

Sol. (c) Here density of bob, $\rho = \frac{4}{3}\times 1000\ \text{kgm}^{-3}$

and density of water, $\sigma = 1000\ \text{kgm}^{-3}$

$\therefore$ In air $t_0 = 2\pi\sqrt{\frac{L}{g}}$ and in water

$$t = 2\pi\sqrt{\frac{L}{g\left(1-\frac{\sigma}{\rho}\right)}}$$

$$= 2\pi\sqrt{\frac{L}{g\left(1-\frac{3}{4}\right)}}$$

$$= 2\times 2\pi\sqrt{\frac{L}{g}} = 2t_0$$

Important Points Related to Simple Pendulum

- The time period of a simple pendulum is

$$T = 2\pi\sqrt{l/g}$$

$$\Rightarrow \quad T \propto \sqrt{l} \quad \text{or} \quad T \propto \frac{1}{\sqrt{g}} \Rightarrow T \propto \sqrt{\frac{l}{g}}$$

Using these relations. We may conclude

(a) The graph between T^2 and l is a straight line.

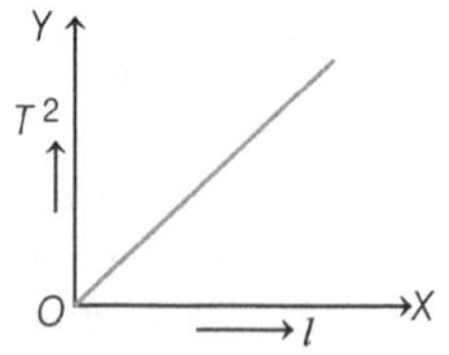

(b) The graph between T and l is a parabola.

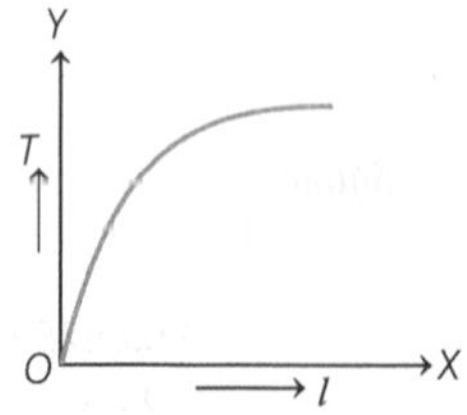

(c) The graph between T^2 and $1/g$ is a straight line.

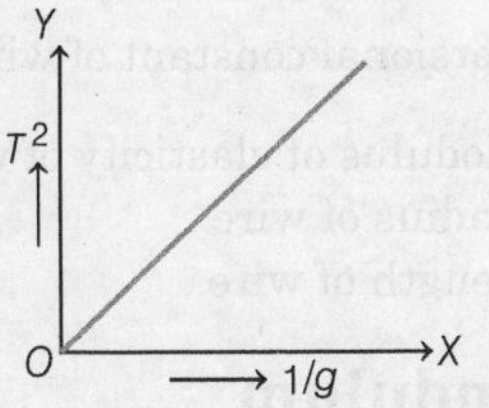

(d) The graph between T^2 and g is a rectangular hyperbola.

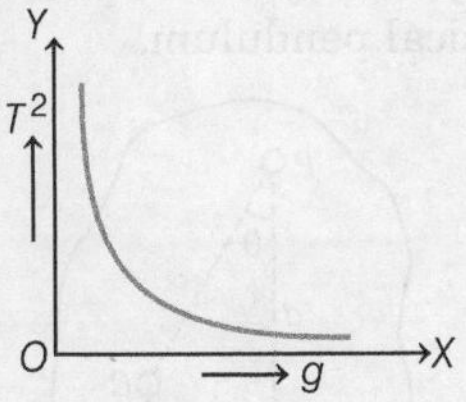

- In the case of water oscillating in a U-tube

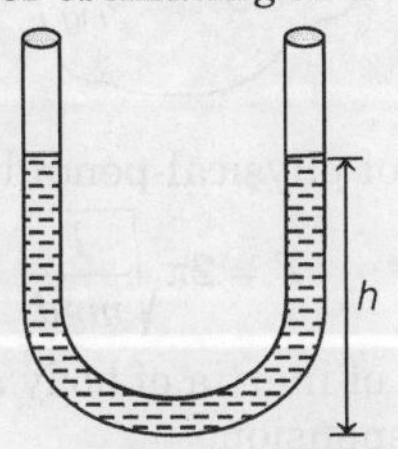

$$T = 2p\sqrt{\left(\frac{h}{g}\right)}$$

where, h is the height of liquid column in each limb.

- When a pendulum is kept in a car which is sliding down, then

$$T = 2\pi\sqrt{\left(\frac{l}{g\cos\theta}\right)}$$

where, θ is the angle of inclination.

- If a simple pendulum oscillates in a non-viscous liquid of density ρ, then its time period is

$$T = 2p\sqrt{\left[\frac{l}{\left(1-\frac{s}{r}\right)g}\right]}$$

where, ρ = density of suspended mass.

- If the mass m attached to a spring oscillates in a non-viscous liquid of density σ, then its time period is

$$T = 2\pi\left[\frac{m}{k}\left(1-\frac{\sigma}{\rho}\right)\right]^{1/2}$$

where, k = force constant.

- For a body executing SHM in a tunnel dug along any chord of earth.

Time period, $T = 2\pi\sqrt{\frac{R_e}{g}} = 84.6$ min

where, R_e is the radius of Earth.

- If the time period of simple pendulum is 2s, then it is called as second's pendulum.
- If the simple pendulum is placed in some non-inertial frame of reference like an accelerated lift, g is replaced by g_{eff} whose value can be computed by considering the inertial force. In these cases, the equilibrium position may also change.
- If the length of simple pendulum is very large, then g can't be taken along vertical direction.

In this case, $T = 2\pi\sqrt{\frac{1}{g\left(\frac{1}{l}+\frac{1}{R}\right)}}$

where, R = Radius of length of the pendulum.

- If temperature of system changes, then time period of simple pendulum changes due to change in length of the simple pendulum.
- If a simple pendulum is in a carriage which is accelerating with an acceleration **a**, then

$$\mathbf{g}_{\text{eff}} = \mathbf{g} - \mathbf{a}$$

e.g., if the acceleration **a** is upwards, then

$$|\mathbf{g}_{\text{eff}}| = g + a \text{ and } T = 2\pi\sqrt{\frac{l}{g+a}}$$

If the acceleration **a** is downwards, then ($g > a$)

$$|\mathbf{g}_{\text{eff}}| = g - a \text{ and } T = 2\pi\sqrt{\frac{l}{g-a}}$$

If the acceleration **a** is in the horizontal direction, then

$$|\mathbf{g}_{\text{eff}}| = \sqrt{a^2 + g^2}$$

In a freely falling lift, $g_{\text{eff}} = 0$ and $T = \infty$, *i.e.* the pendulum will not oscillate.

- If in addition to gravity one additional force **F** (*e.g.* electrostatic force $\mathbf{F}_e$) is also acting on the bob, then in that case

$$\mathbf{g}_{\text{eff}} = \mathbf{g} + \frac{\mathbf{F}}{m}$$

Here, m is the mass of the bob.

Example 26. *The acceleration due to gravity on the surface of the moon is 1.7 ms^{-2}. What is the time period of a simple pendulum on the moon, if its time period on the earth is 3.5 s? (g on the Earth 9.8 ms^{-2}).*

(a) 8.4 s (b) 8.2 s (c) 7.4 s (d) 6.4 s

Sol. (a) Given, $g_m = 1.7$ ms^{-2},

$$g_e = 9.8 \text{ ms}^{-2}, T_m = ?; T = 3.5 \text{ s}$$

As, $$T_e = 2\pi\sqrt{\frac{l}{g_e}}$$

and $$T_m = 2\pi\sqrt{\frac{l}{g_m}}$$

$\therefore$ $$\frac{T_m}{T_e} = \sqrt{\frac{g_e}{g_m}}$$

or $$T_m = T_e\sqrt{\frac{g_e}{g_m}} = 3.5\sqrt{\frac{9.8}{1.7}} = 8.4 \text{ s}$$

Example 27. *A simple pendulum oscillating in air has period T. The bob of the pendulum is completely immersed in a non-viscous liquid. The density of the liquid is $\frac{1}{16}$th of the material of the bob. If the bob is inside liquid all the time, its period of oscillation in this liquid is* [JEE Main 2019]

(a) $2T\sqrt{\frac{1}{10}}$ (b) $2T\sqrt{\frac{1}{14}}$ (c) $4T\sqrt{\frac{1}{14}}$ (d) $4T\sqrt{\frac{1}{15}}$

Sol. (d) We know that,
Time period of a pendulum is given by

$$T = 2\pi\sqrt{L/g_{\text{eff}}} \quad \text{...(i)}$$

Here, L is the length of the pendulum and g_{eff} is the effective acceleration due to gravity in the respective medium in which bob is oscillating.
Initially, when bob is oscillating in air, $g_{\text{eff}} = g$.

So, initial time period, $T = 2\pi\sqrt{\frac{L}{g}}$ ...(ii)

Let ρ_{bob} be the density of the bob.
When this bob is dipped into a liquid whose density is given as

$$\rho_{\text{liquid}} = \frac{\rho_{\text{bob}}}{16} = \frac{\rho}{16} \quad \text{(given)}$$

$\therefore$ Net force on the bob is $F_{\text{net}} = V\rho g - V\cdot\frac{\rho}{16}\cdot g$ ...(iii)

(where, V = volume of the bob = volume of displaced liquid by the bob when immersed in it). If effective value of gravitational acceleration on the bob in this liquid is g_{eff}, then net force on the bob can also be written as

$$F_{\text{net}} = V\rho g_{\text{eff}} \quad \text{...(iv)}$$

Equating Eqs. (iii) and (iv), we have

$$V\rho g_{\text{eff}} = V\rho g - V\rho g/16$$

$$\Rightarrow \quad g_{\text{eff}} = g - g/16 = \frac{15}{16}g \quad \text{...(v)}$$

Substituting the value of g_{eff} from Eq. (v) in Eq. (i), the new time period of the bob will be

$$T' = 2\pi\sqrt{\frac{L}{g_{\text{eff}}}} = 2\pi\sqrt{\frac{16}{15}\frac{L}{g}}$$

$$\Rightarrow \quad T' = \sqrt{\frac{16}{15}}\times 2\pi\sqrt{\frac{L}{g}} = \frac{4}{\sqrt{15}}\times T \quad \text{[using Eq. (ii)]}$$

Torsional pendulum

In a torsional pendulum, an object is suspended from a wire. If such a wire is twisted due to elasticity, it exerts a restoring torque $\tau = C\theta$

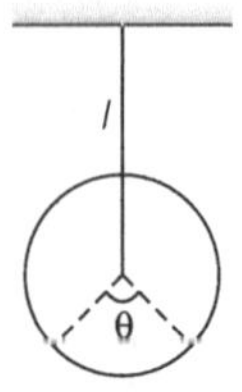

In this case, time period is given by

$$T = 2\pi\sqrt{\frac{I}{C}}$$

where, I = Moment of inertia of the object,
C = Torsional constant of wire = $\frac{\pi\eta r^n}{2l}$,
η = Modulus of elasticity of wire,
r = Radius of wire
and l = Length of wire.

Physical Pendulum

When a rigid body of any shape is capable of oscillating about an axis (may or may not be passing through it), it constitutes a physical pendulum.

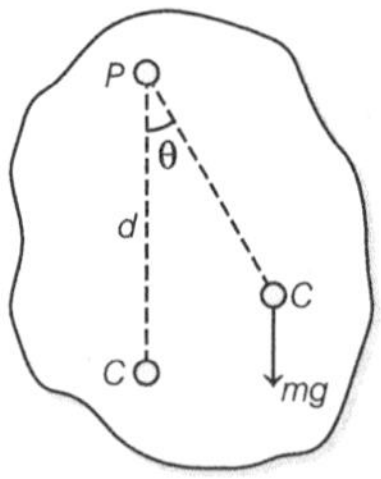

So, the time period of physical-pendulum can be given as

$$\Rightarrow \quad T = 2\pi\sqrt{\frac{I}{mgd}}$$

where, I = moment of inertia of body about axis passing through point of suspension.

d = distance of COM of body from point of suspension.

The simple pendulum whose time period is same as that of a physical pendulum is termed as an *equivalent simple pendulum*.

$$T = 2\pi\sqrt{\frac{I}{mgd}} = 2\pi\sqrt{\frac{l}{g}}$$

The length of an equivalent simple pendulum is given by

$$l = \frac{I}{md}$$

Example 28. *A uniform rod of mass m and length l_0 is pivoted at one end and is hanging in the vertical direction. The period of small angular oscillations of the rod is*

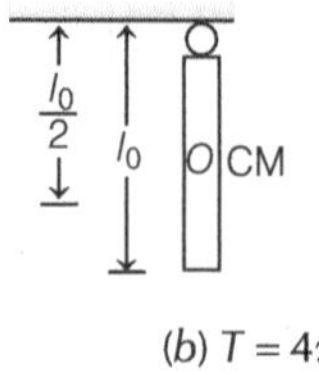

(a) $T = 3\pi\sqrt{\frac{2l_0}{3g}}$ (b) $T = 4\pi\sqrt{\frac{l_0}{3g}}$

(c) $T = 4\pi\sqrt{\frac{2l_0}{3g}}$ (d) $T = 2\pi\sqrt{\frac{2l_0}{3g}}$

Sol. (d) Here the rod is oscillating about an end point O. Hence, moment of inertia of rod about the point of oscillating is

$$I = \frac{1}{3}ml_0^2$$

Moreover, length l of the pendulum = distance from the oscillation axis to centre of mass of rod = $l_0/2$

$\therefore$ Time period of oscillation,

$$T = 2\pi\sqrt{\frac{I}{mgl}} = 2\pi\sqrt{\frac{\frac{1}{3}ml_0^2}{mg\left(\frac{l_0}{2}\right)}} \Rightarrow T = 2\pi\sqrt{\frac{2l_0}{3g}}$$

Example 29. *Two physical pendulums perform small oscillations about the same horizontal axis with frequencies ω_1 and ω_2. Their moments of inertia relative to the given axes are I_1 and I_2 respectively. In the equilibrium positions, they are joined rigidly. The frequency of small oscillations of the combined pendulum is*

(a) $\sqrt{\frac{I_1\omega_1^2 + I_2\omega_2^2}{I_1 + I_2}}$ (b) $\sqrt{\frac{I_1 + I_2}{I_1^2\omega_1 + I_2^2\omega_2}}$

(c) $\sqrt{\frac{I_1\omega_1^2 - I_2\omega_2^2}{I_1 - I_2}}$ (d) $\sqrt{\frac{I_1 - I_2}{I_1^2\omega_1 - I_2^2\omega_2}}$

Sol. (a) When the pendulums are rigidly joined and set to oscillate, each exerts a torque on the other. These torques are equal and opposite, thus

$$I_1\alpha = -\omega_1^2 I_1\theta + G \quad \text{...(i)}$$

$$I_2\alpha = -\omega_2^2 I_2\theta - G \quad \text{...(ii)}$$

Adding Eqs. (i) and (ii), we get

$$\alpha = -\frac{(I_1\omega_1^2 + I_2\omega_2^2)\,\theta}{I_1 + I_2}$$

Comparing with $\alpha = -\omega^2\theta$, we get

Frequency, $$\omega = \sqrt{\frac{I_1\omega_1^2 + I_2\omega_2^2}{I_1 + I_2}}$$

Free, Forced, Damped and Resonant Vibrations

Free Vibrations If a given body is once set into vibrations and then let free to vibrate with its own natural frequency, the vibrations are said to be free vibrations.

Forced Vibrations The vibrations in which a body oscillates under the effect of an external periodic force, whose frequency is different from the natural frequency of oscillating body are called forced vibrations.

Damped Vibrations When a body is set in free vibrations, there is a dissipation of energy due to dissipative causes like viscous drag of a fluid, frictional force, hysteresis, electromagnetic damping force, etc. and as a result, amplitude of vibration regularly decreases with time. Such vibrations of continuously falling amplitudes are called damped vibrations.

If the velocity of an oscillator is v, the damping force

$$F_d = -bv$$

where, b = damping constant.

Resultant force on a damped oscillator is given by

$$F = F_R + F_D = -kx - bv$$

or $$\frac{md^2x}{dt^2} + \frac{bdx}{dt} + kx = 0$$

Displacement of a damped oscillator is given by

$$x = x_m e^{-bt/2m} \sin(\omega' t + \phi)$$

$$\omega' = \sqrt{\omega_0^2 - (b/2m)^2}$$

where, ω' = angular frequency of the damped oscillator.

Mechanical energy of a damped oscillator which decreases exponentially with time can be given as

$$E = \frac{1}{2}kx_m^2 e^{-bt/m}$$

Resonant Vibrations It is a special case of forced vibrations in which frequency of external force is exactly same as the natural frequency of oscillator. As a result, the oscillating body begins to vibrate with a large amplitude leading to the resonance phenomenon to occur. Resonant vibrations play a very important role in music and tuning of station/channel in a radio/TV.

Example 30. *A damped harmonic oscillator has a frequency of 5 oscillations per second. The amplitude drops to half its value for every 10 oscillations. The time it will take to drop to $\frac{1}{1000}$ of the original amplitude is close to* **[JEE Main 2019]**

(a) 20 s (b) 50 s
(c) 100 s (d) 10 s

Sol. (a) Given, frequency of oscillations is

$$f = 5 \text{ osc s}^{-1}$$

$\Rightarrow$ Time period of oscillations is $T = \frac{1}{f} = \frac{1}{5}$ s

So, time for 10 oscillations is $= \frac{10}{5} = 2$s

Now, if A_0 = initial amplitude at $t = 0$ and γ = damping factor, then for damped oscillations, amplitude after t second is given as

$$A = A_0 e^{-\gamma t}$$

$\therefore$ After 2 s,

$$\frac{A_0}{2} = A_0 e^{-\gamma(2)}$$

$$\Rightarrow \quad 2 = e^{2\gamma}$$

$$\Rightarrow \quad \gamma = \frac{\log 2}{2} \quad \text{...(i)}$$

Now, when amplitude is $\frac{1}{1000}$ of initial amplitude, *i.e.*

$$\frac{A_0}{1000} = A_0 e^{-\gamma t}$$

$$\Rightarrow \quad \log(1000) = \gamma t$$

$$\Rightarrow \quad \log(10^3) = \gamma t$$

$$3\log 10 = \gamma t$$

$$\Rightarrow \quad t = \frac{2 \times 3\log 10}{\log 2} \quad \text{[using Eq. (i)]}$$

$$\Rightarrow \quad t = 19.93 \text{ s}$$

or $$t \approx 20 \text{ s}$$

Practice Exercise

ROUND I Topically Divided Problems

Oscillatory Motion and Simple Harmonic Motion

1. The displacement of a damped harmonic oscillator is given by $x(t) = e^{-0.1t}\cos(10\pi t + \phi)$. Here, t is in seconds.
The time taken for its amplitude of vibration to drop to half of its initial value is close to **[JEE Main 2019]**
(a) 27 s
(b) 13 s
(c) 4 s
(d) 7 s

2. The displacement of the particle varies with time according to the relation.
$$y = a\sin\omega t + b\cos\omega t, \text{ then}$$
[NCERT Exemplar]
(a) The motion is oscillating but not SHM
(b) The motion is SHM with amplitude $a + b$
(c) The motion is SHM with amplitude $a^2 + b^2$
(d) The motion is SHM with amplitude $\sqrt{a^2 + b^2}$

3. The displacement of a particle is represented by the equation $y = 3\cos\left(\frac{\pi}{4} - 2\omega t\right)$. The motion of the particle is **[NCERT Exemplar]**
(a) simple harmonic with period $2\pi/\omega$
(b) simple harmonic with period π/ω
(c) periodic but not simple harmonic
(d) non-periodic

4. The displacement of a particle is represented by the equation $y = \sin^3\omega t$. The motion is **[NCERT Exemplar]**
(a) non-periodic
(b) periodic but not simple harmonic
(c) simple harmonic with period $2\pi/\omega$
(d) simple harmonic with period π/ω

5. A simple harmonic oscillator has amplitude a and time period T. The time required by it to travel from $x = a$ to $x = \frac{a}{2}$ is
(a) $\frac{T}{6}$ (b) $\frac{T}{4}$
(c) $\frac{T}{3}$ (d) $\frac{T}{2}$

6. Motion of an oscillating liquid column in a U-tube is **[NCERT Exemplar]**
(a) periodic but not simple harmonic
(b) non-periodic
(c) simple harmonic and time period is independent of the density of the liquid
(d) simple harmonic and time-period is directly proportional to the density of the liquid

7. A particle is acted simultaneously by mutually perpendicular simple hormonic motions $x = a\cos\omega t$ and $y = a\sin\omega t$. The trajectory of motion of the particle will be **[NCERT Exemplar]**
(a) an ellipse (b) a parabola
(c) a circle (d) a straight line

8. The acceleration d^2x/dt^2 of a particle varies with displacement x as $\frac{d^2x}{dt^2} = -kx$ where k is a constant of the motion. The time period T of the motion is equal to
(a) $2\pi k$ (b) $2\pi\sqrt{k}$
(c) $2\pi/\sqrt{k}$ (d) $2\pi/k$

9. A block is resting on a piston which is moving vertically with SHM of period 1.0 s. At what amplitude of motion will the block and piston separate?
(a) 0.2 m (b) 0.25 m
(c) 0.3 m (d) 0.35 m

10. Figure shows the circular motion of a particle. The radius of the circle, the period, sense of revolution and the initial position are indicated on the figure. The simple harmonic motion of the x-projection of the radius vector of the rotating particle P is

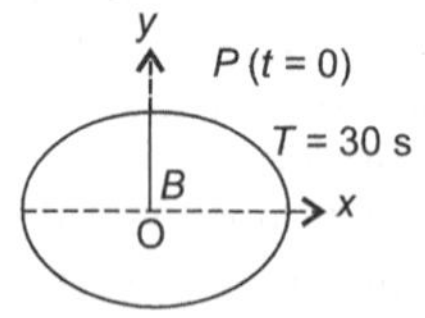

[NCERT Exemplar]
(a) $x(t) = B\sin\left(\frac{2\pi t}{30}\right)$ (b) $x(t) = B\cos\left(\frac{\pi t}{15}\right)$
(c) $x(t) = B\sin\left(\frac{\pi t}{15} + \frac{\pi}{2}\right)$ (d) $x(t) = B\left(\frac{\pi t}{15} + \frac{\pi}{2}\right)$

11. Two pendulums have time period T and $5T/4$. They start SHM at the same time from the mean position. What will be the phase difference between them after the bigger pendulum completed one oscillation?
(a) 45° (b) 90° (c) 60° (d) 30°

12. The displacement of two particles executing SHM are represented by equations

$$y_1 = 2\sin(10t + \theta),\ y_2 = 3\cos 10t.$$

The phase difference between the velocity of these particles is
(a) θ (b) $-\theta$ (c) $\theta + \pi/2$ (d) $\theta - \pi/2$

13. A particle executes simple harmonic motion with an amplitude of 5 cm. When the particle is at 4 cm from the mean position, the magnitude of its velocity in SI units is equal to that of its acceleration. Then, its periodic time (in seconds) is **[JEE Main 2019]**
(a) $\frac{4\pi}{3}$ (b) $\frac{8\pi}{3}$ (c) $\frac{7}{3}\pi$ (d) $\frac{3}{8}\pi$

14. Two pendulums of length 121 cm and 100 cm start vibrating. At some instant both are in the mean position in the same phase. After how many vibrations of the shorter pendulum, both will be again in phase at the mean position?
(a) 10 (b) 11 (c) 20 (d) 21

15. The piston in the cylinder head of a locomotive has a stroke (twice the amplitude) of 1.0 m. If the piston moves with simple harmonic motion with an angular frequency of 200 rad/min, what is its maximum speed? **[NCERT]**
(a) 100 m/min (b) 200 m/min
(c) 300 m/min (d) 50 m/min

16. Two points are located at a distance of 10 m and 15 m from the source of oscillation. The period of oscillation is 0.05s and the velocity of the wave is 300 m/s. What is the phase difference between the oscillations of two points?
(a) π (b) $\frac{\pi}{6}$ (c) $\frac{\pi}{3}$ (d) $\frac{2\pi}{3}$

17. A horizontal plank has a rectangular block placed on it. The plank starts oscillating vertically and simple harmonically with an amplitude of 40 cm. The block just loses contact with the plank when the later is momentarily at rest, then
(a) the block becomes weighless
(b) the block weighs double its weight when the plank is at one of the positions of momentary at rest
(c) the block weighs 1.5 times its weight on the plank at extreme position
(d) the block weighs its true weight on the plank, at one of the positions of momentary at rest

18. A body has a time period T_1 under the action of one force and T_2 under the action of another force, the square of the time period when both the forces are acting in the same direction is
(a) $T_1^2T_2^2$ (b) $T_1^2T_2^2$
(c) $T_1^2 + T_2^2$ (d) $T_1^2T_2^2/(T_1^2 + T_2^2)$

19. Two linear SHMs of equal amplitude A and angular frequencies ω and 2ω are impressed on a particle along the axes x and y respectively. If the initial phase difference between them is $\pi/2$, the resultant path followed by the particle is
(a) $y^2 = x^2(1 - x^2/A^2)$ (b) $y^2 = 2x^2(1 - x^2/A^2)$
(c) $y^2 = 4x^2(1 - x^2/A^2)$ (d) $y^2 = 8x^2(1 - x^2/A^2)$

20. A coin is placed on a horizontal platform, which undergoes horizontal SHM about a mean position O. The coin placed on platform does not slip, coefficient of friction between the coin and the platform is μ. The amplitude of oscillation is gradually increased. The coil will begin to slip on the platform for the first time
(a) at the mean position
(b) at the extreme position of oscillations
(c) for an amplitude of $\mu g/\omega^2$
(d) for an amplitude of $g/\mu\omega^2$

21. A particle in SHM is described by the displacement function $x(t) = A\cos(\omega t + \phi)$, $\omega = 2\pi/T$. If the initial ($t = 0$) position of the particle is 1 cm, its initial velocity is π cm s^{-1} and its angular frequency is πs^{-1}, then the amplitude of its motion is
(a) π cm (b) 2 cm (c) $\sqrt{2}$ cm (d) 1 cm

22. A cylindrical plastic bottle of negligible mass is filled with 310 mL of water and left floating in a pond with still water. If pressed downward slightly and released, it starts performing simple harmonic motion at angular frequency ω. If the radius of the bottle is 2.5 cm, then ω is close to
(Take, density of water = 10^3 kg/m^3) **[JEE Main 2019]**
(a) 2.50 rad s^{-1} (b) 8.00 rad s^{-1}
(c) 1.25 rad s^{-1} (d) 3.75 rad s^{-1}

23. A large horizontal surface moves up and down in SHM with an amplitude of 1 cm. If a mass of 10 kg (which is placed on the surface) is to remain continuously in contact with it, then the maximum frequency of SHM will be
(a) 5 Hz (b) 0.5 Hz
(c) 1.5 Hz (d) 10 Hz

24. A horizontal platform vibrates with simple harmonic motion in the horizontal direction with a period 2 s. A body of mass 0.5 kg is placed on the platform. The coefficient of static friction between the body and platform is 0.3. What is the maximum frictional force on the body when the platform is oscillating with an amplitude 0.2 m? Assume $\pi^2 = 10 = g$.

(a) 0.5 N (b) 1 N
(c) 1.5 N (d) 2 N

25. The composition of two simple harmonic motions of equal periods at right angle to each other and with a phase difference of π results in the displacement of the particle along

(a) circle (b) figure of eight
(c) straight line (d) ellipse

26. The bob of a simple pendulum of length L is released at time $t = 0$ from a position of small angular displacement. Its linear displacement at time t is given by

(a) $x = a \sin 2\pi \sqrt{\frac{L}{g}} \times t$ (b) $x = a \cos 2\pi \sqrt{\frac{g}{L}} \times t$

(c) $x = a \sin \sqrt{\frac{g}{L}} \times t$ (d) $x = a \cos \sqrt{\frac{g}{L}} \times t$

27. Displacement-time equation of a particle executing SHM is , $x = 3 \sin \omega t + 4 \sin (\omega t + \pi/3)$. Here x is in centimetre and t in second. The amplitude of oscillation of the particle is approximately

(a) 5 cm (b) 6 cm
(c) 7 cm (d) 9 cm

28. Which one of the following equations does not represent SHM, x = displacement and t = time? (parameters a, b and c are the constants of motion)

(a) $x = a \sin bt$ (b) $x = a \cos bt + c$
(c) $x = a \sin bt + c \cos bt$ (d) $x = a \sec bt + c \operatorname{cosec} bt$

29. A particle is performing simple harmonic motion along X-axis with amplitude 4 cm and time period 1.2 s. The minimum time taken by the particle to move from $x = +2$ to $x = 4$ cm and back again is given by

(a) 0.4 s (b) 0.3 s (c) 0.2 s (d) 0.6 s

Energy in Simple Harmonic Motion

30. The angular velocity and the amplitude of a simple pendulum is ω and a respectively. At a displacement x from the mean position, if its kinetic energy is T and potential energy is V, then the ratio of T to V is

(a) $(a^2 - x^2\omega^2)/x^2\omega^2$ (b) $x^2\omega^2/(a^2 - x^2\omega^2)$
(c) $(a^2 - x^2)/x^2$ (d) $x^2/(a^2 - x^2)$

31. A particle is vibrating in a simple harmonic motion with an amplitude of 4 cm. At what displacement from the equilibrium position is its energy half potential and half kinetic?

(a) 1 cm (b) $\sqrt{2}$ cm
(c) 3 cm (d) $2\sqrt{2}$ cm

32. When the potential energy of a particle executing simple harmonic motion is one-fourth of its maximum value during the oscillation, the displacement of the particle from the equilibrium position in terms of its amplitude a is

(a) $\frac{a}{4}$ (b) $\frac{a}{3}$ (c) $\frac{a}{2}$ (d) $\frac{2a}{3}$

33. The potential energy of a particle (U_x) executing SHM is given by

(a) $U_x = \frac{k}{2}(x-a)^2$ (b) $U_x = k_1x + k_2x^2 + k_3x^3$
(c) $U_x = Ae^{-bx}$ (d) U_x = constant

34. A particle of mass m is executing oscillations about the origin on the X-axis with amplitude A. Its potential energy is given as $U(x) = \alpha x^4$, where α is positive constant. The x-coordinate of mass where potential energy is one-third of the kinetic energy of particle, is

(a) $\pm\frac{A}{\sqrt{3}}$ (b) $\pm\frac{A}{\sqrt{2}}$ (c) $\pm\frac{A}{3}$ (d) $\pm\frac{A}{2}$

35. A particle starts SHM from the mean position. Its amplitude is a and total energy E. At one instant, its kinetic energy is $3E/4$. Its displacement at this instant is

(a) $y = a/\sqrt{2}$ (b) $y = \frac{a}{2}$
(c) $y = \frac{a}{\sqrt{3/2}}$ (d) $y = a$

36. A point particle of mass 0.1 kg is executing SHM of amplitude 0.1 m. When the particle passes through the mean position, its kinetic energy is 8×10^{-3} J. The equation of motion of this particle, if its initial phase of oscillation is 45°, is

(a) $y = 0.1 \sin\left(\frac{r}{4} + \frac{\pi}{4}\right)$ (b) $y = 0.1 \sin\left(\frac{t}{2} + \frac{\pi}{4}\right)$

(c) $y = 0.1 \sin\left(4t - \frac{\pi}{4}\right)$ (d) $y = 0.1 \sin\left(4t + \frac{\pi}{4}\right)$

37. If a simple pendulum of length l has maximum angular displacement θ, then the maximum kinetic energy of bob of mass m is

(a) $\frac{1}{2} \times \left(\frac{l}{g}\right)$ (b) $\frac{1}{2} \times \frac{mg}{l}$
(c) $mgl \times (1 - \cos\theta)$ (d) $\frac{1}{2} \times mgl \sin\theta$

38. For a particle executing SHM, the kinetic energy K is given by $K = K_0 \cos^2 \omega t$. The equation of its displacement can be

(a) $\left(\frac{K_0}{m\omega^2}\right)^{1/2} \sin \omega t$ (b) $\left(\frac{2K_0}{m\omega^2}\right)^{1/2} \sin \omega t$

(c) $\left(\frac{2\omega^2}{mK_0}\right)^{1/2} \sin \omega t$ (d) $\left(\frac{2K_0}{m\omega}\right)^{1/2} \sin \omega t$

Springs and their Oscillations

39. Two springs of force constants k and $2k$ are connected to a mass as shown below. The frequency of oscillation of the mass is

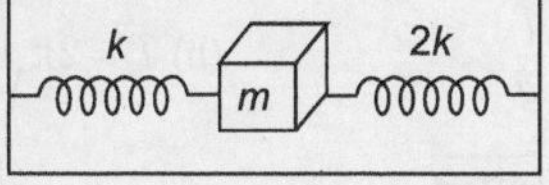

(a) $\frac{1}{2\pi}\sqrt{k/m}$ (b) $\frac{1}{2\pi}\sqrt{2k/m}$

(c) $\frac{1}{2\pi}\sqrt{\frac{3k}{m}}$ (d) $\frac{1}{2\pi}\sqrt{\frac{m}{k}}$

40. A simple spring has length l and force constant k. It is cut into two springs of length l_1 and l_2 such that $l_1 = nl_2$ (n = an integer). The force constant of the spring of length l_2 is

(a) $k(1+n)$ (b) $\frac{(n+1)k}{n}$ (c) k (d) $k/(n+1)$

41. A weightless spring which has a force constant k oscillates with frequency n when a mass m is suspended from it. The spring is cut into two equal halves and a mass 2 m is suspended from one part of spring. The frequency of oscillation will now become

(a) n (b) $2n$ (c) $\frac{n}{\sqrt{2}}$ (d) $n(2)^{1/2}$

42. An object suspended from a spring exhibits oscillations of period T. Now, the spring is cut in two halves and the same object is suspended with two halves as shown in figure. The new time period of oscillation will become

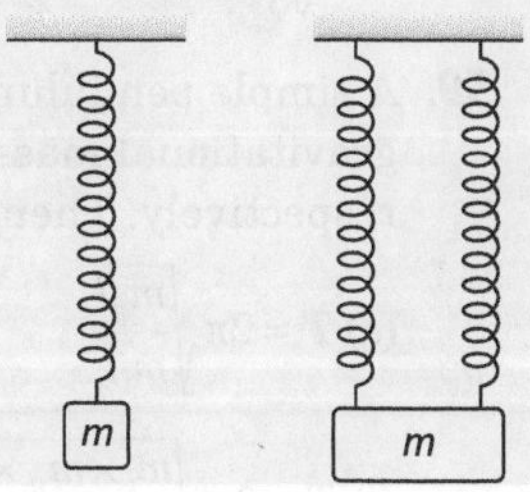

(a) $\frac{T}{2\sqrt{2}}$ (b) $\frac{T}{2}$ (c) $\frac{T}{\sqrt{2}}$ (d) $2T$

43. On a smooth inclined plane, a body of mass M is attached between two springs. The other ends of the springs are fixed to firm support. If each spring has force constant k, then the period of oscillation of the body (assuming the springs as massless) is

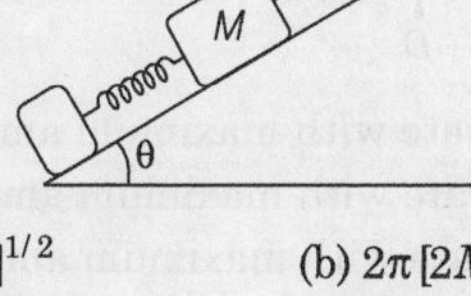

(a) $2\pi[M/2k]^{1/2}$ (b) $2\pi[2M/k]^{1/2}$

(c) $2\pi[Mg\sin\theta/2k]^{1/2}$ (d) $2\pi[2Mg/k]^{1/2}$

44. Time period of mass m suspended by a spring is T. If the spring is cut to one-half and made to oscillate by suspending double mass, the time period of the mass will be

(a) $8T$ (b) $4T$

(c) $\frac{T}{2}$ (d) T

45. What will be the force constant of the spring system shown in figure?

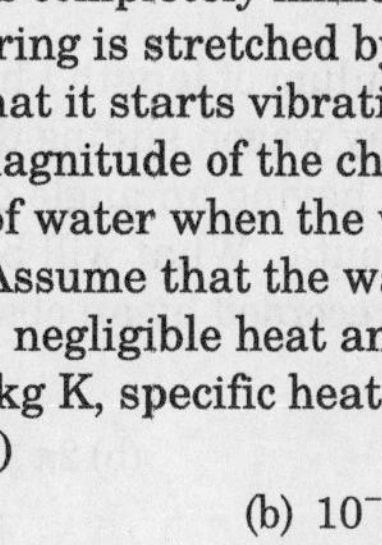

(a) $\frac{k_1}{2}+k_2$ (b) $\left(\frac{1}{2k_1}+\frac{1}{k_2}\right)^{-1}$

(c) $\frac{1}{2k_1}+\frac{1}{k_2}$ (d) $\left(\frac{2}{k_1}+\frac{1}{k_2}\right)^{-1}$

46. A particle of mass m is fixed to one end of a light spring having force constant k and unstretched length l. The other end is fixed. The system is given an angular speed ω about the fixed end of the spring such that it rotates in a circle in gravity free space. Then, the stretch in the spring is **[JEE Main 2020]**

(a) $\frac{ml\omega^2}{k+m\omega}$ (b) $\frac{ml\omega^2}{k+m\omega^2}$

(c) $\frac{ml\omega^2}{k-m\omega^2}$ (d) $\frac{ml\omega^2}{k-m\omega}$

47. A massless spring (k = 800 N/m), attached with a mass (500 g) is completely immersed in 1 kg of water. The spring is stretched by 2 cm and released, so that it starts vibrating. What would be the order of magnitude of the change in the temperature of water when the vibrations stop completely? (Assume that the water container and spring receive negligible heat and specific heat of mass = 400 J/kg K, specific heat of water = 4184 J/kg K) **[JEE Main 2019]**

(a) 10^{-4} K (b) 10^{-3} K

(c) 10^{-1} K (d) 10^{-5} K

48. A mass M, attached to a spring, oscillates with a period of 2 s. If the mass is increased by 4 kg, the time period increases by 1 s, the initial mass M was

(a) 3.2 kg (b) 1 kg

(c) 2 kg (d) 8 kg

49. A mass M is suspended from a light spring. An additional mass m is added and displaces the spring further by a distance X. Now the combined mass will oscillate on the spring with period

(a) $T = 2\pi\sqrt{\frac{mg}{X(M+m)}}$ (b) $T = 2\pi\sqrt{\frac{(M+m)X}{mg}}$

(c) $T = \pi/2\sqrt{\frac{mg}{X(M+m)}}$ (d) $T = 2\pi\sqrt{\frac{(M+m)}{mg}}$

50. Two blocks with masses $m_1 = 1$ kg and $m_2 = 2$ kg are connected by a spring of spring constant $k = 24\ \text{Nm}^{-1}$ and placed on a frictionless horizontal surface. The block m_1 is imparted an initial velocity $v_0 = 12\ \text{cms}^{-1}$ to the right, the amplitude of oscillation is

(a) 1 cm (b) 2 cm (c) 3 cm (d) 4 cm

51. A mass 1 kg suspended from a spring whose force constant is $400\ \text{Nm}^{-1}$, executes simple harmonic oscillation. When the total energy of the oscillator is 2 J, the maximum acceleration experienced by the mass will be

(a) $2\ \text{ms}^{-2}$ (b) $4\ \text{ms}^{-2}$

(c) $40\ \text{ms}^{-2}$ (d) $400\ \text{ms}^{-2}$

Simple Pendulum and Other Oscillating Systems

52. A simple pendulum of length l and mass (bob) m is suspended vertically. The string makes an angle θ with the vertical. The restoring force acting on the pendulum is

(a) $mg\tan\theta$ (b) $-mg\sin\theta$

(c) $mg\sin\theta$ (d) $-mg\cos\theta$

53. A man measures the period of a simple pendulum inside a stationary lift and finds it to be T second. If the lift accelerates upwards with an acceleration $g/4$, then the period of pendulum will be

(a) $2T\sqrt{5}$ (b) T (c) $\frac{2T}{\sqrt{5}}$ (d) $\frac{T}{4}$

54. A simple pendulum of length l has been set up inside a railway wagon sliding down a frictionless inclined plane having an angle of inclination $\theta = 30°$ with the horizontal. What will be its period of oscillation as recorded by an observer inside the wagon?

(a) $2\pi\sqrt{\frac{2l}{\sqrt{3}g}}$ (b) $2\pi\sqrt{2l/g}$

(c) $2\pi\sqrt{l/g}$ (d) $2\pi\sqrt{\frac{\sqrt{3}l}{2g}}$

55. If a simple pendulum is taken to a place where g decreases by 2%, then the time period

(a) increases by 0.5% (b) increases by 1%

(c) increases by 2.0% (d) decreases by 0.5%

56. A heavy sphere of mass m is suspended by string of length l. The sphere is made to revolve about a vertical line passing through the point of suspension in a horizontal circle such that the string always remains inclined to the vertical at an angle θ. What is its period of revolution?

(a) $T = 2\pi\sqrt{\frac{l}{g}}$ (b) $T = 2\pi\sqrt{\frac{l\cos\theta}{g}}$

(c) $T = 2\pi\sqrt{\frac{l\sin\theta}{g}}$ (d) $T = 2\pi\sqrt{\frac{l\tan\theta}{g}}$

57. A ring is hung on a nail. It can oscillate without slipping or sliding

(i) in its plane with a time period T_1 and

(ii) back and forth in a direction perpendicular to its plane, with a period T_2.

The ratio $\frac{T_1}{T_2}$ will be **[JEE Main 2020]**

(a) $\frac{2}{\sqrt{3}}$ (b) $\frac{2}{3}$ (c) $\frac{3}{\sqrt{2}}$ (d) $\frac{\sqrt{2}}{3}$

58. The mass and the diameter of a planet are three times the respective values for the earth. The period of oscillation of a simple pendulum on the earth is 2 s. The period of oscillation of the same pendulum on the planet would be **[JEE Main 2019]**

(a) $\frac{2}{\sqrt{3}}$ s (b) $\frac{3}{2}$ s (c) $2\sqrt{3}$ s (d) $\frac{\sqrt{3}}{2}$ s

59. A simple pendulum has a length l. The inertial and gravitational masses of the bob are m_i and m_g respectively. Then the time period T is given by

(a) $T = 2\pi\sqrt{\frac{m_g l}{m_i g}}$ (b) $T = 2\pi\sqrt{\frac{m_i l}{m_g g}}$

(c) $T = 2\pi\sqrt{\frac{m_i \times m_g \times l}{g}}$ (d) $T = 2\pi\sqrt{\frac{l}{m_i \times m_g \times g}}$

60. Four pendulums A, B, C and D are suspended from the same elastic support as shown in figure. A and C are of the same length, while B is smaller than A and D is larger than A. **[NCERT Exemplar]**

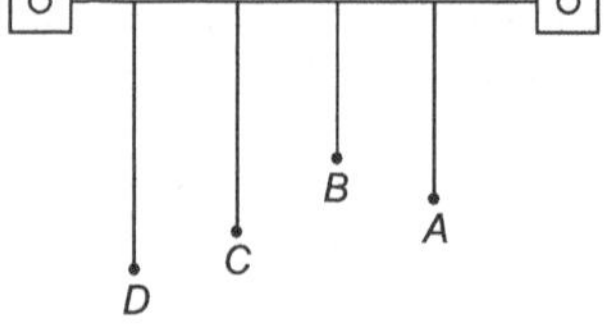

(a) D will vibrate with maximum amplitude

(b) C will vibrate with maximum amplitude

(c) B will vibrate with maximum amplitude

(d) All the four will oscillate with equal amplitude

61. If the length of second's pendulum is increased by 2%. How many seconds it will lose per day?

(a) 3927 s (b) 3427 s

(c) 3737 s (d) 864 s

62. A pendulum bob of mass m is hanging from a fixed point by a light thread of length l. A horizontal speed v_0 is imparted to the bob, so that it takes up horizontal position. If g is the acceleration due to gravity, then v_0 is

(a) mgl (b) $\sqrt{2gl}$

(c) $\sqrt{gl}$ (d) gl

63. The bob of a pendulum of length l is pulled a side from its equilibrium position through an angle θ and then released. The bob will then pass through its equilibrium position with a speed v, where v equals

(a) $\sqrt{2gl(1-\cos\theta)}$

(b) $\sqrt{2gl(1+\sin\theta)}$

(c) $\sqrt{2gl(1-\sin\theta)}$

(d) $\sqrt{2gl(1+\cos\theta)}$

64. A tunnel is made across the earth of radius R, passing through its centre. A ball is dropped from a height h in the tunnel. The motion will be periodic with time period

(a) $2\pi\sqrt{\dfrac{R}{g}}+4\sqrt{\dfrac{h}{g}}$

(b) $2\pi\sqrt{\dfrac{R}{g}}+4\sqrt{\dfrac{2h}{g}}$

(c) $2\pi\sqrt{\dfrac{R}{g}}+\sqrt{\dfrac{h}{g}}$

(d) $2\pi\sqrt{\dfrac{R}{g}}+\sqrt{\dfrac{2h}{g}}$

65. Two simple pendulums of length 0.5 m and 20 m respectively are given small linear displacement in one direction at the same time. They will again be in the phase when the pendulum of shorter length has completed ...... oscillations.

(a) 5 (b) 1

(c) 2 (d) 3

66. In damped oscillation the amplitude of oscillations is reduced to one-third of its initial value a_0 at the end of 100 oscillation. When the oscillation completes 200 oscillations, its amplitude must be

(a) $\dfrac{a_0}{2}$ (b) $\dfrac{a_0}{4}$

(c) $\dfrac{a_0}{6}$ (d) $\dfrac{a_0}{9}$

67. A and B are fixed points and the mass M is tied by strings at A and B. If the mass M is displaced slightly out of this plane and released, it will execute oscillations with period (Given, $AM = BM = L$, $AB = 2d$)

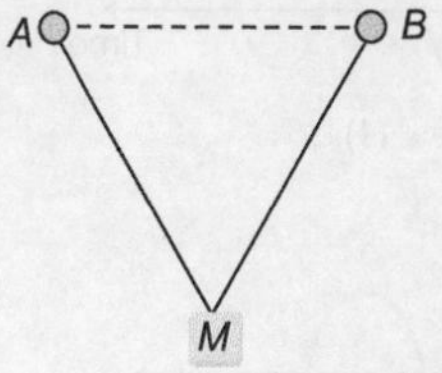

(a) $2\pi\sqrt{\dfrac{L}{g}}$ (b) $2\pi\sqrt{\dfrac{(L^2-d^2)^{1/2}}{g}}$

(c) $2\pi\sqrt{\dfrac{(L^2+d^2)^{1/2}}{g}}$ (d) $2\pi\sqrt{\dfrac{(2d^2)^{3/2}}{g}}$

68. A piece of wood has dimensions a, b and c. Its relative density is d. It is floating in water such that the side c is vertical. It is now pushed down gently and released. The time period is

(a) $T = 2\pi\sqrt{\left(\dfrac{abc}{g}\right)}$ (b) $T = 2\pi\sqrt{\left(\dfrac{ba}{dg}\right)}$

(c) $T = 2\pi\sqrt{\left(\dfrac{g}{dc}\right)}$ (d) $T = 2\pi\sqrt{\left(\dfrac{dc}{g}\right)}$

69. A pendulum clock is placed on the Moon, where object weighs only one-sixth as much as on the Earth. How many seconds the clock tick out in an actual time of 1 min the clock keeps good time on the Earth?

(a) 12.25 (b) 24.5 (c) 2.45 (d) 0.245

70. A uniform cylinder of length L and mass M having cross-sectional area A is suspended with its vertical length, from a fixed point by a massless spring, such that it is half submerged in a liquid of density d at equilibrium position. When released, it starts oscillating vertically with a small amplitude. If the force constant of the spring is k, the frequency of oscillation of the cylinder is

(a) $\dfrac{1}{2\pi}\left(\dfrac{k-Adg}{M}\right)^{1/2}$ (b) $\dfrac{1}{2\pi}\left(\dfrac{k+Adg}{M}\right)^{1/2}$

(c) $\dfrac{1}{2\pi}\left(\dfrac{k-dgL}{M}\right)^{1/2}$ (d) $\dfrac{1}{2\pi}\left(\dfrac{k+AgL}{Adg}\right)^{1/2}$

71. A particle, with restoring force proportional to displacement and resisting force proportional to velocity is subjected to a force, $F = F_0 \sin \omega t$. If the amplitude of the particle is maximum for $\omega = \omega_1$ and the energy of the particle is maximum for $\omega = \omega_2$, then

(a) $\omega_1 = \omega_0$ and $\omega_2 \neq \omega_0$ (b) $\omega_1 = \omega_0$ and $\omega_2 = \omega_0$

(c) $\omega_1 \neq \omega_0$ and $\omega_2 = \omega_0$ (d) $\omega_1 \neq \omega_0$ and $\omega_2 \neq \omega_0$

72. Which of the following figure represent(s) damped simple harmonic motions?

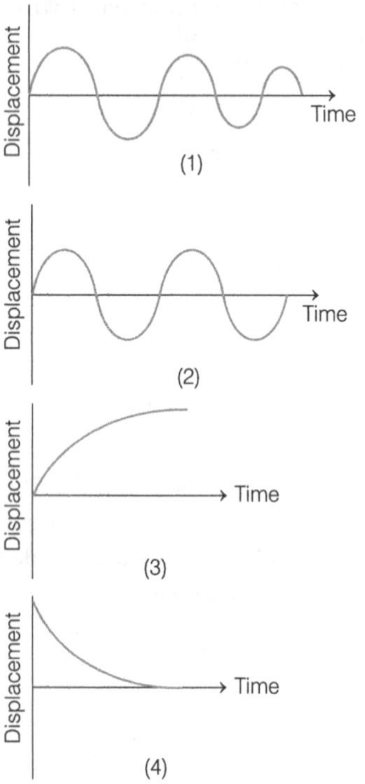

(a) Fig. 1 (b) Fig. 2
(c) Fig. 4 (d) Figs. 3 and 4

73. A particle of mass m is released from rest and follows a parabolic path as shown. Assuming that the displacement of the mass from the origin is small. Which graph correctly depicts the position of the particle as a function of time?

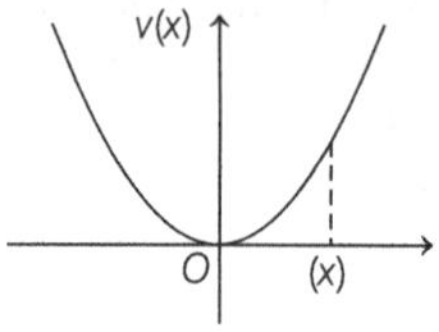

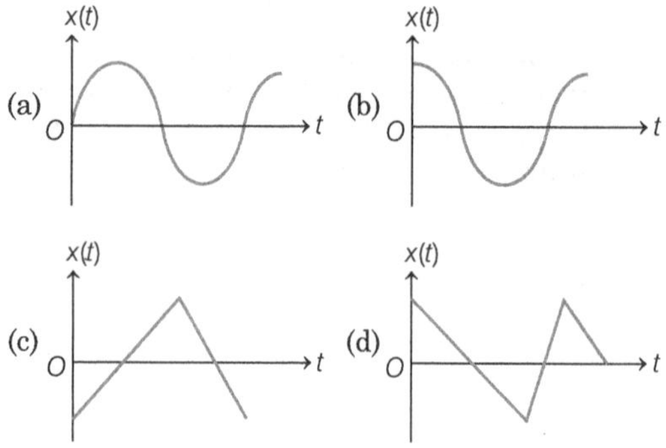

74. The amplitude of damped oscillator becomes (1/3)rd in 2 s. Its amplitude after 6 s is $1/n$ times the original. Then n is equal to

(a) 2^3 (b) 3^2 (c) $3^{1/3}$ (d) 3^3

ROUND II Mixed Bag

Only One Correct Option

1. Motion of a ball bearing inside a smooth curved bowl, when released from a point slightly above the lower point is **[NCERT Exemplar]**

(a) simple harmonic motion (b) non-periodic motion
(c) periodic but not SHM (d) None of the above

2. The time period of a mass suspended from a spring is 5 s. The spring is cut into four equal parts and the same mass is now suspended from one of its parts. The period is now

(a) 5 s (b) 2.5 s (c) 1.25 s (d) $\frac{1}{16}$ s

3. Displacement vs. time curve for a particle executing SHM is shown in figure. Choose the correct statements. **[NCERT Exemplar]**

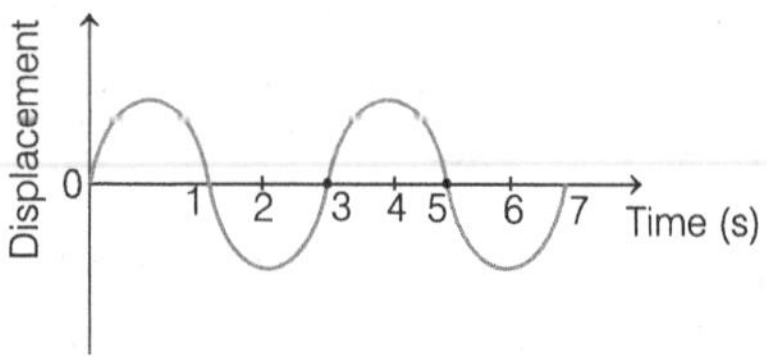

(a) Phase of the oscillator is same at $t=0$ s and $t=2$ s
(b) Phase of the oscillator is different at $t=2$ s and $t=6$ s
(c) Phase of the oscillator is same at $t=1$ s and $t=7$ s
(d) Phase of the oscillator is same at $t=1$ s and $t=5$ s

4. The total energy of a particle executing SHM is 80 J. What is the potential energy when the particle is at a distance of 3/4 of amplitude from the mean position?

(a) 60 J (b) 10 J (c) 40 J (d) 45 J

5. A particle is executing simple harmonic motion with a time period T. At time $t = 0$, it is at its position of equilibrium. The kinetic energy-time graph of the particle will look like **[JEE Main 2017]**

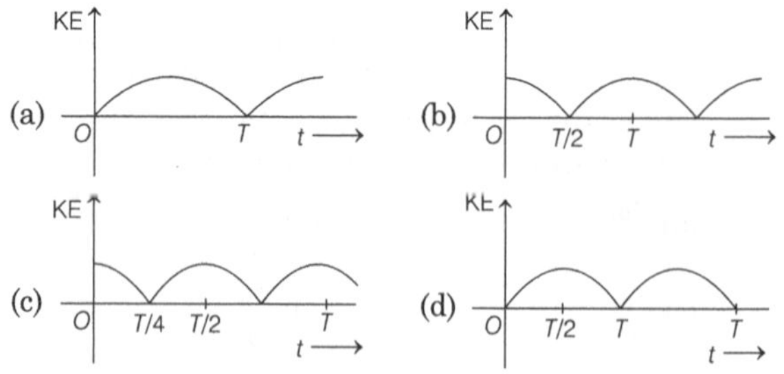

6. For a simple pendulum, a graph is plotted between its Kinetic Energy (KE) and Potential Energy (PE) against its displacement d. Which one of the following represents these correctly? (graphs are schematic and not drawn to scale) **[JEE Main 2015]**

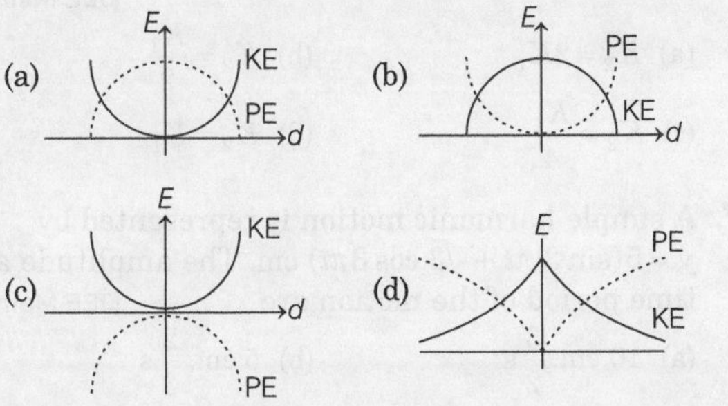

7. Two simple harmonic motions act on a particle. These harmonic motions are
$x = A\sin(\omega t + \alpha)$; $y = A\cos(\omega t + \alpha)$, then path of particle is
(a) an ellipse and the actual motion is counter clockwise
(b) an ellipse and the actual motion is clockwise
(c) a circle and the actual motion is counter clockwise
(d) a circle and the actual motion is clockwise

8. A uniform cylindrical metal rod A of length L and radius R is suspended at its mid-point from a rigid support through a strong metal wire of length l. The rod is given a small angular twist and released so that it oscillates to and fro about its mean position with a time period T_1. The rotational inertia of metal rod about the wire as an axis is

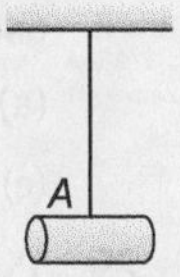

(a) $\frac{ML^2}{12}$ (b) $\frac{MR^2}{2}$
(c) $M\left[\frac{L^2}{12}+\frac{R^2}{2}\right]$ (d) $M\left[\frac{L^2}{12}+\frac{R^2}{4}\right]$

9. A particle is having kinetic energy 1/3 of the maximum value at a distance of 4 cm from the mean position. Find the amplitude of motion.
(a) $2\sqrt{6}$ cm (b) $2/\sqrt{6}$ cm
(c) $\sqrt{2}$ cm (d) $6/\sqrt{2}$

10. If a spring extends by x on loading, then the energy stored in the spring is (if T is the tension and k is the force constant of the spring)
(a) $\frac{T^2}{2x}$ (b) $\frac{T^2}{2k}$ (c) $\frac{2k}{T^2}$ (d) $\frac{2T^2}{k}$

11. The period of oscillation of a mass m suspended from a spring is 2 s. If along with it another mass 2 kg is also suspended, the period of oscillation increases by 1 s. The mass m will be
(a) 2 kg (b) 1 kg (c) 1.6 kg (d) 2.6 kg

12. Two pendulums begin to swing simultaneously. The first pendulum makes 9 full oscillations when the other makes 7. The ratio of lengths of the two pendulums is
(a) 9/7 (b) 7/9 (c) 49/81 (d) 81/49

13. A ball of mass (m) 0.5 kg is attached to the end of a string having length (L) 0.5m. The ball is rotated on a horizontal circular path about vertical axis. The maximum tension that string can bear is 324 N. The maximum possible value of angular velocity of ball (in rad/s) is

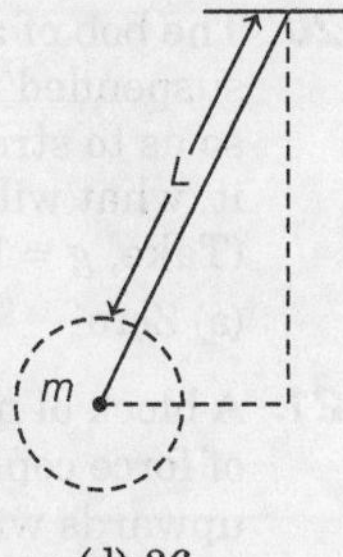

(a) 9 (b) 18 (c) 27 (d) 36

14. A point mass is subjected to two simultaneous sinusoidal displacement in X-direction
$X_1(t) = A\sin\omega t$ and $X_2(t) = A\sin\left(\omega t + \frac{2\pi}{3}\right)$.
Adding a third sinusoidal displacement $X_3(t) = B\sin(\omega t + \phi)$ brings the mass to a complete rest. The values of B and ϕ are
(a) $\sqrt{2}A, \frac{3\pi}{4}$ (b) $A, \frac{4\pi}{3}$ (c) $\sqrt{3}A, \frac{5\pi}{6}$ (d) $A, \frac{\pi}{3}$

15. Average value of kinetic energy and potential energy over entire time period is
(a) $0, \frac{1}{2}m\omega^2A^2$ (b) $\frac{1}{2}m\omega^2A^2, 0$
(c) $\frac{1}{2}m\omega^2A^2, \frac{1}{2}m\omega^2A^2$ (d) $\frac{1}{4}m\omega^2A^2, \frac{1}{4}m\omega^2A^2$

16. A particle in SHM is described by the displacement function $x(t) = A\cos(\omega t + \theta)$. If the initial ($t = 0$) position of the particle is 1 cm and its initial velocity is π cms^{-1}, what is its amplitude? The angular frequency of the particle is π s^{-1}.
(a) 1 cm (b) $\sqrt{2}$ cm (c) 2 cm (d) 2.5 cm

17. The period of particle in SHM is 8 s. At $t = 0$, particle is at the mean position. The ratio of the distances travelled by it in Ist second and 2nd second is
(a) 1.6 : 1 (b) 2.4 : 1 (c) 3.2 : 1 (d) 4.2 : 1

18. A simple pendulum of length l has a bob of mass m, with a charge q on it. A vertical sheet of charge, with surface charge density σ passes through the point of suspension. At equilibrium, the spring makes an angle θ with the vertical. Its time period of oscillations is T in this position. Then
(a) $\tan\theta = \frac{\sigma q}{2\varepsilon_0 mg}$ (b) $\tan\theta = \frac{\sigma q}{\varepsilon_0 mg}$
(c) $T > 2\pi\sqrt{\frac{l}{g}}$ (d) $T = 2\pi\sqrt{\frac{l}{g}}$

19. The time period of a particle in simple harmonic motion is 8 s. At $t = 0$, it is at the mean position. The ratio of the distances travelled by it in the first and second, seconds is

(a) 1 : 2 (b) $1:\sqrt{2}$ (c) $1:(\sqrt{2}-1)$ (d) $1:\sqrt{3}$

20. The bob of a simple pendulum is of mass 10 g. It is suspended with a thread of 1 m. If we hold the bob so as to stretch the string horizontally and release it, what will be the tension at the lowest position? (Take, $g = 10\ \text{ms}^{-2}$)

(a) Zero (b) 0.1 N (c) 0.3 N (d) 1.0 N

21. A block of mass M is suspended from a light spring of force constant k. Another mass m moving upwards with velocity v hits the mass M and gets embedded in it. What will be the amplitude of oscillation of the combined mass?

(a) $\dfrac{mv}{k\sqrt{(M-m)}}$ (b) $\dfrac{mv}{\sqrt{(M-m)k}}$

(c) $\dfrac{mv}{k\sqrt{(M+m)}}$ (d) $\dfrac{mv}{\sqrt{(M+m)k}}$

22. Lissajous figure shown below corresponds to which one of the following?

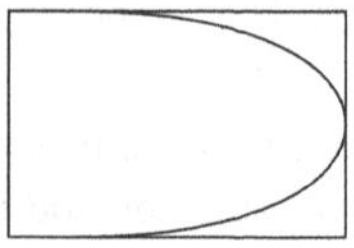

(a) Phase difference $\pi/2$ and period 1 : 2
(b) Phase difference $3\pi/4$ and period 1 : 2
(c) Phase difference $\pi/4$ and period 2 : 1
(d) Phase difference $2\pi/3$ and period 2 : 1

23. The amplitude of a damped oscillator decreases to 0.9 times its original magnitude in 5s. In another 10s, it will decrease to α times its original magnitude, where α equals **[JEE Main 2013]**

(a) 0.7 (b) 0.81 (c) 0.729 (d) 0.6

24. A silver atom in a solid oscillates in simple harmonic motion in some direction with a frequency of 10^{12} per second. What is the force constant of the bonds connecting one atom with the other? (Take, molecular weight of silver = 108 and Avogadro number = $6.02 \times 10^{23}\ \text{g mol}^{-1}$)

(a) 6.4 N/m (b) 7.1 N/m **[JEE Main 2018]**
(c) 2.2 N/m (d) 5.5 N/m

25. A particle moves with simple harmonic motion in a straight line. In first t sec, after starting from rest, it travels a distance a and in next t sec, it travels $2a$ in same direction, then **[JEE Main 2014]**

(a) amplitude of motion is $3a$
(b) time period of oscillations is $5t$
(c) amplitude of motion is $4a$
(d) time period of oscillations is $6t$

26. A pendulum is executing simple harmonic motion and its maximum kinetic energy is K_1. If the length of the pendulum is doubled and it performs simple harmonic motion with the same amplitude as in the first case, its maximum kinetic energy is K_2. Then **[JEE Main 2019]**

(a) $K_2 = 2K_1$ (b) $K_2 = \dfrac{K_1}{2}$

(c) $K_2 = \dfrac{K_1}{4}$ (d) $K_2 = K_1$

27. A simple harmonic motion is represented by $y = 5(\sin 3\pi t + \sqrt{3}\cos 3\pi t)$ cm. The amplitude and time period of the motion are **[JEE Main 2019]**

(a) 10 cm, $\dfrac{3}{2}$ s (b) 5 cm, $\dfrac{2}{3}$ s

(c) 5 cm, $\dfrac{3}{2}$ s (d) 10 cm, $\dfrac{2}{3}$ s

28. A highly rigid cubical block A of small mass M and side L is fixed rigidly on the another cubical block of same dimensions and low modulus of rigidity η such that the lower face of A completely covers the upper face of B. The lower face of B is rigidly held on a horizontal surface. A small force F is applied perpendicular to one of the side faces of Z. After the force is withdrawn, block A executes small oscillations, the time period of which is given by

(a) $2\pi\sqrt{ML\eta}$ (b) $2\pi\sqrt{M\eta/L}$
(c) $2\pi\sqrt{ML/\eta}$ (d) $2\pi\sqrt{M/\eta L}$

29. Two pendulums of length 1 m and 16 m start vibrating one behind the other from the same stand. At some instant, the two are in the mean position in the same phase. The time period of shorter pendulum is T. The minimum time after which the two threads of the pendulum will be one behind the other is

(a) $T/4$ (b) $T/3$ (c) $4T/3$ (d) $4T$

30. A particle is in linear simple harmonic motion between two points A and B, 10 cm apart (Fig). Take the direction from A to B as the +ve direction and choose the correct statements. **[NCERT Exemplar]**

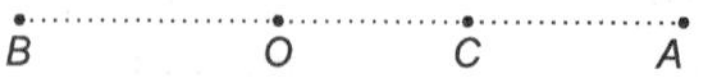

$AO = OB = 5\ \text{cm}, \quad BC = 8\ \text{cm}$

(a) The sign of velocity, acceleration and force on the particle when it is 3 cm away from A going towards B are positive
(b) The sign of velocity of the particle at C going towards O is negative
(c) The sign of velocity, acceleration and force on the particle when it is 5 cm away from B going towards A are positive
(d) The sign of acceleration and force on the particle when it is at point B is positive

31. A particle performs harmonic oscillation along the X-axis about the equilibrium position $x = 0$. The oscillation frequency is $\omega = 4.00\ s^{-1}$. At a certain moment of time the particle has a coordinate $x_0 = 25.0$ cm and its velocity is equal to $v_{x_0} = 100\ cms^{-1}$. Find the equation of motion of the particle.

(a) $y = 13\sqrt{3}\sin\left(4t + \frac{\pi}{4}\right)$ (b) $y = 25\sqrt{2}\sin\left(4t + \frac{\pi}{4}\right)$

(c) $y = 27\sqrt{2}\sin\left(4t + \frac{\pi}{4}\right)$ (d) $y = 27\sqrt{5}\sin\left(t + \frac{\pi}{2}\right)$

32. A metal rod of length L and mass m is pivoted at one end. A thin disc of mass M and radius $R (< L)$ is attached at its centre to the free end of the rod. Consider two ways, the disc is attached (case A), the disc is not free to rotate about its centre and (case B) the disc is free to rotate about its centre. The rod-disc system perform SHM in vertical plane after being released from the same displaced position in each case. Which of the following statements is (are) true?

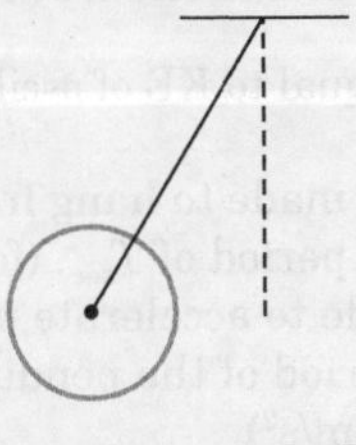

(a) Restoring torque in case A = Restoring torque in case B.

(b) Restoring torque in case A < Restoring torque in case B.

(c) Angular frequency for case A > Angular frequency for case B.

(d) None of the above

33. A mass m is suspended separately by two different springs in successive order, then time periods is t_1 and t_2 respectively. If m is connected by both springs as shown in figure, then time period is t_0, the correct relation is

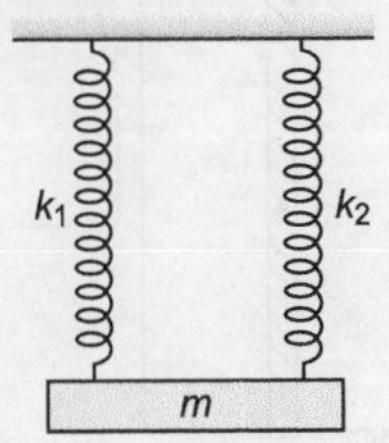

(a) $t_0^2 = t_1^2 + t_2^2$

(b) $t_0^{-2} = t_1^{-2} + t_2^{-2}$

(c) $t_0^{-1} = t_1^{-1} + t_2^{-1}$

(d) $t_0 = t_1 + t_2$

34. A body of mass 4.9 kg hangs from a spring and oscillates with a period 0.5 s on the removal of the body, the spring is shortened by (Take $g = 10\ ms^{-2}$, $\pi^2 = 10$)

(a) 6.3 m (b) 0.63 m

(c) 6.25 cm (d) 63.5 cm

35. A simple pendulum of length l and having a bob of mass M is suspended in a car. The car is moving on a circular track of radius R with a uniform speed v. If the pendulum makes small oscillations in a radial direction about its equilibrium position, what will be its time period?

(a) $2\pi\sqrt{\frac{3l}{g + \frac{v^2}{R}}}$ (b) $2\pi\sqrt{l\Big/\left(g^2 + \frac{v^4}{R^2}\right)}$

(c) $2\pi\sqrt{\frac{2l}{\sqrt{\left(g^2 + \frac{v^2}{R^2}\right)}}}$ (d) $2\pi\sqrt{\frac{2l}{(g^2 + v^2/R)}}$

36. When a body of mass 1.0 kg is suspended from a certain light spring hanging vertically, its length increases by 5 cm. By suspending 2.0 kg block to the spring and if the block is pulled through 10 cm and released, the maximum velocity of it, (in ms^{-1}) is (Take, $g = 10\ ms^{-2}$)

(a) 0.5 (b) 1

(c) 2 (d) 4

37. A simple pendulum of length 1 m is oscillating with an angular frequency 10 rad/s. The support of the pendulum starts oscillating up and down with a small angular frequency of 1 rad/s and an amplitude of 10^{-2} m. The relative change in the angular frequency of the pendulum is best given by **[JEE Main 2019]**

(a) 1 rad/s (b) 10^{-5} rad/s

(c) 10^{-3} rad/s (d) 10^{-1} rad/s

38. A particle performs simple harmonic motion with amplitude A. Its speed is trebled at the instant when it is at a distance $\frac{2}{3}A$ from equilibrium position. The new amplitude of the motion is **[JEE Main 2016]**

(a) $\frac{A}{3}\sqrt{41}$ (b) $3A$ (c) $A\sqrt{3}$ (d) $\frac{7}{3}A$

39. An ideal gas enclosed in a vertical cylindrical container supports a freely moving piston of mass M. The piston and the cylinder have equal cross-sectional area A. When the piston is in equilibrium, the volume of the gas is V_0 and its pressure is p_0. The piston is slightly displaced from the equilibrium position and released. Assuming that the system is completely isolated from its surrounding, the piston executes a simple harmonic motion with frequency [JEE Main 2013]

(a) $\frac{1}{2\pi}\frac{A\gamma p_0}{V_0 M}$

(b) $\frac{1}{2\pi}\frac{V_0 M p_0}{A^2\gamma}$

(c) $\frac{1}{2\pi}\sqrt{\frac{A^2\gamma p_0}{MV_0}}$

(d) $\frac{1}{2\pi}\sqrt{\frac{MV_0}{A\gamma p_0}}$

40. A block of mass 1 kg attached to a spring is made to oscillate with an initial amplitude of 12 cm. After 2 min, the amplitude decrease to 6 cm. Determine the value of the damping constant for this motion. [JEE Main 2021]

(a) $0.69\times10^2 \text{kg s}^{-1}$
(b) $3.3\times10^2 \text{kg s}^{-1}$
(c) $1.16\times10^2 \text{kg s}^{-1}$
(d) $5.7\times10^{-3} \text{kg s}^{-1}$

41. A mass M is attached to a horizontal spring of force constant k fixed on one side to a rigid support as shown in figure. The mass oscillates on a frictionless surface with time period T and amplitude A. When the mass is in equilibrium position, another mass m is gently placed on it. What will be the new amplitude of oscillations?

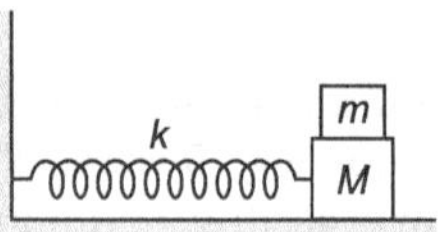

(a) $A\sqrt{\left(\frac{M}{M-m}\right)}$
(b) $A\sqrt{\left(\frac{M-m}{M}\right)}$
(c) $A\sqrt{\left(\frac{M}{M+m}\right)}$
(d) $A\sqrt{\left(\frac{M+m}{M}\right)}$

42. A bottle weighing 200 g and area of cross-section 50 cm² and height 4 cm oscillates on the surface of water in vertical position. Its frequency of oscillation is

(a) 1.5 Hz
(b) 2.5 Hz
(c) 3.5 Hz
(d) 4.5 Hz

43. One end of a U-tube containing mercury is connected to a suction pump and the other end to atmosphere. A small pressure difference is maintained between the two columns. The suction pump is removed, the column of mercury in the U-tube will show [NCERT]

(a) periodic motion
(b) oscillation
(c) simple harmonic motion
(d) None of the above

44. A spring balance has a scale that reads from 0 to 50 kg. The length of the scale is 20 cm. A body suspended from this balance, when displaced and released, oscillates with a period of 0.6 s. What is the weight of the body? [NCERT]

(a) 222.13 N
(b) 200.13 N
(c) 193.13 N
(d) 219.13 N

45. The displacement time graph of a particle executing SHM is shown in figure. Which of the following statement is/are true? [NCERT Exemplar]

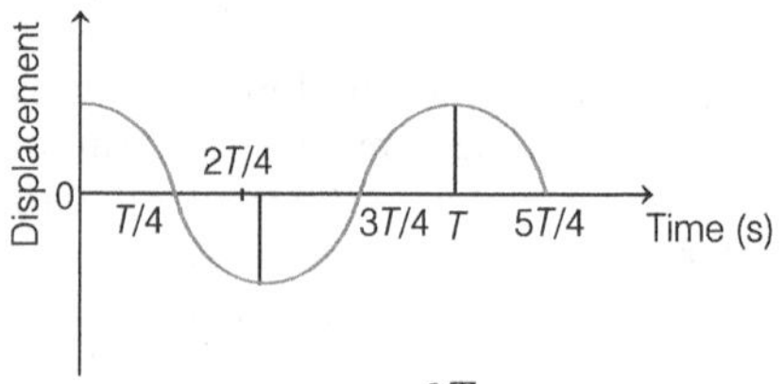

(a) The force is zero at $t=\frac{3T}{4}$.

(b) The acceleration is maximum at $t=\frac{3T}{4}$.

(c) The velocity is maximum at $t=\frac{T}{2}$.

(d) The PE is equal to KE of oscillation at $t=\frac{T}{2}$.

46. A pendulum is made to hang from a ceilling of an elevator.It has period of T_{sec}. (for small angles). The elevator is made to accelerate upwards with 10 m/s².The period of the pendulum now will be (assume g =10 m/s²)

(a) $T\sqrt{2}$
(b) infinite
(c) $T/\sqrt{2}$
(d) zero

47. Two light identical springs of spring constant k are attached horizontally at the two ends of an uniform horizontal rod AB of length l and mass m. The rod is pivoted at its centre 'O' and can rotate freely in horizontal plane. The other ends of the two springs are fixed to rigid supports as shown in figure.

The rod is gently pushed through a small angle and released. The frequency of resulting oscillation is

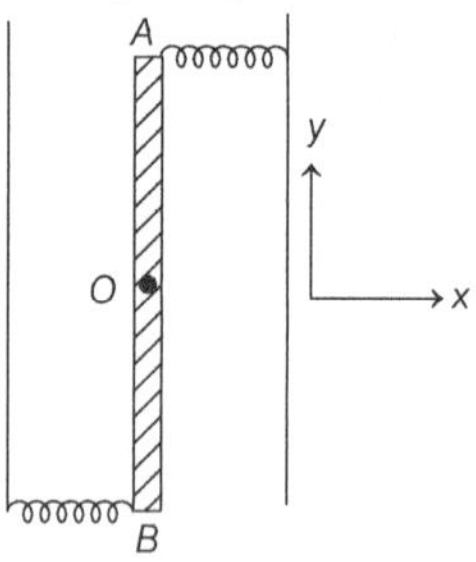

[JEE Main 2019]

(a) $\frac{1}{2\pi}\sqrt{\frac{2k}{m}}$
(b) $\frac{1}{2\pi}\sqrt{\frac{3k}{m}}$
(c) $\frac{1}{2\pi}\sqrt{\frac{6k}{m}}$
(d) $\frac{1}{2\pi}\sqrt{\frac{k}{m}}$

Numerical Value Questions

48. Two point masses of 3.0 kg and 1.0 kg are attached to opposite ends of a horizontal spring whose spring constant is 300 Nm^{-1} as shown in adjacent figure. The natural frequency (in Hz) of vibration of the system is ....

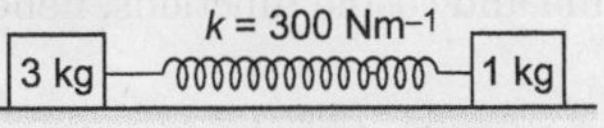

49. A particle performs simple harmonic motion with a period of 2 s. The time taken by the particle to cover a displacement equal to half of its amplitude from the mean position is $(1/a)$ s . The value of a to the nearest integer is .... **[JEE Main 2021]**

50. Two springs of force constants 20 N/m and 10 N/m, have equal highest velocities when executing SHM, then the ratio of their amplitudes (given their masses are equal) will be $\frac{1}{\sqrt{x}}$. Find the value of x.

51. A particle of mass 4 cm is executing oscillations about the origin on the X-axis. Its potential energy is $U(x) = k|x|^3$, where k is a positive constant. If the amplitude of oscillation is 4 cm, then its time period T will be proportional to ....

52. As shown in figure, system consisting of massless pulley, a light spring of force constant $k = 16$ N/m and a block of mass 0.01 kg. If the block is slightly displaced vertically downwards from its equilibrium position and released, then the frequency of vertical oscillations will be $\frac{n}{\pi}$ Hz. Find the value of n.

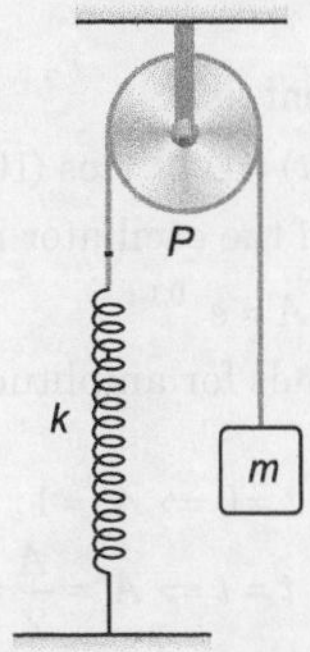

53. Consider two identical springs each of spring constant k and negligible mass compared to the mass M as shown in figure. Fig. 1 shows one of them and Fig. 2 shows their series combination. The ratios of time period of oscillation of the two SHM is $\frac{T_b}{T_a} = \sqrt{x}$, where value of x is .... (Rounded off to the nearest integer) **[JEE Main 2021]**

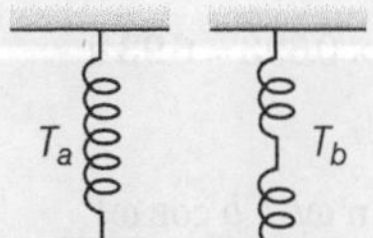

Fig. 1 Fig. 2

54. A particle is executing SHM of amplitude 25 cm and time period 3 s. The minimum time (in s) required for the particle to move between two points 12.5 cm on either side of the mean position will be .... s.

Answers

Round I

1. (d)	2. (d)	3. (b)	4. (b)	5. (a)	6. (c)	7. (c)	8. (c)	9. (b)	10. (a)
11. (b)	12. (d)	13. (b)	14. (b)	15. (a)	16. (d)	17. (b)	18. (d)	19. (c)	20. (c)
21. (c)	22. (b)	23. (a)	24. (b)	25. (c)	26. (d)	27. (b)	28. (d)	29. (a)	30. (c)
31. (d)	32. (c)	33. (a)	34. (b)	35. (b)	36. (d)	37. (c)	38. (d)	39. (c)	40. (b)
41. (a)	42. (b)	43. (a)	44. (d)	45. (b)	46. (c)	47. (d)	48. (a)	49. (b)	50. (b)
51. (c)	52. (b)	53. (c)	54. (a)	55. (b)	56. (b)	57. (a)	58. (c)	59. (b)	60. (b)
61. (d)	62. (b)	63. (a)	64. (b)	65. (a)	66. (d)	67. (b)	68. (d)	69. (b)	70. (b)
71. (c)	72. (a)	73. (a)	74. (d)						

Round II

1 (a)	2. (b)	3. (d)	4. (d)	5. (c)	6. (b)	7. (c)	8. (d)	9. (a)	10. (d)
11. (c)	12. (c)	13. (d)	14. (b)	15. (d)	16. (b)	17. (b)	18. (a)	19. (c)	20. (c)
21. (d)	22. (a)	23. (c)	24. (b)	25. (d)	26. (b)	27. (d)	28. (d)	29. (c)	30. (a)
31. (b)	32. (a)	33. (b)	34. (c)	35. (b)	36. (b)	37. (c)	38. (d)	39. (c)	40. (*)
41. (c)	42. (b)	43. (c)	44. (d)	45. (a)	46. (c)	47. (c)	48. 3	49. 6	50. 2
51. 0.5	52. 20	53. 2	54. 0.5						

Solutions

Round I

1. Given, displacement is

$$x(t) = e^{-0.1t} \cos(10\pi t + \phi)$$

Here, amplitude of the oscillator is

$$A = e^{-0.1t} \quad \dots (i)$$

Let it takes t seconds for amplitude to be dropped by half.

At $t = 0 \Rightarrow A = 1$ [from Eq. (i)]

At $t = t \Rightarrow A' = \frac{A}{2} = \frac{1}{2}$

So, Eq. (i) can be written as

$$e^{-0.1t} = \frac{1}{2}$$

or $e^{0.1t} = 2$

or $0.1\, t = \ln(2)$

or $t = \frac{1}{0.1} \ln(2) = 10 \ln(2)$

Now, $\ln(2) = 0.693$

$\therefore$ $t = 10 \times 0.693 = 6.93$ s

or $t \approx 7$ s

2. Given, $y = a \sin \omega t + b \cos \omega t$

Let $a = A \cos \theta$ and $b = A \sin \theta$ ...(i)

then $y = A \cos \theta \sin \omega t + A \sin \theta \cos \omega t$

$y = A \sin(\omega t + \theta)$

which is in the form of SHM.

From Eq. (i)

$$a^2 + b^2 = A^2 \cos^2 \theta + A^2 \sin^2 \theta$$

$\Rightarrow$ $A = \sqrt{a^2 + b^2}$

3. Given, $y = 3 \cos\left(\frac{\pi}{4} - 2\omega t\right)$ ... (i)

Velocity, $v = \frac{dy}{dt} = 3 \times 2\, \omega \sin\left(\frac{\pi}{4} - 2\, \omega t\right)$

Acceleration,

$$A = \frac{dv}{dt} = -4\, \omega^2 \times 3 \cos\left(\frac{\pi}{4} - 2\omega t\right) = -4\, \omega^2 y$$

As $A \propto y$ and $-$ ve sign shows that it is directed towards equilibrium (or mean position), hence particle will execute SHM. Comparing Eq. (i) with equation

$$y = r \cos(\phi - \omega' t)$$

we have, $\omega' = 2\, \omega$

or $\frac{2\pi}{T'} = 2\, \omega$

or $T' = \frac{\pi}{\omega}$

4. Given, $y = \sin^3 \omega t = \frac{1}{4}(3 \sin \omega t - \sin 3\, \omega t)$

As this motion is not represented by single harmonic function, hence it is not an SHM. As this motion involves sine and cosine functions, hence it is periodic motion.

5. It is required to calculate the time for extreme position.

Hence in this case, equation of displacement of particle can be written as $x = a \sin\left(\omega t + \frac{\pi}{2}\right) = a \cos \omega t$

$\Rightarrow$ $\frac{a}{2} = a \cos \omega t$ (as per question)

$\frac{1}{2} = \cos \omega t$

$\Rightarrow$ $\cos \frac{\pi}{3} = \cos \omega t$ $\left\{\because \cos \frac{\pi}{3} = \frac{1}{2}\right\}$

$\Rightarrow$ $\omega t = \frac{\pi}{3}$

$\Rightarrow$ $\frac{2\pi}{T} \cdot t = \frac{\pi}{3}$

$\Rightarrow$ $t = \frac{T}{6}$

6. Motion of an oscillating liquid column in a U-tube is SHM with period, $T = 2\pi \sqrt{\frac{l}{g}}$, where l is the height of liquid column in one arm of U-tube in equilibrium position of liquid. Therefore, T is independent of density of liquid.

7. $x = a \cos \omega t$ and $y = a \sin \omega t$

$\therefore$ $x^2 + y^2 = a^2(\cos^2 \omega t + \sin^2 \omega t) = a^2$

It is an equation of a circle. Thus, trajectory of motion of the particle will be a circle.

8. As, $d^2x/dt^2 = -kx$

and $T = 2\pi \sqrt{\frac{\text{displacemant}}{\text{acceleration}}}$

So, $T = 2\pi \sqrt{\frac{x}{kx}} = 2\pi \sqrt{\frac{1}{k}}$

9. Weight kept on the system will separate from the piston, when the maximum force just exceeds the weight of the body. Hence,

$$m\omega^2 y = mg$$

or $y = g/\omega^2 = 9.8/(2\pi)^2 = 0.25$ m

10. Given, $T = 30$ s, $OQ = B$. The projection of the radius vector on the diameter of the circle when a particle is moving with uniform angular velocity (ω) on a circle of reference is SHM. Let the particle move from P to Q in time t, then $\angle POQ = \omega t = \angle OQP$.

The projection of radius OQ on X-axis will be $OR = x(t)$ say.

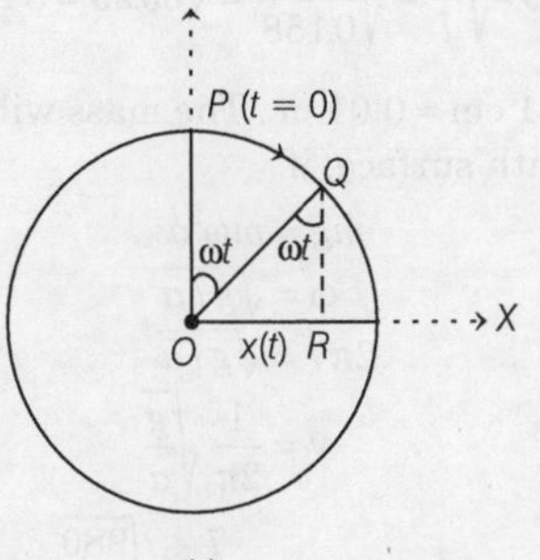

In ΔOQR, $\sin \omega t = \dfrac{x(t)}{B}$

or $x(t) = B\sin \omega t = B\sin \dfrac{2\pi}{T} t = B\sin \dfrac{2\pi}{30} t$

11. When bigger pendulum of time period $(5T/4)$ completes one vibrations, the smaller pendulum will complete $(5/4)$ vibrations. It means the smaller pendulum will be leading the bigger pendulum by phase $T/4$ sec $= \dfrac{\pi}{2}$ rad $= 90°$.

12. We can find the velocities, $v_1 = \dfrac{dy_1}{dt} = 2\times 10\cos(10t+\theta)$

and $v_2 = -3\times 10\sin 10t = 30\cos(10t + \pi/2)$

$\therefore$ Phase difference $= (10t+\theta) - (10t+\pi/2) = \theta - \pi/2$

13. In simple harmonic motion, position (x), velocity (v) and acceleration (a) of the particle are given by

$$x = A\sin \omega t$$
$$v = \omega\sqrt{A^2 - x^2}$$

or $\quad v = A\omega\cos\omega t$

and $\quad a = -\omega^2 x$

or $\quad a = -\omega^2 A\sin\omega t$

Given, amplitude $A = 5$ cm and displacement $x = 4$ cm. At this time (when $x = 4$ cm), velocity and acceleration have same magnitude.

$\Rightarrow \quad |v_{x=4}| = |a_{x=4}|$

or $\quad |\omega\sqrt{5^2 - 4^2}| = |-4\omega^2|$

$\Rightarrow \quad 3\omega = +4\omega^2$

$\Rightarrow \quad \omega = (3/4)$ rad/s

So, time period, $T = \dfrac{2\pi}{\omega}$

$\Rightarrow \quad T = \dfrac{2\pi}{3}\times 4 = \dfrac{8\pi}{3}$ s

14. The time periods, $T_1 = 2\pi\sqrt{\dfrac{121}{g}}$ and $T_2 = 2\pi\sqrt{\dfrac{100}{g}}$

So, $T_1 > T_2$, Let the shorter pendulum makes n vibrations, then the longer pendulum will make less than n vibrations to come in phase again.

So, $\quad nT_2 = (n-1)T_1$

or $\quad n\times 2\pi\sqrt{\dfrac{100}{g}} = (n-1)\times 2\pi\sqrt{\dfrac{121}{g}}$

or $\quad 10n = (n-1)11$

or $\quad n = 11$

15. Given, angular frequency of the piston, $\omega = 200$ rad/min and stroke length = 1 m

$\therefore$ Amplitude of SHM, $A = \dfrac{\text{Stroke length}}{2} = \dfrac{1}{2} = 0.5$ m

Now, $\quad v_{max} = \omega A$

$= 200\times 0.5 = 100$ m/min

16. As, $\quad$ Wavelength = Velocity of wave × Time period

i.e. $\quad \lambda = 300\times 0.05 = 15$ m

According to the problem path difference between two points

$= 15 - 10 = 5$ m

$\therefore$ Phase difference $= \dfrac{2\pi}{\lambda}\times$ path difference

$$\Delta\phi = \frac{2\pi}{\lambda}\times\Delta x = \frac{2\pi}{15}\times 5 = \frac{2\pi}{3}$$

17. At one of the extreme position, weight of block = restoring force. At the other extreme position where it is momentary at rest, weight of block and restoring force both act in downward direction. So the block weighs double than its weight.

18. As, $F_1 = \dfrac{m4\pi^2 a}{T_1^2}$ and $F_2 = \dfrac{m4\pi^2 a}{T_2^2}$

Net force, $\quad F = F_1 + F_2$

$$= \frac{4\pi^2 ma}{T_1^2} + \frac{4\pi^2 ma}{T_2^2}$$

$$= 4\pi^2 ma\left(\frac{1}{T_1^2} + \frac{1}{T_2^2}\right)$$

or $\quad \dfrac{4\pi^2 ma}{T^2} = 4\pi^2 ma\left(\dfrac{1}{T_1^2} + \dfrac{1}{T_2^2}\right)$

or $\quad \dfrac{1}{T^2} = \dfrac{1}{T_1^2} + \dfrac{1}{T_2^2}$

or $\quad \dfrac{1}{T^2} = \dfrac{T_1^2 + T_2^2}{T_1^2 T_2^2}$

or $\quad T^2 = \dfrac{T_1^2 T_2^2}{T_1^2 + T_2^2}$

19. As, $\quad x = A\sin(\omega + \pi/2) = A\cos\omega t$

$\therefore \quad \cos\omega t = x/A$

and $\quad \sin\omega t = \sqrt{1 - (x^2/A^2)}$

$$y = A\sin 2\omega t = \sqrt{1 - (x^2/A^2)}$$
$$y = A\sin 2\omega t = 2A\sin\omega t\cos\omega t$$

or $y^2 = 4A^2 \sin^2 \omega t \cos^2 \omega t$

$$= 4A^2 \times \frac{x^2}{A^2} \times \left(\frac{A^2 - x^2}{A^2}\right)$$

$$= 4x^2\left(1 - \frac{x^2}{A^2}\right)$$

20. Let, O be the position and x be the distance of coin from O. The coin will slip if pseudo force on coin just becomes equal to force of friction, *i.e.* $mx\omega^2 = \mu mg$.

The coin will slip if, x = maximum = amplitude A

$$mA\omega^2 = \mu mg$$

or $A = \mu g / \omega^2$

21. As, $x(t) = A\cos(\omega t + \phi)$ ...(i)

$\therefore$ $1 = A\cos(\pi \times 0 + \phi) = A\cos\phi$ ...(ii)

velocity $= \frac{d[x(t)]}{dt} = -A\omega \sin(\omega t + \phi)$

$$\pi = -A \times \pi \sin(0 + \phi) = -\pi A \sin\phi - 1$$

$= A\sin\phi$ ...(ii)

Squaring and adding Eqs. (ii) and (iii), we have

$$1 + 1 = A^2(\cos^2\omega + \sin^2\omega) = A^2$$

or $A = \sqrt{2}$ cm

22. In equilibrium condition bottle floats in water and its length l inside water is same as the height of water upto which bottle is filled.

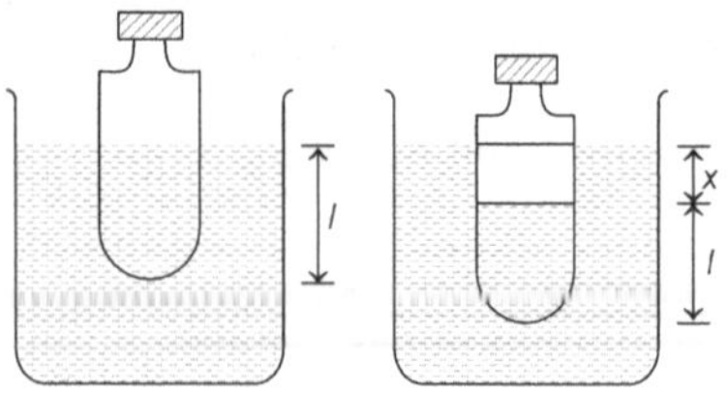

So, l = Volume of water in bottle/Area

$$= \frac{310}{\pi \times (2.5)^2} = 15.8 \text{ cm} = 0.158 \text{ m}$$

When bottle is slightly pushed inside by an amount x then, restoring force acting on the bottle is the upthrust of fluid displaced when bottle goes into liquid by amount x.

So, restoring force,

$F = -(\rho Ax)g$ ...(i)

where ρ = density of water,

A = area of cross-section of bottle

and x = displacement from equilibrium position.

But $F = ma$ ...(ii)

where, m = mass of water and bottle system $- Al\rho$

From Eqs. (i) and (ii), we have

$$Al\rho a = -\rho Axg$$

or $a = -\frac{g}{l}x$

As for SHM, $a = -\omega^2 x$

We have $\omega = \sqrt{\frac{g}{l}} = \sqrt{\frac{10}{0.158}} = \sqrt{63.29} \approx 8 \text{ rad s}^{-1}$

23. Here, $a = 1$ cm $= 0.01$ m. The mass will remain in contact with surface, if

$$mg = m\omega^2 a$$

or $\omega = \sqrt{g/a}$

or $2\pi\nu = \sqrt{g/a}$

or $\nu = \frac{1}{2\pi}\sqrt{\frac{g}{a}}$

$$= \frac{7}{2 \times 22}\sqrt{\frac{980}{1}}$$

$$= 4.9 \text{ Hz} \approx 5 \text{ Hz}$$

24. Maximum force on body while in SHM

$$= m\omega^2 a = 0.5\,(2\pi/2)^2 \times 0.2 = 1 \text{ N}$$

Maximum force of friction $= \mu mg = 0.3 \times 0.5 \times 10 = 1.5$ N

Since, the maximum force on the body due to SHM of the platform is less than the maximum possible frictional force, so the maximum force of friction will be equal to the maximum force acting on body due to SHM of platform, *i.e.* 1 N.

25. As, $x = a\sin\omega t$

and $y = b\sin(\omega t + \pi) = -b\sin\omega t$

$\Rightarrow \frac{x}{a} = -\frac{y}{b} \Rightarrow y = -\frac{b}{a}x$

It is an equation of a straight line.

26. Time period, $T = 2\pi\sqrt{\frac{L}{g}}$ and $\omega = \frac{2\pi}{T} = \sqrt{\frac{g}{L}}$

$\therefore$ Displacement, $x = a\cos\omega t = a\cos\sqrt{\frac{g}{L}}\,t$

27. Given, $x = 3\sin\omega t + 4\sin(\omega t + \pi/3)$

Comparing it with the equation

$$x = r_1 \sin\omega t + r_2 \sin(\omega t + \phi)$$

We have, $r_1 = 3$ cm, $r_2 = 4$ cm and $\phi = \pi/3$

The amplitude of combination is

$$r = \sqrt{r_1^2 + r_2^2 + 2r_1 r_2 \cos\phi}$$

$$= \sqrt{3^2 + 4^2 + 2 \times 3 \times 4 \times \cos\pi/3}$$

$$= \sqrt{37} \approx 6 \text{ cm}$$

28. As, $\sec bt$ is not define for $bt = \pi/2$

and $x = a\sec bt + c\,\text{cosec}\,bt = \frac{a\sin bt + c\cos bt}{\sin bt\cos bt}$

This equation cannot be modified in the form of simple equation of SHM containing sine or cosine function.

So, it cannot represent SHM.

29. When particle is at $x = 2$, the displacement is $y = 4 - 2 = 2$ cm. If t is the time taken by the particle to go from $x = 4$ cm to $x = 2$ cm, then

$$y = a\cos\omega t = a\cos\frac{2\pi t}{T} = a\cos\frac{2\pi t}{1.2}$$

or $\cos\frac{2\pi t}{1.2} = \frac{y}{a} = \frac{2}{4}$

$\cos\frac{2\pi t}{1.2} = \frac{1}{2} = \cos\frac{\pi}{3}$

$\Rightarrow \frac{2\pi t}{1.2} = \frac{\pi}{3}$

or $\frac{2t}{1.2} = \frac{1}{3}$

or $t = \frac{1.2}{6} = 0.2$ s

Time taken to move from $x = +2$ cm to $x = +4$ cm and back again $= 2t = 2\times 0.2$ s $= 0.4$ s.

30. Potential energy, $V = \frac{1}{2}m\omega^2x^2$

and kinetic energy $E, T = \frac{1}{2}m\omega^2(a^2 - x^2)$

$\therefore \frac{T}{V} = \frac{a^2 - x^2}{x^2}$

31. Let x be point, where KE = PE

Hence, $\frac{1}{2}m\omega^2(a^2 - x^2) = \frac{1}{2}m\omega^2x^2$

$\Rightarrow 2x^2 = a^2$

$\Rightarrow x = \frac{a}{\sqrt{2}} = \frac{y}{\sqrt{2}} = 2\sqrt{2}$ cm

32. As, $\frac{U}{U_{\max}} = \frac{\frac{1}{2}m\omega^2y^2}{\frac{1}{2}m\omega^2a^2} = \frac{1}{4}$

$\frac{y^2}{a^2} = \frac{1}{4} \Rightarrow y = \frac{a}{2}$

33. Potential energy of body in SHM at an instant,

$U_x = \frac{1}{2}ky^2$

If the displacement, $y = (a - x)$, then

$U_x = \frac{1}{2}k(a - x)^2 = \frac{1}{2}k(x - a)^2$

34. Total energy of oscillation, $E = \alpha A^4$

(at maximum displacement)

Kinetic energy of mass at $x = x$ is

$K = E - U = \alpha(A^4 - x^4)$

As $K = 3U$

$\alpha(A^4 - x^4) = 3\alpha x^4$

or $x = \pm\frac{A}{\sqrt{2}}$

35. Total energy, $E = \frac{1}{2}m\omega^2a^2$ ...(i)

$\text{KE} = \frac{3E}{4} = \frac{1}{2}m\omega^2(a^2 - y^2)$ ...(ii)

From Eqs. (i) and (ii), we get

$\frac{3}{4} = \frac{a^2 - y^2}{a^2}$

or $y^2 = a^2/4$

or $y = a/2$

36. Kinetic energy at mean position $= \frac{1}{2}m\omega^2a^2 = 8\times10^{-3}$

or $\omega = \left(\frac{2\times8\times10^{-3}}{ma^2}\right)^{1/2} = \left[\frac{2\times8\times10^{-3}}{0.1\times(0.1)^2}\right]^{1/2} = 4$ rad/s

Equation of SHM is,

$y = a\sin(\omega t + \theta) = 0.1\sin\left(4t + \frac{\pi}{4}\right)$

37. From the figure, $OC = l\cos\theta$

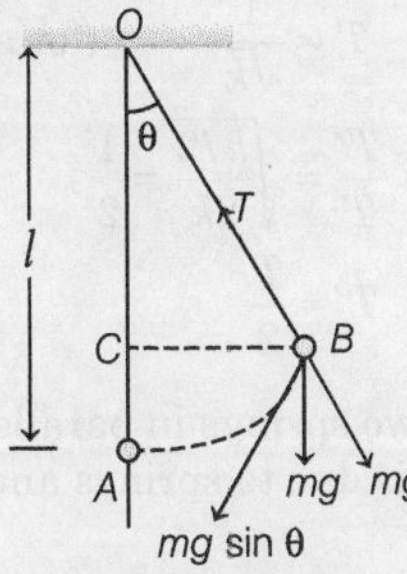

$\therefore AC = OA - OC$

$= l - l\cos\theta$

$= l(1 - \cos\theta)$

Maximum KE of bob at O or A = Maximum PE of bob at B

$= ma\times AC = mgl(1 - \cos\theta)$

38. If m is the mass and r is the amplitude of oscillation, then maximum kinetic energy,

$K_0 = \frac{1}{2}m\omega^2r^2$

or $r = \left(\frac{2K_0}{m\omega^2}\right)^{1/2}$

The displacement equation can be

$y = r\sin\omega t = \left(\frac{2K_0}{m\omega}\right)^{1/2}\sin\omega t$

39. As, $n = \frac{1}{2\pi}\sqrt{\frac{k_{\text{effective}}}{m}} = \frac{1}{2\pi}\sqrt{\frac{k + 2k}{m}} = \frac{1}{2\pi}\sqrt{\frac{3k}{m}}$

(for parallel combination of spring, $k_{eq} = k_1 + k_2$)

40. Let k be the force constant of spring of length l_2. Since, $l_1 = nl_2$, where n is an integer, so the spring is made of $(n+1)$ equal parts in length, each of length l_2.

$\therefore \frac{1}{k} = \frac{(n+1)}{k}$

or $k = (n+1)k$

The spring of length $l_1 = n\,l_2$ will be equivalent to n spring connected in series where spring constant

$k' = \frac{k}{n} = \frac{(n+1)k}{n}$

41. We have, $n = \frac{1}{2\pi}\sqrt{\frac{k}{m}}$;

$$n' = \frac{1}{2\pi}\sqrt{\frac{k'}{2m}} = \frac{1}{2\pi}\sqrt{\frac{2k}{2m}} = n \qquad (\because k' = 2k)$$

42. Let k be the spring constant of each half part of the spring. For a complete spring, the spring constant $k' = k/2$ (spring in series). When two splitted parts of a spring are connected to the body, then spring are in parallel. Their effective spring constant, $k' = k + k = 2k$.

As $$T = 2\pi\sqrt{\frac{m}{k}}$$

or $$T \propto \frac{1}{\sqrt{k}} \qquad \text{(for a fixed value of } m)$$

$\therefore$ $$\frac{T'}{T} = \sqrt{\frac{k/2}{2k}} = \frac{1}{2}$$

or $$T' = \frac{T}{2}$$

43. It is a system of two springs in parallel. The restoring force on the body is due to springs and not due to gravity.

Therefore slope is irrelevant.
Here, the effective spring constant $= k + k = 2k$
Thus time period, $T = 2\pi\sqrt{M/2k}$

44. As, $T = 2\pi\sqrt{\frac{m}{k}}$

and $$T' = 2\pi\sqrt{\frac{2m}{2k}} = 2\pi\sqrt{\frac{m}{k}} = T$$

45. Two spring, each of spring constant k_1 in parallel, give equivalent spring constant of $2k_1$ and this is in series with another spring of spring constant k_2, so equivalent spring constant

$$k = \left(\frac{1}{k_2} + \frac{1}{2k_1}\right)^{-1}$$

46. Initially, it is given that the unstretched length of the spring is l. When it is given an angular speed ω, then let x = stretched length of spring.

Then, total length of the spring system while rotating will be $(l + x)$ as shown in the figure.

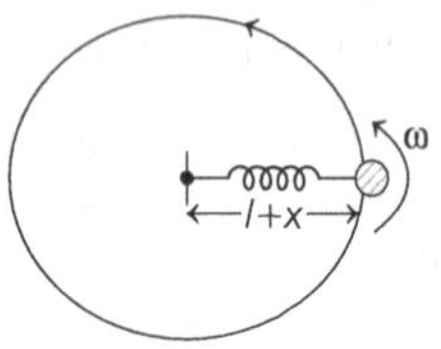

As we know, spring force will give the necessary centripetal force for rotation.

So, $$kx = m(l + x)\omega^2$$

$\Rightarrow$ $$(k - m\omega^2)x = ml\omega^2$$

$\Rightarrow$ $$x = \frac{ml\omega^2}{(k - m\omega^2)}$$

47. The given situation is shown in the figure given below.

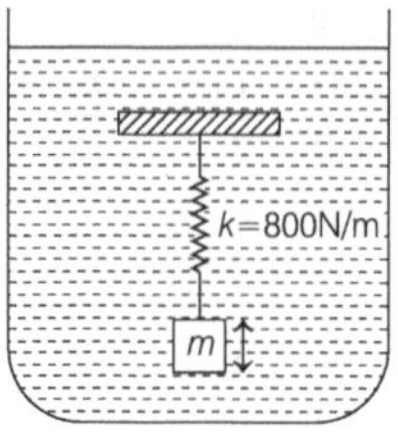

When vibrations of mass are suddenly stopped, oscillation energy (or stored energy of spring) is dissipated as heat, causing rise of temperature.

So, conversation of energy gives

$$\frac{1}{2}kx_m^2 = (m_1 s_1 + m_2 s_2)\Delta T$$

where, x_m = amplitude of oscillation,
s_1 = specific heat of mass,
s_2 = specific heat of water
and ΔT = rise in temperature.

Substituting values given in question, we have

$$\frac{1}{2} \times 800 \times (2 \times 10^{-2})^2 = \left[\left(\frac{500}{1000}\right) \times 400 + 1 \times 4184\right]\Delta T$$

$\Rightarrow$ $$\Delta T = \frac{16 \times 10^{-2}}{4384}$$

$$= 3.65 \times 10^{-5} \text{ K}$$

So, the order of magnitude of change in temperature is approx 10^{-5}.

48. As, $T = 2 = 2\pi\sqrt{\frac{M}{k}}$

and $$2 + 1 = 2\pi\sqrt{\frac{M+4}{h}} \qquad \text{(from question)}$$

or $$3 = 2\pi\sqrt{\frac{k+4}{k}}$$

So, $$\frac{4}{9} = \frac{M}{M+4}$$

or $$4M + 16 = 9M$$

or $$M = \frac{16}{5} = 3.2 \text{ kg}$$

49. A total restoring force, $F = kX = mg$

or $$k = mg/X$$

$\therefore$ $$T = 2\pi\sqrt{\frac{(M+m)}{mg/X}} = 2\pi\sqrt{\frac{(M+m)X}{mg}}$$

50. The amplitude of oscillations will be the maximum when compression in the spring is maximum. At the time of maximum compression, velocities of both the blocks are equal say v, then using law of conservation of momentum,

$$m_1 v_0 = (m_1 + m_2)v$$

or $$1 \times 12 = (1 + 2)v$$

or $$v = 4 \text{ cms}^{-1}$$

Using law of conservation of energy, we have

$$\frac{1}{2}m_1v_0^2 = \frac{1}{2}kx^2 + \frac{1}{2}(m_1+m_2)v^2$$

Putting the value of $v = 4\ \text{cms}^{-1}$ and solving, we get $x = 2$ cm.

51. As, energy stored = work done

$\Rightarrow \quad E = \frac{1}{2}kr^2$ (where, r = displacement)

or $\quad r = \sqrt{\frac{2E}{k}} = \sqrt{\frac{2\times 2}{400}} = \frac{1}{10}\text{m}$

Now, $\quad a = \omega^2 r = \left(\sqrt{\frac{k}{m}}\right)^2 \times \frac{1}{10}$

$$= \left(\frac{400}{1}\right)\times\frac{1}{10} = 40\ \text{ms}^{-2}$$

52.

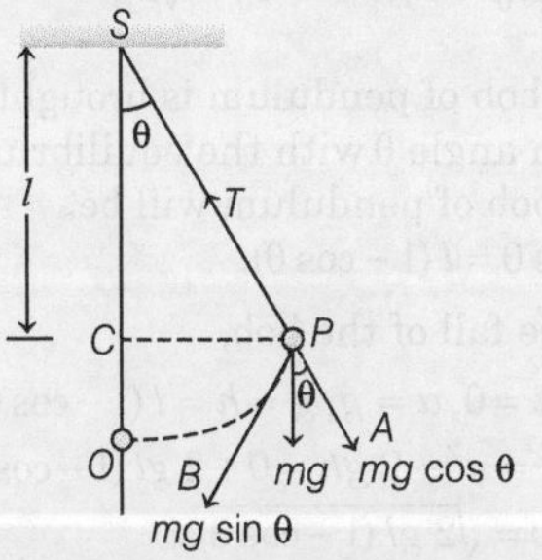

When the bob is displaced to position P, through a small angle θ from the vertical, the various forces acting on the bob at P are

(i) the weight mg of the bob acting vertically downwards

(ii) the tension T in the string acting along PS

Resolving mg into two rectangular components, we get

(a) $mg\cos\theta$ acts along PA, opposite to tension,

(b) $mg\sin\theta$ acts along PB, tangent to the arc OP and directed towards O.

If the string neither slackens nor breaks but remains taut, then

$$T = mg\cos\theta$$

The force $mg\sin\theta$ tends to bring the bob back to its mean position O.

$\therefore$ Restoring force acting on the bob is

$$F = -mg\sin\theta$$

53. As, $T = 2\pi\sqrt{\frac{l}{g}}$. When lift is accelerated upwards with acceleration a $(= g/4)$, the effective acceleration due to gravity inside the lift,

$$g_1 = g + a = g + \frac{g}{4} = \frac{5g}{4}$$

$\therefore \quad T_1 = 2\pi\sqrt{\frac{l}{5g/4}}$

$$= 2\pi\sqrt{\frac{l}{g}}\times\frac{2}{\sqrt{5}} = \frac{2T}{\sqrt{5}}$$

54. On the inclined plane, the effective acceleration due to gravity

$$g' = g\cos 30° = g\times\sqrt{3}/2$$

$$T = 2\pi\sqrt{\frac{l}{g'}} = 2\pi\sqrt{\frac{2l}{\sqrt{3}g}}$$

55. As, $T = 2\pi\sqrt{l/g}$

$\Rightarrow \quad \log T = \log 2 + \log\pi + \frac{1}{2}\log l - \frac{1}{2}\log g$

Differentiating it, we get

$$\frac{dT}{T} = \frac{1}{2}\frac{dl}{l} - \frac{1}{2}\frac{dg}{g} = -\frac{1}{2}\frac{dg}{g} \quad (\because l \text{ is constant})$$

% change in time period

$$= \frac{dT}{T}\times 100 = \frac{1}{2}\frac{dg}{g}\times 100$$

$$= -\frac{1}{2}\left(\frac{-2}{100}\right)\times 100 = 1\% \quad \text{(increase)}$$

56. Resolving tension T in string into two rectangular components, we get

$$T\cos\theta = mg$$

and $\quad T\sin\theta = mr\omega^2$

So, $\quad \frac{T\sin\theta}{T\cos\theta} = \tan\theta = \frac{r\omega^2}{g}$

or $\quad g\tan\theta = r\omega^2 = r\times 4\pi^2/T^2$

or $\quad T = 2\pi\sqrt{\frac{r}{g\tan\theta}} = 2\pi\sqrt{\frac{l\sin\theta}{g\tan\theta}}$

$$= 2\pi\sqrt{\frac{l\cos\theta}{g}}$$

57. Let I_1 be the moment of inertia in case (i) and I_2 be the moment of inertia in case (ii).

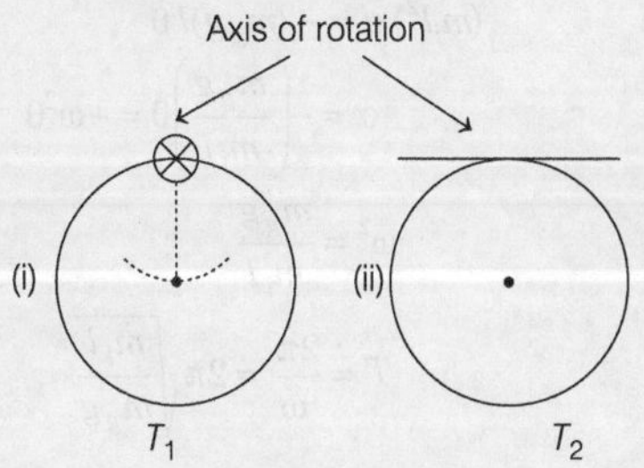

The time period of physical pendulum is given by

$$T = 2\pi\sqrt{\frac{I}{mgd}} \quad \text{(where, } d = R)$$

Here, $\quad I_1 = 2mR^2 \Rightarrow I_2 = \frac{3}{2}mR^2$

$\therefore \quad T_1 = 2\pi\sqrt{\frac{I_1}{mgd}}$

and $\quad T_2 = 2\pi\sqrt{\frac{I_2}{mgd}}$

$\Rightarrow \quad \frac{T_1}{T_2} = \sqrt{\frac{I_1}{I_2}} = \sqrt{\frac{2mR^2}{\frac{3}{2}mR^2}} = \frac{2}{\sqrt{3}}$

58. Period of motion of a pendulum is given by

$$T = 2\pi\sqrt{\frac{l}{g}} \quad \ldots\text{(i)}$$

On the surface of Earth, let period of motion is T_e and acceleration due to gravity is g_e

$$\therefore \quad T_e = 2\pi\sqrt{\frac{l}{g_e}} \quad \ldots\text{(ii)}$$

On the another planet, let period of motion is T_p and gravitational acceleration is g_p

$$\therefore \quad T_p = 2\pi\sqrt{\frac{l}{g_p}} \quad \ldots\text{(iii)}$$

($\because$ Pendulum is same, so l will be same)

From Eqs. (ii) and (iii), we get

$$\frac{T_e}{T_p} = \frac{2\pi\sqrt{\frac{l}{g_e}}}{2\pi\sqrt{\frac{l}{g_p}}} = \sqrt{\frac{g_p}{g_e}} \quad \ldots\text{(iv)}$$

Now, $g_e = \dfrac{GM_e}{R_e^2}$ and $g_p = \dfrac{GM_p}{R_p^2}$

Given, $M_p = 3M_e$ and $R_p = 3R_e$

$$\therefore \quad g_p = \frac{G \times 3M_e}{9R_e^2} = \frac{1}{3} \cdot \frac{GM_e}{R_e^2} = \frac{1}{3} g_e$$

$$\Rightarrow \quad \frac{g_p}{g_e} = \frac{1}{3} \text{ or } \sqrt{\frac{g_p}{g_e}} = \frac{1}{\sqrt{3}} \quad \ldots\text{(v)}$$

From Eqs. (iv) and (v), $T_p = \sqrt{3}\, T_e$

or $T_p = 2\sqrt{3}$ s ($\because T_e = 2$ s)

59. Torque acting on the bob $= I\alpha = -(mg) l \sin\theta$

or $(m_i l^2)\alpha = -(m_g g) l\theta$

or $\alpha = -\left(\dfrac{m_g g}{m_i l}\right)\theta = -\omega^2\theta$

where, $\omega^2 = \dfrac{m_g g}{m_i l}$

$$\therefore \quad T = \frac{2\pi}{\omega} = 2\pi\sqrt{\frac{m_i l}{m_g g}}$$

60. Since length of pendulums A and C is same and $T = 2\pi\sqrt{l/g}$, hence their time period is same and they will have same frequency of vibration. Due to it, a resonance will take place and the pendulum C will vibrate with maximum amplitude.

61. According to question,

$$T' = T\sqrt{\frac{l + 2l/100}{l}}$$

$$= T\left(1 + \frac{2}{100}\right)^{1/2} = 2\left(1 + \frac{1}{100}\right)$$

[$\because$ using binomial theorem, $(1+x)^n = 1 + nx$

$\Rightarrow \left(1 + \dfrac{2}{100}\right)^{1/2} = \left(1 + \dfrac{1}{100}\right)$]

$$\therefore \quad T' - T = \frac{2}{100} = \frac{1}{50}\text{ s}$$

Therefore, loss in seconds per day

$$= \frac{1/50}{2} \times 24 \times 60 \times 60 = 864\text{ s}$$

62. According to the law of conservation of mechanical energy, we get

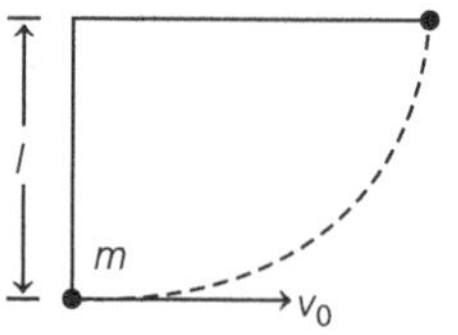

$$\frac{1}{2}mv_0^2 = mgl \Rightarrow v_0 = \sqrt{2gl}$$

63. When the bob of pendulum is brought to a position making an angle θ with the equilibrium position, the height of bob of pendulum will be,
$h = l - l\cos\theta = l(1 - \cos\theta)$.

Taking free fall of the bob,

$$u = 0, a = g, g = h = l(1-\cos\theta), v = ?$$

Now, $v^2 = u^2 + 2gh = 0 + 2gl(1 - \cos\theta) f$

or $v = \sqrt{2gl(1-\cos\theta)}$

64. When the ball of mass m falls from a height h, it reaches the surface of Earth in time, $t = \sqrt{2h/g}$ with velocity $v = \sqrt{2gh}$. It then moves into the tunnel and reaches upto a height h from other side of Earth. The ball then returns back and thus executes periodic motion. Outside the Earth, ball crosses distance h four times. When the ball is in the tunnel at distance x from the centre of the Earth, then gravitational force acting on ball is

$$F = \frac{Gm}{x^2} \times \left(\frac{4}{3}\pi x^3 \rho\right) = G \times \left(\frac{4}{3}\pi\rho\right) mx$$

Mass of the Earth, $M = \dfrac{4}{3}\pi R^3 \rho$

or $\dfrac{4}{3}\pi\rho = \dfrac{M}{R^3}$

$$\therefore \quad F = \frac{GMmx}{R^3}$$

i.e. $F \propto x$

As, this force, F is directed towards the centre of Earth, *i.e.* the mean position. So, the ball will execute periodic motion about the centre of Earth.

Here, Inertia factor = mass of ball = m

Spring factor $= \dfrac{GMm}{R^3} = \dfrac{gm}{R}$

$\therefore$ Time period of oscillation of ball in the tunnel is

$$T' = 2\pi\sqrt{\frac{\text{inertia factor}}{\text{spring factor}}}$$

$$= 2\pi\sqrt{\frac{m}{gm/R}} = 2\pi\sqrt{\frac{R}{g}}$$

Time spent by ball outside the tunnel on both the sides will be $= 4\sqrt{2h/g}$.

Therefore, total time period of oscillation of ball is

$$= 2\pi\sqrt{\frac{R}{g}} + 4\sqrt{\frac{2h}{g}}$$

65. Let T_1 and T_2 be the time period of shorter length and longer length pendulums respectively. As per question, $nT_1 = (n-1)T_2$

So $\quad n\,2\pi\sqrt{\frac{0.5}{g}} = (n-1)2\pi\sqrt{\frac{20}{g}}$

or $\quad n = (n-1)\sqrt{40} \approx (n-1)6$

Hence, $\quad 5n = 6$

Hence, after 5 oscillations they will be in same phase.

66. In damped oscillation, amplitude goes on decaying exponentially,

$$a = a_0\, e^{-bt}$$

where, b = damping coefficient initially

$$\frac{a_0}{3} = a_0 e^{-b\times 100T},$$

T = time of one oscillation

or $\quad \frac{1}{3} = e^{-100bT} \quad$ ...(i)

Finally, $\quad a = a_0 e^{-b\times 200T}$

or $\quad a = a_0 (e^{-100bt})^2$

or $\quad a = a_0 \times \left(\frac{1}{3}\right)^2 \quad$ [$\because$ from Eq. (i)]

$$a = \frac{a_0}{9}$$

67. The motion of M is SHM, with length, $l = \sqrt{L^2 - d^2}$

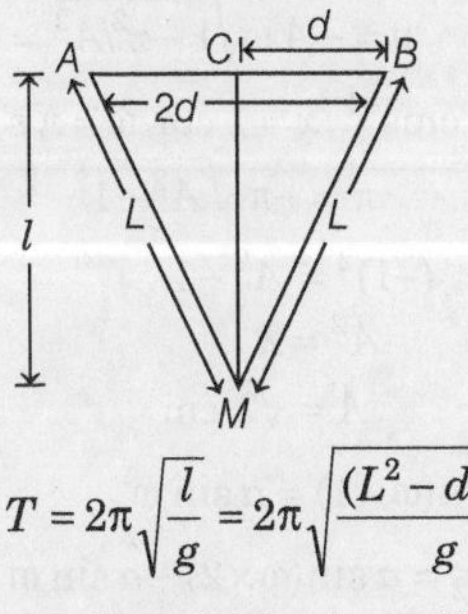

$\therefore$ Time period, $T = 2\pi\sqrt{\frac{l}{g}} = 2\pi\sqrt{\frac{(L^2-d^2)^{1/2}}{g}}$

68. Let the distance x of vertical side c of block be pushed in liquid, when block is floating, the buoyancy force

$= abxd\omega\, g = abxg \quad (\because d\omega = 1)$

The mass of piece of wood $= abcd$

So, $\quad$ acceleration $= -abxg / abcd = -\left(\frac{g}{cd}\right)x$

Hence, time period, $\quad T = 2\pi\sqrt{\frac{dc}{g}}$

69. We have, $T' = 2\pi\sqrt{l/(g/6)} = \sqrt{6}T$

Hence, the clock will tick in one minute,

$$n = 60/\sqrt{6} = 24.5 \text{ times}$$

70. When the cylinder is given a small downward displacement, say y, the additional restoring force is due to (i) additional extension y, which is, $F_1 = ky$ (ii) additional buoyancy which is $F_2 = AYd\,g$.

Total restoring force,

$$-F = F_1 + F_2 = (k + Adg)Y$$

= new force constant

$\therefore \quad n = \frac{1}{2\pi}\sqrt{\frac{k'}{M}}$

$$= \frac{1}{2\pi}\sqrt{\frac{k + Adg}{M}}$$

71. Amplitude resonance takes place at a frequency of external force which is less than the frequency of undamped maximum vibration, i.e. $\omega_1 \neq \omega_0$. Velocity-resonance takes place (*i.e.* maximum energy), when frequency of external periodic force is equal to natural frequency of undamped vibrations, i.e. $\omega_2 = \omega_0$.

72. Fig. (1) alone represents damped SHM as displacement is decreasing regularly with time.

73. Motion given here is SHM starting from rest. Therefore, the graph shown in option (a) is correct.

74. For damped motion, $a = a_0 e^{-bt}$

For first case,

$$\frac{a_0}{3} = a_0 e^{-b\times 2} \quad \text{or} \quad \frac{1}{3} = e^{-2b}$$

For second case, $\quad \frac{a_0}{n} = a_0 e^{-(b\times 6)}$

or $\quad \frac{1}{n} = e^{-6b} = (e^{-2b})^3 = \left(\frac{1}{3}\right)^3 \Rightarrow n = 3^3$

Round II

1. A ball bearing when released a little above the lower limit inside a smooth curved bowl, will execute SHM with a definite period.

2. As, $\quad T = 2\pi\sqrt{\frac{m}{k}}$

and $\quad T' = 2\pi\sqrt{\frac{m}{4k}} = \frac{T}{2} = \frac{5}{2}\text{ s} = 2.5\text{ s}$

3. Phase is the state of a particle as regards with its position and direction of motion w.r.t. mean position. In the given curve phase is same when $t = 1$ s and $t = 5$ s. Also, phase is same when $t = 2$ s and $t = 6$ s.

4. Given, $\quad \frac{1}{2}m\omega^2 r^2 = 80$ J;

$\therefore \quad \text{PE} = \frac{1}{2}m\omega^2 y^2 = \frac{1}{2}m\omega^2 \times \left(\frac{3}{4}r\right)^2$

$$= \frac{9}{16}\left(\frac{1}{2}m\omega^2 r^2\right) = \frac{9}{16}\times 80 = 45\text{ J}$$

5. KE is maximum at mean position and minimum at extreme position $\left(\text{at } t = \frac{T}{4}\right)$, so correct graph is (c).

6. During oscillation, (motion of a simple pendulum) KE is maximum at mean position where PE is minimum. At extreme position, KE is minimum and PE is maximum. Thus, correct graph is depicted in option (b).

7. Given, $x = A\sin(\omega t + \alpha)$ ...(i)

and $y = A\cos(\omega t + \alpha)$ ...(ii)

Squaring and adding Eqs. (i) and (ii), we get

$$x^2 + y^2 = A^2[\sin^2(\omega t + \alpha) + \cos^2(\omega t + \alpha)] = A^2$$

It is an equation of a circle. The given motion is counter-clockwise.

8. The moment of inertia of a cylindrical rod about axis of wire(*i.e.* an axis passing through the centre of rod and perpendicular to its length) is

$$I = M\left(\frac{L^2}{12} + \frac{R^2}{4}\right)$$

9. As, $\frac{1}{2}m\omega^2(r^2 - y^2) = \frac{1}{3}\times\frac{1}{2}m\omega^2 r^2$

or $r^2 - y^2 = \frac{1}{3}r^2$

or $3r^2 - 3y^2 = r^2$

or $2r^2 - 3y^2 = 0$

or $r = \sqrt{\frac{3}{2}}\times y = \sqrt{\frac{3}{2}}\times 4 = 2\sqrt{6}$ cm

10. In equilibrium, $T = mg$

Work done $= mg = mgx = \frac{1}{2}kx^2$

or $x = \frac{2mg}{k} = \frac{2T}{k}$ ...(i)

Energy stored $= mgx = Tx$

$= T\times\frac{2T}{k} = \frac{2T^2}{k}$ [from Eq. (i)]

11. Here, $2 = 2\pi\sqrt{\frac{m}{k}}$

and $3 = 2\pi\sqrt{\frac{m+2}{k}}$

So, $\frac{3}{2} = \sqrt{\frac{m+2}{m}}$

or $9m = 4m + 8$

or $m = 1.6$ kg

12. Since the time period of simple pendulum is

$$T = 2\pi\sqrt{\frac{L}{g}}$$

$$\frac{t_0}{9} = 2\pi\sqrt{\frac{l_1}{g}} \text{ and } \frac{t_0}{7} = 2\pi\sqrt{\frac{l_2}{g}}$$

$\therefore$ $\frac{l_1}{l_2} = \left(\frac{7}{9}\right)^2 = \left(\frac{49}{81}\right)$

13. From the figure, $T\sin\theta = mL\sin\theta\omega^2$

$$324 = 0.5\times 0.5\times\omega^2$$

$$\omega^2 = \frac{324}{0.5\times 0.5}$$

$\Rightarrow$ $\omega = \sqrt{\frac{324}{0.5\times 0.5}} = \frac{18}{0.5} = 36$ rad/s

14.

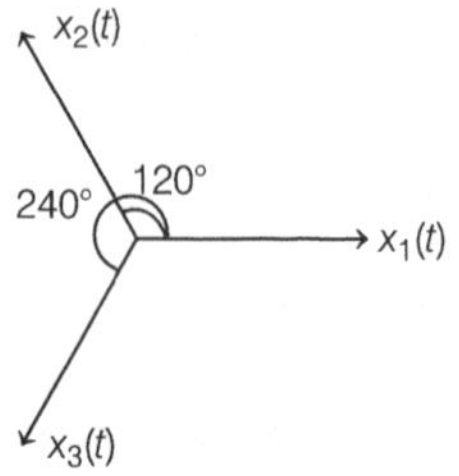

It is clear from figure that, magnitude of $x_3(t)$, must be the resultant of $x_1(t)$ and $x_2(t)$, to bring the mass at rest. So, $B = A$ and $\phi = 240° = \frac{4\pi}{3}$.

15. Maximum KE $= \frac{1}{2}m\omega^2A^2$; minimum KE $= 0$

Average KE $= \frac{0 + \frac{1}{2}m\omega^2A^2}{2} = \frac{1}{4}m\omega^2A^2$

Similarly, average PE $= \left(\frac{0 + \frac{1}{2}m\omega^2A^2}{2}\right) = \frac{1}{4}m\omega^2A^2$

16. Given, $x(t) = A\cos(\omega t + \theta)$

Velocity, $v = \frac{dx}{dt} = -A\omega\sin(\omega t + \theta)$

$= -A\omega\sqrt{1 - \cos^2(\omega t + \theta)}$

$\Rightarrow$ $v = -A\omega\sqrt{1 - x^2/A^2} = -\omega\sqrt{A^2 - x^2}$

Here, $v = \pi$ cms^{-1}, $x = 1$ cm, $\omega = \pi$ s^{-1}

So, $\pi = -\pi\sqrt{A^2 - 1^2}$

or $(-1)^2 = A^2 - 1$

or $A^2 = 2$

or $A = \sqrt{2}$ cm

17. As, $x_1 = a\sin(\omega\times 1) = a\sin\omega$

and $x_2 = a\sin(\omega\times 2) - a\sin\omega$

Now $\frac{x_2}{x_1} = \frac{\sin(2\omega) - \sin\omega}{\sin\omega}$

$= \sin 2\times(2\pi/8) - \sin 2\pi/8$ (from question)

$= \frac{1 - (1/\sqrt{2})}{(1/\sqrt{2})} = \frac{\sqrt{2} - 1}{1}$

or $\frac{x_1}{x_2} = \frac{1}{\sqrt{2} - 1} = \frac{\sqrt{2} + 1}{(\sqrt{2} - 1)(\sqrt{2} + 1)}$

$= \frac{\sqrt{2} + 1}{2 - 1}$

$= 2.414 = 2.4$

18. Electric intensity at B due to sheet of charge,

$$E = \frac{1}{2}\frac{\sigma q}{\varepsilon_0}$$

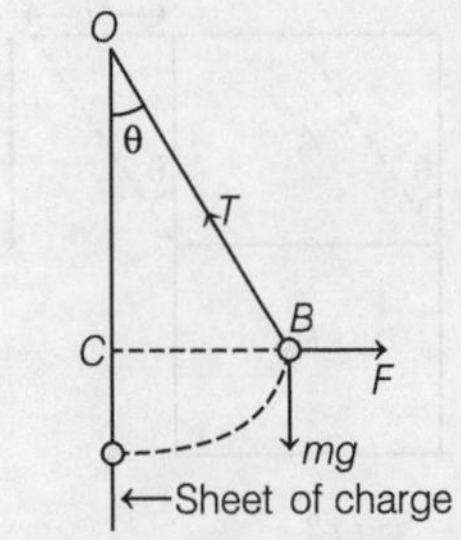

Force on the bob due to sheet of charge,

$$F = qE = \frac{1}{2}\frac{\sigma q}{\varepsilon_0}$$

As, the bob is in equilibrium, so

$$\frac{mg}{OC} = \frac{F}{CB} = \frac{T}{BO}$$

$$\therefore \quad \tan\theta = \frac{CB}{OC} = \frac{F}{mg} = \frac{\frac{1}{2}\sigma q/\varepsilon_0}{mg} = \frac{\sigma q}{2\varepsilon_0 mg}$$

19. When $t = 1\,\text{s}$, $y_1 = r\sin\omega\times 1 = r\sin\omega$

when $t = 2\,\text{s}$, $y_2 = r\sin\omega\times 2 = r\sin 2\omega$

$$\therefore \quad \frac{y_1}{y_2} = \frac{r\sin\omega}{r\sin 2\omega}$$

$$= \frac{1}{2\cos\omega} = \frac{1}{2\cos 2\pi/T}$$

$$= \frac{1}{2\cos 2\pi/8}$$

$$= \frac{1}{2\cos\pi/4} = \frac{1}{2(1/\sqrt{2})} = \frac{1}{\sqrt{2}}$$

$$\therefore \quad y_2 = \sqrt{2}y_1$$

Distance covered in 2nd second $= y_2 - y_1 = (\sqrt{2}-1)y_1$

$\therefore$ Ratio $= 1 : (\sqrt{2}-1)$

20. When the bob falls through a vertical height h, the velocity acquired at the lowest point,

$$v = \sqrt{2gh} = \sqrt{2\times 10\times 1} = \sqrt{20}\ \text{ms}^{-1}$$

Centrifugal force $= \dfrac{mv^2}{r} = \dfrac{0.01\times 20}{1} = 0.20$ N

Net tension = Weight + Centrifugal force

$$= (0.01\times 10 + 0.20) = 0.30\ \text{N}$$

21. v and v' are the velocities of the block of mass M and $(M+m)$ respectively while passing from the mean position when executing SHM.

Using law of conservation of linear momentum, we have

$$mv = (M+m)\,v'$$

or $$v' = mv/(M+m)$$

Also, maximum PE = maximum KE

$$\therefore \quad \frac{1}{2}kA'^2 = \frac{1}{2}(M+m)\,v'^2$$

or $$A' = \left(\frac{M+m}{k}\right)^{1/2}\times\frac{mv}{(M+m)}$$

$$= \frac{mv}{\sqrt{(M+m)k}}$$

22. The Lissajous figure will be parabola, if period ratio is 1 : 2 and phase difference is $\pi/2$.

Let $x = a\sin(2\omega t + \pi/2)$ and $y = b\sin\omega t$

$$\therefore \quad \sin\omega t = y/b$$

Now, $$\frac{x}{a} = \sin(2\omega t + \pi/2) = \cos 2\omega/t$$

$$\Rightarrow \quad \frac{x}{a} = 1 - 2\sin^2\omega t = 1 - \frac{2y^2}{b^2}$$

or $$\frac{2y^2}{b^2} = 1 - \frac{x}{a} = -\left(\frac{x-a}{a}\right)$$

or $$y^2 = -\frac{b^2}{2a}(x-a)$$

It is an equation of a parabola as given in figure. Hence, period ratio is 1 : 2 and phase difference is $\pi/2$.

23. Amplitude of damped oscillator is given by

$$A = A_0\,e^{-\frac{bt}{2m}}$$

After 5 s, $$0.9\,A_0 = A_0\,e^{-\frac{b(5)}{2m}}$$

$$\Rightarrow \quad 0.9 = e^{-\frac{b(5)}{2m}} \quad \text{...(i)}$$

After, 10 s, $$A = A_0 e^{-\frac{b(15)}{2m}}$$

$$\Rightarrow \quad A = A_0\,(e^{-\frac{5b}{2m}})^3 \quad \text{...(ii)}$$

From Eqs. (i) and (ii), we get

$$A = 0.729\,A_0$$

Hence, $$\alpha = 0.729$$

24. For a harmonic oscillator,

$$T = 2\pi\sqrt{\frac{m}{k}}$$

where, k = force constant and $T = \dfrac{1}{f}$.

$$\therefore \quad k = 4\pi^2 f^2 m$$

$$= 4\times\left(\frac{22}{7}\right)^2\times(10^{12})^2\times\frac{108\times 10^{-3}}{6.02\times 10^{23}}$$

$$\Rightarrow \quad k = 7.1\ \text{N/m}$$

25. In SHM, a particle starts from rest, we have

$$x = A\cos\omega t, \text{ at } t = 0,\ x = A$$

When $t = t$, then $x = A - a$

When $t = 2t$, then $x = A - 3a$

So, $$A - a = A\cos\omega t$$

$$A - 3a = A\cos 2\omega t$$

As, $$\cos 2\omega\tau = 2\cos^2\omega t - 1$$

$$\Rightarrow \quad \frac{A-3a}{A} = 2\left(\frac{A-a}{A}\right)^2 - 1$$

$$\Rightarrow \quad \frac{A-3a}{A} = \frac{2A^2+2a^2-4Aa-A^2}{A^2}$$

$$A^2 - 3aA = A^2 + 2a^2 - 4Aa$$

$$a^2 = 2aA \Rightarrow A = 2a$$

Now, $A - a = A\cos\omega t \Rightarrow \cos\omega t = 1/2$

$$\frac{2\pi t}{T} = \frac{\pi}{3}$$

$$T = 6t$$

26. Kinetic energy of a pendulum is maximum at its mean position. Also, maximum kinetic energy of pendulum

$$K_{\max} = \frac{1}{2} m\omega^2 a^2$$

where, angular frequency

$$\omega = \frac{2\pi}{T} = \frac{2\pi}{2\pi\sqrt{\frac{l}{g}}}$$

or $$\omega = \sqrt{\frac{g}{l}} \text{ or } \omega^2 = \frac{g}{l}$$

and a = amplitude.

As amplitude is same in both cases, so

$$K_{\max} \propto \omega^2$$

or $$K_{\max} \propto \frac{1}{l} \quad [\because g \text{ is constant}]$$

According to given data, $K_1 \propto \frac{1}{l}$ and $K_2 \propto \frac{1}{2l}$

$$\therefore \quad \frac{K_1}{K_2} = \left(\frac{1/l}{1/2l}\right) = 2$$

or $$K_1 = 2K_2$$

$$\Rightarrow \quad K_2 = \frac{K_1}{2}$$

27. Equation for SHM is given as

$$y = 5(\sin 3\pi t + \sqrt{3}\cos 3\pi t)$$

$$= 5 \times 2\left(\frac{1}{2} \times \sin 3\pi t + \frac{\sqrt{3}}{2}\cos 3\pi t\right)$$

$$= 5 \times 2\left(\cos\frac{\pi}{3} . \sin 3\pi t + \sin\frac{\pi}{3} . 3\pi t\right)$$

$$= 5 \times 2\sin\left(3\pi t + \frac{\pi}{3}\right)$$

[using, $\sin(a+b) = \sin a\cos b + \cos a \sin b$]

or $$y = 10\sin\left(3\pi t + \frac{\pi}{3}\right)$$

Comparing this equation with the general equation of SHM, *i.e.*

$$y = A\sin\left(\frac{2\pi t}{T} + \phi\right)$$

We get, amplitude, $A = 10$ cm

and $$3\pi = \frac{2\pi}{T}$$

or Time period, $T = \frac{2}{3}$ s

28. When the force F is applied to one side of block A, let the upper face of A be displaced through distance ΔL. Then

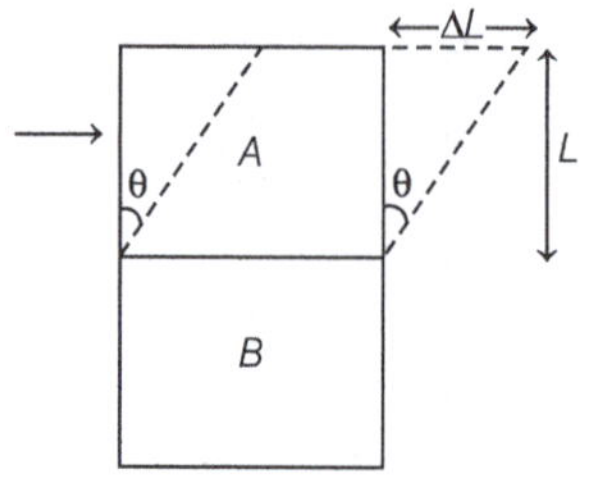

$$\eta = \frac{F/L^2}{\Delta L/L} \text{ or } F = \eta L \Delta L \quad \ldots(i)$$

So, $F \propto \Delta L$ and this force is restoring one. So, if the force is removed, the block will execute SHM.

From Eq. (i) spring factor $= \eta L$

Here, inertia factor $= M$

$\therefore$ Time period, $$T = 2\pi\sqrt{\frac{M}{\eta L}}$$

29. As, $$\frac{T_1}{T} = \sqrt{\frac{l_1}{l}} = \sqrt{\frac{16}{1}} = 4 \text{ or } T_1 = 4T$$

Let after time t, the pendulum be in the same phase. It will be so, then

$$\frac{t}{T_1} = \frac{t}{T} - 1 = \frac{t-T}{T}$$

or $$\frac{t}{4T} = \frac{t-T}{T}$$

or $$t = 4t - 4T$$

or $$3t = 4T$$

or $$t = 4T/3$$

30. As per question, the direction from A to B, *i.e.* from A towards mean position O is positive, therefore if a particle starting from A reaches at D, where $AD = 3$ cm, then its direction of motion is towards the mean position O. Hence, its velocity is positive, acceleration is positive and force is positive.

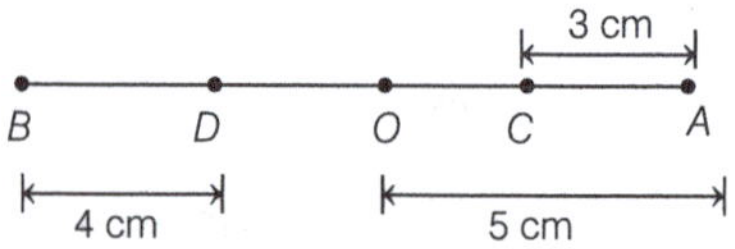

When a particle from B reaches point O, where $BO = 4$ cm, then its direction of motion is towards BA, *i.e.* along BO, then velocity, acceleration and force are negative.

When particle reaches at B, its velocity becomes zero but its acceleration and force are towards BA, *i.e.* negative.

31. Let $x(t) = A\sin(\omega t + \phi_0)$

and $v(t) = A\omega\cos(\omega t + \phi_0)$

We have,

$$x(0) = A\sin\phi_0 = 25$$

$\Rightarrow \quad \sin \phi_0 = \dfrac{25}{A} = \dfrac{1}{\sqrt{2}}$

$\therefore \quad A = 25\sqrt{2}$

Also, $\quad v(0) = A\omega \cos \phi_0$

$\Rightarrow \quad \cos \phi_0 = \dfrac{100}{A\omega} = \dfrac{100}{25\sqrt{2} \times 4} = \dfrac{1}{\sqrt{2}}$

$\therefore \quad \phi_0 = \dfrac{\pi}{4}$

The equation of motion,

$$x(t) = 25\sqrt{2} \sin\left(4t + \frac{\pi}{4}\right)$$

or

$$y = 25\sqrt{2} \sin\left(4t + \frac{\pi}{4}\right)$$

32. Torque is same for both the cases, *i.e.* restoring torque is same for both cases.

$\therefore \quad T = 2\pi\sqrt{\dfrac{I}{mgd}}$

Since, $\quad I_A > I_B$

therefore, $\omega_a < \omega_b$

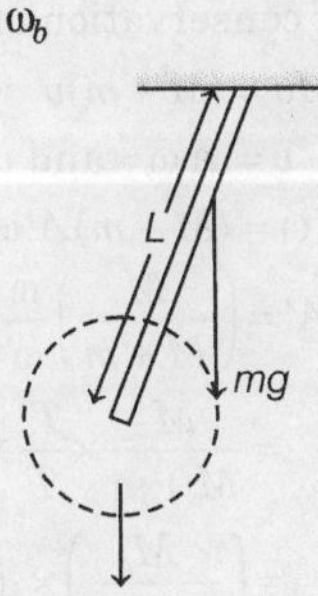

33. As, $t_1 = 2\pi\sqrt{\dfrac{m}{k_1}}$

or $\quad t_1^2 = \dfrac{2\pi^2 m}{k_1}$

or $\quad k_1 = \dfrac{4\pi^2 m}{t_1^2} \quad$ ...(i)

Similarly, $\quad k_2 = \dfrac{4\pi^2 m}{t_2^2} \quad$ ...(ii)

and $\quad (k_1 + k_2) = \dfrac{4\pi^2 m}{t_0^2} \quad$ ...(iii)

From Eqs. (i), (ii) and (iii), we get

$\therefore \quad \dfrac{4\pi^2 m}{t_0^2} = \dfrac{4\pi^2 m}{t_1^2} + \dfrac{4\pi^2 m}{t_2^2}$

or $\quad \dfrac{1}{t_0^2} = \dfrac{1}{t_1^2} + \dfrac{1}{t_2^2}$

34. Time period of oscillation, $T = 2\pi\sqrt{\dfrac{m}{k}}$

where m is the mass of body suspended from a spring and K is spring constant of the spring and $K = \dfrac{4\pi^2 m}{T^2}$.

Substituting the given values, we get

$$K = \frac{4 \times 10 \times 4.9}{(0.5)^2} \text{ Nm}^{-1} \quad \text{...(i)}$$

On the removal of the body the spring is shortened by x

$\therefore \quad mg = kx$

$\Rightarrow \quad x = \dfrac{mg}{k} = \dfrac{4.9 \times 10 \times (0.5)^2}{4 \times 10 \times 4.9} \quad$ [from Eq. (i)]

$= \dfrac{0.25}{4} = 0.0625 \text{ m} = 6.25 \text{ cm}$

35. The bob is subjected to two simultaneous, accelerations perpendicular to each other *viz* acceleration due to gravity g and radial acceleration $a_R = \dfrac{v^2}{R}$ towards the centre of the circular path.

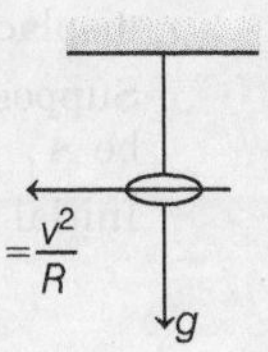

$\therefore$ Effective acceleration, $a_{\text{eff}} = \sqrt{g^2 + \left(\dfrac{v^2}{R}\right)^2}$

$\therefore$ Time period of the simple pendulum, $T = 2\pi\sqrt{\dfrac{l}{a_{\text{eff}}}}$

$$= 2\pi\sqrt{\frac{l}{\sqrt{g^2 + \left(\frac{v^2}{R}\right)^2}}} = 2\pi\sqrt{\frac{l}{\sqrt{g^2 + \frac{v^4}{R^2}}}}$$

36. As, $F = mg = kx$

For first case, $k = \dfrac{mg}{x} = \dfrac{1 \times 10 \text{ N}}{0.05} = 200 \text{ Nm}^{-1}$

For second case, $\omega = \sqrt{\dfrac{k}{m}} = \sqrt{\dfrac{200}{2.0}} = \sqrt{100} = 10\text{Hz}$

$r = \dfrac{m'g}{k} = \dfrac{2 \times 10}{200} = 0.1 \text{ m}$

$\therefore \quad v_{\max} = r\omega = 0.1 \times 10 = 1 \text{ ms}^{-1}$

37. We know that, time period of a pendulum is given by

$$T = 2\pi\sqrt{\frac{l}{g}}$$

So, angular frequency, $\omega = \dfrac{2\pi}{T} = \sqrt{\dfrac{g}{l}} \quad$ ..(i)

Now, differentiate both side w.r.t g

$\therefore \quad \dfrac{d\omega}{dg} = \dfrac{1}{2\sqrt{g}\sqrt{l}}$

$d\omega = \dfrac{dg}{2\sqrt{g}\sqrt{l}} \quad$ ...(ii)

By dividing Eq. (ii) by Eq. (i), we get

$$\frac{d\omega}{\omega} = \frac{dg}{2g}$$

Or we can write

$$\frac{\Delta\omega}{\omega} = \frac{\Delta g}{2g} \quad \text{...(iii)}$$

As Δg is due to oscillation of support.

$\therefore \quad \Delta g = 2\omega^2 A \quad (\omega_1 = 1 \text{ rad/s, support})$

Putting value of Δg in Eq. (iii) we get

$$\frac{\Delta\omega}{\omega} = \frac{1}{2}\cdot\frac{2\,\omega_1^2 A}{g} = \frac{\omega_1^2 A}{g}; \qquad (A = 10^{-2}\text{m})$$

$$\Rightarrow \quad \frac{\Delta\omega}{\omega} = \frac{1\times 10^{-2}}{10} = 10^{-3} \text{ rad/s}$$

38. The velocity of a particle executing SHM at any instant, is defined as the time rate of change of its displacement at that instant.

$$v = \omega\sqrt{A^2 - x^2}$$

where, ω is angular frequency, A is amplitude and x is displacement of a particle.

Suppose that the new amplitude of the motion be A'.

Initial velocity of a particle performing SHM,

$$v^2 = \omega^2\left[A^2 - \left(\frac{2A}{3}\right)^2\right] \qquad \text{... (i)}$$

where, A is initial amplitude and ω is angular frequency.

Final velocity,

$$(3v)^2 = \omega^2\left[A'^2 - \left(\frac{2A}{3}\right)^2\right] \qquad \text{...(ii)}$$

From Eqs. (i) and (ii), we get

$$\frac{1}{9} = \frac{A^2 - \frac{4A^2}{9}}{A'^2 - \frac{4A^2}{9}} \Rightarrow A' = \frac{7A}{3}$$

39. Pressure applied by piston,

$$\frac{Mg}{A} = p_0$$

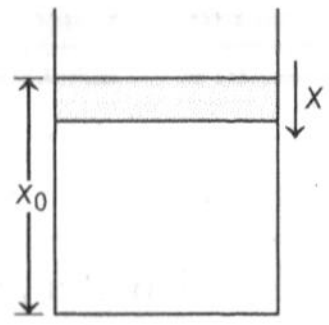

$$Mg = p_0 A \qquad \text{...(i)}$$

As no exchange of heat, so process is adiabatic.

$$p_0 V_0^\gamma = p V^\gamma$$

$$p_0 A x_0^\gamma = pA(x_0 - x)^\gamma$$

$$p = \frac{p_0 x_0^\gamma}{(x_0 - x)^\gamma}$$

Let piston be displaced by x.

$$Mg - \left(\frac{p_0 x_0^\gamma}{(x_0 - x)^\gamma}\right) A = f_{\text{restoring}}$$

$$p_0 A\left(1 - \frac{x_0^\gamma}{(x_0 - x)^\gamma}\right) = F_{\text{restoring}} \qquad (\because x_0 - x \approx x_0)$$

$$F = \frac{\gamma\, p_0\, Ax}{x_0}$$

$$\therefore \quad f = \frac{1}{2\pi}\sqrt{\frac{\gamma p_0 A}{x_0 M}} = \frac{1}{2\pi}\sqrt{\frac{\gamma p_0 A^2}{M V_0}}$$

40. For damped motion, $A = A_0 e^{-rt}$

$$12 = 6e^{-\frac{bt}{2m}}$$

or

$$\ln 2 = \frac{b}{2m}\times 120$$

$$b = \frac{0.693\times 2\times 1}{120}$$

$$= 1.16\times 10^{-2} \text{ kg s}^{-1}$$

41. When a mass m is placed on mass M, the new system is of mass $= (M + m)$, attached to the spring. New time period of oscillation,

$$T' = 2\pi\sqrt{\frac{M+m}{k}}$$

$$T = 2\pi\sqrt{\frac{M}{k}}$$

Let v = velocity of mass M while passing through the mean position.

v' = velocity of mass $(M + m)$, while passing through the mean position.

According to law of conservation of linear momentum,

$$Mv = (M + m)v'$$

At mean position, $v = A\,\omega =$ and $v' = A'\omega'$

$$\therefore \quad MA\,\omega = (M + m)A'\,\omega'$$

or

$$A' = \left(\frac{M}{M+m}\right)\frac{\omega}{\omega'}A$$

$$= \frac{M}{M+m}\times\frac{T'}{T}\times A$$

$$= \left(\frac{M}{M+m}\right)\times\sqrt{\frac{M+m}{M}}\times A$$

$$= A\sqrt{\left(\frac{M}{M+m}\right)}$$

42. Let h be the depth of in water, then

$$A\,h\,\rho\,g = mg$$

or

$$h = \frac{m}{A\rho} = \frac{200}{50\times 1} = 4 \text{ cm} \qquad (\because \rho = 1 \text{ gcm}^{-3})$$

$$T = 2\pi\sqrt{\frac{h}{g}}$$

Now,

$$\nu = \frac{1}{T} = \frac{1}{2\pi}\sqrt{\frac{g}{h}} = \frac{7}{2\times 22}\sqrt{\frac{980}{4}} = 2.5 \text{ Hz}$$

43. Density of mercury column $= \rho$

Acceleration due to gravity $= g$

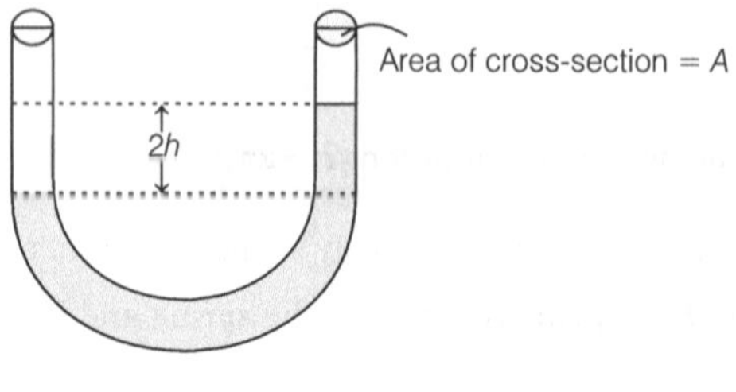

Restoring force,

$F = -$ Weight of mercury column in excess of one arm

$= -$ (Volume $\times$ density $\times g$)

$= -(A \times 2h \times \rho \times g)$

$= -2A\rho gh = -k \times$ Displacement in one arm (h)

Clearly, $2A\rho g = \text{constant} = k$ (say)

As, $F = -kx \Rightarrow F \propto -h$

Hence, motion is SHM.

$\therefore$ Time period, $T = 2\pi\sqrt{\frac{m}{k}} = 2\pi\sqrt{\frac{m}{2A\rho g}}$

where, m = mass of the mercury column of length l.

If ρ_{Hg} is density of mercury, then

$$m = Al\rho$$

$$\therefore \quad T = 2\pi\sqrt{\frac{Al\rho}{2A\rho g}} = 2\pi\sqrt{\frac{l}{2g}}$$

44. As, the length of the scale is 20 cm and it can read upto 50 kg. The maximum extension of 20 cm will correspond to maximum weight of 50 kg $\times$ 9.8 m/s^2.

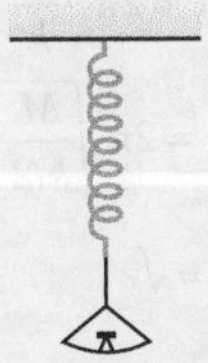

Using, $F = -kx$

$$|\mathbf{F}| = F = kx$$

Here, $x = 20 \times 10^{-2}$ m

$$k = \frac{50 \times 9.8}{20 \times 10^{-2}} = 2450 \text{ N/m}$$

We have for loaded oscillation

$$T = 2\pi\sqrt{\frac{m}{k}}$$

or $T^2 = 4\pi^2 \frac{m}{k}$

$$m = \frac{T^2 k}{4\pi^2} = \frac{(0.60)^2 \times 2450}{4 \times (3.14)^2}$$

$$= 22.36 \text{ kg}$$

$\therefore$ Weight $= mg = 22.36 \times 9.8 = 219.13$ N

45. For the given SHM, the displacement is given by

$y = a\cos\omega t$

Velocity, $v = \frac{dy}{dt} = -\omega\sin\omega t = a\omega\sin(\omega t + \pi)$

Acceleration, $A = \frac{dV}{dt} = a\omega^2\cos\omega t$

Force = mass $\times$ acceleration $= -m\,a\,\omega^2\cos\omega t$

Force is zero, when $\cos\omega t = 0$ or $\omega t = \frac{\pi}{2}$ or $\frac{3\pi}{2}$

i.e. $\frac{2\pi}{T}t = \frac{\pi}{2}$ or $\frac{3\pi}{2}$

If $\frac{2\pi}{T}t = \frac{\pi}{2}$, then, $t = \frac{T}{4}$

If $\frac{2\pi}{T}t = \frac{3\pi}{2}$, then $t = \frac{3T}{4}$ (given)

So for $t = \frac{3T}{4}$, force is zero.

Acceleration is maximum, if $\cos\omega t = 1$ or 2π

or $\frac{2\pi}{T}t = 2\pi$ or $t = T = \frac{4T}{4}$ s (given)

Velocity is maximum, if $\sin(\omega t + \pi) = 1$ or $\omega t + \pi = \pi/2$

or $\omega t = \frac{\pi}{2} - \pi = -\pi/2$

or $\frac{2\pi}{T}t = -\frac{\pi}{2}$ or $t = -\frac{T}{4}$ s

$$\text{PE} = \frac{1}{2}m\,\omega^2 y^2 = \frac{1}{2}m\omega^2 a^2\cos^2\omega t$$

$$\text{KE} = \frac{1}{2}m\,\omega^2 a^2\sin^2\omega t$$

If PE = KE, then $\cos^2\omega t = \sin^2\omega t$ or $\cos\omega t = \sin\omega t$ or $\tan\omega t = 1$

or $\omega t = \frac{\pi}{4}$ or $\frac{2\pi}{T}t = \frac{\pi}{4}$ or $t = \frac{T}{8}$ s

46. $T' = 2\pi\sqrt{\frac{l}{g_{net}}}$

$$g_{net} = g + a = 10 + 10 = 20 \text{ m/s}^2$$

$$T' = \frac{T}{\sqrt{2}}$$

47. When a system oscillates, the magnitude of restoring torque of system is given by

$$\tau = C\theta \quad \text{...(i)}$$

where, C = constant that depends on system.

Also, $\tau = I\alpha$ ...(ii)

where, I = moment of inertia

and α = angular acceleration

From Eqs. (i) and (ii), we get

$$\alpha = \frac{C}{I}\cdot\theta \quad \text{...(iii)}$$

and time period of oscillation of system will be

$$T = 2\pi\sqrt{\frac{I}{C}}$$

In given case, magnitude of torque is

τ = Force $\times$ Perpendicular distance

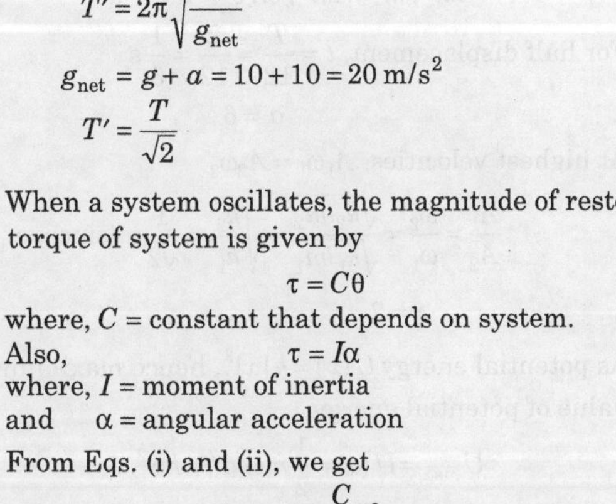

$$\tau = 2kx \times \frac{l}{2}\cos\theta$$

For small deflection, $\tau = \left(\frac{kl^2}{2}\right)\theta$ ...(iv)

$\because$ For small deflections, $\sin\theta = \frac{x}{(l/2)} \approx \theta$

$\Rightarrow \quad x = \frac{l\theta}{2}$

Also, $\cos\theta \approx 1$

Comparing Eqs. (iv) and (i), we get

$$C = \frac{kl^2}{2} \Rightarrow \alpha = \frac{(kl^2/2)}{\left(\frac{1}{12}ml^2\right)}\cdot\theta \Rightarrow \alpha = \frac{6k}{m}\cdot\theta$$

Hence, time period of oscillation is $T = 2\pi\sqrt{\frac{m}{6k}}$

Frequency of oscillation is given by

$$f = \frac{1}{T} = \frac{1}{2\pi}\sqrt{\frac{6k}{m}}$$

48. Here reduced mass of the system, $\mu = \frac{m_1 m_2}{m_1 + m_2} = \frac{3\times1}{3+1}$

$= 0.75\text{kg}$

$\therefore$ Vibrational frequency,

$$\nu = \frac{1}{2\pi}\sqrt{\frac{k}{m}} = \frac{1}{2\pi}\sqrt{\frac{300}{0.75}} = \frac{20}{2\pi} = \frac{10}{\pi} \cong 3\text{ Hz}$$

49. For half displacement, $t = \frac{T}{12} = \frac{2}{12} = \frac{1}{6}$ s

$\therefore \quad a = 6$

50. At highest velocities, $A_1\omega_1 = A_2\omega_2$

$\therefore \quad \frac{A_1}{A_2} = \frac{\omega_2}{\omega_1} = \frac{\sqrt{k_2/m_2}}{\sqrt{k_1/m_1}} = \sqrt{\frac{k_2}{k_1}} = \frac{1}{\sqrt{2}}$ (since, $m_1 = m_2$)

$\therefore \quad x = 2$

51. As potential energy $U(x) = k|x|^3$, hence maximum value of potential energy,

$$U_{\max} = U(a) = \frac{1}{2}m\omega^2 a^2 = ka^2$$

$\Rightarrow \quad \omega \propto \sqrt{a}$

As, $T = \frac{2\pi}{\omega}$, hence $T \propto \frac{1}{\sqrt{4}} = \frac{1}{2} = 0.5$

52. In equilibrium, due to weight mg spring is stretched by y_0 as shown in figure. When further depressed by a small distance y, the restoring force will be

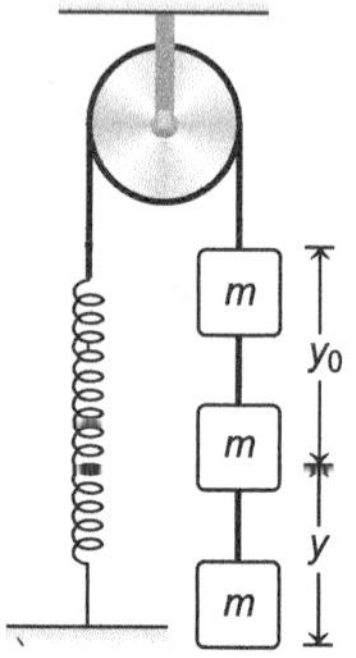

$$F = -[k(y + y_0) - mg]$$
$$= [k(y + y_0) - ky_0] = -ky$$

$\therefore \quad$ Acceleration, $a = \frac{F}{m} = \frac{k}{m}y$

As acceleration is proportional to the displacement y and opposite to y, the motion will be SHM. Frequency of oscillations will be

$$\nu = \frac{1}{2\pi}\sqrt{\frac{k}{m}} = \frac{1}{2\pi}\sqrt{\frac{16}{0.01}} = \frac{20}{\pi}\text{ Hz}$$

$\therefore \quad n = 20$

53. $T_a = 2\pi\sqrt{\frac{M}{k}}$

In series combination, $k' = \frac{k}{2}$

$$T_b = 2\pi\sqrt{\frac{M}{k'}} = 2\pi\sqrt{\frac{M}{k/2}} = 2\pi\sqrt{\frac{2M}{k}}$$

$$\frac{T_b}{T_a} = \sqrt{2}$$

$\therefore \quad x = 2$

54. Here, $A = 25$ cm; $T = 3$ s;

Let the particle be at the locating -12.5 cm at time t_1 and $+12.5$ cm at time t_2.

Using the relation, $x = A\cos\left(\frac{2\pi t}{T} + \phi\right)$

First condition $-12.5 = 25\cos\left(\frac{2\pi t_1}{3} + \phi\right)$ ...(i)

Second condition $12.5 = 25\cos\left(\frac{2\pi t_2}{3} + \phi\right)$ ...(ii)

From Eq. (i), $\cos\left(\frac{2\pi t_1}{3} + \phi\right) = \frac{-12.5}{25} = -\frac{1}{2} = \cos\frac{2\pi}{3}$

$\therefore \quad \frac{2\pi t_1}{3} + \phi = \frac{2\pi}{3}$

or $\quad 2\pi t_1 + 3\phi = 2\pi$ ...(iii)

From Eq. (ii), $\cos\left(\frac{2\pi t_2}{2} + \phi\right) = \frac{12.5}{25} = \frac{1}{2} = \cos\frac{\pi}{3}$

$\therefore \quad \frac{2\pi t_2}{3} + \phi = \frac{\pi}{3}$

or $\quad 2\pi t_2 + 3\phi = \pi$ ...(iv)

Subtracting Eq. (iv) from Eq. (iii), we get

$$2\pi(t_1 - t_2) = \pi$$

or $\quad (t_1 - t_2) = \pi/2\pi = 1/2 = 0.5$ s

17

Waves

Waves occur when a system is disturbed from its equilibrium position and this disturbance propagates from one region to other.

IN THIS CHAPTER ....

Wave Motion

Wave motion involves transfer of disturbance (energy) from one point to another without actual transport of matter between the two points.

In this motion, the disturbance travels through the medium due to the repeated periodic oscillations of the particles of the medium about their mean positions.

Types of Waves

Waves can be classified into two broad categories

(i) **Mechanical waves** These are the waves which requires a material medium for their propagation, *e.g.* sound waves, water waves, etc.
For propagation of mechanical waves, medium must possess inertia, elasticity and minimum frictional force.

(ii) **Non-mechanical waves** These are the waves which can travel with or without vacuum, *e.g.* electromagnetic waves.

Note All electromagnetic waves are transverse in nature.

Mechanical waves are further classified into two categories

(i) Transverse Wave

In this type of wave, the particles of the medium oscillate perpendicular to the direction in which the wave travels. *e.g.* Travelling waves on a tight rope. In this, disturbance travels along the rope in the form of crests (upward peaks) and troughs (downward peaks).

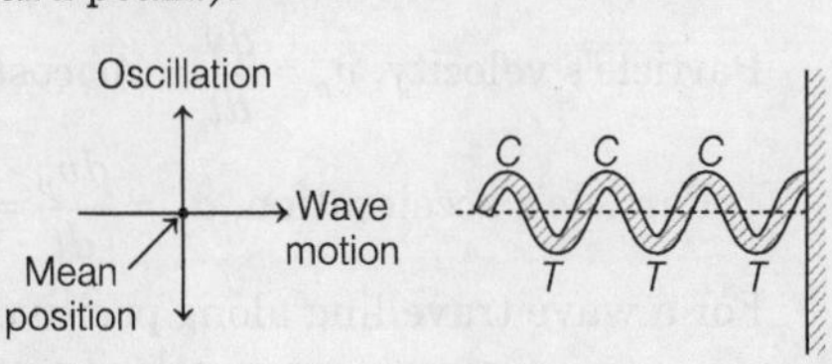

(ii) Longitudinal Wave

In this type of wave, the particles of the medium oscillate about their mean or equilibrium position along the direction of propagation of the wave motion itself. *e.g.* Waves on springs or sound waves in air.

This wave travels in the form of compressions and rarefactions.

Some important terms related to wave motion are as follows

(i) **Wavelength** It is the distance travelled by the disturbance in one time period, *i.e.* the time in which particle of the medium completes one oscillation.

In case of a transverse wave, wavelength (λ) is equal to the distance between two consecutive crests or troughs.

However, in case of a longitudinal wave, wavelength (λ) is equal to the distance between two consecutive compressions or rarefactions.

(ii) **Frequency** (ν) It is the number of waves produced per unit time in the given medium.

If T is period of oscillation of the particle in the medium, then

$$\nu = \frac{1}{T}$$

(iii) **Velocity** Since, in one time period (T), wave travels a distance which is equal to its wavelength (λ).

Hence, velocity of wave is given by

$$v = \frac{\lambda}{T} = \left(\frac{1}{T}\right)\lambda \quad \text{or} \quad v = \nu\lambda$$

Progressive Waves

A wave which travels from one point of the medium to another continuously in the same direction without any change in its amplitude is called a progressive wave or a *travelling wave.* It may be transverse or longitudinal.

General form of a Progressive Wave

For a wave travelling from left to right, *i.e.* along positive direction of X-axis with velocity v, its displacement relation is given by $y(x, t) = f(x - vt)$

Here, x is the distance of wave pulse from origin.

Similarly, for a wave pulse travelling from right to left, *i.e.* along negative direction of X-axis, the wave function will be

$$y(x, t) = f(x + vt)$$

Displacement Relation for Plane Progressive Harmonic Wave

During the propagation of a wave through a medium, if the particles of the medium oscillate simple harmonically about their mean position, then the wave is said to be *plane progressive harmonic wave.*

For such a wave travelling in + x-direction, displacement relation is given as

$$y = A\sin\left\{2\pi\left(\frac{t}{T} - \frac{x}{\lambda}\right) \pm \phi\right\}$$

In terms of velocity of wave (v), displacement relation is

$$y = A\sin\left\{\frac{2\pi}{\lambda}(vt - x) \pm \phi\right\}$$

Different forms of displacement relation for progressive waves are as follows

$$y = A\sin\{(\omega t \pm kx) \pm \phi\}$$
$$= A\sin\{k(vt \mp x) \pm \phi\}$$
$$= A\sin\left\{\omega\left(t \mp \frac{x}{v}\right) \pm \phi\right\}$$

Description of the various terms in the displacement relation of plane progressive waves

- ω is the *angular frequency* of the particle oscillating in SHM.

  $\omega = 2\pi\nu = \frac{2\pi}{T}$, where ν is the natural frequency and T is the time period.
- t is the time elapsed from when the wave begin its motion, *i.e.* time elapsed, since origin of the wave.
- k is the *propagation constant* or wave number.

  $k = \frac{2\pi}{\lambda} = \frac{\omega}{v}$, where v is wave velocity.
- The coefficient of sin or cos function, *i.e.* A gives the amplitude of the wave while its argument ($\omega t \pm kx$) denotes its phase.
- ϕ represents *phase constant* or initial phase.

Important points related to plane progressive wave

- The following equations

  $y = A\sin(kx - \omega t)$ and $y = A\sin(\omega t - kx)$ represent a wave travelling in positive x-direction with speed $v = \frac{\omega}{k}$.

  However, the difference between them is that they are out of phase, *i.e.* phase difference between them is π.
- For a particle with displacement, $y = A\sin(\omega t - kx)$

  Particle's velocity, $v_p = \frac{dy}{dt} = A\omega\cos(\omega t - kx)$

  $\therefore$ Particle's acceleration, $a_p = \frac{dv_p}{dt} = -\omega^2 y$.
- For a wave travelling along positive X-axis as Shown,

  $$v_p = -\ \text{wave velocity} \times \text{slope}$$

i.e. Particle's velocity at a given position and time is equal to negative of the product of wave velocity with the slope of the wave at that point at that instant.

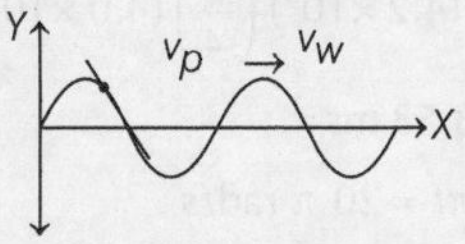

- **Phase difference and path difference** At any instant t, if ϕ_1 and ϕ_2 are the phases of two particles whose distances from the origin are x_1 and x_2 respectively, then

$$\phi_2 - \phi_1 = k(x_2 - x_1)$$

or Phase difference, $\Delta\phi = \frac{2\pi}{\lambda}$ (Path difference Δx)

- **Phase difference and time difference** If the phases of a particle at distance x from the origin is ϕ_1 at time t_1 and ϕ_2 at time t_2, then $\phi_1 - \phi_2 = \omega(t_1 - t_2)$

or Phase difference $(\Delta\phi) = \frac{2\pi}{T}$(Time difference Δt)

Example 1. *A wave travelling along a string is described by*

$$y(x, t) = 0.005 \sin(80x - 3t)$$

in which numerical constants are in SI units (0.005 m, 80 rad m^{-1}, 3 rad s^{-1}). The wavelength λ is then given by

(a) 2.85 cm (b) 7.85 cm (c) 10.83 cm (d) 18.2 cm

Sol. (b) Comparing the given displacement equation with standard equation,

$$y(x, t) = a \sin(kx - \omega t)$$

We have, amplitude of the wave is 0.005 m = 5 mm, angular wave number k and angular frequency ω are $k = 80\ \text{m}^{-1}$ and $\omega = 3\ \text{s}^{-1}$.

Then, from equation

$$\lambda = \frac{2\pi}{k} = \frac{2\pi}{80\ \text{m}^{-1}} = 7.85\ \text{cm}$$

Example 2. *Equation of a transverse wave travelling in a rope is given by $y = 5 \sin(4.0t - 0.02x)$, where y and x are expressed in cm and time in seconds. Then, the velocity of the wave and acceleration of the particle of rope respectively is*

(a) 200 cms^{-1}, 40 cms^{-2} (b) 200 cms^{-1}, 80 cms^{-2}

(c) 250 cms^{-1}, 20 cms^{-2} (d) None of these

Sol. (b) Given, $y = 5\sin(4.0t - 0.02x)$.

Comparing this with the standard equation of wave motion,

$$y = A \sin\left(2\pi ft - \frac{2\pi}{\lambda}x\right)$$

where, A, f and λ are amplitude, frequency and wavelength, respectively. Thus, amplitude, $A = 5$ cm, $2\pi f = 4$

Frequency, $f = \frac{4}{2\pi} = 0.673$ cycle s^{-1}

Again, $\frac{2\pi}{\lambda} = 0.02$

or wavelength, $\lambda = \frac{2\pi}{0.02} = 100\ \pi$ cm

Velocity of the wave, $v = f\lambda = \frac{4}{2\pi} \times \frac{2\pi}{0.02} = 200\ \text{cms}^{-1}$

Transverse velocity of the particle,

$$u = \frac{\partial y}{\partial t} = 5 \times 4 \cos(4.0t - 0.02x)$$

$$= 20 \cos(4.0t - 0.02x)$$

Maximum velocity of the particle = 20 cms^{-1}

Particle acceleration, $a = \frac{\partial y}{\partial t} = 20 \times 4 \cos(4.0t - 0.02x)$

Maximum particle acceleration = 80 cms^{-2}

Energy in Wave Motion

Related to the energy associated with the wave motion. There are three terms namely : energy density (u), power (P) and intensity (I).

Energy density (u) It is defined as the total mechanical energy (kinetic + potential) per unit volume of the medium through which the wave is travelling.

$$\therefore \qquad u = \frac{E}{V} = \frac{\frac{1}{2}m\omega^2A^2}{V} = \frac{1}{2}\rho\omega^2A^2 \qquad \left(\because \rho = \frac{m}{V}\right)$$

Power (P) If we consider a transverse wave on a string, then the instantaneous rate at which energy is transferred along the string is called power. Thus,

$$\text{Power } (P) = \frac{1}{2}\rho\omega^2A^2\Delta v$$

Intensity (I) It is the flow of energy per unit area of cross-section of the string in unit time. Thus,

$$I = \frac{\text{Power}}{\text{Area of cross-section}} = \frac{P}{S}$$

or $$I = \frac{1}{2}\rho\omega^2A^2v$$

Note The intensity of waves emitting in all directions due to a point source varies inversely as the square of the distance (r).

i.e. $$I \propto \frac{1}{r^2}$$

The intensity of waves from a linear source varies inversely as the distance (r) perpendicular to the source, i.e. $I \propto \frac{1}{r}$.

Speed of Transverse Waves

The expression for speed of transverse waves in a stretched string is given as $v = \sqrt{\frac{T}{m}}$

where, T is tension and m is linear mass density of the string.

Example 3. *A steel wire has a length of 12.0 m and a mass of 2.10 kg. What should be the tension in the wire, so that the speed of a transverse wave on the wire equals the speed of sound in dry air at 20°C? (Take, speed of sound in dry air at 20° C = 343 m/s)*

(a) 2.06×10^3 N (b) 2.06×10^4 N

(c) 3.8×10^5 N (d) 3.8×10^6 N

Sol. (*b*) Here, speed of sound in air, $v = 343\ ms^{-1}$;

Length of the wire, $l = 12.0$ m;

Total mass of the wire, $M = 2.10$ kg

Therefore, mass per unit length of the wire,

$$m = \frac{M}{l} = \frac{2.10}{12.0} = 0.175\ kgm^{-1}$$

Now, $$v = \sqrt{\frac{T}{m}}$$

or $$T = v^2 m = (343)^2 \times 0.175 = 20588.6\ N = 2.06 \times 10^4\ N$$

Example 4. *Equation of travelling wave on a stretched string of linear mass density 5 g/m is $y = 0.03 \sin(450t - 9x)$, where distance and time are measured in SI units. The tension in the string is* **[JEE Main 2019]**

(a) 5 N (b) 12.5 N
(c) 7.5 N (d) 10 N

Sol. (*b*) Given equation can be rewritten as

$$y = 0.03 \sin 450 \left(t - \frac{9x}{450}\right) \quad \ldots(i)$$

We know that, the general equation of a travelling wave is given as

$$y = A \sin \omega (t - x/v) \quad \ldots(ii)$$

Comparing Eqs. (i) and (ii), we get

$$\text{Velocity, } v = \frac{450}{9} = 50\ m/s$$

and angular velocity, $\omega = 450$ rad/s

As, the velocity of wave on stretched string with tension (T) is given as $v = \sqrt{T/\mu}$

where, μ is linear density.

$$\therefore \quad T = \mu v^2 = 5 \times 10^{-3} \times 50 \times 50 = 12.5\ N \quad (\because \text{given, } \mu = 5\ g/m = 5 \times 10^{-3} kg/m)$$

Example 5. *A stretched string is forced to transmit transverse waves by means of an oscillator coupled to one end. The string has a diameter of 4 mm. The amplitude of the oscillation is 10^{-4} m and the frequency is 10 Hz. Tension in the string is 100 N and mass density of wire is $4.2 \times 10^3\ kgm^{-3}$. Then,*

(a) *the equation of the waves along the string is $(10^{-4})\ m \sin [(1.44\ m^{-1}) x - (62.83\ rads^{-1})] t$*

(b) *the energy per unit volume of the wave $8.29 \times 10^{-2}\ Jm^{-3}$.*

(c) *the average energy flow per unit time across any section of the string $4.53 \times 10^{-5}\ Js^{-1}$.*

(d) *All of the above*

Sol. (*d*) Speed of transverse wave on the string, $v = \sqrt{\frac{T}{\rho S}}$

Substituting the values, we have

$$v = \sqrt{\frac{100}{(4.2 \times 10^3)\left(\frac{\pi}{4}\right)(4.0 \times 10^{-3})^2}} = 43.53\ ms^{-1}$$

$$\omega = 2\pi f = 20\pi\ rad/s = 62.83\ rad/s$$

$$k = \frac{\omega}{v} = 1.44\ m^{-1}$$

$\therefore$ Equation of the waves along the string,

$$y(x, t) = A \sin (kx - \omega t) = (10^{-4}\ m) \sin [(1.44\ m^{-1})\, x - (62.83\ rad\ s^{-1})\, t]$$

Energy per unit volume of the string,

$$u = \text{energy density} = \frac{1}{2}\rho\omega^2 A^2$$

Substituting the values, we have

$$u = \left(\frac{1}{2}\right)(4.2 \times 10^3)(62.83)^2 (10^{-4})^2 = 8.29 \times 10^{-2}\ Jm^{-3}$$

Average energy flow per unit time,

$$= \left(\frac{1}{2}\rho\omega^2 A^2\right)(Sv) = (u)(Sv)$$

Substituting the values, we have

$$P = (8.29 \times 10^{-2})\left(\frac{\pi}{4}\right)(4.0 \times 10^{-3})^2 (43.53) = 4.53 \times 10^{-5}\ Js^{-1}$$

Sound Waves

These waves are longitudinal in nature. They are classified into three groups according to their range of frequencies, namely, *infrasonic waves* (below 20Hz), *audible waves* (20Hz-20 kHz) and *ultrasonic waves* (above 20 kHz).

Speed of Sound Wave

The speed of longitudinal wave (sound wave) in a medium depends on the elastic and inertial properties of medium.

- Speed of sound wave in a medium is given as
$$v = \sqrt{\frac{B}{\rho}}$$
where, B is bulk modulus and ρ is the density of the medium.
- Speed of longitudinal wave (sound wave) in a solid bar,
$$v = \sqrt{\frac{Y}{\rho}}$$
where, Y is the Young's modulus of the medium.
- Speed of sound wave through a solid of bulk modulus B and modulus of rigidity η is given as
$$v = \sqrt{\frac{B + (4/3)\eta}{\rho}}$$

- **Newton's formula** Speed of longitudinal waves in gases,

$$v = \sqrt{\frac{p}{\rho}}$$

where, p is the pressure exerted by the gas,
ρ is the density of the gas at given temperature.
By this formula, speed of sound at STP is approx. 250 m/s, which is smaller as compared to the experimental value of 331 ms^{-1}.

- **Laplace's correction** After modifying Newton's formula, speed of sound in gases is given as

$$v = \sqrt{\frac{\gamma p}{\rho}}$$

where, γ = ratio of two specific heat capacities (C_p/C_V) of the gas.
By this formula, speed of sound comes out to be 331.3 ms^{-1}.

Factors Affecting Speed of Sound in a Gas

- **Effect of temperature** $v \propto \sqrt{T}$

i.e. with the increase in temperature (in kelvin), velocity of sound in a gas also increases.
If velocity of sound at 0°C is 332 m/s, then its value at t°C will be

$$v_t = 332 + 0.61\,t$$

Thus, velocity of sound in air increases roughly by 0.61 m/s per degree centigrade rise in temperature.

- **Effect of pressure** If the temperature of the gas remains constant, then there is no effect of the pressure change on the speed of sound.
- **Effect of humidity** Speed of sound increases with the increase in humidity. Thus, speed of sound in moist air is slightly greater than that in dry air.

Note (i) Speed of sound is greater in solids, and liquids than in gases even though they are denser than gases.
(ii) Speed of sound in air is independent of its frequency.

Example 6. *The density of air at NTP is 1.29 kgm^{-3}. Assume air to be diatomic with $\gamma = 1.4$. The velocity of sound at 127°C is*

(a) 382.8 ms^{-1} (b) 350 ms^{-1}
(c) 350.6 ms^{-1} (d) 348.6 ms^{-1}

Sol. (a) Velocity of sound in air at NTP

$$= \sqrt{\frac{\gamma p}{\rho}} = \sqrt{\frac{1.4 \times 1.013 \times 10^5\ \text{Nm}^{-2}}{1.29\ \text{kgm}^{-3}}} = 331.6\ \text{ms}^{-1}$$

The velocity of sound is proportional to the square root of absolute temperature

$$\Rightarrow \quad \frac{v_2}{v_1} = \sqrt{\frac{T_2}{T_1}}$$

$$\Rightarrow \quad v_2 = v_1\sqrt{\frac{T_2}{T_1}} = 331.6\sqrt{\frac{273+127}{273+27}} = 382.8\ \text{ms}^{-1}$$

Displacement or Pressure Waves

A sound wave can be considered as either a displacement wave,

$$y = A\sin(\omega t - kx)$$

or a pressure wave,

$$p = p_0\cos(\omega t - kx)$$

i.e. Pressure wave is 90° out of phase with respect to displacement wave.
Also, amplitude of pressure wave,

$$p_m = ABk = Ak\rho v^2 = \rho v A\omega$$

where, B is bulk modulus.

Example 7. *The pressure wave $p = 0.01\sin[1000t - 3x]$ Nm^{-2}, corresponds to the sound produced by a vibrating blade on a day when atmospheric temperature is 0° C. On some other day when temperature is T, the speed of sound produced by the same blade and at the same frequency is found to be 336 ms^{-1}. Approximate value of T is* **[JEE Main 2019]**

(a) 15° C (b) 11° C (c) 12° C (d) 4° C

Sol. (d) Given, $p = 0.01\sin(1000t - 3x)$ N/m^2
Comparing with the general equation of pressure wave of sound, *i.e.* $p_0\sin(\omega t - kx)$, we get

$$\omega = 1000 \text{ and } k = 3$$

Also, $k = \frac{\omega}{v}$

$\Rightarrow$ $v = \omega/k$

$\therefore$ Velocity of sound, $|v_1| = \frac{1000}{3}$

Or

Speed of sound wave can also be calculated as

$$v = -\frac{(\text{coefficient of } t)}{(\text{coefficient of } x)} = -\frac{1000}{(-3)} = \frac{1000}{3}\ \text{m/s}$$

Now, relation between velocity of sound and temperature is

$$v = \sqrt{\frac{\gamma RT}{m}}$$

$\Rightarrow$ $v \propto \sqrt{T}$

or $\frac{v_2}{v_1} = \sqrt{\frac{T_2}{T_1}} \Rightarrow T_2 = \frac{v_2^2}{v_1^2}\cdot T_1$

Here, $v_2 = 336$ m/s, $v_1 = 1000/3$ m/s,
$T_1 = 0°\text{C} = 273$ K

$$\therefore \quad T_2 = \frac{(336)^2}{(1000/3)^2} \times 273 = 277.38\ \text{K}$$

$$\therefore \quad T_2 = 4.38°\text{C} \simeq 4°\text{C}$$

Intensity of Sound Waves

Intensity of sound waves in terms of pressure is given as

$$I = \frac{v(\Delta p)_m^2}{2B}$$

where, Δp_m is the amplitude of pressure variation.

If the intensity of sound waves in watt per square metre is I, then the intensity level β in decibels (dB) is given by

$$\beta = 10 \log_{10} \frac{I}{I_0}$$

where, $I_0 \cong 10^{-12}\ \text{W/m}^2$

For comparison of two different sounds in dB, we can write

$$\beta_2 - \beta_1 = 10 \log_{10} \frac{I_2}{I_1}$$

Note Threshold of hearing is 0 dB and threshold of pain is 120 dB.

Example 8. *Assume that the displacement(s) of air is proportional to the pressure difference Δp created by a sound wave. Displacement further depends on the speed of sound v, density of air ρ and the frequency f. If $\Delta p = 10$Pa, $v = 300$ m/s, $\rho = 1$ kg/m^3 and $f = 1000$ Hz, then displacement(s) will be the order of (Take, the multiplicative constant to be 1)* [JEE Main 2020]

(a) 10 mm (b) 1 mm
(c) $\frac{1}{10}$ mm (d) $\frac{3}{100}$ mm

Sol. (d) The pressure difference is given by

$$\Delta p = \frac{B\omega}{v} \times s$$

$\therefore$ Displacement of air, $s = \frac{\Delta p \times v}{B\omega} = \frac{\Delta p v}{\rho v^2 \omega}$

$$= \frac{\Delta p}{\rho v \omega} = \frac{10}{1 \times 300 \times 1000}$$

$$\approx \frac{3}{100}\ \text{mm}$$

Example 9. *A window whose area is 2 m^2 opens on a street where the street noise results in an intensity level at the window of 60 dB. Now, if a sound absorber is fitted at the window, how much energy from the street will it collect in a day?*

(a) 0.73 J (b) 0.173 J (c) 2.73 J (d) 1.73 J

Sol. (b) By definition, sound level $= 10 \log \frac{I}{I_0} = 60$

or $\frac{I}{I_0} = 10^6$

$I = 10^{-12} \times 10^6 = 1\ \mu\text{ Wm}^{-2}$ $[\because I_0 = 10^{-12}\ \text{Wm}^{-2}]$

Power entering the room $= 1 \times 10^{-6} \times 2 = 2\ \mu\text{W}$

Energy collected in a day $= 2 \times 10^{-6} \times 86400 = 0.173$ J

Example 10. *A small speaker delivers 2 W of audio output. At what distance from the speaker will an observer detect 120 dB intensity sound? (Take, reference intensity of sound as 10^{-12} W/m^2)* [JEE Main 2019]

(a) 40 cm (b) 20 cm (c) 10 cm (d) 30 cm

Sol. (a) Loudness of sound in decibel is given by $\beta = 10 \log_{10}\left(\frac{I}{I_0}\right)$

where, I = intensity of sound in W/m^2,

I_0 = reference intensity ($= 10^{-12}$ W/m^2), chosen because it is near the lower limit of the human hearing range.

Here, $\beta = 120$ dB

So, we have $120 = 10 \log_{10}\left(\frac{I}{10^{-12}}\right)$

$$\Rightarrow \quad 12 = \log_{10}\left(\frac{I}{10^{-12}}\right)$$

Taking antilog, we have

$$\Rightarrow \quad 10^{12} = \frac{I}{10^{-12}}$$

$$\Rightarrow \quad I = 1\ \text{W/m}^2$$

This is the intensity of sound reaching the observer.

Now, intensity, $I = \frac{P}{4\pi r^2}$

where, r = distance from source,
P = power of output source.

Here, $P = 2$ W, we have

$$1 = \frac{2}{4\pi r^2}$$

$$\Rightarrow \quad r^2 = \frac{1}{2\pi}$$

$$\Rightarrow \quad r = \sqrt{\frac{1}{2\pi}}\ \text{m} = 0.398\ \text{m} \approx 40\ \text{cm}$$

Principle of Superposition of Waves

Two or more waves can travel independently in a medium without affecting the motion of one another. Thus, the resultant displacement of each particle of the medium at any instant is equal to the vector sum of displacements produced by the two waves separately. This principle is called principle of superposition of waves.

At a given point for two waves arriving with a phase difference ϕ and with equations

$$y_1 = A_1 \sin(\omega t - kx)$$

$$y_2 = A_2 \sin(\omega t - kx + \phi)$$

By principle of superposition, resultant wave is given by

$$\mathbf{y} = \mathbf{y}_1 + \mathbf{y}_2 = A \sin(\omega t - kx + \phi)$$

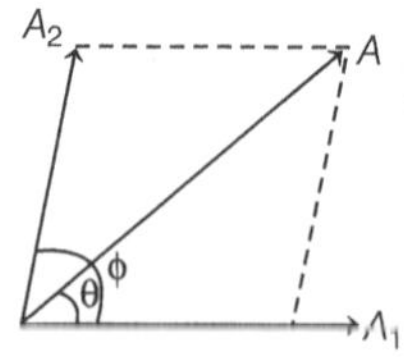

where, $A = \sqrt{A_1^2 + A_2^2 + 2A_1A_2 \cos\phi}$

$$\tan\theta = \frac{A_2 \sin\phi}{A_1 + A_2 \cos\phi}$$

Interference of Waves

When two coherent waves of same frequency propagate in same direction and superimpose on each other, then the intensity of the resultant wave becomes maximum at some points and minimum at some points. This phenomenon of intensity variation is called interference of waves.

If I_1 and I_2 are intensities of the interfering waves and ϕ is the phase difference, then the resultant intensity is given by

$$I = I_1 + I_2 + 2\sqrt{I_1 I_2}\cos\phi$$

If $I_1 = I_2 = I_0$, then $I = 4I_0 \cos^2 \frac{\phi}{2}$

Constructive interference These are those points, where the resultant amplitude/intensity is maximum.

For maximum amplitude,

$$\cos\phi = +1$$

or $\phi = 0, 2\pi, \ldots, 2n\pi$

$\Rightarrow$ $A = A_{max} = A_1 + A_2$

$$I = I_{max} = (\sqrt{I_1} + \sqrt{I_2})^2$$

Destructive interference These are those points, where resultant amplitude/ intensity is minimum.

For minimum amplitude,

$$\cos\phi = -1$$

or $\phi = \pi, 3\pi, \ldots, (2n-1)\pi$

$\Rightarrow$ $A = A_{min} = A_1 - A_2$

$$I = I_{min} = (\sqrt{I_1} - \sqrt{I_2})^2$$

In interference,

$$\frac{I_{max}}{I_{min}} = \frac{(A_1 + A_2)^2}{(A_1 - A_2)^2} = \frac{(\sqrt{I_1} + \sqrt{I_2})^2}{(\sqrt{I_1} - \sqrt{I_2})^2}$$

Example 11. *Two coherent sound sources are at distances $x_1 = 0.20$ m and $x_2 = 0.48$ m from a point. The intensity of the resultant wave at that point, if the frequency of each wave is $f = 400$ Hz and velocity of wave in the medium is $v = 448\ ms^{-1}$ is (Take, the intensity of each wave is $I_0 = 60\ Wm^{-2}$)*

(a) 120 Wm^{-2} (b) 125 Wm^{-2}
(c) 130 Wm^{-2} (d) 135 Wm^{-2}

Sol. (a) Path difference, $\Delta x = x_2 - x_1 = 0.48 - 0.20 = 0.28$ m

Phase difference, $\phi = \frac{2\pi}{\lambda}\Delta x = \left(\frac{2\pi f}{v}\right)\Delta x$

$$= \frac{2\pi(400)(0.28)}{448} = \frac{\pi}{2}$$

$$I = I_1 + I_2 + 2\sqrt{I_1 I_2}\cos\phi$$

or $I = I_0 + I_0 + 2I_0 \cos(\pi/2)$

$$= 2I_0 = 2(60) = 120\ Wm^{-2}$$

Reflection and Transmission of Waves

When sound waves are incident on a boundary separating two media, a part of it is reflected back into the initial medium while the remaining is partly absorbed and partly transmitted into the second medium.

Following are the characteristics of reflection and transmission of waves

(i) In case of reflection from a denser medium or rigid support or fixed end, there is an inversion of the reflected displacement wave, *i.e.* if the incident wave is $y = A_i \sin(\omega t - kx)$, then reflected wave will be

$$y = -A_r \sin(\omega t + kx)$$
$$= A_r \sin(\omega t + kx + \pi)$$

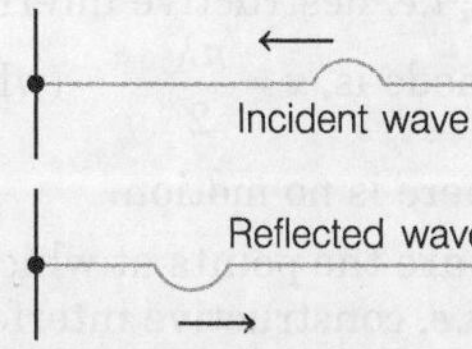

Reflection from fixed end

i.e. In case of reflection from a denser medium, displacement wave change its phase by π. While in case of reflection from a rarer medium or free end no inversion of wave or phase change occurs.

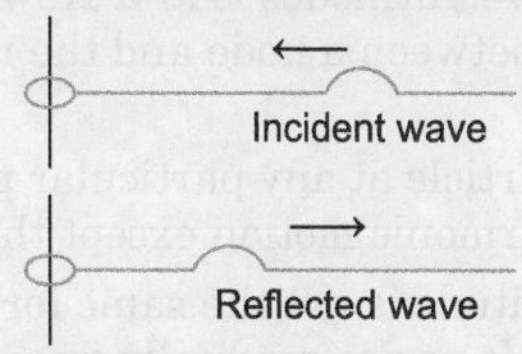

Reflection from free end

(ii) In case of pressure wave, there is no phase change when the wave is reflected from a denser medium or fixed end.

Wave property	Reflection	Transmission (Refraction)
v	does not change	changes
f, T, ω	do not change	do not change
λ, k	do not change	change
A, I	change	change
ϕ	$\Delta\phi = 0$, from a rarer medium $\Delta\phi = \pi$, from a denser medium	$\Delta\phi = 0$

Note (i) The concept of rarer or denser medium for a wave is through speed (and not density of medium). For example, water is rarer for sound and denser for light than air, as for sound $v_w > v_a$, while for light $v_w < v_a$.

(ii) Multiple reflection of sound is called an *echo*. If the distance of reflector from the source is, d and t is the time of echo, then

$$d = \frac{vt}{2}.$$

Standing or Stationary Waves

When two harmonic waves of equal frequency and amplitude travelling through a medium in the opposite directions along the same line superimpose, then we get stationary waves.

For two such waves having equations,

$$y_1 = A\sin(kx - \omega t) \text{ and } y_2 = A\sin(kx + \omega t)$$

The resultant equation after interference will be

$$y = 2A\sin(kx)\cos(\omega t) = 2A\sin\left(\frac{2\pi x}{\lambda}\right)\cos\left(\frac{2\pi t}{T}\right)$$

This is the equation of standing wave.

Following are some important points related to standing waves

(i) **Nodes** are the points at which the amplitude is always zero, *i.e.* destructive interference points.

Position of node is, $x = \frac{n\lambda}{2}$ (where, $n = 0, 1, 2 \ldots$)

At nodes, there is no motion.

(ii) **Antinodes** are the points at which amplitude is maximum, *i.e.* constructive interference points.

Position of antinode is,

$$x = (2n + 1)\frac{\lambda}{4} \quad \text{(where, } n = 0, 1, 2, \ldots)$$

(iii) The distance between two consecutive nodes or two consecutive antinodes is $\lambda/2$. However, the distance between a node and the next antinode is $\lambda/4$.

(iv) All the particle at any particular point x executes simple harmonic motion except those at nodes.

(v) The amplitude is not the same for different particles. It varies gradually from zero at nodes to maximum at antinodes.

(vi) Energy is not transported along the string to the right or to the left, because energy cannot flow past the nodes points in the string which are permanently at rest.

(vii) Due to persistance of vision, these waves appear in the form of loops. All the particles in a loop are in the same phase. But the particles in adjacent loops differ in phase by π.

(viii) Stationary waves may be transverse or longitudinal.

(ix) Two identical waves moving in opposite directions along the string will still produce standing waves even, if their amplitudes are unequal (as shown in figure).

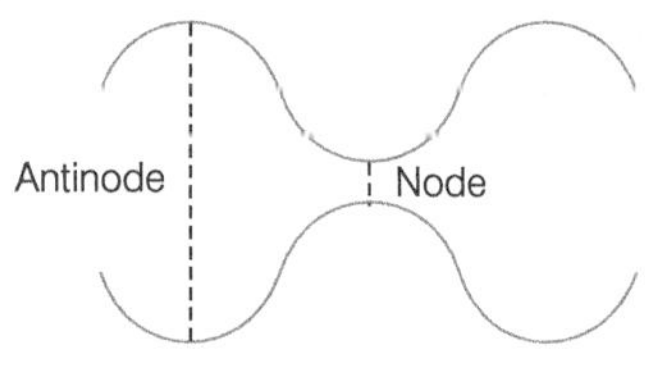

The standing wave ratio (SWR) is defined as

$$= \frac{A_{\max}}{A_{\min}} = \frac{A_i + A_r}{A_i - A_r}$$

For 100% reflection SWR = ∞ and for no reflection SWR = 1

(x) Energy in standing waves in given loop between two nodes oscillates between elastic PE and KE of the particles of the medium. When the particles are at their mean position, KE is maximum while elastic PE is minimum.

When particles are at their extreme positions KE is minimum while elastic PE is maximum.

Vibrations of Air Columns and Strings

Few terms related to standing waves in air columns and strings are given below

(i) When a sound source produces sound waves, it consists of mixture of many frequencies. This mixture of sound is called *note*.

(ii) If produced sound constains only one frequency, then it is called *tone*.

(iii) Tone of minimum frequency is called *fundamental tone*.

(iv) Those tones which are having frequencies greater than fundamental tones are called *overtones*.

(v) Sound produced from a source consists of frequencies which contains fundamental frequency as well as frequencies which are multiple of fundamental frequency, called *harmonics*.

The ratio between the frequencies of two notes is called the *musical interval*.

Following are the names of some musical intervals

(a) Unison $\frac{n_2}{n_1} = 1$

(b) Octave $\frac{n_2}{n_1} = 2$

(c) Major tone $\frac{n_2}{n_1} = \frac{9}{8}$

(d) Minor tone $\frac{n_2}{n_1} = \frac{10}{9}$

(e) Semi tone $\frac{n_2}{n_1} = \frac{16}{15}$

(f) Fifth tone $\frac{n_2}{n_1} = \frac{3}{2}$

Stationary Waves in Strings

Stationary waves in the strings can be produced by two ways, *i.e.*

(i) **When both the ends of the string are fixed**

A string of length L is stretched between two fixed points. When the string is set into vibrations, a transverse progressive wave begins to travel along the string. It is reflected at the other fixed end.

The incident and the reflected waves interfere to produce a stationary transverse wave in which the ends are always nodes.

Various modes of vibrations of a stretched string are shown below

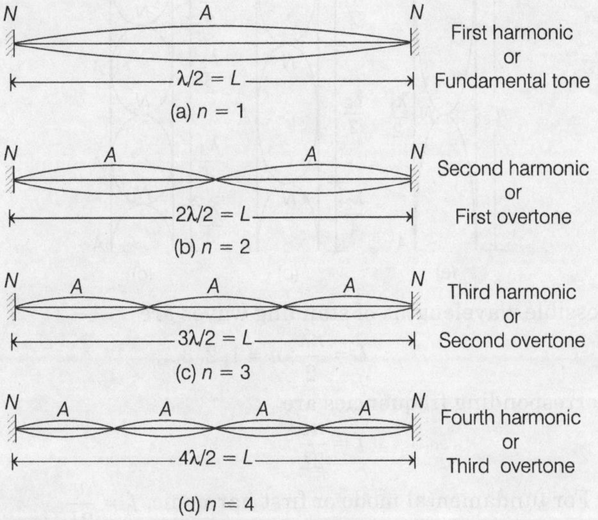

Possible wavelength of stationary wave are

$$\lambda = \frac{2L}{n}; n = 1, 2, 3, \ldots$$

$\therefore$ Natural frequencies of oscillations are

$$f = n\left(\frac{v}{2L}\right) = \frac{n}{2L}\sqrt{\frac{T}{\mu}}$$

So, fundamental frequency or first harmonic is

$$f_1 = \frac{v}{2L}$$

Similarly, $f_2 = \frac{2v}{2L} = 2f_1$

(first overtone or second hermonic)

$$f_3 = \frac{3v}{2L} = 3f_1$$

(second overtone or third harmonic)

(ii) **When one end of the string is fixed and other end is free to move**

Fundamental frequency of vibration or first harmonic is

f_0 or $n_0 = \frac{v}{4l}$

Frequency of third harmonic, $n_1 = \frac{3v}{4l} = 3n_0$

Frequency of fifth harmonic, $n_2 = \frac{5v}{4l} = 5n_0$

$$n_0 : n_1 : n_2 \ldots\ldots = 1:3:5 : \ldots\ldots$$

Laws of Vibration of Stretched String

- **Law of length** For a given wire under a given tension, the frequency of wire varies inversely as its vibrating length, *i.e.* $f \propto \frac{1}{L}$

  or $f_1L_1 = f_2L_2$

  where, T and m are constants.

- **Law of tension** For a uniform wire of given length and material, the frequency of the wire varies directly as the square root of tension, *i.e.*

  $$f \propto \sqrt{T}$$

  or $$\frac{f_1}{f_2} = \sqrt{\left(\frac{T_1}{T_2}\right)}$$

  where, L and m are constants.

- **Law of mass** When L and T are constants, the frequency of vibration of the wire varies inversely as the square root of mass per unit length of the wire, *i.e.*

  $$f \propto \frac{1}{\sqrt{m}}$$

  where, L and T are constants.

So, $$f \propto \frac{1}{L} \propto \frac{1}{\sqrt{m}}$$

Hence, a graph between L and $\sqrt{m}$ is a straight line.

Example 12. *A string 2.0 m long and fixed at its ends is driven by a 240 Hz vibrator. The string vibrates in its third harmonic mode. The speed of the wave and its fundamental frequency is* **[JEE Main 2019]**

(a) 180 m/s, 80 Hz
(b) 320 m/s, 80 Hz
(c) 320 m/s, 120 Hz
(d) 180 m/s, 120 Hz

Sol. (*b*) Frequency of vibration of a string in *n*th harmonic is given by

$$f_n = n \cdot \frac{v}{2L} \quad \ldots \text{(i)}$$

where, v = speed of sound and L = length of string.

Here, $f_3 = 240$ Hz, $L = 2$ m and $n = 3$

Substituting these values in Eq. (i), we get

$$240 = 3 \times \frac{v}{2 \times 2}$$

$$\Rightarrow \quad v = \frac{4 \times 240}{3} = 320 \text{ ms}^{-1}$$

Also, fundamental frequency,

$$f = \frac{f_n}{n} = \frac{f_3}{3}$$

$$= \frac{240}{3} = 80 \text{ Hz}$$

Example 13. *A wire of length 2L, is made by joining two wires A and B of same length but different radii r and 2r and made of the same material. It is vibrating at a frequency such that the joint of the two wires forms a node. If the number of antinodes in wire A is p and that in B is q, then the ratio p : q is* **[JEE Main 2019]**

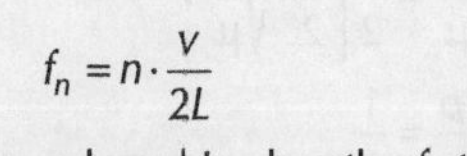

(a) 3 : 5 (b) 4 : 9
(c) 1 : 2 (d) 1 : 4

Sol. (c)

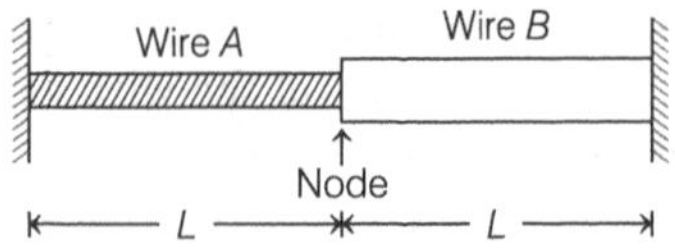

Let mass per unit length of wires are μ_A and μ_B, respectively.

$\because$ For same material, density is also same.

So, $$\mu_A = \frac{\rho \pi r^2 L}{L} = \mu$$

and $$\mu_B = \frac{\rho 4\pi r^2 L}{L} = 4\mu$$

Tension (T) in both connected wires are same.

So, speed of wave in wires are

$$v_A = \sqrt{\frac{T}{\mu_A}} = \sqrt{\frac{T}{\mu}} \qquad (\because \mu_A = \mu \text{ and } \mu_B = 4\mu)$$

and $$v_B = \sqrt{\frac{T}{\mu_B}} = \sqrt{\frac{T}{4\mu}}$$

So, nth harmonic in such wires system is

$$f_{nth} = \frac{pv}{2L}$$

$$\Rightarrow \quad f_A = \frac{pv_A}{2L} = \frac{p}{2L}\sqrt{\frac{T}{\mu}} \qquad \text{(for } p \text{ antinodes)}$$

Similarly, $$f_B = \frac{qv_B}{2L} = \frac{q}{2L}\sqrt{\frac{T}{4\mu}} = \frac{1}{2}\left(\frac{q}{2L}\sqrt{\frac{T}{\mu}}\right) \qquad \text{(for } q \text{ antinodes)}$$

As frequencies f_A and f_B are given equal.

So, $$f_A = f_B$$

$$\Rightarrow \quad \frac{p}{2L}\sqrt{\frac{T}{\mu}} = \frac{q}{2}\left[\frac{1}{2L}\sqrt{\frac{T}{\mu}}\right]$$

$$\frac{p}{q} = \frac{1}{2}$$

$$\Rightarrow \quad p : q = 1 : 2$$

Stationary Waves in Air Column

When longitudinal waves propagate in a fluid in a pipe with finite length, the waves are reflected from the ends. Thus, the superposition of the waves travelling in opposite directions forms a longitudinal standing wave.

- Longitudinal standing waves are described either in terms of the displacement or the pressure variation in the fluid.
- Displacement node and antinode refers to the points, where the particles of the fluid have zero and maximum displacement, respectively.
- Pressure node and antinode refers to the points, where the pressure and density variation is zero and maximum, respectively.

Open Organ Pipe

If both ends of a pipe of length L are open and a system of air is directed against an edge, standing longitudinal waves can be set up in the tube. The open end has a displacement antinode.

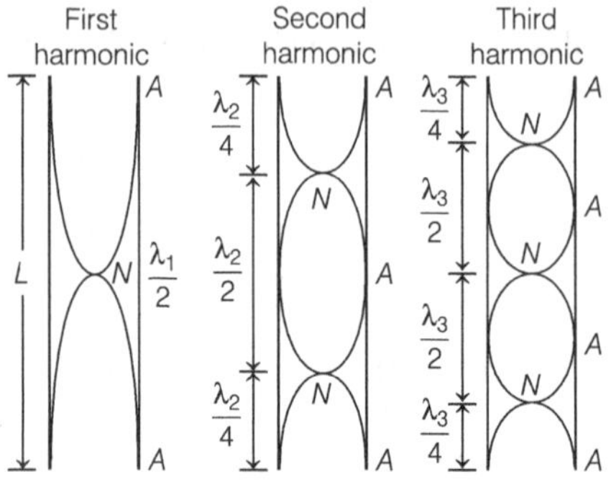

(a) (b) (c)

Possible wavelengths of standing waves are

$$L = \frac{n\lambda}{2};\ n = 1, 2, 3 \ldots$$

Corresponding frequencies are

$$f = \frac{nv}{2L}$$

$\therefore$ For fundamental mode or first harmonic, $f_1 = \dfrac{v}{2L}$

For the second harmonic or first overtone, $f_2 = \dfrac{v}{2L} = 2f_1$

Similarly, for the third harmonic or second overtone,

$$f_3 = \frac{3v}{2L} = 3f_1$$

The above relations, we get

$$f_1 : f_2 : f_3 : \ldots = 1 : 2 : 3 : \ldots$$

i.e. The natural frequencies of oscillations form a harmonic series that includes all integral multiples of fundamental frequency.

Closed Organ Pipe

In this type of organ pipe, one end is closed and the other end is open. A displacement node is formed at the closed end and an antinode at the open end.

For a organ pipe of length L, possible wavelengths of the stationary waves are

$$\lambda = \frac{4L}{n};\ \text{where } n = 1, 3, 5 \ldots$$

$\therefore$ Corresponding frequencies are, $f = \dfrac{nv}{4L}$

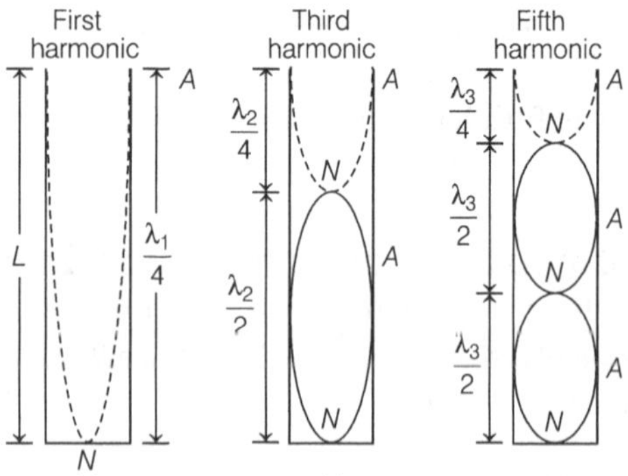

(a) (b) (c)

$\therefore$ Fundamental frequency given as

$$f_1 = \frac{v}{4L}$$

Frequency of third harmonic or first overtone,

$$f_2 = \frac{3v}{4L} = 3f_1$$

Similarly, frequency of fifth harmonic or second overtone,

$$f_3 = \frac{5v}{4L} = = 5f_1$$

From above relations, we get

$$f_1 : f_2 : f_3 \ldots = 1 : 3 : 5 \ldots$$

i.e. The natural frequencies of oscillations form a harmonic series that includes only odd integral multiples of the fundamental frequencies.

Example 14. *A pipe, 30 cm long is open at both ends, harmonic mode of the pipe which resonates at 1.1 kHz source is (Take, speed of sound in air is 330 ms^{-1})*

(a) *first* (b) *second*
(c) *third* (d) *four*

Sol. (b) The first harmonic frequency is

$$f_1 = \frac{v}{\lambda_l} = \frac{v}{2L} \quad \text{(open pipe)}$$

where, L is the length of the pipe. The frequency of its nth harmonic is

$$f_n = \frac{nv}{2L} \quad \text{for } n = 1, 2, 3, \ldots \text{(open pipe)}$$

Given, $L = 30$ cm, $v = 330 \text{ ms}^{-1}$

$$f_n = \frac{n(330 \text{ ms}^{-1})}{0.6 \text{ m}} = 550 \text{ ms}^{-1}$$

The frequencies of 2nd harmonic, 3rd hormonic, 4th hormonic, ... are $2 \times 550 = 1100$ Hz, $3 \times 550 = 1650$ Hz, $4 \times 550 = 2200$ Hz.

Clearly, a source of frequency 1.1 kHz will resonate at f_2, *i.e.* the second harmonic.

Example 15. *An open pipe is suddenly closed at one end with the result that the frequency of third harmonic of the closed pipe is found to be higher by 100 Hz than the fundamental frequency of the open pipe. The fundamental frequency of the open pipe is*

(a) 100 *Hz* (b) 150 *Hz*
(c) 200 *Hz* (d) 250 *Hz*

Sol. (c) For fundamental mode in open pipe,

$$L = \lambda/2$$

$$\Rightarrow \quad \lambda = 2L$$

and

$$f_F = \frac{v}{\lambda} = \frac{v}{2L} \quad \ldots\text{(i)}$$

For third harmonic in closed pipe,

$$L = \frac{3\lambda}{4}$$

$$\Rightarrow \quad \lambda = \frac{4L}{3}$$

$$f_{H_2} = \frac{v}{\lambda} = \frac{v}{\frac{4}{3}L} = \frac{3v}{4L} \quad \ldots\text{(ii)}$$

From Eqs. (i) and (ii), we have

$$\frac{f_H}{f_F} = \frac{3}{2}$$

$$\Rightarrow \quad f_H = f_F \times \frac{3}{2} \quad \ldots\text{(iii)}$$

But $\quad f_H - f_F = 100$

$$\frac{3}{2}f_F - f_F = 100$$

$$\Rightarrow \quad f_F = 200 \text{ Hz}$$

End Correction

It was found that the antinode is not formed exactly at the open end of the *organ pipe* but actually due to finite momentum of the particles the reflection takes place a little above the open end; that is why the antinode is formed a little above the open end.

For this, a correction is applied being known as end correction. This is denoted by *e*. If length of organ pipe is *L* and end correction is *e*, then length of air-column in closed pipe will be $(L + e)$ and in *open pipe*, $(L + 2e)$. Thus, for a closed organ pipe,

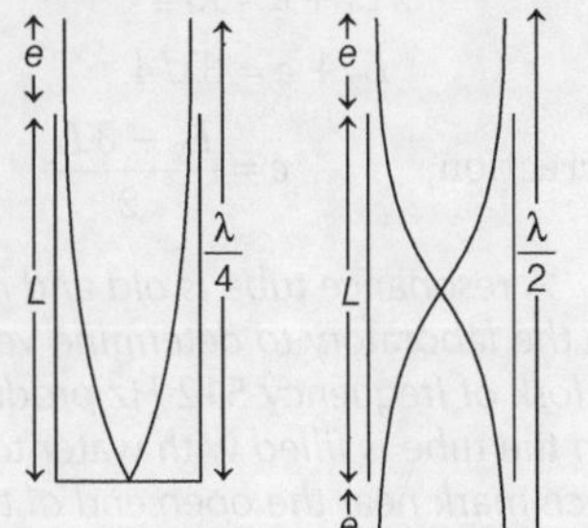

(a) (b)

$$f_1 = \frac{v}{4(L + e)}$$

and for an open organ pipe,

$$f_2 = \frac{v}{2(L + 2e)}$$

The value of end correction *e* is $0.6r$ for closed organ pipe and $1.2r$ for an open organ pipe, where *r* is the radius of the pipe.

Resonance Tube

It is a closed organ pipe in which length of air-column can be increased or decreased. When a vibrating tuning fork is brought at its mouth as shown in figure, then forced vibrations are set up in its air-column. If we adjust the length of air-column as such its any natural frequency equals to the frequency of tuning fork, then the amplitude of forced vibrations of air-column increases very much. This is the state of *resonance*.

When length of air-column is $L = \lambda/4$, then the first resonance occurs. As shown in Fig. (a), antinode is formed at an open end and a node is formed at the water surface.

Now, when length of air-column is $L_2 = 3\lambda/4$, then second resonance occurs. In this condition, two antinodes and two nodes are formed as shown in Fig. (b).

End correction In resonance tube, antinode is not formed exactly at open end but it is formed a little above the open end known as end correction (e). So, in first and second state of air-column, the lengths are $L_1 + e$ and $L_2 + e$.

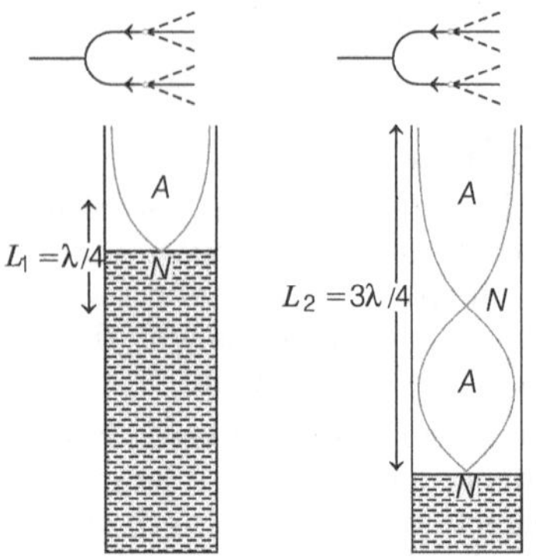

(a) (b)

$$L_1 + e = \lambda/4$$

and $$L_2 + e = 3\lambda/4$$

Hence, end correction, $$e = \frac{L_2 - 3L_1}{2}$$

Example 16. *A resonance tube is old and has jagged end. It is still used in the laboratory to determine velocity of sound in air. A tuning fork of frequency 512 Hz produces first resonance when the tube is filled with water to a mark 11 cm below a reference mark near the open end of the tube. The experiment is repeated with another fork of frequency 256 Hz which produces first resonance when water reaches a mark 27 cm below the reference mark. The velocity of sound in air obtained in the experiment is close to* **[JEE Main 2019]**

(a) $328\ ms^{-1}$ (b) $341\ ms^{-1}$

(c) $322\ ms^{-1}$ (d) $335\ ms^{-1}$

Sol. (a) In first resonance, length of air column $= \frac{\lambda}{4}$.

So, $$L_1 + e = \frac{\lambda}{4}$$

or $$11 \times 4 + 4e = \lambda$$

So, speed of sound,

$$v = f_1\lambda = 512(44 + 4e) \quad \ldots(i)$$

and in second case,

$$L_1' + e = \frac{\lambda'}{4}$$

or $$27 \times 4 + 4e = \lambda'$$

$\Rightarrow$ $$v = f_2\lambda' = 256(108 + 4e) \quad \ldots(ii)$$

Dividing both Eqs. (i) and (ii), we get

$$1 = \frac{512(44 + 4e)}{256(108 + 4e)}$$

$\Rightarrow$ $$e = 5\text{ cm}$$

Substituting value of e in Eq. (i), we get

Speed of sound, $$v = 512(44 + 4e)$$
$$= 512(44 + 4 \times 5)$$
$$= 512 \times 64\text{ cm s}^{-1}$$
$$= 327.68\text{ ms}^{-1}$$
$$\approx 328\text{ ms}^{-1}$$

Beats

When two sound waves of nearly same frequency are produced simultaneously, then the intensity of resultant sound wave increases and decreases with time. This change in the intensity of sound waves is called as the phenomenon of beats.

The time interval between two successive beats is called *beat period* and the number of beats per second is called the *beat frequency*.

If f_1 and f_2 are the frequencies ($f_1 > f_2$) of the two waves, then the beat frequency, $b = f_1 - f_2$

$\therefore$ Beat period, $$T = \frac{1}{f_1 - f_2}$$

Example 17. *The first overtone of an open pipe and the fundamental note of a pipe closed at one end gives 5 beats s^{-1}, when sounded together. If the length of the pipe closed at one end is 25 cm, what are the possible lengths of the open pipe? (Neglect end corrections and take the velocity of sound in air to be 340 ms^{-1})*

(a) 90.5 and 120.5 cm

(b) 98.5 and 101.5 cm

(c) 95.5 and 102.5 cm

(d) 95.5 and 200 cm

Sol. (b) Let the fundamental frequency of the closed end pipe of length 25 cm be f_o. Then,

$$f_o = \frac{v}{4l} = \frac{340 \times 100}{4 \times 25} = 340\text{ Hz}$$

Possible frequencies of first overtone of the required open pipe are 340 ± 5, *i.e.* 345 or 335 Hz.

For the first overtone of an open pipe, the length of the pipe l equals the wavelength of the vibration.

Hence, $$345 = \frac{v}{l}$$

or $$l = \frac{34000}{345} = 98.5\text{ cm}$$

Other possible length l' is given by

$$335 = \frac{v}{l'}$$

$$l' = \frac{34000}{335} = 101.5\text{ cm}$$

Hence, possible lengths of the open pipe are 98.5 and 101.5 cm.

Example 18. *Two tuning forks A and B sounded together give 8 beat s^{-1}. With an air resonance tube closed at one end, the two forks give resonance when the two air columns are 32 cm and 33 cm, respectively. The frequencies of forks are*

(a) 260 Hz, 250 Hz (b) 264 Hz, 256 Hz
(c) 274 Hz, 256 Hz (d) 2709 Hz, 250 Hz

Sol. (b) Let the frequency of the first fork be f_1 and that of second be f_2. Then, we have

$$f_1 = \frac{v}{4\times 32} \text{ and } f_2 = \frac{v}{4\times 33}$$

We also see that $f_1 > f_2$

$$\therefore \quad f_1 - f_2 = 8 \quad \text{...(i)}$$

$$\text{and} \quad \frac{f_1}{f_2} = \frac{33}{32} \quad \text{...(ii)}$$

Solving Eqs. (i) and (ii), we get

$$f_1 = 264 \text{ Hz}$$

$$\text{and} \quad f_2 = 256 \text{ Hz}$$

Doppler's Effect

If a source of wave and the observer/receiver are in relative motion to each other, then the frequency observed (f_a) by the observer is different from the actual source frequency (f_o). This phenomenon is called Doppler's effect.

For different situations of the source and the observer the apparent frequencies observed due to Doppler's effect are tabulated as follows

Observer (o)	Sources (s)	Apparent frequency
$\rightarrow v_o$	Stationary ($v_s = 0$)	$f_a = \left(\frac{v + v_o}{v}\right) f_o$
$v_o \leftarrow$	Stationary ($v_s = 0$)	$f_a = \left(\frac{v - v_o}{v}\right) f_o$
Stationary ($v_o = 0$)	$\rightarrow v_s$	$f_a = \left(\frac{v}{v + v_s}\right) f_o$
Stationary ($v_o = 0$)	$v_s \leftarrow$	$f_a = \left(\frac{v}{v - v_s}\right) f_o$
$\rightarrow v_o$	$\rightarrow v_s$	$f_a = \left(\frac{v + v_o}{v + v_s}\right) f_o$
$v_o \leftarrow$	$\rightarrow v_s$	$f_a = \left(\frac{v - v_o}{v + v_s}\right) f_o$
$\rightarrow v_o$	$v_s \leftarrow$	$f_a = \left(\frac{v + v_o}{v - v_s}\right) f_o$
$v_o \leftarrow$	$v_s \leftarrow$	$f_a = \left(\frac{v - v_o}{v - v_s}\right) f_o$

Here, v = speed of sound, v_s = speed of source, v_o = speed of observer, f_o = original frequency of the source.

Note Doppler's effect is a wave phenomenon, it hold not only for sound waves but also for electromagnetic waves.

Special Cases of Doppler's Effect in Sound

1. Doppler's effect in reflected sound

For a source of sound moving with velocity v_s towards a wall which is itself moving in same direction and an observer too moving in the same direction as shown in figure.

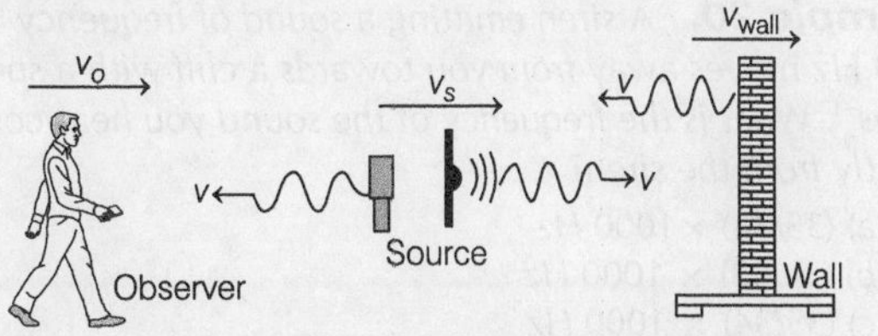

A descriptive explanation of this case of Doppler's effect can be understood by example given just below.

Example 19. *Submarine A travelling at 18 km/h is being chased along the line of its velocity by another submarine B travelling at 27 km/h. B sends a sonar signal of 500 Hz to detect A and receives a reflected sound of frequency ν. The value of ν is close to (Speed of sound in water = 1500 ms^{-1})*

[JEE Main 2019]

(a) 504 Hz (b) 507 Hz
(c) 499 Hz (d) 502 Hz

Sol. (d) Given, velocity of submarine (A),

$$v_A = 18 \text{ km/h} = \frac{18000}{3600} \text{ m/s}$$

$$\text{or} \quad v_A = 5 \text{ m/s} \quad \text{...(i)}$$

and velocity of submarine (B),

$$v_B = 27 \text{ km/h} = \frac{27000}{3600} \text{ m/s}$$

$$\text{or} \quad v_B = 7.5 \text{ m/s} \quad \text{...(ii)}$$

Signal sent by submarine (B) is detected by submarine (A) can be shown as

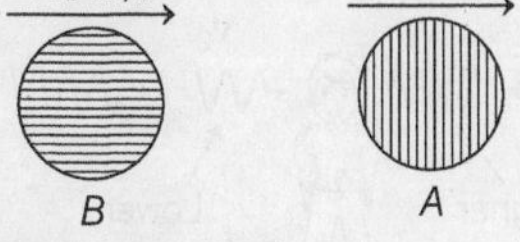

Frequency of the signal, f_o = 500 Hz

So, in this relative motion, frequency received by submarine (A) is

$$f_1 = \left(\frac{v_s - v_A}{v_s - v_B}\right) f_o = \left(\frac{1500 - 5}{1500 - 7.5}\right) 500 \text{ Hz}$$

$$\Rightarrow \quad f_1 = \frac{1495}{1492.5} \times 500 \text{ Hz}$$

The reflected frequency f_1 is now received back by submarine (B). So, frequency received at submarine (B) is

$$f_2 = \left(\frac{v_s + v_B}{v_s + v_A}\right) f_1$$

$$= \left(\frac{1500 + 7.5}{1500 + 5}\right)\left(\frac{1495}{1492.5}\right) 500 \text{ Hz}$$

$$\Rightarrow \quad f_2 = \left(\frac{1507.5}{1505}\right)\left(\frac{1495}{1492.5}\right) 500 \text{ Hz}$$

$$\Rightarrow \quad f_2 = 1.00166 \times 1.00167 \times 500$$

$$\Rightarrow \quad f_2 = 501.67 \text{ Hz}$$

$$\approx 502 \text{ Hz}$$

Example 20. *A siren emitting a sound of frequency 1000 Hz moves away from you towards a cliff with a speed of 10 ms⁻¹. What is the frequency of the sound you hear coming directly from the siren?*

(a) (33/34) × 1000 Hz
(b) (34/33) × 1000 Hz
(c) (35/34) × 1000 Hz
(d) (34/35) × 1000 Hz

Sol. (a) Sound heard directly,

$$f_1 = f_o\left(\frac{v}{v + v_s}\right)$$

$$v_s = 10 \text{ ms}^{-1}$$

$$\therefore \quad f_1 = \left(\frac{330}{330 + 10}\right) \times 1000$$

$$= \frac{33}{34} \times 1000 \text{ Hz}$$

Example 21. *A stationary observer receives sound from two identical tuning forks, one of which approaches and the other one recedes with the same speed (much less than the speed of sound). The observer hears 2 beats/s. The oscillation frequency of each tuning fork is $\nu_0 = 1400$ Hz and the velocity of sound in air is 350 m/s. The speed of each tuning fork is close to* **[JEE Main 2020]**

(a) $\frac{1}{8}$ m/s (b) $\frac{1}{2}$ m/s (c) 1 m/s (d) $\frac{1}{4}$ m/s

Sol. (d) Due to Doppler's effect, frequency of approaching tuning fork is higher and receding tuning fork is lower than emitted frequency.

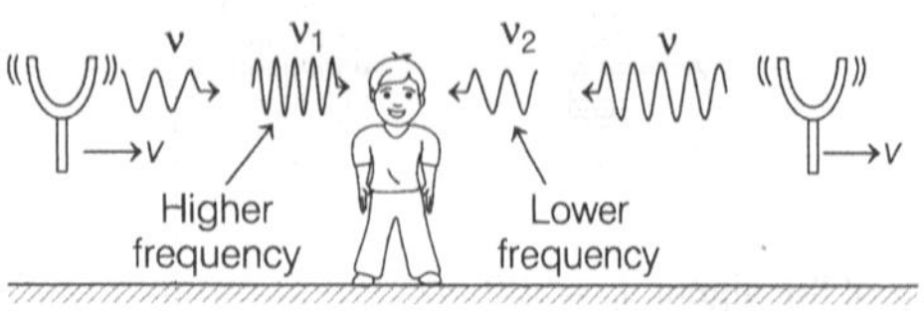

Frequency received from tuning fork moving towards observer,

$$\nu_1 = \nu\left(\frac{v}{v - v_s}\right)$$

where, ν = frequency emitted by tuning fork, v = speed of sound and v_s = speed of tuning fork.

Frequency received from receding tuning fork,

$$\nu_2 = \nu\left(\frac{v}{v + v_s}\right)$$

Beat frequency is the difference of both received frequencies.

So, $\quad \nu_{beat} = \nu_1 - \nu_2$

$$\Rightarrow \quad \nu_{beat} = \nu\left(\frac{v}{v - v_s}\right) - \nu\left(\frac{v}{v + v_s}\right)$$

$$= \frac{2\,\nu v_s v}{v^2 - v_s^2} = \frac{2\nu v_s}{v\left(1 - \frac{v_s^2}{v^2}\right)}$$

If $v_s << v$, then $\nu_{beat} = \frac{2 f v_s}{v}$

Here, $\nu = 1400$ Hz, $v = 350$ ms⁻¹ and $\nu_{beat} = 2$ s⁻¹

So, $\quad 2 = \frac{2 \times 1400 \times v_s}{350}$

$$\Rightarrow \quad v_s = \frac{1}{4} \text{ ms}^{-1}$$

Example 22. *A source S of acoustic wave of the frequency $\nu_0 = 1700$ Hz and a receiver R are located at the same point. At the instant $t = 0$, the source starts from rest to move away from the receiver with a constant acceleration a. The velocity of sound in air is 340 m/s. If $a = 10$ m/s², the apparent frequency that will be recorded by the stationary receiver at $t = 10$ s will be*

(a) 1700 Hz (b) 1.35 Hz (c) 2.89 Hz (d) 1300 Hz

Sol. (b) Source frequency, $\nu_o = 1700$ Hz. Source (coinciding with observer at $t = 0$) moves away with uniform acceleration a. Consider the wave which is received by the observer at instant $t = \tau$. It will have left the source at an earlier instant of time, say $t\,(< \tau)$, when the distance of source was r (say), if u be velocity of source at instant t, then $r = \left(\frac{1}{2}\right) at^2$ and $u = at$. The relation between τ and t is

$$\tau = t + \frac{r}{v} = t + \frac{at^2}{2v}$$

This is a quadratic equation in t, giving the solution

$$at = \frac{-2v + \sqrt{4v^2 + 8vat}}{2}$$

$$u = at = v\left(\sqrt{1 + \frac{2a\tau}{v}} - 1\right)$$

$$u = 340\left(\sqrt{1 + \frac{2 \times 10 \times 10}{340}} - 1\right)$$

$$= 340\left(\frac{\sqrt{27}}{17} - 1\right)$$

Then, apparent frequency is given by

$$\nu_a = \left(\frac{v}{v + u}\right)\nu_o$$

Putting the values $v = 340$ m/s, $\tau = 10$ s, $a = 10$ m/s², we have

$$\nu_a = \left(\frac{340}{340 + u}\right) 1700$$

$$\nu_a = 1700\sqrt{\frac{17}{27}} = 1.35 \text{ Hz}$$

2. Doppler's effect when the source and observer are not in same line of motion

From the figure, the position of a source is S and of observer is O. The component of velocity of source towards the observer is $v\cos\theta$. For this situation, the approach frequency is

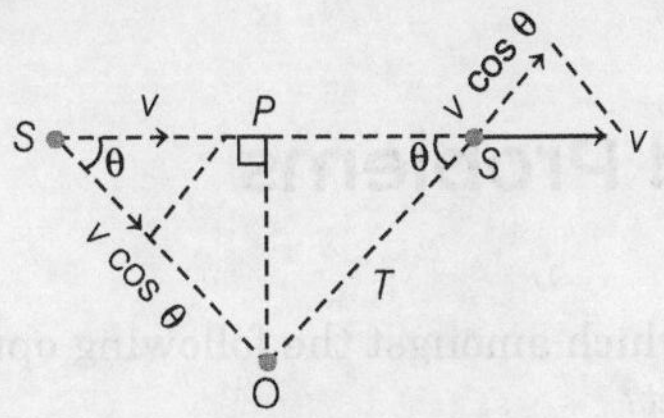

$$f' = \frac{v}{v - v_s\cos\theta} \times f$$

Similarly, if the source is moving away from the observer as shown above with velocity component $v_s\cos\theta$, then

$$f' = \frac{v}{v + v_s\cos\theta} \times f$$

If $\theta = 90°$, then $v_s\cos\theta = 0$ and there is no shift in the frequency. Thus, at point P, Doppler's effect does not occur.

Note If wind blows at a speed v_w from the source to the observer, take $v \to v + v_w$ (both in numerator and denominator) and if in opposite direction (*i.e.* from observer to source), take $v \to v - v_w$. Thus, the modified formula is

$$f' = \left(\frac{v \pm v_w \pm v_o}{v \pm v_w \pm v_s}\right) f$$

Example 23. *A person P is 600 m away from the station, when train is approaching station with 72 km/h, it flows a whistle of frequency 800 Hz when 800 m away from the station. Frequency heard by the person is (Take, speed of sound = 340 ms^{-1})*

(a) 800 Hz (b) 839.5 Hz
(c) 829.5 Hz (d) 834.5 Hz

Sol. (*b*) From Doppler's effect, frequency,

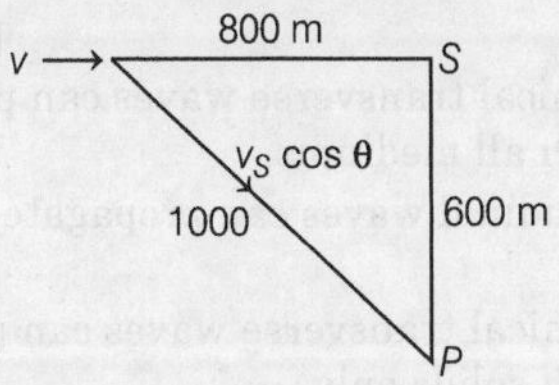

$$f_{app} = \frac{v}{v - v_S\cos\theta} f$$

$$= \frac{340}{340 - 16} \times 800$$

$$= 839.5 \text{ Hz}$$

Practice Exercise

ROUND I Topically Divided Problems

Wave Motion and Progressive Waves

1. Which of the following statements are true for wave motion? [NCERT Exemplar]

(a) Mechanical transverse waves can propagate through all mediums.
(b) Longitudinal waves can propagate through solids only.
(c) Mechanical transverse waves can propagate through solids only.
(d) Longitudinal waves can propagate through vacuum.

2. The figure shows three progressive waves A, B and C moving in the same direction. What can be concluded with respect to wave A from the figure?

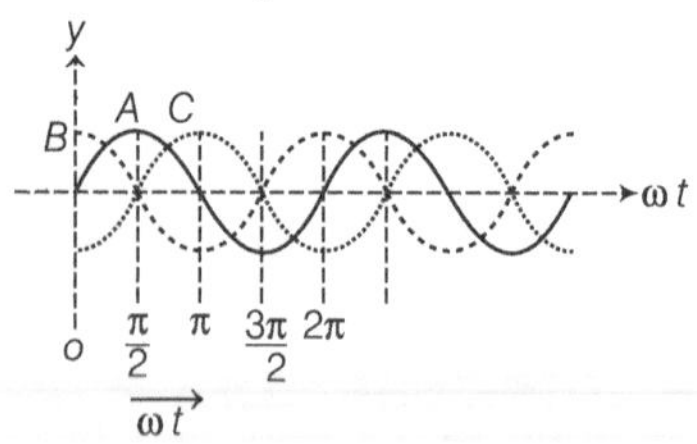

(a) The wave C lags behind by a phase angle of $\pi/2$ and the wave B is ahead by a phase angle $\pi/2$
(b) The wave C is ahead by a phase angle of $\pi/2$ and the wave B lags behind by a phase angle $\pi/2$
(c) The wave C is ahead by a phase angle of π and the wave B lags behind by a phase angle π
(d) The wave C lags behind by a phase angle π and the wave B is ahead by a phase angle π

3. The equation of wave is represented by $y = 10^{-4} \sin\left[100t - \frac{x}{10}\right]$ m, then the velocity of wave will be

(a) 100 ms^{-1}
(b) 4 ms^{-1}
(c) 1000 ms^{-1}
(d) zero

4. A transverse harmonic wave on a string is described by $y(x, t) = 3.0 \sin(36t + 0.018x + \pi/4)$, where x and y are in cm and t is in second. The positive direction of x is from left to right.

Then, which amongst the following option is incorrect?

(a) The wave is travelling from right to left.
(b) The speed of the wave is 20m/s.
(c) Frequency of the wave is 5.7 Hz.
(d) The least distance between two successive crests in the wave is 2.5 cm.

5. A transverse wave is described by the equation $y = y_0 \sin 2\pi\left[ft - \frac{x}{\lambda}\right]$. The maximum particle velocity is equal to four times the wave velocity, if

(a) $\lambda = \pi y_0/4$
(b) $\lambda = 2\pi y_0$
(c) $\lambda = \pi/y_0$
(d) $\lambda = \pi y_0/2$

6. For the wave shown in figure, write the equation of this wave, if its position is shown at $t = 0$. Speed of wave is $v = 300$ m/s.

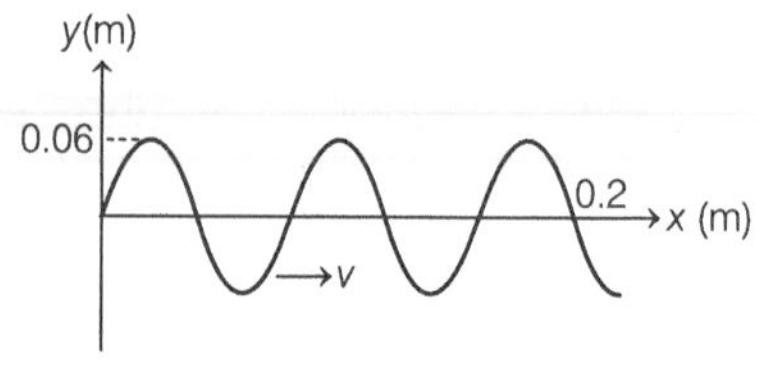

(a) $y = (0.06)\cos[(78.5\text{m}^{-1})x + (23562\text{s}^{-1})t]$ m
(b) $y = (0.06)\sin[(78.5\text{ m}^{-1})x - (23562\text{ s}^{-1})t]$ m
(c) $y = (0.06)\sin[(78.5\text{ m}^{-1})x + (28562\text{ s}^{-1})t]$ m
(d) $y = (0.06)\cos[(78.5\text{m}^{-1})x - (28562\text{s}^{-1})t]$ m

7. A travelling harmonic wave is represented by the equation $y(x, t) = 10^{-3} \sin(50t + 2x)$, where x and y are in metre and t is in second. Which of the following is a correct statement about the wave? [JEE Main 2019]

(a) The wave is propagating along the negative X-axis with speed 25 ms^{-1}.
(b) The wave is propagating along the positive X-axis with speed 25 ms^{-1}.
(c) The wave is propagating along the positive X-axis with speed 100 ms^{-1}.
(d) The wave is propagating along the negative X-axis with speed 100 ms^{-1}.

8. In a transverse wave the distance between a crest and neighbouring trough at the same instant is 4.0 cm and the distance between a crest and trough at the same place is 1.0 cm. The next crest appears at the same place after a time interval of 0.4 s. The maximum speed of the vibrating particles in the medium is **[JEE Main 2013]**

(a) $\frac{3\pi}{2}$ cm/s (b) 5π cm/s
(c) $\frac{\pi}{2}$ cm/s (d) 2π cm/s

9. A transverse mechanical harmonic wave is travelling on a string. Maximum velocity and maximum acceleration of a particle on the string are 3 m/s and 90 m/s^2, respectively. If the wave is travelling with a speed of 20 m/s on the string. The wave function describing the wave is

(a) $0.1 \cos (30t \pm 1.5x)$ (b) $0.1 \sin (30t \pm 1.5x)$
(c) $0.2 \sin (1.5t \pm 30x)$ (d) $0.2 \cos (1.5t \pm 30x)$

10. A transverse sinusoidal wave moves along a string in the positive x-direction at a speed of 10 cm s^{-1}. The wavelength of the wave is 0.5 m and its amplitude is 10 cm. At a particular time t, the snapshot of the wave is shown in the figure. The velocity of the point P when its displacement is 5 cm, is

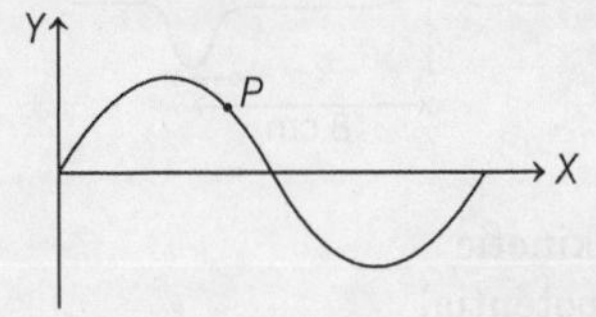

(a) $\frac{\sqrt{3}\pi}{50}\,\hat{j}\,\text{ms}^{-1}$ (b) $-\frac{\sqrt{3}\pi}{50}\,\hat{j}\,\text{ms}^{-1}$
(c) $\frac{\sqrt{3}\pi}{50}\,\hat{i}\,\text{ms}^{-1}$ (d) $-\frac{\sqrt{3}\pi}{50}\,\hat{i}\,\text{ms}^{-1}$

11. From a point source, if amplitude of waves at a distance r is A, its amplitude at a distance $2r$ will be

(a) A (b) $2A$ (c) $A/2$ (d) $A/4$

12. A simple harmonic progressive wave is represented by the equation $y = 8 \sin 2\pi (0.1x - 2t)$ where x and y are in cm and t is in seconds. At any instant, the phase difference between two particles separated by 2.0 cm in the x-direction is

(a) 18° (b) 54° (c) 36° (d) 72°

13. A string of mass 2.5 kg is under a tension of 200 N. The length of the stretched string is 20.0 m. If the transverse jerk is struck at one end of the string, the disturbance will reach the other end in **[NCERT Exemplar]**

(a) 1 s (b) 0.5 s
(c) 2 s (d) Data is insufficient

14. A transverse wave travels on a taut steel wire with a velocity of v when tension in it is 2.06×10^4 N. When the tension is changed to T, the velocity changed to $v/2$. The value of T is close to **[JEE Main 2020]**

(a) 10.2×10^2 N (b) 5.15×10^3 N
(c) 2.50×10^4 N (d) 30.5×10^4 N

15. The amplitude of wave disturbance propagating in positive direction of X-axis is given by $y = \frac{1}{1+x^2}$ at $t = 0$ and by $y = \frac{1}{1+(x-1)^2}$ at $t = 2$ s, where x and y are in metres. The shape of the wave disturbance does not change during propagation. The velocity of the wave is

(a) 0.5 ms^{-1} (b) 2.0 ms^{-1}
(c) 1.0 ms^{-1} (d) 4.0 ms^{-1}

16. A 100 Hz sinusoidal wave is travelling in the positive x-direction along a string with a linear mass density of 3.5×10^{-3} kg/ m and a tension of 35 N. At time $t = 0$, the point $x = 0$, has maximum displacement in the positive y-direction. Next when this point has zero displacement, the slope of the string is $\pi/20$. Which of the following expression represent(s) the displacement of string as a function of x (in metre) and t (in second)?

(a) $y = 0.025 \cos (200\pi t - 2\pi x)$
(b) $y = 0.5 \cos (200\pi t - 2\pi x)$
(c) $y = 0.025 \cos (100\pi t - 10\pi x)$
(d) $y = 0.5 \cos (100\pi t - 10\pi x)$

Sound Waves

17. Speed of sound waves in a fluid depends upon

(a) directly on density of the medium
(b) square of bulk modulus of the medium
(c) inversely of the square root of density
(d) inversely of the square root of bulk modulus of the medium

18. v_1 and v_2 are the velocities of sound at the same temperature in two monoatomic gases of densities ρ_1 and ρ_2, respectively. If $\frac{\rho_1}{\rho_2} = \frac{1}{4}$, then the ratio of velocities v_1 and v_2 will be

(a) 1: 2 (b) 4 :1 (c) 2 : 1 (d) 1: 4

19. If the temperature is raised by 1 K from 300 K, the percentage change in the speed of sound in the gaseous mixture is (R = 8.31 J/mol-K)

(a) 0.167% (b) 0.334% (c) 1% (d) 2%

20. The speed of sound in a mixture of 1 mole of helium and 2 mol of oxygen at 27°C is

(a) 800 ms^{-1} (b) 400.8 ms^{-1}
(c) 600 ms^{-1} (d) 1200 ms^{-1}

21. In brass, the velocity of longitudinal wave is 100 times the velocity of the transverse wave. If $Y = 1 \times 10^{11}$ Nm^{-2}, then stress in the wire is

(a) 1×10^{13} Nm^{-2} (b) 1×10^{9} Nm^{-2}
(c) 1×10^{11} Nm^{-2} (d) 1×10^{7} Nm^{-2}

22. The displacement-time graphs for two sound waves A and B are shown in figure, then the ratio of their intensities I_A / I_B is equal to

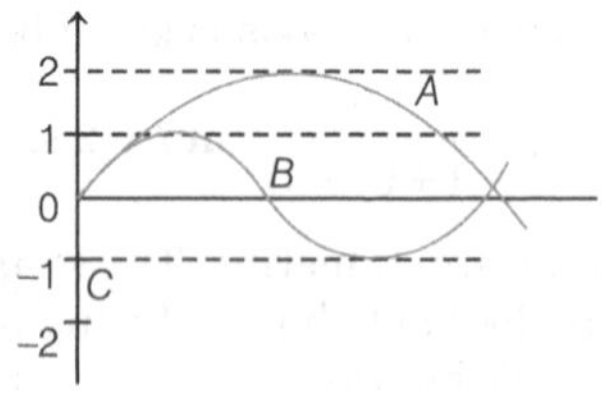

(a) 1 : 4 (b) 1 : 16
(c) 1 : 2 (d) 1 : 1

23. A sound wave of frequency 245 Hz travel with the speed of 300 ms^{-1} along the positive X-axis. Each point of the wave moves to and fro through a total distance of 6 cm. What will be the mathematical expression of this travelling wave? **[JEE Main 2021]**

(a) $y(x, t) = 0.03 \sin[5.1x - (0.2 \times 10^3)t]$
(b) $y(x, t) = 0.06 \sin[5.1x - (1.5 \times 10^3)t]$
(c) $y(x, t) = 0.06 \sin[0.8x - (0.5 \times 10^3)t]$
(d) $y(x, t) = 0.03 \sin[5.1x - (1.5 \times 10^3)t]$

24. A plane longitudinal wave of angular frequency 1000 rad s^{-1} is travelling along negative x-direction in a homogeneous gaseous medium of density $\rho = 1$ kg m^{-3}. Intensity of the wave is $I = 10^{-10}$ Wm^{-2} and the maximum pressure change is 2×10^{-4} Nm^{-2}. Then, displacement equation is

(a) $y = 10^{-9} \sin\left(1000\,t - 5x + \frac{\pi}{2}\right)$
(b) $y = 10^{-9} \cos(1000\,t + 5x)$
(c) $y = 10^{-9} \tan(1000\,t - 5x)$
(d) $y = 10^{-9} \cos(1000\,t - 5x)$

Stationary Waves in Strings and in Organ Pipes

25. Equation of a plane progressive wave is given by $y = 0.6 \sin 2\pi \left(t - \frac{x}{2}\right)$. On reflection from a denser medium, its amplitude becomes 2/3 of the amplitude of the incident wave. The equation of the reflected wave is **[NCERT Exemplar]**

(a) $y = 0.6 \sin 2\pi \left(t + \frac{x}{2}\right)$ (b) $y = -0.4 \sin 2\pi \left(t + \frac{x}{2}\right)$
(c) $y = 0.4 \sin 2\pi \left(t + \frac{x}{2}\right)$ (d) $y = -0.4 \sin 2\pi \left(t - \frac{x}{2}\right)$

26. The displacement of a particle executing periodic motion is given by $y = 4\cos^2(t/2)\sin(1000\,t)$. This expression may be considered to be a result of superposition of

(a) two waves (b) three waves
(c) four waves (d) five waves

27. Four simple harmonic vibrations,

$$y_1 = 8\cos\omega t$$
$$y_2 = 4\cos\left(\omega t + \frac{\pi}{2}\right)$$
$$y_3 = 2\cos(\omega t + \pi)$$
$$y_4 = \cos\left(\omega t + \frac{3\pi}{2}\right)$$

are superimposed on one another. The resulting amplitude and phase are respectively

(a) $\sqrt{45}$ and $\tan^{-1}\left(\frac{1}{2}\right)$ (b) $\sqrt{45}$ and $\tan^{-1}\left(\frac{1}{3}\right)$
(c) $\sqrt{75}$ and $\tan^{-1}(2)$ (d) $\sqrt{75}$ and $\tan^{-1}\left(\frac{1}{3}\right)$

28. Two pulses in a stretched string whose centres are initially 8 cm apart are moving towards each other as shown in the figure. The speed of each pulse is 2 cm s^{-1}. After 2 s, the total energy of the pulses will be

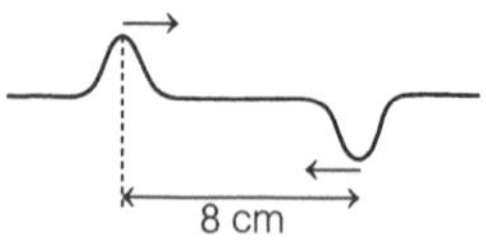

(a) zero
(b) purely kinetic
(c) purely potential
(d) partly kinetic and partly potential

29. When two sound waves travel in the same direction in a medium, the displacement of a particle located at X at time t is given by

$$y_1 = 0.05\cos(0.50\pi x - 100\pi t)$$
$$y_2 = 0.05\cos(0.46\pi x - 92\pi t)$$

where, y_1, y_2 and x are in metres and t in seconds. The speed of sound in the medium is **[JEE Main 2013]**

(a) 92 m/s (b) 200 m/s
(c) 100 m/s (d) 332 m/s

30. The following equations represent progressive transverse waves

$Z_1 = A\cos(\omega t - kX)$, $Z_2 = A\cos(\omega t + kX)$
$Z_3 = A\cos(\omega t - kY)$, $Z_4 = A\cos(2\omega t - 2kY)$

A stationary wave will be formed by superposing

(a) Z_1 and Z_2
(b) Z_1 and Z_4
(c) Z_2 and Z_3
(d) Z_3 and Z_4

31. A wave represented by the given equation $y = a\cos(kx - \omega t)$ is superposed with another wave to form a stationary wave such that the point $x = 0$ is a node. The equation for the other wave is
(a) $y = a\sin(kx + \omega t)$
(b) $y = -a\cos(kx + \omega t)$
(c) $y = -a\cos(kx - \omega t)$
(d) $y = -a\sin(kx - \omega t)$

32. A wave of wavelength 2 m is reflected from a surface, if a node is formed at 3 m from the surface, then at what distance from the surface another node will be formed?
(a) 3 m
(b) 2 m
(c) 3 m
(d) 4 m

33. The displacement of a string is given by $y(x, t) = 0.06\sin(2\pi x/3)\cos(120\pi t)$, where x and y are in metre and t in second. The length of the string is 1.5m and its mass is 3.0×10^{-2} kg.
(a) It represents a progressive wave of frequency 60 Hz.
(b) It represents a stationary wave of frequency 120 Hz.
(c) It is the result of superposition of two waves of wavelength 3 m, frequency 60 Hz each travelling with a speed of 180 m/s in opposite direction.
(d) Amplitude of this wave is constant.

34. Amongst the following statements which is correct for stationary waves?
(a) Nodes and antinodes are formed in case of stationary transverse wave only
(b) In case of longitudinal stationary wave, compressions and rarefactions are obtained in place of nodes and antinodes respectively
(c) Suppose two plane waves, one longitudinal and the other transverse having same frequency and amplitude are travelling in a medium in opposite directions with the same speed, by superposition of these waves, stationary waves cannot be obtained
(d) None of the above

35. Two identical sinusoidal waves travel in opposite direction in a wire 15 m long and produce a standing wave in the wire. If the speed of the waves is 12 ms^{-1} and there are 6 nodes in the standing wave. Find the frequency.
(a) 4 Hz
(b) 2 Hz
(c) 6 Hz
(d) 9 Hz

36. The transverse displacement of a string (clamped at its both ends) is given by

$$y(x, t) = 0.06\sin(2\pi x/3)\cos(120\pi t)$$

All the points on the string between two consecutive nodes vibrate with
I. same frequency
II. same phase
III. same energy
IV. different amplitude
(a) I, II, III
(b) I, III only
(c) I, II, IV
(d) II, IV only

37. Two wires made up of same material are of equal lengths but their radii are in the ratio 1 : 2. On stretching each of these two wires by the same tension, the ratio between the fundamental frequencies is
(a) 1 : 2
(b) 2 : 1
(c) 1 : 4
(d) 4 : 1

38. n waves are produced on a string in one second. When the radius of the string is doubled and the tension is maintained the same, the number of waves produced in one second for the same harmonic will be
(a) $\frac{n}{2}$
(b) $\frac{n}{3}$
(c) $2n$
(d) $\frac{n}{\sqrt{2}}$

39. A string in a musical instrument is 50 cm long and its fundamental frequency is 800 Hz. If a fundamental frequency of 1000 Hz is to be produced, then required length of string is
(a) 62.5 cm
(b) 50 cm
(c) 40 cm
(d) 37.5 cm

40. Two identical strings X and Z made of same material have tensions T_X and T_Z in them. If their fundamental frequencies are 450 Hz and 300 Hz respectively, then the ratio of T_X / T_Z is **[JEE Main 2020]**
(a) 1.25
(b) 2.25
(c) 0.44
(d) 1.5

41. A string of length 1 m and mass 5 g is fixed at both ends. The tension in the string is 8.0 N. The string is set into vibration using an external vibrator of frequency 100 Hz. The separation between successive nodes on the string is close to **[JEE Main 2019]**
(a) 16.6 cm
(b) 33.3 cm
(c) 10.0 cm
(d) 20.0 cm

42. Choose the correct statement from the following options.
(a) Under identical conditions of pressure and density, the speed of sound is highest in a monoatomic gas.
(b) A sound absorber attenuates the sound level by 20 dB. The intensity decreases by a factor of 100.
(c) A sound wave is passing through air column in the form of compression and rarefaction. In consecutive compression and rarefactions, there is no transfer of heat.
(d) All of the above

43. A person blows into the open end of a long pipe. As a result, a high-pressure pulse of air travels down the pipe. When this pulse reaches the other end of the pipe,
(a) a high-pressure pulse starts travelling up the pipe, if the other end of the pipe is open
(b) a low-pressure pulse starts travelling up the pipe, if the other end of the pipe is open
(c) a low-pressure pulse starts travelling up the pipe, if the other end of the pipe is closed
(d) Both (a) and (b)

44. Which of the following is/are correct?

(a)

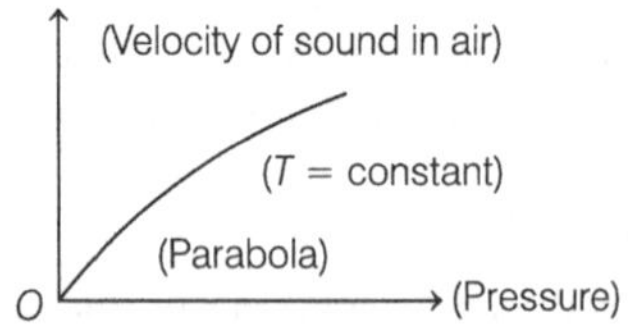

(b)

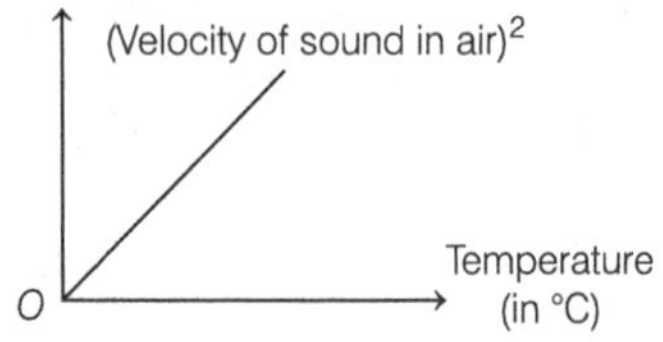

(c)

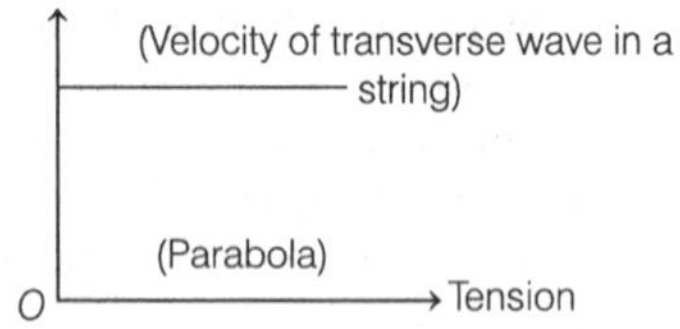

(d)

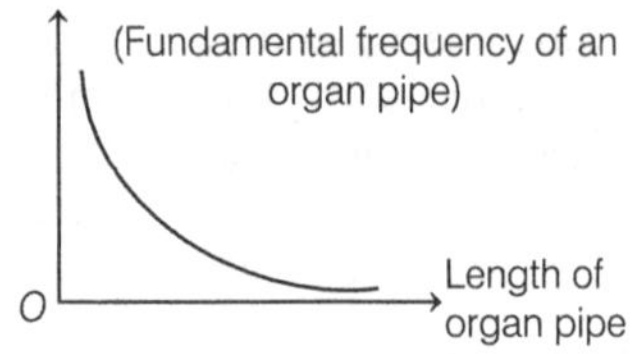

45. If n_1, n_2 and n_3 are the fundamental frequencies of three segments into which a string is divided, then the original fundamental frequency n of the string is given by

(a) $\frac{1}{n} = \frac{1}{n_1} + \frac{1}{n_2} + \frac{1}{n_3}$ (b) $\frac{1}{\sqrt{n}} = \frac{1}{\sqrt{n_1}} + \frac{1}{\sqrt{n_2}} + \frac{1}{\sqrt{n_3}}$

(c) $\sqrt{n} = \sqrt{n_1} + \sqrt{n_2} + \sqrt{n_3}$ (d) $n = n_1 + n_2 + n_3$

46. A wire of length L and mass per unit length 6.0×10^{-3} kgm^{-1} is put under tension of 540 N. Two consecutive frequencies that it resonates at are 420 Hz and 490 Hz, then L in metres is **[JEE Main 2020]**

(a) 8.1 (b) 2.1

(c) 5.1 (d) 1.1

47. A heavy uniform rope hangs vertically from the ceiling with its lower end free. A disturbance on the rope travelling upwards from the lower end has a velocity v at a distance x from the lower end such that

(a) $v \propto x$ (b) $v \propto \sqrt{x}$ (c) $v \propto \frac{1}{x}$ (d) $v \propto \frac{1}{\sqrt{x}}$

48. A string is clamped at both the ends and it is vibrating in its 4th harmonic. The equation of the stationary wave is $y = 0.3 \sin(0.157x) \cos(200\pi t)$. The length of the string is (All quantities are in SI units) **[JEE Main 2019]**

(a) 60 m (b) 40 m (c) 80 m (d) 20 m

49. A granite rod of 60 cm length is clamped at its middle point and is set into longitudinal vibrations. The density of granite is 2.7×10^3 kg/m^3 and its Young's modulus is 9.27×10^{10} Pa. What will be the fundamental frequency of the longitudinal vibrations? **[JEE Main 2018]**

(a) 5 kHz (b) 2.5 kHz

(c) 10 kHz (d) 7.5 kHz

50. An organ pipe open at one end is vibrating in first overtone and is in resonance with another pipe open at both ends and vibrating in third harmonic. The ratio of length of two pipes is

(a) 3 : 8 (b) 8 : 3

(c) 1 : 2 (d) 4 : 1

51. A wave of frequency 100 Hz is sent along a string towards a fixed end. When this wave travels back after reflection, a node is formed at a distance of 10 cm from the fixed end of the string. The speeds of incident (and reflected) waves are

(a) 5 ms^{-1} (b) 10 ms^{-1}

(c) 20 ms^{-1} (d) 40 ms^{-1}

52. The vibrating four air columns are represented in the figure. The ratio of frequencies $n_p : n_q : n_r : n_s$ is

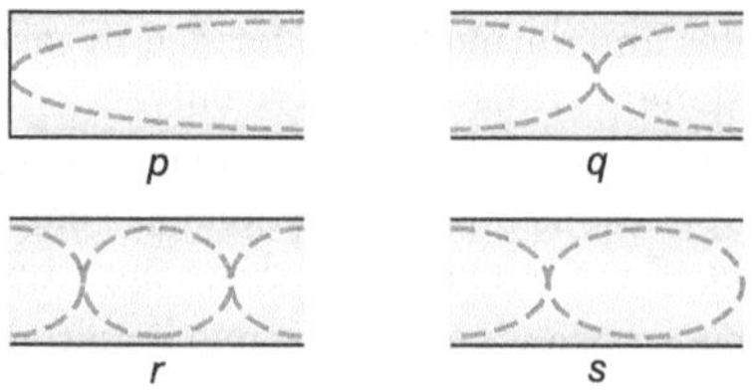

(a) 1 : 2 : 3 : 5

(b) 1 : 2 : 4 : 3

(c) 4 : 2 : 3 : 1

(d) 6 : 2 : 3 : 4

53. An open pipe is in resonance in 2nd harmonic with frequency f_1. Now, one end of the tube is closed and frequency is increased to f_2 such that the resonance again occurs in nth harmonic. Choose the correct option.

(a) $n = 3, f_2 = \frac{3}{4} f_1$ (b) $n = 3, f_2 = \frac{5}{4} f_1$

(c) $n = 5, f_2 = \frac{5}{4} f_1$ (d) $n = 5, f_2 = \frac{3}{4} f_1$

54. Six antinodes are observed in the air column when a standing wave forms in a closed (at both ends) tube. A steel bar (in its fundamental mode) of length 1m and clamped at the middle is in unison with the air column. The speed of sound in steel is 5250 ms^{-1} and in air, it is 343 ms^{-1}. Length of the air column is

(a) 79 cm (b) 49 cm

(c) 39 cm (d) 29 cm

55. A closed organ pipe of length L and an open organ pipe contain gases of densities ρ_1 and ρ_2, respectively. The compressibility of gases are equal in both the pipes. Both the pipes are vibrating in their first overtone with same frequency. The length of the open organ pipe is

(a) $\frac{L}{3}$ (b) $\frac{4L}{3}$

(c) $\frac{4L}{3}\sqrt{\frac{\rho_1}{\rho_2}}$ (d) $\frac{4L}{3}\sqrt{\frac{\rho_2}{\rho_1}}$

56. A glass tube of length 1.0 m is completely filled with water. A vibrating tuning fork of frequency 500 Hz is kept over the mouth of the tube and the water is drained out slowly at the bottom of the tube. If the velocity of sound in air is 330 ms^{-1}, then the total number of resonances that occur will be

(a) 2 (b) 3

(c) 1 (d) 5

57. A metre-long tube open at one end with a movable piston at the other end, shows resonance with a fixed frequency source (a tunning fork of frequency 340 Hz) when the tube length is 25.5 cm or 79.3 cm. Estimate the speed of sound in air at the temperature of the experiment. The edge effect may be neglected. [NCERT]

(a) 336 m/s (b) 331 m/s

(c) 356 m/s (d) 366 m/s

58. A closed organ pipe has a fundamental frequency of 1.5 kHz. The number of overtones that can be distinctly heard by a person with this organ pipe will be (Assume that the highest frequency a person can hear is 20000 Hz) [JEE Main 2019]

(a) 7 (b) 4 (c) 5 (d) 6

59. A pipe open at both ends has a fundamental frequency f in air. The pipe is dipped vertically in water, so that half of it is in water. The fundamental frequency of the air column is now [JEE Main 2016]

(a) $\frac{f}{2}$ (b) $\frac{3f}{4}$ (c) $2f$ (d) f

60. In a resonance column first and second resonance are obtained at depths 22.7 cm and 70.2 cm. The third resonance will be obtained at a depth of

(a) 117.7 cm (b) 92.9 cm

(c) 115.5 cm (d) 113.5 cm

61. A pipe of length 85 cm is closed from one end. Find the number of possible natural oscillations of air column in the pipe whose frequencies lie below 1250 Hz. The velocity of sound in air is 340 m/s. [JEE Main 2014]

(a) 12 (b) 8

(c) 6 (d) 4

62. A tuning fork of frequency 480 Hz is used in an experiment for measuring speed of sound (v) in air by resonance tube method. Resonance is observed to occur at two successive lengths of the air column $l_1 = 30$ cm and $l_2 = 70$ cm. Then, v is equal to [JEE Main 2019]

(a) 332 ms^{-1}

(b) 384 ms^{-1}

(c) 338 ms^{-1}

(d) 379 ms^{-1}

63. Figure shows a stretched string of length L and pipes of length L, $2L$, $L/2$ and $L/2$ in options (a), (b), (c) and (d), respectively. The string's tension is adjusted until the speed of waves on the string equals the speed of sound waves in the air. The fundamental mode of oscillation is then set up on the string. In which pipe will the sound produced by the string cause resonance?

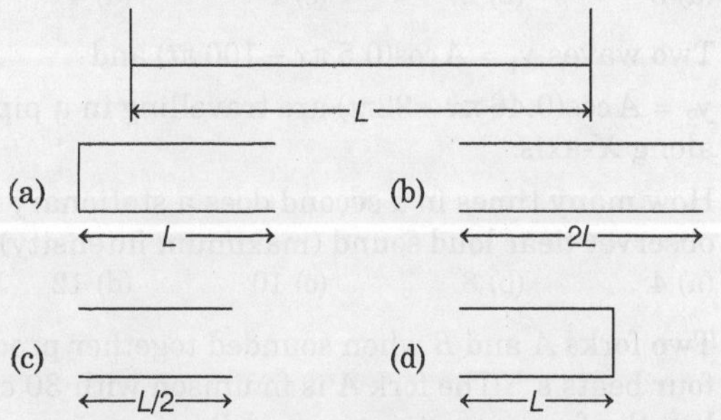

64. In a resonance tube experiment, when the tube is filled with water upto a height of 17.0 cm from bottom, it resonates with a given tuning fork. When the water level is raised, the next resonance with the same tuning fork occurs at a height of 24.5 cm. If the velocity of sound in air is 330 m/s, the tuning fork frequency is [JEE Main 2020]

(a) 1100 Hz (b) 3300 Hz

(c) 2200 Hz (d) 550 Hz

Beats

65. When 2 tuning forks (fork 1 and fork 2) are sounded simultaneously, 4 beats s^{-1} are heard. Now, some tape is attached on the prong of fork 2. When the tuning forks are sounded again, 6 beats s^{-1} are heard. If the frequency of fork 1 is 200 Hz, then what was the original frequency of fork 2 ?

(a) 196 Hz (b) 200 Hz

(c) 202 Hz (d) 204 Hz

66. In two similar wires of tensions 16 N and T, 3 beats are heard, then T is equals to

(a) 49 N (b) 25 N

(c) 64 N (d) None of these

67. A source of frequency n gives 5 beats s^{-1}, when sounded with a source of frequency 200 s^{-1}. The second harmonic ($2n$) gives 10 beats s^{-1}, when sounded with a source of frequency 420 s^{-1}, then n is equal to
(a) 200 s^{-1} (b) 205 s^{-1} (c) 195 s^{-1} (d) 210 s^{-1}

68. Ten tuning forks are arranged in increasing order of frequency in such a way that any two nearest tuning forks produce 4 beats s^{-1}. The highest frequency is twice that of the lowest. Possible highest and lowest frequencies are
(a) 80 and 40 (b) 100 and 50
(c) 44 and 32 (d) 72 and 36

69. Three sound waves of equal amplitudes have frequencies $(\nu - 1)$, ν, $(\nu + 1)$. They superpose to give beats. The number of beats produced per second will be
(a) 3 (b) 2 (c) 1 (d) 4

70. Two waves $y_1 = A\cos(0.5\,\pi x - 100\,\pi t)$ and $y_2 = A\cos(0.46\,\pi x - 92\pi t)$ are travelling in a pipe along X-axis.

How many times in a second does a stationary any observer hear loud sound (maximum intensity)?
(a) 4 (b) 8 (c) 10 (d) 12

71. Two forks A and B when sounded together produce four beats s^{-1}. The fork A is in unison with 30 cm length of a sonometer wire and B is in unison with 25 cm length of the same wire at the same tension. The frequencies of the forks are
(a) 24 Hz, 28 Hz (b) 20 Hz, 24 Hz
(c) 16 Hz, 20 Hz (d) 26 Hz, 30 Hz

72. Two organ pipes, each closed at one end, give 5 beats s^{-1} when emitting their fundamental notes. If their lengths are in the ratio 50 : 51, their fundamental frequencies (in Hz) are
(a) 250, 255 (b) 255, 260
(c) 260, 265 (d) 265, 270

73. A source of sound gives 5 beats s^{-1} when sounded with another source of frequency 100 Hz. The second harmonic of the source together with a source of frequency 205 Hz gives 5 beats s^{-1}. What is the frequency of the source?
(a) 105 Hz (b) 205 Hz
(c) 95 Hz (d) 100 Hz

74. Two uniform wires are vibrating simultaneously in their fundamental notes. The tension, lengths diameters and the densities of the two wires are in the ratio 8 : 1, 36 : 35, 4 : 1 and 1 : 2, respectively. If the note of the higher pitch has a frequency 360 Hz, the number of beats produced per second is
(a) 5 (b) 15 (c) 10 (d) 20

75. A tuning fork of frequency 250 Hz produces a beat frequency of 10 Hz when sounded with a sonometer vibrating as its fundamental frequency. When the tuning fork is filled, the beat frequency decreases. If the length of sonometer wire is 0.5 m, the speed of transverse wave is
(a) 260 ms^{-1} (b) 250 ms^{-1}
(c) 240 ms^{-1} (d) 500 ms^{-1}

76. The correct figure that shows schematically, the wave pattern produced by superposition of two waves of frequencies 9Hz and 11 Hz, is **[JEE Main 2019]**

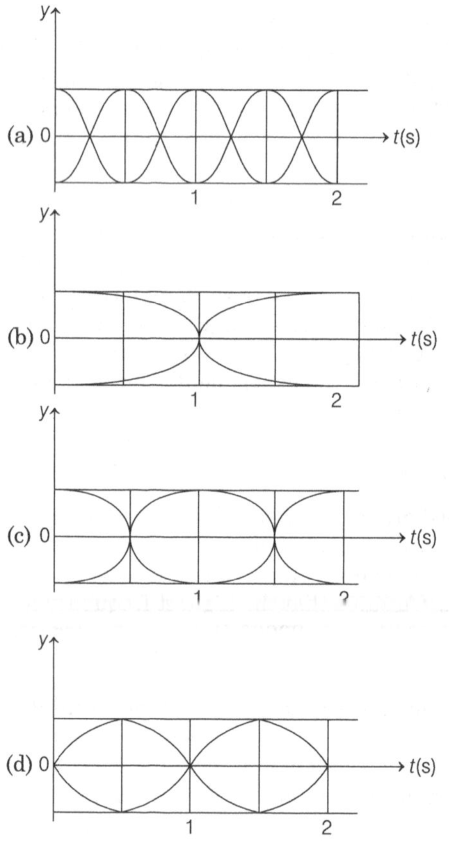

77. Two sounding bodies are producing progressive waves given by $y_1 = 2\sin(400\pi t)$ and $y_2 = 1\sin(404\,\pi t)$, where t is in second, which superpose near the ears of a person? The person will hear
(a) 2 beats/s with intensity ratio 9/4 between maxima and minima
(b) 2 beats/s with intensity ratio 9 between maxima and minima
(c) 4 beats/s with intensity ratio 16 between maxima and minima
(d) 4 beats/s with intensity ratio 16/9 between maxima and minima

Doppler's Effect

78. A source and an observer move away from each other with a velocity of 10 m/s with respect to ground. If the observer finds the frequency of sound coming from the source as 1950 Hz, then actual frequency of the source is (velocity of sound in air = 340 m/s)

(a) 1950 Hz (b) 2068 Hz
(c) 2132 Hz (d) 2486 Hz

79. A train whistling at constant frequency is moving towards a station at a constant speed v. The train goes past a stationary observer on the station. The frequency n' of the sound as heard by the observer is plotted as a function of time t (figure). Identify the expected curve. **[NCERT Exemplar]**

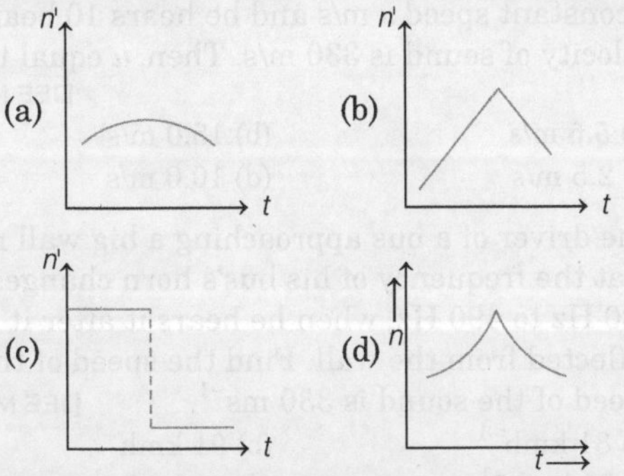

80. An observer is moving with half the speed of light towards a stationary microwave source emitting waves at frequency 10 GHz. What is the frequency of the microwave measured by the observer? (Take, speed of light = 3×10^8 ms^{-1}) **[JEE Main 2017]**

(a) 12.1 GHz (b) 17.3 GHz
(c) 15.3 GHz (d) 10.1 GHz

81. A train is moving on a straight track with speed 20 ms^{-1}. It is blowing its whistle at the frequency of 1000 Hz. The percentage change in the frequency heard by a person standing near the track as the train passes him is close to (Take, speed of sound = 320 ms^{-1}) **[JEE Main 2015]**

(a) 6% (b) 12% (c) 18% (d) 24%

82. Two cars A and B are moving away from each other in opposite directions. Both the cars are moving with a speed of 20 ms^{-1} with respect to the ground. If an observer in car A detects a frequency 2000 Hz of the sound coming from car B, what is the natural frequency of the sound source in car B? (Take, speed of sound in air = 340 ms^{-1}) **[JEE Main 2019]**

(a) 2060 Hz (b) 2250 Hz
(c) 2300 Hz (d) 2150 Hz

83. A detector is released from rest over a source of sound of frequency $f_0 = 10^3$ Hz. The frequency observed by the detector at time t is plotted in the graph. The speed of sound in air is ($g = 10$ m/s^2)

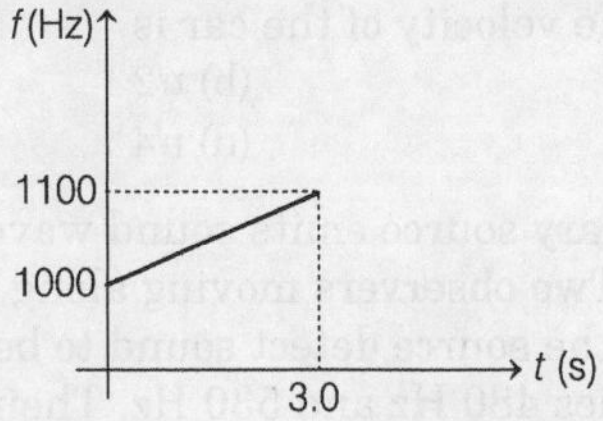

(a) 330 m/s (b) 350 m/s
(c) 300 m/s (d) 310 m/s

84. A motor cycle starts from rest and accelerates along a straight path at 2 ms^{-2}. At the starting point of the motor cycle, there is a stationary electric siren. How far has the motor cycle gone when the driver hears the frequency of the siren at 94% of its value when the motor cycle was at rest? (Take, speed of sound = 330 ms^{-1})

(a) 98 m (b) 147 m
(c) 196 m (d) 49 m

85. A bus is moving with a velocity of 5 ms^{-1} towards a huge wall. The driver sounds a horn of frequency 165 Hz. If the speed of sound in air is 335 ms^{-1}, the number of beats heard per second by the passengers in the bus will be

(a) 3 (b) 4
(c) 5 (d) 6

86. Two sound sources emitting sound each of wavelength λ are fixed at a given distance apart. A listener moves with a velocity u along the line joining the two sources. The number of beats heard by him per second is

(a) $2u/\lambda$ (b) u/λ
(c) $\dfrac{u}{3\lambda}$ (d) $\dfrac{2\lambda}{u}$

87. A sound wave of frequency n travels horizontally to the right. It is reflected from a large vertical plane surface moving to the left with speed v. The speed of the sound in the medium is c, then

(a) the frequency of the reflected wave is $\left[\dfrac{c+v}{c-v}\right]$

(b) the wavelength of the reflected wave is $\left[\dfrac{c}{n}\right]\left[\dfrac{c+v}{c-v}\right]$

(c) the number of waves striking the surface per second is $\left[\dfrac{c+v}{c}\right]n$

(d) the number of beats heard by a stationary listener to the left to the reflecting surface is $\dfrac{nv}{c-v}$

88. A racing car moving towards a cliff sounds its horn. The driver observes that the sound reflected from the cliff has a pitch one octave higher than the actual sound of the horn. If v is the velocity of sound, the velocity of the car is

(a) $v/\sqrt{2}$ (b) $v/2$
(c) $v/3$ (d) $v/4$

89. A stationary source emits sound waves of frequency 500 Hz. Two observers moving along a line passing through the source detect sound to be of frequencies 480 Hz and 530 Hz. Their respective speeds are in ms^{-1} (Take, speed of sound = 300 m/s) **[JEE Main 2019]**

(a) 12, 16 (b) 12, 18
(c) 16, 14 (d) 8, 18

90. A train moves towards a stationary observer with speed 34 m/s. The train sounds a whistle and its frequency registered by the observer is f_1. If the speed of the train is reduced to 17 m/s, the frequency registered is f_2. If speed of sound is 340 m/s, then the ratio $\frac{f_1}{f_2}$ is **[JEE Main 2019]**

(a) $\frac{19}{18}$ (b) $\frac{21}{20}$ (c) $\frac{20}{19}$ (d) $\frac{18}{17}$

91. A source of sound S is moving with a velocity of 50 m/s towards a stationary observer. The observer measures the frequency of the source as 1000 Hz. What will be the apparent frequency of the source when it is moving away from the observer after crossing him? (Take, velocity of sound in air is 350 m/s) **[JEE Main 2019]**

(a) 807 Hz (b) 1143 Hz (c) 750 Hz (d) 857 Hz

92. A source of sound emitting a tone of frequency 200 Hz moves towards an observer with a velocity v equal to the velocity of sound. If the observer also moves away from the source with the same velocity v, the apparent frequency heard by the observer is

(a) 50 Hz (b) 100 Hz (c) 150 Hz (d) 200 Hz

93. A and B are two sources generating sound waves. A listener is situated at C. The frequency of the source at A is 500 Hz. A now moves towards C with a speed 4 m/s. The number of beats heard at C is 6. When A moves away from C with speed 4 m/s, the number of beats heard at C is 18. The speed of sound is 340 m/s. The frequency of the source at B is **[JEE Main 2013]**

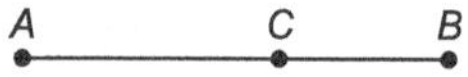

(a) 500 Hz (b) 506 Hz
(c) 512 Hz (d) 494 Hz

94. Two sources of sound S_1 and S_2 produce sound waves of same frequency 660 Hz.
A listener is moving from source S_1 towards S_2 with a constant speed u m/s and he hears 10 beats/s. The velocity of sound is 330 m/s. Then, u equal to **[JEE Main 2019]**

(a) 5.5 m/s (b) 15.0 m/s
(c) 2.5 m/s (d) 10.0 m/s

95. The driver of a bus approaching a big wall notices that the frequency of his bus's horn changes from 420 Hz to 490 Hz, when he hears it after it gets reflected from the wall. Find the speed of the bus, if speed of the sound is 330 ms^{-1}. **[JEE Main 2020]**

(a) 81 kmh^{-1} (b) 91 kmh^{-1}
(c) 71 kmh^{-1} (d) 61 kmh^{-1}

96. A vibrating tuning fork tied to the end of a string 1.988 m long is whirled round a circle. If it makes the revolutions in a second, calculate the ratio of the frequencies of the highest and the lowest notes heard by a stationary observer situated in the plane of rotation of tuning fork at a large distance from the tuning fork. Speed of sound is 350 ms^{-1}. (In both, the observer is assumed to be along the line of velocity sound)

(a) 1.732 (b) 1.154
(c) 1.278 (d) 1.000

ROUND II Mixed Bag

Only One Correct Option

1. A progressive wave travelling along the positive x-direction is represented by $y(x, t) = A \sin(kx - \omega t + \phi)$. Its snapshot at $t = 0$ is given in the figure. [JEE Main 2019]

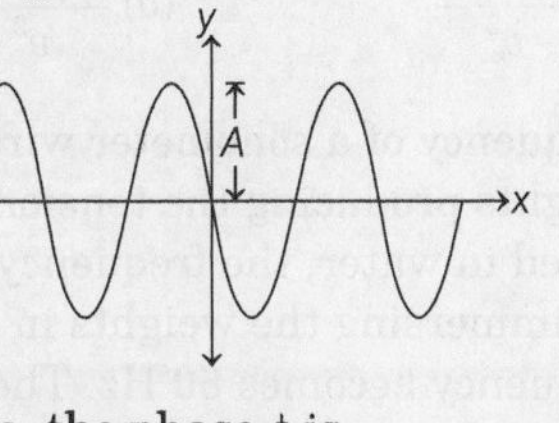

For this wave, the phase ϕ is

(a) $-\frac{\pi}{2}$ (b) π (c) 0 (d) $\frac{\pi}{2}$

2. The frequency of a tuning fork A is 2% more than the frequency of a standard tuning fork. The frequency of the same standard tuning fork is 3 % more than the frequency of tuning fork B. If 6 beats s^{-1} are heard when the two tuning forks A and B are excited, the frequency of A is

(a) 120 Hz (b) 122.4 Hz
(c) 116.4 Hz (d) 130 Hz

3. Two uniform strings A and B made of steel are made to vibrate under the same tension. If the first overtone of A is equal to the second overtone of B and if the radius of A is twice that of B, the ratio of the lengths of the strings is

(a) 2 : 1 (b) 3 : 4 (c) 3 : 2 (d) 1 : 3

4. A closed organ pipe and an open organ pipe of same length produce 2 beats s^{-1} when they are set into vibrations together in fundamental mode. The length of open pipe is now halved and that of closed pipe is doubled. The number of beats produced will be

(a) 7 (b) 4 (c) 8 (d) 2

5. A string is under tension, so that its length is increased by $1/n$ times its original length. The ratio of fundamental frequency of longitudinal vibrations and transverse vibrations will be

(a) $1 : n$ (b) $n^2 : 1$ (c) $\sqrt{n} : 1$ (d) $n : 1$

6. Standing waves are produced by the superposition of two waves

$$y_1 = 0.05 \sin(3\pi t - 2x)$$
$$y_2 = 0.05 \sin(3\pi t + 2x)$$

where, x and y are in metres and t is in second. What is the amplitude of the particle at $x = 0.5$ m? (Take, $\cos 57.3° = 0.54$)

(a) 2.7 cm (b) 5.4 cm
(c) 8.1 cm (d) 10.8 cm

7. The ends of a stretched wire of length L are fixed at $X = 0$ and $X = L$. In one experiment, the displacement of the wire is $Y_1 = A \sin\left(\frac{\pi x}{L}\right) \sin \omega t$ and energy is E_1 and in another experiment, its displacement is $y_2 = A \sin(2\pi x/L) \sin 2\omega t$ and energy is E_2. Then,

(a) $E_2 = E_1$ (b) $E_2 = 2E_1$
(c) $E_2 = 4E_1$ (d) $E_2 = 16E_1$

8. In an experiment, it was found that string vibrates in n loops when a mass M is placed on the pan. What mass should be placed on the pan to make it vibrate in $2n$ loops, with same frequency. Neglect the mass of the pan.

(a) $M/4$ (b) $4M$
(c) $2M$ (d) $M/2$

9. A string of mass 0.2 kg/m has length $L = 0.6$ m. It is fixed at both ends and stretched such that it has a tension of 80 N. The string vibrates in three segments with amplitude = 0.5 cm. The amplitude of transverse velocity is

(a) 9.42 ms^{-1} (b) 3.14 ms^{-1}
(c) 1.57 ms^{-1} (d) 6.28 ms^{-1}

10. When a train approaches a stationary observer, the apparent frequency of the whistle is n' and when the same train recedes away from the observer, the apparent frequency is n''. Then, the apparent frequency n when the observer moves with the train is

(a) $n = \frac{n' + n}{2}$ (b) $n = \sqrt{n'n''}$
(c) $n = \frac{2n'n''}{n' + n''}$ (d) $n = \frac{2n'n''}{n' - n''}$

11. A table is revolving on its axis at 5 revolutions per second. A sound source of frequency 1000 Hz is fixed on the table at 70 cm from the axis. The minimum frequency heard by a listener standing at a distance from the table will be (Take, speed of sound = 352 ms^{-1})

(a) 1000 Hz (b) 1066 Hz
(c) 941 Hz (d) 352 Hz

12. A light pointer fixed to one prong of a tuning fork touches a vertical plate. The fork is set vibrating and the plate is allowed to fall freely. Eight complete oscillations are counted when the plate falls through 10 cm. What is the frequency of the tuning fork?

(a) 112 Hz (b) 56 Hz
(c) $\frac{8}{7}$ Hz (d) $\frac{7}{8}$ Hz

13. An observer starts moving with uniform acceleration a towards a stationary sound source of frequency f_0. As the observer approaches the source, the apparent frequency (f) heard by the observer varies with time (t) as

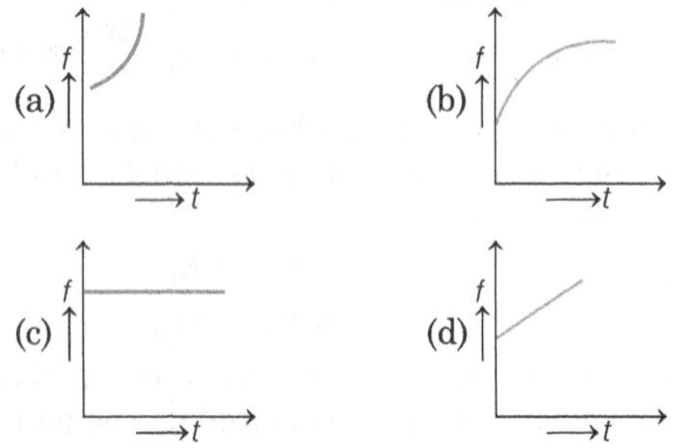

14. A uniform thin rope of length 12 m and mass 6 kg hangs vertically from a rigid support and a block of mass 2 kg is attached to its free end. A transverse short wavetrain of wavelength 6 cm is produced at the lower end of the rope. What is the wavelength of the wavetrain (in cm) when it reaches the top of the rope? [JEE Main 2020]

(a) 3 (b) 6 (c) 12 (d) 9

15. A uniform rope having mass m hangs vertically from a rigid support. A transverse wave pulse is produced at the lower end. The speed (v) of wave pulse varies with height h from the lower end as

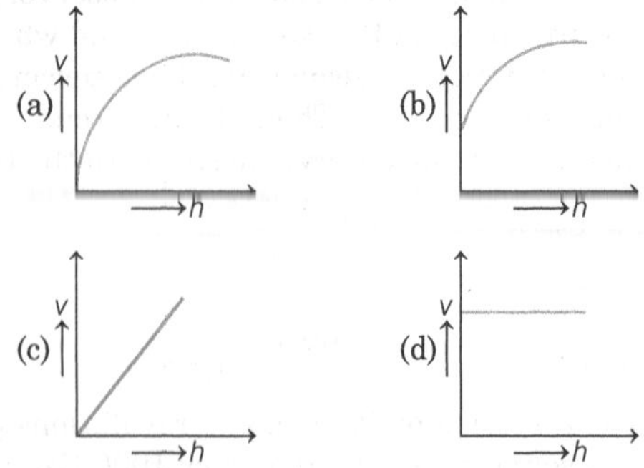

16. Speed of a transverse wave on a straight wire (mass 6.0 g, length 60 cm and area of cross-section 1.0 mm^2) is 90 ms^{-1}. If the Young's modulus of wire is 16×10^{11} Nm^{-2}, the extension of wire over its natural length is [JEE Main 2020]

(a) 0.01 mm (b) 0.04 mm
(c) 0.03 mm (d) 0.02 mm

17. Equations of a stationary wave and a travelling wave are $y_1 = a \sin kx \cos \omega t$ and $y_2 = a \sin(\omega t - kx)$. The phase difference between two points $x_1 = \dfrac{\pi}{3k}$ and $x_2 = \dfrac{3\pi}{2k}$ are ϕ_1 and ϕ_2 respectively for the two waves. The ratio ϕ_1/ϕ_2 is

(a) 1 (b) 5/6 (c) 3/4 (d) 6/7

18. A band playing music at a frequency f is moving towards a wall at a speed v_b. A motorist is following the band with a speed v_m. If v is speed of sound, the expression for the beat frequency heard by the motorist is

(a) $\dfrac{(v+v_m)f}{v+v_b}$ (b) $\dfrac{(v+v_m)f}{v-v_b}$

(c) $\dfrac{2v_b(v+v_m)f}{v^2-v_b^2}$ (d) $\dfrac{2v_m(v+v_b)f}{v^2-v_b^2}$

19. The frequency of a sonometer wire is 100 Hz. When the weights producing the tension are completely immersed in water, the frequency becomes 80 Hz and on immersing the weights in a certain liquid, the frequency becomes 60 Hz. The specific gravity of the liquid is

(a) 1.42 (b) 1.77 (c) 1.21 (d) 1.82

20. A massless rod is suspended by two identical strings AB and CD of equal length. A block of mass m is suspended from point O such that BO is equal to (x) further it is observed that the frequency of 1st harmonic (fundamental frequency) in AB is equal to 2nd harmonic frequency in CD. Then, length of BO is

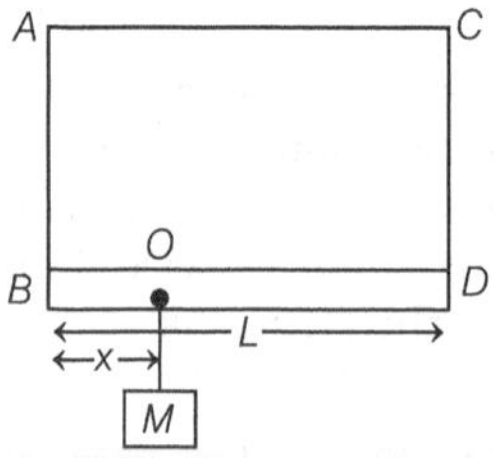

(a) $\dfrac{L}{5}$ (b) $\dfrac{4L}{5}$ (c) $\dfrac{3L}{4}$ (d) $\dfrac{L}{4}$

21. Source and observer, both start moving simultaneously from origin, one along X-axis and the other along Y-axis with speed of source equal to twice the speed of observer. The graph between the apparent frequency (n') observed by observer and time t in figure would be

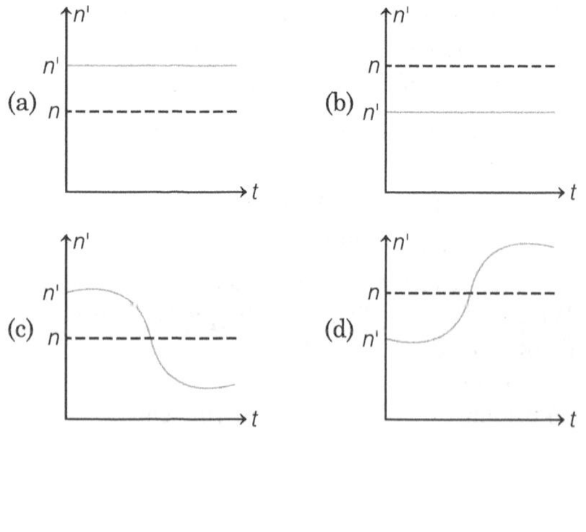

22. Two radio stations broadcast their programmes at the same amplitude A and at slightly different frequencies, ω_1 and ω_2 respectively, where $\omega_2 - \omega_1 = 10^3$ Hz. A detector is receiving signals from the two stations simultaneously. It can only detect signals of intensity $> 2A^2$. The time interval between successive maxima of the intensity of the signal received by the detector is

(a) 10^3 s (b) 10^{-3} s

(c) 10^{-4} s (d) 10^4 s

23. A musician produce the sound of second harmonics from open end flute of 50 cm. The other person moves toward the musician with speed 10 km/h from the second end of room. If the speed of sound 330 m/s, the frequency heard by running person will be **[JEE Main 2019]**

(a) 666 Hz (b) 500 Hz

(c) 753 Hz (d) 333 Hz

24. A sound source S is moving along a straight track with speed v and is emitting sound of frequency ν_o (see figure). An observer is standing at a finite distance, at the point O, from the track. The time variation of frequency heard by the observer is best represented by

(Here, t_0 represents the instant when the distance between the source and observer is minimum.) **[JEE Main 2020]**

(a)
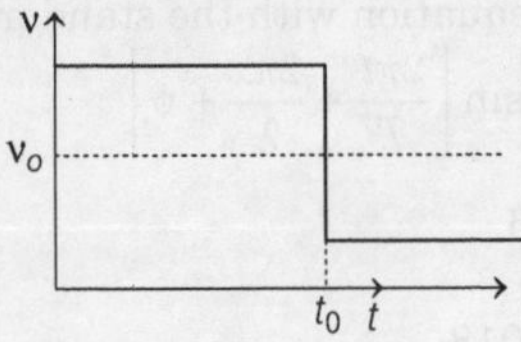

(b)
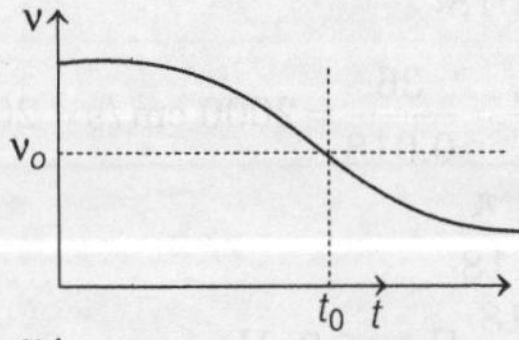

(c)
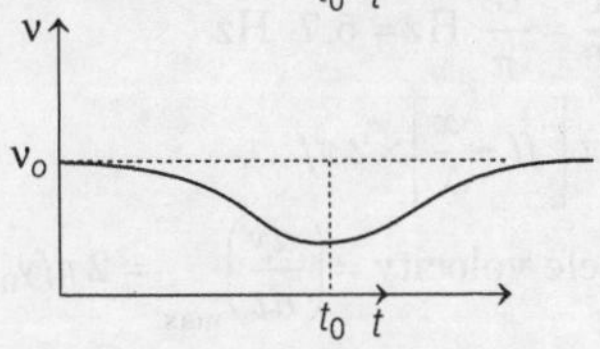

(d)
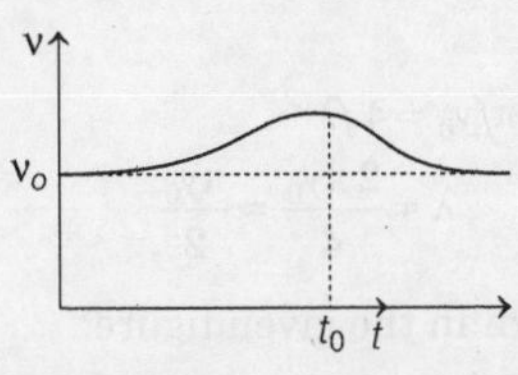

25. For a transverse wave travelling along a straight line, the distance between two peaks (crests) is 5 m, while the distance between one crest and one trough is 1.5 m. The possible wavelengths (in metre) of the waves are **[JEE Main 2020]**

(a) 1, 2, 3,...... (b) $\frac{1}{1}, \frac{1}{3}, \frac{1}{5}, \ldots$

(c) 1, 3, 5,..... (d) $\frac{1}{2}, \frac{1}{4}, \frac{1}{6}, \ldots$

Numerical Value Questions

26. The mass of 1 mol of air is 29×10^{-3} kg, then the speed of sound (in m/s) in air at standard temperature and pressure is ....

27. A whistle of frequency 540 Hz rotates in a circle of radius 2 m at a linear speed of 30 m/s. What is the difference of lowest and highest frequency heard by an observer a long distance away at rest with respect to the centre of circle (in Hz)? (Take, speed of sound in air = 330 m/s)

28. A closed organ pipe of length L and an open organ pipe contain gases of densities ρ_1 and ρ_2, respectively. The compressibility of gases are equal in both the pipes. Both the pipes are vibrating in their first overtone with same frequency.

The length of the open pipe is $\frac{x}{3} L \sqrt{\frac{\rho_1}{\rho_2}}$, where x is .... (Round off to the nearest integer) **[JEE Main 2021]**

29. Four harmonic waves of equal frequencies and equal intensities (I_0) have phase angles 0, $\frac{\pi}{3}$, $\frac{2\pi}{3}$ and π. When they are superposed, the intensity of the resulting wave is nI_0. The value of n is ....

30. A stretched rope having linear mass density 5×10^{-2} kg m^{-1} is under a tension of 80 N. The power (in watt) that has to be supplied to the rope to generate harmonic waves at a frequency of 60 Hz and an amplitude of 6 cm is ....

31. A one metre long (both ends open) organ pipe is kept in a gas that has double the density of air at STP. Assuming the speed of sound in air at STP is 300 m/s, the frequency difference between the fundamental and second harmonic of this pipe is ...... Hz. **[JEE Main 2020]**

32. An aluminium wire is clamped at each end and under zero tension at room temperature. Reducing the temperature, which results in a decrease in the wire's equilibrium length, increases the tension in the wire. The strain $(\Delta L/L)$ that results in a transverse wave speed of 100 m/s is 3.86×10^{-n}, then the value of n is ....

(Take, the cross-sectional area of the wire to be 5.00×10^{-6} m^2, the density to be 2.70×10^3 kg/m^3, and Young's modulus to be 7.00×10^{10} N/m^2)

Answers

Round I

1. (c)	*2. (a)*	*3. (c)*	*4. (d)*	*5. (d)*	*6. (b)*	*7. (a)*	*8. (b)*	*9. (b)*	*10. (a)*
11. (c)	*12. (d)*	*13. (b)*	*14. (b)*	*15. (a)*	*16. (a)*	*17. (c)*	*18. (c)*	*19. (a)*	*20. (b)*
21. (d)	*22. (d)*	*23. (d)*	*24. (b)*	*25. (b)*	*26. (b)*	*27. (a)*	*28. (b)*	*29. (b)*	*30. (a)*
31. (b)	*32. (d)*	*33. (c)*	*34. (c)*	*35. (b)*	*36. (c)*	*37. (b)*	*38. (a)*	*39. (c)*	*40. (b)*
41. (d)	*42. (d)*	*43. (b)*	*44. (d)*	*45. (a)*	*46. (b)*	*47. (b)*	*48. (c)*	*49. (a)*	*50. (c)*
51. (c)	*52. (b)*	*53. (c)*	*54. (c)*	*55. (c)*	*56. (b)*	*57. (d)*	*58. (d)*	*59. (d)*	*60. (a)*
61. (c)	*62. (b)*	*63. (b)*	*64. (c)*	*65. (a)*	*66. (a)*	*67. (b)*	*68. (d)*	*69. (c)*	*70. (a)*
71. (b)	*72. (a)*	*73. (a)*	*74. (c)*	*75. (a)*	*76. (a)*	*77. (b)*	*78. (b)*	*79. (c)*	*80. (b)*
81. (b)	*82. (b)*	*83. (c)*	*84. (a)*	*85. (c)*	*86. (a)*	*87. (c)*	*88. (c)*	*89. (b)*	*90. (a)*
91. (c)	*92. (d)*	*93. (c)*	*94. (c)*	*95. (b)*	*96. (b)*				

Round II

1. (b)	*2. (b)*	*3. (d)*	*4. (a)*	*5. (c)*	*6. (b)*	*7. (c)*	*8. (a)*	*9. (c)*	*10. (c)*
11. (c)	*12. (b)*	*13. (d)*	*14. (c)*	*15. (a)*	*16. (c)*	*17. (d)*	*18. (c)*	*19. (b)*	*20. (a)*
21. (b)	*22. (b)*	*23. (a)*	*24. (b)*	*25. (b)*	*26. 332.5*		*27. 99*		
28. 4	*29. 3*	*30. 518*	*31. 106.06*		*32. 4*				

Solutions

Round I

1. Mechanical transverse waves can propagate through solids only as solids have elasticity of shape.

2. In the figure, C reaches the position, where A already reached, after $\omega t = \frac{\pi}{2}$ and A reaches the position, where B already reached, after $\omega t = \pi/2$.

3. Given, $y = 10^{-4}\sin\left[100t - \frac{x}{10}\right]$

Comparing it with the standard equation of wave motion

$$y = A\sin\left[\frac{2\pi}{T}t - \frac{2\pi}{\lambda}x\right], \text{ we get}$$

$$\frac{2\pi}{T} = 100$$

or $$T = \frac{2\pi}{100} = \frac{\pi}{50}\text{ s}$$

$$\frac{2\pi}{\lambda} = \frac{1}{10}$$

or $$\lambda = 20\pi$$

and velocity, $$v = \frac{\lambda}{T} = \frac{20\pi}{\pi/50} = 1000 \text{ ms}^{-1}$$

4. The given equation is

$$y(x, t) = 3.0\sin\left[36t + 0.018x + \frac{\pi}{4}\right]$$

As positive direction is from left to right and x is positive, therefore wave is travelling from right to left.

Compare the given equation with the standard from

$$y = r\sin\left[\frac{2\pi t}{T} + \frac{2\pi x}{\lambda} + \phi\right]$$

$$\frac{2\pi}{T} = 36$$

$$\frac{2\pi}{\lambda} = 0.018$$

Speed of wave, $v = \frac{\lambda}{T} = \frac{36}{0.018} = 2000 \text{ cm/s} = 20 \text{ m/s}$

Again, $T = \frac{2\pi}{36} = \frac{\pi}{18}$

Frequency, $\nu = \frac{1}{T} = \frac{18}{\pi}\text{ Hz} = 5.7 \text{ Hz}$

5. As, $\frac{dy}{dt} = y_0\cos 2\pi\left[ft - \frac{x}{\lambda}\right] \times 2\pi f$

Maximum particle velocity $= \left(\frac{dy}{dt}\right)_{max} = 2\pi f y_0 \times 1$

Wave velocity $= f\lambda$

As, $$2\pi f y_0 = 4f\lambda$$

$\therefore$ $$\lambda = \frac{2\pi y_0}{4} = \frac{\pi y_0}{2}$$

6. From the wave shown in the given figure.

The amplitude, $A = 0.06$ m

$$\frac{5}{2}\lambda = 0.2 \text{ m}$$

$\therefore$ $$\lambda = 0.08 \text{ m}$$

$$f = \frac{v}{\lambda} = \frac{300}{0.08} = 3750 \text{ Hz}$$

$$k = \frac{2\pi}{\lambda} = 78.5 \text{ m}^{-1}$$

and $\omega = 2\pi f = 23562$ rad/s

At $t = 0, x = 0, \frac{dy}{dx}$ = positive

and the given curve is a sine curve.

Hence, equation of wave travelling in positive x-direction should have the form

$$y(x, t) = A \sin(kx - \omega t)$$

Substituting the values, we have

$$y = (0.06) \sin[(78.5 \text{ m}^{-1})\, x - 23562 \text{ s}^{-1})t] \text{ m}$$

7. Wave equation is given by, $y = 10^{-3} \sin(50\,t + 2x)$

Speed of wave is obtained by differentiating phase of wave.

Now, phase of wave from given equation is

$$\phi = 50\,t + 2x = \text{constant}$$

Differentiating 'ϕ' w.r.t. 't', we get

$$\frac{d}{dx}(50\,t + 2x) = \frac{d}{dt}(\text{constant})$$

$$\Rightarrow \quad 50 + 2\left(\frac{dx}{dt}\right) = 0$$

$$\Rightarrow \quad \frac{dx}{dt} = \frac{-50}{2} = -25 \text{ ms}^{-1}$$

So, wave is propagating in negative x-direction with a speed of 25 ms^{-1}.

Alternate Solution

The general equation of a wave travelling in negative x-direction is given as

$$y = a \sin(\omega t + kx) \quad \text{...(i)}$$

Given equation of wave is

$$y = 10^{-3} \sin(50t + 2x) \quad \text{...(ii)}$$

Comparing Eqs. (i) and (ii), we get

$$\omega = 50 \text{ and } k = 2$$

Velocity of the wave, $v = \frac{\omega}{k} = \frac{50}{2} = 25$ m/s

8. Given, $\frac{\lambda}{4} = 4$ cm

$\therefore \quad \lambda = 16$ cm and $T = 0.4$ s

As, $\quad f\lambda \times T = 2\pi$

$$\Rightarrow \quad f = \frac{2\pi}{16 \times 0.4} = \frac{5\pi}{16} \text{ s}^{-1}$$

Now, $\quad v = f\lambda = \frac{5\pi}{16} \times 16 = 5\pi$ cm/s

9. Maximum particle velocity, $u_{max} = \omega A \quad \text{...(i)}$

Maximum particle acceleration, $a_{max} = \omega^2 A \quad \text{...(ii)}$

Dividing Eq. (ii) by Eq. (i), we get

Angular frequency, $\omega = \frac{a_{max}}{u_{max}} = \frac{90}{3} = 30$ rad/s

From Eq. (i), amplitude, $A = \frac{u_{max}}{\omega} = \frac{3}{30} = 0.1$ m

Propagation constant, $k = \frac{\omega}{v} = \frac{30}{20} = 1.5 \text{ m}^{-1}$

Equation of wave is $y = A \sin(\omega t \pm kx)$ or wave function $y = 0.1 \sin(30t \pm 1.5\,x)$ wave x is in metres and t in seconds.

Positive sign is for wave propagation along negative X-axis and negative sign for wave propagating along positive X-axis.

10. Here, $y = A \sin(\omega t \pm \phi)$

Given, $A = 10$ cm and $y = 5$ cm,

$$\therefore \quad \sin(\omega t \pm \phi) = \frac{1}{2} = \sin 30°$$

$$\Rightarrow \quad \omega t \pm \phi = 30°$$

Here, $\omega = \frac{2\pi}{T} = 2\pi\left(\frac{v}{\lambda}\right) = 2\pi\left(\frac{10 \text{ cm s}^{-1}}{50 \text{ cm}}\right) = \frac{2\pi}{5} \text{ rad s}^{-1}$

$\therefore$ The velocity of the point P is

$$v_P = \frac{dy}{dt} = A\omega \cos(\omega t \pm \phi) = (10 \text{ cm})\left(\frac{2\pi}{5} \text{s}^{-1}\right)(\cos 30°)$$

$$= 2\pi\sqrt{3} \text{ cms}^{-1} = \frac{\pi\sqrt{3}}{50} \text{ m s}^{-1}$$

$$\therefore \mathbf{v}_P = \frac{\pi\sqrt{3}}{50} \hat{\mathbf{j}} \text{ m s}^{-1}$$

11. As, intensity = power/area

From a point source, energy spreads over the surface of a sphere of radius r.

$$\therefore \quad \text{Intensity} = \frac{P}{A} = \frac{P}{4\pi r^2} \propto \frac{1}{r^2}$$

But $\quad$ intensity = (amplitude)²

$$\therefore \quad (\text{Amplitude})^2 \propto \frac{1}{r^2}$$

or $\quad \text{Amplitude} \propto \frac{1}{r}$

Clearly, at distance $2r$, amplitude becomes $A/2$.

12. Given, $y = 8 \sin 2\pi\,(0.1\,x - 2\,t)$

Compare it with the equation of wave motion

$$y = r \sin 2\pi\left(\frac{x}{\lambda} - \frac{t}{T}\right)$$

$$\frac{1}{\lambda} = 0.1 = 10 \text{ cm}$$

From $\quad \phi = \frac{2\pi}{\lambda}\Delta x = \frac{2\pi}{10} \times 2 = 0.4 \times 180° = 72°$

13. Here, $m = \frac{2.5}{20}$ kg/m, $T = 200$N

$$v = \sqrt{\frac{T}{m}} = \sqrt{\frac{200}{2.5/20}}$$

$$v = \sqrt{\frac{200 \times 200}{25}}$$

$$= \frac{200}{5} = 40 \text{ m/s}$$

Time taken, $\quad t = \frac{L}{v} = \frac{20}{40} = 0.5$ s

14. Transverse wave speed over a string is given by

$$v = \sqrt{\frac{T}{\mu}} \quad \dots \text{(i)}$$

where, T = tension in string

and μ = mass per unit length of string.

Here, when velocity is v, then tension, $T_1 = 2.06 \times 10^4$ N

Let when velocity is $\frac{v}{2}$, then tension is T, hence from Eq. (i), we get

$$\frac{v}{\frac{v}{2}} = \sqrt{\frac{T_1}{T}}$$

$$\Rightarrow \quad T = \frac{T_1}{4} = \frac{2.06 \times 10^4}{4}$$

or $\quad T = 5.15 \times 10^3$ N

15. In a wave equation, x and t must be related in the form $(x - vt)$. Therefore, we rewrite the given equation as

$$y = \frac{1}{1 + (x - vt)^2}$$

For $t = 0$, it becomes $y = \frac{1}{1 + x^2}$

and for $t = 2$, it becomes

$$y = \frac{1}{[1 + (x - 2v)^2]} = \frac{1}{1 + (x - 1)^2}$$

$\therefore \quad 2v = 1$

or $\quad v = 0.5 \text{ ms}^{-1}$

16. $\because \quad \omega = 2\pi f = (2\pi)(100) = (200\pi)$ rad/s

$$\therefore \quad v = \frac{\omega}{k}$$

$$\Rightarrow \quad k = \frac{\omega}{v} = \omega\sqrt{\frac{\mu}{T}}$$

$$= (200\pi)\sqrt{\frac{3.5 \times 10^{-3}}{35}} = 2\pi \text{ m}^{-1}$$

At zero displacement,

$$v_P = \omega A = -v\frac{\partial y}{\partial x}$$

$$\therefore \quad |A| = \frac{v\left(\frac{\partial y}{\partial x}\right)}{\omega} = \frac{\left(\sqrt{\frac{T}{\mu}}\right)\text{(slope)}}{\omega}$$

$$= \frac{\left(\sqrt{\frac{35}{3.5 \times 10^{-3}}}\right)(\pi/20)}{200\pi}$$

$$\Rightarrow \quad |A| = 0.025 \text{ m}$$

Hence, $y = 0.025\cos(200\pi t - 2\pi x)$

17. According to Newton's formula for velocity of sound in a fluid,

$$v = \sqrt{\frac{B_a}{\rho}}$$

$$v \propto \sqrt{B_a}$$

and $\quad v \propto \frac{1}{\sqrt{\rho}}$

18. At given temperature and pressure,

$$v \propto \frac{1}{\sqrt{\rho}}$$

$$\Rightarrow \quad \frac{v_1}{v_2} = \sqrt{\frac{\rho_2}{\rho_1}} = \sqrt{\frac{4}{1}} = 2:1$$

19. From $v = \sqrt{\frac{\gamma RT}{M}}$

$$\frac{\Delta v}{v} = \frac{1}{2}\left(\frac{\Delta T}{T}\right)$$

$$\Rightarrow \quad \frac{\Delta v}{v} \times 100 = \frac{1}{2}\left(\frac{1}{T}\right) \times 100$$

$$= \frac{1}{2} \times \frac{1}{300} \times 100 = 0.167\%$$

20. Molecular weight of mixture,

$$M_{\text{mix}} = \frac{n_1 M_1 + n_2 M_2}{n_1 + n_2}$$

$$= \frac{1 \times 4 + 2 \times 32}{1 \times 2} = \frac{68}{3}$$

$$= \frac{68}{3} \times 10^{-3} \text{ kg mol}^{-1}$$

For helium, $C_{V_1} = \frac{3}{2}R$

For oxygen, $C_{V_2} = \frac{5}{2}R$

$$(C_V)_{\text{mix}} = \frac{n_1 C_{V_1} + n_2 C_{V_2}}{n_1 + n_2} = \frac{1 \times \frac{3R}{2} + 2 \times \frac{5R}{2}}{1 + 2} = \frac{13R}{6}$$

Now, $\quad (C_p)_{\text{mix}} = (C_V)_{\text{mix}} + R$

$$= \frac{13R}{6} + R = \frac{19R}{6}$$

$$\Rightarrow \quad \gamma_{\text{mix}} = \frac{(C_p)_{\text{mix}}}{(C_V)_{\text{mix}}} = \frac{19}{13}$$

Speed of sound, $v = \sqrt{\frac{\gamma_{\text{mix}} RT}{M_{\text{mix}}}} = \sqrt{\frac{19}{13} \times \frac{8.31 \times 300}{\frac{68}{3} \times 10^{-3}}}$

$$= 400.8 \text{ ms}^{-1}$$

21. As, $v_L = \sqrt{\frac{Y}{\rho}}$ and $v_T = \sqrt{\frac{T}{m}} = \sqrt{\frac{T}{\pi r^2 \rho}}$

$$\because \quad \frac{v_L}{v_T} = \sqrt{\frac{Y}{\rho} \times \frac{\pi r^2 \rho}{T}} = \sqrt{\frac{Y}{T/\pi r^2}} = \sqrt{\frac{Y}{\text{stress}}}$$

$$\therefore \text{ Stress} = \frac{Y}{(v_L/v_T)^2} = \frac{1 \times 10^{11}}{(100)^2} = 1 \times 10^7 \text{ Nm}^{-2}$$

22. As, intensity $\propto a^2\omega^2$

Here, $\quad \frac{a_A}{a_B} = \frac{2}{7}$ and $\frac{\omega_A}{\omega_B} = \frac{1}{2}$

$$\rightarrow \quad \frac{I_A}{I_B} - \left(\frac{2}{1}\right)^2 \times \left(\frac{1}{2}\right)^2 - \frac{1}{1}$$

23. Amplitude, $A = \frac{6}{2} = 3$ cm $= 0.03$ cm

Angular frequency, $\omega = 2\pi f = 2\pi \times 245$

$= 1538.6 \text{ rad s}^{-1}$

$= 1.5 \times 10^3 \text{ rad s}^{-1}$

Wave number, $k = \frac{\omega}{v} = \frac{1538.6}{300} = 5.1 \text{ radm}^{-1}$

$\therefore \quad y(x, t) = 0.03 \sin[5.1x - (1.5 \times 10^3)t]$

24. $I = \frac{(\Delta p)^2_{max}}{2\rho V}$

$\Rightarrow \quad v = \frac{(\Delta p)^2_{max}}{2\rho I} = (2 \times 10^{-4})^2 / 2 \times 1 \times 10^{-10}$

$= 200 \text{ ms}^{-1}$

Amplitude of wave is

$A = \frac{(\Delta p)_{max}}{\omega \rho v} = \frac{2 \times 10^{-4}}{10^3 \times 1 \times 200} = 10^{-9} \text{ m}$

$\therefore \quad k = \frac{\omega}{v} = \frac{1000}{200} = 5 \text{ m}^{-1}$

As, the wave is travelling in $-x$-direction, the only equation with positive sign between ωt and kx, with suitable ω and k values is option (b).

25. On reflection from a denser medium, there is a phase reversal of 180°.

New amplitude $= \frac{2}{3} \times 0.6 = 0.4$

$\therefore$ Equation of reflected wave,

$$y = 0.4 \sin 2\pi \left[t + \frac{x}{2} + 180^\circ \right]$$

$$= -0.4 \sin 2\pi \, (t + x/2)$$

26. As, $y = 4\cos^2(t/2)\sin(1000t)$

$= 2\,[2\cos^2(t/2)\sin(1000t)]$

$= 2\,[2(1 \cos t)\sin(1000t)]$

$= 2\sin 1000t + 2\sin 1000t \cos t$

$y = 2\sin 1000t + \sin(1001)t + \sin(999t)$

$\therefore$ The given wave equation represents the superposition of three waves.

27. Resultant displacement along X-axis,

$x = y_1 - y_3 = 8 - 2 = 6$

Resultant displacement along Y-axis,

$y = y_2 - y_4 = 4 - 1 = 3$

Net displacement,

$r = \sqrt{x^2 + y^2} = \sqrt{6^2 + 3^2} = \sqrt{45}$

Also, $\tan\theta = \frac{y}{x} = \frac{3}{6} = \frac{1}{2}$

$\theta = \tan^{-1}(1/2)$

28. After 2 s, the pulses will overlap completely. The string becomes straight and therefore, does not have any potential energy. Its entire energy must be purely kinetic.

29. $y_1 = 0.05 \cos\,(0.50\,\pi x - 100\,\pi t)$

and $y_2 = 0.05 \cos\,(0.46\pi x - 92\pi t)$

Comparing these two equations are

$y = A\sin(kx - \omega t)$, we get

$\omega_1 = 100\,\pi$ and $\omega_2 = 92\,\pi$

Now, speeds, $v_1 = \frac{A}{100\,\pi} = \frac{0.05}{100\,\pi}$

and $v_2 = \frac{A}{92\,\pi} = \frac{0.05}{92\pi}$

Now, the resultant speed,

$$v = \sqrt{\left(\frac{0.05}{92\,\pi}\right)^2 + \left(\frac{0.05}{100\,\pi}\right)^2}$$

$= 200$ m/s

30. Since, amplitude at all waves are same but Z_1 and Z_2 are displacements of two waves of same frequency travelling in opposite directions.They will form a stationary wave.

31. Since, the point $x = 0$ is a node and reflection is taking place from point $x = 0$.This means that reflection must be taking place from the fixed end, hence the reflected ray must suffer an additional phase change of π or a path change of $\frac{\lambda}{2}$.

So, if $y_{\text{incident}} = a\cos(kx - \omega t)$

$\Rightarrow \quad y_{\text{reflected}} = a\cos(-kx - \omega t + \pi) = -a\cos(\omega t + kx)$

32. Distance between two consecutive node is $\frac{\lambda}{2}$.

$\therefore \quad \frac{\lambda}{2} = \frac{2}{2} \text{ m} = 1 \text{ m}$

So, the distance of another node from the surface will be

$$3 + \frac{\lambda}{2} = 3 + 1 = 4 \text{ m}$$

33. The given equation is

$$y\,(x, t) = 0.06 \sin\left(\frac{2\pi x}{3}\right) \cos\,(120\pi t)$$

As terms involving x and t are independent of each other, the given equation represents a stationary wave.

Compare the given equation with the standard form of equation of stationary wave

$$y\,(x, t) = 2r \sin kx \cos \omega t$$

$$k = \frac{2\pi}{\lambda} = \frac{2\pi}{3}$$

$\therefore \quad \lambda = 3\text{m}$

$\omega = 120\pi$

$\therefore \quad \nu = \frac{\omega}{2\pi} = \frac{120\pi}{2\pi} = 60 \text{ Hz}$

and $\quad v = \nu\lambda = 60 \times 3 = 180$ m/s

Hence, the given stationary wave is the result of superposition of two waves of wavelength 3 m and frequency 60 Hz each, travelling with a velocity of 180 m/s in opposite directions.

34. Node means a point at which medium particles do not displace from its mean position and antinode mean a point at which particles oscillate with maximum possible amplitude. Nodes and antinodes are obtained for both types of stationary waves, transverse and longitudinal. Hence, options (a) and (b) both are wrong. To obtain a stationary wave, two waves travelling in opposite directions, having same amplitude, same frequency are required. They must have same nature, means either both of the waves should be longitudinal or both of them should be transverse. Hence, option (c) is correct.

35. Given, $L = 15.0$ m, $v = 12\ \text{ms}^{-1}$

Since, there are 6 nodes, with the ends as nodes, there will be five half wavelength in the string.

So, $$\frac{5\lambda}{2} = L = 15$$

$\Rightarrow$ $$\lambda = 6.0 \text{ m}$$

Using $f = \frac{v}{\lambda} = \frac{12}{6} = 2.0$ Hz

36. The given equation is

$$y(x, t) = 0.06 \sin\left(\frac{2\pi x}{3}\right) \cos(120\,\pi t)$$

It represents a stationary wave. Therefore, all the points between two consecutive nodes.

I. vibrate with same frequency

II. in same phase, but

IV. different amplitudes. The amplitude is zero at nodes and maximum at antinodes (between the nodes).

37. Here, $\rho_1 = \rho_2$; $\frac{r_1}{r_2} = \frac{1}{2}$, $T_1 = T_2$

$\Rightarrow$ $$n_1 = \frac{1}{2Lr_1}\sqrt{\frac{T_1}{\pi \rho_1}}$$

and $$n_2 = \frac{1}{2Lr_2}\sqrt{\frac{T_2}{\pi \rho_2}}$$

$\therefore$ $$\frac{n_1}{n_2} = \frac{r_1}{r_2} = 2:1$$

38. From $n = \frac{L}{LD}\sqrt{\frac{T}{\pi\rho}}$

When radius of string is doubled, diameter D becomes twice. As T and ρ are same , n becomes $1/2$, *i.e.* $n/2$.

39. According to the law of length,

$$n_1 l_1 = n_2 l_2$$

$\therefore$ $$l_2 = \frac{n_1 l_1}{n_2} = \frac{800 \times 50}{1000} = 40 \text{ cm}$$

40. Fundamental frequency of oscillation of a taut string is

$$f = \frac{1}{2L}\sqrt{\frac{T}{\mu}}$$

where, L = length, T = tension

and μ = mass per unit length.

Now, given X and Z are identical strings.

$\therefore$ L and μ are same for both X and Z.

$\Rightarrow$ $$f \propto \sqrt{T}$$

$\Rightarrow$ $$\frac{f_X}{f_Z} = \sqrt{\frac{T_X}{T_Z}}$$

$\Rightarrow$ $$\frac{T_X}{T_Z} = \frac{f_X^2}{f_Z^2}$$

$$= \left(\frac{450}{300}\right)^2 = \frac{9}{4} = 2.25$$

41. Velocity 'v' of the wave on the string $= \sqrt{\frac{T}{\mu}}$

where, T = tension and μ = mass per unit length. Substituting the given values, we get

$$v = \sqrt{\frac{8}{5} \times 1000} = 40\ \text{ms}^{-1}$$

Wavelength of the wave on the string, $\lambda = \frac{v}{f}$

where, f = frequency of wave.

$\Rightarrow$ $$\lambda = \frac{40}{100} \text{ m} = 40 \text{ cm}$$

$\therefore$ Separation between two successive nodes,

$$d = \frac{\lambda}{2} = \frac{40}{2} = 20.0 \text{ cm}$$

42. Speed of sound in gas, $v = \sqrt{\frac{\gamma p}{\rho}}$, γ is highest for monoatomic gas.

Since, $\Delta B = 10 \log_{10}\left(\frac{I_1}{I_2}\right)$

$\Rightarrow$ $$20 = 10 \log_{10}\left(\frac{I_1}{I_2}\right) \Rightarrow \frac{I_1}{I_2} = 100$$

Passing of sound wave is considered as adiabatic process.

43. When reflection of a sound wave occurs at a rigid boundary (like the closed end of a pipe), the particles at the boundary are unable to vibrate. The reflected wave thus generated interferes with the incident wave to produce zero displacement (or node). At this displacement node exists the pressure antinode. Thus, reflected pressure wave has the same phase as the incident wave and a high pressure compression pulse gets reflected as a compression pulse.

Similarly, for reflection from the open end of the pipe, the particles vibrate with increased amplitude (displacement antinode) and pressure remains at the average value (pressure node). The reflected pressure wave interferes destructively with the incident wave, so that a phase change of π occurs from the open end. Hence, a high-pressure compression pulse reflects as a low-pressure refraction pulse.

44. (a) Velocity of sound wave in air independent of pressure, if T = constant

(b) $v = \sqrt{\frac{\gamma RT}{M}} = \sqrt{\frac{\gamma R(t + 273)}{M}}$

or $$v^2 \propto (t + 273)$$

(c) $v = \sqrt{T/\mu}$

$\therefore$ $$v \propto \sqrt{T}$$

or $$v^2 \propto T$$

(d) $f_0 = \frac{v}{2L}$ or $\frac{v}{4L}$

$\Rightarrow \quad f_0 \propto \frac{1}{L}$

Hence, option (d) is correct.

45. As, $n \propto \frac{1}{L}$ or $L \propto \frac{1}{n}$

and $\quad L = L_1 + L_2 + L_3$

$$\Rightarrow \quad \frac{1}{n} = \frac{1}{n_1} + \frac{1}{n_2} + \frac{1}{n_3}$$

46. Resonant frequency for a stretched string is given by

$$f_n = nf_1 = \frac{n}{2L}\sqrt{\frac{T}{\mu}}$$

Given, two consecutive resonant frequencies are 420 Hz and 490 Hz.

So, $$\frac{n}{2L}\sqrt{\frac{T}{\mu}} = 420 \quad \text{...(i)}$$

and $$\frac{n+1}{2L}\sqrt{\frac{T}{\mu}} = 490 \quad \text{...(ii)}$$

Subtracting Eq. (ii) from Eq. (i), we get

$$\frac{1}{2L}\sqrt{\frac{T}{\mu}} = 490 - 420 = 70$$

Here, $\quad T = 540\text{ N}, \mu = 6 \times 10^{-3}\text{ kg m}^{-1}$

So, $$L = \frac{1}{2 \times 70}\sqrt{\frac{T}{\mu}} = \frac{1}{2 \times 70} \times \sqrt{\frac{540}{6 \times 10^{-3}}}$$

$$= 2.14\text{ m} \simeq 2.1\text{ m}$$

47. Let m = mass per unit length of rope,

T = tension in the rope at a distance x from the lower end

$\therefore \quad T = (mg)\,x$ = weight of x metre of rope

As $v = \sqrt{\frac{T}{m}}$, therefore $v = \sqrt{\frac{mgx}{m}} = \sqrt{gx}$

i.e. $v \propto \sqrt{x}$

48. Given equation of stationary wave is

$$y = 0.3 \sin (0.157x) \cos (200\pi t)$$

Comparing it with general equation of stationary wave, *i.e.* $y = a \sin kx \cos \omega t$, we get

$$k = \left(\frac{2\pi}{\lambda}\right) = 0.157$$

$$\Rightarrow \quad \lambda = \frac{2\pi}{0.157} = 4\pi^2 \quad \left(\because \frac{1}{2\pi} \approx 0.157\right) \text{...(i)}$$

and $\omega = 200\,\pi = \frac{2\pi}{T} \Rightarrow T = \frac{1}{20}$ s

As the possible wavelength associated with nth harmonic of a vibrating string, *i.e.* fixed at both ends is given as

$$\lambda = \frac{2L}{n} \text{ or } L = n\left(\frac{\lambda}{2}\right)$$

Now, according to question, string is fixed from both ends and oscillates in 4th harmonic, so

$$4\left(\frac{\lambda}{2}\right) = L \Rightarrow 2\lambda = L$$

or $\quad L = 2 \times 4\pi^2 = 8\pi^2 \quad$ [using Eq.(i)]

Now, $\quad \pi^2 \approx 10 \Rightarrow L \approx 80\text{ m}$

49.

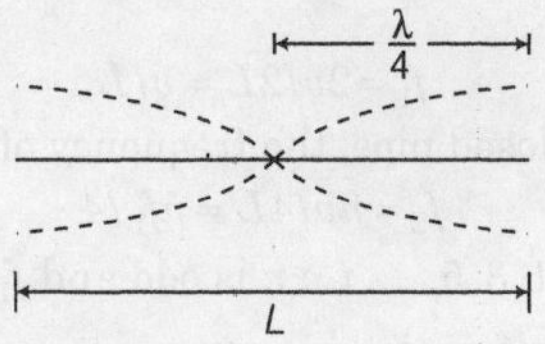

From vibration mode,

$$\frac{\lambda}{2} = L \Rightarrow \lambda = 2L$$

$\therefore$ Wave speed, $v = \sqrt{\frac{Y}{\rho}}$

So, frequency $f = \frac{v}{\lambda}$

$$\Rightarrow \quad f = \frac{1}{2L}\sqrt{\frac{Y}{\rho}}$$

$$= \frac{1}{2 \times 60 \times 10^{-2}}\sqrt{\frac{9.27 \times 10^{10}}{2.7 \times 10^3}} \approx 5000\text{ Hz}$$

$$f = 5\text{ kHz}$$

50. For an organ pipe open at one end,

frequency of 1st overtone, $n_1 = \frac{3v}{4L_1}$

For the organ pipe open at both ends,

frequency of 3rd harmonic, $n_2 = \frac{3v}{2L_2}$

As, $\quad n_1 = n_2$

$$\therefore \quad \frac{3v}{4L_1} = \frac{3v}{2L_2}$$

or $$\frac{L_1}{L_2} = \frac{2}{4} = \frac{1}{2}$$

51. As fixed end is a node, therefore distance between two consecutive nodes $= \frac{\lambda}{2} = 10$ cm

$$\lambda = 20\text{ cm} = 0.2\text{ m}$$

Now, $\quad v = \nu\lambda$

$\therefore \quad v = 100 \times 0.2 = 20\text{ ms}^{-1}$

52. As is clear from figure of question,

$$L = \frac{\lambda_p}{4},\ \lambda_p = 4L,\ n_p = \frac{v}{\lambda_p} = \frac{v}{4L}$$

$$L = \frac{\lambda_q}{2},\ \lambda_q = 2L,\ n_q = \frac{v}{\lambda_q} = \frac{v}{2L}$$

$$L = \lambda r, \lambda_r = 1, n_r = \frac{v}{\lambda_r} = \frac{v}{L}$$

$$L = \frac{3\lambda_s}{4},\ \lambda_s = \frac{4L}{3},\ n_s = \frac{v}{\lambda_s} = \frac{3v}{4L}$$

$$L = \frac{3\lambda s}{4}, \lambda_s = \frac{4L}{3}$$

$$h_s = \frac{v}{\lambda_s} = \frac{3v}{4L}$$

$\therefore \quad n_p : n_q : n_r : n_s = \frac{v}{4L} : \frac{v}{2L} : \frac{v}{L} : \frac{3v}{4L} = 1 : 2 : 4 : 3$

53. In case of open pipe, the frequency of second harmonic is,

$$f_1 = 2v/2L = v/L$$

In case of closed pipe, the frequency of nth harmonic is

$$f_2 = nv/4L = nf_1/4$$

where, $n = 1, 3, 5, \ldots$, *i.e.* n is odd and $f_1 > f_2$

It will be so if $n = 5$

$\therefore \quad f_2 = \frac{5}{4} f_1$

54. $L_0 = \frac{5\lambda_{air}}{2} \Rightarrow \lambda_{air} = \frac{2L_0}{5}$

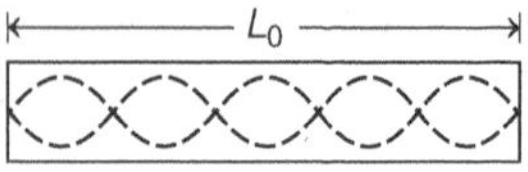

$$f = v_{air}/\lambda_{air} = \frac{343 \times 3}{2L_0} = \frac{1029}{2L_0}$$

$$\frac{\lambda}{4} = 1\text{ m} \Rightarrow \lambda = 4\text{ m}$$

$\therefore \quad f = v_{steel}/\lambda = \frac{5250}{4} = 1312.5\text{ Hz}$

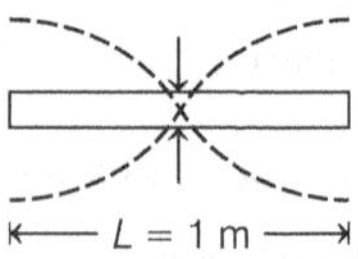

As both of them are in unison, $\frac{1029}{2L_0} = 1312.5\text{ Hz}$

$\therefore \quad L_0 = 0.392\text{ m} \simeq 39\text{ cm}$

55. As, $f_c = f_o$ (both first overtone)

or $3\left(\frac{v_c}{4L}\right) = 2\left(\frac{v_o}{2l_o}\right)$

$\therefore \quad l_o = \frac{4}{3}\left(\frac{v_o}{v_c}\right)L$

$$= \frac{4}{3}\frac{\sqrt{B/\rho_2}}{\sqrt{B/\rho_1}}L$$

$$= \frac{4}{3}\sqrt{\frac{\rho_1}{\rho_2}}L$$

56. As, $\lambda = \frac{v}{n} = \frac{330}{500} = 0.66\text{ m} = 66\text{ cm}$

The successive resonance lengths are at

$$\frac{\lambda}{4}, \frac{3\lambda}{4}, \frac{5\lambda}{4}, \frac{7\lambda}{4} \text{ and so on.}$$

Within one metre length of the tube, total number of resonances is 3(as, $\frac{7\lambda}{4}$ is more than 1.0 m).

57. Length at which first resonance occurs

$$L_1 = 25.5\text{ cm}$$

Length at which second resonance occurs $L_2 = 79.3$ cm

$\therefore$ Wavelength, $\lambda = 2(L_2 - L_1)$

$= 2(79.3 - 25.5)$

$= 2 \times 53.8$

$= 107.6$ cm

$= 1.076$ m

Using $v = \nu\lambda$

$\therefore$ Speed of sound in air $v = 340(107.6)$ m/s ($\because \nu = 340$ Hz)

$= 365.84$ m/s

$\simeq 366\text{ ms}^{-1}$

58. Fundamental frequency of closed organ pipe is given by $f_0 = v/4L$, where v is the velocity of sound in it and L is the length of the pipe.

Also, overtone frequencies are given by

$$f = (2n+1)\frac{v}{4L}$$

or $\quad f = (2n+1) f_0$

Given, $f_0 = 1500$ Hz and $f_{max} = 20000$ Hz

This means, $f_{max} > f$

So, $\quad f_{max} > (2n+1) f_0$

$\Rightarrow \quad 20000 > (2n+1)1500$

$\Rightarrow \quad 2n + 1 < 13.33$

$\Rightarrow \quad 2n < 13.33 - 1$

$\Rightarrow \quad 2n < 12.33$

or $\quad n < 6.16$

or $n = 6$ (integer number)

Hence, total six overtones will be heard.

59. For open ends, fundamental frequency f in air, we have

$$\frac{\lambda}{2} = L \Rightarrow \lambda = 2L$$

$$v = f\lambda \Rightarrow f = \frac{v}{\lambda} = \frac{v}{2L} \quad \ldots\text{(i)}$$

When a pipe is dipped vertically in water, so that half of it is in water, we have

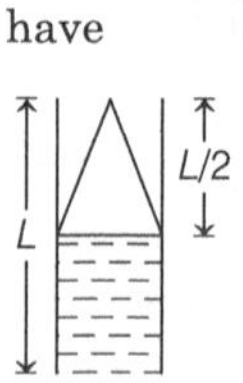

$$\frac{\lambda}{4} = \frac{L}{2}$$

$\Rightarrow \quad \lambda = 2L$

$\Rightarrow \quad v = f'\lambda$

$$\Rightarrow \qquad f' = \frac{v}{\lambda} = \frac{v}{2L} = f \qquad \dots\text{(ii)}$$

Thus, the fundamental frequency of the air column is now,

$$f = f'$$

60. As,
$$L_1 + x = \frac{\lambda}{4} = 22.7$$
$$L_2 + x = \frac{3\lambda}{4} = 70.2$$
and
$$L_3 + x = \frac{5\lambda}{4}$$
$$\therefore \qquad x = \frac{L_2 - 3L_1}{2}$$
$$= \frac{70.2 - 68.1}{2}$$
$$= \frac{2.1}{2} = 1.05 \text{ cm}$$

Now,
$$\frac{L_3 + x}{L_1 + x} = 5$$
$$\Rightarrow \qquad L_3 = 5L_1 + 4x$$
$$= 5 \times 22.7 + 4 \times 1.05$$
$$= 117.7 \text{ cm}$$

61. For closed organ pipe $= \frac{(2n+1)v}{4l}$ $(n = 0, 1, 2 \dots\dots)$

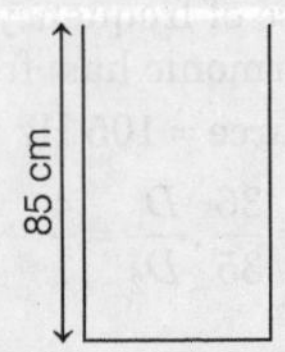

$$\frac{(2n+1)v}{4l} < 1250$$
$$(2n+1) < 1250 \times \frac{4 \times 0.85}{340}$$
$$(2n+1) < 12.5$$
$$\Rightarrow \qquad 2n < 11.50$$
$$n < 5.25$$

So, $n = 0, 1, 2, 3, \dots 5$

So, we have 6 possibilities.

Alternate method In closed organ pipe, fundamental node

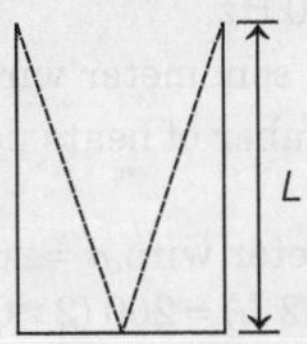

i.e.
$$\frac{\lambda}{4} = 0.85$$
$$\lambda = 4 \times 0.85$$

As, we know, $\nu = \frac{v}{\lambda}$

$$= \frac{340}{4 \times 0.85} = 100 \text{ Hz}$$

$\therefore$ Possible frequencies = 100 Hz, 300 Hz, 500 Hz, 700 Hz, 900 Hz, 1100 Hz below 1250 Hz.

62. In a resonance tube apparatus, first and second resonance occur as shown below

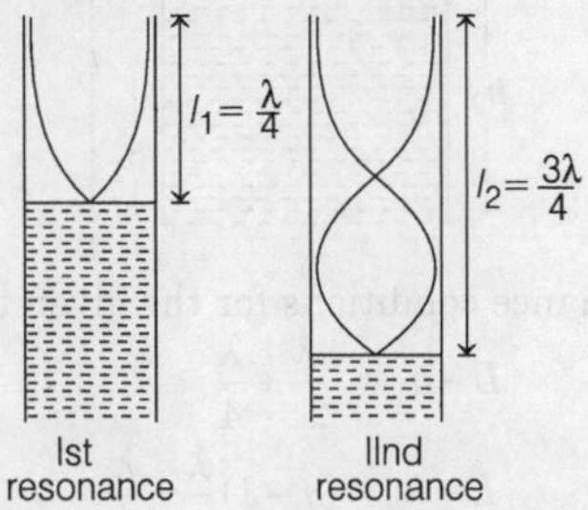

As in a stationary wave, distance between two successive nodes is $\frac{\lambda}{2}$ and distance of a node and an antinode is $\frac{\lambda}{4}$.

$$l_2 - l_1 = \frac{3\lambda}{4} - \frac{\lambda}{4} = \frac{\lambda}{2}$$

So, speed of sound, $v = f\lambda$
$$= f \times 2(l_2 - l_1)$$
$$= 480 \times 2 \times (70 - 30) \times 10^{-2}$$
$$= 384 \text{ ms}^{-1}$$

63. Given,

fundamental frequency of wire (f_{wire}) = $v/2L$.

(a)

$f = \frac{v}{4L}, \frac{3v}{4L}, \frac{5v}{4L}$ cannot match with f_{wire}.

(b)

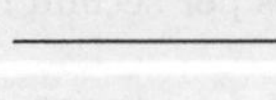

$f = \frac{v}{2(2L)}, \frac{2v}{2(2L)}, \frac{3v}{2(2L)}$ its second harmonic $\frac{2v}{2(2L)}$ matches with f_{wire}.

(c)

$f = \frac{v}{2(L/2)}, \frac{2v}{2(L/2)}$ cannot match with f_{wire}.

(d)

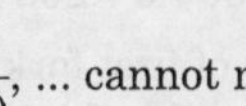

$f = \frac{v}{4(L/2)}, \frac{3v}{4(L/2)}, \dots$ cannot match with f_{wire}.

64. Let λ be the wavelength and L be the length of the tube.

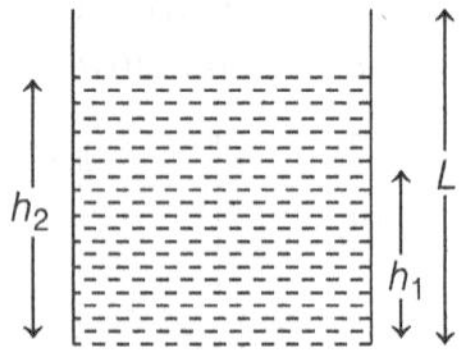

The resonance conditions for the given two cases are

$$L - h_1 = \frac{n\lambda}{2} + \frac{\lambda}{4} \quad \text{...(i)}$$

$$L - h_2 = (n-1)\frac{\lambda}{2} + \frac{\lambda}{4} \quad \text{...(ii)}$$

Subtracting Eq. (i) from Eq. (ii), we get

$$(L - h_1) - (L - h_2) = \frac{n\lambda}{2} + \frac{\lambda}{4} - (n-1)\frac{\lambda}{2} - \frac{\lambda}{4}$$

$$h_2 - h_1 = \frac{\lambda}{2}$$

Here, $h_1 = 17.0$ cm and $h_2 = 24.5$ cm

Substituting these values in above equation, we get

$$24.5 - 17.0 = \frac{\lambda}{2}$$

$$\Rightarrow \quad \frac{\lambda}{2} = 7.5 \text{ cm}$$

$$\Rightarrow \quad \lambda = 15.0 \text{ cm}$$

Now, the tuning fork frequency,

$$f = \frac{v}{\lambda} = \frac{330}{15 \times 10^{-2}} = 2200 \text{ Hz}$$

65. Here, $n_1 = 200$ Hz

Number of beats s^{-1}; $m = 4$

$\therefore \quad n_2 = 200 \pm 4 = 204$ or 196 Hz

On loading fork 2, its frequency decreases. And number of beats per second increases to 6. Therefore, m is negative.

$$n_2 = 200 - 4 = 196 \text{ Hz}$$

66. Here, $T_1 = 16$ N, $T_2 = T = ?$

As per the choice given, $T_2 > T_1$

$\therefore \quad n_2 > n_1$ and given that $(n_2 - n_1) = 3$ ...(i)

Now, $\quad n \propto \sqrt{T}$

$$\therefore \quad \frac{n_2}{n_1} = \sqrt{\frac{T}{16}} = \frac{\sqrt{T}}{4}$$

If n_1 corresponds to 4, then n_2 corresponds to $3 + 4 = 7$, which is $\sqrt{T}$. Therefore, $T = 49$ N

67. Here, $n = 200 \pm 5$ and $2n = 420 \pm 10$. This is possible only when $n = 200 + 5 = 205$ Hz or s^{-1}.

68. If n is frequency of first fork, then frequency of the last (10th fork) $= n + 4(10 - 1) = 2n$

$\therefore \quad n = 36$ and $2n = 72$

69. If we assume that all the three waves are in same phase at $t = 0$, we shall hear only one beat per second.

70. Here, $\omega_1 = 100\pi$ and $\omega_2 = 92\pi$

Hence $\nu_1 = \frac{100\pi}{2\pi} = 50$ Hz

and $\quad \nu_2 = \frac{92\pi}{2\pi} = 46$ Hz

$\therefore$ Number of beats per second $= \nu_1 - \nu_2 = 50 - 46 = 4$

71. $\frac{n_1}{n_2} = \frac{L_2}{L_1} = \frac{25}{30} = \frac{5}{6}$

and $\quad n_2 - n_1 = 4$

On solving, we get

$$n_2 = 24 \text{ Hz}$$
$$n_1 = 20 \text{ Hz}$$

72. As, $\frac{n_1}{n_2} = \frac{L_2}{L_1} = \frac{51}{50}$

and $\quad n_1 - n_2 = 5$

On solving, we get

$$n_2 = 250 \text{ Hz}, \ n_1 = 255 \text{ Hz}$$

73. Two possible frequencies of source are $= 100 \pm 5$ $= 105$ or 95 Hz

Frequencies of 2nd harmonic $= 210$ or 190 Hz

5 beats with source of frequency 205 Hz are possible only when 2nd harmonic has frequency $= 210$ Hz

$\therefore$ Frequency of source $= 105$ Hz

74. Here, $\frac{T_1}{T_2} = \frac{8}{1}, \frac{L_1}{L_2} = \frac{36}{35}, \frac{D_1}{D_2} = \frac{4}{1}$

$$\frac{\rho_1}{\rho_2} = \frac{1}{2}$$

and $\quad n_1 = 360$ Hz, $n_2 = ?$

Now, $\quad \frac{n_2}{n_1} = \frac{L_1 D_1}{L_2 D_2}\sqrt{\frac{T_2 \rho_1}{\rho_2 T_1}}$

$$\frac{n_2}{n_1} = \frac{36}{35} \times \frac{4}{1}\sqrt{\frac{1}{8} \times \frac{1}{2}} = \frac{36}{35}$$

Clearly, $\quad n_2 > n_1$.

When $\quad n_2 = 360$ Hz, $n_1 = 350$ Hz

Number of beats per second $= n_2 - n_1 = 360 - 350 = 10$

75. Frequency of fork $= 250$ Hz.

Possible frequencies of sonometer wire $= (250 \pm 10)$ Hz

On filling the fork, number of beats per second decreases.

$\therefore$ Frequency of sonometer wire, $n = 260$ Hz

$$v = n\lambda = 260\,(2L) = 260\,(2 \times 0.5)$$
$$= 260 \text{ ms}^{-1}$$

76. When waves of nearby frequencies overlaps, beats are produced.

Beat frequency is given by, $f_{beat} = f_1 - f_2$

Here, $f_1 = 11$ Hz and $f_2 = 9$ Hz

$\Rightarrow$ Beat frequency is, $f_{beat} = 11 - 9 = 2$ Hz

Hence, time period of beats or time interval between beats,

$$T = \frac{1}{f_{\text{beat}}}$$

$$\Rightarrow \quad T = \frac{1}{2} = 0.5 \text{ s}$$

So, resultant wave has a time period of 0.5 s which is correctly depicted in option (a) only.

77. $$f_1 = \frac{\omega_1}{2\pi} = \frac{(400\pi)}{2\pi} = 200 \text{ Hz}$$

$$f_2 = \frac{\omega_2}{2\pi} = \frac{404\,\pi}{2\pi} = 202 \text{ Hz}$$

$$\therefore \quad f_b = f_2 - f_1 = 2 \text{ Hz}$$

$$A_1 = 2 \text{ and } A_2 = 1$$

$$\frac{I_{\max}}{I_{\min}} = \left(\frac{A_1 + A_2}{A_1 - A_2}\right)^2 = \frac{9}{1}$$

78. From Doppler's effect,

$$n' = n\left(\frac{v - v_o}{v + v_s}\right) = n\left(\frac{340 - 10}{340 + 10}\right) = 1950 \text{ Hz} \quad \text{(given)}$$

$$n = 2068 \text{ Hz}$$

79. Whistling train is the source of sound, $v_s = V$. Before crossing a stationary observer on station, frequency heard is $n' = \dfrac{vn}{(v - v_s)} = \dfrac{vn}{v - V}$ = constant and $n' > n$.

Here, v is velocity of sound in air and n is actual frequency of whistle.
After crossing the stationary observer, frequency heard is $n' = \dfrac{vn}{(v + v_s)} = \dfrac{vn}{v + V}$ = constant and $n' < n$.

Therefore, the expected curve is given in option (c).

80. As the observer is moving towards the source, so frequency of waves emitted by the source will be given by the formula

$$f_{\text{observed}} = f_{\text{actual}} \cdot \left(\frac{1 + v/c}{1 - v/c}\right)^{1/2}$$

Here, frequency, $\dfrac{v}{c} = \dfrac{1}{2}$

So, $$f_{\text{observed}} = f_{\text{actual}} \left(\frac{3/2}{1/2}\right)^{1/2}$$

$$\therefore \quad f_{\text{observed}} = 10 \times \sqrt{3} = 17.3 \text{ GHz}$$

81. Apparent frequency heard by the person before crossing the trains,

$$f_1 = \left(\frac{c}{c - v_s}\right) f_0 = \left(\frac{320}{320 - 20}\right) 1000$$

Similarly, apparent frequency heard after crossing the train,

$$f_2 = \left(\frac{c}{c + v_s}\right) f_0 = \left(\frac{320}{320 + 20}\right) 1000$$

(c = speed of sound)

$$\Delta f = f_1 - f_2 = \left(\frac{2\,cv_s}{c^2 - v_s^2}\right) f_0$$

or $$\frac{\Delta f}{f_0} \times 100 = \left(\frac{2\,cv_s}{c^2 - v_s^2}\right) \times 100$$

$$= \frac{2 \times 320 \times 20}{300 \times 340} \times 100$$

$$= \frac{2 \times 32 \times 20}{3 \times 34} = 12.54\%$$

$$= 12\%$$

82. The given condition can be shown below as

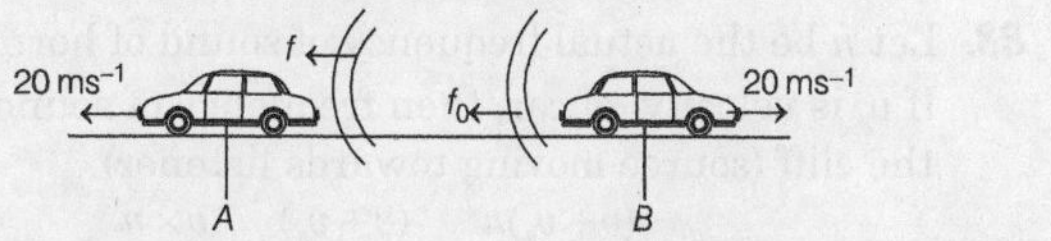

Here, source and observer both are moving away from each other. So, by Doppler's effect, observed frequency is given by

$$f = f_0\left(\frac{v + v_o}{v - v_s}\right) \quad \text{... (i)}$$

where, v = speed of sound = 340 ms^{-1},
v_o = speed of observer = $-$ 20 ms^{-1},
v_s = speed of source = $-$ 20 ms^{-1},
f_0 = true frequency
and f = apparent frequency = 2000 Hz

Substituting the given values in Eq. (i), we get

$$2000 = \left(\frac{340 - 20}{340 + 20}\right) \times f_0$$

$$\Rightarrow \quad f_0 = \frac{2000 \times 360}{320} = 2250 \text{ Hz}$$

83. $$f = f_0\left(\frac{v + v_o}{v}\right) = 10^3\left(1 + \frac{10t}{v}\right) \quad (\text{as } v_o = gt)$$

Hence, f *versus* t graph is a straight line of slope $\dfrac{10^4}{v}$.

$$\therefore \quad \frac{10^4}{v} = \text{slope} = \frac{100}{3}$$

$$\therefore \quad v = 300 \text{ m/s}$$

84. As the listener on motor cycle is moving away from the source (siren), therefore $\dfrac{n'}{n} = \dfrac{v - v_L}{v} = \dfrac{94}{100}$

$$\Rightarrow \quad 1 - \frac{v_L}{v} = \frac{94}{100}$$

$$\Rightarrow \quad \frac{v_L}{v} = 1 - \frac{94}{100} = \frac{6}{100}$$

$$\Rightarrow \quad v_L = \frac{6 \times v}{100} = \frac{6 \times 330}{100} = 19.8 \text{ ms}^{-1}$$

$$\text{Distance covered} = \frac{v_L^2}{2a} = \frac{19.8 \times 19.8}{2 \times 2} = 98 \text{ m}$$

85. ν = 165 Hz and

$$\nu' = \frac{335 + 5}{335} \times \frac{335}{330} \times 165 = 170 \text{ Hz}$$

$\therefore$ Number of beats per second = $\nu' - \nu = 170 - 165 = 5$

86. Number of extra waves received per second

$$= \frac{u}{\lambda} - (-u/\lambda) = \frac{2u}{\lambda}$$

$=$ Number of beats heard per second

87. Large vertical plane acts as listener moving per second.

$$\therefore \qquad n' = \frac{(c+v)n}{c}$$

This is the number of waves striking the surface per second.

88. Let n be the actual frequency of sound of horn.

If v_s is velocity of car, then frequency of sound striking the cliff (source moving towards listener)

$$n' = \frac{(v+v_s)n'}{v} = \frac{(v+v_s)}{v} \times \frac{v \times n}{(v-v_s)}$$

or
$$\frac{n''}{n} = \frac{v+v_s}{v-v_s} = 2$$

$$v + v_s = 2v - 2v_s$$

$$3v_s = v, v_s = \frac{v}{3}$$

89. Given, frequency of sound source $(f_0) = 500$ Hz

Apparent frequency heard by observer 1, $f_1 = 480$ Hz and apparent frequency heard by observer 2, $f_2 = 530$ Hz.

Let v_o be the speed of sound.

When observer moves away from the source,

Apparent frequency, $f_1 = f_o\left(\frac{v - v_o'}{v}\right)$ ... (i)

When observer moves towards the source,

Apparent frequency, $f_2 = f_o\left(\frac{v + v_o''}{v}\right)$ ... (ii)

Substituting the values in Eq. (i), we get

$$480 = 500\left(\frac{300 - v_o'}{300}\right)$$

$$\Rightarrow \qquad 96 \times 3 = 300 - v_o'$$

$$\Rightarrow \qquad v_o' = 12 \text{ m/s}$$

Substituting the values in Eq. (ii), we get

$$530 = 500\left(\frac{330 + v_o''}{300}\right)$$

$$\Rightarrow \qquad 106 \times 3 = 300 + v_o''$$

$$\Rightarrow \qquad v_o'' = 18 \text{ m/s}$$

Thus, their respective speeds (in m/s) is 12 and 18.

90. When a source is moving towards a stationary observer, observed frequency is given by

$$f_{\text{observed}} = f\left(\frac{v}{v+v_s}\right)$$

where, f = frequency of sound from the source, v = speed of sound and v_s = speed of source.

Now, applying above formula to two different conditions given in problem, we get

f_1 = Observed frequency initially

$$= f\left(\frac{340}{340-34}\right) = f\left(\frac{340}{306}\right)$$

and f_2 = Observed frequency when speed of source is reduced

$$= f\left(\frac{340}{340-17}\right) = \frac{340}{323}$$

So, the ratio $f_1 : f_2$ is, $\frac{f_1}{f_2} = \frac{323}{306} = \frac{19}{18}$.

91. Initially,

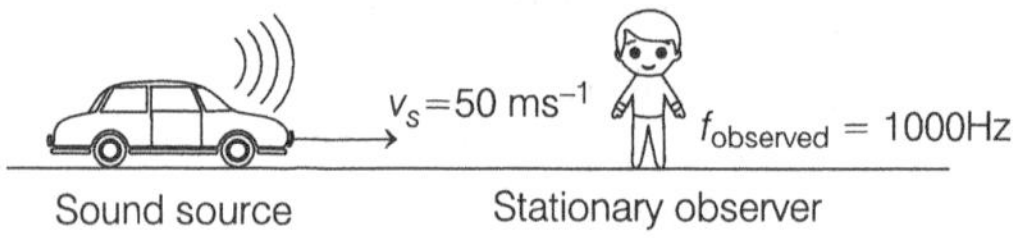

After sometime,

When source is moving towards stationary observer, frequency observed is more than source frequency due to Doppler's effect, it is given by

$$f_{\text{observed}} = f\left(\frac{v}{v - v_s}\right)$$

where, f = source frequency,

f_o = observed frequency = 1000 Hz,

v = speed of sound in air = 350 ms^{-1}

and v_s = speed of source = 50 ms^{-1}

So, $f = \frac{f_{\text{obs}}(v - v_s)}{v} = \frac{1000(350-50)}{350} = \frac{6000}{7}$ Hz

When source moves away from stationary observer, observed frequency will be lower due to Doppler's effect and it is given by

$$f_0 = f\left(\frac{v}{v+v_s}\right) = \frac{6000 \times 350}{7 \times (350+50)} = \frac{6000 \times 350}{7 \times 400} = 750 \text{ Hz}$$

92. As source and observer both are moving in the same direction with the same velocity, their relative velocity is zero. Therefore, $n' = n = 200$ Hz.

93. Here, frequency of source = 500 Hz

Speed of source $A = 4$ m/s $= u$

Then, source is moving towards stationary observer,

$$v' = \frac{v}{v-u} v_0 \quad \text{(where } v = \text{speed of sound)}$$

$$= \frac{340}{340-4} \times 500$$

$$\Rightarrow v' = \frac{340}{336} \times 500 \text{ Hz} = 506 \text{ Hz}$$

Now, when source is reciding from the observer,

$$v' = \frac{v}{v+u} v_o$$

$$\Rightarrow \quad = \frac{340}{344} \times 500 \text{ Hz}$$

$$\therefore \quad v' = 494 \text{ Hz}$$

According to question,

Let frequency of source B is Z Hz.

$\therefore \quad Z = 506 \pm 6 \Rightarrow Z = 500$ or 512

and $\quad Z = 494 \pm 18 \Rightarrow Z = 512$ or 476

Thus, required frequency = 512 Hz

94. When observer moves away from S_1 and towards S_2,

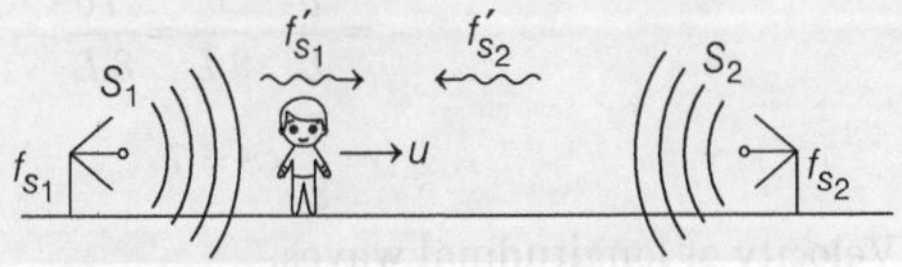

then due to Doppler's effect observed frequencies of sources by observer are

$$f'_{S_1} = \frac{v - v_o}{v} \cdot f_{S_1}$$

(observer moving away from source)

and $$f'_{S_2} = \left(\frac{v + v_o}{v}\right) \cdot f_{S_2}$$

(observer moving towards source)

(where, v = speed of sound, v_o = speed of observer)

So, beat frequency heard by observer is

$$f_b = f'_{S_2} - f'_{S_1}$$

Here, $\quad v_o = u, v = 330 \text{ ms}^{-1}$

$f_b = 10$ Hz, $f_{S_1} = f_{S_2} = 660$ Hz

On putting the values, we get

$$f_b = f'_{S_2} - f'_{S_1}$$

$$= \left(\frac{v + v_o}{v}\right) \cdot f_{S_2} - \left(\frac{v - v_o}{v}\right) f_{S_1}$$

$$= f_{S_1}\left(\frac{v + v_o}{v} - \frac{v - v_o}{v}\right)$$

$$= f_{S_1} \cdot \frac{2v_o}{v}$$

$$\Rightarrow 10 = \frac{660 \times 2u}{330} \quad (\because v_o = u)$$

$$\Rightarrow u = \frac{330 \times 10}{2 \times 660}$$

$$\Rightarrow u = 2.5 \text{ ms}^{-1}$$

95. Let v_B be the speed of the bus. Before reflection, the frequency of the horn is 420 Hz.

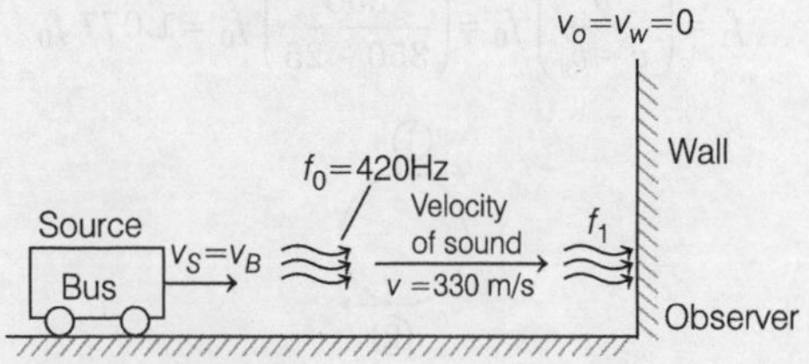

Using Doppler's effect,

$$f_1 = \left(\frac{v}{v - v_s}\right) f_0$$

$$\Rightarrow f_1 = \left(\frac{330}{330 - v_B}\right) \times 420 \quad \text{...(i)}$$

After reflection from the wall, the frequency of the horn becomes 490 Hz.

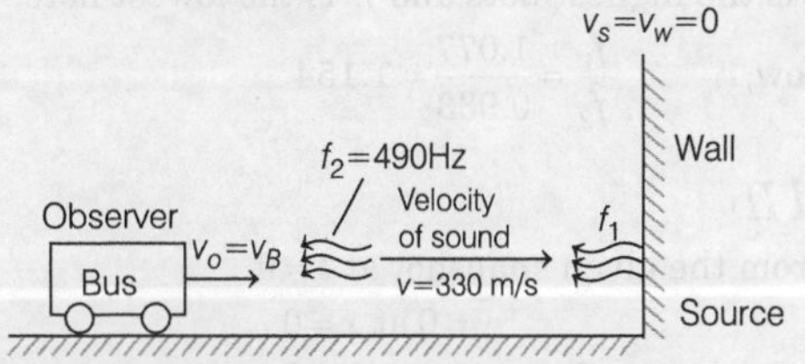

Again using Doppler's effect ,

$$f_2 = \left(\frac{v + v_o}{v}\right) f_1$$

$$\Rightarrow 490 = \left(\frac{330 + v_B}{330}\right) \times \left(\frac{330}{330 - v_B}\right) \times 420$$

$$\Rightarrow \frac{490}{420} = \frac{330 + v_B}{330 - v_B}$$

$$\Rightarrow \frac{7}{6} = \frac{330 + v_B}{330 - v_B}$$

On applying Componendo-Dividendo rule, we get

$$\frac{7+6}{7-6} = \frac{(330 + v_B) + (330 - v_B)}{(330 + v_B) - (330 - v_B)}$$

$$\Rightarrow \frac{13}{1} = \frac{2 \times 330}{2v_B}$$

$$\Rightarrow v_B = \frac{330}{13} \text{ m/s}$$

$$= \frac{330}{13} \times \frac{18}{5} \text{ km/h}$$

$$= 91.38 \text{ km/h} \simeq 91 \text{ km/h}$$

96. Number of revolutions per second = 2

Radius of the circle = 1.988 m

Linear velocity of the tuning fork is

$$v = (2\pi R) \times 2 = 4 \times \frac{22}{7} \times 1.988 = 25 \text{ ms}^{-1}$$

Let the listener is located on the left side at a large distance in the diagram.

$\therefore$ Apparent frequency, when the tuning fork is approaching the listener is

$$f_1 = \left(\frac{v}{v - v_s}\right) f_0 = \left(\frac{350}{350 - 25}\right) f_0 = 1.077\ f_0$$

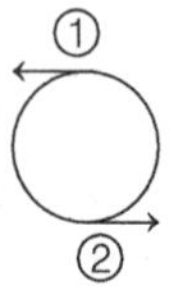

Apparent frequency, when the tuning fork is moving away from the listener is

$$f_2 = \left(\frac{v}{v + v_s}\right) f_0$$

$$f_2 = \left(\frac{350}{350 + 25}\right) \cdot f_0 = 0.933\ f_0$$

f_1 is the highest note and f_2 is the lowest note.

Now, $\frac{f_1}{f_2} = \frac{1.077}{0.933} = 1.154$

Round II

1. From the given snapshot at $t = 0$,

$$y = 0 \text{ at } x = 0$$

and $y = -$ ve when x increases from zero.

Standard expression of any progressive wave is given by $y = A \sin(kx - \omega t + \phi)$

Here, ϕ is the phase difference, we need to get

at $t = 0,\ y = A \sin(kx + \phi)$

Clearly $\phi = \pi$, so that

$$y = A \sin(kx + \pi)$$

$$y = -A \sin(kx)$$

$\Rightarrow$ $y = 0$ at $x = 0$

and $y = -$ve at $x > 0$

which satisfies the given snapshot.

2. Let the frequency of standard tuning fork $= x$

$\therefore$ $n_A = \frac{102}{100} x$ and $n_B = \frac{97}{100} x$

Number of beats per second $= n_A - n_B = 6$

$\because$ $\frac{102}{100} x - \frac{97}{100} x = 6$

$\Rightarrow$ $x = \frac{6 \times 100}{5} = 120$ Hz

So, frequency of $A = \frac{102}{100} \times 120$ Hz $= 122.4$ Hz

3. Frequency of 1st overtone of A,

$$n_1 = \frac{2}{2L_1}\sqrt{\frac{T}{m}} = \frac{2}{L_1 D_1}\sqrt{\frac{T}{\pi\rho}}$$

Frequency of 2nd overtone of B,

$$n_2 = \frac{3}{2L_2}\sqrt{\frac{T}{m}} = \frac{2}{L_2 D_2}\sqrt{\frac{T}{\pi\rho}}$$

As, $n_1 = n_2$ (resonance condition)

$\therefore$ $\frac{2}{L_1 D_1}\sqrt{\frac{T}{m}} = \frac{2}{L_1 D_2}\sqrt{\frac{T}{\pi\rho}}$

$$\frac{L_1 D_1}{L_1 D_2} = \frac{2}{3}$$

$$\frac{L_1}{L_2} = \frac{2D_2}{D_2} = \frac{2}{3} \quad \text{or} \quad \frac{L_1}{L_2} = 1:3$$

4. $n_c = \frac{v}{4L}$ and $n_o = \frac{v}{2L}$

Now, $n_o - n_c = 2$

$\therefore$ $\frac{v}{2L} - \frac{v}{4L} = 2$

or $\frac{v}{L} = 8$

Also, $n'_o = \frac{v}{2L/2} = \frac{v}{L}$

and $n'_c = \frac{v}{4(2L)} = \frac{v}{8L}$

Number of beats per second $= n'_o - n'_c$

$$= \frac{v}{L} - \frac{v}{8L} = \frac{7v}{8L}$$

$$= \frac{7}{8} \times 8 = 7$$

5. Velocity of longitudinal waves,

$$v_1 = \sqrt{\frac{Y}{\rho}}$$

and velocity of transverse waves,

$$v_2 = \sqrt{\frac{T}{m}}$$

If a is area of cross-section of string, then

$$m = \frac{\text{mass}}{\text{length}} = \frac{\text{mass}}{\text{volume}} \times \text{area} = \rho a$$

$\therefore$ $v_2 = \sqrt{\frac{T}{\rho a}}$

$\because$ $\frac{v_1}{v_2} = \sqrt{\frac{Y}{\rho} \cdot \frac{\rho a}{T}} = \sqrt{\frac{Ya}{T}}$

As, $Y = \frac{F}{a\Delta l / l} = \frac{T}{a(\Delta l / l)}$

$\Rightarrow$ $\frac{v_1}{v_2} = \sqrt{\frac{T}{a\left(\frac{\Delta l}{l}\right)} \frac{a}{T}} = \left(\frac{\Delta l}{l}\right)^{-1/2}$

We are given, $\frac{\Delta l}{l} = \frac{1}{n}$

$\therefore$ $\frac{v_1}{v_2} = \left(\frac{1}{n}\right)^{-1/2} = \sqrt{n}$

If f_1 and f_2 are the corresponding fundamental frequencies of longitudinal and transverse vibrations, then

$$v_1 = f_1 \lambda$$

and $v_2 = f_2 \lambda$

$\therefore$ $\frac{v_1}{v_2} = \frac{f_1}{f_2} = \sqrt{n} : 1$

6. Here, $y_1 = 0.05\sin(3\pi t - 2x)$

$$y_2 = 0.05\sin(3\pi t + 2x)$$

According to superposition principle, the resultant displacements is

$$y = y_1 + y_2$$
$$= 0.05[\sin(3\pi t - 2x) + \sin(3\pi t + 2x)]$$
$$y = 0.05 \times 2 \sin 3\pi t \cos 2x$$
$$y = (0.1\cos 2x) \sin 3\pi t = R\sin 3\pi t$$

where, $R = 0.1\cos 2x$ = amplitude of the resultant standing wave.

At $x = 0.5$ m,

$$R = 0.1\cos 2x = 0.1\cos 2\times 0.5$$
$$= 0.1\cos 1(\text{radian}) = 0.1\cos\frac{180^\circ}{\pi}$$
$$= 0.1\cos 57.3^\circ$$

or $R = 0.1 \times 0.54$ m $= 0.054$ m $= 5.4$ cm

7. As, energy $E \propto (\text{amplitude})^2(\text{frequency})^2$

Amplitude is same in both the cases; but frequency 2ω in the second case is two times the frequency (ω) in the first case.

Hence, $E_2 = 4E_1$

8. As the string vibrates in n loops, therefore

$$L = \frac{n\lambda}{2}$$

Therefore, v would become $\frac{1}{2}$ time.

As, $v \propto \sqrt{T}$

Therefore, to make v half-time, T must be made $\frac{1}{4}$ time, *i.e.* $M/4$.

9. As the string is vibrating in three segments, therefore,

$$L = \frac{3\lambda}{2}$$

or $$\lambda = \frac{2L}{3} = \frac{2(0.6)}{3} = 0.4 \text{ m}$$

Now, $$v = \sqrt{\frac{T}{m}}$$

$$v = \sqrt{\frac{80}{2.0}} = 20 \text{ ms}^{-1}$$

$\therefore$ $$n = \frac{v}{\lambda} = \frac{20}{0.4} = 50 \text{ Hz}$$

Amplitude of particle velocity

$$= \left(\frac{dy}{dt}\right)_{\max} = (a_{\max})\,\omega = a_{\max}(2\pi n)$$
$$= (0.5\times 10^{-2}) \times 2\pi \times 50 = 1.57 \text{ ms}^{-1}$$

10. As, $n' = \frac{vn}{v - v_s}$ and $n'' = \frac{vn}{v + v_s}$

$\therefore$ $$\frac{n}{n'} = 1 - \frac{v_s}{v},$$
$$\frac{n}{n''} = 1 + \frac{v_s}{v}$$

Adding the two, we get

$$\frac{n}{n'} + \frac{n}{n''} = 2$$

$\therefore$ $$n = \frac{2n'n''}{n' + n''}$$

11. As, $v_s = r\omega = r \times 2\pi\nu = \frac{70}{100} \times 2 \times \frac{22}{7} \times 5 = 22 \text{ ms}^{-1}$

Frequency is minimum when source is moving away from listenery.

Therefore, from Doppler's effect,

$$\nu' = \frac{u \times v}{u + u_s} = \frac{352 \times 1000}{352 + 22} = 941 \text{ Hz}$$

12. The time taken by the plate falling through a distance y is given by

$$t = \sqrt{(2yg)} = \sqrt{\left(\frac{2\times 10}{980}\right)} = \left(\frac{1}{7}\right) \text{s}$$

The number of oscillations completed in $\frac{1}{7}$ s is 8.

$\therefore$ Frequency = Number of oscillations completed in one second

$$= \frac{8}{1/7} = 56 \text{ Hz}$$

13. Let v be the speed of sound in air, v_L velocity of observer at time t. As, the observer approaches the source, therefore apparent frequency,

$$f = \frac{(v + v_L)}{v} f_0 = \left[\frac{v + (0 + at)}{v}\right] f_0 = f_0 + \left(\frac{f_0 at}{v}\right)$$

This is the equation of a straight line with a positive intercept (f_0) and positive slope $\left(\frac{f_0 a}{v}\right)$. Therefore, option (d) is correct.

14. As wavetrain moves up over string, tension and hence, wave speed also changes. Also, frequency (f) of wavetrain remains constant as it depends only on the source.

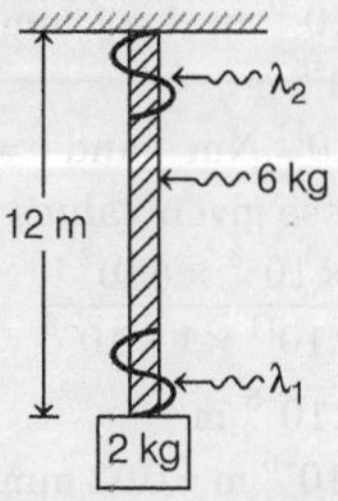

Now, using $v = \sqrt{\frac{T}{\mu}}$

where, v = wave speed, T = tension in string and μ = mass per unit length of string.

Tension T_1 at bottom $= 2g$

Tension T_2 at top $= (2 + 6)\,g = 8\,g$

If λ_1 and λ_2 are wavelengths at bottom and top, then

$$\frac{v_1}{v_2} = \frac{f\lambda_1}{f\lambda_2} = \frac{\sqrt{\frac{T_1}{\mu}}}{\sqrt{\frac{T_2}{\mu}}}$$

$$\Rightarrow \quad \lambda_2 = \lambda_1\sqrt{\frac{T_2}{T_1}}$$

$$= 6 \times \sqrt{\frac{8\,g}{2\,g}} = 12 \text{ cm}$$

15. Let m be the total mass of the rope of length l. Tension in the rope at a height h from lower end = weight of rope of length h,

i.e. $$T = \frac{mg}{L}(h)$$

As, $$v = \sqrt{\frac{T}{(m/L)}}$$

$\therefore$ $$v = \sqrt{\frac{mg(h)}{L(m/L)}} = \sqrt{gh}$$

$$v^2 = gh$$

which is a parabola. Therefore, h *versus* v graph is a parabola, hence option (a) is correct.

16. Speed of transverse wave over a string,

$$v = \sqrt{\frac{T}{\mu}} \quad \text{...(i)}$$

where, T = tension or force on string

and $\mu = \frac{m}{L}$ = mass per unit length.

Also, Young's modulus of string,

$$Y = \frac{TL}{A\Delta L}$$

$$\Rightarrow \quad T = \frac{YA\Delta L}{L} \quad \text{...(ii)}$$

From Eqs. (i) and (ii), we have

$$v^2 = \frac{YA\Delta L}{\mu L}$$

or $$\Delta L = \frac{mv^2}{YA} \quad \text{(iii)}$$

Here, $m = 6\text{ g} = 6 \times 10^{-3}$ kg, $L = 60$ cm
$= 60 \times 10^{-2}$ m, $A = 1\text{ mm}^2$
$= 1 \times 10^{-6}\text{ m}^2$,
$Y = 16 \times 10^{11}\text{ Nm}^{-2}$ and $v = 90\text{ ms}^{-1}$

Substituting these given values in Eq. (iii), we get

$$\Delta L = \frac{6 \times 10^{-3} \times (90)^2}{16 \times 10^{11} \times 1 \times 10^{-6}}$$

$$= 3.03 \times 10^{-5}\text{ m}$$

$$\approx 30 \times 10^{-6}\text{ m} = 0.03\text{ mm}$$

17. Equation of stationary wave is

$$y_1 = a \sin kx \cos \omega t$$

and equation of progressive wave is

$$y_2 = a \sin(\omega t - kx)$$

$$= a(\sin \omega t \cos kx - \cos \omega t \sin kx)$$

At $$x_1 = \frac{\pi}{3k} \text{ and } x_2 = \frac{3\pi}{2k}$$

$\sin kx_1$ or $\sin kx_2$ is zero.

So, neither x_1 nor x_2 is node.

$$\Delta x = x_2 - x_1 = \frac{3\pi}{2k} - \frac{\pi}{3k} = \frac{7\pi}{6k}$$

As, $$\Delta x = \frac{7\pi}{6x}$$

Therefore, $$\frac{2\pi}{k} > \Delta x > \frac{\pi}{k}$$

But $$\frac{2\pi}{k} = \lambda$$

So, $$\lambda > \Delta x > \frac{\lambda}{2}$$

In case of a stationary wave, phase difference between any two points is either zero or π.

$\therefore$ $$\phi_1 = \pi$$

and $$\phi_2 = k\Delta x = k\frac{7\pi}{6k} = \frac{7}{6}\pi$$

$\therefore$ $$\frac{\phi_1}{\phi_2} = \frac{\pi}{\frac{7}{6}\pi} = \frac{6}{7}$$

18. The motorist receives two sound waves, one directly from the band and second reflected from the wall which is shown in figure.

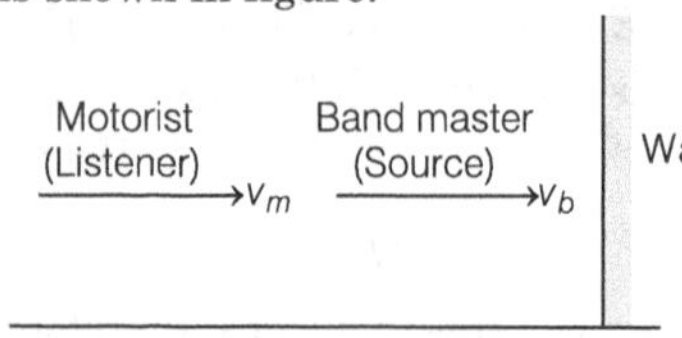

For direct sound waves, apparent frequency,

$$f' = \frac{(v + v_m)f}{v + v_b}$$

For reflected sound waves,
frequency of sound wave, reflected from the wall,

$$f'' = \frac{v \times f}{v - v_b}$$

Frequency of reflected waves as received by the moving motorist,

$$f'' = \frac{(v + v_m)f''}{v} = \frac{(v + v_m)f}{v - v_b}$$

$\therefore$ Beat frequency $= f''' - f'$

$$= \frac{(v + v_m)f}{v - v_b} - \frac{(v + v_m)f}{v + v_b}$$

$$= \frac{2v_b(v + v_m)f}{v^2 - v_b^2}$$

19. As is known, frequency of vibration of a stretched string,

$$n \propto \sqrt{T} \propto \sqrt{mg} \propto \sqrt{g}$$

As, $$n_\omega = \frac{80}{100}n_a = 0.8\,n_a$$

$\therefore$ $$\frac{g'}{g} = \left(\frac{n_\omega}{n_a}\right)^2 = (0.8)^2 = 0.64$$

If ρ_ω = relative density of water (= 1),
ρ_m = relative density of mass,
ρ_l = relative density of liquid, then

$$\frac{g'}{g}=\left(1-\frac{\rho_\omega}{\rho_\mu}\right)=0.64$$

$$\frac{\rho_\omega}{\rho_m}=1-0.64=0.36 \quad \text{...(i)}$$

Similarly, in the liquid,

$$\frac{g'}{g}=\left(\frac{n_L}{n_a}\right)^2=(0.6)^2=0.36$$

$$\frac{g'}{g}=\left(1-\frac{\rho_L}{\rho_m}\right)=0.36$$

$$\frac{\rho_L}{\rho_m}=1-0.36=0.64 \quad \text{...(ii)}$$

Dividing Eq. (i) by Eq. (ii), we get

$$\frac{\rho_L}{\rho_\omega}=\frac{0.64}{0.34}=1.77$$

Hence, specific gravity of liquid = 1.77

20. According to the question,

$$\frac{1}{2l}\sqrt{\frac{T_1}{\mu}}=\frac{1}{l}\sqrt{\frac{T_2}{\mu}}$$

$$\Rightarrow \quad T_2=T_1/4$$

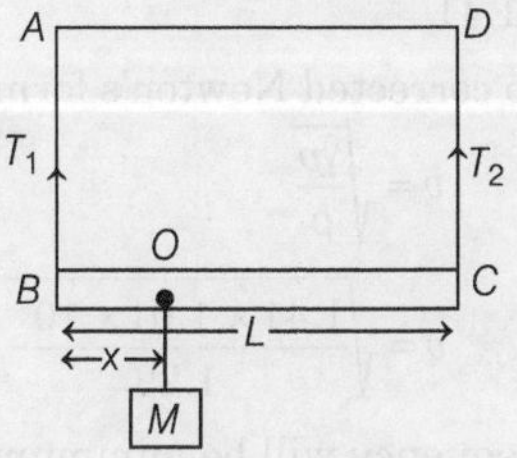

For rotational equilibrium, net torque should be equal to zero

$$\Rightarrow \quad T_1\,x=T_2\,(L-x)$$

$$x=\frac{L}{5}$$

21. Let speed of observer be $v_L = v$ along Y-axis and speed of source be $v_s = 2v_L = 2v$ along X-axis

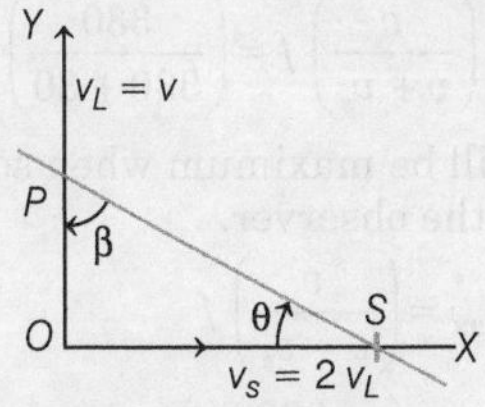

$$\therefore \quad PS=2\,(OL)$$

$$\cos\alpha=\frac{2}{\sqrt5}$$

and $$\cos\beta=\frac{2}{\sqrt5}$$

Now, apparent frequency n' is given by

$$n'=\frac{(v-v_L\cos\beta)n}{(v+v_L\cos\alpha)}$$

where, v is velocity of sound.

$$n'=\frac{(v-v\sqrt5)n}{(v+4v\sqrt5)}$$

Clearly, n' is constant but $n' < n$. This is shown in curve (b).

22. Here, $A_1 = A_2$; $n_1 = \omega$, $n_2 = \omega_2$

$$\therefore \quad y_1=A\sin 2\pi\omega_1 t,$$

and $$y_2=A\sin 2\pi\omega_2 t$$

Now, $$y=y_1+y_2 \quad \text{(from superposition principle)}$$

$$=2A\,\frac{\cos 2\pi\,(\omega_2-\omega_1)\,t}{2}\,\sin\frac{2\pi\,(\omega_2+\omega_1)\,t}{2}$$

$$=A'\sin\pi\,(\omega_2+\omega_1)\,t$$

where, $$A'=2A\cos\pi(\omega_2-\omega_1)t$$

Sound heard will be of maximum intensity ($> 2A^2$)

when $$\cos\pi(\omega_2-\omega_1)\,t=\max=\pm1$$

$$\pi(\omega_2-\omega_1)t=0,\ \pi,\ 2\pi$$

$$t=0,\ \frac{1}{\omega_2-\omega_1};\ \frac{2}{\omega_2-\omega_1},\ldots$$

Time interval between two successive maxima

$$=\frac{1}{\omega_2-\omega_1}=\frac{2}{10^3}=10^{-3}\text{ s}$$

23. According to the question, the musician uses a open flute of length 50 cm and produce second harmonic sound waves.

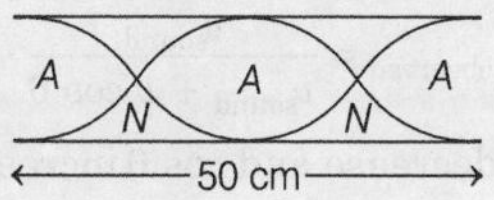

When the flute is open from both ends and produce second harmonic, then

$$L=\lambda_2 \Rightarrow f_2=\frac{v}{L}$$

where, λ_2 = wavelength for second harmonic,

f_2 = frequency for second harmonic

and v = speed of wave.

For given question, $f_2=\dfrac{v}{L}=\dfrac{330}{50\times10^{-2}}$

$f_2 = 660$ Hz (frequency produce by source)

Now, a person runs towards the musician from another end of a hall,

$$v_{\text{observer}}=10\text{ km/h (towards source)}$$

There is apparent change in frequency, which heard by person and given by Doppler's effect formula,

$$\nu'=f'=\nu\left[\frac{v_{\text{sound}}+v_{\text{observer}}}{v_{\text{sound}}}\right]$$

$$f'=f_2\left[\frac{v_s+v_o}{v_{\text{sound}}}\right]$$

$$f'=660\left[\frac{330+\frac{50}{18}}{330}\right]$$

$$f'=666\text{ Hz}$$

24. The situation can be shown for a stationary observer of height D,

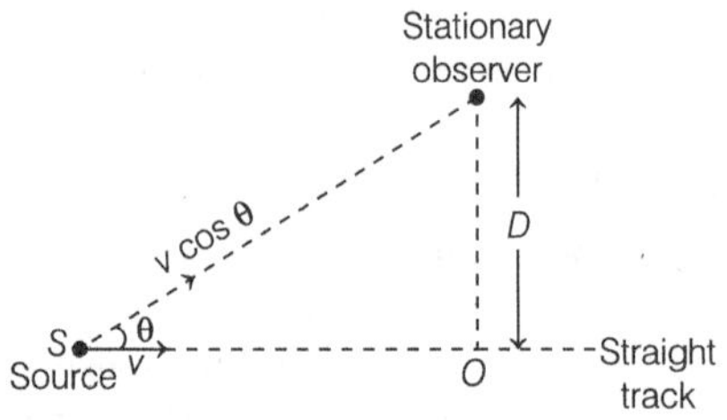

Source frequency = ν_o

Using the concept of Doppler's effect,

Observed frequency,

$$\nu_{observed} = \left(\frac{v_{sound}}{v_{sound} - v\cos\theta}\right)\nu_o$$

When $t < t_0 (0° < \theta < 90°)$

Initially, θ is less but increasing with time, so $\cos\theta$ decreases continuously and $\nu_{observed}$ also decreases.

When $t = t_0 (\theta = 90°)$

$$\cos\theta = \cos 90° = 0, \nu_{observed} = \nu_o$$

When $t > t_0$ $(0° < q < 90°)$

i.e. sound source is moving away from observer.

So, in this case, expression of observed frequency will be

$$\nu_{observed} = \frac{v_{sound}}{v_{sound} + v\cos\theta} \cdot \nu_o$$

With time θ decrease and $\cos\theta$ increases, so $\nu_{observed}$ decreases continuously.

So, correct graph is

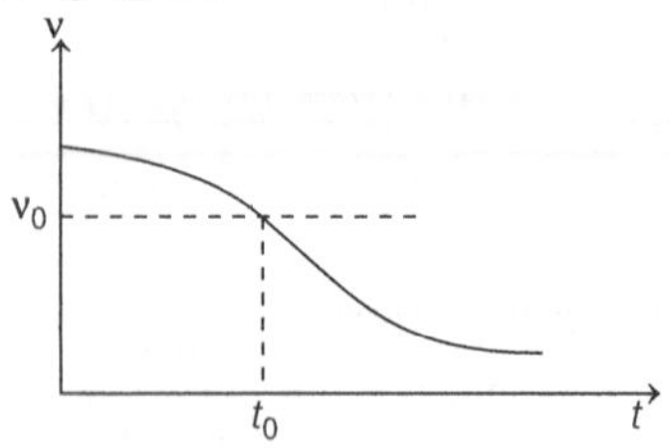

25. Distance between any two crests,

$$n\lambda = 5 \quad (n \text{ is an integer}) \quad \ldots\text{(i)}$$

Distance between any crest and any trough,

$$(2m+1)\frac{\lambda}{2} = 1.5 \quad (m \text{ is an integer})$$

$$\Rightarrow \quad (2m+1)\lambda = 3 \quad \ldots\text{(ii)}$$

Dividing Eq. (i) by Eq. (ii), we get

$$\frac{n\lambda}{(2m+1)\lambda} = \frac{5}{3}$$

$$3n = 5(2m+1)$$

$$3n = 10m + 5 \quad \ldots\text{(iii)}$$

Now, putting the values of m and n in Eq. (iii), which satisfy the equation (only integral values of m and n are acceptable).

(I) $m = 1$ and $n = 5$

Now, on putting these values in Eqs. (i) and (ii), we get

$$\lambda = 1$$

(II) $m = 4$ and $n = 15$

Now, on putting these values in Eqs. (i) and (ii), we get

$$\lambda = \frac{1}{3}$$

(III) $m = 7$ and $n = 25$

Now, on putting these values in Eqs. (i) and (ii), we get

$\lambda = \frac{1}{5}$ and so on.

So, possible values of $\lambda = 1, \frac{1}{3}, \frac{1}{5}, \ldots\ldots$

Hence, correct option is (b).

26. Here, $p = 1$ atmospheric pressure

$$= 1.01 \times 10^5 \text{ N/m}^2$$

$$\text{Density of air, } \rho = \frac{\text{Mass of one mole}}{\text{Volume of one mole}}$$

$$= \frac{29.0 \times 10^{-3} \text{ (kg)}}{22.4 \times 10^{-3} \text{ (m}^3\text{)}}$$

$$\rho = 1.29 \text{ kgm}^{-3}$$

For air, $\gamma = 1.41$

According to corrected Newton's formula,

$$v = \sqrt{\frac{\gamma p}{\rho}}$$

$$v = \sqrt{\frac{1.41 \times 1.01 \times 10^5}{1.29}} = 332.5 \text{ ms}^{-1}$$

27. Apparent frequency will be minimum when the source is at N and moving away from the observer.

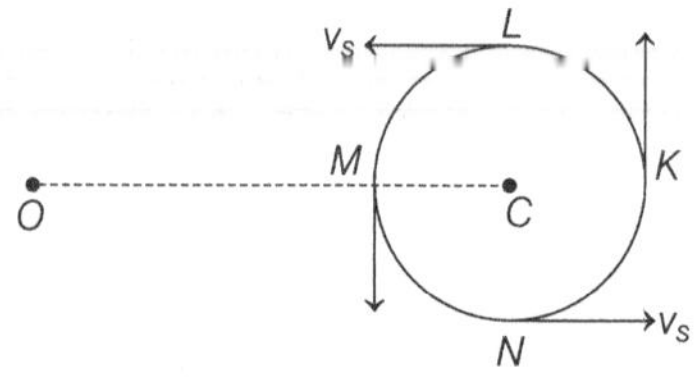

$$f_{min} = \left(\frac{v}{v + v_s}\right) f = \left(\frac{330}{330 + 30}\right)(540) = 495 \text{ Hz}$$

Frequency will be maximum when source is at L and approaching the observer.

$$f_{max} = \left(\frac{v}{v - v_s}\right) f$$

$$= \left(\frac{330}{330 - 30}\right)(540) = 594 \text{ Hz}$$

$$\therefore \quad f_{max} - f_{min} = 594 - 495 = 99 \text{ Hz}$$

28. Here, $f_c = f_o$

$$\frac{3v_c}{4L} = \frac{2v_o}{2L'}$$

$$L' = \frac{4L}{3}\frac{v_o}{v_c} = \frac{4L}{3}\sqrt{\frac{\rho_1}{\rho_2}}$$

$$\therefore \quad x = 4$$

29. The first and the fourth waves have phase difference π, so they interfere destructively leading to zero intensity.

The second and the third waves differ in phase by $\left(\frac{2\pi}{3}-\frac{\pi}{3}\right)=\frac{\pi}{3}$. So the net intensity,

$$I_{net} = I_0 + I_0 + 2\sqrt{I_0}\,\sqrt{I_0}\cos\frac{\pi}{3} = 3I_0$$

$$\therefore \qquad n = 3$$

30.
$$P = \frac{1}{2}\rho\omega^2A^2sv = \frac{1}{2}\mu\omega^2A^2v$$

$$= \frac{1}{2}\mu\omega^2A^2\sqrt{\frac{T}{\mu}}$$

or
$$P = \frac{1}{2}\omega^2A^2\sqrt{\mu T}$$

$$= \frac{1}{2}(2\pi\times 60)^2(6\times 10^{-2})^2\sqrt{5\times 10^{-2}\times 80}$$

$$\simeq 518 \text{ W}$$

31. Speed of sound in air by Laplace formula,

$$v = \sqrt{\frac{\gamma p}{\rho}}$$

So, ratio of speed of sound in a gas to that of air,

$$\frac{v_{gas}}{v_{air}} = \sqrt{\frac{\rho_{air}}{\rho_{gas}}}$$

Here, $v_{air} = 300 \text{ ms}^{-1}$ and $\frac{\rho_{air}}{\rho_{gas}} = \frac{1}{2}$

So, we have

$$\frac{v_{gas}}{300} = \frac{1}{\sqrt{2}}$$

$$\Rightarrow \qquad v_{gas} = \frac{300}{\sqrt{2}} = 150\sqrt{2} \text{ ms}^{-1}$$

Now, for a tube open at both ends.

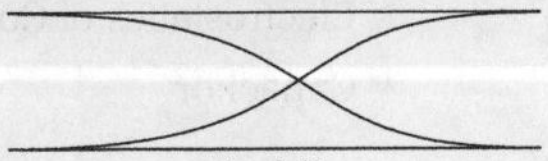

$\therefore$ Fundamental frequency,

$$f_1 = \frac{v}{\lambda} = \frac{v}{2L}$$

Second harmonic frequency f_2 in tube,

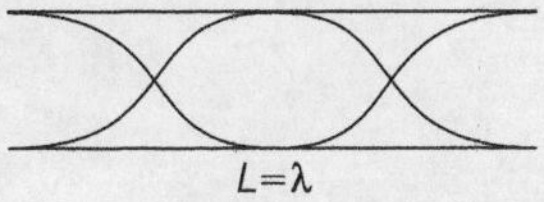

$$f_2 = \frac{v}{\lambda},\ f_2 = v/L$$

So, frequency difference,

$$\Delta f = f_2 - f_1 = \frac{v}{L} - \frac{v}{2L} = \frac{v}{2L} = \frac{150\sqrt{2}}{2\times 1} \qquad (\because L = 1\text{ m})$$

$$= 75\sqrt{2} = 106.06 \text{ Hz}$$

32. The expression for the elastic modulus, $Y = \dfrac{F/A}{\Delta L/L}$

$$\Rightarrow \qquad \frac{\Delta L}{L} = \frac{F/A}{Y} \qquad \ldots\text{(i)}$$

As, the wave speed,

$$v = \sqrt{\frac{T}{\mu}} = \sqrt{\frac{F}{\mu}}$$

where, tension T is the same as stretching force F.

$$\mu = \frac{m}{L} = \frac{\rho(AL)}{L} = \rho A$$

Substituting these values, we get

$$v^2 = \frac{F}{\mu} = \frac{1}{\rho}\left(\frac{F}{A}\right)$$

or
$$\left(\frac{F}{A}\right) = \rho v^2 \qquad \ldots\text{(ii)}$$

From Eqs. (i) and (ii), we get

$$\frac{\Delta L}{L} = \frac{\rho v^2}{Y}$$

$$= \frac{2.70\times 10^3\times 100}{7.00\times 10^{10}}$$

$$= 3.86\times 10^{-4}$$

$$\therefore \qquad n = 4$$

18

Electrostatics

In this branch of Physics, we deals with static free charge, its properties and phenomenon exhibited by it. Charge which is static (not flowing in a conductor) and free (unbound unlike an electron of inner shell or proton in nucleus), produces only an electric field.

IN THIS CHAPTER ....

Electric Charge

Charge is the property associated with subatomic matter particles by virtue of which they attract dissimilar particles or repels similar particles.

Charges are of two types namely

(i) Positive charge (ii) Negative charge

It is measured in coulomb (C).

Note Other units of charge are

1 microcoulomb = 10^{-6} coulomb and 1 picocoulomb = 10^{-12} coulomb

Basic Properties of Electric Charge

- **Additivity of Charges** Net electric charge on a body is equal to the algebraic sum of all electric charges distributed on different parts of body.
- **Charge is Conservative** Charge can neither be created nor be destroyed, but can be transferred from one body to another body.
- **Quantisation of Charge** Charge on a body is always an integral multiple of charge of an electron (e).

  i.e. $q = ne$

  where, $n = 1, 2, 3, \ldots$ and $e = 1.6 \times 10^{-19}$ C.

  The charge on an electron is taken to be negative.

Example 1. *A polythene piece rubbed with wool is found to have a negative charge of 3×10^{-7} C. Estimate the number of electrons transferred from wool to polythene.*

(a) 1.8×10^{13} (b) 1.8×10^{12} (c) 1.2×10^{11} (d) 1.2×10^{10}

Sol. (b) Here, total charge transferred, $q = -3 \times 10^{-7}$ C

Charge on an electron, $e = -1.6 \times 10^{-19}$ C

From quantisation of charge, $q = ne$

Therefore, number of electrons transferred,

$$n = \frac{q}{e} = \frac{-3 \times 10^{-7}}{-1.6 \times 10^{-19}} = 1.8 \times 10^{12}$$

Methods of Statically Charging a Body

To charge a body statically, we either remove electrons from body to make it positively charged or we add electrons to body to make it negatively charged which can be done by the following methods.

Methods of charging
- Charging by friction
- Charging by conduction
- Charging by induction

Coulomb's Law : Force between Two Point Charges

Coulomb's law is a quantitative statement about the force between two point charges. It states that "the force of interaction between any two point charges is directly proportional to the product of the charges and inversely proportional to the square of the distance between them".

Suppose two point charges q_1 and q_2 are separated in vacuum by a distance r, then force between two charges is given by

$$F_e = \frac{k|q_1 q_2|}{r^2}$$

where, k is constant of proportionality. Its value depends upon the system of units and on the nature of medium between the charges.

When two charges are located in vacuum or air

$$k = \frac{1}{4\pi\varepsilon_0} = 9 \times 10^9 \quad \text{(In SI unit)}$$

where ε_0 is called *absolute permittivity* of free space and its value is $8.85 \times 10^{-12} C^{-12} N^2 m^{-2}$.

If there is another medium between the point charges except air or vacuum, then ε_0 is replaced by $\varepsilon_0 K$ or $\varepsilon_0 \varepsilon_r$ or ε.

Here, K or ε_r is called *dielectric constant* or *relative permittivity of the medium*.

$$K = \varepsilon_r = \frac{\varepsilon}{\varepsilon_0}$$

where, ε = permittivity of the medium.

For air or vacuum, $K = 1$

For water, $K = 81$

For metals, $K = \infty$

Coulomb's Law in Vector Form

Consider two point charges q_1 and q_2 separated by distance r in vacuum. Let $\mathbf{F}_{21}$ be the force on q_2 due to q_1 and $\mathbf{F}_{12}$ the force on q_1 due to q_2.

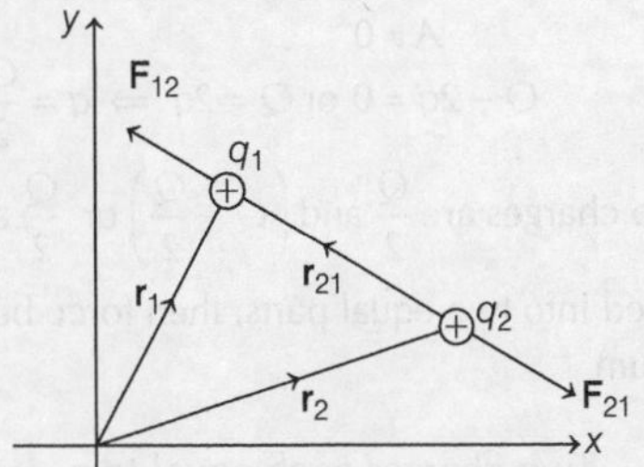

Then, $$\mathbf{F}_{21} = k\frac{q_1 q_2}{r^2}\hat{\mathbf{r}}_{12}$$

where, $\hat{\mathbf{r}}_{12}$ is a unit vector pointing from q_1 to q_2.

and $$\mathbf{F}_{12} = k\frac{q_1 q_2}{r^2}\hat{\mathbf{r}}_{21}$$

where, $\hat{\mathbf{r}}_{21}$ is a unit vector pointing from q_2 to q_1.

$\therefore$ Force on q_1 due to q_2 = – Force on q_2 due to q_1

or $$\mathbf{F}_{12} = -\mathbf{F}_{21}$$

Note The forces due to two point charges are parallel to the line joining the point charges; such forces are called central forces and so electrostatic forces are conservative forces.

Force between Multiple Charges : Superposition Principle

According to this principle, the net force on a given charge due to number of charges is equal to the vector sum of individual forces exerted on it due to the presence of other charges.

Consider a system of n-point charges $q_1, q_2, q_3 \ldots\ldots q_n$ distributed in space. Let the charges be $q_2, q_3 \ldots q_n$, exert forces $\mathbf{F}_{12}, \mathbf{F}_{13}, \ldots\ldots \mathbf{F}_{1n}$ on charge q_1. The total force on charge q_1 is given by

$$\mathbf{F}_1 = \frac{1}{4\pi\varepsilon_0}\left(\frac{q_1 q_2}{r_{12}^2}\hat{\mathbf{r}}_{12} + \frac{q_1 q_2}{r_{13}^2}\hat{\mathbf{r}}_{13} + \ldots\ldots + \frac{q_1 q_2}{r_{1n}^2}\hat{\mathbf{r}}_{1n}\right)$$

It is important to note here the force of two the charges exert on each other is not changed by the presence of a third charge.

Example 2. *A charge Q is divided into two parts and then they are placed at a fixed distance. The force between the two charges is always maximum when the charges are*

(a) $\frac{Q}{2}, \frac{Q}{2}$ (b) $\frac{Q}{3}, \frac{Q}{3}$

(c) $\frac{Q}{4}, \frac{3Q}{4}$ (d) $\frac{Q}{5}, \frac{4Q}{5}$

Sol. (a) Let the two charges be q and $(Q - q)$.

As, $F = A\,q(Q - q)$ (where, A is a constant)

Since, F is maximum when $\frac{dF}{dq} = 0$

$\Rightarrow$ $$\frac{d}{dq}A(qQ - q^2) = 0$$

or $$A(Q - 2q) = 0$$

As $\quad A \neq 0$

$\therefore \quad Q - 2q = 0$ or $Q = 2q \Rightarrow q = \frac{Q}{2}$

Hence, the two charges are $\frac{Q}{2}$ and $\left(Q - \frac{Q}{2}\right)$ or $\frac{Q}{2}$ and $\frac{Q}{2}$, when a charge is divided into two equal parts, then force between them is always maximum.

Example 3. *Two charges each equal to q, are kept at x = − a and x = a on the X-axis. A particle of mass m and charge $q_0 = q/2$ is placed at the origin. If charge q_0 is given, a small displacement (y << a) along the Y-axis, the net force acting on the particle is proportional to* [JEE Main 2013]

(a) y (b) $-y$ (c) $1/y$ (d) $-1/y$

Sol. (b)

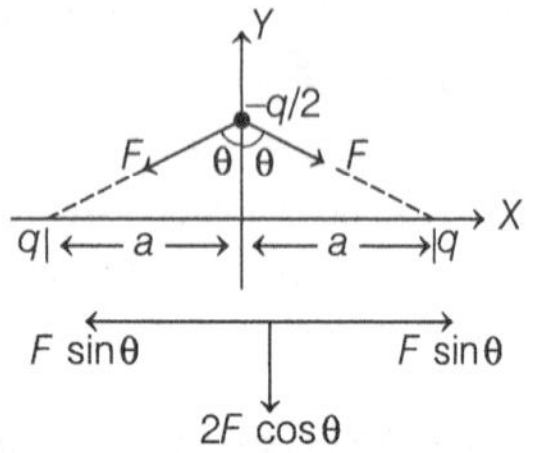

$$F_{net} = 2F\cos\theta$$

$$\therefore \quad F_{net} = -\frac{2kq\left(\frac{q}{2}\right)}{(\sqrt{y^2 + a^2})^2} \cdot \frac{y}{\sqrt{y^2 + a^2}}$$

[negative sign indictate the net force is towards the mean position]

$$F_{net} = -\frac{2kq\left(\frac{q}{2}\right)y}{(y^2 + a^2)^{3/2}} \Rightarrow F \simeq \frac{kq^2y}{a^3} \propto -y$$

Example 4. *Two identical charged spheres suspended from a common point by two massless strings of length l are initially a distance d (d << l) apart because of their mutual repulsion. The charge begins to leak from both the spheres at a constant rate. As a result, charges approach each other with a velocity v. Then, as a function of distance x between them, is* [AIEEE 2011]

(a) $v \propto x^{-1}$ (b) $v \propto x^{1/2}$

(c) $v \propto x$ (d) $v \propto x^{-1/2}$

Sol. (d) At any instant,

$$T\cos\theta = mg \quad \text{...(i)}$$

$$T\sin\theta = F_e = \frac{kq^2}{x^2} \quad \text{...(ii)}$$

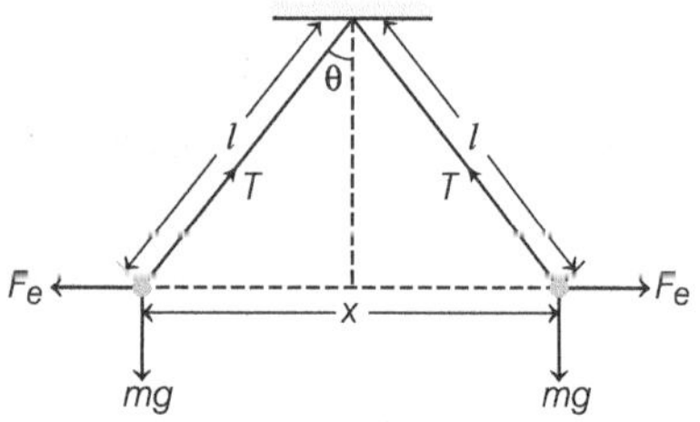

From Eqs. (i) and (ii), we get

$$\frac{kq^2}{x^2} = mg\tan\theta$$

$$\Rightarrow \quad q^2 = \frac{mg}{k} \cdot \frac{x}{2l} \cdot x^2 \qquad \left[\because \tan\theta \approx \frac{x}{2l}\right]$$

$$\Rightarrow \quad q^2 = \frac{mg}{2kl}x^3 \quad \text{...(iii)}$$

$$\Rightarrow \quad 2q\frac{dq}{dt} = \frac{3mg}{2kl}x^2\frac{dx}{dt}$$

$$\Rightarrow \quad 2\left(\frac{mg}{2kl}x^3\right)^{1/2}\frac{dq}{dt} = \frac{3mg}{2kl}x^2v \qquad \left[\because q = \left(\frac{mg}{2kl}x^3\right)^{1/2}\right]$$

$$\Rightarrow \quad vx^{1/2} = \text{constant}$$

$$\Rightarrow \quad v \propto x^{-1/2}$$

Continuous Charge Distribution

There are three types of continuous charge distribution and according to charge distribution, there are three types of charge density as given below

(i) **Linear charge density** (λ) If charge dq is distributed uniformly along a line element dl then the linear charge density, $\lambda = \frac{dq}{dl}$.

Its unit is coulomb metre^{-1} (Cm^{-1}).

(ii) **Surface charge density** (σ) If charge dq is distributed over a surface element ds, then surface charge density, $\sigma = dq/ds$.

Its unit is coulomb metre^{-2} (Cm^{-2}).

(iii) **Volume charge density** (ρ) If charge dq is distributed uniformly over the volume element ΔV, then volume charge density, $\rho = \frac{dq}{dV}$.

Its unit is coulomb metre^{-3} (Cm^{-3}).

Electric Field

The space surrounding an electric charge q in which another charge q_0 experiences a force of attraction or repulsion, is called the electric field of charge q. The charge q is called the **source charge** and the charge q_0 is called the **test charge**. The test charge must be negligibly small, so that it does not modify the electric field of the source charge.

Intensity of Electric Field (*E*)

The intensity of electric field at a point in an electric field is the ratio of the forces acting on the test charge placed at that point to the magnitude of the test charge.

$\mathbf{E} = \lim_{q_0 \to 0}\left(\frac{\mathbf{F}}{q_0}\right)$, where **F** is the force acting on q_0.

It is a vector quantity.

The direction of electric field is same as that of force acting on the positive test charge. Unit of E is NC^{-1} or Vm^{-1}.

$\therefore$ Dimensions of electric field are $[MLT^{-3}A^{-1}]$.

From $\mathbf{E} = \frac{\mathbf{F}}{q}$, we get $\mathbf{F} = q\mathbf{E}$

- If q is positive charge, **F** on it is in the direction of **E**.
- If q is negative charge, **F** on it is opposite to the direction of **E**

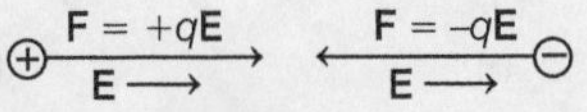

Electric Field due to a Point Charge

Electric field intensity due to a point charge q at a distance r is given by

$$E = \frac{1}{4\pi\varepsilon_0}\frac{q}{r^2}$$

Electric Field due to System of Charges

Electric field **E** at point P due to the systems of charges is given by

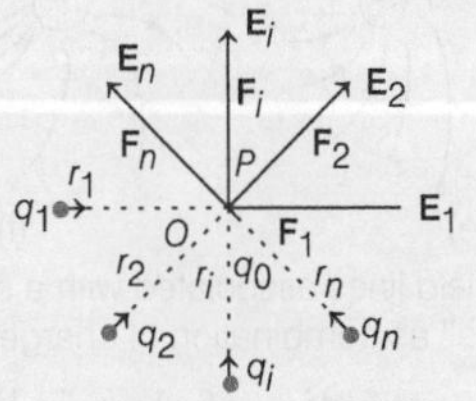

$$\mathbf{E}(r) = \frac{1}{4\pi\varepsilon_0}\sum_{i=1}^{n}\frac{q_i}{r_i^2}\hat{\mathbf{r}}_i$$

Electric Field Due to Continuous Charge Distribution

(i) Electric field due to the line charge distribution at the location of charge q_0 is

$$\mathbf{E}_L = \frac{1}{4\pi\varepsilon_0}\int_L \frac{\lambda}{r^2}\,dL\hat{\mathbf{r}}$$

(ii) Electric field due to the surface charge distribution at the location of charge q_0 is

$$\mathbf{E}_S = \frac{1}{4\pi\varepsilon_0}\int_S \frac{\sigma}{r^2}\,dS\,\hat{\mathbf{r}}$$

(iii) Electric field due to the volume charge distribution at the location of charge q_0 is

$$\mathbf{E}_V = \frac{1}{4\pi\varepsilon_0}\int \frac{\rho}{r^2}\,dV\hat{\mathbf{r}}$$

Example 5. *Two point charges of $+16\ \mu C$ and $-9\ \mu C$ are placed 8 cm apart in air. What is the position of the point from $-9\ \mu C$ charge at which the resultant electric field is zero?*

(a) 9 cm (b) 16 cm
(c) 24 cm (d) 35 cm

Sol. (c) Here, $q_A = +16\mu C = +16\times10^{-6}$ C; $q_B = -9\ \mu C = -9\times10^{-6}$ C; $r = 8$ cm $= 0.08$ m

Now, the electric field cannot be zero between the two charges. It is because, the charges are of opposite signs. Also, the electric field cannot be zero at a point to the left of charge q_A. It is because the magnitude of charge q_A is greater than that of q_B. Suppose that the resultant electric field due to the two charges is zero at point O located to the right of charge q_B as shown in figure. If $OB = x$, then $OA = x + 0.08$.

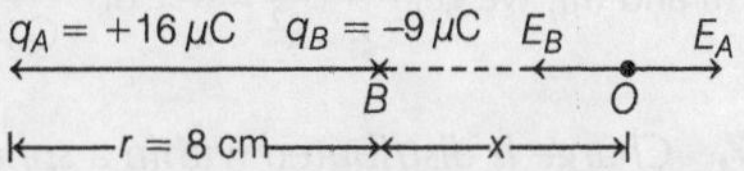

The electric fields E_A and E_B due to the two charges at the point O are in opposite directions. Since, resultant electric field is zero at point O, E_A and E_B are equal in magnitude,

i.e. $\frac{1}{4\pi\varepsilon_0}\cdot\frac{q_A}{OA^2} = \frac{1}{4\pi\varepsilon_0}\frac{q_B}{OB^2}$ or $\frac{16\times10^{-6}}{(x+0.08)^2} = \frac{9\times10^{-6}}{x^2}$

or $\frac{(x+0.08)}{4} = \pm\frac{x}{3}$ or $x = 0.24$ m or $-\frac{0.24}{7}$ m

At, $x = -\frac{0.24}{7}$ m

(at point to the left of point B), the magnitudes of E_A and E_B are equal but the two fields will not cancel each other. It is because at this point, both the fields will be in the same direction. Hence, electric field is zero at a point at distance 0.24 m or 24 cm from the charge of $-9\ \mu C$ as shown in figure.

Example 6. *Charges Q_1 and Q_2 are at points A and B of a right angle triangle OAB (see figure). The resultant electric field at point O is perpendicular to the hypotenuse, then Q_1/Q_2 is proportional to* **[JEE Main 2020]**

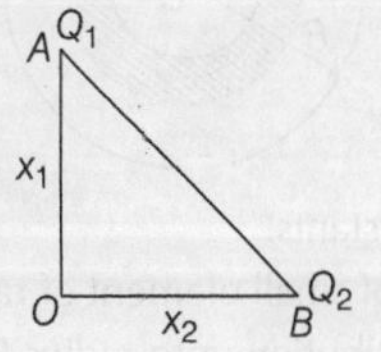

(a) $\frac{x_1^3}{x_2^3}$ (b) $\frac{x_2}{x_1}$

(c) $\frac{x_1}{x_2}$ (d) $\frac{x_2^2}{x_1^2}$

Sol. (c) Let electric field produced by charges Q_1 and Q_2 at point O be E_1 and E_2, respectively.

The direction of fields are shown in the figure below and a perpendicular is also drawn on side AB, that passes through point O.

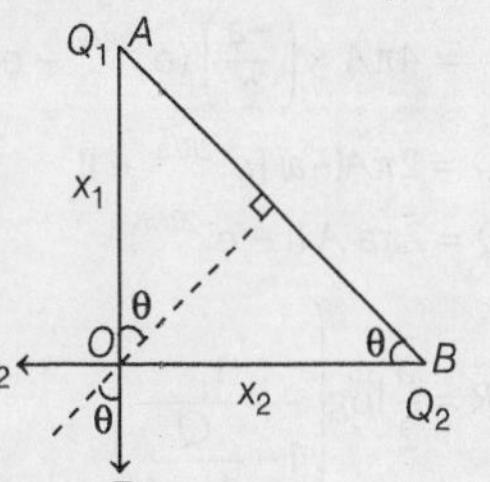

If the resultant electric field at point O is perpendicular to hypotenuse, this means resultant of E_1 and E_2 must be along it.

$$\therefore \qquad \tan\theta = \frac{E_2}{E_1} = \frac{\dfrac{KQ_2}{x_2^2}}{\dfrac{KQ_1}{x_1^2}} = \frac{Q_2}{Q_1}\frac{x_1^2}{x_2^2} \qquad \ldots \text{(i)}$$

In ΔOAB,
$$\tan\theta = \frac{x_1}{x_2} \qquad \ldots \text{(ii)}$$

Equating Eqs. (i) and (ii), we get $\dfrac{Q_2}{Q_1}\cdot\dfrac{x_1^2}{x_2^2} = \dfrac{x_1}{x_2}$ or $\dfrac{Q_1}{Q_2} = \dfrac{x_1}{x_2}$

Example 7. *Charge is distributed within a sphere of radius R with a volume charge density* $\rho(r) = \dfrac{A}{r^2}e^{\frac{-2r}{a}}$, *where A and a are constants. If Q is the total charge of this charge distribution, the radius R is* **[JEE Main 2019]**

(a) $a\log\left(\dfrac{1}{1-\dfrac{Q}{2\pi aA}}\right)$ (b) $a\log\left(1-\dfrac{Q}{2\pi aA}\right)$

(c) $\dfrac{a}{2}\log\left(1-\dfrac{Q}{2\pi aA}\right)$ (d) $\dfrac{a}{2}\log\left(\dfrac{1}{1-\dfrac{Q}{2\pi aA}}\right)$

Sol. (d) Here, volume charge density, $\rho(r) = \dfrac{A}{r^2}\cdot e^{-\frac{2r}{a}}$

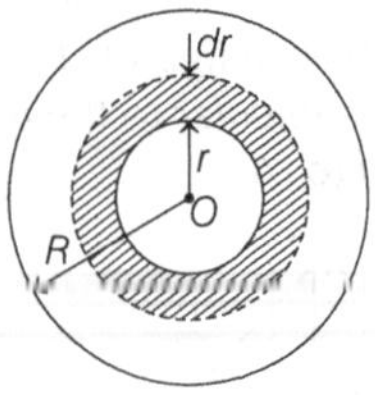

where, a and A are constants.
Let a spherical region of small element of radius r.
If Q is total charge distribution upto radius R, then

$$Q = \int_0^R \rho\cdot dV = \int_0^R \frac{A}{r^2}e^{-2r/a}(4\pi r^2\,dr)$$

(from figure, we observe $dV = A\cdot dr = 4\pi r^2\cdot dr$)

$$= 4\pi A\int_0^R e^{-2r/a}dr$$
$$= 4\pi A\left(\frac{e^{-2r/a}}{-2/a}\right)_0^R$$
$$= 4\pi A\times\left(\frac{-a}{2}\right)(e^{-2R/a} - e^0)$$
$$= 2\pi A(-a)\,[e^{-2R/a} - 1]$$

or
$$Q = 2\pi a\,A\,(1 - e^{-2R/a})$$

or
$$R = \frac{a}{2}\log\left(\frac{1}{1-\dfrac{Q}{2\pi aA}}\right)$$

Electric Field Lines

"An electric field line is an imaginary line or curve drawn through a region of space, so that its tangent at any point is in the direction of the electric field vector at that point.

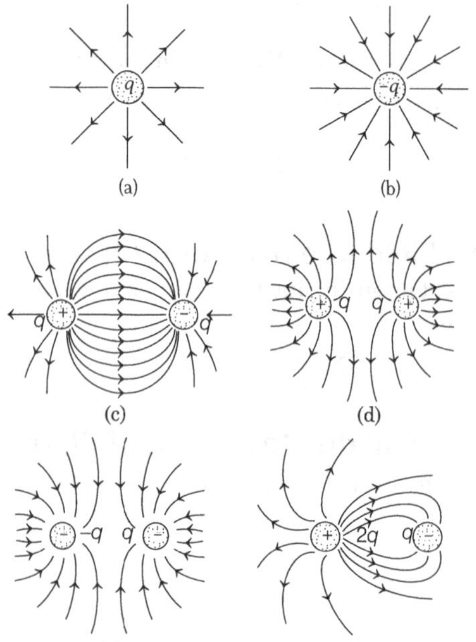

Electric field lines associated with a single as well as combination of charges

Following are the properties of electric lines of forces

(i) It is a curved path on which a unit positive charge will move.

(ii) The lines of force do not pass through a conductor.

(iii) In the region of strong electric field, lines of forces are closely packed while in the region of weak field they are far apart.

(iv) The electric lines of forces emanate (originate and move outwards) from positive charge and terminate on the nearest negative charge.

(v) Two electric lines of forces never intersect each other.

(vi) The electric lines of forces are always normal, *i.e.* perpendicular to the surface of the charged body, *i.e.* at an angle 90° over the surface of the charged body.

Example 8. *A long cylindrical shell carries positive surface charge σ in the upper half and negative surface charge – σ in the lower half. The electric field lines around the cylinder will look like figure given in (figures are schematic and not drawn to scale)* **[JEE Main 2015]**

(a) 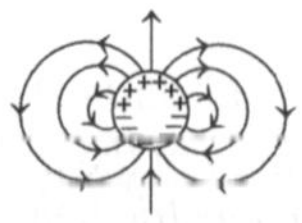

(b) 

(c) 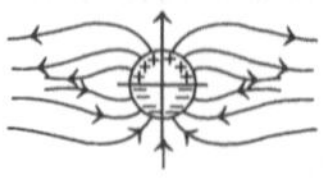

(d) 

Sol. (a) Field lines should originate from positive charge and terminate to negative charge. Thus, (b) and (c) are not possible. Electric field lines cannot form corners as shown in (d). Thus, correct option is (a).

Electric Flux

It is the measure of electric field through a surface.

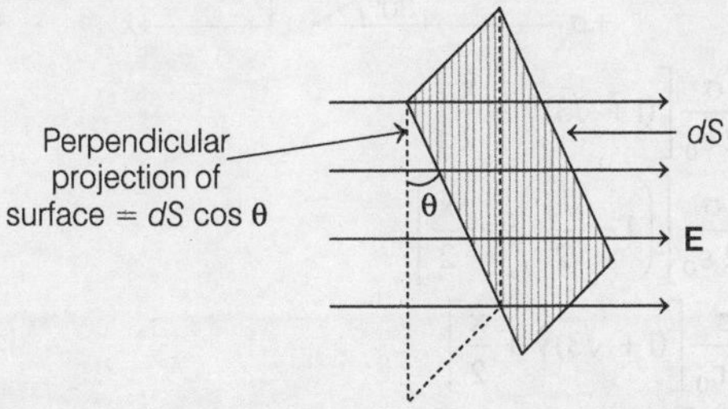

Mathematically, electric flux is the product of an area element $\Delta \mathbf{S}$ and normal component of $\mathbf{E}$ integrated over a surface, $\phi_E = \int \mathbf{E} \cdot \Delta \mathbf{S} = \int \mathbf{E} \cdot \hat{\mathbf{n}}\, \Delta S = \int E \Delta S \cos\theta$.

where, $\hat{\mathbf{n}}$ is the unit vector normal to area element ΔS.

It is a scalar quantity and its SI unit is N-m^2/C m^2 or V-m.

Example 9. *An electric field* $\mathbf{E} = 4x\hat{\mathbf{i}} - (y^2 + 1)\hat{\mathbf{j}}$ *N/C passes through the box shown in figure. The flux of the electric field through surfaces ABCD and BCGF are marked as* ϕ_I *and* ϕ_{II}*, respectively. The difference between* $(\phi_I - \phi_{II})$ *is (in N-m*^{2}*/C)* *[JEE Main 2020]*

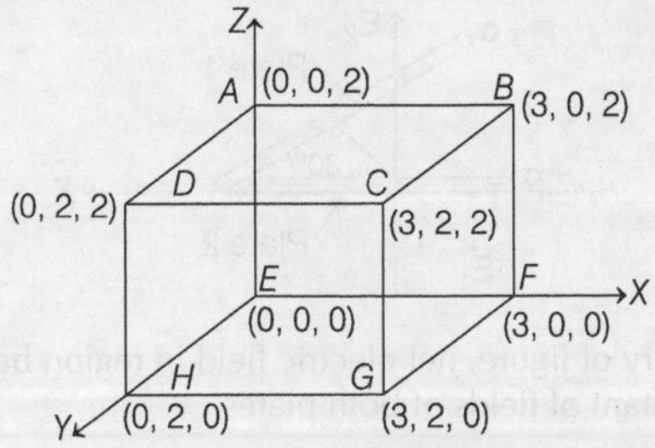

(a) 45 (b) – 45 (c) 48 (d) – 48

Sol. (d)

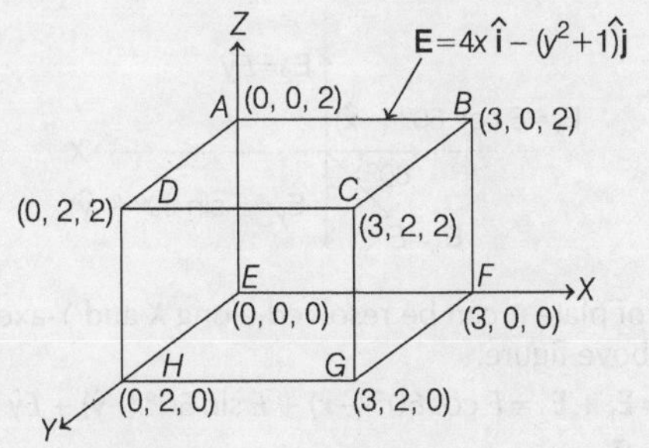

Area vector of face $ABCD$, $\mathbf{A}_1 = 2 \times 3\hat{\mathbf{k}} = 6\hat{\mathbf{k}}$

Area vector of face $BCGF$, $\mathbf{A}_2 = 2 \times 2\hat{\mathbf{i}} = 4\hat{\mathbf{i}}$

So, flux through face $ABCD$,

$$\phi_1 = \mathbf{E} \cdot \mathbf{A}_1 = (4x\hat{\mathbf{i}} - (y^2+1)\hat{\mathbf{j}}) \cdot 6\hat{\mathbf{k}} = 0$$

Flux through face $BCGF$,

$$\phi_2 = \mathbf{E} \cdot \mathbf{A}_2 = (4x\hat{\mathbf{i}} - (y^2+1)\hat{\mathbf{j}}) \cdot 4\hat{\mathbf{i}} = 16x$$

At face $BCGF$, $x = 3$

So, $\phi_2 = 16 \times 3 = 48$ units

$\therefore \quad \phi_1 - \phi_2 = 0 - 48 = -48$ N-m^{2}C^{-1}

Gauss's Law

Net flux over any a closed surface is $\frac{1}{\varepsilon_0}$ times the net charge enclosed within the surface.

i.e. $\phi_E = \oint \mathbf{E} \cdot d\mathbf{S} = \frac{1}{\varepsilon_0} \Sigma q$

where, Σq = net charge enclosed in that surface.

Note If a closed body, not enclosing any charge, is placed in an electric field (either uniform or non-uniform), total flux linked with it will be zero.

Applications of Gauss's Law

To calculate electric field by Gauss's law, we will draw a Gaussian surface (either sphere or cylinder according to situation) in such a way that electric field is perpendicular at each point of surface and its magnitude is same at every point and then apply Gauss's law. Following are few application of this law

(i) **Electric field due to a point charge** Electric field due to a point charge at a distance r is

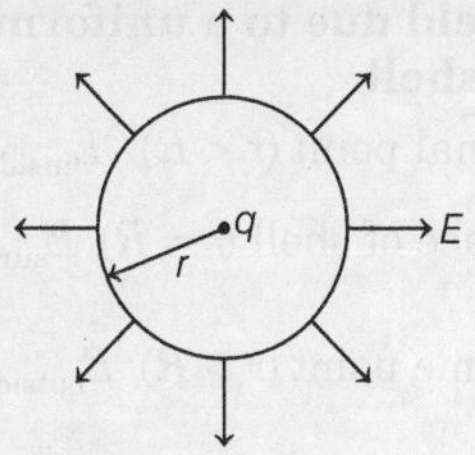

$$E = \frac{1}{4\pi\varepsilon_0} \frac{q}{r}$$

(ii) **Electric field due to infinitely long uniformly charged straight wire** At a distance r from wire electric field is

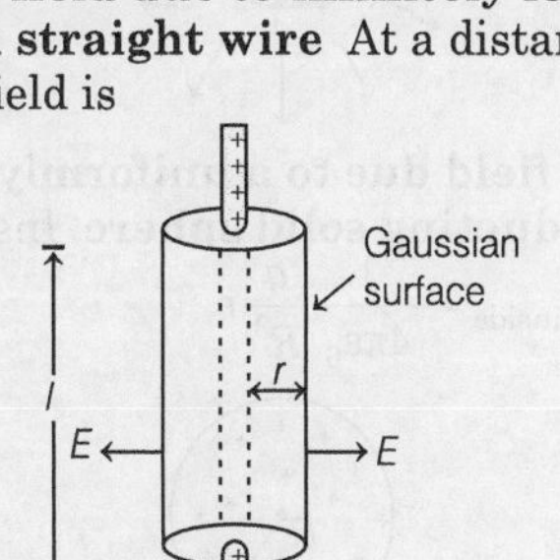

$$E = \frac{\lambda}{2\pi\varepsilon_0 r} = 2\frac{k\lambda}{r} \text{ or } E \propto \frac{1}{r}$$

(iii) **Electric field due to an plane sheet of charge** For a thin sheet with charge density σ, electric field

$$E = \frac{\sigma}{2\varepsilon_0}$$

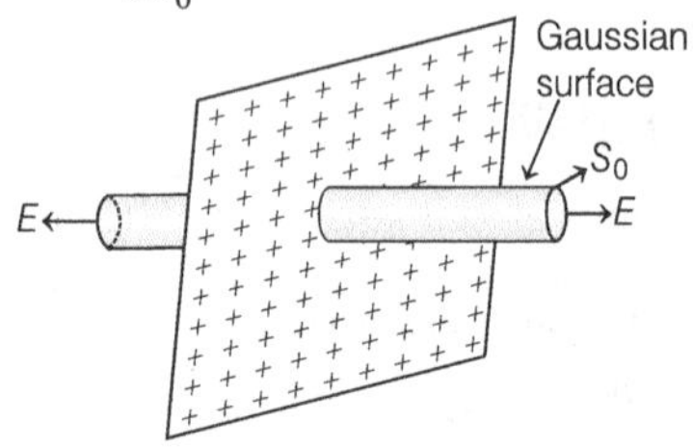

(iv) **Electric field near a charged conducting surface** For a charged conducting surface, field near the surface is

$$E = \frac{\sigma}{\varepsilon_0}$$

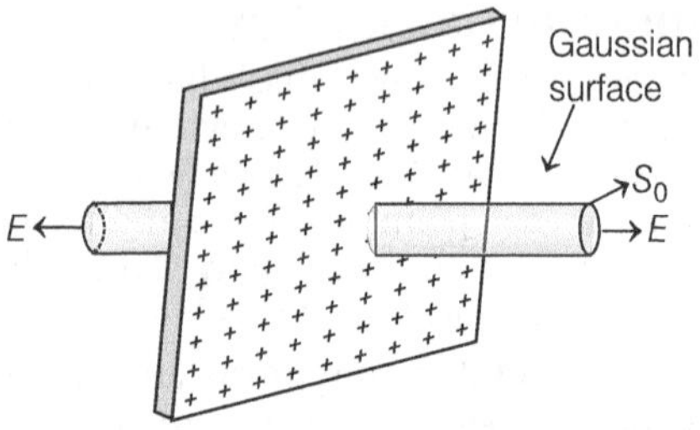

(v) **Electric field due to a uniformly charged spherical shell**

At an internal point $(r < R)$, $E_{\text{inside}} = 0$

At the surface of shell $(r = R)$, $E_{\text{surface}} = \dfrac{1}{4\pi\varepsilon_0}\dfrac{q}{R^2}$

At an extreme point $(r > R)$, $E_{\text{outside}} = \dfrac{1}{4\pi\varepsilon_0}\dfrac{q}{r^2}$

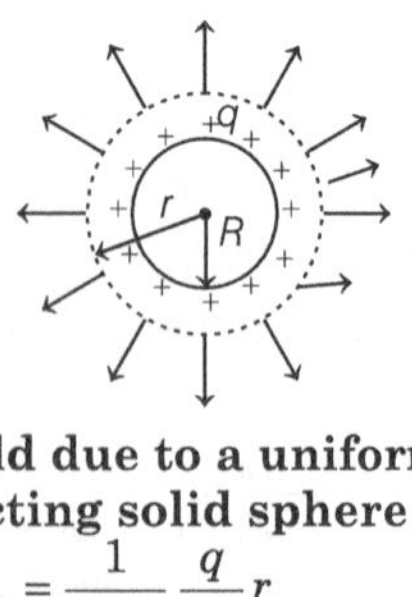

(vi) **Electric field due to a uniformly charged non-conducting solid sphere** Inside the sphere $(r < R)$, $E_{\text{inside}} = \dfrac{1}{4\pi\varepsilon_0}\dfrac{q}{R^3}r$

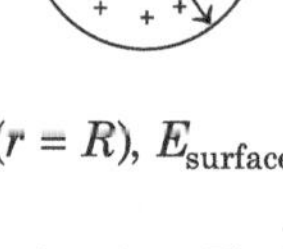

On the surface $(r = R)$, $E_{\text{surface}} = \dfrac{1}{4\pi\varepsilon_0}\dfrac{q}{R^2}$

At an external point $(r > R)$, $E_{\text{outside}} = \dfrac{1}{4\pi\varepsilon_0}\dfrac{q}{r^2}$

Example 10. *Two infinite planes each with uniform surface charged density +σ are kept in such a way that the angle between them is 30°. The electric field in the region shown between them is given by* **[JEE Main 2020]**

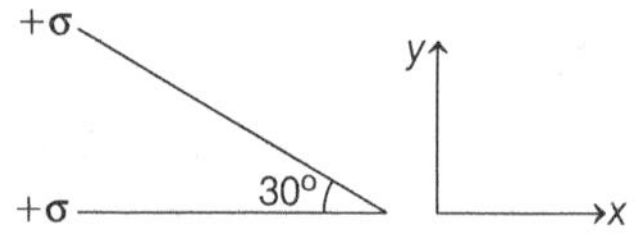

(a) $\dfrac{\sigma}{2\varepsilon_0}\left[(1+\sqrt{3})\hat{\mathbf{y}} - \dfrac{\hat{\mathbf{x}}}{2}\right]$

(b) $\dfrac{\sigma}{2\varepsilon_0}\left[\left(1-\dfrac{\sqrt{3}}{2}\right)\hat{\mathbf{y}} - \dfrac{\hat{\mathbf{x}}}{2}\right]$

(c) $\dfrac{\sigma}{2\varepsilon_0}\left[(1+\sqrt{3})\hat{\mathbf{y}} + \dfrac{\hat{\mathbf{x}}}{2}\right]$

(d) $\dfrac{\sigma}{\varepsilon_0}\left[\left(1+\dfrac{\sqrt{3}}{2}\right)\hat{\mathbf{y}} + \dfrac{\hat{\mathbf{x}}}{2}\right]$

Sol. (b) Electric field of an infinite plate is perpendicular to the plane of plate and its magnitude is $E = \dfrac{\sigma}{2\varepsilon_0}$.

We are given with two positively charged plates with a set of coordinate axes as shown in the figure.

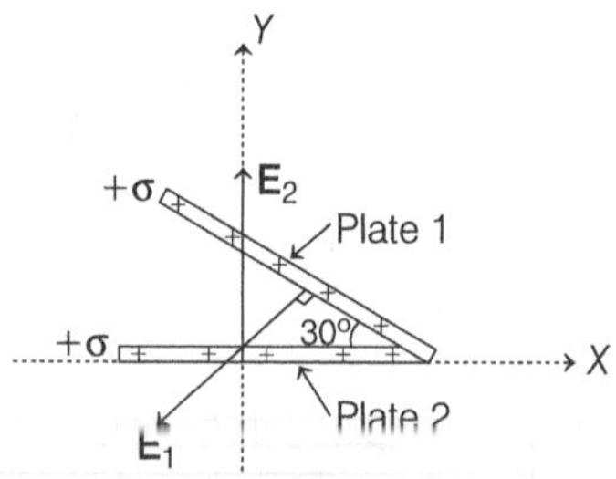

From geometry of figure, net electric field in region between plates is resultant of fields of both plates,

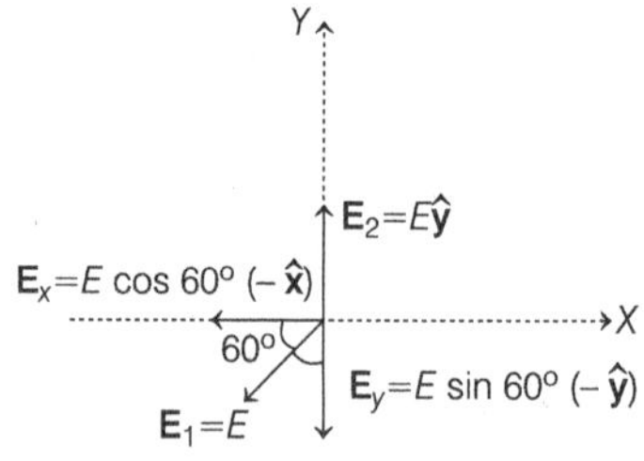

Now, field of plate 1 can be resolved along X and Y-axes as shown in above figure.

Now, $\mathbf{E}_{\text{net}} = \mathbf{E}_1 + \mathbf{E}_2 = E\cos 60^\circ(-\hat{\mathbf{x}}) + E\sin 60^\circ(-\hat{\mathbf{y}}) + E\hat{\mathbf{y}}$

Here, $E = \dfrac{\sigma}{2\varepsilon_0}$ and $\hat{\mathbf{x}}$ & $\hat{\mathbf{y}}$ are unit vectors along X & Y-axes.

$$= \frac{\sigma}{2\varepsilon_0}\left(-\frac{\hat{\mathbf{x}}}{2} - \frac{\sqrt{3}}{2}\hat{\mathbf{y}} + \hat{\mathbf{y}}\right)$$

$$= \frac{\sigma}{2\varepsilon_0}\left[\left(1-\frac{\sqrt{3}}{2}\right)\hat{\mathbf{y}} - \frac{\hat{\mathbf{x}}}{2}\right]$$

Electric Field Intensity of Various Systems

System	Electric Field Intensity
Isolated charge 	$E = \frac{1}{4\pi\varepsilon_0} \cdot \frac{q}{r^2}$
A ring of charge 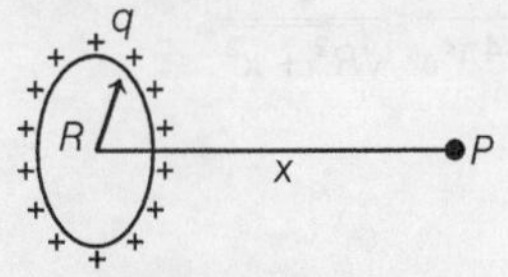	$E = \frac{1}{4\pi\varepsilon_0} \cdot \frac{qx}{(R^2 + x^2)^{3/2}}$
A disc of charge 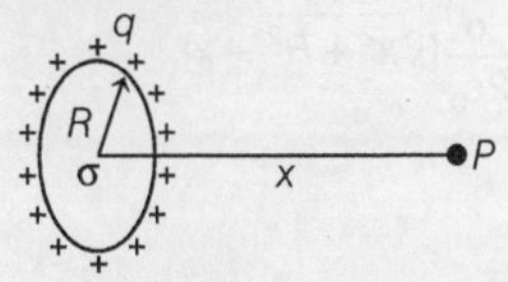	$E = \frac{\sigma}{2\varepsilon_0}\left[1 - \frac{x}{\sqrt{x^2 + R^2}}\right]$
Infinitely long line of charge 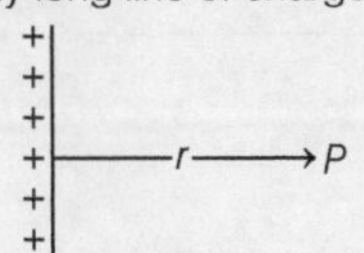	$E = \frac{\lambda}{2\pi\varepsilon_0 r}$
Finite line of charge 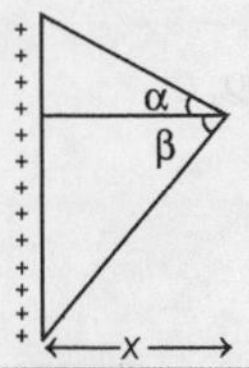	$E_\perp = \frac{\lambda}{4\pi\varepsilon_0 x}(\sin\alpha + \sin\beta)$ $E_{\parallel} = \frac{\lambda}{4\pi\varepsilon_0 x}(\cos\alpha - \cos\beta)$
At centre of semicircular ring 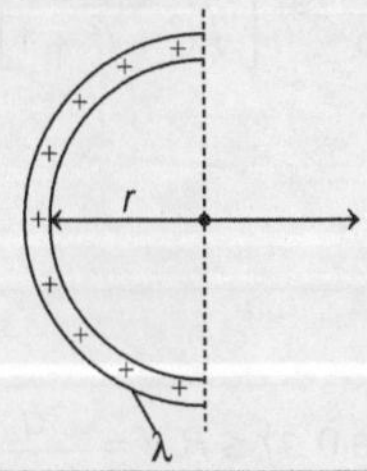	$E = \frac{\lambda}{2\pi\varepsilon_0 r}$
At centre of quarter circular ring 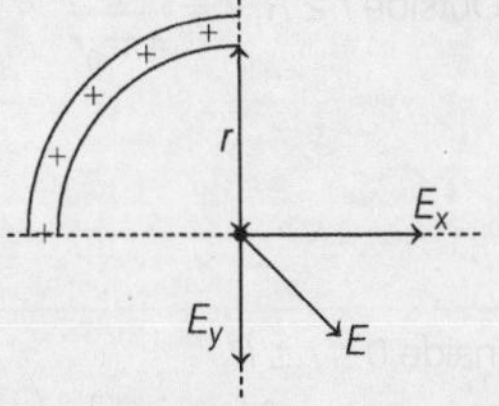	$E = \frac{1}{2\sqrt{2}} \cdot \frac{1}{\pi\varepsilon_0 r}$ $E_x = \frac{\lambda}{4\pi\varepsilon_0 r}$ $E_y = \frac{\lambda}{4\pi\varepsilon_0 r}$

Electric Potential (*V*)

Electric potential at any point is equal to the work done per unit positive charge in carrying it from infinity to that point in electric field.

Electric potential, $V = \frac{W}{q}$

Its SI unit is J/C or volt and its dimensions are $[ML^2T^{-3}A^{-1}]$.

It is a scalar quantity.

Electric Potential Difference

The electric potential difference between two points *A* and *B* is equal to the work done by the external force in moving a unit positive charge against the electrostatic force from point *B* to *A* along any path between these two points.

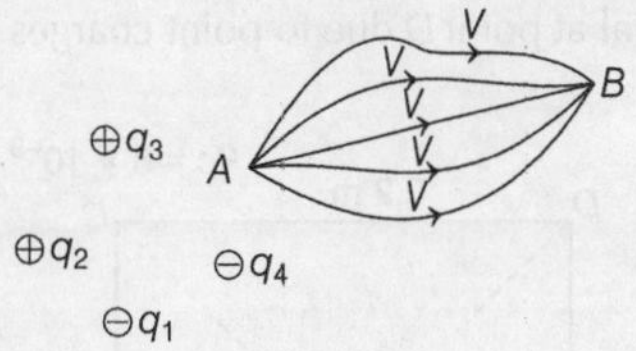

If V_A and V_B be the electric potentials at points *A* and *B* respectively, then $\Delta V = V_A - V_B$

or $$\Delta V = \frac{W_{AB}}{q}$$

The SI unit of potential difference is volt (V).

The dimensional formula for electric potential difference is given by $[ML^2T^{-3}A^{-1}]$.

Electric Potential due to a Point Charge

Electric potential due to a point charge *q* at any point *P* lying at a distance *r* as shown in the figure below, is given by

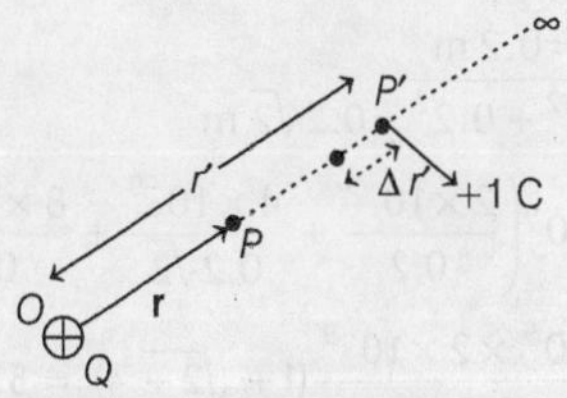

$$V = \frac{1}{4\pi\varepsilon_0} \cdot \frac{q}{r}$$

Electric Potential due to a System of Charges

Electric potential at any point *P* due to system of charges is equal to the algebraic sum of potentials due to individual charges.

i.e. $$V = \frac{1}{4\pi\varepsilon_0} \cdot \left(\frac{q_1}{r_{1P}} + \frac{q_2}{r_{2P}} + \dots \frac{q_n}{r_{nP}}\right)$$

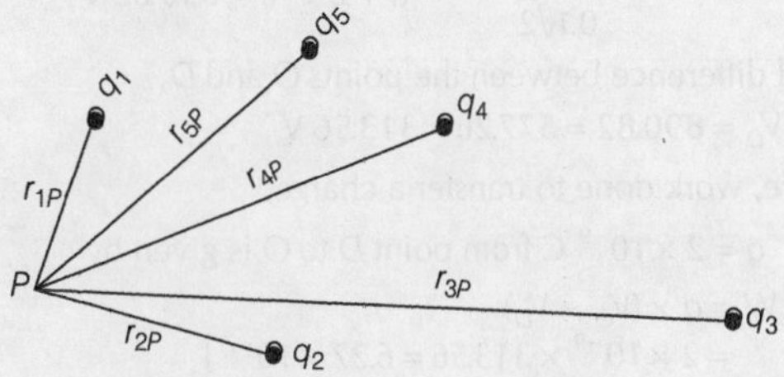

Example 11. *ABCD is a square of side 0.2 m. Charges of $2\times10^{-9}, 4\times10^{-9}$ and 8×10^{-9} C are placed at the corners A, B and C respectively. The work required to transfer a charge of 2×10^{-9} C from D to the centre of the square is*

(a) 6.27×10^{-7} J (b) 4.85×10^{-5} J
(c) 4.8×10^{-4} J (d) 3.2×10^{-2} J

Sol. (a) The charges of $q_A = 2\times10^{-9}$ C; $q_B = 4\times10^{-9}$ C and $q_C = 8\times10^{-9}$ C are placed at the corners A, B and C of the square $ABCD$ of each side of length 0.2 m.

Let V_D be potential at point D due to point charges placed at A, B and C, then

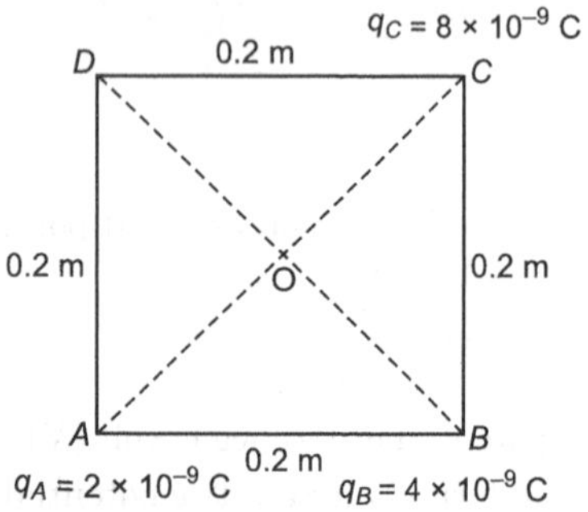

V_D = sum of the potentials due to the charges q_A, q_B and q_C

$$= \frac{1}{4\pi\varepsilon_0}\cdot\frac{q_A}{AD} + \frac{1}{4\pi\varepsilon_0}\cdot\frac{q_B}{BD} + \frac{1}{4\pi\varepsilon_0}\cdot\frac{q_C}{CD}$$

$$= \frac{1}{4\pi\varepsilon_0}\left(\frac{q_A}{AD} + \frac{q_B}{BD} + \frac{q_C}{CD}\right)$$

Here, $AD = CD = 0.2$ m

and $BD = \sqrt{0.2^2 + 0.2^2} = 0.2\sqrt{2}$ m

$$\therefore \quad V_D = 9\times10^9\left(\frac{2\times10^{-9}}{0.2} + \frac{4\times10^{-9}}{0.2\sqrt{2}} + \frac{8\times10^{-9}}{0.2}\right)$$

$$= \frac{9\times10^9\times2\times10^{-9}}{0.2}(1+\sqrt{2}+4) = 577.26 \text{ V}$$

Now, potential at point O due to charges q_A, q_B and q_C,

$$V_0 = \frac{1}{4\pi\varepsilon_0}\left(\frac{q_A}{AO} + \frac{q_B}{BO} + \frac{q_C}{CO}\right)$$

Now, $AO = BO = CO = \frac{1}{2}\times$ diagonal of square

$$= \frac{1}{2}\times0.2\sqrt{2} = 0.1\sqrt{2} \text{ m} \quad (\because BD = 0.2\sqrt{2} \text{ m})$$

$$\therefore \quad V_0 = 9\times10^9\times\left(\frac{2\times10^{-9}}{0.1\sqrt{2}} + \frac{4\times10^{-9}}{0.1\sqrt{2}} + \frac{8\times10^{-9}}{0.1\sqrt{2}}\right)$$

$$= \frac{9\times10^9\times2\times10^{-9}}{0.1\sqrt{2}}(1+2+4) = 890.82 \text{ V}$$

Potential difference between the points O and D,

$V_O - V_D = 890.82 = 577.26 = 313.56$ V

Therefore, work done to transfer a charge,

$q = 2\times10^{-9}$ C from point D to O is given by

$W = q\times(V_O - V_D)$
$= 2\times10^{-9}\times313.56 = 6.27\times10^{-7}$ J

Electric Potential of Various Systems

System	Electrical Potential
Isolated charge 	$V = \frac{q}{4\pi\varepsilon_0 r}$
A ring of charge 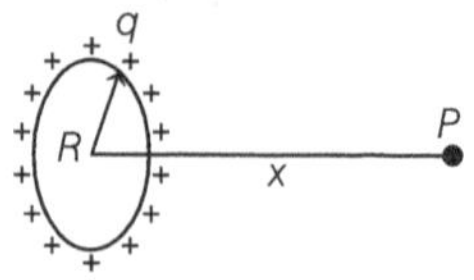	$V = \frac{1}{4\pi\varepsilon_0}\cdot\frac{q}{\sqrt{R^2+x^2}}$
A disc of charge 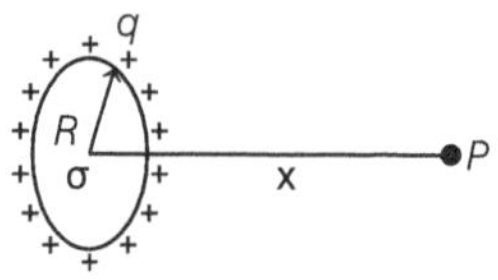	$V = \frac{\sigma}{2\varepsilon_0}[\sqrt{x^2+R^2} - x]$
Infinite sheet of charge 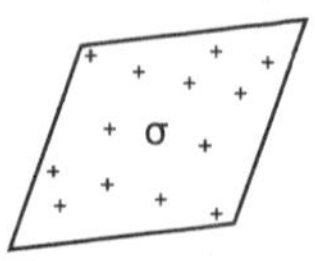	$V = \frac{-\sigma r}{2\varepsilon_0}$
Infinitely long line of charge 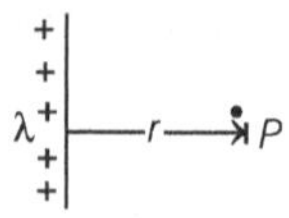	$V = \frac{\lambda}{2\pi\varepsilon_0}\log_e r$
Finite line of charge 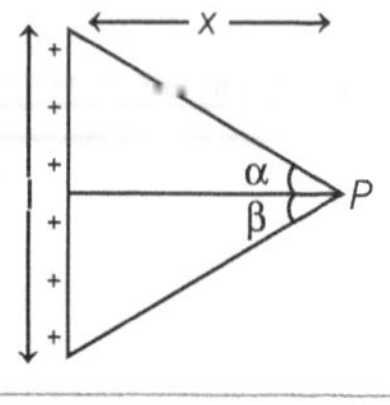	$V = \frac{\lambda}{2\pi\varepsilon_0}\log_e\left[\frac{\sqrt{x^2+l^2}-l}{\sqrt{x^2+l^2}+1}\right]$
Charged spherical shell 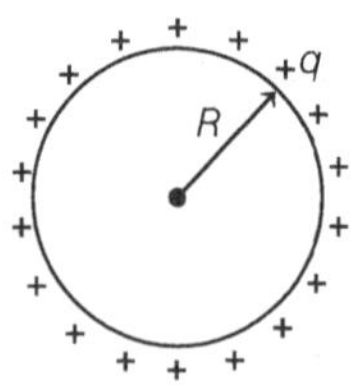	(a) Inside $0 \le r \le R, V = \frac{q}{4\pi\varepsilon_0 R}$ (b) Outside $r \ge R, V = \frac{q}{4\pi\varepsilon_0 r}$
Solid sphere of charge 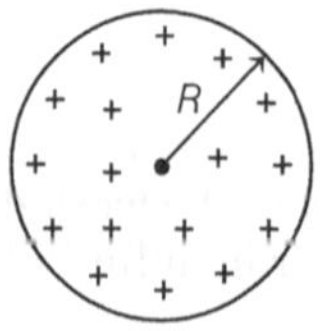	(a) Inside $0 \le r \le R$, $V = \frac{\rho R^2}{6\varepsilon_0}\left[3 - \frac{r^2}{R^2}\right]$ (b) Outside $r > R$, $V = \frac{\rho R^3}{3\varepsilon_0}\left[\frac{1}{r}\right]$

Example 12. *A solid conducting sphere, having a charge Q, is surrounded by an uncharged conducting hollow spherical shell. Let the potential difference between the surface of the solid sphere and that of the outer surface of the hollow shell be V. If the shell is now given a charge of −4 Q, the new potential difference between the same two surfaces is* **[JEE Main 2019]**

(a) −2 V (b) 2 V (c) 4 V (d) V

Sol. (d) Initially when uncharged shell encloses charge Q, charge distribution due to induction will be as shown

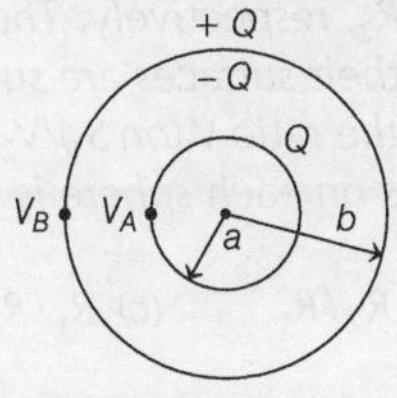

The potential on surface of inner shell is

$$V_A = \frac{kQ}{a} + \frac{k(-Q)}{b} + \frac{kQ}{b} \quad \text{...(i)}$$

Potential on surface of outer shell is

$$V_B = \frac{kQ}{b} + \frac{k(-Q)}{b} + \frac{kQ}{b} \quad \text{...(ii)}$$

Then, potential difference is

$$\Delta V_{AB} = V_A - V_B = kQ\left(\frac{1}{a} - \frac{1}{b}\right)$$

Given, $\Delta V_{AB} = V$

So, $kQ\left(\frac{1}{a} - \frac{1}{b}\right) = V$ ...(iii)

Finally after giving charge (− 4Q) to outer shell, potential difference will be

$$\Delta V_{AB} = V_A - V_B$$

$$= \left(\frac{kQ}{a} + \frac{k(-4Q)}{b}\right) - \left(\frac{kQ}{b} + \frac{k(-4Q)}{b}\right)$$

$$= kQ\left(\frac{1}{a} - \frac{1}{b}\right) = V \quad \text{[from Eq. (iii)]}$$

Hence, we obtain that potential difference does not depend on the charge of outer sphere and so it remains same.

Example 13. *Three concentric spherical metallic spheres A, B and C of radii a, b and c(a < b < c) have surface charge densities σ, − σ and σ, respectively. Find the potentials of the three shells A, B and C.*

(a) $\frac{\sigma}{\varepsilon_0}(a - b + c), \frac{\sigma}{\varepsilon_0}\left[\frac{a^2}{b} - b + c\right], \frac{\sigma}{\varepsilon_0}\left[\frac{a^2 - b^2}{c} + c\right]$

(b) $\frac{\sigma}{\varepsilon_0}\left[\frac{a^2}{b} - b + c\right], \frac{\sigma}{\varepsilon_0}\left[\frac{a^2 - b^2}{c} + c\right], \frac{\sigma}{\varepsilon_0}(a - b + c)$

(c) $\frac{\sigma}{\varepsilon_0}\left[\frac{a^2}{b} - b + c\right], \frac{\sigma}{\varepsilon_0}(a - b + c), \frac{\sigma}{\varepsilon_0}\left[\frac{a^2 - b^2}{c} + c\right]$

(d) None of the above

Sol. (a) In case of charged sphere,

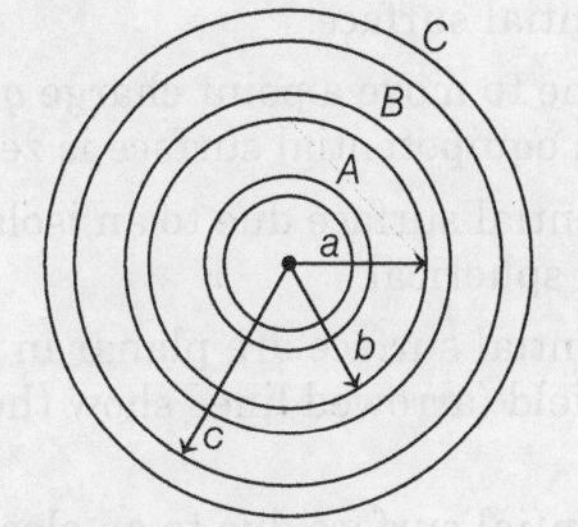

$$V_{out} = \frac{1}{4\pi\varepsilon_0}\frac{q}{r}$$

while $V_{in} = V_s \; \frac{1}{4\pi\varepsilon_0}\frac{q}{R} = \frac{\sigma R}{\varepsilon_0}$ [as, $q = 4\pi R^2\sigma$]

So, (i) $V_A = (V_A)_s + (V_B)_{in} + (V_C)_{in}$

$$V_A = \frac{1}{4\pi\varepsilon_0}\frac{\sigma a}{a} + \frac{-\sigma b}{\varepsilon_0} + \frac{\sigma c}{\varepsilon_0} = \frac{\sigma}{\varepsilon_0}(a - b + c)$$

where, $q_A = 4\pi a^2\sigma$

(ii) $V_B = (V_A)_{out} + (V_B)_s + (V_C)_{in}$

$$= \frac{1}{4\pi\varepsilon_0}\frac{q_A}{b} + \frac{-\sigma b}{\varepsilon_0} + \frac{\sigma c}{\varepsilon_0} = \frac{\sigma}{\varepsilon_0}\left[\frac{a^2}{b} - b + c\right]$$

where, $q_A = 4\pi a^2\sigma$

(iii) $V_C = (V_A)_{out} + (V_B)_{out} + (V_C)_s$

$$= \frac{1}{4\pi\varepsilon_0}\frac{q_A}{c} + \frac{1}{4\pi\varepsilon_0}\frac{q_B}{c} + \frac{\sigma c}{\varepsilon_0}$$

$$= \frac{\sigma}{\varepsilon_0}\left[\frac{a^2 - b^2}{c} + c\right]$$

where $q_A = 4\pi a^2\sigma$ and $q_B = 4\pi b^2(-\sigma)$.

Equipotential Surfaces

Equipotential surface is an imaginary surface joining the points of same potential in an electric field. So, we can say that the potential difference between any two points on an equipotential surface is zero.

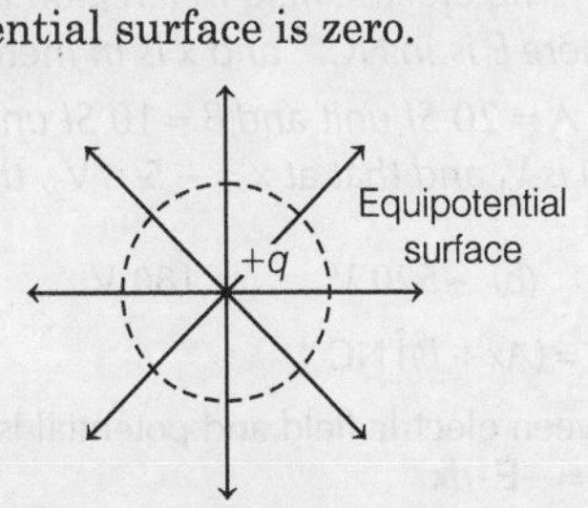

Regarding equipotential surface, following are few important points

(i) Equipotential surface may be planar, etc., but it can never be point size.

(ii) Since, equipotential surface is single valued, so these surfaces never cross each other.

(iii) Electric field is always perpendicular to equipotential surface.

(iv) Work done to move a point charge q between two points on equipotential surface is zero.

(v) Equipotential surface due to an isolated point charge is spherical.

(vi) Equipotential surface are planar in an uniform electric field (arrowed lines show the electric lines of force).

(vii) Equipotential surface due to an electric dipole is shown in the figure.

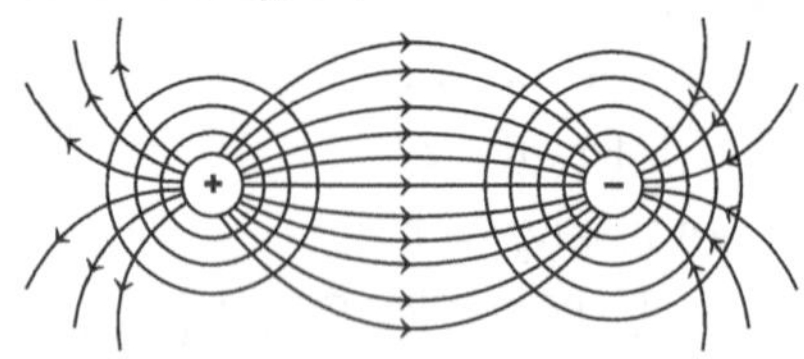

Relation between Electric Field and Electric Potential

Relation between E and V is given as

$$E = \frac{-dV}{dx}$$

where, dx is the perpendicular distance between two equipotential surfaces and dV is the change in V in the direction of E.

For a three-dimensional potential function,

$$|\mathbf{E}| = \frac{\partial V}{\partial x}\hat{\mathbf{i}} + \frac{\partial V}{\partial Y}\hat{\mathbf{j}} + \frac{\partial V}{\partial z}\hat{\mathbf{k}}$$

where, $\frac{\partial V}{\partial x}$ is partial derivation of potential function which is derivative of V with respect to x treating y and z constants.

Thus, the electric field intensity E is the negative gradient of potential. This means that decrease in potential is along the direction of E. The SI unit of E is volt per metre (Vm^{-1}).

Example 14. *The electric field in a region is given by $\mathbf{E} = (Ax + B)\hat{\mathbf{i}}$, where E is in NC^{-1} and x is in metres. The values of constants are $A = 20$ SI unit and $B = 10$ SI unit. If the potential at $x = 1$ is V_1 and that at $x = -5$ is V_2, then $V_1 - V_2$ is* *[JEE Main 2019]*

(a) -48 V (b) -520 V (c) 180 V (d) 320 V

Sol. (c) Given, $\mathbf{E} = (Ax + B)\hat{\mathbf{i}}\ \text{NC}^{-1}$

The relation between electric field and potential is given as

$$dV = -\mathbf{E}\cdot d\mathbf{x}$$

Integrating both sides within the specified limits, we get

$$\therefore \quad \int_1^2 dV = V_2 - V_1 = -\int_{x_1}^{x_2} \mathbf{E}\cdot d\mathbf{x}$$

$$\Rightarrow \quad V_1 - V_2 = \int_{x_1}^{x_2} \mathbf{E}\cdot d\mathbf{x}$$

$$= \int_{x_1}^{x_2} (Ax + B)\hat{\mathbf{i}}\cdot(dx\hat{\mathbf{i}}) = \int_{x_1}^{x_2} (Ax + B)\cdot dx$$

Here, $A = 20$ SI unit, $B = 10$ SI unit,

$$x_1 = 1 \text{ and } x_2 = -5 \Rightarrow V_1 - V_2 = \int_1^{-5} (20x + 10)\cdot dx$$

$$= \left[\frac{20x^2}{2} + 10x\right]_1^{-5} = 10\ [x^2 + x]_1^{-5}$$

$$= 10\ [(-5)^2 + (-5) - (1)^2 - (1)]$$

$$= 10\ (25 - 5 - 2) = 180 \text{ V}$$

Example 15. *Consider two charged metallic spheres S_1 and S_2 of radii R_1 and R_2, respectively. The electric fields E_1 (on S_1) and E_2 (on S_2) on their surfaces are such that $E_1 / E_2 = R_1 / R_2$. Then the ratio V_1(on S_1)/V_2 (on S_2) of the electrostatic potentials on each sphere is* *[JEE Main 2020]*

(a) $\left(\frac{R_1}{R_2}\right)^3$ (b) R_2 / R_1 (c) R_1 / R_2 (d) $(R_1 / R_2)^2$

Sol. (d) Given, $\frac{E_1}{E_2} = \frac{R_1}{R_2}$ ...(i)

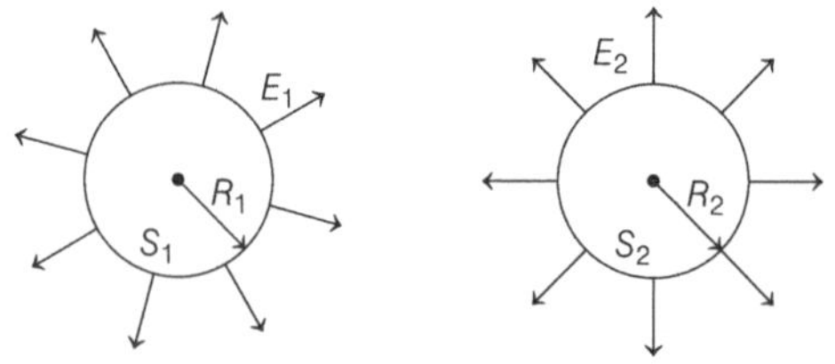

For a charged sphere, potential is given by

$$V = \frac{kQ}{R} = \frac{kQ}{R^2}\cdot R \text{ or } V = E\cdot R$$

So, $$\frac{V_1}{V_2} = \frac{E_1R_1}{E_2R_2} = \frac{R_1}{R_2} \times \frac{R_1}{R_2}$$ [from Eq. (i)]

$$\Rightarrow \quad \frac{V_1}{V_2} = \frac{R_1^2}{R_2^2} = \left(\frac{R_1}{R_2}\right)^2$$

Potential Energy of a System of Charges

Electrostatic potential energy of a system of two point charges is given by

$$U = \frac{1}{4\pi\varepsilon_0}\cdot\frac{q_1q_2}{r}$$

where, q_1 and q_2 are charges and r is the separation between the point charges.

The total work done in assembling the charges at the given locations, as shown in the figure below is given as

$$U = \frac{1}{4\pi\varepsilon_0}\left(\frac{q_1q_2}{r_{12}} + \frac{q_1q_3}{r_{13}} + \frac{q_2q_3}{r_{23}}\right)$$

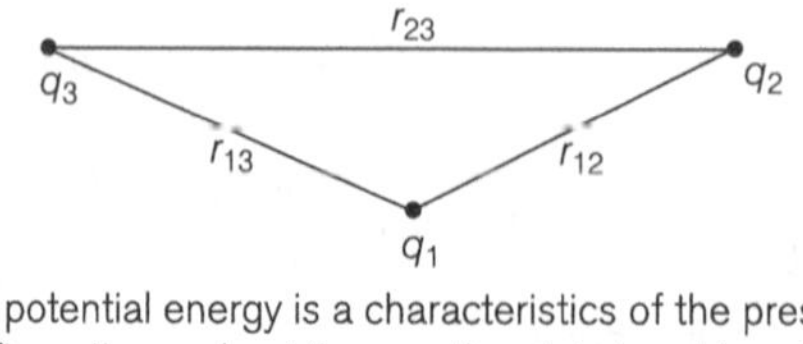

Note The potential energy is a characteristics of the present state of configuration and not the way the state is achieved.

Potential Energy in an External Field

Potential energy of a single charge q at a point with position vector $\mathbf{r}$, in an external field is $qV(\mathbf{r})$, where $V(\mathbf{r})$ is the potential at the point due to external electric field E.

Potential energy of a system of two charges in an external field can be given as

$$U = q_1 V(\mathbf{r}_1) + q_2 V(\mathbf{r}_2) + \frac{q_1 q_2}{4\pi\varepsilon_0 \mathbf{r}_{12}}$$

where, q_1 and q_2 = two point charges at position vectors $\mathbf{r}_1$ and $\mathbf{r}_2$ respectively,

$V(\mathbf{r}_1)$ = potential at $\mathbf{r}_1$ due to the external field

and $V(\mathbf{r}_2)$ = potential at $\mathbf{r}_2$ due to the external field.

Example 16. *Two point charges 4q and – q are fixed on the X-axis at* $x = \frac{-d}{2}$ *and* $x = \frac{d}{2}$*, respectively. If a third point charge q is taken from the origin to x = d along the semi-circle as shown in the figure, the energy of the charge will* *[JEE Main 2020]*

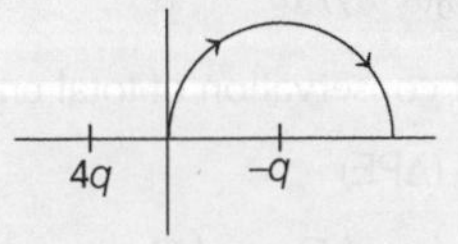

(a) increase by $\frac{2q^2}{3\pi\varepsilon_0 d}$

(b) decrease by $\frac{4q^2}{3\pi\varepsilon_0 d}$

(c) increase by $\frac{3q^2}{4\pi\varepsilon_0 d}$

(d) decrease by $\frac{q^2}{4\pi\varepsilon_0 d}$

Sol. (b)

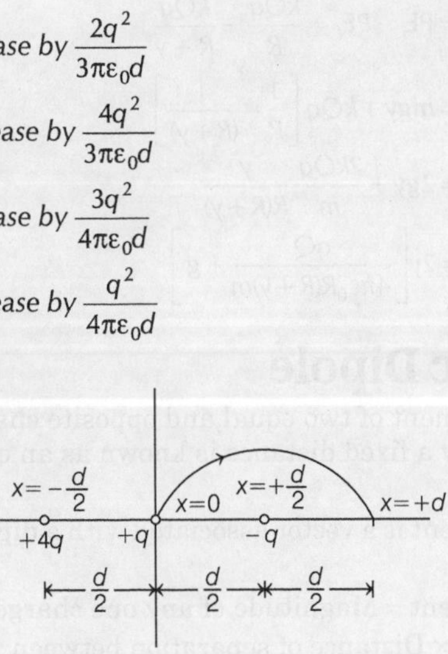

Change in potential energy, $\Delta U = U_f - U_i$

$$\Delta U = \left[\frac{k(4q)(-q)}{d} + \frac{k(4q)(q)}{\frac{3d}{2}} + \frac{k(-q)(q)}{\frac{d}{2}}\right] - \left[\frac{k(4q)(-q)}{d} + \frac{k(4q)(q)}{\frac{d}{2}} + \frac{k(-q)(q)}{\frac{d}{2}}\right]$$

$$= \frac{k(4q)(-q)}{d} + \frac{k(4q)(q)}{\frac{3d}{2}} + \frac{k(-q)(q)}{\frac{d}{2}} - \frac{k(4q)(-q)}{d} - \frac{k(4q)(q)}{\frac{d}{2}} - \frac{k(-q)(q)}{\frac{d}{2}}$$

$$= \frac{k(4q)(q)}{\frac{3d}{2}} - \frac{k(4q)(q)}{\frac{d}{2}} = \frac{8kq^2}{3d} - \frac{8kq^2}{d}$$

$$= -\frac{16kq^2}{3d} = -\frac{16q^2}{3d \times 4\pi\varepsilon_0}$$

$$= -\frac{4q^2}{3\pi\varepsilon_0 d}$$

So, ΔU will decrease by $\frac{4q^2}{3\pi\varepsilon_0 d}$.

Motion of a Charged Particle in an Electric Field

Consider a charged particle having charge q and mass m is initially at rest in an electric field of strength E. The particle will experience an electric force which causes its motion.

The force experienced by the charged particle is F, where

$$F = qE$$

$\therefore$ Acceleration produced by this force,

$$a = \frac{F}{m} = \frac{qE}{m}$$

Suppose at point A particle is at rest and after sometime t, it reaches the point B and attains velocity v.

$$\therefore \quad v = at$$

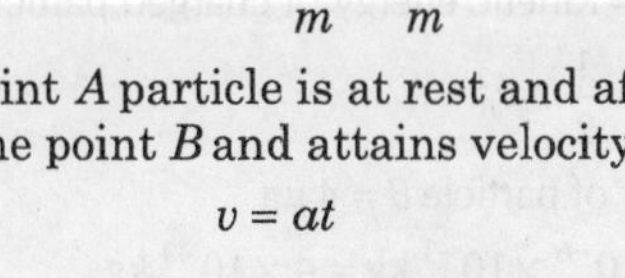

If potential difference between A and B be ΔV and the distance between them is d, then

$$v = \frac{qEt}{m} = \sqrt{\frac{2q\Delta V}{m}}$$

Kinetic energy gained by the particle in time t,

$$K = \frac{1}{2}mv^2 = \frac{1}{2}m\left(\frac{qEt}{m}\right)^2 = q\Delta v$$

Example 17. *In free space, a particle A of charge 1μC is held fixed at a point P. Another particle B of the same charge and mass 4 μg is kept at a distance of 1 mm from P. If B is released, then its velocity at a distance of 9 mm from P is*

$\left[\text{Take, } \frac{1}{4\pi\varepsilon_0} = 9 \times 10^9 \text{ N-m}^2\text{C}^{-2}\right]$ **[JEE Main 2019]**

(a) 1.5×10^2 m/s

(b) 3.0×10^4 m/s

(c) 1.0 m/s

(d) 2.0×10^3 m/s

Sol. (d) Given situation is shown in the figure below

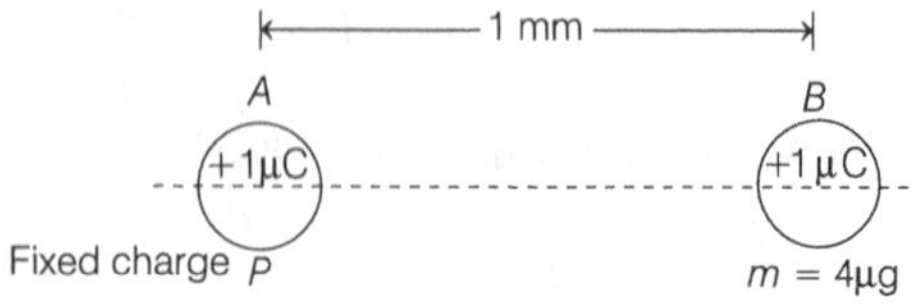

When charged particle B is released due to mutual repulsion, it moves away from A. In this process, potential energy of system of charges reduces and this change of potential energy appears as kinetic energy of B.

Now, potential energy of system of charges at separation of 1 mm,

$$U_1 = \frac{Kq_1q_2}{r}$$

Here, $q_1 = q_2 = 1 \times 10^{-6}$ C

$r = 1 \text{ mm} = 1 \times 10^{-3}$ m

$$\therefore \quad U_1 = \frac{9 \times 10^9 \times 1 \times 10^{-6} \times 1 \times 10^{-6}}{1 \times 10^{-3}} = 9 \text{ J}$$

Potential energy of given system of charges at separation of 9 mm,

$$U_2 = \frac{Kq_1q_2}{r} = \frac{9 \times 10^9 \times (1 \times 10^{-6})^2}{9 \times 10^{-3}} = 1 \text{ J}$$

By energy conservation,

Change in potential energy of system of A and B

= Kinetic energy of charged particle B

$$\Rightarrow \quad U_1 - U_2 = \frac{1}{2} m_B v_B^2$$

where, m_B = mass of particle B = 4 μg

$= 4 \times 10^{-6} \times 10^{-3}$ kg $= 4 \times 10^{-9}$ kg

and v_B = velocity of particle B at separation of 9 mm

$$\Rightarrow \quad 9 - 1 = \frac{1}{2} \times 4 \times 10^{-9} \times v_B^2$$

$$\Rightarrow \quad v_B^2 = 4 \times 10^9$$

$$\Rightarrow \quad v_B = 2 \times 10^3 \text{ ms}^{-1}$$

Example 18. *A solid sphere of radius R carries a charge Q + q distributed uniformly over its volume. A very small point-like piece of it of mass m gets detached from the bottom of the sphere and falls down vertically under gravity. This piece carries charge q. If it acquires a speed v when it has fallen through a vertical height y (see figure), then (Assume the remaining portion to be spherical)* **[JEE Main 2020]**

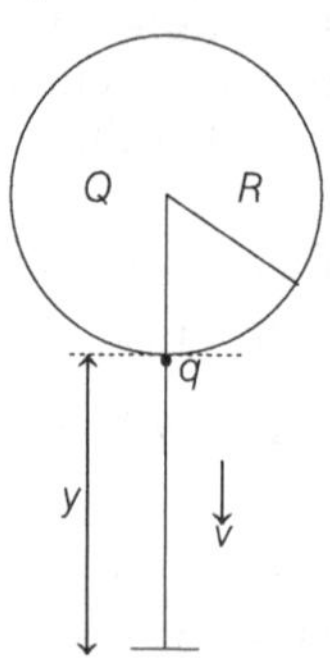

(a) $v^2 = y\left[\frac{qQ}{4\pi\varepsilon_0 R^2 ym} + g\right]$

(b) $v^2 = 2y\left[\frac{qQ}{4\pi\varepsilon_0 R(R+y)m} + g\right]$

(c) $v^2 = y\left[\frac{qQ}{4\pi\varepsilon_0 R(R+y)m} + g\right]$

(d) $v^2 = 2y\left[\frac{qQR}{4\pi\varepsilon_0 (R+y)^3 m} + g\right]$

Sol. (b) Using law of conservation of total energy,

$$\frac{1}{2}mv^2 = mgy + (\Delta\text{PE})$$

Here, $\Delta\text{PE} = \text{PE}_i - \text{PE}_f = \frac{kQq}{R} - \frac{kQq}{R+y}$

$$\Rightarrow \quad \frac{1}{2}mv^2 = mgy + kQq\left[\frac{1}{R} - \frac{1}{(R+y)}\right]$$

$$\Rightarrow \quad v^2 = 2gy + \frac{2kQq}{m}\frac{y}{R(R+y)}$$

$$\Rightarrow \quad v^2 = 2y\left[\frac{qQ}{4\pi\varepsilon_0 R(R+y)m} + g\right] \qquad \left[\because k = \frac{1}{4\pi\varepsilon_0}\right]$$

Electric Dipole

An arrangement of two equal and opposite charges separated by a fixed distance is known as an electric dipole.

Dipole moment is a vector associated with a dipole which is given as

Dipole moment = Magnitude of any one charge × Distance of separation between two charges

or $\mathbf{p} = q \times 2\mathbf{d}$

SI unit of dipole moment is coulomb-metre (C-m).

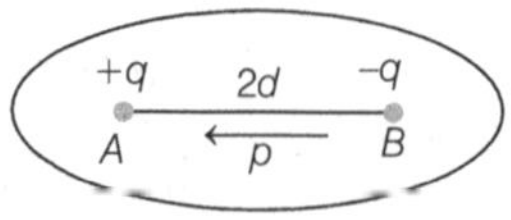

Dimensions of $\mathbf{p} = [\text{M}^0\text{LTA}]$

The dipole moment is always directed from negative charge to the positive charge.

Electric Field Intensity and Potential due to an Electric Dipole

(i) **On Axial Line**

Electric field intensity, $E = \dfrac{1}{4\pi\varepsilon_0}\dfrac{2pr}{(r^2 - a^2)^2}$

If $r >> 2a$, then $E = \dfrac{1}{4\pi\varepsilon_0}\dfrac{2p}{r^3}$.

Electric potential, $V = \dfrac{1}{4\pi\varepsilon_0}\dfrac{p}{(r^2 - a^2)}$

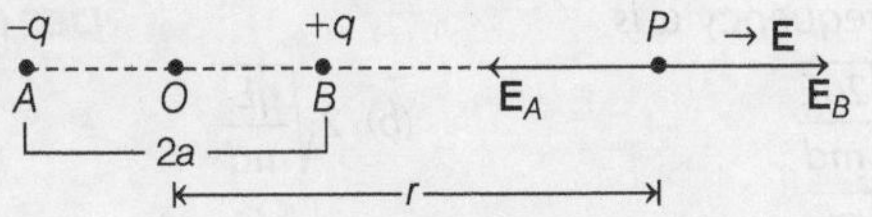

If $r >> 2a$, then $V = \dfrac{1}{4\pi\varepsilon_0}\dfrac{p}{r^2}$.

(ii) **On Equatorial Line**

Electric field intensity, $E = \dfrac{1}{4\pi\varepsilon_0}\dfrac{p}{(r^2 + a^2)^{3/2}}$

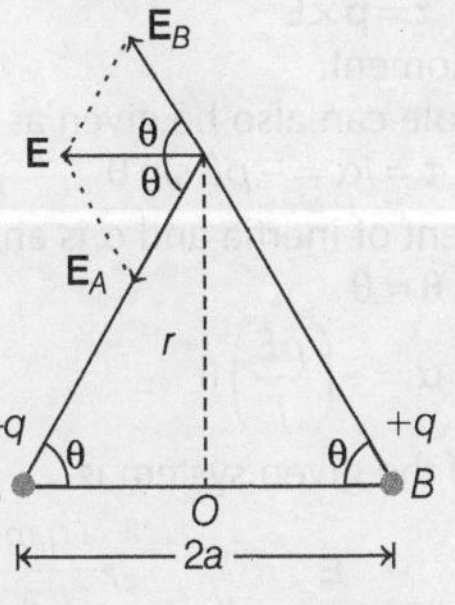

If $r >> 2a$, then

$$E = \frac{1}{4\pi\varepsilon_0}\frac{p}{r^3}$$

Electric potential, $V = 0$.

(iii) **At any Point along a Line Making θ Angle with Dipole Axis**

Electric field intensity,

$$E = \frac{1}{4\pi\varepsilon_0}\frac{p\sqrt{(1 + 3\cos^2\theta)}}{r^3}$$

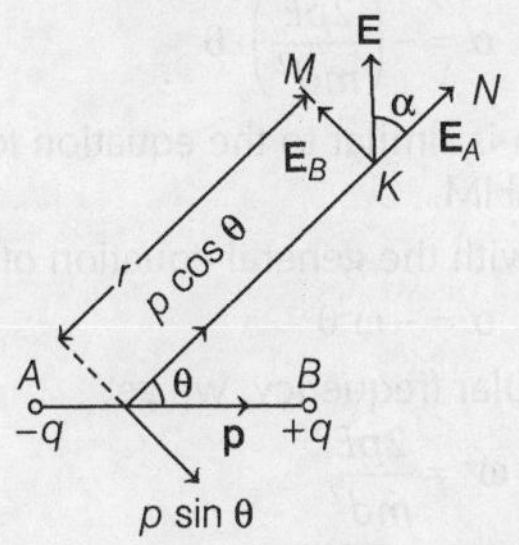

Electric potential, $V = \dfrac{1}{4\pi\varepsilon_0}\dfrac{p\cos\theta}{(r^2 - a^2\cos^2\theta)}$

If $r >> 2a$, then $V = \dfrac{1}{4\pi\varepsilon_0}\dfrac{p\cos\theta}{r^2}$.

Example 19. *Determine the electric dipole moment of the system of three charges, placed on the vertices of an equilateral triangle as shown in the figure.* **[JEE Main 2019]**

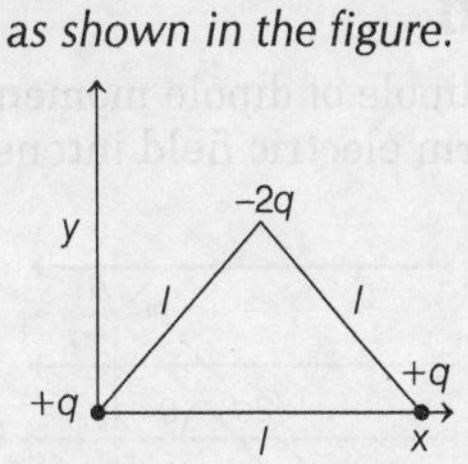

(a) $\sqrt{3}\, ql\, \dfrac{\hat{\mathbf{j}} - \hat{\mathbf{i}}}{\sqrt{2}}$ (b) $2ql\,\hat{\mathbf{j}}$ (c) $-\sqrt{3}\, ql\,\hat{\mathbf{j}}$ (d) $(ql)\dfrac{\hat{\mathbf{i}} + \hat{\mathbf{j}}}{\sqrt{2}}$

Sol. (c) Given system is equivalent to two dipoles inclined at 60° to each other as shown in the figure below

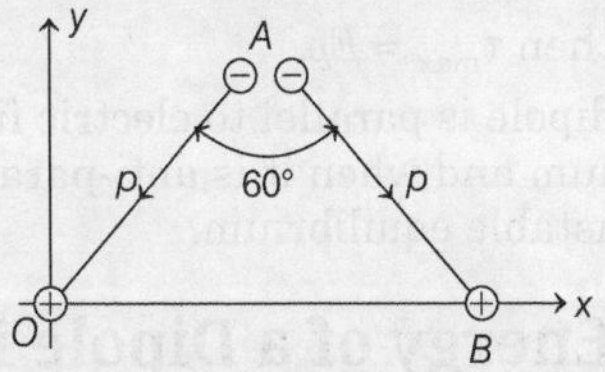

Now, magnitude of resultant of these dipole moments is

$$p_{\text{net}} = \sqrt{p^2 + p^2 + 2p \cdot p\cos 60^\circ} = \sqrt{3}p = \sqrt{3}ql$$

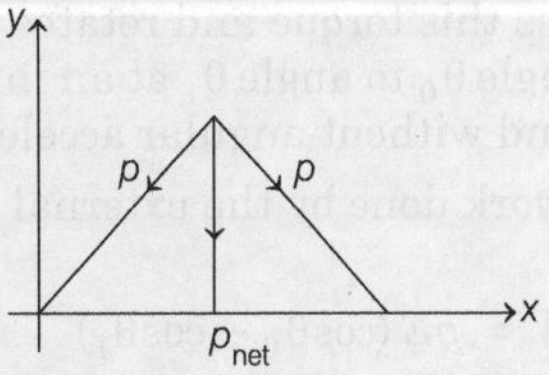

As, resultant is directed along negative y-direction

$p_{\text{net}} = -\sqrt{3}p\hat{\mathbf{j}} = -\sqrt{3}ql\,\hat{\mathbf{j}}$

Example 20. *An electric dipole of moment* $\mathbf{p} = (-\hat{\mathbf{i}} - 3\hat{\mathbf{j}} + 2\hat{\mathbf{k}}) \times 10^{-29}$ *C-m is at the origin (0, 0, 0). The electric field due to this dipole at* $\mathbf{r} = +\hat{\mathbf{i}} + 3\hat{\mathbf{j}} + 5\hat{\mathbf{k}}$ *(note that* $\mathbf{r}\cdot\mathbf{p} = 0$*) is parallel to* **[JEE Main 2020]**

(a) $(+\hat{\mathbf{i}} - 3\hat{\mathbf{j}} - 2\hat{\mathbf{k}})$ (b) $(-\hat{\mathbf{i}} - 3\hat{\mathbf{j}} + 2\hat{\mathbf{k}})$

(c) $(+\hat{\mathbf{i}} + 3\hat{\mathbf{j}} - 2\hat{\mathbf{k}})$ (d) $(-\hat{\mathbf{i}} + 3\hat{\mathbf{j}} - 2\hat{\mathbf{k}})$

Sol. (c) Given, $\mathbf{r}\cdot\mathbf{p} = 0$, so $\mathbf{r} \perp \mathbf{p}$, *i.e.* we have following situation

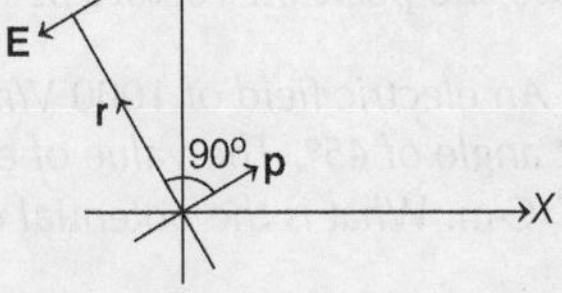

So, we have to find direction of electric field at equatorial line. As **E** is directed opposite to **p** at all equatorial points, so direction of **E** is along $-\mathbf{p}$.

So, $\mathbf{E} = \lambda(-\mathbf{p}) = \lambda[-(-\hat{\mathbf{i}} - 3\hat{\mathbf{j}} + 2\hat{\mathbf{k}})] = \lambda(+\hat{\mathbf{i}} + 3\hat{\mathbf{j}} - 2\hat{\mathbf{k}})$

∴ Electric field due to dipole is parallel to $(+\hat{\mathbf{i}} + 3\hat{\mathbf{j}} - 2\hat{\mathbf{k}})$

Torque on a Dipole in a Uniform Electric Field

When an electric dipole of dipole moment $\mathbf{p} = q \times 2d$ be placed in an uniform electric field intensity it experiences a torque given by

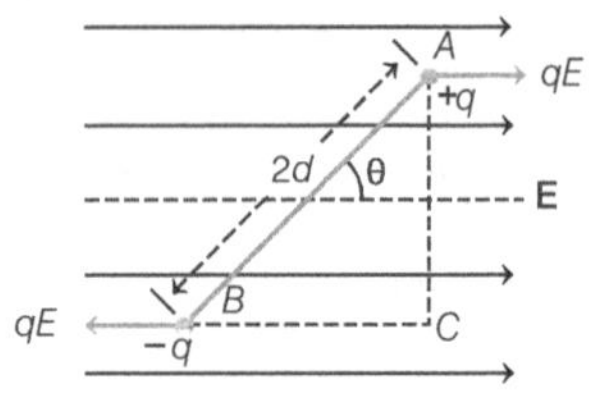

$$\tau = \mathbf{p} \times \mathbf{E}$$

or $$|\tau| = |\mathbf{p}||\mathbf{E}| \sin\theta$$

When $\theta = 90°$, then $\tau_{max} = Ep$.

When electric dipole is parallel to electric field, it is in stable equilibrium and when it is anti-parallel to electric field, it is in unstable equilibrium.

Potential Energy of a Dipole in an External Field

If an external torque τ_{ext} is applied in such a manner that it just neutralises this torque and rotates it in the plane of paper from angle θ_0 to angle θ_1 at an infinitesimal angular speed and without angular acceleration.

The amount of work done by the external torque will be given by

$$W = pE(\cos\theta_0 - \cos\theta_1)$$

If dipole is initially placed parallel to **E** field, then $\theta_1 = 0°$ and when the dipole is rotated by angle θ, then $\theta_2 = \theta$ and work done,

$$W = pE[\cos 0° - \cos\theta] = pE[1 - \cos\theta]$$

Potential energy of a dipole in a uniform electric field,

$$U(\theta) = -\mathbf{p} \cdot \mathbf{E} = -pE\cos\theta$$

The potential energy $U'(\theta)$ with an inclination θ of the dipole in an external field is given by

$$U'(\theta) = q[V(\mathbf{r}_1) - V(\mathbf{r}_2)] - \frac{q^2}{4\pi\varepsilon_0 \times 2a}$$

where, q is magnitude of either charge of electric dipole and $\mathbf{r}_1$ & $\mathbf{r}_2$ denote the position vectors of $+q$ & $-q$.

Example 21. *An electric field of 1000 V/m is applied to an electric dipole at angle of 45°. The value of electric dipole moment is 10^{-29} C-m. What is the potential energy of the electric dipole?* **[JEE Main 2019]**

(a) -9×10^{-20} J (b) -10×10^{-29} J

(c) -20×10^{-18} J (d) -7×10^{-27} J

Sol. (d) Given, $\mathbf{E} = 1000$ V/m, $\theta = 45°$

and $\mathbf{p} = 10^{-29}$ C-m

We know that, electric potential energy stored in an electric dipole kept in uniform electric field is given by the relation

$$U = -\mathbf{p} \cdot \mathbf{E} = -pE\cos\theta$$
$$= -10^{-29} \times 1000 \times \cos 45°$$

$\Rightarrow$ $$U \approx -7 \times 10^{-27} \text{ J}$$

Example 22. *An electric dipole is formed by two equal and opposite charges q with separation d. The charges have same mass m. It is kept in a uniform electric field E. If it is slightly rotated from its equilibrium orientation, then its angular frequency ω is* **[JEE Main 2019]**

(a) $\sqrt{\frac{2qE}{md}}$ (b) $2\sqrt{\frac{qE}{md}}$

(c) $\sqrt{\frac{qE}{md}}$ (d) $\sqrt{\frac{qE}{2md}}$

Sol. (a) When an electric dipole is placed in an electric field **E** at some angle θ, then two forces equal in magnitude but opposite in direction acts on the +ve and −ve charges, respectively. These forces forms a couple which exert a torque, which is given as

$$\tau = \mathbf{p} \times \mathbf{E}$$

where, **p** is dipole moment.

∴ Torque on the dipole can also be given as

$$\tau = I\alpha = -pE\sin\theta$$

where, I is the moment of inertia and α is angular acceleration.

For small angles, $\sin\theta \approx \theta$

∴ $$\alpha = -\left(\frac{pE}{I}\right)\theta \quad \text{...(i)}$$

Moment of inertia of the given system is

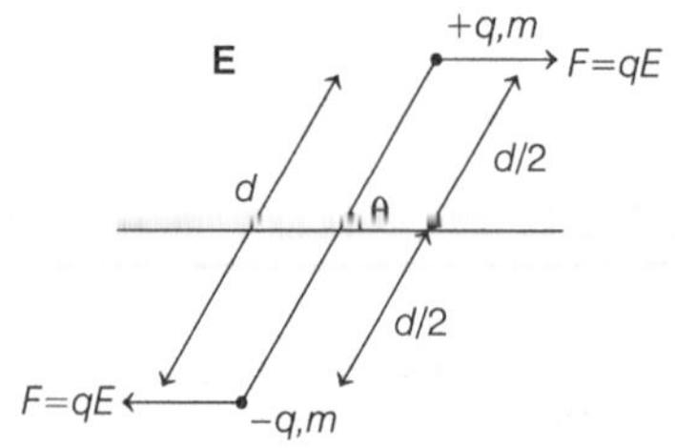

$$I = m\left(\frac{d}{2}\right)^2 + m\left(\frac{d}{2}\right)^2 = \frac{2md^2}{4} = \frac{md^2}{2}$$

Substituting the value of I in Eq. (i), we get

$\Rightarrow$ $$\alpha = -\left(\frac{2pE}{md^2}\right) \cdot \theta \quad \text{...(ii)}$$

The above equation is similar to the equation for a system executing angular SHM.

Comparing Eq. (ii) with the general equation of angular SHM,

$$\alpha = -\omega^2\theta$$

where, ω is the angular frequency, we get

$$\omega^2 = \frac{2pE}{md^2}$$

or $$\omega = \sqrt{\frac{2pE}{md^2}}$$

As, $$p = qd$$

∴ $$\omega = \sqrt{\frac{2qdE}{md^2}} = \sqrt{\frac{2qE}{md}}$$

Dielectric

These are insulating (non-conducting) materials that can produce electric effect without conduction.

Dielectrics are of two types

(i) The **non-polar dielectrics** (like N_2, O_2, benzene, methane, etc.) are made up of non-polar atoms/molecules, in which the centre of positive charge coincides with the centre of negative charge of the atom/molecule.

(ii) The **polar dielectric** (like H_2O, CO_2, NH_3, etc.) are made up of polar atoms/molecules, in which the centre of positive charge does not coincide with the centre of negative charge of the atom.

Dielectric Constant (K)

The ratio of the strength of the applied electric field to the strength of the reduced value of electric field on placing the dielectric between the plates of a capacitor is the dielectric constant. It is denoted by K (or ε_r). It is given by $K = \dfrac{E_0}{E}$.

Polarisation (P) and Electric Susceptibility (χ_e)

The induced dipole moment developed per unit volume in a dielectric slab on placing it in an electric field is called **polarisation**. It is denoted by P.

$$P = \chi_e E$$

where, χ_e is known as **electric susceptibility of the dielectric medium.**

It is a dimensionless constant.It describes the electrical behaviour of a dielectric. It has different values for different dielectrics.

For vacuum, $\chi = 0$.

Relation between dielectric constant and electric susceptibility can be given as

$$K = 1 + \chi$$

Electrostatics of Conductors

In electrostatics, the conductors have following properties

(i) Inside a conductor, electrostatic field is zero.

(ii) The interior of a conductor can have no excess charge in the static situation.

(iii) Electrostatic potential is constant throughout the volume of the conductor and has the same value (as inside) on its surface.

(iv) Electric field at the surface of a charged conductor is given as

$$E = \frac{\sigma}{\varepsilon_0}\hat{n}$$

where, σ is the surface charge density and $\hat{n}$ is a unit vector normal to the surface in the outward direction.

For $\sigma > 0$, electric field is normal to the surface outward and for $\sigma < 0$, the electric field is normal to the surface inward.

(v) For any charge and field configuration outside, any cavity in a conductor remains shielded from outside electric influence. The field inside the cavity is always zero, this is known as **electrostatic shielding.**

Capacitor

It is a device that stores electrical energy. It consists of pair of two conductors of any shape and size carrying charges of equal magnitudes and opposite signs and separated by an insulating medium.

Capacity of a capacitor is the amount of charge required to raise the potential of the capacitor by one unit (in SI one volt).

If q = amount of charge given to one of the conductor and V = potential difference between the conductors, then

Capacity of the capacitor, $C = \dfrac{q}{V}$

Its SI unit is farad (F) or coulomb/volt

Its other units are

$$1\mu F = 10^{-6}\text{ F}$$
$$1pF = 10^{-12}\text{ F}$$

Its dimensions are $[M^{-1}L^{-2}T^4A^2]$.

Note Capacitance C depends only on the geometrical configuration (shape, size and separation) of the system of conductors. It also depends on the nature dielectric separating two conductors.

Spherical Capacitor

It consists of two concentric spheres of radii a and b as shown. The inner sphere is positively charged to potential V and outer sphere is at zero potential.

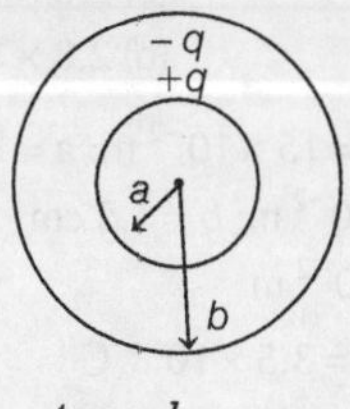

Capacitance, $C = \dfrac{Q}{V} = \dfrac{4\pi\varepsilon_0 ab}{b - a}$

For a dielectric (K) between the spheres, $C = \dfrac{4\pi K\varepsilon_0 ab}{b - a}$

When outer sphere is earthed, $C = \dfrac{4\pi\varepsilon_0 b^2}{(b - a)}$

If both spheres are separated by a distance d, then, the capacitance of the system

$$C = \frac{4\pi\varepsilon_0}{\left(\frac{1}{a} + \frac{1}{b} - \frac{2}{d}\right)}$$

Isolated Spherical Capacitor

For an isolated spherical capacitor of radius R having charge q,

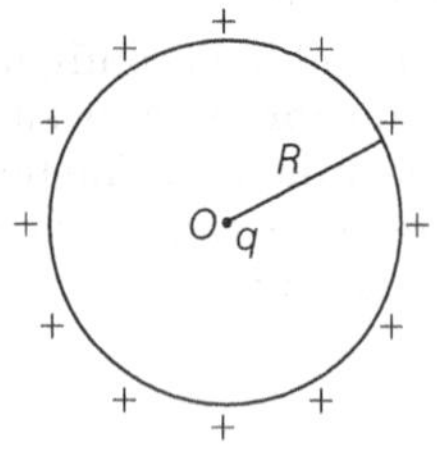

Capacitance in vacuum (or air) is given by

$$C = 4\pi\varepsilon_0 R$$

If sphere is kept in a medium of dielectric constant K or ε_r, then capacity will be

$$C = 4\pi\varepsilon_0\varepsilon_r R = 4\pi\varepsilon_0 KR$$

Cylindrical Capacitor

It consists of two-axial cylinders of radii a and b and length l. The electric field exists in the region between the cylinders. If K is the dielectric constant of the material between the cylinders, then capacitance is given by

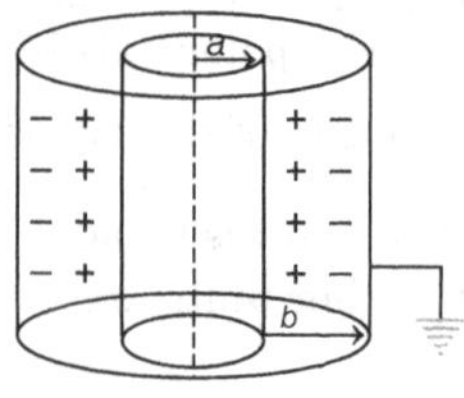

$$C = \frac{2\pi K\varepsilon_0 l}{\log_e\left(\frac{b}{a}\right)}$$

Example 23. *A cylindrical capacitor has two coaxial cylinders of length 15 cm and radii 1.5 cm and 1.4 cm. The outer cylinder is earthed and the inner cylinder is given a charge of 3.5 μC. What will be the potential of the inner cylinder? Neglect and effects (i.e. bending of field lines at the ends).*

(a) 2.8×10^2 V (b) 2.8×10^3 V
(c) 2.8×10^4 V (d) 2.8×10^5 V

Sol. (c) Here, $l = 15\text{ cm} = 15 \times 10^{-2}$ m; $a = 1.4$ cm $= 1.4 \times 10^{-2}$ m; $b = 1.5$ cm $= 1.5 \times 10^{-2}$ m

$$q = 3.5\,\mu\text{C} = 3.5 \times 10^{-6}\text{ C}$$

Now,
$$C = \frac{2\pi\varepsilon_0 l}{2.303\log_{10}\left(\frac{b}{a}\right)}$$

$$= \frac{2\pi \times 8.854 \times 10^{-12} \times 15 \times 10^{-2}}{2.303\log_{10}\frac{1.5 \times 10^{-2}}{1.4 \times 10^{-2}}} = 1.12 \times 10^{-10}\text{ F}$$

Since, outer cylinder is earthed, the potential of the inner cylinder will be equal to the potential difference between them.

Therefore, potential of inner cylinder,

$$V = \frac{q}{C} = \frac{3.5 \times 10^{-6}}{1.21 \times 10^{-10}} = 2.89 \times 10^4\text{ V}$$

Parallel Plate Capacitor

It consists of two metal plates parallel to each other and separated by a distance that is very small as compared to the dimensions of the plates.

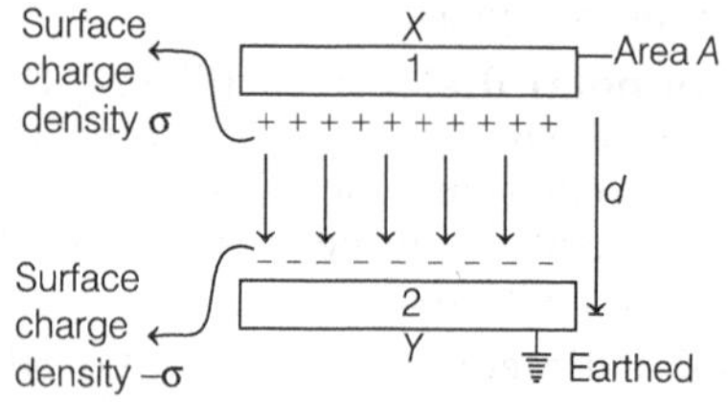

Intensity of the electric field between the plates of a capacitor,

$$E = \frac{\sigma}{\varepsilon_0} = \frac{q}{A\varepsilon_0}$$

Potential difference between the plates,

$$V = Ed = \frac{qd}{A\varepsilon_0}$$

where, d is the distance between the conductor plates.

Its capacity, $C = \dfrac{\varepsilon_0 A}{d}$

Following are few important cases related to parallel plate capacitor

(i) If both plates of parallel plate capacitor are connected by a metallic wire, then

$$C = \frac{q}{V} = \frac{q}{0} = \infty$$

(ii) If both plates are earthed,

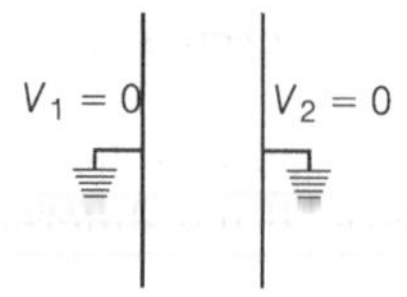

$$V = V_1 - V_2 = 0$$

$$C = \frac{q}{V} = \frac{q}{0} = \infty$$

(iii) If two plates have different charges, then

$$q = \frac{(q_1 - q_2)}{2} \text{ and } V = \frac{(q_1 - q_2)d}{2A\varepsilon_0}$$

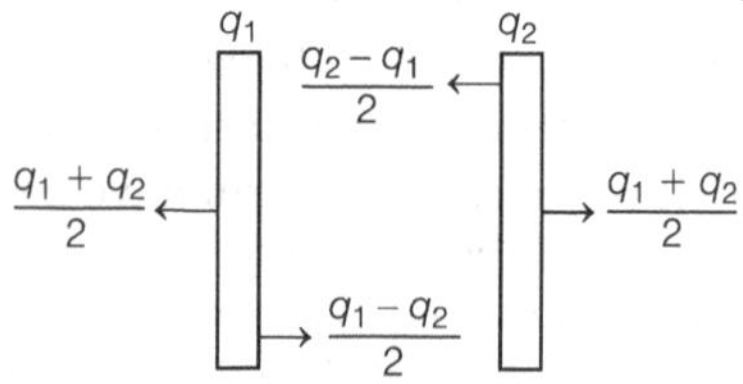

$$\therefore \quad C = \frac{q}{V} = \frac{\varepsilon_0 A}{d}$$

Example 24. *A parallel plate capacitor has 1μF capacitance. One of its two plates is given + 2 μC charge and the other plate + 4μC charge. The potential difference developed across the capacitor is* **[JEE Main 2019]**

(a) 1 V (b) 5 V (c) 2 V (d) 3 V

Sol. (a) Net value of charge on plates of capacitor after steady state is reached is

$$q_{net} = \frac{q_2 - q_1}{2}$$

where, q_2 and q_1 are the charges given to plates.
(Note that this formula is valid for any polarity of charge.)

Here, $q_2 = 4\mu C, q_1 = 2\mu C$

$\therefore$ Charge of capacitor is $q = \Delta q_{net} = \frac{4-2}{2} = 1\mu C$

Potential difference between capacitor plates is

$$V = \frac{Q}{C} = \frac{1\mu C}{1\mu F} = 1\,V$$

Effect of Dielectric on Capacitor

When a dielectric slab of dielectric constant K and thickness t is placed between the two plates.

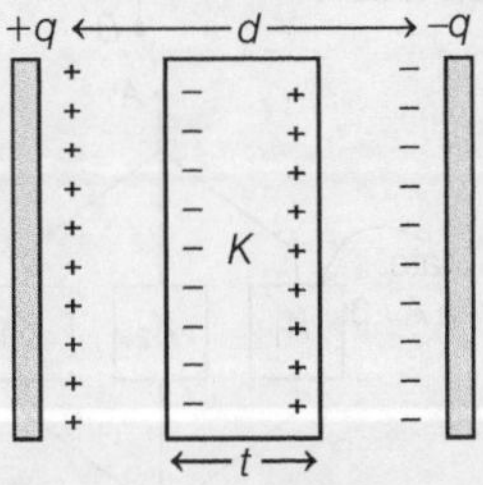

Then, $C = \frac{\varepsilon_0 A}{d - t + \frac{t}{K}} = \frac{\varepsilon_0 A}{d - t\left(1 - \frac{1}{K}\right)}$

(i) If the slab completely fills the space between the plates, then $t = d$, and therefore

$$C = \frac{K\varepsilon_0 A}{d}$$

(ii) If a conducting slab ($K = \infty$) is placed between the plates, then

$$C = \frac{\varepsilon_0 A}{d - t}$$

(iii) If the space between the plates is completely filled with a conductor, then $t = d$ and $K = \infty$.
Then, $C = \infty$

(iv) When more than one dielectric slabs are placed fully between the plates, then

$$C = \frac{A\varepsilon_0}{\left(\frac{t_1}{K_1} + \frac{t_2}{K_2} + \frac{t_3}{K_3} + \ldots + \frac{t_n}{K_n}\right)}$$

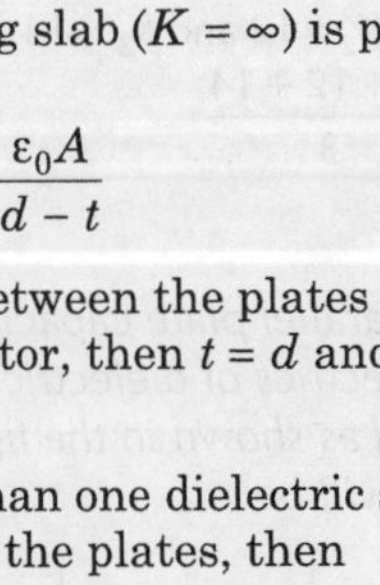

Example 25. *A parallel plate capacitor of capacitance 90 pF is connected to a battery of emf 20 V. If a dielectric material of dielectric constant $K = \frac{5}{3}$ is inserted between the plates, the magnitude of the induced charge will be* **[JEE Main 2018]**

(a) 1.2 nC (b) 0.3 nC (c) 2.4 nC (d) 0.9 nC

Sol. (a) Magnitude of induced charge is given by

$$Q' = (K-1)\,CV_0$$
$$= \left(\frac{5}{3} - 1\right) 90 \times 10^{-12} \times 20 = 1.2 \times 10^{-9}\,C$$
$$\Rightarrow \quad Q' = 1.2\text{ nC}$$

Example 26. *Voltage rating of a parallel plate capacitor is 500 V. Its dielectric can withstand a maximum electric field of 10^6 V/m. The plate area is $10^{-4}\,m^2$. What is the dielectric constant, if the capacitance is 15 pF? (Take, $\varepsilon_0 = 8.86 \times 10^{-12} C^2/N\text{-}m^2$)* **[JEE Main 2019]**

(a) 3.8 (b) 8.5 (c) 4.5 (d) 6.2

Sol. (b) As we know, capacitance of a capacitor filled with dielectric medium,

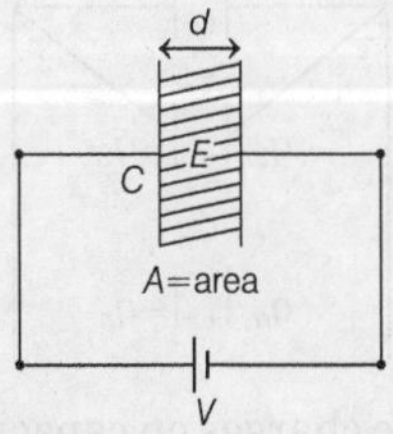

$$C = \frac{\varepsilon_0 KA}{d} \qquad \ldots(i)$$

and electric field between plates is

$$E = \frac{V}{d} \Rightarrow d = \frac{V}{E} \qquad \ldots(ii)$$

So, by combining Eqs. (i) and (ii), we get

$$K = \frac{CV}{\varepsilon_0 AE} \qquad \ldots(iii)$$

Given, $C = 15\text{pF} = 15 \times 10^{-12}$ F,

$V = 500$ V, $E = 10^6$ Vm^{-1},

$A = 10^{-4}\text{m}^2$ and $\varepsilon_0 = 8.85 \times 10^{-12}\,C^2N^{-1}m^{-2}$

Substituting values in Eq. (iii), we get

$$K = \frac{15 \times 10^{-12} \times 500}{8.85 \times 10^{-12} \times 10^{-4} \times 10^6} = 8.47 \approx 8.5$$

Combination of Capacitors

Series Combination In series combination, capacitors are connected one after another as shown in figure. If a source of emf V volt is connected between points X and Y, then charge on each capacitor is same, *i.e.* $+q$ and $-q$ and the applied potential difference of the source are divided.

If $V_1, V_2, V_3, \ldots, V_n$ are potential differences across capacitors $C_1, C_2, C_3, \ldots, C_n$ respectively, then

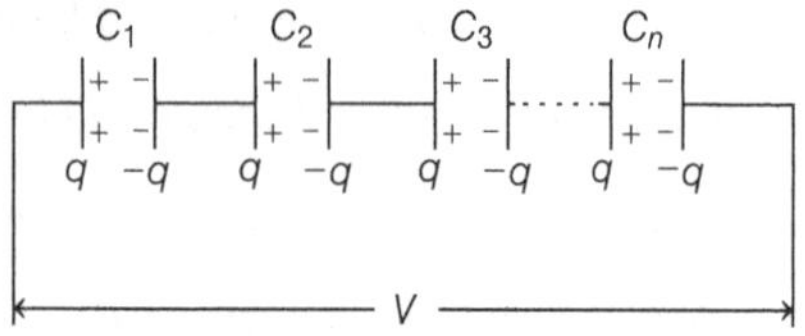

$$V = V_1 + V_2 + V_3 + \ldots + V_n$$

If C_s = equivalent capacity in series combination, then

$$\frac{1}{C_s} = \frac{1}{C_1} + \frac{1}{C_2} + \frac{1}{C_3} + \ldots + \frac{1}{C_n}$$

Parallel Combination In parallel combination, capacitors are connected one upon another as shown in figure. If a source of emf V volt is connected between points X and Y, the potential difference across each capacitors is same but charges on different capacitors are different.

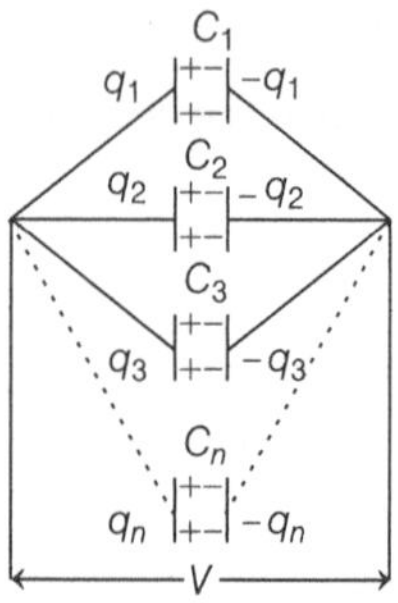

If $q_1, q_2, q_3, \ldots, q_n$ are charges on capacitors $C_1, C_2, C_3, \ldots C_n$ respectively, then charge delivered by source,

$$q = q_1 + q_2 + q_3 + \ldots + q_n$$

If C_p = equivalent capacity in parallel combination, then

$$C_p = C_1 + C_2 + C_3 + \ldots + C_n$$

Mixed Combination If n capacitors each of capacity C are connected in a series to form n row and m such rows are connected in parallel as shown in figure.

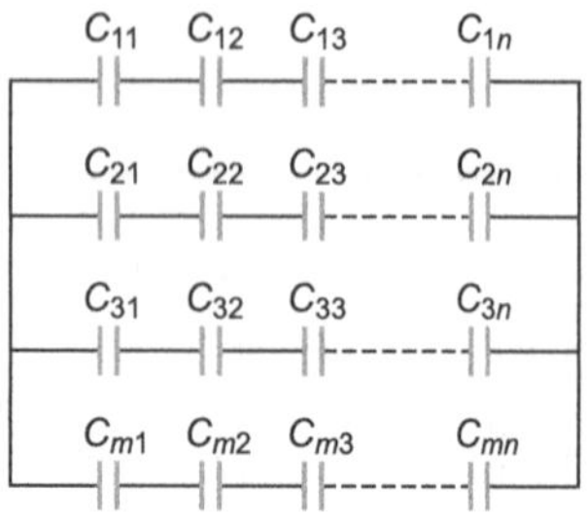

The number of capacitors used, $N = m \times n$

Hence, equivalent capacity of the combination,

$$C_{eq} = m \times \frac{C}{n}$$

Example 27. *A parallel plate capacitor is of area 6 cm² and a separation 3 mm. The gap is filled with three dielectric materials of equal thickness (see figure) with dielectric constants $K_1 = 10$, $K_2 = 12$ and $K_3 = 14$. The dielectric constant of a material which give same capacitance when fully inserted in above capacitor, would be* [JEE Main 2019]

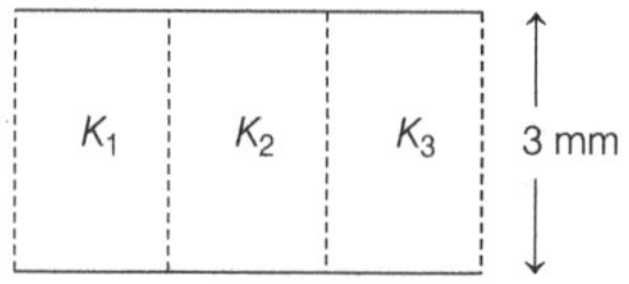

(a) 4 (b) 36
(c) 12 (d) 14

Sol. (c) In the given arrangement, capacitor can be viewed as three-different capacitors connected in parallel as shown below

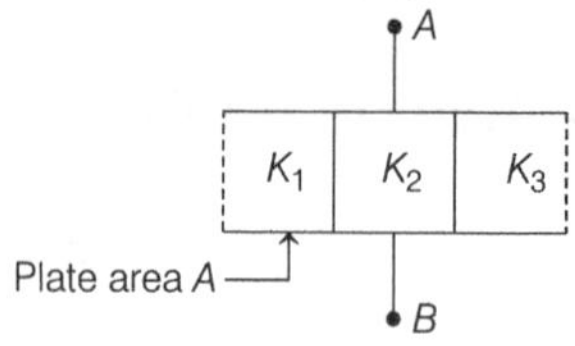

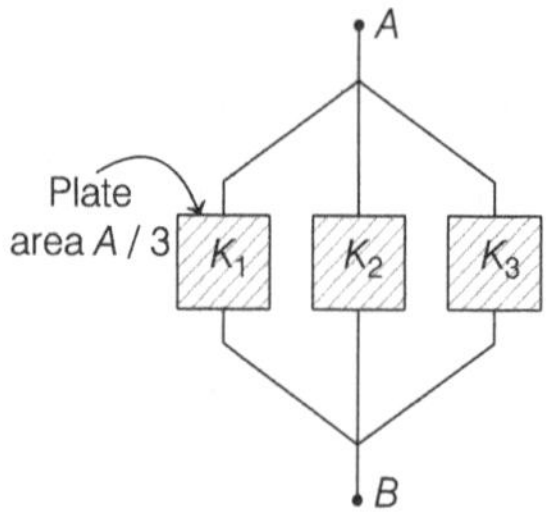

So, equivalent capacity of the system is

$$C_{eq} = C_1 + C_2 + C_3$$

$$\Rightarrow \quad \frac{K\varepsilon_0 A}{d} = \frac{K_1\varepsilon_0 A/3}{d} + \frac{K_2\varepsilon_0 A/3}{d} + \frac{K_3\varepsilon_0 A/3}{d}$$

$$\Rightarrow \quad K = \frac{K_1}{3} + \frac{K_2}{3} + \frac{K_3}{3}$$

Here $K_1 = 10, K_2 = 12$ and $K_3 = 14$

So, $K = \dfrac{10 + 12 + 14}{3}$

$\Rightarrow K = 12$

Example 28. *A parallel plate capacitor with square plates is filled with four dielectrics of dielectric constants K_1, K_2, K_3, K_4 arranged as shown in the figure. The effective dielectric constant K will be* [JEE Main 2019]

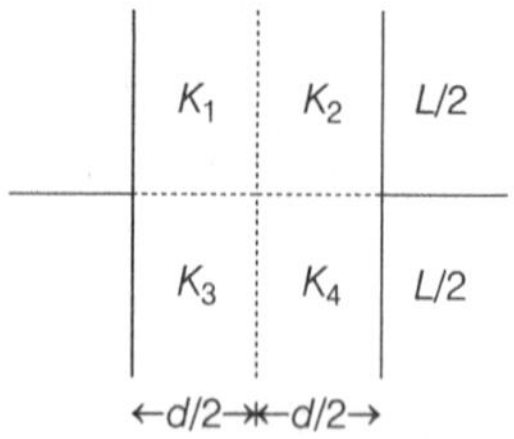

(a) $K = \dfrac{(K_1 + K_2)(K_3 + K_4)}{2(K_1 + K_2 + K_3 + K_4)}$ (b) $K = \dfrac{(K_1 + K_2)(K_3 + K_4)}{K_1 + K_2 + K_3 + K_4}$

(c) $K = \dfrac{(K_1 + K_3)(K_2 + K_4)}{K_1 + K_2 + K_3 + K_4}$ (d) $K = \dfrac{(K_1 + K_4)(K_2 + K_3)}{2(K_1 + K_2 + K_3 + K_4)}$

Sol. (*) This capacitor system can be converted into two parts as shown in the figure

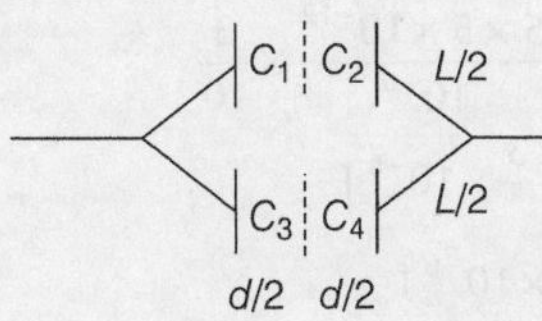

where C_1, C_2, C_3 and C_4 are capacitance of the capacitor having dielectric constants K_1, K_2, K_3 and K_4, respectively.

Here, $C_1 = \dfrac{K_1\varepsilon_0 A/2}{d/2} = \dfrac{K_1\varepsilon_0 A}{d}$

Similarly, $C_2 = \dfrac{K_2\varepsilon_0 A}{d}, C_3 = \dfrac{K_3\varepsilon_0 A}{d}$ and $C_4 = \dfrac{K_4\varepsilon_0 A}{d}$

Since, equivalent capacitance in series combination is

$$C_{eq} = \frac{C_1 \cdot C_2}{C_1 + C_2}$$

Here, C_1, C_2 and C_3, C_4 are in series combination.

$$\therefore \quad (C_{eq})_{12} = \frac{C_1 \cdot C_2}{C_1 + C_2} = \frac{\frac{K_1\varepsilon_0 A}{d} \cdot \frac{K_2\varepsilon_0 A}{d}}{\frac{K_1\varepsilon_0 A}{d} + \frac{K_2\varepsilon_0 A}{d}} = \frac{K_1 \cdot K_2}{K_1 + K_2} \cdot \frac{\varepsilon_0 A}{d}$$

Similarly, $(C_{eq})_{34} = \dfrac{K_3 \cdot K_4}{K_3 + K_4} \cdot \dfrac{\varepsilon_0 A}{d}$

Now, $(C_{eq})_{12}$ and $(C_{eq})_{34}$ are in parallel combination.

$$\therefore \quad C_{net} = (C_{eq})_{12} + (C_{eq})_{34} = \frac{K_1 \cdot K_2}{K_1 + K_2} \cdot \frac{\varepsilon_0 A}{d} + \frac{K_3 \cdot K_4}{K_3 + K_4} \cdot \frac{\varepsilon_0 A}{d}$$

$$\Rightarrow \quad C_{net} = \left(\frac{K_1 \cdot K_2}{K_1 + K_2} + \frac{K_3 \cdot K_4}{K_3 + K_4}\right)\frac{\varepsilon_0 A}{d} \quad \text{...(i)}$$

If K is effective dielectric constant, then $C_{net} = \dfrac{K\,\varepsilon_0 A}{d}$ ...(ii)

From Eqs. (i) and (ii), we get

$$\frac{K\,\varepsilon_0 A}{d} = \left(\frac{K_1 \cdot K_2}{K_1 + K_2} + \frac{K_3 \cdot K_4}{K_3 + K_4}\right)\frac{\varepsilon_0 A}{d}$$

or
$$K = \left(\frac{K_1 \cdot K_2}{K_1 + K_2} + \frac{K_3 \cdot K_4}{K_3 + K_4}\right)$$

Example 29. *In the circuit shown, charge on the 5 μF capacitor is* **[JEE Main 2020]**

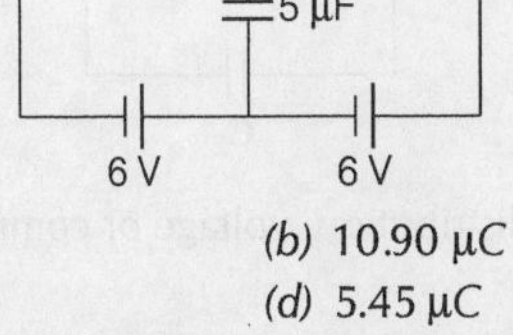

(a) 18.00 μC (b) 10.90 μC
(c) 16.36 μC (d) 5.45 μC

Sol. (c) Let q_2 be the required charge on 5 μF capacitor. The given circuit is shown below

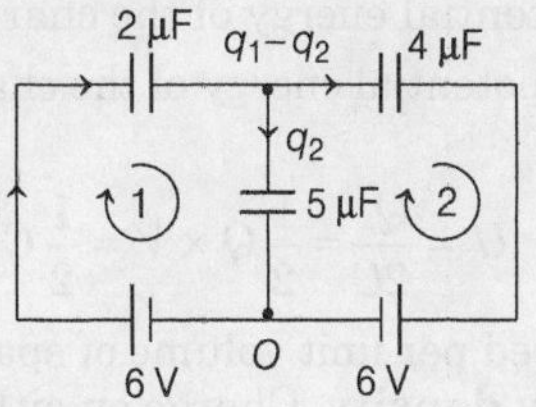

Applying KVL in loop 1,

$$6 - \frac{q_1}{2} - \frac{q_2}{5} = 0 \quad \text{...(i)}$$

Now, applying KVL in loop 2,

$$-\left(\frac{q_1 - q_2}{4}\right) - 6 + \frac{q_2}{5} = 0 \quad \text{...(ii)}$$

Adding Eqs. (i) and (ii), we get

$$-\left(\frac{q_1 - q_2}{4}\right) - \frac{q_1}{2} = 0$$

$$\Rightarrow \quad -\frac{q_1}{4} + \frac{q_2}{4} - \frac{q_1}{2} = 0$$

$$\frac{q_2}{4} = \frac{q_1}{2} + \frac{q_1}{4}$$

$$\Rightarrow \quad \frac{q_2}{4} = \frac{3q_1}{4}$$

$$\Rightarrow \quad q_1 = \frac{q_2}{3} \quad \text{...(iii)}$$

Substituting the value of q_1 in Eq. (i), we get

$$6 - \frac{q_2}{6} - \frac{q_2}{5} = 0 \Rightarrow 6 = \frac{5q_2 + 6q_2}{30}$$

$$\frac{11}{30}q_2 = 6 \Rightarrow q_2 = \frac{180}{11}\,\mu C = 16.36\,\mu C$$

Example 30. *Effective capacitance of parallel combination of two capacitors C_1 and C_2 is 10 μF. When these capacitors are individually connected to a voltage source of 1 V, the energy stored in the capacitor C_2 is 4 times that of C_1. If these capacitors are connected in series, then their effective capacitance will be* **[JEE Main 2020]**

(a) 4.2 μF (b) 3.2 μF (c) 1.6 μF (d) 8.4 μF

Sol. (c) Let capacitances of capacitors are C_1 and C_2.

Given, energy stored in capacitor 1 = 4 × energy stored in capacitor 2

$$\Rightarrow \quad \frac{1}{2}C_1 \times V^2 = 4 \times \frac{1}{2}C_2 \times V^2$$

$$\Rightarrow \quad C_1 = 4C_2$$

Also, equivalent capacity in parallel combination is 10 μF.

$$\Rightarrow \quad C_1 + C_2 = 10$$

$$\Rightarrow \quad 5C_2 = 10 \quad (\because C_1 = 4C_2)$$

$$\Rightarrow \quad C_2 = 2\,\mu F$$

So, $\quad C_1 = 4 \times 2 = 8\,\mu F$

Equivalent capacity in series combination will be

$$C_{eq} = \frac{C_1 C_2}{C_1 + C_2} = \frac{8 \times 2}{8 + 2} = 1.6\,\mu F$$

Energy Stored in a Conductor

The work done in charging a conductor is stored as electrostatic potential energy of the charged conductor.

$\therefore$ Electrostatic potential energy of the charged conductor or capacitor,

$$U = \frac{Q^2}{2C} = \frac{1}{2} Q \times V = \frac{1}{2} CV^2$$

The energy stored per unit volume of space in a capacitor is called **energy density**. Charge on either plate of capacitor is

$$Q = \sigma A = \varepsilon_0 EA$$

Energy stored in the capaitor is

$$U = \frac{Q}{2C} = \frac{(\varepsilon_0 EA)^2}{2 \cdot \varepsilon_0 A/d} = \frac{1}{2} \varepsilon_0 E^2 \cdot Ad$$

Energy density, $u = \dfrac{\text{Energy stored}}{\text{Volume of capacitor}} = \dfrac{U}{Ad}$

$$\therefore \quad u = \frac{1}{2} \varepsilon_0 E^2$$

When the dielectric is introduced in between parallel plate capacitor

(i) and battery is removed

$$C' = KC, \quad Q' = Q$$

$$V' = \frac{V}{K}, \quad E' = \frac{E}{K}$$

$$\text{and} \quad U' = \frac{U}{K}$$

(ii) and battery remains connected

$$C' = KC, \; Q' = KQ, \; V' = V, \; E' = E \text{ and } U' = KU$$

Note Force between the plates of a parallel plate capacitor is

$$|F| = \frac{\sigma A}{2\varepsilon_0} = \frac{Q^2}{2\varepsilon_0 A} = \frac{CV^2}{2d}$$

Example 31. *A capacitor with capacitance 5 μF is charged to 5 μC. If the plates are pulled apart to reduce the capacitance to 2 μF, how much work is done?* **[JEE Main 2019]**

(a) 6.25×10^{-6} J (b) 2.16×10^{-6} J

(c) 2.55×10^{-6} J (d) 3.75×10^{-6} J

Sol. (d) Potential energy stored in a capacitor is

$$U = \frac{1}{2} QV = \frac{1}{2} \frac{Q^2}{C}$$

So, initial energy of the capacitor, $U_i = \frac{1}{2} Q^2 / C_1$

Final energy of the capacitor, $U_f = \frac{1}{2} Q^2 / C_2$

As we know, work done, $W = \Delta U = U_f - U_i = \frac{1}{2} Q^2 \left[\frac{1}{C_2} - \frac{1}{C_1}\right]$

Here, $Q = 5\,\mu\text{C} = 5 \times 10^{-6}$ C,

$C_1 = 5\,\mu\text{F} = 5 \times 10^{-6}$F

and $C_2 = 2\,\mu\text{F} = 2 \times 10^{-6}$ F.

$$\Rightarrow \quad \Delta U = \frac{1}{2} \times (5 \times 10^{-6})^2 \left[\frac{1}{2 \times 10^{-6}} - \frac{1}{5 \times 10^{-6}}\right]$$

$$= \frac{1}{2} \times \frac{5 \times 5 \times 10^{-12}}{10^{-6}} \times \frac{3}{10}$$

$$= \frac{25 \times 3}{20} \times 10^{-6} \text{ J}$$

$$\Rightarrow \quad \Delta U = 3.75 \times 10^{-6} \text{ J}$$

$\therefore$ Work done in reducing the capacitance from 5 μF to 2 μF by pulling plates of capacitor apart is 3.75×10^{-6} J.

Redistribution of Charge

When two isolated charged conductors are connected to each other, then charge is redistributed in the ratio of their capacitances.

Common potential, $V = \dfrac{q_1 + q_2}{C_1 + C_2} = \dfrac{C_1 V_1 + C_2 V_2}{C_1 + C_2}$

$$\text{Energy loss} = \frac{1}{2} \frac{C_1 C_2 (V_1 - V_2)^2}{(C_1 + C_2)}$$

This energy is lost in the form of heat in connecting wires.

When n small drops, each of capacitance C, charged to potential V with charge q, surface charge density σ and potential energy U coalesce to form a single drop.

Then for new drop,

Total charge = nq, total capacitance = $n^{1/3}C$,

total potential = $n^{2/3}$ V, surface charge density = $n^{1/3}$ σ

and total potential energy = $n^{2/3}$ U.

Example 32. *A 10 μF capacitor is fully charged to a potential difference of 50 V. After removing the source voltage, it is connected to an uncharged capacitor in parallel. Now, the potential difference across them becomes 20 V. The capacitance of the second capacitor is* **[JEE Main 2020]**

(a) 15 μF (b) 30 μF

(c) 20 μF (d) 10 μF

Sol. (a) Initially we have a capacitor of capacitance, $C_1 = 10$ μF and $V_1 = 50$ V

Charge on capacitor is

$$Q = C_1 V_1 = 10\,\mu\text{F} \times 50\,\text{V} = 500\,\mu\text{C}$$

Now, another uncharged capacitor of capacity C_2 is connected in parallel with C_1 as shown in figure.

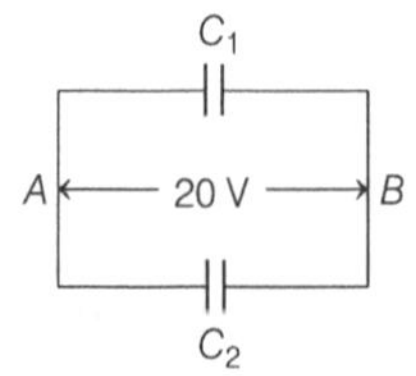

Given that, after redistribution, voltage or common potential is 20 V.

Final charge on C_1 is

$$Q_1 = C_1V = 10\ \mu F \times 20\ V = 200\ \mu C$$

So, another capacitor has a charge of $Q_2 = 500 - 200 = 300\ \mu F$

Potential, $V = 20$ V

$$\therefore \text{Capacity, } C_2 = \frac{Q_2}{V} = \frac{300\ \mu F}{20\ V} = 15\ \mu F$$

Example 33. *Two isolated conducting spheres S_1 and S_2 of radii $\frac{2}{3}R$ and $\frac{1}{3}R$, have charges 12 μC and – 3 μC respectively, and are at a large distance from each other. They are now connected by a conducting wire. A long time after this is done, the charges on S_1 and S_2 respectively, are* **[JEE Main 2020]**

(a) 4.5 μC on both *(b)* + 4.5 μC and – 4.5 μC

(c) 3 μC and 6 μC *(d)* 6 μC and 3 μC

Sol. *(d)* When both spheres are connected by a conducting wire, charge from higher potential sphere flows to lower potential sphere till both spheres reach a common potential V_C which is given by

$$V_C = \frac{Q_1 + Q_2}{C_1 + C_2}$$

For a sphere, capacitance, $C = 4\pi\varepsilon_0 R$

so,
$$V_C = \frac{12\times10^{-6} + (-3\times10^{-6})}{4\pi\varepsilon_0(R_1 + R_2)}$$

$$= \frac{9\times10^{-6}}{4\pi\varepsilon_0\left(\frac{2}{3}R + \frac{1}{3}R\right)} = \frac{9\times10^{-6}}{4\pi\varepsilon_0 R}$$

Values of charges on spheres after redistribution are

Q_1 = charge on first sphere

$$Q_1 = C_1V_C = 4\pi\varepsilon_0 R_1 \cdot \frac{(9\times10^{-6})}{4\pi\varepsilon_0 R}s$$

$$= \frac{\left(4\pi\varepsilon_0 \frac{2}{3}R\right)(9\times10^{-6})}{4\pi\varepsilon_0 R}$$

$$= \frac{2}{3}\times 9\times10^{-6}\ C$$

$$= \frac{2}{3}\times 9\ \mu C = 6\ \mu C$$

and Q_2 = charge on second sphere

$$Q_2 = C_2V_C$$

$$= \frac{4\pi\varepsilon_0\left(\frac{1}{3}R\right)\times 9\times10^{-6}}{4\pi\varepsilon_0 R} = 3\times10^{-6}\ C = 3\ \mu C$$

Example 34. *Two capacitors of capacitances C and 2C are charged to potential differences V and 2V, respectively. These are then connected in parallel in such a manner that the positive terminal of one is connected to the negative terminal of the other. The final energy of this configuration is* **[JEE Main 2020]**

(a) $\frac{9}{2}CV^2$ *(b)* $\frac{3}{2}CV^2$ *(c)* $\frac{25}{6}CV^2$ *(d)* zero

Sol. *(b)* Before connecting the two capacitors in parallel, Charge on capacitor C, $Q_1 = CV$ and charge on capacitor $2C$, $Q_2 = 2C \times 2V = 4CV$

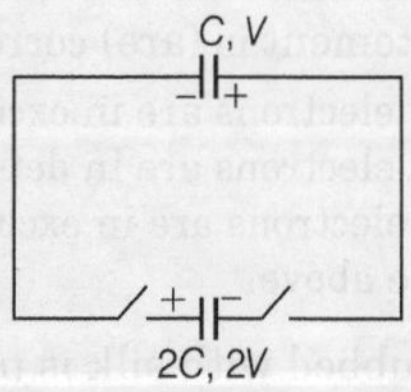

After connecting the two capacitors in parallel,

Using law of conservation of charge,

$$4CV - CV = (C + 2C)V_{\text{common}}$$

$$V_{\text{common}} = \frac{4CV - CV}{3C} = V$$

$$V_{\text{common}} = V$$

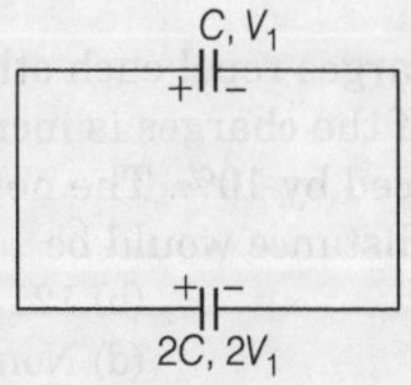

Now, final energy of the configuration,

$$U_f = \frac{1}{2}CV^2 + \frac{1}{2}\times 2CV^2 = \frac{3}{2}CV^2$$

Practice Exercise

ROUND I Topically Divided Problems

Electric Charge, Coulomb's Force and Electric Field

1. Two identical spheres carrying charges $-9\,\mu C$ and $5\,\mu C$ respectively are kept in contact and then separated from each other. Which amongst the following statement is (are) correct?

(a) 1.25×10^{13} electrons are in excess.
(b) 1.25×10^{13} electrons are in deficit.
(c) 4.15×10^{12} electrons are in excess.
(d) None of the above

2. A glass rod rubbed with silk is used to charge a gold leaf electroscope and the leaves are observed to diverse. The electroscope thin charged leaf is exposed to X-rays for short period, then

(a) the leaves will diverge further
(b) the leaves will melt
(c) the leaves will not be affected
(d) None of the above

3. Two point charges repel each other with a force of 100 N. One of the charges is increased by 10% and other is reduced by 10%. The new force of repulsion at the same distance would be

(a) 100 N (b) 121 N
(c) 99 N (d) None of these

4. Three charges each of $+1\mu C$ are placed at the corners of an equilateral triangle. If the force between any two charges be F, then the net force on either charge will be

(a) $\sqrt{2}F$ (b) $F\sqrt{3}$
(c) $2F$ (d) $3F$

5. Three charges $+Q$, q and $+Q$ are placed respectively at distance 0, $\frac{d}{2}$ and d from the origin on the X-axis. If the net force experienced by $+Q$ placed at $x = 0$ is zero, then value of q is [JEE Main 2019]

(a) $\frac{+Q}{2}$ (b) $\frac{+Q}{4}$
(c) $\frac{-Q}{2}$ (d) $\frac{-Q}{4}$

6. If two charges are placed at a distance of 5 cm. If a brass sheet is placed between them, the force between two charges will be

(a) decrease to 0 (b) increase to ∞
(c) increase to 0 (d) decrease to ∞

7. The electrostatic force of repulsion between two positively charged ions carrying equal charge is 3.7×10^{-9} N, when they are separated by a distance of $5\,\text{Å}$. What are the number of electrons are missing from each ion?

(a) 2 (b) 4
(c) 0 (d) 10

8. Two point charges $+3\mu C$ and $+8\mu C$ repel each other with a force of 40 N. If a charge of $-5\mu C$ is added to each of them, then the force between them will become

(a) -10 N (b) $+10$ N
(c) $+20$ N (d) -20 N

9. A charged spherical conductor of radius R carries a charge q_0. A point test charge q_0 is placed at a distance x from the surface of the conductor. The force experienced by the test charge will be proportional to

(a) $(R+x)^2$ (b) $(R-x)^2$
(c) $\frac{1}{(R-x)^2}$ (d) $\frac{1}{(R+x)^2}$

10. Charge $q_1 = +6.0$ nC is on Y-axis at $y = +3$ cm and charge $q_2 = -6.0$ nC is on Y-axis at $y = -3$ cm. Calculate force on a test charge $q_0 = 2$ nC placed on X-axis at $x = 4$ cm.

(a) $-51.8\,\hat{j}\,\mu N$ (b) $+51.8\,\hat{j}\,\mu N$
(c) $-5.18\,\hat{j}\,\mu N$ (d) $5.18\,\hat{j}\,\mu N$

11. Two point charges exert on each other a force F when they are placed r distance apart in air. If they are placed R distance apart in a medium of dielectric constant K, they exert the same force. The distance R equals

(a) $\frac{r}{K}$ (b) rK
(c) $r\sqrt{K}$ (d) $\frac{r}{\sqrt{K}}$

12. In figure, two positive charges q_2 and q_3 fixed along the Y-axis, exert a net electric force in the $+x$-direction on a charge q_1 fixed along the X-axis. If a positive charge Q is added at $(x, 0)$, the force on q_1 **[NCERT Exemplar]**

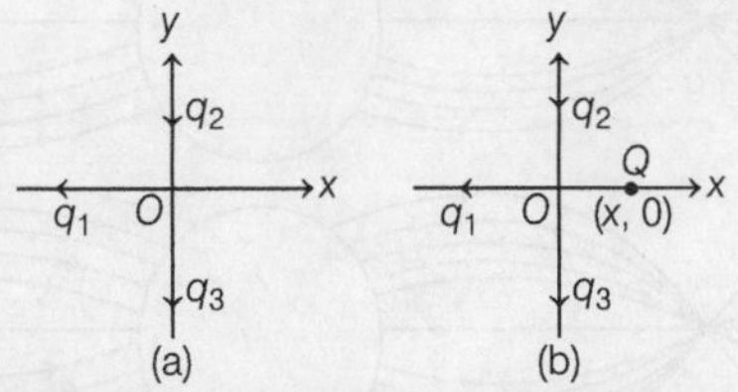

(a) shall increase along the positive X-axis
(b) shall decrease along the positive X-axis
(c) shall point along the negative X-axis
(d) shall increase but the direction changes because of the intersection of Q with q_2 and q_3

13. A pitch ball A of mass 9×10^{-5} kg carries a charge of $5\ \mu C$. What must be the magnitude and sign of the charge on a pitch ball B held 2 cm directly above the pitch ball A, such that the pitch ball A, remains stationary?
(a) 5×10^{-6} C (b) 5×10^{-12} C
(c) 8×10^{-6} C (d) 7.84×10^{-12} C

14. There are two charged identical metal spheres A and B repel each other with a force 3×10^{-5} N. Another identical uncharged sphere C is touched with A and then placed at the mid-point between A and B. Net force on C is
(a) 1×10^{-5} N (b) 2×10^{-5} N
(c) 1.5×10^{-5} N (d) 3×10^{-5} N

15. Two small conducting sphere of equal radii have charges $+10\ \mu C$ and $-20\ \mu C$ respectively and placed at a distance R from each other experience force F_1. If they are brought in contact and separated to the same distance, they experience force F_2. The ratio of F_1 to F_2 is
(a) 1 : 2 (b) −8 : 1
(c) 1 : 8 (d) −2 : 1

16. Figure shows the electric field lines around three point charges A, B and C. Which charge has the largest magnitude? **[NCERT Exemplar]**

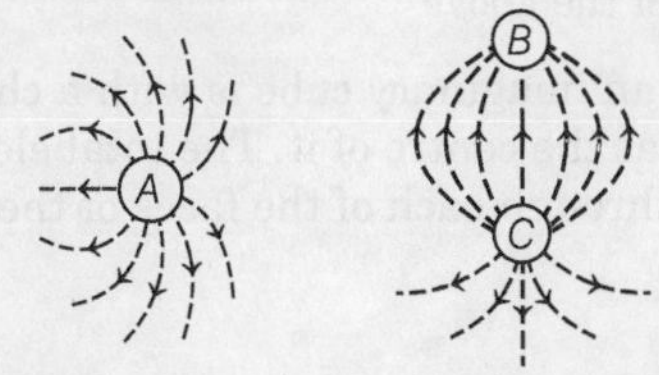

(a) Charge A
(b) Charge B
(c) Charges A and B
(d) Charge C

17. In the electric field shown in figure, the electric lines in the left have twice the separation as that between those on right. If the magnitude of the field at point A is 40 NC^{-1}, the force experienced by a proton placed at point A is

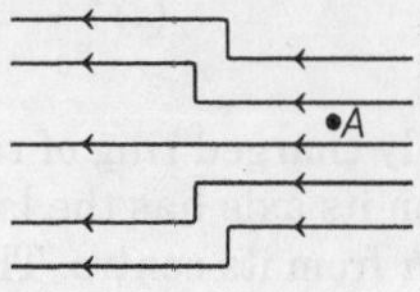

(a) 6.4×10^{-18} N (b) 3.2×10^{-15} N
(c) 5.0×10^{-12} N (d) 1.2×10^{-18} N

18. Two conducting sphere of radii r_1 and r_2 are charged to the same surface charge density. The ratio of electric field near their surface is
(a) r_1^2 / r_2^2 (b) r_2^2 / r_1^2
(c) r_1 / r_2 (d) 1 : 1

19. Figure shows the electric lines of force energy from a charged body. If the electric field at A and B are E_A and E_B, respectively and the displacement between A and B is r, then

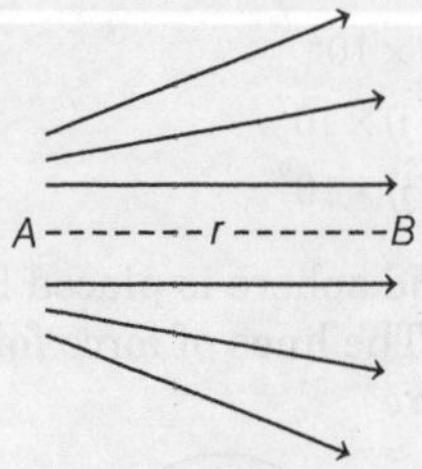

(a) $E_A < E_B$ (b) $E_A > E_B$
(c) $E_A = E_B$ (d) $E_A = 2E_B$

20. Two point charges $-q$ and $+q/2$ are situated at the origin and at the point $(a, 0, 0)$, respectively. The point along the X-axis, whereas the electric field vanished, is
(a) $x = \dfrac{\sqrt{2}a}{\sqrt{2}-1}$ (b) $X = \sqrt{2}a - \sqrt{2} - 1$
(c) $x = (\sqrt{2}-1)\sqrt{2}a$ (d) None of these

21. The maximum field intensity on the axis of a uniformly charged ring of charge q and radius R will be
(a) $\dfrac{1}{4\pi\varepsilon_0} \cdot \dfrac{q}{3\sqrt{3}R^2}$
(b) $\dfrac{1}{4\pi\varepsilon_0} \cdot \dfrac{2q}{3R^2}$
(c) $\dfrac{1}{4\pi\varepsilon_0} \cdot \dfrac{2q}{3\sqrt{3}R^2}$
(d) $\dfrac{1}{4\pi\varepsilon_0} \cdot \dfrac{3q}{2\sqrt{3}R^2}$

22. An insulated sphere of radius R has charge density ρ. The electric field at a distance r from the centre of the sphere $(r < R)$

(a) $\frac{\rho r}{3\varepsilon_0}$ (b) $\frac{\rho R}{3\varepsilon_0}$

(c) $\frac{\rho r}{\varepsilon_0}$ (d) $\frac{\rho R}{\varepsilon_0}$

23. For a uniformly charged ring of radius R, the electric field on its axis has the largest magnitude at a distance h from its centre. Then, value of h is **[JEE Main 2019]**

(a) $\frac{R}{\sqrt{2}}$ (b) $R\sqrt{2}$

(c) R (d) $\frac{R}{\sqrt{5}}$

24. Two point charges q_1 ($\sqrt{10}$ μC) and q_2 (−25 μC) are placed on the X-axis at $x = 1$ m and $x = 4$ m, respectively. The electric field (in V/m) at a point $y = 3$ m on Y-axis is

$\left(\text{Take, } \frac{1}{4\pi\varepsilon_0} = 9 \times 10^9 \text{ N-m}^2\text{C}^{-2}\right)$ **[JEE Main 2019]**

(a) $(63\,\hat{\mathbf{i}} - 27\hat{\mathbf{j}}) \times 10^2$
(b) $(81\,\hat{\mathbf{i}} - 81\hat{\mathbf{j}}) \times 10^2$
(c) $(-81\,\hat{\mathbf{i}} + 81\,\hat{\mathbf{j}}) \times 10^2$
(d) $(-63\,\hat{\mathbf{i}} + 27\hat{\mathbf{j}}) \times 10^2$

25. A metallic solid sphere is placed in a uniform electric field. The lines of force follow the paths shown in figure

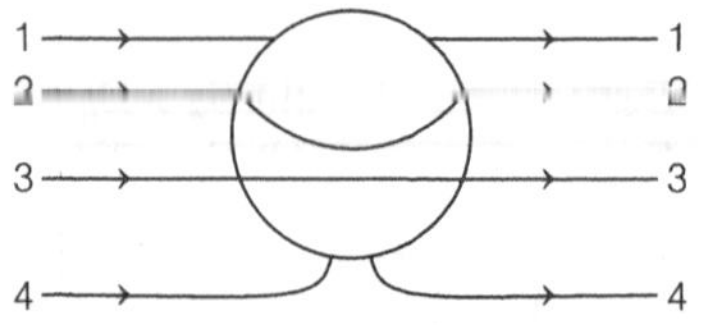

(a) 1 (b) 2
(c) 3 (d) 4

26. A point charge $+q$ is placed at a distance d from an isolated conducting plane. The field at a point P on the other side of the plane is **[NCERT Exemplar]**

(a) directed perpendicular to the plane and away from the plane
(b) directed perpendicular to the plane but towards the plane
(c) directed radially away from the point charge
(d) directed radially towards the point charge

27. A hemisphere is uniformly charged positively. The electric field at a point on a diameter away from the centre is directed **[NCERT Exemplar]**

(a) perpendicular to the diameter
(b) parallel to the diameter
(c) at an angle tilted towards the diameter
(d) at an angle tilted away from the diameter

28. A point positive charge brought near an isolated conducting sphere (figure). The electric field is best given by **[NCERT Exemplar]**

(a) 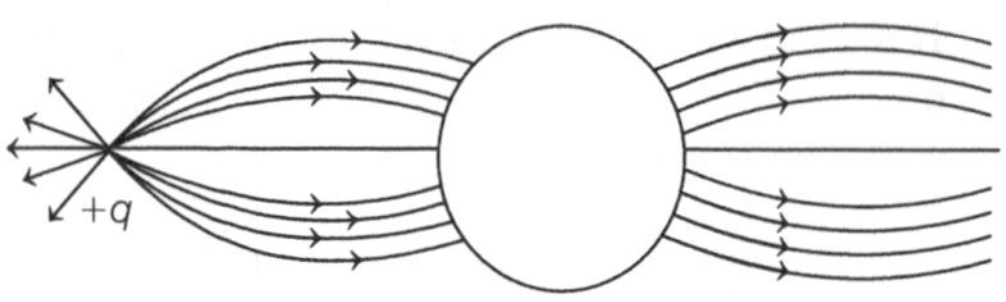

(b) 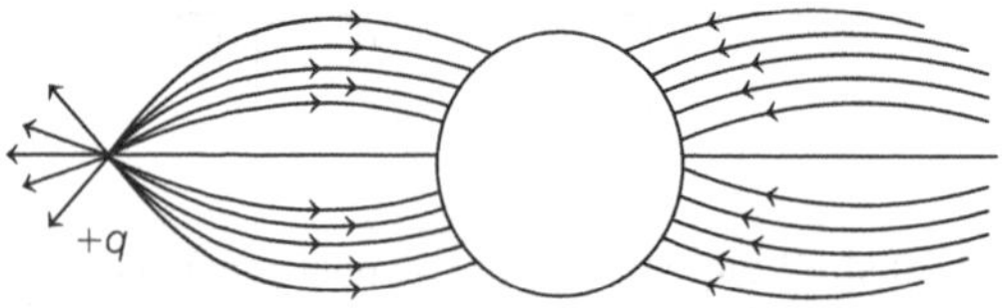

(c) 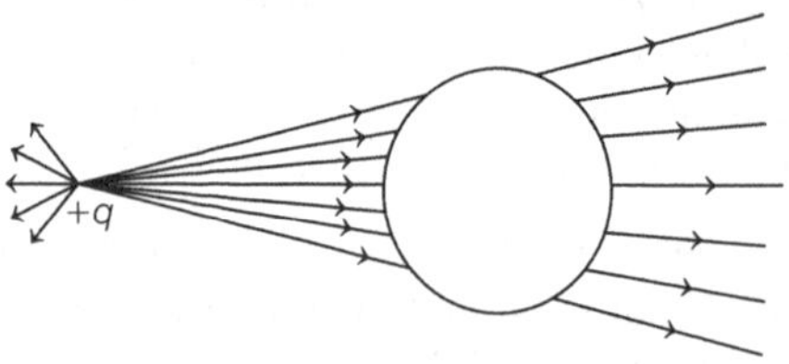

(d) 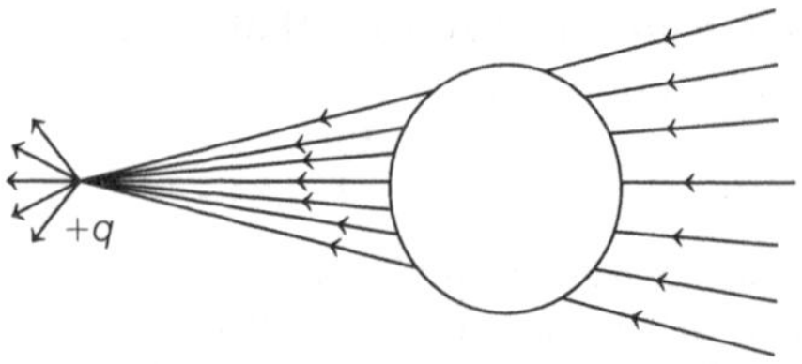

29. In a region of space, the electric field is given by $\mathbf{E} = 8\hat{\mathbf{i}} + 4\hat{\mathbf{j}} + 3\hat{\mathbf{k}}$. The electric flux through a surface of area of 100 units XY-plane is

(a) 800 units
(b) 300 units
(c) 400 units
(d) 1500 units

Electric Flux and Gauss's Theorem

30. A cylinder of radius R and length L is placed in a uniform electric field E parallel to the cylinder axis. The total flux for the surface of the cylinder is given by

(a) zero
(b) $\pi R^2 / E$
(c) $2\pi R^2 E$
(c) None of the above

31. Suppose an imaginary cube is with a charge situated at the centre of it. The total electric flux passing through each of the faces of the cube will be

(a) $\frac{q}{6\varepsilon_0}$

(b) $\frac{q}{2\varepsilon_0}$

(c) $\frac{q}{12\varepsilon_0}$

(c) None of the above

32. In finding the electric field using Gauss's law, the formula $|\mathbf{E}| = \frac{q_{enc}}{\varepsilon_0 |A|}$ is applicable. In the formula, ε_0 is permittivity of free space, A is the area of Gaussian surface and q_{enc} is charge enclosed by the Gaussian surface. This equation can be used in which of the following situation? **[JEE Main 2020]**

(a) Only when the Gaussian surface is an equipotential surface and $|\mathbf{E}|$ is constant on the surface.

(b) Only when the Gaussian surface is an equipotential surface.

(c) For any choice of Gaussian surface.

(d) Only when $|\mathbf{E}|$ = constant on the surface.

33. The electric flux through the surface **[NCERT Exemplar]**

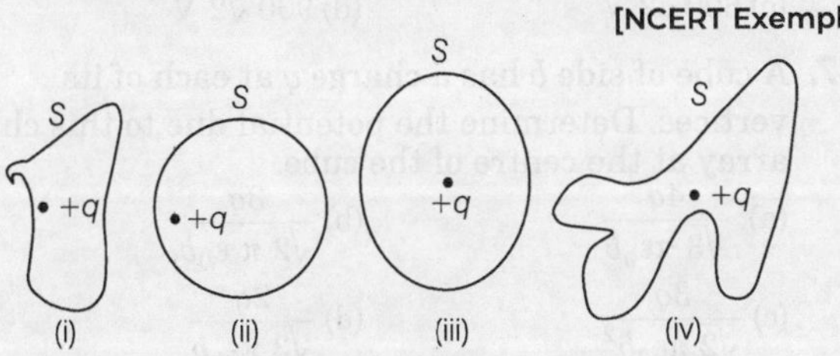

(a) in Fig. (iv) is the largest

(b) in Fig. (iii) is the least

(c) in Fig .(ii) is same as fig. (iii) but is smaller than Fig. (iv)

(d) is the same for all the figures

34. Five charges q_1, q_2, q_3, q_4 and q_5 are fixed at their positions as shown in figure, S is a Gaussian surface. The Gauss's law is given by $\oint \mathbf{E} \cdot d\mathbf{S} = \frac{q}{\varepsilon_0}$.

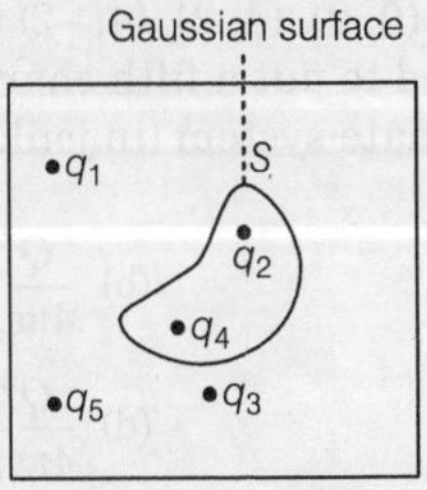

Which of the following statements is correct? **[NCERT Exemplar]**

(a) **E** on the LHS of the above equation will have a contribution from q_1, q_5 and q_3 on the RHS will have a contribution from q_2 and q_4 only.

(b) **E** on the LHS of the above equation will have a contribution from all charges while q on the RHS will have a contribution from q_2 and q_4 only.

(c) **E** on the LHS of the above equation will have a contribution from all charges while q on the RHS will have a contribution from q_1, q_3 and q_5 only.

(d) Both **E** on the LHS and q on the RHS will have contributions from q_2 and q_4 only.

35. The adjacent diagram shows a charge $+Q$ held on an insulating support S and enclosed by a hollow spherical conductor, O represents the centre of the spherical conductor and P is a point such that $OP = x$ and $SP = r$. The electric field at point, P will be

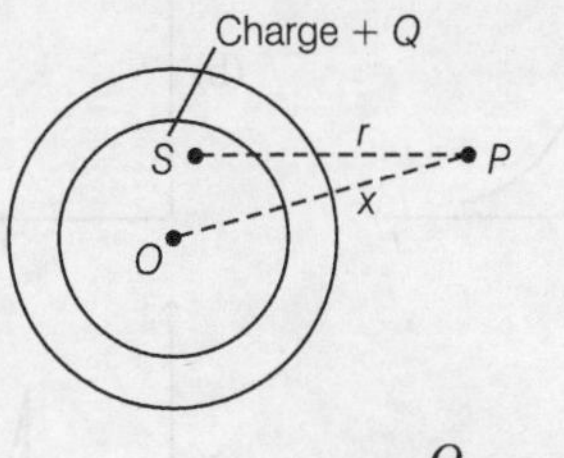

(a) zero (b) $\frac{Q}{4\pi\varepsilon_0 x^2}$

(c) $\frac{Q}{\varepsilon_0 x^2}$ (d) None of these

36. The electrostatic potential inside a charged spherical ball is given by $\phi = ar^2 + b$, where r is the distance from the centre, a and b are constants. Then, the charge density inside the ball is

(a) $-24\pi\, a\, \varepsilon_0 r$

(b) $-6a\varepsilon_0$

(c) $-24\,\pi\,\varepsilon_0$

(d) $-6\,a\varepsilon_0\, r$

37. In infinite parallel plane sheet of a metal is charged to charge density σ coulomb per square metre in a medium of dielectric constant K. Intensity of electric field near the metallic surface will be

(a) $E = \frac{\sigma}{\varepsilon_0 K}$ (b) $E = \frac{K}{3\varepsilon_0}$

(c) $E = \frac{\sigma}{2\varepsilon_0 K}$ (d) $E = \frac{K}{2\varepsilon_0}$

38. A long charged cylinder of linear charged density λ is surrounded by a hollow co-axial conducting cylinder. What is the electric field in the space between the two cylinders? **[NCERT]**

(a) $\frac{\lambda}{2\pi\varepsilon_0 r}$ (b) $\frac{\lambda r}{\sqrt{2}\,\pi\varepsilon_0}$

(c) $\frac{\lambda}{\sqrt{2}\,\pi\varepsilon_0 r}$ (d) None of these

39. A spherical charged conductor has σ as the surface density of charge. The electric field on its surface is E. If the radius of the sphere is doubled, keeping the surface density of the charge unchanged, what will be the electric field on the surface of the new sphere?

(a) $\frac{E}{4}$ (b) $\frac{E}{2}$

(c) E (d) $2E$

40. Which one of the following graphs shows the variation of electric field strength E with distance d from the centre of the hollow conducting sphere?

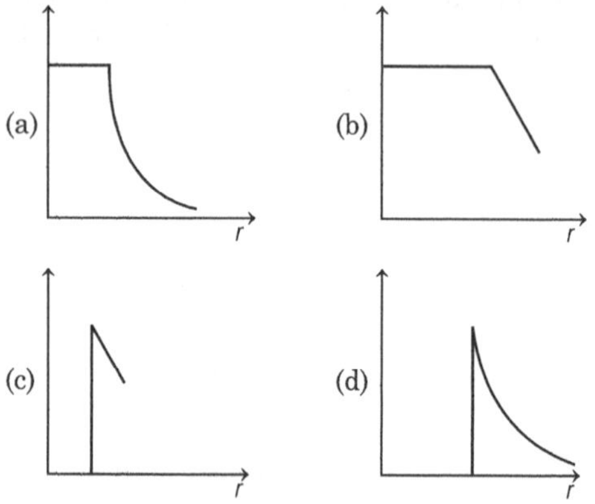

41. A positive point charge q is carried from a point B to a point A in the electric field of a point charge $+Q$ at O. If the permittivity of free space is ε_0, the work done in the process is given by (where, $a = OA$ and $b = OR$)

(a) $\frac{qQ}{4\pi\varepsilon_0}\left(\frac{1}{a}+\frac{1}{b}\right)$ (b) $\frac{qQ}{4\pi\varepsilon_0}\left(\frac{1}{a}-\frac{1}{b}\right)$

(c) $\frac{qQ}{4\pi\varepsilon_0}\left(\frac{1}{a^2}-\frac{1}{b^2}\right)$ (d) $\frac{qQ}{4\pi\varepsilon_0}\left(\frac{1}{a^2}+\frac{1}{b^2}\right)$

42. The electrostatic potential on the surface of a charged conducting sphere is 100 V. Two statements are made in this regard.

S_1 : At any point inside the sphere, electric intensity is zero.

S_2 : At any point inside the sphere, the electrostatic potential is 100 V.

Which of the following is a correct statements. **[NCERT Exemplar]**

(a) S_1 is true but S_2 is false.

(b) Both S_1 and S_2 are false.

(c) S_1 is true, S_2 is also true and S_1 is the cause of S_2.

(d) S_1 is true, S_2 is also true but the statements are independent.

43. A hollow conducting sphere of radius R has a charge $(+Q)$ on its surface. What is the electric potential within the sphere at a distance $r = R/3$ from its centre?

(a) $\frac{1}{4\pi\varepsilon_0}\cdot\frac{Q}{r}$ (b) $\frac{1}{4\pi\varepsilon_0}\cdot\frac{Q}{r^2}$

(c) $\frac{1}{4\pi\varepsilon_0}\cdot\frac{Q}{R}$ (d) Zero

44. n small drops of same size are charged to V volt each. If they coalesce to form a single large drop, then its potential will be

(a) Vn (b) Vn^{-1}

(c) $Vn^{1/3}$ (d) $Vn^{2/3}$

45. A charge $+q$ is fixed at each of the points $x = x_0$, $x = 3x_0$, $x = 5x_0 \ldots \infty$, on the X-axis and a charge $-q$ is fixed at each of the pointsx $= 2x_0$, $x = 4x_0$ $x = 6x_0 \ldots \infty$. Here, x_0 is the constant. Take, the electric potential at a point due to a charge Q at a distance r from it to be $Q/4\pi\varepsilon_0 r$. Then, the potential at the origin due to the above system of charges is

(a) $\frac{q}{4\pi\varepsilon_0 x_0}\log_e 2$ (b) $\frac{q}{8\pi\varepsilon_0 x_0}\log_e 2$

(c) 0 (d) ∞

46. The charge of $+\frac{10}{3}\times 10^{-9}$C are placed at each of the four corners of a square of side 8 cm. The potential at the point of intersection of the diagonals, is

(a) $1500\sqrt{2}$ V (b) $1800\sqrt{2}$ V

(c) $600\sqrt{2}$ V (d) $900\sqrt{2}$ V

47. A cube of side b has a charge q at each of its vertices. Determine the potential due to this charge array at the centre of the cube.

(a) $\frac{4q}{\sqrt{3}\,\pi\varepsilon_0 b}$ (b) $\frac{3q}{\sqrt{2}\,\pi\,\varepsilon_0 b}$

(c) $\frac{3q}{\sqrt{2}\,\pi\varepsilon_0 b^2}$ (d) $\frac{2q}{\sqrt{3}\,\pi\varepsilon_0 b}$

48. The tangential component of electrostatic field is continuous from one side of a charged surface to another is **[NCERT]**

(a) $\frac{1}{4\pi\varepsilon_0}\left(\frac{1}{r_A}+\frac{1}{r_B}-\frac{1}{r_C}\right)$ (b) zero

(c) $\frac{1}{4\pi\varepsilon_0}\left(\frac{1}{r_A}-\frac{1}{r_B}+\frac{1}{r_C}\right)$ (d) $\frac{1}{4\pi\varepsilon_0}\left(\frac{1}{r_A}+\frac{1}{r_B}+\frac{1}{r_C}\right)$

49. Four equal point charges Q each are placed in the XY-plane at (0, 2), (4, 2), (4, –2) and (0, – 2). The work required to put a fifth charge Q at the origin of the coordinate system (in joule) will be **[JEE Main 2019]**

(a) $\frac{Q^2}{4\pi\varepsilon_0}$ (b) $\frac{Q^2}{4\pi\varepsilon_0}\left(1+\frac{1}{\sqrt{3}}\right)$

(c) $\frac{Q^2}{2\sqrt{2}\pi\varepsilon_0}$ (d) $\frac{Q^2}{4\pi\varepsilon_0}\left(1+\frac{1}{\sqrt{5}}\right)$

50. A charge Q is distributed over two concentric conducting thin spherical shells of radii r and R $(R > r)$. If the surface charge densities on the two shells are equal, the electric potential at the common centre is **[JEE Main 2020]**

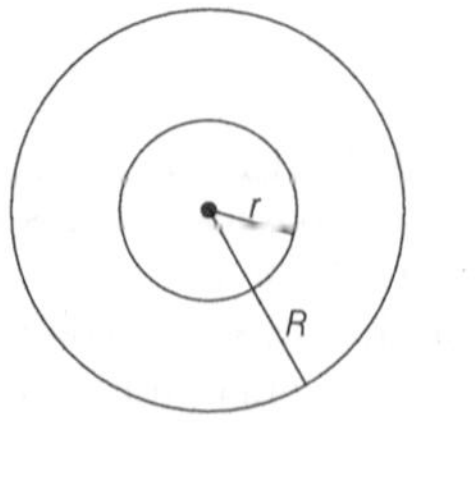

(a) $\frac{1}{4\pi\varepsilon_0}\frac{(R+r)}{2(R^2+r^2)}Q$ (b) $\frac{1}{4\pi\varepsilon_0}\frac{(2R+r)}{(R^2+r^2)}Q$

(c) $\frac{1}{4\pi\varepsilon_0}\frac{(R+2r)Q}{2(R^2+r^2)}$ (d) $\frac{1}{4\pi\varepsilon_0}\frac{(R+r)}{(R^2+r^2)}Q$

51. Ten charges are placed on the circumference of a circle of radius R with constant angular separation between successive charges. Alternate charges 1, 3, 5, 7, 9 have charge $+q$ each, while 2, 4, 6, 8, 10 have charge $-q$ each. The potential V and the electric field E at the centre of the circle respectively, are (Take, $V = 0$ at infinity) **[JEE Main 2020]**

(a) $V = \frac{10q}{4\pi\varepsilon_0 R}$; $E = 0$ (b) $V = 0$; $E = \frac{10q}{4\pi\varepsilon_0 R^2}$

(c) $V = 0$; $E = 0$ (d) $V = \frac{10q}{4\pi\varepsilon_0 R}$; $E = \frac{10q}{4\pi\varepsilon_0 R^2}$

52. A positively charged particle is released from rest in a uniform electric field. The electric potential energy of the charge **[NCERT Exemplar]**

(a) remains a constant because the electric field is uniform

(b) increases because the charge moves along the electric field

(c) decreases because the charge moves along the electric field

(d) decreases because the charge moves opposite to the electric field

53. The work done by electric field done during the displacement of a negatively charged particle towards a fixed positively charged particle is 9 J. As a result, the distance between the charges has been decreased by half. What work is done by the electric field over the first half of this distance?

(a) 3 J (b) 6 J

(c) 1.5 J (d) 9 J

54. A charge 5 μC is placed at a point. What is the work required to carry 1C of charge once round it in a circle of 12 cm radius?

(a) 100 (b) 0 (c) 1 (d) ∞

55. In the electric field of a point charge q at a certain point charge is carried from point A to B, C, D and E as shown in figure. The work done is

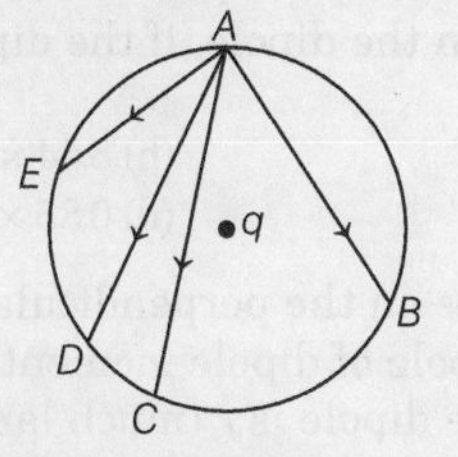

(a) least along the path AE

(b) least along the path AC

(c) zero along any of the paths

(d) least along AB

56. Equipotentials at a great distance from a collection of charges whose total sum is not zero are approximately **[NCERT Exemplar]**

(a) spheres (b) planes

(c) paraboloids (d) ellipsoids

57. An electric field is given by $\mathbf{E} = (y\hat{\mathbf{i}} + x\hat{\mathbf{j}})$ NC^{-1}. The work done (in joule) in moving a 1 C charge from $\mathbf{r}_A = (2\hat{\mathbf{i}} + 2\hat{\mathbf{j}})$ m to $\mathbf{r}_B$ = (in joule $4\hat{\mathbf{i}} + 2\hat{\mathbf{j}}$) m is

(a) 2 y (b) 3 y

(c) zero (d) infinity

58. Figures shows some equipotential lined distributed in space. A charged object is moved from point A to point B. **[NCERT Exemplar]**

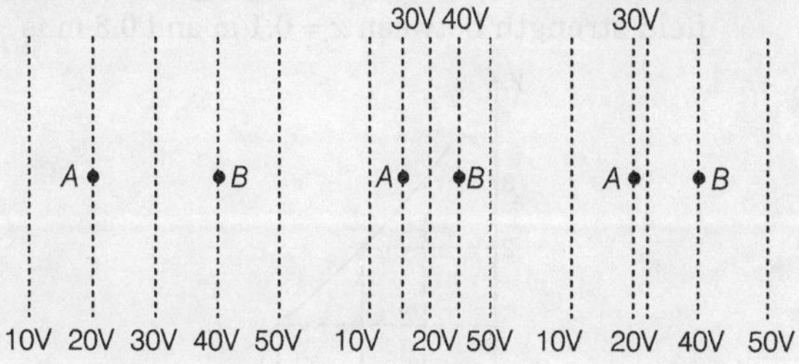

Fig.(i) Fig.(ii) Fig.(iii)

(a) The work done in Fig. (i) is the greatest

(b) The work done in Fig. (ii) is least

(c) The work done in Fig. (i), Fig. (ii) and Fig. (iii) is equal

(d) The work done in Fig. (iii) is greater than Fig. (ii) but equal to that in Fig. (i)

59. The potential at a point x (measured in μm) due to some charges situated on the X-axis is given by $V(x) = 20/(x^2 - 4)$ volt.
The electric field E at $x = 4$ μm is given by **[AIEEE 2007]**

(a) $\frac{5}{3}$ V/μm and in the –ve x-direction

(b) $\frac{5}{3}$ V/μm and in the +ve x-direction

(c) $\frac{10}{9}$ V/μm and in the –ve x-direction

(d) $\frac{10}{9}$ V/μm and in the +ve x-direction

60. A test charge q_0 is moved without acceleration from A to C and covers the path ABC as shown in figure. The potential difference between A and C is

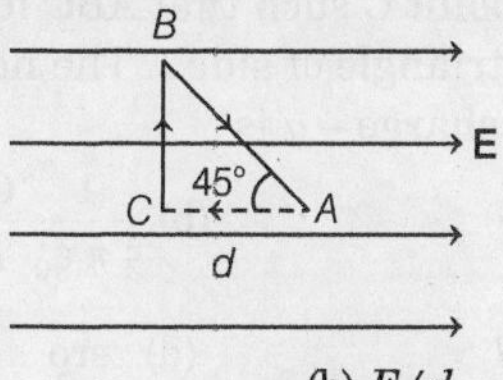

(a) Ed (b) E/d

(c) 2 Ed (d) $Ed/2$

Electric Potential and Electric Potential Energy

61. A non-conducting ring of radius 0.5 m carries total charge of 1.11×10^{-10} C distributed non-uniformly on its circumference producing an electric field everywhere in space.

The value of the line integral $\oint_{l=\infty}^{l=0} -E \cdot dl$ ($l = 0$, being centre of ring) in volt is

(a) + 2 (b) −1
(c) − 2 (d) zero

62. The variation of potential V with distance x from a fixed point charge is shown in figure. The electric field strength between $x = 0.1$ m and 0.3 m is

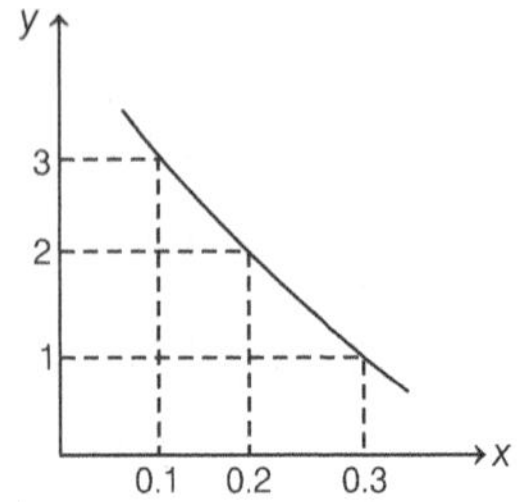

(a) + 0.4 Vm^{-1} (b) − 0.4 Vm^{-1}
(c) + 10 Vm^{-1} (d) − 10 Vm^{-1}

63. The electric potential V at any point (x, y, z) in space is given by $V = 4x^2$. The electric field at (1, 0, 2) m in Vm^{-1} is
(a) 8, along negative X-axis
(b) 8, along positive X-axis
(c) 16, along negative X-axis
(d) 16, along positive Z-axis

64. A ball of mass 1 kg carrying a charge 10^{-8} C moves from a point A at potential 600 V to a point B at zero potential. The change in its kinetic energy is
(a) -6×10^{-6} erg
(b) -6×10^{-6} J
(c) 6×10^{-6} J
(d) 6×10^{-6} erg

65. A charge $-q$ and another charge $+Q$ are kept at two points A and B, respectively. Keeping the charge $+Q$ fixed at B, the charge $-q$ at A is moved to another point C such that ABC forms an equilateral triangle of side l. The net work done in moving the charge $-q$ is

(a) $\frac{1}{4\pi\varepsilon_0}\frac{Qq}{l}$ (b) $\frac{1}{4\pi\varepsilon_0}\frac{Qq}{l^2}$

(c) $\frac{1}{4\pi\varepsilon_0}Qql$ (d) zero

66. A particle of mass m and charge q is released from rest in a uniform electric field. If there is no other force on the particle, the dependence of its speed v on the distance x travelled by it is correctly given by (graphs are schematic and not drawn to scale) **[JEE Main 2020]**

(a)
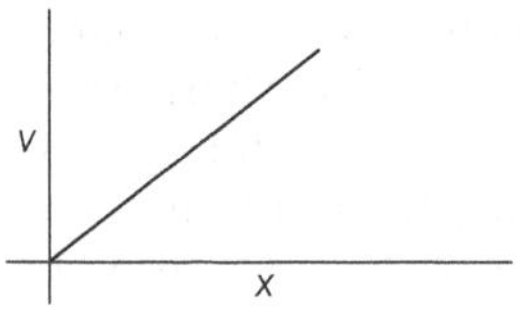

(b)
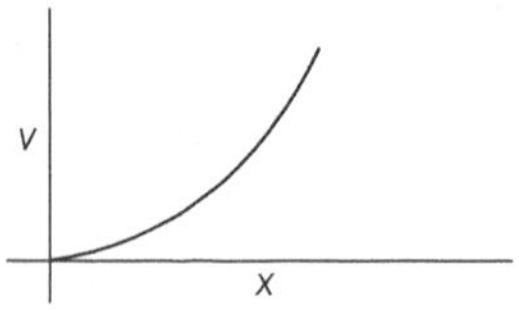

(c)
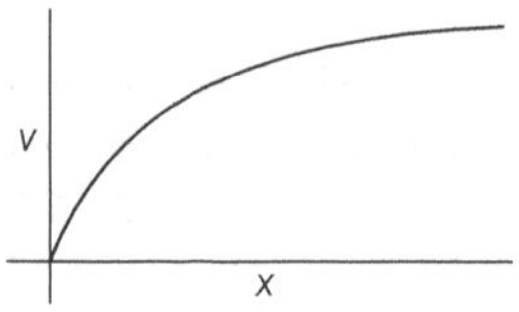

(d)
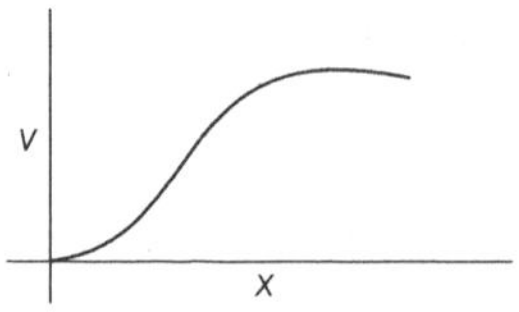

67. Figure shows electric field lines in which an electric dipole **p** is placed as shown. Which of the following statements is correct? **[NCERT Exemplar]**

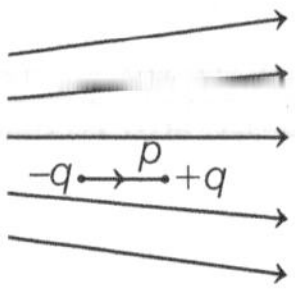

(a) The dipole will not experience any force.
(b) The dipole will experience a force towards left.
(c) The dipole will experience a force towards right.
(d) The dipole will experience a force upwards.

68. An electric dipole is placed at an angle of 60° with an electric field of intensity 10^5 NC^{-1}. It experiences a torque equal to $8\sqrt{3}$ N-m. Calculate the charge on the dipole, if the dipole length is 2 cm.
(a) -8×10^3 C (b) 8.54×10^{-4} C
(c) 8×10^{-3} C (d) 0.85×10^{-6} C

69. A point Q lies on the perpendicular bisector of an electrical dipole of dipole moment p. If the distance of Q from the dipole is r (much larger than the size of the dipole), then the electric field intensity E at Q is proportional to
(a) r^{-2} (b) r^{-4} (c) r^{-1} (d) r^{-3}

70. A given charge situated at a certain distance from an electric dipole in the end on position, experiences a force F. If the distance of charge is doubled, the force acting on the charge will be
(a) $2F$ (b) $F/2$
(c) $F/4$ (d) $F/8$

71. Two point charges of 1 μC and −1 μC are separated by a distance of 100 Å. A point P is at a distance of 10 cm from the mid-point and on the perpendicular bisector of the line joining the two charges. The electric field at P will be
(a) 9 NC^{-1} (b) 0.9 Vm^{-1}
(c) 90 Vm^{-1} (d) 0.09 NC^{-1}

72. A point dipole $\mathbf{p} = -p_0\,\hat{\mathbf{x}}$ is kept at the origin. The potential and electric field due to this dipole on the Y-axis at a distance d are respectively (Take, $V = 0$ at infinity) [JEE Main 2019]
(a) $\frac{|\mathbf{p}|}{4\pi\varepsilon_0 d^2}, \frac{\mathbf{p}}{4\pi\varepsilon_0 d^3}$ (b) $0, \frac{-\mathbf{p}}{4\pi\varepsilon_0 d^3}$
(c) $0, \frac{\mathbf{p}}{4\pi\varepsilon_0 d^3}$ (d) $\frac{|\mathbf{p}|}{4\pi\varepsilon_0 d^2}, \frac{-\mathbf{p}}{4\pi\varepsilon_0 d^3}$

73. Two electric dipoles A, B with respective dipole moments $\mathbf{d}_A = -4qa\,\hat{\mathbf{i}}$ and $\mathbf{d}_B = -2qa\,\hat{\mathbf{i}}$ are placed on the X-axis with a separation R, as shown in the figure

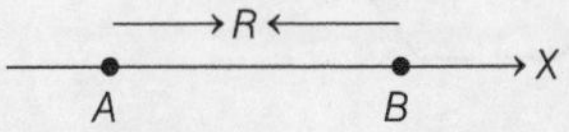

The distance from A at which both of them produce the same potential is [JEE Main 2019]
(a) $\frac{\sqrt{2}R}{\sqrt{2}+1}$ (b) $\frac{\sqrt{2}R}{\sqrt{2}-1}$
(c) $\frac{R}{\sqrt{2}+1}$ (d) $\frac{R}{\sqrt{2}-1}$

74. An electric dipole has a fixed dipole moment $\mathbf{p}$, which makes angle θ with respect to X-axis. When subjected to an electric field $\mathbf{E}_1 = E\hat{\mathbf{i}}$, it experiences a torque $\mathbf{T}_1 = \tau\hat{\mathbf{k}}$. When subjected to another electric field $\mathbf{E}_2 = \sqrt{3}E_1\hat{\mathbf{j}}$, it experiences a torque $\mathbf{T}_2 = -\mathbf{T}_1$. The angle θ is [JEE Main 2017]
(a) 45° (b) 60°
(c) 90° (d) 30°

75. Two electric dipoles of moment p and $64\,p$ are placed in opposite direction on a line at a distance of 25 cm. The electric field will be zero at point between the dipoles whose distance from dipole of moment p is
(a) 10 cm (b) 5 cm
(c) 8 cm (d) 20 cm

76. Two identical electric point dipoles have dipole moments $\mathbf{p}_1 = p\,\hat{\mathbf{i}}$ and $\mathbf{p}_2 = -p\,\hat{\mathbf{i}}$ are held on the X-axis at distance a from each other. When released, they move along the X-axis with the direction of their dipole moments remaining unchanged. If the mass of each dipole is m, their speed when they are infinitely far apart is [JEE Main 2020]
(a) $\frac{p}{a}\sqrt{\frac{1}{\pi\varepsilon_0 ma}}$ (b) $\frac{p}{a}\sqrt{\frac{1}{2\pi\varepsilon_0 ma}}$
(c) $\frac{p}{a}\sqrt{\frac{2}{\pi\varepsilon_0 ma}}$ (d) $\frac{p}{a}\sqrt{\frac{3}{2\pi\varepsilon_0 ma}}$

Capacitors and Capacitance

77. A capacitor connected to a 10 V battery collects a charge of 40 μC with air as dielectric and 100 μC with a given oil as dielectric. The dielectric constant of the oil is
(a) 1.5 (b) 2.0
(c) 2.5 (d) 3.0

78. A slab of copper of thickness b is inserted in between the plates of parallel plate capacitor as shown in figure. The separation between the plates is d, if $b = d/2$, then the ratio of capacities of capacitors after and before inserting the slab will be

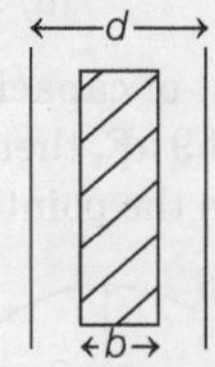

(a) $\sqrt{2}:1$ (b) $2:1$
(c) $1:1$ (d) $1:\sqrt{2}$

79. The capacitance of a spherical condensers is 1 μF. If the spacing between two spheres is 1 mm, the radius of the outer sphere is
(a) 3 m (b) 7 m
(c) 8 m (d) 9 m

80. Two identical metal plates are given positive charges Q_1 and Q_2 ($< Q_1$), respectively. If they are now brought close together to form a parallel plate capacitor with capacitance C, the potential difference between them is
(a) $\frac{Q_1 + Q_2}{2C}$ (b) $\frac{Q_1 + Q_2}{C}$
(c) $\frac{Q_1 - Q_2}{C}$ (d) $\frac{Q_1 - Q_2}{2C}$

81. A parallel plate capacitor is made of two dielectric blocks in series. One of the blocks has thickness d_1 and dielectric constant K_1 and the other has thickness d_2 and dielectric constant K_2 as shown in figure. This arrangement can be thought as a dielectric slab of thickness $d(= d_1 + d_2)$ and effective dielectric constant K. The K is **[NCERT Exemplar]**

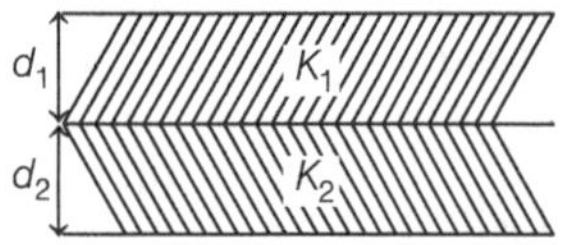

(a) $\dfrac{K_1d_1 + K_2d_2}{d_1 + d_2}$ (b) $\dfrac{K_1d_1 + K_2d_2}{K_1 + K_2}$

(c) $\dfrac{K_1K_2(d_1 + d_2)}{(K_2d_1 + K_1d_2)}$ (d) $\dfrac{2K_1K_2}{K_1 + K_2}$

82. The equivalent capacity between points A and B in figure will be, while capacitance of each capacitor is 3 μF.

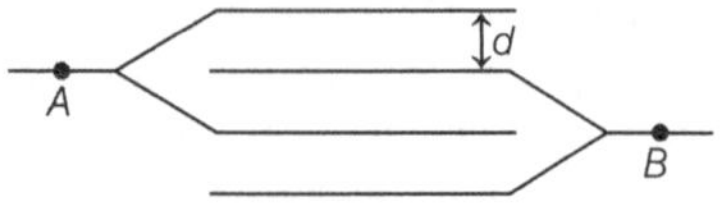

(a) 2 μF (b) 4 μF
(c) 7 μF (d) 9 μF

83. In the arrangement of capacitors shown in figure, each capacitor is of 9 μF, then the equivalent capacitance between the points A and B is

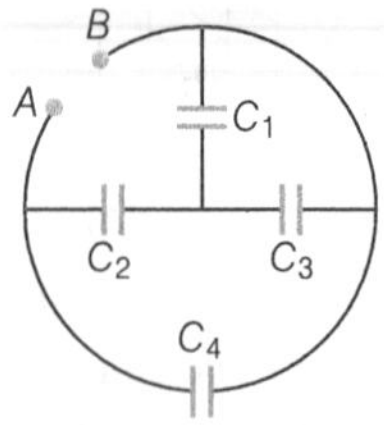

(a) 9 μF
(b) 18 μF
(c) 4.5 μF
(d) 15 μF

84. An electrical technician requires a capacitance of 2 μF in a circuit across a potential difference of 1 kV. A large number of 1 μF capacitors are available to him each of which can withstand a potential difference of not more than 400 V. Suggest a possible arrangement that requires the minimum number of capacitors. **[NCERT]**

(a) Six rows having 3 capacitors in each row
(b) Three rows having 6 capacitors in each row
(c) Nine rows having 2 capacitors in each row
(d) Two rows having 9 capacitors in each row

85. Seven capacitors each of the capacitance 2 μF are connected in a configuration to obtain an effective capacitance of $\frac{10}{11}$ μF. Which of the combination(s) shown in figure will achieve the desired result?

(a) 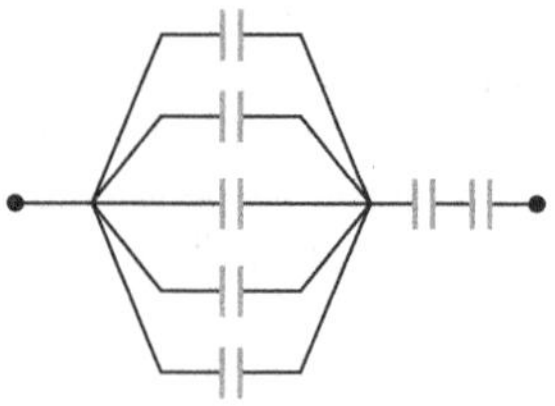

(b) 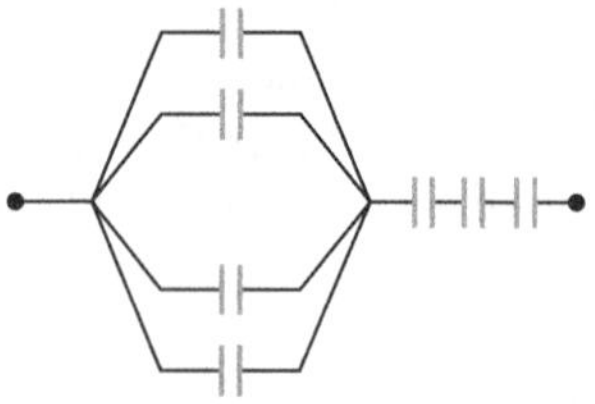

(c) 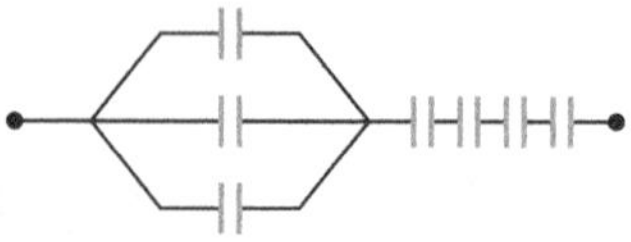

(d) 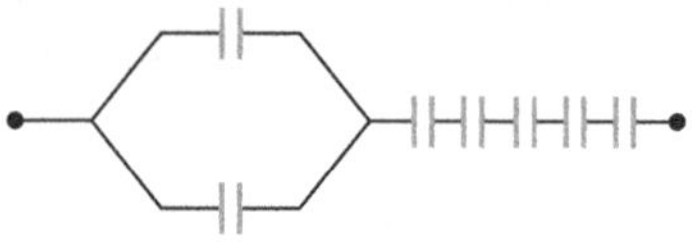

86. Consider the arrangement of three metal plates A, B and C of equal surface area and separation d as shown in figure. The energy stored in the arrangement, when the plates are fully charged, is

(a) $\dfrac{\varepsilon_0 AV^2}{2d}$ (b) $\dfrac{\varepsilon_0 AV^2}{d}$

(c) $\dfrac{2\varepsilon_0 AV^2}{d}$ (d) $\dfrac{3\varepsilon_0 AV^2}{2d}$

87. The equivalent capacitance of the combination of three capacitors, each of capacitance C as shown in figure between points A and B is

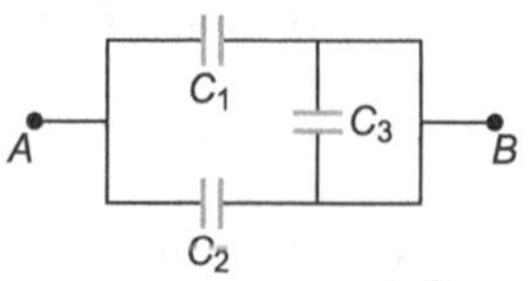

(a) $\dfrac{C}{2}$ (b) $\dfrac{3C}{2}$

(c) $\dfrac{1}{3C}$ (d) $2C$

88. Figure shows charge (q) *versus* voltage (V) graph for series and parallel combination of two given capacitors. The capacitances are [JEE Main 2019]

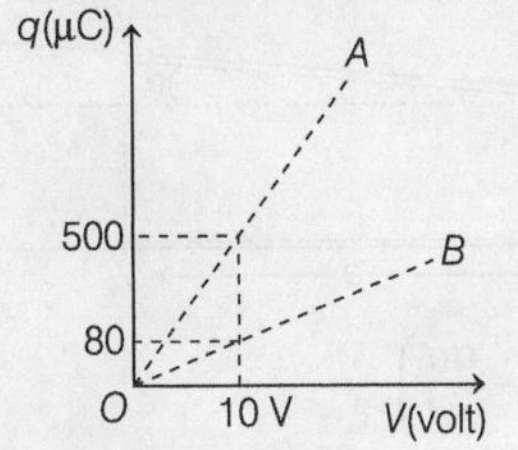

(a) 60 μF and 40 μF
(b) 50 μF and 30 μF
(c) 20 μF and 30 μF
(d) 40 μF and 10 μF

89. In the given circuit, the charge on 4 μF capacitor will be [JEE Main 2019]

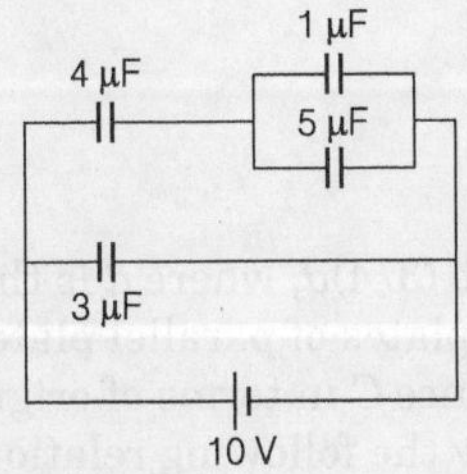

(a) 5.4 μC
(b) 9.6 μC
(c) 13.4 μC
(d) 24 μC

90. Two condenser one of capacity C and other of capacity $C/2$ are connected to 9V battery as shown in figure. The work done in charging fully both condensers is

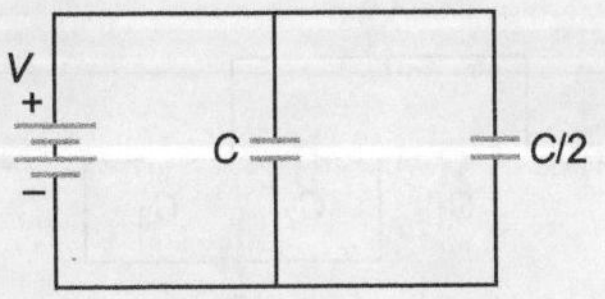

(a) $(1/4) CV^2$
(b) $2 CV^2$
(c) $(3/4) CV^2$
(d) $(1/2) CV^2$

91. Two identical capacitors have the same capacitance C. One of them is charged to potential V_1 and the other to V_2. The negative ends of the capacitors are connected together. When the positive ends are also connected, the decrease in energy of the system is

(a) $\frac{1}{4} C (V_1^2 - V_2^2)$ (b) $\frac{1}{4} C (V_1^2 + V_2^2)$

(c) $\frac{1}{4} C (V_1 - V_2)^2$ (d) $\frac{1}{4} C (V_1 + V_2)^2$

92. The force on each plate of parallel plate capacitor has a magnitude equal to $\frac{1}{2} QE$, where Q is the charge on the capacitor and E is the magnitude of electric field between the plates. Then, [NCERT]

(a) $\frac{E}{2}$ contributes to the force against which the plates are moved
(b) $\frac{E}{3}$ contributes to the force against which the plates are moved
(c) E contributes force against which the plates are moved
(d) None of the above

93. In the circuit shown in the figure, the total charge is 750 μC and the voltage across capacitor C_2 is 20 V. Then, the charge on capacitor C_2 is [JEE Main 2020]

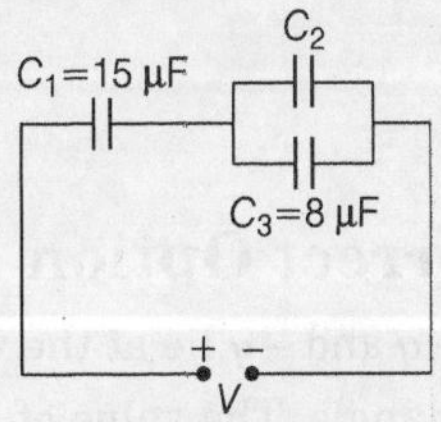

(a) 450 μC
(b) 590 μC
(c) 160 μC
(d) 650 μC

94. A capacitor C is fully charged with voltage V_0. After disconnecting the voltage source, it is connected in parallel with another uncharged capacitor of capacitance $\frac{C}{2}$. The energy loss in the process after the charge is distributed between the two capacitors is [JEE Main 2020]

(a) $\frac{1}{3} CV_0^2$
(b) $\frac{1}{6} CV_0^2$
(c) $\frac{1}{2} CV_0^2$
(d) $\frac{1}{4} CV_0^2$

95. Two identical parallel plate capacitors of capacitance C each, have plates of area A, separated by a distance d. The space between the plates of the two capacitors, is filled with three dielectrics of equal thickness and dielectric constants K_1, K_2 and K_3. The first capacitor is filled as shown in Fig. I, and the second one is filled as shown in Fig. II. If these two modified capacitors are charged by the same potential V, the ratio of the energy stored in the two, would be (E_1 refers to capacitor (I) and E_2 to capacitor (II)) : [JEE Main 2019]

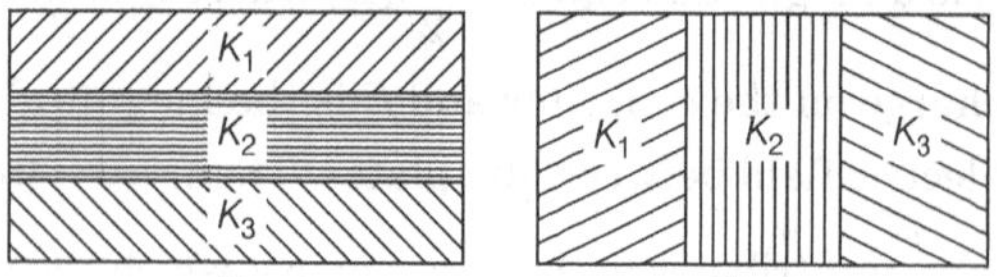

(I) (II)

(a) $\dfrac{E_1}{E_2}=\dfrac{K_1K_2K_3}{(K_1+K_2+K_3)(K_2K_3+K_3K_1+K_1K_2)}$

(b) $\dfrac{E_1}{E_2}=\dfrac{(K_1+K_2+K_3)(K_2K_3+K_3K_1+K_1K_2)}{K_1K_2K_3}$

(c) $\dfrac{E_1}{E_2}=\dfrac{9K_1K_2K_3}{(K_1+K_2+K_3)(K_2K_3+K_3K_1+K_1K_2)}$

(d) $\dfrac{E_1}{E_2}=\dfrac{(K_1+K_2+K_3)(K_2K_3+K_3K_1+K_1K_2)}{9K_1K_2K_3}$

96. A capacitor is made of two square plates each of side a making a very small angle α between them as shown in figure. The capacitance will be close to [JEE Main 2020]

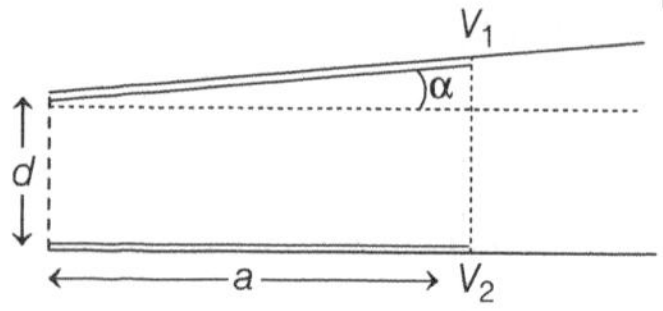

(a) $\dfrac{\varepsilon_0 a^2}{d}\left(1-\dfrac{\alpha a}{4d}\right)$

(b) $\dfrac{\varepsilon_0 a^2}{d}\left(1+\dfrac{\alpha a}{d}\right)$

(c) $\dfrac{\varepsilon_0 a^2}{d}\left(1-\dfrac{\alpha a}{2d}\right)$

(d) $\dfrac{\varepsilon_0 a^2}{d}\left(1-\dfrac{3\alpha a}{2d}\right)$

ROUND II Mixed Bag

Only One Correct Option

1. Charges $2q$, $-q$ and $-q$ lie at the vertices of an equilateral triangle. The value of E and V at the centroid of the triangle will be

(a) $E\neq 0$ and $V\neq 0$ (b) $E=0$ and $V=0$
(c) $E\neq 0$ and $V=0$ (d) $E=0$ and $V\neq 0$

2. If a positively charged pendulum is oscillating in a uniform electric field as shown in figure. Its time period as compared to that when it was uncharged will

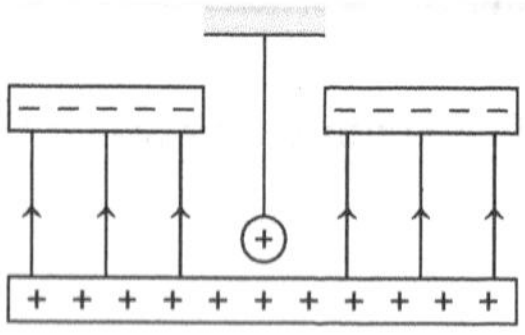

(a) increase
(b) decrease
(c) not change
(d) first increase and then decrease

3. Two particles of equal mass m and charge q are placed at a distance of 16 cm. They do not experience any force. The value of $\dfrac{q}{m}$ is

(a) $\sqrt{\dfrac{\pi\varepsilon_0}{G}}$ (b) $\sqrt{\dfrac{G}{\pi\varepsilon_0}}$

(c) $\sqrt{4\pi\varepsilon_0 G}$ (d) l

4. For changing the capacitance of a given parallel plate capacitor, a dielectric material of dielectric constant K is used, which has the same area as the plates of the capacitor. The thickness of the dielectric slab $(3/4)d$, where d is the separation between the plates of parallel plate capacitor. The new capacitance C in terms of original capacitance C_0 is given by the following relation [JEE Main 2021]

(a) $C'=\dfrac{3+K}{4K}C_0$ (b) $C'=\dfrac{4+K}{3}C_0$

(c) $C'=\dfrac{4K}{K+3}C_0$ (d) $C'=\dfrac{4}{3+K}C_0$

5. The effective capacitance between points X and Y shown in figure is (Assuming, $C_2=10\,\mu F$ and that outer capacitors are all $4\,\mu F$)

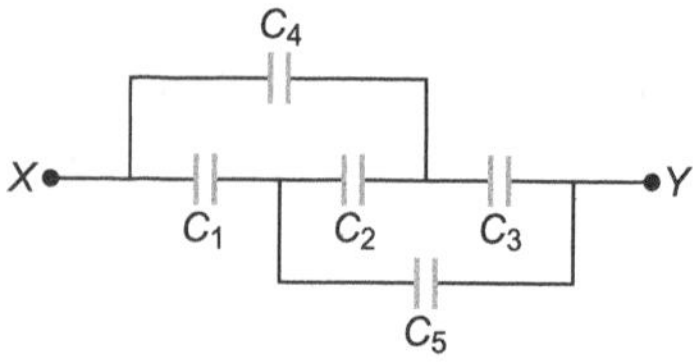

(a) $1\,\mu F$ (b) $3\,\mu F$ (c) $4\,\mu F$ (d) $5\,\mu F$

6. The four capacitors, each of $25\,\mu F$ are connected as shown in figure. The DC voltmeter reads 200 V. The charge on each plate of capacitor is

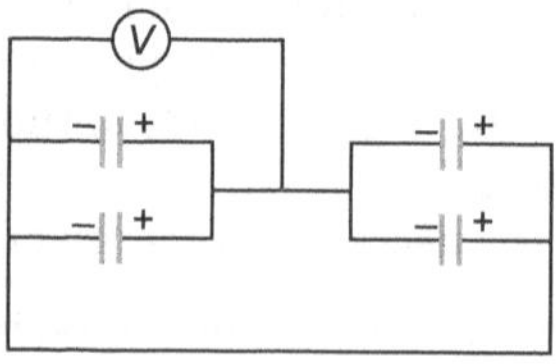

(a) $\pm 2\times 10^{-3}$ C (b) $\pm 5\times 10^{-3}$ C
(c) $\pm 2\times 10^{-2}$ C (d) $\pm 5\times 10^{-2}$ C

7. Three charges $-q_1$, $+q_2$ and $-q_3$ are placed as shown in figure. The x-component of the force on $-q_1$ is proportional to

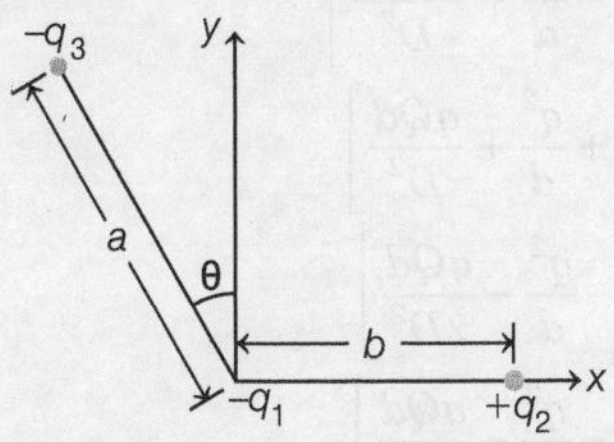

(a) $\frac{q_2}{b^2} - \frac{q_3}{a^2}\sin\theta$ (b) $\frac{q_2}{b^2} - \frac{q_3}{a^2}\cos\theta$

(c) $\frac{q_2}{b^2} + \frac{q_3}{a^2}\sin\theta$ (d) $\frac{q_2}{b^2} + \frac{q_3}{a^2}\cos\theta$

8. Three large parallel plates have uniform surface charge densities as shown in the figure. Find the electric field at point P.

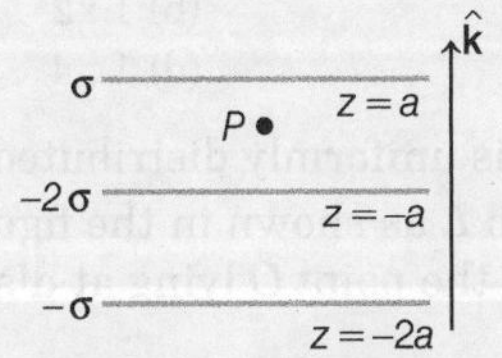

(a) $\frac{-4\sigma}{\varepsilon_0}\hat{k}$ (b) $\frac{4\sigma}{\varepsilon_0}\hat{k}$

(c) $\frac{-2\sigma}{\varepsilon_0}\hat{k}$ (d) $\frac{2\sigma}{\varepsilon_0}\hat{k}$

9. A 100 eV electron is fired directly towards a large metal plate having surface charge density $2\times10^{-6}\,\text{cm}^{-2}$. The distance from where the electrons be projected so that it just fails to strike the plate is

(a) 0.22 mm (b) 0.44 mm

(c) 0.66 mm (d) 0.88 mm

10. A uniform electric field pointing in positive x-direction exists in a region. Let A be the origin, B be the point on the X-axis at $x = +1$ cm and C be the point on the Y-axis at $y = +1$ cm. Then, the potentials at the points A, B and C satisfy the condition

(a) $V_A < V_B$ (b) $V_A > V_B$

(c) $V_A < V_C$ (d) $V_A > V_C$

11. A solid conducting sphere having a charge Q is surrounded by an uncharged concentric conducting hollow spherical shell. Let the potential difference between the surface of the solid sphere and that of the outer surface of the hollow shell be V. If the shell is now given a charge $-3Q$, the new potential difference between the same two surfaces is

(a) V (b) $2V$

(c) $4V$ (d) $-2V$

12. For the circuit shown in figure, which of the following statements is true?

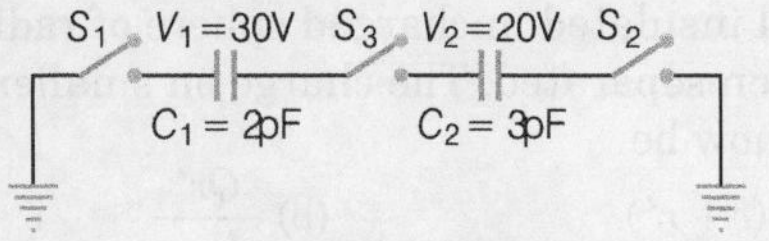

(a) With S_1 closed, $V_1 = 15\text{V}$, $V_2 = 20\text{V}$.

(b) With S_3 closed, $V_1 = V_2$, $V_2 = 20\text{V}$.

(c) With S_1 and S_3 closed, $V_1 = V_2 = 0$.

(d) With S_1 and S_3 closed, $V_1 = 30\text{V}$, $V_2 = 20\text{V}$.

13. Consider the force F on a charge q due to a uniformly charged spherical shell of radius R carrying charge Q distributed uniformly over it. Which one of the following option is correct for F, if q is placed at distance r from the centre of the shell? **[JEE Main 2020]**

(a) $F = \frac{1}{4\pi\varepsilon_0}\frac{Qq}{R^2}$ (for $r < R$)

(b) $\frac{1}{4\pi\varepsilon_0}\frac{Qq}{R^2} > F > 0$ (for $r < R$)

(c) $F = \frac{1}{4\pi\varepsilon_0}\frac{Qq}{r^2}$ (for $r > R$)

(d) $F = \frac{1}{4\pi\varepsilon_0}\frac{Qq}{R^2}$ (for all r)

14. Point charges 4×10^{-6} C and 2×10^{-6} C are placed at the vertices A and B of a right angled triangle ABC, respectively. B is the right angle, $AC = 2\times10^{-2}$ m and $BC = 10^{-2}$ m. Find the magnitude and direction of the resultant electric intensity at C.

(a) $1.73\times10^4\ \text{NC}^{-1}$; 34 .7° (b) $2.38\times10^8\ \text{NC}^{-1}$; 40.9°

(c) $4.28\times10^9\ \text{NC}^{-1}$; 45° (d) $4.9\times10^{10}\ \text{NC}^{-1}$; 34.7°

15. A parallel plate capacitor of area A, plate separation d and capacitance C is filled with three different dielectric materials having dielectric constants K_1, K_2 and K_3 as shown in figure. If a single dielectric material is to be used to have the same capacitance C is this capacitors, then its dielectric constant K is given by

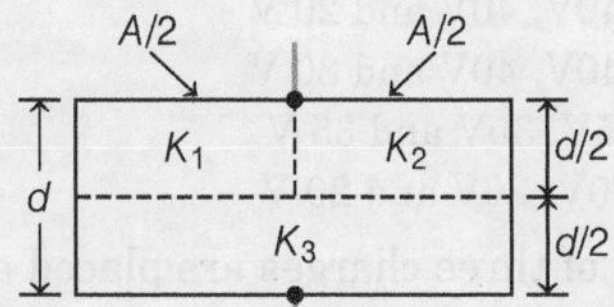

(a) $\frac{1}{K} = \frac{1}{K_1} + \frac{1}{K_2} + \frac{1}{2K_3}$ (b) $\frac{1}{K} = \frac{1}{K_1 + K_2} + \frac{1}{2K_3}$

(c) $K = \frac{K_1 K_2}{K_1 + K_2} + 2K_3$ (d) $K = K_1 + K_2 + 2K_3$

16. A large insulated sphere of radius r charged with Q units of electricity is placed in contact with a small insulated uncharged sphere of radius r' and in then separated. The charge on smaller sphere will now be

(a) $Q(r + r')$ (b) $\frac{Qr'}{r' + r}$

(c) $Q(r + r')$ (d) $\frac{Q}{r' + r}$

17. Two charged thin infinite plane sheets of uniform surface charge densities σ_+ and σ_-, where $|\sigma_+| > |\sigma_-|$, intersect at right angle. Which of the following best represents the electric field lines for this system? [JEE Main 2020]

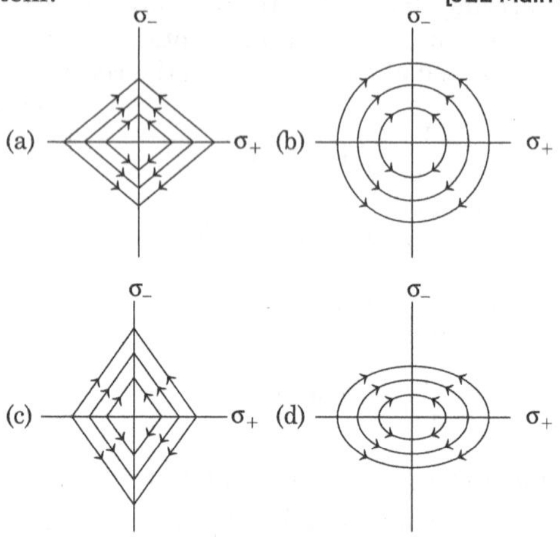

18. In the circuit arrangement shown in figure, the value of $C_1 = C_2 = C_3 = 30$ pF and $C_4 = 120$ pF. If the combination of capacitors is charged with 140V DC supply, the potential differences across the four capacitors will be respectively

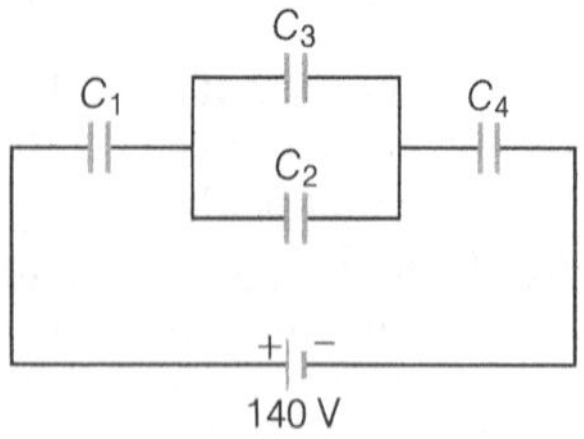

(a) 80 V, 40V, 40V and 20 V
(b) 20V, 40V, 40V and 80 V
(c) 35V, 35V, 35V and 35 V
(d) 80V, 20V, 20V and 20 V

19. A system of three charges are placed as shown in the figure

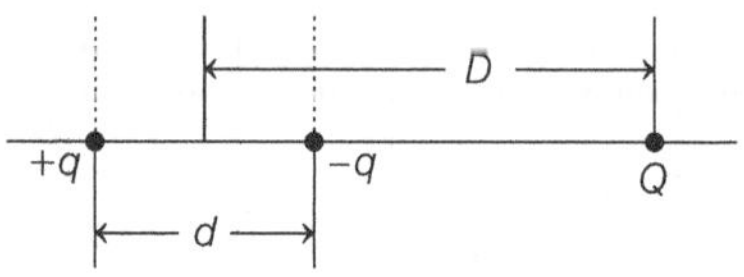

If $D >> d$, the potential energy of the system is best given by [JEE Main 2019]

(a) $\frac{1}{4\pi\varepsilon_0}\left[-\frac{q^2}{d} + \frac{2qQd}{D^2}\right]$

(b) $\frac{1}{4\pi\varepsilon_0}\left[+\frac{q^2}{d} + \frac{qQd}{D^2}\right]$

(c) $\frac{1}{4\pi\varepsilon_0}\left[-\frac{q^2}{d} - \frac{qQd}{2D^2}\right]$

(d) $\frac{1}{4\pi\varepsilon_0}\left[-\frac{q^2}{d} - \frac{qQd}{D^2}\right]$

20. Two spherical conductors A and B of radii 1 mm and 2 mm are separated by a distance of 5 cm and are uniformly charged. If the spheres are connected by a conducting wire, then in equilibrium condition, the ratio of the magnitude of the electric fields at the surfaces of spheres A and B is

(a) 4 : 1 (b) 1 : 2
(c) 2 : 1 (d) 1 : 4

21. A charge Q is uniformly distributed over a long rod AB of length L as shown in the figure. The electric potential at the point O lying at distance L from the end A is [JEE Main 2013]

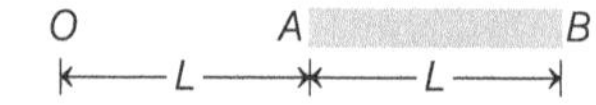

(a) $\frac{Q}{8\pi\varepsilon_0 L}$ (b) $\frac{3Q}{4\pi\varepsilon_0 L}$

(c) $\frac{Q}{4\pi\varepsilon_0 L \ln 2}$ (d) $\frac{Q \ln 2}{4\pi\varepsilon_0 L}$

22. Concentric metallic hollow spheres of radii R and $4R$ hold charges Q_1 and Q_2, respectively. Given that, surface charge densities of the concentric spheres are equal. The potential difference $V(R) - V(4R)$ is [JEE Main 2020]

(a) $\frac{3Q_2}{4\pi\varepsilon_0 R}$ (b) $\frac{3Q_1}{16\pi\varepsilon_0 R}$

(c) $\frac{3Q_1}{4\pi\varepsilon_0 R}$ (d) $\frac{Q_2}{4\pi\varepsilon_0 R}$

23. A 2μF capacitor is charged as shown in figure. The percentage of the stored energy dissipated after the switch S is turned to position 2 is

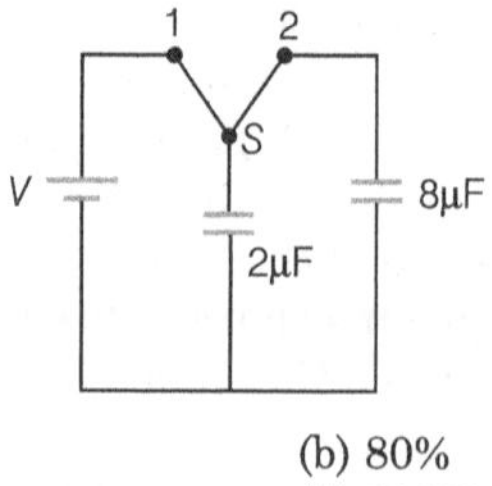

(a) 20% (b) 80%
(c) 10% (d) 100%

24. Two parallel plane sheets 1 and 2 carry uniform charge densities σ_1 and σ_2, as shown in figure. The magnitude of the resultant electric field in the region marked III is $(\sigma_1 > \sigma_2)$

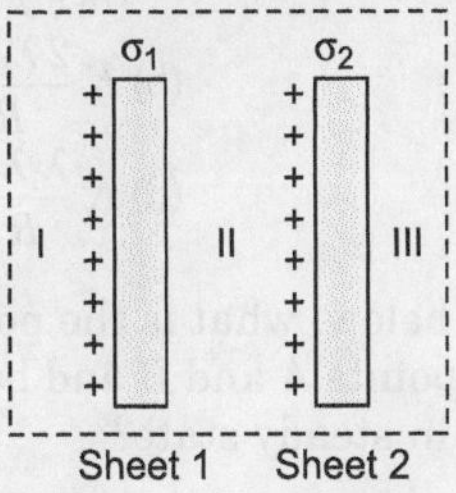

(a) $\dfrac{\sigma_1 + \sigma_2}{2\varepsilon_0}$ (b) $\dfrac{\sigma_1 \sigma_2}{\varepsilon_0}$

(c) $\dfrac{\sigma_1}{2\sigma_2 \varepsilon_0}$ (d) $\dfrac{\sigma_2}{2\varepsilon_0}$

25. A simple pendulum of length L is placed between the plates of a parallel plate capacitor having electric field E, as shown in figure. Its bob has mass m and charge q. The time period of the pendulum is given by **[JEE Main 2019]**

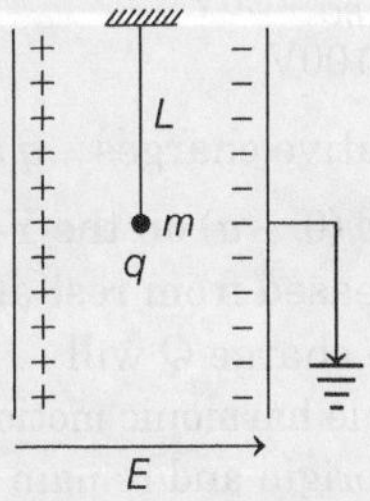

(a) $2\pi\sqrt{\dfrac{L}{\sqrt{g^2 + \left(\dfrac{qE}{m}\right)^2}}}$ (b) $2\pi\sqrt{\dfrac{L}{\sqrt{g^2 - \dfrac{q^2E^2}{m^2}}}}$

(c) $2\pi\sqrt{\dfrac{L}{\left(g + \dfrac{qE}{m}\right)}}$ (d) $2\pi\sqrt{\dfrac{L}{\left(g - \dfrac{qE}{m}\right)}}$

26. Charges $-q$ and $+q$ located at A and B respectively, constitute an electric dipole. Distance $AB = 2a$, O is the mid-point of the dipole and OP is perpendicular to AB. A charge Q is placed at P, where $OP = y$ and $y >> 2a$. The charge Q experiences an electrostatic force F.

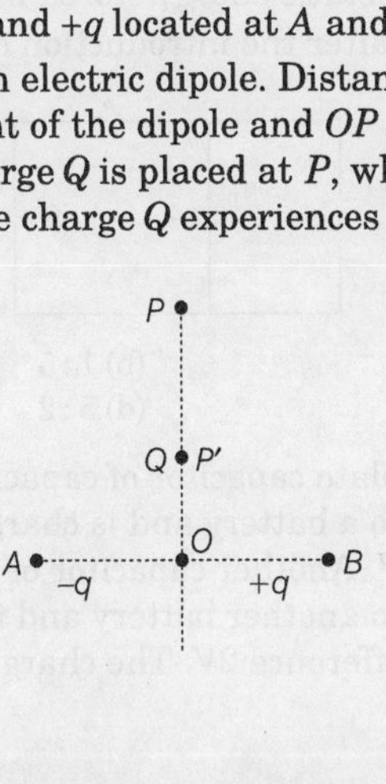

If Q is now moved along the equatorial line to P' such that $OP' = \left(\dfrac{y}{3}\right)$, the force on Q will be close to $\left(\dfrac{y}{3} >> 2a\right)$ **[JEE Main 2019]**

(a) $\dfrac{F}{3}$ (b) $3F$ (c) $9F$ (d) $27F$

27. A parallel plate capacitor is made of two square plates of side a separated by a distance d $(d<<a)$. The lower triangular portions is filled with a dielectric of dielectric constant K, as shown in the figure. Capacitance of this capacitor is **[JEE Main 2019]**

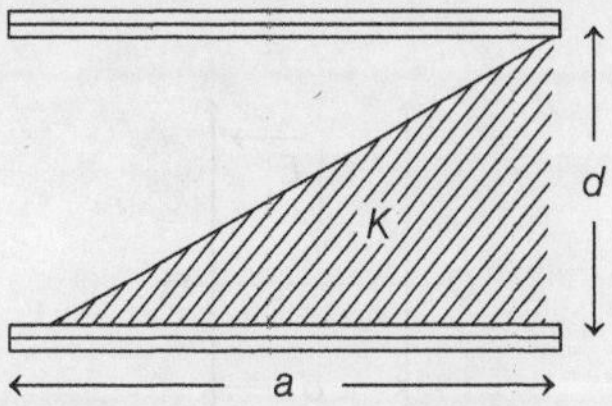

(a) $\dfrac{K\varepsilon_0 a^2}{d} \ln K$ (b) $\dfrac{K\varepsilon_0 a^2}{d(K-1)} \ln K$

(c) $\dfrac{K\varepsilon_0 a^2}{2d(K+1)}$ (d) $\dfrac{1}{2} \cdot \dfrac{K\varepsilon_0 a^2}{d}$

28. Seven capacitors, each of capacitance $2\,\mu F$ are to be connected in a configuration to obtain an effective capacitance of $\left(\dfrac{6}{13}\right)\mu F$. Which of the combinations shown in figures below will achieve the desired value? **[JEE Main 2019]**

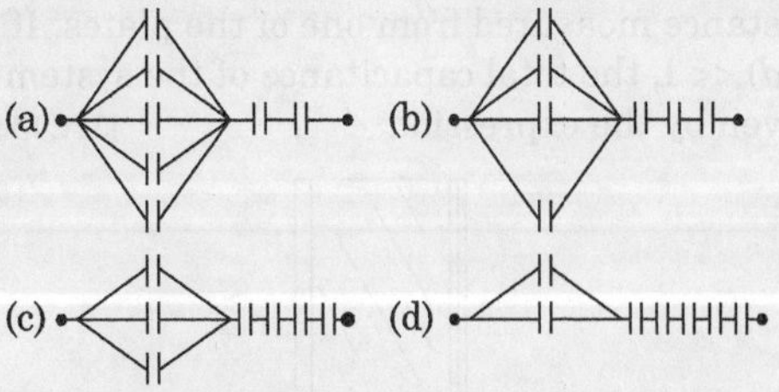

29. A parallel plate capacitor has plate of length l, width w and separation of plates is d. It is connected to a battery of emf V. A dielectric slab of the same thickness d and of dielectric constant $K = 4$ is being inserted between the plates of the capacitor. At what length of the slab inside plates, will the energy stored in the capacitor be two times the initial energy stored? **[JEE Main 2020]**

(a) $\dfrac{2l}{3}$ (b) $\dfrac{l}{3}$ (c) $\dfrac{l}{4}$ (d) $\dfrac{l}{2}$

30. The parallel combination of two air filled parallel plate capacitors of capacitance C and nC is connected to a battery of voltage V. When the capacitors are fully charged, the battery is removed and after that a dielectric material of dielectric constant K is placed between the two plates of the first capacitor. The new potential difference of the combined system is [JEE Main 2019]

(a) $\dfrac{(n+1)V}{(K+n)}$ (b) $\dfrac{nV}{K+n}$

(c) V (d) $\dfrac{V}{K+n}$

31. A charged particle (mass m and charge q) moves along X-axis with velocity v_0. When it passes through the origin, it enters a region having uniform electric field $\mathbf{E} = -E\,\hat{\mathbf{j}}$ which extends upto $x = d$. Equation of path of electron in the region $(x > d)$ is [JEE Main 2020]

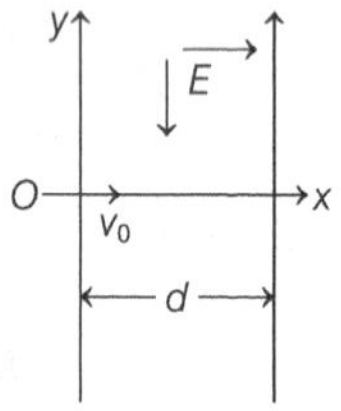

(a) $y = \dfrac{qEd}{mv_0^2}x$ (b) $y = \dfrac{qEd}{mv_0^2}(x-d)$

(c) $y = \dfrac{qEd}{mv_0^2}\left(\dfrac{d}{2}-x\right)$ (d) $y = \dfrac{qEd^2}{mv_0^2}x$

32. A parallel plate capacitor has plates of area A separated by distance d between them. It is filled with a dielectric which has a dielectric constant that varies as $k(x) = K(1+\alpha x)$, where x is the distance measured from one of the plates. If $(\alpha d) << 1$, the total capacitance of the system is best given by the expression [JEE Main 2020]

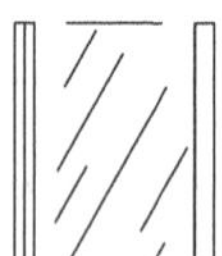

(a) $\dfrac{AK\varepsilon_0}{d}(1+\alpha d)$ (b) $\dfrac{A\varepsilon_0 K}{d}\left(1+\left(\dfrac{\alpha d}{2}\right)^2\right)$

(c) $\dfrac{AK\varepsilon_0}{d}\left(1+\dfrac{\alpha d}{2}\right)$ (d) $\dfrac{A\varepsilon_0 K}{d}\left(1+\dfrac{\alpha^2 d^2}{2}\right)$

33. If on the concentric hollow sphere of radii r and $R(> r)$ the charge Q is distributed such that their surface densities are same, then the potential at their common centre is

(a) $\dfrac{Q(R^2+r^2)}{4\pi\varepsilon_0(R+r)}$ (b) $\dfrac{Q(R+r)}{4\pi\varepsilon_0(R^2+r^2)}$

(c) zero (d) $\dfrac{QR}{R+r}$

34. Two infinitely long parallel wires having linear charge densities λ_1 and λ_2 respectively are placed at a distance of R metres. The force per unit length on either wire will be $\left(K = \dfrac{1}{4\pi\varepsilon_0}\right)$

(a) $K\dfrac{2\lambda_1\lambda_2}{R^2}$ (b) $K\dfrac{2\lambda_1\lambda_2}{R}$

(c) $K\dfrac{\lambda_1\lambda_2}{R^2}$ (d) $K\dfrac{\lambda_1\lambda_2}{R}$

35. In the figure below, what is the potential difference between the points A and B and between B and C respectively, in steady state?

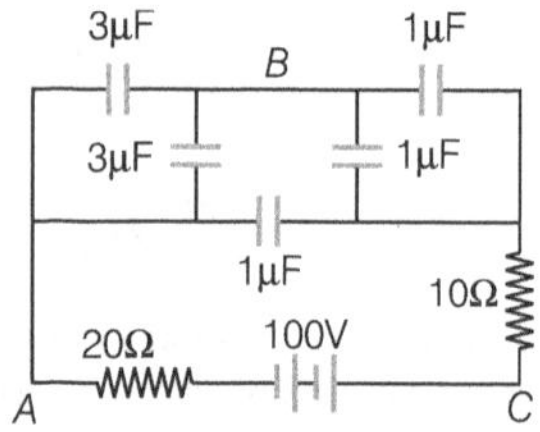

(a) $V_{AB} = V_{BC} = 50\text{V}$
(b) $V_{AB} = 25\text{V}, V_{BC} = 75\text{V}$
(c) $V_{AB} = 75\text{V}, V_{BC} = 25\text{V}$
(d) $V_{BC} = V_{AB} = 100\text{V}$

36. Two equal negative charges $-q$ are fixed at the points $(0, a)$ and $(0, -a)$ on the Y-axis. A positive charge Q is released from rest at the point $(2a, 0)$ on the X-axis. The charge Q will

(a) execute simple harmonic motion about the origin
(b) move to the origin and remain at rest
(c) move to infinity
(d) execute oscillatory but not simple harmonic motion

37. Figure given below shows two identical parallel plate capacitors connected to a battery with switch S closed. The switch is now opened and the free space between the plate of capacitors is filled with a dielectric constant 3. What will be the ratio of total electrostatic energy stored in both capacitors before and after the introduction of the dielectric?

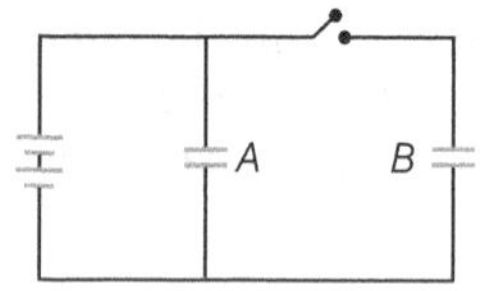

(a) 1 : 2 (b) 1 : 5
(c) 3 : 5 (d) 5 : 2

38. A parallel plate capacitor of capacitance C is connected to a battery and is charged to a potential difference V. Another capacitor of capacitance $2C$ is connected to another battery and is charged to potential difference $2V$. The charging batteries are now disconnected and the capacitors are connected in parallel to each other in such a way that the positive terminal of one is connected to the negative terminal of the other. The final energy of the configuration is

(a) infinite (b) zero

(c) $\frac{3CV^2}{2}$ (d) $\frac{6CV^2}{2}$

39. A point charge q moves from point P to point S along the path $PQRS$ in a uniform electric field $\mathbf{E}$ pointing parallel to the positive direction of the X-axis as shown in figure.

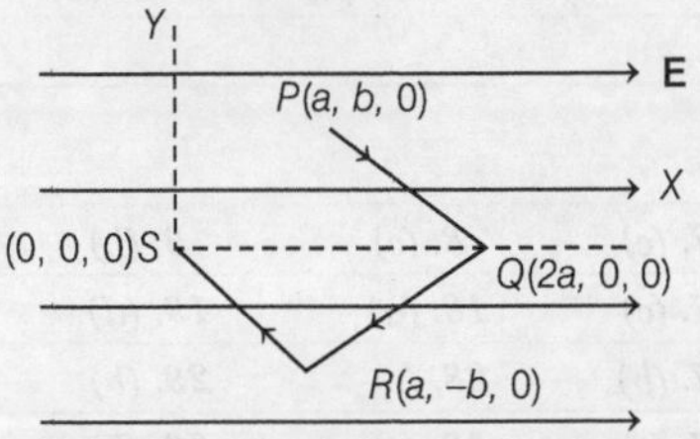

The coordinates of the points P, Q, R and S are $(a, b, 0)$ $(2a, 0, 0)$, $(a, -b, 0)$ and $(0, 0, 0)$, respectively. The work done by the field in the above process is given by the expression

(a) qE (b) $-qaE$

(c) $q(\sqrt{a^2+b^2})+E$ (d) $3qE(\sqrt{a^2+b^2})$

40. The electron is projected from a distance d and with initial velocity u parallel to a uniformly charged flat conducting plate as shown in figure. It strikes the plate after travelling a distance l along the direction. The surface charge density of conducting plate is equal to

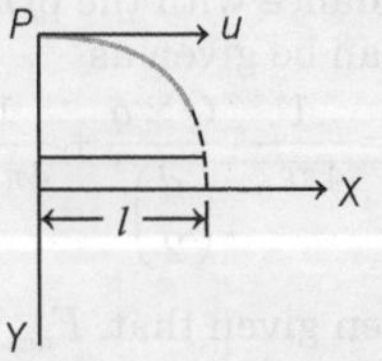

(a) $\frac{2d\varepsilon_0 m u^2}{el^2}$ (b) $\frac{2d\varepsilon_0 m u}{el}$

(c) $\frac{d\varepsilon_0 m u^2}{el}$ (d) $\frac{d\varepsilon_0 m u}{el}$

Numerical Value Questions

41. 27 identical drops of mercury are charged simultaneously to the same potential of 10 V each. Assuming drops to be spherical, if all the charged drops are made to combine to form one larger drop, then the potential (in volt) of larger drop would be ....

42. In a parallel plate capacitor set up, the plate area of capacitor is 2m^2 and the plates are separated by 1 m. If the space between the plates are filled with a dielectric material of thickness 0.5 m and area 2m^2 (see figure) the capacitance of the set up will be .... ε_0. (Dielectric constant of the material = 3.2)

(Round off to the nearest integer)

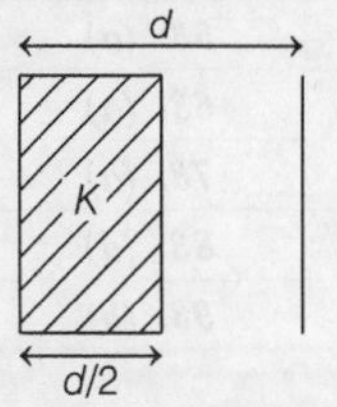

[JEE Main 2021]

43. A 10 μF capacitor and a 20 μF capacitor are connected in series across 200 V supply line. The charged capacitors are then disconnected from the line and reconnected with their positive plates together and negative plates together and no external voltage is applied. What is the potential difference (in volt) across each capacitor?

44. A 5 μF capacitor is charged fully by a 220V supply. It is then disconnected from the supply and is connected in series to another uncharged 2.5 μF capacitor. If the energy change during the charge redistribution is $\frac{X}{100}$ J, then value of X to the nearest integer is .... [JEE Main 2020]

45. A 60 pF capacitor is fully charged by a 20 V supply. It is then disconnected from the supply and is connected to another uncharged 60 pF capacitor in parallel. The electrostatic energy that is lost in this process by the time, the charge is redistributed between them is(in nJ) .... [JEE Main 2020]

46. A parallel plate capacitor whose capacitance C is 14pF is charged by a battery to a potential difference $V = 12\text{V}$ between its plates. The charging battery is now disconnected and a porcelain plate with $K = 7$ is inserted between the plates, then the plate would oscillate back and forth between the plates with a constant mechanical energy of .... pJ. (Assume no friction) [JEE Main 2021]

47. A parallel plate capacitor is made of two circular plates separated by a distance of 5 mm with a dielectric of dielectric constant 2.2 between them. When the electric field in the dielectric is 3×10^4 V/m, the charge density of the positive plate will be close to 6×10^{-x} C/m^2, where the value of x is ....

Answers

Round I

1. (a)	2. (a)	3. (c)	4. (b)	5. (d)	6. (a)	7. (a)	8. (a)	9. (d)	10. (a)
11. (d)	12. (a)	13. (d)	14. (d)	15. (b)	16. (d)	17. (a)	18. (d)	19. (b)	20. (a)
21. (c)	22. (a)	23. (a)	24. (a)	25. (d)	26. (a)	27. (a)	28. (a)	29. (b)	30. (a)
31. (a)	32. (a)	33. (d)	34. (b)	35. (b)	36. (b)	37. (c)	38. (a)	39. (c)	40. (d)
41. (b)	42. (c)	43. (c)	44. (d)	45. (a)	46. (a)	47. (a)	48. (b)	49. (d)	50. (d)
51. (c)	52. (c)	53. (a)	54. (b)	55. (c)	56. (a)	57. (a)	58. (c)	59. (d)	60. (a)
61. (a)	62. (c)	63. (a)	64. (c)	65. (d)	66. (c)	67. (b)	68. (c)	69. (d)	70. (d)
71. (d)	72. (b)	73. (a)	74. (b)	75. (b)	76. (b)	77. (c)	78. (b)	79. (a)	80. (d)
81. (c)	82. (d)	83. (d)	84. (a)	85. (a)	86. (b)	87. (d)	88. (d)	89. (d)	90. (c)
91. (c)	92. (a)	93. (b)	94. (b)	95. (d)	96. (c)				

Round II

1. (c)	2. (a)	3. (c)	4. (c)	5. (c)	6. (b)	7. (c)	8. (c)	9. (b)	10. (b)
11. (a)	12. (d)	13. (c)	14. (b)	15. (b)	16. (b)	17. (c)	18. (a)	19. (d)	20. (c)
21. (d)	22. (b)	23. (b)	24. (a)	25. (a)	26. (d)	27. (b)	28. (c)	29. (b)	30. (a)
31. (c)	32. (c)	33. (b)	34. (b)	35. (b)	36. (d)	37. (c)	38. (c)	39. (b)	40. (a)
41. 90	42. 3	43. 88.89	44. 4	45. 6	46. 864	47. 7			

Solutions

Round I

1. Each sphere having $-2\,\mu$C.

$$\therefore \quad n = \frac{q}{e} = -\frac{2\times10^{-6}}{-1.6\times10^{-19}} = 1.25\times10^{13}$$

Thus, 1.25×10^{13} electrons are in excess.

2. Charge on the glass rod is positive, so charge on the gold leaves will also be positive. Due to X-rays, more electrons from leaf will be emitted, so leaf becomes more positive and diverse further.

3. $\therefore\ 100 = k\dfrac{q_1q_2}{r^2}$

Now, $\quad q_1 = 0.9q_1 \quad$ (decrease by 10%)

and $\quad q_2 = 1.1q_2 \quad$ (increase by 10%)

$$\because \quad F' = \frac{k(0.9\,q_1)(1.1\,q_2)}{r^2}$$

$$\Rightarrow \quad \frac{F'}{100} = \frac{0.9\times1.1}{1}$$

$$\therefore \quad F' = 99\text{ N}$$

4. Angle between two forces due to individual charges is equal to 60°.

$$\therefore \quad R = \sqrt{F^2+F^2+2\,F\,F\cos60^\circ} = F\sqrt{3}$$

5. The given condition is shown in the figure given below

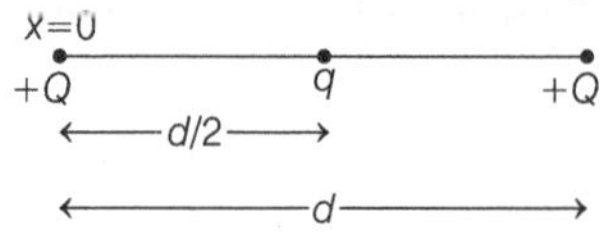

Then, according to the Coulomb's law, the electrostatic force between two charges q_1 and q_2 such that the distance between them is (r) given as

$$F = \frac{1}{4\pi\varepsilon_0}\cdot\frac{q_1q_2}{r^2}$$

$\therefore$ Net force on charge Q placed at origin, *i.e.* at $x=0$ in accordance with the principle of superposition can be given as

$$F_{\text{net}} = \frac{1}{4\pi\varepsilon_0}\cdot\frac{Q\times q}{\left(\frac{d}{2}\right)^2} + \frac{1}{4\pi\varepsilon_0}\cdot\frac{Q\times Q}{(d)^2}$$

Since, it has been given that, $F_{\text{net}} = 0$.

$$\Rightarrow \quad \frac{1}{4\pi\varepsilon_0}\cdot\frac{Q\times q}{\left(\frac{d}{2}\right)^2} + \frac{1}{4\pi\varepsilon_0}\cdot\frac{Q\times Q}{(d)^2} = 0$$

$$\Rightarrow \quad \frac{1}{4\pi\varepsilon_0}\cdot\frac{Q\times q}{\left(\frac{d}{2}\right)^2} = -\frac{1}{4\pi\varepsilon_0}\cdot\frac{Q\times Q}{(d)^2}$$

$$\text{or} \quad q = -\frac{Q}{4}$$

6. From $F = \dfrac{1}{4\pi\varepsilon_0K}\,\dfrac{q_1\,q_2}{r^2}$

For a brass sheet, $K = \infty$

$$\therefore \quad F = \frac{1}{4\pi\varepsilon_0(\infty)}\,\frac{q_1\,q_2}{r^2}$$

$$\Rightarrow \quad F = 0$$

7. From Coulomb's law,

$$F = \frac{1}{4\pi\varepsilon_0}\frac{q_1 q_2}{r^2}$$

$$\Rightarrow \quad 3.7\times10^{-9} = 9\times10^{9}\times\frac{q^2}{(5\times10^{-10})^2} \quad (\because q_1 = q_2 = q)$$

$$\Rightarrow \quad q = 3.2\times10^{-19}\ \text{C}$$

As, $\quad q = ne$

$$\Rightarrow \quad n = \frac{q}{e} = \frac{3.2\times10^{-19}}{1.6\times10^{-19}} = 2$$

8. Here, $F = 40 = \dfrac{k\,(3)\,(8)}{r^2} = \dfrac{24\,k}{r^2}$

and $\quad F' = \dfrac{k\,(3-5)\,(8-5)}{r^2} = \dfrac{k\,(-2)\,(3)}{r^2} = \dfrac{-6k}{r^2}$

$$\because \quad \frac{F'}{F} = \frac{-6k}{r^2}\times\frac{r^2}{24\,k} = -\frac{1}{4}$$

$$\Rightarrow \quad F' = -\frac{F}{4} = -\frac{40}{4} = -10\ \text{N}$$

9. In case of spherical conductor, the whole charge is concentrated at the centre. Now, the distance between two charges will be $(R + x)$. Thus, the force will be proportional to $1/(R+x)^2$.

10. Here, $q = \pm\ 6.0\ \text{nC} = \pm\ 6.0\times10^{-9}\ \text{C}$,

$$2\,a = 6\ \text{cm} = 6\times10^{-2}\ \text{m}$$

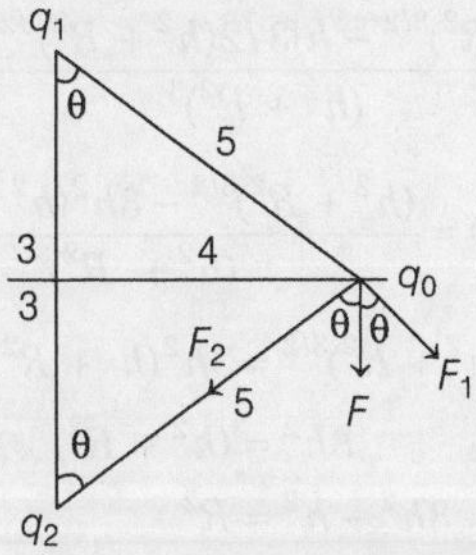

$$r = 4\ \text{cm (on equatorial line)}$$
$$= 4\times10^{-2}\ \text{m}$$

and $\quad q_0 = 2\ \text{nC} = 2\times10^{-9}\ \text{C}, F = ?$

$$\therefore \quad F = F_1\cos\theta + F_2\cos\theta$$

$$= 2\times\frac{1}{4\pi\varepsilon_0}\frac{qq_0}{r^2}\cos\theta$$

$$= 2\times9\times10^{9}\times\frac{6\times10^{-9}\times2\times10^{-9}}{(5\times10^{-2})^2}\times\frac{3}{5}$$

or $\quad F = 5.18\times10^{-5}\ \text{N}$

Clearly, this force is along $\hat{\mathbf{j}}$.

So, $\quad \mathbf{F} = -51.8\ \hat{\mathbf{j}}\ \mu\text{N}$

11. As, $F = \dfrac{q_1 q_2}{4\pi\varepsilon_0 r^2}$

In a medium of dielectric constant, $K = \dfrac{\varepsilon}{\varepsilon_0}$

$$F' = \frac{q_1 q_2}{4\pi\varepsilon R^2}$$

As, $\quad F' = F$

$$\Rightarrow \quad \frac{q_1 q_2}{4\pi\varepsilon R^2} = \frac{q_1 q_2}{4\pi\varepsilon_0 r^2}$$

or $\quad R^2 = \dfrac{\varepsilon_0}{\varepsilon}r^2 = \dfrac{r^2}{K}$

$$\therefore \quad R = \frac{r}{\sqrt{K}}$$

12. As, q_2 and q_3 are positive charges and net force on q_1 is along + x-direction, therefore q_1 must be negative as shown in figure

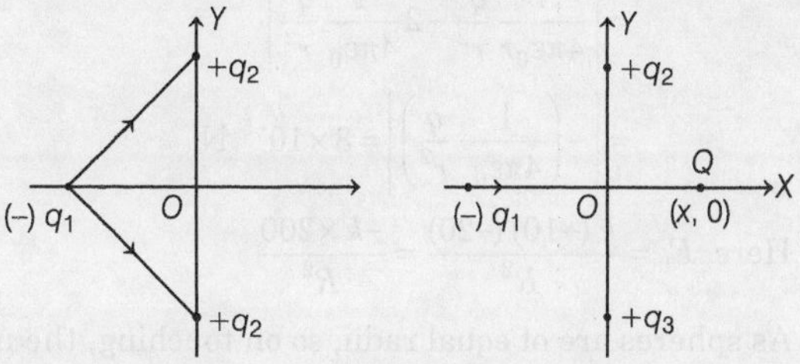

When a positive charge Q is added at $(x, 0)$, it will attract $(-\ q_1)$ along + x-direction, Fig. (b). Therefore, force on q_1 will increase along the positive X-axis.

13. Here, charge on pitch ball A, $q_1 = 5\ \mu\text{C} = 5\times10^{-6}$ C

Mass of pitch ball A, $\quad m_1 = 9\times10^{-5}$ kg

The weight $m_1\ g$ of the pitch ball A acts vertically downwards.

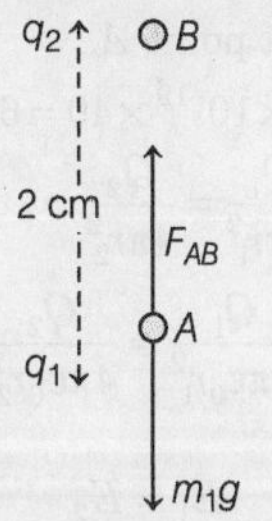

Let q_2 be charge on the pitch ball B held 2 cm above the pitch ball A, so that the pitch ball A remains stationary. It can be possible only, if the charges on two pitch balls are of opposite signs, *i.e.* if charge on pitch ball A is positive, charge on B must be negative. Then, the force on pitch ball A due to B, *i.e.* F_{AB} will act vertically upwards (figure). For charge q_1 to remain stationary,

$$F_{AB} = m_1 g \quad \text{or} \quad \frac{1}{4\pi\varepsilon_0}\cdot\frac{q_1 q_2}{AB^2} = m_1 g$$

Here, $\quad AB = 2\ \text{cm} = 0.02\ \text{m}$

$$\therefore \quad 9\times10^{9}\times\frac{5\times10^{-6}\times q_2}{(0.02)^2} = 9\times10^{-5}\times9.8$$

or $\quad q_2 = 7.84\times10^{-12}$ C

14. If same charges on spheres A and B are q, then

force, $\quad F = \dfrac{1}{4\pi\varepsilon_0}\dfrac{q^2}{r^2} = 3\times10^{-5}$ N

Charge on A and C after touching,

$$q'_A = q_C = \frac{q}{2}$$

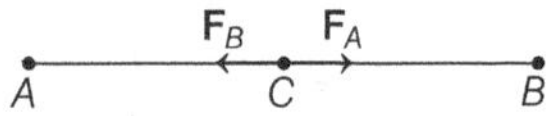

$\therefore$ Net force on C,

$$\mathbf{F} = \mathbf{F}_A + \mathbf{F}_B$$

$$\therefore \quad F = \left| \frac{1}{4\pi\varepsilon_0} \frac{(q/2)(q/2)}{(r/2)^2} - \frac{1}{4\pi\varepsilon_0} \frac{(q/2) \times q}{(r/2)^2} \right|$$

$$= \left| \frac{1}{4\pi\varepsilon_0 r} \frac{q^2}{r^2} - 2 \frac{1}{4\pi\varepsilon_0} \frac{q^2}{r^2} \right|$$

$$= \left| -\left(\frac{1}{4\pi\varepsilon_0} \cdot \frac{q}{r^2} \right) \right| = 3 \times 10^{-5} \text{ N}$$

15. Here, $F_1 = \dfrac{k(+10)(-20)}{R^2} = \dfrac{-k \times 200}{R^2}$

As spheres are of equal radii, so on touching, the net charge $= (+10 - 20)\mu\text{C} = -10\mu\text{C}$, is shared equally between them, *i.e.* each sphere carries $-5\mu\text{C}$ charge.

$$F_2 = \frac{k(-5)(-5)}{R^2} = \frac{k \times 25}{R^2}$$

$$\because \quad \frac{F_1}{F_2} = \frac{-8}{1} \Rightarrow F_1 : F_2 = -8 : 1$$

16. Charge (C) has the largest magnitude, since maximum number of field lines are associated with it.

17. Force on proton at point A,

$$F_A = qE_A = 1.6 \times 10^{-19} \times 40 = 6.4 \times 10^{-18} \text{ N}$$

18. As, $\sigma_1 = \sigma_2 \Rightarrow \dfrac{Q_1}{4\pi r_1^2} = \dfrac{Q_2}{4\pi r_2^2}$

$$\Rightarrow \quad \frac{Q_1}{4\pi\varepsilon_0 r_1^2} = \frac{Q_2}{4\pi\varepsilon_0 r_2^2}$$

$$\Rightarrow \quad E_1 = E_2 \qquad \left[\because E = \frac{Q}{4\pi\varepsilon_0 r^2} \right]$$

or $\quad E_1 / E_2 = 1$

19. The density of lines of force $\propto E$.

Here, the density of lines of force at A is greater than at B.

Thus, $\quad E_A > E_B$

20. According to the question,

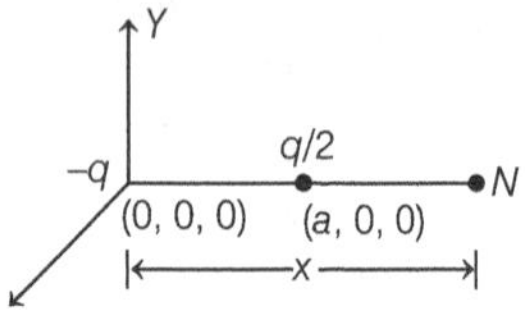

$$\frac{hq}{x^2} = \frac{hq/2}{(x-a)^2}$$

or $\quad 2(x-a)^2 = x^2$

$$\sqrt{2}(x-a) = x$$

or $\quad (\sqrt{2} - 1)x = \sqrt{2}a$

$$\Rightarrow \quad x = \frac{\sqrt{2}a}{\sqrt{2} - 1}$$

21. For a ring, $E = \dfrac{1}{4\pi\varepsilon_0} \cdot \dfrac{qx}{(x^2 + R^2)}$

and E is maximum when $x = \dfrac{R}{\sqrt{2}}$

$$\Rightarrow \quad E_{\text{max}} = \frac{1}{4\pi\varepsilon_0} \cdot \frac{2q}{3\sqrt{3}R^2}$$

22. Charge enclosed in the sphere of radius r,

$$q = \frac{4}{3}\pi r^3 \rho$$

Electric field, $E = \dfrac{q}{4\pi\varepsilon_0 r^2} = \dfrac{\frac{4}{3}\pi r^3 \rho}{4\pi\varepsilon_0 r^2} = \dfrac{r\rho}{3\varepsilon_0}$

23. Electric field at a distance h from the centre of uniformly charged ring of total charge q (say) on its axis is given as

$$E = \frac{1}{4\pi\varepsilon_0} \cdot \frac{qh}{(h^2 + R^2)^{3/2}}$$

For the magnitude to be maximum, then $\dfrac{dE}{dh} = 0$

$$\Rightarrow \quad \frac{dE}{dh} = \frac{q}{4\pi\varepsilon_0}$$

$$\left[\frac{(h^2 + R^2)^{3/2} - h[3/2(h^2 + R^2)^{1/2} 2h]}{(h^2 + R^2)^3} \right] = 0$$

$$\Rightarrow \quad 0 = \frac{(h^2 + R^2)^{3/2} - 3h^2(h^2 + R^2)^{1/2}}{(h^2 + R^2)^3}$$

$$\Rightarrow \quad (h^2 + R^2)^{3/2} = 3h^2(h^2 + R^2)^{1/2}$$

$$\Rightarrow \quad 3h^2 = (h^2 + R^2)$$

$$\Rightarrow \quad 3h^2 - h^2 = R^2$$

$$\Rightarrow \quad 2h^2 = R^2$$

$$\Rightarrow \quad h = \pm \frac{R}{\sqrt{2}}$$

$\therefore$ At $\dfrac{R}{\sqrt{2}}$, the value of electric field associated with a charged ring on its axis has the maximum value.

24. Here, $q_1 = \sqrt{10}\,\mu\text{C} = \sqrt{10} \times 10^{-6}$ C,

$$q_2 = -25\,\mu\text{C} = -25 \times 10^{-6} \text{ C}$$

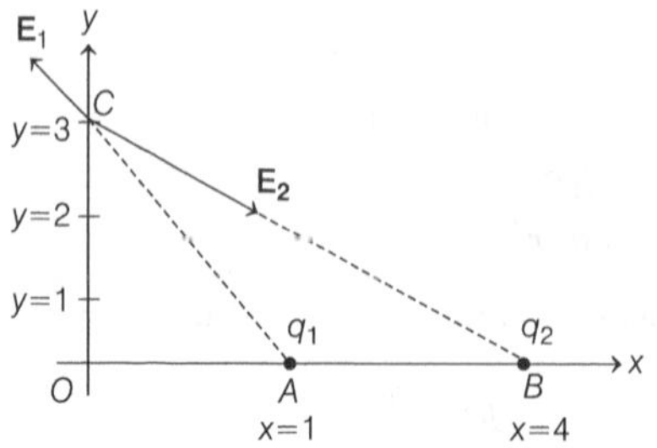

Let $\mathbf{E}_1$ and $\mathbf{E}_2$ are the values of electric field due to q_1 and q_2, respectively.

Here, $E_1 = \frac{1}{4\pi\varepsilon_0} \cdot \frac{q_1}{AC^2} = \frac{1}{4\pi\varepsilon_0} \times \frac{\sqrt{10} \times 10^{-6}}{(1^2 + 3^2)}$

$= 9 \times 10^9 \times \sqrt{10} \times 10^{-7}$

$= 9\sqrt{10} \times 10^2$

$\therefore \quad \mathbf{E}_1 = 9\sqrt{10} \times 10^2 [\cos\theta_1 (-\hat{\mathbf{i}}) + \sin\theta_1 \hat{\mathbf{j}}]$

From ΔOAC,

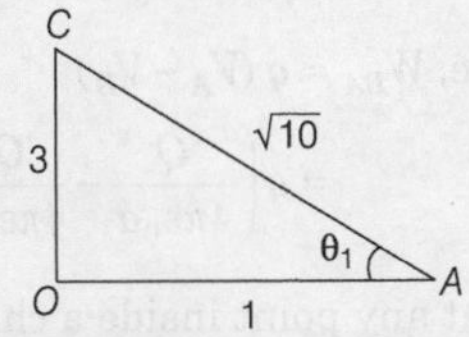

$\sin\theta_1 = \frac{3}{\sqrt{10}}$

and $\quad \cos\theta_1 = \frac{1}{\sqrt{10}}$

$\therefore \quad E_1 = 9\sqrt{10} \times 10^2 \left[\frac{1}{\sqrt{10}}(-\hat{\mathbf{i}}) + \frac{3}{\sqrt{10}}\hat{\mathbf{j}}\right]$

$= 9 \times 10^2 [-\hat{\mathbf{i}} + 3\hat{\mathbf{j}}]$

$= (-9\hat{\mathbf{i}} + 27\hat{\mathbf{j}}) \times 10^2$ V/m

and $\quad E_2 = \frac{1}{4\pi\varepsilon_0} \cdot \frac{-25 \times 10^{-6}}{(4^2 + 3^2)}$

$= 9 \times 10^3$ V/m

From ΔOBC,

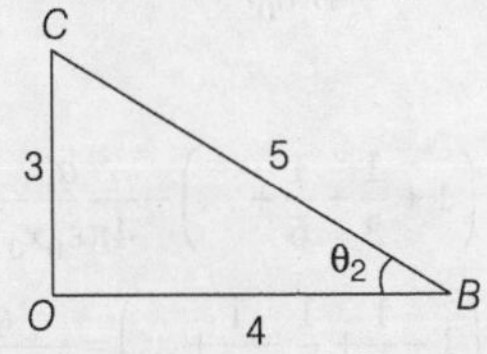

$\sin\theta_2 = \frac{3}{5}$

and $\quad \cos\theta_2 = \frac{4}{5}$

$\therefore \quad \mathbf{E}_2 = 9 \times 10^3 [\cos\theta_2 \hat{\mathbf{i}} - \sin\theta_2 \hat{\mathbf{j}}]$

$\mathbf{E}_2 = 9 \times 10^3 \left[\frac{4}{5}\hat{\mathbf{i}} - \frac{3}{5}\hat{\mathbf{j}}\right]$

$= (72\hat{\mathbf{i}} - 54\hat{\mathbf{j}}) \times 10^2$

$\therefore \quad \mathbf{E} = \mathbf{E}_1 + \mathbf{E}_2$

$= (63\hat{\mathbf{i}} - 27\hat{\mathbf{j}}) \times 10^2$ V/m

25. In a uniform electric field, field lines should be straight but line of force cannot pass through the body of metal sphere and must end/start from the sphere normally. All these conditions are fulfilled only in option (d).

26. When a point charge $+q$ is placed at a distance (d) from an isolated conducting plane, some negative charge developes on the surface of the plane towards the charge and an equal positive charge develops on opposite side of the plane.

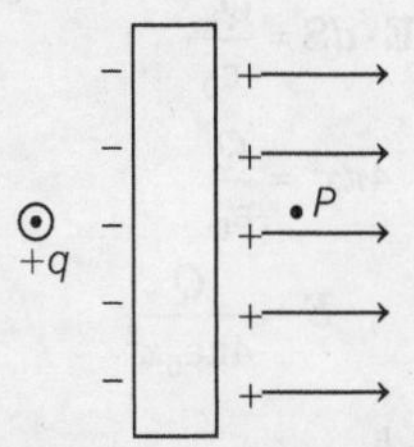

Hence, the field at a point P on the other side of the plane is directed perpendicular to the plane and away from the plane as shown in figure.

27. When the point is situated at a point on diameter away from the centre of hemisphere charged uniformly positively, the electric field is perpendicular to the diameter. The components of electric intensity parallel to the diameter cancel out.

28. Electric field lines always move from positive to negative charge and is always normal to the surface of conductor.

29. Here, $\quad \mathbf{E} = 8\hat{\mathbf{i}} + 4\hat{\mathbf{j}} + 3\hat{\mathbf{k}}$

$\mathbf{S} = 100\,\hat{\mathbf{k}}$

(direction of area is perpendicular to XY-plane)

$\therefore \quad \phi = \mathbf{E} \cdot \mathbf{S} = (8\hat{\mathbf{i}} + 4\hat{\mathbf{j}} + 3\hat{\mathbf{k}}) \cdot 100\,\hat{\mathbf{k}} = 300$ units

30. Flux through surface A,

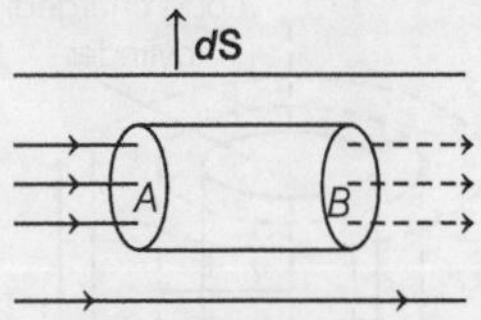

$\phi_A = E \times \pi R^2$

and $\quad \phi_B = -E \times \pi R^2$

Electric flux through the curved surface,

$\phi = \int \mathbf{E} \cdot d\mathbf{S} = \int E\,dS \cos 90° = 0$

$\therefore$ Total flux $= E\pi R^2 - E\pi R^2 + 0 = 0$

32. Equation $|\mathbf{E}| = \frac{q_{\text{enc}}}{\varepsilon_0 |A|}$ gives $\int \mathbf{E} \cdot d\mathbf{A} = \frac{q_{\text{enc}}}{\varepsilon_0}$

Now, in finding the electric field by above equation, the integral is easy to evaluate, if $|\mathbf{E}| =$ constant. Also, if $|\mathbf{E}| =$ constant for the surface, then surface is equipotential.

33. The electric flux through a surface depends only on amount of charge enclosed by that surface. It does not depend on size and shape of the surface, as per Gauss's theorem in electrostatic. Therefore, electric flux through the given surfaces is the same for all the figures.

34. According to Gauss's theorem in electrostatics, $\oint \mathbf{E} \cdot d\mathbf{S} = \frac{q}{\varepsilon_0}$. Here, $\mathbf{E}$ is due to all the charges q_1, q_2, q_3, q_4 and q_5. As, q is charge enclosed by the Gaussian surface, therefore $q = q_2 + q_4$.

35. According to Gauss's theorem,

$$\oint \mathbf{E}\cdot d\mathbf{S} = \frac{Q_{\text{in}}}{\varepsilon_0}$$

$$\Rightarrow \quad E\cdot 4\pi x^2 = \frac{Q}{\varepsilon_0}$$

or $$E = \frac{Q}{4\pi\varepsilon_0 x^2}$$

36. Here, $\phi = ar^2 + b$

As, $$E = -\frac{d\phi}{dr} = -2ar$$

$$\therefore \quad \oint \mathbf{E}\cdot d\mathbf{S} = \frac{q}{\varepsilon_0}$$

$$-2ar\cdot 4\pi r^2 = \frac{q}{\varepsilon_0}$$

$$\Rightarrow \quad q = -8\,\varepsilon_0 a\pi r^3 \quad \ldots\text{(i)}$$

$$\therefore \quad \rho = \frac{q}{\frac{4}{3}\pi r^3}$$

$$\Rightarrow \quad \rho = -6\,a\varepsilon_0 \quad \text{[from Eq. (i)]}$$

37. We have, $E = \dfrac{\sigma}{2\,\varepsilon} = \dfrac{\sigma}{2\,\varepsilon_0 K}$ $\quad \left(\because \dfrac{\varepsilon}{\varepsilon_0} = K\right)$

38. The charge on cylinder A, $q = \lambda l$

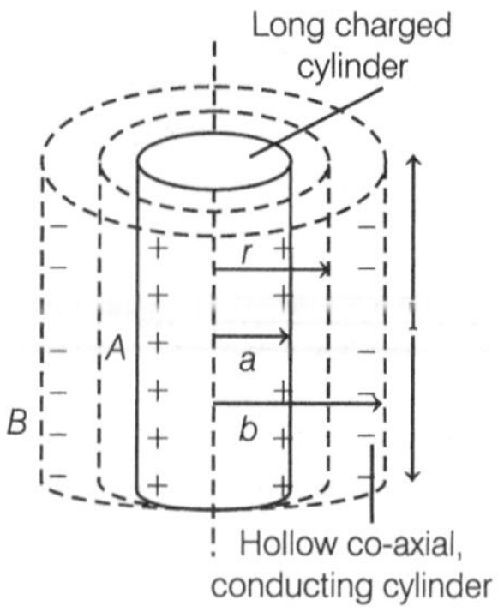

Total charge = Linear charge density × Length

This charge spreads uniformly on A and a charge $-q$ is induced on B. Let **E** be the electric field produced in the space between the two cylinders. Consider a Gaussian cylindrical surface of radius r between the two given cylinders.

Electric flux linked with the Gaussian surface,

$$\phi_E = \int \mathbf{E}\cdot d\mathbf{S} = \int E\cdot dS\cos 0^\circ = E\int dS = E\times 2\pi rl$$

(As angle between the direction of electric field and area vector is zero)

According to Gauss's theorem, $\phi_E = E\times 2\pi rl = \dfrac{q}{\varepsilon_0}$

$$\Rightarrow \quad E\times 2\pi rl = \frac{\lambda l}{\varepsilon_0} \Rightarrow E = \frac{\lambda}{2\pi\varepsilon_0 r}$$

39. As, $E = \dfrac{1}{4\pi\varepsilon_0}\dfrac{Q}{R^2} = \dfrac{1}{4\pi\varepsilon_0}\times\dfrac{4\pi R^2\sigma}{R^2} = \dfrac{\sigma}{\varepsilon_0}$

This is independent of radius and depends on σ. Hence, the electric field on the surface of new sphere will be E.

40. Inside hollow sphere, $E = 0$. On the surface of hollow sphere, E = maximum and outside the sphere, $E \propto 1/r^2$. So the variations are correctly shown in option (d).

41. As, work done, $W_{BA} = q\,(V_A - V_B)$

$$= q\left(\frac{Q}{4\pi\varepsilon_0 a} - \frac{Q}{4\pi\varepsilon_0 b}\right) = \frac{qQ}{4\pi\varepsilon_0}\left(\frac{1}{a} - \frac{1}{b}\right)$$

42. As we know, at any point inside a charged conducting sphere, $E = 0$ and $V = \dfrac{kq}{R}$ = potential on the surface = 100 V

Also, $$E = -\frac{dV}{dr} = -\frac{d}{dr}\text{(constant)} = \text{Zero}$$

43. Electric potential inside the hollow conducting sphere is constant and equal to potential at the surface of the sphere, *i.e.* $\dfrac{Q}{4\pi\varepsilon_0 R}$.

44. As, $$\frac{4}{3}\pi R^3 = n\times\frac{4}{3}\pi r^3$$

$$\therefore \quad R = n^{1/3} r$$

New potential, $V' = \dfrac{nq}{4\pi\varepsilon_0 r} = n^{2/3} V$

45. As we have,

$$V = \frac{q}{4\pi\varepsilon_0 x_0}\left(1 + \frac{1}{3} + \frac{1}{5} + \ldots\right) - \frac{q}{4\pi\varepsilon_0 x_0}\left(\frac{1}{2} + \frac{1}{4} + \frac{1}{6}\ldots\right)$$

$$= \frac{q}{4\pi\varepsilon_0 x_0}\left(1 - \frac{1}{2} + \frac{1}{3} - \frac{1}{4} + \ldots\right) = \frac{q}{4\pi\varepsilon_0 x_0}\log_e 2$$

46. Potential at the centre O,

$$V = 4\times\frac{1}{4\pi\varepsilon_0}\cdot\frac{q}{a\sqrt{2}}$$

where, $q = \dfrac{10}{3}\times 10^{-9}$ C, in magnitude

and $a = 8\text{ cm} = 8\times 10^{-2}\,\text{m}$

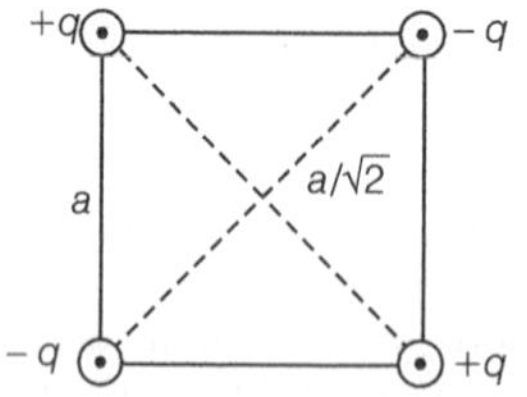

Then, $$V = 4\times 9\times 10^9\times\frac{\frac{10}{3}\times 10^{-9}}{\frac{8\times 10^{-2}}{\sqrt{2}}} = 1500\sqrt{2}\,\text{V}$$

47. Let there is a cube of side b and its centre is O. The charge q is placed at each of the corners. Side of the cube $= b$

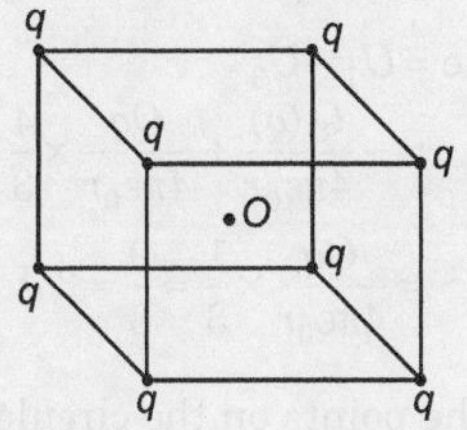

Length of the main diagonal of the cube

$$= \sqrt{b^2 + b^2 + b^2} = \sqrt{3}b$$

Distance of centre O from each of the vertices,

$$r = \frac{b\sqrt{3}}{2} \quad \ldots\text{(i)}$$

Potential at point O due to one charge,

$$V = \frac{1}{4\pi\varepsilon_0} \cdot \frac{q}{r}$$

Potential at point O due to all charges placed at the vertices of the cube,

$$V' = 8V = \frac{8 \times 1 \times q}{4\pi\varepsilon_0 \cdot r}$$

$$= \frac{8q \times 2}{4\pi\varepsilon_0 \cdot b\sqrt{3}} \quad \text{[from Eq. (i)]}$$

$$= \frac{4q}{\sqrt{3}\pi\varepsilon_0 b}$$

48. The tangential component of electrostatic field is continuous from one side of a charged surface to another, we use that the work done by electrostatic field on a closed loop is zero.

Let ABA be a charged surface in the field of a point charge q lying at origin.

Let $\mathbf{r}_A$ and $\mathbf{r}_B$ be its positive vectors at points A and B, respectively.

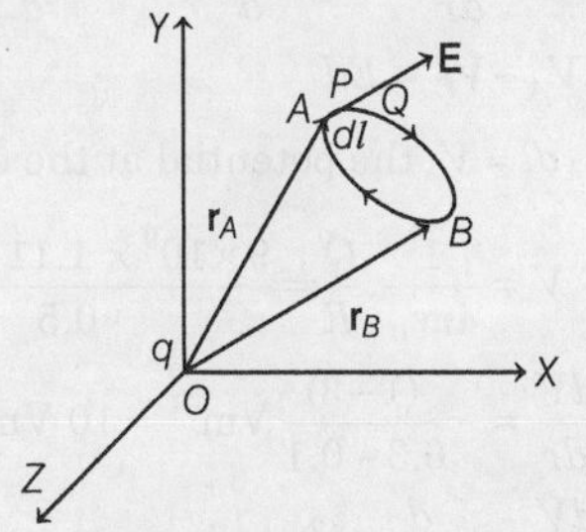

Let $\mathbf{E}$ be the electric field at point P, thus $E\cos\theta$ is the tangential component of electric field $\mathbf{E}$.

To prove that $E\cos\theta$ is continuous from one to another side of the charge surface, we have to find the value of $\oint_{ABA} \mathbf{E}\cdot d\mathbf{l}$. If it comes to be zero, then we can say that tangential component of $\mathbf{E}$ is continuous.

$$\int_A^B \mathbf{E}\cdot d\mathbf{l} = \frac{1}{4\pi\varepsilon_0} q \cdot \left(\frac{1}{r_A} - \frac{1}{r_B}\right)$$

and $$\int_B^A \mathbf{E}\cdot d\mathbf{l} = \frac{1}{4\pi\varepsilon_0} q \cdot \left(\frac{1}{r_B} - \frac{1}{r_A}\right)$$

$$\therefore \quad \oint_{ABA} \mathbf{E}\cdot d\mathbf{l} = \int_A^B \mathbf{E}\cdot d\mathbf{l} + \int_B^A \mathbf{E}\cdot d\mathbf{l}$$

$$= \frac{1}{4\pi\varepsilon_0} \cdot q \cdot \left(\frac{1}{r_A} - \frac{1}{r_B} + \frac{1}{r_B} - \frac{1}{r_A}\right) = 0$$

49. The four charges are shown in the figure below

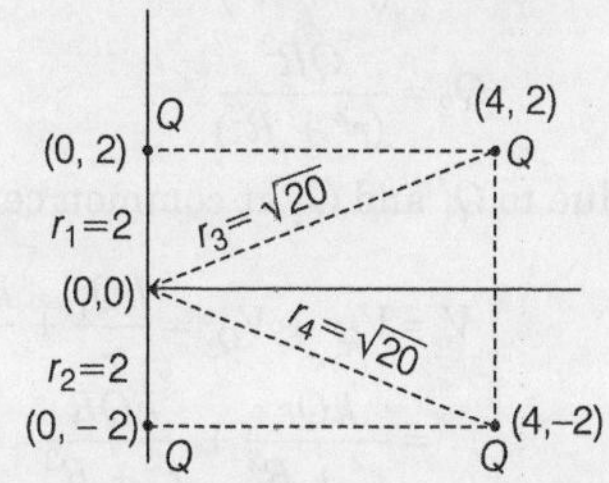

Electric potential at origin (0, 0) due to these charges can be found by scalar addition of electric potentials due to each charge.

$$\therefore \quad V = \frac{kQ}{r_1} + \frac{kQ}{r_2} + \frac{kQ}{r_3} + \frac{kQ}{r_4} \quad \ldots\text{(i)}$$

$$\Rightarrow \quad V = kQ\left[\frac{1}{2} + \frac{1}{2} + \frac{1}{\sqrt{20}} + \frac{1}{\sqrt{20}}\right] = kQ\left[1 + \frac{1}{\sqrt{5}}\right]$$

$$\Rightarrow \quad V = kQ\frac{(\sqrt{5}+1)}{\sqrt{5}} \text{ volt} \quad \ldots\text{(ii)}$$

Now, if another charge Q is placed at origin, then work done to get the charge at origin,

$$W = QV \quad \ldots\text{(iii)}$$

By putting the value of V from Eq. (ii) in Eq. (iii), we get

$$W = kQ^2\frac{(\sqrt{5}+1)}{\sqrt{5}} \text{ J}$$

or $$W = \frac{Q^2}{4\pi\varepsilon_0}\left(1 + \frac{1}{\sqrt{5}}\right) \text{ J}$$

50. Given, total charge Q is uniformly distributed over concentric shells of radii r and R.

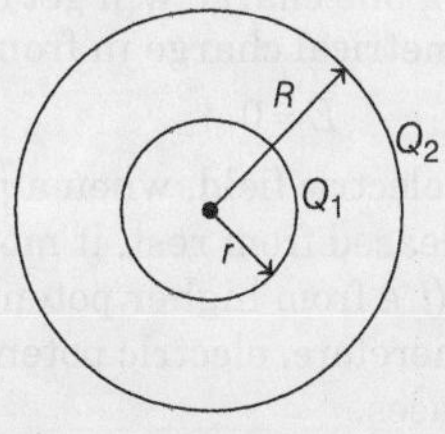

Let Q_1 and Q_2 be the charges over inner and outer shells, such that charge densities are equal on both shells.

Charge density of inner shell, $\sigma = \frac{Q_1}{4\pi r^2}$

Charge density of outer shell, $\sigma = \frac{Q_2}{4\pi R^2}$

$\because \quad \dfrac{Q_1}{4\pi r^2} = \dfrac{Q_2}{4\pi R^2}$

$\Rightarrow \quad \dfrac{Q_1}{r^2} = \dfrac{Q_2}{R^2}$

$\Rightarrow \quad Q_2 = \dfrac{Q_1 R^2}{r^2}$

Also, $\quad Q_1 + Q_2 = Q$

$\Rightarrow \quad Q_1 + \dfrac{Q_1 R^2}{r^2} = Q$

or $\quad Q_1 = \dfrac{Qr^2}{(r^2 + R^2)}$

and $\quad Q_2 = \dfrac{QR^2}{(r^2 + R^2)}$

Potential due to Q_1 and Q_2 at common centre of shells is

$$V = V_{Q_1} + V_{Q_2} = \frac{kQ_1}{r} + \frac{kQ_2}{R}$$
$$= \frac{kQr}{r^2 + R^2} + \frac{kQR}{r^2 + R^2}$$
$$= \frac{1}{4\pi\varepsilon_0} \cdot \frac{Q(r + R)}{r^2 + R^2}$$

51. The arrangement of all the charges is shown in the following figure

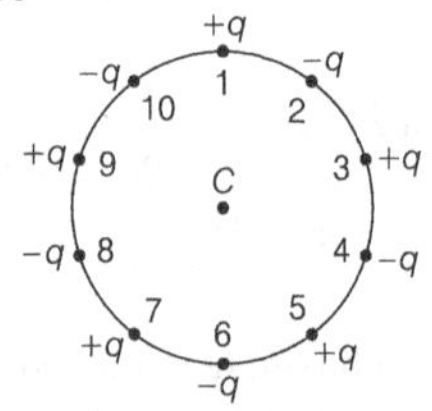

Potential at the centre of the circle C,

$$V = \frac{kq}{R} + \frac{k(-q)}{R} + \frac{kq}{R} + \frac{k(-q)}{R} + \ldots$$
$$= \Sigma \frac{kq}{R} = k\frac{\Sigma(q)}{R} = \frac{k(0)}{R} = 0$$

Electric field at centre of the circle C,

$\mathbf{E}$ = ($\mathbf{E}$ due to 5 + ve charges) + ($\mathbf{E}$ due to 5 – ve charges)

As, all the charges are equidistant from centre. So, the electric field of one charge will get cancelled due to another symmetrical charge in front of it.

$\therefore \quad E = 0$

52. In a uniform electric field, when a positively charged particle is released from rest, it moves along the electric field (*i.e.* from higher potential to lower potential). Therefore, electric potential energy of charge decreases.

53. Here, $U_1 = \dfrac{Q(-q)}{4\pi\varepsilon_0 r}$; $U_2 = \dfrac{Q(-q)}{4\pi\varepsilon_0 (r/2)}$

$$\therefore \quad U_1 - U_2 = \frac{Q(-q)}{4\pi\varepsilon_0}\left[\frac{1}{r} - \frac{2}{r}\right]$$
$$= \frac{Qq}{4\pi\varepsilon_0 r} = 9\text{ J} \qquad \ldots\text{(i)}$$

When negative charge travels first half of distance, *i.e.* $r/4$, potential energy of the system,

$$U_3 = \frac{Q(-q)}{4\pi\varepsilon_0 (3r/4)} = -\frac{Qq}{4\pi\varepsilon_0 r} \times \frac{4}{3}$$

$\therefore \quad$ Work done $= U_1 - U_3$

$$= -\frac{Q(q)}{4\pi\varepsilon_0 r} + \frac{Qq}{4\pi\varepsilon_0 r} \times \frac{4}{3}$$
$$= \frac{Qq}{4\pi\varepsilon_0 r} \times \frac{1}{3} = \frac{9}{3} = 3\text{ J}$$

54. Because, all the points on the circular path are at same potential. So, work done is zero.

55. For charge q placed at the centre of circle, the circular path is an equipotential surface and hence, work done along all paths AB or AC or AD or AE is zero.

56. From a collection of charges, whose total sum is not zero, equipotentials at large distances must be spheres only.

57. Here, displacement would be

$$\Delta\mathbf{r} = (4 - 2)\,\hat{\mathbf{i}} + (2 - 2)\,\hat{\mathbf{j}} = 2\hat{\mathbf{i}}$$

Now, $\quad \Delta\mathbf{V} = \mathbf{E} \cdot \Delta\mathbf{r} = (y\hat{\mathbf{i}} + x\hat{\mathbf{j}}) \cdot 2\hat{\mathbf{i}} = 2y$

$\therefore$ Work done, $W = q(\Delta\mathbf{V}) = q \cdot 2y = 1 \times 2y = 2y$ J

58. We observe that in all the three parts, $V_A = 20$ V and $V_B = 40$ V. Work done in carrying a charge q from A to B is $W = q(V_B - V_A)$ = same in all the three figures.

59. $E = -\dfrac{\partial V}{\partial x}\hat{\mathbf{i}} - \dfrac{\partial V}{\partial y}\hat{\mathbf{j}} - \dfrac{\partial V}{\partial z}\hat{\mathbf{k}}$

$$\Rightarrow \quad E_x = -\frac{\partial V}{\partial x} = -\frac{d}{dx}\left[\frac{20}{x^2 - 4}\right] = \frac{40x}{(x^2 - 4)^2}$$

$$\Rightarrow \quad E_x \text{ at } x = 4\,\mu\text{m} = \frac{10}{9}\text{ V}/\mu\text{m}$$

and is along positive x-direction.

60. As, it is clear from figure

$$\mathbf{AB} + \mathbf{BC} = \mathbf{AC}$$

As, $\quad E = -\dfrac{dV}{dr} = -\dfrac{(V_C - V_A)}{d} = \dfrac{V_A - V_C}{d}$

$\therefore \quad V_A - V_C = Ed$

61. As, $\int_{l=\infty}^{l=0} -E \cdot dl = V$, the potential at the centre of ring,

$$V = \frac{1}{4\pi\varepsilon_0} \cdot \frac{Q}{R} = \frac{9 \times 10^9 \times 1.11 \times 10^{-10}}{0.5} = 2\text{V}$$

62. As, $E = -\dfrac{dV}{dr} = -\dfrac{(1-3)}{0.3 - 0.1}\text{Vm}^{-1} = 10\text{ Vm}^{-1}$

63. As, $E = -\dfrac{dV}{dr} = -\dfrac{d}{dx}(4x^2) = -8x = -8\,(1) = -8\text{Vm}^{-1}$

Negative sign indicates $\mathbf{E}$ is along negative direction of X-axis.

64. As the work has been done by field, so it should be positive.

The work done, $W = q(V_A - V_B)$

We know that, change in kinetic energy = Work done

$$= q(V_A - V_B)$$
$$= 10^{-8}(600 - 0) = 6 \times 10^{-6}\,\text{J}$$

65. As, net work done = final PE – initial PE

$$= \frac{Qq}{4\pi\varepsilon_0 l} - \frac{Qq}{4\pi\varepsilon_0 l} = \text{zero}$$

66.

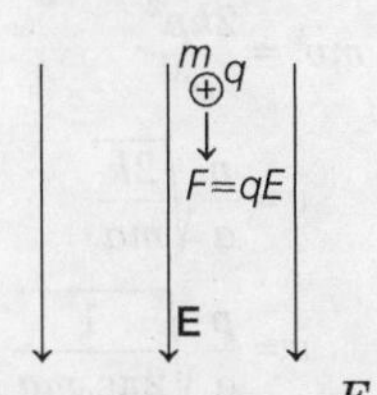

Acceleration of the particle, $a = \frac{F}{m} = \frac{q}{m} \cdot E$

Velocity v and distance x can be related using $v^2 - u^2 = 2ax$

$$\Rightarrow \quad v = \sqrt{2\left(\frac{q}{m}E\right)x} \qquad (\because u = 0)$$

or $$v^2 = 2\left(\frac{q}{m}E\right)x$$

This equation resembles a parabola $y^2 = 4ax$. So, the graph between v and x will be as shown in option (c).

67. In figure, spacing between electric lines of force increases from left to right. Therefore, E on left is greater that E on right. Force on + q charge of dipole is smaller and to the right. Force on – q charge of dipole is bigger and to the left. Hence, the dipole will experience a net force towards the left.

68. Here, $\theta = 60°$, $E = 10^5\ \text{NC}^{-1}$,

$$\tau = 8\sqrt{3}\ \text{N-m}, q = ?, 2a = 2\ \text{cm} = 2 \times 10^{-2}\ \text{m}$$

From, $\tau = pE\sin\theta = q\,(2a)\,E\sin\theta$

or $$q = \frac{\tau}{2\,aE\sin\theta}$$
$$= \frac{8\sqrt{3}}{2 \times 10^{-2} \times 10^5 \times \sin 60°}$$
$$= \frac{8\sqrt{3}}{2 \times 10^3 \times \sqrt{3}/2}$$

or $$q = 8 \times 10^{-3}\ \text{C}$$

69. On equatorial line of electric dipole, $E \propto \frac{1}{r^3}$

70. In case of an electric dipole, $F \propto \frac{1}{r^3}$

$\therefore$ New force $= F/2^3 = F/8$

71. The point lies on equatorial line of a short dipole.

$$\therefore \quad E = \frac{p}{4\pi\varepsilon_0 r^3}$$
$$= \frac{9 \times 10^9\,(10^{-6} \times 10^{-8})}{(10^{-1})^3}$$
$$= 9 \times 10^{-2}\ \text{NC}^{-1} = 0.09\ \text{NC}^{-1}$$

72. The given problem can be shown as clearly potential difference at point P due to dipole is

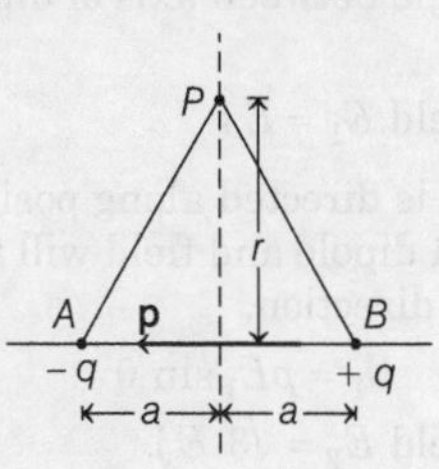

$$V = V_{AP} + V_{BP} \qquad \text{(scalar addition)}$$

$$\Rightarrow \quad V = \frac{k(-q)}{AP} + \frac{k(q)}{BP} \qquad \text{...(i)}$$

Here, $\quad AP = BP = \sqrt{a^2 + r^2}$

$$\therefore \quad V = -\frac{kq}{\sqrt{a^2 + r^2}} + \frac{kq}{\sqrt{a^2 + r^2}} = 0 \qquad \text{...(ii)}$$

Now, electric field at any point on Y-axis, *i.e.* equatorial line of the dipole can be given by

$$\mathbf{E} = -\frac{k\,\mathbf{p}}{r^3} \qquad \text{(standard expression)}$$

$$\Rightarrow \quad \mathbf{E} = -\frac{1}{4\pi\varepsilon_0}\frac{\mathbf{p}}{r^3}$$

Given, $\quad r = d$

$$\therefore \quad \mathbf{E} = -\frac{1}{4\pi\varepsilon_0}\frac{\mathbf{p}}{d^3} \qquad \text{...(iii)}$$

73. As, dipole moments point in same direction

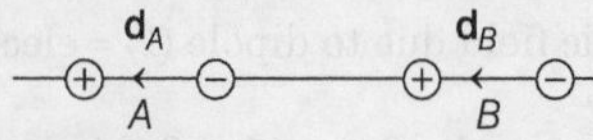

So, potential of both dipoles can be same at some point between A and B.

Let potentials are same at P, distant x from B as shown below

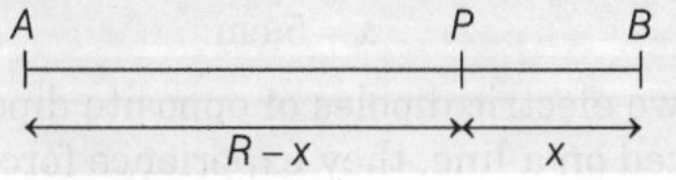

Then, $$\frac{4qa}{(R - x)^2} = \frac{2qa}{(x)^2}$$

$$\Rightarrow \quad 2x^2 = (R - x)^2 \Rightarrow \sqrt{2}\,x = R - x$$

$$\Rightarrow \quad x = \frac{R}{\sqrt{2} + 1}$$

Distance from A is

$$R - x = R - \frac{R}{\sqrt{2} + 1} = \frac{\sqrt{2}R}{\sqrt{2} + 1}$$

74.

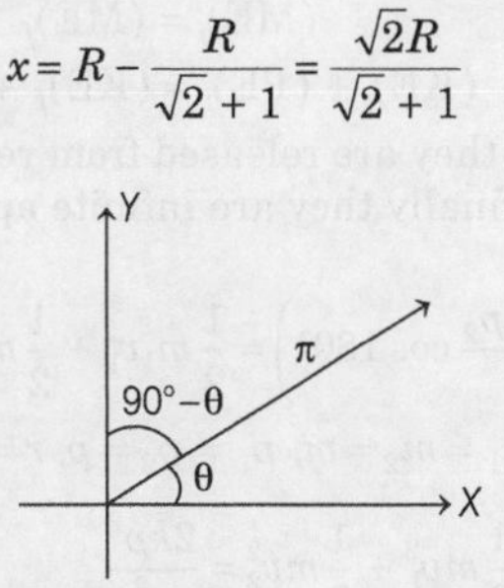

Torque applied on a dipole, $\tau = pE \sin\theta$

where, θ = angle between axis of dipole and electric field.

For electric field $E_1 = E\,\hat{\mathbf{i}}$

It means field is directed along positive x-direction, so angle between dipole and field will remain θ, therefore torque in this direction,

$$E_1 = pE_1 \sin\theta$$

For electric field $E_2 = \sqrt{3}\, E\,\hat{\mathbf{j}}$,

It means field is directed along positive Y-axis, so angle between dipole and field will be $90^\circ - \theta$.

Torque in this direction, $\tau_2 = pE \sin(90^\circ - \theta)$

$$= p\sqrt{3}\, E_1 \cos\theta$$

According to question, $\tau_2 = -\tau_1 \Rightarrow |\tau_2| = |\tau_1|$

$\therefore \quad pE_1 \sin\theta = p\sqrt{3}\, E_1 \cos\theta$

$\Rightarrow \quad \tan\theta = \sqrt{3}$

$\Rightarrow \quad \tan\theta = \tan 60^\circ$

$\therefore \quad \theta = 60^\circ$

75. Let us consider a neutral point O lies at a distance x from the dipole of moment p or at a distance $(25 - x)$ from dipole of $64\,p$.

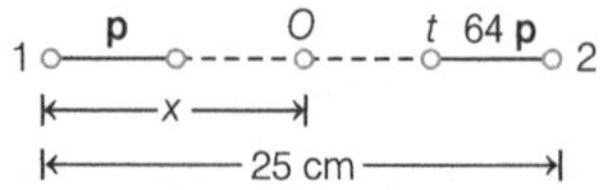

At O, electric field due to dipole (1) = electric field due to dipole (2)

$\Rightarrow \quad \dfrac{1}{4\pi\varepsilon_0}\dfrac{2p}{x^3} = \dfrac{1}{4\pi\varepsilon_0}\dfrac{2(64\,p)}{(25-x)^3}$

$\Rightarrow \quad \dfrac{1}{x^3} = \dfrac{64}{(25-x)^3}$

$\Rightarrow \quad x = 5$ cm

76. When two electric dipoles of opposite dipole moments are placed on a line, they experience force of attraction along the same line as shown below

Considering both dipoles as a system, we find that net external force on system is zero, *i.e.* $\mathbf{F}_{ext} = 0$.

So, total mechanical energy = constant

$$(\text{ME})_i = (\text{ME})_f$$

or $\quad (\text{KE})_i + (\text{PE})_i = (\text{KE})_f + (\text{PE})_f$

As, initially they are released from rest, so initial KE is zero and finally they are infinite apart, so final PE is zero.

$$0 + \left(-\frac{2kp_1p_2}{r^3}\cos 180^\circ\right) - \frac{1}{2}m_1v_1^2 + \frac{1}{2}m_2v_2^2 + 0$$

Here, $\quad m_1 = m_2 = m,\ p_1 = p_2 = p,\ r = a$

$$\frac{1}{2}mv_1^2 + \frac{1}{2}mv_2^2 = \frac{2kp^2}{a^3} \quad \ldots\text{(i)}$$

Using conservation of momentum,

$$\mathbf{p}_i = \mathbf{p}_f$$

$\Rightarrow \quad 0 = m\mathbf{v}_1 + m\mathbf{v}_2$

or $\quad \mathbf{v}_1 = -\mathbf{v}_2$

$\Rightarrow \quad v_1 = v_2 = v \quad \ldots\text{(ii)}$

Putting this value in Eq. (i), we get

$$mv^2 = \frac{2kp^2}{a^3}$$

or $\quad v = \dfrac{p}{a}\sqrt{\dfrac{2k}{ma}}$

or $\quad v = \dfrac{p}{a}\sqrt{\dfrac{1}{2\pi\varepsilon_0 ma}} \quad \left(\because k = \dfrac{1}{4\pi\varepsilon_0}\right)$

77. As, charge on capacitor, $Q = CV$

Since, V is constant, therefore $Q \propto C$

Hence, C becomes $\dfrac{100}{40} = 2.5$ times

$\therefore \quad K = 2.5$

78. As, $C = \dfrac{A\varepsilon_0}{d} \quad \ldots\text{(i)}$

After inserting the slab,

$$C' = \frac{A\varepsilon_0}{(d-b)} = \frac{A\varepsilon_0}{d - \dfrac{d}{2}}$$

or $\quad C' = \dfrac{2A\varepsilon_0}{d} \quad \ldots\text{(ii)}$

From Eqs. (i) and (ii), we get

$$\frac{C'}{C} = \frac{2}{1} = 2:1$$

79. As, $r_b - r_a = 1\text{ mm} = 10^{-3}$ m

From $\quad C = \dfrac{4\pi\varepsilon_0 r_a r_b}{r_b - r_a}$

$\Rightarrow \quad 10^{-6} = \dfrac{1\,(r_b - 10^{-3})\,r_b}{9\times 10^9\,(10^{-3})}$

$\Rightarrow \quad r_b^2 = 9$

or $\quad r_b = 3$ m

80. On bringing the charged metal plates closer, electric field $\mathbf{E}$ in the intervening space is

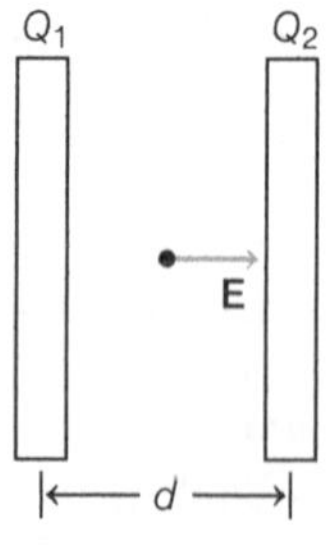

$\therefore \quad E = \dfrac{\sigma_1}{2\varepsilon_0} - \dfrac{\sigma_2}{2\varepsilon_0} = \dfrac{Q_1}{2A\varepsilon_0} - \dfrac{Q_2}{2A\varepsilon_0}$

or $$E = \frac{(Q_1 - Q_2)}{2A\varepsilon_0} = \frac{V}{d}$$

or $$V = \frac{(Q_1 - Q_2)d}{2A\varepsilon_0}$$

$$\Rightarrow \quad V = \frac{Q_1 - Q_2}{2C} \qquad \left(\because C = \frac{\varepsilon_0 A}{d}\right)$$

81. The capacities of two individual condensers are

$$C_1 = \frac{K_1\varepsilon_0 A}{d_1}$$

and $$C_2 = \frac{K_2\varepsilon_0 A}{d_2}$$

The arrangement is equivalent to two capacitors joined in series. Therefore, the combined capacity (C_s) is given by

$$\frac{1}{C_s} = \frac{1}{C_1} + \frac{1}{C_2} = \frac{d_1}{K_1\varepsilon_0 A} + \frac{d_2}{K_2\varepsilon_0 A}$$

$$\Rightarrow \quad \frac{1}{C_s} = \frac{1}{\varepsilon_0 A}\left[\frac{d_1}{K_1} + \frac{d_2}{K_2}\right] = \frac{d_1K_2 + d_2K_1}{\varepsilon_0 AK_1K_2}$$

$$\Rightarrow \quad C_s = \frac{\varepsilon_0 AK_1K_2}{d_1K_2 + d_2K_1} \text{ but } C_s = \frac{K\varepsilon_0 A}{(d_1 + d_2)}$$

$$\therefore \quad K = \frac{K_1K_2(d_1 + d_2)}{(d_1K_2 + d_2K_1)}$$

82. Positive plate of all the three condensers is connected to one point A and negative plate of all the three condensers is connected to point B, *i.e.* they are joined in parallel.

$$\therefore \quad C_p = 3 + 3 + 3 = 9\,\mu\text{F}$$

83. The arrangement can be redrawn as shown in the adjoining figure

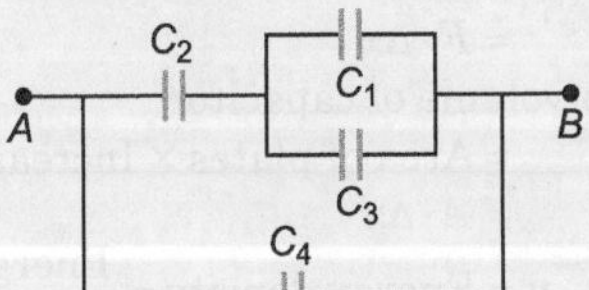

$$\therefore \quad C_{13} = C_1 + C_3 = 9 + 9 = 18\,\mu\text{F}$$

and $$C_{2\text{-}13} = \frac{C_2 \times C_{13}}{C_2 + C_{13}} = \frac{9\,\mu\text{F} \times 18\,\mu\text{F}}{(9+18)\,\mu\text{F}} = 6\,\mu\text{F}$$

$$\therefore \quad C = C_{2\text{-}13} + C_4 = 6\,\mu\text{F} + 9\,\mu\text{F} = 15\,\mu\text{F}$$

84. The required capacitance, $C = 2\,\mu$F

Potential difference,

$$V = 1\text{ kV} = 1000\text{ V}$$

Capacitance of each capacitor $C_1 = 1\,\mu$F and it can withstand a potential difference of $V_1 = 400$ V

Let the n capacitors are connected in series and there are m rows of such capacitors.

As the potential difference across each row is 1000 V.

So, the potential difference across each capacitor $= \frac{1000}{n}$

Minimum number of capacitors that must be connected in series in a row are

$$\frac{1000}{n} = 400$$

$$\Rightarrow \quad n = 2.5$$

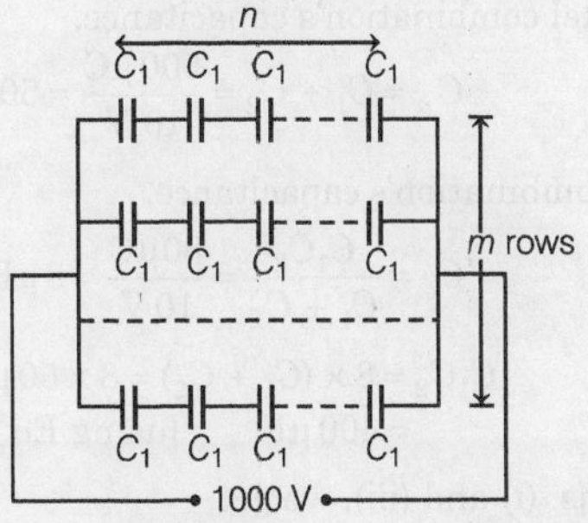

Here, n is the number of capacitors, so it should be a whole number. If we take $n = 2$, then potential difference across each capacitor is 500 V.

Here, according to question, a capacitor can bear only 400 V, so they burst. We take the value of $n = 3$.

So, the capacitance of each row (in series),

$$\frac{1}{C'} = \frac{1}{1} + \frac{1}{1} + \frac{1}{1} = \frac{3}{1}$$

$$\Rightarrow \quad C' = \frac{1}{3}$$

The total capacitance of m rows is $m \times \frac{1}{3} = \frac{m}{3}$

According to question, the total capacitance required is $2\,\mu$F.

So, $$\frac{m}{3} = 2 \Rightarrow m = 6$$

Thus, the total number of capacitor $= m \times n = 3 \times 6 = 18$

So, 1 μF capacitors are connected in 6 rows having 3 capacitors in each row.

85. Net capacity of 5 capacitors joined in parallel $= 5 \times 2 = 10\,\mu$F

Now, it is connected with two capacitors of $2\,\mu$F each in series, thus equivalent capacitance is $\frac{10}{11}\,\mu$F.

86. The arrangement behaves as a combination of 2 capacitors each of capacitance, $C = \frac{\varepsilon_0 A}{d}$.

Thus, equivalent capacity $= 2C$

$\therefore$ Total energy stored, $U = \frac{1}{2} \times (2C)V^2 = CV^2 = \frac{\varepsilon_0 A}{d}V^2$

87. In the arrangement shown both plates of capacitor C_3 are joined to point B. Hence, it does not act as a capacitor and is superfluous. Now, C_1 and C_2 are in parallel, hence

$$C_{AB} = C_1 + C_2 = C + C = 2C$$

88. In the given figure,

slope of OA > slope of OB

Since, we know that,

net capacitance of parallel combination > net capacitance of series combination

$\therefore$ Parallel combination's capacitance,

$$C_p = C_1 + C_2 = \frac{500\,\mu\text{C}}{10\text{ V}} = 50\,\mu\text{F} \quad \dots \text{(i)}$$

Series combination's capacitance,

$$C_s = \frac{C_1 C_2}{C_1 + C_2} = \frac{80\,\mu\text{C}}{10\text{ V}} = 8\,\mu\text{F} \quad \dots \text{(ii)}$$

or $\quad C_1 C_2 = 8 \times (C_1 + C_2) = 8 \times 50\,\mu\text{F}$

$= 400\,\mu\text{F} \quad$ [using Eq. (i)] $\quad \dots$(iii)

From Eqs. (i) and (iii), we get

$$C_1 = 50 - C_2 \text{ and } C_1 C_2 = 400$$

$\Rightarrow \quad C_2 (50 - C_2) = 400$

$\Rightarrow \quad 50\, C_2 - C_2^2 = 400$

or $\quad C_2^2 - 50\, C_2 + 400 = 0$

$\Rightarrow \quad C_2 = \dfrac{+50 \pm \sqrt{2500 - 1600}}{2} = \dfrac{+50 \pm 30}{2}$

$\Rightarrow \quad C_2 = +40\,\mu\text{F}$ or $+10\,\mu\text{F}$

Also, $\quad C_1 = 50 - C_2$

$\Rightarrow \quad C_1 = +10\,\mu\text{F}$ or $+40\,\mu\text{F}$

Hence, capacitance of two given capacitors is $10\,\mu\text{F}$ and $40\,\mu\text{F}$.

89. Given circuit is

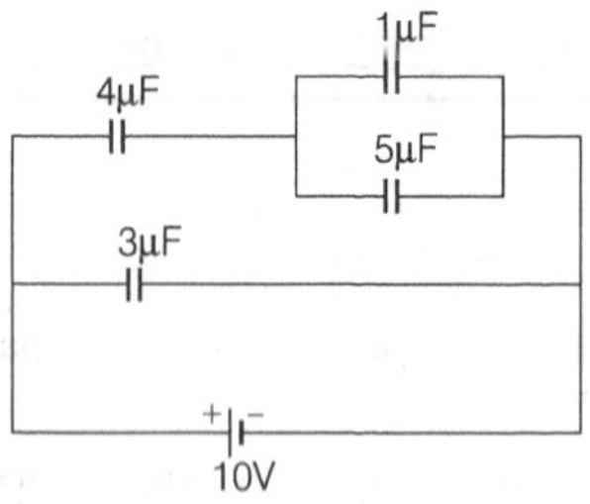

In parallel, $\quad C_{\text{eq}} = 5 + 1 = 6\,\mu\text{F}$

and in series, $C'_{\text{eq}} = \dfrac{6 \times 4}{6 + 4} = 2.4\,\mu\text{F}$

This is equivalent to

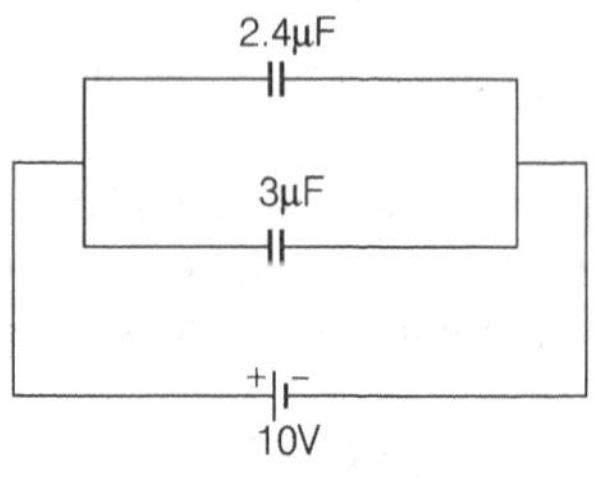

So, potential difference across upper branch = 10 V

Using $Q = C \times V$, charge delivered to upper branch is

$$Q = C'_{\text{eq}} \cdot V$$
$$= 2.4\,\mu\text{F} \times 10\text{V} = 24\,\mu\text{C}$$

As we know, in series connection, same charge is shared by capacitors, so charge on $4\,\mu$F capacitor and $6\,\mu$F capacitor would be same,

i.e. $\quad Q'_{4\mu\text{F}} = 24\,\mu\text{C}$

90. The two condensers are connected in parallel, we have

$$C_p = C + \frac{C}{2} = \frac{3C}{2}$$

$\therefore$ Total work done in charging both the condensers,

$$W = \frac{1}{2} C_p V^2 = \frac{1}{2} \times \frac{3C}{2} V^2 = \frac{3}{4} CV^2$$

91. On sharing of charges loss in electrical energy,

$$\Delta U = \frac{C_1 C_2}{2(C_1 + C_2)} (V_1 - V_2)^2$$

In present case, $C_1 = C_2 = C$

$\therefore \quad \Delta U = \dfrac{C^2}{2(2C)} (V_1 - V_2)^2 = \dfrac{1}{4} C (V_1 - V_2)^2$

92. Let the distance between the plates be increased by a very small distance Δx. The force on each plate is F.

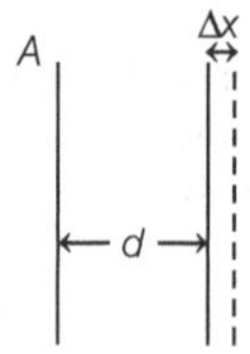

The amount of work done in increasing the separation by Δx

$= $ Force × Increased distance

$= F \cdot \Delta x \quad \dots$(i)

Increase in volume of capacitor

$=$ Area of plates × Increased distance

$= A \cdot \Delta x$

$$u = \text{Energy density} = \frac{\text{Energy}}{\text{Volume}}$$

$$\text{Energy} = u \times \text{Volume} = u \cdot A \cdot \Delta x \quad \dots \text{(ii)}$$

As, $\quad$ energy = work done

$u \cdot A \cdot \Delta x = F \cdot \Delta x \quad$ [from Eqs. (i) and (ii)]

$\Rightarrow \quad F = u \cdot A$

$= \dfrac{1}{2} \varepsilon_0 E^2 \cdot A \quad \left(\because\ u = \dfrac{1}{2} \varepsilon_0 E^2 \text{ and } E = \dfrac{V}{d} \right)$

$= \dfrac{1}{2} \varepsilon_0 \cdot \dfrac{V^2}{d^2} \cdot A$

$= \left(\dfrac{\varepsilon_0 A}{d} \cdot V \right) \dfrac{V}{d} \times \dfrac{1}{2} \quad \left(\because\ C = \dfrac{\varepsilon_0 A}{d},\ CV = q \right)$

$= \dfrac{1}{2} \cdot E \cdot C \cdot V$

$= \dfrac{1}{2} QE$

The factor of $\frac{1}{2}$ in the force can be explained by the fact that the field is zero inside the conductor and outside the conductor, field is E. So, the average value of the field, *i.e.* $\frac{E}{2}$ contributes to the force against which the plates are moved.

93.

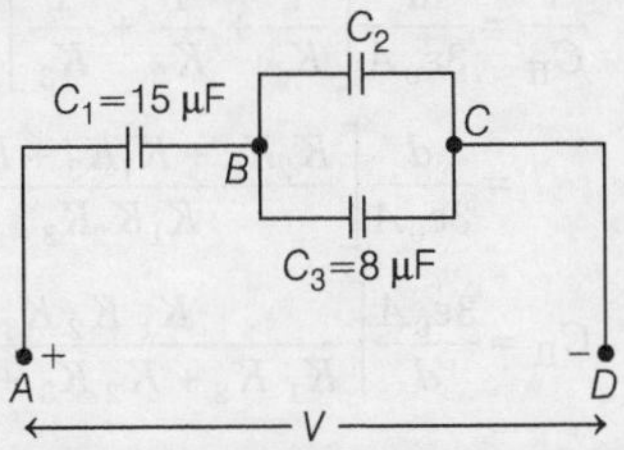

In given arrangement,

V_{BC} = potential drop across C_2 = 20V

As combination of capacitors C_2 and C_3 is in parallel, so potential drop across C_3 is also 20 V.

So, $\quad Q_3$ = charge on $C_3 = C_3 \times V$

$= 8 \times 10^{-6} \times 20 = 160\ \mu C$

Also, total charge given by cell, $Q = 750\ \mu C$

As combination of capacitors C_2 and C_3 is in series with capacitor C_1, so charge on

C_1 = charge on combination of capacitors C_2 and C_3 = total charge given by cell.

Hence, $Q_2 + Q_3 = 750\ \mu C$

$\Rightarrow \quad Q_2 + 160\ \mu C = 750\ \mu C$

$\Rightarrow \quad Q_2 = 750 - 160 = 590\ \mu C$

94.

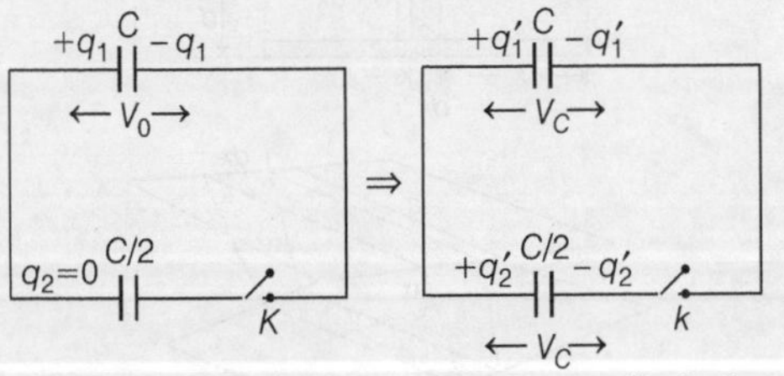

Before charge distribution $\quad$ After charge distribution

Using law of conservation of charge, we have

$$q_1 + q_2 = q_1' + q_2'$$

$$CV_0 + \frac{C}{2}(0) = CV_C + \frac{C}{2}(V_C)$$

$$CV_0 + 0 = \frac{3}{2}CV_C$$

$$\Rightarrow \quad CV_0 = \frac{3}{2}CV_C$$

$$\Rightarrow \quad V_C = \frac{2}{3}V_0 \quad \text{...(i)}$$

Here, V_C = common potential.

Now, initial electrical energy,

$$U_i = \frac{1}{2}CV_0^2 + \frac{1}{2}\left(\frac{C}{2}\right)(0)^2$$

$$= \frac{1}{2}CV_0^2 + 0$$

$$= \frac{1}{2}CV_0^2$$

and final electrical energy,

$$U_f = \frac{1}{2}\left(C + \frac{C}{2}\right)V_C^2$$

$$= \frac{1}{2}\left(\frac{3C}{2}\right)\left(\frac{2}{3}V_0\right)^2$$

$$= \frac{1}{2} \times \frac{3C}{2} \times \frac{4}{9}V_0^2$$

$$= \frac{1}{3}CV_0^2$$

Loss in electrical energy,

$$\Delta U_{\text{loss}} = U_i - U_f$$

$$= \frac{1}{2}CV_0^2 - \frac{1}{3}CV_0^2$$

$$= \frac{1}{6}CV_0^2$$

95. A capacitor filled with dielectrics can be treated/compared as series/parallel combinations of capacitor having individual dielectric. *e.g.*

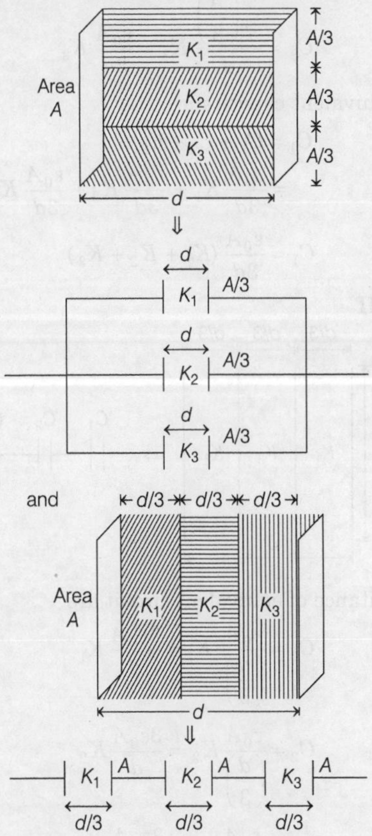

Case I

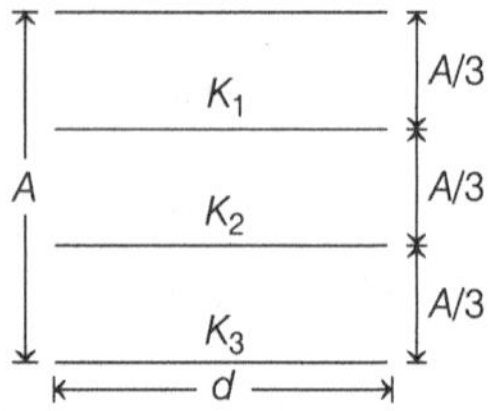

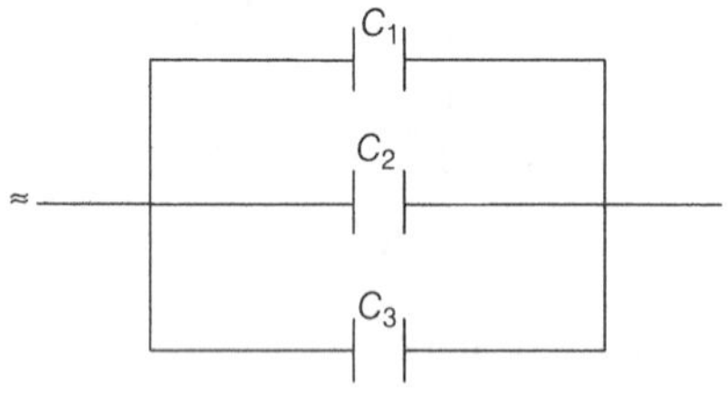

Capacitance in the equivalent circuit are

$$C_1 = \frac{\varepsilon_0\left(\frac{A}{3}\right)}{d} K_1 = \frac{\varepsilon_0 A}{3d} K_1$$

$$C_2 = \frac{\varepsilon_0\left(\frac{A}{3}\right)}{d} K_2 = \frac{\varepsilon_0 A}{3d} K_2$$

and $$C_3 = \frac{\varepsilon_0\left(\frac{A}{3}\right)}{d} K_3 = \frac{\varepsilon_0 A}{3d} K_3$$

So, equivalent capacitance,

$$C_I = C_1 + C_2 + C_3$$

$$= \frac{\varepsilon_0 A}{3d} K_1 + \frac{\varepsilon_0 A}{3d} K_2 + \frac{\varepsilon_0 A}{3d} K_3$$

$$C_I = \frac{\varepsilon_0 A}{3d}(K_1 + K_2 + K_3) \quad \ldots\text{(i)}$$

Case II

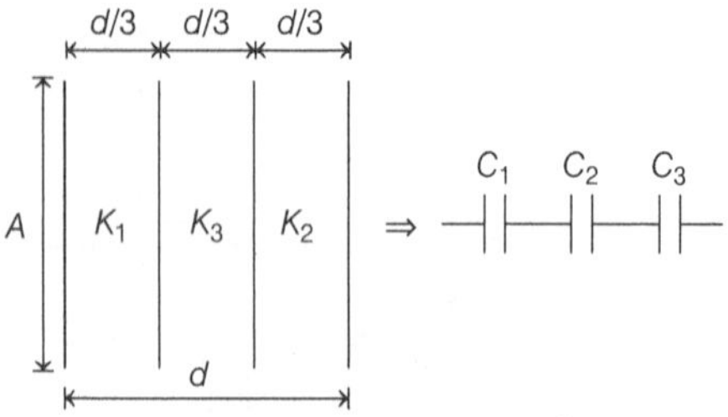

Capacitance of equivalent circuit are

$$C_1 = \frac{\varepsilon_0 A}{\left(\frac{d}{3}\right)} \cdot K_1 = \frac{3\varepsilon_0 A}{d} K_1$$

$$C_2 = \frac{\varepsilon_0 A}{\left(\frac{d}{3}\right)} K_2 = \frac{3\varepsilon_0 A}{d} K_2$$

and $$C_3 = \frac{\varepsilon_0 A}{\left(\frac{d}{3}\right)} K_3 = \frac{3\varepsilon_0 A}{d} K_3$$

So, equivalent capacitance,

$$\frac{1}{C_{II}} = \frac{1}{C_1} + \frac{1}{C_2} + \frac{1}{C_3}$$

$$= \frac{d}{3\varepsilon_0 AK_1} + \frac{d}{3\varepsilon_0 AK_2} + \frac{d}{3\varepsilon_0 AK_3}$$

$$\Rightarrow \quad \frac{1}{C_{II}} = \frac{d}{3\varepsilon_0 A}\left[\frac{1}{K_1} + \frac{1}{K_2} + \frac{1}{K_3}\right]$$

$$= \frac{d}{3\varepsilon_0 A}\left[\frac{K_2K_3 + K_1K_3 + K_1K_2}{K_1K_2K_3}\right]$$

$$C_{II} = \frac{3\varepsilon_0 A}{d}\left[\frac{K_1 K_2 K_3}{K_1 K_2 + K_2 K_3 + K_3 K_1}\right] \quad \ldots\text{(ii)}$$

On dividing Eq. (i) by Eq. (ii), we get

$$\frac{C_I}{C_{II}} = \frac{\varepsilon_0 A}{3d}(K_1 + K_2 + K_3) \times \frac{d(K_1K_2 + K_2K_3 + K_3K_1)}{3\varepsilon_0 A(K_1K_2K_3)}$$

$$= \frac{(K_1 + K_2 + K_3)(K_1K_2 + K_2K_3 + K_3K_1)}{9K_1K_2K_3}$$

Now, energy stored in capacitor, $E = \frac{1}{2}CV^2$

$$\Rightarrow \quad E \propto C$$

$$\therefore \quad \frac{E_I}{E_{II}} = \frac{C_I}{C_{II}}$$

96. For given capacitor, consider a small width dx at a distance x from origin O.

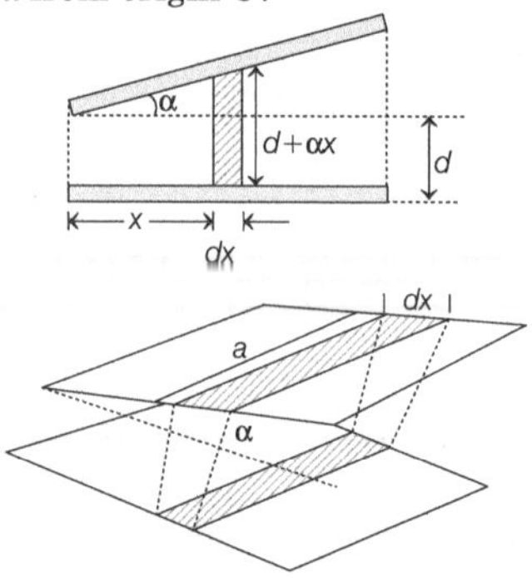

For this elemental capacitor, we have

Area of each plate $= a \times dx$

Distance between plates $= d + \alpha x$

Capacity of this elemental capacitor, $dC = \frac{\varepsilon_0 a dx}{d + \alpha x}$

Total capacity of arrangement is sum of capacitances of all such elemental capacitors because they all are in parallel.

Hence, $C_{\text{system}} = \int_0^a dC = \int_0^a \frac{\varepsilon_0 a}{d + \alpha x} \cdot dx = \frac{\varepsilon_0 a}{\alpha} \ln\left(1 + \frac{\alpha a}{d}\right)$

As, for small values of $\frac{a\alpha}{d}$,

$$\ln\left(1 + \frac{a\alpha}{d}\right) \approx \frac{a\alpha}{d} - \frac{1}{2}\left(\frac{a\alpha}{d}\right)^2$$

So, $$C_{\text{system}} = \frac{\varepsilon_0 a^2}{d}\left(1 - \frac{a\alpha}{2d}\right)$$

Round II

1. In an equilateral triangle, distance of centroid from all the vertices is same (say r).

$$\therefore \quad V = V_1 + V_2 + V_3$$

$$= \frac{1}{4\pi\varepsilon_0}\left[\frac{q_1}{r} + \frac{q_2}{r} + \frac{q_3}{r}\right] = \frac{1}{4\pi\varepsilon_0}\left[\frac{2q}{r} - \frac{q}{r} - \frac{q}{r}\right] = 0$$

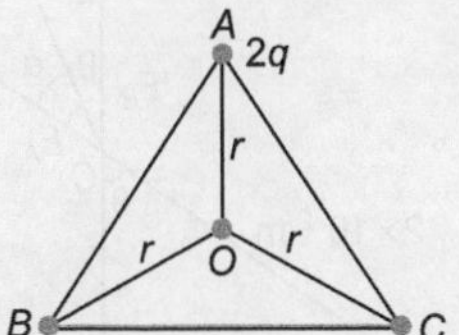

But $\quad \mathbf{E}_A = \frac{1}{4\pi\varepsilon_0}\cdot\frac{2q}{r^2}$ along AO,

$$\mathbf{E}_B = \frac{1}{4\pi\varepsilon_0}\cdot\frac{q}{r^2} \text{ along } OB$$

and $\quad \mathbf{E}_C = \frac{1}{4\pi\varepsilon_0}\cdot\frac{q}{r^2}$ along OC

Obviously, $\mathbf{E}_B + \mathbf{E}_C$ will also be in the direction of AO (extended) and hence, $\mathbf{E}_A$ and $(\mathbf{E}_B + \mathbf{E}_C)$ being in same direction will not give zero resultant.

2. Bob will experience an additional force, $F = qE$ in vertically upward direction and hence, effective acceleration due to gravity is reduced from g to $(g - a)$. Consequently, time period of oscillation will become

$$T = 2\pi\sqrt{\frac{l}{(g-a)}}$$

i.e. time period will increase.

3. They will not experience any force, if $|\mathbf{F}_G| = |\mathbf{F}_e|$

$$\Rightarrow \quad G\frac{m^2}{(16\times10^{-2})^2} = \frac{1}{4\pi\varepsilon_0}\cdot\frac{q^2}{(16\times10^{-2})^2}$$

$$\Rightarrow \quad \frac{q}{m} = \sqrt{4\pi\varepsilon_0 G}$$

4. The given situation can be shown as a combination of two capacitors

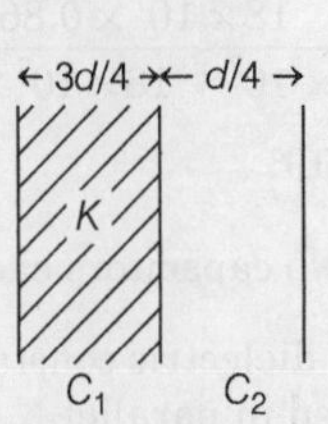

The equivalent capacitance is

$$\frac{1}{C'} = \frac{1}{C_1} + \frac{1}{C_2}$$

$$= \frac{3d/4}{\varepsilon KA} + \frac{d/4}{\varepsilon_0 A}$$

$$= \frac{d}{4\varepsilon_0 A}\left(\frac{3+K}{K}\right)$$

$$C' = \frac{4K}{3+K}\frac{\varepsilon_0 A}{d}$$

$$= \frac{4K}{K+3}C_0 \qquad \left(\because C_0 = \frac{\varepsilon_0 A}{d}\right)$$

5. The arrangement shows a Wheatstone bridge.

As $\frac{C_1}{C_3} = \frac{C_4}{C_5} = 1$, therefore the bridge is balanced.

$$\frac{1}{C_{s_1}} = \frac{1}{4} + \frac{1}{4} = \frac{2}{4}, C_{s_1} = 2\,\mu\text{F}$$

Similarly, $C_{s_2} = 2\,\mu\text{F}$

$\therefore$ Effective capacitance

$$= C_p = C_{s_1} + C_{s_2}$$

$$= 2 + 2 = 4\,\mu\text{F}$$

6. Charge on each plate of capacitor will be

$$Q = \pm CV$$

$$= \pm 25\times10^{-6}\times200$$

$$= \pm 5\times10^{-3}\ \text{C}$$

7. Force on $-q_1$ due to q_2,

$$F_{12} = \frac{kq_1q_2}{b^2}, \text{ along } X\text{-axis}$$

Force on $-q_1$ due to $-q_3$, $F_{13} = \frac{kq_1q_3}{a^2}$,

at $\angle\theta$ with negative direction of Y-axis.

$\therefore$ x-component of force on $-q_1$ is

$$F_x = F_{12} + F_{13}\sin\theta$$

$$= kq_1\left[\frac{q_2}{b^2} + \frac{q_3}{a^2}\sin\theta\right]$$

i.e. $\quad F_x \propto \left[\frac{q_2}{b^2} + \frac{q_3}{a^2}\sin\theta\right]$

8. Here, $\quad \mathbf{E} = \mathbf{E}_1 + \mathbf{E}_2 + \mathbf{E}_3$

$$= \left(+\frac{\sigma}{2\varepsilon_0}\right)(-\hat{\mathbf{k}}) + \left(\frac{2\sigma}{2\varepsilon_0}\right)(-\hat{\mathbf{k}}) + \left(\frac{\sigma}{2\varepsilon_0}\right)(-\hat{\mathbf{k}})$$

$$= -\left(\frac{2\sigma}{\varepsilon_0}\right)\hat{\mathbf{k}}$$

9. Here, KE = 100 eV = $100\times1.6\times10^{-19}$ J = 1.6×10^{-17} J.

This is lost when electron moves through a distance (d) towards the negative plate.

$$\therefore \quad \text{KE} = \text{Work done}$$

$$= F\times s = qE\times s = e\left(\frac{\sigma}{\varepsilon_0}\right)d$$

$$d = \frac{(\text{KE})\varepsilon_0}{e\sigma}$$

$$\Rightarrow \quad d = \frac{1.6\times10^{-17}\times8.86\times10^{-12}}{1.6\times10^{-19}\times2\times10^{-6}}$$

$$= 4.43\times10^{-4}\text{m}$$

$$= 0.44\text{ mm}$$

10. As, electric field is along positive x-axis and $E = -\frac{dV}{dx}$.

Hence, potential at A must be greater than that at B, *i.e.* $V_A > V_B$.

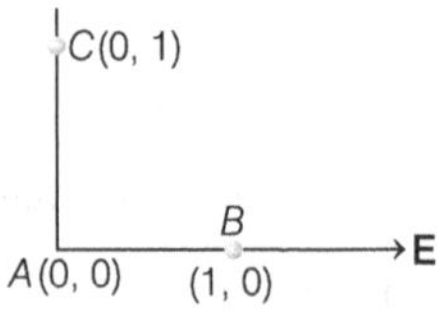

11. Due to additional charge of $-3Q$, given to external spherical shell, the potential difference between conducting sphere and the outer shell will not change because by presence of charge on outer shell, potential everywhere inside and on the surface of the shell, will change by same amount. Therefore, the potential difference between sphere and shell will remain unchanged.

12. With S_1 and S_3 closed, the capacitors C_1 and C_2 are in series arrangement. In series arrangement, potential difference developed across capacitors are in the inverse ratio of their capacities. Hence,

$$\frac{V_1'}{V_2'} = \frac{C_2}{C_1} = \frac{3\text{ pF}}{2\text{ pF}} = \frac{3}{2} \text{ and}$$

$$V_1' + V_2' = V_1 + V_2 = 30 + 20 = 50\text{V}$$

On simplification, we get

$$V_1' = V_1 = 30\text{V}$$

and $$V_2' = V_2 = 20\text{V}$$

13. To calculate force on a point charge q, we need to find electric field due to uniformly charged spherical shell at various points.

If $r < R$, *i.e.* inside the shell, then $E = 0$

$\Rightarrow$ $$F = qE = 0$$

If $r > R$, *i.e.* outside the shell, then $E = \frac{1}{4\pi\varepsilon_0}\frac{Q}{r^2}$

$\Rightarrow$ $$F = qE$$

$\Rightarrow$ $$F = \frac{1}{4\pi\varepsilon_0}\frac{Qq}{r^2}$$

If $r = R$, *i.e.* at surface of shell, then

$$E = \frac{1}{4\pi\varepsilon_0}\frac{Q}{R^2}$$

$\Rightarrow$ $$F = \frac{Qq}{4\pi\varepsilon_0 R^2}$$

If $E = R$, $E = \frac{kQ}{R^2}$ and $F = \frac{kQq}{R^2}$

14. Figure shows the right angled triangle ABC, such that $AC = 2 \times 10^{-2}$ m and $BC = 10^{-2}$ m. The charges of $q_A = 4 \times 10^{-6}$ C and $q_B = 2 \times 10^{-6}$ C are placed at the vertices A and B respectively. Let E_A and E_B be electric field intensity at point C due to charges q_A and q_B respectively, then

$$E_A = \frac{1}{4\pi\varepsilon_0} \times \frac{q_A}{(AC)^2}$$

$$= 9 \times 10^9 \times \frac{4 \times 10^{-6}}{(2 \times 10^{-2})^2}$$

$$= 9 \times 10^7 \text{ NC}^{-1} \quad \text{(along } AC\text{)}$$

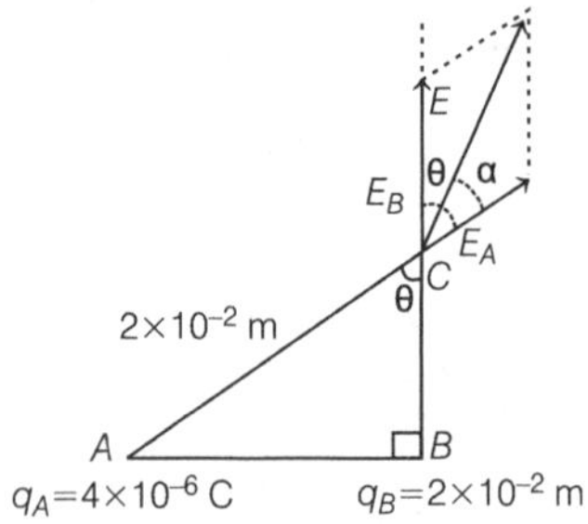

$$E_B = \frac{1}{4\pi\varepsilon_0} \times \frac{q_B}{BC^2} = 9 \times 10^9 \times \frac{2 \times 10^{-6}}{(10^{-2})^2}$$

$$= 18 \times 10^7 \text{ NC}^{-1} \quad \text{(along } BC\text{)}$$

If θ is angle between the directions of E_A and E_B, then

$$E = \sqrt{E_A^2 + E_B^2 + 2E_AE_B\cos\theta}$$

In right angled ΔABC, $\angle ACB = \theta$

$\therefore$ $$\cos\theta = \frac{BC}{AC} = \frac{10^{-2}}{2 \times 10^{-2}} = 0.5 \text{ or } \theta = 60^\circ$$

Hence,

$$E = \sqrt{(9 \times 10^7)^2 + (18 \times 10^7)^2 + 2 \times 9 \times 10^7 \times 18 \times 10^7 \times 0.5}$$

$$= 9 \times 10^7 \sqrt{1 + 4 + 2}$$

$$= 2.38 \times 10^8 \text{ NC}^{-1}$$

Suppose that the resultant electric intensity E makes an angle α with line AC, then

$$\tan\alpha = \frac{E_B \sin\theta}{E_A + E_B\cos\theta}$$

$$= \frac{18 \times 10^7 \sin 60^\circ}{9 \times 10^7 + 18 \times 10^7 \times \cos 60^\circ}$$

$$= \frac{18 \times 10^7 \times 0.866}{9 \times 10^7 + 18 \times 10^7 \times 0.5} = 0.866$$

or $$\alpha = 40.9^\circ$$

15. Capacitance of two capacitors each of area $\frac{A}{2}$, plate separation d but dielectric constants K_1 and K_2, respectively joined in parallel

$$C_1 = \frac{K_1\varepsilon_0\left(\frac{A}{2}\right)}{d/2} + \frac{K_2\varepsilon_0\left(\frac{A}{2}\right)}{d/2} = \frac{(K_1 + K_2)\varepsilon_0 A}{d}$$

It is in series with a capacitor of plate area A, plate separation $d/2$ and dielectric constant K_3, *i.e.* $C_2 = \frac{K_3\varepsilon_0 A}{d/2}$.

If resultant capacitance be taken as $C = \frac{K\varepsilon_0 A}{d}$, then

$$\frac{1}{C} = \frac{1}{C_1} + \frac{1}{C_2}$$

$$\therefore \quad \frac{d}{K\varepsilon_0 A} = \frac{d}{(K_1 + K_2)\varepsilon_0 A} + \frac{d/2}{K_3 \varepsilon_0 A}$$

$$\Rightarrow \quad \frac{1}{K} = \frac{1}{(K_1 + K_2)} + \frac{1}{2K_3}$$

16. Common potential,

$$V = \frac{\text{Total charge}}{\text{Total capacity}} = \frac{Q + 0}{4\pi\varepsilon_0 (r + r')}$$

$\therefore$ Charge on smaller sphere

$$= 4\pi\varepsilon_0 r' \times V = \frac{Qr'}{r + r'}$$

17. Electric field in each quadrant will look like this

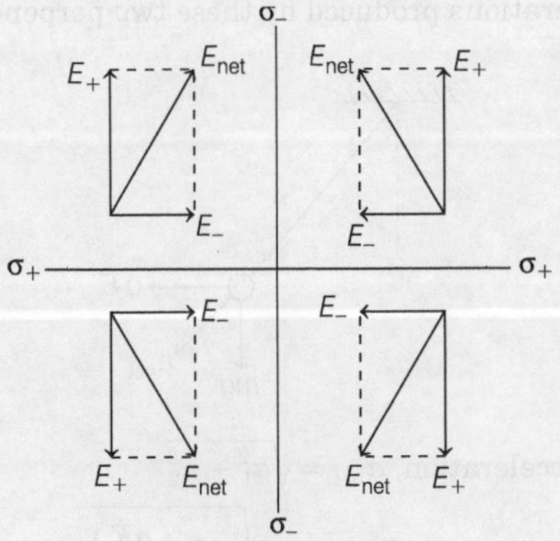

$$\because \quad |\sigma_+| > |\sigma_-|$$

$$\therefore \quad |E_+| > |E_-|$$

So, the final electric field will become as shown below

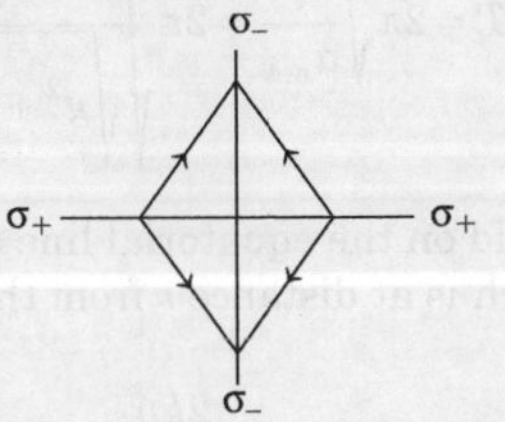

So, the nearest matching option is

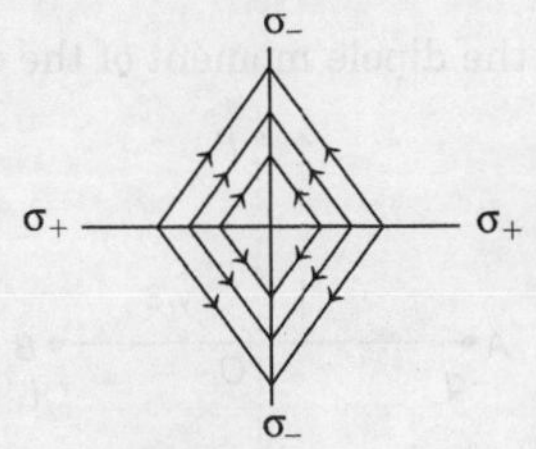

Hence, correct option is (c).

18. Here, $C_{23} = 30 + 30 = 60$ pF.

Total equivalent capacitance is given by

$$\frac{1}{C} = \frac{1}{C_1} + \frac{1}{C_{23}} + \frac{1}{C_4} = \frac{1}{30} + \frac{1}{60} + \frac{1}{120} = \frac{7}{120}$$

$$\Rightarrow \quad C = \frac{120}{7} \text{ pF}$$

$\therefore$ Total charge, $Q = CV = \frac{120}{7} \times 140 \text{ pC} = 2400 \text{ pC}$

$$\therefore \quad V_1 = \frac{Q}{C_1} = \frac{2400 \text{ pC}}{30 \text{ pF}} = 80 \text{ V}$$

$$V_2 = V_3 = V_{23} = \frac{Q}{C_{23}} = \frac{2400 \text{ pC}}{60 \text{ pF}} = 40 \text{ V}$$

and $$V_4 = \frac{Q}{C_4} = \frac{2400 \text{ pC}}{120 \text{ pF}} = 20 \text{ V}$$

19. The system of two charges, *i.e.* $+q$ and $-q$ that are separated by distance d can be considered as a dipole. Thus, the charge Q would be at D distance from the centre of an electric dipole on its axial line.

So, the total potential energy of the system will be due to two components.

(1) Potential energy of dipole's own system,

$$(\text{PE})_1 = \frac{Kq_1q_2}{d} = -\frac{Kq^2}{d} \quad \ldots\text{(i)}$$

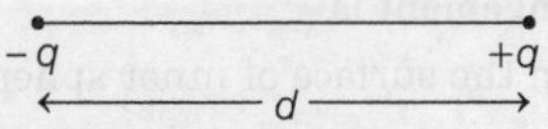

(2) Potential energy of charge Q and dipole system,

$$(\text{PE})_2 = -\frac{KQq}{D^2} \cdot d \quad \ldots\text{(ii)}$$

Hence, total potential energy of the system,

$$(\text{PE})_{\text{total}} = (\text{PE})_1 + (\text{PE})_2 = -\frac{Kq^2}{d} - \frac{KQq}{D^2} \cdot d$$

$$\Rightarrow \quad (\text{PE})_{\text{total}} = -\frac{1}{4\pi\varepsilon_0}\left[\frac{q^2}{d} + \frac{Qqd}{D^2}\right]$$

20. Here, $\frac{r_1}{r_2} = \frac{1 \text{ mm}}{2 \text{ mm}} = \frac{1}{2}$. When the spheres are connected by a conducting wire, then $V_1 = V_2$

or $$\frac{q_1}{4\pi\varepsilon_0 r_1} = \frac{q_2}{4\pi\varepsilon_0 r_2}$$

$$\Rightarrow \quad \frac{q_1}{q_2} = \frac{r_1}{r_2} = \frac{1}{2}$$

Now, $$\frac{E_1}{E_2} = \frac{q_1}{q_2} \cdot \left(\frac{r_2}{r_1}\right)^2 = \frac{1}{2} \times \left(\frac{2}{1}\right)^2 = \frac{4}{2} = 2:1$$

21.

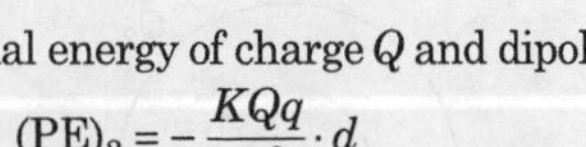

Let an element of length dx, charge dq, at distance x from point O.

$$dV = k\frac{dq}{x}$$

where, $$dq = \frac{Q}{L}\,dx$$

$$\therefore \quad V = \int_L^{2L} \frac{k\,dq}{x} = k\int_L^{2L} \frac{\left(\frac{Q}{L}\right)dx}{x}$$

$$= \frac{Q}{4\pi\varepsilon_0 L} \int_L^{2L} \left(\frac{1}{x}\right) dx$$
$$= \frac{Q}{4\pi\varepsilon_0 L} [\log_e x]_L^{2L}$$
$$= \frac{Q}{4\pi\varepsilon_0 L} [\log_e 2L - \log_e L]$$
$$= \frac{Q}{4\pi\varepsilon_0 L} \left[\log_e^2 \frac{2L}{L}\right]$$
$$= \frac{Q}{4\pi\varepsilon_0 L} \ln(2)$$

22. Potential due to a hollow spherical charge distribution is

$$V = \begin{cases} \frac{kq}{r}, & r > R \text{ (outside points)} \\ \frac{kq}{R}, & r = R \text{ (at surface)} \end{cases}$$
$$\begin{cases} \frac{kq}{R}, & r < R \text{ (inside points)} \end{cases}$$

Given arrangement is

Potential on the surface of inner sphere,

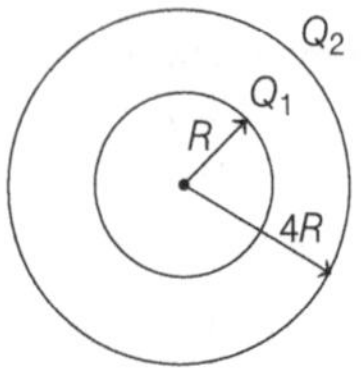

$V(R)$ = potential due to charge Q_1 + potential due to charge Q_2

$$= \frac{kQ_1}{R} + \frac{kQ_2}{4R}$$

Potential on the surface of outer sphere,

$V(4R)$ = potential due to charge Q_1 + potential due to charge Q_2

$$= \frac{kQ_1}{4R} + \frac{kQ_2}{4R}$$

The potential difference $V(R) - V(4R)$

$$= \left(\frac{kQ_1}{R} + \frac{kQ_2}{4R}\right) - \left(\frac{kQ_1}{4R} + \frac{kQ_2}{4R}\right)$$
$$= \frac{kQ_1}{R}\left(\frac{3}{4}\right) = \frac{3}{16\pi\varepsilon_0} \cdot \frac{Q_1}{R}$$

23. We have, $U_i = \frac{1}{2}CV^2 = \frac{1}{2} \times 2 \times V^2$

$\Rightarrow \quad U_i = V^2 \quad \quad$ ...(i)

Final voltage, $V_f = \frac{C_1 V_1}{C_1 + C_2} = \frac{2V}{10} = \frac{V}{5}$

and $\quad U_f = \frac{1}{2}(C_1 + C_2) V_f^2$

$$= \frac{1}{2}(2+8)\left(\frac{V}{5}\right)^2$$
$$= 5\left(\frac{V}{5}\right)^2 = \frac{5V^2}{25} = \frac{V^2}{5} \quad \text{...(ii)}$$

Now, $\frac{U_i - U_f}{U_i} \times 100 = \frac{V^2 - \frac{V^2}{5}}{V^2} \times 100$

$$= \frac{4V^2/5}{V^2} \times 100 = 80\%$$

24. The resultant electric field is given by

$$\mathbf{E} = \mathbf{E}_1 + \mathbf{E}_2$$

or $\quad E = \frac{\sigma_1}{2\varepsilon_0} + \frac{\sigma_2}{2\varepsilon_0} = \frac{\sigma_1 + \sigma_2}{2\varepsilon_0}$

25. When pendulum is oscillating between capacitor plates, it is subjected to two forces

(i) Weight downwards $(ma_1) = mg$

(ii) Electrostatic force acting horizontally $(ma_2) = qE$

So, net acceleration of pendulum bob is resultant of accelerations produced by these two perpendicular forces.

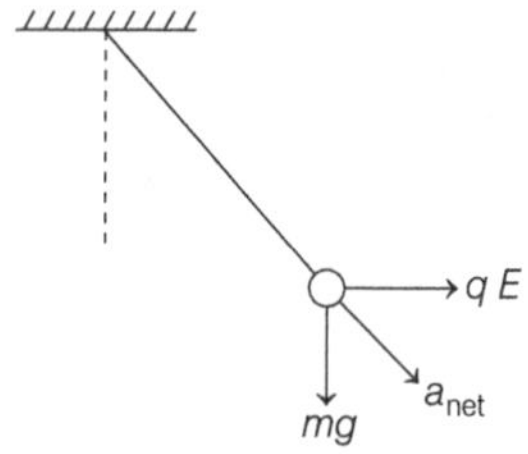

Net acceleration, $a_{net} = \sqrt{a_1^2 + a_2^2}$

$$= \sqrt{g^2 + \left(\frac{qE}{m}\right)^2}$$

So, time period of oscillations of pendulum,

$$T = 2\pi\sqrt{\frac{l}{a_{net}}} = 2\pi\sqrt{\frac{L}{\sqrt{g^2 + \left(\frac{qE}{m}\right)^2}}}$$

26. Electric field on the equatorial line of a dipole at any point, which is at distance r from the centre is given by

$$E = \frac{2kp}{(r^2 + a^2)^{3/2}} \quad \text{... (i)}$$

where, p is the dipole moment of the charges.

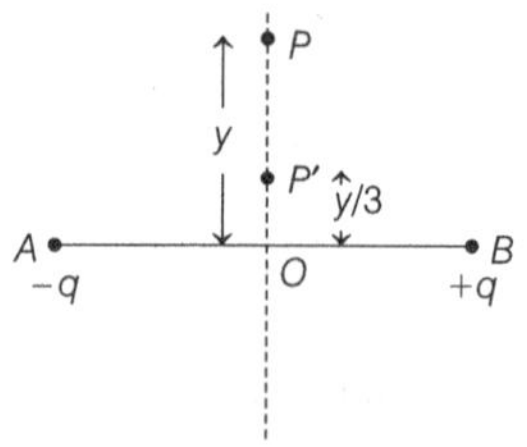

In first case $\quad r = y$

$\Rightarrow \quad E_1 = \frac{2kp}{(y^2 + a^2)^{3/2}}$

Here, $\quad y^2 >> a^2$

$$\Rightarrow \qquad y^2 + a^2 \approx y^2 \text{ or } E_1 = \frac{2kp}{y^3} \qquad \dots \text{(ii)}$$

So, force on the charge in its position at P will be

$$F = QE_1 = \frac{2kpQ}{y^3} \qquad \dots \text{(iii)}$$

In second case $r = y/3$

From Eq. (i), electric field at point P' will be

$$E_2 = \frac{2kp}{\left[\left(\frac{y}{3}\right)^2 + a^2\right]^{3/2}}$$

Again, $\qquad \frac{y}{3} >> a$

$$\Rightarrow \qquad \left(\frac{y}{3}\right)^2 + a^2 \approx \left(\frac{y}{3}\right)^2$$

$$\Rightarrow \qquad E_2 = \frac{2kp}{(y/3)^3} \Rightarrow E_2 = 27 \times \frac{2kp}{y^3}$$

Force on charge in this position,

$$F' = QE_2 = 27 \times \frac{2kpQ}{y^3} \qquad \dots \text{(iv)}$$

From Eqs. (iii) and (iv), we get

$$F' = 27\,F$$

27. Let's consider a strip of thickness dx at a distance of x from the left end as shown in figure. From the figure, ΔABC and ΔADE are similar triangles,

$$\Rightarrow \qquad \frac{y}{x} = \frac{d}{a} \Rightarrow y = \left(\frac{d}{a}\right)x \qquad \dots \text{(i)}$$

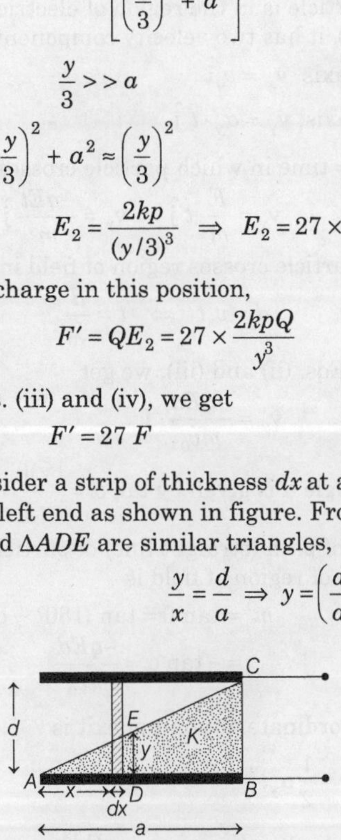

We know that, the capacitance of parallel plate capacitor, $\quad C = \frac{\varepsilon_0 A}{d}$

$$C_1 = \frac{\varepsilon_0 (adx)}{(d-y)} \text{ and } C_2 = \frac{K\varepsilon_0 (adx)}{y}$$

Here, two capacitors are placed in series with variable thickness, therefore

$$C_{eq} = \frac{C_1 C_2}{C_1 + C_2}$$

$$\Rightarrow \qquad C_{eq} = \frac{K\varepsilon_0 adx}{Kd + (1-K)y} \qquad \dots \text{(ii)}$$

Now, integrate it from 0 to a

$$C = \int_0^a \frac{K\varepsilon_0 adx}{Kd + (1-K)y}$$

Using Eq. (i), we get

$$C = \varepsilon_0 a \int_0^a \frac{dx}{d + \left(\frac{1}{K} - 1\right)\frac{d}{a}x}$$

$$\Rightarrow \qquad C = \frac{\varepsilon_0 a}{\left(\frac{1-K}{K}\right)\frac{d}{a}} \ln\left[\frac{1}{K}\right]$$

$$\Rightarrow \qquad C = \frac{\varepsilon_0 a^2 K \ln K}{(K-1)\,d}$$

28. Here, effective capacitance $= \left(\frac{6}{13}\right)\mu F$

In the given options, some capacitors are joined in parallel and some are in series.

For parallel combination, $C_P = C_1 + C_2 + C_3 + \dots$

For series combination, $\frac{1}{C_S} = \frac{1}{C_1} + \frac{1}{C_2} + \frac{1}{C_3} \dots$

Here, capacity of each capacitor is $2\mu F$.

$\therefore$ Three capacitors must be in parallel to get $6\,\mu F$. Now, we consider the combinations with three capacitors in parallel (as shown in the figure below)

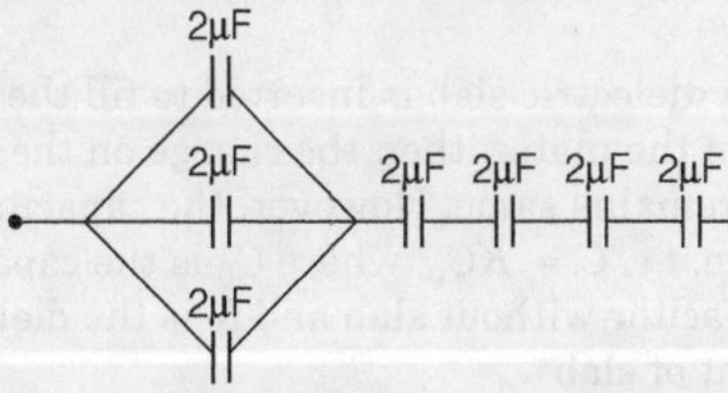

$$\therefore \qquad \frac{1}{C_{eq}} = \frac{1}{6} + \frac{1}{2} + \frac{1}{2} + \frac{1}{2} + \frac{1}{2}$$

$$= \frac{1}{6} + 2 = \frac{13}{6} \text{ or } C_{eq} = \frac{6}{13}\mu F$$

29. Let x be the required length of the slab.

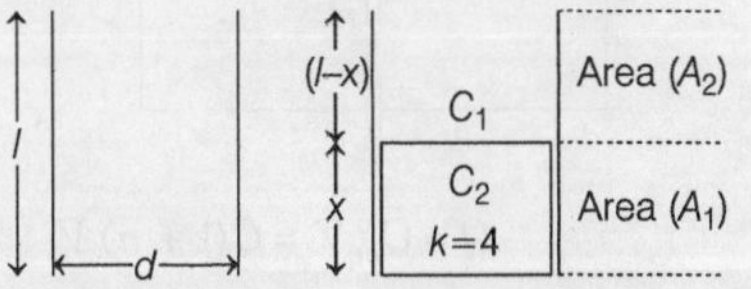

Parallel plate capacitor   Dielectric slab

The capacitance of a parallel plate capacitor is given by

$$C = \frac{\varepsilon_0 A}{d}$$

Capacitance of parallel plate capacitor before inserting the dielectric slab,

$$C_{\text{initial}} = \frac{\varepsilon_0 A}{d}$$

$$\Rightarrow \qquad C_{\text{initial}} = \frac{\varepsilon_0 lw}{d}$$

Now, capacitance of parallel plate capacitor after inserting the dielectric slab,

$$C_{\text{final}} = C_1 + C_2$$

$$= \frac{k\varepsilon_0 A_1}{d} + \frac{\varepsilon_0 A_2}{d}$$

$$= \frac{k\varepsilon_0 wx}{d} + \frac{\varepsilon_0 w(l-x)}{d}$$

$$= \frac{\varepsilon_0 w}{d}[kx + (l - x)]$$

$$= \frac{\varepsilon_0 w}{d}[l + (k-1)x]$$

According to question,

Energy stored in capacitor = 2 × Initial energy stored

$$\Rightarrow \quad \frac{1}{2} C_{\text{final}} V^2 = 2\left(\frac{1}{2} C_{\text{initial}} V^2\right)$$

$$\Rightarrow \quad C_{\text{final}} = 2\,C_{\text{initial}}$$

$$\Rightarrow \quad \frac{\varepsilon_0 w}{d}[l + (4-1)x] = \frac{2\varepsilon_0 wl}{d}$$

$$\Rightarrow \quad \frac{\varepsilon_0 w}{d}[l + 3x] = \frac{2\varepsilon_0 wl}{d}$$

$$\Rightarrow \quad l + 3x = 2l$$

$$\Rightarrow \quad 3x = l$$

$$\Rightarrow \quad x = \frac{l}{3}$$

30. When a dielectric slab is inserted to fill the space between the plates, then the charge on the capacitor plates remains same. However, the capacitance increase, *i.e.* $C = KC_0$, where C_0 is the capacitance of the capacitor without slab and K is the dielectric constant of slab.

When parallel combination is fully charged, charge on the combination is

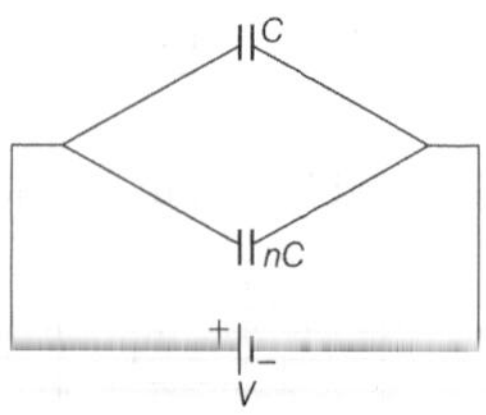

$$Q = C_{\text{eq}} V = C(1+n)\,V$$

When battery is removed and a dielectric slab is placed between two plates of first capacitor, then charge on the system remains same. Now, equivalent capacitance after insertion of dielectric is

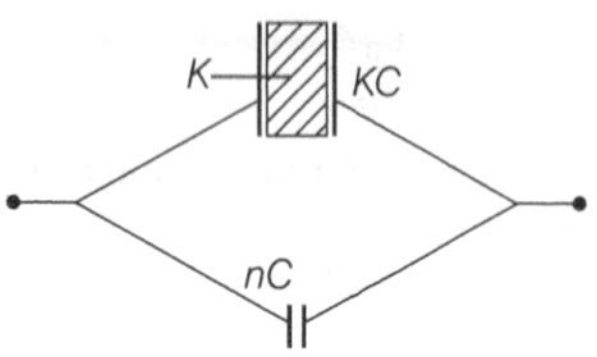

$$C_{\text{eq}} = KC + nC = (n+K)\,C$$

If potential value after insertion of dielectric is V', then charge on system is

$$Q' = C_{\text{eq}} V' = (n+K)\,CV'$$

As $Q = Q'$, we have

$$C(1+n)V = (n+K)\,CV'$$

$$\therefore \quad V' = \frac{(1+n)\,V}{(n+K)}$$

31. The path followed by charged particle is shown in figure below

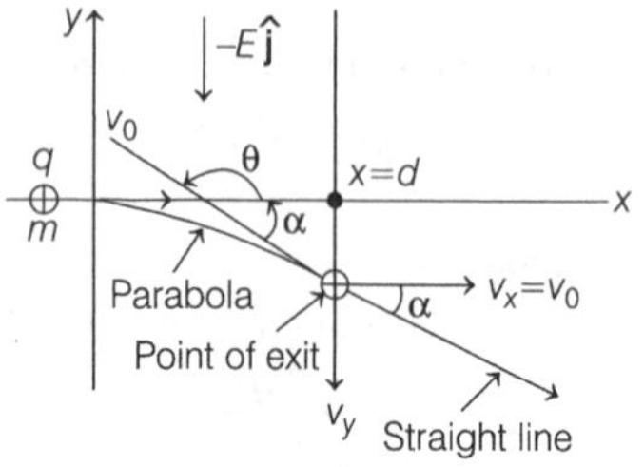

When particle is in the region of electric field $(0 < x \le d)$, it has two velocity components.

Along X-axis, $\mathbf{v}_x = v_0 \hat{\mathbf{i}}$ ...(i)

Along Y-axis, $\mathbf{v}_y = a_y \cdot t\,\hat{\mathbf{j}}$

where, t = time in which particle crosses region of field.

$$\Rightarrow \quad \mathbf{v}_y = \frac{F}{m} \cdot t\,\hat{\mathbf{j}} \Rightarrow \mathbf{v}_y = \frac{-qEt}{m}\,\hat{\mathbf{j}} \qquad \text{...(ii)}$$

Now, if particle crosses region of field in time t, then

$$d = v_x t \quad \Rightarrow \quad t = \frac{d}{v_0} \qquad \text{...(iii)}$$

So, from Eqs. (ii) and (iii), we get

$$\mathbf{v}_y = \frac{-qEd}{mv_0} \cdot \hat{\mathbf{j}}$$

Hence, angle α is given by $\tan\alpha = \frac{|\mathbf{v}_y|}{|\mathbf{v}_x|} = \frac{qEd}{mv_0^2}$

So, slope of path (straight line) of particle when it comes out of region of field is

$$m' = \tan\theta = \tan(180^\circ - \alpha)$$

$$= -\tan\alpha = \frac{-qEd}{mv_0^2} \qquad \text{...(iv)}$$

Now, y-coordinate at point of exit is

$$y = \frac{1}{2} a_y t^2$$

$$\Rightarrow \quad y = \frac{-1}{2}\left(\frac{qE}{m}\right) \cdot t^2 = -\frac{1}{2}\frac{qEd^2}{mv_0^2}$$

So, the intercept length c is $-y$ as shown in figure.

$$c = \left(\frac{1}{2}\frac{qEd^2}{mv_0^2}\right) \qquad \text{...(v)}$$

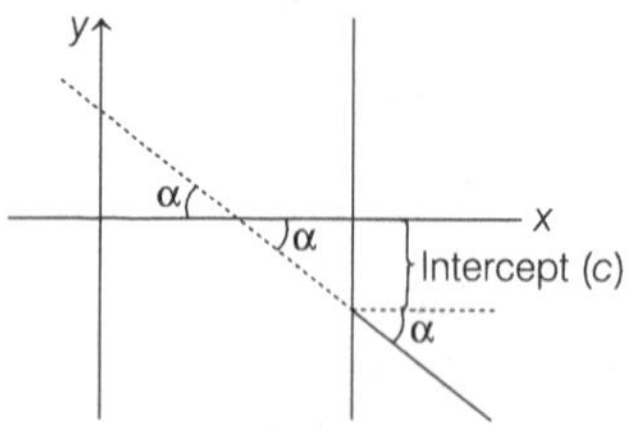

∴ Equation of path will be

$$y = m'x + c \qquad \text{[from Eqs. (iv) and (v)]}$$

$$\Rightarrow \quad y = \frac{-qEdx}{mv_0^2} + \frac{1}{2}\frac{qEd^2}{mv_0^2} = \frac{qEd}{mv_0^2}\left(\frac{d}{2} - x\right)$$

32. Consider a dielectric slab of differentiable thickness dx at a distance x from left hand side plate.

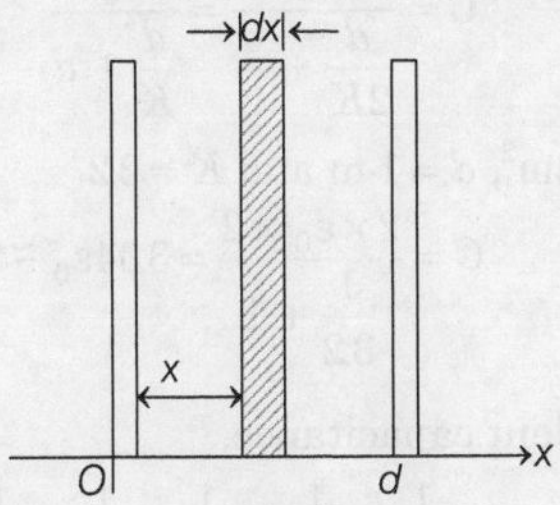

Capacity of this differential thickness capacitor,

$$dC = \frac{\varepsilon_0 K(x) A}{dx} = \frac{\varepsilon_0 K(1+\alpha x) A}{dx}$$

Total capacity of system is equivalent capacitance of series combination of these differential capacitors.

As in series, $\frac{1}{C_{eq}} = \sum_{i=1}^{n} \frac{1}{C_i}$

So, $\frac{1}{C_{eq}} = \int_0^d \frac{1}{dC} = \int_0^d \frac{dx}{\varepsilon_0 K(1+\alpha x) A}$

or $\frac{1}{C_{eq}} = \frac{1}{\varepsilon_0 AK} \int_0^d \frac{dx}{1+\alpha x}$

$$= \frac{1}{\varepsilon_0 A \alpha K} [\ln(1+\alpha x)]_0^d$$

$$= \frac{1}{\varepsilon_0 A \alpha K} [\ln(1+d\alpha) - \ln(1)]$$

$$= \frac{1}{\varepsilon_0 A \alpha K} \ln(1+d\alpha)$$

Now, we use expansion,

$\ln(1+y) = y - \frac{y^2}{2} + \frac{y^3}{3}$ ...., we get

$$\frac{1}{C_{eq}} = \frac{1}{\varepsilon_0 A \alpha K}\left(\alpha d - \frac{(\alpha d)^2}{2} + \frac{(\alpha d)^3}{3} \ldots\ldots\ldots\right)$$

$$= \frac{\alpha d}{\varepsilon_0 A \alpha K}\left[1 - \frac{\alpha d}{2} + \frac{(\alpha d)^2}{3} \ldots\right]$$

or $C_{eq} = \frac{\varepsilon_0 AK}{d}\left(1 - \frac{\alpha d}{2}\right)^{-1}$

From binomial approximation $(1+x)^n = 1 + nx$, we have

$$C_{eq} = \frac{\varepsilon_0 AK}{d}\left(1 + \frac{\alpha d}{2}\right)$$

33. As, $q_1 + q_2 = Q$

Here, $\frac{q_1}{4\pi r^2} = \frac{q_2}{4\pi R^2}$

$\therefore$ $q_1 = \frac{Qr^2}{R^2 + r^2}$

and $q_2 = \frac{QR^2}{R^2 + r^2}$

$\therefore$ Potential at component centre

$$= \frac{1}{4\pi\varepsilon_0}\left(\frac{Qr^2}{(R^2+r^2)r} + \frac{QR^2}{(R^2+r^2)R}\right)$$

$$= \frac{1}{4\pi\varepsilon_0} \frac{Q(R+r)}{(R^2+r^2)}$$

34. The force on l length of the wire 2,

$$f_2 = QE_1 = (\lambda_2 l)\frac{2K\lambda_1}{R}$$

$\Rightarrow$ $\frac{f_2}{l} = \frac{2K\lambda_1\lambda_2}{R}$

Also, $\frac{f_1}{l} = \frac{f_2}{l} = \frac{f_1}{l} = \frac{2K\lambda_1\lambda_2}{R}$

35. We have,

$$C_{eq} = \frac{(3+3)\times(1+1)}{(3+3)\times(1+1)} + 1 = \frac{6\times 2}{6+2} + 1 = \frac{12}{8} + 1$$

$$= \frac{12+8}{8} = \frac{20}{8} = \frac{5}{2}\,\mu\text{F}$$

$\because$ $Q = CV$

$\therefore$ $Q = \frac{5}{2} \times 100 = 250\,\mu\text{C}$

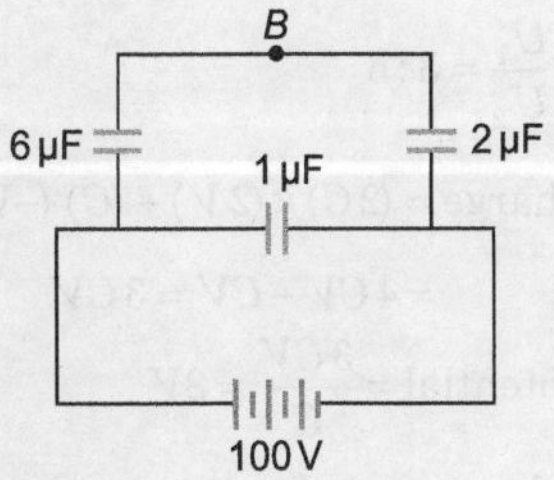

Charge in 6 μF branch

$$= CV = \left(\frac{6\times 2}{6+2}\right) \times 100 = 150\,\mu\text{C}$$

$$V_{AB} = \frac{150}{6} = 25\text{ V}$$

and $V_{BC} = 100 - V_{AB} = 75\text{ V}$ ...(i)

36. By symmetry of problem, the components of force on Q due to charges at A and B along y will cancel each other while along X-axis it will be added up. If at any time, charge Q at a distance x from O, then net force on charge Q,

$$F_{net} = 2F\cos\theta$$

Here, $F = \frac{-qQ}{4\pi\varepsilon_0 (a^2 + x^2)}$

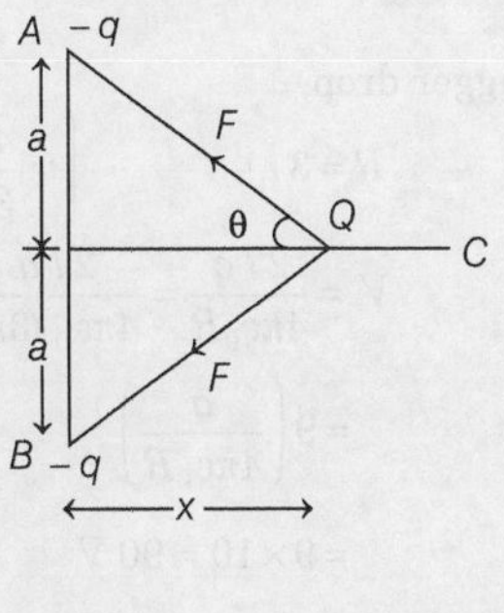

and $\cos\theta = \frac{x}{(a^2 + x^2)^{1/2}}$

$\Rightarrow F_{net} = -\frac{1}{4\pi\varepsilon_0}\frac{2qQx}{(a^2+x^2)^{3/2}}$

As the resulting force F_{net} is not linear. So, the motion will be oscillating, but not simple harmonic.

37. The energy of the system when the both capacitors are same.

$$U_1 = \frac{1}{2}CV^2 + \frac{1}{2}CV^2 = CV^2 \quad \text{...(i)}$$

In 2nd case, when K is opened and dielectric medium is filled between the plates, capacitance of both the capacitors becomes $3C$, while potential difference across A is V and potential difference across B is $V/3$.

Then, $U_2 = \frac{1}{2}(3C)V^2 + \frac{1}{3}(3C)\frac{V^2}{3}$

$$= \frac{10}{6}CV^2 = \frac{5}{3}CV^2 \quad \text{...(ii)}$$

Dividing Eq. (i) by Eq. (ii), we get

$\therefore \frac{U_1}{U_2} = 3:5$

38. The total charge $= (2C)\times(2V) + (C)(-V)$

$$= 4CV - CV = 3CV$$

Common potential $= \frac{3CV}{C} = 3V$

$\therefore$ Energy $= \frac{1}{2}(3C)V^2 = \frac{3}{2}CV^2$

39. As the electrostatic forces are conservative, so work done is independent of path

$\therefore W = \mathbf{F}\cdot d\mathbf{s} = qE\hat{\mathbf{i}}\cdot[(0-a)\hat{\mathbf{i}} + (0-b)\hat{\mathbf{j}}]$

$$= -qEa$$

40. Here, $E = \frac{\sigma}{\varepsilon_0}$ and $t = \frac{l}{u}$

Along Y-axis, $u = 0, a = \frac{F}{m} = \frac{eE}{m}$

$\therefore s = d = \frac{1}{2}at^2 = \frac{1}{2}\frac{eE}{m}t^2 = \frac{1}{2}\frac{e\sigma}{m\varepsilon_0}\frac{l^2}{u^2}$

or $\sigma = \frac{2d\varepsilon_0 mu^2}{el^2}$

41. Radius of bigger drop,

$R = 3r \quad \left[\because \frac{4}{3}\pi R^3 = 27\times\frac{4}{3}\pi r^3\right]$

$\therefore V = \frac{27q}{4\pi\varepsilon_0 R} = \frac{27q}{4\pi\varepsilon_0(3r)}$

$$= 9\left(\frac{q}{4\pi\varepsilon_0 R}\right)$$

$$= 9\times 10 = 90\text{ V}$$

42. The capacitance with dielectric,

$$C = \frac{\varepsilon_0 A}{\frac{d}{2K}+\frac{d}{2}} = \frac{2\varepsilon_0 A}{\frac{d}{K}+d}$$

Here, $A = 2\text{m}^2$, $d = 1$ m and $K = 3.2$

$$C = \frac{2\times\varepsilon_0\times 2}{\frac{1}{3.2}+1} = 3.04\varepsilon_0 \simeq 3\varepsilon_0$$

43. For equivalent capacitance,

$$\frac{1}{C_s} = \frac{1}{C_1} + \frac{1}{C_2} = \frac{1}{10} + \frac{1}{20} = \frac{3}{20}$$

$$C_s = \frac{20}{3}\mu\text{F}$$

$\therefore$ Charge on each capacitor

$$= C_sV = \frac{20}{3}\times 200$$

$$= \frac{4000}{3}\mu\text{C}$$

Common potential $= \frac{\text{Total charge}}{\text{Total capacity}}$

$$= \frac{2\times 4000/3}{10+20} = \frac{800}{9}\text{ V}$$

$$= 88.89\text{ V}$$

44. Initial energy of charged capacitor,

$$U_1 = \frac{1}{2}C_1V_1^2 = \frac{1}{2}\times 5\times 10^{-6}\times(220)^2$$

$$= 121\times 10^{-3}\text{ J}$$

Common potential after redistribution of charges,

$$V = \frac{C_1V_1 + C_2V_2}{C_1 + C_2}$$

$$\simeq \frac{C_1V_1}{C_1+C_2}$$

$$\simeq \frac{5\times 10^{-6}\times 220}{5\times 10^{-6} + 2.5\times 10^{-6}}$$

$$= 220\times\frac{2}{3}\text{ V}$$

Final stored energy,

$$U_2 = \frac{1}{2}(C_1 + C_2)V^2$$

$$= \frac{1}{2}\times 7.5\times 10^{-6}\times\left(220\times\frac{2}{3}\right)^2$$

$$\approx 80\times 10^{-3}\text{ J}$$

Loss of energy

$$\Delta U = (121 - 80)\times 10^{-3}\text{ J}$$

$$= 41\times 10^{-3}\text{ J}$$

$$= \frac{4.1}{100}\text{ J} \approx \frac{4}{100}\text{ J}$$

Given, $\Delta U = \frac{X}{100}$ J

$\therefore X = 4$

45. Initially a 60 pF capacitor is fully charged by a 20 V supply as shown in the figure.

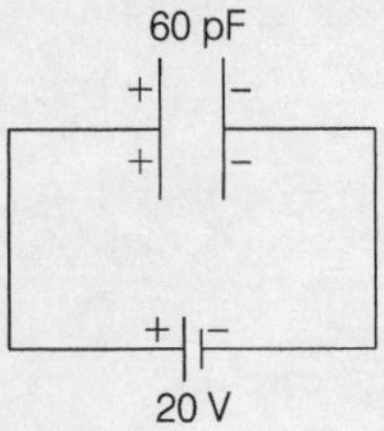

Its energy, $U_1 = \frac{1}{2}CV^2$

$$\Rightarrow \quad U_1 = \frac{1}{2} \times 60 \times 10^{-12} \times (20)^2$$

$$= 12 \times 10^{-9} \text{ J}$$

Then, it is connected to another uncharged 60 pF capacitor as shown in the figure.

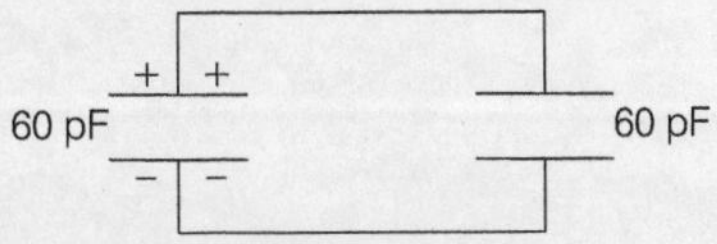

Transfer of charge from charged to uncharged capacitor occurs till both reaches a common potential.

If V = common potential, then

$$V = \frac{\text{total charge}}{\text{total capacitance}}$$

$$\Rightarrow \quad V = \frac{Q_1}{C_1 + C_2} = \frac{C_1V_1}{C_1 + C_2}$$

$$= \frac{60 \times 10^{-12} \times 20}{60 \times 10^{-12} + 60 \times 10^{-12}} = 10 \text{ V}$$

Final energy of system,

$$U_2 = \frac{1}{2}(C_1 + C_2) \cdot V^2$$

$$= \frac{1}{2} \times 2 \times 60 \times 10^{-12} \times (10^2)$$

$$= 60 \times 10^{-10}$$

$$= 6 \times 10^{-9} \text{ J}$$

Loss of energy,

$$\Delta U = U_1 - U_2$$

$$= 12 \times 10^{-9} - 6 \times 10^{-9}$$

$$= 6 \times 10^{-9} \text{ J}$$

$$= 6 \text{ nJ}$$

46. Initial energy of capacitor,

$$U_1 = \frac{1}{2}CV^2$$

$$= \frac{1}{2} \times 14 \times 12 \times 12$$

$$= 1008 \text{ pJ}$$

Final energy, $U_f = \frac{U_i}{K} = \frac{1008}{7} = 144$ pJ $\quad (\because C_m = KC_0)$

Mechanical energy,

$$\Delta U = U_i - U_f = 1008 - 144 = 864 \text{ pJ}$$

47. Electric field inside dielectric, $\frac{\sigma}{K\varepsilon_0} = 3 \times 10^4$

$$\Rightarrow \quad \sigma = 2.2 \times 8.58 \times 10^{-12} \times 3 \times 10^4$$

$$= 6.6 \times 8.85 \times 10^{-8}$$

$$= 5.841 \times 10^{-7}$$

$$= 6 \times 10^{-7} \text{ C/m}^2$$

19

Current Electricity

IN THIS CHAPTER ....

The directional flow of charge in a conductor under a potential difference maintained between the ends of the conductor constitutes an electric current.

Electric Current

It is defined as the amount of charge flowing across any section of wire per unit time. If charge Δq passes through the area in time interval Δt at uniform rate, then current i is defined as

$$i = \frac{\Delta q}{\Delta t}$$

If rate of flow of charge is not steady, then instantaneous current is given by

$$i = \lim_{\Delta t \to 0} \frac{\Delta q}{\Delta t} = \frac{dq}{dt}$$

The SI unit of current is ampere (A). Smaller currents are more conveniently expressed in milliampere (mA) or microampere (μA).

$$1 \text{ mA} = 10^{-3} \text{ A and } 1\,\mu\text{A} = 10^{-6} \text{ A}$$

Note Total charge in time interval t_1 and t_2 is given as

$$q = \int_{t_1}^{t_2} i \cdot dt = \text{Area under the graph } i \text{ versus } t \text{ in interval } t_1 \text{ to } t_2.$$

Current Density

Current density at any point inside a conductor is defined as "the amount of charge flowing per second through a unit area held normal to the direction of the flow of charge at that point."

Current density is a vector quantity and its direction is along the motion of the positive charge.

Current density, $\mathbf{J} = \frac{q/t}{\mathbf{A}} = \frac{i}{\mathbf{A}}$

The SI unit of current density is ampere per square metre, *i.e.* Am^{-2} and its dimensional formula is $[AL^{-2}]$.

Drift Velocity

It is defined as "the average velocity with which the free electrons in a conductor get drifted towards the positive end of the conductor under the influence of an electric field applied across the conductor."

It is given by, $v_d = \frac{eE}{m}\tau$

where, e = charge on electron, E = electric field, m = mass of the electron and τ = relaxation time.

The electric current relates with drift velocity as

$$i = neAv_d$$

Hence, current density is also given by, $J = \frac{i}{A} = nev_d$.

The direction of drift velocity for electrons in a metal is opposite to that of applied electric field.

The drift velocity of electron is very small of the order of $10^{-4}\,ms^{-1}$ as compared to thermal speed ($\approx 10^5$ m/s) of electron at room temperature.

Relaxation Time (τ)

As free electrons move in a conductor, they continuously collide with positive ions. The time interval between two successive collisions of electrons with the positive ions in the metallic lattice is defined as relaxation time.

$$\tau = \frac{\text{mean free path}}{\text{rms velocity of electrons}} = \frac{\lambda}{v_{rms}}$$

With rise in temperature v_{rms} increases, consequently τ decreases.

Mobility (μ)

Drift velocity per unit electric field is called mobility of electron, *i.e.* $\mu = \frac{v_d}{E} = \frac{e\tau}{m}$.

Its unit is m^2/V-s.

Mobility of free electrons is independent of electric field.

Example 1. *The current in a wire varies with time according to the equation $i = 4 + 2t$, where i is in ampere and t is in second. The quantity of charge which has to be passes through a cross-section of the wire during the time t = 2 s to t = 6 s is*

(a) 40 C (b) 48 C
(c) 38 C (d) 43 C

Sol. (b) Let dq be the charge which passes in a small interval of time dt. Then, $dq = idt = (4 + 2t)dt$

$\Rightarrow$ $q = \int_2^6 (4 + 2t)dt = [4t + t^2]_2^6 = 48$ C

Example 2. *Drift speed of electrons, when 1.5 A of current flows in a copper wire of cross-section 5 mm^2 is v. If the electron density in copper is $9 \times 10^{28}\,/\,m^3$, the value of v (in mm/s) is close to (Take, charge of electron to be $= 1.6 \times 10^{-19}$ C)*

[JEE Main 2019]

(a) 0.02 (b) 0.2 (c) 2 (d) 3

Sol. (a) Relation between current (i) flowing through a conducting wire and drift velocity of electrons (v_d) is given as $i = neAv_d$ where, n is the electron density and A is the area of cross-section of wire.

$\Rightarrow$ $v_d = \frac{i}{neA}$

Substituting the given values, we get

$$v = \frac{1.5}{9 \times 10^{28} \times 1.6 \times 10^{-19} \times 5 \times 10^{-6}}$$

$$v = \frac{1.5 \times 10^{-3}}{72}\text{ m/s} = 0.2 \times 10^{-4}\text{ m/s}$$

or $v = 0.02$ mm/s

Ohm's Law

It states that, "the current i flowing through a conductor is always directly proportional to the potential difference V applied across the ends of the conductor", provided that the physical conditions (temperature, mechanical strain, etc.) are kept constant.

i.e. $i \propto V$ or $V \propto i$ or $V = Ri$

or $\frac{V}{i} = R$ = a constant

where, R = electrical resistance of the conductor.

The SI unit of R is Ω (ohm) and its dimensions are $[ML^2T^{-3}A^{-2}]$.

Graph between V and i for a metallic conductor is a straight line as shown below

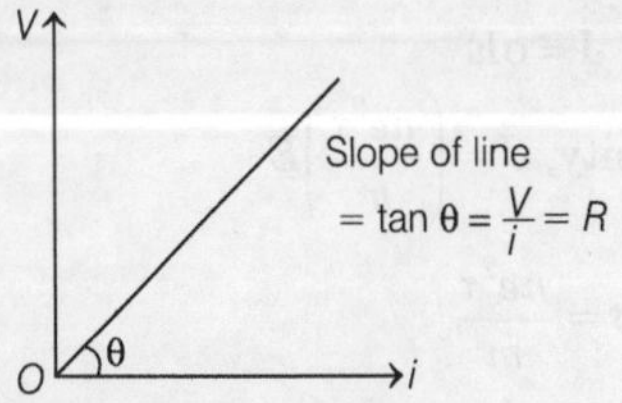

Limitations of Ohm's Law

There are some materials and devices used in electric circuits, where the proportionality of V and i does not hold. The deviations broadly are one or more of the following types:

(i) V ceases to be proportional to i.

(ii) If i is the current for a certain V, then reversing the direction of V keeping its magnitude fixed, does not produce a current of the same magnitude as i in the opposite direction. This happens in a diode.

Electrical Resistance and Resistivity

The property of a substance by virtue of which it opposes the flow of current through it is known as resistance.

It depends on the geometrical factors of the substance [length (l), cross-sectional area (A)] as well as on the nature of the substance from which the resistor is made.

$\therefore$ Electrical reistance, $R = \frac{\rho l}{A}$ ...(iv)

where, ρ is a constant of proportionality called *resistivity* or specific resistance of substance.

Resistivity depends only on the nature of the material of the resistor and its physical conditions such as temperature and pressure. Its unit is ohm-m (Ω-m).

Conductance and Conductivity

Conductance The reciprocal of resistance of a conductor is called its conductance. Thus, conductance (G) of a conductor having resistance R is given by

$$G = \frac{1}{R}$$

Unit of conductance is ohm^{-1}(Ω^{-1}), which is also called *mho*. The unit of conductance in SI system is *siemens* and is denoted by the symbol S.

Conductivity The reciprocal of resistivity of the material of a conductor is called its conductivity. It is denoted by σ. Thus,

$$\sigma = \frac{1}{\rho}$$

Its SI unit is siemen metre^{-1} (Sm^{-1}).

Vector Form of Ohm's Law

Ohm's law in vector form is given as

$$\mathbf{E} = \rho \mathbf{J}$$

or $$\mathbf{J} = \sigma \mathbf{E}$$

$\therefore$ Current density, $J = \left(\frac{ne^2\tau}{m}\right)E$

Conductivity, $\sigma = \frac{ne^2\tau}{m}$

and resistivity, $\rho = \frac{1}{\sigma} = \frac{m}{ne^2\tau}$

Example 3. *A copper wire is stretched to make it 0.5% longer. The percentage change in its electrical resistance, if its volume remains unchanged is* **[JEE Main 2019]**

(a) 2.0% (b) 1.0% (c) 0.5% (d) 2.5%

Sol. (b) Electrical resistance of wire of length l, area of cross-section A and resistivity ρ is given as

$$R = \rho \frac{l}{A} \quad \text{...(i)}$$

Since we know, volume of the wire is $V = A \times l$ ...(ii)

$\therefore$ From Eqs. (i) and (ii), we get

$$R = \rho \frac{l^2}{V} \quad \text{...(iii)}$$

As, the length has been increased to 0.5%.

$\therefore$ New length of the wire,

$$l' = l + 0.5\% \text{ of } l = l + 0.005\, l = 1.005\, l$$

But V and ρ remains unchanged.

So, new resistance, $R' = \frac{\rho[(1.005)\, l]^2}{V}$ ...(iv)

Dividing Eq. (iv) by Eq. (iii), we get

$$\frac{R'}{R} = (1.005)^2$$

$\Rightarrow$ % change in the resistance $= \left(\frac{R'}{R} - 1\right) \times 100$

$$= [(1.005)^2 - 1] \times 100 = 1.0025\% \simeq 1\%$$

Example 4. *Space between two concentric conducting spheres of radii a and b (b > a) is filled with a medium of resistivity ρ. The resistance between the two spheres (in ohm) will be* **[JEE Main 2019]**

(a) $\frac{\rho}{2\pi}\left(\frac{1}{a} + \frac{1}{b}\right)$ (b) $\frac{\rho}{4\pi}\left(\frac{1}{a} - \frac{1}{b}\right)$

(c) $\frac{\rho}{2\pi}\left(\frac{1}{a} - \frac{1}{b}\right)$ (d) $\frac{\rho}{4\pi}\left(\frac{1}{a} + \frac{1}{b}\right)$

Sol. (b) For an elemental shell of radius x and thickness dx,

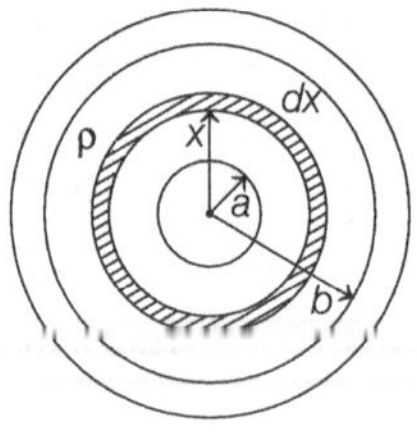

Resistance, $dR = \rho \frac{l}{A} \Rightarrow dR = \rho \frac{dx}{4\pi x^2}$

So, resistance of complete arrangement is

$$R = \int_a^b dR = \int_a^b \rho \frac{dx}{4\pi x^2} = \frac{\rho}{4\pi}\int_a^b x^{-2} dx$$

$$\Rightarrow \quad R = \frac{\rho}{4\pi}\left(\frac{x^{-1}}{-1}\right)_a^b = \frac{\rho}{4\pi}\left(-\frac{1}{x}\right)_a^b$$

$$= \frac{\rho}{4\pi}\left(-\frac{1}{b} - \left(-\frac{1}{a}\right)\right)$$

$$= \frac{\rho}{4\pi}\left(\frac{1}{a} - \frac{1}{b}\right) \text{ ohm}$$

Example 5. *In a conductor, if the number of conduction electrons per unit volume is 8.5×10^{28} m^{-3} and mean free time is 25 fs (femto second), it's approximate resistivity is (Take, $m_e = 9.1 \times 10^{-31}$ kg)* **[JEE Main 2019]**

(a) 10^{-7} Ω-m (b) 10^{-5} Ω-m

(c) 10^{-6} Ω-m (d) 10^{-8} Ω-m

Sol. (d) Resistivity of a conductor is $\rho = \dfrac{m_e}{ne^2\tau}$ ... (i)

where, m_e = mass of electron = 9.1×10^{-31} kg,

n = free charge density = $8.5 \times 10^{28}\ m^{-3}$,

τ = mean free time = 25 fs = 25×10^{-15} s

and e = charge of electron = 1.6×10^{-19} C

Substituting the values in Eq. (i), we get

$$\rho = \frac{9.1 \times 10^{-31}}{8.5 \times 10^{28} \times (1.6 \times 10^{-19})^2 \times 25 \times 10^{-15}}$$

$$= \frac{9.1 \times 10^{-6}}{8.5 \times 2.56 \times 25} = 0.016 \times 10^{-6}$$

$$= 1.6 \times 10^{-8}\ \Omega\text{-m}$$

$$\simeq 10^{-8}\ \Omega\text{-m}$$

Temperature Dependence of Resistivity

Resistivity of a conductor varies approximately linearly with temperature as

$$\rho = \rho_0[1 + \alpha\,(T - T_0)]$$

Here, ρ is resistivity at temperature T and ρ_0 is resistivity at temperature T_0.

Also, α is temperature coefficient of resistivity which is given as $\alpha = \dfrac{\rho - \rho_0}{\rho_0\,(T - T_0)}$

In terms of resistance, $R = R_0\,[1 + \alpha\,(T - T_0)]$

$$\Rightarrow \quad \alpha = \frac{R - R_0}{R_0\,(T - T_0)}$$

Variation of resistance of some electrical material with temperature is as given in table

Material	Temperature coefficient of resistance	Effect of temperature rise on resistance
Metals	$\alpha > 0$	Increases
Solid non-metal	$\alpha \approx 0$	No effect
Semiconductors	$\alpha < 0$	Decreases
Electrolytes	$\alpha < 0$	Decreases
Alloys	$0 < \alpha < 1$	Nearly constant

Example 6. *A heating element using nichrome connected to a 230 V supply draws an initial current of 3.2 A in first 2 seconds, which settles, after a few seconds to a steady value of 2.8 A. What is the steady temperature of the heating element, if the room temperature is 27°C? Temperature coefficient of resistance of nichrome averaged over the temperature range involved is $1.70 \times 10^{-4}\,°C^{-1}$.*

(a) 678.35°C (b) 768.35°C
(c) 867.35°C (d) 976.35°C

Sol. (c) Here, $V = 230$ V, initial current, $i_1 = 3.2$ A, steady current, $i_2 = 2.8$ A

Therefore, resistance of wire at room temperature ($\theta_{12} = 27°$ C),

$$R_1 = \frac{V}{i_1} = \frac{230}{3.2} = 71.875\ \Omega$$

Resistance of wire at steady temperature θ_2° C,

$$R_2 = \frac{V}{i_2} = \frac{230}{2.8} = 82.143\ \Omega$$

Now, $$\alpha = \frac{R_2 - R_1}{R_1(\theta_2 - \theta_1)}$$

$$\therefore \quad 1.70 \times 10^{-4} = \frac{82.143 - 71.875}{71.875 \times (\theta_2 - 27)}$$

or $$\theta_2 = 27 + \frac{82.143 - 71.875}{71.875 \times 1.70 \times 10^{-4}}$$

$$= 27 + 840.35$$

$$= 867.35°\ C$$

Example 7. *In an experiment, the resistance of a material is plotted as a function of temperature (in some range). As shown in the figure, it is a straight line.*

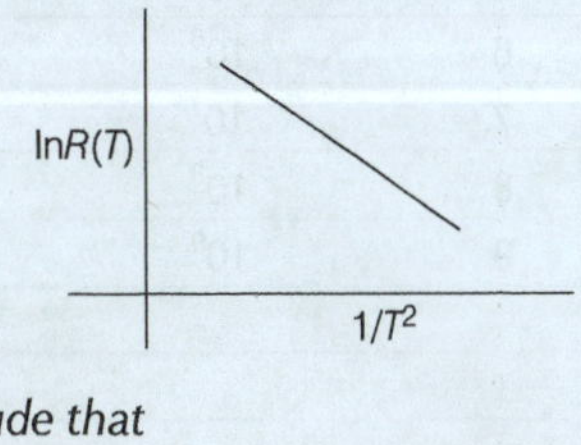

[JEE Main 2019]

One may conclude that

(a) $R(T) = R_0 e^{-T^2/T_0^2}$ (b) $R(T) = R_0 e^{T^2/T_0^2}$
(c) $R(T) = R_0 e^{-T_0^2/T^2}$ (d) $R(T) = \dfrac{R_0}{T^2}$

Sol. (c) From the given graph,

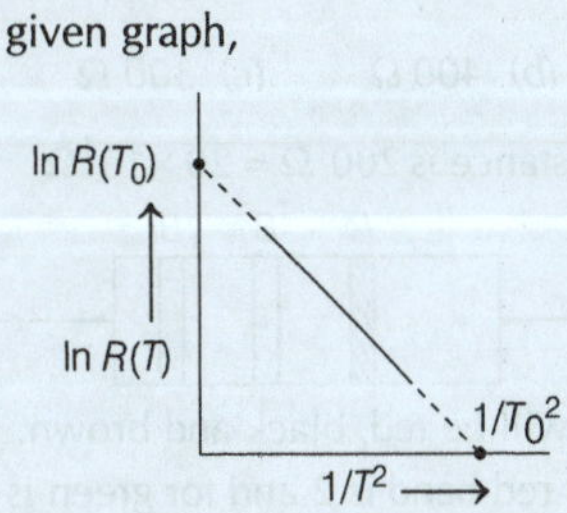

We can say that, $\ln R\,(T) \propto -\dfrac{1}{T^2}$

Negative sign implies that the slope of the graph is negative.

or $$\ln R(T) = \text{constant}\left(-\frac{1}{T^2}\right)$$

$$\Rightarrow \quad R(T) = \frac{\exp\,(\text{const.})}{\exp\left(\dfrac{1}{T^2}\right)}$$

$$\Rightarrow \quad R(T) = R_0\,\exp\left(-\frac{T_0^2}{T^2}\right)$$

Colour Code for Carbon Resistors

The resistance of a carbon resistor can be calculated by the code given on it in the form of coloured strips as shown below

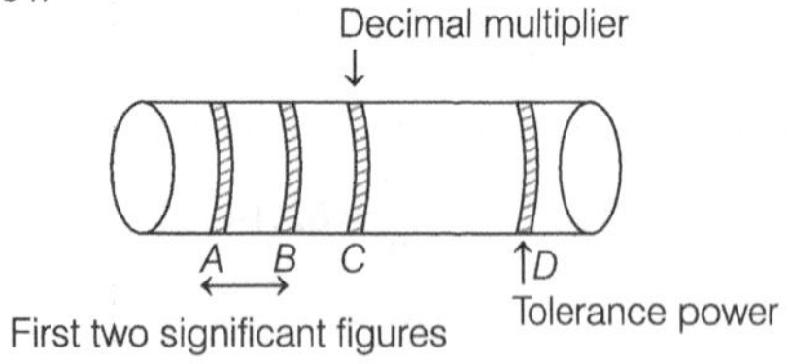

∴ The value of a carbon resistor is given as

$$R = AB \times C \pm D\%$$

The table for colour code of carbon resistor is given below

Color codes	Values (*A*, *B*)	Multiplier (*C*)	Tolerance (*D*) (%)
Black	0	10^0	
Brown	1	10^1	
Red	2	10^2	
Orange	3	10^3	
Yellow	4	10^4	
Green	5	10^5	
Blue	6	10^6	
Violet	7	10^7	
Grey	8	10^8	
White	9	10^9	
Gold			± 5%
Silver			± 10%
No colour			± 20%

Example 8. *A 200 Ω resistor has a certain colour code. If one replaces the red colour by green in the code, the new resistance will be* **[JEE Main 2019]**

(a) 100 Ω (b) 400 Ω (c) 300 Ω (d) 500 Ω

Sol. (d) Given, resistance is $200\ \Omega = 20 \times 10^1 \Omega$

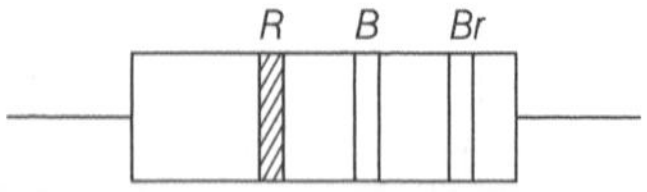

So, colour scheme will be red, black and brown.

Significant figure of red band is 2 and for green is 5. When red (2) is replaced with green (5), new resistance will be

200 ohm ⟶ 500 ohm

Combination of Resistors

There are two types of combination of resistances, *i.e.* series and parallel combination.

Series Combination Combination of resistors in series is as shown below

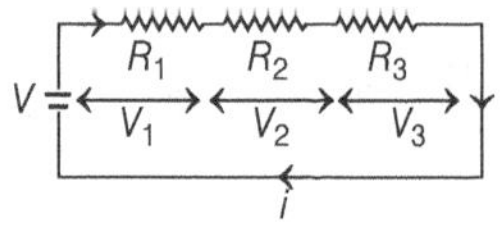

Same current flows through each resistance but potential difference distributes in the ratio of their resistance, *i.e.*, $V \propto R$.

So, the total potential drop is equal to the sum of potential applied across the combination.

Equivalent resistance in series combination is given by

$$R_{eq} = R_1 + R_2 + R_3$$

i.e. Equivalent resistance is greater than the maximum value of resistance in the combination.

If n identical resistances are connected in series, then $R_{eq} = nR$ and potential difference across each resistance, is $V' = V/n$.

Parallel Combination Combination of resistors in parallel is as shown below

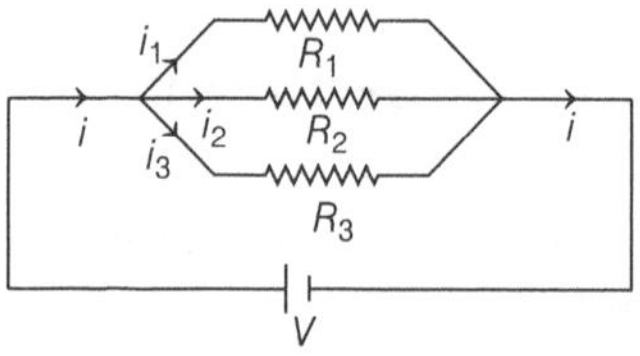

Same potential difference appears across each resistance but current distributes in the reverse ratio of their resistance, *i.e.* $I \propto \frac{V}{R}$. So, the total current is equal to the sum of currents through each resistance.

Equivalent resistance is given by, $\frac{1}{R_{eq}} = \frac{1}{R_1} + \frac{1}{R_2} + \frac{1}{R_3}$

i.e. Equivalent resistance is less than the minimum value of resistance in the combination.

In n identical resistance are connected in parallel, then $R_{eq} = \frac{R}{n}$ and current through each resistance is, $I' = \frac{I}{n}$.

Example 9. *A uniform metallic wire has a resistance of 18 Ω and is bent into an equilateral triangle, then the resistance between any two vertices of the triangle is* **[JEE Main 2019]**

(a) 12 Ω (b) 8 Ω
(c) 2 Ω (d) 4 Ω

Sol. (d) Resistance of each arm of equilateral triangle will be

$$R = \frac{18}{3} = 6\ \Omega$$

So, we have following combination

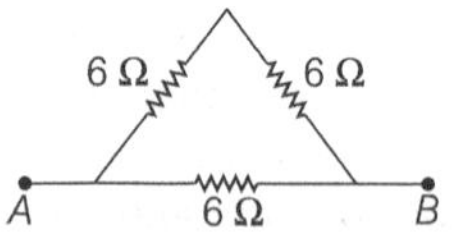

Equivalent resistance,

$$R_{AB} = \frac{12 \times 6}{12 + 6} = \frac{12 \times 6}{18} = 4\ \Omega$$

Example 10. *A metal wire of resistance 3 Ω is elongated to make a uniform wire of double its previous length. This new wire is now bent and the ends joined to make a circle. If two points on this circle make an angle 60° at the centre, the equivalent resistance between these two points will be* **[JEE Main 2019]**

(a) $\frac{7}{2}$ Ω (b) $\frac{5}{2}$ Ω (c) $\frac{12}{5}$ Ω (d) $\frac{5}{3}$ Ω

Sol. (d) Initial resistance of wire is 3 Ω. Let its length is l and area is A.

Then, $R_{\text{initial}} = \rho\frac{l}{A} = 3\ \Omega$ ... (i)

When wire is stretched twice its length, then its area becomes A', on equating volume, we have

$$Al = A'2l \Rightarrow A' = \frac{A}{2}$$

So, after stretching, resistance of wire will be

$$R' = R_{\text{final}} = \rho\frac{l'}{A'} = 4\rho\frac{l}{A} = 12\ \Omega \quad \text{[using Eq. (i)]}$$

Now, this wire is made into a circle and connected across two points A and B (making 60° angle at centre) as

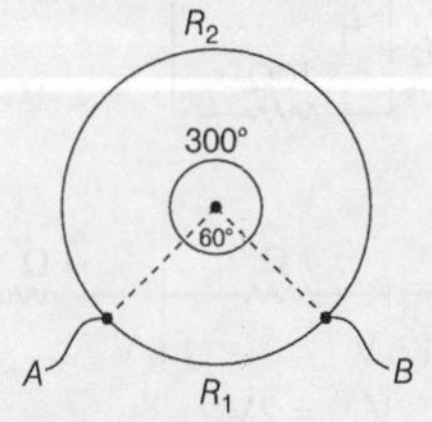

Now, above arrangement is a combination of two resistances in parallel,

$$R_1 = \frac{60 \times R'}{360} = \frac{1}{6}\times 12 = 2\ \Omega$$

and $$R_2 = \frac{300}{360}\times R' = \frac{5}{6}\times 12 = 10\ \Omega$$

Since, R_1 and R_2 are connected in parallel.

So, $$R_{AB} = \frac{R_1R_2}{R_1+R_2} = \frac{10\times 2}{12} = \frac{5}{3}\ \Omega$$

Example 11. *Two resistors 400 Ω and 800 Ω are connected in series across a 6 V battery. The potential difference measured by a voltmeter of 10 kΩ across 400 Ω resistor is close to* **[JEE Main 2020]**

(a) 1.95 V (b) 2 V (c) 1.8 V (d) 2.05 V

Sol. (a) The circuit diagram is

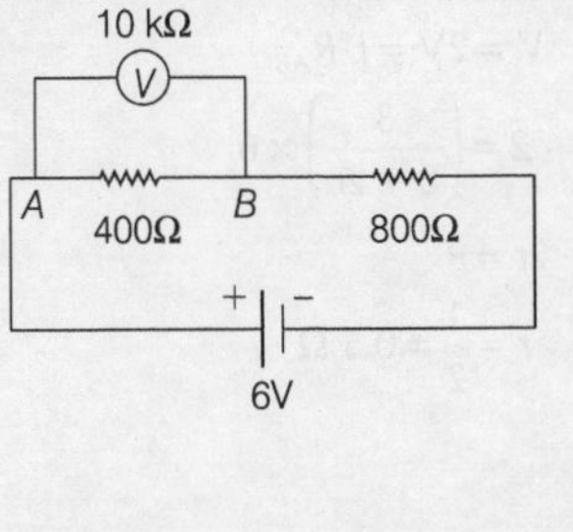

The equivalent resistance across AB,

$$R_{AB} = \frac{10000\times 400}{10000+400} = \frac{10000\times 400}{10400} = \frac{40000}{104} = 384.6\ \Omega \approx 385\ \Omega$$

∴ Total circuit resistance = 385 + 800 = 1185 Ω

Hence, circuit current is $i = \frac{V}{R_{\text{eq}}} = \frac{6}{1185} = 5.06\times 10^{-3}$ A

Now, potential drop across points A and B is

$$V_{AB} = i\cdot R_{AB} = 5.06\times 10^{-3}\times 385 = 1.946\text{V} \simeq 1.95\text{V}$$

Electric Cell

It is a device which maintains a continuous flow of charge (or electric current) in a circuit by a chemical reaction. It converts chemical energy into electrical energy.

Electromotive Force (emf) of a Cell (ε) Electric cell has to do some work in maintaining the current through a circuit. The work done by the cell in moving unit positive charge through the whole circuit (including the cell) is called the *electromotive force* (emf) of the cell.

If during the flow of q coulomb of charge in an electric circuit, the work done by the cell is W, then

$$\text{emf of the cell, } \varepsilon = \frac{W}{q}$$

Its unit is joule/coulomb or volt.

Internal Resistance (r) Internal resistance of a cell is defined as the resistance offered by the electrolyte of the cell to the flow of current through it. It is denoted by r. Its unit is ohm.

Terminal Potential Difference (V) Terminal potential difference of a cell is defined as the potential difference between the two terminals of the cell in a closed circuit (*i.e.* when current is drawn from the cell). It is represented by V and its unit is volt.

Terminal potential difference of a cell is always less than the emf of the cell.

Relation between Terminal Potential Difference, emf of the Cell and Internal Resistance of a Cell

- If no current is drawn from the cell, *i.e.* the cell is in open circuit, so emf of the cell will be equal to the terminal potential difference of the cell.

  $$i = 0 \text{ or } V = \varepsilon$$

- Consider a cell of emf ε and internal resistance r is connected across an external resistance R.

  Current drawn from the cell, $i = \frac{\varepsilon}{R+r}$

where, ε = emf of the cell, R = external resistance

and r = internal resistance of a cell;

and $r = \left(\frac{\varepsilon}{V} - 1\right)R$

Grouping of Cells

Series Grouping Suppose n cells each of emf ε and internal resistance r are connected in series as shown in figure, then net emf = $n\varepsilon$

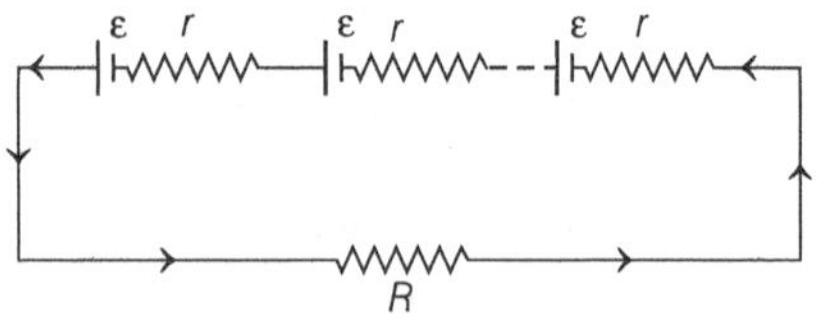

Total resistance = $nr + R$

$\therefore$ Current in the circuit $= \frac{\text{Net emf}}{\text{Total resistance}}$

or $$i = \frac{n\varepsilon}{nr + R}$$

Note If polarity of m cells is reversed, then equivalent emf $\varepsilon_{eq} = (n - 2m)\varepsilon$ while total resistance is still $nr + R$, so $i = \frac{(n-2m)\varepsilon}{nr + R}$.

Parallel Grouping Let n cells each of emf ε and internal resistance r are connected in parallel as shown in figure, then net emf = ε

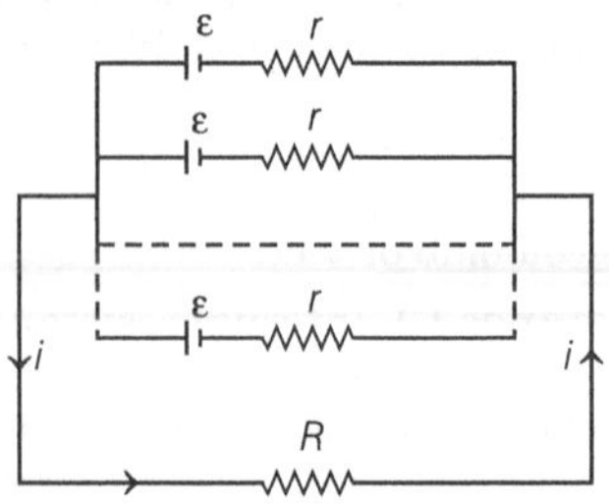

Total resistance $= \frac{r}{n} + R$

$\therefore$ Current in the circuit, $i = \frac{\text{Net emf}}{\text{Total resistance}}$

or $$i = \frac{\varepsilon}{R + r/n}$$

Mixed Grouping If n identical cells are in a row and such m rows are connected in parallel as shown, then net emf = $n\varepsilon$

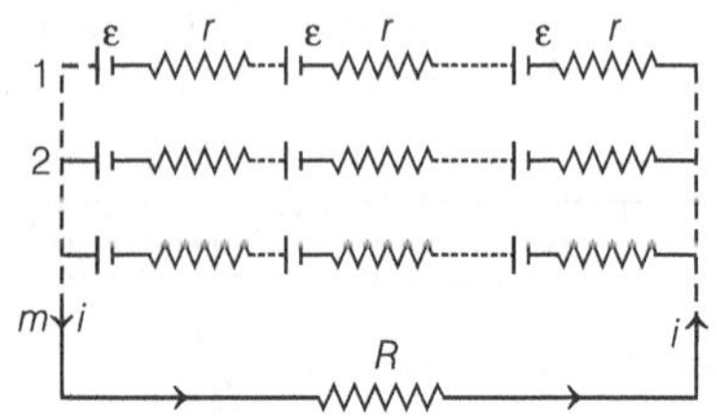

Total resistance, $R_{eq} = R + \frac{nr}{m}$

Hence, $$i = \frac{n\varepsilon}{R + \frac{nr}{m}}$$

Example 12. *In the given circuit, an ideal voltmeter connected across the 10 Ω resistance reads 2 V. The internal resistance r, of each cell is* **[JEE Main 2019]**

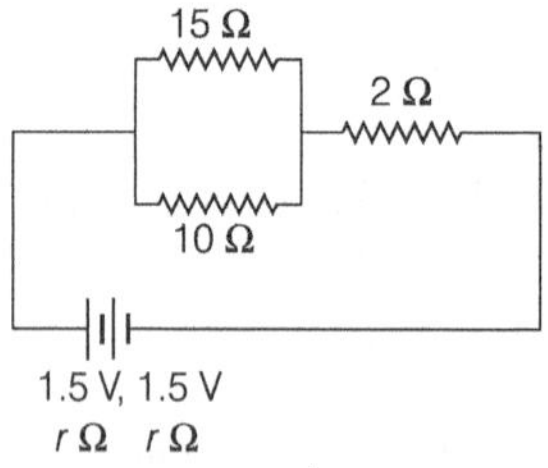

(a) 1.5 Ω (b) 0.5 Ω
(c) 1 Ω (d) 0 Ω

Sol. (b) For the given circuit,

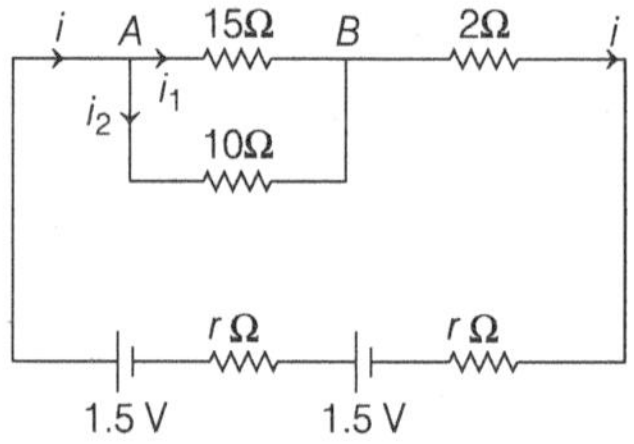

Given, $V_{AB} = 2\text{V}$

Equivalent resistance between A and B is

$$\frac{1}{R_{AB}} = \frac{1}{10} + \frac{1}{15} = \frac{1}{6}$$

$\Rightarrow$ $$R_{AB} = 6\,\Omega$$

$\therefore$ Equivalent resistance of the entire circuit is,

$$R_{eq} = 6\,\Omega + 2\,\Omega + 2r = 8 + 2r$$

Now, current passing through the circuit is given as

$$i = \frac{\varepsilon_{net}}{R + r_{eq}} = \frac{\varepsilon_{net}}{R_{eq}}$$

where, R is external resistance, r_{eq} is net internal resistance and ε_{net} is the emf of the cells.

Here, $\varepsilon_{net} = 1.5 + 1.5 = 3\text{ V}$

$r_{eq} = r + r = 2r$

$\Rightarrow$ $$i = \frac{3}{8 + 2r}$$

Also, reading of the voltmeter,

$$V = 2\text{V} = i \cdot R_{AB}$$

$$2 = \left(\frac{3}{8 + 2r}\right) \times 6$$

$\Rightarrow$ $$8 + 2r = 9$$

or $$r = \frac{1}{2} = 0.5\,\Omega$$

Example 13. *For the circuit shown with $R_1 = 1.0\ \Omega$, $R_2 = 2.0\Omega$, $\varepsilon_1 = 2$ V and $\varepsilon_2 = \varepsilon_3 = 4$ V, the potential difference between the points a and b is approximately (in volt)*

[JEE Main 2019]

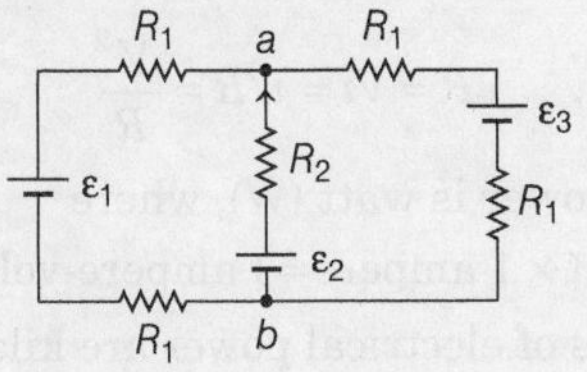

(a) 2.7 (b) 2.3 (c) 3.7 (d) 3.3

Sol. (d) Given circuit is

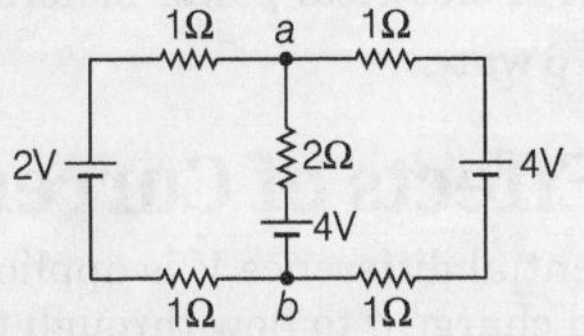

Above circuit can be viewed as

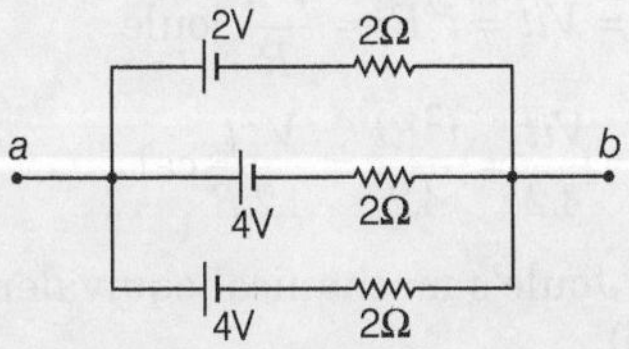

This is a parallel combination of three cells or in other words, a parallel grouping of three cells with internal resistances.

So, $$V_{ab} = \varepsilon_{eq} = \frac{i_{eq}}{r_{eq}} = \frac{\frac{\varepsilon_1}{r_1} + \frac{\varepsilon_2}{r_2} + \frac{\varepsilon_3}{r_3}}{\frac{1}{r_1} + \frac{1}{r_2} + \frac{1}{r_3}}$$

$$= \frac{\frac{2}{2} + \frac{4}{2} + \frac{4}{2}}{\frac{1}{2} + \frac{1}{2} + \frac{1}{2}}$$

$$= \frac{10}{3}\text{ V} \approx 3.3\text{ V}$$

Kirchhoff's Laws

Many complex electrical circuits cannot be reduced to simple series or parallel combinations.

However, such circuits can be analysed by applying two rules, devised by Kirchhoff which are as follows

Junction Rule The sum of the currents meeting at any junction in a closed circuit is zero, *i.e.* $\sum_{\text{junction}} i = 0$

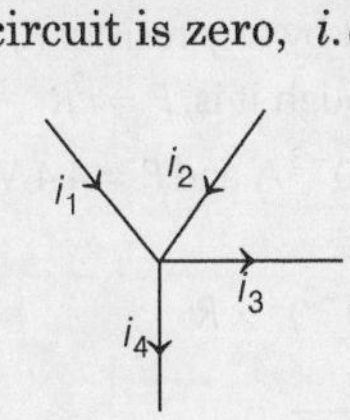

In figure, $$i_1 + i_2 = i_3 + i_4$$

This rule is based on *conservation of electric charge.*

Loop Rule The algebraic sum of the potential differences in any closed circuit is zero.

i.e. $$\sum_{\text{closed loop}} \Delta V = 0$$

This law is based on **conservation of energy.**

Sign Convention in Kirchhoff's Laws

In applying the loop rule, we need sign convention as discussed below

(i) **When we travel through a source in the direction from –ve to +ve, the emf is considered to be positive.**

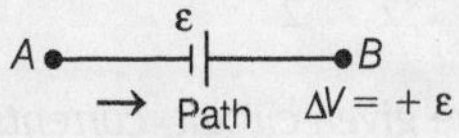

(a)

(ii) **When we travel from +ve to –ve, the emf is considered to be negative.**

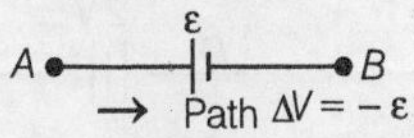

(b)

(iii) **When we travel through a resistor in the same direction as the assumed current, the iR term is negative because the current goes in the direction of decreasing potential.**

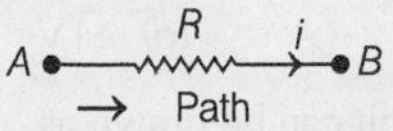

(c)

(iv) **When we travel through a resistor in the direction opposite to the assumed current, the iR term is positive because this represents a rise of potential.**

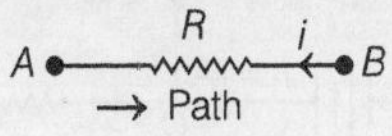

(d)

Example 14. *When the switch S in the circuit shown is closed, then the value of current i will be*

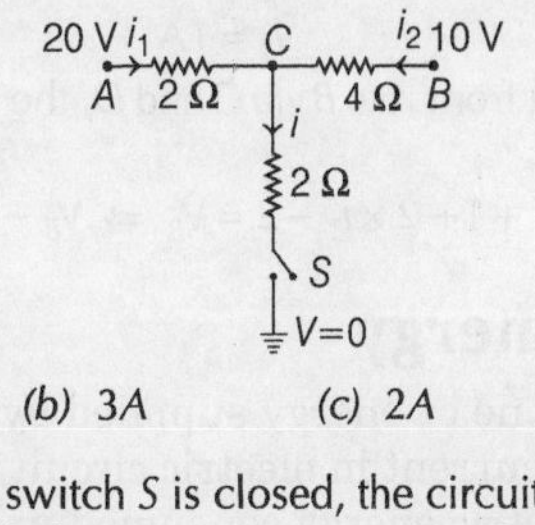

[JEE Main 2019]

(a) 4A (b) 3A (c) 2A (d) 5A

Sol. (d) When the switch S is closed, the circuit hence formed is given in the figure below

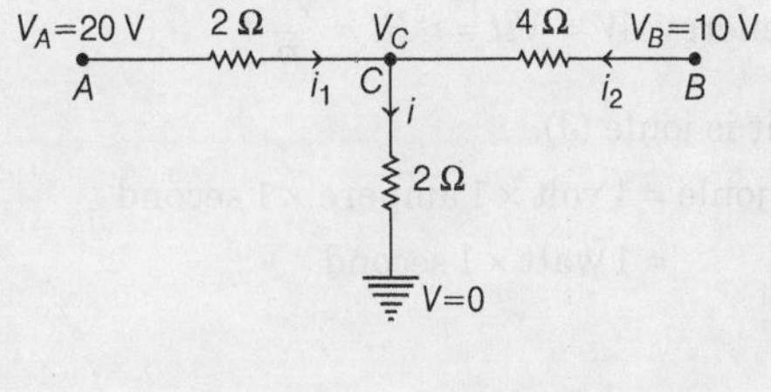

Then, according to Kirchhoff's current law, which states that the sum of all the currents directed towards a point in a circuit is equal to the sum of all the currents directed away from that point. Since, in the above circuit, that point is C.

$\therefore \quad i_1 + i_2 = i$

$\Rightarrow \quad \frac{V_A - V_C}{2} + \frac{V_B - V_C}{4} = \frac{V_C - V}{2}$ (∵ using Ohm's law, $V = iR$)

or $\quad \frac{20 - V_C}{2} + \frac{10 - V_C}{4} = \frac{V_C - 0}{2}$

$\Rightarrow \quad 20 - V_C + (10 - V_C)2 = V_C \Rightarrow 40 = V_C + 3V_C$

$40 = 4V_C$ or $V_C = 10$ V

$\therefore \quad$ Current, $i = \frac{V_C}{2} = \frac{10}{2} = 5$A

Example 15. *In the given circuit, currents in different branches and value of one resistor are shown. Then, potential at point B with respect to the point A is* **[JEE Main 2020]**

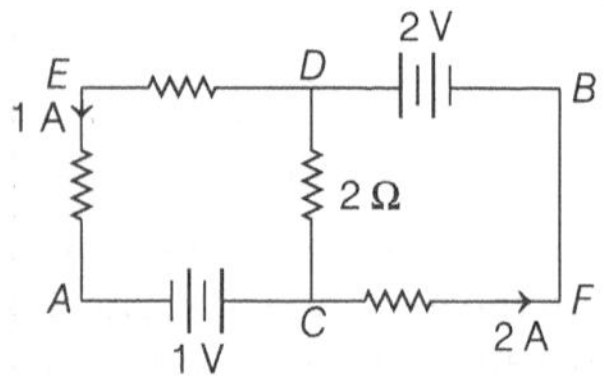

(a) +2 V (b) −2 V

(c) −1V (d) +1V

Sol. (d) The given circuit can be drawn as

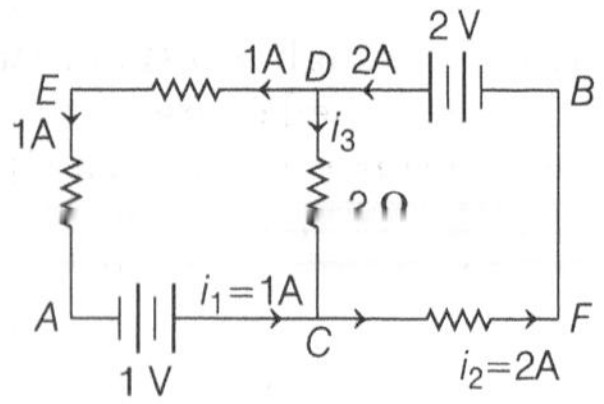

Current in branch DC (using KCL at point C),

$$i_1 + i_3 = i_2$$
$$2 - 1 - i_3 = 0$$
$$i_3 = 1\text{A}$$

Now, while moving from A to B via C and D, the potential along $ACDB$,

$$V_A + 1 + 2 \times i_3 - 2 = V_B \Rightarrow V_B - V_A = 1\text{V}$$

Electrical Energy

The total work done or energy supplied by the source in maintaining the current in electric circuit for a given time is called electric energy consumed in the circuit.

Electric energy, $W = Vit = i^2Rt = \frac{V^2}{R}t$

Its SI unit is joule (J).

where, 1 joule = 1 volt × 1 ampere × 1 second

= 1 watt × 1 second

Electrical Power

It is defined as the rate of electrical energy supplied per unit time to maintain flow of electric current through a conductor.

Mathematically, $\quad P = Vi = i^2R = \frac{V^2}{R}$

The SI unit of power is watt (W), where

1 watt = 1 volt × 1 ampere = 1 ampere-volt.

The bigger units of electrical power are kilowatt (kW) and megawatt (MW)

where, 1kW = 1000 W and 1 MW = 10^6W

Commercial unit of electrical power is horse power (hp) where, 1 hp = 74 6 watt.

Heating Effects of Current

When some potential difference V is applied across a resistance R and charge q to flow through the circuit in time t, the heat absorbed or produced, is given by

$$W = qV = Vit = i^2Rt = \frac{V^2t}{R} \text{ joule}$$

or $$= \frac{Vit}{4.2} = \frac{i^2Rt}{4.2} = \frac{V^2t}{4.2R} \text{ cal}$$

where, J is the Joule's mechanical equivalent of heat (4.21 J/cal).

Electricity Consumption

To measure the electrical energy consumed commercially, joule is not sufficient. So, to express electrical energy consumed commercially a special unit *kilo-watt-hour* is used in place of joule. It is also called 1 unit of electrical energy.

1 kilowatt hour or 1 unit of electrical energy is the amount of energy dissipated in 1 hour in a circuit, when the electric power in the circuit is 1 kilowatt.

1 kilowatt hour (kWh) = 3.6× 10^6 joule (J)

Example 16. *A current of 2 mA was passed through an unknown resistor which dissipated a power of 4.4 W. Dissipated power when an ideal power supply of 11 V is connected across it is* **[JEE Main 2019]**

(a) 11×10^{-4} W (b) 11×10^{-5} W

(c) 11×10^{5} W (d) 11×10^{-3} W

Sol. (b) Power dissipated by any resistor R, when i current flows through it is, $P = i^2R$ ... (i)

Given, $i = 2$ mA $= 2 \times 10^{-3}$A and $P = 4.4$ W

Using Eq. (i), we get

$$4.4 = (2 \times 10^{-3})^2 \times R$$

or $\quad R = \frac{4.4}{4 \times 10^{-6}}$ ...(ii)

When this resistance R is connected with 11 V supply, then power dissipated is,

$$P = \frac{V^2}{R}$$

or $$P = \frac{(11)^2}{4.4} \times 4 \times 10^{-6}$$ [∵ using Eq. (ii)]

$\Rightarrow$ $$P = \frac{11 \times 11 \times 4 \times 10^{-6}}{44 \times 10^{-1}} \text{ W}$$

or $$P = 11 \times 10^{-5} \text{ W}$$

Example 17. *In a building, there are 15 bulbs of 45 W, 15 bulbs of 100 W, 15 small fans of 10 W and 2 heaters of 1 kW. The voltage of electric main is 220 V. The minimum fuse capacity (rated value) of the building will be* ***[JEE Main 2020]***

(a) 25 A (b) 10 A
(c) 20 A (d) 15 A

Sol. (c) Power of 15 bulbs (each of 45 W),

$$P_1 = 15 \times 45 = 675 \text{ W}$$

Power of 15 bulbs (each of 100 W),

$$P_2 = 15 \times 100 = 1500 \text{ W}$$

Power of 15 fans (each of 10 W),

$$P_3 = 15 \times 10 = 150 \text{ W}$$

Power of 2 heaters (each of 1 kW),

$$P_4 = 2 \times 1000 = 2000 \text{ W}$$

Total power usage of building,

$$P = P_1 + P_2 + P_3 + P_4 = 4325 \text{ W}$$

Using $P = V \cdot i$, current drawn from mains supply of 220 V,

$$i = \frac{P}{V} = \frac{4325}{220} = 19.66 \text{ A}$$

∴ Minimum fuse capacity required is 20 A.

Example 18. *A battery of 3.0 V is connected to a resistor dissipating 0.5 W of power. If the terminal voltage of the battery is 2.5 V, the power dissipated within the internal resistance is* ***[JEE Main 2020]***

(a) 0.072 W (b) 0.125 W
(c) 0.50 W (d) 0.10 W

Sol. (d) Using Kirchhoff's loop law,

$$V = \varepsilon - ir$$

$$2.5 = 3 - ir$$

$\Rightarrow$ $$ir = 0.5 \quad \text{...(i)}$$

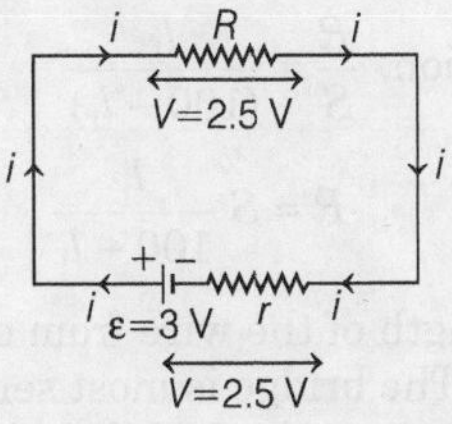

Now, from Ohm's law, $V = iR$

$$2.5 = iR$$

$$iR = 2.5 \quad \text{...(ii)}$$

and given that, $$P_R = 0.5, i^2R = 0.5 \quad \text{...(iii)}$$

Dividing Eq. (i) by Eq. (ii), we get

$$\frac{ir}{iR} = \frac{0.5}{2.5} \Rightarrow \frac{r}{R} = \frac{1}{5} \quad \text{...(iv)}$$

Now, $$\frac{P_r}{P_R} = \frac{i^2r}{i^2R}$$

$\Rightarrow$ $$\frac{P_r}{0.5} = \frac{r}{R}$$

$\Rightarrow$ $$\frac{P_r}{0.5} = \frac{1}{5}$$

$\Rightarrow$ $$P_r = 0.10 \text{ W}$$

Maximum Power Transfer Theorem

It states that the power output across load due to a cell or or battery is maximum, if the load (external) resistance is equal to the effective internal resistance of cell of battery. It means, when the effective internal resistance of cell or a battery is equal to external load resistance in a circuit, the efficiency of battery or cell is maximum.

Example 19. *A cell of internal resistance r drives current through an external resistance R. The power delivered by the cell to the external resistance will be maximum when* ***[JEE Main 2019]***

(a) $R = 2r$ (b) $R = r$
(c) $R = 0.001r$ (d) $R = 1000\,r$

Sol. (b) Given circuit is shown in the figure below

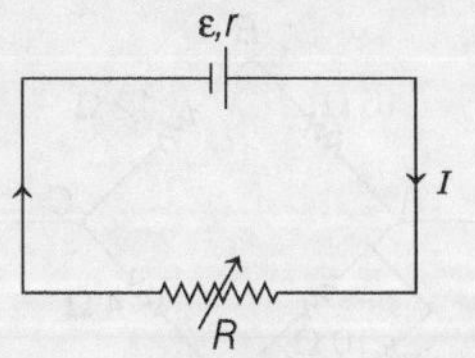

Net current, $$I = \frac{\varepsilon}{R + r} \quad \text{...(i)}$$

Power across R is given as, $$P = I^2R = \left(\frac{\varepsilon}{R + r}\right)^2 \cdot R \quad \text{[using Eq. (i)]}$$

For the maximum power, $$\frac{dP}{dR} = 0$$

$\Rightarrow$ $$\frac{dP}{dR} = \frac{d}{dR}\left(\left(\frac{\varepsilon}{R + r}\right)^2 \cdot R\right) = \varepsilon^2 \frac{d}{dR}\left(\frac{R}{(R + r)^2}\right)$$

$$= \varepsilon^2 \left[\frac{(R + r)^2 \times 1 - 2R \times (R + r)}{(R + r)^4}\right] = 0$$

$\Rightarrow$ $$(R + r)^2 = 2R(R + r)$$

or $$R + r = 2R \Rightarrow r = R$$

∴ The power delivered by the cell to the external resistance is maximum when $R = r$.

Wheatstone Bridge

It is an arrangement of four resistances used to measure one of them in terms of other three.

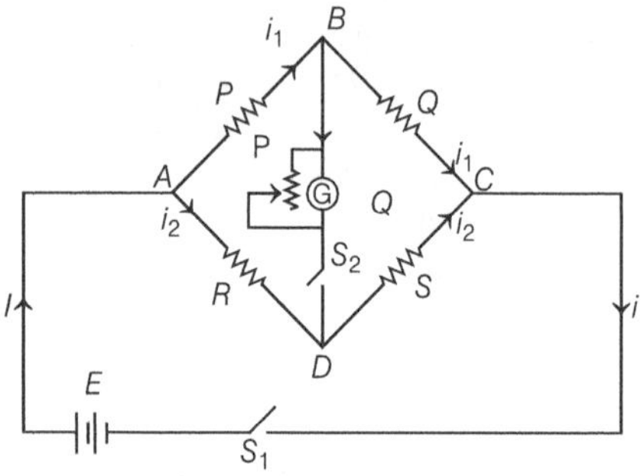

The bridge is said to be balanced when deflection in galvanometer is zero, *i.e.* $i_g = 0$. So, we have the balanced condition as

$$\frac{P}{Q} = \frac{R}{S}$$

Example 20. *Four resistances of 15 Ω, 12 Ω, 4 Ω and 10 Ω respectively in cyclic order to form Wheatstone's network. The resistance that is to be connected in parallel with the resistance of 10 Ω to balance the network is* **[JEE Main 2020]**

(a) 4 Ω (b) 7 Ω
(c) 10 Ω (d) 12 Ω

Sol. (c) Cyclic order and resistance X Ω which is connected to obtain balance condition is as shown in the figure.

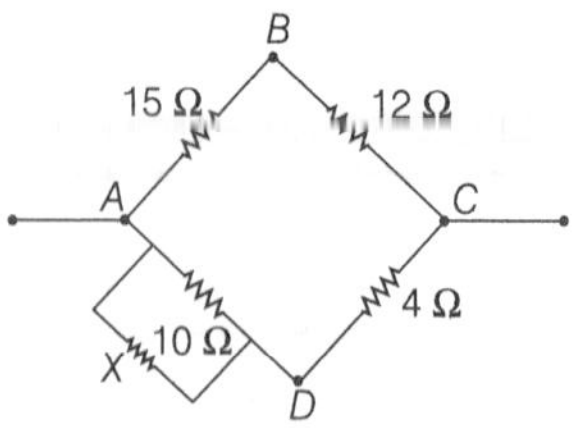

For balance, $\frac{15}{12} = \frac{\left(\frac{10X}{10+X}\right)}{4} \Rightarrow \frac{10X}{10+X} = 5$

$\Rightarrow X = 10\ \Omega$

Example 21. *The Wheatstone bridge shown in figure here, gets balanced when the carbon resistor is used as R_1 has the colour code (orange, red, brown). The resistors R_2 and R_4 are 80 Ω and 40 Ω, respectively.*

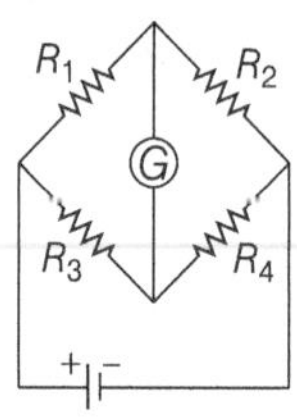

Assuming that the colour code for the carbon resistors gives their accurate values, the colour code for the carbon resistor used as R_3 would be **[JEE Main 2019]**

(a) brown, blue, black
(b) brown, blue, brown
(c) grey, black, brown
(d) red, green, brown

Sol. (b) The value of R_1 (orange, red, brown)

$$= 32 \times 10 = 320\ \Omega$$

Given, $R_2 = 80\ \Omega$ and $R_4 = 40\ \Omega$

In balanced Wheatstone bridge condition,

$$\frac{R_1}{R_2} = \frac{R_3}{R_4}$$

$$\Rightarrow \quad R_3 = R_4 \times \frac{R_1}{R_2}$$

$$\Rightarrow \quad R_3 = \frac{40 \times 320}{80}$$

or $R_3 = 160\ \Omega = 16 \times 10^1$

Comparing the value of R_3 with the colours assigned for the carbon resistor, we get

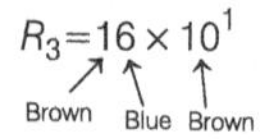

Meter Bridge

A meter bridge is **slide wire bridge** or Carey Foster **bridge**. It is an instrument which work on the principle of Wheatstone bridge.

It consists of a straight and uniform wire along a meter scale (AC) and by varying the taping point B as shown in figure, the bridge is balanced.

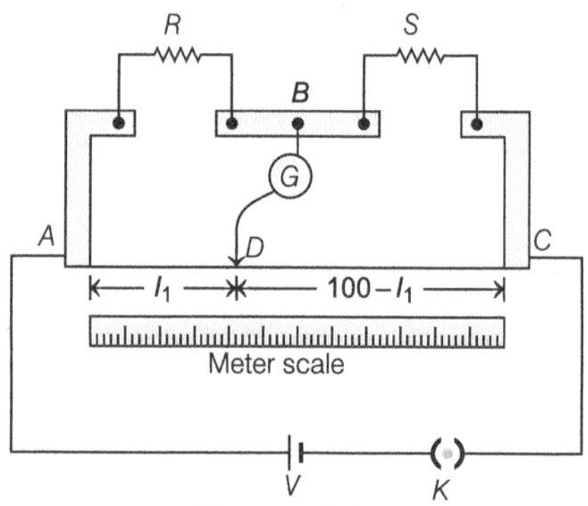

In balanced condition, $\frac{R}{S} = \frac{l_1}{(100 - l_1)}$

or $R = S \frac{l_1}{100 - l_1}$

where, l_1 is the length of the wire from one end, where null point is obtained. The bridge is most sensitive, when null point is somewhere near the middle point of the wire.

Example 22. *In a meter bridge experiment, the circuit diagram and the corresponding observation table are shown in figure* **[JEE Main 2019]**

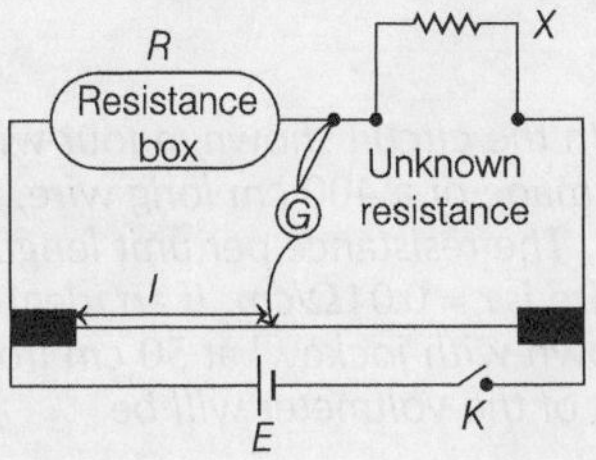

S. No.	R (Ω)	l (cm)
1.	1000	60
2.	100	13
3.	10	1.5
4.	1	1.0

Which of the readings is inconsistent?

(a) 3 (b) 2
(c) 1 (d) 4

Sol. (d) Unknown resistance 'X' in meter bridge experiment is given by

$$X = \left(\frac{100 - l}{l}\right) R$$

Case I When $R = 1000\ \Omega$ and $l = 60$ cm, then

$$X = \frac{(100 - 60)}{60} \times 1000 = \frac{40 \times 1000}{60}$$

$$\Rightarrow \quad X = \frac{2000}{3}\ \Omega \approx 667\ \Omega$$

Case II When $R = 100\ \Omega$ and $l = 13$ cm, then

$$X = \left(\frac{100 - 13}{13}\right) \times 100 = \frac{100 \times 87}{13}$$

$$= \frac{8700}{13}\ \Omega \approx 669\ \Omega$$

Case III When $R = 10\ \Omega$ and $l = 1.5$ cm, then

$$X = \left(\frac{100 - 1.5}{1.5}\right) \times 10 = \frac{98.5}{1.5} \times 10$$

$$= \frac{9850}{15}\ \Omega \approx 656\ \Omega$$

Case IV When $R = 1\ \Omega$ and $l = 1.0$ cm, then

$$X = \left(\frac{100 - 1}{1}\right) \times 1$$

$$\therefore \quad X = 99\ \Omega$$

Thus, from the above cases, it can be concluded that, value calculated in case (4) is inconsistent.

Potentiometer

It is an ideal device used to measure the potential difference between two points. It consists of a long resistance wire AB of uniform cross-section in which a steady direct current is set up by means of a battery.

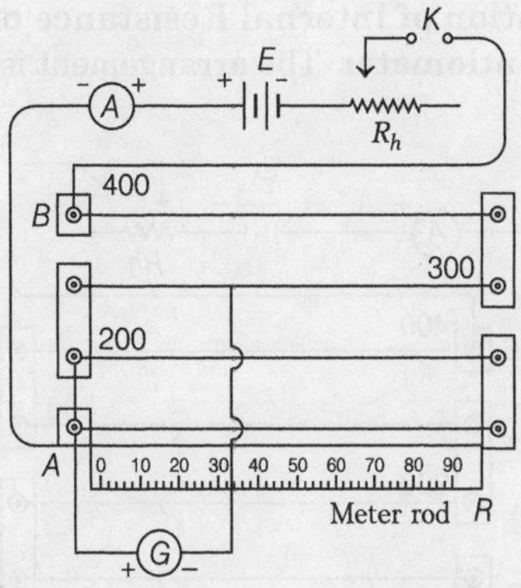

The principle of potentiometer states that, when a constant amount of current flows through a wire of uniform cross-section, then the potential drop across the wire is directly proportional to its length,

i.e. $\quad V \propto l$

$\Rightarrow \quad V = kl$

where, k is known as *potential gradient*.

SI unit of k is Vm^{-1}

Sensitivity of potentiometer is increased by increasing length of potentiometer wire.

Applications of Potentiometer

Potentiometer can be used for following purpose

- **To Compare the emf's of Two Cells using Potentiometer** The arrangement of two cells of emfs ε_1 and ε_2 which are to be compared is shown in the figure below

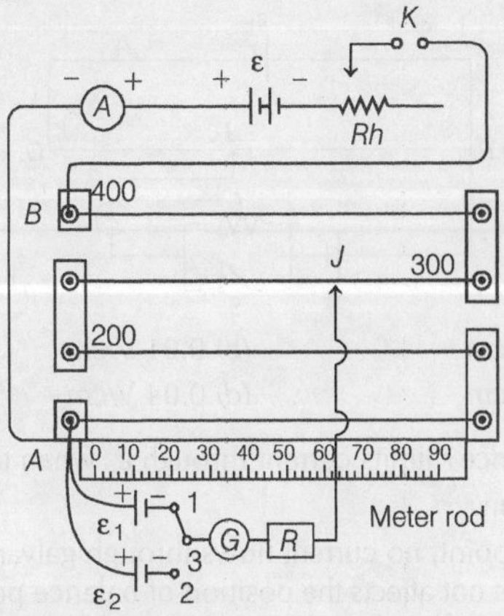

If the plug is put in the gap between 1 and 3, we get

$$\varepsilon_1 = (x\,l_1)\,i \qquad \text{...(i)}$$

where, x = resistance per unit length.

Similarly, when the plug is put in the gap between 2 and 3, we get

$$\varepsilon_2 = (x\,l_2)\,i \qquad \text{...(ii)}$$

From Eqs. (i) and (ii), we get

or $$\frac{\varepsilon_1}{\varepsilon_2} = \frac{l_1}{l_2}$$

- **Determination of Internal Resistance of a Cell using Potentiometer** The arrangement is shown in figure

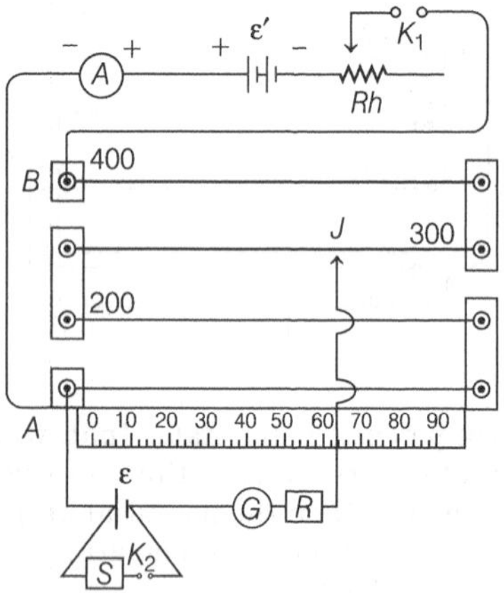

When K_2 is kept out, $\varepsilon = xl_1 i$

But if by inserting key K_2 and introducing some resistance S (say), then potential difference V is balanced by a length l_2, where $V = kl_2$

Internal resistance of cell, $r = \dfrac{\varepsilon - V}{V} R = \dfrac{l_1 - l_2}{l_2} R$

Example 23. *A potentiometer wire PQ of 1 m length is connected to a standard cell ε_1. Another cell ε_2 of emf 1.02 V is connected with a resistance r and switch S (as shown in figure). With switch S open, the null position is obtained at a distance of 49 cm from Q. The potential gradient in the potentiometer wire is* **[JEE Main 2020]**

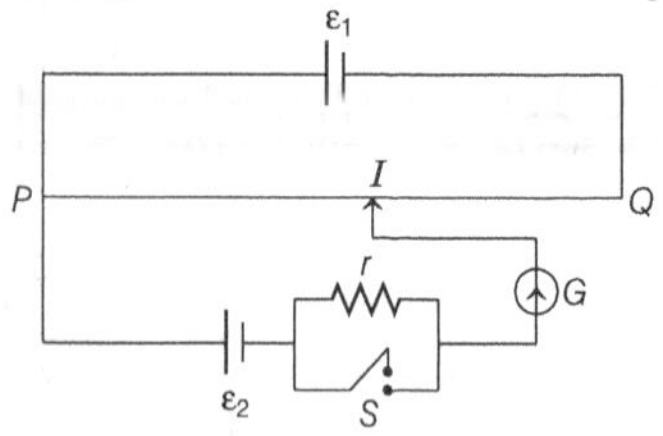

(a) 0.02 V/cm (b) 0.01 V/cm
(c) 0.03 V/cm (d) 0.04 V/cm

Sol. (a) Resistance r limits current through ε_2 when there is no balance situation.

But at balance point, no current flows through galvanometer G and ε_2, so r does not affects the position of balance point as shown in figure.

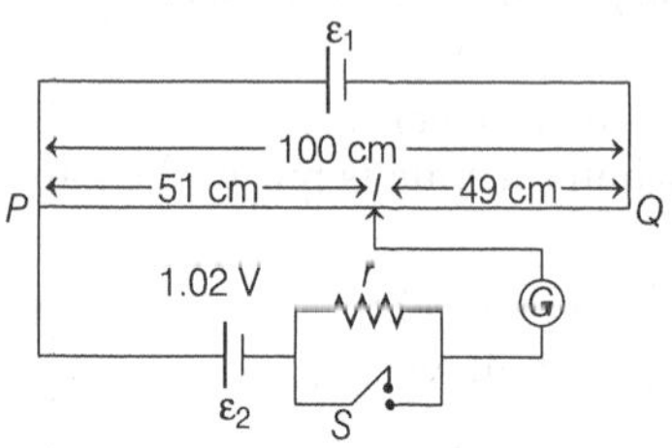

Now, 1.02 V is balanced against 51 cm length, so potential gradient of wire PQ is;

$$\text{Potential gradient} = \text{Fall of potential per unit length} = \frac{1.02 \text{ V}}{51 \text{ cm}}$$
$$= 0.02 \text{ V/cm}$$

Example 24. *In the circuit shown, a four-wire potentiometer is made of a 400 cm long wire, which extends between A and B. The resistance per unit length of the potentiometer wire is $r = 0.01\,\Omega$/cm. If an ideal voltmeter is connected as shown with jockey J at 50 cm from end A, the expected reading of the voltmeter will be* **[JEE Main 2019]**

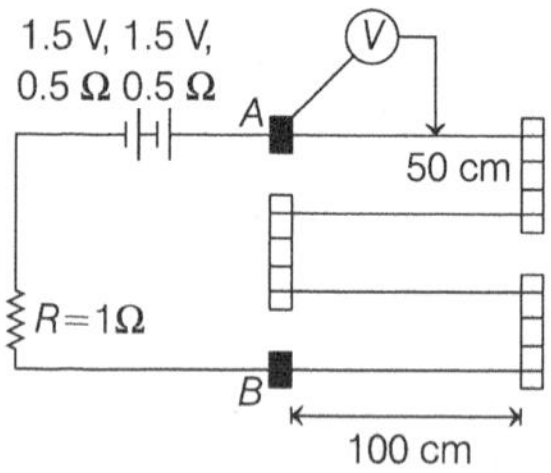

(a) 0.20 V (b) 0.75 V
(c) 0.25 V (d) 0.50 V

Sol. (c) In given potentiometer, resistance per unit length is $x = 0.01\,\Omega\text{cm}^{-1}$.

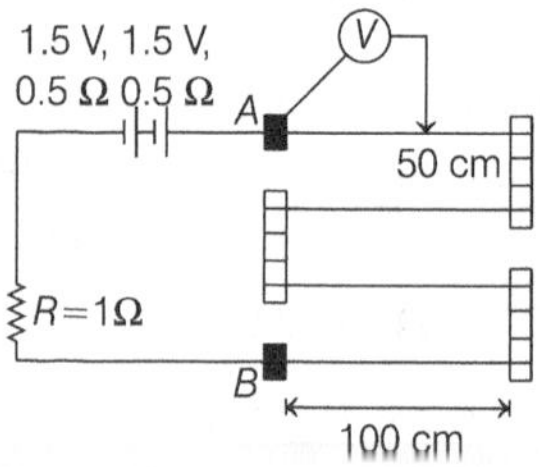

Length of potentiometer wire is $L = 400$ cm

Net resistance of the wire AB is

$$R_{AB} = \text{resistance per unit length} \times \text{length of } AB$$
$$= 0.01 \times 400$$
$$\Rightarrow \quad R_{AB} = 4\ \Omega$$

Net internal resistance of the cells connected in series,

$$r = 0.5 + 0.5 = 1\Omega$$

$\therefore$ Current in given potentiometer circuit is

$$I = \frac{\text{Net emf}}{\text{Total resistance}}$$
$$= \frac{\text{Net emf}}{r + R + R_{AB}}$$
$$= \frac{3}{1 + 1 + 4}$$
$$= 0.5 \text{ A}$$

Reading of voltmeter when the jockey is at 50 cm (l') from one end A,

$$V = IR = I(xl')$$
$$= 0.5 \times 0.01 \times 50$$
$$= 0.25 \text{ V}$$

Practice Exercise

ROUND I Topically Divided Problems

Electric Current and Drift Velocity

1. The current flowing through a wire depends on time as $i = 3t^2 + 2t + 5$. The charge flowing the cross-section of the wire in time from $t = 0$ to $t = 2$ s is

(a) 21 C (b) 10 C (c) 22 C (d) 1 C

2. There is a current of 0.21 A in a copper wire whose area of cross-section is 10^{-6} m^2. If the number of free electrons per m^3 is 8.4×10^{28}, then find the drift velocity. (Take, $e = 1.6 \times 10^{-19}$ C)

(a) 2×10^{-5} ms^{-1} (b) 1.56×10^{-5} ms^{-1}
(c) 1×10^{-5} ms^{-1} (d) 0.64×10^{-5} ms^{-1}

3. Two wires of the same material but of different diameters carry the same current i. If the ratio of their diameters is 2 : 1, then the corresponding ratio of their mean drift velocities will be

(a) 4 : 1 (b) 1 : 1 (c) 1 : 2 (d) 1 : 4

4. Every atom makes one free electron in copper. If 1.1 A current is flowing in the wire of copper having 1 mm diameter, then the drift velocity (approx.) will be (Take, density of copper $= 9 \times 10^3$ kg m^{-3} and atomic weight of copper = 63)

(a) 0.1 mms^{-1} (b) 0.2 mms^{-1}
(c) 0.3 mms^{-1} (d) 0.2 cms^{-1}

5. A straight conductor of uniform cross-section carries a current i. If s is the specific charge of an electron, the momentum of all the free electrons per unit length of the conductor, due to their drift velocity only is

(a) is (b) $\sqrt{i/s}$ (c) i/s (d) $(i/s)^2$

Resistance and Resistivity

6. The current i and voltage V graphs for a given metallic wire at two different temperatures T_1 and T_2 are shown in the figure. It is concluded that

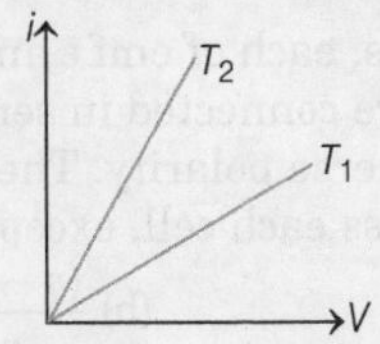

(a) $T_1 > T_2$ (b) $T_1 < T_2$ (c) $T_1 = T_2$ (d) $T_1 = 2T_1$

7. The resistance of a 10 m long wire is 10Ω. Its length is increased by 25% by stretching the wire uniformly. The resistance of wire will change to (approximately)

(a) 12.5 Ω (b) 14.5 Ω (c) 15.6 Ω (d) 16.6 Ω

8. Masses of the three wires of same material are in the ratio of 1 : 2 : 3 and their lengths in the ratio of 3 : 2 : 1. Electrical resistance of these wires will be in the ratio of

(a) 1 : 1 : 1 (b) 1 : 2 : 3
(c) 9 : 4 : 1 (d) 27 : 6 : 1

9. Two conductors are made of the same material and have the same length. Conductor A is a solid wire of diameter 1 mm. Conductor B is a hollow tube of outer diameter 2 mm and inner diameter 1mm. What is the ratio of resistances R_A to R_B?

(a) 1 : 3 (b) 3 : 1
(c) 2 : 3 (d) 3 : 2

10. A metal rod of length 10 cm and a rectangular cross-section of 1 cm $\times \frac{1}{2}$ cm is connected to a battery across opposite faces. The resistance will be

(a) maximum when the battery is connected across 1 cm $\times \frac{1}{2}$ cm
(b) maximum when the battery is connected across 10 cm × 1 cm faces
(c) maximum when the battery is connected across 10 cm $\times \frac{1}{2}$ cm faces
(d) same irrespective of the three faces

11. Consider four conducting materials copper, tungsten, mercury and aluminium with resistivity ρ_C, ρ_T, ρ_M and ρ_A, respectively. Then, **[JEE Main 2020]**

(a) $\rho_A > \rho_T > \rho_C$ (b) $\rho_A > \rho_M > \rho_C$
(c) $\rho_C > \rho_A > \rho_T$ (d) $\rho_M > \rho_A > \rho_C$

12. A resistance is shown in the figure. Its value and tolerance are given respectively by **[JEE Main 2019]**

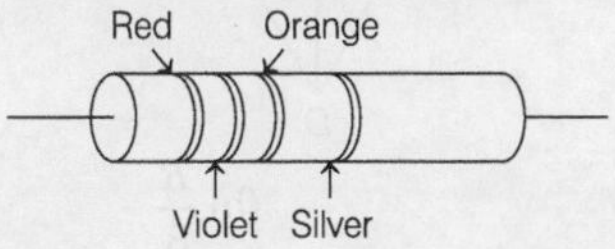

(a) 270 Ω, 5% (b) 27 k Ω, 20%
(c) 27 k Ω, 10% (d) 270 k Ω, 10%

13. Resistance of a resistor at temperature $t°$ C is

$$R_t = R_0 (1 + \alpha t + \beta t^2)$$

Here, R_0 is the resistance at 0°C. The temperature coefficient of resistance at temperature $t°$ C is

(a) $\frac{(1 + \alpha t + \beta t^2)}{\alpha + 2\beta t}$ (b) $(\alpha + 2\beta t)$

(c) $\frac{\alpha + 2\beta t}{(1 + \alpha t + \beta t^2)}$ (d) $\frac{(\alpha + 2\beta t)}{2(1 + \alpha t + \beta t)}$

14. A silver wire has a resistance of 2.1 Ω at 27.5 °C and a resistance of 2.7 Ω at 100 °C. Determine the temperature coefficient of resistivity of silver.

(a) 0.049/°C

(b) 0.0049/°C

(c) 0.0039/°C

(d) 0.039/°C

15. A copper wire of length 1 m and radius 1 mm is joined in series with an iron wire of length 2 m and radius 3 mm and a current is passed through the wires. The ratio of the current density in the copper and iron wires is

(a) 2 : 3 (b) 6 : 1

(c) 9 : 1 (d) 18 : 1

Grouping of Resistors

16. Six equal resistances each of 4Ω are connected to form a figure. The resistance between two corners A and B is

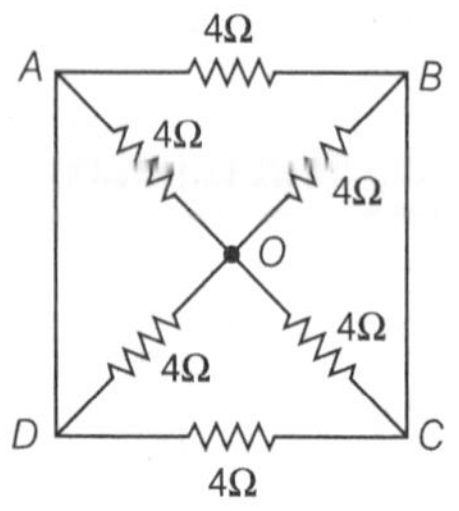

(a) 4 Ω (b) (4/3) Ω

(c) 12 Ω (d) 2 Ω

17. The effective resistance between points A and B is

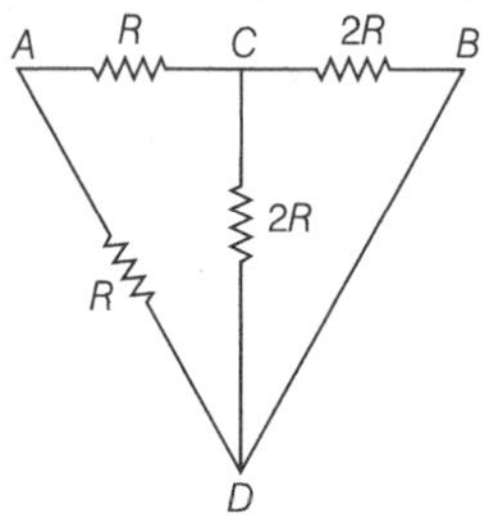

(a) R (b) $\frac{R}{3}$

(c) $\frac{2R}{3}$ (d) $\frac{3R}{5}$

18. The resistance across R and Q in the figure.

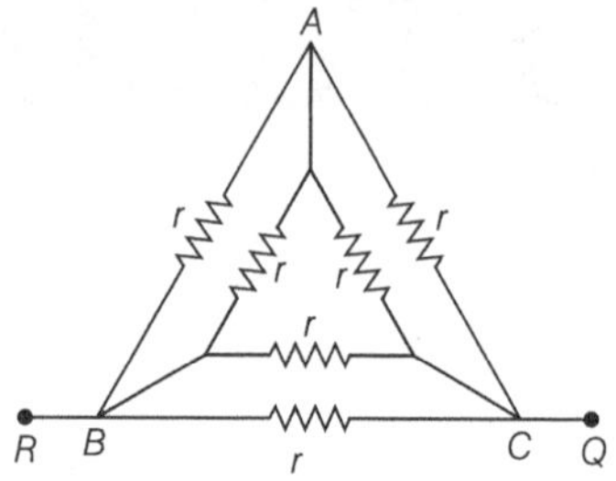

(a) $r/3$ (b) $r/2$ (c) $2r$ (d) $6r$

19. What is the equivalent resistance across the points A and B in the circuit given below ?

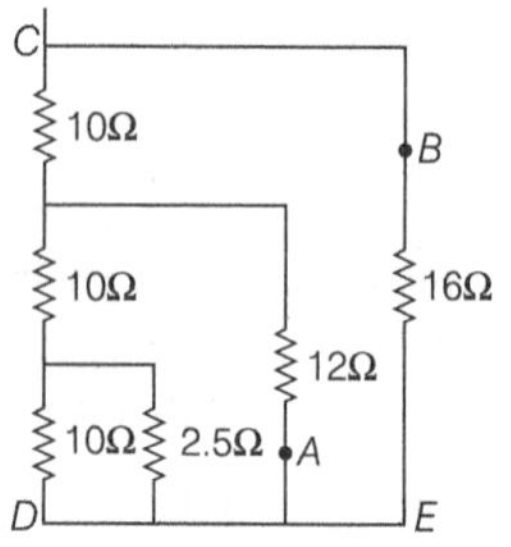

(a) 8 Ω (b) 12 Ω

(c) 16 Ω (d) 32 Ω

20. The current i_1 (in ampere) flowing through 1Ω resistor in the following circuit is

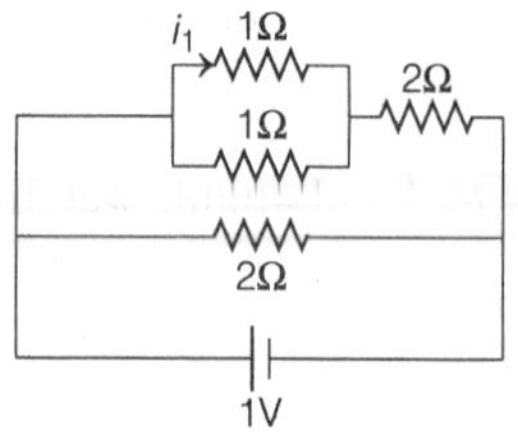

[JEE Main 2020]

(a) 0.25 (b) 0.4

(c) 0.5 (d) 0.2

Grouping of Cells

21. The strength of current in a wire of resistance R will be the same for connection in series or in parallel of n identical cells each of the internal resistance r, when

(a) $R = nr$ (b) $R = r/n$

(c) $R = r$ (d) $R \to \infty, r \to 0$

22. n identical cells, each of emf ε and internal resistance r, are connected in series, a cell A is joined with reverse polarity. The potential difference across each cell, except A is

(a) $\frac{2n\varepsilon}{n-2}$ (b) $\frac{(n-2)\varepsilon}{n}$

(c) $\frac{(n-1)\varepsilon}{n}$ (d) $\frac{2\varepsilon}{n}$

23. Two cells of emf $2E$ and E with internal resistances r_1 and r_2 respectively are connected in series to an external resistor R (see figure). The value of R, at which the potential difference across the terminals of the first cell becomes zero is **[JEE Main 2021]**

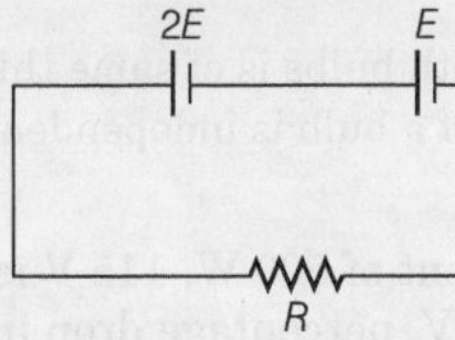

(a) $r_1 + r_2$ (b) $\frac{r_1}{2} - r_2$ (c) $\frac{r_1}{2} + r_2$ (d) $r_1 - r_2$

24. To get a maximum current through a resistance of 2.5 Ω, one can use m rows of cells each row having n cells. The internal resistance of each cell is 0.5Ω. What are the values of m and n, if the total number of cells are 20?

(a) $m = 2, n = 10$ (b) $m = 4, n = 5$
(c) $m = 5, n = 4$ (d) $n = 2, m = 10$

25. The series combination of two batteries, both of the same emf 10 V, but different internal resistance of 20 Ω and 5 Ω, is connected to the parallel combination of two resistors 30 Ω and R Ω. The voltage difference across the battery of internal resistance 20 Ω is zero, the value of R(in Ω) is **[JEE Main 2020]**

(a) 15 (b) 30 (c) 45 (d) 60

Kirchhoff's Law

26. In the given circuit diagram, the currents $i_1 = 0.3$ A, $i_4 = 0.8$ A and $i_5 = 0.4$ A, are flowing as shown. The currents i_2, i_3 and i_6 respectively are **[JEE Main 2019]**

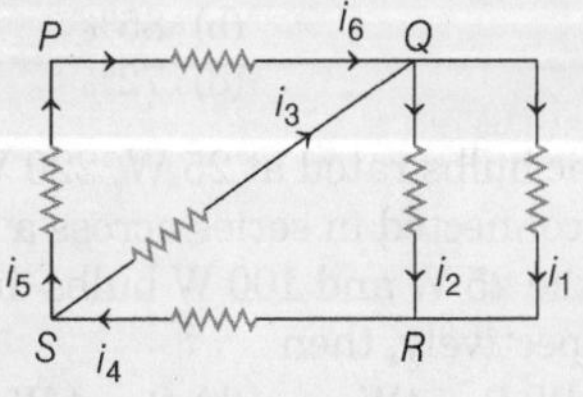

(a) 1.1 A, 0.4 A, 0.4 A (b) 1.1 A, −0.4 A, 0.4 A
(c) 0.4 A, 1.1 A, 0.4 A (d) −0.4 A, 0.4 A, 1.1 A

27. For the given circuit, terminal potential differences of cells are around

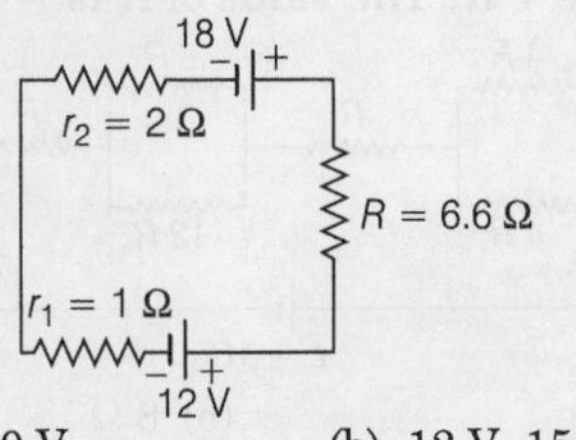

(a) 15 V, 10 V (b) 12 V, 15 V
(c) 6 V, 12 V (d) 17 V, 13 V

28. In the circuit shown the current through 2 Ω resistance is

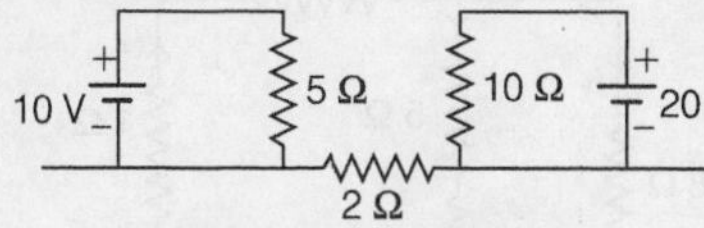

(a) 5 A (b) 2 A (c) 0 (d) 4 A

29. In the given circuit diagram, a wire is joining points B and D. The current in this wire is **[JEE Main 2020]**

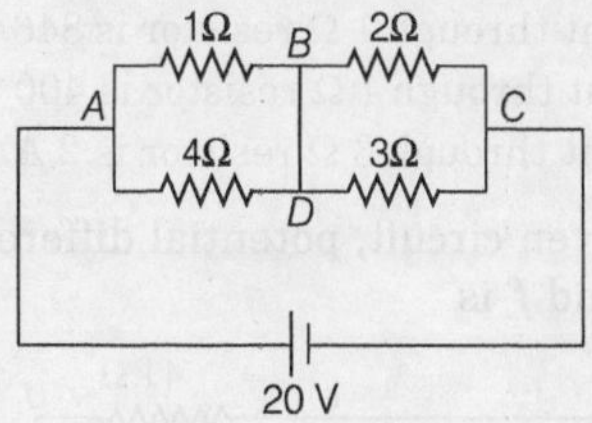

(a) 0.4 A (b) zero (c) 2A (d) 4A

30. For the given circuit,

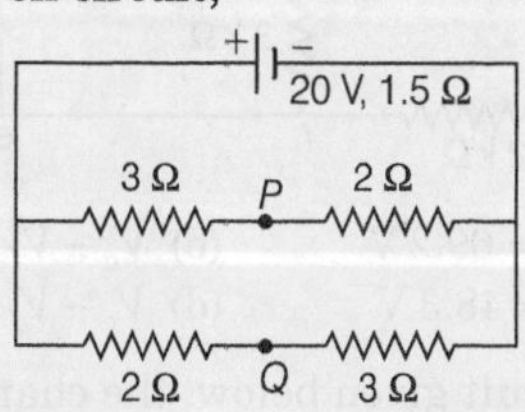

If internal resistance of cell is 1.5 Ω, then

(a) $V_P - V_Q = 0$ (b) $V_P - V_Q = 4$ V
(c) $V_P - V_Q = -4$ V (d) $V_P - V_Q = -2.5$ V

31. In the given circuit, the cells have zero internal resistance. The currents (in ampere) passing through resistances R_1 and R_2 respectively are **[JEE Main 2019]**

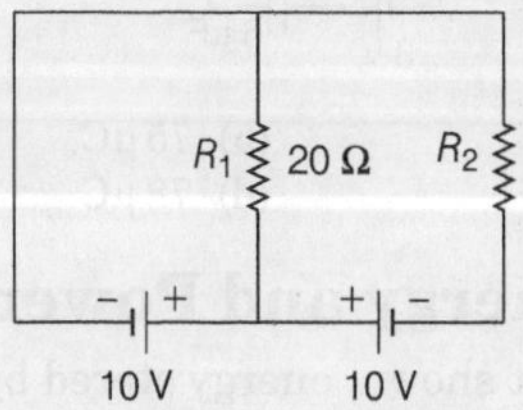

(a) 0.5, 0 (b) 1, 2 (c) 2, 2 (d) 0, 1

32. In the circuit shown in figure, if the ammeter reads 2 A. Then, i_1 and i_2 are

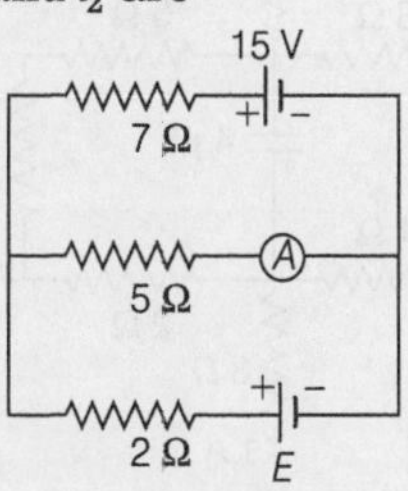

(a) $i_1 = 1.8$ A, $i_2 = 0.2$ A (b) $i_1 = 0.8$ A, $i_2 = 1.2$ A
(c) $i_1 = 0.75$ A, $i_2 = 1.25$ A (d) $i_1 = 0.714$ A, $i_2 = 1.29$ A

33. For the given network, mark the correct statement.

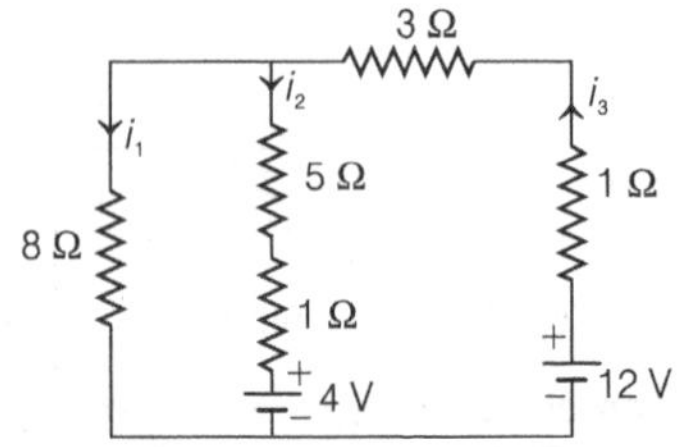

(a) Current through 8 Ω resistor is 846 mA.
(b) Current through 1 Ω resistor is 846 mA.
(c) Current through 1 Ω resistor is 400 mA.
(d) Current through 3 Ω resistor is 2 A.

34. For the given circuit, potential difference between points c and f is

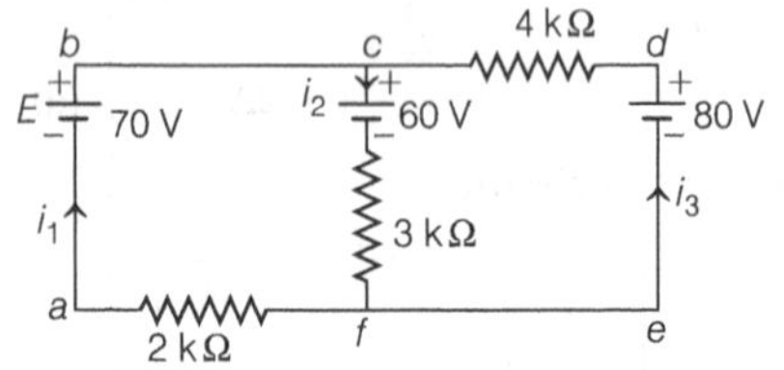

(a) $V_c - V_f = 69.2$ V
(b) $V_c - V_f = -69.2$ V
(c) $V_c - V_f = 48.3$ V
(d) $V_c - V_f = -48.3$ V

35. For the circuit given below, the charge on the capacitor is

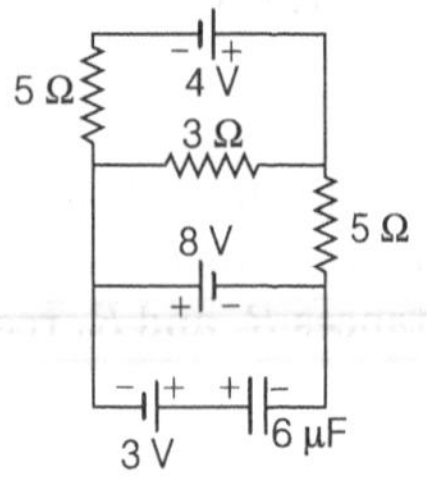

(a) 45 μC
(b) 75 μC
(c) 66 μC
(d) 78 μC

Electrical Energy and Power

36. For the circuit shown, energy stored by the capacitor is

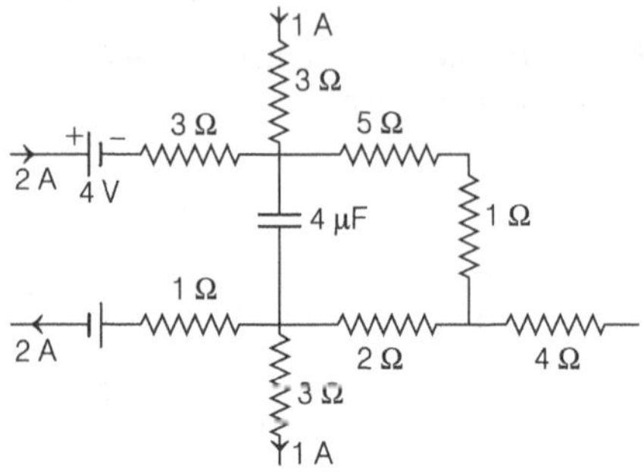

(a) around 0.3 mJ
(b) around 0.5 mJ
(c) around 0.8 mJ
(d) around 0.1 mJ

37. For two incandescent bulbs of rated power P_1 and P_2, if $P_1 > P_2$, then
(a) filament of bulb 1 is more thicker than filament of bulb 2
(b) filament of bulb 1 is thinner than the filament of bulb 2
(c) filament of both bulbs is of same thickness
(d) rated power of a bulb is independent of filament thickness

38. A heating filament of 500 W, 115 V is being operated at 110 V, percentage drop in output power (with comparison to rated output power) is
(a) 10.20% (b) 8.1% (c) 8.6% (d) 7.6%

39. Power of a heater is 500 W at 800°C. What will be its power at 200°C, if $\alpha = 4 \times 10^{-4}$ per°C?
(a) 484 W (b) 672 W (c) 526 W (d) 620 W

40. A wire when connected to 220 V mains supply has power dissipation P_1. Now, the wire is cut into two equal pieces which are connected in parallel to the same supply. Power dissipation in this case is P_2. Then, $P_2 : P_1$ is
(a) 1 : 1
(b) 4 : 1
(c) 2 : 1
(d) 3 : 1

41. In a large building, there are 15 bulbs of 40 W, 5 bulbs of 100 W, 5 fans of 80 W and 1 heater of 1 kW. The voltage of the electric mains is 220 V. The minimum capacity of the main fuse of the building will be **[JEE Main 2014]**
(a) 8 A
(b) 10 A
(c) 12 A
(d) 14 A

42. An electrical power line, having a total resistance of 2 Ω, delivers 1 kW at 220 V. The efficiency of the transmission line is approximately **[JEE Main 2020]**
(a) 91%
(b) 85%
(c) 96%
(d) 72%

43. Two electric bulbs rated at 25 W, 220 V and 100 W, 220 V are connected in series across a 220 V voltage source. If the 25 W and 100 W bulbs draw powers P_1 and P_2 respectively, then **[JEE Main 2019]**
(a) $P_1 = 16$ W, $P_2 = 4$W
(b) $P_1 = 4$ W, $P_2 = 16$W
(c) $P_1 = 9$ W, $P_2 = 16$W
(d) $P_1 = 16$ W, $P_2 = 9$W

44. The resistive network shown below is connected to a DC source of 16 V. The power consumed by the network is 4 W. The value of R is **[JEE Main 2019]**

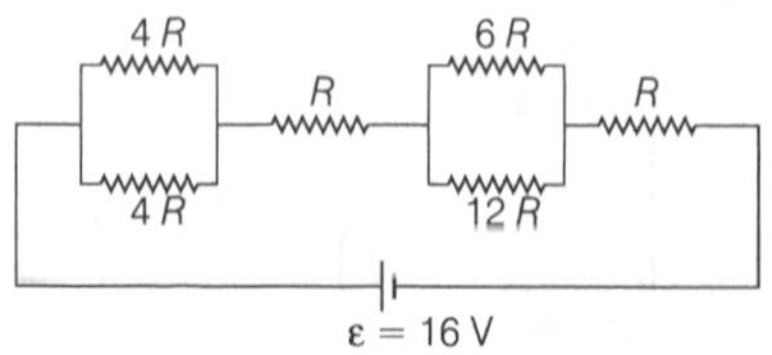

(a) 6 Ω
(b) 8 Ω
(c) 1 Ω
(d) 16 Ω

45. A resistor develops 500 J of thermal energy in 20s when a current of 1.5 A is passed through it. If the current is increased from 1.5 A to 3A, what will be the energy developed in 20 s? [JEE Main 2021]

(a) 1500 J (b) 1000 J (c) 500 J (d) 2000 J

46. A torch battery of length l is to be made up of a thin cylindrical bar of radius a and a concentric thin cylindrical shell of radius b is filled in between with an electrolyte of resistivity ρ (see figure). If the battery is connected to a resistance R, the maximum Joule's heating in R will takes place for

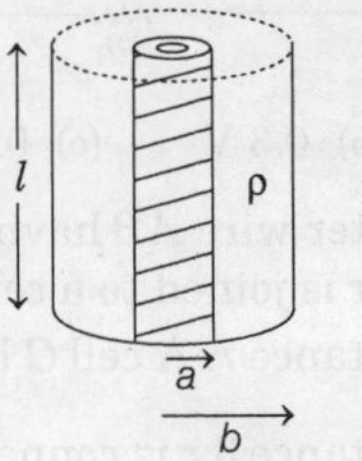

[JEE Main 2020]

(a) $R=\frac{\rho}{2\pi l}\left(\frac{b}{a}\right)$ (b) $R=\frac{\rho}{2\pi l}\ln\left(\frac{b}{a}\right)$

(c) $R=\frac{\rho}{\pi l}\ln\left(\frac{b}{a}\right)$ (d) $R=\frac{2\rho}{\pi l}\ln\left(\frac{b}{a}\right)$

Measuring Instruments

47. Which of the following statements is false? [JEE Main 2017]

(a) In a balanced Wheatstone bridge, if the cell and the galvanometer are exchanged, the null point is disturbed.

(b) A rheostat can be used as a potential divider.

(c) Kirchhoff's second law represents energy conservation.

(d) Wheatstone bridge is the most sensitive when all the four resistances are of the same order of magnitude.

48. The four arms of a Wheatstone bridge have resistances as shown in the figure. A galvanometer of 15 Ω resistance is connected across BD. Calculate the current through the galvanometer when a potential difference of 10 V is maintained across AC. [JEE Main 2021]

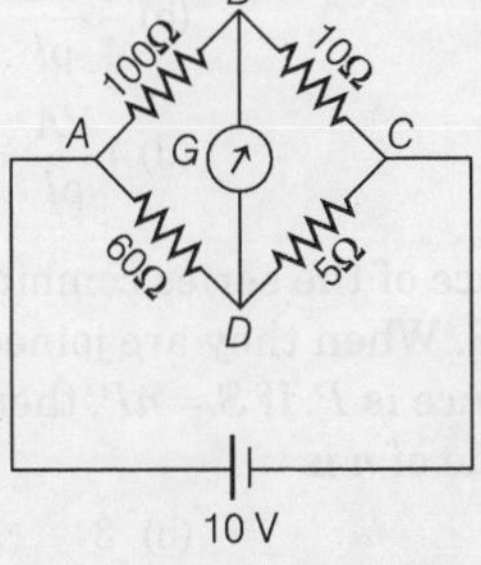

(a) 2.44 μA (b) 2.44 mA (c) 4.87 mA (d) 4.87 μA

49. In a Wheatstone bridge (see figure), resistances P and Q are approximately equal. When $R = 400\,\Omega$, the bridge is balanced. On interchanging P and Q, the value of R for balance is 405 Ω. The value of X is close to

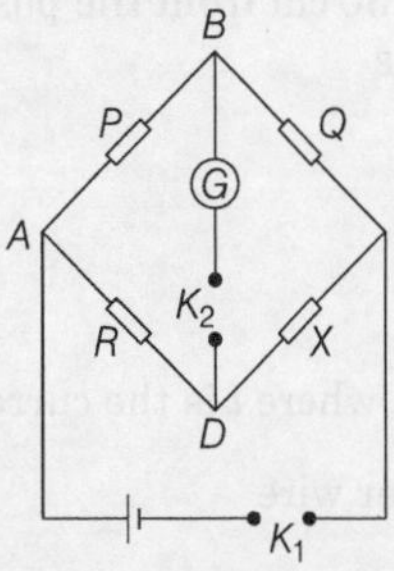

[JEE Main 2019]

(a) 404.5 Ω (b) 401.5 Ω (c) 402.5 Ω (d) 403.5 Ω

50. On interchanging the resistances, the balance point of a meter bridge shifts to the left by 10 cm. The resistance of their series combination is 1 kΩ. How much was the resistance on the left slot before interchanging the resistances? [JEE Main 2018]

(a) 990 Ω (b) 505 Ω (c) 550 Ω (d) 910 Ω

51. In the experimental set up of meter bridge shown in the figure, the null point is obtained at a distance of 40 cm from A. If a 10 Ω resistor is connected in series with R_1, the null point shifts by 10 cm. The resistance that should be connected in parallel with $(R_1 + 10)$ Ω such that the null point shifts back to its initial position is [JEE Main 2019]

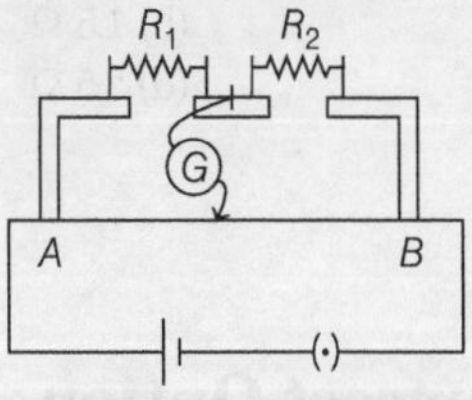

(a) 60 Ω (b) 20 Ω (c) 30 Ω (d) 40 Ω

52. In a meter bridge, the wire of length 1m has a non-uniform cross-section such that the variation $\frac{dR}{dl}$ of its resistance R with length l is $\frac{dR}{dl} \propto \frac{1}{\sqrt{l}}$.

Two equal resistance are connected as shown in the figure. The galvanometer has zero deflection when the jockey is at point P. What is the length AP? [JEE Main 2019]

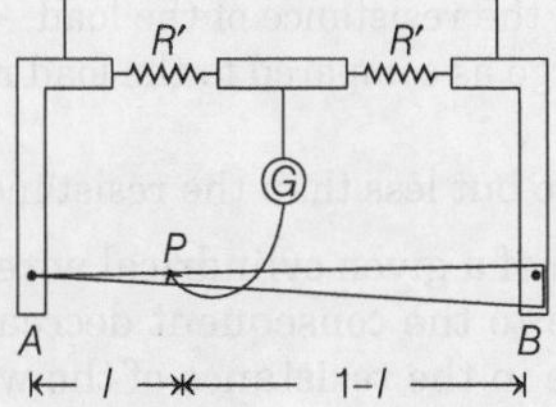

(a) 0.3 m (b) 0.25 m (c) 0.2 m (d) 0.35 m

53. The length of a wire of a potentiometer is 100 cm and the emf of its stand and cell is ε volt. It is employed to measure the emf of a battery whose internal resistance is 0.5 Ω. If the balance point is obtained at $l = 30$ cm from the positive end, the emf of the battery is

(a) $\frac{30\varepsilon}{100.5}$

(b) $\frac{30\varepsilon}{100 - 0.5}$

(c) $\frac{30(\varepsilon - 0.5 i)}{100}$, where i is the current in the potentiometer wire

(d) $\frac{30\varepsilon}{100}$

54. In a potentiometer experiment, the balancing with a cell is at length 240 cm. On shunting the cell with a resistance of 2 Ω, the balancing length becomes 120 cm. The internal resistance of the cell is

(a) 1 Ω (b) 0.5 Ω
(c) 4 Ω (d) 2 Ω

55. In a potentiometer experiment, it is found that no current passes through the galvanometer when the terminals of the cell are connected across 52 cm of the potentiometer wire. If the cell is shunted by a resistance of 5 Ω, a balance is found when the cell is connected across 40 cm of the wire. Find the internal resistance of the cell. [JEE Main 2018]

(a) 1 Ω (b) 1.5 Ω
(c) 2 Ω (d) 2.5 Ω

56. The resistance of the potentiometer wire AB in given figure is 4 Ω. With a cell of emf $\varepsilon = 0.5$ V and rheostat resistance $Rh = 2\,\Omega$. The null point is obtained at some point J. When the cell is replaced by another one of emf $\varepsilon = \varepsilon_2$, the same null point J is found for $Rh = 6\,\Omega$. The emf ε_2 is [JEE Main 2019]

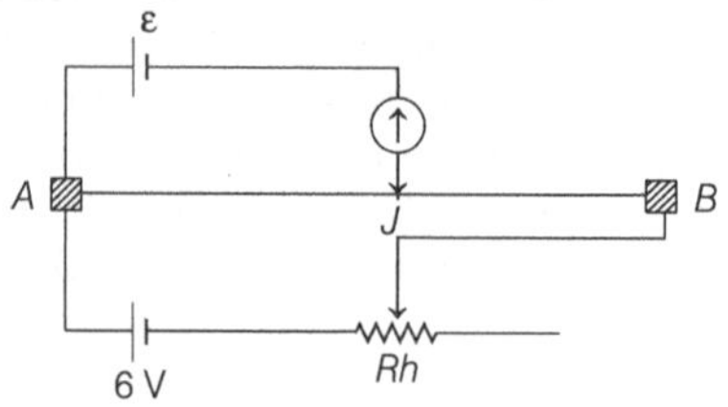

(a) 0.6 V (b) 0.3 V (c) 0.5 V (d) 0.4 V

57. A potentiometer wire AB having length L and resistance $12r$ is joined to a cell D of emf ε and internal resistance r. A cell C having emf $\frac{\varepsilon}{2}$ and internal resistance $3r$ is connected. The length AJ at which the galvanometer as shown in figure shows no deflection is

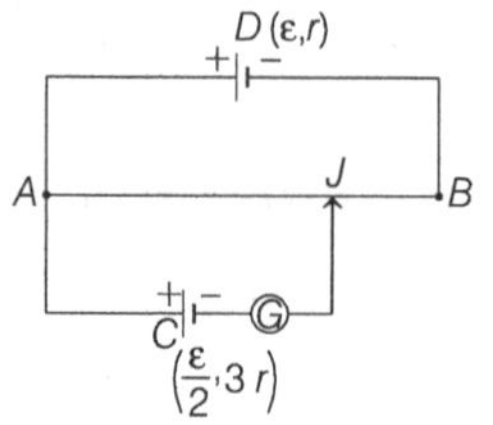

(a) $\frac{5}{12}L$ (b) $\frac{11}{12}L$ (c) $\frac{13}{24}L$ (d) $\frac{11}{24}L$

ROUND II Mixed Bag

Only One Correct Option

1. The temperature dependence of resistances of Cu and undoped Si in the temperature range 300-400 K, is best described by [JEE Main 2016]

(a) linear increase for Cu, linear increase for Si
(b) linear increase for Cu, exponential increase for Si
(c) linear increase for Cu, exponential decrease for Si
(d) linear decrease for Cu, linear decrease for Si

2. An energy source will supply a constant current into the load, if its internal resistance is

(a) equal to the resistance of the load
(b) very large as compared to the load resistance
(c) zero
(d) non-zero but less than the resistance of the load

3. The length of a given cylindrical wire is increased by 100%. Due to the consequent decrease in diameter, the change in the resistance of the wire will be

(a) 200% (b) 100% (c) 50% (d) 300%

4. A conducting wire of length l, area of cross-section A and electric resistivity ρ is connected between the terminals of a battery. A potential difference V is developed between its ends, causing an electric current. If the length of the wire of the same material is doubled and the area of cross-section is halved, resultant current would be [JEE Main 2021]

(a) $\frac{1}{4}\frac{VA}{\rho l}$ (b) $\frac{3}{4}\frac{VA}{\rho l}$

(c) $\frac{1}{4}\frac{\rho l}{VA}$ (d) $4\frac{VA}{\rho l}$

5. The resistance of the series combination of two resistors is S. When they are joined in parallel, the total resistance is P. If $S = nP$, then the minimum possible value of n is

(a) 4 (b) 3
(c) 2 (d) 1

6. A current of 10 A exists in a wire of cross sectional area of 5 mm^2 with a drift velocity of 2×10^{-3} ms^{-1}. The number of free electrons in each cubic meter of the wire is **[JEE Main 2021]**

(a) 2×10^6 (b) 625×10^{25}
(c) 2×10^{25} (d) 1×10^{23}

7. A material B has twice the specific resistance of A. A circular wire made of B has twice the diameter of a wire made of A. Then, for the two wires to have the same resistance, the ratio l_B/l_A of their respective lengths must be

(a) 1 (b) 2/1
(c) 1/4 (d) 2

8. A carbon resistor has a following colour code. What is the value of the resistance? **[JEE Main 2019]**

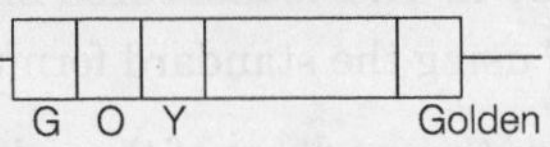

(a) 5.3 MΩ ± 5% (b) 64 kΩ ± 10%
(c) 6.4 MΩ ± 5% (d) 530 kΩ ± 5%

9. The value of current i_1 flowing from A to C in the circuit diagram is **[JEE Main 2020]**

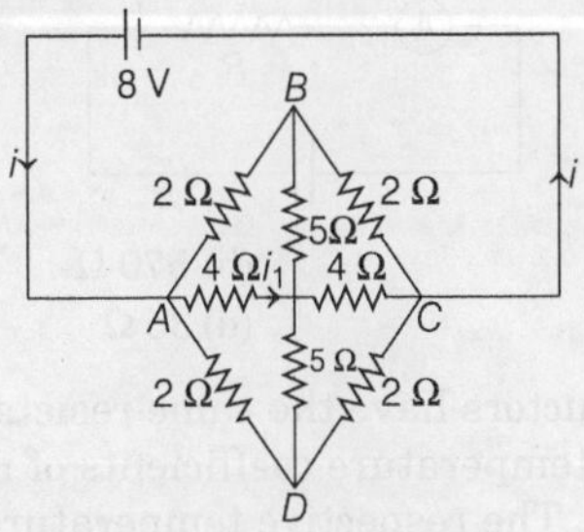

(a) 1 A (b) 5 A
(c) 2 A (d) 4 A

10. In a Wheatstone's bridge, three resistances P, Q and R are connected in the three arms and the fourth arm is formed by two resistances S_1 and S_2 connected in parallel. The condition for the bridge to be balanced will be

(a) $\frac{P}{Q} = \frac{2R}{S_1 + S_2}$

(b) $\frac{P}{Q} = \frac{R(S_1 + S_2)}{S_1 S_2}$

(c) $\frac{P}{Q} = \frac{R(S_1 + S_2)}{2 S_1 S_2}$

(d) $\frac{P}{Q} = \frac{R}{S_1 + S_2}$

11. A 2 W carbon resistor is colour coded with green, black, red and brown, respectively. The maximum current which can be passed through this resistor is **[JEE Main 2019]**

(a) 0.4 mA (b) 63 mA
(c) 20 mA (d) 100 mA

12. A 5 V battery with internal resistance 2 Ω and a 2 V battery with internal resistance 1 Ω are connected to a 10 Ω resistor as shown in the figure

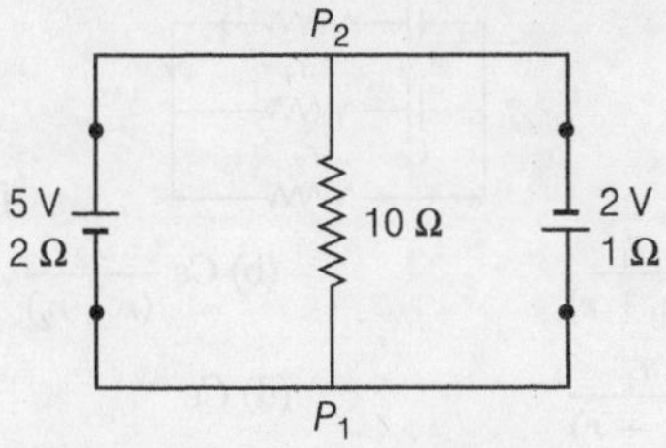

The current in the 10 Ω resistor is

(a) 0.27 A, P_2 to P_1 (b) 0.03 A, P_1 to P_2
(c) 0.03 A, P_2 to P_1 (d) 0.27 A, P_1 to P_2

13. When 5V potential difference is applied across a wire of length 0.1m, the drift speed of electrons is 2.5×10^{-4} ms^{-1}. If the electron density in the wire is 8×10^{28} m^{-3} the resistivity of the material is close to **[JEE Main 2015]**

(a) 1.6×10^{-8} Ω-m (b) 1.6×10^{-7} Ω-m
(c) 1.6×10^{-6} Ω-m (d) 1.6×10^{-5} Ω-m

14. The supply voltage to room is 120 V. The resistance of the lead wires is 6 Ω. A 60 W bulb is already switched on. What is the decrease of voltage across the bulb, when a 240 W heater is switched on in parallel to the bulb? **[JEE Main 2013]**

(a) Zero (b) 2.9 V
(c) 13.3 V (d) 10.04 V

15. Two equal resistances when connected in series to a battery consume electric power of 60 W. If these resistances are now connected in parallel combination to the same battery, the electric power consumed will be **[JEE Main 2019]**

(a) 60 W (b) 30 W
(c) 240 W (d) 120 W

16. Two sources of equal emf are connected to an external resistance R. The internal resistances of the two sources are R_1 and $R_2 (R_2 > R_1)$. If the potential difference across the source having internal resistance R_2 is zero, then

(a) $R = \frac{R_2 \times (R_1 + R_2)}{(R_2 - R_1)}$ (b) $R = R_2 - R_1$

(c) $R = \frac{R_1 R_2}{(R_1 + R_2)}$ (d) $R = \frac{R_1 R_2}{(R_2 - R_1)}$

17. An ideal battery of 4 V and resistance R are connected in series in the primary circuit of a potentiometer of length 1 m and resistance 5 Ω. The value of R to give a potential difference of 5 mV across 10 cm of potentiometer wire is **[JEE Main 2019]**

(a) 395 Ω (b) 495 Ω
(c) 490 Ω (d) 480 Ω

18. In the given circuit diagram, when the current reaches steady state in the circuit, the charge on the capacitor of capacitance C will be

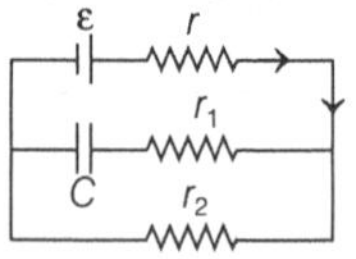

[JEE Main 2017]

(a) $C\varepsilon \dfrac{r_1}{(r_2 + r)}$ (b) $C\varepsilon \dfrac{r_2}{(r + r_2)}$

(c) $C\varepsilon \dfrac{r_1}{(r_1 + r)}$ (d) $C\varepsilon$

19. The total current supplied to the circuit by the battery is

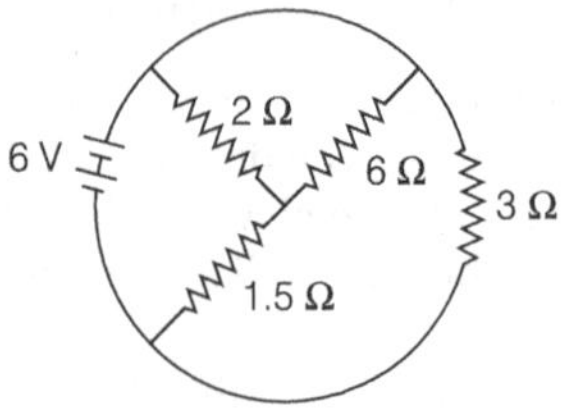

(a) 1 A (b) 2 A (c) 4 A (d) 6 A

20. The current i drawn from the 5 V source will be

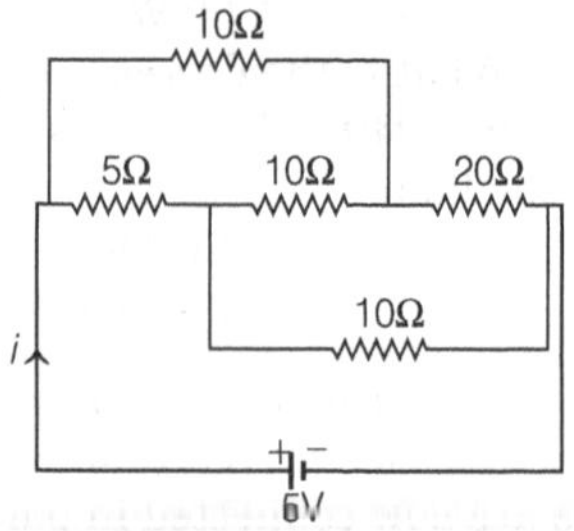

(a) 0.33 A (b) 0.5 A (c) 0.67 A (d) 0.17 A

21. In above figure shown, the current in the 10 V battery is close to [JEE Main 2020]

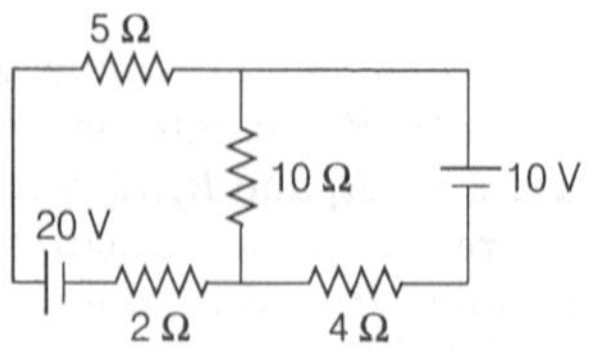

(a) 0.71 A from positive to negative terminal
(b) 0.42 A from positive to negative terminal
(c) 0.21 A from positive to negative terminal
(d) 0.36 A from negative to positive terminal

22. Two batteries with emf 12 V and 13 V are connected in parallel across a load resistor of 10 Ω. The internal resistances of the two batteries are 1 Ω and 2 Ω, respectively. The voltage across the load lies between [JEE Main 2018]

(a) 11.6 V and 11.7 V (b) 11.5 V and 11.6 V
(c) 11.4 V and 11.5 V (d) 11.7 V and 11.8 V

23. In the given circuit, the internal resistance of the 18 V cell is negligible. If $R_1 = 400\ \Omega$, $R_3 = 100\ \Omega$ and $R_4 = 500\ \Omega$ and the reading of an ideal voltmeter across R_4 is 5 V, then the value of R_2 will be [JEE Main 2019]

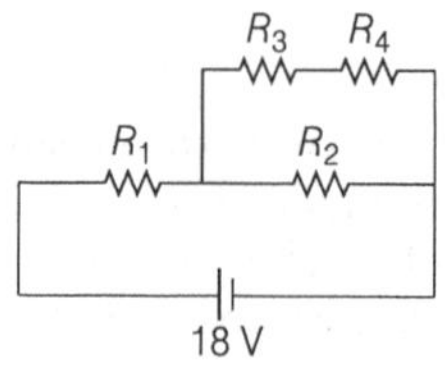

(a) 550 Ω (b) 230 Ω
(c) 300 Ω (d) 450 Ω

24. The actual value of resistance R, shown in the figure is 30 Ω. This is measured in an experiment as shown using the standard formula $R = \dfrac{V}{i}$, where V and i are the readings of the voltmeter and ammeter, respectively. If the measured value of R is 5% less, then the internal resistance of the voltmeter is

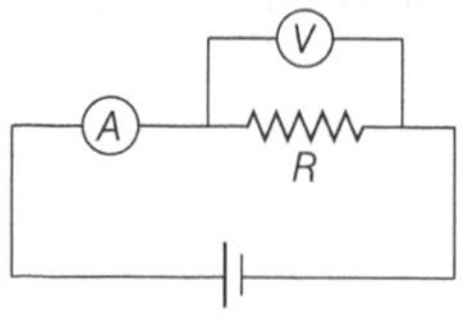

[JEE Main 2019]

(a) 600 Ω (b) 570 Ω
(c) 350 Ω (d) 35 Ω

25. Two conductors have the same resistance at 0°C but their temperature coefficients of resistance are α_1 and α_2. The respective temperature coefficients of their series and parallel combinations are nearly

(a) $\dfrac{\alpha_1 + \alpha_2}{2}, \alpha_1 + \alpha_2$ (b) $\alpha_1 + \alpha_2, \dfrac{\alpha_1 + \alpha_2}{2}$

(c) $\alpha_1 + \alpha_2, \dfrac{\alpha_1\alpha_2}{\alpha_1 + \alpha_2}$ (d) $\dfrac{\alpha_1 + \alpha_2}{2}, \dfrac{\alpha_1 + \alpha_2}{2}$

26. Four resistances 40 Ω, 60 Ω, 90 Ω and 110 Ω make the arms of a quadrilateral $ABCD$. Across AC is a battery of emf 40 V and internal resistance negligible. The potential difference across BD (in volt) is

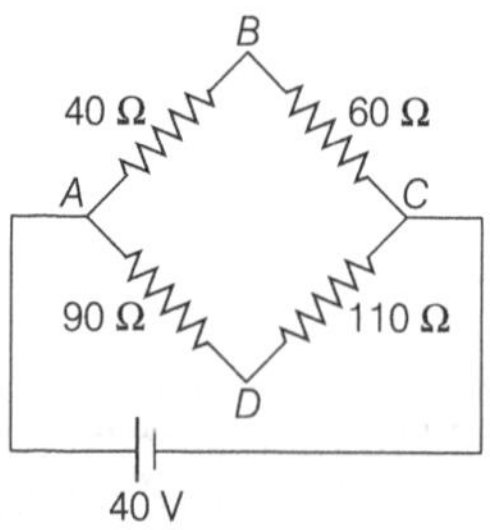

[JEE Main 2020]

(a) 1 (b) 2
(c) 3 (d) 4

27. Fig. (a) shows a meter bridge (which is nothing but a practical Wheatstone bridge) consisting of two resistors X and Y together in parallel with one metre long constant wire of uniform cross-section.

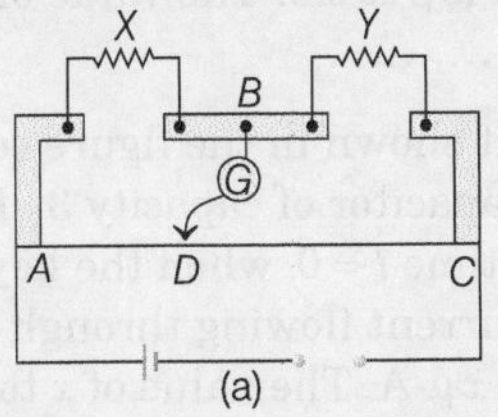

(a)

With the help of a movable contact D, one can change the ratio of the resistances of the two segments of the wire, until a sensitive galvanometer G connected across B and D shows to deflection. The null point is found to be at a distance of 33.7 cm from the end A. The resistance Y is shunted by a resistance Y' of 12.0 Ω [Fig. (b)] and the null point is found to shift by a distance of 18.2 cm. Determine the resistances of X and Y.

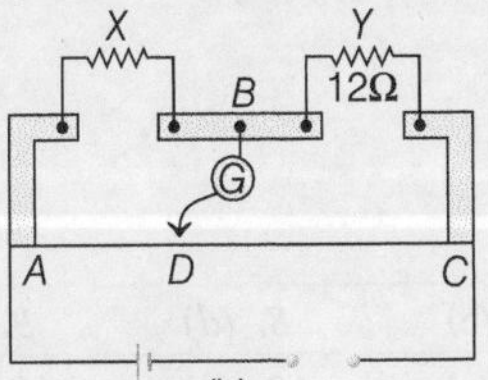

(b)

(a) $Y = 13.5\ \Omega$ and $X = 6.86\ \Omega$
(b) $Y = 13.5\ \Omega$ and $X = 5.86\ \Omega$
(c) $Y = 11.5\ \Omega$ and $X = 6.86\ \Omega$
(d) $Y = 12.5\ \Omega$ and $X = 6.86\ \Omega$

28. A wire of resistance R is bent to form a square $ABCD$ as shown in the figure. The effective resistance between E and C is [E is mid-point of arm CD] [JEE Main 2019]

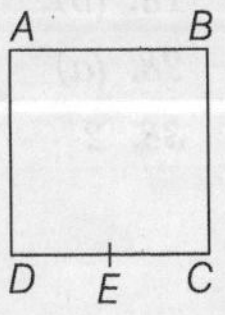

(a) $\frac{7}{64}R$ (b) $\frac{3}{4}R$ (c) R (d) $\frac{1}{16}R$

29. In the figure shown, what is the current (in ampere) drawn from the battery? You are given $R_1 = 15\ \Omega$, $R_2 = 10\ \Omega$, $R_3 = 20\ \Omega$, $R_4 = 5\ \Omega$, $R_5 = 25\ \Omega$, $R_6 = 30\ \Omega$, $\varepsilon = 15$ V [JEE Main 2019]

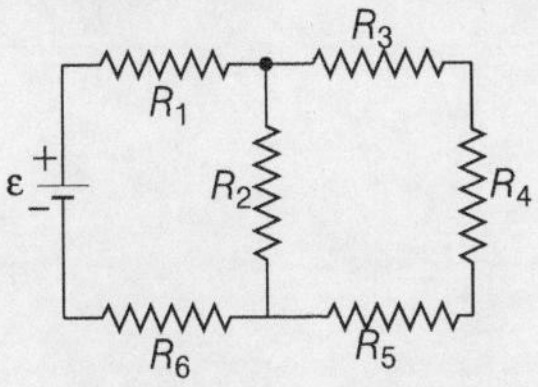

(a) 13/24 (b) 7/18 (c) 20/3 (d) 9/32

30. In the circuit shown, the potential difference between A and B is [JEE Main 2019]

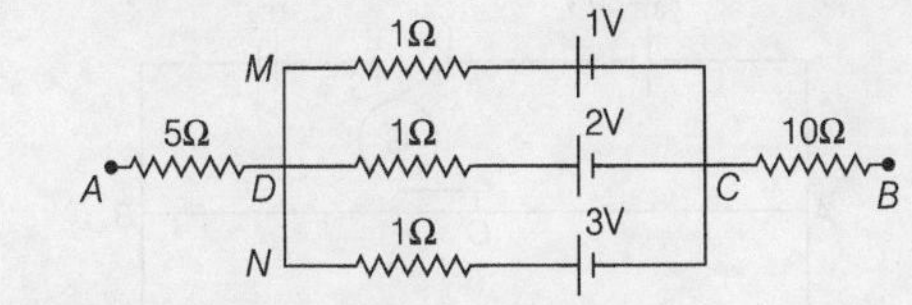

(a) 3 V (b) 1 V (c) 6 V (d) 2 V

31. Two cells of emf's 2V and 1V and of internal resistances 1 Ω and 2Ω respectively, have their positive terminals connected by a wire of 10Ω resistance and their negative terminals by wire of 4Ω resistance. Another coil of 10Ω is connected between the middle points of these wires. The potential difference across the 10Ω coil is
(a) 1.07 V (b) 2.03 V (c) 3.45 V (d) 4.25 Vs

32. Determine the charge on the capacitor in the following circuit.

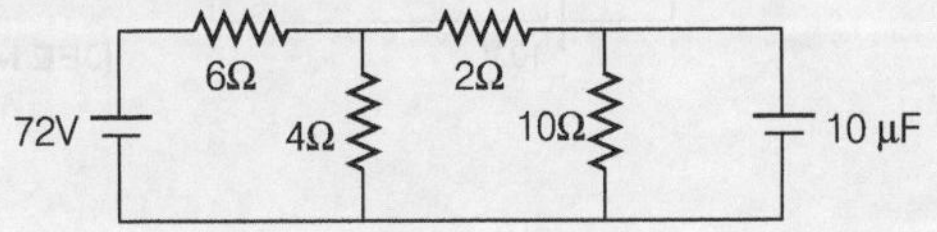

[JEE Main 2019]

(a) 2μC (b) 200μC (c) 10μC (d) 60μC

Numerical Value Questions

33. The number density of free electrons in a copper conductor is estimated as $8.5 \times 10^{28}\ m^{-3}$. How long does an electron take (in min) to drift from one end of a wire 3.0 m long to its other end? The area of cross-section of the wire is $2.0 \times 10^{-6}\ m^2$ and it is carrying a current of 3.0 A.

34. A conducting open pipe has shape of a half cylinder of length L. Its semi-circular cross-section has radius r and thickness of the conducting wall is $t(<< r)$. The resistance of the conductor when the current enters and leaves as shown in Fig. (a) is R_1 and its resistance is R_2 when the current is as shown in Fig. (b). Find the ratio of R_1 / R_2. (Take, $L = 1$ m, $\pi^2 = 10$ and $r = 0.1$ m)

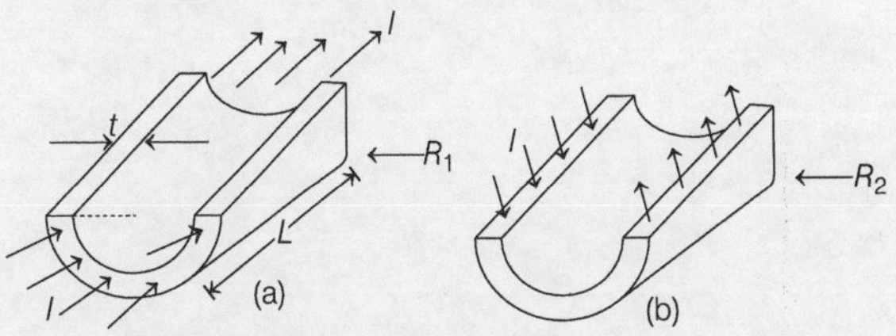

35. In the circuit shown in figure, AB is a uniform wire of length $L = 5$ m. It has a resistance of 2 Ω/m. When $AC = 2.0$ m, it was found that the galvanometer shows zero reading when switch S is placed in either of the two positions 1 or 2.

Find the emf ε_1 (in volts).

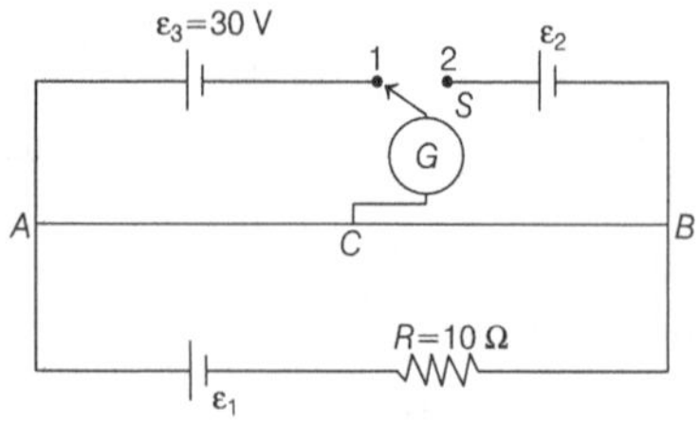

36. An ideal cell of emf 10 V is connected in circuit shown in figure. Each resistance is 2 Ω. The potential difference (in volt) across the capacitor when it is fully charged is

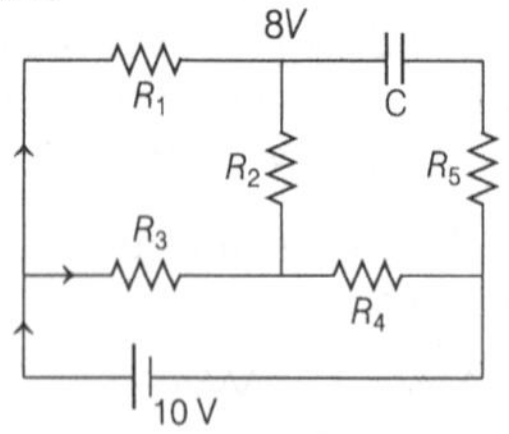

[JEE Main 2020]

37. Two wires of same length and thickness having specific resistance 6Ω cm and 3Ω cm respectively are connected in parallel. The effective resistivity is ρ Ω cm. The value of ρ, to the nearest integer is .... [JEE Main 2021]

38. The circuit shown in the figure consists of a charged capacitor of capacity 3 μF and a charge of 30 μC. At time $t = 0$, when the key is closed, the value of current flowing through the 5 MΩ resistor is x μ-A. The value of x to the nearest integer is ....

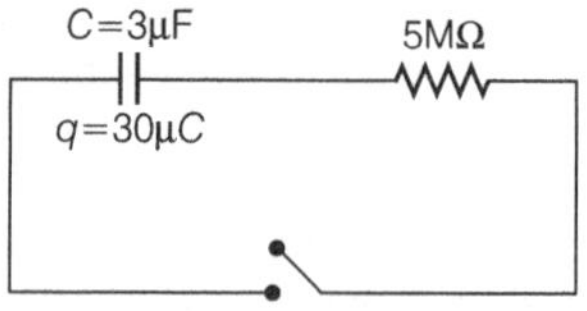

[JEE Main 2021]

Answers

Round I

1. (c)	2. (b)	3. (d)	4. (a)	5. (c)	6. (a)	7. (c)	8. (d)	9. (b)	10. (a)
11. (d)	12. (c)	13. (c)	14. (c)	15. (c)	16. (b)	17. (c)	18. (a)	19. (d)	20. (d)
21. (c)	22. (d)	23. (b)	24. (a)	25. (b)	26. (a)	27. (d)	28. (c)	29. (c)	30. (d)
31. (a)	32. (d)	33. (a)	34. (b)	35. (c)	36. (a)	37. (a)	38. (c)	39. (d)	40. (b)
41. (c)	42. (c)	43. (a)	44. (b)	45. (d)	46. (b)	47. (a)	48. (c)	49. (c)	50. (c)
51. (a)	52. (b)	53. (d)	54. (d)	55. (b)	56. (b)	57. (c)			

Round II

1. (d)	2. (c)	3. (d)	4. (a)	5. (a)	6. (b)	7. (b)	8. (d)	9. (a)	10. (b)
11. (c)	12. (c)	13. (d)	14. (d)	15. (c)	16. (b)	17. (a)	18. (b)	19. (c)	20. (b)
21. (c)	22. (b)	23. (c)	24. (b)	25. (d)	26. (b)	27. (a)	28. (a)	29. (d)	30. (d)
31. (a)	32. (b)	33. 453	34. 10	35. 150	36. 8	37. 4	38. 2		

Solutions

Round I

1. We have, $i = \frac{dq}{dt} = 3t^2 + 2t + 5$

$\Rightarrow \quad dq = (3t^2 + 2t + 5)dt$

$\therefore \quad q = \int_0^2 (3t^2 + 2t + 5)\, dt$

$$= \left(\frac{3t^3}{3} + \frac{2t^2}{2} + 5t\right)\Bigg|_0^2 = \left[\frac{3(2)^3}{3} + 2\frac{(2)^2}{2} + 5 \times 2\right]$$

$$= 8 + 4 + 10 = 22\text{C}$$

2. $\therefore \; v_d = i/nAe = 0.21/(8.4 \times 10^{28} \times 10^{-6} \times 1.6 \times 10^{-19})$

$= 1.56 \times 10^{-5}\ \text{ms}^{-1}$

3. $\therefore \quad v_d = \frac{i}{nAe} = \frac{i \times 4}{n\pi D^2 e} \Rightarrow v_d \propto \frac{1}{D^2}$

$\because \quad \frac{v_{d_1}}{v_{d_2}} = \frac{D_2^2}{D_1^2} = \left(\frac{1}{2}\right)^2 = \frac{1}{4}$

4. $\therefore \quad v_d = i/n\,Ae$

where, $n = N\rho/M$

$= 6.023 \times 10^{26} \times 9 \times 10^3/63$

$= 0.860 \times 10^{29} = 8.6 \times 10^{28}$

and $A = \pi D^2/4 = \frac{22}{7} \times (10^{-3})^2/4\ \text{m}^2$

$= \frac{11}{14} \times 10^{-6}\ \text{m}^2$

$$v_d = \frac{1.1}{8.6 \times 10^{28} \times \frac{11}{14} \times 10^{-6} \times 1.6 \times 10^{-19}}$$

$$= \frac{1}{9.6 \times 10^{+3}} = \frac{100 \times 10^{-4}}{96} = 0.1\ \text{mms}^{-1}$$

5. $\therefore i = nAev_d$

or $\quad v_d = \frac{i}{nAe}$

Total number of free electrons in the unit length of conductor,

$N = nA \times 1$

Total linear momentum of all the free electrons per unit length

$$= (Nm)\, v_d = nAm \times \frac{i}{nAe} = \frac{i}{(e/m)} = \frac{i}{s}$$

6. Slope of the graph will give us reciprocal of resistance. Here, resistance at temperature T_1 is greater than that at T_2. Since, resistance of metallic wire is more at higher temperature, then at lower temperature, hence $T_1 > T_2$.

7. Given, $l_1 = l + \frac{25}{100} l = \frac{5\,l}{4}$. Since, volume of wire remains unchanged on increasing length, hence

$Al = A_1 \times 5\,l/4$

or $\quad A_1 = 4\,A/5$

Given, $\quad R = 10 = \rho l/A$

and $\quad R_1 = \frac{\rho l_1}{A_1} = \frac{\rho\, 5l/4}{4A/5} = \frac{25}{16}\frac{\rho l}{A}$

or $\quad R_1 = \frac{25}{16} \times 10 = \frac{250}{16} = 15.6\ \Omega$

8. Mass, $M = \text{Volume} \times \text{Density} = Al \times d$

or $\quad A = M/ld$

Resistance, $R = \rho l/A = \rho l/(M/ld) = \frac{\rho l^2 d}{M}$

So, $\quad R \propto l^2/M$

Thus, $\quad R_1 : R_2 : R_3 = \frac{l_1^2}{M_1} : \frac{l_2^2}{M_2} : \frac{l_3^2}{M_3}$

$= \frac{3^2}{1} : \frac{2^2}{2} : \frac{1^2}{3}$

$= 27 : 6 : 1$

9. We have, $\quad R_A = \frac{\rho l'}{\pi\,(10^{-3} \times 0.5)^2}$

$R_B = \frac{\rho l}{\pi\,[(10^{-3})^2 - (0.5 \times 10^{-3})^2]}$

$\because \quad \frac{R_A}{R_B} = \frac{(10^{-3})^2 - (0.5 \times 10^{-3})^2}{(0.5 \times 10^{-3})^2} = 3$

$\therefore \quad R_A : R_B = 3:1$

10. We know that, $R = \frac{\rho l}{A}$

(a) When the battery is connected across 1 cm × 1/2 cm faces, then

$l = 10\ \text{cm};\ A = 1 \times 1/2\ \text{cm}^2, R_1 = \frac{\rho \times 10}{1 \times 1/2} = 20\rho\Omega$

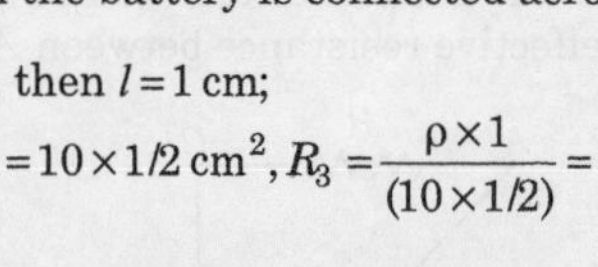

(b) When the battery is connected across 10 cm × 1 cm faces, then $l = \frac{1}{2}$ cm; $A = 10 \times 1\ \text{cm}^2$, $R_2 = \frac{\rho \times 1/2}{10 \times 1} = \frac{\rho}{20}\ \Omega$

(c) When the battery is connected across $10\text{cm} \times \frac{1}{2}$ cm faces, then $l = 1$ cm;

$A = 10 \times 1/2\ \text{cm}^2, R_3 = \frac{\rho \times 1}{(10 \times 1/2)} = \frac{\rho}{5}\ \Omega$

11. Aluminium is more resistive than copper and mercury is most resistive of all.

So, $\quad \rho_M > \rho_A > \rho_C$

12. Using colour code, resistance is

$R = 27 \times 10^3\ \Omega, \pm 10\%$

$= 27\ \text{k}\,\Omega, \pm 10\%$

13. Temperature coefficient of resistance $= \dfrac{1}{R_t}\dfrac{dR}{dt}$

$$= \frac{1}{R_0(1+\alpha t+\beta t^2)} \times \frac{d}{dt}[R_0(1+\alpha t+\beta t^2)]$$

$$= \frac{\alpha + 2\beta t}{1+\alpha t+\beta t^2}$$

14. Given, resistance of silver wire at 27.5 °C, $R_{27.5} = 2.1\ \Omega$

Resistance of silver wire at 100 °C, $R_{100} = 2.7\ \Omega$

Let the temperature coefficient of silver be α.

$$\alpha = \frac{R_{t_2} - R_{t_1}}{R_1\,(t_2 - t_1)}$$

$$\alpha = \frac{R_{100} - R_{27.5}}{R_{27.5}\,(100-27.5)} = \frac{2.7-2.1}{2.1\times 72.5}$$

$$\alpha = 0.0039/\ ^\circ\text{C}$$

15. The ratio, $\dfrac{J_1}{J_2} = \dfrac{i/\pi r_1^2}{i/\pi r_2^2} = \left(\dfrac{r_2}{r_1}\right)^2 = \left(\dfrac{3}{1}\right)^2 = \dfrac{9}{1}$

16. Equivalent circuit of this combination of resistances is as shown in figure. The effective resistance of arm

$$EG = 2\times\left[\frac{4\times 4}{4+4}\right] = 4\ \Omega$$

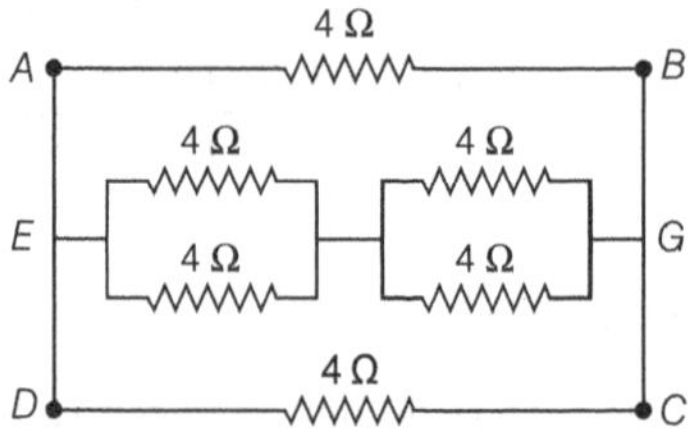

Total resistance between A and D will be

$$\frac{1}{R} = \frac{1}{4}+\frac{1}{4}+\frac{1}{4} = \frac{3}{4}$$

or $$R = \frac{4}{3}\ \Omega$$

17. Here, points B and D are common. So, $2R$ in arm DC and $2R$ in arm CB are in parallel between C and B.

Their effective resistance $= \dfrac{2R\times 2R}{2R+2R} = R$

The modified and simpler circuit will be as shown in figure. The effective resistance between A and B is

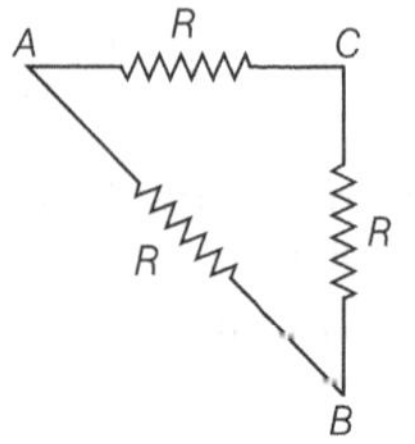

$$R_{\text{eff}} = \frac{R\times(R+R)}{R+(R+R)} = \frac{2}{3}R$$

18. Two resistances of each side of triangle are connected in parallel. Therefore, the effective resistance of each arm of the triangle would be $= \dfrac{r\times r}{r+r} = \dfrac{r}{2}$.

The two arms AB and AC are in series and they together are in parallel with third one.

$\therefore$ $$R' = (r/2)+(r/2) = r$$

Total resistance,

$$\frac{1}{R} = \frac{1}{r}+\frac{2}{r} = \frac{3}{r}$$

$\Rightarrow$ $$R = \frac{r}{3}$$

19. As, $\dfrac{1}{R_1} = \dfrac{1}{10}+\dfrac{1}{2.5} = \dfrac{5}{10} = \dfrac{1}{2} \Rightarrow R_1 = 2\ \Omega$

Now, 2 Ω and 10 Ω are in series,

$$R_2 = 10+2 = 12\ \Omega$$

R_2 and 12 Ω are in parallel, $\dfrac{1}{R_3} = \dfrac{1}{12}+\dfrac{1}{12} \Rightarrow R_3 = 6\ \Omega$

Now, R_3 and 6 Ω are in series, $R_4 = 10+6 = 16\ \Omega$

Now, R_4 and 16 Ω are in parallel.

$\therefore$ $$\frac{1}{R} = \frac{1}{16}+\frac{1}{16}$$

$\Rightarrow$ $$R = 32\ \Omega$$

20. Given circuit is

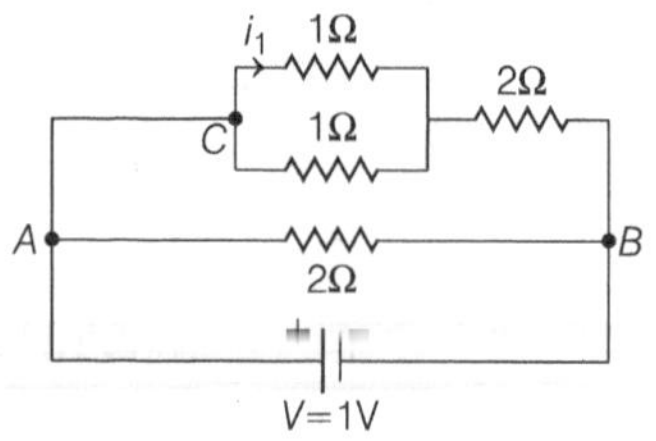

In given circuit $V_{AB} = 1$ V, so upper branch of circuit is as shown in below figure.

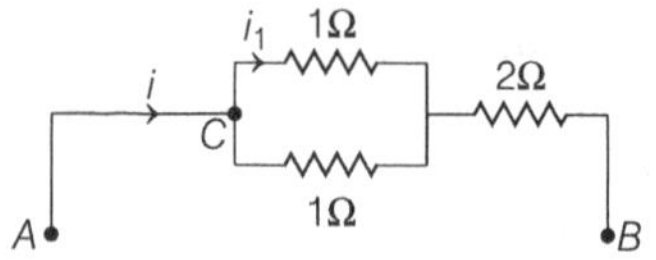

Equivalent resistance of upper branch,

$$R_{\text{eq}} = (1\ \Omega\,||\,1\ \Omega) + 2\ \Omega = \frac{1}{2}+2 = \frac{5}{2}\ \Omega$$

So, current in upper branch,

$$i = \frac{V}{R} = \frac{V_{AB}}{R_{\text{eq}}} = \frac{1}{5/2} = \frac{2}{5}\ \text{A}$$

At point C, this current is equally divided into two parts.

So, $$i_1 = \frac{1}{2}\left(\frac{2}{5}\right) = 0.2\ \text{A}$$

21. In series combination of cells current, $i = \dfrac{n\varepsilon}{nr+R}$

In parallel combination of cells, $i' = \dfrac{\varepsilon}{(r/n)+R}$

If $i = i'$, then $\dfrac{n\varepsilon}{nr+R} = \dfrac{\varepsilon}{(r/n)+R} = \dfrac{n\varepsilon}{r+nR}$

It will be so if $r = R$.

22. When one cell is wrongly connected in series, the emf of cells decreases by 2ε, but internal resistances of cells remains the same for all the cells.

Current in the circuit is $i = \dfrac{(n-2)\varepsilon}{nr} \times r$

Potential difference across each cell is

$$V = \varepsilon - ir = \varepsilon - \frac{(n-2)\varepsilon}{nr} \times r = \frac{2\varepsilon}{n}$$

23.

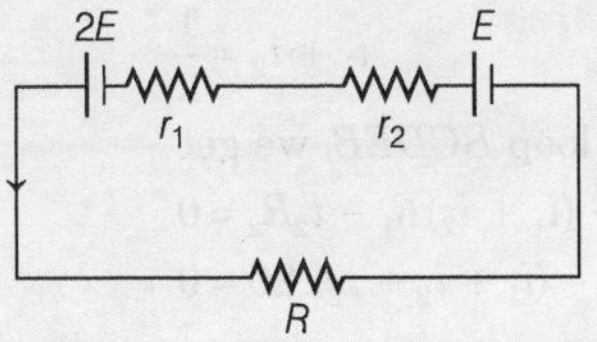

$$i = \frac{3E}{R + r_1 + r_2}$$

Total potential difference $= 2E - ir_1 = 0$

$$2E = ir_1$$

$$2E = \frac{3E \times r_1}{R + r_1 + r_2}$$

$$2R + 2r_1 + 2r_2 = 3r_1$$

$$R = \frac{r_1}{2} - r_2$$

24. Here, $mn = 20$

For maximum current, $R = nr/m$

or $\quad 2.5 = n \times 0.5/m$

or $\quad n = 5\,m$

From Eq. (i), we get

$$m \times 5m = 20$$

or $\quad m^2 = 4$

or $\quad m = 2$

Therefore, $\quad n = 5 \times 2 = 10$

25. Given arrangement of batteries and resistances is shown below

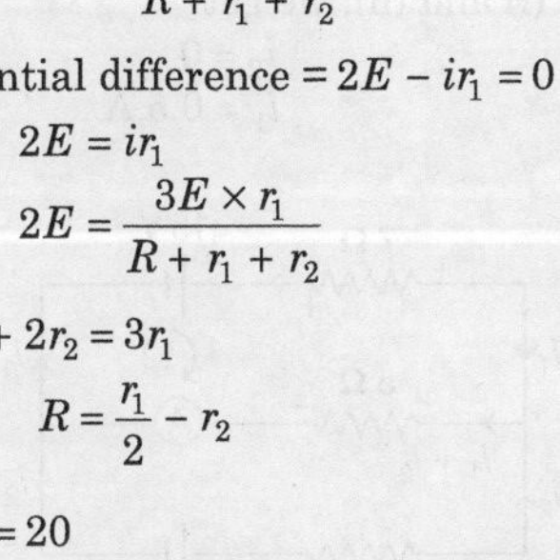

Let i = circuit current, then it is given that potential difference across battery of 10V and 20 Ω is zero.

i.e. $\quad \varepsilon_1 - ir_1 = 0$

$\Rightarrow \quad 10 - i(20) = 0 \Rightarrow i = \dfrac{1}{2} = 0.5\text{A}$

Now, potential drop across combination of resistors,

$$V_{AB} = (\varepsilon_1 + \varepsilon_2) - i(r_1 + r_2)$$
$$= 20 - 0.5 \times (25) = 7.5 \text{ V}$$

Now, at junction B,

$$i = i_1 + i_2$$

$\Rightarrow \quad 0.5 = \dfrac{7.5}{R} + \dfrac{7.5}{30}$

On solving, we get

$$R = 30\,\Omega$$

26. Given circuit with currents is as shown in the figure below, [in the question $i_1 = 0.3$ A is given, due to it, we change the direction of i_1, in this figure]

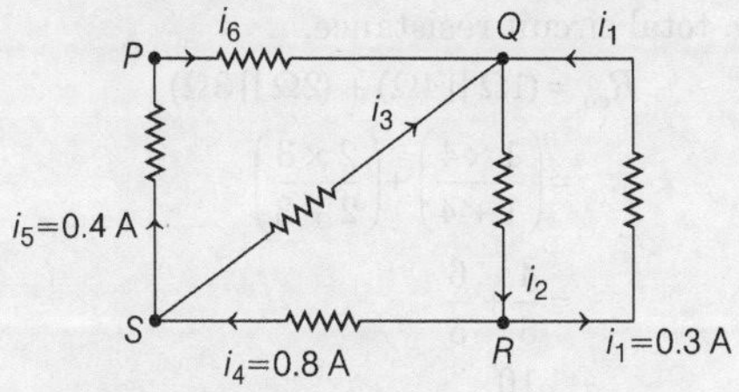

From Kirchhoff's junction rule, $\Sigma i = 0$

At junction S, $\quad i_4 = i_5 + i_3$

$\Rightarrow \quad 0.8 = 0.4 + i_3$

$\Rightarrow \quad i_3 = 0.4$ A

At junction P, $\quad i_5 = i_6 \Rightarrow i_6 = 0.4$ A

At junction Q, $\quad i_2 = i_1 + i_3 + i_6 = 0.3 + 0.4 + 0.4 = 1.1$ A

27. From Kirchhoff's loop rule,

$$-i(2) + 18 - i(6.6) - 12 - i(1) = 0$$

$\Rightarrow \quad i = \dfrac{6}{9.6}$ A

Terminal voltage of 18 V battery,

$$V = \varepsilon - ir_1$$
$$= 18 - i \times 2$$
$$= 18 - 0.625 \times 2$$
$$= 16.75 \approx 17 \text{ V}$$

For 12 V battery,

$$V = \varepsilon + ir_2 = 12 + 1 \times 0.625$$
$$= 12.625 \approx 13 \text{ V}$$

28. In loop 1,

$$i_1 = \frac{5}{10} = \frac{1}{2} \text{ A}$$

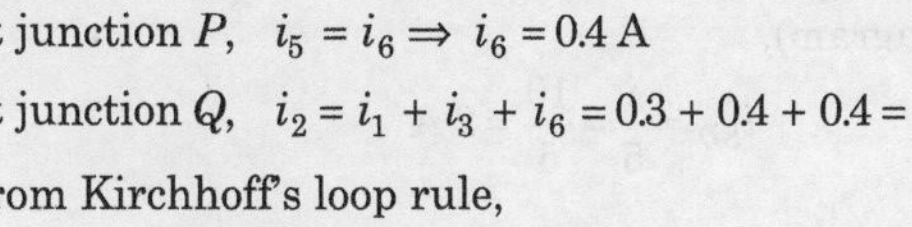

In loop 2, $\quad i_2 = \dfrac{20}{10} = 2$ A

But both currents are confined to separate loops and so current through 2 Ω resistor is zero.

29. In given circuit, current distribution following Kirchhoff's law will be as shown in the figure.

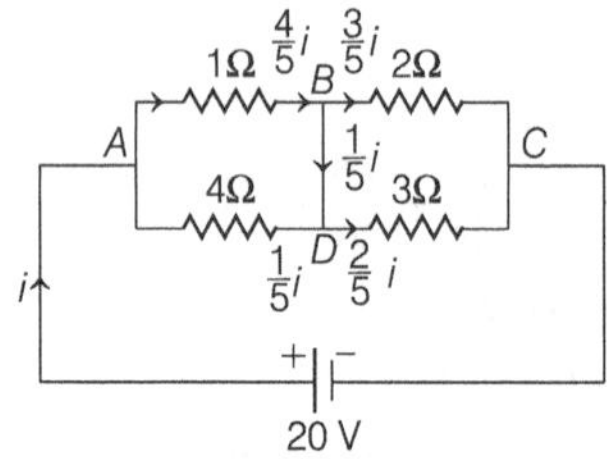

Since, the current flows in inverse ratio of the resistance of branch.

Now, total circuit resistance,

$$R_{eq} = (1\Omega \,||\, 4\Omega) + (2\Omega \,||\, 3\Omega)$$
$$= \left(\frac{1 \times 4}{1+4}\right) + \left(\frac{2 \times 3}{2+3}\right)$$
$$= \frac{4}{5} + \frac{6}{5}$$
$$= \frac{10}{5}\,\Omega = 2\Omega$$

So, current drawn from cell,

$$i = \frac{V}{R_{eq}} = \frac{20V}{2\Omega} = 10\text{ A}$$

Hence, current through BD arm is (refer to circuit diagram),

$$i_{BD} = \frac{i}{5} = \frac{10}{5} = 2\text{ A}$$

30. $R_{eq} = \frac{5}{2} + 1.5 = 4$ A

Current through cell $= \frac{V}{R} = \frac{20}{4} = 5$ A

So, current through each branch $= \frac{i}{2} = 2.5$ A

Now, considering loop

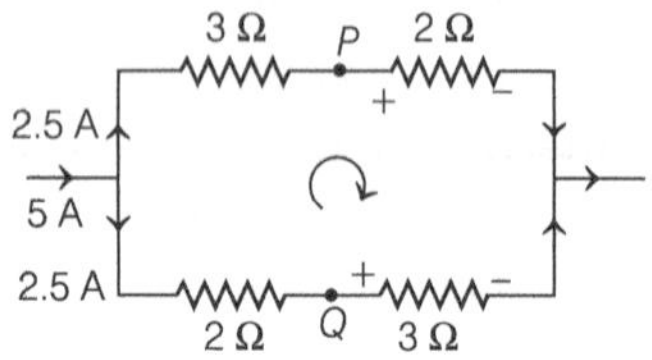

Now, applying Kirchhoff's rule to part loop P to Q,

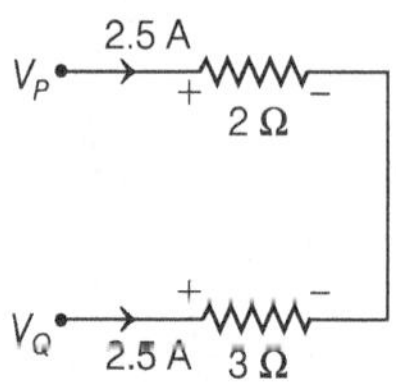

$$V_P - 2 \times 2.5 + 3 \times 2.5 - V_Q = 0$$
$$\Rightarrow \quad V_P - V_Q = -2.5\text{ V}$$

31. By Kirchhoff's loop rule in the given loop $ABEFA$, we get

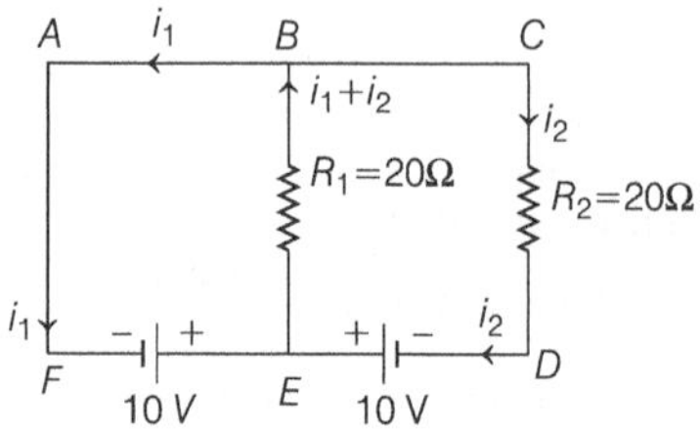

$$10 - (i_1 + i_2)R_1 = 0$$
$$\Rightarrow \quad 10 - (i_1 + i_2)\,20 = 0$$
$$\text{or} \quad i_1 + i_2 = \frac{1}{2} \quad \text{...(i)}$$

and from loop $BCDEB$, we get

$$10 - (i_1 + i_2)R_1 - i_2R_2 = 0$$
$$\Rightarrow \quad 10 - (i_1 + i_2 + i_2)\,20 = 0$$
$$\Rightarrow \quad i_1 + 2i_2 = \frac{1}{2} \quad \text{...(ii)}$$

From Eqs. (i) and (ii), we get

$$i_2 = 0$$
$$\text{and} \quad i_1 = 0.5\text{ A}$$

32. Given,

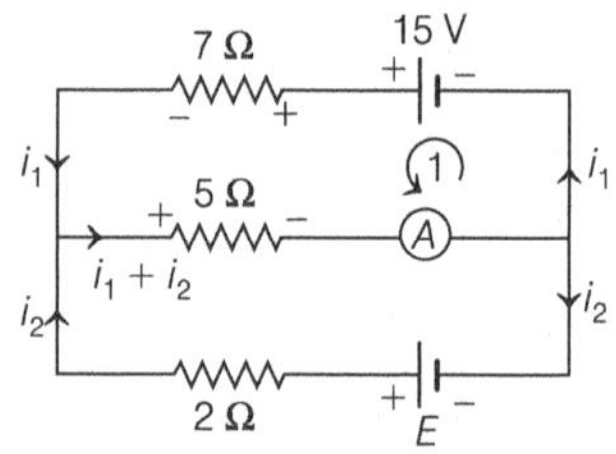

Given, $i_1 + i_2 = 2$ A

In loop 1, by Kirchhoff's rule, $\Sigma V = 0$

$$15 - 7i_1 - 2 \times 5 = 0$$
$$\Rightarrow \quad i_1 = \frac{5}{7}\text{ A}$$
$$\Rightarrow \quad i_1 = 0.714\text{ A}$$

Also, $i_1 + i_2 = 2$ A

$$\Rightarrow \quad i_2 = 2 - \frac{5}{7} = \frac{9}{7}\text{ A}$$
$$\Rightarrow \quad i_2 = 1.29\text{ A}$$

33. By Kirchhoff's rule for loop 1,

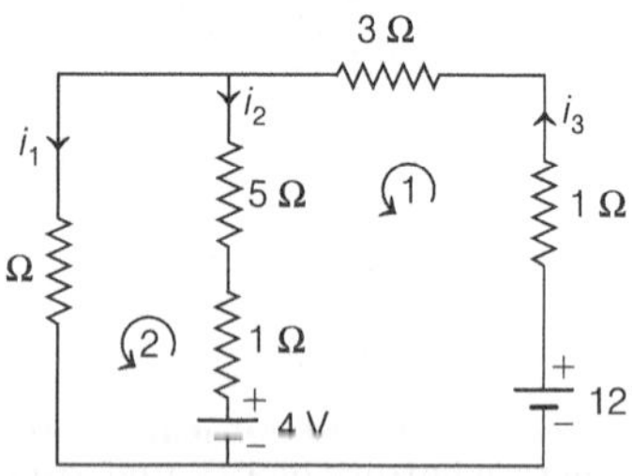

$$12 - 4\,i_3 - 6\,i_2 - 4 = 0$$
$$\Rightarrow \quad 8 = 4i_3 + 6i_2$$

and from loop 2, $-6\,i_2 - 4 + 8\,i_1 = 0$

On solving, we get $\quad i_1 = 846$ mA

and $\quad i_2 = 462$ mA

Hence, current through 8 Ω resistor is 846 mA.

34. Let currents in branches are i_1, i_2 and i_3.

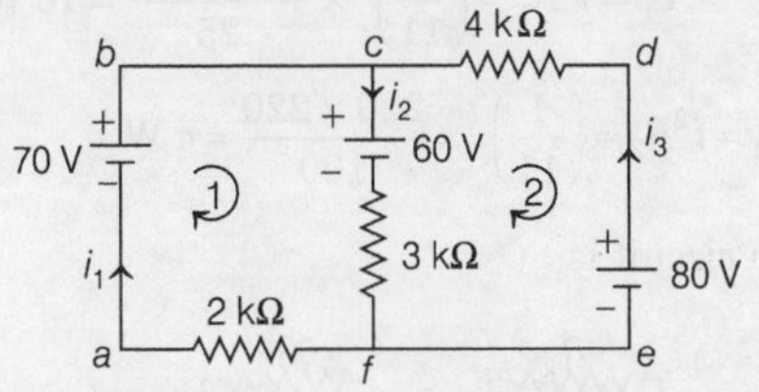

Then, from loops 1 and 2, we have

$$70 - 60 - i_2 \times 3 - i_1 \times 2 = 0 \qquad \text{...(i)}$$

$$80 - i_3 \times 4 - 60 - i_2 \times 3 = 0 \qquad \text{...(ii)}$$

and $$i_2 = i_1 + i_3 \qquad \text{...(iii)}$$

On solving Eqs. (i), (ii) and (iii), we get

$$i_1 = 0.385 \text{ mA}$$

$$i_2 = 3.08 \text{ mA}$$

$$i_3 = 2.69 \text{ mA}$$

and $$V_{cf} = -60 - 3.08 \times 3$$

i.e. $$V_c - V_f = -69.2 \text{ V}$$

Point c is at higher potential.

35. Current distribution in the given circuit is taken as

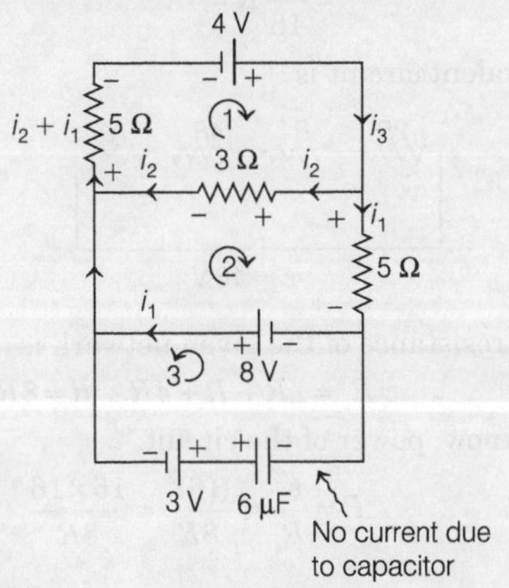

Using equations of Kirchhoff's law as given below

$$i_1 + i_2 = i_3$$

$$4 - 3i_2 - 5i_3 = 0 \qquad \text{(loop 1)}$$

$$3i_2 - 5i_1 + 8 = 0 \qquad \text{(loop 2)}$$

On solving these equations, we get

$$i_2 = -0.364 \text{ A}$$

$$i_1 = 1.38 \text{ A}$$

$$i_3 = 1.02 \text{ A}$$

Now, going anti-clockwise in loop 3,

$$\Sigma V = 0$$

$$\Rightarrow \quad 8 + 3 - \Delta V_{\text{capacitor}} = 0$$

$$\Rightarrow \quad \Delta V_{\text{capacitor}} = 11 \text{ V}$$

So, charge on a capacitor,

$$C = Q\Delta V = 6\,\mu\text{F} \times 11 \text{ V} = 66\,\mu\text{C}$$

36. By Kirchhoff's laws,

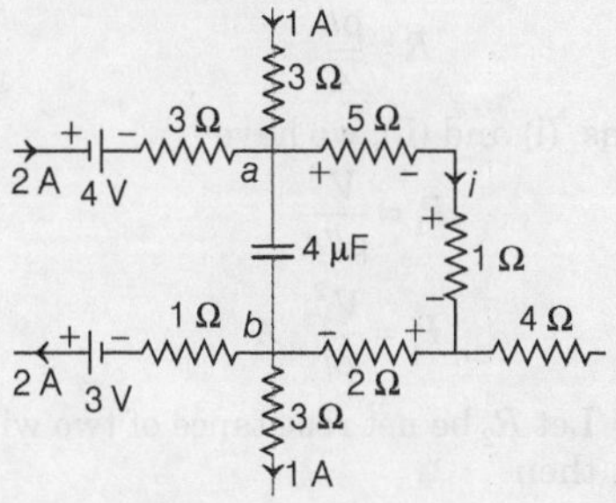

At junction a and b,

$$i = 2\text{A} + 1\text{ A} = 3 \text{ A}$$

and also current through corner 4 Ω resistor is zero.

Now, for path ab,

$$-3 \times 5 - 3 \times 1 + 3 \times 2 = V_{ab} = V_b - V_a$$

$$\Rightarrow \quad V_{ab} = V_b - V_a = -12 \text{ V}$$

So, potential drop across capacitor = 12 V

Hence, energy stored $= \frac{1}{2}CV^2 = \frac{1}{2} \times 4 \times 10^{-6} \times 12^2$ J

$$= 0.288 \text{ mJ} \approx 0.3 \text{ mJ}$$

37. As, bulbs can be compared only when they are to be used with same source, so we take V = constant for both.

Now, $$P_1 > P_2$$

$$\frac{V^2}{R_1} > \frac{V^2}{R_2} \Rightarrow R_1 < R_2 \qquad \left(\text{as, } R = \frac{\rho l}{A}\right)$$

$$\Rightarrow \quad \frac{\rho l}{A_1} < \frac{\rho l}{A_2} \Rightarrow A_1 > A_2$$

Filament of first bulb is thicker.

38. Power consumed $= \left(\frac{\text{Supply voltage}}{\text{Rate voltage}}\right)^2 \times$ Rated power

$$= \left(\frac{110}{115}\right)^2 \times 500 = 456.6 \text{ W}$$

$$\therefore \% \text{ drop in heat output} = \frac{P_{\text{actual}} - P_{\text{consumed}}}{P_{\text{actual}}} \times 100$$

$$= \left(\frac{500 - 456.6}{500}\right) \times 100 = 8.6\%$$

39. $\because \quad P = \dfrac{V^2}{R}$

$$\because \quad \frac{P_{200}}{P_{800}} = \frac{V^2/R_{200}}{V^2/R_{800}} = \frac{R_{800}}{R_{200}}$$

$$= \frac{R_{200}(1 + \alpha\,\Delta T)}{R_{200}}$$

$$\Rightarrow \quad P_{200} = P_{800} \times (1 + 4 \times 10^{-4} \times 600)$$

$$= 500\,(1 + 4 \times 10^{-4} \times 600) = 620 \text{ W}$$

40. ***Case* I** Using the formula, $P = \frac{V^2}{R}$ ...(i)

where, R is resistance of wire, V is voltage across wire and P is power dissipation in wire

and $R = \frac{\rho l}{A}$ ...(ii)

From Eqs. (i) and (ii), we have

$$P_1 = \frac{V^2}{\rho l / A}$$

or $P_1 = \frac{V^2}{\rho l} \cdot A$ ...(iii)

***Case* II** Let R_2 be net resistance of two wires in parallel, then

$$R_2 = \frac{R \times R}{R + R} = \frac{R}{2}$$

where, R is the resistance of half wire

$\Rightarrow$ $R_2 = \frac{\rho \cdot \left(\frac{l}{2}\right)}{A \cdot 2} = \frac{\rho l}{4A}$

or $P_2 = \frac{V^2}{\rho l} \cdot 4A$ ...(iv)

On dividing Eq. (iii) by Eq. (iv), we get

$$\frac{P_1}{P_2} = \frac{1}{4} \Rightarrow \frac{P_2}{P_1} = \frac{4}{1}$$

41. Total power (P) consumed

$$= (15 \times 40) + (5 \times 100) + (5 \times 80) + (1 \times 1000)$$
$$= 2500 \text{ W}$$

As we know power, *i.e.* $P = Vi$

$\Rightarrow$ $i = \frac{2500}{220} \text{ A} = \frac{125}{11} = 11.3 \text{ A}$

Hence, minimum capacity should be 12 A.

42. Given, $P = 1 \text{ kW} = 1000 \text{ W}$,

$$V = 220 \text{ V}, \ R = 2 \, \Omega$$

Current, $i = \frac{P}{V} = \frac{1000}{220} \text{ A}$

Power loss, $P_{\text{loss}} = i^2 R = \left(\frac{1000}{220}\right)^2 \times 2 = 41.32 \text{ W}$

Now, the efficiency of transmission line,

$$\eta = \left(\frac{P}{P + P_{\text{loss}}}\right) \times 100 = \left(\frac{1000}{1000 + 41.32}\right) \times 100$$

$$\eta = 96.03\% \approx 96\%$$

43. Resistance of a bulb of power P and with a voltage source V is given by

$$R = \frac{V^2}{P}$$

Resistance of the given two bulbs are

$$R_1 = \frac{V^2}{P_1} = \frac{(220)^2}{25} \text{ and } R_2 = \frac{V^2}{P_2} = \frac{(220)^2}{100}$$

Since, bulbs are connected in series. This means same amount of current flows through them.

$\therefore$ Current in circuit is

$$i = \frac{V}{R_{\text{total}}} = \frac{220}{\frac{(220)^2}{25} + \frac{(220)^2}{100}} = \frac{1}{11} \text{ A}$$

Power drawn by bulbs are respectively,

$$P_1 = i^2 R_1 = \left(\frac{1}{11}\right)^2 \times \frac{220 \times 220}{25} = 16 \text{ W}$$

and $P_2 = i^2 R_2 = \left(\frac{1}{11}\right)^2 \times \frac{220 \times 220}{100} = 4 \text{ W}$

44. Given circuit is

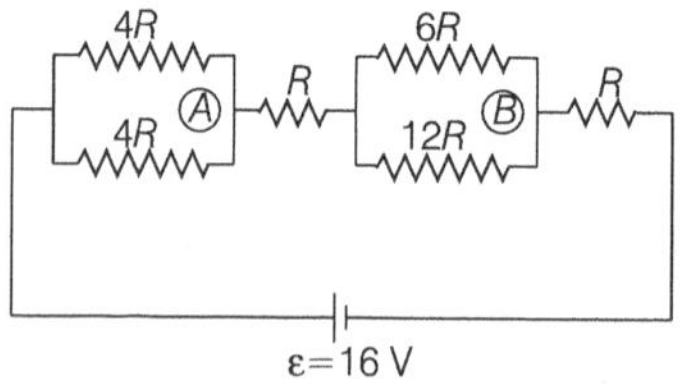

Equivalent resistance of part A,

$$R_A = \frac{4R \times 4R}{4R + 4R} = 2R$$

Equivalent resistance of part B,

$$R_B = \frac{6R \times 12R}{6R + 12R}$$
$$= \frac{72}{18} R = 4R$$

$\therefore$ Equivalent circuit is

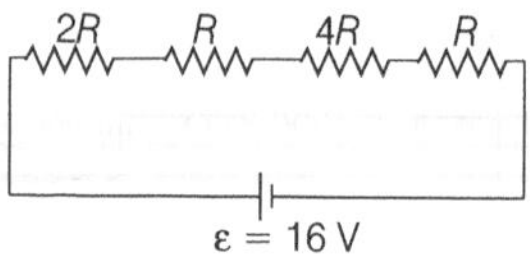

$\therefore$ Total resistance of the given network is

$$R_s = 2R + R + 4R + R = 8R$$

As we know, power of the circuit,

$$P = \frac{\varepsilon^2}{R_s} = \frac{(16)^2}{8R} = \frac{16 \times 16}{8R} \quad \text{...(i)}$$

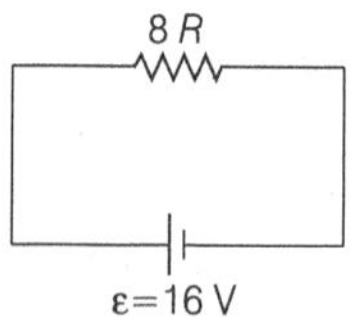

According to question, power consumed by the network, $P = 4 \text{ W}$

From Eq. (i), we get

$\therefore$ $\frac{16 \times 16}{8R} = 4$

$\Rightarrow$ $R = \frac{16 \times 16}{8 \times 4} = 8 \Omega$

45. We have, $H = i^2Rt \Rightarrow 500 = (1.5)^2 \times R \times 20$

$$E = (3)^2 \times R \times 20$$
$$E = 2000 \text{ J}$$

46. By maximum power theorem, maximum Joule's heating in external resistance R takes place when internal resistance of battery is equal to external resistance R.

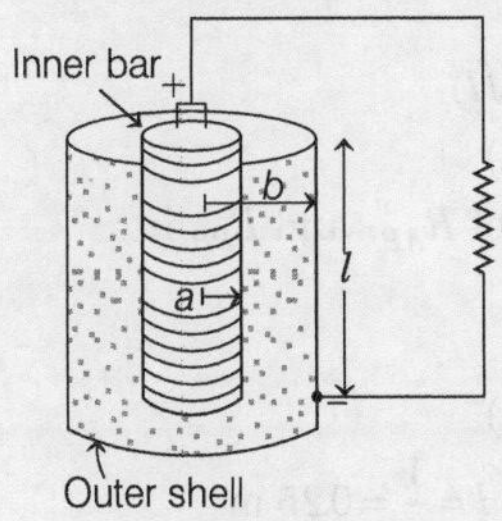

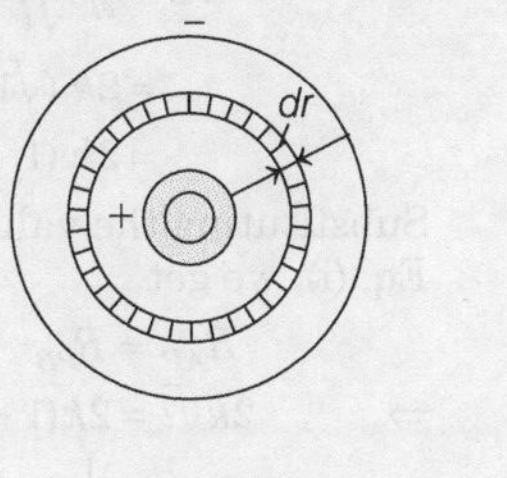

Now, resistance of an elemental cylinder of radius r and thickness dr is

$$(dR)_{\text{internal}} = \frac{\rho \cdot dr}{2\pi rl}$$

$$\Rightarrow \quad R_{\text{internal}} = \int (dR)_{\text{internal}}$$

$$= \int_{r=a}^{r=b} \frac{\rho dr}{2\pi rl} = \frac{\rho}{2\pi l} \int_a^b \frac{dr}{r} = \frac{\rho}{2\pi l} [\ln r]_a^b$$

$$= \frac{\rho}{2\pi l} (\ln b - \ln a)$$

$$= \frac{\rho}{2\pi l} \cdot \ln\left(\frac{b}{a}\right)$$

So, the maximum Joule's heating in R will takes place when its value is

$$= R_{\text{internal}} = \frac{\rho}{2\pi l} \cdot \ln\left(\frac{b}{a}\right)$$

47. In a balanced Wheatstone bridge, there is no effect on position of null point, if we exchange the battery and galvanometer. So, option (a) is incorrect.

48.

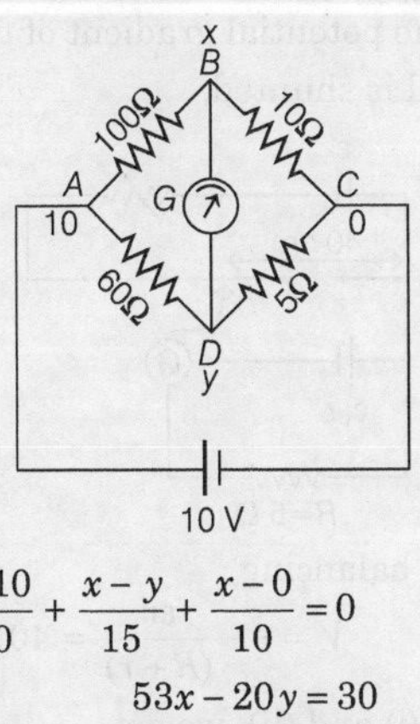

$$\frac{x-10}{100} + \frac{x-y}{15} + \frac{x-0}{10} = 0$$

$$53x - 20y = 30 \quad \text{...(i)}$$

$$\frac{y-10}{60} + \frac{y-x}{15} + \frac{y-0}{5} = 0$$

$$17y - 4x = 10 \quad \text{...(ii)}$$

On solving Eqs. (i) and (ii), we get

$$x = 0.865 \Rightarrow y = 0.792$$
$$\Delta V = x - y = 0.073 \text{ and } R = 15\ \Omega$$
$$i = \frac{\Delta V}{R} = 4.87 \text{ mA}$$

49. For a balanced Wheatstone bridge,

$$\frac{P}{R} = \frac{Q}{X}$$

In first case when $R = 400\ \Omega$, the balancing equation will be

$$\frac{P}{R} = \frac{Q}{X}$$

$$\Rightarrow \quad \frac{P}{400\ \Omega} = \frac{Q}{X}$$

$$\Rightarrow \quad P = \frac{400 \times Q}{X} \quad \text{...(i)}$$

In second case, P and Q are interchanged and $R = 405\ \Omega$

$$\therefore \quad \frac{Q}{R} = \frac{P}{X} \quad \Rightarrow \quad \frac{Q}{405} = \frac{P}{X} \quad \text{...(ii)}$$

Substituting the value of P from Eq. (i) in Eq. (ii), we get

$$\frac{Q}{405} = \frac{Q \times 400}{X^2}$$

$$\Rightarrow \quad X^2 = 400 \times 405$$

$$\Rightarrow \quad X = \sqrt{400 \times 405} = 402.5$$

The value of X is close to 402.5 Ω.

50. We have, $X + Y = 1000\ \Omega$

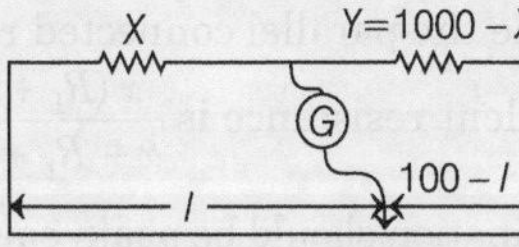

Initially, $$\frac{X}{l} = \frac{1000 - X}{100 - l} \quad \text{...(i)}$$

When X and Y are interchanged, then

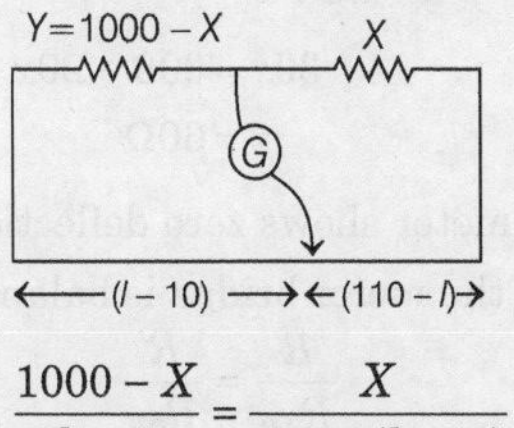

$$\frac{1000 - X}{l - 10} = \frac{X}{100 - (l - 10)}$$

or $$\frac{1000 - X}{l - 10} = \frac{X}{110 - l} \quad \text{...(ii)}$$

From Eqs. (i) and (ii), we get

$$\frac{100 - l}{l} = \frac{l - 10}{110 - l}$$

$$(100 - l)(110 - l) = (l - 10)\,l$$

$$11000 - 100\,l - 110\,l + l^2 = l^2 - 10\,l$$

$$\Rightarrow \quad 11000 = 200\,l$$

$$\therefore \quad l = 55 \text{ cm}$$

Substituting the value of l in Eq. (i), we get

$$\frac{X}{55} = \frac{1000 - 55}{100 - 55}$$

$$\Rightarrow \quad 20X = 11000$$

$$\therefore \quad X = 550\ \Omega$$

51. For meter bridge, if balancing length is l cm, then in first case, $\frac{R_1}{R_2} = \frac{l}{(100 - l)}$

It is given that, $l = 40$ cm

So, $\frac{R_1}{40} = \frac{R_2}{100 - 40}$ or $\frac{R_1}{R_2} = \frac{2}{3}$ ...(i)

In second case, $R_1' = R_1 + 10$, and balancing length is now 50 cm, then

$$\frac{R_1 + 10}{50} = \frac{R_2}{(100 - 50)}$$

or $R_1 + 10 = R_2$ ...(ii)

Substituting the value of R_2 from Eq. (ii) in Eq. (i), we get

or $$\frac{R_1}{10 + R_1} = \frac{2}{3}$$

$\Rightarrow$ $3R_1 = 20 + 2R_1$

or $R_1 = 20\Omega$

$\Rightarrow$ $R_2 = 30\Omega$

Let us assume the parallel connected resistance is x.

Then, equivalent resistance is $\frac{x(R_1 + 10)}{x + R_1 + 100}$.

So, this combination should be again equal to R_1.

$$\frac{(R_1 + 10)x}{R_1 + 10 + x} = R_1$$

$$\Rightarrow \quad \frac{30x}{30 + x} = 20$$

or $30x = 600 + 20x$

or $x = 60\Omega$

52. As, galvanometer shows zero deflection.

This means, the meter bridge is balanced.

$$\frac{R'}{R_{AP}} = \frac{R'}{R_{PB}}$$

$\Rightarrow$ $R_{AP} = R_{PB}$ ...(i)

Now, for meter bridge wire, $\frac{dR}{dl} = \frac{k}{\sqrt{l}}$

where, k is the constant of proportionality.

$$\Rightarrow \quad dR = \frac{k}{\sqrt{l}}\,dl$$

Integrating both sides, we get

$$\Rightarrow \quad R = \int \frac{k}{\sqrt{l}}\,dl$$

So, $$R_{AP} = \int_0^l \frac{k}{\sqrt{l}}\,dl = k(2\sqrt{l})\Big|_0^l = 2k\sqrt{l}$$

and $$R_{PB} = \int_l^1 \frac{k}{\sqrt{l}}\,dl = 2k(\sqrt{l})\Big|_l^1$$

$$= 2k(\sqrt{1} - \sqrt{l})$$

$$= 2k(1 - \sqrt{l})$$

Substituting the values of R_{AP} and R_{PB} in Eq. (i), we get

$$R_{AP} = R_{PB}$$

$$\Rightarrow \quad 2k\sqrt{l} = 2k(1 - \sqrt{l})$$

$$\Rightarrow \quad \sqrt{l} = \frac{1}{2} \quad \text{or} \quad l = \frac{1}{4} = 0.25 \text{ m}$$

53. $\because$ $V \propto l$

$$\therefore \quad \frac{V}{\varepsilon} = \frac{l}{L}$$

where, l = balance point distance

and L = length of potentiometer wire.

or $$V = \frac{l}{L}\varepsilon \text{ or } V = \frac{30 \times \varepsilon}{100} = \frac{30}{100}\varepsilon$$

54. The internal resistance of the cell,

$$r = \left(\frac{l_1 - l_2}{l_2}\right)R = \left(\frac{240 - 120}{120}\right) \times 2 = 2\ \Omega$$

55. With only the cell,

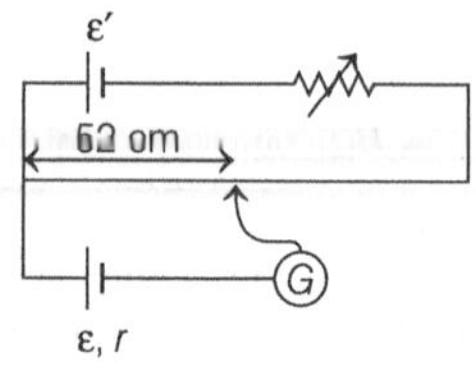

On balancing, $\varepsilon = 52 \times x$ ...(i)

where, x is the potential gradient of the wire.

When the cell is shunted,

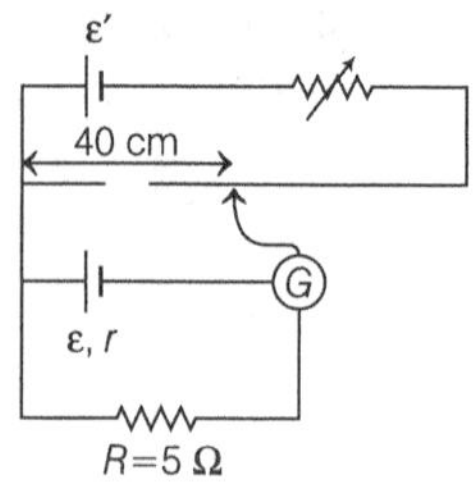

Similarly, on balancing,

$$V = \varepsilon - \frac{\varepsilon r}{(R + r)} = 40 \times x \quad \text{...(ii)}$$

Solving Eqs. (i) and (ii), we get

$$\frac{\varepsilon}{V} = \frac{1}{1 - \frac{r}{R + r}} = \frac{52}{40}$$

$$\Rightarrow \quad \frac{\varepsilon}{V} = \frac{R+r}{R} = \frac{52}{40} \Rightarrow \frac{5+r}{5} = \frac{52}{40}$$

$$\Rightarrow \quad r = \frac{3}{2}\,\Omega \Rightarrow r = 1.5\,\Omega$$

56. Let length of null point J be x and length of the potentiometer wire be L.

In first case, current in the circuit, $i_1 = \frac{6}{4+2} = 1\text{ A}$

$\therefore$ Potential gradient $= i \times R = \frac{1 \times 4}{L}$

$\Rightarrow$ Potential difference in part 'AJ' $= \frac{1 \times 4}{L} \times x = \varepsilon_1$

Given, $\quad \varepsilon_1 = 0.5 = \frac{4x}{L} \quad$ or $\quad \frac{x}{L} = \frac{1}{8} \quad$ ...(i)

In second case, current in the circuit, $i_2 = \frac{6}{4+6} = 0.6\text{ A}$

$\therefore$ Potential gradient $= \frac{0.6 \times 4}{L}$

$\Rightarrow$ Potential difference in part 'AJ'

$$= \frac{0.6 \times 4}{L} \times x = \varepsilon_2$$

$$\Rightarrow \quad \varepsilon_2 = \frac{0.6 \times 4}{L} \times \frac{L}{8} \quad \text{[using Eq. (i)]}$$

$$\Rightarrow \quad \varepsilon_2 = 0.3\text{ V}$$

57. Given, length of potentiometer wire $(AB) = L$

Resistance of potentiometer wire $(AB) = 12\,r$

emf of cell D of potentiometer $= \varepsilon$

Internal resistance of cell $D = r$

emf of cell $C = \frac{\varepsilon}{2}$

Internal resistance of cell $C = 3r$

Current in potentiometer wire

$$i = \frac{\text{emf of cell of potentiometer}}{\text{total resistance of potentiometer circuit}}$$

$$\Rightarrow \quad i = \frac{\varepsilon}{r + 12r} = \frac{\varepsilon}{13r}$$

Potential drop across the balance length AJ of potentiometer wire is $V_{AJ} = i \times R_{AJ}$

$\Rightarrow \quad V_{AJ} = i$ (Resistance per unit length of potentiometer wire $\times$ length AJ)

$$\Rightarrow \quad V_{AJ} = i\left(\frac{12r}{L} \times x\right)$$

where, x = balance length AJ.

As null point occurs at J, so potential drop across balance length AJ = emf of the cell C.

$$\Rightarrow \quad V_{AJ} = \frac{\varepsilon}{2} \Rightarrow i\left(\frac{12r}{L} \times x\right) = \frac{\varepsilon}{2}$$

$$\Rightarrow \quad \frac{\varepsilon}{13r} \times \frac{12r}{L} \times x = \frac{\varepsilon}{2}$$

$$\Rightarrow \quad x = \frac{13}{24}L$$

Round II

1. As, we know Cu is a conductor, so when there is increase in temperature, resistance will increase linearly. Then, Si is semiconductor, so with increase in temperature, resistance will decrease linearly.

2. $\quad i = \frac{\varepsilon}{R+r}$

$$\Rightarrow \quad i = \frac{\varepsilon}{R} = \text{constant}$$

where, R = external resistance

r = internal resistance = 0

3. Given, $l' = l + 100\%\, l = 2\,l$

Initial volume = Final volume

i.e. $\quad \pi r^2 l = \pi r'^2 l'$

or $\quad r'^2 = \frac{r^2 l}{l'} = r^2 \times \frac{l}{2\,l}$

or $\quad r'^2 = \frac{r^2}{2}$

$\therefore \quad R' = \rho\frac{l'}{A'} = \rho\frac{2l}{\pi r'^2} \quad \left(\because R = \frac{\rho l}{A}\right)$

$$= \frac{\rho \cdot 4l}{\pi r^2} = 4R$$

Thus, $\quad \Delta R = R' - R = 4R - R = 3\,R$

$\therefore \quad \%\,\Delta R = \frac{3R}{R} \times 100\% = 300\%$

4. As per the question,

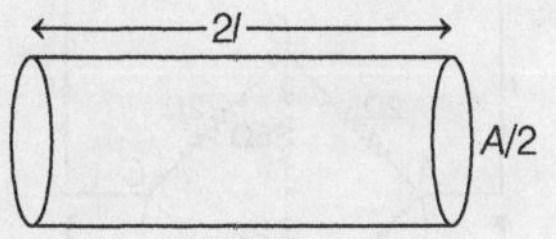

$$\text{Resistance} = \frac{\rho(2l)}{(A/2)} = \frac{4\rho l}{A}$$

$$\Rightarrow \quad \text{Current} = \frac{V}{R} = \frac{VA}{4\rho l}$$

5. Let resistances be R_1 and R_2, then

$$S = R_1 + R_2 \text{ and } P = \frac{R_1 R_2}{R_1 + R_2}$$

$\therefore \quad (R_1 + R_2) = \frac{n \times R_1 R_2}{R_1 + R_2} \quad (\because S = nP)$

or $\quad (R_1 + R_2)^2 = nR_1R_2$

$$\Rightarrow \quad n = \left[\frac{R_1^2 + R_2^2 + 2\,R_1R_2}{R_1R_2}\right] = \left[\frac{R_1}{R_2} + \frac{R_2}{R_1} + 2\right]$$

We know, Arithmetic Mean $\geq$ Geometric Mean

$$\frac{\frac{R_1}{R_2} + \frac{R_2}{R_1}}{2} \geq \sqrt{\frac{R_1}{R_2} \times \frac{R_2}{R_1}} \Rightarrow \frac{R_1}{R_2} + \frac{R_2}{R_1} \geq 2$$

So, n (minimum value) $= 2 + 2 = 4$

6. Given, $i = 10\text{ A},\ A = 5\text{ mm}^2 = 5 \times 10^{-6}\text{ m}^2$

and $v_d = 2 \times 10^{-3}$ m/s

We know that, $i = neAv_d$

$\therefore \quad 10 = n \times 1.6 \times 10^{-19} \times 5 \times 10^{-6} \times 2 \times 10^{-3}$

$\Rightarrow \quad n = 0.625 \times 10^{28} = 625 \times 10^{25}$

7. Let (ρ_A, l_A, r_A, A_A) and (ρ_B, l_B, r_B, A_B) be specific resistances, lengths, radii and areas of wires A and B, respectively.

Resistance of A, $R_A = \dfrac{\rho_A l_A}{A_A} = \dfrac{\rho_A l_A}{\pi r_A^2}$

Resistance of B, $R_B = \dfrac{\rho_B l_B}{A_B} = \dfrac{\rho_B l_B}{\pi r_B^2}$

From given information,

$$\rho_B = 2\rho_A$$
$$r_B = 2r_A$$

and $R_A = R_B$

$\therefore \quad \dfrac{\rho_A l_A}{\pi r_A^2} = \dfrac{\rho_B l_B}{\pi r_B^2}$

or $\quad \dfrac{\rho_A l_A}{\pi r_A^2} = \dfrac{2\rho_A \times l_B}{\pi (2r_A)^2}$

or $\quad \dfrac{l_B}{l_A} = \dfrac{2}{1} = 2:1$

8. Using colour code, we have

$$R = 53 \times 10^4 \pm 5\%\ \Omega$$

or $\quad R = 530 \times 10^3\ \Omega \pm 5\%$

or $\quad R = 530\text{ k}\Omega \pm 5\%$

9.

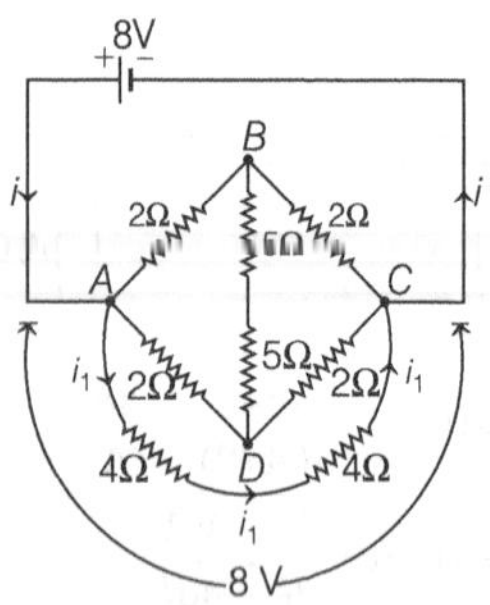

From the given circuit diagram,
Potential drop across $AC, V = 8$ V
Resistance of mentioned wire,

$$R = 4 + 4 = 8\ \Omega$$

So, the current flowing from A to C,

$$i_1 = \frac{V}{R} = \frac{8\text{V}}{8\Omega} = 1\text{ A}$$

10. For balanced Wheatstone's bridge,

$$\frac{P}{Q} = \frac{R}{S}$$

Here, $\quad S = S_1 \,||\, S_2 = \dfrac{S_1 S_2}{S_1 + S_2}$

$\Rightarrow \quad \dfrac{P}{Q} = \dfrac{R(S_1 + S_2)}{S_1 S_2}$

11. Colour code of carbon resistor is shown in the figure below

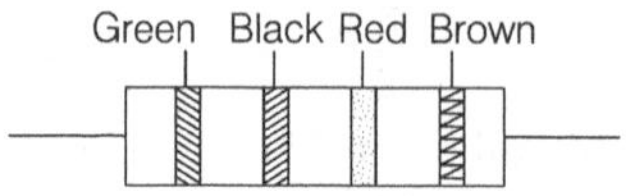

So, resistance value of resistor using colour code is

$$R = 502 \times 10 = 50.2 \times 10^2\ \Omega$$

Here, we must know that for given carbon resistor first three colours give value of resistance and fourth colour gives multiplier value.

Now using power, $P = i^2R$, we get

$\Rightarrow \quad i = \sqrt{\dfrac{P}{R}} = \sqrt{\dfrac{2}{50.2 \times 10^2}}$

$$\simeq 20 \times 10^{-3}\text{ A}$$
$$= 20\text{ mA}$$

12. Let potential at P_1 be 0 V and potential at P_2 be V_0.

Now, apply KCL at P_2,

$$\Sigma i = 0$$

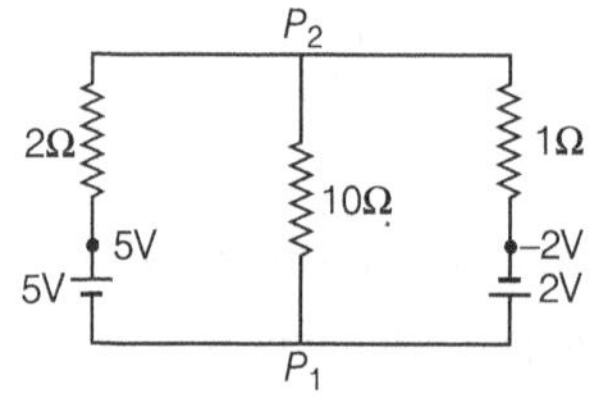

$$\frac{V_0 - 5}{2} + \frac{V_0 - 0}{10} + \frac{V_0 - (-2)}{1} = 0$$

or $\quad V_0 = \dfrac{5}{16}$

So, current through 10 Ω resistor is $\dfrac{V_0}{10}$ ($\simeq 0.03$ A) from P_2 to P_1.

13. According to the question,

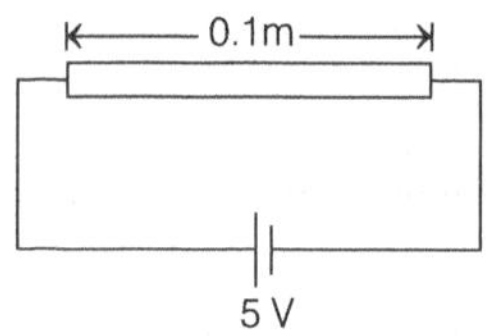

$$v_d = 2.5 \times 10^{-4}\text{ m/s},$$
$$n = 8 \times 10^{28}/\text{m}^3$$

We know that,

$$J = nev_d \text{ or } i = nev_d A$$

where, symbols have their usual meanings.

$\Rightarrow \quad \dfrac{V}{R} = nev_d A$

or $\quad \dfrac{V}{\dfrac{\rho L}{A}} = nev_d A$

or $$\frac{V}{\rho L} = nev_d$$

or $$\rho = \frac{V}{nev_d\, L}$$

$$= \frac{5}{8 \times 10^{28} \times 1.6 \times 10^{-19} \times 2.5 \times 10^{-4} \times 0.1}$$

or $\rho = 1.6 \times 10^{-5}\ \Omega\text{-m}$

14. $$P = \frac{V^2}{R}$$

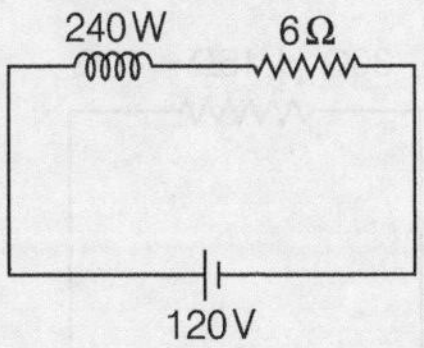

Resistance of the bulb,

$$R = \frac{120 \times 120}{60} = 240\ \Omega$$

$$R_{eq} = 240 + 6 = 246\ \Omega$$

$$\Rightarrow \quad i_1 = \frac{V}{R_{eq}} \quad \frac{120}{246}$$

$$\because \quad V_1 = i_1 R_1 = \frac{120}{246} \times 240 = 117.073\ \text{V}$$

Resistance of the heater $= \dfrac{V^2}{P} = \dfrac{120 \times 120}{240} = 60\ \Omega$

As bulb and heater are connected in parallel.

Net resistance $= \dfrac{240 \times 60}{300} = 48\ \Omega$

Total resistance, $R_2 = 48 + 6 = 54\ \Omega$

Total current, $i_2 = V/R_2 = 120/54$

Potential across heater = Potential across bulb

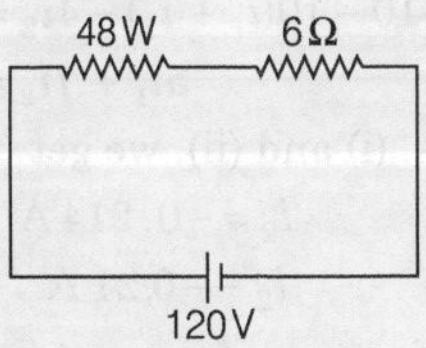

$$V_2 = \frac{120}{54} \times 48 = 106.66\ \text{V}$$

$$\because \quad V_1 - V_2 = 117.073 - 106.66 = 10.04\ \text{V}$$

15. Let P_1 and P_2 be the individual electric powers of the two resistances, respectively.

In series combination, power is

$$P_0 = \frac{P_1 P_2}{P_1 + P_2} = 60\text{W}$$

Since, the resistances are equal and the current through each resistor in series combination is also same. Then,

$$P_1 = P_2 = 120\ \text{W}$$

In parallel combination, power is

$$P = P_1 + P_2 = 120 + 120 = 240\ \text{W}$$

Alternate Solution

Let R be the resistance.

$\therefore$ Net resistance in series $= R + R = 2R$

$$P = \frac{V^2}{2R} = 60\ \text{W} \Rightarrow \frac{V^2}{R} = 120\ \text{W}$$

New resistance in parallel $= \dfrac{R \times R}{R + R} = R/2$

$$P' = \frac{V^2}{R/2} = 2\left(\frac{V^2}{R}\right) = 240\ \text{W}$$

16. As R_1, R_2 and R in series.

$$R_{eq} = R_1 + R_2 + R$$

$\therefore$ Net current, $$i = \frac{2\,\varepsilon}{R_1 + R_2 + R}$$

According to the question,

$$-(V_A - V_B) = \varepsilon - iR_2$$

$$\therefore \quad 0 = \varepsilon - iR_2$$

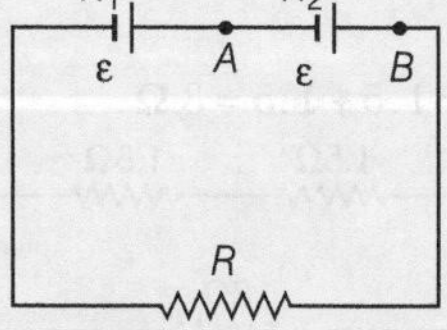

or $$\varepsilon E = iR_2$$

or $$\varepsilon = \frac{2\,\varepsilon}{R_1 + R_2 + R} R_2$$

or $$R_1 + R_2 + R = 2R_2$$

or $$R = R_2 - R_1$$

17. Given, potential difference of 5 mV is across 10 cm length of potentiometer wire. So, potential drop per unit length is

$$= \frac{5 \times 10^{-3}}{10 \times 10^{-2}} = 5 \times 10^{-2}\left(\frac{\text{V}}{\text{m}}\right)$$

Hence, potential drop across 1 m length of potentiometer wire is

$$V_{AB} = 5 \times 10^{-2}\left(\frac{\text{V}}{\text{m}}\right) \times 1 = 5 \times 10^{-2}\ \text{V}$$

Now, potential drop that must occurs across resistance R is

$$V_R = 4 - 5 \times 10^{-2} = \frac{395}{100}\ \text{V}$$

Now, circuit current is $i = \dfrac{V}{R_{\text{total}}} = \dfrac{4}{R + 5}$

Hence, for resistance R, using $V_R = iR$, we get

$$\frac{395}{100} = \frac{4}{R + 5} \times R$$

$$\Rightarrow \quad 395\,(R + 5) = 400R$$

$$\Rightarrow \quad 395 \times 5 = (400 - 395)R \Rightarrow R = 395\ \Omega$$

18. In steady state, no current flows through the capacitor. So, resistance r_p becomes ineffective.

So, the current in circuit, $i = \dfrac{\varepsilon}{r + r_2}$

$\because$ Potential drop across capacitor = Potential drop across $r_2 = ir_2 = \dfrac{\varepsilon r_2}{r + r_2}$

$\therefore$ Stored charge of capacitor, $Q = CV = C\varepsilon \dfrac{r_2}{r + r_2}$

19. The equivalent circuit can be drawn as 6 Ω and 2 Ω are in parallel.

$$R' = \frac{2 \times 6}{8} = 1.5\ \Omega$$

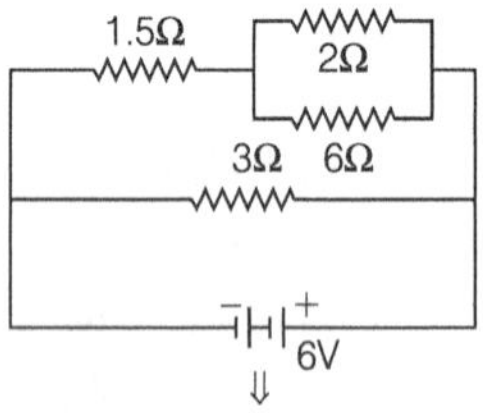

⇓

As, 1.5 Ω and 1.5 Ω are in series. (as shown in above circuit)

$$R'' = 1.5 + 1.5 = 3\ \Omega$$

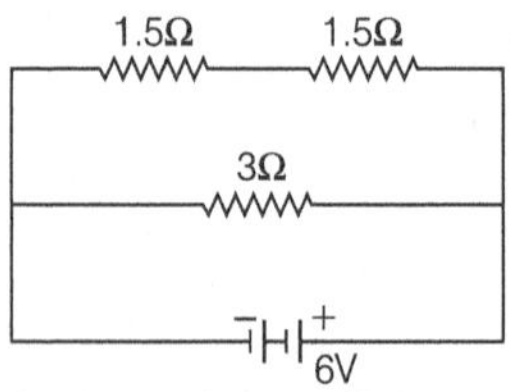

Now, 3 Ω and 3 Ω are in parallel.

$$R_{eq} = \frac{3 \times 3}{6} = 1.5\ \Omega$$

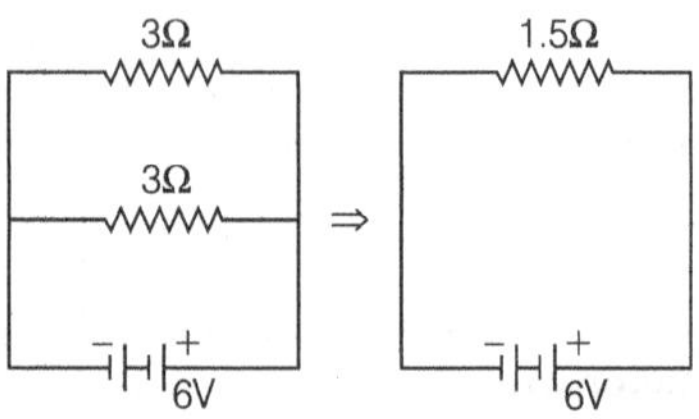

Hence, current supplied by the battery is

$$i = \frac{V}{R_{eq}} = \frac{6}{1.5} = 4\ \text{A}$$

20. The given circuit can be redrawn as

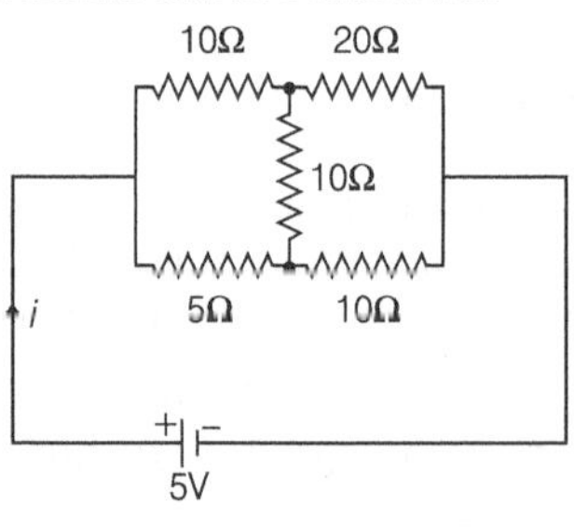

which is a balanced Wheatstone's bridge and hence, no current flows in the middle resistor, so equivalent circuit would be as shown below

⇒

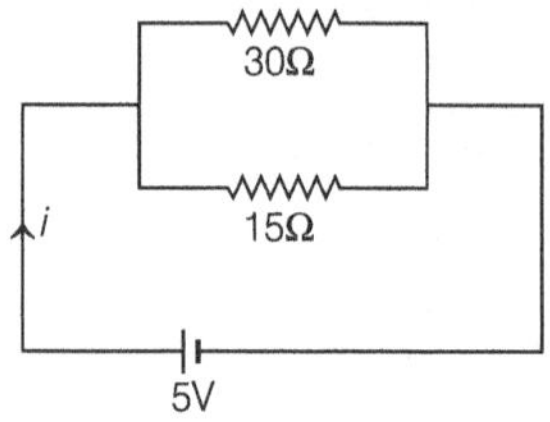

⇒

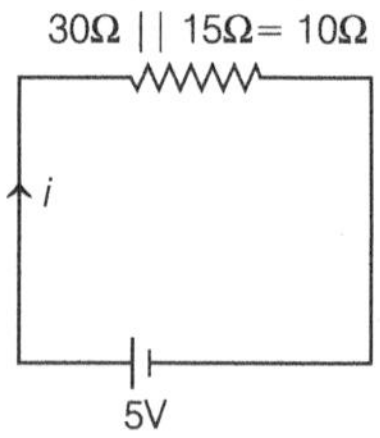

$$i = \frac{V}{R} = \frac{5}{10} = 0.5\ \text{A}$$

21. Assume the current in branches as shown in figure

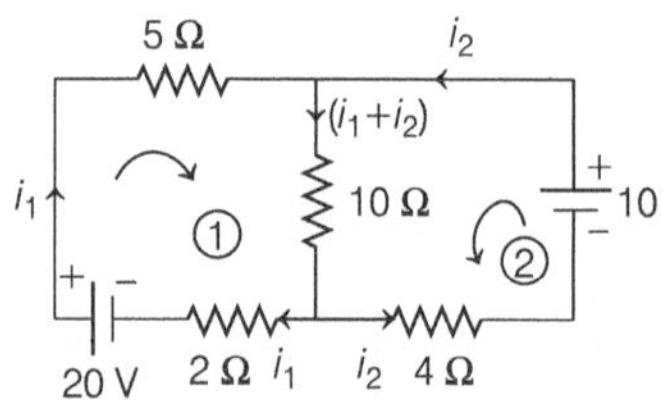

Applying KVL in loop 1,

$$+20 - 5i_1 - 10(i_1 + i_2) - 2i_1 = 0$$

$$17i_1 + 10i_2 = 20 \qquad \text{...(i)}$$

Applying KVL in loop 2,

$$+10 - 10(i_1 + i_2) - 4i_2 = 0$$

$$5i_1 + 7i_2 = 5 \qquad \text{...(ii)}$$

On solving Eqs. (i) and (ii), we get

$$i_2 = -0.214\ \text{A}$$

or $\quad i_2 \simeq -0.21\ \text{A}$

It means the direction of current is from positive to negative terminal of battery inside it.

22. For parallel combination of cells,

$$\varepsilon_{eq} = \frac{\dfrac{\varepsilon_1}{r_1} + \dfrac{\varepsilon_2}{r_2}}{\dfrac{1}{r_1} + \dfrac{1}{r_2}}$$

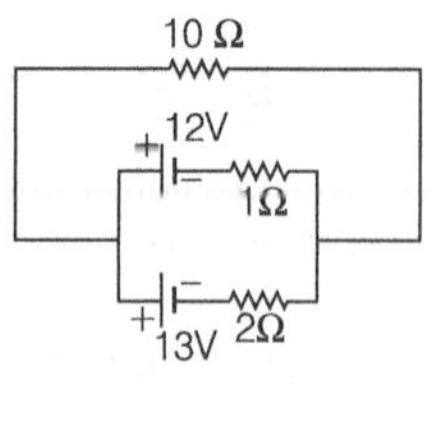

$$\therefore \quad \varepsilon_{eq} = \frac{\frac{12}{1} + \frac{13}{2}}{\frac{1}{1} + \frac{1}{2}} = \frac{37}{3} \text{ V}$$

Potential drop across 10 Ω resistance,

$$V = \left(\frac{\varepsilon}{R_{total}}\right) \times 10$$

$$= \frac{37/3}{\left(10 + \frac{2}{3}\right)} \times 10$$

$$= 11.56 \text{ V}$$

$$\therefore \quad V = 11.56 \text{ V}$$

23. According to question, the voltage across R_4 is 5 V, then the current across it

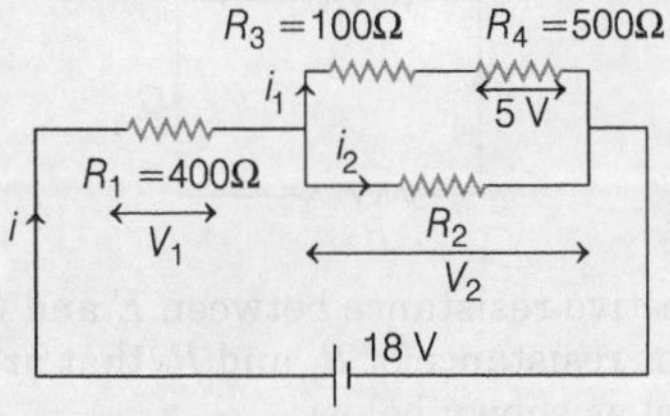

According to Ohm's law,

$$V = iR$$

$$\Rightarrow \quad 5 = i_1 \times R_4$$

$$\Rightarrow \quad 5 = i_1 \times 500$$

$$\Rightarrow \quad i_1 = \frac{5}{500} = \frac{1}{100} \text{ A}$$

The potential difference across series combination of R_3 and R_4,

$$V_2 = (R_3 + R_4)i = 600 \times \frac{1}{100} = 6 \text{ V}$$

So, potential difference (across R_1),

$$V_1 = 18 - 6 = 12 \text{ V}$$

Current through R_1,

$$i = \frac{V_1}{R_1} = \frac{12}{400} = \frac{3}{100} \text{ A}$$

So current through R_2,

$$i_2 = i - i_1 = \frac{3}{100} - \frac{1}{100} \text{ A} = \frac{2}{100} \text{ A}$$

Now, from $V = IR$, we have

$$R_2 = \frac{V_2}{i_2} = \frac{6}{(2/100)} = 300 \ \Omega$$

24. Measured value of R = 5% less than actual value of R.

Actual values of $R = 30 \ \Omega$

So, measured value of R is

$$R' = 30 - (5\% \text{ of } 30)$$

$$= 30 - \frac{5}{100} \times 30$$

$$\Rightarrow \quad R' = 28.5 \ \Omega \qquad \text{...(i)}$$

Now, let us assume that internal resistance of voltmeter R_V. Replacing voltmeter with its internal resistance, we get following circuit.

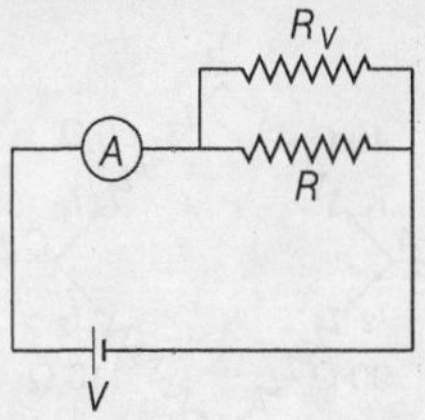

It is clear that the measured value, R' should be equal to parallel combination of R and R_V. Mathematically,

$$R' = \frac{RR_V}{R + R_V} = 28.5 \ \Omega$$

Given, $\quad R = 30 \ \Omega$

$$\Rightarrow \quad \frac{30R_V}{30 + R_V} = 28.5$$

$$\Rightarrow \quad 30R_V = (28.5 \times 30) + 28.5 \ R_V$$

$$\Rightarrow \quad 1.5R_V = 28.5 \times 30$$

$$\Rightarrow \quad R_V = \frac{28.5 \times 30}{1.5} = 19 \times 30 \text{ or } R_V = 570 \ \Omega$$

25. Let R_0 be the initial resistance of both conductors.

$\therefore$ At temperature θ, their resistances will be

$$R_1 = R_0(1 + \alpha_1\theta)$$

and $\quad R_2 = R_0(1 + \alpha_2\theta)$

For series combination,

$$R_s = R_1 + R_2$$

$$R_{s0}(1 + \alpha_s\theta) = R_0(1 + \alpha_1\theta) + R_0(1 + \alpha_2\theta)$$

where, $\quad R_{s0} = R_0 + R_0 = 2R_0$

$$\therefore \quad 2R_0(1 + \alpha_s\theta) = 2R_0 + R_0\theta(\alpha_1 + \alpha_2)$$

or $\quad \alpha_s = \dfrac{\alpha_1 + \alpha_2}{2}$

For parallel combination,

$$R_p = \frac{R_1R_2}{R_1 + R_2}$$

$$R_{p0}(1 + \alpha_p\theta) = \frac{R_0(1 + \alpha_1\theta)R_0(1 + \alpha_2\theta)}{R_0(1 + \alpha_1\theta) + R_0(1 + \alpha_2\theta)}$$

where, $\quad R_{p0} = \dfrac{R_0R_0}{R_0 + R_0} = \dfrac{R_0}{2}$

$$\therefore \quad \frac{R_0}{2}(1 + \alpha_p\theta) = \frac{R_0^2(1 + \alpha_1\theta + \alpha_2\theta + \alpha_1\alpha_2\theta^2)}{R_0(2 + \alpha_1\theta + \alpha_2\theta)}$$

As, α_1 and α_2 are small quantities.

$\therefore$ $\alpha_1 \alpha_2$ is negligible.

or $\quad \alpha_p = \dfrac{\alpha_1 + \alpha_2}{2 + (\alpha_1 + \alpha_2)\theta}$

$$= \frac{\alpha_1 + \alpha_2}{2}\left[1 - \left(\frac{\alpha_1 + \alpha_2}{2}\right)\theta\right]$$

As, $(\alpha_1 + \alpha_2)^2$ is negligible.

$$\therefore \quad \alpha_p = \frac{\alpha_1 + \alpha_2}{2}$$

26. The resistors 40 Ω and 60 Ω are connected in series combination. Similarly, the resistors 90 Ω and 110 Ω are also connected in series combination.

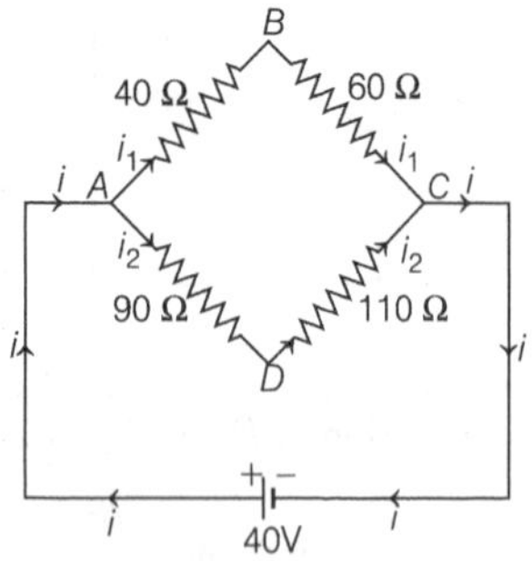

So, $i_1 = \dfrac{V_{AC}}{R_{AB} + R_{BC}}$

$= \dfrac{40}{40+60} = \dfrac{40}{100} = \dfrac{2}{5}$ A

and $i_2 = \dfrac{V_{AC}}{R_{AD} + R_{DC}}$

$= \dfrac{40}{90+110} = \dfrac{40}{200} = \dfrac{1}{5}$ A

For path *BAD*, using KVL (Kirchhoff's voltage law),

$$V_B + i_1 \times 40 - i_2 \times 90 = V_D$$

$\Rightarrow \quad V_B + \dfrac{2}{5} \times 40 - \dfrac{1}{5} \times 90 = V_D$

$\Rightarrow \quad V_B + 16 - 18 = V_D$

$\Rightarrow \quad V_B - 2 = V_D$

$\Rightarrow \quad V_B - V_D = 2$ V

So, the potential difference across *BD* (in volt) is 2.

27. Since, the wire is of uniform cross-section, the resistances of the two segments of the wire *AD* and *DC* are in the ratio of the lengths of *AD* and *DC*. Using the null point conditions of a Wheatstone bridge, we have

$$\left(\frac{X}{Y}\right) = \left(\frac{33.7}{66.3}\right) \quad \ldots \text{(i)}$$

When *Y* is shunted by a resistance of 12.0 Ω, net resistance changes,

$$Y' = 12Y/(Y+12)$$

Since, Y' is less than Y, the ratio X/Y' is greater than $\dfrac{X}{Y}$. Thus, the null point must shift towards the end *C*, *i.e.*

$$\left(\frac{X}{Y'}\right) = \left(\frac{51.9}{48.1}\right)$$

or $\quad X(Y+12)/12Y = (51.9/48.1)$

i.e. $\quad \dfrac{Y+12}{12} = \left(\dfrac{51.9}{48.1}\right) \times \dfrac{66.3}{33.7}$

which give $Y = 13.5$ Ω and $X = 6.86$ Ω using Eq. (i)

28. Let the length of each side of square *ABCD* is a.

$\therefore$ Resistance per unit length of each side $= \dfrac{R}{4a}$

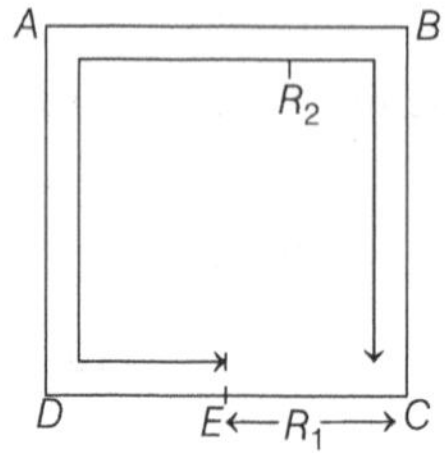

Now, $\quad R_1 = \dfrac{R}{4a}(EC) = \dfrac{R}{4a} \times \dfrac{a}{2} = \dfrac{R}{8}$

Similarly, $R_2 = \dfrac{R}{4a}(EDABC) = \dfrac{R}{4a} \times \dfrac{7a}{2} = \dfrac{7R}{8}$

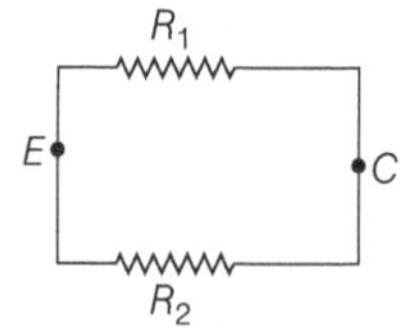

Now, effective resistance between *E* and *C* is the equivalent resistance of R_1 and R_2 that are connected in parallel as shown below.

$$R_{EC} = \frac{R_1 R_2}{R_1 + R_2} = \frac{(R/8) \times (7R/8)}{(R/8) + (7R/8)} = \frac{7R^2}{64} \times \frac{8}{8R} = \frac{7}{64}R$$

29. Given circuit is redrawn and can be simplified as

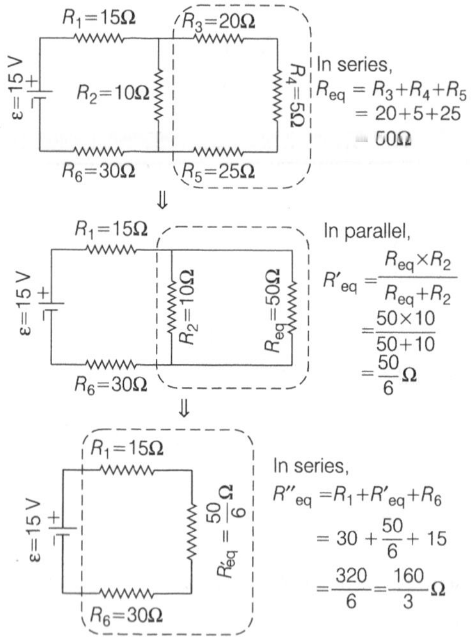

So, current drawn through cell is

$$i = \frac{\text{Voltage}}{\text{Net resistance of the circuit}}$$

$$= \frac{V}{R''_{eq}} = \frac{15}{(160/3)} = \frac{9}{32} \text{ A}$$

30. In the given circuit, let's assume currents in the arms are i_1, i_2 and i_3, respectively.

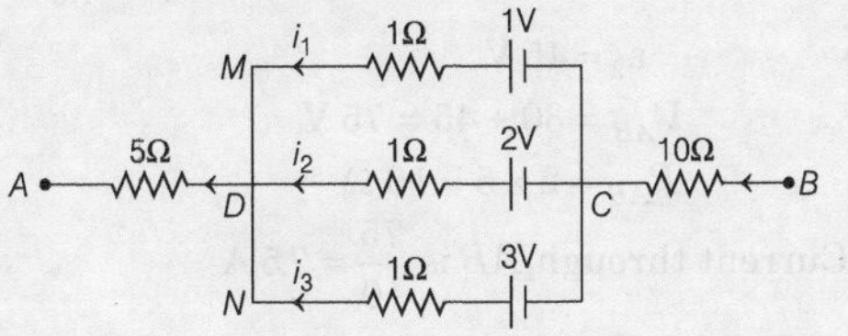

Now, $$i_1 = \frac{V_1}{R_1} = \frac{1}{1} = 1\text{ A}$$

Similarly, $$i_2 = \frac{2}{1} = 2\text{ A and } i_3 = \frac{3}{1} = 3\text{ A}$$

Total current in the arm DA,

$$i = i_1 + i_2 + i_3 = 6\text{ A}$$

As all three resistors between D and C are in parallel.

$\therefore$ Equivalent resistance between terminals D and C,

$$\frac{1}{R_{DC}} = \left(\frac{1}{1} + \frac{1}{1} + \frac{1}{1}\right)$$

$\therefore$ $$R_{DC} = \frac{1}{3}\,\Omega$$

So, potential difference across D and C,

$$V_{DC} = iR_{DC} = 6 \times \frac{1}{3} \Rightarrow V_{DC} = 2\text{ V}$$

Now, V_{AD} and $V_{CB} = 0$ (in case of open circuits, $i = 0$)

So, $V_{AB} = V_{AD} + V_{DC} + V_{CB} = V_{DC}$

So, $V_{AB} = 2\text{V}$

31. The positive terminals of the cells ε_1 and ε_2 are connected to the wire AE of resistance 10Ω and negative terminals to the wire BD of resistance 4Ω. The resistance of 10Ω is connected between the middle points F and C of the wires AE and BD, respectively. Therefore,

$$R_1 = R_2 = \frac{10}{2} = 5\Omega$$

$$R_3 = R_4 = \frac{4}{2} = 2\Omega$$

The distribution of currents in various branches is shown in the figure

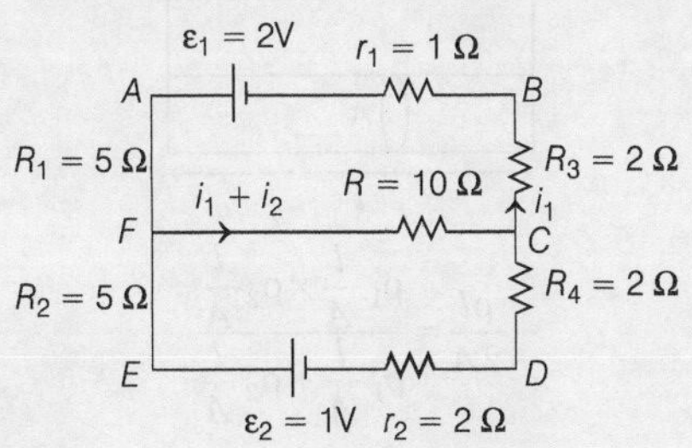

In closed part $ABCFA$ of the circuit,

$$i_1 \times r_1 + i_1 \times R_1 + (i_1 + i_2)R + i_1 \times R_3 = \varepsilon_1$$

$$i_1 \times 1 + i_1 \times 5 + (i_1 + i_2) \times 10 + i_1 \times 2 = 2$$

or $$9i_1 + 5i_2 = i \qquad \text{...(i)}$$

In closed part $CDEFC$ of the circuit,

$$i_2 \times r_2 + i_2 \times R_2 + (i_1 + i_2) \times R + i_2 \times R_4 = \varepsilon_2$$

or $$i_2 \times 2 + i_2 \times 5 + (i_1 + i_2) \times 10 + i_2 \times 2 = 1$$

or $$10i_1 + 19i_2 = 1 \qquad \text{...(ii)}$$

Solving Eqs. (i) and (ii), we have

$$i_1 = \frac{14}{121}\text{ A and } i_2 = -\frac{1}{121}\text{ A}$$

Therefore, current through resistance R,

$$i_1 + i_2 = \frac{14}{121} + \left(-\frac{1}{121}\right) = \frac{13}{121}\text{ A}$$

Potential difference across the resistance,

$$R = (i_1 + i_2)R = \frac{13}{121} \times 10 = 1.07\text{ V}$$

32. Given circuit is

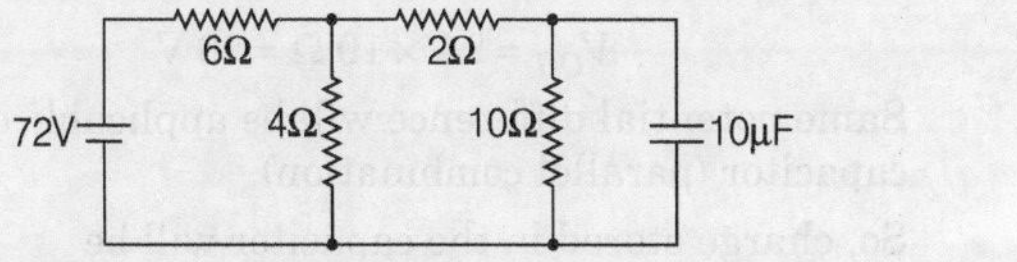

To find charge on capacitor, we need to determine voltage across it. In steady state, capacitor will acts as open circuit and circuit can be reduced as

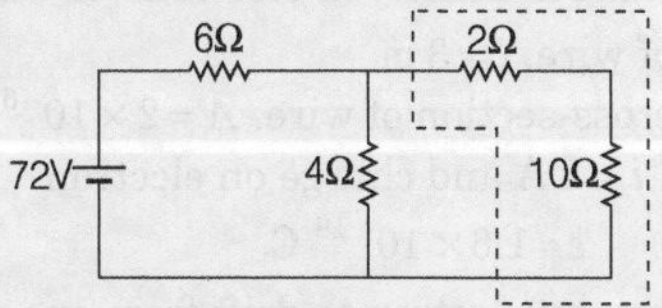

In series, $R_{\text{eq}} = 2\,\Omega + 10\,\Omega = 12\,\Omega$

⇓

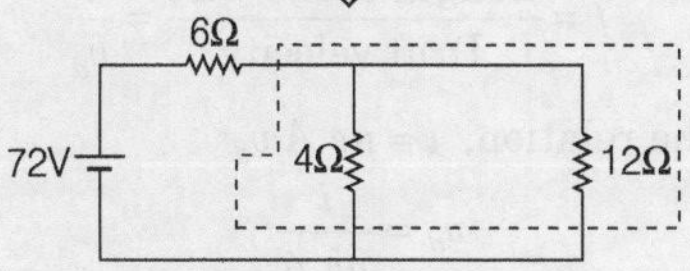

In parallel, $$R_{\text{eq}} = \frac{4 \times 12}{4 + 12} = 3\,\Omega$$

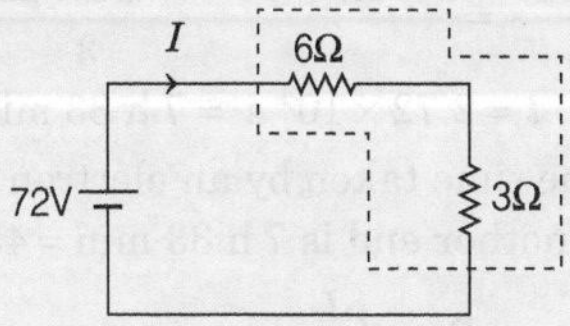

In series, $R_{\text{eq}} = 6\,\Omega + 3\,\Omega = 9\,\Omega$

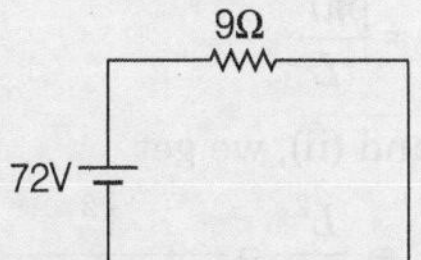

So, current in steady state, $i = \dfrac{V}{R} = \dfrac{72}{9} = 8\text{ A}$

Now, by using current division at point P, current in 6 Ω branch is

$$\frac{72 - V_P}{6\,\Omega} = 8\text{ A} \Rightarrow V_P = 72 - 48 = 24\text{ V}$$

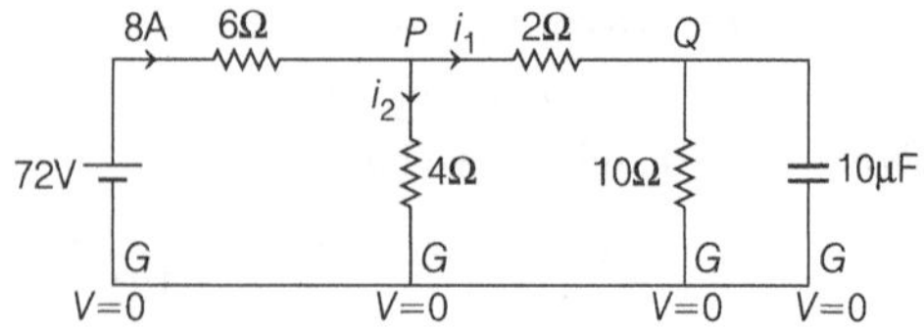

Current in 4 Ω branch,

$$i_2 = \frac{V_P - 0}{4} = \frac{24-0}{4} = 6\text{ A}$$

So, current in 2 Ω resistance,

$$i_1 = 8 - i_2 \qquad (\because i = i_1 + i_2)$$
$$= 8 - 6 = 2\text{A}$$

∴ Potential difference across 10 Ω resistor,

$$V_{QG} = 2\text{A} \times 10\ \Omega = 20\text{ V}$$

Same potential difference will be applicable over the capacitor (parallel combination).

So, charge stored in the capacitor will be

$$Q = CV = 10 \times 10^{-6} \times 20$$
$$\Rightarrow \quad Q = 2 \times 10^{-4}\text{ C} = 200\ \mu\text{C}$$

33. Given, number density of electrons, $n = 8.5 \times 10^{28}/\text{m}^3$

Length of wire, $l = 3$ m

Area of cross-section of wire, $A = 2 \times 10^{-6}\text{m}^2$

Current, $i = 3$ A and charge on electron,

$$e = 1.6 \times 10^{-19}\text{ C}$$

Time taken by electron to drift from one end to another of the wire,

$$t = \frac{\text{Length of the wire}}{\text{Drift velocity}} = \frac{l}{v_d} \qquad \ldots\text{(i)}$$

Using the relation, $i = ne\,A\,v_d$

$$\text{or} \qquad v_d = \frac{i}{ne\,A} \qquad \ldots\text{(ii)}$$

From, Eqs. (i) and (ii), we get

$$t = \frac{l\,ne\,A}{i} = \frac{3 \times 8.5 \times 10^{28} \times 1.6 \times 10^{-19} \times 2 \times 10^{-6}}{3}$$

$$\text{or} \qquad t = 2.72 \times 10^4\text{ s} = 7\text{ h }33\text{ min}$$

Thus, the time taken by an electron to drift from one end to another end is 7 h 33 min = 453 min.

34.
$$R_1 = \frac{\rho L}{\pi r \cdot t} \qquad [\because A_1 = \pi r t] \ldots\text{(i)}$$
$$R_2 = \frac{\rho \pi r}{tL} \qquad [\because A_2 = tL] \ldots\text{(ii)}$$

From Eqs. (i) and (ii), we get

$$\frac{R_1}{R_2} = \frac{L^2}{\pi^2 r^2} = \frac{1^2}{10 \times 0.1 \times 0.1} = 10$$

35. $V_{AC} = \varepsilon_3 = 30$ V ...(i)

$$V_{CB} = \varepsilon_2 \qquad \ldots\text{(ii)}$$

From Eqs. (i) and (ii), we get

$$\frac{\varepsilon_2}{30} = \frac{V_{CB}}{V_{AC}}$$

$$\Rightarrow \quad \frac{\varepsilon_2}{30} = \frac{3}{2} \qquad \left(\because \frac{V_{CB}}{V_{AC}} = \frac{l_{CB}}{l_{AC}} = \frac{3}{2}\right)$$

$$\Rightarrow \quad \varepsilon_2 = 45\text{ V}$$
$$\therefore \quad V_{AB} = 30 + 45 = 75\text{ V}$$
$$\because \quad R_{AB} = 2 \times 5 = 10\ \Omega$$

∴ Current through AB is $\frac{75}{10} = 7.5$ A

$$\therefore \quad \varepsilon_1 = 7.5 \times (R_{AB} + R) = 150\text{ V}$$

36. When capacitor is fully charged, circuit is reduced to as shown below

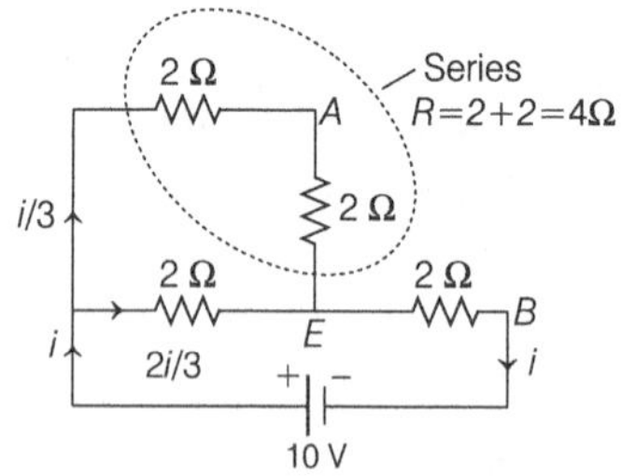

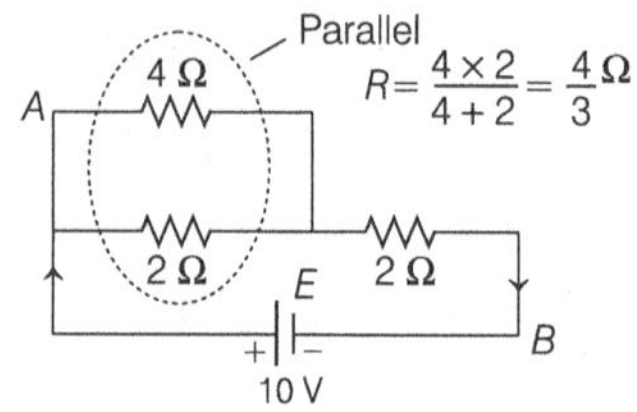

So, total resistance, $R_{\text{eq}} = \frac{4}{3} + 2 = \frac{10}{3}\ \Omega$

Current in circuit, $i = \frac{V}{R_{\text{eq}}} = \frac{10}{10/3} = 3$ A

Hence, potential difference across capacitor

$$= \text{potential difference across } AEB$$
$$= 2\,i/3 + 2 \times i = 2 \times \frac{3}{3} + 2 \times 3 = 8\text{ V}$$

37. $\because$ In parallel, $\quad R_{\text{net}} = \frac{R_1 R_2}{R_1 + R_2}$

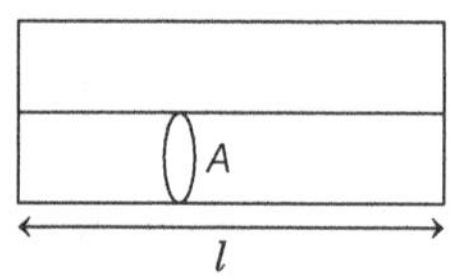

$$\frac{\rho l}{2A} = \frac{\rho_1 \frac{l}{A} \times \rho_2 \frac{l}{A}}{\rho_1 \frac{l}{A} + \rho_2 \frac{l}{A}}$$

$$\frac{\rho}{2} = \frac{6 \times 3}{6+3} = 2$$

$$\rho = 4$$

38. $\because \quad i_0 = \frac{V}{R} = \frac{30/3}{5 \times 10^6} = 2 \times 10^{-6}\text{ A} = 2\,\mu\text{A}$

$$\therefore \quad x = 2$$

20

Magnetic Effects of Current

In the year 1820, Oersted realised that electricity and magnetism were related to each other. He showed experimentally that the electric current through a straight wire causes noticeable deflection of the magnetic compass needle held near the wire.
This shows that the magnetic field is associated with a moving charge or a current carrying conductor. The branch of physics which deals with the magnetism due to electric current is called **electromagnetism**.

IN THIS CHAPTER ....

- Magnetic Field
- Biot-Savart's Law
- Ampere's Circuital Law
- Force on a Moving Charge in a Uniform Magnetic Field
- Lorentz Force
- Cyclotron
- Force on a Current Carrying Conductor in a Magnetic Field
- Moving Coil Galvanometer

Magnetic Field

The space around a magnet, moving charges or current carrying conductor in which its magnetic effect can be experienced, is called magnetic field (**B**). It is a vector quantity. The SI unit for magnetic field is tesla or weber/m^2 and its CGS unit is gauss (1 tesla = 10^4 gauss) or maxwell/cm^2.

Rules to Find Direction of Magnetic Field

- **Right Hand Palm Rule** If we spread our right hand in such a way that thumb is towards the direction of current and fingers are towards that point where we have to find the direction of field, then the direction of field will be perpendicular to the palm.

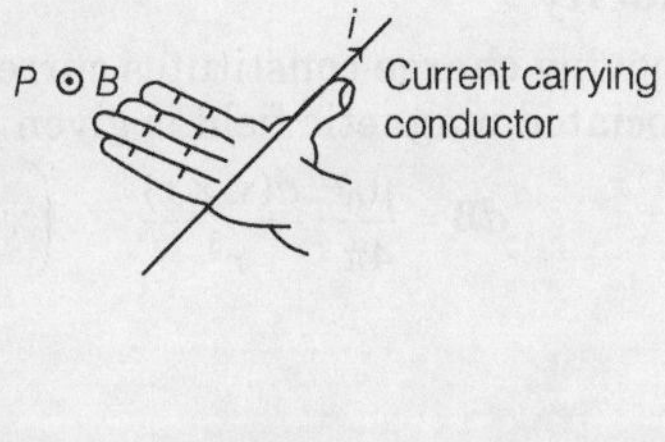

- **Maxwell's Right Handed Screw Rule** If a right handed cork screw is rotated so that its tip moves in the direction of flow of current through the conductor, then the rotation of the head of the screw gives the direction of magnetic field.

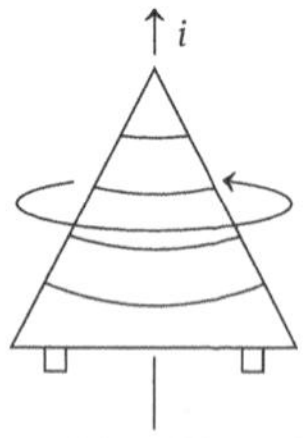

- **Right hand thumb rule** If a straight current carrying conductor is held in the right hand such that the thumb of the hand represents the direction of current flow, then the direction of the folding fingers will represent the direction of magnetic field lines of force.

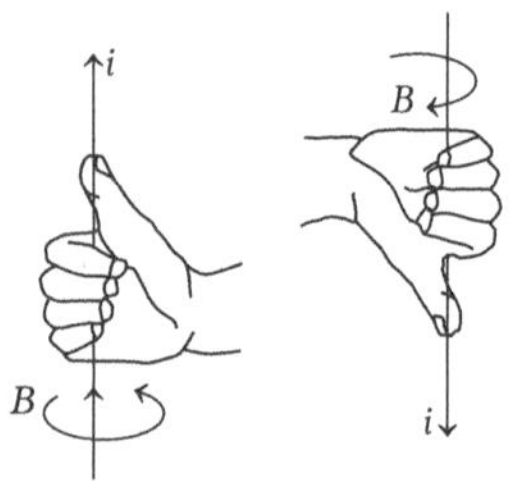

Note (i) If magnetic field is directed perpendicular and into the plane of the paper, it is represented by ⊗(cross) whereas if magnetic field is directed perpendicular and out of the plane of the paper, it is represented by ⊙ (dot).

(ii) **In** Magnetic field is away from the observer or perpendicular inwards.

Out Magnetic field is towards the observer or perpendicular outwards.

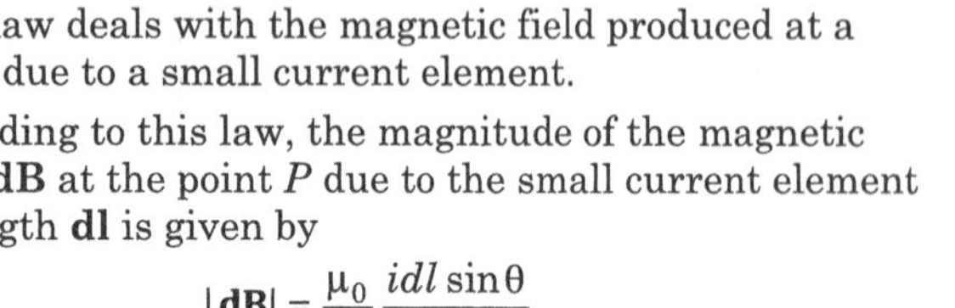

Biot-Savart's Law

This law deals with the magnetic field produced at a point due to a small current element.

According to this law, the magnitude of the magnetic field **dB** at the point P due to the small current element of length **dl** is given by

$$|\mathbf{dB}| = \frac{\mu_0}{4\pi}\frac{idl\sin\theta}{r^2}$$

where, μ_0 is a constant and is called, *magnetic permeability of free space.*

$$\mu_0 = 4\pi \times 10^{-7}\,\text{Wb/A-m}$$

The value of $\frac{\mu_0}{4\pi}$ in CGS is unity.

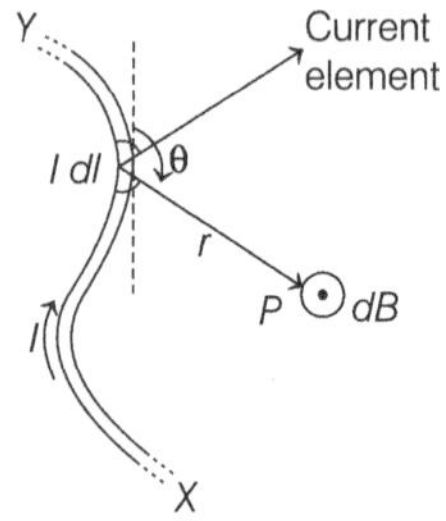

Its direction is perpendicular to the plane containing $d\mathbf{l}$ and $\mathbf{r}$.

In vector form, Biot-Savart's law can be written as

$$d\mathbf{B} \propto \frac{i\,(d\mathbf{l} \times \mathbf{r})}{|\mathbf{r}|^3}$$

$$= \frac{\mu_0}{4\pi}\cdot\frac{id\mathbf{l} \times \mathbf{r}}{r^3}$$

The direction of $d\mathbf{B}$ is represented by the right hand screw rule or right hand thumb rule.

Similarly, magnetic field induction at point P due to current through entire wire,

$$\mathbf{B} = \int \frac{\mu_0}{4\pi}\cdot\frac{i(d\mathbf{l} \times \mathbf{r})}{r^3}$$

or

$$\mathbf{B} = \int \frac{\mu_0}{4\pi}\cdot\frac{id\,l\sin\theta}{r^2}$$

Biot-Savart's Law in a Medium

If the conductor is placed in a medium, then vector form of magnetic field is given as

$$d\mathbf{B} = \frac{\mu}{4\pi}\frac{i(d\mathbf{l} \times \mathbf{r})}{|\mathbf{r}|^3}$$

$$= \frac{\mu_0\mu_r}{4\pi}\frac{i\,(d\mathbf{l} \times \mathbf{r})}{r^3}$$

where, μ_r = relative permeability = $\frac{\mu}{\mu_0}$

and μ = absolute permeability of the medium.

For air or vacuum $\mu_r = 1$.

Biot-Savart's Law in Terms of Current Density *J*

$$d\mathbf{B} = \frac{\mu_0}{4\pi}\cdot\frac{(\mathbf{J} \times \mathbf{r})}{r^3}dV \qquad \left(\because J = \frac{i}{A} = \frac{idA}{Adl} = \frac{idl}{dV}\right)$$

Biot-Savart's Law in Terms of Charge *q* and its Velocity *v*

A moving charge constitutes current and hence, associated magnetic field is given as

$$d\mathbf{B} = \frac{\mu_0}{4\pi}\cdot\frac{q(\mathbf{v} \times \mathbf{r})}{r^3} \qquad \left(\because idl = \frac{q}{dt}dl = q\frac{dl}{dt} = qv\right)$$

Applications of Biot-Savart's Law

Let us consider few applications of Biot-Savart law

(i) **Magnetic field due to a straight current carrying conductor is**

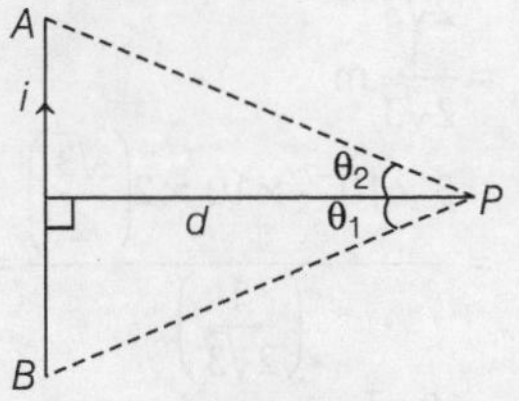

$$B = \frac{\mu_0 i}{4\pi d}(\sin\theta_1 + \sin\theta_2)$$

- For an infinitely long straight wire,

$$\theta_1 = \theta_2 = 90^\circ$$

$$\Rightarrow \quad B = \frac{\mu_0 i}{2\pi d}$$

- When wire is semi-infinite (at the foot of long wire),

$$\theta_1 = 0^\circ \text{ and } \theta_2 = \frac{\pi}{2}$$

$$\Rightarrow \quad B = \frac{\mu_0 i}{4\pi d}$$

- For axial position of wire, *i.e.* point P lies on axial position of current carrying conductor, then B at P is zero.
- For an infinitely long straight wire, $B \propto \frac{1}{d}$

$\Rightarrow$ *B-d* graph is a rectangular hyperbola as shown below

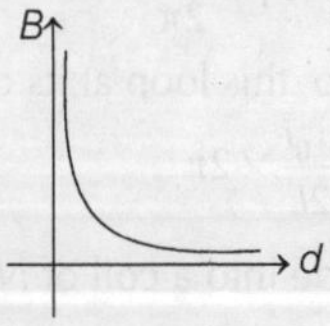

(ii) **Magnetic field at the centre of a circular current carrying coil**

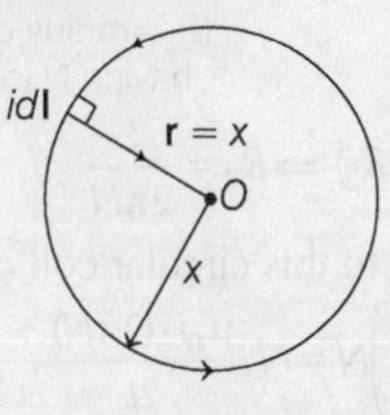

$$B = \frac{\mu_0 i}{2\pi a}$$

- Due to whole circular loop, direction of this field is outward perpendicular to the plane of the paper. If the loop have N turns, then magnetic field, $B = \frac{\mu_0 Ni}{2\pi a}$.

- Magnetic field due to an arc of a circular current carrying coil at the centre is given by

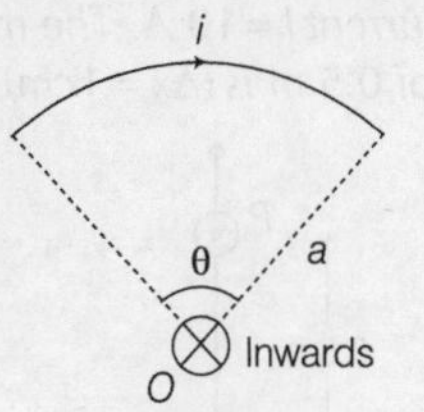

$$B = \left(\frac{\mu_0}{4\pi}\right)\left(\frac{i}{a}\right)\theta$$

Here, θ is to be substituted in radians.

(iii) **Magnetic field on the axis of a circular coil having N turns is**

$$B = \frac{\mu_0 N i R^2}{2(R^2 + x^2)^{3/2}}$$

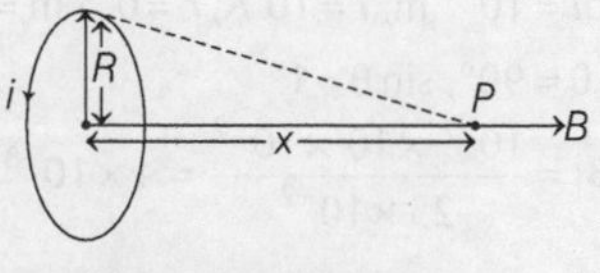

Here, R = radius of the coil,
x = distance of point P from centre
and i = current in the coil.

- At the centre of the loop, $x = 0$

$$\therefore \quad B = \frac{\mu_0 Ni}{2R}$$

- For $x >> R$, $x^2 + R^2 = x^2$

$$\therefore \quad B = \frac{\mu_0 N i R^2}{2x^3} \Rightarrow B \propto \frac{1}{x^3}$$

- *B-r* graph shows the variation of magnetic field at various position on the axis of circular current carrying coil. Magnetic field is maximum at the centre and decreases as we move away from the centre on the axis of the loop.

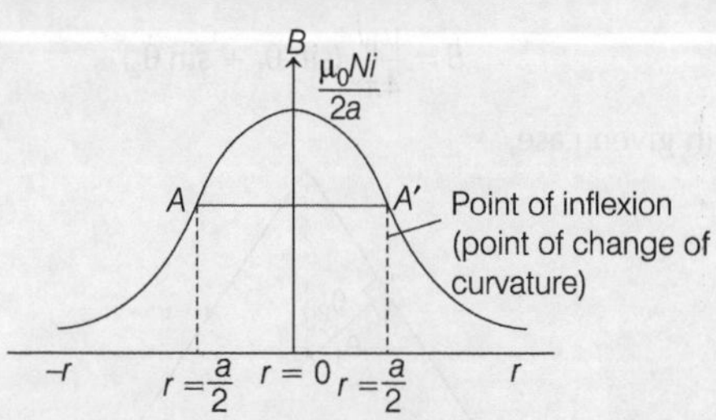

The magnetic field lines due to current carrying circular coil form closed loops and is shown as follows

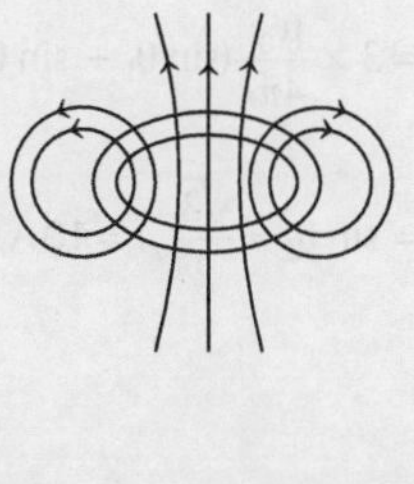

Example 1. *An element $\Delta L = \Delta x \cdot \hat{i}$ is placed at the origin and carries a large current $I = 10$ A. The magnetic field on the Y-axis at a distance of 0.5 m is ($\Delta x = 1$ cm)*

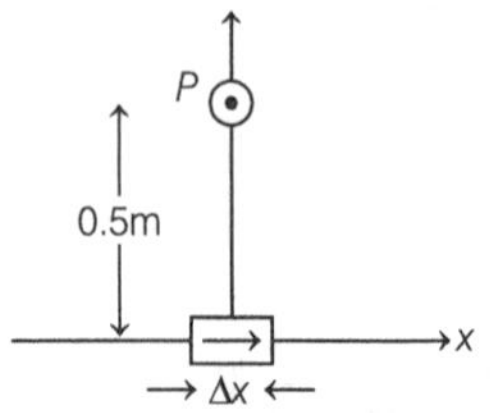

(a) 2×10^{-8} T (b) 6×10^{-8} T
(c) 8×10^{-8} T (d) 4×10^{-8} T

Sol. (d) According to Biot-Savart law,

$$|d\mathbf{B}| = \frac{\mu_0}{4\pi} \frac{idl \sin\theta}{r^2}$$

Given, $\Delta x = dl = 10^{-2}$ m, $I = 10$ A, $r = 0.5$ m $= y$

$$\theta = 90°, \sin\theta = 1$$

$$\therefore \quad |d\mathbf{B}| = \frac{10^{-7} \times 10 \times 10^{-2}}{25 \times 10^{-2}} = 4 \times 10^{-8} \text{ T}$$

Example 2. *The magnitude of the magnetic field at the centre of an equilateral triangular loop of side 1 m which is carrying a current of 10 A is [Take, $\mu_0 = 4\pi \times 10^{-7} NA^{-2}$]* **[JEE Main 2019]**

(a) 9 μT (b) 1 μT
(c) 3 μT (d) 18 μT

Sol. (d) For a current carrying wire, from result obtained by Biot-Savart's law, magnetic field at a distance r is given by

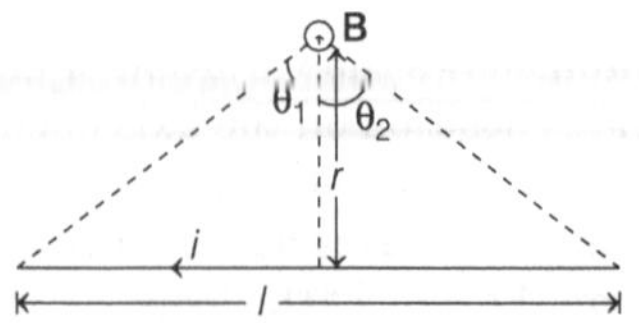

$$B = \frac{\mu_0 i}{4\pi r}(\sin\theta_1 + \sin\theta_2)$$

Now, in given case,

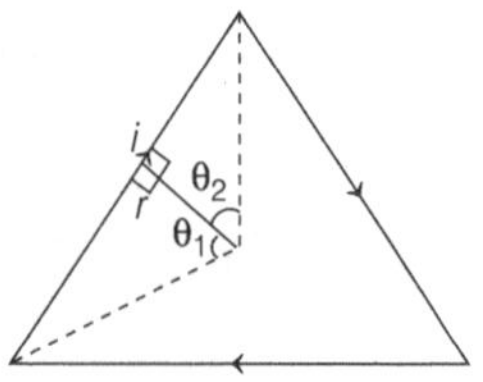

Due to symmetry of arrangement, net field at centre of triangle is

B_{net} = Sum of fields of all wires (sides)

$$= 3 \times \frac{\mu_0 i}{4\pi r}(\sin\theta_1 + \sin\theta_2)$$

Here, $\theta_1 = \theta_2 = 60°$

$$\therefore \quad \sin\theta_1 = \sin\theta_2 = \frac{\sqrt{3}}{2}, i = 10\text{A}, \frac{\mu_0}{4\pi} = 10^{-7}\text{NA}^{-2}$$

and
$$r = \frac{1}{3} \times \text{altitude}$$
$$= \frac{1}{3} \times \frac{\sqrt{3}}{2} \times \text{side length}$$
$$= \frac{1}{2\sqrt{3}} \times 1$$
$$= \frac{1}{2\sqrt{3}} \text{ m}$$

So,
$$B_{net} = \frac{3 \times 10^{-7} \times 10 \times 2\left(\frac{\sqrt{3}}{2}\right)}{\left(\frac{1}{2\sqrt{3}}\right)} = 18 \times 10^{-6} \text{ T}$$

$$\Rightarrow \quad B_{net} = 18 \ \mu\text{T}$$

Example 3. *One of the two identical conducting wires of length L is bent in the form of a circular loop and the other one into a circular coil of N identical turns. If the same current is passed in both, the ratio of the magnetic field at the centre of the loop B_L to that at the centre of the coil B_C, i.e. $\frac{B_L}{B_C}$ will be* **[JEE Main 2019]**

(a) $\frac{1}{N}$ (b) N
(c) $\frac{1}{N^2}$ (d) N^2

Sol. (c) Let consider the length of first wire is L, then according to question, if radius of loop formed is R_1, then

For first wire,

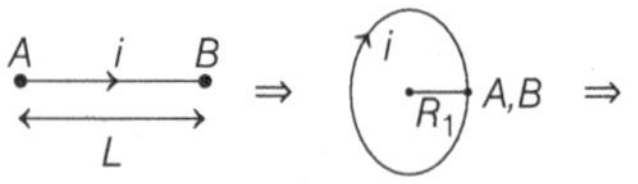

$$L = 2\pi R_1 \Rightarrow R_1 = \frac{L}{2\pi}$$

The magnetic field due to this loop at its centre,

$$B_L = \frac{\mu_0 i}{2R_1} = \frac{\mu_0 i}{2L} \times 2\pi \quad \text{...(i)}$$

Now, second wire is made into a coil of N turns.

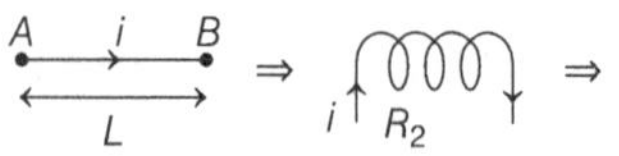

(R_2=radius of coil having N loops)

Thin,
$$L = N(2\pi R_2) \Rightarrow R_2 = \frac{L}{2\pi N}$$

The magnetic field due to this circular coil of N turns,

$$B_C = \left(\frac{\mu_0 i}{2R_2}\right)N = N \cdot \frac{\mu_0 i \cdot (2\pi N)}{2L} \quad \text{...(ii)}$$

Using Eqs. (i) and (ii), the ratio of $\frac{B_L}{B_C}$ is

$$\Rightarrow \quad \frac{B_L}{B_C} = \frac{\frac{\mu_0 i}{2R_1}}{N\frac{\mu_0 i}{2R_2}} = \frac{\frac{\mu_0 i}{2L} \cdot (2\pi)}{\frac{\mu_0 i}{2L} \cdot (2\pi)N^2} = \frac{1}{N^2}$$

Example 4. *A wire A, bent in the shape of an arc of a circle, carrying a current of 2 A and having radius 2 cm and another wire B, also bent in the shape of arc of a circle, carrying a current of 3 A and having radius of 4 cm, are placed as shown in the figure. The ratio of the magnetic fields due to the wires A and B at the common centre O is*

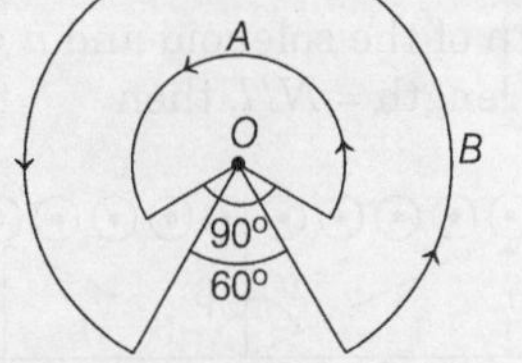

[JEE Main 2020]

(a) 4 : 6 (b) 6 : 4 (c) 6 : 5 (d) 2 : 5

Sol. (c) For wire A,

$$\theta_A = 360° - 90° = 270° = \frac{3\pi}{2} \text{ rad}$$

$$i_A = 2 \text{ A}$$

$$R_A = 2 \text{ cm} = 2 \times 10^{-2} \text{ m}$$

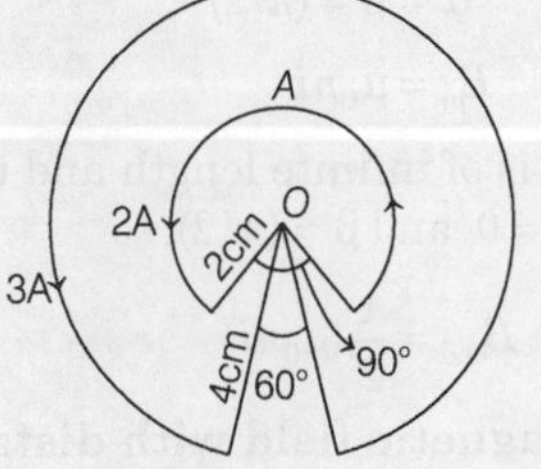

For wire B,

$$\theta_B = 360° - 60° = 300° = \frac{5\pi}{3} \text{ rad}$$

$$i_B = 3 \text{ A}$$

$$R_B = 4 \text{ cm} = 4 \times 10^{-2} \text{ m}$$

So, the ratio of magnetic fields due to wires A and B at the common centre O is

$$\frac{B_A}{B_B} = \frac{\frac{\mu_0 \theta_A i_A}{4\pi R_A}}{\frac{\mu_0 \theta_B i_B}{4\pi R_B}} = \frac{\theta_A i_A R_B}{\theta_B i_B R_A}$$

$$= \frac{\frac{3\pi}{2} \times 2 \times 4 \times 10^{-2}}{\frac{5\pi}{3} \times 3 \times 2 \times 10^{-2}}$$

$$= \frac{12}{10} = \frac{6}{5}$$

So, $B_A : B_B = 6 : 5$

Example 5. *The magnetic field B due to a current carrying circular loop of radius 12 cm at its centre is 0.5×10^{-4} T. Find the magnetic field due to this loop at a point on the axis at a distance of 5.0 cm from the centre.*

(a) 3.9×10^{-5} T (b) 0.5×10^{-4} T
(c) 3.9×10^{4} T (c) 1.0×10^{4} T

Sol. (a) Magnetic field at the centre of a circular loop, $B_1 = \frac{\mu_0 i}{2R}$

and that at an axial point, $B_2 = \frac{\mu_0 i R^2}{2(R^2 + x^2)^{3/2}}$

Thus, $\frac{B_2}{B_1} = \frac{R^3}{(R^2 + x^2)^{3/2}}$ or $B_2 = B_1 \left[\frac{R^3}{(R^2 + x^2)^{3/2}}\right]$

Substituting the values, we have

$$B_2 = (0.5 \times 10^{-4}) \left[\frac{(12)^3}{(144 + 25)^{3/2}}\right]$$

$$= 3.9 \times 10^{-5} \text{ T}$$

Ampere's Circuital Law

It states that, the line integral of **B** around any closed path or circuit is equal to μ_0 times the total current bounded or threaded by that closed path provided the electric field inside the loop remains constant. Thus,

$$\int \mathbf{B} \cdot d\mathbf{l} = \mu_0 (i_{net})$$

Its simplified form is $Bl = \mu_0 i_{net}$

This simplified equation can be used under the following conditions

(i) At every point of the closed path, $\mathbf{B} \parallel d\mathbf{l}$.
(ii) Magnetic field has the same magnitude B at all places on the closed path.
(iii) Ampere circuital law holds for steady currents which do not change with time.

Magnetic Field due to a Cylindrical Wire

- **Outside the cylinder** ($r > R$)

In all the cases, $B_{surface} = \frac{\mu_0 i}{2\pi R}$

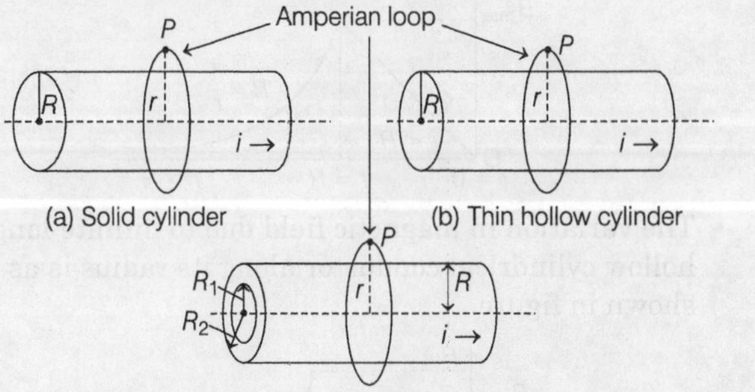

(a) Solid cylinder (b) Thin hollow cylinder

(c) Thick hollow cylinder

- **Inside the hollow cylinder** ($r < R$) Magnetic field inside the hollow cylinder is zero because no current is enclosed by Amperian loops as shown below

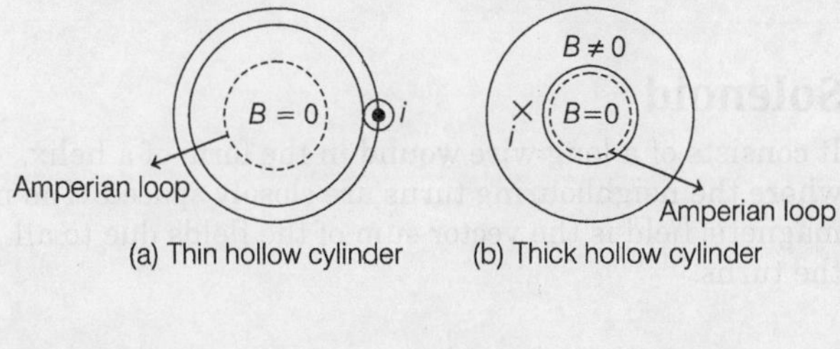

(a) Thin hollow cylinder (b) Thick hollow cylinder

- **Inside the solid cylinder** ($r < R$) Current i' enclosed by Amperian loop is lesser than the total current i.

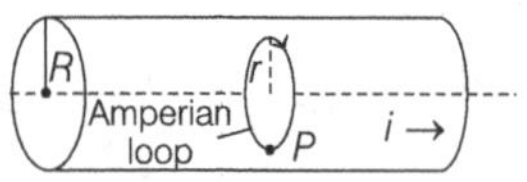

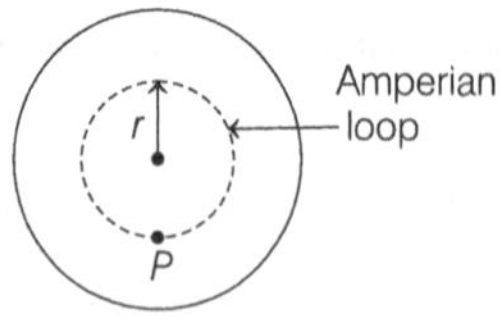

At inside point, $\oint \mathbf{B}_{in} \cdot d\mathbf{l} = \mu_0 i'$

$$\Rightarrow \quad B_{in} = \frac{\mu_0}{2\pi} \cdot \frac{ir}{R^2}$$

- **Inside the thick portion of hollow cylinder** Current enclosed by the Amperian loop of radius r is given by

$$i' = i \times \frac{(r^2 - R_1^2)}{(R_2^2 - R_1^2)}$$

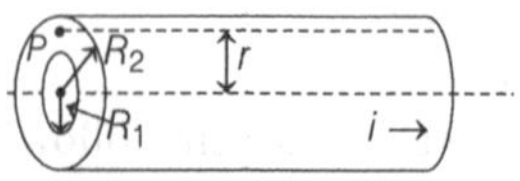

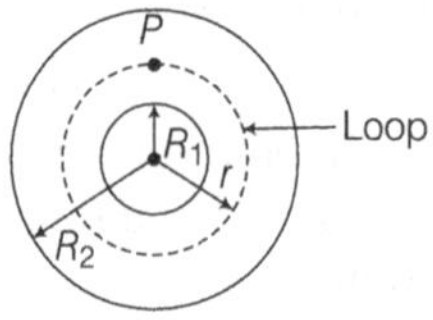

$$B_{in} = \frac{\mu_0 i}{2\pi r} \cdot \frac{(r^2 - R_1^2)}{(R_2^2 - R_1^2)}$$

Variation in Magnetic Field with Radius

- The variation in magnetic field due to infinite long solid cylindrical conductor along its radius is as shown in figure

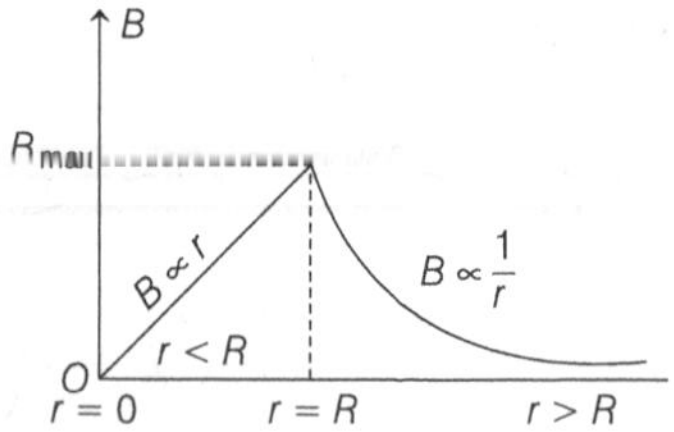

- The variation in magnetic field due to infinite long hollow cylindrical conductor along its radius is as shown in figure

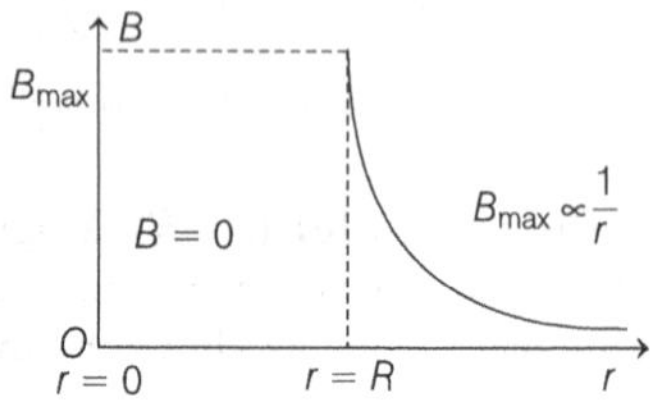

Solenoid

It consists of a long wire wound in the form of a helix, where the neighbouring turns are closely spaced. The net magnetic field is the vector sum of the fields due to all the turns.

Magnetic field due to a straight solenoid can be given as

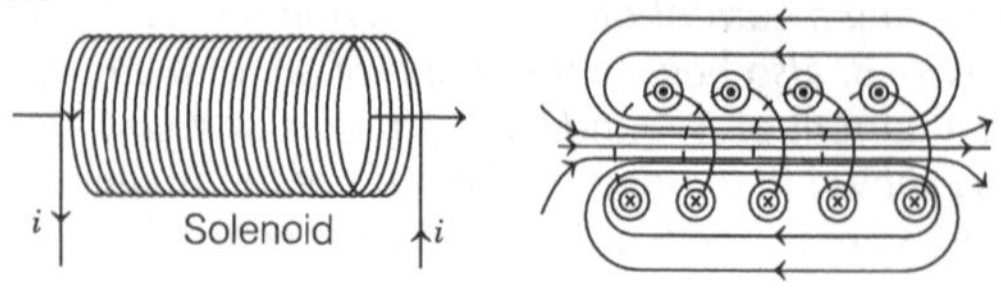

- **Finite Length Solenoid** If N = total number of turns, l = length of the solenoid and n = number of turns per unit length = N/l, then

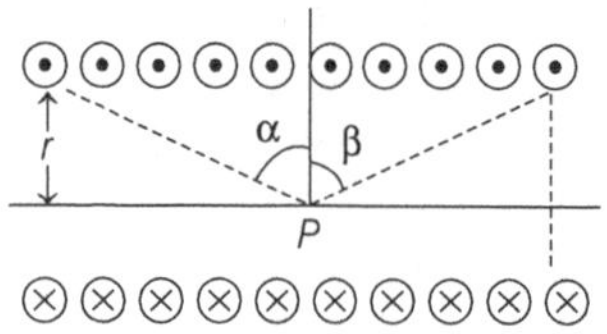

Magnetic field inside the solenoid at point P is given by

$$B = \frac{\mu_0 n i}{2} [\sin \alpha + \sin \beta]$$

- **Infinite Length Solenoid** If the solenoid is of infinite length and the point is well inside it,

i.e. $\qquad \alpha = \beta = (\pi/2)$

So, $\qquad B_{in} = \mu_0 n i$

If the solenoid is of infinite length and the point is near one end, *i.e.* $\alpha = 0$ and $\beta = (\pi/2)$, so

$$B_{end} = \frac{1}{2}(\mu_0 n i) \qquad \left(\because B_{end} = \frac{1}{2} B_{in}\right)$$

Variation in magnetic field with distance *d* from its centre

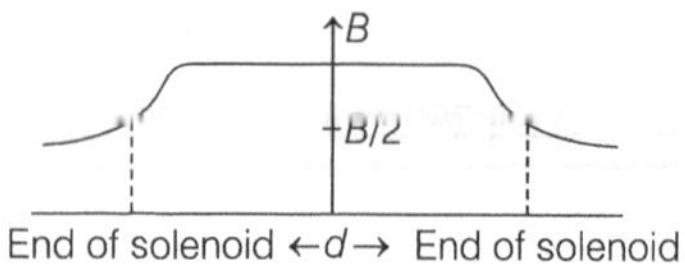

Toroid

It is a hollow circular ring in which a large number of turns of a wire are closely wound.

It can be viewed as a solenoid which has been bent into a circular shape to close on itself as shown below

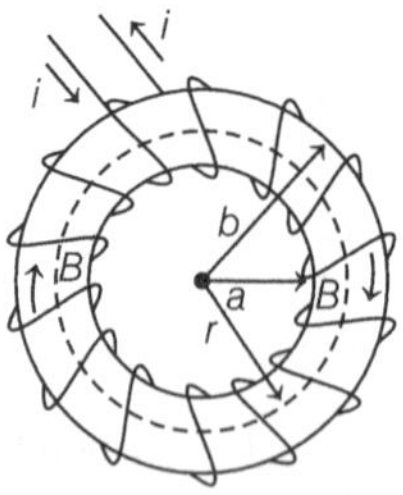

Magnetic field due to toroidal solenoid can be given as

- In the open space, exterior of toroidal solenoid,

$$B = 0$$

- Inside the toroidal solenoid,

$$B = \frac{\mu_0 N i}{2\pi r}$$

where, N = total number of turns = $2\pi rn$.

Then, $B = \mu_0 \, ni$

Note Ampere's law is valid only for steady current. Further, it is useful only for calculating the magnetic fields of current configurations with high degrees of symmetry, just as Gauss's law is useful only for calculating the electric fields of highly symmetric charge distributions.

Example 6. *A long, straight wire of radius a carries a current distributed uniformly over its cross-section. The ratio of the magnetic fields due to the wire at distance $\frac{a}{3}$ and 2a respectively, from the axis of the wire is* **[JEE Main 2020]**

(a) $\frac{3}{2}$ (b) 2 (c) $\frac{2}{3}$ (d) $\frac{1}{2}$

Sol. (c) By Ampere's circuital law, magnetic field inside and outside a wire carrying a uniformly distributed current are

$$B_{\text{out}} = \frac{\mu_0 i}{2\pi r} \quad \text{...(i)}$$

$$B_{\text{in}} = \frac{\mu_0}{2\pi} \cdot \frac{ir}{R^2} \quad \text{...(ii)}$$

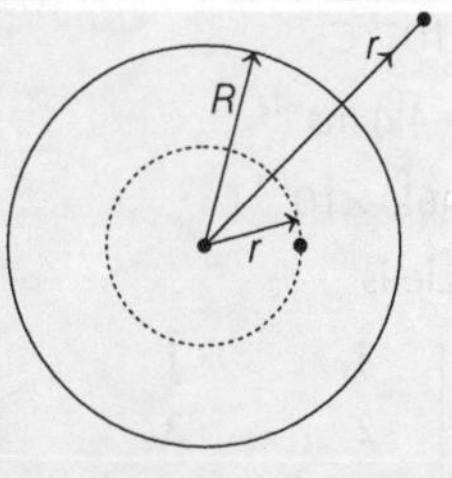

In given case, radius of wire is a. So, magnetic field due to wire at a distance $\frac{a}{3}$ from centre,

$$B_1 = \frac{\mu_0}{2\pi} \frac{i\left(\frac{a}{3}\right)}{a^2} = \frac{\mu_0 i}{6\pi a} \quad \text{[from Eq. (ii)]}$$

and magnetic field at a distance 2a from centre,

$$B_2 = \frac{\mu_0 i}{2\pi(2a)} = \frac{\mu_0 i}{4\pi a} \quad \text{[from Eq. (i)]}$$

Ratio of fields will be

$$\frac{B_1}{B_2} = \frac{\left(\frac{\mu_0 i}{6\pi a}\right)}{\left(\frac{\mu_0 i}{4\pi a}\right)} = \frac{2}{3}$$

Example 7. *A solenoid of length 0.4 m and diameter 0.6 m consists of a single layer of 1000 turns of fine wire carrying a current of 5×10^{-3} A. Calculate the magnetic field on the axis at the middle and at the end of the solenoid.*

(a) $8.7 \times 10^{-6} T, 6.28 \times 10^{-6} T$

(b) $6.28 \times 10^{-6} T, 8.7 \times 10^{-6} T$

(c) $5.7 \times 10^{-6} T, 6.28 \times 10^{-6} T$

(d) $8.7 \times 10^{-6} T, 8.28 \times 10^{-6} T$

Sol. (a) In case of solenoid, the field at a point on the axis as shown in figure is given by $B = \frac{\mu_0}{4\pi} 2\pi ni \, (\sin\alpha + \sin\beta)$

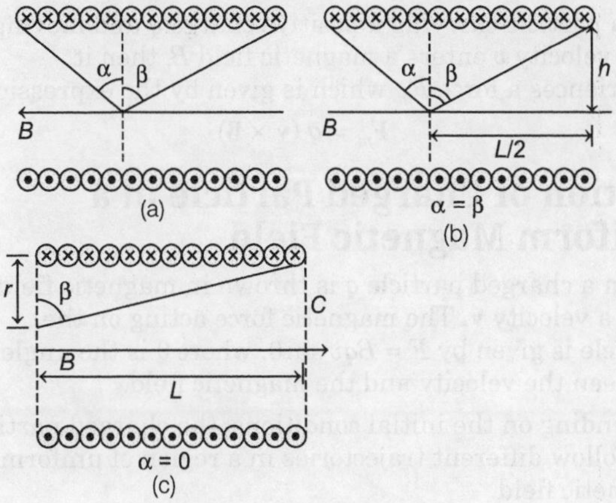

Here, $n = \frac{N}{L} = \frac{1000}{0.4} = 2.5 \times 10^3$ turns m^{-1}

So, $B = 10^{-7} \times 2\pi \times 2.5 \times 10^3 \times 5 \times 10^{-3} (\sin\alpha + \sin\beta)$

i.e. $B = 2.5\pi \times 10^{-6} (\sin\alpha + \sin\beta)$

So, (a) when the point is at the middle on the axis $\alpha = \beta$ with

$$\sin\alpha = \sin\beta = \frac{L}{\sqrt{L^2 + 4r^2}}$$

$$= \frac{0.4}{\sqrt{(0.4)^2 + 4(0.3)^2}} = \frac{4}{7.2}$$

Then, $B = 2.5\pi \times 10^{-6} \times 2 \times \frac{4}{7.2}$

$$= 8.7 \times 10^{-6} \text{T}$$

and (b) when the point is at the end on the axis $\alpha = 0$ with

$$\sin\beta = \frac{L}{\sqrt{L^2 + r^2}} = \frac{0.4}{\sqrt{(0.4)^2 + (0.3)^2}} = \frac{4}{5}$$

Then, $B = 2.5\pi \times 10^{-6} \times \frac{4}{5} = 6.28 \times 10^{-6}$ T

Example 8. *A coil wrapped around a toroid has inner radius of 20.0 cm and an outer radius of 25.0 cm. If the wire wrapping makes 800 turns and carries a current of 12.0 A. Find the ratio of maximum and minimum values of the magnetic field within the toroid.*

(a) 2 (b) 1.25 (c) 3 (d) 4.29

Sol. (b) Let a and b denote the inner and outer radii of the toroid, then maximum value of magnetic field,

$$B_{\max} = \mu_0 ni = \frac{\mu_0 N}{2\pi a} i = \frac{4\pi \times 10^{-7} \times 800 \times 12.0}{2\pi \times 20.0 \times 10^{-2}} \quad \left(\because n = \frac{N}{2\pi r}\right)$$

$$= 9.6 \times 10^{-3} \text{ T} = 9.6 \text{ mT}$$

Minimum value of magnetic field, $B_{\min} = \mu_0 ni = \mu_0 \frac{Ni}{2\pi b}$

$$B_{\min} = \frac{4\pi \times 10^{-7} \times 800 \times 12.0}{2\pi \times 25.0 \times 10^{-2}} = 7.68 \times 10^{-3} \text{T} = 7.68 \text{ mT}$$

$$\Rightarrow B_{\max}/B_{\min} = 9.6/7.68 = 1.25$$

Force on a Moving Charge in a Uniform Magnetic Field

For a particle carrying a positive charge q and moving with velocity v enters a magnetic field B, then it experiences a force F_m which is given by the expression

$$\mathbf{F}_m = q(\mathbf{v} \times \mathbf{B})$$

Motion of Charged Particle in a Uniform Magnetic Field

When a charged particle q is thrown in magnetic field $\mathbf{B}$ with a velocity $\mathbf{v}$. The magnetic force acting on the particle is given by $F = Bqv \sin\theta$, where θ is the angle between the velocity and the magnetic field.

Depending on the initial conditions, the charged particle can follow different trajectories in a region of uniform magnetic field.

***Case* I** When $\theta = 0°$ or $180°$

The magnetic force is $F = 0$.

Hence, path of the charged particle is a straight line (undeviated) when it enters parallel or antiparallel to magnetic field.

***Case* II** When $\theta = 90°$

The magnetic force is $F = Bqv$.

This magnetic force is perpendicular to the velocity at every instant. Hence, path is circle. Then,

- Radius of the circle, $r = \dfrac{mv}{Bq} = \dfrac{\sqrt{2qVm}}{Bq}$

  Here, K = kinetic energy of particle $= \dfrac{p^2}{2m}$ or $p = \sqrt{2Km}$

  where, p = momentum of particle.
- Time period of the circular path, $T = \dfrac{2\pi m}{Bq}$
- Angular speed of the particle,

  $\therefore \qquad \omega = \dfrac{Bq}{m}$
- Frequency of rotation, $f = \dfrac{Bq}{2\pi m}$

***Case* III** When θ is other than $0°$, $180°$ or $90°$,

then velocity of charged particle can be resolved in two components one along $\mathbf{B}$ and another perpendicular to $\mathbf{B}$.

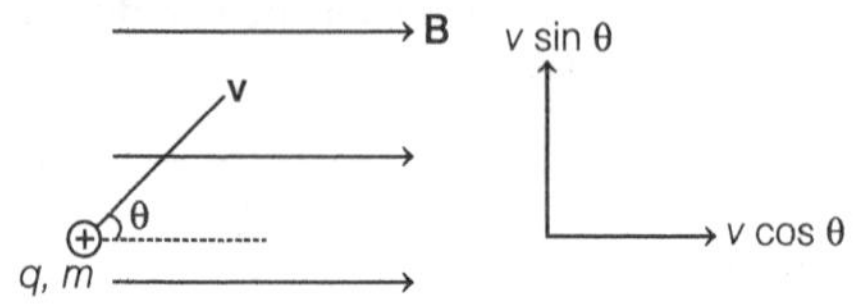

$$v_{\parallel} = v\cos\theta \quad \text{and} \quad v_{\perp} = v\sin\theta$$

Hence, the resultant path will be helical. Then,

- The radius of this helical path,

$$r = \frac{mv_{\perp}}{Bq} = \frac{mv\sin\theta}{Bq}$$

- Time period, $T = \dfrac{2\pi m}{Bq}$
- Frequency, $f = \dfrac{Bq}{2\pi m}$
- **Pitch** is defined as the distance travelled along magnetic field in one complete cycle.

  i.e. $\qquad P = v_{\parallel} T = \dfrac{2\pi m\, v\cos\theta}{Bq}$

Example 9. *A charged particle carrying charge 1 μC is moving with velocity $(2\hat{\mathbf{i}} + 3\hat{\mathbf{j}} + 4\hat{\mathbf{k}})\ ms^{-1}$. If an external magnetic field of $(5\hat{\mathbf{i}} + 3\hat{\mathbf{j}} - 6\hat{\mathbf{k}}) \times 10^{-3}$ T exists in the region, where the particle is moving, then the force on the particle is $\mathbf{F} \times 10^{-9}$ N. The vector* **F** *is* **[JEE Main 2020]**

(a) $-0.30\hat{\mathbf{i}} + 0.32\hat{\mathbf{j}} - 0.09\hat{\mathbf{k}}$
(b) $-30\hat{\mathbf{i}} + 32\hat{\mathbf{j}} - 9\hat{\mathbf{k}}$
(c) $-300\hat{\mathbf{i}} + 320\hat{\mathbf{j}} - 90\hat{\mathbf{k}}$
(d) $-3.0\hat{\mathbf{i}} + 3.2\hat{\mathbf{j}} - 0.9\hat{\mathbf{k}}$

Sol. (b) Force on a charged particle moving in region of a magnetic field is given by

$$\mathbf{F} = q(\mathbf{v} \times \mathbf{B})$$

Here, $q = 1\ \mu C = 1 \times 10^{-6}$ C,

$\mathbf{v} = (2\hat{\mathbf{i}} + 3\hat{\mathbf{j}} + 4\hat{\mathbf{k}})\ ms^{-1}$

and $\mathbf{B} = (5\hat{\mathbf{i}} + 3\hat{\mathbf{j}} - 6\hat{\mathbf{k}}) \times 10^{-3}$ T

So, force on the particle is

$$\mathbf{F} \times 10^{-9} = 1 \times 10^{-6} \begin{vmatrix} \hat{\mathbf{i}} & \hat{\mathbf{j}} & \hat{\mathbf{k}} \\ 2 & 3 & 4 \\ 5 \times 10^{-3} & 3 \times 10^{-3} & -6 \times 10^{-3} \end{vmatrix}$$

$$= 10^{-9} \begin{vmatrix} \hat{\mathbf{i}} & \hat{\mathbf{j}} & \hat{\mathbf{k}} \\ 2 & 3 & 4 \\ 5 & 3 & -6 \end{vmatrix}$$

$$\Rightarrow \qquad \mathbf{F} = -30\hat{\mathbf{i}} + 32\hat{\mathbf{j}} - 9\hat{\mathbf{k}}$$

Example 10. *A proton, an electron and a helium nucleus, have the same energy. They are in circular orbits in a plane due to magnetic field perpendicular to the plane. Let r_p, r_e and r_{He} be their respective radii, then* **[JEE Main 2019]**

(a) $r_e < r_p = r_{He}$
(b) $r_e > r_p = r_{He}$
(c) $r_e < r_p < r_{He}$
(d) $r_e > r_p > r_{He}$

Sol. (c) When a moving charged particle is placed in a magnetic field **B**, then the net magnetic force acting on it is

$$\mathbf{F}_m = q(\mathbf{v} \times \mathbf{B})$$

$$F_m = q\,vB\sin\theta$$

Here, $\theta = 90°$

$$F_m = q\,vB$$

Also, due to this net force, the particle transverses a circular path, whose necessary centripetal force is being provided by F_m, i.e.

$$\frac{mv^2}{r} = qvB$$

$$\Rightarrow \quad r = \frac{mv^2}{qvB} = \frac{mv}{qB} \Rightarrow r \propto m$$

So, **for electron** $\quad r_e = \frac{m_e v}{eB}$ or $r_e \propto m_e$

For proton $\quad r_p = \frac{m_p v}{eB}$ or $r_p \propto m_p$

For He-particle $\quad r_{He} = \frac{4m_p v}{2eB} = \frac{2m_p v}{eB}$

Clearly, $\quad r_{He} > r_p \quad (\because r_{He} = 2r_p)$

and we know that, $m_p > m_e \quad \left(\because m_p \approx 10^{-27} \text{kg}, \; m_e \approx 10^{-31} \text{ kg}\right)$

$\therefore \quad r_p > r_e \Rightarrow r_{He} > r_p > r_e$

Example 11. *A beam of protons with a velocity $4 \times 10^5 ms^{-1}$ enters a uniform magnetic field of 0.3 T at an angle of 60° to the magnetic field. Find the pitch of the helix. (Take, mass of the proton = 1.67×10^{-27} kg)*

(a) 2.35 cm (b) 5.35 cm

(c) 4.35 cm (d) 6.35 cm

Sol. (c) When a charged particle is projected at an angle θ to a magnetic field, the component of velocity parallel to the field is v cos θ while perpendicular to the field is v sin θ, so the particle will move in a circle of radius

$$r = \frac{mv \sin\theta}{qB}$$

$$= \frac{1.67 \times 10^{-27} \times 4 \times 10^5 \times (\sqrt{3}/2)}{1.6 \times 10^{-19} \times 0.3}$$

$$= \frac{2}{\sqrt{3}} \times 10^{-2} \text{m} = 1.2 \text{ cm}$$

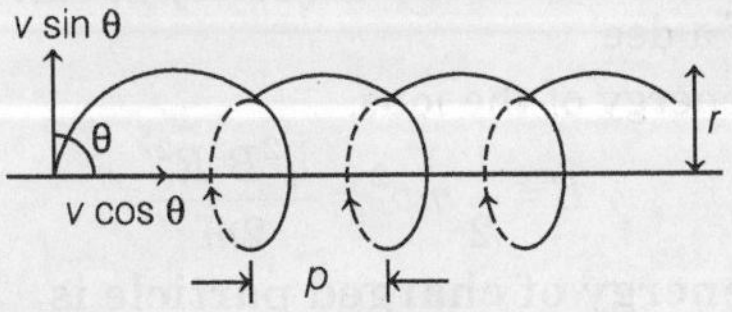

As, $\quad T = \frac{2\pi r}{v \sin\theta} = \frac{2\pi \times 1.2 \times 10^{-2}}{4 \times 10^5 \times (\sqrt{3}/2)} = 2.175 \times 10^{-7}$ s

So, pitch, $\quad P = v \cos\theta \times T = 4 \times 10^5 \times \frac{1}{2} \times 2.175 \times 10^{-7}$

i.e. $\quad P = 4.35 \times 10^{-2}$ m = 4.35 cm

Lorentz Force

Motion of a charged particle in combined electric and magnetic fields.

If a charge q is moving with velocity **v** enters in a region in which electric field **E** and magnetic field **B** both are present, it experiences force due to both fields simultaneously. The force experienced by the charged particle is given by the expression,

$$\mathbf{F} = q(\mathbf{v} \times \mathbf{B}) + q\mathbf{E} = \mathbf{F}_m + \mathbf{F}_e$$

Here, magnetic force $\mathbf{F}_m = q(\mathbf{v} \times \mathbf{B}) = Bqv \sin\theta$ and electric force $\mathbf{F}_e = q\mathbf{E}$.

This force is known as **Lorentz force.**

The direction of magnetic force is same as **v** × **B**, if charge is positive and opposite to **v** × **B**, if charge q is negative.

***Case* I** When **vB** and **B** are all collinear.

In this, magnetic force on the particle will be zero. So,

$$\mathbf{a} = \frac{q\mathbf{E}}{m}$$

The particle will pass through the field following a straight line path with change in its speed.

***Case* II** When **v**, **E** and **B** are mutually perpendicular.

In this, $\quad \mathbf{F} = \mathbf{F}_e + \mathbf{F}_m = 0$

or $\quad \mathbf{a} = 0$

The particle will pass through the field with same velocity without any deviation in it's path.

Thus, $\quad F_e = F_m$

or $\quad v = \frac{E}{B}$

This principle is used in velocity selector to get a charged beam having a specific velocity.

Example 12. *An electron is moving along +x-direction with a velocity of 6×10^6 ms^{-1}. It enters a region of uniform electric field of 300 V/cm pointing along +y-direction. The magnitude and direction of the magnetic field set up in this region such that the electron keeps moving along the x-direction will be* **[JEE Main 2020]**

(a) 3×10^{-4} *T, along + z-direction*

(b) 5×10^{-3} *T, along – z-direction*

(c) 5×10^{-3} *T, along + z-direction*

(d) 3×10^{-4} *T, along – z-direction*

Sol. (c) Given that, $\mathbf{v} = 6 \times 10^6 \,\hat{\mathbf{i}}$ m/s,

$$\mathbf{E} = 300\hat{\mathbf{j}} \text{ V/cm} = 3 \times 10^4 \,\hat{\mathbf{j}} \text{ V/m}$$

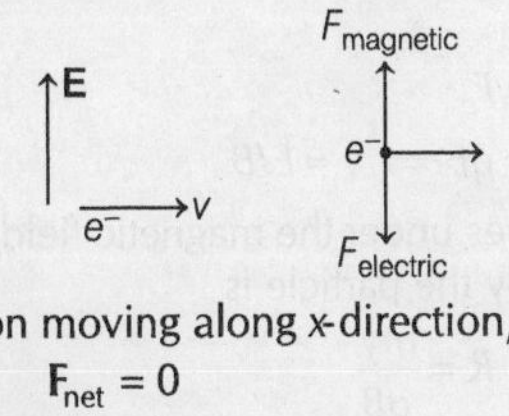

To keep the electron moving along x-direction,

$$\mathbf{F}_{net} = 0$$

or $\quad \mathbf{F}_{electric} + \mathbf{F}_{magnetic} = 0$

$$q\mathbf{E} + q(\mathbf{v} \times \mathbf{B}) = 0$$

or $\quad \mathbf{E} + \mathbf{v} \times \mathbf{B} = 0$

It is possible when magnetic field **B** must be along + z-direction and magnitude is

$$\mathbf{B} = \frac{\mathbf{E}}{\mathbf{v}} = \frac{3 \times 10^4 \hat{\mathbf{j}}}{6 \times 10^6 \hat{\mathbf{i}}} = 5 \times 10^{-3} \hat{\mathbf{k}} \text{T}$$

Example 13. *A particle having the same charge as of electron moves in a circular path of radius 0.5 cm under the influence of a magnetic field of 0.5 T. If an electric field of 100 V/m makes it to move in a straight path, then the mass of the particle is (Take, charge of electron = 1.6×10^{-19} C)* **[JEE Main 2019]**

(a) 1.6×10^{-19} kg
(b) 1.6×10^{-27} kg
(c) 9.1×10^{-31} kg
(d) 2.0×10^{-24} kg

Sol. (d) According to given condition, when a particle having charge same as electron move in a magnetic field on circular path, then the force always acts towards the centre and perpendicular to the velocity.

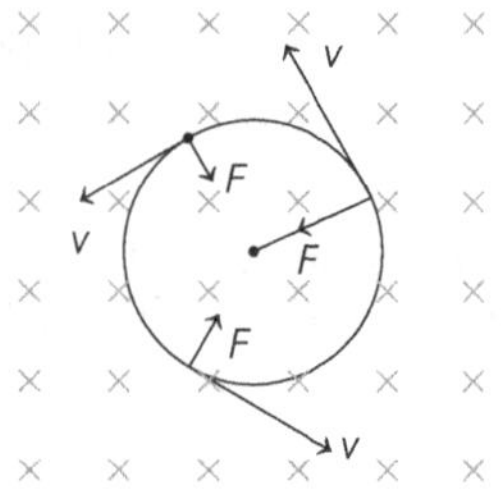

Here, $B = 0.5$ T

R = radius of circular path = 0.5 cm = 0.5×10^{-2} m

Now, the magnetic force is

$$F_m = q(v\times B) = q\,vB\sin 90^\circ$$

$$F_m = qvB \quad \ldots(i)$$

When the electric field applied, then the particle moves in a straight path, then this is the case of velocity selector.

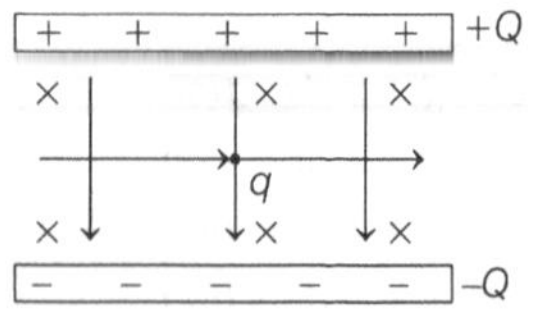

Here, the electric force on charge,

$$F_e = qE \quad \ldots(ii)$$

In velocity selector,

$$F_m = F_e$$

$$\Rightarrow \quad qvB = qE \Rightarrow v = E/B \quad \ldots(iii)$$

Initially particle moves under the magnetic field, so the radius of circular path taken by the particle is

$$R = \frac{mv}{qB} \quad \ldots(iv)$$

From Eqs. (iii) and (iv), we get

$$m = \frac{qB^2R}{E}$$

$$m = \frac{1.6\times 10^{-19}\times 0.25\times 0.5\times 10^{-2}}{10^2}$$

$$m = 2\times 10^{-24}\text{ kg}$$

Cyclotron

It is a device used for accelerating positively charged particle (like α-particles, deutrons, etc.) with the help of uniform magnetic field upto energy of the order of MeV.

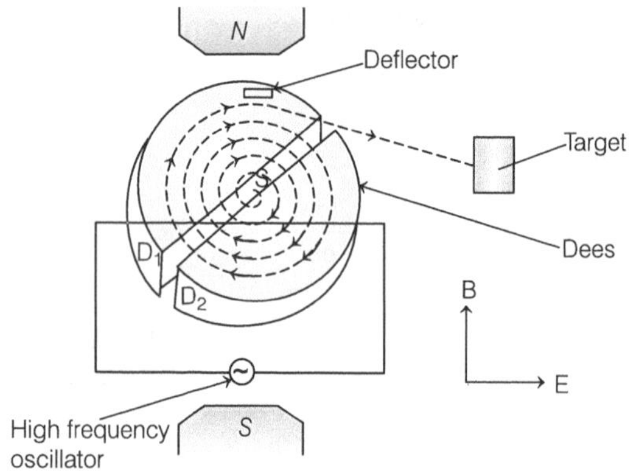

The period of revolution is given by

$$T = \frac{1}{\nu_c} = \frac{2\pi m}{qB}$$

or

$$\nu_c = \frac{qB}{2\pi m}$$

This frequency is called the *cyclotron frequency*.

The frequency ν_a of the applied voltage is adjusted, so that the polarity of the dees is reversed in the same time that it takes the ions to complete one-half of the revolution. The requirement $\nu_a = \nu_c$ is called the *resonance condition*.

The velocity of the charged particle or ions when it leave the system *via* an exit slit,

$$v = \frac{qBR}{m}$$

where, R is the radius of the trajectory at exit and equals the radius of a dee.

The kinetic energy of the ions,

$$E = \frac{1}{2}mv^2 = \frac{q^2B^2R^2}{2m}$$

Maximum energy of charged particle is

$$E_{\max} = \left(\frac{q^2B^2}{2m}\right)r_0^2$$

where, r_0 is maximum radius of the circular path followed by positive ion.

Important Points Related to Cyclotron

- Cyclotron is suitable only for accelerating heavy particles like proton, deutron, α-particle, etc. Electrons cannot be accelerated by the cyclotron because the mass of the electron is small and a small increase in energy of the electron makes the electrons move with a very high speed. As a result of it, the electrons go quickly out of step with oscillating electric field.

- When a positive ion is accelerated by the cyclotron, it moves with greater and greater speed. As the speed of ion becomes comparable with that of light, the mass of the ion increase according to the relation

$$m = \frac{m_0}{\sqrt{\left(1 - \frac{v^2}{c^2}\right)}}$$

where m_0 = the rest mass of the ion,

m = the mass of the ion while moving velocity v

and c = velocity of light.

Now, the time taken by the ion to describe semicircular path is

$$t = \frac{\pi m}{qB} = \frac{\pi}{qB} \cdot \frac{m_0}{\sqrt{\left(1 - \frac{v^2}{c^2}\right)}}$$

From the above relation, we can conclude that, the ion cannot move with a speed beyond a certain limit in a cyclotron.

Force on a Current Carrying Conductor in a Magnetic Field

When a current carrying conductor is placed in a magnetic field, the conductor experience a force in a direction perpendicular to both the direction of magnetic field and the direction of current flowing in the conductor.

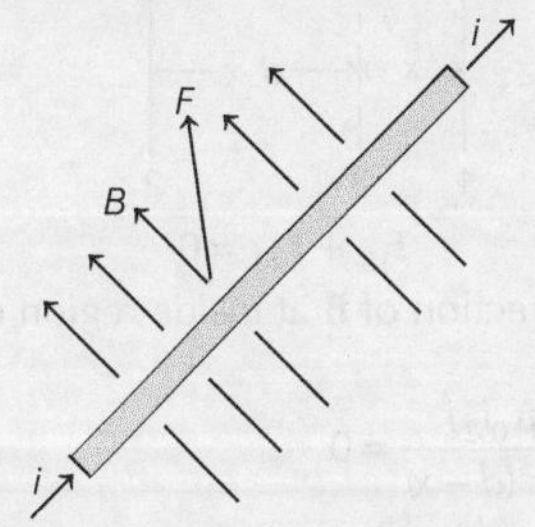

The magnetic force is

$$F = ilB \sin \theta$$

In vector form, $\mathbf{F} = i(\mathbf{l} \times \mathbf{B})$

where, B = intensity of magnetic field,

i = current in the conductor,

l = length of the conductor

and θ = angle between the length of conductor and direction of magnetic field.

There arises two cases

- If $\theta = 90^\circ$ or $\sin\theta = 1$, then $F = ilB$ (maximum).

  Therefore, force will be maximum when the conductor carrying current is perpendicular to magnetic field.
- If $\theta = 0^\circ$ or $\sin\theta = 0$, then $F = ilB \times 0 = 0$.

  Thus, the force will be zero, when the current carrying conductor is parallel to the field.

  The direction of this force can be found out either by Fleming's left hand rule or by right hand palm rule.

Rules to Find the Direction of Force

- **Right hand palm rule** If we stretch the right hand palm such that the fingers and the thumb are mutually perpendicular to each other and the fingers point in the direction of magnetic field and thumb points in the direction of motion of positive charge, then the direction of force will be along the outward normal on the palm.

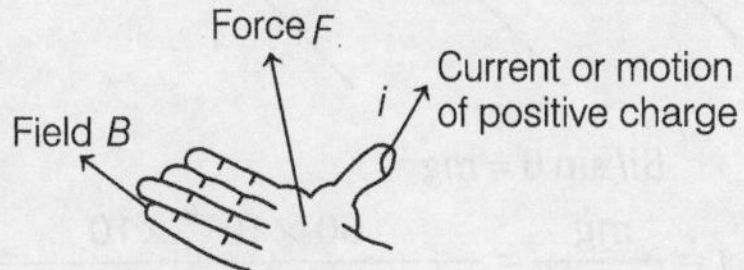

- **Fleming's left hand rule** If we spread the forefinger, central finger and thumb of our left hand in such a way that all three are perpendicular to each other then, if forefinger is in the direction of magnetic field, central finger is in the direction of current, then thumb will represent the direction of force.

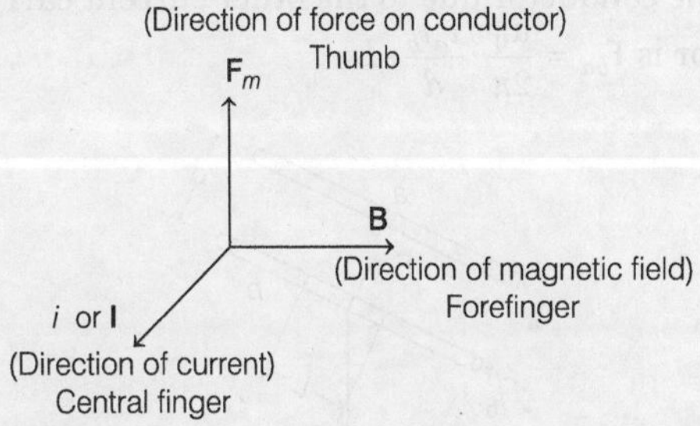

Note To learn this rule, remember the sequence of Father, Mother, Child.

Thumb → Father → Force

Forefinger → Mother → Magnetic field

Central finger → Child → Current of direction of positive charge

Magnetic Force on an Arbitrarily Shaped Wire

If the current carrying conductor in the form or a loop of any arbitrary shape is placed in a uniform field,

$$\mathbf{F} = \oint i d\mathbf{L} \times \mathbf{B} = i[\oint d\mathbf{L} \times \mathbf{B}]$$

and as for a closed loop, the vector sum of $d\mathbf{L}$ is always zero.

So, $$\mathbf{F} = \oint i d\mathbf{L} \times \mathbf{B} = i[\oint d\mathbf{L} \times \mathbf{B}]$$

i.e. the net magnetic force on a current loop in a uniform magnetic field is always zero.

Example 14. *A straight wire of length 30 cm and mass 60 μg lies in a direction 30° east of north. The earth's magnetic field at this site is horizontal and has a magnitude of 0.8 G. What current must be passed through the wire, so that it may float in air?*

(a) 10 A (b) 20 A

(c) 40 A (d) 50 A

Sol. (d) As shown in figure, if a current i is passed through the wire from end A towards B it will experience a force $BiL \sin\theta$ vertically up and hence will float, if

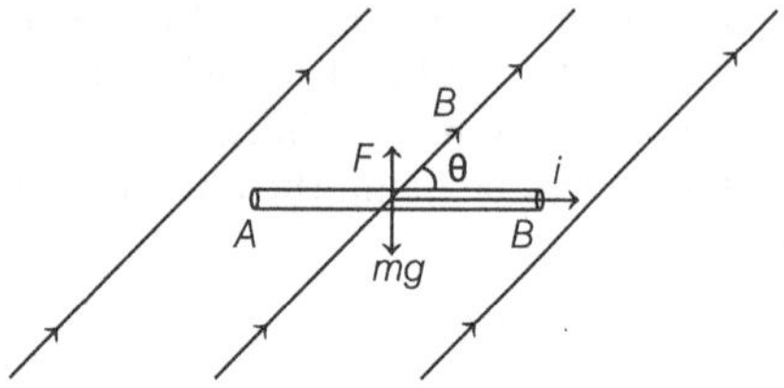

$$Bil \sin\theta = mg$$

i.e. $$i = \frac{mg}{Bl \sin\theta} = \frac{60 \times 10^{-6} \times 10}{0.8 \times 10^{-4} \times 30 \times 10^{-2} \times 1/2} = 50 \text{ A}$$

Force between the Two Parallel Carrying Current Conductors

If two infinitely long parallel conductors carrying the currents i_a and i_b respectively and are separated by a distance d, then magnetic force experienced by length L of any one conductor due to the other current carrying conductor is $\mathbf{F}_{ba} = \frac{\mu_0}{2\pi} \cdot \frac{i_a i_b}{d} \cdot L$.

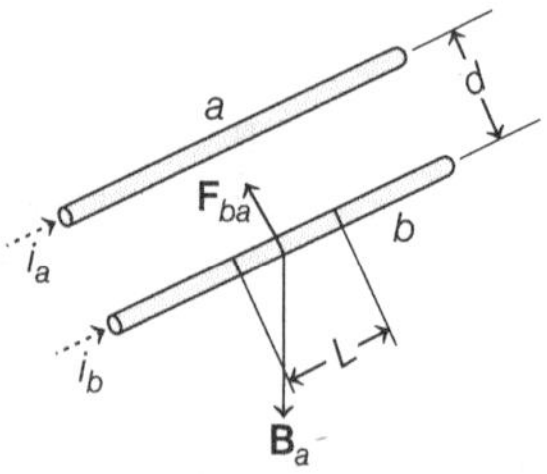

The force $\mathbf{F}_{ba}$ per unit length, $f_{ba} = \frac{\mu_0 i_a i_b}{2\pi d}$

The direction of force depends on the direction of current in them as

(i) attractive, if current flow in same direction.
(ii) repulsive, if current flow in opposite direction.

Definition of Ampere

One ampere is the current which flows through each of the two parallel uniform long linear conductors, which are placed in free space at a distance of 1 m from each other and attract or repel each other with a force of 2×10^{-7} N/m.

Example 15. *The force between two parallel current carrying conductor is F. If the current in each conductor is doubled, then the force between them becomes*

(a) $4F$ (b) $2F$ (c) F (d) $F/4$

Sol. (a) The force between two parallel current carrying conductor is

$$F = \frac{\mu_0}{4\pi} \frac{I_1 I_2}{d} \times l$$

When currents in each conductor are doubled, then force will be

$$F' = \frac{\mu_0}{2\pi} \frac{2I_1 \times 2I_2}{d} \times l = 4 \frac{\mu_0}{2\pi} \frac{I_1 I_2}{d} \times l = 4F$$

Example 16. *Two wires A and B are carrying currents i_1 and i_2 as shown in the figure. The separation between them is d. A third wire C carrying a current i is to be kept parallel to them at a distance x from A such that the net force acting on it is zero. The possible values of x are*

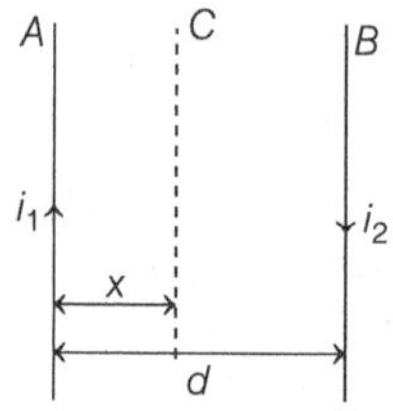

[JEE Main 2019]

(a) $x = \left(\frac{i_2}{i_1 + i_2}\right) d$ and $x = \left(\frac{i_2}{i_1 - i_2}\right) d$

(b) $x = \left(\frac{i_1}{i_1 - i_2}\right) d$ and $x = \left(\frac{i_2}{i_1 + i_2}\right) d$

(c) $x = \left(\frac{i_1}{i_1 + i_2}\right) d$ and $x = \left(\frac{i_2}{i_1 - i_2}\right) d$

(d) $x = \pm \frac{i_1 d}{(i_1 - i_2)}$

Sol. (d) Net force on the third wire, carrying current i in the following first case is

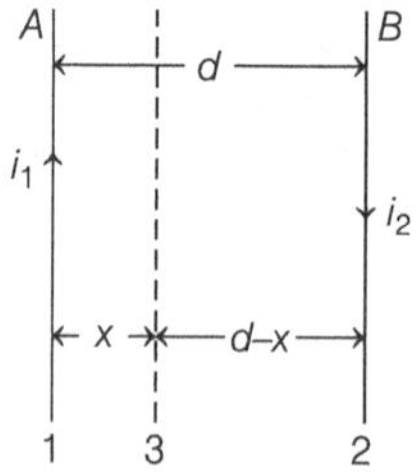

$$F_{13} + F_{23} = 0$$

Using thumb rule, direction of **B** at inside region of wires A and B will be same.

$\therefore$ $$\frac{\mu_0 i_1 i}{2\pi x} + \frac{\mu_0 i_2 i}{2\pi (d - x)} = 0$$

$\Rightarrow$ $$\frac{i_1}{x} + \frac{i_2}{d - x} = 0 \Rightarrow \frac{i_1}{x} = \frac{i_2}{x - d}$$

or $$(x - d)\, i_1 = x\, i_2 \Rightarrow x(i_1 - i_2) = d i_1$$

$\Rightarrow$ $$x = \frac{i_1}{(i_1 - i_2)} \cdot d \quad \text{... (i)}$$

Second case of balanced force can be as shown

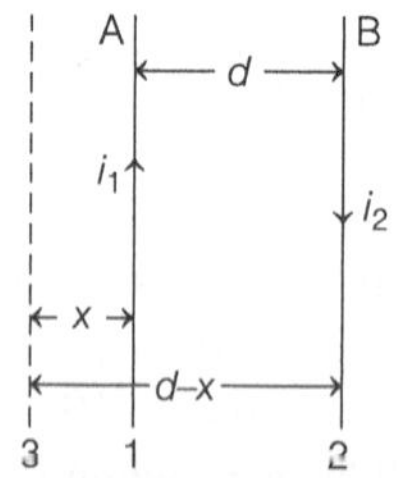

Using thumb rule, directions of **B** at any point on wires A and B will be opposite, so net force,

$$\frac{\mu_0 i_1 i}{2\pi x} - \frac{\mu_0 i_2 i}{2\pi (d + x)} = 0$$

or $\frac{i_1}{x} - \frac{i_2}{(d+x)} = 0 \Rightarrow \frac{i_1}{x} = \frac{i_2}{d+x}$

$\Rightarrow \quad (d+x)\, i_1 = x\, i_2$

$\Rightarrow \quad (i_2 - i_1)\, x = d i_1$

$\Rightarrow \quad x = \frac{i_1}{(i_1 - i_2)} \cdot d \qquad \text{... (ii)}$

From Eqs. (i) and (ii), it is clear that

$$x = \pm \frac{i_1}{(i_1 - i_2)} d$$

Magnetic Force between Two Moving Charges

Consider two charges q_1 and q_2 are moving with velocities v_1 and v_2 respectively and at any instant, the distance between them is r.

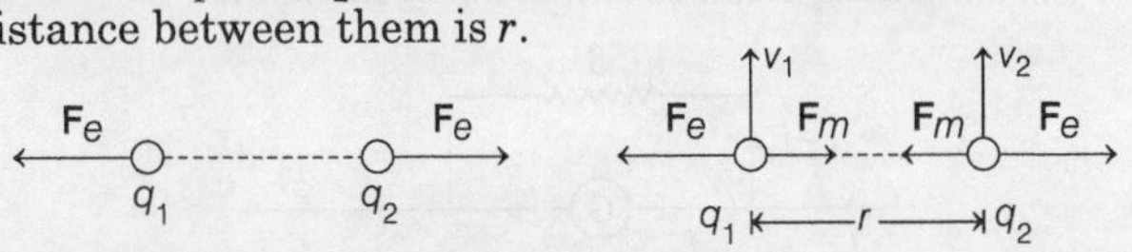

A magnetic force F_m will appear between them alongwith the electric force.

i.e. $$F_m = \frac{\mu_0}{4\pi} \frac{q_1 q_2\, v_1 v_2}{r^2}$$

Torque Acting on a Current Carrying Coil

Torque acting on a current carrying coil placed inside a uniform magnetic field is given by

$$\tau = NBiA \sin\theta$$

where, N = number of turns in the coil,
B = magnetic field intensity,
i = current in the coil,
A = area of cross-section of the coil
and θ = angle between magnetic field and normal to the plane of the coil.

Moving Coil Galvanometer (MCG)

It is a device whose principle is based on the torque on a current carrying loop. When a current flows through the coil of a galvanometer, a torque acts on it. This torque is given by $\tau = NiAB$.

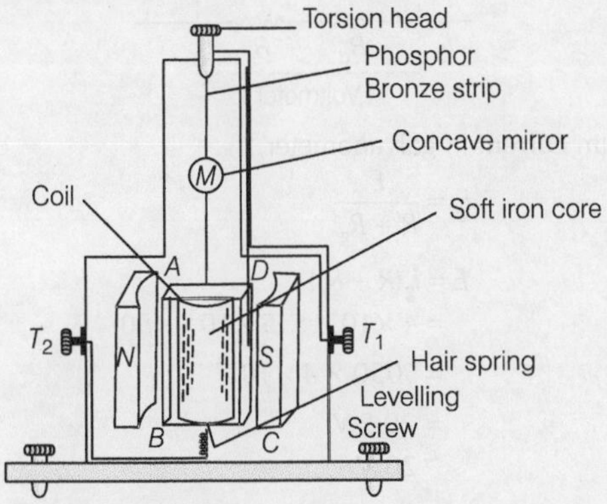

In equilibrium, deflecting torque = restoring force

$$k\phi = NiAB$$

where, k is the torsional constant of the spring.

Due to this torque, a deflection ϕ is indicated on the scale by the pointer attached to the spring as

$$\phi = \left(\frac{NAB}{k}\right) i$$

where, N = number of turns in the coil,
B = magnetic field intensity,
A = area of cross-section of the coil
and θ = angle of twist.

Current Sensitivity The deflection produced per unit current in galvanometer is called its current sensitivity.

Current sensitivity, $i_S = \frac{\theta}{i} = \frac{NBA}{k}$

Voltage Sensitivity The deflection produced per unit voltage applied across the ends of the galvanometer is called its voltage sensitivity.

Voltage sensitivity, $V_s = \frac{\theta}{V} = \frac{NBA}{kr}$

where, R is the resistance of the galvanometer.

Therefore for a sensitive galvanometer

- N should be large,
- B should be large,
- A should be large
- and k should be small.

Example 17. *A moving coil galvanometer has a coil with 175 turns and area 1 cm^2. It uses a torsion band of torsion constant 10^{-6} N-m/rad. The coil is placed in a magnetic field B parallel to its plane. The coil deflects by 1° for a current of 1 mA. The value of B (in tesla) is approximately* **[JEE Main 2019]**

(a) 10^{-3} (b) 10^{-4}
(c) 10^{-1} (d) 10^{-2}

Sol. (a) In a moving coil galvanometer in equilibrium, torque on coil due to current is balanced by torque of torsion band.

As, torque on coil,

$$C = NIAB \sin\alpha$$

where, B = magnetic field strength,
i = current and N = number of turns of coil.

Since, plane of the coil is parallel to the field.

$\therefore \quad \alpha = 90° \Rightarrow \tau = NiBA$

Torque of torsion band, $T = k\theta$

where, k = torsion constant of torsion band
and θ = deflection of coil in radians or angle of twist of restoring torque.

$\therefore \quad BiNA = k\theta$ or $B = \frac{k\theta}{iNA} \qquad \text{... (i)}$

Here, $k = 10^{-6}$ N - m / rad,
$i = 1 \times 10^{-3}$ A,
$N = 175$,
$A = 1\,\text{cm}^2 = 1 \times 10^{-4}\,\text{m}^2$

$$\theta = 1° = \frac{\pi}{180}\,\text{rad}$$

Substituting values in Eq. (i), we get

$$B = \frac{10^{-6} \times 22}{1 \times 10^{-3} \times 175 \times 7 \times 180 \times 10^{-4}}$$
$$= 0.998 \times 10^{-3}$$
$$\simeq 10^{-3}\ \text{T}$$

Conversion of Galvanometer to Ammeter

A current measuring instrument is called an ammeter. A galvanometer can be converted into an ammeter by connecting a small resistance S (called shunt) in parallel with it.

If G is the resistance of a galvanometer and it give full scale deflection for current i_g, then required low resistance S, connected in its parallel for converting it into an ammeter of range i is given by

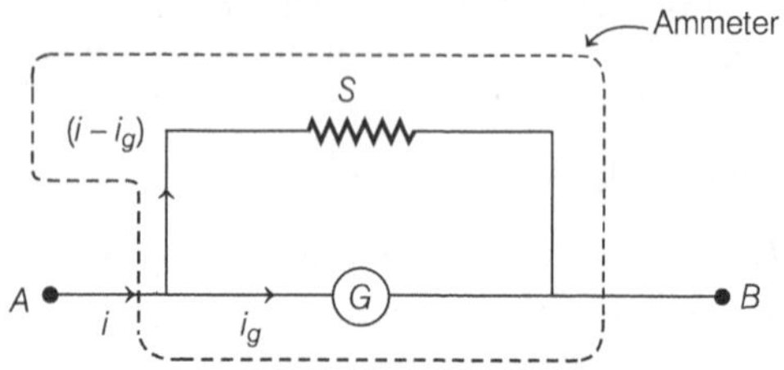

$$i_g \times G = (i - i_g) \times S$$

$$\Rightarrow \quad S = \left(\frac{i_g}{i - i_g}\right) G$$

Note (i) Percentage error in measuring a current through an ammeter is

$$= \left(\frac{R_A}{R + R_A}\right) \times 100$$

(ii) Resistance of ammeter, $R_A = \dfrac{GS}{G + S}$

Conversion of Galvanometer to Voltmeter

A voltage measuring device is called a *voltmeter*. It measures the potential difference between two points. A galvanometer can be converted into voltmeter by connecting a high resistance R in series with it.

If a galvanometer of resistance G shows full scale deflection for current i_g, then required high resistance R, connected in series for converting it into a voltmeter of range V is given by

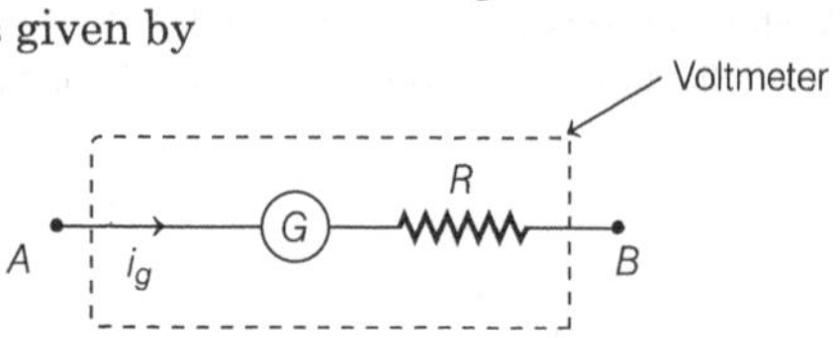

$$V = i_g\,(G + R)$$

$$\Rightarrow \quad R = \frac{V}{i_g} - G$$

Note (i) Percentage error in measuring the potential difference by a

$$\text{voltmeter} = \left(\frac{1}{1 + \dfrac{r}{R_V}}\right) \times 100$$

(ii) Resistance of voltmeter, $R_V = R + G$

Example 18. *The resistance of a galvanometer is 50 Ω and the maximum current which can be passed through it is 0.002 A. What resistance must be connected to it in order to convert it into an ammeter of range 0-0.5 A?* **[JEE Main 2019]**

(a) 0.2 Ω (b) 0.5 Ω
(c) 0.002 Ω (d) 0.02 Ω

Sol. (a) Ammeter circuit is shown in the figure below

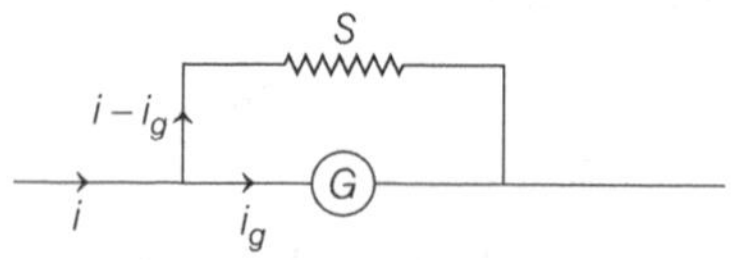

So, $i_g G = (i - i_g)\, S$

Here, $i_g = 0.002$ A,

$i = 0.5$ A,

$G = 50\ \Omega$

So, shunt resistance required,

$$S = \frac{i_g G}{i - i_g} = \frac{0.002 \times 50}{(0.5 - 0.002)} \approx 0.2\ \Omega$$

Example 19. *A moving coil galvanometer has resistance 50 Ω and it indicates full deflection at 4 mA current. A voltmeter is made using this galvanometer and a 5 kΩ resistance. The maximum voltage, that can be measured using this voltmeter, will be close to* **[JEE Main 2019]**

(a) 40 V (b) 10 V (c) 15 V (d) 20 V

Sol. (d) Given, resistance of galvanometer, $R_g = 50\ \Omega$

Current, $i_g = 4\ \text{mA} = 4 \times 10^{-3}$ A

Resistance used in converting a galvanometer in voltmeter,

$$R = 5\ \text{k}\Omega = 5 \times 10^3\ \Omega$$

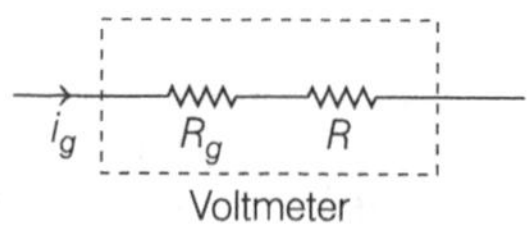

∴ Maximum current in galvanometer,

$$i_g = \frac{E}{R + R_g}$$

$$\therefore \quad E = i_g (R + R_g)$$
$$= 4 \times 10^{-3} \times (5 \times 10^3 + 50)$$
$$= 5050 \times 4 \times 10^{-3}$$
$$= 20.2\ \text{V}$$
$$\simeq 20\ \text{V}$$

Practice Exercise

ROUND I Topically Divided Problems

Magnetic Field, Biot-Savart's Law and its Applications

1. Which of the following graph represents the variation of magnetic flux density B with distance r for a straight long wire carrying an electric current?

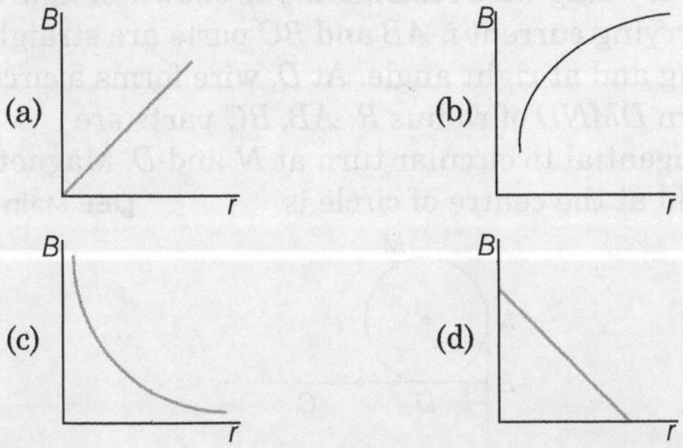

2. Biot-Savart law indicates that the moving electrons (velocity $\mathbf{v}$) produce a magnetic field $\mathbf{B}$ such that **[NCERT Exemplar]**

(a) $\mathbf{B} \perp \mathbf{v}$

(b) $\mathbf{B} \parallel \mathbf{v}$

(c) it obeys inverse cube law

(d) it is along the line joining the electron and point of observation

3. Find the magnitude and direction of magnetic field at point P due to the current carrying wire as shown in figure.

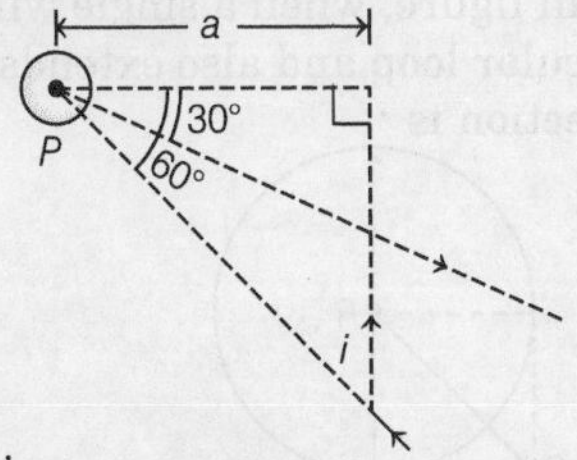

(a) $\mathbf{B} = \dfrac{\mu_0 i}{4\pi a}[-1/2 - \sqrt{3}/2]$

(b) $\mathbf{B} = \dfrac{\mu_0 i}{4\pi a}[1/2 + \sqrt{3}/2]$

(c) $\mathbf{B} = \dfrac{\mu_0 i}{4\pi a}[-1/2 + \sqrt{3}/2]$

(d) $\mathbf{B} = \dfrac{\mu_0 i}{4\pi a}[-1/2 + \sqrt{5}/2]$

4. A horizontal overhead power line carries a current of 90 A in east to west direction. What are the magnitude and direction of the magnetic field due to the current 1.5 m below the line? **[NCERT]**

(a) 1.2×10^{-5} T, perpendicularly outward to the plane of paper

(b) 1.9×10^{-5} T, perpendicularly outward to the plane of paper

(c) 2.6×10^{-5} T, perpendicularly inward to the plane of paper

(d) 2.6×10^{-5} T, perpendicularly inward to the plane of paper

5. Two wires PQ and QR, carry equal currents i as shown in figure. One end of both the wires extends to infinity $\angle PQR = \theta$. The magnitude of the magnetic field at O on the bisector angle of these two wires at a distance r from point Q, is

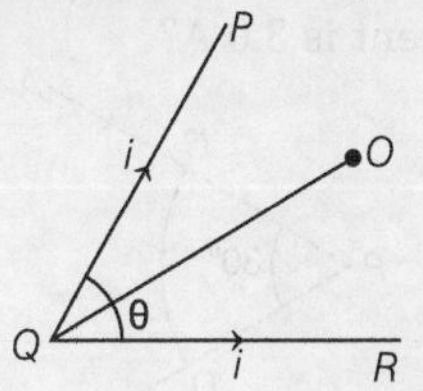

(a) $\dfrac{\mu_0}{4\pi}\dfrac{i}{r}\sin\dfrac{\theta}{2}$ (b) $\dfrac{\mu_0}{4\pi}\dfrac{i}{r}\cot\theta$

(c) $\dfrac{\mu_0}{4\pi}\dfrac{i}{r}\tan\dfrac{\theta}{2}$ (d) $\dfrac{\mu_0}{2\pi}\dfrac{i}{r}\dfrac{(1+\cos\theta/2)}{(\sin\theta/2)}$

6. A pair of stationary and infinite long bent wires are placed in the xy-plane. The wires carrying currents of 10 A each are shown in figure. The segments L and M are parallel to X-axis. The segments P and Q are parallel to Y-axis, such that $OS = OR = 0.02$ m. The magnetic field induction at the origin O is

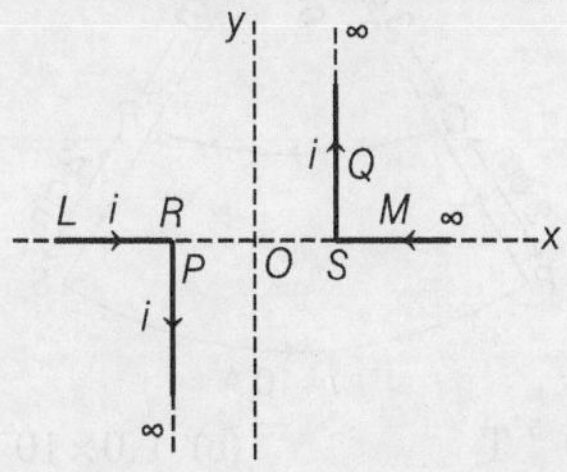

(a) 10^{-3} T (b) 4×10^{-3} T (c) 2×10^{-4} T (d) 10^{-4} T

7. A length l of wire carries a steady current i. It is bent first to form a circular plane coil of one turn. The same length is now bent more sharply to give three loops of smaller radius. The magnetic field at the centre caused by the same current is
(a) one-third of its initial value
(b) unaltered
(c) three times of its initial value
(d) nine times of its initial value

8. Equal current i flows in two segments of a circular loop in the direction shown in figure. Radius of the loop is r. The magnitude of magnetic field induction at the centre of the loop is

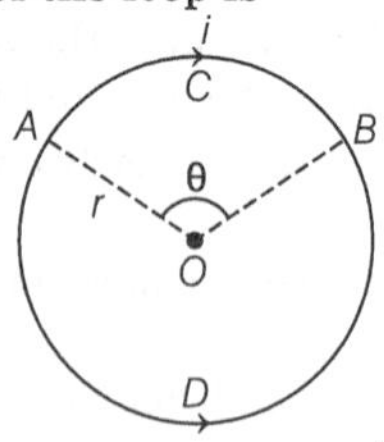

(a) zero (b) $\frac{\mu_0 i\theta}{3\pi r}$
(c) $\frac{\mu_0}{2\pi}\frac{i}{r}(\pi-\theta)$ (d) $\frac{\mu_0}{2\pi}\frac{i}{r}(2\pi-\theta)$

9. A current path shaped as shown in figure produces a magnetic field at P, the centre of the arc. If the arc subtends an angle of 30° and the radius of the arc is 0.6 m. What is the magnitude of the field at P, if the current is 3.0 A?

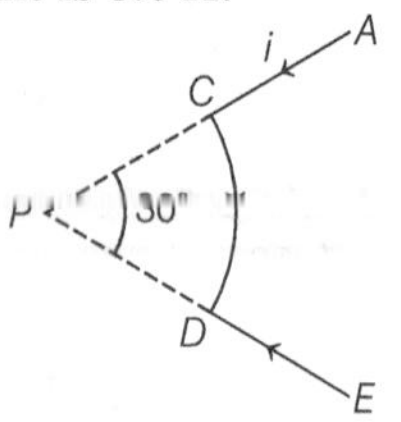

(a) 2.62×10^{-6} T (b) 2.62×10^{-7} T
(c) 3.62×10^{-7} T (d) 2.62×10^{-8} T

10. A current loop, having two circular arcs joined by two radial lines as shown in the figure. It carries a current of 10 A. The magnetic field at point O will be close to **[JEE Main 2019]**

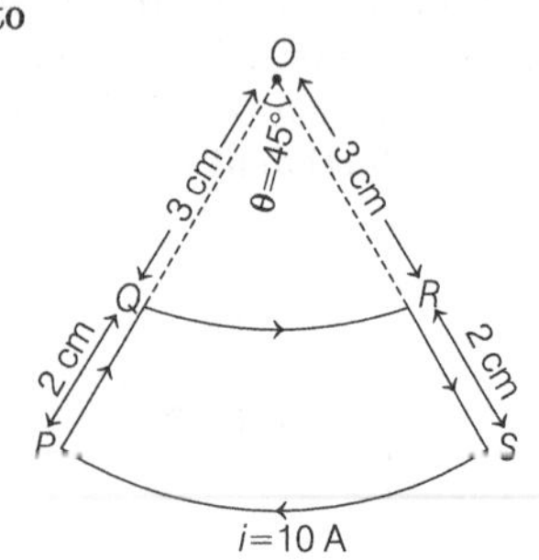

(a) 1.0×10^{-5} T (b) 1.0×10^{-7} T
(c) 1.5×10^{-7} T (d) 1.5×10^{-5} T

11. Magnitude of magnetic field (in SI unit) at the centre of a hexagonal shaped coil of side 10 cm, 50 turns and carrying current i ampere in units of $\frac{\mu_0 i}{\pi}$ is **[JEE Main 2020]**
(a) $250\sqrt{3}$ (b) $50\sqrt{3}$
(c) $500\sqrt{3}$ (d) $5\sqrt{3}$

12. The magnetic field normal to the plane of a wire of n turns and radius r which carries a current i is measured on the axis of the coil at a small distance h from the centre of the coil. This is smaller than the magnetic field at the centre by the fraction
(a) $(2/3)r^2/h^2$
(b) $(3/2)r^2/h^2$
(c) $(2/3)h^2/r^2$
(d) $(3/2)h^2/r^2$

13. A very long wire $ABDMNDC$ is shown in figure carrying current i. AB and BC parts are straight, long and at right angle. At D, wire forms a circular turn $DMND$ of radius R. AB, BC parts are tangential to circular turn at N and D. Magnetic field at the centre of circle is **[JEE Main 2020]**

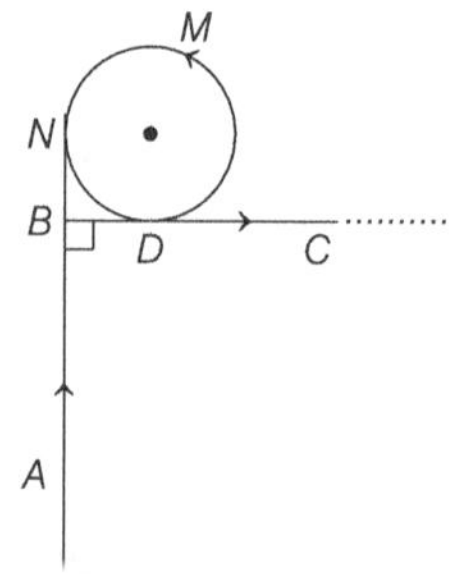

(a) $\frac{\mu_0 i}{2\pi R}\left(\pi+\frac{1}{\sqrt{2}}\right)$ (b) $\frac{\mu_0 i}{2R}$
(c) $\frac{\mu_0 i}{2\pi R}(\pi+1)$ (d) $\frac{\mu_0 i}{2\pi R}\left(\pi-\frac{1}{\sqrt{2}}\right)$

14. The magnetic field at the centre of the circular loop as shown in figure, when a single wire is bent to form a circular loop and also extends to form straight section is

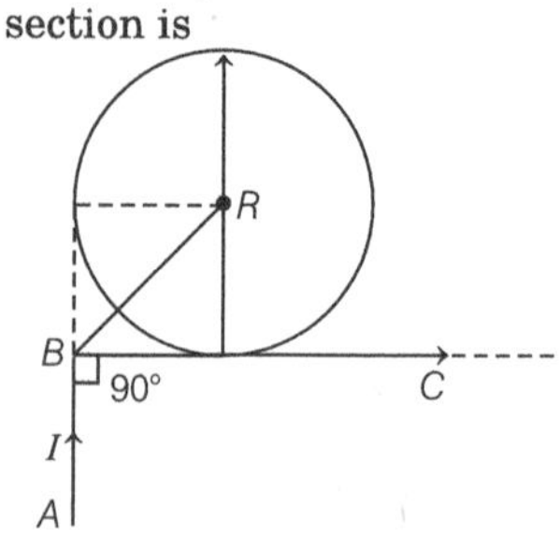

(a) $\frac{\mu_0 i}{2R}$ (b) $\frac{\mu_0 i}{2R}\left(1+\frac{1}{\pi\sqrt{2}}\right)$
(c) $\frac{\mu_0 i}{2R}\left(1-\frac{1}{\pi\sqrt{2}}\right)$ (d) $\frac{\mu_0 i}{R}\left(1-\frac{1}{\pi\sqrt{2}}\right)$

15. Two concentric coils, each of radius equal to 2π cm, are placed at right angle to each other. Currents of 3 A and 4 A respectively, are flowing through the two coils. The magnetic induction (in Wbm^{-2} or T) at the centre of the coils will be (Take, $\mu_0 = 4\pi \times 10^{-7}$ Wb $(Am)^{-1}$)

(a) 5×10^{-5} (b) 7×10^{-5}
(c) 12×10^{-5} (d) 10^{-5}

16. A hair pin like shape as shown in figure is made by bending a long current carrying wire. What is the magnitude of a magnetic field at point P which lies on the centre of the semi-circle? **[JEE Main 2021]**

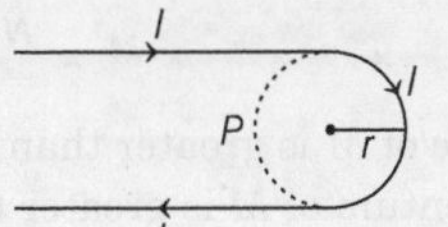

(a) $\frac{\mu_0 I}{4\pi r}(2-\pi)$ (b) $\frac{\mu_0 I}{4\pi r}(2+\pi)$
(c) $\frac{\mu_0 I}{2\pi r}(2+\pi)$ (d) $\frac{\mu_0 I}{2\pi r}(2-\pi)$

17. The magnetic flux density B at a distance r from a long straight rod carrying a steady current varies with r as shown in figure.

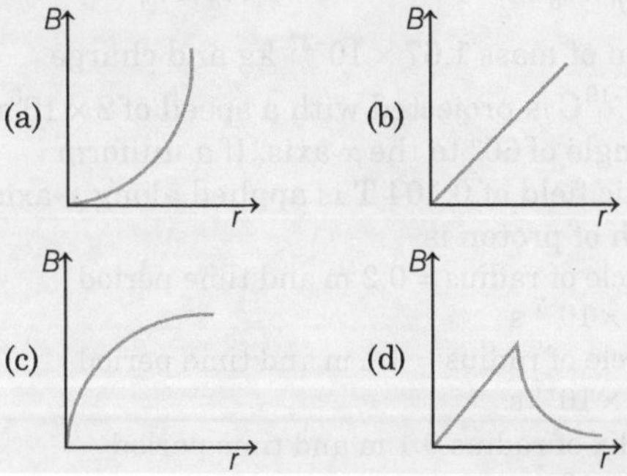

Ampere's Circuital Law and Solenoid

18. A current of i ampere flows along an infinitely long straight thin walled tube, then the magnetic induction at any point inside the tube is

(a) infinite (b) zero
(c) $\frac{\mu_0 2i}{4\pi r}$ T (d) $\frac{\mu_0 i}{2r}$ T

19. A solenoid of 1000 turns per metre has a core with relative permeability 500. Insulated windings of the solenoid carry an electric current of 5A. The magnetic flux density produced by the solenoid is (Take, permeability of free space $= 4\pi \times 10^{-7}$ H/m) **[JEE Main 2021]**

(a) πT (b) $2 \times 10^{-3}\pi$T
(c) $\frac{\pi}{5}$ T (d) $10^{-4}\pi$T

20. A closely wound solenoid 80 cm long has 5 layers of windings of 400 turns each. The diameter of the solenoid is 1.8 cm. If the current carried is 8.0 A, estimate the magnitude of B inside the solenoid near its centre. **[NCERT]**

(a) 1.5×10^{-2}T, opposite to the axis of solenoid
(b) 2.5×10^{-2}T, along the axis of solenoid
(c) 3.5×10^{-2}T, along the axis of solenoid
(d) 1.5×10^{-2}T, opposite to the axis of solenoid

Force on Charged Particle in Electric and Magnetic Fields

21. An electron is projected with uniform velocity along the axis of a current carrying long solenoid. Which of the following is true? **[NCERT Exemplar]**

(a) The electron will be accelerated along the axis
(b) The electron path will be circular about the axis
(c) The electron will experience a force at 45° to the axis and hence execute a helical path
(d) The electron will continue to move with uniform velocity along the axis of the solenoid

22. An electron gun is placed inside a long solenoid of radius R on its axis. The solenoid has n turns/length and carries a current I. The electron gun shoots an electron along the radius of the solenoid with speed v. If the electron does not hit the surface of the solenoid, maximum possible value of v is (all symbols have their standard meaning) **[JEE Main 2020]**

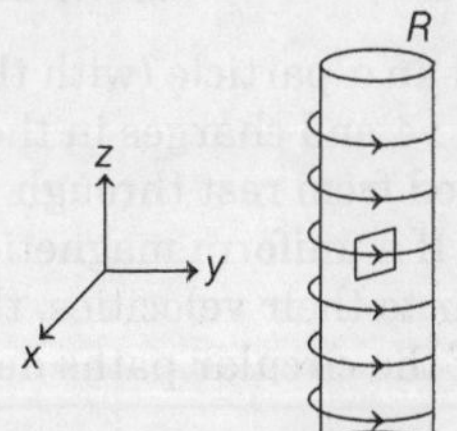

(a) $\frac{2e\mu_0 nIR}{m}$
(b) $\frac{e\mu_0 nIR}{4m}$
(c) $\frac{e\mu_0 nIR}{2m}$
(d) $\frac{e\mu_0 nIR}{m}$

23. A particle of charge q and mass m is moving with a velocity $-v\hat{i}(v \neq 0)$ towards a large screen placed in the YZ-plane at a distance d. If there is a magnetic field $\mathbf{B} = B_0\hat{k}$, the minimum value of v for which the particle will not hit the screen is **[JEE Main 2020]**

(a) $\frac{qdB_0}{3m}$ (b) $\frac{2qdB_0}{m}$
(c) $\frac{qdB_0}{m}$ (d) $\frac{qdB_0}{2m}$

24. An electron moving along the X-axis with an initial energy of 100 eV, enters a region of magnetic field $\mathbf{B} = (1.5 \times 10^{-3}\text{ T})\,\hat{\mathbf{k}}$ at S (see figure). The field extends between $x = 0$ and $x = 2$ cm. The electron is detected at the point Q on a screen placed 8 cm away from the point S. The distance d between P and Q (on the screen) is

(Take, electron's charge $= 1.6 \times 10^{-19}$ C, mass of electron $= 9.1 \times 10^{-31}$ kg) [JEE Main 2019]

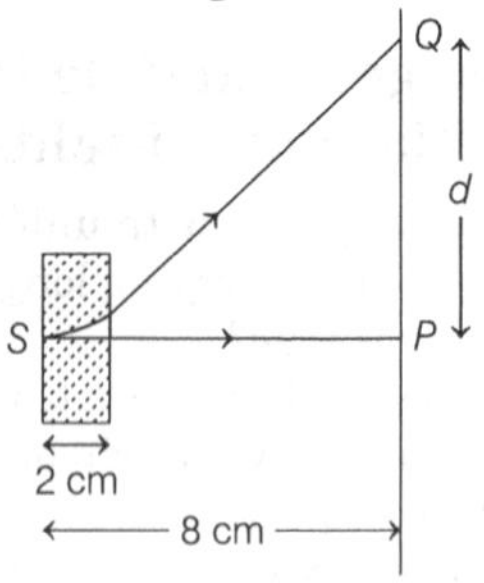

(a) 11.65 cm (b) 12.87 cm
(c) 1.22 cm (d) 2.25 cm

25. In an experiment, electrons are accelerated, from rest by applying a voltage of 500 V. Calculate the radius of the path, if a magnetic field 100 mT is then applied. (Take, charge of the electron $= 1.6 \times 10^{-19}$ C and mass of the electron $= 9.1 \times 10^{-31}$ kg) [JEE Main 2019]

(a) 7.5×10^{-2} m (b) 7.5×10^{-4} m
(c) 7.5×10^{-3} m (d) 7.5 m

26. A proton and an α-particle (with their masses in the ratio of 1 : 4 and charges in the ratio of 1 : 2) are accelerated from rest through a potential difference V. If a uniform magnetic field B is set up perpendicular to their velocities, the ratio of the radii $r_p : r_\alpha$ of the circular paths described by them will be [JEE Main 2019]

(a) $1 : \sqrt{2}$ (b) $1 : \sqrt{3}$
(c) $1 : 3$ (d) $1 : 2$

27. A deuteron of kinetic energy 50 keV is describing a circular orbit of radius 0.5 m, is plane perpendicular to magnetic field B. The kinetic energy of proton that describes a circular orbit of radius 0.5 m in the same plane with the same magnetic field B, is

(a) 200 keV (b) 50 keV
(c) 100 keV (d) 25 keV

28. An electron and a proton enter a magnetic field perpendicularly. Both have same kinetic energy, which of the following statement is true ?

(a) Trajectory of electron is less curved.
(b) Trajectory of proton is less curved.
(c) Both trajectories are equally curved.
(d) Both move on straight line path.

29. A charge Q is moving $d\mathbf{l}$ distance in the magnetic field $\mathbf{B}$. Find the value of work done by $\mathbf{B}$. [JEE Main 2021]

(a) 1 (b) Infinite (c) Zero (d) −1

30. Two charged particles M and N enter a space of uniform magnetic field, with velocities perpendicular to the magnetic field. The paths are as shown in figure. The possible reason (s) is/are

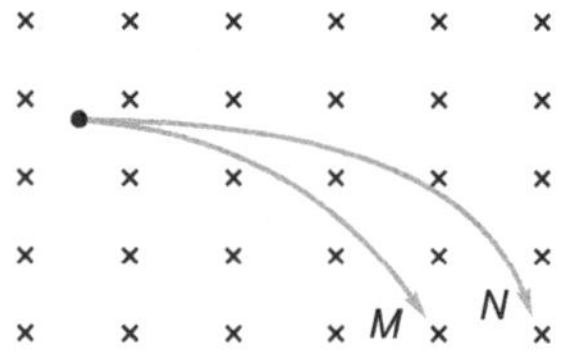

(a) the charge of M is greater than that of N
(b) the momentum of M is greater than that of N
(c) specific charge of M is greater than that of N
(d) the speed of M is greater than that of N

31. An electron, a proton and an alpha particle having the same kinetic energy are moving in circular orbits of radii r_e, r_p, r_α respectively, in a uniform magnetic field B. The relation between r_e, r_p, r_α is [JEE Main 2018]

(a) $r_e > r_p = r_\alpha$ (b) $r_e < r_p = r_\alpha$
(c) $r_e < r_p < r_\alpha$ (d) $r_e < r_\alpha < r_p$

32. A proton of mass 1.67×10^{-27} kg and charge 1.6×10^{-19} C is projected with a speed of $2 \times 10^6\text{ ms}^{-1}$ at an angle of 60° to the x-axis. If a uniform magnetic field of 0.104 T is applied along y-axis, the path of proton is

(a) a circle of radius = 0.2 m and time period $= 2\pi \times 10^{-7}$ s
(b) a circle of radius = 0.1 m and time period $= 2\pi \times 10^{-7}$ s
(c) a helix of radius 0.1 m and time period $= 2\pi \times 10^{-7}$ s
(d) a helix of radius 0.2 m and time period $= 2\pi \times 10^{-7}$ s

33. A beam of protons with speed $4 \times 10^5\text{ ms}^{-1}$ enters a uniform magnetic field of 0.3 T at an angle of 60° to the magnetic field. The pitch of the resulting helical path of protons is close to

(Take, mass of the proton $= 1.67 \times 10^{-27}$ kg and charge of the proton $= 1.69 \times 10^{-19}$ C) [JEE Main 2020]

(a) 2 cm (b) 4 cm
(c) 5 cm (d) 12 cm

34. The figure shows a region of length l with a uniform magnetic field of 0.3 T in it and a proton entering the region with velocity $4 \times 10^5\text{ ms}^{-1}$ making an angle 60° with the field. If the proton completes 10 revolutions by the time it cross the region shown, l is close to (Take, mass of proton $= 1.67 \times 10^{-27}$ kg, charge of the proton $= 1.6 \times 10^{-19}$ C) **[JEE Main 2020]**

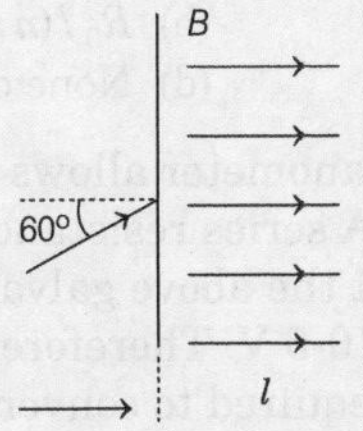

(a) 0.11 m (b) 0.88 m
(c) 0.44 m (d) 0.22 m

35. A particle of mass m and charge q is in an electric and magnetic field is given by

$$\mathbf{E} = 2\hat{\mathbf{i}} + 3\hat{\mathbf{j}},\ \mathbf{B} = 4\hat{\mathbf{j}} + 6\hat{\mathbf{k}}.$$

The charged particle is shifted from the origin to the point $P(x = 1;\ y = 1)$ along a straight path. The magnitude of the total work done is **[JEE Main 2019]**

(a) $0.35\ q$ (b) $(0.15)\ q$
(c) $2.5\ q$ (d) $5\ q$

36. A particle of charge q and mass m starts moving from the origin under the action of an electric field, $E = E_0\ \hat{\mathbf{i}}$ and $B = B_0\ \hat{\mathbf{i}}$ with a velocity, $v = v_0\ \hat{\mathbf{j}}$. The speed of the particle will become $\frac{\sqrt{5}}{2}v_0$ after a time

(a) $\frac{mv_0}{qE}$ (b) $\frac{mv_0}{2qE}$
(c) $\frac{\sqrt{3}mv_0}{2qE}$ (d) $\frac{\sqrt{5}mv_0}{2qE}$

Force on a Current Carrying Conductor/Coil in a Magnetic Field

37. A long horizontal wire P carries a current of 50 A. It is rigidly fixed. Another fine wire Q is placed directly above and parallel to P. The weight of wire Q is 0.075 Nm^{-1} and carries a current of 25 A. Find the position of wire Q from P, so that wire Q remains suspended due to the magnetic repulsion.

(a) $\frac{1}{2} \times 10^{-2}$ m (b) $\frac{1}{3} \times 10^{-2}$ m
(c) $\frac{1}{4} \times 10^{-2}$ m (d) $\frac{1}{5} \times 10^{-2}$ m

38. A metal wire of mass m slides without friction on two rails placed at a distance l apart. The track lies in a uniform vertical magnetic field B. A constant current i flows along the rails across the wire and breakdown the other rail. The acceleration of the wire is

(a) $\frac{Bmi}{l}$ (b) $mBi\,l$
(c) $\frac{Bil}{m}$ (d) $\frac{mil}{B}$

39. A current i_1 carrying wire AB is placed near an another long wire CD carrying current i_2 as shown in figure. If free to move, wire AB will have

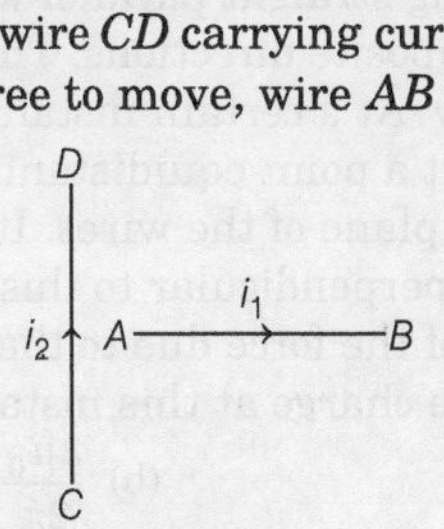

(a) rotational motion only
(b) translational motion only
(c) rotational as well as translational motion
(d) neither rotational nor translational motion

40. Two long and parallel straight wires A and B carrying currents of 8.0 A and 5.0 A in the same direction are separated by a distance of 4.0 cm. Estimate the force on a 10 cm section of wire A.

(a) 1.5×10^{-5} N (b) 2×10^{-5} N
(c) 4×10^{-5} N (d) 3.2×10^{-5} N

41. In the given diagram, two long parallel wires carry equal currents in opposite direction. Point O is situated mid-way between the wires and the XY-plane contains the two wires and the positive Z-axis comes normally out of the plane of paper. The magnetic field B at O is non-zero along

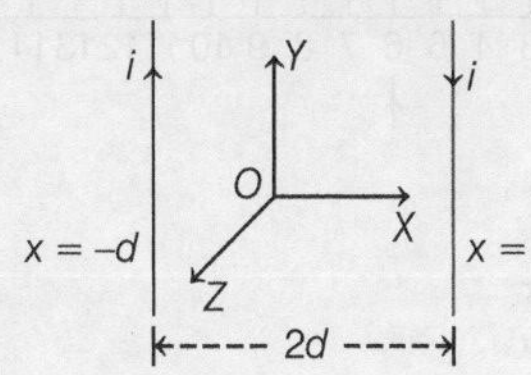

(a) X, Y and Z-axes (b) X-axis (positive)
(c) Y-axis (negative) (d) Z-axis (negative)

42. A rigid square loop of side a and carrying current i_2 is lying on a horizontal surface near a long current i_1 carrying wire in the same plane as shown in figure. The net force on the loop due to the wire will be **[JEE Main 2019]**

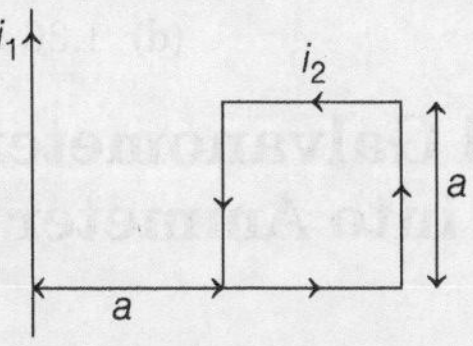

(a) repulsive and equal to $\frac{\mu_0 i_1 i_2}{2\pi}$
(b) attractive and equal to $\frac{\mu_0 i_1 i_2}{3\pi}$
(c) zero
(d) repulsive and equal to $\frac{\mu_0 i_1 i_2}{4\pi}$

43. Two very long straight parallel wires carry currents i and $2i$ in opposite directions. The distance between the wires is r. At a certain instant of time, a point charge q is at a point equidistant from the two wires in the plane of the wires. Its instantaneous velocity v is perpendicular to this plane. The magnitude of the force due to the magnetic field acting on the charge at this instant is

(a) zero (b) $\frac{3\mu_0}{2\pi}\frac{iqv}{r}$

(c) $\frac{\mu_0}{\pi}\frac{iqv}{r}$ (d) $\frac{\mu_0}{2\pi}\frac{iqv}{r}$

44. Two parallel long wires A and B carry currents i_1 and i_2 ($< i_1$). When i_1 and i_2 are in the same direction, the magnetic field at a point mid-way between the wires is $10\mu T$. If i_2 is reversed, the field becomes $30\,\mu T$. The ratio, i_1/i_2 is

(a) 1 (b) 2

(c) 3 (d) 4

45. Two parallel long straight conductors are placed at right angle to the meter scale at the 2 cm and 6 cm marks as shown in the figure. If they carry currents i and $3i$ respectively in the same direction, then they will produce zero magnetic field at

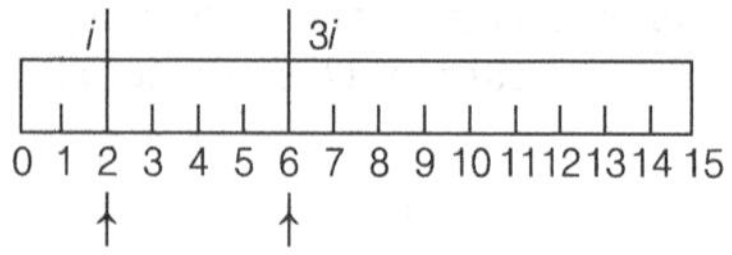

(a) zero mark

(b) 9 cm mark

(c) 3 cm mark

(d) 7 cm mark

46. A square coil of side 10 cm consists of 20 turns and carries a current of 12 A. The coil is suspended vertically and the normal to the plane of the coil makes an angle of 30° with the direction of a uniform horizontal magnetic field of magnitude 0.80 T. What is the magnitude of torque experienced by the coil? [NCERT Exemplar]

(a) 0.96 N-m (b) 2.06 N-m

(c) 0.23 N-m (d) 1.36 N-m

Moving Coil Galvanometer, Its Conversion into Ammeter and Voltmeter

47. A galvanometer having a coil resistance of 100 Ω gives a full scale deflection when a current of 1 mA is passed through it. The value of the resistance which can convert this galvanometer into ammeter giving a full scale deflection for a current of 10 A, is [JEE Main 2016]

(a) 0.01 Ω (b) 2 Ω

(c) 0.1 Ω (d) 3 Ω

48. An ammeter has resistance R_0 and range I. What resistance should be connected in parallel with it to increase its range by nI ?

(a) $R_0/(n-1)$ (b) $R_0/(n+1)$

(c) R_0/n (d) None of these

49. A moving coil galvanometer allows a full scale current of 10^{-4} A. A series resistance of 2 MΩ is required to convert the above galvanometer into a voltmeter of range 0-5 V. Therefore, the value of shunt resistance required to convert the above galvanometer into an ammeter of range 0.10 mA is [JEE Main 2019]

(a) 100 Ω (b) 500 Ω

(c) 200 Ω (d) 10 Ω

50. A galvanometer having a coil resistance 100 Ω gives a full scale deflection when a current of 1 mA is passed through it. What is the value of the resistance which can convert this galvanometer into a voltmeter giving full scale deflection for a potential difference of 10 V? [JEE Main 2020]

(a) 8.9 kΩ (b) 10 kΩ

(c) 9.9 kΩ (d) 7.9 kΩ

51. A voltmeter has resistance of 2000 Ω and it can measure upto 2 V. If we want to increase its range by 8 V, then required resistance in series will be

(a) 4000 Ω (b) 6000 Ω

(c) 7000 Ω (d) 8000 Ω

52. A galvanometer of resistance 100 Ω has 50 divisions on its scale and has sensitivity of 20 μA/division. It is to be converted to a voltmeter with three ranges of 0-2 V, 0-10 V and 0-20 V. The appropriate circuit to do so is [JEE Main 2019]

(a)

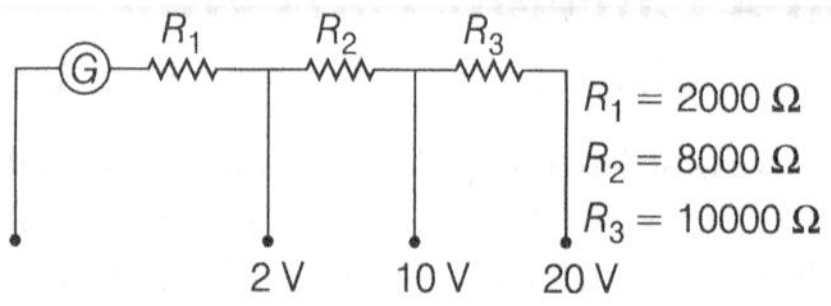

(b)

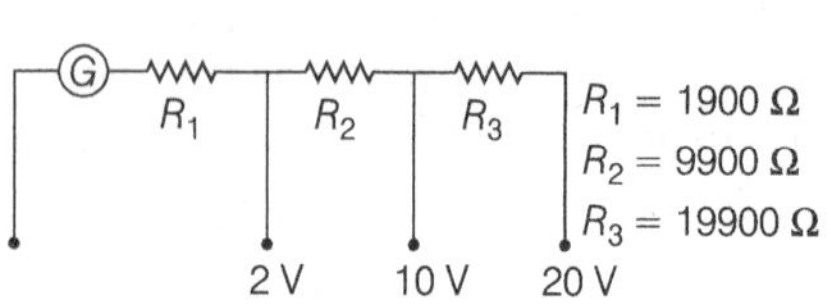

(c)

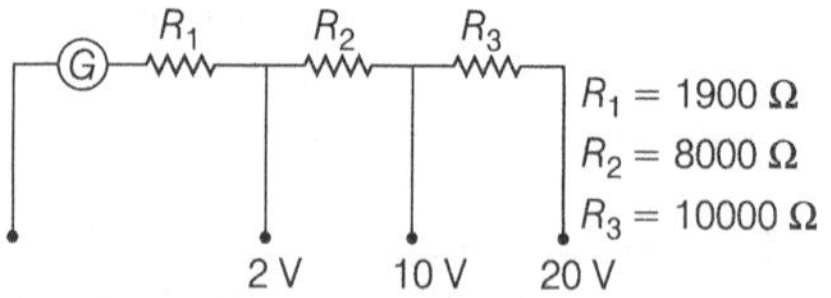

(d)

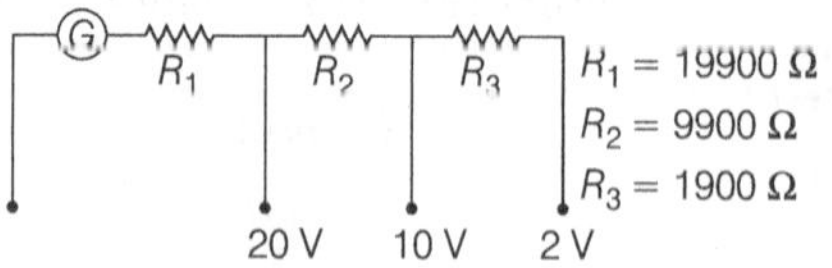

53. A microammeter has a resistance of 100 Ω and full scale range of 50μA. It can be used as a voltmeter or as a higher range ammeter provided a resistance is added to it. Pick the correct range and resistance combinations
(a) 50 V range with 10 kΩ resistance in series
(b) 10 V range with 200 kΩ resistance in series
(c) 10 mA range with 1 Ω resistance in parallel
(d) 10 mA range with 0.1 Ω resistance in parallel

54. A moving coil galvanometer, having a resistance G, produces full scale deflection when a current i_g flows through it. This galvanometer can be converted into (i) an ammeter of range 0 to $i_0 (i_0 > i_g)$ by connecting a shunt resistance R_A to it and (ii) into a voltmeter of range 0 to $V\ (V = Gi_0)$ by connecting a series resistance R_V to it. Then, **[JEE Main 2019]**

(a) $R_A R_V = G^2\left(\frac{i_0 - i_g}{I_g}\right)$ and $\frac{R_A}{R_V} = \left(\frac{i_g}{(i_0 - i_g)}\right)^2$

(b) $R_A R_V = G^2$ and $\frac{R_A}{R_V} = \left(\frac{i_g}{i_0 - i_g}\right)^2$

(c) $R_A R_V = G^2\left(\frac{i_g}{i_0 - i_g}\right)$ and $\frac{R_A}{R_V} = \left(\frac{i_0 - i_g}{i_g}\right)^2$

(d) $R_A R_V = G^2$ and $\frac{R_A}{R_V} = \frac{i_g}{(i_0 - i_g)}$

ROUND II Mixed Bag

Only One Correct Option

1. A current i flows along the length of an infinitely long, straight and thin-walled pipe, then
(a) the magnetic field at all points inside the pipe is the same, but not zero
(b) the magnetic field at any point inside the pipe is zero
(c) the magnetic field as zero only on the axis of the pipe
(d) the magnetic field at different at different points inside the pipe

2. A rectangular loop carrying current is placed near a long straight fixed wire carrying strong current such that long sides are parallel to wire. If the current in the nearer long side of loop is parallel to current in the wire. Then the loop

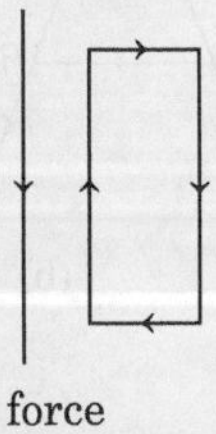

(a) experiences no force
(b) experiences a force towards wire
(c) experiences a force away from wire
(d) experiences a torque but no force

3. Three infinite straight wires A, B and C carry currents as shown in figure. The net force on the wire B is directed

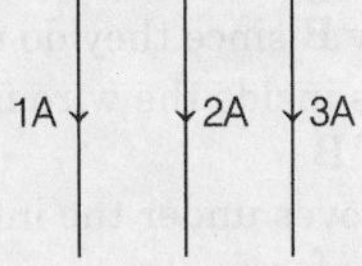

(a) towards A
(b) towards C
(c) normal to plane of paper
(d) zero

4. For the arrangement as shown in the figure, the magnetic induction at the centre is

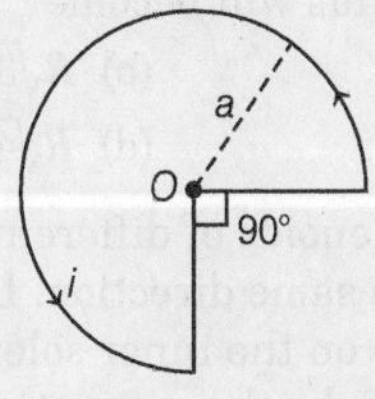

(a) $\frac{3\mu_0 i\pi}{4a}$ (b) $\frac{\mu_0 i}{4\pi a}(1 + \pi)$

(c) $\frac{\mu_0 i}{4\pi a}$ (d) $\frac{3\mu_0 i}{8a}$

5. Four wires each of length 2.0 m are bent into four P, Q, R and S and then suspended into a uniform magnetic field. If same current is passed in each loop, then

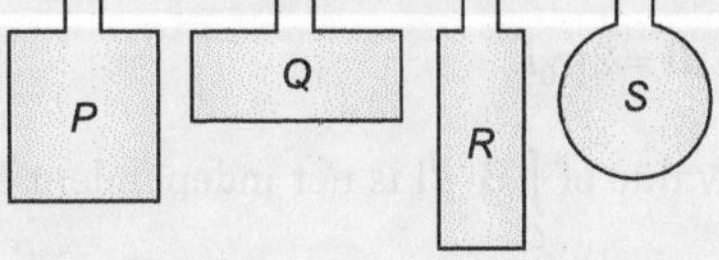

(a) couple on loop P will be maximum
(b) couple on loop Q will be maximum
(c) couple on loop R will be maximum
(d) couple on loop S will be maximum

6. An electron is revolving around a proton in a circular path of diameter 0.1 nm. It produces a magnetic field 14 T at a proton. Then the angular speed of the electron is
(a) 8.8×10^6 rad s^{-1}
(b) 4.4×10^{16} rad s^{-1}
(c) 2.2×10^{16} rad s^{-1}
(d) 1.1×10^{16} rad s^{-1}

7. A pulsar is a neutron star having magnetic field at 10^{12} G at its surface. The maximum magnetic force experienced by an electron moving with velocity $0.9\,c$ is

(a) 43.2 N (b) 4.32×10^{-3} N
(c) 4.32×10^{3} N (d) zero

8. An electron is shot in steady electric and magnetic fields such that its velocity v, electric field E and magnetic field B are mutually perpendicular. The magnitude of E is 1 Vcm^{-1} and that of B is 2 T. Now if it so happens that the Lorentz (magnetic) force cancels the electrostatic force on the electron, then the velocity of the electron is

(a) 50 ms^{-1} (b) 2 cms^{-1}
(c) 0.5 cms^{-1} (d) 200 cms^{-1}

9. An electron having kinetic energy E is moving in a circular orbit of radius R perpendicular to a uniform magnetic field induction B. If kinetic energy is doubled and magnetic field induction is tripled, the radius will become

(a) $R\sqrt{9/4}$ (b) $R\sqrt{3/2}$
(c) $R\sqrt{2/9}$ (d) $R\sqrt{4/3}$

10. Two coaxial solenoids of different radii carry current i in the same direction. Let $\mathbf{F}_1$ be the magnetic force on the inner solenoid due to the outer one and $\mathbf{F}_2$ be the magnetic force on the outer solenoid due to the inner one. Then, [JEE Main 2015]

(a) $\mathbf{F}_1 = \mathbf{F}_2 = 0$
(b) $\mathbf{F}_1$ is radially inwards and $\mathbf{F}_2$ is radially outwards
(c) $\mathbf{F}_1$ is radially inwards and $\mathbf{F}_2 = 0$
(d) $\mathbf{F}_1$ is radially outwards and $\mathbf{F}_2 = 0$

11. Two identical current carrying coaxial loops, carry current i in an opposite sense. A simple amperian loop passes through both of them once. Calling the loop as C, [NCERT Exemplar]

(a) $\int_C \mathbf{B}\cdot d\mathbf{l} = 2\mu_0 i$
(b) the value of $\int_C \mathbf{B}\cdot d\mathbf{l}$ is not independent of sense of C
(c) there may be a point on C where $\mathbf{B}$ and $d\mathbf{l}$ are perpendicular.
(d) $\mathbf{B}$ vanishes everywhere on C

12. A particle of mass m and charge q released from the origin in a region occupied by electric field E and magnetic field B,

$$B = -B_0\,\hat{\mathbf{j}},\ E = E_0\,\hat{\mathbf{i}}$$

The velocity of the particle will be

(a) $\sqrt{\dfrac{2qE_0}{m}}$ (b) $\sqrt{\dfrac{qE_0}{m}}$
(c) $\sqrt{\dfrac{qE_0}{2m}}$ (d) None of these

13. Consider the following statements regarding a charged particle in a magnetic field

(i) starting with zero velocity, it accelerates in a direction perpendicular to the magnetic field.
(ii) while deflecting in the magnetic field, its energy gradually increases.
(iii) only the component of magnetic field perpendicular to the direction of motion of the charged particle is effective in deflecting it.
(iv) direction of deflecting force on the moving charged particle is perpendicular to its velocity.

Amongst these statements,
(a) (ii) and (iii) are correct
(b) (iii) and (iv) are correct
(c) (i), (iii) and (iv) are correct
(d) (i), (ii) and (iii) are correct

14. When a current of 5 mA is passed through a galvanometer having a coil of resistance 15 Ω, it shows full scale deflection. The value of the resistance to be put in series with the galvanometer to convert it into a voltmeter of range 0-10 V is [JEE Main 2017]

(a) $2.045\times 10^3\ \Omega$ (b) $2.535\times 10^3\ \Omega$
(c) $4.005\times 10^3\ \Omega$ (d) $1.985\times 10^3\ \Omega$

15. An equilateral triangle of side l is formed from a piece of wire of uniform resistance. The current i is fed as shown in the figure. The magnitude of the magnetic field at its centre O is

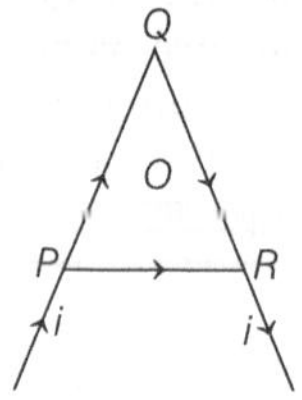

(a) $\dfrac{\sqrt{3}\mu_0 i}{2\pi l}$ (b) $\dfrac{3\sqrt{3}\mu_0 i}{2\pi l}$
(c) $\dfrac{\mu_0 i}{2\pi l}$ (d) zero

16. Consider a wire carrying a steady current, i placed in a uniform magnetic field $\mathbf{B}$ perpendicular to its length. Consider the charges inside the wire. It is known that magnetic forces do no work. This implies that, [NCERT Exemplar]

(a) motion of charges inside the conductor is unaffected by $\mathbf{B}$ since they do not absorb energy
(b) some charges inside the wire move to the surface as a result of $\mathbf{B}$
(c) if the wire moves under the influence of $\mathbf{B}$, no work is done by the force
(d) if the wire moves under the influence of $\mathbf{B}$, then work is done by the magnetic force on the ions, assumed fixed within the wire

17. Two wires A and B carry currents as shown in figure. The magnetic interactions

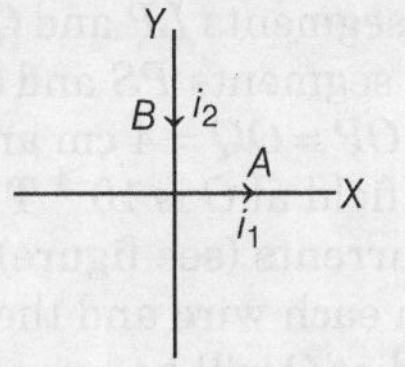

(a) push i_2 away from i_1 (b) pull i_2 closer to i_1
(c) turn i_2 clockwise (d) turn i_2 counter-clockwise

18. A current i carrying circular wire of radius R is placed in a magnetic field B perpendicular to its plane. The tension T along the circumference of wire is

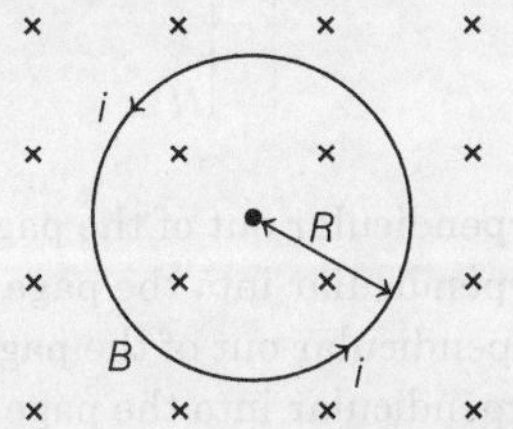

(a) BiR (b) $2\pi BiR$
(c) πBiR (d) $2BiR$

19. Two very long, straight and insulated wires are kept at 90° angle from each other in XY-plane as shown in the figure.

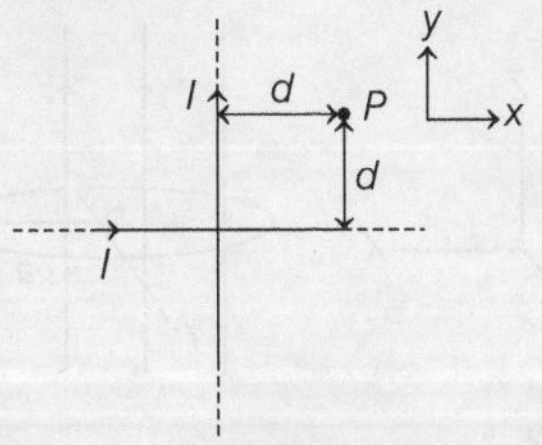

These wires carry currents of equal magnitude i, whose directions are shown in the figure. The net magnetic field at point P will be [JEE Main 2019]

(a) zero (b) $\frac{+\mu_0 i}{\pi d}(\hat{z})$

(c) $-\frac{\mu_0 i}{2\pi d}(\hat{x}+\hat{y})$ (d) $\frac{\mu_0 i}{2\pi d}(\hat{x}+\hat{y})$

20. A straight wire of mass 200g and length 1.5 m carries a current of 2 A. It is suspended in mid-air by a uniform horizontal magnetic field B. The magnitude of the magnetic field is [NCERT]

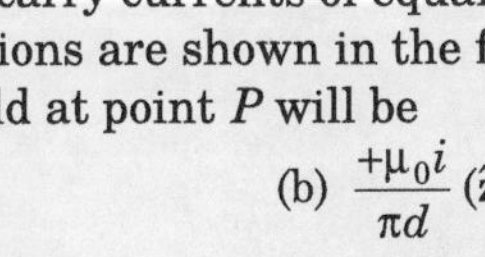

(a) 0.35 T (b) 0.65 T (c) 0.25 T (d) 0.88 T

21. A galvanometer of resistance G is converted into a voltmeter of range 0-1V by connecting a resistance R_1 in series with it. The additional resistance that should be connected in series with R_1 to increase the range of the voltmeter to 0-2V will be [JEE Main 2020]

(a) R_1+G (b) R_1 (c) G (d) R_1-G

22. A uniform electric and magnetic fields are produced pointing in the same direction. If an electron is projected with its velocity pointing in the same direction, [NCERT Exemplar]

(a) the electron velocity will decrease in magnitude
(b) the electron velocity will increase in magnitude
(c) neither (a) nor (b)
(d) None of the above

23. Net magnetic field at the centre of the circle O due to a current through a loop as shown in figure ($\theta < 180°$)

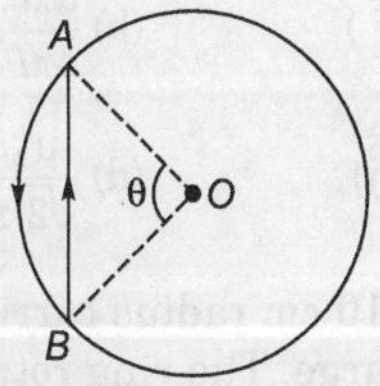

(a) zero
(b) perpendicular to paper inwards
(c) perpendicular to paper outwards
(d) perpendicular to paper outwards if $90° \leq \theta < 180°$

24. A particle of mass m and charge q has an initial velocity $\mathbf{v}=v_0\hat{\mathbf{j}}$. If an electric field $\mathbf{E}=E_0\hat{\mathbf{i}}$ and magnetic field $\mathbf{B}=B_0\hat{\mathbf{i}}$ act on the particle, its speed will double after a time [JEE Main 2020]

(a) $\frac{3mv_0}{qE_0}$ (b) $\frac{\sqrt{3}mv_0}{qE_0}$

(c) $\frac{2mv_0}{qE_0}$ (d) $\frac{\sqrt{2}mv_0}{qE_0}$

25. A conductor lies along the Z-axis at $-1.5 \leq z < 1.5$ m and carries a fixed current of 10.0 A in $-a_z$ direction (see figure). For a field $B = 3.0\times10^{-4}\, e^{-0.2x} a_y$ T, find the power required to move the conductor at constant speed to $x = 2.0$ m, $y = -0$ in 5×10^{-3} s. Assume, parallel motion along the X-axis. [JEE Main 2014]

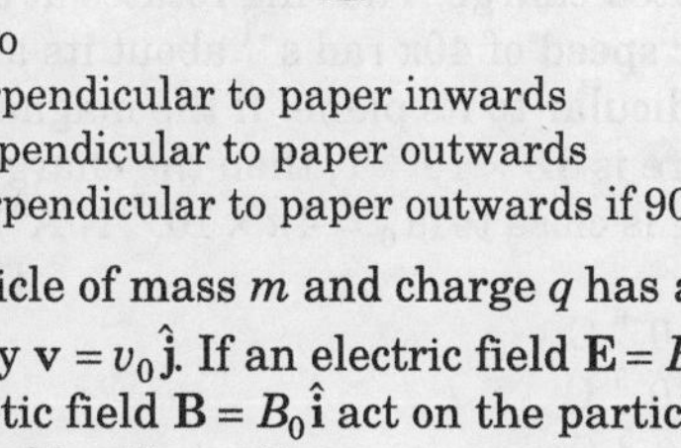

(a) 1.57 W (b) 2.97 W
(c) 14.85 W (d) 29.7 W

26. In the figure, $ABCDEFA$ was a square loop of side l, but is folded in two equal parts so that half of it lies in the XZ-plane. The origin O is centre of the frame also. The loop carries current i, the magnetic field at the centre is

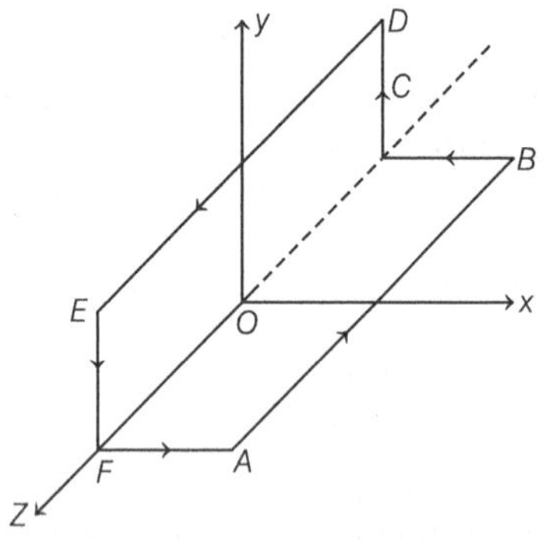

(a) $\frac{\mu_0 i}{2\sqrt{2}\,\pi l}(\hat{\mathbf{i}}-\hat{\mathbf{j}})$ (b) $\frac{\mu_0 i}{4\pi l}(-\hat{\mathbf{i}}+\hat{\mathbf{j}})$

(c) $\frac{\sqrt{2}\,\mu_0 i}{\pi l}(\hat{\mathbf{i}}+\hat{\mathbf{j}})$ (d) $\frac{\mu_0 i}{\sqrt{2}\,\pi l}(\hat{\mathbf{i}}+\hat{\mathbf{j}})$

27. A thin ring of 10 cm radius carries a uniformly distributed charge. The ring rotates at a constant angular speed of 40π rad s^{-1} about its axis, perpendicular to its plane. If the magnetic field at its centre is 3.8×10^{-9} T, then the charge carried by the ring is close to ($\mu_0 = 4\pi \times 10^{-7}$ N/A^2). **[JEE Main 2019]**

(a) 2×10^{-6} C
(b) 3×10^{-5} C
(c) 4×10^{-5} C
(d) 7×10^{-6} C

28. Two long current carrying thin wires, both with current i, are held by insulating threads of length L and are in equilibrium as shown in the figure, with threads making an angle θ with the vertical. If wires have mass λ per unit length, then the value of i is (g = gravitational acceleration)

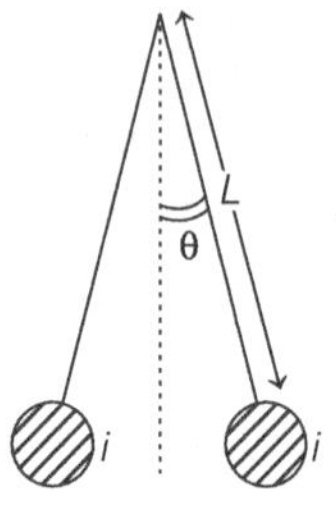

[JEE Main 2015]

(a) $\sin\theta\sqrt{\frac{\pi\lambda g L}{\mu_0 \cos\theta}}$ (b) $2\sin\theta\sqrt{\frac{\pi\lambda g L}{\mu_0 \cos\theta}}$

(c) $2\sqrt{\frac{\pi g L}{\mu_0}\tan\theta}$ (d) $\sqrt{\frac{\pi\lambda g L}{\mu_0}\tan\theta}$

29. As shown in the figure, two infinitely long, identical wires are bent by 90° and placed in such a way that the segments LP and QM are along the X-axis, while segments PS and QN are parallel to the Y-axis. If $OP = OQ = 4$ cm and the magnitude of the magnetic field at O is 10^{-4} T and the two wires carry equal currents (see figure), the magnitude of the current in each wire and the direction of the magnetic field at O will be
(Take, $\mu_0 = 4\pi \times 10^{-7}$ NA^{-2}) **[JEE Main 2019]**

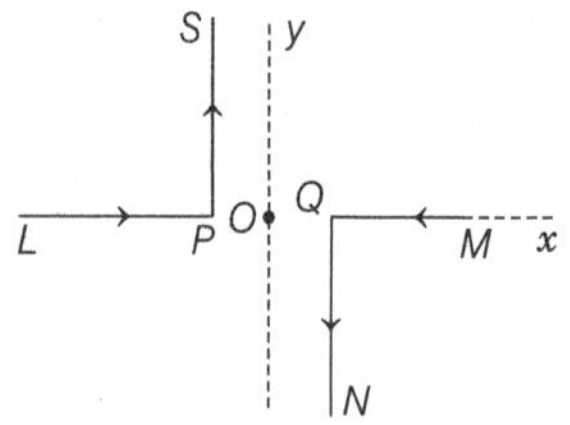

(a) 40 A, perpendicular out of the page
(b) 20 A, perpendicular into the page
(c) 20 A, perpendicular out of the page
(d) 40 A, perpendicular into the page

30. An elliptical loop having resistance R, of semi-major axis a and semi-minor axis b is placed in a magnetic field as shown in the figure. If the loop is rotated about the X-axis with angular frequency ω, then the average power loss in the loop due to joule's heating is **[JEE Main 2020]**

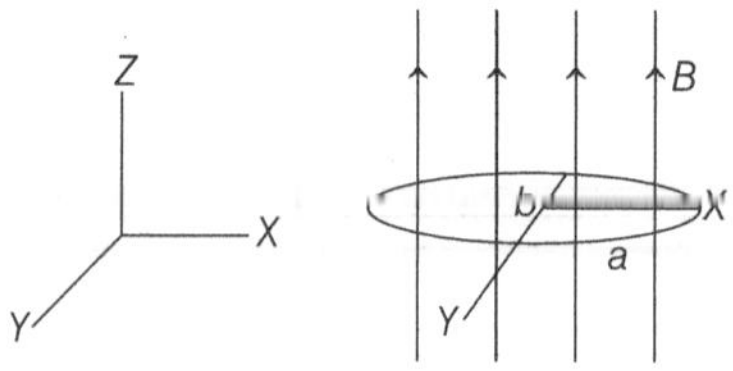

(a) $\frac{\pi^2 a^2 b^2 B^2 \omega^2}{2R}$

(b) zero

(c) $\frac{\pi a b B \omega}{R}$

(d) $\frac{\pi^2 a^2 b^2 B^2 \omega^2}{R}$

31. A square loop of side $2a$ and carrying current i is kept in XY-plane with its centre at origin. A long wire carrying the same current i is placed parallel to the Z-axis and passing through the point $(0, b, 0)$, $(b >> a)$. The magnitude of the torque on the loop about Z-axis is given by **[JEE Main 2020]**

(a) $\frac{2\mu_0 I^2 a^3}{\pi b^2}$ (b) $\frac{2\mu_0 I^2 a^2}{\pi b}$

(c) $\frac{\mu_0 I^2 a^3}{2\pi b^2}$ (d) $\frac{\mu_0 I^2 a^2}{2\pi b}$

32. Two identical wires A and B, each of length l, carry the same current i. Wire A is bent into a circle of radius R and wire B is bent to form a square of side a. If B_A and B_B are the values of magnetic field at the centres of the circle and square respectively, then the ratio $\frac{B_A}{B_B}$ is **[JEE Main 2016]**

(a) $\frac{\pi^2}{8}$ (b) $\frac{\pi^2}{16\sqrt{2}}$

(c) $\frac{\pi^2}{16}$ (d) $\frac{\pi^2}{8\sqrt{2}}$

33. Proton with kinetic energy of 1 MeV moves from south to north. It gets an acceleration of 10^{12} m/s^2 by an applied magnetic field (west to east). The value of magnetic field (rest mass of proton is 1.6×10^{-27} kg) **[JEE Main 2020]**

(a) 71 mT (b) 0.071 mT

(c) 0.71 mT (d) 7.1 mT

34. An infinitely long wire carrying current i is along Y-axis such that its one end is at point $(0, b)$ while the wire extends upto ∞. The magnitude of magnetic field strength at point $P(a, 0)$ is

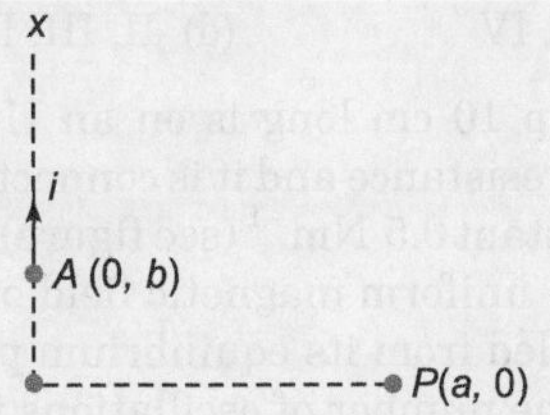

(a) $\frac{\mu_0 i}{4\pi a}\left(1+\frac{b}{\sqrt{a^2+b^2}}\right)$ (b) $\frac{\mu_0 i}{4\pi a}\left(1-\frac{b}{\sqrt{a^2+b^2}}\right)$

(c) $\frac{\mu_0 i}{4\pi a}\left(1-\frac{a}{\sqrt{a^2+b^2}}\right)$ (d) $\frac{\mu_0 i}{4\pi a}\left(1+\frac{a}{\sqrt{a^2+b^2}}\right)$

35. A straight rod of mass m and length L is suspended from the identical springs as shown in figure. The spring is stretched a distance x_0 due to the weight of the wire. The circuit has total resistance R. When the magnetic field perpendicular to the plane of paper is switched ON, springs are observed to extend further by the same distance. The magnetic field strength is

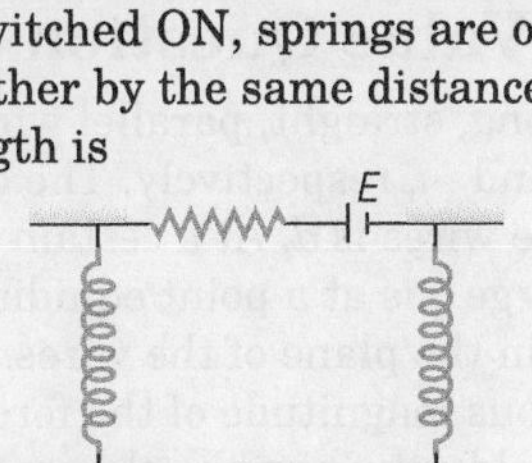

(a) $\frac{2mgR}{LE}$ (b) $\frac{mgR}{LE}$ (c) $\frac{mgR}{2LE}$ (d) $\frac{mgR}{E}$

36. A thin disc having radius r and charge q distributed uniformly over the disc is rotated n rotations per second about its axis. The magnetic field at the centre of the disc is

(a) $\frac{\mu_0 qn}{2r}$ (b) $\frac{\mu_0 qn}{r}$ (c) $\frac{\mu_0 qn}{4r}$ (d) $\frac{3\mu_0 qn}{4r}$

37. A particle of charge $+q$ and mass m moving under the influence of a uniform electric field $E\hat{\mathbf{i}}$ and a uniform magnetic field $B\hat{\mathbf{k}}$ follows a trajectory from P to Q as shown in figure. The velocities at P and Q are $v\hat{\mathbf{i}}$ and $-2v\hat{\mathbf{i}}$, respectively. Which of the following statement(s) is/are correct?

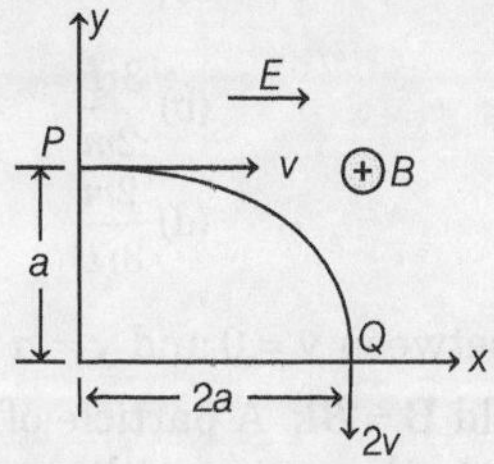

(a) $E = \frac{1}{4}\left(\frac{\mu v^2}{qa}\right)$

(b) Rate of work done by the electric field at P is $\frac{1}{4}\left(\frac{mv^2}{a}\right)$.

(c) Rate of work done by the electric field at P is zero.

(d) Rate of work done by both the fields at Q is zero.

38. A cubical region of space is filled with some uniform electric and magnetic fields. An electron enters the cube across one of its faces with velocity **v** and a positron enters *via* opposite face with velocity − **v**. At this instant, **[NCERT Exemplar]**

(a) the electric forces on both the particles cause identical accelerations

(b) the magnetic forces on both the particles cause unequal accelerations

(c) Both particles gain or loose energy at the same rate

(d) the motion of the centre of mass (CM) is determined by **E** alone

39. A square loop of side $2a$ and carrying current i is kept in XZ-plane with its centre at origin. A long wire carrying the same current i is placed parallel to Z-axis and passing through point $(0, b, 0)$, $(b >> a)$. The magnitude of torque on the loop about Z-axis will be **[JEE Main 2020]**

(a) $\frac{2\mu_0 i^2 a^2}{\pi b}$ (b) $\frac{2\mu_0 i^2 a^2 b}{\pi(a^2+b^2)}$

(c) $\frac{\mu_0 i^2 a^2 b}{2\pi(a^2+b^2)}$ (d) $\frac{\mu_0 i^2 a^2}{2\pi b}$

40. A straight conductor of mass m and carrying a current i is hinged at one end and placed in a plane perpendicular to the magentic field of intensity B as shown in the figure. At any moment if the conductor is left free, then the angular acceleration of the conductor will be (Assume gravity free region) will be

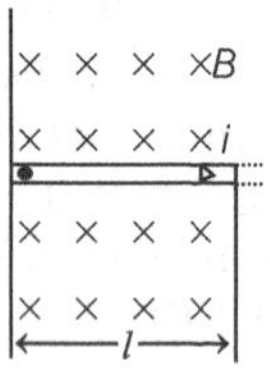

(a) $\frac{iB}{m}$ (b) $\frac{3iB}{2m}$

(c) $\frac{2m}{iB}$ (d) $\frac{2m}{3iB}$

41. The region between $y = 0$ and $y = d$ contains a magnetic field $\mathbf{B} = B\hat{\mathbf{k}}$. A particle of mass m and charge q enters the region with a velocity $\mathbf{v} = v\hat{\mathbf{i}}$. If $d = \frac{mv}{2qB}$ then the acceleration of the charged particle at the point of its emergence at the other side is **[JEE Main 2019]**

(a) $\frac{qvB}{m}\left(\frac{\sqrt{3}}{2}\hat{\mathbf{i}} + \frac{1}{2}\hat{\mathbf{j}}\right)$ (b) $\frac{qvB}{m}\left(\frac{1}{2}\hat{\mathbf{i}} - \frac{\sqrt{3}}{2}\hat{\mathbf{j}}\right)$

(c) $\frac{qvB}{m}\left(\frac{-\hat{\mathbf{j}}+\hat{\mathbf{i}}}{\sqrt{2}}\right)$ (d) $\frac{qvB}{m}\left(\frac{\hat{\mathbf{i}}+\hat{\mathbf{j}}}{\sqrt{2}}\right)$

42. An infinitely long current-carrying wire and a small current carrying loop are in the plane of the paper as shown in figure. The radius of the loop is a and distance of its centre from the wire is $d (d >> a)$. If the loop applies a force F on the wire, then **[JEE Main 2019]**

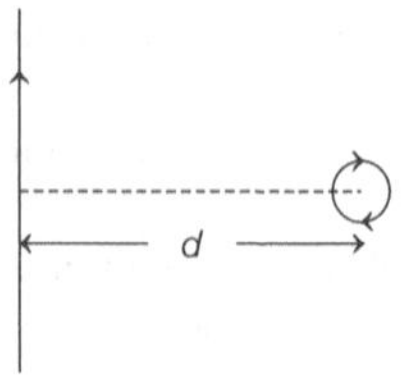

(a) $F \propto \left(\frac{a^2}{d^3}\right)$ (b) $F = 0$

(c) $F \propto \left(\frac{a}{d}\right)$ (d) $F \propto \left(\frac{a}{d}\right)^2$

43. A charged particle of mass m and charge q moving under the influence of uniform electric field $E\hat{\mathbf{i}}$ and a uniform magnetic field $B\hat{\mathbf{k}}$ follows a trajectory from point P to Q as shown in the figure. The velocities at P and Q are respectively, $v\hat{\mathbf{i}}$ and $-2v\hat{\mathbf{j}}$. Then, which of the following statements (A, B, C, D) are the correct? (Trajectory shown is schematic and not to scale)

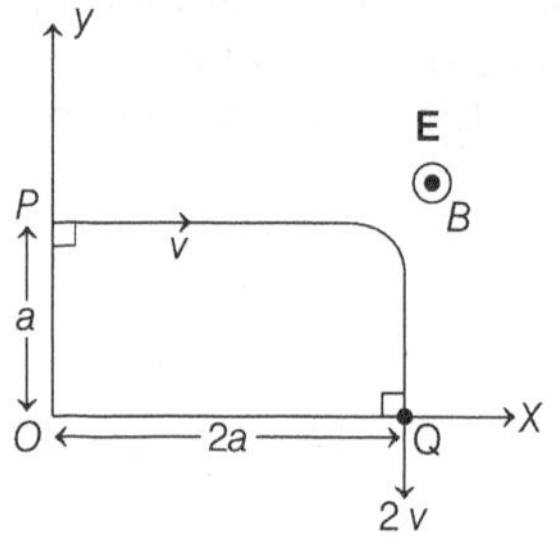

I. $E = \frac{3}{4}\left(\frac{mv^2}{qa}\right)$

II. Rate of work done by the electric field at P is $\frac{3}{4}\left(\frac{mv^3}{a}\right)$.

III. Rate of work done by both the fields at Q is zero.

IV. The difference between the magnitude of angular momentum of the particle at P and Q is $2\,mav$. **[JEE Main 2020]**

(a) I, III,IV (b) I, II, III

(c) I, II, III, IV (d) II, III, IV

44. A thin strip 10 cm long is on an U-shaped wire of negligible resistance and it is connected to a spring of spring constant $0.5\ \text{Nm}^{-1}$ (see figure). The assembly is kept in a uniform magnetic field of 0.1 T. If the strip is pulled from its equilibrium position and released, the number of oscillations it performs before its amplitude decreases by a factor of e is N. If the mass of the strip is 50 g, its resistance $10\ \Omega$ and air drag negligible, N will be close to

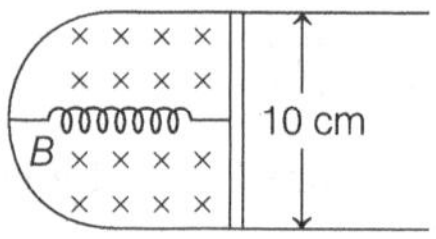

[JEE Main 2019]

(a) 1000 (b) 50000

(c) 5000 (d) 10000

Numerical Value Questions

45. Two very long, straight, parallel wires carry steady currents i and $-i$, respectively. The distance between the wires is d. At a certain instant of time, a point charge q is at a point equidistant from the two wires, in the plane of the wires. Its instantaneous magnitude of the force due to the magnetic field acting on the charge at this instant is ....

46. A galvanometer coil has 500 turns and each turn has an average area of $3 \times 10^{-4}\ \text{m}^2$. If a torque of 1.5 N-m is required to keep this coil parallel to a magnetic field when a current of 0.5 A is flowing through it, the strength of the field (in T) is ....

[JEE Main 2020]

47. In a square loop $PQRS$ made with a wire of cross-section current i enters from point P and leaves from point S. The magnitude of magnetic field induction at the centre O of the square is ....

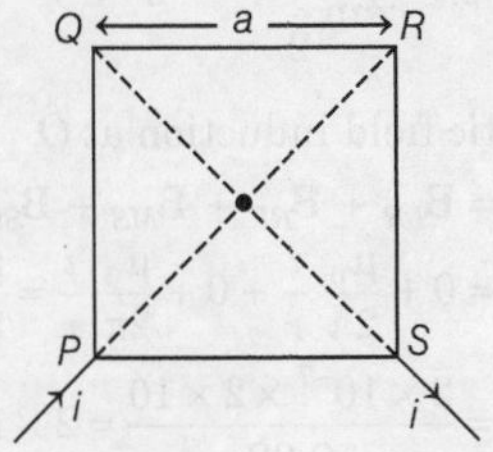

48. In the following figure, the magnetic field at the point P will be $\frac{\mu_0 i}{P\pi a}\sqrt{\pi^2+4}$, where the value of P is ....

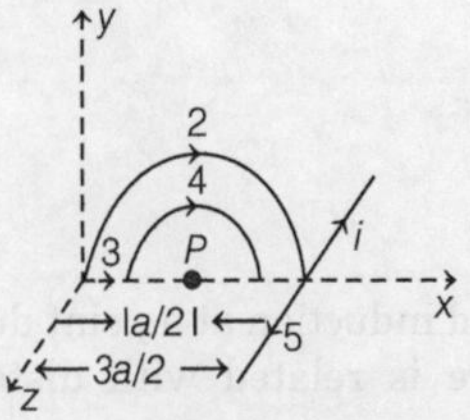

49. An electron moves straight inside a charged parallel plate capacitor of uniform surface charge density σ. The space between the plates is filled with constant magenetic field of induction **B**, as shown in the figure. If the gravity is neglected, then the time of straight line motion of the electron in the capacitor will be $\frac{\varepsilon_0 lB}{x\sigma}$, where the value of x is

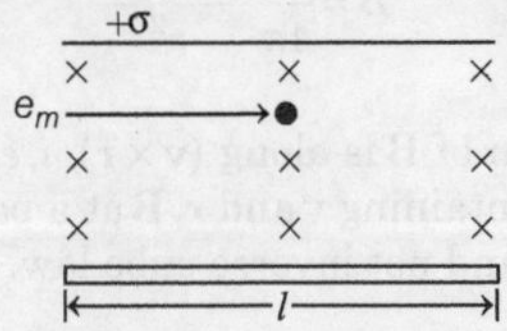

Answers

Round I

1. (c)	2. (a)	3. (c)	4. (a)	5. (d)	6. (c)	7. (d)	8. (c)	9. (b)	10. (a)
11. (c)	12. (d)	13. (a)	14. (b)	15. (a)	16. (b)	17. (d)	18. (b)	19. (a)	20. (b)
21. (d)	22. (c)	23. (c)	24. (*)	25. (b)	26. (a)	27. (c)	28. (b)	29. (c)	30. (c)
31. (b)	32. (c)	33. (b)	34. (c)	35. (d)	36. (b)	37. (b)	38. (c)	39. (c)	40. (b)
41. (d)	42. (d)	43. (a)	44. (b)	45. (c)	46. (a)	47. (a)	48. (c)	49. (*)	50. (c)
51. (d)	52. (c)	53. (b)	54. (b)						

Round II

1. (b)	2. (b)	3. (a)	4. (d)	5. (d)	6. (b)	7. (b)	8. (a)	9. (c)	10. (a)
11. (c)	12. (a)	13. (c)	14. (d)	15. (d)	16. (b)	17. (d)	18. (a)	19. (a)	20. (b)
21. (a)	22. (a)	23. (b)	24. (b)	25. (b)	26. (c)	27. (b)	28. (b)	29. (b)	30. (a)
31. (b)	32. (d)	33. (c)	34. (b)	35. (b)	36. (b)	37. (d)	38. (c)	39. (a)	40. (b)
41. (*)	42. (d)	43. (b)	44. (c)	45. 0	46. 20	47. 0	48. 3	49. 1	

Solutions

Round I

1. Magnetic field induction at a point due to a long current carrying wire is related with distance r by relation $B \propto \frac{1}{r}$.

   Therefore, graph given in option (c) is correct.

2. According to Biot Savart's law, the magnetic field $\mathbf{B}$ at a point distance r from a charge q moving with a velocity $\mathbf{v}$ is given by

$$\mathbf{B} = \frac{\mu_0}{4\pi} \frac{q\,(\mathbf{v} \times \mathbf{r})}{r^3}$$

or
$$B = \frac{\mu_0}{4\pi} \frac{qv \sin\theta}{r^2}$$

   The direction of $\mathbf{B}$ is along $(\mathbf{v} \times \mathbf{r})$, *i.e.* perpendicular to the plane containing $\mathbf{v}$ and $\mathbf{r}$. $\mathbf{B}$ at a point obeys inverse square law and not inverse cube law.

3. As, $\mathbf{B} = \frac{\mu_0 i}{4\pi a} [\sin\theta_1 + \sin\theta_2]$

   Here, $\theta_1 = -30°, \theta_2 = 60°$

$$\sin\theta_1 = -\frac{1}{2} \text{ and } \sin\theta_2 = \frac{\sqrt{3}}{2}$$

   Putting these values, we get

$$\mathbf{B} = \frac{\mu_0 i}{4\pi a} [-1/2 + \sqrt{3}/2]$$

4. Given, $i = 90$ A and $r = 1.5$ m

   Here, point P is below the power line, where we have to find the magnetic field and its direction.

   The magnitude of magnetic field,

$$B = \frac{\mu_0}{4\pi} \cdot \frac{2i}{r}$$

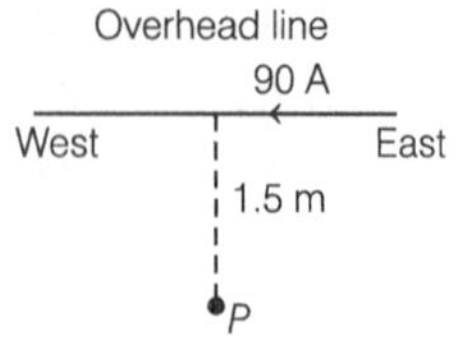

$$= \frac{10^{-7} \times 2 \times 90}{1.5} = 1.2 \times 10^{-5} \text{ T}$$

   The direction of magnetic field is given by Maxwell's right hand rule, so the direction of magnetic field at point P due to the flowing current is perpendicularly outwards to the plane of paper.

5. Perpendicular to O from PQ or QR,

$$a = r \sin\frac{\theta}{2}$$

   Magnetic field induction at O due to current through PQ and QR,

$$B = \frac{\mu_0}{4\pi} \frac{i}{a} [\sin(90° - \theta/2) + \sin 90°] \times 2$$

$$= \frac{\mu_0}{2\pi} \frac{i}{r \sin\frac{\theta}{2}} \left(\cos\frac{\theta}{2} + 1\right) = \frac{\mu_0}{2\pi} \frac{i}{r} \frac{\left(1 + \cos\frac{\theta}{2}\right)}{\sin\frac{\theta}{2}}$$

6. Total magnetic field induction at O,

$$\mathbf{B} = \mathbf{B}_{LR} + \mathbf{B}_{RP} + \mathbf{B}_{MS} + \mathbf{B}_{SQ}$$

$$= 0 + \frac{\mu_0}{2\pi} \frac{i}{r} + 0 + \frac{\mu_0}{2\pi} \frac{i}{r} = \frac{\mu_0}{2\pi} \frac{2i}{r}$$

$$= \frac{2 \times 10^{-7} \times 2 \times 10}{0.02} = 2 \times 10^{-4} \text{ T}$$

7. Magnetic field induction at the centre of circular coil carrying current, $B = \frac{\mu_0}{4\pi} \frac{2\pi n i}{r}$, *i.e.* $B \propto n/r$

   But, $2\pi r = 3 \times 2\pi r_1$ or $r_1 = \frac{r}{3}$

   So, $\frac{B_1}{B} = \frac{n_1}{r_1} \times \frac{r}{n} = \frac{3 \times r}{\left(\frac{r}{3}\right) \times 1} = 9$

$$B_1 = 9B$$

   The magnetic field at the centre caused by the same current is nine times of its initial value.

8. Magnetic field induction at O due to current through ACB is $B_1 = \frac{\mu_0 i\theta}{4\pi r}$

   It is acting perpendicular to the paper downwards. Magnetic field induction at O due to current through ABD is

$$B_2 = \frac{\mu_0}{4\pi} \frac{i(2\pi - \theta)}{r}$$

   It is acting perpendicular to paper upwards.

   $\therefore$ Total magnetic field at O due to current loop is

$$B = B_2 - B_1 = \frac{\mu_0}{4\pi} \frac{i}{r} (2\pi - \theta) - \frac{\mu_0}{4\pi} \frac{i}{r} \theta$$

$$= \frac{\mu_0}{2\pi} \frac{i}{r} (\pi - \theta)$$

9. Magnetic field at centre of the circular loop,

$$B = \frac{\mu_0 N i}{2R}$$

   Magnetic field due to an arc of a circle at the centre,

$$B = \left(\frac{\theta}{2\pi}\right) \frac{\mu_0 i}{2R} = \left(\frac{\mu_0}{4\pi}\right)\left(\frac{i}{R}\right)\theta$$

   Here, $\theta = 30°$ and $i = 3$ A, $R = 0.6$ m

$$B = \left(\frac{\mu_0}{4\pi}\right)\left(\frac{3}{0.6}\right)\left(\frac{\pi}{6}\right)$$

$$= \frac{10^{-7} \times 3 \times \pi}{0.6 \times 6} = 2.6 \times 10^{-7} \text{ T}$$

10. From the given figure as shown below

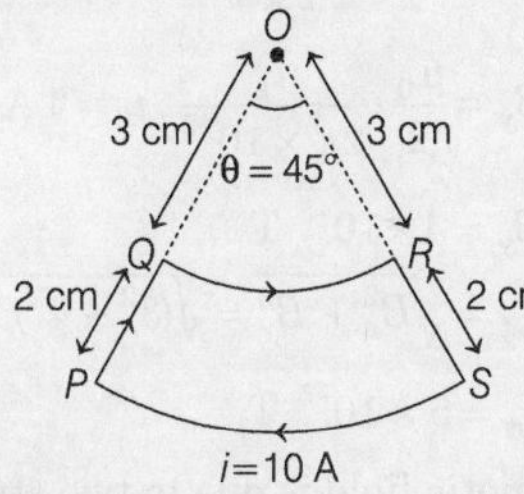

The magnetic field at point O due to wires PQ and RS will be zero.

Magnetic field due to arc QR at point O will be

$$B_1 = \frac{\theta}{2\pi}\left(\frac{\mu_0 i}{2a}\right)$$

Here, $\theta = 45° = \frac{\pi}{4}$ rad, $i = 10$ A

and $a = 3$ cm $= 3 \times 10^{-2}$ m

$$\Rightarrow \quad B_1 = \frac{\pi}{2\pi \times 4}\left(\frac{\mu_0 \times 10}{2 \times 3 \times 10^{-2}}\right)$$

$$= \frac{\mu_0 \times 5}{2 \times 12 \times 10^{-2}} = \frac{5 \times \mu_0 \times 10^2}{24}$$

Direction of field B_1 will be coming out of the plane of figure.

Similarly, field at point O due to arc SP will be

$$B_2 = \frac{\pi}{4}\left(\frac{1}{2\pi}\right)\left[\frac{\mu_0 \times 10}{2 \times (2+3) \times 10^{-2}}\right]$$

$$= \frac{\mu_0 \times 10^2}{8}$$

Direction of B_2 is going into the plane of the figure.

$\therefore$ The resultant field at O is

$$B = B_1 - B_2 = \frac{1}{2}\left(\frac{5 \times \mu_0}{12 \times 10^{-2}} - \frac{\mu_0}{4 \times 10^{-2}}\right)$$

$$= \frac{4\pi \times 10^{-7}}{12 \times 10^{-2}} \cong 1 \times 10^{-5} \text{ T}$$

11. Magnitude of magnetic field due to one side of hexagonal loop at the centre is

$$B_1 = \frac{\mu_0 i}{4\pi r} \cdot 2 \sin \phi$$

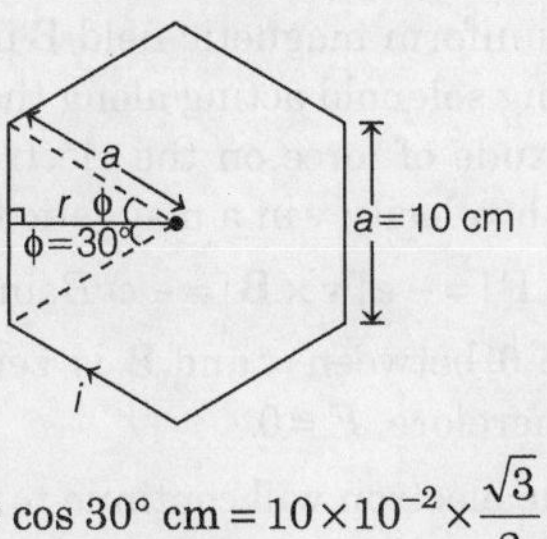

Here, $r = 10 \cos 30°$ cm $= 10 \times 10^{-2} \times \frac{\sqrt{3}}{2}$ m

$$= 5\sqrt{3} \times 10^{-2} \text{ m}$$

So, $B_1 = \frac{\mu_0 i \times 2 \sin 30°}{4\pi (5\sqrt{3} \times 10^{-2})}$ $\quad (\because \phi = 30°)$

So, magnetic field due to complete 1 turn of loop of six sides is

$$B_2 = 6B_1 = \frac{6\,\mu_0 i \times 2 \sin 30°}{4\pi (5\sqrt{3} \times 10^{-2})}$$

Magnetic field of 50 such turns at centre is

$$B_3 = 50B_2 = \frac{50 \times 6 \times \mu_0 i \times 2 \sin 30°}{4\pi (5\sqrt{3} \times 10^{-2})}$$

$$= 500\sqrt{3}\left(\frac{\mu_0 i}{\pi}\right) \text{T}$$

$$= 500\sqrt{3} \text{ in units of } \frac{\mu_0 i}{\pi}$$

12. $B_1 = \frac{\mu_0}{4\pi}\frac{2\pi n i}{r}$ and $B_2 = \frac{\mu_0}{4\pi}\frac{2\pi n i r^2}{(r^2 + h^2)^{3/2}}$

So, $$\frac{B_2}{B_1} = \frac{r^3}{(r^2 + h^2)^{3/2}} = \left(1 + \frac{h^2}{r^2}\right)^{-3/2}$$

Fractional decrease in the magnetic field will be

$$= \frac{B_1 - B_2}{B_1} = \left(1 - \frac{B_2}{B_1}\right)$$

$$= \left[1 - \left(1 + \frac{h^2}{r^2}\right)^{-3/2}\right]$$

$$= 1 - \left(1 - \frac{3h^2}{2r^2}\right) = \frac{3h^2}{2r^2}$$

13. Given arrangement consists of two current carrying wires and one current loop as shown in the question figure.

For wire ABN, magnetic field at O is given by

$$B_1 = \frac{\mu_0}{4\pi} \times \frac{i}{R}(\sin \phi_1 + \sin \phi_2)$$

Here, $\phi_1 = +90°$ and $\phi_2 = -45°$.

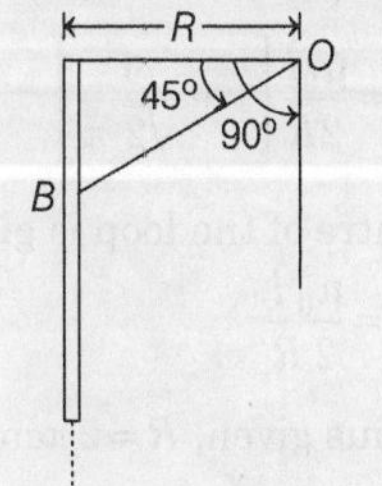

Hence, magnetic field of wire,

$$B_1 = \frac{\mu_0}{4\pi} \times \frac{i}{R}[\sin 90° + \sin(-45°)]$$

$$= \frac{\mu_0 i}{4\pi R}\left(1 - \frac{1}{\sqrt{2}}\right) \otimes$$

Also, field of wire BDC at O,

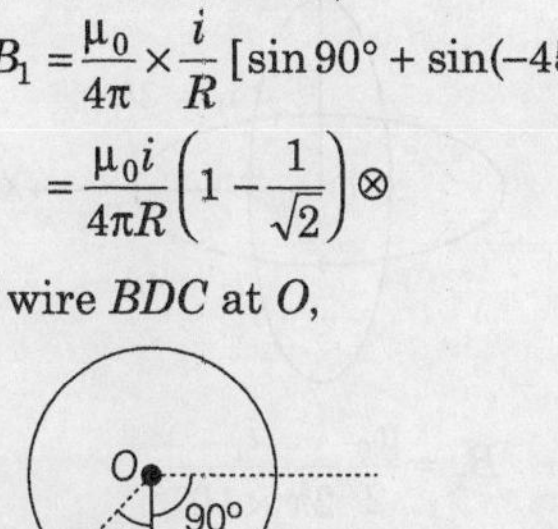

$$B_2 = \frac{\mu_0 i}{4\pi R}(\sin\phi_1 + \sin\phi_2)$$

Here, $\phi_1 = 45°$ and $\phi_2 = 90°$.

$$\therefore \quad B_2 = \frac{\mu_0 i}{4\pi R}\left(\frac{1}{\sqrt{2}} + 1\right) \odot$$

Magnetic field of circular coil at O,

$$B_3 = \frac{\mu_0 i}{2R} \odot$$

Hence, net field at O,

$$B_{net} = B_1 \otimes + B_2 \odot + B_3 \odot$$

$$= \frac{\mu_0 i}{4\pi R}\left(1 - \frac{1}{\sqrt{2}}\right) - \frac{\mu_0 i}{4\pi R}\left(\frac{1}{\sqrt{2}} + 1\right) - \frac{\mu_0 i}{2R}$$

$$= \frac{\mu_0 i}{2R}\left(\frac{1}{2\pi} - \frac{1}{\sqrt{2}\cdot 2\pi} - \frac{1}{\sqrt{2}\cdot 2\pi} - \frac{1}{2\pi} - 1\right)$$

$$= \frac{-\mu_0 i}{2\pi R}\left(\pi + \frac{1}{\sqrt{2}}\right) = \frac{\mu_0 i}{2\pi R}\left(\pi + \frac{1}{\sqrt{2}}\right).$$

Negative sign is taken for outward field.

14. From the figure, magnetic field due to AB,

$$\mathbf{B}_1 = \frac{\mu_0 i}{4\pi R}\left[\sin\frac{\pi}{2} - \sin\frac{\pi}{4}\right] = \frac{\mu_0 i}{4\pi R}\left[1 - \frac{1}{\sqrt{2}}\right]$$

Magnetic field due to circular loop,

$$\mathbf{B}_2 = \frac{\mu_0 i}{2R}$$

Magnetic field due to straight wire BC,

$$\mathbf{B}_3 = \frac{\mu_0 i}{4\pi R}\left[\sin\frac{\pi}{2} + \sin\frac{\pi}{4}\right]$$

$$= \frac{\mu_0 i}{4\pi R}\left[1 + \frac{1}{\sqrt{2}}\right]$$

$\therefore$ Resultant magnetic field, $\mathbf{B} = \mathbf{B}_1 + \mathbf{B}_2 + \mathbf{B}_3$

$$= \left(\frac{\mu_0 i}{2R} + \frac{2\mu_0 i}{4\pi R}\frac{1}{\sqrt{2}}\right)$$

$$= \frac{\mu_0 i}{2R}\left[1 + \frac{1}{\sqrt{2}\,\pi}\right]$$

15. Field at the centre of the loop is given by

$$B = \frac{\mu_0 i}{2R}$$

where, R is radius given, $R = 2\pi\text{cm} = 2\pi \times 10^{-2}$ m

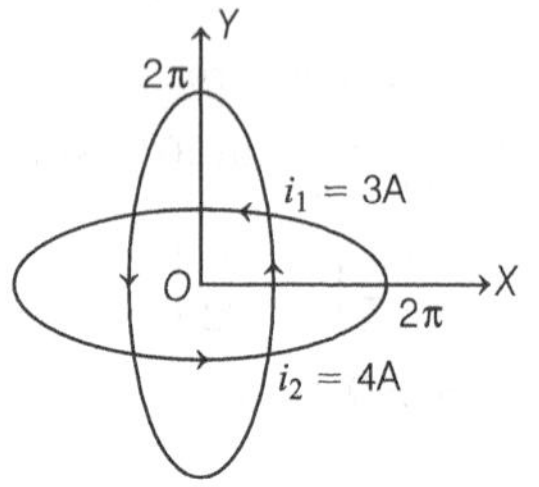

$$B_x = \frac{\mu_0}{2}\frac{i_1}{2\pi \times 10^{-2}}$$

$i_1 = 3$ A

$$\therefore \quad B_x = \frac{\mu_0}{2}\cdot\frac{3 \times 10^2}{2\pi} = 3 \times 10^{-5} \text{ T}$$

$$B_y = \frac{\mu_0}{2}\cdot\frac{i_2}{2\pi \times 10^{-2}}, \; i_2 = 4 \text{ A}$$

$$B_y = 4 \times 10^{-5} \text{ T}$$

$$\therefore \quad B_{net} = \sqrt{B_x^2 + B_y^2} = \sqrt{(3^2 + 4^2) \times 10^{-10}}$$

$$B_{net} = 5 \times 10^{-5} \text{ T}$$

16. As, total magnetic field is due to two straight paths and the circular loop.

$$B_{total} = 2 \times B_{st} + B_{loop}$$

$$= \frac{2\mu_0 I}{4\pi r} + \frac{\mu_0 I}{2r}\left(\frac{\pi}{2\pi}\right)$$

$$= \frac{\mu_0 I}{4\pi r}(2 + \pi)$$

17. Magnetic flux inside rod, $B \propto r$ and outside the rod, $B \propto \frac{i}{r}$. Therefore, the graph shown in option (d) is correct.

18. For a point inside the tube, using Ampere law, $\int B \cdot dl = \mu_0 i$. Here, we have $i = 0$ for inside the tube.

$$\therefore \quad B = 0$$

19. Magnetic field inside solenoid is

$$B = \mu n I = \mu_0 \mu_r n I = 4\pi \times 10^{-7} \times 500 \times 1000 \times 5 = \pi\text{T}$$

20. The length of solenoid, $l = 80$ cm $= 0.8$ m

Number of layers $= 5$

Number of turns per layer $= 400$

Diameter of solenoid $= 1.8$ cm

Current in solenoid, $i = 8$ A

$\therefore$ The total number of turns, $N = 400 \times 5 = 2000$

and number of turns/length, $n = \frac{2000}{0.8} = 2500$

The magnitude of magnetic field inside the solenoid

$$B = \mu_0 n i = 4 \times 3.14 \times 10^{-7} \times 2500 \times 8 = 2.5 \times 10^{-2} \text{ T}$$

The direction of magnetic field is along the axis of solenoid.

21. There is a uniform magnetic field $\mathbf{B}$ inside the current carrying long solenoid acting along the axis of solenoid. The magnitude of force on the electron of charge $(-e)$ moving with velocity $\mathbf{v}$ in a magnetic field $\mathbf{B}$ is

$$|\mathbf{F}| = -e|\mathbf{v} \times \mathbf{B}| = -evB\sin\theta$$

Here angle θ between $\mathbf{v}$ and $\mathbf{B}$ is zero, *i.e.* $\theta = 0°$ and $\sin\theta = 0$. Therefore, $F = 0$.

It means the electron will continue to move with a uniform velocity along the axis of the solenoid.

22. In given arrangement, electron gun shoots electron perpendicularly to the direction of magnetic field.

Here, we are showing cross-section of solenoid.

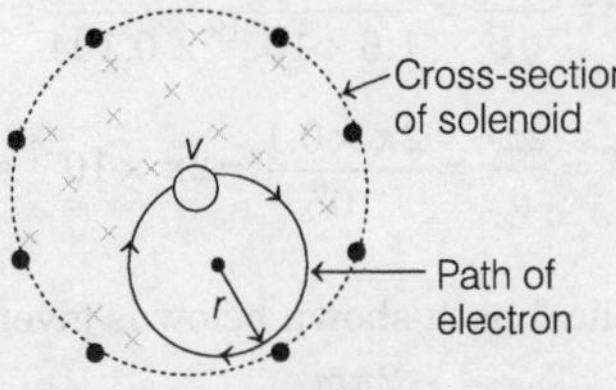

As electron is moving at 90° to magnetic field, its path is a circle. Maximum value of radius of path taken by electron, such that it does not collide with solenoid wall is $\frac{R}{2}$.

So, $$r_{\max} = \frac{R}{2} = \frac{mv_{\max}}{eB}$$

Here, $$B = \mu_0 nI$$

So, $$v_{\max} = \frac{ReB}{2m} = \frac{e\mu_0 nIR}{2m}$$

23. As we know that, if a charge particle moves in a uniform magnetic field, then its path is always circular. Considering the charge positive,the direction of magnetic force acting on it, has been shown by dotted line.

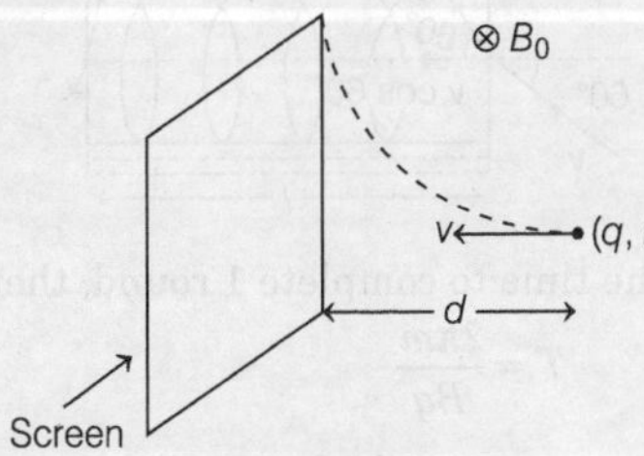

As, charge particle should not hit the screen this means radius of circular path must be less than screen distance d.

i.e. $$R \le d$$

$$\Rightarrow \quad \frac{mv}{qB_0} \le d \qquad \left(\because R = \frac{mv}{qB_0}\right)$$

or $$v \le \frac{qB_0 d}{m}$$

Maximum value of $v = \frac{qB_0 d}{m}$

24.

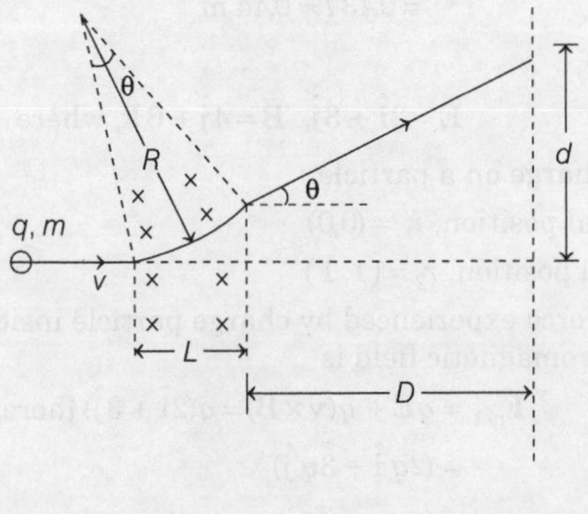

When electron enters the region of magnetic field, it experiences a magnetic force which rotates electron in a circular path of radius R.

So, magnetic force acts like a centripetal force and we have

$$\frac{mv^2}{R} = Bqv$$

where, m = mass of electron, q = charge of electron, v = speed of electron, R = radius of path and B = magnetic field intensity.

Radius of path of electron,

$$R = \frac{mv}{Bq}$$

Now, from geometry of given arrangement, comparing values of $\tan\theta$, we have

$$\tan\theta = \frac{L}{R} = \frac{d}{D}$$

$$\Rightarrow \quad d = \frac{LD}{R} = \frac{Bq\,LD}{mv}$$

$$\Rightarrow \quad d = \frac{BqLD}{\sqrt{2mK}} \qquad (\because mv = \sqrt{2mK})$$

where, K = kinetic energy of electron

Here, $B = 1.5 \times 10^{-3}$ T,

$q = 1.6 \times 10^{-19}$ C, $L = 2 \times 10^{-2}$ m, $D = 6 \times 10^{-2}$ m,

$m = 9.1 \times 10^{-31}$ kg, $K = 100 \times 1.6 \times 10^{-19}$ J

So, $$d = \frac{(1.5 \times 10^{-3} \times 1.6 \times 10^{-19} \times 2 \times 10^{-2} \times 6 \times 10^{-2})}{\sqrt{(2 \times 9.1 \times 10^{-31} \times 100 \times 1.6 \times 10^{-19})}}$$

$$= 5.34 \text{ cm}$$

No option is matching.

25. During the circular motion of accelerated electron in the presence of magnetic field, its radius is given by

$$r = \frac{mv}{Be} = \frac{\sqrt{2meV}}{eB}$$

where, v is velocity and V is voltage.

After substituting the given values, we get

$$= \frac{\sqrt{2 \times 9.1 \times 10^{-31} \times 1.6 \times 10^{-19} \times 500}}{1.6 \times 10^{-19} \times 100 \times 10^{-3}}$$

$$= 10\left[\frac{2 \times 9.1 \times 500}{1.6} \times 10^{-12}\right]^{1/2},$$

$$r = 7.5 \times 10^{-4} \text{ m}$$

26. Radius of path of charged particle q in a uniform magnetic field B of mass m moving with velocity v is

$$r = \frac{mv}{Bq} = \frac{m\sqrt{(2qV/m)}}{Bq}$$

$$\Rightarrow \quad r \propto \frac{\sqrt{m}}{\sqrt{q}}$$

So, required ratio is

$$\frac{r_p}{r_\alpha} = \sqrt{\frac{m_p}{m_\alpha}} \times \sqrt{\frac{q_\alpha}{q_p}} = \sqrt{\frac{1}{4}} \times \sqrt{\frac{2}{1}} = \frac{\sqrt{2}}{2} = \frac{1}{\sqrt{2}}$$

27. As, $r = \frac{\sqrt{2mE}}{Bq} = \frac{\sqrt{2m_1E_1}}{Bq}$

or $E_1 = \frac{mE}{m_1} = \frac{(2m_1)}{m_1} \times 50 \text{ keV} = 100 \text{ keV}$

28. As, $E = \frac{1}{2}mv^2$ or $v = \sqrt{\frac{2E}{m}}$

and $r = \frac{mv}{Bq} = \frac{m}{Bq}\sqrt{2E/m}$

or $r = \frac{\sqrt{2Em}}{Bq}$ or $r \propto \sqrt{m}$

Now, $m_e < m_p$, so $r_e < r_p$.
Therefore, trajectory of proton is less curved.

29. The force on a point charge Q in a magnetic field is

$$\mathbf{F} = Q(\mathbf{v} \times \mathbf{B})$$

Its direction is perpendicular to direction of motion of charge, so work done,

$$W = \mathbf{F} \cdot \mathbf{s} = Fs\cos 90^\circ = 0$$

30. In magnetic field, the radius of circular path,

$$r = \frac{mv}{Bq} = \frac{v}{B\left(\frac{q}{m}\right)}, \text{ i.e. } r \propto 1/(q/m)$$

As, $r_M < r_N$
So, the specific charge of M is greater than that of N.

31. From $Bqv = \frac{mv^2}{r}$, we have

$$r = \frac{mv}{Bq} = \frac{\sqrt{2mK}}{Bq}$$

where, K is the kinetic energy.
As, kinetic energies of particles are same.

$$r \propto \frac{\sqrt{m}}{q}$$

$\Rightarrow$ $r_e : r_p : r_\alpha = \frac{\sqrt{m_e}}{e} : \frac{\sqrt{m_p}}{e} : \frac{\sqrt{4m_p}}{2e}$

Clearly, $r_p = r_\alpha$ and r_e is least $(\because m_e < m_p)$

So, $r_p = r_\alpha > r_e$

32. Since, the proton is entering the magnetic field at some angle other than 90°, its path is helix. Component velocity of proton along X-axis,

$$v_x = v\cos 60^\circ = 2 \times 10^6 \times \frac{1}{2} = 10^6 \text{ ms}^{-1}$$

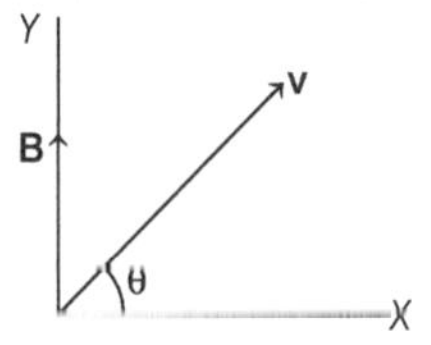

Due to component of velocity v_x, the radius of the helix described is given by the relation,

$$r = \frac{mv_x}{qB} = \frac{1.67 \times 10^{-27} \times 10^6}{1.6 \times 10^{-19} \times 0.104} = 0.1 \text{ m}$$

Now, $T = \frac{2\pi r}{v_x} = \frac{2\pi \times 0.1}{10^6} = 2\pi \times 10^{-7} \text{ s}$

33. Pitch of helical path shown below is given by

$$\text{Pitch} = T \cdot v\cos\theta = \frac{2\pi m}{Bq} \cdot v\cos\theta$$

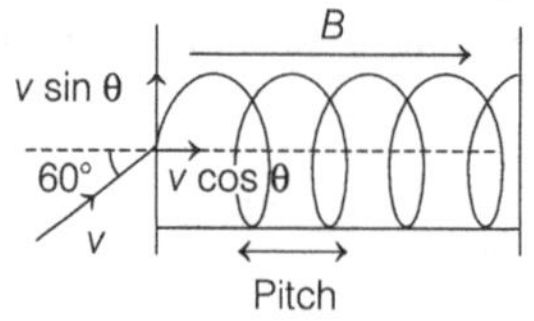

$$\Rightarrow \text{Pitch} = \frac{2 \times 3.14 \times 1.67 \times 10^{-27} \times 4 \times 10^5 \times \cos 60^\circ}{(0.3)(1.69 \times 10^{-19})}$$

$= 0.04 \text{ m} = 4 \text{ cm}$

34. As $\mathbf{v}$ is not perpendicular to $\mathbf{B}$, so path of particle is helical in magnetic field as shown in figure,

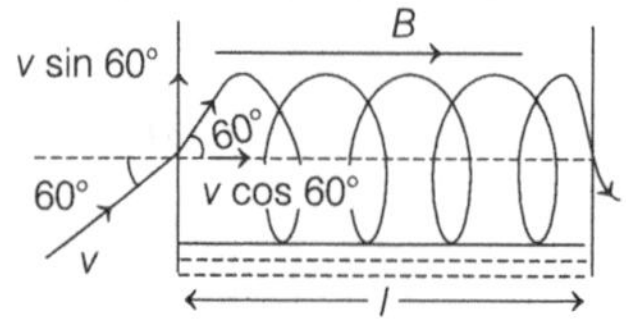

Let T be the time to complete 1 round, then

$$T = \frac{2\pi m}{Bq}$$

So, total time to cross the length l of region of magnetic field is

$$t = 10\,T = 10 \times \frac{2\pi m}{Bq}$$

Hence, length of path,

$$l = \text{speed} \times \text{time} = v\cos\theta \times t$$

$$= v\cos 60^\circ \times 10 \times \frac{2\pi m}{Bq}$$

$\Rightarrow$ $l = \frac{4 \times 10^5 \times \frac{1}{2} \times 10 \times 2\pi \times 1.67 \times 10^{-27}}{0.3 \times 1.6 \times 10^{-19}}$

$= 0.437 \approx 0.44 \text{ m}$

35. Here,

$$\mathbf{E} = 2\hat{\mathbf{i}} + 3\hat{\mathbf{j}}, \ \mathbf{B} = 4\hat{\mathbf{j}} + 6\hat{\mathbf{k}}, \text{ where}$$

q = charge on a particle.
Initial position, $r_1 = (0,0)$
Final position, $r_2 = (1,1)$
Net force experienced by charge particle inside electromagnetic field is

$$\mathbf{F}_{\text{net}} = q\mathbf{E} + q(\mathbf{v} \times \mathbf{B}) = q(2\hat{\mathbf{i}} + 3\hat{\mathbf{j}}) \text{ [here, } \mathbf{v} \times \mathbf{B} = 0]$$

$$= (2q\,\hat{\mathbf{i}} + 3q\,\hat{\mathbf{j}})$$

$\therefore \quad dW = \mathbf{F}_{net} \cdot d\mathbf{r}$

$\Rightarrow \quad \int dW = \int_{r_1}^{r_2} (2q\,\hat{\mathbf{i}} + 3q\,\hat{\mathbf{j}}) \cdot (dx\,\hat{\mathbf{i}} + dy\,\hat{\mathbf{j}})$

[here $d\mathbf{r} = dx\,\hat{\mathbf{i}} + dy\,\hat{\mathbf{j}}$]

$\Rightarrow \quad W = 2q\int_0^1 dx + 3q\int_0^1 dy$

or $\quad W = 2q + 3q$ or $W = 5q$

36. Here, E and B are acting along X-axis and v is acting along Y-axis, *i.e.* perpendicular to both E and B. Therefore, the path of charged particle is a helix with increasing speed. Speed of particle at time t is

$$v = \sqrt{v_x^2 + v_y^2} \quad \text{...(i)}$$

Here, $v_y = v_0$; $v_x = \frac{qE}{m}t$ and $v = \frac{\sqrt{5}}{2}v_0$

Putting values in Eq. (i), we get $t = \frac{mv_0}{2qE}$

37. As force per unit length between two parallel current carrying wires separated by a distance d is given by

$$\frac{dF}{dL} = \frac{\mu_0}{4\pi}\frac{2i_1 i_2}{d}$$

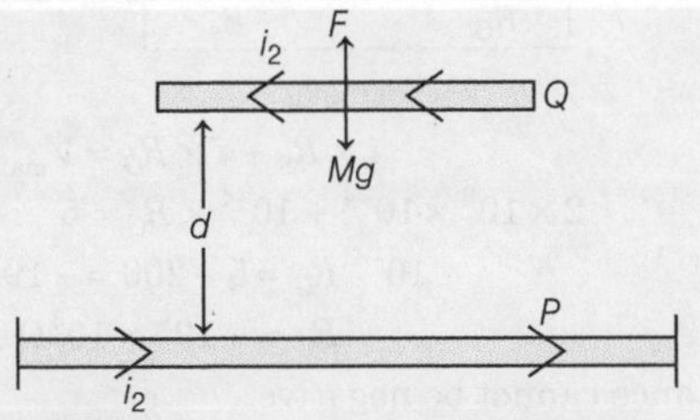

and is repulsive if the current in the wires is in opposite direction (otherwise attractive).

So, in order that wire Q may remain suspended, the force F on it must be repulsive and equal to its weight, *i.e.* the current in the two wires must be opposite directions and

$$F = Mg, \text{ i.e. } \frac{F}{L} = \frac{Mg}{L}$$

or $\quad \frac{\mu_0}{4\pi}\frac{2i_1 i_2}{d} = \frac{Mg}{L}\left[\text{as, } \frac{dF}{dL} = \frac{\mu_0}{4\pi}\frac{2i_1 i_2}{d}\right]$

or $\quad d = 10^{-7} \times \frac{2 \times 50 \times 25}{0.075} = \frac{1}{3} \times 10^{-2}\text{m}$

$$\left[\text{as, } \frac{Mg}{L} = 0.075\ \text{Nm}^{-1}\right]$$

38. Force on wire, $F = Bil \sin 90° = Bil$. It acts perpendicular to the magnetic field as well as the length of wire. The acceleration in the wire,

$$a = \frac{F}{m} = \frac{Bil}{m}$$

39. Since, the magnetic field due to current through wire CD at various locations on wire AB is not uniform. Therefore, the wire AB, carrying current i_1 is subjected to variable magnetic field. Due to which, neither the force nor the torque on the wire AB will be zero. As a result of which, the wire AB will have both translational and rotational motion.

40. Given, $i_1 = 8$ A, $i_2 = 5$ A and $r = 4$ cm $= 0.04$ m

Force per unit length on two parallel wire carrying current,

$$F = \frac{\mu_0}{4\pi}\cdot\frac{2i_1 \cdot i_2}{r}$$

$$= \frac{10^{-7} \times 2 \times 8 \times 5}{0.04}$$

$$= 2 \times 10^{-4}\ \text{N}$$

The force on A of length 10 cm, $F' = F \times 0.1$

($\because$ 1m = 100 cm)

$\Rightarrow \quad F' = 2 \times 10^{-4} \times 0.1 = 2 \times 10^{-5}$ N

41. As the currents are in opposite directions, the magnetic field induction due to current in each wire will add up at O. The direction of magnetic field is perpendicular to XY-plane and is directed inward, *i.e.* along negative Z-axis.

42. Force on a wire 1 in which current i_1 is flowing due to another wire 2 which are separated by a distance r is given as

$$\mathbf{F} = i_1(\mathbf{l} \times \mathbf{B}_2)$$

or $\quad F = \frac{\mu_0 i_1 i_2}{2\pi r} \cdot l \sin\theta \quad \left(\because \mathbf{B}_2 = \frac{\mu_0 i_2}{2\pi r}\right)$

Thus, the given square loop can be drawn as shown below

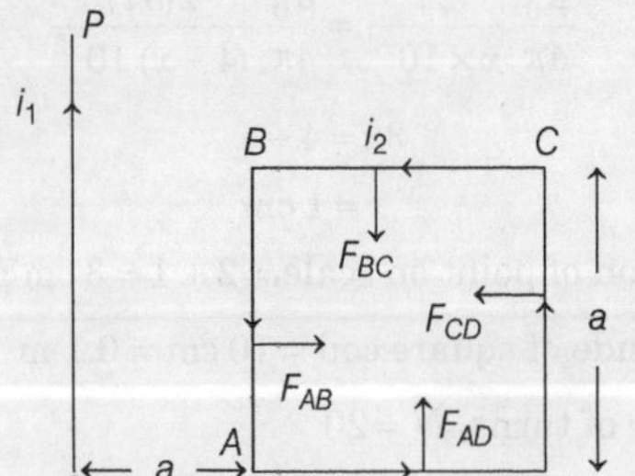

$F_{AB} = \frac{\mu_0 i_1}{2\pi a} \cdot i_2 a$ (away from wire PQ) $(\because \theta = 90°)$

$F_{BC} = F_{AD} = 0 \quad (\because \theta = 0°)$

$F_{CD} = \frac{\mu_0 i_1}{2\pi(2a)} \cdot i_2 a \quad (\because \theta = 90°)$

$= \frac{\mu_0 i_1}{4\pi a} \cdot i_2 a$ (towards the wire PQ)

$\therefore \quad F_{net} = F_{AB} - F_{CD}$

$= \frac{\mu_0 i_1 i_2}{2\pi} - \frac{\mu_0 i_1 i_2}{4\pi}$ (away from wire)

$= \frac{\mu_0 i_1 i_2}{4\pi}$ (repulsive in nature)

43. The magnetic field induction at P due to currents through both the wires is

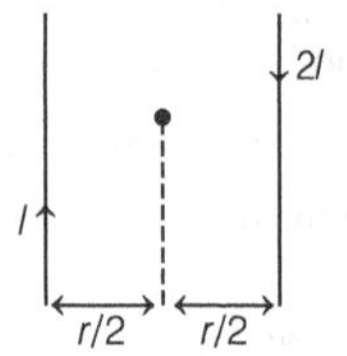

$$B = \frac{\mu_0}{4\pi}\frac{2i}{r/2} + \frac{\mu_0}{4\pi}\frac{2(2i)}{r/2} = \frac{\mu_0}{4\pi}\cdot\frac{12i}{r}$$

$= \frac{\mu_0}{4\pi} \times \frac{12\,i}{r}$ acting perpendicular to plane of wire inwards. Now, B and v are acting in the same direction, *i.e.* $\theta = 0°$.

Force on charged particle is $F = qvB\sin\theta = qvB \times 0 = 0$.

44. $$\frac{\mu_0 i}{4\pi}\frac{2i_i}{r} - \frac{\mu_0}{4\pi}\frac{2i_2}{r} = 10\,\mu\text{T} \quad \ldots\text{(i)}$$

$$\frac{\mu_0}{4\pi}\frac{2\,i_1}{r} + \frac{\mu_0}{4\pi}\frac{2\,i_2}{r} = 30\,\mu\text{T} \quad \ldots\text{(ii)}$$

On solving Eqs. (i) and (ii), we get

$$i_1 = 20\text{ A and } i_2 = 10\text{ A, so } \frac{i_1}{i_2} = 2$$

45. Distance between two linear conductors $= 6 - 2 = 4$ cm.

Let the distance of the point on scale from conductor carrying current i be x cm, where resultant magnetic field is zero. Then, the distance of this point from other conductor is $(4 - x)$ cm. As per question,

$$\frac{\mu_0}{4\pi}\frac{2\,i}{x \times 10^{-2}} = \frac{\mu_0}{4\pi}\frac{2(3\,i)}{(4 - x)\,10^{-2}}$$

or $\quad 3x = 4 - x$

or $\quad x = 1$ cm

$\therefore$ Location of point on scale $= 2 + 1 = 3$ cm mark.

46. Given, side of square coil $= 10$ cm $= 0.1$ m

Number of turns, $N = 20$

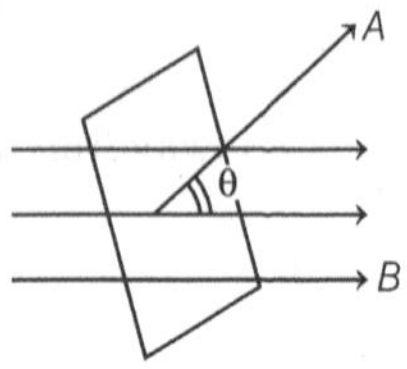

Current in square coil, $I = 12$ A

Angle made by coil, $\theta = 30°$

Magnetic field, $B = 0.80$ T

The magnitude of torque experienced by the coil,

$$\tau = NIAB\sin\theta$$
$$= 20 \times 12 \times (10 \times 10^{-2})^2 \times 0.80 \times \sin 30°$$
$$\tau = 2.4 \times 0.80\sin 30° = \frac{2.4 \times 0.80}{2} = 0.96\text{ N-m}$$

47. Maximum voltage that can be applied across the galvanometer coil $= 100\,\Omega \times 10^{-3}$ A $= 0.1$ V.

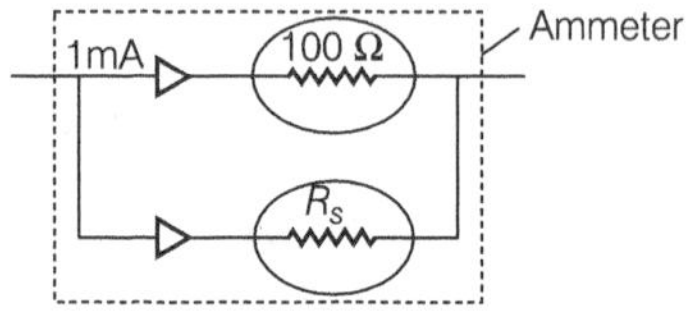

If R_s is the shunt resistance, then

$$R_s \times 10\text{ A} = 0.1\text{ V}$$
$$\Rightarrow \quad R_s = 0.01\,\Omega$$

48. Given, $i_g = i$, $G = R_0$

$$I = n\,i + i = (n + 1)\,i$$
$$\therefore \quad S = \frac{i_g G}{I - i_g} = \frac{iR_0}{(n + 1)i - i} = \frac{R_0}{n}$$

49. Given data,

$$i = 10^{-4}\text{ A},$$
$$R_S = 2\text{ M}\Omega = 2 \times 10^6\,\Omega,$$
$$V_{\text{max}} = 5\text{ V}$$

Let internal resistance of galvanometer is R_G.

Voltmeter

Then, $\quad i \times R_S + i \times R_G = V_{\text{max}}$

$$\Rightarrow \quad 2 \times 10^6 \times 10^{-4} + 10^{-4} \times R_G = 5$$
$$\Rightarrow \quad 10^{-4}\,R_G = 5 - 200 = -195$$
$$\text{or} \quad R_G = -195 \times 10^4\,\Omega$$

Resistance cannot be negative.

$\therefore$ No option is correct.

50. Here, $i_g =$ full scale deflection current

$$= 1\text{mA} = 1 \times 10^{-3}\text{ A}$$

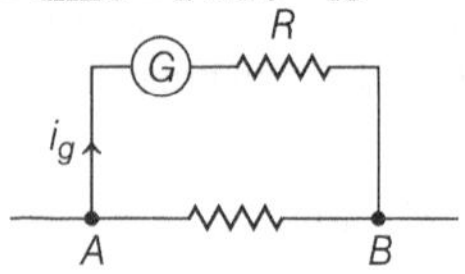

$V_{AB} =$ full scale deflection potential $= 10$ V

$G =$ Galvanometer resistance $= 100\,\Omega$

If $R =$ resistance used in series for converting galvanometer into voltmeter, then from above circuit diagram,

$$i_g\,(G + R) = V$$
$$\Rightarrow \quad 10^{-3}(100 + R) = 10$$
$$\Rightarrow \quad R = 10000 - 100 = 9900\,\Omega = 9.9\text{k}\Omega$$

51. As, $i_g = \frac{2}{2000} = \frac{1}{1000}$ A;

Now for range, $V = 8\text{V} + 2\text{V} = 10\text{V}$

$$\therefore \quad R = \frac{V}{i_g} - G = \frac{10}{\left(\frac{1}{1000}\right)} - 2000 = 8000\,\Omega$$

52. Given, divisions in scale of galvanometer,

$$n = 50$$

Sensitivity of galvanometer,

$$\frac{i_g}{n} = 20\ \mu\text{A/division}$$

$\therefore$ Current in galvanometer,

$$I_g = \frac{I_g}{n} \times n = 20\mu\text{A} \times 50$$

$$\Rightarrow \quad I_g = 1000\ \mu\text{A} = 1\ \text{mA}$$

Now, for R, it should be converted into 2V voltmeter.

$$\therefore \quad V_1 = i_g\ (R_1 + G)$$

$$2 = 10^{-3}(R_1 + 100)$$

$$\Rightarrow \quad 2000 = R_1 + 100$$

$$\therefore \quad R_1 = 1900\ \Omega \quad \text{...(i)}$$

For R_2, it should be converted into 10V voltmeter.

$$\therefore \quad V_2 = i_g\ [(R_1 + R_2) + G]$$

$$\Rightarrow \quad 10 = 10^{-3}\ [(R_1 + R_2) + 100]$$

$$\Rightarrow \quad 10000 = R_1 + R_2 + 100 = 2000 + R_2$$

$$\therefore \quad R_2 = 8000\ \Omega \quad \text{...(ii)}$$

For R_3, it should be converted into 20V voltmeter.

$$\therefore \quad V_3 = I_g[(R_1 + R_2 + R_3 + G]$$

$$\Rightarrow \quad 20 = 10^{-3}\ [1900 + 8000 + R_3 + 100]$$

$$\Rightarrow \quad 20000 = R_3 + 10000$$

$$\therefore \quad R_3 = 10000\ \Omega \quad \text{...(iii)}$$

From Eqs. (i), (ii) and (iii), it is clear that option (c) is correct.

53. To convert a glavanometer into a voltmeter, a resistance $R = \frac{V}{i_g} - G$ is connected in series with it.

To convert galvanometer into an ammeter, a resistance, $S = i_g G/(i - i_g)$ is to be connected in parallel with galvanometer.

Therefore, the combination should have a range of 10 V with connection of resistance 200 kΩ in series.

54. To use galvanometer as an ammeter, a low resistance in parallel is used.

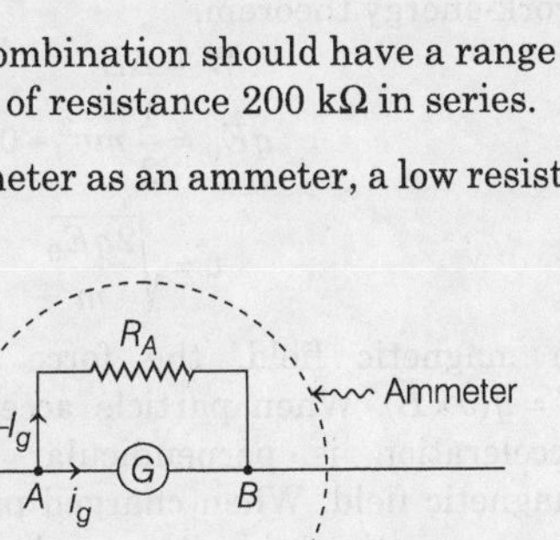

In ammeter, if i_g = full scale deflection current, then equating potential drops across points marked AB, we have

$$V_{AB} = i_g G$$

$$= (i_0 - i_g) R_A$$

$$\Rightarrow \quad R_A = \frac{i_g G}{i_0 - i_g} \quad \text{...(i)}$$

Here, G = resistance of galvanometer coil.

When a galvanometer is used as a voltmeter, a high resistance (R_V) in series is used.

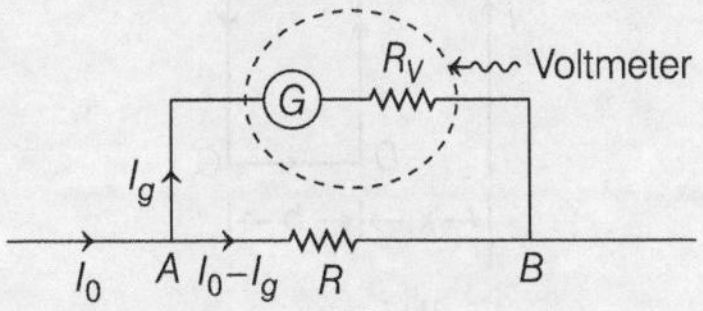

Equating potential across point AB,

$$V_{AB} = (G + R_V)\ i_g$$

But $\quad V_{AB} = i_0 G \quad$ (given)

So, $\quad i_0 G = (G + R_V)\ i_g$

$$\Rightarrow \quad R_V = \frac{(i_0 - i_g)G}{i_g} \quad \text{...(ii)}$$

From Eqs. (i) and (ii), we have

$$\frac{R_A}{R_V} = \frac{\left(\frac{i_g G}{i_0 - i_g}\right)}{\frac{(i_0 - i_g)G}{i_g}}$$

$$= \frac{i_g^2}{(i_0 - i_g)^2}$$

and $\quad R_A \times R_V = \frac{i_g G}{(i_0 - i_g)} \times \frac{(i_0 - i_g)\ G}{i_g} = G^2$

Round II

1. Figure shows infinitely long straight thin walled pipe carrying current I.

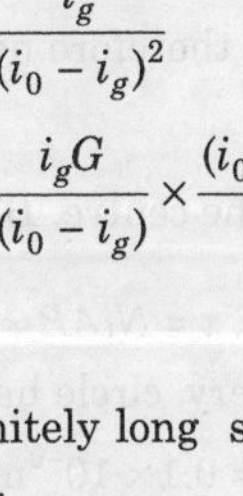

Let P be any point at a distance r from the axis OO_1 of the pipe. Let B be magnetic field at P. Consider a closed circular path passing through point P as shown in figure. From Ampere's circuital law $\oint B \cdot dl = \mu_0 i$

i = current through the closed path. Obviously, $i = 0$

$\therefore \quad 2\pi RB = 0$ or $B = 0$

2. The magnetic force on AB and CD are equal and opposite due to symmetry and opposite currents in these sides. The magnetic force on AD,

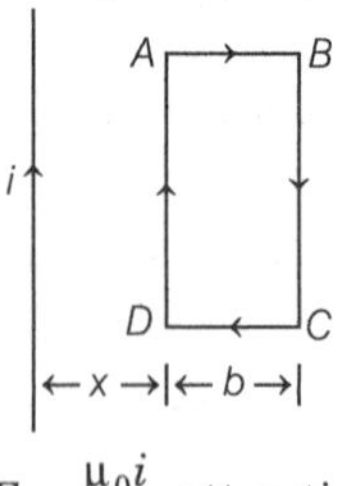

$$F_1 = \frac{\mu_0 i}{2\pi x}, \text{ attractive}$$

Magnetic force on BC

$$F_2 = \frac{\mu_0 i}{2\pi (x + b)}, \text{ repulsive}$$

As $F_1 > F_2$, so magnetic force will be experienced towards wire.

3. Force on wire B due to A,

$$F_{BA} = \frac{\mu_0 \times 1 \times 2}{2\pi r} = \frac{\mu_0}{\pi r} \text{ towards } C$$

Force on wire B due to C,

$$F_{BC} = \frac{\mu_0 \times 2 \times 2}{2\pi r} = \frac{3\mu_0}{\pi r} \text{ towards } A$$

Clearly, $F_{BC} > F_{BA}$, therefore net force on D is directed towards A.

4. Magnetic field at the centre, $B = \frac{\mu_0 i\theta}{4\pi a} = \frac{\mu_0 i}{4\pi a} \cdot \frac{3\pi}{2} = \frac{3\mu_0 i}{8a}$

5. Torque on the loop, $\tau = NiAB \propto A$

For a given periphery, circle has maximum area.

6. Here, $2r = 0.1$ nm $= 0.1 \times 10^{-9}$m $= 10^{-10}$m;

$i = \frac{e}{T} = \frac{e\omega}{2\pi}$ where, ω = angular speed.

Now, $\quad B = \frac{\mu_0}{4\pi} \frac{2\pi n i}{r} = \frac{\mu_0}{4\pi} \frac{n e \omega}{r}$

or $\quad \omega = B \cdot \frac{4\pi}{\mu_0} \cdot \frac{r}{ne}$

$$= 14 \times \frac{1}{10^{-7}} \times \frac{(10^{-10})/2}{1.6 \times 10^{-19}}$$

$$= 4.4 \times 10^{16} \text{ rad/s}$$

7. Maximum value of force, $F_{max} = evB$

$$= (1.6 \times 10^{-19}) \times (0.9 \times 3 \times 10^8) \times (10^8)$$

$$= 4.32 \times 10^{-3} \text{ N}$$

8. As, $eE = evB \Rightarrow v = \frac{E}{B}$

Here, $\quad E = 1 \text{ Vcm}^{-1} = 100 \text{ Vm}^{-1}, B = 2\text{T}$

$\therefore \quad v = \frac{100}{2} = 50 \text{ ms}^{-1}$

9. As, $E = \frac{B^2 q^2 r^2}{2m}$ or $r = \frac{\sqrt{2mE}}{Bq}$

So, $\quad r \propto \sqrt{E}/B$

$\therefore \quad \frac{r_2}{r_1} = \sqrt{\frac{2E}{E}} \cdot \frac{B}{3B} = \sqrt{\frac{2}{9}}$

$\therefore \quad r_2 = \sqrt{\frac{2}{9}} = \sqrt{\frac{2}{9}} R$

10. Consider the two coaxial solenoids. Due to one of the solenoids, magnetic field at the centre of the other can be assumed to be constant.

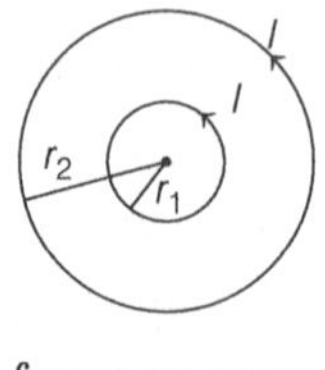

Due to symmetry, forces on upper and lower part of a solenoid will be equal and opposite and hence resultant is zero.

Therefore, $\mathbf{F}_1 = \mathbf{F}_2 = 0$.

11. Consider a simple amperian loop passing once through both the identical current carrying coaxial loops.

(i) According to Ampere circuital law, $\oint_C \mathbf{B} \cdot d\mathbf{l} = \mu_0 (i - i) = 0$.

(ii) As $\oint_C \mathbf{B} \cdot d\mathbf{l} = 0$, therefore $\oint_C \mathbf{B} \cdot d\mathbf{l}$ is independent of sense of C.

(iii) The value of $\mathbf{B}$ does not vanish on various points of C.

12. Since, the magnetic field does not perform any work, therefore whatever has been gain in kinetic energy, it is only because of the work done by electric field. Applying work-energy theorem,

$$W = \Delta E$$

$$qE_0 = \frac{1}{2} mv^2 - 0$$

or $\quad v = \sqrt{\frac{2qE_0}{m}}$

13. In magnetic field, the force on charged particle $F = q(v \times B)$. When particle accelerates from rest, its acceleration is perpendicular to the direction of magnetic field. When charged particle is deflected by the magnetic field, its speed does not change but direction of velocity changes because the deflecting force acts perpendicular to v and B. Only the component of magnetic field perpendicular to the direction of motion is effective in deflecting the particle. Also, while deflecting in the magnetic field, its energy gradually decreases.

14. Suppose a resistance R_s is connected in series with galvanometer to convert it into voltmeter.

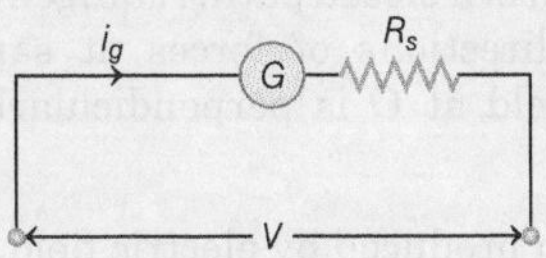

$$i_g (G + R_s) = V$$

$$\Rightarrow \quad R = \frac{V}{i_g} - G$$

$$\Rightarrow \quad R = 1985 = 1.985 \text{ k}\Omega$$

or $$R = 1.985 \times 10^3 \ \Omega$$

15. Magnetic field at O due to PR,

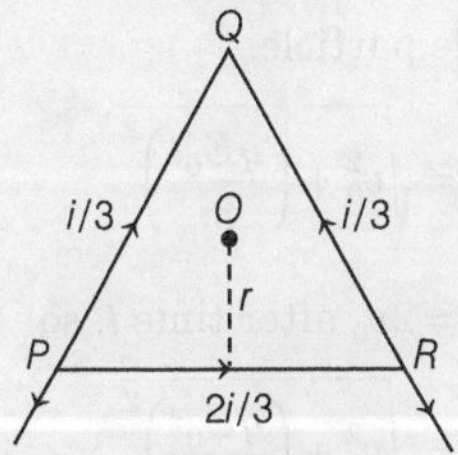

$$B_1 = \frac{\mu_0}{4\pi} \frac{2i/3}{r} [\sin 30° + \sin 30°]$$

$$= \frac{\mu_0}{4\pi} \frac{2i}{3r}$$

It is directed outside the paper. Magnetic field at O, due to PQR (*i.e.*, for the wire PQ and QR)

$$B_2 = 2 \times \frac{\mu_0}{4\pi} \frac{\left(\frac{i}{3}\right)}{\mu} [\sin 30° + \sin 30°]$$

$$= \frac{\mu_0}{4\pi} \frac{2i}{3r}$$

It is directed inside the paper.

$\therefore$ Resultant magnetic field at O

$$B = B_1 - B_2 = 0$$

16. (a) Motion of charges inside the conductor is affected by magnetic field **B**, due to magnetic force **F** given by

$$\mathbf{F} = q\,(\mathbf{v} \times \mathbf{B})$$

(b) Due to magnetic force, some charges inside the wire move to the surface of wire.

(c) The force on wire of length l, carrying current I when subjected to magnetic field **B** is, $\mathbf{F} = I\,(\mathbf{l} \times \mathbf{B})$. It acts perpendicular to the plane containing **l** and **B** and is directed as given by Right Hand rule. If the wire moves under the influence of **B** at an angle θ, where $\theta \neq 90°$, then work done, $W = Fs\cos\theta$, can not be zero.

(d) When wire moves under the influence of **B**, then displacement of the ions is perpendicular to the magnetic force **F**. Therefore **work done is zero.**

17. Magnetic field at point P due to i_1, $B_1 = \frac{\mu_0 i_1}{2\pi x}$ upward perpendicular to plane of paper. Therefore, magnetic force on element P

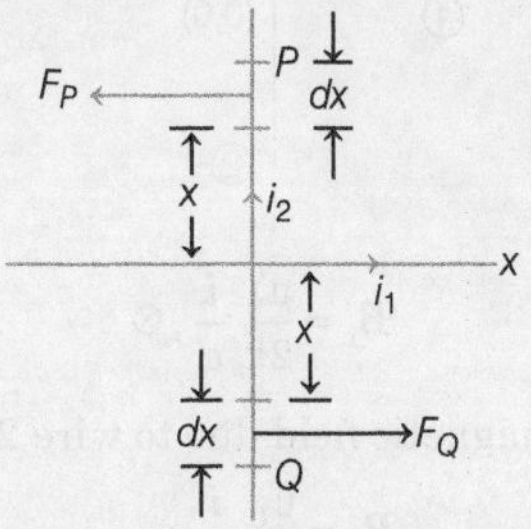

$$F_P = B_1 i_2 dx = \frac{\mu_0 i_1 i_2}{2\pi x} dx, \text{ along } X\text{-axis}$$

Magnetic field at point Q due to i_1,

$B_2 = \frac{\mu_0 i_1}{2\pi x}$ downward perpendicular to plane to paper.

Magnetic force on element Q,

$$F_Q = B_2\, i_2\, dx = \frac{\mu_0 i_1 i_2}{2\pi x} dx, \text{ along } (+)\ X\text{-axis}$$

So, wire i_2 will turn counter-clockwise.

18. For small element of wire,

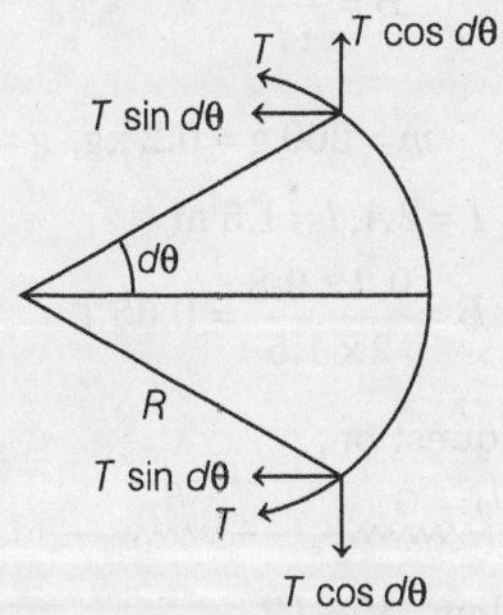

$$2T \sin d\theta = 2R\, d\theta\, iB$$

$$2Td\theta = 2RiBd\theta$$

$$T = iRB$$

19. Magnetic field due to an infinitely long straight wire at point P is given as

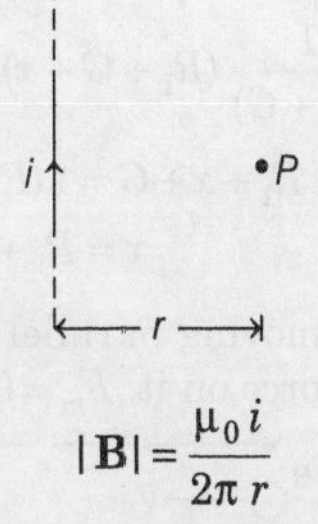

$$|\mathbf{B}| = \frac{\mu_0 i}{2\pi r}$$

Thus, in the given situation, magnetic field due to wire 1 at point P is

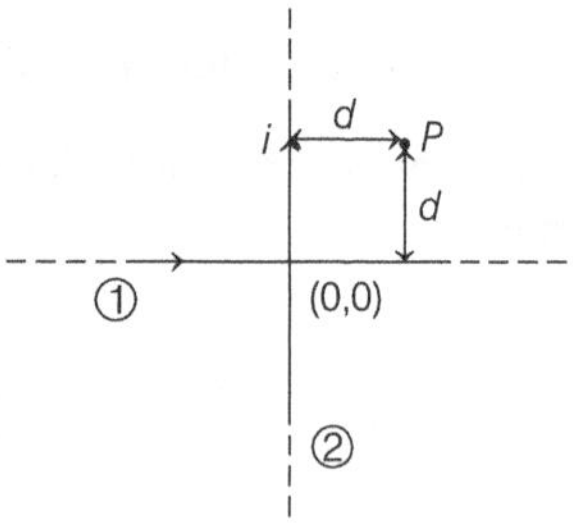

$$\mathbf{B}_1 = \frac{\mu_0}{2\pi}\frac{i}{d}, \otimes$$

Similarly, magnetic field due to wire 2 at point P is

$$\mathbf{B}_2 = \frac{\mu_0}{2\pi}\frac{i}{d}, \odot$$

Resultant field at point P is

$$\mathbf{B}_{net} = \mathbf{B}_1 + \mathbf{B}_2$$

Since, $|\mathbf{B}_1| = |\mathbf{B}_2|$, but they are opposite in direction.

Thus, $\mathbf{B}_{net} = 0$

$\therefore$ Net magnetic field at point P will be zero.

20. Applying Fleming's rule, we find that upward force F of magnitude IlB acts. For mid-air suspension, this must be balanced by the force due to gravity.

$$\therefore \quad mg = i\,lB$$

$$\Rightarrow \quad B = \frac{mg}{i\,l}$$

Given, $\quad m = 200\text{ g} = 0.2\text{ kg},\ g = 9.8\text{ m/s}^2$

$$I = 2A,\ l = 1.5\text{ m}$$

we have, $\quad B = \dfrac{0.2 \times 9.8}{2 \times 1.5} = 0.65\text{ T}$

21. According to question,

From Ohm's law, $V = IR$

$$\therefore \quad i_g(R_1 + G) = 1\text{V} \quad \text{...(i)}$$

Let x be the required resistance.

Now, $\quad i_g(R_1 + G + x) = 2\text{V}$

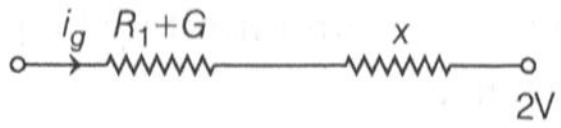

Using Eq. (i) in above equation, we get

$$\frac{1}{(R_1 + G)}(R_1 + G + x) = 2$$

$$R_1 + x + G = 2(R_1 + G)$$

$$x = R_1 + G$$

22. Since election is moving parallel to the magnetic field, hence magnetic force on it. $F_m = 0$

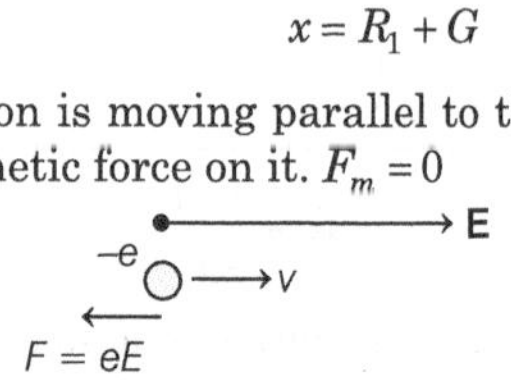

So, the only force which is acting on electron in the direction, is electric force which reduces its speed.

23. The current through loop is in anticlockwise direction. Hence, magnetic field at the points within the loop acts perpendicularly outward to the plane. The magnetic field lines form a closed path. A tangent to the field lines gives the directions of forces at same point. Thus, magnetic field at O is perpendicularly inward to the paper.

24. Acceleration produced by electric field,

$$\mathbf{a} = \frac{qE_0}{m}\hat{\mathbf{i}}$$

After time t, velocity of the particle,

$$\mathbf{v} = \mathbf{u} + \mathbf{a}t$$

$$\Rightarrow \quad \mathbf{v} = v_0\hat{\mathbf{j}} + \left(\frac{qE_0t}{m}\right)\hat{\mathbf{i}} \qquad [\because u = \mathbf{v} = v_0\hat{\mathbf{j}}]$$

So, speed of the particle,

$$|\mathbf{v}| = \sqrt{v_0^2 + \left(\frac{qE_0t}{m}\right)^2}$$

Now, given $|\mathbf{v}| = 2v_0$ after time t, so

$$2v_0 = \sqrt{v_0^2 + \left(\frac{qE_0t}{m}\right)^2}$$

$$\Rightarrow \quad 4v_0^2 = v_0^2 + \left(\frac{qE_0t}{m}\right)^2$$

$$\Rightarrow \quad 3v_0^2 = \left(\frac{qE_0t}{m}\right)^2$$

So, time interval, $t = \dfrac{\sqrt{3}\,mv_0}{qE_0}$

Here, we must note that no change in magnitude of velocity is caused by a perpendicular magnetic field. So, we are not taking effect of magnetic field while calculating change in speed.

25. Force exerted on a current carrying conductor,

$$F_{ext} = B(x)\,iL$$

$$\text{Average power} = \frac{\text{Work done}}{\text{Time taken}}$$

$$P = \frac{1}{t}\int_0^2 F_{ext}\cdot dx$$

$$= \frac{1}{t}\int_0^2 B(x)\,iL\,dx$$

$$= \frac{1}{5 \times 10^{-3}}\int_0^2 3 \times 10^{-4} e^{-0.2x} \times 10 \times 3\,dx$$

$$= 9\,[1 - e^{-0.4}]$$

$$= 9\left[1 - \frac{1}{e^{0.4}}\right]$$

$$= 2.967 \approx 2.97\text{ W}$$

26. From the figure, due to $FABC$, the magnetic field at O is along Y-axis and due to $CDEF$, the magnetic field is along axis. Hence, the field will be of the form $A(\hat{\mathbf{i}}+\hat{\mathbf{j}})$.

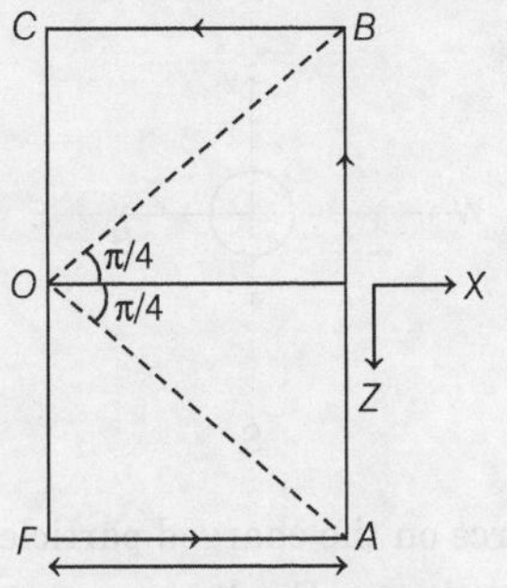

Calculating the field due to $FABC$

Due to AB,

$$\mathbf{B}_{AB}=\frac{\mu_0 i}{4\pi\left(\frac{l}{2}\right)}(\sin 45°+\sin 45°)\hat{\mathbf{i}}=\sqrt{2}\,\frac{\mu_0 i}{2\pi l}$$

Due to BC, $\mathbf{B}_{AB}=\frac{\mu_0 i}{4\pi l\left(\frac{l}{2}\right)}(\sin 0°+\sin 45°)=\frac{\mu_0 i}{2\sqrt{2}\,\pi l}$

$$\mathbf{B}_{AB}=\frac{\mu_0 i}{2\sqrt{2}\,\pi l}\,\hat{\mathbf{i}}$$

Hence, the field due to $FABC$,

$$\mathbf{B}_{FABC}=\frac{\mu_0 i}{\pi l}\left[\frac{1}{2\sqrt{2}}+\frac{1}{2\sqrt{2}}+\frac{\sqrt{2}}{2}\right]\hat{\mathbf{i}}$$

$$\Rightarrow \quad \mathbf{B}_{FABC}=\frac{\sqrt{2}\,\mu_0 i}{\pi l}\,\hat{\mathbf{i}}$$

Similarly, due to $CDEP$, we have

$$\mathbf{B}_{CDEF}=\frac{\sqrt{2}\,\mu_0 i}{\pi l}\,\hat{\mathbf{j}}$$

$$\therefore \quad \mathbf{B}_{\text{net}}=\mathbf{B}_{FABC}+\mathbf{B}_{CDEF}=\frac{\sqrt{2}\,\mu_0 i}{\pi l}(\hat{\mathbf{i}}+\hat{\mathbf{j}})$$

27. Given, $\mu_0=4\pi\times10^{-7}$ N/A^2,

$$\omega=40\pi \text{ rad/s}$$

$$B_{\text{at centre}}=3.8\times10^{-9}\text{ T}$$

and $\quad R=10\text{ cm}=0.1\text{ m}$

Now, we know that, magnetic field at the centre of a current carrying ring is given by

$$B=\frac{\mu_0 i}{2R} \qquad \text{...(i)}$$

Here, i can be determined by flow of charge per rotation, *i.e.*

$$i=\frac{Q}{T} \qquad \text{...(ii)}$$

Here, $\quad T=\frac{2\pi}{\omega}$

$$\Rightarrow \quad i=\frac{Q\omega}{2\pi} \qquad \text{...(iii)}$$

By putting value of i from Eq. (iii) to Eq. (i), we get

$$B=\frac{\mu_0 Q\omega}{2R\times2\pi} \text{ or } Q=\frac{2BR\times2\pi}{\mu_0\omega}$$

$$=\frac{2\times3.8\times10^{-9}\times0.1\times2\pi}{4\pi\times10^{-7}\times40\pi}$$

$$=\frac{2\times3.8\times0.1}{2\times40\pi}\times10^{-2}$$

$$=0.003022\times10^{-2}\text{ C}$$

$$=3.022\times10^{-5}\text{ C}$$

or $\quad Q=3\times10^{-5}$ C

28. Consider free body diagram of the wire.

As the wires are in equilibrium, they must carry current in opposite direction.

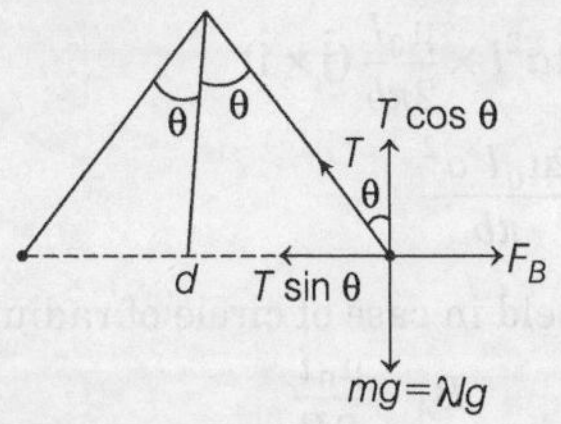

Here, $F_B=\frac{\mu_0 i^2 l}{2\pi d}$, where l is length of each wire and d is separation between wires.

From figure, $d=2L\sin\theta$

$$T=\cos\theta=mg=\lambda lg \quad \text{(in vertical direction)...(i)}$$

$$T\sin\theta=F_B=\frac{\mu_0 i^2 l}{4\pi L\sin\theta} \text{(in horizontal direction) ...(ii)}$$

From Eqs. (i) and (ii),

$$\frac{T\sin\theta}{T\cos\theta}=\frac{\mu_0 i^2 l}{4\pi L\sin\theta\times\lambda\, lg}$$

$$\therefore \quad i=\sqrt{\frac{4\pi\lambda Lg\sin^2\theta}{\mu_0\cos\theta}}=2\sin\theta\sqrt{\frac{\pi\lambda Lg}{\mu_0\cos\theta}}$$

29. There is no magnetic field along axis of a current-carrying wire.

Also, magnetic field near one of end of an infinitely long wire is $\frac{\mu_0 i}{4\pi r}$ tesla.

Hence, magnetic field due to segments LP and MQ at 'O' is zero. Using right hand rule, we can check that magnetic field due to segments PS and QN at O is in same direction perpendicularly into the plane of page.

Hence, $B_O=B_{PS}+B_{QN}=\frac{\mu_0 i}{4\pi r}+\frac{\mu_0 i}{4\pi r}=\frac{\mu_0 i}{2\pi r}$

So, $\quad i=\frac{2\pi r B_0}{\mu_0}$

Here, $r=OP=OQ=4$ cm and $B_0=10^{-4}$ T.

Substituting values, we get

$$\Rightarrow \quad i=\frac{2\pi\times4\times10^{-2}\times10^{-4}}{4\pi\times10^{-7}}$$

$$\Rightarrow \quad i=20\text{ A}$$

Also, magnetic field points perpendicular to the plane of page.

30. Average power developed which eventually gets lost as heat due to resistance of loop is given by

$$P_{avg} = \frac{V_{rms}^2}{R} = \frac{(V_{max}/\sqrt{2})^2}{R} = \frac{V_{max}^2}{2R}$$

Also, $V_{max} = NBA\,\omega$

In given loop, $N = 1, A = \pi ab$

So, $$P_{avg} = \frac{(BA\omega)^2}{2R} = \frac{B^2\pi^2a^2b^2\omega^2}{2R}$$

31. Torque $(\tau) = \mathbf{M} \times \mathbf{B}$

Here, $\mathbf{M} = (2a)^2 \times I = 4a^2I\hat{\mathbf{j}}$

and $\mathbf{B} = \frac{\mu_0 I}{2\pi b}\hat{\mathbf{i}}$

$\therefore$ $$\tau = 4a^2I \times \frac{\mu_0 I}{2\pi b}(\hat{\mathbf{j}} \times \hat{\mathbf{i}})$$

$\Rightarrow$ $$|\tau| = \frac{2\mu_0 I^2 a^2}{\pi b}$$

32. Magnetic field in case of circle of radius R, we have

$$B_A = \frac{\mu_0 i}{2R}$$

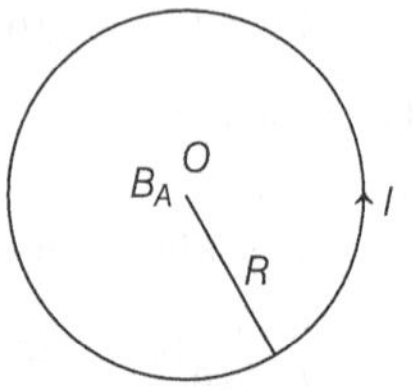

As, $2\pi R = l$, (where l is length of a wire)

$$R = \frac{l}{2\pi}$$

$\Rightarrow$ $$B_A = \frac{\mu_0 i}{2 \times \frac{l}{2\pi}} = \frac{\mu_0 i\pi}{l} \quad \text{...(i)}$$

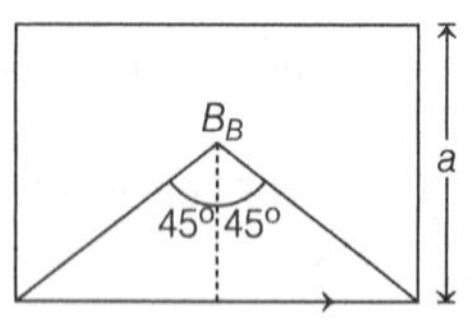

Magnetic field in case of square of side a, we get

$$B_B = 4 \times \frac{\mu_0}{4\pi} \times \frac{i}{\left(\frac{a}{2}\right)}\left(\frac{1}{\sqrt{2}} + \frac{1}{\sqrt{2}}\right)$$

$\Rightarrow$ $$B_B = \frac{4i\mu_0}{\pi a\sqrt{2}} = \frac{\mu_0 2\sqrt{2}i}{a\pi}$$

As, $4a = l$, $a = \frac{l}{4}$

$\Rightarrow$ $$B_B = \frac{8\sqrt{2}\,\mu_0 i}{\pi l} \quad \text{... (ii)}$$

Dividing Eq. (i) by Eq. (ii), we get

$$\frac{B_A}{B_B} = \frac{\pi^2}{8\sqrt{2}}$$

33. Given, proton is moving from south to north and magnetic field is directed from west to east.

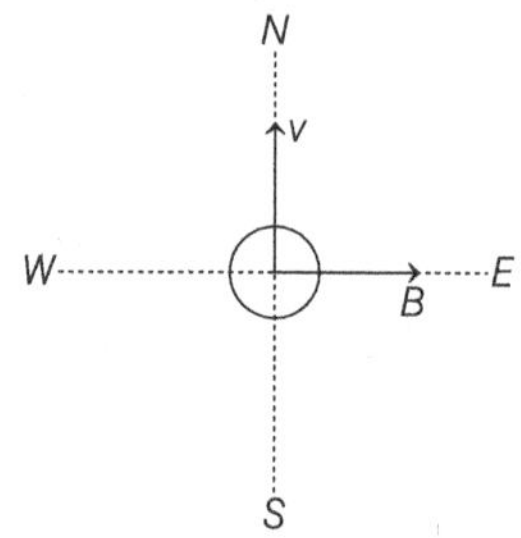

As $\mathbf{v} \perp \mathbf{B}$, force on the charged particle,

$$F = Bqv$$

If m = mass and a = acceleration of the particle, then

$$F = ma$$

So, $$ma = Bqv$$

or $$B = \frac{ma}{qv} \quad \text{...(i)}$$

If K = kinetic energy of the particle,

then $$K = \frac{1}{2}mv^2$$

$\Rightarrow$ $$v = \sqrt{\frac{2K}{m}} \quad \text{...(ii)}$$

From Eqs. (i) and (ii), we get

$$B = \frac{ma}{q\sqrt{\frac{2K}{m}}} = \frac{m^{3/2} \cdot a}{q \times \sqrt{2K}} \quad \text{...(iii)}$$

Here, $m = 1.6 \times 10^{-27}$ kg,

$a = 10^{12}$ ms^{-2},

$q = 1.6 \times 10^{-19}$ C,

$K = 1$ MeV

$= 1 \times 10^6 \times 1.6 \times 10^{-19}$J

$= 1.6 \times 10^{-13}$ J

Substituting these given values in Eq. (iii), we get

$$B = \frac{(1.6 \times 10^{-27})^{3/2} \times 10^{12}}{1.6 \times 10^{-19} \times (2 \times 1.6 \times 10^{-13})^{1/2}}$$

$$= \frac{(1.6)^{3/2} \times 10^{-27 \times \frac{3}{2}} \times 10^{12}}{(1.6)^{3/2} \times 2^{1/2} \times 10^{-19} \times 10^{-13/2}}$$

$$= \frac{1}{\sqrt{2}} \times 10^{-3}$$

$= 0.71$ mT

34. As shown in the figure, consider an element dl at C of wire, where $OC = l$.

Let $PC = r$

and $\angle OPC = \phi$.

According to Biot-Savart's law, magnitude of magnetic field induction at P due to current element at C is

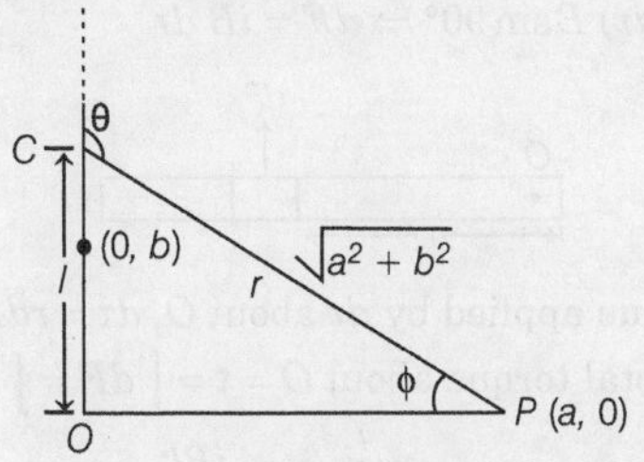

$$dB = \frac{\mu_0}{4\pi}\frac{idl\sin\theta}{r^2}$$

Here, $\theta = 90° + \phi; r = a\sec\phi$

and $l = a\tan\phi$

$\therefore$ $dl = a\sec^2\phi\, d\phi$ (after differentiating)

$\therefore$ $$dB = \frac{\mu_0}{4\pi}\frac{i\,a(\sec^2\phi\, d\phi)\sin(90° + \phi)}{a^2\sec^2\phi}$$

$$= \frac{\mu_0}{4\pi}\frac{i}{a}\cos\phi\, d\phi$$

The magnetic field induction at P is

$$B = \int dB = \int_{90°}^{\phi}\frac{\mu_0}{4\pi}\frac{i}{a}\cos\phi\, d\phi = \frac{\mu_0}{4\pi}\frac{i}{a}[-\sin\phi]_{90°}^{\phi}$$

$$= \frac{\mu_0}{4\pi}\frac{i}{a}(1 - \sin\phi) = \frac{\mu_0 i}{4\pi a}\left(1 - \frac{b}{\sqrt{a^2 + b^2}}\right)$$

35. In the absence of magnetic field,

$$mg = 2kx_0 \quad \text{...(i)}$$

$\therefore$ Magnetic force on the rod is $F_m = BiL = \frac{BLE}{R}$

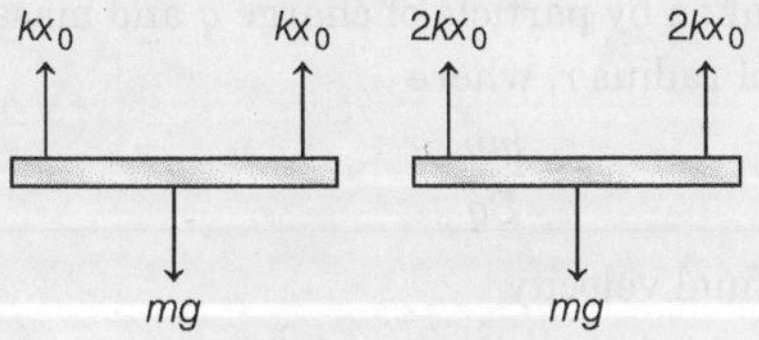

In downward direction,

$\therefore$ $$2kx_0 = mg + \frac{BLE}{R} \quad \text{...(ii)}$$

From Eqs. (i) and (ii); we get, $4kx_0 = 2kx_0 + \frac{BLE}{R}$

$$B = \frac{2kx_0R}{EL} = \frac{mgR}{LE}$$

36. Consider a hypothetical ring of radius x and thickness dx of a disc as shown in figure.

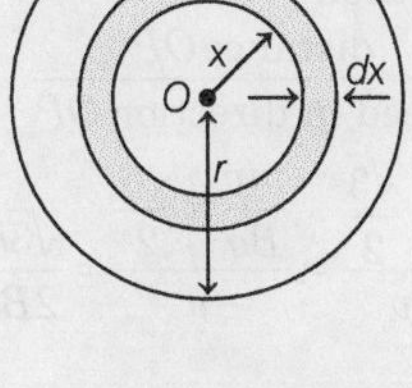

Charge on the ring, $dq = \frac{q}{\pi r^2}x(2\pi x\, dx)$

Current due to rotation of charge on ring,

$$di = \frac{dq}{T} = \frac{dq}{1/n} = ndq = \frac{nq2x\, dx}{r^2}$$

Magnetic field at the centre O due to current of ring element is

$$dB = \frac{\mu_0 di}{2x} = \frac{\mu_0 nq2x\, dx}{r^2(2x)} = \frac{\mu_0 n\, dx}{r^2} \quad \text{... (i)}$$

Integrating Eq. (i), we get

$$\int_0^B dB = \int_0^r \frac{\mu_0 ndx}{r^2}$$

$\Rightarrow$ $$B = \frac{\mu_0 n}{r^2}\int_0^r dx \Rightarrow B = \frac{\mu_0 nr}{r^2}$$

Therefore, total magnetic field induction due to current of whole disc is

$$B = \frac{\mu_0 nq}{r^2}r = \frac{\mu_0 nq}{r}.$$

37. Increase in kinetic energy of particle

$$= \frac{1}{2}m(2v)^2 - \frac{1}{2}mv^2 = \frac{3}{2}mv^2$$

Work done by the uniform electric field, E in going from P to $Q = (qE) \times 2a = 2qEa$

Hence, $2qEa = \frac{3}{2}mv^2$ (by work-energy theorem)

or $$E = \frac{3mv^2}{4qa}$$

Rate of work done (*i.e.* power by the electric field),

$$P = F \cdot v = qE \cdot v = qE\hat{\mathbf{i}} \cdot v\hat{\mathbf{i}} = qEv$$

$$= q \cdot \frac{3mv^2}{4qa} \cdot v = \frac{3}{4}\frac{mv^2}{a}$$

$\therefore$ Power at $Q = qE\,\hat{\mathbf{i}} \cdot (-2v\hat{\mathbf{j}}) = 0$

At Q, rate of work done by both the fields is zero.

38. Acceleration of electron due to electric field is $\mathbf{a}_E = -e\mathbf{E}/m$.

Its direction is opposite to the direction of **E**.

Acceleration of positron due to electric field is $\mathbf{a}'_E = e\mathbf{E}/m$. Its direction is along the direction of **E**.

Thus $\mathbf{a}'_E \neq \mathbf{a}_E$.

Magnetic force on electron, $\mathbf{F} = -e(\mathbf{v} \times \mathbf{B})$

Acceleration of electron, $\mathbf{a}_m = \frac{\mathbf{F}}{m} = -\frac{e(\mathbf{v} \times \mathbf{B})}{m}$

Magnetic force on positron, $\mathbf{F}' = e(-\mathbf{v} \times \mathbf{B}) = -e(\mathbf{v} \times \mathbf{B})$

Acceleration of positron, $\mathbf{a}'_m = \frac{\mathbf{F}'}{m} = -\frac{e(\mathbf{v} \times \mathbf{B})}{m}$

$\therefore$ $$\mathbf{a}_m = \mathbf{a}'_m.$$

As both the particles (electron and positron) are of same mass and same charge in magnitude having same acceleration due to magnetic field, hence they gain or loose the energy at the same rate.

Due to electric field, net electric force on electron-positron pair, $= -e\mathbf{E} + e\mathbf{E} = 0$.

Net magnetic force on electron-positron pair $= -e(\mathbf{v}\times\mathbf{B}) + [-e(\mathbf{v}\times\mathbf{B})] = -2e(\mathbf{v}\times\mathbf{B})$

Therefore, the motion of the centre of mass (CM) is determined by magnetic field alone.

39. According to question, the situation can be drawn as

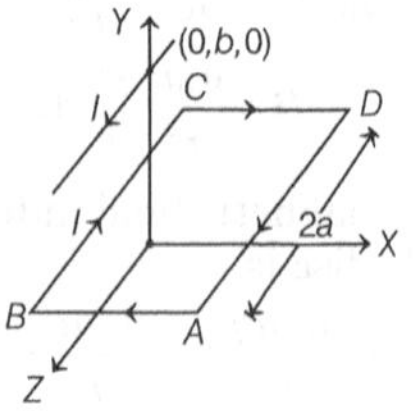

Force of a current carrying conducting wire,

$$F = Bil\sin\theta$$

So, force acting on wire AB and CD are zero

$(\because \theta = 0°$ and $\sin 0° = 0)$.

Only sides BC and AD will experience force due to current carrying wire.

Force on side AD = Force on side $BC = F$ (say)

Net force on loop, $F = Bil_{\text{loop}}\sin 90°$

$$= \left(\frac{\mu_0 i}{2\pi d}\right) i(2a)$$

$$F = \frac{\mu_0 i^2 a}{\pi\sqrt{a^2+b^2}} \qquad (\because d = \sqrt{a^2+b^2})$$

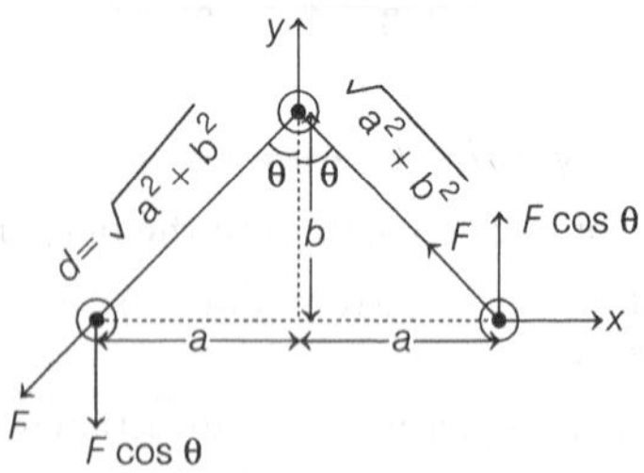

Force on AD and BC acts in two different directions as shown in above figure, so torque is produced in square loop.

$\therefore$ Torque on loop $= F\cos\theta \times 2a$

$$= \frac{\mu_0 i^2 a}{\pi\sqrt{a^2+b^2}} \times \frac{b}{\sqrt{a^2+b^2}} \times 2a$$

$$= \frac{2\mu_0 i^2 a^2 b}{\pi(a^2+b^2)}$$

If $b >> a$, then $a^2 + b^2 \simeq b^2$.

$$\therefore \qquad \tau = \frac{2\mu_0 i^2 a^2}{\pi b}$$

40. The force acting on the elementary portion of the current carrying conductor is given as,

$dF = i\,(dr)\,B\sin 90° \Rightarrow dF = iB\,dr$

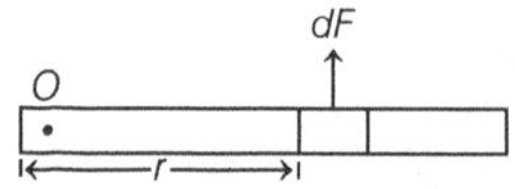

The torque applied by dr about O, $d\tau = r dF$.

$\Rightarrow$ The total torque about $O = \tau = \int dF = \int r\,(iBdr)$

$$\Rightarrow \qquad \tau = iB\int_0^l r dr = \frac{iBl^2}{2}$$

The angular acceleration, $\alpha = \frac{\tau}{I}$ (where, MI = moment of inertia)

$$\Rightarrow \qquad \alpha = \left(\frac{ibl^2}{2}\right) / \left(\frac{ml^2}{3}\right) \Rightarrow \alpha = \frac{3iB}{2m}$$

41. Situation given in question is shown below

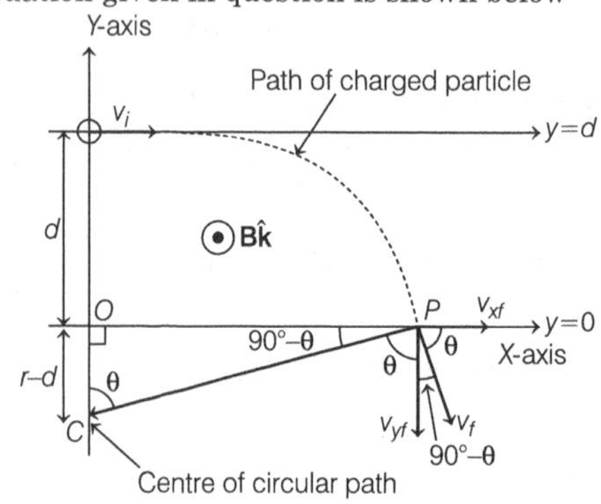

Path taken by particle of charge q and mass m is a circle of radius r, where

$$r = \frac{mv}{Bq}$$

Here, final velocity

$$\mathbf{v}_f = \mathbf{v}_{xf}\hat{\mathbf{i}} + \mathbf{v}_{yf}(-\hat{\mathbf{j}}) = v\cos 60°\,\hat{\mathbf{i}} - v\sin 60°\,\hat{\mathbf{j}}$$

$$= v\left(\frac{1}{2}\hat{\mathbf{i}} - \frac{\sqrt{3}}{2}\hat{\mathbf{j}}\right)$$

So, change of velocity of charged particle is

$$\Delta\mathbf{v} = \mathbf{v}_f - \mathbf{v}_i = v\left(\frac{1}{2}\hat{\mathbf{i}} - \frac{\sqrt{3}}{2}\hat{\mathbf{j}}\right) - v\hat{\mathbf{i}}$$

$$= -v\left(\frac{1}{2}\hat{\mathbf{i}} + \frac{\sqrt{3}}{2}\hat{\mathbf{j}}\right)$$

If t = time taken by charged particle to cross region of magnetic field, then

$$t = \frac{\text{distance } OP}{\text{speed in direction } OP}$$

$$= \frac{r \times \frac{\sqrt{3}}{2}}{v} = \frac{\frac{mv}{Bq} \times \frac{\sqrt{3}}{2}}{v} = \frac{\sqrt{3}m}{2Bq}$$

So, acceleration of charged particle at the point its emergence,

$$\mathbf{a} = \frac{\Delta \mathbf{v}}{\Delta t} = \frac{-v\left(\frac{1}{2}\hat{\mathbf{i}} + \frac{\sqrt{3}}{2}\hat{\mathbf{j}}\right)}{\frac{\sqrt{3}}{2}\frac{m}{Bq}} = \frac{-Bqv}{m}\left(\frac{\hat{\mathbf{i}}}{\sqrt{3}} + \hat{\mathbf{j}}\right) \text{ ms}^{-2}$$

None of the option given matches with answer.

42. Suppose the effective length of this dipole be l. Thus, the top view of the condition can be shown in the figure given below.

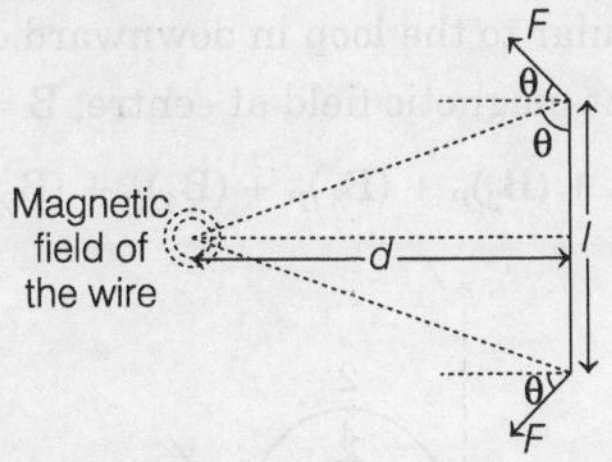

Now, the net force on the loop (*i.e.* at the two poles) due to the wire is given as,

$$F_{\text{net}} = 2F\cos\theta = 2mB\cos\theta$$

where, m is the pole strength.

From the figure, we have

$$\cos\theta = \frac{\frac{l}{2}}{x} = \frac{\frac{l}{2}}{\sqrt{d^2 + \frac{l^2}{4}}} \Rightarrow F_{\text{net}} = \frac{2mB\,l}{2\sqrt{d^2 + \frac{l^2}{4}}} \quad \text{...(i)}$$

Since, the magnetic moment of a loop of radius a is

$$m = iA = i\pi a^2 \quad \text{...(ii)}$$

and magnetic field due to a straight infinitely long current carrying conductor at a distance x is

$$B = \frac{\mu_0 i'}{2\pi b} \quad \text{...(iii)}$$

$\therefore$ Using Eqs. (ii) and (iii), rewriting Eq. (i), we get

$$F_{\text{net}} = \frac{i\pi a^2 \mu_0 i'}{2\pi\left(\sqrt{d^2 + \frac{l^2}{4}}\right)^2}$$

$$\Rightarrow \quad F_{\text{net}} \propto \frac{a^2}{\left(d^2 + \frac{l^2}{4}\right)} \text{ or } F_{\text{net}} \propto \left(\frac{a}{d}\right)^2$$

43. Charged particle follows trajectory as shown in the figure.

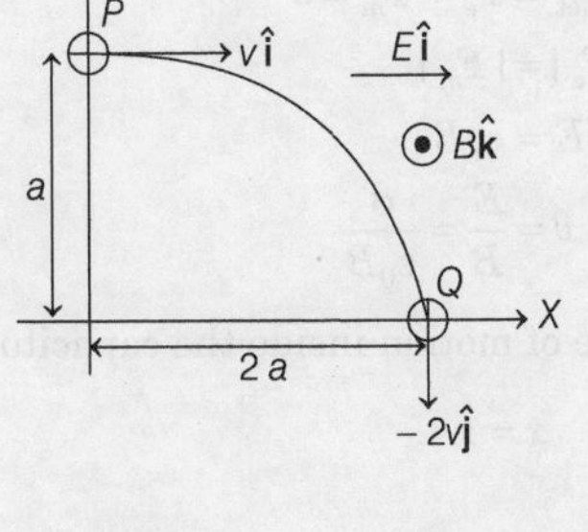

As magnetic field is perpendicular to the velocity of charged particle, work done by magnetic field is zero.

So, work done by electric field = change in kinetic energy of charged particle

$$\Rightarrow \quad qE2a = \frac{1}{2}m(2v)^2 - \frac{1}{2}m(v)^2$$

$$\Rightarrow \quad E = \frac{3}{4}\frac{mv^2}{qa} \qquad \text{(statement I is correct)}$$

Rate of work done = Power = $\mathbf{F}\cdot\mathbf{v} = qE\hat{\mathbf{i}}\cdot v\hat{\mathbf{i}}$

$$= qEv = q\cdot\frac{3}{4}\frac{mv^2}{qa}\cdot v$$

$$= \frac{3}{4}\frac{mv^3}{a} \qquad \text{(statement II is correct)}$$

At point Q, both electric field and magnetic field are perpendicular to velocity of charged particle, so work done by both $\mathbf{E}$ and $\mathbf{B}$ at Q is zero.

(statement III is correct)

Angular momentum about origin at $P = mva\hat{\mathbf{i}}$.

Angular momentum about origin at $Q = -4mva\hat{\mathbf{j}}$.

Difference between magnitudes of angular momentum at P and Q,

$$\Delta L = L_Q - L_P$$

$$= 4mva - mva$$

$$= 3mva \qquad \text{(statement IV is incorrect)}$$

Thus, option (b) is correct.

44. There are two forces on slider.

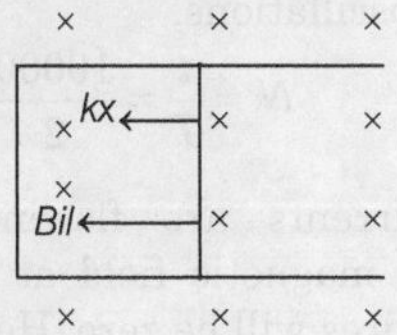

Spring force = kx

where, k = spring constant.

As the slider is kept in a uniform magnetic field $B = 0.1$ T, hence it will experience a force, *i.e.*

Magnetic force = Bil

where, l = length of the strip.

Now, using

$$F_{\text{net}} = ma$$

We have,

$$(-kx) + (-Bil) = ma$$

$$\Rightarrow \quad -kx - Bil - ma = 0$$

$$\Rightarrow \quad -kx - \frac{B^2l^2}{R}\cdot v - ma = 0 \qquad \left(\because i = \frac{Blv}{R}\right)$$

and acceleration, $a = \dfrac{d^2x}{dt^2}$

Hence, the modified equation becomes

$$\Rightarrow \quad \frac{md^2x}{dt^2} + \frac{B^2l^2}{R}\left(\frac{dx}{dt}\right) + kx = 0$$

This is the equation of damped simple harmonic motion.

So, amplitude of oscillation varies with time as

$$A = A_0 e^{-\frac{B^2l^2}{2Rm}\cdot t}$$

Now, when amplitude is $\frac{A_0}{e}$, then

$$\frac{A_0}{e} = \frac{A_0}{e^{\frac{B^2l^2}{2Rm}\cdot t}} \quad \text{(as given)}$$

$$\Rightarrow \left(\frac{B^2l^2}{2Rm}\right)t = 1 \text{ or } t = \frac{2Rm}{B^2l^2}$$

According to the question, magnetic field $B = 0.1$ T, mass of strip $m = 50 \times 10^{-3}$ kg, resistance $R = 10\,\Omega$, $l = 10 \text{ cm} = 10 \times 10^{-2}$ m

$$\therefore \quad t = \frac{2Rm}{B^2l^2} = \frac{2 \times 10 \times 50 \times 10^{-3}}{(0.1)^2 \times (10 \times 10^{-2})^2}$$

$$= \frac{1}{10^{-4}} = 10000 \text{ s}$$

Given, spring constant, $k = 0.5 \text{ Nm}^{-1}$

Also, time period of oscillation is

$$T = 2\pi\sqrt{\frac{m}{k}}$$

$$= 2\pi\sqrt{\frac{50 \times 10^{-3}}{0.5}} = \frac{2\pi}{\sqrt{10}} \approx 2 \text{ s}$$

So, number of oscillations,

$$N = \frac{t}{T} = \frac{10000}{2} = 5000$$

45. Since, the currents are flowing in the opposite directions, the magnetic field at a point equidistant from the two wires will be zero. Hence, the force acting on the charge at this instant will be zero.

46. Torque, $\tau = BINA \sin\theta$

Here, $\theta = 90°$

$\Rightarrow \sin\theta = 1, \tau = 1.5$ N-m

$$A = 3 \times 10^{-4} \text{m}^2, N = 500 \text{ and } i = 0.5 \text{ A}$$

So, $$B = \frac{\tau}{iNA} = \frac{1.5}{0.5 \times 500 \times 3 \times 10^{-4}} = 20 \text{ T}$$

47. The resistance of arm $PQRS$ is 3 times the resistance of arm PS. If resistance of arm $PS = r$, then resistance of arm $PQRS = 3r$.

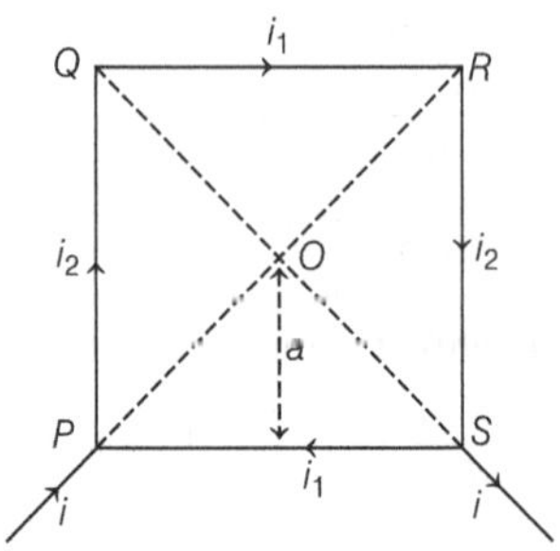

Potential difference across P and $S = i_1 r = i_2 \times 3r$

Magnetic field induction at O due to current through arm PS is $B_1 = \frac{\mu_0}{4\pi}\frac{i_1}{a}[\sin 45° + \sin 45°]$ acting perpendicular to the loop upwards.

Magnetic field due to PQ and RS are equal and opposite.

Therefore magnetic field due to QR,

$$B_2 = \frac{\mu_0}{4\pi}\cdot\frac{I_1}{a}[\sin 45° + \sin 45°]$$

perpendicular to the loop in downward direction.

$\therefore$ Resultant magnetic field at centre, $\mathbf{B} = \mathbf{B}_1 + \mathbf{B}_2 = 0$

48. $\mathbf{B}_P = (\mathbf{B}_1)_P + (\mathbf{B}_2)_P + (\mathbf{B}_3)_P + (\mathbf{B}_4)_P + (\mathbf{B}_5)_P$

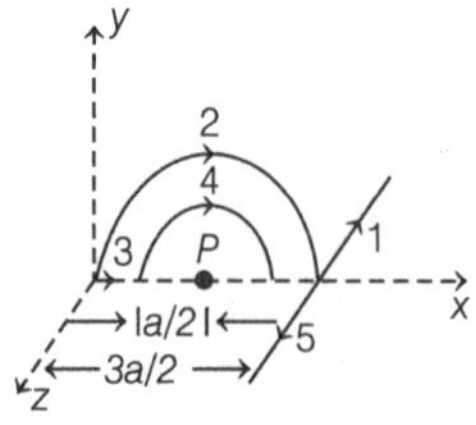

where, $(\mathbf{B}_1)_P = \frac{\mu_0 i}{4\pi\left(\frac{3a}{2}\right)}(\hat{\mathbf{j}})$ (semi-finite wire)

$$(\mathbf{B}_2)_P = \frac{\mu_0 i}{4(3a/2)}(-\hat{\mathbf{k}}), (\mathbf{B}_3)_P = 0$$

$$(\mathbf{B}_4)_P = \frac{\mu_0 i}{4(3a/2)}(-\hat{\mathbf{k}})$$

$$\Rightarrow (\mathbf{B}_5)_P = \frac{\mu_0 i}{4\pi\left(\frac{a}{2}\right)}(-\hat{\mathbf{j}})$$

$$\Rightarrow B_{\text{net}} = \sqrt{(B_4 - B_2)^2 + (B_5 - B_1)^2}$$

$$= \sqrt{\left(\frac{2\mu_0 i}{6a}\right)^2 + \left(\frac{\mu_0 i}{2\pi a} + \frac{\mu_0 i}{6\pi a}\right)^2} = \frac{\mu_0 i}{3\pi a}\sqrt{\pi^2 + 4}$$

$$\Rightarrow P = 3$$

49. The net electric field,

$$\mathbf{E} = \mathbf{E}_1 + \mathbf{E}_2$$

$$\Rightarrow E = \frac{\sigma}{2\varepsilon_0} + \frac{\sigma}{2\varepsilon_0} = \frac{\sigma}{\varepsilon_0}$$

The net force acting on the electron is zero because it moves with cosntant velocity

$$\mathbf{F}_{\text{net}} = \mathbf{F}_e + \mathbf{F}_m = 0$$

$$\Rightarrow |\mathbf{F}_e| = |\mathbf{F}_m|$$

or $$eE = evB$$

or $$v = \frac{E}{B} = \frac{\sigma}{\varepsilon_0 B}$$

$\therefore$ The time of motion inside the capacitor, $t = \frac{l}{v} = \frac{\varepsilon_0 lB}{\sigma}$.

$$\Rightarrow x = 1$$

21

Magnetostatics

Magnetism is a phenomenon by the virtue of which there develops an attracting or repulsive force between two magnetic objects. The ability of an object or a material to produce invisible magnetic field lines in order to develop a force which attracts other magnetic materials such as iron, steel, etc., are known as *magnets*.

IN THIS CHAPTER ....

Bar Magnet and Magnetic Poles

Bar magnet consists of two equal and non-separate magnetic poles. One pole is designated as north pole (N) and the other as south pole (S).

Following are few characteristics of bar magnet

(i) A freely suspended magnet stays always in north-south direction. The end of the magnet which always rests in the geographical north direction is called the north pole and the end which always rests in the geographical south direction is called the south pole. These poles are represented by the letters N and S, respectively.

(ii) Like poles repel each other and unlike poles attract each other. This repulsion or attraction obeys the inverse-square law.

(iii) We cannot isolate the north or south pole of magnet. If a bar magnet is broken into two halves, we get two similar bar magnets with somewhat weaker properties.

Magnetic Field Lines

The magnetic field lines of a magnet are the imaginary lines which continuously represent the direction of that magnetic field or the direction of the magnetic force on a north monopole at any given position.

Important properties of magnetic field lines are given below

(i) These lines forms closed continuous curves.

(ii) The tangent drawn at any point of a field line represents the direction of net magnetic field.

(iii) These lines cannot intersect each other.

(iv) Outside a magnet, they are directed from north to south pole and inside a magnet, they are directed from south to north.

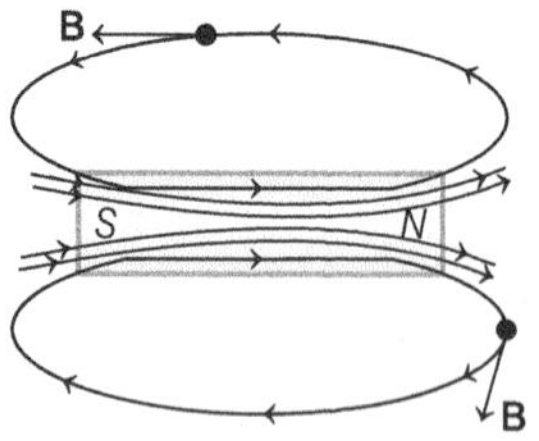

Magnetic Dipole

A magnetic dipole is an arrangement which consists two magnetic poles of equal and opposite strengths separated at a small distance. A bar magnet, a compass needle, etc., are the examples of magnetic dipoles.

Pole Strength (m)

It can be defined as the strength of magnetic pole to attract magnetic material towards itself. It is a scalar quantity and its SI unit is ampere-metre (A-m).

Magnetic Dipole Moment

The product of the distance ($2l$) between the two poles and the pole strength of either pole is called magnetic dipole moment.

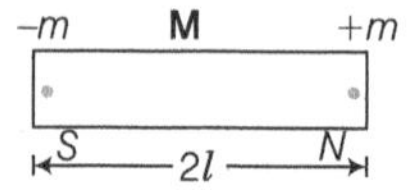

Magnetic dipole moment, $\mathbf{M} = m(2l)$

Its SI unit is joule/tesla (J/T) or ampere-metre^2 (A-m^2).

Its direction is from south pole towards north pole.

Pole Strength and Magnetic Dipole Moment in Special Cases

Special Cases	Figure	Effect on Pole Strength	New Magnetic Dipole Moment
If bar magnet is cut into two equal pieces such that the length of each piece becomes half	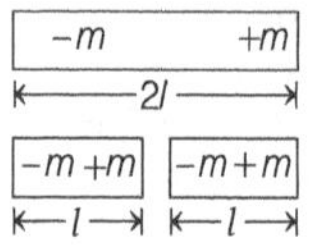	Remains unchanged	$M' = m \cdot \frac{2l}{2} = \frac{M}{2}$ (becomes half)
If bar magnet is cut into two equal pieces such that the width of each piece becomes half	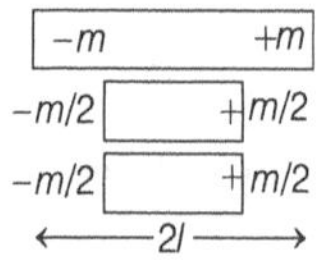	Pole strength of each piece becomes half	$M' = \left(\frac{m}{2}\right)(2l) = \frac{M}{2}$ (becomes half)
If bar magnet is bent in the form of semi-circle	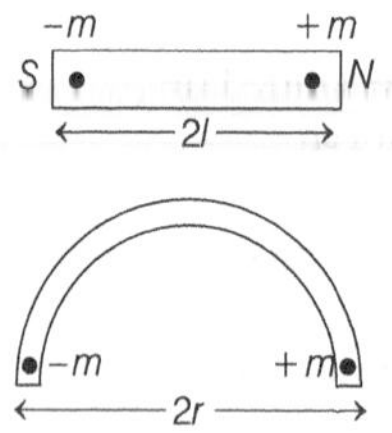	Remains unchanged	$M' = m(2r)$ ($\because 2l = \pi r$) $M' = m \times 2\left(\frac{2l}{\pi}\right) = \frac{2M}{\pi}$ $\left(\text{becomes } \frac{2}{\pi} \text{ times}\right)$
When two identical bar magnets are joined perpendicular to each other	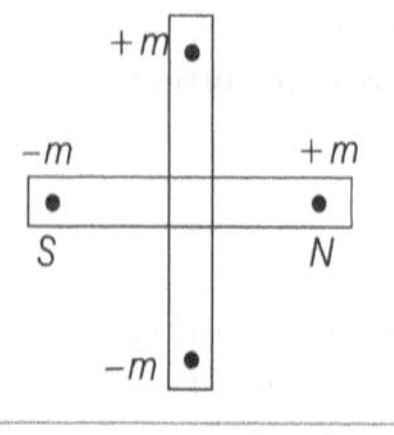	Remains unchanged	$M = \sqrt{M_1^2 + M_2^2} = \sqrt{2}M$
When two bar magnets are inclined at an angle θ.	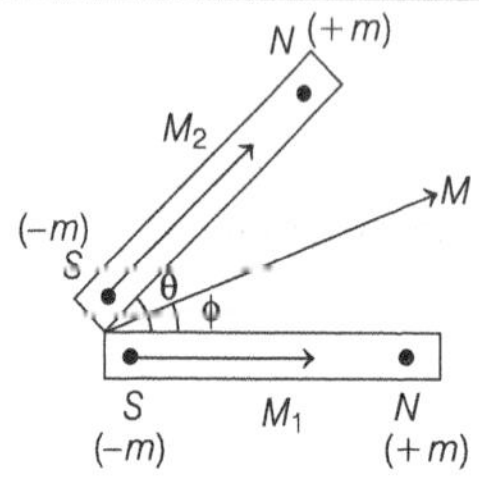	Remains unchanged	Resultant magnetic moment, $M' = \sqrt{M_1^2 + M_2^2 + 2M_1M_2\cos\theta}$ Angle made by resultant magnetic moment (M) with M_1 is given by $\tan\phi = \frac{M_2\sin\theta}{M_1 + M_2\cos\theta}$

Example 1. *A square loop is carrying a steady current I and the magnitude of its magnetic dipole moment is m. If this square loop is changed to a circular loop and it carries the same current, the magnitude of the magnetic dipole moment of circular loop (in A-m) will be* **[JEE Main 2019]**

(a) $\frac{4m}{\pi}$ (b) $\frac{3m}{\pi}$ (c) $\frac{2m}{\pi}$ (d) $\frac{m}{\pi}$

Sol. (a) Let the given square loop has side a, then its magnetic dipole moment will be $m = Ia^2$.

When square is converted into a circular loop of radius r.

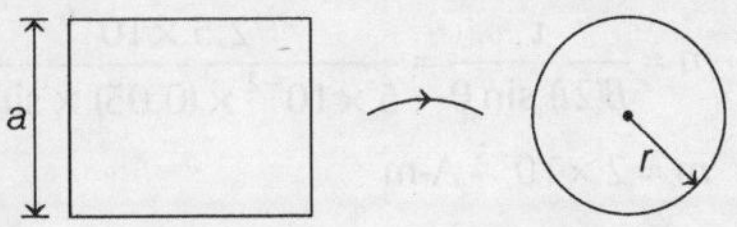

Then, wire length will be same in both area,

$$4a = 2\pi r \Rightarrow r = \frac{4a}{2\pi} = \frac{2a}{\pi}$$

Hence, area of circular loop formed is, $A' = \pi r^2 = \pi\left(\frac{2a}{\pi}\right)^2 = \frac{4a^2}{\pi}$

Magnitude of magnetic dipole moment of circular loop will be

$$m' = IA' = I\frac{4a^2}{\pi}$$

Ratio of magnetic dipole moments of both shapes is,

$$\frac{m'}{m} = \frac{I\cdot\frac{4a^2}{\pi}}{Ia^2} = \frac{4}{\pi} \Rightarrow m' = \frac{4m}{\pi}$$

Magnetic Field Strength at a Point due to Magnetic Dipole

The strength of a magnetic field at any point is defined as the force experienced by a hypothetical (unit magnetic) north pole placed at that point,

i.e. $\mathbf{B} = \frac{\mathbf{F}}{m}$, where m is the pole strength of hypothetical north pole. It is a vector quantity.

It's direction is the direction in which hypothetical north pole would tend to move, if free to do so.

There arises following cases

When Point lies on Axial or End-on Line of a Bar Magnet Magnetic field strength at point P due to bar magnet at a distance r from its centre is

$$B = \frac{\mu_0}{4\pi}\cdot\frac{2Mr}{(r^2 - l^2)^2}$$

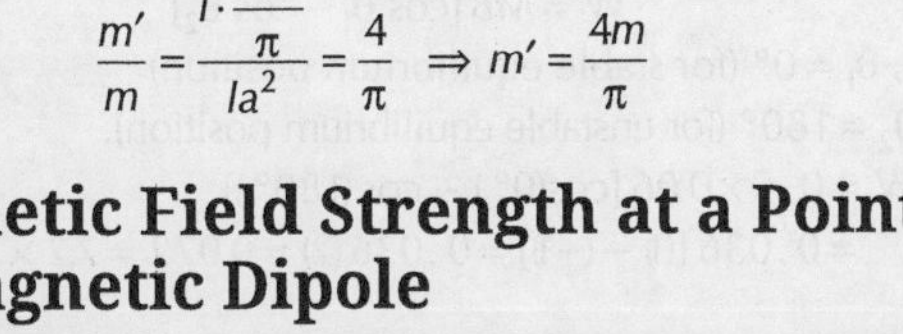

If magnet is short, $l^2 << r^2$, then $B = \frac{\mu_0}{4\pi}\frac{2\,Mr}{r^4} = \frac{\mu_0}{4\pi}\frac{2M}{r^3}$

The direction of magnetic field at point P is along NP.

When Point lies on Equatorial or Broad Side-on Line of a Bar Magnet Magnetic field strength at point P due to bar magnet at a distance r from its centre is

$$B = \frac{\mu_0}{4\pi}\cdot\frac{M}{(r^2 + l^2)^{3/2}}$$

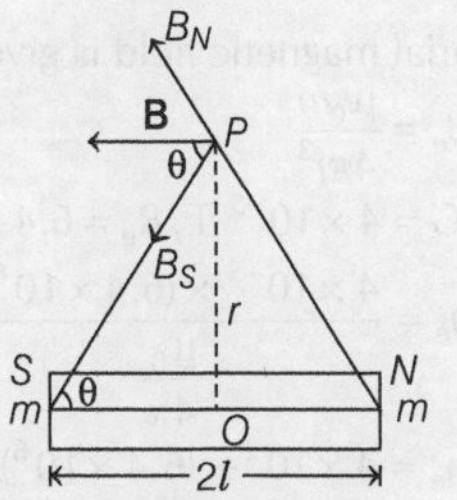

If $r^2 >> l^2$, then $B = \frac{\mu_0}{4\pi}\cdot\frac{M}{(r^2)^{3/2}} = \frac{\mu_0}{4\pi}\frac{M}{r^3}$

The magnetic field at point P is directed parallel to the length of the magnet from its N to S-pole.

When Point makes Angle θ with Axis of a Bar Magnet At an angle θ with the axis of magnet, the magnetic field at point P at a distance r from centre of magnet in vacuum (or air) is

$$B = \frac{\mu_0}{4\pi}\cdot\frac{M}{r^3}\sqrt{1 + 3\cos^2\theta}$$

and $$\tan\alpha = \frac{1}{2}\tan\theta$$

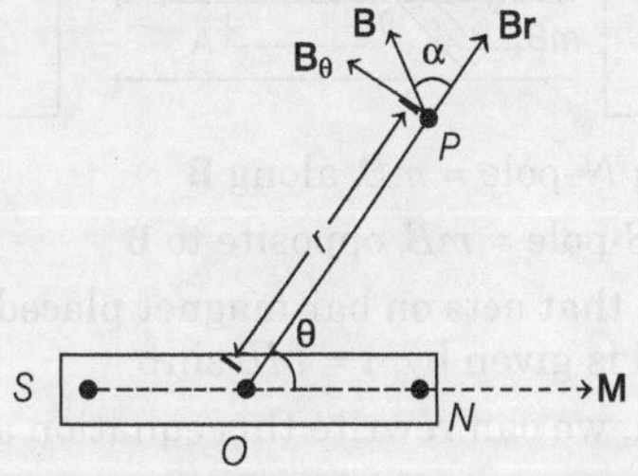

For axial position of point P, θ = 0° and for equitorial position, θ = 90°.

Example 2. *The pole strength of 12 cm long bar magnet is 20 A-m. The magnetic induction at a point 10 cm away from the centre of the magnet on its axial line is*

$\left(\text{Take, } \frac{\mu_0}{4\pi} = 10^{-7}\ Hm^{-1}\right)$

(a) 1.17×10^{-3} T (b) 2.20×10^{-2} T
(c) 1.17×10^{-2} T (d) 2.21×10^{-2} T

Sol. (a) Here, $2l = 12$ cm $= 0.12$ m,

$m = 20$ A-m, $d = 10$ cm $= 0.1$ m

On axial line, $B = \frac{\mu_0}{4\pi}\frac{2Md}{(d^2 - l^2)^2} = \frac{\mu_0}{4\pi}\frac{2m(2l)d}{(d^2 - l^2)^2}$

$$= 10^{-7} \times \frac{2(20)\,(0.12) \times 0.1}{[(0.1)^2 - (0.06)^2]^2}$$

$$= 1.17 \times 10^{-3}\ \text{T}$$

Example 3. *The earth's magnetic field at the equator is approximately 0.4 G. The earth's dipole moment is (Take, $R_e = 6.4 \times 10^6$ m)*

(a) 4×10^2 A-m^2 (b) 1.05×10^{23} A-m^2
(c) 4×10^{-5} A-m^2 (d) 1.05×10^{-20} A-m^2

Sol. (b) The equitorial magnetic field is given by

$$B_e = \frac{\mu_0 m}{4\pi r^3}$$

Given, $B_e \sim 0.4\,G = 4 \times 10^{-5}$ T, $R_e = 6.4 \times 10^6$ m

$$\therefore \quad m_e = \frac{4 \times 10^{-5} \times (6.4 \times 10^6)^3}{\frac{\mu_0}{4\pi}}$$

$$m_e = 4 \times 10^2 \times (6.4 \times 10^6)^3$$
$$= 1.05 \times 10^{23} \text{ A-m}^2$$

Torque on Bar Magnet in Uniform Magnetic Field

Consider a bar magnet of length $2l$ with strength of each pole m placed in a uniform magnetic field **B**. The magnet is held at angle θ with the direction of **B**.

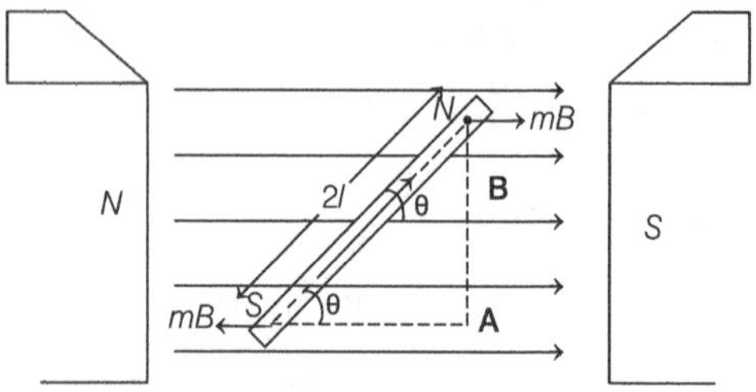

Then, force on N-pole = mB, along **B**

Force on S-pole = mB, opposite to **B**

So, the torque that acts on bar magnet placed in magnetic field is given by, $\tau = MB \sin\theta$

In vector form, we can rewrite this equation as

$$\tau = \mathbf{M} \times \mathbf{B}$$

The direction of τ is perpendicular to the plane containing **M** and **B**.

Potential Energy of a Magnetic Dipole in a Uniform Magnetic Field

The torque acting on the dipole tends to align it in the direction of the field. Work has to be done in rotating the dipole against the action of the magnetic torque. This work done is stored in the form of potential energy of the dipole and is given as $U = W = -MB(\cos\theta - \cos\theta_0)$

In vector form, we may rewrite this equation as

$$U = -\mathbf{M}\cdot\mathbf{B}$$

Note When $\theta = 90°$, $U = 0$, *i.e.* potential energy is zero, when the magnetic dipole is perpendicular to the field. Above equation shows that at $\theta = 0°$, potential energy is minimum ($= -MB$), which is the most stable position. Further at $\theta = 180°$, potential energy is maximum ($= +MB$), which is most unstable position.

Example 4. *A bar magnet when placed at an angle of 30° to the direction of magnetic field induction of 5×10^{-2} T, experiences a moment of couple 2.5×10^{-6} N-m. If the length of the magnet is 5 cm, its pole strength is*

(a) 2×10^2 A-m (b) 2×10^{-3} A-m
(c) 5 A-m (d) 5×10^{-2} A-m

Sol. (b) Here, $\theta = 30°$, $B = 5 \times 10^{-2}$ T,

$$\tau = 2.5 \times 10^{-6} \text{ N-m},$$
$$2l = 5 \text{ cm} = 0.05 \text{ m}, m = ?$$

$\therefore$ Torque, $\tau = MB \sin\theta = m(2l) B \sin\theta$

$$\Rightarrow \quad m = \frac{\tau}{B(2l)\sin\theta} = \frac{2.5 \times 10^{-6}}{5 \times 10^{-2} \times (0.05) \times \sin 30°}$$

$$\Rightarrow \quad m = 2 \times 10^{-3} \text{ A-m}$$

Example 5. *A small bar magnet placed with its axis at 30° with an external field of 0.06 T experiences a torque of 0.018 N-m. The minimum work required to rotate it from its stable to unstable equilibrium position is* **[JEE Main 2020]**

(a) 9.2×10^{-3} J (b) 7.2×10^{-2} J
(c) 6.4×10^{-2} J (d) 11.7×10^{-3} J

Sol. (b) Torque on a bar magnet, $\tau = MB \sin\theta$

$$0.018 = M \times 0.06 \times \sin 30°$$
$$\Rightarrow \quad M = 0.6 \text{ A-m}^2$$

Now, work done to rotate bar magnet,

$$W = MB[\cos\theta_1 - \cos\theta_2]$$

Here, $\theta_1 = 0°$ (for stable equilibrium position)
and $\theta_2 = 180°$ (for unstable equilibrium position).

So, $W = 0.6 \times 0.06 [\cos(0°) - \cos(180°)]$
$= 0.036 [(1) - (-1)] = 0.036(2) = 0.072 = 7.2 \times 10^{-2}$ J

Current Carrying Loop as a Magnetic Dipole

A current carrying loop behaves as a magnetic dipole. If we look the upper face of the loop and current is flowing anti-clockwise, then it has a north polarity and if current is flowing clockwise, then it has a south polarity.

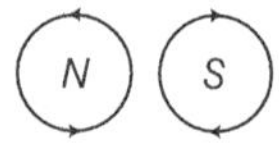

Magnetic dipole moment of a current carrying loop is given by $M = IA$

For N such turns, $M = NIA$

where, I = current and A = area of cross-section of the coil.

Magnetic Dipole Moment of a Revolving Electron in an Atom

The circular motion of an electron around the positively charged nucleus of an atom can be treated as a current carrying loop producing a magnetic field. Hence, it behaves like a magnetic dipole.

Consider m_e be the mass and $-e$ the charge of an electron revolving with speed v in a circular orbit of radius r.

The magnitude of the magnetic dipole moment **M** associated with the revolving electron is

$$\mathbf{M} = IA = \frac{ev}{2\pi r} \times \pi r^2 = \frac{evr}{2}$$

The magnitude of the orbital angular momentum **L** of electron,

$$L = m_e vr \Rightarrow \frac{M}{L} = \frac{e}{2m_e}$$

Thus, **the ratio of the magnitude of the magnetic dipole moment to the magnitude of the angular momentum of the revolving electron is a constant. This ratio is called the gyromagnetic ratio.** Its value for an electron is 8.8×10^{10} C/kg.

The vector form of above equation can be written as

$$\mathbf{M} = -\left(\frac{e}{2m_e}\right)\mathbf{L}$$

$$\frac{M}{nh/2\pi} = \frac{e}{2m_e} \Rightarrow M = \frac{neh}{4\pi m_e}$$

When $n = 1$, $M = \mu$ (the elementary magnetic dipole moment), thus

$$\mu = \frac{eh}{4\pi m_e}$$

The elementary magnetic moment of a revolving electron is also known as Bohr magneton (μ).

Note Since, a current carrying loop acts as an magnetic dipole. So, the torque on the loop in a magnetic field is $\tau = NBiA \sin\theta = \mathbf{M} \times \mathbf{B}$.

Example 6. *A straight wire carrying current i is turned into a circular loop. If the magnitude of magnetic moment associated with it MKS unit is M, the length of wire will be*

(a) $\frac{4\pi}{M}$ (b) $\frac{M\pi}{4i}$

(c) $\sqrt{\frac{4\pi M}{i}}$ (d) $\sqrt{\frac{4\pi i}{M}}$

Sol. (c) Magnetic moment, $M = iA = i(\pi r^2)$, where $l = 2\pi r$

$$r = \sqrt{\frac{M}{\pi i}}$$

$$\therefore \quad l = 2\pi\sqrt{\frac{M}{\pi i}} = \sqrt{\frac{4\pi M}{i}}$$

Example 7. *The electron in hydrogen atom moves with a speed of 2.2×10^6 m/s in an orbit of radius 5.3×10^{-11}cm. The magnetic moment of the orbiting electron is*

(a) 3×10^{-20} A-m^2 (b) 3.3×10^{-10} A-m^2

(c) 9.3×10^{-26} A-m^2 (d) 9.6×10^{-10} A-m^2

Sol. (c) Given, $v = 2.2 \times 10^6$ ms^{-1},

$$r = 5.3 \times 10^{-11} \text{ cm} = 5.3 \times 10^{-13} \text{ m}$$

Frequency of revolution, $f = \frac{v}{2\pi r}$

The moving charge is equivalent to a current loop, given by

$$i = f \times e \text{ or } i = \frac{ev}{2\pi r}$$

Let A be the area of the orbit, then the magnetic moment of the orbiting electron is,

$$M = iA = \left(\frac{ev}{2\pi r}\right)(\pi r^2) = \frac{evr}{2}$$

Putting the numerical values, we have

$$M = \frac{(1.6 \times 10^{-19})(2.2 \times 10^6)(5.3 \times 10^{-13})}{2}$$

$$\Rightarrow \quad M = 9.3 \times 10^{-26} \text{ A-m}^2$$

Example 8. *A circular coil having N turns and radius r carries a current I. It is held in the XZ-plane in a magnetic field $B\hat{\mathbf{i}}$. The torque on the coil due to the magnetic field (in N-m) is* ***[JEE Main 2019]***

(a) $\frac{Br^2I}{\pi N}$ (b) $B\pi r^2 IN$ (c) $\frac{B\pi r^2 I}{N}$ (d) zero

Sol. (b) According to the question, the situation can be drawn as

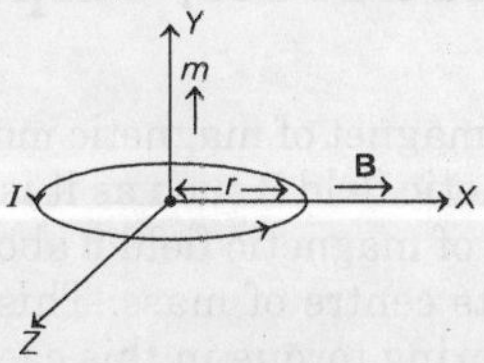

Let the current I is flowing in anti-clockwise direction, then the magnetic moment of the coil is $m = NIA$.

where, N = number of turns in coil

and A = area of each coil = πr^2.

Its direction is perpendicular to the area of coil and is along Y-axis.

Then, torque on the current coil is

$$\tau = \mathbf{m} \times \mathbf{B} = mB \sin 90°$$
$$= NIAB = NI\pi r^2 B \text{ (N-m)}$$

Bar Magnet as an Equivalent Solenoid

The resemblance of magnetic field lines of a bar magnet and a solenoid suggest that a bar magnet may be thought of as a large number of circulating currents in analogy with a solenoid.

- Cutting a bar magnet in half is like cutting a solenoid, we get two smaller solenoids with weaker magnetic properties.
- The field lines remain continuous, emerging from one face of the solenoid and entering into the other face.
- The magnitude of the magnetic moment of the solenoid is M = (total number of turns × current × cross-sectional area) = $n(2l)\, I(\pi a^2)$.
- A bar magnet and a solenoid produce similar magnetic fields. The magnetic moment of a bar magnet is thus equal to the magnetic moment of an equivalent solenoid that produces the same magnetic field.

Example 9. *A closely wound solenoid of 2000 turns and area of cross-section 1.6×10^{-4} m^2, carrying a current of 4 A is suspended through its centre allowing it to turn in a horizontal plane. If a uniform horizontal magnetic field of 7.5×10^{-2} T is set up at angle of 30° with the axis of the solenoid, then the torque acting on the solenoid is*

(a) 0.048 J (b) 0.012 J (c) 48 J (d) 12 J

Sol. (a) Let M = magnetic moment of the solenoid, *i.e.*

$$M = NIA$$

Here, $N = 2000, I = 4\text{A}, A = 1.6 \times 10^{-4}\ \text{m}^2$

$\therefore$ $M = 2000 \times 4 \times 1.6 \times 10^{-4}$

$M = 1.28\ \text{JT}^{-1}$

Torque, $\tau = MB \sin\theta$

Here, $\theta = 30°, B = 7.5 \times 10^{-2}\ \text{T}, M = 1.28\ \text{JT}^{-1}$

$\tau = 1.28 \times 7.5 \times 10^{-2} \times \sin 30°$

$= 1.28 \times 7.5 \times 10^{-2} \times \frac{1}{2} = 0.048\ \text{J}$

Oscillations of a Freely Suspended Magnet

When a small bar magnet of magnetic moment **M** is placed in a uniform magnetic field **B** such as it is free to vibrate in a horizontal plane of magnetic field **B** about a vertical axis passing through its centre of mass. This bar magnet will oscillate. The restoring torque in this case will be

$$\tau = -MB\theta \quad (\because \text{for small oscillation, } \sin\theta \simeq \theta)$$

The deflecting torque on the magnet is

$$\tau = I\alpha = I\frac{d^2\theta}{dt^2}$$

In equilibrium, deflecting torque = restoring torque

or $$\frac{d^2\theta}{dt^2} = \frac{-MB\theta}{I} = -\omega^2\theta, \text{ where } \omega = \sqrt{\frac{MB}{I}}$$

The period of vibration is given by, $T = \frac{2\pi}{\omega} = 2\pi\sqrt{\frac{I}{MB}}$

Magnetic field B can be calculated from above equation and is given as $B = \frac{4\pi^2 I}{MT^2}$

Magnetism and Gauss's Law

The net magnetic flux ϕ_B through any closed surface is always zero, *i.e.* the number of magnetic field lines leaving the surface is balanced by the number of lines entering it, *i.e.*

$\phi_B = \Sigma\, \mathbf{B} \cdot \Delta\, \mathbf{S} = 0$

where, B = magnetic field

and ΔS = area element.

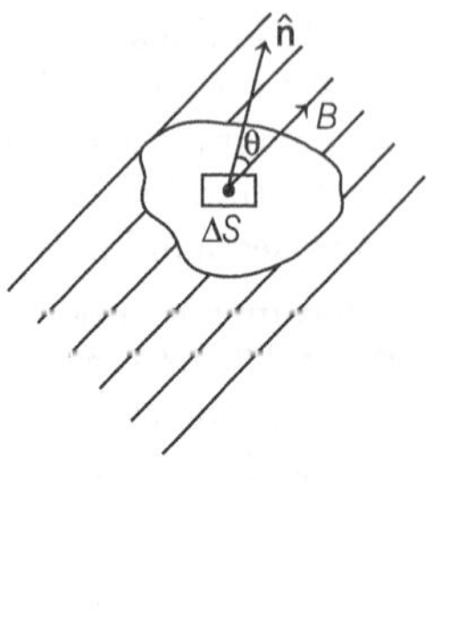

Earth's Magnetism

Our earth behaves as a huge powerful magnet. The value of magnetic field on the surface of earth is a few tenths of a gauss (1 G = 10^{-4} T) and its strength varies from place to place on the earth's surface. The earth's magnetic south pole is located near the geographic north pole and the earth's magnetic north pole is located near the geographic south pole.

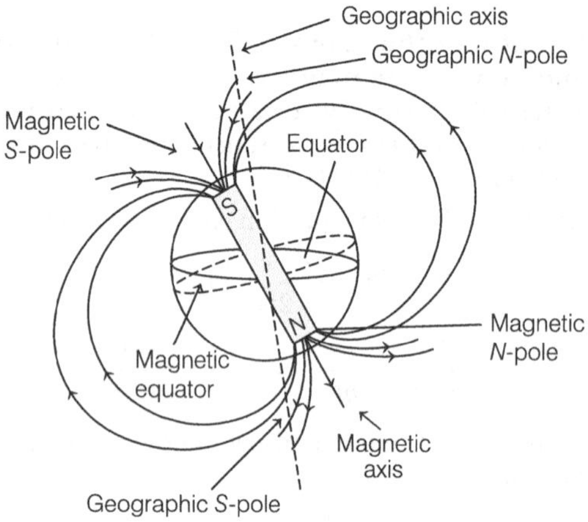

Some definitions related to earth's magnetism are

Magnetic equator The huge circle whose plane is perpendicular to the earth's magnetic axis is called earth's magnetic equator.

Magnetic meridian The line joining the earth's magnetic poles is called the *magnetic axis* and a vertical plane passing through it is called the magnetic meridian.

Geographical meridian The line joining the geographical north and south poles is called the geographic axis and a vertical plane passing through it is called the geographical meridian.

Magnetic Elements of Earth

The earth's magnetic field at a place can be completely described by three parameters which are called elements of earth's magnetic field. These three elements are

Angle of Declination or Magnetic Declination The angle between the magnetic meridian and geographical meridian at a place is called the angle of declination (or simply the declination) at that place.

Angle of Dip or Magnetic Inclination The angle of dip (θ) at a place is the angle between the direction of earth's magnetic field (B) and horizontal line in the magnetic meridian.

Horizontal-Vertical Component of the Earth's Magnetic Field Consider B_e be the net magnetic field at some point. H and V be the horizontal and vertical components of B_e. Let θ is the angle of dip at the some place, then we can see that,

$$H = B_e \cos\theta \quad \ldots\text{(i)}$$

and $\quad V = B_e \sin\theta \quad$ ...(ii)

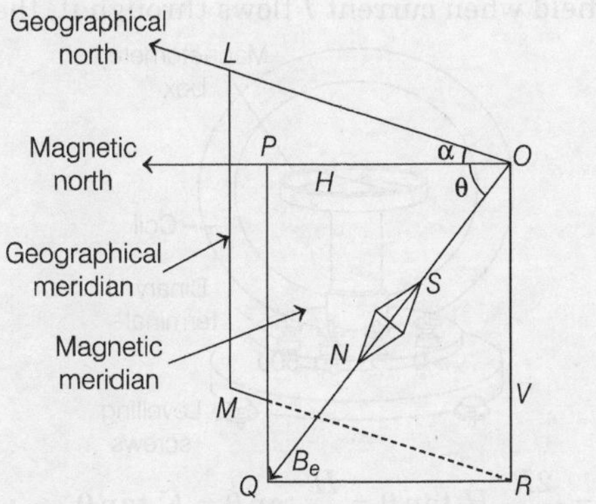

Elements of earth's magnetic field

Squaring and adding Eqs. (i) and (ii), we get

$$B_e = \sqrt{H^2 + V^2}$$

Further, dividing Eqs. (ii) by Eq. (i), we get

$$\theta = \tan^{-1}\left(\frac{V}{H}\right)$$

Neutral Points

The point where resultant magnetic field due to magnet and due to earth is zero are called neutral points. Following are two cases related to neutral points

(i) When a bar magnet is placed along the magnetic meridian with its north pole pointing towards geographic north, then neutral points lie on the equatorial line of the magnet.

At each neutral point,

$$B_{eq} = \frac{\mu_0}{4\pi} \cdot \frac{M}{(r^2 + l^2)^{3/2}} = B_H$$

If $r >> l$, then $\frac{\mu_0}{4\pi} \cdot \frac{M}{r^3} = B_H$

(ii) When a bar magnet is placed along the magnetic meridian with its north pole pointing towards geographic south, two neutral points are obtained on either side of the magnet along its axial line.

At each neutral point,

$$B_{\text{axial}} = \frac{\mu_0}{4\pi} \frac{2Mr}{(r^2 - l^2)^2} = B_H$$

If $r >> l$, then $\frac{\mu_0}{4\pi} \cdot \frac{2M}{r^3} = B_H$

Example 10. *If a magnet is suspended at an angle 30° to the magnetic meridian, the dip needle makes an angle of 45° with the horizontal. The real dip is*

(a) $\tan^{-1}(\sqrt{3}/2)$ (b) $\tan^{-1}(\sqrt{3})$
(c) $\tan^{-1}(\sqrt{3/2})$ (d) $\tan^{-1}(2/\sqrt{3})$

Sol. (d) As, $\tan\theta' = \frac{\tan\theta}{\cos\theta} = \frac{\tan 45°}{\cos 30°} = \frac{1}{\sqrt{3}/2} = \frac{2}{\sqrt{3}}$

$\Rightarrow \quad \theta' = \tan^{-1}(2/\sqrt{3})$

Example 11. *In the magnetic meridian of a certain place, the horizontal component of the earth's magnetic field is 0.26 G and the angle of dip is 60°. The magnetic field of the earth at this location is* **[NCERT]**

(a) 0.26 G (b) 0.42 G (c) 0.52 G (d) 0.80 G

Sol. (c) The earth's magnetic field is B_e and its horizontal and vertical components are H_e and V_e.

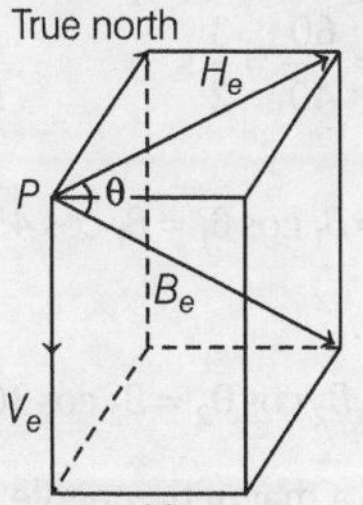

From the above figure, $\cos\theta = \frac{H_e}{B_e}$

Given, $\quad H_e = 0.26$ G, $\theta = 60°$

$\therefore \quad B_e = \frac{H_e}{\cos 60°} = \frac{0.26}{\left(\frac{1}{2}\right)} = 0.52$ G

Example 12. *Two short bar magnets of length 1 cm each have magnetic moments 1.20 A-m² and 1.00 A-m², respectively. They are placed on a horizontal table parallel to each other with their N poles pointing towards the south. They have a common magnetic equator and are separated by a distance of 20.0 cm. The value of the resultant horizontal magnetic induction at the mid-point O of the line joining their centres is close to (Take, horizontal component of the earth's magnetic induction = 3.6×10^{-5} Wb/m²)* **[JEE Main 2013]**

(a) 3.6×10^{-5} Wb/m² (b) 2.56×10^{-4} Wb/m²
(c) 3.50×10^{-4} Wb/m² (d) 5.80×10^{-4} Wb/m²

Sol. (b) Net magnetic field, $B_{net} = B_1 + B_2 + B_H$

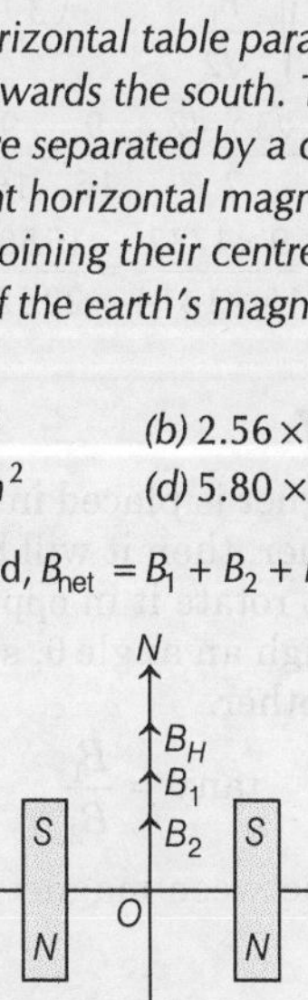

$$B_{net} = \frac{\mu_0}{4\pi} \frac{(M_1 + M_2)}{r^3} + B_H = \frac{10^{-7}(1.2 + 1)}{(0.1)^3} + 3.6 \times 10^{-5}$$

$$= 2.56 \times 10^{-4} \text{ Wb/m}^2$$

Example 13. *A magnetic compass needle oscillates 30 times per minute at a place, where the dip is 45° and 40 times per minute, where the dip is 30°. If B_1 and B_2 are respectively, the total magnetic field due to the earth at the two places, then the ratio $\frac{B_1}{B_2}$ is best given by* **[JEE Main 2019]**

(a) 1.8 T (b) 0.7 T (c) 3.6 T (d) 2.2 T

Sol. (b) Given, at first place, angle of dip, $\theta_1 = 45°$

Time period, $T_1 = \frac{60}{30} = 2\text{ s}$

At second place, angle of dip, $\theta_2 = 30°$

Time period, $T_2 = \frac{60}{40} = \frac{3}{2}\text{ s}$

Now, at first place,

$$B_{H_1} = B_1 \cos\theta_1 = B_1 \cos 45° = \frac{B_1}{\sqrt{2}} \quad \text{...(i)}$$

and at second place,

$$B_{H_2} = B_2 \cos\theta_2 = B_2 \cos 30° = \frac{\sqrt{3}}{2} B_2 \quad \text{...(ii)}$$

Also, time period of a magnetic needle is given by

$$T = 2\pi\sqrt{\frac{I}{MB_H}} \quad \text{...(iii)}$$

$$\therefore \quad T \propto \sqrt{\frac{1}{B_H}} \text{ or } \frac{T_1}{T_2} = \sqrt{\frac{B_{H_2}}{B_{H_1}}} \quad \text{...(iv)}$$

By putting the values from Eqs. (i) and (ii) into Eq. (iv), we get

$$\frac{2}{\frac{3}{2}} = \sqrt{\frac{\sqrt{3}\frac{B_2}{2}}{\frac{B_1}{\sqrt{2}}}} \text{ or } \left(\frac{4}{3}\right)^2 = \frac{\sqrt{3}\times\sqrt{2}\,B_2}{2B_1}$$

$$\Rightarrow \quad \frac{B_1}{B_2} = \frac{\sqrt{3}\times\sqrt{2}}{2}\times\frac{9}{16} = \frac{9\sqrt{3}}{16\sqrt{2}}$$

$$= \frac{9\times 1.732}{16\times 1.414} = \frac{15.588}{22.624} = 0.689 \approx 0.7\text{ T}$$

Tangent Law

It states that, if a magnet is placed in two magnetic fields right angle to each other, then it will be acted upon by two couples tending to rotate it in opposite directions. It will be deflected through an angle θ, such that two couples balance each other.

Also, $$\tan\theta = \frac{B_1}{B_2}$$

where, θ is the angle between magnet and magnetic field B_2.

Tangent Galvanometer

It is a device used to measure very small current. It is a moving magnet type galvanometer and works on the principle of tangent law. In tangent galvanometer, a magnetic compass needle is placed horizontally at the centre of a vertical fixed current carrying coil whose plane is in the magnetic meridian. So, if the needle in equilibrium subtends an angle ϕ with the earth's magnetic field when current I flows through it, then

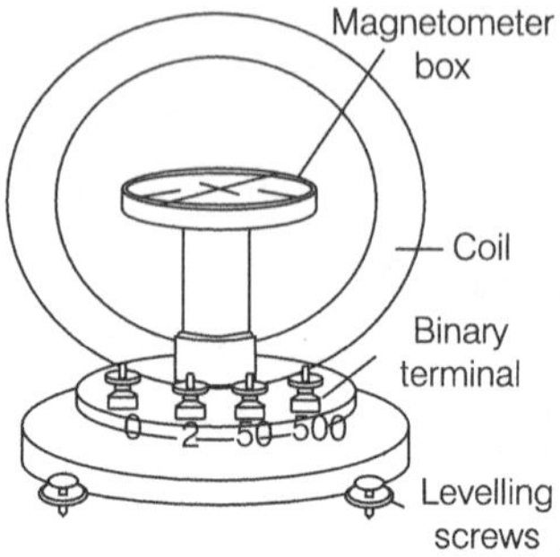

$$I = \frac{2R}{N\mu_0} H \tan\theta = \frac{H}{G}\tan\theta = K\tan\theta$$

where, $G = \frac{N\mu_0}{2R}$ is called *galvanometer constant* and $K = \frac{H}{G}$ is called *reduction factor* of tangent galvanometer.

Here, N is number of turns in the coil and R is radius of the coil.

Tangent galvanometer is also called moving magnet type galvanometer.

Sensitivity It is the measure of change in the deflection produced by a unit current.

It is given by $$\frac{d\theta}{dI} = \frac{1}{k\sec^2\theta} = \frac{1}{k\left(1+\frac{I^2}{k^2}\right)}$$

Vibration Magnetometer

An instrument which is used to compare magnetic moments of two bar magnets or to determine the horizontal component of earth's magnetic field is known as vibration magnetometer. It is based on the principle that when a bar magnet suspended freely in a uniform magnetic field is displaced from its equilibrium position, it starts executing simple harmonic motion about the equilibrium position.

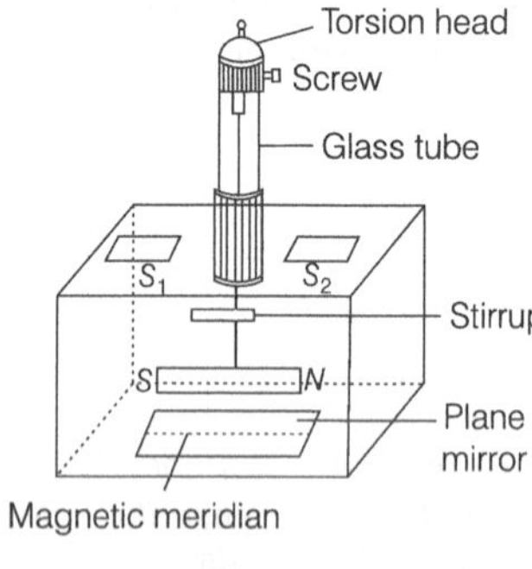

The time period of vibration of the magnet of moment of inertia I and magnetic moment M vibrating in uniform magnetic field of strength B_H is given by $T = 2\pi\sqrt{\frac{I}{MB_H}}$.

Uses of Vibration Magnetometer

(i) **To compare magnetic moments of two bar magnets** For two bar magnets A and B of same size and mass,

$$T_1 = 2\pi\sqrt{\frac{I}{M_1 B_H}}$$

and $$T_2 = 2\pi\sqrt{\frac{I}{M_2 B_H}}$$

where, M_1 and M_2 be the magnetic moments of the magnets A and B respectively and I be the moment of inertia of each magnets.

$$\therefore \quad \frac{M_1}{M_2} = \frac{T_2^2}{T_1^2}$$

(ii) **To compare the magnetic moments by sum and difference method**

(a) When the two magnets are placed with their poles in the same direction.

Net magnetic moment, $M_S = M_1 + M_2$

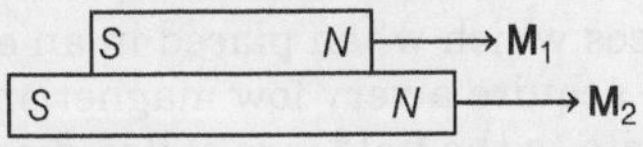

Net moment of inertia, $I_S = I_1 + I_2$

Time period of oscillation of this pair in Earth's magnetic field (B_H)

$$T_S = 2\pi\sqrt{\frac{I_S}{M_S B_H}} = 2\pi\sqrt{\frac{I_1 + I_2}{(M_1 + M_2)B_H}}$$

(b) When the two bar magnets are placed with their unlike poles in the same direction, the ratio of their magnetic moments can be given as

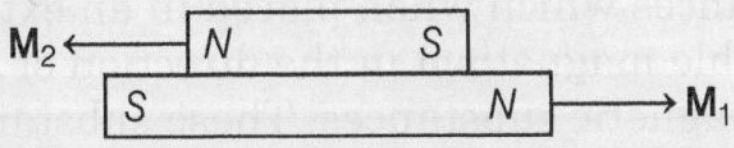

$$T_d = 2\pi\sqrt{\frac{I_1 + I_2}{(M_1 - M_2)H}}$$

and $$\frac{M_1}{M_2} = \frac{T_d^2 + T_S^2}{T_d^2 - T_S^2}$$

(iii) **To compare horizontal components of Earth's magnetic field at two places**

$T = 2\pi\sqrt{\frac{I}{MB_H}}$, since I and M of the magnet are constant, so $T^2 \propto \frac{1}{B}$

$$\therefore \quad \frac{B_H}{B'_H} = \frac{T'^2}{T^2}$$

where, B_H and B'_H be the value of horizontal components of earth's magnetic field at two places A and B, respectively.

Example 14. *In a tangent galvanometer, when a current of 10 mA is passed, the deflection is 31°. By what percentage, the current has to be increased, so as to produce a deflection of 42°?*

(a) 10% (b) 50%
(c) 20% (d) 80%

Sol. (b) Here, $I_1 = 10$ mA, $\theta_1 = 31°$, $I_2 = ?$, $\theta_2 = 42°$

As, $$\frac{I_2}{I_1} = \frac{\tan\theta_2}{\tan\theta_1}$$

$$\Rightarrow \quad I_2 = I_1\frac{\tan\theta_2}{\tan\theta_1} = 10 \times \frac{\tan 42°}{\tan 31°}$$

$$= \frac{10 \times 0.9}{0.6} = 15 \text{ mA}$$

Percentage increase in current,

$$\frac{(I_2 - I_1)}{I_1} \times 100 = \frac{(15 - 10)}{10} \times 100 = 50\%$$

Example 15. *The time period of vibration of two magnets in sum position is 3s. When polarity of weaker magnet is reversed, the combination makes 12 oscillations per minute. The ratio of magnetic moments of two magnets is*

(a) $\frac{16}{17}$ (b) $\frac{17}{8}$
(c) $\frac{3}{5}$ (d) $\frac{4}{5}$

Sol. (b) Here, $T_1 = 3$ s, $T_2 = \frac{1}{12}$ min $= \frac{60 \text{ s}}{12} = 5$ s

$$\frac{M_1}{M_2} = \frac{T_2^2 + T_1^2}{T_2^2 - T_1^2}$$

$$\therefore \quad \frac{M_1}{M_2} = \frac{5^2 + 3^2}{5^2 - 3^2} = \frac{34}{16} = \frac{17}{8}$$

Magnetisation of Materials

To describe the magnetic properties of materials, we have to understand the following terms

Magnetic Induction (B) The number of magnetic lines of force inside a magnetic substance crossing per unit area normal to their direction is called the magnetic induction or **magnetic flux density** inside the substance. It is denoted by **B**. The SI unit of **B** is tesla (T) or weber/metre^2 (Wb/m^2). The CGS unit is gauss (G).

$$1 \text{ Wb/m}^2 = 1 \text{ T} = 10^4 \text{ G}$$

Intensity of Magnetisation (I) The magnetic moment induced in unit volume of a magnetic substance placed in a magnetic field is called the intensity of magnetisation. It is denoted by I.

Thus, $$I = \frac{M}{V} = \frac{m}{A} = \frac{\text{Pole strength}}{\text{Area}}$$

i.e. The intensity of magnetisation may also be defined as the pole strength per unit cross-sectional area.

The unit of I is Am^{-1}.

Magnetic Intensity or Magnetic Field Strength (*H*) The capability of the magnetising field to magnetise the substance is expressed by means of a vector **H**, called the *magnetic intensity* of the field. It is defined through the vector relation,

$$\mathbf{H} = \frac{\mathbf{B}}{\mu_0} - \mathbf{I}$$

The SI unit of **H** is same as that of **I**, *i.e.* ampere/metre (A/m). The CGS unit is oersted.

Magnetic Permeability (μ) It is defined as the ratio of the magnetic induction **B** inside the magnetised substance to the magnetic intensity **H** of the magnetising field, *i.e.*

$$\mu = \frac{\mathbf{B}}{\mathbf{H}}$$

It is basically a measure of conduction of magnetic lines of force through a substance. The SI unit of magnetic permeability is weber/ampere-metre (Wb/A-m) or tesla metre-ampere^{-1} (TmA^{-1}).

Relative Magnetic Permeability (μ_r) It is the ratio of the magnetic permeability μ of the substance to the permeability of free space.

Thus, $$\mu_r = \frac{\mu}{\mu_0}$$

μ_r is a pure ratio, hence dimensionless.

For vacuum, its value is 1.

μ_r can also be defined as the ratio of the magnetic field B inside the substance when placed in magnetic field B_0.

Thus, $\mu_r = \dfrac{B}{B_0}$

Magnetic Susceptibility (χ_m) It is the property of the substance which shows how easily a substance can be magnetised when kept in a magnetising field. It can also be defined as the ratio of intensity of magnetisation (I) in a substance to the magnetic intensity (H) applied to the substance, *i.e.*

$$\chi_m = \frac{I}{H}$$

Relation between μ_r and χ_m We have, $B = \mu_0(I + H)$

or $$B = B_0(\chi_m + 1)$$

$\Rightarrow$ $$\mu_r = \chi_m + 1$$

Example 16. *A paramagnetic substance in the form of a cube with sides 1 cm has a magnetic dipole moment of 20×10^{-6} J/T when a magnetic intensity of 60×10^3 A/m is applied. Its magnetic susceptibility is* **[JEE Main 2019]**

(a) 3.3×10^{-4}

(b) 3.3×10^{-2}

(c) 4.3×10^{-2}

(d) 2.3×10^{-2}

Sol. (a) Given, side of cube = 1 cm = 10^{-2} m

$\therefore$ Volume, $V = 10^{-6} \text{m}^3$

Dipole moment, $M = 20 \times 10^{-6}$ J/T

Applied magnetic intensity, $H = 60 \times 10^3$ A/m

Intensity of magnetisation

$$I = \frac{M}{V} = \frac{20 \times 10^{-6}}{10^{-6}} = 20 \text{ A/m}$$

Now, magnetic susceptibility χ is,

$$\chi = \frac{\text{Intensity of magnetisation}}{\text{Applied magnetic intensity}} = \frac{I}{H} = \frac{20}{60 \times 10^3}$$

$$\Rightarrow \quad \chi = \frac{1}{3} \times 10^{-3} = 3.3 \times 10^{-4}$$

Classification of Magnetic Materials

According to behaviour of magnetic substances, they are classified into three categories as follows

Diamagnetic Substances

Those substances which when placed in an external magnetic field, acquire a very low magnetism in a direction opposite to the field, are called diamagnetic substances. These substances when brought near the end of a strong magnet, get repelled by it.

e.g. Copper (Cu), silver (Ag), bismuth (Bi), zinc (Zn), diamond (C), salt (NaCl), water (H_2O), mercury (Hg), nitrogen (N_2), hydrogen (H_2), magnesium (Mg), gold (Au), etc.

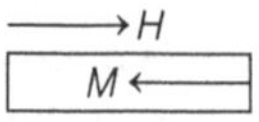

Paramagnetic Substances

Those substances which when placed in an external field, acquire a feeble magnetism in the direction of field, are called paramagnetic substances. These substances when brought near the end of a strong magnet, get attracted towards it.

e.g. Aluminium (Al), sodium (Na), potassium (K), platinum (Pt), manganese (Mn), copper sulphate ($CuSO_4$), oxygen (O_2), etc.

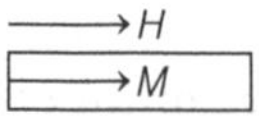

Ferromagnetic Substances

Those substances which when placed in a magnetic field, acquire a strong magnetism in the direction of field, are called ferromagnetic substances. These substances when brought near the end of a strong magnet get radially attracted towards it.

e.g. Iron (Fe), nickel (Ni), cobalt (Co), magnetite (Fe_3O_4) or natural magnet, etc.

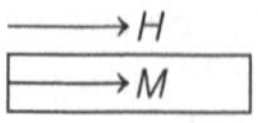

Following are few important graphs related to magnetic materials

- *I-H* curve

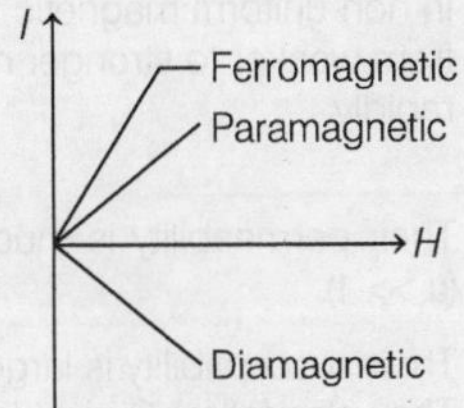

- χ-*T* curve

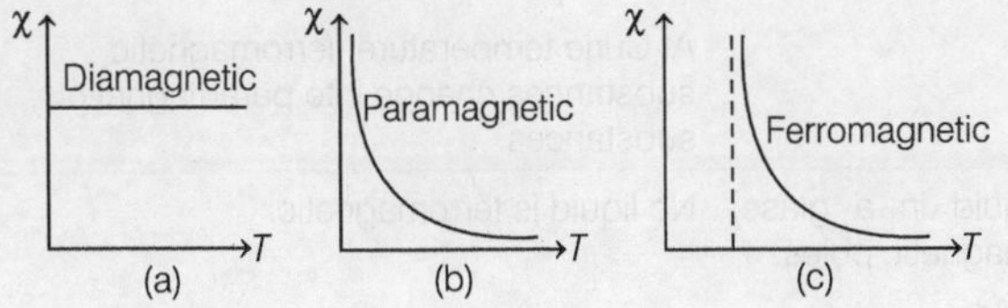

Curie's Law

According to this law, the magnetic susceptibility of paramagnetic substances is inversely proportional to absolute temperature, *i.e.*

$$\chi \propto \frac{1}{T} \Rightarrow \chi T = \text{constant}$$

Here, T = absolute temperature.

At Curie temperature, ferromagnetic substances changes into paramagnetic substances.

Curie-Wiess Law

Above Curie temperature or Curie point, the magnetic susceptibility of ferromagnetic substances is inversely proportional to $(T - T_C)$, *i.e.*

$$\chi \propto \frac{1}{T - T_C}$$

or

$$\chi = \frac{C}{T - T_C}$$

Here, T_C = Curie temperature.

Example 17. *A paramagnetic material has* 10^{28} *atoms/m*^{3}. *Its magnetic susceptibility at temperature 350 K is* 2.8×10^{-4}. *Its susceptibility at 300 K is* **[JEE Main 2019]**

(a) 3.726×10^{-4} (b) 3.672×10^{-4}

(c) 2.672×10^{-4} (d) 3.267×10^{-4}

Sol. (*d*) From Curie's law for paramagnetic substance, we have

Magnetic susceptibility, $\chi \propto \frac{1}{T}$

$$\therefore \quad \frac{\chi_2}{\chi_1} = \frac{T_1}{T_2}$$

$$\Rightarrow \quad \chi_2 = \frac{\chi_1 \cdot T_1}{T_2}$$

$$\Rightarrow \quad \chi_2 = \frac{2.8 \times 10^{-4} \times 350}{300} = 3.267 \times 10^{-4}$$

Comparative Study of Magnetic Materials

Diamagnetic Substances	Paramagnetic Substances	Ferromagnetic Substances
When a diamagnetic solution is poured into a U-tube and one arm is placed between the poles of strong magnet, the level of solution in that arm is lowered. 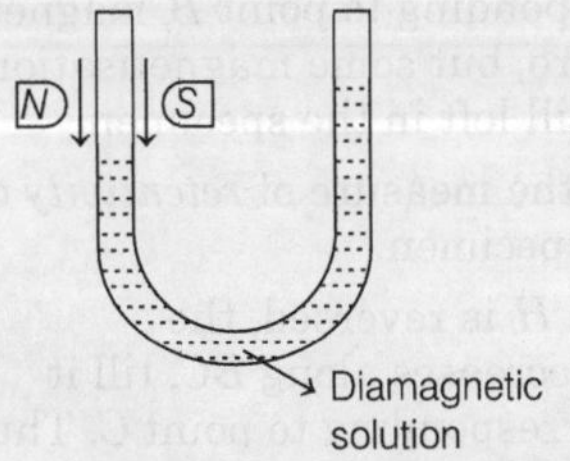	The level of the paramagnetic solution in that arm rises. 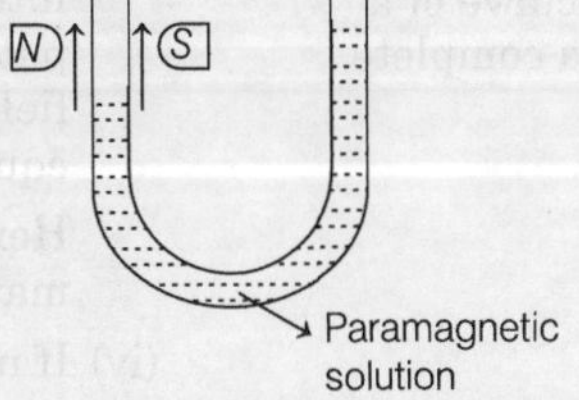	No liquid is ferromagnetic.
If a rod of diamagnetic material is suspended freely between two magnetic poles, its axis becomes perpendicular to the magnetic field. 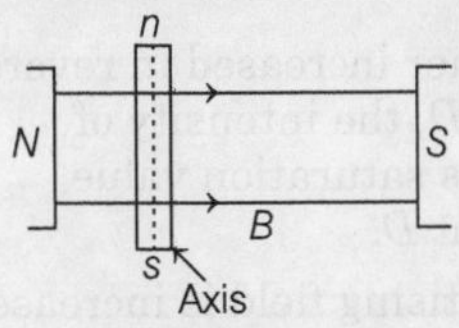	Paramagnetic rod becomes parallel to the magnetic field. 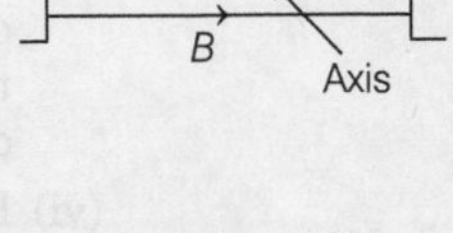	Ferromagnetic rod also becomes parallel to the magnetic field. 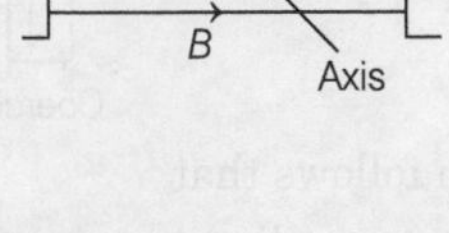

Diamagnetic Substances	Paramagnetic Substances	Ferromagnetic Substances
In non-uniform magnetic field, the diamagnetic substances are attracted towards the weaker fields, *i.e.* they move from stronger to weaker magnetic field.	In non-uniform magnetic field, they move from weaker to stronger part of the magnetic field slowly.	In non-uniform magnetic field, they move from weaker to stronger magnetic field rapidly.
Their permeability is less than one ($\mu < 1$).	Their permeability is slightly greater than one ($\mu > 1$).	Their permeability is much greater than one ($\mu >> 1$).
Their susceptibility is small and negative. Their susceptibility is independent of temperature.	Their susceptibility is small and positive. Their susceptibility is inversely proportional to absolute temperature which is Curie's law. *i.e.* $\chi \propto \frac{1}{T}$.	Their susceptibility is large and positive. They also follow Curie's law. *i.e.* $\chi \propto \frac{1}{T}$. At Curie temperature, ferromagnetic substances change into paramagnetic substances.
Shape of diamagnetic liquid in a glass crucible and kept over two magnetic poles. 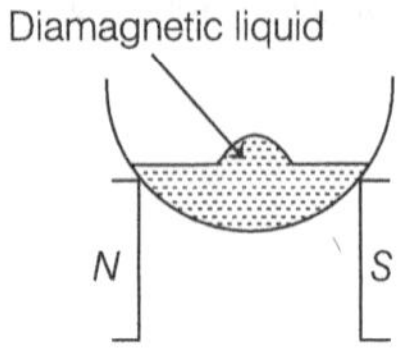	Shape of paramagnetic liquid in a glass crucible and kept over two magnetic poles. 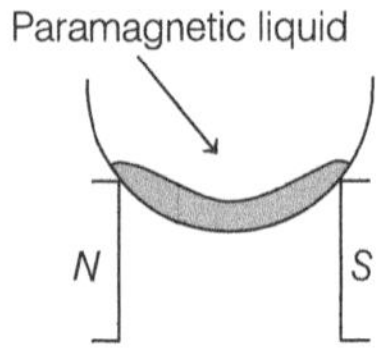	No liquid is ferromagnetic.
In these substances, the magnetic lines of force are farther than in air.	In these substances, the magnetic lines of force are closer than in air.	In these substances, magnetic lines of force are much closer than in air.
The resultant magnetic moment of these substances is zero.	These substances have a permanent magnetic moment.	These substances also have a permanent magnetic moment.

Hysteresis

The lagging of intensity of magnetisation (I) or magnetic induction (B) behind the magnetising field (H) during the process of magnetisation and demagnetisation of a ferromagnetic material is called hysteresis.

The following figure shows the magnetisation curve of a ferromagnetic material, when it is taken over a complete cycle of magnetisation.

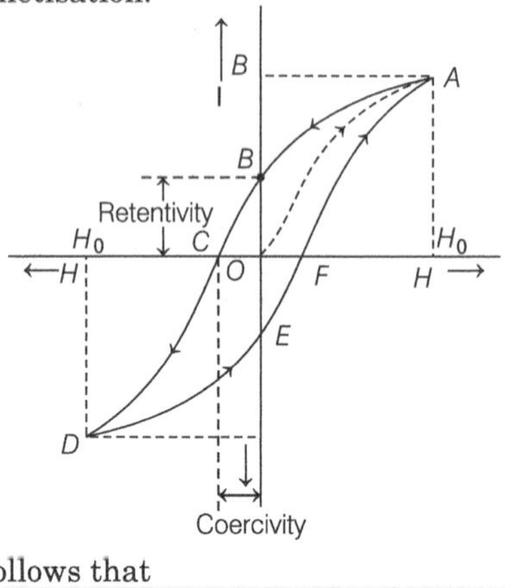

The graph follows that

(i) Corresponding to point O, the magnetisation (H) is zero and likewise intensity of magnetisation (I) is also zero.

(ii) As magnetising field is increased, intensity of magnetisation I also increases along OA and becomes maximum at A. This maximum value is called *saturation value*.

(iii) If magnetic field is now decreased slowly, intensity of magnetisation decreases along the path AB. Corresponding to point B, magnetising field becomes zero, but some magnetisation equal to OB is still left in the specimen.

Here, OB gives the measure of *retentivity* of the material of the specimen.

(iv) If magnetic field H is reversed, the magnetisation decreases along BC, till it becomes zero corresponding to point C. Thus, to make I to be zero, magnetising field equal to OC has to be applied in reverse direction. Here, OC gives the measure of *coercivity* of the material of the specimen.

(v) When field H is further increased in reverse direction, *i.e.* along CD, the intensity of magnetisation attains saturation value corresponding to point D.

(vi) Finally, when magnetising field is increased in original direction, the point A is reached *via* EFA.

If the magnetising field is repeatedly changed between H_0 and $-H_0$, the curve $ABCDEFA$ is retraced. This curve is called the hysteresis loop.

The energy loss in magnetising and demagnetising a specimen is proportional to the area of hysteresis loop.

Note For steel, coercivity is large. However, retentivity is comparatively smaller in case of steel. Due to high value of coercivity and fairly large value of retentivity, steel is used to make permanent magnets. For soft iron, coercivity is very small and area of hysteresis loop is small. Because of these characteristics, soft iron is an ideal material for making electromagnets.

Example 18.

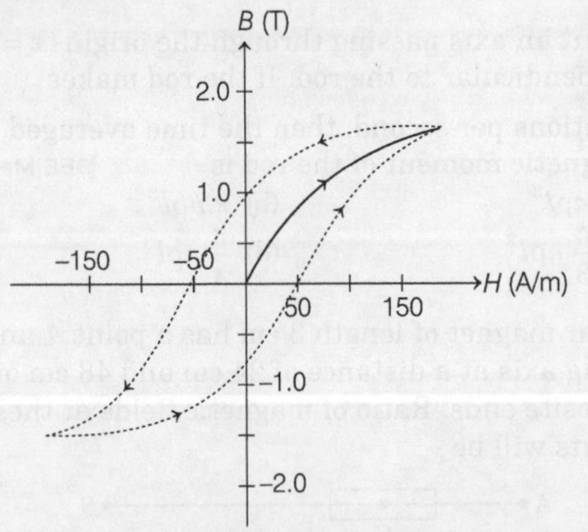

The figure gives experimentally measured B versus H variation in a ferromagnetic material. The retentivity, coercivity and saturation respectively of the material are ***[JEE Main 2020]***

(a) $1.0\ T$, $50\ A/m$ and $1.5\ T$
(b) $150\ A/m$, $1.0\ T$ and $1.5\ T$
(c) $1.5\ T$, $50\ A/m$ and $1.0\ T$
(d) $1.5\ T$, $50\ A/m$ and $1.0\ T$

Sol. (a) We can obtain the values of retentivity (the remaining value of magnetic field B when $H = 0$), coercivity (value of H when $B = 0$) and saturation value of induced magnetic field (B_{max}) as shown in the figure.

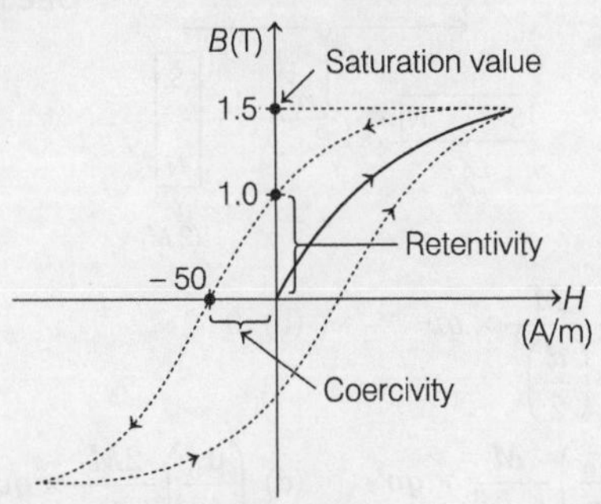

Clearly, retentivity $= 1\text{T}$, coercivity $= 50\ \text{Am}^{-1}$ and saturation value $= 1.5\ \text{T}$

Permanent Magnets

Those substances which at room temperature retain their ferromagnetic property for a long period of time are called permanent magnets. An efficient way to make a permanent magnet is to place a ferromagnetic rod in a solenoid and pass a current. The magnetic field of the solenoid magnetises the rod.

The hysteresis curve allows us to select the suitable materials for permanent magnets. The material should have high retentivity, so that the magnet is strong and high coercivity, so that the magnetisation is not erased by stray magnetic fields, temperature fluctuations or minor mechanical damage.

Electromagnets

As we know that a current carrying solenoid behaves like a bar magnet. If we place a soft iron rod in the solenoid, then the magnetism of the solenoid increases hundreds of times and the solenoid is called an electromagnet. It is a temporary magnet.

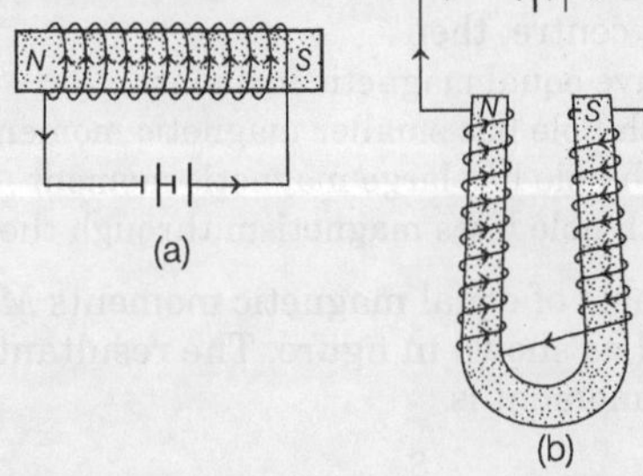

The ferromagnetic property of the iron core causes the internal magnetic domains of the iron to line up with the smaller driving magnetic field produced by the current in the solenoid.

These are widely used in electric and electromechanical devices, including motors and generators. The main advantage of an electromagnet over a permanent magnet is that the magnetic field can be rapidly manipulated over a wide range by controlling the amount of electric current.

Example 19. *Magnetic materials used for making permanent magnets (P) and magnets in a transformer (T) have different properties of the following, which property best matches for the type of magnet required?* ***[JEE Main 2020]***

(a) *T: Large retentivity, small coercivity*
(b) *P : Large retentivity, large coercivity*
(c) *P : Small retentivity, large coercivity*
(d) *T : Large retentivity, large coercivity*

Sol. (b) Permanent magnets (P) must be able to hold large magnetism for a long time. So, they must have large coercivity and large retentivity.

Electromagnet and transformer cores (T) must be able to loose magnetisation quickly when current in their coils are zero. So, they must have small retentivity.

Practice Exercise

ROUND I Topically Divided Problems

Bar Magnet and Magnetic Dipole

1. A magnet of magnetic moment M and pole strength m is divided in two equal parts, the magnetic moment of each part will be

(a) M (b) $M/2$
(c) $M/4$ (d) $2M$

2. Two magnets have the same length and the same pole strength. But one of the magnets has a small hole at its centre, then

(a) both have equal magnetic moment
(b) one with hole has smaller magnetic moment
(c) one with hole has large magnetic moment
(d) one with hole loses magnetism through the hole

3. Two magnets of equal magnetic moments M each are placed as shown in figure. The resultant magnetic moment is

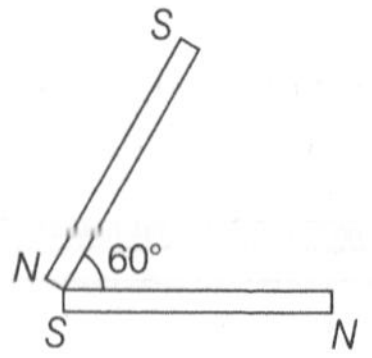

(a) M (b) $\sqrt{3}\,M$
(c) $\sqrt{2}\,M$ (d) $M/2$

4. In which orientation, the resultant magnetic moment of two magnets will be zero, if magnetic moment of each magnets is M in the following figures?

(a)
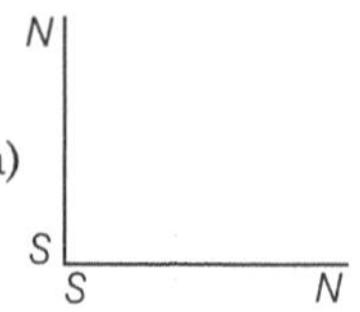

(b)
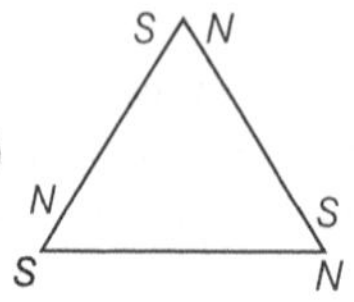

(c)

(d)
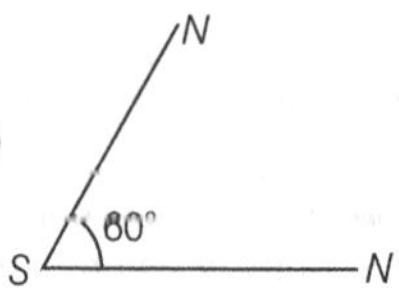

5. An insulating thin rod of length l has a linear charge density $\rho(x) = \rho_0 \dfrac{x}{l}$ on it. The rod is rotated about an axis passing through the origin ($x = 0$) and perpendicular to the rod. If the rod makes rotations per second, then the time averaged magnetic moment of the rod is **[JEE Main 2019]**

(a) $n\rho l^3$ (b) $\pi\, n\rho l^3$
(c) $\dfrac{\pi}{3} n\rho l^3$ (d) $\dfrac{\pi}{4} n\rho l^3$

6. A bar magnet of length 3 cm has a point A and B along axis at a distance of 24 cm and 48 cm on the opposite ends. Ratio of magnetic fields at these points will be

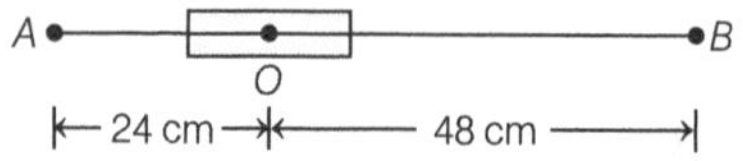

(a) 8 (b) 3
(c) 4 (d) $1/2\sqrt{2}$

7. Two magnetic dipoles X and Y are placed at a separation d, with their axes perpendicular to each other. The dipole moment of Y is twice that of X. A particle of charge q is passing through their mid-point P, at angle $\theta = 45°$ with the horizontal line, as shown in figure. What would be the magnitude of force on the particle at that instant? (d is much larger than the dimensions of the dipole) **[JEE Main 2019]**

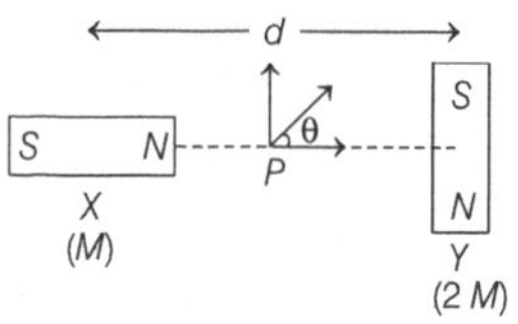

(a) $\left(\dfrac{\mu_0}{4\pi}\right)\dfrac{M}{\left(\dfrac{d}{2}\right)^3} \times qv$ (b) 0

(c) $\sqrt{2}\left(\dfrac{\mu_0}{4\pi}\right)\dfrac{M}{\left(\dfrac{d}{2}\right)^3} \times qv$ (d) $\left(\dfrac{\mu_0}{4\pi}\right)\dfrac{2M}{\left(\dfrac{d}{2}\right)^3} \times qv$

8. The magnetic potential due to a magnetic dipole at a point on its axis distant 40 cm from its centre is found to be 2.4×10^{-5} JA^{-1}m^{-1}. The magnetic moment of the dipole will be

(a) 28.6 A-m^2 (b) 32.2 A-m^2
(c) 38.4 A-m^2 (d) None of these

9. A magnetic needle lying parallel to a magnetic field required W units of work to turn it through 60°. The torque required to maintain the needle in this position will be

(a) $\sqrt{3}\,W$ (b) W (c) $\sqrt{3}\,\dfrac{W}{2}$ (d) 2 W

10. Rate of change of torque τ with deflection q is maximum for a magnet suspended freely in a uniform magnetic field of induction B, when

(a) $\theta = 0°$ (b) $\theta = 45°$ (c) $\theta = 60°$ (d) $\theta = 90°$

11. A rectangular coil (dimension 5 cm × 2.5 cm) with 100 turns, carrying a current of 3A in the clockwise direction, is kept centred at the origin and in the XZ-plane. A magnetic field of 1 T is applied along X-axis. If the coil is tilted through 45° about Z-axis, then the torque on the coil is [JEE Main 2019]

(a) 0.27 N-m (b) 0.38 N-m
(c) 0.42 N-m (d) 0.55 N-m

12. A wire carrying current I is bent in the shape $ABCDEFA$ as shown, where rectangle $ABCDA$ and $ADEFA$ are perpendicular to each other. If the sides of the rectangles are of lengths a and b, then the magnitude and direction of magnetic moment of the loop $ABCDEFA$ is [JEE Main 2020]

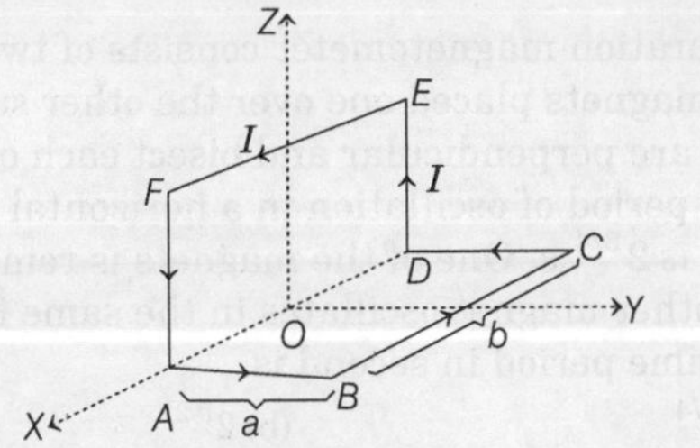

(a) abI, along $\left(\dfrac{\hat{\mathbf{j}}}{\sqrt{2}} + \dfrac{\hat{\mathbf{k}}}{\sqrt{2}}\right)$

(b) $\sqrt{2}\,abI$, along $\left(\dfrac{\hat{\mathbf{j}}}{\sqrt{2}} + \dfrac{\hat{\mathbf{k}}}{\sqrt{2}}\right)$

(c) $\sqrt{2}\,abI$, along $\left(\dfrac{\hat{\mathbf{j}}}{\sqrt{5}} + \dfrac{2\hat{\mathbf{k}}}{\sqrt{5}}\right)$

(d) abI, along $\left(\dfrac{\hat{\mathbf{j}}}{\sqrt{5}} + \dfrac{2\hat{\mathbf{k}}}{\sqrt{5}}\right)$

13. A toroid of n turns, mean radius R and cross-sectional radius a carries current I. It is placed on a horizontal table taken as XY-plane. Its magnetic moment $\mathbf{m}$ [NCERT Exemplar]

(a) is non-zero and points in the z-direction by symmetry
(b) points along the axis of the toroid ($\mathbf{m} = m$)
(c) is zero, otherwise there would be a field falling as $\dfrac{1}{r^3}$ at large distances outside the toroid
(d) is pointing radially outwards

Earth's Magnetism

14. A bar magnet of length 14 cm is placed in the magnetic meridian with its north pole pointing towards the geographic north pole. A neutral point is obtained at a distance of 18 cm from the centre of the magnet. If $B_H = 0.4$ G, the magnetic moment of the magnet is

(a) 2.880×10^3 JT^{-1} (b) 2.880×10^2JT^{-1}
(c) 2.880JT^{-1} (d) 28.80JT^{-1}

15. A circular coil of 16 turns and radius 10 cm carrying a current of 0.75 A rests with its plane normal to an external field of magnitude 5.0×10^{-2} T. The coil is free to turn about an axis in its plane perpendicular to the field direction. When the coil is turned slightly and released, it oscillates about its stable equilibrium with a frequency of 2.0/s. What is the moment of inertia of the coil about its axis of rotation?

(a) 1.2×10^{-4} g-cm^2 (b) 3×10^{-4} kg-m^2
(c) 0.3×10^{-4} kg-m^2 (d) 1.2×10^{-4} kg-m^2

16. Two bar magnets having same geometry with magnetic moments M and $2\,M$ are firstly placed in such a way that their similar poles are at the same side. Time period of oscillations is T_1. Now, the polarity of one of the magnets is reversed and time period of oscillation is T_2

(a) $T_1 < T_2$ (b) $T_1 = T_2$
(c) $T_1 > T_2$ (d) $T_2 = \infty$

17. The period of oscillation of a freely suspended bar magnet is 4 s. If it is cut into two equal parts in length, then the time period of each part will be

(a) 4 s (b) 2 s
(c) 0.5 s (d) 0.25 s

18. A small circular loop of conducting wire has radius a and carries current i. It is placed in a uniform magnetic field B perpendicular to its plane such that when rotated slightly about its diameter and released, it starts performing simple harmonic motion of time period T. If the mass of the loop is m, then [JEE Main 2020]

(a) $T = \sqrt{\dfrac{\pi m}{iB}}$ (b) $T = \sqrt{\dfrac{2m}{iB}}$

(c) $T = \sqrt{\dfrac{2\pi m}{iB}}$ (d) $T = \sqrt{\dfrac{\pi m}{2iB}}$

19. Consider the two idealised systems: (i) a parallel plate capacitor with large plates and small separation and (ii) a long solenoid of length $L >> R$, radius of cross-section. In (i), E is ideally treated as a constant between plates and zero outside. In (ii), magnetic field is constant inside the solenoid and zero outside. These idealised assumptions, however contradict fundamental laws as below **[NCERT Exemplar]**

(a) case (i) contradicts Gauss's law for electrostatic fields.
(b) case (i) contradicts Gauss's law for magnetic fields.
(c) case (i) agrees with $\oint \mathbf{E}\cdot d\mathbf{l} = 0$.
(d) case (ii) contradicts $\oint \mathbf{H}\cdot d\mathbf{l} = I_{en}$.

20. The vertical component of earth's magnetic field is zero at or the earth's magnetic field always has a vertical component except at the **[NCERT Exemplar]**

(a) magnetic poles (b) geographic poles
(c) every place (d) magnetic equator

21. The magnetic field of the earth can be modelled by that of a point dipole placed at the centre of the earth. The dipole axis makes an angle of 11.3° with the axis of the earth. At Mumbai, declination is nearly zero, then **[NCERT Exemplar]**

(a) the declination varies between 11.3° W to 11.3° E
(b) the least declination is 0°
(c) the plane defined by dipole axis and earth axis passes through Greenwich
(d) declination averaged over earth must be always negative

22. At a certain place, the horizontal component of the earth's magnetic field is B_0 and the angle of dip is 45°. The total intensity of the field at that place will be

(a) B_0 (b) $\sqrt{2}\, B_0$
(c) $2\, B_0$ (d) B_0^2

23. The earth's magnetic field at a certain place has a horizontal component of 0.3 G and total strength of 0.5 G. Find angle of dip in $\tan^{-1}$.

(a) $\delta = \tan^{-1}\frac{4}{3}$ (b) $\delta = \tan^{-1}\frac{3}{4}$
(c) $\delta = \tan^{-1}\frac{5}{3}$ (d) $\delta = \tan^{-1}\frac{3}{5}$

Magnetic Materials and Hysteresis

24. The magnetic needle of a tangent galvanometer is deflected at an angle 30° due to a magnet. The horizontal component of earth's magnetic field 0.34×10^{-4} T is along the plane of the coil. The magnetic intensity is

(a) 1.96×10^{-4} T (b) 1.96×10^{4} T
(c) 1.96×10^{-5} T (d) 1.96×10^{5} T

25. Two tangent galvanometers having coils of the same radius are connected in series. A current flowing in them produces deflections of 60° and 45°, respectively.
The ratio of the number of turns in the coils is

(a) 4/3 (b) $(\sqrt{3}+1)/1$
(c) $\frac{\sqrt{3}+1}{\sqrt{3}-1}$ (d) $\frac{\sqrt{3}}{1}$

26. A magnet freely suspended in a vibration magnetometer makes 40 oscillations per minute at a place A and 20 oscillations per minute at a place B. If the horizontal component of earth's magnetic field at A is 36×10^{-6} T, then its value at B is

(a) 36×10^{-6} T (b) 9×10^{-6} T
(c) 144×10^{-6} T (d) 228×10^{-6} T

27. Two magnets held together in earth's magnetic field with same polarity together make 12 vib min^{-1} and when opposite poles together make 4 vib min^{-1}. The ratio of magnetic moments is

(a) 9 : 1 (b) 1 : 3
(c) 1 : 9 (d) 5 : 4

28. A magnet performs 10 oscillations per minute in a horizontal plane at a place, where the angle of dip is 45° and the total intensity is 0.707 CGS units. The number of oscillations per minute at a place, where dip angle is 60° and total intensity is 0.5 CGS units will be

(a) 5 (b) 7
(c) 9 (d) 11

29. A vibration magnetometer consists of two identical bar magnets placed one over the other such that they are perpendicular and bisect each other. The time period of oscillation in a horizontal magnetic field is $2^{5/4}$ s. One of the magnets is removed and if the other magnet oscillates in the same field, then the time period in second is

(a) $2^{1/4}$ (b) $2^{1/2}$
(c) 2 (d) 4

30. The time of vibration of a dip needle vibrating in the vertical plane is 3 s. When magnetic needle is made to vibrate in the horizontal plane, the time of vibration is $3\sqrt{2}$ s. Then, the angle of dip is

(a) 30° (b) 45°
(c) 60° (d) 90°

31. A bar magnet is placed north-south with its north pole due north. The points of zero magnetic field will be in which direction from centre of magnet

(a) north and south
(b) east and west
(c) north-east and south-west
(d) north-east and south-east

32. A short bar magnet with the north pole facing north forms a neutral point P in the horizontal plane. If the magnet is rotated by 90° in the horizontal plane, the net magnetic induction at P is (Horizontal component of earth's magnetic field $= B_H$)

(a) zero (b) $2\,B_H$

(c) $\frac{\sqrt{5}}{2}B_H$ (d) $\sqrt{5}\,B_H$

33. A bar magnet 20 cm in length is placed with its south pole towards geographic north. The neutral points are situated at a distance of 40 cm from centre of the magnet. If horizontal component of earth's field 3.2×10^{-5} T, then pole strength of magnet is

(a) 5 A-m (b) 10 A-m

(c) 45 A-m (d) 20 A-m

34. The horizontal component of flux density of earth's magnetic field is 1.7×10^{-5} T. The value of horizontal component of intensity of earth's magnetic field will be

(a) 24.5 Am^{-1} (b) 13.5 Am^{-1}

(c) 1.53 Am^{-1} (d) 0.35 Am^{-1}

35. A bar magnet is cut into two equal halves by a plane parallel to the magnetic axis. Of the following physical quantities, the one which remains unchanged is

(a) pole strength

(b) magnetic moment

(c) intensity of magnetisation

(d) moment of inertia

36. A rod of ferromagnetic material with dimensions 10 cm × 0.5 cm × 0.2 cm is placed in a magnetic field of strength 0.5×10^{4} Am^{-1} as a result of which a magnetic moment of 5 A-m^2 is produced in the rod. The value of magnetic induction will be

(a) 0.54 T (b) 6.28 T

(c) 0.358 T (d) 2.591 T

37. The space inside a toroid is filled with tungsten whose susceptibility is 6.8×10^{-5}. The percentage increase in the magnetic field will be

(a) 0.0068% (b) 0.068%

(c) 0.68% (d) None of these

38. A paramagnetic sample shows a net magnetisation of 8 Am^{-1} when placed in an external magnetic field of 0.6 T at a temperature of 4 K. When the same sample is placed in an external magnetic field of 0.2 T at a temperature of 16 K, the magnetisation will be **[NCERT Exemplar]**

(a) $\frac{32}{3}$ Am^{-1} (b) $\frac{2}{3}$ Am^{-1}

(c) 6 Am^{-1} (d) 2.4 Am^{-1}

39. The variation of the intensity of magnetisation (I) with respect to the magnetising field (H) in a diamagnetic substance is described by the graph in figure.

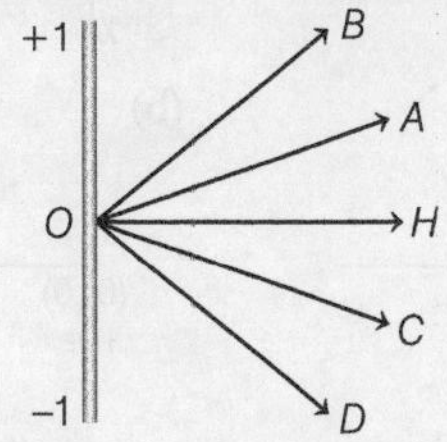

(a) OD (b) OC (c) OB (d) OA

40. The variation of magnetic susceptibility (χ) with temperature for a diamagnetic substance is best represented by figure

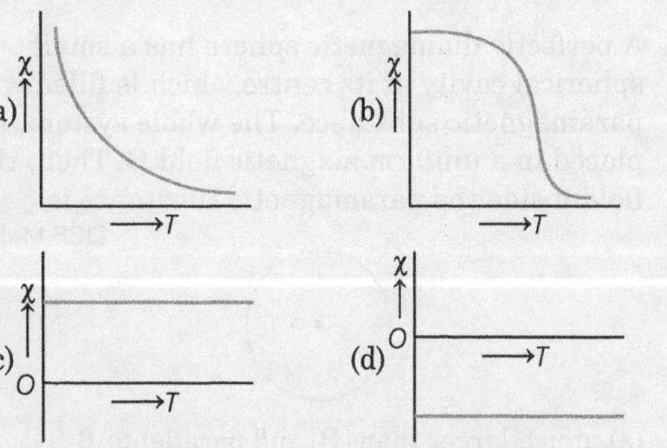

41. A copper rod is suspended in a non-homogeneous magnetic field region. The rod when in equilibrium will align itself

(a) in the region, where magnetic field is strongest

(b) in the region, where magnetic field is weakest and parallel to direction of magnetic field there

(c) in the direction in which it was originally suspended

(d) in the region, where magnetic field is weakest and perpendicular to the direction of magnetic field there

42. A uniform magnetic field parallel to the plane of paper, exsisted in space initially directed from left to right. When a bar of soft iron is placed in the field parallel to it, the lines of force passing through it will be represented by figure.

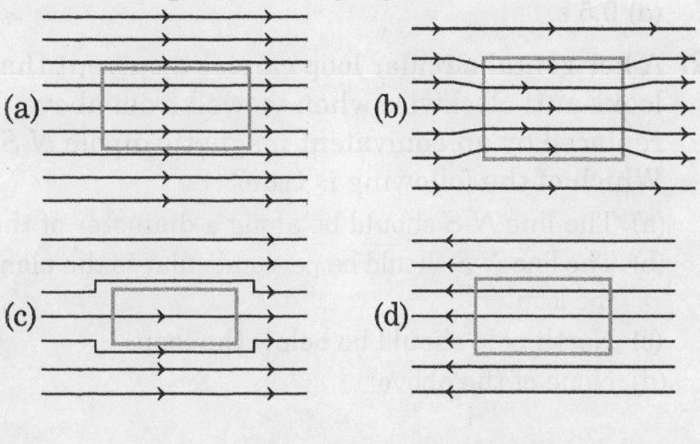

43. The variation of magnetic susceptibility (χ) with absolute temperature T for a ferromagnetic substance is given by

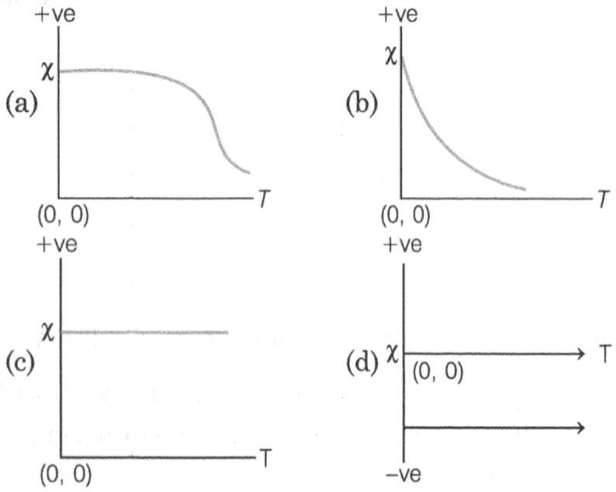

44. A perfectly diamagnetic sphere has a small spherical cavity at its centre,which is filled with a paramagnetic substance. The whole system is placed in a uniform magnetic field **B**. Then, the field inside the paramagnetic substance is [JEE Main 2020]

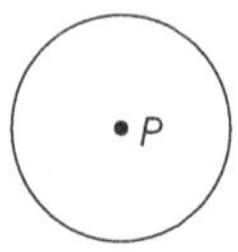

(a) much larger than $|\mathbf{B}|$ and parallel to **B**
(b) **B**
(c) zero
(d) much larger than $|\mathbf{B}|$ but opposite to **B**

45. The relative permeability of a substance X is slightly less than unity and that of substance Y is slightly more than unity, then
(a) X is paramagnetic and Y is ferromagnetic
(b) X is diamagnetic and Y is ferromagnetic
(c) X and Y both are paramagnetic
(d) X is diamagnetic and Y is paramagnetic

46. A bar magnet is demagnetised by inserting it inside a solenoid of length 0.2 m, 100 turns and carrying a current of 5.2 A. The coercivity of the bar magnet is [JEE Main 2019]
(a) 1200 A/m (b) 285 A/m
(c) 2600A/m (d) 520A/m

47. Hysteresis loops for two magnetic materials a and b are as given below

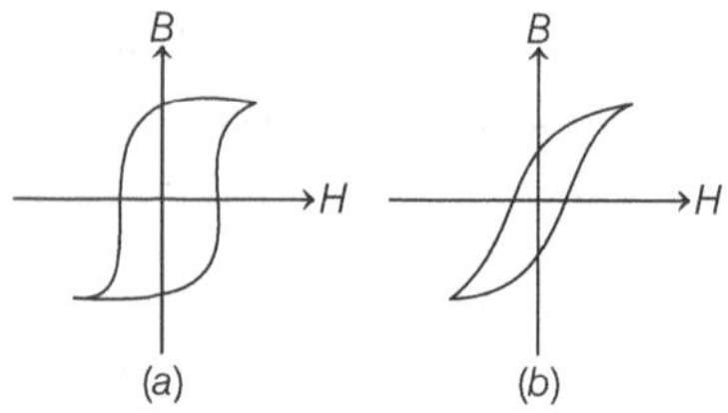

These materials are used to make magnets for electric generators, transformer core and electromagnet core. Then, it is proper to use [JEE Main 2016]
(a) a for electric generators and transformers
(b) a for electromagnets and b for electric generators
(c) a for transformers and b for electric generators
(d) b for electromagnets and transformers

ROUND II Mixed Bag

Only One Correct Option

1. The period of oscillation of a bar magnet in a vibration magnetometer is 2 s. The period of oscillation of another bar magnet whose magnetic moment is 4 times to that of Ist magnet is
(a) 4 s
(b) 1 s
(c) 2 s
(d) 0.5 s

2. A horizontal circular loop carries a current that looks anti-clockwise when viewed from above. It is replaced by an equivalent magnetic dipole N-S. Which of the following is true?
(a) The line N-S should be along a diameter of the loop
(b) The line N-S should be perpendicular to the plane of the loop
(c) North pole should be below the loop
(d) None of the above

3. The points A and B are situated perpendicular to the axis of 2 cm long bar magnet at large distances x and $3x$ from the centre on opposite sides. The ratio of magnetic fields at A and B will be approximately equal to
(a) 27 : 1 (b) 1 : 27
(c) 9 : 1 (d) 1 : 9

4. Which of the following statements are correct?
I. Electric monopoles do not exist whereas magnetic monopoles exist.
II. Magnetic field lines due to a solenoid at its ends and outside cannot be completely straight and confined.
III. Magnetic field lines are completely confined within a toroid.
IV. Magnetic field lines inside a bar magnet are not parallel.
V. $\chi = -1$ is the condition for a perfect diamagnetic material, where χ is its magnetic susceptibility.

Choose the correct answer from the options given below. **[JEE Main 2021]**

(a) III and V only (b) II and IV only
(c) I and II only (d) II and III only

5. A magnetic needle of magnetic moment 6.7×10^{-2} A-m^2 and moment of inertia 7.5×10^{-6} kg -m^2 is performing simple harmonic oscillations in a magnetic field of 0.01 T. Time taken for 10 complete oscillations is **[JEE Main 2017]**

(a) 8.89 s (b) 6.98 s
(c) 8.76 s (d) 6.65 s

6. The coercivity of a small magnet, where the ferromagnet gets demagnetised is 3×10^3 Am^{-1}. The current required to be passed in a solenoid of length 10 cm and number of turns 100, so that the magnet gets demagnetised when kept inside the solenoid is **[JEE Main 2014]**

(a) 30 mA (b) 60 mA
(c) 3 A (d) 6 A

7. The susceptibility of a paramagnetic material is K at 27 °C. At what temperature will its susceptibility be $K/2$?

(a) 600° C (b) 287° C (c) 54° C (d) 327° C

8. Force between two identical short bar magnets whose centres are r metre apart is 8.1 N, when their axes are along the same line. If separation is increased to $3r$ and the axis are rearranged perpendicularly, the force between them would become

(a) 2.4 N (b) 1.2 N (c) 0.1 N (d) 1.15 N

9. A coil of 50 turns and area 1.25×10^{-3} m^2 is pivoted about a vertical diameter in a uniform horizontal magnetic field and carries a current of 2A. When the coil is held with its plane in N-S direction, it experience a couple of 0.04 N-m; and when its plane is E-W, the corresponding couple is 0.03 N-m. The magnetic induction is

(a) 0.2 T (b) 0.3 T (c) 0.4 T (d) 0.5 T

10. A long magnet is placed vertically with its S-pole resting on the table. A neutral point is obtained 10 cm from the pole due geographic north of it. If $H = 3.2 \times 10^{-5}$ T, then the pole strength of magnet is

(a) 8 ab-A-cm^{-1} (b) 16 ab-A-cm^{-1}
(c) 32 ab-A-cm^{-1} (d) 64 ab-A-cm^{-1}

11. A dip needle vibrates in the vertical plane perpendicular to magnetic meridian. The time period of vibration is found to be 2 s. The same needle is then allowed to vibrate in the horizontal plane and time period is again found to be 2 s. Then, the angle of dip is

(a) 0° (b) 30° (c) 45° (d) 90°

12. Two magnets of equal mass are joined at 90° to each other as shown in figure. Magnet N_1S_1 has a magnetic moment $\sqrt{3}$ times that of N_2S_2. The arrangement is pivoted, so that it is free to rotate in horizontal plane. When in equilibrium, what angle should N_1S_1 make with magnetic meridian?

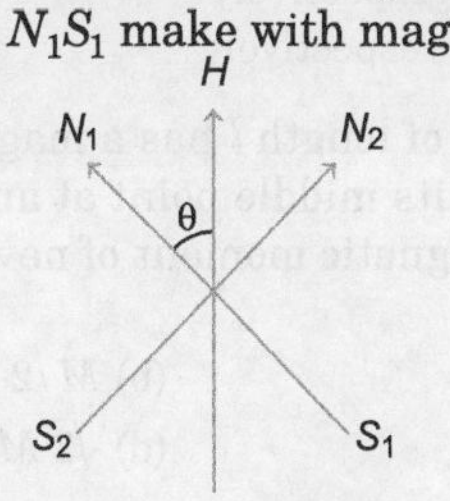

(a) 75° (b) 60° (c) 30° (d) 45°

13. S is the surface of a lump of magnetic material. **[NCERT Exemplar]**

(a) Lines of B are necessarily continuous across S
(b) Some lines of B must be continuous across S
(c) Lines of H are necessarily continuous across S
(d) Both (a) and (c)

14. An iron rod of volume 10^{-3} m^3 and relative permeability 1000 is placed as core in a solenoid with 10 turns/cm. If a current of 0.5 A is passed through the solenoid, then the magnetic moment of the rod will be **[JEE Main 2020]**

(a) 50×10^2 A-m^2 (b) 5×10^2 A-m^2
(c) 500×10^2 A-m^2 (d) 0.5×10^2 A-m^2

15. The dipole moment of a circular loop carrying a current I, is m and the magnetic field at the centre of the loop is B_1. When the dipole moment is doubled by keeping the current constant, the magnetic field at the centre of the loop is B_2. The ratio of $\frac{B_1}{B_2}$ is **[JEE Main 2018]**

(a) 2 (b) $\sqrt{3}$
(c) $\sqrt{2}$ (d) $\frac{1}{\sqrt{2}}$

16. A rectangular loop of sides 10 cm and 5 cm carrying a current I of 12 A is placed in different orientations as shown in the figures below.

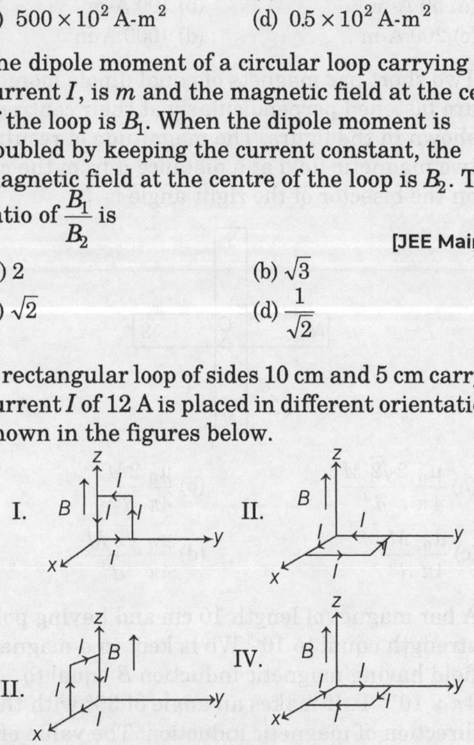

If there is a uniform magnetic field of 0.3 T in the positive z-direction in which orientations the loop would be in (i) stable equilibrium and (ii) unstable equilibrium? [JEE Main 2015]

(a) I and II respectively
(b) I and III respectively
(c) II and IV respectively
(d) II and III respectively

17. A steel wire of length l has a magnetic moment M. It is bent at its middle point at an angle of 60°, then the magnetic moment of new shape of wire will be

(a) $M/\sqrt{2}$ (b) $M/2$
(c) M (d) $\sqrt{2}\,M$

18. The mass of a specimen of a ferromagnetic material is 0.6 kg and the density is 7.8×10^3 kg m^{-3}. If the area of hysteresis loop of alternating magnetising field of frequency 50 Hz is 0.722 MKS units, then hysteresis loss per second will be

(a) 27.77×10^{-5} J (b) 2.777×10^{-5} J
(c) 27.77×10^{-4} J (d) 27.77×10^{-6} J

19. In an experiment with vibration magnetometer, the value of $4\pi^2 I / T^2$ for a short bar magnet is observed as 36×10^{-4}. In the experiment with deflection magnetometer with the same magnet, the value of $4\,\pi d^3 / 2\mu_0$ is observed as $10^8/36$. The magnetic moment of the magnet used is

(a) 50 A-m (b) 100 A-m
(c) 200 A-m (d) 1000 A-m

20. Two short bar magnets of equal dipole moment M are fastened perpendicularly at their centres as shown in the figure. The magnitude of resultant of two magnetic field at a distance d from the centre on the bisector of the right angle is

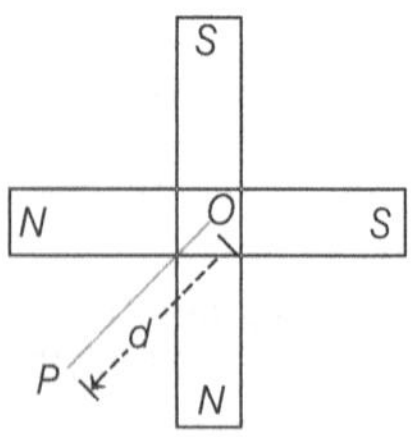

(a) $\frac{\mu_0}{4\pi}\frac{2\sqrt{2}\,M}{d^3}$ (b) $\frac{\mu_0}{4\pi}\frac{2M}{d^3}$

(c) $\frac{\mu_0}{4\pi}\frac{M}{d^3}$ (d) $\frac{\mu_0}{4\pi}\frac{\sqrt{2}\,M}{d^3}$

21. A bar magnet of length 10 cm and having pole strength equal to 10^{-3} Wb is kept in a magnetic field having magnetic induction B equal to $4\pi \times 10^{-3}$ T. It makes an angle of 30° with the direction of magnetic induction. The value of the torque acting on the magnet is

(a) 0.5 N-m (b) $2\pi \times 10^{-5}$ N-m
(c) $\pi \times 10^{-5}$ N-m (d) 0.5×10^{-5} N-m

22. A uniform magnetic needle is suspended from its centre by a thread. Its upper end is now loaded with a mass of 50 mg and the needle becomes horizontal. If the strength of each pole is 98.1 ab-amp-cm and $g = 981$ cms^{-2}, then the vertical component of earth's magnetic induction is

(a) 0.50 G (b) 0.25 G (c) 0.005 G (d) 0.05 G

23. In a vibration magnetometer, the time period of a bar magnet oscillating in horizontal component of earth's magnetic field is 2 s. When a magnet is brought near and parallel to it, the time period reduces to 1 s. The ratio F/H of the fields, F due to magnet and H, the horizontal component will be

(a) $\sqrt{3}$ (b) $\frac{1}{\sqrt{3}}$ (c) $\frac{1}{3}$ (d) 3

24. The magnetic needle of an oscillation magnetometer makes 10 oscillations per min under the action of earth's magnetic field alone. When a bar magnet is placed at some distance along the axis of the needle, it makes 14 oscillations per min. If the bar magnet is turned, so that its poles interchange their positions, then the new frequency of oscillation of the needle is

(a) 10 vib min^{-1} (b) 2 vib min^{-1}
(c) 4 vib min^{-1} (d) 20 vib min^{-1}

25. A bar magnet has a magnetic moment equal to 5×10^{-5} Wb-m. It is suspended in a magnetic field which has a magnetic induction B equal to $8\pi \times 10^{-4}$ T. The magnet vibrates with a period of vibration equal to 15 s. The moment of inertia of magnet is

(a) 4.54×10^{4} kg-m^2 (b) 4.54×10^{-5} kg-m^2
(c) 4.54×10^{-4} kg-m^2 (d) 4.54×10^{5} kg-m^2

26. The plane of a dip circle is set in the geographic meridian and the apparent dip is δ_1. It is then set in a vertical plane perpendicular to the geographic meridian. The apparent dip angle is δ_2. The declination θ at the place is

(a) $\theta = \tan^{-1}(\tan\delta_1 \tan\delta_2)$
(b) $\theta = \tan^{-1}(\tan\delta_1 + \tan\delta_2)$
(c) $\theta = \tan^{-1}\left(\frac{\tan\delta_1}{\tan\delta_2}\right)$
(d) $\theta = \tan^{-1}(\tan\delta_1 - \tan\delta_2)$

27. A deflection magnetometer is adjusted in the usual way. When a magnet is introduced, the deflection observed is θ and the period of oscillation of the needle in the magnetometer is T. When the magnet is removed, the period of oscillation is T_0. The relation between T and T_0 is

(a) $T^2 = T_0^2 \cos\theta$ (b) $T = T_0 \cos\theta$

(c) $T = \frac{T_0}{\cos\theta}$ (d) $T^2 = \frac{T_0^2}{\cos\theta}$

28. Two short magnets with pole strengths of 900 ab-amp-cm and 100 ab-amp-cm are placed with their axes in the same vertical line, with similar poles facing each other. Each magnet has a length of 1 cm. When separation between the nearer poles is 1 cm, the weight of upper magnet is supported by the repulsive force between the magnets. If $g = 1000 \text{ cms}^{-2}$, then the mass of upper magnet is

(a) 100 g (b) 55 g (c) 45 g (d) 77.5 g

29. At a certain location in Africa, a compass points 12° west of the geographic north. The north tip of the magnetic needle of a dip circle placed in the plane of magnetic meridian points 60° above the horizontal. The horizontal component of the earth's field is measured to be 0.16 G. Specify the direction and magnitude of the earth's field at that location.

(a) 32° west of geographical meridian and 3.2×10^{-4} T
(b) 12° west of geographical meridian and 0.32×10^{-4} T
(c) 12° east of geographical meridian and 0.32×10^{-4} T
(d) 32° east of geographical meridian and 3.2×10^{-4} T

30. A circular coil has moment of inertia 0.8 kg-m^2 around any diameter and is carrying current to produce a magnetic moment of 20 A-m^2. The coil is kept initially in a vertical position and it can rotate freely around a horizontal diameter. When a uniform magnetic field of 4 T is applied along the vertical, it starts rotating around its horizontal diameter. The angular speed of the coil, acquired after rotating by 60°, will be [JEE Main 2020]

(a) 10 rad s^{-1} (b) 20π rad s^{-1}
(c) 10π rad s^{-1} (d) 13.16 rad s^{-1}

31. A paramagnetic sample shows a net magnetisation of 6 A/m, when it is placed in an external magnetic field of 0.4 T at a temperature of 4 K. When the sample is placed in an external magnetic field of 0.3 T at a temperature of 24 K, then the magnetisation will be [JEE Main 2020]

(a) 4 A/m (b) 0.75 A/m
(c) 1 A/m (d) 2.25 A/m

32. A charged particle going around in a circle can be considered to be a current loop. A particle of mass m carrying charge q is moving in a plane with speed v under the influence of magnetic field **B**. The magnetic moment of this moving particle is [JEE Main 2020]

(a) $\frac{mv^2\mathbf{B}}{2B^2}$ (b) $-\frac{mv^2\mathbf{B}}{2\pi B^2}$

(c) $-\frac{mv^2\mathbf{B}}{B^2}$ (d) $-\frac{mv^2\mathbf{B}}{2B^2}$

Numerical Value Questions

33. At a given place on the earth's surface, the horizontal component of earth's magnetic field is 3×10^{-5} T and resultant magnetic field is 6×10^{-6} T. So, the angle of dip (in degree) at this place is ....

34. The earth's magnetic field may be considered due to a short magnet placed at the centre of earth and oriented along magnetic south-north direction. The ratio of magnitude of magnetic field on earth's surface at magnetic equator and that at magnetic poles will be $1 : x$, where the value of x is ....

35. If the magnetic susceptibility of a paramagnetic material at −73 °C is 0.0075, then its value at −173° C will be ....

36. The intensity of magnetic field at a point X on the axis of a small magnet is equal to the field intensity at another point Y on its equatorial axis. The ratio of distances of X and Y from the centre of the magnet will be $2^{1/P}$, where the value of P is ....

37. A loop $ABCDEFA$ of straight edges has six corner points $A(0, 0, 0)$, $B(5, 0, 0)$, $C(5, 5, 0)$, $D(0, 5, 0)$, $E(0, 5, 5)$ and $F(0, 0, 5)$. The magnetic field in this region is $\mathbf{B} = (3\hat{i} + 4\hat{k})$T. The quantity of flux through the loop $ABCDEFA$ (in Wb) is .... [JEE Main 2020]

Answers

Round I

1. (b)	2. (b)	3. (a)	4. (b)	5. (d)	6. (a)	7. (b)	8. (c)	9. (a)	10. (a)
11. (a)	12. (b)	13. (c)	14. (c)	15. (d)	16. (a)	17. (b)	18. (c)	19. (b)	20. (d)
21. (a)	22. (b)	23. (a)	24. (c)	25. (d)	26. (b)	27. (d)	28. (b)	29. (c)	30. (c)
31. (b)	32. (d)	33. (c)	34. (b)	35. (c)	36. (b)	37. (a)	38. (b)	39. (b)	40. (d)
41. (d)	42. (b)	43. (a)	44. (c)	45. (d)	46. (c)	47. (d)			

Round II

1. (b)	2. (b)	3. (a)	4. (a)	5. (d)	6. (c)	7. (d)	8. (c)	9. (c)	10. (c)
11. (c)	12. (c)	13. (a)	14. (b)	15. (c)	16. (c)	17. (b)	18. (c)	19. (b)	20. (a)
21. (a)	22. (b)	23. (d)	24. (b)	25. (d)	26. (c)	27. (a)	28. (b)	29. (b)	30. (d)
31. (b)	32. (d)	33. 60	34. 2	35. 0.015	36. 3	37. 175			

Solutions

Round I

1. If we cut along the axis of magnet of length l, then new pole strength, $m' = \frac{m}{2}$ and new length $l' = l$.

$\therefore$ New magnetic moment, $M' = \frac{m}{2} \times l = \frac{ml}{2} = \frac{M}{2}$

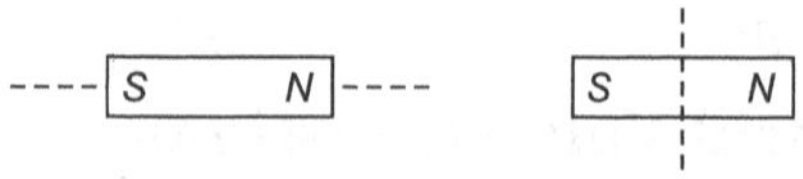

If we cut perpendicular to the axis of magnet, then new pole strength, $m' = m$ and new length, $l' = \frac{l}{2}$

$\therefore$ New magnetic moment,

$$M' = m \times \frac{l}{2} = \frac{ml}{2} = \frac{M}{2}$$

2. Hole reduces the effective length of the magnet and hence, magnetic moment reduces.

3. As, magnetic moments are directed along SN, angle between $\mathbf{M}$ and $\mathbf{M}$ is $\theta = 120^\circ$.

$\therefore$ Resultant magnetic moment

$$= \sqrt{M^2 + M^2 + 2\,M \cdot M \cos 120^\circ}$$

$$= \sqrt{M^2 + M^2 + 2\,M^2(-1/2)} = M$$

4. In Fig. (a), $M' = \sqrt{M^2 + M^2} = \sqrt{2}\,M$

As magnetic moments are in a closed loop in Fig. (b),

$\Rightarrow \quad M' = 0$

In Fig. (c), $M' = M + M = 2\,M$

In Fig. (d), $M' = \sqrt{M^2 + M^2 + 2\,MM \cos 60^\circ} = \sqrt{3}\,M$

5. Let dq be the charge on dx length of rod at a distance x from origin as shown in the figure below

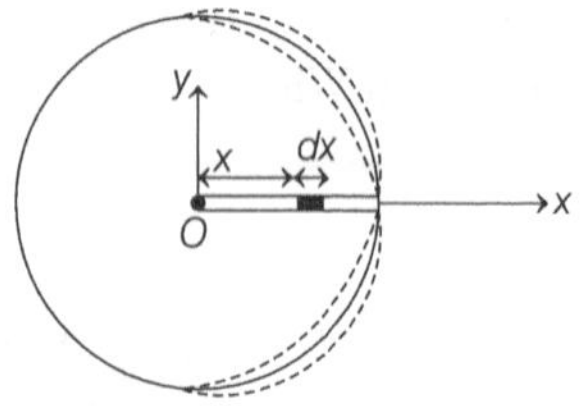

The magnetic moment dm of this portion dx is given as

$$dm = (dI)\,A$$

$$dm = ndqA \qquad (\because I = qn, \therefore dI = n \cdot dq)$$

$$= n\,\rho\,dx\,A$$

where, $\quad \rho = \text{charge density of rod} = \rho_0 \frac{x}{l}$.

So, $$dm = \frac{n\rho_0 x\,dx\,\pi x^2}{l}$$

$$= \frac{\pi\,n\,\rho_0}{l} \cdot x^3 \cdot dx$$

So, magnetic moment associated with complete rod is

$$M = \int_{x=0}^{x=l} dm = \int_0^l \frac{\pi\,n\,\rho_0}{l} \cdot x^3\,dx$$

$$= \frac{\pi\,n\,\rho_0}{l} \cdot \int_0^l x^3\,dx$$

$$= \frac{\pi\,n\,\rho_0}{l} \left[\frac{x^4}{4}\right]_0^l$$

$$= \frac{\pi\,n\,\rho_0\,l^3}{4}$$

At $\quad x = l, \rho = \rho_0$

$\therefore \quad M = \frac{\pi}{4} n \rho l^3$

6. Here, $2l = 3$ cm, $d_1 = 24$ cm and $d_2 = 48$ cm

As, the magnet is short, $\frac{B_1}{B_2} = \frac{d_2^3}{d_1^3} = \left(\frac{48\text{ cm}}{24\text{ cm}}\right)^3 = 8$

7. Let $2l_1$ and $2l_2$ be the length of dipole X and Y respectively.

For dipole X, point P lies on its axial line. So, magnetic field strength at P due to X is

$$\mathbf{B}_X = \frac{\mu_0}{4\pi} \cdot \frac{2Mr}{(r^2 - l_1^2)^2}, \text{ along } OP$$

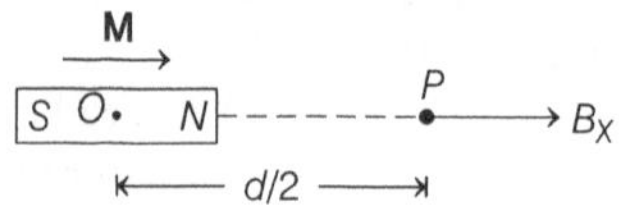

Here, $\quad r = \frac{d}{2}$

Also, $\quad d >> l_1$

$$\Rightarrow \quad |\mathbf{B}_X| = \frac{\mu_0}{4\pi} \cdot \frac{2M(d/2)}{(d/2)^4} = \frac{\mu_0}{4\pi} \frac{2M}{(d/2)^3}$$

Similarly, for dipole Y, point P lies on its equatorial line. So, magnetic field strength at P due to Y is

$$\mathbf{B}_Y = \frac{\mu_0}{4\pi} \cdot \frac{2M}{(r^2 + l_2^2)^{3/2}},$$

along a line perpendicular to $O'P$

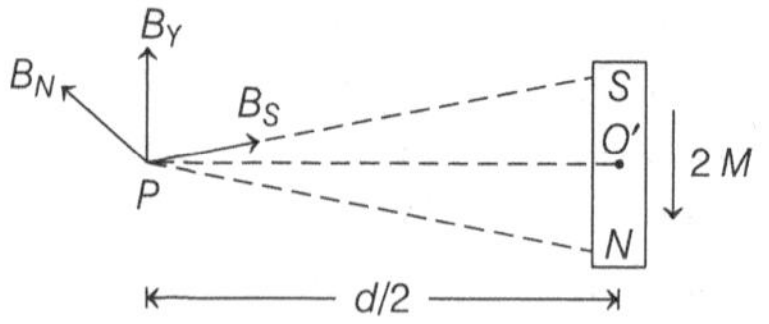

Here, $\quad r = \frac{d}{2}$

Also, $\quad d >> l_2$

$$\Rightarrow \quad |\mathbf{B}_Y| = \frac{\mu_0}{4\pi} \frac{2M}{(d/2)^3}$$

Thus, the resultant magnetic field due to X and Y at P is

$$\mathbf{B}_{\text{net}} = \mathbf{B}_X + \mathbf{B}_Y$$

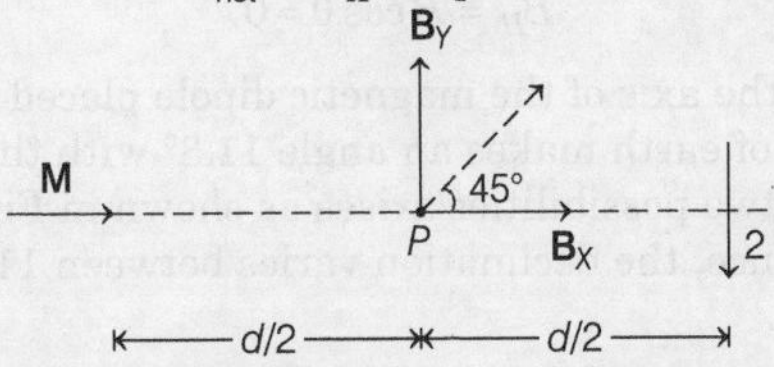

Since, $\quad |\mathbf{B}_Y| = |\mathbf{B}_X|$

Thus, the resultant magnetic field ($\mathbf{B}_{\text{net}}$) at P will be at 45° with the horizontal.

This means, direction of $\mathbf{B}_{\text{net}}$ and velocity of the charged particle is same.

$\therefore$ Force on the charged particle moving with velocity v in the presence of magnetic field which is

$$\mathbf{B} = q(\mathbf{v} \times \mathbf{B}) = q\,|\mathbf{v}|\,|\mathbf{B}| \sin\theta$$

where, θ is the angle between $\mathbf{B}$ and $\mathbf{v}$.

According to the above analysis, we get

$$\theta = 0, \text{ therefore } F = 0$$

Thus, magnitude of force on the particle at that instant is zero.

8. Here, $r = 40$ cm $= 0.4$ m,

$\theta = 0°$ (an axial line)

$V = 2.4 \times 10^{-5}$ J/A-m, $M = ?$

As, $\quad V = \dfrac{\mu_0}{4\pi} \dfrac{M\cos\theta}{r^2}$

$\Rightarrow \quad 2.4 \times 10^{-5} = 10^{-7} \dfrac{M \times 1}{(0.4)^2}$

or $\quad M = 38.4 \text{ A-m}^2$

9. As, $W = BM\cos 60° = \dfrac{MB}{2}$ or $MB = 2W$

$\therefore \quad \tau = MB\sin\theta = (2W)\sin 60° = 2W\dfrac{\sqrt{3}}{2} = W\sqrt{3}$

10. As, $\tau = MB\sin\theta$ or $\dfrac{d\tau}{d\theta} = MB\cos\theta$. It will be maximum, when $\theta = 0°$

11. Given, area of the rectangular coil,

$A = 5 \text{ cm} \times 2.5 \text{ cm}$

$\Rightarrow \quad A = 12.5 \text{ cm}^2 = 12.5 \times 10^{-4} \text{ m}^2$

Number of turns, $N = 100$ turns

Current through the coil, $I = 3$ A

Magnetic field applied, $B = 1$ T

Angle between the magnetic field and area vector of the coil, $\theta = 45°$

As we know that, when a coil is tilted by an angle θ in the presence of some external magnetic field, then the net torque experienced by the coil is,

$$\tau = \mathbf{M} \times \mathbf{B} = NI(\mathbf{A} \times \mathbf{B})$$
$$= NIAB\sin\theta$$

Substituting the given values, we get

$$\tau = 100 \times 3 \times 12.5 \times 10^{-4} \times 1 \times \sin 45°$$
$$\tau = 0.707 \times 100 \times 3 \times 12.5 \times 10^{-4} \text{ N-m}$$
$$= 2.651 \times 10^{-1} \text{ N-m} \approx 0.27 \text{ N-m}$$

12. Given rectangular loop can be viewed as a combination of two loops $ABCDA$ and $ADEFA$.

So, $\mathbf{m}$ (loop) $= \mathbf{m}$ ($ABCDA$) $+ \mathbf{m}$ ($ADEFA$)

The area vector of $ABCDA$ is along Z-axis and that of $ADEFA$ is along Y-axis as shown in figure

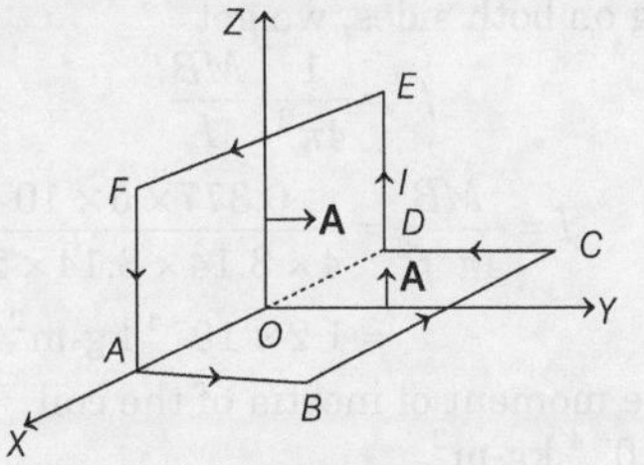

Now, $\quad \mathbf{m}\,(ABCDA) = IA\hat{\mathbf{k}} = I(ab)\hat{\mathbf{k}}$

and $\quad \mathbf{m}\,(ADEFA) = IA\hat{\mathbf{j}} = I(ab)\hat{\mathbf{j}}$

$\therefore \quad \mathbf{m}\text{ (loop)} = Iab\hat{\mathbf{k}} + Iab\hat{\mathbf{j}} = Iab(\hat{\mathbf{k}} + \hat{\mathbf{j}})$

$$= (Iab\sqrt{2})\left(\frac{\hat{\mathbf{k}}}{\sqrt{2}} + \frac{\hat{\mathbf{j}}}{\sqrt{2}}\right)$$

So, magnitude of $\mathbf{m} = \sqrt{2}Iab$ and direction of $\mathbf{m}$ is along $\dfrac{1}{\sqrt{2}}(\hat{\mathbf{j}} + \hat{\mathbf{k}})$.

13. In case of a toroid, the magnetic field is only confined inside the body of toroid in the form of concentric magnetic lines of force and there is no magnetic field outside the body of toroid. Thus, the magnetic moment of toroid is zero. Hence, option (c) is correct.

14.

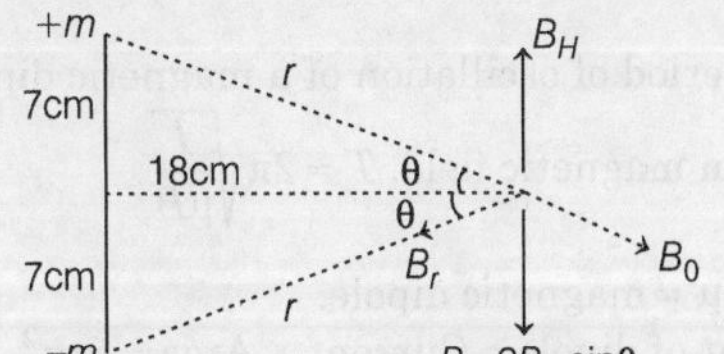

i.e. $\quad \dfrac{2\mu_0}{4\pi}\dfrac{m}{r^2} \times \dfrac{7}{r} = 0.4 \times 10^{-4}$

$$\Rightarrow 2 \times 10^{-7} \times \frac{m \times 7}{(7^2 + 18^2)^{3/2}} \times 10^4 = 0.4 \times 10^{-4}$$

$$m = \frac{4 \times 10^{-2} \times (373)^{3/2}}{14}$$

$$M = m \times 14 \text{ cm} = m \times \frac{14}{100}$$

$$= \frac{0.04 \times (373)^{32}}{14} \times \frac{14}{100}$$

$$= 4 \times 10^{-4} \times 7203.82 = 2.88 \text{ J/T}$$

15. Given, number of turns of circular coil, $n = 16$

Radius of circular coil, $r = 10$ cm $= 0.1$ m

Current, $I = 0.75$ A

Magnetic field, $B = 5.0 \times 10^{-2}$ T

Frequency, $f = 2 \text{ s}^{-1}$

Magnetic moment of the coil, $M = nIA$

$= 16 \times 0.75 \times \pi\ (0.1)^2$

$= 16 \times 0.75 \times 3.14 \times 0.1 \times 0.1$

$= 0.377$ J/T

Frequency of oscillation of the coil, $f = \frac{1}{2\pi}\sqrt{\frac{M \times B}{I}}$

where, I = moment of inertia of the coil.

Squaring on both sides, we get

$$f^2 = \frac{1}{4\pi^2} \cdot \frac{MB}{I}$$

$$\Rightarrow \quad I = \frac{MB}{4\pi^2 f^2} = \frac{0.377 \times 5 \times 10^{-2}}{4 \times 3.14 \times 3.14 \times 2 \times 2}$$

$$= 1.2 \times 10^{-4}\ \text{kg-m}^2$$

Thus, the moment of inertia of the coil is 1.2×10^{-4} kg-m^2.

16. When polarity is reversed, net magnetic moment $2M - M = M$, decreases. Therefore, time period of oscillation increases, *i.e.* $T_2 > T_1$.

17. From $T = 2\pi\sqrt{\frac{I}{MB}}$

When it is cut into two equal parts in length, mass of each part becomes 1/2, $I = \text{mass}\ \frac{(\text{length})^2}{12}$ becomes $\frac{1}{8}$th and M becomes $\frac{1}{2}$.

$$T' = 2\pi\sqrt{\frac{I/8}{(M/2)B}}$$

$$= \frac{1}{2}\left(2\pi\sqrt{\frac{I}{MB}}\right) = \frac{1}{2}T = 2\,\text{s}$$

18. Time period of oscillation of a magnetic dipole in a uniform magnetic field, $T = 2\pi\sqrt{\frac{I}{\mu B}}$

where, μ = magnetic dipole.

Moment of dipole = Current × Area = $i\pi a^2$

I = moment of inertia of a coil $= \frac{ma^2}{2}$

So, $$T = 2\pi\sqrt{\frac{\frac{ma^2}{2}}{\pi a^2 iB}} = 2\pi\sqrt{\frac{m}{2\pi iB}} = \sqrt{\frac{2\pi m}{iB}}$$

19. According to Gauss's law for electrostatic field,

$$\oint \mathbf{E} \cdot d\mathbf{S} = Q/\varepsilon_0$$

It does not contradict for electrostatic fields as the electric field lines do not form a continuous closed path.

According to Gauss's law in magnetic form, $\oint \mathbf{B} \cdot d\mathbf{S} = 0$

It contradicts for magnetic field because there is a magnetic field inside the solenoid and no field is present outside the solenoid carrying current but the magnetic field lines form the closed paths.

Hence, the correct option is (b).

20. At magnetic equator, the angle of dip is 0°. Then, the horizontal component,

$$B_H = B\cos\theta = 0$$

21. Since, the axis of the magnetic dipole placed at the centre of earth makes an angle 11.3° with the axis of earth, two possibilities arises as shown in Fig. (a) and (b). Hence, the declination varies between 11.3°W to 11.3°E.

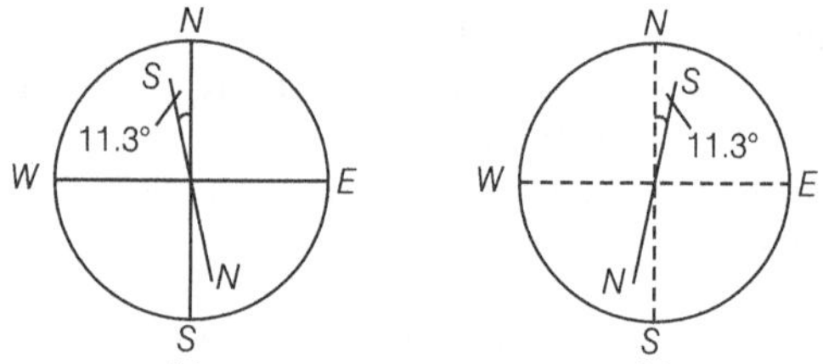

(a) (b)

22. For $H = R\cos\delta$,

$$\Rightarrow \quad R = \frac{H}{\cos\delta} = \frac{B_0}{\cos 45°} \qquad (\because H = B_0)$$

$$= \sqrt{2}\, B_0$$

23. Horizontal component of earth's magnetic field is given by

$$B_H = B\cos\delta$$

or $$\cos\delta = \frac{B_H}{B} = \frac{0.3}{0.5} = \frac{3}{5}$$

$\therefore$ $$\frac{1}{\cos\delta} = \sec\delta = \frac{5}{3}$$

Now, $$\tan\delta = \sqrt{\sec^2\delta - 1} = \sqrt{\left(\frac{5}{3}\right)^2 - 1}$$

$$= \sqrt{\frac{25}{9} - 1} = \frac{4}{3}$$

$\therefore$ $$\delta = \tan^{-1}\frac{4}{3}$$

24. In a tangent galvanometer, applying tangent law,

$$B = H\tan\theta$$

$$\Rightarrow \quad B = 0.34 \times 10^{-4}\tan 30° = 0.34 \times 10^{-4} \times \frac{1}{\sqrt{3}}$$

$$B = 1.96 \times 10^{-5}\ \text{T}$$

25. In series, same current flows through two tangent galvanometers.

i.e. $$i = \frac{2rH}{\mu_0 n_1}\tan\theta_1 = \frac{2rH}{\mu_0 n_2}\tan\theta_2$$

$\therefore$ $$\frac{n_1}{n_2} = \frac{\tan\theta_1}{\tan\theta_2} = \frac{\tan 60°}{\tan 45°} = \sqrt{3} : 1$$

26. Here, $n_1 = 40, n_2 = 20, H_1 = 36 \times 10^{-6}$ T, $H_2 = ?$

$\therefore$ $$\frac{H_2}{H_1} = \frac{n_2^2}{n_1^2} = \frac{(20)^2}{(40)^2} = \frac{1}{4}$$

or $$H_2 = \frac{36 \times 10^{-6}}{4} = 9 \times 10^{-6}\ \text{T}$$

27. As, $\dfrac{M_1}{M_2} = \dfrac{T_2^2 + T_1^2}{T_2^2 - T_1^2} = \dfrac{n_2^2 + n_1^2}{n_1^2 - n_2^2}$

$$\Rightarrow \quad \frac{M_1}{M_2} = \frac{4^2 + 12^2}{12^2 - 4^2} = \frac{160}{128} = 5:4$$

28. Here, $n_1 = 10$ oscillations per min,

$\delta_1 = 45°, R_1 = 0.707$ CGS units,

$n_2 = ?, \delta_2 = 60°$, $R_2 = 0.5$ CGS units

$$\therefore \quad \frac{n_2}{n_1} = \sqrt{\frac{H_2}{H_1}} = \sqrt{\frac{R_2 \cos \delta_2}{R_1 \cos \delta_1}}$$

$$\Rightarrow \quad \frac{n_2}{10} = \sqrt{\frac{0.5 \cos 60°}{0.707 \cos 45°}} = \sqrt{\frac{0.5 \times 1/2}{0.5 \times \sqrt{2} \times 1/\sqrt{2}}} = \frac{1}{\sqrt{2}}$$

or, $\quad n_2 = \dfrac{10}{\sqrt{2}} = 7.07 \approx 7$

29. When two identical bar magnets are held perpendicular to each other.

$$\therefore \quad M_1 = \sqrt{M^2 + M^2} = M\sqrt{2}, I_1 = I$$

and $\quad T_1 = 2^{5/4}$ s, $T_2 = ?$

and $M_2 = M$ (as one magnet is removed)

$$I_2 = I_1 / 2$$

$$T_2 = T_1 \times \frac{1}{2^{1/4}} = 2^{5/4} \times \frac{1}{2^{1/4}} = 2 \text{ s}$$

30. As $t_1 = 3 = 2\pi\sqrt{\dfrac{I}{MR}}$, where R resultant intensity of earth's field.

$$\therefore \quad t_2 = 3\sqrt{2} = 2\pi\sqrt{\frac{I}{MH}}$$

Dividing $\dfrac{t_1}{t_2}$, we get

$$\frac{1}{\sqrt{2}} = \sqrt{\frac{H}{R}} = \sqrt{\frac{R \cos \delta}{R}} = \sqrt{\cos \delta}$$

$$\Rightarrow \quad \cos \delta = \frac{1}{2}$$

$$\Rightarrow \quad \delta = 60°$$

31. Points of zero magnetic field, *i.e.* neutral points lie on equatorial line of magnet, *i.e.* along east and west.

32. In Fig. (a), at neutral point P,

$$B_H = \frac{\mu_0}{4\pi}\left(\frac{M}{d^3}\right)$$

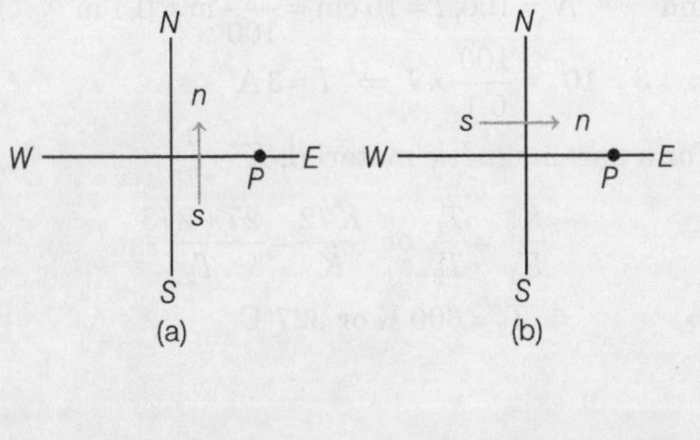

(a) (b)

In Fig. (b),

Net magnetic induction at P = resultant of

$\dfrac{\mu_0}{4\pi}\dfrac{2M}{d^3} = 2B_H$ along horizontal and B_H along vertical

$\Rightarrow$ Net magnetic induction at P

$$= \sqrt{(2B_H)^2 + (B_H)^2}$$

$$= \sqrt{5}\, B_H$$

33. Here, $\quad 2l = 20$ cm

$$\Rightarrow \quad l = 10 \text{ cm}, d = 40 \text{ cm}$$

As, neutral point, $H = B = \dfrac{\mu_0}{4\pi}\dfrac{2Md}{(d^2 - l^2)^2}$

$$\Rightarrow \quad 3.2 \times 10^{-5} = \frac{10^{-7} \times 2M(0.4)}{15 \times 15 \times 10^{-4}}$$

$$\therefore \quad M = \frac{3.2 \times 15 \times 15 \times 10^{-4} \times 10^{-5}}{0.8 \times 10^{-7}} = 9$$

$$\therefore \quad m = \frac{M}{2l} = \frac{9}{0.2} = 45 \text{ A-m}$$

34. Here, $B = 1.7 \times 10^{-5}$ T, $H = ?$

Now, $H = \dfrac{B}{\mu_0} = \dfrac{1.7 \times 10^{-5}}{4\pi \times 10^{-7}} = 13.5 \text{ Am}^{-1}$

35. For each half, pole strength m becomes half, *i.e.* $M = m \times 2l$ and volume $V = a \times 2l$ also becomes half. Therefore, $I = M/V$, remains constant.

36. Here, $\quad V = (10 \times 0.5 \times 0.2) \text{ cm}^3 = 1 \text{ cm}^3 = 10^{-6} \text{ m}^3$,

$H = 0.5 \times 10^4 \text{ Am}^{-1}, M = 5 \text{ A-m}^2, B = ?$

$$\therefore \quad I = \frac{M}{V} = \frac{5}{10^{-6}} = 5 \times 10^6 \text{ Am}^{-1}$$

From $B = \mu_0(I + H)$

$$B = 4\pi \times 10^{-7}(5 \times 10^6 + 0.5 + 10^4) = 6.28 \text{ T}$$

37. When space inside the toroid is filled with air,

$$B_0 = \mu_0 H$$

When filled with tungsten,

$$B = \mu H = \mu_0 \mu_r H$$

$$= \mu_0(1 + \chi_m) H$$

Percentage increase in magnetic field/induction

$$= \frac{(B - B_0) \times 100}{B_0}$$

$$= \frac{\mu_0 \chi_m H \times 100}{\mu_0 H} = \chi_m \times 100$$

$$= 6.8 \times 10^{-5} \times 100$$

$$= 0.0068\%$$

38. Here $I_1 = 8 \text{ Am}^{-1}, B_1 = 0.6$ T, $t_1 = 4$ K,

$I_2 = ?, B_2 = 0.2$ T, $t_2 = 16$ K

According to Curie law,

$$I \propto \frac{B \text{ (magnetic field induction)}}{t \text{(in kelvin)}}$$

$\therefore \quad \frac{I_2}{I_1} = \frac{B_2}{B_1} \times \frac{t_1}{t_2} = \frac{0.2}{0.6} \times \frac{4}{16} = \frac{1}{12}$

or $\quad I_2 = I_1 \times \frac{1}{12} = \frac{8}{12} = \frac{2}{3}\text{ Am}^{-1}$

39. For a diamagnetic substance, I is negative and $-I \propto H$. Therefore, the variation is represented by OC or OD. As magnetisation is small, so OC is the better choice.

40. For diamagnetic substances, the magnetic susceptibility is negative and it is independent of temperature.

41. Copper is a diamagnetic material, therefore its rod align itself, where magnetic field is weakest and perpendicular to the direction of magnetic field at that position.

42. Because of large permeability of soft iron, magnetic lines of force prefer to pass through it. Concentration of lines in soft iron bar increases as shown in Fig. (b).

43. As temperature of a ferromagnetic material is raised, its susceptibility χ remains constant first and then decreases as shown in Fig. (a).

44. As, the sphere is perfectly diamagnetic, it completely expels field as shown below

Therefore, there will be no magnetic field inside the cavity of sphere.

So, the paramagnetic sample at the spherical cavity at the centre of sphere remains uneffected by external field and field inside paramagnetic sample is zero.

45. As, $\mu_r < 1$ for substance X, it must be diamagnetic and $\mu_r > 1$ for substance Y, so it is must be paramagnetic.

46. Coercivity of a bar magnet is the value of magnetic field intensity (H) that is needed to reduce magnetisation to zero.

Since, for a solenoid, magnetic induction is given as,

$$B = \mu_0 nI \quad \text{...(i)}$$

where, n is the number of turns (N) per unit length (l) and I is the current.

Also, $$B = \mu_0 H \quad \text{...(ii)}$$

$\therefore$ From Eqs. (i) and (ii), we get

$$\mu_0 nI = \mu_0 H$$

or $$H = nI = \frac{N}{l} I$$

Substituting the given values, we get

$$H = \frac{100}{0.2} \times 5.2 = 2600 \text{ A/m}$$

Thus, the value of coercivity of the bar magnet is 2600 A/m.

47. Area of hysteresis loop is proportional to the net energy absorbed per unit volume by the material, as it is taken over a complete cycle of magnetisation.

For electromagnets and transformers, energy loss should be low.

i.e. Thin hysteresis curves.

Also, $|B| \to 0$ when $H = 0$ and $|H|$ should be small when $B \to 0$.

Hence, option (d) is correct.

Round II

1. As, $T \propto \frac{1}{\sqrt{M}}$; when M becomes 4 times, then T becomes its half. Therefore, new $T = 1$ s.

2. Field due to circular loop carrying current is perpendicular to the plane of the loop. As, current is anti-clockwise. So, N pole lies above the loop and south pole lies below the loop.

3. On equatorial line, magnetic field due to magnet varies inversely as cube of the distance, therefore

$$\frac{B_1}{B_2} = \left(\frac{3x}{x}\right)^3 = 27:1$$

4. Statement III is correct, because the magnetic field outside the toroid is zero and they form closed loops inside the toroid itself.

Statement V is correct because we know that superconductors are materials inside which the net magnetic field is always zero and they are perfect diamagnetic.

$$\mu_r = 1 + \chi$$
$$\chi = -1$$
$$\mu_r = 0$$

5. Time period of oscillation is

$$T = 2\pi \sqrt{\frac{I}{MB}}$$

$$\Rightarrow \quad T = 2\pi \sqrt{\frac{7.5 \times 10^{-6}}{6.7 \times 10^{-2} \times 0.01}} = 0.665 \text{ s}$$

Hence, time for 10 oscillations is $t = 6.65$ s.

6. For solenoid, the magnetic field needed to magnetise, the magnet is given by

$$B = \mu_0 nI$$

where, n is number of turns per unit length $= N/l$

and $\quad N = 100,\ l = 10 \text{ cm} = \frac{10}{100} \text{ m} = 0.1 \text{ m}$

$\Rightarrow \quad 3 \times 10^3 = \frac{100}{0.1} \times I \Rightarrow I = 3 \text{ A}$

7. For a paramagnetic material, $K \propto \frac{1}{T}$

$\therefore \quad \frac{K_2}{K_1} = \frac{T_1}{T_2}$ or $\frac{K/2}{K} = \frac{27 + 273}{T_2}$

$\Rightarrow \quad T_2 = 600 \text{ K or } 327°\text{C}$

8. When axes are in the same line,

$$F = \frac{\mu_0}{4\pi}\frac{6M_1M_2}{r^4}$$

i.e. $F \propto \frac{1}{r^4}$

When r becomes thrice, F becomes $\frac{1}{(3)^4}$ times, i.e. $\frac{1}{81}$ times. Therefore, $F' = \frac{8.1}{81} = 0.1$ N.

9. As, $M = niA = 50 \times 2 \times 1.25 \times 10^{-3} = 0.125$ A-m^2

If normal to the face of the coil makes an angle θ with the magnetic induction B, then in first case,

torque = $MB\cos\theta = 0.04$

and in second case,

torque = $MB\sin\theta = 0.03$

$\therefore$ $MB = \sqrt{(0.04)^2 + (0.03)^2} = 0.05$

$\Rightarrow$ $B = \frac{0.05}{M} = \frac{0.05}{0.125} = 0.4$ T

10. As, the magnet is long, we assume that the upper north pole produces no effect. But due to south pole strength of the magnet which is equal and opposite to horizontal component of earth's magnetic field, neutral point is obtained. i.e.

$$B = \left(\frac{\mu_0}{4\pi}\right)\frac{m}{r^2} = H$$

In CGS system, $\frac{\mu_0}{4\pi} = 1$

$\therefore$ $1 \times \frac{m}{10^2} = 3.2 \times 10^{-5} \times 10^4$ (gauss)

$m = 32$ ab-A-cm^{-1}

11. As, $T_1 = 2\pi\sqrt{\frac{I}{MV}}$

and $T_2 = 2\pi\sqrt{\frac{I}{MH}}$

$\therefore$ $\frac{T_2}{T_1} = \sqrt{\frac{V}{H}} = \sqrt{\tan\theta}$

or $\tan\theta = \left(\frac{T_2}{T_1}\right)^2 = \left(\frac{2}{2}\right)^2 = 1$

$\Rightarrow$ $\theta = 45°$

12. In equilibrium, the resultant magnetic moment will be along magnetic meridian. Let N_1S_1 makes an angle θ with this resultant,

$\therefore$ $\tan\theta = \frac{M_2}{M_1} = \frac{M}{\sqrt{3}M} = \frac{1}{\sqrt{3}}$

$\therefore$ $\theta = 30°$

13. The lines of magnetic field induction **B** are necessarily continuous across the surface S of a lump of magnetic material. Outside the lump of magnetic material, $H = B/\mu_0$ and inside the lump of magnetic material, $H = B/\mu_0\mu_r$, where μ_r is the relative permeability of material. Thus, the lines of H cannot all be continuous across surface S.

14. Magnetic moment of an iron core solenoid,

$$M = \mu_r NiA$$

where, μ_r = relative permeability = 1000,

N = number of turns = 10 turns/cm,

i = current = 0.5 A

and A = area of cross-section.

Now, $M = \mu_r Ni\frac{V}{l}$ $\left(\because A = \frac{V}{l}\right)$

Substituting all the given values in above equation, we get

$$M = 1000 \times \frac{10}{10^{-2}} \times 0.5 \times 10^{-3} \quad (\because V = 10^{-3}\text{ m}^3)$$

$$= 500 \text{ A-m}^2 = 5 \times 10^2 \text{ A-m}^2$$

15. Initially, $m = I\pi R^2$ and $B_1 = \frac{\mu_0 I}{2R_1}$

Finally, $m' = 2m = I\pi R_2^2$

$\Rightarrow$ $2I\pi R_1^2 = I\pi R_2^2$ or $R_2 = \sqrt{2}R_1$

So, $B_2 = \frac{\mu_0 I}{2(R_2)} = \frac{\mu_0 I}{2\sqrt{2}R_1}$

Hence, $\frac{B_1}{B_2} = \frac{\left(\frac{\mu_0 I}{2R_1}\right)}{\left(\frac{\mu_0 I}{2\sqrt{2}R_1}\right)} = \sqrt{2}$

16. Since, **B** is uniform only torque acts on a current carrying loop.

As, $\tau = \mathbf{M} \times \mathbf{B}$

$\Rightarrow$ $|\tau| = |M||B|\sin\theta$

For orientation shown in (II) $\theta = 0^0$, $\tau = 0$ (stable equilibrium) and for (IV) $\theta = \pi$, $\tau = 0$ (unstable equilibrium)

17. When the wire is bent at its middle point O at 60°, then

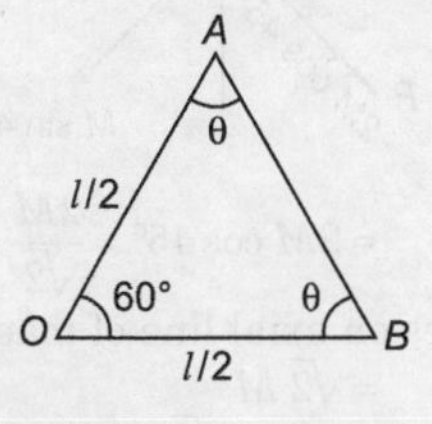

$60° + \theta + \theta = 180°$

or $2\theta = 180° - 60° = 120°$

$\Rightarrow$ $\theta = 60°$

$\because$ ΔOAB is an equilateral triangle.

$\therefore$ $AB = 2l' = l/2$

New magnetic moment,

$$M' = m(2l') = \frac{ml}{2} = \frac{M}{2}$$

18. If V = volume of the material = mass/density,
A = area of hysteresis loop,
ν = frequency of alternate magnetic field applied
and t = time for which field is applied,
then energy loss in the material in t second is

$$E = VA\nu t = \left(\frac{m}{d}\right) A\nu t$$

So, energy loss in 1 s $= \left(\frac{m}{d}\right) A\nu$

$$= \frac{0.6}{7.8\times10^3}\times0.722\times50$$
$$= 2.777\times10^3 = 27.77\times10^{-4}\text{ J}$$

19. In a vibration magnetometer,

$$T = 2\pi\sqrt{\frac{I}{MH}}$$

$\therefore$ $$4\pi^2\frac{1}{T^2} = MH = 36\times10^{-4} \quad \text{...(i)}$$

In a deflection magnetometer,

$$H = \frac{\mu_0}{4\pi}\frac{2M}{d^3}$$
$$\frac{4\pi d^3}{2\mu_0} = \frac{M}{H} = \frac{10^8}{36} \quad \text{...(ii)}$$

Multiplying Eq. (i) and Eq. (ii), we get

$$M^2 = 36\times10^{-4}\times\frac{10^8}{36} = 10^4$$

or $$M = 10^2 = 100\text{ A-m}$$

20. Resolving the magnetic moments along OP and perpendicular to OP. From figure, we find that component of OP perpendicular to it cancel out. Resultant magnetic moment along OP is
$= M\cos45°+M\cos45°$.

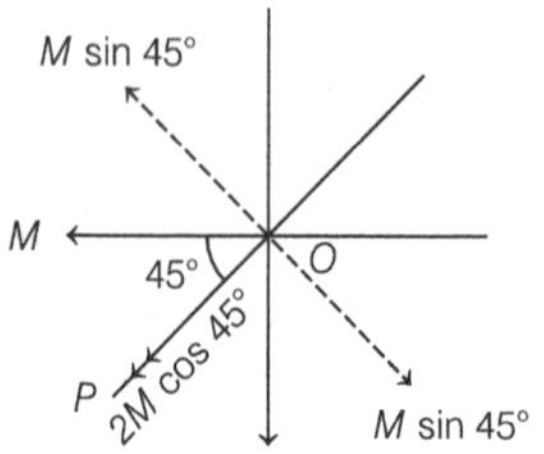

$$= 2M\cos45° = \frac{2M}{\sqrt{2}} = \sqrt{2}\,M$$

The point P lies on axial line of magnetic of moment
$= \sqrt{2}\,M$

$\therefore$ $$B = \frac{\mu_0}{4\pi}\frac{2(\sqrt{2}\,M)}{d^3}$$

21. In SI, the unit of pole strength is A-m. Here, the pole strength is given in weber, which is the unit of $(\mu_0 m)$ and

$$\mu_0 m = 10^{-3}\text{ Wb}$$

or $$m = \frac{10^{-3}}{\mu_0}$$

Magnetic moment of magnet,

$$M = m\times2l = \frac{10^{-3}}{\mu}(0.1) = \frac{10^{-4}}{\mu_0}$$

$\therefore$ Torque, $$\tau = MB\sin\theta$$
$$= \left(\frac{10^{-4}}{\mu_0}\right)(4\pi\times10^{-3})\sin30°$$
$\Rightarrow$ $$= \frac{10^{-4}}{4\pi\times10^{-7}}(4\pi\times10^{-3})\times\frac{1}{2} = 0.5\text{ N-m}$$

22. As shown in figure,

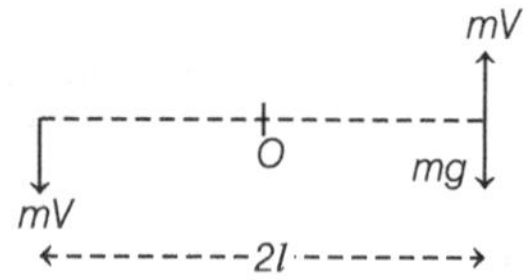

Mass, $M = 50$, $mg = 50\times10^{-3}\ g$
Strength of each pole, $m = 98.1$ ab-amp-cm,
$g = 981\text{ cms}^{-2}$, $V = ?$

At equilibrium, $mV\times2l = Mg\times l$

or $$V = \frac{Mg}{2m}$$

or $$V = \frac{50\times10^{-3}\times981}{2\times98.1} = 0.25\text{ G}$$

23. As, $$T = 2\pi\sqrt{\frac{I}{MH}}$$

$\therefore$ $$2 = 2\pi\sqrt{\frac{I}{MH}} \quad \text{...(i)}$$

When an external magnet is brought near and parallel to H and the time period reduces to 1 s, net field must be $(F+H)$.

$\therefore$ $$1 = 2\pi\sqrt{\frac{1}{M(F+H)}} \quad \text{...(ii)}$$

Dividing Eq. (i) by Eq. (ii), we get

$$\frac{2}{1} = \sqrt{\frac{F+H}{H}} = \sqrt{\frac{F}{H}+1}$$

$\Rightarrow$ $$\frac{F}{H}+1 = 4$$

$\Rightarrow$ $$\frac{F}{H} = 4-1 = 3$$

24. In the first case, $T_1 = \frac{60}{10} = 6$ s

$\Rightarrow$ $$T_1 = 2\pi\sqrt{\frac{I}{MH}} = 6 \quad \text{...(i)}$$

In the second case, $T_2 = \frac{60}{14} = \frac{30}{7}$ s

If B magnetic induction due to external magnet, then

$$T_2 = 2\pi\sqrt{\frac{I}{M(H+B)}} = \frac{30}{7} \quad \text{...(ii)}$$

Dividing Eq. (i) by Eq. (ii), we get

$$\frac{6}{30/7} = \sqrt{\frac{H+B}{H}} = \sqrt{1+\frac{B}{H}}$$

or $$\left(\frac{7}{5}\right)^2 = 1+\frac{B}{H}$$

or $$\frac{B}{H} = \frac{49}{25} - 1 = \frac{24}{25}$$

If n is number of vibrations/min in the third case when polarity of external magnet is reversed, then

$$T_3 = \frac{60}{n} = 2\pi\sqrt{\frac{I}{M(H-B)}} \quad \text{...(iii)}$$

Dividing Eq. (i) by Eq. (iii), we get

$$\frac{6}{60/n} = \sqrt{\frac{H-B}{H}} = \sqrt{1-\frac{B}{H}} = \sqrt{1-\frac{24}{25}}$$

$$\Rightarrow \quad \frac{n}{10} = \frac{1}{5} \text{ or } n = 2 \text{ vib per min}$$

25. Here, magnetic moment is given in weber-metre, which is the unit of $\mu_0 M$.

$$\therefore \quad \mu_0 M = 5\times10^{-5} \text{ Wb-m}$$

or $$M = \frac{5\times10^{-5}}{\mu_0} \text{ A-m}^2$$

Also, $$B = 8\pi\times10^{-4} = \mu_0 H$$

$$\therefore \quad H = \frac{8\pi\times10^{-4}}{\mu_0}$$

Now, $$T = 2\pi\sqrt{\frac{I}{MH}}$$

$$\therefore \quad I = \frac{MHT^2}{4\pi^2} = \frac{5\times10^{-5}\times8\pi\times10^{-4}\times15^2}{4\pi^2\mu_0^2}$$

$$= \frac{5\times10^{-5}\times8\pi\times10^{-4}\times225}{4\pi^2(4\pi\times10^{-7})^2}$$

or $$I = \frac{2250\times10^{-9}}{\pi(16\pi^2)\times10^{-14}}$$

$$= 4.54\times10^5 \text{ kg-m}^2$$

26. Let θ be the declination at the place. As it is clear from figure.

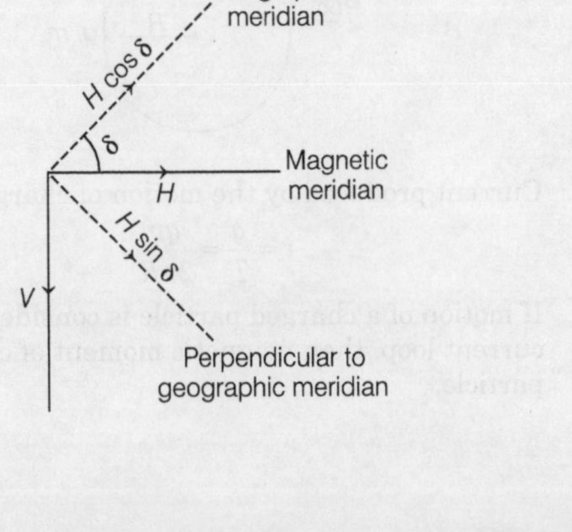

$$\tan\delta_1 = \frac{V}{H\cos\theta}$$

and $$\tan\delta_2 = \frac{V}{H\cos(90^\circ-\theta)} = \frac{V}{H\sin\theta}$$

$$\because \quad \frac{\tan\delta_1}{\tan\delta_2} = \frac{\sin\theta}{\cos\theta} = \tan\theta$$

$$\Rightarrow \quad \theta = \tan^{-1}\left(\frac{\tan\delta_1}{\tan\delta_2}\right)$$

27. In the usual setting of deflecting magnetometer, field due to magnet (F) and horizontal component (H) of earth's field are perpendicular to each other. Therefore, the net field on the magnetic needle is $\sqrt{F^2+H^2}$.

$$\therefore \quad T = 2\pi\sqrt{\frac{I}{M\sqrt{F^2+H^2}}} \quad \text{...(i)}$$

When magnet is removed,

$$T_0 = 2\pi\sqrt{\frac{I}{MH}} \quad \text{...(ii)}$$

Also, $$\frac{F}{H} = \tan\theta$$

Dividing Eq. (i) by Eq. (ii), we get

$$\frac{T}{T_0} = \sqrt{\frac{H}{\sqrt{F^2+H^2}}}$$

$$= \sqrt{\frac{H}{\sqrt{H^2\tan^2\theta + H^2}}}$$

$$= \sqrt{\frac{H}{H\sqrt{\sec^2\theta}}}$$

$$= \sqrt{\cos\theta}$$

or $$\frac{T_2^2}{T_0^2} = \cos\theta$$

$$\Rightarrow \quad T^2 = T_0^2\cos\theta$$

28. In CGS system, $\frac{\mu_0}{4\pi} = 1$

In equilibrium, net repulsion due to magnetic interaction = weight of upper magnet. Therefore, it is clear from figure.

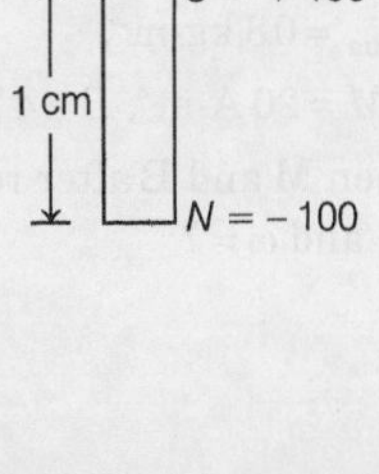

$$\frac{900\,(100)}{1^2}+\frac{900\,(-100)}{2^2}-\frac{900\,(100)}{2^2}-\frac{900\,(-100)}{3^2}=m\times g$$

$$\Rightarrow \quad 900\times 100\left(\frac{1}{1^2}-\frac{1}{2^2}-\frac{1}{2^2}+\frac{1}{3^2}\right)=m\times 1000$$

$$\Rightarrow \quad 90\left(\frac{11}{18}\right)=m$$

$$\Rightarrow \quad m=55\text{ g}$$

29. Given, angle of declination,

$$\theta = 12^\circ \text{ west}$$

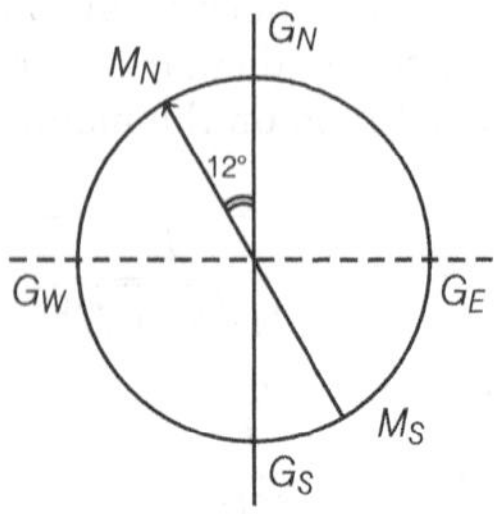

Angle of dip, $\delta = 60^\circ$

Horizontal component of earth's magnetic field,

$$H = 0.16\text{ G}$$

Let the magnitude of earth's magnetic field at that place be R.

Using the formula, $H = R\cos\delta$

or $$R=\frac{H}{\cos\delta}=\frac{0.16}{\cos 60^\circ}=\frac{0.16\times 2}{1}=0.32\text{ G}=0.32\times 10^{-4}\text{ T}$$

The earth's magnetic field lies in a vertical plane 12° west of geographical meridian at an angle 60° above the horizontal.

30.

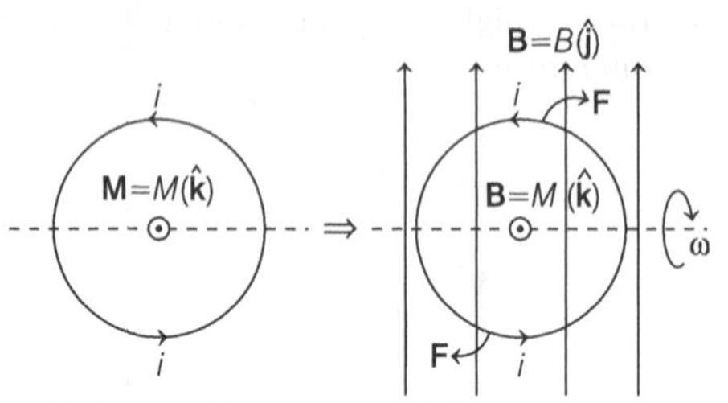

Before applying the magnetic field — After applying the magnetic field

Given that, $I_{dia}=0.8\text{ kg-m}^2$,

$$M = 20\text{ A-m}^2,\ B = 4\text{ T}$$

$\theta=$ angle between **M** and **B** after rotation of 60° by coil $=90^\circ-60^\circ=30^\circ$ and $\omega=?$

Applying the law of conservation of energy,

$$(\text{TE})_{\text{initial}}=(\text{TE})_{\text{final}}$$

$$U_i+K_i=U_f+K_f$$

$$0+0=-MB\cos\theta+\frac{1}{2}I_{\text{dia}}\omega^2$$

$$0=-20\times 4\cos 30^\circ+\frac{1}{2}\times 0.8\times\omega^2$$

$$0=-20\times 4\times\frac{\sqrt{3}}{2}+0.4\,\omega^2$$

$$0=-40\sqrt{3}+0.4\,\omega^2$$

$$\Rightarrow \quad 40\sqrt{3}=0.4\,\omega^2$$

$$\Rightarrow \quad \omega^2=\frac{40\sqrt{3}}{0.4}=100\sqrt{3}=100\times 1.732=173.2$$

$$\omega=\sqrt{173.2}=13.16\text{ rad/s}$$

31. From Curie's law for paramagnetic materials,

$$\chi_m\propto\frac{1}{T}$$

$$\Rightarrow \quad \frac{I}{H}\propto\frac{1}{T}$$

$$\Rightarrow \quad \frac{I}{(B_0/\mu_0)}\propto\frac{1}{T}\ \Rightarrow\ \frac{\mu_0 I}{B_0}\propto\frac{1}{T}$$

$$\Rightarrow \quad I\propto\frac{B_0}{\mu_0 T}\ \Rightarrow\ I\propto\frac{B_0}{T}\qquad (\because \mu_0=\text{constant})$$

$$\Rightarrow \quad \frac{I_1}{I_2}=\frac{(B_0)_1}{(B_0)_2}\times\frac{T_2}{T_1}$$

On putting the given values in above equation, we get

$$\frac{6}{I_2}=\frac{0.4}{0.3}\times\frac{24}{4}$$

$$\Rightarrow \quad \frac{6}{I_2}=\frac{4}{3}\times\frac{6}{1}$$

$$\Rightarrow \quad \frac{6}{I_2}=8\ \Rightarrow\ I_2=\frac{6}{8}$$

$$\Rightarrow \quad I_2=\frac{3}{4}\text{ A/m}$$

$$\Rightarrow \quad I_2=0.75\text{ A/m}$$

32. Let us consider the direction of magnetic field be perpendicular to the plane in inward direction as shown below

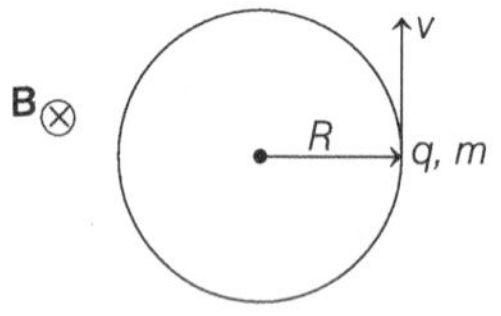

Current produced by the motion of charge particle,

$$i=\frac{q}{T}=\frac{qv}{2\pi R}$$

If motion of a charged particle is considered as a current loop, then magnetic moment of charged particle,

$$M = iA = \frac{qv}{2\pi R} \times \pi R^2$$

$$\Rightarrow \quad M = \frac{1}{2} qvR \qquad \ldots(i)$$

As we know that, while moving in a circle, necessary centripetal force is provided by magnetic force.

i.e. $\frac{mv^2}{R} = qvB$ or $R = \frac{mv}{qB}$

Putting this value in Eq. (i), we get

$$M = \frac{1}{2} qv\left(\frac{mv}{qB}\right)$$

or $$M = \frac{mv^2}{2B}$$

Here, the direction of **M** is determined by right hand thumb rule which is perpendicular to plane but in outward direction, *i.e.* opposite to that of **B**.

$$\therefore \quad \mathbf{M} = \frac{mv^2}{2B}(-\hat{\mathbf{B}})$$

or $$\mathbf{M} = -\frac{mv^2\mathbf{B}}{2B^2}$$

33. Horizontal component of earth's magnetic field,

$$B_H = B_E \cos\theta$$

$$\Rightarrow \quad \cos\theta = \frac{B_H}{B_E} = \frac{3\times10^{-6}}{6\times10^{-5}} = \frac{1}{2} \Rightarrow \theta = 60^\circ$$

34. Any point on magnetic equator is on equatorial line of magnet and at poles, it is on axial line.

$$\therefore \quad \frac{B_{\text{equatorial}}}{B_{\text{axial}}} = \frac{1/2}{1} = \frac{1}{2}$$

$$\therefore \quad x = 2$$

35. Magnetic susceptibility,

$$\chi_{m_1} = 0.0075, T_1 = -73^\circ\text{ C}$$

$$= (-73 + 273)\text{ K} = 200\text{ K}$$

$$\chi_{m_2} = ?, T_2 = -173^\circ\text{ C} = (-173 + 273)\text{ K} = 100\text{ K}$$

As, $\chi_m \propto \frac{1}{T}$

$$\Rightarrow \quad \frac{\chi_{m_2}}{\chi_{m_1}} = \frac{T_1}{T_2} = \frac{200}{100} = 2$$

$$\therefore \quad \chi_{m_2} = 2\chi_{m_1} = 2\times 0.0075 = 0.015$$

36. If d_1 is distance of point X on axial line and d_2 is distance of point Y on equatorial line,

then $$B_1 = \frac{\mu_0}{4\pi}\frac{2M}{d_1^3}, B_2 = \frac{\mu_0}{4\pi}\frac{M}{d_2^3}$$

As, $$B_1 = B_2$$

$$\therefore \quad \frac{\mu_0}{4\pi}\frac{2M}{d_1^3} = \frac{\mu_0}{4\pi}\frac{M}{d_2^3}$$

$$d_1^3 = d_2^3$$

$$\Rightarrow \quad \frac{d_1}{d_2} = 2^{1/3}$$

37. If given situation, loop and magnetic field are given as shown in the figure

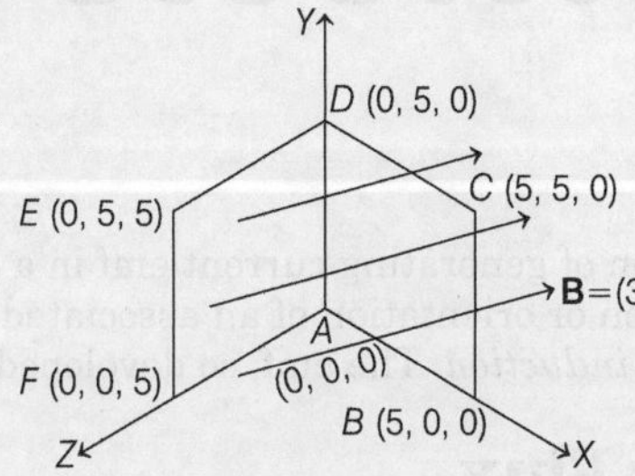

Flux through loop $ABCDEFA$ = Flux through part $ABCDA$ + Flux through part $ADEFA$

$$= \mathbf{B}\cdot\mathbf{A}_1 + \mathbf{B}\cdot\mathbf{A}_2$$

Here, from figure, $\mathbf{A}_1 = 25\hat{\mathbf{k}}$ and $\mathbf{A}_2 = 25\hat{\mathbf{i}}$

So, flux associated with complete loop

$$= (3\hat{\mathbf{i}} + 4\hat{\mathbf{k}})\cdot 25\hat{\mathbf{k}} + (3\hat{\mathbf{i}} + 4\hat{\mathbf{k}})\cdot 25\hat{\mathbf{i}}$$

$$= 100 + 75$$

$$= 175\text{ Wb}$$

22

Electromagnetic Induction

IN THIS CHAPTER ....

The phenomenon of generating current/emf in a conducting circuit by change in strength, position or orientation of an associated external magnetic field is called *electromagnetic induction*. The emf, so developed is called *induced emf.*

Magnetic Flux

Magnetic flux is the volume flow rate of magnetic field linked with an area.

It is also equals to total number of field lines passing through any surface normally when it is placed in magnetic field.

For an element of area $d\mathbf{A}$ on an arbitrary shaped surface as shown in below figure. If the magnetic field at this element is **B**, then the magnetic flux through the element is

$$d\phi_B = \mathbf{B} \cdot d\mathbf{A} = BdA \cos\theta$$

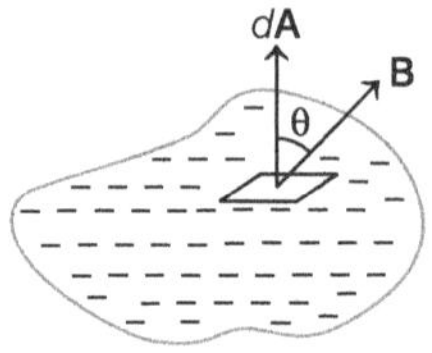

Here, $d\mathbf{A}$ is a vector that is perpendicular to the surface and has a magnitude equal to the area dA and θ is the angle between **B** and dA at that element.

In general, total magnetic flux through the surface is given as

$$\phi_B = \int BdA \cos\theta = \int \mathbf{B} \cdot d\mathbf{A}$$

It is a scalar quantity. The SI unit of magnetic flux is tesla-metre^2 (1 T-m^2). This unit is called *weber* (1Wb).

$$1 \text{ Wb} = 1 \text{ T-m}^2 = 1 \text{ N-m/A}$$

The CGS unit of flux is maxwell (Mx). 1 Wb = 10^8 Mx

The dimensional formula of magnetic flux is $[ML^2T^{-2}A^{-1}]$.

Specific conditions

For $\theta = 0°$	For $\theta = 90°$	For $\theta = 180°$
B is parallel to surface vector, *i.e.* **A**. $\Rightarrow \ \phi_B = BA\cos 0° = BA$ $\Rightarrow \ \phi_B =$ maximum	**B** is perpendicular to surface vector, *i.e.* **A**. $\Rightarrow \phi_B = BA\cos 90° = 0$	**B** is anti-parallel to surface vector, *i.e.* **A**. $\Rightarrow \phi_B = BA\cos 180° = BA(-1) = -BA$

Faraday's Laws of EMI

Experiments revealed that when magnetic flux linked with a conductor varies, an electromotive force (emf) is produced in the conductor. If the conductor forms part of a closed circuit, then emf produced causes an electric current in the circuit. This phenomena is known as **electromagnetic induction** (EMI).

Magnitude and direction of emf and current induced are found by Faraday's and Lenz's laws.

According to Faraday's law, "the induced emf in a conductor is proportional to the time rate of change of magnetic flux linked with the conductor."

$$e \propto \frac{d\phi_B}{dt} \quad \text{or} \quad e = -\frac{d\phi_B}{dt}$$

In SI system, proportionality constant is unity.

If a circuit is a coil consisting of N loops all of the same area and if ϕ_B is the flux through one loop, an emf is induced in every loop, thus the total induced emf in the coil is given by the expression,

$$e = -N\frac{d\phi_B}{dt}$$

Induced current produced in the circuit is given by

$$i = \frac{e}{R} = -\frac{N}{R}\cdot\frac{d\phi_B}{dt}$$

where, R = resistance of loop.

This induced emf and current exists only when flux linked with the conductor is changing.

Induced charge is given as $dq = \frac{1}{R}(d\phi_B)$

Note Integral form of Faraday's law of electromagnetic induction is $\oint \mathbf{E}\cdot d\mathbf{l} = \frac{-d\phi}{dt}$

Example 1. *In a coil of resistance 100 Ω, a current is induced by changing the magnetic flux through it as shown in the figure. The magnitude of change in flux through the coil is* **[JEE Main 2017]**

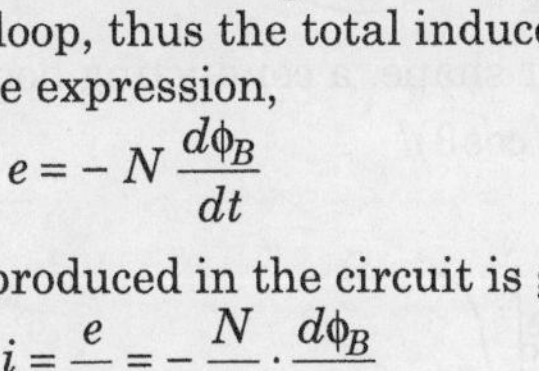

(a) 225 Wb (b) 250 Wb (c) 275 Wb (d) 200 Wb

Sol. (b) Induced current, $i = \frac{e}{R}$

Here, e = induced emf $= -\frac{d\phi_B}{dt}$

$$i = -\left(\frac{d\phi_B}{dt}\right)\cdot\frac{1}{R}$$

$$d\phi_B = -iRdt$$

$$\phi_B = -\int iRdt$$

Here, R is constant.

$\therefore \quad |\phi_B| = R\int idt$

$\int i\cdot dt$ = Area under i-t graph

$$= \frac{1}{2}\times 10\times 0.5 = 2.5$$

$\therefore \quad \phi_B = R\times 2.5 = 100\times 2.5 = 250$ Wb

Example 2. *A uniform magnetic field B exists in a direction perpendicular to the plane of a square loop made of a metal wire. The wire has a diameter of 4 mm and a total length of 30 cm. The magnetic field changes with time at a steady rate $dB/dt = 0.032\ Ts^{-1}$. The induced current in the loop is close to (Take, resistivity of the metal wire $= 1.23\times 10^{-8}\ \Omega m$)* **[JEE Main 2020]**

(a) 0.61 *A* (b) 0.43 *A* (c) 0.53 *A* (d) 0.34 *A*

Sol. (a) Induced emf in square loop,

$$E = A\left(\frac{dB}{dt}\right) = l^2\left(\frac{dB}{dt}\right) = \left(\frac{L}{4}\right)^2\left(\frac{dB}{dt}\right) \qquad (\because \phi_B = BA)$$

Here, $L = 30\ \text{cm} = 30\times 10^{-2}$ m

$\frac{dB}{dt} = 0.032\ \text{Ts}^{-1}$

$\therefore \quad E = 1.8\times 10^{-4}$ V

Induced current, $I = \frac{E}{R} = \frac{E}{\rho\frac{L}{A}} = \frac{EA}{\rho L}$

Here, $\rho = 1.23\times 10^{-8}\ \Omega$-m

$L = 30\times 10^{-2}$ m

$A = \pi r^2 = \pi(2\times 10^{-3})^2 = 12.56\times 10^{-6}\ \text{m}^2$

So, $I = \frac{1.8\times 10^{-4}\times 1.256\times 10^{-5}}{1.23\times 10^{-8}\times 30\times 10^{-2}} = 0.61$ A

Lenz's Law and Conservation of Energy

The negative sign in Faraday's equation of electromagnetic induction describes the direction in which induced emf drives the current. This direction is easily determined with the help of Lenz's law, an associated principle with Faraday's law.

Lenz's law states that, "the direction of any magnetic induction effect is such as to oppose the cause of effect."

This law is based upon law of conservation of energy.

Direction of induced current with the help of Lenz's law

Direction of induced current can be determined by checking whether the flux through a conducting loop or circuit is increasing or decreasing.

- If flux is decreasing, the magnetic field due to induced current will be along the existing magnetic field.
- If flux is increasing, the magnetic field due to induced current will be opposite to existing magnetic field.

Note Also to apply Lenz's law, you can remember RIN or ⊗ IN (when the loop lies on the plane of paper), where

(i) **RIN** In RIN, R stands for right, I stands for increasing and N stands for north pole (anti-clockwise). It means, if a loop is placed on the right side of a straight current-carrying conductor and the current in the conductor is increasing, then induced current in the loop is anti-clockwise (↺).

(ii) ⊗ **IN** In ⊗ IN, suppose the magnetic field in the loop is perpendicular to paper inwards ⊗ and this field is increasing, then induced current in the loop is anti-clockwise (↺).

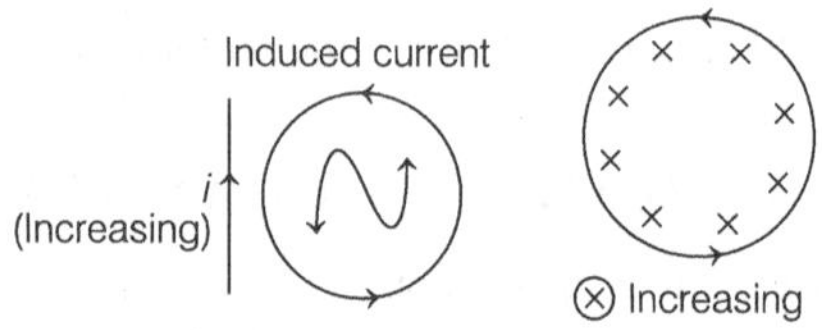

Motional Electromotive Force

Consider a straight conductor of length (as shown below) which is moving through a uniform magnetic field directed into the page.

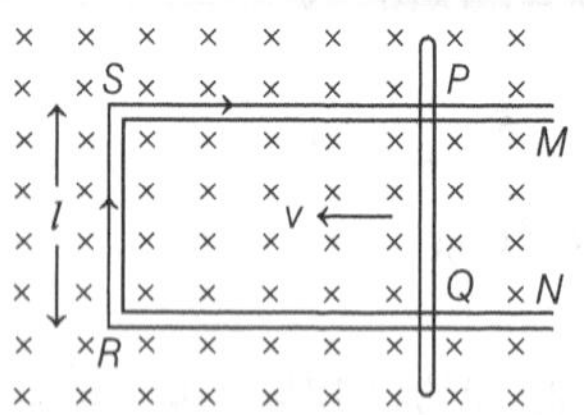

Due to the motion of conductor in **B** a potential difference is maintained between the ends of the conductor as long as the conductor continuous to move through the uniform magnetic field.

This potential difference is called *motional electromagnetic force*, which is denoted by e and is given by

$$e = Bvl$$

If R is the resistance of the closed circuit (as shown above), then current will be written as

$$i = \frac{e}{R} = \frac{Bvl}{R}$$

For a conductor of any shape, moving in a uniform or non-uniform magnetic field, motional emf is given as

$$e = \oint (\mathbf{v} \times \mathbf{B}) \cdot d\mathbf{l} \text{ (for closed conducting loop)}$$

Some important points related to motional emf are given below

- For a semi-circular conducting loop of radius R with the centre at O and moving with velocity v, then the emf is given by

$$e = Bv(2R)$$

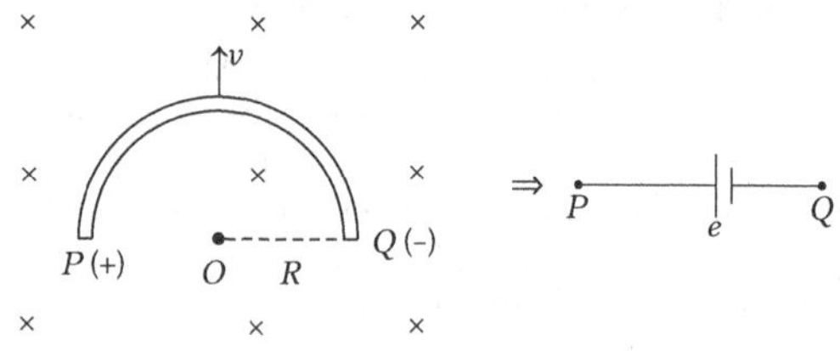

$$V_P - V_Q = 2BvR$$

- For a circular loop of radius R moving with velocity v, the emf is given by

$$e = Bv(2R) = 2BvR$$

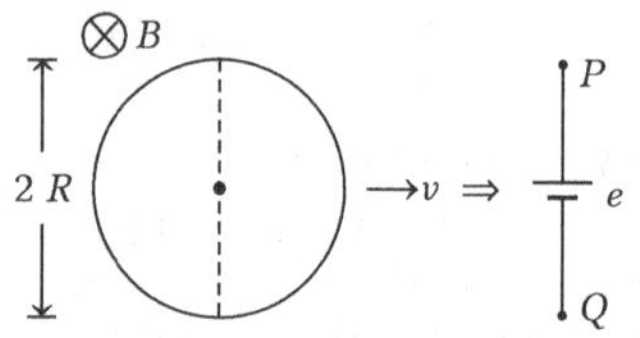

- For a irregular shape, a conducting body of length l. Then, $e = B(v\cos\theta)l$

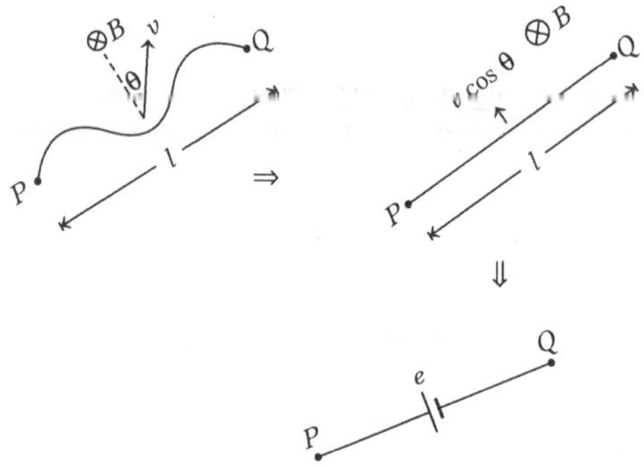

- A moving conductor (PQ) is equivalent to a cell or battery as shown below

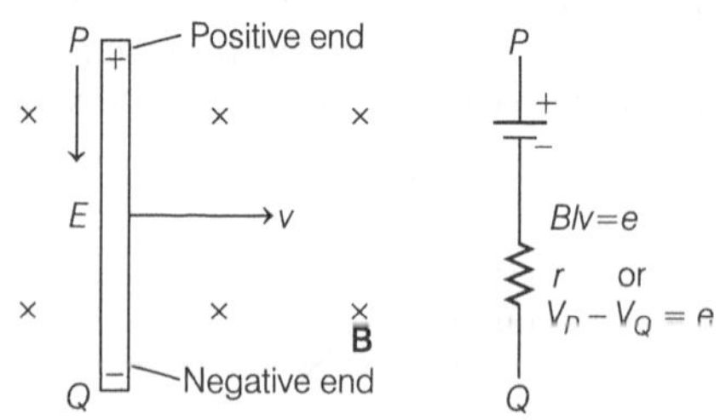

- For a conductor moving at some angle 'θ' with direction of magnetic field, velocity component perpendicular to magnetic field is taken,

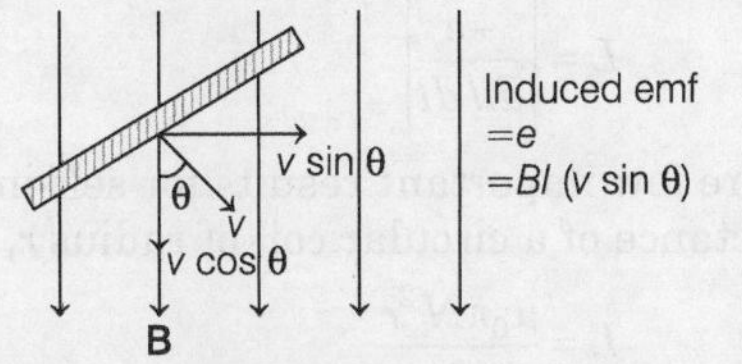

- An emf is induced between the ends of rod, when a conducting rod of length l rotates about an axis passing through one of its ends (that end may be fixed) with an angular velocity ω in a plane perpendicular to the magnetic field B and is given by

$$|\varepsilon| = \frac{1}{2} Bl^2\omega$$

- If a conducting disc of radius r rotates with constant angular velocity ω about its axis in a uniform magnetic field parallel to its axis of rotation

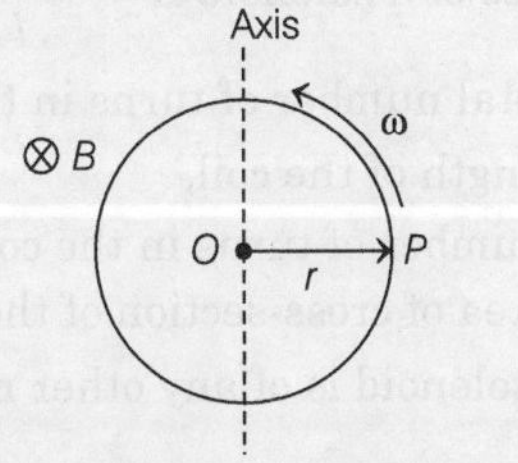

The induced emf is given by

$$e = \frac{1}{2} B\omega r^2$$

That means, disc is equivalent to rod.

- Due to presence of current and magnetic field , arm PQ (as shown in the above figure) experiences a magnetic force given by

$$F = BIl = \frac{B^2 l^2 v}{R}$$

- The force given above is directed opposite to the velocity of the rod and hence the power required to maintain constant speed v is given as

$$P = Fv = \frac{B^2 l^2 v^2}{R}$$

- The direction of induced current is given by Fleming's right hand rule.

Fleming's Right Hand Rule

If we stretch the thumb, the forefinger and the central finger of right hand in such a way that all three are perpendicular to each other, then if thumb represent the direction of motion, the forefinger represent the direction of magnetic field, then central finger will represent the direction of induced current.

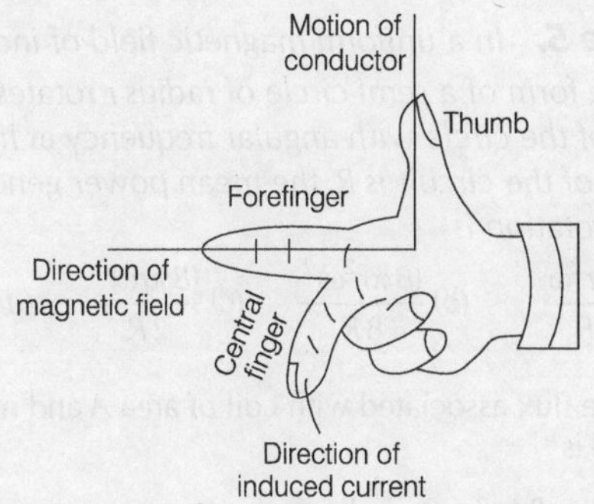

Example 3. *A 10 m long horizontal wire extends from North-East to South-West. It is falling with a speed of 5.0 ms^{-1}, at right angles to the horizontal component of the earth's magnetic field of 0.3×10^{-4} Wb/m². The value of the induced emf in wire is* **[JEE Main 2019]**

(a) 1.5×10^{-3} V (b) 1.1×10^{-3} V
(c) 0.3×10^{-3} V (d) 2.5×10^{-3} V

Sol. (a) Wire falls perpendicularly to horizontal component of earth's magnetic field, so induced emf (e) = Blv

Substituting the given values, we get

$$e = 0.3 \times 10^{-4} \times 10 \times 5 = 1.5 \times 10^{-3} \text{ V}$$

Example 4. *A metallic rod of length l is tied to a string of length $2l$ and made to rotate with angular speed ω on a horizontal table with one end of the string fixed. If there is a vertical magnetic field B in the region, the emf induced across the ends of the rod is* **[JEE Main 2013]**

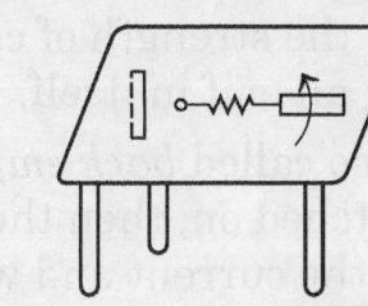

(a) $\frac{2B\omega l^2}{2}$ (b) $\frac{3B\omega l^2}{2}$
(c) $\frac{4B\omega l^2}{2}$ (d) $\frac{5B\omega l^2}{2}$

Sol. (d) ∵ Induced emf is rate of change of magnetic flux.

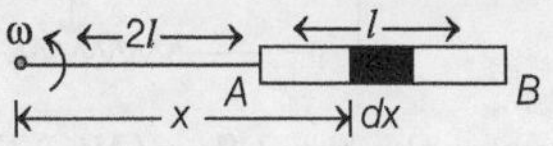

$$\therefore \quad e = \int Bv\, dx$$

$$\Rightarrow \quad e = \int_{2l}^{3l} (\omega x) \cdot B \cdot dx \qquad [\because v = \omega x]$$

Using $\int x^n dx = \frac{x^{n+1}}{n+1}$

$$\Rightarrow \quad e = B\omega \left[\frac{x^2}{2}\right]_{2l}^{3l}$$

$$e = B\omega \frac{[(3l)^2 - (2l)^2]}{2} = \frac{5Bl^2\omega}{2}$$

Example 5. *In a uniform magnetic field of induction B, a wire in the form of a semi-circle of radius r rotates about the diameter of the circle with angular frequency ω. If the total resistance of the circuit is R, the mean power generated per period of rotation is*

(a) $\frac{B\pi r^2\omega}{2R}$ (b) $\frac{(B\pi r^2\omega)^2}{8R}$ (c) $\frac{(B\pi r\omega)^2}{2R}$ (d) $\frac{(B\pi r\omega^2)^2}{8R}$

Sol. (b) The flux associated with coil of area A and magnetic induction B is

$$\phi = BA\cos\theta = \frac{1}{2}B\pi r^2\cos\omega t \quad \left[\because A = \frac{1}{2}\pi r^2, \text{ for semi-circle}\right]$$

$$\therefore \quad e_{\text{induced}} = -\frac{d\phi}{dt} = -\frac{d}{dt}\left(\frac{1}{2}B\pi r^2\cos\omega t\right) = \frac{1}{2}B\pi r^2\omega\sin\omega t$$

$$\therefore \text{ Power, } P = \frac{e_{\text{induced}}^2}{R} \quad [\because P = V^2/R]$$

$$= \frac{B^2\pi^2r^4\omega^2\sin^2\omega t}{4R}$$

Hence, $P_{\text{mean}} = \langle P \rangle$

$$= \frac{B^2\pi^2r^4\omega^2}{4R}\cdot\frac{1}{2} \quad \left[\because \langle\sin^2\omega t\rangle = \frac{1}{2}\right]$$

$$= \frac{(B\pi r^2\omega)^2}{8R}$$

Self-inductance

It is the property of a coil by virtue of which the coil opposes any change in the strength of current flowing through it by inducing an emf in itself.

This induced emf is also called *back emf*. When the current in a coil is switched on, then the self-induction opposes the growth of the current and when it is switched off, then the self-induction opposes the decay of the current. Hence, self-inductance is also known as *inertia of electricity*.

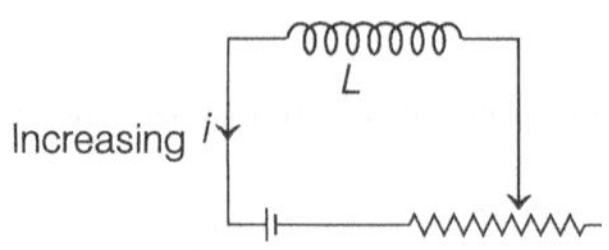

For a coil of N turns, the total flux ($N\phi_B$) linked with the coil is directly proportional to the current (i) in the coil. *i.e.*

$$N\phi_B \propto i \quad \text{or} \quad N\phi_B = Li \quad \text{or} \quad L = \frac{N\phi_B}{i}$$

L is coefficient of self-inductance of coil.

The induced emf e produced in the coil is directly proportional to the rate of change of current. Thus,

$$e = -L\frac{di}{dt}$$

The minus sign here is a reflection of Lenz's law. It says that the self induced emf in a circuit opposes any change in the current in that circuit. From the above equation,

$$L = \left|\frac{-e}{di/dt}\right|$$

Following are few important results for self-inductance

- Self-inductance of a circular coil of radius r,

$$L = \frac{\mu_0\pi N^2 r}{2}$$

$$\Rightarrow \quad L \propto N^2$$

$$\Rightarrow \quad \frac{L_1}{L_2} = \left(\frac{N_1}{N_2}\right) \quad \text{(For constant } r\text{)}$$

- Self-inductance for a square coil of side a,

$$L = \frac{2\sqrt{2}\mu_0 N^2 a}{\pi}$$

$$\Rightarrow \quad L \propto N^2$$

- Self-inductance of a solenoid $L = \frac{\mu_0 N^2 A}{l} = \mu_0 n^2 Al$

  where, N = total number of turns in the solenoid,
  l = length of the coil,
  n = number of turns in the coil
  and A = area of cross-section of the coil.

- If core of the solenoid is of any other magnetic material, then

$$L = \frac{\mu_0\mu_r N^2 A}{l}$$

- Self-inductance of a toroid, $L = \frac{\mu_0 N^2 A}{2\pi r}$

  where, r = radius of the toroid.

- Self-inductance of a toroid with radius r,

$$L = \frac{\mu_0 N^2 r}{2}$$

Kirchhoff's Second Law with an Inductor

According to Kirchhoff's second law (loop rule), when we go through an inductor in the same direction as the assumed current, we encounter a voltage drop equal to $L\,di/dt$, where di/dt is to be substituted with sign.

e.g. In the loop shown in figure, Kirchhoff's second law gives the equation,

$$\Rightarrow \quad E - iR - L\frac{di}{dt} = 0$$

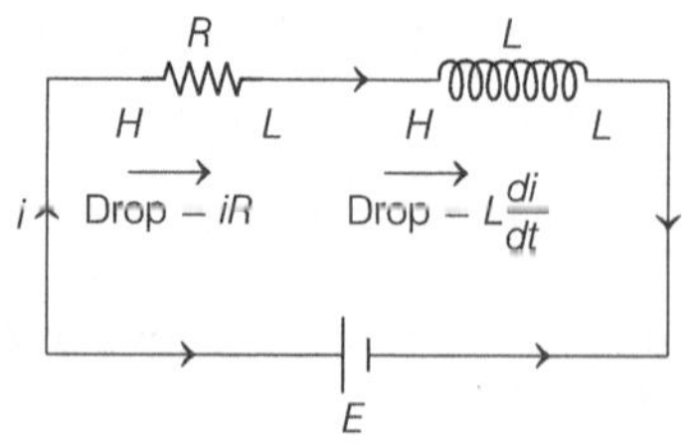

Example 6. *In the circuit diagram shown in figure, $R = 10\ \Omega$, $L = 5H$, $E = 20\ V$, $i = 2\ A$. This current is decreasing at a rate of $-1.0\ A/s$. Find V_{ab} at this instant.*

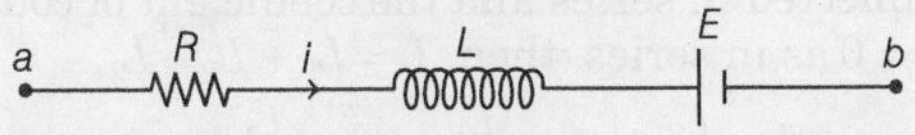

(a) 20 V (b) 35 V

(c) 40 V (d) 25 V

Sol. (b) PD across inductor, $V_L = L\dfrac{di}{dt} = (5)(-1.0) = -5\ V$

Now, using Kirchhoff's law, $V_a - iR - V_L - E = V_b$

$\therefore \quad V_{ab} = V_a - V_b = E + iR + V_L$
$= 20 + (2)(10) - 5 = 35\ V$

Note As the current is decreasing the inductor can be replaced by a source of emf, $e = \left|L \cdot \dfrac{di}{dt}\right| = 5\ V$ in such a manner that this emf supports the decreasing current or it sends the current in the circuit in the same direction as the existing current. So, positive terminal of this source is towards *b*. Thus, the given circuit can be drawn as

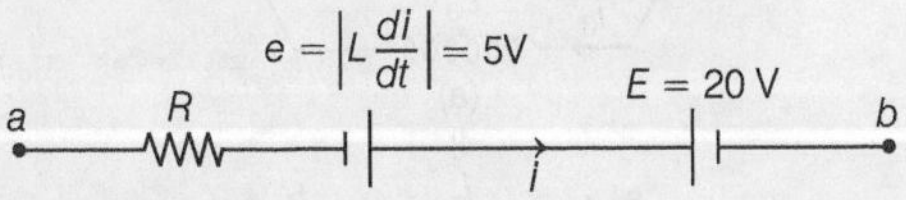

Now, we can find V_{ab}.

Example 7. *A copper wire is wound on a wooden frame, whose shape is that of an equilateral triangle. If the linear dimension of each side of the frame is increased by a factor of 3, keeping the number of turns of the coil per unit length of the frame the same, then the self-inductance of the coil*

(a) increases by a factor of 3 **[JEE Main 2019]**

(b) decreases by a factor of $9\sqrt{3}$

(c) increases by a factor of 27

(d) decreases by a factor of 9

Sol. (c) Self-inductance of a coil is given by the relation

$$L = \mu_0 n^2 A \cdot l$$

where, *n* is number of turns per unit length. Shape of the wooden frame is equilateral triangle.

$\therefore$ Area of equilateral triangle, $A = \dfrac{\sqrt{3}}{4}a^2$

(where, a is side of equilateral triangle)

$\therefore$ Self-inductance, $L = \mu_0 n^2 \left[\dfrac{\sqrt{3}}{4}a^2\right] \cdot l$

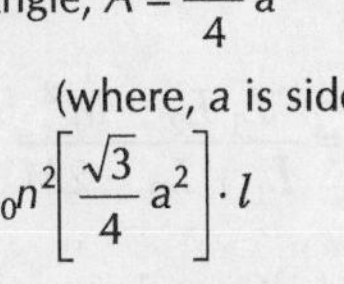

Here, $l = 3a \times N$ (where, *N* is total turns)

$$\therefore \quad L = \mu_0 n^2 \left[\frac{\sqrt{3}}{4}a^2\right] \times 3aN \text{ or } L \propto a^3$$

When each side of frame is increased by a factor 3 keeping the number of turns per unit length of the frame constant,

then $\quad a' = 3a$

$\therefore \quad L' \propto (a')^3$ or $L' \propto (3a)^3$ or $L' \propto 27a^3$ or $L' = 27L$

So, self-inductance will becomes 27 times.

Mutual Inductance

The phenomenon according to which an opposing emf is produced in a coil as a result of change in current or magnetic flux linked with a neighbouring coil is called *mutual inductance.*

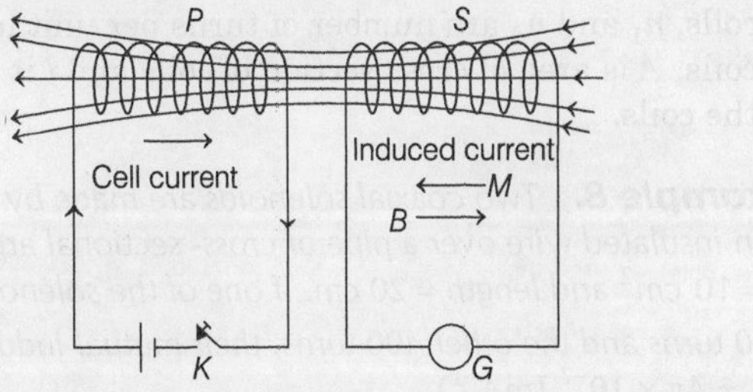

Suppose the circuit 1 has a current i_1 flowing in it. Then total flux $N_2\phi_{B_2}$ linked with circuit 2 is proportional to the current in circuit 1. Thus,

$$N_2 \phi_{B_2} \propto i_1$$

or $$N_2\phi_{B_2} = Mi_1$$

Here, the proportionality constant *M* is known as the mutual inductance *M* of the two circuits.

Thus, $$M = \frac{N_2\phi_{B_2}}{i_1}$$

From the above expression, *M* can be defined as the total flux linked with circuit 2 per unit current in circuit 1. If we change the current in circuit 1 at a rate di_1/dt, an induced emf e_2 is developed in circuit 1, which is proportional to the rate di_1/dt.

Thus, $$e_2 \propto di_1/dt$$

or $$e_2 = -Mdi_1/dt$$

or $$M = \left|\frac{-e_2}{di_1/dt}\right|$$

So, the mutual inductance of two circuits is the magnitude of induced emf e_2 per unit rate of change of current di_1/dt.

The SI unit of mutual inductance is henry (H).

M depends upon closeness of the two circuits, their orientations and sizes and the number of turns, etc.

Reciprocity theorem $M_{21} = M_{12} = M$

As, $\quad e_2 = -M(di_1/dt)$

and $e_1 = -M(di_2/dt)$

$$M_{12} = \frac{N_2 \phi_{B_2}}{i_1} \quad \text{and} \quad M_{21} = \frac{N_1 \phi_{B_1}}{i_2}$$

Following are few important results for mutual inductance

- Mutual inductance for two concentric coils having N_1 and N_2 turns with radius r and $R (R >> r)$, respectively is

$$M = \frac{\pi \mu_0 N_1 N_2 r^2}{2R} \Rightarrow M \propto \frac{r^2}{R}$$

- Mutual inductance for two long coaxial solenoids is

$$M = \frac{\mu_0 N_1 N_2 A}{l} = \mu_0 n_1 n_2 Al$$

where, N_1 and N_2 are total number of turns in both coils, n_1 and n_2 are number of turns per unit length in coils, A is area of cross-section of coils and l is length of the coils.

Example 8. *Two coaxial solenoids are made by winding thin insulated wire over a pipe of cross- sectional area $A = 10\ cm^2$ and length = 20 cm. If one of the solenoids has 300 turns and the other 400 turns, their mutual inductance is ($\mu_0 = 4\pi \times 10^{-7}\ TmA^{-1}$)*

(a) $2.4\pi \times 10^{-5}$ H (b) $4.8\pi \times 10^{-4}$ H

(c) $4.8\pi \times 10^{-5}$ H (d) $2.4\pi \times 10^{-4}$ H

Sol. (d) $M = \dfrac{\mu_0 N_1 \times N_2 \times A}{l}$

where, $N_1 = 300$ turns, $N_2 = 400$ turns,
$A = 10\ cm^2 = 10 \times 10^{-4} m^2$ and $l = 20\ cm = 20 \times 10^{-2} m$

Substituting the values in the given formula, we get

$$M = 2.4\pi \times 10^{-4}\ H$$

Coefficient of Coupling

The coefficient of coupling of two coils gives a measure of the manner in which the two coils are coupled together. If L_1 and L_2 are the self-inductances of two coils and M is their mutual inductance, then their coefficient of coupling is given by

$$K = \frac{M}{\sqrt{L_1 L_2}}$$

It is also defined as

$$K = \frac{\text{magnetic flux linked in secondary coil}}{\text{magnetic flux linked in primary coil}}$$

where, $0 \le K \le 1$.

Coefficient of coupling is maximum ($K = 1$) in case (a), when coils are coaxial and minimum in case (b), when coils are placed a right angles.

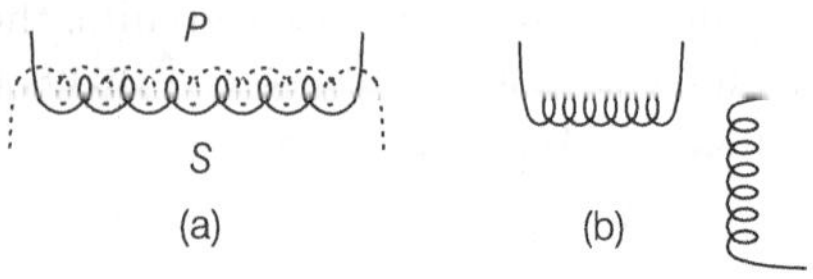

Combination of Inductances

If coefficient of coupling $k = 0$, then

(i) When three coils of inductances L_1, L_2 and L_3 are connected in series and the coefficient of coupling $K = 0$ as in series, then $L = L_1 + L_2 + L_3$

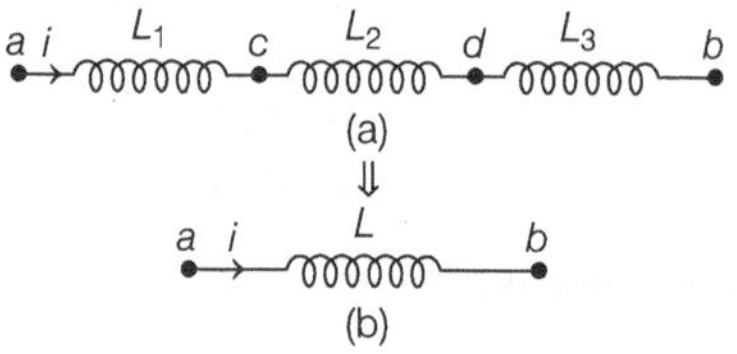

(ii) When three coils of inductances L_1, L_2 and L_3 are connected in parallel and the coefficient of coupling $K = 0$ as in parallel, then

$$\frac{1}{L} = \frac{1}{L_1} + \frac{1}{L_2} + \frac{1}{L_3}$$

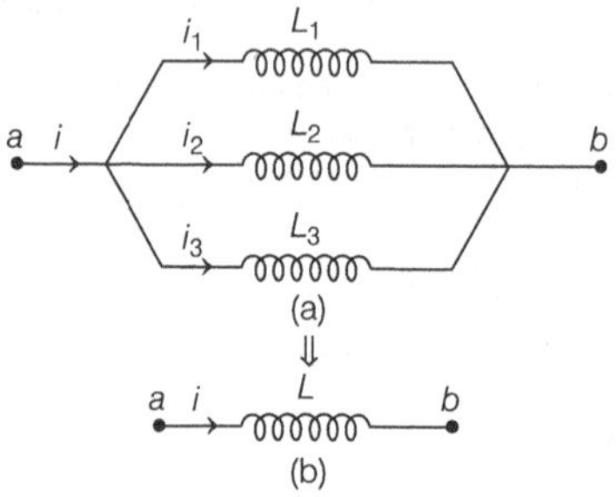

If coefficient of coupling $K = 1$, then

(i) **In series**

(a) If current in two coils are in the same direction, then

$$L = L_1 + L_2 + 2M$$

(b) If current in two coils are in opposite directions, then

$$L = L_1 + L_2 - 2M$$

(ii) **In parallel**

(a) If current in two coils are in same direction, then

$$L = \frac{L_1 L_2 - M^2}{L_1 + L_2 + 2M}$$

(b) If current in two coils are in opposite directions, then

$$L = \frac{L_1 L_2 - M^2}{L_1 + L_2 - 2M}$$

Example 9. *The inductance between A and D is*

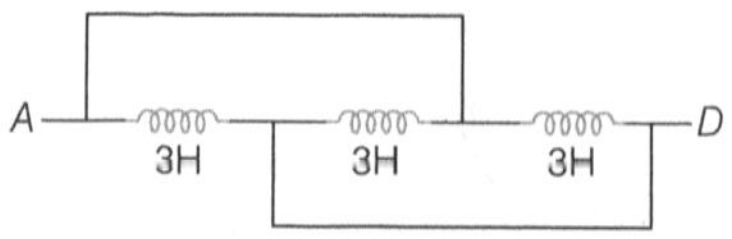

(a) 3.66 H (b) 9 H

(c) 0.66 H (d) 1 H

Sol. (d)

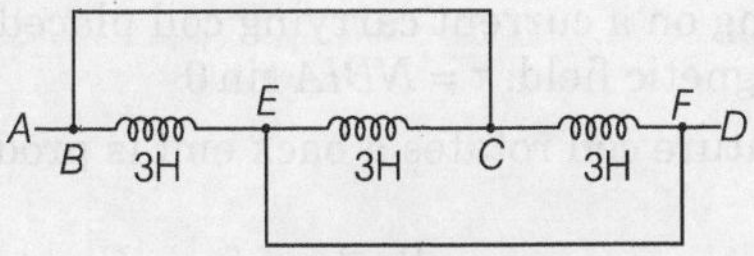

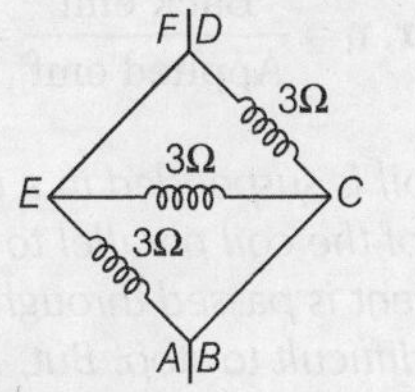

Here, inductors are in parallel.

$\therefore \quad \frac{1}{L} = \frac{1}{3} + \frac{1}{3} + \frac{1}{3}$ or $L = 1$ H

Energy Stored in an Inductor

The total energy stored by the inductor when the current in the circuit increases from zero to final value i is given as

$$U = \int_0^i Lidi = \frac{1}{2}Li^2;\text{ Also } U = \frac{1}{2}(Li)i = \frac{N\phi i}{2}$$

Also, energy density is given by $u_B = \frac{1}{2}\frac{B^2}{\mu_0}$

Energy density is property of magnetic field.

Example 10. *The self-induced emf of a coil is 25 V. When the current in it is changed at uniform rate from 10 A to 25 A in 1s, the change in the energy of the inductance is* **[JEE Main 2019]**

(a) 437.5 J *(b)* 740 J *(c)* 637.5 J *(d)* 540 J

Sol. (a) Energy stored in an inductor of inductance L and current I is given by

$$E = \frac{1}{2}LI^2$$

When current is being changed from I_1 to I_2, change in energy will be

$$\Delta E = E_2 - E_1 = \frac{1}{2}LI_2^2 - \frac{1}{2}LI_1^2 \quad \text{...(i)}$$

As I_1 and I_2 are given, we need to find value of L.

Now, induced emf in a coil is

$$\varepsilon = L\frac{dI}{dt}$$

Here, $\varepsilon = 25$ V,

$dI = I_2 - I_1 = (25 - 10) = 15$ A and $dt = 1$s

$\Rightarrow \quad 25 = L \times \frac{15}{1}$ or $L = \frac{25}{15} = (5/3)$ H

Putting values of L, I_1 and I_2 in Eq. (i), we get

$$\Delta E = \frac{1}{2} \times \frac{5}{3} \times [25^2 - 10^2] = \frac{1}{2} \times \frac{5}{3} \times 525$$

$$\Delta E = 437.5 \text{ J}$$

Example 11. *A 10 H inductor is used in an induction cooker plate operating on a current of 20 A. How much ice at 0°C could be melted by the cooker if whole of energy is used to heat the ice? (Take, latent heat of ice = 22.6 × 10^3 J/kg)*

(a) 1 kg *(b)* 2.2 kg

(c) 0.88 kg *(d)* 0.9 kg

Sol. (c) Energy stored by a inductor $= \frac{1}{2}Li^2$

This energy is completely used in melting the ice.

Hence, $\frac{1}{2}Li^2 = mL_f$

where, L_f = latent heat of fusion.

Hence, mass of ice melted, $m = \frac{Li^2}{2L_f}$

Substituting the values, we have

$$m = \frac{(10)\,(20)^2}{2\,(2.26 \times 10^3)} = 0.88 \text{ kg}$$

Eddy Currents

The currents induced in bulk pieces of conductors when the magnetic flux linked with the conductor changes are known as eddy currents. These currents are always produced in a plane which is perpendicular to the direction of magnetic field. Eddy current shows both heating and magnetic effects.

Eddy currents can be reduced by

(i) Reducing the area of metallic core of transformers, electric motor and other such devices.

(ii) Using laminations of metal to make a metal core. The plane of the laminations must be arranged parallel to the magnetic field.

Various applications of eddy currents are as follows

(i) **Magnetic Breaking in Trains** Strong electromagnets are situated above the rails in some electrically powered trains. When electromagnets situated above rails are activated, the eddy currents induced in the rails opposes the motion of train.

(ii) **Electromagnetic Damping** Certain galvanometer have a fixed core made of non-magnetic metallic material. When the coil oscillates, the eddy currents generated in the core opposes the motion and bring the coil to rest quickly.

(iii) **Induction Furnace** It can be used to produce high temperatures and can be utilised to prepare alloys, by melting the constituent metals.

(iv) **Electric Power Meters** Electric currents are induced in the shiny metal disc by magnetic fields produced by sinusoidally varying currents in a coil.

AC Generator or Dynamo

It is a device which converts mechanical energy into alternating current energy. Its working is based on electromagnetic induction.

The induced emf produced by the AC generator is given by

$$e = NBA\omega \sin \omega t = e_0 \sin \omega t$$

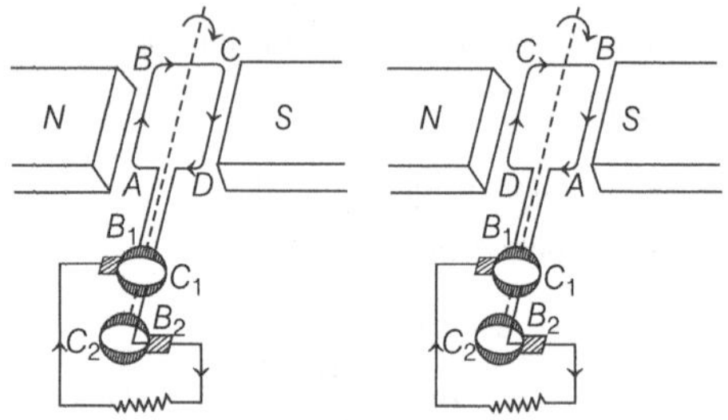

Working of AC dynamo

There are four main parts of an AC generator

(i) **Armature** It is rectangular coil of insulated copper wire having a large number of turns.

(ii) **Field Magnets** These are two pole pieces of a strong electromagnet.

(iii) **Slip Rings** These are two hollow metallic rings.

(iv) **Brushes** These are two flexible metals or carbon rods, which remains slightly in contact with slip rings.

DC Motor

It is a device which converts electrical energy into mechanical energy. Its working is based on the fact that when a current carrying coil is placed in uniform magnetic field a torque acts on it.

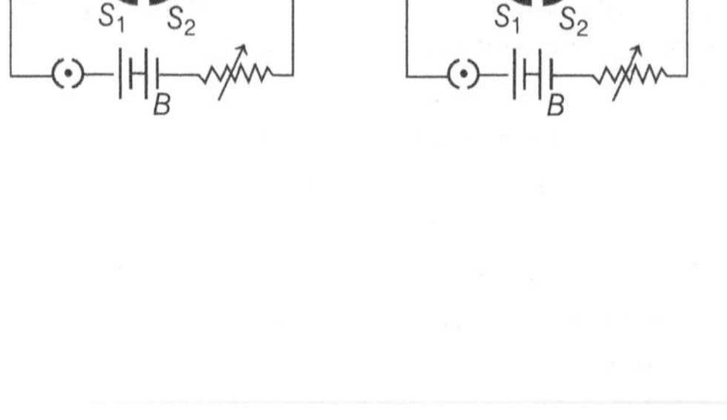

Torque acting on a current carrying coil placed in uniform magnetic field, $\tau = NBIA \sin \theta$

When armature coil rotates a back emf is produced in the coil,

$$\text{Efficiency of a motor, } \eta = \frac{\text{Back emf}}{\text{Applied emf}} = \frac{E}{V}$$

Example 12. *A coil is suspended in a uniform magnetic field, with the plane of the coil parallel to the magnetic lines of force. When a current is passed through the coil it starts oscillating; it is very difficult to stop. But, if an aluminium plate is placed near to the coil, it stops. This is due to*

(a) *development of air current when the plate is placed*

(b) *induction of electrical charge on the plate*

(c) *shielding of magnetic lines of force as aluminium is a paramagnetic material*

(d) *electromagnetic induction in the aluminium plate giving rise to electromagnetic damping*

Sol. (d) According to Lenz's law, electromagnetic induction takes place in the aluminium plate due to which eddy current is developed which oppose the motion or vibrations of coil.

This causes loss in energy which results in damping of oscillatory motion of the coil.

Example 13. *A planar loop of wire rotates in a uniform magnetic field. Initially at $t = 0$, the plane of the loop is perpendicular to the magnetic field. If it rotates with a period of 10s about an axis in its plane, then the magnitude of induced emf will be maximum and minimum respectively at* **[JEE Main 2020]**

(a) 2.5 s and 7.5 s (b) 2.5 s and 5.0 s

(c) 5.0 s and 10.0 s (d) 5.0 s and 7.5 s

Sol. (b) Induced emf in the wire loop, $e = BA\omega \sin \omega t$

Induced emf is maximum, when $\sin \omega t = \pm 1$

$$\Rightarrow \quad \omega t = \frac{\pi}{2}$$

$$\Rightarrow \quad \frac{2\pi}{T} \cdot t = \frac{\pi}{2}$$

Here, $T = 10$ s

So, $t = \frac{10}{4} = 2.5$ s

and induced emf is minimum, when $\sin \omega t = 0$

$$\Rightarrow \quad \omega t = \pi \Rightarrow \frac{2\pi}{T} \cdot t = \pi \Rightarrow t = \frac{10}{2} = 5 \text{ s}$$

Practice Exercise

ROUND I Topically Divided Problems

Magnetic Flux, Faraday's Law and Lenz's Law

1. A square of side L metres lies in the XY-plane in a region, where the magnetic field is given by $\mathbf{B} = B_0\,(2\hat{\mathbf{i}} + 3\hat{\mathbf{j}} + 4\hat{\mathbf{k}})$T, where B_0 is constant. The magnitude of flux passing through the square is **[NCERT Exemplar]**

(a) $2\,B_0\,L^2$ Wb (b) $3\,B_0\,L^2$ Wb
(c) $4\,B_0\,L^2$ Wb (d) $\sqrt{29}\,B_0\,L^2$ Wb

2. Consider a circular coil of wire carrying constant current I, forming a magnetic dipole. The magnetic flux through an infinite plane that contains the circular coil and excluding the circular area is given by ϕ_i. The magnetic flux through the area of the circular coil area is given by ϕ_o. Which of the following option is correct? **[JEE Main 2020]**

(a) $\phi_i > \phi_o$ (b) $\phi_i < \phi_o$
(c) $\phi_i = \phi_o$ (d) $\phi_i = -\phi_o$

3. The magnetic flux through each turn of a 100 turn coil is $(t^3 - 2t) \times 10^{-3}$ Wb, where t is in second. The induced emf at $t = 2$s is

(a) -4 V (b) -1 V
(c) $+1$ V (d) $+4$ V

4. A square loop of wire of side 5 cm is lying on a horizontal table. An electromagnet above and to one side of the loop is turned on, causing a uniform magnetic field downwards at an angle of 60° to the vertical as shown in figure. The magnetic induction is 0.50 T. The average induced emf in the loop, if the field increases from zero to its final value in 0.2 s is

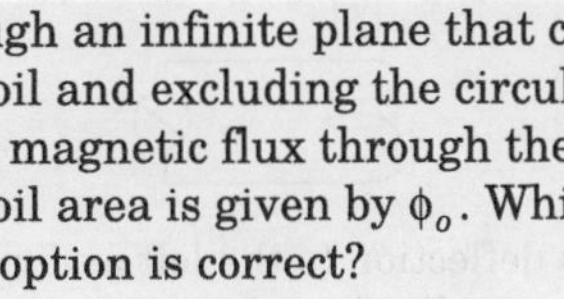

(a) 5.4×10^{-3} V (b) 3.12×10^{-3} V
(c) 0 (d) 0.25×10^{-3} V

5. A coil having n turns and resistance $R\ \Omega$ is connected with a galvanometer of resistance $4R\ \Omega$. This combination is moved in time t second from a magnetic field W_1 weber to W_2 weber. The induced current in the circuit is

(a) $\dfrac{W_2 - W_1}{5\,Rnt}$ (b) $-\dfrac{n(W_2 - W_1)}{5\,Rt}$
(c) $-\dfrac{(W_2 - W_1)}{Rnt}$ (d) $-\dfrac{n(W_2 - W_1)}{Rt}$

6. Some magnetic flux is changed from a coil of resistance 10Ω. As a result an induced current is developed in it, which varies with time as shown in figure. The magnitude of change in flux through the coil (in Wb) is

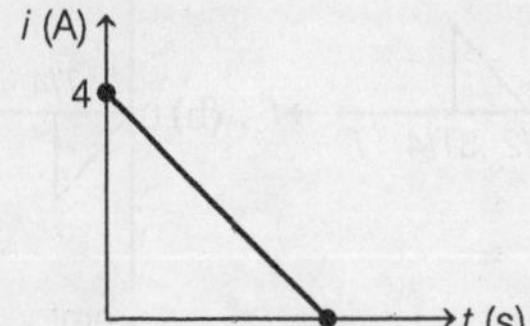

(a) 2 (b) 4 (c) 6 (d) 0

7. The magnetic flux through a circuit of resistance R changes by an amount $\Delta\phi$ in a time Δt, then the total quantity of electric charge Q that passes any point in the circuit during the time Δt is represented by

(a) $\dfrac{\Delta\phi}{R}$ (b) $\dfrac{\Delta\phi}{\Delta t}$ (c) $R.\dfrac{\Delta\phi}{\Delta t}$ (d) $\dfrac{1}{R}.\dfrac{\Delta\phi}{\Delta t}$

8. A circular ring of diameter 20 cm has a resistance of 0.01 Ω. The charge that will flow through the ring, if it is turned from a position perpendicular to a uniform magnetic field of 2.0 T to a position parallel to the field is about

(a) 63 C (b) 0.63 C (c) 6.3 C (d) 0.063 C

9. The charge which will flow through a 200 Ω galvanometer connected to a 400 Ω circular coil of 1000 turns wound on a wooden stick 20 mm in diameter, if a magnetic field $B = 0.012$ T parallel to the axis of the stick decreased suddenly to zero is

(a) 6.3 μC (b) 63 μC (c) 0.63 μC (d) 630 μC

10. The variation of induced emf (E) with time (t) in a coil, if a short bar magnet is moved along its axis with a constant velocity is best represented as

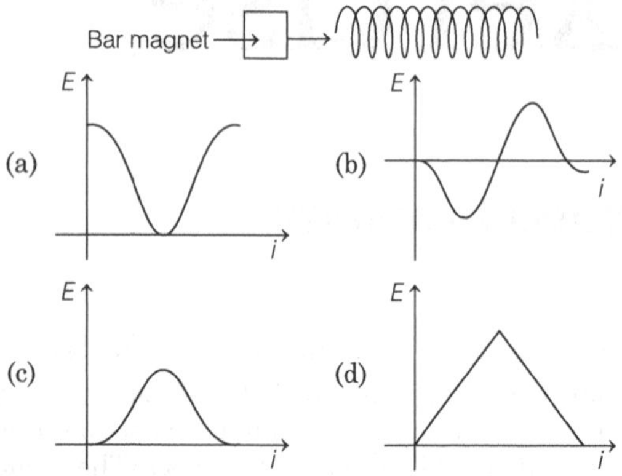

11. The current i in a coil varies with time as shown in the figure. The variation of induced emf with time would be

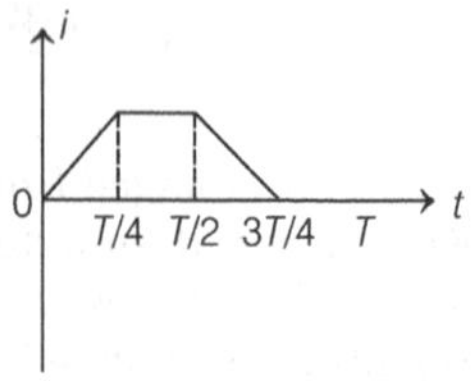

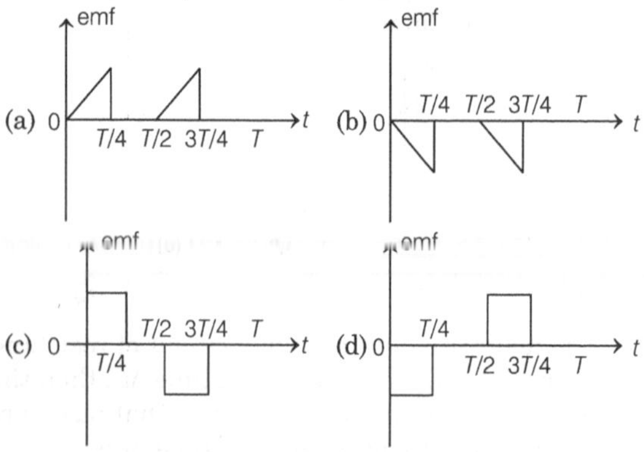

12. Plane figure made of thin wires of resistance $R = 50$ milli Ω/m are located in a uniform magnetic field perpendicular into the plane of the figures and which decrease at the rate $dB/dt = 0.1$ m T/s. Then currents in the inner and outer boundaries are (The inner radius a = 10 cm and outer radius b = 20 cm)

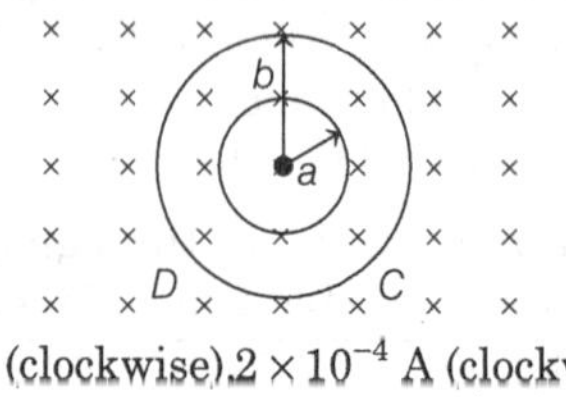

(a) 10^{-4} A (clockwise),2×10^{-4} A (clockwise)
(b) 10^{-4} A (anti-clockwise), 2×10^{-4} A (clockwise)
(c) 2×10^{-4} A (clockwise), 10^{-4} A (anti-clockwise)
(d) 2×10^{-4} A (anti-clockwise), 10^{-4} A (anti-clockwise)

13. There are two coils A and B as shown in figure. A current starts flowing in B as shown, when A is moved towards B and stops when A stops moving. The current in A is counter clockwise. B is kept stationary when A moves. We can infer that

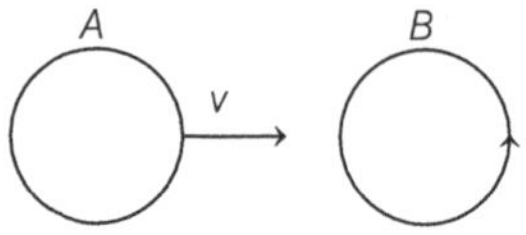

[NCERT Exemplar]

(a) there is a constant current in the clockwise direction in A
(b) there is a varying curent in A
(c) there is no current in A
(d) there is a constant current in the counter clockwise direction in A

14. A magnet N-S is suspended from a spring and while it oscillates, the magnet moves in and out of the coil C. The coil is connected to a galvanometer G, then as the magnet oscillates,

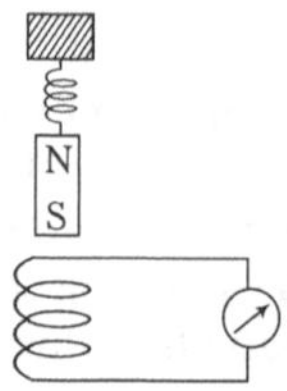

(a) G shows deflection to the left and right with constant amplitude
(b) G shows deflection on one side
(c) G shows no deflection
(d) G shows deflection to the left and right but the amplitude steadily decreases

Motional EMF

15. The figure shows four wire loops, with edge lengths of either L or $2L$. All four loops will move through a region of uniform magnetic field **B** (directed out of the page) at the same constant velocity. Rank the four loops according to the maximum magnitude of the emf induced as they move through the field, greatest first

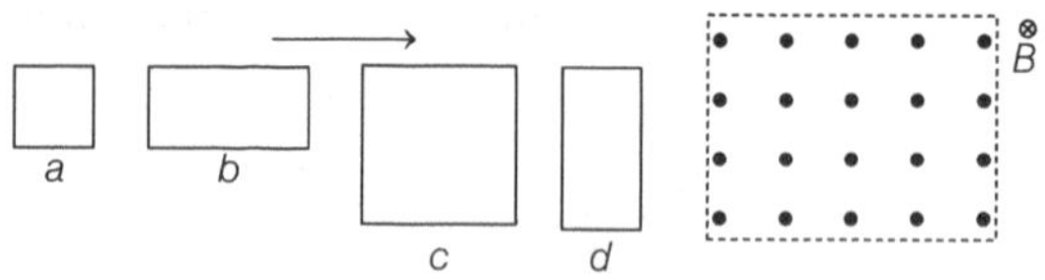

(a) $(e_c = e_d) < (e_a = e_b)$ (b) $(e_c = e_d) > (e_a = e_b)$
(c) $e_c > e_d > e_b > e_a$ (d) $e_c < e_d < e_b < e_a$

16. A thin semi-circular conducting ring of radius R is falling with its plane vertical in a horizontal magnetic induction **B**. At the position MNQ, the speed of the ring is v. The potential difference developed across the ring is

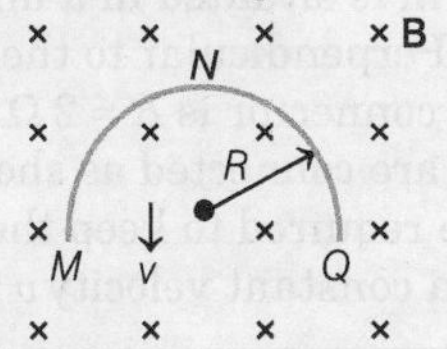

(a) zero

(b) $\frac{1}{2}Bv\pi R^2$ and M is at a higher potential

(c) πRBv and Q is at a higher potential

(d) $2RBv$ and Q is at a higher potential

17. One conducting U-tube can slide inside another as shown in figure, maintaining electrical contacts between the tubes. The magnetic field B is perpendicular to the plane of the figure. If each tube moves towards the other at a constant speed v, then the emf induced in the circuit in terms of B, l and v, where l is the width of each tube, will be

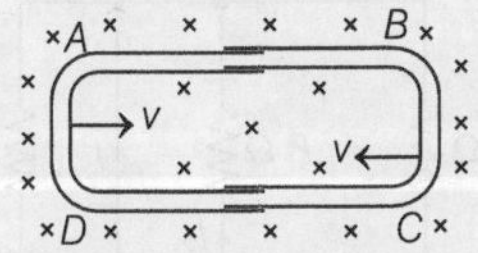

(a) Blv (b) $-Blv$ (c) zero (d) $2\,Blv$

18. An airplane with 20 m wing spread is flying at 250 ms^{-1} straight south parallel to the earth's surface. The earth's magnetic field has a horizontal component of 2×10^{-5} Wbm^{-2} and the dip angle is 60°. Calculate the induced emf between the plane tips.

(a) 0.174 V (b) 0.173 V (c) 1.173 V (d) 0.163 V

19. A square loop of side 5 cm enters a magnetic field with 1 cms^{-1}. The front edge enters the magnetic field at $t = 0$, then which graph best depicts emf variation correctly?

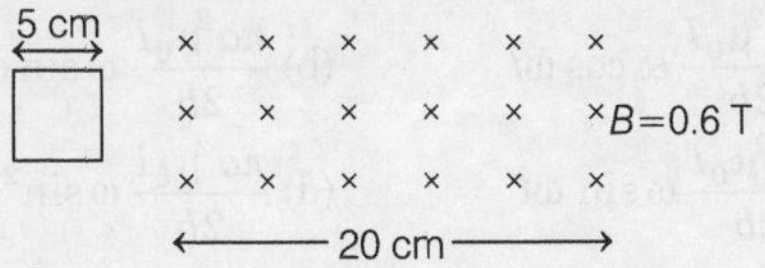

(a)

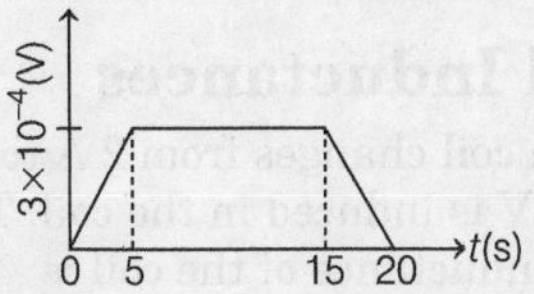

(b)

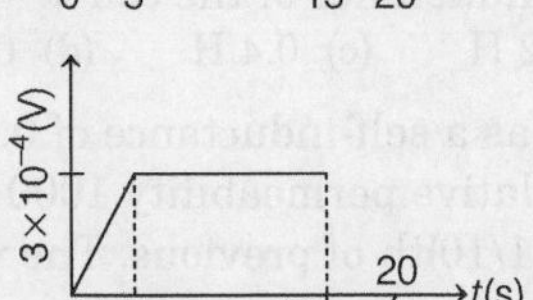

(c)

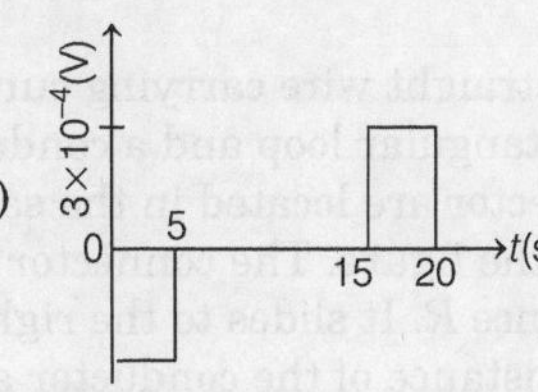

(d)

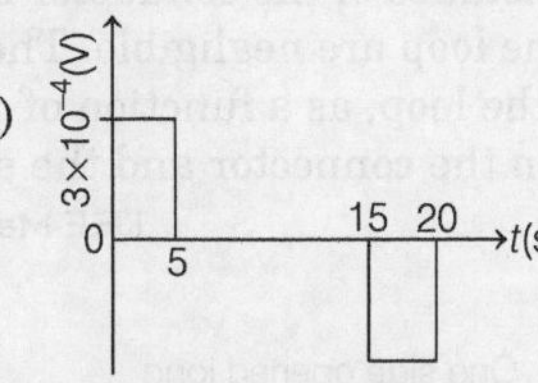

20. The magnetic field in a region is given by $B = B_0\left(\frac{x}{a}\right)\hat{\mathbf{k}}$. A square loop of side d is placed with its edges along the X and Y-axes. The loop is moved with a constant velocity $\mathbf{v} = v_0\hat{\mathbf{i}}$, then emf induced in the loop is **[JEE Main 2021]**

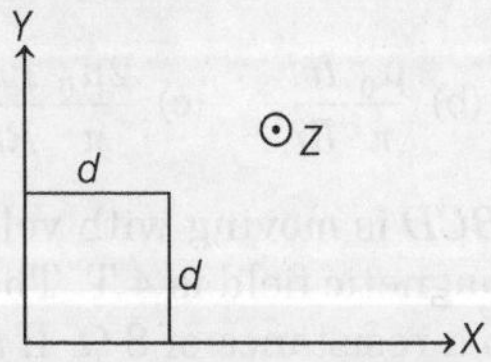

(a) $\frac{B_0 v_0^2 d}{2a}$ (b) $\frac{B_0 v_0 d}{2a}$

(c) $\frac{B_0 v_0 d^2}{a}$ (d) $\frac{B_0 v_0 d^2}{2a}$

21. The rails of a railway track insulated from each other and the ground are connected to a millivoltmeter. Find the reading of voltmeter, when a train travels with a speed of 180 km/h along the track. Given that, the vertical component of earth magnetic field is 0.2×10^{-4} Wb/m^2 and the rails are separated by 1m.

(a) 10^{-4} V (b) 10^{-2} V (c) 10^{-3} V (d) 1 V

22. A conducting rod PQ of length $L = 1.0$ m is moving with a uniform speed $v = 2.0\ ms^{-1}$ in a uniform magnetic field $B = 4.0$ T directed into the paper. A capacitor of capacity $C = 10\ \mu F$ is connected as shown in figure, then

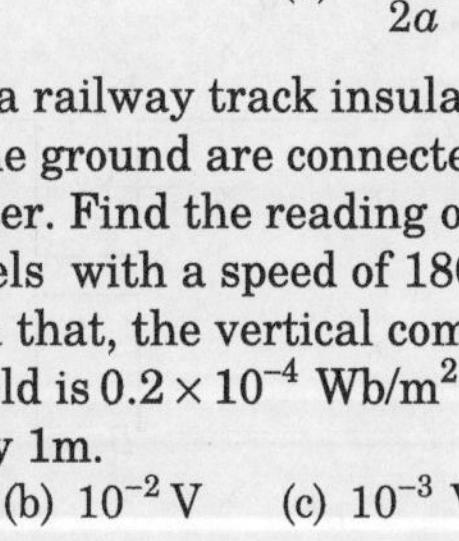

(a) $q_A = -80\ \mu C$ and $q_B = +80\ \mu C$

(b) $q_A = +80\ \mu C$ and $q_B = -80\ \mu C$

(c) $q_A = 0 = q_c$

(d) charge stored in the capacitor increases exponentially with time

23. An infinitely long straight wire carrying current I, one side opened rectangular loop and a conductor C with a sliding connector are located in the same plane, as shown in the figure. The connector has length l and resistance R. It slides to the right with a velocity v. The resistance of the conductor and the self-inductance of the loop are negligible. The induced current in the loop, as a function of separation r between the connector and the straight wire is **[JEE Main 2020]**

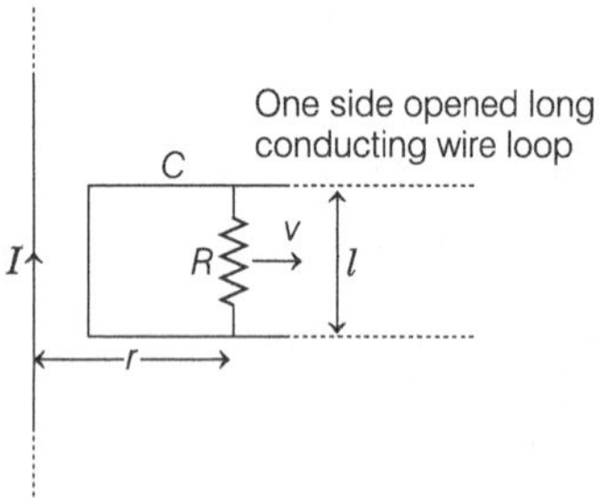

(a) $\frac{\mu_0}{4\pi}\frac{Ivl}{Rr}$ (b) $\frac{\mu_0}{\pi}\frac{Ivl}{Rr}$ (c) $\frac{2\mu_0}{\pi}\frac{Ivl}{Rr}$ (d) $\frac{\mu_0}{2\pi}\frac{Ivl}{Rr}$

24. The loop $ABCD$ is moving with velocity v towards right. The magnetic field is 4 T. The loop is connected to a resistance of 8 Ω. If steady current of 2 A flows in the loop, then value of v, if loop has resistance of 4 Ω, is (Given, $AB = 30$ cm, $AD = 30$ cm, $\sin 37° = \frac{3}{5}$)

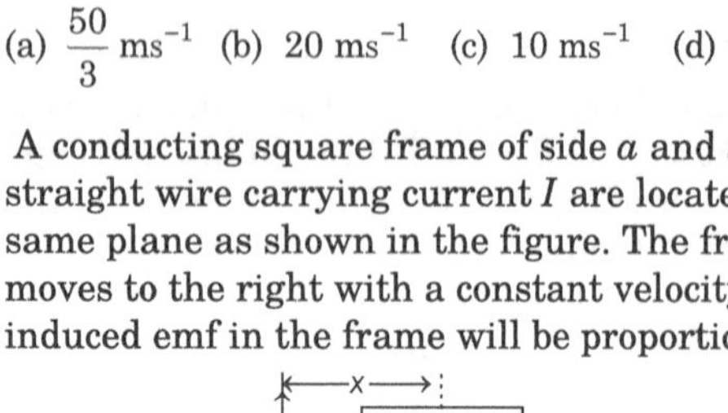

(a) $\frac{50}{3}$ ms^{-1} (b) 20 ms^{-1} (c) 10 ms^{-1} (d) $\frac{100}{3}$ ms^{-1}

25. A conducting square frame of side a and a long straight wire carrying current I are located in the same plane as shown in the figure. The frame moves to the right with a constant velocity v. The induced emf in the frame will be proportional to

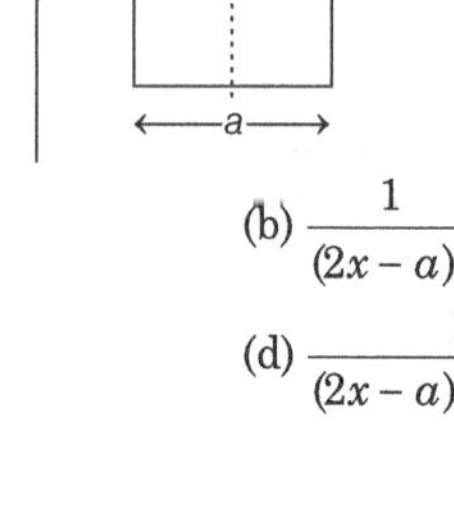

(a) $\frac{1}{x^2}$ (b) $\frac{1}{(2x-a)^2}$

(c) $\frac{1}{(2x+a)^2}$ (d) $\frac{1}{(2x-a)(2x+a)}$

26. A rectangular loop with a sliding connector of length $l = 1.0$ m is situated in a uniform magnetic field $B = 2$ T. Perpendicular to the plane of loop. Resistance of connector is $R = 2\ \Omega$. Two resistance of 6 Ω and 3Ω are connected as shown in figure. The external force required to keep the connector moving with a constant velocity $v = 2\ \text{ms}^{-1}$ is

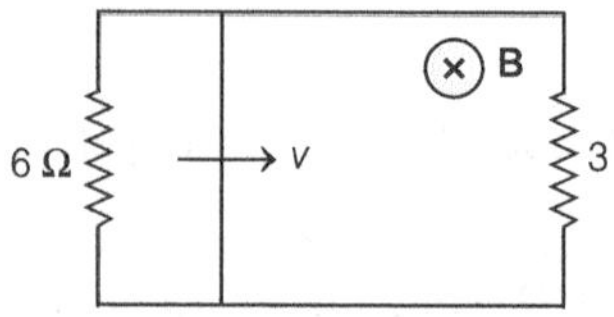

(a) 2 N (b) 1 N (c) 4 N (d) 6 N

27. A rectangular loop has a sliding connector PQ of length l and resistance R Ω and it is moving with a speed v as shown. The set up is placed in a uniform magnetic field going into the plane of the paper. The three currents I_1, I_2 and I are

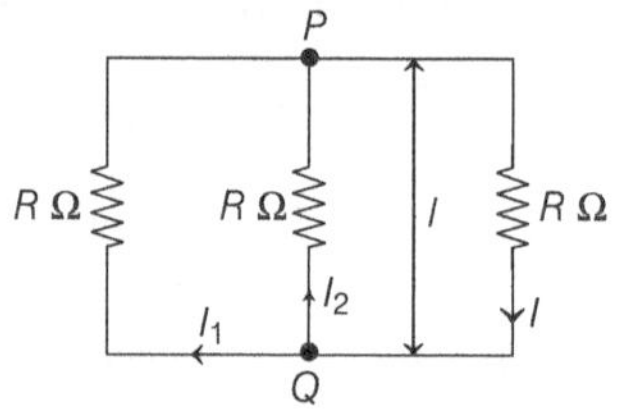

(a) $I_1 = I_2 = \frac{Blv}{6R}, I = \frac{Blv}{3R}$ (b) $I_1 = -I_2 = \frac{Blv}{R}, I = \frac{2Blv}{R}$

(c) $I_1 = I_2 = \frac{Blv}{3R}, I = \frac{2Blv}{3R}$ (d) $I_1 = I_2 = I = \frac{Blv}{R}$

28. A very small circular loop of radius a is initially (at $t = 0$) co-planar and concentric with a much larger fixed circular loop of radius b. A constant current I flows in the larger loop. The smaller loop is rotated with a constant angular speed ω about the common diameter. The emf induced in the smaller loop as a function of time t is

(a) $\frac{\pi a^2 \mu_0 I}{2b}\omega\cos\omega t$ (b) $\frac{\pi a^2 \mu_0 I}{2b}\omega\sin\omega^2 t^2$

(c) $\frac{\pi a^2 \mu_0 I}{2b}\omega\sin\omega t$ (d) $\frac{\pi a^2 \mu_0 I}{2b}\omega\sin^2\omega t$

Self and Mutual Inductances

29. When current in a coil changes from 2 A to −2 A in 0.05s, an emf of 8 V is induced in the coil. The coefficient of self-inductance of the coil is

(a) 0.1 H (b) 0.2 H (c) 0.4 H (d) 0.8 H

30. An air cored coil has a self-inductance of 0.1 H. A soft iron core of relative permeability 1000 has number of coil as (1/10)th of previous. The value of self-inductance now becomes

(a) 1 mH (b) 10 mH (c) 1 H (d) 10 H

31. A coil is wound on a core of rectangular cross-section. If all the linear dimensions of core are increased by a factor 2 and number of turns per unit length of coil remains same, the self-inductance increases by a factor of

(a) 16 (b) 8 (c) 4 (d) 2

32. The self-inductance L of a solenoid of length l and area of cross-section A, with a fixed number of turns N increases as **[NCERT Exemplar]**

(a) Both l and A increase
(b) l decreases and A increases
(c) l increases and A decreases
(d) Both l and A decrease

33. A coil of wire of certain radius has 100 turns and a self-inductance of 15 mH. The self-inductance of a second similar coil of 500 turns will be

(a) 75 mH (b) 375 mH (c) 15 mH (d) 400 mH

34. Two coils have mutual inductance 0.005 H. The current changes in the first coil according to equation $i = i_0 \sin \omega t$, where $i_0 = 10$ A and $\omega = 100\pi$ rads^{-1}. The maximum value of emf in second coil is

(a) 2π (b) 5π (c) π (d) 4π

35. A current of 10 A in the primary coil of a circuit is reduced to zero. If the coefficient of mutual inductance is 3H and emf induced in secondary coil is 30 kV, time taken for the change of current is

(a) 10^3 s (b) 10^2 s (c) 10^{-3} s (d) 10^{-2} s

36. A pair of adjacent coils has a mutual inductance of 1.5 H. If the current in one coil changes from 0 to 20 A in 0.5 s, what is the change of flux linkage with the other coil?

(a) 30 Wb (b) 33 Wb (c) 23 Wb (d) 42 Wb

37. Two inductors L_1 and L_2 are connected in parallel and a time varying current flows as shown in figure. The ratio of current, $\frac{i_1}{i_2}$ at any time t is

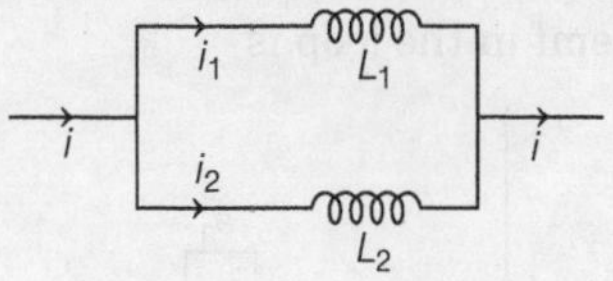

(a) $\frac{L_2}{L_1}$ (b) $\frac{L_1}{L_2}$

(c) $\frac{L_2^2}{(L_1 + L_2)^2}$ (d) $\frac{L_1^2}{(L_1 + L_2)^2}$

38. A rectangular coil $ABCD$ which is rotated at a constant angular velocity about an horizontal as shown in the figure. The axis of rotation of the coil as well as the magnetic field B are horizontal. Maximum current will flow in the circuit when the plane of the coil is

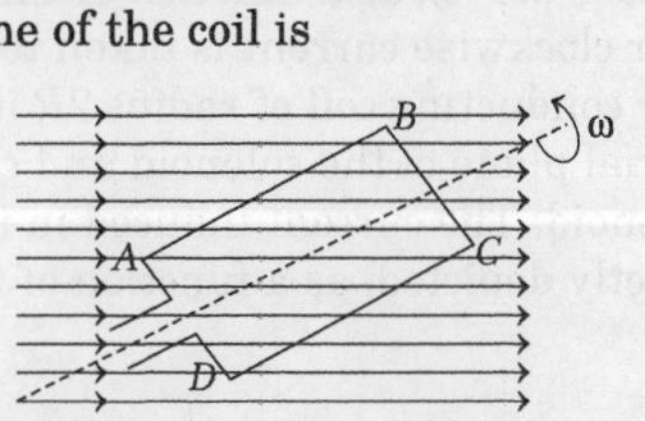

(a) inclined at 30° to the magnetic field
(b) perpendicular to the magnetic field
(c) inclined at 45° to the magnetic field
(d) parallel to the magnetic field

ROUND II Mixed Bag

Only One Correct Option

1. When a sheet of metal is placed in a magnetic field, which changes from zero to a maximum value, the induced currents are set up in the direction shown in figure. What is the direction of magnetic field?

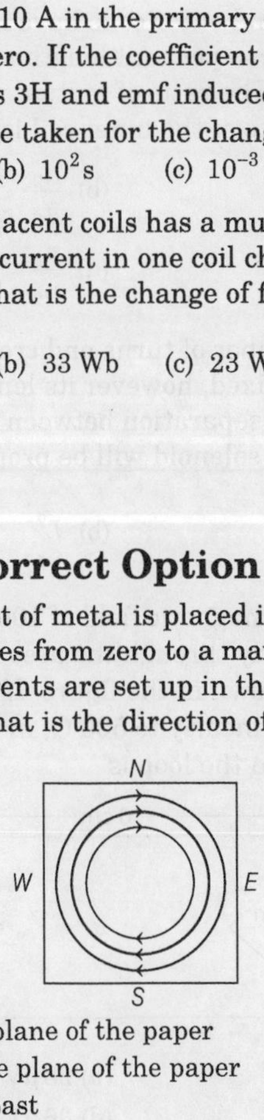

(a) Into the plane of the paper
(b) Out of the plane of the paper
(c) West to East
(d) South to North

2. A copper rod of mass m slides under gravity on two smooth parallel rails l distance apart and set at an angle θ to the horizontal. At the bottom, the rails are joined by a resistance R (figure). There is a uniform magnetic field B perpendicular to the plane of the rails. The terminal velocity of the rod is

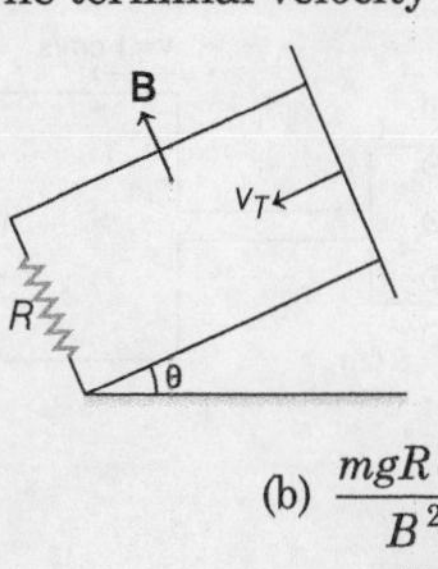

(a) $\frac{mgR \tan \theta}{B^2 l^2}$ (b) $\frac{mgR \cot \theta}{B^2 l^2}$

(c) $\frac{mgR \sin \theta}{B^2 l^2}$ (d) $\frac{mgR \cos \theta}{B^2 l^2}$

3. A square loop of side a placed in the same plane as a long straight wire carrying a current i. The centre of the loop is at a distance r from the wire, where $r >> a_1$ (see the figure). The loop is moved away from the wire with a constant velocity v. The induced emf in the loop is

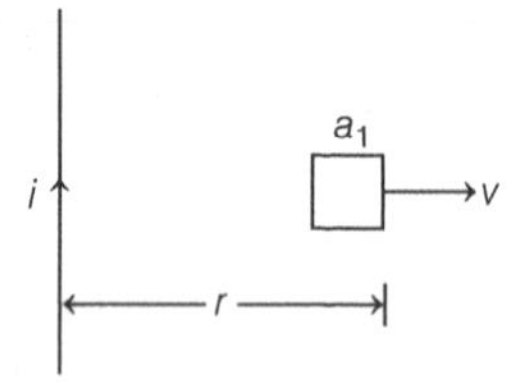

(a) $\frac{\mu_0\, iav}{2\pi r}$ (b) $\mu_0\, ia^3 v$

(c) $\frac{\mu_0\, iv}{2\pi}$ (d) $\frac{\mu_0\, ia^2 v}{2\pi r^2}$

4. A very long solenoid of radius R is carrying current $I(t) = kte^{-\alpha t}(k > 0)$, as a function of time $(t \geq 0)$. Counter clockwise current is taken to be positive. A circular conducting coil of radius $2R$ is placed in the equatorial plane of the solenoid and concentric with the solenoid. The current induced in the outer coil is correctly depicted, as a function of time, by **[JEE Main 2019]**

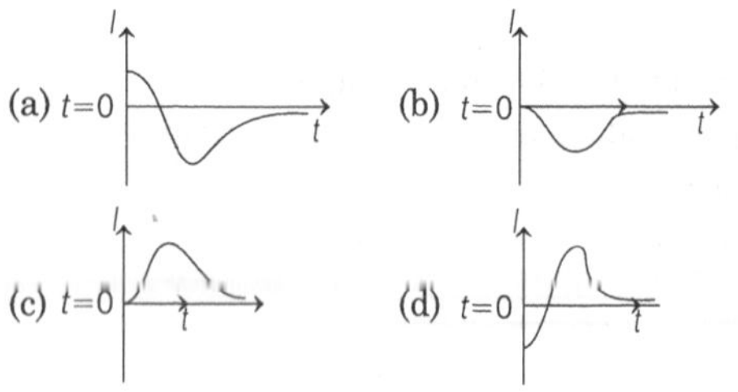

5. The figure shows a square loop L of side 5 cm which is connected to a network of resistances. The whole setup is moving towards right with a constant speed of 1 cm s^{-1}. At some instant, a part of L is in a uniform magnetic field of 1 T, perpendicular to the plane of the loop. If the resistance of L is 1.7 Ω, the current in the loop at that instant will be close to **[JEE Main 2019]**

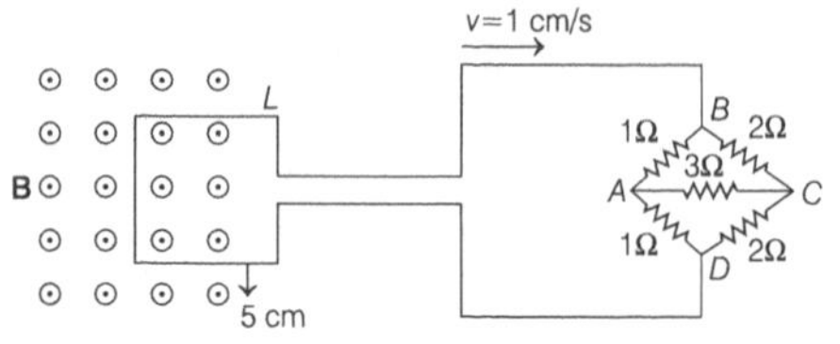

(a) 60 μA
(b) 170 μA
(c) 150 μA
(d) 115 μA

6. A long solenoid of radius R carries a time (t)-dependent current $I(t) = I_0 t(1-t)$. A ring of radius $2R$ is placed coaxially near its middle. During the time interval $0 \leq t \leq 1$, the induced current (I_R) and the induced emf (V_R) in the ring change as **[JEE Main 2020]**

(a) direction of I_R remains unchanged and V_R is maximum at $t = 0.5$ s
(b) at $t = 0.25$ s, direction of I_R reverses and V_R is maximum
(c) direction of I_R remains unchanged and V_R is zero at $t = 0.25$ s
(d) at $t = 0.5$ s, direction of I_R reverses and V_R is zero

7. Two coils P and Q are separated by some distance. When a current of 3 A flows through coil P, a magnetic flux of 10^{-3} Wb passes through Q and no current is passed through Q. When no current passes through P and a current of 2 A passes through Q, the flux through P is **[JEE Main 2019]**

(a) 6.67×10^{-3} Wb (b) 6.67×10^{-4} Wb
(c) 3.67×10^{-3} Wb (d) 3.67×10^{-4} Wb

8. There are two long coaxial solenoids of same length l. The inner and outer coils have radii r_1 and r_2 and number of turns per unit length n_1 and n_2, respectively. The ratio of mutual inductance to the self-inductance of the inner coil is **[JEE Main 2019]**

(a) $\frac{n_2}{n_1} \cdot \frac{r_1}{r_2}$ (b) $\frac{n_2}{n_1} \cdot \frac{r_2^2}{r_1^2}$

(c) $\frac{n_2}{n_1}$ (d) $\frac{n_1}{n_2}$

9. The total number of turns and cross-section area in a solenoid is fixed, however its length L is varied by adjusting the separation between windings. The inductance of solenoid will be proportional to **[JEE Main 2019]**

(a) $1/L$ (b) L^2
(c) L (d) $1/L^2$

10. At time $t = 0$, magnetic field of 1000 G is passing perpendicularly through the area defined by the closed loop shown in the figure. If the magnetic field reduces linearly to 500 G, in the next 5 s, then induced emf in the loop is **[JEE Main 2020]**

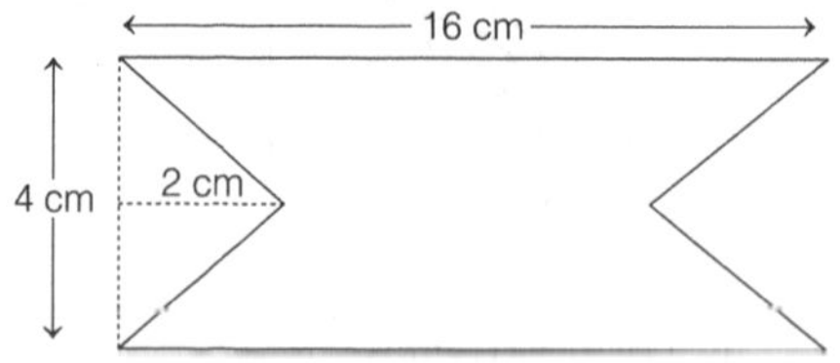

(a) 48 μV (b) 28 μV
(c) 56 μV (d) 36 μV

11. A conducting circular loop is made of a thin wire has area $3.5 \times 10^{-3} \text{m}^2$ and resistance 10 Ω. It is placed perpendicular to a time dependent magnetic field $B(t) = (0.4\text{T}) \sin(0.5\pi t)$. The field is uniform in space, then the net charge flowing through the loop during $t = 0$ s to $t = 10$ ms is close to **[JEE Main 2019]**

(a) 6 mC
(b) 21 mC
(c) 7 mC
(d) 14 mC

Numerical Value Questions

12. In a fluorescent lamp choke (a small transformer), 100 V of reverse voltage is produced when the choke current changes uniformly from 0.25A to 0 in a duration of 0.025 ms. The self-inductance of the choke (in mH) is estimated to be .... **[JEE Main 2020]**

13. A part of a complete circuit is shown in the figure. At some instant, the value of current I is 1 A and it is decreasing at the rate of 10^2 As^{-1}. The value of the potential difference $V_P - V_Q$ (in volt) at that instant, is ....

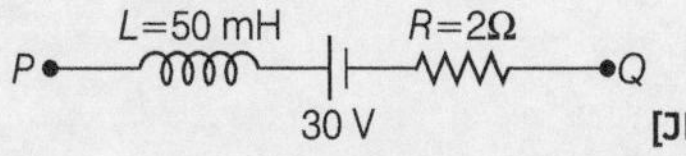

[JEE Main 2020]

14. Two concentric circular coils C_1 and C_2 are placed in the XY-plane. C_1 has 500 turns and radius of 1 cm. C_2 has 200 turns and radius of 20 cm. C_2 carries a time dependent current $I(t) = (5t^2 - 2t + 3)$A, where t is in second. The emf induced in C_1 (in mV), at the instant $t = 1$ s is $\frac{4}{x}$. The value of x is .... **[JEE Main 2020]**

15. A circular coil of radius 10 cm is placed in a uniform magnetic field of 3.0×10^{-5} T with its plane perpendicular to the field initially. It is rotated at constant angular speed about an axis along the diameter of coil and perpendicular to magnetic field, so that it undergoes half of rotation in 0.2s. The maximum value of emf induced (in μV) in the coil will be close to the integer .... **[JEE Main 2020]**

Answers

Round I

1. (c)	2. (d)	3. (b)	4. (b)	5. (b)	6. (a)	7. (a)	8. (c)	9. (a)	10. (b)
11. (d)	12. (a)	13. (d)	14. (d)	15. (b)	16. (d)	17. (d)	18. (b)	19. (c)	20. (c)
21. (c)	22. (b)	23. (d)	24. (d)	25. (d)	26. (a)	27. (c)	28. (c)	29. (b)	30. (c)
31. (b)	32. (b)	33. (b)	34. (b)	35. (c)	36. (a)	37. (a)	38. (d)		

Round II

1. (b)	2. (c)	3. (d)	4. (d)	5. (b)	6. (d)	7. (b)	8. (c)	9. (a)	10. (c)
11. (d)	12. 10	13. 33	14. 5	15. 15					

Solutions

Round I

1. Here, $\mathbf{A} = L^2\,\hat{\mathbf{k}}$ and $\mathbf{B} = B_0\,(2\hat{\mathbf{i}} + 3\hat{\mathbf{j}} + 4\hat{\mathbf{k}})$ T, $\phi = ?$

As, $\quad \phi = \mathbf{B}\cdot\mathbf{A} = B_0\,(2\hat{\mathbf{i}} + 3\hat{\mathbf{j}} + 4\hat{\mathbf{k}})\cdot L^2\hat{\mathbf{k}}$

$\therefore \quad \phi = 4\,B_0 L^2$ Wb

2. We are given with following situation

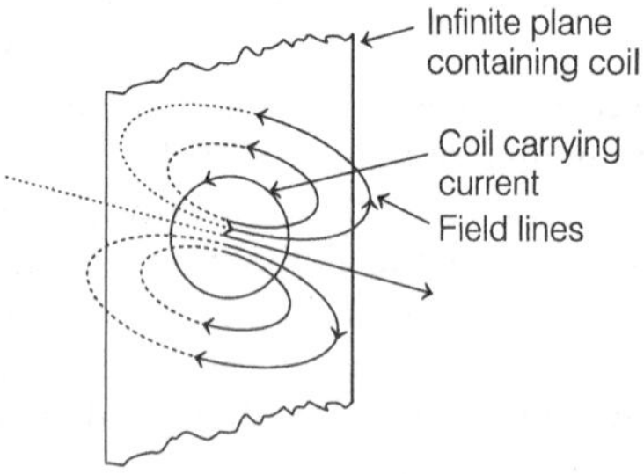

From the diagram of field lines, we can observe that whatever be the number of field lines emitted from coil, all of them goes back into the infinite plane only.

So, magnetic flux emanating from coil is equal and opposite to the flux linked with infinite plane.

So, $\quad \phi_i = -\phi_o$

3. Magnetic flux, $\phi = (t^3 - 2t) \times 10^{-3}$

$$\frac{d\phi}{dt} = (3t^2 - 2) \times 10^{-3}$$

$$\left.\frac{d\phi}{dt}\right|_{t=2} = (3\times 4 - 2)\times 10^{-3}\ \text{Wbs}^{-1}$$

$$= 10^{-2}\ \text{Wbs}^{-1}$$

$$e = -N\frac{d\phi}{dt} = -100\times 10^{-2} = -1\,\text{V}$$

4. $e = \dfrac{d\phi}{dt} = \dfrac{(NBA\cos\theta - 0)}{t}$

$$= \frac{1\times 0.5\times 25\times 10^{-4}\times\cos 60^\circ - 0}{0.2}$$

$e = 3.12\times 10^{-3}$ V

5. The rate of change of flux or induced emf in the coil is

$$e = -n\frac{d\phi}{dt}$$

$\therefore$ Induced current, $I = \dfrac{e}{R'} = -\dfrac{n}{R'}\dfrac{d\phi}{dt}$ ...(i)

Given, $\quad R' = R + 4R = 5R,\ d\phi = W_2 - W_1,\ dt = t$

[here, W_1 and W_2 are flux associated with one turn]

Putting the given values in Eq. (i), we get

$$I = -\frac{n}{5R}\frac{(W_2 - W_1)}{t}$$

6. $|dq| = \dfrac{d\phi}{R} = i\,dt =$ Area under i-t graph

$\therefore d\phi =$ (Area under i-t graph) R

$$= \frac{1}{2}\times 4\times 0.1\times(10) = 2\ \text{Wb}$$

7. $|e| = \dfrac{\Delta\phi}{\Delta t}$

$$i = \frac{|e|}{R} = \frac{1}{R}\frac{\Delta\phi}{\Delta t}$$

$$Q = i\cdot\Delta t = \frac{\Delta\phi}{R\Delta t}\cdot\Delta t = \frac{\Delta\phi}{R}$$

8. As, $q = \dfrac{-d\phi}{R} = \dfrac{BA(\cos 0^\circ - \cos 90^\circ)}{R}$

$$= \frac{B\pi r^2(1-0)}{R} = \frac{B\pi r^2}{R} = \frac{2\times 3.143\times(10^{-1})^2}{0.01}$$

$$= 6.286\ \text{C} \simeq 6.3\ \text{C}$$

9. $q = \dfrac{d\phi}{R} = \dfrac{NA(B_2 - B_1)}{R} = \dfrac{N\pi r^2(B_2 - B_1)}{R}$

$$= \frac{1000\times\pi\times 10^{-4}\times(0.012 - 0)}{(200 + 400)}$$

$$= 6.3\times 10^{-6}\text{C} = 6.3\,\mu\text{C}$$

10. As the magnet moves towards the coil, the magnetic flux increases (non-linearly). Also, there is a change in polarity of induced emf when the magnet passes on to the other side of the coil. It is shown in option (b).

11. $e = -L\dfrac{di}{dt}$

During 0 to $\dfrac{T}{4}$, $\dfrac{di}{dt} =$ constant

So, $e = -$ ve (negative)

For $\dfrac{T}{4}$ to $\dfrac{T}{2}$, $\dfrac{di}{dt} = 0$

$e = 0$

For $\dfrac{T}{2}$ to $\dfrac{3T}{4}$, $\dfrac{di}{dt} =$ constant

$e = +$ ve (positive)

12. Current in the inner coil, $i = \dfrac{e}{R} = \dfrac{A_1}{R_1}\dfrac{dB}{dt}$

Length of the inner coil $= 2\pi a$

So, its resistance, $R_1 = 50\times 10^{-3}\times 2\pi(a)$

$$\therefore i_1 = \frac{\pi a^2}{50\times 10^{-3}\times 2\pi(a)}\times 0.1\times 10^{-3}\ \text{A} = 10^{-4}\ \text{A}$$

According to Lenz's law, direction of i_1 is clockwise.

Induced current in outer coil, $i_2 = \dfrac{e_2}{R_2} = \dfrac{A_2}{R_2}\dfrac{dB}{dt}$

$\Rightarrow \quad i_2 = \dfrac{\pi b^2}{50 \times 10^{-3} \times (2\pi b)} \times 0.1 \times 10^{-3}$

$= 2 \times 10^{-4}$ A (clockwise)

13. Coil A must be carrying a constant current in counter clockwise direction. That is why when A moves towards B, current induced in B is in counter clockwise direction, as per Lenz's law. The current in B would stop when A stops moving.

14. A magnet oscillates the magnetic flux through the coil changes and thus current or emf is produced in it, but in opposite direction, whose magnitude and direction changes with time. So, G shows deflection to the left and right but the amplitude decreases.

15. Emf induces across the length of the wire which cuts the magnetic field. (Length of c = Length of d) > (Length of a = Length of b).

So, $(e_c = e_d) > (e_a = e_b)$.

16. The emf induced in a conductor does not depend on its shape, but only on its end points, M and Q in this case. Thus, the conductor is equivalent to an imaginary straight conductor of $l = MQ = 2R$. Therefore, potential difference developed across the ring $= Blv = B(2R)v$ and the direction of induced current is from Q to M. Therefore, Q is at higher potential.

17. Relative velocity $= v - (-v) = 2v = \dfrac{dl}{dt}$

Now, $\quad e = \dfrac{d\phi}{dt}$

$\Rightarrow \quad e = \dfrac{Bl\,dl}{dt} \quad [\because \phi = BA]$

Induced emf, $e = 2Blv \quad \left[\because \dfrac{dl}{dt} = 2v\right]$

18. As the plane is flying horizontally it will cut the vertical component of earth's field B_V, so the emf induced between its tips

$$e = B_V vl$$

But as by definition of angle of dip,

$$\tan\theta = \frac{B_V}{B_H}$$

i.e. $\quad B_V = B_H \tan\theta$

So, $e = (B_H \tan\theta)vl = 2 \times 10^{-5} \times \sqrt{3} \times 250 \times 20$

i.e. $\quad e = (\sqrt{3}) \times 10^{-1}$ V $= 0.173$ V

19. When loop is entering in the field, magnetic flux (i.e. ×) linked with the loop increases so induced emf in it $e = Bvl = 0.6 \times 1 \times 10^{-2} \times 5 \times 10^{-2} = 3 \times 10^{-4}$ V (negative).

When loop completely entered in the field (after 5 s) flux linked with the loop remains constant, so $e = 0$.

After 15 s, loop begins to exit out, linked magnetic flux decreases, so induced emf, $e = 3 \times 10^{-4}$ V (positive).

20. Let at time t, the loop is at places shown in figure

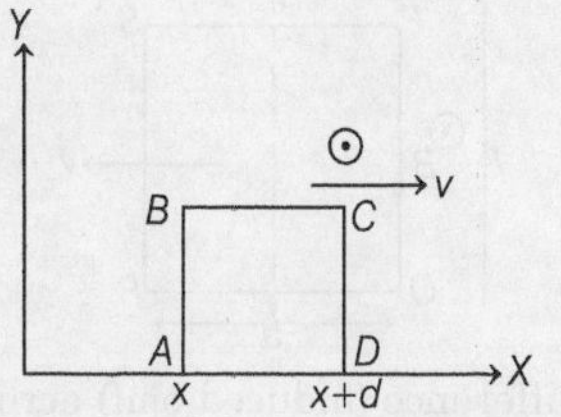

The emf induced in AB, $\varepsilon_{AB} = \dfrac{B_0(x)}{a} v_0 d$

and the emf induced in CD, $\varepsilon_{CD} = \dfrac{B_0(x+d)}{a} v_0 d$

$$\varepsilon_{net} = \varepsilon_{CD} - \varepsilon_{AB} = \frac{B_0 v_0 d^2}{a}$$

21. Given, $B_V = 0.2 \times 10^{-4}$ Wb/m^2

We know that, $v = 180$ km/h

$e = B_V v l$

$= 0.2 \times 10^{-4} \times 180 \times \dfrac{5}{18} \times 1$

$= 2 \times 10^{-5} \times 50$

$= 1 \times 10^{-3}$ V $= 10^{-3}$ V

22. Motional emf across PQ,

$$V = Blv = 4(1)(2) = 8 \text{ V}$$

This is the potential to which the capacitor is charged.

As, $\quad q = CV$

$\therefore \quad q = (10 \times 10^{-6}) \times 8$

$= 8 \times 10^{-5}$ C

$= 80 \mu$C

As magnetic force on electron in the conducting rod PQ is towards Q, therefore A is positively charged and B is negatively charged.

i.e. $\quad q_A = +80 \mu$C

and $\quad q_B = -80 \mu$C

23. Magnetic field due to wire at a distance r is given by

$$B = \frac{\mu_0 I}{2\pi r}$$

Induced emf, $e_{induced} = Bvl = \dfrac{\mu_0 I\, vl}{2\pi r}$

Now, induced current in the loop,

$$I_{induced} = \frac{e_{induced}}{R} = \frac{\mu_0 I}{2\pi} \frac{vl}{Rr}$$

24. Current, $I = \dfrac{\text{Potential difference}}{\text{Resistance}}$

$\therefore \quad$ Potential difference $= 2 \times 12 = 24$ V $= Bvl$

Here, $\quad l = AD \sin 37° = 0.3 \times \dfrac{3}{5} = 0.18$ m

$\therefore$ Velocity, $v = \dfrac{24}{Bl} = \dfrac{24}{4 \times 0.18} = \dfrac{100}{3}$ ms^{-1}

25.

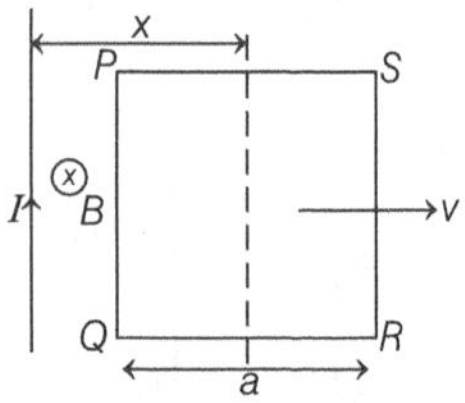

Potential difference (induced emf) across PQ,

$$V_P - V_Q = B_1 av = \frac{\mu_0 I}{2\pi\left(x - \frac{a}{2}\right)} av$$

Potential difference (induced emf) across side RS of frame,

$$V_S - V_R = B_2 av = \frac{\mu_0 I}{2\pi\left(x + \frac{a}{2}\right)} av$$

Hence, the net potential difference (induced emf) in the loop will be

$$V_{\text{net}} = (V_P - V_Q) - (V_S - V_R) = \frac{\mu_0 Iav}{2\pi}\left[\frac{1}{\left(x - \frac{a}{2}\right)} - \frac{1}{\left(x + \frac{a}{2}\right)}\right]$$

$$= \frac{\mu_0 Iav}{2\pi}\left[\frac{a}{\left(x - \frac{a}{2}\right)\left(x + \frac{a}{2}\right)}\right]$$

Thus, $V_{\text{net}} \propto \frac{1}{(2x - a)(2x + a)}$

26. Motional emf induced in the connector,

$$e = Blv = 2(1)(2) = 4 \text{ V}$$

This acts as a cell of emf 4 V and internal resistance 2Ω, 6Ω and 3Ω resistors are in parallel.

$$\therefore \quad \frac{1}{R_p} = \frac{1}{6} + \frac{1}{3} = \frac{1+2}{6} = \frac{3}{6} = \frac{1}{2}$$

$$R_p = 2\,\Omega$$

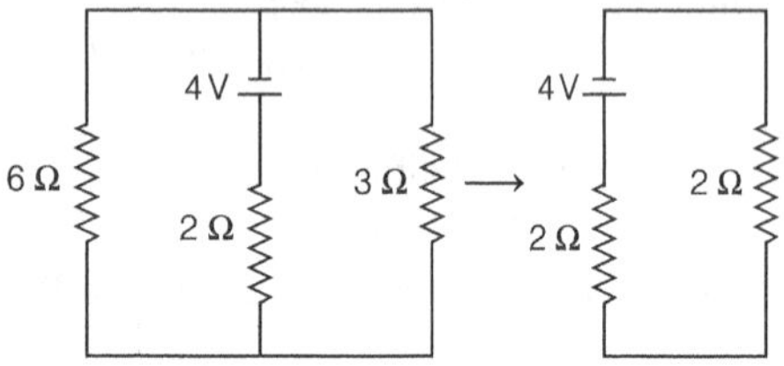

$\therefore$ Current through the connector, i

$$= \frac{E}{R_p + r} = \frac{4}{2+2} = 1 \text{ A}$$

Magnetic force on the connector $= Bil = 2(1)(1) = 2$ N

Therefore, to keep the connector moving with a constant velocity, a force of 2 N has to be applied to the right side.

27. Circuit can be reduced as

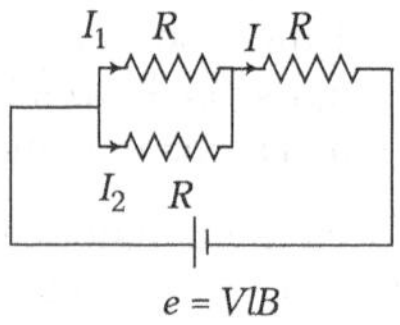

$$I = \frac{e}{3R/2} = \frac{2\,vlB}{3R}$$

$$\Rightarrow \quad I_1 = I_2 = \frac{I}{2} = \frac{vlB}{3R}$$

28. We know that, $e = NBA\,\omega \sin \omega t$

where, N = number of loops = 1,

$$B = \frac{\mu_0 I}{2b} \text{ N/A-m}$$

and $\quad A = \pi a^2 \text{ m}^2$.

Putting the values in Eq. (i), we get

$\therefore$ Induced emf, $e = \frac{\mu_0 I}{2b}(\pi a^2)\,\omega \sin \omega t$

$$= \frac{\pi a^2 \mu_0 I}{2b} \cdot \omega \sin \omega t$$

29. From $e = -L di/dt, L = \left|\frac{-edt}{di}\right| = \frac{8 \times 0.05}{2} = 0.2$ H

30. From $L = \frac{\mu_0 N^2 A}{l} = \frac{\mu_0 \mu_r N^2 A}{l}$

When $\mu_r = 1000$ and N becomes $\frac{1}{10}$,

L becomes $1000 \times \left(\frac{1}{10}\right)^2 = 10$ times

i.e. $\quad L = 10 \times 0.1 = 1$ H

31. As, $\quad L = \mu_0 n^2 Al$

$\therefore \quad L \to 2 \times 2 \times 2$ times $= 8$ times

32. The self-inductance L of a solenoid of length l and area of cross-section A with fixed number of turns N is

$$L = \frac{\mu_0 N^2 A}{l}$$

Obviously, L increases, when l decreases and A increases.

33. $\frac{L_2}{L_1} = \frac{N_2^2}{N_1^2}$

$$\therefore \quad L_2 = L_1 \frac{N_2^2}{N_1^2} = 1.5\left(\frac{500}{100}\right)^2 = 375 \text{ mH}$$

34. As, $e = -M\dfrac{di}{dt} = -M\dfrac{d}{dt}(i_0 \sin \omega t)$

$\therefore \quad e = Mi_0 \cos \omega t(\omega)$

$$e_{max} = Mi_0 \times 1 \times \omega$$
$$= 0.005 \times 10 \times 100\pi = 5\pi$$

35. As, $e = -M\dfrac{di}{dt}$

$\therefore \quad 30 \times 10^3 = -3 \times \dfrac{(0-10)}{dt}$

$$dt = \frac{30}{30 \times 10^3} = 10^{-3} \text{ s}$$

36. Given, mutual inductance of coil, $M = 1.5$ H

Current change in coil, $dI = (0-20) = -20$ A

Time taken in change, $dt = 0.5$ s

Induced emf in the coil, $e = -M\dfrac{dI}{dt} = \dfrac{d\phi}{dt}$

or $\quad d\phi = -M \cdot dI = 1.5 \times 20$

$$d\phi = 30 \text{ Wb}$$

Thus, the change of flux linkage is 30 Wb.

37. As the inductors are in parallel, therefore induced emf across the two inductors is the same, *i.e.* $e_1 = e_2$.

$\therefore \quad L_1\left(\dfrac{di_1}{dt}\right) = L_2\left(\dfrac{di_2}{dt}\right)$

Integrating both sides w.r.t. t, we get

$$L_1 i_1 = L_2 i_2$$

$\therefore \quad \dfrac{i_1}{i_2} = \dfrac{L_2}{L_1}$

38. As the coil is rotated, angle θ (angle which is normal to the coil makes with **B** at any instant t) changes, therefore magnetic flux ϕ linked with the coil changes and hence an emf is induced in the coil. At this instant t, if e is the emf induced in the coil, then

$$e = -\frac{d\phi}{dt} = -\frac{d}{dt}(NSB \cos \omega t)$$

where, N is number of turns in the coil.

or $\quad e = -NSB\dfrac{d}{dt}(\cos \omega t) = -NSB(-\sin \omega t)\,\omega$

or $\quad e = NSB\omega \sin \omega t \quad$ ...(i)

The induced emf will be maximum,

when $\sin \omega t$, *i.e.* maximum

$\therefore \quad e_{max} = e_0 = NSB\omega \times 1 \quad$ or $\quad e = e_0 \sin \omega t$

Therefore, e would be maximum, hence current is maximum (as $i_0 = e_0/R$), when $\theta = 90°$, i.e. normal to plane of coil is perpendicular to the field or plane of coil is parallel to magnetic field.

Round II

1. Induced current are clockwise, therefore induced magnetic field is into the plane of the paper. As it opposes the increasing inducing field, the inducing field must be out of the plane of the paper.

2. Terminal velocity of the rod is attained when magnetic force on the rod (Bil) balances the component of weight of the rod ($mg \sin \theta$) as shown in figure.

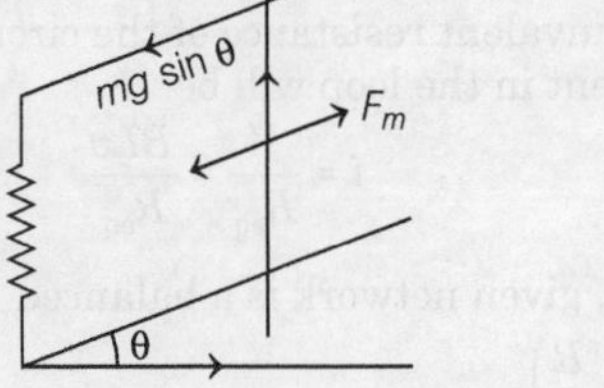

i.e. $\quad Bil = mg \sin \theta$

$$B\left(\frac{e}{R}\right) l = mg \sin \theta$$

$$\frac{Bl}{R}(e) = mg \sin \theta$$

$$\frac{Bl}{R}(Blv_T) = mg \sin \theta$$

$$v_T = \frac{mgR \sin \theta}{B^2 l^2}$$

3. Magnetic field intensity at a distance r from the straight wire carrying current is

$$B = \frac{\mu_0 i}{2\pi r}$$

As area of loop, $A = a^2$

and magnetic flux, $\phi = BA$

$\therefore \quad \phi = \dfrac{\mu_0 i a^2}{2\pi r}$

The induced emf in the loop is

$$e = \left|\frac{d\phi}{dt}\right| = \left|\frac{d}{dt}\frac{\mu_0 i a^2}{2\pi r}\right|$$

$$e = \frac{\mu_0 i a^2}{2\pi r^2}\frac{dr}{dt} = \frac{\mu_0 i a^2 v}{2\pi r^2}$$

where, $v = \dfrac{dr}{dt}$ = velocity.

4. Magnetic flux associated with the outer coil is

$$\phi_{outer} = \mu_0 \pi NR \cdot I$$
$$= \mu_0 N\pi R(kte^{-\alpha t}) = Cte^{-\alpha t}$$

where, $C = \mu_0 N\pi Rk$ = constant.

Induced emf,

$$e = \frac{-d\phi_{outer}}{dt}$$
$$= Ce^{-\alpha t} + (-\alpha t C e^{-\alpha t})$$
$$= Ce^{-\alpha t}(1 - \alpha t)$$

$\therefore$ Induced current, $I = \dfrac{e}{\text{Resistance}}$

At $t = 0$, $I = -$ ve

$\therefore$ The correct graph representing this condition is given in option (d).

5. Induced emf in the conductor of length L moving with velocity of 1 cm/s in the magnetic field of 1T is given by

$$V = BLv \quad \text{...(i)}$$

If equivalent resistance of the circuit is R_{eq}, then current in the loop will be

$$i = \frac{V}{R_{eq}} = \frac{BLv}{R_{eq}} \quad \text{...(ii)}$$

Now, given network is a balanced Wheatstone bridge $\left(\frac{P}{Q} = \frac{R}{S}\right)$.

So, equivalent resistance of the Wheatstone bridge is

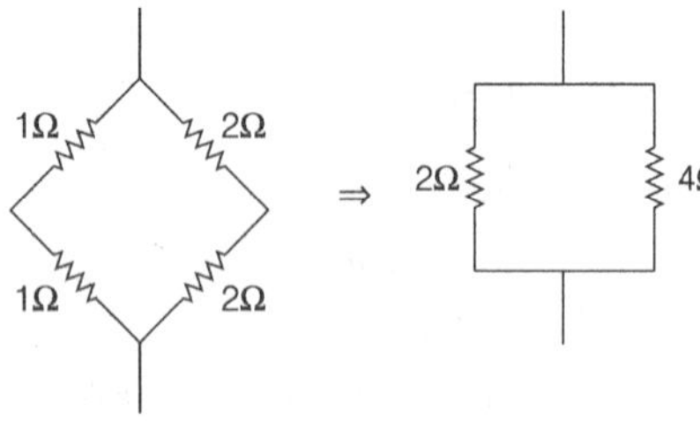

$$R_W = \frac{2\times4}{2+4} = \frac{8}{6} = \frac{4}{3}\,\Omega$$

Again, resistance of conductor is 1.7 Ω.

So, effective resistance will be

$$R_{eq} = \frac{4}{3} + 1.7 = \frac{4}{3} + \frac{17}{10}$$
$$= \frac{40+51}{30} = \frac{91}{30} \simeq 3\Omega$$

By putting given values of R_{eq} B and v in Eq. (ii), we have

$$i = \frac{(1)\,(5\times10^{-2})\times10^{-2}}{3}$$

[Here, $L = 5\times10^{-2}$ m, $v = 1$ cm/s $= 10^{-2}$ m/s]

$$i = \frac{5\times10^{-4}}{3} = 1.67\times10^{-4}\,\text{A}$$
$$i = 167\,\mu\text{A} \approx 170\,\mu\text{A}$$

6. We are given following situation

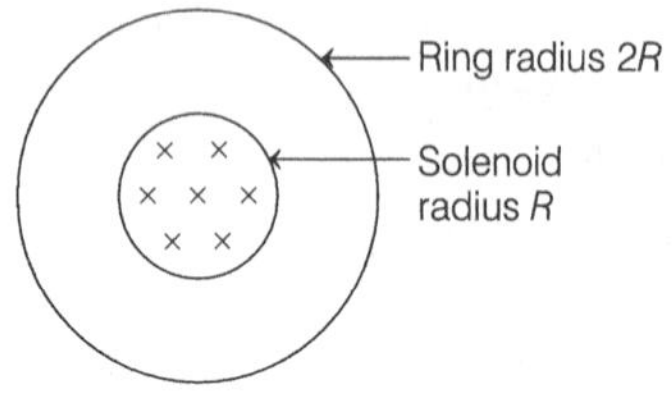

Flux linked with the ring,

$$\phi = (B_{\text{solenoid}})(A_{\text{solenoid}}) = \mu_0 n I_R \times \pi R^2$$

Here, $I_R = I_0 t(1-t)$

So, $\phi = \mu_0 \pi R^2 \times n I_0 \cdot t \cdot (1-t)$

Magnitude of induced emf in the ring is

$$V_R = |E| = \frac{d\phi}{dt}$$
$$= \frac{d}{dt}\mu_0 \pi R^2 \times n I_0 t(1-t)$$
$$= \mu_0 \pi R^2\, n I_0 (1-2t)$$

Clearly, induced emf is zero, when $1 - 2t = 0$

or $\quad t = \frac{1}{2}\,\text{s} = 0.5\,\text{s}$

So, induced emf and induced current are zero at $t = 0.5$ s and their direction also reverses.

7. As, coefficient of mutual induction is same for both coils. i.e.

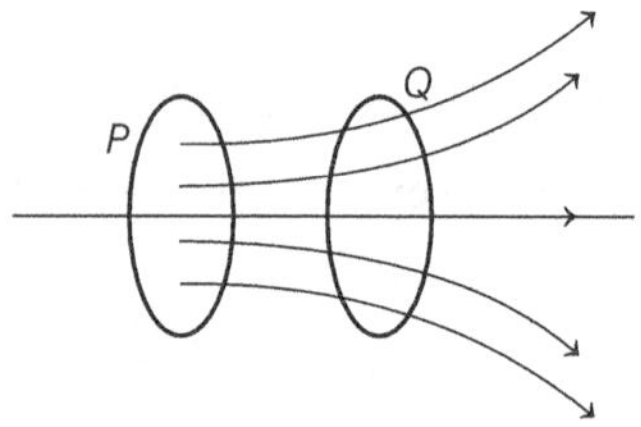

$\therefore \quad M_{PQ} = M_{QP}$

$$\Rightarrow \frac{N_P \phi_{PQ}}{I_Q} = \frac{N_Q \phi_{QP}}{I_P} \quad \text{... (i)}$$

Here, $N_P = N_Q = 1$,

$\phi_{PQ} = ?,\ \phi_{QP} = 10^{-3}$ Wb

$I_Q = 2\text{A},\ I_P = 3\text{A}$

Substituting values in Eq (i), we get

$$\phi_{PQ} = \frac{N_Q \phi_{QP} \cdot I_Q}{N_p \cdot I_p} = \frac{1\times10^{-3}\times2}{1\times3} = \frac{2}{3}\times10^{-3}$$
$$= 0.667\times10^{-3} = 6.67\times10^{-4}\ \text{Wb}$$

8. Mutual inductance for a coaxial solenoid of radius r_1 and r_2 and number of turns n_1 and n_2, respectively is given as,

$M = \mu_0 n_1 n_2\, \pi\, r_1^2\, l$ (for internal coil of radius r_1)

Self-inductance for the internal coil,

$$L = \mu_0 n_1^2 \pi\, r_1^2 l$$
$$\therefore \quad \frac{M}{L} = \frac{n_1 n_2}{n_1^2} = \frac{n_2}{n_1}$$

9. Self-inductance L_{sol} of a solenoid is given by

$$L_{sol} = \mu_0 n^2 \pi r^2 L$$

(Here, $n = N/L$ and L = length of solenoid)

or $\quad L_{sol} = \frac{\mu_0 N^2 \pi r^2}{L}$

Clearly, $\quad L_{sol} \propto \frac{1}{L}$

($\because$ all other parameters are fixed)

10. Induced emf in the loop,

$$e = -\frac{\Delta\phi}{\Delta t} = -\left(\frac{\Delta B}{\Delta t}\right)(A) \quad [\because \phi = BA]$$
$$= -\left(\frac{B_2 - B_1}{\Delta t}\right)A \quad \text{...(i)}$$

Here, $B_2 = 500\ \text{G} = 500\times10^{-4}$ T,

$B_1 = 1000 \text{ G} = 1000 \times 10^{-4}$ T,

$\Delta t = 5\text{s}$

$\therefore$ A = area of loop

= Area of rectangle − Area of two triangles

$= \left(16 \times 4 - 2 \times \frac{1}{2} \times 4 \times 2\right) \text{cm}^2$

$= 56 \times 10^{-4} \text{ m}^2$

Using Eq. (i), we get

$$e = \frac{(1000 - 500) \times 10^{-4} \times 56 \times 10^{-4}}{5}$$

$$= 56 \times 10^{-6} \text{ V} = 56 \, \mu\text{V}$$

11. Since, the magnetic field is dependent on time, so the net charge flowing through the loop will be given as

$$Q = \frac{\text{change in magnetic flux, } \Delta\phi_B}{\text{resistance, } R}$$

As, $\Delta\phi_B = B\,A = BA \cos\theta$

where, A is the surface area of the loop and θ is an angle between B and A.

Here, $\theta = 0 \Rightarrow \Delta\phi_B = BA$

$\therefore$ For the time interval, $t = 0$ s to $t = 10$ ms,

$$Q = \frac{\Delta\phi_B}{R} = \frac{A}{R}(B_{f \text{ at } 0.01\text{s}} - B_{i \text{ at } 0\text{s}})$$

Substituting the given values, we get

$$= \frac{3.5 \times 10^{-3}}{10} [0.4 \sin(0.5\pi) - 0.4 \sin 0]$$

$$= 3.5 \times 10^{-4} \,(0.4 \sin \pi/2)$$

$$= 1.4 \times 10^{-4} \text{C}$$

$$= 14 \text{ mC}$$

12. From an inductor (choke), $E_{\text{induced}} = -L\frac{dI}{dt}$

$\Rightarrow$ $L = \left|\frac{-E \times \Delta t}{\Delta I}\right|$

Here, $E = 100$ V, $\Delta I = 0.25$ A

and $\Delta t = 0.025 \text{ ms} = 0.025 \times 10^{-3}$ s

So, $L = \left|\frac{-100 \times 0.025 \times 10^{-3}}{0.25}\right|$

$$= 100 \times 10^{-4} = 10 \text{ mH}$$

13. Given that, at any instant, $I = 1$ A

$\frac{dI}{dt} = -10^2$ A/s (as, current is decreasing)

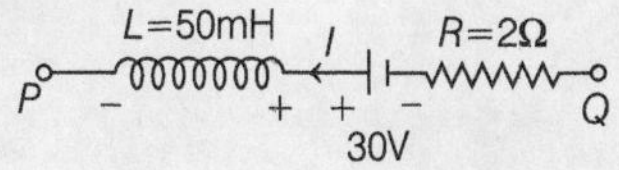

Applying KVL from point P to point Q,

$$V_P + \frac{LdI}{dt} - 30 + IR = V_Q$$

$$V_P - V_Q = 30 - 1 \times 2 - 50 \times 10^{-3} \times (-10^2) = 33 \text{ V}$$

14. Given, $I(t) = 5t^2 - 2t + 3$

$$\frac{dI}{dt} = 10t - 2$$

At $t = 1$ s, $\frac{dI}{dt} = 8$ A/s

Magnetic flux, $\phi = \left(\frac{\mu_0 I n_2}{2R}\right)(\pi r^2 n_1)$

Induced emf, $e = \left|\frac{d\phi}{dt}\right| = \left(\frac{\mu_0 n_2}{2R}\right)\pi r^2 n_1 \frac{dI}{dt}$

Given, $n_1 = 200$, $r = 1 \text{ cm} = 10^{-2}$ m, $n_2 = 500$, $R = 20 \text{ cm} = 20 \times 10^{-2}$ m

and $\mu_0 = 4\pi \times 10^{-7}$

$\therefore$ $e = \frac{4\pi \times 10^{-7} \times 200 \times \pi \times 10^{-4} \times 500}{2 \times 20 \times 10^{-2}} \times 8$

$$= 8 \times 10^{-4} \text{ V} = 0.8 \text{ mV}$$

According to question, $0.8 \text{ mV} = \frac{4}{x} \Rightarrow x = \frac{4}{0.8} = 5$

15. Flux linked with the coil is

$$\phi = \mathbf{B} \cdot \mathbf{A} = BA \cos \omega t$$

Magnitude of emf induced in coil

$$= \left|\frac{d}{dt}\phi\right| = |BA\omega \sin \omega t|$$

Maximum value of induced emf $= BA\omega = \frac{BA2\pi}{T}$

$$= \frac{3 \times 10^{-5} \times \pi \times (0.1)^2 \times 2\pi}{0.4} = 14.7 \times 10^{-6} \text{ V}$$

$$\approx 15 \times 10^{-6} \text{ V} = 15 \, \mu\text{V}$$

23

Alternating Current

IN THIS CHAPTER ....

Types of Current

There are two types of current which flows in any of the electrical appliances. These are as follows

Direct Current

Current whose direction does not change with time in a circuit is known as direct current (DC).

Alternating Current

Currents whose magnitude and direction changes continuously (periodically) with time is known as alternating currents and corresponding voltage as alternating voltage or emf.

The instantaneous value of time varying alternating voltage and current are given by

$$V = V_0 \sin \omega t$$

and
$$i = i_0 \sin \omega t$$

where, V_0 and i_0 = peak value of voltage and current respectively

and ω is called angular frequency of AC.

where, $\omega = \frac{2\pi}{T} = 2\pi\nu$, here T is the time period or period of AC.

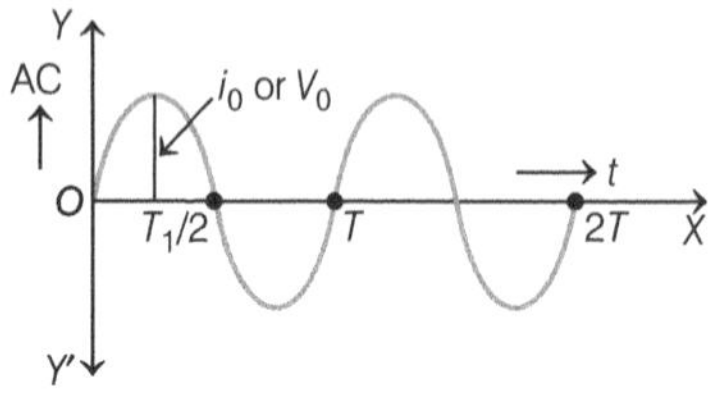

Mean or Average Value of AC

Mean or average value of an alternating current (AC) is the total charge flow for one complete cycle, divided by the time taken to complete the cycle. *i.e.* time period T.

The direction of AC changes after every half cycle. Hence, for a full cycle, the mean or average value is zero. Thus, we find average value for half cycle only.

It is given by $I_{\text{mean}} = \dfrac{1}{T/2}\int_0^{\pi/2} I\,dt$

Mean or average value of alternating current for first half cycle, $\quad I_m = \dfrac{2I_0}{\pi} = 0.637\, I_0$

Mean or average value of alternating current for next half cycle, $\quad I'_m = -\dfrac{2I_0}{\pi} = -\,0.637\, I_0$

So, mean or average value of alternating current for one complete cycle = 0.

In the same way, mean value of alternating voltage,

$$V_m = 0.637\, V_0$$

Root Mean Square Value of AC

The steady current, which when passes through a resistance for a given time will produce the same amount of heat as the alternating current does in the same resistance and in the same time, is called rms value of alternating current. It is given by

$$i_{\text{rms}} = \frac{i_0}{\sqrt{2}} = 0.707\, i_0$$

Similarly, rms value of alternating emf,

$$V_{\text{rms}} = \frac{V_0}{\sqrt{2}} = 0.707\, V_0$$

The different values of i_0, i_{rms} and i_{av} are shown in figure

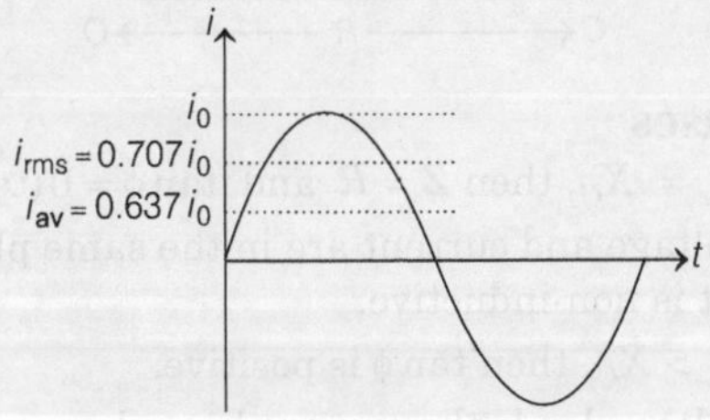

Resistive Circuit or *R*-Circuit

When an AC voltage source (*i.e.* $V = V_0 \sin \omega t$) is connected across a resistor in a circuit, then current in the circuit,

$$i = i_m \sin \omega t$$

where, $i_m = \dfrac{V_m}{R}$.

The voltage and current are in phase with each other.

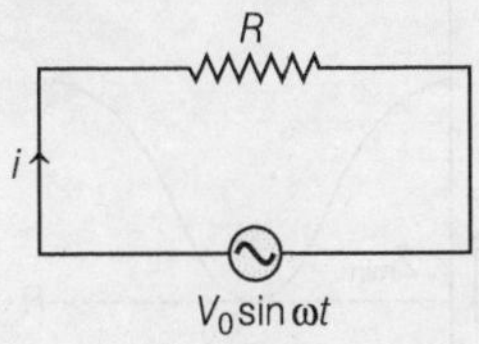

The **phasor diagram** and ***V* & *I* versus** ωt graph are as shown below.

A diagram representing alternating current and alternating emf (of same frequency) as rotating vectors (phasors) with the phase angle between them is called *phasor diagram*.

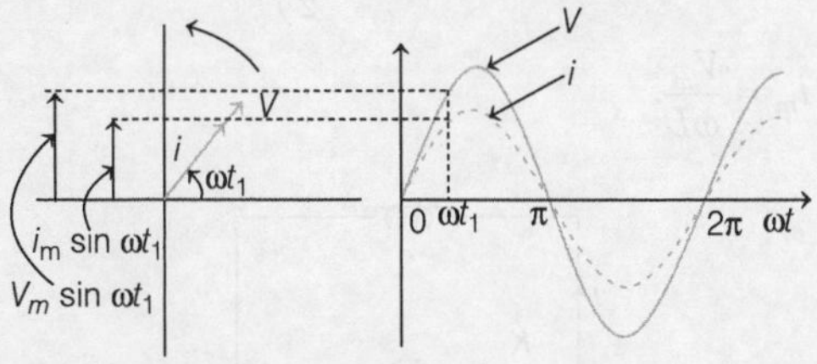

Capacitive Circuit or *C*-Circuit

When a capacitor is connected to an AC voltage source (*i.e.* $V = V_0 \sin \omega t$) in a circuit, then the current in the circuit,

$$i = i_m \sin\left(\omega t + \frac{\pi}{2}\right)$$

where, $\quad i_m = \dfrac{V_m}{1/\omega C}$.

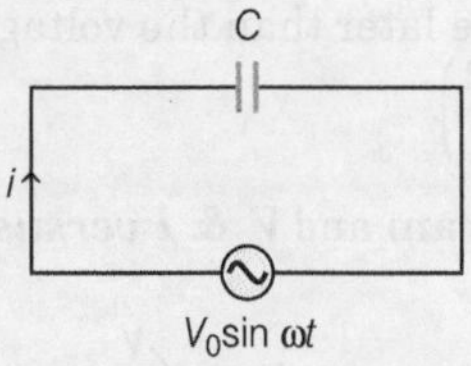

The term $\left(\dfrac{1}{\omega C}\right)$ is also analogous to resistance and it is called *capacitive reactance*. It is denoted by X_C, *i.e.*

$$X_C = \frac{1}{\omega C} = \frac{1}{2\pi \nu C}$$

The dimension of capacitive reactance is the same as that of resistance and its SI unit is ohm (Ω).

The current in this circuit is ahead of voltage in phase by $\dfrac{\pi}{2}$ or one quarter cycle. It means, the current reaches its maximum value earlier than the voltage by one-fourth of a period.

The **phasor diagram** and ***V* & *I* *versus*** ωt graph are as shown below

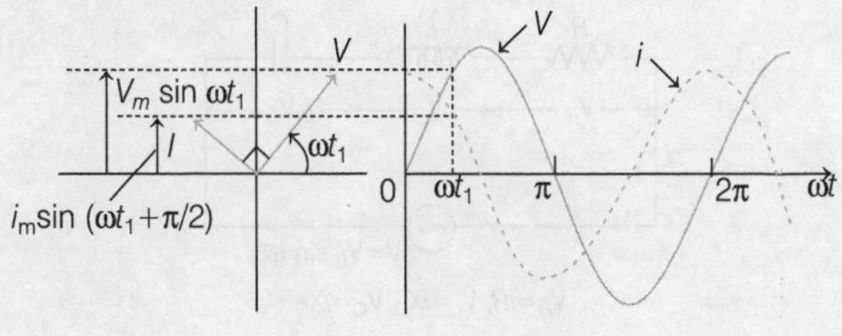

Inductive Circuit or *L*-Circuit

When an AC voltage source (*i.e.* $V = V_0 \sin \omega t$) is connected to a purely inductive circuit, then current in the circuit,

$$i = i_m \sin\left(\omega t - \frac{\pi}{2}\right)$$

where, $i_m = \dfrac{V_m}{\omega L}$.

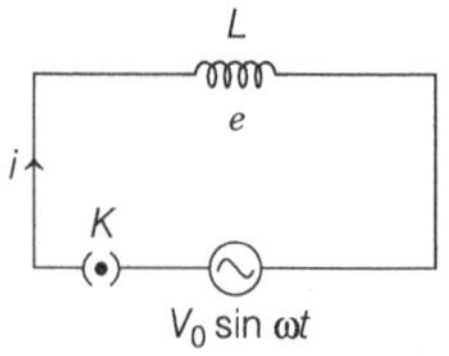

The quantity ωL is analogous to the resistance and is called *inductive reactance*. It is denoted by X_L, *i.e.*

$$X_L = \omega L = (2\pi\nu)L$$

The dimension of inductive reactance is same as that of resistance and its SI unit is ohm (Ω).

The current in this type of circuit lags the voltage in phase by $\dfrac{\pi}{2}$ or one-quarter $\left(\dfrac{1}{4}\right)$ cycle. It means the current reaches its maximum value later than the voltage by one-fourth of a period $\left(\dfrac{T}{4} = \dfrac{\pi/2}{\omega}\right)$.

The **phasor diagram** and ***V* & *I* versus** ωt graph are as shown below

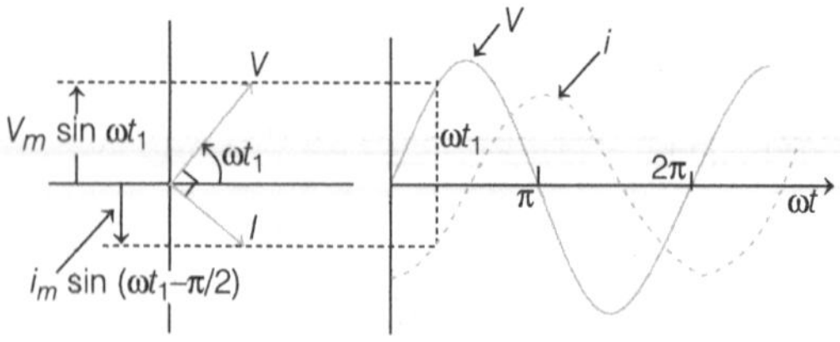

Note Capacitor blocks DC and acts as an open circuit while passing AC of high frequency, while inductor passes DC and blocks AC of very high frequency.

AC Voltage Applied to Series *R-L-C* Circuit

When a resistor, capacitor and an inductor are connected in series with an alternating voltage source as shown below

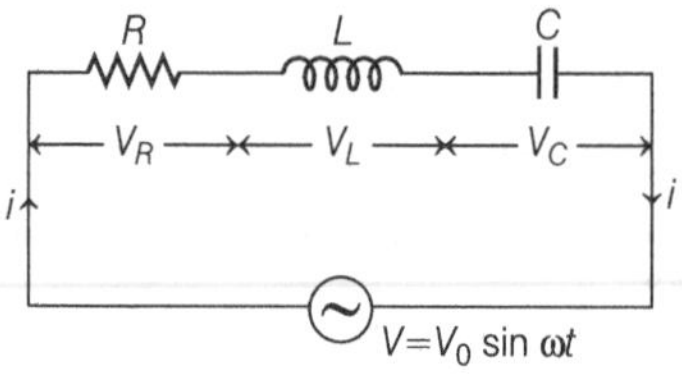

$V_R = iR,\ V_L = iX_L,\ V_C = iX_C$

- **Current in the circuit** $i = i_0 \sin(\omega t \pm \phi)$, where $i_0 = \dfrac{V_0}{Z}$
- **Voltage across the circuit**
$$V = \sqrt{V_R^2 + (V_L - V_C)^2}$$
- **Impedance (*Z*)** The resistance offered by an AC circuit to the flow of alternating current through it, is called impedance of a given circuit.
It is given as $Z = \sqrt{R^2 + (X_L - X_C)^2}$
$$= \sqrt{R^2 + \left(\omega L - \frac{1}{\omega C}\right)^2}$$
- **Phase angle** ϕ is the angle between total voltage V and total electric current and can be given as
$$\tan \phi = \frac{V_L - V_C}{V_R} = \frac{X_L - X_C}{R}$$
$$= \frac{\omega L - \dfrac{1}{\omega C}}{R}$$
$$= \frac{2\pi\nu L - \dfrac{1}{2\pi\nu C}}{R}$$

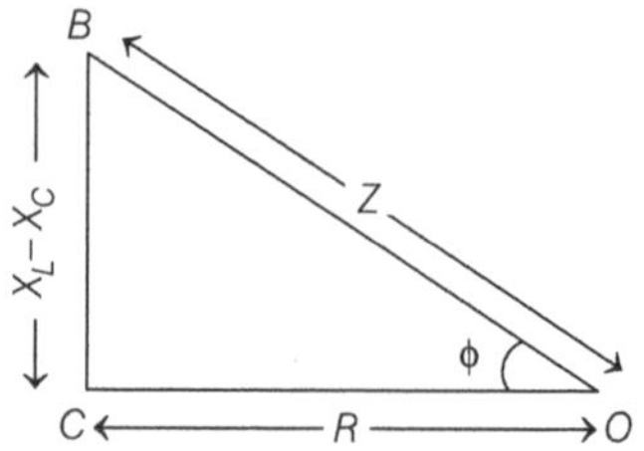

Special Cases

- When $X_L = X_C$, then $Z = R$ and $\tan \phi = 0 \ (\because \phi = 0^\circ)$.
Hence, voltage and current are in the same phase. The AC circuit is non-inductive.
- When $X_L > X_C$, then $\tan \phi$ is positive.
Hence, voltage leads the current by a phase angle ϕ. The AC circuit is inductance dominated circuit.
- When $X_C > X_L$, then $\tan \phi$ is negative.
Hence, voltage lags behind the current by a phase angle ϕ. The AC circuit is capacitance dominated circuit.

The variation of *Z* with frequency ω is shown in figure below. At $\omega = \omega_0$, $X_L = X_C$, $Z = R =$ minimum.

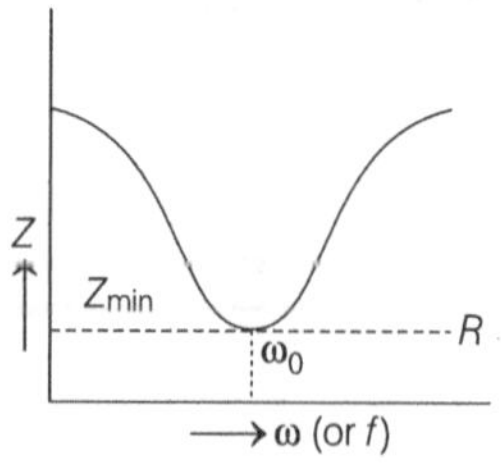

Series R-L-C circuit can be further reduced into R-L, R-C and L-C circuits, whose circuit characteristics are given in the tabular form below

Characteristics	R-L circuit	R-C circuit	L-C circuit
Circuit	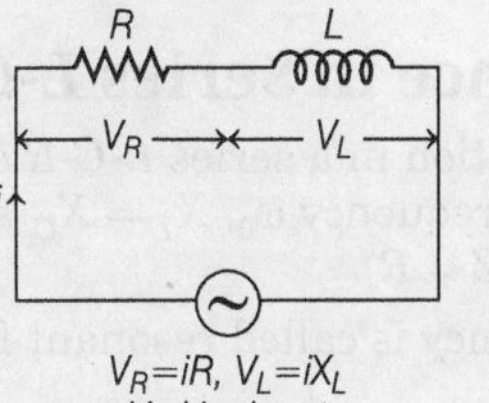 $V_R = iR$, $V_L = iX_L$ $V = V_0 \sin \omega t$	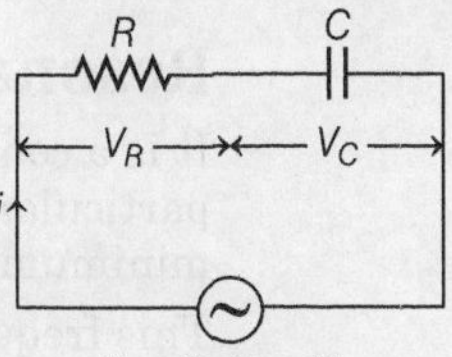 $V_R = iR$, $V_C = iX_C$ $V = V_0 \sin \omega t$	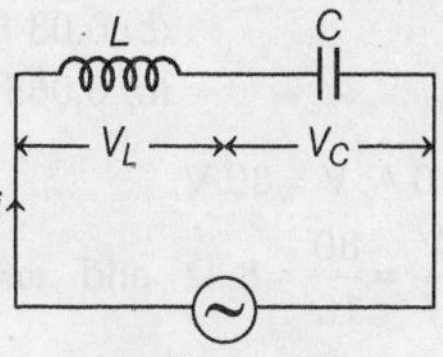 $V_L = iX_L$, $V_C = iX_C$ $V = V_0 \sin \omega t$
Current	$i = i_0 \sin(\omega t - \phi)$	$i = i_0 \sin(\omega t + \phi)$	$i = i_0 \sin\left(\omega t \pm \frac{\pi}{2}\right)$
Peak current	$i_0 = \frac{V_0}{Z} = \frac{V_0}{\sqrt{R^2 + X_L^2}}$ $= \frac{V_0}{\sqrt{R^2 + 4\pi^2 \nu^2 L^2}}$	$i_0 = \frac{V_0}{Z} = \frac{V_0}{\sqrt{R^2 + X_C^2}}$ $= \frac{V_0}{\sqrt{R^2 + \frac{1}{4\pi^2 \nu^2 C^2}}}$	$i_0 = \frac{V_0}{Z} = \frac{V_0}{X_L - X_C}$ $= \frac{V_0}{\omega L - \frac{1}{\omega C}}$
Phasor diagram	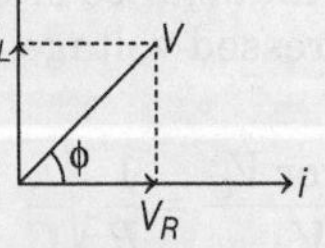	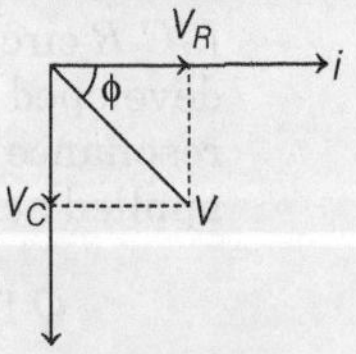	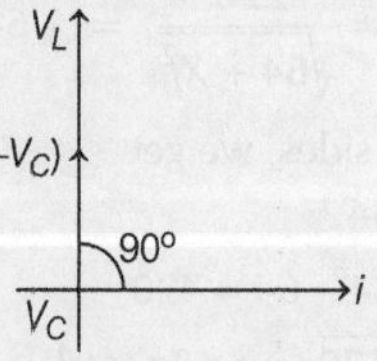
Applied voltage	$V = \sqrt{V_R^2 + V_L^2}$	$V = \sqrt{V_R^2 + V_C^2}$	$V = V_L - V_C$
Impedance	$Z = \sqrt{R^2 + X_L^2} = \sqrt{R^2 + \omega^2 L^2}$ $= \sqrt{R^2 + 4\pi^2 \nu^2 L^2}$	$Z = \sqrt{R^2 + X_C^2} = \sqrt{R^2 + \left(\frac{1}{\omega C}\right)^2}$	$Z = X_L - X_C = X$
Phase difference	$\phi = \tan^{-1} \frac{X_L}{R}$ $= \tan^{-1} \frac{\omega L}{R}$	$\phi = \tan^{-1} \frac{X_C}{R}$ $= \tan^{-1} \frac{1}{\omega CR}$	$\phi = 90^\circ$
Leading quantity	Voltage	Current	Either voltage or current

Example 1. *An alternating voltage $V(t) = 220 \sin 100\pi t$ volt is applied to a purely resistive load of 50 Ω. The time taken for the current to rise from half of the peak value to the peak value is* **[JEE Main 2019]**

(a) 5 ms (b) 2.2 ms (c) 7.2 ms (d) 3.3 ms

Sol. (*d*) In an AC resistive circuit, current and voltage are in phase.

So, $I = \frac{V}{R} \Rightarrow I = \frac{220}{50} \sin(100\pi t)$

∴ Time period of one complete cycle of current is

$$T = \frac{2\pi}{\omega} = \frac{2\pi}{100\pi} = \frac{1}{50} \text{ s}$$

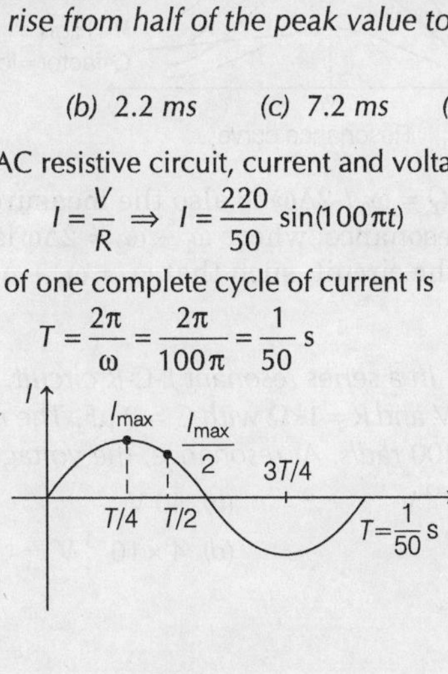

So, current reaches its maximum value at

$$t_1 = \frac{T}{4} = \frac{1}{200} \text{ s}$$

When current is half of its maximum value, then from Eq. (i), we have

$$I = \frac{I_{max}}{2} = I_{max} \sin(100\pi t_2)$$

$$\Rightarrow \quad \sin(100\pi t_2) = \frac{1}{2}$$

$$\Rightarrow \quad 100\pi t_2 = \frac{5\pi}{6}$$

So, instantaneous time at which current is half of maximum value is $t_2 = \frac{1}{120}$ s

Hence, time duration in which current reaches half of its maximum value after reaching maximum value is

$$\Delta t = t_2 - t_1 = \frac{1}{120} - \frac{1}{200} = \frac{1}{300} \text{ s} = 3.3 \text{ ms}$$

Example 2. *An arc lamp requires a direct current of 10 A at 80 V to function. If it is connected to a 220 V (rms), 50 Hz AC supply, the series inductor needed for it to work is close to* **[JEE Main 2016]**

(a) 80 H (b) 0.08 H
(c) 0.044 H (d) 0.065 H

Sol. (d) Given, $I = 10$ A, $V = 80$ V,

$$R = \frac{V}{I} = \frac{80}{10} = 8\ \Omega \text{ and } \omega = 50 \text{ Hz}$$

For AC circuit, we have

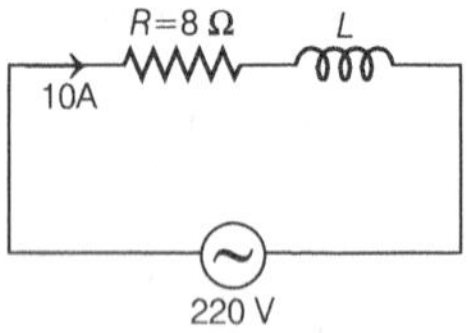

$$I = \frac{V}{\sqrt{8^2 + X_L^2}}$$

$$\Rightarrow \quad 10 = \frac{220}{\sqrt{64 + X_L^2}} \Rightarrow \sqrt{64 + X_L^2} = 22$$

Squaring on both sides, we get

$$64 + X_L^2 = 484$$

$$\Rightarrow \quad X_L^2 = 484 - 64 = 420$$

$$\Rightarrow \quad X_L = \sqrt{420} \Rightarrow 2\pi \times \omega L = \sqrt{420}$$

Series inductor on an arc lamp, $L = \frac{\sqrt{420}}{(2\pi \times 50)} = 0.065$ H

Example 3. *When an alternating voltage of 220 V is applied across a device X, a current of 0.5 A flows through the circuit and is in phase with the applied voltage. When the same voltage is applied across another device Y, the same current again flows through the circuit, but it lags behind the applied voltage by (π /2) rad. Calculate the current flowing in the circuit when same voltage is applied across the series combination of X and Y.*

(a) 0.5 A (b) 0.3 A
(c) 2.5 A (d) 4.5 A

Sol. (b) The current and voltage are in phase with each other, when alternating voltage is applied across a resistor. Hence, the device X is resistor.

Obviously, $R = \frac{V_{rms}}{i_{rms}} = \frac{220}{0.5} = 440\ \Omega$

The current lags behind the voltage by phase angle $\pi/2$, when alternating voltage is applied across an inductor. Hence, the device Y is an inductor.

Obviously, $X_L = \frac{V_{rms}}{I_{rms}} = \frac{220}{0.5} = 440\ \Omega$

Here, $V_{rms} = 220$ V, $R = 440\ \Omega$, $X_L = 440\ \Omega$

If Z is impedance of L-R circuit, then $Z = \sqrt{R^2 + X_L^2}$

$$= \sqrt{440^2 + 440^2} = 400\sqrt{2}\ \Omega$$

Therefore, current in the L-R circuit,

$$i_{rms} = \frac{E_V}{Z} = \frac{220}{440\sqrt{2}} = 0.3535 \text{ A}$$

Resonance in Series *L-C-R* Circuit

It is a condition in a series L-C-R circuit, when at particular frequency ω_0, $X_L = X_C$ and the impedance is minimum ($Z = R$).

This frequency is called resonant frequency and given as

$$\omega_0 = \frac{1}{\sqrt{LC}} \text{ or } f_0 = \frac{1}{2\pi\sqrt{LC}}$$

Resonant frequency is independent of resistance.

At resonating frequency,

$$Z = R = Z_{min} \text{ and } I = \frac{V}{Z} = I_{max}$$

Q-Factor of Series Resonant Circuit

It is the measure of sharpness of the resonance of an L-C-R circuit. It is defined as the ratio of voltage developed across the inductance or capacitance at resonance to the impressed voltage, which is the voltage applied across R.

$$Q\text{-factor} = \frac{V_L \text{ or } V_C}{V_R} = \frac{1}{R}\sqrt{\frac{L}{C}}$$

$$= \frac{\omega_0 L}{R} = \frac{1}{\omega_0 CR}$$

It is also defined as

$$Q = 2\pi \times \frac{\text{Maximum energy stored}}{\text{Energy dissipation}}$$

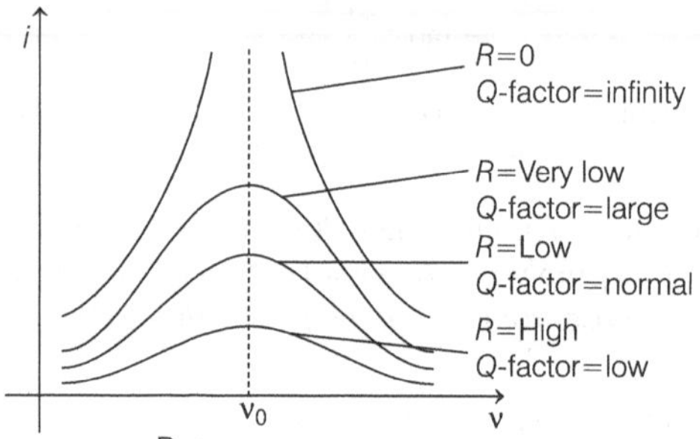

Resonance curve

The quantity ($Q = \omega_0 / 2\Delta\omega$) is also the measure of sharpness of resonance, where $\omega_1 - \omega_2 = 2\Delta\omega$ is called the bandwidth of the circuit, such that $\omega_1 = \omega_0 + \Delta\omega$ and $\omega_2 = \omega_0 - \Delta\omega$.

Example 4. *In a series resonant L-C-R circuit, the voltage across R is 100 V and R = 1kΩ with C = 2 μF. The resonant frequency ω is 200 rad/s. At resonance, the voltage across L is*

(a) 2.5×10^{-2} V (b) 40 V
(c) 250 V (d) 4×10^{-3} V

Sol. (c) At resonance, $\omega L = 1/\omega C$

Current flowing through the circuit, $I = \frac{V_R}{R} = \frac{100}{1000} = 0.1$ A

So, voltage across L is given by $V_L = I X_L = I\omega L$.

But $\omega L = 1/\omega C$

$$\therefore \quad V_L = \frac{I}{\omega C} = V_C = \frac{0.1}{200 \times 2 \times 10^{-6}} = 250 \text{ V}$$

Parallel Circuit (Rejector Circuit)

Consider an alternating source connected across an inductance L in parallel with a capacitor C.

The resistance R is in series with the inductance.

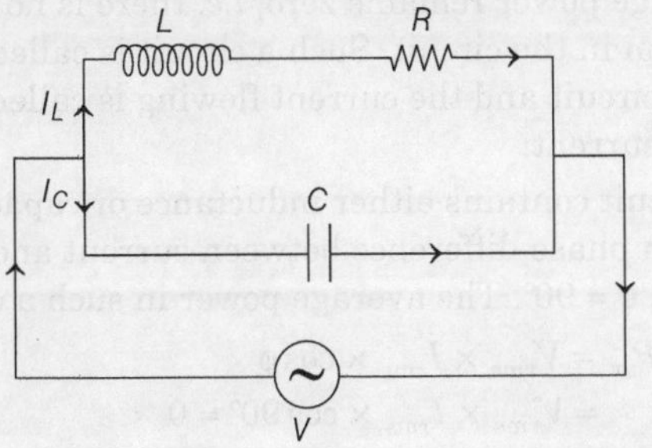

Let the instantaneous value of emf applied be V and the corresponding current be I. Then,

$$I = I_L + I_C$$

$$\Rightarrow \quad \frac{1}{Z} = \frac{1}{R + j\omega L} + j\omega C$$

$\frac{1}{Z}$ is known as admittance (Y). Admittance is defined as the reciprocal of the impedance. Its unit is mho.

$\therefore$ The magnitude of the admittance,

$$|Y| = \left|\frac{1}{Z}\right| = \frac{\sqrt{R^2 + (\omega CR^2 + \omega^3 L^2 C - \omega L)^2}}{R^2 + \omega^2 L^2}$$

The admittance will be minimum, when

$$\omega = \sqrt{\frac{1}{LC} - \frac{R^2}{L^2}}$$

It gives the condition of resonance and the corresponding frequency, is known as *resonance frequency*. It is given as

$$f = \frac{\omega}{2\pi} = \frac{1}{2\pi}\sqrt{\frac{1}{LC} - \frac{R^2}{L^2}}$$

At resonance frequency, admittance is minimum or the impedance is maximum.
Thus, the parallel circuit does not allow this frequency voltage or current from the source to pass in the circuit. Due to this reason the circuit with such a frequency is known as *rejector circuit.*

At resonance, the reactive component of **Y** vanishes or **Y** is real. The reciprocal of the admittance is called the *parallel resistor* or the *dynamic resistance*.

The dynamic resistance is thus, reciprocal of the real part of the admittance.

$$\therefore \quad \text{Dynamic resistance} = \frac{R^2 + \omega^2 L^2}{R}$$

$$\therefore \quad \text{Peak current through the supply} = \frac{V_0}{L/CR} = \frac{V_0 CR}{L}$$

$$\text{The peak current through capacitor} = \frac{V_0}{1/\omega C} = \omega C V_0.$$

The ratio of the peak current through capacitor and through the supply is known as ***Q*-factor**.

$$\text{Thus, } Q\text{-factor} = \frac{V_0 \omega C}{V_0 CR/L} = \frac{\omega L}{R}$$

This is basically the measure of current magnification. The rejector circuit at resonance exhibits current magnification of $\omega L/R$, similar to the voltage magnification of the same ratio exhibited by the series acceptor circuit at resonance.

Fig. (a) shows the variation of current I with angular frequency ω in parallel LC circuit (*i.e.* $R = 0$) and Fig. (b) shows the variation of impedance Z with ω.

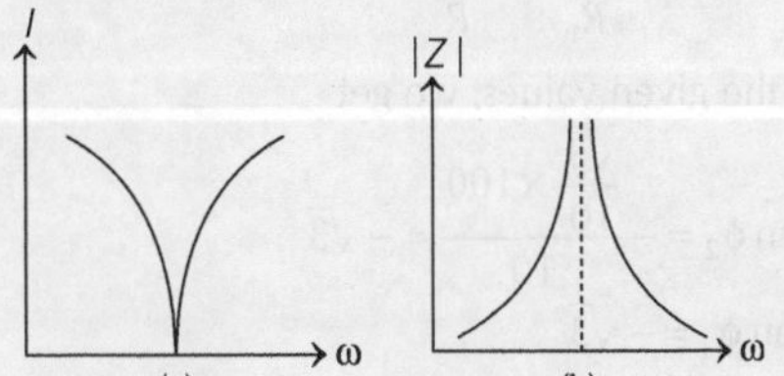

(a) (b)

Note At resonance, the current through the supply and voltage are in phase, while the current through the capacitor leads the voltage by 90°.

Example 5. *An AC circuit has $R = 100\ \Omega$, $C = 2\ \mu F$ and $L = 80$ mH connected in series. The quality factor of the circuit is* **[JEE Main 2020]**

(a) 2 (b) 0.5

(c) 20 (d) 400

Sol. (a) Given, $R = 100\ \Omega$, $C = 2\ \mu$F, $L = 80$ mH

For a series L - C - R AC circuit,

$$\text{Quality factor, } \phi = \frac{1}{R}\sqrt{\frac{L}{C}}$$

$$= \frac{1}{100}\sqrt{\frac{80 \times 10^{-3}}{2 \times 10^{-6}}}$$

$$\phi = 2$$

Example 6. *In the below circuit, $C = \frac{\sqrt{3}}{2}\ \mu F$, $R_2 = 20\ \Omega$, $L = \frac{\sqrt{3}}{10}$ H and $R_1 = 10\ \Omega$. Current in L - R_1 path is I_1 and in C - R_2 path is I_2. The voltage of AC source is given by $V = 200\sqrt{2}\sin(100t)$ volts. The phase difference between I_1 and I_2 is*

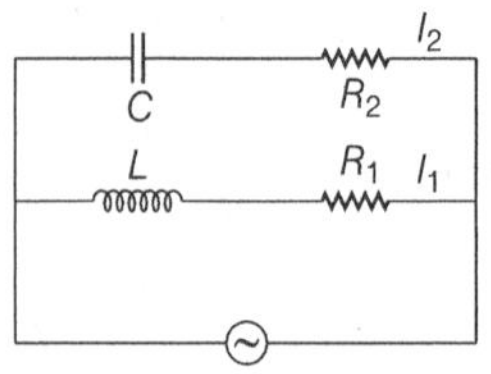

[JEE Main 2019]

(a) 30° (b) 60° (c) 0° (d) 90°

Sol. (a) Phase difference between I_2 and V, i.e. C-R_2 circuit is given by

$$\tan \phi = \frac{X_C}{R_2}$$

$$\Rightarrow \quad \tan \phi = \frac{1}{C\omega R_2}$$

Substituting the given values, we get

$$\tan \phi = \frac{1}{\frac{\sqrt{3}}{2} \times 10^{-6} \times 100 \times 20} = \frac{10^3}{\sqrt{3}}$$

$\therefore \phi_1$ is nearly 90°.

Phase difference between I_1 and V, i.e. in L-R_1 circuit is given by

$$\tan \phi_2 = -\frac{X_L}{R_1} = -\frac{L\omega}{R}$$

Substituting the given values, we get

$$\tan \phi_2 = -\frac{\frac{\sqrt{3}}{10} \times 100}{10} = -\sqrt{3}$$

As, $\tan \phi_2 = -\sqrt{3}$

$\therefore \quad \phi_2 = 120°$

Now, phase difference between I_1 and I_2 is

$$\Delta\phi = \phi_2 - \phi_1 = 120° - 90° = 30°$$

Power in an AC Circuit

It is defined as the rate at which work is being done in the circuit.

Instantaneous power of AC is given by

$$P = Vi = V_0 i_0 \sin \omega t \sin (\omega t + \phi)$$

Average power in an AC circuit,

$$P_{av} = V_{rms} i_{rms} \cos \phi = \frac{V_0}{\sqrt{2}} \frac{i_0}{\sqrt{2}} \cos \phi$$

where, $\cos \phi = \dfrac{\text{resistance } (R)}{\text{impedance } (Z)}$ is called the power factor of AC circuit.

Power factor is defined as cosine of the angle of lag or lead (*i.e.* $\cos \phi$). It is also defined as the ratio of resistance and impedance.

$$\cos \phi = \frac{P_{av}}{V_{rms} I_{rms}} = \frac{\text{True power}}{\text{Apparent power}}$$

The product of V_{rms} and I_{rms} gives the apparent power (or virtual power). While the true power is obtained by multiplying the apparent power by the power factor $\cos \phi$.

Thus, apparent power = $V_{rms} \times I_{rms}$

and true power = apparent power × power factor

Wattless Current

The current which consumes no power for its maintenance in the circuit is called wattless current or idle current.

or

If the resistance in an AC circuit is zero, its power factor will be zero. Although the current flows in the circuit, yet the average power remains zero, *i.e.* there is no energy dissipation in the circuit. Such a circuit is called the wattless circuit and the current flowing is called the wattless current.

If the circuit contains either inductance or capacitance only, then phase difference between current and voltage is 90°, *i.e.* $\phi = 90°$. The average power in such a circuit is

$$P_{av} = V_{rms} \times I_{rms} \times \cos \phi = V_{rms} \times I_{rms} \times \cos 90° = 0$$

The concept of wattless current is similar to that of a frictionless pendulum, where the total work done by gravity upon the pendulum in a cycle is zero.

Special Cases

***Case* I** When AC circuit contains R only. In this case, $\phi = 0$

$$\therefore \quad P_{av} = V_{rms} \times i_{rms} = \frac{V_{rms}^2}{R}$$

***Case* II** When AC circuit contains only capacitor.

In this case, $\phi = -\dfrac{\pi}{2}$

$$\therefore \quad P_{av} = 0$$

***Case* III** When AC circuit contains only inductance.

In this case, $\phi = \dfrac{\pi}{2}$

$$\therefore \quad P_{av} = 0$$

***Case* IV** When AC circuit contains resistance and capacitance both, then

$$\tan \phi = \frac{1/\omega C}{R}$$

and

$$\cos \phi = \frac{R}{\sqrt{R^2 + \frac{1}{\omega^2 C^2}}}$$

$$P_{av} = \frac{V_{rms}^2 R}{\left(R^2 + \frac{1}{\omega^2 C^2}\right)}$$

***Case* V** When AC circuit contains resistance and inductance both.

Now, $\tan\theta = \dfrac{\omega L}{R}$ and $\cos\theta = \dfrac{R}{\sqrt{R^2 + \omega^2 L^2}}$

$$\therefore \quad P_{av} = \frac{V_{rms}^2 R}{(R^2 + \omega L^2)}$$

***Case* VI** When AC circuit contains inductance, capacitance and resistance, then

$$\tan\theta = \frac{\omega L - \dfrac{1}{\omega C}}{R}$$

and $$\cos\theta = \frac{R}{\sqrt{R^2 + \left(\omega L - \dfrac{1}{\omega C}\right)^2}}$$

$$\therefore \quad P_{av} = V_{rms} \times i_{rms} \times \frac{R}{\sqrt{R^2 + \left(\omega L - \dfrac{1}{\omega C}\right)^2}}$$

Example 7. *In a series L-C-R circuit R = 200 Ω and the voltage and the frequency of the main supply is 220 V and 50 Hz, respectively. On taking out the capacitance from the circuit, the current lags behind the voltage by 30°. On taking out the inductor from the circuit, the current leads the voltage by 30°. The power dissipated in the L-C-R circuit is*

(a) 305 W (b) 210 W
(c) zero (d) 242 W

Sol. (d) The given circuit is under resonance as $X_L = X_C$.

(∵ same phase change in *L-R* and *C-R* circuits)

Hence, power dissipated in the circuit is

$$P = \frac{V^2}{R} = 242 \text{ W}$$

Half Power Points in Series *L-C-R* Circuit

The values of ω at which the power input is half of its maximum value are called half power points. At these points,

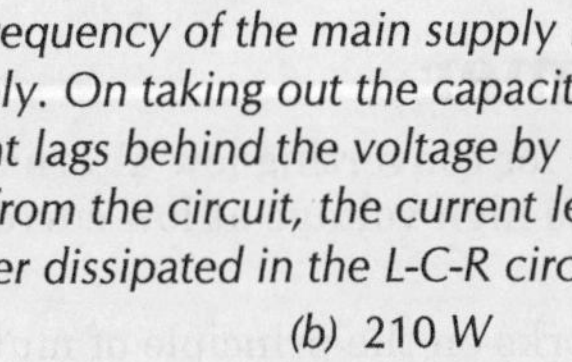

$$\omega = \omega_0 \pm \Delta\omega = \omega_0 \pm \frac{1}{2}\left(\frac{2\Delta\omega}{\omega_0}\right)\omega_0 = \omega_0\left(1 \pm \frac{1}{2Q}\right)$$

The amplitude of the current falls by a factor of $\dfrac{1}{\sqrt{2}}$ and hence, the power ($P \propto I^2$) goes down by half.

The separation between these two half power points is $2\Delta\omega = \dfrac{\omega_0}{Q}$. This separation is called the *full-width* at half maximum and is $\dfrac{1}{Q}$ th fraction of the resonant frequency.

Example 8. *In an AC circuit, the instantaneous emf and current are given by*

$$e = 100 \sin 30t, \; i = 20 \sin\left(30t - \frac{\pi}{4}\right)$$

In one cycle of AC, the average power (in W) consumed by the circuit and the wattless current (in A) are, respectively

[JEE Main 2018]

(a) 50 , 10 (b) $\dfrac{1000}{\sqrt{2}}$, 10 (c) $\dfrac{50}{\sqrt{2}}$, 0 (d) 50 , 0

Sol. (b) Given, $e = 100 \sin 30t$ and $i = 20 \sin\left(30t - \dfrac{\pi}{4}\right)$

∴ Average power, $P_{av} = V_{rms} I_{rms} \cos\phi$

$$= \frac{100}{\sqrt{2}} \times \frac{20}{\sqrt{2}} \times \cos\frac{\pi}{4} = \frac{1000}{\sqrt{2}} \text{ W}$$

Wattless current, $I = I_{rms} \sin\phi = \dfrac{20}{\sqrt{2}} \times \sin\dfrac{\pi}{4} = \dfrac{20}{2} = 10$ A

$$\therefore \quad P_{av} = \frac{1000}{\sqrt{2}} \text{ watt and } I_{wattless} = 10 \text{ A}$$

Example 9. *The power factor of an AC circuit having resistance R and inductance L (connected in series) and an angular velocity ω is*

(a) $\dfrac{R}{\omega L}$ (b) $\dfrac{R}{(R^2 + \omega^2 L^2)^{1/2}}$

(c) $\dfrac{\omega L}{R}$ (d) $\dfrac{R}{(R^2 - \omega^2 L^2)^{1/2}}$

Sol. (b) From the relation, $\tan\phi = \dfrac{\omega L}{R}$

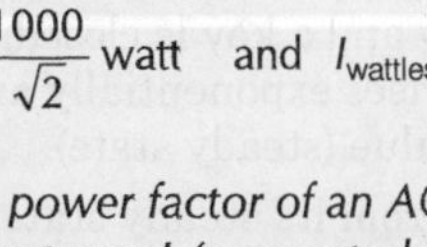

Power factor, $\cos\phi = \dfrac{1}{\sqrt{1 + \tan^2\phi}}$

$$\Rightarrow \quad \cos\phi = \frac{R}{\sqrt{R^2 + \omega^2 L^2}}$$

$$= \frac{1}{\sqrt{1 + (\omega L/R)^2}}$$

$$= \frac{1}{\sqrt{1 + \tan^2\phi}}$$

$$= \frac{1}{\sqrt{1 + (\omega L/R)^2}} = \frac{R}{\sqrt{R^2 + \omega^2 L^2}}$$

L-C Oscillations

When a charged capacitor is allowed to discharge through a non-resistive inductor, electrical oscillations of constant amplitude and frequency are produced. These oscillations are called *L-C* oscillations. The equation of *L-C* oscillations is given by

$$\frac{d^2q}{dt^2} + \frac{1}{LC}q = 0$$

and the charge oscillates with a frequency,

$$\nu = \frac{\omega}{2\pi} = \frac{1}{2\pi\sqrt{LC}}$$

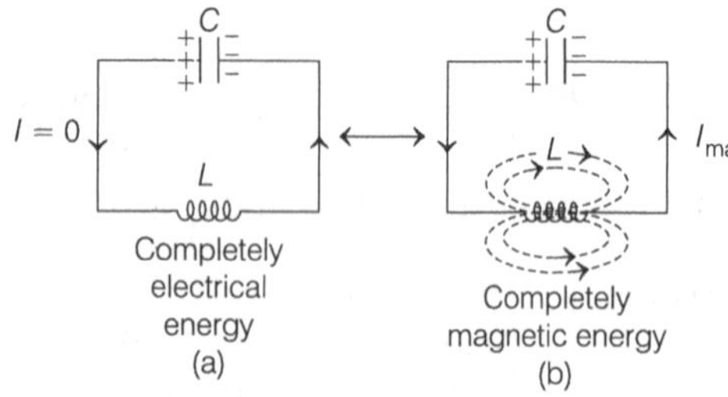

Growth and Decay of Current in *L-R* Circuit

If a circuit containing a pure inductor L and a resistor R in series with a battery and a key is closed, then the current through the circuit rises exponentially and reaches upto a certain maximum value (steady state).

If circuit is opened from its steady state condition, battery is removed and it is closed again, then current through the circuit decreases exponentially.

In case of growth of current at time $t = 0$, inductor offers infinite resistance and at time $t = \infty$ (steady state) inductor offers zero resistance. This type of circuit is called *L-R* circuit.

Growth of Current in *L-R* Circuit

Growth of current in *L-R* circuit at any instant of time t is given by $I = I_0(1 - e^{-Rt/L})$

or $\quad I = I_0(1 - e^{-t/5})$

where, I_0 = maximum current, L = self-inductance of the inductor and R = resistance of the circuit.

Here, $\frac{L}{R} = \tau$ is called *time constant* of an *L-R* circuit.

Time constant of an *L-R* circuit is the time in which current in the circuit grows to 63.2% of the maximum value of current.

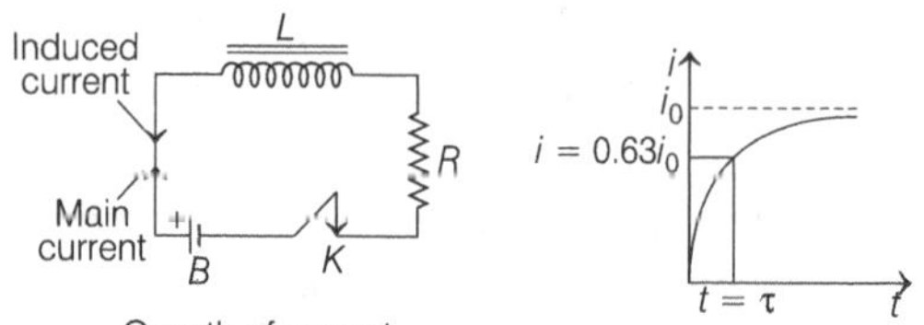

Growth of current

Decay of Current in *L-R* Circuit

Decay of current in *L-R* circuit at any time t is given by

$$I = I_0\, e^{-Rt/L} \text{ or } I = I_0 e^{-t/\tau}$$

Time constant (τ) of an *L-R* circuit is the time in which current decays to 36.8% of the maximum value of current (I_0).

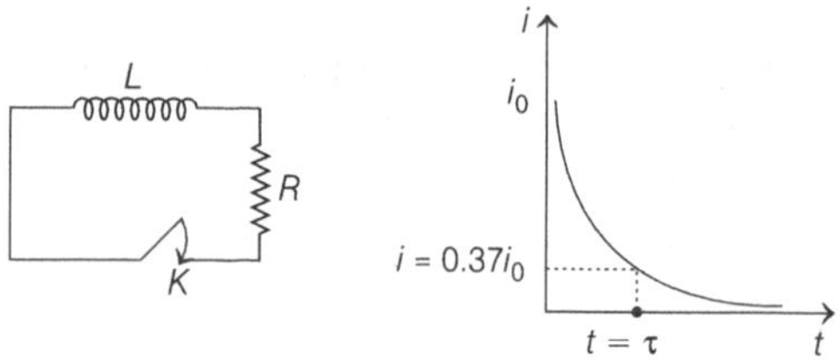

Choke Coil

It is a device having high inductance and negligible resistance. It is used in AC circuits for the purpose of adjusting current to any required value in such a way that power loss in the circuit can be minimised. It is used in fluorescent tubes.

It is based on the principle of wattless current.

Transformer

The device used for converting low alternating voltage at high current into high voltage at low current and *vice-versa*.

Principle It works on the principle of mutual induction, *i.e.* if two coils are inductively coupled and when current or magnetic flux is changed through one of the two coils, then induced emf is produced in the other coil.

Types of Transformers

There are two types of transformers

Step-up Transformer The transformer used to change low voltage alternating emf to high voltage alternating emf (of same frequency) is called as step-up transformer.

In this transformer, $n_s > n_p, V_s > V_p$ and $i_p > i_s$

Step-down Transformer The transformer used to change high voltage alternating emf to low voltage alternating emf of same frequency is called a step-down transformer. In this transformer,

$$n_p > n_s, v_p > v_s \text{ and } i_p < i_s$$

where, n_p = number of turns in primary coil,

n_s = number of turns in secondary coil,

V_p = voltage drop across primary coil

and $\quad V_s$ = voltage induced across secondary coil.

Transformation ratio

If the number of turns in the primary coil is n_p, the number of turns in the secondary coil is n_s and the magnetic flux linked with each turns is ϕ, then for an ideal transformer when there is no leakage of flux, the magnetic flux linked with each turn of primary and secondary coil will be ϕ. Then,

$$\frac{V_s}{V_p} = \frac{n_s}{n_p}$$

This is known as transformation ratio.

Efficiency of Transformer

Efficiency of a transformer

$$= \frac{\text{Energy obtained from the secondary coil}}{\text{Energy given to the primary coil}}$$

$$= \frac{\text{Output power}}{\text{Input power}}$$

$$= \frac{V_s i_s}{V_p i_p}$$

Example 10. *A power transmission line feeds input power at 2300 V to a step-down transformer with its primary windings having 4000 turns. The output power is delivered at 230 V by the transformer. If the current in the primary of the transformer is 5A and its efficiency is 90%, the output current would be* **[JEE Main 2019]**

(a) 45 A *(b) 50 A*

(c) 25 A *(d) 35 A*

Sol. (a) For a transformer, there are two circuits which have N_p and N_s (number of coil turns), I_p and I_s (currents) respectively as shown below

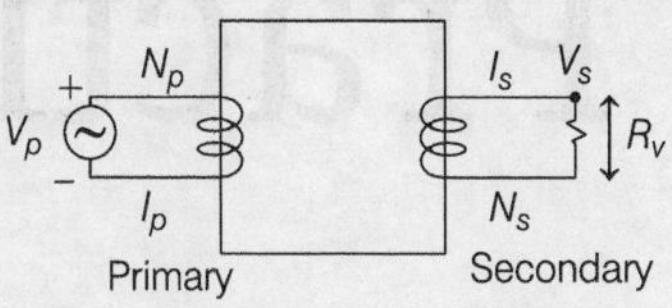

Here, input voltage, $V_p = 2300$ V

Number of turns in primary coil, $N_p = 4000$

Output voltage, $V_s = 230$ V

Output power, $P_s = V_s \cdot I_s$

Input power, $P_p = V_p I_p$

$\therefore$ The efficiency of the transformer is

$$\eta = \frac{\text{Output (secondary) power}}{\text{Input (primary) power}}$$

$$= \frac{V_s \cdot I_s}{V_p \cdot I_p} \times 100$$

$$\Rightarrow \quad \eta = \frac{(230)\,(I_s)}{(2300)\,(5)} \times 100$$

$$90 = \frac{230\, I_s}{(2300) \times 5} \times 100$$

$$\Rightarrow \quad I_s = 45 \text{ A}$$

Practice Exercise

ROUND I Topically Divided Problems

Peak Value, RMS Value and Average Value of AC

1. The voltage of an AC supply varies with time (t) as $V = 120 \sin 100\pi t \cos 100\pi t$. The maximum voltage and frequency respectively are

(a) 120 V 100 Hz (b) $\frac{120}{\sqrt{2}}$ V, 100 Hz

(c) 60 V, 200 Hz (d) 60 V, 100 Hz

2. An alternating current having peak value 14 A used to heat a metal wire. To produce the same heating effect, a constant current I can be used, where I is

(a) 14 A (b) 20 A

(c) 7 A (d) 10 A

3. If the rms current in a 50 Hz AC circuit is 5 A, the value of the current (1/300) s after its value becomes zero is **[NCERT Exemplar]**

(a) $5\sqrt{2}$ A (b) $5\sqrt{3/2}$ A

(c) (5/6) A (d) $(5/\sqrt{2})$ A

4. The output sinusoidal current *versus* time curve of a rectifier is shown in the figure. The average value of output current in this case is

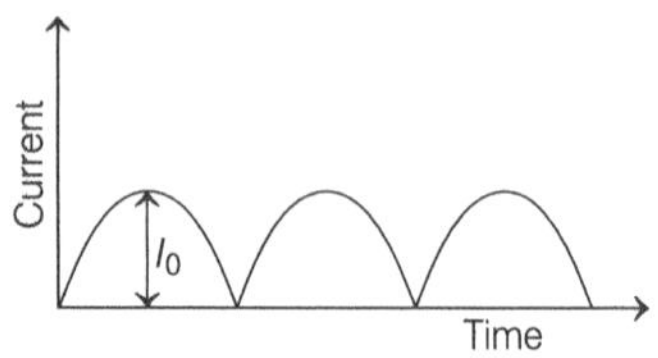

(a) 0 (b) $\frac{I_0}{2}$

(c) $\frac{2I_0}{\pi}$ (d) I_0

5. An alternating current is given by $I = I_1 \cos \omega t + I_2 \sin \omega t$. The root mean square current is given by

(a) $\frac{(I_1 + I_2)}{\sqrt{2}}$ (b) $\frac{(I_1 + I_2)^2}{2}$

(c) $\sqrt{\frac{I_1^2 + I_2^2}{2}}$ (d) $\frac{\sqrt{I_1^2 - I_2^2}}{2}$

6. An AC voltage source has an output of $V = (200 \text{ V}) \sin 2\pi ft$. This source is connected to a 100 Ω resistor. Rms current in the resistance is

(a) 1.41 A (b) 2.41 A

(c) 3.41 A (d) 0.71 A

AC Circuits

7. A 60 μF capacitor is connected to a 110 V, 60 Hz AC supply. Determine the rms value of the current in the circuit. **[NCERT]**

(a) 2.5 A (b) 2.1 A (c) 3.1 A (d) 3.5 A

8. A group of electric lamps having a total power rating of 1000 W is connected by an AC voltage $E = 200 \sin(310t + 60°)$, then the rms value of the circuit current is

(a) 10 A (b) $10\sqrt{2}$ A (c) 20 A (d) $20\sqrt{2}$ A

9. A circuit when connected to an AC source of 12 V gives a current of 0.2 A. The same circuit when connected to a DC source of 12 V, gives a current of 0.4 A. The circuit is

(a) series L-R (b) series R-C

(c) series L-C (d) series L-C-R

10. In AC series circuit, the resistance, inductive reactance and capacitive reactance are 3 Ω, 10 Ω and 14 Ω, respectively. The impedance of the circuit is

(a) 5 Ω (b) 4 Ω (c) 7 Ω (d) 10 Ω

11. In an L-C-R circuit, $R = 100\ \Omega$. When capacitance C is removed, the current lags behind the voltage by $\pi/3$. When inductance L is removed, the current leads the voltage by $\pi/3$. The impedance of the circuit is

(a) 50 Ω (b) 100 Ω

(c) 200 Ω (d) 400 Ω

12. For a series R-L-C circuit, $R = X_L = 2X_C$. The impedance of the circuits and phase difference (between) V and i will be

(a) $\frac{\sqrt{5}R}{2}, \tan^{-1}(2)$ (b) $\frac{\sqrt{5}R}{2}, \tan^{-1}\left(\frac{1}{2}\right)$

(c) $\sqrt{5}X_C, \tan^{-1}(2)$ (d) $\sqrt{5}R, \tan^{-1}\left(\frac{1}{2}\right)$

13. An inductance coil has a reactance of 100 Ω. When an AC signal of frequency 1000 Hz is applied to the coil, the applied voltage leads the current by 45°. The self-inductance of the coil is **[JEE Main 2020]**

(a) 1.1×10^{-2} H
(b) 1.1×10^{-1} H
(c) 5.5×10^{-5} H
(d) 6.7×10^{-7} H

14. A 100 μF capacitor in series with a 40 Ω resistance is connected to a 110 V-60 Hz supply. Calculate the maximum current in the circuit and the phase lag between the current maximum and voltage maximum?

(a) 1.5 A, 33° 33′
(b) 1.5 A, 60°
(c) 3.2 A, 33° 33′
(d) 3.2 A, 60°

15. An AC source of angular frequency ω is fed across a resistor R and a capacitor C in series. The current registered is I. If now the frequency of source is changed to ω/ 3 (but maintaining the same voltage), the current in then circuit is found to be halved. Calculate the ratio of reactance to resistance at the original frequency ω.

(a) $\sqrt{\frac{3}{5}}$ (b) $\sqrt{\frac{2}{5}}$
(c) $\sqrt{\frac{1}{5}}$ (d) $\sqrt{\frac{4}{5}}$

16. A series of R-C circuit is connected to AC voltage source. Consider two cases : (A) when C is without a dielectric medium and (B) when C is filled with dielectric of constant 4. The current I_R through the resistor and voltage V_C across the capacitor are compared in the two cases. Which of the following is/are true? **[IIT JEE 2011]**

(a) $I_R^A > I_R^B$ (b) $I_R^A < I_R^B$
(c) $V_C^A = V_C^B$ (d) $V_C^A < V_C^B$

17. In a series L-C-R circuit, the inductive reactance X_L is 10 Ω and the capacitive reactance X_C is 4 Ω. The resistance R in the circuit is 6 Ω. The power factor of the circuit is **[JEE Main 2021]**

(a) $\frac{1}{2}$ (b) $\frac{1}{2\sqrt{2}}$
(c) $\frac{1}{\sqrt{2}}$ (d) $\frac{\sqrt{3}}{2}$

18. In L-C circuit, the inductance L = 40 mH and capacitance C = 100μF. If a voltage $V(t) = 10\sin(314t)$ is applied to the circuit, the current in the circuit is given as **[JEE Main 2020]**

(a) $0.52\cos 314\,t$ (b) $0.52\sin 314\,t$
(c) $10\cos 314\,t$ (d) $5.2\cos 314\,t$

19. The phase difference between the alternating current and emf is π/2. Which of the following cannot be the constituent of the circuit?

(a) C alone (b) R, L (c) L, C (d) L alone

20. In the given circuit, the AC source has ω = 100 rad/s Considering the inductor and capacitor to be ideal. The correct choice (s) is/are

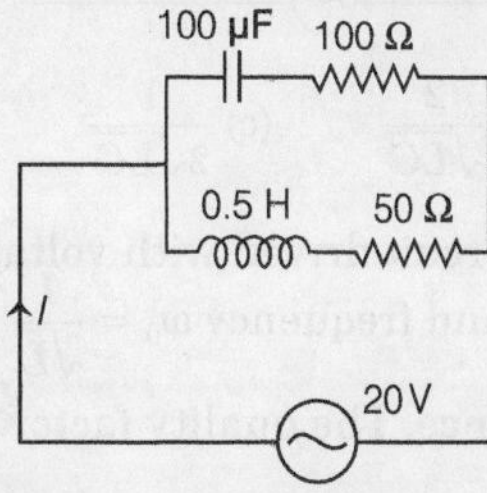

(a) The current through the circuit I is 0.3 A
(b) The current through the circuit I is $0.3\sqrt{2}$ A
(c) The voltage across 100 Ω resistor/s = $10\sqrt{2}$ V
(d) The voltage across 50 Ω resistor/s = 10 V

Resonance and Quality Factor

21. In the circuit shown in the figure, the AC source gives a voltage $V = 20\cos(2000t)$. Neglecting source resistance, the voltmeter and ammeter reading will be

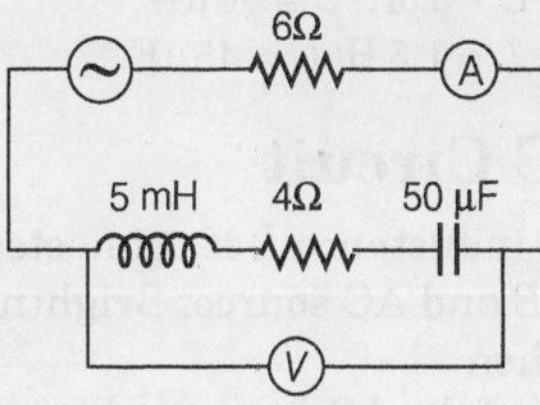

(a) 0 V, 0.47 A (b) 1.68 V, 0.47 A
(c) 0 V, 1.4 A (d) 5.6 V, 1.4 A

22. In a series L-C-R resonance circuit, if we change the resistance only, from a lower to higher value **[JEE Main 2021]**

(a) the bandwidth of resonance circuit will increase
(b) the resonance frequency will increase
(c) the quality factor will increase
(d) the quality factor and the resonance frequency will remain constant

23. In an L-C-R circuit, capacitance is changed from C to $2C$. For the resonant frequency to remains unchanged, the inductance should be changed from L to

(a) $4L$ (b) $2L$ (c) $L/2$ (d) $L/4$

24. The self-inductance of the motor of an electric fan is 10 H. In order to impart maximum power at 50 Hz, it should be connected to a capacitance of

(a) 4 μF (b) 8 μF
(c) 1 μF (d) 2 μF

25. Find resonance frequency in the given circuit.

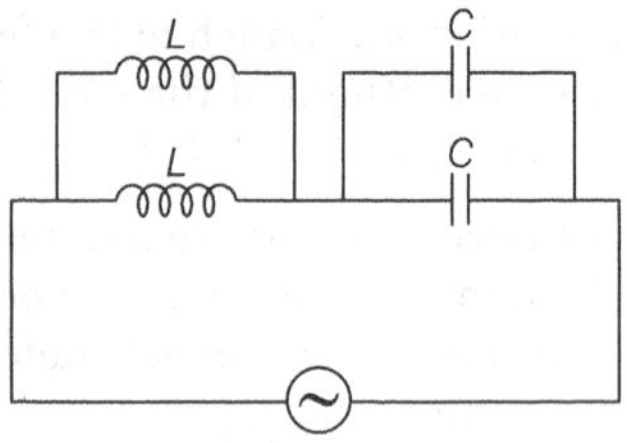

(a) $\frac{1}{\sqrt{LC}}$ (b) $\frac{2}{\sqrt{LC}}$ (c) $\frac{1}{2\sqrt{LC}}$ (d) $\frac{4}{\sqrt{LC}}$

26. For an *R-L-C* circuit driven with voltage of amplitude V_m and frequency $\omega_0 = \frac{1}{\sqrt{L}}$, the current exhibits resonance. The quality factor Q is given by **[JEE Main 2018]**

(a) $\frac{\omega_0 L}{R}$ (b) $\frac{\omega_0 R}{L}$

(c) $\frac{R}{\omega_0 C}$ (d) $\frac{CR}{\omega_0}$

27. Which of the following combinations should be selected for better tuning of an *L-C-R* circuit used for communication? **[NCERT Exemplar]**

(a) $R = 20\ \Omega$, $L = 1.5\text{H}$, $C = 35\ \mu\text{F}$
(b) $R = 25\ \Omega$, $L = 2.5\ \text{H}$, $C = 45\ \mu\text{F}$
(c) $R = 15\ \Omega$, $L = 3.5\text{H}$, $C = 30\ \mu\text{F}$
(d) $R = 25\ \Omega$, $L = 1.5\ \text{H}$, $C = 45\ \mu\text{F}$

Power in AC Circuit

28. A coil of self-inductance L is connected in series with a bulb B and AC source. Brightness of the bulb decreases when

(a) frequency of the AC source is decreased
(b) number of turns in the coil is reduced
(c) a capacitance of reactance $X_C = X_L$ is included in the same circuit
(d) an iron rod is inserted in the coil

29. In an *L-R* circuit, the inductive reactance is equal to the resistance R of the circuit. An emf $E = E_0 \cos(\omega t)$ applied to the circuit. The power consumed in the circuit is

(a) $\frac{E_0^2}{R}$ (b) $\frac{E_0^2}{2R}$

(c) $\frac{E_0^2}{4R}$ (d) $\frac{E_0^2}{8R}$

30. The self-inductance of a choke coil is 10 mH. When it is connected with a 10 V DC source, then the loss of power is 20 W. When it is connected with 10 V AC source loss of power is 10 W, the frequency of AC source will be

(a) 50 Hz (b) 60 Hz
(c) 80 Hz (d) 100 Hz

31. $\frac{2.5}{\pi}\ \mu\text{F}$ capacitor and 3000 Ω resistance are joined in series to an AC source of 200 V and 50 s^{-1} frequency. The power factor of the circuit and the power dissipated in it will respectively

(a) 0.6, 0.06 W (b) 0.06, 0.6 W
(c) 0.6, 4.8 W (d) 4.8, 0.6 W

32. A resistance R draws power P when connected to an AC source. If an inductance is now placed in series with the resistance, such that the impedance of the circuit becomes Z, the power drawn will be

(a) $P\left(\frac{R}{Z}\right)$ (b) $P\sqrt{\frac{R}{Z}}$ (c) $P\ (R/Z)^2$ (d) P

33. In a series *L-R* circuit, power of 400W is dissipated from a source of 250 V, 50 Hz. The power factor of the circuit is 0.8. In order to bring the power factor to unity, a capacitor of value C is added in series to the L and R. Taking the value of C as $\left(\frac{n}{3\pi}\right)\mu\text{F}$, then value of n is **[JEE Main 2020]**

(a) 100 (b) 200
(c) 300 (d) 400

Growth and Decay of Current

34. An inductor of inductance $L = 400$ mH and resistors of resistances $R_1 = 4\ \Omega$ and $R_2 = 2\ \Omega$ are connected to battery of emf 12 V as shown in the figure. The internal resistance of the battery is negligible. The switch S is closed at $t = 0$. The potential drop across L as a function of time is

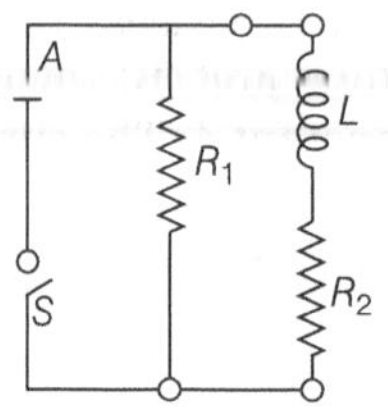

(a) $6e^{-5t}$ V (b) $\frac{12}{t}e^{-3t}$ V

(c) $6(1 - e^{-t/0.2})$ V (d) $12e^{-5t}$ V

35. An emf of 15 V is applied in a circuit coil containing 5 H inductance and 10 Ω resistance, the ratio of currents at time $t = \infty$ and $t = 1$ s is

(a) $\frac{e^{1/2}}{e^{1/2} - 1}$ (b) $\frac{e^2}{e^2 - 1}$

(c) $1 - e^{-1}$ (d) e^{-1}

36. A coil of inductance 300 mH and resistance 2 Ω is connected to a source of voltage 2 V. The current reaches half of its steady state value in

(a) 0.05 s (b) 0.1 s
(c) 0.15 s (d) 0.3 s

37. A coil of self-inductance 10 mH and resistance 0.1 Ω is connected through a switch to a battery of internal resistance 0.9 Ω. After the switch is closed, the time taken for the current to attain 80% of the saturation value is (Take, ln 5 = 1.6) **[JEE Main 2019]**

(a) 0.002 s (b) 0.324 s
(c) 0.103 s (d) 0.016 s

38. Consider the L-R circuit shown in the figure. If the switch S is closed at $t = 0$, then the amount of charge that passes through the battery between $t = 0$ and $t = \dfrac{L}{R}$ is **[JEE Main 2019]**

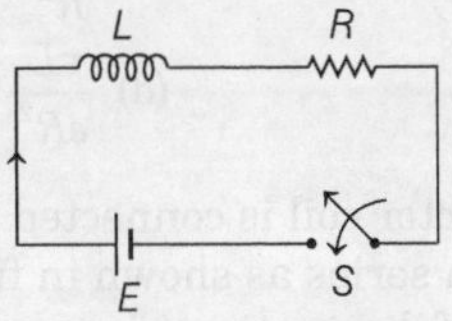

(a) $\dfrac{2.7\,EL}{R^2}$ (b) $\dfrac{EL}{2.7\,R^2}$
(c) $\dfrac{7.3\,EL}{R^2}$ (d) $\dfrac{EL}{7.3\,R^2}$

39. The figure shows an experimental plot discharging of a capacitor in an R-C circuit. The time constant τ of this circuit lies between

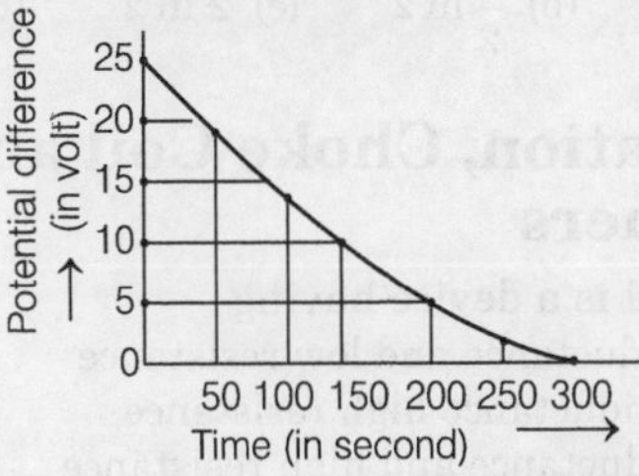

(a) 150 s and 200 s (b) 0 and 50 s
(c) 50 s and 100 s (d) 100 s and 150 s

40. An L-C-R circuit is equivalent to a damped pendulum. In an L-C-R circuit, the capacitor is charged to Q_0 and then connected to the L and R as shown below

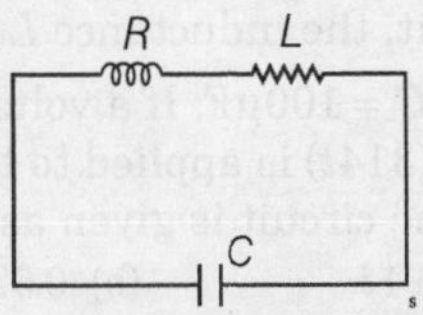

If a student plots graphs of the square of maximum charge (Q^2_{max}) on the capacitor with time (t) for two different values L_1 and L_2 ($L_1 > L_2$) of L, then which of the following represents this graph correctly? (Plots are schematic and not drawn to scale) **[JEE Main 2015]**

(a) 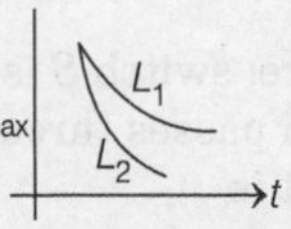

(b) 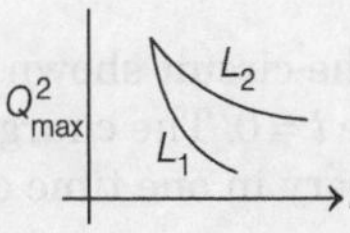

(c) 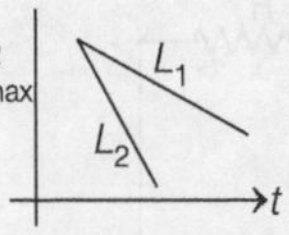

(d) 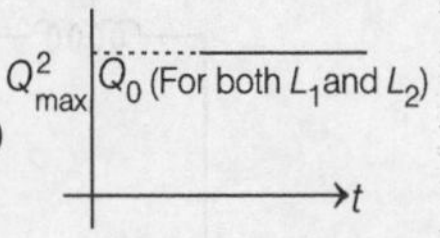

41. An inductor ($L = 0.03$ H) and a resistor ($R = 0.15$ kΩ) are connected in series to a battery of 15V emf in a circuit shown below. The key K_1 has been kept closed for a long time, then at $t = 0$, K_1 is opened and key K_2 is closed simultaneously. At $t = 1$ ms, the current in the circuit will be ($e^5 \cong 150$) **[JEE Main 2015]**

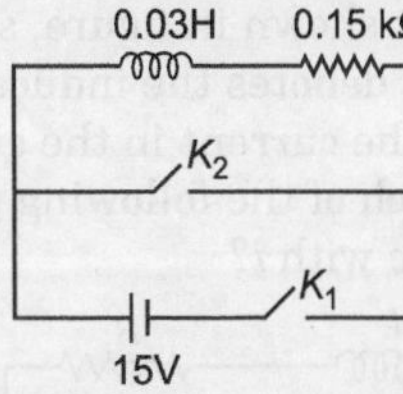

(a) 100 mA (b) 67 mA (c) 6.7 mA (d) 0.67mA

42. A series L-R circuit is connected to a battery of emf V. If the circuit is switched ON at $t = 0$, then the time at which the energy stored in the inductor reaches $\left(\dfrac{1}{n}\right)$ times of its maximum value, is **[JEE Main 2020]**

(a) $\dfrac{L}{R}\ln\left(\dfrac{\sqrt{n}+1}{\sqrt{n}-1}\right)$ (b) $\dfrac{L}{R}\ln\left(\dfrac{\sqrt{n}}{\sqrt{n}+1}\right)$

(c) $\dfrac{L}{R}\ln\left(\dfrac{\sqrt{n}-1}{\sqrt{n}}\right)$ (d) $\dfrac{L}{R}\ln\left(\dfrac{\sqrt{n}}{\sqrt{n}-1}\right)$

43. An emf of 20 V is applied at time $t = 0$ to a circuit containing in series 10 mH inductor and 5Ω resistor. The ratio of the currents at time $t = \infty$ and at $t = 40$ s is close to (Take, $e^2 = 7.389$) **[JEE Main 2020]**

(a) 1.06 (b) 1.15 (c) 1.46 (d) 0.84

44. In the figure shown, a circuit contains two identical resistors with resistance $R = 5\,\Omega$ and an inductance with $L = 2$ mH. An ideal battery of 15 V is connected in the circuit. **[JEE Main 2019]**

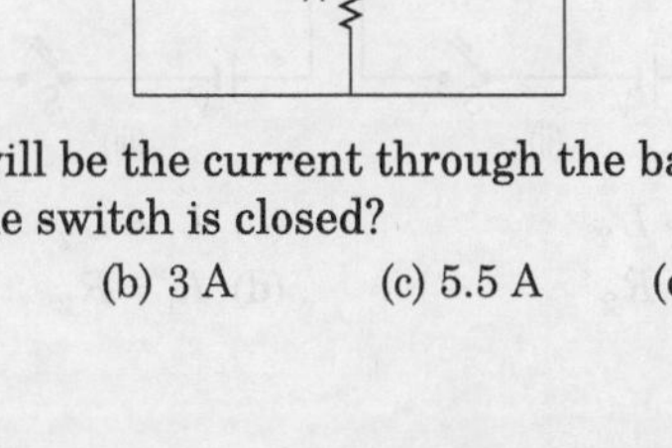

What will be the current through the battery long after the switch is closed?

(a) 6 A (b) 3 A (c) 5.5 A (d) 7.5 A

45. In the circuit shown in figure, switch S is closed at time $t = 0$. The charge which passes through the battery in one time constant is

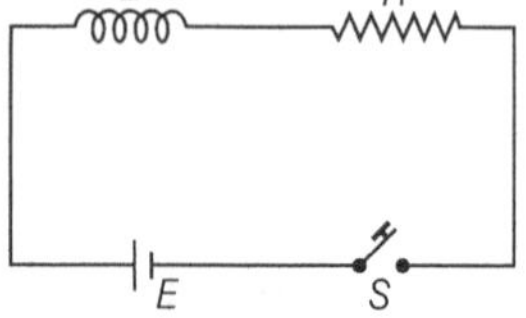

(a) $\frac{EL}{eR^2}$
(b) $\frac{eL}{ER}$
(c) $\frac{eR^2E}{L}$
(d) $E\left(\frac{L}{R}\right)$

46. In an L-R circuit shown in figure, switch S is closed at time $t = 0$. If e denotes the induced emf across inductor and i, the current in the circuit at any time t, then which of the following graphs, shows the variation of e with i?

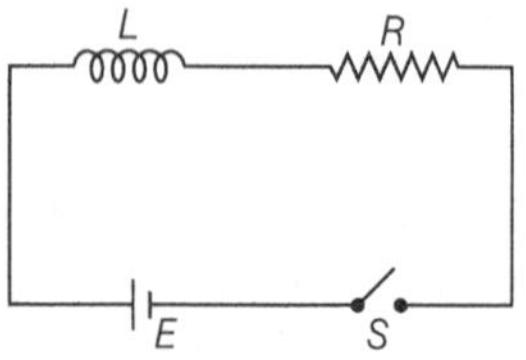

(a)
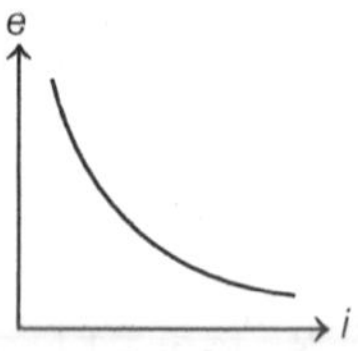

(b)
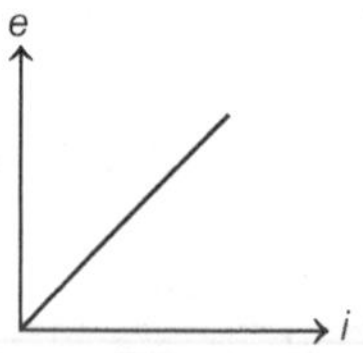

(c)
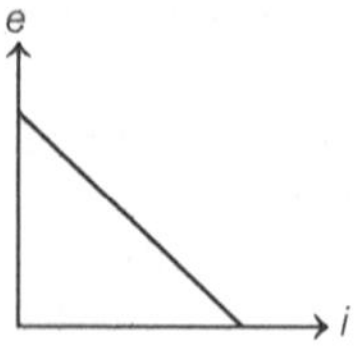

(d)
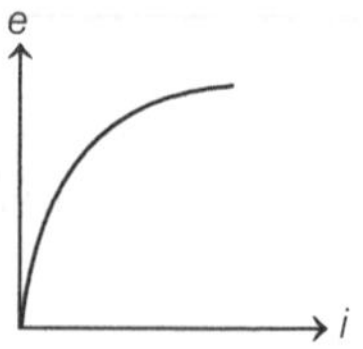

47. Current growth in two L-R circuits (ii) and (iii) is as shown in Fig (i). Let L_1, L_2, R_1 and R_2 be the corresponding values in two circuits, then

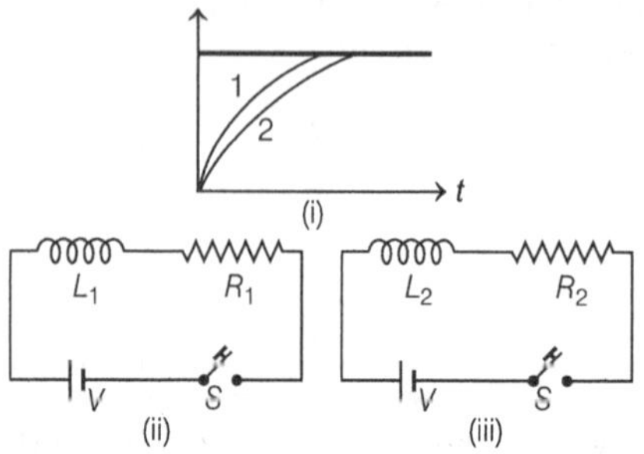

(a) $L_1 > L_2$
(b) $L_1 < L_2$
(c) $R_1 > R_2$
(d) $R_1 = R_2$

48. As shown in the figure, a battery of emf ε is connected to an inductor L and resistance R in series. The switch is closed at $t = 0$. The total charge that flows from the battery, between $t = 0$ and $t = t_c$ (t_c is the time constant of the circuit) is **[JEE Main 2020]**

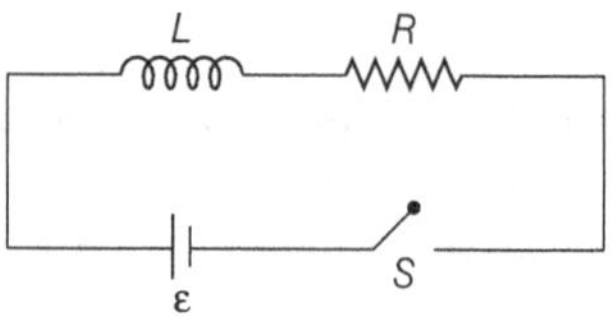

(a) $\frac{\varepsilon L}{R^2}\left(1 - \frac{1}{e}\right)$
(b) $\frac{\varepsilon L}{R^2}$
(c) $\frac{\varepsilon R}{eL^2}$
(d) $\frac{\varepsilon L}{eR^2}$

49. A 20 H inductor coil is connected to a 10 Ω resistance in series as shown in figure. The time at which rate of dissipation of energy (Joule's heat) across resistance is equal to the rate at which magnetic energy is stored in the inductor, is **[JEE Main 2019]**

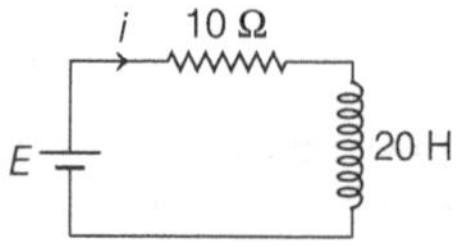

(a) $\frac{2}{\ln 2}$
(b) $\frac{1}{2}\ln 2$
(c) $2 \ln 2$
(d) $\ln 2$

L-C Oscillation, Choke Coil and Transformers

50. Choke coil is a device having
(a) low inductance and low resistance
(b) high inductance high resistance
(c) low inductance and high resistance
(d) high inductance and low resistance

51. A charged 40 μF capacitor is connected to a 16 mH inductor. What is the angular frequency of free oscillations of the circuit?
(a) 1.1 s
(b) $1.25 \times 10^3\ s^{-1}$
(c) $2 \times 10^3\ s^{-1}$
(d) $2.5 \times 10^3\ s^{-1}$

52. In L-C circuit, the inductance $L = 40$ mH and capacitance $C = 100\mu F$. If a voltage $V(t) = 10\sin(314t)$ is applied to the circuit, the current in the circuit is given as **[JEE Main 2020]**
(a) $0.52 \cos 314\, t$
(b) $0.52 \sin 314\, t$
(c) $10 \cos 314\, t$
(d) $5.2 \cos 314\, t$

53. A 1 μF capacitor is charged to 25 V of potential. The battery is then disconnected and a pure 100 mH coil is connected across the capacitor, so that L-C oscillations are set up. The maximum current in the coil is
(a) 0.25 A
(b) 0.01 A
(c) 2.5 A
(d) 1.6 A

54. If maximum energy is stored in a capacitor at $t = 0$, then find the time after which, current in the circuit will be maximum?

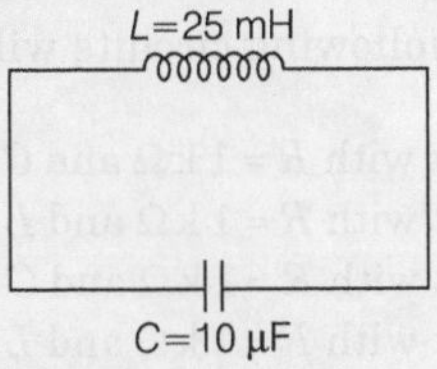

(a) $\frac{\pi}{2}$ ms (b) $\frac{\pi}{4}$ ms (c) π ms (d) 2 ms

55. The core of any transformer is laminated so as to
(a) reduce the energy loss due to eddy currents
(b) make it light weight
(c) make it robust and strong
(d) increase the secondary voltage

56. The turns ratio of transformer is given as 2 : 3. If the current passing through the primary coil is 3 A. Find the current through the load resistance.
(a) 4.5 A (b) 1.5 A (c) 2 A (d) 1 A

57. The number of turns in the primary coil of a transformer is 200 and the number of turns in secondary coil is 10. If 240 V AC is applied to the primary, the output from secondary will be
(a) 48 V (b) 24 V (c) 12 V (d) 6 V

58. In an induction coil, the coefficient of mutual inductance is 4H. If current of 5A in the primary coil is cut-off for 1/1500 s, the emf at the terminals of the secondary coil will be
(a) 15 kV (b) 60 kV
(c) 10 kV (d) 30 kV

59. A transformer consisting of 300 turns in the primary and 150 turns in the secondary gives output power of 2.2 kW. If the current in the secondary coil is 10 A, then the input voltage and current in the primary coil are **[JEE Main 2019]**
(a) 440 V and 5 A
(b) 220 V and 20 A
(c) 220 V and 10 A
(d) 440 V and 20 A

ROUND II Mixed Bag

Only One Correct Option

1. An alternating emf is applied across a parallel combination of a resistance R, capacitance C and an inductance L. If I_R, I_L and I_C are the currents through R, L and C respectively, then the diagram which correctly represents, the phase relationship among I_R, I_L, I_C and source emf E, is given by

(a)
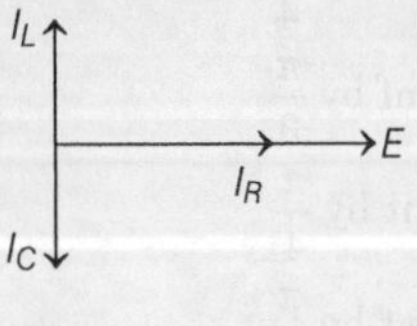

(b)
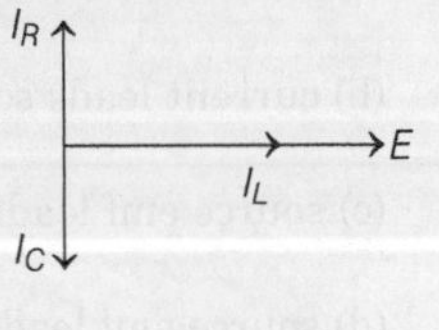

(c)
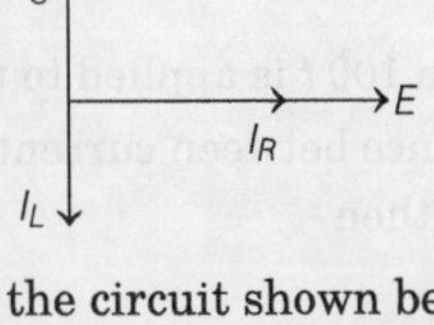

(d)
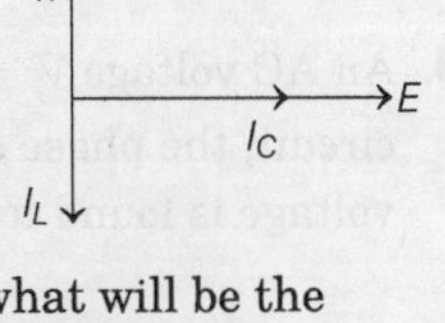

2. In the circuit shown below, what will be the readings of the voltmeter and ammeter?

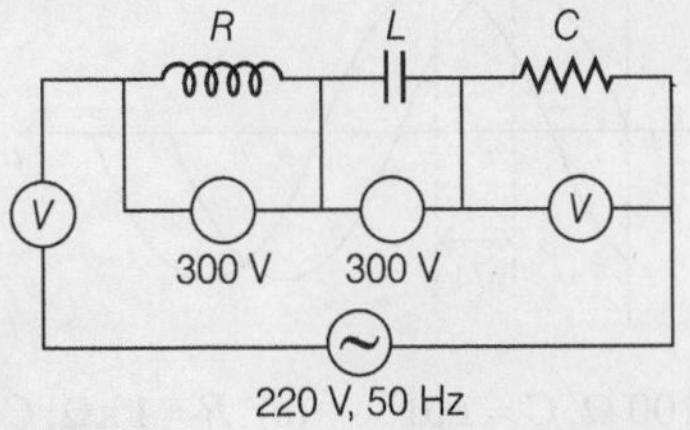

(a) 800 V, 2A (b) 300 V, 2A
(c) 220 V, 2.2 A (d) 100 V, 2A

3. An AC source rated 220 V, 50 Hz is connected to a resistor. The time taken by the current to change from its maximum to the rms value is **[JEE Main 2021]**
(a) 2.5 ms (b) 25 ms
(c) 2.5 s (d) 0.25 ms

4. In the circuit shown in figure neglecting source resistance, the voltmeter and ammeter readings will be respectively,

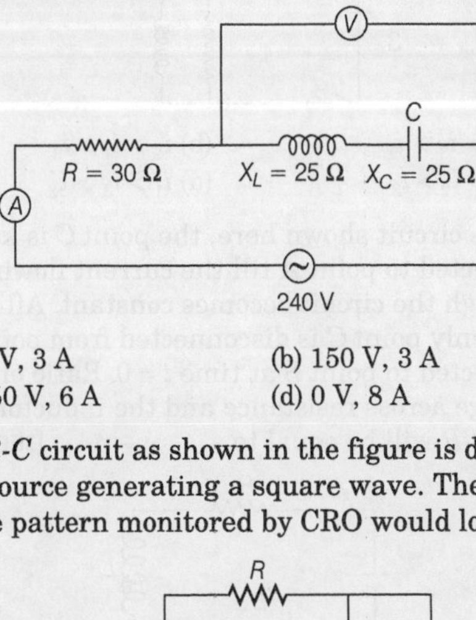

(a) 0 V, 3 A (b) 150 V, 3 A
(c) 150 V, 6 A (d) 0 V, 8 A

5. An R-C circuit as shown in the figure is driven by a AC source generating a square wave. The output wave pattern monitored by CRO would look close to

(a)  (b) 

(c)  (d) 

6. If a circuit made up of a resistance 1 Ω and inductance 0.01 H; and alternating emf 200 V at 50 Hz is connected, then the phase difference between the current and the emf in the circuit is

(a) $\tan^{-1}(\pi)$ (b) $\tan^{-1}\left(\frac{\pi}{2}\right)$

(c) $\tan^{-1}\left(\frac{\pi}{4}\right)$ (d) $\tan^{-1}\left(\frac{\pi}{3}\right)$

7. In the circuit shown, what is the energy stored in the coil at steady state?

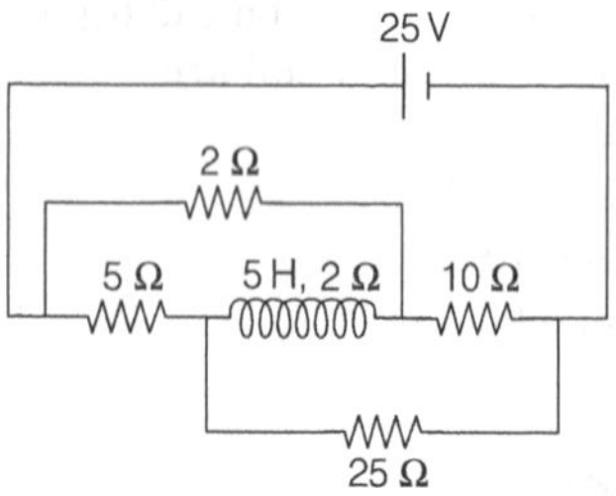

(a) 21.3 J (b) 42.6 J (c) Zero (d) 213 J

8. The figure shows three circuits with identical batteries, inductors and resistance. Rank the circuits according to the currents through the battery just after the switch is closed, greatest first.

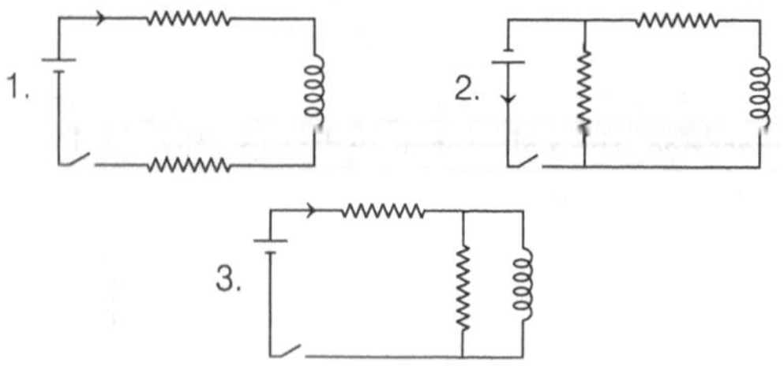

(a) $i_2 > i_3 > i_1$ (b) $i_2 > i_1 > i_3$
(c) $i_1 > i_2 > i_3$ (d) $i_1 > i_3 > i_2$

9. In the circuit shown here, the point C is kept connected to point A till the current flowing through the circuit becomes constant. Afterward, suddenly point C is disconnected from point A and connected to point B at time $t = 0$. Ratio of the voltage across resistance and the inductor at $t = L/R$ will be equal to **[JEE Main 2014]**

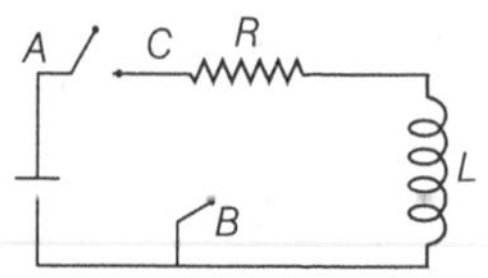

(a) $\frac{e}{1-e}$ (b) 1 (c) −1 (d) $\frac{1-e}{e}$

10. A circuit connected to an AC source of emf $e = e_0 \sin(100t)$ with t in seconds, gives a phase difference of $\frac{\pi}{4}$ between the emf e and current i. Which of the following circuits will exhibit this? **[JEE Main 2019]**

(a) R-C circuit with $R = 1\text{ k}\Omega$ and $C = 1\,\mu\text{F}$
(b) R-L circuit with $R = 1\text{ k}\Omega$ and $L = 1\text{ mH}$
(c) R-C circuit with $R = 1\text{ k}\Omega$ and $C = 10\,\mu\text{F}$
(d) R-L circuit with $R = 1\text{ k}\Omega$ and $L = 10\text{ mH}$

11. A 750 Hz, 20 V (rms) source is connected to a resistance of 100 Ω, an inductance of 0.1803 H and a capacitance of 10 μF all in series combination. The time in which the resistance (heat capacity 2 J/°C) will get heated by 10°C is close to. (Assume no loss of heat to the surroundings) **[JEE Main 2020]**

(a) 418 s (b) 245 s
(c) 365 s (d) 348 s

12. In the given AC circuit, when switch S is at position 1, the source emf leads current by $\pi/6$. Now, if the switch is at position 2, then

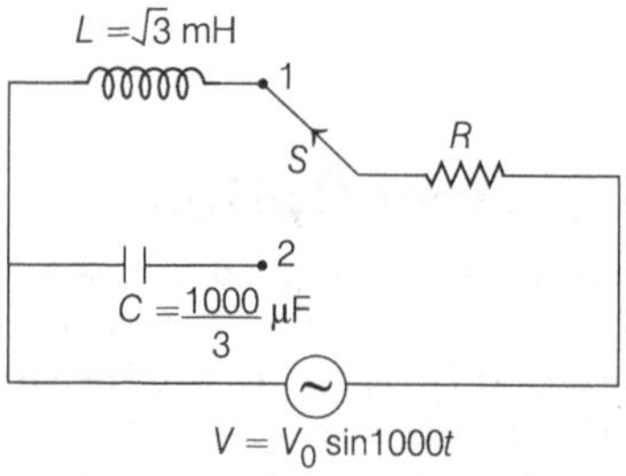

(a) current leads source emf by $\frac{\pi}{4}$

(b) current leads source emf by $\frac{\pi}{3}$

(c) source emf leads current by $\frac{\pi}{4}$

(d) source emf leads current by $\frac{\pi}{3}$

13. An AC voltage $V = V_0 \sin 100\,t$ is applied to the circuit, the phase difference between current and voltage is found to be $\frac{\pi}{4}$, then

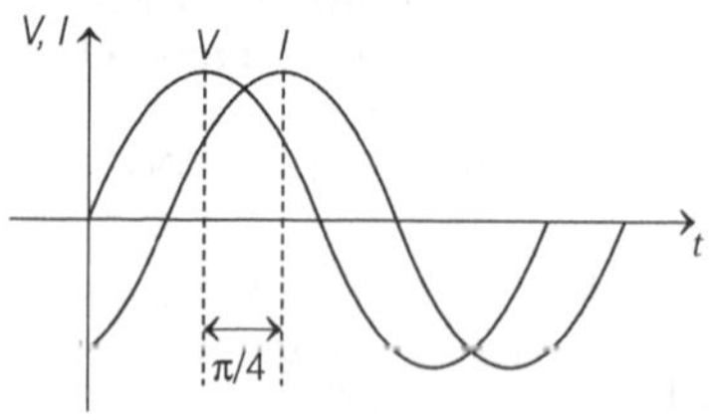

(a) $R = 100\,\Omega, C = 1\,\mu\text{F}$ (b) $R = 1\text{ k}\Omega, C = 10\,\mu\text{F}$
(c) $R = 10\text{ k}\Omega, L = 1\text{ H}$ (d) $R = 1\text{ k}\Omega, L = 10\text{ H}$

14. Two identical incandescent light bulbs are connected as shown in figure. When the circuit is an AC voltage source of frequency f, which of the following observations will be correct?

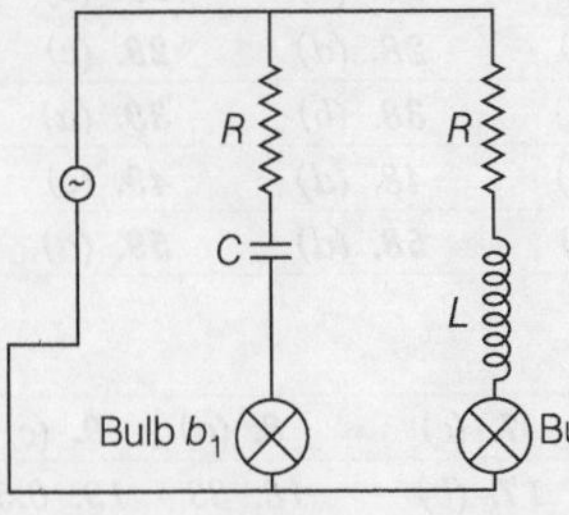

(a) Both bulbs will glow alternatively

(b) Both bulbs will glow with same brightness provided $f = \frac{1}{2\pi}\sqrt{(1/LC)}$

(c) Bulb b_1 will light up initially and goes OFF, bulb b_2 will be ON constantly

(d) Bulb b_1 will blink and bulb b_2 will be ON constantly

15. An inductor ($X_L = 2\,\Omega$), a capacitor ($X_C = 8\,\Omega$) and a resistance ($R = 8\,\Omega$) are connected in series with an AC source. The voltage output of AC source is given by $V = 10\cos(100\pi t)$.

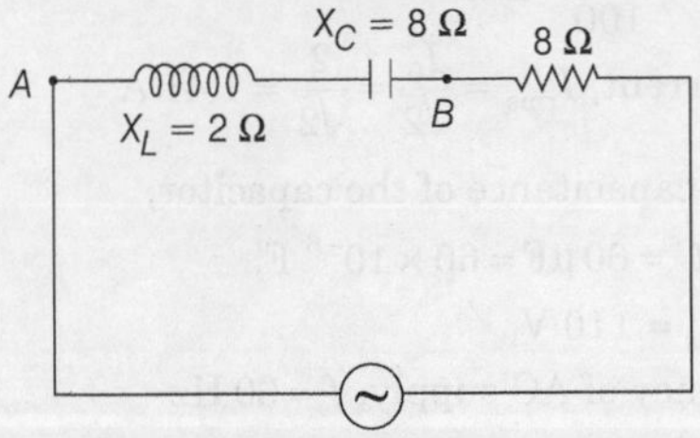

The instantaneous potential difference between points A and B, when the applied voltage is (3/5)th of the maximum value of applied voltage is

(a) 0 V

(b) 6 V

(c) 8 V

(d) None of the above

16. A series AC circuit containing an inductor (20 mH), a capacitor (120 μF) and a resistor (60 Ω) is driven by an AC source of 24 V/50 Hz. The energy dissipated in the circuit in 60 s is [JEE Main 2019]

(a) 3.39×10^3 J

(b) 5.65×10^2 J

(c) 2.26×10^3 J

(d) 5.17×10^2 J

17. An L-C-R circuit behaves like a damped harmonic oscillator. Comparing it with a physical spring-mass damped oscillator having damping constant b, the correct equivalence would be [JEE Main 2020]

(a) $L \leftrightarrow \frac{1}{b}, C \leftrightarrow \frac{1}{m}, R \leftrightarrow \frac{1}{k}$

(b) $L \leftrightarrow m, C \leftrightarrow \frac{1}{k}, R \leftrightarrow b$

(c) $L \leftrightarrow m, C \leftrightarrow k, R \leftrightarrow b$

(d) $L \leftrightarrow k, C \leftrightarrow b, R \leftrightarrow m$

Numerical Based Questions

18. A resistance R, an inductance $L = 0.01$ H and a capacitance C are connected in series. When a voltage $V = 400\cos(300t - 10°)$ V is applied to the series combination, the current flowing is $10\sqrt{2}\cos(3000t - 55°)$ A. Find the capacitance of capacitor in microfarads.

19. When 100 V DC is supplied across a solenoid, a current of 1.0 A flows in it. When 100 V AC is applied across the same coil, the current drops to 0.5 A. If the frequency of AC source is 50 Hz, then inductance (in H) of the solenoid is ....

20. An L-C-R series circuit with a resistance of 100 Ω is connected to an AC source of 200 V (rms) and angular frequency 300 rad/s. When only the capacitor is removed, the current lags behind the voltage by 60°. When only the inductor is removed, the current leads the voltage by 60°. The average power dissipated is .... W.

21. In a series circuit, $C = 2\,\mu$F, $L = 1$ mH and $R = 10\,\Omega$. When the current in the circuit is maximum, at that time, the ratio of the energies stored in the capacitor and the inductor will be $x:5$, then the value of x is ....

22. A sinusoidal voltage of peak value 250 V is applied to a series L-C-R circuit, in which $R = 8\Omega$, $L = 24$ mH and $C = 60\,\mu$F. The value of power dissipated at resonant condition is x kW. The value of x to the nearest integer is .... [JEE Main 2021]

23. Seawater at a frequency $f = 9 \times 10^2$ Hz, has permittivity $\varepsilon = 80\varepsilon_0$ and resistivity $\rho = 0.25\,\Omega$-m. Imagine a parallel plate capacitor is immersed in seawater and is driven by an alternating voltage source $V(t) = V_0 \sin(2\pi ft)$, then the conduction current density becomes 10^x times the displacement current density after time $t = \frac{1}{800}$ s. The value of x is .... $\left(\text{Given, } \frac{1}{4\pi\varepsilon_0} = 9 \times 10^9\ \text{Nm}^2\text{C}^{-2}\right)$

[JEE Main 2021]

Answers

Round I

1. (d)	2. (d)	3. (b)	4. (c)	5. (c)	6. (a)	7. (a)	8. (b)	9. (a)	10. (a)
11. (b)	12. (b)	13. (a)	14. (c)	15. (a)	16. (b)	17. (c)	18. (a)	19. (b)	20. (a)
21. (d)	22. (a)	23. (c)	24. (c)	25. (a)	26. (a)	27. (c)	28. (d)	29. (c)	30. (c)
31. (c)	32. (a)	33. (d)	34. (d)	35. (b)	36. (b)	37. (d)	38. (b)	39. (a)	40. (a)
41. (d)	42. (d)	43. (a)	44. (a)	45. (a)	46. (c)	47. (b)	48. (d)	49. (c)	50. (d)
51. (b)	52. (a)	53. (a)	54. (b)	55. (a)	56. (a)	57. (c)	58. (d)	59. (a)	

Round II

1. (c)	2. (c)	3. (a)	4. (d)	5. (c)	6. (a)	7. (c)	8. (a)	9. (c)	10. (c)
11. (d)	12. (a)	13. (b)	14. (a)	15. (b)	16. (d)	17. (b)	18. 33	19. 0.55	
20. 400	21. 1	22. 4	23. 6						

Solutions

Round I

1. Here, $V = \left(\frac{120}{2}\right) 2\sin 100\pi t \cos 100\pi t$

$$V = 60\sin(200\,\pi t)$$

Maximum voltage, $V_0 = 60$ V and $\omega = 200\,\pi$ rad s^{-1}

or frequency, $\nu = \frac{\omega}{2\pi} = \frac{200\pi}{2\pi} = 100$ Hz

2. $\therefore\ I_{\text{rms}} = \frac{I_0}{\sqrt{2}} = \frac{14}{\sqrt{2}} \approx 10$ A

3. Here, $\nu = 50$ Hz, $I_v = 5$ A, $I = ?$, $t = \frac{1}{300}$ s

$\therefore\quad I_0 = \sqrt{2}\, I_v = \sqrt{2} \times 5$ A

From $I = I_0 \sin \omega t$

$= 5\sqrt{2} \sin 100\,\pi \times \frac{1}{300} = 5\sqrt{2} \sin \frac{\pi}{3} = 5\sqrt{2} \times \frac{\sqrt{3}}{2} = 5\sqrt{\frac{3}{2}}$ A

4. Average value of output current is given by

$$I_{\text{av}} = \frac{\int_0^{T/2} I\,dt}{\int_0^{T/2} dt} = \frac{\int_0^{T/2} I_0 \sin \omega t\, dt}{T/2}$$

$$= \frac{2I_0}{T}\left[-\frac{\cos \omega t}{\omega}\right]_0^{T/2}$$

$$= \frac{2I_0}{T}\left[-\frac{\cos(\omega T/2)}{\omega} + \frac{\cos 0^\circ}{\omega}\right]$$

$$= \frac{2I_0}{\omega T}(-\cos \pi + \cos 0^\circ)$$

$$= \frac{2I_0}{2\pi}(1+1) = \frac{2I_0}{\pi}$$

5. The equation of AC is $I = I_1 \cos \omega t + I_2 \sin \omega t$

The resultant current is given by

$$I_0 = \sqrt{I_1^2 + I_2^2} \qquad \text{...(i)}$$

Hence, the rms current from relation is

$$I_{\text{rms}} = \frac{I_0}{\sqrt{2}} = \frac{\sqrt{I_1^2 + I_2^2}}{\sqrt{2}} \qquad \text{[from Eq. (i)]}$$

$$= \sqrt{\frac{I_1^2 + I_2^2}{2}}$$

6. $I_0 = \frac{V_0}{R} = \frac{200}{100} = 2$ A

rms current, $I_{\text{rms}} = \frac{I_0}{\sqrt{2}} = \frac{2}{\sqrt{2}} = 1.41$ A

7. Given, capacitance of the capacitor,

$$C = 60\,\mu\text{F} = 60 \times 10^{-6}\text{ F},$$

$$V_{\text{rms}} = 110\text{ V}$$

Frequency of AC supply, $f = 60$ Hz

Capacitive reactance,

$$X_C = \frac{1}{2\pi f C} = \frac{1}{2 \times 3.14 \times 60 \times 60 \times 10^{-6}} = 44.23\ \Omega$$

The rms value of the current in the circuit,

$$I_{\text{rms}} = \frac{V_{\text{rms}}}{X_C} = \frac{110}{44.23} = 2.49\text{ A} \simeq 2.5\text{ A}$$

8. $$P = \frac{1}{2} V_0 i_0 \cos \phi$$

$$\Rightarrow\ 1000 = \frac{1}{2} \times 200 \times i_0 \cos 60^\circ$$

$$\Rightarrow\ i_0 = 20\text{ A}$$

$$\Rightarrow\ i_{\text{rms}} = \frac{i_0}{\sqrt{2}} = \frac{20}{\sqrt{2}} = 10\sqrt{2}\text{ A}$$

9. When the circuit is connected to an AC source,

Voltage, $V = 12$ V

Current, $I = 0.2$ A

$\therefore$ Impedance, $Z = \frac{V}{I} = \frac{12}{0.2} = 60\ \Omega$

When it is connected to DC source,

Voltage, $V = 12$ V

Current, $I = 0.4$ A

$\Rightarrow$ Resistance, $R = \frac{V}{I} = \frac{12}{0.4} = 30\,\Omega$

As in case of DC supply, the capacitor act as an open circuit and no current flows through the circuit. So, the given circuit will not have capacitor in series combination. Therefore, the circuit should be a series *L-R* circuit.

10. Given, resistance, $R = 3\,\Omega$

Inductive reactance, $X_L = 10\,\Omega$

Capacitive reactance, $X_C = 14\,\Omega$

The impedance of the series *L-C-R* circuit,

$$Z = \sqrt{R^2 + (X_C - X_L)^2} = \sqrt{(3)^2 + (14 - 10)^2}$$

$$Z = 5\,\Omega$$

11. When C is removed circuit becomes *R-L* circuit, hence

$$\tan\frac{\pi}{3} = \frac{X_L}{R} \quad \text{... (i)}$$

When L is removed circuit becomes *R-C* circuit, hence

$$\tan\frac{\pi}{3} = \frac{X_C}{R} \quad \text{... (ii)}$$

From Eqs. (i) and (ii), we obtain $X_L = X_C$.

This is condition of resonance and in resonance, $Z = R = 100\,\Omega$.

12. Here, $X_L = R, X_C = R/2$

$$\therefore \quad \tan\phi = \frac{X_L - X_C}{R} = \frac{R - \frac{R}{2}}{R} = \frac{1}{2}$$

$$\Rightarrow \quad \phi = \tan^{-1}(1/2)$$

Also, $Z = \sqrt{R^2 + (X_L - X_C)^2} = \sqrt{R^2 + \frac{R^2}{4}} = \frac{\sqrt{5}}{2}R$

13. Given, reactance, $X_L = 100\,\Omega$

Frequency of AC, $f = 1000$ Hz

As voltage leads current by 45°, so there must be some resistance in the coil otherwise $\Delta\phi = 90°$.

Using $\tan\phi = \frac{X_L}{R}$, we have

$$\tan 45° = \frac{X_L}{R} \text{ or } X_L = R$$

$\therefore$ Reactance of circuit containing resistance and inductance is

$$\sqrt{X_L^2 + R^2} = 100\,\Omega$$

$$\Rightarrow \sqrt{X_L^2 + X_L^2} = 100\,\Omega$$

or $X_L = 50\sqrt{2}\,\Omega$

But $X_L = L\omega$

$$\Rightarrow \quad L = \frac{X_L}{\omega} = \frac{X_L}{2\pi f} = \frac{50\sqrt{2}}{2\pi \times 1000}$$

$$\Rightarrow \quad L = 1.1 \times 10^{-2}\text{ H}$$

14. Here, $C = 100\mu\text{F} = 10^{-4}\text{F}, R = 40\,\Omega, V_{\text{rms}} = 110$ V, $f = 60$ Hz

(i) As, $i_{\text{rms}} = \frac{V_{\text{rms}}}{\sqrt{R^2 + \frac{1}{\omega^2C^2}}} = \frac{V_{\text{rms}}}{\sqrt{R^2 + \frac{1}{(2\pi fC)^2}}}$

$$= \frac{110}{\sqrt{40^2 + \frac{1}{(2\pi \times 60 \times 10^{-4})^2}}}$$

$$= \frac{100}{\sqrt{1600 + 703.62}}$$

$$= \frac{100}{48} = 2.292\text{ A}$$

Now, $i_{\text{rms}} = \sqrt{2}i_0 = \sqrt{2} \times 2.292 = 3.24$ A

(ii) For *C-R* circuit,

$$\tan\theta = \frac{1/\omega C}{R} = \frac{1}{2\pi fCR}$$

$$= \frac{1}{2\pi \times 60 \times 10^{-4} \times 40}$$

$$= 0.6631$$

$\theta = 33°33'$ (emf lags behind the current)

15. At angular frequency ω, the current in *R-C* circuit is given by

$$i_{\text{rms}} = \frac{V_{\text{rms}}}{\sqrt{R^2 + \left(\frac{1}{\omega C}\right)^2}} \quad \text{... (i)}$$

Also, $\frac{i_{\text{rms}}}{2} = \frac{V_{\text{rms}}}{\sqrt{R^2 + \left(\frac{1}{\frac{\omega}{3}C}\right)^2}} = \frac{V_{\text{rms}}}{\sqrt{R^2 + \frac{9}{\omega^2C^2}}}$ ... (ii)

From Eqs. (i) and (ii), we get

$$3R^2 = \frac{5}{\omega^2C^2}$$

$$\Rightarrow \quad \frac{\frac{1}{\omega C}}{R} = \sqrt{\frac{3}{5}} \Rightarrow \frac{X_C}{R} = \sqrt{\frac{3}{5}}$$

16. As, $Z = \sqrt{R^2 + \frac{1}{(\omega C)^2}}$ and $Z' = \sqrt{R^2 + \frac{1}{(\omega 4C)^2}}$

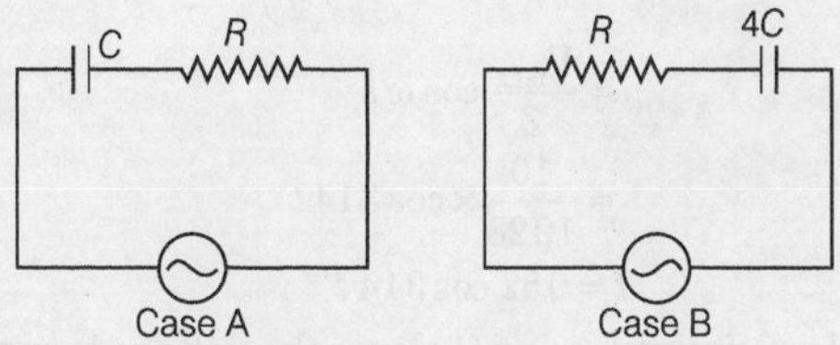

$$I_R^A = \frac{V}{Z}$$

$$I_R^B = \frac{V}{Z'}$$

As, $Z' < Z$

$\Rightarrow$ $I_R^A < I_R^B$

17. 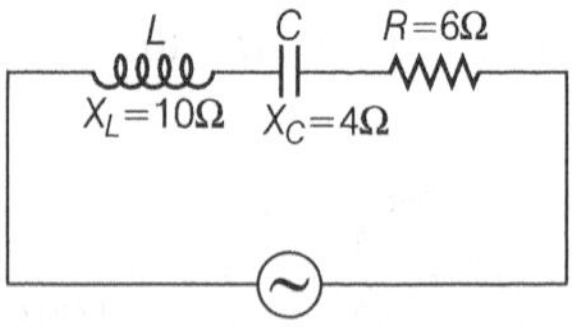

We know that, power factor is $\cos\phi$.

i.e. $$\cos\phi = \frac{R}{Z} \quad \ldots\text{(i)}$$

$$Z = \sqrt{R^2 + (X_L - X_C)^2} \quad \ldots\text{(ii)}$$

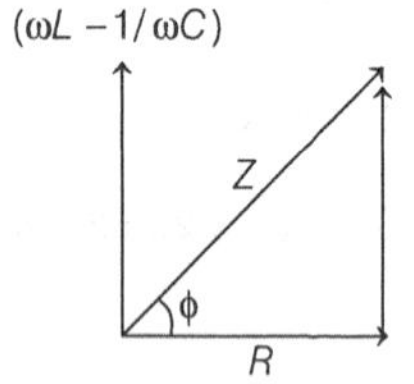

$\Rightarrow$ $$Z = \sqrt{6^2 + (10-4)^2}$$

$\Rightarrow$ $$Z = 6\sqrt{2} \mid \cos\phi = \frac{6}{6\sqrt{2}}$$

$$\cos\phi = \frac{1}{\sqrt{2}}$$

18. Impedance of given circuit,

$$Z = \sqrt{(X_C - X_L)^2} = X_C - X_L$$

$$= \frac{1}{\omega C} - \omega L$$

$$= \frac{1}{314 \times 100 \times 10^{-6}} - 314 \times 40 \times 10^{-3}$$

$$= 19.28\,\Omega$$

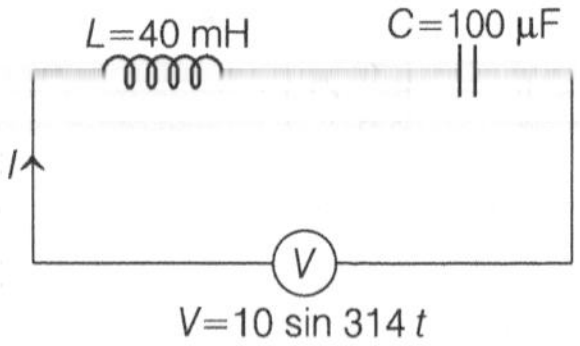

As $X_C > X_L$, circuit is capacitive, hence current in circuit leads emf by $\frac{\pi}{2}$ rad.

Current in circuit is given by

$$I = I_{\max} \sin\left(\omega t + \frac{\pi}{2}\right)$$

$$= \frac{V_{\max}}{Z} \cos\omega t$$

$$= \frac{10}{19.28} \times \cos 314\,t$$

or $$I = 0.52 \cos 314\,t$$

19. (a) In a circuit having C alone, the voltage lags the current by $\frac{\pi}{2}$.

(b) In circuit containing R and L, the phase difference between current and voltage can have any value between 0 to $\pi/2$.

(c) In L-C circuit, the phase difference between current and voltage has a value of 0 or $\frac{\pi}{2}$ depending on the values of L and C.

(d) In a circuit containing L alone, the voltage leads the current by $\pi/2$.

20. We have, $X_L = \omega L = 100 \times 0.5 = 50\,\Omega$

$$X_1' = \frac{1}{\omega C} = \frac{1}{100 \times 100 \times 10^{-6}} = 100\,\Omega$$

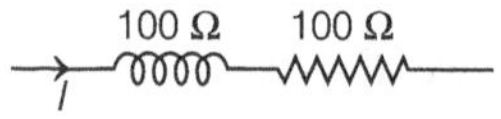

$$Z_1 = 100\sqrt{2},$$

$$I_1 = \frac{20}{100\sqrt{2}} = \frac{1}{5\sqrt{2}}$$

$$V \text{ across } 100\,\Omega = \frac{1}{5\sqrt{2}} \times 100$$

$$= \frac{20}{\sqrt{2}} \times \frac{\sqrt{3}}{\sqrt{2}} = 10\sqrt{2}$$

Phase difference between I_1 and V,

$$\cos\phi_1 = \frac{R_1}{Z_1} = \frac{100}{100\sqrt{2}},$$

$$\phi = \frac{\pi}{4}$$

I_1 and V,

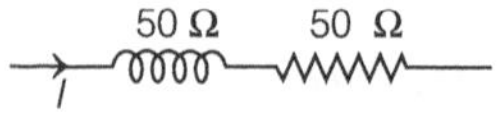

$$Z_2 = 50\sqrt{2},\ I_2 = \frac{20}{50\sqrt{2}} = \frac{2}{5\sqrt{2}}$$

$$V \text{ across } 50\,\Omega = \frac{2}{50\sqrt{2}} \times 50 = \frac{20}{\sqrt{2}} = 10\sqrt{2}$$

$$\phi_2 = \frac{\pi}{4}\ I_2 \text{lag} V \text{ by } \frac{\pi}{4}$$

$$I = I_1 = I_2$$

$$I_{\text{net}} = \sqrt{I_1^2 + I_2^2}$$

$$I = \sqrt{\frac{4}{25 \times 2} + \frac{1}{25 \times 2}}$$

$$= \sqrt{\frac{5}{50}} = \frac{1}{\sqrt{10}} = 0.316\text{ A}$$

21. $Z = \sqrt{(R)^2 + (X_L - X_C)^2}$

$$R = 10\Omega,\ X_L = \omega L = 2000 \times 5 \times 10^{-3} = 10\,\Omega$$

$\therefore$ $$X_C = \frac{1}{\omega C} = \frac{1}{2000 \times 50 \times 10^{-6}} = 10\Omega, \text{ i.e. } Z = 10\Omega$$

Maximum current, $i_0 = \frac{V_0}{Z} = \frac{20}{10} = 2\text{ A}$

Hence, $i_{\text{rms}} = \frac{2}{\sqrt{2}} = 1.41\text{ A}$

and $V_{\text{rms}} = 4 \times 1.41 = 5.64\text{ V}$

22. Bandwidth = R/L

Bandwidth $\propto R$

So, bandwidth of resonance circuit will increase.

23. In the condition of resonance,

$$X_L = X_C \text{ or } \omega L = \frac{1}{\omega C} \quad \ldots\text{(i)}$$

Since, resonance frequency remains unchanged, so

$\sqrt{LC}$ = constant or LC = constant

$\therefore \quad L_1C_1 = L_2C_2$

or $\quad L \times C = L_2 \times 2C$ or $L_2 = L/2$

24. Given, $L = 10$ H, $f = 50$ Hz

For maximum power,

$$X_C = X_L \quad (\because \text{resonance condition})$$

or $\quad \frac{1}{\omega C} = \omega L$ or $C = \frac{1}{\omega^2 L}$

$$\therefore \quad C = \frac{1}{4\pi^2 \times 50 \times 50 \times 10}$$

or $\quad C = 0.1 \times 10^{-5}$ F $= 1\,\mu$F

25. In the given circuit diagram, both inductors of inductance L are parallel to each other, hence

$$L_{eg} = \frac{L_1 L_2}{L_1 + L_2} = \frac{L \times L}{L + L} = \frac{L}{2}$$

Similarly,

$$C_{eg} = C_1 + C_2 = C + C = 2C$$

$\therefore$ Resonance frequency

$$\omega_r = \frac{1}{\sqrt{L_{eg} \cdot C_{eg}}} = \sqrt{\frac{1}{\frac{L}{2} 2C}} = \sqrt{\frac{1}{LC}} = \frac{1}{\sqrt{LC}}$$

26. Sharpness of resonance of a resonant L-C-R circuit is determined by the ratio of resonant frequency with the selectivity of circuit. This ratio is also called "quality factor" or Q-factor.

$$Q\text{-factor} = \frac{\omega_0}{2\Delta\omega} = \frac{\omega_0 L}{R}$$

27. The L-C-R circuit used for communication should possess high quality factor (Q-factor) of resonance, which is given by

$$Q = \frac{1}{R}\sqrt{\frac{L}{C}}$$

Putting values in above equation,

For option (a), $Q = 10.52$

For option (b), $Q = 9.42$

For option (c), $Q = 0.77$

For option (d), $Q = 7.30$

Hence, option (c) is the best representation.

28. As, the iron rod is inserted, the magnetic field inside the coil magnetises, the iron increasing the magnetic field inside it. Hence, the inductance of the coil increases. Consequently, the inductive reactance of the coil increases. As a result, a larger fraction of the applied AC voltage appears across the inductor, leaving less voltage across the bulb. Therefore, the brightness of the light bulb decreases.

29. $P = E_{\text{rms}} i_{\text{rms}} \cos\phi = \frac{E_0}{\sqrt{2}} \times \frac{i_0}{\sqrt{2}} \times \frac{R}{Z}$

$$\Rightarrow P = \frac{E_0}{\sqrt{2}} \times \frac{E_0}{Z\sqrt{2}} \times \frac{R}{Z} \Rightarrow P = \frac{E_0^2 R}{2Z^2}$$

Given $X_L = R$, so $Z = \sqrt{2}R \Rightarrow P = \frac{E_0^2}{4R}$

30. With DC, $P = \frac{V^2}{R} \Rightarrow R = \frac{(10)^2}{20} = 5\,\Omega$

With AC, $P = \frac{V_{\text{rms}}^2 R}{Z^2} \Rightarrow Z^2 = \frac{(10)^2 \times 5}{10} = 50\,\Omega^2$

Also, $\quad Z^2 = R^2 + 4\pi^2\nu^2L^2$

$\Rightarrow \quad 50 = (5)^2 + 4(3.14)^2\nu^2(10 \times 10^{-3})^2 \Rightarrow \nu = 80$ Hz

31. $Z = \sqrt{R^2 + \left(\frac{1}{2\pi\nu C}\right)^2} = \sqrt{(3000)^2 + \frac{1}{\left(2\pi \times 50 \times \frac{2.5}{\pi} \times 10^{-6}\right)^2}}$

$$\Rightarrow Z = \sqrt{(3000)^2 + (4000)^2} = 5 \times 10^3\ \Omega$$

So, power factor, $\cos\phi = \frac{R}{Z} = \frac{3000}{5 \times 10^3} = 0.6$

and power, $P = V_{\text{rms}}\, i_{\text{rms}} \cos\phi = \frac{V_{\text{rms}}^2 \cos\phi}{Z}$

$$\Rightarrow \quad P = \frac{(200)^2 \times 0.6}{5 \times 10^3} = 4.8 \text{ W}$$

32. When a resistor is connected to an AC source, the power drawn will be of

$$P = V_{\text{rms}} \cdot I_{\text{rms}} = V_{\text{rms}} \cdot \frac{V_{\text{rms}}}{R}$$

$$\Rightarrow \quad V_{\text{rms}}^2 = PR$$

When an inductor is connected in series with the resistor, then the power drawn will be

$$P' = V_{\text{rms}} \cdot I_{\text{rms}} \cos\phi$$

where, ϕ = phase difference.

$$\therefore \quad P' = \frac{V_{\text{rms}}^2}{R} \cdot \frac{R}{Z} = \frac{P \cdot R}{R} \cdot \frac{R}{Z}$$

$$\Rightarrow \quad P' = \frac{P \cdot R}{Z} = P\left(\frac{R}{Z}\right)$$

33. Given, for L-R circuit,

$$V_{\text{rms}} = 250 \text{ V}, f = 50 \text{ Hz}, P = 400 \text{ W}$$

and $\cos\phi = 0.8$

As we know, power is given by $P = V_{\text{rms}} i_{\text{rms}} \cos\phi$

$$\Rightarrow \quad P = V_{\text{rms}} \frac{V_{\text{rms}}}{Z} \cos\phi \Rightarrow Z = \frac{V_{\text{rms}}^2}{P} \cos\phi$$

$$\Rightarrow \quad Z = \frac{(250)^2}{400} \times 0.8 = 125\ \Omega$$

Also, $\cos\phi = \frac{R}{Z} \Rightarrow 0.8 = \frac{R}{125} \Rightarrow R = 100\ \Omega$

Impedance, $Z = \sqrt{R^2 + X_L^2} \Rightarrow 125 = \sqrt{(100)^2 + X_L^2}$

$$X_L = 75\ \Omega$$

Now, to obtain power factor unity, X_C must be equal to X_L.

i.e. $X_C = X_L = 75\ \Omega$

$\Rightarrow \quad \frac{1}{\omega C} = 75$

$\Rightarrow \quad C = \frac{1}{\omega \times 75} = \frac{1}{2\pi \times 50 \times 75}$

$\Rightarrow \quad C = \frac{1}{7500\pi}\text{ F} \quad \text{or} \quad C = \frac{400}{3\pi}\ \mu\text{F}$

Given, $C = \frac{n}{3\pi}\ \mu\text{F}$

On comparing the both, we get

$$n = 400$$

34. In the R-L circuit, growth of current is given by

$$I = \frac{E}{R_2}\left[1 - e^{\frac{-R_2 t}{L}}\right]$$

Potential drop across L,

$$V_L = L\frac{dI}{dt} = L\left(\frac{-E}{R_2}\right)e^{\frac{R_2 t}{L}}\left(-\frac{R_2}{L}\right)$$

$$V_L = Ee^{-\frac{R_2 T}{L}} = 12e^{-2t/0.4} = 12e^{-5t}\ \text{V}$$

35. Here, $i = i_0$ at $t = \infty$. Let i be the current at $t = 1$ s

From $i = i_0(1 - e^{-\frac{R}{L}t}) = i_0\left(1 - e^{\frac{-10}{5}\times 1}\right) = i_0\left(1 - \frac{1}{e^2}\right)$

$\therefore \quad \frac{i_0}{i} = \frac{e^2}{e^2 - 1}$

36. The current at any instant is given by

$$I = I_0(1 - e^{-Rt/L})$$

$\Rightarrow \quad \frac{I_0}{2} = I_0(1 - e^{-Rt/L})$

or $\quad \frac{1}{2} = (1 - e^{-Rt/L})$

or $\quad e^{-Rt/L} = \frac{1}{2}$ or $\frac{Rt}{L} = \ln 2$

$\therefore \quad t = \frac{L}{R}\ln 2 = \frac{300 \times 10^{-3}}{2} \times 0.693$

$= 150 \times 0.693 \times 10^{-3}$

$= 0.10395$ s

$= 0.1$ s

37. In an L-R circuit, current during charging of inductor is given by

$$i = i_0(1 - e^{-\frac{R}{L}t})$$

where, i_0 = saturation current.

In given circuit,

inductance of circuit, $L = 10\text{ mH} = 10 \times 10^{-3}$ H

Resistance of circuit, $R = (R_s + r) = 0.1 + 0.9 = 1\ \Omega$

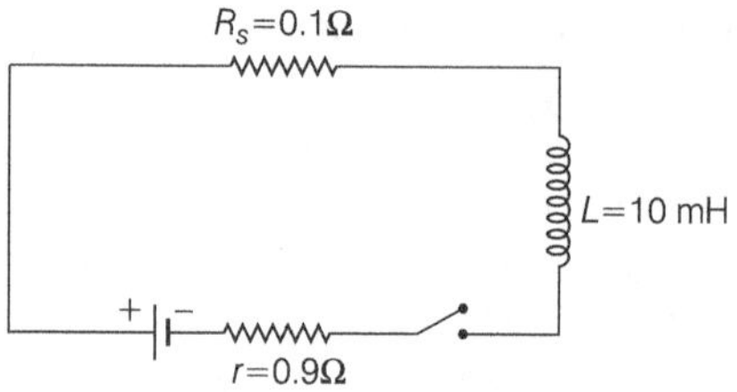

Now, from $\quad i = i_0(1 - e^{-\frac{R}{L}t}) \quad$ ...(i)

Given, $\quad i = 80\%$ of i_0

$\Rightarrow \quad i = \frac{80\, i_0}{100} = 0.8\, i_0$

Substituting the value of i in Eq. (i), we get

$$0.8 = 1 - e^{-\frac{R}{L}t}$$

$\Rightarrow \quad e^{-\frac{R}{L}t} = 0.2 \Rightarrow e^{\frac{R}{L}t} = 5$

$\Rightarrow \quad \ln (e)^{\frac{R}{L}t} = \ln 5$

$\Rightarrow \quad \frac{R}{L}t = \ln 5$

$\Rightarrow \quad t = \frac{L}{R}\cdot \ln(5) = \frac{10 \times 10^{-3}}{1} \times \ln(5)$

$= 10 \times 10^{-3} \times 1.6$

$= 1.6 \times 10^{-2}\text{ s} = 0.016$ s

38. In an L-R circuit, current during charging is given by

$$I = I_0\left(1 - e^{-\frac{R}{L}t}\right)$$

where, $I_0 = \frac{E}{R}$ = saturation current.

So, we have $\frac{dq}{dt} = I = I_0\left(1 - e^{-\frac{R}{L}t}\right)$

$\Rightarrow \quad dq = I_0\left(1 - e^{-\frac{R}{L}t}\right)dt$

So, charge q that passes through battery from time $t = 0$ to $t = \frac{L}{R}$ is obtained by integrating the above equation within the specified limits, *i.e.*

$$q = \int_0^Q dq = \int_{t=0}^{t=\frac{L}{R}} I_0\left(1 - e^{-\frac{R}{L}t}\right)dt$$

$$= I_0\left[\left(t - \frac{1}{\left(-\frac{R}{L}\right)}\cdot e^{-\frac{R}{L}t}\right)\right]_0^{\frac{L}{R}}$$

$$=\frac{E}{R}\left[\left\{\frac{L}{R}+\frac{L}{Re'}\right\}-\left\{0+\frac{L}{R}\right\}\right]$$

$$=\frac{E}{R}\times\frac{L}{Re}=\frac{EL}{R^2e}$$

$$\Rightarrow \quad Q=\frac{EL}{2.7\,R^2} \qquad (\because e\approx 2.72)$$

39. Time constant τ is the duration when the value of potential drops by 63% of its initial maximum value (*i.e.* V_0/e). Here, 37% of 25 V = 9.25 V which lies between 150 s to 200 s in the graph.

40. Consider the *L-C-R* circuit at any time *t*,

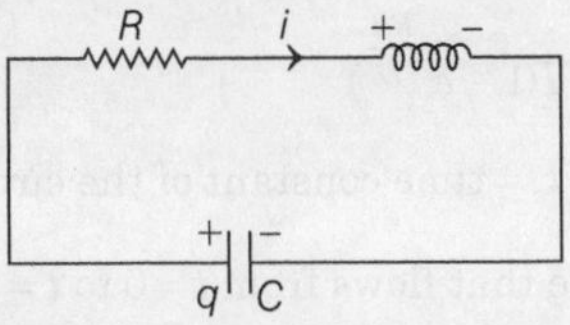

Now, applying KVL, we have

$$\frac{q}{C}-iR-\frac{Ldi}{dt}=0$$

As current is decreasing with time, we can write

$$i=-\frac{dq}{dt}$$

$$\Rightarrow \quad \frac{q}{C}+\frac{dq}{dt}R+\frac{Ld^2q}{dt^2}=0$$

or

$$\frac{d^2q}{dt^2}+\frac{R}{L}\frac{dq}{dt}+\frac{q}{LC}=0$$

This equation is equivalent to that of a damped oscillator.

Thus, we can write the solution as

$$Q_{\max}(t)=Q_0\cdot e^{-Rt/2L} \quad \text{or} \quad Q^2_{\max}=Q_0^2\,e^{-\frac{Rt}{L}}$$

As $L_1 > L_2$ damping is faster for L_2

Hence, the correct graph is shown in option (a).

41. After long time, inductor behaves as short-circuit. At $t=0$, the inductor behaves as short-circuited. The current,

$$I_0=\frac{E_0}{R}=\frac{15\text{ V}}{0.15\text{ k}\Omega}=100\text{ mA}$$

As K_2 is closed, current through the inductor starts decay, which is given at any time *t* as

$$I=I_0\,e^{-\frac{tR}{L}}=(100\text{ mA})\,e^{\frac{-t\times 15000}{3}}$$

At $t=1$ ms,

$$I=(100\text{ mA})\,e^{-\frac{1\times10^{-3}\times15\times10^{3}}{3}}$$

$$I=(100\text{ mA})\,e^{-5}=0.6737\text{ mA}$$

42. Let *t* be the required time at which the energy stored in inductor becomes $\left(\frac{1}{n}\right)$ times of its maximum value.

According to question, for given *L-R* circuit,

$$U=\frac{1}{n}U_{\max}$$

$$\frac{1}{2}LI^2=\frac{1}{n}\times\frac{1}{2}LI^2_{\max}$$

$$\frac{1}{2}L\left[I_{\max}\left(1-e^{\frac{-Rt}{L}}\right)\right]^2=\frac{1}{n}\times\frac{1}{2}LI^2_{\max}$$

$$\frac{1}{2}LI^2_{\max}\left(1-e^{\frac{-Rt}{L}}\right)^2=\frac{1}{n}\times\frac{1}{2}L\,I^2_{\max}$$

$$\left(1-e^{\frac{-Rt}{L}}\right)^2=\frac{1}{n}$$

$$\Rightarrow \quad 1-e^{\frac{-Rt}{L}}=\left(\frac{1}{n}\right)^{\frac{1}{2}}$$

$$e^{\frac{-Rt}{L}}=1-\frac{1}{\sqrt{n}}$$

$$\Rightarrow \quad e^{\frac{-Rt}{L}}=\frac{\sqrt{n}-1}{\sqrt{n}}$$

Taking natural log on both sides, we get

$$\ln\left(e^{\frac{-Rt}{L}}\right)=\ln\left(\frac{\sqrt{n}-1}{\sqrt{n}}\right)$$

or

$$-\frac{Rt}{L}\ln(e)=\ln\left(\frac{\sqrt{n}-1}{\sqrt{n}}\right)$$

$$=-\frac{Rt}{L}\times 1=\ln\left(\frac{\sqrt{n}-1}{\sqrt{n}}\right)$$

$$\Rightarrow \quad t=\frac{-L}{R}\ln\left(\frac{\sqrt{n}-1}{\sqrt{n}}\right)$$

$$t=\frac{L}{R}\ln\left(\frac{\sqrt{n}-1}{\sqrt{n}}\right)^{-1}$$

$$\Rightarrow \quad t=\frac{L}{R}\ln\left(\frac{\sqrt{n}}{\sqrt{n}-1}\right)$$

43. In an *L-R* circuit, current growth occurs as

$$I=I_0\left(1-e^{-\frac{R}{L}t}\right) \qquad \text{...(i)}$$

where, I = instantaneous current,

I_0 = maximum current = current at $t=\infty=\frac{E}{R}$,

R = resistance of circuit,

L = inductance of circuit

and t = instantaneous time.

Here, $R=5\,\Omega$, $L=10\text{ mH}=10\times10^{-3}\text{ H}$,

$E=20\text{ V}$, $t=40\text{ s}$

So, substituting these values in Eq. (i), we get

$$\frac{I}{I_0}=\left(1-e^{-\frac{5}{10\times10^{-3}}\times 40}\right)=1-e^{-2\times10^4}$$

$\Rightarrow \qquad \dfrac{I_0}{I} = \dfrac{1}{1 - e^{-2\times 10^4}} \approx 1$

$\therefore \qquad \dfrac{I_0}{I} \approx 1.06$ (nearest option)

44. After a sufficiently long time, in steady state, resistance offered by inductor is zero. So, circuit is reduced to

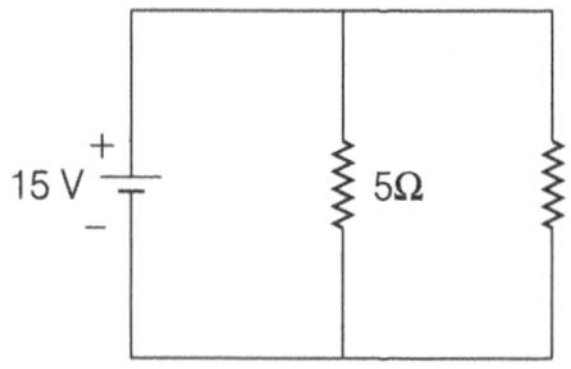

$\therefore$ Current in circuit, $I = \dfrac{E}{R_{eq}} = \dfrac{15}{\left(\dfrac{5\times 5}{5+5}\right)} = \dfrac{15\times 2}{5} = 6\text{ A}$

45. In *L-R* circuit, the growing current at time *t* is given by

$$i = i_0[1 - e^{-t/\tau}], \text{ where } i_0 = \frac{E}{R} \text{ and } \tau = \frac{L}{R}$$

$\therefore$ Charge passed through the battery in one time constant is

$$q = \int_0^{\tau} i\,dt = \int_0^{\tau} i_0(1 - e^{-t/\tau})dt$$

$$q = i_0\tau - \left[\frac{i_0 e^{-t/\tau}}{-2/\tau}\right]_0^{\tau} = i_0\tau + i_0\tau[e^{-1} - 1]$$

$$= i_0\tau - i_0\tau + \frac{i_0\tau}{e}$$

$$q = \frac{i_0\tau}{e} = \frac{(E/R)(L/R)}{e} = \frac{EL}{eR^2}$$

46. In *L-R* circuit, current at any time *t* is given by

$$i = \frac{E}{R}\left(1 - e^{-\frac{R}{L}t}\right) = \frac{E}{R} - \frac{E}{R}e^{-\frac{R}{L}t} \qquad \ldots\text{(i)}$$

$$\frac{di}{dt} = \frac{E}{R}e^{-\frac{R}{L}t}\left(\frac{R}{L}\right) = \frac{E}{L}e^{-\frac{R}{L}t}$$

$$\text{Induced emf} = L\frac{di}{dt} = Ee^{-\frac{R}{L}t} \qquad \ldots\text{(ii)}$$

From Eq. (i), $\quad iR = E - Ee^{-\frac{R}{L}t}$

Using Eq. (ii), $iR = E - e$

or $\qquad e = E - iR$

Therefore, graph between *e* and *i* is a straight line with negative slope and positive intercept.

47. As is clear for Fig. (i), steady state current for t = both the circuits is same. Therefore,

$$\frac{V}{R_1} = \frac{V}{R_2} \text{ or } R_1 = R_2$$

Again, from the same figure, we observe that

$$\tau_1 < \tau_2$$

$$\therefore \qquad \frac{L_1}{R_1} < \frac{L_2}{R_2}$$

As $R_1 = R_2$, therefore $L_1 < L_2$

48. In an *L-R* circuit as shown in the figure, instantaneous current at time *t* is given by

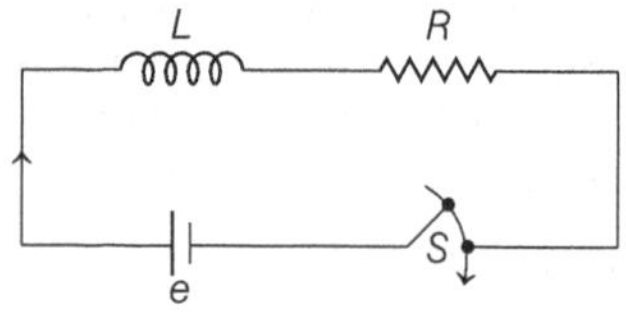

$$i = I(1 - e^{-\frac{R.t}{L}})$$

where, $\dfrac{L}{R} = t_c$ = time constant of the circuit.

Now, charge that flows from $t = 0$ to $t = t_c$ from battery,

$$q = \int_0^{t_c} i\,dt = I\int_0^{t_c}(1 - e^{-\frac{R}{L}.t})dt$$

$$= I\left[t - \frac{1}{-\frac{R}{L}}.e^{\frac{-R}{L}.t}\right]_0^{t_c} = I\left[t + \frac{L}{R}.e^{\frac{-R}{L}.t}\right]_0^{t_c}$$

$$= I\left[\left(t_c + \frac{L}{R}e^{-\frac{R}{L}t_c}\right) - \left(\frac{L}{R}e^0\right)\right]$$

$$= I\left[\frac{L}{R} + \frac{L}{R}e^{-1} - \frac{L}{R}\right] = I.\frac{L}{R}.e^{-1}$$

Here, I is maximum current which occurs at $t = \infty$ and its value,

$$I = \frac{\varepsilon}{R}$$

So, $\qquad q = \dfrac{\varepsilon}{R}.\dfrac{L}{R}.\dfrac{1}{e} = \dfrac{\varepsilon L}{eR^2}$

49. Given circuit is a series *L-R* circuit

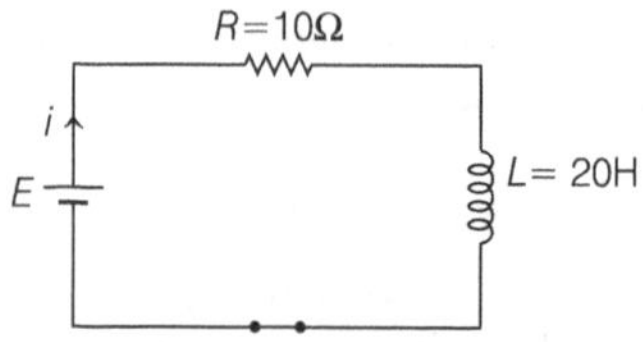

In an *L - R* circuit, current increases is $i = \dfrac{E}{R}(1 - e^{-\frac{R}{L}t})$

Now, energy stored in inductor is $U_L = \dfrac{1}{2}Li^2$

where, L = self-inductance of the coil

and energy dissipated by resistor is $U_R = i^2R$.

Given, rate of energy stored in inductor is equal to the rate of energy dissipated in resistor. So, after differentiating, we get

$$iL\frac{di}{dt} = i^2R \Rightarrow \frac{di}{dt} = \frac{R}{L}i$$

$$\Rightarrow \quad \frac{E}{R}\cdot\frac{R}{L}e^{-\frac{R}{L}t} = \frac{R}{L}\cdot\frac{E}{R}\left(1 - e^{-\frac{R}{L}t}\right)$$

$$\Rightarrow \quad 2e^{-\frac{R}{L}t} = 1 \Rightarrow e^{-\frac{R}{L}t} = \frac{1}{2}$$

Taking log on both sides, we have

$$\Rightarrow \quad \frac{-R}{L}t = \ln\left(\frac{1}{2}\right) \Rightarrow \frac{R}{L}t = \ln 2$$

$$\Rightarrow \quad t = \frac{L}{R}\ln 2 = \frac{20}{10}\ln 2$$

$$\Rightarrow \quad t = 2\ln 2$$

50. Choke coil is a device having high value of inductance and low resistance which is used to control AC.

51. Given, $C = 40\,\mu\text{F} = 40 \times 10^{-6}\,\text{F}$,

and $L = 16\text{ mH} = 16 \times 10^{-3}\text{ H}$

Angular frequency of oscillating circuit,

$$\omega = \frac{1}{\sqrt{LC}} = \frac{1}{\sqrt{(16 \times 10^{-3})(40 \times 10^{-6})}}$$

$$= \frac{10^4}{8} = 1.25 \times 10^3\text{ s}^{-1}$$

52.

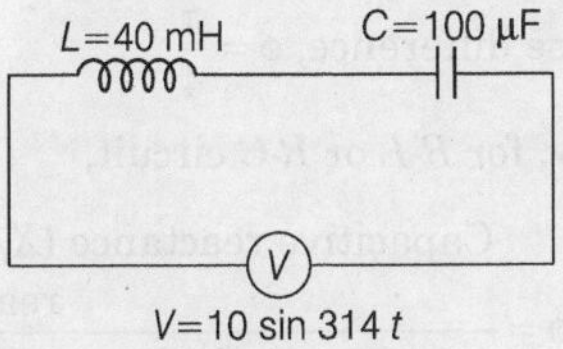

Impedance of given ciruit,

$$Z = \sqrt{(X_C - X_L)^2} = X_C - X_L$$

$$= \frac{1}{\omega C} - \omega L = \frac{1}{314 \times 100 \times 10^{-6}} - 314 \times 40 \times 10^{-3}$$

$$= 19.28\,\Omega$$

As $X_C > X_L$, circuit is capacitive, hence current in circuit leads emf by $\frac{\pi}{2}$ rad.

Current in circuit is given by

$$I = I_{\max}\sin\left(\omega t + \frac{\pi}{2}\right) = \frac{V_{\max}}{Z}\cos\omega t = \frac{10}{19.28} \times \cos 314t$$

or $\quad I = 0.52\cos 314t$

53. For L-C oscillations,

Energy stored in inductor = Energy stored in capacitor

$$\frac{1}{2}Li_m^2 = \frac{1}{2}CV_m^2$$

Given, $V_m = 25\text{ V}$, $C = 10\,\mu\text{F} = 10^{-5}\text{ F}$

and $L = 100\text{ mH} = 10^{-1}\text{ H}$

or $\quad i_m = V_m\sqrt{\frac{C}{L}} = 25\sqrt{\frac{10^{-5}}{10^{-1}}}$

$$= 25 \times 10^{-2}\text{ A} = 0.25\text{ A}$$

54. Given, $\quad L = 25\text{ mH} = 25 \times 10^{-3}\text{ H}$

$$C = 10\,\mu\text{F} = 10^{-5}\text{ F}$$

If T be the time period in L-C oscillation, then

$$T = 2\pi\sqrt{LC}$$

$$= 2\pi\sqrt{25 \times 10^{-3} \times 10^{-5}}$$

$$= \pi \times 10^{-3}\text{ s} = \pi\text{ m-s}$$

Current in the circuit will be maximum, when

$$t = \frac{T}{4} = \frac{\pi}{4}\text{ ms}$$

55. The core of transformer is laminated to reduce energy loss due to eddy currents because induction is reduced by laminating.

56. We have, $N_sI_s = N_pI_p$

$$\Rightarrow \quad \frac{N_S}{N_P} = \frac{I_P}{I_S}$$

$$\Rightarrow \quad \frac{2}{3} = \frac{3}{I_s}$$

$$\Rightarrow \quad I_s = \frac{3 \times 3}{2} = \frac{9}{2} = 4.5\text{ A}$$

57. $E_s = \frac{n_s}{n_p}E_p = \frac{10}{200} \times 240 = 12\text{ V}$

58. $e = \frac{Ldi}{dt} = 4 \times \frac{5}{1/1500}$

$$= 30000\text{ V} = 30\text{ kV}$$

59. Given, number of turns in primary, $N_1 = 300$

Number of turns in secondary, $N_2 = 150$

Output power, $P_2 = 2.2\text{ kW} = 2.2 \times 10^3\text{ W}$

Current in secondary coil, $I_2 = 10\text{ A}$

Output power, $P_2 = I_2V_2$

$$\Rightarrow \quad V_2 = \frac{P_2}{I_2} = \frac{2.2 \times 10^3}{10} = 220\text{ V} \qquad \text{... (i)}$$

We know that,

$$\frac{N_1}{N_2} = \frac{\text{Input voltage}}{\text{Output voltage}} = \frac{V_1}{V_2}$$

$$\Rightarrow \quad V_1 = \left(\frac{N_1}{N_2}\right)V_2$$

$$\Rightarrow \quad V_1 = \left(\frac{300}{150}\right) \times (220\text{ V}) \qquad \text{[using Eq. (i)]}$$

$$V_1 = 440\text{ V} \qquad \text{... (ii)}$$

Again, $\quad \frac{V_1}{V_2} = \frac{I_2}{I_1}$

$$\Rightarrow \quad I_1 = \left(\frac{V_2}{V_1}\right)I_2 = \frac{220}{440} \times 10$$

[using Eqs. (i) and (ii)]

$$\Rightarrow \quad I_1 = 5\text{ A}$$

Round II

1. Current in inductance I_L lags behind the source voltage, current in resistance I_R is in phase with voltage, while current in capacitance I_C leads by a phase of $\pi/2$ with voltage. Hence, correct option is (c).

2. $V^2 = V_R^2 + (V_L - V_C)^2 \Rightarrow V_R = V = 220\text{ V}$

Also, $i = \dfrac{220}{100} = 2.2\text{ A}$

3.
$$i = i_0 \cos(\omega t)$$
$$i = i_0 \text{ at } t = 0$$
$$i = \frac{i_0}{\sqrt{2}} \text{ at } \omega t = \frac{\pi}{4}$$
$$t = \frac{\pi}{4\omega} = \frac{\pi}{4(2\pi f)} = \frac{1}{8f}$$
$$t = \frac{1}{400} = 2.5\text{ ms}$$

4. The voltages V_L and V_C are equal and opposite, so voltmeter reading will be zero.

Also, $R = 30\ \Omega$,

$X_L = X_C = 25\ \Omega$

So,
$$i = \frac{V}{\sqrt{R^2 + (X_L - X_C)^2}} = \frac{V}{R} = \frac{240}{30} = 8\text{ A}$$

5. 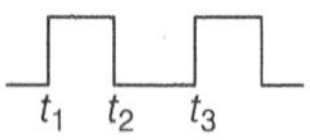

For t_1-t_2 Charging graph

t_2-t_3 Discharging graph

6. The phase difference, $\tan\phi = \dfrac{X_L}{R}$

and $X_L = \omega L = 2\pi f L$

$= 2\pi \times 50 \times 0.01 = \pi\ \Omega$

Also, $R = 1\ \Omega \Rightarrow \phi = \tan^{-1}(\pi)$

7. Four resistances form a balanced Wheatstone bridge.

So, current flowing in the coil will be zero and hence energy stored in the coil is zero.

8. In circuit (1), on closing the switch, the current in the inductor is zero due to self-induction, i.e. $i_1 = 0$.

In circuit (2), on closing the switch, the current in the inductor is zero due to self-induction.

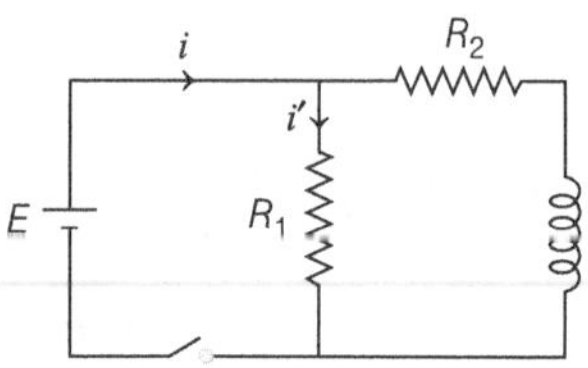

Therefore, $i_2 = i' = \dfrac{E}{R_1}$

In circuit (3), on closing the switch, the current in the inductor is again zero due to the same reason.

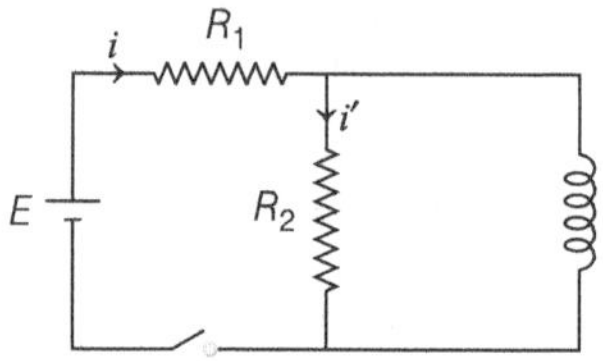

Therefore, $i_3 = i' = \dfrac{E}{R_1 + R_2}$

Thus, it is obvious that, $i_2 > i_3 > i_1 (= 0)$.

9. After connecting C to B hanging the switch, the circuit will act like L-R discharging circuit.

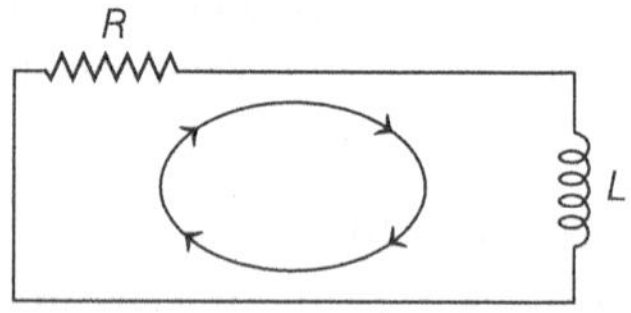

Applying Kirchhoff's loop equation,
$$V_R + V_L = 0 \Rightarrow V_R = -V_L$$
$$\therefore \quad \frac{V_R}{V_L} = -1$$

10. Given, phase difference, $\phi = \dfrac{\pi}{4}$

As we know, for R-L or R-C circuit,
$$\tan\phi = \frac{\text{Capacitive reactance } (X_C) \text{ or inductive reactance } (X_L)}{\text{Resistance } (R)}$$
$$\tan\frac{\pi}{4} = \frac{X_C \text{ or } X_L}{R}$$
$$1 = \frac{X_C \text{ or } X_L}{R}$$
$$\Rightarrow \quad R = X_C \text{ or } X_L$$

Also, given $e = e_0 \sin(100\,t)$

Comparing the above equation with general equation of emf, *i.e.* $e = e_0 \sin\omega t$, we get
$$\omega = 100\text{ rad/s} = 10^2\text{ rad/s}$$

Now, checking option wise,

(a) For R-C circuit, with $R = 1\text{ k}\Omega = 10^3\ \Omega$ and $C = 1\ \mu\text{F} = 10^{-6}\text{ F}$

So, $X_C = \dfrac{1}{\omega C} = \dfrac{1}{10^2 \times 10^{-6}} = 10^4\ \Omega \Rightarrow R \neq X_C$

(b) For R-L circuit, with
$$R = 1\text{ k}\Omega = 10^3\ \Omega$$
and $L = 1\text{ mH} = 10^{-3}\text{ H}$

So, $X_L = \omega L = 10^2 \times 10^{-3} = 10^{-1}\ \Omega \Rightarrow R \neq X_L$

(c) For R-C circuit, with $R = 1\text{ k}\Omega = 10^3\ \Omega$

and $C = 10\ \mu\text{F} = 10 \times 10^{-6}\text{ F} = 10^{-5}\text{ F}$

So, $X_C = \dfrac{1}{10^2 \times 10^{-5}} = 10^3\,\Omega \Rightarrow R = X_C$

(d) For R-L circuit, with $R = 1\,\text{k}\Omega = 10^3\,\Omega$

and $L = 10\ \text{mH} = 10\times 10^{-3}\,\text{H} = 10^{-2}\text{H}$

$X_L = 10^2 \times 10^{-2} = 1\,\Omega \Rightarrow R \neq X_L$

11. Given, frequency, $f = 750$ Hz,

$V_{\text{rms}} = 20\text{V}, R = 100\,\Omega, L = 0.1803$ H

and $C = 10\ \mu\text{F}$

So, impedance of L-C-R circuit is

$$Z = \sqrt{R^2 + \left(2\pi f L - \frac{1}{2\pi f C}\right)^2}$$

$$= \sqrt{(100)^2 + \left(2\pi \times 750 \times 0.1803 - \frac{1}{2\pi \times 750 \times 10 \times 10^{-6}}\right)^2}$$

$$\approx 834\,\Omega$$

Now, power lost by L-C-R circuit, which occurs in resistance is given by

$$P = V_{\text{rms}} I_{\text{rms}} \cos\phi$$

As, $\cos\phi = \dfrac{R}{Z}$ and $I_{\text{rms}} = \dfrac{V_{\text{rms}}}{Z}$

$\therefore \quad P = \dfrac{V_{\text{rms}}^2 \cdot R}{Z^2} = \dfrac{(20)^2 \times 100}{(834)^2} = 0.0575\ \text{Js}^{-1}$

Heat developed in resistor, $H = Pt = S(\Delta\theta)$

where, t = time, S = heat capacity = 2 J/°C and

$\Delta\theta$ = temperature change = 10 °C

$\Rightarrow \quad t = \dfrac{S\Delta\theta}{P} = \dfrac{2 \times 10}{0.0575} = 348$ s

12. In position-1

$$\tan\frac{\pi}{6} = \frac{\omega L}{R} = \frac{(1000)(\sqrt{3} \times 10^{-3})}{R}$$

or $\dfrac{1}{\sqrt{3}} = \dfrac{\sqrt{3}}{R}$

$\therefore \quad R = 3\,\Omega$

In position-2

$$\tan\phi = \frac{X_C}{R} = \frac{(1/\omega C)}{R}$$

$$= \frac{1\Big/\left(1000 \times \frac{1000}{3} \times 10^{-6}\right)}{3} = 1$$

$\therefore \quad \phi = \dfrac{\pi}{4}$

Since, the circuit is C-R, hence current leads emf by $\pi/4$.

13. In the given graph current is leading the voltage by 45°. Therefore circuit is C-R.

$$\tan\phi = \frac{X_C}{R} \text{ and } \phi = \frac{\pi}{4}$$

$\therefore \quad X_C = R$

$\Rightarrow \quad \omega CR = 1$

$\Rightarrow \quad CR = \dfrac{1}{\omega} = \dfrac{1}{100}$

When $R = 1\,\text{k}\Omega = 10^3\,\Omega$, then

$$C = \frac{1}{10^5} = 10^{-5}\ \text{F} = 10\,\mu\text{F}$$

14. This is a parallel circuit, for oscillation, the energy in L and C will be alternatively maximum.

15. $Z = \sqrt{R^2 + (X_C - X_L)^2} = 10\,\Omega$

$$\cos\phi = \frac{R}{Z} = \frac{4}{5}$$

$\therefore \quad \phi = 37°$

$$I_0 = \frac{V_0}{Z} = \frac{10}{10} = 1\,\text{A}$$

$\therefore \quad I = 1\cos(100\pi t + 37°)$

$V = 10\cos(100\pi t)$

The applied voltage is $\dfrac{3}{5}$th of the maximum applied voltage when, or $100\,\pi t = 53°$

$\therefore \quad I = 1\cos(53° + 37°) = 0$

$\therefore \quad V_R = 0$

$\therefore \quad V_{\text{AB}} = V_{\text{applied}} = \dfrac{3}{5} \times 10 = 6$ V

16. The given series R-L-C circuit is shown in the figure below

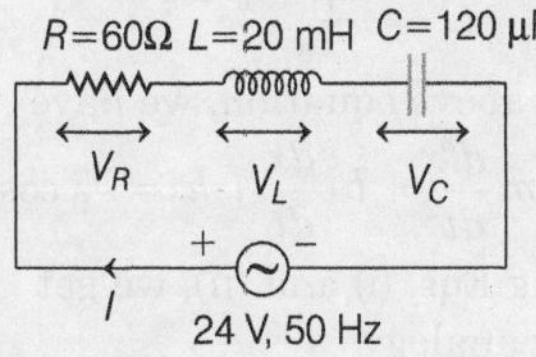

Here, V_R = potential across resistance (R),

V_L = potential across inductor (L)

and V_C = potential across capacitor (C).

Impedance of this series circuit is,

$$Z = \sqrt{R^2 + (X_L - X_C)^2} \quad \text{...(i)}$$

$\because \quad X_L = \omega L = (2\pi f)(L) = 2\pi \times 50 \times 20 \times 10^{-3}\,\Omega$

$X_L = 6.28\,\Omega$ ...(ii)

and $X_C = \dfrac{1}{\omega C} = \dfrac{1}{2\pi f C} = \dfrac{1}{2\pi \times 50 \times 120 \times 10^{-6}} = \dfrac{250}{3\pi}\,\Omega$ ...(iii)

and $X_L - X_C = \left(6.28 - \dfrac{250}{3\pi}\right) = -20.23\,\Omega$ ...(iv)

RMS value of current in circuit,

$$I_{\text{rms}} = \frac{V_{\text{rms}}}{Z} = \frac{24}{\sqrt{R^2 + (X_L - X_C)^2}}$$

$$I_{\text{rms}} = \frac{24}{\sqrt{60^2 + (-20.23)^2}} = \frac{24}{63.18}$$

$$I_{\text{rms}} = 0.379\ \text{A}$$

Therefore, energy dissipated is

$$= I_{rms}^2 \times R \times t$$

$$E = (0.379)^2 \times 60 \times 60$$

or $$= 517.10 = 5.17 \times 10^2 \text{ J}$$

17. For an *L-C-R* circuit,

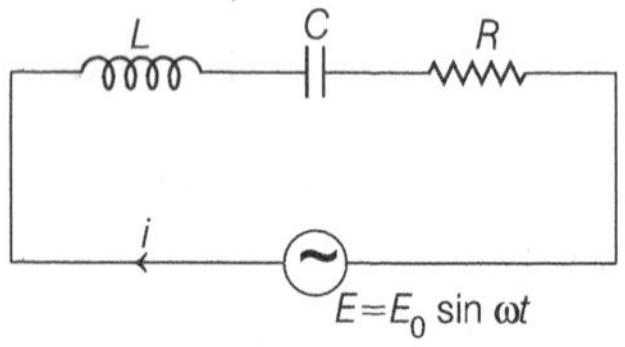

By KVL, we have

$$-L\frac{di}{dt} - \frac{q}{C} - iR + E = 0$$

Above can be rearranged as

$$L\frac{d^2q}{dt^2} + R\cdot\frac{dq}{dt} + \frac{1}{C}q = E_0 \sin \omega t \qquad \ldots\text{(i)}$$

Now, for a damped harmonic oscillator, we have

$$ma = -kx - bv + F_0 \cos \omega t$$

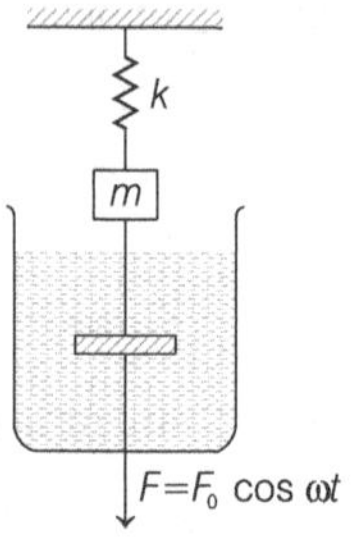

Rearranging above equation, we have

$$m\frac{d^2x}{dt^2} + b\cdot\frac{dx}{dt} + kx = F_0 \cos \omega t \qquad \ldots\text{(ii)}$$

On comparing Eqs. (i) and (ii), we get the following analogy

$$L \equiv m, R \equiv b \text{ and } \frac{1}{C} \equiv k$$

or $$L \leftrightarrow m, C \leftrightarrow \frac{1}{k} \text{ and } R \leftrightarrow b$$

18. The phase difference between the applied voltage and circuit current is $(55° - 10°) = 45°$ with current lagging. The angular frequency is $\omega = 3000$ rad/s. Since, current lags, $X_L > X_C$.

Net reactance, $X = (X_L - X_C)$.

Also, $X_L = \omega L = 3000 \times 0.01 = 30\ \Omega$

$$\tan \phi = X/R \text{ or } \tan 45° = X/R$$

$\therefore \quad X = R$

Now, $Z = \dfrac{V_m}{I_m} = \dfrac{400}{10\sqrt{2}} = 28.3\ \Omega$

$$Z^2 = R^2 + X^2 = 2R^2$$

$\therefore \quad R - Z/\sqrt{2} - 28.3/\sqrt{2} - 20\ \Omega;$

$$X = X_L - X_C = 30 - X_C = 20$$

$$X_C = 10\ \Omega \text{ or } \frac{1}{\omega C} = 10$$

$\Rightarrow \quad C = \dfrac{1}{3000} \qquad (\because \omega = 300\text{ s}^{-1})$

$\Rightarrow \quad C = 33\mu\text{F}$

19. For DC, $R = \dfrac{V}{i} = \dfrac{100}{1} = 100\ \Omega$

For AC, $Z = \dfrac{V}{i} = \dfrac{100}{0.5} = 200\ \Omega$

$$Z = \sqrt{R^2 + (\omega L)^2}$$

$\Rightarrow \quad 200 = \sqrt{(100)^2 + 4\pi^2(50)^2L^2}$

$\therefore \quad L = 0.55$ H

20. $\tan \phi = \dfrac{X_L}{R} = \dfrac{X_C}{R}$

$\Rightarrow \quad \tan 60° = \dfrac{X_L}{R} = \dfrac{X_C}{R}$

$\Rightarrow \quad X_L = X_C = \sqrt{3}R \quad (\because$ the circuit is in resonance)

i.e. $\quad Z = \sqrt{R^2 + (X_L - X_C)^2} = R$

So, average power, $P = \dfrac{V^2}{R} = \dfrac{200 \times 200}{100} = 400$ W

21. Current will be maximum in the condition of resonance, so $i_{max} = \dfrac{V}{R} = \dfrac{V}{10}$ A

Energy stored in the coil,

$$W_L = \frac{1}{2}Li_{max}^2 = \frac{1}{2}L\left(\frac{E}{10}\right)^2$$

$$= \frac{1}{2} \times 10^{-3}\left(\frac{E^2}{100}\right) = \frac{1}{2} \times 10^{-5}E^2 \text{ J}$$

$\therefore$ Energy stored in the capacitor,

$$W_C = \frac{1}{2}CE^2 = \frac{1}{2} \times 2 \times 10^{-6} E^2 = 10^{-6}E^2 \text{ J}$$

$\therefore \quad \dfrac{W_C}{W_L} = \dfrac{1}{5}$

22. At resonance, power,

$$P = \frac{(V_{rms})^2}{R}$$

$$P = \frac{(250/\sqrt{2})^2}{8} = 3906.25 \text{ W} \approx 4 \text{ kW}$$

$\therefore \quad x = 4$

23. $\quad J_c = \dfrac{E}{\rho} = \dfrac{V}{\rho d}$

$$J_d = \frac{1}{A}\frac{dq}{dt} = \frac{C}{A}\frac{dV_c}{dt} = \frac{\varepsilon}{d}\frac{dV_c}{dt}$$

$$\frac{J_c}{J_d} = \frac{\varepsilon_0 \tan\left(2\pi \times \dfrac{900}{800}\right)}{80\varepsilon_0\rho 2\pi f}$$

$$= \frac{2 \times 9 \times 10^9 \tan\left(\dfrac{9\pi}{4}\right)}{80 \times 0.25 \times 900} = 10^6$$

$\therefore \quad x = 6$

24

Electromagnetic Waves

In 1865, Maxwell predicted the existence of electromagnetic waves on the basis of his equations. He concluded that variation of electric and magnetic field vectors perpendicular to each other leads to the generation of electromagnetic disturbances in space. These disturbances possesses the properties of waves and can propagate in space without any medium. Thus, these waves are known as **electromagnetic waves**.

IN THIS CHAPTER ....

- Displacement Current
- Maxwell's Equations
- Electromagnetic Waves
- Electromagnetic Spectrum

Displacement Current

It is a current which is produced due to the time rate of change of electric flux.

It is given by

$$i_D = \varepsilon_0 \frac{d\phi_E}{dt}$$

where, ε_0 is the absolute permittivity of free space and $d\phi_E / dt$ is rate of change of electric flux.

It is also known as **Maxwell's displacement current**.

So, total electric current through a parallel plate capacitor connected with a battery is given as $i = i_C + i_D = i_C + \varepsilon_0 \cdot \dfrac{d\phi_E}{dt}$

where, i_C is the **conduction current** which arises due to the flow of charges.

Important points related to displacement and conduction current are as follows

(i) The conduction and displacement current are individually discontinuous. But the currents together possess the property of continuity through any closed electric circuit.

(ii) Displacement current is precisely equal to the conduction current but they are present in different parts of the circuit.

(iii) The source of magnetic field is not just the conduction current but it is also due to the displacement current.

(iv) For a capacitor, inside the plates, $i_C = 0$ and $i = i_D$. But outside the plates, $i_D = 0$ and $i = i_C$.

Example 1. *The instantaneous displacement current of 1.0 A current in the space between the parallel plates of 1μF capacitor can be established by changing the potential difference of*

(a) $10^5 Vs^{-1}$ (b) $10^6 Vs^{-1}$

(c) $10^{-6} Vs^{-1}$ (d) $10^7 Vs^{-1}$

Sol. (b) As, $i_d = \varepsilon_0 \frac{d\phi_E}{dt} = \varepsilon_0 \frac{d}{dt}(EA) = \varepsilon_0 A \frac{dE}{dt} = \varepsilon_0 A \frac{d}{dt}\left(\frac{V}{d}\right)$

or $$i_d = \frac{\varepsilon_0 A}{d} \times \frac{dV}{dt} = C\frac{dV}{dt}$$

or $$\frac{dV}{dt} = \frac{i_d}{C} = \frac{1.0}{10^{-6}} = 10^6 \text{ Vs}^{-1}$$

Thus, an instantaneous displacement current of 1.0 A can be set-up by changing the potential difference across the parallel plates of capacitor at the rate of 10^6 Vs^{-1}.

Maxwell's Equations

Maxwell gave the basic laws of electricity and magnetism in the form of four fundamental equations, which are known as Maxwell's equations.

These equations give complete description of all electromagnetic interactions.

These equations are as follows

(i) $\oint_S \mathbf{E} \cdot d\mathbf{S} = \frac{q}{\varepsilon_0}$ (Gauss's law for electrostatics)

(ii) $\oint_S \mathbf{B} \cdot d\mathbf{S} = 0$ (Gauss's law for magnetism)

(iii) $\oint \mathbf{E} \cdot d\mathbf{l} = -\frac{d\phi_B}{dt}$

(Faraday's law of electromagnetic induction)

(iv) $\oint \mathbf{E} \cdot d\mathbf{l} = \mu_0\left(i_C + \varepsilon_0 \frac{d\phi_E}{dt}\right)$ (Ampere-Maxwell's law)

Electromagnetic Waves

Maxwell found that the **accelerated** or **oscillated** charges radiate electromagnetic waves. These charges produce an oscillating electric field in space, which produces an oscillating magnetic field, which in turn is a source of oscillating electric field and so on. Thus, oscillating electric and magnetic fields regenerate each other and electromagnetic waves propagates through the space.

Equation of electric field and magnetic field component of EM wave is given as

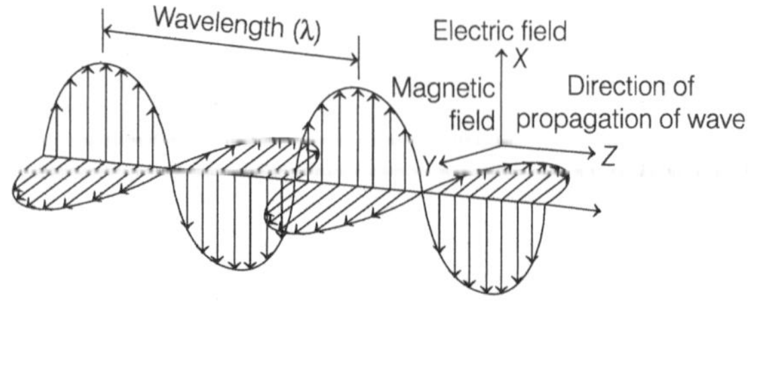

Here, $$E = E_x = E_0 \sin(\omega t - kZ)$$
$$B = B_y = B_0 \sin(\omega t - kZ)$$

where, propagation vector of wave vector, $k = \frac{2\pi}{\lambda}$ (λ is wavelength), angular frequency, $\omega = 2\pi\nu$ (ν is frequency), E_0 and B_0 are the amplitude of varying electric and magnetic field, respectively.

Characteristics of EM Waves

- These waves are self-sustaining oscillations of **E** and **B** in free space or vacuum.
- No material medium is required for their propagation.
- Electromagnetic waves could be diffracted, refracted and polarised.
- These waves are transverse in nature and travel in free space with the speed of light,

$$c \simeq 3 \times 10^8 \text{ m/s}$$

where, $$c = \frac{1}{\sqrt{\mu_0 \varepsilon_0}} = \frac{E_0}{B_0}$$

In isotropic medium, $v = \frac{1}{\sqrt{\mu\varepsilon}} = \frac{1}{\sqrt{\mu_0 \mu_r \varepsilon_0 \varepsilon_r}}$

or $$v = \frac{1}{\sqrt{\mu_r \varepsilon_r}} \cdot \frac{1}{\sqrt{\mu_0 \varepsilon_0}} = \frac{c}{n}$$

Here, $n = \sqrt{\mu_r \varepsilon_r}$ = refractive index of medium,

μ_r = relative permeability of medium

and ε_r = relative permittivity of medium.

- The average energy density (energy per unit volume) of electric field,

$$u_E = \frac{1}{2}\varepsilon_0 E^2 = \frac{1}{2}\varepsilon_0 (E_0/\sqrt{2})^2 = \frac{1}{4}\varepsilon_0 E_0^2$$

The average energy density of magnetic field,

$$u_B = \frac{B^2}{2\mu_0} = \frac{(B_0/\sqrt{2})^2}{2\mu_0} = \frac{B_0^2}{4\mu_0}$$

$\therefore$ $$u_E = u_B \qquad \left(\because c = \frac{E_0}{B_0} = \frac{1}{\sqrt{\mu_0 \varepsilon_0}}\right)$$

Total average energy density,

$$u_{av} = u_E + u_B = 2u_E = 2u_B = \frac{1}{2}\varepsilon_0 E_0^2 = \frac{1}{2}\frac{B_0^2}{\mu_0}$$

The units of u_E and u_B are Jm^{-3}.

- Energy in electromagnetic waves is divided equally between electric and magnetic fields.
- The energy crossing per unit area per unit time, perpendicular to the direction of propagation of EM wave is called **intensity** of EM wave

i.e. Intensity, $I = \frac{\text{Total EM energy}}{\text{Surface area} \times \text{Time}}$

$$= u_{av} \times c = \frac{1}{2}\varepsilon_0 E_0^2 c = \frac{1}{2}\frac{B_0^2 c}{\mu_0} \frac{\text{W}}{\text{m}^2}$$

- If a portion of electromagnetic wave of energy U is propagating with speed c, then **linear momentum** of electromagnetic wave is given by

$$p = \frac{U}{c} \quad \text{(complete absorption)}$$

If the wave is incident on a totally reflecting surface, then momentum delivered,

$$p = \frac{2U}{c}$$

- **Radiation pressure** exerted on the surface is defined as force per unit area (F/A). Thus,

$$p = \frac{F}{A} = \frac{1}{A}\left(\frac{dp}{dt}\right) = \frac{1}{A}\frac{d}{dt}\left(\frac{U}{c}\right) = \frac{1}{c}\frac{(dU/dt)}{A}$$

where, $\frac{dU/dt}{A}$ is the magnitude of the Poynting vector.

Thus, the radiation pressure p exerted on the perfectly absorbing surface, $p = \frac{S}{c}$.

If the surface is perfect reflector and incidence is normal, then $p = \frac{2S}{c}$.

- The rate of flow of energy crossing a unit area in an electromagnetic wave is described by the vector $\mathbf{S}$ called the **Poynting vector,** which is defined by the expression,

$$\mathbf{S} = \frac{1}{\mu_0}\mathbf{E} \times \mathbf{B}$$

Its magnitude S is related to the rate at which energy is transported by a wave across a unit area at any instant.

or $$S = \frac{EB}{\mu_0}$$

The SI unit of poynting vector $\mathbf{S}$ is $Js^{-1}m^{-2}$ or Wm^{-2}.

Example 2. *If the magnetic field of a plane electromagnetic wave is given by*

$$B = 100 \times 10^{-6} \sin\left[2\pi \times 2 \times 10^{15}\left(t - \frac{x}{c}\right)\right],$$

then the maximum electric field associated with it is (Take, speed of light = 3×10^8 m/s) **[JEE Main 2019]**

(a) 6×10^4 N/C
(b) 4×10^4 N/C
(c) 3×10^4 N/C
(d) 4.5×10^4 N/C

Sol. (c) Given, instantaneous value of magnetic field,

$$B = 100 \times 10^{-6} \sin\left[2\pi \times 2 \times 10^{15}\left(t - \frac{x}{c}\right)\right]$$

and speed of light, $c = 3 \times 10^8$ ms^{-1}

For an electromagnetic wave, $E_{max} = B_{max} \times c$

where, E_{max} = maximum value of the electric field.

We get,

$$E_{max} = 100 \times 10^{-6} \times 3 \times 10^8 = 3 \times 10^4 \text{ N/C}$$

Example 3. *The electric field of a plane polarised electromagnetic wave in free space at time $t = 0$ is given by an expression*

$$\mathbf{E}(x, y) = 10\,\hat{\mathbf{j}} \cos(6x + 8z)$$

The magnetic field $\mathbf{B}(x, z, t)$ is given by (where, c is the velocity of light) **[JEE Main 2019]**

(a) $\frac{1}{c}(6\hat{\mathbf{k}} - 8\hat{\mathbf{i}}) \cos[(6x + 8z + 10ct)]$

(b) $\frac{1}{c}(6\hat{\mathbf{k}} - 8\hat{\mathbf{i}}) \cos[(6x + 8z - 10ct)]$

(c) $\frac{1}{c}(6\hat{\mathbf{k}} + 8\hat{\mathbf{i}}) \cos[(6x - 8z + 10ct)]$

(d) $\frac{1}{c}(6\hat{\mathbf{k}} + 8\hat{\mathbf{i}}) \cos[(6x + 8z - 10ct)]$

Sol. (b) Given, electric field is $E(x, y)$, *i.e.* electric field is in XY-plane which is given as

$$\mathbf{E} = 10\,\hat{\mathbf{j}} \cos(6x + 8z)$$

Since, the magnetic field given is $\mathbf{B}(x, z, t)$, this means $\mathbf{B}$ is in XZ-plane.

$\therefore$ Propagation of wave is in y-direction.

($\because$ for an electromagnetic wave, $\mathbf{E} \perp \mathbf{B} \perp$ propagation direction)

As, Poynting vector suggests that $\mathbf{E} \times \mathbf{B}$ is parallel to $(6\,\hat{\mathbf{i}} + 8\,\hat{\mathbf{k}})$.

Let $\hat{\mathbf{B}} = (x\,\hat{\mathbf{i}} + z\,\hat{\mathbf{k}})$, then

$$\hat{\mathbf{E}} \times \hat{\mathbf{B}} = \hat{\mathbf{j}} \times (x\,\hat{\mathbf{i}} + z\,\hat{\mathbf{k}}) = 6\,\hat{\mathbf{i}} + 8\,\hat{\mathbf{k}}$$

or $-x\hat{\mathbf{k}} + z\,\hat{\mathbf{i}} = 6\,\hat{\mathbf{i}} + 8\,\hat{\mathbf{k}}$ or $x = -8$ and $z = 6$

$$\therefore \quad \mathbf{B} = \frac{1}{c}(6\,\hat{\mathbf{k}} - 8\,\hat{\mathbf{i}}) \cos(6x + 8z - 10ct) \quad \left(\because \frac{|\mathbf{E}|}{|\mathbf{B}|} = c\right)$$

Example 4. *The electric fields of two plane electromagnetic plane waves in vacuum are given by*

$$\mathbf{E}_1 = E_0\,\hat{\mathbf{j}} \cos(\omega t - kx)$$

and $$\mathbf{E}_2 = E_0\,\hat{\mathbf{k}} \cos(\omega t - ky)$$

At $t = 0$, a particle of charge q is at origin with a velocity $\mathbf{v} = 0.8c\,\hat{\mathbf{j}}$ (c is the speed of light in vacuum). The instantaneous force experienced by the particle is **[JEE Main 2020]**

(a) $E_0 q\,(0.8\,\hat{\mathbf{i}} - \hat{\mathbf{j}} + 0.4\,\hat{\mathbf{k}})$
(b) $E_0 q\,(0.4\hat{\mathbf{i}} - 3\hat{\mathbf{j}} + 0.8\hat{\mathbf{k}})$
(c) $E_0 q(-0.8\,\hat{\mathbf{i}} + \hat{\mathbf{j}} + \hat{\mathbf{k}})$
(d) $E_0 q\,(0.8\,\hat{\mathbf{i}} + \hat{\mathbf{j}} + 0.2\,\hat{\mathbf{k}})$

Sol. (d) For first electromagnetic wave,

$$\mathbf{E}_1 = E_0\hat{\mathbf{j}}\cos(\omega t - kx)$$

So, magnetic field vector,

$$\mathbf{B}_1 = B_0\hat{\mathbf{k}}\cos(\omega t - kx) = \frac{E_0}{c}\hat{\mathbf{k}}\cos(\omega t - kx)$$

Also, for second electromagnetic wave,

$$\mathbf{E}_2 = E_0\hat{\mathbf{k}}\cos(\omega t - ky)$$

$$\Rightarrow \quad \mathbf{B}_2 = \frac{E_0}{c}\hat{\mathbf{i}}\cos(\omega t - ky)$$

Force on charged particle under influence of both electric and magnetic fields,

$$\mathbf{F} = q(\mathbf{v}\times\mathbf{B} + \mathbf{E}) \qquad \text{...(i)}$$

Here, $\mathbf{v} = 0.8\,c\,\hat{\mathbf{j}}$

$$\because \quad \mathbf{B} = \mathbf{B}_1 + \mathbf{B}_2 = \frac{E_0}{c}\hat{\mathbf{k}}\cos(\omega t - kx) + \frac{E_0}{c}\hat{\mathbf{i}}\cos(\omega t - ky)$$

and at $t = 0$, at origin ($x = 0, y = 0$ and $z = 0$), we have

$$\mathbf{B} = \frac{E_0}{c}(\hat{\mathbf{i}} + \hat{\mathbf{k}})$$

and $\quad \mathbf{E} = \mathbf{E}_1 + \mathbf{E}_2 = E_0\hat{\mathbf{j}}\cos(\omega t - kx) + E_0\hat{\mathbf{k}}\cos(\omega t - ky)$

At $t = 0$ and at origin, $\mathbf{E} = E_0(\hat{\mathbf{j}} + \hat{\mathbf{k}})$

So, from Eq. (i), we get

$$\mathbf{F} = q\left[0.8c\hat{\mathbf{j}} \times \frac{E_0}{c}(\hat{\mathbf{i}} + \hat{\mathbf{k}}) + E_0(\hat{\mathbf{j}} + \hat{\mathbf{k}})\right]$$

$$= E_0 q\,[0.8\hat{\mathbf{j}} \times (\hat{\mathbf{i}} + \hat{\mathbf{k}}) + (\hat{\mathbf{j}} + \hat{\mathbf{k}})]$$

$$= E_0 q\,[0.8(-\hat{\mathbf{k}} + \hat{\mathbf{i}}) + (\hat{\mathbf{j}} + \hat{\mathbf{k}})]$$

$$= E_0 q\,(0.8\hat{\mathbf{i}} + \hat{\mathbf{j}} + 0.2\hat{\mathbf{k}})$$

Example 5. *A plane electromagnetic wave has frequency of 2.0×10^{10} Hz and its energy density is 1.02×10^{-8} J / m^3 in vacuum. The amplitude of the magnetic field of the wave is close to (Take, $\frac{1}{4\pi\varepsilon_0} = 9 \times 10^9 \frac{N\text{-}m^2}{C^2}$ and speed of light $= 3 \times 10^8$ ms^{-1})* **[JEE Main 2020]**

(a) 190 nT (b) 160 nT

(c) 180 nT (d) 150 nT

Sol. (b) Energy density of magnetic field of an electromagnetic wave is given by

$$u_B = \frac{1}{2}\left(\frac{B_0^2}{\mu_0}\right)$$

Here, $\mu_0 = 4\pi \times 10^{-7}$ units and $u_B = 1.02 \times 10^{-8}$ Jm^{-3}

So, amplitude of magnetic field,

$$B_0 = \sqrt{2\mu_0 u_B}$$

$$= \sqrt{2 \times 4\pi \times 10^{-7} \times 1.02 \times 10^{-8}}$$

$$= 16 \times 10^{-8}\,\text{T}$$

$$= 160\,\text{nT}$$

Example 6. *The mean intensity of radiation on the surface of the sun is about 10^8 W/m^2. The rms value of the corresponding magnetic field is closest to* **[JEE Main 2019]**

(a) 1 T (b) 10^2 T

(c) 10^{-4} T (d) 10^{-2} T

Sol. (c) Mean radiation intensity is

$$I = \varepsilon_0 c E_{rms}^2 = \varepsilon_0 c (cB_{rms})^2 \qquad \left(\because \frac{E_{rms}}{B_{rms}} = c\right)$$

$$= \varepsilon_0 c^3 B_{rms}^2$$

$$\Rightarrow \quad B_{rms} = \sqrt{\frac{I}{\varepsilon_0 c^3}}$$

Substituting the given values, we get

$$= \sqrt{\left(\frac{10^8}{8.85 \times 10^{-12} \times (3 \times 10^8)^3}\right)}$$

$$= \sqrt{\left(\frac{10^8}{8.85 \times 27 \times 10^{12}}\right)}$$

$$\approx \sqrt{(10^{-8})} \approx 10^{-4}\,\text{T}$$

Example 7. *If the earth receives 2 cal min^{-1} cm^{-2} solar energy, then the amplitudes of electric field of radiation is (V/m)*

(a) 102.3 (b) 377

(c) 150 (d) 200

Sol. (a) From Poynting vector, we know that,

$$\mathbf{S} = \mathbf{E} \times \mathbf{H} = EH \sin 90° = EH$$

Given, $\quad S = 2$ cal min^{-1} cm^{-2}

$$= \frac{2 \times 4.18 \times 10^4}{60}\,\text{Jm}^{-2}\,\text{s}^{-1}$$

As S represents energy flux per unit area per second, we have

$$EH = \frac{2 \times 4.18 \times 10^4}{60} = 1400$$

$$EH = \sqrt{\frac{\mu_0}{\varepsilon_0}} = 377$$

$$\Rightarrow \quad EH \times \frac{E}{H} = 1400 \times 377$$

$$\therefore \quad E = \sqrt{1400 \times 377} = 726.5\,\text{V/m}$$

Amplitude of electric field,

$$E_0 = E\sqrt{2} = 102.3\,\text{V/m}$$

Electromagnetic Spectrum

It is the orderly arrangement of the electromagnetic waves in increasing or decreasing wavelength λ or frequency ν. These waves includes gamma rays, X-rays, ultraviolet rays, visible light, infrared waves, microwaves and radio waves.

Tabular form of various electromagnetic waves of electromagnetic spectrum with their features and uses are as follows

Name of Wave	Frequency Range	Wavelength Range	Production	Detection	Uses
Radio waves	500 kHz to 1000 MHz	> 0.1 m	Rapid acceleration and decelerations of electrons in aerials.	Receiver's aerials	• These are used in AM (Amplitude Modulation) from 530 kHz to 1710 kHz and ground wave propagation. • These are used in TV waves ranging from 54 MHz to 890 MHz. • These are used in FM (Frequency Modulation) ranging from 88 MHz to 08 MHz.
Microwaves	1 GHz to 300 GHz	0.1 m to 1 mm	Klystron valve or magnetron valve.	Point contact diodes	• These are used in RADAR systems for aircraft navigation. • These are used in microwave oven for cooking purpose. • These are used in study of atomic and molecular structures.
Infrared waves (Heat waves)	3×10^{11} Hz to 4×10^{14} Hz	1 mm to 700 nm	Vibration of atoms and molecules	Thermopiles bolometer and infrared photographic film.	• These are used in physical therapy. • These are used in satellites for army purpose. • These are used in weather forecasting.
Light rays (Visible rays)	4×10^{14} Hz to 7×10^{14} Hz	700 nm to 400 nm	Electrons in atoms emit light, when they move from a higher energy level to a lower energy level.	Eye, photocells and photographic film.	• These are used by the optical organs of humans and animals for three primary purposes given below (i) To see things, avoid bumping into them and escape danger. (ii) To look for food. (iii) To find other living things with which to copulate, so as to prolong the species.
Ultraviolet rays	10^{14} Hz to 10^{16} Hz	400 nm to 1 nm	Inner shell electrons in atoms moving from higher energy level to a lower energy level.	Photocells and photographic film.	• These are used in burglar alarm. • These are used in checking mineral sample. • These are used to study molecular structure. • To kill germs in water purifiers. • Used in LASER eye surgery.
X-rays	3×10^{16} Hz to 3×10^{21} Hz	1 nm to 10^{-3} nm	X-ray tubes or inner shell electrons, bombarding metals by high energy electrons.	Photographic film, Geiger tubes and ionisation chamber.	• These are used in medicine to detect the fracture, diseased organs, stones in the body, etc. • These are used in engineering to detect fault, cracks in bridges and testing of welds. • These are used at metro stations to detect metals or explosive material.
Gamma (γ) rays	3×10^{18} Hz to 5×10^{22} Hz	$< 10^{-3}$ nm	Radioactive decay of the nucleus.	Photographic film and ionisation chamber.	• These are used to produce nuclear reactions. • These are used in radio therapy for the treatment of tumour and cancer. • These are used in food industry to kill pathogenic micro-organism.

Practice Exercise

ROUND I Topically Divided Problems

Displacement Current and Maxwell's Equations

1. The space between plates of parallel capacitor is filled with poorly conducting material. If an AC source is connected across the capacitor, current through material is
(a) only conduction current
(b) only displacement current
(c) displacement current as well as conduction current
(d) None of the above

2. The charge of a parallel plate capacitor is varying as $q = q_0 \sin 2\pi ft$. The plates are very large and close together (area = A, separation = d). Neglecting edge effects, the displacement current through the capacitor is
(a) $\frac{d}{A\varepsilon_0}$ (b) $\frac{d}{\varepsilon_0}\sin 2\pi ft$
(c) $2\pi f q_0 \cos 2\pi ft$ (d) $\frac{2\pi f q_0}{\varepsilon_0}\cos 2\pi ft$

3. A capacitor having a capacity of 2pF. Electric field across the capacitor is changing with a value of $10^{12}\ \text{Vs}^{-1}$. The displacement current is
(a) 2 A (b) 4 A (c) 6 A (d) 10 A

4. A large parallel plate capacitor, whose plates have an area of $1\ \text{m}^2$ and are separated from each other by 1 mm, is being charged at a rate of $25\ \text{Vs}^{-1}$. If the dielectric between the plates has the dielectric constant 10, then the displacement current at this instant is
(a) 25 μA (b) 11 μA (c) 2.2 μA (d) 1.1 μA

5. A parallel plate capacitor consists of two circular plates with radius $R = 10$ cm separated by distance $d = 0.5$ mm. The capacitor is being charged at a uniform rate by applying a changing potential difference between the two plates. Calculate the displacement current for the capacitor. Assume that, the electric field is due to the displacement current only and rate at which the electric field between the plates changes is $5\times 10^{13}\ \text{Vms}^{-1}$.
(a) 12.8 A (b) 12.6 A
(c) 13.9 A (d) 10.5 A

6. A parallel plate capacitor consists of two circular plates each of radius 2 cm, separated by a distance of 0.1 mm. If the potential difference across the plates is varying at the rate of $5\times 10^{6}\ \text{Vs}^{-1}$, then the value of displacement current is
(a) 5.56 A (b) 5.56 mA
(c) 0.556 mA (d) 2.28 mA

7. Match the Column I with Column II and mark the correct option from the codes given below.

	Column I		Column II
A.	$\oint \mathbf{B}\cdot d\mathbf{l} = \mu_0 I$	1.	Gauss's law
B.	$\oint \mathbf{B}\cdot d\mathbf{l} = \mu_0\left(I + \epsilon_0 \frac{d\phi_E}{dt}\right)$	2.	Faraday's laws of electromagnetic induction
C.	$\oint \mathbf{E}\cdot d\mathbf{l} = -\frac{d}{dt}\oint_s \mathbf{B}\cdot d\mathbf{s}$	3.	Ampere-Maxwell's law
D.	$\oint \mathbf{E}\cdot d\mathbf{s} = \frac{1}{\varepsilon_0}\oint_s \rho\, ds$	4.	Ampere's circuital law

Codes

	A	B	C	D
(a)	1	2	3	4
(b)	4	3	2	1
(c)	4	2	1	3
(d)	3	1	2	4

Electromagnetic Waves and its Characteristics

8. Select the correct statement from the following. **[JEE Main 2013]**
(a) Electromagnetic waves cannot travel in vacuum.
(b) Electromagnetic waves are longitudinal waves.
(c) Electromagnetic waves are produced by charges moving with uniform velocity.
(d) Electromagnetic waves carry both energy and momentum as they propagate through space.

9. Which of the following is/are the property/properties of a monochromatic electromagnetic wave propagating in the free space?
I. Electric and magnetic fields will have a phase difference $\frac{\pi}{2}$.
II. The energy of the wave is distributed equally between electric and magnetic fields.
III. The pressure exerted by the wave is the product of its speed and energy density.

IV. The speed of the wave is equal to the ratio of the magnetic field to the electric field.

(a) I and III (b) Only II

(c) II and III (d) Only IV

10. If an electromagnetic wave is propagating in a medium with permittivity ε and permeability μ, then $\sqrt{\frac{\mu}{\varepsilon}}$ is the

(a) intrinsic impedance of the medium

(b) square of the refractive index of the medium

(c) refractive index of the medium

(d) energy density of the medium

11. A radio wave of frequency 90 MHz enters a ferrite rod. If $\varepsilon_r = 10^3$ and $\mu_r = 10$, then the velocity and wavelength of the wave in ferrite are

(a) $3\times10^8 \text{ms}^{-1}; 3.33\times10^{-2}\text{m}$

(b) $3\times10^6 \text{ms}^{-1}; 3.33\times10^{-2}\text{m}$

(c) $3\times10^8 \text{ms}^{-1}; 3.33\times10^{-1}\text{m}$

(d) $3\times10^7 \text{ms}^{-1}; 3.33\times10^{-3}\text{m}$

12. An electromagnetic wave of frequency $\nu = 3.0$ MHz passes from vacuum into a dielectric medium with permittivity $\varepsilon = 4.0$, then

(a) wavelength is doubled and the frequency remains unchanged

(b) wavelength is doubled and frequency becomes half

(c) wavelength is halved and frequency remains unchanged

(d) wavelength and frequency both remains unchanged

13. The amplitude of an electromagnetic wave in vacuum is doubled with no other changes made to the wave. As a result of this doubling of the amplitude, which of the following statement is true?

(a) The speed of wave propagation changes only.

(b) The frequency of the wave changes only.

(c) The wavelength of the wave changes only.

(d) None of the above

14. A plane electromagnetic wave is incident on a material surface. If the wave delivers momentum p and energy E, then

(a) $p=0, E=0$ (b) $p\neq0, E\neq0$

(c) $p\neq0, E=0$ (d) $p=0, E\neq0$

15. An electromagnetic wave going through vacuum is described by $E = E_0 \sin(kx - \omega t)$; $B = B_0 \sin(kx - \omega t)$. Which of the following equation is true?

(a) $E_0 k = B_0 \omega$ (b) $E_0 \omega = B_0 k$

(c) $E_0 B_0 = \omega k$ (d) None of these

16. An electromagnetic wave propagating along North has its electric field vector upwards. Its magnetic field vector point towards

(a) North (b) East (c) West (d) Downwards

17. In a plane electromagnetic wave, the electric field oscillates sinusoidally at a frequency of 2×10^{10} Hz and its amplitude is 54 V. Then, which of the following statement is correct?

(a) The amplitude of oscillating magnetic field will be 18×10^{-8} Wbm^{-2}.

(b) The amplitude of oscillating magnetic field will be 18×10^{-7} Wbm^{-2}.

(c) The wavelength of electromagnetic wave is 1.5 m.

(d) Both (a) and (c)

18. A plane electromagnetic wave propagating along y-direction can have the following pair of electric field (E) and magnetic field (B) components **[JEE Main 2021]**

(a) E_y, B_y or E_z, B_z (b) E_y, B_x or E_x, B_y

(c) E_x, B_z or E_z, B_x (d) E_x, B_y or E_y, B_x

19. A plane electromagnetic wave of frequency 25 GHz is propagating in vacuum along the z-direction. At a particular point in space and time, the magnetic field is given by $\mathbf{B} = 5\times10^{-8}\,\hat{\mathbf{j}}$ T. The corresponding electric field $\mathbf{E}$ is (speed of light, $c = 3\times10^8\ \text{ms}^{-1}$) **[JEE Main 2020]**

(a) $-1.66\times10^{-16}\hat{\mathbf{i}}$ V/m (b) $1.66\times10^{-16}\hat{\mathbf{i}}$ V/m

(c) $-15\,\hat{\mathbf{i}}$ V/m (d) $15\,\hat{\mathbf{i}}$ V/m

20. An electromagnetic wave in vacuum has the electric and magnetic fields $\mathbf{E}$ and $\mathbf{B}$, which are mutually perpendicular to each other. The direction of polarisation is given by $\mathbf{X}$ and that of wave propagation by $\hat{\mathbf{k}}$, then

(a) $\mathbf{X} \parallel \mathbf{B}$ and $\hat{\mathbf{k}} \parallel \mathbf{B}\times\mathbf{E}$

(b) $\mathbf{X} \parallel \mathbf{E}$ and $\hat{\mathbf{k}} \parallel \mathbf{E}\times\mathbf{B}$

(c) $\mathbf{X} \parallel \mathbf{B}$ and $\hat{\mathbf{k}} \parallel \mathbf{E}\times\mathbf{B}$

(d) $\mathbf{X} \parallel \mathbf{E}$ and $\hat{\mathbf{k}} \parallel \mathbf{B}\times\mathbf{E}$

21. In a plane electromagnetic wave, the directions of electric field and magnetic field are represented by $\hat{\mathbf{k}}$ and $2\hat{\mathbf{i}} - 2\hat{\mathbf{j}}$, respectively. What is the unit vector along direction of propagation of the wave? **[JEE Main 2020]**

(a) $\frac{1}{\sqrt{2}}(\hat{\mathbf{i}} + \hat{\mathbf{j}})$ (b) $\frac{1}{\sqrt{2}}(\hat{\mathbf{j}} + \hat{\mathbf{k}})$

(c) $\frac{1}{\sqrt{5}}(\hat{\mathbf{i}} + 2\hat{\mathbf{j}})$ (d) $\frac{1}{\sqrt{5}}(2\hat{\mathbf{i}} + \hat{\mathbf{j}})$

22. The electromagnetic wave travelling along Z-axis is given as $\mathbf{E} = E_0 \cos(kz - \omega t)$. Choose the incorrect options from the following.

(a) The associated magnetic field is given as $\mathbf{B} = \frac{1}{c}\hat{\mathbf{k}}\times\mathbf{E}$

(b) The electromagnetic field can be written in terms of the associated magnetic field as $\mathbf{E} = c(\mathbf{B}\times\hat{\mathbf{k}})$

(c) $\hat{\mathbf{k}}\cdot\mathbf{E} = 0, \hat{\mathbf{k}}\cdot\mathbf{B} = 0$

(d) $\hat{\mathbf{k}}\times\mathbf{E} = 0, \hat{\mathbf{k}}\times\mathbf{B} = 0$

23. The electric field of a plane electromagnetic wave propagating along the x-direction in vacuum is $\mathbf{E} = E_0 \hat{\mathbf{j}} \cos(\omega t - kx)$. The magnetic field $\mathbf{B}$, at the moment $t = 0$ is **[JEE Main 2020]**

(a) $\mathbf{B} = E_0 \sqrt{\mu_0 \varepsilon_0} \cos(kx)\,\hat{\mathbf{j}}$

(b) $\mathbf{B} = \dfrac{E_0}{\sqrt{\mu_0 \varepsilon_0}} \cos(kx)\hat{\mathbf{j}}$

(c) $\mathbf{B} = \dfrac{E_0}{\sqrt{\mu_0 \varepsilon_0}} \cos(kx)\hat{\mathbf{k}}$

(d) $\mathbf{B} = E_0 \sqrt{\mu_0 \varepsilon_0} \cos(kx)\hat{\mathbf{k}}$

24. A plane electromagnetic waves travelling along the x-direction has a wavelength of 3 mm. The variation in the electric field occurs in the y-direction with an amplitude 66 Vm^{-1}. The equation for the electric and magnetic fields as a function of x and t are respectively

(a) $E_y = 33 \cos \pi \times 10^{11} \left(t - \dfrac{x}{c}\right)$ and $B_z = 1.1 \times 10^{-7} \cos \pi \times 10^{11} \left(t - \dfrac{x}{c}\right)$

(b) $E_y = 11 \cos 2\pi \times 10^{11} \left(t - \dfrac{x}{c}\right)$ and $B_y = 11 \times 10^{-7} \cos 2\pi \times 10^{11} \left(t - \dfrac{x}{c}\right)$

(c) $E_x = 33 \cos \pi \times 10^{11} \left(t - \dfrac{x}{c}\right)$ and $B_y = 11 \times 10^{-7} \cos \pi \times 10^{11} \left(t - \dfrac{x}{c}\right)$

(d) $E_y = 66 \cos 2\pi \times 10^{11} \left(t - \dfrac{x}{c}\right)$ and $B_z = 2.2 \times 10^{-7} \cos 2\pi \times 10^{11} \left(t - \dfrac{x}{c}\right)$

25. The magnetic field of a plane electromagnetic wave is $\mathbf{B} = 3 \times 10^{-8} \sin[200\pi(y + ct)]\hat{\mathbf{i}}$ T, where $c = 3 \times 10^{8}\ \text{ms}^{-1}$ is the speed of light. The corresponding electric field is **[JEE Main 2020]**

(a) $\mathbf{E} = 9 \sin[200\pi(y + ct)]\,\hat{\mathbf{k}}$ V/m

(b) $\mathbf{E} = -10^{-6} \sin[200\pi(y + ct)]\,\hat{\mathbf{k}}$ V/m

(c) $\mathbf{E} = 3 \times 10^{-8} \sin[200\pi(y + ct)]\,\hat{\mathbf{k}}$ V/m

(d) $\mathbf{E} = -9 \sin[200\pi(y + ct)]\,\hat{\mathbf{k}}$ V/m

26. A linearly polarised electromagnetic wave given as $\mathbf{E}_r = -E_0 \hat{\mathbf{i}} \cos(kz - \omega t)$ is incident normally on a perfectly reflecting infinite wall at $z = a$. Assuming that the material of the wall is optically inactive, the reflected wave will be given as **[NCERT Exemplar]**

(a) $\mathbf{E}_r = -E_0 \hat{\mathbf{i}} \cos(kz + \omega t)$

(b) $\mathbf{E}_r = E_0 \hat{\mathbf{i}} \cos(kz + \omega t)$

(c) $\mathbf{E}_r = -E_0 \hat{\mathbf{i}} \cos(kz - \omega t)$

(d) $\mathbf{E}_r = -E_0 \hat{\mathbf{i}} \sin(kz - \omega t)$

27. For a plane electromagnetic wave, the magnetic field at a point x and time t is
$\mathbf{B}(x, t) = [1.2 \times 10^{-7} \sin(0.5 \times 10^{3} x + 1.5 \times 10^{11} t)\hat{\mathbf{k}}]$ T
The instantaneous electric field $\mathbf{E}$ corresponding to $\mathbf{B}$ is
(Given, speed of light, $c = 3 \times 10^{8}$ m/s) **[JEE Main 2020]**

(a) $\mathbf{E}(x, t) = [-36 \sin(0.5 \times 10^{3} x + 1.5 \times 10^{11} t)\hat{\mathbf{j}}]$ V/m

(b) $\mathbf{E}(x, t) = [36 \sin(1 \times 10^{3} x + 0.5 \times 10^{11} t)\hat{\mathbf{j}}]$ V/m

(c) $\mathbf{E}(x, t) = [36 \sin(0.5 \times 10^{3} x + 1.5 \times 10^{11} t)\hat{\mathbf{k}}]$ V/m

(d) $\mathbf{E}(x, t) = [36 \sin(1 \times 10^{3} x + 1.5 \times 10^{11} t)\hat{\mathbf{j}}]$ V/m

28. A plane electromagnetic wave is propagating along the direction $\dfrac{\hat{\mathbf{i}} + \hat{\mathbf{j}}}{\sqrt{2}}$ with its polarisation along the direction $\hat{\mathbf{k}}$. The correct form of the magnetic field of the wave would be (here B_0 is an appropriate constant) **[JEE Main 2020]**

(a) $B_0 \hat{\mathbf{k}} \cos\left(\omega t - k\dfrac{\hat{\mathbf{i}} + \hat{\mathbf{j}}}{\sqrt{2}}\right)$

(b) $B_0 \dfrac{\hat{\mathbf{j}} - \hat{\mathbf{i}}}{\sqrt{2}} \cos\left(\omega t + k\dfrac{\hat{\mathbf{i}} + \hat{\mathbf{j}}}{\sqrt{2}}\right)$

(c) $B_0 \dfrac{\hat{\mathbf{i}} - \hat{\mathbf{j}}}{\sqrt{2}} \cos\left(\omega t - k\dfrac{\hat{\mathbf{i}} + \hat{\mathbf{j}}}{\sqrt{2}}\right)$

(d) $B_0 \dfrac{\hat{\mathbf{i}} + \hat{\mathbf{j}}}{\sqrt{2}} \cos\left(\omega t - k\dfrac{\hat{\mathbf{i}} + \hat{\mathbf{j}}}{\sqrt{2}}\right)$

29. An electromagnetic wave is represented by the electric field $\mathbf{E} = E_0\, \hat{\mathbf{n}} \sin[\omega t + (6y - 8z)]$.
Taking unit vectors in x, y and z-directions to be $\hat{\mathbf{i}}, \hat{\mathbf{j}}, \hat{\mathbf{k}}$, the direction of propagation $\hat{\mathbf{s}}$, is **[JEE Main 2019]**

(a) $\hat{\mathbf{s}} = \dfrac{3\hat{\mathbf{i}} - 4\hat{\mathbf{j}}}{5}$

(b) $\hat{\mathbf{s}} = \dfrac{-4\hat{\mathbf{k}} + 3\hat{\mathbf{j}}}{5}$

(c) $\hat{\mathbf{s}} = \left(\dfrac{-3\hat{\mathbf{j}} + 4\hat{\mathbf{k}}}{5}\right)$

(d) $\hat{\mathbf{s}} = \dfrac{4\hat{\mathbf{j}} - 3\hat{\mathbf{k}}}{5}$

30. A plane electromagnetic wave having a frequency $\nu = 23.9$ GHz propagates along the positive z-direction in free space. The peak value of the electric field is 60 V/m. Which among the following is the acceptable magnetic field component in the electromagnetic wave? **[JEE Main 2019]**

(a) $\mathbf{B} = 2 \times 10^{7} \sin(0.5 \times 10^{3} z + 1.5 \times 10^{11} t)\,\hat{\mathbf{i}}$

(b) $\mathbf{B} = 2 \times 10^{-7} \sin(0.5 \times 10^{3} z - 1.5 \times 10^{11} t)\,\hat{\mathbf{i}}$

(c) $\mathbf{B} = 60 \sin(0.5 \times 10^{3} x + 1.5 \times 10^{11} t)\hat{\mathbf{k}}$

(d) $\mathbf{B} = 2 \times 10^{-7} \sin(1.5 \times 10^{2} x + 0.5 \times 10^{11} t)\,\hat{\mathbf{j}}$

31. The magnetic field of a plane electromagnetic wave is given by $\mathbf{B} = B_0[\cos(kz - \omega t)]\hat{\mathbf{i}} + B_1 \cos(kz + \omega t)\hat{\mathbf{j}}$, where $B_0 = 3 \times 10^{-5}$ T and $B_1 = 2 \times 10^{-6}$ T. The rms value of the force experienced by a stationary charge $Q = 10^{-4}$ C at $z = 0$ is closest to **[JEE Main 2019]**

(a) 0.1 N (b) 3×10^{-2} N
(c) 0.6 N (d) 0.9 N

32. In free space, the energy of electromagnetic wave in electric field is U_E and in magnetic field is U_B. Then, **[JEE Main 2019]**

(a) $U_E = U_B$ (b) $U_E > U_B$
(c) $U_E < U_B$ (d) $U_E = \frac{U_B}{2}$

33. Assume that, a lamp radiates power P uniformly in all directions. What is the magnitude of electric field strength at a distance r from the lamp?

(a) $\frac{P}{\pi c \varepsilon_0 r^2}$ (b) $\frac{P}{2\pi c \varepsilon r^2}$
(c) $\sqrt{\frac{P}{2\pi\varepsilon_0 r^2 c}}$ (d) $\sqrt{\frac{P}{\pi\varepsilon_0 c r^2}}$

34. The amplitude of the electric field in a parallel beam of plane electromagnetic waves of intensity 53.1 Wm^{-2} is (Permittivity of free space $= 8.85 \times 10^{-12}\ C^2N^{-1}m^{-2}$)

(a) 400 NC^{-1} (b) 50 NC^{-1}
(c) 100 NC^{-1} (d) 200 NC^{-1}

35. An earth orbiting statellite has solar energy collecting panel with total area 5 m^2. If solar radiations are perpendicular and completely absorbed, the average force associated with the radiation pressure is
(Solar constant = 1.4 kWm^{-2})

(a) 2.33×10^{-3} N
(b) 2.33×10^{-4} N
(c) 2.33×10^{-5} N
(d) 2.33×10^{-6} N

36. An electromagnetic wave with Poynting vector 6 Wm^{-2} is absorbed by a surface area 12 m^2. The force on the surface is

(a) 24×10^{-8} N
(b) 30×10^{-5} N
(c) 2.4×10^{-12} N
(d) 24×10^{8} N

37. A 27 mW laser beam has a cross-sectional area of 10 mm^2. The magnitude of the maximum electric field in this electromagnetic wave is given by (Take, permittivity of space, $\varepsilon_0 = 9 \times 10^{-12}$ SI units and speed of light, $c = 3 \times 10^8$ m / s) **[JEE Main 2019]**

(a) 1 kV/m (b) 0.7 kV/m
(c) 2 kV/m (d) 1.4 kV/m

38. The electric field intensity produced by the radiations coming from 100 W bulb at a 3 m distance is E. The electric field intensity produced by the radiations coming from 50 W bulb at the same distance is **[NCERT Exemplar]**

(a) $E/2$ (b) $2E$
(c) $E\sqrt{2}$ (d) $\sqrt{2}E$

39. A plane electromagnetic wave of wave intensity 6 Wm^{-2} strikes a small mirror of area 30 cm^2, held perpendicular to the approaching wave. The momentum transferred (in $kg\text{-}ms^{-1}$) by the wave to the mirror each second will be

(a) 1.2×10^{-10} (b) 2.4×10^{-9}
(c) 3.6×10^{-8} (d) 4.8×10^{-7}

40. In a region of free space, the electric field at some instant of time is $\mathbf{E} = (80\hat{\mathbf{i}} + 32\hat{\mathbf{j}} - 64\hat{\mathbf{k}})$ V/m and the magnetic field is $\mathbf{B} = (0.2\,\hat{\mathbf{i}} + 0.08\,\hat{\mathbf{j}} + 0.29\,\hat{\mathbf{k}})\ \mu$T. The poynting vector for these fields is

(a) $-11.45\,\hat{\mathbf{i}} + 28.6\,\hat{\mathbf{j}}$
(b) $-28.6\,\hat{\mathbf{i}} + 11.45\,\hat{\mathbf{i}}$
(c) $28.6\,\hat{\mathbf{i}} - 11.45\,\hat{\mathbf{j}}$
(d) $11.45\,\hat{\mathbf{i}} - 28.6\,\hat{\mathbf{j}}$

41. A lamp emits monochromatic green light uniformly in all directions. The lamp is 3% efficient in converting electric power to electromagnetic waves and consume 100 W of power. The amplitude of the electric field associated with the electromagnetic radiation at a distance of 5 m from the lamp will be

(a) 2.68 Vm^{-1} (b) 3.15 Vm^{-1}
(c) 2.01 Vm^{-1} (d) 0

42. A cube of edge a has its edges parallel to X, Y and Z-axis of rectangular coordinate system. A uniform electric field $\mathbf{E}$ is parallel to Y-axis and a uniform magnetic field is $\mathbf{B}$ parallel to X-axis. The rate at which energy flows through each face of the cube is

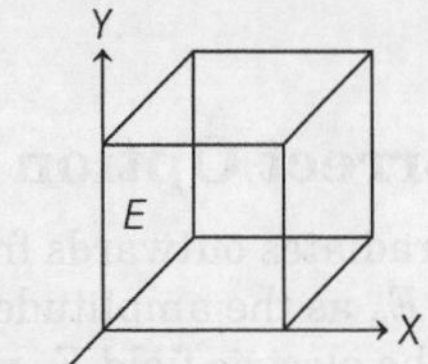

(a) $\frac{a^2EB}{2\mu_0}$ parallel to XY-plane and zero in others
(b) $\frac{a^2EB}{\mu_0}$ parallel to XY-plane and zero in others
(c) $\frac{a^2EB}{2\mu_0}$ from all faces
(d) $\frac{a^2EB}{2\mu_0}$ parallel to yz faces and zero in others

Electromagnetic Spectrum

43. Radio waves diffract around building although light waves do not. The reason is that radio waves
(a) travel with speed target than c
(b) have much larger wavelength than light
(c) carry news
(d) are not electromagnetic waves

44. An electromagnetic wave is generated due to oscillation of a charged particle with a frequency 10^9 Hz.
(a) The wavelength of electromagnetic wave in vacuum is 60 cm
(b) This wave is radio wave
(c) This wave is microwave
(d) This wave is light wave

45. This question has Statement I and Statement II. Of the four choices given after the Statement, choose the one that best describes the two Statements. [JEE Main 2015]

I. Out of radio waves and microwaves, the radio waves undergo more diffraction.

II. Radio waves have greater frequency compared to microwaves.

(a) I is true, II is true and II is the correct explanation of I.
(b) I is false, II is true.
(c) I is true, II is false.
(d) I is true, II is true but II is not the correct explanation of I.

46. Choose the correct option relating wavelengths of different parts of electromagnetic wave spectrum. [JEE Main 2020]
(a) $\lambda_{\text{X-rays}} < \lambda_{\text{microwaves}} < \lambda_{\text{radio waves}} < \lambda_{\text{visible}}$
(b) $\lambda_{\text{visible}} > \lambda_{\text{X-rays}} > \lambda_{\text{radio waves}} > \lambda_{\text{microwaves}}$
(c) $\lambda_{\text{radio waves}} > \lambda_{\text{microwaves}} > \lambda_{\text{visible}} > \lambda_{\text{X-rays}}$
(d) $\lambda_{\text{visible}} < \lambda_{\text{microwaves}} < \lambda_{\text{radio waves}} < \lambda_{\text{X-rays}}$

47. Match Table I (Electromagnetic wave type) with Table II (Its association/application) and select the correct option from the codes given below the tables. [JEE Main 2014]

	Table I		Table II
A.	Infrared waves	1.	To treat muscular strain
B.	Radio waves	2.	For broadcasting
C.	X-rays	3.	To detect fracture of bones
D.	Ultraviolet rays	4.	Absorbed by the ozone layer of the atmosphere

Codes

	A	B	C	D		A	B	C	D
(a)	4	3	2	1	(b)	1	2	4	3
(c)	3	2	1	4	(d)	1	2	3	4

48. The correct sequence of the increasing wavelength of the given radiation sources is
(a) radioactive sources, X-ray tube, crystal oscillator, sodium vapour lamp
(b) radioactive source, X-ray tube, sodium vapour lamp, crystal oscillator
(c) X-ray tube, radioactive source, crystal oscillator, sodium vapour lamp
(d) X-ray tube, crystal oscillator, radioactive source, sodium vapour lamp

49. The correct match between the entries in table are : [JEE Main 2020]

	Table I (Radiation)		Table II (Wavelength)
A.	Microwaves	1.	100 m
B.	γ-rays	2.	10^{-15} m
C.	AM radio waves	3.	10^{-10} m
D.	X-rays	4.	10^{-3} m

Codes

	A	B	C	D		A	B	C	D
(a)	2	1	4	3	(b)	1	3	4	2
(c)	3	2	1	4	(d)	4	2	1	3

ROUND II Mixed Bag

Only One Correct Option

1. An EM wave radiates outwards from a dipole antenna with E_0 as the amplitude of its electric field vector. The electric field E_0 which transports significant energy from the source falls off as
(a) $\frac{1}{r^3}$ (b) $\frac{1}{r^2}$
(c) $1/r$ (d) remains constant

2. A sinusoidal voltage is applied directly across 8 μF capacitor. The frequency of the source is 3.00 kHz and the voltage amplitude is 30.0 V. Find the maximum value of displacement current between the plates of the capacitor.
(a) 42.5 A (b) 4.25 μA (c) 4.52 A (d) 4.52 μA

3. An EM wave of intensity I falls on a surface kept in vacuum and exerts radiation pressure p on it . Which of the following is false?
(a) Radiation pressure is I/c, if the wave is totally absorbed
(b) Radiation pressure is I/c, if the wave is totally reflected
(c) Radiation pressure is $2I/c$, if the wave is totally reflected
(d) Radiation pressure is in the range $I/c < p < 2I/c$ for the real surfaces

4. An EM wave from air enters a medium. The electric fields are $\mathbf{E}_1 = E_{01}\,\hat{\mathbf{x}} \cos\left[2\pi\nu\left(\frac{z}{c} - t\right)\right]$ in air and $\mathbf{E}_2 = E_{02}\,\hat{\mathbf{x}} \cos[k(2z - ct)]$ in medium, where the wave number k and frequency ν refer to their values in air. The medium is non-magnetic.

If ε_{r_1} and ε_{r_2} refer to relative permittivities of air and medium respectively, which of the following options is correct? [JEE Main 2018]

(a) $\frac{\varepsilon_{r_1}}{\varepsilon_{r_2}} = 4$ (b) $\frac{\varepsilon_{r_1}}{\varepsilon_{r_2}} = 2$ (c) $\frac{\varepsilon_{r_1}}{\varepsilon_{r_2}} = \frac{1}{4}$ (d) $\frac{\varepsilon_{r_1}}{\varepsilon_{r_2}} = \frac{1}{2}$

5. An electromagnetic wave travels in vacuum along z-direction: $E = (E_1\hat{i} + E_2\hat{j})\cos(kz - \omega t)$. Choose the correct options from the following.

(a) The associated magnetic field is given as $\mathbf{B} = \frac{1}{c}(E_1\hat{\mathbf{i}} + E_2\hat{\mathbf{j}})\cos(kz - \omega t)$

(b) The associated magnetic field is given as $\mathbf{B} = \frac{1}{c}(E_1\hat{\mathbf{i}} - E_2\hat{\mathbf{j}})\cos(kz - \omega t)$

(c) The given electromagnetic field is circularly polarised

(d) None of the above

6. The electric field of a plane electromagnetic wave is given by $\mathbf{E} = E_0\hat{\mathbf{i}}\cos(kz)\cos(\omega t)$

The corresponding magnetic field $\mathbf{B}$ is then given by [JEE Main 2019]

(a) $\mathbf{B} = \frac{E_0}{c}\hat{\mathbf{j}}\sin(kz)\sin(\omega t)$

(b) $\mathbf{B} = \frac{E_0}{c}\hat{\mathbf{j}}\sin(kz)\cos(\omega t)$

(c) $\mathbf{B} = \frac{E_0}{c}\hat{\mathbf{k}}\sin(kz)\cos(\omega t)$

(d) $\mathbf{B} = \frac{E_0}{c}\hat{\mathbf{j}}\cos(kz)\sin(\omega t)$

7. The magnetic field of an electromagnetic wave is given by

$\mathbf{B} = 1.6 \times 10^{-6}\cos(2\times10^7 z + 6\times10^{15} t)(2\hat{\mathbf{i}} + \hat{\mathbf{j}})\ \text{Wbm}^{-2}$

The associated electric field will be [JEE Main 2019]

(a) $\mathbf{E} = 4.8\times10^2\cos(2\times10^7 z - 6\times10^{15} t)(-2\hat{\mathbf{j}} + \hat{\mathbf{i}})\ \text{Vm}^{-1}$

(b) $\mathbf{E} = 4.8\times10^2\cos(2\times10^7 z - 6\times10^{15} t)(2\hat{\mathbf{j}} + \hat{\mathbf{i}})\ \text{Vm}^{-1}$

(c) $\mathbf{E} = 4.8\times10^2\cos(2\times10^7 z + 6\times10^{15} t)(\hat{\mathbf{i}} - 2\hat{\mathbf{j}})\ \text{Vm}^{-1}$

(d) $\mathbf{E} = 4.8\times10^2\cos(2\times10^7 z + 6\times10^{15} t)(-\hat{\mathbf{i}} + 2\hat{\mathbf{j}})\text{Vm}^{-1}$

8. The electric field of a plane electromagnetic wave is given by $\mathbf{E} = E_0\frac{\hat{\mathbf{i}} + \hat{\mathbf{j}}}{\sqrt{2}}\cos(kz + \omega t)$

At $t = 0$, a positively charged particle is at the point $(x, y, z) = \left(0, 0, \frac{\pi}{k}\right)$.

If its instantaneous velocity at $(t = 0)$ is $v_0\hat{\mathbf{k}}$, the force acting on it due to the wave is [JEE Main 2020]

(a) zero (b) anti-parallel to $\frac{\hat{\mathbf{i}} + \hat{\mathbf{j}}}{\sqrt{2}}$

(c) parallel to $\frac{\hat{\mathbf{i}} + \hat{\mathbf{j}}}{\sqrt{2}}$ (d) parallel to $\hat{\mathbf{k}}$

9. The electric field of a plane electromagnetic wave is given by

$$\mathbf{E} = E_0(\hat{\mathbf{x}} + \hat{\mathbf{y}})\sin(kz - \omega t)$$

Its magnetic field will be given by [JEE Main 2020]

(a) $\frac{E_0}{c}(\hat{\mathbf{x}} + \hat{\mathbf{y}})\sin(kz - \omega t)$ (b) $\frac{E_0}{c}(-\hat{\mathbf{x}} + \hat{\mathbf{y}})\sin(kz - \omega t)$

(c) $\frac{E_0}{c}(\hat{\mathbf{x}} - \hat{\mathbf{y}})\sin(kz - \omega t)$ (d) $\frac{E_0}{c}(\hat{\mathbf{x}} - \hat{\mathbf{y}})\cos(kz - \omega t)$

10. An electromagnetic wave of intensity 50 Wm^{-2} enters in a medium of refractive index n without any loss. The ratio of the magnitudes of electric fields and the ratio of the magnitudes of magnetic fields of the wave before and after entering into the medium are respectively, given by [JEE Main 2019]

(a) $\left(\frac{1}{\sqrt{n}}, \sqrt{n}\right)$ (b) $(\sqrt{n}, \sqrt{n})$

(c) $\left(\frac{1}{\sqrt{n}}, \frac{1}{\sqrt{n}}\right)$ (d) $\left(\sqrt{n}, \frac{1}{\sqrt{n}}\right)$

11. An electron is constrained to move along the Y-axis with a speed of $0.1\,c$ (c is the speed of light) in the presence of electromagnetic wave, whose electric field is

$$\mathbf{E} = 30\hat{\mathbf{j}}\sin(1.5\times10^7 t - 5\times10^{-2}x)\ \text{V/m}.$$

The maximum magnetic force experienced by the electron will be (Given, $c = 3\times10^8\ \text{ms}^{-1}$ and electron charge $= 1.6\times10^{-19}$ C) [JEE Main 2020]

(a) 2.4×10^{-18} N (b) 4.8×10^{-19} N

(c) 1.6×10^{-19} N (d) 3.2×10^{-18} N

Numerical Value Questions

12. In a plane electromagnetic wave, the electric field oscillates with a frequency $2\times10^{10}\ \text{s}^{-1}$ and amplitude 40 Vm^{-1}, then the energy density due to electric field is $p\times10^{-9}\ \text{J/m}^3$. The value of p is .... ($\varepsilon_0 = 8.85\times10^{-12}\ \text{Fm}^{-1}$)

13. An electromagnetic wave of frequency 2 MHz propagates from vacuum to a non-magnetic medium of relative permittivity 9. Then, its wavelength decrease by .... m.

14. If 2.5×10^{-6} N average force is exerted by a light wave on a non-reflecting surface of 30 cm^2 area during to minutes of time span, the energy flux of light just before it falls on the surface is ...... W/cm^2. (Round off to the nearest integer)

(Assume complete absorption and normal incidence conditions are there) [JEE Main 2021]

15. Suppose that, intensity of a laser is $\left(\frac{315}{\pi}\right)\text{W/m}^2$.

The rms electric field (in V/m) associated with this source is close to the nearest integer is .... (Take, $\varepsilon_0 = 8.86\times10^{-12}\ \text{C}^2\text{N}^{-1}\text{m}^{-2}$ and $c = 3\times10^8\ \text{ms}^{-1}$) [JEE Main 2020]

Answers

Round I

1. (c)	2. (c)	3. (a)	4. (c)	5. (c)	6. (c)	7. (b)	8. (d)	9. (b)	10. (a)
11. (b)	12. (c)	13. (d)	14. (b)	15. (a)	16. (b)	17. (a)	18. (c)	19. (d)	20. (b)
21. (a)	22. (d)	23. (d)	24. (d)	25. (d)	26. (b)	27. (a)	28. (c)	29. (c)	30. (b)
31. (c)	32. (a)	33. (c)	34. (d)	35. (c)	36. (a)	37. (d)	38. (a)	39. (a)	40. (d)
41. (a)	42. (b)	43. (b)	44. (b)	45. (c)	46. (c)	47. (d)	48. (b)	49. (d)	

Round II

1. (c)	2. (c)	3. (b)	4. (c)	5. (a)	6. (a)	7. (c)	8. (b)	9. (b)	10. (d)
11. (b)	12. 3.54	13. 100	14. 25	15. 194					

Solutions

1. Due to conductivity, conduction current is present and due to variation of electric field, displacement current is also present.

2. As, $i = \frac{dq}{dt} = \frac{d}{dt}(q_0 \sin 2\pi ft) = q_0 2\pi f \cos 2\pi ft$

3. As, $i = \frac{dQ}{dt} = \frac{d}{dt}(CV) = C\frac{dV}{dt} = 2 \times 10^{-12} \times 10^{12} = 2$ A

4. As, $C = \frac{\varepsilon_0 KA}{d} = \frac{(8.85 \times 10^{-12}) \times 10 \times 1}{10^{-3}}$

$= 8.85 \times 10^{-8}$ F

Now, $i = \frac{dQ}{dt} = \frac{d}{dt}(CV) = C\frac{dV}{dt} = 8.85 \times 10^{-8} \times 25$

$= 2.2 \times 10^{-6}\text{A} = 2.2\,\mu\text{A}$

5. Given, radius of plate, $R = 10$ cm $= 0.1$ m and cross-section area of a capacitor,

$$A = \pi R^2 = \pi(0.1)^2 = 3.14 \times 10^{-2}\ \text{m}^2$$

Rate of change of electric field,

$$\frac{dE}{dt} = 5 \times 10^{13}\ \text{Vms}^{-1}$$

Thus, displacement current,

$$i_d = \varepsilon_0 \frac{d\phi_E}{dt} = \varepsilon_0 A \frac{dE}{dt} \qquad [\because \phi_E = EA]$$

$$= (8.85 \times 10^{-12})(3.14 \times 10^{-2})(5 \times 10^{13})$$

$\therefore$ $i_d = 13.9$ A

6. According to the question, a parallel plate capacitor is shown in the figure below.

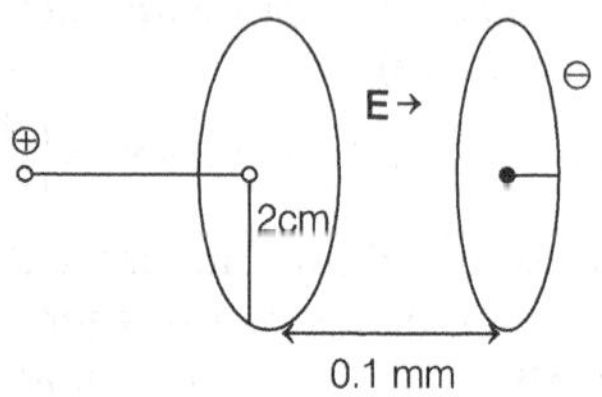

Given, radius = 2 cm, distance between plates, $d = 0.1$ mm and $\frac{dV}{dt} = 5 \times 10^6\ \text{Vs}^{-1}$

Displacement current in the capacitor,

$$I_d = \varepsilon_0 \frac{d\phi}{dt} = \varepsilon_0 \frac{dE}{dt} A$$

$$I_d = \varepsilon_0 A \frac{dE}{dt}$$

$$\Rightarrow \quad I_d = \varepsilon_0 \frac{A}{d}\frac{dV}{dt} \qquad \left(\because E = \frac{V}{d}\right)$$

Putting the given values, we get

$$\Rightarrow \quad I_d = \frac{8.85 \times 10^{-12} \times \pi \times (4 \times 10^{-4}) \times 5 \times 10^6}{1 \times 10^{-4}}\ \text{A}$$

$$\Rightarrow \quad I_d = 556.28 \times 10^{-6}\ \text{A} = 0.556\ \text{mA}$$

7. According to Gauss's law.

$$\oint \mathbf{E} \cdot \mathbf{dS} = \oint \frac{\rho}{\varepsilon_0} \cdot dS$$

$$\left[\because E = \frac{\rho}{\varepsilon_0}, \rho = \text{surface charge density}\right]$$

According to Faraday's law of electromagnetic induction,

$$\oint \mathbf{E} \cdot \mathbf{dl} = \frac{-d\phi}{dt} = \frac{-d}{dt}\oint \mathbf{B} \cdot \mathbf{dS}$$

$$\therefore \quad \phi = \oint \mathbf{B} \cdot \mathbf{dS}$$

According to Ampere-Maxwell's law,

$$\oint \mathbf{B} \cdot \mathbf{d}l = \mu_0\left(I + \varepsilon_0 \frac{d\phi_E}{dt}\right)$$

According to Ampere's circuital law,

$$\oint \mathbf{B} \cdot \mathbf{dl} = \mu_0 I$$

8. As electromagnetic wave contains both electric field and magnetic field. It carry both energy and momentum. Also, it can travel in vacuum and it is produced by oscillating charges.

9. The energy of electromagnetic wave is equally distributed between electric and magnetic fields. Some more properties of electromagnetic waves are as given below

(i) Electric and magnetic fields will have zero phase difference.

(ii) The pressure exerted by electromagnetic wave is the ratio of energy density to the speed of light.

(iii) The speed of wave is equal to the ratio of electric field to the magnetic field.

Hence, the correct option is (b).

10. As, $\sqrt{\frac{\mu}{\varepsilon}}$ = has the dimensions of resistance, hence it is called the intrinsic impedance of the medium.

11. As, $v_{\text{ferrite}} = \frac{c}{\sqrt{\mu_r \varepsilon_r}} = \frac{3\times10^8}{\sqrt{10\times10^3}} = 3\times10^6 \text{ ms}^{-1}$

and $\lambda_{\text{ferrite}} = \frac{v_{\text{ferrite}}}{\nu} = \frac{3\times10^6}{90\times10^6} = 3.33\times10^{-2}$ m

12. In vacuum, $\varepsilon_0 = 1$

In medium, $\varepsilon = 4$

$\therefore$ Refractive index, $\mu = \sqrt{\frac{\varepsilon}{\varepsilon_0}} = \sqrt{\frac{4}{1}} = 2$

Wavelength, $\lambda' = \frac{\lambda}{\mu} = \frac{\lambda}{2}$

and wave velocity, $v = \frac{c}{\mu} = \frac{c}{2}$ $\left(\because \mu = \frac{c}{v}\right)$

Hence, it is clear that wavelength and velocity will become half but frequency remains unchanged, when the wave is passing through any medium.

13. Velocity of an electromagnetic wave,

$c = \frac{1}{\sqrt{\mu_0\varepsilon_0}} = 3\times10^8 \text{ ms}^{-1}$, which is independent of amplitude of electromagnetic wave, its frequency and wavelength.

14. Electromagnetic wave carry momentum and hence, it can exert pressure on surfaces. Thus, they also transfer energy to the surface. To $p \neq 0$ and $E \neq 0$.

15. As, $\frac{E_0}{B_0} = c$

Also, $k = \frac{2\pi}{\lambda}$ and $\omega = 2\pi\nu$

Thus, we get $E_0 k = B_0 \omega$

16. In electromagnetic wave, **E** and **B** are perpendicular to each other and to the direction of propagation, so the magnetic field vector points towards East.

17. As, $B_0 = \frac{E_0}{c} = \frac{54}{3\times10^8} = 18\times10^{-8}$ T or Wb m^{-2}

$\therefore \quad \lambda = \frac{c}{\nu} = \frac{3\times10^8}{2\times10^{10}} = 1.5\times10^{-2}$ m = 1.5 cm

18. The electromagnetic waves are produced due to sinusoidal variation of electric field vector and magnetic field vector perpendicular to each other as well as perpendicular to the direction of propagation of waves. Since, electromagnetic wave is propagating along y-direction, therefore **E** and **B** should be either along X-axis and Z-axis or along Z and X-axis, respectively.

19. Given, $\mathbf{B} = 5\times10^{-8}\hat{\mathbf{j}}$ and $\mathbf{v} = 3\times10^8\hat{\mathbf{k}}$

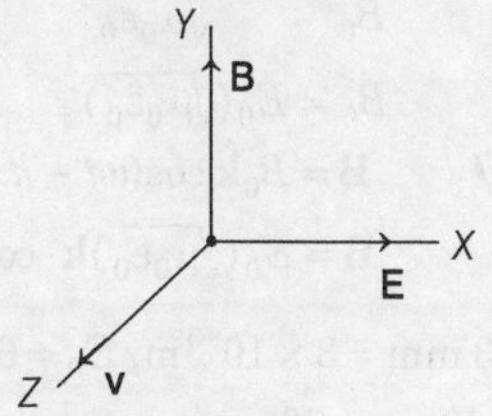

Using $\mathbf{E} = \mathbf{B}\times\mathbf{v}$, we have

$\mathbf{E} = (5\times10^{-8}\hat{\mathbf{j}})\times(3\times10^8\hat{\mathbf{k}}) = 15\hat{\mathbf{i}}$ V/m

20. In an electromagnetic wave, the direction of propagation of wave, electric field and magnetic field are mutually perpendicular to each other, *i.e.* the wave propagates perpendicular to E and B or along $\mathbf{E}\times\mathbf{B}$ i.e. $\hat{\mathbf{k}} \parallel \mathbf{E}\times\mathbf{B}$. While polarisation of wave takes place parallel to electric field vector i.e. $\mathbf{X} \parallel \mathbf{E}$.

21. Direction of propagation of an electromagnetic wave is given by $\mathbf{E}\times\mathbf{B}$.

$\therefore$ A unit vector in the direction of propagation

$$= \frac{\mathbf{E}\times\mathbf{B}}{|\mathbf{E}\times\mathbf{B}|} = \frac{\hat{\mathbf{k}}\times(2\hat{\mathbf{i}}-2\hat{\mathbf{j}})}{|\mathbf{E}\times\mathbf{B}|}$$

$$\left[\begin{aligned} \because \mathbf{E}\times\mathbf{B} &= \hat{\mathbf{k}}\times(2\hat{\mathbf{i}}-2\hat{\mathbf{j}}) \\ &= 2(\hat{\mathbf{k}}\times\hat{\mathbf{i}}) - 2(\hat{\mathbf{k}}\times\hat{\mathbf{j}}) \\ &= 2(\hat{\mathbf{j}}) - 2(-\hat{\mathbf{i}}) \\ &= 2\hat{\mathbf{j}} + 2\hat{\mathbf{i}} \\ \therefore |\mathbf{E}\times\mathbf{B}| &= \sqrt{2^2+2^2} = 2\sqrt{2} \end{aligned}\right]$$

$$= \frac{2\hat{\mathbf{j}}+2\hat{\mathbf{i}}}{2\sqrt{2}} = \frac{\hat{\mathbf{i}}+\hat{\mathbf{j}}}{\sqrt{2}}$$

22. Suppose an electromagnetic wave is travelling along negative z-direction. Its electric field is given by $\mathbf{E} = E_0 \cos(kz - \omega t)$

which is perpendicular to Z-axis. It acts along negative y-direction.

The associated magnetic field **B** in electromagnetic wave is along X-axis, *i.e.* along $\hat{\mathbf{k}}\times\mathbf{E}$.

As, $B_0 = \frac{E_0}{c}$

$\therefore \quad \mathbf{B} = \frac{1}{c}(\hat{\mathbf{k}}\times\mathbf{E})$

The associated electric field can be written in terms of magnetic field as $\mathbf{E} = c(\mathbf{B}\times\hat{\mathbf{k}})$.

23. Given, $\mathbf{E} = E_0\hat{\mathbf{j}}\cos(\omega t - kx)$

This equation represents electric field vector of a wave which travels in positive x-direction, *i.e.* along $\hat{\mathbf{i}}$.

As, $\mathbf{E}\times\mathbf{B}$ = direction of wave propagation, we have

$$\hat{\mathbf{j}}\times\mathbf{B} = \hat{\mathbf{i}}$$

$\therefore$ $\mathbf{B}$ is along $\hat{\mathbf{k}}$.

Also, $$\frac{E_0}{B_0} = c = \frac{1}{\sqrt{\mu_0\varepsilon_0}}$$

so $$B_0 = E_0(\sqrt{\mu_0\varepsilon_0})$$

So, $$\mathbf{B} = B_0\hat{\mathbf{k}}\cos(\omega t - kx)$$

or $$\mathbf{B} = E_0(\sqrt{\mu_0\varepsilon_0})\hat{\mathbf{k}}\cdot\cos kx$$

24. Here, $\lambda = 3\text{ mm} = 3\times10^{-3}\text{m}$, $E_0 = 66\text{ Vm}^{-1}$

$$\therefore \quad B_0 = \frac{E_0}{c} = \frac{66}{3\times10^8} = 2.2\times10^{-7}\text{ T}$$

As, electromagnetic wave is propagating along X-axis and electric field oscillation is along y-direction, the magnetic field oscillation is along z-direction using the relation for harmonic wave,

$$E_y = E_0\cos\frac{2\pi}{\lambda}(ct - x)$$

$$E_y = E_0\cos\frac{2\pi c}{\lambda}(t - x/c)$$

$$\therefore \quad E_y = 66\cos\frac{2\pi\times3\times10^8}{3\times10^{-3}}(t - x/c)$$

$$= 66\cos 2\pi\times10^{11}(t - x/c)$$

and $$B_z = B_0\cos\frac{2\pi c}{\lambda}(t - x/c)$$

$$= 2.2\times10^{-7}\cos 2\pi\times10^{11}(t - x/c)$$

25. We have the following facts

(i) An electromagnetic wave propagates in direction of vector $\mathbf{E}\times\mathbf{B}$.

(ii) Phase gives velocity of wave.

(iii) Electric field vector of electromagnetic wave is given by

$$\mathbf{E} = c\mathbf{B}$$

Here, $$\mathbf{B} = B_0\sin[200\pi(y + ct)]\hat{\mathbf{i}}$$

$$= 3\times10^{-8}\sin[200\pi(y + ct)]\hat{\mathbf{i}}$$

$$\Rightarrow \quad E_0 = cB_0 = 3\times10^8\times3\times10^{-8} = 9\text{ V/m}$$

Phase = $y + ct$ = constant

$$\Rightarrow \quad \frac{dy}{dt} = -c$$

i.e. Wave propagates in negative y-direction.

$\Rightarrow$ $\mathbf{E}\times\mathbf{B}$ is parallel to $-\hat{\mathbf{j}}$.

$$\Rightarrow \quad \mathbf{E}\times\hat{\mathbf{i}} = \lambda(-\hat{\mathbf{j}})$$

So, $$\hat{\mathbf{E}} = -\hat{\mathbf{k}}$$

Hence, electric field vector is given by

$$\mathbf{E} = E_0\sin(ky + \omega t)(-\hat{\mathbf{k}})$$

$$= -9\sin[200\pi(y + ct)]\hat{\mathbf{k}}\text{ V/m}$$

26. When a wave is reflected from denser medium, the reflected wave is without change in type of wave but with a change in phase by 180° or π radian. Therefore, for the reflected wave, we use $z = -z$, $\hat{\mathbf{i}} = -\hat{\mathbf{i}}$ and additional phase of π in the incident wave. The incident electromagnetic wave is $E_r = E_0\hat{\mathbf{i}}\cos(kz - \omega t)$.

The reflected electromagnetic wave is

$$\mathbf{E}_r = E_0(-\hat{\mathbf{i}})\cos[k(-z) - \omega t + \pi]$$

$$= -E_0\hat{\mathbf{i}}\cos[-(kz + \omega t) + \pi]$$

$$= E_0\hat{\mathbf{i}}\cos[-(kz + \omega t)] \quad [\because \cos(\theta + \pi) = -\cos\theta]$$

$$= E_0\hat{\mathbf{i}}\cos(kz + \omega t) \quad [\because \cos(-\theta) = \cos\theta]$$

27. Given,

$\mathbf{B}(x, t) = [1.2\times10^{-7}\sin(0.5\times10^3 x + 1.5\times10^{11}t)\hat{\mathbf{k}}]\text{T}$

Speed of light, $c = 3\times10^8$ m/s

Maximum value of electric field,

E_0 = (Maximum value of B_0) $\times$ (Speed of light)

$= B_0\times c = 1.2\times10^{-7}\times3\times10^8$

$= 36$ V/m

Since, wave is propagating in negative x-direction and magnetic field is in z-direction. So, electric field vector must be in negative y-direction as direction of wave propagation is in the direction of $\mathbf{E}\times\mathbf{B}$.

So, electric field at a point x and time t is given by

$\mathbf{E}(x, t) = -36\sin(0.5\times10^3 x + 1.5\times10^{11}t)\hat{\mathbf{j}}$ V/m

Here, phase of $\mathbf{E}$ will be same that of $\mathbf{B}$.

28. Given, direction of propagation of electromagnetic wave is along $\hat{\mathbf{i}} = \frac{\hat{\mathbf{i}} + \hat{\mathbf{j}}}{\sqrt{2}}$, also direction of its electric field vector $\mathbf{E}$ (or direction of its polarisation) is $\hat{\mathbf{k}}$.

As, direction of propagation of wave is along $\mathbf{E}\times\mathbf{B}$, so direction of $\mathbf{B}$ is such that

$$\hat{\mathbf{k}}\times\hat{\mathbf{B}} = \frac{\hat{\mathbf{i}} + \hat{\mathbf{j}}}{\sqrt{2}}$$

$$\Rightarrow \quad \hat{\mathbf{B}} = \left(\frac{\hat{\mathbf{i}} + \hat{\mathbf{j}}}{\sqrt{2}}\right)\hat{\mathbf{k}}$$

$$= \frac{(\hat{\mathbf{i}}\times\hat{\mathbf{k}}) + (\hat{\mathbf{j}}\times\hat{\mathbf{k}})}{\sqrt{2}}$$

$$= \frac{(\hat{\mathbf{i}} - \hat{\mathbf{j}})}{\sqrt{2}}$$

So, correct expression of $\mathbf{B}$ is

$$\mathbf{B} = B_0\hat{\mathbf{B}}\cos(\omega t - k\hat{\mathbf{i}})$$

$$= B_0\left(\frac{\hat{\mathbf{i}} - \hat{\mathbf{j}}}{\sqrt{2}}\right)\cdot\cos\left\{\omega t - k\left(\frac{\hat{\mathbf{i}} + \hat{\mathbf{j}}}{\sqrt{2}}\right)\right\}$$

29. Standard expression of electromagnetic wave is given by

$$\mathbf{E} = E_0\,\hat{\mathbf{n}}[\sin(\omega t - \mathbf{k}\cdot\hat{\mathbf{r}})] \quad \ldots\text{(i)}$$

Here, $\mathbf{k}$ is the propagation vector.

Direction of propagation in this case is $\hat{\mathbf{k}}$.

Given expression of electromagnetic wave,

$$\mathbf{E} = E_0 \hat{\mathbf{n}} \sin [\omega t + (6y - 8z)]$$

$$\Rightarrow \quad \mathbf{E} = E_0 \hat{\mathbf{n}} \sin [\omega t - (8z - 6y)] \quad \ldots\text{(ii)}$$

Comparing Eq. (ii) with Eq. (i), we get

$$\mathbf{k} \cdot \hat{\mathbf{r}} = 8z - 6y \quad \ldots\text{(iii)}$$

Here, $\hat{\mathbf{r}} = x\hat{\mathbf{i}} + y\hat{\mathbf{j}} + z\hat{\mathbf{k}}$

and $\mathbf{k} = k_x \hat{\mathbf{i}} + k_y \hat{\mathbf{j}} + k_z \hat{\mathbf{k}}$

$$\therefore \quad \mathbf{k} \cdot \hat{\mathbf{r}} = xk_x + yk_y + zk_z \quad \ldots\text{(iv)}$$

From Eqs. (iii) and (iv), we get

$$xk_x = \text{zero} \Rightarrow k_x = 0$$

$$yk_y = -6y \Rightarrow k_y = -6$$

$$zk_z = 8z \Rightarrow k_z = 8$$

Hence, $\mathbf{k} = -6\hat{\mathbf{j}} + 8\hat{\mathbf{k}}$

So, direction of propagation,

$$\hat{\mathbf{s}} = \hat{\mathbf{k}} = \frac{\mathbf{k}}{|\mathbf{k}|} = \frac{-6\hat{\mathbf{j}} + 8\hat{\mathbf{k}}}{\sqrt{6^2 + 8^2}} = \frac{-6\hat{\mathbf{j}} + 8\hat{\mathbf{k}}}{10} = \frac{-3\hat{\mathbf{j}} + 4\hat{\mathbf{k}}}{5}$$

30. In an electromagnetic wave, magnetic field and electric field are perpendicular to each other and both are also perpendicular to the direction of propagation of wave.

Now, given direction of propagation is along z-direction. So, magnetic field is in either x or y-direction. Also, angular wave number for wave is

$$k = \frac{2\pi}{\lambda} = \frac{2\pi\nu}{c} = \frac{2\pi \times 23.9 \times 10^9}{3 \times 10^8} \approx 0.5 \times 10^3 \text{ m}^{-1}$$

and angular frequency ω for wave is

$$\omega = 2\pi\nu = 2\pi \times 23.9 \times 10^9 \text{ Hz} = 1.5 \times 10^{11} \text{ Hz}$$

Magnitude of magnetic field is

$$B_0 = \frac{E_0}{c} = \frac{60}{3 \times 10^8} = 2 \times 10^{-7} \text{T}$$

As the general equation of magnetic field of an electromagnetic wave propagating in + z- direction is given as

$$\mathbf{B} = B_0 \sin (kz - \omega t)\, \hat{\mathbf{i}} \text{ or } \hat{\mathbf{j}}$$

Thus, substituting the values of B_0, k and ω, we get

$$\Rightarrow \mathbf{B} = 2 \times 10^{-7} \sin(0.5 \times 10^3 z - 1.5 \times 10^{11} t)\hat{\mathbf{i}} \text{ or } \hat{\mathbf{j}}$$

31. Given, magnetic field of an electromagnetic wave is

$$\mathbf{B} = B_0 [\cos(kz - \omega t]\, \hat{\mathbf{i}} + B_1 [\cos(kz + \omega t]\, \hat{\mathbf{j}}$$

Here, $B_0 = 3 \times 10^{-5}$ T and $B_1 = 2 \times 10^{-6}$ T

Also, stationary charge, $Q = 10^{-4}$C at $z = 0$.

As charge is released from the rest at $z = 0$, in this condition.

Maximum electric field, $E_0 = cB_0$ and $E_1 = cB_1$

So, $E_0 = c \times 3 \times 10^{-5}$

and $E_1 = c \times 2 \times 10^{-6}$

Now,the direction of electric field of an electromagnetic wave is perpendicular to **B** and to the direction of propagation of wave (**E** × **B**) which is $\hat{\mathbf{k}}$.

So, for E_0,

$$\mathbf{E}_0 \times \mathbf{B}_0 = \hat{\mathbf{k}}$$

$$\Rightarrow \mathbf{E}_0 \times \hat{\mathbf{i}} = \hat{\mathbf{k}}$$

$$\Rightarrow \mathbf{E}_0 = -\hat{\mathbf{j}}$$

Similarly,for E_1, $\mathbf{E}_1 \times \mathbf{B}_1 = \mathbf{k}$

$$\Rightarrow \mathbf{E}_1 \times \hat{\mathbf{j}} = \hat{\mathbf{k}}$$

$$\Rightarrow \mathbf{E}_1 = \hat{\mathbf{i}}$$

$$\therefore \mathbf{E}_0 = c \times 3 \times 10^{-5} (-\hat{\mathbf{j}}) \text{ NC}^{-1}$$

$$\mathbf{E}_1 = c \times 2 \times 10^{-6} (+\hat{\mathbf{i}}) \text{ NC}^{-1}$$

$\therefore$ Maximum force experienced by stationary charge is

$$\mathbf{F}_{\text{max}} = Q\,\mathbf{E} = Q(\mathbf{E}_0 + \mathbf{E}_1) = Q \times c\, [-3 \times 10^{-5} \hat{\mathbf{j}} + 2 \times 10^{-6} \hat{\mathbf{i}}]$$

$$\Rightarrow |\mathbf{F}_{\text{max}}| = 10^{-4} \times 3 \times 10^8 \times \sqrt{(3 \times 10^{-5})^2 + (2 \times 10^{-6})^2} = 3 \times 10^4 \times 10^{-6} \sqrt{900 + 4} = 3 \times 10^{-2} \times \sqrt{904} \approx 0.9 \text{ N}$$

$\therefore$ rms value of experienced force is

$$F_{\text{rms}} = \frac{F_{\text{max}}}{\sqrt{2}} = \frac{0.9}{\sqrt{2}} = 0.707 \times 0.9 = 0.6363 \text{ N} \approx 0.6 \text{ N}$$

32. Energy density of an electomagnetic wave in electric field,

$$U_E = \frac{1}{2} \varepsilon_0 \cdot E^2 \quad \ldots\text{(i)}$$

Energy density of an electromagnetic wave in magnetic field,

$$U_B = \frac{B^2}{2\mu_0} \quad \ldots\text{(ii)}$$

where, E = electric field,

B = magnetic field,

ε_0 = permittivity of free space

and μ_0 = magnetic permeability of free space.

From the theory of electromagnetic waves, the relation between μ_0 and ε_0 is

$$c = \frac{1}{\sqrt{\mu_0 \varepsilon_0}} \quad \ldots\text{(iii)}$$

where, c = velocity of light

$$= 3 \times 10^8 \text{ m/s}$$

and $$\frac{E}{B} = c \quad \text{...(iv)}$$

On dividing Eq. (i) by Eq. (ii), we get

$$\frac{U_E}{U_B} = \frac{\frac{1}{2}\varepsilon_0 E^2}{\frac{1}{2}B^2 \times \frac{1}{\mu_0}} = \frac{\mu_0 \varepsilon_0 E^2}{B^2} \quad \text{...(v)}$$

Using Eqs. (iii), (iv) and (v), we get

$$\frac{U_E}{U_B} = \frac{c^2}{c^2} = 1$$

Therefore, $U_E = U_B$

33. As, intensity, $I = \frac{P}{4\pi r^2}$

$$= \text{Average energy density} \times \text{velocity}$$

$$= \frac{1}{2}\varepsilon_0 E_0^2 c$$

$$\therefore \quad E_0 = \sqrt{\frac{2P}{4\pi\varepsilon_0 r^2 c}} = \sqrt{\frac{P}{2\pi\varepsilon_0 r^2 c}}$$

34. Given, intensity of electromagnetic wave ,

$$I = 53.1 \text{ Wm}^{-2}$$

Permittivity of free space,

$$\varepsilon_0 = 8.85 \times 10^{-12} \text{ C}^2\text{N}^{-1}\text{m}^{-2}$$

Intensity of electromagnetic wave, $I = \frac{1}{2}\varepsilon_0 E_0^2 c$

where, E_0 = amplitude of the electric field.

$$E_0 = \sqrt{\frac{2I}{\varepsilon_0 c}} = \sqrt{\frac{2 \times 53.1}{8.85 \times 10^{-12} \times 3 \times 10^8}}$$

$$E_0 = 200 \text{ NC}^{-1}$$

35. As, power = $I \times$ area = $(1.4 \times 10^3) \times 5$

Force, $$F = \frac{\text{Power}}{c} = \frac{\text{Solar constant} \times \text{Area}}{\text{Speed of light}}$$

$$= \frac{1.4 \times 10^3 \times 5}{3 \times 10^8}$$

$$= 2.33 \times 10^{-5} \text{ N}$$

36. As, $F = PA$, but $P = \frac{1}{c}$

So, $$F = \frac{IA}{c} = \frac{6 \times 12}{3 \times 10^8} = 24 \times 10^{-8} \text{ N}$$

37. Given, power of laser beam,

$$P = 27\text{mW} = 27 \times 10^{-3} \text{ W}$$

Area of cross-section, A

$$= 10 \text{ mm}^2 = 10 \times 10^{-6} \text{m}^2$$

Permittivity of free space, $\varepsilon_0 = 9 \times 10^{-12}$ SI unit

Speed of light, $c = 3 \times 10^8$ m/s

Intensity of electromagnetic wave is given by the relation,

$$I = \frac{1}{2} n c \varepsilon_0 E^2$$

where, n is refractive index, for air $n = 1$.

$$\therefore \quad I = \frac{1}{2} c \cdot \varepsilon_0 E^2 \quad \text{...(i)}$$

Also, $$I = \frac{P}{A} \quad \text{...(ii)}$$

From Eqs. (i) and (ii), we get

$$\frac{1}{2} c \varepsilon_0 E^2 = \frac{P}{A}$$

or $$E^2 = \frac{2P}{Ac\varepsilon_0}$$

or $$E = \sqrt{\frac{2 \times 27 \times 10^{-3}}{10 \times 10^{-6} \times 3 \times 10^8 \times 9 \times 10^{-12}}}$$

$$\simeq 1.4 \times 10^3 \text{ V/m}$$

$$= 1.4 \text{ kV/m}$$

38. Electric field intensity on a surface due to incident radiation is $E = \frac{U}{At} = \frac{P}{A}$, where $\frac{U}{t} = P$ = power

$\therefore E \propto P$ (for the given area of the surface)

Hence, $$\frac{E'}{E} = \frac{P'}{P} = \frac{50}{100} = \frac{1}{2} \text{ or } E' = \frac{E}{2}$$

39. Momentum transferred in one second by electromagnetic wave to the mirror is

$$p = \frac{2S_{av}A}{c} = \frac{2 \times 6 \times (30 \times 10^{-4})}{3 \times 10^8}$$

$$= 1.2 \times 10^{-10} \text{ kg-ms}^{-1}$$

40. The poynting vector, $\mathbf{S} = \frac{1}{\mu_0}(\mathbf{E} \times \mathbf{B})$

Substituting the given values, we get

$$\mathbf{S} = \frac{10^{-6}}{4\pi \times 10^{-7}} \begin{vmatrix} \hat{\mathbf{i}} & \hat{\mathbf{j}} & \hat{\mathbf{k}} \\ 80 & 32 & -64 \\ 0.2 & 0.08 & 0.29 \end{vmatrix} = 11.45\,\hat{\mathbf{i}} - 28.6\,\hat{\mathbf{j}}$$

41. Average intensity $= \frac{P}{4\pi R^2} = \frac{1}{2}\varepsilon_0 E_0^2 c$

or $$E_0 = \sqrt{\frac{P}{2\pi R^2 \varepsilon_0 c}}$$

$$= \sqrt{\frac{3}{2 \times 3.14 \times 25 \times 8.85 \times 10^{-12} \times 3 \times 10^8}}$$

$$= 2.68 \text{ Vm}^{-1}$$

42. Energy flowing per second per unit area from a face is $= \frac{1}{\mu_0}[\mathbf{E} \times \mathbf{B}]$. It will be in the negative z-direction. It shows that the energy will be flowing in faces parallel to XY-plane and is zero in all other faces. Total energy flowing per second from a face in XY-plane

$$= \frac{1}{\mu_0}(EB\sin 90^\circ)a^2 = \frac{EBa^2}{\mu_0}$$

43. Diffraction takes places when the wavelength of wave is comparable with the size of the obstacle in path. The wavelength of radio waves is greater than the wavelength of light waves. Therefore, radio waves are diffracted around building.

44. $\lambda = \frac{c}{f} = \frac{3\times 10^8}{10^9} = 3\times 10^{-1} = 0.3 \text{ m} = 30 \text{ cm}$

This wave is lying in radio wave region.

45. As, frequency of radio waves $= 3\times 10^8$ Hz

and frequency of microwaves $= 10^{10}$ Hz

$\therefore$ The frequency of radio waves is less than the frequency of microwaves.

Also, radio wave undergo more diffraction than microwaves.

46.

$$\lambda_{\text{X-rays}} = 0.01 \text{ Å to } 100 \text{ Å}$$

$$\lambda_{\text{visible}} = 3800 \text{ Å to } 7000 \text{ Å}$$

$$\lambda_{\text{microwaves}} = 10^7 \text{ Å to } 3\times 10^9 \text{ Å}$$

$$\lambda_{\text{radio waves}} = 10^7 \text{ Å to } 10^{15} \text{ Å}$$

So, it is clear from the values that

$$\lambda_{\text{radio waves}} > \lambda_{\text{microwaves}} > \lambda_{\text{visible}} > \lambda_{\text{X-rays}}$$

47. (i) Infrared waves are used to treat muscular strain.

(ii) Radio waves are used for broadcasting purposes.

(iii) X-rays are used to detect fracture of bones.

(iv) Ultraviolet rays are absorbed by the ozone layer of the atmosphere.

48. Radioactive source $\Rightarrow$ γ-rays

X-ray tube $\Rightarrow$ X-rays

Sodium vapour lamp $\Rightarrow$ Visible light

Crystal oscillator $\Rightarrow$ Microwaves

$\therefore$ Radioactive source < X-ray tube < Sodium vapour lamp < Crystal oscillator

49. Energy is inversely proportional to wavelength, *i.e.*

$$E = \frac{hc}{\lambda}$$

$$\Rightarrow \quad E \propto \frac{1}{\lambda}$$

Energies of given radiations have the following order

$$E_{\gamma\text{-rays}} > E_{\text{X-rays}} > E_{\text{microwaves}} > E_{\text{AM radio waves}}$$

$$\therefore \quad \lambda_{\gamma\text{-rays}} < \lambda_{\text{X-rays}} < \lambda_{\text{microwaves}} < \lambda_{\text{AM radio waves}}$$

Thus, microwaves $\rightarrow 10^{-3}$ m

γ-rays $\rightarrow 10^{-15}$ m

AM radio waves $\rightarrow 100$ m

X-rays $\rightarrow 10^{-10}$ m

Hence, (A) $\rightarrow$ (4), (B) $\rightarrow$ (2), (C) $\rightarrow$ (1), (D) $\rightarrow$ (3).

Round II

1. From a diode antenna, the electromagnetic waves are radiated outwards. The amplitude of electric field vector (E_0) which transports significant energy from the source falls off intensity inversely as the distance (r) from the antenna, *i.e.* $E_0 \propto 1/r$.

2. Given, $C = 8\,\mu\text{F} = 8\times 10^{-6}$ F, $\nu = 3.00$ kHz

and $V_0 = 30.0$ V

Clearly, $\omega = 2\pi\nu = 2\pi \times (3.00\times 10^3 \text{ s}^{-1}) = 6\pi \times 10^3 \text{ s}^{-1}$

Voltage across the capacitor, $V = V_0 \sin \omega t = (30.0) \sin (6\pi \times 10^3\, t)$

Displacement current,

$$i_d = i_c = \frac{dq}{dt} = \frac{d}{dt}(q) = \frac{d}{dt}(CV) = C\frac{dV}{dt}$$

$$= (8\times 10^{-6})\frac{d}{dt}[30.0 \sin (6\pi \times 10^3\, t)]$$

$$= (8\times 10^{-6})(30.0)\frac{d}{dt}[\sin (6\pi \times 10^3\, t)]$$

$$= (8\times 10^{-6})(30.0)(6\pi\times 10^3)\cos (6\pi \times 10^3\, t)$$

$$= 4.52 \cos (6\pi \times 10^3\, t)\text{A}$$

Hence, the displacement current varies sinusoidally with time and has a maximum value of 4.52 A.

3. Radiation pressure (p) is the force exerted by electromagnetic wave on unit area of the surface, *i.e.* rate of change of momentum per unit area of the surface.

Momentum per unit time per unit area

$$= \frac{\text{Intensity}}{\text{Speed of wave}} = \frac{I}{c}$$

Change in momentum per unit time per unit area $= \Delta I/c =$ radiation pressure (p), *i.e.* $p = \Delta I/c$.

Momentum of incident wave per unit time per unit area $= I/c$.

When wave is totally absorbed by the surface, the momentum of the reflected wave per unit time per unit area = 0.

Radiation pressure (p) = Change in momentum per unit time per unit area $= \frac{\Delta I}{c} = \frac{I}{c} - 0 = \frac{I}{c}$

When wave is totally reflected, then momentum of the reflected wave per unit time per unit area $= -I/c$.

$$\text{Radiation pressure } (p) = \frac{I}{c} - \left(-\frac{I}{c}\right) = \frac{2I}{c}$$

Here, p lies between I/c.

4. As electric field in air,

$$E_1 = E_{01}\,\hat{\mathbf{x}} \cos\left(\frac{2\pi\nu z}{c} - 2\pi\nu t\right)$$

$$\therefore \quad \text{Speed in air} = \frac{2\pi\nu}{\left(\frac{2\pi\nu}{c}\right)} = c$$

Also, $$c = \frac{1}{\sqrt{\mu_0 \varepsilon_{r_1} \varepsilon_0}} \quad \ldots(i)$$

In medium, $E_2 = E_{02}\,\hat{\mathbf{x}} \cos(2kz - kct)$

$\therefore$ Speed in medium $= \frac{kc}{2k} = \frac{c}{2}$

Also, $$\frac{c}{2} = \frac{1}{\sqrt{\mu_0 \varepsilon_{r_2} \varepsilon_0}} \quad \ldots(ii)$$

As, medium is non-magnetic medium, $\mu_{medium} = \mu_{air}$

On dividing Eq. (i) by Eq. (ii), we have

$$2 = \sqrt{\frac{\varepsilon_{r_2}}{\varepsilon_{r_1}}} \Rightarrow \frac{\varepsilon_{r_1}}{\varepsilon_{r_2}} = \frac{1}{4}$$

5. In electromagnetic wave, the electric field vector is given as

$$\mathbf{E} = (E_1\hat{\mathbf{i}} + E_2\hat{\mathbf{j}})\cos(kz - \omega t)$$

In electromagnetic wave, the associated magnetic field vector,

$$\mathbf{B} = \frac{\mathbf{E}}{c} = \frac{(E_1\hat{\mathbf{i}} + E_2\hat{\mathbf{j}})}{c}\cos(kz - \omega t)$$

As, **E** and **B** are perpendicular to each other and the propagation of electromagnetic wave is perpendicular to **E** as well as **B**, so the given electromagnetic wave is plane polarised.

6. We know that, $\mathbf{E} \times \mathbf{B}$ represents direction of propagation of an electromagnetic wave

$$\Rightarrow (\mathbf{E} \times \mathbf{B}) \,||\, v$$

$\therefore$ From the given electric field, we can state that direction of propagation is along Z-axis and direction of **E** is along X-axis.

Thus, from the above discussion, direction of **B** must be Y-axis.

From Maxwell's equation,

$$\nabla \times \mathbf{E} = -\frac{\partial \mathbf{B}}{\partial t}$$

Here, $$\frac{\partial \mathbf{E}}{\partial Z} = -\frac{\partial \mathbf{B}}{\partial t} \quad \ldots(i)$$

and $$B_0 = E_0 / c \quad \ldots(ii)$$

Given, $\mathbf{E} = E_0\,\hat{\mathbf{i}} \cos kz \cos \omega t$

$$\Rightarrow \frac{-\partial \mathbf{E}}{\partial Z} = kE_0 \sin kz \cos \omega t$$

$\therefore$ Using Eq. (i), we get

$$\frac{\partial \mathbf{B}}{\partial t} = kE_0 \sin kz \cos \omega t$$

Integrating both sides of the above equation w.r.t. t, we get

$$\mathbf{B} = \frac{k}{\omega} E_0 \sin kz \sin \omega t$$

$$= \frac{E_0}{c} \sin kz \sin \omega t$$

$$\Rightarrow \mathbf{B} = \frac{E_0}{c} \sin(kz) \sin(\omega t)\,\hat{\mathbf{j}}$$

7. Given, $\mathbf{B} = 1.6 \times 10^{-6} \cos(2 \times 10^7 z + 6 \times 10^{15} t)$

$$(2\hat{\mathbf{i}} + \hat{\mathbf{j}})\ \text{Wbm}^{-2}$$

From the given equation, it can be said that the electromagnetic wave is propagating along negative z-direction, *i.e.* $-\hat{\mathbf{k}}$.

Equation of associated electric field will be

$$\mathbf{E} = (|\mathbf{B}_0| c)\cos(kz + \omega t)\cdot \hat{\mathbf{n}}$$

where, $\hat{\mathbf{n}}$ = a vector perpendicular to **B**.

So,

$$|\mathbf{E}_0| = |\mathbf{B}_0| \cdot c$$
$$= 1.6 \times 10^{-6} \times 3 \times 10^8 = 4.8 \times 10^2\ \text{V/m}$$

Since, we know that for an electromagnetic wave, **E** and **B** are mutually perpendicular to each other.

So, $$\mathbf{E} \cdot \mathbf{B} = 0$$

From the given options, when $\hat{\mathbf{n}} = \hat{\mathbf{i}} - 2\hat{\mathbf{j}}$

$$\mathbf{E} \cdot \mathbf{B} = (2\hat{\mathbf{i}} + \hat{\mathbf{j}}) \cdot (\hat{\mathbf{i}} - 2\hat{\mathbf{j}}) = 0$$

Also, when $\hat{\mathbf{n}} = -\hat{\mathbf{i}} + 2\hat{\mathbf{j}}$

$$\mathbf{E} \cdot \mathbf{B} = (2\hat{\mathbf{i}} + \hat{\mathbf{j}}) \cdot (-\hat{\mathbf{i}} + 2\hat{\mathbf{j}}) = 0$$

But, we also know that the direction of propagation of electromagnetic wave is perpendicular to both **E** and **B**, *i.e.* it is in the direction of $\mathbf{E} \times \mathbf{B}$.

Again, when $\hat{\mathbf{n}} = \hat{\mathbf{i}} - 2\hat{\mathbf{j}}$

$$\mathbf{E} \times \mathbf{B} = (2\hat{\mathbf{i}} + \hat{\mathbf{j}}) \times (\hat{\mathbf{i}} - 2\hat{\mathbf{j}}) = -\hat{\mathbf{k}}$$

and when $\hat{\mathbf{n}} = -\hat{\mathbf{i}} + 2\hat{\mathbf{j}}$

$$\mathbf{E} \times \mathbf{B} = (2\hat{\mathbf{i}} + \hat{\mathbf{j}}) \times (-\hat{\mathbf{i}} + 2\hat{\mathbf{j}}) = \hat{\mathbf{k}}$$

But, it is been given in the question that the direction of propagation of wave is in $-\hat{\mathbf{k}}$.

Thus, associated electric field will be

$$\mathbf{E} = 4.8 \times 10^2 \cos(2 \times 10^7 z + 6 \times 10^{15} t)(\hat{\mathbf{i}} - 2\hat{\mathbf{j}})\ \text{Vm}^{-1}$$

8. We are given with following situation

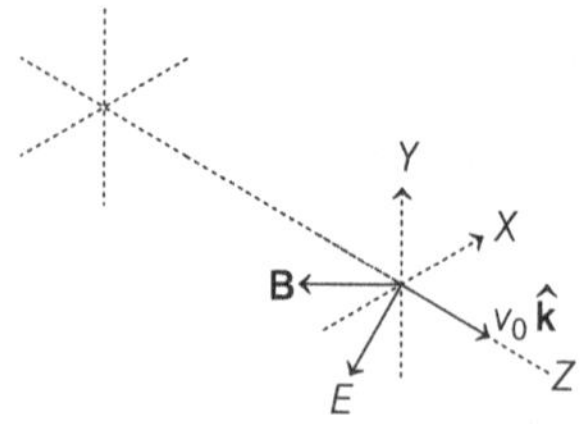

Given, $$\mathbf{E} = E_0\left(\frac{\hat{\mathbf{i}} + \hat{\mathbf{j}}}{\sqrt{2}}\right) \cdot \cos(kz + \omega t)$$

So, electromagnetic wave is propagating along Z-axis and its electric vector **E** is perpendicular to Z-axis; and is along angle bisector of IIIrd quadrant of XY-plane as shown in the figure. So, magnetic vector **B** is along angle bisector of IInd quadrant of XY-plane and is perpendicular to both **E** and **v** vectors.

Now, force due to electric field is along **E** vector and force due to magnetic field is obtained by Fleming's left hand rule.

Thus, direction of magnetic force is also along $\mathbf{E}$. So, force on charged particle will be along $\mathbf{E}$ or along $-\left(\frac{\hat{\mathbf{i}}+\hat{\mathbf{j}}}{\sqrt{2}}\right)$ vector.

Hence, force on charged particle is anti-parallel to $\left(\frac{\hat{\mathbf{i}}+\hat{\mathbf{j}}}{\sqrt{2}}\right)$ vector.

9. Electric field vector of the given EM wave is

$$\mathbf{E} = E_0\,(\hat{\mathbf{x}}+\hat{\mathbf{y}})\sin(kz-\omega t)$$

$\therefore$ Amplitude of electric field $= E_0$

and unit vector of electric field,

$$\hat{\mathbf{E}} = \frac{\hat{\mathbf{x}}+\hat{\mathbf{y}}}{\sqrt{1^2+1^2}} = \frac{\hat{\mathbf{x}}+\hat{\mathbf{y}}}{\sqrt{2}}$$

The coefficient of k in the given equation is z. So, wave travels along positive z-direction. So, unit vector into the direction of propagation of EM wave $\hat{\mathbf{c}} = \hat{\mathbf{z}}$.

Now, for magnetic field vector of the given EM wave, we can conclude

Amplitude of magnetic field, $B_0 = \frac{E_0}{c}$

and unit vector of magnetic field $= \hat{\mathbf{B}}$

So, the expected equation of magnetic field vector will be

$$\mathbf{B} = B_0(\hat{\mathbf{B}})\sin(kz-\omega t)$$

$$= \frac{E_0}{c}(\hat{\mathbf{B}})\sin(kz-\omega t) \qquad \text{...(i)}$$

Also, $\quad \hat{\mathbf{E}}\times\hat{\mathbf{B}} = \hat{\mathbf{c}} = \hat{\mathbf{z}} \qquad \text{...(ii)}$

Now, let's check the given alternatives

(a) $\mathbf{B} = \frac{E_0}{c}(\hat{\mathbf{x}}+\hat{\mathbf{y}})\sin(kz-\omega t)$

Here, $\quad \hat{\mathbf{B}} = \frac{\hat{\mathbf{x}}+\hat{\mathbf{y}}}{\sqrt{1^2+1^2}} = \frac{\hat{\mathbf{x}}+\hat{\mathbf{y}}}{\sqrt{2}}$

So, $\hat{\mathbf{E}}\times\hat{\mathbf{B}} = \left(\frac{\hat{\mathbf{x}}+\hat{\mathbf{y}}}{\sqrt{2}}\right)\times\left(\frac{\hat{\mathbf{x}}+\hat{\mathbf{y}}}{\sqrt{2}}\right)$

$$= \frac{1}{2}[(\hat{\mathbf{x}}\times\hat{\mathbf{x}}) + (\hat{\mathbf{x}}\times\hat{\mathbf{y}}) + (\hat{\mathbf{y}}\times\hat{\mathbf{x}}) + (\hat{\mathbf{y}}\times\hat{\mathbf{y}})]$$

$$= \frac{1}{2}[(0) + (\hat{\mathbf{z}}) + (-\hat{\mathbf{z}}) + (0)]$$

$$= \frac{1}{2}[\hat{\mathbf{z}} - \hat{\mathbf{z}}] = 0$$

which is impossible for an EM wave.

So, this option is incorrect.

(b) $\mathbf{B} = \frac{E_0}{c}(-\hat{\mathbf{x}}+\hat{\mathbf{y}})\sin(kz-\omega t)$

Here, $\quad \hat{\mathbf{B}} = \frac{-\hat{\mathbf{x}}+\hat{\mathbf{y}}}{\sqrt{1^2+1^2}} = \frac{-\hat{\mathbf{x}}+\hat{\mathbf{y}}}{\sqrt{2}}$

So, $\hat{\mathbf{E}}\times\hat{\mathbf{B}} = \left(\frac{\hat{\mathbf{x}}+\hat{\mathbf{y}}}{\sqrt{2}}\right)\times\left(-\frac{\hat{\mathbf{x}}+\hat{\mathbf{y}}}{\sqrt{2}}\right)$

$$= \frac{1}{2}[-(\hat{\mathbf{x}}\times\hat{\mathbf{x}}) + (\hat{\mathbf{x}}\times\hat{\mathbf{y}}) - (\hat{\mathbf{y}}\times\hat{\mathbf{x}}) + (\hat{\mathbf{y}}\times\hat{\mathbf{y}})]$$

$$= \frac{1}{2}[-(0) + (\hat{\mathbf{z}}) - (-\hat{\mathbf{z}}) + (0)]$$

$$= \frac{1}{2}[\hat{\mathbf{z}} + \hat{\mathbf{z}}] = \frac{1}{2}(2\,\hat{\mathbf{z}}) = \hat{\mathbf{z}}$$

which satisfies for Eq. (ii).

So, this option is correct.

(c) $\mathbf{B} = \frac{E_0}{c}(\hat{\mathbf{x}}-\hat{\mathbf{y}})\sin(kz-\omega t)$

Here, $\hat{\mathbf{B}} = \frac{\hat{\mathbf{x}}-\hat{\mathbf{y}}}{\sqrt{1^2+1^2}} = \frac{\hat{\mathbf{x}}-\hat{\mathbf{y}}}{\sqrt{2}}$

So, $\hat{\mathbf{E}}\times\hat{\mathbf{B}} = \left(\frac{\hat{\mathbf{x}}+\hat{\mathbf{y}}}{\sqrt{2}}\right)\times\left(\frac{\hat{\mathbf{x}}-\hat{\mathbf{y}}}{\sqrt{2}}\right)$

$$= \frac{1}{2}[(\hat{\mathbf{x}}\times\hat{\mathbf{x}}) - (\hat{\mathbf{x}}\times\hat{\mathbf{y}}) + (\hat{\mathbf{y}}\times\hat{\mathbf{x}}) - (\hat{\mathbf{y}}\times\hat{\mathbf{y}})]$$

$$= \frac{1}{2}[(0) - (\hat{\mathbf{z}}) + (-\hat{\mathbf{z}}) - (0)]$$

$$= \frac{1}{2}[-\hat{\mathbf{z}} - \hat{\mathbf{z}}]$$

$$= \frac{1}{2}(-2\,\hat{\mathbf{z}}) = -\hat{\mathbf{z}}$$

which does not satisfy for Eq. (ii).

So, this option is incorrect.

(d) $\mathbf{B} = \frac{E_0}{c}(\hat{\mathbf{x}}-\hat{\mathbf{y}})\cos(kz-\omega t)$

This option is incorrect, as the function in Eq. (i) is a "sine" function, not a "cosine" function.

Hence, option (b) is correct.

10. In the free space, the speed of electromagnetic wave is given as

$$c = \frac{1}{\sqrt{\mu_0\varepsilon_0}} = \frac{E_0}{B_0} \qquad \text{...(i)}$$

where, E_0 and B_0 are the amplitudes of varying electric and magnetic fields, respectively.

Now, when it enters in a medium of refractive index n, its speed is given as

$$v = \frac{1}{\sqrt{\mu\varepsilon}} = \frac{1}{\sqrt{K\varepsilon_0\mu}} = \frac{E}{B} \qquad \text{...(ii)}$$

where, K is dielectric strength of the medium.

Using Eqs. (i) and (ii), we get

$$\frac{v}{c} = \frac{1}{\sqrt{K}} \qquad \text{...(iii)}$$

($\because$ For a transparent medium, $\mu_0 \approx \mu$)

Also, refractive index of medium is n and is given as

$$\frac{c}{v} = n \quad \text{or} \quad \frac{v}{c} = \frac{1}{n} \qquad \text{...(iv)}$$

$\therefore$ From Eqs. (iii) and (iv), we get

$$n = \sqrt{K} \quad \text{or} \quad K = n^2 \qquad \text{...(v)}$$

The intensity of an EM wave is given as

$$I = \frac{1}{2}\varepsilon_0 E_0^2 c$$

and in the medium, it is given as $I' = \frac{1}{2} K\varepsilon_0 E^2 v$

It is given that, $I = I'$

$$\Rightarrow \quad \frac{1}{2}\varepsilon_0 E_0^2 c = \frac{1}{2} K\varepsilon_0 E^2 v$$

$$\text{or} \quad \left(\frac{E_0}{E}\right)^2 = \frac{Kv}{c} \quad \ldots\text{(vi)}$$

From Eqs. (iv), (v) and (vi), we get

$$\left(\frac{E_0}{E}\right)^2 = \frac{n^2}{n} = n \quad \text{or} \quad E_0 / E = \sqrt{n}$$

Similarly, $\frac{1}{2} \cdot \frac{B_0^2}{\mu_0} c = \frac{1}{2} \cdot \frac{B^2}{\mu_0} v$

$$\Rightarrow \quad \frac{B_0}{B} = \frac{1}{\sqrt{n}}$$

11. The moving electron is shown in the following figure.

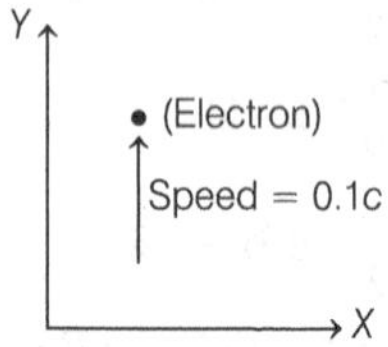

In electromagnetic wave, $B_{max} = \frac{E_{max}}{c}$

We know that, magnetic force on a moving charge (electron) is given by

$$\mathbf{F} = q\mathbf{v} \times \mathbf{B}$$

Here, angle between $\mathbf{v}$ and $\mathbf{B}$ is 90°.

So, $F_{max} = qvB_{max} \sin\theta$

$$= qvB_{max} \sin 90° = \frac{qvE_{max}}{c}$$

Given, $c = 3 \times 10^8$ m/s, $v = 0.1\, c$

Charge on electron, $q = 1.6 \times 10^{-19}$ C

Substituting these values in above equation, we get

$$F_{max} = \frac{1.6 \times 10^{-19} \times 0.1 \times 3 \times 10^8 \times 30}{3 \times 10^8}$$

$$= 4.8 \times 10^{-19} \text{ N}$$

12. Energy density due to electric field,

$$u = \frac{1}{2}\varepsilon_0 E^2$$

Here, $E = \frac{E_0}{\sqrt{2}} = \frac{40}{\sqrt{2}}$

$$\therefore \quad u = \frac{1}{2} \times 8.85 \times 10^{-12} \times \frac{40 \times 40}{2}$$

$$= 3.54 \times 10^{-9} \text{ J/m}^3$$

$\therefore$ Hence, p is 3.54.

13. In vacuum, wavelength,

$$\lambda_0 = \frac{c}{f_0} = \frac{3 \times 10^8}{2 \times 10^6} = 1.5 \times 10^2 = 150 \text{ m}$$

In medium, speed of wave,

$$v = \frac{c}{\sqrt{\varepsilon_r \mu_r}} \approx \frac{c}{\sqrt{\varepsilon_r}}$$

(As medium is non-magnetic)

$$\therefore \quad v = \frac{c}{\sqrt{9}} = \frac{c}{3} = 1 \times 10^8 \text{ms}^{-1}$$

Frequency remaining same, wavelength in medium is

$$\lambda = \frac{v}{f_0} = \frac{10^8}{2 \times 10^6} = 50 \text{ m}$$

$\therefore$ Wavelength decreases by 100 m.

14. As, $F = pA = \frac{UA}{c}$

$$\Rightarrow \quad U = \frac{Fc}{A}$$

$$= \frac{2.5 \times 10^{-6} \times 3 \times 10^8}{30} = 25 \text{ W/ cm}^2$$

15. Given, intensity, $I = \left(\frac{315}{\pi}\right)$ W/m^2

Absolute permittivity, $\varepsilon_0 = 8.86 \times 10^{-12}\, \text{C}^2\text{N}^{-1}\text{m}^{-2}$

Speed of light, $c = 3 \times 10^8$ m/s

As, intensity, $I = \frac{1}{2}\varepsilon_0 E_0^2 c$

$$\text{or} \quad E_0 = \sqrt{\frac{2I}{\varepsilon_0 c}} = \sqrt{\frac{2 \times \left(\frac{315}{\pi}\right)}{8.86 \times 10^{-12} \times 3 \times 10^8}}$$

$$E_0 \approx 275 \text{ V/m}$$

$$\therefore \quad E_{rms} = \frac{E_0}{\sqrt{2}} = \frac{275}{\sqrt{2}} \approx 194 \text{ V/m}$$

25

Ray Optics and Optical Instruments

IN THIS CHAPTER ....

Light is a form of energy which makes objects visible to our eyes. The branch of physics, which deals with nature of light, its sources, properties, measurement, effects and vision is called **optics**. This branch can be divided into two sub-branches namely (i) **geometrical optics** or **ray optics**, and (ii) **wave optics**.

Ray optics deals with the propagation of light in terms of rays, whereas wave optics deals with the wave phenomenon.

Reflection of Light

It is the phenomenon of bouncing back of light rays in the same medium on striking a smooth surface.

After reflection, the frequency, speed and wavelength do not change, but a phase change may occur depending on the nature of reflecting surface.

Laws of Reflection

(i) The incident ray, reflected ray and the normal to the reflecting surface at the point of incidence lie in the same plane.

(ii) The angle of incidence, $\angle i$ = angle of reflection, $\angle r$.

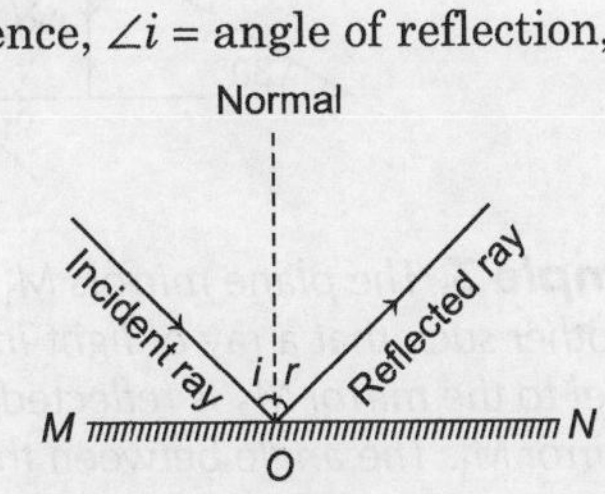

Note In geometrical optics, angles are measured with normal.

Image

If rays emanating from a point actually meet at another point after reflection and/or refraction, that point is called the **image of the first point**.

(i) The image is **real**, if the rays actually converge to the point.

(ii) The image is **virtual**, if the rays do not actually meet but appear to diverge from the point when produced backwards.

Reflection from Plane Mirror (Surface)

In case of reflection from plane surface such as plane mirror

- The image is always erect, virtual, of same size and at the same distance from the mirror as the distance of the object from mirror.
- The image is laterally inverted.
- If keeping the incident ray fixed, the plane mirror is rotated through an angle θ, the reflected ray turns through an angle 2θ in that direction only.
- If the object is fixed and the mirror moves relative to the object with a speed v, the image moves with a speed $2v$ relative to the object.
- If the mirror is fixed and the object moves relative to the mirror with a speed v, the image also moves with the same speed v relative to the mirror.
- Deviation produced by plane mirror is given by $\delta = 180° - 2i$.
- The minimum size of a mirror required to see the full image of a person, is half the height of the person.
- If a plane mirror is rotated about an axis perpendicular to plane of mirror, then reflected ray image do not rotate.
- The minimum height of a plane mirror to see object's full height in it is $H/2$, where H is the height of object.
- Minimum height of the plane mirror fixed on the wall of a room in which an observer at the centre of the room can see the full image of the wall behind him, is one-third the height of wall.
- If plane mirror moves a distance x towards or away from the object, the image will move a distance $2x$ towards or away from the mirror/object.
- If both plane mirror and object are moved by a distance x, each in opposite directions, the image will be displaced by a distance $3x$ in the direction of the displacement of the mirror.
- If a luminous object is placed in front of a thick glass mirror, multiple images are formed due to multiple reflections. The second image formed by the first reflection by the polished surface is much brighter than the others. The intensity of the other images rapidly fade away.
- If an object is placed between two mirrors facing each other at an angle θ, the number of images is given by $\frac{360°}{\theta} = N$ and actual number of images are n, where

  (a) $n = N - 1$, if N is even integer,

  (b) $n = N$, if N is odd integer and object is not on the bisector of mirrors,

  (c) $n = N - 1$, if N is odd integer and object is on the bisector of mirrors and

  (d) If $\frac{360°}{\theta} = N$ is a fraction, the number of images will be equal to its lower boundary of integral part.

Example 1. *Find velocity of image when object and mirror both are moving toward each other with velocity $2\ ms^{-1}$ and $3\ ms^{-1}$, respectively.*

(a) $8\ ms^{-1}$ (b) $-8\ ms^{-1}$

(c) $-5\ ms^{-1}$ (d) $5\ ms^{-1}$

Sol. (a) Here,
$$v_{OM} = v_{IM}$$
$$v_O - v_M = (v_I - v_M)$$
$$\Rightarrow (+2ms^{-1}) - (-3ms^{-1}) = v_I + (-3\ ms^{-1})$$
$$\Rightarrow v_I = 8\ ms^{-1}$$

Example 2. *Two plane mirrors are inclined at 30° as shown in figure. A light ray is incident at angle 45°. Find total deviation produced by combination of mirror after two successive reflections.*

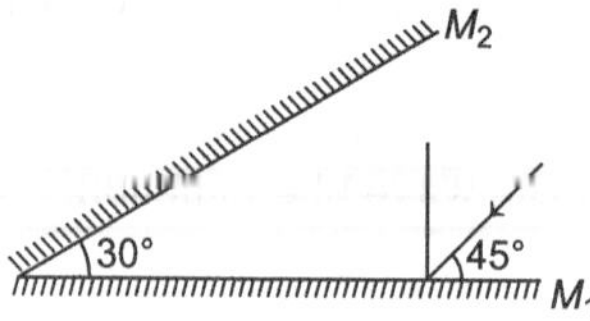

(a) 60° ↑ (b) 58° ↑

(c) 50° ↑ (d) 68° ↑

Sol. (a) Deviation at mirror M_1, $\delta_1 = 180° - 2 \times 45° = 90°$ ↑

Deviation at mirror M_2, $\delta_2 = 180° - 2 \times 15° = 150°$ ↑

Total deviation $\delta = \delta_2 - \delta_1 = 150° - 90° = 60°$ ↑

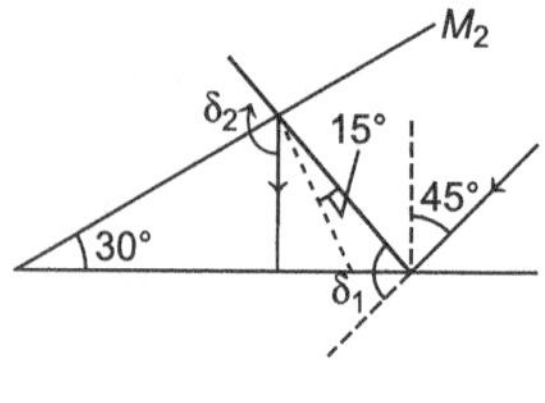

Example 3. *The plane mirrors M_1 and M_2 are inclined to each other such that a ray of light incident on mirror M_1 and parallel to the mirror M_2 is reflected from mirror M_2 parallel to the mirror M_1. The angle between the two mirror is*

[JEE Main 2019]

(a) 45° (b) 75°

(c) 90° (d) 60°

Sol. (*d*) The given condition is shown in the figure given below, where two plane mirrors inclined to each other such that a ray of light incident on the first mirror M_1 and parallel to the second mirror M_2 is finally reflected from second mirror M_2 parallel to the first mirror.

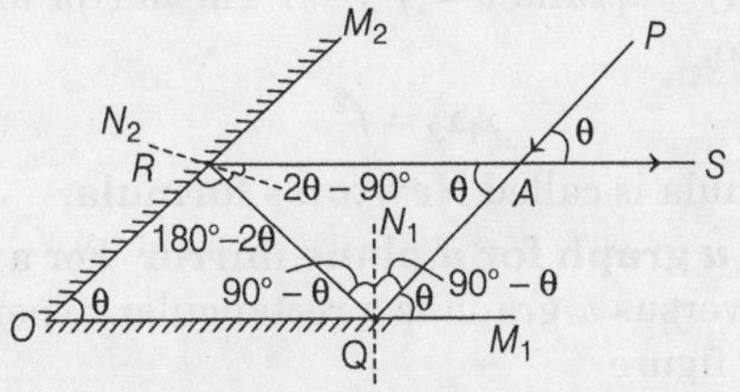

where, PQ = incident ray parallel to the mirror M_2, QR = reflected ray from the mirror M_1, RS = reflected ray from the mirror M_2 which is parallel to the M_1 and θ = angle between M_1 and M_2.

According to geometry,

$$\angle PAS = \angle PQM_1 = \theta \text{ (angle on same line)}$$

$$\angle AQN_1 = \text{angle of incident} = 90° - \theta$$

$$\angle N_1QR = \text{angle of reflection} = (90° - \theta).$$

Therefore, for triangle ΔORQ, (according to geometry)

$$\angle\theta + \angle\theta + \angle ORQ = 180°$$

$$\angle ORQ = 180° - 2\theta$$

For normal N_2, angle of incidence (i) = angle of reflection (r)

$$= 2\theta - 90°$$

$$\therefore \angle ORA = i + r = 2\theta - 90° + 2\theta - 90° = 4\theta - 180°$$

Therefore, for the triangle ΔRAQ,

$$4\theta - 180° + 180° - 2\theta + \theta = 180° \Rightarrow 3\theta = 180° \Rightarrow \theta = 60°$$

Spherical Mirrors

Mirrors whose reflecting surface is a part of a hollow sphere is called spherical mirror. It is of two types namely; concave mirror and convex mirror.

Some terms related to spherical mirrors are as follows

Pole (P) of the spherical mirror is geometric centre of its reflecting surface.

Centre of curvature (C) of a mirror is the centre of the sphere of which the mirror is a part.

Radius of curvature (R) is the distance between the pole and centre of curvature.

Principal axis of a spherical mirror is the line joining the pole and centre of curvature.

Principal focus (F) is a point on the principal axis of the mirror at which the light rays coming parallel to principal axis actually meet or appear to meet after reflection.

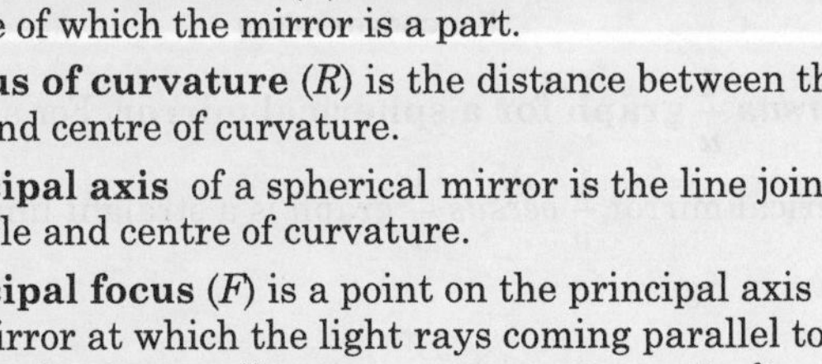

(a) Concave mirror (b) Convex mirror

Note For concave mirror focus, is infront of the mirror, while for convex mirror, focus is behind the mirror. Focus of concave mirror is real, while focus of convex mirror is virtual.

Focal length (f) is the distance between pole and focus of a spherical mirror.

i.e. $$f = \frac{R}{2}$$

Rules for Image Formation in Spherical Mirrors

- A ray going through centre of curvature which strikes the mirror normally is reflected back along the same path.
- A ray parallel to principal axis after reflection either actually passes through the principal focus F or appear to diverge from it.
- A ray passing through the principal focus F or a ray which appears to converge at F is reflected parallel to the principal axis.
- A ray striking at pole P is reflected symmetrically back in the opposite side.

Sign Convention for Mirrors

According to cartesian sign convention,

- All distances are measured from the pole (P).
- Distances measured in the direction of incident ray are taken as positive, while in the direction opposite of incident ray are taken negative.
- Height measured upwards with respect to X-axis and normal to the principal axis (X-axis) of the mirror are taken as positive. Heights measured downwards are taken as negative.

Image Formation by Concave Mirror

Position of Object	Ray Diagram	Details of Image
At infinity	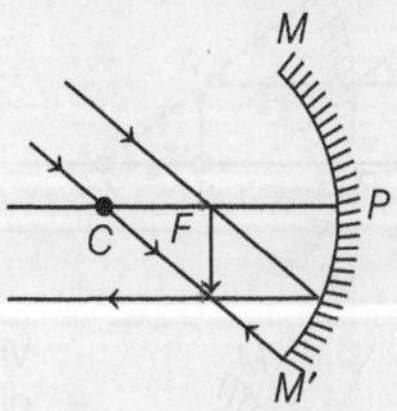	Real, inverted, very small, at F
Between infinity and C	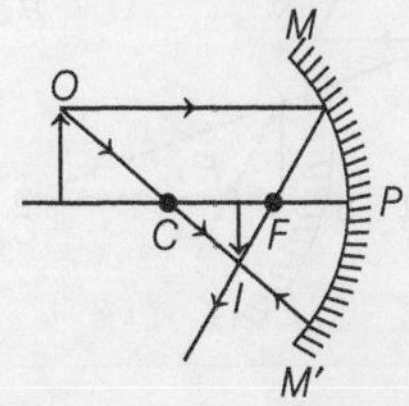	Real, inverted, diminished, between F and C
At C	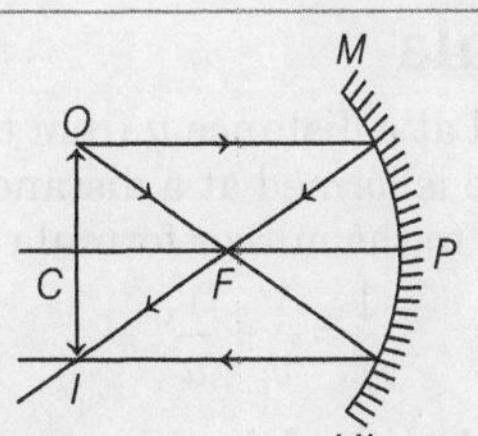	Real, inverted, equal in size, at C

Position of Object	Ray Diagram	Details of Image
Between F and C	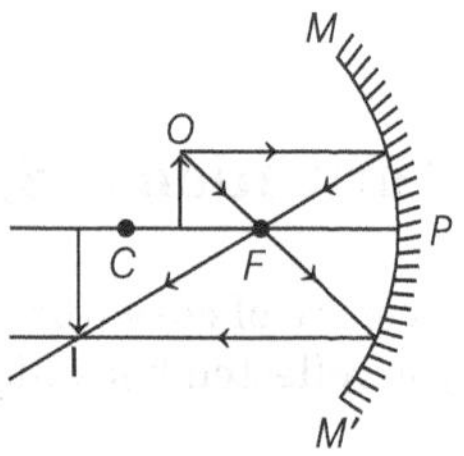	Real, inverted and very large, between $2F$ and infinity
At F	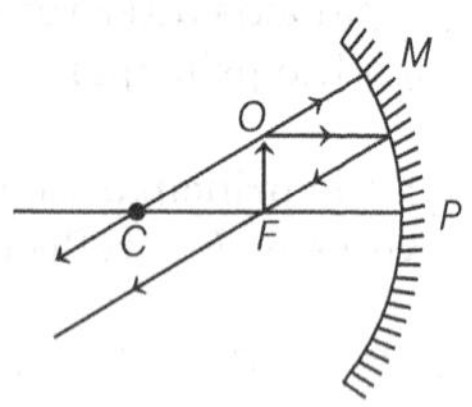	Real, inverted, very large, at infiinity
Between F and P	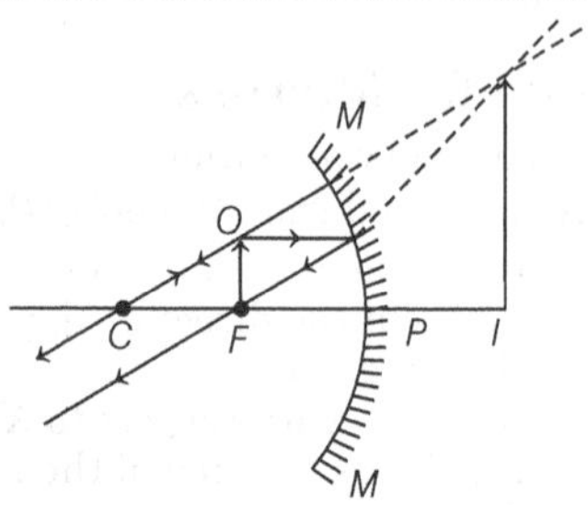	Virtual, erect, large in size, behind the mirror

Image Formation by Convex Mirror

Position of Object	Ray Diagram	Details of Image
At infinity	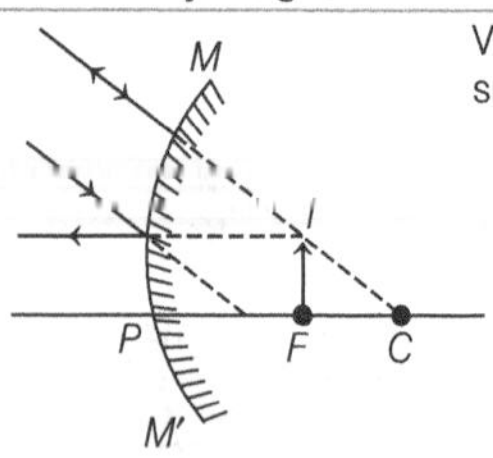	Virtual, erect, very small, at F
In front of mirror	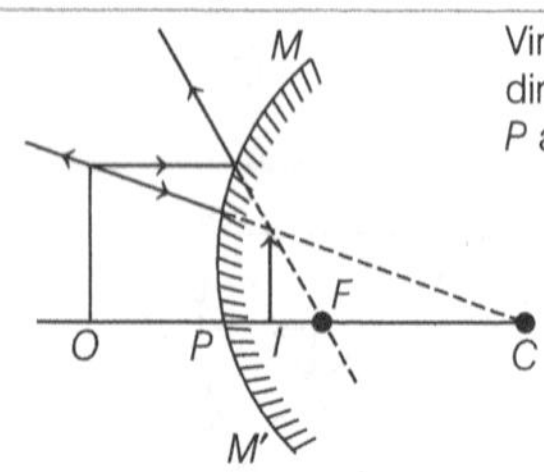	Virtual, erect, diminished, between P and F

Mirror Formula

If an object is placed at a distance u from the pole of a mirror and its image is formed at a distance v from the pole, then according to the mirror formula

$$\frac{1}{f} = \frac{1}{v} + \frac{1}{u}$$

where, f is the focal length of the mirror.

Following are few important points related to mirror formula

- In case of spherical mirrors, if object distance x_1 and image distance x_2 are measured from focus instead of pole, $u = (f + x_1)$ and $v = (f + x_2)$. The mirror formula reduces to,

$$x_1 x_2 = f^2$$

  This formula is called **Newton's formula**.

- ***v* versus *u* graph for a plane mirror** For a spherical mirror v versus u graph is a rectangular hyperbola as shown in figure

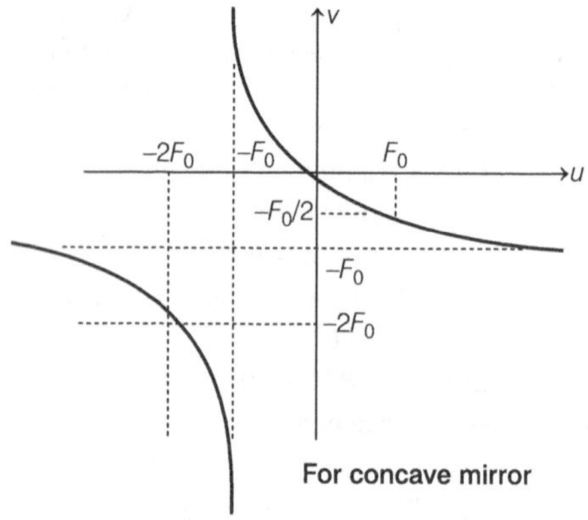

For concave mirror

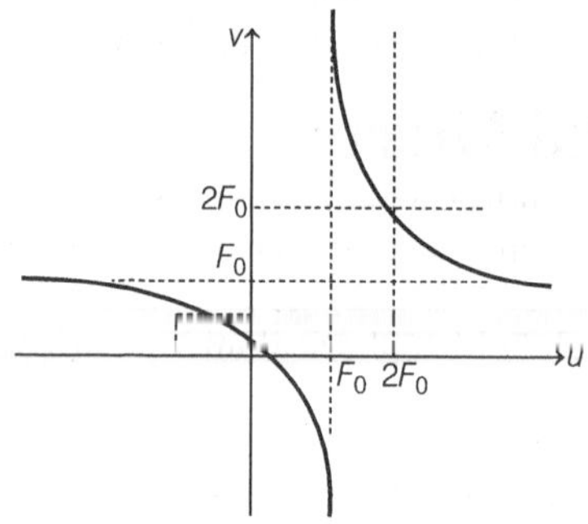

For convex mirror

- $\frac{1}{v}$ ***versus*** $\frac{1}{u}$ **graph for a spherical mirror** For a spherical mirror $\frac{1}{v}$ *versus* $\frac{1}{u}$ graph is a straight line as shown in figure

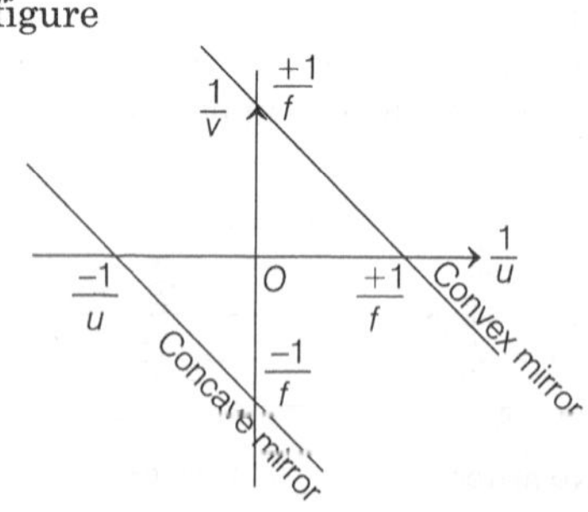

$\frac{1}{v}$ *versus* $\frac{1}{u}$ graph for a spherical mirror

Magnification

It is the ratio of the height of the image (I) to height of the object (O).

Lateral (linear or transverse) **magnification,**

$$m = \frac{I}{O} = -\frac{v}{u} = \frac{f}{f-u} = \frac{f-v}{f}$$

Axial (longitudinal) **magnification,**

$$m_{ax} = -\frac{dv}{du} = \frac{v^2}{u^2} = \left(\frac{f}{f-u}\right)^2 = \left(\frac{f-v}{f}\right)^2$$

Also, length of image,

$$= \left(\frac{v}{u}\right)^2 \times \text{length of object } (L_O)$$

$$\Rightarrow \quad L_I = \left(\frac{f}{u-f}\right)^2 \cdot L_O$$

Areal magnification,

$$m_{ar} = \frac{A_I}{A_O} = \frac{v^2}{u^2}$$

$$= \left(\frac{f}{f-u}\right)^2 = \left(\frac{f-v}{f}\right)^2$$

where, A_I = area of image and A_O = area of object.

Example 4. *Lower half of concave mirror's reflecting surface is covered with an opaque (non-reflective) material. The intensity of an image of an object placed in front of the mirror is*

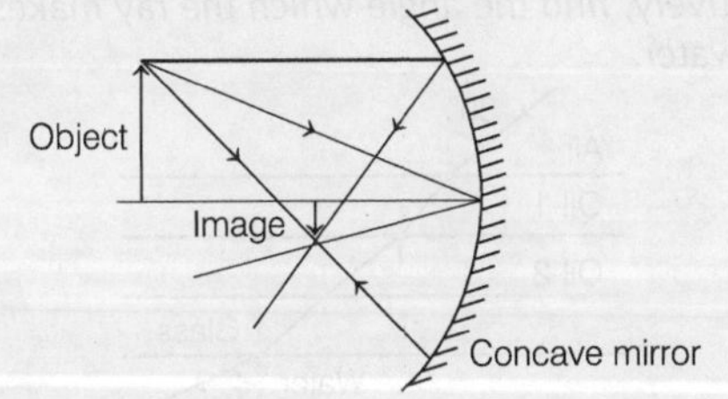

(a) *twice the original image*
(b) *half the original image*
(c) *same as the original image*
(d) *one-fourth the original image*

Sol. (b) One may think that the image will now show only half of the object, but according to the laws of reflection to be true for all points of the remaining part of the mirror, the image will be that of the whole object. However, as the area of the reflecting surface has been reduced, the intensity of the image will be low (in this case half).

Example 5. *An object of length 2.5 cm is placed at 1.5f from a concave mirror, where f is the focal length of the mirror. The length of the object is perpendicular to the principal axis. Find the length of the image. Is the image erect or inverted?*

(a) – 5 *cm* (b) 5 *cm*
(c) 6 *cm* (d) – 6 *cm*

Sol. (a) The focal length, $F = -f$

and $u = -1.5f$

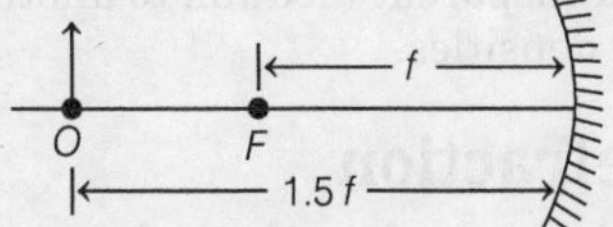

From mirror formula, we have

$$\frac{1}{u} + \frac{1}{v} = \frac{1}{f}$$

or $$\frac{1}{-1.5f} + \frac{1}{v} = -\frac{1}{f}$$

$$\frac{1}{v} = \frac{1}{1.5f} - \frac{1}{f} = -\frac{1}{3f}$$

Now, $$m = \frac{v}{u} = \frac{3f}{1.5f} = 2$$

or $$\frac{h_2}{h_1} = -2$$

or $$h_2 = -2h_1 = -5 \text{ cm}$$

The image is 5 cm long. The minus sign shows that it is inverted.

Example 6. *A concave mirror of focal length 10 cm and a convex mirror of focal length 15 cm are placed facing each other 40 cm apart. A point object is placed between the mirrors, on their common axis and 15 cm from the concave mirror. Find the position of the image produced by the successive reflections, first at concave mirror and then at convex mirror.*

(a) –30 *cm*, 6 *cm* (b) –30 *cm*, +10 *cm*
(c) +30 *cm*, + 15 *cm* (d) + 6 *cm*, + 30 *cm*

Sol. (a) According to given problem, for concave mirror.

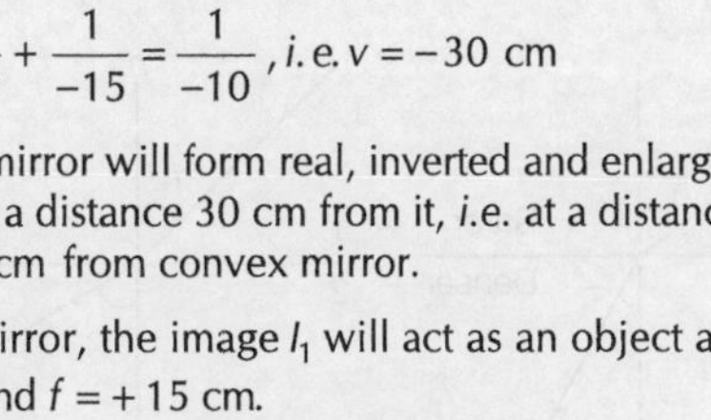

Given, $u = -15$ cm

and $f = -10$ cm

So, $$\frac{1}{v} + \frac{1}{-15} = \frac{1}{-10}, \text{ i.e. } v = -30 \text{ cm}$$

i.e. Concave mirror will form real, inverted and enlarged image I_1 of object O at a distance 30 cm from it, *i.e.* at a distance $40 - 30 = 10$ cm from convex mirror.

For convex mirror, the image I_1 will act as an object and so for it $u = -10$ cm and $f = +15$ cm.

$$\frac{1}{v} + \frac{1}{-10} = \frac{1}{15}, \text{ i.e. } v = +6 \text{ cm}$$

So, final image I_2 is formed at a distance 6 cm behind the convex mirror and is virtual as shown in above figure.

Refraction of Light

It is the phenomena of bending of ray of light when they pass from one transparent medium to another depending on their optical densities.

Laws of Refraction

(i) The incident ray, the refracted ray and the normal to the interface at the point of incidence, all lie in the same plane.

(ii) The ratio of the sine of angle of incidence i to the sine of angle of refraction r is constant, *i.e.*

$$\frac{\sin i}{\sin r} = n_{21} \text{ or } {}^1n_2$$

where, n_{21} is a constant, called the **refractive index** of the second medium with respect to the first medium.

This is known as **Snell's law of refraction**.

Note n_{21} is a characteristic of the pair of media (and also depend on the wavelength of light) but is independent of the angle of incidence.

Following are few important points related to refraction of light.

- When light travels from vacuum (or air) to any transparent medium, then refractive index of medium with respect to vacuum (or air) is called it's absolute refractive index, *i.e.* $\mu = \dfrac{\text{speed of light in vacuum}}{\text{speed of light in medium}} = \dfrac{c}{v}$
- If $n_{21} > 1, r < i$, then the refracted ray bend towards the normal. In such a case, medium 2 is said to be **optically denser than medium 1**.
- If $n_{21} < 1, r > i$, the refracted ray bend away from the normal. This is the case when incident ray in a denser medium refracts into a rarer medium.
- We can also write Snell's law as $\mu \sin i =$ constant

  For two media, $\mu_1 \sin i_1 = \mu_2 \sin i_2$

  Also, $${}_1\mu_2 = \frac{\sin i_1}{\sin i_2} = \frac{v_1}{v_2} = \frac{\lambda_1}{\lambda_2} = \frac{\mu_1}{\mu_2}$$

  where, v_1 is the speed of light in medium 1 and v_2 in medium 2.

  Similarly, λ_1 and λ_2 are the corresponding wavelengths.

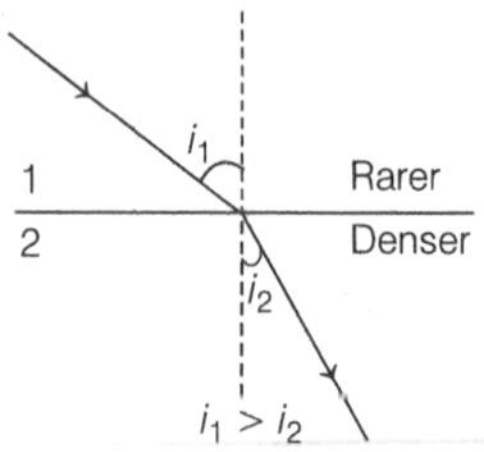

$i_1 > i_2$
$v_2 < v_1$
$\mu_2 > \mu_1$
$\lambda_2 < \lambda_1$

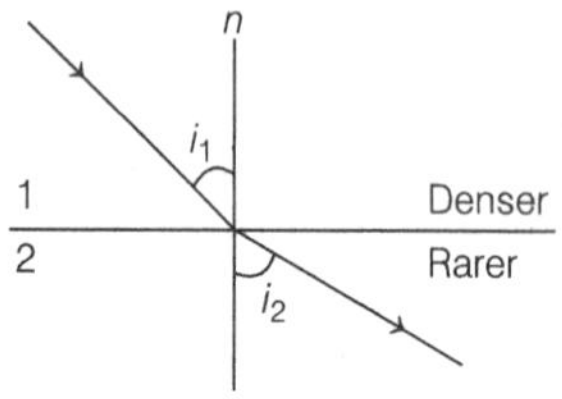

$i_1 < i_2$
$v_2 > v_1$
$\mu_2 < \mu_1$
$\lambda_2 > \lambda_1$

If $\mu_2 > \mu_1$, then $v_1 > v_2$ and $\lambda_1 > \lambda_2$, *i.e.*, in a rarer medium, speed and hence, wavelength of light is more.

- For n mediums, ${}_1\mu_2 \times {}_2\mu_3 \times {}_3\mu_4 \times \ldots \times (n-1)\mu_n = {}_1\mu_n$

  It is called **reversibility of light**. According to this principal, when a light ray, after suffering n number of reflections and refractions, will has its final path reversed, *i.e.*, it travels back along its entire initial path.
- Experiments show that, if the boundaries of the media are **parallel** the emergent ray CD, although laterally displaced. If reaches in the same medium, is parallel to the incident ray, however it is laterally displaced.

Example 7. *Refractive index of glass with respect to water is 9/8. Refractive index of glass with respect to air is 3/2. Find the refractive index of water with respect to air.*

(a) 3/4 (b) 4/3 (c) 9/8 (d) 8/9

Sol. (b) Given, ${}_w\mu_g = 9/8$ and ${}_a\mu_g = 3/2$

As, $\quad {}_a\mu_g \times {}_g\mu_w \times {}_w\mu_a = 1$

$$\therefore \quad \frac{1}{{}_w\mu_a} = {}_a\mu_w = {}_a\mu_g \times {}_g\mu_w = \frac{{}_a\mu_g}{{}_w\mu_g}$$

$\therefore$ Refractive index of water with respect to air,

$${}_a\mu_w = \frac{3/2}{9/8} = \frac{4}{3}$$

Example 8. *Light is incident from air on oil at an angle of 30°. After moving through oil-1, oil-2 and glass it enters water. If the refractive indices of glass and water are 1.5 and 1.3, respectively, find the angle which the ray makes with normal in water.*

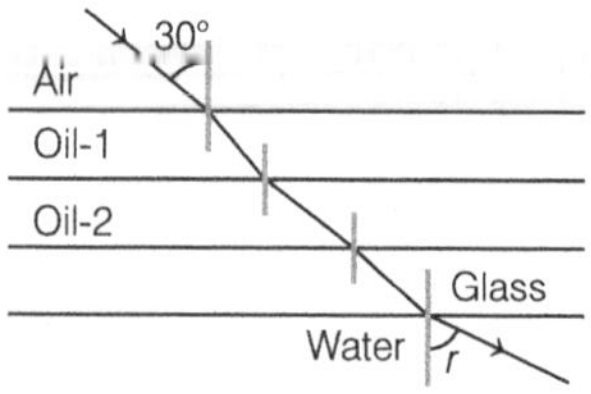

(a) $\sin^{-1}\left(\dfrac{1}{2.6}\right)$ (b) $\sin^{-1}\left(\dfrac{3}{2.6}\right)$

(c) $\sin^{-1}\left(\dfrac{1}{3.6}\right)$ (d) $\sin^{-1}(2.6)$

Sol. (a) As we know, $\mu \sin i =$ constant

$$\Rightarrow \quad \mu_{air} \sin i_{air} = \mu_{glass} \sin r_{glass}$$

$$\sin r_{glass} = \frac{\mu_{air}}{\mu_{glass}} \sin i_{air} \quad \ldots(i)$$

Again, $\quad \mu_{glass} \sin i_{glass} = \mu_{water} \sin i_{water} \quad \ldots(ii)$

From Eqs. (i) and (ii), we get

$$\sin 30° = 1.3 \sin r$$

$$\sin r = \frac{1}{2 \times 1.3} = \frac{1}{2.6}, \; r = \sin^{-1}\left(\frac{1}{2.6}\right)$$

Example 9. *A searchlight on a yacht is being used at night to illuminate a sunken box, as shown in figure. At what angle of incidence, θ_1 should the light be aimed?*

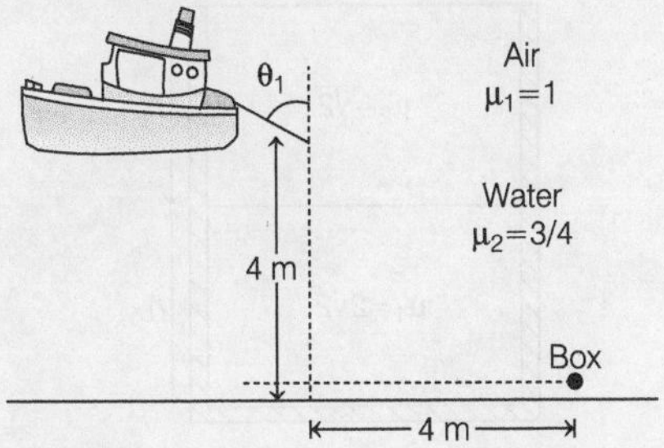

(a) $\sin^{-1}\left(\frac{2\sqrt{2}}{3}\right)$ (b) $\sin^{-1}(\sqrt{2})$

(c) $\cos^{-1}\left(\frac{3}{2\sqrt{2}}\right)$ (d) $\cos^{-1}(\sqrt{2})$

Sol. (a) Here, $\tan\theta_2 = \frac{4}{4}, \Rightarrow \theta_2 = 45°$

Given, $\mu_1 = 1$ (for air) and $\mu_2 = 3/4$ (for water)

From Snell's law, $\mu_1 \cdot \sin\theta_1 = \mu_2 \cdot \sin\theta_2$

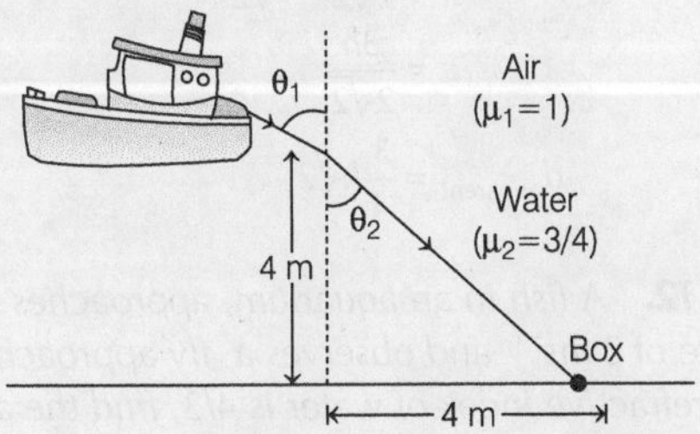

$$\Rightarrow \quad 1\cdot\sin\theta_1 = \frac{4}{3}\cdot\sin 45° \Rightarrow \sin\theta_1 = \frac{4}{3}\cdot\frac{1}{\sqrt{2}}$$

$$\Rightarrow \quad \sin\theta_1 = \frac{2\sqrt{2}}{3} \Rightarrow \theta_1 = \sin^{-1}\left(\frac{2\sqrt{2}}{3}\right)$$

Apparent Shift of an Object due to Refraction

Due to bending of light at the interface of two different media, the image formation due to refraction creates an illusion of shifting of the object position.

***Case* I** The object O is in denser medium (glass, water, etc.) and is seen from rarer medium (air) normally through plane surface.

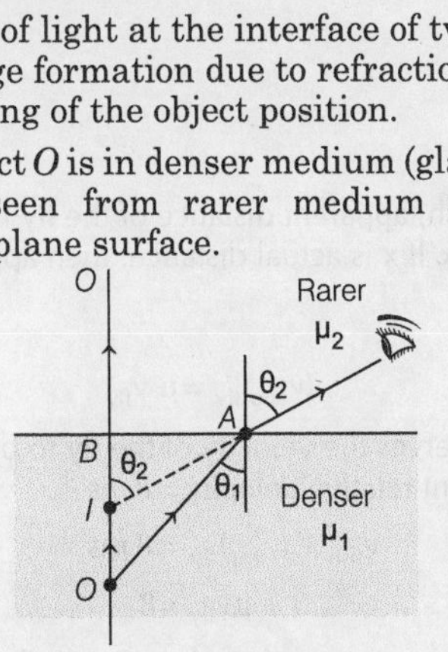

$$AI = d_{\text{apparent}} = \frac{d}{\mu}$$

$$\therefore \quad \text{Shift} = OI = AO - AI = d\left(1 - \frac{1}{\mu}\right)$$

where, d is actual depth of the object in the denser medium of refractive index μ.

Lateral magnification If immiscible liquids of refractive indices μ_1, μ_2 and μ_3 ...... (with $\mu_3 > \mu_2 > \mu_1$) are filled in a vessel. Their depths are d_1, d_2 and d_3...... and so on respectively. Then, the apparent depth (for normal incidence) when seen from top of the first liquid will be

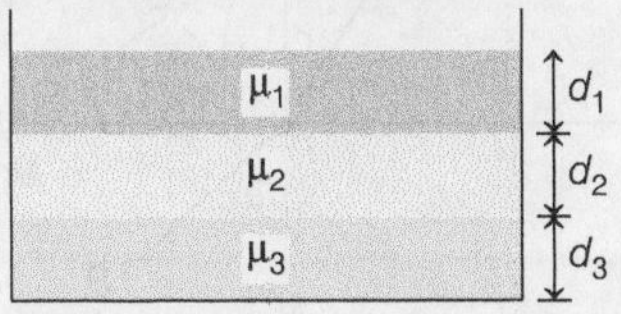

$$d_{\text{app}} = \frac{d_1}{\mu_1} + \frac{d_2}{\mu_2} + \frac{d_3}{\mu_3} + \ldots\ldots$$

Apparent shift

$$= d_1\left(1 - \frac{1}{\mu_1}\right) + d_2\left(1 - \frac{1}{\mu_2}\right) + \ldots + d_n\left(1 - \frac{1}{\mu_n}\right)$$

***Case* II** The object O is in rarer medium (air) and is seen from denser medium (glass, water, etc.) *i.e.* normally through plane surface.

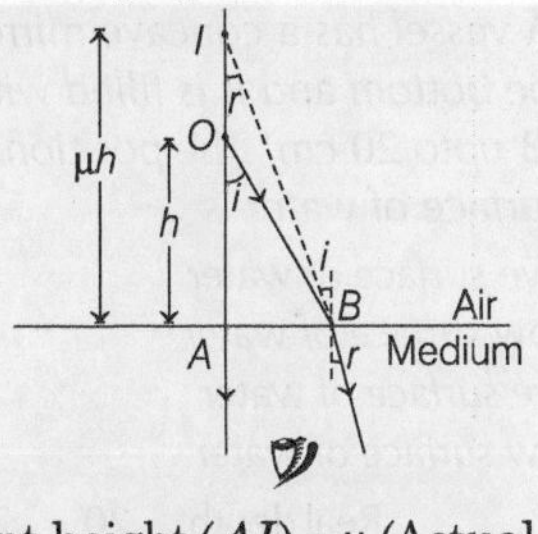

$\therefore$ Apparent height $(AI) = \mu$ (Actual height)
and Shift $= (\mu - 1)h$

Refraction through a Glass Slab

Normal shift If a glass slab is placed in the path of a converging or diverging beam of light, then point of convergence or point of divergence appears to be shifted as shown in figure.

Normal shift, $OO' = x = \left(1 - \frac{1}{\mu}\right)t$

Normal shift by a glass slab

Note Due to slab, if the $\mu_{\text{slab}} > \mu_{\text{medium}}$, then the shift is along the direction of incident ray and if, $\mu_{\text{slab}} > \mu_{\text{medium}}$, then the shift will be opposite direction.

Lateral Shift Refraction of a ray AO incident on the slab at an angle of incidence i trought the glass slab $EFGH$ is as shown below

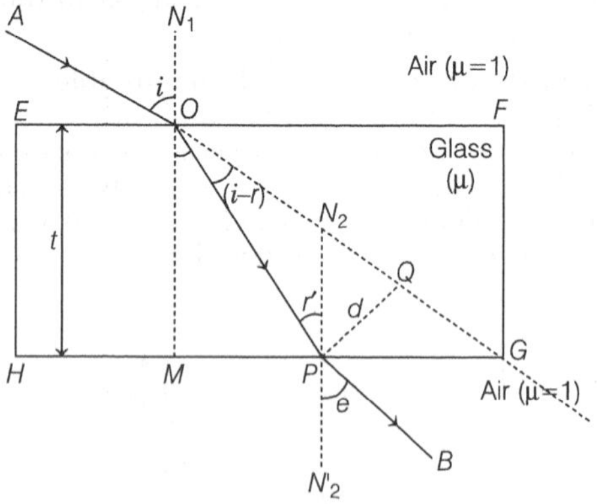

- If e is angle of emergence, then from Snell's law at faces EF and HG, $e = i$. *i.e.* the emergent ray is parallel to the incident ray.
- If PQ be the perpendicular dropped from P on incident ray produced, then the lateral displacement caused by the plate $= \dfrac{t\sin(i-r)}{\cos r}$.

Example 10. *A vessel has a concave mirror of focal length 30 cm placed at the bottom and it is filled with water of refractive index 4/3 upto 20 cm. The position of the image of the sun from the surface of water is*

(a) 7.5 cm above surface of water
(b) 7.5 cm below surface of water
(c) 15 cm above surface of water
(d) 15 cm below surface of water

Sol. (a) Apparent depth $= \dfrac{\text{Real depth}}{\mu} = \dfrac{20}{4/3} = \dfrac{20 \times 3}{4} = 15$ cm

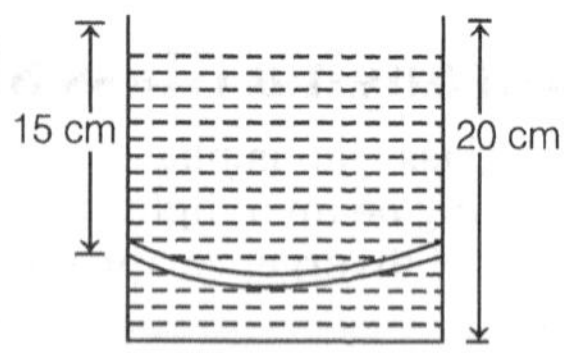

Shift in position $= 20 - 15 = 5$ cm

Apparent, $R = 60 - 5 = 55$ cm

Apparent, $f = 27.5$ cm

$\therefore$ Image of the sun is formed $(27.5 - 20 = 7.5$ cm) above the surface of water.

Example 11. *A vessel of depth 2h is half filled with a liquid of refractive index $2\sqrt{2}$ and the upper half with another liquid of refractive index $\sqrt{2}$. The liquids are immiscible. The apparent depth of the inner surface of the bottom of vessel will be* ***[JEE Main 2020]***

(a) $\dfrac{h}{\sqrt{2}}$ (b) $\dfrac{3}{4}h\sqrt{2}$ (c) $\dfrac{h}{3\sqrt{2}}$ (d) $\dfrac{h}{2(\sqrt{2}+1)}$

Sol. (b)

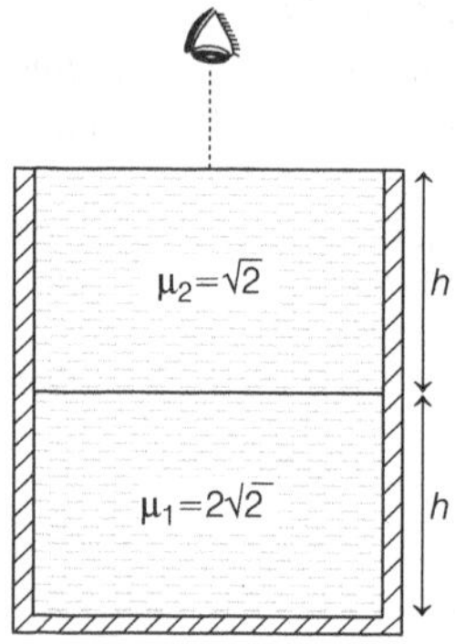

When a vessel contains immiscible liquids, then apparent depth will be

$$d_{\text{apparent}} = \frac{d_1}{\mu_1} + \frac{d_2}{\mu_2} + \dots$$

In given case,

$$d_{\text{apparent}} = \frac{h}{\mu_1} + \frac{h}{\mu_2} = \frac{h}{2\sqrt{2}} + \frac{h}{\sqrt{2}} = \frac{3h}{2\sqrt{2}}$$

or

$$d_{\text{apparent}} = \frac{3}{4}h\sqrt{2}$$

Example 12. *A fish in an aquarium, approaches the left wall at a rate of 3 ms^{-1} and observes a fly approaching it at 8 ms^{-1}. If the refractive index of water is 4/3, find the actual velocity of the fly.*

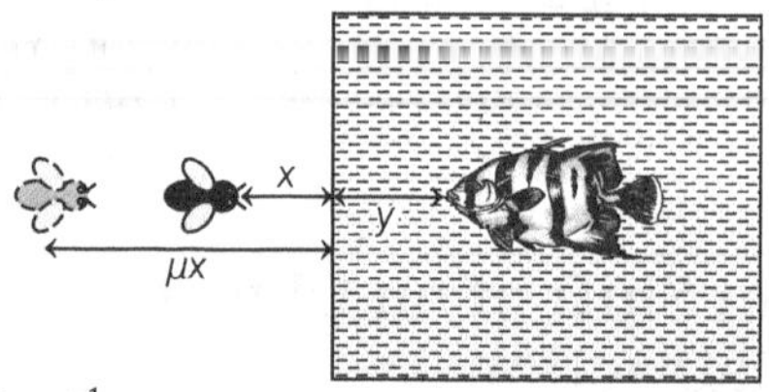

(a) 3.75 ms^{-1}
(b) 2.75 ms^{-1}
(c) 0.75 ms^{-1}
(d) 4.75 ms^{-1}

Sol. (a) For the fish, apparent distance of the fly from the wall of the aquarium is μx. If x is actual distance, then apparent velocity will be $= \dfrac{d(\mu x)}{dt}$.

$$(v_{\text{app}})_{\text{fly}} = \mu\, v_{\text{fly}}$$

Now, the fish observes the velocity of the fly to be 8 ms^{-1}. Therefore, apparent relative velocity = 8 ms^{-1}

$$v_{\text{fish}} + (v_{\text{app}})_{\text{fly}} = 8 \text{ ms}^{-1}$$

$$\Rightarrow \quad 3 + \mu v_{\text{fly}} = 8$$

$$v_{\text{fly}} = 5 \times \frac{3}{4} = 3.75 \text{ ms}^{-1}$$

Total Internal Reflection

The angle of incidence ($\angle i$) in denser medium for which the angle of refraction ($\angle r$) in rarer medium is 90° is called the critical angle ($\angle C$) for the given pair of media. *i.e.*

$$C = \sin^{-1}\left(\frac{\mu_r}{\mu_d}\right)$$

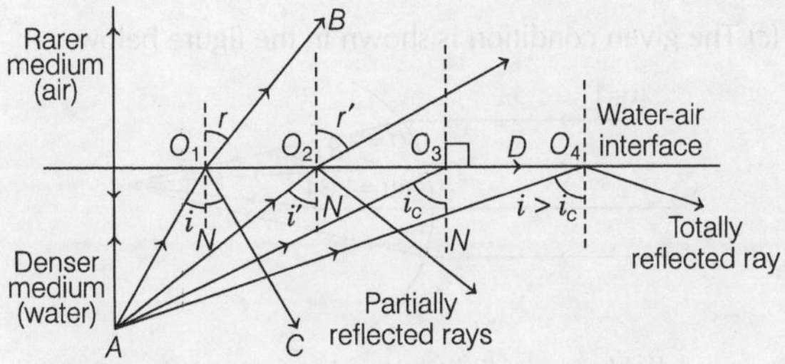

Refraction and internal reflection of rays from a point A in the denser medium (water) incident at different angles at the interface with a rarer medium (air)

Now, if the angle of incidence ($\angle i$) in the rarer medium is greater than the critical angle ($\angle C$), then the ray instead of suffering refraction is reflected back in the same (denser) medium. This phenomenon is called **total internal reflection**.

For total internal reflection to take place following set of conditions must be obeyed

(i) The ray must travel from denser medium to rarer medium.

(ii) The angle of incidence ($\angle i$) must be greater than critical angle ($\angle C$).

Applications of Total Internal Reflection

- **Mirage** It is the phenomena, in which an inverted image of distant tall objects cause an optical illusion of water. This type of mirage is especially common in hot deserts.
- **Diamond** The critical angle for diamond-air interface (≅ 24.4°) is very small, therefore once light enters a diamond, it is very likely to undergo total internal reflection inside it. Due to this, diamond shines brilliantly.
- **Prism** Few prisms are designed to bend light by 90° or by 180°, making use of total internal reflection. Such prisms are also used to invert images, without changing their size.
- **Optical Fibres** These fibres are fabricated with high quality composite glass/quartz. Each fibre consists of a core and cladding such that refractive index of the material of the core is higher than that of the material of cladding.

  When a signal in the form of light is directed at one end of the fibre at a suitable angle, it undergoes repeated total internal reflections along the length of the fibre and finally comes out of the other end.

  Thus, these are extensively used for transmitting audio and video signals through long distances.

Example 13. *An isotropic point source (bulb) is placed at a depth h below the water surface. A floating opaque disc is placed on the surface of water, so that the bulb is not visible from the surface. What is the minimum radius of the disc? (Take, refractive index of water = μ)*

(a) $\frac{2h}{\mu}$ (b) $\frac{h}{\sqrt{\mu-1}}$ (c) $\frac{h}{2\mu-1}$ (d) $\frac{\mu}{2h}$

Sol. (b) As shown in figure, light from bulb will not emerge out of the water, if at the edge of disc.

$$i > C$$

$$\sin i > \sin C \quad \text{...(i)}$$

Now, if R is the radius of disc and h is the depth of bulb from it

$$\sin i = \frac{R}{\sqrt{R^2+h^2}} \text{ and } \sin C = \frac{1}{\mu}$$

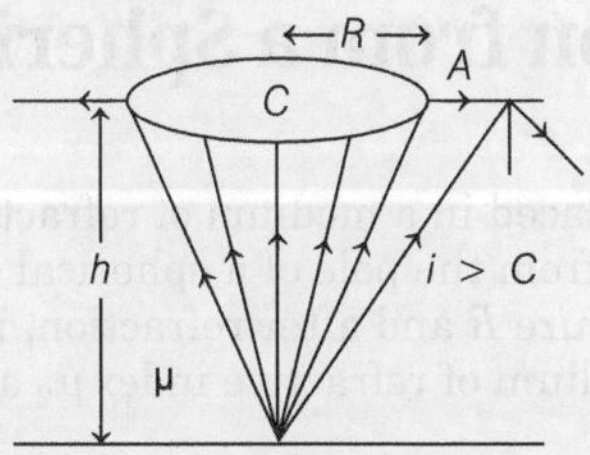

So, Eq. (i) becomes, $\frac{R}{\sqrt{R^2+h^2}} > \frac{1}{\mu}$ or $R > \frac{h}{\sqrt{\mu-1}}$

Example 14. *In figure, the optical fibre is l = 2 m long and has a diameter of d = 20 μm. If a ray of light is incident on one end of the fibre at angle θ_1 = 40°, the number of reflections it makes before emerging from the other end is close to (Take, refractive index of fibre is 1.31 and sin 40° = 0.64)* **[JEE Main 2019]**

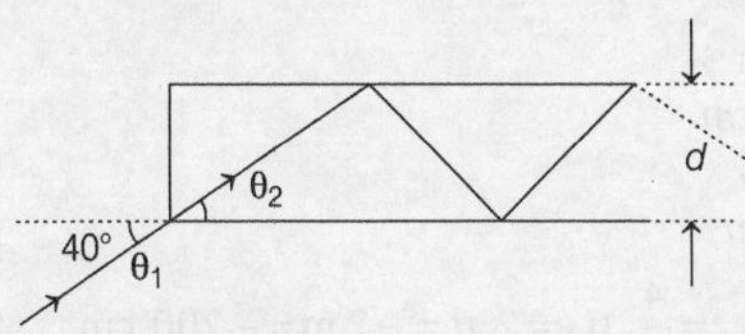

(a) 55000 (b) 66000 (c) 45000 (d) 57000

Sol. (d) Total internal reflection occurs through given glass rod as shown in figure

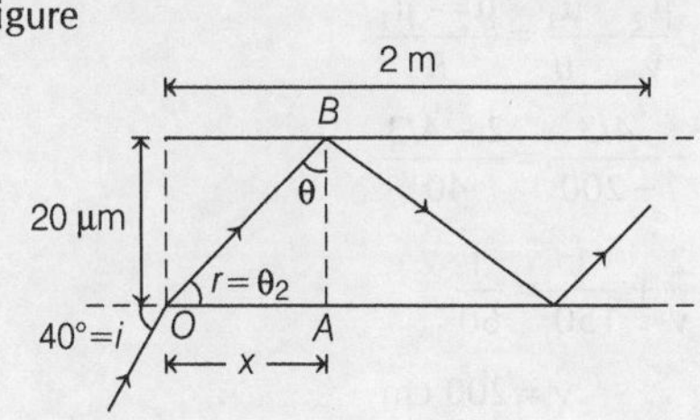

From Snell's law, $n_1 \sin i = n_2 \sin r$

where, $n_1 = 1, n_2 = 1.31$ and $i = 40°$.

So, we get

$$1 \sin 40° = 1.31 \sin r$$

$$\Rightarrow \quad \sin r = \frac{0.64}{1.31} = 0.49 \approx 0.5$$

So, $\quad r = 30°$

From ΔOAB, $\quad \theta = 90° - r = 60°$

Now, $\quad \tan \theta = \frac{x}{20 \ \mu m}$

$$\Rightarrow \quad x = 20\sqrt{3} \ \mu m \quad [\because \tan 60° = \sqrt{3}]$$

One reflection occurs in $20\sqrt{3}$ μm.

$\therefore$ Total number of reflections occurring in 2m,

$$n = \frac{2}{20\sqrt{3} \times 10^{-6}}$$

$$= 57735 \text{ reflections}$$

$$\approx 57000 \text{ reflections}$$

Refraction from a Spherical Surface

If an object is placed in a medium of refractive index μ_1, at a distance u from the pole of a spherical surface of radius of curvature R and after refraction, its image is formed in a medium of refractive index μ_2 at a distance v, then

$$\frac{\mu_2}{v} - \frac{\mu_1}{u} = \frac{\mu_2 - \mu_1}{R}$$

Note This equation holds for any curved spherical surface.

Example 15. *A convex refracting surface of radius of curvature 40 cm separates two media of refractive indices 4/3 and 2. An object is placed in the first medium* $\left(\mu = \frac{4}{3}\right)$ *at a distance of 2 m from the refracting surface. The position of image is*

(a) 100 cm
(b) – 150 cm
(c) 200 cm
(d) – 80 cm

Sol. (c) Here, $\mu_1 = \frac{4}{3}, \mu_2 = 2, u = -2 \text{ m} = -200$ cm

and $\quad R = +40$ cm

As, $\quad \frac{\mu_2}{v} - \frac{\mu_1}{u} = \frac{\mu_2 - \mu_1}{R}$

$$\Rightarrow \quad \frac{2}{v} - \frac{4/3}{-200} = \frac{2 - 4/3}{40}$$

$$\Rightarrow \quad \frac{2}{v} + \frac{1}{150} = \frac{1}{60}$$

$$\Rightarrow \quad v = 200 \text{ cm}$$

Example 16. *The eye can be regarded as a single refracting surface. The radius of curvature of this surface is equal to that of cornea (7.8 mm) and this surface separates two media of refractive indices 1 and 1.34. Calculate the distance from the refracting surface at which a parallel beam of light will come to focus.* ***[JEE Main 2019]***

(a) 4.0 cm (b) 2 cm
(c) 3.1 cm (d) 1 cm

Sol. (c) The given condition is shown in the figure below

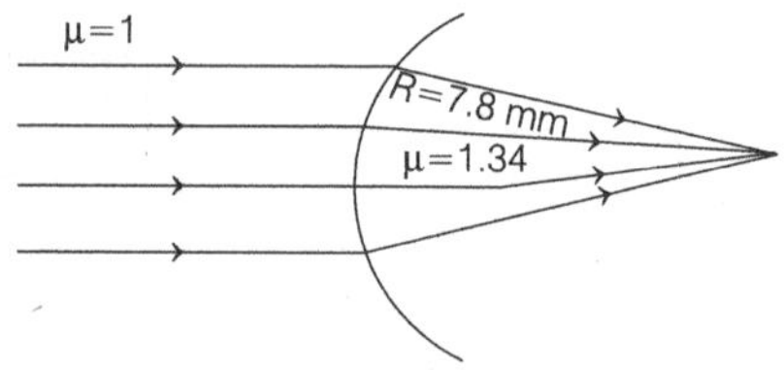

where, a parallel beam of light is coming from air ($\mu = 1$) to a spherical surface (eye) of refractive index 1.34.

Radius of curvature of this surface = 7.8 mm

From the image formation formula for spherical surface, *i.e.* relation between object, image and radius of curvature

$$\frac{\mu_R}{v} - \frac{\mu_I}{u} = \frac{\mu_R - \mu_I}{R} \quad \dots (i)$$

Given, $\mu_R = 1.34, \mu_I = 1, u = \infty$ (– ve) and $R = 7.8$

Substituting the given values, we get

$$\frac{1.34}{v} + \frac{1}{\infty} = \frac{1.34 - 1}{7.8}$$

or $\quad \frac{1.34}{v} = \frac{0.34}{7.8}$

$$\Rightarrow \quad v = \frac{1.34 \times 7.8}{0.34} \text{ mm}$$

$$\Rightarrow \quad v = \frac{4}{3} \times 3 \times 7.8 \text{ mm}$$

$(\because$ approximately $1.34 = 4/3$ and $0.34 = 1/3)$

$$\Rightarrow \quad v = 31.2 \text{ mm}$$

$$= 3.12 \text{ cm} \simeq 3.1 \text{ cm}$$

Lens

It is a transparent medium bounded by two curved surfaces at least one of which should be spherical.

Lenses are of two types namely; Convex or convergent lens and Concave or divergent lens.

Some of the terms related to lenses are as follows

Optical centre is a point within or outside the lens, at which incident rays refract without deviation in its path.

Principal axis is the straight line joining the centres of curvature of the two surfaces of the lens.

Principal focus Lens has two principal foci

(i) **First principal focus** It is a point on the principal axis of lens, the rays starting from which (convex lens) or appear to converge at which (concave lens) become parallel to principal axis after refraction.

(ii) **Second principal focus** It is the point on the principal axis at which the rays coming parallel to the principal axis converge (convex lens) or appear to diverge (concave lens) after refraction from the lens.

Both the foci of convex lens are real, while that of concave lens are virtual.

Focal length is the distance between focus and optical centre of lens.

Rules for Image Formation in Lens

- The rays coming parallel to principal axis of lens passes through the focus or appear to diverge after refraction.
- The rays coming through the first focus of lens go parallel to the principal axis of lens after refraction.
- The rays of light passing through optical centre go straight after refraction without changing their path.

Image Formation by a Convex Lens

Position of Object	Ray Diagram	Position of Image	Nature and Size of Image
At infinity	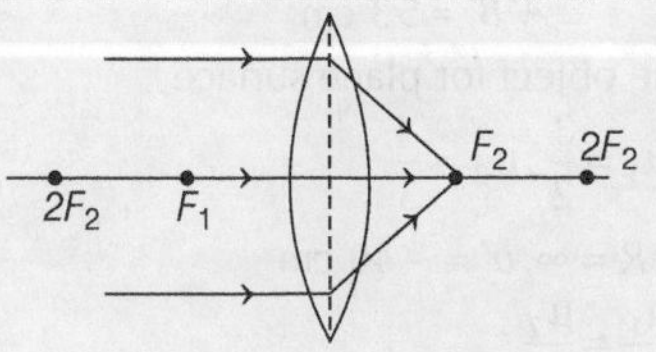	At the principal focus F_2 or in the focal plane	Real, inverted and extremely diminished
Beyond $2F_1$	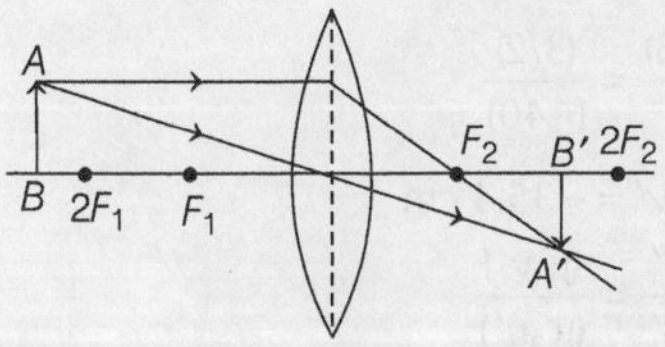	Between F_2 and $2F_2$	Real, inverted and diminished
At $2F_1$	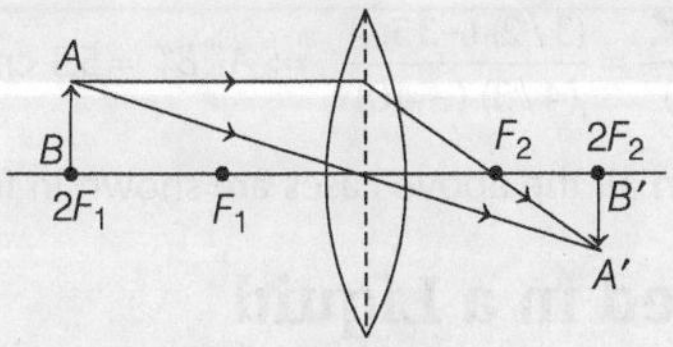	At $2F_2$	Real, inverted and of same size as the object
Between F_1 and $2F_1$	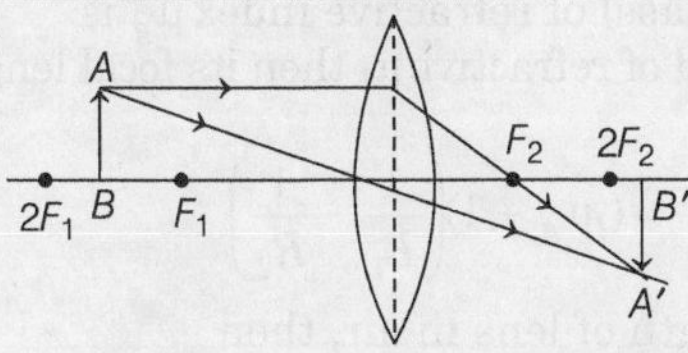	Beyond $2F_2$	Real, inverted and magnified
At F_1	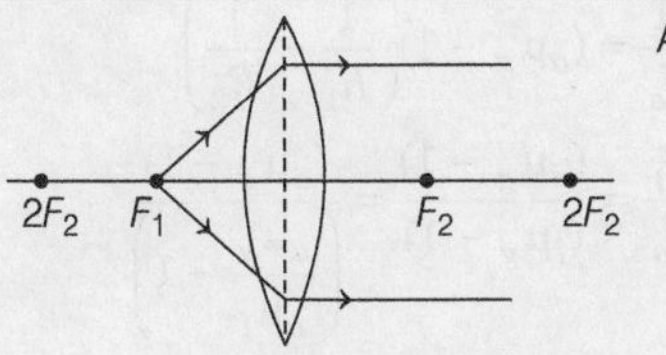	At infinity	Real, inverted and highly magnified
Between F_1 and optical centre	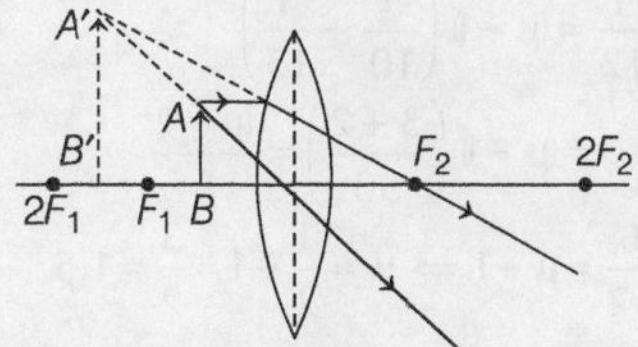	On the same side as the object	Virtual, erect and magnified

Image Formation by Concave Lens

Position of Object	Ray Diagram	Position of Image	Nature and Size of Image
At infinity	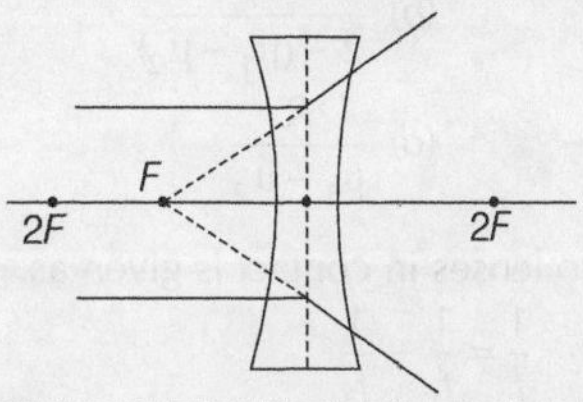	At the focus	Virtual erect and point size
Anywhere on the principal axis	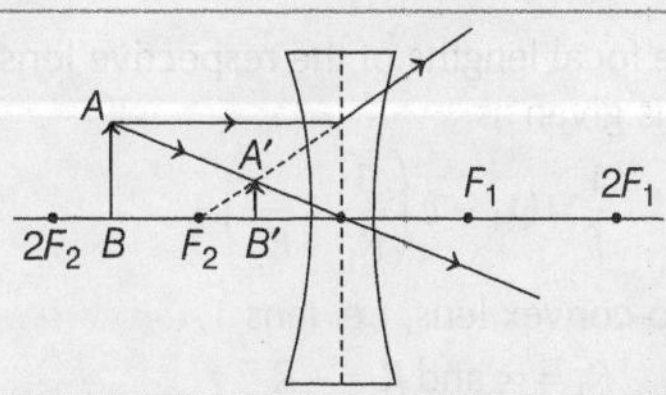	Between optical centre and F_2	Virtual, erect, diminished

Lens Maker's Formula

If R_1 and R_2 are the radii of curvature of first and second refracting surfaces of a thin lens of focal length f and refractive index μ (with respect to surrounding medium), the relation between f, μ, R_1 and R_2 is known as lens maker's formula and is given by,

$$\frac{1}{f} = (\mu - 1)\left(\frac{1}{R_1} - \frac{1}{R_2}\right)$$

If the lens is thin, then lens maker's formula is given as

$$\frac{1}{v} - \frac{1}{u} = \frac{1}{f}$$

This is usually known as **thin lens formula.**

Example 17. *The radii of curvature of the faces of a double convex lens are 10 cm and 15 cm and its focal length is 12 cm. The refractive index of glass lens is*

(a) 1.33 (b) 1.05
(c) 1.5 (d) 1.0

Sol. (c) Given, $f = +12\text{cm}, R_1 = +10\text{cm}, R_2 = -15\text{cm}$.

Refractive index of air is taken as unity.

From the Lens maker's formula, we have

$$\frac{1}{f} = (\mu - 1)\left(\frac{1}{R_1} - \frac{1}{R_2}\right)$$

Putting the values in above expression, we get

$$\frac{1}{12} = (\mu - 1)\left(\frac{1}{10} - \frac{1}{-15}\right)$$

$$\Rightarrow \quad \frac{1}{12} = (\mu - 1)\left(\frac{1}{10} + \frac{1}{15}\right)$$

$$= (\mu - 1)\left(\frac{3+2}{30}\right) = \frac{(\mu - 1)}{6}$$

$$\Rightarrow \quad \frac{6}{12} = \mu - 1 \Rightarrow \mu = \frac{1}{2} + 1 = \frac{3}{2} = 1.5$$

Example 18. *One plano-convex and one plano-concave lens of same radius of curvature R but of different materials are joined side-by-side as shown in the figure. If the refractive index of the material of 1 is μ_1 and that of 2 is μ_2, then the focal length of the combination is* **[JEE Main 2019]**

(a) $\frac{2R}{\mu_1 - \mu_2}$ (b) $\frac{R}{2 - (\mu_1 - \mu_2)}$

(c) $\frac{R}{2(\mu_1 - \mu_2)}$ (d) $\frac{R}{\mu_1 - \mu_2}$

Sol. (d) Focal length of two lenses in contact is given as

$$\frac{1}{f} = \frac{1}{f_1} + \frac{1}{f_2}$$

where, f_1 and f_2 are the focal lengths of the respective lenses. Focal length of a lens is given as

$$\frac{1}{f} = (\mu_1 - 1)\left(\frac{1}{R_1} - \frac{1}{R_2}\right)$$

$\therefore$ Focal length of plano-convex lens, *i.e.* lens 1,

$$R_1 = \infty \text{ and } R = -R$$

$$\Rightarrow \quad \frac{1}{f_1} = \frac{\mu_1 - 1}{R}$$

$$\text{or} \quad f_1 = \frac{R}{(\mu_1 - 1)} \quad \text{...(i)}$$

Similarly, focal length of plano-concave lens, *i.e.* lens 2,

$$R_1 = -R \text{ and } R_2 = \infty$$

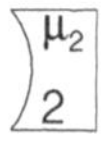

$$\Rightarrow \quad \frac{1}{f_2} = -\frac{(\mu_2 - 1)}{R}$$

$$\text{or} \quad f_2 = \frac{-R}{(\mu_2 - 1)} \quad \text{... (ii)}$$

From Eqs. (i) and (ii), we get

$$\frac{1}{f} = \frac{\mu_1 - 1}{R} - \frac{\mu_2 - 1}{R} = \frac{\mu_1 - \mu_2}{R} \Rightarrow f = \frac{R}{\mu_1 - \mu_2}$$

Example 19. *A linear object of length 4 cm is placed at 30 cm from the plane surface of hemispherical glass of radius 10 cm. The hemispherical glass is surrounded by water. The final position of image and its size respectively, are*

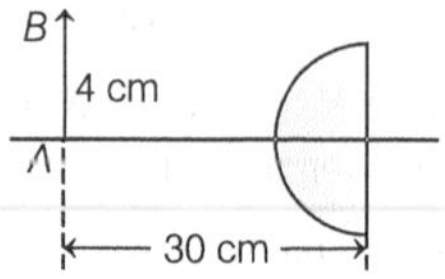

(a) – 30 cm, 5.3 cm (b) – 20 cm, 4.3 cm

(c) –15 cm, 5.3 cm (d) –10 cm, 3.6 cm

Sol. (a) For 1st surface,

$$\mu_1 = \frac{4}{3}, \mu_2 = \frac{3}{2}, u = -20 \text{ cm}$$

and $R = +10$ cm

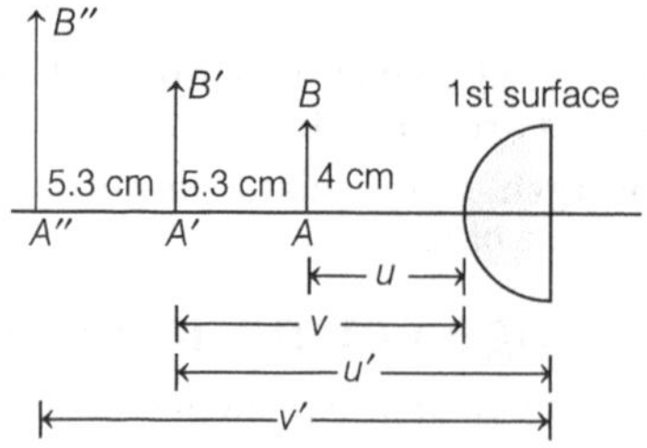

Using, $$\frac{\mu_2}{v} - \frac{\mu_1}{u} = \frac{(\mu_2 - \mu_1)}{R}$$

$$\Rightarrow \quad \frac{(3/2)}{v} - \frac{(4/3)}{(-20)} = \frac{(3/2 - 4/3)}{10}$$

$$\Rightarrow \quad v = -30 \text{ cm}$$

Using, $$\frac{A'B'}{AB} = \frac{\mu_1 v}{\mu_2 u}$$

$$\Rightarrow \quad \frac{A'B'}{(4\text{cm})} = \frac{(4/3)(-30)}{(3/2)(-20)}$$

$$\Rightarrow \quad A'B' = 5.3 \text{ cm}$$

$A'B'$ behaves as the object for plane surface

$$\mu_1 = \frac{3}{2}, \mu_2 = \frac{4}{3}$$

and $R = \infty, u' = -40$ cm

$$\Rightarrow \quad \frac{\mu_1}{v'} = \frac{\mu_2}{u'}$$

$$\Rightarrow \quad \frac{(4/3)}{v'} = \frac{(3/2)}{(-40)}$$

Solving it, we get $v' = -35.4$ cm

Now, using, $$\frac{A''B''}{A'B'} = \frac{(\mu_1 v')}{(\mu_2 u')}$$

$$\frac{A''B''}{(5.3)} = \frac{(3/2)(-35.4)}{(4/3)(-40)} \Rightarrow A''B'' = 5.3 \text{ cm}$$

The final images in all the above cases are shown in figure.

Lens Immersed in a Liquid

If a lens (made of glass) of refractive index μ_g is immersed in a liquid of refractive μ_l then its focal length f_l is given by

$$\frac{1}{f_l} = ({}_l\mu_g - 1)\left(\frac{1}{R_1} - \frac{1}{R_2}\right)$$

If f_a is the focal length of lens in air, then

$$\frac{1}{f_a} = ({}_a\mu_g - 1)\left(\frac{1}{R_1} - \frac{1}{R_2}\right)$$

$$\Rightarrow \quad \frac{f_l}{f_a} = \frac{({}_a\mu_g - 1)}{({}_l\mu_g - 1)} = \frac{({}_a\mu_g - 1)}{\left(\frac{{}_a\mu_g}{{}_a\mu_l} - 1\right)}$$

There are three cases based on this

(i) If $\mu_g > \mu_l$, *i.e.* lens is immersed in a liquid whose refractive index is less than that of material of lens, then f_l and f_a are of same sign and $f_l > f_a$, *i.e.* the nature of lens remains unchanged, but its focal length increases.

(ii) If $\mu_g = \mu_l$, *i.e.* lens is immersed in a liquid whose refractive index is same as that of material of lens, then $f_l = \infty$. It means lens behaves as a plane glass plate and becomes invisible in the medium.

(iii) If $\mu_g < \mu_l$, *i.e.* lens is immersed in a liquid whose refractive is greater than that of material of lens, then f_l and f_a have opposite signs and the nature of lens changes. *i.e.* A convex lens diverges the light rays and concave lens converges the light rays *i.e.* convex lens behaves like a concave lens and concave lens behaves like a convex lens.

Note **Newton's formula**

(i) In case of thin convex lens, if an object is placed at a distance x_1, from first focus and its image is formed at distance x_2, from the second focus from $x_1 x_2 = f^2$.

(ii) If some portion of a lens is covered with black paper, full image will be formed by brightness will be reduced.

Example 20. *A thin lens made of glass (refractive index = 1.5) of focal length f = 16 cm in immersed in a liquid of refractive index 1.42. If its focal length in liquid is f_1, then the ratio f_1 / f is closest to the integer* **[JEE Main 2020]**

(a) 1 (b) 5
(c) 9 (d) 17

Sol. (c) Focal length of lens of refractive index n_2 placed in a medium of refractive index n_1 is

$$\frac{1}{f} = \left(\frac{n_2}{n_1} - 1\right)\left(\frac{1}{R_1} - \frac{1}{R_2}\right)$$

For a glass-lens in air, focal length,

$$\frac{1}{f_{\text{air}}} = \frac{1}{f} = \left(\frac{n_g}{n_a} - 1\right)\left(\frac{1}{R_1} - \frac{1}{R_2}\right) \quad \dots \text{(i)}$$

When lens is dipped in a liquid, its focal length,

$$\frac{1}{f_{\text{liquid}}} = \frac{1}{f_1} = \left(\frac{n_g}{n_l} - 1\right)\left(\frac{1}{R_1} - \frac{1}{R_2}\right) \quad \dots \text{(ii)}$$

From Eqs. (i) and (ii), we have

$$\frac{f_1}{f} = \frac{\left(\frac{n_g}{n_a} - 1\right)}{\left(\frac{n_g}{n_l} - 1\right)}$$

As, $f = 16$ cm, $n_g = 1.5$ and $n_l = 1.42$, we have

$$\frac{f_1}{f} = \frac{\left(\frac{1.5}{1} - 1\right)}{\left(\frac{1.5}{1.42} - 1\right)}$$

$$\Rightarrow \quad \frac{f_1}{f} = \frac{\left(\frac{3}{2} - 1\right)}{\left(\frac{150}{142} - 1\right)} = \frac{\left(\frac{1}{2}\right)}{\left(\frac{4}{71}\right)} = \frac{71}{8} \approx 8.9$$

So, $\frac{f_1}{f}$ is closest to integer 9.

Magnification Produced by a Lens

It is defined as, the ratio of the size of the image to that of the object. Mathematically,

$$m = \frac{\text{size of image}}{\text{size of object}} = \frac{I}{O} = \frac{v}{u}$$

or

$$m = \frac{f}{f + u} = \frac{f - u}{f}$$

For erect (and virtual image), m is positive and for an inverted (and real) image, m is negative.

Power of a Lens

It is the measure of the convergence or divergence which a lens introduces in the light falling on it. Mathematically,

$$P = \frac{1}{\text{Focal length (in m)}} \quad \text{[Dioptre]}$$

or

$$P = \frac{100}{f\text{ (in cm)}}$$

It is also defined as, the tangent of the angle by which it converges or diverges a beam of light falling at unit distance from the optical centre.

i.e. $$\tan \delta = \frac{h}{f}$$

Note Power of lens is positive for a converging lens and negative for a diverging lens.

Example 21. *A biconvex lens has a radius of curvature of magnitude 20 cm, which one of the following options describe best the image formed of an object height 2 cm placed 30 cm from the lens? (Take, refractive index of lens μ = 3/2)*

(a) *Real, inverted, height = 4 cm*
(b) *Virtual, upright, height = 1 cm*
(c) *Real, inverted, height = 1 cm*
(d) *Virtual, inverted, height = 0.3 cm*

Sol. (a) Given, $R_1 = +20$ cm, $R_2 = -20$ cm and $\mu = \frac{3}{2}$

Using lens Maker formula,

$$\frac{1}{f} = (\mu - 1)\left(\frac{1}{R_1} - \frac{1}{R_2}\right) = \left(\frac{3}{2} - 1\right)\left(\frac{1}{20} - \frac{1}{20}\right) = \frac{1}{20}$$

$$\Rightarrow \quad f = 20 \text{ cm}$$

Here, $u = -30$ cm, $f = 20$ cm, $v = ?$

Using lens formula, $\frac{1}{f} = \frac{1}{v} - \frac{1}{u}$ or $\frac{1}{20} = \frac{1}{v} - \frac{1}{-30}$

$$\Rightarrow \quad \frac{1}{v} = \frac{1}{60} \quad \text{or} \quad v = 60 \text{ cm}$$

Magnification, $m = \frac{v}{u} = \frac{h_i}{h_o}$

or $\frac{60}{-30} = \frac{h_i}{2}$

$\Rightarrow$ $h_i = -4$ cm

It means the image is real, inverted and of height $(h) = 4$ cm.

Example 22. *A convex lens is made of glass of refractive index 1.5. If radius of curvature of each of its two surfaces is 20 cm, find the ratio of the power of the lens, when placed in air to its power, when immersed inside a liquid of refractive index 1.25.*

(a) $\frac{2}{5}$ (b) $\frac{5}{2}$ (c) $\frac{4}{3}$ (d) $\frac{4}{9}$

Sol. (b) Given, $R_1 = 20$ cm $= 0.20$ m; $R_2 = -20$ cm $= -0.2$ m

When lens is placed in air If P_1 is power of the lens when placed in air, then

$$P_1 = (\mu - 1)\left(\frac{1}{R_1} + \frac{1}{R_2}\right) = (1.5 - 1)\left(\frac{1}{0.2} - \frac{1}{-0.2}\right)$$
$$= 0.5(5 + 5) = 5 \text{ D}$$

When lens is placed inside the liquid Let μ' be the refractive index of the material of the lens w.r.t. the liquid, then power of the lens, when placed inside the liquid is given by

$$P_2 = (\mu' - 1)\left(\frac{2}{R_1} - \frac{1}{R_2}\right)$$

Here, $\mu' = \frac{\mu_{\text{glass}}}{\mu_{\text{liquid}}} = \frac{1.5}{1.25} = 1.2$

$\therefore$ $P_2 = (1.2 - 1)\left(\frac{1}{0.2} - \frac{1}{-0.2}\right) = 0.2\ (5 + 5) = 2$ D

Hence, $\frac{P_1}{P_2} = \frac{5}{2}$

Displacement Method to Determine the Focal Length of a Convex Lens

The focal length of convex lens can be determined by displacement method. If an object and a screen placed at a distance $D (> 4f)$ apart, for two different position of lens two images (I_1 and I_2) of an object are formed at the screen.

Let distance between the two positions of the lens is x, then the focal length of the lens is given by

$$f = \frac{D^2 - x^2}{4D} = \frac{x}{m_1 - m_2}.$$

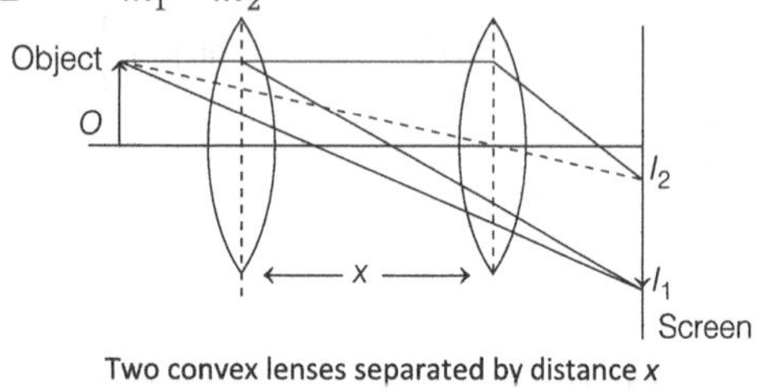

Two convex lenses separated by distance x

where, $m_1 = \frac{I_1}{O}$, $m_2 = \frac{I_2}{O}$ and $m_1 m_2 = 1$.

Also, size of object, $O = \sqrt{I_1 I_2}$.

Combination of Thin lenses

If two or more lenses are placed in contact, then equivalent focal length of the combination,

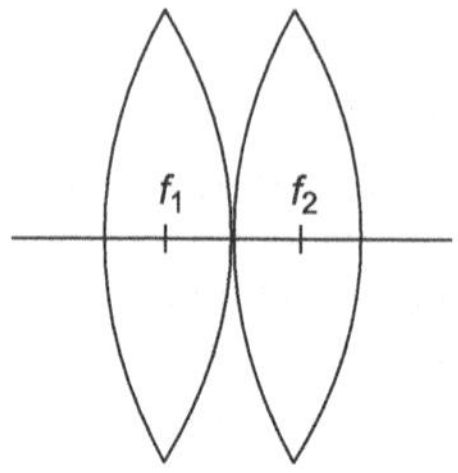

$$\frac{1}{f} = \frac{1}{f_1} + \frac{1}{f_2} + \ldots = \sum_{i=1}^{n} \frac{1}{f_i}$$

Power of combination, $P = P_1 + P_2 + \ldots = \sum_{i=1}^{n} P_i$

Magnification of combination,

$$M = m_1 \times m_2 \times \ldots = \sum_{i=1}^{n} m$$

Similarly, if two lenses of focal lengths f_1 and f_2 are separated by a distance x, then its equivalent focal length,

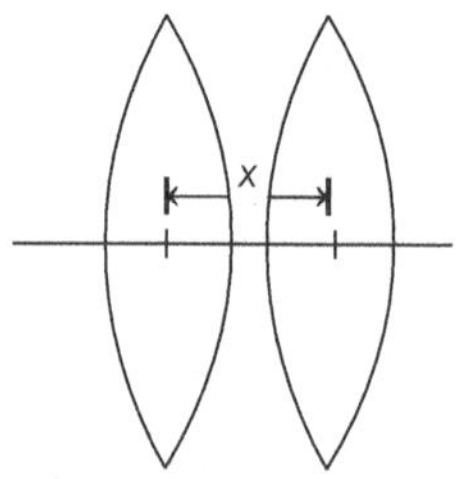

$$\frac{1}{F} = \frac{1}{f_1} + \frac{1}{f_2} - \frac{x}{f_1 f_2}$$

Power of combination, $P = P_1 + P_2 - x P_1 P_2$

Total magnification remains unchanged, *i.e.*

$$m = m_1 \times m_2$$

Note If a lens is made of a number of layers of different refractive indices, then number of images of an object formed by the lens is equal to number of different media.

Example 23. *An object is placed at a distance of 10 cm to the left on the axis of a convex lens L_1 of focal length 20 cm. A second convex lens L_2 of focal length 10 cm is placed co-axially to the right of the lens L_1 at a distance of 5 cm from it. Find the position of the final image and its magnification.*

(a) $16\frac{2}{3}$ *cm on the right of the second lens, 3.33*

(b) $16\frac{2}{3}$ *cm on the right of the second lens, 1.33*

(c) $16\frac{2}{3}$ *cm on the right of the first lens, 1.33*

(d) *None of the above*

Sol. (b) Here, **for 1st lens** $u_1 = -10\,\text{cm}$

and $f_1 = 20$ cm

$$\frac{1}{v_1} - \frac{1}{u_1} = \frac{1}{f_1}$$

$$\Rightarrow \quad \frac{1}{v_1} = \frac{1}{20} - \frac{1}{10}$$

$$\Rightarrow \quad v_1 = -20\,\text{cm}$$

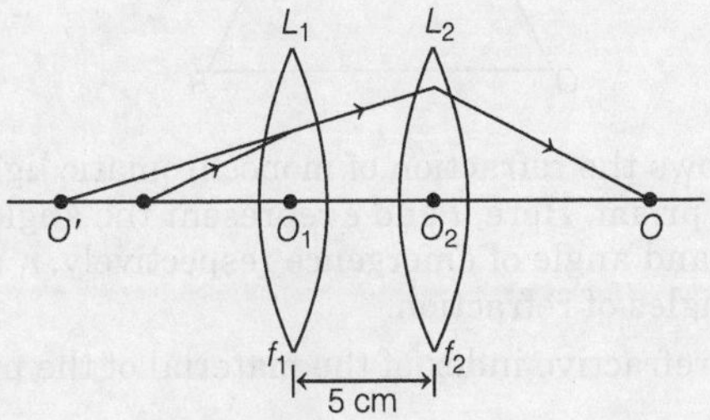

i.e. the image is virtual and hence lies on the same side of the object. This will behave as an object for the second lens.

For 2nd lens, $\frac{1}{v_2} - \frac{1}{u_2} = \frac{1}{f_2}$

Here, $u_2 = -(20+5)$, $f_2 = 10$ cm

$$\frac{1}{v_2} + \frac{1}{25} = \frac{1}{10}$$

$$\Rightarrow \quad v_2 = \frac{50}{3} = 16\frac{2}{3}\text{ cm}$$

i.e. Final image is at a distance of $16\frac{2}{3}$ cm on the right of the second lens.

The magnification of the image is given by

$$m = \frac{v_1}{u_1}\frac{v_2}{u_2}$$

$$= \frac{20}{10}\frac{50}{3\times 25}$$

$$= \frac{4}{5} = 1.33$$

Cutting of a Lens

- If a symmetrical convex lens of focal length f is cut into two parts along its optical axis, then focal length of each part (a plano-convex lens) is $2f$. However, if the two parts are joined as shown in figure, the focal length of combination is again f.

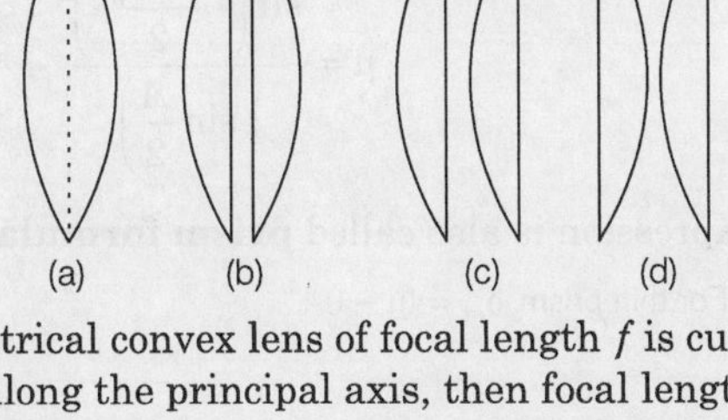

(a) (b) (c) (d)

- If a symmetrical convex lens of focal length f is cut into two parts along the principal axis, then focal length of each part remains changed at f. If these two parts are joined with curved ends on one side, focal length of the combination is $\frac{f}{2}$. But on joining two parts in opposite sense, the net focal length becomes ∞ (or net power = 0).

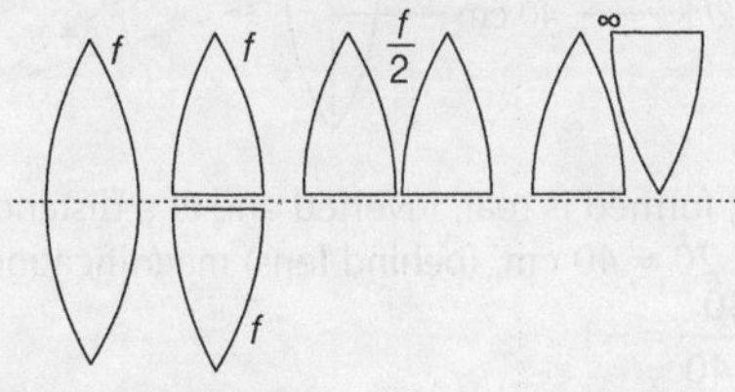

(a) (b) (c) (d)

Silvering of a Lens

- Let a plano-convex lens is having a curved surface of radius of curvature R and has refractive index μ. If its plane surface is silvered, it behaves as a concave mirror of focal length,

$$f = \frac{R}{2(\mu-1)}$$

- If the curved surface of plano-convex lens is silvered, then it behaves as a concave mirror of focal length,

$$f = \frac{R}{2\mu}$$

- If one surface of a symmetrical double convex lens is silvered, then the lens behaves as a concave mirror of focal length,

$$f = \frac{R}{2(2\mu-1)}$$

Combination of Lenses and Mirrors

- When several lenses or mirrors are combined coaxially, image by first serves as an object for second and so on. The net magnification is given by

$$m = m_1 \times m_2 \times m_3 \times \ldots$$

- In some cases, object and image are formed at same place after interacting the (lens + mirror) combination. For this, after refraction from lens, rays must retrace their path, *i.e.* must be incident normally on the mirror after refraction from the lens. In other word, we can say that image by lens should be formed at centre of curvature C of mirror.

Example 24. *An upright object is placed at a distance of 40 cm in front of a convergent lens of focal length 20 cm. A convergent mirror of focal length 10 cm is placed at a distance of 60 cm on the other side of the lens. The position and size of the final image will be* ***[JEE Main 2019]***

(a) 20 cm from the convergent mirror, same size as the object
(b) 40 cm from the convergent mirror, same size as the object
(c) 40 cm from the convergent lens, twice the size of the object
(d) 20 cm from the convergent mirror,twice size of the object

Sol. (*) In given system of lens and mirror, position of object O in front of lens is at a distance $2f$.
i.e. $u = 2f = 40$ cm

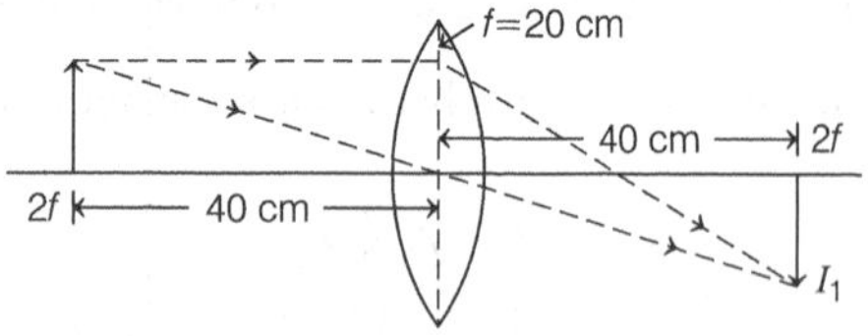

So, image I_1 formed is real, inverted and at a distance, $v = 2f = 2 \times 20 = 40$ cm, (behind lens) magnification,

$$m_1 = \frac{v}{u} = \frac{40}{-40} = -1$$

Thus, size of image is same as that of object.
This image I_1 acts like a real object for mirror.

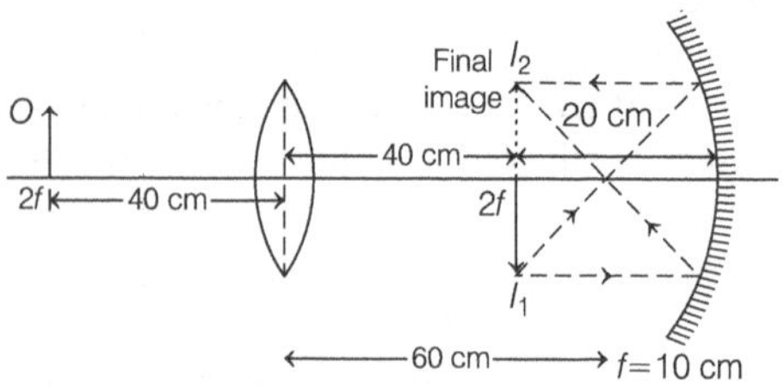

As, object distance for mirror is $u = C = 2f = -20$ cm
where, C = centre of curvature.
So, image I_2 formed by mirror is at $2f$.
$\therefore$ For mirror $v = 2f = 2(-10) = -20$ cm

$$\text{Magnification, } m_2 = -\frac{v}{u} = -\frac{(-20)}{(-20)} = -1$$

Thus, image size is same as that of object.
The image I_2 formed by the mirror will act like an object for lens.

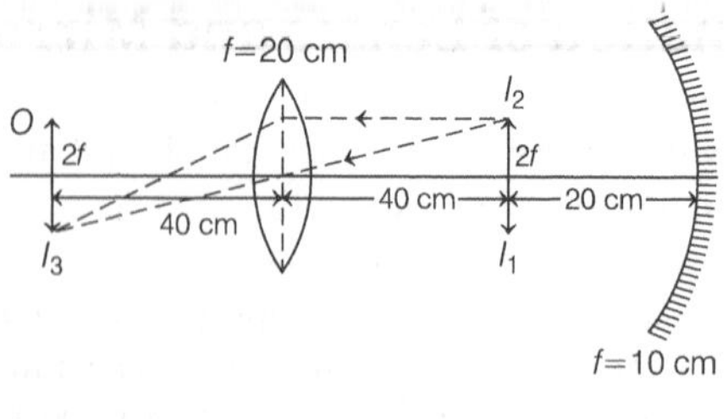

As the object is at $2f$ distance from lens, so image I_3 will be formed at a distance $2f$ or 40 cm. Thus, magnification,

$$m_3 = \frac{v}{u} = \frac{40}{-40} = -1$$

So, final magnification,

$$m = m_1 \times m_2 \times m_3 = -1$$

Hence, final image I_3 is real, inverted of same size as that of object and coinciding with object.

Refraction Through a Prism

A prism is a homogeneous, transparent medium bounded by two plane surfaces inclined at an angle A with each other.

These surfaces are called as **refracting surfaces** and the angle between them is called **angle of prism A.**

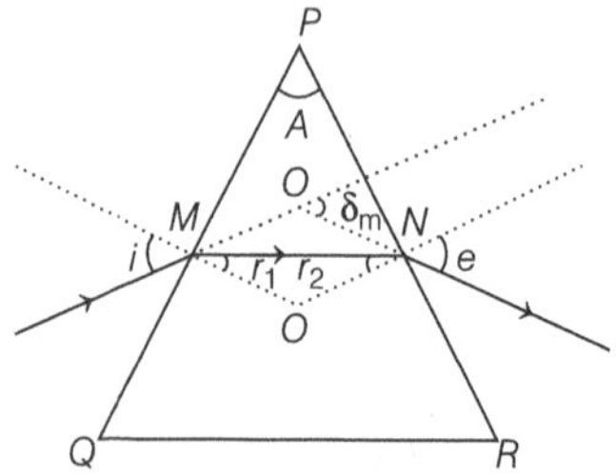

Figure shows the refraction of monochromatic light through a prism. Here, i and e represent the angle of incidence and angle of emergence respectively, r_1 and r_2 are two angles of refraction.

If μ is the refractive index of the material of the prism, then

$$\mu = \frac{\sin i}{\sin r_1} = \frac{\sin e}{\sin r_2}$$

The angle between the incident ray and the emergent ray is known as the **angle of deviation** δ.

For refraction through a prism, it is found that

$$i + e = A + \delta$$

and $$r_1 + r_2 = A$$

Minimum Deviation

The variation of angle of deviation δ with the angle of incidence i of the ray incident on the first refracting face of the prism is shown in figure

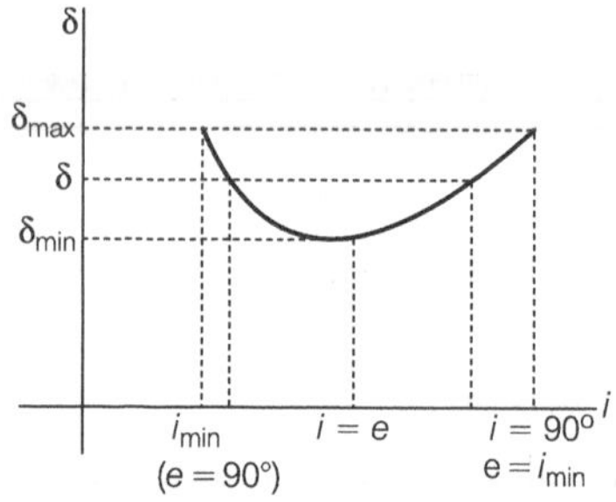

At minimum deviation δ_m, the refracted ray inside the prism becomes parallel to its base. At this value,

$$i = e$$

Further at δ_m, refractive index of prism is given as,

$$\mu = \frac{\sin\left(\frac{A + \delta_m}{2}\right)}{\left(\sin\frac{A}{2}\right)}$$

This expression is also called **prism formula**.

Note For thin prism, $\delta_m = (\mu - 1)A$.

Example 25. *A prism is made of glass of unknown refractive index. A parallel beam of light is incident on a face of the prism. The angle of minimum deviation is measured to be 40°. The refracting angle of the prism is 60°. If the prism is placed in water (refractive index 1.33), then new angle of minimum deviation of a parallel beam of light is*

(a) 20° (b) 10°20′ (c) 30° (d) 60°23′

Sol. (b) Using the relation, $\mu = \dfrac{\sin\left(\dfrac{A+\delta_m}{2}\right)}{\sin\left(\dfrac{A}{2}\right)}$

$$= \frac{\sin\left(\dfrac{60° + 40°}{2}\right)}{\sin\left(\dfrac{60°}{2}\right)} = \frac{\sin 50°}{\sin 30°}$$

$$= \frac{0.7660}{0.50} = 1.532$$

$$\approx 1.53$$

Let δ'_m be the new angle of minimum deviation of a parallel beam of light when the prism is placed in water ${}_a\mu_w = 1.33$, ${}_a\mu_g = 1.532$

$$\therefore \quad {}_w\mu_g = \frac{{}_a\mu_g}{{}_a\mu_w} = \frac{1.532}{1.33} = 1.152$$

$${}_w\mu_g = \frac{\sin\left(\dfrac{A+\delta'_m}{2}\right)}{\sin\left(\dfrac{A}{2}\right)}$$

or $$1.152 = \frac{\sin\left(\dfrac{60+\delta'_m}{2}\right)}{\sin\dfrac{60}{2}}$$

or $$\sin\left(30° + \frac{\delta'_m}{2}\right) = 1.152 \times \frac{1}{2} = 0.5759 = \sin 35.17°$$

or $$30° + \frac{\delta'_m}{2} = 35.17°$$

or $$\frac{\delta'_m}{2} = 35.17° - 30 = 5.17°$$

Angle of minimum deviation, $\delta'_m = 2 \times 5.17° = 10.34° = 10°20'$

Dispersion of Light

The phenomenon of splitting of light into different colours is known as dispersion. The pattern of colour components of light thus obtained is called the **spectrum of light.**

Dispersion takes place because the refractive index of medium for different frequencies (colours) is different. *e.g.* Bending of red light is least while it is most for the violet. Equivalently, red light travels faster than violet light in a glass prism.

Rainbow is an example of dispersion of sunlight by water drops in atmosphere. It is a phenomenon based on the combined effect of dispersion, refraction and reflection of sunlight by spherical water droplets of rain.

Dispersion of Light by a Prism

Dispersion of light is the phenomenon of splitting of white light into its constituent wavelengths on passing through a dispersive medium, *e.g.* prism. The reason for dispersion is the variation of refractive index of prism with wavelength. As, $\lambda_V > \lambda_R$, hence $\mu_V > \mu_R$ and consequently, $\delta_V > \delta_R$.

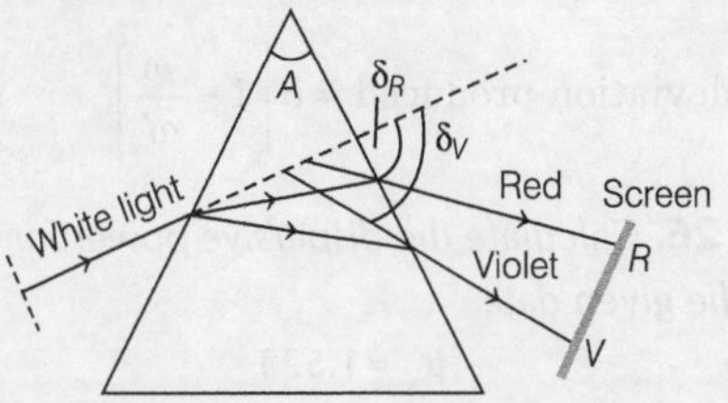

Angular Dispersion It is the angular separation between the two extreme rays.

Angular dispersion, $\theta = \delta_V - \delta_R$

$$= (\mu_V - \mu_R)A$$

Dispersive Power The dispersive power of a prism material is measured by the ratio of angular dispersion to the mean deviation suffered by light beam.

$$\therefore \text{ Dispersive power,} \quad \omega = \frac{\delta_V - \delta_R}{\delta} = \frac{\mu_V - \mu_R}{\mu - 1}$$

where, μ is the mean value of refractive index of prism.

The dispersive power of a prism depends only on its material and is independent of angle of prism and angle of incidence or size of the prism.

Dispersion without Deviation (Direct Vision Prism)

To produce dispersion without mean deviation, the combination of two prisms of different materials such that

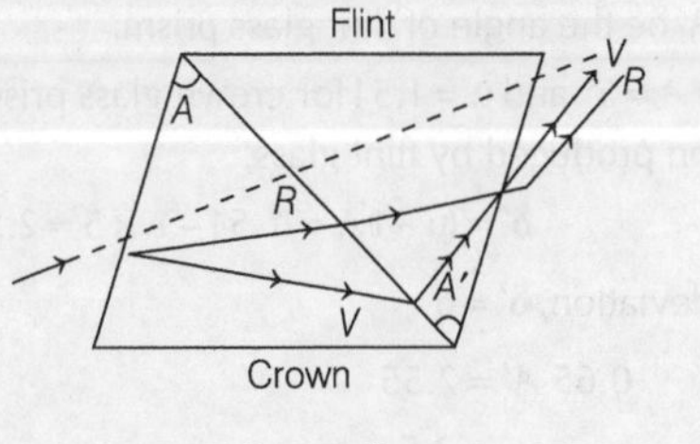

$$A' = \left(\frac{\mu - 1}{\mu' - 1}\right)A$$

$$\text{Net dispersion caused} = (\mu_V - \mu_R)\,A + (\mu'_V - \mu'_R)A'$$
$$= \delta\,(\omega - \omega')$$

Deviation without Dispersion (Achromatic Prism)

To produce deviation without dispersion, the combination of two prisms of different materials is as shown below such that $A' = \dfrac{[\mu_V - \mu_R]}{[\mu'_V - \mu'_R]}A$

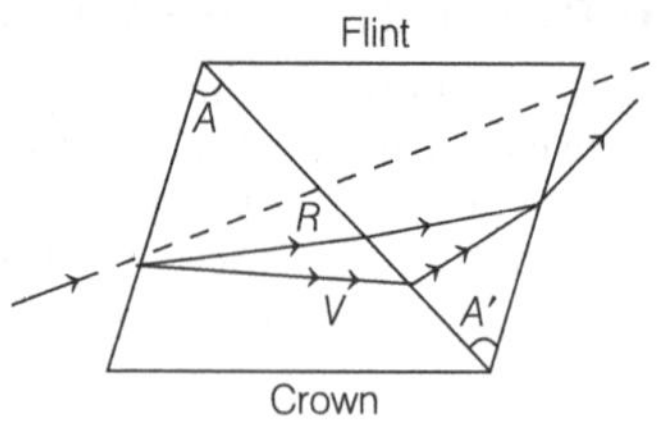

Resultant deviation produced $= \delta\left[1 - \dfrac{\omega}{\omega'}\right]$.

Example 26. *Calculate the dispersive power for crown glass from the given data*

$$\mu_V = 1.523$$

and $$\mu_R = 1.5145$$

(a) 0.01639 (b) 1.05639
(c) 0.05639 (d) 2.05639

Sol. (a) Here, $\mu_V = 1.523$ and $\mu_R = 1.145$

Mean refractive index, $\mu = \dfrac{1.523 + 1.5145}{2} = 1.51875$

Dispersive power is given by

$$\omega = \frac{(\mu_V - \mu_R)}{(\mu - 1)}$$

$$= \frac{1.523 - 1.5145}{(1.51875 - 1)} = 0.01639$$

Example 27. *A prism of crown glass with refracting angle of 5° and mean refractive index = 1.51 is combined with a flint glass prism of refractive index = 1.65 to produce no deviation. Find the angle of flint glass.*

(a) 3.92° (b) 4.68°
(c) 5.32° (d) 7.28°

Sol. (a) Let A' be the angle of flint glass prism.

Here, $A = 5°$ and $\mu = 1.51$ for crown glass prism.

Deviation produced by flint glass,

$$\delta' = (\mu - 1)A = (1.51 - 1) \times 5 = 2.55°$$

For no deviation, $\delta' = \delta$

or $$0.65\,A' = 2.55$$

$$A' = \frac{2.55}{0.65} = 3.92°$$

Defects of Images

Actual image formed by an optical system is usually imperfect, the defects of images are called **aberrations.**

There are of two types namely; chromatic aberration and monochromatic aberration.

(i) **Chromatic aberration** The image of an object formed by a lens is usually coloured and blurred. This defect of image is called chromatic aberration. This defect arises due to the fact that focal length of a lens is different for different colours.

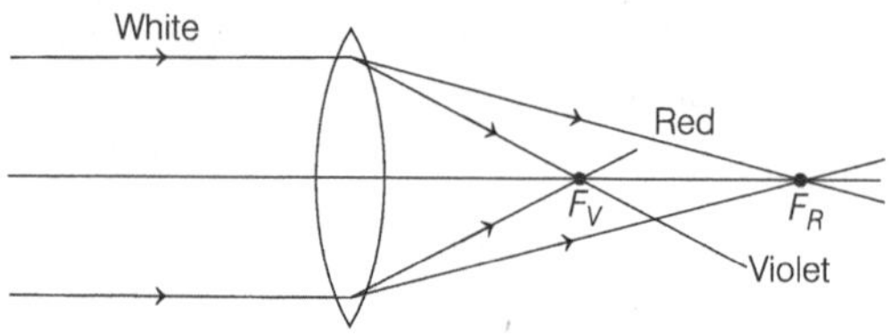

Chromatic aberration

The difference between f_R and f_V is a measure of longitudinal chromatic aberration (LCA), thus

$$\text{LCA} = f_R - f_V = \omega f_Y$$

where, $$\omega = \frac{\mu_V - \mu_R}{\mu_Y - 1}$$

Here, $$\mu_Y = \frac{\mu_V + \mu_R}{2}$$

To get achromatism, we use a pair of two lenses in contact. The combination will be free from chromatic aberration.

$$\therefore \quad \frac{\omega_1}{f_1} + \frac{\omega_2}{f_2} = 0$$

This is the **condition of achromatism.**

(ii) **Monochromatic aberration** This is the defect in image due to optical system. Monochromatic aberration is of many types such as; spherical, coma, distortion, curvature and antigmatism.

Spherical aberration arises due to spherical nature of lens (or mirror). It is the inability of the lens to form a point image of an axial point object.

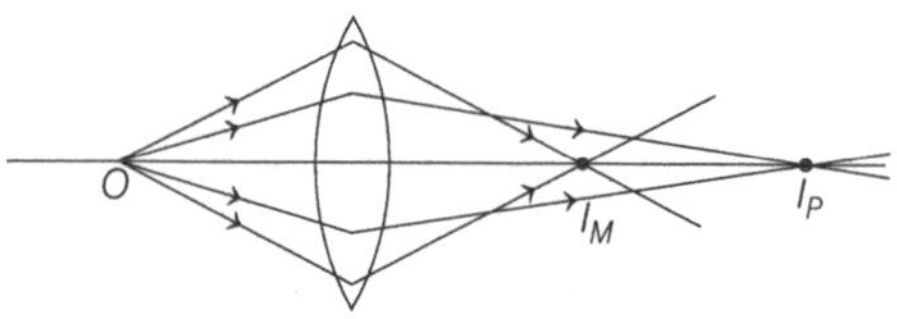

Spherical aberration

Scattering of Light

As sunlight travels through the earth's atmosphere, it gets scattered (changes its direction) by the atmospheric particles. Light of shorter wavelength is scattered much more than light of larger wavelengths.

Rayleigh Scattering

According to this, the amount of scattering is inversely proportional to the fourth power of the wavelength.

Some phenomena based on scattering are bluish colour that predominates in a clear sky, sun appears reddish to us, etc.

Optical Instruments

These are those instruments which are used to assist the eye in viewing an object, *e.g.* microscope, telescope, etc.

Microscope

It is an optical instrument which forms a magnified image of a small nearby object and thus increases the visual angle subtended by the image at the eye, so that the object is seen to be bigger and distinct.

Simple Microscope

It is a convex lens of short focal length which is fixed in a frame provided with handle.

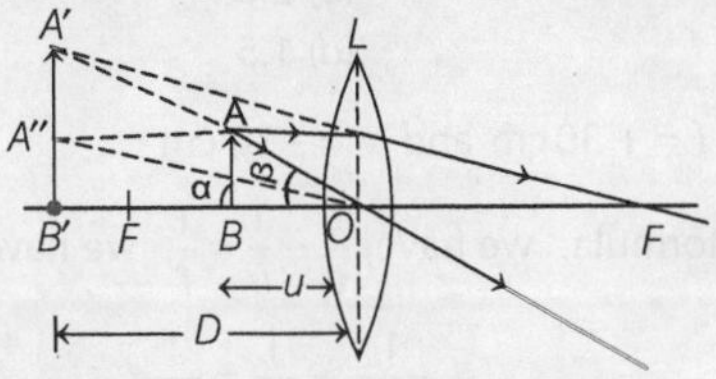

Magnification of simple microscope

(a) When final image is formed at least distance of distinct vision,

$$M = 1 + \frac{D}{f}$$

(b) For relaxed eye, $M = \dfrac{D}{f}$

where, D = least distance of distinct vision.

Compound Microscope

Figure shows a simplified version of a compound microscope. It consists of two converging lenses arranged coaxially. The one facing the object is called **objective** and the one close to eye is called **eyepiece**. The objective has a smaller aperture and smaller focal length than those of the eyepiece.

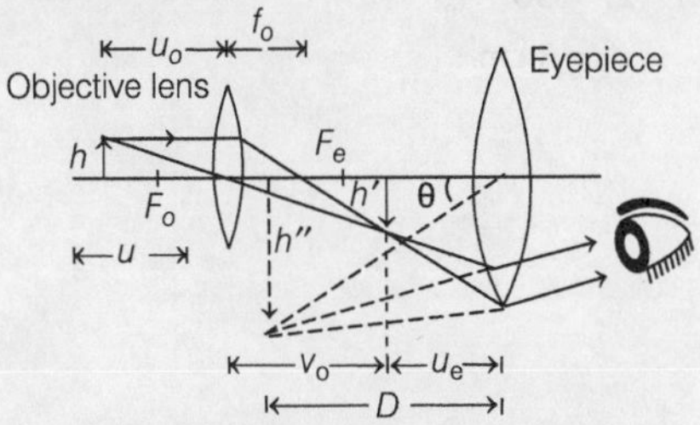

Magnification of compound microscope

(a) For relaxed eye, $M_\infty = -\dfrac{v_o}{u_o}\left(\dfrac{D}{f_e}\right)$

In this position, length of microscope,

$$L_\infty = v_o + f_e$$

(b) When final image is formed at least distance of distinct vision.

$$M_D = -\frac{v_o}{u_o}\left(1 + \frac{D}{f_e}\right)$$

Length of microscope,

$$L_D = v_o + u_e$$

where, v_o = distance of first image from objective lens,

u_o = distance of object from objective lens

and f_e = focal length of eyepiece.

Telescope

It is an optical instrument which increases the visual angle at the eye by forming the image of a distant object at the least distance of distinct vision, so that the object is seen distinct and bigger.

Astronomical Telescope

It consists of two converging lenses placed coaxially. The one facing the distant object is called the **objective** and has a large aperture and large focal length. The other is called the **eyepiece**, which is placed closed to eye. The eyepiece tube can slide within the objective tube, so that the separation between the objective and the eyepiece may be varied.

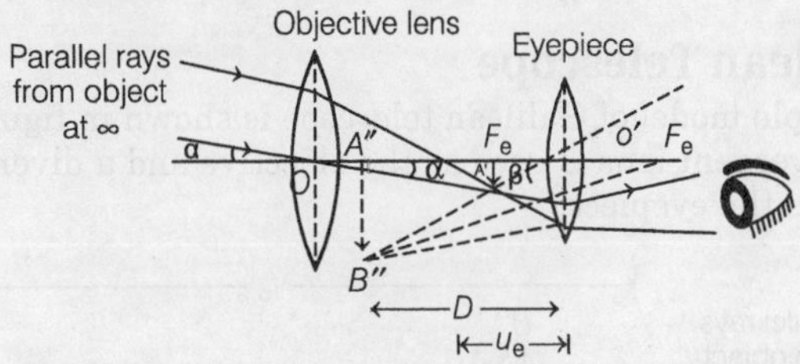

Magnification of astronomical telescope

(a) For relaxed eye, $M_\infty = -\dfrac{f_o}{f_e}$

In this position, length of telescope,

$$L_\infty = f_o + f_e$$

(b) When final image is formed at least distance of distinct vision

$$M_D = -\frac{f_o}{f_e}\left(1 + \frac{f_e}{D}\right)$$

Length of telescope,

$$L_D = f_o + u_e$$

where, f_o = focal length of objective lens

and f_e = focal length of eyepiece.

Terrestrial Telescope

In an astronomical telescope, the final image is inverted with respect to the object. To remove this difficulty, a convex lens of focal length f is included between the objective and the eyepiece in such a way that the focal plane of the objective is a distance $2f$ away from this lens.

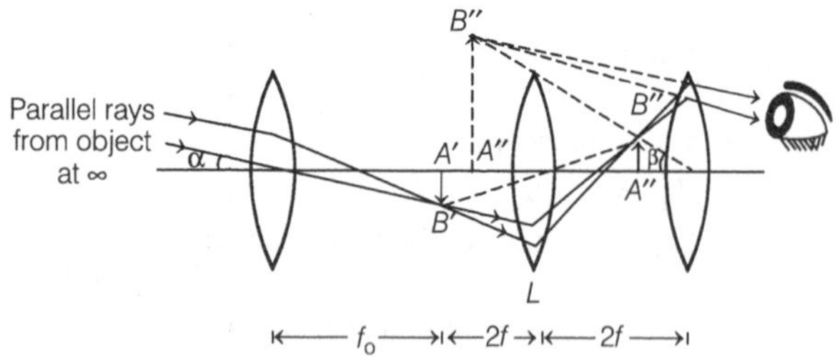

Magnification of terrestrial telescope,

(a) For relaxed eye, $M_\infty = f_o / f_e$

In this position, length of telescope,

$$L_\infty = f_o + 4f + f_e$$

(b) When final image is formed at least distance of distinct vision,

$$M_o = \frac{f_o}{f_e}\left(1 + \frac{f_e}{D}\right)$$

Length of telescope,

$$L_\infty = f_o + 4f + u_e$$

f_o = focal length of objective lens

and f_e = focal length of eyepiece.

Galilean Telescope

A simple model of Galilean telescope is shown in figure. A convergent lens is used as the objective and a divergent lens as the eyepiece.

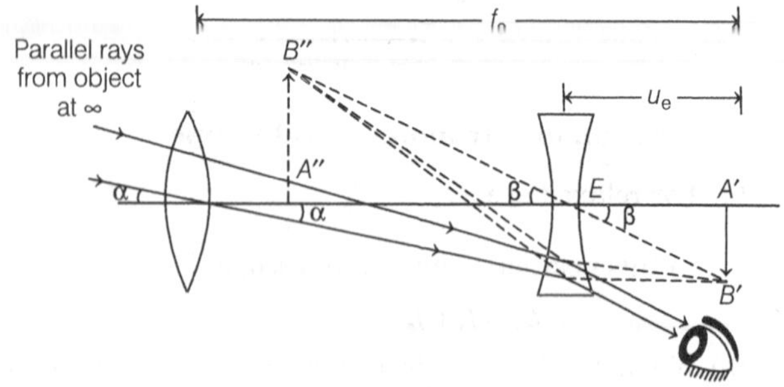

Magnification of Galilean telescope

(a) For relaxed eye, $M_\infty = \frac{f_o}{f_e}$

In this position, length of telescope,

$$L_\infty = f_o - f_e$$

(b) When final image is formed at least distance of distinct vision, $M_D = \frac{f_o}{f_e}\left(1 - \frac{f_e}{D}\right)$

Length of telescope, $L_D = f_o - u_e$

Example 28. *An object is seen through a simple microscope of focal length 20 cm. Find the angular magnification produced, if the image is formed at 30 cm from the lens.*

(a) 2.08 (b) 2.05
(c) 3.08 (d) 1.5

Sol. (a) Given, $f = +30$ cm and $v = -30$ cm

Using the formula, we have, $\frac{1}{v} - \frac{1}{u} = \frac{1}{f}$, we have

$$\frac{1}{-30} - \frac{1}{u_o} = \frac{1}{20}$$

$$u_o = 12\text{cm}$$

The angular magnification, $M = \frac{D}{u_o} = \frac{25}{12} = 2.08$

Example 29. *A Galilean telescope is 27 cm long when focussed to form an image at infinity. If the objective has a focal length of 30 cm, what is the focal length of the eyepiece?*

(a) 3 cm (b) −3 cm
(c) 2 cm (d) −2 cm

Sol. (a) Given, $f_o = +30$ cm

Length of telescope when final image forms at infinity,

$$L_\infty = 27 \text{ cm}$$

$$\therefore \quad L_\infty = f_o - f_e$$

$$27 = 30 - f_e$$

$$\Rightarrow \quad f_e = 3 \text{ cm}$$

Practice Exercise

ROUND I Topically Divided Problems

Reflection from Plane and Spherical Mirror

1. An object is placed a symmetrically between two plane mirrors inclined at an angle of 72°. The number of image formed is

(a) 5 (b) 4
(c) 2 (d) infinite

2. Two plane mirrors are inclined to each other at an angle θ. A ray of light is reflected first at one mirror and then at the other. The total deviation of the ray is

(a) 2θ (b) $240° - 2\theta$
(c) $360° - 2\theta$ (d) $180° - \theta$

3. A candle is placed before a thick plane mirror. When looked obliquely in the mirror, a number of images are seen from the surfaces of the plane mirror. Then

(a) first image is brightest
(b) second image is brightest
(c) third image is brightest
(d) all images beyond second are brighter

4. An object is approaching a plane mirror at 10 cms^{-1}. A stationary observer sees the image. At what speed will the image approach the stationary observer?

(a) 10 cms^{-1}
(b) 5 cms^{-1}
(c) 20 cms^{-1}
(d) 15 cms^{-1}

5. A soldier directs a laser beam on an enemy by reflecting the beam from a mirror. If the mirror is rotated by an angle θ, by what angle will reflected beam rotate?

(a) $\theta/2$ (b) θ
(c) 2θ (d) None of these

6. It is necessary to illuminate the bottom of a well by reflected solar beam when the light is incident at an angle of $\alpha = 40°$ to the vertical. At what angle β to the horizontal should a plane mirror be placed?

(a) 70° (b) 20°
(c) 50° (d) 40°

7. A car is moving with at a constant speed of 60 kmh^{-1} on a straight road. Looking at the rear view mirror, the driver finds that the car following him is at distance of 100 m and is approaching with a speed of 5 km h^{-1}. In order to keep track of the car in the rear, the driver begins to glance alternatively at the rear and side mirror of his car after every 2 s till the other car overtakes. If the two cars were maintaining their speeds, which of the following statement (s) is/are correct? **[NCERT Exemplar]**

(a) The speed of the car in the rear is 65 km h^{-1}.
(b) In the side mirror, the car in the rear would appear to approach with a speed of 5 km h^{-1} to the driver of the leading car.
(c) In the rear view mirror, the speed of the approaching car would appear to decrease as the distance between the cars decreases.
(d) In the side mirror, the speed of the approaching car would appear to increase as the distance between the cars decreases.

8. A point source of light S is placed at a distance L in front of the centre of plane mirror of width d, which is hanging vertically on a wall. A man walks in front of the mirror along a line parallel to the mirror, at a distance $2L$ as shown below.

The distance over which the man can see the image of the light source in the mirror is **[JEE Main 2019]**

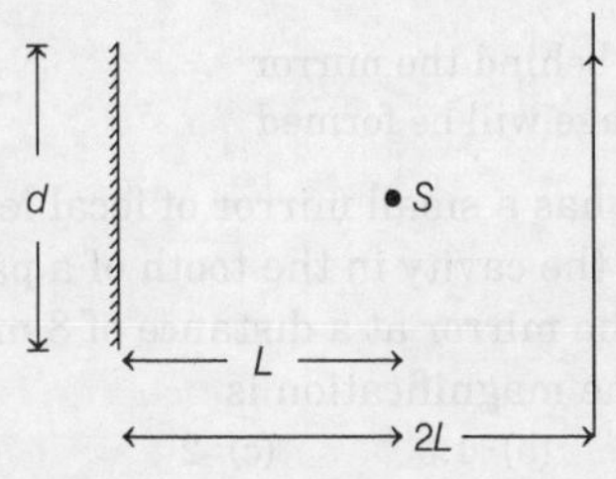

(a) $\frac{d}{2}$
(b) d
(c) $3d$
(d) $2d$

9. From a spherical mirror, the graph of $1/v$ *versus* $1/u$ is given by

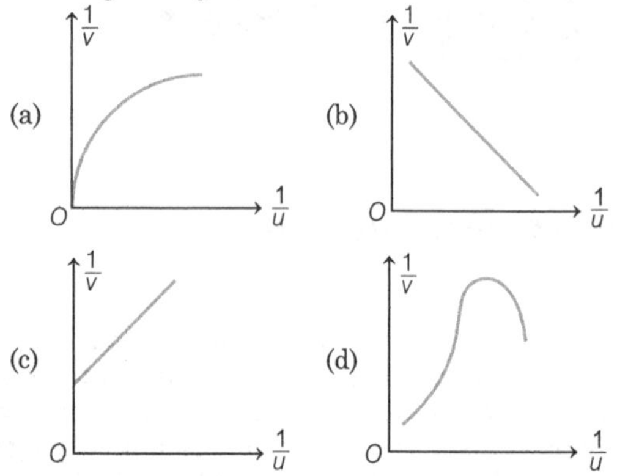

10. A convex mirror forms an image one-fourth the size of the object. If object is at a distance of 0.5 m from the mirror, the focal length of mirror is

(a) 0.17 m (b) – 1.5 m (c) 0.4 m (d) – 0.4 m

11. With a concave mirror, an object is placed at a distance x_1 from the principal focus, on the principal axis. The image is formed at a distance x_2 from the principal focus. The focal length of the mirror is

(a) $x_1 x_2$ (b) $\frac{x_1 + x_2}{2}$

(c) $\sqrt{\frac{x_1}{x_2}}$ (d) $\sqrt{x_1 x_2}$

12. A person of 6 feet in length can see his full size erect image in a mirror 2 feet in height. This mirror has to be

(a) plane or convex (b) plane or concave

(c) necessarily convex (d) necessarily concave

13. A point object is placed at a distance of 30 cm from a convex mirror of a focal length 30 cm. The image will form at

(a) infinity

(b) pole

(c) 15 cm behind the mirror

(d) no image will be formed

14. A dentist has a small mirror of focal length 16 mm. He views the cavity in the tooth of a patient by holding the mirror at a distance of 8 mm from the cavity. The magnification is

(a) 1 (b) 1.5 (c) 2 (d) 3

15. A car is fitted with a convex mirror of focal length 20 cm. A second car 2 m broad and 1.6 m height is 6 m away from the first car. The position of the second car as seen in the mirror of the first car is

(a) 19.35 cm (b) 17.45 cm

(c) 21.48 cm (d) 15.49 cm

16. A plane mirror is reflecting a ray of incident light is rotated through an angle of about an axis through the point of incidence in the plane of the mirror perpendicular to the plane of incident, then

(a) the reflected ray rotates through an angle 2θ

(b) the reflected ray rotates through an angle of θ

(c) the reflected ray does not rotate

(d) None of the above

17. A man has a concave shaving mirror or focal length 0.2 m. How far should the mirror be held from his face in order to give an image of two fold magnification?

(a) – 0.1 m (b) 0.2 m

(c) 0.3 m (d) 0.4 m

18. A spherical mirror forms diminished virtual image of magnification 1/3. Focal length is 18 cm. The distance of the object is

(a) 18 cm (b) – 36 cm

(c) 48 cm (d) infinite

19. Sun subtends an angle of 0.5° at the centre of curvature of a concave mirror of radius of curvature 15 m. The diameter of the image of the sun formed by the mirror is

(a) 8.55 cm (b) 7.55 cm

(c) 6.55 cm (d) 6.55 cm

20. A short linear object of length b lies along the axis of a concave mirror of focal length f at a distance u from the pole of the mirror. The size of the image is equal to

(a) $b\left(\frac{u-f}{f}\right)^{1/2}$ (b) $b\left(\frac{f}{u-f}\right)^{1/2}$

(c) $b\left(\frac{u-f}{f}\right)$ (d) $b\left(\frac{f}{u-f}\right)^{2}$

21. A convex mirror of radius of curvature 1.6 m has an object placed at a distance of 1 m from it. The image is formed at a distance of

(a) (8/13) m in front of the mirror

(b) (8/13) m behind the mirror

(c) (4/9) m in front of the mirror

(d) (4/9) m behind the mirror

22. A small object is placed 10 cm in front of a plane mirror. If you stand behind the object, 30 cm from the mirror and look at its image, for what distance must you focus your eyes?

(a) 20 cm (b) 60 cm

(c) 80 cm (d) 40 cm

23. A spherical mirror forms an image of magnification $m = \pm 3$. The object distance, if focal length of mirror is 24 cm, may be

(a) 32 cm, 24 cm (b) 32 cm, 16 cm

(c) 32 cm only (d) 16 cm only

24. A 4.5 cm needle is placed 12 cm away from a convex mirror of focal length 15 cm. Then, choose the correct option.
(a) Location of the image is 6.7 cm.
(b) Magnification is 5/9.
(c) If the needle is moved farther from the mirror, the image goes on decreasing.
(d) All of the above

25. When an object is at distances of u_1 and u_2 from the poles of a concave mirror, images of the same size are formed. The focal length of the mirror is
(a) $|u_1 + u_2|$ (b) $|u_1 - u_2|$
(c) $\left|\frac{u_1 + u_2}{2}\right|$ (d) $\left|\frac{u_1 - u_2}{2}\right|$

26. A concave mirror for face viewing has focal length of 0.4 m. The distance at which you hold the mirror from your face in order to see your image upright with a magnification of 5 is **[JEE Main 2019]**
(a) 0.16 m (b) 1.60 m (c) 0.32 m (d) 0.24 m

27. The direction of ray of light incident on a concave mirror is shown by PQ while directions in which the ray would travel after reflection is shown by four rays marked 1, 2, 3 and 4 (figure). Which of the four rays correctly shows the direction of reflected ray?

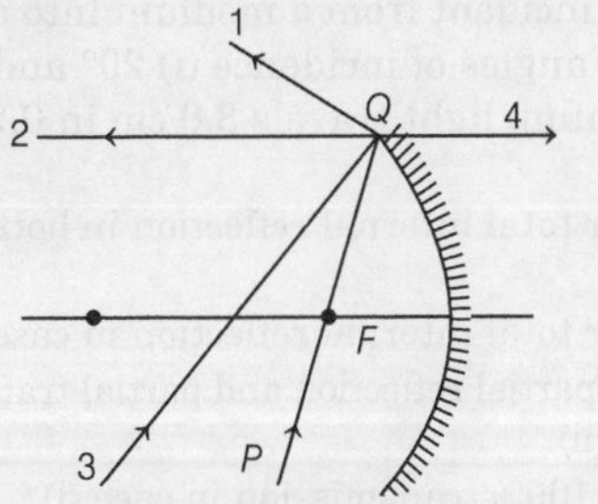

(a) 1 (b) 2 (c) 3 (d) 4

28. A spherical mirror is obtained as shown in the above figure from a hollow glass sphere. If an object is in front of the mirror, what will be the nature and magnification of the image of the object? (Figure drawn as schematic and not to scale)

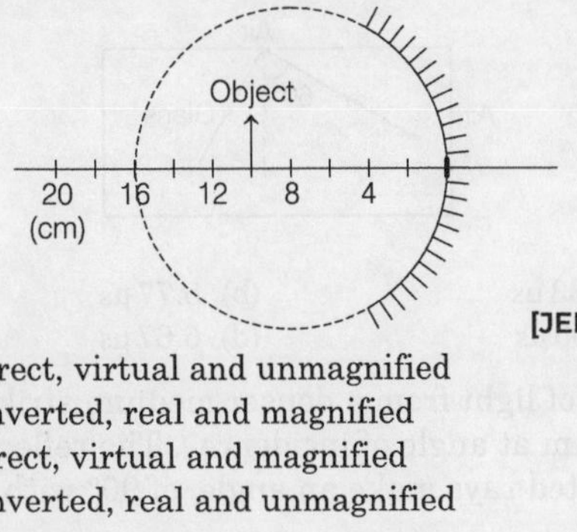

[JEE Main 2020]
(a) Erect, virtual and unmagnified
(b) Inverted, real and magnified
(c) Erect, virtual and magnified
(d) Inverted, real and unmagnified

Refraction of Light

29. There are certain materials developed in laboratories which have a negative refractive index (figure). A ray incident from air (medium 1) into such a medium (medium 2) shall follow a path given by **[NCERT Exemplar]**

(a) 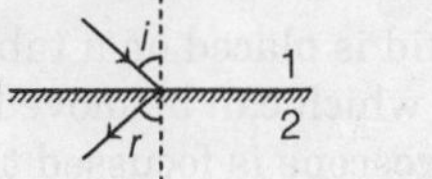

(b) 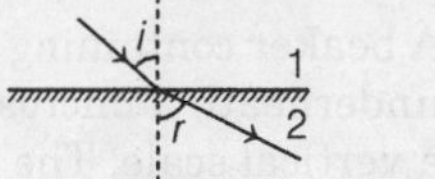

(c) 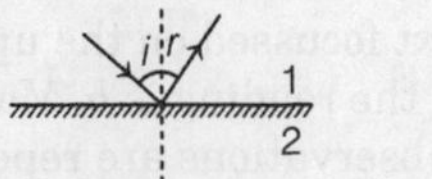

(d) 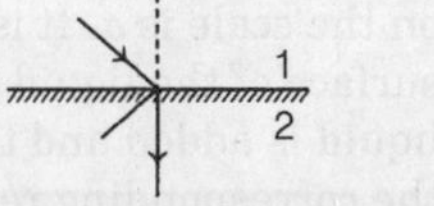

30. Monochromatic light of wavelength λ_1 travelling in medium of refractive index n_1 enters a denser medium of refractive index n_2. The wavelength in the second medium is
(a) $\lambda_1\left(\frac{n_1}{n_2}\right)$ (b) $\lambda_1\left(\frac{n_2}{n_1}\right)$
(c) λ_1 (d) $\lambda_1\left(\frac{n_2 - n_1}{n_1}\right)$

31. Light is incident from a medium X at an angle of incident i and is refracted into a medium Y at angle of refraction r. The graph sin i *versus* sin r is shown in figure. Which of the following conclusions would fit the situation?

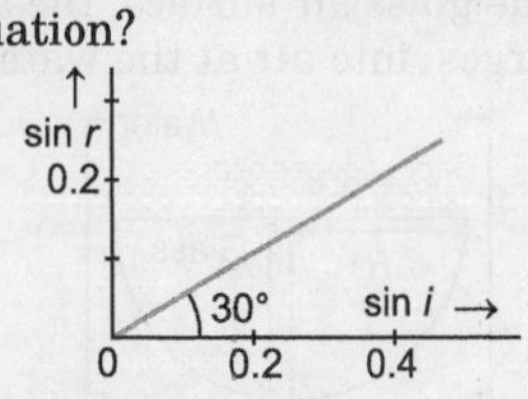

1. Speed of light in medium Y is $\sqrt{3}$ times that in medium X.
2. Speed of light in medium Y is $1/\sqrt{3}$ times that in medium X.

(a) 1 only (b) 1 and 2
(c) 2 only (d) Neither 1 nor 2

32. A ray of light AO in vacuum is incident on a glass slab at angle 60° and refracted at angle 30° along OB as shown in the figure. The optical path length of light ray from A to B is **[JEE Main 2019]**

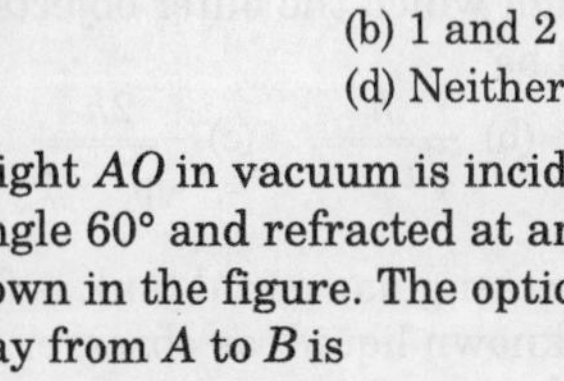

(a) $\frac{2\sqrt{3}}{a} + 2b$ (b) $2a + \frac{2b}{3}$
(c) $2a + 2b$ (d) $2a + \frac{2b}{\sqrt{3}}$

33. A printed page is pressed by a glass of water. The refractive index of the glass and water is 1.5 and 1.33, respectively. If the thickness of the bottom of glass is 1 cm and depth of water is 5 cm, how much the page will appear to be shifted, if viewed from the top? (JEE Main 2013)

(a) 1.033 cm (b) 3.581 cm (c) 1.3533 cm (d) 1.90 cm

34. A beaker containing liquid is placed on a table underneath a microsope which can be moved along a vertical scale. The microscope is focussed through the liquid, on a mark on the table and the reading on the scale is a. It is next focussed on the upper surface of the liquid and the reading is b. More liquid is added and the observations are repeated, the corresponding reading being c and d. The refractive index of the liquid is

(a) $\frac{d-b}{d-c-b+a}$ (b) $\frac{b-d}{d-c-b+a}$

(c) $\frac{d-c-b+a}{d-b}$ (d) $\frac{d-b}{d+c-b-a}$

Total Internal Reflection

35. A ray of light travelling in glass $\left(\mu = \frac{3}{2}\right)$ is incident on a horizontal glass air surface at the critical angle θ_C. If thin layer of water $\left(\mu = \frac{4}{3}\right)$ is now poured on the glass air surface, the angle at which the ray emerges into air at the water-air surface is

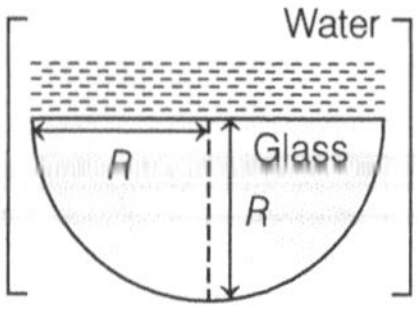

(a) 60° (b) 45° (c) 90° (d) 180°

36. If eye is kept at a depth h inside water of refractive index and viewed outside, then the diameter of the circle through which the outer objects become visible, will be

(a) $\frac{h}{\sqrt{\mu^2-1}}$ (b) $\frac{h}{\sqrt{\mu_2+1}}$ (c) $\frac{2h}{\sqrt{\mu^2-1}}$ (d) $\frac{h}{\sqrt{\mu^2}}$

37. When the rectangular metal tank is filled to the top with an unknown liquid, as observer with eyes level with the top of the tank can just see the corner E; a ray that refracts towards the observer at the top surface of the liquid is shown. The refractive index of the liquid will be

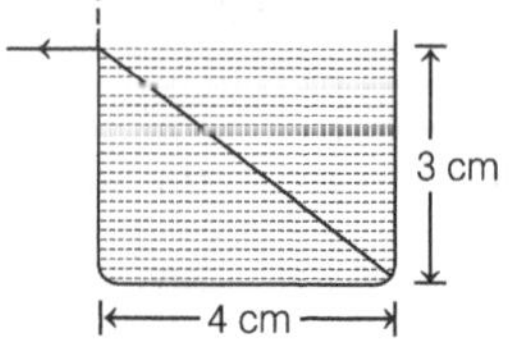

(a) 1.2 (b) 1.4

(c) 1.6 (d) 1.9

38. A rectangular block of glass $ABCD$ has a refractive index 1.6. A pin is placed midway on the face AB (figure). When observed from the face AD, the pin shall

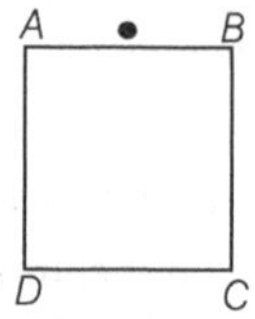

(a) appear to be near A (b) appear to be near D

(c) not be seen at all (d) Both (a) and (c)

39. The critical angle of a medium for a specific wavelength, if the medium has relative permittivity 3 and relative permeability $\frac{4}{3}$ for this wavelength, will be [JEE Main 2020]

(a) 45° (b) 60°

(c) 15° (d) 30°

40. When a glass slab is placed on a cross made on a sheet, the cross appears raised by 1 cm. The thickness of the glass is 3 cm. The critical angle for glass is

(a) $\sin^{-1}(0.33)$ (b) $\sin^{-1}(0.5)$

(c) $\sin^{-1}(0.67)$ (d) $\sin^{-1}(\sqrt{3}/2)$

41. Light is incident from a medium into air at two possible angles of incidence (i) 20° and (ii) 40°. In the medium, light travels 3.0 cm in 0.2 ns. The ray will

(a) suffer total internal reflection in both cases (i) and (ii)

(b) suffer total internal reflection in case (ii) only

(c) have partial reflection and partial transmission in case (ii)

(d) have 100% transmission in case (i)

42. A light ray from air is incident as shown in figure at one end of the glass fibre (refractive index $\mu = 1.5$) making an incidence angle of 60° on the lateral surface, so that it undergoes a total internal reflection. How much time would it take to traverse the straight fibre of light 1 km?

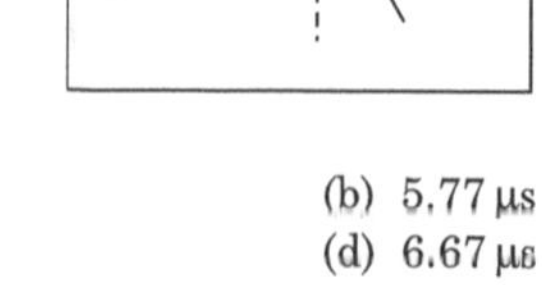

(a) 3.33 μs (b) 5.77 μs

(c) 3.85 μs (d) 6.67 μs

43. A ray of light from a denser medium strikes a rarer medium at angle of incidence i. The reflected and refracted rays make an angle of 90° with each

other. The angles of reflection and refraction are r and r', respectively. The critical angle is
(a) $\sin^{-1}(\tan r')$ (b) $\sin^{-1}(\tan r)$
(c) $\tan^{-1}(\tan r')$ (d) $\tan^{-1}(\tan i)$

44. A glass prism ABC (refractive index 1.5), immersed in water (refractive index 4/3). A ray of light is incident normally on face AB. If it is totally reflected at face AC, then

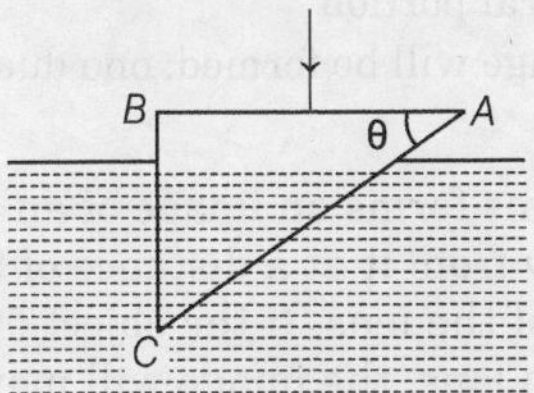

(a) $\sin\theta \geq \frac{8}{9}$ (b) $\sin\theta \geq \frac{2}{3}$
(c) $\sin\theta = \frac{\sqrt{3}}{2}$ (d) $\frac{2}{3} < \sin\theta < \frac{8}{9}$

45. A point source of light is held at a depth h below the surface of water. If C is critical angle of air-water interface, the diameter of circle of light coming from water surface would be
(a) $2h\tan C$ (b) $h\tan C$
(c) $h\sin C$ (d) $h/\sin C$

46. You are given four sources of light each one providing a light of a single colour, blue, green and yellow. Suppose the angle of refraction for a beam of yellow light corresponding to a particular angle of incidence at the interface of two media is 90°. Which of the following statement is correct if the source of yellow light is replaced with that of other lights without changing the angle of incidence? **[NCERT Exemplar]**
(a) The beam of red light would undergo total internal reflection
(b) The beam of red light would bend towards normal while it gets refracted through the second medium
(c) The beam of blue light would undergo total internal reflection
(d) The beam of green light would bend away from the normal as it gets refracted through the second medium

47. A small bulb is placed at the bottom of a tank containing water to a depth of 80 cm. What is the area of the surface of water through which light from the bulb can emerge out? Refractive index of water is 1.33. (Consider the bulb to be a point source.) **[NCERT]**
(a) 4.6 m^2 (b) 3.2 m
(c) 5.6 m^2 (d) 2.6 m^2

48. A transparent cube of side d, made of a material of refractive index μ_2, is immersed in a liquid of refractive index $\mu_1(\mu_1 < \mu_2)$. A ray is incident on the face AB at an angle θ (shown in the figure). Total internal reflection takes place at point E on the face BC. Then, θ must satisfy **[JEE Main 2019]**

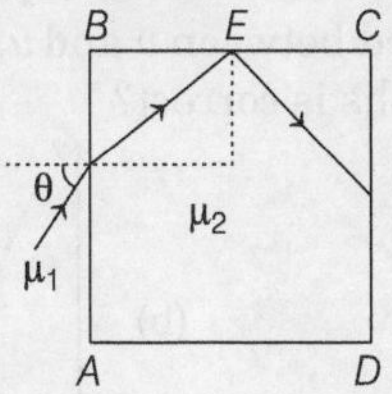

(a) $\theta < \sin^{-1}\frac{\mu_1}{\mu_2}$ (b) $\theta > \sin^{-1}\sqrt{\frac{\mu_2^2}{\mu_1^2} - 1}$
(c) $\theta < \sin^{-1}\sqrt{\frac{\mu_2^2}{\mu_1^2} - 1}$ (d) $\theta > \sin^{-1}\frac{\mu_1}{\mu_2}$

49. Parallel beam containing light of $\lambda = 400$ nm and 500 nm is incident on a prism as shown in figure. The refractive index μ of the prism is given by the relation,

$$\mu(\lambda) = 1.20 + \frac{0.8 \times 10^{-14}}{\lambda^2}$$

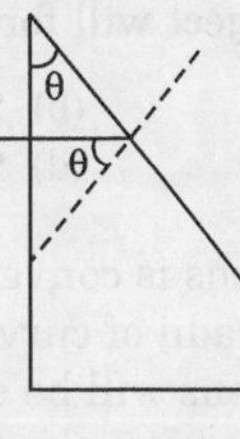

Which of the following statement is correct?
(a) Light of $\lambda = 400$ nm undergoes total internal reflection.
(b) Light of $\lambda = 500$ nm undergoes total internal reflection.
(c) Neither of the two wavelengths undergoes total internal reflection.
(d) Both wavelengths undergoes total internal reflection.

Refraction from Spherical Surfaces and Lenses

50. Parallel rays of light are falling on convex spherical surface of radius of curvature $R = 20$ cm as shown. Refractive index of the medium is $\mu = 1.5$. After refraction from the spherical surface parallel rays

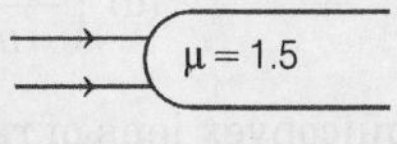

(a) actually meet at some point
(b) appears to meet after extending the refracted rays backwards

(c) meet (or appears to meet) at a distance of 30 cm from the spherical surface
(d) meet (or appears to meet) at a distance of 90 cm from the spherical surface

51. The distance v of the real image formed by a convex lens is measured for various object distance u. A graph is plotted between v and u. Which one of the following graphs is correct?

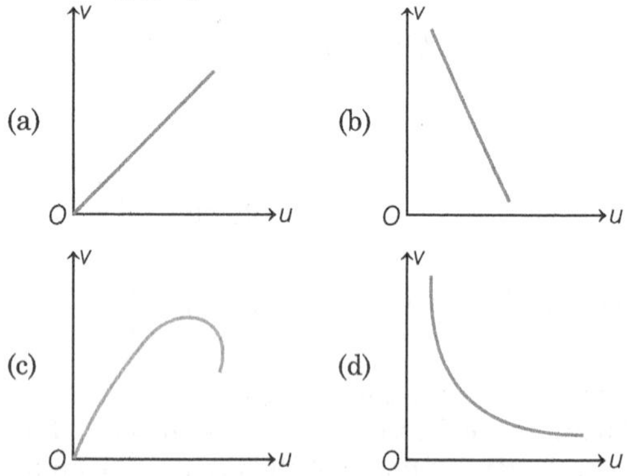

52. A layered lens as shown in figure is made of two types of transparent materials indicated by different shades. A point object is placed on its axis. The object will form

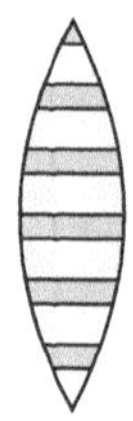

(a) 1 image (b) 2 images
(c) 3 images (d) 9 images

53. One surface of a lens is convex and the other is concave. If the radii of curvature are r_1 and r_2 respectively, the lens will be convex, if
(a) $r_1 > r_2$ (b) $r_1 = r_2$
(c) $r_1 < r_2$ (d) $r_1 = 1/r_2$

54. A concave lens of focal length 20 cm produces an image half in size of the real object. The distance of the real object is
(a) 20 cm (b) 30 cm (c) 10 cm (d) 60 cm

55. An object 15 cm high is placed 10 cm from the optical centre of a thin lens. Its image is formed 25 cm from the optical centre in the same side of the lens as the object. The height of the image is
(a) 2.5 cm (b) 0.2 cm (c) 16.7 cm (d) 37.5 cm

56. A convex lens of focal length f produces a virtual image n times the size of the object. Then, the distance of the object from the lens is
(a) $(n-1)f$ (b) $(n+1)f$
(c) $\left(\frac{n-1}{n}\right)f$ (d) $\left(\frac{n+1}{n}\right)f$

57. Consider an equiconvex lens of radius of curvature R and focal length f. If $f > R$, the refractive index μ of the material of the lens is greater than
(a) zero but less than 1.5 (b) 1.5 but less than 2.0
(c) one but less than 1.5 (d) None of these

58. Which amongst the following statement is correct regarding the image formed by a convex lens, if the central portion of the lens is wrapped in blank paper as shown in the figure.

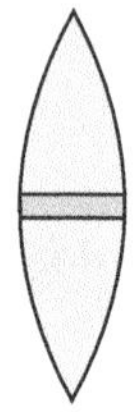

(a) No image will be formed
(b) Full image will be formed but is less bright
(c) Full image will be formed but without the central portion
(d) Two image will be formed, one due to each exposed half

59. A convex lens forms an image of an object placed 20 cm away from it at a distance of 20 cm on the other side of the lens. If the object is moved 5 cm towards the lens, the image will move
(a) 5 cm towards the lens
(b) 5 cm away from the lens
(c) 10 cm towards the lens
(d) 10 cm away from the lens

60. The radius of curvature of the curved surface of a plano-convex lens is 20 cm. If the refractive index of the material of the lens be 1.5, it will **[NCERT Exemplar]**
(a) act as a convex lens only for the objects that lie on its curved side
(b) act as a concave lens for the objects that lie on its curved side
(c) act as a convex lens irrespective of the side on which the object lies
(d) act as a concave lens irrespective of side on which the object lies

61. An object approaches a convergent lens from the left of the lens with a uniform speed 5 m/s and stops at the focus. The image **[NCERT Exemplar]**
(a) moves away from the lens with a uniform speed 5 m/s
(b) moves away from the lens with a uniform acceleration
(c) moves away from the lens with a non-uniform acceleration
(d) moves towards the lens with a non-uniform acceleration

62. What is the relation between refractive indices μ_1, μ_2 and μ_3, if the behaviour of light rays is as shown in figure?

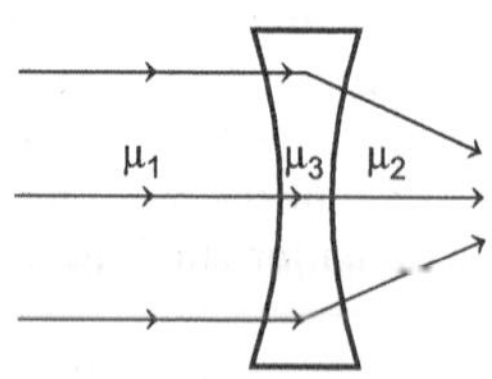

(a) $\mu_3 < \mu_2, \mu_3 = \mu_1$ (b) $\mu_2 < \mu_1, \mu_2 = \mu_3$
(c) $\mu_3 < \mu_2 < \mu_1$ (d) $\mu_3 > \mu_2 > \mu_1$

63. A concave lens with unequal radii of curvature made of glass ($\mu_g = 1.5$) has focal length of 40 cm. If it is immersed in a liquid of refractive index $\mu = 2$, then
(a) it behaves like a convex lens of 80 cm focal length
(b) it behaves like a concave lens of 20 cm focal length
(c) its focal length becomes 60 cm
(d) nothing can be said

64. An object is placed 30 cm to the left of a diverging lens whose focal length is of magnitude 20 cm. Which one of the following correctly states the nature and position of the virtual image formed?

Nature of image	Distance from lens
(a) inverted enlarged	60 cm to the right
(b) erect, diminished	12 cm to the left
(c) inverted, enlarged	60 cm to the left
(d) erect, diminished	12 cm to the right

65. A 16 cm long image of an object is formed by a convex lens on a screen. On moving the lens towards the screen, without changing the position of the object and the screen, 9 cm long image is formed again on the screen. The size of the object is
(a) 9 cm (b) 11 cm (c) 12 cm (d) 13 cm

66. A thin lens has focal length f_1 and its aperture has diameter d. It forms an image of intensity I. Now, the central part of the aperture upto diameter $\frac{d}{2}$ is blocked by an opaque paper. The focal length and image intensity will be change to
(a) f and $\frac{I}{4}$ (b) f and $\frac{3I}{4}$
(c) $\frac{f}{2}$ and $\frac{I}{2}$ (d) $\frac{3f}{4}$ and $\frac{I}{2}$

67. A convex lens of focal length f is placed somewhere in between an object and a screen. The distance between object and screen is x. If numerical value of magnification produced by lens is m, focal length of lens is
(a) $\frac{mx}{(m+1)^2}$ (b) $\frac{mx}{(m-1)^2}$
(c) $\frac{(m+1)^2}{m}x$ (d) $\frac{(m-1)^2}{m}x$

68. A point object is placed at a distance of 25 cm from a convex lens of focal length 20 cm. If a glass slab of thickness t and refractive index 1.5 is inserted between the lens and the object, the image is formed at infinity. The thickness t is
(a) 15 cm (b) 5 cm (c) 10 cm (d) 20 cm

69. The focal lengths for violet, green and red light rays are f_V, f_G and f_R, respectively. Which of the following is the true relationship?
(a) $f_R < f_G < f_V$ (b) $f_V < f_G < f_R$
(c) $f_G < f_R < f_V$ (d) $f_G < f_V < f_R$

70. A beam of parallel rays is brought to a focus by a plano-convex lens. A thin concave lens of the same focal length is joined to the first lens. The effect of this is
(a) the focal points shifts away from the lens by a small distance
(b) the focus remains undisturbed
(c) the focus shifts to infinity
(d) the focal points shifts towards the lens by a small distance

71. The power of a thin convex lens ($_a n_g = 1.5$) is + 5.0 D. When it is placed in a liquid of refractive index $_a n_e$, then it behaves as a concave lens of focal length 100 cm. The refractive index of the liquid $_a n_l$ will be
(a) 5/3 (b) 4/3
(c) $\sqrt{3}$ (d) 5/4

72. For a concave lens of focal length f, the relation between object and image distances u and v respectively, from its pole can best be represented by ($u = v$ is the reference line) **[JEE Main 2020]**

(a)
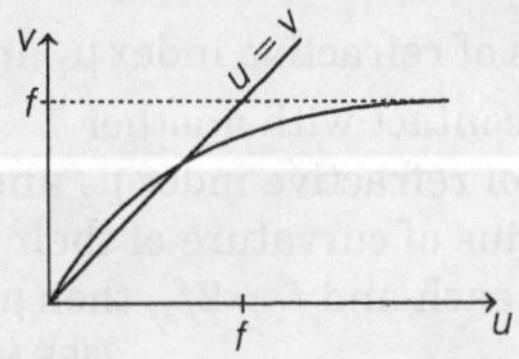

(b)
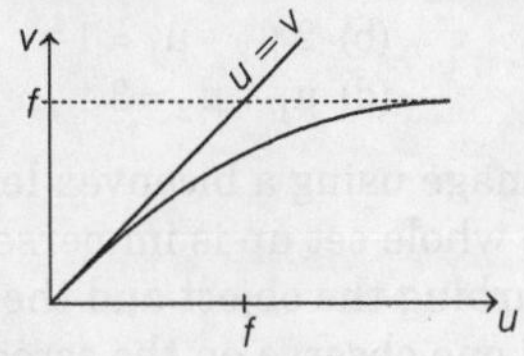

(c)
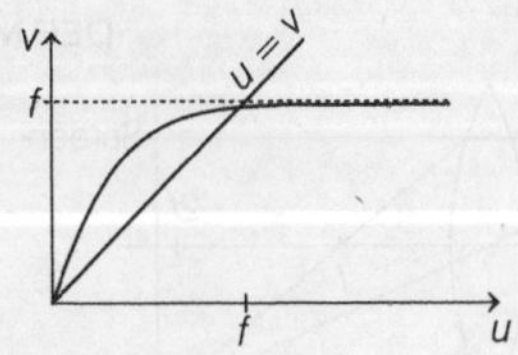

(d)
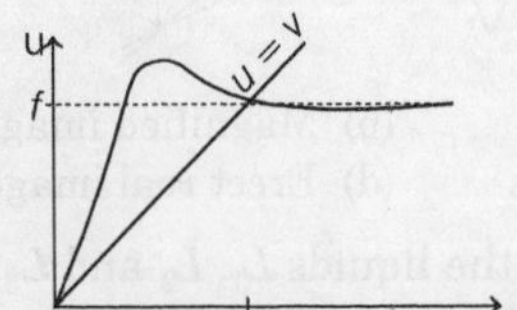

73. An object is at a distance of 20 m from a convex lens of focal length 0.3 m. The lens forms an image of the object. If the object moves away from the lens at a speed of 5 m/s, the speed and direction of the image will be **[JEE Main 2019]**
(a) 3.22×10^{-3} m/s towards the lens
(b) 0.92×10^{-3} m/s away from the lens
(c) 2.26×10^{-3} m/s away from the lens
(d) 1.16×10^{-3} m/s towards the lens

74. A plano-convex lens (focal length f_2, refractive index μ_2, radius of curvature R) fits exactly into a plano-concave lens (focal length f_1, refractive index μ_1, radius of curvature R). Their plane surfaces are parallel to each other. Then, the focal length of the combination will be **[JEE Main 2019]**

(a) $f_1 - f_2$ (b) $\frac{R}{\mu_2 - \mu_1}$ (c) $f_1 + f_2$ (d) $\frac{2f_1 f_2}{f_1 + f_2}$

75. Diameter of a plano-convex lens is 6 cm and thickness at the centre is 3 mm. If speed of light in material of lens is 2×10^8 m/s, the focal length of the lens is **[JEE Main 2013]**

(a) 15 cm (b) 20 cm (c) 30 cm (d) 10 cm

76. A diminished image of an object is to be obtained on a screen 1.0 m away from it. This can be achieved by approximately placing,

(a) a convex mirror of suitable focal length
(b) a convex mirror of suitable focal length
(c) a convex lens of focal length less than 0.25 m
(d) a concave lens of suitable focal length

77. A plano-convex lens of refractive index μ_1 and focal length f_1 is kept in contact with another plano-concave lens of refractive index μ_2 and focal length f_2. If the radius of curvature of their spherical faces is R each and $f_1 = 2f_2$, then μ_1 and μ_2 are related as **[JEE Main 2019]**

(a) $3\mu_2 - 2\mu_1 = 1$ (b) $2\mu_2 - \mu_1 = 1$
(c) $2\mu_1 - \mu_2 = 1$ (d) $\mu_1 + \mu_2 = 3$

78. Formation of real image using a biconvex lens is shown below. If the whole set up is immersed in water without disturbing the object and the screen positions, what will one observe on the screen? **[JEE Main 2019]**

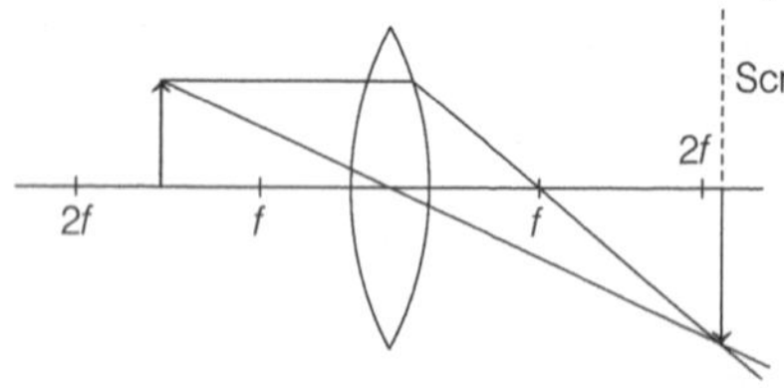

(a) No change (b) Magnified image
(c) Image disappears (d) Erect real image

79. As shown in figure, the liquids L_1, L_2 and L_3 have refractive indices 1.55, 1.50 and 1.20, respectively. Therefore, the arrangement corresponds to

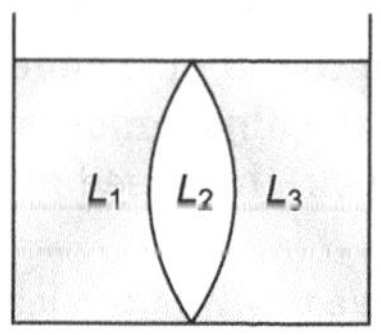

(a) biconvex lens (b) biconcave lens
(c) concavo-convex lens (d) convexo-concave lens

80. The refractive index of a converging lens is 1.4. What will be the focal length of this lens if it is placed in a medium of same refractive index? (Assume the radii of curvature of the faces of lens are R_1 and R_2 respectively) **[JEE Main 2021]**

(a) 1 (b) Infinite
(c) $\frac{R_1 R_2}{R_1 - R_2}$ (d) Zero

81. The thickness at the centre of a plano-convex lens is 3 mm and the diameter is 6 cm. If the speed of light in the material of the lens is 2×10^8 ms^{-1}. The focal length of the lens is **[JEE Main 2021]**

(a) 0.30 cm (b) 15 cm
(c) 1.5 cm (d) 30 cm

82. A convex lens A of focal length 20 cm and a concave lens B of focal length 56 cm are kept along the same axis with the distance d between them. If a parallel beam of light falling on A leaves B as a parallel beam, then distance d (in cm) will be

(a) 25 (b) 36
(c) 30 (d) 50

83. A convex lens is placed in contact with a mirror as shown in figure. If the space between them is filled with water, its power will

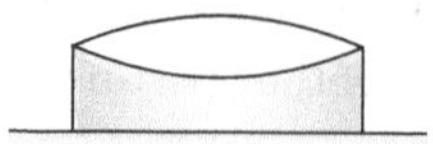

(a) decrease
(b) increase
(c) remain unchanged
(d) increase or decrease depending on the focal length

84. A diverging lens with magnitude of focal length 25 cm is placed at a distance of 15 cm from a converging lens of magnitude of focal length 20 cm. A beam of parallel light falls on the diverging lens. The final image formed is **[JEE Main 2017]**

(a) virtual and at a distance of 40 cm from convergent lens
(b) real and at a distance of 40 cm from the divergent lens
(c) real and at a distance of 6 cm from the convergent lens
(d) real and at a distance of 40 cm from convergent lens

85. Two lenses, one concave and the other convex of same power are placed such that their principal axes coincide. If the separation between the lenses is x, then

(a) real image is formed for $x = 0$ only
(b) real image is formed for all values of x
(c) system will behave like a glass plate for $x = 0$
(d) virtual image is formed for all values of x other than zero

86. Two similar plano-convex lenses are combined together in three different ways as shown in figure. The ratio of focal lengths in three cases will be

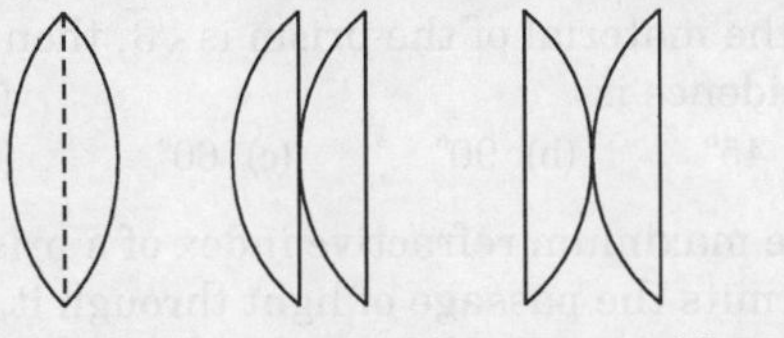

(a) 2 : 1 : 1 (b) 1 : 2 : 2 (c) 2 : 2 : 1 (d) 1 : 1 : 1

87. What is the position and nature of image formed by lens combination shown in figure? (where, f_1 and f_2 are focal lengths) [JEE Main 2019]

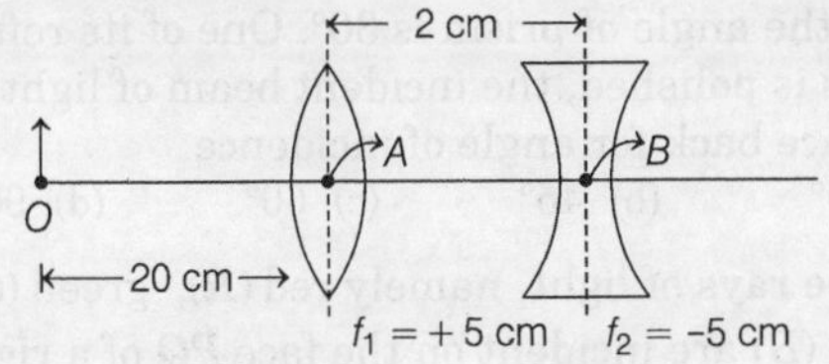

(a) $\frac{20}{3}$ cm from point B at right, real
(b) 70 cm from point B at right, real
(c) 40 cm from point B at right, real
(d) 70 cm from point B at left, virtual

88. A plano-convex lens has a thickness of 4 cm. When placed on a horizontal table, with the curved surface in contact with it, the apparent depth of the bottom most point of the lens is found to be 3 cm. If the lens is inverted such that the plane face is in contact with the table, the apparent depth of the centre of the plane face is found to be 25/8 cm. Find the focal length of the lens. Assume thickness to be negligible.

(a) 85 cm (b) 59 cm (c) 75 cm (d) 7.5 cm

89. A point like object is placed at a distance of 1 m in front of a convex lens of focal length 0.5 m. A plane mirror is placed at a distance of 2 m behind the lens. The position and nature of the final image formed by the system is [JEE Main 2020]

(a) 2.6 m from the mirror, real
(b) 1 m from the mirror, virtual
(c) 1 m from the mirror, real
(d) 2.6 m from the mirror, virtual

90. A convex lens is put 10 cm from a light source and it makes a sharp image on a screen, kept 10 cm from the lens. Now, a glass block (refractive index is 1.5) of 1.5 cm thickness is placed in contact with the light source.

To get the sharp image again, the screen is shifted by a distance d. Then, d is [JEE Main 2019]

(a) 0
(b) 1.1 cm away from the lens
(c) 0.55 cm away from the lens
(d) 0.55 cm towards the lens

91. A thin convex lens L (refractive index $= 1.5$) is placed on a plane mirror M. When a pin is placed at A, such that $OA = 18$ cm, its real inverted image is formed at A itself as shown in figure. When a liquid of refractive index μ_l is put between the lens and the mirror, the pin has to be moved to A', such that $OA' = 27$ cm, to get its inverted real image at A' itself. The value of μ_l will be [JEE Main 2019]

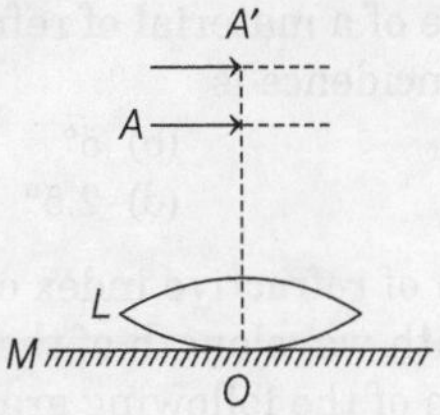

(a) $\sqrt{3}$ (b) $\sqrt{2}$ (c) $\frac{4}{3}$ (d) $\frac{3}{2}$

92. The graph shows how the magnification m produced by a thin lens varies with image distance v. What is the focal length of the lens used? [JEE Main 2019]

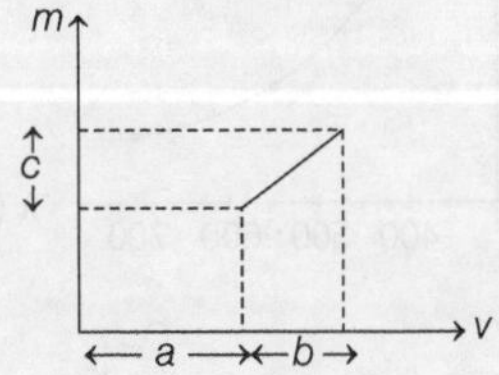

(a) $\frac{b^2c}{a}$ (b) $\frac{b^2}{ac}$ (c) $\frac{a}{c}$ (d) $\frac{b}{c}$

93. A double convex lens has power P and same radii of curvature R of both the surfaces. The radius of curvature of a surface of a plano-convex lens made of the same material with power 1.5 P is [JEE Main 2020]

(a) $2R$ (b) $\frac{R}{2}$ (c) $\frac{3R}{2}$ (d) $\frac{R}{3}$

Refraction and Dispersion Through Prism

94. What is the angle of incidence for an equilateral prism of refractive index $\sqrt{3}$, so that the ray is parallel to the base inside the prism?

(a) 30° (b) 45°
(c) 60° (d) Either 30° or 60°

95. At what angle should a ray of light be incident on the face of a prism of refracting angle 60° so that it just suffers total internal reflection at the other face if, the refractive index of the material of the prism is 1.524? [NCERT]

(a) 16°15′ (b) 29°75′
(c) 45°75′ (d) 58°15′

96. The refractive index of a prism for a monochromatic wave is $\sqrt{2}$ and its refracting angle is 60°. For minimum deviation, the angle of incidence will be

(a) 30° (b) 45°
(c) 60° (d) 75°

97. A ray of light incident at an angle θ on a refracting face of a prism emerges from the other face normally. If the angle of the prism is 5° and the prism is made of a material of refractive index 1.5, the angle of incidence is **[NCERT Exemplar]**

(a) 7.5° (b) 5°
(c) 15° (d) 2.5°

98. The variation of refractive index of a crown glass thin prism with wavelength of the incident light is shown. Which of the following graphs is the correct one, if D_m is the angle of minimum deviation? **[JEE Main 2019]**

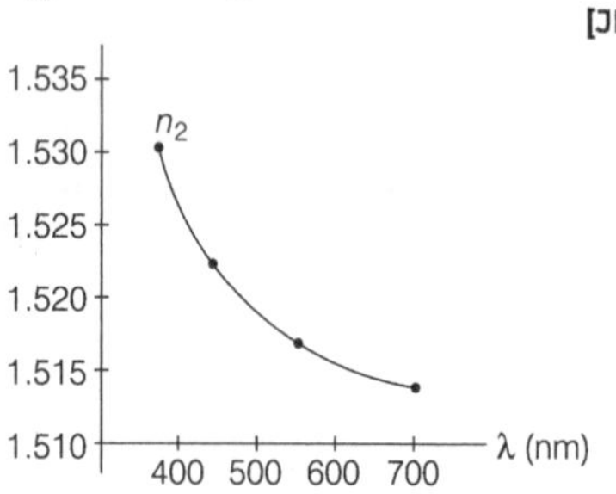

(a)
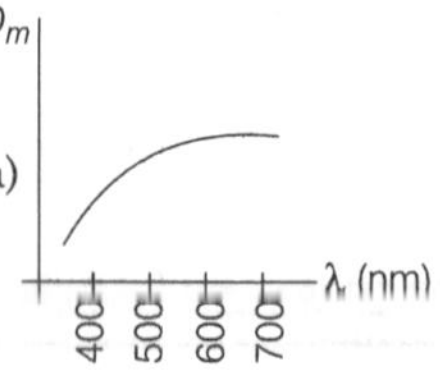

(b)
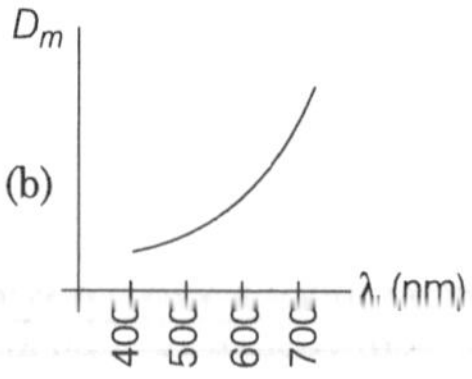

(c)
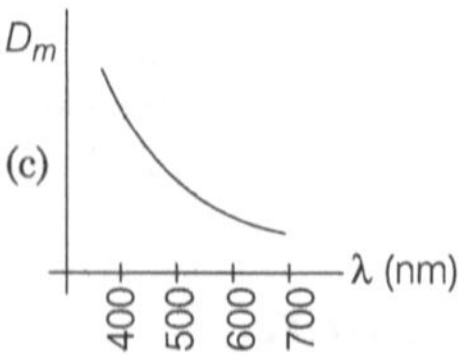

(d)
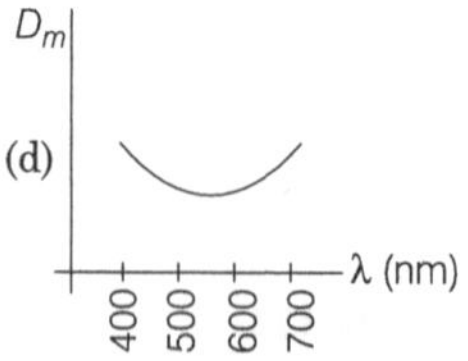

99. Two parallel light rays are incident at one surface of a prism of refractive index 1.5 as shown in figure. The angle between the emergent rays is nearly

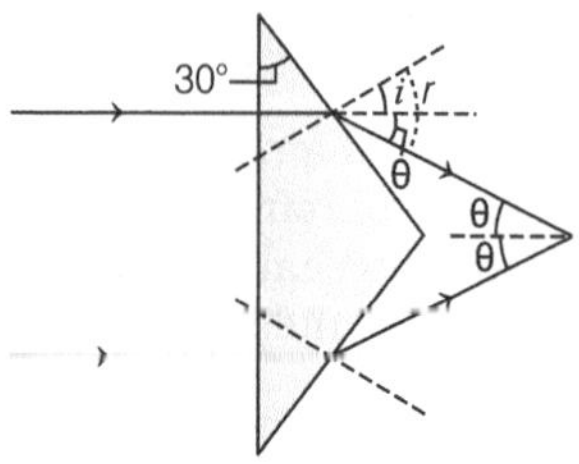

(a) 19° (b) 37° (c) 45° (d) 49°

100. A monochromatic light is incident at a certain angle on an equilateral triangular prism and suffers minimum deviation. If the refractive index of the material of the prism is $\sqrt{3}$, then the angle of incidence is **[JEE Main 2019]**

(a) 45° (b) 90° (c) 60° (d) 30°

101. The maximum refractive index of a prism which permits the passage of light through it, when the refracting angle of the prism is 90°, is

(a) $\sqrt{3}$ (b) $\sqrt{2}$ (c) $\frac{\sqrt{3}}{2}$ (d) $\frac{3}{2}$

102. The refractive index of the material of a prism is $\sqrt{2}$ and the angle of prism is 30°. One of its refracting faces is polished, the incident beam of light will retrace back for angle of incidence

(a) 0° (b) 45° (c) 60° (d) 90°

103. Three rays of light, namely red (R), green (G) and Blue (B) are incident on the face PQ of a right angled prism PQ as shown in figure.

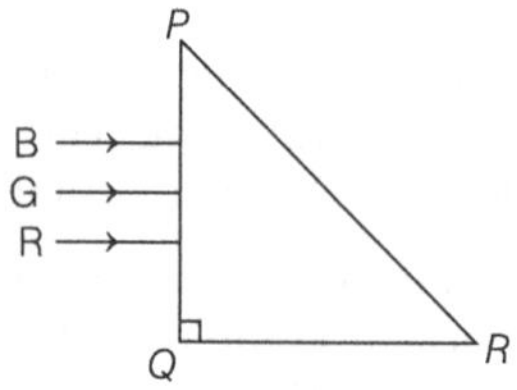

The refractive indices of the material of the prism for red, green and blue wavelength are 1.27, 1.42 and 1.49 respectively. The colour of the ray (s) emerging out of the face PR is **[JEE Main 2021]**

(a) green (b) red
(c) blue and green (d) blue

104. Monochromatic light is incident on a glass prism of angle A. If the refractive index of the material of the prism is μ, a ray incident at an angle θ, on the face AB would get transmitted through the face AC of the prism provided

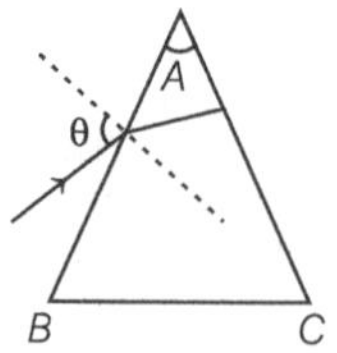

[JEE Main 2015]

(a) $\theta > \sin^{-1}\left[\mu \sin\left(A - \sin^{-1}\left(\frac{1}{\mu}\right)\right)\right]$

(b) $\theta < \sin^{-1}\left[\mu \sin\left(A - \sin^{-1}\left(\frac{1}{\mu}\right)\right)\right]$

(c) $\theta > \cos^{-1}\left[\mu \sin\left(A + \sin^{-1}\left(\frac{1}{\mu}\right)\right)\right]$

(d) $\theta < \cos^{-1}\left[\mu \sin\left(A + \sin^{-1}\left(\frac{1}{\mu}\right)\right)\right]$

105. A thin prism P_1 with angle 4° and made from glass of refractive index 1.54 is combined with another thin prism P_2 made from glass of refractive index 1.72 to produce dispersion without deviation. The angle of prism P_2 is
(a) 3° (b) 4° (c) 5.33° (d) 2.6°

106. An achromatic convergent doublet of two lenses in contact has a power of + 2 D. The convex lens has power +5D. What is the ratio of the dispersive powers of the convergent and divergent lenses?
(a) 2 : 5 (b) 3 : 5 (c) 5 : 2 (d) 5 : 3

107. It is desired to make a converging achromatic combination of mean focal length 50 cm by using two lenses of materials *A* and *B*. If the dispersive powers of *A* and *B* are in ratio 1 : 2, the focal lengths of the convex and the concave lenses are respectively
(a) 25 cm and 50 cm (b) 50 cm and 25 cm
(c) 50 cm and 100 cm (d) 100 cm and 50 cm

Optical Instruments

108. A simple microscope consists of a concave lens of power – 10D and a convex lens of power + 20D in contact. If the image is formed at infinity, then the magnifying power when $D = 25$ cm is
(a) 2.5 (b) 3.5 (c) 2.0 (d) 3.0

109. In compound microscope, magnifying power is 95 and the distance of object from objective lens is $\frac{1}{3.8}$ cm. The focal length of objective lens is $\frac{1}{4}$ cm. What is the magnification of eyepiece?
(a) 5 (b) 10 (c) 100 (d) 200

110. The focal lengths of the objective and eyelenses of a microscope are 1.6 cm and 2.5 cm, respectively. The distance between the two lenses is 21.7 cm. If the final image is formed at infinity, the distance between the object and objective lens is
(a) 1.8 cm (b) 1.70 cm
(c) 1.65 cm (d) 1.75 cm

111. The focal length of the objective and the eyepiece of a microscope are 4 mm and 25 mm, respectively. If the final image is formed at infinity and the length of the tube is 16 cm, then the magnifying power of microscope will be
(a) – 337.5 (b) –3.75
(c) 3.375 (d) 33.75

112. An astronomical telescope has a converging eyepiece of focal length 5 cm and objective of focal length 80 cm. When the final image is formed at the least distance of distinct vision (25 cm), the separation between the two lenses is
(a) 75.0 cm (b) 80.0 cm
(c) 84.2 cm (d) 85.0 cm

113. The focal length of objective and eye lens of an astronomical telescope are respectively 2cm and 5cm. Final image is formed at (1) least distance of distinct vision (2) infinity. Magnifying powers in two cases will be
(a) –48, –40 (b) – 40, 48
(c) – 40, + 48 (d) – 48, + 40

114. The magnifying power of a telescope is 9. When it is adjusted for parallel rays, the distance between the objective and the eyepiece is found to be 20 cm. The focal lengths of the lenses are
(a) 18 cm, 2 cm (b) 11 cm, 9 cm
(c) 10 cm, 10 cm (d) 15 cm, 5 cm

115. The magnifying power of a telescope with tube length 60 cm is 5. What is the focal length of its eyepiece? [JEE Main 2020]
(a) 10 cm (b) 20 cm (c) 30 cm (d) 40 cm

116. If we need a magnification of 375 from a compound microscope of tube length 150 mm and an objective of focal length 5 mm, the focal length of the eyepiece should be close to [JEE Main 2020]
(a) 22 mm (b) 2 mm (c) 12 mm (d) 33 mm

117. A telescope has an objective of focal length 50 cm and an eyepiece of focal length 5 cm. The least distance of distinct vision is 25 cm. The telescope is focussed for distinct vision on a scale 200 cm away. The separation between the objective and eyepiece is
(a) 74 cm (b) 75 cm (c) 60 cm (d) 71 cm

118. A Cassegrain telescope uses two mirrors as shown in figure. Such a telescope is built with the mirrors 20 mm apart. If the radius of curvature of the large mirror is 220 mm and the small mirror is 140 mm, where will the final image of an object at infinity be? [NCERT]

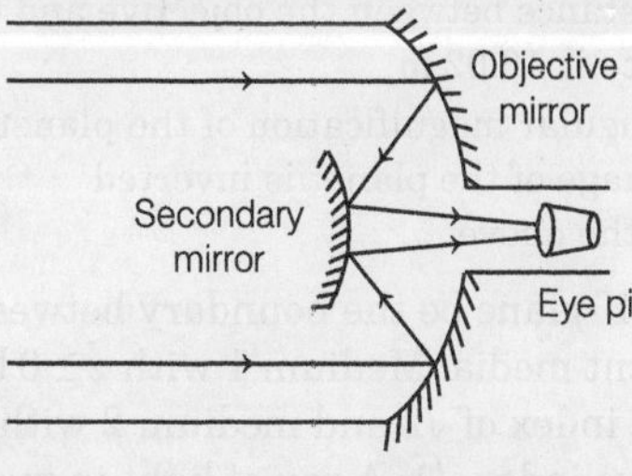

(a) 235 mm away from small mirror
(b) 285 mm away from small mirror
(c) 305 mm away from small mirror
(d) 315 mm away from small mirror

119. A hypermetropic person having near point at a distance of 0.75 m puts on spectacles of power 2.5 D. The near point now is at
(a) 0.75 m (b) 0.83 m
(c) 0.36 m (d) 0.26 m

ROUND II Mixed Bag

Only One Correct Option

1. A convex lens forms a real image of a point object placed on its principal axis. If the upper half of the lens is painted black, then
(a) the image will be shifted upward
(b) the image will be shifted downward
(c) the intensity of the image will decrease
(d) None of the above

2. A magnifying glass is used, as the object to be viewed can be brougth closer to the eye than the normal near point. This results in
(a) a larger angle to be subtended by the object at the eye and hence viewed in greater detail
(b) the formation of a real image
(c) increase in the field of view
(d) infinite magnification at the near point

3. Consider an extended object immersed in water contained in a plane trough.
When the object is seen close to the edge of the trough, then the object looked distorted because,
(a) the apparent depth of the points close to the edge are nearer the surface to the water compared to the points away from the edge.
(b) the angle subtended by the image of the object at the eye is smaller than the actual angle subtended by the object in air
(c) some of the points of the object far away from the edge may not be visible because of total internal reflection
(d) All of the above

4. A planet is observed by an astronomical refracting telescope having an objective of focal length 16 m and an eyepiece of focal length 2 cm, then
(a) The distance between the objective and the eyepiece is 16.02 m
(b) The angular magnification of the planet is −800
(c) The image of the planet is inverted
(d) All of the above

5. Let the XY-plane be the boundary between two transparent media. Medium 1 with $z \geq 0$ has a refractive index of $\sqrt{2}$ and medium 2 with $z < 0$ has a refractive index $\sqrt{3}$. A ray of light in medium 1 given by the vector $\mathbf{A} = 6\sqrt{3}\,\hat{\mathbf{i}} + 8\sqrt{3}\,\hat{\mathbf{j}} - 10\hat{\mathbf{k}}$ is incident on the plane of separation. The angle of refraction in medium 2 is **[AIEEE 2011]**
(a) 45° (b) 60°
(c) 75° (d) 30°

6. A car is fitted with a convex side-view mirror of focal length 20 cm. A second car 2.8 m behind the first car is overtaking the first car at a relative speed of 15m/s. The speed of the image of he second car as seen in the mirror of the first one is **[AIEEE 2011]**
(a) $\frac{1}{15}$ m/s (b) 10 m/s (c) 15 m/s (d) $\frac{1}{10}$ m/s

7. A ray of light passes through four transparent medium with refractive indices μ_1, μ_2, μ_3 and μ_4 as shown in the figure. The surfaces of all media are parallel. If the emergent ray CD is parallel to the incident ray AB, we must have

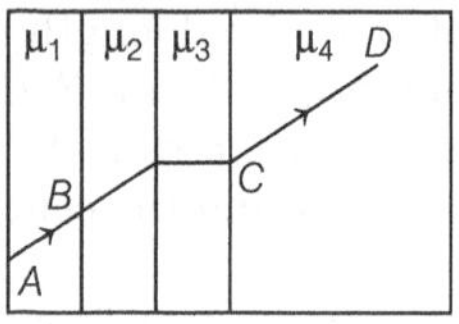

(a) $\mu_1 = \mu_2$ (b) $\mu_1 = \mu_3$ (c) $\mu_3 = \mu_4$ (d) $\mu_3 = \mu_1$

8. For a optical arrangement shown in the figure, find the position and nature of image.

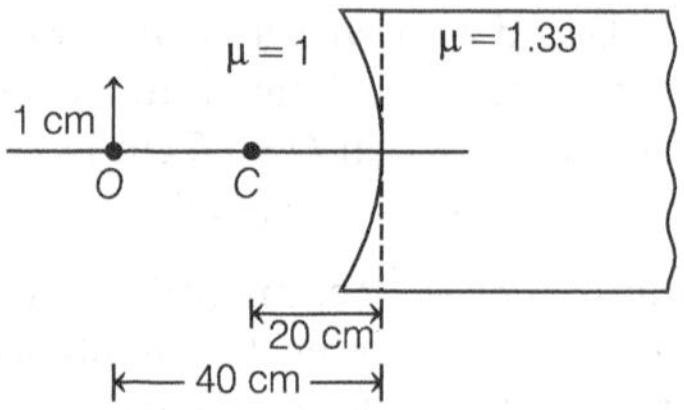

(a) 32 cm (b) 0.6 cm (c) 6 cm (d) 0.5 cm

9. A ray of light makes an angle of 10° with the horizontal above it and strikes a plane mirror which is inclined at an angle θ to the horizontal. The angle θ for which the reflected ray becomes vertical is
(a) 50° (b) 80°
(c) 100° (d) 40°

10. A 4 cm thick layer of water covers a 6 cm thick glass slab. A coin is placed at the bottom of the slab and is being observed from the air side along the normal to the surface. Find the apparent position of the coin from the surface.

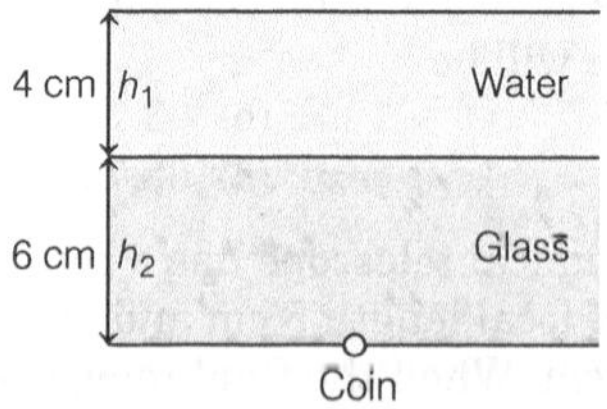

(a) 7.0 cm (b) 8.0 cm
(c) 10 cm (d) 5 cm

11. A plane mirror as placed at the bottom of a tank containing a liquid of refractive index μ and P is a small object at a height h above the mirror. An observer O vertically above P outside the liquid sees P and its image in a mirror. The apparent distance between these two will be

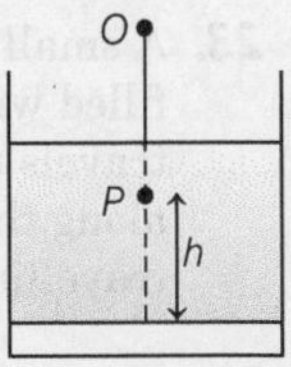

(a) $2\mu h$ (b) $\frac{2h}{\mu}$ (c) $\frac{2h}{\mu - 1}$ (d) $h\left(1+\frac{1}{\mu}\right)$

12. An object 2.4 m in front of a lens forms a sharp image on a film 12 cm behind the lens. A glass plate 1 cm thick, of refractive index 1.50 is interposed between lens and film with its plane faces parallel to film. At what distance (from lens) should object shifted to be in sharp focus on film? **[AIEEE 2012]**

(a) 7.2 (b) 2.4 (c) 3.2 (d) 5.6

13. An object is gradually moving away from the focal point of a concave mirror along the axis of the mirror. The graphical representation of the magnitude of linear magnification m *versus* distance of the object from the mirror x is correctly given by (graphs are drawn schematically and are not to scale) **[JEE Main 2020]**

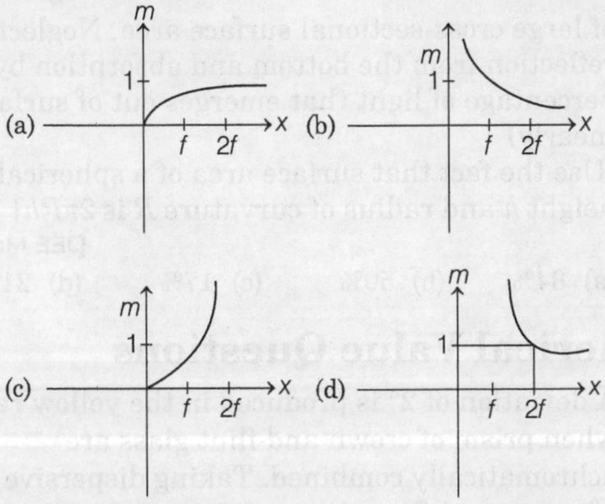

14. A transparent solid cylinder rod has a refractive index of $\frac{2}{\sqrt{3}}$. It is surrounded by air. A light ray is incident at the mid point of one end of the rod as shown in the figure.

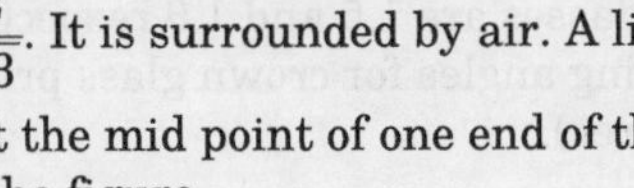

The incident angle θ for which the light ray grazes along the wall of the rod is

(a) $\sin^{-1}\left(\frac{1}{2}\right)$ (b) $\sin^{-1}\left(\frac{\sqrt{3}}{2}\right)$

(c) $\sin^{-1}\left(\frac{2}{\sqrt{3}}\right)$ (d) $\sin^{-1}\left(\frac{1}{\sqrt{3}}\right)$

15. A rectangular glass slab $ABCD$ of refractive index μ_1, is immersed in water of refractive index $\mu_2 (\mu_1 > \mu_2)$. A ray of light is incident at the surface AB of the slab as shown in figure. The maximum value of the angle of incidence α_{max}, such that the ray comes out only from the other surface CD is given by

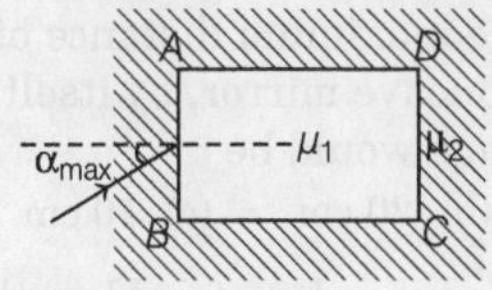

(a) $\sin^{-1}\left[\left(\frac{\mu_1}{\mu_2}\right)\cos\left(\sin^{-1}\frac{\mu_2}{\mu_1}\right)\right]$

(b) $\sin^{-1}\left(\frac{\alpha_2}{\alpha_1}\right)$

(c) $\sin^{-1}\left[\alpha_1 \cdot \cos\left(\sin^{-1}\frac{1}{\alpha_2}\right)\right]$

(d) $\left[\sin^{-1}\frac{\alpha_1}{\alpha_2}\right]$

16. A ray OP of monochromatic light is incident on the face AB of prism $ABCD$ near vertex B at an incident angle of 60° (see figure). If the refractive index of the material of the prism is $\sqrt{3}$, which of the following is (are) correct?

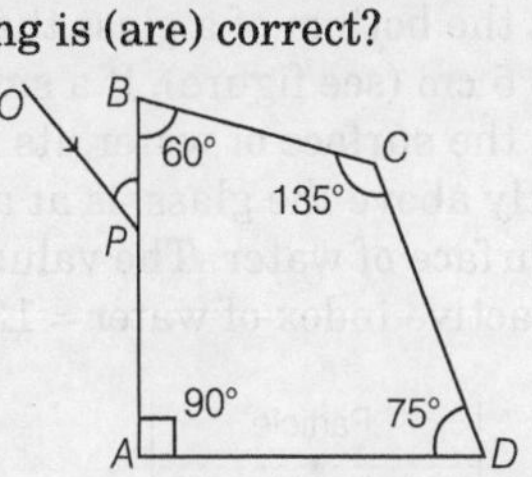

(a) The ray gets totally internally reflected at face CD

(b) The ray comes out through face AD

(c) The angle between the incident ray and the emergent ray is 90°

(d) All of the above

17. A convex lens of focal length 20 cm produces images of the same magnification 2 when an object is kept at two distances x_1 and x_2 $(x_1 > x_2)$ from the lens. The ratio of x_1 and x_2 is **[JEE Main 2019]**

(a) 5 : 3 (b) 2 : 1 (c) 4 : 3 (d) 3 : 1

18. P is a point on the axis of a concave mirror. The image of P formed by the mirror, coincides with P. A rectangular glass slab of thickness t and refractive index μ is now introduced between P and the mirror. For image of P to coincide with P again, the mirror must be moved

(a) towards P by $(\mu - 1)t$ (b) away from P by $(\mu - 1)t$

(c) towards P by $t\left(1-\frac{1}{\mu}\right)$ (d) away from P $t\left(1-\frac{1}{\mu}\right)$

19. A convex lens (of focal length 20 cm) and a concave mirror, having their principal axes along the same lines, are kept 80 cm apart from each other. The concave mirror is to the right of the convex lens. When an object is kept at a distance of 30 cm to the left of the convex lens, its image remains at the same position even if the concave mirror is removed. The maximum distance of the object for which this concave mirror, by itself would produce a virtual image would be **[JEE Main 2019]**

(a) 25 cm (b) 20 cm (c) 10 cm (d) 30 cm

20. A thin plano-convex lens focal length f is split into two halves. One of the halves is shifted along the optical axis and the separation between object and image plane is 1.8 m. The magnification of the image formed by one of the half lens is 2. Find the focal length of the lens and separation between the two halves.

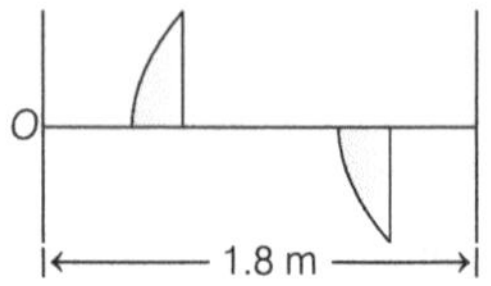

(a) 0.1 m (b) 0.4 m (c) 0.9 m (d) 1 m

21. A concave mirror has radius of curvature of 40 cm and it is at the bottom of a glass that has water filled up to 5 cm (see figure). If a small particle is floating on the surface of water, its image as seen, from directly above the glass, is at a distance d from the surface of water. The value of d is close to [Take, refractive index of water = 1.33] **[JEE Main 2019]**

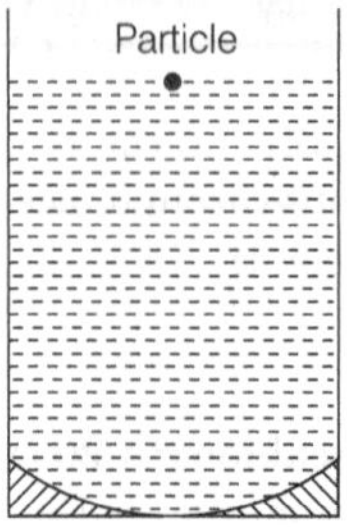

(a) 6.7 cm (b) 13.4 cm
(c) 8.8 cm (d) 11.7 cm

22. For a normal eye, the far point is at infinity and the near point of distinct vision is about 25 cm in front of the eye. The cornea of the eye provides a converging power of about 40 D and the least converging power of the eye lens be, hind the cornea is about 20 D. From this rough data, estimate the range of accommodation (*i.e.* the range of converging power of the eye lens) of a normal eye.

(a) 10 D to14 D (b) 20 D to 24 D
(c) 28 D to 32 D (d) 14 D to 18 D

23. A small coin is resting on the bottom of the beaker filled with a liquid. A ray of light from the coin travels upto the surface of the liquid and moves along the surface (figure). How fast is light travelling in the liquid?

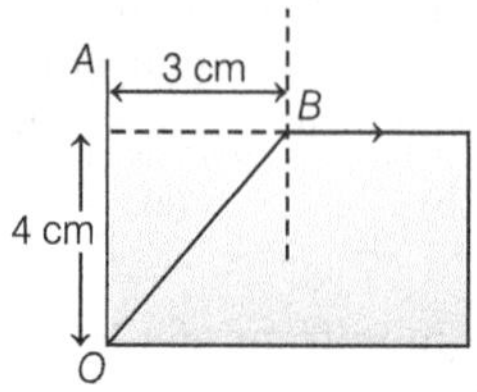

(a) 2.4×10^8 m/s (b) 1.8×10^8 m/s
(c) 3.0×10^8 m/s (d) 5.0×10^5 m/s

24. A simple telescope, consisting of an objective of focal length 60 cm and a single eyelens of focal length 60 cm and a single eyelens of focal length 5 cm is focussed on a distant object is such a way that parallel rays come out from the eyelens. If the object subtends an angle 2° at the objective, the angular width of the image will be

(a) 10° (b) 24°
(c) 35° (d) 48°

25. There is a small source of light at some depth below the surface of water (refractive index = $\frac{4}{3}$) in a tank of large cross-sectional surface area. Neglecting any reflection from the bottom and absorption by water, percentage of light that emerges out of surface is (nearly)
[Use the fact that surface area of a spherical cap of height h and radius of curvature R is $2\pi Rh$] **[JEE Main 2020]**

(a) 34% (b) 50% (c) 17% (d) 21%

Numerical Value Questions

26. A deviation of 2° is produced in the yellow ray when prism of crown and flint glass are achromatically combined. Taking dispersive powers of crown and flint glass are 0.02 and 0.03 respectively and refractive index for yellow light for these glasses are 1.5 and 1.6 respectively. The refracting angles for crown glass prism will be ...... (in degree). **[JEE Main 2021]**

(Round off to the nearest integer)

27. A prism of angle $A = 1°$ has a refractive index $\mu = 1.5$. A good estimate for the minimum angle of deviation (in degree) is close to $N/10$. Value of N is .... **[JEE Main 2020]**

28. A point object in air is in front of the curved surface of a plano-convex lens. The radius of curvature of the curved surface is 30 cm and the refractive index of the lens material is 1.5, then the focal length of the lens (in cm) is .... **[JEE Main 2020]**

29. When an object is kept at a distance of 30 cm from a concave mirror, the image is formed at a distance of 10 cm from the mirror. If the object is moved with a speed of 9 cms^{-1}, the speed (in cms^{-1}) with which image moves at that instant is .... **[JEE Main 2020]**

30. A compound microscope has an objective and eyepiece as thin lenses of focal lengths 1 cm and 5 cm, respectively. The distance between the objective and the eyepiece is 20 cm. The distance (in cm) at which the object must be placed in front of the objective, if the final image is located at 25 cm from the eyepiece, is numerically ....

31. In a compound microscope, the magnified virtual image is formed at a distance of 25 cm from the eyepiece. The focal length of its objective lens is 1 cm. If the magnification is 100 and the tube length of the microscope is 20 cm, then the focal length of the eyepiece lens (in cm) is .... **[JEE Main 2020]**

32. A compound microscope consists of an objective lens of focal length 1 cm and an eyepiece of focal length 5 cm with a separation of 10 cm. The distance between an object and the objective lens, at which the strain on the eye is minimum is $\frac{n}{40}$ cm. The value of n is .... **[JEE Main 2020]**

33. The distance between an object and a screen is 100 cm. A lens can produce real image of the object on the screen for two different positions between the screen and the object. The distance between these two positions is 40 cm. If the power of the lens is close to $\left(\frac{N}{100}\right)$ D, where N is an integer. Then, the value of N is .... **[JEE Main 2020]**

34. A light ray enters a solid glass sphere of refractive index $\mu = \sqrt{3}$ at an angle of incidence 60°. The ray is both reflected and refracted at the farther surface of the sphere. The angle (in degree) between the reflected and refracted rays at this surface is .... **[JEE Main 2020]**

35. An observer can see through a small hole on the side of a jar (radius 15 cm) at a point at height of 15 cm from the bottom (see figure). The hole is at a height of 45 cm. When the jar is filled with a liquid up to a height of 30 cm, the same observer can see the edge at the bottom of the jar. If the refractive index of the liquid is $N/100$, where N is an integer, the value of N is .... **[JEE Main 2020]**

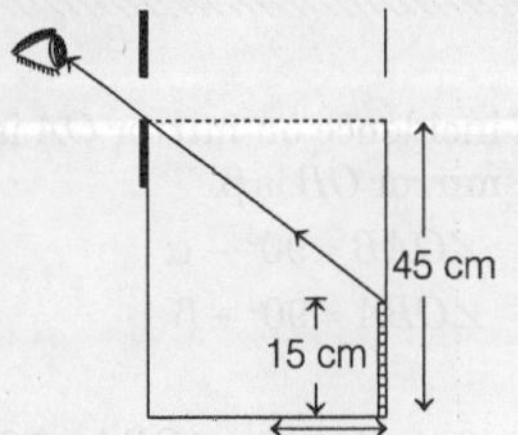

Answers

Round I

1. (a)	2. (c)	3. (b)	4. (a)	5. (c)	6. (a)	7. (d)	8. (c)	9. (b)	10. (a)
11. (d)	12. (c)	13. (c)	14. (c)	15. (a)	16. (a)	17. (a)	18. (b)	19. (c)	20. (d)
21. (d)	22. (d)	23. (b)	24. (d)	25. (c)	26. (c)	27. (b)	28. (d)	29. (a)	30. (a)
31. (c)	32. (c)	33. (d)	34. (a)	35. (c)	36. (c)	37. (a)	38. (d)	39. (d)	40. (c)
41. (b)	42. (c)	43. (b)	44. (a)	45. (a)	46. (c)	47. (d)	48. (c)	49. (a)	50. (a)
51. (d)	52. (b)	53. (c)	54. (a)	55. (d)	56. (c)	57. (c)	58. (b)	59. (d)	60. (c)
61. (c)	62. (a)	63. (a)	64. (b)	65. (c)	66. (b)	67. (a)	68. (a)	69. (b)	70. (c)
71. (a)	72. (b)	73. (d)	74. (b)	75. (c)	76. (c)	77. (c)	78. (c)	79. (c)	80. (b)
81. (d)	82. (b)	83. (a)	84. (d)	85. (c)	86. (d)	87. (b)	88. (c)	89. (a)	90. (c)
91. (c)	92. (d)	93. (d)	94. (c)	95. (b)	96. (b)	97. (a)	98. (c)	99. (b)	100. (c)
101. (b)	102. (b)	103. (b)	104 (a)	105. (a)	106. (b)	107. (a)	108. (a)	109. (a)	110. (d)
111. (a)	112. (c)	113. (a)	114. (a)	115. (a)	116. (a)	117. (d)	118. (d)	119. (d)	

Round II

1. (c)	2. (a)	3. (d)	4. (d)	5. (a)	6. (a)	7. (a)	8. (b)	9. (d)	10. (a)
11. (b)	12. (d)	13. (d)	14. (d)	15. (a)	16. (d)	17. (d)	18. (c)	19. (c)	20. (b)
21. (c)	22. (b)	23. (b)	24. (b)	25. (c)	26. 12	27. 5	28. 60	29. 1	30. 1.067
31. 6	32. 50	33. 476	34. 90	35. 158					

Solutions

Round I

1. The number of images formed, if two plane mirrors are kept at an angle θ is given by $\frac{360^\circ}{\theta}$, if $\frac{360^\circ}{\theta}$ is odd and it is given by $\frac{360^\circ}{\theta} - 1$, if $\frac{360^\circ}{\theta}$ is even.

For $\theta = 72^\circ$, number of images,

$$n = \frac{360^\circ}{72^\circ} = 5$$

As, $\frac{360^\circ}{72^\circ} = 5$ is an odd number.

2. The ray diagram for given situation is shown below.

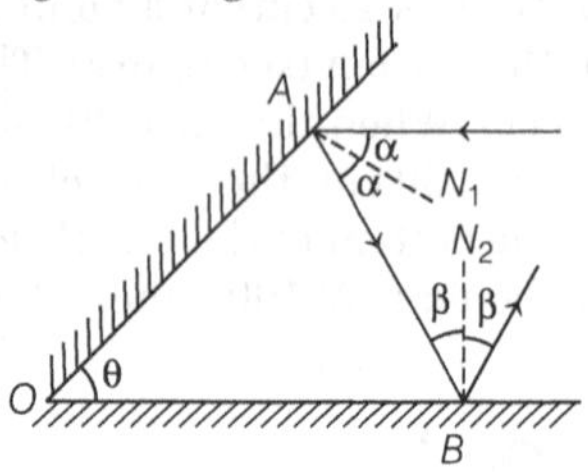

Let, angle of incidence on mirror OA is α and angle of incidence on mirror OB is β.

$\therefore \quad \angle OAB = 90^\circ - \alpha$

and $\quad \angle OBA = 90^\circ - \beta$

In ΔOAB,

$$\theta + \angle OAB + \angle OBA = 180^\circ$$

$$\Rightarrow \quad \theta + 90^\circ - \alpha + 90^\circ - \beta = 180^\circ$$

$$\Rightarrow \quad \theta = \alpha + \beta$$

$$\Rightarrow \quad \alpha + \beta = \theta \qquad \text{...(i)}$$

$\therefore$ Total deviation of the ray

$$\delta = (180^\circ - 2\alpha) + (180^\circ - 2\beta)$$
$$= 360^\circ - 2(\alpha + \beta)$$
$$= 360^\circ - 2\theta \qquad \text{[from Eq. (i)]}$$

3. The first images is due to reflection from the front surface, *i.e.* in unpolished surface of the mirror. So, only a small fraction is the incident light energy is reflected.

The second image is due to reflection from polished surface. So, a major portion of light is reflected. Thus, the second image is the brightest.

4. It is clear from figure, the distance of image with reference to observer reduces by 10 cm in one second. Thus, required speed, $20 - 10 = 10 \text{ cms}^{-1}$.

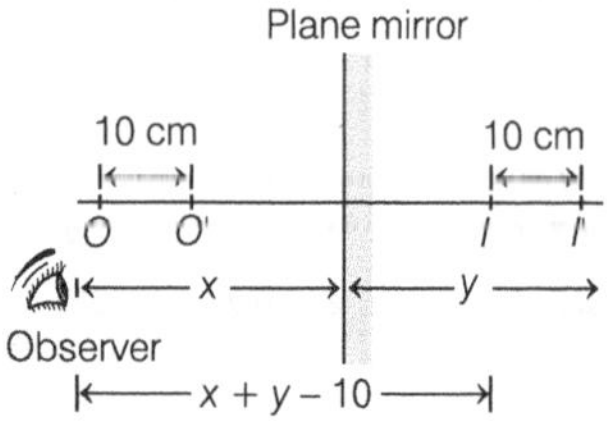

5. If the plane mirror is rotated through an angle θ, the reflected ray turns through an angle 2θ in that direction keeping the incident ray fixed.

6. Clearly, $i + r = i + i = 140^\circ$

or $\quad i = 70^\circ$

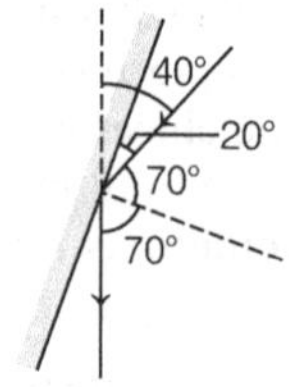

Clearly, the plane mirror makes an angle of 20° with vertical and 70° with horizontal. Thus, β is 70°.

7. As we known in the side mirror, the speed of approaching car would appear to increase, since the distance between the cars decreases.

8. Light from mirror is reflected in a straight line and it is appear to come from its image formed at same distance (as that of source) behind the mirror as shown in the ray diagram below

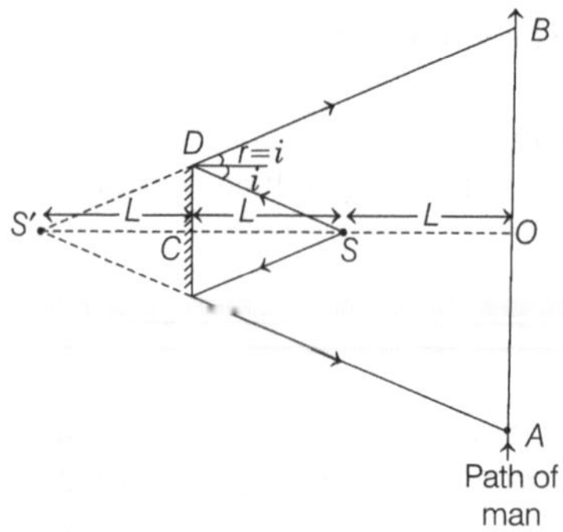

From ray diagram in similar triangles $\Delta S'CD$ and $\Delta S'OB$, we have

$$\frac{S'C}{S'O} = \frac{CD}{OB}$$

So, $\quad OB = \frac{CD \times S'O}{S'C} = \frac{\frac{d}{2} \times 3L}{L} = \frac{3d}{2}$

Also, $OA = \frac{3d}{2}$

So, distance over which man can see the image S' is

$$\frac{3d}{2} + \frac{3d}{2} = 3d$$

9. We have, $\quad \frac{1}{u} + \frac{1}{v} = \frac{1}{f}$

or $\quad \frac{1}{v} = -\frac{1}{u} + \frac{1}{f}$

Now, compare it with equation, $y = mx + c$.
Therefore, graph is a straight line having negative slope.

10. As, $\frac{f}{f-u} = \frac{1}{4} = \frac{f}{f-(-0.5)}$

or $4f = f + 0.5$ or $3f = 0.5$

or $f = \frac{0.5}{3}$ m $= 0.17$ m

11. As, $u = f - x_1$ and $v = f - x_2$

Now, $\frac{1}{u} + \frac{1}{v} = \frac{1}{f}$

$\Rightarrow \frac{1}{f-x_1} + \frac{1}{f-x_2} = \frac{1}{f}$

or $\frac{f-x_2+f-x_1}{(f-x_1)(f-x_2)} = \frac{1}{f}$

or $f^2 - fx_2 - fx_1 + x_1x_2 = 2f^2 - f(x_1+x_2)$

or $f^2 = x_1x_2$

or $f = \sqrt{x_1x_2}$

This is Newton's mirror formula.

12. The image is erect and diminished. So, the mirror is necessarily convex.

13. We have, $\frac{1}{-30} + \frac{1}{v} = \frac{1}{30}$

or $\frac{1}{v} = \frac{2}{30} = \frac{1}{15}$

or $v = 15$ cm

The distance is behind the mirror.

14. We have, $m = \frac{f}{f-u}$

$= \frac{-16}{-16-(-8)} = \frac{-16}{-8} = 2$

15. From the mirror formula,

$\frac{1}{v} + \frac{1}{(-600)} = \frac{1}{20}$

The given situation is shown in the figure.

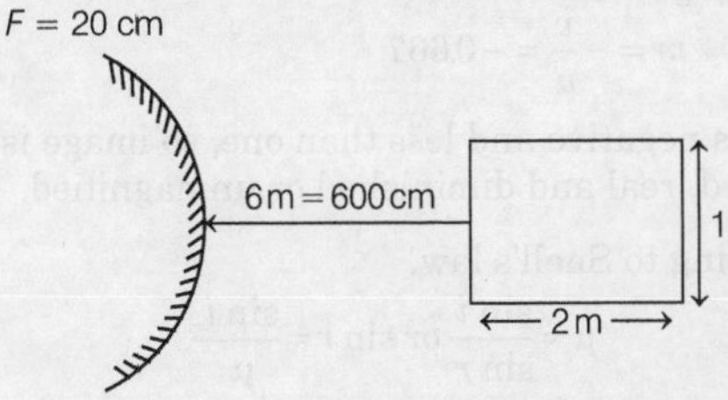

or $\frac{1}{v} = \frac{1}{20} + \frac{1}{600}$

or $\frac{1}{v} = \frac{31}{600}$

or $v = \frac{600}{31}$ cm $= 19.35$ cm

16. By keeping the incident ray is fixed, if plane mirror rotates through an angle θ reflected ray rotates through an angle 2θ.

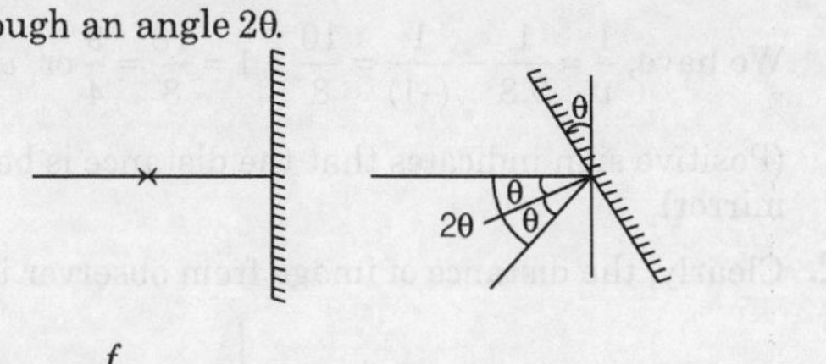

17. As, $m = \frac{f}{f-u}$

$\Rightarrow 2 = \frac{-0.2}{-0.2-u}$

or $2 = \frac{0.2}{0.2+u}$

or $0.4 + 2u = 0.2$

or $2u = 0.2 - 0.4 = -0.2$

or $u = -0.1$ m

18. Clearly, the given mirror is a convex mirror.

As $m = \frac{f}{f-u}$

$\Rightarrow \frac{1}{3} = \frac{18}{18-u}$

or $3 \times 18 = 18 - u$

or $u = -2 \times 18$ cm

or $u = -36$ cm

19. We have, $\theta = \frac{1}{2} \times \frac{\pi}{180}$ rad

As, $\frac{\text{diameter of image}}{\text{focal length}} = \theta$

or diameter of image

$= \frac{1}{2} \times \frac{\pi}{180} \times \frac{15}{2} \times 100$ cm

$= 6.55$ cm

20. Using mirror formula,

$\frac{1}{v} + \frac{1}{(-u)} = \frac{1}{-f}$

$\Rightarrow v = \frac{fu}{f-u}$

$\Rightarrow \frac{\delta V}{\delta x} = \frac{\delta}{\delta x}\left(\frac{fu}{u-f}\right)$

$\Rightarrow \frac{\partial V}{\partial x} = \frac{f^2}{(u-f)^2} \cdot \frac{\partial u}{\partial x}$

$\Rightarrow d_v = \frac{f^2}{(u-f)^2} \cdot d_u$

Now, $d_u = b$ (given)

$\therefore d_v = \frac{bf^2}{(u-f)^2} = b\left(\frac{f}{u-f}\right)^2$

= size of image.

21. Here, $f = \frac{1.6}{2}$ m $= 0.8$ m, $u = -1$ m

We have, $\frac{1}{v} = \frac{1}{0.8} - \frac{1}{(-1)} = \frac{10}{8} + 1 = \frac{18}{8} = \frac{9}{4}$ or $v = \frac{4}{9}$ m

(Positive sign indicates that the distance is behind the mirror)

22. Clearly, the distance of image from observer is 40 cm.

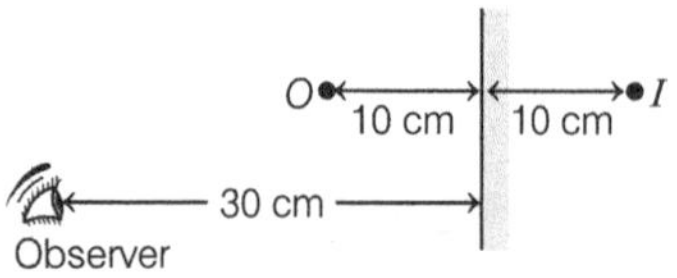

23. As, magnification, $m = \frac{f}{f-u}$

If $f = -24$ cm and $m = +3$, then

$$3 = \frac{-24}{-24-u}$$

or $\quad -24 - u = -8$

or $\quad u + 24 = 8$

or $\quad u = -16$ cm

Now, if $m = -3$, then

$$-3 = \frac{-24}{-24-u}$$

or $\quad u + 24 = -8$

or $\quad u = -32$ cm

24. Given, focal length of convex mirror, $f = +15$ cm (focal length of convex mirror is taken as positive)

Distance of object, $u = -12$ cm

Size of object, $O = 4.5$ cm

Using the mirror formula,

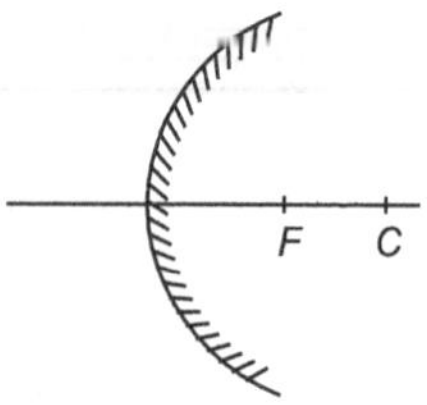

$$\frac{1}{f} = \frac{1}{v} + \frac{1}{u}$$

$$\frac{1}{15} = \frac{1}{v} - \frac{1}{12}$$

$$\Rightarrow \quad \frac{1}{v} = \frac{9}{60}$$

Distance of image from the mirror, $v = 6.7$ cm.

The positive sign shows that the image is formed behind the mirror.

Using the formula of magnification,

$$m = -\frac{v}{u} = \frac{I}{O}$$

$$\frac{-6.7}{-12} = \frac{I}{4.5}$$

Size of image, $I = 2.5$ cm

As I is positive, so image is erect and virtual.

Magnification m is given by

$$m = \frac{I}{O} = \frac{2.5}{4.5} = \frac{25}{45} = \frac{5}{9}$$

As the needle moves away from the mirror, the image also moves away from the mirror (as $u \to \infty$, $v \to f$) and the size of image goes on decreasing.

25. One image will be real and the other will be virtual. Since, they are of the same size, one will have magnification m and the other $-m$.

$$\frac{1}{u_1} + \frac{1}{u_1 m} = \frac{1}{f}$$

or $$\frac{1}{u_1}\left(1 + \frac{1}{m}\right) = \frac{1}{f} \quad \text{... (i)}$$

and $$\frac{1}{u_2} - \frac{1}{u_2 m} = \frac{1}{f}$$

or $$\frac{1}{u_2}\left(1 - \frac{1}{m}\right) = \frac{1}{f} \quad \text{... (ii)}$$

From Eqs. (i) and (ii), we get

$$\frac{u_1 + u_2}{f} = 2$$

or $$f = \frac{u_1 + u_2}{2}$$

26. Magnification produced by a mirror can also be given as

$$m = \frac{f}{f-u}$$

Substituting the given values, we get

$$5 = \frac{-0.4}{-0.4-u}$$

or $\quad u = -0.32$ m

27. In question figure, PQ is a ray of light passing through focus, and falling on the surface of a concave mirror. On reflection from the mirror, the ray becomes parallel to principal axis of the mirror.

28. Radius of curvature of given concave mirror, $R = -8$ cm

So, focal length, $f = \frac{R}{2} = -4$ cm

Object distance, $u = -10$ cm

Using mirror formula,

$$\frac{1}{v} = \frac{1}{f} - \frac{1}{u} = -\frac{1}{4} - \frac{1}{(-10)} = -\frac{3}{20}$$

$\Rightarrow \quad v = -6.67$ cm (real)

Also, magnification,

$$m = -\frac{v}{u} = -0.667$$

As, m is negative and less than one, so image is inverted, real and diminished or unmagnified.

29. According to Snell's law,

$$\mu = \frac{\sin i}{\sin r} \text{ or } \sin r = \frac{\sin i}{\mu}$$

As μ is negative, $\sin r$ is negative.

Therefore, r is negative.

If incident ray from air (medium 1) incident on those material, the ray refract or bend same side of the normal as in option (a).

30. As, $$n_1 = \frac{c}{v_1} = \frac{\nu\lambda}{\nu\lambda_1} = \frac{\lambda}{\lambda_1}$$

and $\quad n_2 = \frac{c}{v_2} = \frac{v\lambda}{v\lambda_2} = \frac{\lambda}{\lambda_2}$

Now, $\quad \frac{n_1}{n_2} = \frac{\lambda_2}{\lambda_1}$ or $\lambda_2 = \left(\frac{n_1}{n_2}\right)\lambda_1$

31. We have, $\frac{\sin r}{\sin i} = \tan 30° = \frac{1}{\sqrt{3}}$

or $\quad \frac{\sin i}{\sin r} = \sqrt{3}$

$\Rightarrow \quad \frac{\text{Speed of light in medium } X}{\text{Speed of light in medium } Y} = \sqrt{3}$

$\Rightarrow$ Speed of light in medium $Y = \frac{1}{\sqrt{3}} \times$ Speed of light in medium X

32. From the figure,

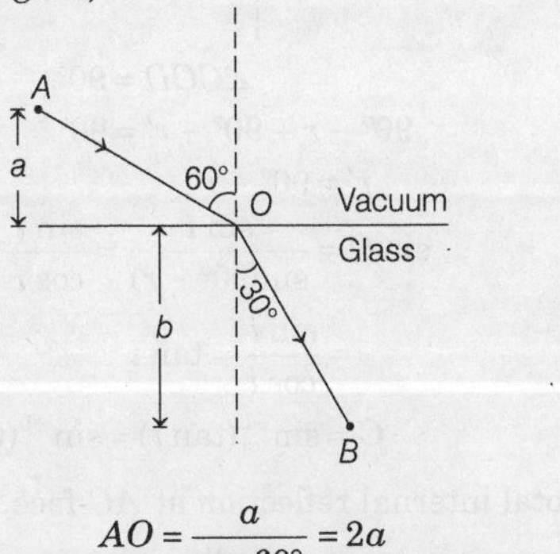

$$AO = \frac{a}{\cos 60°} = 2a$$

and $\quad BO = \frac{b}{\cos 30°} = \frac{2}{\sqrt{3}}b$

From Snell's law,

$$\mu = \frac{\sin 60°}{\sin 30°} = \frac{\sqrt{3}/2}{1/2} = \sqrt{3}$$

$\therefore$ Optical path $= AO + \mu\,(BO)$

$$= 2a + (\sqrt{3} \times \frac{2}{\sqrt{3}} b) = 2a + 2b$$

33. Given, $\mu = 1.5$, $t_1 = 5$ cm, $\mu_2 = 1.33$ and $t_2 = 1$ cm

Change in path $= \Delta t_1 + \Delta t_2$

$$= \left(1 - \frac{1}{\mu_1}\right) \times t_1 + \left(1 - \frac{1}{\mu_2}\right) \times t_2$$

$$= \left(1 - \frac{1}{1.5}\right) \times 5 + \left(1 - \frac{1}{1.33}\right) \times 1 \simeq 1.90 \text{ cm}$$

34. We have, $\mu_g \sin \theta_C = \mu_1 \sin 90°$

or $\quad \mu_g \sin \theta_C = 1$

When water is poured, then

$\quad \mu_w \sin r = \mu_s \sin \theta_C$

or $\quad \mu_w \sin r = 1$

Again, $\quad \mu_a \sin \theta = \mu_w \sin r$

or $\quad \mu_a \sin \theta = 1$

or $\quad \sin \theta = 1$

or $\quad \theta = 90°$

35. Let r be the radius of circle through which other objects become visible. The rays of light must be incident at critical angle C.

Now, $\quad \sin C = \frac{1}{\mu} = \frac{r}{\sqrt{r^2 + h^2}}$

or $\quad \mu^2 r^2 = r^2 + h^2$

or $\quad (\mu^2 - 1)r^2 = h^2$

$\Rightarrow \quad r = \frac{h}{\sqrt{\mu^2 - 1}}$

Diameter, $\quad 2r = \frac{2h}{\sqrt{\mu^2 - 1}}$

36. Light ray is going from liquid (denser) to air (rarer) and angle of refraction is 90°, so angle of incidence must be equal to critical angle, therefore $\sin C = \frac{4}{5}$

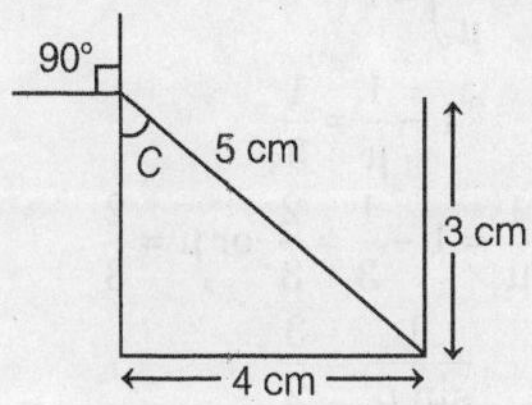

Also, $\quad \mu = \frac{1}{\sin C} = \frac{5}{4} = 1.2$

37. In figure, a pin is held at L, mid-point of AB. When seen from face AD (so long $i < C$) image of L appears to be at L', close to A.

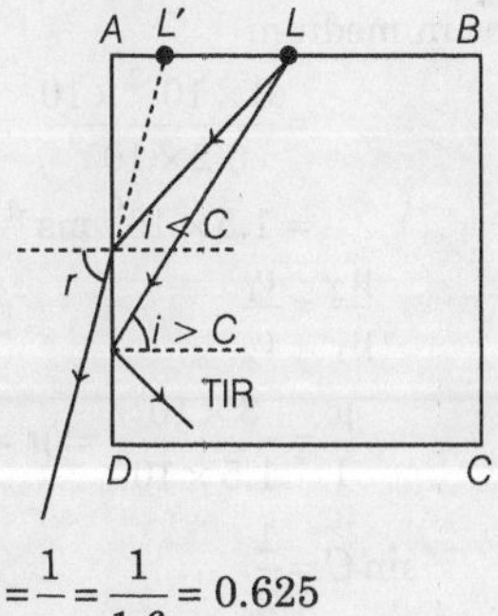

From $\sin C = \frac{1}{\mu} = \frac{1}{1.6} = 0.625$

$\Rightarrow \quad C = \sin^{-1}(0.625) = 38.7°$

So when angle of incidence becomes greater than $C (= 38.7°)$, the rays starting from L will undergo total internal reflection and pin shall not be seen at all.

38. The real depth = Refractive index (Apparent depth)

In first case, the real depth $h_1 = \mu\,(d - c)$.

Similarly, in the second case, the real depth, $h_2 = \mu\,(d - c)$

Since $h_2 \geq h_1$, the difference of the depths $= h_2 - h_1$

$$= \mu(d - c - b + a)$$

Since, the liquid is added in second case, $h_2 - h_1 = d - b$

$$\mu = \frac{d - b}{d - c - b + a}$$

39. Speed of light in a medium is given by

$$v = \frac{1}{\sqrt{\varepsilon\mu}} = \frac{1}{\sqrt{\varepsilon_0\mu_0}} \times \frac{1}{\sqrt{\varepsilon_r\mu_r}} = \frac{c}{\sqrt{\varepsilon_r\mu_r}}$$

Here, $\varepsilon_r = 3$ and $\mu_r = \frac{4}{3}$

$$\therefore \quad \frac{v}{c} = \frac{1}{\sqrt{\varepsilon_r\mu_r}} = \frac{1}{\sqrt{3 \times \frac{4}{3}}} = \frac{1}{2}$$

So, refractive index of medium w.r.t. vacuum or air,

$$n = \frac{c}{v} = 2$$

Hence, critical angle,

$$i_c = \sin^{-1}\left(\frac{1}{n}\right) = \sin^{-1}\frac{1}{2} = 30°$$

40. We have $3\left(1 - \frac{1}{\mu}\right) = 1$

or $\quad 1 - \frac{1}{\mu} = \frac{1}{3}$

or $\quad \frac{1}{\mu} = 1 - \frac{1}{3} = \frac{2}{3}$ or $\mu = \frac{2}{3}$

Now, $\quad \frac{1}{\sin i_C} = \frac{3}{2}$

or $\quad \sin i_C = \frac{2}{3}$

or $\quad i_C = \sin^{-1}\left(\frac{2}{3}\right)$

or $\quad i_C = \sin^{-1}(0.67)$

41. Speed of light in medium

$$= \frac{3 \times 10^{-2} \times 10}{0.2 \times 10^{-9}}$$

$$= 1.5 \times 10^{9}\ \text{ms}^{-1}$$

As, $\quad \frac{\mu_2}{\mu_1} = \frac{v_1}{v_2}$

$$\frac{\mu}{1} = \frac{3 \times 10^{8}}{1.5 \times 10^{8}} \Rightarrow \mu = 2$$

We have, $\quad \sin C = \frac{1}{2}$

$\Rightarrow \quad C = \sin^{-1}\left(\frac{1}{2}\right) = 30°$

Hence, it suffers internal reflection in the case when the angle of incidence is 40°.

42. Given, $\angle i = 60°$ and $x = 1\ \text{km} = 10^3\ \text{m}$

When the total internal reflection just takes place from lateral surface, $i = C$, i.e. $60° = C$

$\Rightarrow \quad \sin C = \frac{1}{\mu}$

$\Rightarrow \quad \sin 60° = \frac{1}{\mu} = \frac{\sqrt{3}}{2}$

$\Rightarrow \quad \mu = \frac{2}{\sqrt{3}}$

The time taken by light to travel some distance in a medium,

$$t = \frac{\mu x}{c} = \frac{\frac{2}{\sqrt{3}} \times 10^3}{3 \times 10^8} = 3.85\,\mu\text{s}$$

43. At the critical angle, $\sin C = \frac{1}{\mu} = \frac{1}{\sin r'/\sin i} = \frac{\sin i}{\sin r'}$

As is clear as shown in figure

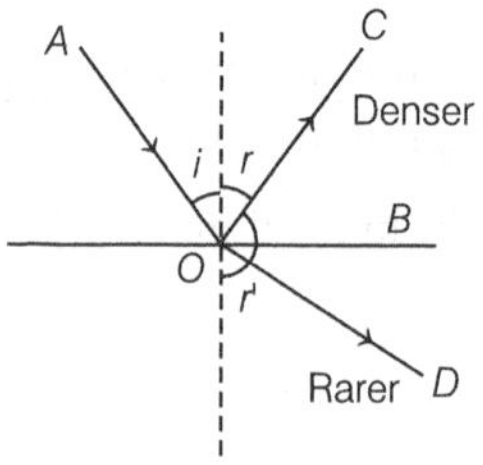

$$\angle COD = 90°$$

$\therefore \quad 90° - r + 90° - r' = 90°$

or $\quad r' = 90° - r$

$\therefore \quad \sin C = \frac{\sin i}{\sin(90° - r)} = \frac{\sin i}{\cos r}$

$$= \frac{\sin i}{\cos i} = \tan i \qquad (\because i = r)$$

$$C = \sin^{-1}(\tan i) = \sin^{-1}(\tan r).$$

44. For total internal reflection at AC-face

$$\sin i \geq \frac{\mu_w}{\mu_g}$$

$$\sin\theta \geq \frac{4}{3 \times 1.5}$$

$$\sin\theta \geq \frac{8}{9}$$

45. As shown in figure

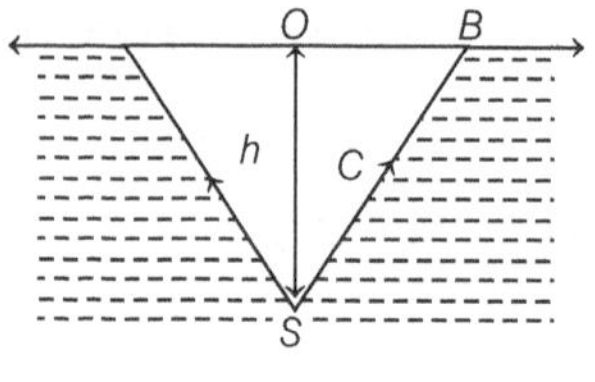

$$OS = h$$

When angle of incidence is slightly greater than C, light undergoes total internal reflection.

$\therefore$ Diameter of circle of light coming from water surface $= 2r = 2(OB) = 2OS\tan C = 2h\tan C$

46. Here, for yellow light, $r = 90°$ when $i = C$. As, i is kept same, C must be smaller for total internal reflection. From $\mu = \frac{1}{\sin C}$, C will be smaller when μ is larger. Out of given colours, μ is largest for blue colour. Critical angle will be smallest for blue colour. Therefore, blue light would undergo total internal reflection.

47. Let the bulb is placed at point O,

$$AB = AC = r$$

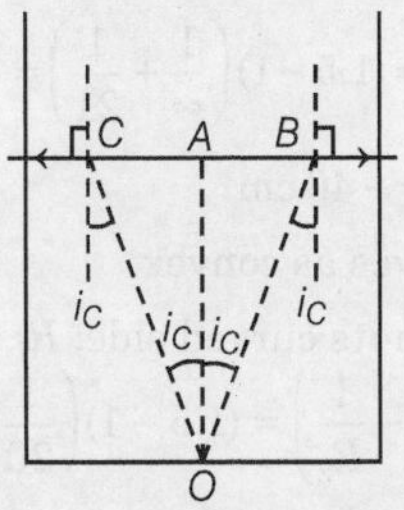

If the light falls at an angle of incidence equal to critical angle i_C, then only a circular area is formed because if angle of incidence is less than the critical angle, it will refract into air and when angle of incidence is greater than critical angle, then it will be reflected back in water.

The source of light is 80 cm below the surface of water, *i.e.* $AO = 80 \text{ cm}, \mu_w = 1.33$

Using the formula for critical angle,

$$\sin i_C = \frac{1}{\mu_w}$$

$$\sin i_C = \frac{1}{1.33} = 0.75$$

$$\Rightarrow \quad i_C = 48.6°$$

In ΔOAB, $$\tan i_C = \frac{AB}{AO}$$

$$\Rightarrow \quad \tan i_C = \frac{r}{l}$$

$$\Rightarrow \quad r = l \tan i_C = 80 \tan 48.6$$

$$r = 80 \times 1.1345 = 90.7 \text{ cm}$$

Area of circular surface of water through which light will emerge,

$$A = \pi r^2$$

$$A = 3.14 \times (90.7)^2 = 25865.36 \text{ cm}^2$$

$$A = 2.58 \text{ m}^2$$

$$\simeq 2.6 \text{ m}^2$$

48. We know that, $\sin C = \frac{\mu_1}{\mu_2}$

$$\because \quad \mu_1 \sin\theta = \mu_2 \sin(90° - C)$$

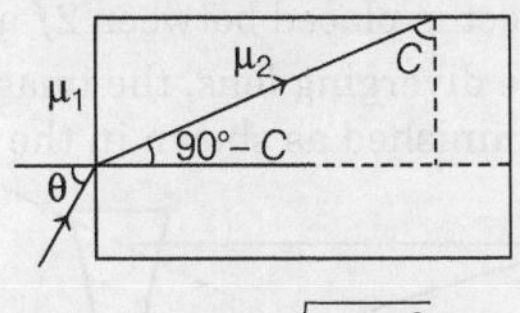

$$\Rightarrow \quad \sin\theta = \frac{\mu_2\sqrt{1 - \frac{\mu_1^2}{\mu_2^2}}}{\mu_1}$$

$$\Rightarrow \quad \theta = \sin^{-1}\sqrt{\frac{\mu_2^2 - \mu_1^2}{\mu_1^2}}$$

For TIR, $\theta < \sin^{-1}\sqrt{\frac{\mu_2^2}{\mu_1^2} - 1}$

49. We have, $$\mu_1 = 1.20 + \frac{0.8 \times 10^{-14}}{(400 \times 10^{-9})^2}$$

or $$\mu_1 = 1.25$$

or $$\sin i_C = \frac{1}{1.25} = 0.8$$

$$\Rightarrow \quad i_C = 53.13°$$

Again, $$\mu_2 = 1.20 + \frac{0.8 \times 10^{-14}}{(500 \times 10^{-9})^2}$$

$$\because \quad \mu_2 = 1.232$$

or $$\sin i_C = \frac{1}{1.232} = 0.81$$

$$\Rightarrow \quad i_C = \sin^{-1} 0.81$$

$$= 54.26°$$

Now, $$\sin\theta = 0.8$$

or $$\theta = 53.13°$$

This angle is clearly greater than critical angle corresponding to wavelength 400 nm. So, light of 400 nm wavelength undergoes total internal reflection.

50. Using $\frac{\mu_2}{v} - \frac{\mu_1}{u} = \frac{\mu_2 - \mu_1}{R}$, we get

$$\frac{1.5}{v} - \frac{1.0}{\infty} = \frac{1.5 - 1.0}{20}$$

or $$v = 60 \text{ cm}$$

Hence, parallel rays of light actually meet at some point *i.e.* at a distance of 60 cm.

51. From a lens formula,

$$\frac{1}{f} = \frac{1}{v} - \frac{1}{u}$$

If $u = \infty$, then $v = f$ and if $u = f$, then $v = \infty$.

Hence, graph between u and v will be a rectangular parabola.

Therefore, option (d) is correct.

52. Since two refractive indices are involved. Therefore, two images will be formed.

53. As, $\frac{1}{f} = (\mu - 1)\left(\frac{1}{r_1} - \frac{1}{r_2}\right)$ (Lens maker formula)

For lens to be concave, $\left(\frac{1}{r_1} - \frac{1}{r_2}\right) > 0$

or $$\frac{1}{r_1} > \frac{1}{r_2}$$

or $$r_1 < r_2$$

54. As, $m = \frac{f}{f + u}$

$$\Rightarrow \quad \frac{1}{2} = \frac{-20}{-20 + u}$$

or $$-20 + u = -40$$

or $$u = -40 + 20$$

or $$u = -20 \text{ cm}$$

55. As, $\frac{I}{O} = \frac{v}{u} =$ magnification

or $\frac{I}{15} = \frac{-35}{-10}$

$\therefore \quad I = 15 \times 2.5 \text{ cm} = 37.5 \text{ cm}$

56. As, $n = \frac{f}{f+u}$

$\Rightarrow \quad f + u = \frac{f}{n}$

$\Rightarrow \quad u = \frac{f}{n} - f = \left(\frac{1-n}{n}\right)f$

$\Rightarrow \quad u = -\left(\frac{n-1}{n}\right)f$

$\therefore \quad |u| = \left(\frac{n-1}{n}\right)f$

57. We have, $\frac{1}{f} = (\mu - 1)\left(\frac{2}{R}\right)$ (By lens maker's formula)

or $f = \frac{R}{2(\mu - 1)}$

Now, $f > R$

$\therefore \quad \frac{R}{2(\mu-1)} > R$

or $\frac{1}{2(\mu-1)} > 1$

or $2(\mu - 1) < 1$

or $\mu - 1 < \frac{1}{2}$

or $\mu < \left(1 + \frac{1}{2}\right)$

or $\mu < 1.5$

58. As the image formed by a large number of rays from the object. So, if the central portion of the lens is wrapped in blank paper, then full image of object will be formed, but due to decrease of number of rays on lens, brightness of image will decrease.

59. Clearly, $2f = 20$ cm or $f = 10$ cm

Now, $u = -15$ cm, $v = ?$

$\because \quad \frac{1}{v} - \frac{1}{(-15)} = \frac{1}{10}$

or $\frac{1}{v} + \frac{1}{15} = \frac{1}{10}$

or $\frac{1}{v} = \frac{1}{10} - \frac{1}{15}$

or $\frac{1}{v} = \frac{3-2}{30} = \frac{1}{30}$

or $v = 30$ cm

The change in image distance is (30 − 20) cm, *i.e.* 10 cm and away from the lens.

60. Here, $\mu = 1.5$

If object lies on its plane side; $R_1 = \infty, R_2 = -20$ cm

$$\frac{1}{f} = (\mu - 1)\left(\frac{1}{R_1} - \frac{1}{R_2}\right)$$

$$= (1.5 - 1)\left(\frac{1}{\infty} + \frac{1}{21}\right) = \frac{1}{40}$$

$\Rightarrow \quad f = +40$ cm

The lens behaves as convex.

If object lies on its curved side; $R_1 = 20$ cm, $R_2 = \infty$

$$\frac{1}{f'} = (\mu - 1)\left(\frac{1}{R_1} - \frac{1}{R_2}\right) = (1.5 - 1)\left(\frac{1}{20} - \frac{1}{\infty}\right) = \frac{1}{40}$$

$\Rightarrow f' = 40$ cm

The lens behaves as convex.

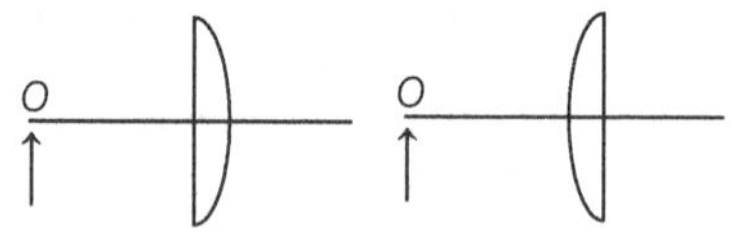

61. When an object approaches a convergent lens from the left of the lens with a uniform speed of 5 m/s, the image moves away from the lens with a non-uniform acceleration. For example, if $f = 20$ m and $v = -50$ m ; -45 m , -40 m and -35 m; we get $v = 33.3$ m ; 36 m ; 40 m and 46.7 m. Clearly, image moves away from the lens with a non-uniform acceleration.

62. The central ray goes undeviated. So, $\mu_1 = \mu_3$. Also, $\mu_3 < \mu_2$.

63. We have, $-\frac{1}{40} = (1.5 - 1)\left(\frac{1}{R_1} - \frac{1}{R_2}\right)$

(from lens maker formula)

$\Rightarrow \quad \frac{1}{R_1} - \frac{1}{R_2} = -\frac{1}{20}$

Now, $\frac{1}{f} = \left(\frac{1.5}{2} - 1\right)\left(-\frac{1}{20}\right)$

or $\frac{1}{f} = -\frac{0.5}{2}\left(-\frac{1}{20}\right)$

or $\frac{1}{f} = \frac{1}{80}$

or $f = 80$ cm

So, it behaves like a convex lens of focal length 80 cm.

64. When an object is placed between $2f$ and f (focal length) of the diverging lens, the image is virtual, erect and diminished as shown in the figure.

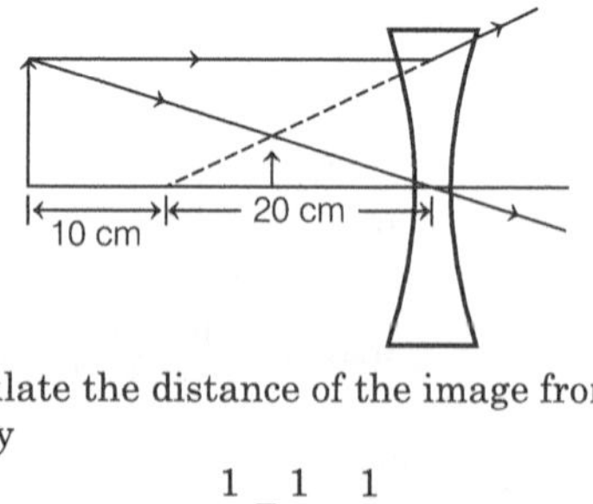

To calculate the distance of the image from the lens, we apply

$$\frac{1}{f} = \frac{1}{v} - \frac{1}{u}$$

$\Rightarrow \quad \frac{1}{-20} = \frac{1}{v} + \frac{1}{30}$ (as, $\mu = -30$ cm)

$\Rightarrow \quad v = -\frac{(20)(30)}{20+30}$

Thus, distance = − 12 cm

(to the left of the diverging lens)

65. As, $y = \sqrt{y_1 \times y_2} = \sqrt{16 \times 9} = 4 \times 3 = 12$ cm

66. As $I \propto A^2$, where I = intensity and A = amplitude

$$\Rightarrow \frac{I_2}{I_1} = \left(\frac{A_2}{A_1}\right)^2 = \frac{\pi r^2 - \frac{\pi r^2}{4}}{\pi r^2} = \frac{4\pi r^2 - \pi r^2}{4\pi r^2} = \frac{3\pi r^2}{4\pi r^2} = \frac{3}{4}$$

$$\Rightarrow \frac{I_2}{I_1} = \frac{3}{4}$$

$\Rightarrow \quad I_2 = \frac{3}{4} I_1$ and the focal length remains unchanged.

67. Here, $x = u + v$ ...(i)

As, $\quad m = \frac{f}{f+u} = \frac{f-v}{f}$

and image is real, magnification is negative.

$\therefore \quad -m = \frac{f}{f+u}, u = \frac{-(m+1)f}{m}$

From $\quad -m = \frac{f-v}{f}$

$\Rightarrow \quad v = (m+1)f$

Put in Eq. (i), we get

$$x = \frac{-(m+1)}{m} f + (m+1)f$$

Solving, we get

$$f = \frac{mx}{(m+1)^2}$$

68. In the situation given, the image will be formed at infinity, if the object is at focus of the lens, *i.e.* at 20 cm from the lens. Hence, shift in position of object,

$$x = 25 - 20 = \left(1 - \frac{1}{\mu}\right)t$$

$$5 = \left(1 - \frac{1}{1.5}\right)t$$

$\Rightarrow \quad t = 15$ cm

69. As, $f \propto \frac{1}{\mu - 1}$

$\Rightarrow f \propto \frac{1}{\mu}$

Since, $\mu_V > \mu_G > \mu_R$

$\therefore \quad f_V < f_G < f_R$

70. As, $\frac{1}{F} = \frac{1}{f_1} + \frac{1}{f_2}$

$\Rightarrow \quad f_1 = -f_2$

$F = \infty$

71. We have, from lens maker formula,

$$5 = (1.5 - 1)\left(\frac{2}{R}\right) \quad \text{...(i)}$$

and $\quad -1 = \left(\frac{1.5}{n} - 1\right)\left(\frac{1}{R}\right)$ ...(ii)

Dividing Eq. (i) by Eq. (ii), we get

$$-5 = \frac{0.5n}{1.5 - n}$$

or $\quad -7.5 + 5n = 0.5n$

or $\quad -7.5 = -4.5n$

$\therefore \quad a^n l = n = \frac{75}{45} = \frac{5}{3}$

72. Using lens formula,

$$\frac{1}{f} = \frac{1}{v} - \frac{1}{u}$$

$\Rightarrow \quad \frac{1}{v} = \frac{1}{f} + \frac{1}{u}$

$\Rightarrow \quad v = \frac{uf}{u+f}$

Now, considering two cases

Case I If $v = u$

$\Rightarrow \quad u = \frac{uf}{u+f}$

$\Rightarrow \quad u + f = u$

$\Rightarrow \quad u = 0$

So, the curve will pass through origin.

Case II If $u = \infty \Rightarrow v = f$

So, as u tends towards infinity, v tends towards focus and all other distances will be less than focal length f and it will never intersect $u = v$ line.

Hence, correct option is (b).

73. Lens formula is given as

$$\frac{1}{f} = \frac{1}{v} - \frac{1}{u} \quad \text{... (i)}$$

$\Rightarrow \quad \frac{1}{v} = \frac{1}{f} + \frac{1}{u}$

$\Rightarrow \quad \frac{uf}{u+f} = v \Rightarrow \frac{v}{u} = \frac{f}{u+f}$ ...(ii)

Now, by differentiating Eq. (i), we get

$$0 = -\frac{1}{v^2} \cdot \frac{dv}{dt} + \frac{1}{u^2} \cdot \frac{du}{dt}$$

[$\because f$ (focal length of a lens is constant)]

or $\quad \frac{dv}{dt} = \frac{v^2}{u^2} du/dt$

$\Rightarrow \quad \frac{dv}{dt} = \left(\frac{f}{u+f}\right)^2 \frac{du}{dt}$ [using Eq. (ii)]

Given, $f = 0.3$ m, $u = -20$ m, $du/dt = 5$ m/s

$\therefore \quad \frac{dv}{dt} = \left(\frac{0.3}{0.3-20}\right)^2 \times 5 = \left(\frac{3}{197}\right)^2 \times 5$

$= 1.16 \times 10^{-3}$ m/s

Thus, the image is moved with a speed of 1.16×10^{-3} m/s towards the lens.

74. Given combination is as shown below

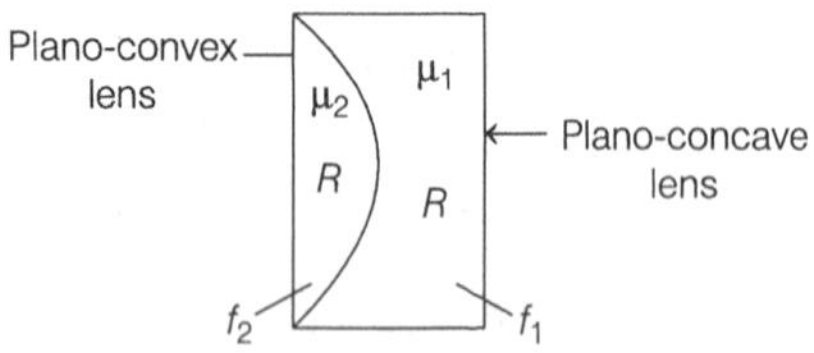

As lenses are in contact, equivalent focal length of combination is

$$\frac{1}{f_{eq}} = \frac{1}{f_1} + \frac{1}{f_2}$$

Using lens Maker's formula,

$$\frac{1}{f} = (\mu - 1)\left(\frac{1}{R_1} - \frac{1}{R_2}\right)$$

Here, $\frac{1}{f_1} = (\mu_1 - 1)\left(\frac{1}{-R} - \frac{1}{\infty}\right) = \frac{(1-\mu_1)}{R}$

and $\frac{1}{f_2} = (\mu_2 - 1)\left(\frac{1}{\infty} - \frac{1}{(-R)}\right) = \frac{(\mu_2 - 1)}{R}$

$\therefore \quad \frac{1}{f_{eq}} = \left(\frac{1-\mu_1}{R}\right) + \left(\frac{\mu_2 - 1}{R}\right)$

$$= \frac{\mu_2 - 1 + 1 - \mu_1}{R} = \frac{\mu_2 - \mu_1}{R}$$

So, $f_{eq} = \frac{R}{\mu_2 - \mu_1}$

75. By Pythagoras theorem,

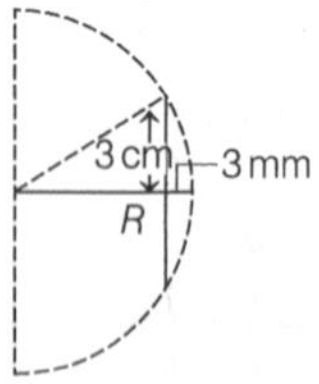

$$R^2 = (3)^2 + (R - 0.3)^2$$

$\Rightarrow \quad R \approx 15$ cm

Refractive index of material of lens, $\mu = \frac{c}{v}$

Here, c = speed of light in vacuum $= 3 \times 10^8$ m/s

v = speed of light in material of lens $= 2 \times 10^8$ m/s

$$= \frac{3 \times 10^8}{2 \times 10^8} = \frac{3}{2}$$

From lens maker's formula,

$$\frac{1}{f} = (\mu - 1)\left(\frac{1}{R_1} - \frac{1}{R_2}\right)$$

Here, $R_1 = R$ and $R_2 = \infty$ (for plane surface)

$$\frac{1}{f} = \left(\frac{3}{2} - 1\right)\left(\frac{1}{15}\right)$$

$\Rightarrow \quad f = 30$ cm

76. Image can be formed on the screen, if it is real. Real image of reduced size can be formed by a concave mirror of a convex lens.

Let $u = 2f + x$, then

mirror equation, $\frac{1}{u} + \frac{1}{v} = \frac{1}{f}$

$\Rightarrow \quad \frac{1}{2f + x} + \frac{1}{v} = \frac{1}{f}$

$\Rightarrow \quad \frac{1}{v} = \frac{1}{f} - \frac{1}{2f + x} = \frac{f + x}{f(2f + x)}$

$\Rightarrow \quad v = \frac{f(2f + x)}{f + x}$

It is given that, $u + v = 1.0$ m

$$2f + x + \frac{f(2f + x)}{f + x} = (2f + x)\left[1 + \frac{f}{f + x}\right] < 1.0 \text{ m}$$

or $\quad \frac{(2f + x)^2}{f + x} < 1.0$ m

or $\quad (2f + x)^2 < (f + x)$

This will be true only, when $f < 0.25$ m

77. Given, $f_1 = 2f_2$

$\Rightarrow \quad \frac{1}{|f_1|} = \frac{1}{|2f_2|}$ ...(i)

Using lens maker's formula, we get

$$\frac{1}{f_1} = (\mu_1 - 1)\left(\frac{1}{R_1} - \frac{1}{R_2}\right)$$

and $\frac{1}{f_2} = (\mu_2 - 1)\left(\frac{1}{R_1'} - \frac{1}{R_2'}\right)$

$\Rightarrow \left|(\mu_1 - 1)\left(\frac{1}{\infty} - \frac{1}{(-R)}\right)\right| = \left|\frac{1}{2}(\mu_2 - 1)\left(\frac{1}{-R} - \frac{1}{\infty}\right)\right|$

[$\because$ using Eq. (i)]

$\Rightarrow \quad \frac{\mu_1 - 1}{R} = \frac{\mu_2 - 1}{2R}$

$\Rightarrow \quad 2\mu_1 - \mu_2 = 1$

78. When given set up is immersed in water, focal length of lens increases,

Note

$$\frac{f_{\text{liquid}}}{f_{\text{air}}} = \frac{n_{ga} - 1}{n_{gl} - 1} = \frac{n_{ga} - 1}{\left(\frac{n_{ga}}{n_{la}} - 1\right)}$$

Now, $n_{ga} = \frac{3}{2}$ and $n_{la} = \frac{4}{3}$

$$\therefore f_{\text{liquid}} = f_{\text{air}}\left(\frac{\frac{1}{2}}{\frac{1}{8}}\right) = 4 f_{\text{air}}$$

As, focal length increases, there is no focussing of image on screen.

$\therefore$ Image will disappear

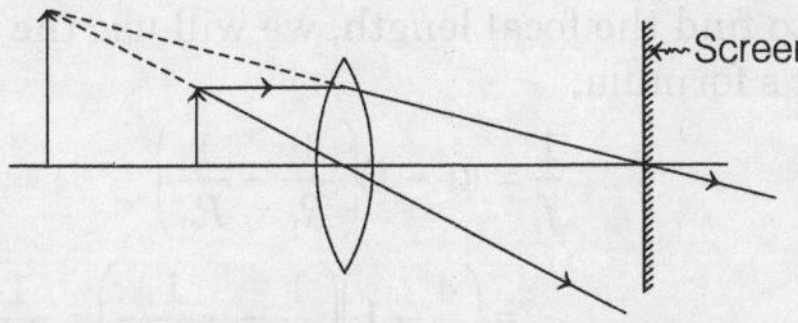

Actually, a virtual image is formed on same side of object.

79. As seen from a relatively rarer medium (L_2 or L_3), the interface $L_1 L_2$ is concave and $L_2 L_3$ is convex. The divergence produced by concave surface is much smaller than the convergence due to the convex surface. Hence, the arrangement corresponds to concavo-convex lens.

80. $\dfrac{1}{F} = \left[\dfrac{\mu_L}{\mu_S} - 1\right]\left[\dfrac{1}{R_1} - \dfrac{1}{R_2}\right]$

If $\mu_L = \mu_S \Rightarrow \dfrac{1}{F} = 0 \Rightarrow F = \infty$

81. $$R^2 = r^2 + (R - t)^2$$
$$R^2 = r^2 + R^2 + t^2 - 2Rt$$

Neglecting t^2, we get

$$R = \frac{r^2}{2t}$$

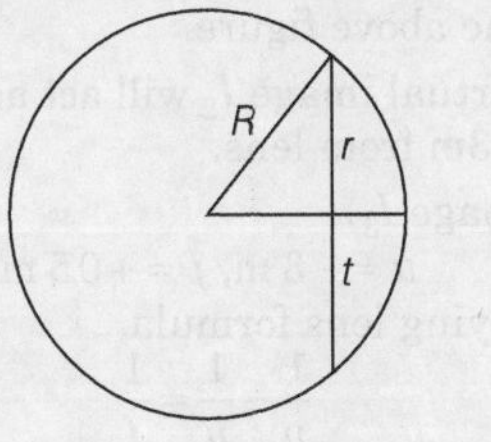

$\therefore$ $$\frac{1}{f} = (\mu - 1)\left(\frac{1}{R} - \frac{1}{\infty}\right) = \frac{\mu - 1}{R}$$

$\therefore$ $$f = \frac{R}{\mu - 1} = \frac{r^2}{2t(\mu - 1)}$$

$$= \frac{(3 \times 10^{-2})^2}{2 \times 3 \times 10^{-3} \times \left(\frac{3}{2} - 1\right)} = 0.3 \text{ m} = 30 \text{ cm}$$

82. As, power of combination,

$$P = \frac{1}{f_1} + \frac{1}{f_2} - \frac{d}{f_1 f_2} = 0$$

$\therefore$ $$\frac{1}{f_1} + \frac{1}{f_2} = \frac{d}{f_1 f_2}$$

$\Rightarrow$ $$\frac{1}{20} - \frac{1}{56} = \frac{d}{20(-56)}$$

$\Rightarrow$ $$\frac{56 - 20}{20 \times 56} = \frac{d}{-20 \times 56}$$

$\Rightarrow$ $$d = -36 \text{ cm}$$

83. The power of the given system is a combination of the positive power of the convex lens, negative power of the plano-concave lens of water and zero power of the plane mirror. Clearly, the power of the system decreases.

84. Focal length of diverging lens is 25 cm.

As the rays are coming parallel, so the image (I_1) will be formed at the focus of diverging lens, *i.e.* at 25 cm towards left of diverging lens.

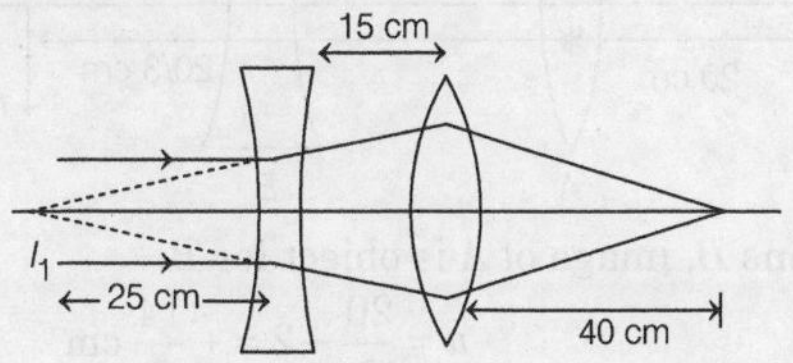

Now, the image (I_1) will work as object for converging lens.

For converging lens, distance of object u

(*i.e.* distance of I_1) $= -(25 + 15)$

$= -40$ cm

$f = 20$ cm

$\therefore$ From lens formula, $\dfrac{1}{f} = \dfrac{1}{v} - \dfrac{1}{u}$

$$\frac{1}{20} = \frac{1}{v} - \frac{1}{(-40)}$$

$\Rightarrow$ $$\frac{1}{v} = \frac{1}{20} - \frac{1}{40}$$

$\Rightarrow$ $$\frac{1}{v} = \frac{1}{90}$$

$\Rightarrow$ $$v = 40 \text{ cm}$$

v is positive, so image will be real and will form at right side of converging lens at 40 cm.

85. As, $\dfrac{1}{f} = \dfrac{1}{f_1} + \dfrac{1}{f_2} - \dfrac{x}{f_1 f_2} = \dfrac{1}{f_1} - \dfrac{1}{f_1} + \dfrac{x}{f_1^2}$

$\Rightarrow$ $$\frac{1}{f} = \frac{x}{f_1^2}$$

$\Rightarrow f > 0$ for every x.

For $x = 0$, $f = \infty$

Hence, for $x = 0$, system will behave like a glass plate.

86. In every condition, two plano-convex lenses are placed, closed to each other, *i.e.* distance between is assumed to be zero.

i.e. $$\frac{1}{F} = \frac{1}{f_1} + \frac{1}{f_2}$$

Hence, focal length of combination will be same in each cases.

87. For lens A, object distance, $u = -20$ cm

Focal length, $f = +5$ cm

From lens formula, $\dfrac{1}{v} - \dfrac{1}{u} = \dfrac{1}{f}$

We have, $\frac{1}{v} = \frac{1}{5} - \frac{1}{20}$

$\Rightarrow \quad \frac{1}{v} = \frac{15}{100}$

$\Rightarrow \quad v = \frac{20}{3}$ cm

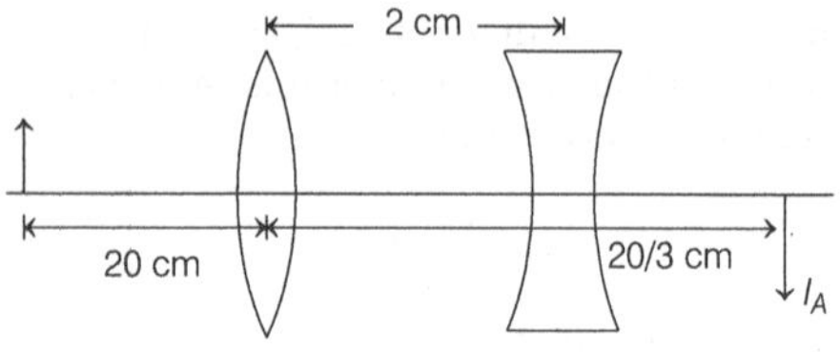

For lens B, image of A is object for B.

$$\therefore \quad u = \frac{20}{3} - 2 = +\frac{14}{3} \text{ cm}$$

$$f = -5 \text{ cm}$$

Now, from lens formula, we have

$$\frac{1}{v} - \frac{1}{u} = \frac{1}{f}$$

$$\Rightarrow \quad \frac{1}{v} = \frac{1}{-5} + \frac{3}{14}$$

$$\Rightarrow \quad \frac{1}{v} = \frac{1}{70}$$

$$\Rightarrow \quad v = 70 \text{ cm}$$

Hence, image is on right of lens B and is real in nature.

88. As shown in Fig. (a),

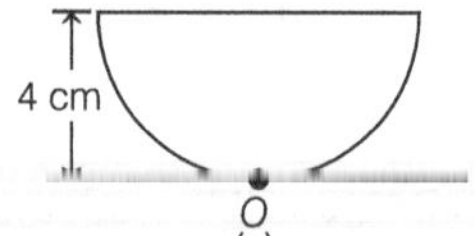

(a)

In this case, refraction of the rays starting from t_0 takes place from a plane surface. So, we can use,

$$d_{\text{app}} = \frac{d_{\text{actual}}}{\mu}$$

or $\quad 3 = \frac{4}{\mu}$

or $\quad \mu = \frac{4}{3}$

As shown in Fig. (b)

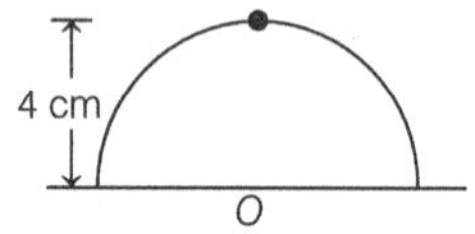

(b)

In this case, refraction takes place from a spherical surface. Hence, applying

$$\frac{\mu_2}{v} - \frac{\mu_1}{u} = \frac{\mu_2 - \mu_1}{R}, \text{ we have}$$

$$\frac{1}{(-25/8)} - \frac{4/3}{(-4)} = \frac{1-4/3}{-R}$$

or $\quad \frac{1}{3R} = \frac{1}{3} - \frac{8}{25} = \frac{1}{75}$

$\therefore \quad R = 25$ cm

Now, to find the focal length, we will use the lens maker's formula,

$$\frac{1}{f} = (\mu - 1)\left(\frac{1}{R_1} - \frac{1}{R_2}\right)$$

$$= \left(\frac{4}{3} - 1\right)\left(\frac{1}{\infty} - \frac{1}{(-25)}\right) = \frac{1}{75}$$

$\therefore \quad f = 75$ cm

89. Given that, for convex lens,

$$u_1 = -1 \text{ m}, f = +0.5 \text{ m}$$

Applying lens formula,

$$\frac{1}{v_1} - \frac{1}{u_1} = \frac{1}{f}$$

$$\Rightarrow \quad \frac{1}{v_1} - \frac{1}{(-1)} = \frac{1}{0.5}$$

$$\Rightarrow \quad v_1 = +1 \text{ m}$$

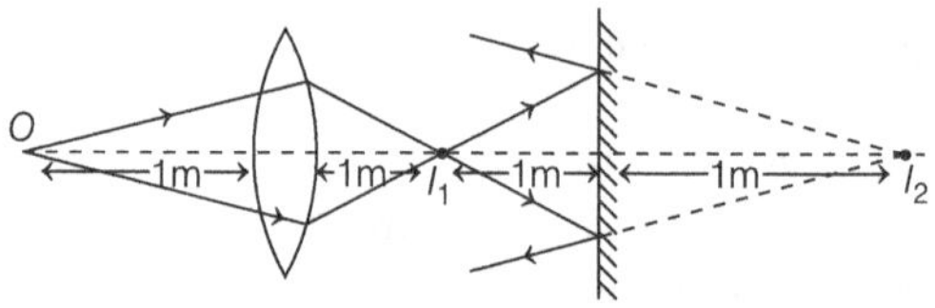

The image I_1 will behave as an object for plane mirror. So, image will be formed 1 m behind the mirror as shown in the above figure.

Now, the virtual image I_2 will act as an object at a distance of 3m from lens.

For final image I_3,

$$u = -3 \text{ m}, f = +0.5 \text{ m}$$

Again, applying lens formula,

$$\frac{1}{v} - \frac{1}{u} = \frac{1}{f}$$

$$\Rightarrow \quad \frac{1}{v} - \frac{1}{(-3)} = \frac{1}{0.5}$$

$$\Rightarrow \quad v = \frac{3}{5} = +0.6 \text{ m}$$

So, total distance of final image from mirror $= 2 + 0.6 = 2.6$ m. The image will be real (but in answer it was given virtual).

90. Initially, when a light source (i.e. an object) is placed at 10 cm from the convex mirror and an image is form on the screen as shown in the figure below,

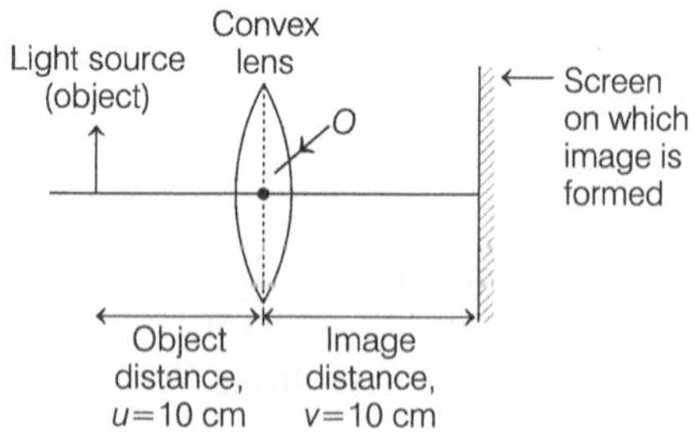

The distance from the optical centre (O) of the lens and $2f$ is $2 \times$ focal length (f) = object distance

$$\Rightarrow \quad 2 \times f = 10$$

$$\text{or} \quad f = 5 \text{ cm}$$

Now, when a glass block is placed in contact with the light source *i.e.*, object, then the situation is shown in the figure given below

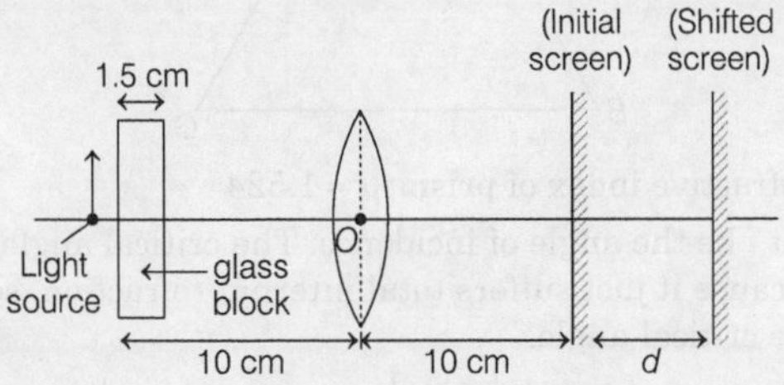

The shift in the position of the object is given as

$$x = \left(1 - \frac{1}{\mu}\right) t$$

where, μ is the refractive index of the block and t is its thickness.

$$\Rightarrow \quad x = \left(1 - \frac{1}{1.5}\right) 1.5 = \left(1 - \frac{2}{3}\right)\frac{3}{2}$$

$$= \frac{1}{3} \times \frac{3}{2} = \frac{1}{2} = 0.5 \text{ cm}$$

$\therefore$ The new object distance of the light source in front of the lens will be

$$u' = 10 - 0.5 = 9.5 \text{ cm}$$

Since, the focal length of the lens is 5 cm.

Therefore, the image distance of the light source now can be given as,

$$\frac{1}{v'} = \frac{1}{f} + \frac{1}{u'} \quad \text{(using lens formula)}$$

Substituting the values, we get

$$\frac{1}{v'} = \frac{1}{5} + \left(\frac{1}{-9.5}\right) = \frac{+9.5 - 5}{47.5} = \frac{4.5}{47.5}$$

$$\text{or} \quad v' = 10.55 \text{ cm}$$

$\therefore$ The value of $d = v' - v = 10.55 - 10$

$$= 0.55 \text{ cm, away from the lens}$$

91. Light from plane mirror is reflected back on it's path, so that image of A coincides with A itself.

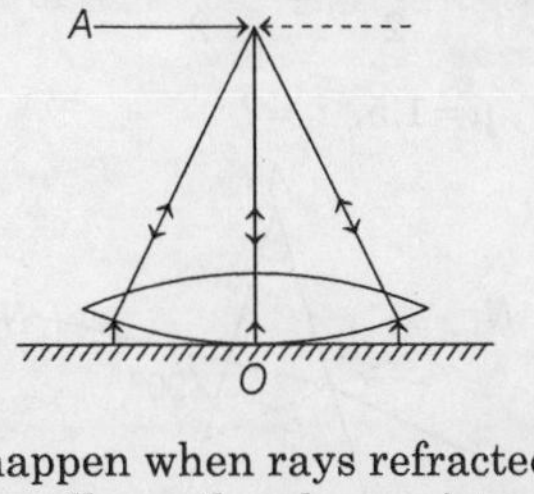

This would happen when rays refracted by the convex lens falls normally on the plane mirror, *i.e.* the refracted rays form a beam parallel to principal axis of the lens. Hence, the object would then be considered at the focus of convex lens.

$\therefore$ Focal length of curvature of convex lens,

$$f_1 = 18 \text{ cm}$$

With liquid between lens and mirror, image is again coincides with object, so the second measurement is focal length of combination of liquid lens and convex lens.

$$\therefore \quad \frac{1}{f_1} + \frac{1}{f_2} = \frac{1}{f_{eq}}$$

$$\Rightarrow \quad \frac{1}{18} + \frac{1}{f_2} = \frac{1}{27}$$

$$\Rightarrow \quad f_2 = -54 \text{ cm}$$

For convex lens by lens maker's formula, we have

$$\frac{1}{f} = (\mu - 1)\left(\frac{2}{R}\right)$$

$$\Rightarrow \quad \frac{1}{18} = 0.5 \times \frac{2}{R}$$

$$\Rightarrow \quad R = 18 \text{ cm}$$

and for plano-convex liquid lens, we have

$$\frac{1}{f} = (\mu_l - 1)\left(\frac{-1}{R}\right)$$

$$\Rightarrow \quad -\frac{1}{54} = (\mu_l - 1)\left(\frac{-1}{18}\right)$$

$$\Rightarrow \quad \mu_l = 1 + \frac{1}{3} = \frac{4}{3}$$

92. First we find relation of magnification m and focal length f. By lens equation,

$$\frac{1}{v} - \frac{1}{u} = \frac{1}{f}$$

$$\Rightarrow \quad 1 - \frac{v}{u} = \frac{v}{f}$$

$$\Rightarrow \quad 1 - m = \frac{v}{f} \quad \left(\because \frac{v}{u} = m\right)$$

$$\Rightarrow \quad m = 1 - \frac{v}{f} \quad \text{...(i)}$$

Now, from given graph and from Eq. (i),

At $v = a$, magnification is

$$m_1 = 1 - \frac{a}{f} \quad \text{...(ii)}$$

At $v = a + b$, magnification is

$$m_2 = 1 - \frac{a+b}{f} \quad \text{...(iii)}$$

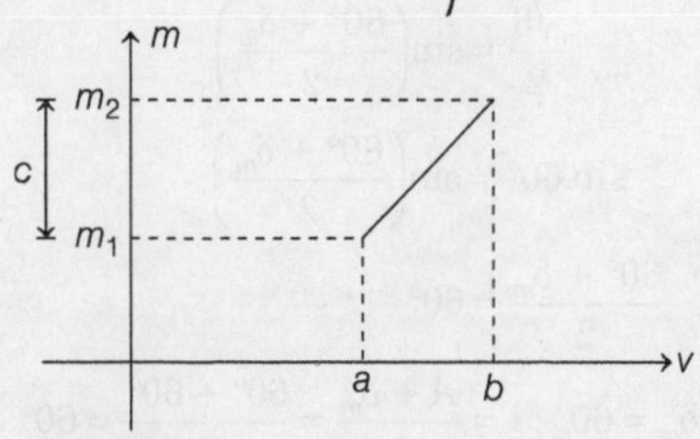

From graph, we can also say that

$$m_2 - m_1 = c \quad \text{...(iv)}$$

So, from Eqs. (ii), (iii) and (iv), we have

$$\left(1 - \frac{(a+b)}{f}\right) - \left(1 - \frac{a}{f}\right) = c$$

$$\Rightarrow \quad \frac{a - (a+b)}{f} = c$$

$$\Rightarrow \quad f = -\frac{b}{c} \text{ or } |f| = \frac{b}{c}$$

93. For double convex lens,

$$R_1 = R_2 = R \quad \text{(given)}$$

$$\frac{1}{f} = (\mu - 1)\left(\frac{1}{R} - \frac{1}{(-R)}\right)$$

$$\frac{1}{f} = (\mu - 1)\left(\frac{2}{R}\right)$$

$$\text{Power,} \quad P = \frac{1}{f} = \frac{2(\mu - 1)}{R} \quad \text{...(i)}$$

For plano-convex lens,

$$R_1 = R', R_2 = \infty$$

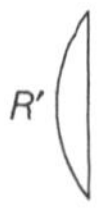

$$\frac{1}{f'} = (\mu - 1)\left(\frac{1}{R'} - \frac{1}{\infty}\right)$$

$$\frac{1}{f'} = \frac{\mu - 1}{R'}$$

$$\text{Power,} \; P' = \frac{1}{f'} = \frac{(\mu - 1)}{R'}$$

Given, $P' = 1.5P$

$$\Rightarrow \quad 1.5P = \frac{\mu - 1}{R'} \quad \text{...(ii)}$$

Dividing Eq. (i) by Eq. (ii), we get

$$\frac{P}{1.5P} = \frac{\frac{2(\mu - 1)}{R}}{\frac{(\mu - 1)}{R'}}$$

$$\Rightarrow \quad \frac{2}{3} = \frac{2R'}{R}$$

$$\text{or} \quad R' = \frac{R}{3}$$

94. We have, $\mu = \sqrt{3} = \dfrac{\sin\left(\dfrac{60° - \delta_m}{2}\right)}{\sin\left(\dfrac{60°}{2}\right)}$

where, μ = refractive index.

$$\Rightarrow \quad \frac{\sqrt{3}}{2} = \sin\left(\frac{60° + \delta_m}{2}\right)$$

$$\sin 60° = \sin\left(\frac{60° + \delta_m}{2}\right)$$

$$\text{or} \quad \frac{60° + \delta_m}{2} = 60°$$

$$\text{or} \quad \delta_m = 60° \; ; \; i = \frac{A + \delta_m}{2} = \frac{60° + 60°}{2} = 60°$$

95. Angle of prism, $A = 60°$

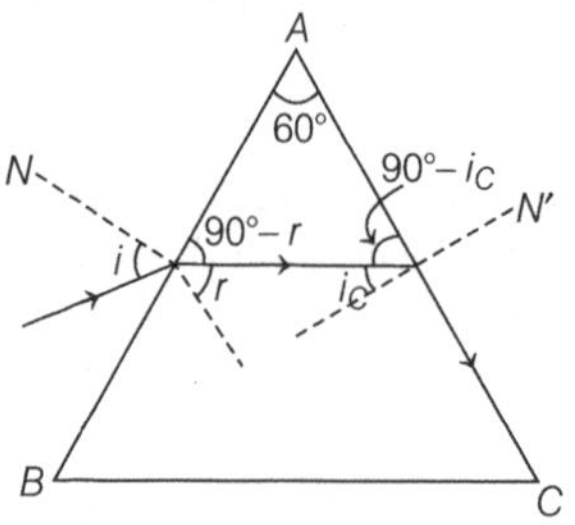

Refractive index of prism, $\mu = 1.524$

Let i be the angle of incidence. The critical angle is i_c because it just suffers total internal refraction, so we use critical angle.

$$\sin i_C = \frac{1}{\mu} = \frac{1}{1.524} = 0.6561$$

$$i_C = 41°$$

For a prism $r_1 + r_2 = A$, here $r_2 = i_c$

$$\therefore \quad r_1 + i_C = A$$

$$r_1 + 41° = 60°$$

$$\Rightarrow \quad r_1 = 19°$$

Using the formula, $\mu = \dfrac{\sin i_1}{\sin r_1}$

$$\text{or} \quad \sin i_1 = 1.524 \sin 19° = 1.524 \times 0.3256$$

$$\text{or} \quad i_1 = \sin^{-1}(0.4962)$$

$$i_1 = 29° \; 75'$$

Thus, the angle should be 29° 75′.

96. As, $\mu = \dfrac{\sin\left(\dfrac{A + \delta_m}{2}\right)}{\sin\dfrac{A}{2}}$

$$\Rightarrow \quad \sqrt{2} = \frac{\sin\left(\dfrac{60° + \delta_m}{2}\right)}{\sin\dfrac{60°}{2}} \quad (\text{as, } \mu = \sqrt{2})$$

$$\frac{1}{\sqrt{2}} = \sin\left(\frac{60° + \delta_m}{2}\right)$$

$$\text{or} \quad \sin 45° = \sin\left(\frac{60° + \delta_m}{2}\right)$$

$$\Rightarrow \quad \delta_m = 30°$$

$\therefore$ Angle of incidence,

$$i = \frac{A + \delta_m}{2} = \frac{60 + 30}{2} = 45°$$

97. Here, $A = 5°$, $\mu = 1.5$, $i_1 = ?$

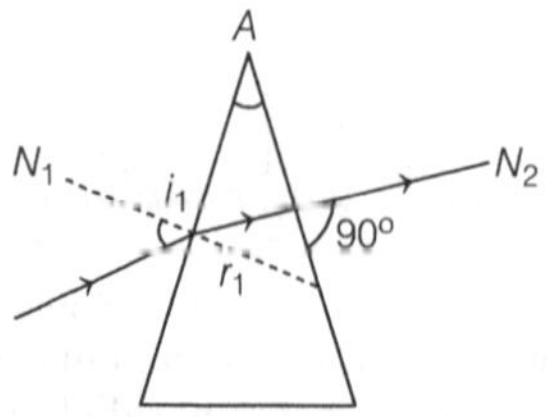

As the ray emerges from the other face of prism normally,

$$i_2 = 0^\circ \quad \therefore \quad r_2 = 0^\circ$$

As, $r_1 + r_2 = A$

$\Rightarrow$ $r_1 = A - r_2 = 5 - 0 = 5^\circ$

From $\mu = \dfrac{\sin i_1}{\sin r_1}$

$\Rightarrow$ $\sin i_1 = \mu \sin r_1$

$= 1.5 \times \sin 5^\circ$

$= 1.5 \times 0.087$

$= 0.1305$

$\Rightarrow$ $i_1 = \sin^{-1}(0.1325) = 7.5^\circ$

98. For a crown glass thin prism, *i.e.* prism with small angle, the angle of minimum deviation is given as

$$D_m = (n-1)A$$

where, A is prism angle and n is refractive index.

$\Rightarrow$ $D_m \propto n$ ... (i)

Since, from the given graph, the value of n decrease with the increase in λ. Thus, from relation (i), we can say that, D_m will also decrease with the increase in λ. Hence, option (c) is correct.

99. Following arguments lead us easily to the right choice.

(i) Angle between any two lines is the same as the angle between their perpendiculars.

$\therefore$ $i = 30^\circ$

(ii) $\dfrac{1}{1.5} = \dfrac{\sin 30^\circ}{\sin r}$ (by Snell's law)

or $\sin r = \dfrac{1.5}{2} = 0.75$

or $r = 48.6^\circ$

(iii) $\theta = r - i = 18.6^\circ$

$\therefore$ Required angle $= 2 \times 18.6 = 37.2^\circ$

100. $n = \sqrt{3}$, prism angle, $A = 60^\circ$

Method 1

Using prism formula, $n = \dfrac{\sin\left(\dfrac{A+\delta}{2}\right)}{\sin\left(\dfrac{A}{2}\right)}$

$\Rightarrow$ $\sqrt{3} = \dfrac{\sin\left(\dfrac{60^\circ+\delta}{2}\right)}{\sin 30^\circ}$

$\Rightarrow$ $\sin\left(\dfrac{60^\circ+\delta}{2}\right) = \dfrac{\sqrt{3}}{2}$

$\Rightarrow$ $\sin\left(\dfrac{60^\circ+\delta}{2}\right) = \sin 60^\circ$

or $\dfrac{60^\circ+\delta}{2} = 60^\circ$

or angle of minimum deviation, $\delta = 60^\circ$

Incident angle, $i = \dfrac{60+\delta}{2} = 60^\circ$

Method 2 For minimum deviation, ray should pass symmetrically (*i.e.* parallel to the base of the equilateral prism)

From geometry of given figure, we have

$$r = 30^\circ$$

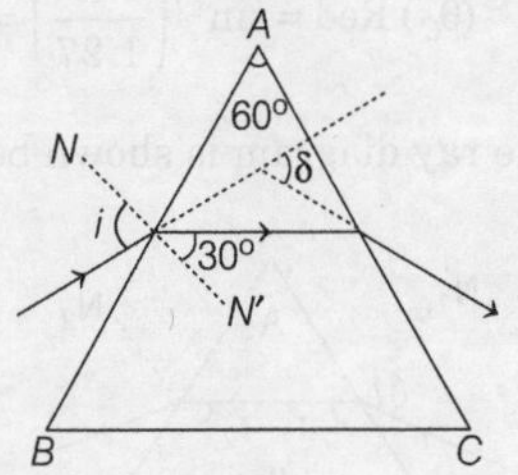

Using Snell's law, $n = \dfrac{\sin i}{\sin r}$

$\Rightarrow$ $\sin i = n \sin r = \sqrt{3} \sin 30^\circ$

$\Rightarrow$ $\sin i = \dfrac{\sqrt{3}}{2}$ or $i = 60^\circ$

101. For passing the ray from prism,

$$\mu \le \operatorname{cosec}\frac{A}{2}$$

$$\mu \le \operatorname{cosec}\left(\frac{90^\circ}{2}\right)$$

$$\mu \le \sqrt{2}$$

$\therefore$ $\mu_{max} = \sqrt{2}$

102. As, $A = r_1 + r_2$

$\therefore$ $30^\circ = r_1 + 0^\circ$

or $r_1 = 30^\circ$

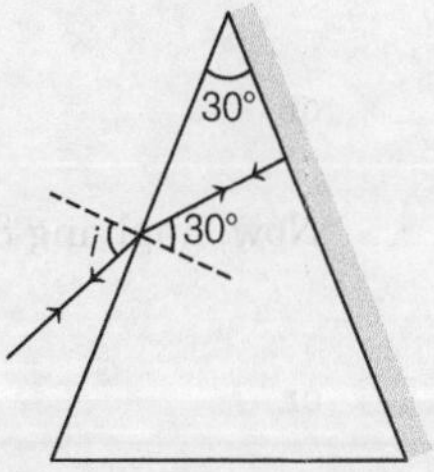

Now, $\dfrac{\sin i}{\sin 30^\circ} = \sqrt{2}$

or $\sin i = \sqrt{2} \times \dfrac{1}{2}$

or $\sin i = \dfrac{1}{\sqrt{2}}$

or $i = 45^\circ$

103. Assuming that the right angled prism is an isoceles prism, so the other angles will be 45° each. Each incident ray will make an angle of 45° with the normal at face PR.

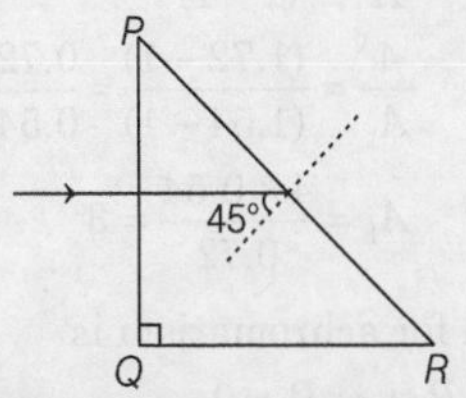

The wavelength corresponding to which the incidence angle is less than the critical angle, will pass through PR.

$\Rightarrow \quad \theta_C =$ critical angle

$$\Rightarrow \quad \theta_C = \sin^{-1}\left(\frac{1}{\mu}\right)$$

If $\theta_C \geq 45°$, then light will pass

$$\Rightarrow \quad (\theta_C)\text{ Red} = \sin^{-1}\left(\frac{1}{1.27}\right) = 51.94°$$

104. Consider the ray diagram is shown below

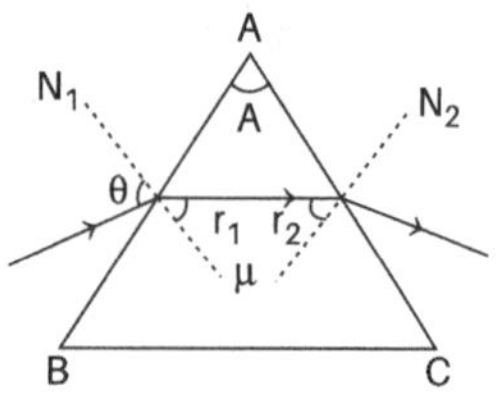

A ray of light incident on face AB at an angle θ.
Here, r_1 = angle of refraction on face AB
and r_2 = angle of incidence at face AC.
For transmission of light through face AC,

$$i_{AC} < i_C$$

or $\quad A - r_1 < i_C$

or $\quad \sin(A - r_1) < \sin i_C$

or $\quad \sin(A - r_1) < \dfrac{1}{\mu}$

or $\quad A - r_1 < \sin^{-1}\left(\dfrac{1}{\mu}\right)$

or $\quad \sin r_1 > \sin\left[A - \sin^{-1}\left(\dfrac{1}{\mu}\right)\right]$

Now, applying Snell's law at the face AB,

$$1 \times \sin\theta = \mu \sin r_1$$

or $\quad \sin r_1 = \dfrac{\sin\theta}{\mu}$

$\Rightarrow \quad \dfrac{\sin\theta}{\mu} > \sin\left[A - \sin^{-1}\left(\dfrac{1}{\mu}\right)\right]$

or $\quad \theta > \sin^{-1}\left[\mu \sin\left\{A - \sin^{-1}\left(\dfrac{1}{\mu}\right)\right\}\right]$

105. For dispersion without deviation,

$$\frac{A}{A'} = \frac{(\mu' - 1)}{(\mu - 1)}$$

$\Rightarrow \quad \dfrac{4}{A_1} = \dfrac{(1.72 - 1)}{(1.54 - 1)} = \dfrac{0.72}{0.54}$

$\Rightarrow \quad A_1 = \dfrac{4 \times 0.54}{0.72} = 3°$

106. The condition for achromatism is

$$\omega_1 P_1 + \omega_2 P_2 = 0$$

$$\omega_1 P_1 = -\omega_2 P_2$$

$\Rightarrow \quad \dfrac{\omega_1}{\omega_2} = -\dfrac{P_2}{P_1}$

Now, $\quad P_1 + P_2 = 2\text{D}$

or $\quad 5 + P_2 = 2$

or $\quad P_2 = -3\text{D}$

$\therefore \quad \dfrac{\omega_1}{\omega_2} = -\dfrac{-3}{5} = \dfrac{3}{5}$ (in magnitude)

107. We have, ratio of dispersive powers,

$$\frac{\omega_1}{\omega_2} = \frac{1}{2}$$

Now, $\quad \dfrac{f_1}{f_2} = -\dfrac{\omega_1}{\omega_2} = -\dfrac{1}{2}$

or $\quad f_2 = -2f_1$

Now, $\quad \dfrac{1}{F} = \dfrac{1}{f_1} + \dfrac{1}{f_2}$ (for the combination of lenses)

$\Rightarrow \quad \dfrac{1}{50} = \dfrac{1}{f_1} + \dfrac{1}{(-2f_1)}$

or $\quad 50f = \dfrac{-2+1}{-2f_1} = \dfrac{1}{2f_1}$

or $\quad 2f_1 = 50$

or $\quad f_1 = 25$ cm

Again, $\quad f_2 = -2 \times 25 \text{ cm} = -50$ cm

108. Power of combination, $P = P_1 + P_2 = +20 - 10 = +10\text{D}$

$\therefore \quad f = \dfrac{1}{P} = \dfrac{1}{10}\text{ m} = 10$ cm

For image at infinity, $M = \dfrac{D}{f} = \dfrac{25}{10} = 2.5$

109. For the objective,

$\therefore \quad \dfrac{1}{v_o} - \dfrac{1}{-1/3.8} = \dfrac{1}{1/4}$

or $\quad \dfrac{1}{v_o} + 3.8 = 4$

or $\quad \dfrac{1}{v_o} = 0.2 = \dfrac{1}{5}$

or $\quad v_o = 5$ cm

Now, $\quad M_o = \dfrac{5}{-\dfrac{1}{3.8}} = -19$

Again, $\quad M = M_o \times M_e$

$$-95 = -19 \times M_e$$

or $\quad M_e = \dfrac{95}{19} = 5$

110. As, $L = v_o + f_e$

$\Rightarrow \quad v_o = L - f_e$

or $\quad v_o = 19.2$ cm

$\therefore$ From lens formula, $\dfrac{1}{19.2} - \dfrac{1}{u_o} = \dfrac{1}{1.6}$

or $\quad -\dfrac{1}{u_o} = \dfrac{10}{16} - \dfrac{10}{192}$

or $$-\frac{1}{u_o}=\frac{120-10}{192}=\frac{110}{192}$$

or $$u_o=-\frac{192}{110}\text{ cm}=-1.75\text{ cm}$$

111. When final image is formed at infinite, then

$$M=\frac{v_o}{u_o}\left(\frac{d}{f_e}\right)=\frac{v_o}{f_o}\left(\frac{d}{f_e}\right)$$

Now, $$v_o=16-f_e=16-2.5=13.5\text{ cm}$$

$\therefore$ $$M=\frac{13.5}{-0.4}\times\frac{25}{2.5}=-337.5$$

112. Separation $=f_o+\dfrac{f_e D}{f_e+D}=80+\dfrac{5\times 25}{5+25}=80+\dfrac{125}{30}$

$=84.16\text{ cm}=84.2\text{ cm}$

113. **Case I** As, $M=-\dfrac{f_o}{f_e}\left(1+\dfrac{f_e}{D}\right)$

$$M=-\frac{200}{5}\left(1+\frac{5}{25}\right)$$

$$M=-40\left(1+\frac{1}{5}\right)=-40\times\frac{6}{5}=-48$$

Case II $M=\dfrac{f_o}{f_e}=-\dfrac{200}{5}=-40$

114. As, $M=\dfrac{f_o}{f_e}$

or $$9=\frac{f_o}{f_e}\text{ or }f_o=9f_e$$

Also, $$L=f_o+f_e$$

or $$20=f_o+f_e$$

or $$20=9f_e+f_e$$

or $$20=10f_e$$

or $$f_e=2\text{ cm}$$

$\therefore$ $$f_o=9\times 2\text{ cm}=18\text{ cm}$$

115. For a telescope in normal setting,

$f_o+f_e=L$ (length of the tube of telescope)

and $\dfrac{f_o}{f_e}=m$ (magnification)

where, f_o and f_e is the focal length of the objective and eyepiece, respectively.

According to the given values in the question, we have

$$f_o+f_e=60\text{ cm and }\frac{f_o}{f_e}=5\Rightarrow f_o=5f_e$$

Solving the above equations, we have

$$f_e=10\text{ cm}$$

116. In normal setting, magnification obtained by a microscope is given by

$$m=\frac{L\times D}{f_o\times f_e}$$

where, L = tube length = 150 mm,

D = distance of distinct vision = 25 cm = 250 mm,

f_o = focal length of objective = 5 mm,

f_e = focal length of eyepiece

and m = magnification = 375.

So, we have

$$375=\frac{150\times 250}{5\times f_e}\text{ or }f_e=20\text{ mm}$$

So, focal length of eyepiece required is closest to 22 mm.

117. Here, $f_o=50$ cm, $f_e=5$ cm, $D=25$ cm and $u_o=200$ cm

$\therefore$ Separation between the objective and eyepiece lens is,

$$L=\frac{u_o f_o}{u_o-f_o}+\frac{f_e D}{(f_e+D)}$$

$$=\frac{200\times 50}{(200-50)}+\frac{5\times 25}{(5+25)}=71\text{ cm}$$

118. Given, distance between objective mirror and another mirror,

$$d=20\text{ mm}$$

Radius of curvature of objective mirror, $R_1=220$ mm

$\therefore$ Focal length of objective mirror, $f_1=\dfrac{220}{2}=110$ mm

Radius of curvature of small mirror, $R_2=140$ mm

$\therefore$ Focal length of small mirror, $f_2=\dfrac{140}{2}=70$ mm

The image of an object placed at infinity, formed by the objective mirror, will act as a virtual object for small mirror.

So, the object distance for small mirror, $u=f_1-d$

i.e. $$u=110-20=90\text{ mm}$$

Using mirror formula, $\dfrac{1}{v}+\dfrac{1}{u}=\dfrac{1}{f_2}$

$\Rightarrow$ $$\frac{1}{v}=\frac{1}{f_2}-\frac{1}{u}=\frac{1}{70}-\frac{1}{90}=\frac{2}{630}$$

$\Rightarrow$ $$v=315\text{ mm or }v=31.5\text{ cm}$$

Thus, the final image is formed at 315 mm away from small mirror.

119. As, $\dfrac{1}{f}=\dfrac{1}{v}-\dfrac{1}{u}$ (lens formula)

$\Rightarrow$ $$2.5=\frac{1}{-0.75}-\frac{1}{u}$$

or $$\frac{1}{u}=-\frac{100}{75}-\frac{25}{10}\text{ or }\frac{1}{u}=-\frac{4}{3}-\frac{5}{2}$$

or $$\frac{1}{u}=\frac{-8-15}{6}=-\frac{23}{6}$$

or $$u=-\frac{6}{23}\text{ m}=-0.26\text{ m}$$

Round II

1. By painting black the upper half of the lens, intensity of image will reduce but its position will not be shifted.

2. A magnifying glass is used as the object to be viewed can be brought closer to the eye than the normal near point. This results in a larger angle to be subtended by the object at the eye and hence viewed in greater detail. Also, it results in the formation of a virtual, erect image.

3. Here, an extended object lies immersed in water contained in a plane trough. When seen from close to the edge of the trough, the object looks distorted on account of refraction of light from denser to rarer medium. Therefore, apparent depths of the points close to the edge and nearer to the surface of water is more compared to points away from the edge.Further, the angle subtended by the image of the object at the eye is smaller than the actual angle subtended by the object in air. Again, some of the points of the object, far away from the edge may not be visible because of total internal reflection.

4. Distance between objective and eyepiece = Tube length

$$= f_o + f_e$$
$$= 16 + 0.02 = 16.02 \text{ m}$$

Angular magnification $= \frac{f_o}{f_e} = -\frac{16}{0.02} = -800$

The final image of the planet is inverted as magnification is negative.

5. As, refractive index for $z > 0$ and $z \leq 0$ is different XY-plane should be boundary between two media.

Angle of incidence, $\cos i = \left|\frac{A_z}{\sqrt{A_x^2 + A_y^2 + A_z^2}}\right|$

$$= \frac{10}{\sqrt{(6\sqrt{3})^2 + (8\sqrt{3})^2 + 100^2}} = \frac{1}{2}$$

$\therefore$ $i = 60°$

From Snell's law, $\frac{\sin i}{\sin r} = \frac{\sqrt{3}}{\sqrt{2}}$

$\Rightarrow$ $\sin r = \frac{\sqrt{2}}{\sqrt{3}} \sin 60° = \frac{1}{\sqrt{2}}$

$$r = 45°$$

6. As, $\frac{1}{u} + \frac{1}{v} = \frac{1}{f}$

$\Rightarrow$ $-\frac{1}{u^2}\frac{du}{dt} - \frac{1}{v^2}\frac{dv}{dt} = 0$

$\Rightarrow$ $\frac{dv}{dt} = \frac{-v^2}{u^2}\left(\frac{du}{dt}\right)$

But $\frac{v}{u} = \frac{f}{u-f}$

$\therefore$ $\frac{dv}{dt} = -\left(\frac{f}{u-f}\right)^2\left(\frac{du}{dt}\right)$

$$= \left(\frac{0.2}{-2.8 - 0.2}\right)^2 \times 15$$

$$= \frac{1}{15} \text{ ms}^{-1}$$

7. As, there is no deflection between mediums 1 and 2. Therefore, $\mu_1 = \mu_2$.

8. According to Cartesian sign convention,

$$u = -40 \text{ cm}, R = -20 \text{ cm}$$

Given, $\mu_1 = 1$, and $\mu_2 = 1.33$

Applying equation for refraction through spherical surface, we get

$$\frac{\mu_2}{v} - \frac{\mu_2}{u} = \frac{\mu_2 - \mu_1}{R}$$

$$\frac{1.33}{v} - \frac{1}{-40} = \frac{1.33}{-20}$$

After solving $v = -32$ cm

The magnification is, $m = \frac{h_2}{h_1} = \frac{\mu_1 v}{\mu_2 u}$

$\therefore$ $\frac{h_2}{1} = -\frac{1\,(32)}{1.33(-40)}$

or $h_2 = 0.6$ cm

The positive sign shows that the image is erect.

9. From figure, we see that

$$\theta + \theta + 10° = 90°$$

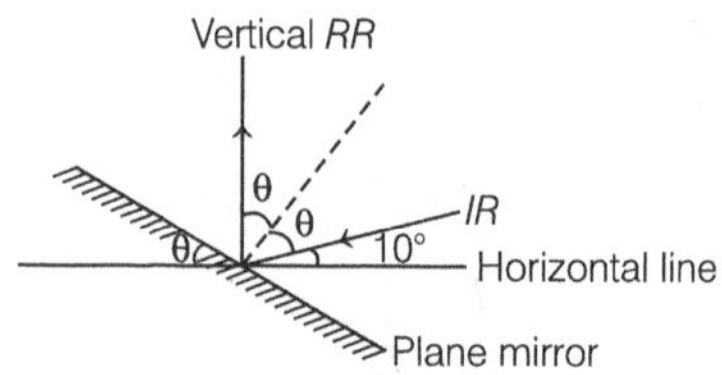

$\Rightarrow$ $2\theta + 10 = 90$

$\Rightarrow$ $2\theta = 80$

$\Rightarrow$ $\theta = 40°$

10. Using equation, the total apparent shift,

$$S = h_1\left(1 - \frac{1}{\mu_1}\right) + h_2\left(1 - \frac{1}{\mu_2}\right)$$

or $S = 4\left(1 - \frac{1}{4/3}\right) + 6\left(1 - \frac{1}{3/2}\right)$

$$= 3.0 \text{ cm}$$

$\therefore$ Apparent position of the coin from the surface,

Thus, $h = h_1 + h_2 - S = 4 + 6 - 3$

$$= 7.0 \text{ cm}$$

11. Image formation by a mirror does not depend on the medium. As, P is at a height h above the mirror, image of P will be at a depth h below the mirror.

If d is depth of liquid in the tank, apparent depth of P,

$$x_1 = \frac{d-h}{\mu}$$

$\therefore$ Apparent distance between P and its image

$$= x_2 - x_1 = \frac{d+h}{\mu} - \frac{d-h}{\mu} = \frac{2h}{\mu}$$

12. Shift in image position due to glass plate,

$$S = \left(1 - \frac{1}{\mu}\right)t = \left(1 - \frac{1}{1.5}\right) \times 1 \text{ cm} = \frac{1}{3} \text{ cm}$$

For focal length of the lens,

$$\frac{1}{f} = \frac{1}{v} - \frac{1}{u} = \frac{1}{12} - \frac{1}{-240}$$

or $\quad \frac{1}{f} = \frac{20+1}{240}$

$\Rightarrow \quad f = \frac{240}{21}$ cm

Now, to get back image on the film, lens has to form image at $\left(12 - \frac{1}{3}\right)$ cm $= \frac{35}{3}$ cm such that the glass plate will shift the image on the film.

As $\quad \frac{1}{f} = \frac{1}{v} - \frac{1}{u}$

$\Rightarrow \quad \frac{1}{u} = \frac{1}{v} - \frac{1}{f} = \frac{3}{35} - \frac{21}{240}$

$\quad \frac{1}{u} = \frac{48 \times 3 - 7 \times 21}{1680} = -\frac{1}{560}$

$\Rightarrow \quad u = -5.6$ m

13. In case of a concave mirror, when object is at f, image is formed at infinity.
So, magnification when object is at f,

$$m = \frac{-v}{u} = \infty$$

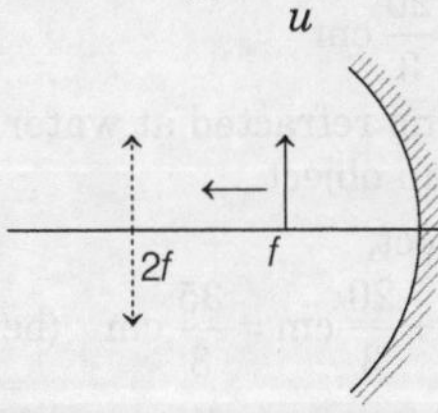

Also, if an object is at of $2f$ from mirror, then the image is also formed at $2f$. So, magnification,

$$m = \frac{-v}{u} = \frac{-(-2f)}{(-2f)} = -1$$

or $\quad |m| = 1$

So, these values are correctly depicted in option (d) only.

14. At interface 1,

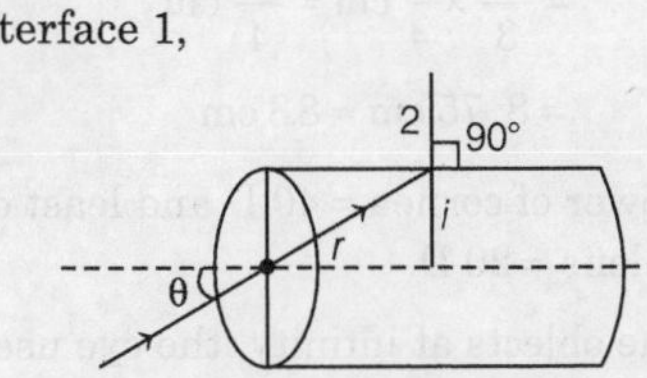

$$\mu = \frac{\sin \theta}{\sin r}$$

$\Rightarrow \quad \sin \theta = \mu \sin r \quad \ldots \text{(i)}$

At interface 2, $(90° - r) = C$

$\Rightarrow \quad \sin(90° - r) = \sin C$

$\Rightarrow \quad \cos r = \frac{1}{\mu} \quad \left(\because \sin C = \frac{1}{\mu}\right)$

$\Rightarrow \quad \cos r = \frac{1}{2/\sqrt{3}} = \frac{\sqrt{3}}{2}$

$\Rightarrow \quad r = 30°$

From Eq. (i), we see that

$$\sin \theta = \frac{2}{\sqrt{3}} \sin 30° = \frac{1}{\sqrt{3}}$$

$\therefore \quad \theta = \sin^{-1}\left(\frac{1}{\sqrt{3}}\right)$

15. The ray comes out from CD, means rays after refraction from AB get, total internal reflected at AD.

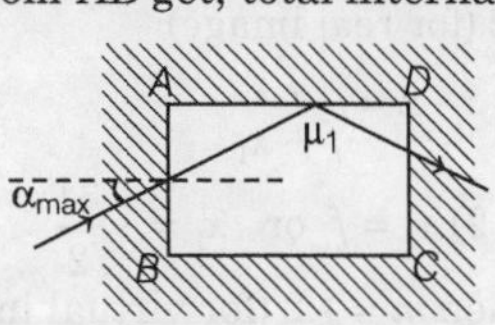

According to Snell's law,

$$\frac{\mu_1}{\mu_2} = \frac{\sin \alpha_{\max}}{\sin r_1}$$

$\Rightarrow \quad \alpha_{\max} = \sin^{-1}\left[\frac{\mu_1}{\mu_2} \sin r_1\right] \quad \ldots \text{(i)}$

Also, $\quad r_1 + r_2 = 90°$

$\Rightarrow \quad r_1 = 90° - r_2 = 90 - C$

$\Rightarrow \quad r_1 = 90° - \sin^{-1}\left(\frac{1}{2\mu_1}\right)$

$\Rightarrow \quad r_1 = 90° - \sin^{-1}\left(\frac{\mu_2}{\mu_1}\right) \quad \ldots \text{(ii)}$

Then, from Eqs. (i) and (ii), we get

$$\alpha_{\max} = \sin^{-1}\left[\frac{\mu_1}{\mu_2} \sin\left(90° - \sin^{-1}\frac{\mu_2}{\mu_1}\right)\right]$$

$$= \sin^{-1}\left[\frac{\mu_1}{\mu_2} \cos\left(\sin^{-1}\frac{\mu_2}{\mu_1}\right)\right]$$

16. The given situation is shown below.

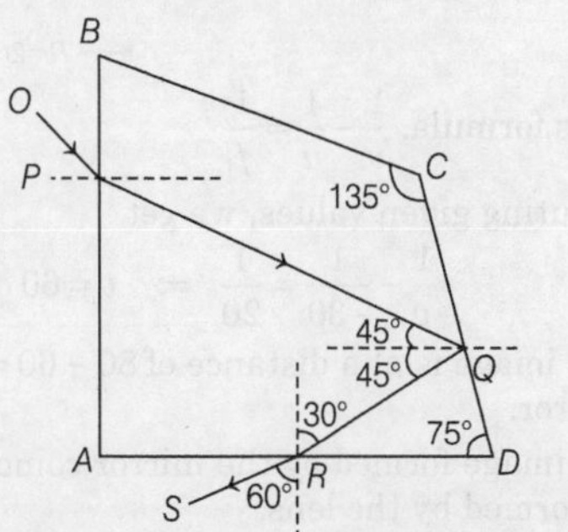

Ray OP goes parallel to the side BC. Hence, the angle of incidence of this ray on the side CD becomes 45°.

According to Snell's law,

$$\sqrt{3}\sin 45° = \sin r > 1$$

Hence, this is a case of total internal reflection. Thus, the angle of incidence on Q is equal to angle of reflection. This forms 30° as angle of incidence on side AD. By Snell's law

$$\sqrt{3}\sin 30° = \sin r = \frac{\sqrt{3}}{2}$$

$$\Rightarrow \qquad r = 60°$$

Thus, the angle between rays OP and RS is 90°.

17. Magnification for a lens can also be written as

$$m = \left(\frac{f}{f+u}\right)$$

When $m = -2$ (for real image)

$$-2 = \frac{f}{f+x_1}$$

$$\Rightarrow \quad -2f + (-2)x_1 = f \text{ or } x_1 = -\frac{3f}{2}$$

Similarly, when $m = +2$ (for virtual image),

$$+2 = \frac{f}{f+x_2}$$

$$\Rightarrow \quad 2f + 2x_2 = f \text{ or } x_2 = \frac{-f}{2}$$

Now, the ratio of x_1 and $x_2 = \dfrac{\frac{-3f}{2}}{-f/2} = \dfrac{3}{1}$

18. When a slab of thickness, t is introduced between P and the mirror, the apparent position of P shifts towards the mirror by $\left(t - \frac{t}{\mu}\right)$. Hence, the mirror must be moved in the same direction through the same distance.

19. The given situation can be drawn as shown below

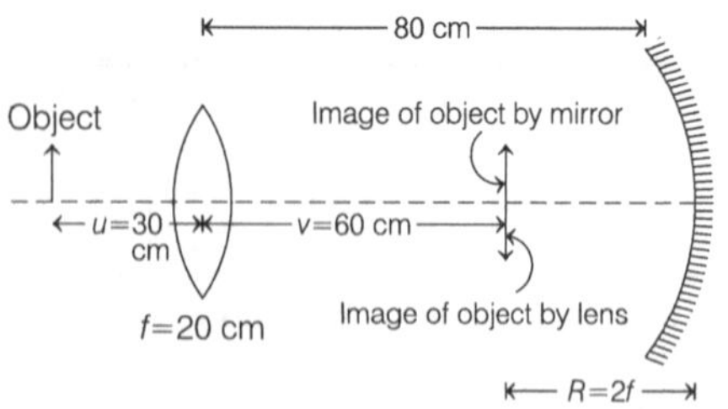

For lens formula, $\frac{1}{v} - \frac{1}{u} = \frac{1}{f}$

Substituting given values, we get

$$\frac{1}{v} - \frac{1}{-30} = \frac{1}{20} \Rightarrow v = 60 \text{ cm}$$

So, this image is at a distance of $80 - 60 = 20$ cm from the mirror.

As, the image formed by the mirror coincides with image formed by the lens.

This condition is only possible, if any object that has been placed in front of concave mirror is at centre of curvature, *i.e.* at $2f$.

So, radius of curvature of mirror is $R = 20$ cm.

$\therefore$ Focal length of mirror, $f = \frac{R}{2} = 10$ cm

As, for virtual image, the object is to kept between pole and focus of the mirror.

$\therefore$ The maximum distance of the object for which this concave mirror by itself produce a virtual image would be 10 cm.

20. This is a modified displacement method problem.

Here, $a = 1.8$ m and $\frac{a+d}{a-d} = \frac{2}{1}$

Solving, this, we get $\quad d = 0.6$ m

$$\therefore \qquad f = \frac{a^2 - d^2}{4a} = 0.4 \text{ m}$$

21. In the given case, $u = -5$ cm

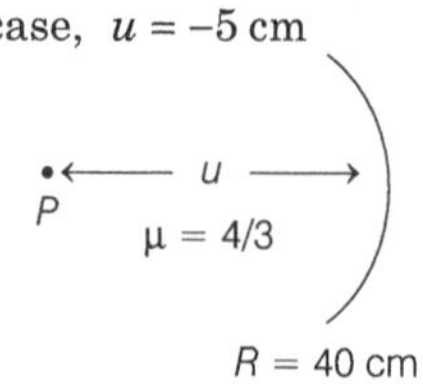

Focal length,

$$f = \frac{-R}{2} = \frac{-40}{2} = -20 \text{ cm}$$

Now, using mirror formula,

$$\frac{1}{v} + \frac{1}{u} = \frac{1}{f} \Rightarrow \frac{1}{v} = \frac{1}{f} - \frac{1}{u} = -\frac{1}{20} + \frac{1}{5} = +\frac{3}{20}$$

$$\Rightarrow \qquad v = +\frac{20}{3} \text{ cm}$$

For the light getting refracted at water surface, this image will act as an object.

So, distance of object,

$$d = 5 \text{ cm} + \frac{20}{3} \text{ cm} = \frac{35}{3} \text{ cm} \quad \text{(below the surface)}$$

Let's assume final image at distance d after refraction.

$$\frac{d'}{d} = \frac{\mu_2}{\mu_1}$$

$$\Rightarrow \qquad d' = d\left(\frac{\mu_2}{\mu_1}\right) = \left(\frac{35}{3}\text{ cm}\right)\left(\frac{1}{\frac{4}{3}}\right)$$

$$= \frac{35}{3} \times \frac{3}{4} \text{ cm} = \frac{35}{4} \text{ cm}$$

$$= 8.75 \text{ cm} \approx 8.8 \text{ cm}$$

22. Given, the power of cornea = 40 D and least converging power of eye lens = 20 D.

To observe the objects at infinity, the eye uses its least converging power means power is maximum, *i.e.* $40 + 20 = 60$ D.

The distance between cornea = focal length of eye lens

$$f = \frac{100}{P} = \frac{100}{60} = \frac{5}{3} \text{ cm}$$

To focus objects at the near point on the retina

$$u = -25 \text{ cm}, \; v = \frac{5}{3} \text{ cm}$$

Using lens formula,

$$\frac{1}{f} = \frac{1}{v} - \frac{1}{u} = \frac{1 \times 3}{5} + \frac{1}{25} = \frac{15+1}{25} = \frac{16}{25}$$

$$\Rightarrow \quad f = \frac{25}{16} \text{ cm}$$

$$\text{Power of lens} = \frac{1}{f} = \frac{100 \times 16}{25} = 64 \text{ D}$$

$\therefore$ Power of eye lens $= 64 - 40 = 24$ D

Thus, the range of accommodation of the eye lens is 20 D to 24 D.

23. From figure, $OA = 4$ cm, $AB = 3$ cm

$$\therefore \quad OB = \sqrt{4^2 + 3^2} = 5 \text{ cm}$$

$$\text{Now,} \quad \mu = \frac{1}{\sin C} = \frac{1}{AB/OB} = \frac{OB}{AB} = \frac{5}{3}$$

$$\text{From,} \quad \mu = \frac{c}{c/v} = \frac{c}{\mu} = \frac{3 \times 10^8}{5/3} = 1.8 \times 10^8 \text{ m/s}$$

24. As, $m = \dfrac{f_o}{f_e}$

We know that, $m = \dfrac{\text{Angle subtended by an image}}{\text{Angle subtended by an object}}$

$$\therefore \quad \frac{f_o}{f_e} = \frac{\alpha}{\beta}$$

$$\Rightarrow \quad \alpha = \frac{\beta \times f_o}{f_e} = \frac{6002 \times 60}{5} = 2 \times 12 = 24°$$

25. Due to total internal reflection, light comes out from a cone of angle $2C$ as shown in the figure (C = angle of critical incidence).

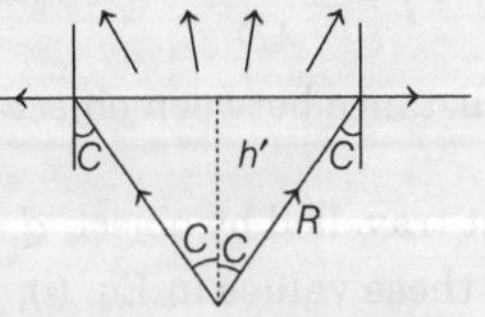

Now, it is given that surface area of spherical cap of height h and radius R is $2\pi Rh$.

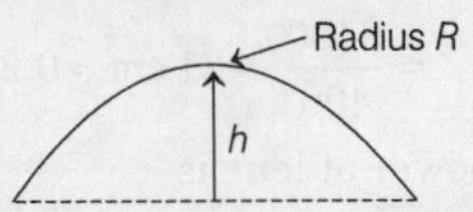

Here, we are given with following situation

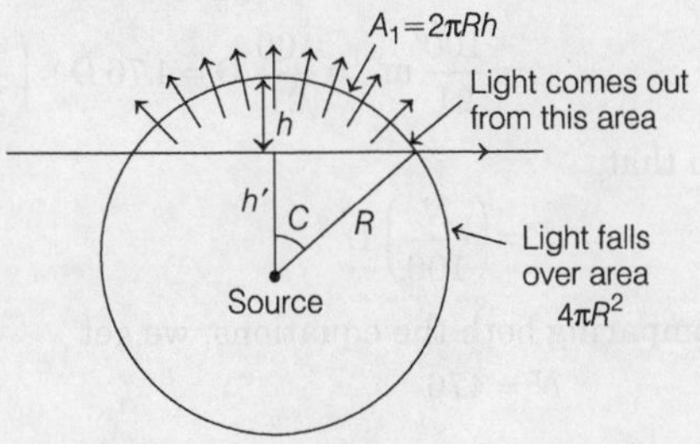

So, percentage of light that emerges out of surface is given by

% of light coming out

$$= \frac{\text{Area through which light comes out}}{\text{Total area over which light falls}} \times 100$$

$$= \frac{2\pi Rh}{4\pi R^2} \times 100$$

$$= \frac{1}{2} \times \left(\frac{R - h'}{R} \right) \times 100$$

$$\text{Now,} \quad \frac{h'}{R} = \frac{\text{base}}{\text{hypotenuse}} = \cos C$$

$$\text{As,} \quad \sin C = \frac{1}{n} = \frac{1}{\frac{4}{3}} = \frac{3}{4}$$

$$\Rightarrow \quad \cos C = \frac{\sqrt{7}}{4} = \frac{h'}{R}$$

$\therefore$ % of light coming out

$$= \frac{1}{2} \times \left(\frac{R - h'}{R} \right) \times 100$$

$$= \frac{1}{2} \times \left(1 - \frac{h'}{R} \right) \times 100$$

$$= \frac{1}{2} \times \frac{\sqrt{7}}{4} \times 100$$

$$= 16.9\%$$

$$\simeq 17\%$$

Nearest answer is 17 %.

26. $\omega_1 = 0.02$; $\mu_1 = 1.5$; $\omega_2 = 0.03$; $\mu_2 = 1.6$

Achromatic combination,

$$\therefore \quad \theta_{net} = 0$$

$$\theta_1 - \theta_2 = 0$$

$$\theta_1 = \theta_2$$

$$\omega_1 \delta_1 = \omega_2 \delta_2$$

$$\text{and} \quad \delta_{net} = \delta_1 - \delta_2 = 2°$$

$$\delta_1 - \frac{\omega_1 \delta_1}{\omega_2} = 2°$$

$$\delta_1 \left(1 - \frac{\omega_1}{\omega_2} \right) = 2°$$

$$\delta_1 \left(1 - \frac{2}{3} \right) = 2°$$

$$\delta_1 = 6°$$

$$\delta_1 = (\mu_1 - 1) A_1$$

$$6° = (1.5 - 1) A_1$$

$$A_1 = 12°$$

27. Deviation for small-angled prism is given by

$$\delta = (\mu - 1) A$$

Given, $A = 1°$, $\mu = 1.5$

Substituting these values in above equation, we get

$$\delta = (1.5 - 1)1 \Rightarrow \delta = 0.5$$

According to question, $\delta = \frac{N}{10}$

$$\Rightarrow \quad 0.5 = \frac{N}{10} \Rightarrow N = 5$$

Hence, the value of N is 5.

28. Focal length of lens by lens Maker's formula,

$$\frac{1}{f} = (n_{21} - 1)\left(\frac{1}{R_1} - \frac{1}{R_2}\right)$$

For a plano-convex lens,

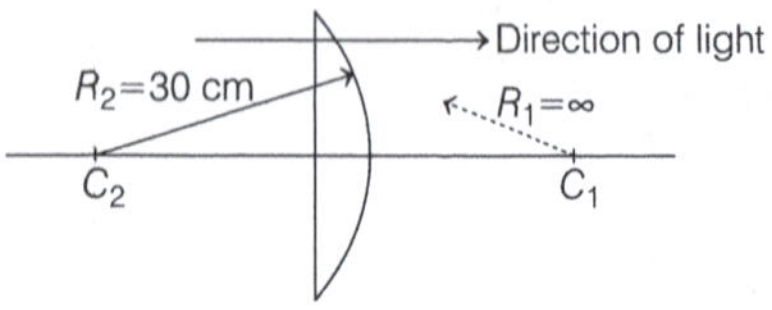

$$R_1 = \infty, R_2 = -30 \text{ cm}, n_{21} = 1.5$$

$$\therefore \quad \frac{1}{f} = (1.5 - 1)\left(\frac{1}{\infty} - \frac{1}{(-30)}\right)$$

$$\Rightarrow \quad \frac{1}{f} = \frac{1}{2} \times \frac{1}{30} \Rightarrow f = 60 \text{ cm}$$

29. Given, object distance, $u = -30$ cm,

Image distance, $v = -10$ cm

and speed of object, $v_o = 9 \text{ cms}^{-1}$

Let v_i be the speed of image.

Magnification of concave mirror,

$$m = -\frac{v}{u} = -\frac{(-10)}{(-30)} = -\frac{1}{3}$$

Now, $v_i = v_o(m^2) = 9 \times \left(\frac{-1}{3}\right)^2 = 1 \text{ cms}^{-1}$

30. For the eyepiece,

$$v_e = -25 \text{ cm}, f_e = 5 \text{ cm}$$

or $$\frac{1}{u_e} = -\frac{1}{25} - \frac{1}{5}$$

or $$\frac{1}{u_e} = -\frac{-1-5}{25}$$

or $$u_e = -\frac{25}{6}$$

Now, $$v_o = L - |u_e| = 20 - \frac{25}{6}$$

$$= \frac{120 - 25}{6} \text{ cm} = \frac{95}{6} \text{ cm}$$

Now, $$\frac{1}{95/6} - \frac{1}{u_o} = \frac{1}{1}$$

or $$\frac{1}{u_o} = \frac{6}{95} - 1$$

or $$u_o = -\frac{95}{89} \text{ cm}$$

$$\therefore \quad |u_o| = \frac{95}{89} \text{ cm} = 1.067 \text{ cm}$$

31. When the final image is formed at least distance of distinct vision (*i.e.* at 25 cm) from eyepiece of a compound microscope, then the magnification is given by

$$m = -\frac{v_0}{u_0}\left(1 + \frac{D}{f_e}\right)$$

$$\Rightarrow \quad m = \frac{-L}{f_0}\left(1 + \frac{D}{f_e}\right)$$

$$\Rightarrow \quad -100 = -\frac{20}{1}\left(1 + \frac{25}{f_e}\right)$$

$$\Rightarrow \quad 5 = 1 + \frac{25}{f_e}$$

$$4 = \frac{25}{f_e} \Rightarrow f_e = \frac{25}{4} = 6.25 \simeq 6 \text{ cm}$$

32. Strain on the eye is minimum, so the intermediate image is formed at focal length of eyepiece.

For objective, $v = 5$ cm, $f = 1$ cm

$\therefore$ Using lens formula, $\frac{1}{v} - \frac{1}{u} = \frac{1}{f}$

$$\Rightarrow \quad \frac{1}{5} - \frac{1}{u} = \frac{1}{1} \Rightarrow u = \frac{-5}{4} \text{ cm}$$

$$\Rightarrow \quad |u| = \frac{5}{4} \text{ cm} \Rightarrow \frac{n}{40} = \frac{5}{4}$$

$$\Rightarrow \quad n = \frac{5}{4} \times 40 \Rightarrow n = 50$$

33. Using displacement method, the focal length of a lens is given by

$$f = \frac{D^2 - d^2}{4D} \quad \ldots\text{(i)}$$

Given that, distance between object and screen, $D = 100$ cm

Displacement travelled by object, $d = 40$ cm

Substituting these values in Eq. (i), we get

$$f = \frac{(100)^2 - (40)^2}{4(100)} = \frac{10000 - 1600}{400}$$

$$= \frac{8400}{400} = 21 \text{ cm} = 0.21 \text{ m}$$

Now, optical power of lens is

$$P = \frac{1}{f} = \frac{1}{0.21 \text{ m}} = \frac{100}{21 \text{ m}}$$

$$= \frac{100}{21} \text{m}^{-1} = \frac{100}{21} \text{ D} = 4.76 \text{ D} = \left(\frac{476}{100}\right) \text{D}$$

Given that,

$$P = \left(\frac{N}{100}\right) \text{D}$$

On comparing both the equations, we get

$$N = 476$$

34. We are given following situation as shown in figure,

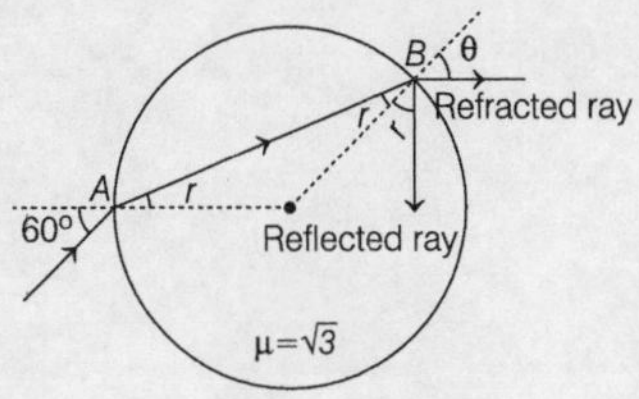

Using Snell's law at point A,

$$n_1 \sin i = n_2 \sin r$$

$$1 \times \sin 60^\circ = \sqrt{3} \times \sin r$$

$$\Rightarrow \quad \sin r = \frac{\sqrt{3}/2}{\sqrt{3}} = \frac{1}{2} \quad \Rightarrow \quad r = 30^\circ$$

Now, at point B, angle of incidence, $r = 30^\circ$

Again, using Snell's law at point B,

$$\sqrt{3} \sin 30^\circ = 1 \times \sin \theta$$

$$\Rightarrow \quad \sin \theta = \frac{\sqrt{3}}{2} \quad \Rightarrow \quad \theta = 60^\circ$$

and angle of reflection, $r = 30^\circ$

$\therefore$ Angle between reflected ray and refracted ray at point $B = 180^\circ - \theta - r$

$$= 180^\circ - 60^\circ - 30^\circ = 90^\circ$$

35. Initially, when the jar is empty.

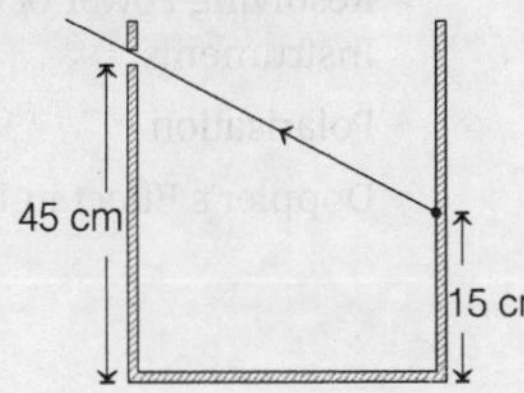

Finally, when the jar is filled with liquid.

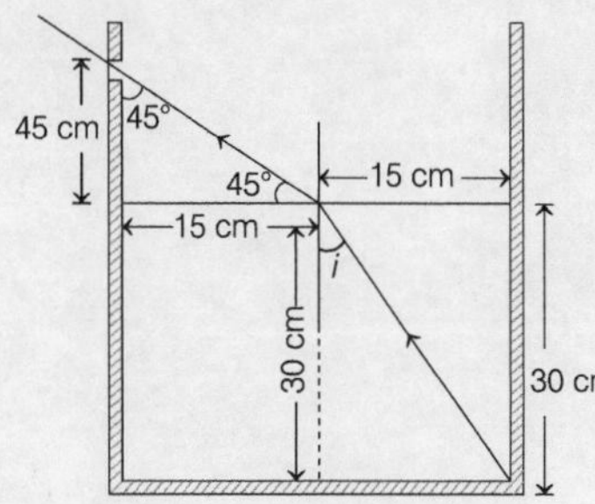

If i be the angle of incidence, then

$$\tan i = \frac{15}{30} = \frac{1}{2}$$

$$\Rightarrow \quad \sin i = \frac{\text{perpendicular}}{\text{hypotenuse}} = \frac{1}{\sqrt{5}}$$

According to law of refraction,

$$n_1 \sin i = n_2 \sin r$$

$$\Rightarrow \quad \mu\left(\frac{1}{\sqrt{5}}\right) = 1 \times \sin 45^\circ$$

$$\Rightarrow \quad \mu = \frac{\sqrt{5}}{\sqrt{2}} = 1.581$$

Also, given that,

$$\frac{N}{100} = \mu$$

$$\Rightarrow \quad N = 100\mu$$

$$= 100 \times 1.581 = 158.1$$

So, nearest integer value of N is 158.

26

Wave Optics

IN THIS CHAPTER ....

- Huygens' Wave Theory
- Interference of Light
- Young's Double Slit Experiment
- Diffraction of Light
- Resolving Power of Optical Instruments
- Polarisation
- Doppler's Effect in Light

Ray optics uses the geometry of straight lines to explain the phenomenon of rectilinear propagation of light, reflection and refraction of light, etc. However, wave optics deals light on the basis of waves, where light can bend around objects, can diffract and interfere, etc.

Huygens' Wave Theory

In 1678, a Dutch scientist, Christian Huygen propounded the wave theory of light. He assumed that, light travels in the form of waves. These waves travel in all the directions with the velocity of light. He also introduced the concept of wavefront.

Wavefront

It is the locus of points (wavelets) having same phase (a surface of constant phase) of oscillations. As the wave travels forward, the wavefront also moves forward. The perpendicular to the surface of a wavefront gives the direction of light ray. The energy transfers along the direction of rays.

Types of Wavefronts

- **Light emerging from a point source** Wavefronts are spherical and eccentric with point source at their centre as shown in figure by circles 1, 2, 3 and 4. Rays are radial as shown in arrows.

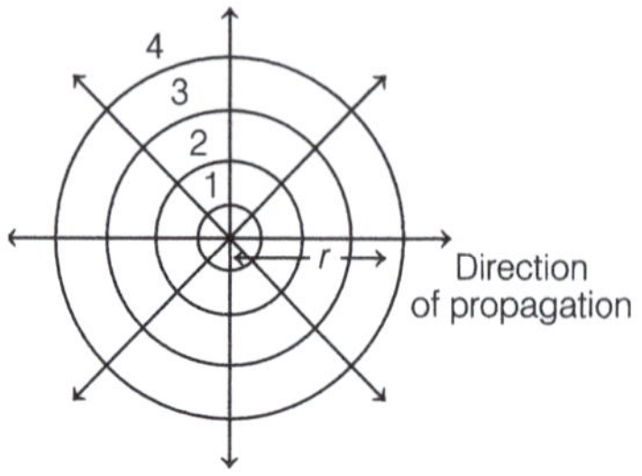

In spherical wavefront,

$$\text{amplitude} \propto \frac{1}{r} \quad \text{and} \quad \text{intensity} \propto \frac{1}{r^2}$$

- **Light coming out from a line source** Wavefronts 1, 2 and 3 are cylindrical and coaxial with the straight source as their common axis.

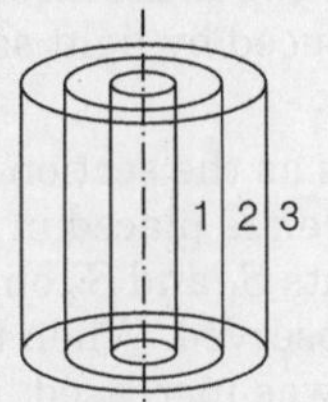

- **When light sources are emitting parallel rays or when the light is coming from a very far-off source** Wavefronts will be planes as shown in following figure.

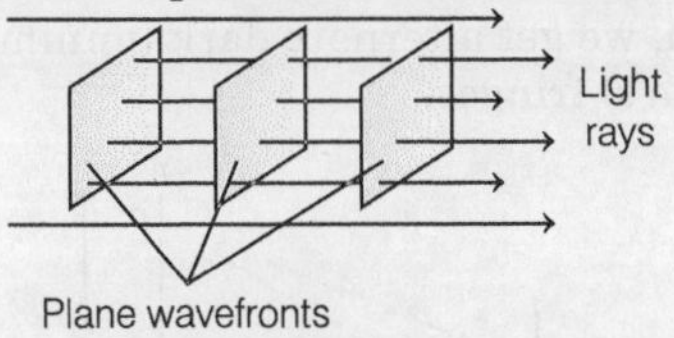

Plane wavefronts

Note Wavefronts parallel to XY-plane can be represented by Z = constant. Similarly, wavefronts parallel to YZ-plane and ZX-plane are X = constant and Y = constant, respectively.

Huygens' Principle of Secondary Wavelets

In 1678, in order to explain the propagation of wave in a medium, Huygen propounded a principle known as Huygen's principle of secondary wavelets by which at any instant, we can geometrically obtain the position of wavefront. He made following three assumptions:

(i) Every point on a given wavefront (called **primary wavefront**) can be regarded as fresh source of new disturbance and sends out its own spherical wavelets called **secondary wavelets.**

(ii) The secondary wavelets spread in all directions with the velocity of wave (*i.e.* velocity of light).

(iii) A surface touching these secondary wavelets, tangentially in the forward direction at any instant gives the position and shape of the new wavefront at the instant. This is called **secondary wavefront.**

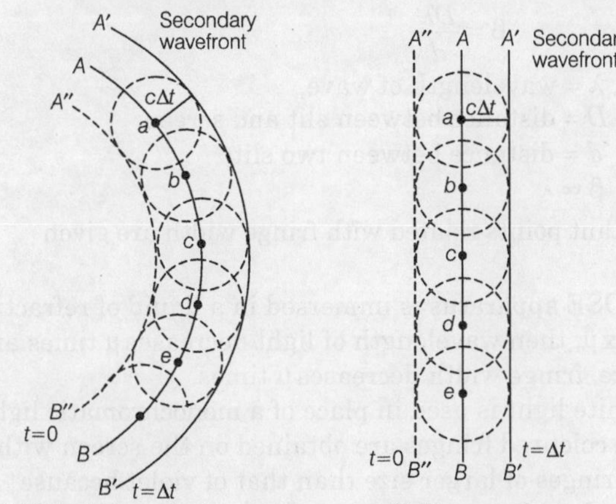

(a) Spherical wavefront (b) Plane wavefront

Principle of Superposition of Waves

According to superposition principle, at a particular point in the medium, the resultant displacement y produced by a number of waves is the vector sum of the displacements $(y_1, y_2, \ldots)$ produced by each of the waves.

i.e. $y = y_1 + y_2 + y_3 + y_4 + \ldots$

Note The superposition principle which was stated first for mechanical waves is equally applicable to the electromagnetic (light) waves.

Coherent and Incoherent Sources

Two sources are said to be **coherent**, if they have the same frequency and have constant phase difference with time.

However, two such sources of light, which do not emit light waves with constant phase difference are called **incoherent sources.**

Interference of Light

When two light waves of exactly equal frequency having phase difference which is constant with respect to time travelled in same direction and overlap each other, then at some points, the intensity of light is maximum, while at some points, the intensity of light is minimum.

Thus, formation of maximum intensity at some points and minimum intensity at some other points by the two identical light waves travelling in same direction is called the *interference of light*.

The interference at the points where the two waves meet in same phase, *i.e.* the intensity of light is maximum, is called the *constructive interference* while at the points where the two waves meet in opposite phase, *i.e.* the intensity of light is minimum is called the *destructive interference.*

Conditions of Maxima and Minima

For two waves from sources S_1 and S_2 interfering at a certain point such that $y_1 = A_1 \sin \omega t$ and $y_2 = A_2 \sin(\omega t + \phi)$

where, A_1 and A_2 are the respective amplitudes of the two waves and ϕ is the constant phase angle by which second wave leads the first wave.

By the principle of superposition, the magnitude of resultant displacement of the waves is

$$y = A \sin(\omega t + \theta)$$

where , $$A = \sqrt{A_1^2 + A_2^2 + 2A_1 A_2 \cos\phi}$$

and $$\theta = \tan^{-1}\left[\frac{A_2 \sin\phi}{A_1 + A_2 \cos\phi}\right]$$

As, resultant intensity I is $I \propto A^2$

$$\Rightarrow \qquad I = I_1 + I_2 + 2\sqrt{I_1 I_2} \cos\phi$$

For constructive interference (maximum intensity) The intensity I is maximum, when $\cos\phi = +1$, *i.e.* when phase difference $\phi = 2n\pi$; $n = 0, 1, 2, \ldots$, etc.

If Δx is the path difference between the interfering waves, then $\Delta x = n\lambda$.

So, $I_{max} = A_1^2 + A_2^2 + 2A_1A_2 = (A_1 + A_2)^2$

$= I_1 + I_2 + 2\sqrt{I_1 I_2} = (\sqrt{I_1} + \sqrt{I_2})^2$

When $I_1 = I_2 = I_0$, then resultant intensity,

$$I_{max} = 4I_0 \cos^2\left(\frac{\phi}{2}\right)$$

For destructive interference (minimum intensity)

The intensity I is minimum, when $\cos\phi = -1$, *i.e.* when phase difference,

$$\phi = (2n - 1)\pi;\ n = 1, 2, \ldots, \text{etc.}$$

Here, path difference, $\Delta x = (2n - 1)\frac{\lambda}{2}$

So, $I_{min} = A_1^2 + A_2^2 - 2A_1A_2 = (A_1 - A_2)^2$

$= I_1 + I_2 - 2\sqrt{I_1 I_2} = (\sqrt{I_1} - \sqrt{I_2})^2$

When $I_1 = I_2 = I_0$, then resultant intensity,

$$I_{min} = 0$$

$\therefore$ We can write,

$$\frac{I_{max}}{I_{min}} = \left(\frac{\sqrt{I_1} + \sqrt{I_2}}{\sqrt{I_1} - \sqrt{I_2}}\right)^2 = \left(\frac{A_1 + A_2}{A_1 - A_2}\right)^2$$

Note Relation between path difference, phase difference and time difference

$$\frac{2\pi}{\text{Phase diff.}} = \frac{\lambda}{\text{Path diff.}} = \frac{T}{\text{Time diff.}}$$

Example 1. *Two coherent sources produce waves of different intensities which interfere. After interference, the ratio of the maximum intensity to the minimum intensity is 16. The intensity of the waves are in the ratio* **[JEE Main 2019]**

(a) 16 : 9 (b) 5 : 3
(c) 25 : 9 (d) 4 : 1

Sol. (c) Let the intensity of two coherent sources be I_1 and I_2, respectively.

It is given that, $\frac{\text{maximum intensity}}{\text{minimum intensity}} = \frac{I_{max}}{I_{min}} = \frac{16}{1}$

Since, we know, $I_{max} = (\sqrt{I_1} + \sqrt{I_2})^2$

and $I_{min} = (\sqrt{I_1} - \sqrt{I_2})^2$

$\therefore$ We can write, $\frac{I_{max}}{I_{min}} = \frac{(\sqrt{I_1} + \sqrt{I_2})^2}{(\sqrt{I_1} - \sqrt{I_2})^2} = 16$

$\Rightarrow \frac{\sqrt{I_1} + \sqrt{I_2}}{\sqrt{I_1} - \sqrt{I_2}} = \frac{\sqrt{16}}{1} = \frac{4}{1}$

$\sqrt{I_1} + \sqrt{I_2} = 4\sqrt{I_1} - 4\sqrt{I_2}$

$\Rightarrow 5\sqrt{I_2} = 3\sqrt{I_1}$, $\frac{\sqrt{I_1}}{\sqrt{I_2}} = \frac{5}{3}$

Squaring both the sides, we get $\frac{I_1}{I_2} = \frac{25}{9}$

Young's Double Slit Experiment

Young's experimental arrangement for double slit experiment consists of two monochromatic narrow slits S_1 and S_2 which are produced by light source S_1 as shown in the figure given below.

These S_1 and S_2 serves as the section of two coherent sources of light. A screen is placed in the path of light emerging out of the slits S_1 and S_2 on which alternate coloured fringes are observed. When the distance between the two slits was increased, it was observed that the bands or fringes disappeared. Therefore, an optimum distance between them is required to observe the well defined fringes pattern.

In this pattern, we get alternate dark (minima's) and bright (maxima's) fringes.

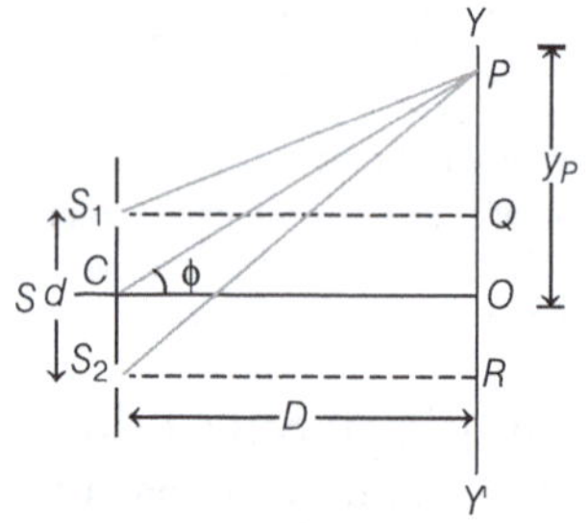

For light waves reaching at a point P, situated at a distance x from central point O, the path difference Δy is given by

$$\Delta y = S_2P - S_1P = \frac{xd}{D}$$

If $\frac{xd}{D} = n\lambda$, then we get nth bright fringe, where $n = 1, 2, 3, 4$

Similarly, if $\frac{xd}{D} = (2n - 1)\frac{\lambda}{2}$, then we get nth dark fringe, where $n = 1, 2, 3, 4\ldots$.

Fringe Width

The separation between any two consecutive bright or dark fringes is called *fringe width* (β) and it is given by

$$\beta = \frac{D\lambda}{d}$$

where, λ = wavelength of wave,
D = distance between slit and screen
and d = distance between two slits.
or $\beta \propto \lambda$

Important points related with fringe width are given below

- If YDSE apparatus is immersed in a liquid of refractive index μ, then wavelength of light decreases μ times and hence, fringe width decreases μ times.
- If white light is used in place of a monochromatic light, then coloured fringes are obtained on the screen with red fringes of larger size than that of violet because $\lambda_{red} > \lambda_{violet}$.

But note that centre is still white because path difference there is zero for all colours. Hence, all the wavelengths interfere constructively.

Angular width of a fringe

It is given as, $\theta = \theta_{n+1} - \theta_n = \frac{\lambda}{d}$ or $\theta = \frac{\lambda}{d} = \frac{\beta}{D}$

It is independent of n, *i.e.* angular widths of all fringes are same.

Distribution of intensity

The distribution of intensity in Young's double slit experiment is shown below.

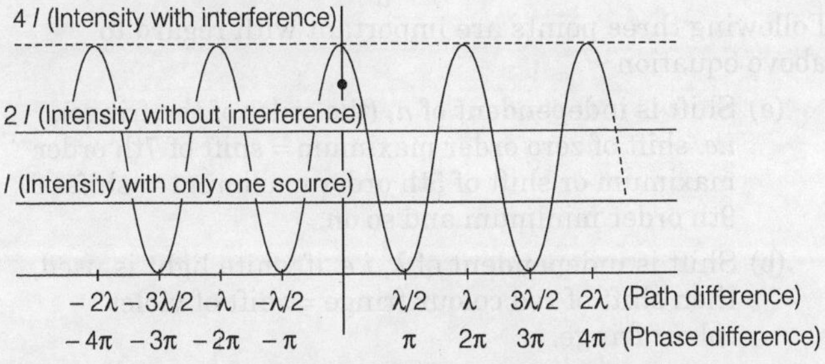

Distribution of intensity in YDSE

Example 2. *Consider a Young's double slit experiment as shown in figure.*

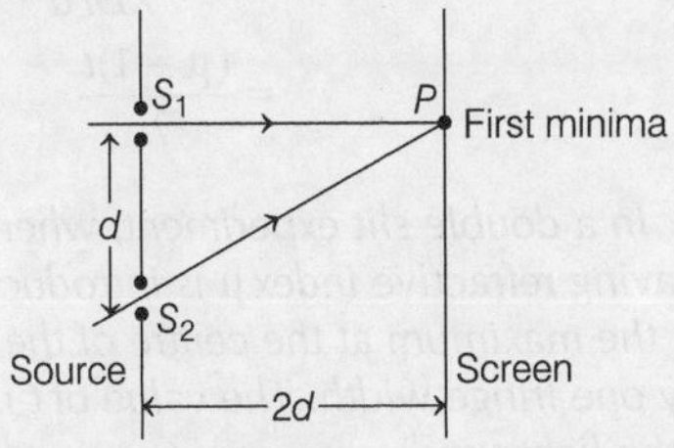

What should be the slit separation d in terms of wavelength λ such that the first minima occurs directly in front of the slit S_1? **[JEE Main 2019]**

(a) $\frac{\lambda}{2(5-\sqrt{2})}$ (b) $\frac{\lambda}{(5-\sqrt{2})}$

(c) $\frac{\lambda}{2(\sqrt{5}-2)}$ (d) $\frac{\lambda}{(\sqrt{5}-2)}$

Sol. (c) In the given case, figure for first minima will be as shown below.

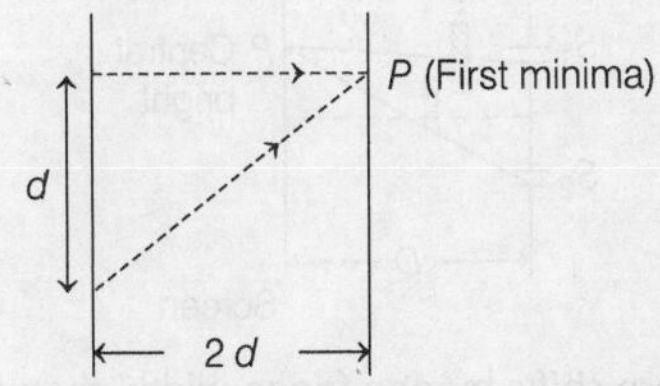

We know that condition for minima in Young's double slit experiment, path difference is

$$\Delta x = (2n-1)\lambda/2$$

For first minima, $n = 1$

$$\Rightarrow \quad \Delta x = \lambda/2 \quad \dots \text{(i)}$$

Path difference between the rays coming from virtual sources S_1 and S_2 at point P will be

$$\Delta x = S_2P - S_1P \quad \dots \text{(ii)}$$

From triangle S_1S_2P,

$$S_1P = 2d \quad \dots \text{(iii)}$$

and $(S_2P)^2 = (S_1S_2)^2 + (S_1P)^2$ (from Pythagoras theorem)

$$= d^2 + (2d)^2$$

$\Rightarrow \quad (S_2P)^2 = 5d^2$

or $S_2P = \sqrt{5}\,d$ ...(iv)

Substituting the values from Eqs. (iii) and (iv) in Eq. (ii), we get

$$\Delta x = \sqrt{5}\,d - 2d \quad \dots \text{(v)}$$

From Eqs. (i) and (v), we get

$$\sqrt{5}\,d - 2d = \lambda/2$$

$$\Rightarrow \quad d = \frac{\lambda}{2(\sqrt{5}-2)}$$

Example 3. *In a Young's double slit experiment, the separation between the slits is 0.15 mm. In the experiment, a source of light of wavelength 589 nm is used and the interference pattern is observed on a screen kept 1.5 m away. The separation between the successive bright fringes on the screen is* **[JEE Main 2020]**

(a) 4.9 mm (b) 3.9 mm (c) 6.9 mm (d) 5.9 mm

Sol. (d) In YDSE, separation between successive bright fringes is width of one dark fringe.

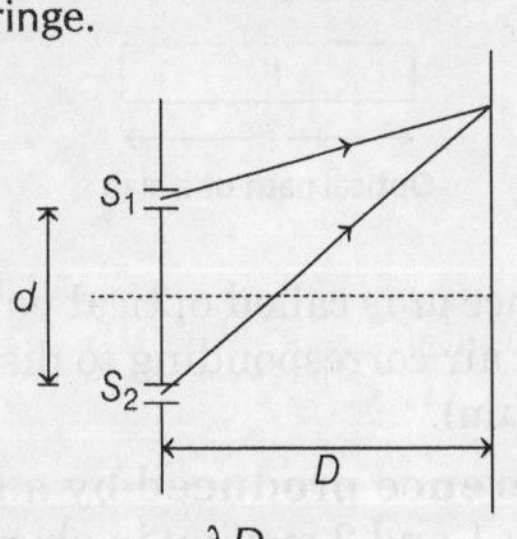

As, fringe width in YDSE, $\beta = \frac{\lambda D}{d}$ ...(i)

Here, $\lambda = 589 \text{ nm} = 589 \times 10^{-9}$ m,

$D = 1.5$ m, $d = 0.15$ mm $= 0.15 \times 10^{-3}$ m

Substituting the given values in Eq. (i), we get

$$\beta = \frac{589 \times 10^{-9} \times 1.5}{0.15 \times 10^{-3}} = 589 \times 10^{-5} \text{ m}$$

$$= 5.89 \text{ mm or } 5.9 \text{ mm}$$

Example 4. *In a YDSE, if D = 2 m, d = 6 mm, λ = 6000Å, then the fringe width and the position of the 3rd maxima are*

(a) 0.2 mm, 0.6 mm (b) 0.8 mm, 0.1 mm

(c) 1.2 mm, 0.2 mm (d) None of these

Sol. (a) We know that,

Fringe width, $\beta = \frac{\lambda D}{d} = \frac{6000 \times 10^{-10} \times 2}{6 \times 10^{-3}} = 0.2$ mm

Position of 3rd maxima, $y_3 = \frac{3\lambda D}{d} = 3\beta = 3 \times 0.2 = 0.6$ mm

Example 5. *In Young's double slit experiment interference fringes 1° apart are produced on the screen, the slit separation is*

(a) 0.546 mm (b) 0.0337 mm
(c) 0.246 mm (d) 0.0927 mm

Sol. (b) The fringe width, $\beta = \frac{\lambda D}{d}$

The angular separation of the fringes is given by

$$\theta = \frac{\beta}{D} = \frac{\lambda}{d}$$

Given, $\theta = 1^\circ = \frac{\pi}{180}$ rad

$\lambda = 589$ nm

$$\therefore \quad d = \frac{\lambda}{\theta} = \frac{589 \times 180 \times 10^{-9}}{\pi}$$

$= 0.0337$ mm

Insertion of Transparent Slab in YDSE

If a thin transparent plate (or slab) of thickness t and refractive index μ is placed in the path of one of interfering waves in YDSE, then fringe pattern shifts. First we will discuss the optical path, path difference due to the slab and then shifting in pattern as follows:

(i) **Optical path of a slab** If light travels a distance t in a slab (medium) of refractive index μ in a given time t_0, then in this same time, it travels a distance μt in vacuum.

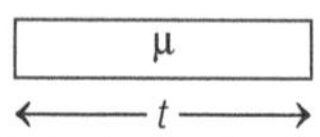

Optical path of a slab

This distance μt is called optical path of the medium in air corresponding to distance t in the slab (medium).

(ii) **Path difference produced by a slab** For two light waves 1 and 2 moving in air parallel to each other, if a slab of refractive index μ and thickness t is inserted between the path of one of the rays, then a path difference

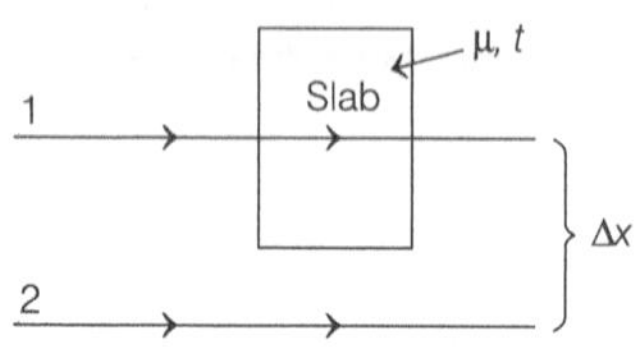

Path difference between rays 1 and 2

$$\Delta x = (\mu - 1)\, t$$

is produced among them.

(iii) **Displacement of fringes** If a thin transparent plate of thickness t and refractive index μ is placed in the path of one of interfering waves (say in path S_1P), then effective path in air is increased by an amount $(\mu - 1)\,t$ due to introduction of plate.

Effective path difference in air is

$$\Delta x = \frac{yd}{D} - (\mu - 1)t$$

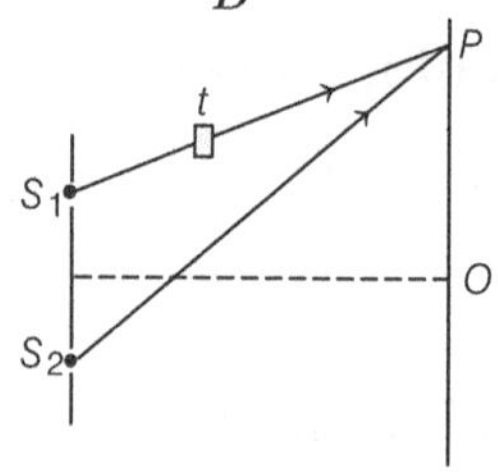

$$\therefore \quad \text{Shift} = \frac{(\mu - 1)\, t D}{d}$$

Following three points are important with regard to above equation

(a) Shift is independent of n, (the order of the fringe), *i.e.* shift of zero order maximum = shift of 7th order maximum or shift of 5th order maximum = shift of 9th order minimum and so on.

(b) Shift is independent of λ, *i.e.* if white light is used, then shift of red colour fringe = shift of violet colour fringe.

(c) Number of fringes shifted $= \frac{\text{Shift}}{\text{Fringe width}}$

$$= \frac{(\mu - 1)\, t\, D/d}{\lambda D/d} = \frac{(\mu - 1)t}{\lambda}$$

Example 6. *In a double slit experiment, when a thin film of thickness t having refractive index μ is introduced in front of one of the slits, the maximum at the centre of the fringe pattern shifts by one fringe width. The value of t is (λ is the wavelength of the light used)* **[JEE Main 2019]**

(a) $\frac{2\lambda}{(\mu - 1)}$ (b) $\frac{\lambda}{2(\mu - 1)}$

(c) $\frac{\lambda}{(\mu - 1)}$ (d) $\frac{\lambda}{(2\mu - 1)}$

Sol. (c) As we know,

Path difference introduced by thin film,

$$\Delta x = (\mu - 1)\, t \quad \text{...(i)}$$

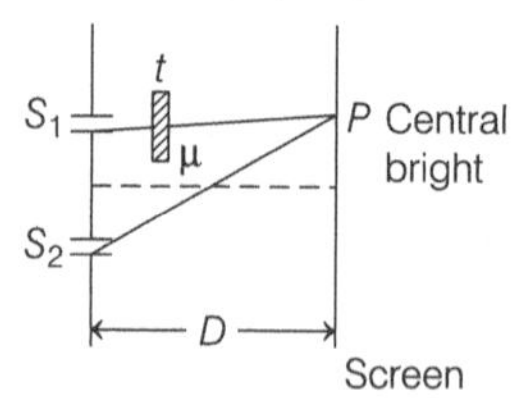

and if fringe pattern shifts by one fringe width, then path difference,

$$\Delta x = 1 \times \lambda = \lambda \quad \text{...(ii)}$$

So, from Eqs. (i) and (ii), we get

$$(\mu - 1)\, t = \lambda \Rightarrow t = \frac{\lambda}{\mu - 1}$$

Example 7. *In a double slit arrangement fringes are produced using light of wavelength 4800Å. One slit is covered by a thin plate of glass of refractive index 1.4 and the other with another glass plate of same thickness but of refractive index 1.7. By doing so, the central bright shifts to original 4th bright fringe from centre. What is the thickness of the glass plate?*

(a) 6.8 m (b) 9.2 m (c) 6.4 μm (d) 9.2 μm

Sol. (c) As, we know that, shift due to insertion of a slab is

$$\Delta x = \frac{\beta}{\lambda}(\mu - 1)t$$

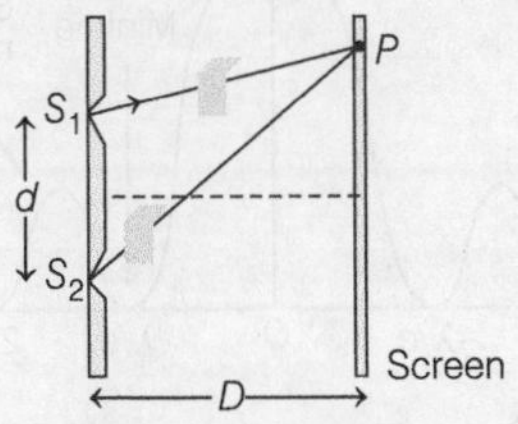

Shift due to one plate, $\Delta x_1 = \frac{\beta}{\lambda}(\mu_1 - 1)t$

Shift due to another path, $\Delta x_2 = \frac{\beta}{\lambda}(\mu_2 - 1)t$

Net shift, $\Delta x = \Delta x_2 - \Delta x_1 = \frac{\beta}{\lambda}(\mu_2 - \mu_1)t$ ...(i)

Here, it is given that, $\Delta x = 4\beta$ ...(ii)

Hence, $4\beta = \frac{\beta}{\lambda}(\mu_2 - \mu_1)t$

$$\Rightarrow \quad t = \frac{4\lambda}{(\mu_2 - \mu_1)} = \frac{4 \times 4800 \times 10^{-10}}{(1.7 - 1.4)}$$

$$= 6.4 \times 10^{-6} \text{m} = 6.4 \mu\text{m}$$

Interference in thin Films

Interference effects are commonly observed in thin films, such as thin layers of oil on water or the thin surface of a soap bubble. The varied colours observed when white light is incident on such films as the result of the interference of waves reflected from the two surfaces of the film. A film of uniform thickness t and index of refraction μ, is shown in the following figure.

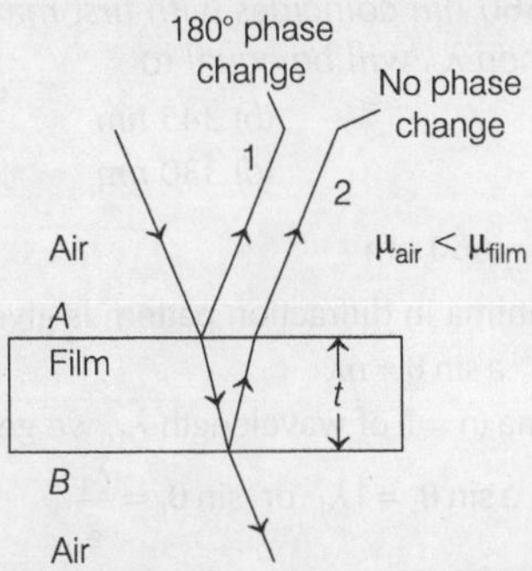

Interference in light reflected from a thin film is due to a combination of rays reflected from the upper and lower surfaces of the film.

If a wave is reflected from a denser medium, then it undergoes a phase change of 180°. The path difference between the two rays 1 and 2 is $2t$, while the phase difference between them is 180°. Hence, condition of constructive interference will be

$$2\mu t = \left(n - \frac{1}{2}\right)\lambda \qquad \left(\because \lambda_\mu = \frac{\lambda}{\mu}\right)$$

Similarly, condition of destructive interference will be

$$2\mu t = n\lambda \qquad (\text{where, } n = 0, 1, 2, \ldots)$$

Example 8. *In solar cells, a silicon solar cell (μ = 3.5) is coated with a thin film of silicon monoxide SiO (μ = 1.45) to minimise reflective losses from the surface. What is the minimum thickness of SiO that produces the least reflection at a wavelength of 550 nm, near the centre of the visible spectrum?*

(a) 94.8 nm (b) 6.28 nm (c) 9.26 m (d) 8.26 μm

Sol. (a) The reflected light is a minimum when rays 1 and 2 (shown in figure) meet the condition of destructive interference. Both rays undergo a 180° phase change upon reflection. The net change in phase due to reflection is therefore zero and the condition for a reflection to be minimum requires a path difference of $\lambda_\mu/2$.

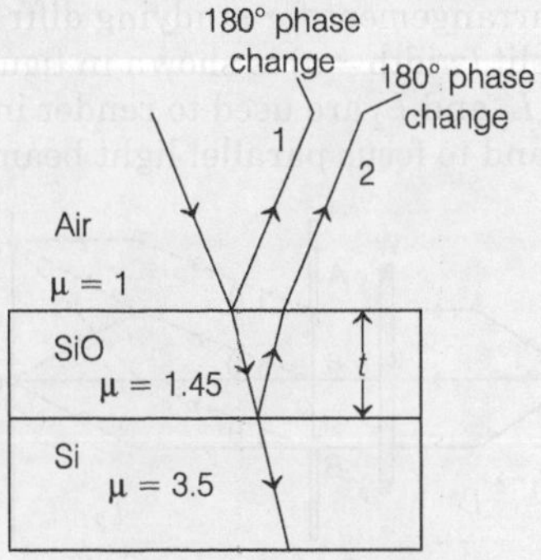

Hence, $2t = \frac{\lambda}{2\mu}$

Thus, the minimum thickness of silicon monoxide,

$$t = \frac{\lambda}{4\mu} = \frac{550}{4\,(1.45)} = 94.8 \text{ nm}$$

Fresnel's Biprism

It is a combination of two prisms of very small refracting angles placed base to base. It is used to obtain two coherent sources from a single light source.

Diffraction of Light

It is the phenomenon of bending of light around corners of an obstacle or aperture in the path of light. This bending light penetrates into the geometrical shadow of the obstacle. The light thus deviates from its linear path. This deviation is more effective when the dimensions of the aperture or the obstacle are comparable to the wavelength of light.

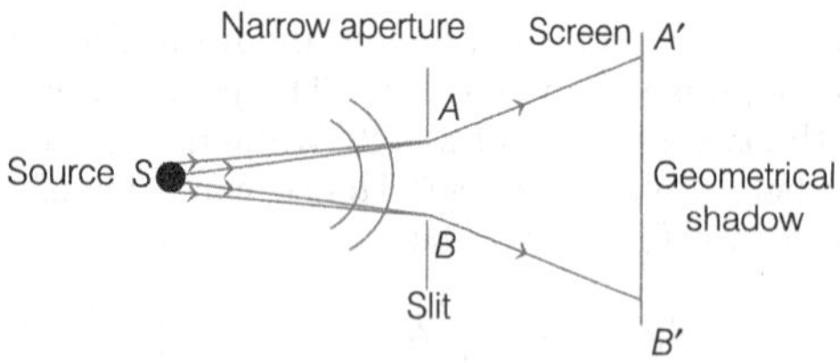

The phenomenon of diffraction is divided mainly in the following two classes

(a) Fresnel class (b) Fraunhoffer class

Fresnel class	Fraunhoffer class
The source is at a finite distance.	The source is at infinite distance.
No opticals are required.	Opticals in the form of collimating lens and focusing lens are required.
Fringes are not sharp and well defined.	Fringes are sharp and well defined.

Diffraction Due to Single Slit

Fraunhoffer's arrangement for studying diffraction at a single narrow slit (width = a) is shown in figure given below. Lenses, L_1 and L_2 are used to render incident light beam parallel and to focus parallel light beam.

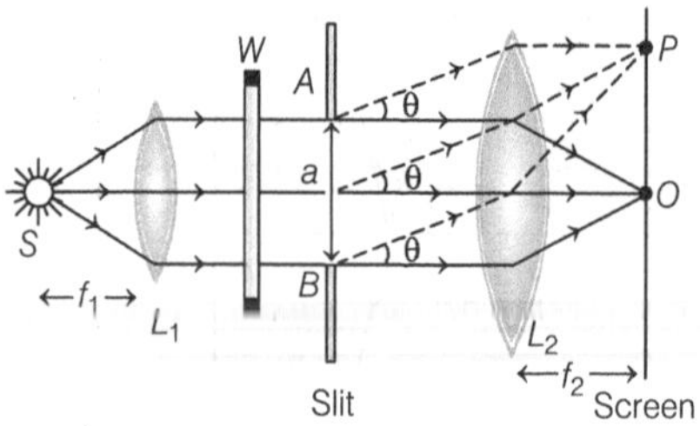

Following are few important points related to diffraction.

- **nth order secondary minima** is obtained, when

$$a \sin\theta = n\lambda, \quad \text{where } n = 1, 2, 3, \dots$$

- **nth order secondary maxima** is obtained, when

$$a \sin\theta = \left(n + \frac{1}{2}\right)\lambda, \text{ where } n = 0, 1, 2, 3, \dots$$

- **Angular separation** of nth order minima,

$$\theta_n = \frac{n\lambda}{a}, \text{ where } n = 1, 2, 3, \dots$$

- **Linear separation** of nth order secondary minima,

$$x_n = \frac{Dn\lambda}{a}$$

- **Angular position** of nth order secondary maxima,

$$\theta_n = \left(n + \frac{1}{2}\right)\frac{\lambda}{a}$$

- **Width of central maximum** $= 2y = \dfrac{2D\lambda}{a}$

As the slit width a increases, width of central maximum decreases.

- **Angular width of central maxima,** $2\theta = \dfrac{2\lambda}{a}$.
- **Intensity of central (principal) maxima** is maximum and intensity of secondary maxima decreases with the increase of their order. The diffraction pattern is graphically shown below.

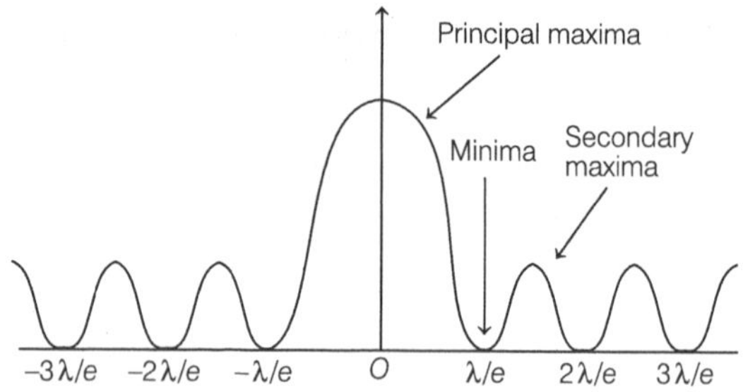

Fresnel's distance

When a slit or hole of size a is illuminated by a parallel beam, it is diffracted into an angle of $\approx \dfrac{\lambda}{a}$.

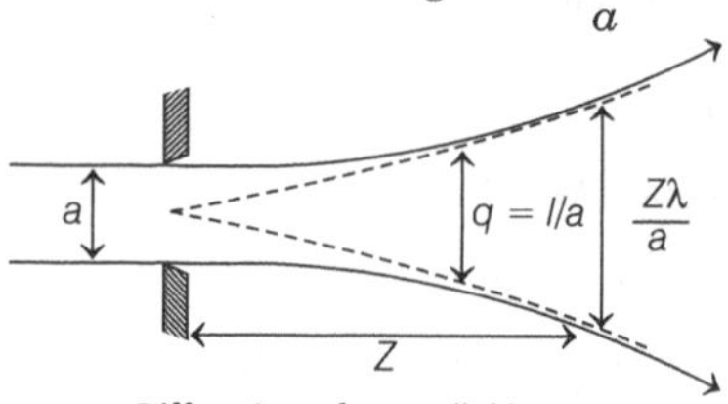

Diffraction of a parallel beam

In travelling a distance Z, size of beam is $Z\lambda / a$.

$$\Rightarrow \qquad Z \geq \frac{a^2}{\lambda}$$

Now, distance Z_F is called *Fresnel's distance,*

$$Z_F = a^2/\lambda$$

So, image formation can be explained by ray optics for distance less than Z_F.

Example 9. *In a single slit diffraction experiment, first minima for $\lambda_1 = 660$ nm coincides with first maxima for wavelength λ_2, then λ_2 will be equal to*

(a) 240 nm (b) 345 nm
(c) 440 nm (d) 330 nm

Sol. (c) Given, $\lambda = 660$ nm

Position of minima in diffraction pattern is given by

$$a \sin\theta = n\lambda$$

For first minima ($n = 1$) of wavelength λ_1, we get

$$a \sin\theta_1 = 1\lambda_1 \text{ or } \sin\theta_1 = \frac{\lambda_1}{a}$$

For the first maxima approximately of wavelength λ_2,

$$a \sin\theta_2 = \frac{3}{2}\lambda_2, \ \sin\theta_2 = \frac{3\lambda_2}{2a}$$

The two will be considered if,

$$\theta_1 = \theta_2 \text{ or } \sin\theta_1 = \sin\theta_2$$

$$\frac{\lambda_1}{a} = \frac{3\lambda_2}{2a}$$

or $$\lambda_2 = \frac{2}{3}\lambda_1 = \frac{2}{3} \times 660 \text{ nm}$$

$$\lambda_2 = 440 \text{ nm}$$

Example 10. *The angular width of the central maximum in a single slit diffraction pattern is 60°. The width of the slit is 1 μm. The slit is illuminated by monochromatic plane waves. If another slit of same width is made near it, Young's fringes can be observed on a screen placed at a distance 50 cm from the slits. If the observed fringe width is 1 cm, what is slit separation distance? (i.e. distance between the centres of each slit.)* **[JEE Main 2018]**

(a) 25 μm *(b)* 50 μm *(c)* 75 μm *(d)* 100 μm

Sol. (a) Angular width of diffraction pattern = 60°

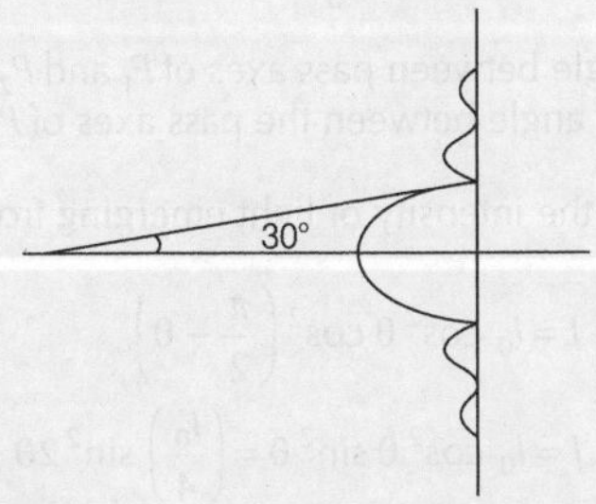

For first minima,

$$\frac{a}{2}\sin\theta = \frac{\lambda}{2}, \text{ [here, } a = 10^{-6} \text{ m, } \theta = 30°]$$

$$\Rightarrow \quad \lambda = 10^{-6} \times \sin 30° \Rightarrow \lambda = \frac{10^{-6}}{2} \text{ m}$$

Now, in case of interference caused by bringing second slit,

$\therefore$ Fringe width, $\beta = \frac{\lambda D}{d}$

$$\left[\text{here, } \lambda = \frac{10^{-6}}{2} \text{ m}, \beta = 1 \text{ cm} = \frac{1}{100} \text{ m}, d = ? \text{ and } D = 50 \text{ cm} = \frac{50}{100} \text{ m}\right]$$

So, $$d = \frac{\lambda D}{\beta} = \frac{10^{-6} \times 50}{2 \times \frac{1}{100} \times 100} = 25 \times 10^{-6} \text{ m}$$

or $$d = 25 \text{ μm}$$

Resolving Power of Optical Instruments

It is the ability of the instrument to produce distinctly separate images of two close objects.

The least separation between two objects, so that they appear just separated is given by

$$d = \frac{\lambda}{2\mu \sin\theta}$$

where, μ is the refractive index of the medium between the objective of the microscope and the object. This distance is called **limit of resolution** of the microscope.

$$\therefore \text{ Resolving power of a microscope} = \frac{1}{d} = \frac{2\mu \sin\theta}{\lambda}$$

where, θ = half angle of the cone of light from the point object, μ sin θ = numerical aperture.

Similarly, the smallest angular separation between two objects, so that they appear just separated is found to be

$$d\theta = \frac{1.22\,\lambda}{D}$$

where, D is the diameter of objective.

$$\therefore \text{Resolving power of telescope} = \frac{1}{d\theta} = \frac{D}{1.22\,\lambda}$$

If d is the width of white strip (gap) which separates the two regions and D be the distance of screen from eye, then d / D is the limit of resolution of our eye.

Example 11. *A telescope is used to resolve two stars separated by 4.6×10^{-6} rad. If the wavelength of light is 5460Å, what should be the aperture of the objective of the telescope?*

(a) 0.0448 *m* *(b)* 0.1448 *m*

(c) 1.1448 *m* *(d)* 0.011 *m*

Sol. *(b)* Here, $d\theta = 4.6 \times 10^{-6}$ rad

$$\lambda = 5460 \text{ Å} = 5460 \times 10^{-10} \text{ m}$$

The aperture (*D*) of the telescope is given as

$$D = \frac{1.22\,\lambda}{d\theta} = \frac{1.22 \times 5460 \times 10^{-10}}{4.6 \times 10^{-6}} = 0.1448 \text{ m}$$

Example 12. *The aperture of a telescope is 5m. The separation between the moon and the earth is 4×10^5 km. With light of wavelength of 5500 Å, the minimum separation between objects on the surface of moon, so that they are just resolved, is close to* **[JEE Main 2020]**

(a) 600 *m* *(b)* 20 *m* *(c)* 60 *m* *(d)* 200 *m*

Sol. *(c)* From Rayleigh's criteria, two objects will be just resolved when their angular separation, $\theta_R = 1.22\frac{\lambda}{a}$

where, λ = wavelength and a = aperture size.

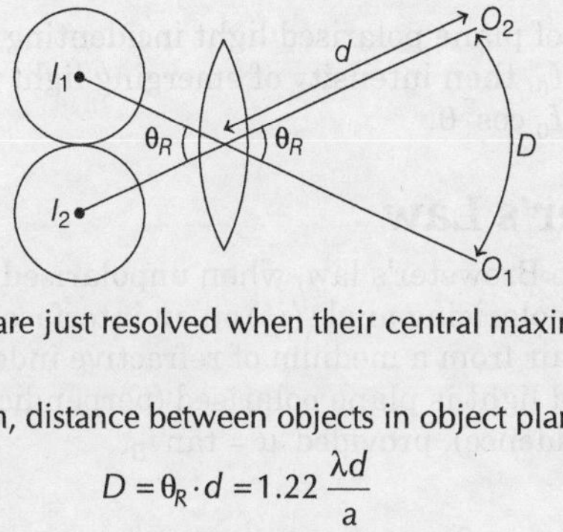

Two objects are just resolved when their central maximas does not overlap.

From diagram, distance between objects in object plane,

$$D = \theta_R \cdot d = 1.22\frac{\lambda d}{a}$$

Here, $a = 5$ m, $d = 4 \times 10^5$ km $= 4 \times 10^8$ m,

$\lambda = 5500$ Å $= 5500 \times 10^{-10}$ m

So, $D = \frac{(1.22 \times 5500 \times 10^{-10} \times 4 \times 10^8)}{5} \approx 54$ m

From options, minimum separation distance is 60 m.

Polarisation

The phenomenon of restricting the vibrations of light (electric vector) in a particular direction, perpendicular to the direction of wave motion is called *polarisation of light*.The plane perpendicular to the plane of oscillation is called *plane of polarisation*.

Polaroids

These are thin and large sheets of crystalline polarising material (made artificially) capable of producing plane polarised beams of large cross-section.

A polaroid has a characteristic plane called *transmission plane*. When unpolarised light falls on a polaroid, only the vibrations parallel to the transmission plane get transmitted.

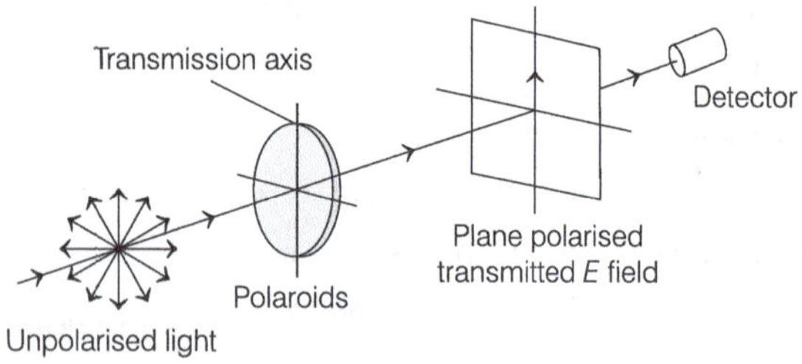

The crystal or polaroid on which unpolarised light is incident is called *polariser*. Crystal or polaroid on which polarised light is incident is called *analyser*.

Malus' Law

This law states that, when a beam of completely plane polarised light is incident on an analyser, the resultant intensity of light (I) transmitted from the analyser varies directly as the square of cosine of angle (θ) between plane of transmission of analyser and polariser.

i.e. $$I \propto \cos^2\theta$$

If intensity of plane polarised light incidenting on analyser is I_0, then intensity of emerging light from analyser is $I_0 \cos^2\theta$.

Brewster's Law

According to Brewster's law, when unpolarised light is incident at polarising angle (i_B) on an interface separating air from a medium of refractive index μ, then the reflected light is plane polarised (perpendicular to the plane of incidence), provided $\mu = \tan i_B$.

Here, the angle of incidence at which the reflected light is completely plane polarised is called **polarising angle** or **Brewster's angle** i_B.

At polarising angle, $i_B + r = \frac{\pi}{2}$

i.e. Reflected plane polarised light is at right angle to the refracted light.

Example 13. *When a polaroid sheet is rotated between two crossed polaroids, the intensity of the transmitted light will be maximum, when angle θ between pass axes is*

(a) $\frac{\pi}{2}$ (b) $\frac{3\pi}{4}$

(c) $\frac{\pi}{4}$ (d) $\frac{2\pi}{3}$

Sol. (c) Let I_0 be the intensity of polarised light after passing through the first polariser P_1. Then the intensity of light after passing through second polariser P_2 will be

$$I = I_0 \cos^2\theta$$

where, θ is the angle between pass axes of P_1 and P_2. Since, P_1 and P_3 are crossed the angle between the pass axes of P_2 and P_3 will be $\left(\frac{\pi}{2} - \theta\right)$. Hence, the intensity of light emerging from P_2 will be

$$I = I_0 \cos^2\theta \cos^2\left(\frac{\pi}{2} - \theta\right)$$

$$I = I_0 \cos^2\theta \sin^2\theta = \left(\frac{I_0}{4}\right)\sin^2 2\theta$$

Intensity of transmitted light is maximum,

when $\sin^2 2\theta = 1$

$\Rightarrow$ $\sin 2\theta = 1 = \sin\frac{\pi}{2}$

$\Rightarrow$ $2\theta = \frac{\pi}{2} \Rightarrow \theta = \frac{\pi}{4}$

Example 14. *A polariser-analyser set is adjusted such that the intensity of light coming out of the analyser is just 10% of the original intensity. Assuming that the polariser-analyser set does not absorb any light, the angle by which the analyser need to be rotated further to reduce the output intensity to be zero, is* **[JEE Main 2020]**

(a) 71.6° (b) 90°

(c) 45° (d) 18.4°

Sol. (d) A polariser-analyser set is as shown below.

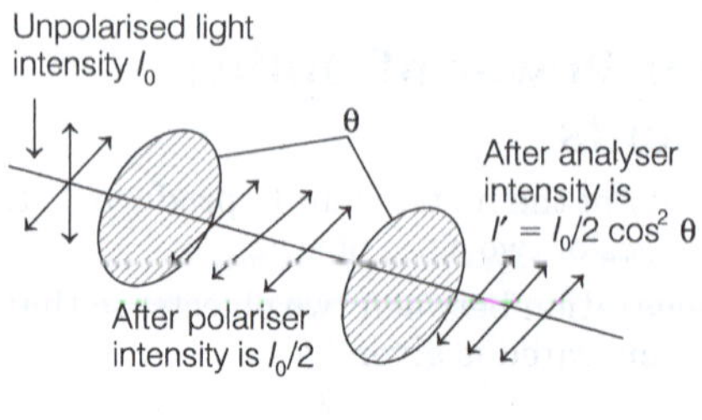

By Malus' law,

Output intensity obtained from analyser, $I' = \frac{I_0}{2}\cos^2\theta$ ...(i)

where, θ = angle between polariser and analyser.

Now, given $I' = 10\%$ of initial intensity

$$\Rightarrow \quad I' = \frac{1}{10} I_0 \quad ...(ii)$$

From Eqs. (i) and (ii), we get

$$\Rightarrow \quad \cos^2\theta = \frac{1}{5}$$

$$\Rightarrow \quad \cos\theta = \frac{1}{\sqrt{5}}$$

$$\Rightarrow \quad \theta = \cos^{-1}\left(\frac{1}{\sqrt{5}}\right) \approx 63^\circ$$

Now, output intensity is zero, when angle between polariser and analyser is 90°.

So, analyser must be rotated by 90° − 63° = 27° to reduce output intensity to zero.

From given options, we can see that output intensity is least when analyser rotation is 18.4°.

Example 15. *Light reflected from the surface of glass plate of refractive index 1.57 is linearly polarised. What is the angle of refraction of glass ?*

(a) 32.5° (b) 42.5° (c) 30.2° (d) 44.5°

Sol. (a) Given that, $\mu = 1.57$

According to Brewster's law,

$$\mu = \tan i_p = 1.57, i_p = \tan^{-1}(1.57) = 57.5^\circ$$

As, $r = 90^\circ - i_p$ or $r = 90^\circ - 57.5^\circ = 32.5^\circ$

Doppler's Effect in Light

The phenomenon of apparent change in frequency (or wavelength) of the light due to the relative motion between the source of light and the observer is called Doppler's effect in light.

A stationary observer will receive ν wave per second. In case, the observer moves towards or away from the source of light with velocity v along the direction of propagation of light, then the number of waves received per second, *i.e. apparent frequency* of light will be

$$\nu' = \nu\left(1 \pm \frac{v}{c}\right)$$

Doppler shift is given as

$$\Delta\nu = \pm\frac{v}{c}\nu$$

The positive sign is considered, when the source and the observer approach each other and the negative sign considered is, when both source and observer move away from each other.

Doppler shift in terms of the wavelength of the light is given by

$$\Delta\lambda = \pm\frac{v}{c}\lambda$$

Example 16. *The spectral line for a given element in light received from a distant star is shifted towards the longer wavelength by 0.32%. Deduce the velocity of star in the line of sight.*

(a) 9.6×10^4 m/s (b) 11.6×10^{-4} m/s

(c) 12.6×10^{-4} m/s (d) None of the above

Sol. (a) Here, $\frac{\Delta\lambda}{\lambda} = \frac{0.032}{100}$ (position, as the shift is towards longer wavelength)

Now, Doppler shift is given by

$$\Delta\lambda = -\frac{V}{c}\lambda$$

$$V = -\frac{\Delta\lambda}{\lambda}c = -\frac{0.032}{100} \times 3 \times 10^8$$

$$= -9.6 \times 10^4 \text{ ms}^{-1}$$

The negative sign indicates that the star is receding.

Practice Exercise

ROUND I Topically Divided Problems

Wavefront and Huygens' Principle

1. For light diverging from a point source
(a) the wavefront is spherical
(b) the intensity increases in proportion to the distance squared
(c) the wavefront is parabolic
(d) the intensity at the wavefront does not depend on the distance

2. Consider a point at the focal point of a convergent lens. Another convergent lens of short focal length is placed on the other side. What is the nature of the wavefront emerging from the final image?
(a) Cylindrical (b) Elliptical
(c) Spherical (d) Square

3. Light waves travel in vacuum along the Y-axis. Which of the following may represent the wavefront?
(a) $y = \text{constant}$ (b) $x = \text{constant}$
(c) $z = \text{constant}$ (d) $x + y + z = \text{constant}$

4. On a hot summer night, the refractive index of air is smallest near the ground and increases with height from the ground. When a light beam is directed horizontally, the Huygens' principle leads us to conclude that as it travels, the light beam
(a) become narrower [JEE Main 2015]
(b) goes horizontally without any deflection
(c) bends downwards
(d) bends upwards

Coherent and Incoherent Sources : Interference and YDSE

5. Two non-coherent sources emit light beams of intensities I and $4I$. The maximum and minimum intensities in the resulting beam are
(a) $9I$ and I (b) $9I$ and $3I$
(c) $5I$ and I (d) $5I$ and $3I$

6. Two beams of light having intensities I and $4I$ interfere to produce a fringe pattern on a screen. The phase difference between the beams is $\pi/2$ at point A and π at point B, then the difference between the resultant intensities at points A and B is
(a) $2I$ (b) $4I$
(c) $5I$ (d) $7I$

7. The maximum intensity in the case of n identical incoherent waves, each of intensity 2 Wm^{-2} is 32 Wm^{-2}. The value of n is
(a) 4 (b) 16 (c) 32 (d) 64

8. Three waves of equal frequency having amplitudes $10\mu m$, $4\mu m$, $7\mu m$ arrive at a given point with successive phase difference of $\frac{\pi}{2}$, the amplitude of the resulting wave (in μm) is given by
(a) 4 (b) 5 (c) 6 (d) 7

9. A parallel beam of light of intensity I_0 is incident on a glass plate, 25% of light is reflected by upper surface and 50% of light is reflected from lower surface. The ratio of maximum to minimum intensity in interference region of reflected rays is
(a) $\left(\frac{\frac{1}{2}+\sqrt{\frac{3}{8}}}{\frac{1}{2}-\sqrt{\frac{3}{8}}}\right)^2$ (b) $\left(\frac{\frac{1}{4}+\sqrt{\frac{3}{8}}}{\frac{1}{2}-\sqrt{\frac{3}{8}}}\right)^2$
(c) $\frac{5}{8}$ (d) $\frac{8}{5}$

10. Two waves originating from sources S_1 and S_2 having zero phase difference and common wavelength λ will show complete destructive interference at a point P, if $(S_1P - S_2P) =$
(a) 5λ (b) $\frac{3\lambda}{4}$ (c) $\frac{4\lambda}{2}$ (d) $\frac{11\lambda}{2}$

11. In the setup shown in figure, the two slits S_1 and S_2 are not equidistant from the slit S. The central fringe at O, then

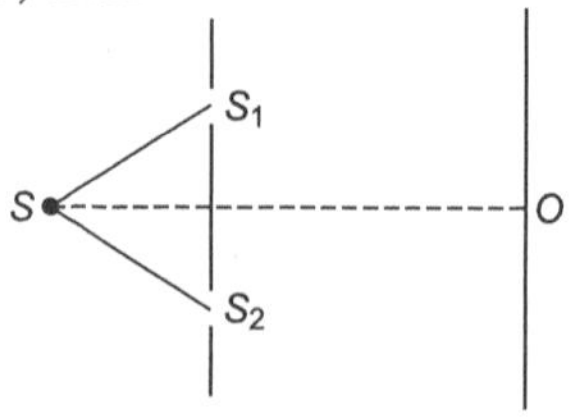

(a) always bright
(b) always dark
(c) Either dark or bright depending on the position of S
(d) Neither dark nor bright

12. Two coherent sources of sound S_1 and S_2, produce sound waves of the same wavelength $\lambda = 1$ m, in phase. S_1 and S_2 are placed 1.5 m apart (see figure). A listener, located at L, directly in front of S_2 finds that the intensity is at a minimum when he is 2 m away from S_2. The listener moves away from S_1, keeping his distance from S_2 fixed. The adjacent maximum of intensity is observed when the listener is at a distance d from S_1. Then, d is **[JEE Main 2020]**

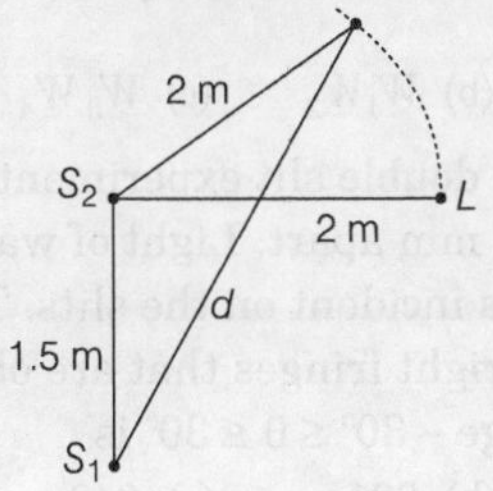

(a) 12 m (b) 5 m
(c) 2 m (d) 3 m

13. Two coherent point sources S_1 and S_2 are separated by a small distance d as shown. The fringes obtained on the screen will be **[JEE Main 2013]**

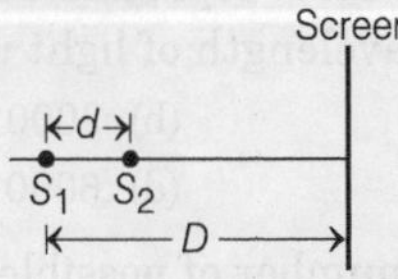

(a) points (b) straight lines
(c) semi-circle (d) concentric circles

14. A ray of light of intensity I is incident on a parallel glass slab at point A as shown in diagram. It undergoes partial reflection and refraction. At each reflection, 25% of incident energy is reflected. The rays AB and $A'B'$ undergo interference. The ratio of I_{max} and I_{min} is

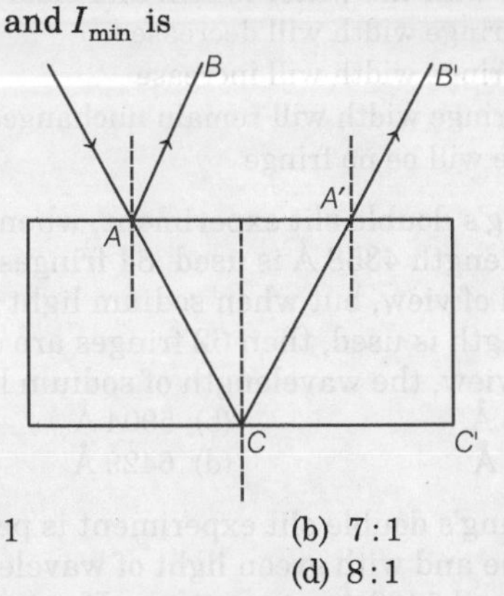

(a) 49 : 1 (b) 7 : 1
(c) 4 : 1 (d) 8 : 1

15. In Young's experiment, the wavelength of red light is 7.8×10^{-5} cm and that of blue light is 5.2×10^{-5} cm. The value of n for which $(n + 1)$th blue light band coincides with nth red bond is
(a) 4 (b) 2 (c) 3 (d) 1

16. In a Young's double slit experiment, the source is white light. One of the holes is covered by a red filter and another by a blue filter. In this case **[NCERT Exemplar]**
(a) there shall be alternate interference patterns of red and blue
(b) there shall be an interference pattern for red distinct from that for blue
(c) there shall be no interference fringes
(d) there shall be an interference pattern for red mixing with one for blue

17. In an experiment, the two slits are 0.5 mm apart and the fringes are observed to 100 cm from the plane of the slits. The distance of the 11th bright fringe from the 1st bright fringe is 9.72 mm. The wavelength is
(a) 4.86×10^{-5} cm
(b) 5.72×10^{-4} cm
(c) 5.87×10^{-4} cm
(d) 3.25×10^{-4} cm

18. In a Young's double slit experiment, the slits are separated by 0.28 mm and the screen is placed 1.4 m away. The distance between the central bright fringe and the fourth bright fringe is measured to be 1.2 cm. Determine the wavelength of light used in the experiment. **[NCERT]**
(a) 4×10^{-9} m (b) 5×10^{-10} m
(c) 3×10^{-7} m (d) 6×10^{-7} m

19. Interference fringes are observed on a screen by illuminating two thin slits 1 mm apart with a light source ($\lambda = 632.8$ nm). The distance between the screen and the slits is 100 cm. If a bright fringe is observed on a screen at a distance of 1.27 mm from the central bright fringe, then the path difference between the waves, which are reaching this point from the slits is close to **[JEE Main 2020]**
(a) 2 nm (b) 1.27 μm
(c) 2.87 nm (d) 2.05 μm

20. In Young's double slit experiment, the 8th maximum with wavelength λ_1 is at a distance, d_1 from the central maximum and the 6th maximum with wavelength λ_2 is at a distance, d_2. Then, d_1/d_2 is equal to
(a) $\frac{4}{3}\left(\frac{\lambda_2}{\lambda_1}\right)$ (b) $\frac{4}{3}\left(\frac{\lambda_1}{\lambda_2}\right)$
(c) $\frac{3}{4}\left(\frac{\lambda_2}{\lambda_1}\right)$ (d) $\frac{3}{4}\left(\frac{\lambda_1}{\lambda_2}\right)$

21. In a Young's double slit experiment, slits are separated by 0.5 mm and the screen is placed 150 cm away. A beam of light consisting of two wavelengths, 650 nm and 520 nm, is used to obtain interference fringes on the screen. The least distance from the common central maximum to the point where the bright fringes due to both the wavelengths coincide, is **[JEE Main 2017]**

(a) 7.8 mm (b) 9.75 mm
(c) 15.6 mm (d) 1.56 mm

22. In a Young's double slit experiment with slit separation 0.1 mm, one observes a bright fringe at angle $\frac{1}{40}$ rad by using light of wavelength λ_1. When the light of the wavelength λ_2 is used a bright fringe is seen at the same angle in the same set up. Given that λ_1 and λ_2 are in visible range (380 nm to 740 nm), their values are **[JEE Main 2019]**

(a) 380 nm, 525 nm
(b) 400 nm, 500 nm
(c) 380 nm, 500 nm
(d) 625 nm, 500 nm

23. The correct curve between fringe width β and distance between the slits (d) in figure is

(a) 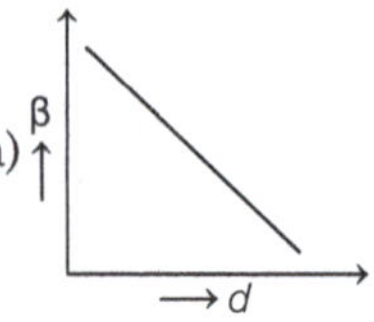

(b) 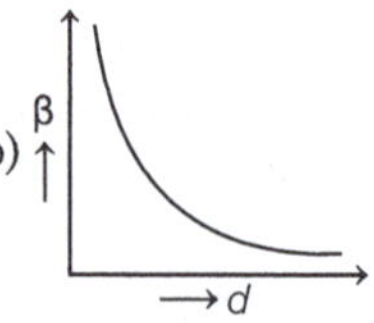

(c) 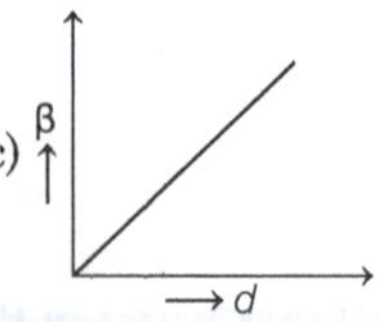

(d) 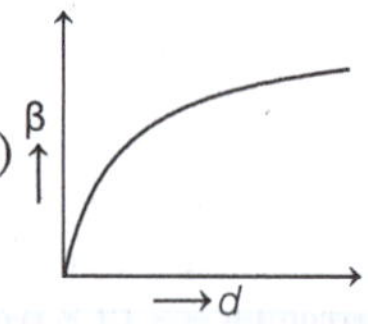

24. In Young's double slit experiment, the separation between slit is halved and the distance between the slits and screen is doubled. The fringe width is

(a) unchanged (b) halved
(c) double (d) quardrupled

25. In Young's double slit arrangement, slits are separated by a gap of 0.5 mm, and the screen is placed at a distance of 0.5 m from them. The distance between the first and the third bright fringe formed when the slits are illuminated by a monochromatic light of 5890 Å is **[JEE Main 2021]**

(a) 1178×10^{-9} m (b) 1178×10^{-6} m
(c) 1178×10^{-12} m (d) 5890×10^{-7} m

26. In a Young's experiment, two coherent sources are placed 0.90 mm apart and the fringes are observed one metre away. If it produces the second dark fringe at a distance of 1 mm from the central fringe, the wavelength of monochromatic light used will be

(a) 60×10^{-4} cm (b) 10×10^{-4} cm
(c) 10×10^{-5} cm (d) 6×10^{-5} cm

27. Interference fringes are being produced on screen XY by the slits S_1 and S_2. In figure, the correct fringe locus is

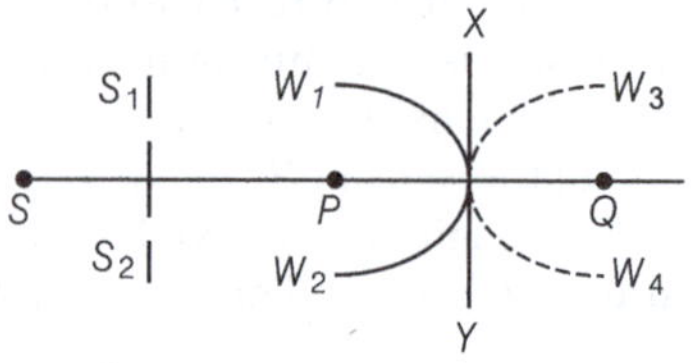

(a) PQ (b) W_1W_2 (c) W_3W_4 (d) XY

28. In a Young's double slit experiment, the slits are placed 0.320 mm apart. Light of wavelength $\lambda = 500$ nm is incident on the slits. The total number of bright fringes that are observed in the angular range $-30° \le \theta \le 30°$ is **[JEE Main 2019]**

(a) 320 (b) 321 (c) 640 (d) 641

29. In a two slits experiment with monochromatic light, fringes are obtained on a screen placed at some distance from the slits. If the screen is moved by 5×10^{-2} m towards the slits, the change in fringe width is 3×10^{-5} m. If separation between the slits is 10^{-3} m, the wavelength of light used is

(a) 4500 Å (b) 3000 Å
(c) 5000 Å (d) 6000 Å

30. The maximum number of possible interference maxima for slit separation equal to twice the wavelength in Young's double slit experiment is

(a) infinite (b) 5
(c) 3 (d) zero

31. If Young's double slit experiment, is performed in water
(assume that the water is still and clear)

(a) the fringe width will decrease
(b) the fringe width will increase
(c) the fringe width will remain unchanged
(d) there will be no fringe

32. In Young's double slit experiment, when violet light of wavelength 4358 Å is used, 84 fringes are seen in the field of view, but when sodium light of certain wavelength is used, then 62 fringes are seen in the field of view, the wavelength of sodium light is

(a) 6893 Å (b) 5904 Å
(c) 5523 Å (d) 6429 Å

33. The Young's double slit experiment is performed with blue and with green light of wavelengths 4360 Å and 5460 Å, respectively. If, x is the distance of 4th maxima from the central one, then

(a) x (blue) = x (green)
(b) x (blue) > x (green)
(c) x (blue) < x (green)
(d) x (blue)/x (green) = 5400/4360

34. Figure shows a standard two slit arrangement with slits S_1, S_2. P_1 and P_2 are the two minima points on either side of P.

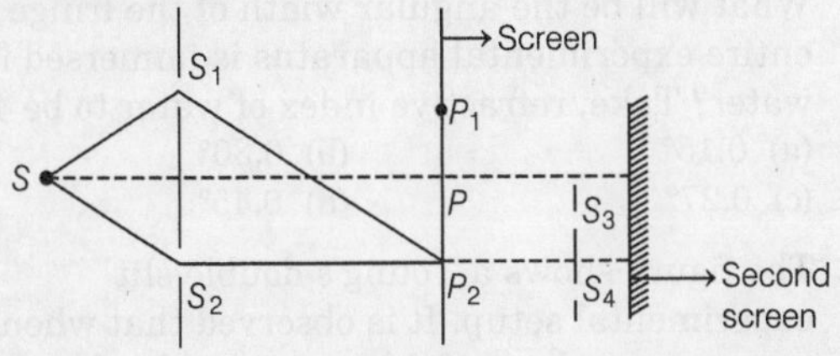

At P_2 on the screen, there is a hole and behind P_2 is a second 2-slit arrangement with slits S_3, S_4 and a second screen behind them. **[NCERT Exemplar]**

(a) There would be no interference pattern on the second screen but it would be lighted.
(b) The second screen would be totally dark.
(c) There would be a single bright point on the second screen.
(d) There would be a regular two slit pattern on the second screen.

35. In a Young's double slit experiment, the path difference at a certain point on the screen between two interfering waves is $\frac{1}{8}$th of wavelength. The ratio of the intensity at this point to that at the centre of a bright fringe is close to **[JEE Main 2019]**

(a) 0.80 (b) 0.74
(c) 0.94 (d) 0.85

36. In the Young's double slit experiment, the interference pattern is found to have an intensity ratio between bright and dark fringes as 9. This implies that

(a) the intensities at the screen due to two slits are 5 units and 4 units respectively
(b) the intensities at the screen due to two slits are 4 units and 1 unit respectively
(c) the amplitude ratio is 3
(d) the amplitude ratio is 2

37. A mixture of light, consisting of wavelength 590 nm and an unknown wavelength, illuminates Young's double slit and gives rise to two overlapping interference patterns on the screen. The central maximum of both lights coincide. Further, it is observed that the third bright fringe of known light coincides with the 4th bright fringe of the unknown light. From this data, the wavelength of the unknown light is **[AIEEE 2009]**

(a) 393.4 nm (b) 885.0 nm
(c) 442.5 nm (d) 776.8 nm

38. Two identical radiators have a separation of $d = \lambda/4$, where λ is the wavelength of the waves emitted by either source. The initial phase difference between the source is $\pi/4$. Then the intensity on the screen at a distant point situated at an angle, $\theta = 30°$ from the radiators is (here, I_0 is intensity at that point due to one radiator alone)

(a) I_0 (b) $2I_0$
(c) $3I_0$ (d) $4I_0$

39. In a Young's double slit experiment, the ratio of the slit's width is 4 : 1. The ratio of the intensity of maxima to minima, close to the central fringe on the screen, will be **[JEE Main 2019]**

(a) 4 : 1
(b) 25 : 9
(c) 9 : 1
(d) $(\sqrt{3}+1)^4 : 16$

40. In Young's double slit experiment, the two slits act as coherent sources of equal amplitude A and wavelength λ. In another experiment with the same setup, the two slits are sources of equal amplitude A and wavelength λ but are incoherent. The ratio of the intensity of light at the mid-point of the screen in the first case to that in the second case is

(a) 2 : 1 (b) 1 : 2
(c) 3 : 4 (d) 4 : 3

41. A thin glass plate of thickness $\frac{2500}{3}\lambda$ (where, λ is wavelength of light used) and refractive index $\mu = 1.5$ is inserted between one of the slits and the screen in Young's double slit experiment. At a point on the screen equidistant from the slits, the ratio of the intensities before and after the introduction of the glass plate is **[JEE Main 2013]**

(a) 2:1 (b) 1:4
(c) 4:1 (d) 4:3

42. In a Young's double slit experiment, 16 fringes are observed in a certain segment of the screen when light of wavelength 700 nm is used. If the wavelength of light is changed to 400 nm, the number of fringes observed in the same segment of the screen would be **[JEE Main 2020]**

(a) 24 (b) 30
(c) 18 (d) 28

43. In an interference pattern, the position of zeroth order maxima is 4.8 mm from a certain point P on the screen. The fringe width is 0.2 mm. The position of second maxima from point P is

(a) 5.1 mm (b) 5 mm
(c) 40 mm (d) 5.2 mm

44. In a Young's experiment, one of the slits is covered with a transparent sheet of thickness 3.6×10^{-3} cm due to which position of central fringe shifts to a position originally occupied by 30th fringe. The refractive index of the sheet, if $\lambda = 6000$ Å, is

(a) 1.5 (b) 1.2
(c) 1.3 (d) 1.7

45. In Young's double slit experiment, distance between two sources is 0.1 mm. The distance of screen from the sources is 20 cm. Wavelength of light used is 5460 Å. Then angular position of first dark fringe is

(a) 0.08° (b) 0.16°
(c) 0.20° (d) 032°

46. In Young's double slit experiment, light of 500 nm is used to produce an interference pattern. When the distance between the slits is 0.05 mm, the angular width (in degree) of the fringes formed on the distance screen is close to [JEE Main 2020]

(a) 0.17° (b) 0.57°
(c) 1.7° (d) 0.07°

47. In Young's double slit experiment, the spacing between the slits is d and wavelength of light used is 6000 Å. If the angular width of a fringe formed on a distance screen is 1°, then value of d is

(a) 1 mm (b) 0.05 mm
(c) 0.03 mm (d) 0.01 mm

48. In double-slit experiment using light of wavelength 600 nm, the angular width of a fringe formed on a distant screen is 0.1°. What is the spacing between the two slits?

(a) 3.44×10^{-4} m (b) 3.03×10^{-4} m
(c) 4.03×10^{-4} m (d) 2.68×10^{-4} m

49. In the given figure, C is middle point of lines S_1S_2. A monochromatic light of wavelength λ is incident on slits. The ratio of intensities of S_3 and S_4 is

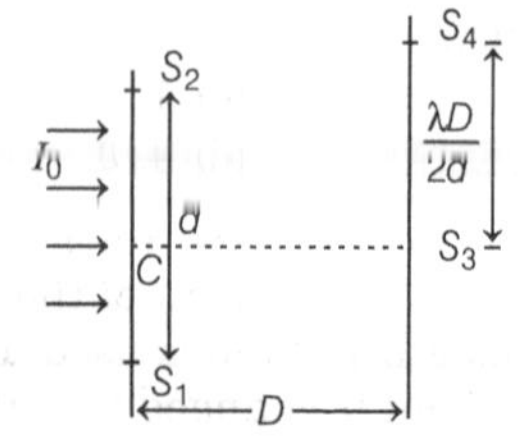

(a) zero (b) ∞
(c) 4 : 1 (d) 1:4

50. The maximum intensity of fringes in Young's experiment is I. If one of the slit is closed, then the intensity at that place becomes I_0. Which of the following relation is true?

(a) $I = I_0$ (b) $I = 2I_0$
(c) $I = 4I_0$ (d) $I = 0$

51. In a double slit experiment, at a certain point on the screen the path difference between the two interfering waves is $\frac{1}{8}$ th of a wavelength. The ratio of the intensity of light at that point to that at the centre of a bright fringe is [JEE Main 2020]

(a) 0.568 (b) 0.853
(c) 0.760 (d) 0.672

52. In a double slit experiment, the angular width of a fringe is found to be 0.2° on a screen placed 1 m away. The wavelength of light used is 600 nm. What will be the angular width of the fringe if the entire experimental apparatus is immersed in water? Take, refractive index of water to be 4/3.

(a) 0.15° (b) 0.30°
(c) 0.27° (d) 0.45°

53. The figure shows a Young's double slit experimental setup. It is observed that when a thin transparent sheet of thickness t and refractive index μ is put in front of one of the slits, the central maximum gets shifted by a distance equal to n fringe widths. If the wavelength of light used is λ, t will be [JEE Main 2019]

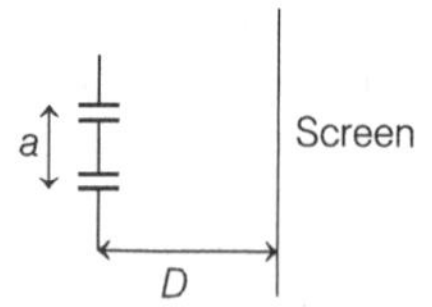

(a) $\frac{2nD\lambda}{a(\mu-1)}$ (b) $\frac{2D\lambda}{a(\mu-1)}$ (c) $\frac{D\lambda}{a(\mu-1)}$ (d) $\frac{nD\lambda}{a(\mu-1)}$

54. In a biprism experiment, 5th dark fringe is obtained at a point. If a thin transparent film is placed in the path of one of waves, then 7th bright fringes is obtained at the same point. The thickness of the film in terms of wavelength l and refractive index, μ will be

(a) $\frac{1.5\lambda}{(\mu-1)}$ (b) $1.5(\mu-1)\lambda$
(c) $2.5(\mu-1)\lambda$ (d) $\frac{2.5\lambda}{(\mu-1)}$

Diffraction of Light

55. The main difference between the phenomena of interference and diffraction is that

(a) diffraction is caused by reflected waves from a source whereas interference is caused due to refraction of waves from a source
(b) diffraction is due to interaction of waves derived from the same source, whereas interference is that bending of light from the same wavefront
(c) diffraction is due to interaction of light from wavefront, whereas the interference is the interaction of two waves derived from the same source
(d) diffraction is due to interaction of light from the same wavefront whereas interference is the interaction of waves from two isolated sources

56. When monochromatic light is replaced by white light in Fresnel's biprism arrangement, the central fringe is

(a) coloured (b) white
(c) dark (d) None of these

57. Consider sunlight incident on a slit of width 10^4 Å. The image seen through the slit shall **[NCERT Exemplar]**

(a) be a fine sharp slit white in colour at the centre
(b) a bright slit white at the centre diffusing to zero intensities at the edges
(c) a bright slit white at the centre diffusing to regions of different colours
(d) Only be a diffused slit white in colour

58. Estimate the distance for which ray optics is good approximation for an aperture of 4 mm and wavelength 400 nm. **[NCERT]**

(a) 40 m (b) 43 m
(c) 47 m (d) 38 m

59. A parallel beam of light of wavelength 3141.59 Å is incident on a small aperture. After passing through the aperture, the beam is no longer parallel but diverges at 1° to the incident direction. What is the diameter of the aperture?

(a) 180 m (b) 18 μm
(c) 1.8 m (d) 0.18 m

60. Plane microwaves are incident on a long slit having a width of 5 cm. The wavelength of the microwaves if the first minimum is formed at 30° is

(a) 2.5 cm (b) 2 cm
(c) 25 cm (d) 2 mm

61. A plane wave of wavelength 6250 Å is incident normally on a slit of width 2×10^{-2} cm. The width of the principal maximum on a screen distant 50 cm will be

(a) 312.5×10^{-3} cm
(b) 312.5×10^{-4} cm
(c) 312 cm
(d) 312.5×10^{-5} cm

62. The distance between the first and the sixth minima in the diffraction pattern of a single slit is 0.5 mm. The screen is 0.5 m away from the slit. If the wavelength of light used is 5000 Å. Then the slit width will be

(a) 5 mm
(b) 0.5 mm
(c) 1.25 mm
(d) 1.0 mm

63. The Fraunhoffer diffraction pattern of a single slit is formed in the focal plane of a lens of focal length 1 m. The width of slit is 0.3 mm. If third minimum is formed at a distance of 5 mm from central maximum, then wavelength of light will be

(a) 5000 Å
(b) 2500 Å
(c) 7500 Å
(d) 8500 Å

64. Light of wavelength λ is incident on a slit width d. The resulting diffraction pattern is observed on a screen at a distance D. The linear width of the principal maximum is equal to the width of the slit, if D equals

(a) $\frac{d^2}{2\lambda}$ (b) $\frac{d}{\lambda}$
(c) $\frac{2\lambda^2}{d}$ (d) $\frac{2\lambda}{d}$

65. The ratio of intensities of successive maxima in the diffraction pattern due to single slit is

(a) 1:4:9 (b) 1:2:3
(c) $1:\frac{4}{9\pi^2}:\frac{4}{25\pi^2}$ (d) $1:\frac{4}{\pi^2}:\frac{9}{\pi^2}$

66. Visible light of wavelength 6000×10^{-8} cm falls normally on a single slit and produces a diffraction pattern. It is found that the second diffraction minimum is at 60° from the central maximum. If the first minimum is produced at θ_1, then θ_1 is close to **[JEE Main 2020]**

(a) 25° (b) 30°
(c) 20° (d) 45°

Resolving Power of Optical Instruments

67. Diameter of the objective lens of a telescope is 250 cm. For light of wavelength 600 nm coming from a distant object, the limit of resolution of the telescope is close to **[JEE Main 2019]**

(a) 3.0×10^{-7} rad (b) 2.0×10^{-7} rad
(c) 1.5×10^{-7} rad (d) 4.5×10^{-7} rad

68. Calculate the limit of resolution of a telescope objective having a diameter of 200 cm, if it has to detect light of wavelength 500 nm coming from a star. **[JEE Main 2019]**

(a) 610×10^{-9} rad (b) 305×10^{-9} rad
(c) 457.5×10^{-9} rad (d) 152.5×10^{-9} rad

69. The value of numerical aperture of the objective lens of a microscope is 1.25. If light of wavelength 5000Å is used, the minimum separation between two points, to be seen as distinct, will be **[JEE Main 2019]**

(a) 0.24 μm (b) 0.38 μm
(c) 0.12 μm (d) 0.48 μm

70. Assuming human pupil to have a radius of 0.25 cm and a comfortable viewing distance of 25 cm, the minimum separation between two objects that human eye can resolve at 500 nm wavelength is **[JEE Main 2015]**

(a) 1 μm (b) 30 μm
(c) 100 μm (d) 300 μm

Polarisation of Light

71. Consider a light beam incident from air to a glass slab at Brewster's angle as shown in the figure. [NCERT Exemplar]

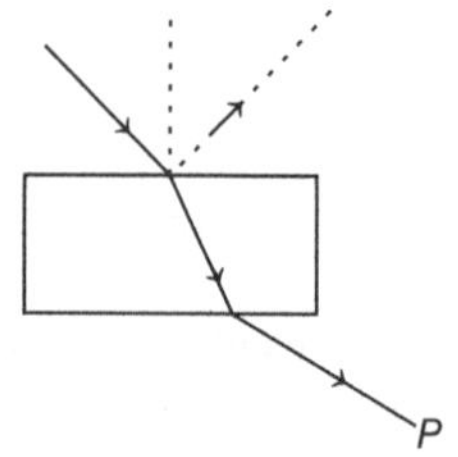

A polaroid is placed in the path of the emergent ray at a point P and rotated about an axis passing through the centre and perpendicular to the plane of the polaroid.

(a) For a particular orientation, there shall be darkness as observed through the polaroid.

(b) The intensity of light as seen through the polaroid shall be independent of the rotation.

(c) The intensity of light as seen through the polaroid shall go through a minimum but not zero for two orientations of the polaroid.

(d) The intensity of light as seen through the polaroid shall go through a minimum for four orientations of the polaroid.

72. Consider a tank made of glass (refractive index is 1.5) with a thick bottom. It is filled with a liquid of refractive index μ. A student finds that, irrespective of what the incident angle i (see figure) is for a beam of light entering the liquid, the light reflected from the liquid glass interface is never completely polarised.

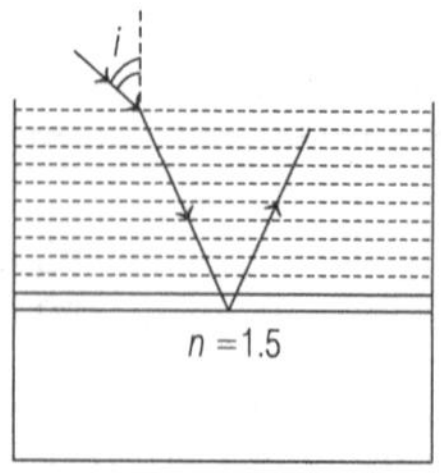

For this to happen, the minimum value of μ is [JEE Main 2019]

(a) $\dfrac{3}{\sqrt{5}}$ (b) $\dfrac{5}{\sqrt{3}}$

(c) $\dfrac{4}{3}$ (d) $\sqrt{\dfrac{5}{3}}$

73. λ_a and λ_m are the wavelengths of a beam of light in air and medium respectively. If θ is the polarising angle, the correct relation between λ_a, λ_m and θ is

(a) $\lambda_a = \lambda_m \tan^2 \theta$ (b) $\lambda_m = \lambda_a \tan^2 \theta$

(c) $\lambda_a = \lambda_m \cot \theta$ (d) $\lambda_m = \lambda_a \cot \theta$

74. A beam of unpolarised light of intensity I_0 is passed through a polaroid A and then through another polaroid B which is oriented so that its principal plane makes an angle of 45° relative to that of A. The intensity of the emergent light is

(a) I_0 (b) $I_0/2$ (c) $I_0/4$ (d) $I_0/8$

75. An unpolarised beam of intensity $2a^2$ passes through a thin polaroid. Assuming zero absorption in the polaroid, the instensity of emergent plane polarised light is

(a) $2a^2$ (b) a^2 (c) $\sqrt{2}\, a^2$ (d) $\dfrac{a^2}{2}$

76. A beam of ordinary unpolarised light passes through a tourmaline crystal C_1 and then it passes through another tourmaline crystal C_2, which is oriented such that its principal plane is parallel to that of C_2. The intensity of emergent light is I_0. Now C_2 is rotated by 60° about the ray. The emergent ray will have an intensity

(a) $2I_0$ (b) $I_0/\sqrt{2}$ (c) $I_0/4$ (d) $I_0/\sqrt{2}$

77. Two Nicol prisms are first crossed and then one of them is rotated through 60°. The percentage of incident light transmitted is

(a) 1.25 (b) 25.0

(c) 37.5 (d) 50

78. Unpolarised light of intensity I passes through an ideal polariser A. Another identical polariser B is placed behind A. The intensity of light beyond B is found to be $\dfrac{I}{2}$. Now, another identical polariser C is placed between A and B. The intensity beyond B is now found to be $\dfrac{1}{8}$. The angle between polariser A and C is [JEE Main 2018]

(a) 0° (b) 30°

(c) 45° (d) 60°

79. A system of three polarisers P_1, P_2, P_3 is set up such that the pass axis of P_3 is crossed with respect to that of P_1. The pass axis of P_2 is inclined at 60° to the pass axis of P_3. When a beam of unpolarised light of intensity I_0 is incident on P_1, the intensity of light transmitted by the three polarisers is I. The ratio (I_0 / I) equals (nearly) [JEE Main 2019]

(a) 5.33 (b) 16.00

(c) 10.67 (d) 1.80

80. The 6563 Å H_α sign line emitted by hydrogen in a star is found to be red-shifted by 15 Å. Estimate the speed with which the star is receding from the earth.

(a) 6.9 m/s approaching the earth

(b) 6.86 m/s receding the earth

(c) 7.9 m/s receding the earth

(d) 8.9 m/s receding the earth

ROUND II Mixed Bag

Only One Correct Option

1. Interference was observed in interference chamber when air was present, now the chamber is evacuated and if, the same light is used, a careful observer will see

(a) interference in which width of the fringe will be slightly increased
(b) interference with bright band
(c) interference with dark band
(d) All of the above

2. A beam of light of wavelength 600 nm from a distant source falls on a single slit 1.00 mm wide and the resulting diffraction pattern is observed on a screen 2 m away. The distance between the first dark fringes on either side of the central bright fringe is

(a) 1.2 cm (b) 1.2 mm (c) 2.4 cm (d) 2.4 mm

3. Find the thickness of a plate which will produce a change in optical path equal to half the wavelength λ of the light passing through it normally. The refractive index of the plate μ is equal to

(a) $\frac{\lambda}{4(\mu-1)}$ (b) $\frac{2\lambda}{4(\mu-1)}$ (c) $\frac{\lambda}{(\mu-1)}$ (d) $\frac{\lambda}{2(\mu-1)}$

4. Red light differs from blue light as they have [JEE Main 2021]

(a) different frequencies and different wavelengths
(b) different frequencies and same wavelengths
(c) same frequencies and same wavelengths
(d) same frequencies and different wavelengths

5. Consider the diffraction pattern for a small pinhole. As the size of the hole is increased,

(a) the size decreases
(b) the intensity increases
(c) the size increases
(d) Both (a) and (b)

6. Two slits separated by a distance of 1 mm are illuminated with red light of wavelength 6.5×10^{-7} m. The interference fringes are observed on a screen placed at 1 m from the slits. The distance between the third dark fringe and the fifth bright fringe is equal to

(a) 0.65 mm (b) 1.63 mm
(c) 3.25 mm (d) 4.88 mm

7. In a double slit experiment, green light (5303 Å) falls on a double slit having a separation of 19.44 μm and a width of 4.05 μm. The number of bright fringes between the first and the second diffraction minima is [JEE Main 2019]

(a) 5 (b) 10 (c) 9 (d) 4

8. A beam of unpolarized light having flux 10^{-3} W falls normally on a polarizer of cross-sectional area 3×10^{-4} m^2. The polarizer rotates with an angular frequency of 31.4 rads^{-1}. The energy of light passing through the polarizer per revolution will be

(a) 10^{-4} J (b) 10^{-3} J (c) 10^{-2} J (d) 10^{-1} J

9. White light is used to illuminate the two slits in a Young's double slit experiment. The separation between slits is b and the screen is at a distance $d(>>b)$ from the slits. At a point on the screen directly in front of one of the slits, certain wavelengths are missing, figure. Some of these missing wavelengths are

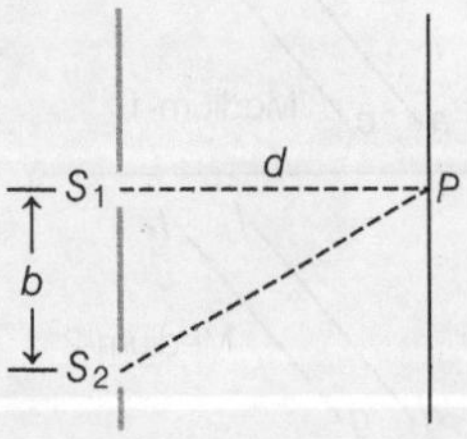

(a) $\lambda = \frac{b^2}{d}, \frac{2b^2}{3d}$ (b) $\lambda = \frac{b^2}{2d}, \frac{3b^2}{2d}$

(c) $\lambda = \frac{2b^2}{3d}$ (d) $\lambda = \frac{3b^2}{4d}$

10. Two sources S_1 and S_2 of intensity I_1 and I_2 are placed in front of a screen [Fig.(a)]. Then, which amongst the following is incorrect about the pattern of intensity distribution seen in the central portion is given by Fig. (b)? [NCERT Exemplar]

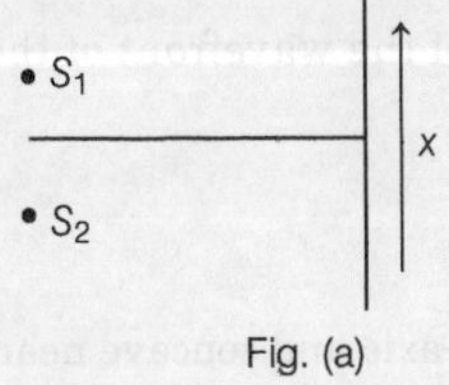

Fig. (a)

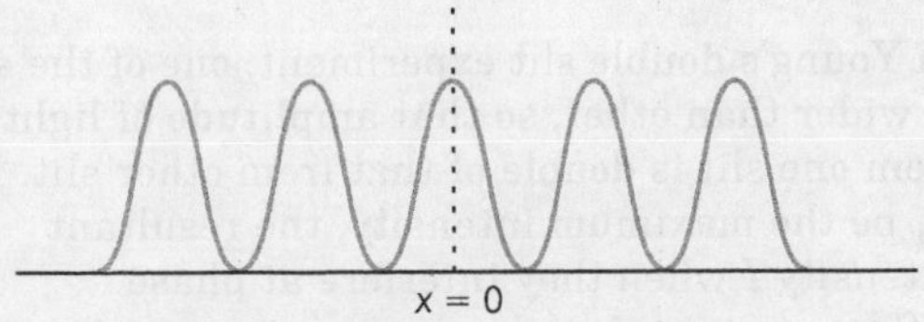

Fig. (b)

(a) S_1 and S_2 have the same intensities
(b) S_1 and S_2 have a constant phase difference
(c) S_1 and S_2 have the same phase
(d) S_1 and S_2 have the same wavelength

11. In Young's double slit experiment $\frac{d}{D} = 10^{-4}$ (d = distance between slits, D = distance of screen from the slits). At a point P on the screen, resultant intensity is equal to the intensity due to individual slit I_0. Then the distance of point P from the central maximum is ($\lambda = 6000$ Å)

(a) 0.5 mm (b) 2 mm
(c) 1 mm (d) 4 mm

12. The figure shows a surface XY separating two transparent media, medium-1 and medium-2. The lines ab and cd represent wavefronts of a light wave travelling in medium-1 and incident on XY. The lines ef and gh represent wavefronts of the light wave in medium-2 after refraction.

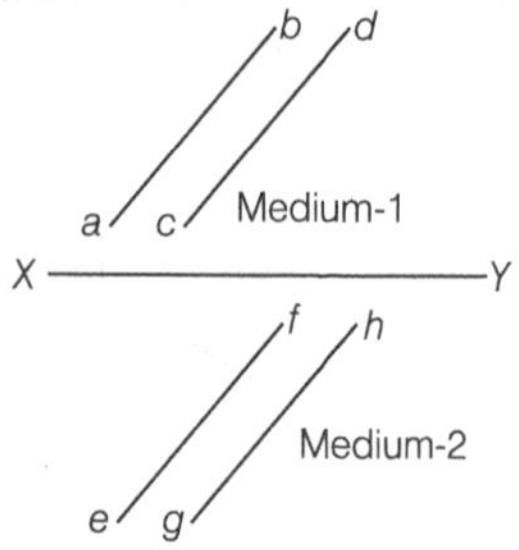

The phases of the light wave at c, d, e and f are ϕ_c, ϕ_d, ϕ_e and ϕ_f respectively. It is given that $\phi_c \neq \phi_f$

(a) ϕ_c cannot be equal to ϕ_d
(b) ϕ_d can be equal to ϕ_e
(c) $(\phi_d - \phi_f)$ is equal to $(\phi_c - \phi_e)$
(d) $(\phi_d - \phi_c)$ is not equal to $(\phi_f - \phi_e)$

13. An initially parallel cylindrical beam travels in a medium of refractive index $\mu(I) = \mu_0 + \mu_2 I$, where μ_0 and μ_2 are positive constants and I is the intensity of the light beam. The intensity of the beam is decreasing with increasing radius.

The initial shape of the wavefront of the beam is **[AIEEE 2010]**

(a) planar
(b) convex
(c) concave
(d) convex near the axis and concave near the periphery

14. In Young's double slit experiment, one of the slit is wider than other, so that amplitude of light from one slit is double of that from other slit. If, I_m be the maximum intensity, the resultant intensity I when they interfere at phase difference ϕ, is given by **[AIEEE 2012]**

(a) $\frac{I_m}{9}(4 + 5\cos\phi)$ (b) $\frac{I_m}{3}\left(1 + \cos^2\frac{\phi}{2}\right)$
(c) $\frac{I_m}{5}\left(1 + 4\cos^2\frac{\phi}{2}\right)$ (d) $\frac{I_m}{9}\left(1 + 8\cos^2\frac{\phi}{2}\right)$

15. A beam of plane polarised light of large cross-sectional area and uniform intensity of 3.3 Wm^{-2} falls normally on a polariser, which rotates about its axis with an angular speed of 31.4 rad/s. The energy of light passing through the polariser per revolution is close to
(Take, cross-sectional area $= 3 \times 10^{-4}$ m^2) **[JEE Main 2020]**

(a) 1.5×10^{-4} J
(b) 5.0×10^{-4} J
(c) 1.0×10^{-4} J
(d) 1.0×10^{-5} J

16. In an interference arrangement similar to Young's double slit experiment, the slits S_1 and S_2 are illuminated with coherent microwave sources each of frequency 10^6 Hz. The sources are synchronized to have zero phase difference. The slits are separated by distance $d = 150$ m. The intensity $I(\theta)$ is measured as a function of θ, where θ is defined as shown. If I_0 is maximum intensity, then $I(\theta)$ for $0 \leq \theta \leq 90°$ is given by

(a) $I(\theta) = I_0$ for $\theta = 30°$
(b) $I(\theta) = \frac{I_0}{2}$ for $\theta = 30°$
(c) $I(\theta) = \frac{I_0}{4}$ for $\theta = 90°$
(d) $I(\theta)$ is constant for all values of θ

17. In the given figure, P and Q are two equally intense coherent sources emitting radiation of wavelength 20 m. The separation between P and Q is 5 m and the phase of P is ahead of that of Q by 90°. A, B and C are three distinct points of observation, each equidistant from the mid-point of PQ. The intensities of radiation at A, B and C will be in the ratio

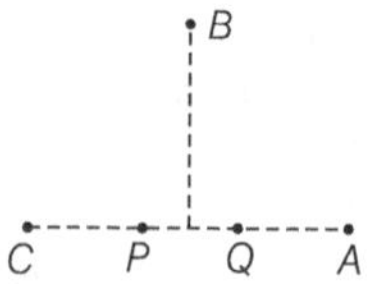

[JEE Main 2020]

(a) 0 : 1 : 4 (b) 2 : 1 : 0
(c) 0 : 1 : 2 (d) 4 : 1 : 0

18. Two light waves having the same wavelength λ in vacuum are in phase initially. Then, the first wave travels a path L_1 through a medium of refractive index n_1 while the second wave travels a path of length L_2 through a medium of refractive index n_2. After this, the phase difference between the two waves is **[JEE Main 2020]**

(a) $\frac{2\pi}{\lambda}(n_1 L_1 - n_2 L_2)$ (b) $\frac{2\pi}{\lambda}\left(\frac{L_1}{n_1} - \frac{L_1}{n_2}\right)$
(c) $\frac{2\pi}{\lambda}\left(\frac{L_2}{n_1} - \frac{L_1}{n_2}\right)$ (d) $\frac{2\pi}{\lambda}(n_2 L_1 - n_1 L_2)$

Numerical Value Questions

19. If amplitude ratio of two sources producing interference is 3:5. The ratio of intensities of maxima and minima is $x:1$, where the value of x is ……… .

20. In Young's double slit experiment, an interference pattern is obtained for $\lambda = 6000 Å$, coming from two coherent sources S_1 and S_2. At certain point P on the screen, third dark fringe is formed. Then the path difference $S_1P - S_2P$ (in micron) is ……… .

21. In a Young's double slit experiment, 15 fringes are observed on a small portion of the screen when light of wavelength 500 nm is used. Ten fringes are observed on the same section of the screen when another light source of wavelength λ is used. Then, the value of λ is (in nm) …… **[JEE Main 2020]**

22. Young's double slit experiment is carried out using microwaves of wavelength $\lambda = 3$ cm. Distance between the slits is $d = 5$ cm and the distance between the plane of slits and the screen is $D = 100$ cm, then what is the number of maximas on the screen?

23. Orange light of wavelength 6000×10^{-10} m illuminates a single slit of width 0.6×10^{-4} m. The maximum possible number of diffraction minima produced on both sides of the central maximum is ……… . **[JEE Main 2020]**

24. A Young's double slit experiment is performed using monochromatic light of wavelength λ. The intensity of light at a point on the screen, where the path difference is λ, is K units. The intensity of light at a point where the path difference is $\frac{\lambda}{6}$ is given by $\frac{nK}{12}$, where n is an integer. The value of n is ……… **[JEE Main 2020]**

Answers

Round I

1. (a)	2. (c)	3. (a)	4. (d)	5. (a)	6. (b)	7. (b)	8. (b)	9. (a)	10. (d)
11. (c)	12. (d)	13. (d)	14. (a)	15. (b)	16. (c)	17. (a)	18. (d)	19. (b)	20. (b)
21. (a)	22. (d)	23. (b)	24. (d)	25. (b)	26. (d)	27. (c)	28. (d)	29. (d)	30. (b)
31. (a)	32. (b)	33. (c)	34. (d)	35. (d)	36. (b)	37. (c)	38. (b)	39. (c)	40. (a)
41. (c)	42. (d)	43. (a)	44. (a)	45. (b)	46. (b)	47. (c)	48. (a)	49. (b)	50. (c)
51. (b)	52. (a)	53. (*)	54. (d)	55. (d)	56. (b)	57. (a)	58. (a)	59. (b)	60. (a)
61. (a)	62. (b)	63. (a)	64. (a)	65. (c)	66. (a)	67. (a)	68. (b)	69. (a)	70. (b)
71. (c)	72. (a)	73. (d)	74. (c)	75. (b)	76. (c)	77. (c)	78. (c)	79. (c)	80. (b)

Round II

1. (a)	2. (d)	3. (d)	4. (a)	5. (d)	6. (b)	7. (a)	8. (a)	9. (a)	10. (c)
11. (b)	12. (c)	13. (a)	14. (d)	15. (c)	16. (b)	17. (b)	18. (a)	19. 16	20. 1.5
21. 750	22. 3	23. 198	24. 9						

Solutions

Round I

1. For light diverging from a point source, the wavefront is (diverging) spherical. The intensity varies inversely as the area of the wavefront $(=4\pi r^2)$, *i.e.* intensity decreases in proportion to the distance squared.

2. The given situation is shown below

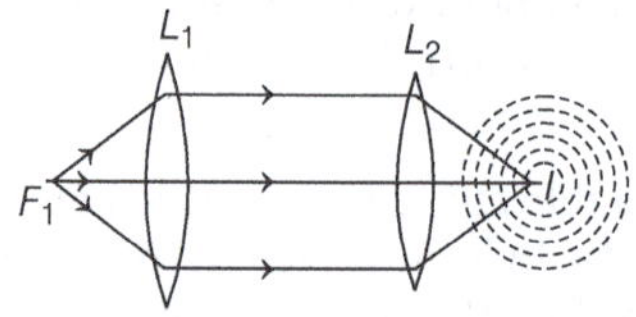

Consider a point F on focus of converging lens L_1. The light rays from F, becomes parallel after refraction through L_1. When these parallel rays falls on converging lens L_2 placed co-axial on the other side of F of lens L_1, lens L_2 converges the rays at its focus at I. It now behaves like a point source of rays and hence form a spherical wavefront.

3. As, velocity of light is perpendicular to the wavefront and light is travelling in vacuum along the Y-axis, therefore the wavefront is represented by $y=$ constant.

4. According to Snell's law,

$$\mu \sin\theta = \text{constant}$$
$$\sin\theta \propto \frac{1}{\mu}$$

As, μ increases, θ decreases.
Hence, the light beam will bend upward.

5. Resultant intensity is given by

$$I_R = I_1 + I_2 + 2\sqrt{I_1 I_2}\cos\phi$$

Here, $I_1 = I$ and $I_2 = 4I$

For maximum intensity, $\cos\phi = 1$,
i.e. $\phi = 0°$

$\therefore$ $I_{max} = I_1 + I_2 + 2\sqrt{I_1 I_2} = I + 4I + 2\sqrt{I\cdot 4I}$

$\Rightarrow$ $I_{max} = 9I$

For minimum intensity, $\cos\phi = -1$, *i.e.* $\phi = 180°$

$$\therefore \quad I_{min} = I_1 + I_2 - 2\sqrt{I_1 I_2} = I + 4I - 2\sqrt{I\cdot 4I} = 5I - 4I = I$$

6. Here, $I_1 = I, I_2 = 4I, \theta_1 = \pi/2, \theta_2 = \pi$

Resultant intensity at point A,

$$I_{\theta_1} = I_1 + I_2 + 2\sqrt{I_1 I_2}\cos\theta_1 = I + 4I + 2\sqrt{I \times 4I}\cos\pi/2 = 5I$$

Resultant intensity at point B,

$$I_{\theta_2} = I_1 + I_2 + 2\sqrt{I_1 I_2}\cos\theta_2 = I + 4I + 2\sqrt{I\times 4I}\cos\pi = 5I - 4I = I$$

$$\therefore \quad I_{\theta_1} - I_{\theta_2} = 5I - I = 4I$$

7. For incoherent waves,

$$I_{max} = nI$$
$$\therefore \quad n = \frac{I_{max}}{I} = \frac{32}{2} = 16$$

8. The amplitude of the waves are $a_1 = 10\,\mu\text{m}$, $a_2 = 4\,\mu\text{m}$ and $a_3 = 7\,\mu\text{m}$ and phase difference between Ist and IInd wave is $\frac{\pi}{2}$ and that between 2nd and 3rd wave is $\frac{\pi}{2}$. Then, phase difference between 1st and 3rd is π.

Combining Ist with 3rd, their resultant amplitude is given by

$$A_1^2 = a_1^2 + a_3^2 + 2a_1 a_3\cos\phi$$

or
$$A_1 = \sqrt{10^2 + 7^2 + 2\times 10\times 7\cos\pi} = \sqrt{100+49-140} = \sqrt{9} = 3\,\mu\text{m}$$

Now, combining this with 2nd wave, we have the resultant amplitude

$$A^2 = A_1^2 + a_2^2 + 2A_1 a_2\cos\frac{\pi}{2}$$

or
$$A = \sqrt{3^2 + 4^2 + 2\times 3\times 4\times\cos 90°} = \sqrt{9+16} = \sqrt{25} = 5\,\mu\text{m}$$

9. The intensity of light reflected from upper surface is

$$I_1 = I_0 \times 25\% = I_0 \times \frac{25}{100} = \frac{I_0}{4}$$

The intensity of transmitted light from upper surface is

$$I = I_0 - \frac{I_0}{4} = \frac{3I_0}{4}$$

$\therefore$ The intensity of reflected light from lower surface is

$$I_2 = \frac{3I_0}{4}\times\frac{50}{100} = \frac{3I_0}{8}$$

$$\therefore \quad \frac{I_{max}}{I_{min}} = \frac{(\sqrt{I_1}+\sqrt{I_2})^2}{(\sqrt{I_1}-\sqrt{I_2})^2}$$

$$\therefore \quad \frac{I_{max}}{I_{min}} = \frac{\left(\sqrt{\frac{I_0}{4}}+\sqrt{\frac{3I_0}{8}}\right)^2}{\left(\sqrt{\frac{I_0}{4}}-\sqrt{\frac{3I_0}{8}}\right)^2} = \frac{\left(\frac{1}{2}+\sqrt{\frac{3}{8}}\right)^2}{\left(\frac{1}{2}-\sqrt{\frac{3}{8}}\right)^2}$$

10. Hence, path difference, $S_1P - S_2P =$ odd integral multiple of $\lambda/2 = 11\,(\lambda/2)$

11. Path difference, $x = (SS_1 + S_1O) - (SS_2 + S_2O)$. If, $x = n\lambda$, then central fringe at O will be bright. If, $x = (2n-1)\lambda/2$, the central fringe at O will be dark.

12. In the given figure, join the points S_1 and L.

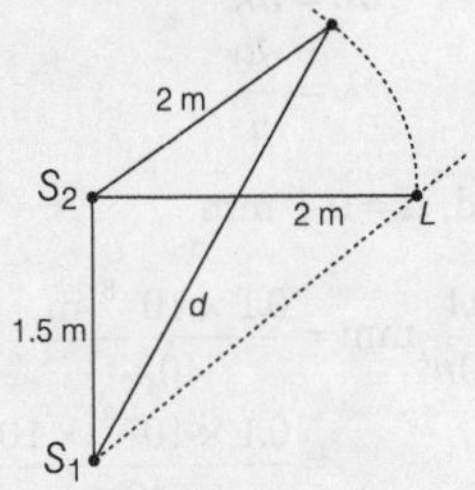

From figure, $S_2L = 2$ m

$$S_1L = \sqrt{(2)^2 + (1.5)^2} = \sqrt{4 + 2.25} = 2.5 \text{ m}$$

Initial path difference where minima is observed,

$$\Delta x_1 = S_1L - S_2L = 2.5 - 2 = 0.5 \text{ m}$$

Path difference at new position, where maxima is observed,

$$\Delta x_2 = d - 2$$

Path difference between adjacent maxima and minima is $\frac{\lambda}{2}$.

i.e. $\Delta x_2 - \Delta x_1 = \frac{\lambda}{2}$

$\Rightarrow$ $d - 2 - 0.5 = \frac{1}{2}$ $(\because \lambda = 1 \text{ m})$

$\Rightarrow$ $d = 3$ m

13. After interference, in accordance to superposition principle, the fringes on the screen will be concentric circles.

14. A ray of light of intensity I is incident on a parallel glass slab, undergoes partial refraction and reflection. According to figure given in question, the rays AB and $A'B'$ undergo interference.

Intensity of reflected ray AB,

$$I_{AB} = 25\% \text{ of } I = \frac{I}{4}$$

$\therefore$ Intensity of ray AC,

$$I_{AC} = I - I_{AB} = I - \frac{I}{4} = \frac{3I}{4}$$

Similarly, intensity of ray $A'C$ is

$$I_{A'C} = 25\% \text{ of } I_{AC} = \frac{25}{100} \times \frac{3I}{4} = \frac{3I}{16}$$

Similarly, $I_{A'B'} = 75\%$ of $I_{A'C} = \frac{75}{100} \times \frac{3I}{16} = \frac{9I}{64}$

$$\therefore \quad \frac{I_{max}}{I_{min}} = \left(\frac{\sqrt{\frac{I_{A'B'}}{I_{AB}}} + 1}{\sqrt{\frac{I_{A'B'}}{I_{AB}}} - 1}\right)^2 = \left(\frac{\sqrt{\frac{9}{16}} + 1}{\sqrt{\frac{9}{16}} - 1}\right)^2 = \frac{49}{1}$$

$\therefore$ $I_{max} : I_{min} = 49 : 1$

15. Here, $n_1\lambda_1 = n_2\lambda_2$

$$n(7.8 \times 10^{-5}) = (n + 1)(5.2 \times 10^{-5})$$

$$\frac{n}{n+1} = \frac{5.2 \times 10^{-5}}{7.8 \times 10^{-5}}$$

$\therefore$ $n = 2$

16. As light from two slits of YDSE is of different colours/wavelength/frequencies, therefore there shall be no interference fringes.

17. Here, $2d = 0.5\text{mm} = 5 \times 10^{-2}\text{cm}$ and $D = 100$ cm

$$x_n = x_{11} - x_1 = 9.72 \text{ mm}$$

i.e. $n = 10$

We know that, $x_{11} - x_1 = \frac{n\lambda D}{2d}$

$$\therefore \quad \lambda = \frac{(x_n)(2d)}{nD}$$

$$= \frac{(0.972) \times (5 \times 10^{-2})}{10 \times 100}$$

$$= 4.86 \times 10^{-5} \text{ cm}$$

18. Given, separation between slits, $d = 0.28$ mm

$$= 0.28 \times 10^{-3} \text{ m}$$

Distance between screen and slit, $D = 1.4$ m,

Distance between central bright fringe and fourth bright fringe,

$$x = 1.2 \text{ cm} = 1.2 \times 10^{-2} \text{ m}$$

Number of fringes, $n = 4$

For constructive interference, $x = n\frac{D\lambda}{d}$

$$1.2 \times 10^{-2} = \frac{4 \times 1.4 \times \lambda}{0.28 \times 10^{-3}}$$

Wavelength, $\lambda = \frac{1.2 \times 10^{-2} \times 0.28 \times 10^{-3}}{4 \times 1.4}$

$$= 6 \times 10^{-7} \text{ m}$$

The wavelength of light is 6×10^{-7} m.

19. Given, setup of Young's double slit experiment is as shown

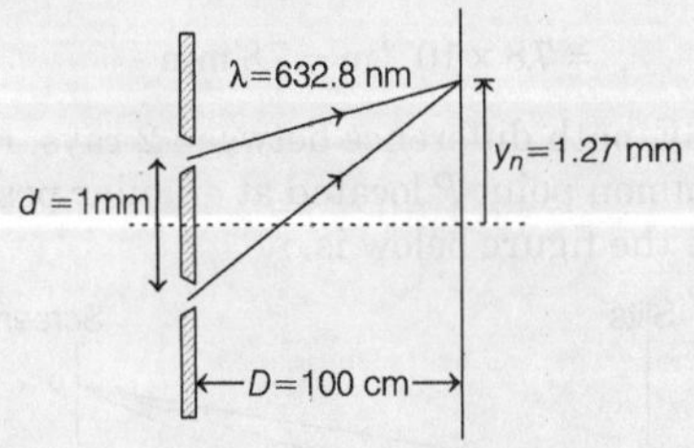

Distance of a bright fringe from central bright fringe is

$$y_n = \frac{n\lambda D}{d}$$

where, $n\lambda$ = path difference between rays from lower and upper slit $= \frac{y_n d}{D} = \frac{1.27 \times 10^{-3} \times 10^{-3}}{100 \times 10^{-2}}$

$$= 1.27 \times 10^{-6} \text{m} = 1.27 \, \mu\text{m}$$

20. Position of nth maxima from central maxima is given by $x_n = \frac{n\lambda D}{d}$

For 8th maxima, $x_8 = \frac{8\lambda_1 D}{d_1}$

and for 6th maxima, $x_6 = \frac{6\lambda_2 D}{d_2}$

Now, $x_6 = x_8$

$$\Rightarrow \quad \frac{d_1}{d_2} = \frac{8\lambda_1}{6\lambda_2} = \frac{4}{3}\left(\frac{\lambda_1}{\lambda_2}\right)$$

21. Given, $d = 0.5 \text{ mm} = 5 \times 10^{-4}\text{m}$

$$D = 150 \text{ cm} = 1.5 \text{ m}$$
$$\lambda_1 = 650 \text{ nm} = 6.5 \times 10^{-7} \text{ m}$$
$$\lambda_2 = 520 \text{ nm} = 5.2 \times 10^{-7}\text{m}$$

Fringe width pattern due to wavelength λ_1 only,

$$\beta_1 = \frac{D\lambda_1}{d} \quad \ldots\text{(i)}$$

Fringe width pattern due to wavelength λ_2 only,

$$\beta_2 = \frac{D\lambda_2}{d} \quad \ldots\text{(ii)}$$

If n_1 and n_2 are number of bright fringes due to fringe patterns using λ_1 and λ_2 respectively, then

$$n_1\beta_1 = n_2\beta_2$$
$$\Rightarrow \quad n_1\frac{D\lambda_1}{d} = n_2\frac{D\lambda_2}{d}$$
$$\Rightarrow \quad \frac{n_1}{n_2} = \frac{\lambda_2}{\lambda_1} = \frac{5.2 \times 10^{-7}}{6.5 \times 10^{-7}}$$
$$\Rightarrow \quad \frac{n_1}{n_2} = \frac{4}{5}$$
$$\Rightarrow \quad n_1 = 4, n_2 = 5$$

If x be the least distance where both the bright fringes coincides, then

$$\frac{xd}{D} = n_1\lambda_1$$
$$\Rightarrow \quad x = \frac{n_1\lambda_1 D}{d} = \frac{4 \times 6.5 \times 10^{-7} \times 1.5}{5 \times 10^{-4}}$$
$$= 7.8 \times 10^{-3}\text{m} = 7.8 \text{ mm}$$

22. In a YDSE, path difference between 2 rays, reaching at some common point P located at angular position θ as shown in the figure below is

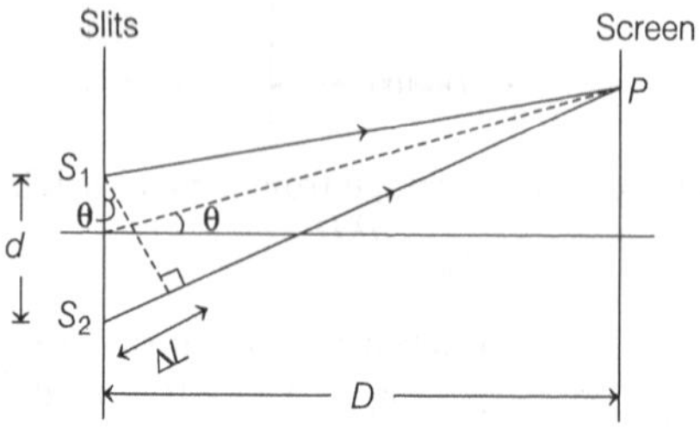

$$\Delta L = d\sin\theta$$

For small value of θ, $\sin\theta \approx \theta$

So, path difference $= \Delta L = d\theta$

For a bright fringe at same angular position θ, both of the rays from slits S_1 and S_2 are in phase.

Hence, path difference is an integral multiple of wavelength of light used.

i.e. $\quad \Delta L = n\lambda$

or $\quad d\theta = n\lambda$

$$\Rightarrow \quad \lambda = \frac{d\theta}{n}$$

Here, $\theta = \frac{1}{40}$ rad, $d = 0.1$ mm

Hence, $\lambda = \frac{0.1}{40n}\text{ mm} = \frac{0.1 \times 10^{-3} \text{ m}}{40n}$

$$= \frac{0.1 \times 10^{-3} \times 10^{9}}{40n} \text{ nm}$$
$$\Rightarrow \quad \lambda = \frac{2500}{n} \text{ nm}$$

So, with light of wavelength λ_1, we have

$$\lambda_1 = \frac{2500}{n_1} \text{ nm}$$

and with light of wavelength λ_2, we have

$$\lambda_2 = \frac{2500}{n_2} \text{ nm}$$

Now, choosing different integral values for n_1 and n_2, (i.e. n_1, $n_2 = 1, 2, 3 \ldots$, etc) we find that for $n_1 = 4$,

$$\lambda_1 = \frac{2500}{4} = 625 \text{ nm}$$

and for $n_2 = 5$,

$$\lambda_2 = \frac{2500}{5} = 500 \text{ nm}$$

These values lie in given interval 500 nm to 625 nm.

23. As, $\beta = \frac{\lambda D}{d}$

$$\therefore \quad \beta \propto 1/d$$

This means that fringe width (β) will decrease with increase in distance between the slits (d) or *vice-versa*. This is clearly depicted in the graph given in option (b).

24. In Young's double slit experiment,

fringe width, $\beta = \frac{D\lambda}{d} \quad \ldots\text{(i)}$

when $d' = \frac{d}{2}$ and $D' = 2D$, then

fringe width, $\beta' = \frac{D'\lambda}{d'} = \frac{2D \cdot \lambda}{d/2}$

$$= \frac{4D\lambda}{d} = 4\beta \quad \text{[from Eq. (i)]}$$

Hence, fringe width is quardrupled.

25. Fringe width, $\beta = \frac{\lambda D}{d} = \frac{5890 \times 10^{-10} \times 0.5}{0.5 \times 10^{-3}} = 589 \times 10^{-6}$ m

Distance between first and third bright fringe is

$$2\beta = 2 \times 589 \times 10^{-6} \text{ m}$$
$$= 1178 \times 10^{-6} \text{ m}$$

26. As, $x = (2n - 1)\frac{\lambda}{2}\frac{D}{d}$

$$\therefore \quad \lambda = \frac{2xd}{(2n-1)D}$$

$$= \frac{2\times10^{-3}\times0.9\times10^{-3}}{(2\times2-1)\times1}$$
$$= 6\times10^{-7}\ \text{m} = 6\times10^{-5}\ \text{cm}$$

27. Interference fringes are the bands on screen XY running parallel to the lengths of slits. Therefore, the locus of fringes is represented correctly by W_3W_4.

28. In Young's double slit experiment, the condition of bright fringe and dark fringe are discussed below.
For bright fringes (maxima), path difference = $n\lambda$
$$d\sin\theta = n\lambda$$
for dark fringes (minima),
path difference = $(2n-1)\dfrac{\lambda}{2}$
$$d\sin\theta = (2n-1)\frac{\lambda}{2}$$

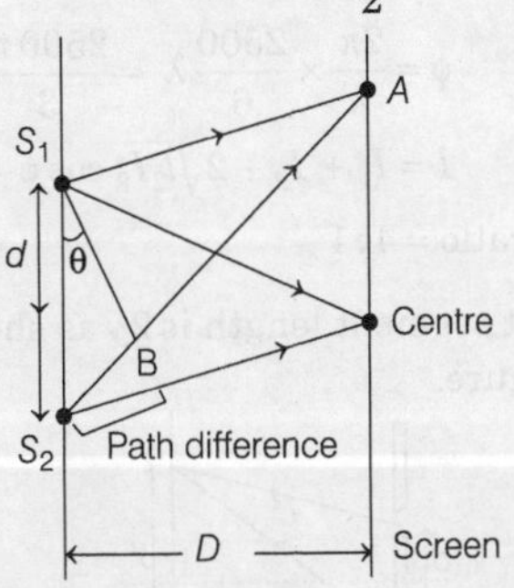

For the given question, distance between slits $(d) = 0.320$ mm
Wavelength of light used $(\lambda) = 500$ nm
Angular range for bright fringe
$$(\theta) = -30^\circ \text{ to } 30^\circ$$
Hence, for bright fringe,
$$n\lambda = d\sin\theta$$
$$n = \frac{d\sin\theta}{\lambda} = \frac{0.320\times10^{-3}\times\sin30^\circ}{500\times10^{-9}}$$
$$n_{max} = 320$$
$\therefore$ Total number of maxima between the two lines are
$$n = (n_{max}\times2)+1$$
Here, $\quad n = (320\times2)+1$
$$n = 641$$

29. As, $\beta = \dfrac{\lambda D}{d}$

and $\quad \beta' = \dfrac{\lambda D'}{d}$

$\therefore \quad \beta - \beta' = \dfrac{\lambda(D-D')}{d}$
$$3\times10^{-5} = \frac{\lambda\times5\times10^{-2}}{10^{-3}}$$
or $\quad \lambda = \dfrac{3\times10^{-5}}{50} = 6\times10^{-7}\ \text{m} = 6000\ \text{Å}$

30. For maxima on the screen,
$$d\sin\theta = n\lambda \qquad \text{...(i)}$$
Given, $\quad d = 2\lambda$
$\therefore \quad 2\lambda\sin\theta = n\lambda$
$$2\sin\theta = n$$
The maximum value of $\sin\theta = +1$
$\therefore \quad n = 2$
$\therefore$ Eq. (i) is satisfied by $-2, -1, 0, 1, 2$.

31. As we know,
$$\beta = \frac{D}{d}\lambda \qquad \text{...(i)}$$
and $\quad \lambda \propto \dfrac{1}{\mu} \qquad$ ...(ii)
From Eqs. (i) and (ii), we get
$$\beta \propto \lambda \propto \frac{1}{\mu}$$
$\therefore \quad \beta \propto \dfrac{1}{\mu}$
The refractive index of water is greater than air, therefore fringe width will decrease.

32. As, field of view is same in both cases
$$n_1\beta_1 = n_2\beta_2$$
or $\quad n_1\left(\dfrac{D\lambda_1}{d}\right) = n_2\left(\dfrac{D\lambda_2}{d}\right)$ or $\lambda_2 = \left(\dfrac{n_1}{n_2}\right)\lambda_1$
$\therefore \quad \lambda_2 = \left(\dfrac{84}{62}\right)\times4358$
$$\lambda_2 = 5904\ \text{Å}$$

33. Distance of nth maxima, $x = n\lambda\dfrac{D}{d} \propto \lambda$
$\Rightarrow \quad \lambda_b < \lambda_g$
$\therefore \quad x_{blue} < x_{green}$

34. As there is a hole at minima point P_2, the hole will act as a source of fresh light for the slits S_3 and S_4. Therefore, there will be a regular two slit pattern on the second screen. Hence, option (d) is correct.

35. Let intensity of each wave be I_0, then intensity at the centre of bright fringe will be $4I_0$.
Given, path difference, $\Delta x = \lambda/8$
$\therefore$ Phase difference, $\phi = \Delta x\times\dfrac{2\pi}{\lambda}$
$\Rightarrow \quad \phi = \dfrac{\lambda}{8}\times\dfrac{2\pi}{\lambda}$ or $\phi = \pi/4$
Intensity of light at this point,
$$I' = I_0 + I_0 + 2I_0\cos(\pi/4)$$
$$= 2I_0 + \sqrt{2}\,I_0 = 3.41\,I_0$$
Now, $\quad \dfrac{I'}{4I_0} = \dfrac{3.41I_0}{4I_0} = \dfrac{3.41}{4} = 0.85$

36. As, $\dfrac{I_{max}}{I_{min}} = \dfrac{(a+b)^2}{(a-b)^2} = 9$ or $\dfrac{a+b}{a-b} = 3$
or $\quad 3a - 3b = a + b$
or $\quad 2a = 4b$
$\Rightarrow \quad a = 2b$
$\therefore \quad \dfrac{I_1}{I_2} = \dfrac{a^2}{b^2} = \dfrac{4b^2}{b^2} = 4:1$

37. $$3\lambda_1 = 4\lambda_2 \qquad \left[\because \frac{3D\lambda_1}{d} = 4\frac{D\lambda_2}{d}\right]$$

$$\Rightarrow \quad \lambda_2 = \frac{3}{4}\lambda_1 = \frac{3}{4}\times 590$$

$$= \frac{1770}{4} = 442.5 \text{ nm}$$

38. The intensity at a point on screen is given by

$$I = 4I_0 \cos^2(\phi/2)$$

where, ϕ is the phase difference. In this problem, ϕ arises (i) due to initial phase difference of $\frac{\pi}{4}$ and (ii) due to path difference for the observation point situated at $\theta = 30°$.

Thus, $$\phi = \frac{\pi}{4} + \frac{2\pi}{\lambda}(d\sin\theta) = \frac{\pi}{4} + \frac{2\pi}{\lambda}\times\frac{\lambda}{4}(\sin 30°)$$

$$= \frac{\pi}{4} + \frac{\pi}{4} = \frac{\pi}{2}$$

Hence, $$\frac{\phi}{2} = \frac{\pi}{4}$$

and $$I = 4I_0\cos^2\left(\frac{\pi}{4}\right) = 2I_0$$

39. Amplitude of light is directly proportional to area through which light is passing.

For same length of slits,

$$\text{amplitude} \propto (\text{width})^{1/2}$$

Also, $$\text{intensity} \propto (\text{amplitude})^2$$

In YDSE, ratio of intensities of maxima and minima is given by

$$\frac{I_{max}}{I_{min}} = \frac{(\sqrt{I_1} + \sqrt{I_2})^2}{(\sqrt{I_1} - \sqrt{I_2})^2}$$

where, I_1 and I_2 are the intensities obtained from two slits, respectively.

$$\Rightarrow \quad \frac{I_{max}}{I_{min}} = \frac{(a_1 + a_2)^2}{(a_1 - a_2)^2}$$

where, a_1 and a_2 are light amplitudes from slits 1 and 2, respectively.

$$\Rightarrow \quad \frac{I_{max}}{I_{min}} = \frac{(\sqrt{W_1} + \sqrt{W_2})^2}{(\sqrt{W_1} - \sqrt{W_2})^2}$$

where, W_1 and W_2 are the widths of slits, respectively.

Here, $$\left(\frac{W_1}{W_2}\right) = \left(\frac{a_1}{a_2}\right)^2 = \frac{4}{1}$$

$$\Rightarrow \quad \sqrt{\frac{W_1}{W_2}} = \frac{2}{1}$$

So, we have

$$\frac{I_{max}}{I_{min}} = \left(\frac{\sqrt{W_1} + \sqrt{W_2}}{\sqrt{W_1} - \sqrt{W_2}}\right)^2 = \left(\frac{\sqrt{\frac{W_1}{W_2}} + 1}{\sqrt{\frac{W_1}{W_2}} - 1}\right)^2$$

$$= \left(\frac{2+1}{2-1}\right)^2 = 9:1$$

40. When sources are coherent, then

$$I_R = I_1 + I_2 + 2\sqrt{I_1 I_2}\cos\phi$$

At middle point of the screen $\phi = 0°$, then

$$I_R = I + I + 2\sqrt{I\cdot I}\cos 0° = 4I$$

When sources are incoherent, then

$$I_R' = I_1 + I_2 = I + I = 2I$$

$$\therefore \quad \frac{I_R}{I_R'} = \frac{4I}{2I} = 2$$

41. Here, path difference,

$$\Delta x = (l_1 - 1)t = (1.5 - 1)\times\frac{2500}{3}\lambda$$

$$\therefore \quad \phi = \frac{2\pi}{\lambda}\Delta x$$

$$\Rightarrow \quad \phi = \frac{2\pi}{\lambda}\times\frac{2500}{6}\lambda = \frac{2500\pi}{3}$$

$$I = I_1 + I_2 + 2\sqrt{l_2 I_2}\cos\phi$$

$\therefore$ Intensity ratio $= 4:1$

42. In YDSE, let segment length is $2y$ as shown in the following figure.

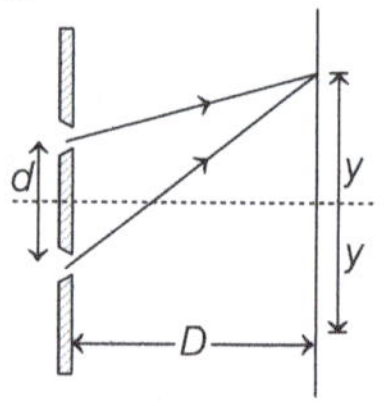

Then, $$2y = 2\times N\times\frac{\lambda D}{d}$$

where, $N =$ number of fringes in segment length $2y$.

$$\Rightarrow \quad N\lambda = \frac{yd}{D}$$

So, for same segment length $N\lambda$ is constant.

$$\Rightarrow \quad N_1\lambda_1 = N_2\lambda_2$$

Given, $N_1 = 16$, $\lambda_1 = 700$ nm

$N_2 = ?$, $\lambda_2 = 400$ nm

$$\Rightarrow \quad N_2 = \frac{N_1\lambda_1}{\lambda_2} = 16\times\frac{700}{400} = 28$$

$\therefore$ 28 fringes are observed in same segment length.

43. The distance between zeroth order maxima and second order minima is

$$y_1 = \frac{\beta}{2} + \beta = \frac{3}{2}\beta = \frac{3}{2}\times 0.2 \text{ mm} = 0.3 \text{ mm}$$

$\therefore$ The distance of second maxima from point P is

$$y = (4.8 + 0.3) \text{ mm} = 5.1 \text{ mm}$$

44. The position of 30th bright fringe,

$$y_{30} = \frac{30\lambda D}{d}$$

Now, position shift of central fringe is

$$y_0 = \frac{30\lambda D}{d}$$

But we know, $y_0 = \frac{D}{d}(\mu - 1)t$

$$\frac{30\lambda D}{d} = \frac{D}{d}(\mu - 1)t$$

$$\Rightarrow \quad (\mu - 1) = \frac{30\lambda}{t} = \frac{30 \times (6000 \times 10^{-10})}{(3.6 \times 10^{-5})} = 0.5$$

$$\therefore \quad \mu = 1.5$$

45. Here, $d = 0.1 \text{ mm} = 10^{-4} \text{m}, D = 20 \text{ cm} = \frac{1}{5} \text{m}$

$$\lambda = 5460 \text{ Å} = 5.46 \times 10^{-7} \text{ m}$$

Angular position of first dark ($n = 1$) fringe is

$$\theta = (2n - 1)\frac{\lambda}{2d} = (2 \times 1 - 1)\frac{\lambda}{2d} = \frac{\lambda}{2d} = \frac{5.46 \times 10^{-7}}{2 \times 10^{-4}}$$

$$= 2.73 \times 10^{-3} \text{ rad}$$

$$= 2.73 \times 10^{-3} \times \frac{180^\circ}{\pi}$$

$$= 0.156^\circ \simeq 0.16^\circ$$

46.

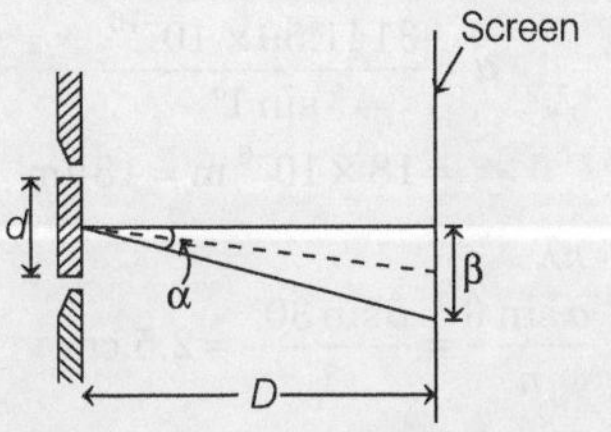

Angular width of central maxima in YDSE is given by

$$\alpha = \frac{\beta}{D} = \frac{\lambda}{d} \text{ (in radian)}$$

$$= \frac{\lambda}{d} \times \frac{180^\circ}{\pi} \text{ (in degree)}$$

Here, $\lambda = 500 \text{ nm} = 500 \times 10^{-9} \text{m}$

and $d = 0.05 \text{ mm} = 0.05 \times 10^{-3} \text{ m}$.

Hence, angular width is

$$\alpha = \frac{500 \times 10^{-9}}{0.05 \times 10^{-3}} \times \frac{180^\circ}{\pi} = 0.57^\circ$$

47. Here, $\sin\theta = \left(\frac{y}{D}\right)$

So, $\Delta\theta = \frac{\Delta y}{D}$

Angular fringe width, $\theta_0 = \Delta\theta$ (width $\Delta y = \beta$)

$$\Rightarrow \quad \theta_0 = \frac{\beta}{D} = \frac{D\lambda}{d} \times \frac{1}{D} = \frac{\lambda}{d}$$

Now, $\theta_0 = 1^\circ = \pi/180$ rad

and $\lambda = 6 \times 10^{-7}$ m

$$\therefore \quad d = \frac{\lambda}{\theta_0} = \frac{180}{\pi} \times 6 \times 10^{-7}$$

$$= 3.44 \times 10^{-5} \text{ m}$$

$$\approx 0.03 \text{ mm}$$

48. Given, wavelength of light, $\lambda = 600 \text{ nm} = 600 \times 10^{-9}$ m

Angular width of fringe, $\theta = 0.1^\circ = \frac{0.1\,\pi}{180}$ rad

Using the formula, $\theta = \frac{\lambda}{d}$

Spacing between the slits $d = \frac{\lambda}{\theta} = \frac{600 \times 10^{-9} \times 180}{0.1 \times \pi}$

$$d = 3.44 \times 10^{-4} \text{ m}$$

Thus, the spacing between the two slits is 3.44×10^{-4} m.

49. Path difference, $\Delta x = S_1S_3 - S_2S_3 = 0$

$$\therefore \quad \phi = \frac{2\pi}{\lambda}\Delta x = 0$$

$$\therefore \quad I_3 = I_0 + I_0 + 2\sqrt{I_0 \cdot I_0}(\cos 0^\circ)$$

$$\therefore \quad I_3 = 4I_0$$

The path difference at S_4 is

$$\Delta x' = S_1S_4 - S_2S_4 = \frac{xd}{D} \quad \left(\text{Here, } x = \frac{\lambda D}{2d}\right)$$

$$= \frac{d}{D} \times \frac{\lambda D}{2d} = \frac{\lambda}{2}$$

$$\therefore \quad \phi' = \frac{2\pi}{\lambda} \cdot \frac{\lambda}{2} = \pi$$

$$\therefore \quad I_4 = I_0 + I_0 + 2I_0 \cos\pi = 0$$

$$\therefore \quad \text{Ratio} = \frac{I_3}{I_4} = \frac{4I_0}{0} = \infty$$

50. Consider that the slit widths are equal, so they produce waves of equal intensity say, I'. Resultant intensity at any point $I_R = 4I'\cos^2\phi$, where ϕ is the phase difference between the wave at the point observation.

For maximum intensity,

$$\phi = 0^\circ \Rightarrow I_{\max} = 4I' = I \quad \text{...(i)}$$

If one of the slits is closed, resultant intensity at the same point will be I' only, *i.e.*

$$I' = I_0 \quad \text{...(ii)}$$

Comparing Eqs. (i) and (ii), we get

$$I = 4I_0$$

51. In YDSE, intensity at a point is given by

$$I_{\text{net}} = 2I_0(1 + \cos\phi) \quad \text{.... (i)}$$

where, ϕ = phase difference between interfering waves.

Here, $\phi = \frac{1}{8}\lambda = \frac{1}{8} \times 2\pi$

(a path difference of one wavelength is equivalent to a phase difference of 2π radians)

$$\Rightarrow \quad \phi = \frac{\pi}{4} \Rightarrow \cos\phi = \frac{1}{\sqrt{2}} \approx 0.7$$

So, I_{net} at given point P is

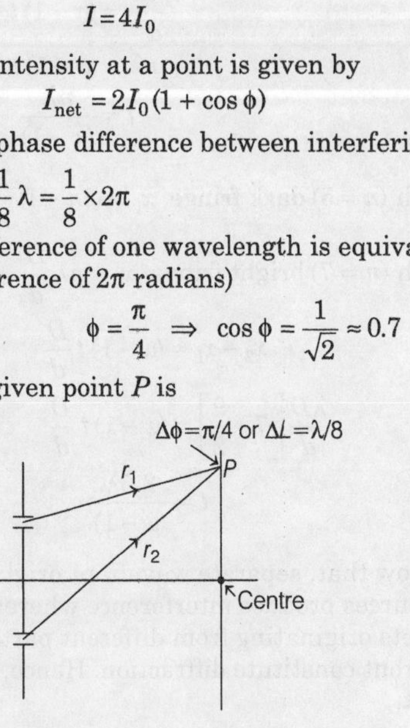

Also, intensity maxima at centre of a bright fringe is obtained by substituting $\phi = 0, 2\pi \dots n\pi$ in Eq. (i).

So, $I_{net} = I_2 = 2I_0(1 + \cos 0°) = 4I_0$

Hence, required ratio is

$$\frac{I_1}{I_2} = \frac{2 \times 1.7 I_0}{4I_0} = \frac{1.7}{2} = 0.853$$

52. Given, angular width, $\theta = 0.2°$

Distance between screen and slit, $D = 1$ m

Wavelength of light, $\lambda = 600$ nm $= 600 \times 10^{-9}$ m

Refractive index of water, $\mu_w = \frac{4}{3}$

Using the formula of angular width,

$$\theta = \frac{\lambda}{D} \quad \dots(i)$$

and $$\theta' = \frac{\lambda'}{D} \quad \left(\text{where, } \lambda' = \frac{\lambda}{\mu}\right) \quad \dots(ii)$$

Dividing Eq.(ii) from Eq.(i), we get

$$\frac{\theta'}{\theta} = \frac{\lambda'}{\lambda} = \frac{\lambda}{\mu\lambda}$$

or $$\theta' = \frac{\theta}{\mu} = \frac{0.2 \times 3}{4} = 0.15° \quad \left(\because \mu = \frac{4}{3}\right)$$

Thus, the angular fringe width is 0.15° as the apparatus is immersed in water.

53. Path difference introduced by a slab of thickness t and refractive index μ is given by

$$\Delta x = (\mu - 1)t$$

Position of the fringe is $x = \frac{\Delta x D}{d} = \frac{(\mu - 1)tD}{d}$

Also, fringe width is given by

$$\beta = \frac{\lambda D}{d}$$

According to the question, $n\beta = x$

$$\Rightarrow \quad \frac{n\lambda D}{d} = (\mu - 1)t\frac{D}{d}$$

$$\Rightarrow \quad n\lambda = (\mu - 1)t$$

or $$t = \frac{n\lambda}{(\mu - 1)}$$

54. For 5th ($n = 5$) dark fringe, $x_1 = (2n - 1)\frac{\lambda}{2}\frac{D}{d} = \frac{9\lambda D}{2d}$

For 7th ($n = 7$) bright fringe, $x_2 = n\lambda\frac{D}{d} = \frac{7\lambda D}{d}$

Now, $$x_2 - x_1 = (\mu - 1)t\frac{D}{d}$$

$$\Rightarrow \quad \frac{\lambda D}{d}\left[7 - \frac{9}{2}\right] = (\mu - 1)t\frac{D}{d}$$

or, $$t = \frac{2.5\lambda}{(\mu - 1)}$$

55. We know that, separate wavefront originating from two sources produce interference whereas secondary wavelets originating from different parts of same wavefront constitute diffraction. Hence, option (d) is correct.

56. At the centre, all colours meet in phase, hence central fringe is white.

57. Here, the width of the slit is 10^4 Å, *i.e.* 10000 Å. The wavelength of (visible) sunlight varies from 4000 Å to 8000 Å. As width of slit $a > \lambda$ (wavelength of light), therefore no diffraction occurs. The image seen through the slit shall be a fine sharp slit white in colour at the centre.

58. Given, aperture, $a = 4$ mm $= 4 \times 10^{-3}$ m

Wavelength, $\lambda = 400$ nm $= 400 \times 10^{-9}$ m

Ray optics is good approximation upto a distance equal to Fresnel's distance (Z_F).

$$Z_F = \frac{a^2}{\lambda} = \frac{4 \times 10^{-3} \times 4 \times 10^{-3}}{400 \times 10^{-9}}$$

$$Z_F = 40 \text{ m}$$

59. From diffraction at a single slit,

size of aperture, $a = \frac{\lambda}{\sin\theta}$

or $$a = \frac{3141.59 \times 10^{-10}}{\sin 1°}$$

$$= 18 \times 10^{-6} \text{ m} = 18 \mu\text{m}$$

60. As, $a \sin\theta = n\lambda$

$$\Rightarrow \quad \lambda = \frac{a\sin\theta}{n} = \frac{5\sin 30°}{1} = 2.5 \text{ cm}$$

61. Here, $\lambda = 6250$ Å $= 6250 \times 10^{-10}$ m

$a = 2 \times 10^{-2}$ cm $= 2 \times 10^{-4}$ m

$D = 50$ cm $= 0.5$ m

$\therefore$ Width of central maxima $= \frac{2\lambda D}{a}$

$$= \frac{2 \times 6250 \times 10^{-10} \times 0.5}{2 \times 10^{-4}}$$

$$= 312.5 \times 10^{-3} \text{ cm}$$

62. Distance between first and sixth minima,

$$x = \frac{5\lambda D}{d}$$

For 1st maxima,

$$\therefore \quad d = \frac{n\lambda D}{x} = \frac{1 \times 5000 \times 10^{-10} \times 0.5}{0.5 \times 10^{-3}}$$

$$d = \frac{5000 \times 10^{-10} \times 0.5}{0.5 \times 10^{-3}}$$

$$\Rightarrow \quad d = 5 \times 10^{-4} = 0.5 \times 10^{-3} = 0.5 \text{ mm}$$

63. As, $a \sin\theta = n\lambda$

$$\Rightarrow \quad \frac{ax}{D} = 3\lambda \quad \text{(Hence, focal length is used as } D\text{)}$$

or $$\lambda = \frac{ax}{3D}$$

$$= \frac{0.3 \times 10^{-3} \times 5 \times 10^{-3}}{3 \times 1}$$

$$= 5 \times 10^{-7} \text{ m} = 5000 \text{ Å}$$

64. The linear width of central principal maximum $= \frac{2\lambda D}{d}$

If it is equal to width of slit (d), then

$$\frac{2\lambda D}{d} = d \text{ or } D = \frac{d^2}{2\lambda}$$

65. The intensity of maxima in the diffraction pattern due to single slit is given by

$$I = \frac{I_0}{\left[\left(\frac{2n+1}{2}\right)\pi\right]^2} \quad \text{...(i)}$$

$\therefore$ For I_1, put $n = 1$ in Eq. (i), we have

$$I_1 = \frac{I_0}{\left[\left(\frac{2\times1+1}{2}\right)\pi\right]^2} = \frac{I_0}{\frac{9\pi^2}{4}} = \frac{4I_0}{9\pi^2}$$

Similarly, for I_2, put $n = 2$ in Eq. (ii), we have

$$I_2 = \frac{I_0}{\left[\left(\frac{2\times2+1}{2}\right)\pi\right]^2} = \frac{I_0}{\frac{25\pi^2}{4}} = \frac{4}{25\pi^2} I_0$$

$$\therefore\ I_0 : I_1 : I_2 = I_0 : \frac{4I_0}{9\pi^2} : \frac{4}{25\pi^2} I_0 = 1 : \frac{4}{9\pi^2} : \frac{4}{25\pi^2}$$

66. In case of a single slit diffraction pattern as shown below

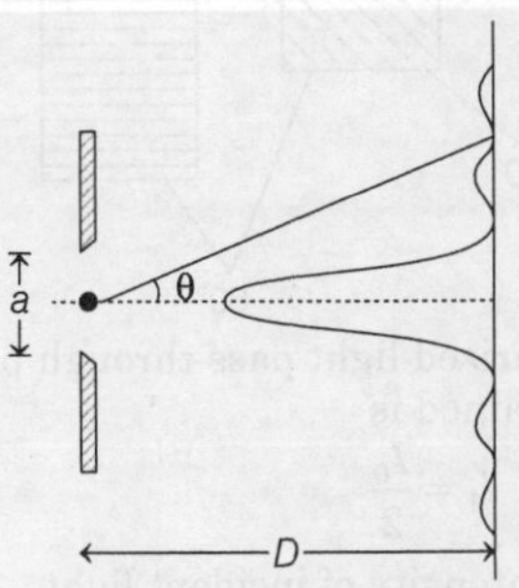

angular position of nth minima is given by

$$\sin\theta_n = \frac{n\lambda}{D} \quad \text{...(i)}$$

Given, $\theta_2 = 60°$ and $\lambda = 6000 \times 10^{-8}$ cm

As, $\sin\theta_2 = \frac{2\lambda}{D}$

We have, $\sin 60° = \frac{2\lambda}{D}$

or $\frac{\lambda}{D} = \frac{\sin 60°}{2} = 0.43 \approx \sin 25°$

[**Note** $\sin 30° = 0.5$, $\sin 25° = 0.422$ and $\sin 20° = 0.34$]

Now, if θ_1 = angular position of first minima, then

$\sin\theta_1 = \frac{\lambda}{D}$ [from Eq. (i), $n = 1$]

$\Rightarrow$ $\sin\theta_1 = 0.43 \approx \sin 25°$ or $\theta_1 \approx 25°$

67. Limit of resolution for a telescope from Rayleigh's criteria is

$$\theta_R = \frac{1.22\,\lambda}{D}$$

Here, $D = 250 \text{ cm} = 250 \times 10^{-2}$ m

and $\lambda = 600 \text{ nm} = 600 \times 10^{-9}$m

So, limit of resolution is

$$\theta_R = \frac{1.22 \times 600 \times 10^{-9}}{250 \times 10^{-2}} = 2.93 \times 10^{-7} \text{ rad}$$

$$\approx 3\times 10^{-7}\text{rad}$$

68. For a telescope, limit of resolution is given by

$$\Delta\theta = \frac{1.22\lambda}{D}$$

Here, $\lambda = 500 \text{ nm} = 500 \times 10^{-9}$ m,

$D = 200 \text{ cm} = 200 \times 10^{-2}$ m

So, limit of resolution is $\Delta\theta = \frac{1.22 \times 500 \times 10^{-9}}{200 \times 10^{-2}}$

$$= 305 \times 10^{-9} \text{ rad}$$

69. Given, $\lambda = 5000\text{Å} = 5000\times10^{-10}$ m and

NA = 1.25

Now, $d = \frac{0.61\,\lambda}{\text{NA}} = \frac{0.61\times5000\times10^{-10}}{1.25}$

or $d = \frac{3.05}{1.25}\times10^{-7}$ m

$= 2.4\times10^{-7}$ m

or $d = 0.24\,\mu\text{m}$

70. We can write resolving angle of necked eye as

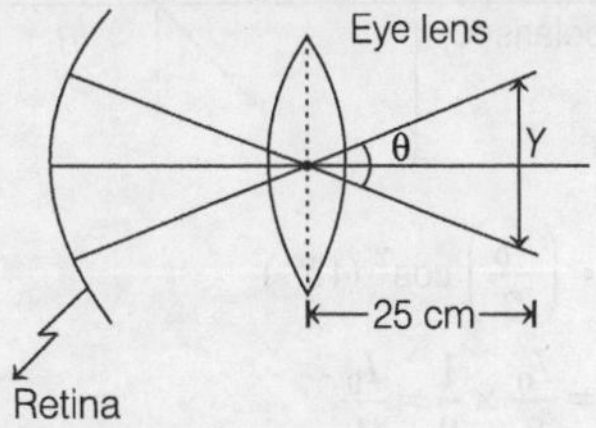

$$\theta = 1.22\frac{\lambda}{D}$$

where, D is the diameter of eye lens.

or $\frac{Y}{25\times10^{-2}} = \frac{1.22\times500\times10^{-9}}{0.25\times2\times10^{-2}}$

$$Y = 30\times10^{-6} \text{ m} = 30\,\mu\text{m}$$

71. According to Brewster's law, the light reflected from the top of glass slab gets polarised. The light refracted into the glass slab and the light emerging from the glass slab is only partially polarised. Therefore, when a polaroid is held in the path of emergent light at P, and rotated about an axis passing through the centre and perpendicular to plane of polaroid, the intensity of light shall go through a minimum but not zero for two orientations of the polaroid.

72. When a beam of unpolarised light is reflected from a transparent medium of refractive index μ, then the reflected light is completely plane polarised at a certain angle of incidence i_B, which is known as Brewster's angle.

In the given condition, the light reflected irrespective of an angle of incidence is never completely polarised.
So, $i_C > i_B$
where, i_C is the critical angle.

$$\Rightarrow \quad \sin i_C < \sin i_B \quad \text{...(i)}$$

From Brewster's law, we know that

$$\tan i_B = {}^w\mu_g = \frac{\mu_{\text{glass}}}{\mu_{\text{water}}} = \frac{1.5}{\mu} \quad \text{...(ii)}$$

From Eqs. (i) and (ii), we get

$$\frac{1}{\mu} < \frac{1.5}{\sqrt{(1.5)^2 + (\mu)^2}}$$

$$\Rightarrow \quad \sqrt{(1.5)^2 + \mu^2} < 1.5\,\mu$$

$$\mu^2 + (1.5)^2 < (1.5\,\mu)^2 \text{ or } \mu < \frac{3}{\sqrt{5}}$$

$\therefore$ The minimum value of μ should be $\frac{3}{\sqrt{5}}$.

73. We know, $\lambda_m = \frac{\lambda_a}{\mu}$

and $\mu = \tan\theta$

$$\therefore \quad \lambda_m = \frac{\lambda_a}{\tan\theta} = \lambda_a \cot\theta$$

74. Relation between intensities is

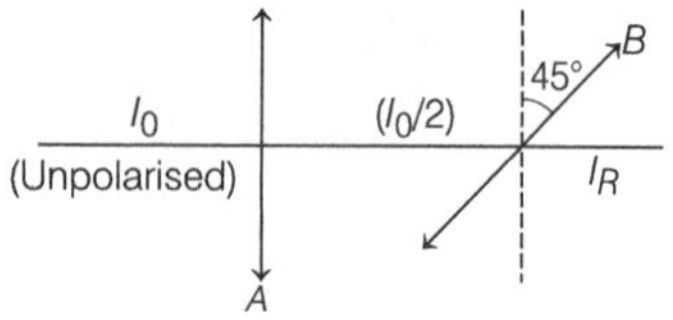

$$I_R = \left(\frac{I_0}{2}\right)\cos^2(45^\circ)$$

$$= \frac{I_0}{2} \times \frac{1}{2} = \frac{I_0}{4}$$

75. The intensity of plane polarised light $= 2a^2$.

$\therefore$ Intensity of polarised light from thin polaroid

$$= \frac{I_0}{2} = \frac{1}{2} \times 2a^2 = a^2$$

76. Intensity of light from $C_2 = I_0$

On rotating the crystal through 60°,

$$I = I_0 \cos^2 60^\circ \quad \text{(Malus' law)}$$

$$= I_0\left(\frac{1}{2}\right)^2 = I_0/4$$

77. If I_0 is intensity of unpolarized light,

then from 1st Nicol prism, $I_1 = \frac{I_0}{2}$

From 2nd Nicol prism, $I_2 = I_1 \cos^2(90^\circ - 60^\circ)$

$$= \frac{I_0}{2}\left(\frac{\sqrt{3}}{2}\right)^2 = \frac{3}{8} I_0$$

$$\Rightarrow \quad \frac{I_2}{I_0} = 0.375 = 37.5\,\%$$

78.

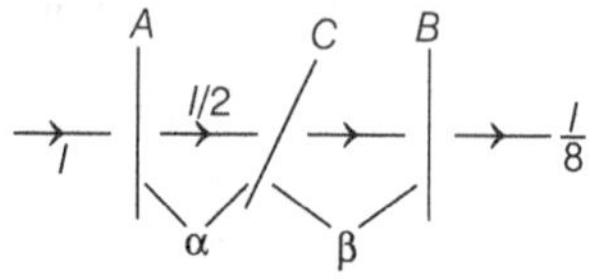

Using Malus's law, intensity available after polariser C

$$I' = \frac{I}{2} \times \cos^2\alpha$$

and intensity available after polariser

$$B = I'\cos^2\beta = \frac{I}{2}\cos^2\alpha \times \cos^2\beta = \frac{I}{8} \quad \text{(given)}$$

$$\text{So, } \frac{I}{2} \times \cos^2\alpha \cdot \cos^2\beta = \frac{I}{8}$$

$$\Rightarrow \quad \cos^2\alpha \cdot \cos^2\beta = \frac{1}{4}$$

This is satisfied with $\alpha = 45^\circ$ and $\beta = 45^\circ$

So, angle between A and C is 45°.

79.

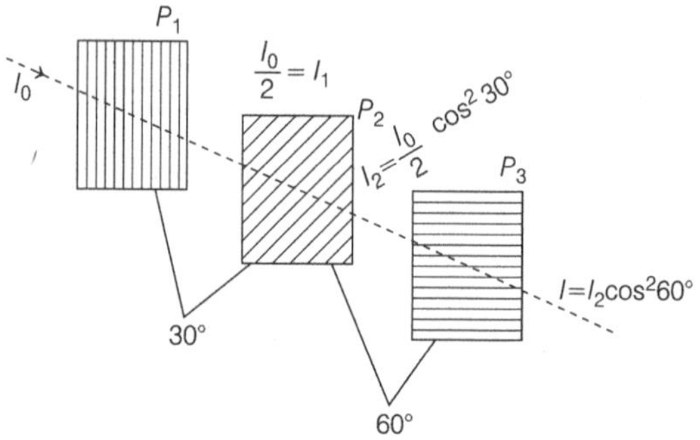

When unpolarised light pass through polaroid P_1, intensity obtained is

$$I_1 = \frac{I_0}{2}$$

where, I_0 = intensity of incident light.

Now, this transmitted light is polarised and it passes through polariser P_2. So, intensity I_2 transmitted is obtained by Malus' law.

$$\Rightarrow \quad I_2 = I_1 \cos^2\theta$$

As angle of pass axis of P_1 and P_3 is 90° and angle of pass axis of P_2 and P_3 is 60°, so angle between pass axis of P_1 and P_2 is $(90^\circ - 60^\circ) = 30^\circ$.

$$\text{So,} \quad I_2 = \frac{I_0}{2}\cos^2 30^\circ$$

$$= \frac{I_0}{2} \times \left(\frac{\sqrt{3}}{2}\right)^2 = \frac{3}{8} I_0$$

When this light pass through third polariser P_3, intensity I transmitted is again obtained by Malus' law.

$$\text{So,} \quad I = I_2 \cos^2 60^\circ = \left(\frac{3}{8} I_0\right)\cos^2 60^\circ$$

$$= \frac{3}{8} I_0 \times \left(\frac{1}{2}\right)^2 = \frac{3}{32} I_0$$

So, ratio $\frac{I_0}{I} = \frac{32}{3} = 10.67$.

80. Given, wavelength of H_α, $\lambda = 6563\ \text{Å} = 6563 \times 10^{-10}$ m

Red- shift $\Delta\lambda = 15\ \text{Å} = 15 \times 10^{-10}$m

Since, the star is found to be red-shifted, hence star is receding away from earth and doppler's shift is negative.

$$\Delta\lambda = -v\frac{\lambda}{c}$$

$$v = -\frac{\Delta\lambda \cdot c}{\lambda} = -\frac{15 \times 10^{-10} \times 3 \times 10^{8}}{6563 \times 10^{-10}}$$

$$v = -6.86 \times 10^{5} \text{ m/s}$$

Negative sign shows that the star is receding (away) from earth.

Round II

1. The refractive index of air is slightly more than 1. When chamber is evacuated, refractive index decreases and hence, the wavelength increases and fringe width also increases.

2. Given, $\lambda = 600 \text{ nm} = 6 \times 10^{-7}$ m,

$$a = 1 \text{ mm} = 10^{-3} \text{ m}, D = 2 \text{m}$$

Distance between the first dark fringes on either side of central bright fringe = width of central maximum

$$= \frac{2\lambda D}{a} = \frac{2 \times 6 \times 10^{-7} \times 2}{10^{-3}}$$

$$= 24 \times 10^{-4} \text{ m} = 2.4 \text{ mm}$$

3. From, $(\mu - 1)\, t = n\lambda$, we get,

$$t = \frac{n\lambda}{(\mu - 1)}$$

When $n\lambda = \frac{\lambda}{2}$,

then $t = \frac{\lambda}{2\,(\mu - 1)}$

4. Red light and blue light have different wavelengths and different frequencies.

5. As size of hole (a) is increased, width of central maximum of diffraction pattern of hole $\left(= \frac{2f\lambda}{a}\right)$ decreases. As the same amount of light emerge is now distributed over a smaller area, the intensity increases.

6. Here, $d = 1 \text{ mm} = 10^{-3}$ m,

$$\lambda = 6.5 \times 10^{-7} \text{ m}$$

$$D = 1 \text{ m}$$

Now, $x_5 = n\lambda\frac{D}{d} = 5 \times 6.5 \times 10^{-7} \times \frac{1}{10^{-3}}$

$$= 32.5 \times 10^{-4} \text{ m}$$

and $x_3 = (2n - 1)\frac{\lambda}{2}\frac{D}{d}$

$$= \frac{(2 \times 3 - 1) \times 6.5 \times 10^{-7}}{2 \times 10^{-3}}$$

$$= 16.25 \times 10^{-4} \text{ m}$$

$\therefore$ $x_5 - x_3 = (32.5 - 16.25)10^{-4}$ m $= 16.25 \times 10^{-4}$ m $= 1.63$ mm

7. Here, wavelength of light (λ) = 5303 Å.

Distance between two slit (d) = 19.44 μm

Width of single slit (a) = 4.05 μm

Here, angular width between first and second diffraction minima, $\theta \simeq \frac{\lambda}{a}$

and angular width of a fringe due to double slit is

$$\theta' = \frac{\lambda}{d}$$

$\therefore$ Number of fringes between first and second diffraction minima, $n = \frac{\theta}{\theta'} = \frac{\frac{\lambda}{a}}{\frac{\lambda}{d}}$

$$= \frac{d}{a} = \frac{19.44}{4.05} = 4.81$$

or $n \simeq 5$

$\therefore$ 5 interfering bright fringes lie between first and second diffraction minima.

8. Here, $\omega = 31.4$ rads^{-1}

$\therefore$ Time period of revolution,

$$T = \frac{2\pi}{\omega} = \frac{2 \times 3.14}{31.4} = 0.2 \text{ s}$$

Energy transmitted/revolution,

$$= (IA)T = \left(\frac{I_0}{2}A\right)T$$

$$= \frac{\phi_0 T}{2} = \frac{10^{-3} \times 0.2}{2} = 10^{-4} \text{ J}$$

9. Distance, $S_2P = (d^2 + b^2)^{1/2} = d\left(1 + \frac{b^2}{d^2}\right)^{1/2}$

$$= d\left(1 + \frac{b^2}{2d^2}\right)$$

$$= d + \frac{b^2}{2d}$$

and $S_1P = d$

$\therefore$ Path difference $= S_2P - S_1P$

$$x = d + \frac{b^2}{2d} - d = \frac{b^2}{2d}$$

For missing wavelengths, $(2n - 1)\frac{\lambda}{2} = x = \frac{b^2}{2d}$

For $n = 1$, $\lambda = \frac{b^2}{d}$

For $n = 2$, $\lambda = \frac{2b^2}{3d}$

10. From the pattern of intensity distribution seen in the central portion, we find that

(a) as intensity of successive maxima is the same, S_1 and S_2 have the same intensity.

(b) as width of successive maxima appears to increase slightly, S_1, S_2 must have a constant phase difference.
(d) as minimum intensity is zero, S_1, S_2 must have the same wavelength.

11. The resultant intensity at any point P is

$$I = 4I_0 \cos^2\left(\frac{\phi}{2}\right)$$

$\therefore \quad I_0 = 4I_0 \cos^2 \phi/2$

or $\quad \cos\frac{\phi}{2} = \frac{1}{2}$

$\therefore \quad \frac{\phi}{2} = \frac{\pi}{3}$ or $\phi = \frac{2\pi}{3}$

If Δx is the corresponding value of path difference at P, then

$$\phi = \frac{2\pi}{\lambda}(\Delta x)$$

$\Rightarrow \quad \frac{2\pi}{3} = \frac{2\pi}{\lambda}\Delta x$

As, $\quad \Delta x = \frac{xd}{D}$

$\therefore \quad \frac{1}{3} = \frac{1}{\lambda}\frac{xd}{D}$

or $\quad x = \frac{\lambda}{3d/D} = \frac{6\times10^{-7}}{3\times10^{-4}} = 2\times10^{-3}$ m

$x = 2$ mm

This is the distance of point P from central maximum.

12. Points c and d are on same wavefront.

i.e. $\quad \phi_c = \phi_d$

Also, $\quad \phi_e = \phi_f$

Thus, $\quad \phi_d - \phi_f = \phi_c - \phi_e$

13. Parallel cylindrical beam gives planar wavefront.

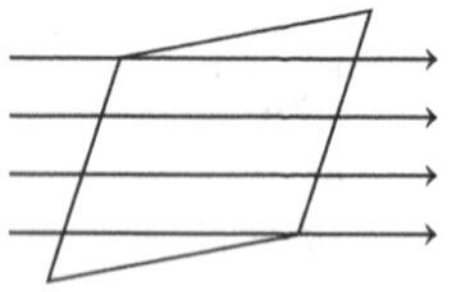

14. Here, $a_1 = 2a_2 \Rightarrow I_1 = 4I_2 = 4I_0$

As, $\quad I_m = (\sqrt{I_1} + \sqrt{I_2})^2 = (3\sqrt{I_2})^2 = 9I_2 = 9I_0$

$\therefore \quad I_0 = \frac{I_m}{9} \quad$ ...(i)

Now, $\quad I = I_1 + I_2 + 2\sqrt{I_1 I_2}\cos\phi$

$= 4I_0 + I_0 + 2\sqrt{4I_0 I_0}\cos\phi$

$= 5I_0 + 4I_0\cos\phi$

$= \frac{I_m}{9}(5 + 4\cos\phi) \quad$ [from Eq. (i)]

$= \frac{I_m}{9}[1 + 4(1 + \cos\phi)]$

$= \frac{I_m}{9}\left(1 + 8\cos^2\frac{\phi}{2}\right)$

15. Intensity falling on polariser, $I_0 = 3.3$ W/m^2

Area of polariser, $A = 3\times10^{-4}$ m^2

Angular speed of rotation of polariser,

$$\omega = 31.4 \text{ rad/s}$$

Now, average energy is given by

$< E > = < IAt >$

$< E > = < I_0 \cos^2\theta \times A \times t >$

$< E > = I_0 At < \cos^2\theta >$

$< E > = I_0 At\left(\frac{1}{2}\right) \quad \left[\because < \cos^2\theta > = \frac{1}{2}\right]$

Now, the calculation is about one revolution, then

Time taken = Time period

$$t = T = \frac{2\pi}{\omega} = \frac{2\pi}{31.4} = \frac{2\pi}{10\pi} = \frac{1}{5} \text{ s}$$

Now, $\quad < E > = I_0 At\left(\frac{1}{2}\right)$

Substituting all the values in it, we get

$< E > = 3.3\times3\times10^{-4}\times\frac{1}{5}\times\left(\frac{1}{2}\right)$

$= 0.99\times10^{-4}$ J

$\simeq 1\times10^{-4}$ J

16. For microwave, $\lambda = \frac{c}{v} = \frac{3\times10^8}{10^6} = 300$ m

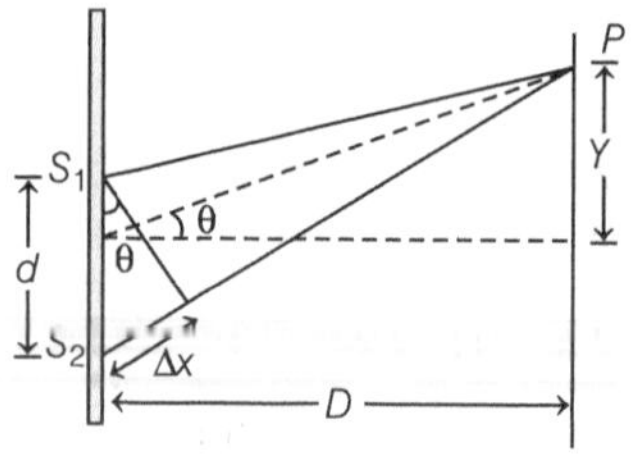

As, $\quad \Delta x = d\sin\theta$

Phase difference, $\phi = \frac{2\pi}{\lambda}$ (Path difference)

$= \frac{2\pi}{\lambda}(d\sin\theta)$

$= \frac{2\pi}{300}(150\sin\theta) = \pi\sin\theta$

$\therefore \quad I_R = I_1 + I_2 + 2\sqrt{I_1 I_2}\cos\phi$

Here, $I_1 = I_2$ and $\phi = \pi\sin\theta$

$\therefore \quad I_R = 2I_1[1 + \cos(\pi\sin\theta)] = 4I_1\cos^2\left(\frac{\pi\sin\theta}{2}\right)$

I_R will be maximum, when $\cos^2\left(\frac{\pi\sin\theta}{2}\right) = 1$

$\therefore \quad (I_R)_{max} = 4I_1 = I_0$

Hence, $\quad I = I_0\cos^2\left(\frac{\pi\sin\theta}{2}\right)$

If $\theta = 0$, then $I = I_0\cos^2 0° = I_0$

If $\theta = 30°$, then $I = I_0\cos^2(\pi/4) = I_0/2$

If $\theta = 90°$, then $I = I_0\cos^2(\pi/2) = 0$

17. **For point A**

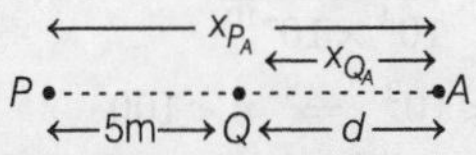

Path difference, $\Delta x = x_{P_A} - x_{Q_A} = 5$ m

Phase difference due to path difference,

$$\Delta\phi = \frac{2\pi}{\lambda} \times \Delta x = \frac{2\pi}{20} \times 5$$

$\Rightarrow$ $$\Delta\phi = \frac{\pi}{2}$$

Also, phase of P is ahead of Q by 90° or $\frac{\pi}{2}$.

So, total phase difference at point A,

$$\Delta\phi = \frac{\pi}{2} - \frac{\pi}{2} = 0°$$

$\therefore$ Intensity at point A,

$$I_A = I_1 + I_2 + 2\sqrt{I_1}\sqrt{I_2}\cos\Delta\phi$$
$$= I_0 + I_0 + 2I_0 \cos 0°$$
$$= 4I_0$$

For point B

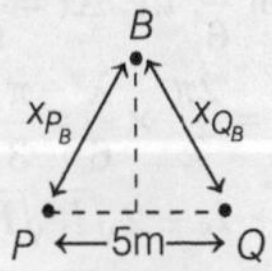

$$\Delta x = x_{P_B} - x_{Q_B} = 0$$

$\therefore$ $\Delta\phi = 0$ (due to path difference)

So, total phase difference at point B,

$$\Delta\phi = \frac{\pi}{2} + 0 = 90°$$

$\therefore$ Intensity at point B,

$$I_B = I_1 + I_2 + 2\sqrt{I_1}\sqrt{I_2}\cos\Delta\phi$$
$$= I_0 + I_0 + 2I_0 \cos 90° = 2I_0$$

For point C

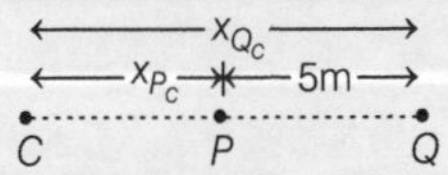

$$\Delta x = x_{Q_C} - x_{P_C} = 5 \text{ m}$$

$\Delta\phi$ due to path difference,

$$\Delta\phi = \pi/2$$

Total phase difference at point C,

$$\Delta\phi = \frac{\pi}{2} + \frac{\pi}{2} = \pi = 180°$$

($\because$ phase of P is ahead of Q by 90°)

$\therefore$ Intensity at point C,

$$I_C = I_1 + I_2 + 2\sqrt{I_1}\sqrt{I_2}\cos\Delta\phi$$
$$= I_0 + I_0 + 2I_0 \cos 180° = 0$$

Ratio of intensities,

$$I_A : I_B : I_C = 4I_0 : 2I_0 : 0$$
$$I_A : I_B : I_C = 2 : 1 : 0$$

18.

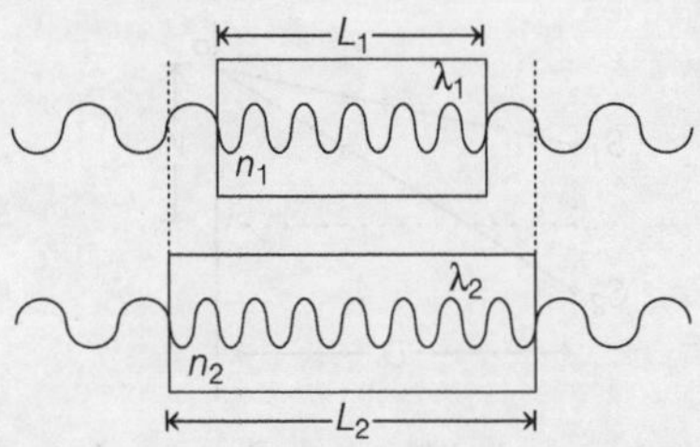

Let λ_1 be the wavelength inside medium of refractive index n_1 and it travels a length L_1. Let λ_2, n_2 and L_2 be the parameters for second wave.

So, $$\frac{\lambda_1}{v} = \frac{\lambda}{c}$$

$\Rightarrow$ $$\lambda_1 = \frac{\lambda}{n_1} \text{ and } \lambda_2 = \frac{\lambda}{n_2}$$

Let there are N_1 number of wavelengths in medium 1 of length L_1, then

$$N_1 = \frac{L_1}{\lambda_1} = \frac{L_1 n_1}{\lambda} \text{ and } N_2 = \frac{L_2}{\lambda_2} = \frac{L_2 n_2}{\lambda}$$

So, difference in number of waves, ΔN

$$= N_1 - N_2 = \frac{L_1 n_1}{\lambda} - \frac{L_2 n_2}{\lambda} = \frac{L_1 n_1 - L_2 n_2}{\lambda}$$

As each wavelength is equivalent to a phase of 2π, we can write

$$\Delta Q = 2\pi \Delta N = 2\pi(N_1 - N_2)$$
$$= 2\frac{\pi}{\lambda}(L_1 n_1 - L_2 n_2)$$

19. We know in interference,

$$I_{\max} = (\sqrt{I_1} + \sqrt{I_2})^2 \text{ while } I_{\min} = (\sqrt{I_1} - \sqrt{I_2})^2$$

$$\frac{I_{\max}}{I_{\min}} = \frac{(\sqrt{I_1} + \sqrt{I_2})^2}{(\sqrt{I_1} - \sqrt{I_2})^2} = \frac{(A_1 + A_2)^2}{(A_1 - A_2)^2}$$

According to problem,

$$\frac{A_1}{A_2} = \frac{3}{5}, \frac{I_{\max}}{I_{\min}} = \frac{(3+5)^2}{(3-5)^2} = \frac{16}{1}$$

$$\left(\because \frac{I_{\max}}{I_{\min}} = \frac{x}{1}, \text{ given}\right)$$

$\Rightarrow$ $$x = 16$$

20. For dark dringe at P, $S_1P - S_2P = \frac{(2n-1)\lambda}{2}$

$$S_1P - S_2P = \frac{5\lambda}{2}$$

$$= 5 \times \frac{6000}{2} = 15000 = 1.5 \text{ micron}$$

21. Given, in a YDSE,

Fringe width of 15 fringes for wavelength of light of 500 nm

Fringe width of 10 fringes for wavelength of light of λ nm

$\Rightarrow$ $$\frac{15 \times 500 \times D}{d} = \frac{10 \times \lambda \times D}{d}$$

$\Rightarrow$ $$\lambda = \frac{15 \times 500}{10} = 750 \text{ nm}$$

22.

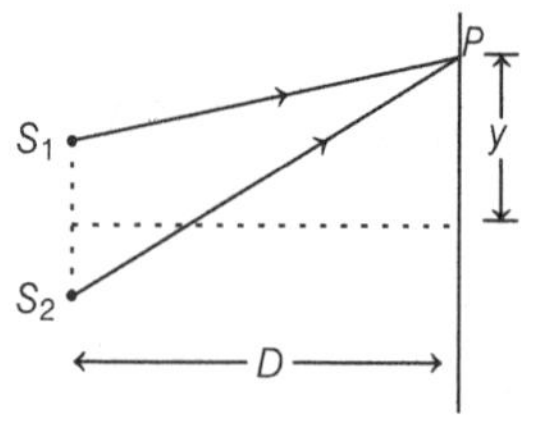

The maximum path difference that can be produced = distance between the sources, *i.e.* 5 cm.

Hence, in this condition, we can have only 3 maximas, *i.e.* one central maxima and two on its either sides for a path difference of λ or 3 cm.

Now, for maximum intensity at P

$$S_2P - S_1P = \lambda$$

or $\sqrt{(y + d/2)^2 + D^2} - \sqrt{(y - d/2)^2 + D^2} = \lambda$

Putting, $d = 5$ cm, $D = 100$ cm

and $\lambda = 3$ cm, we get

$$y = \pm 75 \text{ cm}$$

Thus, the three maximas will be at

$$y = 0 \text{ and } y = \pm 75 \text{ cm}$$

23. In single slit diffraction, if width of slit is a, wavelength of light is λ and angular position on screen is θ, then for nth minima, the condition is

$$a\sin(\theta) = n\lambda$$

$\Rightarrow$ $\sin(\theta) = \dfrac{n\lambda}{a}$ ...(i)

$\because$ $\sin(\theta) < 1$

$\Rightarrow$ $\dfrac{n\lambda}{a} < 1$

$\Rightarrow$ $n < \dfrac{a}{\lambda}$

$\Rightarrow$ $n < \dfrac{0.6 \times 10^{-4}}{6000 \times 10^{-10}}$

$\Rightarrow$ $n < \dfrac{10^{-4}}{10^4 \times 10^{-10}}$

$\Rightarrow$ $n < 10^2 \Rightarrow n < 100$

So, maximum possible number of minima on one side of central maximum = 99.

Therefore, maximum possible number of minima on both sides of central maximum

$$= 2 \times 99 = 198$$

24. For path difference of λ, *i.e.* $\Delta x = \lambda$

Phase difference,

$$\Delta\phi = \frac{2\pi}{\lambda} \times \Delta x = \frac{2\pi}{\lambda} \times \lambda = 2\pi$$

Intensity, $I = I_1 + I_2 + 2\sqrt{I_1}\sqrt{I_2}\cos\Delta\phi$

As, light is monochromatic, so

$$I_1 = I_2 = I_0 \text{ (say)}$$

and $I = K$ (given)

$\Rightarrow$ $K = I_0 + I_0 + 2\sqrt{I_0}\sqrt{I_0}\cos 2\pi$

$$K = 4I_0 \text{ or } I_0 = \frac{K}{4}$$

For path difference of $\dfrac{\lambda}{6}$, *i.e.* $\Delta x = \dfrac{\lambda}{6}$.

Phase difference, $\Delta\phi = \dfrac{2\pi}{\lambda} \times \dfrac{\lambda}{6} = \dfrac{\pi}{3}$

So, intensity, $I' = I_1 + I_2 + 2\sqrt{I_1}\sqrt{I_2}\cos\Delta\phi$

$$= I_0 + I_0 + 2\sqrt{I_0}\sqrt{I_0}\cos\frac{\pi}{3}$$

$$= \frac{K}{4} + \frac{K}{4} + 2 \times \frac{K}{4} \times \frac{1}{2}$$

$$= \frac{3K}{4} = \frac{9K}{12}$$

Given, $I' = \dfrac{nK}{12}$

On comparing the both, we get

$$n = 9$$

27

Dual Nature of Radiation and Matter

Electrons are regarded as particles because they possess charge, mass and behave according to the laws of particle mechanics. However, it can be considered that a moving electron can be assumed as a wave as it is interpreted as a particle. We considered electromagnetic waves as waves because under suitable circumstances, they exhibit diffraction, interference and polarisation. Similarly, under other circumstances, they behave as a streams of particles. Rather, we can say they have the *dual nature*.

IN THIS CHAPTER ....

- Emission of Electrons
- Photoelectric Emission
- Photoelectric Effect and Wave Theory of Light
- Particle Nature of Light : Photon
- Wave Nature of Particle
- Davisson and Germer Experiment

Emission of Electrons

At room temperature, the free electrons move randomly within the conductor, but they cannot leave the surface of the conductor due to attraction of positive charges. Some external energy is required to emit these electrons from a metal surface.

Thus, a sufficient minimum energy which is required to emit these electrons from the surface of the conductor is called the *work function* (denoted by W or ϕ) of the conductor. It is the property of the metallic surface.

Photoelectric Emission

When light of suitable frequency illuminates a metal surface, electrons are emitted from the metal surface. This is called photoelectric emission. These photo (light) generated electrons are *photoelectrons* and the current, so produced is called *photoelectric current*.

Note ϕ is measured in eV (electron-volt), $1\text{ eV} = 1.602 \times 10^{-19}$ J.

Experimental Study of Photoelectric Effect

On the basis of the experimental arrangement used for studying the photoelectric effect, the variations of photo current with intensity of radiation, frequency of radiation and the potential difference between the plates A and C are as follows.

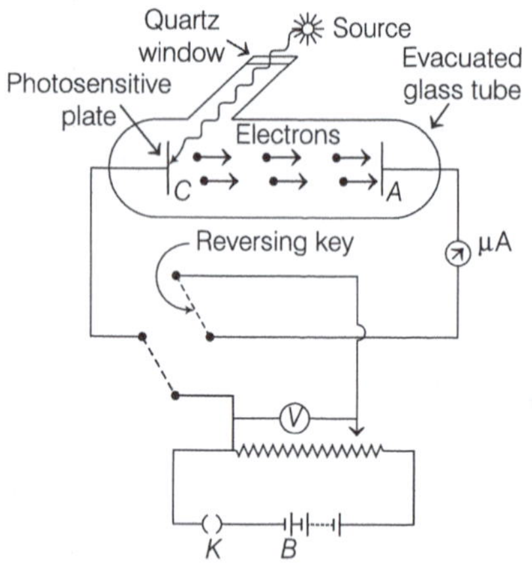

Effect of Potential on Photoelectric Current

For a fixed intensity and frequency of incident radiation, the variation of photoelectric current with potential difference between cathode and anode is as shown below,

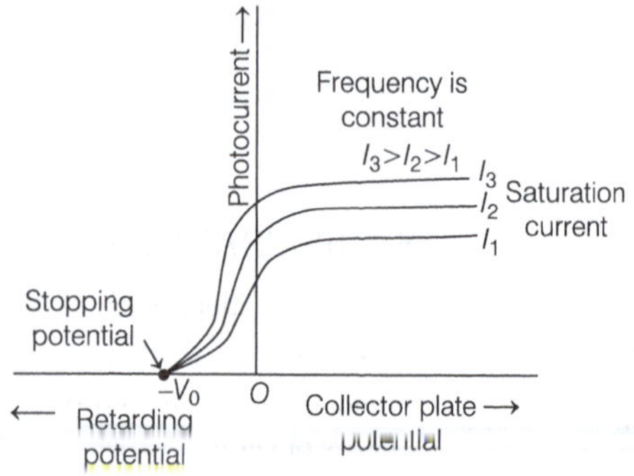

From the above graph, we can observe that,

(i) For a fixed frequency and intensity of incident radiation, the photocurrent increases with increase in the potential applied to the collector. At some stage, the photoelectric current becomes maximum or saturates.

(ii) Maximum value of photoelectric current is called *saturation current*. It corresponds to the case when all the photoelectrons emitted by the emitter plate reaches the collector plate.

(iii) For a particular frequency of incident radiation, the minimum negative (retarding) potential V_0 given to the plate A for which the photocurrent stops or becomes zero is called the *cut-off* or *stopping potential*.

(iv) Photoelectric current is zero when the stopping potential is sufficient to repel even the most energetic photoelectrons with the maximum kinetic energy K_{max}, so that $K_{max} = eV_0$.

(v) Thus, for a given frequency of the incident radiation, the stopping potential is independent of its intensity.

In other words, the maximum kinetic energy of photoelectrons depends on the light source and the emitter plate material, but it is independent of intensity of incident radiation.

Effect of Frequency of Incident Radiation on Stopping Potential

For a fixed frequency of incident radiation, the variation of photoelectric current against the potential difference between the plates is as shown below.

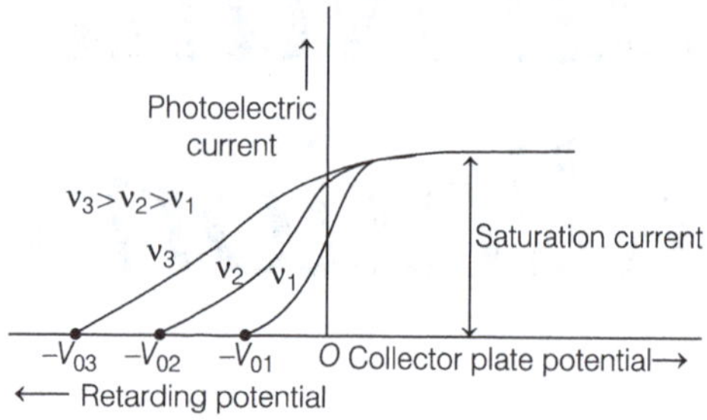

From the above graph, we observe that,

(i) Stopping potentials are in the order of $V_{03} > V_{02} > V_{01}$, if the frequencies are in the order of $\nu_3 > \nu_2 > \nu_1$.

(ii) Thus, greater the frequency of incident light, greater is the maximum kinetic energy of photoelectrons. Consequently, greater retarding potential is required to stop them completely.

The graph between frequency (ν) and stopping potential (V_0) is found to be a straight line which is not passing through the origin.

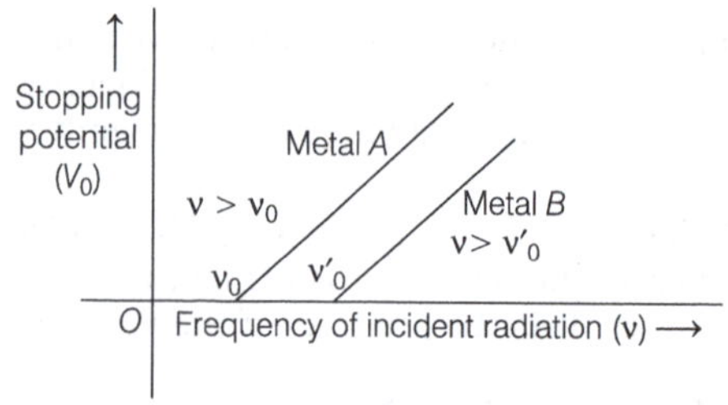

This graph shows that,

(i) The stopping potential V_0 varies linearly with the frequency of incident radiation for a given photosensitive material.

(ii) There exist a certain minimum cut-off frequency (ν_0) for which the stopping potential is zero.

Note The stopping potential is more negative for higher frequencies of incident radiations.

Implication of the Experimental Study of Photoelectric Effect

(i) Maximum KE of photoelectrons varies linearly with ν of incident radiation, but it is independent of its intensity.

(ii) For a given material, there exists a certain minimum frequency of the incident radiation below which no emissions of photoelectrons take place. This is called *threshold frequency*.

(iii) The photoelectric emission is an instantaneous process. The time lag between the incidence of radiations and emission of photoelectrons is very small, less than 10^{-9}s, even when incident radiation is made exceedingly dim.

Note **Threshold wavelength** (λ_0) is the maximum wavelength of incident radiation above which no photoelectric emission takes place.

Photoelectric Effect and Wave Theory of Light

The wave theory of light could not explain photoelectric effect due to following main reasons.

(i) Wave theory suggests that the kinetic energy of the photoelectrons should increase with the increase in intensity of light. However, $K_{max} = eV_0$ suggests that it is independent of the intensity of light.

(ii) According to wave theory, the photoelectric effect should occur for any frequency of the light, provided that the light is intense enough. However, $E \geq \phi$ for $\nu \geq \nu_0$ or $\lambda \leq \lambda_0$.

(iii) Wave theory suggests that absorption of energy by electron takes place continuously, so it can take hours or more for a single electron to pick up sufficient energy to overcome the ϕ_0 and come out of the metal. However, photoemission is an instantaneous process.

Einstein's Photoelectric Equation

Albert Einstein explained the implications of photoelectric effect on the basis of Planck's quantum theory. According to this, light radiations consist of packets of energy called quanta and each quanta of light has energy $h\nu$, where h is Planck 's constant and ν is the frequency of incident radiation.

∴ Einstein's photoelectric equation is given as

$$K_{max} = \frac{1}{2}mv_{max}^2 = h\nu - h\nu_0 = h\nu - \phi_0 = hc\left(\frac{1}{\lambda} - \frac{1}{\lambda_0}\right)$$

Here $h = 6.626 \times 10^{-34}$ J-s $= 4.136 \times 10^{-15}$ eV-s

Also, $K_{max} = eV_0$

$$eV_0 = h\nu - \phi_0 \text{ or } V_0 = \left(\frac{h}{e}\right)\nu - \frac{\phi_0}{e}$$

The above equation implies that V_0 *versus* ν graph is a straight line with slope $\left(\frac{h}{e}\right)$, which is independent of the nature of the material. Also, the intercept on the V_0-axis is $\frac{-h\nu_0}{e}$.

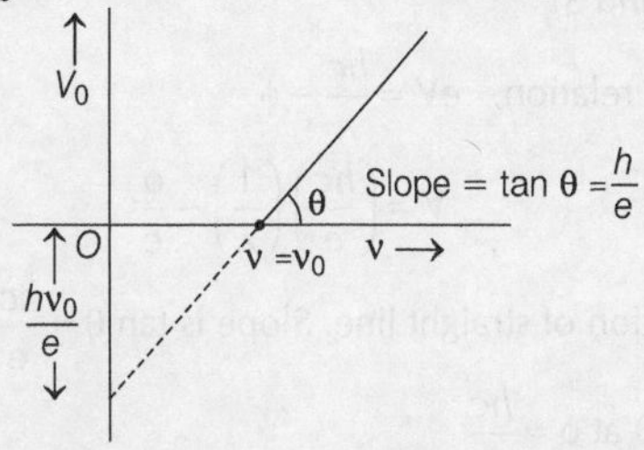

Example 1. *When the wavelength of radiation falling on a metal is changed from 500 nm to 200 nm, the maximum kinetic energy of the photoelectrons becomes three times larger. The work function of the metal is close to* *[JEE Main 2020]*

(a) 0.81 eV *(b) 1.02 eV* *(c) 0.52 eV* *(d) 0.61 eV*

Sol. (d) Maximum kinetic energy of emitted electrons is given by,

$$K_{max} = \frac{hc}{\lambda} - \phi_0$$

where, $hc \approx 1240$ eV- nm and ϕ_0 = work function.

Here, as per question, $K = \frac{1240}{500} - \phi_0$ ...(i)

and $3K = \frac{1240}{200} - \phi_0$ ...(ii)

From Eqs. (i) and (ii), we get

$$3\left(\frac{1240}{500} - \phi_0\right) = \frac{1240}{200} - \phi_0$$

$$\Rightarrow \quad 2\phi_0 = \frac{3 \times 1240}{500} - \frac{1240}{200}$$

$$\Rightarrow \quad 2\phi_0 = \frac{1240}{100}\left(\frac{3}{5} - \frac{1}{2}\right) = \frac{1240}{100}\left(\frac{1}{10}\right)$$

$$\Rightarrow \quad \phi_0 = \frac{1240}{2 \times 1000} = 0.62 \text{ eV}$$

which is close to 0.61 eV.

Example 2. *The graph between 1/λ and stopping potential (V) of three metals having work-functions ϕ_1, ϕ_2 and ϕ_3 in an experiment of photoelectric effect is plotted as shown in the figure given below. Which of the following statement(s) is/are correct? (Here, λ is the wavelength of the incident ray).*

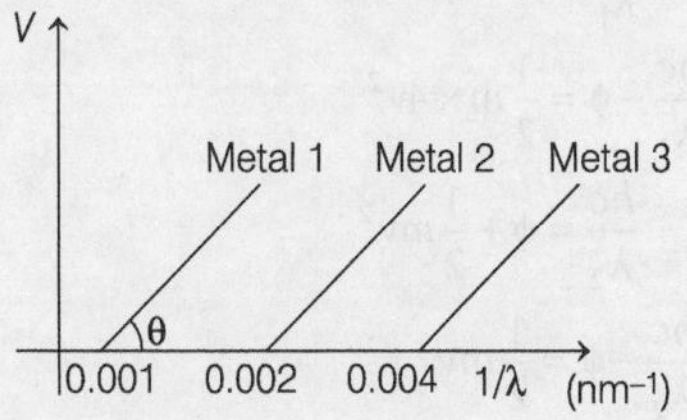

(a) *Ratio of work-functions* $\phi_1 : \phi_2 : \phi_3 = 1 : 3 : 4$
(b) *Ratio of work-functions* $\phi_1 : \phi_2 : \phi_3 = 4 : 2 : 1$
(c) $\tan\theta$ *is directly proportional to* hc/e, *where h is Planck constant and c is the speed of light*
(d) *The violet colour light can eject photoelectrons from metals 2 and 3*

Sol. (c) From the relation, $eV = \frac{hc}{\lambda} - \phi$

or $$V = \left(\frac{hc}{e}\right)\left(\frac{1}{\lambda}\right) - \frac{\phi}{e}$$

This is equation of straight line. Slope is $\tan\theta = \frac{hc}{e}$.

Further $V = 0$ at $\phi = \frac{hc}{\lambda}$

$$\therefore \quad \phi_1 : \phi_2 : \phi_3 = \frac{hc}{\lambda_{01}} : \frac{hc}{\lambda_{02}} : \frac{hc}{\lambda_{03}}$$
$$= \frac{1}{\lambda_{01}} : \frac{1}{\lambda_{02}} : \frac{1}{\lambda_{03}}$$
$$= 1 : 2 : 4$$

Violet colour has wavelength 4000 Å.

So, violet colour can eject photoelectrons from metal 1 and metal 2.

Example 3. *Surface of certain metal is first illuminated with light of wavelength* $\lambda_1 = 350$ *nm and then by light of wavelength* $\lambda_2 = 540$ *nm. It is found that the maximum speed of the photoelectrons in the two cases differ by a factor of 2. The work function of the metal (in eV) is close to*

(energy of photon $= \frac{1240}{\lambda(\text{in nm})}$ *eV)*

(a) 5.6
(b) 2.5
(c) 1.8
(d) 1.4

Sol. (c) Let velocity of one is twice in factor with second then.
Let $v_1 = 2v$ and $v_2 = v$.
We know that from Einstein's photoelectric equation, energy of incident radiation = work function + KE

or $$\frac{hc}{\lambda} = \phi + \frac{1}{2}mv^2$$

Let when $\lambda_1 = 350$ nm, then $v_1 = 2v$
and when $\lambda_1 = 540$ nm, then $v_2 = v$
$\therefore$ Above equation becomes

$$\frac{hc}{\lambda_1} = \phi + \frac{1}{2}m(2v^2)$$

or $$\frac{hc}{\lambda_1} - \phi = \frac{1}{2}m \times 4v^2 \quad \ldots\text{(i)}$$

and $$\frac{hc}{\lambda_2} - \phi + \frac{1}{2}mv^2$$

or $$\frac{hc}{\lambda_2} - \phi = \frac{1}{2}mv^2 \quad \ldots\text{(ii)}$$

Now, we divide Eq. (i) by (ii) Eq.

$$\frac{\frac{hc}{\lambda_1} - \phi}{\frac{hc}{\lambda_2} - \phi} = \frac{\frac{1}{2}m \times 4v^2}{\frac{1}{2}mv^2} = 4$$

or $$\frac{hc}{\lambda_1} - \phi = \frac{4hc}{\lambda_2} - 4\phi$$

or $$\phi = \frac{1}{3}hc\left(\frac{4}{\lambda_2} - \frac{1}{\lambda_1}\right)$$
$$= \frac{1}{3} \times 1240 \left(\frac{4 \times 350 - 540}{350 \times 540}\right)$$

or $$\phi = 1.8 \text{ eV}$$

Example 4. *The magnetic field associated with a light wave is given at the origin, by* $B = B_0$ $[\sin(3.14 \times 10^7)ct + \sin(6.28 \times 10^7)ct]$.
If this light falls on a silver plate having a work function of 4.7 eV, what will be the maximum kinetic energy of the photoelectrons?
(Take, $c = 3 \times 10^8 \text{ ms}^{-1}$ *and* $h = 6.6 \times 10^{-34}$ *J-s)*

(a) 7.72 eV
(b) 6.82 eV
(c) 8.52 eV
(d) 12.5 eV

Sol. (a) According to question, the wave equation of the magnetic field which produce photoelectric effect

$$B = B_0[(\sin(3.14 \times 10^7 ct) + \sin(6.28 \times 10^7 ct)]$$

Here, the photoelectric effect produced by the angular frequency

$$\omega = 6.28 \times 10^7 c$$

$\Rightarrow$ $$\omega = 6.28 \times 10^7 \times 3 \times 10^8$$
$$\omega = 2\pi \times 10^7 \times 3 \times 10^8 \text{ rad/s} \quad \ldots\text{(i)}$$

Using Eqs. (i)

$$h\nu = \frac{h\omega}{2\pi} = \frac{h \times 2\pi \times 10^7 \times 3 \times 10^8}{2\pi}$$
$$h\nu = 12.4 \text{ eV}$$

Therefore, according to Einstein equation for photoelectric effect

$$E = h\nu = \phi + KE_{max}$$

$\Rightarrow$ $$KE_{max} = E - \phi \quad \text{(where, } \phi = \text{work-function} = 4.7 \text{ eV)}$$
$$KE_{max} = 12.4 - 4.7 = 7.7 \text{ eV}$$

or $$KE_{max} = 7.7 \text{ eV}$$

Particle Nature of Light : Photon

Photoelectric effect gave evidence that light in interaction with matter behaved as if, it was made up of quanta or packets of energy, each having energy $h\nu$. This packet of energy is associated with a particle called *photon*.

Characteristic Properties of Photons

(i) In interaction of radiation with matter, radiation behaves as if it is made up of particles called **photons**.

(ii) A photon in vacuum travels with a speed of light, c (*i.e.* $3 \times 10^8 \, ms^{-1}$) in straight line.

(iii) It has zero rest mass, *i.e.* the photon cannot exist at rest.

(iv) The inertial mass of a photon is given by

$$m = \frac{E}{c^2} = \frac{h}{c\lambda} = \frac{h\nu}{c^2}$$

(v) Irrespective of the intensity of radiation, all the photons of a particular frequency (ν) or wavelength λ have the same energy

$$E\left(= h\nu = \frac{hc}{\lambda}\right) \text{ and momentum } p\left(= \frac{h\nu}{c} = \frac{h}{\lambda}\right).$$

(vi) Energy of a photon depends upon frequency of the photon, so the energy of the photon does not change when photon travels from one medium to another.

(vii) Photons are not deflected by electric and magnetic fields. This shows that photons are electrically neutral.

(viii) Total energy and total momentum are conserved in photon-particle collision. However, the number of photons may not be conserved in collision. As, the photon may be absorbed or a new photon may be created in a collision.

Intensity of Light

As a moving photon carries momentum with it. So, pressure and intensity are also associated with photons.

For area A through which light (photons) of total energy E and intensity I is passing normally,

Total energy of photons, $E = Nh\nu = N\left(\frac{hc}{\lambda}\right)$

$$\therefore \quad \text{Power of source, } P = \frac{E}{t} = \frac{Nh\nu}{t} = nh\nu = \frac{nhc}{\lambda} \quad \left[\because n = \frac{N}{t}\right]$$

$$\Rightarrow \quad n = P/E = \frac{P}{h\nu} = \frac{P\lambda}{hc}$$

$$\Rightarrow \quad \text{Intensity, } I = \frac{P}{A} = \frac{E}{tA} = \frac{nh\nu}{A} = n\frac{hc}{A\lambda}$$

Note At a distance r from a point source of power P, intensity is given by $I = \frac{P}{4\pi r^2} \Rightarrow I \propto \frac{1}{r^2}$.

For a line source, $I \propto \frac{P}{2\pi r l} \Rightarrow I \propto \frac{1}{r}$.

Force and Pressure Exerted by Incident Light

When light is incident on a surface, it exerts a force (or pressure) on the surface.

The expression for force and pressure being exerted on perfectly reflecting and perfectly absorbing surfaces are as follows.

Force and Pressure Exerted on Perfectly Reflecting Surface

For N photons incident on a perfectly reflected surface in time t as shown in the figure.

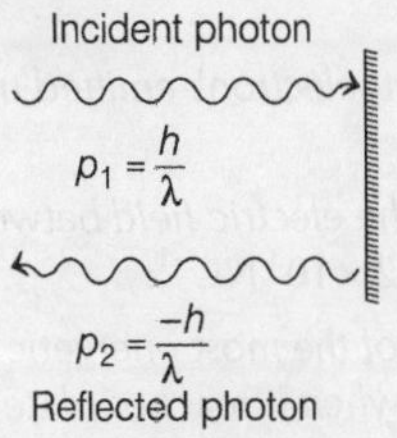

Force on the surface,

$$F = \frac{\Delta p}{\Delta t} = \frac{2Nh}{\lambda t} = n\left(\frac{2h}{\lambda}\right) \quad \left[\because n = \frac{N}{t}\right]$$

$$\Rightarrow \quad F = \left(\frac{2h}{\lambda}\right) \times \left(\frac{P\lambda}{hc}\right) = \frac{2P}{c} \quad \left[\because n = \frac{P\lambda}{hc}\right]$$

$$\therefore \text{ Pressure on the surface} = \frac{F}{A} = \frac{2P}{cA} = \frac{2I}{c}$$

where, c is speed of light.

Special case When a beam of light is incident at angle ϕ on perfectly reflecting surface

$$F = \frac{2P}{c}\cos\phi = n\left(\frac{2h}{\lambda}\right)\cos\phi$$

$$\text{Pressure} = \frac{F}{A} = \frac{2P}{cA}\cos\phi = \frac{2I\cos\phi}{c}$$

Force and Pressure on Perfectly Absorbing Surface

For N photons incident on a perfectly absorbing surface, then in time t.

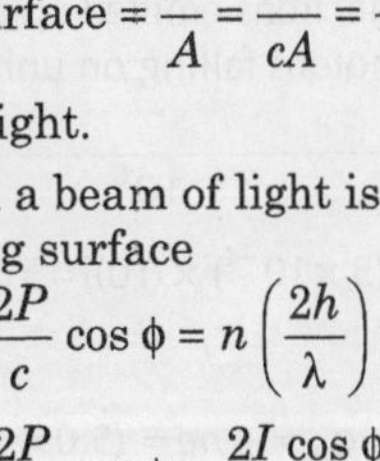

$$\therefore \quad F = \left(\frac{P\lambda}{hc}\right)\left(\frac{h}{\lambda}\right) = \frac{P}{c}$$

$$\text{Pressure on the surface} = \frac{F}{A} = \frac{P}{Ac} = \frac{I}{c}$$

Photocell

It is a device which converts light energy into electrical energy. It is also called an *electric eye.*

It works on the principle of photoelectric emission.

Example 5. *Two metallic plates A and B each of area $5\times10^{-4}m^2$, are placed parallel to each other at a separation of 1 cm. Plate B carries a positive charge of 33.7×10^{-12} C. A monochromatic beam of light, with photons of energy 5 eV each, starts falling on plate A at $t=0$ so that 10^{16} photons fall on it per square metre per second. Assume that one photoelectron is emitted for every 10^6 incident photons. Also assume that all the emitted photoelectrons are collected by plate B and the work-function of plate A remains constant at the value 2 eV. Then*

(a) *the number of photoelectrons emitted upto $t=10$ s is 6×10^7.*

(b) *the magnitude of the electric field between the plates A and B at $t=10$ s is 2×10^3 NC^{-1}.*

(c) *the kinetic energy of the most energetic photoelectrons emitted at $t=10$ s when it reaches plate B is 23 eV. Neglect the time taken by the photoelectron to reach plate B. (Take, $\varepsilon_0=8.85\times10^{-12}C^2/N\text{-}m^2$).*

(d) *Both (b) and (c)*

Sol. (d) Area of plates, $A=5\times10^{-4}$ m^2

Distance between the plates, $d=1$ cm $=10^{-2}$ m

(a) Number of photoelectrons emitted upto $t=10$ s are

$$n=\frac{\text{(number of photons falling on unit area in unit time)}\times\text{(area}\times\text{time)}}{10^6}$$

$$=\frac{1}{10^6}[(10)^{16}\times(5\times10^{-4})\times(10)]=5.0\times10^7$$

(b) At time $t=10$ s,

Charge on plate A, $q_A=+ne=(5.0\times10^7)(1.6\times10^{-19})$

$$=8.0\times10^{-12}\text{ C}$$

and charge on plate B,

$$q_B=(33.7\times10^{-12}-8.0\times10^{-12})$$

$$=25.7\times10^{-12}\text{ C}$$

$\therefore$ Electric field between the plates, $E=\dfrac{(q_B-q_A)}{2A\varepsilon_0}$

or $$E=\frac{(25.7-8.0)\times10^{-12}}{2\times(5\times10^{-4})(8.85\times10^{-12})}=2\times10^3\text{ N/C}$$

(c) Energy of most energetic photoelectrons at plate A,

$$=E-W=(5-2)\text{ eV}=3\text{ eV}$$

Increase in energy of photoelectrons

$$=(eEd)\text{ joule}=(Ed)\text{ eV}$$

$$=(2\times10^3)(10^{-2})\text{ eV}=20\text{ eV}$$

Energy of photoelectrons at plate B = (20 + 3) eV = 23 eV

Example 6. *A beam of light has three wavelengths 4144 Å, 4972 Å and 6216 Å with a total intensity of 3.6×10^{-3} Wm^{-2} equally distributed amongst the three wavelengths. The beam falls normally on an area 1.0 cm^2 of a clean metallic surface of work-function 2.3 eV. Assume that there is no loss of light by reflection and that each energetically capable photon ejects one electron. The number of photoelectrons liberated in two seconds.*

(a) 1.1×10^{12} (b) 2×10^6

(c) 4×10^8 (d) 9×10^6

Sol. (a) Energy of photon having wavelength 4144 Å,

$$E_1=\frac{12375}{4144}\text{ eV}$$

$$=2.99\text{ eV}$$

Similarly, $$E_2=\frac{12375}{4972}\text{ eV}$$

$$=2.49\text{ eV}$$

and $$E_3=\frac{12375}{6216}\text{ eV}$$

$$=1.99\text{ eV}$$

Since, only E_1 and E_2 are greater than the work-function $W=2.3$ eV, only first two wavelengths are capable for ejecting photoelectrons. Given intensity is equally distributed in all wavelengths. Therefore, intensity corresponding to each wavelength is

$$\frac{3.6\times10^{-3}}{3}=1.2\times10^{-3}\text{ W/m}^2$$

Or energy incident per second in the given area ($A=1.0$ cm^2 $=10^{-4}$ m^2) is

$$P=1.2\times10^{-3}\times10^{-4}$$

$$=1.2\times10^{-7}\text{ J/s}$$

Let n_1 be the number of photons incident per unit time in the given area corresponding to first wavelength, then

$$n_1=\frac{P}{E_1}$$

$$=\frac{1.2\times10^{-7}}{2.99\times1.6\times10^{-19}}$$

$$=2.5\times10^{11}$$

Similarly, $$n_2=\frac{P}{E_2}$$

$$=\frac{1.2\times10^{-7}}{2.49\times1.6\times10^{-19}}$$

$$=3.0\times10^{11}$$

Since, each energetically capable photon ejects one electron, total number of photoelectrons liberated in 2 s.

$$=2(n_1+n_2)$$

$$=2(2.5+3.0)\times10^{11}$$

$$=1.1\times10^{12}$$

Wave Nature of Particle

In case of light, some phenomenon like diffraction and interference can be explained on the basis of its wave character. However, certain other phenomenon such as black body radiation and photoelectric effect can be explained only on the basis of its particle nature. Thus, light is said to have a dual character. Such studies on light wave were made by Einstein in 1905.
Louis-de-Broglie, in 1942 extended the idea of photons to material particles such as electron and he proposed that matter also has a dual character as wave and as particle.

Matter Waves : de-Broglie Relation

According to de-Broglie, moving particles of matter should display wave-like properties under suitable conditions. The waves associated with such particles are called *matter waves* or *de-Broglie waves*. These then proposed that the wavelength λ associated with moving particles can expressed as,

$$\lambda = \frac{h}{mv} = \frac{h}{p}$$

where, m and v are the mass and velocity of the particle, respectively, p is momentum and h is Planck's constant.

de-Broglie wavelength associated with different moving particles are as follows.

(i) For a photon, $\lambda = \frac{h}{mc} = \frac{h}{p}$

where, $p = mc$ is the momentum of a photon.

(ii) For a charged particle, $\lambda = \frac{h}{p} = \frac{h}{\sqrt{2mE}} = \frac{h}{\sqrt{2mqV}}$

where V is the potential through which particle is accelerated and E is kinetic energy.

Specifically for an electron $\lambda = \frac{1.227}{\sqrt{V}}$ nm.

(iii) For a gas molecule, $\lambda = \frac{h}{\sqrt{3mkT}}$

where, T = absolute temperature

and k = Boltzmann's constant = 1.38×10^{-23} J/K

(iv) For neutron, $\lambda = \frac{0.286}{\sqrt{E \text{ (in eV)}}}$ Å.

Example 7. *The de-Broglie wavelength associated with an electron moving with a speed of 5.4×10^6 m/s is λ_e and for a ball of mass 150 g travelling at 30 m/s is λ_b, then which of the following relations between λ_e and λ_b is true?*

(a) $\lambda_e < \lambda_b$ (b) $\lambda_b < \lambda_e$
(c) $\lambda_e = \lambda_b$ (d) $2\lambda_b = \lambda_e$

Sol. (b) We know that de-Broglie wavelength is given by

$$\lambda = \frac{h}{p}$$

where, h is Planck's constant and p is momentum.

For the electron,

$$m = 9.11 \times 10^{-31} \text{ kg},$$
$$v = 5.4 \times 10^6 \text{ m/s}$$
$$p = mv = 9.11 \times 10^{-31} \times 5.4 \times 10^6$$
$$= 4.92 \times 10^{-24} \text{ kg-m/s}$$
$$\therefore \quad \lambda = \frac{h}{p} = \frac{6.63 \times 10^{-34}}{4.92 \times 10^{-24}}$$
$$\lambda_e = 0.135 \text{ nm}$$

For the ball, $m' = 0.150$ kg

$$v' = 30 \text{ m/s}$$
$$p' = m'v'$$
$$= 0.150 \times 30$$
$$= 4.50 \text{ kg-m/s}$$
$$\therefore \quad \lambda_b = \frac{h}{p'} = \frac{6.63 \times 10^{-34} \text{ J-s}}{4.50 \text{ kg-m/s}}$$
$$= 1.47 \times 10^{-34} \text{ m}$$
$$\therefore \quad \lambda_b < \lambda_e$$

Example 8. *A particle is moving 5 times as fast as an electron. The ratio of the de-Broglie wavelength of the particle to that of the electron is 1.878×10^{-4}. The mass of the particle is close to* **[JEE Main 2020]**

(a) 4.8×10^{-27} kg (b) 9.1×10^{-31} kg
(c) 1.2×10^{-28} kg (d) 9.7×10^{-28} kg

Sol. (d) de-Broglie wavelength of a particle is given by

$$\lambda = \frac{h}{p} = \frac{h}{mv}$$

Given, $$\frac{\lambda_{\text{particle}}}{\lambda_{\text{electron}}} = 1.878 \times 10^{-4}$$

$$\Rightarrow \quad \frac{m_e v_e}{m_p v_p} = 1.878 \times 10^{-4}$$

$$\Rightarrow \quad \frac{m_e \times v_e}{m_p \times 5v_e} = 1.878 \times 10^{-4} \quad (\because v_p = 5v_e)$$

$$\Rightarrow \quad m_p = \frac{m_e}{5 \times 1.878 \times 10^{-4}}$$

$$= \frac{9.1 \times 10^{-31}}{5 \times 1.878 \times 10^{-4}}$$

$$= 9.7 \times 10^{-28} \text{ kg}$$

Davisson and Germer Experiment

The wave nature of the material particles as predicted by de-Broglie wave confirmed by **Davisson** and **Germer** (1927) in United States and by **GP Thomson** (1928) **Scotland**.

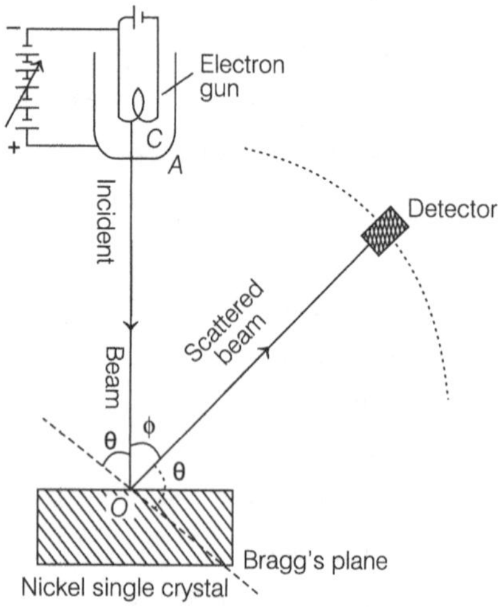

They found that the intensity of scattered beam of electrons was not the same but different at different angles of scattering.

The beam of electron was allowed to fall normally on the surface of nickel crystal. It is observed that below 44 V, the graph of electron intensity is smooth, at 44 V a bump begins to appear and continues to move upwards reaching a pronounced maximum at 54 V. Beyond 54 V, the bump diminishes with increasing potential and almost vanish at around 68 V.

Thus, for angles of incidence 50°, scattering angle relative to the set of Bragg's plane is 65°.

For most of the crystals, the spacing between atomic planes is about 1 Å, therefore the Bragg's equation for maxima in the diffraction pattern becomes

$$2d \sin\theta = \lambda$$

$$\lambda = 2 \times 1 \times \sin 65^\circ = 1.67 \text{ Å}$$

$\therefore$ The de-Broglie wavelength of diffracted electrons is 1.67 Å. Now, for 54 V electron, the de-Broglie wavelength is

$$\lambda = \frac{12.27}{\sqrt{V}} \text{ Å} = \frac{12.27}{\sqrt{54}} \text{ Å} = 1.67 \text{ Å}$$

Practice Exercise

ROUND I Topically Divided Problems

Emission of Electrons and Photoelectric Effect

1. When a point source of light is 1 m away from a photoelectric cell, the photoelectric current is found to be I mA. If the same source is placed at 4 m from the same photoelectric cells, the photoelectric current (in mA) will be

(a) $\frac{I}{16}$ (b) $\frac{I}{4}$ (c) $4I$ (d) $16I$

2. The figures shows the variation of photo-current i with anode potential V for three differential radiations. Let I_a, I_b and I_c be the intensities and f_a, f_b and f_c be the frequencies for the curves a, b and c respectively. Then

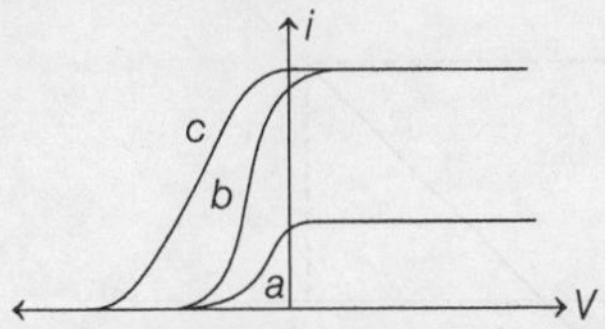

(a) $f_a = f_b$ and $I_a \neq I_b$ (b) $f_a = f_c$ and $I_a = I_c$
(c) $f_a = f_b$ and $I_a = I_b$ (d) $f_b = f_c$ and $I_b = I_c$

3. Which one of the following statements regarding photoemission of electrons is correct?

(a) Kinetic energy of electrons increases with the frequency of incident light.
(b) Electrons are emitted when the wavelength of the incident light is above a certain threshold wavelength.
(c) Photoelectric emission is instantaneous with the incidence of light.
(d) Photo electrons are emitted whenever a gas is irradiated with ultraviolet light.

4. For the graph as shown below

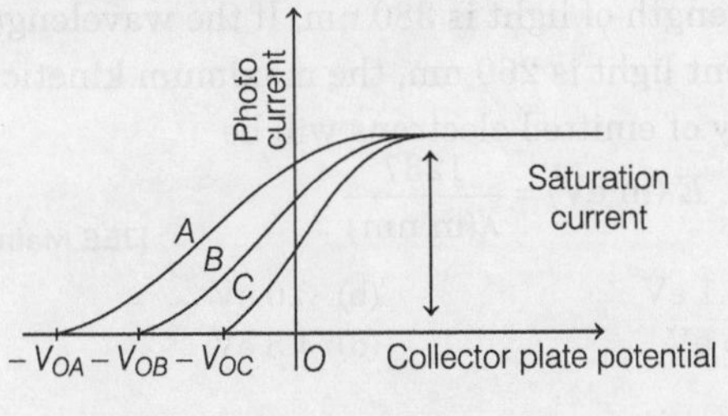

Which of the following statement(s) is/are correct?

I. Stopping potential are in the order $V_{OA} > V_{OB} > V_{OC}$, if the frequencies are in the order $\nu_A > \nu_B > \nu_C$.
II. Greater the frequency of incident radiation, lesser would be the maximum kinetic energy of photoelectrons.

(a) Only I
(b) Only II
(c) Both I and II
(d) Neither I nor II

5. Consider a beam of electrons (each electron with energy E_0) incident on a metal surface kept in an evacuated chamber. Then, **[NCERT Exemplar]**

(a) no electrons will be emitted as only photons can emit electrons
(b) electrons can be emitted but all with an energy, E_0
(c) electons can be emitted with any energy, with a maximum of $E_0 - \phi$ (ϕ is the work function)
(d) electrons can be emitted with any energy, with a maximum of E_0

6. The maximum KE (E_k) of the emitted photoelectrons against frequency of the incident radiation is plotted as shown in figure. This graph help in determining the following quantities.

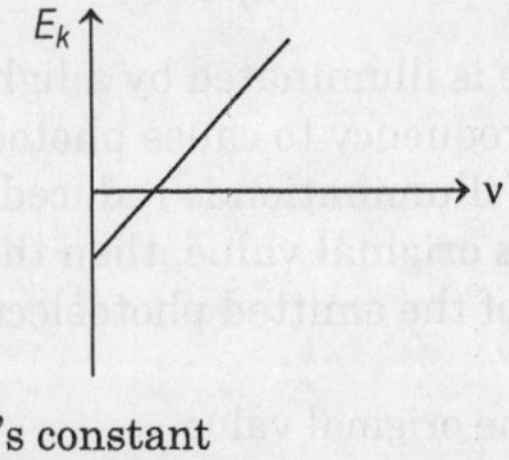

(a) Planck's constant
(b) charge on an electron
(c) threshold frequency
(d) work function of cathode metal

7. The work function of tungsten and sodium are 4.5 eV and 2.3 eV respectively. If the threshold wavelength, λ for sodium is 5460 Å, the value of λ for tungsten is

(a) 2791 Å (b) 3260 Å
(c) 1925 Å (d) 1000 Å

8. Radiations of two photon's energy, twice and ten times the work function of metal are incident on the metal surface successively. The ratio of maximum velocities of photoelectrons emitted in two cases is

(a) 1 : 2 (b) 1 : 3
(c) 1 : 4 (d) 1 : 1

9. Consider a metal exposed to light of wavelength 600 nm. The maximum energy of the electron doubles when light of wavelength 400 nm is used. The work function in eV is **[NCERT Exemplar]**

(a) 1.50 eV (b) 1.02 eV
(c) 1.94 eV (d) 2.76 eV

10. The wavelength of the photoelectric threshold for silver is λ_0. The energy of the electron ejected from the surface of silver by an incident light of wavelength λ $(\lambda < \lambda_0)$ will be

(a) $hc(\lambda_0 - \lambda)$ (b) $\dfrac{hc}{\lambda_0 - \lambda}$

(c) $\dfrac{h}{c}\left(\dfrac{1}{\lambda} - \dfrac{1}{\lambda_0}\right)$ (d) $hc\left(\dfrac{\lambda_0 - \lambda}{\lambda_0 \lambda}\right)$

11. In a Frank-Hertz experiment, an electron of energy 5.6 eV passes through mercury vapour and emerges with an energy 0.7 eV. The minimum wavelength of photons emitted by mercury atoms is close to **[JEE Main 2019]**

(a) 250 nm (b) 2020 nm
(c) 1700 nm (d) 220 nm

12. An isolated lead ball is charged upon continuous irradiation by EM radiation of wavelength, $\lambda = 221$ nm. The maximum potential attained by the lead ball, if its work function is 4.14 eV is (Take, $h = 6.63 \times 10^{-34}$ J-s, $c = 3 \times 10^8$ m/s and $e = 1.6 \times 10^{-19}$ C)

(a) 1.49 V (b) 2.67 V
(c) 3.14V (d) 0.51V

13. A metal surface is illuminated by a light of given intensity and frequency to cause photoemission. If the intensity of illumination is reduced to one-fourth of its original value, then the maximum kinetic energy of the emitted photoelectrons would become

(a) four times the original value
(b) twice the original value
(c) (1/6)th of the original value
(d) unchanged

14. The threshold wavelength for photoelectron emission from a materials is 5200 Å. Photo electrons will be emitted when this material is illuminated with monochromatic radiation from a

(a) 50 W infrared lamp (b) 1 W infrared lamp
(c) 1 W ultraviolet lamp (d) None of these

15. The work function for the surface of Al is 4.2 eV. How much potential difference will be required to just stop the emission of maximum energy electrons emitted by light of 2000 Å ?

(a) 1.51 V (b) 1.99 V
(c) 2.99 V (d) None of these

16. Light is incident on the cathode of a photocell and the stopping voltages are measured for light of two different wavelengths. From the data given below, determine the work functions of the metal of the cathode in eV.

Wavelength (Å)	Stopping voltage (volt)
4000	0.9
4500	1.3

(a) 1.2 eV (b) 2.3 eV
(c) 4.2 eV (d) 6 eV

17. Following graph represents kinetic energy of most energetic photoelectrons K_{max} (in eV) and frequency ν for a metal used as cathode in photoelectric experiment. The threshold frequency of light for the photoelectric emission from the metal is

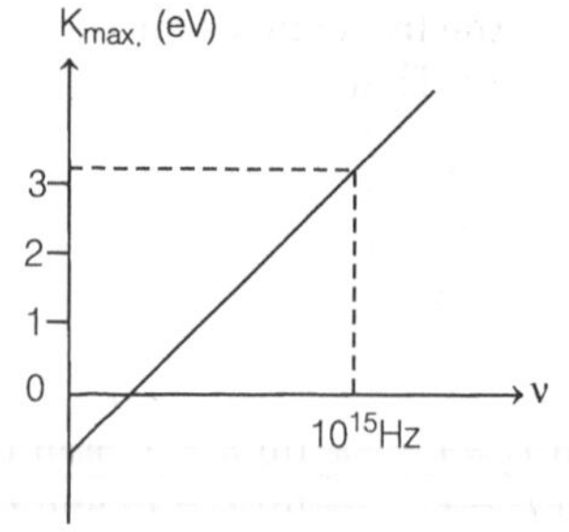

(a) 4×10^{14} Hz (b) 3.5×10^{14} Hz
(c) 2.0×10^{14} Hz (d) 2.7×10^{14} Hz

18. Light of wavelength λ strikes a photo sensitive surface and electrons are ejected with kinetic energy E. If the KE is to be increased to $2E$, the wavelength must be changed to λ' where

(a) $\lambda' = \dfrac{\lambda}{2}$ (b) $\lambda' = 2\lambda$

(c) $\dfrac{\lambda}{2} < \lambda' < \lambda$ (d) $\lambda' > \lambda$

19. In a photoelectric effect experiment, the threshold wavelength of light is 380 nm. If the wavelength of incident light is 260 nm, the maximum kinetic energy of emitted electrons will be

Given, E (in eV) $= \dfrac{1237}{\lambda(\text{in nm})}$ **[JEE Main 2019]**

(a) 15.1 eV (b) 3.0 eV
(c) 1.5 eV (d) 4.5 eV

20. When a certain photosensitive surface is illuminated with monochromatic light of frequency ν, the stopping potential for the photocurrent is $-V_0/2$. When the surface is illuminated by monochromatic light of frequency $\nu/2$, the stopping potential is $-V_0$. The threshold frequency for photoelectric emission is **[JEE Main 2019]**

(a) $\frac{4}{3}\nu$ (b) 2ν
(c) $\frac{3\nu}{2}$ (d) $\frac{5\nu}{3}$

21. In a photoelectric experiment, the wavelength of the light incident on a metal is changed from 300 n-m to 400 n-m. The decrease in the stopping potential is close to $\left(\frac{hc}{e} = 1240 \text{ n-mV}\right)$ **[JEE Main 2019]**

(a) 0.5 V (b) 2.0 V
(c) 1.5 V (d) 1.0 V

22. Cut off potentials for a metal surface for light of wavelengths λ_1, λ_2 and λ_3 are V_1, V_2 and V_3 respectively. If V_1, V_2 and V_3 are in arithmatic progression, then λ_1, λ_2 and λ_3 are in

(a) arithmetic progression
(b) geometric progression
(c) harmonic progression
(d) None of the above

23. Radiation of wavelength λ is incident on a photocell. The fastest emitted electron has speed v. If the wavelength is changed to $\frac{3\lambda}{4}$, the speed of the fastest emitted electron will be

(a) $> v\left(\frac{4}{3}\right)^{1/2}$ (b) $< v\left(\frac{4}{3}\right)^{1/2}$
(c) $= v\left(\frac{4}{3}\right)^{1/2}$ (d) $= v\left(\frac{3}{4}\right)^{1/2}$

24. The following figure shows few data points in a photoelectric effect experiment for a certain metal. The minimum energy for ejection of electron from its surface is (Take, Planck's constant, $h = 6.62 \times 10^{-34}$ J-s) **[JEE Main 2020]**

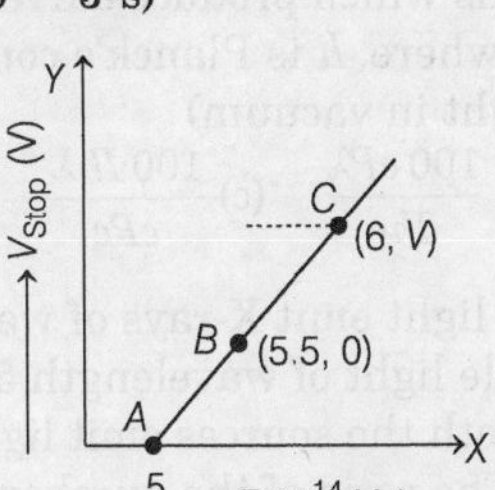

(a) 1.93 eV (b) 2.59 eV
(c) 2.27 eV (d) 2.10 eV

25. In a photoelectric effect experiment, the graph of stopping potential V *versus* reciprocal of wavelength $(1/\lambda)$ obtained is shown in the figure. As the intensity of incident radiation is increased,

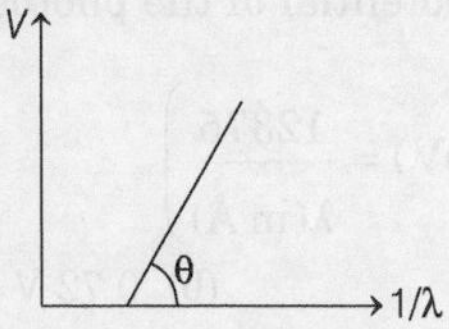

[JEE Main 2020]

(a) graph does not change
(b) straight line shifts to left
(c) slope of the straight line get more steep
(d) straight line shifts to right

26. The stopping potential V_0 (in volt) as a function of frequency (ν) for a sodium emitter, is shown in the figure. The work function of sodium, from the data plotted in the figure, will be (Take, Planck's constant $(h) = 6.63 \times 10^{-34}$ J-s, electron charge, $e = 1.6 \times 10^{-19}$ C] **[JEE Main 2019]**

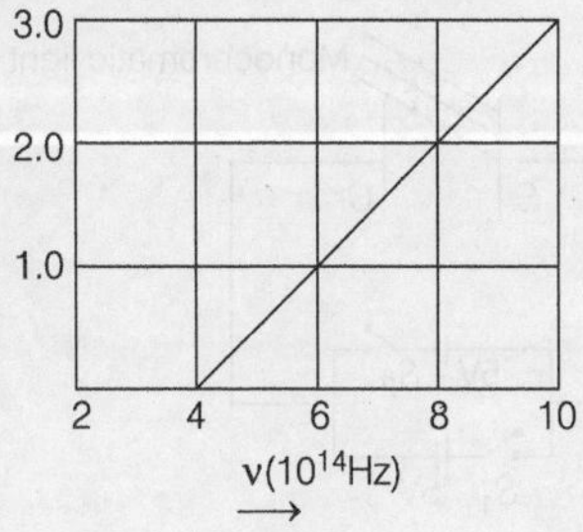

(a) 1.82 eV
(b) 1.66 eV
(c) 1.95 eV
(d) 2.12 eV

27. A photon of energy E ejects a photoelectron from a metal surface whose work function is W_0. If this electron enters into a uniform magnetic field of induction B in a direction perpendicular to the field and describes a circular path of radius r, then the radius, r is given by

(a) $\sqrt{\frac{2m(W_0 - E)}{eB}}$ (b) $\sqrt{\frac{2e(E - W_0)}{mB}}$
(c) $\frac{\sqrt{2m(E - W_0)}}{eB}$ (d) $\sqrt{\frac{2mW_0}{eB}}$

28. All electrons ejected from a metallic surface by incident light of wavelength 400 nm travelled 1 m in the direction of uniform electric field of 2 NC^{-1} and came to rest. The work function of the surface is

(a) 1.1 eV (b) 2.2 eV
(c) 3.1 eV (d) 5.1 eV

29. The electric field of light wave is given as $E = 10^{-3}\cos\left(\dfrac{2\pi x}{5\times10^{-7}} - 2\pi\times6\times10^{14}t\right)\hat{x}\ NC^{-1}$. This light falls on a metal plate of work function 2eV. The stopping potential of the photoelectrons is **[JEE Main 2019]**

$$\left(\text{Given, } E\text{ (in eV)} = \frac{12375}{\lambda(\text{in Å})}\right)$$

(a) 0.48 V (b) 0.72 V
(c) 2.0 V (d) 2.48 V

30. In a photoelectric experiment, a monochromatic light is incident on the emitter plate E, as shown in the figure. When switch S_1 is closed and switch S_2 is open, the photoelectrons strike the collector plate C with a maximum kinetic energy of 1 eV. If switch S_1 is open and switch S_2 is closed and the frequency of the incident light is doubled the photoelectrons strike the collector plate with a maximum kinetic energy of 20 eV. The threshold wavelength of the emitter plate is

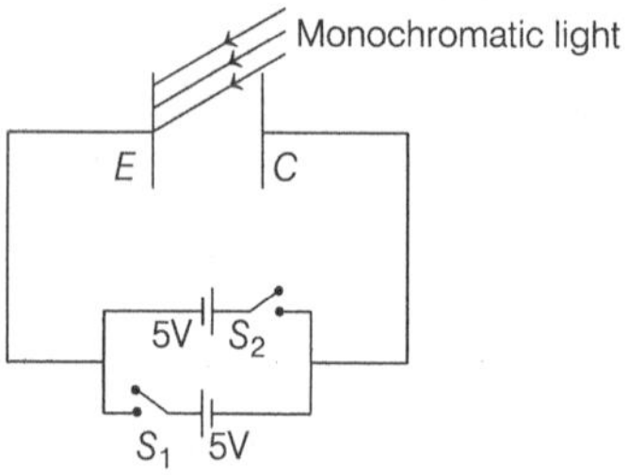

(a) 5233.3 Å (b) 4133.3 Å
(c) 4166.7 Å (d) 5336.7 Å

31. A small piece of cesium metal (ϕ = 1.9 eV) is kept at a distance of 17.7 cm from a large metal plate having a charge density of $1.0\times10^{-9}\,C/m^2$ on the surface facing the cesium piece. A monochromatic light of wavelength 400 nm is incident on the cesium piece. Find the minimum and maximum kinetic energy of the photoelectrons reaching the large metal plate. Neglect any change in electric field due to the small piece of cesium present.

(a) 20 eV, 21.2 eV (b) 21.2 eV, 60 eV
(c) 0.2 eV, 0.31 eV (d) 6 eV, 8 eV

Particle Nature of Light

32. There are two sources of light each emitting with a power of 100 W. One emits X-rays of wavelength 1 nm and the other visible light of wavelength 500 nm. Find the ratio of number of photons of X-rays and photons of visible light of the given wavelength?

(a) 1 : 500 (b) 1 : 250
(c) 1 : 20 (d) 100

33. Two monochromatic beams A and B of equal intensity I, hit a screen. The number of photons hitting the screen by beam A is twice that by beam B. Then, what can inference you about their frequencies ?

(a) The frequency of beam B is twice that of A
(b) The frequency of beam B is half that of A
(c) The frequency of beam A is twice of B
(d) None of the above

34. A parallel beam of light is incident normally on a plane surface absorbing 40% of the light and reflecting the rest. If the incident beam carries 60 W of power, the force exerted by it on the surface is

(a) 3.2×10^{-8} N (b) 3.2×10^{-7} N
(c) 5.12×10^{-7} N (d) 5.12×10^{-8} N

35. Ultraviolet light of wavelength 300 nm and intensity 1.0 Wm^{-2} falls on the surface of a photosensitive material. If one percent of the incident photons produce photoelectrons, then the number of photoelectrons emitted from an area of 1.0 cm^2 of the surface is nearly

(a) $9.61\times10^{14}\ s^{-1}$ (b) $4.12\times10^{13}\ s^{-1}$
(c) $1.51\times10^{12}\ s^{-1}$ (d) $2.13\times10^{11}\ s^{-1}$

36. What will be the number of photons emitted per second by a 10 W sodium vapour lamp assuming that 90% of the consumed energy is converted into light? [Wavelength of sodium light is 590 nm, $h = 6.63\times10^{-34}$ J-s]

(a) 0.267×10^{18} (b) 0.267×10^{19}
(c) 0.267×10^{20} (d) 0.267×10^{17}

37. A 2 mW laser operates at a wavelength of 500 nm. The number of photons that will be emitted per second is (Given, Planck's constant $h = 6.6\times10^{-34}$ Js, speed of light $c = 3.0\times10^8$ m/s)

(a) 1×10^{16} (b) 5×10^{15} **[JEE Main 2019]**
(c) 1.5×10^{16} (d) 2×10^{16}

38. Photons of wavelength λ emitted by a source of power P incident on a photo cell. If the current produced in the cell is I, then the percentage of incident photons which produce current in the photo cell is. (where, h is Planck's constant and c is the speed of light in vacuum)

(a) $\dfrac{100\,ePc}{Ih\lambda}$ (b) $\dfrac{100\,eP\lambda}{Ihc}$ (c) $\dfrac{100\,Ih\lambda}{ePc}$ (d) $\dfrac{100Ihc}{eP\lambda}$

39. Two sources of light emit X-rays of wavelength 1 nm and visible light of wavelength 500 nm, respectively. Both the sources emit light of the same power 200 W. The ratio of the number density of photons of X-rays to the number density of photons of the visible light of the given wavelengths is **[JEE Main 2020]**

(a) 1/250 (b) 500 (c) 250 (d) 1/500

40. A cobalt (Co) plate is placed at a distance of 1 m from a point source of power 1 W. Assume a circular area of the plate of radius, $r = 1$ Å is exposed to the radiation and ejects photoelectrons. The light energy is considered to be spread uniformly and the work function of cobalt is 5 eV. The minimum time the target should be exposed to the light source to eject a photoelectron (Assuming no reflection loses) is
(a) 320 s (b) 450 s
(c) 860 s (d) 100 s

41. A photodiode sensor is used to measure the output of a 300 W lamp kept 10 m away. The sensor has an opening of 2 cm in diameter. How many photons enter the sensor if the wavelength of the light is 660 nm and the exposure time is 100 ms. (Assume that all the energy of the lamp is given off as light and $h = 6.6 \times 10^{-34}$ Js)
(a) 3.6×10^{13} (b) 2.8×10^{13}
(c) 2.5×10^{13} (d) 1.8×10^{13}

42. Photons absorbed in matter are converted to heat. A source emitting n photon/s of frequency ν is used for converting of ice at 0°C to water at 0°C. Then, which amongst the following is incorrect regarding the time, T taken for the conversion?
(a) Decreases with increasing n, with υ fixed
(b) Decreases with n fixed, υ increasing
(c) Remains constant with n and υ changing such that $n\upsilon$ = constant
(d) Increases when the product $n\nu$ increases

43. A beam of light consists of five wavelengths 4000 Å, 4800Å, 6000Å, 7000Å and 7800Å. The light beam is falling normally over a metal surface of area 10^{-4} m^2 with a work function of 1.9 eV. Intensity of light beam is 7.5×10^{-3} Wm^{-2}, which is equally divided among the constituent wavelengths. If there is no loss of light energy, number of photoelectrons emitted per second is
(a) 1.12×10^{12} (b) 3.15×10^{12}
(c) 1.77×10^{12} (d) 4.06×10^{12}

44. Light is incident normally on a completely absorbing surface with an energy flux of 25 W cm^{-2}. If the surface has an area of 25 cm^2, the momentum transferred to the surface in 40 min time duration will be **[JEE Main 2019]**
(a) 3.5×10^{-6} N-s
(b) 6.3×10^{-4} N-s
(c) 1.4×10^{-6} N-s
(d) 5.0×10^{-3} N-s

45. A metal plate of area 1×10^{-4} m^2 is illuminated by a radiation of intensity 16 mW/m^2. The work function of the metal is 5 eV. The energy of the incident photons is 10 eV and only 10% of it produces photoelectrons. The number of emitted photoelectrons per second and their maximum energy, respectively will be (Take, 1 eV = 1.6×10^{-19} J) **[JEE Main 2019]**
(a) 10^{11} and 5 eV (b) 10^{12} and 5 eV
(c) 10^{10} and 5 eV (d) 10^{14} and 10 eV

Wave Nature of Particle

46. A particle of mass 1 mg has the same wavelength as an electron moving with a velocity of 3×10^6 ms^{-1}. The velocity of the particle is
(a) 3×10^{-31} ms^{-1} (b) 2.7×10^{-21} ms^{-1}
(c) 2.7×10^{-18} ms^{-1} (d) 9×10^{-2} ms^{-1}

47. An electron and a proton have the same de-Broglie wavelength. Then the kinetic energy of the electron is
(a) zero
(b) infinity
(c) equal to kinetic energy of the proton
(d) greater than the kinetic energy of proton

48. The de-Broglie wavelength L associated with an elementary particle of linear momentum p is best represented by the graph

(a) 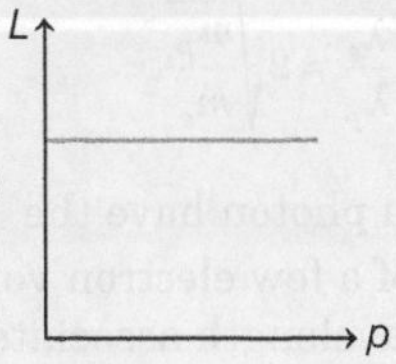
(b) 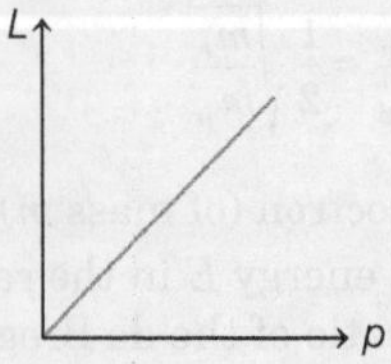
(c) 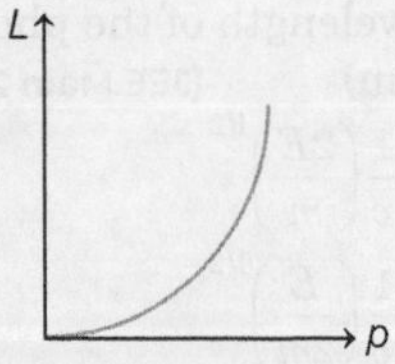
(d) 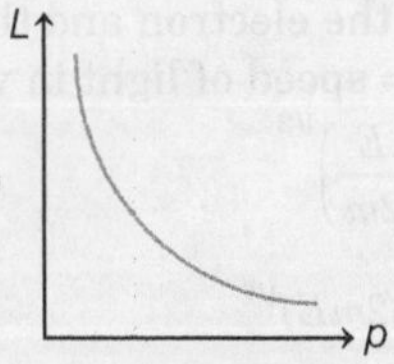

49. The de-Broglie wavelength associated with an electron and a proton were calculated by accelerating them through same potential of 100 V. What should nearly be the ratio of their wavelengths? ($m_p = 1.00727$ u, $m_e = 0.00055$ u)
(a) 1860 : 1 (b) $(1860)^2 : 1$ **[JEE Main 2021]**
(b) 41.4 : 1 (d) 43 : 1

50. The energy that should be added to an electron to reduce its de-Broglie wavelength from 10^{-10} m to 0.5×10^{-10} m, will be
(a) four times the initial energy
(b) thrice the initial energy
(c) equal to the initial energy
(d) twice the initial energy

51. A proton, a neutron, an electron and an α-particle have same energy, then their de-Broglie wavelengths compare as **[NCERT Exemplar]**
(a) $\lambda_p = \lambda_n > \lambda_e > \lambda_\alpha$ (b) $\lambda_e > \lambda_p = \lambda_n > \lambda_\alpha$
(c) $\lambda_e = \lambda_p > \lambda_n > \lambda_\alpha$ (d) $\lambda_e = \lambda_p > \lambda_n > \lambda_\alpha$

52. A particle moving with kinetic energy E has de Broglie wavelength λ. If energy ΔE is added to its energy, the wavelength become $\lambda/2$. Value of ΔE is [JEE Main 2020]

(a) $2E$ (b) $4E$ (c) $3E$ (d) E

53. An electron, a doubly ionised helium ion He $^{++}$) and a proton are having the same kinetic energy. The relation between their respective de-Broglie wavelengths λ_e, $\lambda_{He^{++}}$ and λ_p is [JEE Main 2020]

(a) $\lambda_e > \lambda_{He^{++}} > \lambda_p$ (b) $\lambda_e < \lambda_{He^{++}} = \lambda_p$

(c) $\lambda_e > \lambda_p > \lambda_{He^{++}}$ (d) $\lambda_e < \lambda_p < \lambda_{He^{++}}$

54. Electrons are accelerated through a potential difference V_0 and protons are accelerated through a potential difference 4 V. The de-Broglie wavelength are λ_e and λ_p for electrons and protons respectively. The ratio of $\frac{\lambda_e}{\lambda_p}$ is given by

(Given m_e is mass of electrons and m_p is mass of proton).

(a) $\frac{\lambda_e}{\lambda_p} = \sqrt{\frac{m_p}{m_e}}$ (b) $\frac{\lambda_e}{\lambda_p} = \sqrt{\frac{m_e}{m_p}}$

(c) $\frac{\lambda_e}{\lambda_p} = \frac{1}{2}\sqrt{\frac{m_e}{m_p}}$ (d) $\frac{\lambda_e}{\lambda_p} = 2\sqrt{\frac{m_p}{m_e}}$

55. An electron (of mass m) and a photon have the same energy E in the range of a few electron volt. The ratio of the de Broglie wavelength associated with the electron and the wavelength of the photon is (c = speed of light in vacuum) [JEE Main 2020]

(a) $\left(\frac{E}{2m}\right)^{1/2}$ (b) $\frac{1}{c}\left(\frac{2E}{m}\right)^{1/2}$

(c) $c\,(2mE)^{1/2}$ (d) $\frac{1}{c}\left(\frac{E}{2m}\right)^{1/2}$

56. A particle A of mass m and charge q is accelerated by a potential difference of 50 V. Another particle B of mass $4m$ and charge q is accelerated by a potential difference of 2500 V. The ratio of de-Broglie wavelengths $\frac{\lambda_A}{\lambda_B}$ is close to [JEE Main 2019]

(a) 4.47 (b) 10.00 (c) 0.07 (d) 14.14

57. When photon of energy 4.0 eV strikes the surface of a metal A, the ejected photoelectrons have maximum kinetic energy T_A eV and de-Broglie wavelength λ_A. The maximum kinetic energy of photoelectrons liberated from another metal B by photon of energy 4.50 eV is $T_B = (T_A - 1.5)$ eV.

If the de-Broglie wavelength of these photoelectrons $\lambda_B = 2\lambda_A$, then the work function of metal B is [JEE Main 2020]

(a) 4 eV (b) 2 eV (c) 1.5 eV (d) 3 eV

58. A particle is dropped from a height H. The de-Broglie wavelength of the particle as a function of height is proportional to [NCERT Exemplar]

(a) H (b) $H^{1/2}$

(c) H^0 (d) $H^{-1/2}$

59. An electron of mass m and magnitude of charge $|e|$ initially at rest gets accelerated by a constant electric field $\mathbf{E}$. The rate of change of de Broglie wavelength of this electron at time t ignoring relativistic effects is [JEE Main 2020]

(a) $\frac{h}{|e|\mathbf{E}\sqrt{t}}$ (b) $-\frac{h}{|e|\mathbf{E}t}$

(c) $\frac{-h}{|e|\mathbf{E}t^2}$ (d) $\frac{|e|\mathbf{E}t}{h}$

60. An electron is moving with an initial velocity $\mathbf{v} = v_0\hat{\mathbf{i}}$ and is in a magnetic field $\mathbf{B} = B_0\hat{\mathbf{j}}$. Then it's de-Broglie wavelength [NCERT Exemplar]

(a) remains constant

(b) increases with time

(c) decreases with time

(d) increases and decreases periodically

61. A particle A of mass m and initial velocity v collides with a particle B of mass $\frac{m}{2}$ which is at rest. The collision is head on, and elastic. The ratio of the de-Broglie wavelengths λ_A to λ_B after the collision is [JEE Main 2017]

(a) $\frac{\lambda_A}{\lambda_B} = 2$ (b) $\frac{\lambda_A}{\lambda_B} = \frac{2}{3}$

(c) $\frac{\lambda_A}{\lambda_B} = \frac{1}{2}$ (d) $\frac{\lambda_A}{\lambda_B} = \frac{1}{3}$

62. Particle A of mass $m_A = \frac{m}{2}$ moving along the X-axis with velocity v_0 collides elastically with another particle B at rest having mass $m_B = \frac{m}{3}$. If both particles move along the X-axis after the collision, the change $\Delta\lambda$ in de-Broglie wavelength of particle A, in terms of its de-Broglie wavelength λ_0 before collision is [JEE Main 2020]

(a) $\Delta\lambda = \frac{5}{2}\lambda_0$ (b) $\Delta\lambda = 4\lambda_0$

(c) $\Delta\lambda = 2\lambda_0$ (d) $\Delta\lambda = \frac{3}{2}\lambda_0$

63. Two particles move at right angle to each other. Their de-Broglie wavelengths are λ_1 and λ_2, respectively. The particles suffer perfectly inelastic collision. The de-Broglie wavelength λ of the final particle, is given by [JEE Main 2019]

(a) $\frac{1}{\lambda^2} = \frac{1}{\lambda_1^2} + \frac{1}{\lambda_2^2}$ (b) $\lambda = \sqrt{\lambda_1\lambda_2}$

(c) $\lambda = \frac{\lambda_1 + \lambda_2}{2}$ (d) $\frac{2}{\lambda} = \frac{1}{\lambda_1} + \frac{1}{\lambda_2}$

64. An electron (mass m) with initial velocity $\mathbf{v} = v_0\hat{\mathbf{i}} + v_0\hat{\mathbf{j}}$ is in an electric field $\mathbf{E} = -E_0\hat{\mathbf{k}}$. If λ_0 is initial de-Broglie wavelength of electron, then its de Broglie wavelength at time t is given by **[JEE Main 2020]**

(a) $\dfrac{\lambda_0}{\sqrt{1+\dfrac{e^2E_0^2t^2}{m^2v_0^2}}}$ (b) $\dfrac{\lambda_0}{\sqrt{2+\dfrac{e^2E_0^2t^2}{m^2v_0^2}}}$

(c) $\dfrac{\lambda_0}{\sqrt{1+\dfrac{e^2E_0^2t^2}{2m^2v_0^2}}}$ (d) $\dfrac{\lambda_0\sqrt{2}}{\sqrt{1+\dfrac{e^2E_0^2t^2}{m^2v_0^2}}}$

65. An electron microscope uses electrons accelerated by a voltage of 50 kV. If other factors (such as numerical aperture, etc.) are taken to be roughly the same, then

(a) de-Broglie wavelength associated with e^- is 5.5×10^{-12} m
(b) the resolving power of an electron microscope is about 10^5 times greater than that of an optical microscope which uses yellow light
(c) de-Broglie wavelength associated with e^- is 5.9×10^{-7} m
(d) Both (a) and (b)

66. Consider the figure given below. Suppose the voltage applied to A is increased. The diffracted beam will have the maxima at a value of θ that **[NCERT Exemplar]**

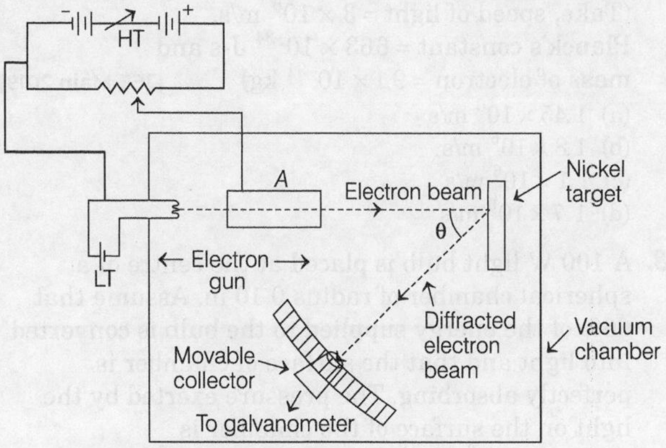

(a) will be larger than the earlier value
(b) will be the same as the earlier value
(c) will be less than the earlier value
(d) will depend on the target

ROUND II Mixed Bag

Only One Correct Option

1. Two identical metal plates shown photoelectric effect by a light of wavelength λ Å falls on plate A and λ_B on plate $B(\lambda_A = 2\lambda_B)$. The maximum kinetic energy is

(a) $2\,K_A = K_B$ (b) $K_A < \dfrac{K_B}{2}$
(c) $K_A = 2K_B$ (d) $K_A = \dfrac{K_B}{2}$

2. Given that a photon of light of wavelength 10,000 Å has an energy equal to 1.23 eV. When light of wavelength 5000 Å and intensity I_0 falls on a photoelectric cell, the surface current is 0.40×10^{-6} A and the stopping potential is 1.36 V, then the work function is

(a) 0.43 eV (b) 0.55 eV
(c) 1.10 eV (d) 1.53 eV

3. An electron of mass m when accelerated through a potential difference has de-Broglie wavelength λ. The de-Broglie wavelength associated with a proton of mass M accelerated through the same potential difference will be

(a) $\lambda\dfrac{m}{M}$ (b) $\lambda\sqrt{\dfrac{m}{M}}$
(c) $\lambda\dfrac{M}{m}$ (d) $\lambda\sqrt{\dfrac{M}{m}}$

4. The stopping potential in the context of photoelectric effect depends on the following property of incident electromagnetic radiation: **[JEE Main 2021]**

(a) Phase (b) Intensity
(c) Amplitude (d) Frequency

5. The potential energy of a particle of mass m is given by

$$U(x) = \begin{cases} E_0; & 0 \le x \le 1 \\ 0\;; & x > 1 \end{cases}$$

λ_1 and λ_2 are the de-Broglie wavelengths of the particle when $0 \le x \le 1$ and $x > 1$ respectively. If the total energy of particle is $2E_0$ the ratio $\dfrac{\lambda_1}{\lambda_2}$ will be

(a) 2 (b) 1 (c) $\sqrt{2}$ (d) $\dfrac{1}{\sqrt{2}}$

6. 50 W/m^2 energy density of sunlight is normally incident on the surface of a solar panel. Some part of incident energy (25%) is reflected from the surface and the rest is absorbed. The force exerted on 1 m^2 surface area will be close to (Take, $c = 3 \times 10^8$ m/s) **[JEE Main 2019]**

(a) 20×10^{-8} N
(b) 35×10^{-8} N
(c) 15×10^{-8} N
(d) 10×10^{-8} N

7. If the de-Broglie wavelength of an electron is equal to 10^{-3} times, the wavelength of a photon of frequency 6×10^{14} Hz, then the speed of electron is equal to
(Take, speed of light = 3×10^8 m/s,
Planck's constant = 6.63×10^{-34} J-s and
mass of electron = 9.1×10^{-31} kg) **[JEE Main 2019]**
(a) 1.45×10^6 m/s
(b) 1.8×10^6 m/s
(c) 1.1×10^6 m/s
(d) 1.7×10^6 m/s

8. A 100 W light bulb is placed at the centre of a spherical chamber of radius 0.10 m. Assume that 66% of the energy supplied to the bulb is converted into light and that the surface of chamber is perfectly absorbing. The pressure exerted by the light on the surface of the chamber is
(a) 0.87×10^{-6} Pa (b) 1.75×10^{-6} Pa
(c) 3.50×10^{-6} Pa (d) None of these

9. In an electron microscope, the resolution that can be achieved is of the order of the wavelength of electrons used. To resolve a width of 7.5×10^{-12} m, the minimum electron energy required is close to **[JEE Main 2019]**
(a) 500 keV (b) 1 keV
(c) 100 keV (d) 25 keV

10. Relativistic corrections become neccssary when the expression for the kinetic energy $\frac{1}{2}mv^2$, becomes comparable with mc^2 where m is the mass of the particle. At what de-Broglie wavelength will relativistic corrections become important for an electron?
(a) $\lambda = 10$ nm (b) $\lambda = 10^{-1}$ nm
(c) $\lambda = 10^{-4}$ nm (d) $\lambda = 10^{-6}$ nm

11. A photon and electron have same de-Broglie wavelength. Give that v is the speed of electron and c is the velocity of light. E_e, E_p are the kinetic energy of electron and photon respectively. p_e, p_h are the momentum of electron and photon respectively. Then which of the following relation is correct?
(a) $\frac{E_e}{E_p} = \frac{v}{2c}$ (b) $\frac{E_e}{E_p} = \frac{2c}{v}$
(c) $\frac{p_e}{p_h} = \frac{c}{2v}$ (d) $\frac{p_e}{p_h} = \frac{2c}{v}$

12. The de-Broglie wavelength of a photon is twice the de-Broglie wavelength of an electron. The speed of the electron is $v_e = \frac{c}{100}$. Then
(a) $\frac{E_e}{E_p} = 10^{-4}$ (b) $\frac{E_e}{E_p} = 10^{-3}$
(c) $\frac{p_e}{m_e c} = 10^{-2}$ (d) $\frac{p_e}{m_e c} = 10^{-4}$

13. Radiation with wavelength 6561 Å falls on a metal surface to produce photoelectrons. The electrons are made to enter a uniform magnetic field of 3×10^{-4}T. If the radius of the largest circular path followed by the electrons is 10 mm, the work function of the metal is close to **[JEE Main 2020]**
(a) 0.8 eV (b) 1.1 eV
(c) 1.8 eV (d) 1.6 eV

14. A particle p is formed due to a completely inelastic collision of particles x and y having de-Broglie wavelengths λ_x and λ_y, respectively. If x and y were moving in opposite directions, then the de-Broglie wavelength of p is **[JEE Main 2019]**
(a) $\lambda_x - \lambda_y$ (b) $\frac{\lambda_x \lambda_y}{\lambda_x - \lambda_y}$
(c) $\frac{\lambda_x \lambda_y}{\lambda_x + \lambda_y}$ (d) $\lambda_x + \lambda_y$

Numerical Value Questions

15. At an incident radiation frequency of ν_1, which is greater than the threshold frequency, the stoopping potential for a certain metal is V_1. At frequency $2\nu_1$, the stopping potential is $3V_1$. If the stopping potential at frequency $4\nu_1$ is nV_1, then n is ....

16. 1.5 mW of 400 nm light is directed at a photoelectric cell. If 0.1 % of the incident photons produce photoelectrons, find the current (in μA) in the cell.

17. Light of wavelength 180 nm ejects photoelectrons from a plate of metal whose work-function is 2 eV. If a uniform magnetic field of 5×10^{-5} T be applied parallel to the plate, what would be the radius (in m) of the path followed by electrons ejected normally from the plate with maximum energy?

18. The surface of a metal is illuminated alternately with photons of energies $E_1 = 4$ eV and $E_2 = 2.5$ eV respectively. The ratio of maximum speeds of the photoelectrons emitted in the two cases is 2. The work function of the metal (in eV) is .... **[JEE Main 2020]**

19. When radiation of wavelength λ is used to illuminate a metallic surface, the stopping potential is V. When the same surface is illuminated with radiation of wavelength 3λ, the stopping potential is $V/4$. If the threshold wavelength for the metallic surface is $n\lambda$, then value of n will be .... **[JEE Main 2020]**

20. The energy flux of sunlight reaching the surface of the earth is 1.388×10^3 W/m^2. The photons (nearly) per square metre are incident on the earth per second is 3.838×10^n, where the value of n is ....

Assume that, the photons in the sunlight have an average wavelength of 550 nm. **[NCERT]**

21. A beam of electromagnetic radiation of intensity 6.4×10^{-5} W/cm^2 is comprised of wavelength $\lambda = 310$ nm. It falls normally on a metal (work function $\phi = 2$ eV) of surface area of 1 cm^2. If one in 10^3 photons ejects an electron, total number of electrons ejected in 1 s is 10^x. ($hc = 1240$ eVnm, 1 eV $= 1.6 \times 10^{-19}$ J), then x is .... **[JEE Main 2020]**

22. A beam of electrons of energy E scatters from a target having atomic spacing of 1Å. The first maximum intensity occurs at $\theta = 60°$, then E (in eV) is ....
(Given, Planck's constant, $h = 6.64 \times 10^{-34}$ Js,
1 eV $= 1.6 \times 10^{-19}$ J and electron mass,
$m = 9.1 \times 10^{-31}$ kg. **[JEE Main 2020]**

23. A photon with energy of 4.9 eV ejects photoelectrons from tungsten. When the ejected electron enters a constant magnetic field of strength $B = 2.5$ mT at an angle of 60° with the field direction, the maximum pitch of the helix described by the electron is found to be 2.7 mm. Find the work-function of the metal in electron-volt. Given that, specific charge of electron is 1.76×10^{11} C / kg.

24. The maximum kinetic energy (in eV) of a photoelectron liberated from the surface of lithium with work function 2.35 eV by electromagnetic radiation whose electric component varies with time as : $E = a[1 + \cos(2\pi f_1 t)] \cos 2\pi f_2 t$ (where a is a constant) is ...... ($f_1 = 3.6 \times 10^{15}$ Hz, $f_2 = 1.2 \times 10^{15}$ Hz and Planck's constant $h = 6.6 \times 10^{-34}$ Js) **[JEE Main 2021]**

Answers

Round I

1. (a)	2. (a)	3. (a)	4. (a)	5. (d)	6. (a)	7. (a)	8. (b)	9. (b)	10. (d)
11. (a)	12. (a)	13. (d)	14. (c)	15. (b)	16. (b)	17. (d)	18. (c)	19. (c)	20. (c)
21. (d)	22. (c)	23. (a)	24. (c)	25. (a)	26. (b)	27. (c)	28. (a)	29. (a)	30. (b)
31. (a)	32. (a)	33. (a)	34. (b)	35. (c)	36. (c)	37. (b)	38. (d)	39. (b)	40. (a)
41. (c)	42. (d)	43. (a)	44. (d)	45. (a)	46. (c)	47. (d)	48. (d)	49. (d)	50. (b)
51. (b)	52. (c)	53. (c)	54. (d)	55. (d)	56. (d)	57. (a)	58. (d)	59. (c)	60. (a)
61. (a)	62. (b)	63. (a)	64. (c)	65. (d)	66. (c)				

Round II

1. (b)	2. (c)	3. (b)	4. (d)	5. (c)	6. (a)	7. (a)	8. (b)	9. (d)	10. (d)
11. (a)	12. (c)	13. (b)	14. (b)	15. 7	16. 0.48	17. 0.148	18. 2	19. 9	20. 21
21. 11	22. 50	23. 4.5	24. 17.45						

Solutions

Round I

1. Photoelectric current $(I) \propto$ Intensity of incident light

and intensity $\propto \dfrac{1}{(\text{distance})^2}$

So, $I \propto \dfrac{1}{(\text{distance})^2}$

Hence $I' = I\left(\dfrac{1}{4}\right)^2 = \dfrac{I}{16}$

2. Stopping potentials for b and a are same

$\therefore \quad f_a = f_b$

and saturation current for b and a are different

$\therefore \quad I_b \neq I_a$

3. KE of photoelectrons increases with increase in frequency of the incident light and is independent of the intensity of incident light.

4. The statement I is correct but rest is incorrect and it can be corrected as,

From the given graph, we can say that we have obtained different values of stopping potential but the same value of the saturation current for incident radiation of different frequencies. Since, the energy of emitted electron depends on the frequency of the incident radiations. This means, stopping potential is more negative for higher frequencies of incident radiation. Thus, we can conclude that,

I. stopping potential are in the order $V_{OA} > V_{OB} > V_{OC}$, if the frequencies are in the order $\nu_A > \nu_B > \nu_C$.

II. This further implies that greater the frequency of incident radiation, greater is the maximum kinetic energy of the photoelectrons. Consequently, we need greater retarding potential to stop them completely.

5. When a beam of electrons of energy E_0 is incident on a metal surface kept in an evacuated chamber, electrons can be emitted with maximum energy E_0 (due to elastic collision) and with any energy less than E_0, when part of incident energy of electron is used in liberating the electrons from the surface of metal.

6. According to graph between frequency (ν) of incident photon and maximum kinetic energy (E_K) of emitted electrons is given as,

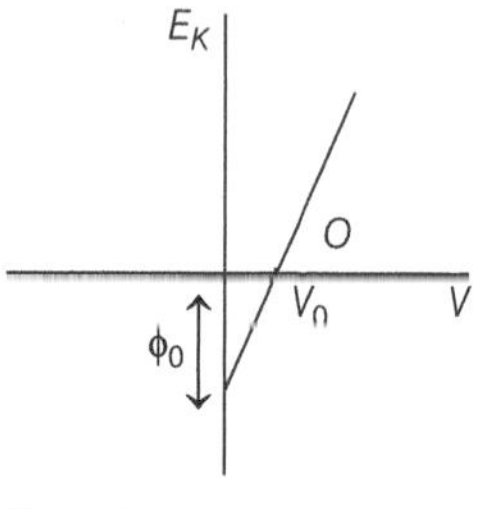

$E_K = h\nu - \phi_0$

Comparing above equation with $y = mx + C$, we get

Slope $m = h$ (Planck's constant)

7. As, $W_0 = \dfrac{hc}{\lambda_0}$; $\dfrac{(W_0)_T}{(W_0)_{Na}} = \dfrac{\lambda_{Na}}{\lambda_T}$

or $\lambda_T = \dfrac{\lambda_{Na} \times (W_0)_{Na}}{(W_0)_T} = \dfrac{5460 \times 2.3}{4.5} = 2791\ \text{Å}$

8. We have, $\dfrac{1}{2}mv_1^2 = 2\phi_0 - \phi_0 = \phi_0$

and $\dfrac{1}{2}mv_2^2 = 10\phi_0 - \phi_0 = 9\phi_0$

$\therefore \quad \dfrac{v_1}{v_2} = \sqrt{\dfrac{\phi_0}{9\phi_0}} = \dfrac{1}{3}$

9. The maximum energy $= h\nu - \phi$

$$K_1 = \frac{K_2}{2}$$

$$\left(\frac{hc}{\lambda_1} - \phi\right) = \frac{1}{2}\left(\frac{hc}{\lambda_2} - \phi\right)$$

$$\Rightarrow \left(\frac{1240}{600} - \phi\right) = \frac{1}{2}\left(\frac{1240}{400} - \phi\right)$$

$(\because hc = 1240\ \text{eV-nm})$

$$\Rightarrow \phi = \frac{1240}{1200} = 1.03 \approx 1.02\ \text{eV}$$

10. As, $E_k = \dfrac{hc}{\lambda} - \dfrac{hc}{\lambda_0} = hc\left(\dfrac{\lambda_0 - \lambda}{\lambda_0 \lambda}\right)$

11. Minimum wavelength occurs when mercury atom de-excites from highest energy level.

$\therefore$ Maximum possible energy absorbed by mercury atom $= \Delta E = 5.6 - 0.7 = 4.9$ eV

Wavelength of photon emitted in deexcitation is

$$\lambda = \frac{hc}{E} \approx \frac{1240\ \text{eV-nm}}{4.9\ \text{eV}} \approx 250\ \text{nm}$$

12. Given, $\lambda = 2.21 \times 10^{-7}$ m

So, energy of imparted by radiation in the lead ball,

$$E = \frac{hc}{\lambda} = \frac{6.62 \times 10^{-34} \times 3 \times 10^8}{2.21 \times 10^{-7}} = 9 \times 10^{-19}\ \text{J}$$

or $E = \dfrac{9 \times 10^{-19}}{1.6 \times 10^{-19}}$ eV $= 5.63$ eV

Work function of lead ball $= 4.14$ eV

So, maximum energy attained by ball

$= 5.63 - 4.14 = 1.49$ eV

Equivalent maximum potential is 1.49 V.

13. The maximum KE of the emitted photoelectrons is independent of the intensity of the incident light but depends upon the frequency of the incident light. Therefore, when the intensity of illumination is reduced to one-fourth of its original value, then the maximum kinetic energy of photoelectrons would be unchanged.

14. Here, $\lambda_{th} = 5200$ Å,

Thus, wavelength greater than 5200 Å cannot produce the photoelectric effect. The wavelength of UV-rays is 100-4000Å, so photoelectrons will be emitted by using 1W UV lamp.

15. As, $E_k = h\nu - W_0 = \frac{hc}{\lambda} - W_0$

$$W_0 = 4.2 \text{ eV} = 4.2 \times 1.6 \times 10^{-19}$$
$$= 6.72 \times 10^{-19} \text{ J}$$

So, $E_k = \frac{6.62 \times 10^{-34} \times 3 \times 10^8}{2000 \times 10^{-10}} - 6.72 \times 10^{-19}$

$\Rightarrow$ $E_k = 9.9 \times 10^{-19} - 6.72 \times 10^{-19}$

$\Rightarrow$ $E_k = 3.18 \times 10^{-19}$ J

Hence, stopping potential

$$V_0 = \frac{E_k}{e} = \frac{3.18 \times 10^{-19}}{1.6 \times 10^{-19}} = 1.99 \text{ V}$$

16. As, $K_{max} = hf - W_0$

or $V_0 = \frac{hc}{e\lambda} - \frac{W_0}{e}$

$\therefore$ $\Delta V_0 = (V_0)_2 - (V_0)_1$

$$= \left[\frac{hc}{e\lambda_2} - \frac{W_0}{e}\right] - \left[\frac{hc}{e\lambda_1} - \frac{W_0}{e}\right]$$

$$= \frac{hc}{e}\left[\frac{1}{\lambda_2} - \frac{1}{\lambda_1}\right]$$

$$\Delta V_0 = \frac{hc}{e}\left[\frac{\lambda_1 - \lambda_2}{\lambda_1\lambda_2}\right]$$

$\Rightarrow$ $\frac{hc}{e} = \frac{\Delta V_0 \cdot \lambda_1\lambda_2}{\lambda_1 - \lambda_2}$

Substituting given values, we get

$$= \frac{(1.3 - 0.9) \times 4000 \times 10^{-10} \times 4500 \times 10^{-10}}{500 \times 10^{-10}}$$

$$= 1.44 \times 10^{-6} \text{ Vm}$$

Also, $V_0 = \frac{hc}{e\lambda} - \frac{W_0}{e}$

$$\frac{W_0}{e} = \frac{hc}{e\lambda} - V_0$$

$$= \frac{1.44 \times 10^{-6}}{4000 \times 10^{-10}} - 1.3 = 2.3 \text{ V}$$

$\Rightarrow$ $W_0 = 2.3$ eV

17. From graph, $\nu = 10^{15}$ Hz

$$K_{max} = 3\text{eV} = 3 \times 1.6 \times 10^{-19} \text{ J}$$

As, $K_{max} = h\nu - h\nu_0$

or $\nu_0 = \nu - \frac{K_{max}}{h}$

$$= 10^{15} - \frac{3 \times 1.6 \times 10^{-19}}{6.6 \times 10^{-34}}$$
$$= (10 - 7.3) \times 10^{14}$$
$$= 2.7 \times 10^{14} \text{ Hz}$$

18. As, $E = \frac{hc}{\lambda} - W_0$ and $2E = \frac{hc}{\lambda'} - W_0$

$\Rightarrow$ $\frac{\lambda'}{\lambda} = \frac{E + W_0}{2E + W_0}$

$\Rightarrow$ $\lambda' = \lambda \frac{\left(1 + \frac{W_0}{E}\right)}{\left(2 + \frac{W_0}{E}\right)}$

Since, $\frac{\left(1 + \frac{W_0}{E}\right)}{\left(2 + \frac{W_0}{E}\right)} > \frac{1}{2}$

So, $\lambda' > \frac{\lambda}{2}$

Hence, $\frac{\lambda}{2} < \lambda' < \lambda.$

19. Given, threshold wavelength, $\lambda_0 = 380$ nm

Wavelength of incident light, $\lambda = 260$ nm

Using Einstein's relation of photoelectric effect,

$$(\text{KE})_{max} = eV_0 = h\nu - h\nu_0 \quad \text{...(i)}$$

But $h\nu = E = \frac{1237}{\lambda(\text{nm})}$ eV (Given)

$\therefore$ $E_0 = \frac{1237}{\lambda_0(\text{nm})}$ eV ...(ii)

From Eqs. (i) and (ii), we get

$$(\text{KE})_{max} = E - E_0\left(\frac{1237}{\lambda} - \frac{1237}{\lambda_0}\right) \text{eV}$$

$$= 1237\left[\frac{1}{\lambda} - \frac{1}{\lambda_0}\right] \text{eV } (\lambda \text{ in nm}) \quad \text{... (iii)}$$

By putting values of λ and λ_0 in Eq. (iii), we get

$$(\text{KE})_{max} = 1237\left(\frac{1}{260} - \frac{1}{380}\right) \text{eV}$$

$$= 1237 \times \left[\frac{380 - 260}{380 \times 260}\right] \text{eV}$$

$\Rightarrow$ $(\text{KE})_{max} = 1.5$ eV

20. Relation between stopping potential and incident light's frequency is $eV_0 = hf - \phi_0$, where V_0 is the stopping potential and ϕ_0 is the the work function of the photosensitive surface.

So, from given data, we have

$$-e\frac{V_0}{2} = h\nu - \phi_0 \quad \text{...(i)}$$

and $-eV_0 = \frac{h\nu}{2} - \phi_0$ ...(ii)

Subtracting Eqs. (i) from (ii), we have

$$-eV_0-\left(-\frac{eV_0}{2}\right)=\frac{h\nu}{2}-h\nu$$

$$\Rightarrow \quad -\frac{eV_0}{2}=-\frac{h\nu}{2}$$

$$\Rightarrow \quad eV_0=h\nu$$

Substituting this in Eq. (i), we get

$$-\frac{eV_0}{2}=eV_0-\phi_0$$

$$\Rightarrow \quad -\left(\frac{3}{2}eV_0\right)=-\phi_0 \text{ or } \frac{3}{2}h\nu=\phi_0$$

If threshold frequency is ν_0, then

$$h\nu_0=\frac{3}{2}h\nu \Rightarrow \nu_0=\frac{3}{2}\nu$$

21. Given, $\lambda_1=300$ nm; $\lambda_2=400$ nm

$$\frac{hc}{e}=1240 \text{ nm}$$

Using Einstein equation for photoelectric effect,

$$E=h\nu=\phi+eV_0 \quad \text{...(i)}$$

(here, ϕ is work function of the metal and V_0 is stopping potential)

For λ_1 wavelength's wave,

$$E_1=h\nu_1=\phi+eV_{01}$$

or $$\frac{hc}{\lambda_1}=\phi+eV_{01} \quad \text{...(ii)}$$

Similarly, $$\frac{hc}{\lambda_2}=\phi+eV_{02} \quad \text{...(iii)}$$

From Eqs. (ii) and (iii), we get

$$hc\left[\frac{1}{\lambda_1}-\frac{1}{\lambda_2}\right]=e(V_{01}-V_{02}) \text{ or } \frac{hc}{e}\left[\frac{1}{\lambda_1}-\frac{1}{\lambda_2}\right]=\Delta V$$

By using given values,

$$\Delta V=1240\left[\frac{1}{300}-\frac{1}{400}\right]\frac{\text{nm-V}}{\text{nm}}$$

$$=1240\times\frac{1}{1200}\text{ V}$$

$$\Rightarrow \quad \Delta V=1.03\text{ V}\approx 1\text{V}$$

22. $eV_1=\frac{hc}{\lambda_1}-\phi_0,\ eV_2=\frac{hc}{\lambda_2}-\phi_0,\ eV_3=\frac{hc}{\lambda_3}-\phi_0$

$$\Rightarrow \quad V_1=\frac{hc}{e\lambda_1}-\frac{\phi_0}{e}$$

$$V_2=\frac{hc}{e\lambda_2}-\frac{\phi_0}{e}$$

$$V_3=\frac{hc}{e\lambda_3}-\frac{\phi_0}{e}$$

$\because$ V_1, V_2 and V_3 are in AP.

$$\Rightarrow \quad 2V_2=V_1+V_3$$

$$\Rightarrow \quad \frac{2hc}{e\lambda_2}-\frac{2\phi_0}{e}=\frac{hc}{e\lambda_1}+\frac{hc}{e\lambda_3}-\frac{2\phi_0}{e}$$

$$\Rightarrow \quad \frac{2}{\lambda_2}=\frac{1}{\lambda_1}+\frac{1}{\lambda_3}$$

$\Rightarrow$ λ_1, λ_2 and λ_3 are in harmonic progression.

23. According to the law of conservation of energy, *i.e.* Energy of a photon $(h\nu)$ = Work function (ϕ) + Kinetic energy of the photoelectron $\left(\frac{1}{2}mv^2_{max}\right)$

According to Einstein's photoelectric emission of light

$$\Rightarrow \quad (\text{KE})_{\max}=\frac{hc}{\lambda}-\phi$$

$$\Rightarrow \quad \frac{1}{2}mv^2=\frac{hc}{\lambda}-\phi \quad \text{(i)}$$

If the wavelength of radiation is changed to $\frac{3\lambda}{4}$, then

$$\frac{1}{2}mv_2^2=\frac{hc}{3\lambda/4}-\phi=\frac{4hc}{3\lambda}-\phi$$

$$\Rightarrow \quad \frac{1}{2}mv_2^2=\frac{4hc}{3\lambda}-\phi \quad \text{(ii)}$$

Clearly, from Eqs. (i) and (ii), we get

$$\frac{1}{2}mv_2^2=\frac{4}{3}\left(\frac{1}{2}mv^2+\phi\right)-\phi$$

$$\frac{1}{2}mv_2^2=\frac{2}{3}mv^2+\frac{\phi}{3}$$

$$\Rightarrow \quad \frac{1}{2}mv_2^2>\frac{2}{3}mv^2$$

$$v_2^2>\frac{4}{3}v^2$$

$$\Rightarrow \quad v_2>v\left(\frac{4}{3}\right)^{1/2}$$

i.e. $$v'>v\left(\frac{4}{3}\right)^{1/2}$$

24. From Einstein's photoelectric equation,

$$\phi=\phi_0+eV_0 \quad [\text{here, } V_0-\text{stopping potential}]$$

$$\Rightarrow hf=\phi_0+eV_0 \quad \text{...(i)}$$

From graph, take point B $(5.5\times10^{14}$ Hz, 0 V)

On satisfying Eq. (i) with point B, we get

$$6.62\times10^{-34}\times5.5\times10^{14}=\phi_0+1.6\times10^{-19}\times0$$

$$36.41\times10^{-20}=\phi_0+0$$

$$\phi_0=36.41\times10^{-20}\text{ J}$$

$$=\frac{36.41\times10^{-20}}{1.6\times10^{-19}}\text{ eV}$$

$$=22.75\times10^{-1}\text{ eV}$$

$$=2.275\text{ eV}$$

$$\simeq 2.27\text{eV}$$

25. From photoelectric equation,

$$\phi=\phi_0+eV$$

$$\Rightarrow \quad \frac{hc}{\lambda}=\phi_0+eV$$

$$eV=\frac{hc}{\lambda}-\phi_0$$

$$\Rightarrow \quad V=\frac{hc}{\lambda e}-\frac{\phi_0}{e}$$

$$V = \left(\frac{hc}{e}\right)\left(\frac{1}{\lambda}\right) - \frac{\phi_0}{e}$$

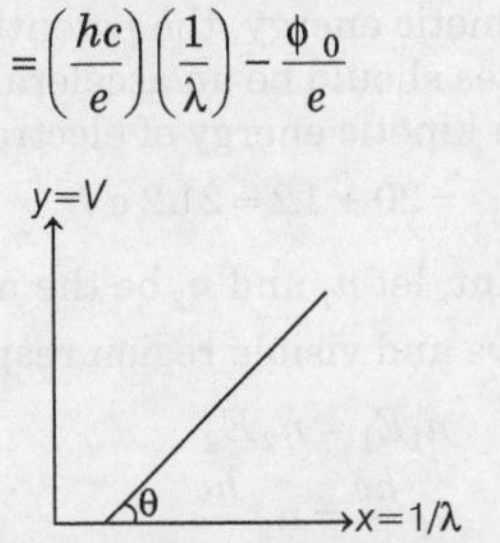

On comparing the above equation with equation of straight line, *i.e.* $y = mx + c$, we get

Slope, $\quad m = \frac{hc}{e}$

y-intercept $c = -\frac{\phi_0}{e}$

Now, if we change intensity of incident radiation, then h, c, e and ϕ_0 will not change. Therefore, neither slope of the graph, nor the y-intercept cut by graph will change on changing the intensity of incident radiation. So, graph will not change.

26. Given, Planck's constant,

$$h = 6.63 \times 10^{-34} \text{ J-s}, \; e = 1.6 \times 10^{-19} \text{ C}$$

and there is a graph between stopping potential and frequency.

We need to determine work function W.

Using Einstein's relation of photoelectric effect,

$$(\text{KE})_{\max} = eV_0 = h\nu - h\nu_0 = h\nu - W \qquad [\because W = h\nu_0]$$

or $\quad V_0 = \frac{h}{e}\nu - \frac{W}{e}$

From graph at $V_0 = 0$ and $\nu = 4 \times 10^{14}$ Hz

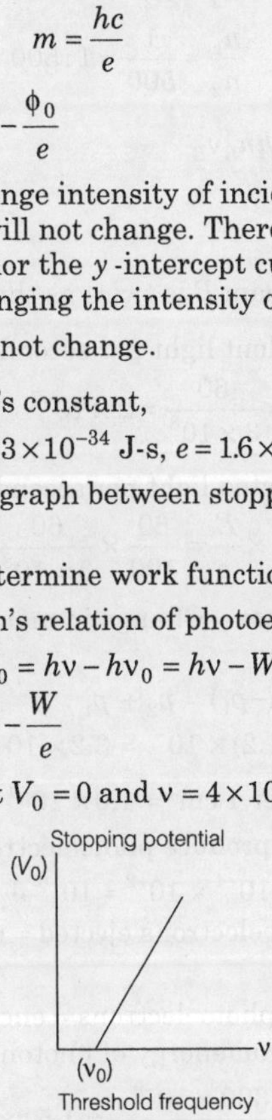

$\therefore \quad 0 = \frac{6.63 \times 10^{-34}}{e} \times 4 \times 10^{14} - \frac{W}{e}$

$\Rightarrow \quad \frac{W}{e} = \frac{6.63 \times 10^{-34} \times 4 \times 10^{14}}{e}$ J

or $\quad W = 6.63 \times 4 \times 10^{-20}$ J

or $\quad W = \frac{6.63 \times 4 \times 10^{-20}}{1.6 \times 10^{-19}}$ eV $= 1.657$ eV

$\therefore \quad W = 1.66$ eV

27. We know that, according to Einstein equation,

$$E = W_0 + \frac{1}{2}mv^2$$

or $\quad \sqrt{\frac{2(E - W_0)}{m}} = v$

When emitted electron enters into magnetic field B in ⊥ direction, it moves on circular path and radius of circular path is given as,

$\Rightarrow \quad r = \frac{mv}{eB} = \frac{\sqrt{2m(E - W_0)}}{eB}$

28. Given, wavelength of incident radiation,

$$\lambda = 400 \text{ nm} = 4 \times 10^{-7} \text{ m}$$

electric field, $E = 2$ N/C and distance, $s = 1$ m

$\therefore$ Energy of the incident light,

$$E = \frac{hc}{\lambda} = \frac{6.6 \times 10^{-34} \times 3 \times 10^8}{4 \times 10^{-7}}$$

$$= 4.95 \times 10^{-19} \text{ J}$$

$$= \frac{4.95 \times 10^{19}}{1.6 \times 10^{-19}} \text{ J}$$

$$= 3.09 \simeq 3.1 \text{ eV}$$

If a be the retardation of emitted electrons in the electric field,

then, $\quad a = \frac{qE}{m}$

initial speed u of emitted electron is calculated as

$$v^2 = u^2 - 2as$$

$\Rightarrow \quad 0 = u^2 - 2a \times 1$

$\Rightarrow \quad u^2 = 2a = \frac{2qE}{m}$

$\therefore$ Maximum kinetic energy of the electron,

$$K_{\max} = \frac{1}{2}mu^2$$

$$= \frac{1}{2}m \cdot \frac{2qE}{m} = qE$$

$$= 1.6 \times 10^{-19} \times 2$$

$$K_{\max} = 3.2 \times 10^{-19} \text{ J} = 2 \text{ eV}$$

$\therefore$ Work function of the surface,

$$W_0 = E - K_{\max} = 3.1 - 2 = 1.1 \text{ eV}$$

29. Given, $\mathbf{E} = 10^{-3} \cos\left(\frac{2\pi x}{5 \times 10^{-7}} - 2\pi \times 6 \times 10^{14} t\right) \hat{\mathbf{x}} \text{NC}^{-1}$

By comparing it with the general equation of electric field of light, *i.e.*

$$E = E_0 \cos(kx - \omega t)\,\hat{\mathbf{x}}, \text{ we get}$$

$$k = \frac{2\pi}{5 \times 10^{-7}} = 2\pi/\lambda \quad (\text{from definition, } k = 2\pi/\lambda)$$

$\Rightarrow \quad \lambda = 5 \times 10^{-7}$ m $= 5000$ Å $\qquad$ ...(i)

The value of λ can also be calculated as, after comparing the given equation of **E** with standard equation, we get

$$\omega = 6 \times 10^{14} \times 2\pi$$

$\Rightarrow \quad \nu = 6 \times 10^{14} \qquad [\because 2\pi\nu = \omega]$

As, $\quad c = \nu\lambda$

$\Rightarrow \quad \lambda = \frac{c}{\nu} = \frac{3 \times 10^8}{6 \times 10^{14}} = 5 \times 10^{-7}$ m $= 5000$ Å

According to Einstein's equation for photoelectric effect, *i.e.*

$$\frac{hc}{\lambda} - \phi = (\text{KE})_{max} = eV_0 \quad \text{...(ii)}$$

For photon, substituting the given values,

$$E = \frac{hc}{\lambda} = \frac{12375 \text{ eV}}{\lambda} \quad \text{[given]}$$

or $$\frac{hc}{\lambda} = \frac{12375}{5000} \text{ eV} \quad \text{[using Eq. (i)] ...(iii)}$$

Now, substituting the values from Eq. (iii) in Eq. (ii), we get

$$\frac{12375}{5000} \text{ eV} - 2\text{eV} = eV_0$$

$\Rightarrow$ $2.475 \text{ eV} - 2 \text{ eV} = eV_0$

or $V_0 = 2.475 \text{ V} - 2 \text{ V} = 0.475 \text{ V}$

$\Rightarrow$ $V_0 \approx 0.48 \text{ V}$

30. Let threshold frequency of emitter plate be ν_0

Energy of photon in first case is E.

When switch S_1 is closed and switch S_2 is open,

So, $$E = h\nu_0 + (5+1) \text{ eV} \quad \text{...(i)}$$

For second case, when switch S_1 is open and switch S_2 is closed and frequency of incident light is doubled.

then, $$2E = h\nu_0 + (20-5) \text{ eV} \quad \text{...(ii)}$$

From Eqs. (i) and (ii), we get

$\Rightarrow$ $2(h\nu_0 + 6\text{eV}) = h\nu_0 + 15\text{eV}$

$\Rightarrow$ $2h\nu_0 + 12\text{eV} = h\nu_0 + 15\text{eV}$

$\Rightarrow$ $h\nu_0 = 3 \text{ eV}$

$\Rightarrow$ $$\nu_0 = \frac{3 \times 1.6 \times 10^{-19}}{6.62 \times 10^{-34}} = 7.25 \times 10^{14} \text{ Hz}$$

$\therefore$ $$\lambda_0 = \frac{c}{\nu_0} = \frac{3 \times 10^8}{7.25 \times 10^{14}}$$

$$\lambda_0 = 4133.3 \text{ Å}$$

Hence, the threshold wavelength of the emitter plate is 4133.3 Å.

31. Electric field between the plate, $E = \frac{\sigma}{\varepsilon_0}$

The potential difference between the plates,

$$V = E \cdot d = \frac{\sigma}{\varepsilon_0} \cdot d$$

$\Rightarrow$ $$V = \frac{1.0 \times 10^{-9} \times 17.7}{8.85 \times 10^{-12} \times 100} = 20 \text{ V}$$

From Einstein photoelectric equation,

$$K_{max} = \frac{hc}{\lambda} - W_0 = \frac{12400}{4000} - 1.9 = 1.2 \text{ eV}$$

But $K_{max} = eV_0$

$\Rightarrow$ $1.2 \text{ eV} = eV_0$

$\Rightarrow$ $V_0 = 1.2 \text{ V}$

As V_0 is less than the potential difference between the plates (20 V). Hence, the minimum energy required to reach the charged plate should be $K_{min} = 20$ eV.

For maximum kinetic energy, the potential difference between the plates should be an accelerating one. Hence maximum kinetic energy of electron should be

$$K_{max} = 20 + 1.2 = 21.2 \text{ eV}$$

32. Total E is constant, let n_1 and n_2 be the number of photons of X-rays and visible region respectively.

Then, $$n_1 E_1 = n_2 E_2$$

$\Rightarrow$ $$n_1 \frac{hc}{\lambda_1} = n_2 \frac{hc}{\lambda_2}$$

$\Rightarrow$ $$\frac{n_1}{n_2} = \frac{\lambda_1}{\lambda_2}$$

$\Rightarrow$ $$\frac{n_1}{n_2} = \frac{1}{500} = 1:500$$

33. As, $E = hn_A\nu_n = hn_B\nu_B$

$\Rightarrow$ $$\frac{n_A}{n_B} = 2 = \frac{\nu_B}{\nu_A}$$

The frequency of beam B is twice as that of A.

34. Momentum of incident light per second

$$p_1 = \frac{E}{c} = \frac{60}{3 \times 10^8} = 2 \times 10^{-7}$$

Momentum of reflected light per second

$$p_2 = \frac{60}{100} \times \frac{E}{c} = \frac{60}{100} \times \frac{60}{3 \times 10^8} = 1.2 \times 10^{-7}$$

$\therefore$ Force on the surface = Change in momentum per second

$$= p_2 - (-p_1) = p_2 + p_1$$
$$= (2 + 1.2) \times 10^{-7} = 3.2 \times 10^{-7} \text{ N}$$

35. Energy incident over 1 cm^2 = 1.0×10^{-4} J ;

Energy required to produce photoelectrons

$$= 1.0 \times 10^{-4} \times 10^{-2} = 10^{-6} \text{ J}$$

As, number of photoelectrons ejected = number of photons

which can produce photoelectrons = energy required for producing electron/energy of photon.

$$= \frac{10^{-6}}{hc/\lambda} = \frac{10^{-6} \times 300 \times 10^{-9}}{6.6 \times 10^{-34} \times 3 \times 10^8} = 1.51 \times 10^{12} \text{ s}^{-1}$$

36. Energy of photon,

$$E = \frac{hc}{\lambda} = \frac{6.63 \times 10^{-34} \times 3 \times 10^8}{590 \times 10^{-9}} = \frac{6.63 \times 3}{59} \times 10^{-18}$$

Light energy produced per second $= \frac{90}{100} \times 10 = 9$ W

$\therefore$ Number of photons emitted per sec

$$= \frac{9 \times 59}{6.63 \times 3 \times 10^{-18}} = 2.67 \times 10^{19} \approx 0.267 \times 10^{20}$$

37. Power of laser is given as $P = \frac{\text{Energy}}{\text{Time}}$

$$= \frac{\text{Number of photons emitted} \times \text{Energy of one photon}}{\text{Time}}$$

$$\Rightarrow \quad P = \frac{NE}{t} = \left(\frac{N}{t}\right) \cdot E$$

So, number of photons emitted per second

$$= \frac{N}{t} = \frac{P}{E}$$

$$= \frac{P}{hc/\lambda} = \frac{P\lambda}{hc} \qquad \left[\because E = h\nu = \frac{hc}{\lambda}\right]$$

Here, $h = 6.6 \times 10^{-34}$ J-s, $\lambda = 500$ nm

$= 500 \times 10^{-9}$ m; $c = 3 \times 10^8$ ms^{-1}

$P = 2$ mW $= 2 \times 10^{-3}$ W

$$\therefore \quad \frac{N}{t} = \frac{2 \times 10^{-3} \times 500 \times 10^{-9}}{6.6 \times 10^{-34} \times 3 \times 10^8}$$

$$= 5.56 \times 10^{15}$$

$$\approx 5 \times 10^{15} \text{ photons per second}$$

38. In photoemission, only one electron can produce one photon.

So, if only x% of incident photons can produce electrons, then

$$\text{then, } x\% = \frac{\text{Energy of incident photons which emit electrons}}{\text{Total energy of source}} \times 100$$

$$= \frac{hc/\lambda}{P \times t} \times 100 = \frac{hc \times 100}{P\lambda t} = \frac{100hc}{P\lambda\left(\frac{e}{I}\right)} \qquad [\because e = It]$$

$$x\% = 100hcI / P\lambda e$$

39. Power of source $= \dfrac{\text{Energy emitted}}{\text{Time}}$

$= \dfrac{\text{Number of photons} \times \text{Energy of 1 photon}}{\text{Time}}$

$=$ Number density of photons $\times$ Energy of 1 photon

$= n \cdot \dfrac{hc}{\lambda}$

Now, $\quad P_{\text{X-rays}} = P_{\text{visible light}}$

$$\Rightarrow \quad n_X\left(\frac{hc}{\lambda_X}\right) = n_V\left(\frac{hc}{\lambda_V}\right)$$

$$\Rightarrow \quad \frac{n_V}{n_X} = \frac{\lambda_V}{\lambda_X} = \frac{500}{1} = 500$$

40. Given,

source power, $P = 1$W

Intensity of light source at a distance of 1 m is given by

$$I = \frac{P}{4\pi r^2} = \frac{1}{4\pi . 1^2} = \frac{1}{4\pi} \text{ W/m}^2$$

Power absorbed by the circular plates of radius 1 Å is

$P = I \times$ Area of plate,

$$= \frac{1}{4\pi} \times \pi r^2 \qquad [\because r = 1 \text{ Å} = 10^{-10} \text{m}]$$

$$P = \frac{1}{4}(10^{-10})^2 = \frac{10^{-20}}{4}$$

Work function of cobalt

$$W = 5 \text{ eV}$$

$$= 5 \times 1.6 \times 10^{-19} \text{ J}$$

$$P = \frac{W}{t}$$

$$t = \frac{W}{P}$$

$$\therefore \quad t = \frac{5 \times 1.6 \times 10^{-19}}{\frac{10^{-20}}{4}} = 320 \text{ s}$$

41. Given that, power of lamp, $P = 300$ W

distance of lamp from sensor, $d = 10$ m,

radius of sensor opening, $r = 1$ cm, wavelength of light emitted from photodiode, $\lambda = 660$ nm and exposure time, $t = 100$ ms $= 100 \times 10^{-3}$ s.

Engergy of photons, $E = \dfrac{hc}{\lambda}$

Putting the given values, we get

$$\Rightarrow \quad E = \frac{6.6 \times 10^{-34} \times 3 \times 10^8}{660 \text{ nm}} = 3 \times 10^{-19}$$

The area of of exposure of lamp at a radius of 10 m

$$A_0 = 4\pi r^2 = 4\pi \times 10^2 \text{m}^2$$

Similarly, The area of sensor

$$A'_s = \pi r^2 = \pi \times (10^{-2})^2 = \pi \times 10^{-4} \text{m}^2$$

Energy emitted by lamp in exposure time,

$E_0 =$ Power of lamp (p) $\times$ Exposure time (t)

$$E_0 = 300 \times 100 \times 10^{-3} = 30 \text{ J}$$

So, the number of photons entering to sensor is given by

$$N = \frac{E_0}{E} \times \frac{A'_s}{A_0} = \frac{30}{3 \times 10^{-19}} \times \frac{\pi \times 10^{-4}}{4\pi \times 10^2}$$

$$\Rightarrow \quad N = 2.5 \times 10^{13}$$

42. Energy spent in conversion $= mL = 80000$ cal

Energy used by photon $= nT \times E = nT \times h\nu$

$$\Rightarrow \quad nTh\nu = mL$$

$$\text{or} \quad T = mL/nh\nu$$

$\therefore \ T \propto \dfrac{1}{n}$, when ν is fixed, $T \propto \dfrac{1}{\nu}$ when n is fixed.

43. $E_1 = \dfrac{12400}{4000} = 3.1$ eV

$$E_2 = \frac{12400}{4800} = 2.58 \text{ eV}$$

$$E_3 = \frac{12400}{6000} = 2.06 \text{ eV}$$

$$E_4 = \frac{12400}{7000} = 1.77 \text{ eV}$$

$$E_5 = \frac{12400}{7800} = 1.58 \text{ eV}$$

Clearly 4th and 5th wavelengths are not emitting any electron because energy of photons corresponding to these wavelengths is less than work function.

Now, number of photoelectrons emitted per second

= Number of photons incident per second

$$= \frac{I_1A_1}{E_1} + \frac{I_2A_2}{E_2} + \frac{I_3A_3}{E_3}$$

$$= IA\left(\frac{1}{E_1} + \frac{1}{E_2} + \frac{1}{E_3}\right) \quad (\because I_1A_1 = I_2A_2 = I_3A_3 = IA)$$

$$= \frac{7.5\times10^{-3}}{5} \times 10^{-4} \times \frac{1}{1.6\times10^{-19}} \times \left(\frac{1}{3.1} + \frac{1}{2.58} + \frac{1}{2.06}\right)$$

$= 1.12\times10^{12}$ electrons per second

44. Radiation pressure over an absorbing surface is,

$$p = \frac{I}{c}$$

where, I = intensity or energy flux

and c = speed of light.

If A = area of surface, then force due to radiation on the surface is

$$F = p \times A = \frac{IA}{c}$$

If force F acts for a duration of Δt seconds, then momentum transferred to the surface is

$$\Delta p = F \times \Delta t = \frac{IA}{c} \times \Delta t$$

Here, $I = 25$ W cm^{-2}, $A = 25$ cm^2,

$c = 3\times10^8$ ms^{-1},

$\Delta t = 40$ min $= 2400$ s

So, momentum transferred to the surface,

$$\Delta p = \frac{25\times25\times2400}{3\times10^8} = 5\times10^{-3} \text{ N-s}$$

45. We know that, intensity of a radiation I with energy 'E' incident on a plate per second per unit area is given as

$$\Rightarrow \quad I = \frac{dE}{dA\times dt}$$

$$\Rightarrow \quad \frac{dE}{dt} = IdA \text{ or } IA$$

i.e. energy incident per unit time = IA

Substituting the given values, we get

$$\frac{dE}{dt} = 16\times10^{-3}\times1\times10^{-4}$$

$$\frac{dE}{dt} = 16\times10^{-7} \text{ W} \quad \ldots\text{(i)}$$

Using Einstein's photoelectric equation, we can find kinetic energy of the incident radiation as

$$E = \frac{1}{2}mv^2 + \phi$$

(Here, ϕ is work function of metal)

or $E = \text{KE} + \phi$

$\text{KE} = E - \phi = 10 \text{ eV} - 5 \text{ eV}$

$\Rightarrow \text{KE} = 5 \text{ eV} \quad \ldots\text{(ii)}$

Now, energy per unit time for incident photons will be

$\because \quad E = Nh\nu$

$$\therefore \quad \frac{dE}{dt} = h\nu\frac{dN}{dt} \text{ or } h\nu\dot{N} \quad \ldots\text{(iii)}$$

From Eqs. (i) and (iii), we get

$$h\nu\dot{N} = 16\times10^{-7} \text{ or } E\dot{N} = 16\times10^{-7}$$

But $E = 10$ eV, so

$$\dot{N}(10\times1.6\times10^{-19}) = 16\times10^{-7} \Rightarrow \dot{N} = 10^{12}$$

$\because$ Only 10% of incident photons emit electrons.

So, emitted electrons per second are

$$\frac{10}{100}\times10^{12} = 10^{11}$$

46. de-Broglie wavelength, $\lambda = \dfrac{h}{mv}$

As both particle and electrons have same wavelength, therefore their momentum will be equal to

$\Rightarrow \quad m_pv_p = m_ev_e$

$$\Rightarrow \quad v_p = \frac{m_ev_e}{m_p} = \frac{9.1\times10^{-31}\times3\times10^6}{10^{-6}}$$

$\Rightarrow \quad v_p = 2.7\times10^{-18}$ m/s

47. As, $E_K = \dfrac{1}{2}mv^2$ or $mv = \sqrt{2mE_K}$

As per question, $m_pv_p = m_ev_e$

or $\sqrt{2m_pE_{K_p}} = \sqrt{2m_eE_{K_e}}$

or $\dfrac{E_{K_e}}{E_{K_p}} = \dfrac{m_p}{m_e} > 1$

or $E_{K_e} > E_{K_p}$

48. As, $\lambda = \dfrac{h}{p}$ or $L = \dfrac{h}{p}$, *i.e.* $L \propto \dfrac{1}{p}$. The curve (d) is correct.

49. $\lambda = \dfrac{h}{mv} = \dfrac{h}{\sqrt{2mK}} = \dfrac{h}{\sqrt{2mqV}}$

$$\frac{\lambda_1}{\lambda_2} = \sqrt{\frac{m_2}{m_1}}$$

$$\frac{\lambda_e}{\lambda_p} = \sqrt{\frac{m_p}{m_e}} = \sqrt{1831.4} = 42.79 \simeq 43$$

50. As, $\lambda = \dfrac{h}{\sqrt{2mE}}$

$$\Rightarrow \quad \lambda \propto \frac{1}{\sqrt{E}}$$

$$\Rightarrow \quad \frac{\lambda_1}{\lambda_2} = \sqrt{\frac{E_2}{E_1}}$$

$$\Rightarrow \quad \frac{10^{-10}}{0.5\times10^{-10}} = \sqrt{\frac{E_2}{E_1}}$$

$\Rightarrow \quad E_2 = 4E_1$

Therefore, added energy $= E_2 - E_1 = 3E_1$

51. Kinetic energy of particle $K = \dfrac{1}{2}mv^2$

or $\quad mv = \sqrt{2mK}$

de-Broglie wavelength $\lambda = \dfrac{h}{mv} = \dfrac{h}{\sqrt{2mK}}$

For the given value of $K, \lambda \propto 1/\sqrt{m}$.

$\therefore \quad \lambda_p : \lambda_n : \lambda_e : \lambda_\alpha = \dfrac{1}{\sqrt{m_p}} : \dfrac{1}{\sqrt{m_n}} : \dfrac{1}{\sqrt{m_e}} : \dfrac{1}{\sqrt{m_\alpha}}$

Since $m_p = m_n$, hence $\lambda_p = \lambda_n$

As $m_\alpha > m_p$, therefore $\lambda_\alpha < \lambda_p$,

As $m_e < m_n$, therefore $\lambda_e > \lambda_n$,

Hence, $\quad \lambda_\alpha < \lambda_p = \lambda_n < \lambda_e$

or $\quad \lambda_e > \lambda_p = \lambda_n > \lambda_\alpha$

52. de-Broglie wavelength of a moving particle is given by

$$\lambda = \frac{h}{p} = \frac{h}{\sqrt{2mK}}$$

where, K = kinetic energy of the particle.

Now, in given condition,

$$\lambda = \frac{h}{\sqrt{2mE}} \quad \text{...(i)}$$

and $$\frac{\lambda}{2} = \frac{h}{\sqrt{2m(E + \Delta E)}} \quad \text{...(ii)}$$

On dividing Eq. (i) by Eq. (ii), we get

$$2 = \frac{\sqrt{E + \Delta E}}{\sqrt{E}}$$

$$\Rightarrow \quad \Delta E = 3E$$

53. de-Broglie wavelength of a particle having kinetic energy, is given by

$$\lambda = \frac{h}{p} = \frac{h}{\sqrt{2m\,(\text{KE})}}$$

For same kinetic energy, $\lambda \propto \dfrac{1}{\sqrt{m}}$

$\because \quad m_{\text{He}^{++}} > m_p > m_e$

$\Rightarrow \quad \lambda_{\text{He}^{++}} < \lambda_p < \lambda_e$

54. We have, $E = qV$, we know that $E = \dfrac{1}{2}mv^2$

$$\Rightarrow \quad v = \sqrt{\frac{2E}{m}}$$

As, $$\lambda = \frac{h}{mv} = \frac{h}{m\sqrt{\dfrac{2E}{m}}}$$

$$\Rightarrow \quad \lambda = \frac{h}{\sqrt{2mqV}} \quad \text{...(i)}$$

For electron, $$\lambda_e = \frac{h}{\sqrt{2m_e qV}} \quad \text{...(ii)}$$

For protron, $$\lambda_p = \frac{h}{\sqrt{2m_p qV}}$$

$$\therefore \quad \lambda_p = \frac{h}{\sqrt{2m_p q \cdot 4V}} \quad (\because V = 4V) \text{ ...(iii)}$$

Dividing Eq. (ii) by Eq. (iii), we get

$$\frac{\lambda_e}{\lambda_p} = 2\sqrt{\frac{m_p}{m_e}}$$

55. Here, energy of electron $= E$

$$\Rightarrow \quad \frac{1}{2}mv^2 = E$$

$$\Rightarrow \quad m^2v^2 = 2mE \Rightarrow p^2 = 2mE$$

or momentum of electron, $p = \sqrt{2mE}$

So, de Broglie wavelength associated with electron,

$$\lambda_e = \frac{h}{p} = \frac{h}{\sqrt{2mE}}$$

Now, photon energy $= E$

$$\Rightarrow \quad hf = E \Rightarrow \frac{hc}{\lambda_p} = E$$

or wavelength of photon, $\lambda_p = \dfrac{hc}{E}$

Hence, ratio $\dfrac{\lambda_e}{\lambda_p} = \dfrac{\left(\dfrac{h}{\sqrt{2mE}}\right)}{\left(\dfrac{hc}{E}\right)} = \dfrac{1}{c}\left(\dfrac{E}{2m}\right)^{1/2}$

56. de-Broglie wavelength associated with a moving charged particle of charge q is

$$\lambda = \frac{h}{p} = \frac{h}{\sqrt{2mqV}}$$

where, V = accelerating potential

Ratio of de-Broglie wavelength for particle A and B is,

$$\frac{\lambda_A}{\lambda_B} = \frac{\sqrt{m_B q_B V_B}}{\sqrt{m_A q_A V_A}} = \sqrt{\frac{m_B}{m_A}} \cdot \sqrt{\frac{q_B}{q_A}} \cdot \sqrt{\frac{V_B}{V_A}}$$

Substituting the given values, we get

$$= \sqrt{\frac{4m}{m}} \cdot \sqrt{\frac{q}{q}} \cdot \sqrt{\frac{2500}{50}} = 2 \times 1 \times 5 \times 1.414 = 14.14$$

57. Kinetic energy T of electron and its de Broglie wavelength λ are related as

$$\lambda = \frac{h}{p} = \frac{h}{\sqrt{m^2v^2}} = \frac{h}{\sqrt{2m \times \dfrac{1}{2}mv^2}} = \frac{h}{\sqrt{2mT}}$$

and given, $\lambda_B = 2\lambda_A$

$$\Rightarrow \quad \frac{h}{\sqrt{2mT_B}} = 2 \cdot \frac{h}{\sqrt{2mT_A}} \Rightarrow T_A = 4T_B$$

Also given,

$$T_B = T_A - 1.5 \Rightarrow T_B = 4T_B - 1.5$$

$$\Rightarrow \quad 3T_B = 1.5 \Rightarrow T_B = 0.5 \text{ eV}$$

Now, for metal B,

incident energy = kinetic energy of electron + work function

$$\Rightarrow \quad 4.5 = 0.5 + \phi_0 \Rightarrow \phi_0 = 4 \text{ eV}$$

58. Velocity acquired by a particle while falling from a height H is, $v = \sqrt{2gH}$

$$\lambda = \frac{h}{mv} = \frac{h}{m\sqrt{2gH}} \text{ or } \lambda \propto H^{-\frac{1}{2}}$$

59. Initial acceleration of electron,

$$a = \frac{F}{m} = \frac{|e|\mathbf{E}}{m}$$

Velocity of electron after time t,

$$v = u + at = \frac{|e|\mathbf{E}}{m} \cdot t$$

Momentum of electron, $p = mv = |e|\mathbf{E} \cdot t$

de Broglie wavelength λ associated with electron

$$\lambda = \frac{h}{p} = \frac{h}{|e|\mathbf{E} \cdot t}$$

Rate of change of de Broglie wavelength with time,

$$\frac{d\lambda}{dt} = \frac{-h}{|e|\mathbf{E}t^2}$$

Here, negative sign shows wavelength decreases with time.

60. Here, $\mathbf{v} = v_0\hat{\mathbf{i}}, \mathbf{B} = B_0\hat{\mathbf{j}}$

Force on moving electron due to magnetic field is

$$\mathbf{F} = -e(\mathbf{v} \times \mathbf{B}) = -e[v_0\hat{\mathbf{i}} \times B_0\hat{\mathbf{j}}] = -eV_0 B_0 \hat{\mathbf{k}}$$

As this force is perpendicular to $\mathbf{v}$ and $\mathbf{B}$, so the magnitude of $\mathbf{v}$ will not change, *i.e.* momentum $(= mv)$ will remain constant in magnitude. Hence, de-Broglie wavelength $\lambda = h/mv$ remains constant.

61. For elastic collision,

$$p_{\text{before collision}} = p_{\text{after collision}}.$$

$$mv = mv_A + \frac{m}{2} v_B$$

$$2v = 2v_A + v_B \quad \cdots \text{(i)}$$

Now, coefficient of restitution,

$$e = \frac{v_B - v_A}{u_A - u_B}$$

Here, $u_B = 0$ (Particle at rest) and for elastic collision $e = 1$

$$\therefore \quad 1 = \frac{v_B - v_A}{v}$$

$$\Rightarrow \quad v = v_B - v_A \quad \ldots \text{(ii)}$$

From Eqs. (i) and (ii), we get

$$v_A = \frac{v}{3}$$

and

$$v_B = \frac{4v}{3}$$

Hence,

$$\frac{\lambda_A}{\lambda_B} = \frac{\left(\dfrac{h}{mv_A}\right)}{\dfrac{h}{\dfrac{m}{2} \cdot v_B}} = \frac{v_B}{2v_A} = \frac{4/3}{2/3} = 2$$

62.

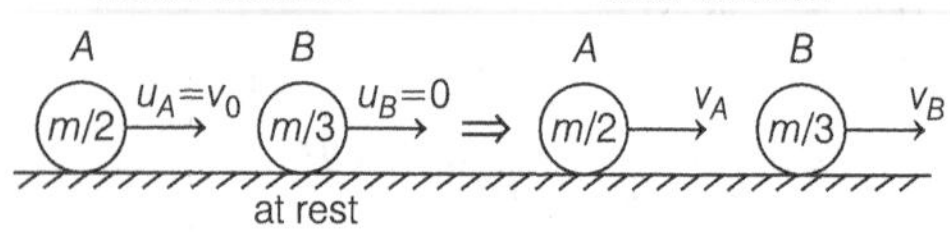

Using law of conservation of linear momentum, we have

$$m_A u_A + m_B u_B = m_A v_A + m_B v_B$$

$$\frac{m}{2} \times v_0 + \frac{m}{3} \times 0 = \frac{m}{2} \times v_A + \frac{m}{3} \times v_B$$

$$\frac{v_0}{2} = \frac{v_A}{2} + \frac{v_B}{3}$$

$$\Rightarrow \quad \frac{v_0}{2} = \frac{3v_A + 2v_B}{6}$$

$$3v_A + 2v_B = 3v_0 \quad \ldots \text{(i)}$$

By the definition of coefficient of restitution,

$$e = \frac{v_B - v_A}{u_A - u_B} \Rightarrow 1 = \frac{v_B - v_A}{v_0 - 0}$$

($\because$ for elastic collision, $e = 1$)

$$\Rightarrow \quad -v_A + v_B = v_0$$

$$\Rightarrow \quad -3v_A + 3v_B = 3v_0 \quad \ldots \text{(ii)}$$

On adding Eqs. (i) and (ii), we get

$$\Rightarrow \quad 5v_B = 6v_0$$

$$\Rightarrow \quad v_B = \frac{6v_0}{5}$$

Substituting this value in eq. (i), we get

$$3v_A + 2\left(\frac{6v_0}{5}\right) = 3v_0 \Rightarrow 3v_A = 3v_0 - \frac{12}{5}v_0$$

$$3v_A = \frac{3}{5}v_0 \Rightarrow v_A = \frac{v_0}{5}$$

Now, de-Broglie wavelength of particle A before collision,

$$(\lambda_A)_i = \frac{h}{(p_A)_i}$$

$$\lambda_0 = \frac{h}{m_A u_A} \quad [\because (\lambda_A)_i = \lambda_0 \text{ (given)}]$$

$$= \frac{h}{\dfrac{m}{2} \times v_0} = \frac{2h}{mv_0} \quad \ldots \text{(iii)}$$

and de-Broglie wavelength of particle A after collision,

$$(\lambda_A)_f = \frac{h}{(p_A)_f} = \frac{h}{m_A v_A}$$

$$= \frac{h}{\dfrac{m}{2} \times \dfrac{v_0}{5}} = \frac{10h}{mv_0}$$

$$= 5 \times \frac{2h}{mv_0} = 5\lambda_0$$

So, change in de-Broglie wavelength of particle A is

$$(\Delta\lambda)_A = (\lambda_A)_f - (\lambda_A)_i = 5\lambda_0 - \lambda_0 = 4\lambda_0$$

63. Given, de-Broglie wavelengths for particles are λ_1 and λ_2.

So,

$$\lambda_1 = \frac{h}{p_1} \text{ and } \lambda_2 = \frac{h}{p_2}$$

and momentum of particles are

$$p_1 = \frac{h}{\lambda_1} \text{ and } p_2 = \frac{h}{\lambda_2}$$

Given that, particles are moving perpendicular to each other and collide inelastically.

So, they move as a single particle.

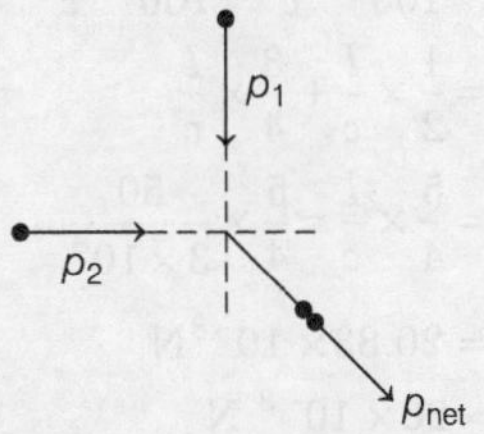

So, by conservation of momentum and vector addition law, net momentum after collision,

$$p_{net} = \sqrt{p_1^2 + p_2^2 + 2p_1p_2\cos 90^\circ}$$

$$= \sqrt{p_1^2 + p_2^2}$$

Since, $p_1 = \frac{h}{\lambda_1}$ and $p_2 = \frac{h}{\lambda_2}$

So, $$p_{net} = \sqrt{\frac{h^2}{\lambda_1^2} + \frac{h^2}{\lambda_2^2}} \quad ...(i)$$

Let the de-Broglie wavelength after the collision is λ_{net}, then

$$p_{net} = \frac{h}{\lambda_{net}} \quad ...(ii)$$

From Eqs. (i) and (ii), we get

$$\frac{h}{\lambda_{net}} = \sqrt{\frac{h^2}{\lambda_1^2} + \frac{h^2}{\lambda_2^2}}$$

$$\Rightarrow \quad \frac{1}{\lambda_{net}^2} = \frac{1}{\lambda_1^2} + \frac{1}{\lambda_2^2}$$

$$\Rightarrow \quad \frac{1}{\lambda^2} = \frac{1}{\lambda_1^2} + \frac{1}{\lambda_2^2} \quad (\because \lambda_{net} = \lambda)$$

64. Initial velocity of electron, $\mathbf{v} = v_0\hat{\mathbf{i}} + v_0\hat{\mathbf{j}}$

$$\Rightarrow \quad |\mathbf{u}| = \sqrt{v_0^2 + v_0^2} = v_0\sqrt{2}$$

Initial de Broglie wavelength associated with electron,

$$\lambda_0 = \frac{h}{p} = \frac{h}{mv_0\sqrt{2}} \quad ...(i)$$

After t seconds velocity of electron,

$$\mathbf{v} = \mathbf{u} + \mathbf{a}t$$

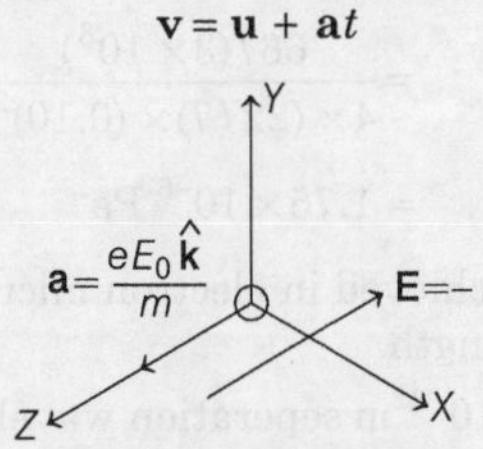

or $$\mathbf{v} = (v_0\hat{\mathbf{i}} + v_0\hat{\mathbf{j}}) + \frac{eE_0}{m}\cdot t\cdot\hat{\mathbf{k}}$$

Note that electron is accelerated opposite to the field.

Magnitude of velocity after time t,

$$|\mathbf{v}| = \sqrt{v_0^2 + v_0^2 + \left(\frac{eE_0}{m}t\right)^2}$$

$$= \sqrt{2v_0^2 + \frac{e^2E_0^2t^2}{m^2}}$$

$$= v_0\sqrt{2 + \frac{e^2E_0^2t^2}{m^2v_0^2}}$$

$\therefore$ de-Broglie wavelength of electron at time t,

$$\lambda = \frac{h}{p} = \frac{h}{m|\mathbf{v}|}$$

$$\Rightarrow \quad \lambda = \frac{h}{mv_0\sqrt{2 + \frac{e^2E_0^2t^2}{m^2v_0^2}}}$$

$$= \frac{\lambda_0\sqrt{2}}{\sqrt{2 + \frac{e^2E_0^2t^2}{m^2v_0^2}}} = \frac{\lambda_0}{\sqrt{1 + \frac{e^2E_0^2t^2}{2m^2v_0^2}}}$$

65. Here, $V = 50\text{ kV} = 5\times10^4\text{ V}$, $m_e = 9.11\times10^{-31}\text{ kg}$

KE of an electron,

$$K = 50\text{ eV} = 1.6\times10^{-19}\times5\times10^4\text{ J} = 8\times10^{-15}\text{ J}$$

$\therefore$ de-Broglie wavelength of electrons is

$$\lambda = \frac{h}{\sqrt{2mK}}$$

$$= \frac{6.63\times10^{-34}}{\sqrt{2\times9.11\times10^{-31}\times8\times10^{-15}}}\text{ m}$$

$$= \frac{6.63\times10^{-11}}{12.07}\text{ m} = 5.5\times10^{-12}\text{ m}$$

Wavelength of yellow light, $\lambda_y = 5.9\times10^{-7}$ m

Resolving power of a microscope $\propto 1/\lambda$

$$\therefore \quad \frac{\text{Resolving power of electron microscope}}{\text{Resolving power of optical microscope}} = \frac{\lambda_y}{\lambda}$$

$$\frac{5.9\times10^{-7}}{5.5\times10^{-12}} \cong 10^5$$

66. In Davisson-Germer experiment, the de-Broglie wavelength associated with electron is

$$\lambda = \frac{12.27}{\sqrt{V}}\text{ Å} \quad ...(i)$$

where, V is the applied voltage.

If there is a maxima of the diffracted electrons at an angle θ, then

$$2d\sin\theta = \lambda \quad ...(ii)$$

From Eq. (i), we note that if V is inversely proportional to the wavelength λ.

i.e. V will increase with the decrease in the λ.

From Eq. (ii), we note that wavelength λ is directly proportional to $\sin\theta$ and hence, θ.

So, with the decrease in λ, θ will also decrease.

Thus, when the voltage applied to A is increased, the diffracted beam will have the maxima at a value of θ that will be less than the earlier value.

Round II

1. We know, $K_A = \frac{hc}{\lambda_A} - \phi_0$

and $$K_B = \frac{hc}{\lambda_B} - \phi_0$$

$\therefore$ $$\frac{K_A}{K_B} = \frac{\frac{hc}{2\lambda_B}}{\frac{hc}{\lambda_B}} < \frac{1}{2}$$

or $$K_A < K_B/2$$

2. As, $E = \frac{hc}{\lambda}$ or $E \propto \frac{1}{\lambda}$

$\therefore$ $$\frac{E_2}{E_1} = \frac{\lambda_1}{\lambda_2}$$

or $$E_2 = E_1 \times \frac{\lambda_1}{\lambda_2} = 1.23 \times \frac{10000}{5000} = 2.46 \text{ eV}$$

Now, $h\nu - \phi_0 = \frac{1}{2} mv_{\text{max}}^2 = eV_s$

$$\phi_0 = h\nu_2 - eV_s = E_s - eV_s$$

or $$= 2.46 - 1.36 = 1.10 \text{ eV}$$

3. As, $\lambda = \frac{h}{\sqrt{2m_0 E}} = \frac{h}{\sqrt{2m_0 qV}} \Rightarrow \lambda \propto \frac{1}{\sqrt{m_0}}$

where, m_0 = mass of the charge

$$\Rightarrow \quad \frac{\lambda_1}{\lambda_2} = \sqrt{\frac{M}{m}}$$

$$\Rightarrow \quad \lambda_2 = \lambda_1 \sqrt{\frac{m}{M}}$$

or $$\lambda_2 = \lambda \sqrt{\frac{m}{M}} \quad (\because \lambda_1 = \lambda)$$

4. Stopping potential changes linearly with frequency of incident radiation.

5. From Einstein's photoelectric equation,

$$\text{KE} = 2E_0 - E_0 = E_0 \quad (\text{for } 0 \le x \le 1)$$

$$\Rightarrow \quad \lambda_1 = \frac{h}{\sqrt{2mE}} \quad \text{...(i)}$$

Similarly, KE $= 2E_0$ (for $x > 1$)

$$\Rightarrow \quad \lambda_2 = \frac{h}{\sqrt{4mE_0}}$$

From Eqs. (i) and (ii), we get

$$\frac{\lambda_1}{\lambda_2} = \sqrt{2}$$

6. Radiation pressure or momentum imparted per second per unit area when light falls is

$$p = \begin{cases} \frac{2I}{c} & ; \text{ for reflection of radiation} \\ \frac{I}{c} & ; \text{ for absorption of radiation} \end{cases}$$

where, I is the intensity of the light.

In given case, there is 25% reflection and 75% absorption, so radiation pressure = force per unit area

$$= \frac{25}{100} \times \frac{2I}{c} + \frac{75}{100} \times \frac{I}{c}$$

$$= \frac{1}{2} \times \frac{I}{c} + \frac{3}{4} \times \frac{I}{c}$$

$$= \frac{5}{4} \times \frac{I}{c} = \frac{5}{4} \times \frac{50}{3 \times 10^8}$$

$$= 20.83 \times 10^{-8} \text{N}$$

$$\approx 20 \times 10^{-8} \text{ N}$$

7. Wavelength of the given photon is given as,

$$\lambda_p = \frac{c}{\nu_p} = \frac{3 \times 10^8}{6 \times 10^{14}} \text{ m} = 5 \times 10^{-7} \text{m} \quad \text{...(i)}$$

As, it is given that, de-Broglie wavelength of the electron is

$$\lambda_e = 10^{-3} \times \lambda_p \quad [\because \text{ using Eq. (i)}]$$

$$= 5 \times 10^{-10} \text{m}$$

Also, the de-Broglie wavelength of an electron is given as,

$$\lambda_e = \frac{h}{p} = \frac{h}{mv_e} \Rightarrow v_e = \frac{h}{\lambda_e m_e}$$

Substituting the given values, we get

$$= \frac{6.63 \times 10^{-34}}{5 \times 10^{-10} \times 9.1 \times 10^{-31}} \text{ m/s}$$

$$= 1.45 \times 10^6 \text{ m/s}$$

8. Light falling per second on the surface of sphere

$$E = \frac{66}{100} \times 100 = 66 \text{ W}$$

Momentum of the light falling per second on the surface of sphere $= \frac{E}{c}$

Momentum of the reflected light = 0; as the light is completely absorbed.

Force exerted by light, $F = \frac{E}{c} - 0 = \frac{E}{c}$

Pressure on surface, $p = \frac{F}{4\pi r^2} = \frac{E/c}{4\pi r^2}$

$$= \frac{66/(3 \times 10^8)}{4 \times (22/7) \times (0.10)^2}$$

$$= 1.75 \times 10^{-6} \text{ Pa}$$

9. Given, resolution achieved in electron microscope is of the order of wavelength.

So, to resolve 7.5×10^{-12}m seperation wavelength associated with electrons is

$$\lambda = 7.5 \times 10^{-12} \text{m}$$

$\therefore$ Momentum of electrons required is

$$p = \frac{h}{\lambda}$$

or kinetic energy of electron must be

$$\text{KE} = \frac{p^2}{2m} = \frac{(h/\lambda)^2}{2m}$$

Substituting the given values, we get

$$= \frac{\left(\dfrac{6.6\times10^{-34}}{7.5\times10^{-12}}\right)^2}{2\times9.1\times10^{-31}}\text{ J}$$

$$= \frac{(6.6\times10^{-34})^2}{2\times9.1\times10^{-31}\times(7.5\times10^{-12})^2\times(1.6\times10^{-19})}\text{ eV}$$

$$(\because 1\text{ eV} = 1.6\times10^{-19}\text{ J})$$

$$= 26593.4 \simeq 26.6\times10^3\text{ eV} \simeq 26\text{ keV}$$

which is nearest to 25 keV.

10. Velocity of electron, $v = h/(m\lambda)$

Let $h = 6.6\times10^{-9}$ Js and $m = 9\times10^{-31}$ kg.

(a) When $\lambda_1 = 10\text{ nm} = 10\times10^{-9}\text{ m} = 10^{-8}$ kg

$$v_1 = \frac{6.6\times10^{-34}}{(9\times10^{-31})\times10^{-8}} = \frac{2.2}{3}\times10^5 = 10^5\text{ m/s}$$

(b) When $\lambda_2 = 10^{-1}\text{ nm} = 10^{-1}\times10^{-9}\text{ m} = 10^{-10}$ kg.

$$v_2 = \frac{6.6\times10^{-34}}{(9\times10^{-31})\times10^{-10}} \approx 10^7\text{ m/s}$$

(c) When $\lambda_3 = 10^{-4}\text{ nm} = 10^{-4}\times10^{-9}\text{ m} = 10^{-13}$ m

$$v_3 = \frac{6.6\times10^{-34}}{(9\times10^{-31})\times10^{-13}} \approx 10^{10}\text{ m/s}$$

(d) When $\lambda_4 = 10^{-4}\text{ nm} = 10^{-4}\times10^{-9}\text{ m} = 10^{-13}$ m

$$v_4 = \frac{6.6\times10^{-34}}{(9\times10^{-31})\times10^{-15}} \approx 10^{12}\text{ m/s}$$

As v_3 and v_4 are greater than velocity of light ($= 3\times10^8$ m/s), hence relativistic correction is needed for $\lambda = 10^{-4}$ nm and $\lambda = 10^{-6}$ nm.

11. As, $E_e = \frac{1}{2}mv^2 = \frac{1}{2}(mv)v = \frac{1}{2}\left(\frac{h}{\lambda}\right)v$

and $E_p = \frac{hc}{\lambda}$;

$$\therefore \quad \frac{E_e}{E_p} = \frac{v}{2c}$$

Now, $p_e = mv = \frac{h}{\lambda}$ and $p_h = \frac{h}{\lambda}$

$$\therefore \quad \frac{p_e}{p_h} = 1$$

12. For electron, $\lambda_e = \frac{h}{m_e v_e} = \frac{h}{m_e(c/100)} = \frac{100h}{m_e c}$ ...(i)

Kinetic energy, $E_e = \frac{1}{2}m_e v_e^2$ or $m_e v_e = \sqrt{2E_e m_e}$

$$\lambda_e = \frac{h}{m_e v_e} = \frac{h}{\sqrt{2m_e E_e}} \text{ or } E_e = \frac{h_2}{\lambda_e^2 m_e} \quad \text{...(ii)}$$

For photon of wavelength λ_p, energy $= E_p = \frac{hc}{\lambda_p} = \frac{hc}{2\lambda_e}$

$$(\because \lambda_p = 2\lambda_e)$$

$$\therefore \quad \frac{E_p}{E_e} = \frac{hc}{2\lambda_e} \times \frac{2\lambda_e^2 m_e}{h_2} = \frac{\lambda_e m_e c}{h} = \frac{100h}{m_e c} \times \frac{m_e c}{h} = 100$$

$$\therefore \quad \frac{E_e}{E_p} = \frac{1}{100} = 10^{-2}$$

For electron, $p_e = m_e v_e = m_e \times c/100$

$$\therefore \quad \frac{p_e}{m_e c} = \frac{1}{100} = 10^{-2}$$

13. Radius of largest circular path of electron in perpendicular magnetic field,

$$r = \frac{mv_{\max}}{Be}$$

∵ Velocity of most energetic electron emitted from metal surface,

$$v_{\max} = \frac{Ber}{m}$$

Kinetic energy with which electron is emitted,

$$K_{\max} = \frac{1}{2}mv_{\max}^2 = \frac{1}{2}m\cdot\frac{B^2e^2r^2}{m^2}$$

$$\Rightarrow \quad K_{\max} = \frac{1}{2}\left(\frac{B^2e^2r^2}{m}\right)\text{ joule}$$

Kinetic energy of emitted electron (in electron volts),

$$K_{\max}(\text{eV}) = \frac{K_{\max}\text{ (in joule)}}{e}$$

$$\therefore \quad K_{\max}(\text{eV}) = \frac{1}{2}\left(\frac{B^2er^2}{m}\right)$$

Here given, $B = 3\times10^{-4}$ T, $e = 1.6\times10^{-19}$ C,

$r = 10\text{ mm} = 10^{-2}$ m

and $m = 9.1\times10^{-31}$ kg

So, $$K_{\max} = \frac{(3\times10^{-4})^2\times(1.6\times10^{-19})\times(10^{-2})^2}{2\times(9.1\times10^{-31})}$$

$$= \frac{9\times1.6}{2\times9.1}\times10^{31-31} \approx 0.8\text{ eV}$$

∴ Energy of incident photon $= hf = \frac{hc}{\lambda}$

$$= \frac{12400\text{ eV-Å}}{6561\text{ Å}} \approx 1.9\text{ eV}$$

(**Note** $hc = 1240$ eV-nm $= 12400$ eV-Å)

So, work function of metal,

ϕ_0 = incident energy – kinetic energy of emitted electron

$$= 1.9 - 0.8 = 1.1\text{ eV}$$

14. Initially,

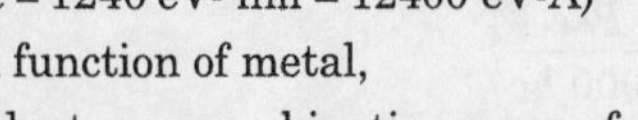

We have, de-Broglie wavelengths associated with particles are

$$\lambda_x = \frac{h}{p_x} \text{ and } \lambda_y = \frac{h}{p_y}$$

$$\Rightarrow \quad p_x = \frac{h}{\lambda_x} \text{ and } p_y = \frac{h}{\lambda_y}$$

Finally, particles collided to form a single particle.

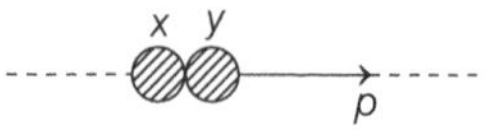

As we know that, linear momentum is conserved in collision.

So, $\mathbf{p}_p = |\mathbf{p}_x - \mathbf{p}_y| \Rightarrow \mathbf{p}_p = \left|\frac{h}{\lambda_x} - \frac{h}{\lambda_y}\right|$

So, de-Broglie wavelength of combined particle is

$$\lambda_p = \frac{h}{|\mathbf{p}_p|} = \frac{h}{\left|\frac{h}{\lambda_x} - \frac{h}{\lambda_y}\right|} = \frac{h}{\left|\frac{h\lambda_y - h\lambda_x}{\lambda_x\lambda_y}\right|}$$

$$= \frac{\lambda_x\lambda_y}{|\lambda_x - \lambda_y|}$$

15. According to Einstein's equation, $h\nu = h\nu_0 + eV$

$\Rightarrow \quad V = \frac{h}{e}(\nu - \nu_0)$

Given, $\quad V_1 = \frac{h}{e}(\nu_1 - \nu_0)$ and $3V_1 = \frac{h}{e}(2\nu_1 - \nu_0)$

$\Rightarrow \quad \frac{3V_1}{V_1} = \frac{2\nu_1 - \nu_0}{\nu_1 - \nu_0} \Rightarrow \nu_1 = 2\nu_0$

$\therefore \quad V_1 = \frac{h}{e}(2\nu_0 - \nu_0) = \frac{h}{e}\nu_0$

At frequency $4\nu_1$, $nV_1 = \frac{h}{e}(4\nu_1 - \nu_0)$

$$= \frac{h}{e}(8\nu_0 - \nu_0) = 7 \times \frac{h}{e}\nu_0$$

$\Rightarrow \quad nV_1 = 7V_1 \Rightarrow n = 7$

16. Number of photons incident per second

$$= \frac{\text{Power}}{\text{Energy of one photon}} = \frac{P}{(hc/\lambda)} = \frac{P\lambda}{hc}$$

Number of electrons emitted per second = 0.1% of

$$\frac{P\lambda}{hc} = \frac{P\lambda}{1000\,hc}$$

$\therefore$ Current = Charge (on photoelectrons per second)

$$= \frac{P\lambda e}{1000\,hc}$$

$$= \frac{(1.5\times 10^{-3})(400\times 10^{-9})(1.6\times 10^{-19})}{(1000)(6.63\times 10^{-34})(3\times 10^{8})}$$

$$= 0.48\times 10^{-6}\ \text{A} = 0.48\ \mu\text{A}$$

17. $E = \frac{12375}{1800} = 6.875$ eV

K_{max} or $K = E - W = 4.875$ eV

$$r = \frac{\sqrt{2Km}}{Bq}$$

$$= \frac{\sqrt{2\times 4.875\times 1.6\times 10^{-19}\times 9.1\times 10^{-31}}}{5\times 10^{-5}\times 1.6\times 10^{-19}} = 0.148\ \text{m}$$

18. Let the work function of the metal be ϕ.

$$\text{KE} = \frac{hc}{\lambda} - \phi$$

For case I, $\quad \frac{1}{2}mv_1^2 = 4 - \phi \quad$ ...(i)

For case II, $\quad \frac{1}{2}mv_2^2 = 2.5 - \phi \quad$ ...(ii)

Given, $\quad \frac{v_1}{v_2} = 2$

From Eqs. (i) and (ii), we get

$$\frac{v_1^2}{v_2^2} = \frac{4-\phi}{2.5-\phi} \Rightarrow (2)^2 = \frac{4-\phi}{2.5-\phi}$$

$\Rightarrow \quad 4(2.5-\phi) = 4-\phi \Rightarrow 10 - 4\phi = 4 - \phi$

$\Rightarrow \quad 3\phi = 6 \Rightarrow \phi = \frac{6}{3} = 2$ eV

So, the work function of the metal is 2 eV.

19. With radiation wavelength λ and work function ϕ_0, stopping potential is V.

$\Rightarrow \quad \frac{hc}{\lambda} - \phi_0 = eV \quad$ ...(i)

When wavelength is 3λ, stopping potential is $\frac{V}{4}$.

$\Rightarrow \quad \frac{hc}{3\lambda} - \phi_0 = \frac{eV}{4} \quad$ ...(ii)

Now, multiply Eq. (ii) by 3 and subtracting Eq (i) from it, we get

$\Rightarrow \left(\frac{hc}{\lambda} - 3\phi_0\right) - \left(\frac{hc}{\lambda} - \phi_0\right) = \frac{3\,\text{eV}}{4} - \text{eV}$

$\Rightarrow \quad 2\phi_0 = \frac{1}{4}eV \Rightarrow eV = 8\phi_0$

Putting this value in Eq. (i), we get

$$\frac{hc}{\lambda} - \phi_0 = 8\phi_0 \Rightarrow \frac{hc}{\lambda} = 9\phi_0 \Rightarrow \frac{hc}{\lambda} = 9\frac{hc}{\lambda_0}$$

{As $\phi_0 = \frac{hc}{\lambda_0}$, where λ_0 = threshold wavelength}

$\Rightarrow \quad \lambda_0 = 9\lambda \quad \therefore \quad n = 9$

20. Given, energy per unit area per second,

$$P = 1.388\times 10^3\ \text{W/m}^2$$

Let n be the number of photons incident on the earth per square metre. Wavelength of each photon $= 550$ nm $= 550\times 10^{-9}$ m.

Energy of each photon, $E = \frac{hc}{\lambda}$

(where h is the Planck's constant)

$$= \frac{6.63\times 10^{-34}\times 3\times 10^8}{550\times 10^{-9}} = 3.616\times 10^{-19}\ \text{J}$$

Number of photons incident on the earth's surface

$$N = \frac{P}{E} = \frac{1.388\times 10^3}{3.616\times 10^{-19}} = 3.838\times 10^{21}$$

$$= 3.838\times 10^{21}\ \text{photon/m}^2\text{-s}$$

$\Rightarrow \quad n = 21$

21. Given, intensity of falling radiation,

$$I = 6.4 \times 10^{-5}\ \text{Wcm}^{-2}$$

Wavelength of radiation, $\lambda = 310$ nm

Work function of metal, $\phi = 2$ eV

Area of metal surface, $A = 1\ \text{cm}^2$

Energy of one photon of radiation,

$$E_1 = \frac{hc}{\lambda} = \frac{1240\ \text{eVnm}}{\lambda\ (\text{nm})}$$

or $$E_1 = \frac{1240}{310}\ \text{eV}$$

$$= \frac{1240}{310} \times 1.6 \times 10^{-19}\ \text{J}$$

$$= 6.4 \times 10^{-19}\ \text{J}$$

So, photon flux of radiation falling over surface = number of photons falling over surface per second

$$= \frac{\text{Total incident energy}}{\text{Energy of one photon}} = \frac{I \times A}{E_1}$$

$$= \frac{6.4 \times 10^{-5} \times 1}{6.4 \times 10^{-19}}$$

$$= 10^{14}\ \text{photons/second}$$

As only one in 10^3 photons are able to emit an electron, number of ejected electrons in 1 second

$$= \frac{10^{14}}{10^3} = 10^{11}$$

$$\Rightarrow \quad x = 11$$

22. From Bragg's law,

$$2d \sin\theta = n\lambda$$

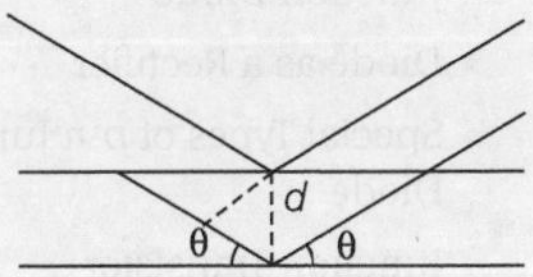

For first maxima, $n = 1$

$$\Rightarrow \quad 2d\sin\theta = \lambda$$

$$\Rightarrow \quad 2d\sin\theta = \frac{h}{\sqrt{2Em}} \qquad \left\{\begin{array}{l} \because \frac{1}{2}mv^2 = E \\ p = \sqrt{2Em} \\ \lambda = \frac{h}{p} = \frac{h}{\sqrt{2mE}} \end{array}\right.$$

$$\Rightarrow \quad 2Em = \frac{h^2}{4d^2\sin^2\theta}$$

$$\Rightarrow \quad E = \frac{h^2}{8md^2\sin^2\theta}$$

Given, $d = 1\ \text{Å} = 10^{-10}$ m, $\theta = 60°$

$$h = 6.64 \times 10^{-34}\ \text{Js},\ m = 9.1 \times 10^{-31}\ \text{kg}$$

Substituting these values in above equation, we get

$$E = \frac{(6.64 \times 10^{-34})^2}{8 \times 9.1 \times 10^{-31} \times (10^{-10})^2 \sin^2 60°}$$

$$= \frac{44.0896 \times 10^{-68}}{8 \times 9.1 \times 10^{-51} \times \left(\frac{3}{4}\right)}$$

$$= 0.8075 \times 10^{-17}\ \text{J}$$

$$= \frac{0.8075 \times 10^{-17}}{1.6 \times 10^{-19}}\ \text{eV}$$

$$= 50.47\ \text{eV} \approx 50\ \text{eV}$$

23. Pitch of helical path, $p = (v\cos\theta)T = \frac{vT}{2}$. (as $\theta = 60°$)

$$T = \frac{2\pi m}{Bq} = \frac{2\pi}{B\alpha} \qquad \left(\alpha = \frac{q}{m}\right)$$

$$\therefore \quad p = \frac{\pi v}{B\alpha}$$

or $$v = \frac{B\alpha p}{\pi} \qquad \text{...(i)}$$

$$\text{KE} = \frac{1}{2}mv^2 = E - W$$

$$\therefore \quad W = E - \frac{1}{2}mv^2 \qquad \text{...(ii)}$$

Substituting value of v from Eq. (i) in Eq. (ii), we get

$$W = 4.9 - \frac{1}{2}$$

$$9.1 \times 10^{-31} \times (2.5 \times 10^{-3})^2\,(1.76 \times 10^{11})^2 \times \frac{(2.7 \times 10^{-3})^2}{\pi^2 \times 1.6 \times 10^{-19}}$$

$$= (4.9 - 0.4)\ \text{eV} = 4.5\ \text{eV}$$

24. Given, work function, $W_0 = 2.35$ eV and the electric component of electromagnetic radiation

$$E = a[1 + \cos(2\pi f_1 t)]\cos(2\pi f_2 t)$$

$$\Rightarrow E = [a\cos(2\pi f_2 t) + a\cos(2\pi f_1 t)\cos(2\pi f_2 t)]$$

$$\left(\because \cos A\cos B = \frac{1}{2}[\cos(A+B) - \cos(A-B)]\right)$$

$$\Rightarrow E = a\cos(2\pi f_2 t) + \frac{a}{2}\cos 2\pi(f_1 + f_2)t - \frac{a}{2}\cos 2\pi(f_1 - f_2)t$$

So, the electric component has 3 sub-components with frequencies are,

$$f_2,\ (f_1 + f_2)\ \text{and}\ (f_1 - f_2)$$

So, for maximum kinetic energy of photoelectron, we take photon of maximum frequency. Hence,

$$E_{\max} = \frac{h\nu_{\max}}{e} = \frac{6.6 \times 10^{-34} \times (3.6 \times 10^{15} + 1.2 \times 10^{15})}{1.6 \times 10^{-19}}$$

$$= 19.8\ \text{eV}$$

Hence, the maximum kinetic energy,

$$\text{KE}_{\max} = E_{\max} - W_0 = 19.8 - 2.35 = 17.45\ \text{eV}$$

28

Electronic Devices

We know that, the solids can be classified in three types as per their electrical conductivity (i) Conductors, (ii) Insulators and (iii) Semiconductors.

Conductors are those in which electric charges can flow easily, while semiconductor are those whose conductivity is intermediate between conductors and insulators. These can be divided on the basis of conductivity and energy band.

IN THIS CHAPTER ....

Energy Bands in Solids

Inside the crystal of a solid, each electron has a unique position and no two electrons see exactly the same pattern of surrounding charges. Because of this, each electron will have a different energy level. These different energy levels with continuous energy variation are called *energy bands.*

These bands are of two types

(i) **Valence band** is the range of energies possessed by the valence electrons.

(ii) **Conduction band** is the range of energies possessed by the conduction electrons.

Note With no external energy, all the valence electrons will reside in the valence band.

Energy Band Gap

It is the gap between the top of the valence band and the bottom of the conduction band.

i.e. $$E_g = E_c - E_v$$

It may be zero, small or large depending upon the material. Also, no free electrons are present in this gap.

As the temperature increases, energy band gap decreases.

Difference between Conductor, Insulator and Semiconductor on the basis of Conductivity and Energy Bands

Conductor (Metal)	Insulator	Semiconductor
Conductivity (σ) lies in the range of 10^4-$10^7 \Omega^{-1} m^{-1}$.	Conductivity (σ) lies in the range of 10^{-20}-$10^{-10} \Omega^{-1} m^{-1}$.	Conductivity (σ) lies in the range of 10^{-6}-$10^4 \Omega^{-1} m^{-1}$.
In conductor, either there is no energy gap between the conduction band which is partially filled with electrons and valence band or the conduction band and valence band overlap each other. 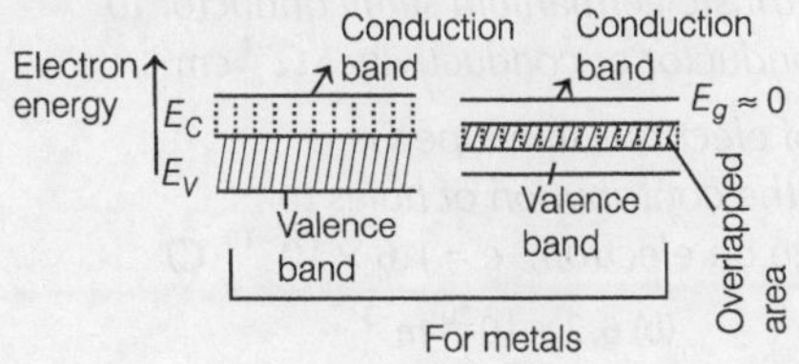	In insulator, the valence band is completely filled, the conduction band is completely empty. In this, energy gap is quite large and even energy from any other source cannot help electrons to overcome it. 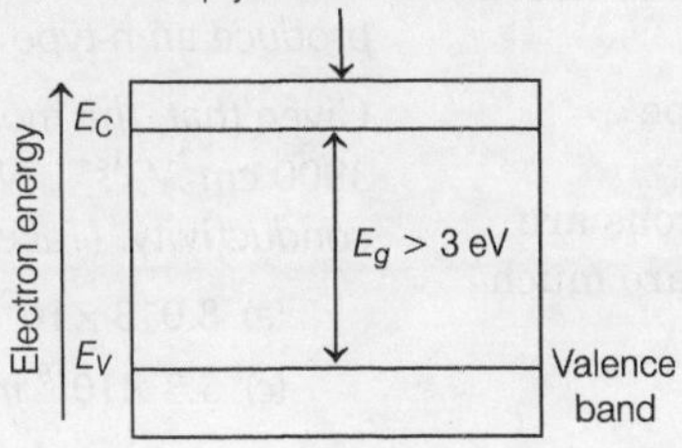	In semiconductor, the valence band is totally filled and the conduction band is empty but the energy gap between conduction band and valence band is very small. Thus, at room temperature, some electrons in the valence band acquire thermal energy greater than energy band gap and jump over to the conduction band where they are free to move under the influence of even a small electric field and acquire small conductivity. 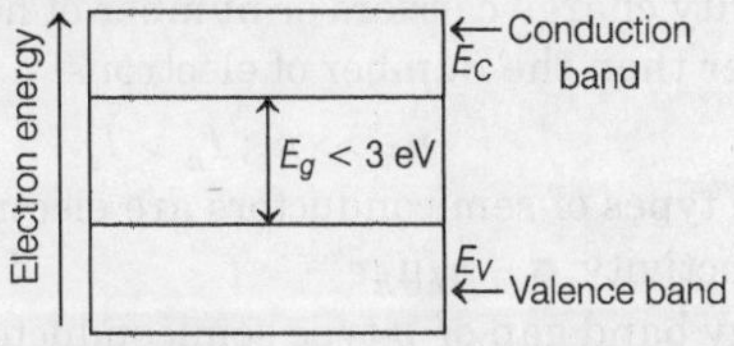
Current carrier Free electrons	—	**Current carriers** Free electrons and holes (a vacancy created when an e^- leaves an carries positive charge)

Types of Semiconductors

On the basis of purity, semiconductors are classified into two types

1. Intrinsic Semiconductors

These are pure semiconductor without any impurity, *e.g.* silicon, germanium, etc. At room temperature, intrinsic semiconductor has equal number of holes and electrons.

i.e. $$n_e = n_h = n_i$$

where, n_i is called *intrinsic carriers concentration.*

The conductivity of intrinsic semiconductors is due to both holes and electrons. The free electrons moves completely independently as conduction electron and give rise to an electron current I_e, under applied field. Similarly, under the effect of field, holes moves towards negative potential giving rise to hole current I_h.

So, total current, $I = I_e + I_h$

At equilibrium, $n_e n_h = n_i^2$

This is known as **mass-action law**.

At $T = 0$ K, it behaves as an insulator. However, at $T > 0$ K, some electrons excite to the conduction band, leaving equal number of holes there.

2. Extrinsic Semiconductors

When a small quantity of impurity is added in a pure or intrinsic semiconductor (nearly 1 atom of impurity in 10^7 atoms of pure semiconductor), the conductivity of semiconductor increases. Such an impure semiconductor is called *extrinsic* or *doped semiconductor.*

Depending upon the nature of impurity added in intrinsic semiconductor, the extrinsic semiconductors are of two types

(i) *n*-type and (ii) *p*-type

n-type Semiconductors

When a small quantity of pentavalent impurity such as antimony (Sb) or arsenic (As), etc., is introduced in pure germanium (or silicon) crystal, then *n*-type semiconductor is formed. The pentavalent impurity is also called **donor atoms** as they donate conduction electrons to crystal.

Following are important points regarding *n*-type semiconductors

- Electrons are majority charge carriers and holes are minority charge carriers or number of electrons are much greater than the number of holes, *i.e.* $n_e >> n_h$; $I_e >> I_h$
- These type of semiconductors are electrically neutral.
- Conductivity, $\sigma \simeq n_e \mu_e e$
- Energy band gap of *n*-type semiconductor at $T > 0$ K is

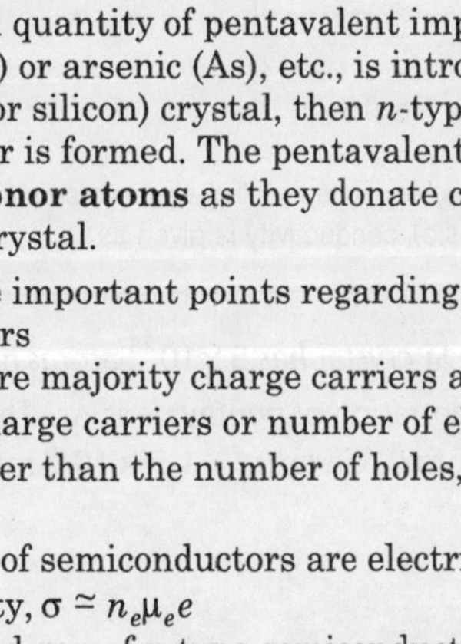

p-type Semiconductors

When a small quantity of a trivalent impurity such as indium (In), boron (B) aluminium (Al), etc., having three valence electrons is introduced into the pure germanium, then such type of semiconductors are called *p-type* or *acceptor type semiconductors.*

This hole is equivalent to a positively charged particle. Since, positive holes are responsible to increase the conductivity in this crystal, the crystal so obtained is called the *p-type crystal* and the impurity atom is called *acceptor impurity.*

Following are important points regarding p-type semiconductors

- Holes are majority charge carriers and electrons are minority charge carriers or number of holes are much greater than the number of electrons.

$$n_h >> n_e; I_h > I_e$$

- These types of semiconductors are electrically neutral.
- Conductivity, $\sigma \simeq n_h \mu_h e$
- Energy band gap of p-type semiconductors at $T > 0$ K is

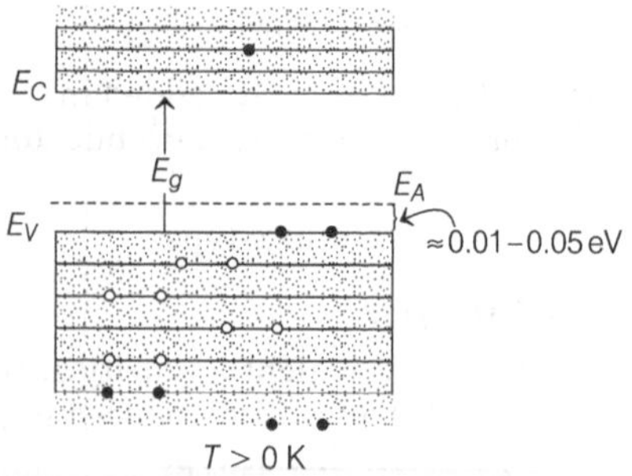

Note Generally, when a battery is connected across a semiconductor (whether intrinsic or extrinsic), conductivity is given as $\sigma = n_e e\mu_e + n_h e\mu_h$.

Example 1. *A pure Si crystal has 5×10^{28} atoms m^{-3}. It is doped by 1 ppm concentration of pentavalent As. The number of holes is (approx) (Given, $n_i = 1.5 \times 10^{16}\ m^{-3}$)* **[NCERT]**

(a) $2.2 \times 10^6\ m^{-3}$
(b) $4.5 \times 10^9\ m^{-3}$
(c) $6.2 \times 10^6\ m^{-3}$
(d) $8.1 \times 10^9\ m^{-3}$

Sol. (b) 1 ppm concentration of As means 1 atom of Si is doped out of 10^6 atoms.

⇒ Number of doped atoms in 5×10^{28} atoms

$$= \frac{5 \times 10^{28}}{10^6} = 5 \times 10^{22} \text{ atoms}$$

Since, 1 As atom donates one electron, hence number of excess electrons $= 5 \times 10^{22} = n_e$

i.e. $$n_e = 5 \times 10^{22}$$

According to law of mass action,

$$n_e \cdot n_h = n_i^2$$

$$\Rightarrow \quad n_h = \frac{n_i^2}{n_e} = \frac{(1.5 \times 10^{16})^2}{5 \times 10^{22}}$$

$$= 4.5 \times 10^9\ m^{-3}$$

∴ Number of holes formed $= 4.5 \times 10^9\ m^{-3}$

Example 2. *Determine the number of donor atoms which have to be added to an intrinsic germanium semiconductor to produce an n-type semiconductor of conductivity $5\ \Omega^{-1}\ cm^{-1}$.*

Given that, the mobility of electrons in n-type Ge is $3900\ cm^2V^{-1}s^{-1}$. Neglect the contribution of holes to conductivity. (Take, charge on electron, $e = 1.6 \times 10^{-19}$ C)

(a) $8.013 \times 10^{21}\ m^{-3}$ (b) $6.2 \times 10^{20}\ m^{-3}$
(c) $5.3 \times 10^{19}\ m^{-3}$ (d) $4.8 \times 10^{18}\ m^{-3}$

Sol. (a) Here, $e = 1.6 \times 10^{-19}$ C,

$$\sigma = 5\ \Omega^{-1}\ cm^{-1} = 500\ \Omega^{-1}\ m^{-1},$$

$$\mu_e = 3900\ cm^2\ V^{-1}s^{-1} = 0.39\ m^2V^{-1}s^{-1}$$

Now, $\sigma = en_e\mu_e$ (neglecting contribution of holes)

$$\therefore \quad n_e = \frac{\sigma}{e\mu_e} = \frac{500}{1.6 \times 10^{-19} \times 0.39}$$

$$= 8.013 \times 10^{21}\ m^{-3}$$

Semiconductor Diode or p-n Junction Diode

When a p-type semiconductor is brought into a close contact with n-type semiconductor crystal, the resulting arrangement is called a semiconductor diode or p n junction diode.

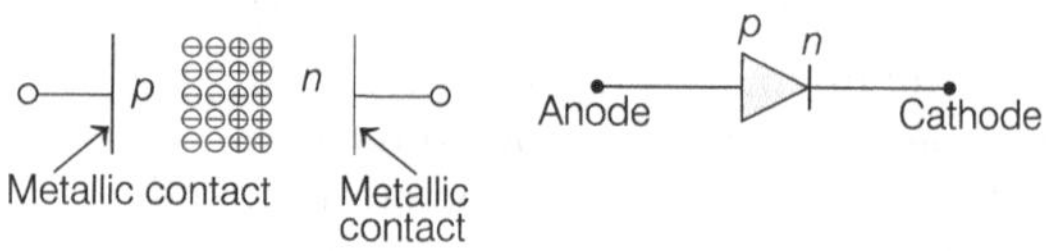

(a) Semiconductor diode (b) Symbol for p-n junction diode

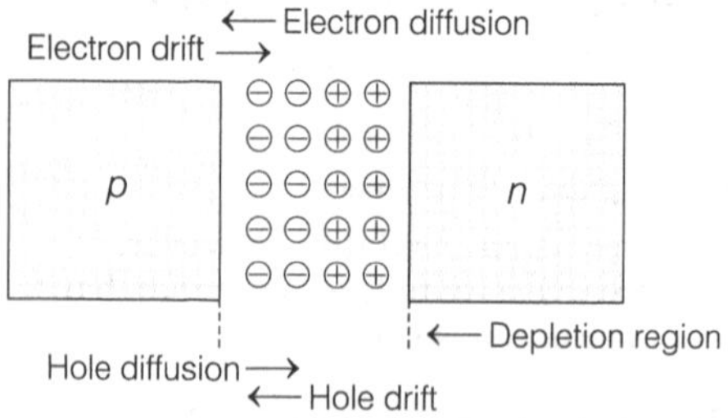

p-n junction formation process

On account of difference in concentration of charge carriers in the two sections of p-n junction, the electrons from n-region diffuse through the junction into p-region and the holes from p-region diffuse into n-region.

Due to this diffusion, the neutrality of the concentration of charge carriers on p and n-type semiconductors is disturbed. Due to which a layer of negative charge ions appear near the junction of p-type crystal and a layer of positive ions appears near the junction of n-type crystal. This layer is called **depletion region**. Also, this causes a difference of potential across the junction of two regions. This potential tends to prevent the movement of electrons from n-region to p-region and is often called **barrier potential**.

Few important graphs related to potential barriers are as follows

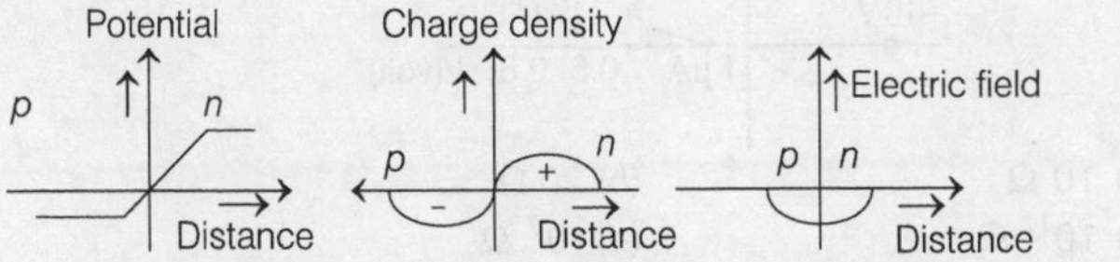

(a) Potential *versus* distance (b) Charge density *versus* distance (c) Electric field *versus* distance across the depletion region

Diffusion and Drift Currents

Current across the p-n junction are of two types

(i) **Diffusion Current (I_{df})** It is the current flowing from p-side to n-side due to the diffusion of electrons and holes because of concentration difference.

(ii) **Drift Current (I_{dr})** It is the current flowing due to electron-hole pairs under the influence of an external electic field.

Biasing of Junction Diode

Biasing is the method of connecting external battery or emf source to a p-n junction diode. The diode can be connected to an external battery in following two ways

(i) **Forward Biasing** A p-n junction is said to be forward biased, if the positive terminal of the external battery B is connected to p-side and the negative terminal to the n-side of p-n junction.

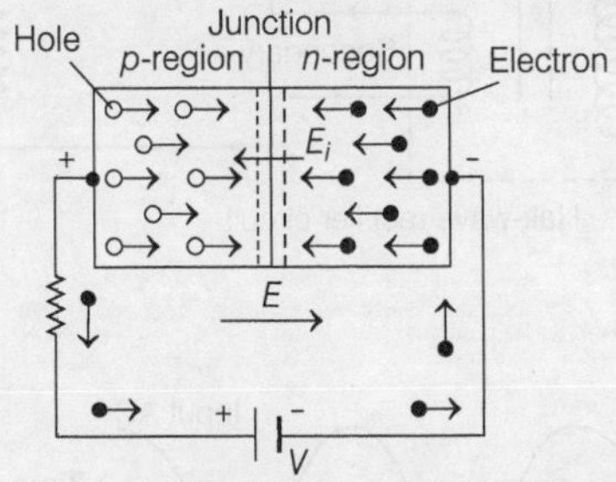

Forward biased p-n junction diode

Following are few important points related to forward biasing

- An electric current flow due to migration of majority carriers across the p-n junction; which is called **forward current**. Since, the small increase in forward voltage shows the large increase in forward current hence, the resistance of p-n junction is low to the flow of current when forward biased.
- The effective barrier height under forward bias is $V_0 - V$.
- Width of the depletion layer decreases.
- Total diode forward current is the sum of hole diffusion current and convential current of electron diffusion. The magnitude of current is in mA.
- The voltage at which current starts to increase is called **cut in (knee) voltage**. For Ge, it is 0.3 V and for Si, it is 0.7 V.

(ii) **Reverse Biasing** A p-n junction is said to be reverse biased, if the positive terminal of the external battery B is connected to n-side and the negative terminal to p-side of the p-n junction.

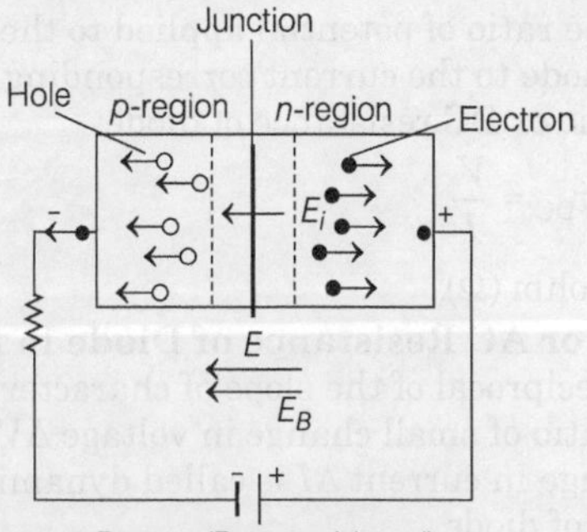

Battery (Reverse biased)

Following are few important points related to reverse biasing.

- A few minority carriers (holes in n-section and electrons in p-section) of p-n junction diode cross the junction after being accelerated by high reverse bias voltage. They constitute a current that flows in the opposite direction, this is called **reverse current of leakage current**. Since, the large increase in reverse voltage shows small increase in reverse current, hence the resistance of p-n junction is high to the flow of current when reverse biased.
- The direction of applied voltage is same as the direction of barrier potential. So, effective barrier height will be $(V_0 + V)$.
- Width of the depletion layer increases.
- The current under reverse bias is essentially voltage independent upto a critical reverse bias voltage, known as *breakdown voltage* (V_{br}).

  When $V = V_{br}$, the reverse current increases sharply.

I-V Characteristics of a Diode

The voltage-current characteristics curves of a p-n junction diode are graphs showing the change in current flowing through the junction with change in an applied voltage when junction is forward and reverse biased. It is as shown below.

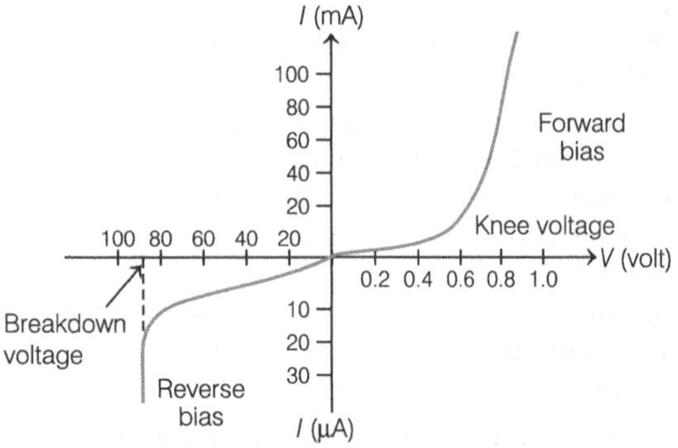

Diode Resistance

(i) **Static or DC Resistance of Diode** In forward biasing, the ratio of potential applied to the junction diode to the current corresponding to it, is called static or DC resistance of diode.

Thus, $r_{DC} = \frac{V}{I}$

Its unit is ohm (Ω).

(ii) **Dynamic or AC Resistance of Diode** In forward bias, the reciprocal of the slope of characteristic curve or ratio of small change in voltage ΔV to a small change in current ΔI is called dynamic or AC resistance of diode.

Thus, $r_{AC} = \frac{\Delta V}{\Delta I}$

Its unit is ohm (Ω).

Example 3. *Figure shows a circuit in which three identical diodes are used. Each diode has forward resistance of 20 Ω and infinite backward resistance. Resistors, $R_1 = R_2 = R_3 = 50\ \Omega$ and battery voltage is 6 V. The current through R_3 is*

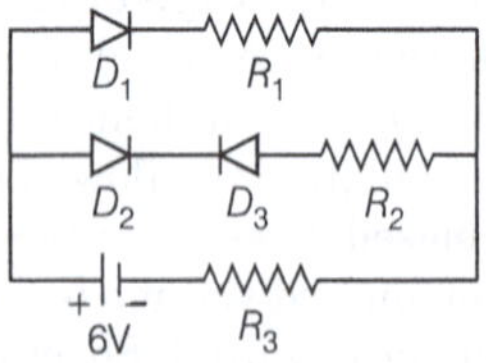

(a) 50 mA (b) 100 mA (c) 60 mA (d) 25 mA

Sol. (a) As diode is conducting in forward bias condition and not conducting in reverse bias condition. Diode D_1 is in forward bias, and diode D_2 is in forward bias but D_3 is reverse bias. So the figure can be drawn as

Here, 20 Ω, R_1Ω and R_3Ω are in series.

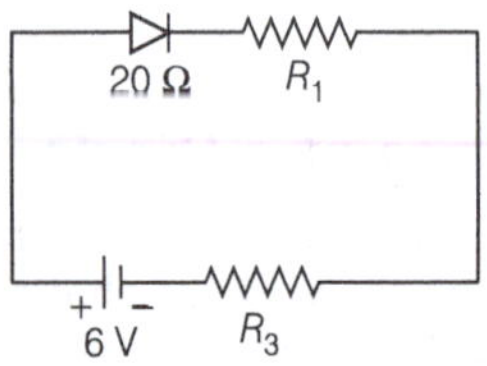

Equivalent resistance = 50 + 50 + 20 = 120 Ω

$$\therefore \quad I = \frac{V}{R} = \frac{6}{120} = \frac{1}{20} \Rightarrow I = 50 \text{ mA}$$

Example 4. *The V-I characteristic of silicon diode is shown in the figure. The resistance of the diode at $V_D = -10$ V is*

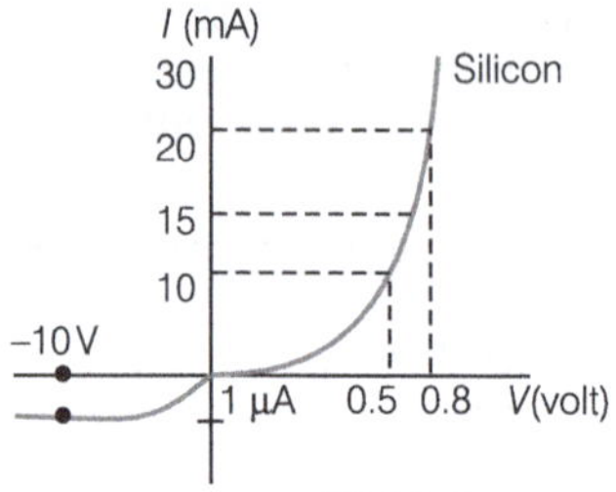

(a) 10 Ω (b) 20 Ω
(c) 10^{12} Ω (d) 10^{7} Ω

Sol. (d) From the curve, at $V = V_D = -10$ V, $I = -1\mu A$

$$\therefore \quad R = \frac{10 \text{ V}}{1\mu A} = 1 \times 10^7\ \Omega = 10^7\ \Omega$$

Diode as a Rectifier

Junction diode allows current to pass only when it is forward biased. So, if an alternating voltage is applied across a diode, the current flows only in that part of the cycle, when the diode is forward biased. This property is used to rectify alternating voltages and the circuit used for this purpose is called a **rectifier** and the process is known as **rectification**.

Half-Wave Rectifier

It is a circuit in which voltage is applied across a diode in series with a load. A pulsating voltage will appear across the load during the half cycle of the AC input when diode is forward biased.

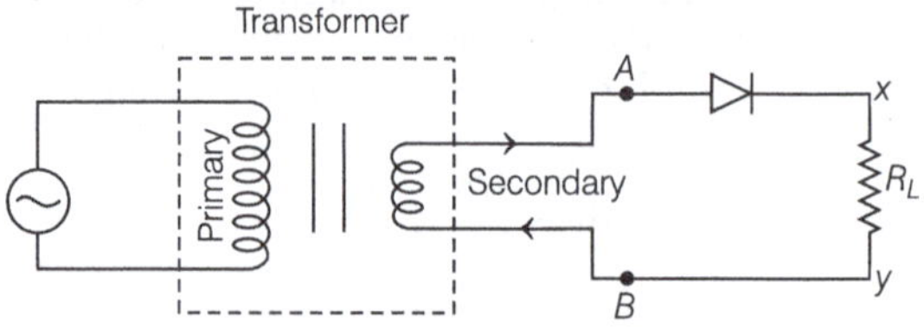

Half-wave rectifier circuit

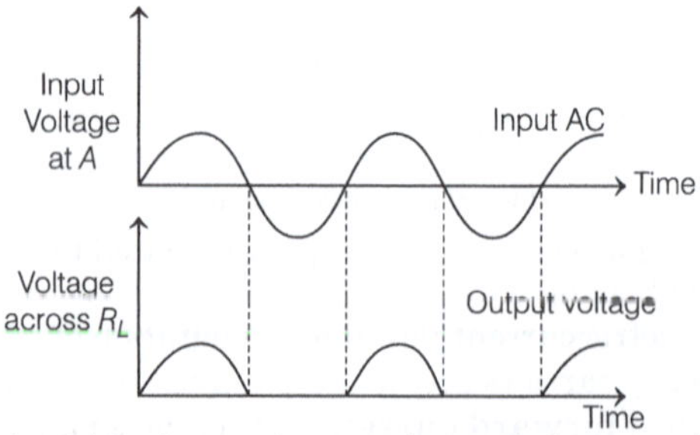

Input AC voltage and output voltage waveforms from the half-wave rectifier circuit

Important Points Related to Half-Wave Rectifier

- **Average output** in one cycle

$$I_{DC} = \frac{I_0}{\pi} \text{ and } V_{DC} = \frac{V_0}{\pi};\ I_0 = \frac{V_0}{r_f + R_L},$$

$$V_{DC} = \frac{V_0}{\pi\left(1 + \frac{r_f}{R_L}\right)} \quad (r_f = \text{forward biased resistance})$$

- **rms output** $I_{rms} = \frac{I_0}{2}, V_{rms} = \frac{V_0}{2}$
- The ratio of the effective alternating component of the output voltage or current to the DC component is known as **ripple factor**.

$$r = \frac{I_{AC}}{I_{DC}} = \left[\left(\frac{I_{rms}}{I_{DC}}\right)^2 - 1\right]^{1/2} = 1.21$$

The smaller the ripple factor, more effective will be the rectifier.

- **Peak inverse voltage** (PIV) The maximum reverse biased voltage that can be applied before commencement of Zener region is called the peak inverse voltage. When diode is not conducting, PIV across it = V_0.
- **Efficiency** It is given by

$$\eta = \frac{P_{out}}{P_{in}} \times 100\% = \frac{40.6\%}{1 + r_f/R_L}$$

If $R_L >> r_f$, then $\eta = 40.6\%$ and if $R_L = r_f$, then $\eta = 20.3\%$.

- The ratio of i_{rms} and i_{DC} is known as **form factor**.

$$\text{Form factor} = \frac{I_{rms}}{I_{DC}} = \frac{\pi}{2} = 1.57$$

- The ripple frequency (ω) for half-wave rectifier is same as that of AC.

Full Wave Rectifier

It is a circuit which gives output rectified voltage corresponding to both the positive as well as negative half of the AC cycle. These can be of two types namely: centre-tap full wave rectifier and full wave bridge rectifier.

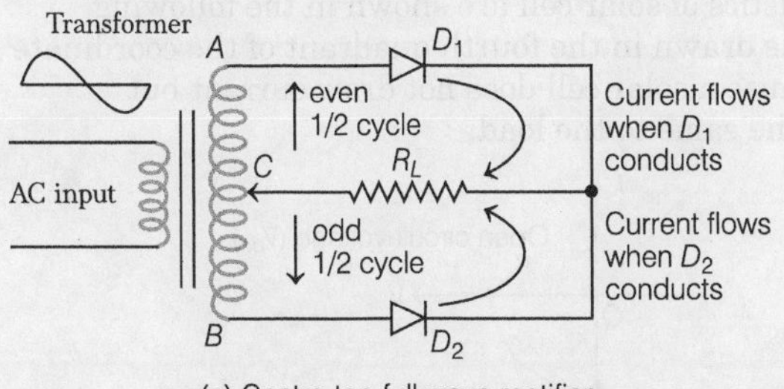

(a) Centre-tap full wave rectifier

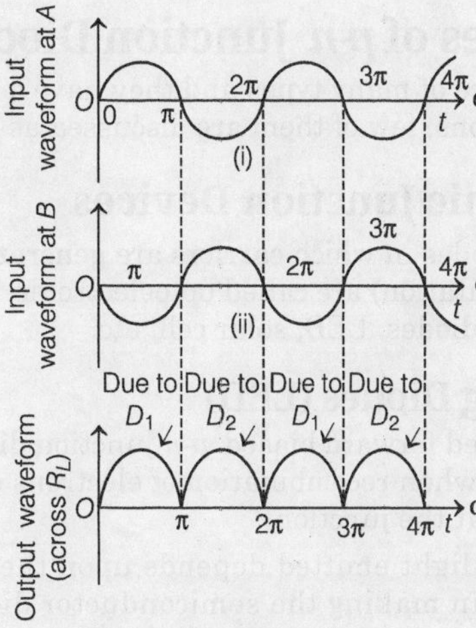

Input/Output waveform

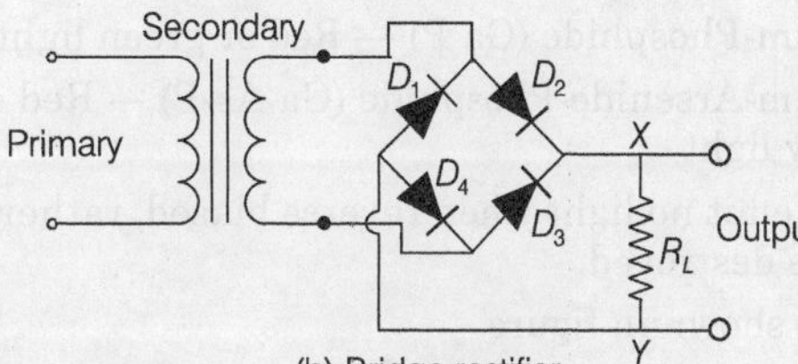

(b) Bridge rectifier

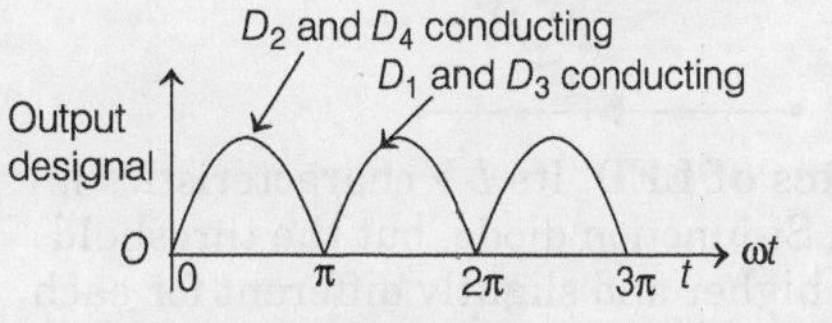

(c) Output waveforms for a bridge rectifier

Important Points Related to Full Wave Rectifier

- Output voltage is obtained across the load resistance R_L. It is not constant but pulsating in nature.
- **Average output**

$$V_{av} = \frac{2V_0}{\pi}, I_{av} = \frac{2I_0}{\pi}, V_{DC} = \frac{2V_0}{\pi\left(1 + \frac{r_f}{R_L}\right)}$$

- **rms output** $V_{rms} = \frac{V_0}{\sqrt{2}}, I_{rms} = \frac{I_0}{\sqrt{2}}$
- **Ripple factor** $r = 0.48 = 48\%$
- **Ripple frequency** The ripple frequency of full wave rectifier

$$= 2 \times \text{Frequency of input AC}$$

- **Peak inverse voltage (PIV)** Its value is $2V_0$.
- **Efficiency** $\eta_{\%} = \frac{81.2}{1 + \frac{r_f}{R_L}}$ for $r_f << R_L, \eta = 81.2\%$
- **Form factor**, $f = \frac{I_{rms}}{I_{DC}} = \frac{\pi}{2\sqrt{2}}$

Special Types of *p-n* Junction Diode

Junction diodes are of many types and they have a wide range of applications, few of them are discussed as below.

Optoelectronic Junction Devices

Semiconductor diodes in which carriers are generated by photons (photoexcitation) are called optoelectronic devices, *e.g.* photodiodes, LED, solar cell, etc.

Light Emitting Diodes (LED)

It is a heavily doped forward biased *p-n* junction diode which emits light when recombination of electrons and holes takes place at the junction.

The colour of the light emitted depends upon the types of material used in making the semiconductor diode.

(i) Gallium-Arsenide (Ga-As) — Infrared radiation

(ii) Gallium-Phosphide (Ga-P) — Red or green light

(iii) Gallium-Arsenide-Phosphide (Ga-As-P) — Red or Yellow light

LEDs emit no light when reverse biased, rather it will be destroyed.

Its symbol is shown in figure

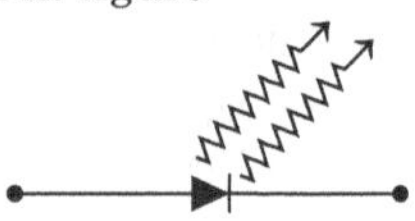

I-V Characteristics of LED Its I-V characteristics is similar to that of a Si-junction diode, but the threshold voltages are much higher and slightly different for each colour.

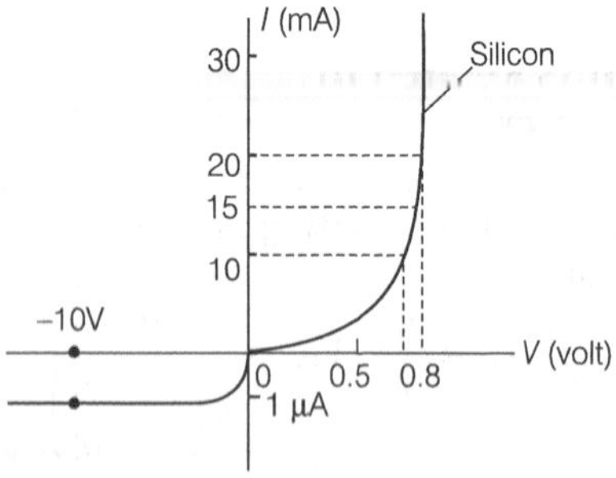

Photodiode

It is a junction diode fabricated with a transparent window to allow light to fall on the diode. It is operated under reverse bias.

Its symbol is given below.

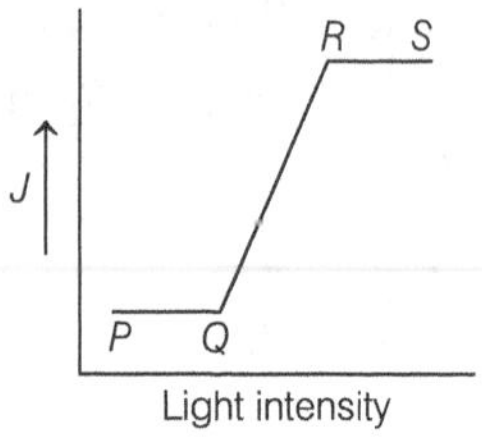

If the photodiode is in forward bias, some current flows in the circuit, shown by the PQ portion of the graph, this current PQ flows when no light is incident on it, is called **dark current**. When the intensity of light is increased, the current goes on increasing, shown by QR portion of the graph. A stage is reached when current does not increase with increase of intensity of light, is called **saturated current**, shown in RS portion of the graph.

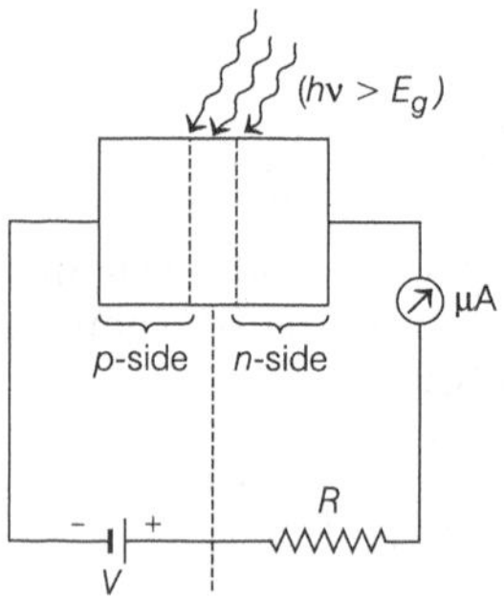

A reverse biased photodiode illuminated with light

I-V Characteristics of Photodiode The characteristics of the photodiode at different intensities are as follows

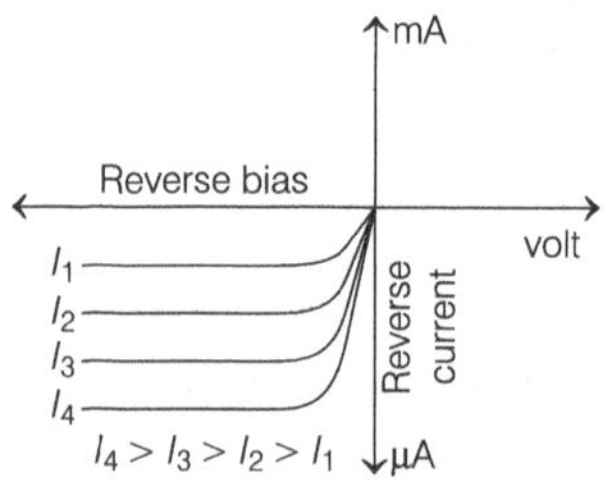

V-I characteristics of photodiode at different intensites

Solar Cell

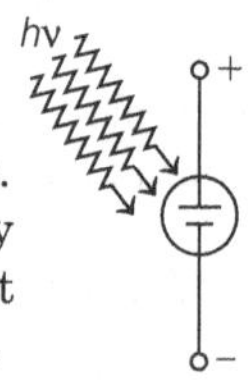

It is basically a *p-n* junction diode which generates emf when solar radiations falls on it. In this diode, either p or n section is made very thin (so that the light energy falling on it is not greatly absorbed before reaching the junction). It is used to convert light energy into electric energy. Its symbol is as shown in figure

I-V Characteristics of Solar Cell The I-V characteristics of solar cell are shown in the following figure. It is drawn in the fourth quadrant of the coordinate axes, because a solar cell does not draw current but supplies the same to the load.

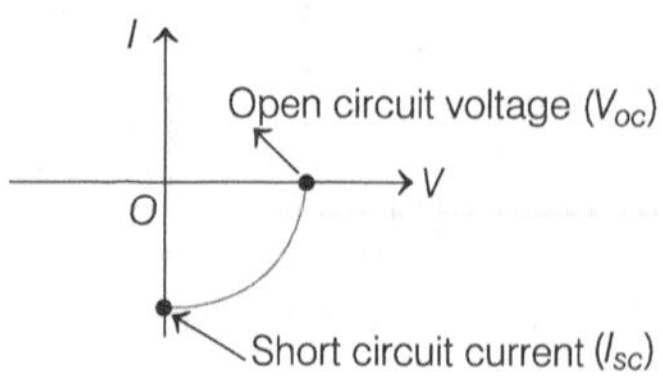

Zener Diode

The specially designed junction diodes, which can operate in the reverse breakdown voltage region continuously without being damaged are called *zener diodes*. Symbol for this diode is given below.

***I-V* Characteristics of Zener Diode** The *I-V* characteristics of Zener diode is shown below and we observe that when the applied reverse voltage (V) reaches the breakdown voltage (V_Z) of the Zener diode, there is a large change in the current.

But after the breakdown voltage V_Z, a large change in the current can be produced by almost insignificant change in the reverse bias voltage.

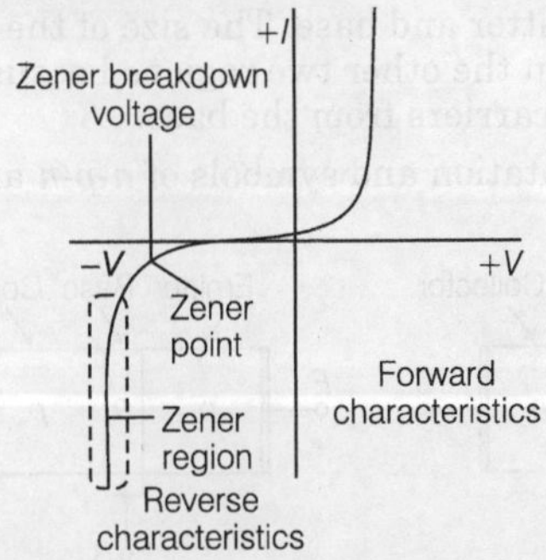

Zener Diode as Voltage Regulator Zener diode can be used as a voltage regulator, in which the constant output voltage is taken across a load resistance connected in parallel with it.

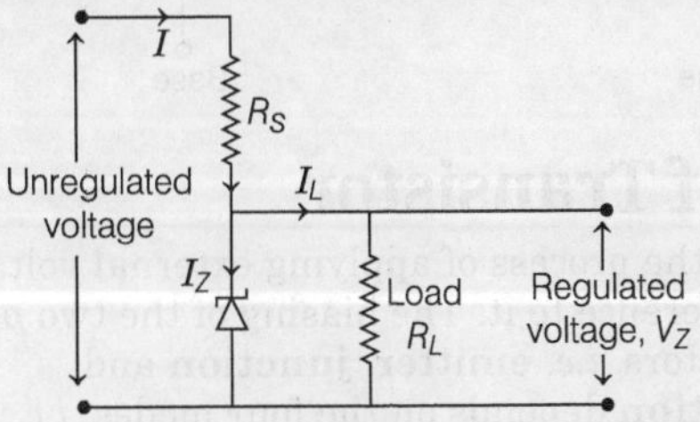

Example 5. *In figure, the input is across the terminals A and C and the output is across B and D. Then the output is*

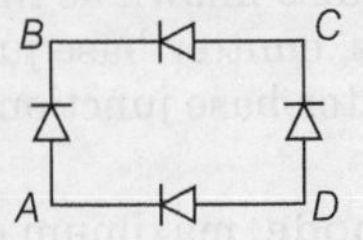

(a) *zero*
(b) same as the input
(c) *half-wave rectified*
(d) *full wave rectified*

Sol. (*d*) AC input is given across the points *A* and C and the output is taken across the points *B* and *D*, which is shown in the following figure.

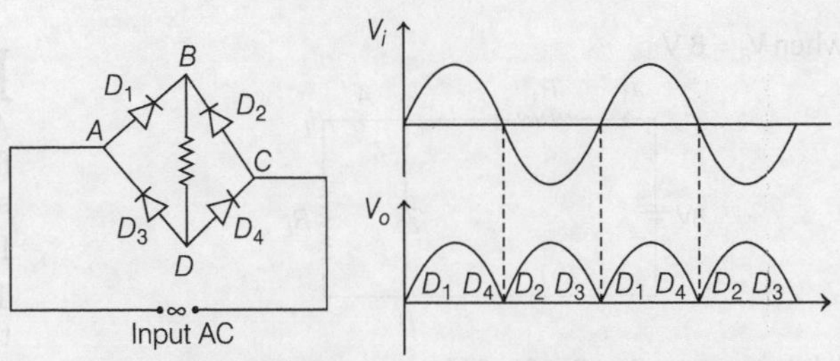

For positive half cycle, diodes D_1 and D_4 are in forward biased while diodes D_2 and D_3 are in reverse biased.
Similarly, for negative half cycle, diodes D_2 and D_3 are in forward biased and diodes D_1 and D_4 are in reverse biased.

Hence, given circuit is full wave bridge rectifier, which gives full wave rectified output.

Example 6. *A photodiode is a special purpose p-n junction diode fabricated with a transparent window to allow light to fall on the diode. It is operated under* **[NCERT]**

(a) *forward bias* (b) *reverse bias*
(c) *Both reverse and forward bias*
(d) *No biasing is required.*

Sol. (*b*) The photodiode is reverse biased for operating in the photoconductive mode. As the photodiode is in reverse bias, the width of the depletion region increases. This reduces the junction capacitance and thereby the response time. In effect, the reverse bias causes faster response time for the photodiode.

Example 7. *The figure represents a voltage regulator circuit using a Zener diode. The breakdown voltage of the Zener diode is 6 V and the load resistance is $R_L = 4\ k\Omega$. The series resistance of the circuit is $R_i = 1\,k\Omega$. If the battery voltage V_B varies from 8V to 16V, what are the minimum and maximum values of the current through Zener diode?* **[JEE Main 2019]**

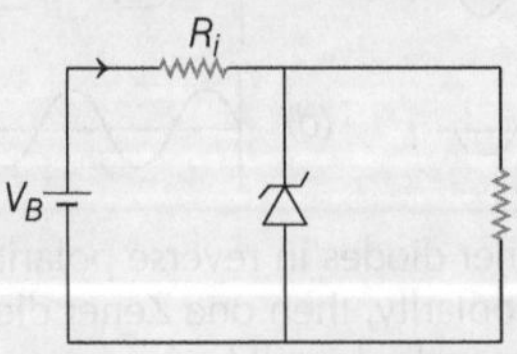

(a) *1.5 mA, 8.5 mA* (b) *1 mA, 8.5 mA*
(c) *0.5 mA, 8.5 mA* (d) *0.5 mA, 6 mA*

Sol. (c) In given voltage regulator circuit

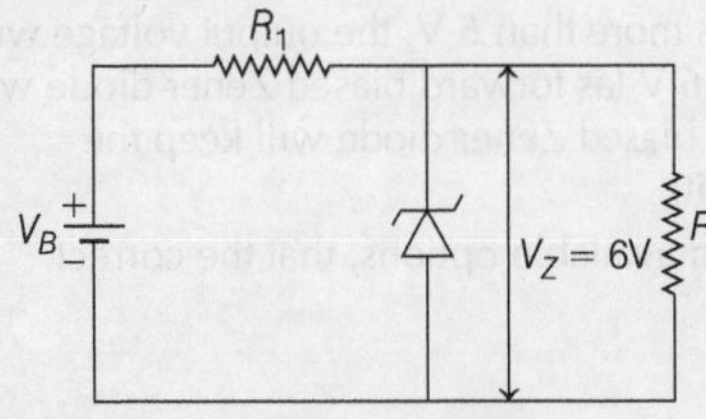

Zener breakdown voltage, $V_Z = 6$ V

So, across R_L, potential drop is always 6 V.

So, current through load resistance is

$$i_L = \frac{V_Z}{R_L} = \frac{6}{4 \times 10^3} = 1.5 \times 10^{-3}\ \text{A}$$

Now, when $V_B = 8$ V

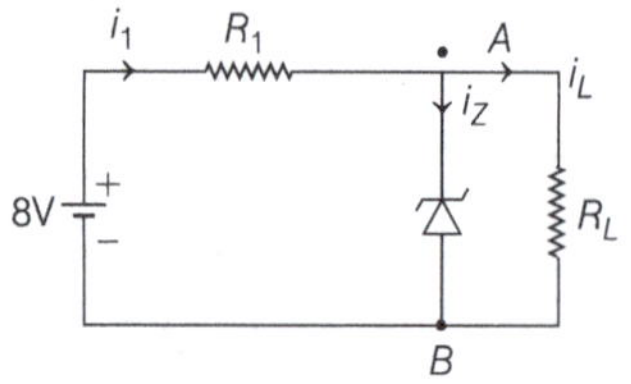

Potential drop across $R_1 = 8 - 6 = 2$ V

So, current through R_1 is $i_1 = \frac{V}{R_1} = \frac{2}{1 \times 10^3} = 2 \times 10^{-3}$ A

So, current through Zener diode is

$$i_Z = i_1 - i_L = 2 \times 10^{-3} - 1.5 \times 10^{-3}$$
$$= 0.5 \times 10^{-3} \text{ A} = 0.5 \text{ mA}$$

Similarly, when $V_B = 16$ V

$$V_{R_1} = 16 - 6 = 10 \text{ V}$$

$$\therefore \quad i_1 = \frac{10}{1 \times 10^3} = 10 \times 10^{-3} \text{ A}$$

Hence, $$i_Z = i_1 - i_L = 10 \times 10^{-3} - 1.5 \times 10^{-3}$$
$$= 8.5 \times 10^{-3} \text{ A} = 8.5 \text{ mA}$$

Example 8. *Take the breakdown voltage of the Zener diode used in the given circuit as 6V. For the input voltage shown in figure below, the time variation of the output voltage is (Graphs are drawn schematically and on not to scale)*

[JEE Main 2020]

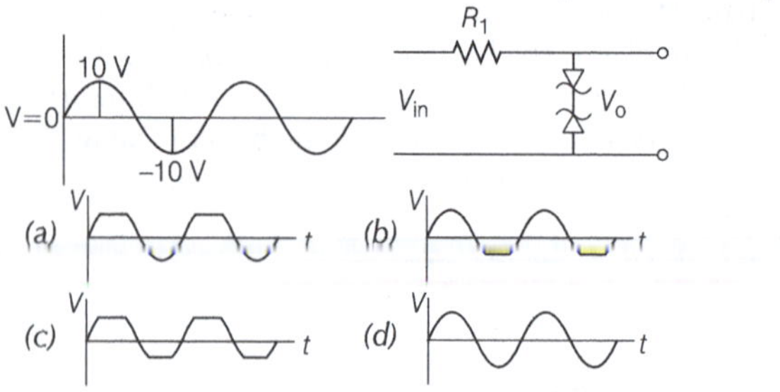

Sol. (c) There are two Zener diodes in reverse polarity. So, if the input voltage changes its polarity, then one Zener diode will be in forward bias and other zener diode will be in reverse bias. This happens at each kind of polarity of input voltage.

Now, if the input voltage is less than 6 V, then output will follow the input voltage, which means it will remain sinusoidal.

And if the input voltage is more than 6 V, the output voltage will remain constant equal to 6 V (as forward biased Zener diode will not be active and reverse biased Zener diode will keep the potential constant across it).

So, we can conclude from available options, that the correct graph is

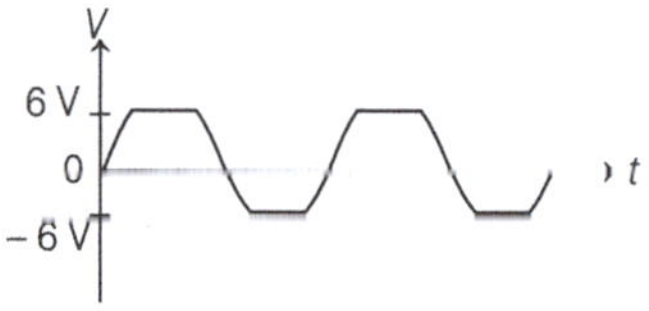

Hence, correct option is (c).

Junction Transistor

A junction transistor is a three terminal device which is formed by sandwiching a thin wafer of one type of semiconductors between two layers of another type.

It is of two types : *n-p-n* and *p-n-p*

The there terminals of transistor are emitter, base and the collector.

Emitter (*E*) It is heavily doped and of moderate size as it emits the majority charge carriers into the base by which current flows in the transistor.

Base (*B*) The base region is very lightly doped and thin. Most of the charge carriers emitted by the emitter pass through it to the collector.

Collector (*C*) The doping level of collector is neither very high nor very low, however it lies between the doping levels of emitter and base. The size of the collector region is larger than the other two regions because it collects the charge carriers from the base.

Schematic representation and symbols of *n-p-n* and *p-n-p* is as shown below.

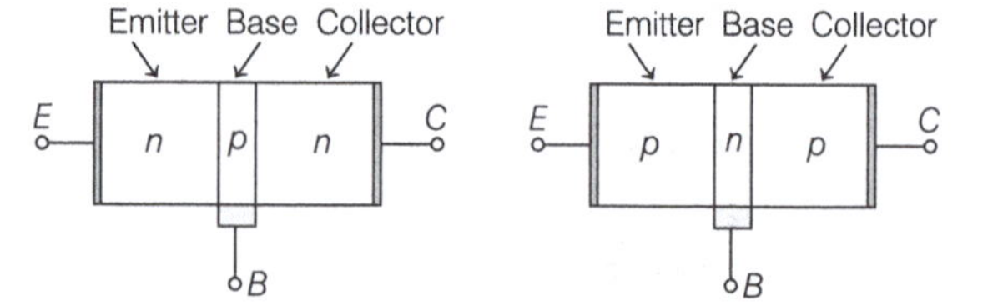

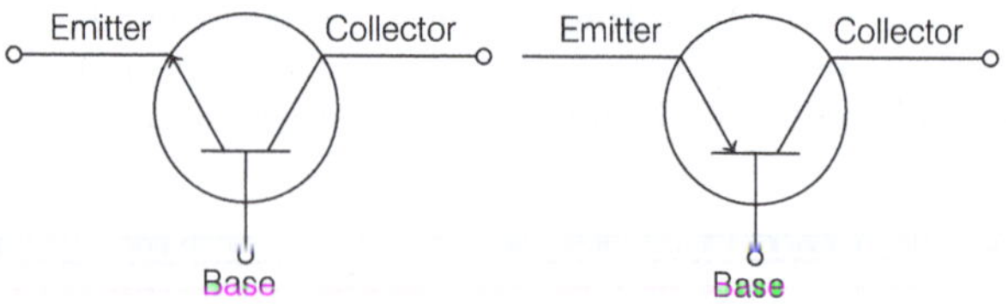

Biasing of Transistor

It is defined as the process of applying external voltages or potential difference to it. The biasing of the two *p-n* junction transistors, *i.e.* **emitter junction** and **collector junction** depends on the four modes. *i.e.* active mode, saturation mode, cut-off mode and inverse mode.

(i) **Active mode** is also known as *linear mode operation*. In this, emitter-base junction is forward biased and collector-base junction is reverse biased.

(ii) In **saturation mode**, maximum collector current flows and transistor acts as a closed switch from collector to emitter terminal.

(iii) In **cut-off mode**, the circuit behaves as open switch where only leakage current flows.

(iv) The emitter and collector are interchanged in **inverse mode**.

Different modes of operation of a transistor is given below.

Operating mode	Emitter-base bias	Collector-base bias
Active	Forward	Reverse
Saturation	Forward	Forward
Cut off	Reverse	Reverse
Inverse	Reverse	Forward

Note In *p-n-p* and *n-p-n* transistors,

Emitter current = Base current + Collector current

Transistor Circuit Configurations

In a transistor, one terminal out of the three serves as a reference point for the entire circuit. This terminal should be common to the input and output circuits and is connected to the ground. So, a transistor can be used in the following three configurations.

(i) Common Emitter (CE) Configuration
(ii) Common Base (CB) Configuration
(iii) Common Collector (CC) Configuration

Common Emitter Characteristics

The graphs drawn between voltages and currents, when emitter of a transistor is common to input and output circuits are known as CE characteristics of a transistor. The circuit of *n-p-n* transistor in CE configuration is as follows

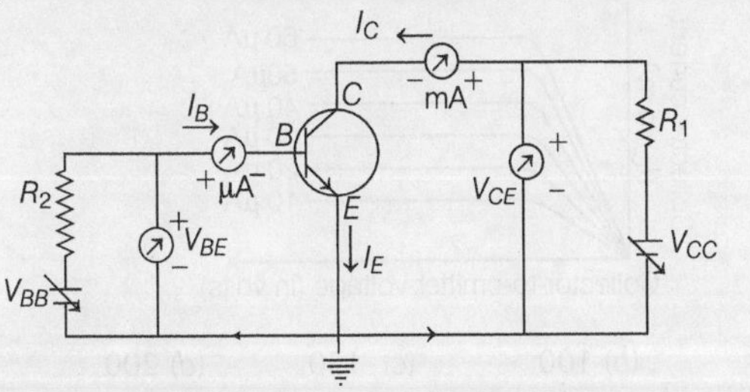

Input Characteristics It represents the variation of base current I_B with the base-emitter voltage V_{BE}. It is as shown below

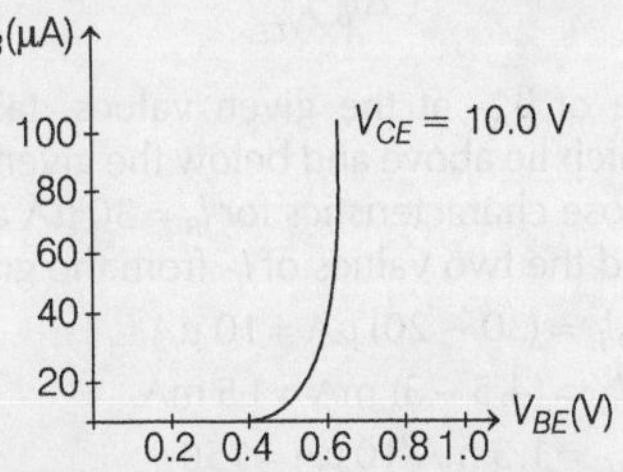

Here, increase in V_{CE} appear as increase in V_{CB}, its effect on I_B is negligible. As a consequence, input characteristics for various values of V_{CE} will give identical curves.

Hence, it is enough to determine only one input characteristics.

Input resistance is reciprocal of slope of I_B-V_{BE} curve.

$$\text{Input resistance, } R_i = \left(\frac{\Delta V_{BE}}{\Delta I_B}\right)_{V_{CE} = \text{constant}}$$

Output Characteristics It represents variation of the collector current I_C with the collector-emitter voltage V_{CE}. It is as shown below.

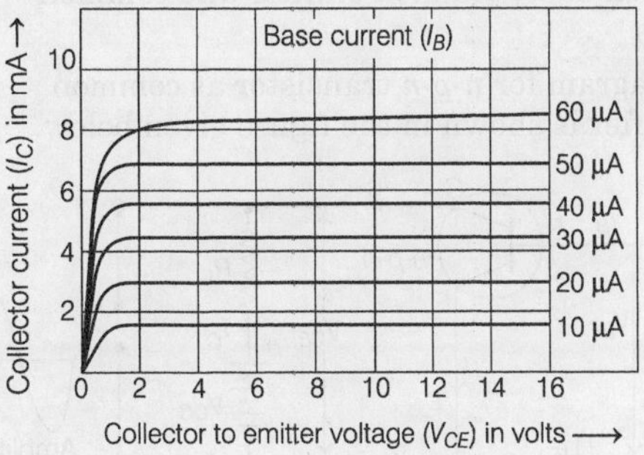

$$\text{Output resistance, } r_o = \left(\frac{\Delta V_{CE}}{\Delta I_C}\right)_{I_B = \text{constant}}$$

$$= \text{Reciprocal of slope of } I_C\text{-}V_{CE} \text{ curve}$$

Transfer Characteristics It is the graph between the collector current I_C and the base current I_B at constant values of collector-emitter voltage V_{CE}.

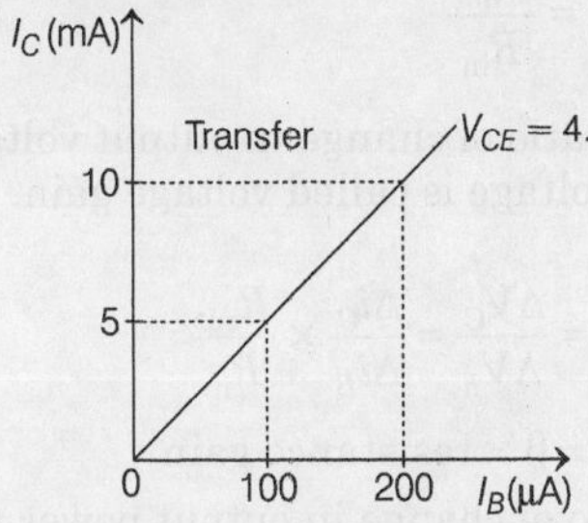

Current Amplification Factor

Current amplification factor β of a transistor in CE configuration is given by $\beta_{AC} = \left(\frac{\Delta I_C}{\Delta I_B}\right)_{V_{CE} = \text{constant}}$

Its value is very large ($\beta_{AC} >> 1$).

In case of common base configuration,

It is the ratio of small change in the collector current to the small change in the emitter current at constant collector-base voltage. It is denoted by α and is given by

$$\alpha = \left[\frac{\Delta I_C}{\Delta I_E}\right]_{V_{CB} = \text{constant}}$$

Relation between α and β

The relation between α and β can be given as follows

$$\alpha = \frac{\beta}{1+\beta} \quad \text{and} \quad \beta = \frac{\alpha}{1-\alpha}$$

Note α and β are independent of current, if the emitter-base junction is forward biased and the collector-base junction is reverse biased.

Transistor as an Amplifier

An amplifier is a device which is used for increasing the strength of amplitude of input signal. A transistor can be used for amplifying a weak signal. For a transistor to be operated as amplifier, in any of the configurations namely common base, common emitter and common collector.

The circuit diagram for *n-p-n* transistor as common emitter amplifier is shown in the figure given below.

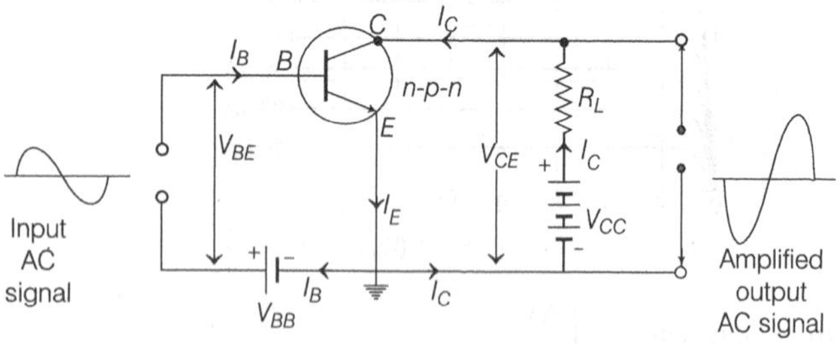

From the above circuit, we get

$$I_E = I_B + I_C \text{ and } V_{CE} = V_{CC} - I_C R_L$$

Resistance gain The ratio of output (load) resistance to the input resistance is called resistance gain.

$$\therefore \quad \text{Resistance gain} = \frac{R_{out}}{R_{in}}$$

Voltage gain The ratio of change in output voltage to the change in input voltage is called voltage gain. It is denoted by A_V.

$$\text{Voltage gain,} \quad A_V = \frac{\Delta V_C}{\Delta V_b} = \frac{\Delta i_C}{\Delta i_B} \times \frac{R_{out}}{R_{in}}$$

$$\Rightarrow \quad A_V = \beta \times \text{resistance gain}$$

Power gain The ratio of change in output power to the change in input power is called power gain.

$$\therefore \quad \text{Power gain} = \frac{P_{out}}{P_{in}}$$

Also, power gain = $\beta^2 \times$ resistance gain

Transconductance (g_m)

It is defined as the ratio of the change in the collector current to the change in the base-to-emitter voltage at constant collector-to-emitter voltage and is denoted by g_m.

$$\text{Thus,} \quad g_m = \left(\frac{\Delta I_C}{\Delta V_{BE}}\right)_{V_{CE}} \text{ or } g_m = \frac{\beta_{AC}}{R_{in}}$$

The unit of g_m is Ω^{-1} (ohm^{-1}) or S (siemen).

Transistor as an Oscillator

Transistor can be used as an oscillator which produces electrical oscillation of constant frequency and amplitude without any external input signal.

External input is necessary to sustain AC signal in the output of an amplifier. In case of oscillator, an AC output is received without any external input signal.

The feedback in the oscillators can be achieved by inductive coupling on *LC* or *RC* networks. The circuit used for an oscillator is an shown below.

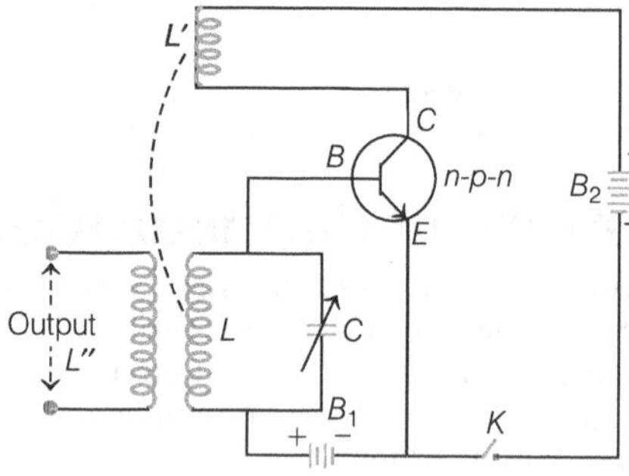

The frequency of oscillations is given by

$$\nu = \frac{1}{2\pi\sqrt{LC}}$$

Example 9. *When a transistor is used in common emitter configuration, the following output characteristics are obtained. When V_{CE} is 10 V and I_C is 4 mA, then the value of β_{AC} (current amplification factor) is*

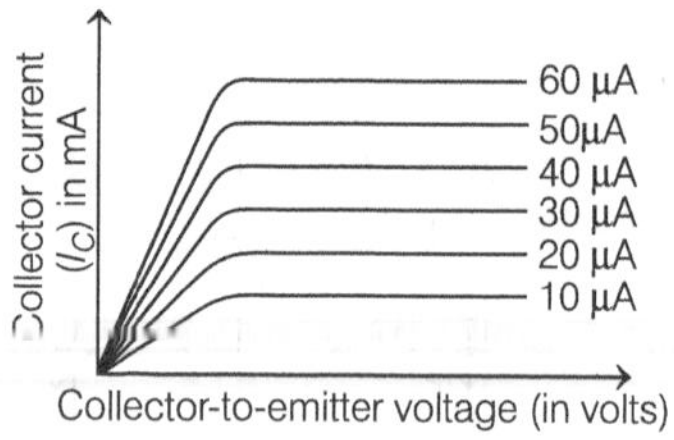

(a) 50 (b) 100 (c) 150 (d) 200

Sol. (c) Current amplification factor (β) is

$$\beta_{AC} = \left(\frac{\Delta I_C}{\Delta I_B}\right)_{V_{CE}}$$

For finding the value of β_{AC} at the given values, take any two characteristics of I_B which lie above and below the given value of I_C. Here, I_C = 4 mA (choose characteristics for $I_B = 30\,\mu$A and $20\,\mu$ A). At V_{CE} = 10 V, we read the two values of I_C from the graph, then

$$\Delta I_B = (30 - 20)\,\mu A = 10\,\mu A$$

$$\Delta I_C = (4.5 - 3)\text{ mA} = 1.5\text{ mA}$$

$$\therefore \quad \beta_{AC} = 1.5\text{ mA}/10\,\mu A = 150$$

Example 10. *In the given figure, the V_{BB} supply can be varied from 0 V to 5V. The Si transistor has $\beta_{DC} = 250$, $R_B = 100\,k\Omega$, $R_C = 1\,k\Omega$ and $V_{CC} = 5$ V. Assume that, when the transistor is saturated, $V_{CE} = 0$ V and $V_{BE} = 0.8$ V. The minimum base current for which the transistor will reach saturation is*

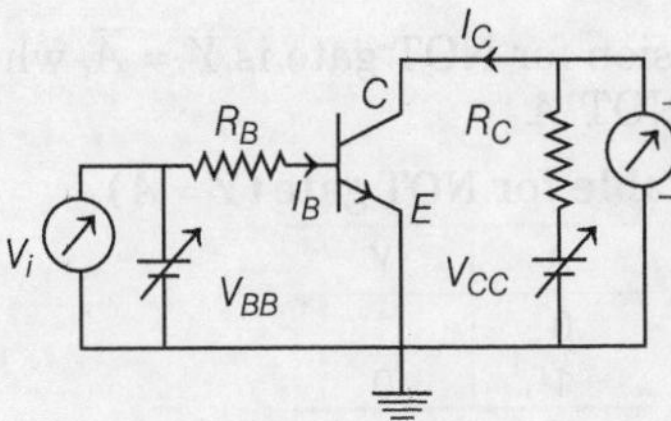

(a) 10 μA (b) 20 μA (c) 30 μA (d) 40 μA

Sol. (b) At saturation, $V_{CE} = 0$ V

$$V_{BE} = 0.8 \text{ V}$$

$$\therefore \quad V_{CE} = V_{CC} - I_C R_C$$

$$I_C = \frac{V_{CC}}{R_C} = \frac{5 \text{ V}}{1 \text{k}\Omega} = 5 \text{ mA}$$

$$\text{and} \quad I_B = \frac{I_C}{\beta} = \frac{5 \text{ mA}}{250} = 20 \text{ μA}$$

Example 11. *For a CE transistor amplifier, the audio signal voltage across the collector resistance of 2 kΩ is 2 V. Suppose the current amplification factor of the transistor is 100. The value of R_B in series with V_{BB} supply of 2V, if the DC base current has to be 10 times the signal current is* ***[NCERT]***

(a) 4 k Ω (b) 14 k Ω (c) 28 k Ω (d) 54 k Ω

Sol. (b) The output AC voltage is 2 V, so the AC collector current, $I_c = \frac{2}{2000} = 1 \text{mA}$

The signal current through the base is given by

$$I_B = \frac{I_c}{\beta} = \frac{1 \text{mA}}{100} = 0.010 \text{ mA}$$

The DC base current has to be $10 \times 0.010 = 0.10$ mA

$$\therefore \quad R_B = \frac{(V_{BB} - V_{BE})}{I_B}, V_{BE} = 0.6 \text{ V}$$

$$\therefore \quad R_B = \frac{(2 - 0.6)}{0.10} = 14 \text{ k}\Omega$$

Example 12. *In a CE transistor amplifier, the output resistance in 500 kΩ and current gain β = 49. If the power gain of the amplifier is 5×10^6, the input resistance is*

(a) 240 Ω (b) 165 Ω (c) 180 Ω (d) 290 Ω

Sol. (a) Given, 500 kΩ, β = 49 and $P = 5 \times 10^6$

We have,

$$P = \beta^2 \frac{R_o}{R_i}$$

$$5 \times 10^6 = \frac{(49)^2 \times 500 \times 10^3}{R_i}$$

$$\therefore \quad R_i = \frac{49 \times 49 \times 500 \times 10^3}{5 \times 10^6} = 240 \ \Omega$$

Example 13. *An n-p-n transistor is used in common emitter configuration as an amplifier with 1 kΩ load resistance. Signal voltage of 10 mV is applied across the base-emitter. This produces a 3 mA change in the collector current and 15 μA change in the base current of the amplifier. The input resistance and voltage gain are* ***[JEE Main 2019]***

(a) 0.67 kΩ, 200 (b) 0.33 kΩ, 1.5
(c) 0.67 kΩ, 300 (d) 0.33 kΩ, 300

Sol. (c) Given, load resistance, $R_L = 1 \text{k}\Omega$

Input voltage, $V_{in} = 10 \text{ mV} = 10 \times 10^{-3}$ V

Base current, $\Delta I_B = 15 \text{ μA} = 15 \times 10^{-6}$ A

Collector current, $\Delta I_C = 3$ mA

$$\text{Input resistance,} \quad R_{in} = \frac{V_{in}}{\Delta I_B} = \frac{10 \times 10^{-3}}{15 \times 10^{-6}} = 0.67 \text{k}\Omega$$

$$\text{and} \quad \text{voltage gain} = \beta \times \frac{R_L}{R_{in}} = \frac{\Delta I_C \times R_L}{\Delta I_B \times R_{in}}$$

$$= \left(\frac{3 \text{ mA}}{15 \text{ μA}}\right) \times \left(\frac{1 \text{k}\Omega}{0.67 \text{ k}\Omega}\right)$$

$$= \left(\frac{3 \times 10^{-3}}{15 \times 10^{-6}}\right)\left(\frac{1 \times 10^3}{0.67 \times 10^3}\right)$$

$$= \frac{1000 \times 3 \times 3}{15 \times 2} = 300 \quad (\because 0.67 \cong 2/3)$$

Logic Gates

A logic gate is an elementary building block of a digital circuit. Most logic gates have two inputs and one output. At any given moment, every terminal is in one of the two binary conditions, i.e. low (0) or high (1), represented by different voltage levels.

The basic logic gates are of three types

OR Gate

It is a device which has two or more inputs and one output. The logic symbol of OR gate is

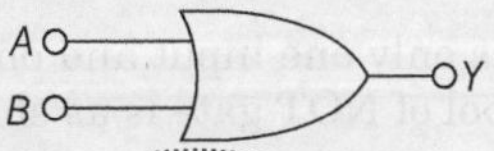

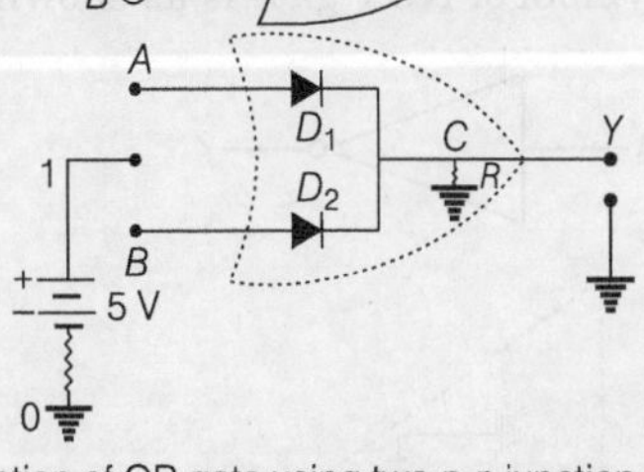

Realisation of OR gate using two *p-n* junction diodes

The Boolean expression for OR gate is $Y = A + B$

This indicates Y equals A OR B.

Truth table for OR gate ($Y = A + B$)

A	B	Y
0	0	0
0	1	1
1	0	1
1	1	1

The output of an OR gate assumes 1, if one or more inputs assume 1. The output is high when either of inputs A or B is high, but not if both A and B are low.

AND Gate

It is a device which has also two or more inputs and one output. The logic symbol of AND gate is given as

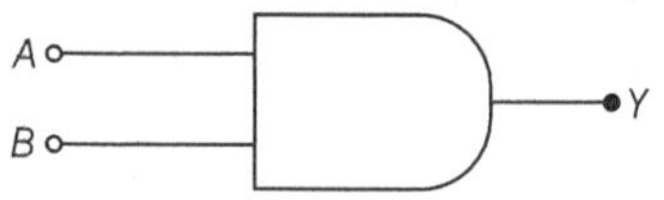

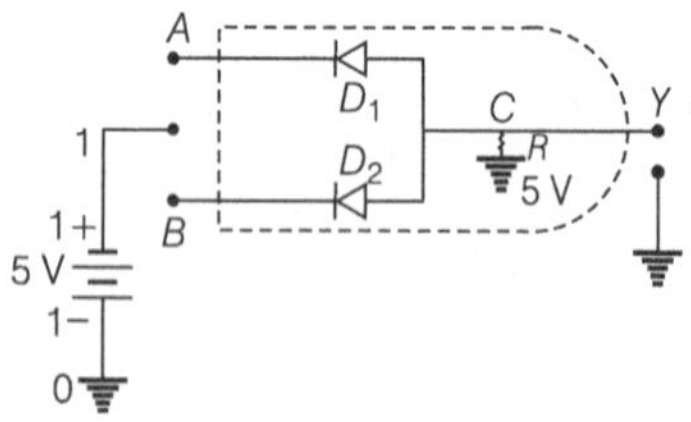

Realisation of an AND gate

The Boolean expression is $Y = A \cdot B$. This indicates Y equals to A AND B.

Truth table for AND gate ($Y = A \cdot B$)

A	B	Y
0	0	0
0	1	0
1	0	0
1	1	1

The output of an AND gate is 1 only, when all the inputs assume 1.

NOT Gate

It is a device which has only one input and only one output. The logic symbol of NOT gate is as shown in the following figure.

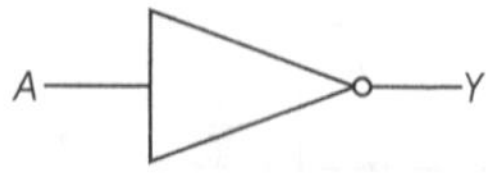

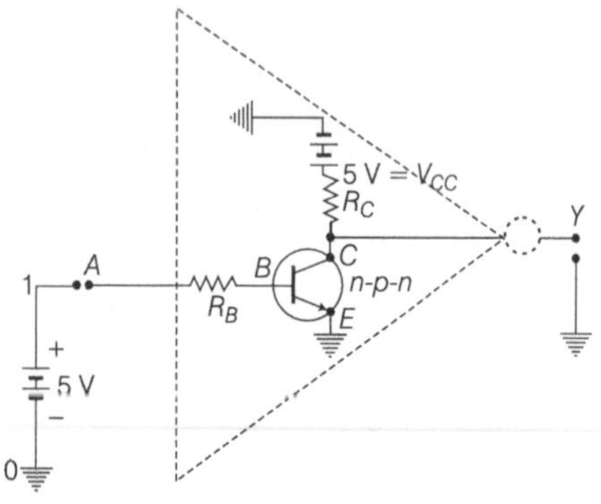

Realisation of NOT gate

The Boolean expression for NOT gate is $Y = \overline{A}$, which indicates Y equals NOT A.

Truth Table for NOT gate ($Y = \overline{A}$)

A	Y
0	1
1	0

The output of a NOT gate assumes 1, if input is 0 and *vice-versa*.

Combination of Logic Gates

Few combinations of basic gates are given below

NAND Gate

In this type of gate, the output of AND gate is fed to input of a NOT gate and final output is obtained at output of NOT gate.

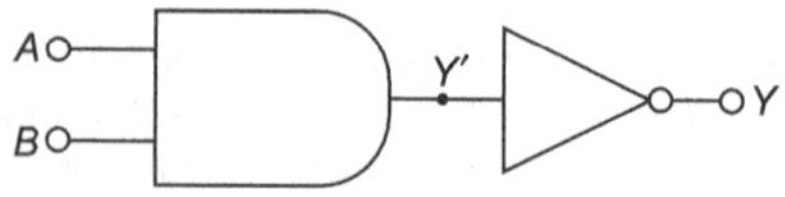

The logic symbol of NAND gate is shown as

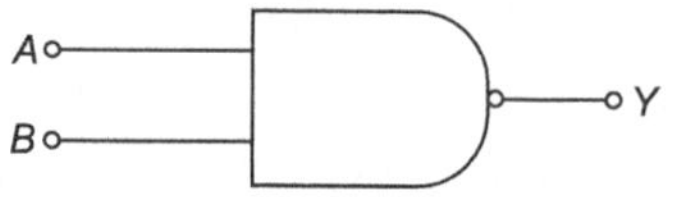

The Boolean expression of NAND gate is $Y = \overline{A \cdot B}$, which indicates A and B are negated.

Truth table for NAND gate

A	B	Y′	Y
0	0	0	1
1	0	0	1
0	1	0	1
1	1	1	0

NOR Gate

In this type of gate, the output of OR gate is fed to input of the NOT gate and final output is obtained at output of the NOT gate.

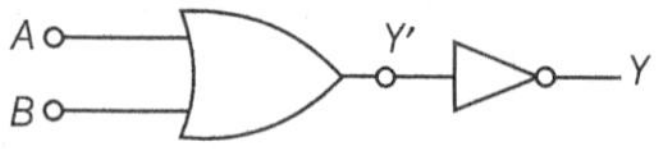

The logic symbol of NOR gate is shown as

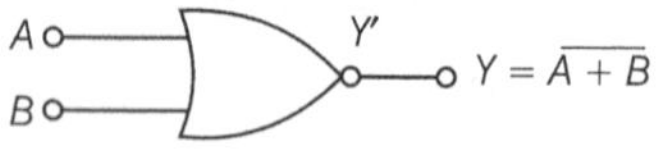

The Boolean expression for NOR gate is $Y = \overline{A + B}$, which indicates that A OR B are negated.

Truth table for NOR gate

A	B	Y′	Y
0	0	0	1
1	0	1	0
0	1	1	0
1	1	1	0

The NOR gate and NAND gate can be said to be **universal gates,** since combinations of them can be used to accomplish any of the basic operations and can thus produce an inverter, an OR gate or an AND gate.

XOR Gate

The XOR gate can be obtained by the combination of OR, AND and NOT gates as shown below.

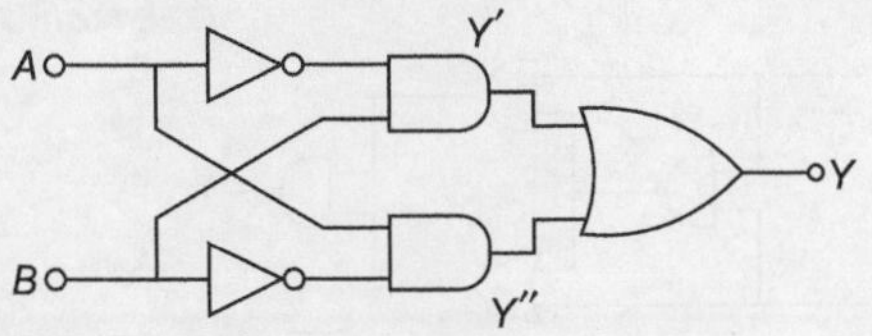

The logic symbol of XOR gate is shown as

The Boolean expression for XOR gate is

$$Y = A\overline{B} + \overline{A}B = A \oplus B$$

Truth Table for XOR gate

A	B	Y
0	0	0
0	1	1
1	0	1
1	1	0

From truth table, we can conclude that if A and B are not indentical, output is 1 and when A and B are indentical, output is 0.

XNOR Gate (exclusive-NOR) Gate

If XOR gate is combined with NOT gate, we get XNOR gate, *i.e.* the XOR gate followed by an inverter. Its output is "true" if the inputs are the same, and "false" if the inputs are different.

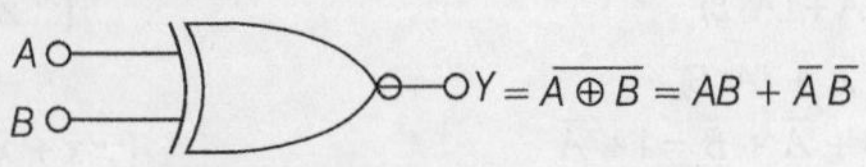

XNOR gate

Truth table for XNOR gate

Inputs		Output
A	B	A XNOR B
0	0	1
0	1	0
1	0	0
1	1	1

Some Useful Boolean Identities

- **Commutative laws**
  (i) $A + B = B + A$; (ii) $A \cdot B = B \cdot A$
- **Associative laws**
  (i) $A + (B + C) = (A + B) + C$;
  (ii) $A \cdot (B \cdot C) = (A + B) \cdot C$
- **Distributive laws**
  (i) $A \cdot (B + C) = A \cdot B + A \cdot C$;
  (ii) $(A + B) \cdot (A + C) = A + B \cdot C$
- **Absorption laws**
  (i) $A + A \cdot B = A$; (ii) $A \cdot (A + B) = A$;
  (iii) $\overline{A} \cdot (A + B) = \overline{A} \cdot B$
- **Double complement function**
  $\overline{\overline{A}} = A$, $\overline{\overline{A + B}} = A + B$; $\overline{\overline{A . B}} = A \cdot B$
- **Boolean indentities**
  (i) $A \cdot (\overline{A} + B) = A \cdot B$
  (ii) $A + \overline{A} \cdot B = A + B$
  (iii) $A + B \cdot C = (A + B) \cdot (A + C)$
  (iv) $(\overline{A} + B) \cdot (A + C) = \overline{A} \cdot C + A \cdot B$
- **De-Morgan's Theorem** It states that the complement of the whole sum is equal to the product of individual complements and *vice-versa. i.e.*
  (i) $\overline{A + B} = \overline{A} \cdot \overline{B}$ (ii) $\overline{A \cdot B} = \overline{A} + \overline{B}$
- **Basic OR and AND relations**

	OR	AND
(i)	$A + 0 = A$	$A \cdot 0 = 0$
(ii)	$A + 1 = 1$	$A \cdot 1 = A$
(iii)	$A = A = A$	$A \cdot A = A$
(iv)	$A + \overline{A} = 1$	
	$A \cdot \overline{A} = 0$	

- $A \oplus B = A\overline{B} + B\overline{A}$ (A AND NOT B OR B AND NOT A).
- $A \oplus B = (A + B)\,(\overline{AB})$ (A OR B AND NOT A AND B).

It can be implemented by the gate arrangements as shown below.

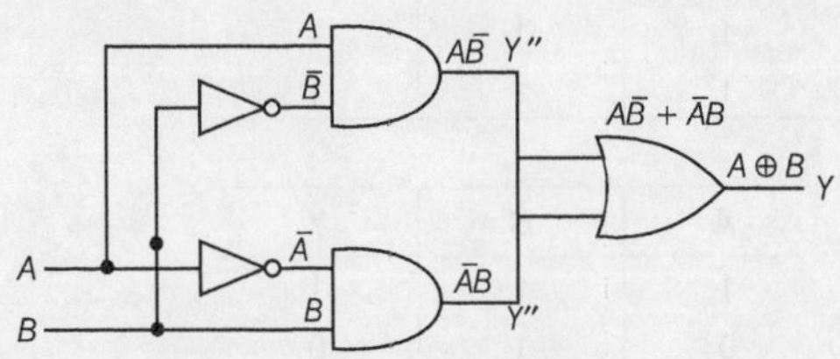

Example 14. *The logic gate equivalent to the given logic circuit is* **[JEE Main 2019]**

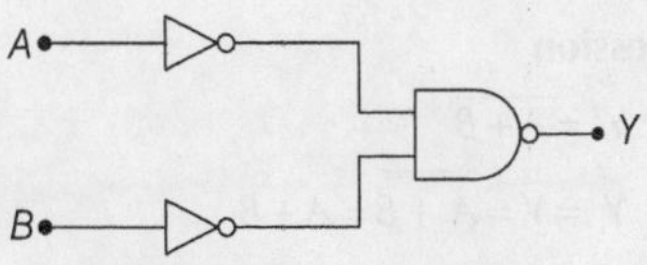

(a) NOR (b) NAND (c) OR (d) AND

Sol. (c) The given logic gate circuit can be drawn as shown below

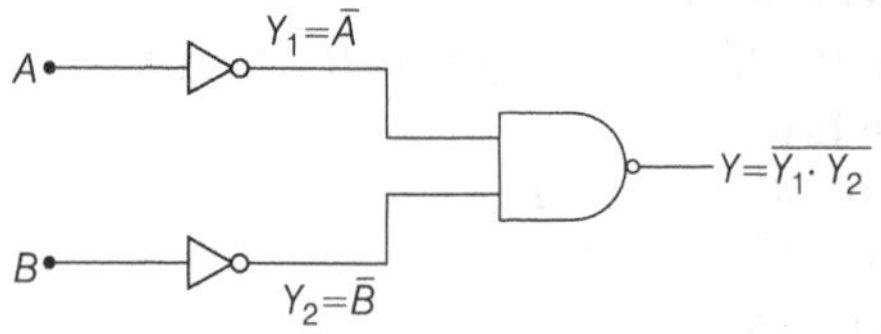

Here, $Y = \overline{Y_1 \cdot Y_2} = \overline{\bar{A} \cdot \bar{B}}$

Using de-Morgan's theorem, *i.e.*

$$\overline{x \cdot y} = \bar{x} + \bar{y}$$

$\therefore$ $Y = \bar{\bar{A}} + \bar{\bar{B}} = A + B$ $[\because \bar{\bar{x}} = x]$

This represents the boolean expression for OR gate.

Example 15. *In the following circuit, the output 1 for all possible inputs A and B is expressed by which of the truth table given below*

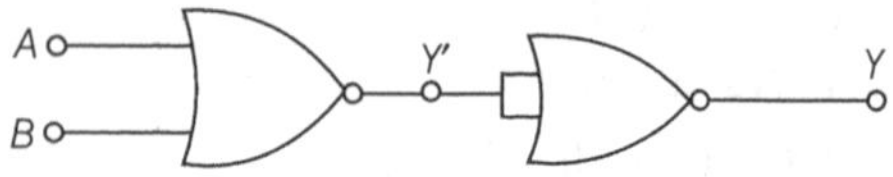

(a)

A	B	Y
0	0	0
0	1	1
1	0	1
1	1	1

(b)

A	B	Y
0	0	1
0	1	0
1	0	1
1	0	1

(c)

A	B	Y
0	0	0
0	1	1
1	1	1
1	1	1

(d)

A	B	Y
1	1	1
0	1	0
1	1	0
0	1	1

Sol. (a) **Boolean expression**

$$Y' = \overline{A + B}$$

and $Y = \bar{Y'} = \overline{\overline{A + B}} = A + B$

Truth Table

A	B	Y'	$\bar{Y'} = Y$
0	0	1	0
1	0	0	1
0	1	0	1
1	1	0	1

Example 16. *The output of the given logic circuit is* **[JEE Main 2019]**

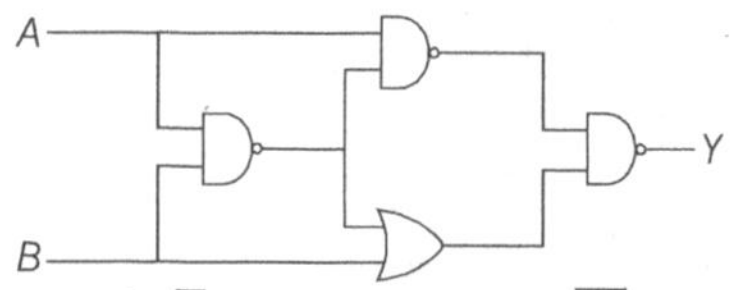

(a) $A\bar{B}$ (b) $\bar{A}B$ (c) $AB + \overline{AB}$ (d) $A\bar{B} + \bar{A}B$

Sol. (a) Truth table for given circuit is

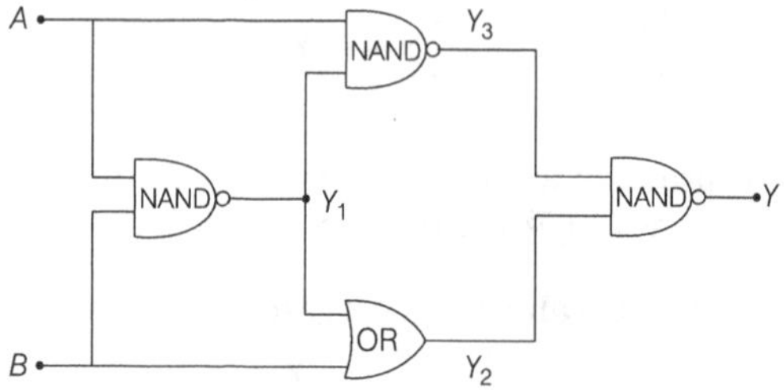

Using Boolean algebra, output of the given logic circuit can be given as

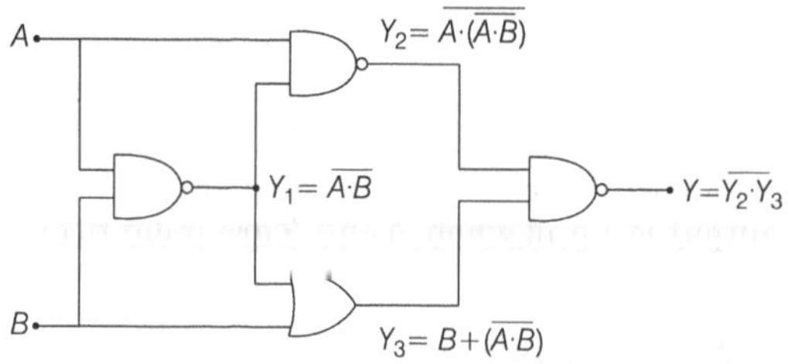

Here $Y_2 = \overline{A \cdot \overline{(A \cdot B)}}$

Using de-Morgan's principle,

$$\overline{x \cdot y} = \bar{x} + \bar{y} \text{ and } \overline{x + y} = \bar{x} \cdot \bar{y}$$

$\Rightarrow$ $Y_2 = \bar{A} + \overline{\overline{(A \cdot B)}}$ $[\because \bar{\bar{x}} = x]$

$= \bar{A} + (A \cdot B)$ ...(i)

Similarly, $Y_3 = B + \bar{A} + \bar{B} = 1 + \bar{A}$ $[\because x + \bar{x} = 1]$

$\Rightarrow$ $Y_3 = 1$ ...(ii) $[\because \bar{x} + 1 = 1]$

As, $Y = \overline{Y_2 \cdot Y_3}$

Using Eqs. (i) and (ii), we get

$$Y = \overline{(\bar{A} + A \cdot B)(1)}$$

$= \overline{(\bar{A} + A \cdot B)} + \bar{1} = \bar{\bar{A}} \cdot \overline{(A \cdot B)} + 0$

$= A \cdot (\bar{A} + \bar{B})$ $[\because x + 0 = x]$

$= A \cdot \bar{A} + A \cdot \bar{B}$ $[\because x\bar{x} = 0]$

$= A\bar{B}$

Practice Exercise

ROUND I Topically Divided Problems

Semiconductors

1. The forbidden energy band gap in conductors, semiconductors and insulators are EG_1, EG_2 and EG_3, respectively. The relation among them, is **[NCERT]**

(a) $EG_1 < EG_2 < EG_3$ (b) $EG_1 > EG_2 > EG_3$
(c) $EG_2 > EG_1 < EG_3$ (d) $EG_3 > EG_1 > EG_2$

2. Hole is **[NCERT Exemplar]**

(a) an anti-particle of electron
(b) a vacancy created when an electron leaves a covalent bond
(c) absence of free electrons
(d) an artifically created particle

3. In an n-type semiconductor, which of the following statements is true? **[NCERT]**

(a) Electrons are majority carriers and trivalent atoms are the dopants.
(b) Electrons are minority carriers and pentavalent atoms are the dopants.
(c) Holes are minority carriers and pentavalent atoms are the dopants.
(d) Holes are majority carriers and trivalent atoms are the dopants.

4. The conductivity of a semiconductor increases with increase in temperature because **[NCERT Exemplar]**

(a) number density of free current carriers increases
(b) relaxation time increases
(c) Both number density of carriers and relaxation time increase
(d) number density of carriers increases, relaxation time decreases but effect of decreases in relaxation time is much less than increase in number density

5. The energy band diagrams for three semiconductor samples of silicon are as shown. We can then assert that,

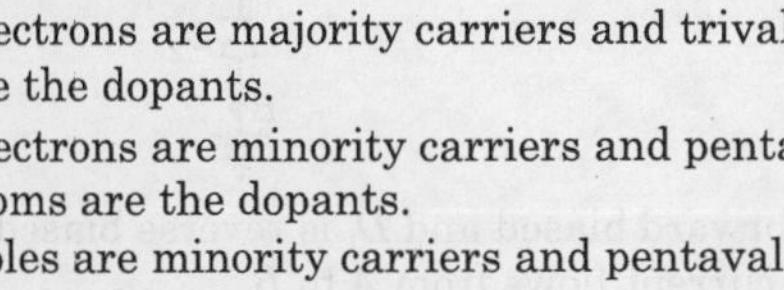

(a) sample X is undoped while samples Y and Z have been doped with a third group impurity respectively
(b) sample X is undoped while both samples Y and Z have been doped with a fifth group impurity
(c) sample X has been doped with equal amounts of third and fifth group impurities while samples Y and Z are undoped
(d) sample X is undoped while samples Y and Z have been doped with a fifth group and a third group impurity respectively

6. The ratio of electron and hole current in a semiconductor is 7/4 and the ratio of drift velocities of electrons and holes is 5/4, then ratio of concentrations of electrons and holes will be

(a) 5/7 (b) 7/5
(c) 25/49 (d) 49/25

7. Mobility of electrons in a semiconductor is defined as the ratio of their drift velocity to the applied electric field. If for an n-type semiconductor, the density of electrons is 10^{19} m^{-3} and their mobility is 1.6 m^2 /V-s, then the resistivity of the semiconductor (since, it is an n-type semiconductor contribution of holes is ignored) is close to **[JEE Main 2019]**

(a) 2 Ω-m (b) 0.2 Ω-m
(c) 0.4 Ω-m (d) 4 Ω-m

8. A silicon specimen is made into a p-type semiconductor by doping, on an average, one indium atom per 5×10^7 silicon atoms. If the number density of atoms in the silicon specimen is 5×10^{28} atoms m^{-3}, then the number of acceptor atoms in silicon per cubic centimetre will be

(a) 2.5×10^{30}
(b) 2.5×10^{35}
(c) 1.0×10^{13}
(d) 1.0×10^{15}

9. The length of germanium rod is 0.928 cm and its area of cross-section is 1 mm^2. If for germanium $n_i = 2.5 \times 10^{19}$ m^{-3}, $\mu_h = 0.19$ m^{2}V^{-1}s^{-1} $\mu_e = 0.39$ m^{2}V^{-1}s^{-1}, then its resistance will be

(a) 2.5 kΩ (b) 4.0 kΩ (c) 5.0 kΩ (d) 10.0 kΩ

10. A pure germanium plate of area 3.5×10^{-4} m^2 and of thickness 1.5×10^{-3} m is connected across a battery of potential 5 V. Find the amount of heat generated in the plate in 120 s. Given that, the concentration of carriers in germanium at room temperature is 1.6×10^{6} per cubic metre. The mobilities of electrons and holes are $0.4\ \text{m}^2\text{V}^{-1}\text{s}^{-1}$ and $0.2\ \text{m}^2\ \text{V}^{-1}\ \text{s}^{-1}$, respectively.

(a) 10.74×10^{-11} (b) 9×10^{-6}
(c) 10.48×10^{-9} (d) 6.26×10^{-10}

Junction Diode

11. Two identical *p-n* junction may be connected in series with a battery in three ways as shown in the adjoining figure. The potential drop across the *p-n* junctions are equal in

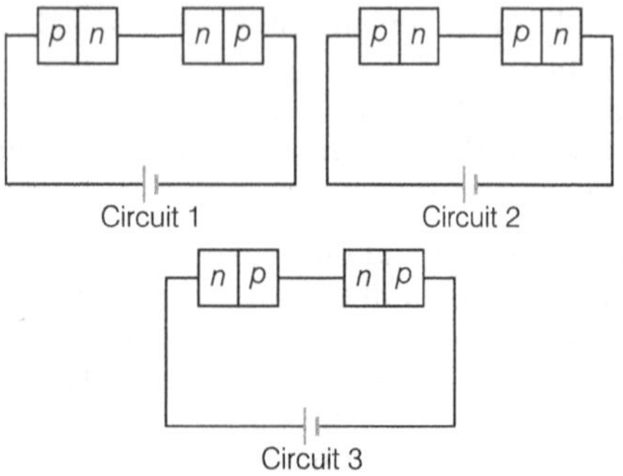

(a) circuit 1 and circuit 2
(b) circuit 2 and circuit 3
(c) circuit 3 and circuit 1
(d) circuit 1 only

12. The correct curve between potential (V) and distance (d) near *p-n* junction is

(a) 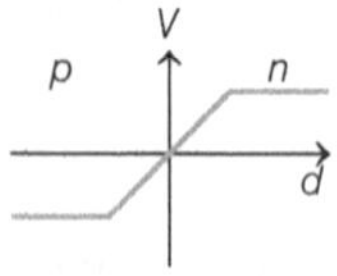
(b) 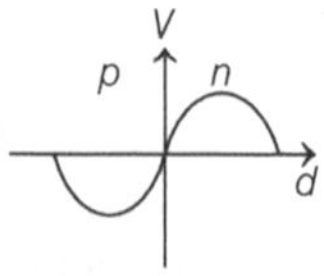
(c) 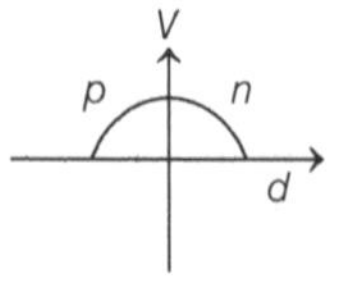
(d) 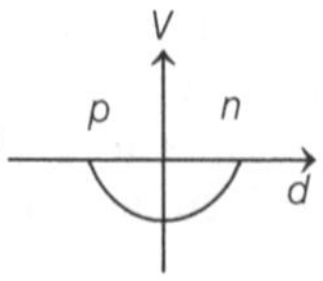

13. The forward biased diode connection is [JEE Main 2014]

(a) 

(b) 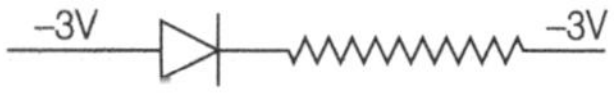

(c) 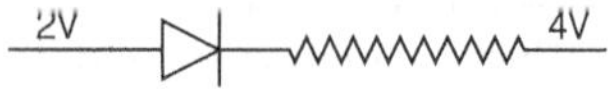

(d) 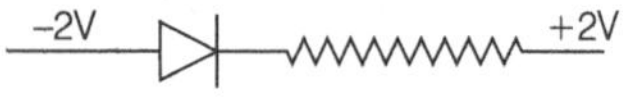

14. In figure, V_0 is the potential barrier across a *p-n* junction, when no battery is connected across the junction [NCERT Exemplar]

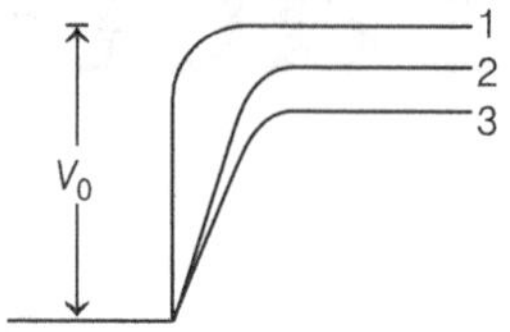

(a) 1 and 3 both correspond to forward bias of junction
(b) 3 corresponds to forward bias of junction and 1 corresponds to reverse bias of junction
(c) 1 corresponds to forward bias and 3 corresponds to reverse bias of junction
(d) 3 and 1 both correspond to reverse bias of junction

15. The reading of the ammeter for a silicon diode in the given circuit is [JEE Main 2018]

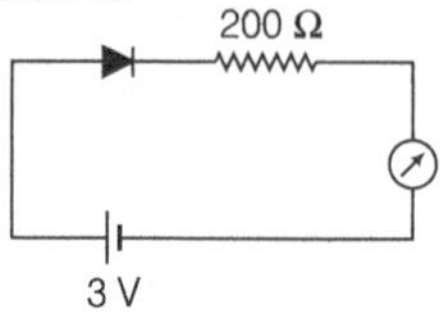

(a) 0 (b) 15 mA (c) 11.5 mA (d) 13.5 mA

16. In figure assuming the diodes to be ideal, [NCERT Exemplar]

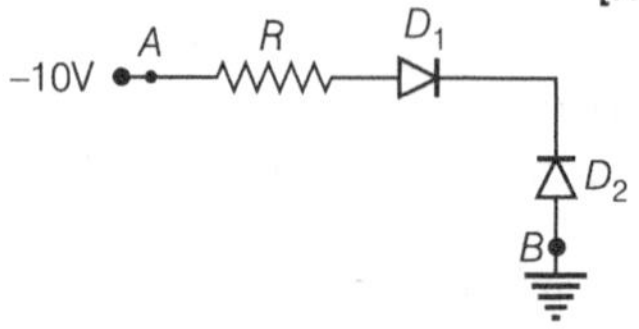

(a) D_1 is forward biased and D_2 is reverse biased and hence current flows from A to B
(b) D_2 is forward biased and D_1 is reverse biased and hence no current flows from B and A and *vice-versa*
(c) D_1 and D_2 are both forward biased and hence current flows from A to B
(d) D_1 and D_2 are both reverse biased and hence no current flows from A to B and *vice-versa*

17. The reverse bias in a junction diode is changed from 8 V to 13 V, then the value of the current changes from 40 μ A to 60 μA. The resistance of junction diode will be

(a) $2\times 10^5\ \Omega$ (b) $2.5\times 10^5\ \Omega$
(c) $3\times 10^5\ \Omega$ (d) $4\times 10^5\ \Omega$

18. The breakdown in a reverse biased *p-n* junction diode is more likely to occur due to

(a) large velocity of the minority charge carriers, if the doping concentration is small
(b) large velocity of the minority charge carriers, if the doping concentration is large
(c) strong electric field in a depletion region, if the doping concentration is small
(d) Both (a) and (c)

19. In the circuit shown in figure. If the diode forward voltage drop is 0.3 V, the voltage difference between A and B is **[NCERT Exemplar]**

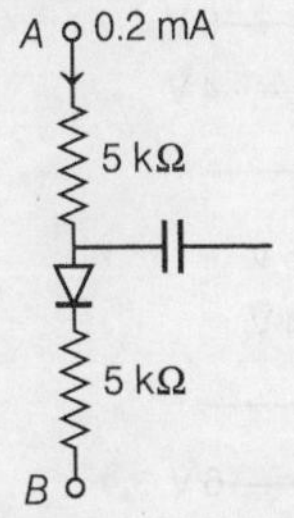

(a) 1.3 V (b) 2.3 V (c) 0 (d) 0.5 V

20. At 0.3 V and 0.7 V, the diodes Ge and Si become conductor respectively. In given figure, if ends of diode Ge overturned, the change in potential V_0 will be **[JEE Main 2019]**

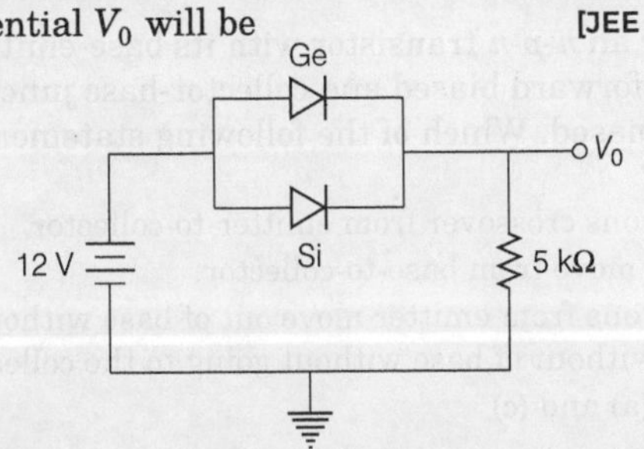

(a) 0.2 V (b) 0.6V (c) 0.4 V (d) 0.8V

21. The circuit shown below contains two ideal diodes, each with a forward resistance of 50 Ω. If the battery voltage is 6 V, the current through the 100 Ω resistance (in ampere) is **[JEE Main 2019]**

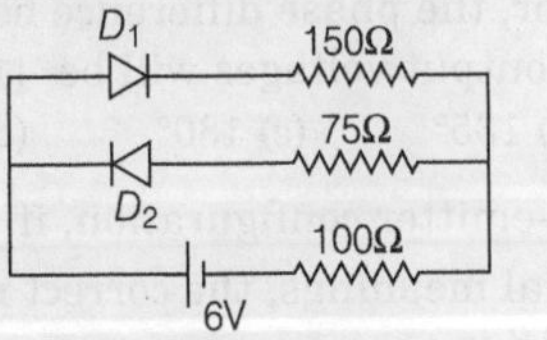

(a) 0.027 (b) 0.020 (c) 0.030 (d) 0.036

Rectifier and Special Purpose Diode

22. The output of the given circuit in figure **[NCERT Exemplar]**

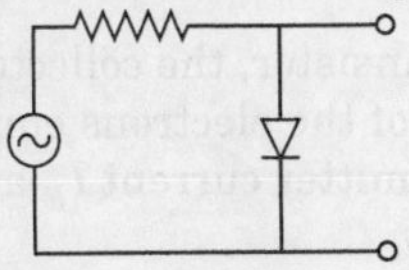

(a) would be zero at all times

(b) would be like a half-wave rectifier with positive cycle in output

(c) would be like a half-wave rectifier with negative cycle in output

(d) would be like that of a full wave rectifier

23. A semiconductor diode is used as a half wave rectifier having internal resistance 200 Ω. The voltage applied is given by $V = (50 \sin \omega t)$ volt and load resistance is 650 Ω. Then

(a) maximum output current is 58 mA

(b) DC output current is 18.5 mA

(c) DC output power is 0.22 W

(d) All of the above

24. The I-V characteristic of an LED is **[JEE Main 2013]**

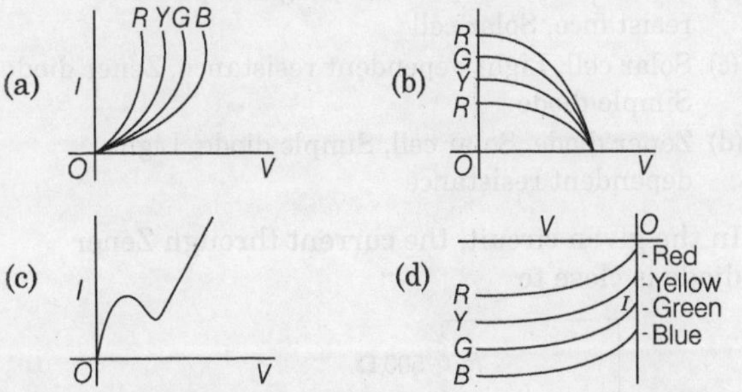

25. If a semiconductor photodiode can detect a photon with a maximum wavelength of 400 nm, then its band gap energy is

(Take, Planck's constant, $h = 6.63 \times 10^{-34}$ J-s and speed of light, $c = 3 \times 10^{8}$ m/s) **[JEE Main 2020]**

(a) 1.5 eV (b) 1.1 eV

(c) 3.1 eV (d) 2.0 eV

26. With increasing biasing voltage of a photodiode, the photocurrent magnitude **[JEE Main 2020]**

(a) increases initially and after attaining certain value, it decreases

(b) increases initially and saturates finally

(c) remains constant

(d) increases linearly

27. Three photodiodes D_1, D_2 and D_3 are made of semiconductors having band gaps of 2.5 eV, 2 eV and 3 eV, respectively. Which one will be able to detect light of wavelength of 6000 Å?

(a) D_1 (b) D_2

(c) D_3 (d) Both D_1 and D_2

28. Identify the semiconductor devices whose characteristics are as given below, in the order (i), (ii), (iii) and (iv).

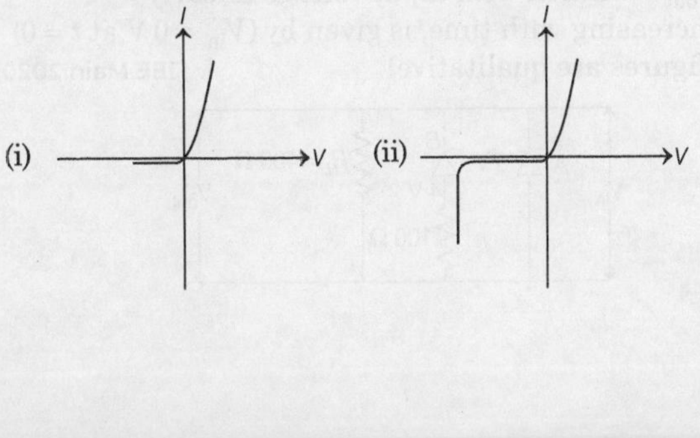

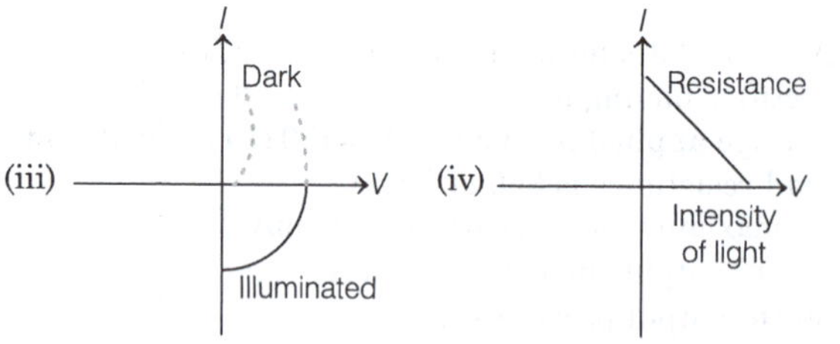

(a) Simple diode, Zener diode, Solar cell, Light dependent resistance
(b) Zener diode, Simple diode, Light dependent resistance, Solar cell
(c) Solar cell, Light dependent resistance, Zener diode, Simple diode
(d) Zener diode, Solar cell, Simple diode, Light dependent resistance

29. In the given circuit, the current through Zener diode is close to

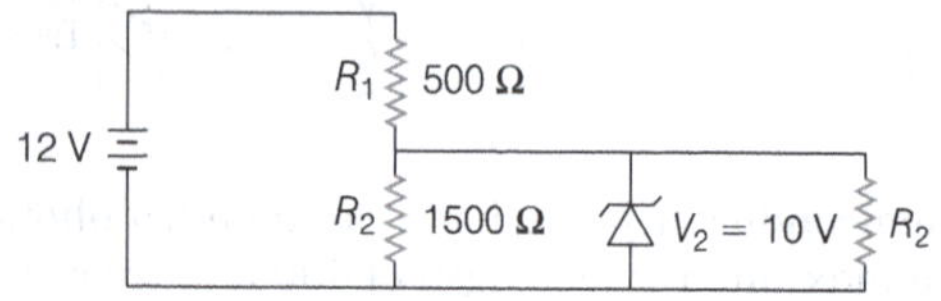

[JEE Main 2019]

(a) 6.0 mA
(b) 6.7 mA
(c) 0
(d) 4.0 mA

30. For the circuit shown below, the current through the Zener diode is [JEE Main 2019]

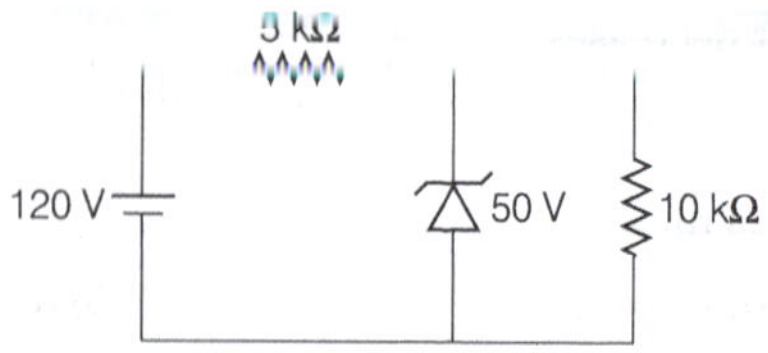

(a) 14 mA
(b) zero
(c) 5 mA
(d) 9 mA

31. Two Zener diodes A and B having breakdown voltages of 6 V and 4 V respectively, are connected as shown in the circuit below. The output voltage V_{out} variation with input voltage linearly increasing with time, is given by (V_{in} = 0 V at t = 0) (figures are qualitative) [JEE Main 2020]

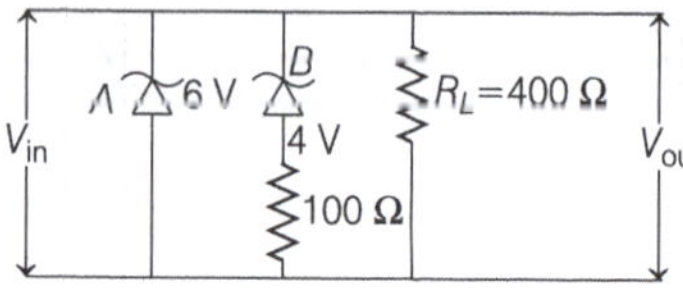

(a)
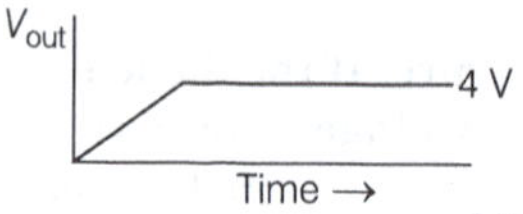

(b)
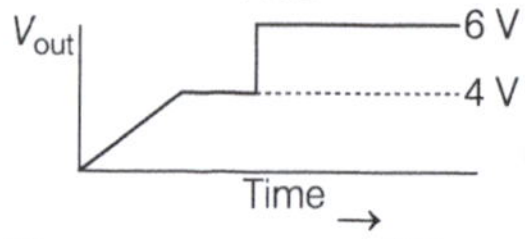

(c)
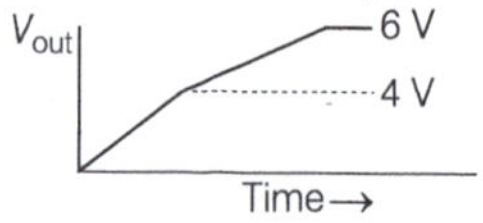

(d)
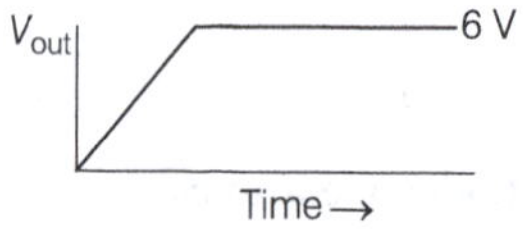

Junction Transistor

32. Consider an n-p-n transistor with its base-emitter junction forward biased and collector-base junction reverse biased. Which of the following statements are true?

(a) Electrons crossover from emitter-to-collector.
(b) Holes move from base-to-collector.
(c) Electrons from emitter move out of base without of base without of base without going to the collector
(d) Both (a) and (c)

33. The current gain of a transistor in a common emitter configuration is 40. If the emitter current is 8.2 mA, then base current is

(a) 0.02 mA (b) 0.2 mA (c) 2.0 mA (d) 0.4 mA

34. In a common emitter amplifier circuit using an n-p-n transistor, the phase difference between the input and the output voltages will be [JEE Main 2017]

(a) 90° (b) 135° (c) 180° (d) 45°

35. For a common-emitter configuration, if α and β have their usual meanings, the correct relationship between α and β is

(a) $\frac{1}{\alpha} = \frac{1}{\beta} + 2$ (b) $\alpha = \frac{\beta}{1-\beta}$

(c) $\alpha = \frac{\beta}{1+\beta}$ (d) $\alpha = \frac{\beta^2}{1+\beta^2}$

36. In an n-p-n transistor, the collector current is 10 mA, if 90% of the electrons emitted reach the collector, the emitter current I_E and base current I_B are given by

(a) $I_E = -1$ mA, $I_b = 9$ mA (b) $I_E = 9$ mA, $I_B = -1$ mA
(c) $I_E = 1$ mA, $I_B = 11$ mA (d) $I_E = 11$ mA, $I_B = 1$ mA

37. The transfer ratio of the transistor is 50. The input resistance of the transistor when used in the CE configuration is 1 kΩ. The peak value for an AC input voltage of 0.01 V of collector current is

(a) 500 μA (b) 0.25 mA (c) 400 mA (d) 0.01 mA

38. In a *n-p-n* transistor, 10^{10} electrons enter the emitter in 10^{-6} s and 4% of the electrons are lost in base. The current transfer ratio will be
(a) 0.98 (b) 0.97 (c) 0.96 (d) 0.94

39. The input signal given to a CE amplifier having a voltage gain of 150 is $V_i = 2\cos\left(15t + \frac{\pi}{3}\right)$. The corresponding output signal will be
(a) $300\cos\left(15t + \frac{\pi}{3}\right)$ (b) $75\cos\left(15t + \frac{2\pi}{3}\right)$
(c) $2\cos\left(15t + \frac{5\pi}{3}\right)$ (d) $300\cos\left(15t + \frac{4\pi}{3}\right)$

40. In a *n-p-n* transistor circuit, the collector current is 10 mA. If 95 per cent of the electrons emitted reach the collector, which of the following statements are true? **[NCERT Exemplar]**
(a) The emitter current will be 8 mA.
(b) The emitter current will be 10.53 mA.
(c) The base current will be 5.53 mA.
(d) The base current will be 2 mA.

41. Figure shows the transfer characteristics of a base biased CE transistor. Which of the following statements are false?

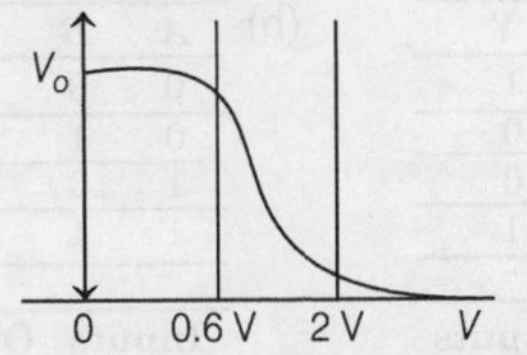

(a) At $V_i = 0.4$ V, transistor is in active state.
(b) At $V_i = 1$ V, it can be used as an amplifier.
(c) At $V_i = 0.5$ V, It can be used as a switch turned OFF.
(d) At $V_i = 2.5$ V, it can be used as a switch turned ON.

42. An *n-p-n* transistor operates as a common emitter amplifier, with a power gain of 60 dB. The input circuit resistance is 100 Ω and the output load resistance is 10 kΩ. The common emitter current gain β is **[JEE Main 2019]**
(a) 10^2 (b) 6×10^2 (c) 10^4 (d) 60

43. The transfer characteristic curve of a transistor, having input and output resistances 100 Ω and 100 kΩ respectively, is shown in the figure. The voltage and power gains are respectively **[JEE Main 2019]**

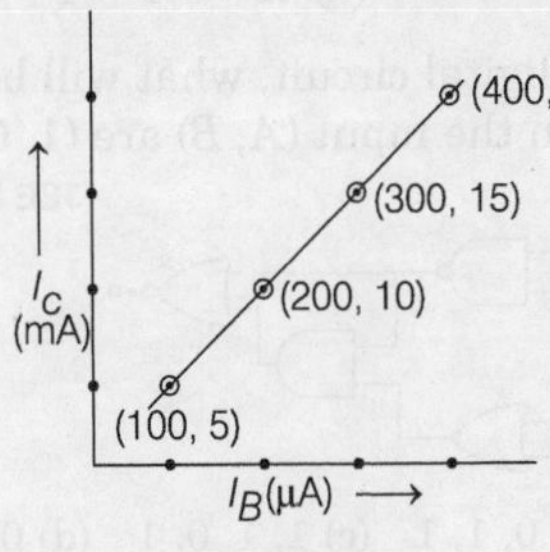

(a) $2.5 \times 10^4, 2.5 \times 10^6$
(b) $5 \times 10^4, 5 \times 10^6$
(c) $5 \times 10^4, 5 \times 10^5$
(d) $5 \times 10^4, 2.5 \times 10^6$

44. A common emitter amplifier circuit, built using an *n-p-n* transistor, is shown in the figure. Its DC current gain is 250, $R_C = 1\,k\Omega$ and $V_{CC} = 10$ V. What is the minimum base current for V_{CE} to reach saturation? **[JEE Main 2019]**

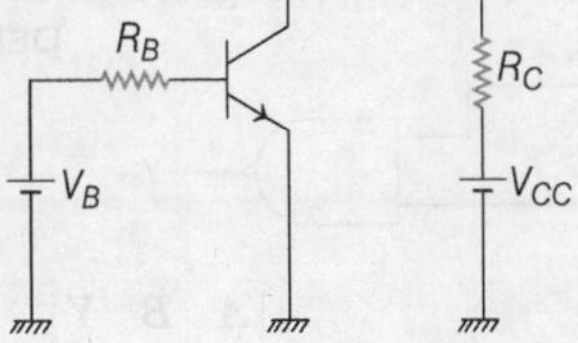

(a) 40 μA (b) 10 μA
(c) 100 μA (d) 7 μA

45. For tuned collector oscillator, using an *n-p-n* transistor its circuit diagram and rise & fall (or built up) of I_C, I_E current graphs, respectively are as shown below. Initially when switch S_1 is ON, it can be concluded

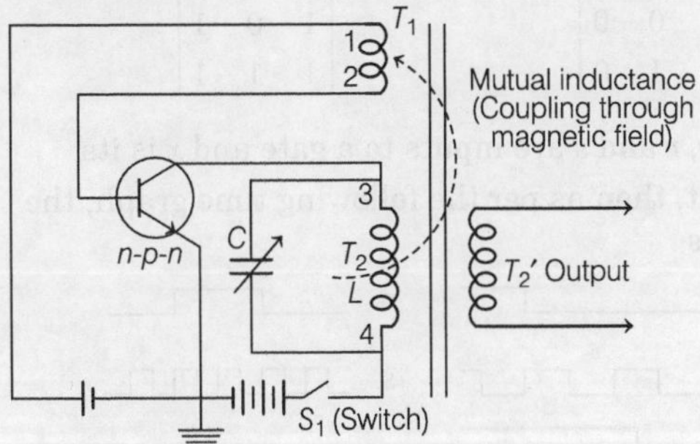

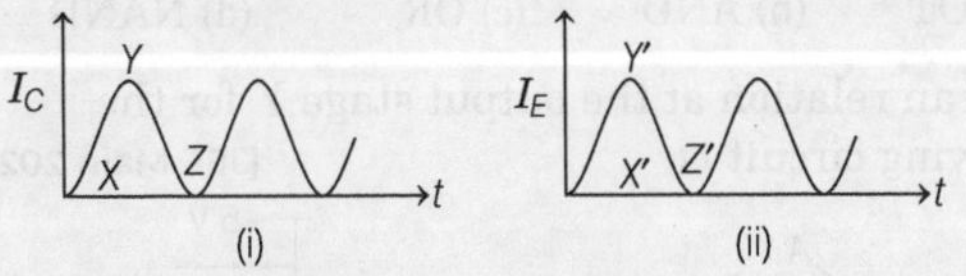

(a) Both I_C and I_E increase initially
(b) Both I_C and I_E decrease
(c) initially I_C increases but I_E decreases
(d) initially I_C decreases but I_E increases

Logic Gates

46. Which of the following gives a reversible operation? **[JEE Main 2020]**

(a) 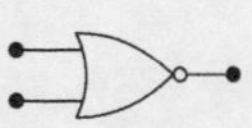 (b) 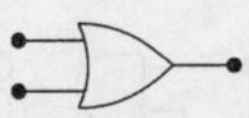
(c) 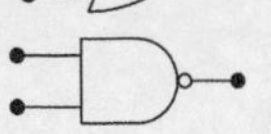 (d) 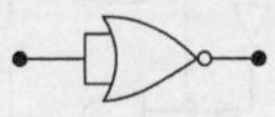

47. Which of the following logic gate is represented by the combination of logic gates?

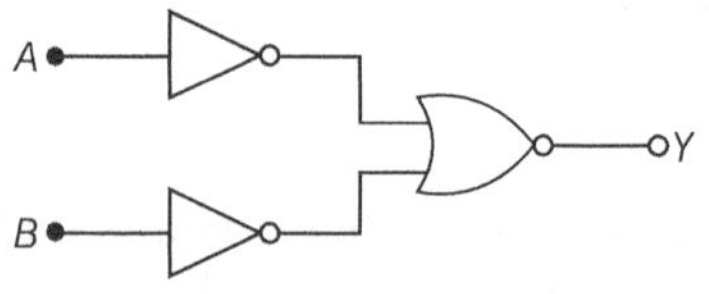

(a) NAND gate (b) NOR gate
(c) AND gate (d) OR gate

48. The truth table for the circuit given in the figure is **[JEE Main 2019]**

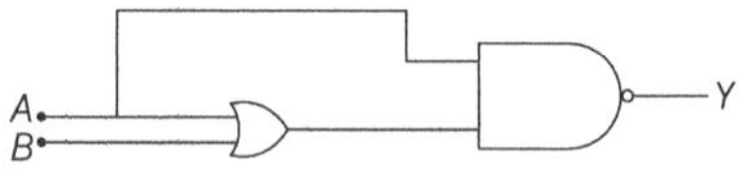

(a)

A	B	Y
0	0	1
0	1	1
1	0	1
1	1	1

(b)

A	B	Y
0	0	1
0	1	0
1	0	0
1	1	0

(c)

A	B	Y
0	0	1
0	1	1
1	0	0
1	1	0

(d)

A	B	Y
0	0	0
0	1	0
1	0	1
1	1	1

49. If p, q, r and s are inputs to a gate and x is its output, then as per the following time graph, the gate is

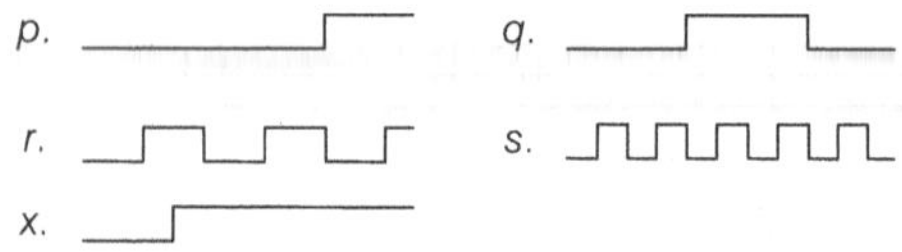

(a) NOT (b) AND (c) OR (d) NAND

50. Boolean relation at the output stage Y for the following circuit is **[JEE Main 2020]**

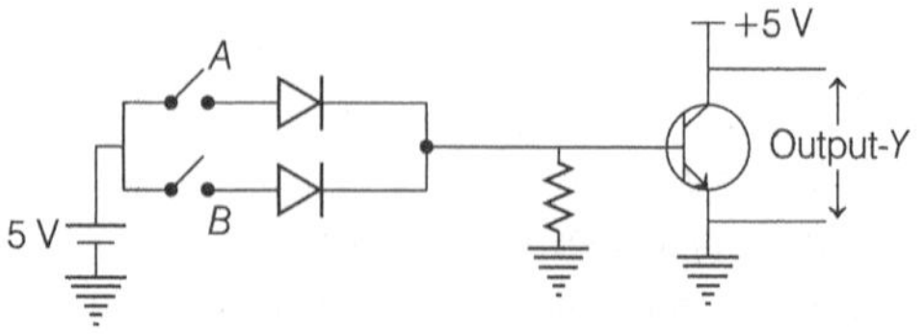

(a) $A \cdot B$ (b) $A + B$
(c) $\overline{A} + \overline{B}$ (d) $\overline{A} \cdot \overline{B}$

51. Truth table for the given circuit. **[NCERT Exemplar]**

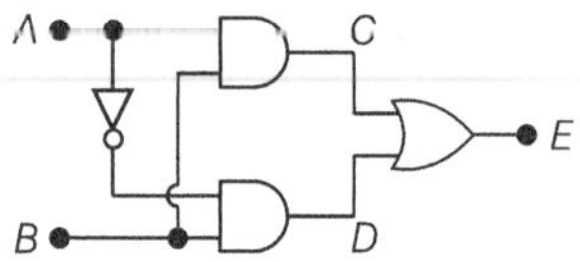

(a)

A	B	E
0	0	1
0	1	0
1	0	1
1	1	0

(b)

A	B	E
0	0	1
0	1	0
1	0	0
1	1	1

(c)

A	B	E
0	0	0
0	1	1
1	0	0
1	1	1

(d)

A	B	E
0	0	0
0	1	1
1	0	1
1	1	0

52. A system of four gates is set up as shown. The truth table corresponding to this system is **[JEE Main 2013]**

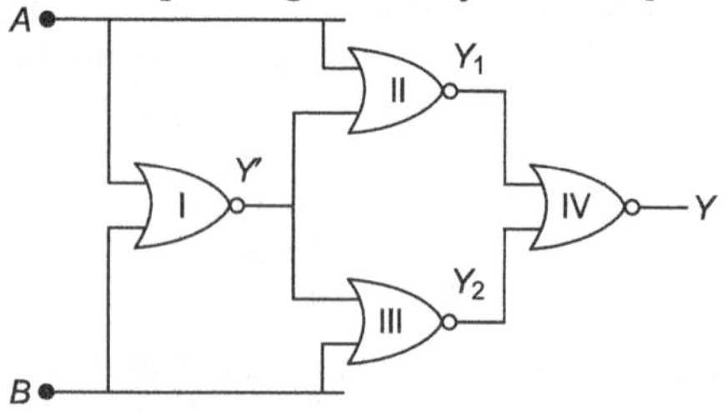

(a)

Inputs		Outputs
A	B	Y
0	0	1
0	1	0
1	0	0
1	1	1

(b)

Inputs		Outputs
A	B	Y
0	0	0
0	1	0
1	0	1
1	1	1

(c)

Inputs		Outputs
A	B	Y
0	0	1
0	1	1
1	0	1
1	1	0

(d)

Inputs		Outputs
A	B	Y
0	0	0
0	1	1
1	0	1
1	1	0

53. To get output 1 at R, for the given logic gate circuit, the input values must be **[JEE Main 2019]**

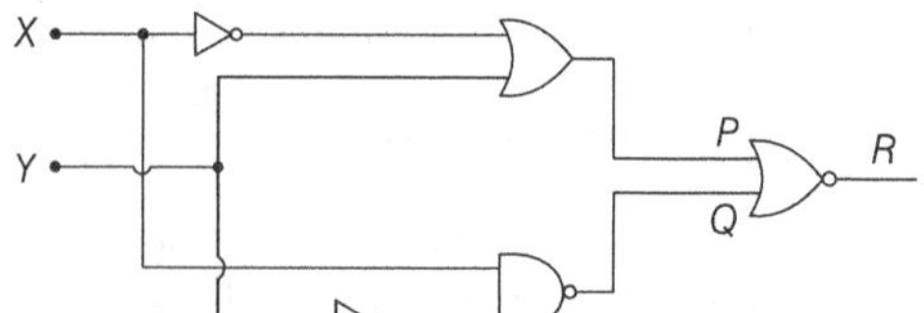

(a) $X = 0, Y = 0$ (b) $X = 1, Y = 0$
(c) $X = 1, Y = 1$ (d) $X = 0, Y = 1$

54. In the following digital circuit, what will be the output at Z, when the input (A, B) are (1, 0), (0, 0), (1, 1) and (0, 1)? **[JEE Main 2020]**

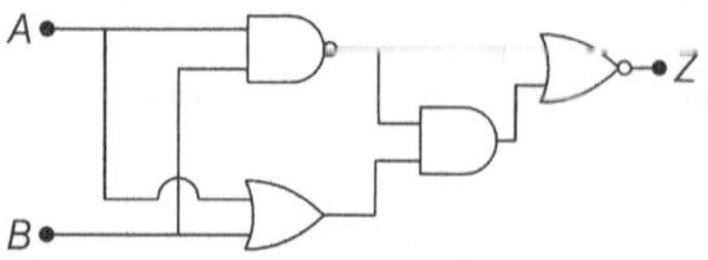

(a) 0, 0, 1, 0 (b) 1, 0, 1, 1 (c) 1, 1, 0, 1 (d) 0, 1, 0, 0

55. Identify the correct output signal Y in the given combination of gates (as shown) for the given inputs A and B. **[JEE Main 2020]**

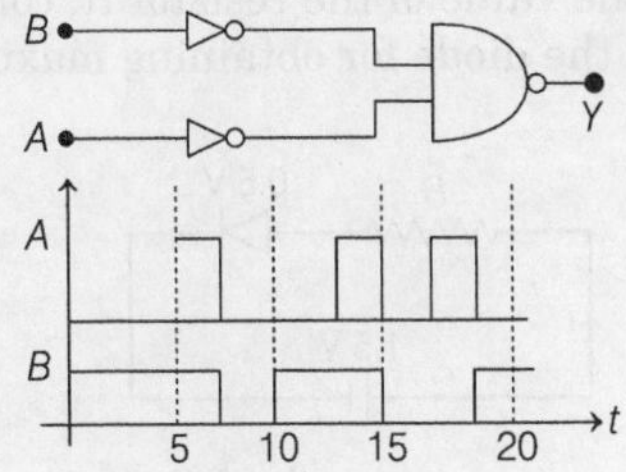

(a) 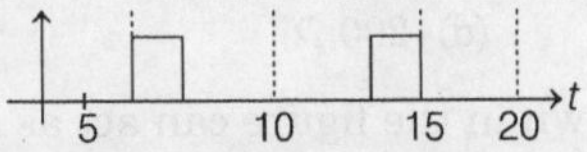

(b) 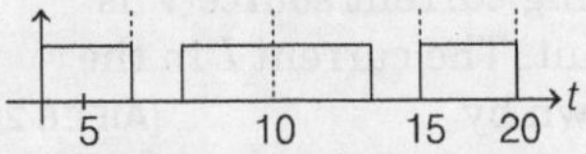

(c) 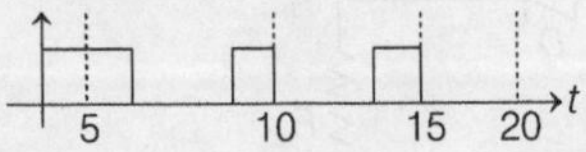

(d) 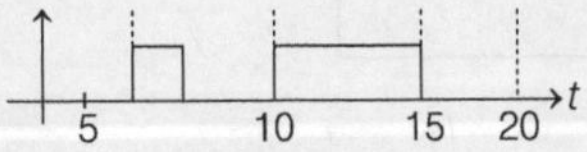

56. Consider two *n-p-n* transistors as shown in figure. If 0 V corresponds to false and 5 V corresponds to true, then the output at C corresponds to

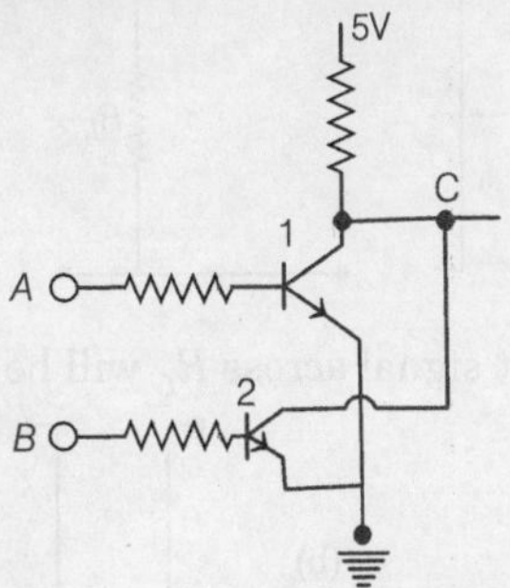

(a) A NAND B (b) A OR B
(c) A AND B (d) A NOR B

57. The following logic gate is equivalent to

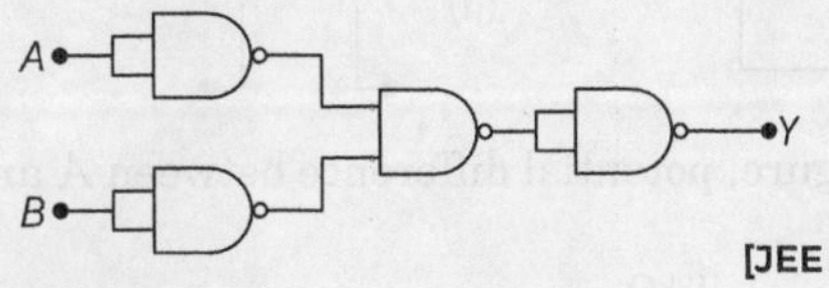

[JEE Main 2021]

(a) NOR gate (b) OR gate
(c) AND gate (d) NAND gate

ROUND II Mixed Bag

Only One Correct Option

1. The output of an OR gate is connected to both the inputs of NAND gate. The combination will serve as a **[IIT JEE 2011]**

(a) OR gate (b) NOT gate
(c) NOR gate (d) AND gate

2. When A is the internal stage gain of an amplifier and β is the feedback ratio, then the amplitude becomes as oscillator if

(a) β is negative and magnitude of $\beta = \frac{A}{2}$

(b) β is negative and magnitude of $\beta = \frac{1}{A}$

(c) β is negative and magnitude of $\beta = A$

(d) β is positive and magnitude of $\beta = \frac{1}{A}$

3. A 220 V AC supply is connected between points A and B (figure). What will be the potential difference V across the capacitor? **[NCERT Exemplar]**

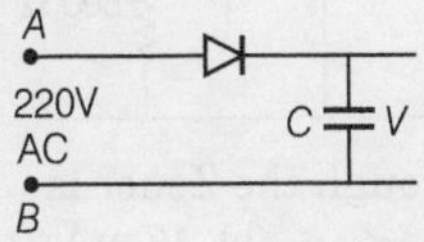

(a) 220 V (b) 110 V
(c) 0 V (d) $220\sqrt{2}$ V

4. When a diode is forward biased, it has a voltage drop of 0.5 V. The safe limit of current through the diode is 10 mA. If a battery of emf 1.5 V is used in the circuit, the value of minimum resistance to be connected in series with the diode, so that the current does not exceed the safe limit is **[JEE Main 2020]**

(a) 300 Ω (b) 50 Ω (c) 100 Ω (d) 200 Ω

5. A Si and a Ge diode has identical physical dimensions and the band gap in Si is larger than that in Ge. An identical reverse bias is applied across the diodes, then **[WB JEE 2008]**

(a) the reverse current in Ge is larger than that in Si
(b) the reverse current in Si is larger than that in Ge
(c) the reverse current is identical in the two diodes
(d) the relative magnitude of the reverse currents cannot be determined from the given data only

6. The output of the given combination gates represents **[JEE Main 2021]**

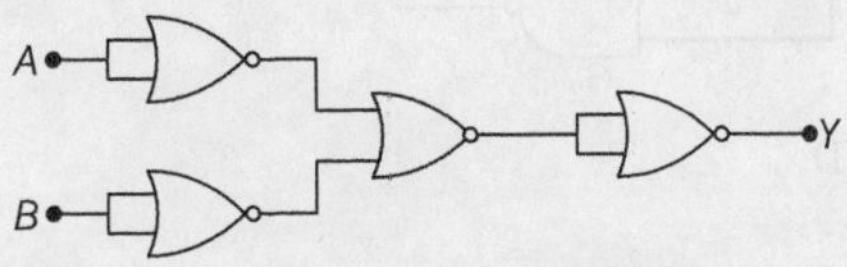

(a) XOR gate (b) NAND gate
(c) AND gate (d) NOR gate

7. If in a *p-n* junction diode, a square input signal of 10 V is applied as shown

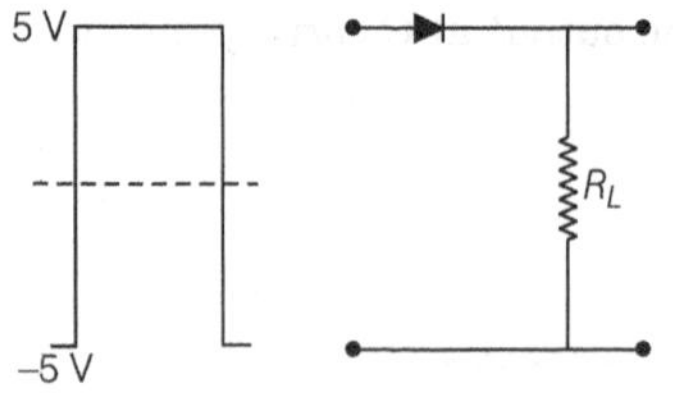

Then the output signal across R_L will be [AIEEE 2009]

(a) 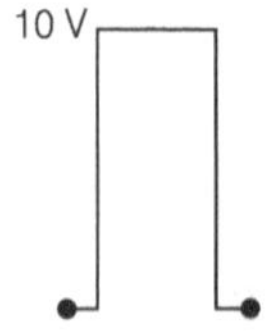

(b) 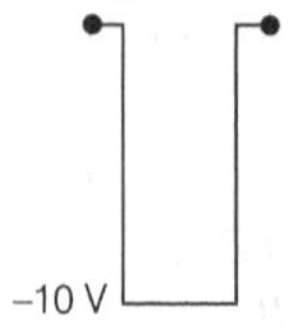

(c) 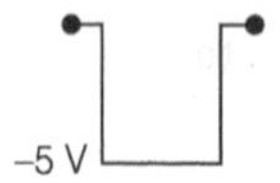

(d) 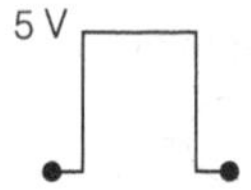

8. In the figure, potential difference between A and B is

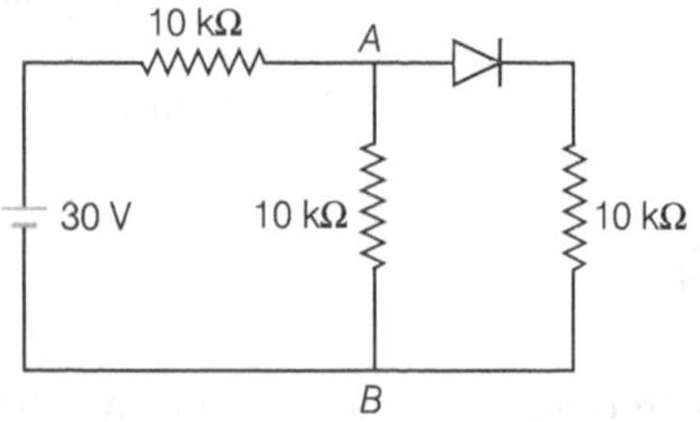

(a) zero (b) 5 V
(c) 10 V (d) 15 V

9. The input resistance of a common emitter transistor amplifier, if the output resistance is 500 kΩ, the current gain, $\alpha = 0.98$ and power gain is 6.0625×10^6, is

(a) 198 Ω
(b) 300 Ω
(c) 100 Ω
(d) 400 Ω

10. The following configuration of gate is equivalent to figure.

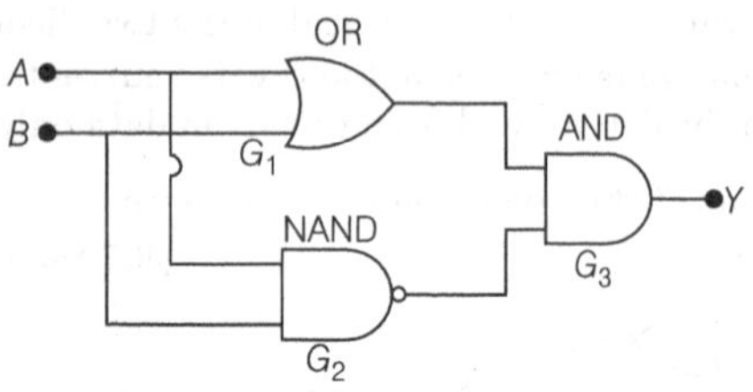

(a) NAND
(b) XOR
(c) OR
(d) None of the above

11. The diode used in the circuit shown in the figure has a constant voltage drop of 0.5 V at all currents and a maximum power rating of 100 mW. What should be the value of the resistor R, connected in series with the diode for obtaining maximum current ?

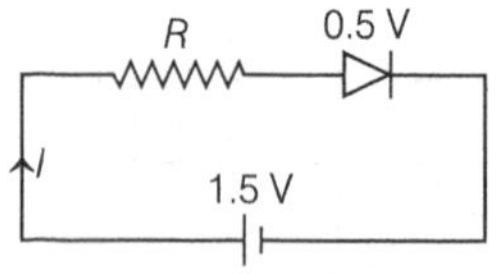

(a) 1.5 Ω (b) 5 Ω
(c) 6.67 Ω (d) 200 Ω

12. A *p-n* junction D shown in the figure can act as a rectifier. An alternating current source V is connected in the circuit. The current I in the resistor R can be shown by [AIEEE 2009]

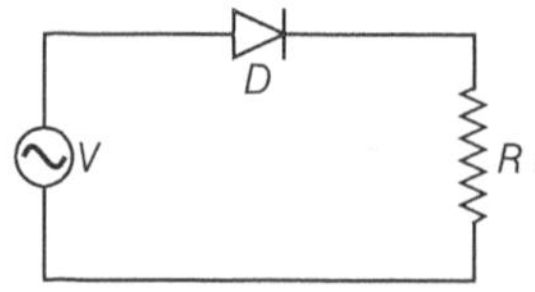

(a) 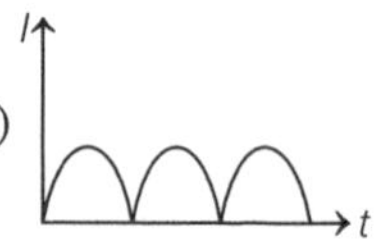

(b) 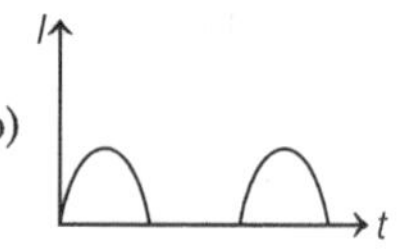

(c) 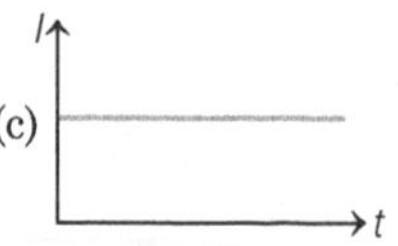

(d) 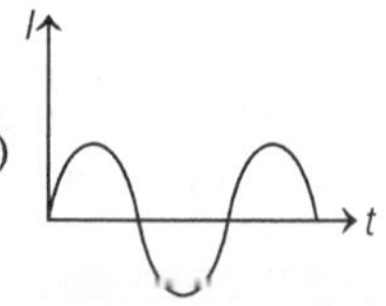

13. The output Y of the logic circuit shown in figure is best represented as

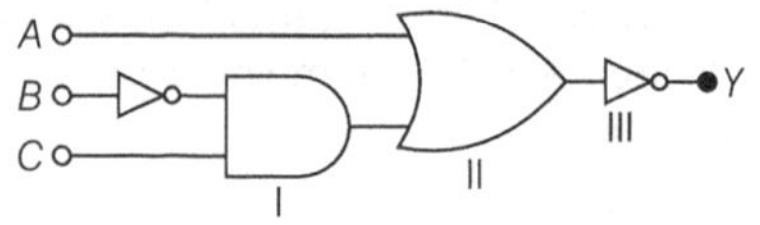

(a) $\overline{A} + \overline{A \cdot C}$ (b) $\overline{A + (\overline{B} \cdot C)}$
(c) $\overline{A + B \cdot C}$ (d) $\overline{A + \overline{B} + C}$

14. The reverse breakdown voltage of a Zener diode is 5.6 V in the given circuit. [JEE Main 2019]

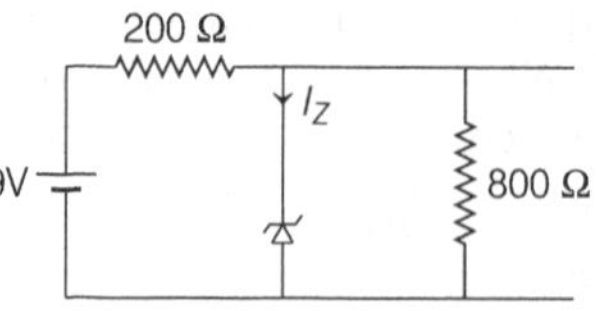

The current I_Z through the Zener is
(a) 10 mA (b) 17 mA
(c) 15 mA (d) 7 mA

15. For the given combination of gates, if the logic states of inputs A, B, C are as follows $A = B = C = 0$ and $A = B = 1, C = 0$, then the logic states of output D are

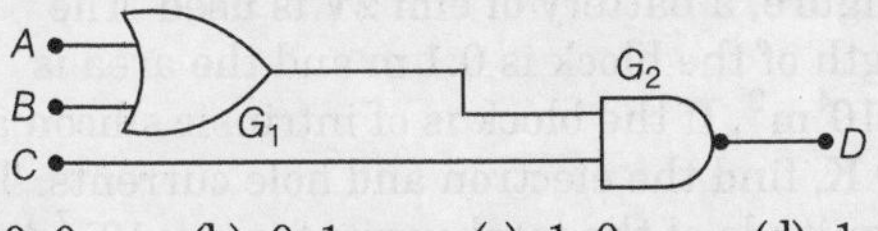

(a) 0, 0 (b) 0, 1 (c) 1, 0 (d) 1, 1

16. The graph given below represents the I-V characteristics of a Zener diode. Which part of the characteristics curve is most relevant for its operation as a voltage regulator? **[AMU Engg. 2009]**

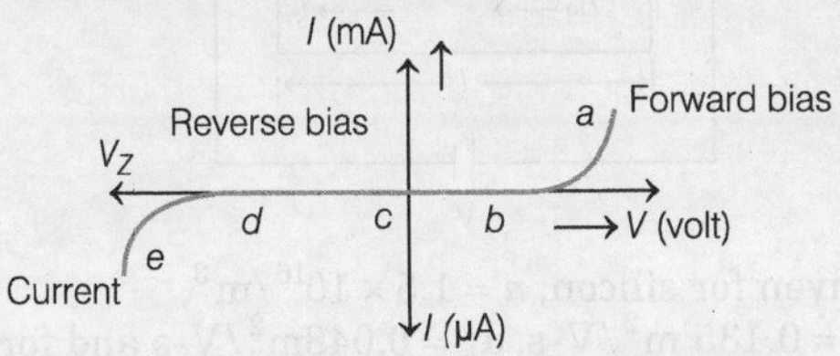

(a) *ab* (b) *bc* (c) *cd* (d) *de*

17. In a common emitter amplifier, using output resistance of 5000 Ω and input resistance of 2000 Ω, if the peak value of input signal voltage is 10 mV and $\beta = 50$, then peak value of output voltage is

(a) 5×10^{-6} V (b) 12.50×10^{-6} V
(c) 1.25 V (d) 125.0 V

18. The junction diode in the following circuit requires a minimum current of 1 mA to be above the knee point (0.7 V) of its I-V characteristic curve. The voltage across the junction diode is independent of current above the knee point, if $V_B = 4$ V, then the maximum value of R so that the voltage is above knee point will be

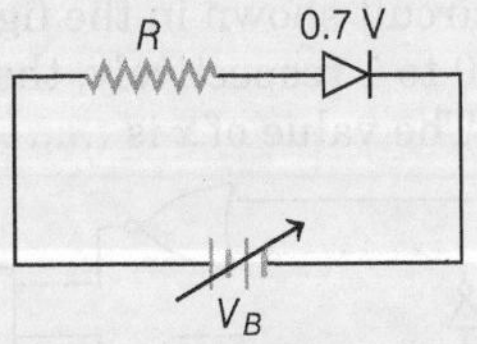

(a) 3.3 kΩ (b) 4.0 kΩ (c) 4.7 kΩ (d) 6.6 kΩ

19. Statement I. In a full wave rectifier circuit operating from 50 Hz mains frequency, the fundamental frequency in the ripple would be 100 Hz.

Statement II. In a semiconducting material, the mobilities of electron and hole are μ_e and μ_h respectively, then $\mu_e < \mu_h$.

Statement III. For an *n-p-n* transistor amplifier in CE configuration, the current gain in the transistor is 100. If the collector changes by 1 mA, then the change in emitter current is 1.01 mA.

Choose the correct statement and mark the correct option given below.

(a) Both I and II (b) Only II
(c) Both I and III (d) All of these

20. Figure shows a DC voltage regulator circuit, with a Zener diode of breakdown voltage = 6 V. If the unregulated input voltage varies between 10 V to 16 V, then what is the maximum Zener current? **[JEE Main 2019]**

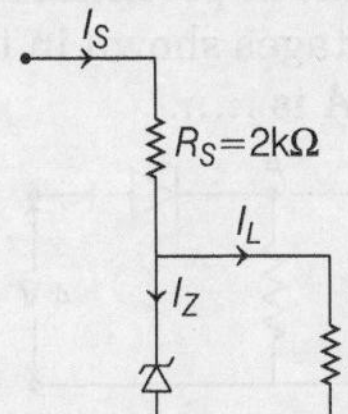

(a) 2.5 mA (b) 1.5 mA
(c) 7.5 mA (d) 3.5 mA

21. Identify the operation performed by the circuit given below. **[JEE Main 2020]**

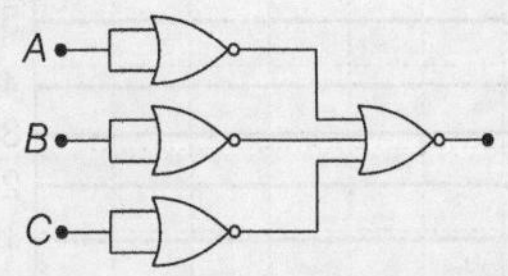

(a) AND (b) NAND
(c) NOT (d) OR

22. In the given circuit, value of Y is **[JEE Main 2020]**

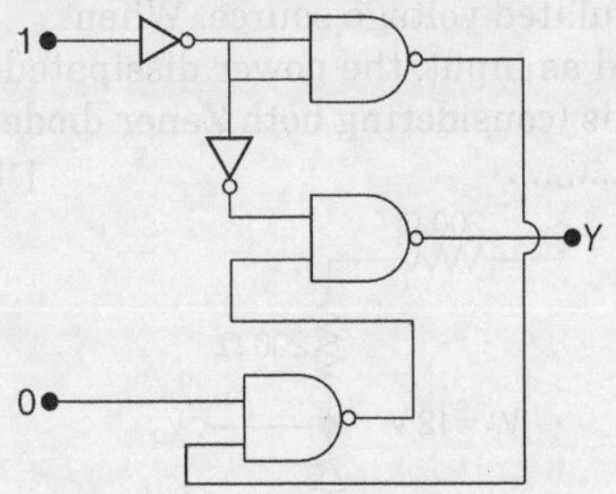

(a) toggles between 0 and 1 (b) 1
(c) 0 (d) will not execute

23. In the figure, given that V_{BB} supply can vary from 0 to 5.0 V, $V_{CC} = 5$ V, $\beta_{DC} = 200$, $R_B = 100$ k Ω, $R_C = 1$ kΩ and $V_{BE} = 1.0$ V. The minimum base current and the input voltage at which the transistor will go to saturation, will be, respectively **[JEE Main 2019]**

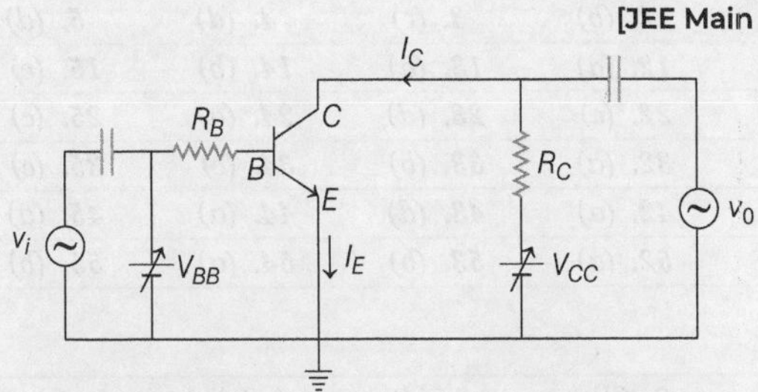

(a) 25 μA and 2.8 V (b) 25 μA and 3.5 V
(c) 20 μA and 3.5 V (d) 20 μA and 2.8 V

Numerical Value Questions

24. Both the diodes used in the circuit shown are assumed to be ideal and have negligible resistance when these are forward biased. Built in potential in each diode is 0.7V. For the input voltages shown in the figure, the voltage (in V) at point A is .... **[JEE Main 2020]**

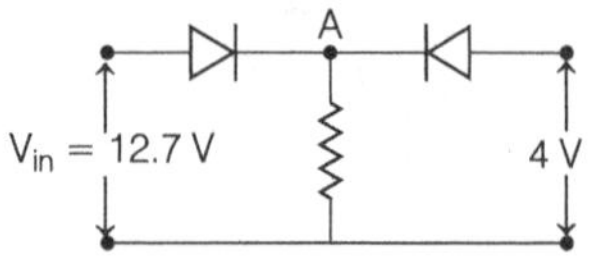

25. The output characteristics of a transistor is shown in the figure. When V_{CE} is 10V and $I_C = 4.0$ mA, then value of β_{AC} is .... **[JEE Main 2020]**

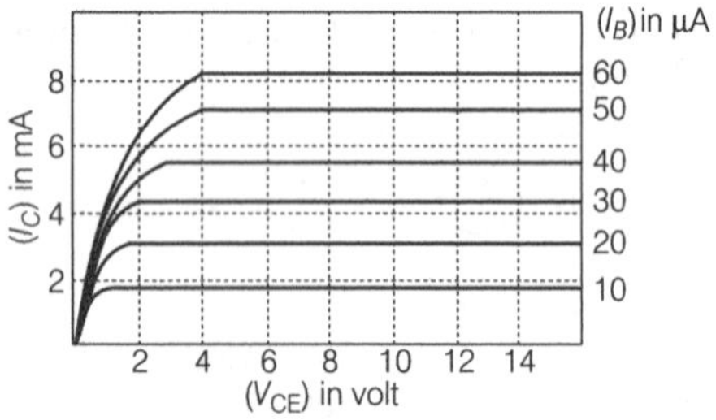

26. The circuit shown below is working as a 8 V DC regulated voltage source. When 12 V is used as input, the power dissipated (in mW) in each diode is (considering both Zener diodes are identical) .... **[JEE Main 2020]**

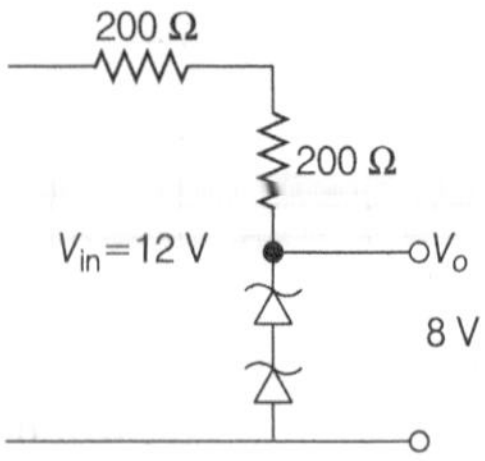

27. For a CE transister amplifier, the audio signal voltage across the collector resistance of 2 kΩ is 2 V Suppose the current amplification factor of the transistor is 100. Find the base current (in μA), if the base resistance is 1 kΩ.

28. In figure, a battery of emf 2V is used. The length of the block is 0.1 m and the area is $1\times10^4 \text{m}^2$. If the block is of intrinsic silicon at 300 K, find the electron and hole currents. The magnitude of the total current is $p\times10^{-7}$ A , where the value of p is ....

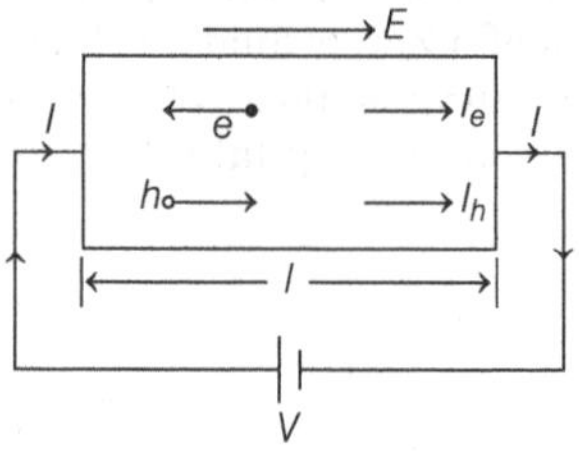

[Given for silicon, $n = 1.5\times10^{16}/\text{m}^3$, $\mu_e = 0.135 \text{ m}^2/\text{V-s}$, $\mu_h = 0.048\text{m}^2/\text{V-s}$ and for germanium, $n = 2.4\times10^{19}/\text{m}^3$, $\mu_e = 0.39 \text{ m}^2/\text{V-s}$ and $\mu_h = 0.19 \text{ m}^2/\text{V-s}$]

29. The value of power dissipated across the zener diode ($V_Z = 15$ V) connected in the circuit as shown in figure is $x\times10^{-1}$ watt.

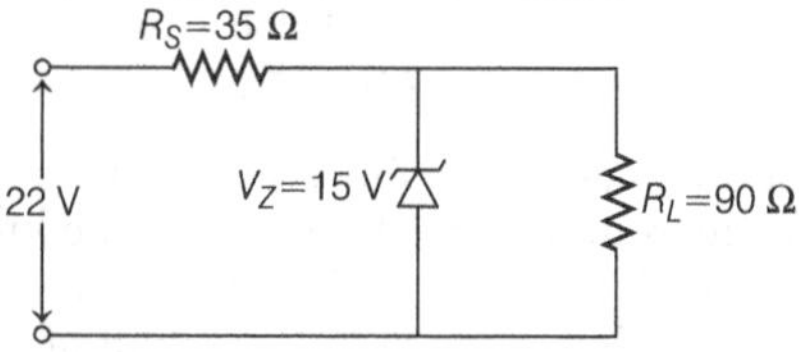

The value of x, to the nearest integer, is .... **[JEE Main 2021]**

30. In the logic circuit shown in the figure, if input A and B are 0 to 1 respectively, the output at Y would be 'x'. The value of x is .... **[JEE Main 2021]**

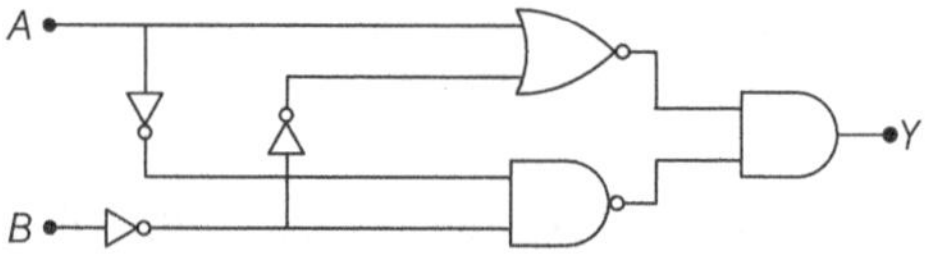

Answers

Round I

1. (a)	2. (b)	3. (c)	4. (d)	5. (d)	6. (b)	7. (c)	8. (d)	9. (b)	10. (a)
11. (b)	12. (a)	13. (a)	14. (b)	15. (c)	16. (b)	17. (b)	18. (b)	19. (b)	20. (c)
21. (b)	22. (c)	23. (d)	24. (a)	25. (c)	26. (b)	27. (b)	28. (a)	29. (c)	30. (d)
31. (c)	32. (a)	33. (b)	34. (c)	35. (c)	36. (d)	37. (a)	38. (c)	39. (d)	40. (b)
41. (a)	42. (a)	43. (d)	44. (a)	45. (a)	46. (d)	47. (c)	48. (c)	49. (c)	50. (d)
51. (c)	52. (a)	53. (b)	54. (a)	55. (b)	56. (c)	57. (a)			

Round II

1. (c)	2. (d)	3. (d)	4. (c)	5. (c)	6. (b)	7. (c)	8. (c)	9. (d)	10. (b)
11. (b)	12. (b)	13. (b)	14. (a)	15. (d)	16. (d)	17. (c)	18. (a)	19. (c)	20. (d)
21. (a)	22. (c)	23. (b)	24. 12	25. 150	26. 40	27. 10	28. 8.784	29. 5	30. 0

Solutions

Round I

1. In insulators, the forbidden energy gap is very large, in case of semiconductor, it is moderate and in conductors the energy-gap is zero.

Hence, $EG_1 < EG_2 < EG_3$.

2. Holes are formed when electrons in atoms move out of the valence band into the conduction band. In this situation, a vacancy is created in covalent band when electron leaves a covalent band, which is called a hole.

3. n-type semiconductor is obtained by doping the Ge or Si with pentavalent (dopants) atoms. In n-type semiconductor, electrons are majority carriers and holes are minority carriers.

4. On increasing the temperature of semiconductor, more number of covalent bonds break and large number of charge carriers are produced. Hence, number density of charge carriers increases. On increassing the temperature, relaxation time also decreases but effect of decrease in relaxation time is much less than increase in number density.

5. In sample X, no impurity energy level is seen, so it is undoped. In sample Y, impurity energy level lies below the conduction band, so it is doped with fifth group impurity. In sample Z, impurity energy level lies above the valence band, so it is doped with third group impurity.

6. As, $I = nAev_d$

or $I \propto nv_d$

$$\therefore \quad \frac{I_e}{I_h} = \frac{n_e v_e}{n_h v_h} \text{ or } \frac{n_e}{n_h} = \frac{I_e}{I_h} \times \frac{v_h}{v_e} = \frac{7}{4} \times \frac{4}{5} = \frac{7}{5}$$

7. Since, it is an n-type semiconductor and concentration of the holes has been ignored. So, its conductivity is given as $\sigma = n_e e \mu_e$, where n_e is the number density of electron, e is the charge on electron and μ_e is its mobility.

Substituting the given values, we get

$$\sigma = 10^{19} \times 1.6 \times 10^{-19} \times 1.6 = 2.56$$

As, resistivity, $\rho = \frac{1}{\sigma} = \frac{1}{2.56}$ or $\rho = 0.39 \simeq 0.4\ \Omega\text{-m}$

8. Number density of atoms in silicon specimen $= 5 \times 10^{28}$ atoms/m$^2 = 5 \times 10^7$ silicon atoms, so total number of indium atoms doped per cm^3 of silicon will be $n = 5 \times 10^{22} / 5 \times 10^7 = 10^{15}$ atoms cm^{-3}.

9. We have, $R = \frac{\rho l}{A} = \frac{L}{n_i e(\mu_e + \mu_h)A}$

$$= \frac{0.928 \times 10^{-2}}{2.5 \times 10^{19} \times 1.6 \times 10^{-19} (0.39 + 0.19) \times 10^{-6}}$$

$$= 4000\ \Omega = 4\ \text{k}\Omega$$

10. Conductivity of semiconductor is given by

$$\sigma = n_e e \mu_e + n_h e \mu_h$$

Since semiconductor is intrinsic

$\Rightarrow \quad n_e = n_h = n_i$

or $\quad \sigma = n_i e(\mu_e + \mu_h)$

$$= 1.6 \times 10^6 \times 1.6 \times 10^{-19}\ (0.4 + 0.2)$$

$$= 1.536 \times 10^{-13}\ \Omega^{-1}\text{m}^{-1}$$

Current flowing, $\quad I = jA \quad$ ...(i)

where, j = current density $= \sigma E = \sigma \left(\frac{V}{d}\right)$

From Eq. (i), we get

$$I = 1.536 \times 10^{-13} \left(\frac{5}{1.5 \times 10^{-3}}\right) (3.5 \times 10^{-4})$$

$$= 1.79 \times 10^{-13}\ \text{A}$$

Heat produced, $H = VIt = 5 \times 1.79 \times 10^{-13} \times 120$

$$= 10.74 \times 10^{-11}\ \text{J}$$

11. In circuit 1, n is connected with n, which is not a series combination of p-n junction. In circuit 2, each p-n junction is forward biased, same current flows, giving same potential difference across junction. In circuit 3, each p-n junction is reverse biased, same leakage current will flow, giving equal potential difference across each p-n junction diode.

12. V-d curve near the junction will be as shown by curve (a).

13. For forward bias, p-side must be a higher potential than n-side.

So,

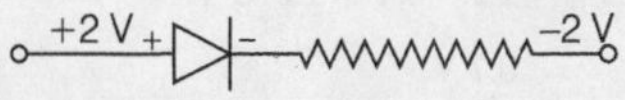

is forward biased.

14. When p-n junction is forward biased, it opposes the potential barrier across junction. When p-n junction is reverse biased, it supports the potential barrier junction, resulting increases in potential barrier across the junction. Thus, option (b) is correct.

15. Potential drop in a silicon diode in forward bias is around 0.7 V.

In given circuit, potential drop across 200 Ω resistor is

$$I = \frac{\Delta V_{\text{net}}}{R} = \frac{3 - 0.7}{200}$$

$\Rightarrow \quad I = 0.0115\ \text{A} \Rightarrow I = 11.5\ \text{mA}$

16. In the given circuit, p side of p-n junction D_1 is connected to lower voltage and n-side of D_1 to higher voltage. Therefore, D_1 is reverse biased. The p-side of p-n junction D_2 is at higher potential and n-side of D_2 is at lower potential. Therefore, D_2 is forward biased.

Due to reverse biasing of diode D_1, circuit becomes open. Hence, no current flows through the junction from B to A, *i.e.* option (b) is correct.

17. As, $R_e = \dfrac{\Delta V}{\Delta I} = \dfrac{(13-8)}{(60-40)\times 10^{-6}} = 2.5\times 10^5\ \Omega$

18. In reverse biasing, the minority charge carriers will be accelerated due to reverse biasing, which on striking with atoms cause ionisation resulting secondary electrons and thus more number of charge carriers.

When doping concentration is large, there will be large number of ions in the depletion region, which will give rise to a strong electric field.

19. Let V be the potential difference between A and B, then

$$V - 0.3 = (5+5)\times 10^3 \times (0.2\times 10^{-3}) = 2$$

or $\quad V = 2 + 0.3 = 2.3\ \text{V}$

20. Initially Ge and Si are both forward biased, so current will effectively pass through Ge diode with a voltage drop of 0.3 V.

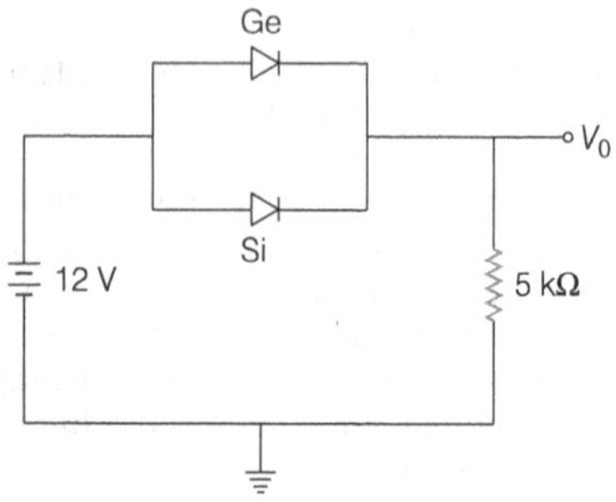

$\therefore$ Initial output voltage,

$$V_0 = 12 - 0.3 = 11.7\ \text{V}$$

If Ge is reversed biased, then only Si diode will work. In this condition, output voltage

$$V_0 = 12 - 0.7 = 11.3\ \text{V}$$

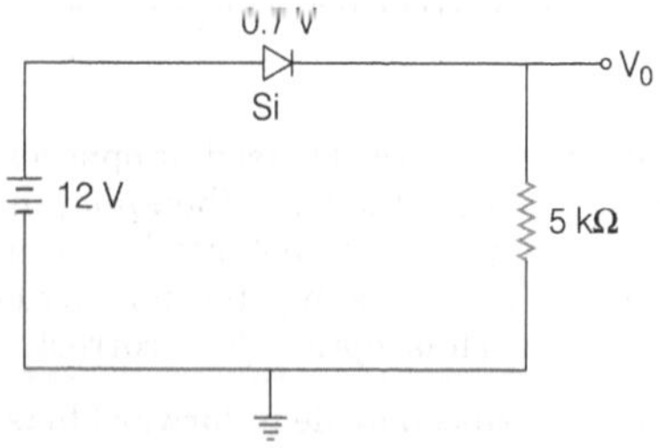

$\therefore$ Change in output voltage

$$= 11.7 - 11.3 = 0.4\ \text{V}$$

21. In this circuit, D_1 is forward biased and D_2 is reversed biased.

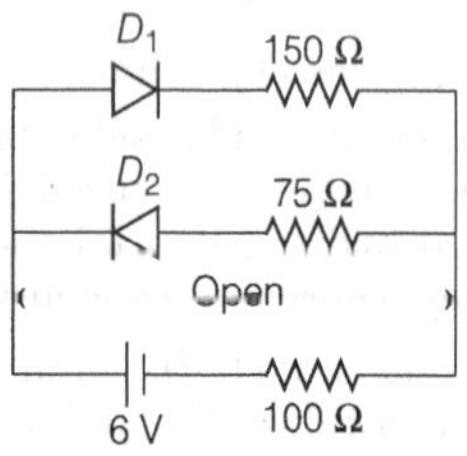

Resistance of D_1 is 50 Ω.

$\therefore$ Net resistance of the circuit,

$$R_{\text{net}} = 50 + 150 + 100 = 300\ \Omega$$

$\therefore$ Current through the 100 Ω resistance

$$= \frac{V}{R_{\text{net}}} = \frac{6}{300} = 0.020\ \text{A}$$

22. During positive half cycle of input AC voltage, the p-n junction is forward biased. The resistance of p-n junction is low. The current in the circuit is maximum. In this situation, a maximum potential difference will appear across resistance connected in series of circuit. Due to it, there is no output voltage across p-n junction.

During the negative half-cycle of input AC voltage, the p-n junction is reverse biased. The resistance of p-n junction becomes high which will be more than resistance in series. Due to it, there will be voltage across p-n junction with negative cycle in output.

23. Here, $V = (50 \sin \omega t)$ volt

Comparing it with general equation $V = V_0 \sin\omega t$, we get

$$V_0 = 50\ \text{V},\ R_L = 650\ \Omega,\ r_d = 200\ \Omega$$

Maximum output current, $I_0 = \dfrac{V_0}{r_d + R_L} = \dfrac{50}{(200+650)}$

$$= 58\ \text{mA}$$

DC output current, $I_{\text{DC}} = \dfrac{I_0}{\pi} = \dfrac{58\times 7}{22} = 18.5\ \text{mA}$

DC output power $= I_{\text{DC}}^2 \cdot R_L = (18.5)^2 \times 650 = 0.22\ \text{W}$

24. For same value of current, higher value of voltage is required for higher frequency.

Since, $\nu_B > \nu_G > \nu_Y > \nu_R$

Hence, $V_B > V_G > V_Y > V_R$

25. Photons are detected by photodiode, so band gap energy is given by

$$\Delta E = \frac{hc}{\lambda} \Rightarrow \Delta E = \frac{1240\ \text{eV-nm}}{400\ \text{nm}} = 3.1\ \text{eV}$$

Note $hc \approx 1240$ eV-nm

26. The I-V characteristic curve for a photodiode is shown below

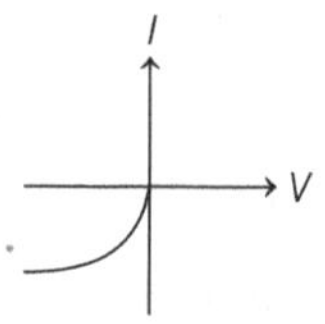

With increasing biasing voltage of a photodiode, the photocurrent first increases and then becomes saturated, *i.e.* after certain potential, electron-hole pair production will be nearly constant.

Hence, correct option is (b).

27. Energy of incident light (photon),

$$h\nu = \frac{hc}{\lambda} = \frac{6.6\times 10^{-34}\times 3\times 10^8}{6\times 10^{-7}\times 1.6\times 10^{-19}} = 2.06\ \text{eV}$$

For the incident radiation to be detected by the photodiode, energy of incident radiation (photon) should be greater than the band gap. This is true only for D_2.

28. Zener diode works in breakdown region.

So, Simple diode $\to$ (a)

Zener diode $\to$ (b)

Solar cell $\to$ (c)

Light dependent resistance $\to$ (d)

29. When the applied reverse voltage V reaches the breakdown voltage of the Zener diode, then only a large amount of current is flown through it, otherwise it is approximately zero.

In the given situation, if we consider that Zener diode is at breakdown, then potential drop across 1500 Ω resistances will be 10 V. So, potential drop at 500 Ω resistor will be 2 V.

$\therefore$ Current in $R_1 = \dfrac{2}{500} = 4\text{mA} = I_1$ (say)

Current in each $R_2 = \dfrac{10}{750} = \dfrac{2}{150} = 13.33\text{ mA} = I_2$ (say)

$\Rightarrow \quad I_1 < I_2$ which is not possible.

So, Zener diode will never reach to its breakdown.

$\therefore$ Current flowing through a reverse biased Zener diode $= 0$.

30. In the circuit, let the current in branches is as shown in figure below

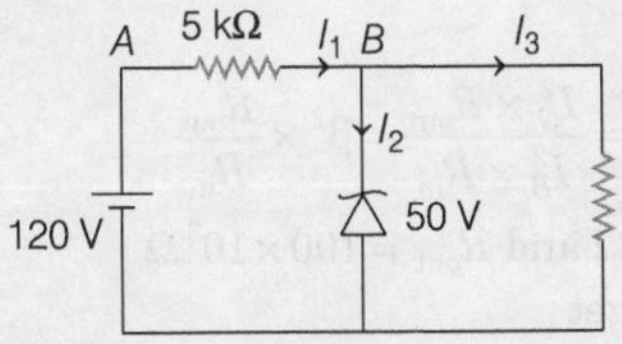

By Kirchhoff's node law,

$$I_1 = I_2 + I_3 \quad \text{... (i)}$$

Now, when diode conducts, voltage difference between points A and B will be

$$V_{AB} = 120 - 50 = 70\text{ V}$$

So, $\quad$ current, $I_1 = \dfrac{V_{AB}}{5\text{ k}\Omega} = \dfrac{70}{5 \times 10^3}$

$$I_1 = 14\text{ mA} \quad \text{...(ii)}$$

Since, diode and 10 kΩ resistor are in parallel combination, so voltage across 10 k Ω resistor will be 50 V only.

$$\Rightarrow \quad I_3 = \frac{50}{10\text{ k}\Omega} = \frac{50}{10 \times 10^3}$$

$$\Rightarrow \quad I_3 = 5\text{ mA} \quad \text{... (iii)}$$

$\therefore$ From Eqs. (i), (ii) and (iii), we get

$$14\text{ mA} = I_2 + 5\text{ mA}$$

or current through diode,

$$I_2 = 14\text{ mA} - 5\text{mA} = 9\text{ mA}$$

31. For $V < 4\text{V}$, both diodes are reverse biased but not working in Zener region. So, output voltage (V_{out}) will be equal to input voltage.

For voltages higher than 4 V, diode A will not conduct current but diode B will work in Zener region and large current will flow through it. So, potential will increase but some voltage drop will occur in connecting wires. So, slope of V_{out} *versus* t graph will get reduced.

Once both the diodes start working in Zener region, the output voltage will become 6 V and after that, it will not change.

Hence, correct option is (c).

32. In an *n-p-n* transistor when base-emitter junction is forward biased and collector-base junction is reverse biased, the majority charge carriers electrons move from emitter-to-base. In base, few electrons get neutralised by electron-hole combination and the remaining electrons crossover the collector.

33. As, $\beta = \dfrac{I_C}{I_B} = \dfrac{I_E - I_B}{I_B} = \dfrac{I_E}{I_B} - 1$ or $\dfrac{I_E}{I_B} = 1 + \beta$

or $I_B = \dfrac{I_E}{1+\beta} = \dfrac{8.2}{1+40} = \dfrac{8.2}{41} = 0.20\text{ mA}$

34.

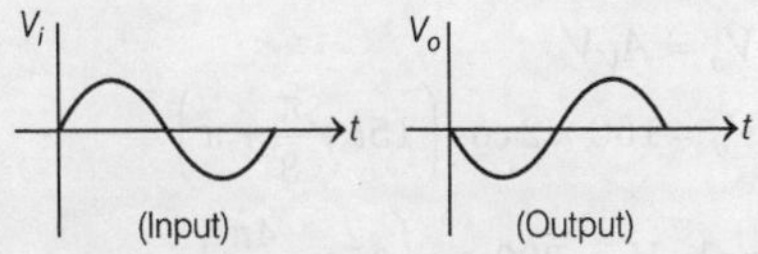

In a CE *n-p-n* transistor amplifier, output voltage is 180° out of phase with input voltage.

35. As we know,

In case of a common emitter configuration, DC current gain, $\alpha = \dfrac{I_C}{I_E}$.

where, I_C is collector current and I_E is emitter current.

AC current gain, $\beta = \dfrac{I_C}{I_B} \quad \text{...(i)}$

where, I_B is base current.

Also, $\quad I_E = I_B + I_C \quad \text{...(ii)}$

Dividing Eq. (ii) by I_C, we get

$\Rightarrow \quad \dfrac{I_E}{I_C} = \dfrac{I_B}{I_C} + 1 \quad$ [from Eq. (i)]

$\Rightarrow \quad \dfrac{1}{\alpha} = \dfrac{1}{\beta} + 1 \Rightarrow \alpha = \dfrac{\beta}{1+\beta}$

36. As, collector current, $I_C = \dfrac{90}{100} \times I_E$

where, $I_E =$ emitter current.

$\Rightarrow \quad 10 = 0.9 \times I_E$

$\Rightarrow \quad I_E = 11\text{ mA}$

Also, $\quad I_E = I_B + I_C$

Base current, $I_B = I_E - I_C$

$= 11 - 10 = 1\text{ mA}$

37. Given, $\beta = 50$, $R_i = 1000\Omega$, $V_i = 0.01$ V

$$\beta = \frac{I_C}{I_B} \quad \text{... (i)}$$

and $$I_B = \frac{V_i}{R_i} = \frac{0.01}{10^3} = 10^{-5} \text{ A}$$

$$I_C = 50 \times 10^{-5} \text{ A} = 500 \, \mu\text{A} \quad \text{[from Eq. (i)]}$$

38. Number of electrons reaching the collector,

$$n_C = \frac{96}{100} \times 10^{10} = 0.96 \times 10^{10}$$

Emitter current, $I_E = \frac{n_E \times e}{t}$

Collector current, $I_C = \frac{n_C \times e}{t}$

Thus, current transfer ratio,

$$\alpha = \frac{I_C}{I_E} = \frac{n_C}{n_E} = \frac{0.96 \times 10^{10}}{10^{10}} = 0.96$$

39. Input signal of a CE amplifier, $V_i = 2\cos\left(15t + \frac{\pi}{3}\right)$

Voltage gain, $A_V = 150$

As CE amplifier gives phase difference of π between input and output signals.

So, $$A_V = \frac{V_o}{V_i}$$

$$\Rightarrow \quad V_o = A_V V_i$$

$$V_o = 150 \times 2\cos\left(15t + \frac{\pi}{3} + \pi\right)$$

Output signal, $V_o = 300\cos\left(15t + \frac{4\pi}{3}\right)$

40. Here, $I_C = 10$ mA

$$\therefore \quad I_C = \frac{95}{100} I_E \Rightarrow 10 \text{ mA} = \frac{95}{100} I_E$$

or $$I_E = \frac{100 \times 10}{95} = 10.53 \text{ mA}$$

$$\Rightarrow \quad i_C = \frac{2}{2 \times 10^3} = 10^{-3} \text{ A}$$

$\therefore$ Current gain, $\beta = \frac{i_C}{i_B}$

$$100 = \frac{10^{-3}}{i_B} \Rightarrow i_B = 10^{-5} \text{ A}$$

41. From the given transfer characteristics of a base biased common emitter transistor, we note that

(a) When $V_i = 0.4$ V, there is no collector current. The transistor circuit is in cut-off state.

(b) When $V_i = 1$ V (which is in between 0.6 V to 2 V), the transistor circuit is in active state and it used as an amplifier.

(c) When $V_i = 0.5$ V, there is no collector current. The transistor is in cut-off state. The transistor circuit can be used as a switch turned OFF.

(d) When $V_i = 2.5$ V, the collector current becomes maximum and transistor is in saturation state and can be used as switch turned ON state.

42. Given, $A_P = 60$ dB (in decibel)

Power gain in decibel can be given as

$$A_P = 10\log_{10}\left(\frac{\text{Output power}}{\text{Input power}}\right)$$

$$\Rightarrow \quad 60 = 10\log_{10}\left(\frac{P_{out}}{P_{in}}\right) \Rightarrow \log_{10}\left(\frac{P_{out}}{P_{in}}\right) = 6$$

$$\Rightarrow \quad \frac{P_{out}}{P_{in}} = 10^6 = A_P \quad \text{... (i)}$$

Also, given $R_{out} = 10$ kΩ, $R_{in} = 100\,\Omega$

$\therefore$ Power gain of a transistor is given by

$$A_P = \beta^2\left(\frac{R_{out}}{R_{in}}\right)$$

where, β is current gain.

$$\Rightarrow \quad \beta^2 = A_P \times \frac{R_{in}}{R_{out}} = 10^6 \times \frac{100}{10 \times 10^3}$$

$$\Rightarrow \quad \beta^2 = 10^4 \text{ or } \beta = 10^2$$

43. Given curve is between I_C and I_B as output and input currents, respectively.

So, it is transfer characteristics curve of a common emitter (CE) configuration.

In CE configuration,

Current gain, $$\beta = \frac{I_{out}}{I_{in}} = \frac{I_C}{I_B} \quad \text{...(i)}$$

Voltage gain,

$$A_V = \frac{V_{out}}{V_{in}} = \frac{I_C \times R_{out}}{I_B \times R_{in}} = \beta \times \frac{R_{out}}{R_{in}} \quad \text{...(ii)}$$

and power gain,

$$A_P = \frac{P_{out}}{P_{in}} = \frac{I_C^2 \times R_{out}}{I_B^2 \times R_{in}} = \beta^2 \times \frac{R_{out}}{R_{in}} \quad \text{...(iii)}$$

Given, $R_{in} = 100\,\Omega$ and $R_{out} = 100 \times 10^3\,\Omega$

From Eq. (i), we get

$$\beta = \frac{5\text{ mA}}{100\,\mu\text{A}}\left(\text{or } \frac{10\text{ mA}}{200\,\mu\text{A}} \text{ or } \frac{15\text{ mA}}{300\,\mu\text{A}} \text{ or } \frac{20\text{ mA}}{400\,\mu\text{A}}\right)$$

$$\Rightarrow \quad \beta = \frac{5 \times 10^{-3}}{100 \times 10^{-6}} = 50 \quad \text{...(iv)}$$

From Eqs. (ii) and (iv), we get

Voltage gain, $A_V = \beta \times \frac{R_{out}}{R_{in}} = 50 \times \frac{100 \times 10^3}{100}$

$$\Rightarrow \quad A_V = 50000 = 5 \times 10^4 \quad \text{...(v)}$$

From Eqs. (iii) and (iv), we get

Power gain, $A_P = \beta^2 \times \frac{R_{out}}{R_{in}} = (50)^2 \times \frac{100 \times 10^3}{100}$

$$= 2500 \times 1000$$

$$\Rightarrow \quad A_P = 2.5 \times 10^6$$

44. For a common emitter n-p-n transistor, DC current gain,

$$\beta_{DC} = \frac{I_C}{I_B}$$

At saturation state, V_{CE} becomes zero.

$$\therefore \quad V_{CC} - I_C R_C = 0$$

$$I_C \approx \frac{V_{CC}}{R_C} = \frac{10}{1000} = 10^{-2}\text{A}$$

Hence, saturation base current,

$$I_B = \frac{I_C}{\beta_{DC}} = \frac{10^{-2}}{250} = 40\mu\text{A}$$

45. When S_1 is put ON to apply proper bias for the first time in the given circuit. Obviously, a surge of collector current flows in the transistor. This current flows through the coil T_2 where terminals are numbered 3 and 4 in the circuit. This current does not reach full amplitude instantaneously but increases from X to Y, as shown in Fig (i).

The inductive coupling between coil T_2 and coil T_1 now causes a current to flow in the emitter circuit (**Note** that this actually is the 'feedback' from input to output). As a result of this positive feedback, this current (in T_1; emitter current) also increases from X' to Y' as shown in Fig (ii). Thus, initially, both I_C and I_E increase.

46. Given gates and their truth tables are as below

1. 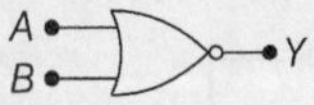

NOR gate

A	B	Y
0	0	1
1	0	0
0	1	0
1	1	0

2. 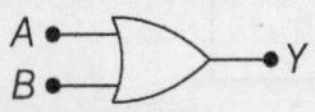

OR gate

A	B	Y
0	0	0
1	0	1
0	1	1
1	1	1

3. 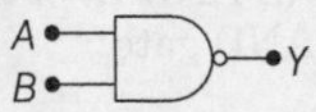

NAND gate

A	B	Y
0	0	1
1	0	1
0	1	1
1	1	0

4. 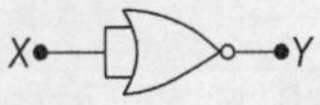

One input NOR gate

X	Y
0	1
1	0

Clearly, 4th gate converts a low voltage (0) into a high voltage (1) and *vice-versa*.

So, 4th gate or one input NOR gate gives a reversible operation.

47. From the figure, the truth table

A	B	Y
0	0	0
0	1	0
1	0	0
1	1	1

i.e. AND gate

48. Given circuit is

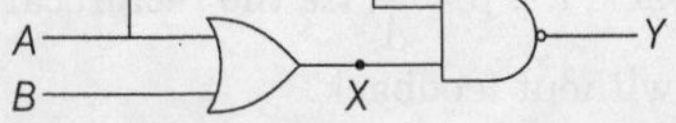

Let the intermediate state X of OR gate is shown in figure.

Clearly, $\quad Y = \overline{AX} \quad$ ...(i)

Here, $\quad X = A + B \quad$ ...(ii)

$$\therefore \quad Y = \overline{A\,(A+B)} = \overline{AA + AB}$$

$$= \overline{A + AB} \qquad (\because AA = A)$$

$$= \overline{A\,(1+B)}$$

$$\Rightarrow \quad Y = \overline{A} \qquad (\because 1 + B = 1)$$

So, truth table shown in option (c) is correct.

49. Output of OR gate is 0 when all inputs are 0 and output is 1 when atleast one of the input is 1.

Observing output *x* It is 0 when all inputs are 0 and it is 1 when atleast one of the input is 1.

∴ The gate is OR.

50. Given circuit is a combination of OR gate and NOT gate.

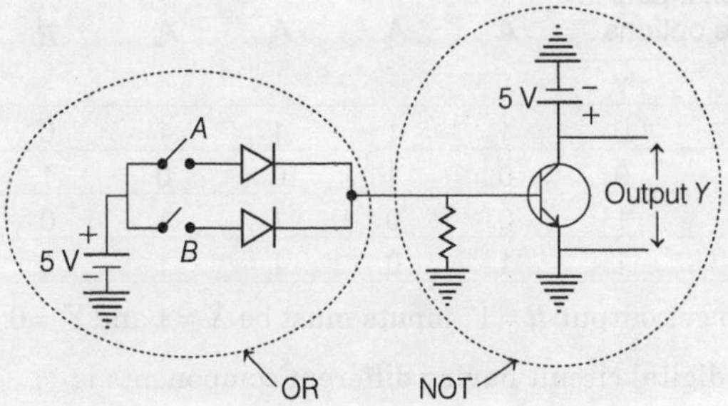

So, given combination is a NOR gate, its truth table is

A	B	Y
0	0	1
1	0	0
0	1	0
1	1	0

Its Boolean relation,

$$Y = \overline{A + B} = \overline{A} \cdot \overline{B} \qquad \text{(by de Morgan's rule)}$$

51. In circuit, $C = A \cdot B$ and $D = \overline{A} \cdot B$

$$E = C + D = (A \cdot B) + (\overline{A} + B)$$

Truth table of this arrangement of gates can be as given below

A	B	$\overline{A}$	$C = A \cdot B$	$D = \overline{A} \cdot B$	$E = (C + D)$
0	0	1	0	0	0
0	1	1	0	1	1
1	0	0	0	0	0
1	1	0	1	0	1

52. Output of gate I is $Y' = \overline{A + B}$

Output of gate II is $Y_1 = \overline{A + Y'}$

Output of gate III is $Y_2 = \overline{B + Y'}$

Output of gate IV is $Y = \overline{Y_1 + Y_2}$

$$\Rightarrow Y = \overline{\overline{A + Y'} + \overline{B + Y'}} = \overline{\overline{(A + Y')}}\;\overline{\overline{(B + Y')}}$$

$$= (A + Y')\,(B + Y')$$

$$\Rightarrow Y = [A + \overline{A + B}]\,[B + \overline{A + B}] = (A + \overline{A} \cdot \overline{B})\,(B + \overline{A} \cdot \overline{B})$$

$$\Rightarrow Y = AB + (A\overline{A})\overline{B} + \overline{A}\,\overline{B}B + (\overline{A}\,\overline{B}) \cdot (\overline{A}\,\overline{B})$$

$$\Rightarrow Y = AB + 0 + 0 + \overline{A}\,\overline{B} + AB + \overline{A}\,\overline{B} = \text{output of XNOR gate.}$$

The truth table of the combination is in option (a).

53. The given circuit can be drawn as shown in the figure given below

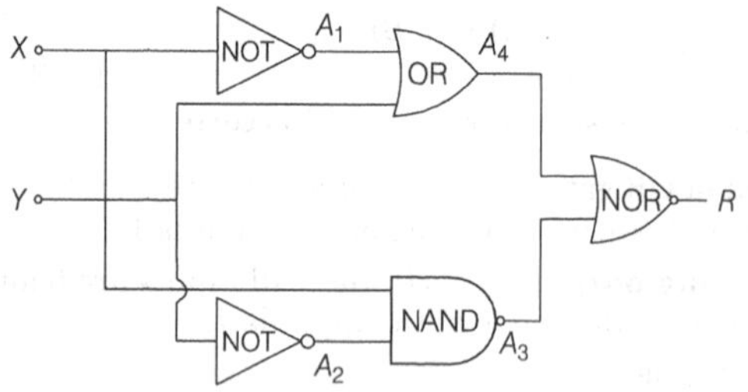

Truth table for this given logic gate is

Given inputs in the options		A_1	A_2	A_3	A_4	R
X	Y					
0	0	1	1	1	1	0
1	0	0	1	0	0	1
1	1	0	0	1	1	0
0	1	1	0	1	1	0

So to get output $R = 1$, inputs must be $X = 1$ and $Y = 0$.

54. The digital circuit having different components is shown below,

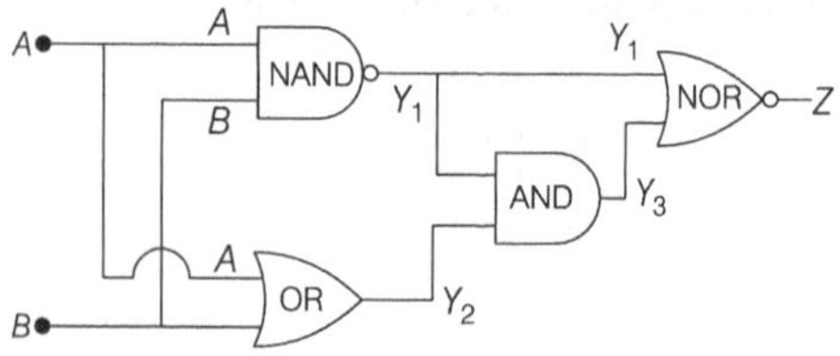

Truth table of above circuit is

A	B	$Y_1 = \overline{A \cdot B}$	$Y_2 = A + B$	$Y_3 = Y_1 \cdot Y_2$	$Z = \overline{Y_1 + Y_3}$
1	0	1	1	1	0
0	0	1	0	0	0
1	1	0	1	0	1
0	1	1	1	1	0

Hence, correct option is (a).

55. Given circuit,

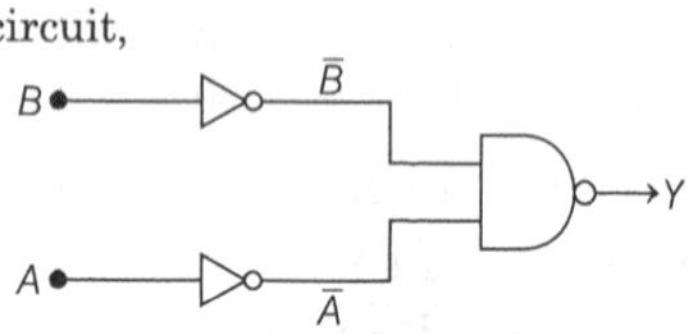

Output, $Y = \overline{\overline{A} \cdot \overline{B}} = \overline{\overline{A}} + \overline{\overline{B}} = A + B$

Truth Table

A	B	$Y = A + B$
0	0	0
0	1	1
1	0	1
1	1	1

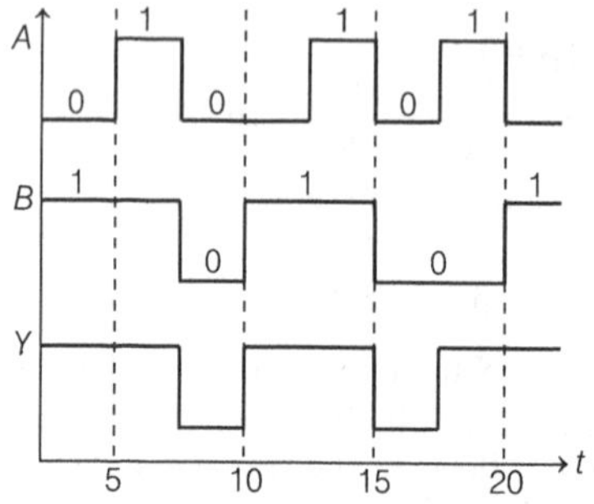

So, most appropriate option is (b).

56. From the figure of AND gate

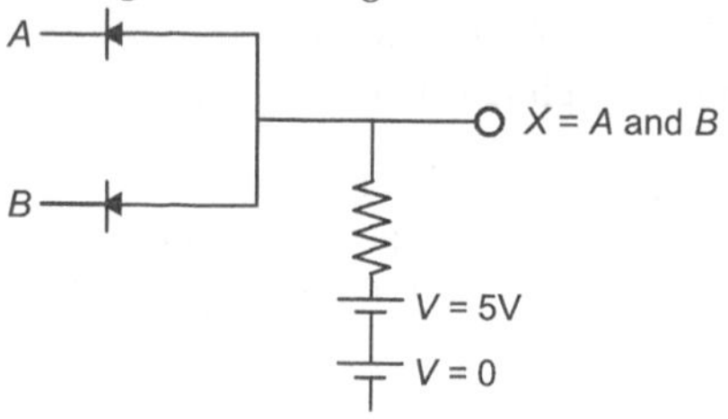

and NOT gate

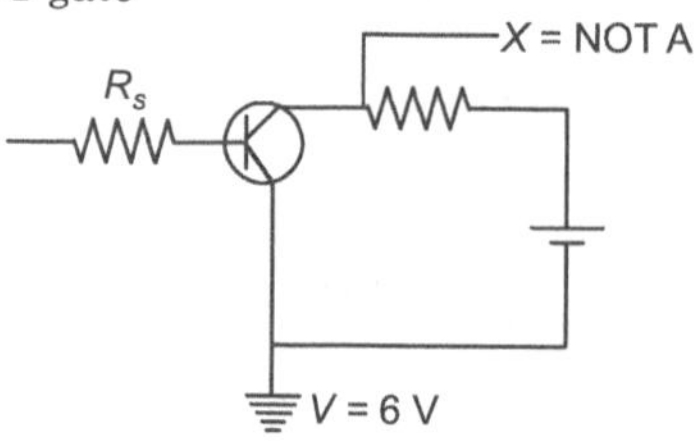

Clearly, the function X = NOT (A AND B) of the logical variables A AND B is called NAND gate.

57. Truth table for the given logic gate

A	B	Y
0	0	1
0	1	0
1	0	0
1	1	0

The truth table is similar to that of a NOR gate.

Round II

1. The output, $Y' = A + B$

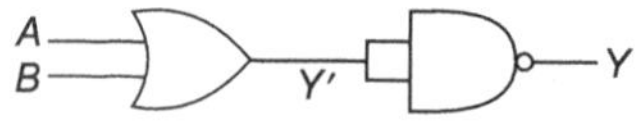

and $Y = \overline{Y'} = \overline{A + B}$

i.e. Output of a NOR gate.

2. The conditions for a circuit to oscillate are (i) the feedback is positive (ii) the fraction of the output voltage feedback, *i.e.* $\beta = \frac{1}{A}$, *i.e.* the reciprocal of the voltage gain without feedback.

3. During positive half cycle, p-n junction conducts. Potential difference across C = peak voltage of the given AC voltage $= V_0 = V_{rms} = \sqrt{2} = 220\sqrt{2}$ V.

4. Given circuit is

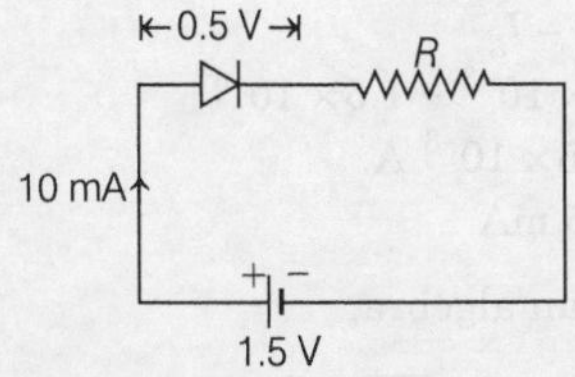

Now, by Kirchhoff 's loop rule,

$$1.5 - 0.5 - R \times 10 \times 10^{-3} = 0 \Rightarrow R = 100\,\Omega$$

5. When diode is reverse biased, the applied voltage supports the barrier voltage. Due to it, the reverse current is weak. It will be identical in two diodes.

6. By De-Morgan's theorem, we have

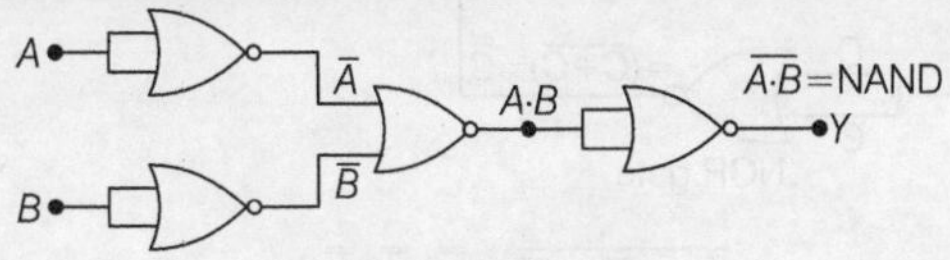

7. For $V_i < 0$,

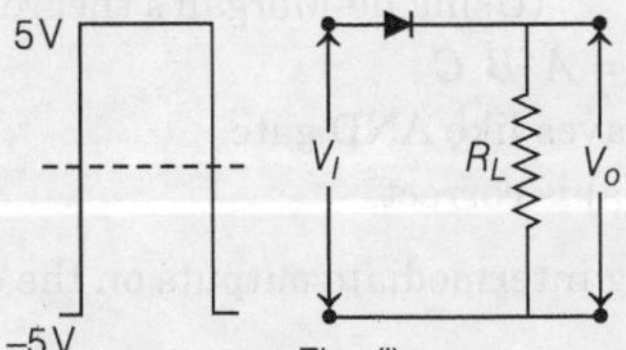

Fig. (i)

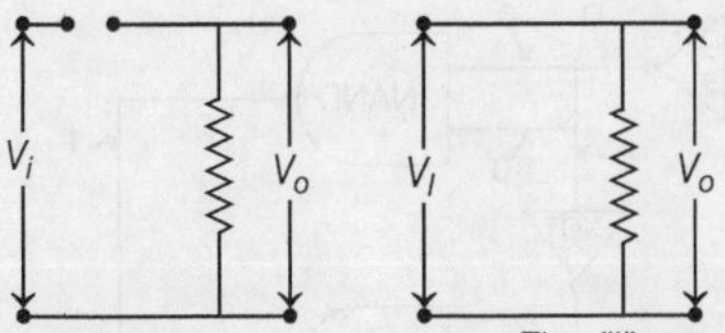

Fig. (ii) Fig. (iii)

The diode is reverse biased and hence offer infinite resistance, so, circuit would be like as shown in Fig. (ii) and $V_o = 0$. For $V_i > 0$, the diode is forward biased and circuit would be as shown in Fig. (iii) and $V_o = V_i$.

8. Here *p-n* junction is forward biased. If *p-n* junction is ideal, its resistance is zero. The effective resistance across *A* and *B*

$$= \frac{10 \times 10}{10 + 10} = 5\ \text{k}\Omega$$

Current in the circuit, $I = \dfrac{V}{R} = \dfrac{30}{15 \times 10^3} = \dfrac{2}{10^3}$ A

Current in arm *AB*, $I = \dfrac{2}{10^3}$

Potential difference across *A* and *B*

$$= \frac{2}{10^3} \times 5 \times 10^3 = 10\ \text{V}$$

9. Voltage gain, $A_V = \beta \dfrac{R_2}{R_1}$

Also, current gain, $\beta = \dfrac{\alpha}{1 - \alpha} = \dfrac{0.98}{1 - 0.98} = 49$

$$A_V = 49\left(\frac{500 \times 10^3}{R_1}\right) \quad \text{...(i)}$$

$\therefore$ Power gain $= 6.0625 \times 10^6$ ...(ii)

From Eqs. (i) and (ii), we get

$$\frac{49 \times 500 \times 10^3}{R_1} = 6.0625 \times 10^6$$

$$\Rightarrow \quad R_1 = \frac{49 \times 5 \times 10^5}{6.0625 \times 10^6}$$

$$\therefore \quad R_1 = 404.37400$$

10. Output of $G_1 = (A + B)$

Output of $G_2 = \overline{A \cdot B}$

Output of $G_3 = (A + B) \cdot \overline{A \cdot B} = (A + B)(\overline{A} + \overline{B})$

$= A\overline{A} + A\overline{B} + B\overline{A} + B\overline{B}$

$= O + A\overline{B} + B\overline{A} + 0 = A\overline{B} + \overline{A}B$

which give XOR gate.

11. Current through circuit,

$$I = \frac{P}{V} = \frac{100 \times 10^{-3}}{0.5} = 0.2\ \text{A}$$

Voltage drop across $R = 1.5 - 0.5 = 1.0$ V

Hence, $R = 1/0.2 = 5\,\Omega$

12. The given circuit works as half-wave rectifier. In this circuit, we will get current through *R* when *p-n* junction diode is forward biased and there is no current when *p-n* junction is reversed biased. Thus, the current through resistance will be shown by graph (b).

13. At logic gate I, the Boolean expression is

$$\overline{B} \cdot C = Y'$$

At logic gate II, the Boolean expression is

$$A + (\overline{B} \cdot C) = Y''$$

At logic gate III, the Boolean expression is

$$\overline{A + (\overline{B} \cdot C)} = Y$$

14. Given circuit is Zener diode circuit.

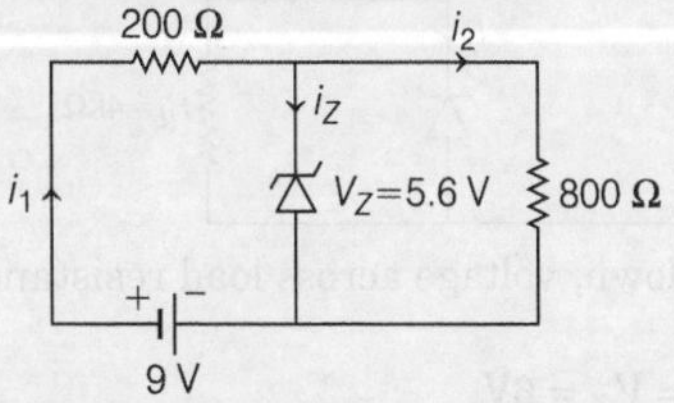

where,potential drop across 800 Ω resistance = potential drop across Zener diode = 5.6 V.

So, current, $i_2 = \dfrac{V}{R} = \dfrac{5.6}{800} = 7$mA

Now, potential drop across 200 Ω resistance

$$= 9 - 5.6 = 3.4\ \text{V}$$

Current, $i_1 = \dfrac{V}{R} = \dfrac{3.4}{200} = 17$ mA

So, current, $i_Z = i_1 - i_2 = 17 - 7 = 10$ mA

Current in 100 Ω $= \dfrac{6}{300} = 0.02$ A

15. Here Boolean expression is $Y = \overline{(A+B)\cdot C}$. So,
$Y = \overline{A \cdot B} = \overline{0.1} = 1$ as the Boolean expression of it is $Y = \overline{A \cdot B}$ (for NAND gate).

Hence, when $A = B = 1$, $C = \overline{AB} = \overline{1 \cdot 1} = \bar{1} = 0$

16. Zener diode is used to regulate voltage with any amount of increase in current while the change in voltage is negligible. This is well defined in the region *de*.

17. $A_V = \dfrac{\Delta V_o}{\Delta V_i} = \beta \dfrac{R_o}{R_i}$

So, $\Delta V_o = \Delta V_i \times \beta \dfrac{R_o}{R_i}$

$\therefore \quad \Delta V_o = 10 \times 50 \times \dfrac{5000}{2000} = 1250 \text{ mV} = 1.25 \text{ V}$

18. As, $V_B = V_{\text{knee}} + IR$

or $\quad 4 = 0.7 + 10^{-3} R$

or $\quad R = 3.3/10^{-3} = 3.3 \times 10^3 \, \Omega = 3.3 \text{ k}\Omega$

19. $f_{\text{out}} = 2 f_{\text{input}} = 2 \times 50 = 100 \text{ Hz}$

Mobility of electrons is 2 to 3 times larger than that of hole. Hence, $\mu_e > \mu_h$.

Current gain $\beta = 100$ and $\alpha = \dfrac{\beta}{1+\beta} = \dfrac{100}{101}$;

Given, $\quad \dfrac{\Delta i_C}{\Delta i_E} = \dfrac{100}{101} = \dfrac{1 \text{ mA}}{\Delta i_E}$

$\therefore \quad \Delta i_E = 1.01 \text{ mA}$

20. In given voltage regulator circuit, breakdown of Zener occurs at 6 V.

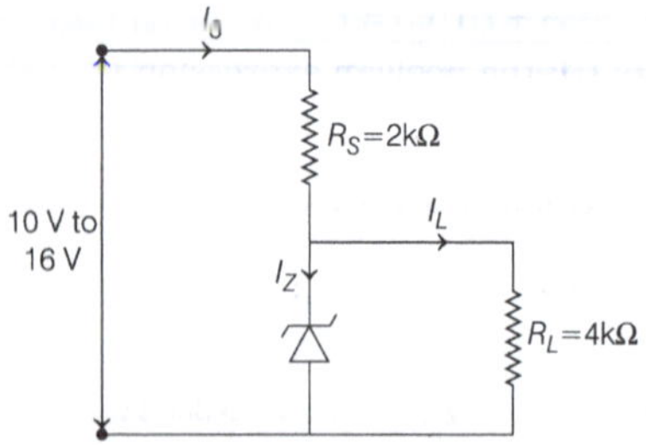

After breakdown, voltage across load resistance ($R_L = 4\text{k}\Omega$),

$$V = V_Z = 6\text{V}$$

$\therefore$ Load current after breakdown,

$$I_L = \frac{V_Z}{R_L} = \frac{6}{4000} = 1.5 \times 10^{-3} \text{ A}$$

When unregulated supply is of 16 V, potential drop occurring across series resistance ($R_S = 2 \text{ k}\Omega$),

$$V_S = V - V_Z$$
$$= 16 - 6 = 10 \text{ V}$$

So, current across series resistance,

$$I_S = \frac{V_S}{R_S} = \frac{10}{2 \times 10^3} = 5 \times 10^{-3} \text{ A}$$

So, current across Zener diode is

$$I_Z = I_S - I_L$$
$$= 5 \times 10^{-3} - 1.5 \times 10^{-3}$$
$$= 3.5 \times 10^{-3} \text{ A}$$
$$= 3.5 \text{ mA}$$

21. Applying Boolean algebra,

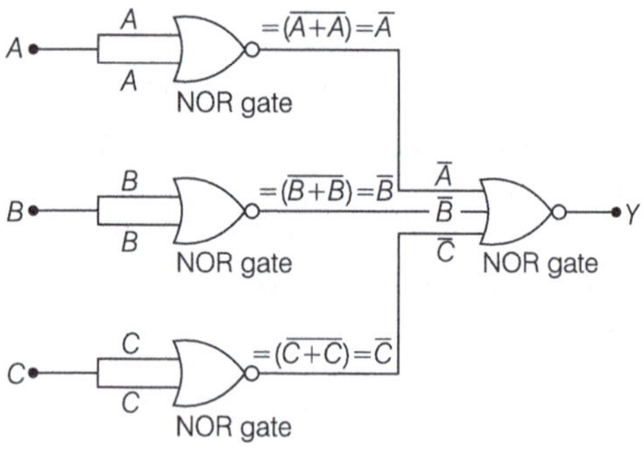

$$Y = \overline{\bar{A} + \bar{B} + \bar{C}} = \bar{\bar{A}} \cdot \bar{\bar{B}} \cdot \bar{\bar{C}}$$

(Using de-Morgan's theorem)

$$= A \cdot B \cdot C$$

So, system behaves like AND gate.
Hence, option (a) is correct.

22. We are showing intermediate outputs on the circuit as given below

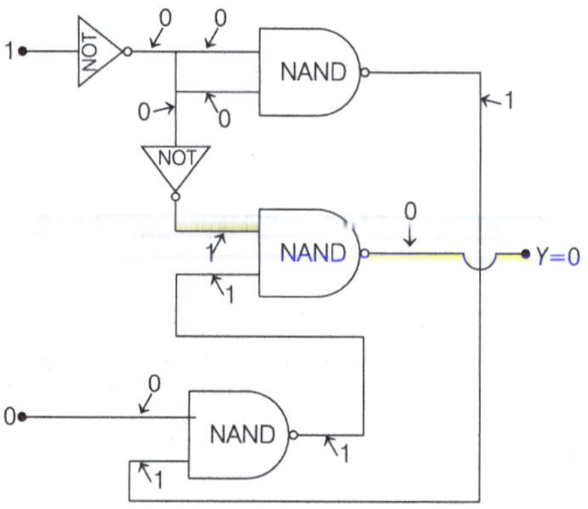

Output of above circuit at given inputs is zero.
To calculate output, we uses truth tables of NAND and NOT gates.

NAND gate

A	B	Y
0	0	1
1	0	1
0	1	1
1	1	0

NOT gate

Y_{in}	Y_{out}
0	1
1	0

23. Transistor saturation occurs when $V_{CE} = 0$.

Now, for closed loop of collector and emitter by Kirchhoff's voltage rule, we have

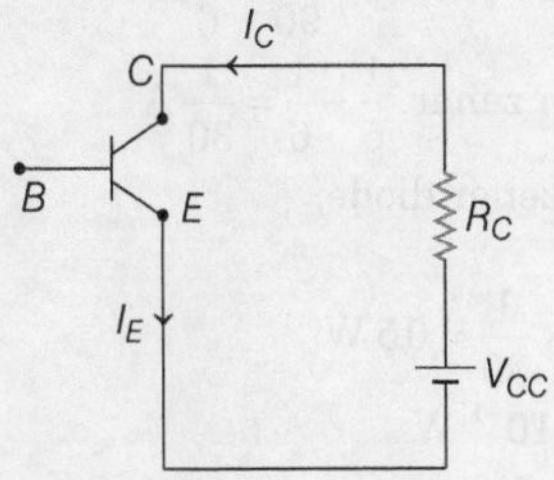

$$V_{CE} = V_{CC} - I_C R_C$$

$$\Rightarrow \quad 0 = V_{CC} - I_C R_c$$

$$\Rightarrow \quad I_C = \frac{V_{CC}}{R_C} = \frac{5}{1 \times 10^3} = 5 \times 10^{-3} \text{ A}$$

Now, $\beta_{DC} = 200$ (given)

$$\Rightarrow \quad \frac{I_C}{I_B} = \beta_{DC} = 200$$

$$\Rightarrow \quad I_B = \frac{I_C}{200} = \frac{5 \times 10^{-3}}{200}$$

$$\Rightarrow \quad I_B = 2.5 \times 10^{-5} = 25\, \mu\text{A}$$

Now, we apply Kirchhoff's voltage rule in base-emitter closed loop, we get

$$V_{BB} = I_B R_B + V_{BE}$$

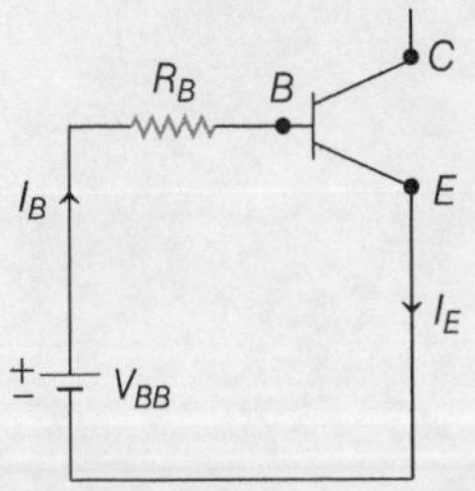

$$\Rightarrow \quad V_{BB} = (25 \times 10^{-6} \times 100 \times 10^3) + 1$$

$$= 3.5 \text{ V}$$

24. We have given circuit,

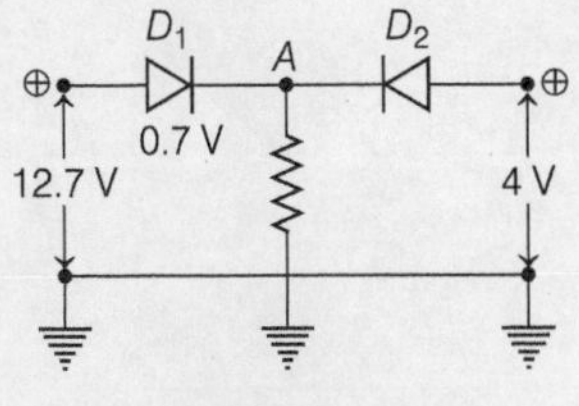

Diode D_1 is in forward bias and D_2 is in reverse bias, so D_1 conducts and D_2 resists.

$\therefore$ Potential at $A = 12.7 - 0.7 = 12$ V

25. For a transistor, $\beta_{AC} = \left(\frac{\Delta I_C}{\Delta I_B}\right)_{V_{CE} = \text{constant}}$

From given figure,

$$\Delta I_B = 30 - 20 = 10\, \mu\text{A}$$

$$\Delta I_C = 4.5 - 3 = 1.5 \text{mA}$$

$$\therefore \quad \beta = \frac{1.5 \times 10^{-3}}{10 \times 10^{-6}} = 150$$

26. As input = 12 V and output = 8 V, this means 4V potential drop occurs in external resistance of 400 Ω.

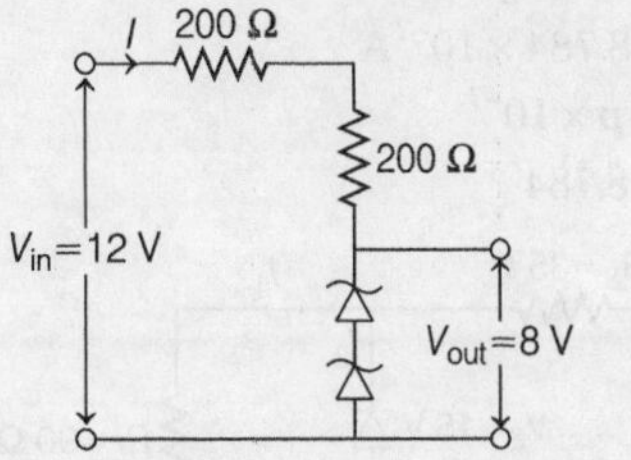

So, current I across resistors, $= \frac{V}{R} = \frac{4}{400} = 10^{-2}$ A

As, there are two identical diodes and potential drop across diodes 8V, *i.e.* potential drop across each diode is 4V.

Hence, power dissipation across each diode,

$$P = VI = 4 \times 10^{-2} \text{ W} = 40 \text{ mW}$$

27. Given, collector resistance, $R_{\text{output}} = 2 \text{ k}\Omega = 2000\ \Omega$

Current amplification factor of the transistor, $\beta_{AC} = 100$

Audio signal voltage, $V_{\text{output}} = 2$ V

Input (base) resistance, $R_{\text{input}} = 1 \text{ k}\Omega = 1000\ \Omega$

$$\because \quad \text{Voltage gain, } A_V = \frac{V_{\text{output}}}{V_{\text{input}}} = \beta_{AC} \frac{R_{\text{output}}}{R_{\text{input}}}$$

$$\therefore \text{ Input signal voltage, } V_{\text{input}} = \frac{V_{\text{output}}}{\beta_{AC}(R_{\text{output}}/R_{\text{input}})}$$

$$= \frac{2}{100\,(2000/1000)}$$

$$= 0.01 \text{ V}$$

$$\text{Base (input) current, } I_B = \frac{V_{\text{input}}}{R_{\text{input}}}$$

$$= \frac{0.01}{1000}$$

$$= 10 \times 10^{-6} \text{ A}$$

$$= 10\, \mu\text{A}$$

28. Here, emf of the battery, $V = 2$ V, $l = 0.1$ m,

$$A = 1 \times 10^{-4} \text{m}^2$$

In case of silicon, $n_e = n_h = n = 1.5 \times 10^{16}/\text{m}^3$

Electric field, $E = \frac{V}{l}$, where emf of the battery $V = 2$ V and length of the block, $l = 0.1$ m

$$\Rightarrow \quad E = \frac{2\text{V}}{0.1 \text{ m}} = 20 \text{ V/m}$$

Clearly, $v_e = E\mu_e = 20 \times 0.135 \text{ m/s} = 2.70 \text{ m/s}$

$$v_h = E\mu_h = 20 \times 0.048 \text{ m/s} = 0.96 \text{ m/s}$$

Electron current,

$$I_e = eAn_e v_e = (1.6 \times 10^{-19})\,(10^{-4})\,(1.5 \times 10^{16})\,(2.70)$$
$$= 6.48 \times 10^{-7}\,\text{A}$$

Hole current,

$$I_h = eAn_h v_h = (1.6 \times 10^{-19})\,(10^{-4})\,(1.5 \times 10^{16})\,(0.96)$$
$$= 2.304 \times 10^{-7}\,\text{A}$$

Total current,

$$I = I_e + I_h = (6.48 \times 10^{-7}\,\text{A}) + (2.304 \times 10^{-7}\,\text{A})$$
$$= 8.784 \times 10^{-7}\,\text{A}$$
$$= p \times 10^{-7}$$
$$\therefore \quad p = 8.784$$

29.

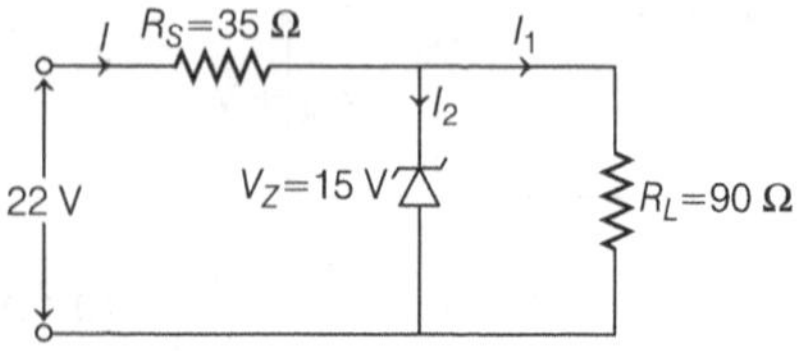

Voltage across $R_S = 22 - 15 = 7\,\text{V}$

Current through R_S, $I = \frac{7}{35} = \frac{1}{5}\,\text{A}$

Current through 90 Ω, $I_2 = \frac{15}{90} = \frac{1}{6}\,\text{A}$

Current through zener, $\frac{1}{5} - \frac{1}{6} = \frac{1}{30}\,\text{A}$

Power through zener diode,

$$P = VI$$
$$P = 15 \times \frac{1}{30} = 0.5\,\text{W}$$
$$P = 5 \times 10^{-1}\,\text{W}$$

The value of x is 5.

30.

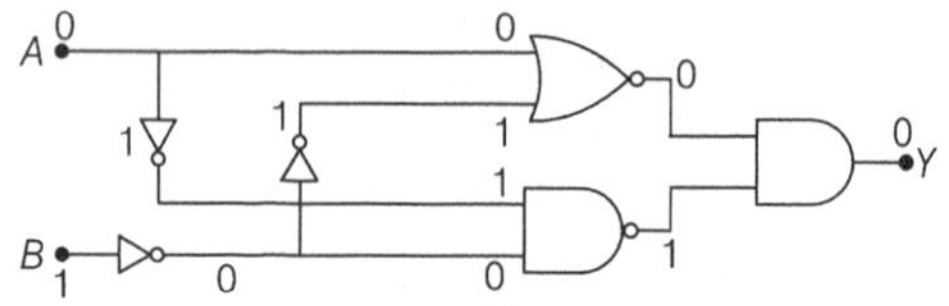

The value of x is 0.

29

Atoms and Nuclei

Atoms are the basic building blocks of matter and defining structure of elements. The first model of an atom was proposed by JJ Thomson in 1898. The schematic arrangement of the Rutherford's α-particle scattering experiment (also known as Geiger-Marsden experiment) is shown below

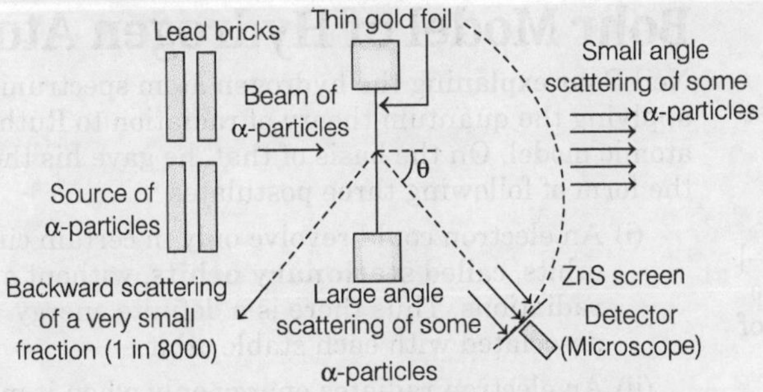

He suggested that atoms are just positively charged lumps of matter with electrons embedded in them like resins in a fruit cake. This model is also called the **plum pudding model**.

IN THIS CHAPTER ....

- α-Particle Scattering Experiment
- Rutherford's Atomic Model
- Bohr Model of Hydrogen Atom
- Hydrogen Spectrum
- Composition and Size of Nucleus
- Equivalence of Mass and Energy
- Nuclear Force
- Radioactivity
- Nuclear Reaction
- Nuclear Energy

α-Particle Scattering Experiment

In α-particle scattering experiment, following facts are observed

- Most of the α-particles pass through the foil without deflection.
- About 0.14% of the incident α-particles are scattered by more than 1%.
- A very small number of α-particles (about 1 in 8000) practically retraced their paths or suffered deflection of nearly 180°.

Rutherford's Atomic Model

Based on the above observations, Rutherford proposed a nuclear model of atom. According to this model,

- entire positive charge and most of the mass of the atom is concentrated in a small volume called **nucleus**, with electrons revolving around the nucleus as planets revolve around the sun.
- size of nucleus is to be about 10^{-15} m to 10^{-14} m. From kinetic theory, the size of an atom is known to be 10^{-10} m. Thus, most of an atom has empty space.

- atomic electrons are so light that they does not affect the path of α-particles.

Following are few important points related to Rutherford's model

- The number of α-particles scattered per unit area $N(\theta)$ at a scattering angle θ varies as

$$N \propto \frac{1}{\sin^4\left(\frac{\theta}{2}\right)}$$

The following graph represents the variation of $N(\theta)$ with θ

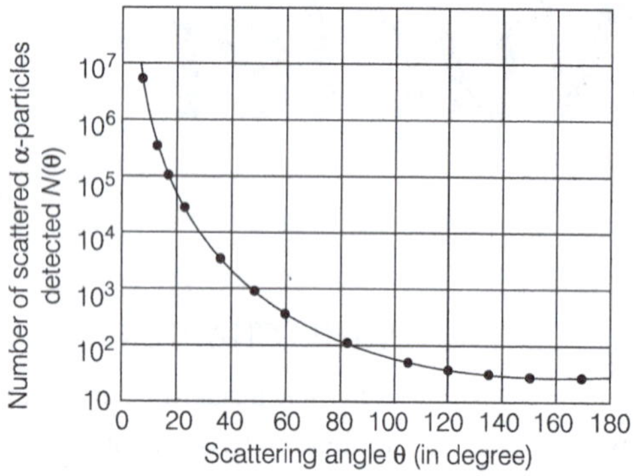

Number of α-particle scattered *versus* scattering angle (θ)

- The minimum distance from the nucleus upto which the α-particle approaches is called **distance of closest approach**. At this distance, whole KE of α-particle converts into electrostatic PE. It is given as

$$r_0 = \frac{2k\,Ze^2}{mv^2} \qquad \left[\text{where, } k = \frac{1}{4\pi\varepsilon_0}\right]$$

- **Impact parameter** (b) is the perpendicular distance of the initial velocity vector of the α-particle from the centre of the nucleus. Mathematically, it is given as

$$b = \frac{1}{4\pi\varepsilon_0}\frac{Ze^2 \cot\frac{\theta}{2}}{\text{KE}}$$

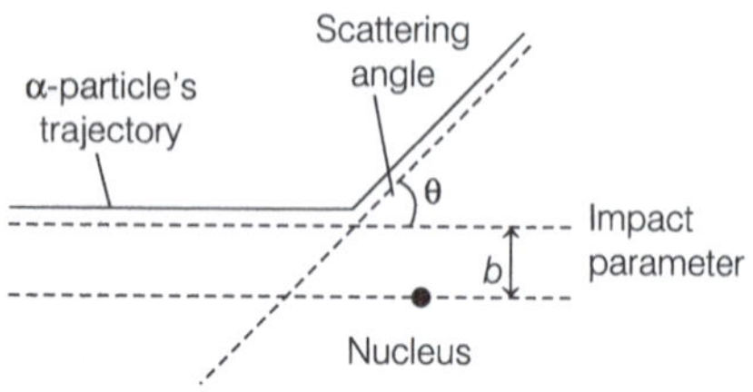

Note Magnitude and direction of the force on the α-particle continuously changes as it approaches nucleus and recedes away from it.

Example 1. *The number of particles scattered at 60° is 100 per minute in α particle scattering experiment, using gold foil. Calculate the number of particles per minute scattered at 90°.*

(a) 100 particle/min (b) 25 particle/min
(c) 65 particle/min (d) 90 particle/min

Sol. (b) Number of α-particles scattered at angle θ,

$$N \propto \frac{1}{\sin^4(\theta/2)}$$

i.e. $$N = \frac{K}{\sin^4(\theta/2)}$$

where, K is a proportionality constant.

$$\therefore \quad \frac{N_{90°}}{N_{60°}} = \frac{\sin^4(60°/2)}{\sin^4(90°/2)}$$

or $$N_{90°} = \frac{\sin^4 30°}{\sin^4 45°} \times N_{60°} = \left[\frac{1/2}{1/\sqrt{2}}\right]^4 \times 100$$

$$= \frac{100}{4} = 25 \text{ particle min}^{-1}$$

Drawbacks of Rutherford's Model

According to classical electromagnetic theory, a accelerated charged particle, must radiate energy in the form of electromagnetic radiation.

The energy of accelerating electron should therefore, continuously decrease. Thus, the electron would spiral inward and eventually fall into nucleus. Hence, such an atom can never be stable.

Rutherford's model was unable to explain line spectrum.

Bohr Model of Hydrogen Atom

Neil Bohr explaning the hydrogen atom spectrum by applying the quantum theory of radiation to Rutherford's atomic model. On the basis of that, he gave his theory in the form of following three postulates

(i) An electron could revolve only in certain circular orbits, called **stationary orbits**, without emitting radiations. Thus there is a definite energy associated with each stable orbit.

(ii) An electron radiates energy only when it makes a transition from one of these orbits to another. The energy is radiated in the form of a photon with energy given by

$$\Delta E = h\nu = E_i - E_f \qquad \text{...(i)}$$

$\therefore$ Frequency of emitted photon is given by

$$\nu = \frac{E_i - E_f}{h}$$

where, E_i and E_f ($E_i > E_f$) are the energies of the initial and final states.

(iii) The electron revolves around the nucleus only in those orbits for which the angular momentum is the integral multiple of $\frac{h}{2\pi}$, where h is Planck's constant ($\simeq 6.6 \times 10^{-34}$ J-s).

$\therefore$ Angular momentum L of the orbiting electron is quantised, i.e.

$$L = mvr = \frac{nh}{2\pi}$$

where, n is an integer (n = 1, 2, 3, ...) and is called **principal quantum number.**

Following are few important points related to Bohr's atomic model

For hydrogen like atoms with atomic number Z,

- **Orbital radius** of an electron in nth orbit is

$$r_n = \frac{\varepsilon_0 n^2 h^2}{\pi m e^2 Z}$$

- **Bohr's radius** is the radius of the first orbit ($n = 1$) of H-atom ($Z = 1$),

$$r_1 = a_0 = \frac{h^2 \varepsilon_0}{\pi m e^2} = 0.53\ \text{Å}$$

Also, $r_n = 0.53 \frac{n^2}{Z}\ \text{Å} = \frac{n^2}{Z} r_1$

$\Rightarrow \quad r_n \propto n^2$

- **Orbital velocity** of an electron in nth orbit,

$$v_n = \frac{Ze^2}{2\varepsilon_0 nh} \text{ or } v_n = \alpha\left(\frac{cZ}{n}\right)$$

where, $\alpha = \frac{e^2}{2h\varepsilon_0 c}$ is the Summerfield's fine structure constant (a pure number) whose value is equal to $\frac{1}{137}$.

$\Rightarrow \quad v_n \propto \frac{1}{n}$

- **Velocity of electron** in first orbit ($n = 1$) of H-atom ($Z = 1$),

$$v_1 = \frac{e^2}{2h\varepsilon_0} = \frac{c}{137} \simeq 2.19 \times 10^6 \text{ m/s}$$

- **Orbital frequency** of an electron in nth orbit,

$$\nu_n = \frac{kZe^2}{nhr_n} \qquad \left[\text{where, } k = \frac{1}{4\pi\varepsilon_0}\right]$$

As, $r_n \propto \frac{n^2}{Z} \Rightarrow \nu_n \propto \frac{Z^2}{n^3}$

$\therefore$ Time period T of an electron is

$$T_n = \frac{1}{\nu_n} = \frac{nhr_n}{k\,Ze^2}$$

$\Rightarrow \quad T_n \propto \frac{n^3}{Z^2}$

Also, $\omega_n = 2\pi\nu_n \Rightarrow \omega_n \propto \frac{Z^2}{n^3}$

Example 2. *Calculate the radius of the third Bohr orbit of the electron of the hydrogen atom.*
(Take, $h = 6.625 \times 10^{-34}$ Js, $e = 1.6 \times 10^{-19}$ C, $m = 9.1 \times 10^{-31}$ kg, $c = 3 \times 10^8$ ms^{-1}, $\varepsilon_0 = 8.85 \times 10^{-12}$ C^{2}N^{-1}m^{-2})

(a) 2.48 Å (b) 3.68 Å (c) 2.68 Å (d) 4.77 Å

Sol. (d) Here, $h = 6.625 \times 10^{-34}$ Js, $e = 1.6 \times 10^{-19}$ C, $m = 9.1 \times 10^{-31}$ kg, $c = 3 \times 10^8$ ms^{-1}, $\varepsilon_0 = 8.85 \times 10^{-12}$ C^{2}N^{-1}m^{-2}, $n = 3$

Radius of the nth Bohr orbit of the hydrogen atom is given by

$$r_n = 4\pi\varepsilon_0 \frac{n^2h^2}{4\pi^2 me^2}$$

$$r_3 = 4\pi \times 8.85 \times 10^{-12} \times \frac{(3)^2 \times (6.625 \times 10^{-34})^2}{4\pi^2 \times 9.1 \times 10^{-31} \times (1.6 \times 10^{-19})^2}$$

$$= 4.777 \times 10^{-10} \text{ m}$$

$$= 4.777 \text{Å}$$

Example 3. *A 10 kg satellite circles earth once every 2 h in an orbit having a radius of 8000 km. Assuming that Bohr's angular momentum postulate applies to satellites just as it does to an electron in the hydrogen atom, then the quantum number of the orbit of the satellite is* **[NCERT]**

(a) 10×10^{40} (b) 5.3×10^{45}
(c) 4×10^{10} (d) 3.2×10^{25}

Sol. (b) From Bohr's postulate, we have $mv_n r_n = \frac{nh}{2\pi}$

Given, $m = 10$ kg, $r_n = 8 \times 10^6$ m,

$$T = 2\text{ h} = 2 \times 60 \times 60 = 7200 \text{ s}$$

$$v_n = \frac{2\pi r_n}{T}$$

The quantum number of the orbit of satellite

$$n = (2\pi r_n)^2 \times \frac{m}{T \times h}$$

Putting the values, we have

$$n = (2\pi \times 8 \times 10^6)^2 \times \frac{10}{(7200 \times 6.64 \times 10^{-34})}$$

$\Rightarrow \quad n = 5.3 \times 10^{45}$

Example 4. *If the average life time of an excited state of hydrogen is of the order of 10^{-8} s, estimate how many orbits an electron makes when it is in the state $n = 2$ and before it suffers a transition to state $n = 1$.*
(Take, Bohr radius, $a_0 = 5.3 \times 10^{-11}$ m)

(a) 6×10^2 (b) 8×10^6
(c) 9×10^8 (d) 6×10^5

Sol. (b) The angular momentum of an election in nth orbit is $nh/2\pi$.

By Bohr's hypothesis, $mvr = nh/2\pi$, where r is the radius of the orbit.

or $$v = \frac{nh}{2\pi mr} \qquad \text{... (i)}$$

The time period of completing an orbit,

$$T = \frac{2\pi r}{v} = \frac{2\pi r(2\pi mr)}{nh}$$

or $$T = \frac{4\pi^2 mr^2}{nh}$$

Since, the radius of the orbit is proportional to n^2, hence $r = a_0 n^2$

$\therefore$ $$T = \frac{4\pi^2 ma_0^2 n^4}{nh} = \frac{4\pi^2 ma_0^2 n^3}{h}$$

Number of orbits completed in 10^{-8} s

$$= \frac{10^{-8}}{T} = \frac{10^{-8} \times h}{4\pi^2 m a_0^2 n^3}$$

$$= \frac{10^{-8} \times (6.6 \times 10^{-34})}{4(3.14)^2 (9.1 \times 10^{-31})(5.3 \times 10^{-11})^2 (2)^3}$$

$$= 8 \times 10^6$$

Energy of an Electron in *n*th Orbit

Kinetic energy (K_n) of an electron in nth orbit,

$$K_n = \frac{1}{2} m v_n^2 = \frac{me^4 Z^2}{8\varepsilon_0{}^2 n^2 h^2} \qquad \left(\because v_n = \frac{Ze^2}{2\varepsilon_0 nh}\right)$$

In terms of Rydberg's, $K_n = \dfrac{RhcZ^2}{n^2}$

where, $R = me^4/8\varepsilon_0^2 ch^3$ = Rydberg's constant

$= 1.09 \times 10^7 \text{ m}^{-1}$

Potential energy U_n of an electron in nth orbit,

$$U_n = -\frac{1}{4\pi\varepsilon_0}\frac{Ze^2}{r_n} = -\frac{mc^4 Z^2}{4\varepsilon_0{}^2 n^2 h^2} \qquad \left(\because r_n = \frac{\varepsilon_0 n^2 h^2}{\pi m e^2 Z}\right)$$

So, $\qquad U_n = \dfrac{-2RhcZ^2}{n^2}$

The **total energy E_n** is the sum of the kinetic and potential energies.

$$\therefore E_n = K_n + U_n = -\frac{me^4 Z^2}{8\varepsilon_0{}^2 n^2 h^2}$$

In simplified form, $E_n = \dfrac{-RhcZ^2}{n^2}$

From the above relations, we get

$$|U_n| = 2|K_n|$$

and $E_n = -K_n$

Substituting, values of m, e, ε_0 and h with $n = 1$, we get the least energy of the atom in first orbit, which is −13.6 eV.

Hence, $\qquad E_1 = -13.6$ eV

and $\qquad E_n = \dfrac{E_1}{n^2} = -\dfrac{13.6}{n^2}$ eV

Substituting, $n = 2, 3, 4 \ldots$, etc., we get energies of electron in different orbits, *i.e.*

$$E_2 = -3.40 \text{ eV}, E_3 = -1.51 \text{ eV}, \ldots E_\infty = 0$$

Note Rhc = 1 Rydberg = 13.6 eV

Energy Levels

The energy of an atom is least, when it is revolving in an orbit closest to the nucleus, *i.e.* for which $n = 1$.

$\therefore \qquad E_1 = -13.6$ eV

This lowest state (minimum energy level) of an atom is called the **ground state**.

However, when a hydrogen atom receives energy by processes such as electron collisions, the atom may acquire sufficient energy to raise the electron to higher states, the atom is then said to be in **excited state**.

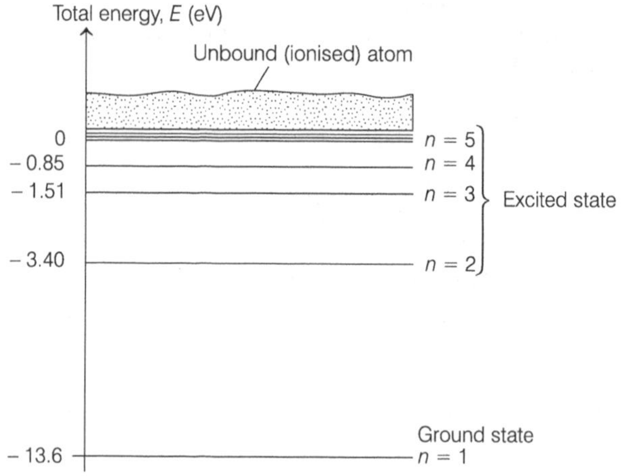

The energy level diagram for the hydrogen atom

Thus, when an atom makes a transition from the higher energy state with quantum number n_i to the lower energy state with quantum number n_f ($n_f < n_i$) in hydrogen like atom, the difference in the energy is carried by a photon of frequency ν_{if} such that,

$$\nu_{if} = \frac{me^4 Z^2}{8\varepsilon_0^2 h^3}\left(\frac{1}{n_f^2} - \frac{1}{n_i^2}\right) \text{ or } \frac{1}{\lambda} = \frac{Z^2 me^4}{8h^2\varepsilon_0^2 c}\left(\frac{1}{n_f^2} - \frac{1}{n_i^2}\right)$$

Note The reciprocal of wavelength is called wave number.

$$\Rightarrow \qquad \bar{\nu} = \frac{1}{\lambda}$$

Ionisation Energy and Potential

It is the energy required to remove the electron completely from its parent atom. In ground state, ($n = 1$) energy of atom is −13.6 eV and energy corresponding to $n = \infty$ is zero. Hence, energy required to remove the electron from ground state is 13.6 eV.

Similarly, the potential difference through which an electron should be accelerated to get ionisation energy is called **ionisation potential**.

$$\therefore \text{Ionisation potential} = \frac{E_{\text{ionisation}}}{e} = \frac{13.6\, Z^2}{n^2} \text{ V}$$

Note The energy required to separate the constituents of a nucleus to infinity is called **binding energy.**

Example 5. *A hydrogen atom, initially in the ground state is excited by absorbing a photon of wavelength 980 Å. The radius of the atom in the excited state in terms of Bohr radius a_0 will be (Take, hc = 12500 eV-Å)* **[JEE Main 2019]**

(a) $4\,a_0$ (b) $9\,a_0$ (c) $16\,a_0$ (d) $25\,a_0$

Sol. (c) We know that, net change in energy of a photon in a transition with wavelength λ is $\Delta E = hc/\lambda$.

Here, $hc = 12500$ eV-Å and $\lambda = 980$ Å

$\therefore \quad \Delta E = 12500/980 = 12.76 \text{ eV} \Rightarrow E_n - E_1 = 12.76$ eV

Since, the energy associated with an electron in nth Bohr's orbit is given as

$$E_n = \frac{-13.6}{n^2} \text{ eV} \quad \text{...(i)}$$

$$\Rightarrow \quad E_n = E_1 + 12.76 \text{ eV} \quad \text{...(ii)}$$

$$= \frac{-13.6}{(1)^2} + 12.76 = -0.84$$

Putting this value in Eq. (i)

$$\Rightarrow \quad n^2 = \frac{-13.6}{-0.84} = 16 \Rightarrow n = 4$$

and radius of nth orbit, $r_n = n^2 a_0 \Rightarrow r_n = 16\, a_0$

Example 6. *A particle of mass m moves in a circular orbit in a central potential field $U(r) = \frac{1}{2}kr^2$. If Bohr's quantisation conditions are applied, radii of possible orbitals and energy levels vary with quantum number n as* **[JEE Main 2019]**

(a) $r_n \propto n, E_n \propto n$ (b) $r_n \propto n^2, E_n \propto \frac{1}{n^2}$

(c) $r_n \propto \sqrt{n}, E_n \propto n$ (d) $r_n \propto \sqrt{n}, E_n \propto \frac{1}{n}$

Sol. (c) As, for conservative fields, $\mathbf{F} = -\left(\frac{d\mathbf{U}}{dr}\right)$

$\therefore$ Magnitude of force on particle is

$$\Rightarrow \quad F = \frac{dU}{dr} = \frac{d}{dr}\left(\frac{1}{2}kr^2\right) \Rightarrow F = kr$$

This force is acting like centripetal force.

$$\therefore \quad \frac{mv^2}{r} = kr \quad \text{...(i)}$$

So, for nth orbit,

$$\Rightarrow \quad m^2v_n^2 = mkr_n^2$$

$$\Rightarrow \quad \frac{n^2h^2}{4\pi^2 r^2} = mkr_n^2 \quad \left[\because v_n = \frac{nh}{2\pi mr}\right]$$

Therefore, $r_n^4 \propto n^2 \Rightarrow r_n^2 \propto n$

So, $r_n \propto \sqrt{n}$

Energy of particle,

$$E_n = \text{PE} + \text{KE} = \frac{1}{2}kr_n^2 + \frac{1}{2}mv_n^2$$

$$= \frac{1}{2}kr_n^2 + \frac{1}{2}kr_n^2 \quad \text{[using Eq. (i)]}$$

$$= kr_n^2$$

So, energy, $E_n \propto r_n^2 \Rightarrow E_n \propto n$

Example 7. *A moving hydrogen atom makes a head on collision with a stationary hydrogen atom. Before collision both atoms are in ground state and after collision they move together. What is the minimum value of the kinetic energy of the moving hydrogen atom, such that one of the atoms reaches one of the excitation state?*

(a) 20.4 eV (b) 6.2 eV (c) 9 eV (d) 0 eV

Sol. (a) Let K be the kinetic energy of the moving hydrogen atom and K' be the kinetic energy of combined mass after collision.

From conservation of linear momentum,

$$p = p'$$

$$\text{or} \quad \sqrt{2Km} = \sqrt{2K'(2m)}$$

$$\text{or} \quad K = 2K' \quad \text{...(i)}$$

From conservation of energy, $\quad K = K' + \Delta E \quad$ ...(ii)

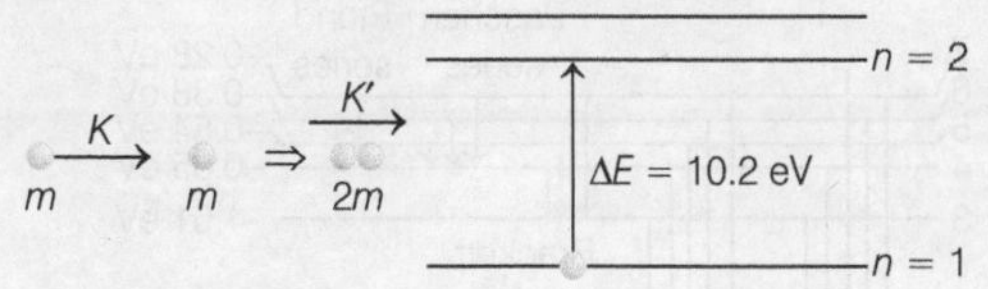

Solving Eqs. (i) and (ii), we get

$$\Delta E = \frac{K}{2}$$

Now, minimum value of ΔE for hydrogen atom is 10.2 eV.

$$\text{or} \quad \Delta E \geq 310.2 \text{ eV}$$

$$\therefore \quad \frac{K}{2} \geq 310.2$$

$$\therefore \quad K \geq 20.4 \text{ eV}$$

Therefore, the minimum kinetic energy of moving hydrogen is 20.4 eV.

Atomic Spectra

When an atomic gas or vapour is excited at low pressure, usually by passing an electric current through it, the emitted radiations has a spectrum containing specific wavelength only. These type of spectrum consists of bright lines on a dark background and is termed as **emission line spectrum**.

When white light passes through a gas and the transmitted light consists of some dark lines in the spectrum. These dark lines correspond precisely to those wavelengths which were found in the emission spectrum of gas. This is called **absorption spectrum** of the gas.

Note Number of lines in emission spectrum or number of transition possible for an electron in nth orbit is $\frac{n(n-1)}{2}$.

Hydrogen Spectrum

In hydrogen spectrum or line spectra of hydrogen atom there is no resemblance of order or regularity in spectral lines but the spacing between the lines within certain sets of spectrum decreases in regular way. Each of these sets is called a **spectral series**.

The wavelength corresponding to various series of spectra for hydrogen atom are as follows

Lyman series, $\quad \frac{1}{\lambda} = R\left(\frac{1}{1^2} - \frac{1}{n^2}\right), n = 2, 3, 4, \ldots$

Balmer series, $\frac{1}{\lambda} = R\left(\frac{1}{2^2} - \frac{1}{n^2}\right)$, $n = 3, 4, 5, \ldots$

Paschen series, $\frac{1}{\lambda} = R\left(\frac{1}{3^2} - \frac{1}{n^2}\right)$, $n = 4, 5, 6, \ldots$

Brackett series, $\frac{1}{\lambda} = R\left(\frac{1}{4^2} - \frac{1}{n^2}\right)$, $n = 5, 6, 7, \ldots$

Pfund series, $\frac{1}{\lambda} = R\left(\frac{1}{5^2} - \frac{1}{n^2}\right)$, $n = 6, 7, 8, \ldots$

Note In Balmer series, the line with largest wavelength 656.3 nm in red is called H_α, the next line with wavelength 486.1 nm in blue is called H_β and so on.

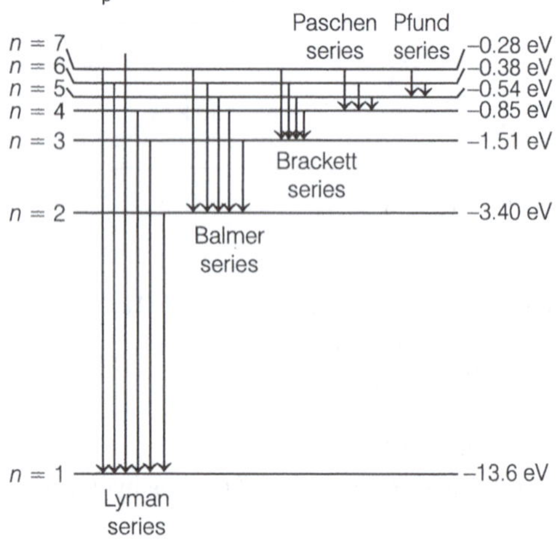

Series Limit

The transition corresponding to maximum energy or smallest wavelength of a given series of hydrogen spectrum is called **series limit**. In general, series limit, $\frac{1}{\lambda} = \frac{R}{n^2}$.

Example 8. *Calculate the frequency of the H_β line of the Balmer series for hydrogen.*

(a) 4.5×10^{11} Hz (b) 5.2×10^{12} Hz

(c) 5.7×10^{13} Hz (d) 6.12×10^{14} Hz

Sol. (d) H_β line of Balmer series corresponds to the transition from $n_2 = 4$ to $n_1 = 2$ level.

$$\frac{1}{\lambda} = R\left[\frac{1}{n_1^2} - \frac{1}{n_2^2}\right]$$

$$\frac{1}{\lambda} = (1.097 \times 10^7)\left(\frac{1}{2^2} - \frac{1}{4^2}\right)$$

$$= 0.2056 \times 10^7$$

$$\therefore \quad \lambda = 4.9 \times 10^{-7} \text{ m}$$

$$\therefore \quad f = \frac{c}{\lambda} = \frac{3.0 \times 10^8}{4.9 \times 10^{-7}}$$

$$= 6.12 \times 10^{14} \text{ Hz}$$

de-Broglie's Explanation of Bohr's Second Postulate of Quantisation

According to de-Broglie, a stationary orbit is that which contains an integral number of de-Broglie standing waves associated with the revolving atom. So, for permissible orbit,

$$2\pi r_n = n\lambda = \frac{nh}{mv_n}$$

or

$$mv_n r_n = \frac{nh}{2\pi}$$

i.e. angular momentum of electron revolving in nth orbit must be an integral multiple of $\frac{h}{2\pi}$, which is quantum condition proposed by Bohr in second postulate.

Example 9. *Orbits of a particle moving in a circle are such that the perimeter of the orbit equals an integer number of de-Broglie wavelengths of the particle. For a charged particle moving in a plane perpendicular to a magnetic field, the radius of the nth orbital will be proportional to*

(a) n^2 (b) n

(c) $n^{1/2}$ (d) $n^{1/4}$

Sol. (c) As, $2\pi r = n\lambda$

$$r = \frac{n\lambda}{2\pi}$$

Now, de-Broglie equation $\lambda = \frac{h}{p}$

$$\Rightarrow \quad mv_n = \frac{h}{\lambda} = \frac{h}{\frac{2\pi r_n}{n}} = \frac{nh}{2\pi r_n}$$

Also, for charged particle moving in a magnetic field

$$r_n = \frac{mv_n}{qB} = \frac{nh}{(2\pi r_n)\, qB}$$

$$\Rightarrow \quad r_n^2 = \frac{nh}{2\pi qB}$$

$$\Rightarrow \quad r_n^2 \propto n \Rightarrow r_n \propto n^{1/2}$$

Composition and Size of Nucleus

The entire positive charge and nearly the entire mass of atom is concentrated in a very small space called the **nucleus** of an atom. It consists of **protons** and **neutrons** and they are together called **nucleons**.

In a nucleus, the number of protons is equal to the atomic number (Z) of that element and total number of nucleons is called the mass number (A), *i.e.*

Number of protons = Atomic number Z

and number of neutrons

= Mass number (A) − Atomic number (Z) = $A - Z$

Thus, the nucleus of an atom represented as $_Z X^A$.

Properties of Nucleus

The nuclear properties are described below

- **Nuclear size**
  (i) Size of the nucleus is of the order of fermi (1 fermi = 10^{15} m)
  (ii) The radius of the nucleus is given by $R = R_0 A^{1/3}$, where $R_0 = 1.3$ fermi and A is the mass number.
  (iii) The size of the atom is of the order of 10^{-19} m.
- **Volume** The volume of nucleus is

$$V = \frac{4}{3}\pi R^3 = \frac{4}{3}\pi (R_0 A^{1/3})^3$$

- **Density**
  (i) Density $= \dfrac{\text{Mass of nucleus}}{\text{Volume of the nucleus}}$

$$= \frac{m_p A}{\frac{4}{3}\pi R^3} = \frac{m_p A}{\frac{4}{3}\pi \cdot (R_0 A^{1/3})^3} = \frac{m_p}{\frac{4}{3}\pi R_0^3}$$

  where, $m_p = 1.6 \times 10^{-27}$ kg = mass of proton and $R_0 = 1.3$ fermi.
  (ii) Density of nuclear matter is of the order of 10^{17} kg/m^3.
  (iii) Density of nuclear matter is independent of the mass number.

Isotopes, Isobars and Isotones

The atoms of same element which have same atomic number but different mass number are called **isotopes**.

For example,

$({}_1H^1, {}_1H^2 \text{ and } {}_1H^3)$, $({}_8O^{16}, {}_8O^{17} \text{ and } {}_8O^{18})$

and $({}_{10}Ne^{20}, {}_{10}Ne^{21} \text{ and } {}_{10}Ne^{22})$, etc., are isotopes.

The atoms of different elements which have same mass number but different atomic number are called **isobars**.

For example,

$({}_8O^{17} \text{ and } {}_9F^{17})$ and $({}_{11}Na^{24} \text{ and } {}_{12}Mg^{24})$ are isobars.

The atoms of different elements of which the nuclei have the same number of neutrons but different number of protons are called the **isotones**.

For example,

$({}_{11}Na^{23} \text{ and } {}_{12}Mg^{24})$ and $({}_{19}K^{39} \text{ and } {}_{20}Ar^{40})$ are isotones.

Equivalence of Mass and Energy

Einstein suggested that energy mass are equivalent. He predicted that if the energy of a body changes by an amount E, its mass changes by an amount m given by the equation, $E = mc^2$

where, c is the speed of light.

This relation is called **mass-energy equivalence** relation.

Example 10. *The energy equivalent of 1g of substance is*
(a) 3×10^{30} J (b) 6×10^{13} J (c) 3×10^{13} J (d) 9×10^{13} J

Sol. *(d) From Einstein's mass energy relation, we have*

$$E = mc^2$$

Given, $m = 1\text{g} = 10^{-3}\text{kg}, c = 3 \times 10^8 \text{m/s}$

$$\therefore \quad E = mc^2 = 10^{-3} \times 9 \times 10^{16} = 9 \times 10^{13} \text{J}$$

Mass Defect

It is found that mass of the nucleus is less than the total mass of protons and neutrons, this mass difference is called **mass defect.**

Since, number of protons inside the nucleus of ${}_ZX^A$ atom is Z and number of neutrons is $(A - Z)$ hence, if ${}_ZM^A$ is the mass of nucleus, then mass defect

$$\Delta M = [Zm_p + (A - Z)m_n] - {}_ZM^A$$

where, m_p is the mass of proton and m_n is the mass of neutron.

Note Energy equivalent to mass defect of 1 amu is equal to 931.5 MeV.

Nuclear Binding Energy

Binding energy (E_b) of a nucleus is defined as the minimum energy required to separate its nucleons and place them at rest at infinite distance apart. In other words, it may be defined as the energy equivalence to the mass defect of the nucleus.

$\therefore$ $E_b = \Delta mc^2$, where Δm is mass defect.

or $E_b = \Delta m \times 931$ MeV

It can also be written as, $E_b = [Zm_p + (A - Z)m_n - m_x]\, c^2$

where, m_p is mass of proton, m_n is mass of neutron and m_x is mass of nucleus.

Binding Energy per Nucleon It is the ratio of binding energy E_b of a nucleus to the number of nucleons A in that nucleus,

$$E_{bn} = \frac{E_b}{A}$$

Example 11. *If the binding energy per nucleon of deuteron is 1.115 MeV, then its mass defect (in amu) is*
(a) 0.056 (b) 0.0072 (c) 0.0024 (d) 0.0017

Sol. (c) Binding energy per nucleon,

$$E_{bn} = \frac{E_b}{A} \quad \text{...(i)}$$

For deuteron, atomic mass $A = 2$

$\therefore$ From Eq. (i), $1.115 = \dfrac{E_b}{2}$

$\Rightarrow$ $E_b = 2.23$ MeV

$\therefore$ Mass defect (in amu) $= \dfrac{E_b}{931.5} = \dfrac{2.23}{931.5} = 0.0024$

Binding Energy Curve

A plot of binding energy per nucleon E_b/A as a function of mass number A for various stable nuclei is shown in figure.

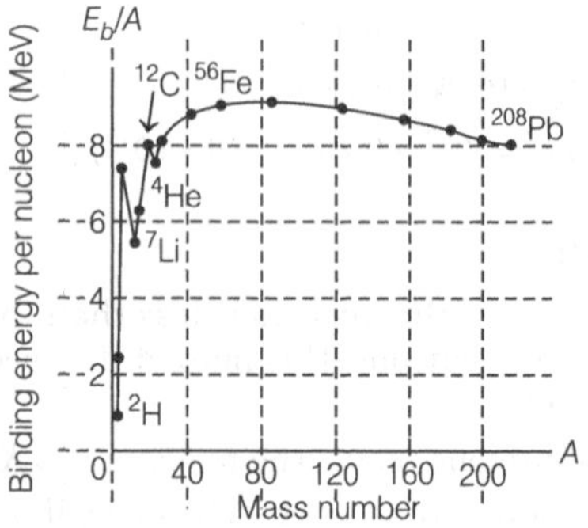

The binding energy per nucleon, E_b / A, as a function of the mass number A

Following conclusions can be drawn from the above graph

- The greater the binding energy per nucleon, the more stable is the nucleus. The curve reaches a maximum of about 8.75 MeV in the vicinity of $^{56}_{26}Fe$ and then gradually falls to 7.6 MeV for $^{238}_{92}U$.
- Force is attractive and sufficiently strong to produce E_{bn} of a few MeV per nucleon.
- The constancy of E_{bn} in the range $30 < A < 170$ is the consequence of the fact that the nuclear force is short-ranged.
- A very heavy nucleus (say $A = 240$) can be split into two lighter nuclei (say $A = 120$) near the flat maximum of the curve, such that E_{bn} will increase. Hence, energy will be released in this process of fission.
- Two very light nuclei ($A \le 10$) joined together to form a heavier nucleus. E_{bn} of the fused nuclei is more than that of lighter nuclei. Again, energy would be released in such as process of fusion.

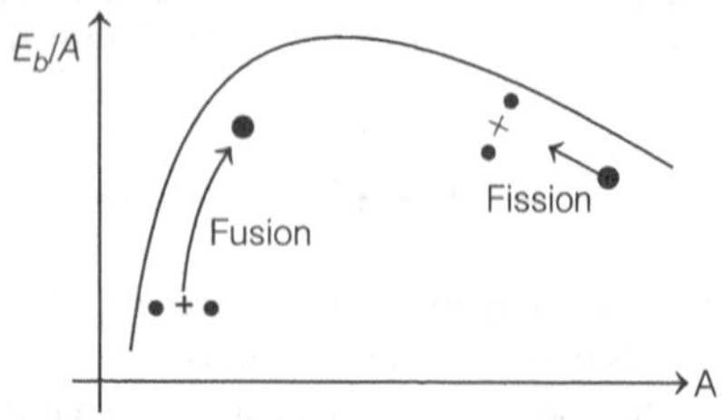

Nuclear Force

The forces that holds the nucleons together inside the nucleus of an atom are called **nuclear forces.**

Following are few important properties of the nuclear forces

- It is much stronger than the Coulomb force acting between the charges or the gravitational force between masses.
- Nuclear force between two nucleons falls rapidly to zero as their distance is more than few femtometres (fm). This leads to saturation of forces or a large sized nucleus, which is the reason for constancy of binding energy per nucleon E_{bn}.
- Nuclear force between neutron-neutron, proton-neutron and proton-proton is approximately the same. It does not depend on the electric charge.

Nuclear Stability

The stability of nucleus is also determined by its neutron to proton ratio. A plot of number of neutrons and number of protons is shown in the figure below

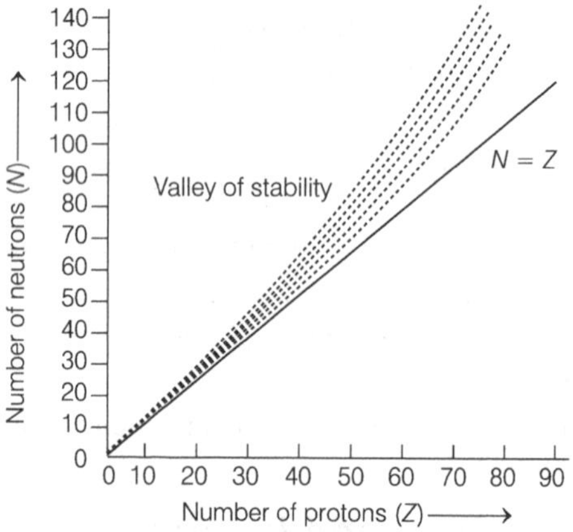

Graphical representation of number of neutrons *versus* number of protons

In the figure, the solid line shows the nuclei with equal number of protons and neutrons ($N = Z$). Only light nuclei are on this line, *i.e.* they are stable if they contain approximately same number of protons and neutrons.

Heavy nuclei are stable only when they have more neutrons than protons.

Radioactivity

It is the phenomenon in which an unstable nucleus undergo a decay with emission of some particles (α, β) and electromagnetic radiation (γ-rays). It is also referred as **radioactive decay**.

Three types of decays are as follows

α-Decay An α-particle is a helium nucleus. Thus, a nucleus emitting an α-particle loses two protons and two neutrons. Therefore, the atomic number Z decreases by 2, the mass number A decreases by 4 and the neutron number N decreases by 2. The decay can be written as

$$^{A}_{Z}X = {}^{A-4}_{Z-2}Y + {}^{4}_{2}He$$

where, X is the parent nucleus and Y is the daughter nucleus.

β-Decay This decay can involve the emission of either electrons or positrons.

In β^- **decay**, a neutron in the nucleus is transformed into a proton, an electron and an antineutrino.

$$n \longrightarrow P + e^- + \bar{\nu}$$

Thus, a parent nucleus with atomic number Z and mass number A decays by β^- emission into a daughter with atomic number $Z + 1$ and the same mass number A.

$${}^A_Z X \xrightarrow{\beta^-} {}_{Z+1}^{\;\;A}Y$$

β^- decay occurs in nuclei that have too many neutrons. An example of β^- decay is the decay of carbon 14 into nitrogen,

$${}^{14}_{6}\text{C} \longrightarrow {}^{14}_{7}\text{N} + e^- + \bar{\nu}$$

In β^+ **decay**, a proton changes into a neutron with the emission of a positron (and a neutrino)

$$p \longrightarrow n + e^+ + \nu$$

Positron (e^+) emission from a nucleus decreases the atomic number Z by 1, while keeping the same mass number A.

$${}^A_Z X \xrightarrow{\beta^+} {}_{Z-1}^{\;\;A}Y$$

β^+ decay occurs in nuclei that have two few neutrons.

γ-Decay In gamma decay, a nucleus changes from a higher energy state to a lower energy state through the emission electromagnetic radiations (photons).

So, no change occurs in atomic number and atomic mass of the parent and daughter atoms in gamma decay.

Example 12. *In a radioactive decay chain, the initial nucleus is ${}^{232}_{90}Th$. At the end, there are 6 α-particles and 4 β-particles which are emitted. If the end nucleus is ${}^A_Z X$, A and Z are given by* ***[JEE Main 2019]***

(a) $A = 202, Z = 80$ (b) $A = 208, Z = 82$
(c) $A = 200, Z = 81$ (d) $A = 208, Z = 80$

Sol. (b) The decay of α-particle reduces, mass number by 4 and atomic number by 2.

∴ Decay of 6 α-particles results, ${}^{232}_{90}\text{Th} \xrightarrow{6\alpha} {}^{232-24}_{90-12}Y = {}^{208}_{78}Y$

A β-decay does not produces any change in mass number but it increases atomic number by 1.

∴ Decay of 4 β-particles results, ${}^{208}_{78}Y \xrightarrow{4\beta} {}^{208}_{82}X$

∴ In the end nucleus $A = 208$, $Z = 82$

Pair Production

When an energetic γ-ray photon falls on a heavy substance, it is absorbed by some nucleus of the substance and an electron and a positron are produced.

This phenomenon is called **pair production** and may be represented by the equation

$$h\nu \longrightarrow {}_{+1}\beta^0 + {}_{-1}\beta^0$$

Pair Annihilation

The converse phenomenon of pair production is called **pair annihilation**. When an electron and a positron come very close to each other, they annihilate each other by combining together and two γ-photons are produced. This phenomenon can be represented by the following equation

$${}_{+1}\beta^0 + {}_{-1}\beta^0 \longrightarrow h\nu + h\nu$$

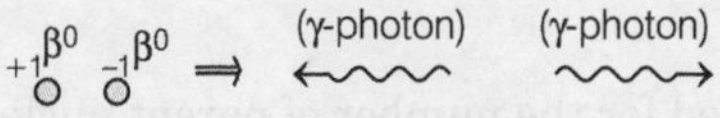

The Radioactive Decay Law

According to this law, the rate of decay of radioactive nuclei at any instant is proportional to the number of nuclei present at that instant. *i.e.* $N = N_0 e^{-\lambda t}$

where, λ is called the **decay constant** and N is the number of active nucleus present in the sample at any instant t and N_0 is the number of nuclear present in the sample initially.

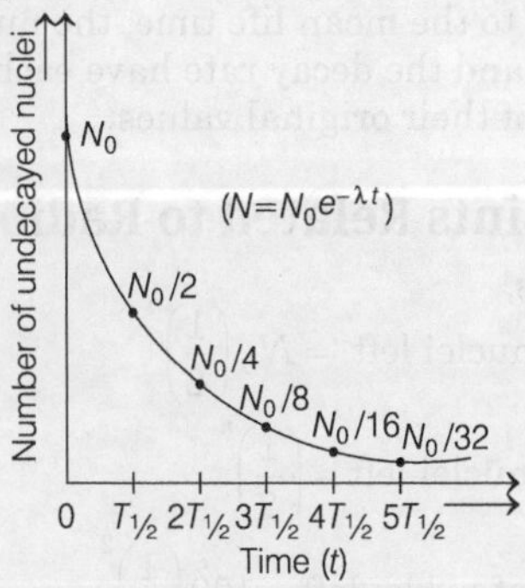

Exponential decay of a radioactive substance

Activity of a Radioactive Substance

The decay rate R of a radioactive substance is the number of decays per second.

It is given as

$$R = -\frac{dN}{dt} \text{ or } R = R_0 e^{-\lambda t}$$

where, $R_0 = \lambda N_0$ is the activity of the radioactive substance at time $t = 0$.

The activity *versus* time graph is shown below.

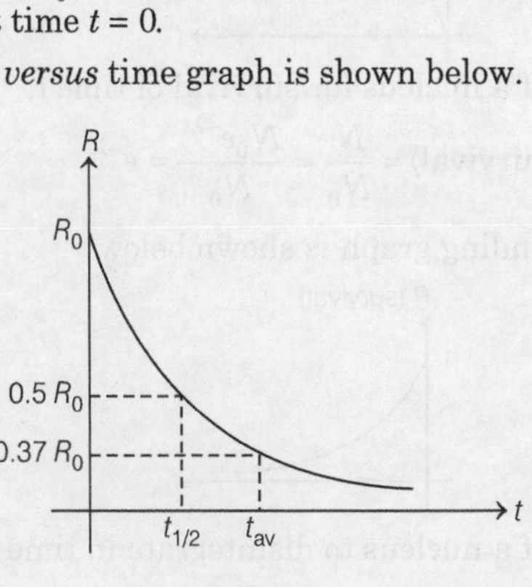

Thus, the number of nuclei and hence the activity of the radioactive substance decrease exponentially with time.

Units of activity The SI unit for the decay rate is the becquerel (Bq), but the curie (Ci) and rutherford (Rd) are often used in practice.

$$1 \text{ Bq} = 1 \text{ decay/s},$$
$$1 \text{ Ci} = 3.7 \times 10^{10} \text{ Bq}$$
and $$1 \text{ Rd} = 10^6 \text{ Bq}$$

Half-life

The time required for the number of parent nuclei to fall 50% of its initial value is called **half-life**. It is denoted by $t_{1/2}$. It is given as

$$t_{1/2} = \frac{\ln(2)}{\lambda} = \frac{0.693}{\lambda}$$

Mean life

The average or mean life t_{av} is the reciprocal of the decay constant, *i.e.* $t_{av} = \frac{1}{\lambda}$

After a time equal to the mean life time, the number of radioactive nuclei and the decay rate have each decreased to 37% of their original values.

Important Points Related to Radioactivity

- After n half lives,
  (a) Number of nuclei left $= N_0\left(\frac{1}{2}\right)^n$
  (b) Fraction of nuclei left $= \left(\frac{1}{2}\right)^n$
  (c) Percentage of nuclei left $= 100\left(\frac{1}{2}\right)^2$
- Number of nuclei decayed after time t,
  $$= N_0 - N = N_0 - N_0e^{-\lambda t} = N_0(1 - e^{-\lambda t})$$
  The corresponding graph is shown below

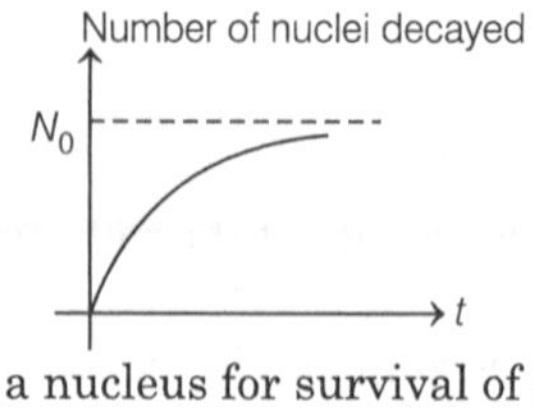

- Probability of a nucleus for survival of time t,
  $$P\text{ (survival)} = \frac{N}{N_0} = \frac{N_0e^{-\lambda t}}{N_0} = e^{-\lambda t}$$
  The corresponding graph is shown below

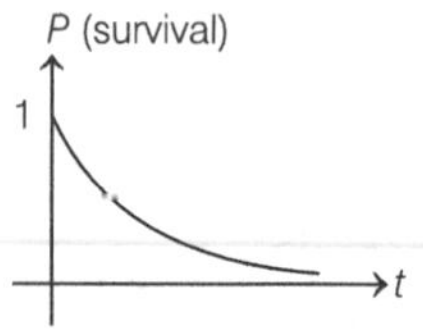

- Probability of a nucleus to disintegrate in time t is
  $$P\text{ (disintegration)} = 1 - P\text{ (survival)} = 1 - e^{-\lambda t}$$
  The corresponding graph is shown below

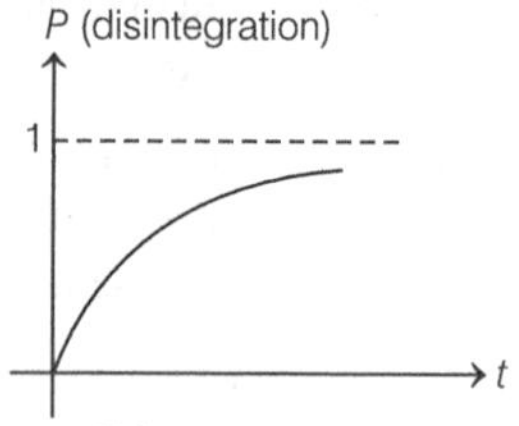

- Half-life and mean life are related to each other by the relation,
  $$t_{1/2} = 0.693\, t_{av}$$
  or $$t_{av} = 1.44\, t_{1/2}$$

Example 13. *The half-life of $^{238}_{92}U$ undergoing α-decay is 4.5×10^9 yr. The activity of 1g sample of $^{238}_{92}U$ is [Avogadro number = 6.022×10^{23}]*

(a) 12.3×10^{10} Bq (b) 1.23×10^4 Bq
(c) 14.2×10^3 Bq (d) 1.42×10^{17} Bq

Sol. (b) Given, $t_{1/2} = 4.5 \times 10^9$ yr
$$= 4.5 \times 10^9 \times 3.16 \times 10^7 \text{ s}$$
$$t_{1/2} = 1.42 \times 10^{17} \text{ s}$$

The number of atoms in 1 g of uranium,
$$N = \frac{6.023 \times 10^{23}}{238}$$

The decay rate (activity) R is
$$R = \lambda N$$
$$= \frac{0.693}{t_{1/2}} N$$
$$= \frac{0.693 \times 6.023 \times 10^{23}}{1.42 \times 10^{17} \times 238}$$
$$= 1.23 \times 10^4 \text{ Bq}$$

Example 14. *In given time t = 0, activity of two radioactive substances A and B are equal. After time t, the ratio of their activities $\frac{R_B}{R_A}$ decreases according to e^{-3t}. If the half-life of A is ln 2, the half-life of B will be* **[JEE Main 2019]**

(a) 4ln2 (b) $\frac{\ln 2}{4}$ (c) $\frac{\ln 2}{2}$ (d) 2ln2

Sol. (b) Activity of radioactive material is given as
$$R = \lambda N$$
where, λ is the decay constant and N is the number of nuclei in the radioactive material.

For substance A,
$$R_A = \lambda_A N_A = \lambda_A N_{0A} \quad (\text{initially, } N_A = N_{0A})$$

For substance B,
$$R_B = \lambda_B N_B = \lambda_B N_{0B} \quad (\text{initially, } N_B = N_{0B})$$

At $t = 0$, activity is equal.

Therefore, $$\lambda_A N_{0A} = \lambda_B N_{0B} \quad \text{...(i)}$$

The half-life is given by $t_{1/2} = \frac{0.693}{\lambda} = \frac{\ln 2}{\lambda}$

So, for substance A,

$$(t_{1/2})_A = \frac{\ln 2}{\lambda_A} \Rightarrow \ln 2 = \frac{\ln 2}{\lambda_A}$$

$$\lambda_A = 1 \quad \text{...(ii)}$$

According to the given question,

at time t, $$\frac{R_B}{R_A} = e^{-3t} \quad \text{...(iii)}$$

Using Eqs. (i), (ii) and (iii), we get

$$\frac{R_B}{R_A} = e^{-3t} = \frac{\lambda_B N_{0B} e^{-\lambda_B t}}{\lambda_A N_{0A} e^{-\lambda_A t}}$$

$$\Rightarrow \quad e^{-3t} = e^{(\lambda_A - \lambda_B)t} \Rightarrow -3 = \lambda_A - \lambda_B$$

$$\Rightarrow \quad \lambda_B = \lambda_A + 3 \Rightarrow \lambda_B = 1 + 3 = 4 \quad \text{...(iv)}$$

The half-life of substance B is $(t_{1/2})_B = \frac{\ln 2}{\lambda_B} = \frac{\ln 2}{4}$

Example 15. *Half lives of two radioactive nuclei A and B are 10 min and 20 min, respectively. If initially a sample has equal number of nuclei, then after 60 min, the ratio of decayed numbers of nuclei A and B will be* ***[JEE Main 2019]***

(a) 3 : 8 (b) 1 : 8 (c) 8 : 1 (d) 9 : 8

Sol. (d) For substance A, half-life is 10 min, so it decays as

$$N_{0_A} \xrightarrow{10\text{min}} \frac{N_{0_A}}{2} \xrightarrow{10\text{min}} \frac{N_{0_A}}{4} \xrightarrow{10\text{min}} \frac{N_{0_A}}{8}$$

(Initial number of nuclei at $t = 0$) (Active nuclei remained after 10 min)

$$\xrightarrow{10\text{min}} \frac{N_{0_A}}{16} \xrightarrow{10\text{min}} \frac{N_{0_A}}{32} \xrightarrow{10\text{min}} \frac{N_{0_A}}{64}$$

$\therefore$ For substance A, number of nuclei decayed in 60 min is

$$N_{0_A} - \frac{N_{0_A}}{64} = \frac{63 N_{0_A}}{64}$$

Similarly, for substance B, half-life is 20 min, so it's decay scheme is

$$N_{0_B} \xrightarrow{20\text{min}} \frac{N_{0_B}}{2} \xrightarrow{20\text{min}} \frac{N_{0_B}}{4} \xrightarrow{20\text{min}} \frac{N_{0_B}}{8}$$

So, number of nuclei of B decayed in 60 min is

$$N_{0_B} - \frac{N_{0_B}}{8} = \frac{7}{8} N_{0_B}$$

Hence, ratio of decayed nuclei of A and B in 60 min is

$$\frac{\frac{63}{64} N_{0_A}}{\frac{7}{8} N_{0_B}} = \frac{9}{8} \quad [\because N_{0_A} = N_{0_B}]$$

Example 16. *Two radioactive materials A and B have decay constants 10 λ and λ, respectively. If initially they have the same number of nuclei, then the ratio of the number of nuclei of A to that of B will be 1/e after a time* ***[JEE Main 2019]***

(a) $\frac{1}{11\lambda}$ (b) $\frac{11}{10\lambda}$ (c) $\frac{1}{9\lambda}$ (d) $\frac{1}{10\lambda}$

Sol. (c) Given, $\lambda_A = 10\lambda$ and $\lambda_B = \lambda$

Number of nuclei (at any instant) present in material is

$$N = N_0 e^{-\lambda t}$$

So, for materials A and B, we can write

$$\frac{N_A}{N_B} = \frac{e^{-\lambda_A t}}{e^{-\lambda_B t}} = e^{-(\lambda_A - \lambda_B)t} \quad \text{... (i)}$$

Given, $$\frac{N_A}{N_B} = \frac{1}{e} \quad \text{... (ii)}$$

Equating Eqs. (i) and (ii), we get

$$\frac{1}{e} = e^{-(\lambda_A - \lambda_B)t} \Rightarrow e^{-1} = e^{-(\lambda_A - \lambda_B)t}$$

Comparing the power of e on both sides, we get

or $$(\lambda_A - \lambda_B)\, t = 1 \Rightarrow t = \frac{1}{\lambda_A - \lambda_B}$$

By putting values of λ_A and λ_B in the above equation, we get

$$t = \frac{1}{10\lambda - \lambda} \Rightarrow t = \frac{1}{9\lambda}$$

Nuclear Reaction

The process by which the identity of a nucleus is changed when it is bombarded by an energetic particle is called nuclear reaction.

The general expression for the nuclear reaction is as follows

$$\underset{\text{(Parent nucleus)}}{X} + \underset{\text{(Incident particle)}}{a} \longrightarrow \underset{\text{(Compound nucleus)}}{C}$$

$$\longrightarrow \underset{\text{(Product nucleus)}}{Y} + \underset{\text{(Daughter nucleus)}}{b} + \underset{\text{energy}}{Q}$$

Here, X and a are the reactants, Y and b are the products and Q is the kinetic energy released in the nuclear reaction.

Q-value

Q-value means the difference between the rest mass energy of initial constituents and the rest mass energy of final constituents of a nuclear reaction.

Example 17. *Given the masses of various atomic particles $m_p = 1.0072\,u, m_n = 1.0087\,u, m_e = 0.000548\,u, m_{\bar{\nu}} = 0$, $m_d = 2.0141u$, where p = proton, n = neutron, e = electron, $\bar{\nu}$ = antineutrino and d = deuteron. Which of the following process is allowed by momentum and energy conservation?* ***[JEE Main 2020]***

(a) $n + n \rightarrow$ deuterium atom (electron bound to the nucleus)

(b) $p \rightarrow n + e^+ + \bar{\nu}$

(c) $n + p \rightarrow d + \bar{\nu}$

(d) $e^+ + e^- \rightarrow \gamma$

Sol. (c) In nuclear reaction, momentum and energy conservation laws are valid only, if total mass on reactant side is greater than that of the product side.

i.e. $\Sigma m_{\text{reactant}} > \Sigma m_{\text{product}}$

because liberated energy appears due to mass defect.

If we check the above condition from given data, then it is found that among all options only option (c) satisfy the condition.

Mass of n + Mass of p = (1.0087 + 1.0072) u
= 2.0159 u

Mass of d + Mass of $\overline{\nu}$ = (2.0141 + 0) u
= 2.0141 u

Hence, only statement in option (c) is true.

Example 18. *Consider the nuclear fission*

$$Ne^{20} \longrightarrow 2He^4 + C^{12}$$

Given that the binding energy/nucleon of Ne^{20}, He^4 and C^{12} are respectively, 8.03 MeV, 7.07 MeV and 7.86 MeV, identify the correct statement.

(a) *Energy of 9.72 MeV will be released.*
(b) *Energy of 12.4 MeV will be supplied.*
(c) *8.3 MeV energy will be released.*
(d) *Energy of 11.9 MeV has to be supplied.*

Sol. (a) Energy absorbed or released in a nuclear reaction is given by

ΔQ = Binding energy of products − Binding energy of reactants

If energy is absorbed, ΔQ is negative and if it is positive, then energy is released.

Also, binding energy = binding energy per nucleon × number of nucleons.

Here, binding energy of products

$$= 2 \times (\text{BE of } He^4) + (\text{BE of } C^{12})$$
$$= 2\ (4 \times 7.07) + (12 \times 7.86) = 150.88 \text{ MeV}$$

and binding energy of reactants

$$= 20 \times 8.03 = 160.6 \text{ MeV}$$

So, $\Delta Q = (BE)_{products} - (BE)_{reactants}$
$= 150.88 - 160.6$
$= -9.72$ MeV

As, ΔQ is negative.

∴ Energy of 9.72 MeV is released in the reaction.

Nuclear Energy

It is the energy released during the transformation of nuclei with less total binding energy to nuclei with greater binding energy.

Two distinct ways of obtaining energy from nucleus are as follows

(i) Nuclear fission (ii) Nuclear fusion

Nuclear Fission

It is the process of splitting of a heavy nucleus (${}^{235}_{92}U$ or ${}^{239}_{94}U$) into two lighter nuclei of comparable masses alongwith the release of a large amount of energy after bombarded by slow neutrons.

When a uranium isotope ${}^{235}_{92}U$ bombarded with a neutron breakes into two intermediate mass nuclear fragments.

$${}^{1}_{0}n + {}^{235}_{92}U \longrightarrow {}^{236}_{92}U \longrightarrow {}^{144}_{51}Ba + {}^{89}_{36}Kr + 3{}^{1}_{0}n + Q \text{ (energy released)}$$

Chain Reaction In nuclear fission many neutrons are produced along with the release of large energy. Under favourable conditions these neutrons can cause further fission of other nuclei, producing large number of neutrons.

Thus, a chain of nuclear fissions is established which continues until the whole of the uranium is consumed.

Nuclear Reactor

It is a device in which nuclear fission can be carried out through a sustained and controlled chain reaction.

It is also called an **atomic pile**. It is thus a source of controlled energy which is utilised for many useful purposes.

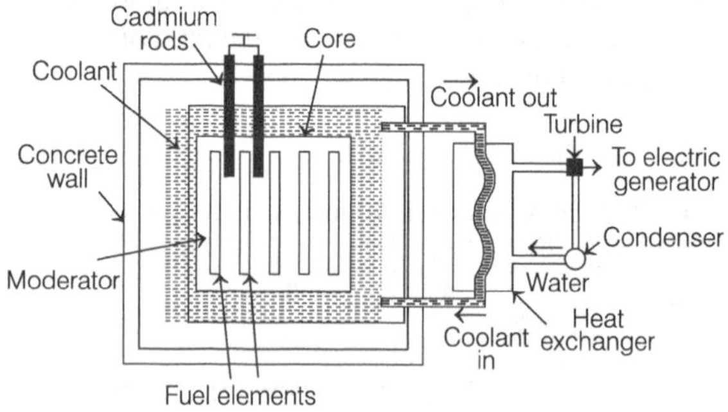

Following are the important terms related to reactor

(i) **Core** is the site of nuclear fission. It is usually consists of ${}^{235}_{92}U$.
(ii) Core contains a **moderator** which is used to slow down fast moving neutrons, *e.g.* water, heavy water (D_2O) and graphite.
(iii) **Reflector** is used to reduce leakage.
(iv) The reactor can be shut down by means of **control rods** (made of cadmium) that have high absorption rate of neutrons.
(v) The **coolant** transfers heat to a working fluid which in turn may produce steam. The steam drives turbines and generates electricity.

Note Multiplication factor k is the ratio of the number of fission produced by a given generation of neutrons to the number of fission of the proceeding generation.

Example 19. *What is the power output of a ${}_{92}U^{235}$ reactor if it takes 30 days to use up 2 kg of fuel, and if each fission gives 185 MeV of usable energy?*

(a) 2.6×10^3 (b) 9.26×10^4
(c) 5.85×10^7 (d) 6.2×10^6

Sol. (c) 235 amu of uranium gives 185 MeV energy. Therefore, the energy given by 1 amu of ${}_{92}U^{235}$

$$= \frac{185}{2.35} \text{ MeV}$$

$$= \frac{185}{235} \times 1.6 \times 10^{-13} \text{J}$$

But 1 amu $= 1.66 \times 10^{-27}$ kg. Therefore, energy released by

$$1.66 \times 10^{-27} \text{ kg of } {}_{92}U^{235} = \frac{185 \times 1.6 \times 10^{-13}}{235}$$

Hence, energy released by 2 kg of ${}_{92}U^{235}$

$$W = \frac{185 \times 1.6 \times 10^{-13} \times 2}{235 \times 1.66 \times 10^{-27}} = 1.517 \times 10^{14} \text{J}$$

Therefore, power output of reactor

$$= \frac{W}{t} = \frac{1.517 \times 10^{14}}{30 \text{ days}}$$

$$P = \frac{1.517 \times 10^{14}}{30 \times 24 \times 60 \times 60} = 5.85 \times 10^{7} \text{ W}$$

Nuclear Fusion

When two lighter nuclei combine to form a heavier nucleus, the process is called **nuclear fusion**. The union of lighter nuclei into heavier nuclei also lead to a transfer of mass and a consequent liberation of energy.

Such a reaction has been achieved in **hydrogen bomb** and it is believed to be the principal source of the sun's energy.

A reaction with heavy hydrogen or deuterium which yields 3.3 MeV per fusion is

$${}_1^2H + {}_1^2H \rightarrow {}_2^3He + {}_0^1n$$

X-rays

When cathode rays strike on a heavy metal of high melting point, then a very small fraction of its energy converts into a new type of waves called X-rays. X-rays were discovered by **Roentgen**.

Properties of X-rays

(i) X-rays are electromagnetic waves of wavelengths ranging from 0.1 Å to 100 Å and frequencies ranging from 10^{16} Hz to 10^{18} Hz.

(ii) Soft X-rays have greater wavelength and lower frequency.

(iii) Hard X-rays have lower wavelength and higher frequency.

(iv) X-rays are produced in coolidge tube.

(v) Molybdenum and tungsten provide suitable targets. These elements have large atomic number and high melting point for the purpose.

(vi) The intensity of X-rays depends on the heating voltage or filament current.

(vii) The kinetic energy of X-ray photons depends upon the voltage applied across the ends of coolidge tube.

(viii) If total energy of fast moving electron transfer to X-ray photon, then its energy $eV = h\nu = \frac{hc}{\lambda}$.

(ix) Wavelength of emitted X-rays is given by $\lambda = \frac{hc}{eV}$, where h = Planck's constant, c = speed of light, e = electronic charge and V = potential difference applied across the ends of the tube.

(x) Absorption of X-rays $I = I_0 e^{-\mu x}$, where I_0 = initial intensity of X-rays, I = final intensity of emergent X-rays, x = thickness of material and μ = absorption coefficient.

Continuous X-rays

The continuous X-rays (or bremsstrahlung X-rays) produced at a given accelerating potential V vary in wavelength, but none has a wavelength shorter than a certain value λ_{min}. This minimum wavelength corresponds to the maximum energy of the X-rays which in turn is equal to the maximum kinetic energy qV or eV of the striking electrons. Thus,

$$\frac{hc}{\lambda_{min}} = eV$$

or

$$\lambda_{min} = \frac{hc}{eV}$$

After substituting values of h, c and e, we obtain the following simple formula for λ_{min}.

$$\lambda_{min} \text{ (in Å)} = \frac{12375}{V} \quad \text{...(i)}$$

If V is increased, then λ_{min} decreases. This wavelength is also known as the **cut-off wavelength** or the **threshold wavelength**.

X-rays Spectrum

The energy spectrum of X-rays is a line spectrum, containing following series

(i) ***K*-series** When electrons of any higher orbit (n = 2, 3, 4, ...) jump to first orbit (n = 1), then K-series of X-rays are produced.

(ii) ***L*-series** When electrons of higher orbit (n = 3, 4, 5, ...) jump to second orbit (n = 2), then L-series of X-rays are produced.

(iii) ***M*-series** When electrons of higher orbit (n = 4, 5, 6, ...) jump to third orbit (n = 3), then M-series of X-rays are produced.

First lines of these series are called K_α, L_α, M_α and second lines of these series are called K_β, L_β, M_β.

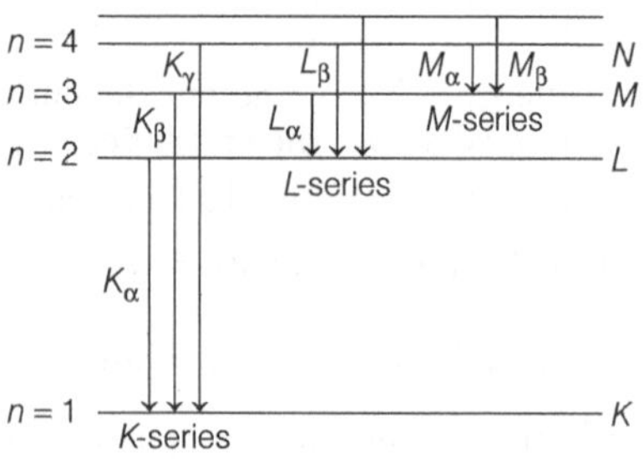

The energy of X-ray radiation as

$$\Delta E = Rhc(Z - b)^2 \left[\frac{1}{n_1^2} - \frac{1}{n_2^2}\right]$$

$$\nu = Rc(Z - b)^2 \left[\frac{1}{n_1^2} - \frac{1}{n_2^2}\right]$$

Moseley's Law

The frequency of X-rays is given by

$$\nu = a\,(Z - b)^2$$

where, a and b are constants and Z is atomic number of element.

$$\nu \propto Z^2$$

Example 20. *Use Moseley's law with $b = 1$ to find the frequency of the K_α X-rays of La ($Z = 57$) if the frequency of the K_α X-rays of Cu ($Z = 29$) is known to be 1.88×10^{18} Hz.*

(a) 7.52×10^{18} Hz (b) 3.25×10^{16} Hz

(c) 8.51×10^{19} Hz (d) 9.1×10^{15} Hz

Sol. (a) Using the equation,

$$\sqrt{f} = a\,(Z - b) \qquad (b = 1)$$

$$\frac{f_{La}}{f_{Cu}} = \left(\frac{Z_{La} - 1}{Z_{Cu} - 1}\right)^2$$

or

$$f_{La} = f_{Cu}\left(\frac{Z_{La} - 1}{Z_{Cu} - 1}\right)^2$$

$$= 1.88 \times 10^{18}\left(\frac{57 - 1}{29 - 1}\right)^2$$

$$= 7.52 \times 10^{18} \text{ Hz}$$

Example 21. *An X-ray tube with a copper target is found to be emitting lines other than those due to copper. The K_α line of copper is known to have a wavelength 1.5405 Å and the other two K_α lines observed have wavelengths 0.7090 Å and 1.6578 Å. Identify the impurities.*

(a) Nickel (b) Molybdenum

(c) Both (a) and (b) (d) Copper

Sol. (c) According to Moseley's equation for K_α radiation,

$$\frac{1}{\lambda} = R(Z - 1)^2 \left[\frac{1}{1^2} - \frac{1}{2^2}\right] \quad (\lambda = \text{wavelength corresponding to Cu})$$

Let λ_1 and λ_2 be the two other unknown wavelengths, then

$$\frac{\lambda_1}{\lambda} = \frac{(Z - 1)^2}{(Z_1 - 1)^2} = \frac{0.7090}{1.5405}$$

For copper $Z = 29$, therefore $(Z_1 - 1) = 28\sqrt{\frac{1.5405}{0.7092}} = 41$

or $Z_1 = 42$ (molybdenum)

Similarly, $\frac{\lambda_2}{\lambda} = \frac{(28)^2}{(Z_2 - 1)^2} = \frac{1.6578}{1.5405}$

$$(Z_2 - 1) = 28\sqrt{\left(\frac{1.5405}{1.6578}\right)}$$

or $(Z_2 - 1) = 27$

or $Z_2 = 28$ (nickel)

So, the impurities are molybdenum and nickel.

Practice Exercise

ROUND I Topically Divided Problems

Rutherford α-Particle Scattering Experiment and Bohr's Model

1. A beam of fast moving α-particles was directed towards a thin gold foil. The parts A', B' and C' of the transmitted and reflected beams corresponding to incident parts A, B and C of the beam are as shown in the figure below.

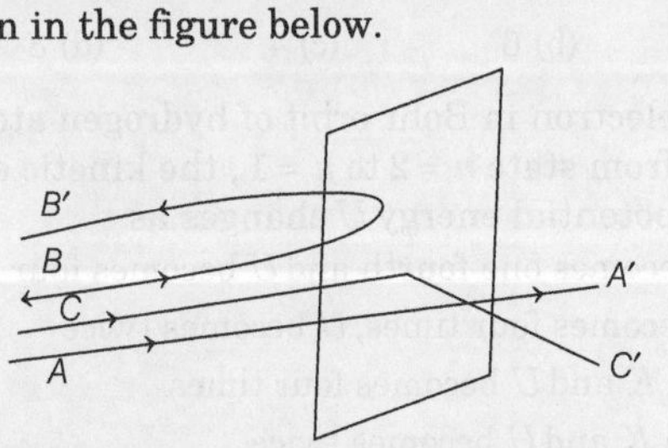

Then, the number of α-particles in part
(a) B' will be minimum and C' is maximum
(b) A' will be minimum and B' is maximum
(c) C' will be minimum and A' is maximum
(d) A' will be maximum and B' will be minimum

2. The graph which depicts the results of Rutherford gold foil experiment with α-particles is
θ : scattering angle
Y : number of scattered α-particles detected
(plots are schematic and not to scale) **[JEE Main 2020]**

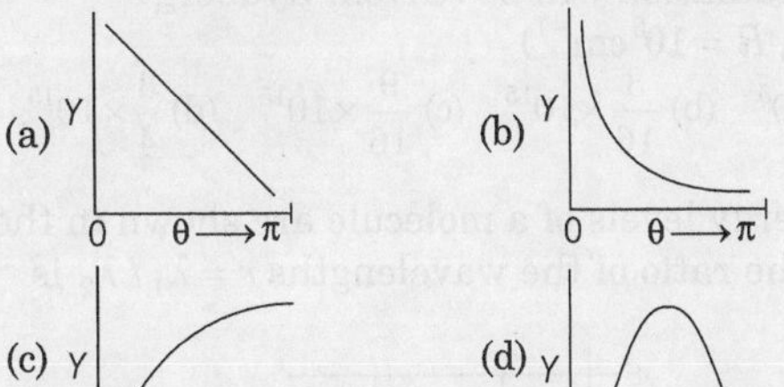

3. An α-particle of energy 5 MeV is scattered through 180° by a fixed uranium nucleus. The distance of closest approach is of the order of
(a) 1 Å (b) 10^{-10} cm
(c) 10^{-12} cm (d) 10–15 cm

4. In some of Rutherford's experiments, a silver foil is used. Let kinetic energy of α-particles is 4.8 MeV, mass of α-particles is 6.64×10^{-27} kg, charge on silver nucleus is $47e$.
For a head-on collision impact parameter for an α-particle, which is scattered at an angle of 45° is
(a) 2.8×10^{-14} m (b) 3.36×10^{-14} m
(c) 1.41×10^{-14} m (d) 5.60×10^{-14}m

5. In α-scattering experiment, in a head-on-collision between an α-particle and a gold nucleus, the minimum distance of separation or closest approach is 4×10^{-14} m. The energy of α-particle is (Take, atomic number of gold = 79)
(a) 5.68 MeV (b) 8 MeV
(c) 4.47 MeV (d) 7.24 MeV

6. An ionised H-molecule consist of an electron and two protons. One proton are separated by a small distance of the order of angstrom. In the ground state
(a) the electron would not move in circular orbits
(b) the energy would be $(z)^4$ times that of a H-atom
(c) the molecule will soon decay in the proton and a H-atom
(d) None of the above

7. For the ground state the electron in the H-atom has an angular momentum = h. According to the simple Bohr model, angular momentum is a vector and hence there will be infinitely many orbits with the vector pointing in all possible directions. In actuality, this is not true, **[NCERT Exemplar]**
(a) because Bohr model gives incorrect values of angular momentum.
(b) because only one of these would have a minimum energy.
(c) angular momentum must be in the direction of spin of electron.
(d) because electrons go around only in horizontal orbits.

8. Taking the Bohr radius as $a_0 = 53$ pm, the radius of Li^{++} ion in its ground state, on the basis of Bohr's model, will be about **[NCERT Exemplar]**
(a) 53 pm (b) 27 pm (c) 18 pm (d) 13 pm

9. In the Bohr's model of hydrogen like atom the force between the nucleus and the electron is modified as $F = \frac{e^2}{4\pi\varepsilon_0}\left(\frac{1}{r^2} + \frac{\beta}{r^3}\right)$, where β is a constant. For this atom, the radius of the nth orbit in terms of the Bohr radius $\left(a_0 = \frac{\varepsilon_0 h^2}{m\pi e^2}\right)$ is

(a) $r = a_0 n - \beta$ (b) $r = a_0 n^2 + \beta$
(c) $r = a_0 n^2 - \beta$ (d) $r = a_0 n + \beta$

10. The electron in a hydrogen atom makes transition from M-shell to L-shell. The ratio of magnitudes of initial to final centripetal acceleration of electron is

(a) 9 : 4 (b) 81 : 16 (c) 4 : 9 (d) 16 : 81

11. Consider an electron in the nth orbit of a hydrogen atom in the Bohr model. The circumference of the orbit can be expressed in terms of the de-Broglie wavelength of that electron as

(a) $(0.529)\, n\lambda$ (b) $\sqrt{n}\, \lambda$
(c) $(13.6)\, \lambda$ (d) $n\lambda$

12. The time period of revolution of electron in its ground state orbit in a hydrogen atom is 1.6×10^{-16} s. The frequency of revolution of the electron in its first excited state (in Hz) is [JEE Main 2020]

(a) 1.6×10^{14} (b) 5.6×10^{12}
(c) 6.2×10^{15} (d) 7.8×10^{14}

13. Consider an electron in a hydrogen atom, revolving in its second excited state (having radius 4.65 Å). The de-Broglie wavelength of this electron is [JEE Main 2019]

(a) 3.5 Å (b) 6.6 Å (c) 12.9 Å (d) 9.7 Å

Energy of Electron and Hydrogen Spectrum

14. As an electron makes a transition from an excited state to the ground state of a hydrogen like atom/ion

(a) its kinetic energy increases but potential energy and total energy decrease
(b) kinetic energy, potential energy and total energy decrease
(c) kinetic energy decreases, potential energy increases but total energy remains same
(d) kinetic energy and total energy decrease but potential energy increases

15. In the Bohr model of the hydrogen atom, let R, V and E represent the radius of the orbit, the speed of electron and the total energy of the electron respectively. Which of the following quantities is proportional to quantum number n?

(a) $\frac{R}{E}$ (b) $\frac{E}{V}$ (c) RE (d) VR

16. Two H-atoms in the ground state collide inelastically. The maximum amount by which their combined kinetic energy is reduced is [NCERT Exemplar]

(a) 10.20 eV (b) 20.40 eV
(c) 13.6 eV (d) 27.2 eV

17. The energy, the magnitude of linear momentum and orbital radius of an electron in a hydrogen atom corresponding to the quantum number n are E, P and r respectively. Then according to Bohr's theory of hydrogen atom, choose the incorrect statement.

(a) Pr is proportional to n
(b) P/E is proportional to n
(c) Er is not constant for all orbits
(d) EPr is proportional to $1/n$

18. Which level of the single ionised carbon has the same energy as the ground state energy of hydrogen atom? [JEE Main 2021]

(a) 1 (b) 6 (c) 4 (d) 8

19. As the electron in Bohr orbit of hydrogen atom passes from state $n = 2$ to $n = 1$, the kinetic energy K and potential energy U changes as

(a) K becomes one-fourth and U becomes four times
(b) K becomes four times, U becomes twice
(c) both K and U becomes four times
(d) both K and U becomes twice

20. The energy of an electron in nth orbit of the hydrogen atom is given by

$$E_n = \frac{-13.6}{n^2}\text{eV}$$

The energy required to raise an electron from the first orbit to the second orbit will be

(a) 10.2 eV (b) 12.1 eV
(c) 13.6 eV (d) 3.4 eV

21. An electron jumps from the 4th orbit to 2nd orbit of hydrogen atom. The frequency (in Hz) of the emitted radiation will be (Given, Rydberg's constant, $R = 10^5 \text{cm}^{-1}$)

(a) $\frac{3}{16} \times 10^5$ (b) $\frac{3}{16} \times 10^{15}$ (c) $\frac{9}{16} \times 10^{15}$ (d) $\frac{3}{4} \times 10^{15}$

22. Some energy levels of a molecule are shown in the figure. The ratio of the wavelengths $r = \lambda_1 / \lambda_2$ is given by

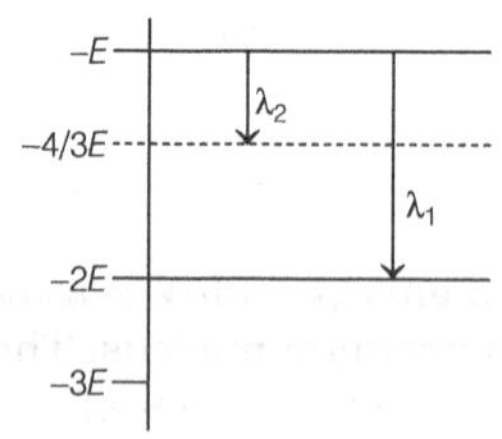

(a) $r = \frac{2}{3}$ (b) $r = \frac{3}{4}$ (c) $r = \frac{1}{3}$ (d) $r = \frac{4}{3}$

23. The ratio of the energies of the hydrogen atom in its first to second excited states is
(a) 9/4 (b) 4/1
(c) 8/1 (d) 1/8

24. A hydrogen atom initially in the ground level absorbs a photon, which excites it to the $n = 4$ level. Determine the wavelength and frequency of photon. [NCERT]
(a) 9.7×10^{-8} m and 3.1×10^{15} Hz
(b) 7.6×10^{-9} m and 2.6×10^{14} Hz
(c) 2.9×10^{-10} m and 4.9×10^{12} Hz
(d) 8.6×10^{-9} m and 3.1×10^{14} Hz

25. The wave number of the energy emitted when electron comes from fourth orbit to second orbit in hydrogen is 20397 cm^{-1}. The wave number of the energy for the same transition in He^+ is
(a) 5099 cm^{-1} (b) 20497 cm^{-1}
(c) 14400 Å (d) 81588 cm^{-1}

26. Hydrogen atom from excited state comes to the ground state by emitting a photon of wavelength λ. If R is the Rydberg constant, then the principal quantum number n of the excited state is
(a) $\sqrt{\dfrac{\lambda R}{\lambda R-1}}$ (b) $\sqrt{\dfrac{\lambda}{\lambda R-1}}$
(c) $\sqrt{\dfrac{\lambda R^2}{\lambda R-1}}$ (d) $\sqrt{\dfrac{\lambda R}{\lambda-1}}$

27. Energy levels A, B and C of a certain atom corresponding to increasing values of energy, *i.e.* $E_A < E_B < E_C$. If λ_1, λ_2 and λ_3 are wavelengths of photon corresponding to transitions shown below.

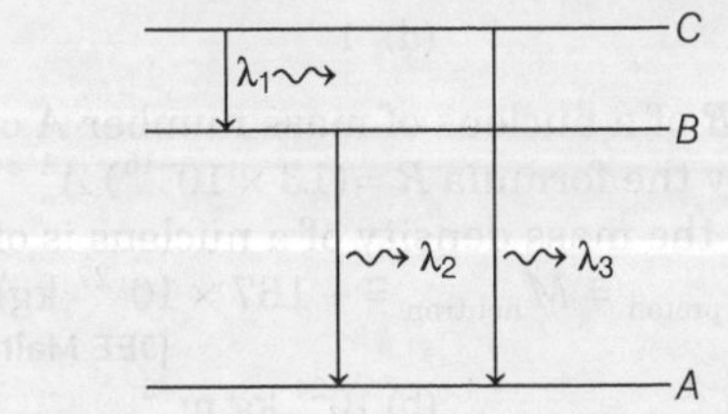

Then,
(a) $\lambda_3 = \lambda_1 + \lambda_2$ (b) $\lambda_3 = \lambda_1\lambda_2 / \lambda_1 + \lambda_2$
(c) $\lambda_1 + \lambda_2 + \lambda_3 = 0$ (d) $\lambda_3^2 = \lambda_1^2 + \lambda_2^2$

28. Imagine an atom made up of proton and a hypothetical particle of double the mass of the electron but having the same charge as the electron. Apply the Bohr atom model and consider all possible transitions of this hypothetical particle to the first excited level. The longest wavelength photon that will be emitted has wavelength λ (given in terms of the Rydberg constant R for the hydrogen atom) equal to (JEE 2000)
(a) $9/5R$ (b) $36/5R$
(c) $18/5R$ (d) $4/R$

29. Hydrogen $({}_1H^1)$, deuterium $({}_1H^2)$, singly ionised helium $({}_2He^4)^+$ and doubly ionised lithium $({}_3Li^8)^{++}$ all have one electron around the nucleus. Consider an electron transition from $n = 2$ to $n = 1$. If the wavelengths of emitted radiation are λ_1, λ_2, λ_3 and λ_4 respectively for four elements, then approximately which one of the following is correct?
(a) $4\lambda_1 = 2\lambda_2 = 2\lambda_3 = \lambda_4$ (b) $\lambda_1 = 2\lambda_2 = 2\lambda_3 = \lambda_4$
(c) $\lambda_1 = \lambda_2 = 4\lambda_3 = 9\lambda_4$ (d) $\lambda_1 = 2\lambda_2 = 3\lambda_3 = 4\lambda_4$

30. The electron in a hydrogen atom first jumps from the third excited state to the second excited state and subsequently to the first excited state. The ratio of the respective wavelengths λ_1 / λ_2 of the photons emitted in this process is [JEE Main 2019]
(a) 20/7 (b) 27/5
(c) 7/5 (d) 9/7

31. In a hydrogen like atom, when an electron jumps from the M-shell to the L-shell, the wavelength of emitted radiation is λ. If an electron jumps from N-shell to the L-shell, the wavelength of emitted radiation will be [JEE Main 2019]
(a) $\dfrac{27}{20}\lambda$ (b) $\dfrac{25}{16}\lambda$
(c) $\dfrac{20}{27}\lambda$ (d) $\dfrac{16}{25}\lambda$

32. The energy required to ionise a hydrogen like ion in its ground state is 9 Rydbergs. What is the wavelength of the radiation emitted when the electron in this ion jumps from the second excited state to the ground state? [JEE Main 2020]
(a) 8.6 nm (b) 24.2 nm
(c) 11.4 nm (d) 35.8 nm

33. An excited He^+ ion emits two photons in succession, with wavelengths 108.5 nm and 30.4 nm, in making a transition to ground state. The quantum number n corresponding to its initial excited state is [for photon of wavelength λ, energy $E = \dfrac{1240 \text{ eV}}{\lambda \text{ (in nm)}}$] [JEE Main 2019]
(a) $n = 4$ (b) $n = 5$
(c) $n = 7$ (d) $n = 6$

34. Radiation coming from transitions $n = 2$ to $n = 1$ of hydrogen atoms fall on He^+ ions in $n = 1$ and $n = 2$ states. The possible transition of helium ions as they absorb energy from the radiation is [JEE Main 2019]
(a) $n = 2$ to $n = 3$ (b) $n = 1$ to $n = 4$
(c) $n = 2$ to $n = 5$ (d) $n = 2$ to $n = 4$

35. A He^+ ion is in its first excited state. Its ionisation energy is [JEE Main 2019]
(a) 54.40 eV (b) 13.6 eV
(c) 48.36 eV (d) 6.04 eV

36. Imagine that the electron in a hydrogen atom is replaced by a muon (μ). The mass of muon particle is 207 times that of an electron and charge is equal to the charge of an electron. The ionisation potential of this hydrogen atom will be [JEE Main 2021]

(a) 13.6 eV (b) 2815.2 eV
(c) 331.2 eV (d) 27.2 eV

37. A muon is an unstable elementary particle whose mass is 207 m_e and whose charge is either $+e$ or $-e$. A negative muon (μ^-) can be captured by a hydrogen nucleus (or proton) to form a muon atom. Then,

(a) the radius of the first Bohr orbit of this atom is 2.85×10^{-13} m
(b) the ionisation energy of the atom is 2.53 keV
(c) radius of first orbit of this atom is 2.86×10^{-10} m
(d) Both (a) and (b)

38. Ionisation potential of hydrogen atom is 13.6 eV. Hydrogen atoms in the ground state are excited by monochromatic radiation of photon energy 12.1 eV. The spectral lines emitted by hydrogen atom according to Bohr's theory will be

(a) one (b) two
(c) three (d) four

39. If the shortest wavelength in the Lyman series is 911.6 Å, the longest wavelength in the same series will be

(a) 1600 Å (b) 2430 Å (c) 1215 Å (d) ∞

40. The first line of Balmer series has wavelength 6563 Å. What will be the wavelength of the first member of Lyman series?

(a) 1215.4 Å (b) 2500 Å (c) 7500 Å (d) 600 Å

41. If the series limit frequency of the Lyman series is ν_L, then the series limit frequency of the Pfund series ν_P is [JEE Main 2018]

(a) $25\,\nu_L$ (b) $16\,\nu_L$ (c) $\frac{\nu_L}{16}$ (d) $\frac{\nu_L}{25}$

42. ν_1 is the frequency of the series limit of Lyman series, ν_2 is the frequency of the first line of Lyman series and ν_3 is the frequency of the series limit of the Balmer series, then [Karnataka CET 2010]

(a) $\frac{1}{\nu} = \frac{1}{\nu_2} + \frac{1}{\nu_3}$ (b) $\frac{1}{\nu_2} = \frac{1}{\nu_1} + \frac{1}{\nu_3}$
(c) $\nu_1 = \nu_2 - \nu_3$ (d) $\nu_1 - \nu_2 = \nu_3$

43. In accordance with the Bohr's model the quantum number that characterises the earth's revolution around the sun in an orbit of radius 1.5×10^{11} m with orbital speed 3×10^4 m/s (Mass of earth $= 6.0 \times 10^{24}$ kg). They belong to

(a) Balmer series (b) Lyman series
(c) Humbnery series (d) None of these

44. Taking the wavelength of first Balmer line in hydrogen spectrum ($n = 3$ to $n = 2$) as 660 nm, the wavelength of the 2nd Balmer line ($n = 4$ to $n = 2$) will be [JEE Main 2019]

(a) 889.2 nm (b) 388.9 nm
(c) 642.7 nm (d) 488.9 nm

45. In Li^{++}, electron in first Bohr orbit is excited to a level by a radiation of wavelength λ. When the ion gets de-excited to the ground state in all possible ways (including intermediate emissions), a total of six spectral lines are observed. What is the value of λ? [Take, $h = 6.63 \times 10^{-34}$ Js; $c = 3 \times 10^8$ ms^{-1}] [JEE Main 2019]

(a) 9.4 nm (b) 12.3 nm (c) 10.8 nm (d) 11.4 nm

Nucleus, Mass-Energy Relation and Binding Energy

46. If Avogadro number is 6×10^{23}, then number of protons, neutrons and electrons is 14 g of $_6C^{14}$ are respectively

(a) 36×10^{23}, 48×10^{23}, 36×10^{23}
(b) 36×10^{23}, 36×10^{23}, 36×10^{23}
(c) 48×10^{23}, 36×10^{23}, 48×10^{23}
(d) 48×10^{23}, 48×10^{23}, 36×10^{23}

47. r_1 and r_2 are the radii of atomic nuclei of mass numbers 64 and 27, respectively. The ratio (r_1/r_2) is

(a) 64/27 (b) 27/64
(c) 4/3 (d) 1

48. The ratio of mass densities of nuclei of ^{40}Ca and ^{16}O is close to [JEE Main 2019]

(a) 5 (b) 2
(c) 0.1 (d) 1

49. The radius R of a nucleus of mass number A can be estimated by the formula $R = (1.3 \times 10^{-15})\,A^{1/3}$ m. It follows that the mass density of a nucleus is of the order of ($M_{proton} \cong M_{neutron} \cong - 1.67 \times 10^{-27}$ kg) [JEE Main 2020]

(a) 10^{17} kg m^{-3} (b) 10^{24} kg m^{-3}
(c) 10^{10} kg m^{-3} (d) 10^{3} kg m^{-3}

50. Highly energetic electrons are bombarded on a target of an element containing 30 neutrons. The ratio of radii of nucleus to that of Helium nucleus is $(14)^{1/3}$. The atomic number of nucleus will be

(a) 25 (b) 26
(c) 56 (d) 30

51. The curve of binding energy per nucleon as function of atomic mass number has a sharp peak for lithium nucleus. This implies that lithium

(a) can be used as fissionable material
(b) is highly radioactive
(c) can be easily broken up
(d) is very stable

52. Consider the binding energy of $_{17}Cl^{35}$ and $_{15}P^{31}$ are 287.67 MeV and 262.48 MeV, respectively. Then,
(a) $_{17}Cl^{35}$ is more stable than $_{15}P^{31}$
(b) $_{15}P^{31}$ is more stable than $_{17}Cl^{35}$
(c) Stability of both the elements is equivalent
(d) cannot be estimated from the given data

53. The binding energies per nucleon of Li^7 and He^4 are 5.6 MeV and 7.06 MeV respectively, then the energy of the reaction $_3Li^7 + p \rightarrow 2\ [_2He^4]$ will be
(a) 17.28 MeV
(b) 39.2 MeV
(c) 28.24 MeV
(d) 1.46 MeV

54. The gravitational force between a H-atom and another particle of mass m will be given by Newton's law : $F = G\dfrac{M \cdot m}{r^2}$, where r is (in km) and **[NCERT Exemplar]**
(a) $M = m_{\text{proton}} + m_{\text{electron}}$
(b) $M = m_{\text{proton}} + m_{\text{electron}} - \dfrac{B}{c^2}$ ($B = 13.6$ eV)
(c) M is not related to the mass of the hydrogen atom.
(d) $M = m_{\text{proton}} + m_{\text{electron}} - \dfrac{|V|}{c^2}$ ($|V|$ = magnitude of the potential energy of electron in the H-atom).

55. Consider the plot of N (neutron number) *versus* Z (photon number) for the different nuclei. Let three nuclides A, B and C are at the positions as shown in the figure.

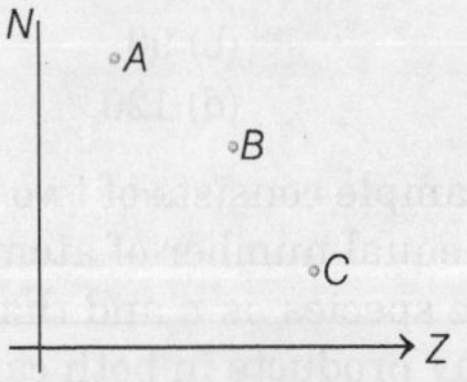

The order of their stability may be
(a) $A > B > C$
(b) $A < B < C$
(c) $B > A > C$
(d) $C < A < B$

56. Find the binding energy per nucleon for $^{120}_{50}$Sn. Mass of proton $m_p = 1.00783$ u, mass of neutron $m_n = 1.00867$ u and mass of tin nucleus $m_{Sn} = 119.902199$ u. (Take, 1 u = 931 MeV) **[JEE Main 2020]**
(a) 9.0 MeV
(b) 7.5 MeV
(c) 8.0 MeV
(d) 8.5 MeV

57. Two nucleons are at a separation of 1 fm. The net force between them is F_1, if both are neutrons, F_2 if both are protons and F_3 if one is a proton and the other is a neutron, then
(a) $F_1 > F_2 > F_3$
(b) $F_2 > F_1 > F_3$
(c) $F_1 = F_3 > F_2$
(d) $F_1 = F_2 > F_3$

58. O_2 molecule consists of two oxygen atoms. In the molecule, nuclear force between the nuclei of the two atoms **[NCERT Exemplar]**
(a) is not important because nuclear forces are short-ranged
(b) is as important as electrostatic force for binding the two atoms
(c) cancels the repulsive electrostatic force between the nuclei
(d) is not important because oxygen nucleus have equal number of neutrons and protons

59. Binding energy for nuclei P, Q and R are E_P, E_Q and E_R, respectively.
In the fusion processes,
$$3P \longrightarrow Q + \text{Energy}(E_1)$$
$$2Q \longrightarrow R + \text{Energy}(E_2)$$
then total energy released in the fusion process
$$6P \longrightarrow R + \text{Energy}(E_3) \text{ is}$$
(a) $E_1 + E_2$ (b) $E_1 - E_2$ (c) $E_1 - 2E_2$ (d) $2E_1 + E_2$

Radioactivity

60. In the uranium radioactive series, the initial nucleus is $_{92}U^{238}$ and that the final nucleus is $_{82}Pb^{206}$. When uranium nucleus decays to lead, the number of α-particle and β-particles emitted are
(a) 8α, 6β (b) 6α, 7β (c) 6α, 8β (d) 4α, 3β

61. For the radioactive nuclei that undergo either α or β-decay, which one of the following cannot occur?
(a) Isobar of original nucleus
(b) Isotope of original nucleus
(c) Nuclei with higher atomic number than that of the original nucleus
(d) Nuclei with lower atomic number than that of the original nucleus

62. The half-life of Au^{198} is 2.7 days. The activity of 1.50 mg of Au^{198} if its atomic weight is 198 g mol^{-1} is, ($N_A = 6 \times 10^{23}$/mol) **[JEE Main 2021]**
(a) 240 Ci
(b) 357 Ci
(c) 535 Ci
(d) 252 Ci

63. Tritium is an isotope of hydrogen whose nucleus triton contains 2 neutrons and 1 proton. Free neutrons decay into $p + e^- + \bar{\nu}$. If one of the neutrons in triton decays, it would transform into He^3 nucleus. This does not happen. This is because **[NCERT Exemplar]**
(a) triton energy is less than that of a He^3 nucleus
(b) the electron created in the beta decay process cannot remain in the nucleus
(c) both the neutrons in trition have to decay simultaneously resulting in a nucleus with 3 protons, which is not a He^3 nucleus
(d) because free neutrons decay due to external perturbations which is absent in a triton nucleus

64. In the following figure, which of the following, sample A or B has shorter mean life?

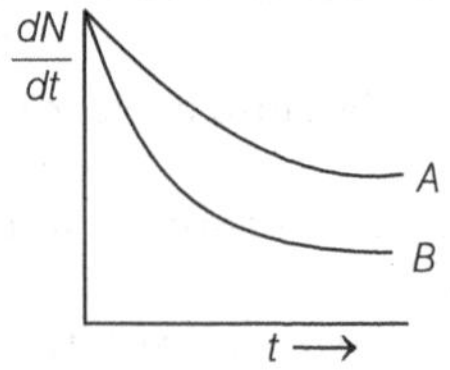

(a) B (b) A

(c) Both A and B (d) Neither A nor B

65. The half-life of radioactive Radon is 3.8 days. The time at the end of which (1/20)th of the Radon sample will remain undecayed is (Given, $\log_{10} e = 0.4343$)

(a) 13.8 days (b) 16.5 days

(c) 33 days (d) 76 days

66. Suppose we consider a large number of containers each containing initially 10000 atoms of a radioactive material with a half-life of 1 yr. After 1 yr, **[NCERT Exemplar]**

(a) all the containers will have 5000 atoms of the materials

(b) all the containers will contain the same number of atoms of the material but that number will only be approximately 5000

(c) the containers will in general have different numbers of the the atoms of the material but their average will be close to 5000

(d) None of the containers can have more than 5000 atoms

67. The half-life period of a radioactive element x is same as the mean life time of another radioactive element y. Initially, both of them have the same number of atoms. Then, **(JEE 1999)**

(a) x and y have the same decay rate initially

(b) x and y decay at the same rate always

(c) y will decay at a faster rate than x

(d) x will decay at a faster rate than y

68. The count rate for 10g of radioactive material was measured at different times and this has been shown in figure with scale given. The half-life of the material and the total count in the first half value period, respectively are

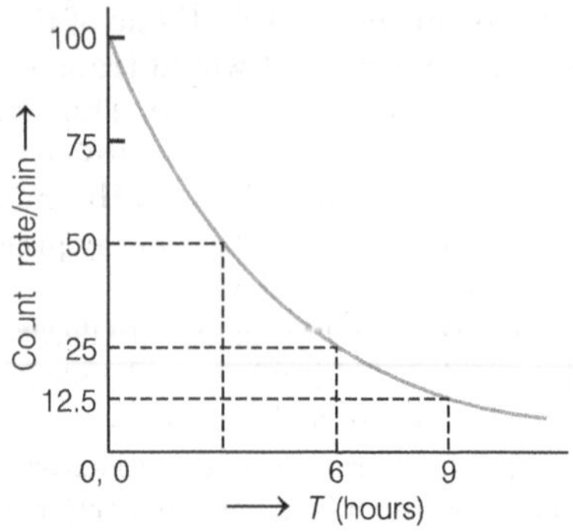

(a) 4 h and 9000 (approximately)

(b) 3 h and 14100 (approximately)

(c) 3 h and 235 (approximately)

(d) 10 h and 157 (approximately)

69. The half-life period of radium is 1600 yr. The fraction of a sample of radium that would remain after 6400 yr is **[NCERT Exemplar]**

(a) $\frac{1}{4}$ (b) $\frac{1}{2}$ (c) $\frac{1}{8}$ (d) $\frac{1}{16}$

70. In a radioactive material, fraction of active material remaining after time t is 9/16. The fraction that was remaining after time $t/2$ is **[JEE Main 2020]**

(a) $\frac{4}{5}$ (b) $\frac{3}{5}$

(c) $\frac{3}{4}$ (d) $\frac{7}{8}$

71. The activity of a radioactive sample falls from 700s^{-1} to 500s^{-1} in 30 min. Its half-life is close to **[JEE Main 2020]**

(a) 62 min (b) 66 min

(c) 72 min (d) 52 min

72. An accident in a nuclear laboratory resulted in depositions of a certain amount of a radioactive material of half-life 18 days inside the laboratory. Tests revealed that the radiation was 64 times the permissible level allowed for safe operation of the laboratory. What is the minimum number of days after which the laboratory can be considered safe for use?

(a) 64 (b) 90

(c) 108 (d) 120

73. A radioactive sample consists of two distinct species having equal number of atoms initially. The mean life of one species is τ and that of the other is 5τ. The decay products in both cases are stable. A plot is made of the total number of radioactive nuclei as a function of time. Which of the following figure best represents the form of this plot?

(a)
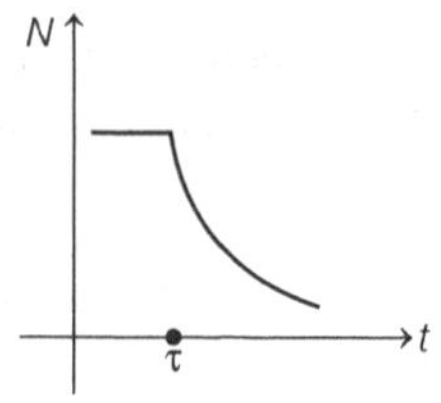

(b)
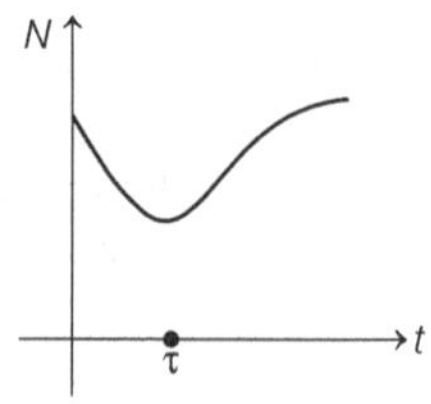

(c)
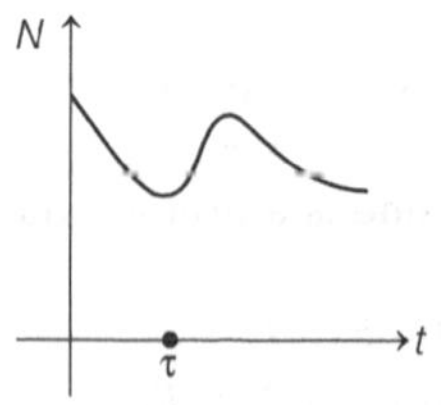

(d)
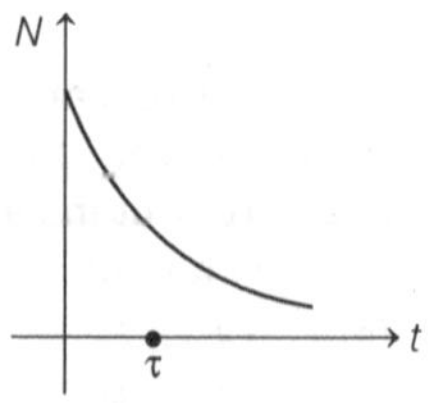

74. Using a nuclear counter, the count rate of emitted particles from a radioactive source is measured. At $t = 0$, it was 1600 counts per second and $t = 8$ s, it was 100 counts per second. The count rate observed as counts per second at $t = 6$ s is close to **[JEE Main 2019]**

(a) 400 (b) 200
(c) 150 (d) 360

75. Samples of two radioactive nuclides A and B are taken. λ_A and λ_B are the disintegration constants of A and B respectively. In which of the following cases, the two samples can simultaneously have the same decay rate at any time? **[NCERT Exemplar]**

(a) Initial rate of decay of A is twice the initial rate of decay of B and $\lambda_A = \lambda_B$
(b) Initial rate of decay of A is twice the initial rate of decay of B and $\lambda_A > \lambda_B$
(c) Initial rate of decay of B is same as the rate of decay of A at $t = 2$ h and $\lambda_B < \lambda_A$
(d) Both (b) and (c)

76. A radioactive nucleus A with a half-life T, decays into a nucleus B. At $t = 0$, there is no nucleus B. After sometime t, the ratio of the number of B to that of A is 0.3. Then, t is given by

(a) $t = T\dfrac{\log 1.3}{\log_e 2}$ (b) $t = T \log 1.3$
(c) $t = \dfrac{T}{\log 1.3}$ (d) $t = \dfrac{T \log_e 2}{2 \log 1.3}$

77. Half-lives of two radioactive elements A and B are 20 min and 40 min, respectively. Initially, the samples have equal number of nuclei. After 80 min, the ratio of decayed numbers of A and B nuclei will be

(a) 1 : 16 (b) 4 : 1
(c) 1 : 4 (d) 5 : 4

78. A sample of radioactive material A, that has an activity of 10 mCi (1 Ci $= 3.7 \times 10^{10}$ decays/s) has twice the number of nuclei as another sample of a different radioactive material B which has an activity of 20 mCi. The correct choices for half-lives of A and B would, then be respectively **[JEE Main 2019]**

(a) 20 days and 10 days
(b) 5 days and 10 days
(c) 10 days and 40 days
(d) 20 days and 5 days

79. A radioactive element decays by β-emission. A detector records n β-particles in 2 s and in next 2 s it records $0.75\,n$ β-particles. Find mean life (in s) correct to nearest whole number. (Given, $\ln |2| = 0.6931$ and $\ln |3| = 1.0986$)

(a) 6 (b) 4
(c) 7 (d) 9

80. A radioactive nucleus decays by two different processes. The half-life for the first process is 10 s and that for the second is 100 s. The effective half-life of the nucleus is close to **[JEE Main 2020]**

(a) 9 s (b) 6 s
(c) 55 s (d) 12 s

81. A radioactive nucleus X decays to a nucleus Y with a decay constant $\lambda_X = 0.1\,\text{s}^{-1}$, Y further decays to a stable nucleus Z with a decay constant $\lambda_Y = 1/30\,\text{s}^{-1}$. Initially, there are only X nuclei and their number is $N_0 = 10^{20}$. Set up the rate equations for the populations of X, Y and Z. The population of Y nucleus as a function of time is given by $N_Y(t) = \{N_0 \lambda_X / (\lambda_X - \lambda_Y)\} [\exp(-\lambda_Y t) - \exp(-\lambda_X t)]$. N_Y is maximum

(a) 16.48s (b) 4.26s
(c) 9s (d) 9.28s

82. A source contains two phosphorous radio nuclides ${}^{32}_{15}P\,(T_{1/2} = 14.3 \text{ days})$ and ${}^{33}_{15}P\,(T_{1/2} = 25.3 \text{ days})$. Initially, 10% of the decay come from ${}^{33}_{15}P$. How long one must wait until 90% do so?

(a) 250 days (b) 295 days
(c) 305 days (d) 208 days

Nuclear Reactions, Nuclear Fission and Nuclear Fusion

83. M_x and M_y denote the atomic masses of the parent and the daughter nuclei respectively in a radioactive decay. The Q-value of a β^- decay is Q_1 and that for a β^+ decay is Q_2. If m_e denotes the mass of an electron, then which of the following statement is correct? **[NCERT Exemplar]**

(a) $Q_1 = (M_x - M_y)c^2$ and $Q_2 = (M_x - M_y - 2m_e)c^2$
(b) $Q_1 = (M_x - M_y)c^2$ and $Q_2 = (M_x - M_y)c^2$
(c) $Q_1 = (M_x - M_y - 2m_e)c^2$ and $Q_2 = (M_x - M_y + 2m_e)c^2$
(d) $Q_1 = (M_x - M_y + 2m_e)c^2$ and $Q_2 = (M_x - M_y + 2m_e)c^2$

84. You are given that mass of ${}^7_3\text{Li} = 7.0160$ u, mass of ${}^4_2\text{He} = 4.0026$ u and mass of ${}^1_1\text{H} = 1.0079$ u. When 20g of ${}^7_3\text{Li}$ is converted into ${}^4_2\text{He}$ by proton capture, the energy liberated (in kWh), is [Mass of nucleon $= 1\,\text{GeV}/c^2$] **[JEE Main 2020]**

(a) 4.5×10^5 (b) 8×10^6
(c) 6.82×10^5 (d) 1.33×10^6

85. The electrostatic energy of Z protons uniformly distributed throughout a spherical nucleus of radius R is given by $E = \dfrac{3}{5}\dfrac{Z(Z-1)e^2}{4\pi\varepsilon_0 R}$

The measured masses of a neutron, ${}^1_1\text{H}$, ${}^{15}_7\text{N}$ and ${}^{15}_8\text{O}$ are 1.008665 u, 1.007825 u, 15.000109 u and 15.003065 u, respectively. Given that the radii

of both the $^{15}_{7}O$ nuclei are the same;
$1\ u = 931.5\ MeV/c^2$, where c is the speed of light and $e^2/4\pi\varepsilon_0 = 1.44$ MeV-fm. Assuming that, the difference between the binding energies of $^{15}_{7}N$ and $^{15}_{8}O$ is purely due to their electrostatic energy, the radius of either of the nuclei is

(a) 2.85 fm (b) 3.03 fm
(c) 3.42 fm (d) 3.80 fm

86. The number of neutrons released during the fission reaction is $^{1}_{0}n + {}^{235}_{92}U \longrightarrow {}^{133}_{51}Sb + {}^{99}_{41}Nb +$ neutrons

(a) 1 (b) 92
(c) 3 (d) 4

87. When uranium is bombarded with neutrons, it undergoes fission. The fission reaction can be written as $_{92}U^{235} + {}_0n^1 \to {}_{56}Ba^{141} + {}_{36}Kr^{92} + 3X + Q$ (energy) where three particles names X are produced and energy Q is released. What is the name of the particle X?

(a) electron (b) α-particle
(c) neutron (d) neutrino

88. Energy released in the fission of a single nucleus is 200 MeV. The fission rate of a $^{235}_{92}U$ filled reactor operating at a power level of 5W is

(a) $1.56 \times 10^{-10}\ s^{-1}$ (b) $1.56 \times 10^{11}\ s^{-1}$
(c) $1.56 \times 10^{-16}\ s^{-1}$ (d) $1.56 \times 10^{-17}\ s^{-1}$

89. Fusion process, like combining two deuterons to form a He nucleus are impossible at ordinary temperatures and pressure. This reasons for this can be traced to the fact

(a) nuclear forces have short range
(b) the original nuclei must be completely ionized before fussion can take place
(c) the original nuclei must first break up before combining with each other.
(d) All of the above

90. It is proposed to use the nuclear fusion reaction, $^{2}_{1}H + {}^{2}_{1}H \to {}^{4}_{2}He$ in a nuclear reactor 200 MW rating. If the energy from the above reaction is used with a 25 per cent efficiency in the reactor, how many grams of deuterium fuel will be needed per day? (The masses of $^{2}_{1}H$ and $^{4}_{2}He$ are 2.0141 amu and 4.0026 amu, respectively.)

(a) 120.26 g (b) 60.46 g
(c) 80 g (d) 90.26 g

91. In a reactor, 2kg of $_{92}U^{235}$ fuel is fully used up in 30 days. The energy released per fission is 200 MeV. Given that, the Avogadro number, $N = 6.023 \times 10^{26}\ K^{-1}\ mol^{-1}$ and $1\ eV = 1.6 \times 10^{-19}$ J. The power output of the reactor is close to **[JEE Main 2020]**

(a) 35 MW (b) 125 MW
(c) 60 MW (d) 54 MW

92. Schematic diagram of a nuclear reactor based on thermal neutron fission is as shown below

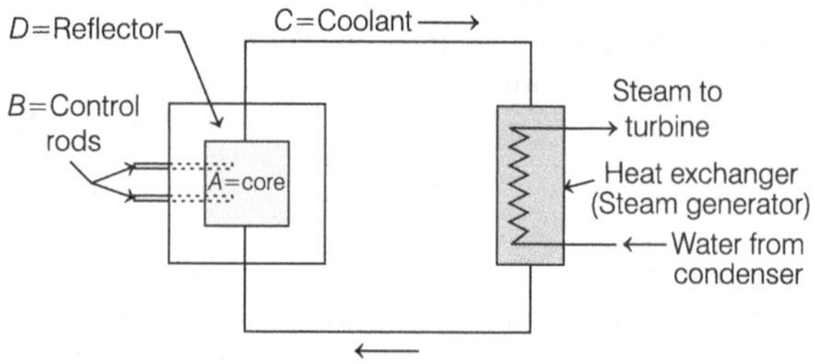

Here, the correct purpose of the parts specified by A, B, C and D is given in

(a) $A \to$ reduces leakage.
(b) $B \to$ can shut down the reactor
(c) $C \to$ site of nuclear fission and contains $^{235}_{92}U$
(d) $D \to$ transfers heat to a working fluid which in turn may produce steam

X-Rays

93. Electrons with de-Broglie wavelength λ fall on the target in an X-ray tube. The cut-off wavelength λ_0 of the emitted X-rays is **[IIT JEE 2007]**

(a) $\lambda_0 = \dfrac{2mc\lambda^2}{h}$ (b) $\lambda_0 = \dfrac{2h}{mc}$
(c) $\lambda_b = \dfrac{2m^2c^2\lambda^2}{h^2}$ (d) $\lambda_0 = \lambda$

94. If λ_1 and λ_2 are the wavelength of characteristic X-rays and γ-rays respectively, then the relation between them is

(a) $\lambda_1 > \lambda_2$ (b) $\lambda_1 < \lambda_2$
(c) $\lambda_1 = \lambda_2$ (d) $\lambda_1\lambda_2 = 1$

95. Which one of the following statement is wrong in the context of X-rays generated from a X-rays tube? **[IIT JEE 2008]**

(a) Wavelength of characteristic X-rays decreases when the atomic number of the target increases
(b) Cut-off wavelength of the continuous X-rays depends on the atomic number of the target
(c) Intentsity of the characteristic X-rays depends on the electrical power given to the X-rays tube
(d) Cut-off wavelength of the continuous X-rays depends on the energy of the electrons in the X-ray tube

96. The binding energy of the innermost electron in tungsten is 40 k eV. To produce characteristic X-rays using a tungsten target in an X-rays tube, the potential difference between the cathode and anti cathode should be

(a) $V < 40$ kV
(b) $V \le 40$ kV
(c) $V > 40$ kV
(d) $V = 40$ kV

97. The intensity of X-rays from a coolidge tube is plotted against wavelength λ as shown in figure. The minimum wavelength found is λ_c and the wavelength of K_α line is λ_K. As the accelerating voltage is increased, then

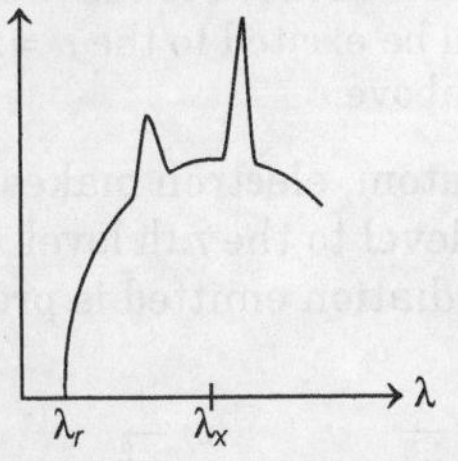

(a) $\lambda_K - \lambda_c$ increases (b) $\lambda_K - \lambda_c$ decreases
(c) λ_K increases (d) λ_K decreases

98. A potential difference of 10^4 V is applied across an X-ray tube. The ratio of the de-Broglie wavelength of X-rays produced is
($\frac{e}{m}$ for electron= 1.8×10^{11} Ckg^{-1})

(a) $\frac{1}{20}$ (b) $\frac{1}{10}$ (c) 1 (d) $\frac{1}{100}$

99. An electron beam is accelerated by a potential difference V to hit a metallic target to produce X-rays. It produces continuous as well as characteristic X-rays. If λ_{min} is the smallest possible wavelength of X-rays in the spectrum, the variation of $\log \lambda_{min}$ with $\log V$ is correctly represented in

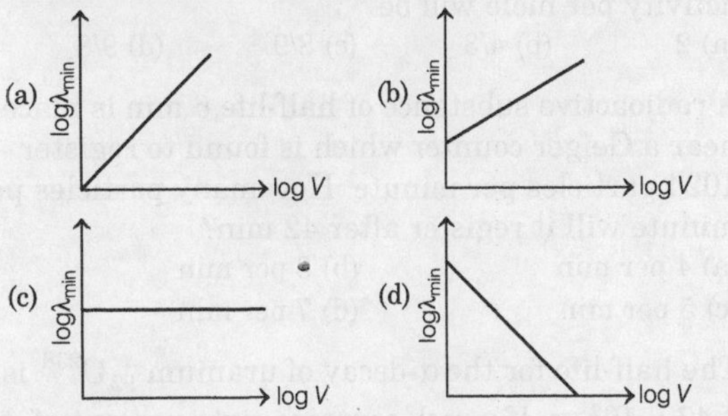

100. An element with atomic number $Z = 11$ emits K_α X-ray of wavelength λ. The atomic number of element which emits K_αX -ray of wavelength 4λ **[IIT Screening 2006]**

(a) 6 (b) 4 (c) 11 (d) 44

101. Find the constants a and b, respectively in Moseley's equation $\sqrt{v} = a(Z - b)$ from the following data.

Element	Z	Wavelength of K_α X-ray
Mo	42	71 pm
Co	27	178.5 pm

(a) 1.37, 5×107 (b) 6.27, 13.4
(c) 5×10^7, 1.37 (d) 9.26, 7×10^7

ROUND II Mixed Bag

Only One correct Option

1. The energy level diagram for an hydrogen like atom is shown in the figure. The radius of its first Bohr orbit is

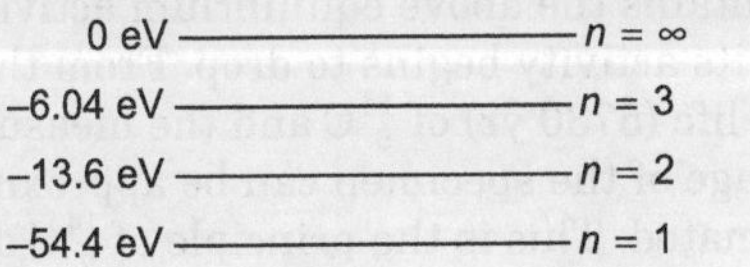

(a) 0.265 Å
(b) 0.53 Å
(c) 0.132 Å
(d) None of the above

2. A hydrogen atom emits a photon corresponding to an electron transition from $n = 5$ to $n = 1$. The recoil speed of hydrogen atom is almost (mass of proton $\approx 1.6 \times 10^{-27}$ kg)

(a) 10 ms^{-1}
(b) 2×10^2 ms^{-1}
(c) 4 ms^{-1}
(d) 8×10^2 ms^{-1}

3. A hydrogen atom and a Li^{2+} ion are both in second excited state. If L_H and L_{Li} are their respective electronic angular momenta and E_H and E_{Li} their respective energies, then

(a) $l_H > l_{Li}$ and $|E_H| > |E_{Li}|$
(b) $L_H = L_{Li}$ and $|E_H| < |E_{Li}|$
(c) $l_H > l_{Li}$ and $|E_H| > |E_{Li}|$
(d) $l_H > l_{Li}$ and $|E_H| << |E_{Li}|$

4. A sample of an element is 10.38 g. If half-life of element is 3.8 days, then after 19 days, how much quantity of element remains?

(a) 0.151 g (b) 0.32 g
(c) 1.51 g (d) 0.16 g

5. The radioactivity of a given sample of whisky due to tritium (half-life 12.3 yr) was found to be only 3% of that measured in a recently purchased bottle marked 7 yr old. The sample must have been prepared about

(a) 220 yr back (b) 300 yr back
(c) 400 yr back (d) 70 yr back

6. The ratio of molecular mass of two radioactive substances is 3/2 and the ratio of their decay constants is 4/3. Then, the ratio of their initial activity per mole will be
(a) 2 (b) 4/3 (c) 8/9 (d) 9/8

7. A radioactive substance of half-life 6 min is placed near a Geiger counter which is found to register 1024 particles per minute. How many particles per minute will it register after 42 min?
(a) 4 per min (b) 8 per min
(c) 5 per min (d) 7 per min

8. The half-life for the α-decay of uranium $_{92}U^{238}$ is 4.47×10^9 yr. If a rock contains sixty percent of its original $_{92}U^{238}$ atoms, its age is
[log 6 = 0.778; log 2 = 0.3]
(a) 3.3×10^9 yr (b) 6.6×10^9 yr
(c) 1.2×10^8 yr (d) 5.4×10^7 yr

9. The electron in a hydrogen atom makes a transition $n_1 \to n_2$ where n_1 and n_2 are the principal quantum numbers of the two states. Assume the Bohr model to be valid. The time period of electron in the initial state is 8 times that in the final state. The possible values of n_1 and n_2 are
(a) $n_1 = 6, n_2 = 3$ (b) $n_1 = 8, n_2 = 2$
(c) $n_1 = n_2 = 1$ (d) $n_1 = 8, n_2 = 1$

10. There are two radioactive substances A and B. Decay constant of B is two times that of A. Initially, both have equal number of nuclei. After n half-lives of A, rate of disintegration of both are equal. The value of n is
(a) 4 (b) 2 (c) 1 (d) 5

11. The binding energy of an electron in the ground state of He is equal to 24.6 eV. The energy required to remove both the electrons is
(a) 49.2 eV (b) 24.6 eV
(c) 38.2 eV (d) 79.0 eV

12. Calculate the time interval between 33% decay and 67% decay, if half-life of a substance is 20 minutes.
(a) 60 minutes (b) 20 minutes
(c) 40 minutes (d) 13 minutes

13. The highly excited states ($n >> 1$) for hydrogen-like atoms (also called Rydberg states) with nuclear charge Ze are defined by their principal quantum number n. Which of the following statement(s) is/are false?
(a) The relative change in the radii of two consecutive orbitals does not depend on Z.
(b) The relative change in the radii of two consecutive orbitals varies as $1/n$.
(c) The relative change in the energy of two consecutive orbitals varies as $1/n^3$.
(d) The relative change in the angular momentum of two consecutive orbitals varies as $1/n$.

14. Let $E_n = \dfrac{-1}{8\varepsilon_0^2\, n^2 h^2}$ be the energy of the nth level of H-atom. If all the H-atoms are in the ground state and radiation of frequency $(E_2 - E_1)/h$ falls on it then
(a) it will not be absorbed at all
(b) some of atoms will move to the first excited state
(c) all atoms will be excited to the $n = 2$ state
(d) None of the above

15. In a hydrogen atom, electron makes a transition from $(n+1)$th level to the nth level. If $n >> 1$, the frequency of radiation emitted is proportional to **[JEE Main 2020]**
(a) $\frac{1}{n}$ (b) $\frac{1}{n^3}$ (c) $\frac{1}{n^2}$ (d) $\frac{1}{n^4}$

16. An electron in hydrogen atom first jumps from second excited state to first excited state and then from first excited state to ground state. Let the ratio of wavelength, momentum and energy of photons emitted in these two cases be a, b and c respectively. Then,
(a) $a = \frac{9}{4}$ (b) $b = \frac{6}{27}$ (c) $c = \frac{7}{27}$ (d) $c = \frac{1}{a}$

17. After absorbing a slowly moving neutron of mass m_N (momentum ~0) a nucleus of mass M breaks into two nuclei of masses m_1 and $5m_1$ $(6m_1 = M + m_N)$, respectively. If the de-Broglie wavelength of the nucleus with mass m_1 is λ, then de-Broglie wavelength of the other nucleus will be **[AIEEE 2011]**
(a) 25λ (b) 5λ (c) $\lambda/5$ (d) λ

18. The normal activity of living carbon containing matter is found to be about 15 decays/min for every gram of carbon. This activity arises from the small proportion of radioactive $^{14}_{6}C$ present with the stable carbon isotope $^{12}_{6}C$. When the organism is dead, its interaction with the atmosphere (which maintains the above equilibrium activity) ceases and its activity begins to drop. From the known half-life (5730 yr) of $^{14}_{6}C$ and the measured activity, the age of the specimen can be approximately estimated. This is the principle of $^{14}_{6}C$ dating used in archaeology. Suppose a specimen from Mohenjodaro gives an activity of 9 decays/min^{-1}g^{-1} of carbon. Estimate the approximate age of the Indus-Valley civilisation.
(a) 5224 yr (b) 4224 yr (c) 8264 yr (d) 6268 yr

19. A radioactive sample disintegrates via two independent decay processes having half lives $T_{1/2}^{(1)}$ and $T_{1/2}^{(2)}$, respectively. The effective half-life $T_{1/2}$ of the nuclei is **[JEE Main 2021]**
(a) $T_{1/2} = \dfrac{T_{1/2}^{(1)} + T_{1/2}^{(2)}}{T_{1/2}^{(1)} - T_{1/2}^{(2)}}$ (b) $T_{1/2} = T_{1/2}^{(1)} + T_{1/2}^{(2)}$
(c) $T_{1/2} = \dfrac{T_{1/2}^{(1)}\, T_{1/2}^{(2)}}{T_{1/2}^{(1)} + T_{1/2}^{(2)}}$ (d) None of these

20. Activities of three radioactive substances A, B and C are represented by the curves A, B and C in the figure. Then, their half-lives $T_{1/2}(A):T_{1/2}(B):T_{1/2}(C)$ are in the ratio **[JEE Main 2020]**

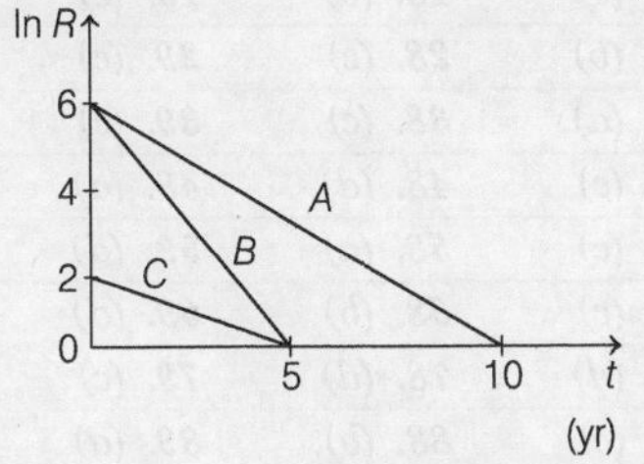

(a) 4 : 3 : 1 (b) 3 : 2 : 1
(c) 2 : 1 : 1 (d) 2 : 1 : 3

21. An electron from various excited states of hydrogen atom emit radiation to come to the ground state. Let λ_n, λ_g be the de-Broglie wavelength of the electron in the nth state and the ground state, respectively. Let Λ_n be the wavelength of the emitted photon in the transition from the nth state to the ground state. For large n, (A, B are constants)

(a) $\Lambda_n \approx A + \dfrac{B}{\lambda_n^2}$ (b) $\Lambda_n \approx A + B\lambda_n^2$
(c) $\Lambda_n^2 \approx A + B\lambda_n^2$ (d) $\Lambda_n^2 \approx \lambda$

22. Photoelectrons are emitted when 400 nm radiation is incident on a surface of work-function 1.9 eV. These photoelectrons pass through a region containing α-particles. A maximum energy electron combines with an α-particle to form a He^+ ion, emitting a single photon in this process. He^+ ions thus formed are in their fourth excited state. Find the energies (in eV) of the photons lying in the 2 to 4 eV range, that are likely to be emitted during and after the combination.
[Take, $h = 4.14 \times 10^{-15}$ eV-s]

(a) 3.4 eV, 3.86 eV (b) 3.86 eV, 3.4 eV
(c) 2.4 eV, 6 eV (d) 9.6 eV, 2.4 eV

Numerical Value Questions

23. The disintegration rate of a certain radioactive sample at any instant is 4750 disintegrations per minute. 5 min later, the rate becomes 2700 min^{-1}. Then, the half-life of the sample (in min) is ....

24. In the fusion reaction $^2_1H + ^2_1H \longrightarrow ^3_2He + ^1_0n$, the masses of deuteron, helium and neutron expressed in amu are 2.015, 3.017 and 1.009, respectively. If 1 g of deuterium undergoes complete fusion, then the amount of total energy released is 9×10^n kJ. Find the value of n.

25. In the chemical analysis of a rock the mass ratio of two radioactive isotopes is found to be 100 : 1. The mean lives of the two isotopes are 4×10^9 yr and 2×10^9 yr, respectively. If it is assumed that at the time of formation the atoms of both the isotopes were in equal proportion, the age of the rock $p \times 10^{10}$ yr, where the value of p is .... (Take, ratio of the atomic weights of the two isotopes is 1.02 : 1.)

26. A hydrogen-like atom of atomic number Z is in an excited state of quantum number $2n$. It can emit a maximum energy photon of 204 eV. If it makes a transition to quantum state n, a photon of energy 40.8 eV is emitted. Calculate the minimum energy (in eV) that can be emitted by this atom during de-excitation. (Ground state energy of hydrogen atom is 13.6 eV.)

27. In an experiment on two radioactive isotopes of an element (which do not decay into one another), their mass ratio at a given instant was formed to be 3. The decaying isotope has a larger mass and activity of 1.0 curie initially. The half-lives of the two radioactive isotopes are known to be 12 h and 16 h. Activity of the each isotope and their mass ratio after 2 days was studied. If the ratio of number of atoms of first isotope to that of the other isotope is found to be $3/x$, then the value of x is ....

28. The first member of the Balmer series of hydrogen atom has a wavelength of 6561 Å. The wavelength of the second member of the Balmer series (in nm) is .... **[JEE Main 2020]**

29. In the line spectra of hydrogen atom, difference between the largest and the shortest wavelengths of the Lyman series is 304 Å. The corresponding difference for the Paschan series (in Å) is .... **[JEE Main 2020]**

30. A particle of mass 200 MeV/c^2 collides with a hydrogen atom at rest. Soon after the collision, the particle comes to rest and the atom recoils and goes to its first excited state. The initial kinetic energy of the particle (in eV) is $\dfrac{N}{4}$. The value of N is .... (Given, the mass of the hydrogen atom to be 1 GeV/c^2) **[JEE Main 2020]**

31. An electron in hydrogen-like atom makes a transition from nth orbit and emits radiation corresponding to Lyman series. If de-Broglie wavelength of electron in nth orbit is equal to the wavelength of radiation emitted, find the value of n. The atomic number of atom is 11.

32. A particle of mass m moves in a circular orbit in a central potential field $U(r) = U_0 r^4$. If Bohr's quantization conditions are applied, radii of possible orbitals r_n vary with $n^{1/\alpha}$, where α is .... **[JEE Main 2021]**

Answers

Round I

1. (d)	2. (b)	3. (c)	4. (b)	5. (a)	6. (a)	7. (a)	8. (c)	9. (c)	10. (d)
11. (d)	12. (d)	13. (d)	14. (a)	15. (d)	16. (a)	17. (c)	18. (b)	19. (c)	20. (a)
21. (c)	22. (c)	23. (a)	24. (a)	25. (d)	26. (a)	27. (b)	28. (c)	29. (c)	30. (a)
31. (c)	32. (c)	33. (b)	34. (d)	35. (b)	36. (b)	37. (a)	38. (c)	39. (c)	40. (a)
41. (d)	42. (d)	43. (b)	44. (d)	45. (c)	46. (a)	47. (c)	48. (d)	49. (a)	50. (b)
51. (d)	52. (b)	53. (a)	54. (b)	55. (c)	56. (d)	57. (c)	58. (a)	59. (d)	60. (a)
61. (b)	62. (a)	63. (a)	64. (b)	65. (b)	66. (c)	67. (c)	68. (b)	69. (d)	70. (c)
71. (a)	72. (c)	73. (d)	74. (b)	75. (c)	76. (a)	77. (d)	78. (d)	79. (c)	80. (a)
81. (a)	82. (d)	83. (a)	84. (d)	85. (c)	86. (d)	87. (c)	88. (b)	89. (a)	90. (a)
91. (c)	92. (b)	93. (a)	94. (a)	95. (a)	96. (c)	97. (a)	98. (b)	99. (d)	100. (a)
101. (a)									

Round II

1. (a)	2. (c)	3. (b)	4. (b)	5. (d)	6. (b)	7. (b)	8. (a)	9. (a)	10. (c)
11. (d)	12. (b)	13. (c)	14. (b)	15. (b)	16. (d)	17. (d)	18. (b)	19. (c)	20. (d)
21. (a)	22. (a)	23. 6.132	24. 9	25. 1.834	26. 10.58	27. 2	28. 486	29. 10553	30. 51
31. 25	32. 3								

Solutions

Round I

1. In α-scattering experiment,

(i) most of the α-particles pass through the foil, which can corresponds to part AA' in the figure.

(ii) only 0.14% are scatter by more than 1°, which can corresponds to part CC' in the figure.

(iii) 1 out of 8000 is deflect by more than 90°, which can corresponds to part BB' in the figure.

Thus, in the given figure, the number of α-particles in part A' will be maximum and part B' will be minimum.

2. In Rutherford's experiment, number of particles scattered at large angles is very less and most of the particles are scattered at small angles.

Hence, graph of Y = number of α- particles and θ = scattering angle is as shown in the figure below.

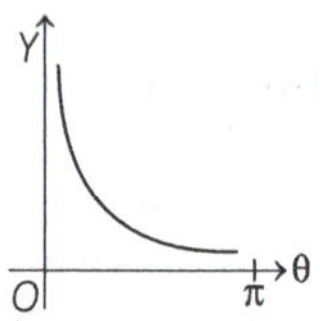

3. As, $r_0 = \frac{(Ze)(2e)}{4\pi\varepsilon_0(E)} = \frac{2\times 92\,(1.6\times 10^{-19})^2\times 9\times 10^9}{5\times 10^6\times 1.6\times 10^{-19}}$

$= 5.3\times 10^{-14}\text{ m} \approx 10^{-12}\text{ cm}$

4. Impact parameter is

$$b = \frac{Z_1 Z_2 e^2}{4\pi\varepsilon_0}\times\frac{1}{2K}\times(\cot\theta/2)$$

For a head on collision, distance of closest approach

$$d_0 = \frac{Z_1 Z_2 e^2}{4\pi\,\varepsilon_0}\times\frac{1}{K}$$

So, from above expressions, we have

$$b = d_0\times\frac{1}{2\tan(\theta/2)}$$

Now, $d_0 = \frac{1}{4\pi\varepsilon_0}\times Z_1 Z_2\times e^2\times\frac{1}{K}$

$$= \frac{9\times 10^9\times 2\times 47\times(1.6\times 10^{-19})^2}{4.8\times 10^6\times 1.6\times 10^{-19}}$$

$$= 2.8\times 10^{-14}\text{ m}$$

$\therefore \quad b = \frac{2.8\times 10^{-14}}{2\times(\sqrt{2}-1)} = 3.36\times 10^{-14}\text{ m}$

5. At distance of minimum separation,

Kinetic Energy [KE] = Potential Energy [PE]

$\Rightarrow \quad KE = \frac{k\,Ze\cdot 2e}{r}$

Here, $k = 9\times 10^9\text{ N-m}^2\text{C}^{-2}$

$Z = 79,\ e = 1.6\times 10^{-19}\text{ C}$

and $r = 4\times 10^{-14}\text{ m}$

$\therefore \quad KE = \frac{9\times 10^9\times 79\times 2\times(1.6\times 10^{-19})^2}{4\times 10^{-14}}\text{ J}$

$$= \frac{9\times 10^9\times 79\times 2\,(1.6\times 10^{-19})^2}{4\times 10^{-14}\times 1.6\times 10^{-19}}\text{ eV}$$

$= 5.68\text{ MeV}$

6. In an ionised hydrogen molecule, as there are two protons and one electron, therefore, electron's orbit would go around the two protons separated by a small distance (~ Å). This orbit shall not be a circular orbit.

7. Simple Bohr model, infact does not give correct values of angular momentum of revolving electron. It gives only the magnitude of angular momentum, which is a vector.

8. On the basis of Bohr's model, $r = \frac{\varepsilon_0 n^2 h^2}{\pi m Z e^2} = a_0 \frac{n^2}{Z}$

Let Li^{++} ion, $Z = 3$; $n = 1$ for ground state

Given, $a_0 = 53$ pm

$\therefore \quad r = \frac{53 \times 1^2}{3} = 18$ pm

9. As, force between nucleus and electron is

$$F = \frac{1}{4\pi\varepsilon_0} \frac{Ze^2}{r^2}$$

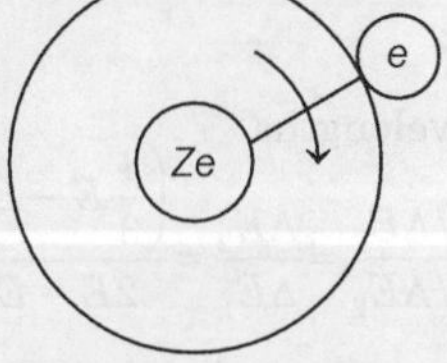

where, r = atomic radius

and e = electronic charge.

From question,

$$F_e = \frac{e^2}{4\pi\varepsilon_0}\left(\frac{1}{r^2} + \frac{\beta}{r^3}\right), \text{ where } \beta = \text{constant.}$$

$$\Rightarrow \quad \frac{mv^2}{r} = \frac{e^2}{4\pi\varepsilon_0}\left(\frac{1}{r^2} + \frac{\beta}{r^3}\right)$$

$$\Rightarrow \quad \frac{m\left(\frac{nh}{2\pi mr}\right)^2}{r} = \frac{e^2}{4\pi\varepsilon_0}\left(\frac{1}{r^2} + \frac{\beta}{r^3}\right)$$

$$\left[\because mvr = \frac{nh}{2\pi} \text{ or } v = \frac{nh}{2\pi mr}\right]$$

$$\Rightarrow \quad \frac{1}{r^2} + \frac{\beta}{r^3} = \frac{mn^2h^2 \cdot 4\pi\varepsilon_0}{4\pi^2 m^2 e^2 r^3} = a_0 \frac{n^2}{r^3}$$

$$\Rightarrow \quad \frac{a_0 n^2}{r^3} = \frac{1}{r^2} + \frac{\beta}{r^3}$$

$$r = a_0 n^2 - \beta$$

10. For an electron in nth state of H-atom,

$$v_n = 2.2 \times 10^6 \frac{Z}{n} \text{ ms}^{-1}$$

$$r_n = 0.53 \times \frac{n^2}{Z} \text{ Å}$$

So, centripetal acceleration of electron $= v_n^2/r_n$

$$= \frac{(2.2 \times 10^6 \, Z/n)^2}{0.53 \times 10^{-10}\left(\frac{n^2}{Z}\right)}$$

$$= \frac{4.84 \times 10^{12}}{0.53 \times 10^{-10}} \cdot \left(\frac{Z^3}{n^4}\right)$$

$$= 9.1 \times 10^{22} \cdot \left(\frac{Z^3}{n^4}\right)$$

So, ratio of centripetal acceleration of 3rd and 2nd orbits is

$$\frac{a_3}{a_2} = \frac{9.1 \times 10^{22} \times \frac{Z^3}{3^4}}{9.1 \times 10^{22} \times \frac{Z^3}{2^4}}$$

$$\Rightarrow \quad \frac{a_3}{a_2} = \frac{2^4}{3^4} = \frac{16}{81}.$$

11. According to de-Broglie hypothesis,

$$mvr = \frac{nh}{2\pi}$$

$$\Rightarrow \quad 2\pi r = n\left(\frac{h}{mv}\right)$$

$$\Rightarrow \quad 2\pi r = n\lambda$$

$\Rightarrow$ Circumference $= n\lambda$

12. Time period of revolution of an electron in its nth state (or nth orbit) of H-atom,

$$T_n = \frac{2\pi r_n}{v_n} = \frac{2\pi r_1 \cdot n^2}{(v_1/n)}$$

where, r_1 and v_1 are radius and velocity of electron in ground state.

So, $\quad T_n = \frac{2\pi r_1}{v_1} \cdot n^3$ or $T_n = T_1 \cdot n^3$

where, T_1 = time period of revolution in ground state.

Now, in first excited state $n = 2$, so time period,

$$T_2 = T_1 \times 2^3 = 8 \times 1.6 \times 10^{-16} \text{ s}$$

As, frequency $= \frac{1}{\text{time period}}$

So, frequency of revolution of electron in first excited state,

$$f_2 = \frac{1}{T_2} = \frac{1}{8 \times 1.6 \times 10^{-16}} = 7.8 \times 10^{14} \text{ Hz}$$

13. By Bohr's second postulate, for revolving electron,

Angular momentum $= \frac{nh}{2\pi}$

$$\Rightarrow \quad mvr_n = \frac{nh}{2\pi}$$

$\Rightarrow$ Momentum of electron, $p = mv = \frac{nh}{2\pi r_n}$

de-Broglie wavelength associated with electron is

$$\lambda_n = \frac{h}{p} = \frac{2\pi r_n}{n}$$

Given, $\quad n = 3$, $r_n = 4.65$ Å

$$\therefore \quad \lambda_n = \frac{(2 \times \pi \times 4.65)}{3} \approx 9.7 \text{ Å}$$

14. As, we know that, kinetic energy of an electron is

$$\text{KE} \propto \left(\frac{Z}{n}\right)^2$$

When the electron makes transition from an excited state to the ground state, then n decreases and KE increases. We know that, PE is lowest for ground state. As, TE = – KE. Hence, TE also decreases.

15. As, $R \propto n^2; V \propto \frac{1}{n}$ and $E \propto \frac{1}{n^2}$

$$\therefore \quad VR \propto \left(\frac{1}{n} \times n^2\right), \text{ i.e. } VR \propto n$$

16. Initial KE of each of two H-atoms in ground state = 13.6 eV

∴ KE of both H-atoms before collision = 2 × 13.6 eV = 27.2 eV

As, the collision is inelastic, linear momentum is conserved, but some KE is lost.

If one H-atom goes over to first excited state and other remains in ground state, then their combined

$$\text{KE after collision} = \left(\frac{13.6}{2^2}\right) + \left(\frac{13.6}{1^2}\right) = 17.0 \text{ eV}$$

∴ Reduction in combined KE = (27.2 – 17.0) eV = 10.2 eV

17. We know that, $E \propto \frac{1}{n^2}; P \propto \frac{1}{n}$ and $r \propto n^2$

$$Pr \propto \frac{1}{n}(n^2), \text{ i.e. } Pr \propto n$$

$$\frac{P}{E} \propto \frac{1/n}{1/n^2}, \text{ i.e. } \frac{P}{E} \propto n$$

$$Er \propto \frac{1}{n^2} \times n^2, \text{ i.e. } Er = \text{constant for all orbits.}$$

$$EPr \propto \frac{1}{n^2} \cdot \frac{1}{n} \cdot n^2, \text{ i.e. } EPr \text{ is proportional to } 1/n.$$

18. Energy of H-atom is $E = -13.6\, Z^2/n^2$

for H-atom $Z = 1$ and for ground state, $n = 1$.

$$\Rightarrow \quad E = -13.6 \times \frac{1^2}{1^2} = -13.6 \text{ eV}$$

Now for carbon atom (single ionised), $Z = 6$

$$E = -13.6 \frac{Z^2}{n^2} = -13.6 \text{ (given)}$$

$$\Rightarrow \quad n^2 = 6^2 \Rightarrow n = 6$$

19. As, we know that, potential energy $U = 2E$, kinetic energy $K = -E$

and total energy of electron, $E = -\frac{13.6}{n^2}$

For state $n = 1$, $E_1 = \frac{-13.6}{(1)^2} = -13.6$

For state $n = 2$, $E_2 = \frac{-13.6}{(2)^2} = \frac{E_1}{4} \Rightarrow E_1 = 4E_2$

But $\quad K = -E$, so $K_1 = 4K_2$

Similarly, for state $n = 1$, potential energy,

$$U_1 = 2E_1 = 2 \times \frac{-13.6}{(1)^2} = -2 \times 13.6$$

For state $n = 2$, potential energy,

$$U_2 = 2E_2 = 2 \times \frac{-13.6}{(2)^2} = \frac{-2 \times 13.6}{4}$$

$$U_2 = \frac{U_1}{4}$$

$$\Rightarrow \quad U_1 = 4U_2$$

Thus, both kinetic energy K and potential energy U becomes four times.

20. As, $E = E_2 - E_1 = -\frac{13.6}{2^2} - \left(-\frac{13.6}{1^2}\right) = 10.2$ eV

21. As, $v = \frac{c}{\lambda} = c \cdot R\left(\frac{1}{n_1^2} - \frac{1}{n_2^2}\right)$

$$= 3 \times 10^8 \times 10^7 \left(\frac{1}{2^2} - \frac{1}{4^2}\right) = \frac{9}{16} \times 10^{15} \text{ Hz}$$

22. We have, $\lambda = \frac{hc}{\Delta E}$

∴ So, ratio of wavelengths

$$\frac{\lambda_1}{\lambda_2} = \frac{hc/\Delta E_1}{hc/\Delta E_2} = \frac{\Delta E_2}{\Delta E_1} = \frac{\left(\frac{4}{3}E - E\right)}{2E - E} = \frac{1}{3}$$

23. Ist excited state corresponds to $n_1 = 2$

2nd excited state corresponds to $n_2 = 3$

As, $\quad E \propto \frac{1}{n^2}$

$$\frac{E_1}{E_2} = \frac{n_2^2}{n_1^2} = \frac{3^2}{2^2} = \frac{9}{4}$$

24. For ground state $n_1 = 1$ to $n_2 = 1$

Energy absorbed by photon, $E = E_2 - E_1$

$$= +13.6\left(\frac{1}{n_1^2} - \frac{1}{n_2^2}\right) \times 1.6 \times 10^{-19} \text{ J}$$

$$= 13.6\left(\frac{1}{1} - \frac{1}{4^2}\right) \times 1.6 \times 10^{-19}$$

$$= 13.6 \times 1.6 \times 10^{-19}\left(\frac{15}{16}\right)$$

$$= 20.4 \times 10^{-19}$$

or $\quad E = hv = 20.4 \times 10^{-19}$

Frequency, $v = \frac{20.4 \times 10^{-19}}{h} = \frac{20.4 \times 10^{-19}}{6.63 \times 10^{-34}}$

$$= 3.076 \times 10^{15}$$

$$\simeq 3.1 \times 10^{15} \text{ Hz}$$

Wavelength of photon, $\lambda = \frac{c}{v} = \frac{3 \times 10^8}{3.076 \times 10^{15}}$

$$= 9.74 \times 10^{-8} \text{ m}$$

$$\simeq 9.7 \times 10^{-8} \text{ m}$$

25. $\bar{\nu} = R\left[\frac{1}{2^2} - \frac{1}{4^2}\right] = \frac{3R}{4} = 20397 \text{ cm}^{-1}$

For the same transition in He^+-atom ($Z = 2$)

$$\bar{\nu} = RZ^2\left(\frac{1}{2^2} - \frac{1}{4^2}\right) = \frac{3R \times 2^2}{4}$$

$$= 20397 \times 4$$
$$= 81588 \text{ cm}^{-1}$$

26. According to Rydberg's formula, $\frac{1}{\lambda} = R\left(\frac{1}{n_f^2} - \frac{1}{n_i^2}\right)$

Here, $n_f = 1$, $n_i = n$

$$\frac{1}{\lambda} = R\left(\frac{1}{1^2} - \frac{1}{n^2}\right)$$

$$\frac{1}{\lambda} = R\left(1 - \frac{1}{n^2}\right) \quad \text{...(i)}$$

$$\Rightarrow \quad \frac{R}{n^2} = R - \frac{1}{\lambda}$$

$$\Rightarrow \quad n = \sqrt{\frac{\lambda R}{\lambda R - 1}}$$

27. Energy at A, B and C state is E_A, E_B and E_C, then

$(E_C - E_B) + (E_B - E_A) = E_C - E_A$

or $$\frac{hc}{\lambda_1} + \frac{hc}{\lambda_2} = \frac{hc}{\lambda_3}$$

$$\Rightarrow \quad \frac{1}{\lambda_1} + \frac{1}{\lambda_2} = \frac{1}{\lambda_3}$$

$$\Rightarrow \quad \lambda_3 = \frac{\lambda_1\lambda_2}{\lambda_1 + \lambda_2}$$

28. In hydrogen atom, $E_n = -\frac{Rhc}{n^2}$

Also, $\quad E_n \propto m$

where, m is the mass of the electron.

Here, the electron has been replaced by a particle whose mass is double of an electron. Therefore, for this hypothetical atom energy in nth orbit will be given by

$$E_n = -\frac{2Rhc}{n^2}$$

The longest wavelength λ_{max} (or minimum energy) photon will correspond to the transition of particle from $n = 3$ to $n = 2$.

$$\therefore \quad \frac{hc}{\lambda_{max}} = E_3 - E_2 = 2Rhc\left(\frac{1}{2^2} - \frac{1}{3^2}\right)$$

This gives, $\lambda_{max} = 18/5R$

29. For hydrogen atom, we get

$$\frac{1}{\lambda} = RZ^2\left(\frac{1}{1^2} - \frac{1}{2^2}\right)$$

$$\frac{1}{\lambda_1} = R(1)^2\left(\frac{3}{4}\right)$$

$$\Rightarrow \quad \lambda_1 = \frac{4}{3R} \quad \text{...(i)}$$

For deuterium ($Z = 1$), we get

$$\frac{1}{\lambda_2} = R(1)^2\left(\frac{3}{4}\right)$$

$$\Rightarrow \quad \lambda_2 = \frac{4}{3R} \quad \text{...(ii)}$$

For singly ionised helium ($Z = 2$), we get

$$\frac{1}{\lambda_3} = R(2)^2\left(\frac{3}{4}\right)$$

$$\Rightarrow \quad \lambda_3 = \frac{1}{4} \cdot \frac{4}{3R} = \frac{\lambda_1}{4}$$

$$\Rightarrow \quad \lambda_1 = 4\lambda_3 \quad \text{...(iii)}$$

For doubly ionised lithium ($Z = 3$), we get

$$\frac{1}{\lambda_4} = R(3)^2\left(\frac{3}{4}\right)$$

$$\Rightarrow \quad \lambda_4 = \frac{1}{9} \cdot \frac{4}{3R} = \frac{\lambda_1}{9}$$

$$\Rightarrow \quad \lambda_1 = 9\lambda_4 \quad \text{...(iv)}$$

From Eqs. (i), (ii), (iii) and (iv), we get

$$\lambda_1 = \lambda_2 = 4\lambda_3 = 9\lambda_4$$

30. When an electron in a hydrogen atom jumps from an initial energy level n_i to some final energy level n_f, then if λ be the wavelength of emitted photon, then by Balmer's formula,

$$\frac{1}{\lambda} = R\left(\frac{1}{n_f^2} - \frac{1}{n_i^2}\right)$$

where, R = Rydberg constant.

In transition from $n = 4$ to $n = 3$, we have

$$\frac{1}{\lambda_1} = R\left(\frac{1}{3^2} - \frac{1}{4^2}\right) = R\left(\frac{7}{9 \times 16}\right) \quad \text{...(i)}$$

In transition from $n = 3$ to $n = 2$, we have

$$\frac{1}{\lambda_2} = R\left(\frac{1}{2^2} - \frac{1}{3^2}\right) = R\left(\frac{5}{9 \times 4}\right) \quad \text{...(ii)}$$

So, from Eqs. (i) and (ii), the ratio of $\frac{\lambda_1}{\lambda_2}$ is

$$\frac{\lambda_1}{\lambda_2} = \frac{\left(\frac{9 \times 16}{7R}\right)}{\left(\frac{9 \times 4}{5R}\right)} = \frac{20}{7}$$

31. For hydrogen or hydrogen like atoms, we know that

$$\frac{1}{\lambda} = RZ^2\left[\frac{1}{n_1^2} - \frac{1}{n_2^2}\right] \quad \text{...(i)}$$

where, R is Rydberg constant and Z is atomic number. When electron jumps from M - shell to the L - shell, then

$n_1 = 2$ (for L-shell)
$n_2 = 3$ (for M-shell)

$\therefore$ Eq (i) becomes,

$$\frac{1}{\lambda} = RZ^2\left[\frac{1}{2^2} - \frac{1}{3^2}\right] = \frac{5}{36} RZ^2 \quad \text{...(ii)}$$

Now, electron jumps from N-shell to the L - shell, for this

$$n_1 = 2 \qquad \text{(for } L \text{ - shell)}$$
$$n_2 = 4 \qquad \text{(for } N \text{ - shell)}$$

$\therefore$ Eq (i) becomes,

$$\frac{1}{\lambda'} = RZ^2\left[\frac{1}{2^2} - \frac{1}{4^2}\right] = \frac{3}{16}RZ^2 \qquad \text{...(iii)}$$

Now, we divide Eq. (ii) by Eq. (iii), we get

$$\frac{\lambda'}{\lambda} = \left(\frac{5}{36}RZ^2\right) \div \left(\frac{3}{16}RZ^2\right) = \frac{20}{27}$$

or $\quad \lambda' = \frac{20}{27}\lambda$

32. As, ionisation energy of ground state

$$= -\text{(ground state energy)}$$

$\Rightarrow \quad 9 \times 13.6 = -E_1 \qquad (\because 1 \text{ Rydberg} = 13.6 \text{ eV})$

or $\quad E_1 = -9 \times 13.6 \text{ eV}$

and energy of nth state for the given atom will be

$$E_n = \frac{E_1}{n^2} = \frac{-9 \times 13.6}{n^2} \text{ eV}$$

In transition from second excited state ($n = 3$) to ground state ($n = 1$), energy released will be

$$\Delta E = E(n = 3) - E(n = 1)$$
$$= \left(\frac{-9 \times 13.6}{3^2}\right) - \left(\frac{-9 \times 13.6}{1^2}\right)$$

or $\quad \Delta E = 9 \times 13.6\left(1 - \frac{1}{9}\right) = 8 \times 13.6 \text{ eV}$

So, wavelength of radiation emitted will be

$$\lambda = \frac{hc}{\Delta E} = \frac{1240 \text{ (eV-nm)}}{\Delta E \text{ (eV)}}$$

$\Rightarrow \quad \lambda = \frac{1240}{8 \times 13.6} = 11.39 \text{ nm} \approx 11.4 \text{ nm}$

33. Change in energy in transition from n to m stage is given by ($n > m$),

$$E_n = -\frac{E_0Z^2}{n^2}$$

Here, $\quad Z = 2$

$$\Delta E_n = +13.6 \times 4\left[\frac{1}{m^2} - \frac{1}{n^2}\right] = \frac{hc}{\lambda} \qquad \text{...(i)}$$

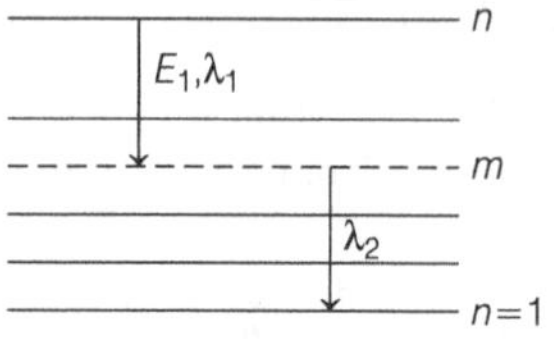

Let it start from n to m and then m to ground.

So, in first case,

$$13.6 \times 4 \times \left(\frac{1}{m^2} - \frac{1}{n^2}\right) = \frac{hc}{108.5 \text{ nm}} \qquad \text{...(ii)}$$

and in second case,

$$13.6 \times 4 \times \left(\frac{1}{1^2} - \frac{1}{m^2}\right) = \frac{hc}{30.4 \text{ nm}}$$

$\Rightarrow \quad \left(1 - \frac{1}{m^2}\right) = \frac{1240 \text{ eV}}{30.4 \times 13.6 \times 4} \quad \left(\because \text{given, } E = \frac{1240 \text{ eV}}{\lambda \text{(in nm)}}\right)$

$\Rightarrow \quad \left(1 - \frac{1}{m^2}\right) = 0.74980 \approx 0.75$

or $\quad \frac{1}{m^2} = 1 - 0.75 = 0.25 \Rightarrow m^2 = \frac{1}{0.25} = 4$

Hence, $m = 2$

So, by putting the value of m in Eq. (ii), we get

$$13.6 \times 4 \times \left(\frac{1}{2^2} - \frac{1}{n^2}\right) = \frac{1240}{108.5} \text{ eV}$$

$\Rightarrow \quad \left(\frac{1}{4} - \frac{1}{n^2}\right) = \frac{1240}{108.5 \times 13.6 \times 4}$

$\Rightarrow \quad \frac{1}{4} - \frac{1}{n^2} = 0.21$

or $\quad \frac{1}{n^2} = 0.25 - 0.21 = 0.04$

$$n^2 = \frac{1}{0.04} = 25 \Rightarrow n^2 = 25 \Rightarrow n = 5$$

34. De-excitation energy of hydrogen electron in transition $n = 2$ to $n = 1$ is

$$E = 13.6 \times \left(\frac{1}{n_f^2} - \frac{1}{n_i^2}\right) \text{eV} = 13.6\left(\frac{1}{1^2} - \frac{1}{2^2}\right)$$
$$= 10.2 \text{ eV}$$

Now, energy levels of helium ion's (He^+) electron are (For helium, $Z = 2$)

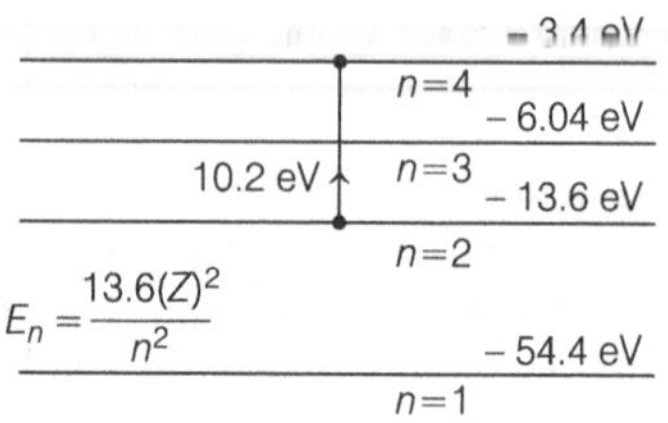

So, a photon of energy 10.2 eV can cause a transition $n = 2$ to $n = 4$ in a He^+ ion.

35. Energy of a hydrogen atom like ion by Bohr's model is $E_n = -13.6\frac{Z^2}{n^2}$

where, Z = atomic number

and $\quad n$ = principal quantum number.

For a He^+ ion in first excited state,

$$n = 2, Z = 2$$

$\therefore \quad E_2 = -13.6 \times \frac{4}{4} = -13.6 \text{ eV}$

$\therefore$ Ionisation energy of He^+ ion in first excited state

$$E = -E_2 = -(-13.6 \text{ eV}) = 13.6 \text{ eV}$$

36. $E \propto \frac{1}{r}, \; r \propto \frac{1}{m}, \; E \propto m$

$$\therefore \text{Ionisation potential} = 13.6 \times \frac{(\text{Mass}_\mu)\text{ eV}}{(\text{Mass}_e)}$$
$$= 13.6 \times 207 \text{ eV}$$
$$= 2815.2 \text{ eV}$$

37. Here, mass of muon, $m = 207m_e$

and mass of proton, $M = 1836m_e$; so the reduced mass is

$$\mu = \frac{mM}{m+M}$$
$$= \frac{(207m_e)(1836m_e)}{207m_e + 1836m_e} = 186m_e$$

According to equation $\lambda = h/p$, the orbit radius corresponding to $n = 1$ is $r_1 = \frac{h^2\varepsilon_0}{\pi m_e e^2} = 5.29 \times 10^{-11}$ m

Hence, the radius r' that corresponds to the reduced mass μ is

$$r_1' = \left(\frac{m}{\mu}\right) r_1 = \left(\frac{m_e}{186m_e}\right) r_1$$
$$= \frac{5.29 \times 10^{-11}}{186}$$
$$= 2.85 \times 10^{-13} \text{ m}$$

The muon is 186 times closer to the proton than an electron would be in the same orbit.

We have, $E_1 = -13.6$ eV (for $n = 1$)

$$E_1' = \left(\frac{\mu}{m}\right) E_1 = 186 \times (-13.6) \text{ eV}$$

$\Rightarrow \quad E_1 = -2.53 \times 10^3 \text{ eV} = -2.53$ keV

The ionisation energy is, therefore, 2.53 keV, 186 times that for an ordinary hydrogen atom.

38. Total energy of electron in excited state $= -13.6 + 12.1 = -1.5$ eV, which corresponds to third orbit. The possible spectral lines are when electron jumps from orbit 3rd to 2nd; 3rd to 1st and 2nd to 1st.

39. As, $$\frac{\lambda_l}{\lambda_s} = \frac{R\left(\frac{1}{1^2} - \frac{1}{\infty}\right)}{R\left(\frac{1}{1^2} - \frac{1}{2^2}\right)} = \frac{4}{3}$$

$$\lambda_l = \frac{4}{3}\lambda_s = \frac{4}{3} \times 911.6$$
$$= 1215.4 \text{ Å}$$
$$\simeq 1215 \text{Å}$$

40. As, $$\frac{\lambda_B}{\lambda_L} = \frac{\left(\frac{1}{1^2} - \frac{1}{2^2}\right)}{\left(\frac{1}{2^2} - \frac{1}{3^2}\right)} = \frac{\frac{3}{4}}{\frac{5}{36}} = \frac{27}{5}$$

$$\lambda_L = \frac{5}{27}\lambda_B = \frac{5}{27} \times 6563 = 1215.4 \text{ Å}$$

41. Series limit occurs in the transition $n_2 = \infty$ to $n_1 = 1$ in Lyman series and $n_2 = \infty$ to $n_1 = 5$ in Pfund series. For Lyman series,

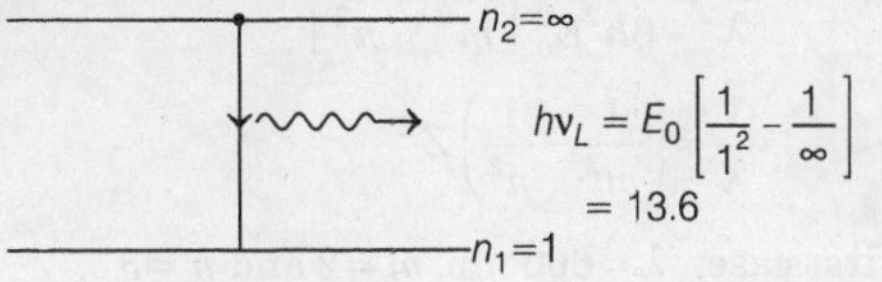

$$h\nu_L = 13.6 \quad \text{...(i)}$$

For Pfund series,

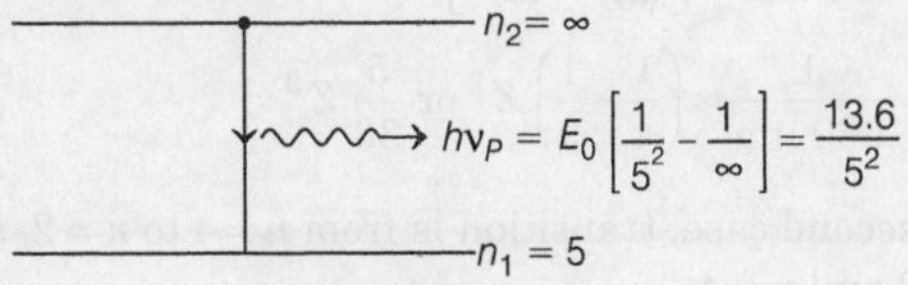

$$h\nu_P = \frac{13.6}{5^2} \quad \text{...(ii)}$$

From Eqs. (i) and (ii), we get

$$25h\nu_P = h\nu_L \Rightarrow \nu_P = \frac{\nu_L}{25}$$

42. As, $\nu = RC\left[\frac{1}{n_f^2} - \frac{1}{n_i^2}\right]$, we get

$$\nu_1 = RC\left(\frac{1}{1} - \frac{1}{\infty}\right) = RC,$$
$$\nu_2 = RC\left(\frac{1}{1} - \frac{1}{4}\right) = \frac{3}{4}RC$$

and $$\nu_3 = RC\left(\frac{1}{4} - \frac{1}{\infty}\right) = \frac{1}{4}RC$$

Clearly, $\nu_1 - \nu_2 = \frac{RC}{4} = \nu_3$, *i.e.* $\nu_1 - \nu_2 = \nu_3$

43. Given, radius of orbit, $r = 1.5 \times 10^{11}$ m

Orbital speed, $v = 3 \times 10^4$ m/s

Mass of earth, $m = 6 \times 10^{24}$ kg

Angular momentum, $mvr = \frac{nh}{2\pi}$

or $$n = \frac{2\pi vrm}{h}$$

[where, n = quantum number of the orbit]

$$= \frac{2 \times 3.14 \times 3 \times 10^4 \times 1.5 \times 10^{11} \times 6 \times 10^{24}}{6.63 \times 10^{-34}}$$
$$= 2.57 \times 10^{74}$$

or $\quad n = 2.6 \times 10^{74}$

Thus, the quantum number is 2.6×10^{74} which is too large.

The electron would jump from $n = 1$ to $n = 3$.

$$E_3 = \frac{-13.6}{3^2} = -1.5 \text{ eV}$$

So, they belong to Lyman series.

44. Expression for the energy of the hydrogenic electron states for atoms of atomic number Z is given by

$$E = h\nu = \frac{Z^2me^4}{8h^2E_0^2}\left[\frac{1}{m^2} - \frac{1}{n^2}\right] \quad \text{(Here, } m < n\text{)}$$

or $$\frac{hc}{\lambda} = \frac{Z^2me^4}{8h^2E_0^2}\left[\frac{1}{m^2} - \frac{1}{n^2}\right]$$

$$\Rightarrow \quad \frac{1}{\lambda} \propto \left(\frac{1}{m^2} - \frac{1}{n^2}\right)Z^2$$

For first case, $\lambda = 660$ nm, $m = 2$ and $n = 3$

$$\therefore \quad \frac{1}{660\text{ nm}} \propto \left[\frac{1}{(2)^2} - \frac{1}{(3)^2}\right]Z^2$$

$$\Rightarrow \quad \frac{1}{660\text{ nm}} \propto \left(\frac{1}{4} - \frac{1}{9}\right)Z^2 \text{ or } \frac{5}{36}Z^2 \quad \text{...(i)}$$

For second case, transition is from $n = 4$ to $n = 2$, *i.e.* $m = 2$ and $n = 4$

$$\therefore \quad \frac{1}{\lambda} \propto \left(\frac{1}{(2)^2} - \frac{1}{(4)^2}\right)Z^2$$

$$\Rightarrow \quad \frac{1}{\lambda} \propto \left(\frac{1}{4} - \frac{1}{16}\right)Z^2$$

or $$\frac{1}{\lambda} \propto \frac{3}{16}Z^2 \quad \text{...(ii)}$$

From Eqs. (i) and (ii), we get

$$\frac{\lambda}{660\text{ nm}} = \frac{5}{36} \times \frac{16}{3}$$

$$\Rightarrow \quad \lambda = \frac{80}{108} \times 660\text{ nm} = 488.9\text{ nm}$$

45. Number of spectral lines produced as an excited electron falls to ground state ($n = 1$) is

$$N = \frac{n(n-1)}{2}$$

In given case, $N = 6$

$$\therefore \quad 6 = \frac{n(n-1)}{2} \Rightarrow n = 4$$

So, L^{++} electron is in it's 3rd excited state.

Now, using the expression of energy of an electron in nth energy level,

$$E_n = -\frac{13.6Z^2}{n^2}\text{ eV}$$

where, Z is the atomic number.

$\therefore$ Energy levels of L^{++} electron are as shown

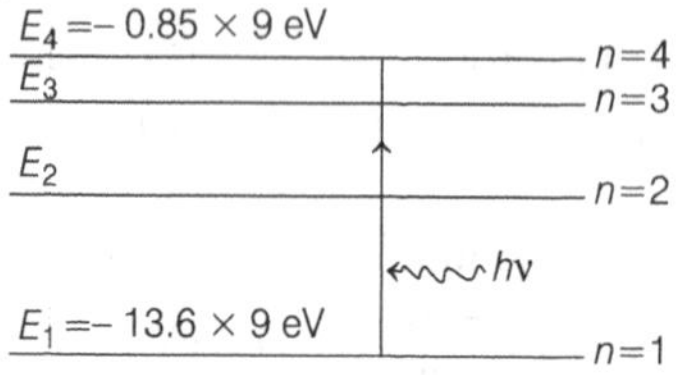

So, energy absorbed by electron from incident photon of wavelength λ is

$$\Delta E = \frac{hc}{\lambda}$$

$$\Rightarrow (13.6 \times 9 - 0.85 \times 9) = \frac{hc}{\lambda}$$

$$\Rightarrow \quad \lambda = \frac{hc}{9(13.6 - 0.85)}$$

$$\Rightarrow \quad \lambda = \frac{1240\text{ eV-nm}}{9 \times 12.75\text{ eV}} = 10.8\text{ nm}$$

46. Each atom of $_6C^{14}$ contains 6 p, 6 e and 8 n.

$\therefore$ In 14 g of $_6C^{14}$,

$$p = 6 \times 6 \times 10^{23} = 36 \times 10^{23}$$
$$n = 8 \times 6 \times 10^{23} = 48 \times 10^{23}$$
$$e = p = 36 \times 10^{23}$$

47. As, $\frac{r_1}{r_2} = \left(\frac{A_1}{A_2}\right)^{1/3} = \left(\frac{64}{27}\right)^{1/3} = \frac{4}{3}$

48. Mass density of nuclear matter is a constant quantity for all elements. It does not depends on element's mass number or atomic radius.

$\therefore$ The ratio of mass densities of ^{40}Ca and ^{16}O is 1 : 1.

49. Mass of nucleus = Mass number × Mass of a nucleon

$$= A \times 1.67 \times 10^{-27}\text{ kg}$$

Volume of nucleus $= \frac{4}{3}\pi R^3 = \frac{4}{3}\pi \times (1.3 \times 10^{-15} \times A^{1/3})^3$

$$= \frac{4}{3}\pi \times (1.3)^3 \times 10^{-45} \times A$$

$\therefore$ Density of nucleus $= \frac{\text{Mass}}{\text{Volume}}$

$$= \frac{A \times 1.67 \times 10^{-27}}{\frac{4}{3}\pi \times (1.3)^3 \times 10^{-45} \times A}$$

$$\approx 10^{17}\text{ kg m}^{-3}$$

50. As, $$R = R_0A^{1/3}$$

$$\Rightarrow \quad \frac{R_1}{R_2} = \left(\frac{A_1}{A_2}\right)^{1/3}$$

$$\Rightarrow \quad \frac{R}{R_{He}} = \left(\frac{A}{4}\right)^{1/3}$$

$$\Rightarrow \quad (14)^{1/3} = \left(\frac{A}{4}\right)^{1/3}$$

$$\Rightarrow \quad A = 56$$

or $$Z = A - N$$

$$\therefore \quad Z = 56 - 30 = 26$$

51. The elements higher on the BE *versus* mass number plot, *i.e.* representing the high peaks on the plot are very tightly bound and hence are stable. However, the elements on lower end of this plot, are less tightly bound due to less binding energy per nucleon and hence are unstable.

As, lithium nucleus shows a peak on this plot and having large value of binding energy per nucleon, so it implies that it is very stable.

52. Given, E_{bn} of chlorine = 287.67 MeV

and E_{bn} of phosphorus = 262.48 MeV

Binding energy per nucleon of ${}_{17}Cl^{35}$

$$= \frac{\text{Binding energy}}{\text{Mass number}}$$

$$= \frac{287.67}{35} = 8.22 \text{ MeV}$$

Similarly,

Binding energy per nucleon of ${}_{15}P^{31} = \frac{262.48}{31}$

$$= 8.47 \text{ MeV}$$

Since, BE per nucleon of ${}_{15}P^{31}$ is more than BE per nucleon of ${}_{17}Cl^{35}$, hence ${}_{15}P^{31}$ is more stable.

53. The reaction is ${}_3Li^7 + {}_1p^1 \longrightarrow 2\,({}_2He^4)$

$$\therefore \quad E_p = 2E\,({}_2He^4) - E_{(Li)}$$

$$= 2\,(4 \times 7.06) - 7 \times 5.6$$

$$= 56.48 - 39.2 = 17.28 \text{ MeV}$$

54. During formation of H-atom, some mass of nucleons convert into energy by $E = mc^2$, this energy is used to bind the nucleons along with nucleus.

$$\therefore \quad E = B = mc^2$$

$$\Rightarrow \quad m = \frac{B}{c^2}$$

So, mass of atom becomes slightly less than sum of actual masses of nucleons and electrons. Actual mass (m') of H-atom

$$m' = m_{\text{proton}} + m_{\text{electron}} - \frac{B}{c^2}$$

where, $B = 13.6$ eV (binding energy) per atom.

55. Most stable, $N = Z(B)$. Nucleus with more neutron number (N) is slightly more stable than nucleus which have more atomic number (Z).

So, order of stability is $B > A > C$.

56. $m_{\text{theoretical}} = Zm_p + (A - Z)m_n$

$$= 50(1.00783 \text{ u}) + (120 - 50)\,(1.00867 \text{ u})$$

$$= 50.391500 \text{ u} + 70.606900 \text{ u}$$

$$= 120.998400 \text{ u}$$

$m_{\text{experimental}} = 119.902199$ u

So, mass defect, $\Delta m = m_{\text{theoretical}} - m_{\text{experimental}}$

$$= 120.998400 \text{ u} - 119.902199 \text{ u}$$

$$= 1.096201 \text{ u}$$

Now, total binding energy, BE = $\Delta m \times 931$ MeV

$$= 1.096201 \times 931 \text{ MeV}$$

$$= 1020.563131 \text{ MeV}$$

Therefore, binding energy per nucleon,

$$(\text{BE})_{\text{per nucleon}} = \frac{\text{BE}}{\text{Total number of nucleons}}$$

$$= \frac{1020.563131}{120}$$

$$= 8.50469275833 \text{ MeV}$$

$$\simeq 8.5 \text{ MeV}$$

57. Nuclear force of attraction between any two nucleons (n-n, p-n, p-n) is same. The difference comes up only due to electrostatic force of repulsion between two protons.

$$\therefore \quad F_1 = F_3 \neq F_2$$

Since, nuclear force is greater than electrostatic force, hence

$$\therefore \quad F_1 = F_3 > F_2$$

58. In the given oxygen molecule, nuclear force between the nuclei of two atoms is not important because nuclear forces being short-ranged are confined only within one particular nucleus.The distance between the nuclei of two atoms may be large. So the nuclear forces between two nuclei may not be operative/effective.

59. (d) As, $\quad 3P \longrightarrow Q + E_1$

$\Rightarrow \quad Q - 3P = E_1$

$\Rightarrow \quad E_1 = E_Q - 3E_P \quad$ ...(i)

Aslo, $\quad 2Q \longrightarrow R + E_2$

$\Rightarrow \quad E_2 = E_R - 2E_Q \quad$ ...(ii)

And $\quad R - 6P = E_3$

$\Rightarrow \quad E_R - 6\,E_P = E_3 \quad$ ...(iii)

Now, $\quad 2E_1 + E_2 = 2(E_Q - 3E_P) + (E_R - 2E_Q)$

$$= E_R - 6\,E_P = E_3$$

60. Let number of α-particle emitted be x and number of β-particles emitted be y.

Difference in mass number $4x = 238 - 206 = 32$

$$x = 8$$

Difference in charge number $2x - 1\,y = 92 - 82 = 10$

$$16 - y = 10 \Rightarrow y = 6$$

61. Isotopes are the atoms of same element having same number of protons but different number of neutrons, *i.e.* Z remains same.

However, in α or β-decay, the nuclei hence formed always have Z either decreased by 2 or increased by 1. Thus, isotopes of the original nuclei cannot be occured in these decay.

62. $\lambda = \frac{0.693}{T_{1/2}} = \frac{0.693}{2.7 \times 24 \times 60 \times 60}$

$$\lambda = 2.9 \times 10^{-6} \text{ s}^{-1}$$

$$N = 6 \times 10^{23} \times \frac{1.0 \text{ mg}}{198 \text{ g}} = 3.03 \times 10^{18}$$

Activity $A = \lambda N$

$$= 2.9 \times 10^{-6} \times 3.03 \times 10^{18}$$

$$= 8.8 \times 10^{12} \text{ disintegration/s}$$

$$= \frac{8.8 \times 10^{12}}{3.7 \times 10^{10}} = 240 \text{ Ci}$$

63. Tritium $\rightarrow {}_1H^3$. The nucleus contains 1 proton and 2 neutrons. If one neutron decays $n \rightarrow p + e^- + \bar{\nu}$, the nucleus may have 2 protons and one neutron, *i.e.* tritium will transform into ${}_2He^3$ (with 2 protons and one neutron). But this does not happen because triton energy is less than that of ${}_2He^3$ nucleus, *i.e.* transformation is not allowed energetically.

64. A has shorter mean life as λ is greater for B.

65. $t_{1/2} = 3.8$ days

$$\therefore \quad \lambda = \frac{0.693}{t_{1/2}} = \frac{0.693}{3.8} = 0.182$$

We know that, $t = \frac{2.303}{\lambda} \log \frac{N_0}{N} = \frac{2.303}{0.182} \log \frac{N_0}{N_0/20}$

$$= \frac{2.303}{0.182} \log 20 = 16.46$$

$$\simeq 16.5 \text{ days}$$

66. Radioactivity is spontaneous self disruptive activity of the radioactive material. In $t = 1$ yr = half-life of material; on the average, half the number of atoms will decay. Therefore the containers will in general have different number of atoms of the material, but their average will be close to 5000.

67. $(t_{1/2})_x = (t_{\text{mean}})_y$

or $$\frac{0.693}{\lambda_x} = \frac{1}{\lambda_y}$$

$$\therefore \quad \lambda_x = 0.693\, \lambda_y$$

$$\lambda_x < \lambda_y$$

or rate of decay $= \lambda N$

Initially, number of atoms (N) of both are equal but since $\lambda_y > \lambda_x$, therefore y will decay at a faster rate than x.

68. Taking average count per minute in the first half value period as $(100 + 50)/2$, *i.e.* 75

Total number of counts during this period
$= 75 \times 3 \times 60 = 13500$

which is closest to the given result (14100).

69. Fraction $= \frac{N}{N_0} = \left(\frac{1}{2}\right) \Rightarrow \frac{6400}{1600} = 4$

Fraction of radium after 4 half-life is

$$= \frac{1}{2^4} = \frac{1}{16}$$

70. By radioactive disintegration (Rutherford and Soddy) law, amount of active (undecayed) radioactive material remained in a sample after time t is given by

$$M(t) = M_0 e^{-\lambda t}$$

where, M_0 – initial sample of mass taken at $t = 0$ and λ = disintegration constant.

Fraction of active material remained after time t is

$$\frac{M(t)}{M_0} = e^{-\lambda t} \quad \text{...(i)}$$

Given that, after time t, $\frac{M(t)}{M_0} = \frac{9}{16}$

So, $$e^{-\lambda t} = \frac{9}{16} \quad \text{[from Eq. (i)]}$$

Similarly, fraction of active material remained after time $t/2$ is

$$\frac{M\left(\frac{t}{2}\right)}{M_0} = e^{-\lambda \frac{t}{2}} = (e^{-\lambda t})^{\frac{1}{2}} = \left(\frac{9}{16}\right)^{\frac{1}{2}} = \frac{3}{4}$$

71. Activity of a sample after time t is related to initial activity as

$$A = A_0 \left(\frac{1}{2}\right)^n \quad \text{... (i)}$$

where, A = present activity
and A_0 = initial activity.

Also, $$n = t / t_{1/2} \quad \text{... (ii)}$$

where, t = time interval
and $t_{1/2}$ = half-life time.

Here, $A = 500\ \text{s}^{-1}$, $A_0 = 700\ \text{s}^{-1}$, $t = 30$ min

So, from Eq. (i), we get

$$500 = 700 \left(\frac{1}{2}\right)^n \Rightarrow 2^n = 1.4 \approx \sqrt{2}$$

So, $n = \frac{1}{2}$

Now, from Eq. (ii), we get

$$\frac{t}{t_{1/2}} = \frac{1}{2} \Rightarrow t_{1/2} = 2 \times t = 60 \text{ min}$$

which is close to 62 min.

72. Activity, $A = A_0 \left(\frac{1}{2}\right)^{t/T}$

or $$\frac{A_0}{64} = A_0 \left(\frac{1}{2}\right)^{t/T}$$

or $$\left(\frac{1}{2}\right)^6 = \left(\frac{1}{2}\right)^{t/\,18 \text{ days}}$$

$$\Rightarrow \quad 6 = \frac{t}{18} \Rightarrow t = 108 \text{ days.}$$

73. The total number of atoms can neither remain constant (as in option a) nor can ever increase (as in options b and c). They will continuously decrease with time. Therefore, (d) is the appropriate option.

74. Here given, at $t = 0$, count rate or initial activity is $A_0 = 1600\ \text{s}^{-1}$

At $t = 8$ s, count rate or activity is $A = 100\ \text{s}^{-1}$

So, decay scheme for given sample is

$$1600 \xrightarrow{t_{1/2}} 800 \xrightarrow{t_{1/2}} 400 \xrightarrow{t_{1/2}} 200 \xrightarrow{t_{1/2}} 100$$

So, $$8\text{s} = 4\, t_{1/2}$$

where, $t_{1/2}$ = half-life time.

$\Rightarrow \quad t_{1/2} = 2$ s

$\therefore$ From above decay scheme, we see that activity after 6 s is 200 counts per second.

75. The two samples of two radioactive nuclides A and B can simultaneously have the same decay rate at any time if initial rate of decay of A is twice the initial rate of decay of B and $\lambda_A > \lambda_B$. Also, when initial rate of decay of B is same as rate of decay of A at $t = 2$ h and $\lambda_B < \lambda_A$.

76. Decay scheme is

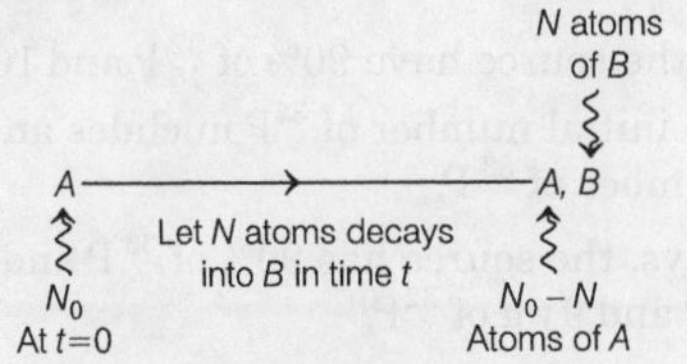

Given, $\dfrac{N_B}{N_A} = 0.3 = \dfrac{3}{10}$

$\Rightarrow \quad \dfrac{N_B}{N_A} = \dfrac{30}{100}$

So, $\quad N_0 = 100 + 30 = 130$ atoms

By using $\quad N = N_0 e^{-\lambda t}$, we have

$$100 = 130\, e^{-\lambda t}$$

$$\Rightarrow \quad \frac{1}{1.3} = e^{-\lambda t}$$

$$\Rightarrow \quad \log 1.3 = \lambda t$$

If T is half-life, then $\lambda = \dfrac{\log_e 2}{T}$

$$\Rightarrow \quad \log 1.3 = \frac{\log_e 2}{T} \cdot t$$

$$\therefore \quad t = \frac{T \cdot \log(1.3)}{\log_e 2}$$

77. Given, 80 min = 4 half-lives of A = 2 half-lives of B.

Let the initial number of nuclei in each sample be N.

For radioactive element A,

$$N_A \text{ after 80 min} = \frac{N}{2^4}$$

$\Rightarrow$ Number of A nuclides decayed $= N - \dfrac{N}{16} = \dfrac{15}{16} N$

For radioactive element B,

$$N_B \text{ after 80 min} = \frac{N}{2^2}$$

$\Rightarrow$ Number of B nuclides decayed

$$= N - \frac{N}{4} = \frac{3}{4} N$$

$\therefore$ Ratio of decayed numbers of A and B nuclei will be

$$\frac{(15/16)N}{(3/4)N} = \frac{5}{4}$$

78. Activity of a radioactive material is given as

$$R = \lambda N$$

where, λ is the decay constant and N is the number of nuclei in the radioactive material.

For substance A,

$$R_A = \lambda_A N_A = 10 \text{ mCi}$$

For substance B,

$$R_B = \lambda_B N_B = 20 \text{ mCi} \quad \text{...(i)}$$

As given in the question,

$$N_A = 2N_B \Rightarrow R_A = \lambda_A (2N_B) = 10 \text{ mCi} \quad \text{...(ii)}$$

$\therefore$ Dividing Eq. (ii) by Eq.(i), we get

$$\frac{R_A}{R_B} = \frac{\lambda_A (2N_B)}{\lambda_B (N_B)} = \frac{10}{20} \text{ or } \frac{\lambda_A}{\lambda_B} = \frac{1}{4} \quad \text{...(iii)}$$

As, half-life of a radioactive material is given as

$$T_{1/2} = \frac{0.693}{\lambda}$$

$\therefore$ For material A and B, we can write

$$\frac{(T_{1/2})_A}{(T_{1/2})_B} = \frac{\dfrac{0.693}{\lambda_A}}{\dfrac{0.693}{\lambda_B}} = \frac{\lambda_B}{\lambda_A}$$

Using Eq. (iii), we get

$$\frac{(T_{1/2})_A}{(T_{1/2})_B} = \frac{4}{1}$$

Hence, from the given options, only option (d) satisfies this ratio.

Therefore, $(T_{1/2})_A = 20$ days and $(T_{1/2})_B = 5$ days.

79. Let n_0 be the number of radioactive nuclei at time $t = 0$. Number of nuclei decayed in time t are given by $n_0\,(1 - e^{-\lambda t})$, which is also equal to the number of β-particles emitted during the same interval of time.

For the given condition,

$$n = n_0\,(1 - e^{-2\lambda}) \quad \text{...(i)}$$

$$(n + 0.75n) = n_0\,(1 - e^{-4\lambda}) \quad \text{...(ii)}$$

Dividing Eq. (ii) by Eq. (i), we get

$$1.75 = \frac{1 - e^{-4\lambda}}{1 - e^{-2\lambda}}$$

or $\quad 1.75 - 1.75\, e^{-2\lambda} = 1 - e^{-4\lambda}$

$\therefore \quad 1.75\, e^{-2\lambda} - e^{-4\lambda} = 0.75 \quad \text{...(iii)}$

Let us take $e^{-2\lambda} = x \quad \text{...(iv)}$

Then, the above equation is

$$x^2 - 1.75x + 0.75 = 0$$

or $\quad x = \dfrac{1.75 \pm \sqrt{(1.75)^2 - (4)\,(0.75)}}{2}$

or $\quad x = 1$ and $\dfrac{3}{4}$

$\therefore$ From Eq. (iv) either

$$e^{-2\lambda} = 1$$

or $\quad e^{-2\lambda} = \dfrac{3}{4}$

but $e^{-2\lambda} = 1$ is not acceptable because which means $\lambda = 0$.

Hence, $e^{-2\lambda} = \frac{3}{4}$

or $-2\lambda \ln(e) = \ln(3) - \ln(4) = \ln(3) - 2\ln(2)$

$\therefore$ $\lambda = \ln(2) - \frac{1}{2}\ln(3)$

Substituting the given values,

$$\lambda = 0.6931 - \frac{1}{2} \times (1.0986) = 0.14395 \text{ s}^{-1}$$

$\therefore$ Mean life, $t_{mean} = \frac{1}{\lambda} = 6.947 \text{ s} \simeq 7 \text{ s}$

80. If a radioactive sample undergoes two different processes simultaneously, then

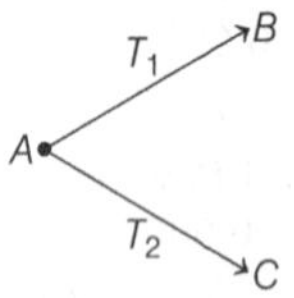

$$\frac{dN}{dT} = -\lambda_1 N - \lambda_2 N \Rightarrow \frac{dN}{dT} = -(\lambda_1 + \lambda_2)N$$

$$\frac{dN}{dT} = -\lambda_{eff} N$$

Here, $\lambda_{eff} = \lambda_1 + \lambda_2$

Let T_{eff} be the effective half-life of the nucleus.

$$\Rightarrow \quad \frac{\ln 2}{T_{eff}} = \frac{\ln 2}{T_1} + \frac{\ln 2}{T_2}$$

$$\frac{1}{T_{eff}} = \frac{1}{T_1} + \frac{1}{T_2} = \frac{1}{10} + \frac{1}{100} = \frac{11}{100}$$

$$\Rightarrow \quad T_{eff} = \frac{100}{11} = 9.09 \text{ s} \approx 9 \text{ s}$$

81. Let at time $t = t$, number of nuclei of Y and Z are N_Y and N_Z. Then, rate equations of the populations of X, Y and Z are

$$\left(\frac{dN_X}{dt}\right) = -\lambda_X N_X \quad \text{...(i)}$$

$$\left(\frac{dN_Y}{dt}\right) = \lambda_X N_X - \lambda_Y N_Y \quad \text{...(ii)}$$

and $$\left(\frac{dN_Z}{dt}\right) = \lambda_Y N_Y \quad \text{...(iii)}$$

Given, $N_Y(t) = \frac{N_0 \lambda_X}{\lambda_X - \lambda_Y}[e^{-\lambda_Y t} - e^{-\lambda_X t}]$

For N_Y to be maximum

$$\frac{dN_Y(t)}{dt} = 0$$

i.e. $\lambda_X N_X = \lambda_Y N_Y$ ...(iv) [from Eq. (ii)]

or $$\lambda_X (N_0 e^{-\lambda_X t}) = \lambda_Y \frac{N_0 \lambda_X}{\lambda_X - \lambda_Y}[e^{-\lambda_Y t} - e^{-\lambda_X t}]$$

or $$\frac{\lambda_X - \lambda_Y}{\lambda_Y} = \frac{e^{-\lambda_Y t}}{e^{-\lambda_X t}} - 1$$

$$\frac{\lambda_X}{\lambda_Y} = e^{(\lambda_X - \lambda_Y)t}$$

or $(\lambda_X - \lambda_Y)\, t \ln(e) = \ln\left(\frac{\lambda_X}{\lambda_Y}\right)$

or $$t = \frac{1}{\lambda_X - \lambda_Y} \ln\left(\frac{\lambda_X}{\lambda_Y}\right)$$

Substituting the values of λ_X and λ_Y, we have

$$t = \frac{1}{(0.1 - 1/30)} \ln\left(\frac{0.1}{1/30}\right) = 15 \ln(3)$$

or $t = 16.48$ s

82. Initially, the source have 90% of ${}^{32}_{15}\text{P}$ and 10% of ${}^{33}_{15}\text{P}$.

Let x g be initial number of ${}^{32}\text{P}$ nuclides and $9x$ g be initial number of ${}^{33}\text{P}$.

After t days, the source has 90% of ${}^{33}\text{P}$ and 10% of ${}^{32}\text{P}$, *i.e.* of ${}^{33}\text{P}$ and $9y$ g of ${}^{32}\text{P}$.

Using the equation,

$$\frac{N}{N_0} = e^{-\lambda t} = \left(\frac{1}{2}\right)^{t/T_{1/2}}$$

$$N = N_0 \left(\frac{1}{2}\right)^{t/T_{1/2}}$$

For P^{33}, $y = 9x \cdot 2^{-t/14.3}$ ...(i)

For P^{32}, $9y = x \cdot 2^{-t/25.3}$ ...(ii)

Dividing Eq. (i) by Eq. (ii),

$$\frac{y}{9y} = \frac{9x}{x} \cdot \frac{2^{-t/14.3}}{2^{-t/25.3}}$$

or $$\frac{1}{9} = 9 \times 2^{(t/25.3 - t/14.3)}$$

or $$\frac{1}{81} = 2^{-11t/25.3 \times 14.3}$$

Taking log on both the sides,

$$\log 1 - \log 81 = -\frac{11t}{25.3 \times 14.3} \log 2$$

or $$-1.9085 = \frac{-11 \times t}{25.3 \times 14.3} \times 0.3010$$

or $$t = \frac{25.3 \times 14.3 \times 1.9085}{11 \times 0.3010}$$

$$= 208.5 \simeq 208 \text{ days}$$

So, we must wait for 208 days to do so.

83. β^- decay is represented as

$$_Z X^A \rightarrow {}_{Z+1}A^Y + {}_{-1}e^0 + \bar{\nu} + Q_1$$

$\therefore$ $Q_1 = [m_N({}_Z X^A) - m_N({}_{Z+1}Y^A) - m_e]c^2$

$= [m_N({}_Z X^A) + Zm_e - m_N({}_{Z+1}Y^A) - (Z+1)m_e]c^2$

$= [m({}_Z X^A) - m({}_{Z+1}Y^A)]\, c^2 = (M_x - M_y)c^2$

β^+ decay is represented as

$$_Z X^A = {}_{Z-1}Y^A + {}_{+1}e^0 + \bar{\nu} + Q_2$$

$\therefore$ $Q_2 - [m_N({}_Z X^A) - m_N({}_{Z-1}Y^A) - m_e]c^2$

$= [m_N({}_Z X^A) + Zm_e - m_N({}_{Z+1}Y^A) - (Z-1)m_e - 2m_e]c^2$

$= [m({}_Z X^A) - m({}_{Z-1}Y^A) - 2m_e]\, c^2$

$= (M_x - M_y - 2m_e)c^2$

84. Given that, $m({}^7_3\text{Li}) = 7.0160$ u

$$m({}^4_2\text{He}) = 4.0026 \text{ u}$$
$$m({}^1_1\text{H}) = 1.0079 \text{ u}$$

Nuclear reaction, ${}^7_3\text{Li} + {}^1_1\text{H} \longrightarrow {}^4_2\text{He} + {}^4_2\text{He}$

Energy released in one nuclear reaction,

$$Q = \Delta mc^2 = \Delta m \times 931 \text{ MeV}$$
$$= (7.0160 + 1.0079 - 2 \times 4.0026) \times 931$$
$$= 17.41 \text{ MeV}$$

Number of atoms in 20 g of Li

$$= \frac{20}{7} \times 6.023 \times 10^{23} = 1.72 \times 10^{24}$$

$\therefore$ Total energy liberated

$$= Q \times \text{Total number of Li atoms}$$
$$= (17.41 \text{ MeV}) \times (1.72 \times 10^{24})$$
$$= 4.79 \times 10^{12} \text{ J}$$

$$\text{Energy (in kWh)} = \frac{4.79 \times 10^{12}}{3.6 \times 10^6} \quad (\because 1 \text{ kWh} = 3.6 \times 10^6 \text{ J})$$
$$= 1.33 \times 10^6 \text{ kWh}$$

85. Consider the reaction, ${}^1_0n + {}^{15}_8\text{O} \rightarrow {}^{15}_7\text{N} + {}^1_1\text{H} + Q$

So, $Q = (\Delta M)c^2 = (M_n + M_\text{O} - M_\text{N} - M_\text{H})c^2$

$$= 0.003796 \times 931.5 = 3.5359 \text{ MeV} \quad \ldots\text{(i)}$$

From the given relation,

$$\Delta Q = \Delta E = E_{{}^{15}_{8}\text{O}} - E_{{}^{15}_{7}\text{N}}$$
$$= \frac{3}{5}\frac{e^2}{4\pi\varepsilon_0 R}(8 \times 7 - 7 \times 6)$$
$$= \frac{3}{5} \times 1.44 \text{ MeV-fm} \times \frac{14}{R} \quad \ldots\text{(ii)}$$

Solving Eqs. (i) and (ii), we get

$$R = 3.42 \text{ fm}$$

86. ${}_0n^1 + {}_{92}\text{U}^{235} \longrightarrow {}_{51}\text{Sb}^{133} + {}_{41}\text{Nb}^{99}$ + neutrons

Charge number is conserved $(92 = 51 + 41)$

Applying principle of conservation of mass number

$$133 + 99 + x = 235 + 1$$
$$x = 236 - 232 = 4$$

$\therefore$ Number of neutrons $({}_0n^1) = 4$

87. The fission of ${}_{92}\text{U}^{235}$ is represented by

$${}_{92}\text{U}^{235} + {}_0n^1 \rightarrow {}_{56}\text{Ba}^{141} + {}_{36}\text{Kr}^{92} + 3\,{}_0n^1 + Q$$

The name of the particle X is neutron $({}_0n^1)$.

88. $\text{Fission rate} = \dfrac{\text{Total power}}{\text{Energy/Fission}}$

$$= \frac{5}{200 \times 1.6 \times 10^{-13}} = 1.56 \times 10^{11} \text{ s}^{-1}$$

89. Nuclear forces are short range forces because they arise from small separation of charges inside neutral atom.

90. Mass defect in the given nuclear reaction,

$$\Delta m = 2\,(\text{mass of deuterium}) - (\text{mass of helium})$$
$$= 2\,(2.0141) - (4.0026) = 0.0256$$

Therefore, energy released

$$\Delta E = (\Delta m)\,(931.48) \text{ MeV} = 23.85 \text{ MeV}$$
$$= 23.85 \times 1.6 \times 10^{-13} \text{ J} = 3.82 \times 10^{-12} \text{ J}$$

Efficiency is only 25%, therefore

$$25\% \text{ of } \Delta E = \left(\frac{25}{100}\right)(3.82 \times 10^{-12}) \text{ J}$$
$$= 9.55 \times 10^{-13} \text{ J}$$

i.e. by the fusion of two deuterium nuclei, 9.55×10^{-13} J energy is available to the nuclear reactor.

Total energy required in one day to run the reactor with a given power of 200 MW,

$$E_\text{Total} = 200 \times 10^6 \times 24 \times 3600 = 1.728 \times 10^{13} \text{ J}$$

$\therefore$ Total number of deuterium nuclei required for this purpose,

$$n = \frac{E_\text{Total}}{\Delta E/2} = \frac{2 \times 1.728 \times 10^{13}}{9.55 \times 10^{-13}}$$
$$= 0.362 \times 10^{26}$$

$\therefore$ Mass of deuterium required = (Number of g-moles of deuterium required) × 2 g

$$= \left(\frac{0.362 \times 10^{26}}{6.02 \times 10^{23}}\right) \times 2 = 120.26 \text{ g}$$

91. Number of moles of U^{235} in given sample,

$$n = \frac{\text{Sample mass}}{\text{Molar mass}} = \frac{2}{235} \text{ (kilomoles)}$$

Number of atoms of U^{235} in sample

$$= \text{Number of moles} \times N = \frac{2}{235} \times 6.023 \times 10^{26}$$

Total energy obtained

$$= \text{Number of atoms} \times \text{Energy from one atom}$$
$$= \frac{2 \times 6.023 \times 10^{26}}{235} \times 200 \times 10^6 \times 1.6 \times 10^{-19} \text{ J}$$

$$\text{Power} = \frac{\text{Energy}}{\text{Time}}$$
$$= \frac{2 \times 6.023 \times 10^{26} \times 200 \times 10^6 \times 1.6 \times 10^{-19}}{235 \times 30 \times 24 \times 3600}\left(\frac{\text{J}}{\text{s}}\right)$$
$$= 63.2 \times 10^6 \text{ W} = 63.2 \text{ MW}$$

It is close to 60 MW.

92. Here,

$A \rightarrow$ is the core of the reactor, which is the site of nuclear fission. It contains elements in suitably fabricated form. The fuel may be enriched uranium say ${}^{235}_{92}\text{U}$.

$D \rightarrow$ is a reflector surrounding the core which is used to reduce the leakage.

$C \rightarrow$ is coolant which helps in removing the energy (heat) released in fission. It transfers heat to working fluid which in turn may produce steam.

$B \rightarrow$ is control rods that shut down the reactor as they have high absorption of neutrons.

Hence, only option (b) shows correct purpose of given part.

93. Let E_k be the KE of the incident electron. Its linear momentum, $p = \sqrt{2mE_k}$

de-Broglie wavelength related to electron is

$$\lambda = \frac{h}{p} = \frac{h}{\sqrt{2mE_k}} \quad \text{or} \quad E_k = \frac{h^2}{2m\lambda^2}$$

The cut-off wavelength of the emitted X-rays is related to the KE of the incident electron as

$$\frac{hv}{\lambda_0} = E_k = \frac{h^2}{2m\lambda^2}$$

or $$\lambda_0 = \frac{2mc\lambda^2}{h}$$

94. In general, X-rays have larger wavelength than that of γ-rays.

95. In X-ray tube, the cut-off wavelength is given by $\lambda_{min} = \frac{hc}{eV}$. The cut-off wavelength depends on the energy eV of the accelerated electrons and is independent of the atomic number of target.

96. Applied potential difference (V) must be greater than binding energy.

i.e. $V > 40$ kV

97. As accelerating voltage V across X-rays tube increases, the value of minimum wavelength of X-rays, $\lambda_c = \frac{hc}{eV}$; decreases ; So, the separation between λ_K and λ_c increases.

98. de-Broglie wavelength, $\lambda_1 = \frac{h}{\sqrt{2meV}}$

X-ray wavelengths, $\lambda_2 = \frac{hc}{eV}$

$$\therefore \quad \frac{\lambda_1}{\lambda_2} = \frac{eV}{c\sqrt{2meV}} = \frac{1}{c}\sqrt{\frac{1}{2}\left(\frac{e}{m}\right)V}$$

$$= \frac{1}{3 \times 10^8}\sqrt{\frac{1}{2} \times 1.8 \times 10^{11} \times 10^4} = \frac{1}{10}$$

99. $\lambda_{min} = \frac{hc}{eV}$

$$\log(\lambda_{min}) = \log\left(\frac{hc}{e}\right) - \log V$$

$$y = c - mx$$

So, the required graph is given in option (d).

100. According to Moseley's law, $\sqrt{v} = a(Z - b)$

or $$v = a^2(Z - b)^2 \text{ or } \frac{c}{\lambda} = a^2(Z - b)^2$$

$$\therefore \quad \frac{\lambda_1}{\lambda_2} = \frac{(Z_2 - 1)^2}{(Z_1 - 1)^2}$$

Here, $\lambda_1 = \lambda$, $\lambda_2 = 4\lambda$, $Z_1 = 11$ and $Z_2 = ?$

$$\therefore \quad \frac{\lambda}{4\lambda} = \frac{(Z_2 - 1)^2}{(11 - 1)^2}$$

or $$(Z_2 - 1)^2 = 25 \text{ or } Z_2 = 6$$

101. Moseley's equation is $\sqrt{v} = a(Z - b)$

Thus, $$\sqrt{\frac{c}{\lambda_1}} = a(Z_1 - b) \quad \ldots\text{(i)}$$

and $$\sqrt{\frac{c}{\lambda_2}} = a(Z_2 - b) \quad \ldots\text{(ii)}$$

From Eqs. (i) and (ii), $\sqrt{c}\left(\frac{1}{\sqrt{\lambda_1}} - \frac{1}{\sqrt{\lambda_2}}\right) = a(Z_1 - Z_2)$

or $$a = \frac{\sqrt{c}}{(Z_1 - Z_2)}\left(\frac{1}{\sqrt{\lambda_1}} - \frac{1}{\sqrt{\lambda_2}}\right)$$

$$= \frac{(3 \times 10^8)^{1/2}}{42 - 27}\left[\frac{1}{(71 \times 10^{-12})^{1/2}} - \frac{1}{(178.5 \times 10^{-12})^{1/2}}\right]$$

$$= 5.0 \times 10^7 \,(\text{Hz})^{1/2}$$

Dividing Eq. (i) by Eq. (ii),

$$\sqrt{\frac{\lambda_2}{\lambda_1}} = \frac{Z_1 - b}{Z_2 - b}$$

or $$\sqrt{\frac{178.5}{71}} = \frac{42 - b}{27 - b} \text{ or } b = 1.37$$

Round II

1. We know that, $E_n = -13.6\frac{Z^2}{n^2}$ eV and $r_n = 0.53\frac{n^2}{Z}$ (Å)

Here, for $n = 1$, $E_1 = -54.4$ eV

Therefore, $-54.4 = -13.6\frac{Z^2}{1^2} \Rightarrow Z = 2$

Hence, radius of first Bohr orbit $r = \frac{0.53\,(1)^2}{2} = 0.265$ Å

2. The hydrogen atom before the transition was at rest. Therefore, from conservation of momentum,

$$p_{\text{H-atom}} = p_{\text{photon}} = \frac{E_{\text{radiated}}}{c} = \frac{13.6\left(\frac{1}{n_1^2} - \frac{1}{n_2^2}\right)\text{eV}}{c}$$

$$1.6 \times 10^{-27} \times v = \frac{13.6\left(\frac{1}{1^2} - \frac{1}{5^2}\right) \times 1.6 \times 10^{-19}}{3 \times 10^8}$$

$$\Rightarrow \quad v = 4.352 \text{ m/s} \approx 4 \text{ m/s}$$

3. In the second excited state, $n = 3$

$$\therefore \quad L_H = L_{Li} = 3\left(\frac{h}{2\pi}\right)$$

As $E \propto Z^2$ and $Z_H = 1$ and $Z_{Li} = 3$

$$\therefore \quad |E_{Li}| = 9|E_H| \text{ or } |E_H| < |E_{Li}|$$

4. $$\frac{N}{N_0} = \left(\frac{1}{2}\right)^{t/T}$$

$$\Rightarrow \quad \frac{N}{N_0} = \left(\frac{1}{2}\right)^{19/3.8} = \left(\frac{1}{2}\right)^5 = \frac{1}{32}$$

$$\Rightarrow \quad N = \frac{N_0}{32} = \frac{10.38}{32} = 0.32 \text{ g}$$

5. After one half-life period, the activity of tritium becomes 50%.

After 2 half-life period 25%

After 3 half-life period 12.5%

After 4 half-life period 6.25%

After 5 half-life period 3.12% = 3%

It is 5×12.5 yr + 7 yr, *i.e.* approximately 70 yr back only.

6. Activity, $A = \frac{-N}{dt} = \lambda N$

As the number of nuclei (N) per mole are equal for both the substances, irrespective of their molecular mass, therefore, $A \propto \lambda$

$$\frac{A_1}{A_2} = \frac{\lambda_1}{\lambda_2} = \frac{4}{3}$$

7. $n = \frac{t}{T} = \frac{42}{6} = 7$

$$\frac{N}{N_0} = \left(\frac{1}{2}\right)^n = \left(\frac{1}{2}\right)^7 = \frac{1}{128}$$

As rate of disintegration $\propto N$

$$\therefore \quad \frac{R}{R_0} = \frac{1}{128}; R = \frac{R_0}{128} = \frac{1024}{128} = 8\,\text{min}^{-1}$$

8. Here, $T = 4.47 \times 10^9$ yr

$$\frac{N}{N_0} = \frac{60}{100} = \left(\frac{1}{2}\right)^n \text{ or } 2^n = \frac{10}{6}$$

$$n \log 2 = \log 10 - \log 6 = 1 - 0.778 = 0.222$$

$$n = \frac{0.222}{\log 2} = \frac{0.222}{0.3} = 0.74$$

$$t = nT = 0.74 \times 4.47 \times 10^9 \text{ yr}$$

$$= 3.3 \times 10^9 \text{ yr}$$

9. As, $T = \frac{2\pi r}{V}$ or $V = \frac{rh}{2\pi mr}$

$$\therefore \quad T = \frac{2\pi r}{nh/2\pi mr} = \frac{mr^2}{nh} \propto \frac{r^2}{n}$$

But $r \propto n^2 \therefore T \propto n^3$

$$\text{or} \quad \frac{T_1}{T_2} = \left(\frac{n_1}{n_2}\right)^3$$

As, $T_1 = 8T_2$

$$\therefore \quad \left(\frac{n_1}{n_2}\right)^3 = 8, \frac{n_1}{n_2} = 2$$

Therefore, in given values $n_1 = 6, n_2 = 3$

10. Let $\lambda_A = \lambda$

$$\therefore \quad \lambda_B = 2\lambda$$

If N_0 is total number of atoms in A and B at $t = 0$, then initial rate of disintegration of $A = \lambda N_0$ and initial rate of disintegration of $B = 2\lambda N_0$

$$\text{As,} \quad \lambda_B = 2\lambda A$$

$$\therefore \quad T_B = \frac{1}{2} T_A$$

i.e. half-life of B is half the half-life of A.

After one half-life of A

$$\left(-\frac{dN}{dt}\right)_A = \frac{\lambda N_0}{2}$$

Equivalently, after two half lives of B

$$\left(-\frac{dN}{dt}\right)_B = \frac{2\lambda N_0}{4} = \frac{\lambda N_0}{2}$$

$$\text{Clearly,} \quad \left(-\frac{dN}{dt}\right)_A = -\left(\frac{dN}{dt}\right)_B$$

After $n = 1$, *i.e.* one half-life of A.

11. Helium atom has 2 electrons. When one electron is removed, the remaining atom is hydrogen like atom, whose energy in first orbit is

$$E_1 = -(2)^2 (13.6 \text{ eV}) = -54.4 \text{ eV}$$

Therefore, to remove the second electron from the atom, the additional energy of 54.4 eV is required. Hence, total energy required to remove both the electrons

$$= 24.6 + 54.4 = 79.0 \text{ eV}$$

12. $N_1 = N_0 e^{-\lambda t_1}$

$$\frac{N_1}{N_0} = e^{-\lambda t_1}$$

$$0.67 = e^{-\lambda t_1}$$

$$\ln(0.67) = -\lambda t_1 \quad \text{...(i)}$$

$$N_2 = N_0 e^{-\lambda t_2}$$

$$\frac{N_2}{N_0} = e^{-\lambda t_2}$$

$$0.33 = e^{-\lambda t_2}$$

$$\ln(0.33) = -\lambda t_2 \quad \text{...(ii)}$$

From Eqs. (i) and (ii), we get

$$\ln(0.67) - \ln(0.33) = \lambda t_1 - \lambda t_2$$

$$\lambda(t_1 - t_2) = \ln\left(\frac{0.67}{0.33}\right)$$

$$\lambda(t_1 - t_2) \cong \ln 2$$

$$t_1 - t_2 \simeq \frac{\ln 2}{\lambda} = t_{1/2}$$

Half life $= t_{1/2} = 20$ minutes.

13. As, radius of nth orbit,

$$r = r_0 \frac{n^2}{Z}$$

$$\Rightarrow \quad r_n \propto \frac{n^2}{Z}$$

$$\therefore \quad \frac{\Delta r_n}{r_n} = \frac{r_{n+1} - r_n}{r_n} \propto \frac{\frac{(n+1)^2}{Z} - \frac{n^2}{Z}}{\frac{n^2}{Z}}$$

$$= \frac{2n+1}{n^2} = \frac{2}{n} + \frac{1}{n^2} \simeq \frac{2}{n} \quad \left[n >> 1, \frac{1}{n^2} \to 0\right]$$

$$\Rightarrow \quad \frac{\Delta r_n}{r_n} \propto \frac{1}{n}$$

Angular momentum, $L = \frac{nh}{2\pi}$

$$\therefore \quad \frac{\Delta L}{L} = \frac{(n+1)\frac{h}{2\pi} - n\frac{h}{2\pi}}{n\frac{h}{2\pi}} = \frac{1}{n}$$

$$\therefore \quad \frac{\Delta L}{L} \propto \frac{1}{n}$$

Total energy, $E_n = -\frac{13.6\, Z^2}{n^2}$

$$\Rightarrow \quad \frac{\Delta E_n}{E_n} = \frac{E_{n+1} - E_n}{E_n} \propto \frac{\frac{Z^2}{(n+1)^2} - \frac{Z^2}{n^2}}{\frac{Z^2}{n^2}}$$

$$\Rightarrow \quad \frac{\Delta E_n}{E_n} \propto \frac{2n+1}{n^2} \propto \frac{2}{n} + \frac{1}{n^2} \simeq \frac{1}{n} \qquad [\text{As, } n >> 1]$$

14. Here, $E_n = \frac{-me^4}{8\varepsilon_0^2 n^2 h^2}$

is the energy of nth level of hydrogen atom. If all the H-atom are in ground state, ($n = 1$), then the radiation of frequency $(E_2 - E_1)/h$ falling on it may be absorbed by some of the atoms and move them to the first excited state ($n = 2$). All atoms may not be excited to $n = 2$ state. Further, as $(E_2 - E_1)/h$ is sufficient only to take the atom from $n = 1$ state to $n = 2$ state, no atoms shall make a tansition to $n = 3$ state.

15. Frequency of emitted radiation during transition from $(n+1)$th level to the nth level in hydrogen atom is

$$f = \frac{\Delta E}{h} \qquad (\because \Delta E = hf)$$

$$= \frac{1}{h}[E_{n+1} - E_n]$$

$$= \frac{1}{h}\left(\frac{-13.6}{(n+1)^2} - \left(\frac{-13.6}{n^2}\right)\right)$$

$$= \frac{1}{h} \times 13.6 \times \left(\frac{1}{n^2} - \frac{1}{(n+1)^2}\right)$$

$$= \frac{13.6}{h}\left\{\frac{2n+1}{n^2(n+1)^2}\right\}$$

$$\Rightarrow \quad f = \frac{13.6}{h}\left\{\frac{n\left(2 + \frac{1}{n}\right)}{n^4\left(1 + \frac{1}{n^2}\right)^2}\right\} \qquad \ldots\text{(i)}$$

If $n >> 1$, $\frac{1}{n}$ and $\frac{1}{n^2} \approx 0$

So, neglecting $\frac{1}{n}$ and $\frac{1}{n^2}$ in Eq (i), we have

$$f = \frac{13.6}{h} \times \left(\frac{2n}{n^4}\right) = \frac{2 \times 13.6}{h} \times \frac{1}{n^3}$$

$$\Rightarrow \quad f \propto \frac{1}{n^3}$$

16. First transition is from $n = 3$ to $n = 2$. Second transition is from $n = 2$ to $n = 1$

$$\therefore \quad \frac{E_1}{E_2} = c = \frac{1/2^2 - 1/3^2}{1/1^2 - 1/2^2}$$

As, $\quad p = \frac{E}{c}$, therefore,

$$\frac{p_1}{p_2} = b = \frac{E_1}{E_2} = c$$

i.e. $\quad b = c = \frac{5}{27}$

As, $\quad E = \frac{hc}{\lambda}$

$$\therefore \quad \lambda \propto \frac{1}{E}$$

or $\quad a = \frac{\lambda_1}{\lambda_2} = \frac{E_2}{E_1} = \frac{27}{5} = \frac{1}{c}$ or $c = \frac{1}{a}$

17. de-Broglie wavelength, $\lambda = \frac{h}{mv} = \frac{h}{p}$

where, p = momentum.

By conservation of momentum, $\mathbf{p}_1 + \mathbf{p}_2 = 0$

or $\quad p_1 = p_2$

$$\therefore \quad \lambda_1 = \lambda_2 = \lambda$$

18. Given, normal activity, $A_0 = 15$ decays/min

Present activity, $A = 9$ decays/min

$$T_{1/2} = 5730 \text{ yr}$$

Using the formula, $\frac{A}{A_0} = e^{-\lambda t}$

$$\frac{9}{15} = e^{-\lambda t}$$

or $\quad \frac{3}{5} = e^{-\lambda t}$ or $e^{\lambda t} = \frac{5}{3}$

Taking log on both the sides, we get

$$\lambda t \log_e e = \log_e 5 - \log_e 3$$

or $\quad \lambda t = 2.303\,(0.69 - 0.47)$

$$\lambda t = 0.5109 \qquad \left(\because \lambda = \frac{0.693}{T_{1/2}}\right)$$

$$t = \frac{0.5066 \times T_{1/2}}{0.693}$$

$$= \frac{0.5066 \times 5730}{0.693}$$

$$= 4224.47 \text{ yr}$$

Thus, the approximate age of Indus-Valley civilisation is 4224 yr.

19. $\lambda_{\text{eq.}} = \lambda_1 + \lambda_2$

$$\frac{1}{T_{1/2}} = \frac{1}{T_{1/2}^{(1)}} + \frac{1}{T_{1/2}^{(2)}}$$

$$T_{1/2} = \frac{T_{1/2}^{(1)}\, T_{1/2}^{(2)}}{T_{1/2}^{(1)} + T_{1/2}^{(2)}}$$

20. According to law of disintegration for radioactive substances,

$$R = R_0 e^{-\lambda t} \quad \Rightarrow \ln R = -\lambda t + \ln R_0$$

Slope of $\ln R$ *versus* t graph will give λ.

Also, decay constant, $\lambda = \dfrac{\ln 2}{T_{1/2}} \Rightarrow T_{1/2} = \dfrac{1}{\lambda} \ln 2$

$\therefore$ Slope of curve $A = \lambda_A = \dfrac{6}{10} \Rightarrow T_{1/2}(A) = \dfrac{10}{6} \ln 2$

Similarly, slope of curve $B = \lambda_B = \dfrac{6}{5}$

$$\Rightarrow \quad T_{1/2}(B) = \frac{5}{6} \ln 2$$

and slope of curve $C = \lambda_C = \dfrac{2}{5} \Rightarrow T_{1/2}(C) = \dfrac{5}{2} \ln 2$

So, $T_{1/2}(A) : T_{1/2}(B) : T_{1/2}(C) = \dfrac{10}{6} : \dfrac{5}{6} : \dfrac{5}{2}$

$$= 20 : 10 : 30 = 2 : 1 : 3$$

21. If wavelength of emitted photon in de-excitation is Λ_n;

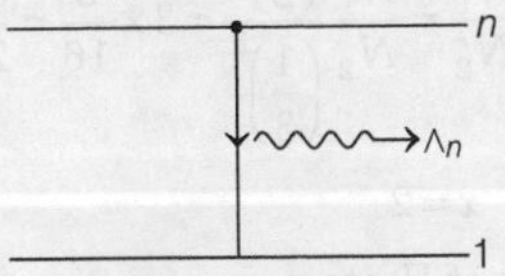

Then, $\dfrac{hc}{\Lambda_n} = E_n - E_g \qquad \left[\because E = \dfrac{p^2}{2m}\right]$

$$\frac{hc}{\Lambda_n} = \frac{p_n^2}{2m} - \frac{p_g^2}{2m}$$

As energies are negative, we get

$$\frac{hc}{\Lambda_n} = \frac{p_g^2}{2m} - \frac{p_n^2}{2m}$$

$$\Rightarrow \quad = \frac{p_g^2}{2m}\left(1 - \left(\frac{p_n}{p_g}\right)^2\right) = \frac{h^2}{2m\lambda_g^2}\left(1 - \frac{\lambda_g^2}{\lambda_n^2}\right)$$

$$\left[\because p \propto \lambda^{-1}, p = \frac{h}{\lambda}\right]$$

$$\Rightarrow \quad \Lambda_n = \frac{2m\lambda_g^2 c}{h}\left(1 - \frac{\lambda_g^2}{\lambda_n^2}\right)^{-1}$$

$$\Rightarrow \quad \Lambda_n = \frac{2m\lambda_g^2 c}{h}\left(1 + \frac{\lambda_g^2}{\lambda_n^2}\right) \qquad [\because (1 - x)^{-n} = 1 + nx]$$

$$\Rightarrow \quad \Lambda_n \simeq A + \frac{B}{\lambda_n^2}$$

where, $A = \left[\dfrac{2mc\lambda_g^2}{h}\right]$ and $B = \left[\dfrac{2mc\lambda_g^4}{h}\right]$ are constants.

22. Given, work-function, $W = 1.9$ eV

Wavelength of incident light, $\lambda = 400$ nm

$\therefore$ Energy of incident light, $E = \dfrac{hc}{\lambda} = 3.1$ eV

(Substituting the values of h, c and λ)

Therefore, maximum kinetic energy of photoelectrons

$$K_{\max} = E - W = (3.1 - 1.9) = 1.2 \text{ eV}$$

Now, the situation is as shown in figure.

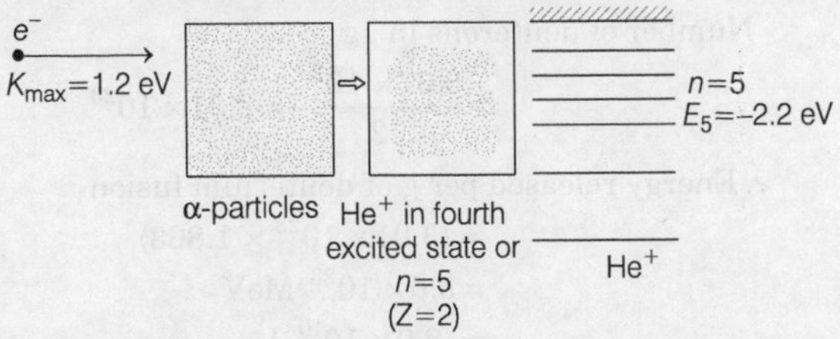

Energy of electron in 4th excited state of He^+ $(n = 5)$ will be

$$E_5 = -13.6 \frac{Z^2}{n^2} \text{ eV}$$

$$\Rightarrow \quad E_5 = -(13.6)\frac{(2)^2}{(5)^2} = -2.2 \text{ eV}$$

Therefore, energy released during the combination

$$= 1.2 - (-2.2) = 3.4 \text{ eV}$$

Similarly, energies in other energy states of He^+ will be

$$E_4 = -13.6 \frac{(2)^2}{(4)^2} = -3.4 \text{ eV}$$

$$E_3 = -13.6 \frac{(2)^2}{(3)^2} = -6.04 \text{ eV}$$

$$E_2 = -13.6 \frac{(2)^2}{(2)^2} = -13.6 \text{ eV}$$

The possible transitions are

$$\Delta E_{5 \to 4} = E_5 - E_4 = 1.2 \text{ eV} < 2 \text{ eV}$$

$$\Delta E_{5 \to 3} = E_5 - E_3 = 3.84 \text{ eV}$$

$$\Delta E_{5 \to 2} = E_5 - E_2 = 11.4 \text{ eV} > 4 \text{ eV}$$

$$\Delta E_{4 \to 3} = E_4 - E_3 = 2.64 \text{ eV}$$

$$\Delta E_{4 \to 2} = E_4 - E_2 = 10.2 \text{ eV} > 4 \text{ eV}$$

Hence, the energy of emitted photons in the range of 2 eV and 4 eV are 3.4 eV during combination and 3.84 eV and 2.64 after combination.

23. From the relation,

$$R = R_0 \, e^{-\lambda t} \qquad \text{(Here, } R = \text{activity of sample)}$$

Substituting the values, we have

$$2700 = 4750 \, e^{-5\lambda}$$

$$\therefore \quad 5\lambda = \ln\left(\frac{4750}{2700}\right) = 0.56$$

$$\therefore \quad \lambda = 0.113 \text{ min}^{-1}$$

Half-life of the sample,

$$t_{1/2} = \frac{\ln 2}{\lambda} = \frac{0.693}{0.113} = 6.132 \text{ min}$$

24. $\Delta m = 2\,(2.015) - (3.017 + 1.009) = 0.004$ amu

$\therefore$ Energy released = (30.004× 931.5) MeV

$$= 3.726 \text{ MeV}$$

Energy released per deuteron

$$=\frac{3.726}{2}=1.863\text{ MeV}$$

Number of deuterons in 1g

$$=\frac{6.02\times 10^{23}}{2}=3.01\times 10^{23}$$

$\therefore$ Energy released per g of deuterium fusion

$$=(3.01\times 10^{23}\times 1.863)$$
$$=5.6\times 10^{23}\text{ MeV}$$
$$\approx 9.0\times 10^{10}\text{ J}$$
$$\simeq 9\times 10^{9}\text{ kJ}$$
$$\simeq 9\times 10^{n}\text{ kJ (given)}$$

$\therefore \quad n=9$

25. At the time of observation ($t=t$),

$$\frac{m_1}{m_2}=\frac{100}{1} \qquad \text{(given)}$$

Further, it is given that $\frac{A_1}{A_2}=\frac{1.02}{1}$

Number of atoms $N=\frac{m}{A}$

$$\therefore \quad \frac{N_1}{N_2}=\frac{m_1}{m_2}\times\frac{A_2}{A_1}=\frac{100}{1.02} \qquad \ldots\text{(i)}$$

Let N_0 be the number of atoms of both the isotopes at the time of formation, then

$$\frac{N_1}{N_2}=\frac{N_0\,e^{-\lambda_1 t}}{N_0\,e^{-\lambda_2 t}}=e^{(\lambda_2-\lambda_1)t} \qquad \ldots\text{(ii)}$$

Equating Eqs. (i) and (ii), we have

$$e^{(\lambda_2-\lambda_1)t}=\frac{100}{1.02}$$

or $\quad (\lambda_2-\lambda_1)t=\ln(100)-\ln(1.02)$

$$\therefore \quad t=\frac{\ln(100)-\ln(1.02)}{\left(\frac{1}{2\times 10^{9}}-\frac{1}{4\times 10^{9}}\right)}$$

Substituting the values, we have

$$t=1.834\times 10^{10}\text{ yr}$$

26. Given, $\quad E_{2n}-E_1=204\text{ eV}$

$$\therefore \quad (13.6)Z^2\left(1-\frac{1}{4n^2}\right)=204 \qquad \ldots\text{(i)}$$

$$E_{2n}-E_n=40.8\text{ eV}$$

$$\therefore \quad 13.6\,Z^2\left(\frac{1}{n^2}-\frac{1}{4n^2}\right)=40.8 \qquad \ldots\text{(ii)}$$

Solving Eqs. (i) and (ii), we get

$$n=2 \quad \text{and} \quad Z=4$$
$$E_1=(-13.6)\,Z^2\text{ eV}$$
$$=(-13.6)\,(4)^2\text{ eV}=-217.6\text{ eV}$$

During de-excitation, minimum energy emitted is

$$E_{\min}=E_{2n}-E_{2n-1}=E_4-E_3$$
$$=\frac{-217.6}{4^2}-\left(\frac{-217.6}{3^2}\right)$$
$$=10.58\text{ eV}$$

27. As, $A_1=\lambda_1 N_1$ and $A_2=\lambda_2 N_2$

$$\lambda_1 N_1=\lambda_2 N_2 \qquad \ldots\text{(i)}$$

$$\Rightarrow \quad \frac{N_1}{N_2}=3 \qquad (\because N_1>N_2)$$

$$T_1=12\text{ h and }T_2=16\text{ h}$$

After integer, *i.e.* $2\times 24=48$ h

$$N_1'=N_1\left(\frac{1}{2}\right)^{t/T_1} \qquad \left(\text{But, }\frac{t}{T_1}=\frac{48}{12}=4\right)$$

$$\therefore \quad N_1'=N_1\left(\frac{1}{2}\right)^4 \qquad \ldots\text{(ii)}$$

and $\quad N_2'=N_2\left(\frac{1}{2}\right)^{t/T_2} \qquad \left(\text{But, }\frac{t}{T_2}=\frac{48}{16}=3\right)$

$$\therefore \quad N_2'=N_2\left(\frac{1}{2}\right)^3 \qquad \ldots\text{(iii)}$$

Now, dividing Eq. (ii) by Eq. (iii), we get

$$\frac{N_1'}{N_2'}=\frac{N_1}{N_2}\frac{\left(\frac{1}{2}\right)^4}{\left(\frac{1}{3}\right)^3}=3\times\frac{8}{16}=\frac{3}{2}=\frac{3}{x}\text{ (given)}$$

$\therefore \quad x=2$

28. Balmer series of H-atom spectrum occurs when an excited electron jumps to second level or $n=2$ state.

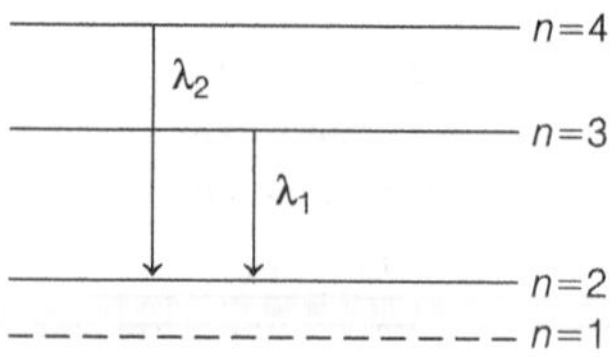

Emitted wavelengths are given by

$$\frac{1}{\lambda}=R\left(\frac{1}{2^2}-\frac{1}{n^2}\right)$$

where, $n=3,4,5\ldots\infty$.

First member is obtained in transition $n=3$ to $n=2$.

$$\Rightarrow \quad \frac{1}{\lambda_1}=R\left(\frac{1}{2^2}-\frac{1}{3^2}\right) \qquad \ldots\text{(i)}$$

Second member is obtained in transition $n=4$ to $n=2$.

$$\Rightarrow \quad \frac{1}{\lambda_2}=R\left(\frac{1}{2^2}-\frac{1}{4^2}\right) \qquad \ldots\text{(ii)}$$

From Eqs. (i) and (ii), we have

$$\frac{\frac{1}{\lambda_1}}{\frac{1}{\lambda_2}}=\frac{\left(\frac{1}{2^2}-\frac{1}{3^2}\right)}{\left(\frac{1}{2^2}-\frac{1}{4^2}\right)} \Rightarrow \frac{\lambda_2}{\lambda_1}=\frac{\left(\frac{1}{4}-\frac{1}{9}\right)}{\left(\frac{1}{4}-\frac{1}{16}\right)}=\frac{5}{36}\times\frac{64}{12}$$

$$\Rightarrow \quad \frac{\lambda_2}{\lambda_1}=\frac{5}{9}\times\frac{4}{3}=\frac{20}{27} \Rightarrow \lambda_2=\lambda_1\times\frac{20}{27}$$

$$=6561\times\frac{20}{27}=4860\text{ Å}=486\text{ nm}$$

29. As, $\frac{1}{\lambda} = R\left(\frac{1}{n_f^2} - \frac{1}{n_i^2}\right)$

$$\Rightarrow \quad \lambda = \frac{1}{R\left(\frac{1}{n_f^2} - \frac{1}{n_i^2}\right)} \quad \text{...(i)}$$

For Lyman series,

Largest wavelength ($n_i = 2$ and $n_f = 1$)

$$\lambda_1 = \frac{1}{R\left(\frac{1}{1^2} - \frac{1}{2^2}\right)} = \frac{1}{R\left(\frac{3}{4}\right)} = \frac{4}{3R}$$

Shortest wavelength ($n_i = \infty$ and $n_f = 1$)

$$\lambda_2 = \frac{1}{R\left(\frac{1}{1^2} - \frac{1}{\infty^2}\right)} = \frac{1}{R(1)} = \frac{1}{R}$$

Now, it is given that the difference between the largest and the shortest wavelengths of the Lyman series is 304 Å.

So, $\Delta\lambda = \lambda_1 - \lambda_2$

$$304 = \frac{4}{3R} - \frac{1}{R} \Rightarrow 304 = \frac{1}{3R}$$

$$R = \frac{1}{3 \times 304} \Rightarrow R = \frac{1}{912} \quad \text{...(ii)}$$

For Paschan series,

Largest wavelength ($n_i = 4$ and $n_f = 3$)

$$\lambda_3 = \frac{1}{R\left(\frac{1}{3^2} - \frac{1}{4^2}\right)} = \frac{1}{R\left(\frac{7}{144}\right)} = \frac{144}{7R}$$

Shortest wavelength ($n_i = \infty$ and $n_f = 3$)

$$\lambda_4 = \frac{1}{R\left(\frac{1}{3^2} - \frac{1}{\infty^2}\right)} = \frac{1}{R\left(\frac{1}{9}\right)} = \frac{9}{R}$$

Now, the difference between the largest and the shortest wavelengths of the Paschan series,

$$\Delta\lambda = \lambda_3 - \lambda_4$$

$$= \frac{144}{7R} - \frac{9}{R} = \frac{144 - 63}{7R} = \frac{81}{7R} = \frac{81}{7\left(\frac{1}{912}\right)}$$

[from Eq. (i)]

$$= \frac{81 \times 912}{7} = 10553.14 \text{ Å} \simeq 10553 \text{ Å}$$

30. Here, mass of hydrogen atom is 5 times mass of colliding particle.

Using law of conservation of linear momentum,

$$mv + 0 = 0 + 5mv' \Rightarrow v' = \frac{v}{5}$$

When the atom goes to its first excited state,

$$\text{Energy dissipated} = \left[\frac{-13.6}{(2)^2}\right] \text{eV} - \left[\frac{-13.6}{(1)^2}\right] \text{eV}$$

$$= -3.4 + 13.6 = 10.2 \text{ eV} \quad \text{...(i)}$$

$$\text{Loss of kinetic energy} = \frac{1}{2}mv^2 - \frac{1}{2}(5m)\left(\frac{v}{5}\right)^2$$

$$= \frac{1}{2}mv^2\left(1 - \frac{1}{5}\right) \text{s} = \frac{4}{5}\left(\frac{mv^2}{2}\right) = \frac{4}{5}K \quad \text{...(ii)}$$

From Eqs. (i) and (ii), we get

$$\frac{4}{5}K = 10.2 \text{ eV}$$

$$\Rightarrow \quad K = \frac{10.2 \times 5}{4} \text{ eV} = \frac{51.0}{4} \text{ eV}$$

$$\Rightarrow \quad \frac{N}{4} = \frac{51}{4} \text{ eV} \Rightarrow N = 51$$

31. If λ is de-Broglie wavelength, then for nth stationary orbit

$$2\pi r_n = n\lambda$$

where r_n is radius of nth orbit.

$$r_n = \frac{\varepsilon_0 h^2 n^2}{\pi m Z e^2}$$

$$\therefore \quad 2\pi\left(\frac{\varepsilon_0 h^2 n^2}{\pi m Z e^2}\right) = n\lambda \Rightarrow \frac{1}{\lambda} = \frac{mZe^2}{2\varepsilon_0 h^2 n} \quad \text{...(i)}$$

For Lyman series of hydrogen like atom, wavelength is given as,

$$\frac{1}{\lambda} = Z^2 R\left(\frac{1}{1^2} - \frac{1}{n^2}\right) \quad \text{...(ii)}$$

From Eqs. (i) and (ii), $Z^2 R\left(1 - \frac{1}{n^2}\right) = \frac{mZe^2}{2\varepsilon_0 h^2 n}$

Rydberg constant, $R = \frac{me^4}{8\varepsilon_0^2 ch^3}$

$$\therefore \quad \frac{Z^2 me^4}{8\varepsilon_0^2 ch^3}\left(1 - \frac{1}{n^2}\right) = \frac{mZe^2}{2\varepsilon_0 h^2 n}$$

$$\text{or} \quad \left(1 - \frac{1}{n^2}\right) = \frac{4\varepsilon_0 ch}{m^2 Z}$$

Substituting given values, we get

$$= \frac{4 \times (8.85 \times 10^{-12}) \times (3 \times 10^8) \times (6.62 \times 10^{-34})}{n \times (1.6 \times 10^{-19})^2 \times 11} = \frac{25}{n}$$

$$\therefore \quad n^2 - 1 = 25n$$

$$\text{or} \quad n^2 - 25n - 1 = 0$$

$$n - \frac{25 \pm \sqrt{(-25)^2 + 4 \times 1 \times 1}}{2} \approx \frac{25 \pm \sqrt{625}}{2} \approx 25$$

As negative n is not possible, $n \approx 25$.

32. $F = \frac{-dU}{dr} = -4U_0 r^3 = \frac{mv^2}{r}$

$$\therefore \quad mv^2 = 4U_0 r^4$$

$$\text{or} \quad v \propto r^2$$

we have $mvr = \frac{nh}{2\pi}$

$$r^3 \propto n$$

$$r \propto n^{1/3}$$

$$\therefore \quad \alpha = 3$$

30

Communication System

In this chapter, we introduce the concepts of communication, namely the mode of communication, the need for modulation, production and deduction of amplitude modulation.

IN THIS CHAPTER ....

- Basic Elements of Communication System
- Antenna
- Propagation of EM Waves
- Modulation and Demodulation
- Bandwidth

Basic Elements of Communication System

A communication system is a set up used to transmit information from one point to another. The essential parts of a communication system are transmitter, communication channel and receiver as shown in the following block diagram

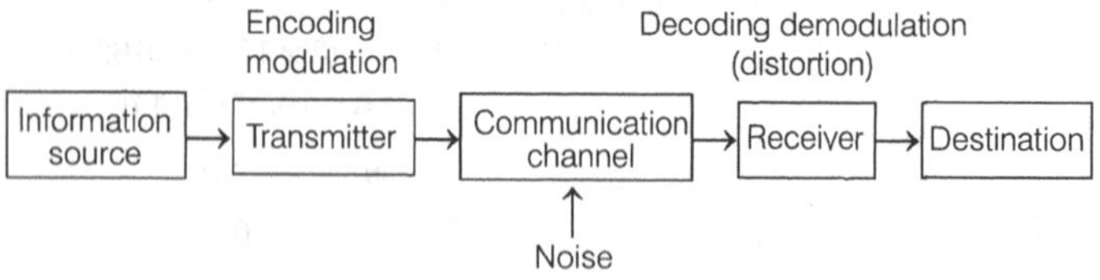

Antenna

An antenna is basically a conductor of small length which is used to radiate or receive electromagnetic waves.

It is of two types, *i.e.* dipole antenna and dish antenna.

Types of antenna are shown in following table relating its length with wavelength

Types of Antenna	Relation of Length (l) with Wavelength (λ)
Hertz antenna	$l = \lambda / 2$
Marconi antenna	$l = \lambda / 4$

Power Radiated by Antenna

The power radiated from a linear antenna of length l is given by

$$P \propto \left(\frac{1}{\lambda}\right)^2$$

For good transmission, for given antenna, length, wavelength λ should be small or frequency ν should be high.

Example 1. *The physical sizes of the transmitter and receiver antenna in a communication system are* **[JEE Main 2019]**

(a) proportional to carrier frequency
(b) inversely proportional to modulation frequency
(c) independent of both carrier and modulation frequency
(d) inversely proportional to carrier frequency

Sol. (d) Size of antenna is directly proportional to the wavelength of the signal.

Also, the speed at which signal moves = carrier frequency × wavelength

$$\Rightarrow \quad f\lambda = c \Rightarrow \lambda \propto \frac{1}{f}$$

$\therefore$ Size of antenna $\propto \frac{1}{f}$.

Note Minimum size of the antenna is $\lambda / 4$.

Propagation of EM Waves

In communication using radio waves, an antenna at the transmitter radiates the electromagnetic waves,which travel through the space and reach the receiving antenna at the other end.

Different layers of the atmosphere and their interaction with the propagating electromagnetic waves are given in following table

Name of the stratum (layer)	Approximate height over earth's surface	Exists during	Frequencies most affected
Troposphere	10 km	Day and night	VHF(upto several GHz)
D (part of stratosphere)	65-75 km	Day only	Reflects LF absorbs MF and HF to some degree
E (part of stratosphere)	100 km	Day only	Helps surface waves, reflects HF
F_1 (part of mesosphere)	170-190 km	Day time merges with F_2 at night	Partially absorbs HF waves yet allowing them to reach F_2
F_2 (thermosphere)	300 km at night 250-400 km during day time	Day and night	Efficiently reflects HF waves, particularly at night.

Different types of propagation of EM wave depending upon properties of medium are given below

Ground Wave Propagation

In this type of communication, transmitting and receiving antenna are close to surface of the earth. This type of propagation can be sustained only at low frequency (≈ 500 kHz to 1500 kHz).

Space Wave Propagation

The transmitted signal is received by the direct interception of the signal by receiving antenna. The frequency range is (100 MHz to 220 MHz). The maximum range of this transmission depends upon the height of transmitting antenna and is given by, $d = \sqrt{2hR}, R >> h$.

where, h = height of the antenna

and R = radius of the earth.

Maximum line of sight distance between two antenna (transmitting and receiving) as per diagram,

$$d_{\max} = d_T + d_R = \sqrt{2Rh_T} + \sqrt{2Rh_R}$$

where, d_T = radio horizon of transmitting antenna,

d_R = radio horizon of receiving antenna

and $d_{\max}$ = maximum line of sight distance between two antennas.

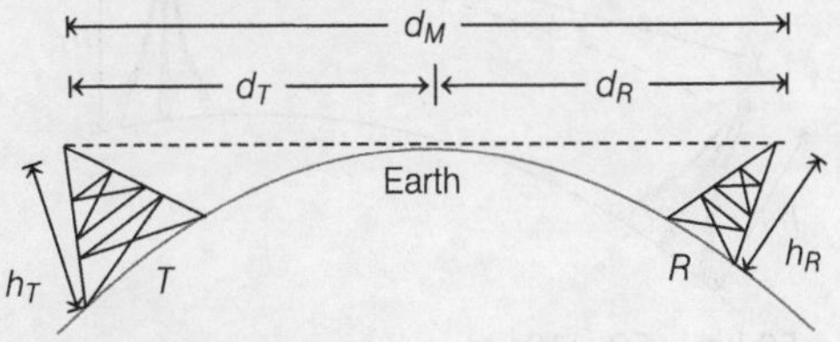

Sky Wave Propagation

Sky wave is the radio wave which is directed towards the sky and reflected back by the ionosphere towards the desired location on the earth. Radio waves of frequencies 2 MHz to 20 MHz can be reflected by the ionosphere by the phenomenon of total internal reflection.

Terms related to sky wave propagation along with their definitions are shown below.

Terms	Definitions	Expressions
Critical Frequency	Maximum frequency of radio waves reflected from ionosphere	$\nu_c \approx 9(N_{\max})^{1/2}$ where, $N_{\max}$ = maximum electron density of ionosphere.
MUF (Maximum Usable Frequency)	The highest radio frequency that can be used for transmission.	MUF = ν_c sec θ where, θ = angle between normal and the direction of incident wave.
Skip distance	Shortest distance from a transmitter, measured along the earth's surface at which a sky wave of fixed frequency will be first received after reflection from ionosphere.	$D_{\text{skip}} = 2h\sqrt{\left(\frac{\nu_{\max}}{\nu_c}\right)^2 - 1}$

Example 2. *On a particular day, the maximum frequency reflected from the ionosphere is 10 MHz. On another day, it was found to increase to 11 MHz. Calculate the ratio of maximum electron density of the ionosphere on the two days.*

(a) 1.21 (b) 0.82 (c) 0.50 (d) 0.25

Sol. (a) As, $f_c = 9\sqrt{N_{max}}$

$$\therefore \quad N_{max} = \frac{f_c^2}{81} \propto f_c^2$$

$$\Rightarrow \quad \frac{(N_2)_{max}}{(N_1)_{max}} = \left(\frac{11}{10}\right)^2 = \frac{121}{100} = 1.21$$

Example 3. *In a line of sight radio communication, a distance of about 50 km is kept between the transmitting and receiving antennas. If the height of the receiving antenna is 70 m, then the minimum height of the transmitting antenna should be (Take, radius of the earth = 6.4×10^6 m)* **[JEE Main 2019]**

(a) 20 m (b) 32 m (c) 40 m (d) 51 m

Sol. (b) Here in figure, h_R and h_T is the height of receiving and transmitting antenna, respectively.

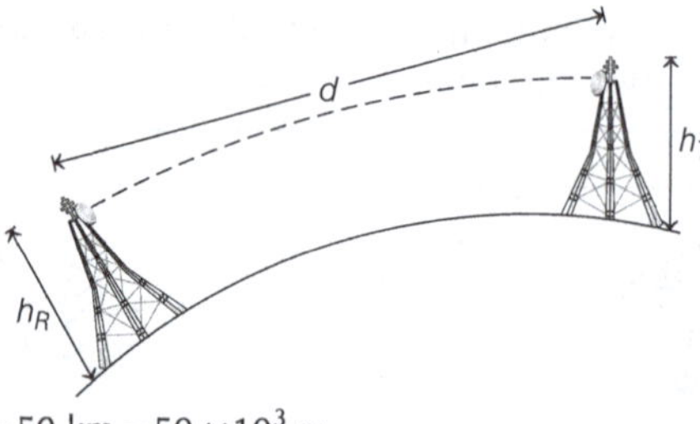

Given, $d = 50 \text{ km} = 50 \times 10^3$ m,

$h_R = 70$ m, $R = 6.4 \times 10^6$ m

Then, distance between transmitting and receiving antenna, i.e.

$$d = \sqrt{2Rh_T} + \sqrt{2Rh_R}$$

$$50 \times 10^3 = \sqrt{2R}(\sqrt{h_T} + \sqrt{h_R}) = \sqrt{2 \times 6.4 \times 10^6}(\sqrt{h_T} + \sqrt{70})$$

$$\Rightarrow \quad \sqrt{h_T} \approx \frac{50 \times 10^3}{3577.7} - 8.37 = 13.98 - 8.37 = 5.61$$

or $\quad h_T = 31.5 \approx 32$ m

Modulation and Demodulation

Modulation and demodulation are explained below.

Modulation

Modulation is process in which characteristics (amplitude, frequency and phase) of carrier wave changes according to the message (or base band) signal.

Need for Modulation

Some important needs for modulation are

(i) **Frequency of signal** The audio frequency signals (20 Hz to 20 kHz) cannot be transmitted without distortion over long distances due to less energy carried by low frequency audio waves.

(ii) **Height of antenna** For efficient radiation and reception, the height of transmitting and receiving antennas should be comparable to a quarter wavelength of the frequency used.

(iii) **Number of channels** Modulation increases the number of channels, *i.e.*

$$\text{Number of channels} = \frac{\text{Total bandwidth of channel}}{\text{Bandwidth per channel}}$$

Example 4. *A telephonic communication service is working at carrier frequency of 10 GHz. Only 10% of it is utilised for transmission. How many telephonic channels can be transmitted simultaneously, if each channel requires a bandwidth of 5 kHz?* **[JEE Main 2018]**

(a) 2×10^3 (b) 2×10^4
(c) 2×10^5 (d) 2×10^6

Sol. (c) Only 10% of 10 GHz is utilised for transmission.

$\therefore$ Band available for transmission

$$= \frac{10}{100} \times 10 \times 10^9 \text{ Hz} = 10^9 \text{ Hz}$$

Now, if there are n channels each using 5 kHz, then

$$n \times 5 \times 10^3 = 10^9$$

$$\Rightarrow \quad n = 2 \times 10^5$$

Frequency Modulation

The phenomenon of superposition of information signal over a high frequency carrier wave such that the frequency of the carrier wave is modified according to frequency of information signal, keeping the amplitude of the carrier wave constant is called amplitude modulation. Frequency modulation is preferred for music or TV transmission.

Amplitude Modulation

In amplitude modulation, the amplitude of the carrier is varied in accordance with the amplitude of information or message signal. Let $c(t) = A_c \sin \omega_c t$ represent carrier wave and $m(t) = A_m \sin \omega_m t$ represent the message or the modulating signal, where, $\omega_m = 2\pi f_m$ is the angular frequency of the message signal. The modulated signal $c_m(t)$ can be written as

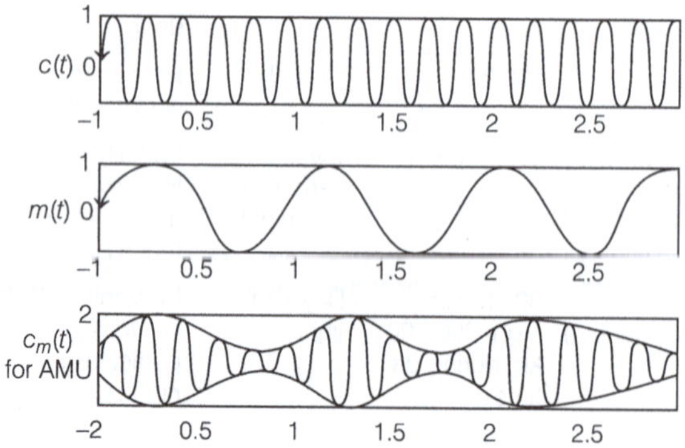

$$c_m(t) = (A_c + A_m \sin\omega_m t)\sin\omega_c t$$
$$= A_c\left(1 + \frac{A_m}{A_c}\sin\omega_m t\right)\sin\omega_c t$$

or
$$c_m(t) = A_c \sin\omega_c t + \mu A_c \sin\omega_m t \sin\omega_c t$$

where, $\mu = \dfrac{A_m}{A_c}$ is the modulation index in practice μ is kept ≤ 1 to avoid distortion.

It can be further written as

$$c_m(t) = A_c \sin\omega_c t + \frac{\mu A_c}{2}\cos(\omega_c - \omega_m)t - \frac{\mu A_c}{2}\cos(\omega_c + \omega_m)t$$

Here, $(\omega_c - \omega_m)$ and $(\omega_c + \omega_m)$ are respectively called the lower side and upper side frequencies. The frequency spectrum of the amplitude modulated signal is as follows

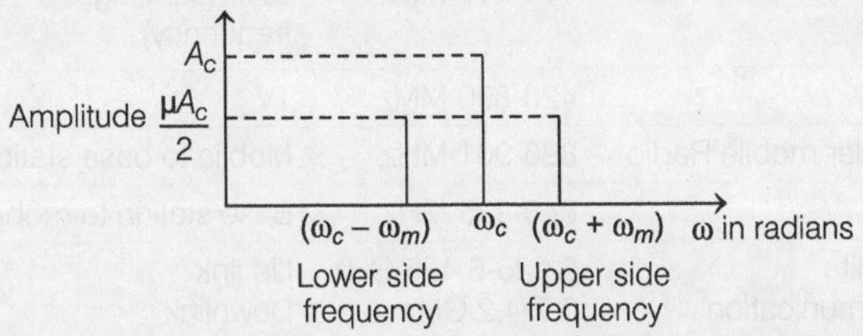

Example 5. *A message signal of frequency 10 kHz and peak voltage of 10 V is used to modulate a carrier of frequency 1 MHz and peak voltage of 20 V. The modulation index is* **[NCERT]**

(a) 1 (b) 0.5 (c) 0.2 (d) 1.2

Sol. (b) Modulation index is given by, $m_a = \dfrac{A_m}{A_c}$

where, A_m is amplitude of the message signal and A_c represents the amplitude of carrier signal.

Given carrier signal amplitude, $A_c = 20$

and message signal amplitude, $A_m = 10$

$$\therefore \quad \mu = \frac{10}{20} = 0.5$$

Modulation Index in Amplitude Modulation

It can be defined as the ratio of the change in amplitude of the carrier wave to the amplitude of the original carrier wave. It is also called modulation index. It used to determine the quality of transmitted signal.

$$m_a = \frac{A_m}{A_c} = \frac{E_{\max} - E_{\min}}{E_{\max} + E_{\min}}$$

where,

$A_{\max}$ = maximum voltage of amplitude modulation wave.

$A_{\min}$ = minimum voltage of amplitude modulation wave.

Modulation Index in Frequency Modulation

Ratio of frequency deviation to the modulating frequency is called modulation index.

$$m_f = \frac{\Delta_f}{f_m}$$

Sideband Frequencies of Amplitude Modulation Wave

- Original carrier wave frequency, $f_c = \dfrac{\omega_c}{2\pi}$ and amplitude $= A_c$.
- The sum of carrier frequency and modulating frequency is called USB (Upper Side Band) frequency $= f_c + f_m$.
- The difference of carrier frequency and modulating frequency is called LSB (Lower Side Band) frequency $= f_c - f_m$.

Power and Current Relations in Amplitude Modulation Wave

Average power in the unmodulated wave, $P_c = \dfrac{A_c^2}{2R}$

Total power in the modulated wave is $P_t = P_c\left(1 + \dfrac{m^2}{2}\right)$

or
$$\frac{P_t}{P_c} = 1 + \frac{m^2}{2}$$

where, m = modulation index

As, $P_t = I_t^2 R$ and $P_c = I_c^2 R$

$$\therefore \quad \frac{I_t}{I_c} = \sqrt{1 + \frac{m^2}{2}}$$

Demodulation

The process of recovering the original audio signal from the modulated wave is called demodulation.

Example 6. *An amplitude modulated wave is represented by the expression $V_m = 5(1 + 0.6\cos 6280t)\sin(211 \times 10^4 t)$ volt. The minimum and maximum amplitudes of the amplitude modulated wave are, respectively* **[JEE Main 2020]**

(a) 3 V, 5 V (b) $\frac{5}{2}$ V, 8V (c) $\frac{3}{2}$ V, 5V (d) 5 V, 8 V

Sol. (b) Equation of amplitude modulated wave given as

$$V = 5(1 + 0.6\cos 6280\, t)\sin(211 \times 10^4 t)$$
$$= (5 + 3\cos 6280\, t)\sin(211 \times 10^4 t)$$

Comparing above equation with standard equation,

$$V = (A_c + A_m \cos\omega_m t)\cdot \sin\omega_c t$$

We get, amplitude of carrier signal $(A_c) = 5$ V

Amplitude of message signal $(A_m) = 3$ V

Maximum amplitude of modulated signal

$$= A_c + A_m = 5 + 3 = 8\text{ V}$$

Minimum amplitude of modulated signal

$$= A_c - A_m = 5 - 3 = 2\text{ V}$$

The answer is close to option (b).

Note Due to phase shift minimum/maximum amplitudes may differ slightly.

Example 7. *A signal A cos ωt is transmitted using $v_0 \sin \omega_0 t$ as carrier wave. The correct amplitude modulated (AM) signal is* **[JEE Main 2019]**

(a) $(v_0 \sin \omega_0 t + A \cos \omega t$
(b) $(v_0 + A) \cos \omega t \sin \omega_0 t$
(c) $v_0 \sin[\omega_0(1 + 0.01A \sin \omega t)t]$
(d) $v_0 \sin \omega_0 t + \frac{A}{2} \sin(\omega_0 - \omega)t + \frac{A}{2} \sin(\omega_0 + \omega)t$

Sol. (d) Given, modulating signal, $A_m = A \cos \omega t$

Carrier wave, $A_c = v_0 \sin \omega_0 t$
In amplitude modulation, modulated wave is given by

$$Y_m = (A_0 + A_m) \sin \omega_0 t$$

where, A_0 is amplitude of the carrier wave (given as v_0)

$$\therefore \quad Y_m = (v_0 + A \cos \omega t) \sin \omega_0 t$$
$$= v_0 \sin \omega_0 t + A \sin \omega_0 t \cos \omega t$$
$$= v_0 \sin \omega_0 t + \frac{A}{2}[\sin(\omega_0 + \omega)t + \sin(\omega_0 - \omega)t]$$
$$= v_0 \sin \omega_0 t + \frac{A}{2} \sin(\omega_0 - \omega)t + \frac{A}{2} \sin(\omega_0 + \omega)t$$

Example 8. *The modulation frequency of an AM radio station is 250 kHz, which is 10% of the carrier wave. If another AM station approaches you for license, what broadcast frequency will you allot?* **[JEE Main 2019]**

(a) 2000 kHz (b) 2250 kHz
(c) 2900 kHz (d) 2750 kHz

Sol. (a) For a given carrier wave of frequency f_c with modulation frequency f_m, the bandwidth is calculated by $f_{upper} = f_c + f_m$

$$f_{lower} = f_c - f_m \quad \text{... (i)}$$

To avoid overlapping of bandwidths, next broadcast frequencies can be

$$f_1 = f_c \pm 2f_m,\ f_2 = f_c \pm 3f_m$$

So, next immediate available broadcast frequency is $f_1 = f_c + 2f_m$

and $\quad f_1' = f_c - 2f_m$

Given, $\quad f_m = 250$ kHz

and also that f_m is 10% of f_c, i.e. $f_c = 2500$ kHz.

So, $\quad f_1 = 2500 + (2 \times 250) = 3000$ kHz

and $\quad f'_1 = 2500 - (2 \times 250) = 2000$ kHz

Bandwidth

It is defined as the difference between the upper side and lower side frequency present in the amplitude modulation wave.

Bandwidth $= (f_c + f_m) - (f_c - f_m) = 2f_m$

Bandwidth of Transmission Medium

The commonly used transmission media are wire, free space and fibre optics cable. Coaxial cable offers a bandwidth of approximately 750 MHz. Such cables are normally operated below 18 GHz.

Service	Frequency bands	Comments
Standard AM broadcast	540-600 kHz	
FM broadcast	88-108 MHz	[VHF (Very high frequency)]
Television	54-72 MHz	VHF (very high frequency)
	76-88 MHz	TV
	174-216 MHz	UHF (ultra high frequency)
	420-890 MMz	TV
Cellular mobile Radio	896-901 MHz	Mobile to base station
	840-935 MHz	Base station to mobile
Satellite Communication	5.925-6.425 GHz 3.7-4.2 GHz	Up link Downlink

Example 9. *Given below in the left column are different modes of communication using the kinds of waves given in the right column.* **[JEE Main 2019]**

A.	Optical fibre communication	P.	Ultrasound
B.	Radar	Q.	Infrared light
C.	Sonar	R.	Microwaves
D.	Mobile phones	S.	Radio waves

From the options given below, find the most appropriate match between entries in the left and the right column.

(a) A-Q, B-S, C-R, D-P
(b) A-S, B-Q, C-R, D-P
(c) A-Q, B-S, C-P, D-R
(d) A-R, B-P, C-S, D-Q

Sol. (c) In optical fibre communication, infrared light is used to transmit information from one point to another.

RADAR (Radio detection and ranging) is a detection system that uses radio waves to determine range, angle or velocity of objects.

SONAR (Sound navigation and ranging) is also a detection system that uses ultrasound to detect under water objects, submarines, etc.

Mobile phone is a portable telephone that can make and receive calls, which make use of microwave.

∴ Correct sequence is

$$A \rightarrow Q,\ B \rightarrow S,\ C \rightarrow P \text{ and } D \rightarrow R$$

Practice Exercise

ROUND I Topically Divided Problems

Basics of Communication Systems

1. Range of frequencies allotted for commercial UHF, TV broadcast is

(a) 470-960 kHz (b) 47-960 MHz
(c) 470-960 MHz (d) 47-960 kHz

2. A 100 m long antenna is mounted on a 500 m tall building. The complex can become a transmission tower for waves with λ **[NCERT Exemplar]**

(a) ~ 400 m (b) ~ 25 m
(c) ~ 150 m (d) ~ 2400 m

3. Digital signals **[NCERT]**

(i) do not provide a continuous set of values
(ii) represent values as discrete steps
(iii) can utilise binary system and
(iv) can utilise decimal as well as binary systems

Which of the above statements are true?

(a) (i) and (ii)
(b) (ii) and (iii)
(c) (i), (ii) and (iii)
(d) All of (i), (ii), (iii) and (iv)

4. In satellite communication, the communication satellite

(a) acts as a reflector for a beam of modulated microwave from transmitter sent directly towards it
(b) acts as a repeater for a signal reaching there, without any change in frequency
(c) receives the coming modulated microwave signal, amplifies it and returns it to earth at a different frequency
(d) None of the above

5. The radio waves of frequency 30 MHz to 300 MHz belong to

(a) high frequency band
(b) very high frequency band
(c) ultra high frequency band
(d) super high frequency band

6. Frequency range for microwaves is

(a) 3×10^4 to 3×10^9 Hz
(b) 3×10^4 to 3×10^{12} Hz
(c) 3×10^5 to 3×10^8 Hz
(d) 1×10^9 to 3×10^{11} Hz

7. Ozone layer above earth's atmosphere will not

(a) prevent infrared radiations from sun reaching on earth
(b) prevent infrared radiations originated from earth from escaping earth's atmosphere
(c) prevent ultraviolet rays from sun
(d) reflect back radio waves

8. Which one of the following is correct?

(a) A single geostationary satellite can cover the whole part of the earth for microwave communication
(b) Atleast three geostationary satellites in the same orbit around earth's can cover the whole part of the earth for microwave communication
(c) The first Indian communication satellite is Apple
(d) The satellite communication is not like the line of sight microwave communication

9. A communication between a fixed base station and several mobile units, located on ships or aircraft utilising two way radio communication in the VHF and UHF is of frequency band

(a) 3 to 30 MHz (b) 30 to 300 MHz
(c) 30 to 470 MHz (d) 30 to 600 MHz

10. The diameter of an optical fibre is

(a) 10^{-5} m (b) 10^{-4} cm
(c) 10^{-3} cm (d) 10^{-2} cm

11. Optical fibre communication is generally preferred over general communication system because

(a) it is more efficient
(b) it has signal security
(c) it cannot be jammed as easily as radio waves
(d) All of the above

12. A laser is a coherent source because it contains

(a) many wavelengths
(b) uncoordinate waves of particular wavelength
(c) coordinated waves of many wavelength
(d) coordinated waves of particular wavelength

13. Quality of transmission depends upon

(a) nature of medium only
(b) nature of signal only
(c) Both (a) and (b)
(d) Neither (a) nor (b)

Propagation of EM Waves

14. Point to point communication requires the use of
(a) a guided medium only
(b) unguided medium only
(c) any medium
(d) None of the above

15. Three waves A, B and C of frequencies 1600 kHz, 5 MHz and 60 MHz respectively are to be transmitted from one place to another. Which of the following is the most appropriate mode of communication? [NCERT Exemplar]
(a) A is transmitted *via* space wave while B and C are transmitted *via* sky wave
(b) A is transmitted *via* ground wave, B *via* sky wave and C *via* space wave
(c) B and C are transmitted *via* ground wave while A is transmitted *via* sky wave
(d) B is transmitted *via* ground wave while A and C are transmitted *via* space wave

16. Is it necessary for a transmitting antenna to be at the same height as that of the receiving antenna for line of sight communication? A TV transmitting antenna is 81 m tall. How much service area can it cover, if the receiving antenna is at the ground level? [NCERT]
(a) 3800 km ^{2} (b) 3260 km ^{2}
(c) 7400 km ^{2} (d) 3320 km ^{2}

17. A 1 kW signal is transmitted using a communication channel which provides attenuation at the rate of − 2dB per km. If the communication channel has a total length of 5 km, the power of the signal received is $\left[\text{Gain in dB} = 10\log\left(\frac{P_0}{P_i}\right)\right]$ [NCERT Exemplar]
(a) 900 W (b) 100 W (c) 990 W (d) 1010 W

18. A signal emitted by an antenna from a certain point can be received at another point of the surface in the form of
(a) sky wave (b) ground wave
(c) sea wave (d) Both (a) and (b)

19. Which of the following frequencies will be suitable for beyond the horizon communication using sky waves? [NCERT Exemplar]
(a) 10 kHz (b) 10 MHz
(c) 1 GHz (d) 1000 GHz

20. Two identical antennas mounted on identical towers are separated from each other by a distance of 45 km. What should nearly be the minimum height of receiving antenna to receive the signals in line of sight? (Assume radius of earth is 6400 km) [JEE Main 2021]
(a) 19.77 m (b) 39.55 m
(c) 79.1 m (d) 158.2 m

21. In earth's atmosphere, for F_1-layer; the virtual height and critical frequency are
(a) 150 km and 3 MHz (b) 160 km and 3.5 MHz
(c) 170 km and 4.5 MHz (d) 180 km and 5 MHz

22. When electromagnetic waves enter the ionised layer of ionosphere, then the relative permittivity, *i.e.* dielectric constant of the ionised layer
(a) does not change
(b) appears to increase
(c) appears to decrease
(d) sometimes appears to increase and sometimes to decrease

23. The highest frequency of radio waves which when sent at some angle towards the ionosphere, gets reflected from that and returns to the earth is called
(a) critical frequency
(b) maximum unusable frequency
(c) polarisation of waves
(d) None of the above

24. In earth's atmosphere, for E-layer, the virtual height and critical frequency are
(a) 80 km and 3 MHz (b) 90 km and 3.5 MHz
(c) 120 km and 4.5 MHz (d) 110 km and 4 MHz

25. In space communication, the information can be passed from one place to another at a distance of 100 km in
(a) 1 s (b) 0.5 s
(c) 0.003 s (d) None of these

26. A TV transmission tower antenna is at a height of 20 m. Suppose that the receiving antenna is at
(i) ground level (ii) a height of 5 m
The increase in antenna range in case (ii) relative to case (i) is n%. The value of n, to the nearest integer, is
(a) 40 (b) 50 (c) 25 (d) 60

27. Which mode of communication is/are employed for the transmission of TV signals?
(a) Ground wave propagation
(b) Sky wave propagation
(c) Space wave propagation
(d) None of the above

28. Frequencies in the UHF range normally propagate by means of
(a) ground waves (b) sky waves
(c) surface waves (d) space waves

29. A TV tower has a height of 75 m. What is the maximum area upto which this TV communication can be possible?
(a) 1509 km^2 (b) 3018 km^2
(c) 2254 km^2 (d) 6036 km^2

30. The maximum distance upto which TV transmission from a TV tower of height h can be received is proportional to
(a) $h^{3/2}$ (b) h (c) $h^{1/2}$ (d) h^2

31. Maximum usable frequency (MUF) infregion layer is x. When the critical frequency is 60 MHz and the angle of incidence is 70°, then n is
(a) 140 MHz (b) 175 MHz
(c) 180 MHz (d) 160 MHz

Modulation and Demodulation

32. In amplitude modulation, carrier wave frequencies are
(a) lower compared to those in frequency modulation
(b) higher compared to those in frequency modulation
(c) same as in frequency modulation
(d) lower sometimes and higher sometimes to those in frequency modulation

33. Sinusoidal carrier voltage of frequency 1.5 MHz and amplitude 50 V is amplitude modulated by sinusoidal voltage of frequency 10 kHz producing 50% modulation. The lower and upper side band frequencies in kHz are
(a) 1490, 1510 (b) 1820, 1012
(c) $\frac{1}{1510}, \frac{1}{1490}$ (d) 2490, 1510

34. 1000 Hz carrier wave is amplitude modulated by the signal frequencies 200-4000 Hz. The channel width of this case is
(a) 3 kHz (b) 8 kHz (c) 10 kHz (d) 14 kHz

35. A male voice after modulation-transmission sounds like that of a female to the receiver. The problem is due to **[NCERT Exemplar]**
(a) poor selection of modulation index (selected $0 < m < 1$)
(b) poor bandwidth selection of amplifiers
(c) poor selection of carrier frequency
(d) loss of energy in transmission

36. When the modulation percentage is 75, an AM transmitter produces 10 kW. How much of this is carrier power?
(a) 10 kW (b) 13.33 kW
(c) 7.5 kW (d) 7.81 kW

37. For an amplitude modulated wave, the maximum amplitude is found to be 10 V while the minimum amplitude is found to be 2 V. Determine the modulation index μ. **[NCERT]**
(a) $\frac{2}{3}, 1$ (b) $1, \frac{2}{3}$ (c) $\frac{4}{5}, 1$ (d) $1, \frac{4}{5}$

38. The speech signal of 3 kHz is used to modulate a carrier signal of frequency 1 MHz, using amplitude modulation. The frequencies of the side bands will be **[NCERT Exemplar]**
(a) 1.003 MHz and 0.997 MHz
(b) 1.954 MHz and 2.0 MHz
(c) 40.5 MHz and 27.6 MHz
(d) 5 MHz and 0.997 MHz

39. A message signal of frequency ω_m is superposed on a carrier wave of frequency ω_c to get an amplitude modulated wave (AM). The frequency of the AM wave will be **[NCERT Exemplar]**
(a) ω_m (b) ω_c
(c) $\frac{\omega_c + \omega_m}{2}$ (d) $\frac{\omega_c - \omega_m}{2}$

40. A carrier wave of peak voltage 12 V is used to transmit a message signal. What should be the peak voltage of the modulating signal in order to have a modulation index of 75%? **[NCERT]**
(a) 7V (b) 6V (c) 9V (d) 8V

41. A carrier wave is modulated by a number of sine waves with modulation indices $m_1, m_2, m_3, \ldots,$. The total modulation index (m) of the wave is
(a) $m_1 + m_2 + m_3 + \ldots$ (b) $m_1 - m_2 + m_3 + \ldots$
(c) $\sqrt{m_1^2 + m_2^2 + m_3^2 + \ldots}$ (d) $\sqrt{\frac{m_1^2 + m_2^2 + m_3^2 + \ldots}{n}}$

42. The antenna current of an AM broadcast transmitter modulated by 50% is 11. Find the carrier current.
(a) 9.25 A (b) 10 A
(c) 10.35 A (d) 5.5 A

43. The total power content of AM wave is 900 W. For 100% modulation, the power transmitted by each size band is
(a) 150 W (b) 140 W
(c) 100 W (d) 90 W

44. The antenna current of an AM transmitter is 8 A when only the carrier is sent, but it increases to 8.93 A when the carrier is sinusoidal modulated. The percentage modulation is nearly
(a) 90% (b) 80%
(c) 75% (d) 70%

45. A carrier is simultaneously modulated by two sine waves having modulation indices of 0.3 and 0.4. The total modulation index will be
(a) 0.1 (b) 0.5
(c) 0.7 (d) 0.35

46. How many 600 kHz waves can be on a 5 mile transmission line simultaneously?
(a) 11 (b) 12
(c) 14 (d) 16

47. A 600 W carrier is modulated to a depth of 75% by a 400 Hz sine. Find the total antenna power.
(a) 769 W (b) 796 W
(c) 679 W (d) 637.5 W

48. A transmitter radiates 12 kW power at 80% modulation index. The power in carrier is nearly
(a) 8.12 kW (b) 6.66 kW
(c) 7.4 kW (d) 9.09 kW

49. When only carrier is transmitted antenna current observed is 8A. When it is modulated with 500 Hz sine wave antenna current becomes 9.6 A. Find % modulation.
(a) 80% (b) 20% (c) 94.26% (d) 83.76%

50. If minimum voltage in an AM wave was found to be 2V and maximum voltage 10V. Find % modulation index.
(a) 80% (b) 66.67%
(c) 64.25% (d) 76.25%

ROUND II Mixed Bag

Only One Correct Option

1. With reference to ionospheric propagation, an electromagnetic wave with a critical frequency of 15 MHz and incident at an angle of 45° will have MUF of
(a) 15 MHz (b) $15\sqrt{2}$ MHz
(c) $15\sqrt{2}$ MHz (d) None of these

2. A TV tower has a height of 150 m. The area of the region covered, the TV broadcast is
(Take, radius of earth = 6.4×10^6 m)
(a) $9.6\pi \times 10^8$ km^2 (b) $19.2\pi \times 10^8$ km^2
(c) $19.2\pi \times 10^7$ km^2 (d) $1.92\pi \times 10^3$ km^2

3. A sky wave with a frequency 55 MHz is incident on the *D*-region of earth's atmosphere at 30°. The angle of refraction is (electron density for *D*-region is 400 electron/cc)
(a) 60° (b) 15° (c) 45° (d) 30°

4. If an electromagnetic wave is transmitted to the height equal to 150 km with maximum frequency 300 kHz and critical frequency 100 kHz, the skip distance is
(a) 426 km (b) 636 km (c) 824 km (d) 849 km

5. What is the value of frequency at which electromagnetic wave must be propagated for the *D*-region of atmosphere to have a refractive index of 0.5, if electron density for *D*-region is 400 electrons/cc.
(a) 200 kHz (b) 104.2 kHz
(c) 208.4 kHz (d) 312.6 kHz

6. The height of a TV antenna is 200 m. The population density is 4000 km^{-2}. Find the population benefitted.
(a) 3.2×10^8 (b) 3.2×10^7
(c) 3.2×10^6 (d) 3.2×10^5

7. A ground receiver station is receiving a signal at 5 MHz and transmitted from a ground transmitter at a height of 300 m, located at a distance of 100 km from the receiver station. The signal is coming *via* (Radius of earth = 6.4×10^6 m). N_{max} is ionosphere = 10^{12} m^3
(a) Space wave propagation
(b) Gromel wave propagation
(c) Sky wave propagation
(d) None of the above

8. Calculate the phase velocity of electromagnetic wave having electron density and frequency of *D*-layer, $N=$ 400 electron cc^{-1}, $\nu = 200$ kHz.
(a) 3×10^8 ms^{-1} (b) 3.5×10^8 ms^{-1}
(c) 6.9×10^8 ms^{-1} (d) 1.1×10^9 ms^{-1}

9. A radar has a power of 1 kW and is operating at a frequency of 10 GHz. It is located on a mountain top of height 500 m. The maximum distance upto which it can detect object located on the surface of the earth (Take, radius of earth = 6.4×10^6 m) is **[AIEEE 2012]**
(a) 80 km (b) 16 km (c) 40 km (d) 64 km

10. Which of the following four alternatives is not correct? We need modulation **[AIEEE 2011]**
(a) to increase the selectivity
(b) to reduce the time lag between transmission and reception of the information signal
(c) to reduce the size of antenna
(d) to reduce the fractional bandwidth, that is the ratio of the signal bandwidth to the centre frequency

11. Moon cannot be used as communication satellite of the earth because
(a) the distance between earth and moon is not proper
(b) the period of revolution of moon is 27.3 days
(c) the moon is not revolving in the equitorial plane of the earth
(d) None of the above

12. A signal wave of frequency 12kHz is modulated with a carrier wave of frequency 2.51 MHz. The upper and lower side band frequencies are respectively
(a) 2512 kHz and 2508 kHz
(b) 2522 kHz and 2488 kHz
(c) 2502 kHz and 2498 kHz
(d) 2522 kHz and 2498 kHz

13. In process of amplitude modulation of signal to be transmitted.
Signal to be modulated is given by $m(t) = A_m \sin \omega_m t$, carrier wave is given by $c(t) = A_c \sin \omega_c t$, modulated signal $c_m(t)$ is given by
(a) $c_m(t) = A_c \sin \omega_c t + A_m \sin \omega_m t$
(b) $c_m(t) = (A_c + A_m) \sin \omega_c t$
(c) $c_m(t) = [A_c + m(t)] \sin \omega_c t$
(d) None of the above

14. The wavelength of the carrier waves in a modern optical fibre communication network is close to **[JEE Main 2019]**

(a) 2400 nm (b) 1500 nm
(c) 600 nm (d) 900 nm

15. To double the covering range of a TV transmission tower, its height should be multiplied by **[JEE Main 2019]**

(a) $\sqrt{2}$ (b) 4
(c) 2 (d) $\frac{1}{\sqrt{2}}$

16. A 100 V carrier wave is made to vary between 160 V and 40 V by a modulating signal. What is the modulation index? **[JEE Main 2019]**

(a) 0.4 (b) 0.5
(c) 0.6 (d) 0.3

17. In communication system, only one per cent frequency of signal of wavelength 800 nm can be used as bandwidth. How many channel of 6MHz bandwidth can be broadcast this? ($c = 3 \times 10^8$ m / s, $h = 6.6 \times 10^{-34}$ J- s) **[JEE Main 2019]**

(a) 3.75×10^6 (b) 3.86×10^6
(c) 6.25×10^5 (d) 4.87×10^5

18. A TV transmission tower has a height of 140 m and the height of the receiving antenna is 40 m. What is the maximum distance upto which signals can be broadcasted from this tower in LOS (Line of Sight) mode? (Take, radius of earth = 6.4×10^6 m) **[JEE Main 2019]**

(a) 65 km (b) 80 km
(c) 40 km (d) 48 km

19. A message signal of frequency 100 MHz and peak voltage 100 V is used to execute amplitude modulation on a carrier wave of frequency 300 GHz and peak voltage 400 V. The modulation index and difference between the two side band frequencies are **[JEE Main 2019]**

(a) 0.25 ; 1×10^8 Hz (b) 4 ; 1×10^8 Hz
(c) 0.25 ; 2×10^8 Hz (d) 4 ; 2×10^8 Hz

20. In an amplitude modulator circuit, the carrier wave is given by $C(t) = 4\sin(20000\,\pi t)$ while modulating signal is given by, $m(t) = 2\sin(2000\,\pi t)$. The values of modulation index and lower side band frequency are **[JEE Main 2019]**

(a) 0.5 and 10 kHz (b) 0.4 and 10 kHz
(c) 0.3 and 9 kHz (d) 0.5 and 9 kHz

21. An amplitude modulates signal is given by $v(t) = 10[1 + 0.3\cos(2.2 \times 10^4 t)]\sin(5.5 \times 10^5 t)$. Here, t is in seconds. The side band frequencies (in kHz) are $\left(\text{Take}, \pi = \frac{22}{7}\right)$ **[JEE Main 2019]**

(a) 892.5 and 857.5 (b) 89.25 and 85.75
(c) 178.5 and 171.5 (d) 1785 and 1715

22. An amplitude modulated signal is plotted below

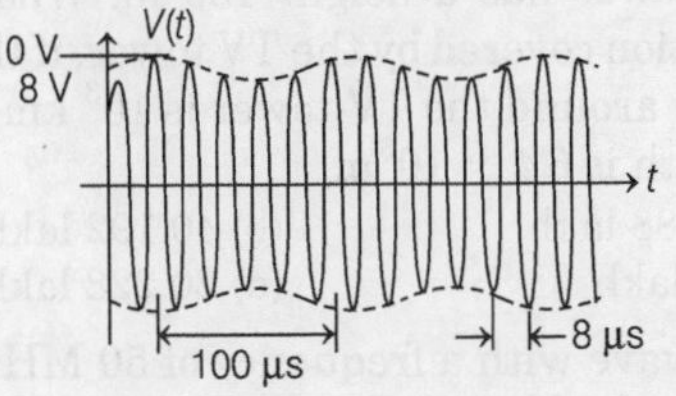

Which one of the following best describes the above signal? **[JEE Main 2019]**

(a) $[1 + 9\sin(2\pi \times 10^4 t)]\sin(2.5\pi \times 10^5 t)$V
(b) $[9 + \sin(2\pi \times 10^4 t)]\sin(2.5\pi \times 10^5 t)$V
(c) $[9 + \sin(4\pi \times 10^4 t)]\sin(5\pi \times 10^5 t)$V
(d) $[9 + \sin(2.5\pi \times 10^5 t)]\sin(2\pi \times 10^4 t)$V

23. An audio signal of 15 kHz frequency cannot be transmitted over long distances without modulation because

(a) the size of the required antenna would be at least 5 km which is not convenient
(b) the audio signal can be transmitted through sky waves
(c) the size of the required antenna would be at least 20 km, which is not convenient
(d) effective power transmitted would be very low, if the size of the antenna is less than 15 km

24. Which of the following statements is/are correct?

(a) The space waves are the radio waves of frequency between 30 MHz to 300 MHz.
(b) The space waves cannot travel through atmosphere from transmitter antenna to receiver antenna either directly or after reflection from ground in the earth troposphere region.
(c) The space wave propagation is not a line of sight propagation.
(d) The ground wave communication is utilised in television communication, radar communication and microwave communication.

25. For sky wave propagation of 10 MHz signal, what should be the minimum electron density in ionosphere? **[AMU]**

(a) ~ 1.2×10^{12} m^{-3} (b) ~ 10^6 m^{-3}
(c) ~ 2.3×10^4 m^{-3} (d) ~ 10^{22}m^{-3}

26. The antenna current of an AM transmitter is 8 A when only the carrier is sent, but it increases to 8.93A when the carrier is modulated by single sine wave. Find the percentage modulation. **[VITEEE]**

(a) 60.1% (b) 70.1% (c) 80.1% (d) 50.1%

27. A TV transmitting antenna is 128 m tall. If the receiving antenna is at the ground level, the maximum distance between them for satisfactory communication in line of sight mode is (radius of the earth = 6.4×10^6 km) **[EAMCET]**

(a) $64 \times \sqrt{10}$ km (b) $\frac{128}{\sqrt{10}}$ km
(c) $128 \times \sqrt{10}$ km (d) $\frac{64}{\sqrt{10}}$ km

28. A TV tower has a height 150 m. What is the total population covered by the TV tower, if the population density around the TV tower is 10^3 km^{-2}? Radius of the earth is 6.4×10^6m. **[Guj CET]**

(a) 60.288 lakh (b) 40.192 lakh
(c) 100 lakh (d) 20.228 lakh

29. If sky wave with a frequency of 50 MHz is incident on D region at an angle of 30°, then angle of refraction is **[BVP]**

(a) 15° (b) 30°
(c) 60° (d) 45°

30. In optical communication system operating at 1200 nm, only 2% of the source frequency is available for TV transmission having a bandwidth of 5 MHz. The number of TV channels that can be transmitted is

(a) 2 million (b) 10 million
(c) 0.1 million (d) 1 million

31. The maximum electron density in the ionosphere in the morning is 10^{10} m^{-3}. At noon time, it increases to 2×10^{10} m^{-3}. Find the ratio of critical frequency at noon and the critical frequency in the morning. **[Guj CET]**

(a) 2.00 (b) 2.82
(c) 4.00 (d) 1.414

32. Match the following Column I with Column II.

Column I (Space communications)	Column II (Frequencies)
I. Ground wave propagation	A. 30 MHz to 300 MHz
II. Sky wave propagation	B. 80 MHz to 200 MHz
III. Space wave propagation	C. 2 MHz to 30 MHz
IV. Television signal propagation	D. 500 kHz to 1500 kHz

Codes

(a) I-D, II-C, III-A, IV-B (b) I-A, II-B, III-C, IV-D
(c) I-B, II-A, III-D, IV-C (d) I-C, II-D, III-B, IV-A

33. I. Sky wave signals are used for long distance radio communication. These signals are in general, less stable than ground wave signals.

II. The state of ionosphere varies from hour to hour, day to day and season to season. **[AIEEE 2011]**

(a) I is true, II is true, II is the correct explanation of I
(b) I is true, II is true, II is not the correct explanation of I
(c) I is false, II is true
(d) I is true, II is false

34. A carrier signal $C(t) = 25\sin(2.512 \times 10^{10} t)$ is amplitude modulated by a message signal $m(t) = 5\sin(1.57 \times 10^8 t)$ and transmitted through an antenna. What will be the bandwidth of the modulated signal? **[JEE Main 2021]**

(a) 8 GHz
(b) 2.01 GHz
(c) 1987.5 MHz
(d) 50 MHz

Numerical Value Questions

35. The value of radio frequency at which the electromagnetic wave must be propagated for the D-layer of atmosphere to have a refractive index of 0.5 is .... kHz (Take, $N = 4 \times 10^8$ m^{-3})

36. There is a transmitter at a height ($h = 400$ m) from the ground at place A, from which the radiowaves can be transmitted. If the station A sends the TV signals, then the maximum area on the earth upto which TV transmission can be received on the earth is ... $\times 10^7$. Coverage population density around the tower is 1000 km^{-2}.

37. For VHF signal broadcasting, ... km^2 of maximum service area will be covered by an antenna tower of height 30 m, if the receiving antenna is placed at ground. Let radius of the earth be 6400 km. (Round off to the nearest integer) (Take π as 3.14) **[JEE Main 2021]**

Answers

Round I

1. (c)	2. (a)	3. (c)	4. (c)	5. (b)	6. (d)	7. (d)	8. (b)	9. (c)	10. (b)
11. (d)	12. (d)	13. (c)	14. (a)	15. (b)	16. (b)	17. (b)	18. (d)	19. (b)	20. (b)
21. (d)	22. (c)	23. (b)	24. (d)	25. (d)	26. (b)	27. (c)	28. (d)	29. (b)	30. (c)
31. (b)	32. (a)	33. (a)	34. (b)	35. (b)	36. (d)	37. (a)	38. (a)	39. (b)	40. (c)
41. (c)	42. (c)	43. (a)	44. (d)	45. (b)	46. (d)	47. (a)	48. (d)	49. (c)	50. (b)

Round II

1. (c)	2. (d)	3. (d)	4. (d)	5. (c)	6. (b)	7. (c)	8. (c)	9. (a)	10. (b)
11. (d)	12. (d)	13. (c)	14. (b)	15. (b)	16. (c)	17. (c)	18. (a)	19. (c)	20. (d)
21. (b)	22. (b)	23. (a)	24. (a)	25. (a)	26. (b)	27. (b)	28. (a)	29. (b)	30. (d)
31. (d)	32. (a)	33. (b)	34. (d)	35. 208	36. 1.6	37. 1206			

Solutions

Round I

1. For commercial UHF, TV broadcasts, the range allotted is 470-960 MHz.

2. For given buildings complex, $l = 100\,\text{m}, \lambda = ?$

As, $l = \frac{\lambda}{4}$ or $\lambda = 4\,l = 4 \times 100 = 400\,\text{m}$

3. A digital signal is a discontinuous function of time in contrast to an analog signal. The digital signals can be stored as digital data and cannot be transmitted along the telephone lines. Digital signal cannot utilise decimal signals.

4. A communication satellite receives the coming modulated microwave signal, amplifies it and return it to earth at a different frequency.

5. Radio waves of frequency 30 MHz to 300 MHz belong to very high frequency band.

6. Frequency range for microwaves is 1×10^9 to 3×10^{11} Hz.

7. Ozone layer will absorb UV rays; reflects the infrared radiation and does not reflect back radio waves.

8. At least three geostationary satellites in the same orbit around earth's can cover the whole part of the earth for microwave communication.

9. A communication link between a fixed base station and mobile units on a ship or aircraft works on 30 to 470 MHz.

10. The diameter of optical fibre is 10^{-4} cm.

11. An optical fibre is very efficient, has signal security and cannot be jammed as easily as radio waves.

12. Laser is a coherent source of coordinated waves of particular wavelength.

13. Quality of transmission is governed both by nature of signal and nature of communication channel/medium.

14. A guided medium alone can provide point to point communication.

15. For ground wave propagation, the frequency range is 530 kHz to 1710 kHz. For sky wave propagation, the frequency range is 1710 kHz to 40 MHz. For space wave propagation, the frequency range is 54 MHz to 4.2 GHz. Thus, option (b) is correct.

16. Given, height of antenna, $h = 81$ m

Radius of earth, $R = 6.4 \times 10^6$ m

No, it is not necessary for line of sight communication, the two antennas may not be at the same height.

$$\text{Area} = \pi d^2$$

$$\because \quad \text{Range, } d = \sqrt{2hR}$$

$$\therefore \quad \text{Service area} = \pi \times 2hR = \frac{22}{7} \times 2 \times 81 \times 6.4 \times 10^6$$

$$= 3258.5 \times 10^6\ \text{m}^2 = 3258.5\ \text{km}^2$$

$$\simeq 3260\ \text{km}^2$$

17. Here, $P_i = 1\ \text{kW} = 1000\ \text{W}$, length of path = 5 km; $P_0 = ?$

Loss of power at a distance of 5 km $= -2 \times 5 = -10$ dB.

As gain in dB $= 101 \log \frac{P_0}{P_i}$

$$\therefore \quad 10 = 10 \log \frac{P_0}{P_i} = 10 \times \log \frac{P_i}{P_0}$$

or $\quad \log \frac{P_i}{P_0} = 1 = \log 10 \quad$ or $\quad \frac{P_i}{P_0} = 10$

or $\quad P_0 = \frac{P_i}{10} = \frac{1000}{10} = 100$ W

18. A signal emitted by an antenna from a certain point can be received at another point of the surface in the form of sky wave and ground wave depending upon the frequency of signal.

19. 10 kHz frequency cannot be radiated due to large antenna size, 1 GHz and 1000 GHz will be penetrated the ionosphere. Hence, option (b) is correct.

20. $d = 2\sqrt{2Rh}$

$$h = \frac{d^2}{8R} = \frac{(45)^2}{8 \times 6400}\ \text{km} \equiv 39.55\ \text{m}$$

21. Virtual height is 180 km while critical frequency is 5 MHz.

22. When electromagnetic waves enter the ionised layer of ionosphere, then the relative permittivity of the ionised layer appears to decrease.

23. The highest frequency of radio waves that can be reflected by the ionosphere is called maximum unusable frequency.

24. The virtual height and critical frequency of E-layer is 110 km and 4 MHz.

25. In space communication, the information can be passed from one place to another with the speed of light ($c = 3 \times 10^8\ \text{ms}^{-1}$). Hence, time taken for a distance of 100 km $= \frac{100 \times 10^3}{3 \times 10^8} = 3.3 \times 10^{-4}$ s

26. Range $= \sqrt{2Rh}$

Range (i) $= \sqrt{2Rh}$

Range (ii) $= \sqrt{2Rh} + \sqrt{2Rh'}$

where, $\quad h = 20$ m and $h' = 5$ m

$$= \frac{\sqrt{2Rh'}}{\sqrt{2Rh}} \times 100\% = \frac{\sqrt{5}}{\sqrt{10}} \times 100\% = 50\%$$

27. The mode of communication used for TV signals transmission is either space wave propagation or line of sight propagation of satellite communication.

28. The frequencies in UHF range normally propagate by means of space waves.

29. Area covered $= \pi \times 2Rh$

$$= \frac{22}{7} \times 2 \times 6.4 \times 10^6 \times 75$$

$$= 3017.1 \text{ km}^2 \simeq 3018 \text{ km}^2$$

30. As, $h = d^2/2R$ or $d^2 = 2hR$

$\therefore \quad d = \sqrt{(2hR)}$

or $\quad d \propto h^{1/2}$

31. Maximum usable frequency (MUF) $= \nu_c \sec i$

$$= (60 \times 10^6) \sec 70^\circ$$

$$= \frac{60 \times 10^6}{0.342} = 175.43 \times 10^6 \text{ Hz} = 175 \text{ MHz}$$

32. In AM, carrier wave frequencies are lower (0.5-20 MHz) compared to those in frequency modulation (80 MHz onwards).

33. Here, $f_c = 1.5 \text{ MHz} = 1500 \text{ kHz}$, $f_m = 10 \text{ kHz}$

$\therefore$ Low side band frequency

$$= f_c + f_m = 1500 \text{ kHz} - 10 \text{ kHz} = 1490 \text{ kHz}$$

Upper side band frequency

$$= f_c + f_m = 1500 \text{ kHz} + 10 \text{ kHz}$$
$$= 1510 \text{ kHz}$$

34. Bandwidth is equal to twice the frequency of modulating signals.

$\therefore$ Bandwidth $= 2\nu_m = 2 \times 4000 \text{ Hz} = 8000 \text{ Hz} = 8 \text{ kHz}$

35. The frequency of male voice is less than that of a female voice. In the given problem, the frequency of modulated signal received becomes more, which is possible with the poor band width selection of amplifiers. It is so because bandwidth in amplitude modulation = twice the frequency of modulating signal.

36. As, $P_t = P_c(1 + m^2/2)$

So, $$P_c = \frac{P_t}{1 + m^2/2} = \frac{10}{1 + \frac{1}{2}\left(\frac{3}{4}\right)^2} = \frac{10 \times 32}{41}$$

$\Rightarrow \quad P_c = 7.81 \text{ kW}$

37. Given, maximum amplitude, $A_{max} = 10 \text{ V}$

Minimum amplitude, $A_{min} = 2 \text{ V}$

Let A_c and A_m be the amplitudes of carrier wave and signal wave.

$\therefore \quad A_{max} = A_c + A_m = 10 \quad \ldots\text{(i)}$

and $\quad A_{min} = A_c - A_m = 2 \quad \ldots\text{(ii)}$

Adding the Eqs. (i) and (ii), we get

$$2A_c = 12$$

or $\quad A_c = 6 \text{ V}$

and $\quad A_m = 10 - 6 = 4 \text{ V}$

Modulation index, $\mu = \frac{A_m}{A_c} = \frac{4}{6} = \frac{2}{3}$

When the minimum amplitude is zero, *i.e.* $A_{min} = 0$

$A_c + A_m = 10 \quad \ldots\text{(iii)}$

$A_c - A_m = 0 \quad \ldots\text{(iv)}$

By solving Eqs. (iii) and (iv), we get

$$2A_m = 10$$

or $\quad A_m = 5$

and $\quad A_c = 5$

Modulation index, $\mu = \frac{A_m}{A_c} = \frac{5}{5} = 1$

38. The frequencies of the side bands will be $f_m + f_c = 1.003$ MHz and $f_c - f_m = 0.997$ MHz.

39. In amplitude modulation, the frequency of modulated wave is equal to the frequency of carrier wave.

40. Given, peak voltage, $V_0 = 12 \text{ V}$

Modulation index, $\mu = 75\% = \frac{75}{100}$

We know that, modulation index

$$(\mu) = \frac{\text{peak voltage of modulating signal } (V_m)}{\text{peak voltage } (V_0)}$$

So, peak voltage of modulating signal,

$$V_m = \mu \times \text{Peak voltage}$$
$$= \frac{75}{100} \times 12 = 9 \text{ V}$$

41. Total modulation index of the wave,

$$m = \sqrt{m_1^2 + m_2^2 + m_3^2 + \ldots}$$

42. $$I_{carrier} = \frac{I_{total}}{(1 + m_a^2/2)^{1/2}} = \frac{11}{\sqrt{1 + (0.5)^{2/2}}} = 10.35 \text{ A}$$

43. $$P_c = P_t = \left(\frac{2}{2 + m^2}\right) = 900\left(\frac{2}{2+1}\right) = \frac{900 \times 2}{3} = 600 \text{ W}$$

Now, $$P_{LSB} = \frac{m^2}{4} \times P_c = \frac{1}{4} \times 600 = 150 \text{ W}$$

44. As, $$\sqrt{1 + \frac{m^2}{2}} = \frac{I_t}{I_c} = \frac{8.93}{3} = 1.11625$$

$\Rightarrow \quad \frac{m^2}{2} = (1.11625)^2 - 1$

$\Rightarrow \quad m = 0.70 = 70\%$

$\Rightarrow \quad m = 70\%$

45. Here, $m_1 = 0.3$ and $m_2 = 0.4$

$\because \quad m = \sqrt{m_1^2 + m_2^2} = \sqrt{0.3^2 + 0.4^2}$

$$= \sqrt{0.09 + 0.16} = \sqrt{0.25} = 0.5$$

46. Given, $f = 600$ kHz, $d = 5$ mile

First determine the wavelength of the 600 kHz, signal using

$$f\lambda = c$$

Use 186000 mile per second as the speed of light, since the line length is given in mile and no figure is given for the velocity factor.

Substituting numerical values,

$$600\times10^3\ \lambda = 186000$$

$$\lambda = \frac{186000}{600\times10^3} = 0.31 \text{ mile}$$

Knowing the wavelength of the signal and the length of the line, the number of cycle on the line can be found from

$$n = \frac{d}{\lambda} = \frac{5}{0.31} = 16.13 \simeq 16$$

47. Total power in the modulated wave,

$$P_{\text{total}} = P_{\text{carrier}}\left(1 + \frac{m_a^2}{2}\right)$$

$$= 600\left(1 + \frac{(0.75)^2}{2}\right) = 768.75\text{W} \simeq 769 \text{ W}$$

48. Power of carrier wave,

$$P_{\text{carrier}} = \frac{P_{\text{total}}}{1 + \dfrac{m_a^2}{2}}$$

$$= \frac{12}{\left[1 + \dfrac{(0.80)^2}{2}\right]} = \frac{12}{1.32} = 9.09 \text{ kW}$$

49. $\left(\dfrac{I_t}{I_C}\right) = \sqrt{1 + \dfrac{\mu^2}{2}}$

or $\quad \mu = \sqrt{2\left[\left(\dfrac{I_t}{I_c}\right)^2 - 1\right]} = \sqrt{2[(1.20)^2 - 1]}$

$= 94.26\%$

50. Modulation index,

$$m_a = \frac{E_{\max} - E_{\min}}{E_{\max} + E_{\min}}$$

$$= \frac{10-2}{10+2} = \frac{2}{3}$$

$= 66.67\%$

Round II

1. MUF $= \nu \sec\theta = 15 \sec 45° = 15\sqrt{2}$ MHz

2. Area of region covered $= \pi d^2 = \pi\ (2\ hR)$

$$= \pi\times2\times150\times6.4\times10^6$$

$$= 1.92\ \pi\times10^9\ \text{m}^2 = 1.92\ \pi\times10^3\ \text{km}^2$$

3. For D-region, $\nu = 55\times10^6$ Hz; $i = 30°$

and $\quad N = 400\times10^6\ \text{m}^{-3}$

$$\mu = \sqrt{\frac{1 - 81.45\ N}{\nu^2}}$$

$$\mu = \sqrt{1 - \frac{81.45\times400\times10^6}{(55\times10^6)}} \approx 1$$

Also $\quad \mu = \dfrac{\sin i}{\sin r}$

or $\quad \sin i = \mu \sin r = \sin r$

or $\quad i = r = 30°$

4. As, $h = 150$ km, $\nu = 300$ kHz

$$\nu_c = 100 \text{ kHz}$$

Therefore, $D_{\text{skip}} = 2h\sqrt{\left(\dfrac{\nu}{\nu_c}\right)^2 - 1}$

$$= 2\times150\times10^5\sqrt{\left(\frac{300\times10^8}{100\times10^8}\right) - 1}$$

$$= 849 \text{ Hz}$$

5. Here, $N = 400$ electrons/cc $= 400\times10^6$ electrons m^{-3}

$$\mu = 0.5 = \sqrt{1 - \frac{81.56\ N}{\nu^2}}$$

$$= \sqrt{\frac{1 - 81.56\times400\times10^6}{\nu^2}}$$

On solving, $\quad \nu = 208.4$ kHz

6. As, $d = \sqrt{2\ Rh}$

Population covered $= \pi d^2 \times$ Population density

$$= \pi\ (2\ Rh)\times\rho$$

$$= \frac{22}{7}\times2\times6400\times0.2\times4000$$

$$= 3.2\times10^7$$

7. Maximum distance covered by space wave communication, $\sqrt{2\ Rh} = 62$ km

Critical frequency, $\nu_c = 9\ (N_{\max})^{1/2} = 9$ MHz

5 MHz $< \nu_c$, sky wave propagation, *i.e.* ionospheric propagation.

8. $\mu = \sqrt{1 - \dfrac{81.45\ N}{\nu^2}} = \sqrt{1 - \dfrac{81.45\times400\times10^6}{(200\times10^3)^2}} = 0.43$

$$\mu = \frac{c}{\nu}$$

$\therefore$ Phase velocity, $v = \dfrac{c}{\mu} = \dfrac{3\times10^8}{0.43} = 6.9\times10^8\ \text{ms}^{-1}$

9. Range of radar on the earth surface optional distance for space wave, *i.e.* line of view.

$$\text{Range} = \sqrt{(R+h)^2 - R^2} = \sqrt{2\ Rh + h^2}$$

$$= \sqrt{2\ Rh} = \sqrt{2\times6400\times\frac{1}{2}}\ \text{km} = 80 \text{ km}$$

10. Modulation does not change time lag between transmission and reception of the information or message signal.

11. For a communication satellite, the distance between the satellite and earth should be about 36000 km. The period of revolution is 24 days and it is revolving in the equatorial plane of the earth.

12. Upper Side Band (USB) frequency $= f_c + f_m$

$$= (2.51)\text{ MHz} + (12)\text{ kHz}$$
$$= (2510 + 12)\text{ kHz} = 2522\text{ kHz}$$

Lower Side Band (LSB) frequency $= f_c - f_m$

$$= (2.51)\text{ MHz} - (12)\text{ kHz}$$
$$= (2510 - 12)\text{ kHz} = 2498\text{ kHz}$$

13. In amplitude modulation, amplitude of carrier wave varies with signal to be modulated. So, modulated signal $c_m(t)$ is given by

$$c_m(t) = [A_c + m(t)]\sin\omega_c t$$
$$\Rightarrow \quad c_m(t) = (A_c + A_m\sin\omega_m t)\sin\omega_c t$$

As, $m(t) = A_m\sin\omega_m t$ (given)

14. In optical fibre communication network, the signals are transmitted by laser light operating in range of 1310 nm-1550 nm.

So, the closest value is 1500 nm.

15. Range of TV transmitting tower, $d = \sqrt{2hR}$

where, h is the height of the transmission tower. When range is doubled, then

$$2d = 2\sqrt{2hR} = \sqrt{2(4h)R}$$

So, height must be multiplied with 4.

16. Modulation index is given by

$$\mu = \frac{A_{\max} - A_{\min}}{A_{\max} + A_{\min}} = \frac{V_{\max} - V_{\min}}{V_{\max} + V_{\min}}$$
$$= \frac{160 - 40}{160 + 40} = \frac{120}{200} = 0.6$$

17. Here, signal wavelength, $\lambda = 800\text{ nm} = 8 \times 10^{-7}$ m

Frequency of source is

$$f = \frac{c}{\lambda} = \frac{3 \times 10^8}{8 \times 10^{-7}} = 3.75 \times 10^{14}\text{ Hz}$$

$\therefore$ Total bandwidth used for communication

$$= 1\% \text{ of } 3.75 \times 10^{14}$$
$$= 3.75 \times 10^{12}\text{ Hz} \quad \text{...(i)}$$

So, number of channel for signals

$$= \frac{\text{total bandwidth available for communication}}{\text{bandwidth of TV signal}}$$
$$= \frac{3.75 \times 10^{12}}{6 \times 10^6} = 0.625 \times 10^6 = 6.25 \times 10^5$$

18. Maximum distance of transmission is given by

$$d = \sqrt{2Rh_T} + \sqrt{2Rh_R}$$

where, h_T = height of transmitter, = 140 m,

h_R = height of receiver = 40 m and

R = radius of earth $= 6.4 \times 10^6$ m.

Substituting the values, we get

$$d = \sqrt{2 \times 6.4 \times 10^6}\,(\sqrt{140} + \sqrt{40}) = 65\text{ km}$$

19. Given, $f_m = 100\text{ MHz} = 100 \times 10^6\text{ Hz}$, $A_m = 100$ V,

$$A_c = 400\text{ V}$$

Range of frequency in case of amplitude modulation is $(f_c - f_m)$ to $(f_c + f_m)$.

$\therefore$ Bandwidth $= 2f_m = 2 \times 100 \times 10^6$ Hz

$$= 2 \times 10^8\text{Hz}$$

and modulation index, $\text{MI} = \dfrac{A_m}{A_c} = \dfrac{100}{400} = 0.25$

20. Given, carrier wave, $C(t) = 4\sin(20000\,\pi t)$

Modulating signal, $m(t) = 2\sin(2000\,\pi t)$

So, carrier wave's amplitude and frequency are

$$A_c = 4\text{ V},\ \omega_c = 20000\,\pi = 2\pi \times 10^4\text{ rad/s}$$
$$\Rightarrow \quad f_c = \frac{\omega_c}{2\pi} = 10^4\text{ Hz} = 10\text{ kHz}$$

and modulating signal's amplitude and frequency are

$$A_m = 2\text{ V},\ \omega_m = 2000\pi = 2\pi \times 10^3\text{ rad/s}$$
$$\Rightarrow \quad f_m = \frac{\omega_m}{2\pi} = 10^3\text{ Hz} = 1\text{ kHz}$$

So, modulating index is $m = \dfrac{A_m}{A_c} = \dfrac{2}{4} = 0.5$

and lower side band frequency is,

$$f_{\text{LSB}} = f_c - f_m = 10 - 1 = 9\text{ kHz}$$

21. $v(t) = 10\,[1 + 0.3\cos(2.2 \times 10^4 t)]\,[\sin(5.5 \times 10^5 t)]$

Upper band angular frequency,

$$\omega_U = (2.2 \times 10^4 + 5.5 \times 10^5)\text{ rad/s}$$
$$= 572 \times 10^3\text{ rad/s}$$

Similarly, lower band angular frequency,

$$\omega_L = (5.5 \times 10^5 - 2.2 \times 10^4)\text{ rad/s}$$
$$= 528 \times 10^3\text{ rad/s}$$

$\therefore$ Side band frequency are,

$$f_u = \frac{\omega_U}{2\pi} = \frac{572}{2\pi}\text{ kHz} \simeq 91\text{ kHz}$$

and $$f_L = \frac{\omega_L}{2\pi} = \frac{528}{2\pi}\text{ kHz} \simeq 84\text{ kHz}$$

Thus, the approximate side band frequencies are 89.25 and 85.75.

22. Equation of an amplitude modulated wave is given by the relation, $C_m = (A_c + A_m\sin\omega_m t)\cdot\sin\omega_c\cdot t$ ...(i)

For the given graph,

maximum amplitude, $A_c + A_m = 10$ ...(ii)

and minimum amplitude, $A_c - A_m = 8$ ...(iii)

From Eqs. (ii) and (iii), we get

$$A_c = 9\text{V and } A_m = 1\text{ V} \quad \text{...(iv)}$$

$\because$ For angular frequency of message signal and carrier wave, we use a relation,

$$\omega_c = \frac{2\pi}{T_c} = \frac{2\pi}{8 \times 10^{-6}}$$

(as from given graph, $T_c = 8 \times 10^{-6}$ s)

$$= 2.5\pi \times 10^5\text{s}^{-1} \quad \text{...(v)}$$

and $\omega_m = \frac{2\pi}{T_m} = \frac{2\pi}{100\times10^{-6}} = 2\pi\times10^4 s^{-1}$

(as from given graph, $T_m = 100\times10^{-6}$s)

When we put values of A_c, A_m, ω_c and ω_m in Eq. (i), we get

$C_m = [9 + \sin(2\pi\times10^4 t)]\sin(2.5\pi\times10^5 t)$ V

23. Here, $\nu_m = 15$ kHz $= 15\times10^3$ Hz

$$\lambda_m = \frac{c}{\nu_m} = \frac{3\times10^8}{15\times10^3} = \frac{1}{5}\times10^5 \text{ m}$$

Size of the antenna required,

$$l = \frac{\lambda}{4} = \frac{1}{4}\left(\frac{1}{4}\times10^5\right) = 5\times10^3 \text{ m} = 5 \text{ km}$$

The audio signals are of low frequency waves. They cannot be transmitted through sky waves as they are absorbed by atmosphere.

If the size of the antenna is less than 5 km, the effective power transmission would be very low because of deviation from resonance wavelength of wave and antenna length.

24. The statement (a) is correct but statements (b), (c) and (d) are wrong as space wave propagation is a line of sight propagation.

25. We know that, $f_c = 9\,(N_{max})^{1/2}$

$$10\times10^6 = 9(N_{max})^{1/2}$$

$$\frac{10^7}{9} = (N_{max})^{1/2} \Rightarrow \frac{10^{14}}{81} = N_{max}$$

$$\frac{100}{81}10^{12} = N_{max}$$

$$N_{max} \approx 1.2\times10^{12} \text{ m}^{-3}$$

26. The power dissipated in any circuit is a function of the square of voltage across the circuit and the effective resistance of the circuit. Equation of AM wave reveals that it has three components of amplitude E_c, $mE_c/2$ and $mE_c/2$. Clearly, power output must be distributed among these components.

Carrier power, $P_c = \frac{(E_c/\sqrt{2})^2}{R} = \frac{E_c^2}{2R}$

Total power of side bands,

$$P_s = \frac{(mE_c/2\sqrt{2})^2}{R} + \frac{(mE_c/2\sqrt{2})^2}{R} = \frac{m^2E_c^2}{4R}$$

$\therefore$ $\frac{P_s}{P_c} = \frac{1}{2}m^2$ and

$$P_t = P_c + P_s = P_c\left(1+\frac{m^2}{2}\right)$$

$\therefore$ $\frac{P_t}{P_c} = 1+\frac{m^2}{2} \Rightarrow \left(\frac{I_t}{I_c}\right)^2 = 1+\frac{m^2}{2}$

Given that, $I_T = 8.93$ A, $I_C = 8$ A, $m = ?$

$\therefore$ $\left(\frac{8.93}{8}\right)^2 = 1+\frac{m^2}{2}$

or $1.246 = 1+\frac{m^2}{2}$ or $\frac{m^2}{2} = 0.246$

or $m = \sqrt{2\times0.246} = 0.701 = 70.1\%$

27. Height of transmitting antenna,

$$d = \sqrt{2R_e h_T}$$

Given, $R_e = 6.4\times10^6$ m

$h_T = 128$ m

$d = ?$

The maximum distance,

$$d = \sqrt{2\times6.4\times10^6\times128} = \sqrt{128\times10^5\times128}$$

$$= \frac{128\times10^2\sqrt{10}}{\sqrt{10}}\times\sqrt{10} = \frac{128\times10^3}{\sqrt{10}} \text{ m} = \frac{128}{\sqrt{10}} \text{ km}$$

28. Given, $h = 150$ m, $R = 6.4\times10^6$ m

Average population density $= 10^3$ km

$$= 10^3\times(10^3)^{-2} = 10^{-3} \text{ m}$$

Distance upto which the transmission would be view

$$d = \sqrt{2hR}$$

Total area over which transmission could be

$$= \pi d^2 = 2\pi hR$$

Population covered $= 10^{-3}\times2\pi hR$

$$= 10^{-3}\times2\times3.14\times150\times6.4\times10^{-6}$$

$$= 60.288 \text{ lakh}$$

29. For *D*-region, $N = 10^9$ m^{-3}

$$\mu = \sqrt{1-\frac{81.45\,N}{\nu^2}} = \sqrt{1-\frac{81.45\times10^9}{(50\times10^6)^2}} \approx 1$$

$$\mu = \frac{\sin i}{\sin r} = 1$$

or $\sin r = \sin i$ or $r = i = 30°$

30. The frequency optical communication,

$$\nu = \frac{c}{\lambda} \text{ or } \nu = \frac{3\times10^8}{1200\times10^{-9}}$$

$$\nu = 25\times10^{13} \text{ Hz}$$

But only 2% of the source frequency is available for TV transmission,

$\nu' = 2.5\times10^{14}\times2\,\%$ or $\nu' = 2.5\times10^{14}\times\frac{2}{100}$

or $\nu' = 5\times10^{12}$ Hz

$\therefore$ Number of channels $= \frac{\nu'}{\text{Bandwidth}}$

or Number of channels $= \frac{5\times10^{12}}{5\times10^6} = 10^6 = 1$ million

31. Given, the maximum electron density in the morning,

$$N_{max} = 10^{10} \text{ m}^{-3}$$

$$f_c' = 9(N_{max})^{1/2}$$

$$= 9\,(10^{10})^{1/2}$$

Critical frequency, $f_c' = 9 \times 10^5$ Hz

The maximum electron density at noon,

$$N_{max} = 2 \times 10^{10} \text{m}^{-3}$$

Critical frequency, $f_c'' = 9 \times (N_{max})^{1/2}$

$$= 9 \times (2 \times 10^{10})^{1/2}$$

$$= 9 \times \sqrt{2} \times 10^5$$

The ratio of critical frequency at noon and critical frequency in the morning,

$$\frac{f_c''}{f_c'} = \frac{9 \times \sqrt{2} \times 10^5}{9 \times 10^5}$$

$$\Rightarrow \quad \frac{f_c''}{f_c'} = \sqrt{2}$$

$$= 1.414$$

32. I. Ground wave propagation $\rightarrow$ 500 kHz to 1500 kHz
II. Sky wave propagation $\rightarrow$ 2 MHz to 30 MHz
III. Space wave propagation $\rightarrow$ 30 MHz to 300 MHz
IV. Television signal propagation $\rightarrow$ 80 MHz to 200 MHz

33. Here both statements are correct but Statement II does not support Statement I.

34. Bandwidth $= 2f_m$

$$\omega_m = 1.57 \times 10^8 = 2\pi f_m$$

$$\text{BW} = 2f_m = \frac{10^8}{2} \text{ Hz} = 50 \text{ MHz}$$

35. Refractive index, $\mu = \mu_0 \sqrt{\dfrac{1 - 81.45\,N}{\nu^2}}$

$$\therefore \quad 0.5 = 1\sqrt{1 - \frac{81.45 \times 4 \times 10^8}{\nu^2}}$$

On solving, we get $\nu = 208.42$ kHz $= 208$ kHz

36. Population covered $= \rho\pi d^2 = \rho\pi\,(2\,hR)$

$$= (1000 \times 10^{-6}) \times \frac{22}{7} \times (2 \times 400 \times 6.4 \times 10^6)$$

$$= 1.6 \times 10^7$$

37. $d = \sqrt{2Rh}$ ...(i)

$$A = \pi d^2$$

Putting the value of d from Eq. (i), we get

$$A = \pi 2Rh = 3.14 \times 2 \times 6400 \times \frac{30}{1000}$$

$$A = 1205.76 \text{ km}^2$$

$$A \simeq 1206 \text{ km}^2$$

31

Experimental Physics

Experiment-1

Object

To measure internal and external diameter and depth of a vessel using vernier callipers.

Apparatus

Vernier callipers, a beaker.

Diagram

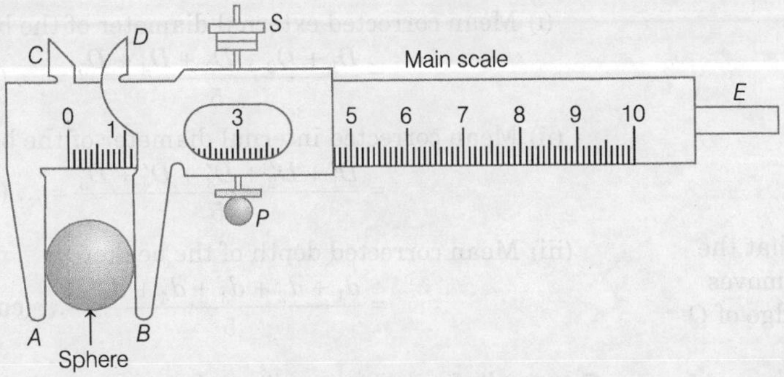

Theory

If the body is kept between jaws and zero of vernier scale lies ahead of Nth division of the main scale, then MSR (main scale reading) = N.

If nth division of vernier scale coincides with any division of main scale, then vernier scale reading (VSR) = $n \times$ (LC)

where, LC = least count of vernier callipers.

Total reading = MSR + VSR = $N + n \times$ (LC)

Procedure

(i) **Measurement of external diameter**

- Determine the least count and zero error of the vernier callipers by bringing the movable jaw in contact with the fixed jaw.
- Place the beaker between the jaws and fix the jaw *BD* in such a way that it grips the beaker.
- Record the main scale reading (MSR) and vernier scale reading (VSR).
- Note down 5 readings for different positions of the beaker.
- Find total reading and apply zero correction.
- Take mean of different values of diameter.

(ii) **Measurement of internal diameter**

- Put the jaws *C* and *D* inside the vessel and open them till each of them touches the inner wall of the vessel without putting any pressure on the wall.
- Note down the main scale reading and vernier scale reading.

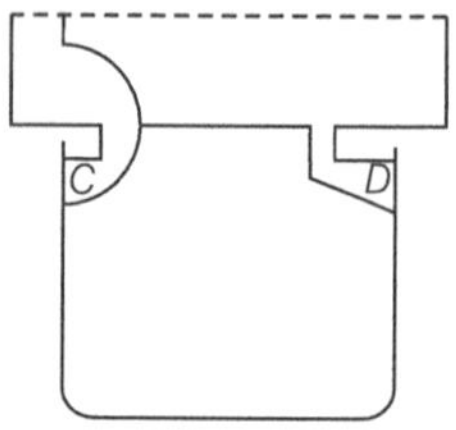

- Rotate the vernier callipers in perpendicular direction and take readings again for its different position.
- Find the total reading, take its mean and apply zero correction.

(iii) **Measurement of depth**

- Keep the right edge of main scale strip *P* on the upper edge of the vessel.

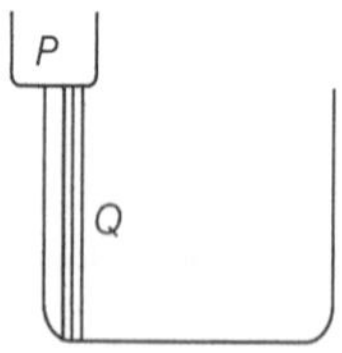

- Start pressing the jaw *BD* downward so that the thin metallic strip *Q* on the backside of *P* moves downward and continue it till the outer edge of *Q* touches the bottoms of the beaker.
- Note down MSR and VSR repeat the previous step for different positions inside the vessel.
- Find the total reading. Take its mean and apply zero correction.

Observations

LC of vernier callipers = ...

Zero error = ...

Table for external diameter of the beaker (*D*)

S.No.	MSR (cm)	VSR = $n \times$ LC (cm)	TR (Total reading) = MSR + VSR (cm)	
			Observed	Corrected
1.				D_1
2.				D_2
3.				D_3
4.				D_4
5.				D_5

Table for internal diameter of the beaker (*D*′)

S.No.	MSR (cm)	VSR = $n \times$ LC (cm)	TR (Total reading) = MSR + VSR (cm)	
			Observed	Corrected
1.				D_1'
2.				D_2'
3.				D_3'
4.				D_4'
5.				D_5'

Table for depth of the beaker (*d*)

S.No.	MSR (cm)	VSR = $n \times$ LC (cm)	TR (Total reading) = MSR + VSR (cm)	
			Observed	Corrected
1.				d_1
2.				d_2
3.				d_3
4.				d_4
5.				d_5

Calculations

(i) Mean corrected external diameter of the beaker

$$= \frac{D_1 + D_2 + D_3 + D_4 + D_5}{5} = \ldots \text{ (cm)}$$

(ii) Mean corrected internal diameter of the beaker

$$= \frac{D_1' + D_2' + D_3' + D_4' + D_5'}{5} = \ldots \text{ (cm)}$$

(iii) Mean corrected depth of the beaker

$$= \frac{d_1 + d_2 + d_3 + d_4 + d_5}{5} = \ldots \text{ (cm)}$$

Example 1. *A vernier callipers has 1 mm marks on the main scale. It has 20 equal divisions on the vernier scale which match with 16 main scale divisions. For this vernier callipers, the least count is*

(a) 0.02 mm

(b) 0.05 mm

(c) 0.1 mm

(d) 0.2 mm

Sol. (d) $\because$ 20 divisions of vernier scale = 16 divisions of main scale

$$\therefore \quad 1\text{ VSD} = \frac{16}{20}\text{MSD} = 0.8\text{ MSD} = 0.8\text{ mm}$$

We know that, $\text{LC} = 1\text{MSD} - 1\text{VSD}$

$$= 1\text{mm} - 0.8\text{ mm}$$
$$= 0.2\text{ mm}$$

Example 2. *The least count of vernier callipers is 0.1 mm. The main scale reading before the zero of the vernier scale is 10 and the zeroth division of the vernier scale coincides with the main scale division. Given that each main scale division is 1 mm, then the measured value should be equal to*

(a) 1 cm (b) 2 cm (c) 0.5 cm (d) 0.1 cm

Sol. (a) Given, least count (LC) = 0.1 mm

Vernier scale reading = 0

Main scale reading = 10 mm

Length measured with the vernier callipers = Number of vernier divisions coinciding with any main division + Reading before the zero of vernier scale × Least count

$$= 10\text{ mm} + 0 \times 0.1\text{ mm}$$
$$= 10\text{ mm} = 1\text{ cm}$$

Example 3. *1 cm on the main scale of vernier callipers is divided into ten equal parts. If 20 divisions of vernier scale coincide with 8 small divisions of the main scale. What will be the least count of vernier callipers?*

(a) 0.01 cm (b) 0.05 cm

(c) 0.001 cm (d) 0.06 cm

Sol. (d) $\because$ 20 divisions of vernier scale = 8 divisions of main scale

$$\therefore \quad 1\text{ VSD} = \left(\frac{8}{20}\right)\text{MSD} = \left(\frac{2}{5}\right)\text{MSD}$$

$$\text{LC} = 1\text{MSD} - 1\text{VSD}$$
$$= 1\text{MSD} - \left(\frac{2}{5}\right)\text{MSD} = \left(1 - \frac{2}{5}\right)\text{MSD}$$
$$= \frac{3}{5}\text{MSD} = \frac{3}{5} \times 0.1\text{ cm} = 0.06\text{ cm}$$
$$\left(\because 1\text{MSD} = \frac{1}{10}\text{ cm} = 0.1\text{ cm}\right)$$

Note For objective questions LC = MSD − VSD

$$= \left(\frac{b-a}{b}\right)\text{MSD} = \left(\frac{20-8}{20}\right)\left(\frac{1}{10}\right)\text{cm}$$
$$= \frac{3}{50}\text{ cm} = 0.06\text{ cm}$$

Example 4. *The least count of the main scale of a vernier callipers is 1 mm. Its vernier scale is divided into 10th division and coincide with 9th division of the main scale. When jaws are touching each other, the 7th division of vernier scale coincides with a division of main scale and the zero of vernier scale is lying right side of the zero of main scale. When this vernier is used to measure length of a cylinder, the zero of the vernier scale lies between 3.1 cm and 3.2 cm and 4th VSD coincides with a main scale division. The length of the cylinder is (where, VSD = vernier scale division)*

[JEE Main 2020]

(a) 3.2 cm (b) 2.99 cm

(c) 3.07 cm (d) 3.21 cm

Sol. (c) Given, least count of main scale = 1 main scale division

(MSD) = 1 mm

Also, 10 vernier scale divisions = 9 main scale divisions

$$\Rightarrow 1\text{ vernier scale division (VSD)} = \frac{9}{10}\text{MSD}$$
$$= 0.9\text{ MSD} = 0.9\text{ mm}$$

Vernier constant or least count of instrument,

$$\text{LC} = 1\text{ MSD} - 1\text{ VSD}$$
$$= 1\text{ MSD} - \frac{9}{10}\text{MSD}$$
$$= \left(\frac{10-9}{10}\right)\text{MSD} = \frac{1}{10}\text{MSD}$$
$$= \frac{1\text{ mm}}{10} = 0.1\text{ mm or } 0.01\text{ cm}$$

The zero of vernier scale lies to the right of main scale, so error is positive.

Zero error = 7 × LC = 7 × 0.01 cm = 0.07 cm

Given, main scale reading = 3.1 cm

Vernier scale reading = 4 divisions

$$\therefore \quad \text{Observed value} = \text{MSR} + \text{VSR} \times \text{LC} - \text{Zero error}$$
$$= 3.1 + 4 \times 0.01 - 0.07$$
$$= 3.14\text{ cm} - 0.07\text{ cm}$$
$$= 3.07\text{ cm}$$

Experiment-2

Object

To determine the thickness/diameter of a thin sheet/wire using screw gauge.

Apparatus

Screw gauge, wire/sheet.

Diagram

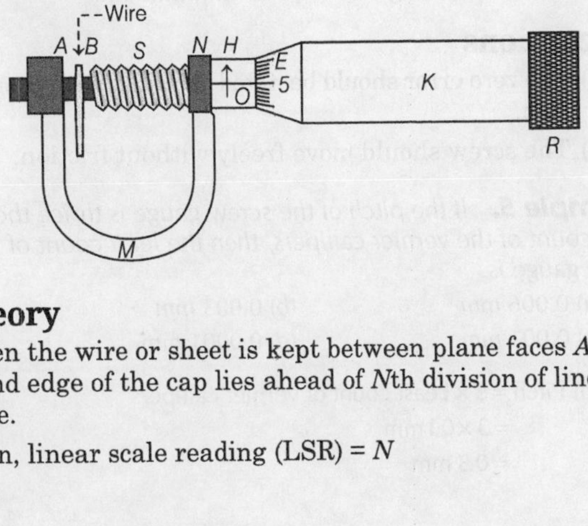

Theory

When the wire or sheet is kept between plane faces A and B and edge of the cap lies ahead of Nth division of linear scale.

Then, linear scale reading (LSR) = N

If nth division of circular scale lies over reference line.
Then, circular scale reading (CSR) = $n \times$ LC
where, LC = least count of screw gauge.

Procedure

(i) Determine the least count and pitch of the screw gauge.

(ii) Bring the faces A and B in close contact and find zero error if any.

(iii) Place the wire between A and B and move the face B towards A using the head R. Stop when R turns without moving the screw.

(iv) Note down linear scale reading (LSR) and circular scale reading (CSR).

(v) Take five readings for different positions of wire.

(vi) Find total reading, take its mean and apply zero correction.

Observations

Least count of screw gauge = ... mm = ... cm

Zero error = ... mm

Zero correction = ... mm

Table for measuring diameter (d)

S.No.	LSR(mm)	CSR = $n \times$ LC (mm)	TR = LSR + CSR(mm)	
			Observed	Corrected
1.				d_1
2.				d_2
3.				d_3
4.				d_4
5.				d_5

Calculations

Mean diameter of the wire

$$= \frac{d_1 + d_2 + d_3 + d_4 + d_5}{5} = \ldots\ldots \text{ mm} = \ldots\ldots \text{ cm}$$

Result

The diameter of the given wire is .... cm.

Precautions

(i) The zero error should be noted carefully with proper sign.

(ii) The screw should move freely without friction.

Example 5. *If the pitch of the screw gauge is thrice the least count of the vernier callipers, then the least count of screw gauge is*

(a) 0.006 mm (b) 0.003 mm
(c) 0.001 mm (d) 0.0001 mm

Sol. (b) Pitch = 3 × Least count of vernier callipers

$= 3 \times 0.1$ mm

$= 0.3$ mm

∴ Least count of screw gauge

$$= \frac{\text{Pitch}}{\text{Number of divisions on circular scale}}$$

$$= \frac{0.3}{100} = 0.003 \text{ mm}$$

Example 6. *Using screw gauge of pitch 0.1 cm and 50 divisions on its circular scale, the thickness of an object is measured. It should correctly be recorded as* **[JEE Main 2020]**

(a) 2.121 cm (b) 2.124 cm
(c) 2.125 cm (d) 2.123 cm

Sol. (b) Given, pitch = 0.1 cm

Number of divisions on circular scale = 50

Least count of screw gauge

$$= \frac{\text{Pitch}}{\text{Number of divisions on circular scale}}$$

$$= \frac{0.1}{50} = 0.002 \text{ cm}$$

Now, correctly recorded measurement is an integral multiple of LC.

So, amongst the given options, only 2.124 cm is the integral multiple of LC. Thus, it is the correctly recorded value.

Hence, option (b) is correct.

Experiment-3

Object

To study dissipation of energy of a simple pendulum by plotting a graph between square of amplitude and time.

Apparatus

A simple pendulum having for its string a 2 m long thin wire with upper end screwed in a torsion screw fixed to rigid support in ceiling and a 2 kg rectangular metallic block as its bob suspended from its lower end, ticker timer, paper tape, spring balance of range 2.5 kg, meter scale and thread.

Diagram

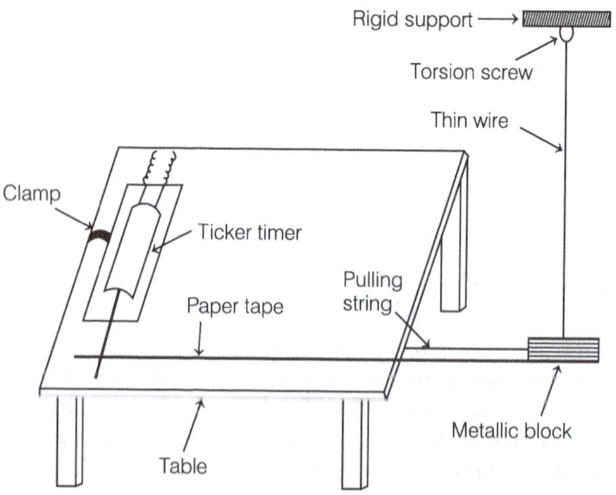

Theory

When a simple pendulum oscillates, the whole of its energy remains in the form of kinetic energy at mean position P and in the form of potential energy at extreme positions Q and R. If we consider any position between P and Q or P and R, say A and B, total energy is sum of potential and kinetic energies.

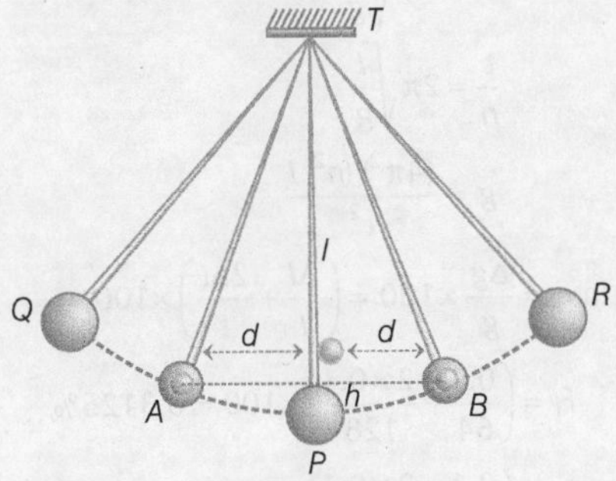

At point (A or B)

Kinetic energy of the bob, KE $= \frac{1}{2} mv^2$

Potential energy of the bob, PE $= mgh = \frac{mgd^2}{2l}$

Total energy of the bob, TE $=$ KE $+$ PE $= \frac{1}{2} mv^2 + \frac{mgd^2}{2l}$

Procedure

(i) Using spring balance, find the mass of the metallic block.

(ii) Clamp one end of a thin metallic wire to a torsion screw fixed with a rigid support.

(iii) From lower end of metallic wire, suspend the metallic block.

(iv) Under the carbon disc, pass the paper tape and attach its end to the centre of the bottom of the metallic block.

(v) Tie a thread at the level of its centre of gravity round the metallic block.

(vi) To displacement the block and the wire, pull the thread towards ticker timer.

(vii) Start the ticker timer carefully and leave the pulled thread.

(viii) As metallic block starts oscillating, it pulls the paper tape and dots can be obtained on the tape to record the position of the metallic block.

(ix) Switch off the ticker timer as soon as the metallic block reaches its outer extreme.

(x) Tear off dotted paper. You can observe symmetrical dots about a central dot.

(xi) Mark P as central dot which corresponds to mean position and mark Q and R as extreme dots.

(xii) Measure distances of the different dots from central dot. This gives values of Q.

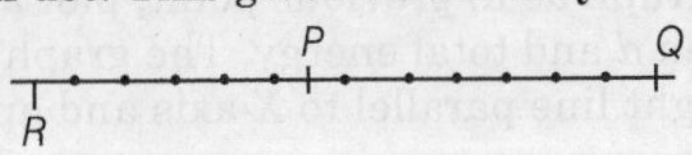

(xiii) Also note the position of a dot with respect to central dot. This gives instant of time for that distance.

Observations

Least count of spring balance = ... kg

Zero error of spring balance = ... kg

Observed mass of metallic block = ... kg

Corrected mass of metallic block = ... kg

Length of simple pendulum l = ... m

Time period of ticker timer = ... s

Table for d and T

Side from central dot	S. No. of the dot	Distance from central dot d(m)	Time interval T(s)	Velocity v(m/s)	v^2	$h = \frac{d^2}{2l}$(m)	KE $= \frac{1}{2}mv^2$ (J)	TE = KE + PE
Right	1.		0.01					
	2.		0.02					
	3.		0.03					
	4.		0.04					
	5.		0.05					
Left	1.		0.01					
	2.		0.02					
	3.		0.03					
	4.		0.04					
	5.		0.05					

Calculations

(i) Plot a graph between d and T taking T along X-axis and d along Y-axis.

(ii) Find slope of the graph which gives value of velocity

$$h = \frac{d^2}{2l}$$

$$\text{KE} = \frac{1}{2} mv^2$$

$$\text{PE} = \frac{mgd^2}{2l}$$

$$\text{TE} = \text{KE} + \text{PE}$$

[Calculate for each value of d]

(iii) Plot a graph between (d) and $\text{KE} = \frac{1}{2} mv^2$. Also plot a graph between d and $\text{PE} = mgh$, taking d along X-axis and mgh along Y-axis.

(vi) Add KE and PE for each value of d and on the same graph as in previous point, plot a graph between d and total energy. The graph comes to be a straight line parallel to X-axis and appears as

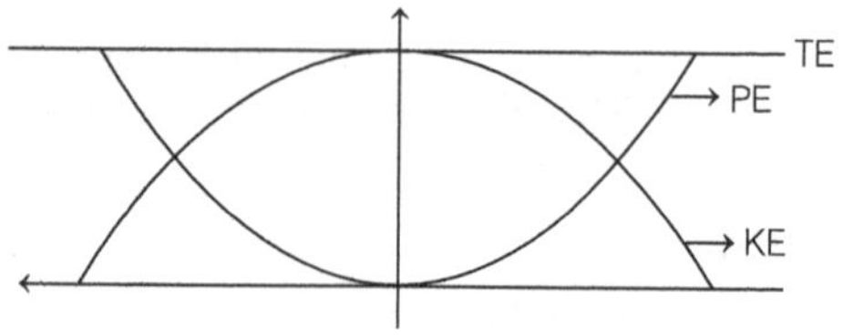

Result

At any point, during oscillation of a simple pendulum, the sum of KE and PE is always constant. This verifies the principle of conservation of energy.

Precautions

(i) Pendulum's length should be sufficient say about 2 m.

(ii) The amplitude should be kept small.

Example 7. *Students I, II and III perform an experiment for measuring the acceleration due to gravity using a simple pendulum. They use different lengths of the pendulum and/or record time for different number of oscillations. The observations are shown in the table. Least count for length = 0.1 cm and least count for time = 0.1s.*

Student	Length of the pendulum (cm)	Number of oscillations (n)	Total time for n oscillations (s)	Time period (s)
I	64.0	8	128.0	16.0
II	64.0	4	64.0	16.0
III	20.0	4	36.0	9.0

If E_I, E_{II} and E_{III} are the percentage errors in g, i.e. $\left(\frac{\Delta g}{g} \times 100\right)$, for students I, II and III respectively.

(a) $E_I = 0$ (b) E_I is minimum
(c) $E_I = E_{II}$ (d) E_{II} is maximum

Sol. (b) Time period $T = 2\pi\sqrt{\frac{l}{g}}$

or $\frac{t}{n} = 2\pi\sqrt{\frac{l}{g}}$

$\therefore$ $g = \frac{(4\pi^2)(n^2)\,l}{t^2}$

$$\%\text{ error in } g = \frac{\Delta g}{g} \times 100 = \left(\frac{\Delta l}{l} + \frac{2\Delta t}{t}\right) \times 100$$

$$E_I = \left(\frac{0.1}{64} + \frac{2 \times 0.1}{128}\right) \times 100 = 0.3125\%,$$

$$E_{II} = \left(\frac{0.1}{64} + \frac{2 \times 0.1}{64}\right) \times 100 = 0.46875\%$$

and $$E_{III} = \left(\frac{0.1}{20} + \frac{2 \times 0.1}{36}\right) \times 100 = 1.055\%$$

Hence, E_I is minimum.

Example 8. *In an experiment to determine acceleration due to gravity, the length of the pendulum is measured as 98 cm by a meter scale of least count of 1 cm. The period of swing/oscillations is measured with the help of a stop watch having a least count of 1s. The time period of 50 oscillations is found to be 98 s. Express value of g with proper error limits.*

(a) $(10.1 \pm 0.5)\ ms^{-2}$ (b) $(10.1 \pm 0.2)\ ms^{-2}$
(c) $(10.1 \pm 0.3)\ ms^{-2}$ (d) $10.1\ ms^{-2}$

Sol. (c) As, $T = 2\pi\sqrt{\frac{l}{g}}$ ...(i)

Now, time period of 50 oscillations is 98 s.

$\therefore$ Time period of one oscillation $= \frac{98}{50} = 1.96$ s

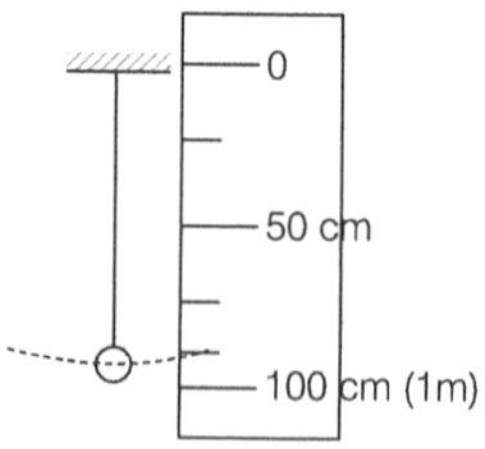

$\Rightarrow$ $g = \frac{4\pi^2 l}{T^2}$ [from Eq. (i)]

$$= 4 \times (3.14)^2 \times \frac{0.98}{1.96 \times 1.96}$$

$$= 10.06\ \text{m/s}^2$$

Now, % error in $g = \frac{\Delta g}{g} \times 100 = \left(\frac{\Delta l}{l} + \frac{2\Delta T}{T}\right) \times 100$

$\Rightarrow \quad \Delta g = 10 \times \left(\frac{1}{98} + \frac{2 \times 1}{98}\right)$

($\because$ least count of meter scale is 1 cm and least count of stop watch is 1s)

$\therefore \quad \Delta g = 0.3 \text{ m/s}^2$

So, final result can be expressed as $(10.1 \pm 0.3)\text{ ms}^{-2}$.

Experiment-4

Object

To determine the mass of a given body using a meter scale by principle of moments.

Apparatus

A meter scale, a broad heavy wedge with sharp edge, a weight box, a body of unknown mass.

Diagram

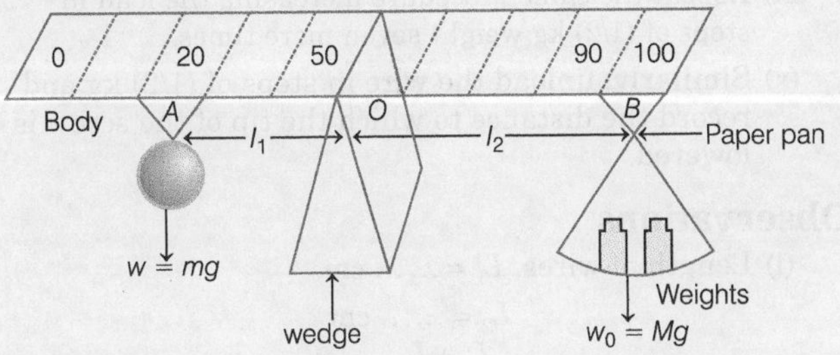

Theory

If a body of mass m be suspended at one end A of a meter scale and a body of unknown mass M be suspended at the end B and then position of point O is adjusted, so that the meter scale is in equilibrium.

By the principle of moments,

$$(mg) \times OA = (Mg) \times OB$$

$$OA = l_1 \text{ and } OB = l_2$$

$$mg \times l_1 = Mg \times l_2$$

$$m = M\left(\frac{l_2}{l_1}\right)$$

Procedure

(i) Place the sharp wedge on a wooden block carefully that is placed on the table.

(ii) Place the meter scale on the wedge such that graduated side facing you.

(iii) Find its centre of gravity by balancing it on the edge of the wedge.

(iv) Suspended the known mass on right hand side of the wedge (at point B) and the body of unknown mass (*i.e.* $w = mg$) on the left side of the wedge (at point A) by means of two threads of equal lengths.

(v) Adjust the position of unknown mass at point A to maintain scale in equilibrium.

(vi) Note the position of m, M and centre of gravity O when the meter scale is in equilibrium.

(vii) Repeat the procedure for three more observations changing the positions of m.

Observations

1. Position of centre of gravity O on the scale =...... cm
2. Magnitude of known mass, M = ... g
3. Known weight, $w = Mg$ = ... g-wt

Table for determination of unknown mass

S.No.	Position of mA(cm)	Position of M B(cm)	$OA = l_1$ (cm)	$OB = l_2$ (cm)	$m = M\left(\frac{l_2}{l_1}\right)$
1.					m_1
2.					m_2
3.					m_3
4.					m_4

Calculation

The mean value of unknown mass,

$$m = \frac{m_1 + m_2 + m_3 + m_4}{4} = \ldots\ldots \text{ g}$$

Result

The value of unknown mass is .... g.

Precautions

(i) Centre of gravity must be determined accurately and after every reading, check that the edge of the wedge remains under this position only.

(ii) The meter scale should be made horizontal in equilibrium position.

Example 9. *A meter rule of length 2m is pivoted at its middle point . If a weight of 50N is hanged at a point 1m left from middle point, then amount of mass required to be applied at 0.5m mark on the right side from the middle point to keep it in a balanced position is (Take, g = 10 m/s²)*

(a) 5 kg (b) 15 kg (c) 10 kg (d) 20 kg

Sol. (c) The given situation is shown below.

If m be the required mass hanging at point C, then according to principle of moments,

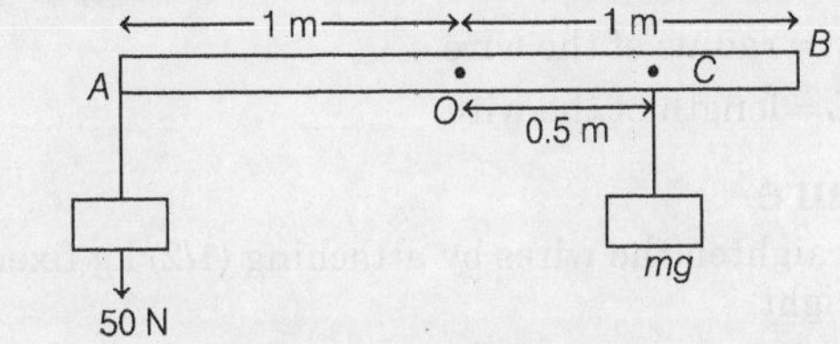

$$50 \times OA = mg \times OC$$

$$\Rightarrow \quad m = \frac{50 \times OA}{g \times OC} = \frac{50 \times 1}{10 \times 0.5} = 10 \text{ kg}$$

Experiment-5

Object

To determine Young's modulus of elasticity of the material of a metallic wire.

Apparatus

Searle's apparatus for Young's modulus, two long wires approximately 2 m length , a (1/2) kg fixed weight, a (1/2) kg hanger along with (1/2) kg slotted weights, a meter scale, a plumb line with long thread, a screw gauge.

Diagram

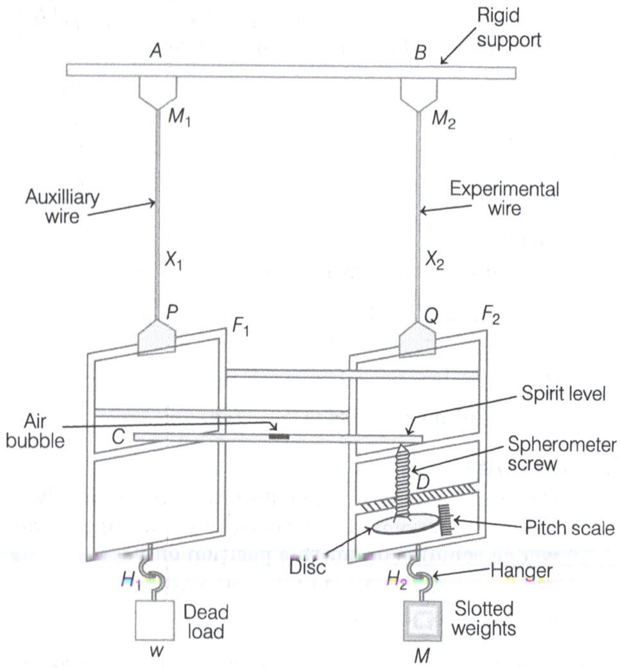

Theory

Young's modulus is defined as the ratio of longitudinal stress and longitudinal strain.

i.e.
$$Y = \frac{\frac{F}{A}}{\frac{\Delta L}{L}} = \frac{mgL}{\pi r^2 \Delta L}$$

where, ΔL = extension of the wire under load mg,

r = radius of the wire

and L = length of the wire.

Procedure

(i) Straighten the wires by attaching (1/2) kg fixed weight.

(ii) Suspend the plumb line along the suspension wire so that its lower end coincides with binding post P and the upper end of the thread A. Measure lengths AP and BQ and take its mean to find L.

(iii) Measure the diameter of the wire by means of screw gauge.

(iv) Note down the value of breaking stress for the wire from the table of constants multiplying with the area of cross-section of the given wire.

(v) Note down the least count of the micrometer screw attached with the Searle's apparatus.

(vi) Ensure that air bubble in the spirit level should be in middle. If not, adjust it with the help of screw.

(vii) Slip a (1/2) kg weight on the hanger due to which bubble in spirit level will move towards F_1. By raising the tip of the screw, bring it back to the centre.

(viii) Record the height to which the tip of the screw is raised. It gives elongation of the wire under (1/2) kg weight.

(ix) Repeat the same procedure increasing the load in steps of (1/2) kg weight seven more times.

(x) Similarly, unload the wire in steps of (1/2) kg and record the distance to which the tip of the screw is lowered.

Observations

(i) Length of wires, L_1 = ... cm,

L_2 = ... cm

Mean length, $L = \frac{L_1 + L_2}{2}$ = ... cm

= ...... m

(ii) Measurement of diameter

- Least count of screw gauge = ... mm
- Zero error = ... mm
- Diameter of the wire, d = ... mm
- Radius, $\frac{d}{2}$ = ... mm
- Area of cross-section of the wire,

$A = \pi r^2$

= ...... mm^2

(iii) Measurement of extension with load

- Breaking stress for the wire,

F = ... kgf/mm^2

Maximum load $= \frac{1}{3} FA$ = ... kgf

- Least count of micrometer screw

$$= \frac{\text{Pitch of micrometer screw}}{\text{Number of division of circular scale}}$$

= ...... cm

Table for extension under different loads

S. No.	Load on hanger H_2		Micrometer reading			Extension	Extension for 2 kg weight
	kg-wt	N	Increasing load L (cm)	Decreasing Load L' (cm)	Mean $L_m = \frac{L+L'}{2}$ (cm)	$L_m - L$ (cm)	$L = L_5 - L$ (cm)
1.	0	0				$L = 0$	
2.	0.5	4.9					$L_1 =$
3.	1.0	9.8					$L_2 =$
4.	1.5	14.7					$L_3 =$
5.	2.0	19.6					$L_4 =$
6.	2.5	24.5					
7.	3.0	29.4					
8.	3.5	24.3					

Calculations

(i) Mean extension, $\Delta L = \dfrac{L_1 + L_2 + L_3 + L_4}{4}$

$= \ldots\ldots \text{ cm} = \ldots\ldots \text{ m}$

(ii) $Y = \dfrac{mgL}{\pi r^2 \Delta L} = \ldots\ldots$

(iii) Draw a graph between load and extension. Calculate slope of the graph.

Result

(i) Young's modulus of the material of given wire is $\ldots\ldots\ldots \text{Nm}^{-2}$.

(ii) The load extension graph is a straight line. It verifies the Hooke's law.

Precautions

(i) Wires must be similar in all respects and must be hanged from the same firm and rigid support.

(ii) While changing the weight, wait for 2 min so that wires reach at equilibrium.

(iii) Add and remove the weights gently.

(iv) Do not load the wire beyond maximum allowed load.

Example 10. *In an experiment for measurement of Young's modulus, following readings are taken : Load = 3.00 kg, length = 2.820 m, diameter = 0.041 cm and extension = 0.87. The percentage error in measurement of Y is around*

(a) 6% (b) 8% (c) 1% (d) 3%

Sol. (a) If Y = Young's modulus of wire, M = mass of wire, g = acceleration due to gravity, x = extension in the wire, A = area of cross-section of the wire and l = length of the wire.

$$Y = \frac{Mgx}{Al} \Rightarrow \frac{\Delta Y}{Y} = \frac{\Delta M}{M} + \frac{\Delta x}{x} + \frac{\Delta A}{A} + \frac{\Delta l}{l}$$

$$\Rightarrow \quad \frac{\Delta Y}{Y} = \frac{0.01}{3.00} + \frac{0.01}{0.87} + \frac{2 \times 0.001}{0.041} + \frac{0.001}{2.820}$$

$$= 0.064 \Rightarrow \frac{\Delta Y}{Y} \times 100 = \pm 6.4\% \simeq 6\%$$

Experiment-6

Object

To determine the surface tension of water by capillary rise method.

Apparatus

Three glass capillaries having bores of different diameters, adjustable stand, travelling microscope, a petridish, a glass slide, an iron stand with clamp, a thermometer, rubber band, dil HCl and NaOH solution, clean water in a beaker.

Diagram

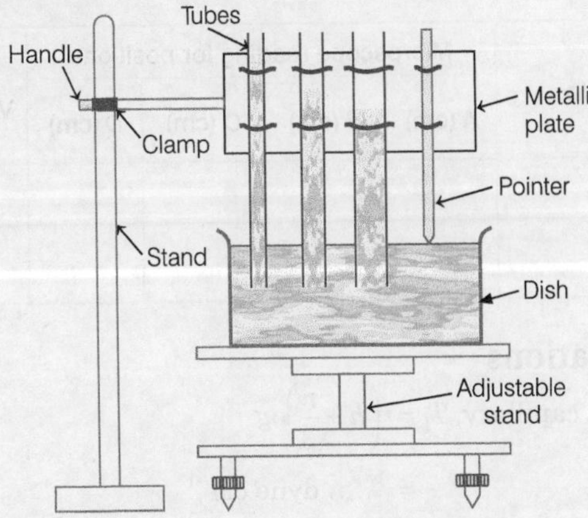

Measurement of surface tension by capillary rise method,

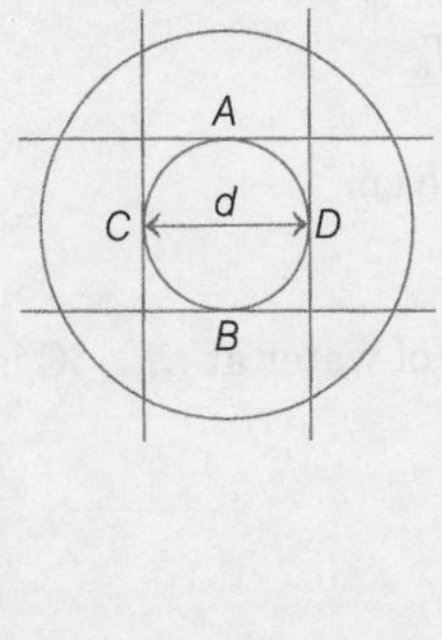

Theory

When a capillary of radius r is immersed in a liquid of density ρ, the height h to which the liquid rises is given by surface tension T.

$$T = \frac{r\left(h + \frac{r}{3}\right)\rho g}{2\cos\theta}$$

For water, $T = \frac{r\left(h + \frac{r}{3}\right)\rho g}{2}$ as θ is small.

Procedure

(i) Clean the tube and petridish first with caustic soda, then with nitric acid and then with water.

(ii) Place the adjustable height stand on the table and make its base horizontal by level.

(iii) Place the petridish containing water on adjustable stand.

(iv) Clamp the capillary tubes of different radii in a metallic plate. Also clamp a pointer after third capillary tube.

(v) Hold the glass plate containing capillaries and pin in the clamp on an iron stand and hold them just above the petridish vertically such that lowest edges of the capillaries are slightly above the water level in the dish.

(vi) Adjust the position of the pointer such that its tip just touches the water surface.

(vii) Record the least count of the microscope for the horizontal scale and the vertical scale.

(viii) Adjust the microscope to a suitable height, keeping its axis horizontal and pointed towards the capillary tube and bring the microscope in front of the first capillary tube.

(ix) Fix the microscope by moving with the help of screw in such a position such that horizontal cross wire is tangential on the meniscus in capillary 1. Record the position of meniscus in capillary 1.

(x) Now move the microscope horizontally and bring it in front of the second capillary tube.

Lower the microscope and repeat the same procedure as in previous step. Similarly, read the position of meniscus in capillary 3.

(xi) Lower the stand so that pointer tip becomes visible.

(xii) Place the first capillary on the stand. Focus the microscope on the end dipped in water. Two concentric circles will be seen.

Observations

Temperature of water (t) = ...°C

Density of water at this temperature = ... gcm^{-3}

Table for internal diameter of the capillary tube

S. No.	Microscope reading for positions				Internal diameter			Internal radius
	A (cm)	B (cm)	C (cm)	D (cm)	Vertical AB (cm)	Horizontal CD (cm)	Mean $= \frac{AB + CD}{2} = d$ (cm)	$\frac{d}{2} = r$ (cm)
1.								
2.								
3.								

Calculations

For first capillary, $T_1 = r\left(h + \frac{r}{3}\right)\rho g$

$= \ldots\ldots$ dyne cm^{-1}

For second capillary, $T_2 = \ldots\ldots$ dyne cm^{-1}

For third capillary, $T_3 = \ldots\ldots$ dyne cm^{-1}

Mean, $T = \frac{T_1 + T_2 + T_3}{3}$

$= \ldots\ldots$ dyne cm^{-1}

Result

The surface tension of water at ...... °C is ......dyne cm^{-1}.

Precautions

(i) Capillary tube and pin should be set vertical.

(ii) Select capillaries of uniform bore.

Example 11. *A capillary tube whose inside radius is 0.5 mm is dipped in water having surface tension 7.0×10^{-2} Nm^{-1}. To what height is the water raised above the normal water level? Angle of contact of water with glass is 0°. (Take, density of water is 10^3 $kg\,m^{-3}$ and $g = 9.8$ ms^{-2})*

(a) 3.25 cm
(b) 1.25 cm
(c) 2.86 cm
(d) 13.76 cm

Sol. (c) Height raised, $h = \frac{2T\cos\theta}{r\rho g}$

Substituting the proper values, we have

$$h = \frac{(2)(7.0\times 10^{-2})\cos 0°}{(0.5\times 10^{-3})(10^3)(9.8)}$$
$$= 2.86\times 10^{-2}\text{ m}$$
$$= 2.86\text{ cm}$$

Experiment-7

Object

To determine the coefficient of viscosity of a given viscous liquid by measuring terminal velocity of a given spherical body.

Apparatus

An iron stand with clamps, a cylindrical glass pipe, two large cork pieces, a He funnel, experimental liquid (glycerine), meter scale, stop watch, thread, spherical metal balls of different sizes, thermometer and screw gauge.

Diagram

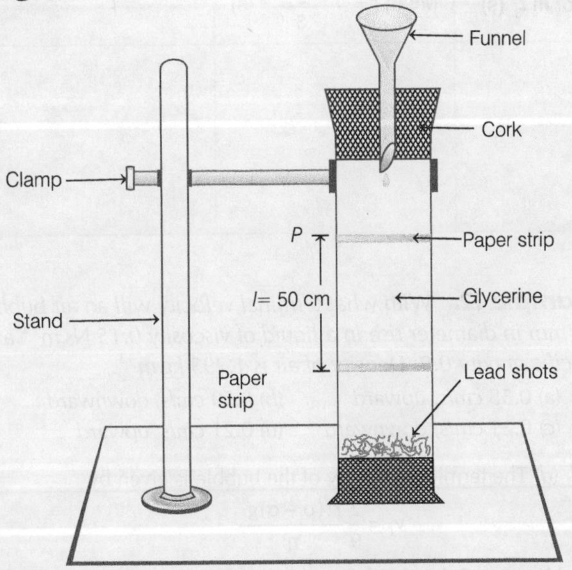

Theory

When a solid falls in a liquid, it experiences a resistive force due to viscosity of the liquid. This is given by

$$F = 6\pi r\eta v$$

where, r = radius of spherical body,

η = coefficient of viscosity of fluid

and v = velocity.

Due to this resistive force, acceleration of the body decreases and after sometime, it attains a constant velocity called terminal velocity which is given by

$$v_t = \frac{2r^2(\rho-\sigma)g}{9\eta}$$

Thus,
$$\eta = \frac{2}{9}\frac{r^2(\rho-\sigma)g}{v_t}$$

where, ρ = density of the material of the sphere,

σ = density of fluid and v_t = terminal velocity.

Procedure

(i) Put the cork at one end of glass pipe and seal it leak proof with wax.

(ii) Clamp the pipe vertically in the iron stand such that its open end is at the top fill it up with glycerine.

(iii) Bore a hole at the centre of the cork and in the bore, insert the stem of the He funnel.

(iv) Put the cork in opening of the pipe such that it closes the opening without leakage of glycerine.

(v) Now put some glycerine in the funnel, so that it fills its stem completely.

(vi) Mark two points separated by a distance = 50 cm, nearly in the middle. Tie up two cotton threads around the tube at these points. Measure the distance l between these points A and B.

(vii) Record the temperature of glycerine.

(viii) Choose lead shots of five different sizes taking at least 3 of each size. Measure its radius.

(ix) Place the lead shots gently on glycerine in funnel. Start stop water as it reaches the point A and stop and it passes point B. Record the time.

(x) Repeat the previous step for two more lead shots of the same radius. Calculate terminal velocity for each of them as $v_t = \frac{l}{t}$ cms^{-1}. Calculate mean value of terminal velocity.

(xi) Now take lead shots of different radii and determine corresponding terminal velocity as in previous steps.

(xii) Record the temperature of glycerine at the end of the experiment.

Observations

(i) Least count of thermometer = ... °C

(ii) • Temperature of glycerine in the beginning of the experiment, θ_1 = ... °C

• Temperature of glycerine at the end of the experiment, θ_2 = ... °C

Mean value of temperature,

$$\theta = \frac{\theta_1+\theta_2}{2} = \text{......}°\text{C}$$

(iii) Density of lead shots = ... gcm^{-3}

(iv) Least count of screw gauge = ... cm

Zero error of screw gauge = ... cm

Table for radius of lead shot

S. No.	Diameter of lead shot			Mean radius $r=\frac{d}{2}$ (cm)	r^2 = ... (cm)^2
	Along one diameter d_1 (cm)	Along perpendicular diameter d_2 (cm)	Mean diameter $d=\frac{d_1+d_2}{2}$ (cm)		
1.					
2.					
3.					
4.					
5.					

Table for terminal velocity of lead shot

Distance between A and B l (cm)	Diameter of lead shot				$v_t=\frac{l}{t}$ (cm s^{-1})
	Lead shot I t_1 (s)	Lead shot II t_2 (s)	Lead shot III t_3 (s)	Mean $t=\frac{t_1+t_2+t_3}{3}$ (s)	
1.					
2.					
3.					
4.					
5.					

Calculations

(i) Plot a graph between r^2 and v_t taking r^2 along X-axis and v_t along Y-axis. It comes out to be a straight line.

(ii) The slope of graph $=\frac{v_t}{r^2}=$......

(iii) $\eta=\frac{2}{9}\frac{(\rho-\sigma)g}{v_t/r^2}=$...... poise

Result

(i) The graph between terminal velocity and square of the radius is a straight line, which implies that $v_t \propto r^2$.

(ii) The coefficient of viscosity of glycerine at ...... °C is ...... poise.

Precautions

(i) Lead shots must be thoroughly immersed in glycerine before dropping in the liquid so that air bubbles are not accumulated.

(ii) The bore in the cork should be exactly at the centre.

(iii) Lead shots should be of uniform diameter.

(iv) Do measure the temperature of glycerine in the beginning and at the end of the experiment.

Example 12. *With what terminal velocity will an air bubble 0.8 mm in diameter rise in a liquid of viscosity 0.15 Ns m^{-2} and specific gravity 0.9. Density of air is 1.293 kg m^{-3}.*

(a) 0.55 cm/s, upward (b) 0.59 cm/s, downward
(c) 0.21 cm/s, downward (d) 0.21 cm/s, upward

Sol. (d) The terminal velocity of the bubble is given by

$$v_t=\frac{2}{9}\frac{r^2(\rho-\sigma)g}{\eta}$$

Here, $r=0.4\times 10^{-3}$ m

$\sigma=0.9\times 10^3$ kgm^{-3}, $\rho=1.293$ kgm^{-3}

$\eta=0.15$ Nsm^{-2} and $g=9.8$ m s^{-2}

Substituting the values, we have

$$v_t=\frac{2}{9}\times\frac{(0.4\times10^{-3})^2(1.293-0.9\times10^3)\times 9.8}{0.15}$$

$$=-0.0021 \text{ ms}^{-1} \text{ or } v_t=-0.21 \text{ cms}^{-1}$$

Note Here, negative sign implies that the bubble will rise up.

Example 13. *Two spherical raindrops of equal size are falling vertically through air with a terminal velocity of 1 ms^{-1}. What would be the terminal speed, if these two drops were to coalesce to form a large spherical drop?*

(a) 1.125 m/s (b) 2.225 m/s (c) 1.587 m/s (d) 3.125 m/s

Sol. (c) As, $v_t \propto r^2$ ...(i)

Let r be the radius of small raindrops and R is the radius of large drop.

Equating the volumes, we have

$$\frac{4}{3}\pi R^3 = 2\left(\frac{4}{3}\pi r^3\right)$$

$\therefore$ $R = (2)^{1/3} \cdot r$

or $\frac{R}{r} = (2)^{1/3}$

$\therefore$ $\frac{v_t'}{v_t} = \left(\frac{R}{r}\right)^2 = (2)^{2/3}$

$\therefore$ $v_t' = (2)^{2/3} v_t = (2)^{2/3} (1.0)\ ms^{-1}$

$= 1.587\ ms^{-1}$

Experiment-8

Object

To study the relationship between the temperature of a hot body and time by plotting a cooling curve.

Apparatus

Two celcius thermometers, Newton's law of cooling apparatus, a clamp stand, a stop watch, a beaker, burner, wire gauge, tripod stand and water.

Diagram

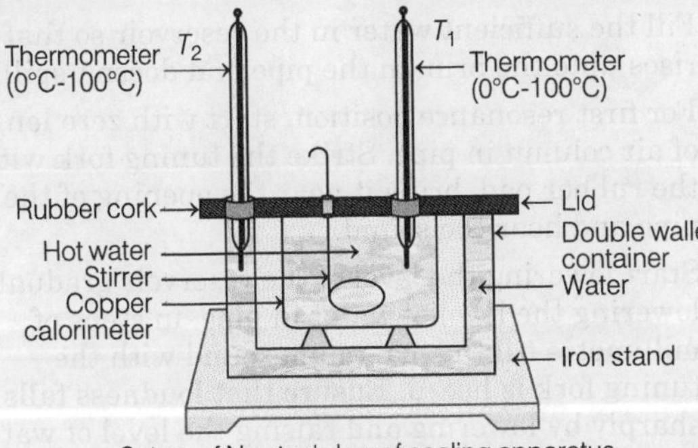

Set up of Newton's law of cooling apparatus

Theory

Newton's law of cooling states that rate of cooling of a body is directly proportional to the difference in temperature of the body and its surroundings provided this difference is not very large.

i.e. $$-\frac{d\theta}{dt} \propto (\theta - \theta_0)$$

$\therefore$ $$(\theta - \theta_0) = (\theta_m - \theta_0)e^{-kt}$$

where, θ = temperature of the body,

θ_0 = temperature of surroundings,

and θ_m = temperature of the body at 0°C.

$$\log(\theta - \theta_0) = -kt + c$$

Thus, graph $\log(\theta - \theta_0)$ *versus* t will be a straight line.

Procedure

(i) Clean the apparatus and place it on the table. Fill water between the two walls of the vessel.

(ii) Put the thermometer in water contained in beaker. Mark them as T_1 and T_2, Record the difference of temperature in two thermometers as $(\theta_2 - \theta_1)$.

(iii) Insert thermometer T_1 in calorimeter with the help of clamp stand and insert T_2 in water between double walls.

(iv) Heat water upto 70°C and pour it in the 'calorimeter. Stir it after placing the lid.

(v) Start noting down temperature in T_1 and where it is about 65°C, start stop watch and record the temperature after every 2 min till the temperature reaches 40°C.

(vi) Record the temperature in T_2 at the end of the experiment.

Observations

Temperature of water as recorded by thermometer

$T_1\ \theta_1 = \ldots\ldots$ °C

Temperature of water as recorded by thermometer

$T_2\ \theta_2 = \ldots\ldots$ °C

Correction applied in

$T_2 = \theta_2 - \theta_1 = \ldots\ldots$ °C

Initial corrected temperature between the walls of the apparatus $= \theta_0 = \frac{\theta_1' + \theta_2'}{2} = \ldots\ldots$ °C

Final corrected temperature $= \theta_2' = \ldots\ldots$ °C

Mean corrected temperature of surroundings of the calorimeter $= \theta_0 = \frac{\theta_1' + \theta_2'}{2} = \ldots\ldots$ °C

Table for measurement of temperature of hot water

S.No.	Time t (min)	Temperature of hot water θ(°C)	$\theta - \theta_0$ (°C)	$\log(\theta - \theta_0)$
1.				
2.				
3.				
4.				
...				
...				
...				
20.				

Calculations

(i) Plot a graph between $(\theta - \theta_0)$ and t taking t along X-axis and $(\theta - \theta_0)$ along Y-axis.

(ii) Plot a graph between $\log(\theta - \theta_0)$ and t taking t along X-axis and $\log(\theta - \theta_0)$ along Y-axis.

Result

(i) The cooling curve for water between $(\theta - \theta_0)$ and t is an exponential curve.

(ii) $\log(\theta - \theta_0)$ *versus* t graph is a straight line. This confirms Newton's law of cooling.

Precautions

(i) The initial temperature of hot water should not be more than 40°C.

(ii) The hot water in the calorimeter should be stirred continuously and gently.

Example 14. *A body cools in 7 min from 60°C to 40°C. What will be its temperature after the next 7 min? The temperature of the surroundings is 10°C.*

(a) 40°C (b) 30°C

(c) 15°C (d) 28°C

Sol. (*d*) In first case, $T_1 = 60°\text{ C}, T_2 = 40°\text{ C}$

$T_0 = 10°\text{ C}, t = 7\text{ min} = 420\text{ s}$

According to Newton's law of cooling, we get

$$mc\frac{T_1 - T_2}{t} = k\left(\frac{T_1 + T_2}{2} - 10\right)$$

$$mc\frac{(60-40)}{420} = k\left(\frac{60+40}{2} - 10\right)$$

$$mc \times \frac{20}{420} = k \times 40 \quad \text{...(i)}$$

In second case, $T_1 = 40°\text{ C}, T_2 = ?, T_0 = 10°\text{ C}$

and $t = 7\text{ min} = 420\text{ s}$

$$mc \times \frac{40 - T_2}{420} = k\left(\frac{40 + T_2}{2} - 10\right) \quad \text{...(ii)}$$

On dividing Eq. (ii) by Eq. (i), we get

$$\frac{20}{40 - T_2} = \frac{40}{\frac{40 + T_2}{2} - 10}$$

$$20 + \frac{T_2}{2} - 10 = 80 - 2T_2$$

On solving, we get $T_2 = 28°\text{ C}$

Experiment-9

Object

To find the speed of sound in air at room temperature using a resonance tube by two resonance positions.

Apparatus

Resonance tube apparatus, two tuning forks of different frequencies, a rubber pad, a plumb line, water in a beaker, a thermometer and a set square.

Diagram

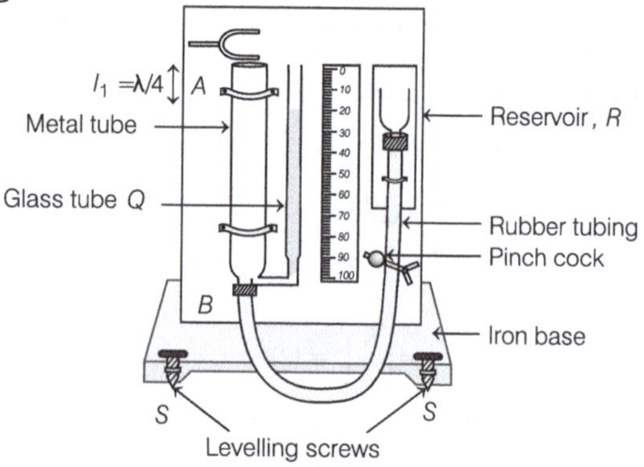

Theory

Speed of sound in air,

$$v = 2\nu(l_2 - l_1)$$

where, ν = frequency of tuning fork,

l_1 = length of the air column at first resonance

and l_2 = length of the air column in second resonance with tuning fork.

Procedure

(i) Using levelling screws and spirit level, set the resonance tube vertical.

(ii) Fill the sufficient water in the reservoir so that it rises upto the brim in the pipe and doesn't spill.

(iii) For first resonance position, start with zero length of air column in pipe. Strike the tuning fork with the rubber pad, bring it near the opening of the pipe and hear the sound.

(iv) Start lowering the level of the reservoir gradually lowering the level of water in pipe in steps of millimetre till the maximum sound with the tuning fork is heard. Ensure that loudness falls sharply by lowering and raising the level of water by one millimetre.

(v) Record the length y_1 of air column at first resonance.

(vi) Increase the length of air column by few centimetre and repeat the same procedure to find resonating length of air column y_2.

(vii) Increase the length of air column to about three times and find the length of air column corresponding to second resonance.

(viii) Repeat the same procedure with tuning fork of different frequencies.

Observations

(i) • Room temperature in the beginning of the experiment, $\theta_1 = \dots\dots ^\circ C$
• Room temperature at the end of the experiment, $\theta_2 = \dots\dots ^\circ C$

Mean value of room temperature,

$$\theta = \frac{\theta_1 + \theta_2}{2} = \dots\dots ^\circ C$$

(ii) Position of the top end A of the resonance tube, $y_0 = \dots\dots$ cm

Table for resonating length with different frequencies

Frequency of tuning fork (Hz)	Resonance	Position of water level at resonance			Length of air column $l = y - y_0$ (cm)
		Level falling y_1 (cm)	Level rising y_2 (cm)	Mean $y = \frac{y_1 + y_2}{2}$ (cm)	
$\nu_1 = 480$	Ist				$l_1 =$
	IInd				$l_2 =$
$\nu_2 = 512$	Ist				$l_1' =$
	IInd				$l_2' =$

Calculations

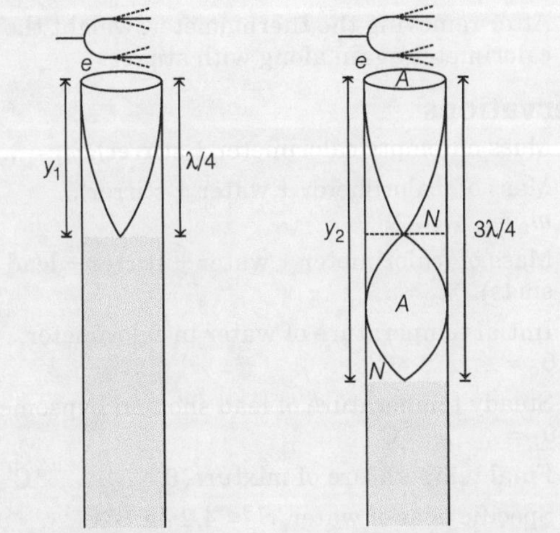

Position of Ist resonance Position of IInd resonance

(i) For tuning fork of 480 Hz

$$v_1 = 2 \times 480\,(l_2 - l_1) = \dots\dots\dots \text{ ms}^{-1}$$

(ii) For tuning fork of 512 Hz

$$v_2 = 2 \times 512\,(l_2' - l_1') = \dots\dots\dots \text{ ms}^{-1}$$

Mean value of velocity of sound in air is

$$\frac{v_1 + v_2}{2} = \dots\dots\dots \text{ ms}^{-1}$$

Result

Velocity of sound in air at ......0°C is...... ms^{-1}.

Precautions

(i) The tuning fork should be struck gently against the rubber pad.

(ii) Ensure that there is no leakage of water from the apparatus.

Example 15. *The internal radius of a 1 m long resonance tube is measured as 3.0 cm. A tuning fork of frequency 2000 Hz is used. The first resonating length is measured as 4.6 cm and the second resonating length is measured as 14.0 cm. Percentage error in calculation of end correction e is*

(a) 100% *(b) 80%* *(c) 94%* *(d) 75%*

Sol. (c) Maximum percentage error in measurement of e, as given by Rayleigh's formula.

(Given, error in measurement of radius is 0.1 cm)

$$\Delta e = 0.6\,\Delta R = 0.6 \times 0.1 = 0.06 \text{ cm}$$

Percentage error is $\frac{\Delta e}{e} \times 100 = \frac{0.06}{0.6 \times 3} \times 100 = 3.33\%$

End correction obtained in the experiment,

$$e = \frac{l_2 - 3l_1}{2}$$

$$= \frac{14.0 - 3 \times 4.6}{2} = 0.1 \text{ cm}$$

So, percentage error in the calculation of e with respect to theoretical value. Percentage error

$$= \frac{0.6 \times 3 - 0.1}{0.6 \times 3} \times 100$$

$$= 94.44\% \simeq 94\%$$

Experiment-10

Object

To determine the specific heat of a given liquid by method of mixtures.

Apparatus

Hypsometer, solid (lead shots), calorimeter with stirrer and lid, celcius thermometers, weight box, heating arrangement, beaker and experimental fluid (turpentine oil).

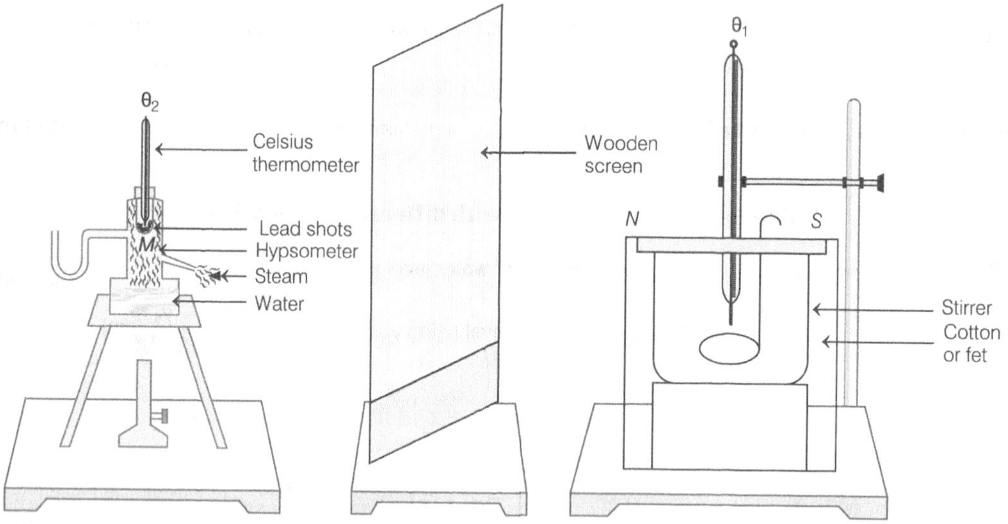

Heating of lead shot in steam using hypsometer

Calorimeter in jacket containing

Theory

The basis of this method is principle of calorimetry, according to which when two bodies at different temperatures are brought close in contact, heat flows from body at higher temperature to body at lower temperature till both bodies acquire the same temperature, specific heat of a given liquid is defined as the amount of heat required to raise the temperature of 1 kg of liquid through 1°C.

Procedure

(i) Fill the hypsometer's tank with sufficient quantity of water and place it on heating arrangement that is kept on one side of insulating wooden screen.

(ii) Tie a thread around the metal of known specific heat and drop it in tube M of the hypsometer.

(iii) On the other side of screen, place a calorimeter in a wooden box.

(iv) To make the calorimeter thermally insulated, fill the empty space with cotton wool around it.

(v) Weigh the calorimeter taking it out and weigh it along with stirrer. Now fill it with turpentine oil and weight again.

(vi) Now insert thermometer T_1 in calorimeter and T_2 in hypsometer tube M which should be surrounded by lead shots.

(vii) Start heating the water in hypsometer till it boils. Continue heating till thermometer T_2 shows a steady temperature θ_2. Record this value.

(viii) Record the initial temperature of the liquid in calorimeter.

(ix) Now shift gently the metal from tube M into the calorimeter after removing its lid.

(x) After closing the lid, stir well and note the highest temperature attains d by T_1.

(ix) After removing the thermometer, weight the calorimeter again along with stirrer.

Observations

(i) Mass of (empty calorimeter + stirrer), $m_1 = \ldots\ldots\ldots$ g

(ii) Mass of (calorimeter + water + stirrer), $m_2 = \ldots\ldots\ldots$ g

(iii) Mass of (calorimeter + water + stirrer + lead shots), $m_3 = \ldots\ldots\ldots$ g

(iv) Initial temperature of water in calorimeter, $\theta_0 = \ldots\ldots\ldots\,°C$

(v) Steady temperature of lead shots in hypsometer $\theta_l = \ldots\ldots\ldots\,°C$

(vi) Final temperature of mixture, $\theta = \ldots\ldots\ldots\,°C$

(vii) Specific heat of water, $S_1 = 4.2\ \text{Jg}^{-1}\text{°C}^{-1}$

(viii) Specific heat of material of calorimeter, $S_2 = \text{Jg}^{-1}\text{°C}^{-1}$

Calculations

(i) Water equivalent of (calorimeter + stirrer),

$$w_1 = \frac{m_1 S_2}{S_1} = \ldots\ldots \text{g}$$

(ii) Mass of water in calorimeter $= (m_2 - m_1) = \ldots\ldots$ g

(iii) Mass of lead shots $= (m_3 - m_2) = \ldots\ldots$ g

(iv) Rise in temperature of water and calorimeter

$$= (\theta - \theta_0) = \ldots\ldots\,°C$$

(v) Fall in temperature of lead shots

$$= (\theta_l - \theta) = \ldots\ldots\,°C$$

According to principle of calorimeter,

Heat lost by lead shots

= heat gained by calorimeter and water

$$(m_3 - m_2) S_L (\theta_l - \theta) = (m_2 - m_1 + w_1) \times 4.2 (\theta - \theta_0)$$

where, S_L = specific heat of given solid.

$$\therefore \quad S_L = \frac{(m_2 - m_1 \times w_1) \times 4.2 \times (\theta - \theta_0)}{(m_3 - m_2)(\theta_l - \theta)}$$

$$= \ldots\ldots \text{ Jg}^{-1} \cdot {}^\circ\text{C}^{-1}$$

Example 16. *The temperature of equal masses of three different liquids A, B and C are 12°C, 19°C and 28°C, respectively. The temperature when A and B are mixed is 16°C and when B and C are mixed is 23°C. The temperature when A and C are mixed, is*

(a) 18.2°C (b) 22°C
(c) 20.2°C (d) 25.2°C

Sol. (c) By applying calorimetry method, we get

Heat gain = Heat lost

$$c_A(16-12) = c_B(19-16) \Rightarrow \frac{c_A}{c_B} = \frac{3}{4}$$

and $$c_B(23-19) = c_C(28-23)$$

$$\Rightarrow \quad \frac{c_B}{c_C} = \frac{5}{4}$$

$$\Rightarrow \quad \frac{c_A}{c_C} = \frac{c_A}{c_B} \times \frac{c_B}{c_C} = \frac{3}{4} \times \frac{5}{4} = \frac{15}{16} \quad \ldots(i)$$

If θ is the temperature when A and C are mixed, then

$$c_A(\theta - 12) = c_C(28 - \theta)$$

$$\Rightarrow \quad \frac{c_A}{c_C} = \frac{28-\theta}{\theta-12} \quad \ldots(ii)$$

On solving Eqs. (i) and (ii), we have

$$\frac{28-\theta}{\theta-12} = \frac{15}{16}$$

$$\Rightarrow \quad 448 - 16\theta = 15\theta - 180$$

$$\therefore \quad \theta = 20.2^\circ \text{ C}$$

Experiment-11

Object

To find resistance of a given wire using meter bridge.

Apparatus

A meter bridge, a galvanometer, a leclanche cell, a resistance box, a jockey, a one way key, a resistance wire, a meter scale and connecting wires.

Circuit Diagram

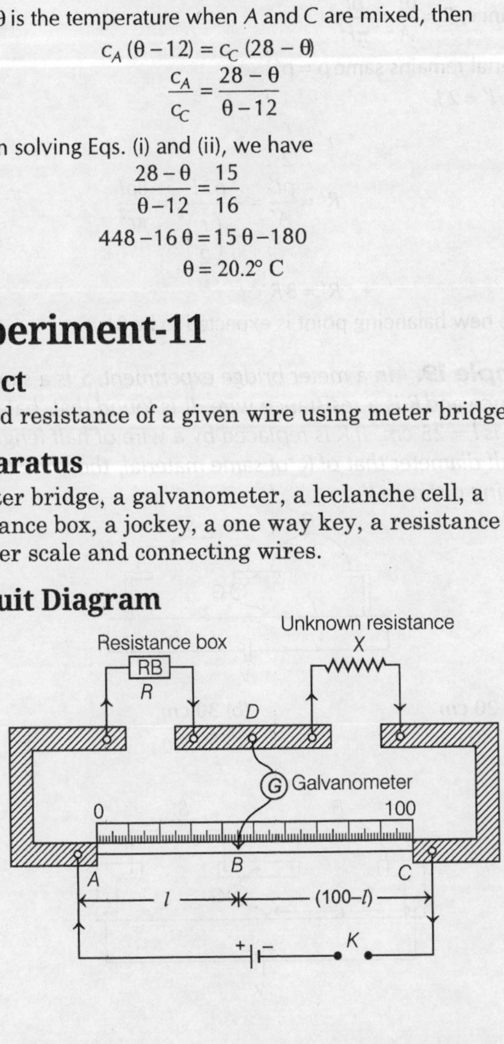

Description to Meter Bridge The practical form of Wheatstone bridge is the slide wire bridge or meter bridge. Usually, ratio arms of fixed resistance are P and Q, and R is variable resistance of known value. X is an unknown resistance as shown in figure. As the bridge uses 1 m long wire, it is called meter bridge and as the jockey is slided over the wire, it is called slide wire bridge.

Theory

The unknown resistance X is given by

$$X = \frac{(100 - l)}{l} R$$

where, R is known resistance placed in the left gap and unknown resistance X in the right gap of the meter bridge, l is the balancing length on the meter bridge wire.

Procedure

(i) Arrange the apparatus according to the arrangement diagram as shown.

(ii) Connect the resistance wire whose resistance is to be determined in the right gap between C and D.

(iii) Connect resistance box of low range in the left hand gap between A and D.

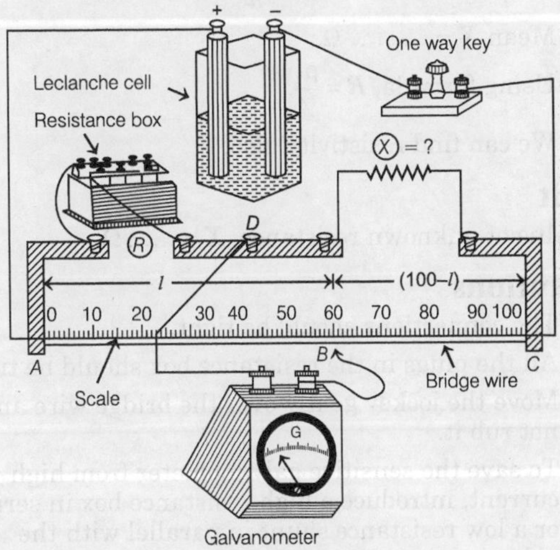

(iv) Take out some resistance (say 4Ω) from the resistance box, plug the key K.

(v) Touch the jockey gently first at left end and then at right end of the bridge wire.

(vi) If the galvanometer shows deflections in opposite directions, the connections are correct.

(vii) Now, move the jockey gently along the wire from left to right till galvanometer shows no deflections. The point where the jockey is touching the wire, when galvanometer shows no deflection is null point B.

(viii) Choose an appropriate value of R from the resistance box such that there is no deflection in the galvanometer when the jockey is nearly in the middle of the wire.

(ix) Take atleast four sets of observations in the same way by changing the value of R by 1Ω in each step.

(x) Record your observations as given ahead in table.

Observations

Table for length (L) and unknown resistance (X)

S. No.	Resistance from the resistance box R (ohm)	Length $AB = l$ (cm)	Length $BC = (100 - l)$ (cm)	Unknown resistance $X = \frac{(100 - l)}{l} R$ (ohm)
1.	4	......	......	$X_1 =$ ......
2.	5	......	......	$X_2 =$ ......
3.	6	......	......	$X_3 =$ ......
4.	7	......	......	$X_4 =$ ......

Calculations

(i) From position of B, find l cm and write in table.

(ii) Find length $(100 - l)$ cm and write in table.

(iii) Calculate X by the given formula and write in table.

(iv) Take mean value of X recorded in table.

$$\text{Mean, } X = \frac{X_1 + X_2 + X_3 + X_4 + \ldots + X_R}{n}$$

Mean $X =$ .... Ω.

Using formula, $R = \frac{\rho l}{A}$.

We can find resistivity also.

Result

The value of unknown resistance, $X =$ ...... Ω.

Precautions

(i) The connections should be tight.

(ii) All the plugs in the resistance box should be tight.

(iii) Move the jockey gently over the bridge wire and do not rub it.

(iv) To save the sensitive galvanometer from high current, introduce a high resistance box in series or a low resistance shunt in parallel with the galvanometer.

(v) Null point should be brought between 40 cm and 60 cm.

Example 17. *The following figure shows a meter bridge set up with null deflection in the galvanometer, the value of the unknown resistance R is*

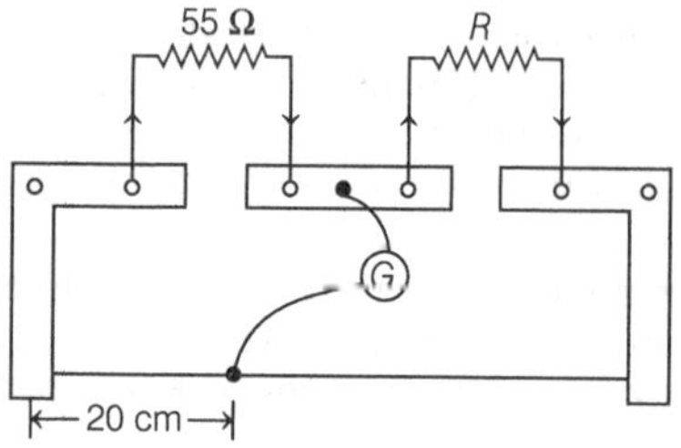

(a) 13.75 Ω (b) 220 Ω (c) 110 Ω (d) 55 Ω

Sol. (b) From the principle of Wheatstone bridge, we have

$$\frac{P}{Q} = \frac{R}{S}$$

Putting $P = 55\ \Omega$, $Q =$ unknown resistance $= R$

$R = 20\ \Omega$, $S = 80\ \Omega$, we have

$$\therefore \quad \frac{55}{R} = \frac{20}{80}$$

$$\Rightarrow \quad R = 220\ \Omega$$

Example 18. *A meter bridge is used to determine the resistance of an unknown wire by measuring the balance point length l. If the wire is replaced by another wire of same material but with double the length and half the thickness, the balancing point is expected to be*

(a) $\frac{1}{8l}$ (b) $\frac{1}{4l}$ (c) $8l$ (d) $16\,l$

Sol. (c) In a meter bridge, the ratio of two resistances is

$$\frac{R}{R'} = \frac{l}{l'}$$

where, l and l' are balancing lengths.

Resistance, $R = \frac{\rho l}{A} = \frac{\rho l}{\pi r^2}$

If material remains same $\rho = \rho'$

Given, $l' = 2\,l$,

$$r' = \frac{r}{2}$$

$$\because \quad R' = \frac{\rho l'}{A'} = \frac{\rho 2\,l}{\pi \left(\frac{r}{2}\right)^2} = \frac{8\rho l}{\pi r^2}$$

$$R' = 8R$$

Hence, new balancing point is expected to be $8l$.

Example 19. *In a meter bridge experiment, S is a standard resistance and R is a resistance wire. It is found that, balancing length is l = 25 cm. If R is replaced by a wire of half length and half diameter that of R of same material, then find the balancing distance l′.*

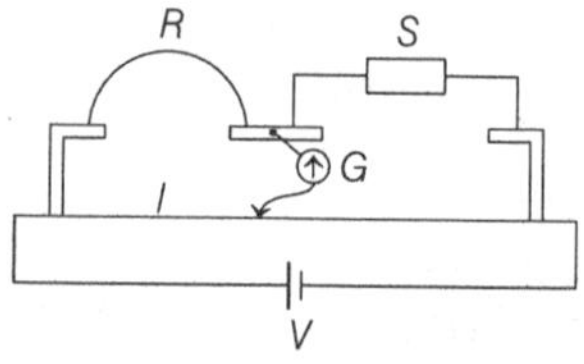

(a) 20 cm (b) 30 cm

(c) 40 cm (d) 50 cm

Sol. (c)

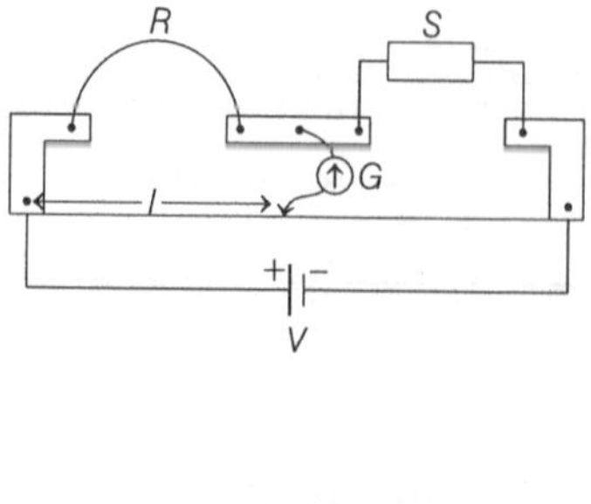

Given, balance length, $l = 25$ cm

At balance point, $\frac{R}{l} = \frac{S}{100 - l}$

$$\Rightarrow \quad \frac{R}{25} = \frac{S}{75} \Rightarrow R = \frac{S}{3}\,\Omega \quad \text{....(i)}$$

When R is replaced by some other resistance of half length and half diameter, then resistance in left gap,

$$R' = \frac{\rho \frac{l}{2}}{\frac{\pi}{4}\left(\frac{d}{2}\right)^2} = 2\left\{\frac{\rho l}{\left(\frac{\pi}{4}d^2\right)}\right\}$$

$$\Rightarrow \quad R' = 2R$$

Now, if balance is obtained at length l', then

$$\frac{R'}{l'} = \frac{S}{(100 - l')}$$

$$\Rightarrow \quad \frac{2R}{l'} = \frac{S}{100 - l'}$$

Substituting the value of S from Eq. (i), we get

$$\frac{2R}{l'} = \frac{3R}{100 - l'}$$

$$\Rightarrow \quad 200 - 2l' = 3l'$$

or $\quad l' = 40$ cm

Experiment-12

Object

To determine resistance per cm of a given wire using Ohm's law or by plotting a graph of potential difference *versus* current.

Apparatus

A resistance wire, a voltmeter, an ammeter of appropriate range, a rheostat, a meter scale, a battery, one way key and connecting wires.

Circuit Diagram

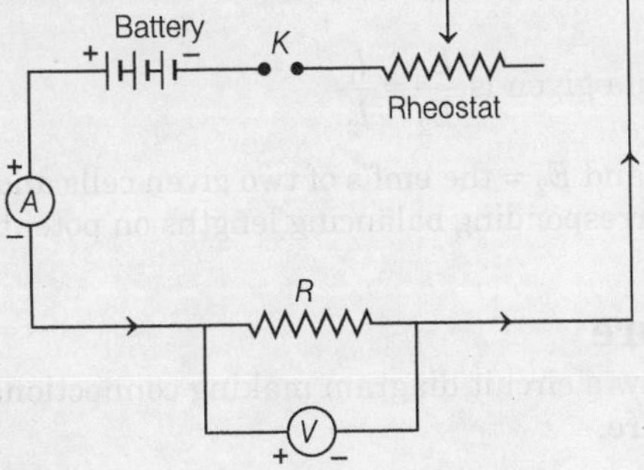

Theory

If I be the current flowing through a conductor and V be the potential difference across its ends. According to Ohm's law,

$$V \propto I$$

or

$$V = RI$$

where, R is the constant of proportionality. It is known as resistance of the conductor.

or $\quad \frac{V}{I} = R$

Procedure

(i) Arrange the apparatus in the same manner as given in arrangement diagram given ahead.

(ii) Make neat, clean and tight connections according to the circuit diagram.

(iii) Determine the least count of voltmeter and ammeter, and also note the zero error, if any.

(iv) Insert the key K, slide the rheostat contact and see that ammeter and voltmeter are working properly.

(v) Adjust the sliding contact of the rheostat such that a small current passes through the resistance wire.

(vi) Note down the value of potential difference V from voltmeter and current I from ammeter.

(vii) Shift the rheostat contact slightly so that both ammeter and voltmeter show full division reading and not in fraction.

(viii) Record the readings of the voltmeter and ammeter.

(ix) Take atleast six sets of independent observations.

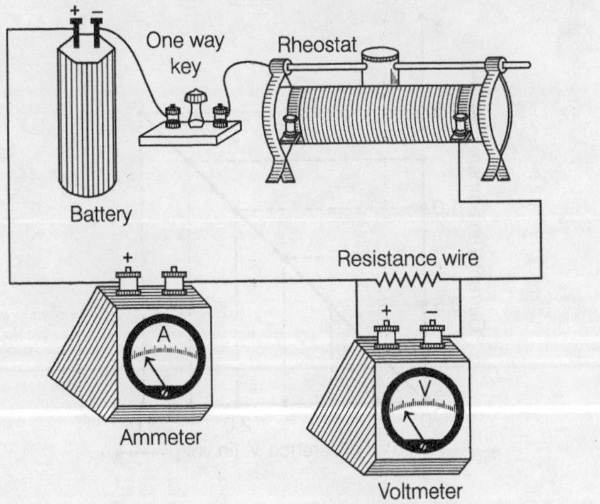

Experimental setup for verification of Ohm's law

Observations

Table for resistance (R)

S. No.	Ammeter reading I (A)	Voltmeter reading V (V)	$\frac{V}{I} = R$ (ohm)
1.	......	......	......
2.	......	......	......
3.	......	......	......
4.	......	......	......
5.	......	......	......
6.	......	......	......

Calculations

(i) Find ratio of V and I for each set of observations.

(ii) Plot a graph between potential difference V and current I taking V along X-axis and I along Y-axis. The graph comes to be a straight line, as shown in figure.

(iii) Constant ratio $\frac{V}{I}$ gives resistance of the wire.

(iv) Resistance of the wire per cm = ... $\Omega\ cm^{-1}$.

Result

Resistance of the wire = ... $\Omega\ cm^{-1}$.

Precautions

(i) The connections should be neat, clean and tight.

(ii) Voltmeter and ammeter should be of proper range.

(iii) A low resistance rheostat should not be used.

(iv) The unknown resistance should not be too low.

(v) The key should be inserted only while taking observations to avoid heating of wire.

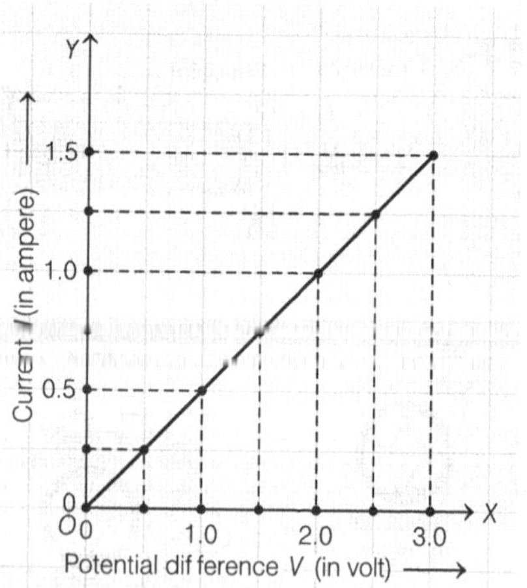

Example 20. *The temperature coefficient of resistance of a wire is 0.00145°C^{-1}. At 100 °C, its resistance is 2 Ω. At what temperature the resistance of the wire will be 3 Ω?*

(a) 667.8 *K* (b) 567.5 *K*

(c) 928.3 *K* (d) 727.56 *K*

Sol. (a) Using the relation, $R = R_0(1 + \alpha T)$

$$R_1 = R_0(1 + \alpha t_1)$$

$\therefore$ $$2 = R_0(1 + 0.00145 \times 100) \quad ...(i)$$

and $$R_2 = R_0(1 + \alpha t_2)$$

$\therefore$ $$3 = R_0(1 + 0.00145 \times t_2) \quad ...(ii)$$

On dividing Eq. (ii) by Eq. (i), we get

$$\frac{3}{2} = \frac{1 + 0.00145 t_2}{1 + 0.00145 \times 100}$$

or $$t_2 = 394.8° \text{ C}$$
$$= 667.8 \text{K}$$

Experiment-13(i)

Object

To compare the emf's of two given primary cells using potentiometer.

Apparatus

Two cells (a leclanche cell, a daniel cell), an ammeter, a voltmeter (0 to 5V), a galvanometer, a potentiometer, a battery, a rheostat of low resistance, a resistance box, a one way key, a two way key, a jockey and connecting wires.

Circuit Diagram

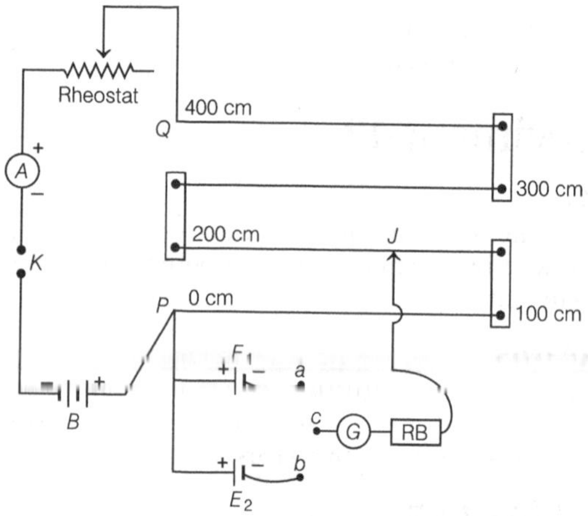

Theory

The formula given is $\frac{E_1}{E_2} = \frac{l_1}{l_2}$

where, E_1 and E_2 = the emf's of two given cells and l_1 and l_2 = the corresponding balancing lengths on potentiometer wire.

Procedure

(i) Draw a circuit diagram making connections as in figure.

(ii) Connect the positive pole of the battery (a battery of constant emf to the zero end (P) of the potentiometer and the negative pole through a one way key, an ammeter and a low resistance rheostat to the other end (Q) of the potentiometer.

(iii) Connect the positive poles of the cells E_1 and E_2 to the terminal at the zero end (P) and the negative poles to the terminals a and b of the two way key.

(iv) Connect the common terminal of the two way key through a galvanometer (G) and a resistance box (RB) to the jockey (J).

(v) Take maximum current from the battery making rheostat resistance zero.

(vi) Insert the plug in the one way key and also between the terminals a and c of the two way key to connect cell E_1 with the circuit.

(vii) Take out a 2000 Ω plug from the resistance box (RB).

(viii) The jockey at the zero end and note the direction of deflection in the galvanometer.

(ix) Touch the jockey at the other end of the potentiometer wire. If the direction of deflection is opposite to that in the previous case, the connections are correct.

(x) Slide the jockey along potentiometer wire so as to obtain a point where galvanometer shows no deflection.

(xi) Put the 2000 Ω plug back in the resistance box and obtain the null point position accurately.

(xii) Note the length l_1 of the wire for the cell E_1. Also, note the current as indicated by the ammeter.

(xiii) Disconnect the cell E_1 by removing the plug from gap ac of two way key and connect the cell E_2 by isserting plug into gap bc of two way key.

(xiv) Take out a 2000 Ω plug from resistance box and slide the jockey along potentiometer wire so as to obtain no deflection position. Put the 2000 Ω plug back in the resistance box and obtain accurate position of null point for second cell E_2.

(xv) Note the length l_2 of wire in this position for the cell E_2. However, make sure that ammeter reading is same as in step 12.

(xvi) Increase the current by adjusting the rheostat and obtain at least three sets of similar observations,

Observations

Table for lengths

S. No.	Corrected ammeter reading (A)	Balance point when E_1 (Leclanche cell) in the circuit L_1 (cm)			Balance point when E_2 (Daniel cell) in the circuit L_2 (cm)			$\frac{E_1}{E_2} = \frac{L_1}{L_2}$
		1	2	Mean l_1	1	2	Mean l_2	
1.	...	...	...	...	...	...	...	...
2.	...	...	...	...	...	...	...	...
3.	...	...	...	...	...	...	...	...

(i) Range of voltmeter = 5.0 V
Least count of voltmeter = 0.1 V
Emf of battery (or battery elimination), E = 2.5 V
Emf of leclanche cell, E_1 = 1.4 V
Emf of daniel cell, E_2 = 1.1 V

(ii) Least count of the ammeter = 0.02 A
Zero error of the ammeter = Nil

Calculations

(i) For each observation find mean l_1 and mean l_2 and record in table above.

(ii) Find $\frac{E_1}{E_2}$ for each set.

(iii) Find mean $\frac{E_1}{E_2}$ of all sets.

Result

The ratio of emf's, $\frac{E_1}{E_2} \cong$ ......

Precautions

(i) The plugs should be introduced in the keys only when the observations are to be taken.

(ii) The positive poles of the battery E, E_1 and E_2 should all be connected to the terminal at the zero of the wires.

(iii) The emf of the battery should be greater than the emf's of the either of the two cells.

(iv) The ammeter reading should remain constant for a particular set of observation.

Experiment-13(ii)

Object

To determine the internal resistance of a given primary cell using potentiometer.

Apparatus

A potentiometer, a battery, two one way keys, a rheostat of low resistance, a galvanometer, a high resistance box, an ammeter, a voltmeter (0-5V), a leclanche cell, a jockey and connecting wires.

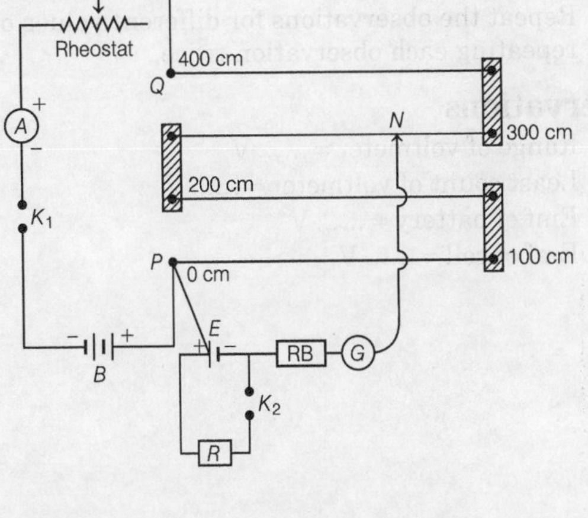

Theory

The internal resistance of a cell is given by

$$r = \left(\frac{l_1}{l_2} - 1\right)R$$

where, l_1 and l_2 are the balancing lengths without shunt and with shunt, respectively and R is the shunt resistance in parallel with the given cell.

Procedure

(i) Draw the circuit diagram showing the scheme of connections as in figure.

(ii) Tight the plugs of the resistance box.

(iii) Check the emf's of the battery and cell and see that emf of the battery is more than that of the given cell, otherwise null or balance point will not be obtained.

(iv) Take maximum current from the battery, making rheostat resistance zero.

(v) Take out 2000 Ω resistance plug from the resistance box. Place the jockey first at the end P of the wire and then at the end Q. If the galvanometer shows deflection in opposite directions in the two cases, the connections are correct.

(vi) Without inserting the plug in the key (K_2) adjust the rheostat, so that a null point is obtained on the fourth wire of potentiometer.

(vii) Insert the 2000 Ω plug back in its position in resistance box and obtain the null point position.

(viii) Measure the balancing length l_1 between this point and the end P of the wire.

(xi) Take out the 2000 Ω plug again from the resistance box (RB). Introduce the plugs in key (K_1) as well as in key (K_2). Take out a small resistance (1 Ω to 5Ω) from the resistance box (R) connected in parallel with the cell.

(x) Slide the jockey along the potentiometer wire and the obtain null point.

(xi) Insert the 2000 Ω plug back in its position in resistance box and again obtain the null point.

(xii) Measure the balancing length l_2 from end P.

(xiii) Repeat the observations for different values of R repeating each observation twice.

Observations

Range of voltmeter = ... V

Least count of voltmeter = ... V

Emf of battery = ... V

Emf of cell = ... V

Table for Lengths

S.No.	Ammeter reading (A)	Postion of null point (cm)						Shunt resistance R (ohm)	Internal resistance $r = \left(\frac{l_1 - l_2}{l_2}\right)R$ (ohm)
		Without shunt R			With shunt R				
		(i)	(ii)	Mean l_1	(i)	(ii)	Mean l_2		
1.	...	...	...	...	...	...	...	...	...
2.	...	...	...	...	...	...	...	...	...
3.	...	...	...	....	...	...	...	...	...
4.	...	...	...	...	...	...	...	...	...
5.	...	...	...	...	...	...	...	...	...
6.	...	...	...	...	...	...	...	...	...

Calculations

(i) For each set of observation, find mean l_1 and l_2.

(ii) Calculate value of r for each set and write it in table.

(iii) Take mean of value of r recorded in table.

Result

The internal resistance of the given cell is ... Ω.

Precautions

(i) The emf of the battery should be greater than that of the cell.

(ii) For one set of observation the ammeter reading should remains constant.

(iii) Current should be passed for short time while finding the null point.

(iv) Cell should not be disturbed during experiment.

To find resistance of a given wire using meter bridge.

Example 21. *The length of a wire of a potentiometer is 100 cm, and the emf of its standard cell is E volt. It is employed to measure the emf a battery whose internal resistance is 0.5 Ω. If the balance point is obtained at $l = 30$ cm from the positive end, then the emf of the battery is*

(a) $\frac{30E}{100}$ (b) $\frac{20}{50}E$

(c) $\frac{100}{30}E$ (d) $\frac{50}{20}E$

Sol. (a) From the formula of potentiometer, we have

$$\frac{E_1}{E_2} = \frac{l_1}{l_2}$$

where, E_1 and E_2 are the emf's of two given cells and l_1 and l_2 are the corresponding balancing lengths on potentiometer wire.

Given, $E_1 = E$ volt, $l_2 = 30$ cm, $l_1 = 100$ cm

$$\therefore \quad E_2 = E_1 \cdot \frac{l_2}{l_1} = E \cdot \frac{30}{100}$$

Example 22. *The potentiometer wire AB shown in the figure is 50 cm long. When AD = 30 cm, no deflection occurs in the galvanometer. The resistance R(in Ω) is*

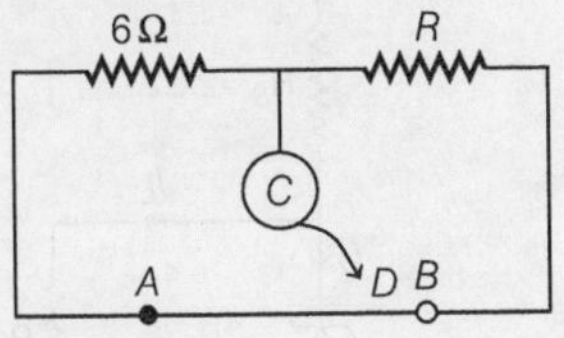

(a) 2 Ω (b) 4 Ω (c) 6 Ω (d) 8 Ω

Sol. (b) In case of no deflection, balance condition is achieved, hence we have

$$\frac{R_1}{R_2} = \frac{l_1}{l_2}$$

Given, $R_1 = 6\ \Omega, R_2 = R\ \Omega, l_1 = 30$ cm

$$l_2 = 50 - 30 = 20 \text{ cm}$$

$$\therefore \quad \frac{6}{R} = \frac{30}{20}$$

$$\Rightarrow \quad R = \frac{6 \times 20}{30} = 4\ \Omega$$

Example 23. *The balancing length for a cell is 560 cm in a potentiometer experiment. When an external resistance of 10Ω is connected in parallel to the cell, the balancing length changes by 60 cm. If the internal resistance of the cell is $\frac{N}{10}$ Ω, where N is an integer, then find the value of N.*

(a) 2 (b) 7
(c) 10 (d) 12

Sol. (d) In experiment of finding internal resistance,

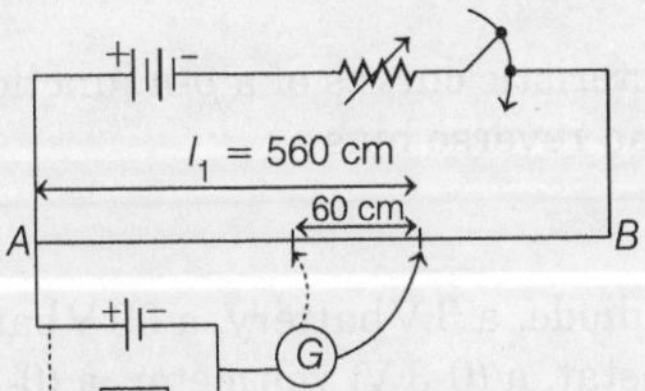

Balancing length with only cell, $l_1 = 560$ cm.
Balancing length with cell and resistance, $l_2 = 560$ cm $-$ 60 cm $= 500$ cm

External resistance, $R = 10\ \Omega$

Now, using $\quad r = R\left(\frac{E}{V} - 1\right) = R\left(\frac{l_1}{l_2} - 1\right)$, we get

$$r = 10\left(\frac{560}{500} - 1\right) = \frac{6}{5}\ \Omega$$

Now, given $\quad r = \frac{N}{10}$

$$\Rightarrow \quad \frac{6}{5} = \frac{N}{10}$$

$$\Rightarrow \quad N = \frac{60}{5} = 12$$

Experiment-14

Object

To draw the characteristic curve of a zener diode and to determine its reverse breakdown voltage.

Apparatus

A zener diode, (V_Z = 6V), a 10 V battery, a high resistance rheostat, two (0 – 10V) voltmeters, an (0 – 100 mA) ammeter, a 20 Ω resistance, a one way key and connecting wires.

Diagram

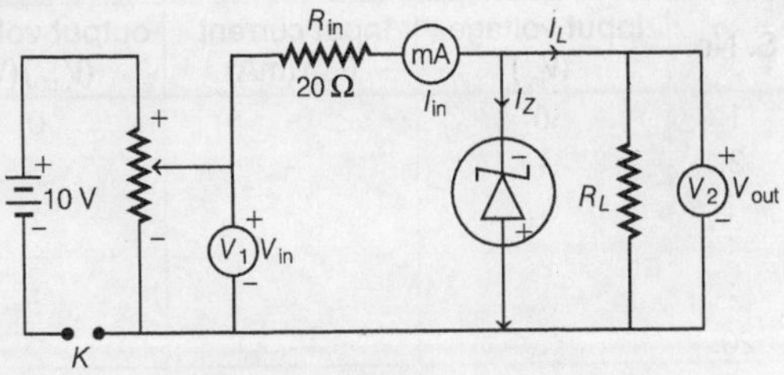

Theory

Zener Diode is a semiconductor diode, having n-type and the p-type sections heavily doped. This heavy doping results in a low value of reverse breakdown voltage.

The reverse breakdown voltage of a zener diode is called zener voltage (V_Z). The reverse current that results after the breakdown, is called zener current (I_Z).

(i) **Relations**

$$I_L = I_{in} - I_Z \quad \text{...(i)}$$

$$V_{out} = V_{in} - R_{in} I_{in} \quad \text{...(ii)}$$

$$V_{out} = R_L I_L \quad \text{...(iii)}$$

Initially, as V_{in} is increased, I_{in} increases a little, then V_{out} increases.

At breakdown, increase of V_{in}, I_{in} increases by large amount, so that $V_{out} = V_{in} - R_{in} I_{in}$ becomes constant. This constant value of V_{out} which is the reverse breakdown voltage, is called zener voltage.

(ii) **Formula used**

$$V_{out} = V_{in} - R_{in} I_{in}$$

Constant value of V_{out} gives reverse breakdown voltage.

Procedure

(i) Draw circuit diagram as shown in figure.

(ii) Bring moving contact of potential divider (rheostat) near negative end.

(iii) Move the contact a little towards positive end to apply some reverse bias voltage (V_{in}). Milliammeter reading remains zero. Voltmeters give equal readings.

i.e. $\quad V_{out} = V_{in} \quad (\because I_{in} = 0)$

(iv) As V_{in} is further increased, I_{in} starts flowing. Then V_{out} becomes less than V_{in}. Note the values of V_{in}, I_{in} and V_{out}.

(v) Go on increasing V_{in} in small steps of 0.5 V. Note corresponding values of I_{in} and V_{out} which will be found to have increased.

(vi) At one stage, as V_{in} is increased further, I_{in} increases by large amount and V_{out} does not increase. This is reverse breakdown situation.

Observations

Table for V_{in}, I_{in} and V_{out}

S. No.	Input voltage (V_{in})	Input current (I_{in}) (mA)	output voltage (V_{out}) (V)
1.	0	-	0
2.	-	-	-
⋮	...	-	⋮
⋮	...	-	⋮
⋮	...	-	⋮
20.	...	-	

Calculations

Plot a graph between input voltage and output voltage along X-axis and along Y-axis respectively. The graph comes as shown below.

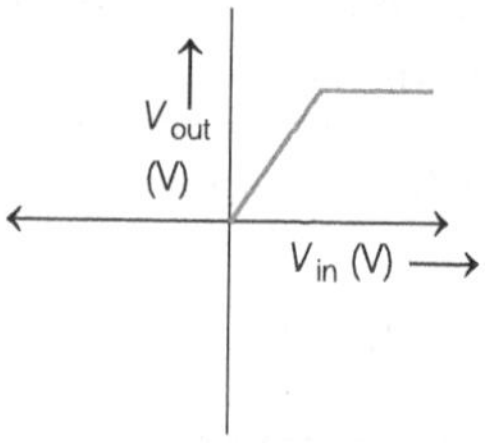

Result

The reverse breakdown voltage of given zener diode is ......V.

Precaution

Key should be used in circuit and opened when the circuit is not being used.

Example 24. *Figure shows a DC voltage regulator circuit, with a Zener diode of breakdown voltage = 6 V. If the unregulated input voltage varies between 10 V to 16 V, then what is the maximum zener current?*

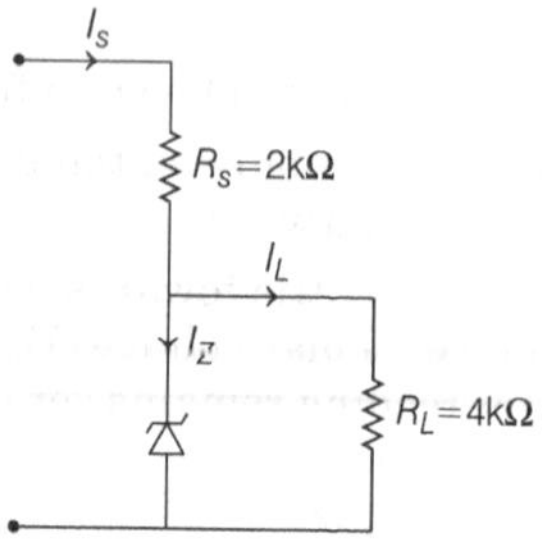

[JEE Main 2019]

(a) 2.5 mA (b) 1.5 mA (c) 7.5 mA (d) 3.5 mA

Sol. (*d*) In given voltage regulator circuit, breakdown of zener occurs at 6 V.

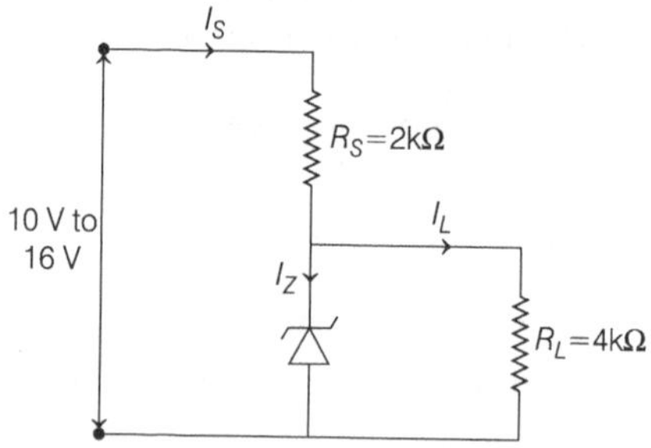

After breakdown, voltage across load resistance ($R_L = 4k\Omega$),

$$V = V_Z = 6\text{ V}$$

$\therefore$ Load current after breakdown,

$$I_L = \frac{V_Z}{R_L} = \frac{6}{4000} = 1.5 \times 10^{-3}\text{ A}$$

When unregulated supply is of 16 V, potential drop occurring across series resistance ($R_S = 2\text{ k}\Omega$),

$$V_S = V - V_Z = 16 - 6 = 10\text{ V}$$

So, current across series resistance,

$$I_S = \frac{V_S}{R_S} = \frac{10}{2 \times 10^3} = 5 \times 10^{-3}\text{ A}$$

So, current across zener diode,

$$I_Z = I_S - I_L = 5 \times 10^{-3} - 1.5 \times 10^{-3}$$
$$= 3.5 \times 10^{-3}\text{ A}$$
$$= 3.5\text{ mA}$$

Experiment-15

Object

To draw characteristic curves of a *p-n* junction diode in forward bias and reverse bias.

Apparatus

A *p-n* junction diode, a 3 V battery, a 30 V battery, a high resistance rheostat, a (0-3 V) voltmeter, a (0-30 V) voltmeter, an (0-100 mA) ammeter, an (0-100 μA) ammeter, a one way key and connecting wires.

Circuit Diagram

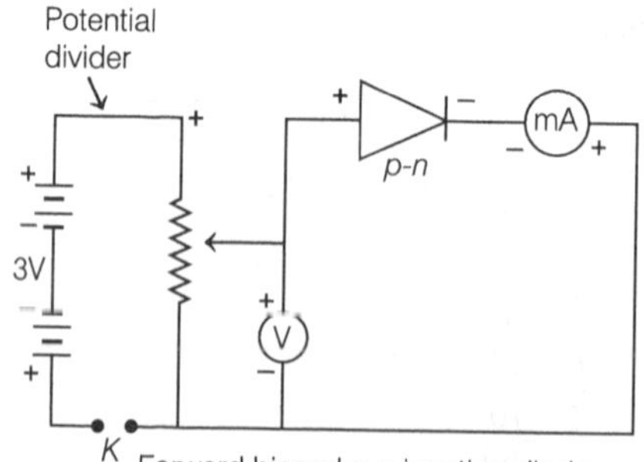

Forward biased *p-n* junction diode

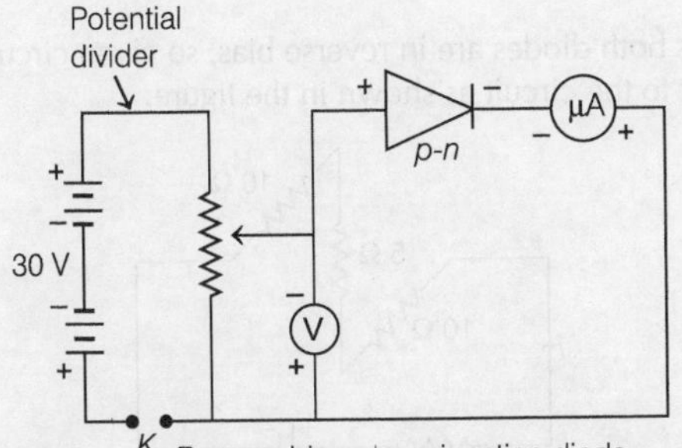

Reverse biased p-n junction diode

Theory

(i) In forward biased p-n junction diode, with increase in bias voltage the forward current increases slowly in the beginning and then rapidly. At about 2.4 V, the current increases suddenly.

(ii) In reverse biased p-n junction diode, in starting no appreciable reverse current flows. At about 5 V, a feeble current starts flowing. With increase in bias voltage, the current slowly increases. At about 25 V, the reverse current increases suddenly.

Procedure

(i) Forward biased *p-n* junction

- Make circuit diagram as shown in figure.
- Find least count and zero error of voltmeter (V) and milliammeter (mA).
- Bring contact of potential divider (rheostat) near negative end and insert the key K. Voltmeter (V) and milliammeter (mA) will give zero reading.
- Move the contact a little towards positive end to apply a forward bias voltage of (V_F) of 0.1 V. Current remains zero.
- Increase V_F to 0.4 V. Milliammeter records a small current.
- Increase V_F in steps of 0.2 V and note the corresponding current. Current increases first slowly and then rapidly, till V_F becomes 2 V.
- Make V_F = 2.2 V. The current will rise by large amount.
- Make V_F = 2.4 V. The current increases suddenly representing forward breakdown stage. Note the current and take out the key at once.

(ii) Reverse biased *p-n* junction

- Make circuit diagram as shown in figure.
- Note least count and zero error of voltmeter (V) and microammeter (μA).
- Bring contact of potential divider (rheostat) near positive end and insert the key (K). Voltmeter (V) and microammeter (μA). will give zero reading.
- Move the contact towards negative end to apply a reverse bias voltage (V_R) of 0.5 V, a feebly reverse current starts flowing.
- Increase in steps of 0.2 V. Current increases first slowly and then rapidly till V_R becomes 20 V.
- Make V_R = 25 V. The current increases suddenly representing reverse breakdown stage. Note the current and take out the key at once.

Observations

(i) For forward bias

Range of voltmeter = 3 V

Least count of voltmeter = 0.1 V

Zero error of voltmeter = Nil

Range of milliammeter = 30 mA

Least count of milliammeter = 0.5 mA

Zero error of milliammeter = Nil

Table for forward bias voltage and forward current

S. No.	Forward bias voltage V_F (V)	Forward current I_F (mA)
1.	...	...
2.	...	...
3.	...	...
4.	...	...
5.	...	...

(ii) For reverse bias

Range of voltmeter = 30 V

Least count of voltmeter = 1 V

Zero error of voltmeter = Nil

Range of microammeter = 30 μA

Least count of microammeter = 0.5 μA

Zero error of microammeter = Nil

Table for reverse bias voltage and reverse current

S. No.	Reverse bias V_R (V)	Reverse current I_R (μA)
1.	...	...
2.	...	...
3.	...	...
4.	...	...
5.	...	...

Calculations

If the voltage is applied along X-axis and current along Y-axis, then it is called a V-I characteristic curve. In the forward region, the voltage where the current starts to increase abruptaly is called knee voltage.

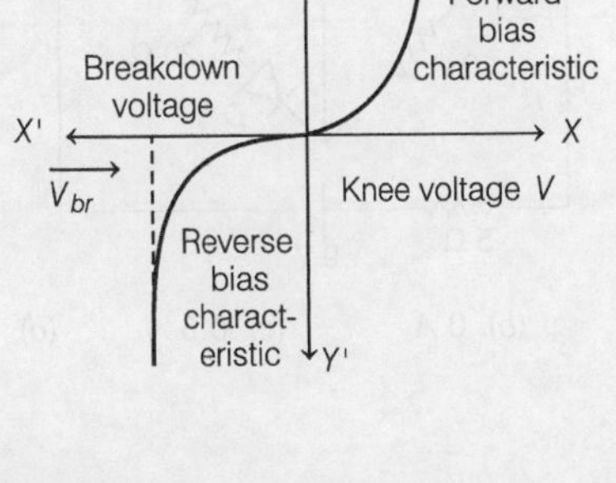

Result

Junction resistance for forward bias $= \frac{\Delta V_F}{\Delta I_F} = \ldots \Omega$

Junction resistance for reverse bias $= \frac{\Delta V_R}{\Delta I_R} = \ldots \Omega$

Precaution

Forward bias and reverse bias voltage should not be applied **beyond breakdown.**

Example 25. *In the figure, potential difference between A and B is* *[JEE Mlain 2020]*

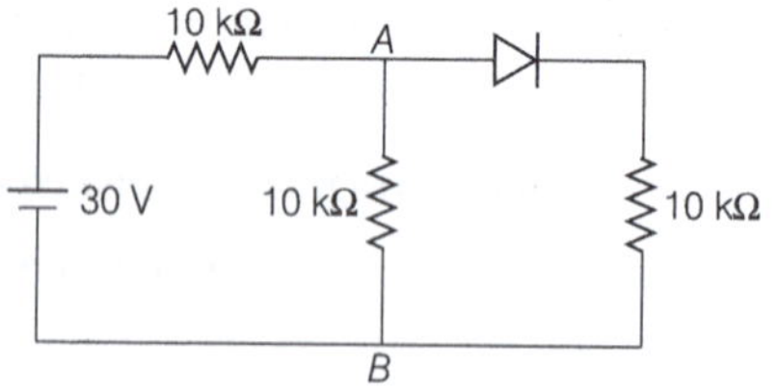

(a) zero (b) 10 V (c) 5 V (d) 15 V

Sol. () In ideal condition, a forward biased *p-n* junction acts like a conductive path of zero resistance. So, given circuit is equivalent to circuit shown below.

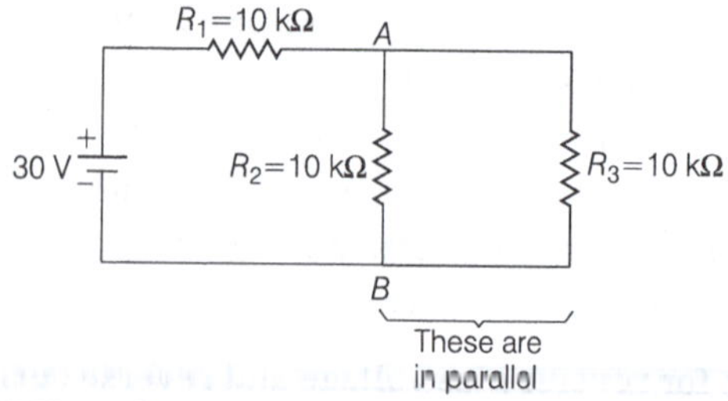

Total resistance of circuit,

R = (R_3 in parallel with R_2) and these are in series with R_1

$$= \frac{10 \times 10}{10 + 10} + 10 = 15 \text{ k}\Omega$$

So, total current in the circuit,

$$I = \frac{V}{R} = \frac{30}{15 \text{ k}\Omega} = 2 \times 10^{-3} \text{ A}$$

Example 26. *The current i in the network is*

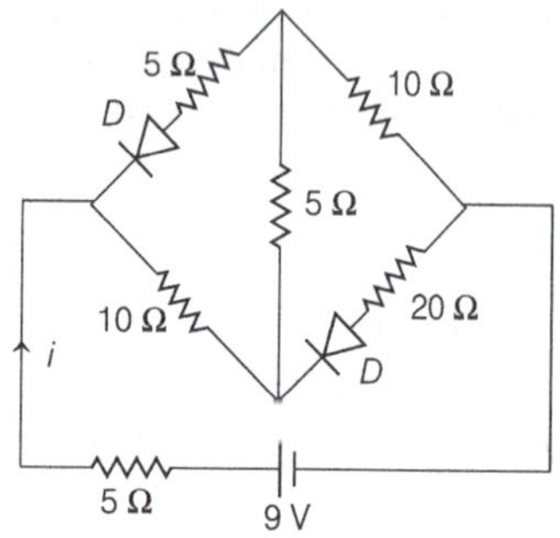

(a) 0.2*A* (b) 0 *A* (c) 0.6 *A* (d) 0.3 *A*

Sol. (*d*) As both diodes are in reverse bias, so given circuit is equivalent to the circuit as shown in the figure.

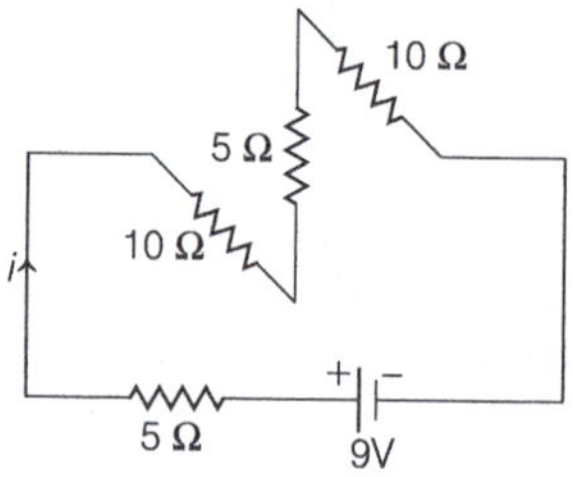

Equivalent resistance of circuit,

$$R = 5 + 10 + 5 + 10 = 30 \Omega$$

$\therefore$ Circuit current, $I = \frac{V}{R} = \frac{9}{30} = 0.3 \text{A}$

Experiment-16

Object

To study the characteristics of a common-emitter *p-n-p* (or *n-p-n*) transistor and to find out the values of current gain and voltage gain.

Apparatus

A *p-n-p* transistor (BC 157 or AC 127), a milliammeter, a microammeter, a voltmeter, a millivoltmeter, two batteries, two rheostats, two one way keys and connecting wires.

Circuit Diagram

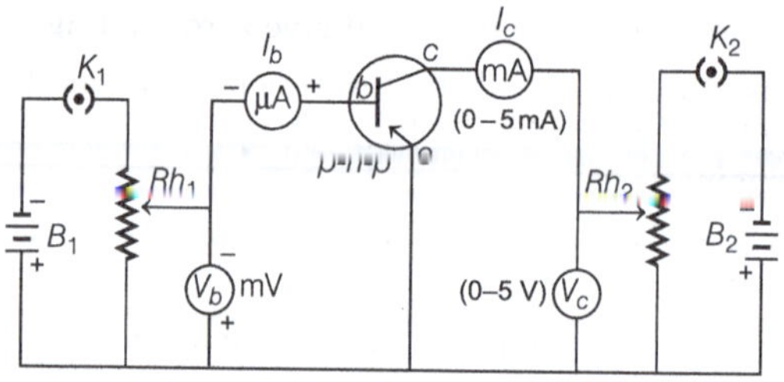

Theory

The *p-n-p* transistor consists of a very thin slice of *n*-type semiconductor sandwitched between two small blocks of *p*-type semiconductor. The middle slice is called the base while the left and right blocks are the emitter and the collector respectively. The emitter-base (*p-n*) junction on the left is under forward bias (low resistance), while the base-collector (*n-p*) junction on right is under reverse bias (high resistance).

Characteristics are the graphical form, helpful in understanding the performance of a transistor. The basic parameters of the transistor are emitter voltage (V_e), emitter current (I_e), collector voltage (V_c), collector current (I_c), and base current (I_b).

The relation between input and output currents and voltages may be represented graphically known as characteristic curves. Different characteristic curves are drawn depending upon which of the three transistor points is common. This common point is taken as the reference point and all measurements are taken w.r.t. this point. Thus, the transistor circuits are named as common-base, common-emitter and common-collector depending upon the common point.

In common-emitter arrangement, the following characteristics are drawn

(i) I_c-V_c characteristics Output characteristics are drawn by noting down the collector current (I_c), for different collector voltage (V_c) for a constant base current.

(ii) I_c-I_b characteristics Collector current versus base current graph is plotted for a constant collector voltage.

(iii) I_b-V_b characteristics Input characteristics are drawn by noting down the base current (I_b) for different base voltage (V_b) when the collector voltage is fixed at a particular value.

Formulae Used

Input resistance, $R_{in} = \dfrac{\Delta V_b}{\Delta I_b}$

Output resistance, $R_{out} = \dfrac{\Delta V_c}{\Delta I_c}$

Resistance gain, $R_{out} = \dfrac{R_{out}}{R_{in}}$

Current gain, $\beta = \dfrac{\Delta V_c}{\Delta I_b}$

Voltage gain = Current gain × Resistance gain

i.e. $A_V = \beta \cdot \dfrac{R_{out}}{R_{in}}$

Procedure

(i) **I_c-V_c Characteristics** (output)

- Make the electrical connections as shown in figure.

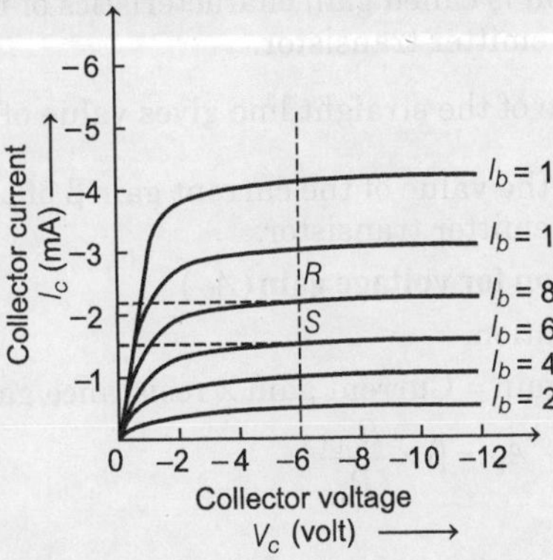

- Close the keys K_1 and K_2 adjust the base current (I_b) to 20 μA by means of the rheostat Rh_1 and keep it constant during this part of the experiment.
- Adjust the collector voltage (V_c) to a suitable value (say-8 V) by means of the rheostat Rh_2 and note the corresponding collector current (I_c).
- Increase the collector voltage (V_c) in equal steps of 1 V and note the corresponding collector current (I_c).
- Now, change the base current I_b(≈ 40 μA, 60μA etc.) by means of the rheostat Rh_1 and repeat the above procedure [steps (3) and (4)] for each value of base current.
- Finally, plot the curves between collector current (I_c) and collector voltage (V_c) on a graph paper as shown in figure.

(ii) **I_c - I_b Characteristics**

- Make the same electrical connections as shown in figure.

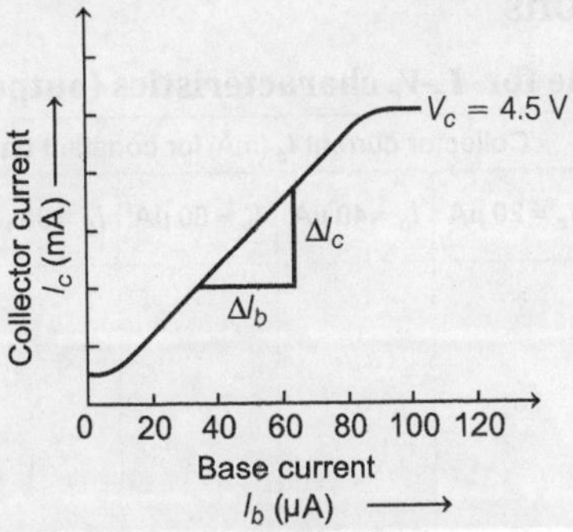

- Adjust the collector voltage (V_c) to –2V with the help of the rheostat Rh_2 and keep it constant throughout the experiment.
- Adjust the base current (I_b) 20 μA by means of the rheostat Rh_1 and note the corresponding collector current (I_c).
- Increase the base current (I_b) in equal steps and note the corresponding collector current (I_c) until it reaches about 5 mA (say).
- Finally, plot the curve between collector current (I_c) against the base current (I_b) as shown in figure.

(iii) **Input characteristics** (I_b-V_b) (Perform this part, if necessary)

- Make the connections as shown in figure. Adjust collector voltage to 5 V by means of rheostat Rh_2 and keep it constant.

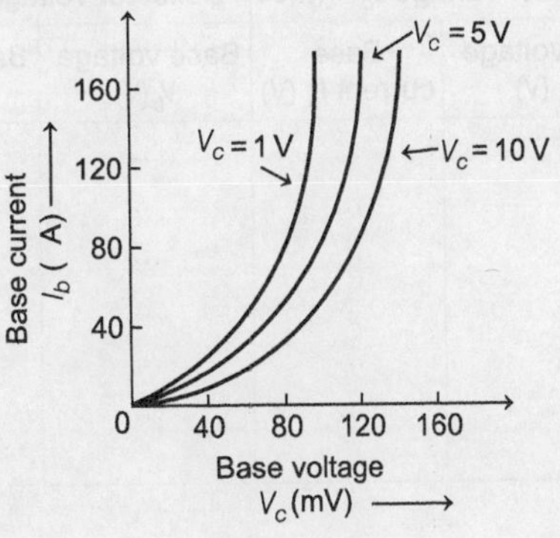

- Set the base voltage (V_b) to (say 0 V) by means of the rheostat Rh_1 and note the corresponding base current (I_b).
- Increase the base voltage (V_b) from zero in equal steps of 1 mV (or so) and note the corresponding base current (I_b).
- Now, change the collector voltage (V_c) by means of rheostat Rh_2 and repeat the above procedure [steps (2) and (3)] for each value of collector voltage and plot the curves between base current (I_b) and base voltage (V_b).

Observations

Table for I_c-V_c characteristics (output)

S. No.	Collector Voltage, V_c (V)	Collector current I_c (mA) for constant base current				
		$I_b = 20\ \mu A$	$I_b = 40\ \mu A$	$I_b = 60\ \mu A$	$I_b = 80\ \mu A$	$I_b = 100\ \mu A$
1.	...	...	...	...	...	...
2.	...	...	...	...	...	...
3.	...	...	...	...	...	...
4.	...	...	...	...	...	...
5.	...	...	...	...	...	...

Table for I_c-I_b characteristics

S. No.	Collector voltage $V_c = -2V$		Collector voltage, $V_c = -3V$	
	Base current I_b (μA)	Collector current I_c (mA)	Base current I_b (mA)	Collector current I_c (mA)
1.	...	...	...	...
2.	...	...	...	...
3.	...	...	...	...
4.	...	...	...	...
5.	...	...	...	...
6.	...	...	...	...

Table for I_b-V_b characteristics (input)
(Take these observations if required)

S. No.	Collector voltage V_c = ... V		Collector voltage, V_c = ... V	
	Base voltage V_b (V)	Base current I_b (V)	Base voltage V_b(V)	Base current I_b (μA)
1.	...	...	...	...
2.	...	...	...	...
3.	...	...	...	...
4.	...	...	...	...
5.	...	...	...	...
6.	...	...	...	...
7.	...	...	...	...

Calculations

(i) **Calculation for input resistance (R_{in})**

Plot a graph between base voltage V_b (table 3) and base current I_b (table 3) for zero collector voltage V_c, taking V_b along X-axis and I_b along Y-axis.

Plot graphs for different values of V_c. The graphs come as shown in Fig.(c).

- These graphs are called input characteristics of the transistor.
- The slope of graphs becomes large at the ends.
- The slope given value of $\frac{\Delta I_b}{\Delta V_b}$. Its reciprocal $\frac{\Delta V_b}{\Delta I_b}$ gives input resistance R_{in}. As graphs run parallel near the ends, all gives same value of R_{in}.

(ii) **Calculation for output resistance (R_{out})**

Plot a graph between collector voltage V_c (table 1) and collector current I_c (table 1) for 20 μA as current I_b, taking V_c along X-axis and I_c along Y-axis.

Plot graphs for different values of I_b. The graphs come as shown in Fig. (a).

These graphs are called output characteristics of the transistor.

The slope of graphs become almost zero at ends. The slope gives value of $\frac{\Delta I_C}{\Delta V_C}$. Its reciprocal $\frac{\Delta V_C}{\Delta I_C}$ gives output resistance R_{out}. As graphs run parallel near the ends, all give same value of R_{out}.

(iii) **Calculation for current gain (β)**

Plot a graph between base current I_b (table 2) and corresponding collector current I_c for collector voltage V_c, taking I_b along X-axis and I_c along Y-axis. The graph comes to be a straight line as shown in Fig. (b).

The graph is called gain characteristics of the common emitter transistor.

The slope of the straight line gives value of $\frac{\Delta I_c}{\Delta I_b}$ which is the value of the current gain β of the common emitter transistor.

(iv) **Calculation for voltage gain (A_V)**

From relation,

Voltage gain = Current gain × resistance gain

$$A_V = \beta \times \frac{R_{out}}{R_{in}}$$

Result

For the given common emitter transistor,

Current gain, β = ...

Voltage gain, A_V = ...

Precautions

(i) Battery with correct polarity should be used in the circuit.

(ii) Overheating of the transistor should be avoided.

(iii) Voltages applied in various parts of the circuit should not exceed the recommended value.

Example 27. *The correct relation between α (ratio of collector current to emitter current) and β (ratio of collector current to base current) of a transistor is*

(a) $\beta = \dfrac{\alpha}{1+\beta}$ (b) $\alpha = \dfrac{\beta}{1-\alpha}$

(c) $\beta = \dfrac{1}{1-\alpha}$ (d) $\alpha = \dfrac{\beta}{1+\beta}$

Sol. (d) We know that, $\alpha = \dfrac{I_C}{I_E}, \beta = \dfrac{I_C}{I_B}$

$$I_E = I_B + I_C$$

$$\alpha = \frac{I_C}{I_B + I_C}$$

$$= \frac{1}{\frac{I_B}{I_C} + 1}$$

$$\alpha = \frac{1}{\frac{1}{\beta} + 1}$$

$$\alpha = \frac{\beta}{1+\beta}$$

Exmaple 28. *If $I_C = 4\ mA$ and $\beta = 45$, then I_E is*

(a) 4.1 *mA* (b) 4.02 *mA*

(c) 4.4 *mA* (d) 0.04 *mA*

Sol. (b) For a transistor, emitter current is sum of collector current and base current, hence

$$I_E = I_C + I_B$$

Also, $I_B = \dfrac{I_C}{\beta} = \dfrac{4}{95} = 0.042\ \text{mA}$

$\therefore$ $I_E = 4 + 0.042$
$= 4.042$

Practice Exercise

1. **On measuring the diameter of a spherical body using vernier callipers, main scale reading = 1.3 cm, 5th vernier scale division is coinciding with any main scale division and zero error is – 0.03 cm, what will be corrected reading?**

(a) 1.38 cm (b) 1.32 cm

(c) 1.35 cm (d) – 1.38 cm

2. **You are given two different vernier callipers *A* and *B* having 10 divisions on vernier scale that coincide with 9 divisions on the main scale each. If 1 cm of main scale *A* is divided into 10 parts and that of *B* in 20 parts, then least count of *A* and *B* are**

(a) 0.001 cm and 0.005 cm

(b) 0.01 cm and 0.05 cm

(c) 0.01 cm and 0.005 cm

(d) 0.01 cm and 0.001 cm

3. **The vernier scale used for measurement has a positive zero error of 0.2 mm. If while taking a measurement, it was noted that 0 on the vernier scale lies between 8.5 cm and 8.6 cm. Vernier coincidence is 6, then the correct value of measurement is ...... cm. (Least count = 0.01 cm)** **[JEE Main 2021]**

(a) 8.36 cm (b) 8.54 cm

(c) 8.58 cm (d) 8.56 cm

4. **One main scale division of a vernier callipers is *a* cm and *n*th division of the vernier scale coincide with (*n* – 1)th division of the main scale. The least count of the callipers (in mm) is** **[JEE Main 2021]**

(a) $\dfrac{10\,na}{(n-1)}$ (b) $\dfrac{10a}{(n-1)}$ (c) $\left(\dfrac{n-1}{10n}\right)a$ (d) $\dfrac{10\,a}{n}$

5. ***N*** **divisions on the main scale of a vernier callipers coincide with *N* + 1 divisions of the vernier scale. If each division of main scale is *a* units, then least count of the instrument is**

(a) $\dfrac{a}{N+1}$ unit (b) a unit

(c) $\dfrac{N}{N+1} \times a$ unit (d) $\dfrac{a}{N}$ unit

6. **A student constructed a vernier callipers as shown. He used two identical inclines and tried to measure the length of line *PQ*. For this instrument determine the least count.**

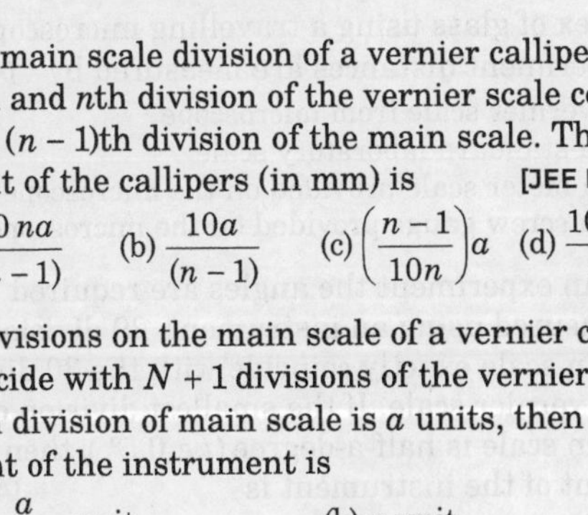

(a) $\frac{l(1-\cos\theta)}{\cos\theta}$ unit (b) $\frac{l}{\cos\theta}$ unit

(c) $l(1-\cos\theta)$ unit (d) $\frac{1-\cos\theta}{l}$ unit

7. For the following diagram [used to measure the length of a small metal piece by using vernier callipers], determine the length of the metal piece. [Least count of the vernier callipers is 0.1 mm]

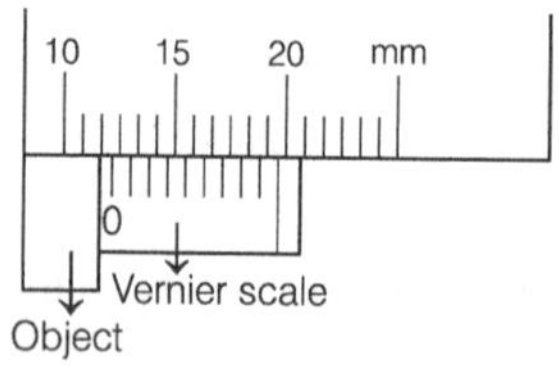

(a) 18 mm (b) 15.7 mm

(c) 12.6 mm (d) None of the above

8. A spectrometer gives the following reading when used to measure the angle of a prism.
Main scale reading : 58.5 degree
Vernier scale reading : 09 divisions
Given that 1 division on main scale corresponds to 0.5 degree. Total divisions on the vernier scale is 30 and match with 29 divisions of the main scale. The angle of the prism from the above data is **[AIEEE 2012]**

(a) 58.59° (b) 58.77° (c) 58.65° (d) 59°

9. An experiment is performed to find the refractive index of glass using a travelling microscope. In this experiment distances are measured by **[AIEEE 2008]**

(a) vernier scale from microscope
(b) a standard laboratory scale
(c) a meter scale provided on the microscope
(d) a screw gauge provided on the microscope

10. In an experiment the angles are required to be measured using an instrument. 29 divisions of the main scale exactly coincide with the 30 divisions of the vernier scale. If the smallest division of the main scale is half-a-degree (*i.e.* 0.5°), then the least count of the instrument is **[AIEEE 2009]**

(a) one minute (b) half minute

(c) one degree (d) half degree

11. On measuring diameter of a wire with help of screw gauge, main scale reading is 1 mm and 6th division of circular scale lying over reference line. On measuring zero error, it is found that zero of circular scale has advanced from reference line by 3 divisions on circular scale, then corrected diameter is

(a) 1.09 mm (b) 1.06 mm (c) 1.03 mm (d) 1.60 mm

12. Two screw gauges A and B have equal number of divisions on circular scale. A has pitch 1 mm and B has pitch 0.5 mm. Which one is more accurate?

(a) A (b) B

(c) Both (d) Can't say

13. On measuring the thickness of a given sheet using screw gauge, observed readings are

S. No.	LSR (mm)	Circular scale division coinciding
1.	1	4
2.	1	6
3.	1	8

If zero error is + 0.06 mm, then corrected thickness of the sheet (in cm) is

(a) 0.1 (b) 1.0

(c) 0.01 (d) 0.11

14. A screw gauge has 50 divisions on its circular scale. The circular scale is 4 units ahead of the pitch scale marking, prior to use. Upon one complete rotation of the circular scale, a displacement of 0.5 mm is noticed on the pitch scale. The nature of zero error involved and the least count of the screw gauge, are respectively **[JEE Main 2020]**

(a) negative, 2 μm (b) positive, 10 μm

(c) positive, 0.1 mm (d) positive, 0.1 μm

15. If the screw on a screw gauge is given six rotations, it moves by 3 mm on the main scale. If there are 50 divisions on the circular scale, the least count of the screw gauge is **[JEE Main 2020]**

(a) 0.001 cm (b) 0.01 cm

(c) 0.02 cm (d) 0.001 cm

16. A screw gauge gives the following reading when used to measure the diameter of a wire.
Main scale reading : 0 mm
Circular scale reading : 52 divisions
Given that 1 mm on main scale corresponds to 100 divisions of the circular scale.
The diameter of wire from the above data is **[AIEEE 2011]**

(a) 0.052 cm (b) 0.026 cm

(c) 0.005 cm (d) 0.52 cm

17. For a simple pendulum, when a graph is plotted between displacement d, KE $= \frac{1}{2}mv^2$ and PE $= mgh$, taking d along X-axis and $\frac{1}{2}mv^2$ and mgh along Y-axis, the graph comes as

(a)
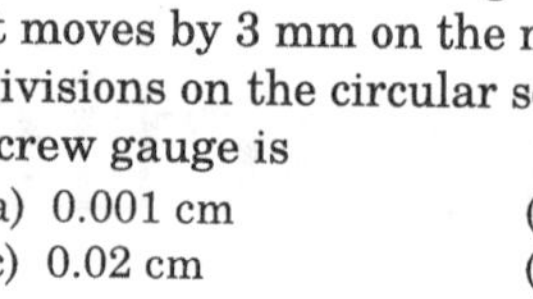

(b)
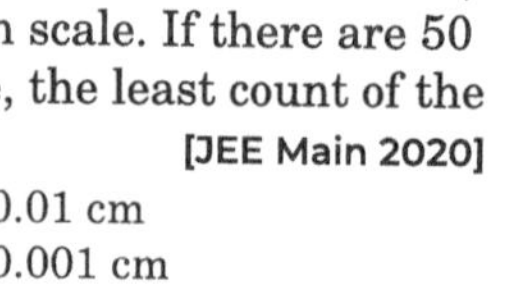

(c)
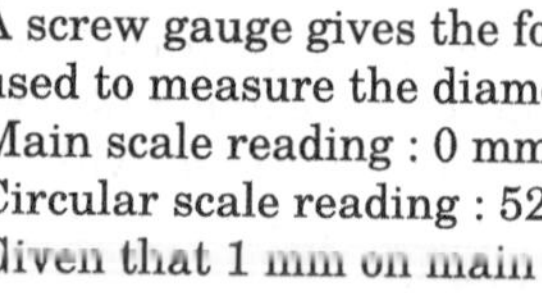

(d)
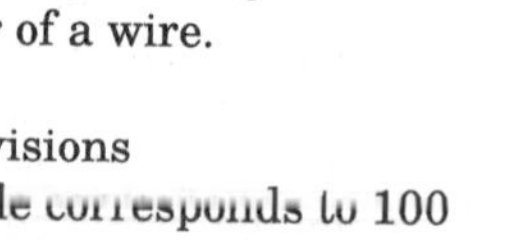

18. The time period of a simple pendulum is given by $T = 2\pi\sqrt{\frac{l}{g}}$. The measured value of the length of pendulum is 10 cm known to a 1 mm accuracy. The time for 200 oscillations of the pendulum is found to be 100 s using a clock of 1s resolution. The percentage accuracy in the determination of g using this pendulum is x. The value of x to the nearest integer is **[JEE Main 2021]**

(a) 2% (b) 3%
(c) 5% (d) 4%

19. A simple pendulum is being used to determine the value of gravitational acceleration g at a certain place. The length of the pendulum is 25.0 cm and a stop watch with 1s resolution measures the time taken for 40 oscillations to be 50 s. The accuracy in g is **[JEE Main 2020]**

(a) 2.40% (b) 5.40%
(c) 4.40% (d) 3.40%

20. If m and M are the masses of two bodies that are tied at two ends of a meter scale that is balanced on a sharp edge of a heavy broad wedge. If $M = 20$ g, its distance from centre = 30 cm and distance of mass m from centre is 25 cm when metre scale is balanced, then m is

(a) 23 g (b) 24 g (c) 25 g (d) 26 g

21. If two masses M and m are tied to two ends of a meter scale. If a balanced point is obtained at point P and if $M > m$, then

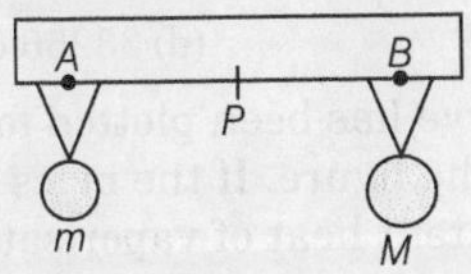

(a) $PA = PB$ (b) $PA > PB$ (c) $PB > PA$ (d) $PA = 3PB$

22. In measurement of mass of a given object by the principle of moments, the meter scale is hung from its mid-point. A known weight of mass 2 kg is hung at one end of meter scale and unknown weight of mass m kg is hung at 20 cm from the centre on other side. The value of m is

(a) 2 kg (b) 5 kg (c) 2.5 kg (d) 0.8 kg

23. In the above question mass of scale is 1 kg and instead of mid-point it is hung at 60 cm from the end where known mass of 2 kg has hung. A mass of 5 kg has to hung at a distance of x cm from pivot to carry out the experiment, then value of x is

(a) 20 cm (b) 10 cm (c) 26 cm (d) 5 cm

24. In order to determine the Young's modulus of a wire of radius 0.2 cm (measured using a scale of least count = 0.001 cm) and length 1 m (measured using a scale of least count = 1 mm), a weight of mass 1 kg (measured using a scale of least count = 1 g) was hanged to get the elongation of 0.5 cm (measured using a scale of least count 0.001 cm). What will be the fractional error in the value of Young's modulus determined by this experiment? **[JEE Main 2021]**

(a) 0.14% (b) 0.9%
(c) 9% (d) 1.4%

25. Two wires A and B have same lengths and made of the same material but A is thicker than B. Both are subjected to the same extending load. Which will extend more?

(a) A (b) B
(c) Same extension (d) Can't predict

26. In experiment for measuring surface tension by capillary rise method, readings for positions A, B, C and D for internal diameter of capillary tube are given as under. Mean internal radius of capillary is

A (cm) = 1,000
B (cm) = 1,000
C (cm) = 1,000
D (cm) = 1,000

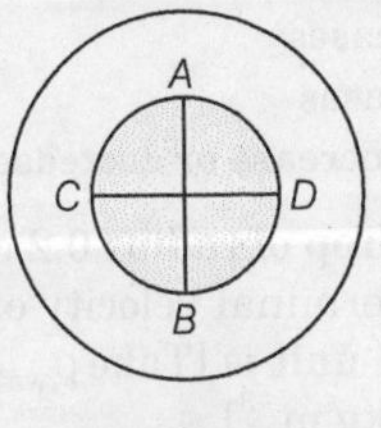

(a) 0.002 cm (b) 0.003 cm
(c) 0.004 cm (d) 0.005 cm

27. When a glass capillary tube of radius 0.015 cm is dipped in water, the water rises to a height of 15 cm within it. Assuming contact angle between water and glass to be 0°, the surface tension of water is

$[\rho_{water} = 1000 \text{ kg m}^{-3}, g = 9.81 \text{ ms}^{-2}]$

(a) 0.11 Nm^{-1} (b) 0.7 Nm^{-1}
(c) 0.072 Nm^{-1} (d) None of these

28. While measuring surface tension of water using capillary rise method, height of the lower meniscus from free surface of water is 3 cm while inner radius of capillary tube is found to be 0.5 cm. Then compute surface tension of water using this data. [Take, contact angle between glass and water as 0° and $g = 9.81 \text{ms}^{-2}$]

(a) 0.72 Nm^{-1}
(b) 0.77 Nm^{-1}
(c) 1.67 Nm^{-1}
(d) None of the above

29. In previous question, if we add some detergent to water, then

(a) liquid level in capillary tube is less than 3 cm
(b) liquid level in capillary tube is greater than 3 cm
(c) liquid level in capillary tube is equal to 3 cm
(d) anything may happen

30. While measuring surface tension of water using capillary rise method, the necessary precaution to be taken is/are
(a) capillary tube should be clean while water should have some grease
(b) both capillary tube and water should be clean
(c) no need to take care of temperature of water
(d) None of the above

31. In an experiment for determining coefficient of viscosity, one lead shot A is having radius r_1 and the other B is having $\frac{r_1}{2}$. Which one will fall fast?
(a) A
(b) B
(c) Both with same speed
(d) Can't say

32. In the above question, as air bubble moves up, its radius
(a) remains constant
(b) decreases
(c) increases
(d) can increase or decrease

33. A rain drop of radius 0.2 cm is falling through air with a terminal velocity of 8.7 ms^{-1}. The viscosity of air in SI unit is [Take ρ_{water} = 1000 kg m^{-3} and ρ_{air} = 1 kg m^{-3}]
(a) 10^{-4} poise
(b) 1×10^{-3} poise
(c) 8.6×10^{-3} poise
(d) 1.02×10^{-3} poise

34. A wide jar is filled with water, in which a steel ball of radius 0.25 cm has been dropped to measure the viscosity of water by using terminal velocity concept.
(a) This method is appropriate
(b) This method is not appropriate
(c) If we take a jar of length 2 m, it will work
(d) None of the above

35. A wide jar is filled with glycerine having specific gravity 1.26, in this jar, a steel ball of radius 0.25 cm has been dropped. After some time it has been observed that ball is taking equal interval of time (1.8 s) to cover equal successive distances, of 20 cm.
[Take, $\rho_{steel} = 7.8 \times 10^3$ kg-m^3, $g = 9.81$ms^{-2}]. The viscosity of glycerine (in N-sm^{-2}) is
(a) 0.802
(b) 1.67
(c) 0.76
(d) 0.963

36. While measuring viscosity of caster oil using terminal velocity concept the following observation table has been taken by a student. Which one is the first correct reading which he should consider for the computation of terminal velocity?

S.No.	Distance	Time
1.	20 cm	1 s
2.	20 cm	1.4 s
3.	20 cm	1.8 s
4.	20 cm	1.81 s
5.	20 cm	1.82 s

(a) 1
(b) 2
(c) 3
(d) 4

37. In an experiment to determine the specific heat of a given liquid by method of mixtures. If room temperature recorded by one thermometer is 29°C and that by second thermometer is 27.5°C. If steady temperature of metal in hypsometer is 62°C, what will be corrected temperature of metal?
(a) 60°C
(b) 60.5°C
(c) 61°C
(d) 63.5°C

38. In the experiment of measuring speed of sound by resonance tube, it is observed that for tuning fork of frequency ν = 480 Hz, length of air column l_1 = 30 cm, l_2 = 70 cm, then v_1 is equal to
(a) 338 ms^{-1}
(b) 379 ms^{-1}
(c) 384 ms^{-1}
(d) 332 ms^{-1}

39 If velocity of sound at room temperature is 35078 cm^{-1}, then velocity of sound at 0°C
(a) 33200 cms^{-1}
(b) 33286 cms^{-1}
(c) 33296 cms^{-1}
(d) 33256 cms^{-1}

40. A heating curve has been plotted for a solid object as shown in the figure. If the mass of the object is 200 g, then latent heat of vaporisation for the material of the objects, is (Power supplied to the object is constant and equal to 1 kW)

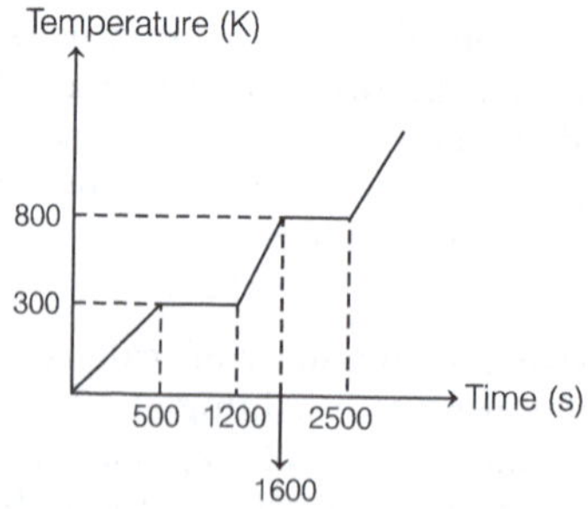

(a) 4.5×10^6 Jkg^{-1}
(b) 4.5×10^6 calg^{-1}
(c) 4.5×10^8 Jkg^{-1}
(d) 4.5×10^4 calg^{-1}

41. The meter bridge shown in the balance position

with $\frac{P}{Q}=\frac{l_1}{l_2}$. If we now interchange the positions of galvanometer and cell, will the bridge work? If yes, that will be the balanced condition?

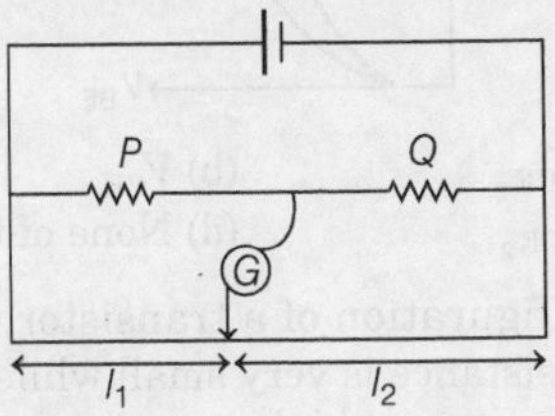

(a) yes, $\frac{P}{Q}=\frac{l_2-l_1}{l_2+l_1}$ (b) no, no null point

(c) yes, $\frac{P}{Q}=\frac{l_2}{l_1}$ (d) yes, $\frac{P}{Q}=\frac{l_1}{l_2}$

42. In the measurement of resistance of a wire using Ohm's law, the plot between V and I is drawn as shown.

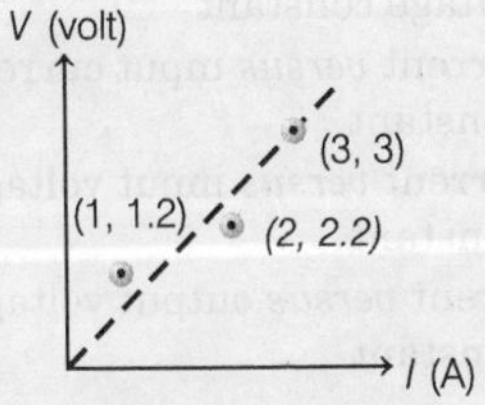

The resistance of the wire is

(a) 0.833 Ω (b) 0.9 Ω

(c) 1 Ω (d) None of these

43. In the experiment of Ohm's law, a potential difference of 5.0 V is applied across the end of a conductor of length 10.0 cm and diameter of 5.00 mm. The measured current in the conductor is 2.00 A. The maximum permissible percentage error in the resistivity of the conductor is **[JEE Main 2021]**

(a) 3.9 (b) 8.4

(c) 7.5 (d) 3.0

44. In comparison of emf's of two cells using potentiometer, the balanced length for batteries having emfs E_1 and E_2 are 60 cm and 20 cm, respectively. Then

(a) $\frac{E_1}{E_2}=3$ (b) $\frac{E_1}{E_2}=\frac{1}{3}$ (c) $\frac{E_1}{E_2}=60$ (d) $\frac{E_1}{E_2}=20$

45. The figure below shows a 2.0 V potentiometer used for the determination of internal resistance of a 2.5 V cell. The balance point of the cell in the open circuit is 75 cm. When a resistor of 10 Ω is used in the external circuit of the cell, the balance point shifts to 65 cm length of potentiometer wire. The internal resistance of the cell is **[JEE Main 2018]**

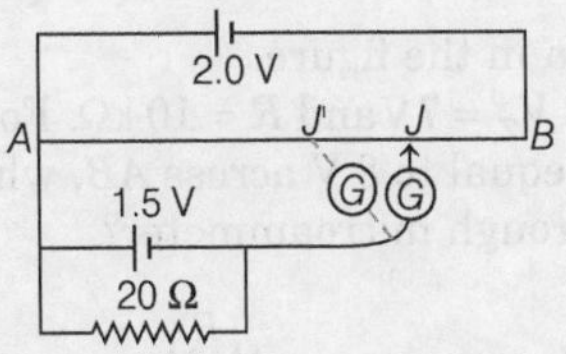

(a) 2.5 Ω (b) 2.0 Ω

(c) 1.54 Ω (d) 1.0 Ω

46. A resistance of R Ω draws current from a potentiometer. Potentiometer has a total resistance R_0 Ω as shown in figure.

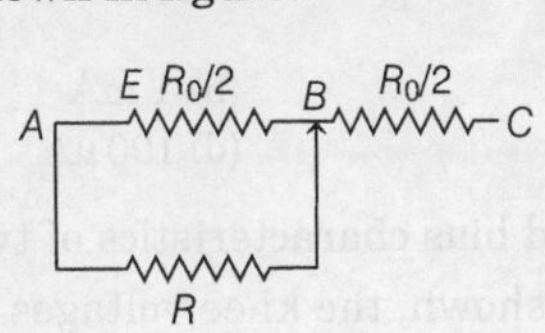

A voltage V is supplied to the potentiometer, then for the voltage across R when the sliding contact is in the middle of the potentiometer is

(a) $\frac{2\,VR}{R_0+4R}$ (b) $\frac{2\,VR}{R_0}$

(c) $\frac{4\,VR}{R_0+2\,R}$ (d) $\frac{VR}{R_0+2\,R}$

47. The readings corresponding to zener diode are given below in the table. From given table, determine the reverse breakdown voltage of the zener diode.

Forward bias		Reverse bias	
V (volt)	I (mA)	V (volt)	I (μA)
0.5	5	0.5	2.0
0.7	20	1.0	2.0
0.8	40	3.0	2.0
1.0	250	5.0	2.0
		5.5	100.0
		5.5	120.0

(a) It is lying between 1.0 V to 5.0 V

(b) 1.0 V

(c) Approx. 5.3 V

(d) None of the above

48. The zener diode normally operates under reverse bias condition, the major use of this fact is in the applications where we require

(a) large value of current

(b) a constant voltage

(c) a current that is increasing without any change in applied voltage

(d) All of the above

49. A zener diode is operating in its normal region *i.e.*, the breakdown region for which the circuit diagram

is as shown in the figure.
Here, take $V_Z = 7\text{V}$ and $R = 10\text{ k}\Omega$. For potential difference equal to 8 V across AB, what is the current through microammeter?

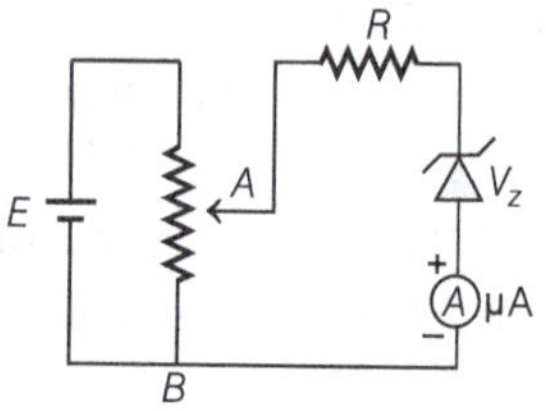

(a) 1000 μA (b) 1 mA
(c) 10 μA (d) 100 μA

50. The forward bias characteristics of two diodes D_1 and D_2 are shown, the knee voltages for D_1 and D_2 are respectively (approx.)

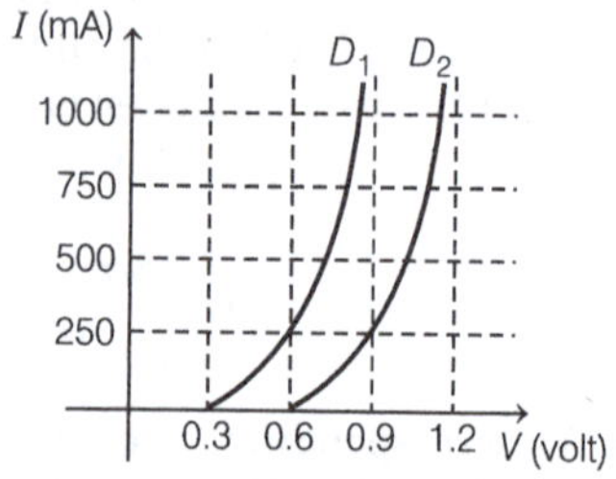

(a) 0.4 V and 0.7 V (b) 0.6 V and 0.9 V
(c) 0.6 V and 0.8 V (d) 0.4 V and 0.9 V

51. The circuit diagram below shows n-p-n transistor in CE configuration. For this configuration, mark the correct statement(s).

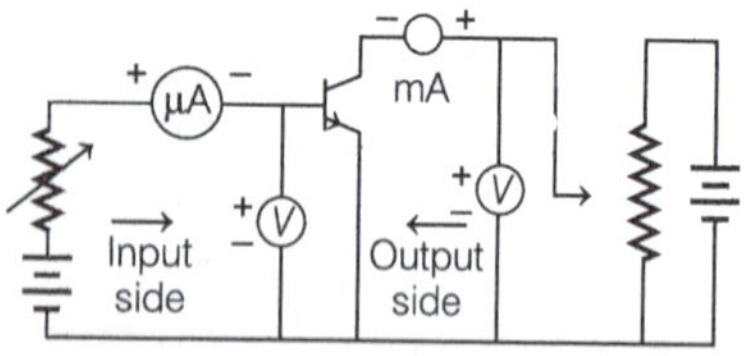

(a) The potential divider on input side is used to keep V_{CE} constant while drawing input characteristics
(b) The potential divider on output side is used to keep V_{CE} constant while drawing output characteristics
(c) The potential divider on input side is used to keep base current constant while drawing output characteristics
(d) Both (b) and (c)

52. Input characteristics are shown for CE configuration of n-p-n transistor for different output voltages. Here,

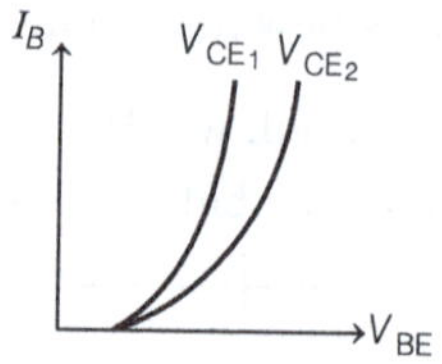

(a) $V_{CE_1} > V_{CE_2}$ (b) V_{CE_1}
(c) $V_{CE_1} < V_{CE_2}$ (d) None of these

53. For CE configuration of a transistor,
(a) input resistance is very small while output resistance is very high
(b) input resistance is very large while output resistance is very small
(c) both input and output resistances are very small
(d) both input and output resistances are very large

54. Transfer characteristic for a transistor is plotted between
(a) output current *versus* input current keeping output voltage constant
(b) output current *versus* input current keeping input voltage constant
(c) output current *versus* input voltage keeping output voltage constant
(d) input current *versus* output voltage keeping input voltage constant

55. Output characteristic of n-p-n transistor in CE configuration is shown. From the characteristic curve determine the current gain at $V_{CE} = 1\text{V}$.

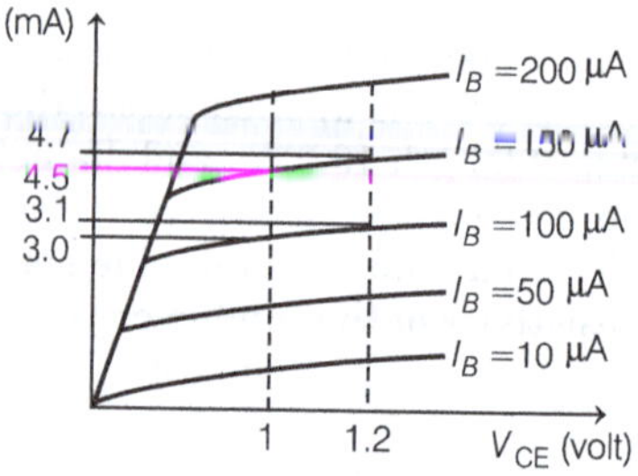

(a) 30 (b) 32 (c) 28 (d) 40

56. Consider the transistor shown in figure, its terminals are marked as 1, 2 and 3. Using multimeter one try to identify the base of transistor, he proceed in the way as follows

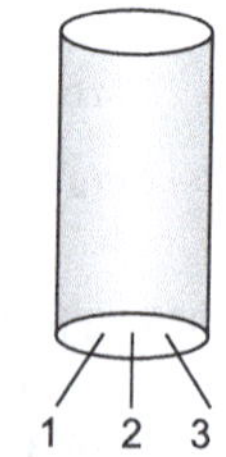

Experiment I He touches the common lead of the multimeter to 2, then on touching other lead of multimeter to 1 he hasn't got any beep (indication of conduction) but when connected to 3 got the beep.

Experiment II He connects the common lead of multimeter to 1 and other lead to 2 and 3 one by one then in this case he got beep for both connections.

From this, we conclude that

(a) 1 is base (b) 2 is base

(c) 3 is base (d) None of these

57. In previous question, the transistor is

(a) *n-p-n*

(b) *p-n-p*

(c) Can't say anything

(d) None of the above

58. To identify whether the transistor is working or not, using multimeter, which statement serves the purpose?

(a) The common lead of multimeter is connected to base and other lead to first emitter and then to collector, only Ist connection shows the continuity

(b) The common lead of multimeter is connected to base and other lead to first emitter and then to collector, both the connections show the continuity

(c) The common lead of multimeter is connected to base and other lead to first emitter and then to collector, none of the connections shows the continuity

(d) All of the above

59. A working transistor with its three legs marked P, Q and R is tested using a multimeter. No conduction is found between P and Q. By connecting the common (– ve) terminal of the multimeter to R and the other (positive) terminal to P or Q, some conduction is seen on the multimeter. Which of the following is true for the transistor? [AIEEE 2008]

(a) It is an *n-p-n* transistor with R as base

(b) It is an *p-n-p* transistor with R as collection

(c) It is a *p-n-p* transistor with R as emitter

(d) It is an *n-p-n* transistor with R as collector

Numerical Value Questions

60. The main scale of a vernier callipers reads in millimeter and its vernier is divided into 10 divisions which coincides with 9 divisions of the main scale. The reading for shown situation is found to be $(x / 10)$ mm. Find the value of x.

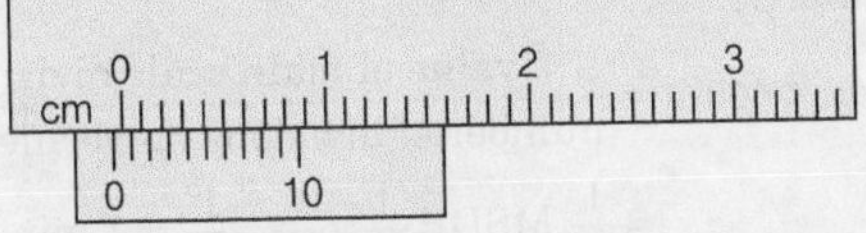

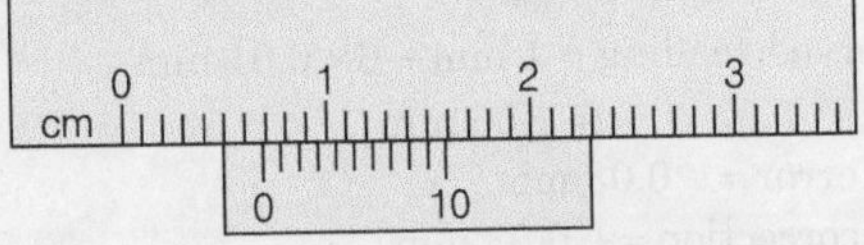

61. An object covers (16.0 ± 0.4) m distance in (4.0 ± 0.2) s. The error in its speed is ± .... m/s.

62. Students I_1, J_1, J_3 and I_2 perform an experiment for measuring the acceleration due to gravity (g) using a simple pendulum. They use different lengths of the pendulum and record time for different number of oscillations. The observations are shown in the table. Least count for length = 0.1 cm and least count for time = 1 s.

Students	Length of Pendulum (cm)	No. of Oscillations (n)	Time period of Pendulum (s)
I_1	100.0	20	20
J_1	400.0	10	40
J_3	100.0	10	20
I_2	400.0	20	40

If P_1, P_2, P_3 and P_4 are the % error in g for students I_1, J_1, J_3 and I_2 respectively, then find ratio $\frac{P_2}{P_1}$.

63. The length of a cylinder is measured with a meter rod having least count 0.1 cm. Its diameter is measured with vernier callipers having least count 0.01 cm. Given that, the length is 5.0 cm and diameter is 2.00 cm. Find the percentage error in the calculated value of volume.

64. In an experiment to determine an unknown resistance, a 100 cm long resistance wire is used. The unknown resistance is kept in the left gap and a known resistance is put into the right gap. The scale used to measure length has a least count 1 mm. The null point B is obtained at 40.0 cm from the left gap. We shall determine the percentage error in the computation of unknown resistance.

65. The value of power dissipated across the zener diode ($V_Z = 15$ V) connected in the circuit as shown in the figure is $x \times 10^{-1}$ watt.

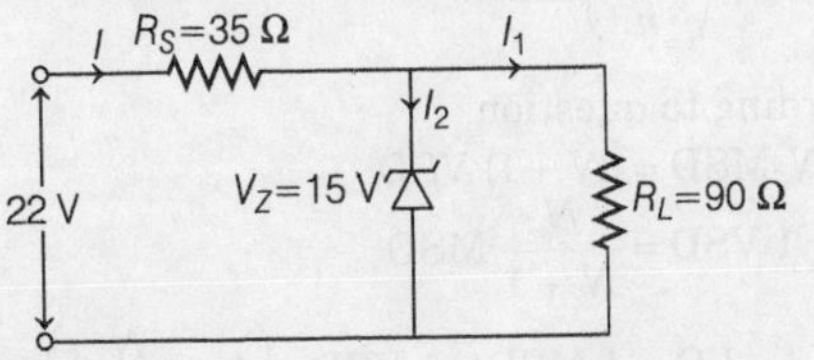

The value of x, to the nearest integer, is .... [JEE Main 2021]

Answers

1. (a)	*2.* (c)	*3.* (b)	*4.* (d)	*5.* (a)	*6.* (a)	*7.* (c)	*8.* (c)	*9.* (a)	*10.* (a)
11. (a)	*12.* (b)	*13.* (a)	*14.* (b)	*15.* (a)	*16.* (a)	*17.* (c)	*18.* (b)	*19.* (c)	*20.* (b)
21. (b)	*22.* (b)	*23.* (c)	*24.* (d)	*25.* (b)	*26.* (a)	*27.* (a)	*28.* (b)	*29.* (a)	*30.* (b)
31. (a)	*32.* (c)	*33.* (b)	*34.* (b)	*35.* (a)	*36.* (c)	*37.* (d)	*38.* (c)	*39.* (c)	*40.* (a)
41. (d)	*42.* (c)	*43.* (a)	*44.* (a)	*45.* (c)	*46.* (a)	*47.* (c)	*48.* (b)	*49.* (d)	*50.* (a)
51. (c)	*52.* (a)	*53.* (a)	*54.* (a)	*55.* (a)	*56.* (b)	*57.* (b)	*58.* (d)	*59.* (a)	*60.* 69
61. 0.3	*62.* 0.7	*63.* 3	*64.* 0.42	*65.* 5					

Solutions

1. As, $TR = MSR + n \times LC$

$$= 1.3 + 5 \times 0.01$$
$$= 1.35 \text{ cm}$$

Zero error $= -0.03$ cm

Zero correction $= +0.03$ cm

Corrected reading $= 1.35 + 0.03 = 1.38$ cm

2. Least count of $A = \dfrac{(1/10)}{10} = 0.01$ cm

Least count of $B = \dfrac{(1/20)}{10} = 0.005$ cm

3. Positive zero error $= 0.2$ mm

Main scale reading $= 8.5$ cm

Vernier scale reading $= 6 \times 0.01 = 0.06$ cm

Final reading $= 8.5 + 0.06 - 0.02 = 8.54$ cm

4. $(n-1)a = n(a')$

$$a' = \frac{(n-1)a}{n}$$

$\therefore \quad LC = 1 \text{ MSD} - 1 \text{ VSD}$

$$= (a - a') \text{ cm}$$
$$= a - \frac{(n-1)a}{n} = \frac{na - na + a}{n} = \frac{a}{n} \text{ cm}$$
$$= \left(\frac{10a}{n}\right) \text{ mm}$$

5. According to question,

$$N \text{ MSD} = (N+1) \text{ VSD}$$
$$\Rightarrow \quad 1 \text{ VSD} = \frac{N}{N+1} \text{ MSD}$$

Now, $\quad LC = 1 \text{ MSD} - 1 \text{ VSD} = \left(1 - \dfrac{N}{N+1}\right) 1 \text{ MSD}$

$$= \frac{a}{N+1} \text{ unit}$$

6. Let θ be the angle of incline. Here, the incline kept horizontally is working as main scale while the other incline kept on horizontally placed incline is treated as vernier scale.

From the given figure, it is clear that,

$1 \text{ MSD} = \dfrac{l}{\cos\theta}$ unit and $1 \text{ VSD} = l$ unit,

So, LC of instrument, $\quad DC = 1\text{MSD} - 1 \text{ VSD}$

$$= \left(\frac{l}{\cos\theta} - l\right) = \frac{l(1-\cos\theta)}{\cos\theta} \text{ unit}$$

7. As, 6th VSD is coinciding with one of the MSD, so reading of the vernier callipers is

$$12 + 6 \times 0.1 = 12.6 \text{ mm}$$

8. 1 Vernier scale division $= \dfrac{29}{30}$ main scale division

$$\Rightarrow \quad 1 \text{ VSD} = \frac{29}{30} \times 0.5° = \left(\frac{29}{60}\right)^{\circ}$$

$\therefore$ The least count $= 1 \text{ MSD} - 1 \text{ VSD}$

$$= \left(\frac{1}{2}\right)^{\circ} - \left(\frac{29}{60}\right)^{\circ} = \left(\frac{1}{60}\right)^{\circ}$$

So, reading = main scale reading + vernier scale reading

$$= MSR + n \times LC$$
$$= 58.5° + 9 \times \left(\frac{1}{60}\right)^{\circ} = 58.65°$$

9. During the experiment performed by travelling microscope of measurement of refractive index, distances are measured by vernier scale from microscope.

10. As, least count $= \dfrac{\text{value of main scale division}}{\text{number of divisions on vernier scale}}$

$$= \frac{1}{30} \text{ MSD} = \frac{1}{30} \times \frac{1°}{2} = \frac{1°}{60} = 1 \text{ min}$$

11. Observed reading $= 1 \text{ mm} + 6 \times 0.01$ mm

$$= 1.06 \text{ mm}$$

Zero error $= -0.03$ mm

Zero correction $= +0.03$ mm

Corrected reading $= 1.06 + 0.03 = 1.09$ mm

12. B's least count is half that of A, hence it is more accurate.

13. Corrected thickness of the sheet

$$= \frac{(0.14 - 0.06) + (0.16 - 0.06) + (0.18 - 0.06)}{3}$$

$$= \frac{0.08 + 0.10 + 0.12}{3}$$

$$= \frac{0.30}{3} = 0.10 \text{ cm}$$

14. Given that, number of divisions on circular scale $= 50$

Pitch $= 0.5$ mm $= 5 \times 10^{-4}$ m

Least count of screw gauge

$$= \frac{\text{Pitch}}{\text{Number of divisions on circular scale}}$$

$$= \frac{5 \times 10^{-4}}{50} = 1 \times 10^{-5} \text{ m} = 10 \mu\text{m}$$

Since, circular scale is 4 units ahead of the pitch scale marking which means screw gauge has positive zero error.

15. Pitch of screw of a screw gauge

$$= \frac{\text{Distance moved by screw}}{\text{Number of rotations}}$$

$$= \frac{3 \text{ mm}}{6} = 0.5 \text{ mm}$$

and least count of screw gauge

$$= \frac{\text{Pitch}}{\text{Number of circular scale divisions}}$$

$$= \frac{0.5 \text{ mm}}{50} = 0.01 \text{ mm} = 0.001 \text{ cm}$$

16. Diameter of wire,

$$d = \text{MSR} + \text{CSR} \times \text{LC}$$

$$= 0 + 52 \times \frac{1}{100} = 0.52 \text{ mm} = 0.052 \text{ cm}$$

17. For a simple pendulum, KE is maximum at mean position and PE is maximum at extreme position. Therefore, curve (c) is correct.

18. $g = \dfrac{4\pi^2 l}{T^2}$

$$\frac{\Delta g}{g} = \frac{\Delta l}{l} + 2\frac{\Delta T}{T} = \frac{0.1}{10} + 2\left(\frac{\frac{1}{200}}{0.5}\right)$$

$$\frac{\Delta g}{g} = \frac{1}{100} + \frac{1}{50}$$

$$\frac{\Delta g}{g} \times 100 = 3\%$$

19. Given, length of pendulum, $l = 25.0$ cm So, there is an uncertainty of 0.1 cm in measurement of length. Resolution of stop watch is 1s, so uncertainty in measurement of time is 1s.

Now using, $T = 2\pi\sqrt{\dfrac{l}{g}}$ or $g = \dfrac{4\pi^2 l}{T^2}$

We have, $\dfrac{\Delta g}{g} = \dfrac{\Delta l}{l} + \dfrac{2\Delta T}{T}$

$$\Rightarrow \quad \frac{\Delta g}{g} \times 100\% = \left(\frac{\Delta l}{l} + \frac{2\Delta T}{T}\right) \times 100\%$$

Accuracy in measurement of g is

$$\frac{\Delta g}{g} \times 100\% = \left(\frac{0.1}{25} + \frac{2 \times 1}{50}\right) \times 100\%$$

$$= (0.004 + 0.04) \times 100\ \% = 4.40\%$$

20. As, $ma = MA$ [principle of moments]

$$\Rightarrow \quad m \times 25 = 20 \times 30$$

$$m = \frac{20 \times 30}{25} = 24 \text{ g}$$

21. In equilibrium, $m \times PA = M \times PB$

$$\frac{PA}{PB} = \frac{M}{m} > 1$$

$$\therefore \quad PA > PB$$

22. Taking moments about the point O.

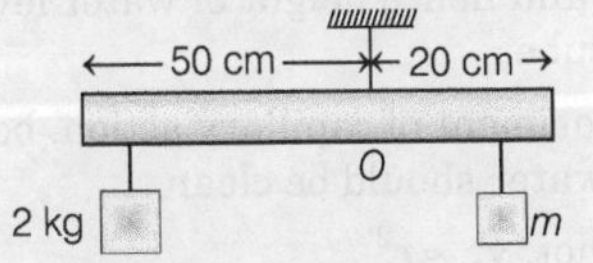

$$2g \times 0.5 = mg \times 0.2$$

$$\Rightarrow \quad m = 5 \text{ kg}$$

23. Taking torque about hanging point.

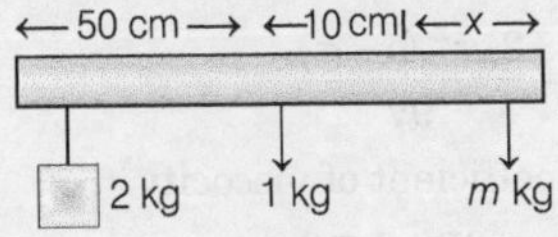

$$2g \times 60 + 1g \times 10 = mg \times x$$

$$\Rightarrow \quad x = \frac{2 \times 60 + 1 \times 10}{5} = 26 \text{ cm}$$

24. $Y = \dfrac{\text{Stress}}{\text{Strain}} = \dfrac{FL}{Al} = \dfrac{mg.L}{\pi R^2.l}$

$$\frac{\Delta Y}{Y} = \frac{\Delta m}{m} + \frac{\Delta L}{L} + 2.\frac{\Delta R}{R} + \frac{\Delta l}{l}$$

$$\frac{\Delta Y}{Y} \times 100 = 100\left[\frac{1}{1000} + \frac{1}{1000} + 2\left(\frac{0.001}{0.2}\right) + \frac{0.001}{0.5}\right]$$

$$= \frac{1}{10} + \frac{1}{10} + 1 + \frac{1}{5} = \frac{14}{10} = 1.4\%$$

25. As, Young's modulus, $Y = \dfrac{F \cdot l}{A\Delta l} \Rightarrow \Delta l = \dfrac{F \cdot l}{YA}$

$$\Rightarrow \quad \Delta l \propto \frac{1}{A}$$

$$\therefore \quad \frac{\Delta l_A}{\Delta l_B} = \frac{A_B}{A_A}$$

As, $\quad A_A > A_B \quad \therefore \quad \Delta l_B > \Delta l_A$

26. Here, $AB = 1.009 - 1.006 = 0.003$ cm

and $CD = 1.009 - 1.004 = 0.005$ cm

$$\therefore \quad d = \frac{AB + CD}{2} = \frac{0.008}{2} = 0.004 \text{ cm}$$

$$r = d/2 = 0.002 \text{ cm}$$

27. For the liquid of meniscus, $2\pi r \times T \cos\theta = \pi r^2 h\rho g$

$$\Rightarrow \quad T = \frac{rh\rho g}{2}$$

$$= \frac{0.015 \times 10^{-2} \times 15 \times 10^{-2} \times 1000 \times 9.8}{2}$$

$$= 0.11 \text{ Nm}^{-1}$$

28. As, $T = \dfrac{r\left(h + \dfrac{r}{3}\right)\rho g}{2\cos\theta}$

$$= \frac{0.5 \times 10^{-2}\left(3 + \dfrac{0.5}{3}\right) \times 10^{-2} \times 10^3 \times 9.81}{2}$$

$$= 0.77 \text{ Nm}^{-1}$$

29. As, detergent is added to water, its surface tension decreases and hence height of water level falls in capillary tube.

30. In the experiment of capillary action, both capillary tube and water should be clean.

31. For lead shot, $v_t \propto r^2$

Hence, A will fall faster.

32. As, air bubble moves up, pressure outside the liquid decreases. As, there is excess of pressure inside the bubble expands due to surface tension.

33. We have, $\eta = \dfrac{2gr^2(\rho - \sigma)}{9v}$

where, η = coefficient of viscocity.

Here, $\rho = \rho_{water}$ and $\sigma = \rho_{air}$

$$= \frac{2 \times 9.81 \times (0.2 \times 10^{-2})^2 \times 999}{9 \times 8.7}$$

$$= 1 \times 10^{-3} \text{ poise}$$

34. As, water is having low viscosity, the terminal velocity will not be acquired by steel ball very soon, so to serve the purpose a very long jar of approximately 1000 m, is needed which is not suitable to perform the experiment.

35. From the expression,

$$\eta = \frac{2r^2(\rho_{steel} - \rho_{glycerine})}{9 \times v} \times g$$

where, v is terminal speed.

Here, $\quad v = \dfrac{0.2}{1.8}$ ms^{-1}

$$\therefore \quad \eta = \frac{2 \times (0.25 \times 10^{-2})^2 \times (7.8 - 1.26) \times 10^3 \times 9.81}{9 \times \dfrac{0.2}{1.8}}$$

$$= 0.802 \text{ N-sm}^{-2}$$

36. After 2nd reading, *i.e.* from 3rd reading the velocity of the ball is almost constant, so this is the first correct reading which shows that terminal speed has been acquired by ball.

37. Corrected temperature of metal

$$= 62° \text{ C} + (29° \text{ C} - 27.5° \text{ C}) = 63.5° \text{ C}$$

38. $v = 2\nu\,(l_2 - l_1)$

$$= 2 \times 480 \times \frac{(70 - 30)}{100} \quad \text{(Convert cm in m)}$$

$$= 2 \times 480 \frac{40}{100} = 384 \text{ ms}^{-1}$$

39. As, $\quad v_0 = v\sqrt{\dfrac{273}{273 + t}}$

$$= 33296 \text{ cms}^{-1} \quad (\text{where, } t = 273 + 25 = 298)$$

40. As, $\quad \displaystyle\int_{1600}^{2500} P\,dt = mL_v$

where, P = power developed,

m = mass

and $\quad L_v$ = latent heat of vaporisation.

$\Rightarrow 10^3 \times 900 = 0.2 \times L_v$

($\because$ time duration is 1600 s to 2500 s)

$$\Rightarrow \quad L_v = 4.5 \times 10^6 \text{ Jkg}^{-1}$$

41. For balanced position in a meter bridge.

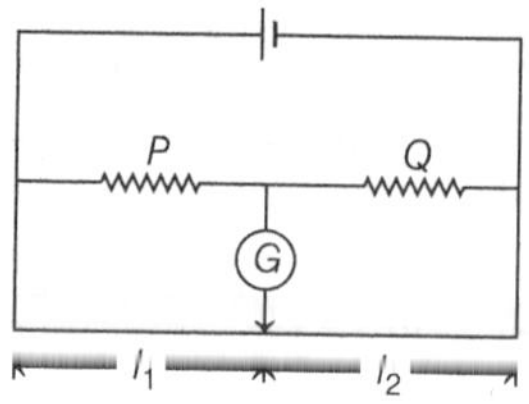

$$\frac{P}{Q} = \frac{l_1}{l_2}$$

Now, if position of G and cell is interchanged,

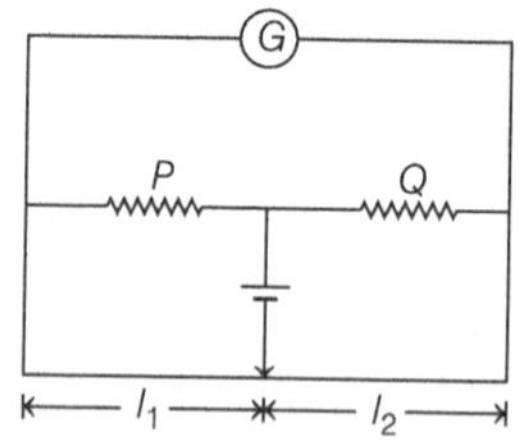

The balance condition still remains the same, if the jockey points at the same point as given in the initial condition, for which there is no deflection in the galvanometer or no current will be drawn from the cell. Thus, the bridge will work as usual and balance condition remains same, *i.e.* $P/Q = l_1/l_2$.

42. We know that, V-I curve for a linear device is a straight line passing through origin. Due to some errors/carelessness on the part of experiment all points may not come on the same line. In this situation, we draw the most appropriate curve.

From the diagram, $R = \frac{3}{3} = 1\ \Omega$.

43. $R = \frac{\rho l}{A} = \frac{V}{I}$

$$\rho = \frac{AV}{Il} = \frac{\pi d^2 V}{4Il} \qquad \left(A = \frac{\pi d^2}{4}\right)$$

$$\therefore \quad \frac{\Delta\rho}{\rho} = \frac{2\Delta d}{d} + \frac{\Delta V}{V} + \frac{\Delta I}{I} + \frac{\Delta l}{l}$$

$$\frac{\Delta\rho}{\rho} = 2\left(\frac{0.01}{5.00}\right) + \frac{0.1}{5.0} + \frac{0.01}{2.00} + \frac{0.1}{10.0}$$

$$\frac{\Delta\rho}{\rho} = 0.004 + 0.02 + 0.005 + 0.01$$

$$\frac{\Delta\rho}{\rho} = 0.039$$

$$\%\ \text{error} = \frac{\Delta\rho}{\rho} \times 100 = 0.039 \times 100 = 3.90\%$$

44. As, for potentiometer, $E \propto l$

where, l is the balanced length of the potentiometer wire.

So, $\frac{E_1}{E_2} = \frac{60}{20} = 3$

45. For a potentiometer, the internal resistance r is given by $r = R\left(\frac{l_1}{l_2} - 1\right)$

Given, $R = 10\ \Omega$, $l_1 = 75$ cm, $l_2 = 65$ cm

$$\therefore \quad r = 10\left(\frac{75}{65} - 1\right) = 10 \times 0.154 = 1.54\ \Omega$$

46. The circuit can be drawn as

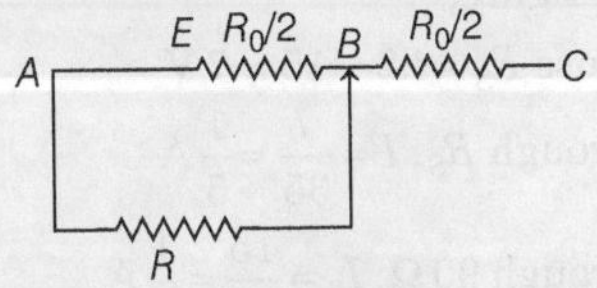

Equivalent resistance across AB,

$$\frac{1}{R_1} = \frac{1}{R} + \frac{1}{(R_0/2)}$$

$$\Rightarrow \quad R_1 = \frac{R_0 R}{R_0 + 2R} \qquad \ldots\text{(i)}$$

The total resistance between A & C will be the sum of resistances between A & B and B & C, *i.e.* $R_1 + R_0/2$.

So, the current flowing through the potentiometer will be

$$I = \frac{V}{R_1 + R_0/2} = \frac{2V}{2R_1 + R_0}$$

The voltage V_1 taken from the potentiometer will be the product of current I and resistance R_1,

$$V_1 = IR_1 = \left(\frac{2V}{2R_1 + R_0}\right) \times R_1$$

$$= \frac{2V}{2\left(\frac{R_0 R}{R_0 + 2R}\right) + R_0} \times \frac{R_0 R}{R_0 + 2R}$$

$$\Rightarrow \quad V_1 = \frac{2VR}{R_0 + 4R} \qquad \text{[using Eq. (i)]}$$

47. Current changes by large amount when we change reverse bias voltage from 5 V to 5.5 V, so reverse breakdown voltage is somewhere there only and thereafter current increases even though no change in voltage occurs.

48. The circuit used for working zener diode is shown. Once the diode attains the breakdown voltage, then there is no change in voltage across the diode even if we change the current in circuit by changing the position of rheostat and that is why the voltage across zener diode is constant.

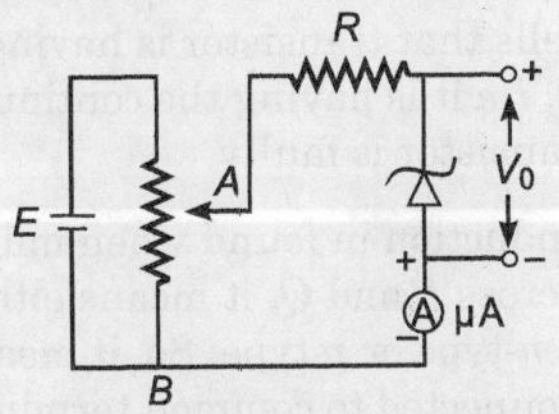

49. Write KVL equation,

$$-V_{BA} + IR + V_Z = 0$$

[As diode is operating in breakdown region]

$$\Rightarrow \quad IR = V_{BA} - V_Z = 8 - 7$$

$$\text{or,} \quad I = \frac{1\text{V}}{10\ \text{k}\Omega} = 100\ \mu\text{A}$$

50. The forward voltage when current in circuit starts increasing abruptly, is the knee voltage.

51. Input characteristic is plotted between I_B *versus* V_{BE} for different values of V_{CE}.
Output characteristic is plotted between I_C *versus* I_{CE} for different values of I_B.

52. Clearly, $V_{CE_1} > V_{CE_2}$

53. As, input resistance,

$$R_i = \frac{\Delta V_{BE}}{\Delta I_C} = \frac{1}{\text{Slope of input characteristic curve}}$$

and output resistance,

$$R_o = \frac{\Delta V_{CB}}{\Delta I_C} = \frac{1}{\text{Slope of output characteristic curve}}$$

So, R_i is very small while R_o is very large.

54. Output current *versus* input current keeps the output voltage constant.

55. Current gain is defined as $\beta = \dfrac{\Delta I_C}{\Delta I_B}$ at constant V_{CE}

$\therefore$ Current gain at $V_{CE} = 1$ V is $\dfrac{(4.5-3)\text{ mA}}{(50)\ \mu\text{A}}$

$$= \frac{1.5 \times 10^3}{50} = 30$$

56. To identify the base of transistor, the multimeter has to show conduction between emitter and base as well as between collector and base keeping one lead of the multimeter common in both cases, then the terminal of the transistor to which the lead of multimeter is common is the base of transistor.

57. If common lead of the multimeter is connected to base and then it shows conduction for other two connections as mentioned in above question, the transistor is *p-n-p* otherwise *n-p-n*.

58. Option (a) tells that transistor is not working, as there is no connection between base and collector *i.e.*, there is some fault in this part of the transistor.

Option (b) tells that transistor is having no open circuit fault, *i.e.* it is having the continuity. Option (c) tells that transistor is faulty.

59. Since, no conduction in found when multimeter is connected across P and Q, it means either one of P and Q are *n*-type or *p*-type. So, it means R is zero. When R is connected to common terminal and conductor is seen when other terminal is connected to P or Q, so it mean transistor is *n-p-n* with R as the base.

60. Least count $= \dfrac{1\text{ mm}}{10} = 0.1$ mm

Zero error $= -(10-6) \times 0.1 = -0.4$ mm

Reading $= 6 + 5 \times (0.1) - (-0.4) = 6.9$ mm

$\Rightarrow \quad 6.9 = \dfrac{x}{10}$ (given)

$x = 6.9 \times 10 = 69$

61. Speed, $v = \dfrac{\text{Distance}}{\text{Time}} = \dfrac{16.0}{4.0} = 4.0$ m/s

Error in speed, $\Delta v = \pm\left(\dfrac{\Delta s}{s} + \dfrac{\Delta t}{t}\right)v$

$$= \left(\frac{0.4}{16.0} + \frac{0.2}{4.0}\right)(4.0)$$

$$= \pm 0.3 \text{ m/s}$$

62. $\therefore T = 2\pi\sqrt{\dfrac{l}{g}}$

So, $\dfrac{\Delta g}{g} = \dfrac{\Delta l}{l} + \dfrac{2\Delta T}{T}$

Therefore, $P = \left(\dfrac{\Delta l}{l} + \dfrac{2\Delta T}{T}\right) \times 100$

$$\Rightarrow \quad P_1 = \left(\frac{0.1}{100} + \frac{2(1)}{400}\right) \times 100 = 0.6\%,$$

$$P_2 = \left(\frac{0.1}{400} + \frac{2(1)}{400}\right) \times 100 = 0.42\%$$

$$\therefore \quad \frac{P_2}{P_1} = \frac{0.42}{0.6} = 0.7$$

63. $V = \pi r^2 h = \dfrac{\pi D^2 h}{4}$

$$\Rightarrow \quad \frac{\Delta V}{V} = \frac{2\Delta D}{D} + \frac{\Delta h}{h}$$

$$\Rightarrow \quad \frac{\Delta V}{V} \times 100 = \left[2 \times \left(\frac{0.01}{2.00}\right) + \left(\frac{0.1}{5.0}\right)\right] \times 100 = 3\%$$

64. As shown in the figure, $\dfrac{P}{Q} = \dfrac{l}{100-l}$, $P \propto \dfrac{l}{100-l}$

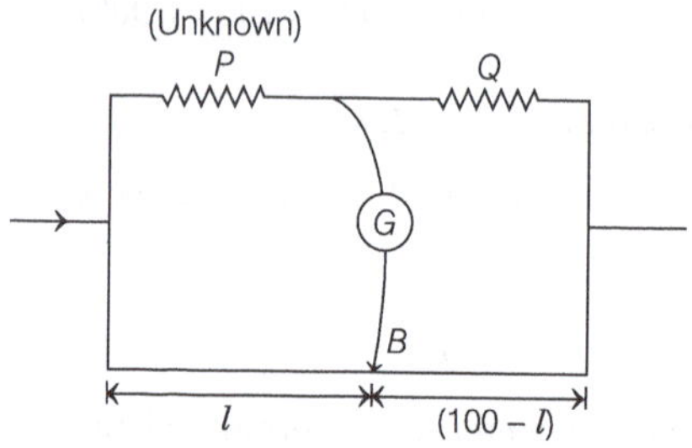

$$\Rightarrow \quad \frac{\Delta P}{P} = \frac{\Delta l}{l} + \frac{\Delta(100-l)}{100-l}$$

$$= \frac{\Delta l}{l} + \frac{\Delta l}{100-l} = \frac{0.1}{400} + \frac{0.1}{600}$$

$$\Rightarrow \quad \frac{\Delta P}{P} \times 100 = 0.42\%$$

65. From given figure,

Voltage across $R_S = 22 - 15 = 7$ V

Current through R_S, $I = \dfrac{7}{35} = \dfrac{1}{5}$ A

Current through 90 Ω, $I_2 = \dfrac{15}{90} = \dfrac{1}{6}$ A

Current through zener $= \dfrac{1}{5} - \dfrac{1}{6} = \dfrac{1}{30}$ A

Power through zener diode,

$P = VI$

$P = 15 \times \dfrac{1}{30} = 0.5$ W

$P = 5 \times 10^{-1}$ W

$\because \quad P = x \times 10^{-1}$ W (given)

$\therefore \quad x = 5$